The History of MANNED SPACE FLIGHT

The History of MANNED SPACE FLIGHT

David Baker, Ph. D.

New Cavendish Books
LONDON

For Nicholas,
so that he may know how it all began

This book is published in commemoration of the personnel, past and present, at the NASA Lyndon B. Johnson Space Center, the only facility worldwide developed exclusively for the pursuit of manned space flight. Formerly known as the Manned Spacecraft Center, and before that the Space Task Group based at the NASA Langley Research Center, JSC epitomizes the spirit of exploration and discovery and from its site outside the Texas city of Houston sent forth the men and machines that successfully planted human bootprints on the surface of the Moon.

Specification:
544 pages, 328 b & w illustrations
and 32 pages full colour

The Publishers express their sincere thanks to the many individuals, companies, agencies and institutes throughout the world without whose assistance this work would not have been possible.

First edition published in Great Britain by New Cavendish Books – 1981

Design and Layout – John B Cooper
Consultant Editors – Geraldine Christy and Narisa Levy
Editorial direction – Allen Levy

Printed and bound by Robert Hartnoll Limited, Bodmin
Phototypeset by Western Printing Services Limited, Bristol

New Cavendish Books, 11 New Fetter Lane, London EC4P 4EE
Distribution: North Way, Andover, Hampshire SP10 5BE

ISBN 0 904568 30 X

Contents

Front cover:
Astronaut Edward H. White II, floats outside his Gemini IV spacecraft during America's first space-walk, 3 June, 1965.

Frontispiece:
Walter M. Schirra photographed in the Apollo command module, October, 1968, during the flight of Apollo 7.

Acknowledgements

It is an impossible task to record by name the many people all over the world who have helped through two decades of cooperation to compile for the author information that led to the publication of this book. Of prime assistance, the US National Aeronautics and Space Administration deserves very special mention, especially personnel in Washington, Houston and at Cape Canaveral. At NASA Headquarters, William J. O'Donnell spent many hours dealing with the author's queries, questions and probings, taking telephone calls for bunches of data, letters for documentary assistance and passing to appropriate offices enquiries on numerous topics. Always, Bill's spring-loaded readiness to assist was a spur to the author's project. At Houston, John McLeaish provided volumes of technical documentation and at the Cape, Dick Young sent gnomes scurrying for information.

Industry was especially helpful and the author would like to acknowledge personnel at Aerojet-General, Astrodata Inc, Avco Corp, Beckman Instruments, Beech Aircraft, Bell Aerosystems, Bendix Corp, Collins, Control Data Corp, Cosmodyne Corp, Dalmo-Victor, McDonnell Douglas Corp, Eagle Picher, Electro-Optical Systems Inc, Garrett Corp, General Motors Corp, Honeywell Inc, ITT Kellog, Ling-Temco-Vought, Lockheed Corp, Marquardt Corp, Motorola Inc, Northrop Corp, RCA, Thiokol, United Aircraft, United Technologies Corp, Westinghouse, and Weston Instruments, for considerate and helpful contributions.

Three companies deserve special mention, for they liased with the author in research on their coveted products that form the basis for this book: McDonnell-Douglas, for work on the Mercury and Gemini programs and the Skylab space station; Rockwell International, on Apollo and ASTP hardware; Grumman Corporation for the Lunar Module. Throughout, the author found a frank and open readiness to both assist impartial and historical analysis and to allow publication of hitherto unknown episodes in the history of manned space flight. The author would like to extend a heartfelt 'thank-you' to all those who contributed to this book and who helped to tell the real story behind their remarkable products.

Other companies whose contribution is also recorded with gratitude include Boeing, Chrysler, Fairchild, General Dynamics, Hughes, Martin Marietta, Rocket Research Cor, Singer, and TRW.

The Pentagon was helpful and contributed significant information for research on the military programs, and the Executive Office of the President of the United States, Office of Management & Budget, contributed valuable budget data. The author is indebted to archivist Lee Saegesser for providing papers from the White House and to historian Jim Grimwood, late of the Johnson Space Center, Houston.

The extent of the contributions received from these and other sources, too numerous to mention by name, were limitless. In preparing material for this book the author was provided, over a period of ten years, with the mission commentary tapes and transcripts which enabled him to read, or hear, every word between the ground and the spacecraft on every manned space flight launched by the United States. Thus, his historical judgements and perspectives had that added dimension rarely afforded the historian, of being able to experience the sensations of the event first hand.

Because of this, and the author's own participation in the glorious days of discovery and endeavour that so characterize the first two decades of Man's flight into space, any injustice or misinterpretation must rest on his shoulders alone. Always, the evidence and the facts were made to write the story and the author has noted several discrepancies between incidents and events recorded in this book and other publications. It is hoped that the facts can set the record straight and that the reader can hold the author responsible for any errors this book contains.

Three very special people, deeply concerned with the production of this book, deserve the author's thanks. Allen Levy, the publisher, provided an almost unlimited canvas on which to paint the succession of historical events and gave the author inspiration to match the precedent of an established high standard. John Cooper, who gave much blood and energy to create the finished product, fabricated from raw clay the sculpture of words and pictures. And Geraldine Christy, with enthusiasm, shaped the flow of sentence and paragraph to the benefit of the text. To all three, the author has profound respect for a thoroughly professional job. It is to them that any credit for this book should go.

Introduction

It is in the very nature of man to explore and to seek vistas beyond the visible frontier. In evolving from one social being to another, the corridors of time have deeply etched his wanderings and his discoveries. For if man were to sit down and rest for even a generation he would be extinct as a traveller among nomadic life forms on Earth. It was inevitable that time and circumstance would match the will and the wisdom for man to leave the mother planet that gave him birth on flights across the darkness of space to set foot upon another world. That stupendous event happened in our generation of time and it is a beginning that can surely know no end.

Not for exploration did man unravel the gravity bonds of Earth; nor for an irresistible urge to push back old frontiers. But rather because it was the only route left for some of his kind to express themselves to others. In the vacuum of insecurity and a feeling of hurt pride, man leapt spaceward to challenge the opponents of his own achievement. But no matter, for in the final analysis the beginning will be seen as gloriously effective in which to start the saga of a new age.

Like Columbus, who sailed to prove himself right, and the Polo brothers who sought trade with the orient, men of space in the 20th century of the Christian era set sail for the Moon to stand upon its dust and dance beneath the planet from whence they came. But the age of Moon exploration has not yet begun and the events of the 1960s brought fleeting prelude to the convoys that will some day slip their celestial berths for harbors deep in space.

We know so little today about the origins and causes of past endeavors that lack of perspective should invigorate our concern for present events. The great stories of man's first twenty years in space are merely an account of first efforts and their chronicle is important now as new technology puts aside the schemes and plans of yesterday. It would be all too easy to forget the high drama of what even now seem simple activities on the road to tomorrow.

Few can remember with ease the morning Yuri Gagarin first rocketed beyond the atmosphere, or the faltering countdowns for John Glenn's orbital flight. The tensions and the expectations dim in the light of greater achievements such as the first dash for home when trouble struck an early Gemini flight, or the first spacewalk outside a protective cabin.

Easily ejected too are memories of men who gave their lives, never willingly, for the unique experience of rolling back the frontiers of human endeavor; men who perished in flaming spacecraft on the ground; men who died because of ironic failure in well-tried aircraft; men who died because of human error and ineptitude. And who recalls readily the emotional drama of voices from around the Moon on Christmas Eve when three men became the first humans to escape the gravity of Earth?

Perhaps the landmark in manned flight came when Neil Armstrong and Edwin Aldrin touched the Moon itself, an hour few who lived those days can surely ever forget. But events dimmed again as lunar landings became an almost routine parade, the players always dressed in the same strange costumes, until long duration missions around the Earth seized new favour. And then, as climax to an age of competition and achievement, three Americans met two Russians in orbit – and shook hands.

As effectively as any one event can seal the days and years of a single era, the joint US-Soviet docking flight of 1975 wrote finis to the beginning of manned space flight. Both nations are today developing new forms of travel designed to completely replace the expendable, throw-away, spacecraft typified by Apollo and Soyuz. For America, the change is a transformation since a deliberate and calculated halt was made in flight operations after the joint cooperative space mission so that funds which would otherwise have gone on continued Apollo activity could be spent on the reusable Shuttle.

For that reason it is a fitting time to pause and review the first twenty years of manned space flight, for the achievements of the next twenty years will be so different from those of the first. The story of the Shuttle is a chronicle of struggles that began before World War Two and as such constitutes a record in itself which the author intends to prepare as a separate history. For that reason, emphasis is here placed on the expendable era beginning with the launch of Russia's Vostok 1 in April, 1961.

In fact that period stands as a landmark in space flight history for within a matter of weeks after Gagarin's flight an American astronaut had also flown into space and the incumbent United States President had set a manned Moon landing as America's goal for the decade. This book stands to commemorate the twentieth anniversary of those days when the future for man in space was assured.

Readers will note the considerably greater emphasis placed on US operations to the apparent detriment of the Soviet story. The reasons for that are three-fold: the Russian authorities are unwilling to open their files for scrutiny and objective analysis, believing their security interests to be threatened by such inspections; considerable speculation surrounds many historical events associated with the Soviet manned program; and the diverse and varied achievements of the United States have not been mirrored by a similar wide-ranging participation in exploration and discovery by the Soviet Union.

In considering the wealth of historical material available for the US story, the author considered it important to set down the details of American manned flight with some precision and deliberation; major events such as the decision to land men on the Moon warrant searching analysis. The examination of many optional routes through the massive quantities of material published on the Soviet programs is more properly performed by the reader as a separate study, for much of it conflicts with official views and contradicts the historical image the Soviet Union wishes to provide for itself.

This book is intended as a complete history of manned space flight development and operations but should serve as a starting point for students of the subject wishing to pursue specific aspects. For that reason it is important to consider the need for timely updates as events overtake the publishing schedules.

Readers in the United States can do no better than seek additional information from the American Astronautical Society, who publish detailed records and reports on technical, engineering and scientific activities associated with manned space flight. For inspecting actual hardware, Washington's National Air and Space Museum is the best depository of spacecraft and rockets set in a historical context. Visitors to the Kennedy Space Center can also tour many of the launch facilities mentioned in this book.

In the United Kingdom, the British Interplanetary Society at 27/29 South Lambeth Road, London SW8 1SZ, publishes magazines and journals devoted to space flight at both general and technical levels and is the best source in Europe for regular information about activities originating both in America and Russia. Visitors to the Science Museum in London will find a liberal display of space-related exhibits and may find actual spacecraft on loan from NASA. The interest of these organizations and museums in attracting participants and visitors ensures a generous value for students of the subject and those with a general enthusiasm. Further research beyond the content of this book is to be encouraged and the reader should consider it a stimulus for continued study.

CHAPTER ONE

Dreams and Inventions

Prophetic design for a two-stage rocket emerged from the work of 16th century artillery chief, Conrad Haas, from Romania.

There can have been few things more significant about the 20th century than that it was the selected period in all history when man first left his planet and looked back upon it from the surface of another world. For generations of man the dream seemed impossibly far, yet the theories and thoughts of sages and scientists inevitably gave way to reasoned debate. Gradually, through the centuries, rockets emerged as the only feasible means of propulsion in space. Devoid of air, the vacuum of space contains no oxygen with which to combust a fuel. But the rocket has both fuel *and* oxidizer, opening endless possibilities for travel beyond Earth.

Rocketry is a form of reaction described by Isaac Newton in the 17th century, but the principles of reactive flight had been known for at least 2,000 years, During the 4th century BC, Aulus Gellius described a mechanical pigeon made to move by jets of steam escaping from its tail. It was mimicked 400 years later by Hero of Alexandŕia, a Greek who recorded an aeolipile device, or sphere, made to rotate by jets of steam from opposing ports.

But these were poor attempts at unique forms of propulsion. The first use of reactive thrust for moving an object rapidly through the air was probably a product of chemical experiments conducted in China at a time when the Christian era was emerging in Europe. No one knows for sure just when alchemy moved from fire baskets to fire arrows, but the use of black powder as a combustible was certainly common by the 6th century AD. Fire baskets were employed to hurl hot materials on the enemy while fire arrows were developed as reactive devices designed specifically to fly under their own reaction. Black powder was mixed and prepared first as an explosive chemistry suited to battle; the display of noise and commotion attending the release of several pyrotechnic contraptions did much to embellish the attack and if contributing little in a practical way, certainly provided a well placed psychological advantage. Firecrackers, as these devices were called, seemed to have originated in the 2nd century BC. Prepared with ingredients of sulfur, saltpeter and charcoal the uses were endless.

Published in 1054 AD, Wu-ching Tsung-yoa's *Complete Compendium of Military Classics* is a valuable record of contemporary developments, indicating the availability of fire arrows for a period pre-dating this great work by at least two centuries. There is little doubt that between the 10th and 13th centuries China exploited the rocket and rocket-propelled fire arrows to good effect. In 994, it is said, an army of 100,000 laid siege to the city of Tzu T'ung and that in addition to the use of war machines the application of rocket projectiles by the defendants drove the invaders to retreat. By the 13th century of the Christian era, Mongol hordes had swept across Asia Minor with a technology that would be applied in a score of nations and states. Adaptation and innovation were key developments in Europe from this time on.

A particular important step was made by one Conrad Haas, chief of the artillery arsenal in Sibiu, Romania, between 1529 and 1569.

Haas prepared a document setting out the principles of rocketry and containing in it a chapter describing the advantages of a multi-stage projectile. Recognizing the dead weight carried along by a rocket as it neared the end of combustion. Haas did away with the unnecessary appendages and converted the basic rocket into a two-stage device. The charge at the rear would be ignited first, consuming paper walls that supported the dense bulk of propellant, so reducing the inert weight and continually improving the thrust to weight ratio. As the charge was consumed, moving from the rear end of the rocket, the heat generated by combustion ignited the second stage which then conveyed the projectile further along its trajectory. Haas prepared drawings to explain his idea and placed a small load, called 'the payload,' at the front of the rocket. His designs included a three-stage missile, a military missile armed with a small powder keg in the nose, and various adaptations, all aimed at reducing the inert mass and improving the performance.

Kasimir Simienowicz, a Polish general under King Wladyslaw IV, published the theory of multi-stage rocketry in a book called *Artis Magnae Artileriae*.

The next major development in solid propellant rocketry came with the designs of William Congreve, a Colonel at Woolwich Arsenal, who set to work on designing an efficient solid propellant rocket for use by the Army after publishing in 1804 a record of early progress. The period between the 16th and 19th centuries produced neither new concepts or principles upon which the science of rocketry could move forward; Congreve, however, was able to bring together the many different technologies then being applied to theoretical rocketry and combine the most promising proposals into an efficient weapon. By the end of the 19th century, however, rockets had given way to artillery. Proved in several major wars during the latter decades of that century, the field piece was seen to be in several ways more efficient than the best Congreve projectile. Having reached the peak of its popularity in Britain, the rocket was to die a slow and unyielding death while an obscure Russian school teacher opened vistas that gave a brave new promise to the future.

That man was Konstantin Eduardovitch Tsiolkovsky, born in 1857 and hailed within his lifetime as the 'father of space travel.' Tsiolkovsky approached problems with a rationality that quickly took his agile mind to the source of a dilemma and nowhere is that ability seen to greater effect than in the way he dealt with the overpowering passion of his entire life: the way that Man might one day fly through space, and the means with which he would build a space-dependant society. Tsiolkovsky knew that the only way to travel through space was by reactive force; careful to recognize that a rocket flew as the reactive product of the active flow of gases through one end, he did not make the mistake many did even decades later, of misinterpreting Newtonian physics by believing the rocket to move forward by pushing on the dense blanket of atmosphere.

Tsiolkovsky also knew that to design a rocket-propelled vehicle effectively, precise information would have to be obtained about the rocket's thrust; this was determined to require a value known as specific impulse, a measure of the efficiency of a given set of propellants: oxidizer and fuel. Because Tsiolkovsky recognized the need to achieve a very efficient concept, he became aware of the value in using high-energy propellants with which to operate the rocket. Accordingly, Tsiolkovsky proposed a new and exceedingly bold solution: use two fluids – liquid hydrogen and liquid oxygen – brought together in a combustion chamber so that when they burn the resulting gases are driven from the chamber through a nozzle, much like the exhausted products of a powder rocket. Tsiolkovsky knew that to be efficient his rocket must be powered by a motor delivering a large thrust at high exhaust velocity; his two-fluid solution was a major step on the road to improving rocket efficiency to a point where it could conceivably support space flight.

Nobody had yet built a liquid propellant rocket motor, and of all the fluids Tsiolkovsky could have selected, he chose the most difficult to handle and the most exotic to prepare: liquid oxygen boils at −183°c, liquid hydrogen at −253°c. Liquid propellant engines promised a considerable improvement in theoretical efficiency over the solid propellant rocket, but of all available chemical combustion propellants, the hydrogen-oxygen combination was the best of all possible choices, the very high exhaust velocity of its combusted gases making it an ideal choice except for the extremely difficult tasks associated with preparing, handling, and operating the super-cold (cryogenic) fluids. That it had to be kept in a liquid state was a necessity resulting from the extremely low mean density of the propellant. Stored as a gas, the resulting volume of the propellant would be prohibitively large.

This liquid propellant engine, conceived by Tsiolkovsky, was chosen because it brought together all the ideal theoretical requirements: reactive forces providing high propellant efficiency in a system which would benefit from the concept of staging. Others were looking to powder rockets – solid propellant rockets – to provide increased power. Tsiolkovsky went straight for the dilemma and sought a solution, albeit in this event one completely radical in its approach and in its engineering. Tsiolkovsky was 15 when he turned his attention to the problems of rocketry, 21 when he proposed the use of rockets for space travel, and 45 when, in 1902, he developed the concept of a liquid hydrogen/liquid oxygen rocket motor. A year later he published *Exploration of Outer Space with Reactive Devices* containing scientific discussion of his ideas and drawings of the liquid propellant rocket. Additional works were published in 1910, 1911, 1912 and 1914.

By this time Tsiolkovsky had been joined historically by Friedrich Arturovitch Tsander, born in 1887, who went on to propose many potential applications of the rocket for space travel and exploration. He will be considered in the next chapter for his contributions he made to the development of manned space flight, but mention should be made here of his proposal that rocket flight in general could benefit from the application of a concept whereby redundant components of the rocket structure could be consumed during flight. In this way the thrust to weight ratio would be improved although the method employed was an extreme case, made perhaps more difficult from an engineering standpoint than application of a basic staging concept.

Yet for all the unique and astonishingly viable designs associated with rocket propulsion that emerged from Tsiolkovsky's pen during the first two decades of this century, it was to another man, in a far off country, that the credit for liquid propellant rocket flight rightfully belongs. To a man called Robert Hutchings Goddard, born at Worcester, Massachusetts, in the USA, on 5 October, 1882. From an early age Goddard was captive to the stirring fiction of H. G. Wells and Jules Verne and at the age of 14 wrote and submitted for publication an article on the navigation of space. Like the Russian mathematician Tsiolkovsky, Goddard was concerned with imminent developments that promised early results and linked the basic Newtonian laws to this need and so became convinced that liquid propellant rockets would be an essential part of space flight. He would later recall how, at the age of 17, while high in a cherry tree removing dead branches he 'imagined how wonderful it would be to make some device which had even the *possibility* of ascending to Mars and how it would look on a small scale if sent up from the meadow at my feet . . .'

Goddard was convinced of the need to develop liquid propulsion, following the award in 1911 of a Ph.D. with activity at Clark University toward that objective. In 1919 he prepared a paper for the Smithsonian Institution which proposed a test flight to the Moon, although Goddard knew it would require considerable effort to realize this event. Outwardly, Goddard retained a calm, cool professionalism that belied his passion for space travel. Humiliated by the press and pre-judged a crank, he knew the price for public affirmation of radical concepts. Nevertheless, working first with a few colleagues from Clark and then with a grant from the Smithsonian, in 1926 the physics professor prepared the world's first liquid-propelled rocket, a simple device with the combustion chamber and exhaust nozzle placed above the propellant tanks and held at that location by pipes that carried fuel and oxidizer.

Goddard knew, as Tsiolkovsky proposed, that hydrogen and oxygen were the ideal propellants for a liquid motor, but for the early development models he was content to work with oxygen and gasoline, easier to handle and an acceptable combination for proving the concept. In fact it would be another 30 years before serious consideration to cryogenic propellants produced the first attempt at high-energy rocket motors burning hydrogen and oxygen. On 16 March, 1926, Goddard successfully flew his simple rocket for 2½ seconds from a farm near Auborn, Massachusetts, where he had set up a rudimentary launch frame with his wife and a few assistants. Esther Goddard was unable to record the historic event; eager to catch every significant step she started a movie camera at ignition, but with just 7 seconds of film available in the camera it ran out while the motor burned for a further 13 seconds before thrust exceeded mass and the rocket sped away to fall 56 meters from the launch frame.

Almost five years later, on 14 March, 1931, Johannes Winkler became the first man outside America to fire a liquid rocket when he successfully flew his device on oxygen and methane to a height of 60 meters. Exactly two years later to the day, the American Interplanetary Society (later the American Rocket Society and, from 1963, the American Institute of Aeronautics and Astronautics) fired its own rocket on oxygen and gasoline. Meanwhile, in Germany, enthusiasts formed a society for the pursuit of rocketry, with space travel as the ultimate objective. Called the Verein fur Raumschiffahrt (German Society for Space Travel) it was disbanded in 1934; practical tests with small rocket motors had absorbed the VfR's limited funds and plunged it into debt. Several leading protagonists of rocketry were taken in by Walter Dornberger from the Army Weapons Department and employed at a research establishment at Kummersdorf. One of those who joined the team, by far the most vocal advocate – a veritable one-man delegation for the pursuit of rocket flight–was

Dr. Robert H. Goddard's liquid propellant rocket tests presaged a new age of military designs, research epitomized here by an experimental flight on 17 July, 1929.

Wernher von Braun. Born in 1912, the young von Braun had an insatiable desire to build rockets and inaugurate the world's first space transportation system. But whereas Tsiolkovsky was the great thinker-philosopher (who never actually pursued practical developments) and Robert Goddard was the practical experimenter and a great inventor, von Braun was the polar head of enthusiasts and protagonists, the voice of stirring hope ready to marshal resources in common assault on technical barriers.

By 1935 the German research group was well on the way toward development of the world's first military missile with designs for precursor rocket vehicles required to test and evaluate the many innovative applications essential for large rocket vehicles. Far away, in Soviet Russia, experimenters there too perfected the principle to the level where a simple rocket flight became possible; on 17 August, 1933, rocket 09 had flown on the thrust from oxygen and gasoline, the first in a long line of liquid rocket developments. To assemble and launch a simple liquid propelled rocket was one thing, to design and build a ballistic missile capable of carrying a 1 tonne warhead across several hundred kilometers was quite another. Yet this was the task Dornberger's team addressed in work that continued on through the 1930s and early 1940s until, on 3 October, 1942, their 13 tonne A-4 missile ascended from its concrete test pad at Peenemunde on the shores of the Baltic. Propelled by a motor burning liquid oxygen and alcohol, the first A-4 flew precisely as planned on the 25 tonnes of thrust it generated, for one minute. The missile was 14 meters tall and was to become the inspiration for rocket engineers on two continents.

At war with the world, Germany turned too late to missile projects like the A-4 for their capabilities to reach full realization. Although able to hurl an amatol warhead across 300 km, the A-4, called V-2 for Vergeltungswaffe Zwei (Revenge Weapon 2), was unable to turn the tide of fate against advancing allied armies rolling across the European mainland toward Germany. Not until September, 1944, did the first V-2 rise in anger and vindicate the beliefs and promises of its inventors, who for a decade had pressed for support that all too often came late. Seven months later, the V-2 flew its last hostile mission and shortly after that the European war was over. At one point von Braun had been arrested and accused of treason, for daring to discuss in private the hopes of many for a peaceful age in which rocketry could be made to serve man's better instincts. During the closing months of war, the V-2 operation had been consumed by the elite SS under Himmler, who for several years had sought control of the missile project in numerous attempts at wresting authority from the non-political Army machine. Now, with Soviet armies pouring across Europe, and combined Anglo-American forces moving from the English Channel through France and the Low Countries, the rocket engineers faced crucial decisions. Where to go, and to whom should they surrender?

Again, von Braun was the spokesman and with an impassioned dedication to continuing his life's work, decided that the fragmented team should give themselves up to the Americans. Eventually, more than 100 ex-Peenemunde engineers migrated to the United States with sufficient equipment for the assembly of about 100 V-2 rockets, snatched from under the Russian guns by a special force. By the end of 1945 Wernher von Braun and his colleagues were installed at Fort Bliss, Texas, there to await assignment. Their first task was to set up a firing range at White Sands Proving Ground, New Mexico. Goddard died in August 1945, and the test area was not far from Roswell where the great inventor had spent many years flying improved versions of his early liquid-propelled rocket. Under the command of Colonel Holger N. Toftoy, the German team were instructed to prepare the ballistic missiles for flight trials.

Seen here with arm in plaster following a motor accident, German rocket engineer Wernher von Braun surrendered to the US Army in 1945.

The first US V-2 firing came on 16 April, 1946, and soon a variety of experimental projects were proposed and authorized. Rockets were deliberately blown up on the pad to measure the effect of blast. Others were launched from ships to determine if sea-going missiles were practicable. Still more were instrumented to see if they could serve effectively as weather probes, and some carried living things in the compartment designed to carry a warhead. In one project, called Operation Bumper, a research program was carried out to test the mechanical functions of the V-2 against changing environments. In this, some were deliberately fired at low angle for maximum range while others went for high altitude, but all carried a small liquid propellant rocket called the Wac Corporal installed on top as a second stage. In this configuration, one Bumper-Wac rocket reached a height of 393 km. In 1949 the Army decided to move the missile program from Fort Bliss to Redstone Arsenal in Huntsville, Alabama.

It was to be the start of a new beginning for the many Germans that came to work for the Army. Before long, communities would develop their own 'German' part of America, whole streets of towns would speak the European language and rocket development from that year forth would be inextricably woven with normal life – at least for the inhabitants. But it was the Korean War that gave the first real impetus to post-war American rocketry. The Redstone missile grew out of specifications laid down in 1950 for a weapon capable of reaching targets 800 km from the firing pad. It was to have a performance about twice that of the V-2 and be used for tactical assault. As time went by the specification was amended until it emerged in 1953 as a 300 km missile, very similar in several respects to the V-2. It was certainly more reliable, a worthy product of several years effort by the ex-Peenemunde team. Redstone was just a little longer than the V-2, but was powered by a liquid oxygen/alcohol engine of similar thrust and adopted the exhaust vane method of guidance and stabilization, with small rudders on stub fins. Goddard had never been sure that aerodynamic stability could be achieved with fins; the Germans found their design a major stumbling block during tests of V-2 models in the early 1940s. But Redstone was far from the only missile to emerge in the decade after World War Two.

Working with new ideas laid down by Karel Bossart, General Dynamics tried to interest the government in a long range missile project capable of sending a warhead across a distance of 8,000 km! Called MX-774, the design was stymied while government departments cancelled and then reinstated the project. It was named Atlas in 1951 and moved slowly ahead as a long range strategic missile. By the Korean War it was still gestating. But Atlas was big, and unlike Redstone and every other missile or rocket, was designed with a thin skin supported in circumferential rings, the whole assembly kept rigid by pressurization or by the mass of propellant contained within the tanks. The walls of the rocket were, in fact, the walls of the propellant tanks and this novel approach by Bossart was the ultimate attempt to reduce the inert weight of the rocket and dramatically improve the propellant/mass ratio, the fraction of the rocket's mass taken up with propellant.

The engines for Atlas were to come from liquid propellant motors developed by Rocketdyne for the Navaho project, a winged missile with a triple propulsion system. Each Rocketdyne engine, developed early after the war by close examination of the V-2 engine, delivered about 60 tonnes of thrust, or twice that of its German predecessor. Atlas would carry three such motors of two different types: a single sustainer engine in the center, flanked by a single booster engine each side. The booster engines would be connected together on a skirt designed to slip down over the sustainer engine 145 sec into the flight, falling away and leaving the rocket to ascend on its single sustainer, lightened by the release of the boosters. Again, it was a novel approach and characterized Atlas. Unlike Redstone, Atlas was an Air Force project and

Derived from research formerly conducted for the V-2, Germany's rocket scientists provided America's first medium-range ballistic missile, called Redstone, from the military arsenal in which it was developed.

was joined in 1955 by Titan. Contracted to the Martin Company, Titan was intended as a hedge against trouble with Atlas, the latter being expected to enter service as an Inter Continental Ballistic Missile (ICBM) and form the first line of defense as a strategic weapon.

Titan went on to replace Atlas in that role, but by the mid-1950s both projects ran concurrently, with Atlas several years ahead. Atlas and Titan were the first of the post-war 'heavyweights.' Compared with the V-2 they were veritable giants. In 1955 reports began to appear criticizing the US missile plans. Intelligence rumor spoke of large-scale Russian plans to build an ICBM fleet. Devoid of a big bomber fleet it was a logical conclusion, but one that was to a large extent ill-founded. Soviet rocket developments had moved ahead after 1945, first with a few captured German scientists and then with the products of an indigenous Soviet research capability. By the mid-1950s work was well under way on a large liquid propellant rocket as equally unique in its design as Atlas.

Called SS-6 Sapwood by NATO code namers when it appeared, the rocket comprised a central sustainer section supporting four main combustion chambers and four small steerable motors called verniers, delivering a combined thrust of 96 tonnes. The sustainer supported four booster sections strapped to its sides. Each booster carried four main combustion chambers and two verniers for a total thrust of 102 tonnes per booster. The single sustainer and each booster had a single turbopump. With all four boosters and the sustainer firing at lift-off, the total thrust was about 504 tonnes. It was a veritable block-buster, a heavyweight missile to lift the early, and heavy, nuclear warheads. The ascent of such a behemoth required the simultaneous firing of 32 rocket motors, the boosters being shut down and separated on the way up to improve the thrust/mass ratio.

In 1955, Sapwood was still an unknown quantity, but in the light of trace elements that contained truth the wild speculation that followed was quite ridiculous, affecting assessments of air power as well as missiles. Nevertheless, the President's special adviser on science and technology, James R. Killian, Jr., headed a committee to look at the question of US defense and the force posture between the two global hemispheres. The Secretary of Defense, Charles Wilson, took the initiative on hearing from the committee evidence that seemed beyond question and which spoke alarmingly of a potential Soviet threat. As a result, a joint Army-Navy project was authorized called Jupiter. It would provide the Army with a long range missile placed in range between the Redstone and the Air Force Atlas and it would give the Navy a shipboard missile for sea duty. Eventually, the Navy would separate their interests and go for a unique Navy project called Polaris. Another project placed before the Army late in 1955, was the Jupiter C, actually a modified Redstone but one which would have small solid propellant upper stages to test warheads. The Air Force had a parallel to the Jupiter in the form of a project called Thor. With a range, at 2,600 km, similar to that laid down for Jupiter, it ensured an Air Force presence further down the range scale. As defined at the end of the year, the Army would have Redstone and Jupiter, short and medium range missiles, and the Air Force would have Thor, Atlas and Titan, medium and intercontinental range weapons.

America had finally woken up to recognize the enormous potential in missile technology, ten years after the war and nearly 30 years after Robert Goddard flew his first liquid propellant rocket. The shocked concern that Soviet Russia could develop and perhaps surpass the strategic capabilities of the United States was the stimulus that forced from government coffers the funds necessary to get the missile programs off the ground. No one yet really thought seriously about any other application. It was still too soon after the major global war, and too close to the Korean conflict, to instill any feeling of crusade or scientific adventure. Yet, by the end of 1955, America and Russia were well on the way to completing the design and fabrication of rockets capable of lifting several tonnes into Earth orbit, and a few hundred kilogrammes to the Moon. Having established the two most important criteria for flight into space – a knowledge of the planets and the stars, and a technology capable of transporting Man into the cosmic vacuum – it only remained for science and technology to unite, and for a will, a motivation, to bring them together.

CHAPTER TWO

Steps to Reality

When thoughts first turned seriously to the prospect of space flight, it was generally assumed that the rocket itself would form an integral part of the crew compartment – the habitable section where life could be sustained through the provision of an atmospheric environment, a controlled temperature and necessary food and sleep facilities. It was further assumed that the motive force, or engine, would be retained throughout the flight and returned to the surface of the Earth at the end of the journey. In reality, events would overtake the orderly progression from simple rocket to complex spaceship and force an adaptation of existing technology rooted in military missile development.

One of the earliest proposals for a manned spaceship came from the rocket pioneer Conrad Haas, attributed in the last chapter to work on the multi-stage projectile. At some time during the middle of the 16th century, Haas prepared an illustration of the device he thought could most effectively represent the ultimate potential of reactive flight: a tiny house placed on top of a massive solid propellant rocket. Haas gave no thought to the shape of this proposed ship, but its arrangement certainly makes it the earliest space station to be seriously considered for human habitation. Haas' work was undoubtedly a genuine attempt to predict what he thought might result from rocketry; others were less inclined to believe in the facts of space flight, preferring the fantasy and its free licence to the impossible.

Space fiction goes back centuries. One of the earliest references comes from *Vera Historia* (True History) written by Lucian of Samosata at the time of Plutarch, a work in which the author advises his reader 'that I mean to speak not a word of truth throughout,' and then proceeds to concoct a fantastic journey to a populated Moon. In a magnificent poem whose 60,000 verses took a lifetime to write, Firdausi told his 11th century readers of ancient kingdoms and lordly demons from space, exciting the intellect with accounts of space travel and celestial conflict. Some scientists went far ahead of accepted teachings and resorted to fiction in order to express their beliefs. One such was Johannes Kepler in his *Somnium* (Dream). Transported to the Moon by nocturnal demons – Kepler knew that Earth-bound animal life would be incapable of making the trip – Earthlings are exposed to the disordered life style of lunar inhabitants.

In the 17th century, Domingo Gonsales (actually Francis Godwin, the Bishop of Hereford) wrote the classic *The Man in the Moon: or a Discourse of a Voyage Thither*. Purportedly an account of personal events, Domingo Gonsales tells of his flight to the Moon in a carriage drawn by 25 geese and describes remarkable forms of plant and vegetable life and beings twice the height of men; Gonsales returned when he noticed the geese grew concerned 'for want of their wanted migration.' Geese were presumed not to migrate beyond the Earth and were, of necessity, required to return perchance they found themselves on the 'Moone.' Yet even as science grew more aware of the true state of nature and the real proportions of the Universe, so did science fiction follow ever more closely the possibilities for space flight and the exploration of other worlds. In no small degree were the works of 19th century writers responsible for stimulating inventors and inspiring pioneers. Sometimes the two sides of scientific literature became blurred and fact mixed itself with fantasy, as did the deliberate Moon hoax of 1835.

Mindful of the time required to travel and communicate across the Atlantic Ocean, Richard Locke Adams issued in the *New York Sun* a report claiming great discoveries from the telescope of Sir John Herschel. (Sir John was the son of William Herschel the famous astronomer.) In an age when very large telescopes were springing up all over Europe and major discoveries were an everyday event, public knowledge at a time of poor education was ripe for manipulation. Adams reported that Herschel saw winged beings on the Moon, demon-like forms with copper-colored hair, giant animals and weird forms of life. That people wanted to believe these fantastic tales is borne out by the popularity of science fiction in the decades following Adams' hoax. It stems from a deep-seated need to liberate the human mind from logic and considered thought; it can be found today expressed in various forms, things seen in the night or the sky. True science fiction has no need of hoax, however, and should, by definition, remain within the gamut of plausibility.

Such was the work of Jules Verne whose *De la Terre à la Lune* (From Earth to the Moon) described a spaceship called 'Columbiad' launched by cannon. The method was quite impossible, but the book touches key aspects of manned space flight and serves as a prelude to the recognition of mechanical principles extolled by Konstantin Tsiolkovsky 50 years later. Verne explains that his cannon-launched travellers were stunned by the force of acceleration as their bullet-shaped projectile shot from the barrel; the force of acceleration would in reality have killed the crew, but Verne recognized the effect and embraced it within his fiction, allowing them to pass into unconsciousness momentarily. Verne provided a shock absorber in the form of a bed of water which served to attenuate the shock of acceleration; (shock attenuators are an integral part of spaceship design and in reality are essential to the physiological well being of the crew).

As they traveled to the Moon, the crew found themselves exposed to the heat of the cabin walls, which increased in temperature due to the friction of moving quickly through Earth's atmosphere; (a spaceship will require protection during the ascent into space and during its flight to other worlds because of the unshielded exposure to solar radiation). Verne's first Moon novel ends with the arrival of the spaceship in the vicinity of the lunar sphere, but in his second work, *Autour de la lune* (Around the Moon), travelers attempted another trip similar to the first. In this, they experienced meteorites and other translunar phenomena before passing around the far side of the Moon. They fired rockets to brake the acceleration of the spaceship and this provided a miss distance of only 47 km. As the ship reached the high point of its far side pass, attempts were again made to reduce its speed and effect a landing. Unable to do so the ship returned to Earth, the landing aborted. It is a prophetic description of a similar flight that would take place nearly one century later.

Through these books Verne stimulated a new generation of science fiction stories, tales that embraced the reality of scientific law if not altogether remaining within the realm of known facts. Verne did not have his ships actually land on the moon, a puzzling reversal that may have been caused by his desire to retain a degree of mysticism, but the seriousness with which he describes in detail the stresses and difficulties of flying beyond Earth are a refreshing change from the synthetic fantasy of former years. But Verne's impact on the world of speculative travel beyond the Earth was more in the form of a transition than an end product in its own right. Verne, and H. G. Wells a few years later, provided motive force by which to move their crews from Earth to the Moon and in this they were more responsible than previous writers, but the new age of science fiction came at the turn of the century when rocket propulsion was seen to be a viable means of transportation. Where Verne used a gun and Wells a mysterious substance called cavorite, 20th century authors were to adopt rocket engines and transform the projectile into a true spaceship. And at that point serious consideration of space travel led to carefully considered plans.

For 2,000 years, speculation on celestial flight remained at large from the development of rocket propulsion. Only when rocketry took on a more promising shape could the dreamers combine with the scientists to define a feasible scheme by which human beings could liberate themselves from the bonds of Earth. And the one single individual who stands proud from the rest is Konstantin Tsiolkovsky, the Russian teacher who did so much more than anyone else to predict the most likely patterns of future development. Tsiolkovsky's contribution to theoretical concepts

In what is regarded as one of the first serious proposals for a space station, Condar Haas presented this 16th century concept of a house set upon a rocket 'for more peace, not war.'

of liquid propellant flight have been outlined already. It is appropriate here to consider his contribution toward the problems of manned space flight, for he was equally proficient in both fields and thus served to bind space travel to rocket propulsion. It seems that as early as 1876, at the age of 21 years, Tsiolkovsky first drew up notes and diagrams to explain and to conceptually evolve the requirements for a space station.

The first prepared work on this was written in 1883, when Konstantin described the effects of weightlessness, and two years later went further in suggesting that workmen would assemble, in Earth orbit, machines for turning solar energy into electrical power 'sufficient to sustain life for 20 billions of inhabitants.' In *Musings on Earth and Heaven and Effects of Universal Gravity*, Tsiolkovsky explained that artificial gravity in the weightlessness of orbital flight could be induced by spinning a large space station around its center of mass so that centrifugal force would press objects to the outer walls. He also foresaw the possibility of graduating the effect of gravity, in effect inducing fractions or multiples of gravity, by slowing or speeding up the rate of rotation.

Perhaps his greatest work was *Exploring Universal Expanses with Jet Instruments*, published in 1902. In this he asserted, 'To set foot on the soil of the asteroids, to lift by hand a rock from the Moon, to place travelling stations in space, to form living rings around the Earth, Moon and Sun, to observe Mars at distances of several tens of versts, and to land on its satellite or even on its surface, what could be more mind dazzling: From the moment jet propulsion instruments are used a grand new era in astronomy will begin – the epoch of orbital study of the heavens.' Tsiolkovsky went on to publish supplements in 1911–1912 at which time he had fully prepared the strategy for inter-planetary exploration. Earth-orbiting space stations would, he said, be developed first, then exploration of the nearer planets could begin. The space stations could be employed as astronomical and astrophysical laboratories, observatories free from the limiting constraints of viewing space through a dense blanket of atmosphere. They could provide facilities where unique physical processes would be put to work; chemical and pharmaceutical laboratories benefiting from the weightlessness of orbital flight. Gradually, from long experience working in the unique environment, cities or space colonies could be fabricated.

In 1923 Tsiolkovsky amplified this strategy: 'We take off in a space ship and stay at a distance of 2,000–3,000 versts (*1,870–2,810 km*) from Earth. Little by little appear colonies with implements, materials, machines and structures brought from Earth. Gradually, independent production, though limited at first, will develop. When life and technical industry have been consolidated in space and settlements have been formed around Earth or in the asteroid belt, in a word, where a surplus of diverse materials can be found for construction – then there will no longer be the need to take monstrous reserves of propellant from Earth.' Tsiolkovsky thereby proposed that orbital colonies would be rendered self-sufficient by converting solar energy into electrical power; uncoupled from the finite fuels of Earth, the expanding requirements of the space habitat would be independently met from space. Several asteroids could be maneuvered into positon around Earth to serve as mineral and fuel sources. By this time the very materials out of which spaceships and rockets would have to be constructed could, thought Tsiolkovsky, be obtained from outside the Earth and when demonstrated this could provide a construction base on which to build the interplanetary rockets.

Concurrent with this activity, mused Tsiolkovsky, other projects would build a massive ring around the Sun at the distance of Earth. Several artificial planets could be joined together to capture solar energy and so provide the electrical power to build more stations. In this way, the expanding space-borne society would be self-sufficient. Tsiolkovsky had the rare quality of tending to the details of his grandiose proposals. He considered the environment into which human life would evolve and designed artificial ecosystems, self-contained life support equipment utilizing plants and bacteria that exude oxygen. For food, Tsiolkovsky stressed the need for organic compounds 'grown' from chemical elements sent from Earth. To wash, he even resolved the need for a shower, with special hot-house provision for tropical plants. Although Tsiolkovsky went far ahead of his day, with plans for massive orbital colonies and cities in space, he substantiated the need for such enterprise by developing a rationale based on the industrialization of the space environment. He produced powerful engineering arguments to justify the exploitation of a space station launched from the surface of the Earth. But his main theme was that, if industrialized at all, space must be made self-sufficient to prevent Earth from being purged of its resources. Several Tsiolkovsky designs were included in the motion film *Space Voyage*, made in 1935.

A contemporary of Tsiolkovsky, F. A. Tsander, prepared several papers on the problems of interplanetary flight, whereas his more distinguished peer sought to present a philosophy for the future development of man in space. Tsander's work can be dated to June, 1907, and notes he made for consideration of the basic problems standing in the way of manned space flight. These were, he thought, a means of propulsion, substances that could provide a habitable environment in the confines of a cabin, the removal of carbon dioxide exhaled by the crew, adequate methods of navigation, etc. Tsander was also concerned about the mode of propulsion and performed valuable work by laying down basic principles for interplanetary flight. He was not so much concerned with great industrial undertakings, rather he searched for solutions to apparent obstacles preventing trips to the Moon and the planets.

Another great Russian space prophet made a major study of landing on other worlds. Yuri Vasilyevich Kondratyuk (1897–1942) repeated many of the pioneering works of Tsiolkovsky and Tsander because he was unaware until the 1920s of their activity. In 1919, however, he set about the task of defining the possible ways in which space flight could be accomplished. Satisfying himself that reactive flight in a vacuum was not only possible but almost exclusive to rocketry, he argued for the gradual progression from vertical ascent to orbital flight, then launches to deep space orbit several thousand kilometers from Earth, circumlunar flight around the Moon and back, and finally a manned lunar landing. There are parallels between Kondratyuk's sequence and that probably attempted by the Soviet Union in the 1960s. But coming so soon after World War One his words were prophetic indeed. Especially in the matter of how to descend to the surface of another world.

It might be thought quite sensible to allow a hypothetical spaceship to be drawn toward the Moon by lunar gravity so that simple braking rockets could be employed to slow the vehicle to a comfortable landing speed. But there is no air on the Moon and no possibility of using aerodynamic braking such as can be employed when landing on Earth. The energy taken out of the descending vehicle, moving faster and faster as it approaches the surface, must wholly depend on slowing it physically by on-board propulsion. Kondratyuk knew that it was energy-expensive to adopt a direct descent mode. So he proposed a unique means of reaching the surface: 'Landing on some other celestial body in no way differs from a takeoff and landing on the Earth, except for the magnitude and the potential. In order to avoid too much consumption of the active substance (propellant), it is possible not to land the whole rocket, but only to reduce its velocity to such a degree that it would revolve uniformly around and as near as possible to the body on which landing must be made. Then, the non-active part (descent module) should be detached with such an amount of active substance (propellant) needed for the non-active part to make a landing and subsequently to return to the rest of the rocket. It is more advantageous not to land the whole rocket on the other planet, but to turn it into a satellite (of the planet), while the landing should be made with such part of the rocket as is required to land on the planet and to return back and join the rocket.'

Kondratyuk proposed a means of reaching the Moon that would dramatically cut the size of the rocket needed to do the job. By making the spaceship a dual vehicle, the complete assembly could first be placed in orbit about the Moon followed by separation of the two modules and descent by the second to the lunar surface. This second vehicle would leave the first in lunar orbit and rendezvous with it when the time came to return to Earth. In effect, it was like running the rocket staging sequence in reverse. Because a multi-stage rocket benefits from discarding excess structure the same would hold good for a descent to the Moon. Only the weight necessary for supporting life during the brief period on the Moon need be taken to the surface and less weight means less thrust needed to land the module. When the time came to lift off the surface, here too the thrust of the rocket motor need be sufficient only to return the part carrying the crew.

Kondratyuk remained relatively obscure until quite recently, the dominant profile of Tsiolkovsky's contribution looming above other, perhaps equally valuable, protagonists. Nevertheless, Kondratyuk's work is the first recorded proposal for serious consideration of using a modular spaceship for landing on another world. But Russian proposals were not the only written testimony to an increasing interest in the marriage of rocketry and space travel. Robert Goddard's contribution to reactive propulsion as the first man to design, build and launch a liquid rocket is well known. What is not so well appreciated is his determination to press ahead in the cause of a more protracted evolution toward manned space flight. Goddard died in 1945, long before the first practical steps toward orbital flight could be taken. Nevertheless, as early as 1932 in a letter to H. G. Wells, he wrote: 'How many more years I shall be able to work on the problem, I do not know; I hope, as long as I live. There can be no thought of finishing, for "aiming at the stars," both literally and figuratively, is a problem to occupy generations.'

Goddard was very much aware that any prophecies he made regarding space flight would be treated with the same humorous contempt that accompanied his paper of 1919 where he tentatively suggested a test flight to the Moon. Yet his private papers tell of the commitment he had to furthering the cause of space travel. Setting his mind to the problems associated with high speed flight through space he proposed a tangential approach path when returning to Earth so that, entering the atmosphere at a shallow angle, the braking effect could be gradually introduced, slowing the vehicle to a point where it could fall vertical to the surface. To protect the spaceship from the fierce heat of re-entry, Goddard proposed the use of ablative compounds, materials that char and partially burn, removing excess heat in the process. In a controlled cycle a thin covering of ablative material could be made to ensure survival of the spaceship, he thought. Both tangential re-entry and ablative compounds were used in later projects. Robert Goddard was sensitive to the ridicule of sceptics and his thoughts on space travel remained obscure for long after his death.

During the period Goddard worked toward the first liquid rocket flight, achieved in 1926, Germany was growing aware of the serious implications of rocketry and space travel. Defeated by internal unrest and the pressure of European and American armies, the peace of 1919 was an unreal sore that festered in the heat of political agitation. Diverse causes stirred the hearts of Germans everywhere and the epitomy of greatness fashioned itself in the promises of new frontiers, new empires and expansion. Space flight was a natural target on which to focus national aspirations. It was a very Germanic concept. The first exponent of literary prophecies published *Die Rakete zu den Planetenräumen* (The Rocket into Interplanetary Space) in 1923. Its author, Hermann Oberth, would inspire the men that a decade hence would begin the German rocket development program. Oberth was born in Hermannstadt, Transylvania, in June, 1894, and in common with Robert Goddard was inspired by the works of Jules Verne. Throughout the Great War of 1914–1918, Oberth pursued his idea that rocketry could serve a useful military function.

After the war he turned his thoughts to space flight and in 1922 wrote to Goddard requesting a copy of his 1919 report 'A method of Reaching Extreme Altitude.' It was this direct contact that would stir thoughts of German aspirations several years later, moving Goddard to warn about the dangers of transferring too much information to that European country. Oberth's book covered several key aspects of space travel, discussed orbital flight and presented preliminary design details of a rocket theoretically capable of probing the upper atmosphere. It was a two-stage device and reflects the level of Oberth's thinking. In this important, albeit small, publication, Oberth proposed a flight around the Moon to study the far side, the launch of spaceships into Earth orbit from where rendezvous and docking operations could be conducted, enabling propellant to be transferred from one to the other, and planetary missions performed on the principles laid down by Kondratyuk in 1919. Oberth had no knowledge of the Russian's work, however, and can be assumed to have derived his solution to this problem quite independently of others. In prophetic similarity to events that would actually evolve, Oberth suggested the placing in Earth orbit of a space station whose crews and equipment could be exchanged on a regular basis by smaller rocket-powered spaceships launched from Earth.

In 1928 a book appeared under the title *Das Problem der Befahrung des Weltraums* (The Problem of Space Flight), written by Captain Potocnik of the Austrian Imperial Army under the pseudonym Hermann Noordung. In it, Potocnik discussed the value of a space station shaped like a wheel and debated with his reader the use that could be made of such a facility placed at a great height so that the time taken to go once round the Earth would equal the time taken for the Earth to spin on its axis: synchronous orbit. Wheeled stations and synchronous orbits would be preferred concepts for space flight proponents after World War Two but in the 1920s they were a radical departure from existing ideas. The value of a synchronous orbit is that an object placed at that location, about 36,000 km from Earth, will appear to remain stationary in the sky, thus enabling a constant watch to be kept on the same area of the planet on a continuous basis. (The use of synchronous orbit was adopted by Arthur C. Clarke when he proposed, after World War Two, placing three satellites 120° apart so that continuous communications relay could be effected between any two points on Earth.)

The Potocnik space station comprised three separate sections: a wheel-shaped living quarters, 50 meters in diameter; a bowl-shaped power supply station; and a cylindrical observatory 24 meters in length, attached to the center of the wheel-shaped assembly. Spun at 1 revolution every 8 seconds, the wheel and bowl-shaped sections provided a 1 g Earth-like environment pressurized with an oxygen/nitrogen atmosphere. Use was made of the vacuum of space by providing thermal control based on the reflective or absorbent qualities of different colors: black was used where temperatures were required to reach a level necessitating preservation of heat and reflective coatings for radiation. Water would be reclaimed by distillation, a process of collecting moisture from the atmosphere and of separating used water from contaminants, with solar rays for boiling and the process of radiation for condensing. Mirrors were to be provided for focusing light into the station and for controlling the internal temperature. The observatory section was made non-rotating so that stars could be held within a fixed field of view while the spinning section was made to rotate by three large electric motors.

A year after Potocnik's book on the proposed station, a design concept which incidentally incorporated several Oberth ideas, the German theoretician published a second edition with 423 pages. Appearing in 1929 the work included detailed description of advanced space industrialization; titled *Wege zur Raumschiffahrt* (The Road to Space Travel), it proposed large orbiting space stations, ferry vehicles from Earth and special propulsive units for moving groups of construction workers from one station to the next. Oberth was an entrepreneur deeply involved with publicizing space travel and rocketry. He got involved with the Ufa Film Company under the director Fritz Lang making a movie called *Frau im Mond* (Girl in the Moon) and built a small test motor which never flew; not for Oberth the meticulous attention to engineering detail that characterized Goddard's work, but certainly the showmanship that put rocketry on the German road.

In 1929 Oberth became president of the German Society for

Space Travel, the organization that built and flew small rockets in the early 1930s, a breeding ground for the cadre of rocket pioneers soon to produce the V-2. Within 5 years the climate for individual effort and propositions for future space exploration had grown cold and inhospitable. By 1935 the German rocket effort was securely wrapped up in the Army Weapons Department, projects that were concerned with ballistic missiles rather than spaceships and space stations. But not all the inter-war activity was theoretical. In fact, the first attempt at sending a man on a rocket flight was made in 1913, when F. Rodman Law sat on a seat attached to a 3 m tall solid propellant rocket. When ignited the occupant was flung 6 m into the air, far short of the anticipated 1,050 m, as the black powder exploded. In 1931 Professor John Q. Steward of Princeton University, estimated a manned Moon flight would cost $2,000 million, although he did say he thought it would not take place for 100 years.

Others seemed to think differently. In 1933 Rudolf Nebel secured a $4,000 bank loan to build a liquid propellant rocket that would lift a man 1 km into the air, after which he would descend by parachute. The manned rocket would have been 10 m high and a precursor model was built for the Magdeburg air show in 1933. Adopting a 250 kg thrust motor built by Nebel in association with Klaus Riedel, the 10-L, or Magedeburg Startgerät, was flown on test between June and September, 1933. It was excess indulgence in practical problems of rocketry that not only prevented the *Piloten-Rakete* (Piloted Rocket) from ever being built but also the German Society for Space Travel from further activity; in 1934 the organization was disbanded by the Gestapo because of their financial difficulties and several key figures, including Wernher von Braun and Klaus Riedel, went to work for the Army's nascent rocket program.

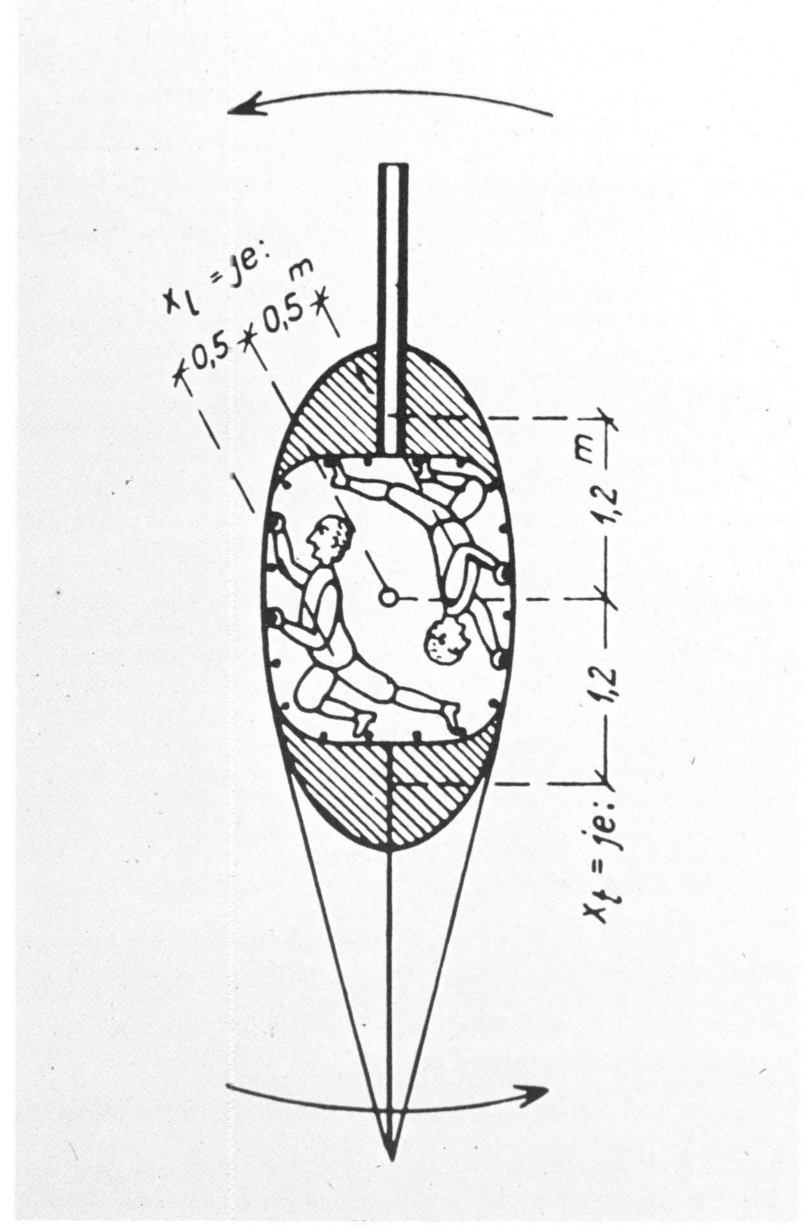

In his book *Die Erreichbarkeit der Himmelskorper* (1925) W. Hohmann suggested the application of body movement to induce rotational control of a manned spaceship.

Further east, in Czechoslovakia, rocket pioneer Ludvik Očenášek was the subject of a newspaper hoax published in a Christmas 1929 edition where mention was made of a proposed Moon flight using a rocket with eight engines. Papers throughout Europe and the United States took up the story and several hundred people wrote to Očenášek volunteering their services in what was probably the first widespread astronaut application on record. One potential crewmember, Miss Salley Gallant (*sic*) of Pennysylvania wrote that she was, 'five feet four inches tall, weight 138 pounds, am blonde, speak Polish and English, I work as a nurse, I am 20 years old, and would like to fly with you to the moon.'

But if the first two decades of the 20th century were dominated by Russian theory and modest experimentation with solid propellant rockets, and if the third was the preserve of American experiments with liquid propellant rockets and German prophecies about manned space flight, the fourth must surely be the decade in which the two opposing concepts for space flight were brought into direct competition: ballistic rockets and winged rocket flying vehicles. The outcome would dictate the way manned space flight could begin. In the 1930s two principal modes of transportation were incorporated by designers and engineers intent on providing a viable means of flying beyond the atmosphere. In Britain, a group of brilliant proponents of space flight designed a manned rocket ship adopting ballistic launch and re-entry characteristics as the first way in which man could attain orbital flight or Moon landings; in Europe an alternative was sought in which manned flight would rely on a winged flying machine capable of ascending and returning in horizontal fashion like the aeroplane it appeared to be. The contest was to be long, extending throughout the evolution and development of manned space vehicles and, putting aside the development of military rockets, the design options for manned space flight separated in the 1930s.

Formed in October 1933, the British Interplanetary Society set up a Technical Committee three years later charged with responsibility for designing a spaceship in which three men could fly to the Moon, remain on its surface for 14 days, and fly back to Earth. Work started early in 1937 and when the constitution was passed in February, 1937, it became the first committee ever to assume such extra-terrestrial responsibility. The members primarily responsible for contributing toward the design included H. Bramhill, A. C. Clarke, A. V. Cleaver, M. K. Hanson, Arthur Janser, S. Klemantaski, H. E. Ross and R. A. Smith. Written in 1938, the report on the project is a remarkable document far ahead of its day and yet appropriate to extant technology. In fact that was an important aspect of the design study; since future technology could not be predicted with certainty, materials and equipment had to be within the state of the art.

The BIS chose to employ solid propellants as the mode of propulsion and arranged 2,490 solids in six stages. The first five stages each carried 168 rockets arranged so as to form a hexagonal pattern in cross-section. The sixth stage was also hexagonal but carried 450 medium size rockets and two tiers of 600 small solids each. On top of the six stages was a habitable cabin section, a pressurized compartment circular in cross-section with a hemispherical top. In the hexagonal corners beneath the cabin and at the sides of the sixth stage, six groups of hydrogen-peroxide liquid propellant motors were attached for controlling the direction of the assembly and for vernier operation during descent to the Moon. Vernier motors are used for adding spurts of thrust to 'balance' an assembly, or for fine-tuning the precise speed of the vehicle.

Directly beneath the cabin section, 6 sets of tangentially mounted liquid propellant rocket motors would control spin during the flight and stop the spin prior to lunar touchdown. Six telescopic legs would be extended for landing on the Moon. In all, the 1,100 tonne assembly would have stood 32 meters tall and possessed a diameter of 6 meters. At launch the crew would be installed in their cabin, 56.6 cubic meters in volume, on three couches contoured to fit the profile of their bodies and in so doing prevent undue stress during the acceleration of powered flight. Several mistaken assumptions were made in the BIS spaceship design, none of them serious but all interesting reflections on the beliefs of the day. It was inconceivable to the designers that weightlessness would not have an adverse effect, so the spaceship was given a rotational spin on the launch pad that would ensure it remained on course (in much the fashion of a spinning rifle bullet) while providing artificial gravity. Also, a protective carapace was placed over the hemispherical dome of the cabin to protect it from the excessive heat built up through friction with the atmosphere on the way up; rockets would be used on the return to reduce the speed of the vehicle below that likely to cause excessive heating when it entered the atmosphere.

Because the spaceship was meant to spin all the way to the Moon and all the way back, only being 'de-spun' for the lunar touchdown, a device was needed to enable the crew to navigate by the stars. This was developed by the Committee and called the coelostat. It consisted of a system of mirrors within a box, transferring the image to the eye via a sort of optical reduction gear so that the spinning star fields appeared to remain stationary thus enabling the crew to navigate. Launch was to be effected from a site as high as possible, to take advantage of the altitude, and as close as possible to the equator, thereby making the most use of Earth's eastward spin to give the rocket a flying start. The latter has been applied throughout the space program. Situated within a flooded caisson, the rocket would have been ejected into the air by high-pressure steam, whereupon 126 of the first stage rockets would ignite for a 1 g acceleration building to 3 g at shutdown of the fifth stage. The sixth stage and cabin would be left alone, flying toward the Moon.

The cabin wall was a double hull to improve thermal insulation and to serve as a meteoroid barrier. Inside, water and oxygen was to be provided from hydrogen peroxide, food was to consist of items high in carbohydrates with cocoa and coffee for drinking – and a little alcohol to celebrate the Moon landing! A battery was to provide electrical power and waste materials were to be ejected through the airlock. Four spacesuits would be carried, one as a spare, made from leather or rubber, goggles would be provided together with sunburn lotion for protection from solar energy, and helmets would be fitted to the suits, fed with oxygen probably stored as liquid. Once out on the lunar surface, crewmen would signal their arrival to waiting colleagues on Earth by high-intensity light flashes; it was even mooted that the high point of the flight *might* include a direct voice link with the BBC! But one very important item was a tent which would be put up on the lunar surface to protect tired Moon walkers from the heat of the Sun during rest between exploratory excursions far from the cabin.

Having landed on the Moon by stopping the spin, turning round, and using the rockets in the sixth stage to decelerate to a gentle touchdown, return to Earth would be facilitated by other rockets in the same stage. Because the Moon possesses only one-sixth the gravity of the Earth at its surface, an object which on Earth weighed 1,000 kg would on the lunar surface have a mass of only 167 kg. In this way, the smaller rocket thrust of the sixth stage effectively provides the necessary energy to return to Earth. The sixth stage expends the last rockets in slowing the cabin down to enter the atmosphere where a combination of rocket and atmospheric braking reduces the speed to a value where the single parachute could be deployed, gently lowering the cabin and its crew – plus valuable Moon samples – to the ground.

The design was far ahead of anything in its day, but it made some naive assumptions about weights which would probably have rendered the entire project untenable; for instance, a three-man compartment supporting life for the three weeks involved in flying to and from the Moon and exploring its surface would weigh considerably more than 1 tonne. Also, it failed to take account of the need for additional gases during prolonged use of the airlock. Nevertheless, it represented the ballistic method of flying in space and, as such, the first comparable effort to a second mode of transportation, one that would not have carried men to the Moon but one which would have made possible the repeated, and hence economic, movement of people through space, albeit much closer to Earth. That project came about because a Viennese engineer believed the future of rocket-propelled flight lay first with an aeroplane powered by rocket motors and then with an orbital aerospace transporter, similarly lifted into space by rocket motors. Amid the profusion of projects, proposals and design concepts, the BIS spaceship and the aerospace transporter stand out as representatives of their respective schools of thought.

Born in 1905, Eugene Sänger was fired with the imaginative concepts in space travel when as a teenager he read Oberth's book *Die Rakete zu den Planetenraumen*. Studying then at the Technische Hochschule in Vienna, Sänger resolved to pursue a study of space engineering which, in the 1920s, was not the most rational decision regarding career orientation. In fact, Sänger was studying to be a construction engineer and found an easier route by first turning to aeronautics. And this is where he became the catalyst for original concepts in aerospace transportation from the theoretical products of Max Valier and Franz Edler von Hoefft. Valier was a member of the German Society for Space Travel when it formed in 1927 but in the years following the end of World War One became seriously interested in rocketry; in 1914 he attached small fireworks to a model aeroplane to propel it through the air. During 1924, Valier entered into correspondence with Oberth about his first book published the year before and from this was introduced to von Hoefft who teamed with him in objecting to Oberth's belief in ballistic rocketry as the road to space travel.

Valier and von Hoefft agreed that winged lifting devices held greater promise than the ballistic, multi-stage missiles proposed by Oberth and nearly every other enthusiast at that time. In 1927, von Hoefft took a small model rocket aeroplane for wind tunnel tests at the Technische Hochschule's Aerodynamic Institute and there was introduced to Eugene Sänger. A year later Sänger applied to join the Scientific Society for High Altitude Flight and in 1929 wrote a dissertation for his doctoral degree. The most important point about Sänger's work after this was that he never once contemplated space flight without a human participation, believing devoutly that the way to space travel was first via a rocket-powered 'stratospheric' aeroplane.

By 1933 Sänger had defined the criteria around which his rocket aeroplanes would evolve. 'One will choose for the body the shape of a projectile,' he said, 'pointed in front and blunt at the back end to give room for the exhaust velocity. The profile of the wings has to be as thin as possible, with sharp leading edges. The wing span can then be kept low because of the negligible resistence of the wing edges.' Sänger was aware of the need to make use of the aeroplane's lifting capabilities but, moving a stage beyond merely atmospheric flight, he said that the performance of the rocket plane was limited only by the amount of propellant it could carry. 'It is this restricted fuel load which prevents these planes from increasing their flight velocity up to the orbital velocity of about 29,000 km per hr. . .; the wings then don't need to provide lift anymore, and the plane circles the Earth continuously, like a moon in a free inertial orbit without needing any driving power.'

The preface to Sänger's book, published in 1933 under the title *Raketenflugtechnik* (Rocket Flight Technique), asserted that, 'This kind of rocket flight is the following fundamental step in the phase of development from the troposphere flight established during the last thirty years. It is the preliminary stage of space flight, the most powerful technical problem of our time. This preliminary stage, and the development to the construction of an orbiting space station of the Earth, is the most noble aim of rocket flight, though its realization still lies in the future.' So did Sänger's orbital rocket 'planes.'

In 1936 he was retained by the German Research Institute for Aviation to direct their rocket programs, mediocre efforts by Sänger's standard, directed at solid propellant rocket assist packages for aircraft and other minor projects. A year later construction started at Trauen of a new research facility for rocket propulsion. In five years beginning August, 1937, Sänger pursued development work toward a 100 tonne thrust rocket motor that could be used for his aerospace transporter, while simultaneously pressing ahead with studies on the aerodynamic problems associated with supersonic flight. In 1938 he prepared a conceptual study of his *Silbervogel* (Silver Bird) and adopted an aerodynamic shape with slab-sided fuselage supporting a wing each side in a structure 28 meters long, spanning 15 meters. With a single 100 tonne thrust rocket motor at the rear, burning propellants stored inside the fuselage, the Silver Bird would have been accelerated to a speed of 500 meters/second by a rocket propelled sled to which it would be attached before lifting into the air.

The sled was to be propelled by a motor generating 610 tonnes of thrust for 11 seconds. The Silver Bird's engine would only be ignited when the 'plane became airborne, burning 90 tonnes of propellant to impart a speed of 22,100 km/hr – sufficient to send the Silver Bird a distance of 23,490 km. Soon after tests began on a device designed to simulate the friction surface of the high-speed sled, war broke out in Europe and priorities changed. To this time, Sänger and his assistants had been insulated from the gradual descent toward war. Isolated geographically, the team

The British Interplanetary Society proposed in 1937 this conceptual design of a lunar spaceship propelled by solid rockets, twenty years before the first artificial satellite.

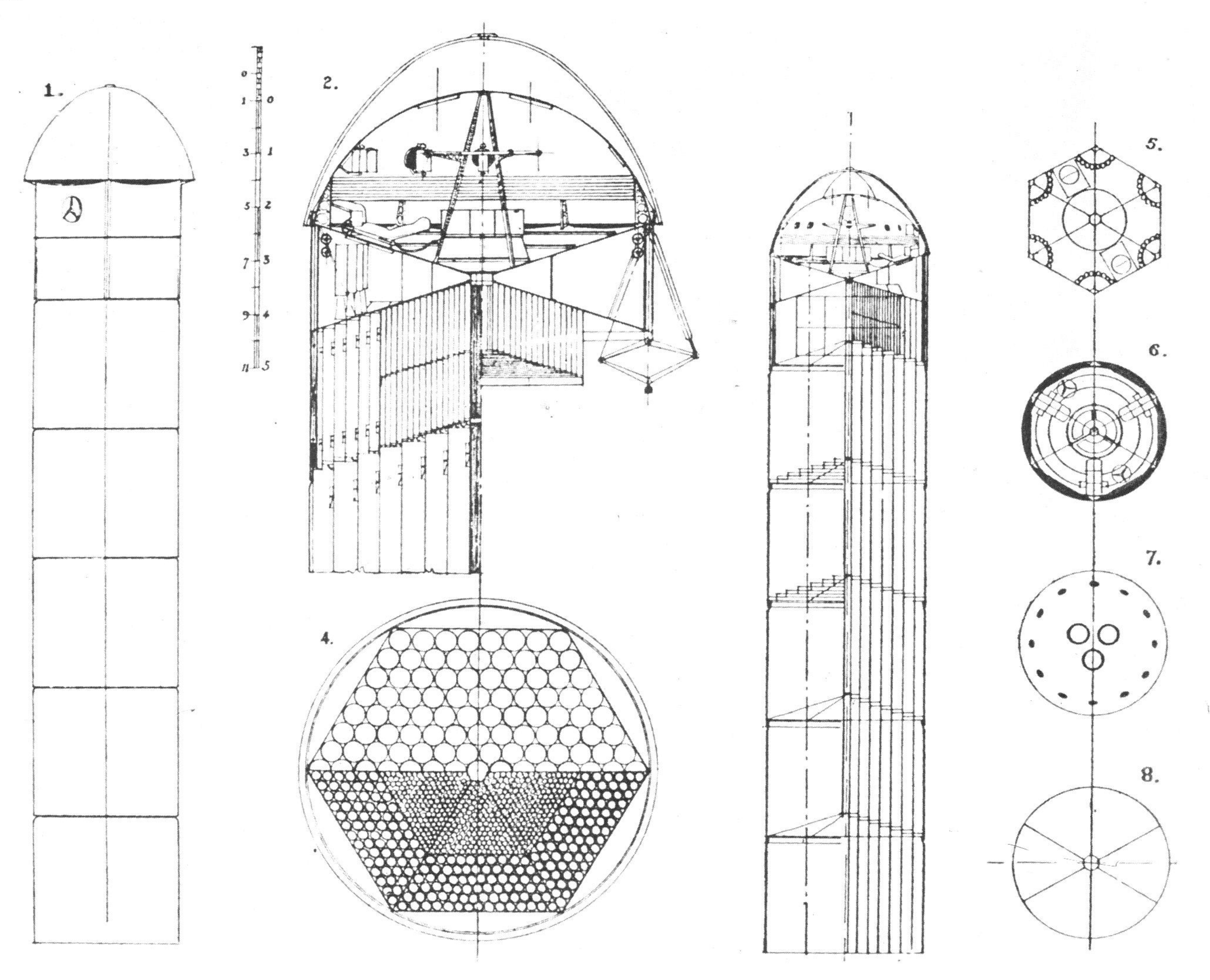

would often spend off-duty hours walking the Oertze Valley on Lüneberg Heath discussing the prospects for rocket-propelled flying machines. The sudden intrusion of politics was a stern reminder that they were in the pay of a military machine that now had very different schedules, projects that would be far removed from aerospace transporters.

In an attempt to retain the interest of the authorities, Sänger prepared a report turning his Silver Bird into a Raketenbomber, inciting the acronym Rabo by which the project was henceforth known, or 'antipodal bomber' by which it is known more appropriately. With the needs of the military very much in mind, Sänger assessed the capabilities of his Rabo to send a 1 tonne payload into Earth orbit for two and one half revolutions, 4 tonnes to a single revolution of the Earth, an altitude of up to 300 km in a ballistic flight profile, or the transportation of 8 tonnes up to half way round the world – the antipodal point. Another report, *Ubereinen Raketenantrieb fur Fernbomber* (On a Rocket Propulsion Engine for Long Distance Bombers), attempted to secure continued financial support for the 100 tonne thrust motor. It failed in this and by the end of 1942 the project was officially dead.

What had begun a decade earlier as the marriage of rocket propulsion to an orbital transportation system, decayed as a stymied project in war-torn Germany. With his wife Irene Bredt, a professional engineer constantly involved with his work, Eugene Sänger went to Paris after the war to work as a consultant for the Arsenal de l'Aeronautique. If nothing else, Sänger's ideas had given form to the theories of men like Valier and von Hoefft who, in challenging the traditional lines of thought epitomized by Hermann Oberth and his ballistic rockets, recognized the greater potential of winged vehicles for manned space flight.

But the apparent rewards of manned space flight were insufficient to stimulate politicians or military decision makers to authorize a major development effort. While a recognized asset to weapons' application, the ballistic rocket would be the way of getting men into space and only when space flight itself was seen to justify a major commerical investment would winged space vehicles again come to be funded. Rocket research in Germany was firmly in the hands of the Army under men trained and schooled in the ways of artillerymen. They saw in the rocket a quick and ready access to long range strategic bombardment. Germany was too heavily committed to other more demanding projects to take an even greater gamble and develop a long range strategic bomber. Too late did Hitler realize the error in emphasizing short range blitzkrieg techniques. And so it was that winged rocket flight died ingloriously in the fires of a defeated nation, not to be resurrected for nearly three decades. It was nearly very different. Although impressed by the performance of the V-2, Wernher von Braun was not blind to the limitations it possessed: too much weight was in structure instead of propellant; using comparatively low energy propellants the combination of fuel and oxidizer, being alcohol and liquid oxygen, was not as efficient as, for instance, hydrogen and oxygen; of purely ballistic capability, the energy imparted by the large rocket motor could be made to extend the flight path if the descending missile flew a lifting trajectory.

Von Braun was a conservative engineer, preferring the 'iron' end of engineering rather than the sophisticated innovation of 'alloy' technology – in other words, he chose to develop large rocket motors for rigid missiles strengthened by redundant support beams. But even von Braun was quick to adopt evolutionary concepts, choosing to set up a potentially rewarding development path. Although schooled in the teachings of Hermann Oberth he leaned heavily on the use of wings for extending the range of ballistic missile designs. Interrogated after the war, von Braun spoke of advanced projects under consideration by the Peenemunde team during the first four years of the 1940s. He spoke of an A-9 missile with a propulsion system similar to the V-2 but with wings that promised a near 100% increase in range to 600 km. Also, 'As a further development it was intended to design the A-9 winged rocket to carry a crew. For that purpose the rocket was to be equipped with a retractable under-carriage, a pressurized cabin for the pilot, manually operated steering gear for use when landing, and special aerodynamic aids to landing. The landing speed . . . would have been as low as 160 km/hr, as it would have contained very little fuel on landing, and would consequently have been light. This piloted A-9 rocket would cover a distance of 600 km in approximately 17 minutes.'

Von Braun indicated a conceptual path that went very close to Sänger's proposal for his Silver Bird. It was determined, said von Braun, that, 'the range of the A-9 . . . could be increased considerably if the propulsion unit were switched on only after the rocket had reached a certain initial velocity,' and that there were two ways of achieving this, one making use, 'of a long catapult with only a slight gradient, which would have given the rocket an initial velocity of approximately 350 m/sec.' Von Braun explained that the second option required, 'Development of a large assisted takeoff rocket of 200 tonne thrust, on which the A-9 rocket would be mounted, and which would give the latter an initial velocity of 1,200 m/sec.' Called the A-10, the large rocket would form the first stage to the A-9 and ascend vertically before pitching over to set the second stage on course. Separated from its carrier, the A-9 would take over to achieve a range of 5,500 km.

But it was in his summary of future possibilities that von Braun came full circle to Sänger's concept. When asked about the next step in rocketry he said that this centred on, 'Development of long-range commerical planes and long-range bombers. . . . The flight duration of a fast rocket aircraft going from Europe to America would be approximately 40 minutes. . . . Construction of multi-staged piloted rockets, which would reach a maximum speed over 7,500 m/sec outside the Earth's atmosphere. At such speeds the rocket would not return to Earth (but) . . . complete one circuit . . . in any time between 1½ hr and several days.' Clearly, this was the way manned space flight was envisaged in the 1940s; these were the topics that occupied hours of off-duty discussion at Peenemunde, talk that sent von Braun and an associate to an SS cell until his boss Walter Dornberger made a personal appeal for his release. How strange that a future world of progress and liberation should be anathema to Nazism! Needless to say, none of the proposed missile developments were pursued, or indeed studied, when the SS took over Army ballistic missile operations toward the end of the war. But it had been a long and fruitful road that began for Germany when Hermann Oberth published his first book in 1923 and inspired two decades of research and engineering that led to the world's first long range missile.

Russian developments were considerable in the field of liquid propellant motor research. The exigencies of war played their part in inhibiting actual progress however; not until late 1944 did German research and development experience the territorial intrusion that hampered Soviet activities. But efforts were made to apply rocketry to aeroplane activity, although that story is outside the realm of potential space flight operations. Not surprising, it was the German rocket achievements that led to several applications after the end of hostilities in 1945. In fact the war was not even over when, in the February, 1945, issue of *Wireless World*, BIS member Arthur C. Clarke proposed the V-2 be used to place a satellite in orbit. 'A rocket which can reach a speed of 8 km/sec parallel to the Earth's surface would continue to circle it for ever in a closed orbit; it would become an "artificial satellite,"' said Clarke, thereby coining the phrase. 'V-2 can only reach a third of this speed under the most favourable conditions, but if its payload consisted of a small 1-ton rocket, this upper component could reach the required velocity with a payload of about 100 pounds (45.4 kg).'

Clarke knew that the V-2 carried a 1 tonne warhead and his calculations showed that a rocket of this mass if placed as a second stage could effect orbital flight. The British Interplanetary Society had had close associations with the V-2. Harry E. Ross, one of the design team of the 1938 spaceship proposal, recalled with humor that toward the end of the war, Val Cleaver was reading aloud at a BIS meeting a letter from Willy Ley in America in which he doubted the validity of reports that Germany was developing large rockets when the building shook with the force of a V-2 exploding nearby! It was Harry Ross's comment that the V-2 was big enough to carry a man that sparked a BIS design study in 1946 on the pc .sibility of adapting a redundant German missile for a ballistic manned space flight. Assisting Ross in the Megaroc concept, as it became known, was R. A. Smith.

To begin with, the designers decided to strengthen the basic V-2 and extend its burning time to 110 seconds at full thrust with a further 38 sec at a reduced thrust level to maintain a constant 3 g acceleration. The additional propellant for the extended burn would be carried in tanks increased in diameter from the standard 164 cm to 218 cm. With the instrument section and warhead

By the late 1940s, rocket designs closely followed the concept of a self-propelled artillery shell.

removed, a light alloy, 586 kg, cabin would be attached to the forward end, surmounted by a fairing to smooth the airflow across the nose. This extended the V-2 in length from 14 meters to 17.5 meters, and increased the weight from 13 tonnes to 20.9 tonnes. The cabin contained a seat or couch on which the pilot would lay supine with his back to the ground. A small persicope-like device, actually a modified coelostat from the 1938 spaceship, was provided to give the pilot a view of the back of the cabin.

For launch, the Megaroc would be placed on a tower inclined 2° from vertical. Ascending for 110 sec at full thrust the rocket would reach a height of 45 km before tracking a constant 3 g acceleration to a maximum altitude of 305 km; several flights were proposed to successively greater altitudes. It is a measure of the change in technical developments that the 1946 Megaroc proposal incorporated radio communication systems and a radio beacon for recovery. The cabin would be separated from the modified V-2 during ascent, after the engine had been shut down and while the assembly was moving upward to the apex of the trajectory. The panels over the cabin would be released when the rocket reached the rarefied layers of the atmosphere. Separated by a charge of compressed air, the cabin would be stabilized by hydrogen peroxide thruster jets during its period of weightless flight until at a height of about 113 km a drag parachute would be deployed to keep deceleration at a constant 3.3 g.

The Megaroc cabin was designed for land or sea recovery, impact on the former being attenuated by a crushable section at the base of the capsule. No provision was made for a separate environmental system since the duration of the flight, measured in minutes, was insufficient to require a change in the atmosphere within the cabin. The first public airing of Megaroc came on 6 November, 1946, when the *Daily Express* enthusiastically published a report. Chapman Pincher took it up and at his prompting the concept was presented to the Ministry of Supply on the day before Christmas Eve, 1946. It got nowhere. R. A. Smith read a paper to the BIS on 7 January, 1948 and the *Journal of the BIS* published it in full four months later.

Smith and Ross were also occupied with a somewhat more ambitious proposal for a large space station in work conducted during 1946. Inspired by the Noordung space station of 1928, it was a legacy from the late pre-war years, a somewhat less relevant digression from the more plausible Megaroc. But Megaroc was limited in performance because it adopted an existing rocket as propulsion for a purely ballistic flight. In this it was precursor to a very similar project that actually took an American astronaut on the first US space flight but one which in 1946 was hardly thought relevant to space flight as then understood. In 1947 a modified proposal was issued by the BIS based on the 1938 lunar landing spaceship study by the first Technical Committee. In this, a lunar spacecraft would be assembled in Earth orbit and fired to the Moon. It reflected the comparatively advanced nature of liquid propellant rocketry by incorporating this type of motor as replacement for the solid propellant units carried by the earlier design. Conical in shape, the spacecraft would decelerate to the lunar surface on the thrust of its rocket motors before settling on four legs.

Another proposal, read at a BIS meeting in 1949, suggested the use of rendezvous techniques. Three spacecraft would be launched into Earth orbit simultaneously, the third providing propellant for the other two. One of these would fly to the Moon, go into lunar orbit and detach propellant tanks before descending to the surface. Ascending to a rendezvous with the tanks, the ship would coast home and go into Earth orbit to rendezvous with the ship left at that location at the start of the mission. Transferring to this ship, the crew would return home, leaving their lunar ship in Earth orbit. Details of the flight were worked out to take advantage of the 'staging' concept, where excess weight is discarded wherever possible. It is, in fact, the ultimate sophistication to the basic multi-stage rocket concept. In that sequence, rocket stages are discarded on the ascent but staging a manned space vehicle means that elements of the ship are left suspended in space and time until needed again for some critical portion of the flight. The BIS prosposal, described by Harry Ross, would require three ships totalling 1,300 tonnes. A direct ascent to the Moon followed by direct return to Earth would require a single ship three times that weight.

In America, engineers were inspired by the apparent progress made by German rocketeers in the war years. When von Braun and his colleagues arrived in the United States there were plenty of government and industry scientists fighting for the chance to interview the engineer, and soon offers of work were being put to the men. But it was not easy for the new arrivals. Ensconced at Fort Bliss, Texas, they were treated like aliens, made to stay on the camp and given escorts when they visited nearby towns; many American parents had lost sons, and wives their husbands, and the authorities were not completely sure of the Germans' professed disconcern for politics. Soon, however, the enormous wealth of knowledge acquired by a decade of rocket research was allowed to filter to the American aerospace industry. Von Braun stayed with the Army until his department was absorbed into the National Aeronautics and Space Administration (NASA) in 1960; Eberhard Rees remained steadfastly at von Braun's right arm throughout; Walter Dornberger went to Bell Aircraft in 1952 (after a spell in British prisons for his 'crime' of directing German research); Krafft A. Ehricke, who went from Panzer divisions on the Russian Front to Peenemunde in 1942, came to America and grew restless with von Braun's conventional concepts before moving to Bell in 1952 and Convair to work with Bossart in 1954; H. H. Koelle, ex-fighter pilot, helped form the German Space Society in 1948 and was brought to the US by von Braun in 1955.

In July, 1945, the Bureau of Aeronautics (BuA) received reports on the German V-2 and intelligence records on von Braun. A month later, Lt. Robert Haviland had prepared a memorandum calling for Navy participation in a major space station project. Haviland was but a single representative of many in the armed forces who saw at first hand the enormous potential inherent in the acquired technology; his superior, Comdr. J. A. Chambers, gave it his blessing and passed the recommendation along. In October, a Committee for Evaluating the Feasibility of Space Rocketry was set up under Capt. R. S. Hatcher. Haviland was one of its members. For the rest of the year, committee meetings held council on the configuration of the propulsion system for a hypothetical space rocket and on the possible combinations of propellant that would be most suitable. The high exhaust velocity obtained from a combination of hydrogen and oxygen was a tempting prospect and several projects using this mixture were to be proposed in the years ahead.

The Jet Propulsion Laboratory of the California Institute of Technology, a lead centre for rocket research, performed studies during the first half of 1946 aimed at presenting a viable design for a single stage vehicle burning liquid hydrogen/liquid oxygen with a high exhaust velocity – 3,240 m/sec at sea level and 4,320 m/sec in space. The Air Force, meanwhile, had been approached by the Bureau of Aeronautics for financial support to bolster flagging interest. Comdr. Harvey Hall, an active protagonist for early satellite development, approached Lt. Gen. Curtiss E. LeMay, Army Air Force (AAF) Deputy Chief, Research and Development, but received a rebuff in March, 1946, when he was told firmly of the lack of interest in a Navy proposal.

Inter-service rivalry was fierce and the Army Air Force was in no mood to cooperate with the Navy or for that matter anybody else; if satellites were on the agenda it would be an Army Air Force show with little or no glory sharing. With the Bureau of Aeronautics so obviously enthusiastic, the Army Air Force asked Rand to study satellite feasibility. Rand was set up at the suggestion of Frank Collbohm, a Douglas Aircraft Company employee, as an independent group of experts contracted by the government to evaluate defense matters and issues pertaining to national security. Douglas was given an initial sum of $10 million on 2 March, 1946; the satellite study took precedence over all other

German engineer Krafft Ehricke explains before a Senate preparedness subcommittee his ideas on how to put a man in space.

Rand work. It was completed in May, 1946, under the title, 'Preliminary Design of an Experimental World-circling Space Ship,' and presented a liquid hydrogen/liquid oxygen concept paralleled by a similar but less efficient design using more conventional propellants; the former needing a two or three stage configuration, the latter at least four.

The Aeronautical Board of the War Department examined both Army Air Force and BuA/Navy proposals and judged more work to be needed; the AAF contracted North American Aviation and the BuA/Navy team brought in the Glenn L. Martin Company. Interest in the prospect of launching an artificial Earth satellite declined in the wake of political troubles encountered by the Truman administration. By January 1947, however, Navy interest in extending the satellite project to embrace civilian as well as military activity spurred Adm. Leslie Stevens of the BuA to propose that the Joint Research and Development Board transfer the effort from the Aeronautical Board to a separate agency specifically assigned control of satellite projects. The Navy effort was getting too public and it was deemed more appropriate for a quasi-civilian organization. Then, in July, 1947, President Truman signed the Act that abolished the Departments of War and Navy and set up the Departments of Army, Navy and Air Force as separate arms of a new National Military Establishment.

By 1949 all hope of an imminent authority to construct a satellite program was gone. The Navy and the Air Force were understandably committed to space project studies. They had a tradition of aviation that went back decades and space flight was seen as an extension of this arm. For its part, the Army had the von Braun team, with ideas in abundance and research tasks extending to massive booster rockets using clustered engines, a legacy from similar work at Peenemunde. But as the 1940s gave way to a new decade it was as if the hopes for a commitment to space research were set far down the list of priorities that faced a world struggling to emerge from the dark days of war; with little or no interest in high places, military studies remained just that, and few could justify the enormous development cost of a project that seemed to vie in size with that conducted for the Atomic Bomb.

Nevertheless, with a deep concern for a new aeronautical technology, the services put money into high speed research ventures that led to rapid increases in aircraft performance. By the end of the war, it had been apparent that the reciprocating engine was outmoded by reaction propulsion adopted by the jet engine and technologists looking far ahead of existing performance capabilities recognized the need for a research tool to examine the characteristics of supersonic flight. It was in March, 1944, that the National Advisory Committee for Aeronautics (NACA) asked for money and support to develop an aircraft capable of flying through the speed of sound in level flight and researching the transonic regime. NACA was the forerunner of the space agency, NASA, and as such warrants description here.

In the early years of this century aeronautical developments were seen to be pursued more vigorously in Europe than in the United States, that historically claims to have performed the first manned flight of a heavier-than-air machine in 1903. This did not go unnoticed by proponents in America and among others Alexander Graham Bell fought for a government body to conduct research in aeronautical science and engineering, much like Britain had with its Advisory Committee for Aeronautics. The lack of American eminence in aviation during these early years can be judged from the sparse number of military aeroplanes in existence: 23, compared with more than 3,500 in Britain, Europe and Russia, at one counting in 1914. Attempts were made to rally support in the form of a rider to the Naval Appropriation Act of 1915, with Charles D. Walcott of the Smithsonian and others pressing for a national body.

In that year the Advisory Committee for Aeronautics was appointed with 12 members selected by the President all of whom were expected to volunteer their services without pay. With a first year budget of only $5,000 and a permanent roll call of 1 member, the NACA (the word National having been appended at the first meeting on 23 April) could do little to 'supervise and direct the scientific study of the problems of flight,' as intended. With America's participation in the Great War came more positive action, increased budgets, and the first research facility. Called the Langley Memorial Aeronautical Laboratory it was the only NACA complex until joined by the Ames Aeronautical Laboratory, in 1941, and the Lewis Propulsion Laboratory, in 1942. These too were authorized by the gathering war clouds that drew America into the fray during 1942.

NACA expanded rapidly, the number of permanent positions growing from 650 in 1940 to nearly 7,000 by 1945, with the budget going from just over $4 million in 1940 to more than $40 million by war's end. In 1945, NACA added a Pilotless Aircraft Research Station at Wallops Island, Virginia, for the purpose of launching small rockets on atmospheric research flights. Two years later a High Speed Flight Station was set up at Edwards Air Force Base in California to support the research tests of the Bell X-1, the product of NACA's appeal for military support of a high-speed test aircraft. Powered by a small liquid propellant rocket motor in the rear fuselage, the fat, straight-winged X-1 flew into history on 14 October, 1947, as the first aircraft to fly

Assembly of rocket stages in orbit was considered an important step in the exploration of space.

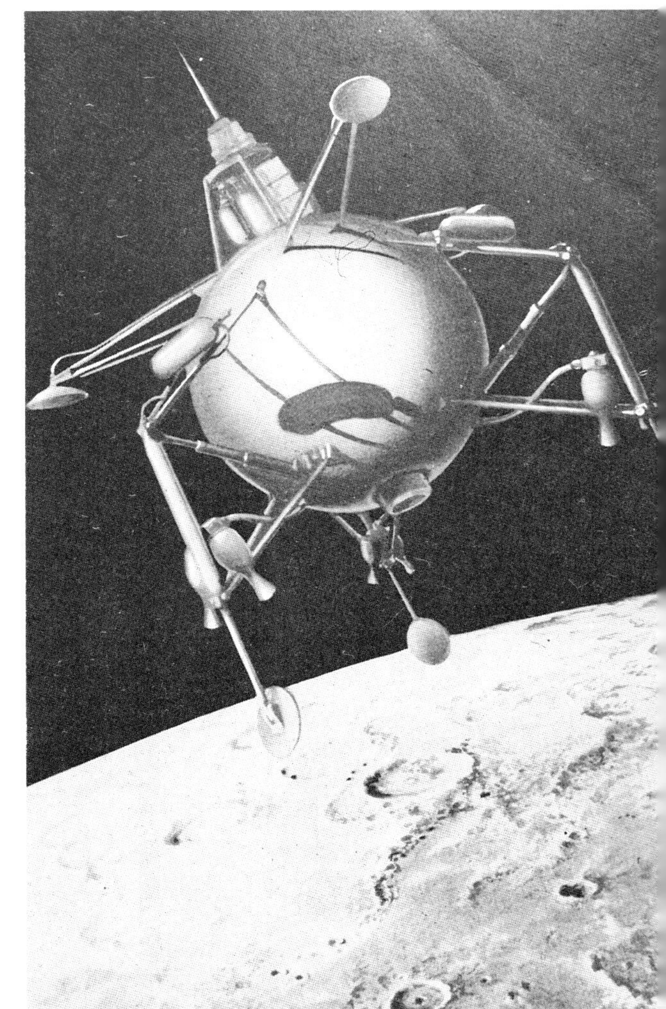

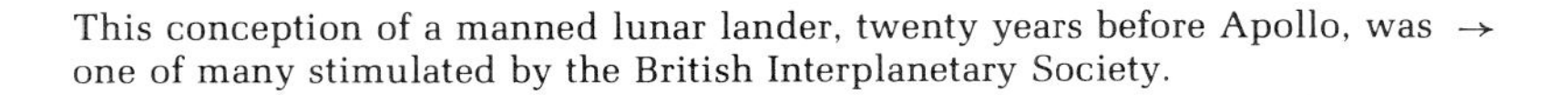

This conception of a manned lunar lander, twenty years before Apollo, was → one of many stimulated by the British Interplanetary Society.

faster than the speed of sound in level flight. The then Captain Charles E. Yeager was at the controls.

Success with the basic X-1 led NACA to proceed with several variations of the basic design, improved models ultimately known as X-1A, -B, -D, and -E models. For its part, the Navy developed the Skyrocket while the Air Force funded the X-2, a catastrophic venture that took several lives before achieving modest success. Skyrocket became the first aircraft to fly through Mach 2 (twice the speed of sound) and achieved that distinction on 20 November, 1953. By that date, NACA were looking considerably beyond the capabilities of the X-1, X-2 and Skyrocket family to a new tool for hypersonic research, rocket-powered aircraft that could fly at five times the speed of sound and to an altitude exceeding 60 km; in 1953 the altitude record stood at just 25.4 km! With public awareness of both high speed/high altitude flights and the rocket tests with redundant V-2s, carrying payloads to more than 200 km, public recognition of the validity in talk about orbital flight and space travel reached a new peak.

If the period between 1945 and 1949 can be categorized as the initial, albeit abortive, attempt at getting a satellite project funded, the first half of the 1950s must be seen as an entrepreneurial bonanza. Late in 1949 Willy Ley's book *The Conquest of Space* appeared with aesthetically attractive illustrations by space artist Chesley Bonestell and descriptions of manned lunar landings and return. In 1950, a color film appeared called *Destination Moon* in which four men traveled to the lunar surface, further popularizing the concept of space travel. The following year Arthur C. Clarke published his book *The Exploration of Space*. In 1952 it appeared in America and added further to the gathering preoccupation with flight beyond the atmosphere. It is an interesting reflection on the times that a major revival in flying saucer sightings came in the period 1947–1952. The public were excited by reports of fantastic new aircraft projects, rocket flights to the stratosphere and missiles that could carry instruments to the fringe of space; it was an expected titillation when beings from other worlds were assumed to parallel developments on Earth with trips from hostile planets.

But serious study of the direction of space activity was given in 1951 when the First Symposium on Space Flight was held at

Conceived in the early 1950s, this artistic rendition of lunar spaceships assembling in earth orbit before leaving for the Moon closely followed the plans formulated by von Braun and his German colleagues.

Winged ferry vehicles (shuttles) were presumed to be the only economic means of assembling in orbit all the many essential elements of a Moon mission.

the Hayden Planetarium, New York, on 12 October. It was in truth the first time proponents in America openly met with their former German enemies to discuss the prospects for space colonization, space law, and the patterns of future exploration, in an equitable atmosphere of public cooperation. Many people resented the presence of the Peenemunde engineers and demonstrations of unity and common purpose helped heal sore wounds. Within the ranks of the military, and at the establishments where they worked, the mood had been one of cooperation from the beginning; it was time for a public affirmation of their 'Americanization,' a recognition of their contribution to America's emerging missile program.

Attending the Symposium was editor Cornelius Ryan from the *Collier Magazine* which was sponsoring the event, with Willy Ley, Fred L. Whipple, Oscar Schacter, Joseph Kaplan and Wernher von Braun. From the meeting came a strategy for lunar exploration the like of which had never been seriously considered before. The operation would, said the participants, require three

Once begun, Moon exploration was believed to be an inevitable product of man's outgoing urge to cross new frontiers.

essential elements: a transporter for ferrying supplies from the surface of the Earth to a close Earth orbit; a toroidal space station where assembly workers and construction crews could live, eat and sleep; and a lunar landing spaceship that would carry men and supplies to the Moon for a preliminary survey. Supplementary equipment would include propulsion units for moving equipment from ferry vehicles to space station and vice versa, and a reconnaissance vehicle for flying around the Moon prior to the first attempted landing flight with men aboard.

The Earth-to-Earth-orbit ferry vehicle was conceived as a three-stage structure 81 meters tall, 19.8 meters in diameter at the base and weighing 6,350 tonnes at lift-off. The first stage would carry 51 liquid propellant rocket motors providing a total thrust of 12,700 tonnes! The first two stages were conventional non-lifting structures but the third stage was provided with wings and two small canard surfaces in the nose–lifting surfaces with a low surface area which are used for trimming pitch control in flight. Approved in the same year that the Symposium held its meeting, the Atlas – America's largest missile project at the time – had an anticipated thrust of less than 300 tonnes! The ferry vehicle concept was a veritable behemoth. After launch, the first two stages would separate in turn and fall to a watery splashdown on parachutes designed to reduce the speed of impact and so provide a reusability that promised lower operating costs. The third, winged, stage would carry on into orbit where it would deliver its cargo before returning through the atmosphere and controlled flight to a conventional landing.

The toroidal space station was to be 76.2 meters in diameter made up from 22 plastic inflatable sections to which would be attached a reflecting mirror and a power generator. Supplementary activities, like Earth observation and astronomical study, were assumed to be possible although the main purpose of the structure would be to support assembly of the Moon landing vehicles. Each of these were to be built from propellant tanks, support beams and rocket motors in configurations totally unlike the conventional rockets designed to fly through the atmosphere. Because they would be assembled in space and used only for operation in a vacuum (the Moon has no atmosphere) they could be designed around the optimum configuration for landing on and ascending from the surface of the Moon. Consequently, they were not streamlined nor did they carry protective covering in the form of skin walls.

Each Moon ship was a collection of three sets of propellant tanks: one set each for leaving Earth orbit, landing on the Moon, and ascending from the surface. Each tank was a large sphere, with a separate sphere on top serving as the pressurized compartment for crewmembers living and working on five decks, where systems equipment would be installed within easy reach. At the bottom of the sphere an airlock afforded access for five men to the exterior without depressurizing the main five-deck structure. Four landing legs were attached, one to each side of the Moon ship, with a fifth at the centre. In all, each ship would be 49 meters tall, form a square-shaped structure 33.5 meters across at the base, and weigh 3,965 tonnes. The main engines at the base of each ship would generate a thrust of 369 tonnes. For the initial expedition to the Moon, two ships would carry 20 men each with a third carrying 10 men and 27 tonnes of supplies.

Before leaving Earth orbit, even while the Moon ships were under construction, a reconnaissance survey would be made with a smaller manned spaceship flying a looping path around the moon and back to Earth orbit, going within 80 km of the lunar surface on the 5 day flight. A manned reconnaissance plan highlights the apparent lack of automation in concepts emerging in the early 1950s. Sophisticated electronics were still a generation away and few realized how precise unmanned space vehicles could be made to perform. Man was assumed to be an integral part of any space exploration plan; manned versus unmanned controversies were relegated to aeronautical parallels in which man was a necessary part of powered flight.

To get to the Moon, the three ships would leave Earth orbit by firing their respective rocket motors and descend to the lunar surface by using propellant from the second set of tanks to reduce speed. The cargo ship would provide equipment to build a lunar shelter capable of supporting the men for six weeks and mechanical shovels, bulldozers and cranes would be used to move about the Moon on geological expeditions. At the end of the six week period, crewmembers would remove from the two return ships the propellant tanks used for landing, thereby lightening the structure for improved thrust/mass ratio. The 50 crewmen would assemble in the two ships, which, with landing legs extended, would stand more than 58 meters tall, and ascend for a direct flight back to Earth on the propellant carried in the last set of tanks. These would contain sufficient fuel and oxidizer to place each ship in Earth orbit so that the winged ferry vehicles – the third stages of the three stage structures – could transfer the men to the surface.

By discarding excess weight, and consuming large quantities of propellant, the original 3,965 tonne lunar landing ships would weigh just 35 tonnes as they orbited the Earth at the end of the flight. By placing them in Earth orbit, crewmembers could return with assembly workers to refurbish the shells and fit new equipment and propellant tanks for another flight. Throughout the planned sequence, very little equipment was expendable: the first two stages of the Earth-orbit ferry vehicle would be reused, as would the winged third stage; the toroidal space station would remain in orbit, presumably for decades, supporting construction tasks and planetary expeditions; the Moon landing ships would discard their propellant tanks but the remainder of the structure, including crew sphere, main engines and support assembly, would be saved for further use.

It was a concept typical of the conventional practices advocated by the von Braun school; massive launch vehicles with little attention to the sophisticated engineering that characterized, for example, the revolutionary Atlas launch vehicle with its very thin walls that required internal pressurization to prevent the structure collapsing. Von Braun had little time for sophistication, preferring instead the sheer power of large engines and conventional propellants. If it did anything useful at all, the 1951 Symposium emphasized the dominating influence of the German approach to rocket engineering, reflected as it was in musings about the future of manned space flight.

The early 1950s were a time for gathering world wide support for expeditions beyond Earth's atmosphere. With development on guided missiles gathering momentum, projects like the Atlas ICBM coming along in the background, and rumours of an increasing commitment to rocket research in the Soviet Union, the time seemed to be approaching when an artificial Earth satellite might be feasible. At the instigation of France's Groupement Astronautique and the British Interplanetary Society, an idea originating from Germany's Association for Space Research germinated into the First International Astronautical Congress in 1959, formation of the International Astronautical Federation a year later and annual meetings in a different host country thereafter. As a means of bringing together great international names in space research it was invaluable; for the international cooperation and friendliness it fostered the IAF was unique.

Accounts of the 1951 Symposium published in *Collier's Magazine* during the following year attracted so much public attention that Walt Disney made a cartoon film embellished with improvements to the basic concept of Moon exploration worked out in the meeting. This came to the attention of S. Fred Singer, a physicist at the University of Maryland, who got in touch with Wernher von Braun. In 1953, at the fourth International Astronautical Congress held in Zurich, Switzerland, Singer put forward his proposal for a Minimum Orbital Unmanned Satellite of the Earth – MOUSE for short. This was inspired by a study conducted by the British Interplanetary Society in 1951, based on original interest in using the V-2 first encountered during the Megaroc project of 1946.

Serious application of a satellite to scientific tasks was proposed in a paper presented by Harry Wexler of the US Weather bureau in 1954 called 'Observing the Weather from a Satellite Vehicle.' George Hoover and Alexander Satin of the Air Branch, Office of Naval Research, worked out the technical feasibility of a satellite launcher based on technology then in existence. Hoover took the argument to the Army at Huntsville where von Braun was Technical Director of the Guided Missile Development Division, Redstone Arsenal. The Army team had only recently completed the design of Redstone, America's first ballistic missile based on V-2 experience but with a more efficient propulsion system. It was this missile that stood as prime contender for the proposed satellite launcher. Coming several years after the abortive attempt at Navy/Air Force cooperation the approach now being made by Navy representatives to the Army stood better chance of success if only because of the available rocketry.

Von Braun was mindful of Redstone's capacity for placing an artificial satellite in orbit and the Hoover plan meshed with his own ideas. Frederick C. Durant III, a former president of the American Rocket Society, arranged a meeting in Washington between von Braun, Satin, David Young, Singer and Fred Whipple, at which the prospect for mounting a satellite launching was discussed. They all agreed that Redstone would place a small satellite in low Earth orbit if supplemented with clusters of small solid propellant rockets. Satin, as the Air Branch's Chief Engineer, drew heavily on a book written by Kenneth W. Gatland, A. M. Kunesch and A. E. Dixon called *The Artificial Satellite*. In this, and in a paper titled 'Minimum Satellite Vehicles' prepared by the same authors and published in the *Journal of the British Interplanetary Society*, November, 1951, the feasibility and technical direction of a satellite launcher design concept were discussed. Satin visited Arthur C. Clarke at his London home in 1952 and there became aware of the Gatland/Kunesch/Dixon report. It played an instrumental part in directing thoughts about satellite launchers and focused attention on the practicality of using small available rockets suitably supplemented with solid propellant upper stages.

Called Project Orbiter, the Redstone-based proposal was put before the National Science Foundation, who agreed with its principles but did little to foster support, and in November, 1954, it was sent to the Office of Naval Research for official comment. This did lead to study contracts which continued into 1955, but by that time other propositions were looming that were, albeit indirectly, to play a leading part in getting America into space. In 1952 the International Council of Scientific Unions agreed to a proposal that the interval between major geophysical surveys of the Earth should be reduced from once every 50 years to an interval of 25 years, placing the next event in the period 1957–1958. Called the International Geophysical Year, or IGY, it would last for 18 months and embrace an international study of the Earth involving scientists from around the world.

Important to the scale of America's participation was an understanding of the amount of money that would be available. President Eisenhower provided generous support in the form of a proposed research and development budget for fiscal year 1955 exceeding $2,000 million, with Congressional approval of a $13 million sum earmarked for the IGY. As a focus for attention on possible satellite observations it was unparalleled. Matters came to a head early in October, 1954, when the Special Committee for the International Geophysical Year met in Rome, Italy, to discuss candidate projects and elicit support from participating nations. Lloyd Berkner, head of the Brookhaven National Laboratory, president of the Radio Union and vice president of the IGY Committee, called leading proponents of satellite projects to his room at the Hotel Majestic on the night before the meeting. Whatever came out of that informal gathering would determine the fate of schemes like Project Orbiter.

Despite reservations, there was unanimous approval for a formal statement of intent to launch an artificial Earth satellite as a part of the US contribution to the IGY. When it was announced, Soviet attendants at the Special Committee session the following day sat quiet and listened, offering no comment or question. On 10 January, 1955, Radio Moscow announced the imminence of a Soviet Earth satellite. But 1955 was the year in which grave misgivings were voiced over the capabilities of the Soviet Union to wage strategic war. It was to result in the several new missile projects authorized for Army and Air Force development and led to a re-examination of the satellite project within the military service structure.

By mid-year, the Army realized their Orbiter proposal was unlikely to be accepted as the IGY satellite project. Several members of Homer J. Stewart's satellite planning committee were unhappy that an American space project would ride into orbit on a rocket designed by engineers hitherto employed in Nazi Germany, a rocket bearing considerable resemblance to the V-2. By year's end, competing proposals had been rejected in favor of a Naval Research Laboratory concept based on the use of the Viking sounding rocket as the first stage of a multi-stage satellite launcher. Until the lines of demarcation were laid down in 1956, allocating responsibility for missiles with a range of less than 320 km to the Army and those with a greater range to the Air Force, military interest in rocket development was a free for all in which outstanding personalities vied for favor with Pentagon officials. Neither the Army nor the Air Force had any thought of opting out of the new missile era and with its Orbiter project moribund, von Braun immediately set about developing an almost identical missile called Jupiter C.

This was essentially the Redstone originally developed by the Huntsville team with the solid propellant upper stages dictated by Orbiter, but justified now as a test missile for nose cones designed to carry warheads. It would prove to be a valuable contribution toward the US satellite effort. As it was, Vanguard, the name given to the Navy proposal, would be developed as the IGY satellite project. For a while at least, the Army would have little or no role in the evolving space program of the United States.

But the Air Force had a more tenacious hold on its proposals. Prevented from moving ahead with its Atlas satellite launch concept as an IGY contender in direct competition with the Army's Orbiter, the Air Force turned its attention to longer range issues – projects that would not mature for at least the next five years but endeavors nevertheless which could revolutionize military flying: a successor to the hypersonic research aircraft known as the X-15, designed to succeed the X-1, X-2 and Skyrocket family.

Conceived as a joint venture between the services and the NACA, the X-15 was formally approved in December 1954 in a Memorandum of Understanding that called for NACA, Navy and Air Force cooperation. The specification had been 2½ years in the making and led, in September, 1955, to the selection of the then North American Aviation as contractor charged with design and construction of three aeroplanes. Five months later Reaction Motors was selected to build the rocket motor that would propel the X-15 to Mach 6 (six times the speed of sound) and altitudes in excess of 106 km. In 1956, the speed record stood at Mach 3.2 and no pilot had flown higher than 38.5 km.

During that year many novel concepts were mooted and several were adopted to give the X-15 a hybrid identity by seeming to place it between aircraft and spacecraft. In reality, the small jet thrusters that spacecraft would adopt for attitude changes in the vacuum of space, the side-arm hand controllers designed to provide a pilot full stick control under high 'g' loads, and the ablative coatings developed to protect the structure from excess heat, although finding permanent application with manned space vehicles were in fact developed quite apart from the teams that would soon be at work building America's manned spaceship. X-15 was a niche in aviation history that owed much to advanced aeronautical science of the day, a true aeroplane – probably the ultimate in performance.

But in 1956, the Air Force was thinking far ahead. It would be two years yet before the X-15 rolled out from its North American Aviation assembly plant, and another beyond that before the first flight; X-15 was designed to be carried beneath the inboard wing of a B-52 to an altitude at which it could be safely dropped prior to ignition of the rocket motor for a high speed or high altitude run. Yet the future role for Air Force interests seemed to lie in space. At a time when increasing emphasis was being placed on the medium and long range strategic missile, many Air Force officials foresaw an age when the manned strategic bomber was no more, the only airborne role for military pilots being a tactical or strike support function, backing up the heavyweight punch from automated systems. It was as if the space frontier was one in which the Air Force should naturally expand, the high ground for the coming decades.

H. Julian Allen played a key role in refining the principle of blunt body dynamics, an important feature of the first US manned spacecraft.

Increased concern over the Soviet developments, politically and industrially, led the Air Force to reassess its priorities and to determine a leading role in plans for manned space flight. The turning point can be dated to 15 February, 1956, when staff met at the headquarters of the Air Research and Development Command (ARDC), Baltimore. It was at this meeting that the ARDC commander, Gen. Thomas S. Power, expressed concern at the apparent lack of post-X-15 proposals; X-15 was, after all, a research tool partially funded by the services but controlled by the National Advisory Committee for Aeronautics and as such was only a proof-of-concept model for evaluating high speed and high altitude flight. The lessons from X-15 should be fed up the line to a more advanced, operational, combat aircraft, and to Gen. Power it was inconceivable that no application could be found for a vehicle beyond this test bed. Power stated his belief that manned ballistic rockets might usefully serve as long range transport carriers and called for a multi-disciplinary study of potential applications.

From this testy challenge to their ingenuity, ARDC officials came up with two proposals: a 'Manned Glide Rocket Research System,' comprising a glider and a conventional rocket booster; and a 'Manned Ballistic Rocket Research System,' being a ballistic capsule launched by a military missile. Only the latter seemed within grasp using technology available in 1956. The manned glider would be boosted to a speed of Mach 21 and an altitude of 122 km before returning to a skip-glide re-entry. It was a conceptual product of work first developed by Eugene Sänger and his Silver Bird in Germany nearly two decades before. Taken up by Walter Dornberger, the artillery officer in charge of German Army missile research, when he went to Bell Aircraft in 1952, the boost-glider was evaluated in Air Force funded studies carried out by Bell.

One design, called Project Bomi, envisaged a manned glider boosted to a speed of 16,400 km/hr and an altitude of 64 km over a target on which it was expected to release a nuclear bomb; the glider would coast to this speed and altitude after release from the booster that despatched it from the launch pad. The glider would make a 180-degree turn and use engines carried in the tail to boost itself back up to the same altitude and speed for the return flight. Tests indicated, however, that the glider would burn up during the turn and that the need for a second propulsion system for the return boost would require an initial rocket of enormous proportions. Dornberger modified the plan to allow the glider to ascend after releasing its bomb and continue on around the Earth to arrive back at its base.

Yet for all the time spent studying the boost-glider there were many unknowns associated with high-speed flight using a delta planform. Greater success had been achieved with ballistic nose cones for warheads carried on missiles under development for the Army and the Air Force and it was with this technology that the 'Manned Ballistic Rocket Research System' found great promise. It also signalled a change in terminology. To this date, manned space vehicle designs were traditionally known as spaceships, a legacy from science fiction and one that found application in the large scale concepts proposed by the von Braun cadre during the 1950s. With ballistic re-entry bodies similar to missile nose cones seriously proposed as manned vehicles the term 'capsule' was the only appropriate name. From the mid 1950s, space vehicles of this type were usually referred to as space-capsules or, more applicably, spacecraft.

It seemed as though the old Oberth-Sänger argument was emerging all over again: ballistic rocket or winged glider. For a while it did. But the Sänger designs were based on theory available in the 1930s and much had yet to be learnt about supersonic and hypersonic flight. New and unanticipated problems arose when shapes were hypothetically flown at speeds beyond Mach 7. Ballistic re-entry dynamics were better understood in 1956 because considerable effort had been expended at NACA's Ames Aeronautical Laboratory on problems associated with missile nose cones slamming into the atmosphere at high Mach numbers.

Serious study of the thermal 'barrier' began in 1952 during NACA research into X-15 shapes and led to a transformation in nose cone design that got the missile industry out of a dilemma. In selecting the best shape for the Atlas nose cone, Convair engineers chose a pointed configuration that seemed the best computer solution. The design team envisaged separation of the nose cone from the main body of the missile soon after the engines shut down but were faced with an atmospheric entry speed of about 24,000 km/hr from peak altitudes of between 800 km and 1,500 km. The build up of heat on the nose of the warhead would reach stagnation temperature of about 6,600°c, far beyond the survival capacity of any known material. This was the dilemma: the best computer solutions – essentially extrapolations into a totally unknown environment from as much meager factual data as existed at the time – dictated a pointed nose cone facing forward; supersonic wind tunnel tests at the Ames Laboratory revealed serious thermal problems with this concept.

A major drawback at the time was that everyone thought in terms of streamlined shapes. Aerodynamic experience had moved rapidly through the speed of sound and by 1956 X-series research aircraft had investigated flight regimes through Mach 2 and Mach 3. Nothing the engineers seemed to see could change the view that for extremely high Mach numbers, a pointed, streamlined configuration was the best. In actual fact, it was not. And ramifications of the solution would extend through any concept of space travel, for a spacecraft returning from Earth orbit encounters a more severe environment than a ballistic missile nose cone; one returning from the vicinity of the Moon would encounter an intolerable regime.

Harry Julian Allen, head of the High Speed Research

Division at Ames, found an answer. He reasoned that the kinetic energy of an incoming nose cone is converted into heat, and that this consists of two types: heat generated by compression of the shock wave in front of the cone, and heat generated at the boundary by compression but, to a greater degree, by friction with the skin – viscous shear. Allen knew that the heat from shock wave compression lay outside the boundary layer and was of no consequence to the temperature of the cone's surface because it could not move across that layer. Pointed cone shapes were heated to excess because the viscous layer generates most of the heat and because it was attached to the surface of the cone the material increased in temperature beyond a survivable level.

In other words, Allen simplified the picture to one in which a strong boundary layer generated heat which was directly conducted to the surface of the cone, with heat in the shock wave unrelated to the temperature of the falling body. Against all convention, he turned the problem around and found a solution. By making the body a blunt-shaped configuration, the shock wave would be greatly strengthened, and since the amount of energy in the falling body depended on velocity and angle of entry, more heat generated in the shock wave would mean less in the boundary layer. Moreover, with a blunt shape, the shock wave would at no point be attached to the entry body as it is where a pointed cone moves quickly through the atmosphere. This is why the tip of a pointed missile is hotter than the skin further back. Also, a blunt-faced entry body would more evenly distribute the heat reaching the surface and render it less prone to local hot spots.

Assisted by Alfred J. Eggers, also of Ames, Allen proved that the blunt body shape generated no more heat than a pointed cone, the quantity generated depending in fact on the rate of deceleration and not the shape, or indeed, its mass. This had profound significance for space transportation in that bodies returning from the Moon would experience the same thermal environment irrespective of their size or weight, so long as they adopted the same shape. There were some modifications to this, when it was found that the rate at which heat built up in the entry body could vary according to the mass and this led to a more pointed profile for large warheads.

Much of the work defining the blunt body entry concept was carried out between 1952 and 1955. In fact it was need for a full size test vehicle to check calculations and wind tunnel experiments that spurred the von Braun team to propose and build the Jupiter-C (Composite test vehicle) from the basic Redstone. It was flying by 1956, carrying models of warheads that would lead to new designs for Atlas, Titan, Thor and Jupiter missiles. But switching the source of heat generation from boundary layer to a 'stand-off' shock wave did not completely solve the problem. It presupposed the availability of thermal protection materials only marginally capable of surviving the environment.

Two approaches were available: heat-sink, or ablation. In the former, a heavy conductive material like copper or beryllium was made to accept and absorb the heat without losing structural integrity; in the latter, material that would partially burn and char carried heat away as soon as it was absorbed, in effect providing a controlled burning cycle at a predictable rate. Heat-sink concepts were, by virtue of their density, much heavier than the comparatively light ablative compounds. Through the period 1955–1957, considerable effort was expended to seek the better method until repeated tests favored the ablative concept for military missile nose cones. So it was that when the Air Force examined manned gliders and manned ballistic re-entry vehicles in selecting an operational successor to the experimental X-15, experience and know-how favored the ballistic re-entry vehicle with a blunt body shape.

But there were mission requirements that seemed to change the picture in favor of the boosted glide vehicle. Air Research and Development Command was interested in moving men and cargo long distances in short periods of time, and of using a similar system for placing a warhead on target across the other side of the globe. Lifting vehicles, re-entry bodies with delta wings, would fly farther for a given boost than the purely ballistic, high-drag, capsule. Glide vehicles could more effectively control the point at which they descended through the atmosphere, provide greater accuracy in landing at the end of a flight and modulate the rate at which they heated up by controlling the rate of descent, a key factor in the amount of heat generated. Principles of blunt body dynamics would still be applied to winged glide vehicles operating at the fringe of space; when they re-entered the Earth's atmosphere they could pitch up to present the large underbody to the direction of travel, effectively moving the shock wave comfortingly ahead of the structure and minimizing the heat load.

Such vehicles would operate at the fringe of orbital conditions, some mission roles dictating a semi-ballistic flight profile where the vehicle boosted itself to a high fraction of orbital speed but without going into orbit, others requiring the glider to ascend into space, above the atmosphere, to attain a speed that kept it circling the Earth for days before returning back through the atmosphere. Orbital conditions were not difficult to achieve in concept, but in reality they demanded precise control over the launch rocket and careful tracking, factors which were met with increasing difficulty by the Vanguard team readying America's first artificial satellite. In principle, orbital mechanics follow closely the laws laid down by Johannes Kepler in the 17th century and rely on forces identical to those that regulate the motion of planetary bodies around the Sun.

To simplify the reasons, a satellite stays in orbit because it is continually falling around the curvature of the planet – in this case, Earth. Objects close to the surface are pulled toward the center of the Earth at an acceleration of 1 g, or 9.8 meters per second (9.8 m/sec^2). But gravity declines on the inverse square law, so an object twice as far out as another will experience only

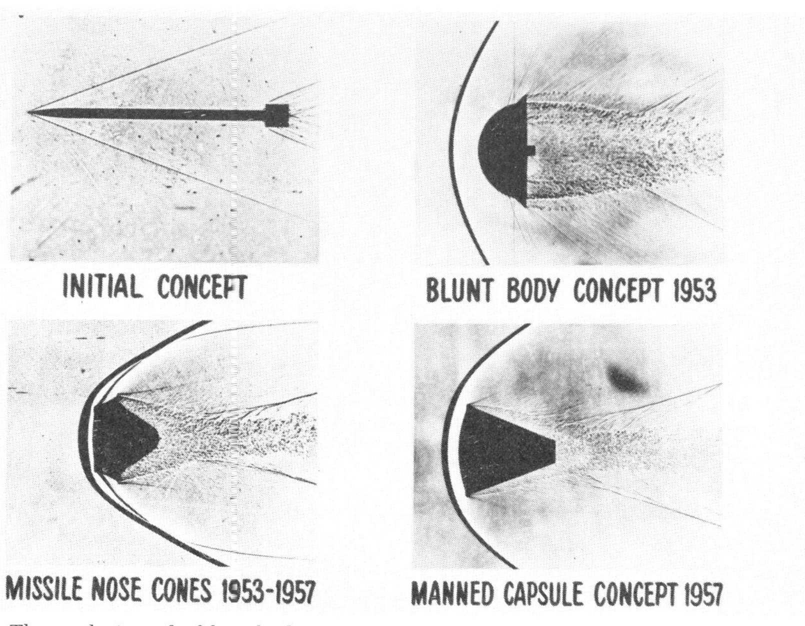

The evolution of a blunt body concept allowed the hot shock wave during re-entry to be removed from the face of the incoming mass, thus reducing to meaningful level the temperature on the exterior skin.

Al Eggers supported H Julian Allen in proving that temperatures affecting a → re-entry body would be reduced by separating the boundary layer from the structure.

one-quarter the force experienced by its closer partner. Therefore, the force of attraction decreases on the inverse square of the distance. But returning to a low orbit case, a satellite will be pulled toward the Earth at a rate of 4.9 meters in the first second. Now it happens that the curvature of the Earth is such that the surface falls 4.9 meters in 8 kilometers of apparently horizontal distance. So, if an object comparatively close to the surface of the Earth was made to move at a speed of 8 kilometers per second (28,800 km/hr), it would fall 4.9 meters due to the attraction of gravity yet be no nearer the actual surface than it was a second before – and so on, and so on, until it made one orbit of the Earth, there to start several orbits unhindered.

In reality, the atmosphere acts as a perturbing influence, extending several thousand kilometres out into space. In measurable terms, however, the atmosphere is negligible beyond an altitude of about 130 km. A minimum orbit could be considered to have a mean height of, for instance, 133 km. At this altitude, orbital speed would be 28,184 km/hr and an object would make one revolution of the Earth in 97 minutes. Because gravity declines on the inverse square of the distance, synchronous altitude is 35,800 km where average velocity would be just 11,060 km/hr and the period of revolution, at 24 hrs, is equal to the rotation rate of the Earth. Hence, an object placed here would seem to keep track of a fixed point on the surface if the object were in an orbit in the plane of the Earth's equator.

So, in considering a trans-orbital boost-glider, the ARDC envisaged a winged re-entry vehicle flying to a high fraction of orbital speed, or actually achieving orbital speed, and reaching altitudes above the sensible atmosphere, or in the vacuum of space proper. In March, 1956, the Air Force set up two study projects – one each for the boost-glider and the ballistic re-entry capsule – with the promise of financial support. The 'Manned Ballistic Rocket Research System' study led to proposals from two companies by December, 1956, and eventually 9 more contractors responded. It was the beginning of a two-year effort that did much to contribute toward the eventual authorization of a US manned space program. Money was not forthcoming, as had been hoped, and the ARDC spent much of its time encouraging private investment from companies intent on securing a piece of the action, once there was action in which to participate.

During 1957, the participating Commands encouraged interest up the Air Force infrastructure, citing contractor studies as proof of concept, and bringing medical evidence to prove that human physiology was capable of enduring the rigors of high-speed/high altitude flight. ARDC's Director of Human Factors, Brig. Gen. Don D. Flickinger, testified to the available data on human response and endorsed the technical claims. Considerable attention had been given to aviation medicine and many thought space flight would merely be an extension of high performance flight in the atmosphere. In 1934 the Aero Medical Laboratory had been set up at Wright-Patterson Field, Ohio, where research was conducted into physiological response to adverse flying conditions and extreme environments. A leading figure in experiments concerning high acceleration was Lt. Col. John P. Stapp who rode rocket-propelled sleds modeled on the sled-booster designed by Eugene Sänger before World War Two.

On 19 March, 1954, Stapp reached a speed of 677 km/hr, increasing this to 808 km/hr on 19 August, and 1,017 km/hr on 10 December. Wired with medical monitoring sensors to record important physiological data, Stapp experienced nearly 40 g momentarily as the sled was brought to a rapid halt at the end of the 1.1 km run. Two months later, a chimpanzee experienced a 247 g load for a fraction of a second but in 1958 the human record was snatched by Capt. Eli L. Beeding with a deceleration of 83 g for 0.04 sec at a rate of 16,000 g/sec. This meant that for 0.04 seconds, the 76 kg man weighed 6.3 tonnes! Tests like these, and careful observation of the physical response, led to a general belief that a human being could withstand 12 g for 4 seconds, 8 g for nearly one minute, 5 g for two minutes and 3 g for five minutes, without any serious impairment to response. Tests with centrifuges demonstrated that trained service personnel could probably withstand 16.5 g for several seconds without repercussion.

While, between mid-1956 and the end of 1957, the Air Force moved slowly ahead with conceptual studies of boost-glider and ballistic capsule, the Army began studies of a clustered booster capable of generating thrusts of about 680 tonnes from several liquid propellant rocket motors at the base. The emerging generation of Atlas and Titan Inter-Continental Ballistic Missiles had opened new possibilities which the Air Force was pursuing with its dual study tasks; the Army considered its role to be closely associated with the colonization of the Moon, or indeed the planets, and moved actively ahead with schemes rooted deep in the formative work of German rocket engineers.

Clustered motors, grouping together in one stage several small engines which collectively could deliver several hundred tonnes of thrust, had been studied in detail during the war. Von Braun had pursued a concept where six modified V-2 motors would form the first stage to a multi-stage launcher but the exigencies of war prevented further analysis. Since 1955, the Rocketdyne Division of North American Aviation had been looking at the feasibility of high-thrust liquid motors, determining as early as 1955 that a single engine of 680 tonne thrust was possible. In February, 1956, the Redstone facility had been re-named the Army Ballistic Missile Agency (ABMA), with von Braun leading conceptual designs that took in both the clustered approach and the high thrust engine. In this way, the Army sought to preserve a role for itself by leapfrogging the coming generation of Air Force ICBM's and their 160–190 tonne thrust capability, to provide a generation of massive booster rockets applicable to a vigorous space program.

Army funds were allocated for research and development, but to the von Braun group it was the initial push that could set off an avalanche. First, Jupiter and Thor type engines would be clustered together to provide a total thrust of 680 tonnes. Then, with the availability of the single 680 tonne thrust engine, that too could be clustered to provide launchers with thrust capability measured in thousands of tonnes. There was certainly a momentum of sorts at work during 1957. For much of that year the Army worked on its National Integrated Missile and Space Vehicle Development Program with projected introduction of the clustered booster in 1963 in a test project embracing 30 flights carrying manned and unmanned capsules. The heavyweight rockets using the 680 tonne thrust engine would be central to 'space exploration and warfare.'

But 1957 was a year of change, for both the direction of ballistic missile activity and for what would rapidly become the nascent space program. In that year, the Army began tests with its Jupiter, the Air Force got under way with Thor and Atlas flights, and the Soviet Union first flew its new ICBM – the SS-6 Sapwood. If 1955 had been a record year for setting up new missile projects, 1957 was certainly the year of maiden flights. It was, more significantly, the year in which the Soviet Union beat America into space by sending into Earth orbit a satellite weighing 83.5 kg called Sputnik 1. Before the year's end, a second satellite, this time weighing 508.5 kg joined it in orbit carrying a dog named Laika.

Although the Russians had been sending animals on ballistic rocket flights for some time, the first successful space flight for live creatures came on 20 September, 1951, when a monkey and 11 mice rode on a straight-up/straight-down flight atop an Aerobee sounding rocket. The first animal to ride in a rocket was Albert I, launched in June 1948, but this died of suffocation in the nose of a V-2 in which it was strapped. Another flight, with Albert II, was made in June 1949, but the monkey died on impact. Another failed to get off the ground on 16 September, when the V-2 blew up on the ascent. The next attempt was made in December 1949, again with a monkey that died when the rocket hit the ground. On 31 October, 1949, a mouse had been sent on a ballistic trajectory but this too perished. A parachute failure on 18 April, 1951, killed another monkey and several mice, before the successful flight in September.

All these attempts were made with one purpose: to study the effect of exposure to solar radiation from high altitude, and to determine the effects, if any, of weightless flight; a ballistic trajectory would offer a few minutes weightlessness. When it came, the Soviet dog flight into orbit was a significant indication of the direction of the Russian activity. Manned space flight was obviously a key feature of future plans, and the weight of Sputnik II reflected the lifting power available to Soviet engineers. Vanguard was capable of lifting a small 10 kg satellite, but for its first flight would be restricted to a payload weighing less than 2 kg. Manned space flight in the United States had been an adopted concept. The administrative machinery simply did not exist in which to promote a serious attempt at manned spacecraft design.

The Air Force adopted the concept and found a military need, but this was more for the sake of keeping up with rapidly changing technology than for any intrinsic value space flight was seen to possess. For its part, the Army believed, largely at the persuasion of von Braun, that space flight would inevitably emerge from some none too clear military requirement and that big boosters and large rocket motors would be needed ultimately. Throughout the 1950s, up to the launch of Sputnik I on 4 October, 1957, studies were performed as a by-product of work essential to perfecting the much needed nose cones for ballistic missiles. There was certainly a serious intent to pursue a manned hypersonic vehicle – lifting or ballistic – and to that end the Air Force was encouraging private support and contractor tests. But the Air Force officially turned away from sanction, or indeed financial provision essential to continued study; the Weapons Systems Plans Office at ARDC never did get the $200,000 it was promised for internal work on the glider versus ballistic capsule controversy.

There is little evidence to suggest that Russian scientists, engineers, or military personnel were any more successful at impressing their political masters with notions of space flight or hypersonic gliders. On the contrary, there is every indication that serious thoughts of manned space flight arose from the obvious political advantage in Soviet lifting capacity. Having prepared and built larger rockets than their American contemporaries, the Soviet politicians recognized the sizable lead they commanded. Sapwood had a thrust of more than 500 tonnes compared to 160 tonnes for Atlas; the payload difference was enormous. With Sapwood capable of accepting an upper stage that would, in turn, permit a payload sufficiently large to carry a man, the emerging space program could be an effective challenge to American technology.

US space policy late in 1957 reflected the mood of the incumbent President, Dwight D. Eisenhower, who, in ordering an increase in missile development as a result of the 1955 intelligence reports, was careful not to overdo the job of putting America in a safe military posture. Eisenhower's strategy was one of cautious reciprocation, building a military capability equal to but not exceeding the task of defending prime US interests. He was concerned to keep apart the increasing trend in the mid-1950s to lump space projects and military interests in one financial research and development bag. When the President gave his formal approval to plans for the US satellite, inspired by the International Geophysical Year, it was made clear that whichever plan succeeded, it would have to be quite apart from missile development and prime service responsibility for the defense of the nation and its interests.

Vanguard was a project from the Naval Research Laboratory, but one which retained a strong scientific bias and which gave no succour to service chiefs who champed at the bit for money to spend on big space operations. The Navy was the only service not openly committed to a major missile development program, although it was tied in with the Army Jupiter project for a while and would shortly develop its own strategic nuclear strike capability in the form of undersea missile arsenals contained in submarines. In 1955, however, the Navy was safely removed from visible involvement with the strategic interests that could, in the light of inter-service rivalry, be seen as an encumbrance. In any event, the satellite effort was a comparatively small affair and hardly impressed the Air Force and Army planners who thought in terms of orbiting space stations and military colonization of the Moon.

Eisenhower had a regard for science that went back to a response in July 1954 to the Office of Defense Mobilization (ODM) Science Advisory Committee request for a searching analysis of defense technology and the future patterns of research and development. The technical study was chaired by James R. Killian, Jr. Former President Truman had brought Killian into the ODM Science Advisory Committee in 1951 and the Technological Capabilities Panel he was to chair for Eisenhower provided stimulus to direct science advisory functions in the White House. The ODM-SAC had been appointed as an ad hoc committee in August, 1948, by Dr. Vannevar Bush, with Dr. Irwin Stewart as chairman. It was to be a temporary forum for discussing how best to meet technical challenges to the United States should that country once again be faced with an emergency such as the one that confronted it at Pearl Harbor. In the event, the ODM-SAC was retained as a permanent watchdog but with little authority to approach the President direct. In fact, the White House strongly resisted attempts to get the Science Advisory Committee within its orbit and when permanently set up in April, 1951, the SAC was kept at arm's length from the administrative functions of the Executive Offices.

Its first really challenging task was given to it by Eisenhower when he set up the Technological Capabilities Panel. Under Eisenhower, expert advice was sought and obtained in a way that would have been impossible in Truman's administration. When the TCP report appeared in early 1955, it warned of the price for vacillation and pointed to several areas of deficiency within which the Soviet Union could exploit to their own advantage a structured attempt to outdo the United States. But if Eisenhower's response to rapidly changing events was to sit back and look at the total situation, preferring not to be diverted from a carefully considered assault, this much lauded stability of purpose was to be the very instigator of the vacillation the TCP report sought to guard against.

One example, that had direct effect on the post-Sputnik attitude to space, was the leak of a report in 1957 concerning US civil defense in the event of a nuclear war. Earlier in the year, the Civilian Defense Administration recommended expenditure of $40,000 million for bomb shelters and other civil defense items. At the behest of the President and the National Security Council, the Science Advisory Committee was requested to look into the civil defense posture by forming a task force that would, in effect, report to the President on America's civil preparedness for total war. Known as the Gaither report (H. Rowan Gaither, Jr., had been appointed co-director by James Killian but fell ill although he did participate in presenting the report to Eisenhower), it portrayed a disturbing picture of almost total lack of defense for private citizens caught up in a nuclear holocaust. In fact it did little more than confirm the official line, that a nuclear war was untenable and therefore not probable because it would be unwinnable and hence unsurvivable; the Russians had a very different view, but that is outside the scope of space developments.

The important link was that the Gaither report appeared in leaked form in the *Washington Post* which incited Congressional demands for publication in full of the original text. The National Security Council had been briefed on the report on 7 November 1957, and the *Post* printed a summary on 20 December. Sputnik's I and II had been launched on 4 October and 3 November; America's Vanguard had blown up on the launch pad at a satellite launch attempt on 6 December. As a note for technical digression, the Army's Jupiter C (ex-Orbiter, ex-Redstone) missile was approved as a back-up to Vanguard and the von Braun team were urged to lose no haste in preparing a different satellite for launch, openly competing with Vanguard. The launcher would be known as Juno I, although it was identical to Jupiter C. But the important connection is that Eisenhower was challenged by internal and external forces which pointed to public conclusions unavoidably siding with critics of the administration. Favorably disposed to maintaining the status quo, the President was forced to reassess his policy in the light of pressure building up within his own administration for a change in orientation.

Openly, the President maintained a cool, calm attitude toward Sputnik, pretending that he found it difficult to understand the general furore over Soviet success. No one in the American administration had recognized that the apparently simple act of launching small satellites into orbit would create such international surprise. But there was good reason to believe that it was such a shock only because the Russians had done it first. Ravaged by a war that took more than 20 million lives, the Russians were thought to be culturally defeated and industrially stagnant. If Vanguard had ascended into orbit first, as everyone expected it to, the shock of a new space capability would have been diluted.

Eisenhower had a profound awareness of being at the helm and knew that when all around was in disarray, he was expected to keep cool. This may have been responsible for his seemingly negative attitude toward Sputnik. In reality, the pent up emotions of two Soviet successes, an American failure, and a secret civil defense report leaked to the press, caused Eisenhower great personal annoyance. And this was reflected in his attitude toward confidants at the White House where he felt able to uncouple his emotions from public exposure. Few who were close to him in late 1957 are able to find a parallel for his anger and disappointment over the direction in which events seemed to be moving. But if publicly calm and unconcerned by claims that Sputnik represented a 'threat,' politically he needed to make a move that would at one stroke appease the critics and not unduly upset the balance of international affairs.

Four days after Sputnik I bleeped its way across the sky, Eisenhower met with advisers and then with representatives of the science community outside the White House, including Dr. Detlev Bronk, president of the National Academy of Sciences. Dr. Bronk assured the President that it would be wrong to respond in like kind to every Soviet initiative and that a degree of measured competition was perhaps desirable. He also brought up the issue of a full time science adviser within the White House and recommended the President consult with I. I. Rabi, then chairman of the ODM Science Advisory Committee. Eisenhower met the ODM-SAC on 15 October in pursuit of a regularly scheduled appointment, a meeting which was, nevertheless, deeply concerned with matters raised by Sputnik.

The President heard of the Committee's concern over the pace of developments in the Soviet Union, over dangers that might exist in proceeding along a path uninfluenced by competitive events, and of their feeling that the President himself was too remote from science and scientists. This, once more, opened the question of a personal adviser. From this meeting came a clarification in the President's mind about the steps he must now take to orchestrate a response to Sputnik and, rather than using it as a prime cause of dramatic change, set up an administrative link with the scientific community to prevent the United States falling out of step with global developments in science and technology. On 7 November, 1957, the President announced that he had asked James Killian to fill a new office formed from the ODM-SAC structure, a group that would henceforth be known as the President's Science Advisory Committee, as Special Assistant for Science and Technology. Killian would occupy an office within the White House, a transfer effective from 29 November. The PSAC was to have 17 members drawn from wide fields of experience in science and administration, a group who were unanimous in choosing Killian as their chairman.

On the other side of the political fence, things moved differently. If there was one man who consistently influenced the direction of America's space program more than any other person it would be Lyndon Baines Johnson – not just an American, but a Texan as well. From his ranch in Texas on the night Sputnik I first flew over, Johnson came to recognize the tides of change: 'That sky had always been so friendly, and had brought us beautiful stars and moonlight and comfort and pleasure; all at once it seemed to have some question marks all over it because of this new development,' he said when recalling thoughts that came to mind of that night.

Johnson – LBJ to many – was born in Gillespie County, Texas, on 27 August, 1908, eldest of five children to parents descended from old trail pioneers. The family were comparatively poor and the young Lyndon had little incentive to search out a higher education; oldest in the group of children he had responsibilities that inevitably led to a lack of adequate study time. With friends he made his way to California, spent some time eking out a living, returned to Texas without funds and went to work on road construction sites. At 19 he was persuaded by his mother to enroll for a high-school teaching certificate at Southwest Texas State Teachers College from where he emerged with a diploma three years later. The astonishing lack of formal education seemed not to hinder the dynamic young man nor to deter him from a career in politics.

Johnson married Claudia Alta Taylor in 1934 while serving in Washington as an aid to Representative Richard M. Kleberg, a post he was requested to fill following energetic contributions to the election campaign. He stood for election to Congress in 1937 to fill a vacancy made possible by the death of the incumbent and was returned for a usual term the following year. He stood for a seat in the Senate four years later, lost it to his rival, and joined the Navy in 1941 when Japan attacked Pearl Harbor. After the war, Johnson held positions on the Select Committee on Postwar Military Policy and on the Joint Atomic Energy Committee before again running for the Senate. In a close battle with Governor Stevenson, Johnson squeezed in with a majority of 87 votes and went to the House of Senators in 1949.

Over the next two years Johnson played at the feet of leading Senate figures, securing favored status to become the Democratic whip in 1951 and party leader in 1953. That year, Eisenhower brought in a Republican majority which relegated Johnson to minority leader. His credibility within the party can be judged from the expertise with which he marshalled total Democratic support for a censure motion restraining Sen. Joseph McCarthy. In 1955 the Democrats returned a majority, propelling Johnson to Majority Leader in the Senate. He was to remain in this position until elected Vice-President in 1961, but in October, 1957, he bristled with indignation over the Soviet spectacle. Not for him the considered and measured response of the Republican President.

After watching Sputnik I move rapidly overhead he returned to his ranch house and telephoned several advisers and Senate conferees to discuss strategy. The following day his charged minions were scurrying to and fro gathering information on the Russian program. Back in Washington, Johnson talked with the senior Republican on the Armed Services Committee, Styles Bridges, and the chairman of the same, Richard Russell. Within a week of the Soviet launch, Johnson was chairing the 'Inquiry into Satellite and Missile Programs' he had personally set up to be conducted by the Senate Armed Services Committee Preparedness Subcommittee. For several years Johnson had had his sights on the White House. As leader of the majority party in Congress, as head of the Democratic opposition, as chairman of armed services review committees, LBJ was not about to pass up the biggest political windfall yet to come his way.

With all the resources at his disposal he set up office in conflict with the administration and, imbued with a sense of purpose motivated by a desire to put pressure on the Republicans, resolved to push for an active space program. Toward that end he set out on a public campaign, speaking in many parts of the United States in support of a strong and independent approach, while the Senate inquiry sailed on through 20 meetings between the end of November and the end of January, 1958. Soon, Johnson became fired with space to the point where it became a personal campaign. His attitude was genuine and to a large extent a product of everything he saw for America's future. It also appealed to his sense of crusade and by early 1958 Johnson was rallying support for the Democratic cause, propelling himself forward on a wave of public recognition that would, he hoped, lead directly to the party nomination for President.

For its part, the military found support from the Russian initiative. The Army had its Huntsville facility under Maj. Gen. J. B. Medaris, commander of the Army Ballistic Missile Agency, with strong interests in preserving the 3,000 engineers marshalled by ex-Peenemunde rocket technicians. The Air Force was wrestling with the Army for control of the Jupiter and Thor projects and turning more forcefully to the post-X-15 program which, even by late 1957, was still a paper performance package. On 5 December, 1957, it was announced that an Advanced Research Projects Agency (ARPA) would be set up within the Department of Defense to look after military space interests by conducting pertinent research. It would have authority over coordination for major projects like the big booster studies under way in the von Braun camp, empowered to bypass the Army Staff in directing work at the ABMA.

The ARPA was effective from 7 February, 1958, one week after a Juno I launcher sent Explorer I, the first American satellite, into Earth orbit. It was a contest of roles. No-one really bothered about function or implication. Russia had put a satellite into *orbit* and only a US satellite in *orbit* would restore the credibility of American technology. Because of this a significant launch went almost unnoticed and even today is ignored in the annals of history concerned with early space achievements. On 16 October, 1957, just 12 days after Sputnik I ascended, an Aerobee rocket was launched from Alamogordo, New Mexico. Accelerated to a speed of 54,000 km/hr, the payload of small artificial meteors left Earth's gravity field forever, becoming the first objects to escape the pull of the Earth and go into orbit about the Sun.

Outside both military and political fields, however, the National Advisory Committee for Aeronautics had its own idea of coming trends. As an essentially research-orientated organization, responsible for non-competitive work in liaison with other government agencies and industrial contractors, NACA was strongly supported by the civil and military machines in aviation. During the 1950s, NACA's work embraced missile research in increasingly demanding work packages. The staff was split between pro and anti space proponents. Many viewed the preoccupation with space as a timely move toward new frontiers in science and technology, seeing the NACA role as one in which research on high speed and high altitude flight could be fully

NASA's first management team. Dr. T. Keith Glennan (centre), Administrator; Dr. Hugh L. Dryden (left), Deputy Administrator; and Richard E. Horner (right), Associate Administrator.

justified. Others believed it to be just that, a preoccupation away from the central issues of the day in research and engineering. Many staff members were wedded to the days of piston-engined aircraft, only reluctantly recognizing the advantages of a reaction device like the jet engine; rocket propulsion was anathema to several key figures at NACA.

Before the Vanguard debacle of 6 December, NACA appointed H. Guyford Stever, a physicist from the Massachusetts Institute of Technology, to chair a Special Committee on Space Technology which would examine the current posture regarding space and report back on the role it recommended for the NACA. Members included, among others, Wernher von Braun from the ABMA, Dr. James A. Van Allen from Iowa University, and William H. Pickering, director of the Jet Propulsion Laboratory. It was not a NACA oriented committee and was expected to respond objectively to the situation it found, or at least as objectively as it could with von Braun around. On 21 November, 1957, the same day the Stever Committee was set up, James Van Allen issued a report on 'A National Mission to Explore Outer Space' from the Rocket and Satellite Research Panel he chaired. In supporting a radical new approach for the administration of space programs, it 'proposed that a national project be established with the mission of carrying out the scientific exploration and eventual habitation of outer space. . . To carry out the objective. . . it is recommended that a National Space Establishment be created. . .'

In December, the NACA director Dr. Hugh L. Dryden gathered key personnel together from laboratories across the country. The Lewis people were enthusiastic about establishing a more positive attitude toward space research; the Langley Laboratory was somewhat diffident about the whole idea; the Ames group were flatly opposed to it. A major drive was exercised at the meeting by Abe Silverstein from Lewis, when he presented a cogent argument constructed by Bruce Lundin from his staff. In it, the two enthusiasts asserted the bold nature of future decisions on space, claiming it as a measure of national capability, essential 'to our national survival . . . our traditional role of leadership . . .' and an example of the need for 'consolidating present Governmental research on space problems in to a single agency.' With Dryden, the chairman of the NACA, James Doolittle, invited representatives of middle management to a dinner in Washington on 18 December to sound out at a lower level the arguments put at the earlier meeting. They got an enthusiastic response and heard some old controversies but the outcome resulted in consolidation of the advisory committee – a common NACA practice in times of great decision making – set up under Stever.

By the new year NACA top management had made up its mind to promote the concept of 'A National Research Program for Space Technology' in a report issued 14 January, 1958. It called for new and improved facilities, increased budgets, continued liaison with defense agencies on large satellite projects, but above all a continued posture of research and development. NACA was playing a cautious tune in calling for changes to the administration of a space program. By 16 January, NACA had defined the shape of its proposal and recommended a joint venture with the Department of Defense, the National Academy of Sciences and the National Science Foundation. In this the NACA failed, as had Eisenhower, to recognize the enormous value in public attitudes world wide that would be influenced by a visible and active space program. In adopting a scientific basis for space, or a security asset with defense projects, the Committee retained the measured response to Soviet initiatives without allowing its own program to change on the whim of political ploys – in effect, the Eisenhower approach.

The NACA report 'A Program for Expansion of NACA Research in Space Flight Technology' proposed development of large rocket motors, a facilities' expansion program costing $380 million over the next five years and an annual budget increase of $100 million on the $80 million it already received. The NACA had many friends close to the White House. Doolittle was a member of the President's Science Advisory Committee; Killian was a personal friend of Hugh Dryden, the NACA director, and when Sen. Clinton Anderson sent a bill to Congress proposing the Atomic Energy Commission take a large slice of the national space program, that relationship paid off. The administration felt obliged to reject the bill by opposing it in session but were expected to prepare an alternative, which they did by requesting on 4 February that James Killian find a solution.

Killian sought help from the Space Sciences Panel of the PSAC and conferred with William Finan at the Bureau of the Budget over the impact any organizational change might have on the government infrastructure. Finan endorsed Killian's view that the NACA should form the nucleus of a National Aeronautics and Space Agency. By forming it from an existing organization, the Agency would emerge with strong management posture and a traditional, well established, liaison with civil and military interests. Very soon the Executive Office of the President decided this was the only way to retain the administration's desire for a separate interest in space in order to conduct peaceful exploration. Killian went with Finan and his associates to report this conclusion to the Committee on Government Organization which endorsed the finding and sent a recommendation to the President that 'leadership of the civil space effort be lodged in a strengthened and redesignated National Advisory Committee for Aeronautics.' Eisenhower approved the move on 5 March, 1958, on the basis that NACA had already moved into space research in a modest way, would have its continued existence brought to question if it was legislatively excluded from the space programs of coming years, and was a visibly civilian organization already.

But time was running out. Eisenhower knew the Defense Department would contest any transfer of space administration to a purely civilian sector and the current session of Congress only had a few weeks until recess. Accordingly, Killian and the Budget Bureau hastily prepared draft documents outlining the legislative language for Congress, sent it to the Defense Department for comment only at the end of March and to the Congress several days later on 2 April in time for speedy enactment. The machinery of government was well equipped and fully oiled up for smooth running when the National Aeronautics and Space Act rode into Congress. The Senate had created the Special Committee on Space and Astronautics on 6 February; the House of Representatives had set up the Select Committee on Astronautics and Space Exploration the day Eisenhower approved the Killian recommendation. But there were ugly sounds from the Pentagon that Defense Department officials had been given insufficient time to comment on the proposed Act when, as Lyndon Johnson quipped, it 'whizzed through . . . on a motorcycle!'

The administration knew that if the Act lost momentum it would be deferred for discussion that would give the Pentagon time to stall it in debate about language before further debate in Congress on the Act itself. But Lyndon Johnson, chairman of the Senate Space Committee, had done a fine job rallying support and greasing the trackways that led to Senate ratification. Johnson was in the upper House and had to rely on majority leader John W. McCormack to boost it through the lower division. Hearings began on 15 April. Points of contest arose from the clause that dictated civilian control of civilian space projects, 'except insofar as such activities may be peculiar to or primarily associated with weapons systems of military operations, in which case the agency may act in cooperation with or on behalf of the Department of Defense.'

To the DoD this smacked of a penalty clause whereby the Pentagon would give up military space interests if it bordered on projects already approved for the civilian agency. The House tried to inhibit the control of space activity by military interests; the Senate supported a strong defense orientated decision making structure. At a lower level, departments in the Pentagon felt the Act would influence adversely the flexibility for service decisions; the Air Force felt its posture would be compromised (it was already conducting a massive undercover effort to secure manned space flight before the new NASA could get hold, as will be discussed in the following chapter), and the Army believed its opportunities would be inhibited by excessive legislation. Basically, the Army was in technology development for a massive space endeavor while the Air Force concentrated on the immediate gains from a hypersonic/near space operational commitment. The DoD finally agreed to accept a wholly civilian agency but with language securing the right for Pentagon administration of 'defense objectives.'

The Senate too gave ground to the House in agreeing to a civilian base for Space Agency operations but was unwilling to accept the view that this could be secured by cooperation with the DoD, preferring instead to clearly define by legal language the boundary separating the two. This was crucial to the future development of manned space operations and, although seeming to appease contestants on both sides, nevertheless effectively prevented military interests from easily securing a foothold in space. When the bill came to the Senate, Johnson played a key role in upgrading the future examination of space projects by opposing the proposed National Aeronautics and Space Board that would, he said, be little more than a re-structured NACA committee with no weight and little influence.

Basing his argument on the proven structure of the National Security Council, Johnson sought support for a National Aeronautics and Space Council, a considerably more powerful group which would be responsible for structuring policy in national interests as an entity in its own right, and not one that would be appended to existing interests and subordinate to national policy. The incumbent administration wanted to keep the space issue in perspective, the PSAC wanted that, and even the House balked at such a colossus to administer what seemed in 1958 to be a small scale program with no direction other than that dictated by long term political goals. Nevertheless, Johnson held his ground, believing that space would grow rapidly to become a major issue in national politics and international affairs. It certainly would if he had anything to do with it and the criticism of Eisenhower's handling of technology in the new space era was moving Johnson strongly toward the position of rallying point for pro-space interests.

On 7 July, 1958, Johnson got an appointment with Eisenhower in an attempt to change his mind about opposing the Council. The President was adamant that not only would a Council over-magnify the whole issue of space policy, but that it would become an instrument carrying a momentum difficult to hold back. This was just what Johnson wanted and Eisenhower knew it. To the President, Johnson was setting up just the very thing the public had clamoured for but that he, as Chief Executive, had opposed. Sensing that Eisenhower's real objections centered on the independence from Executive control such a Council would have, Johnson proposed a solution by designating the President himself chairman of the group. This broke the stalemate. Eisenhower called Killian after Johnson left and met him in his car the next morning for a discussion while the President drove to the airport. Eisenhower stated his approval for the Council structure if he was allowed to chair it, clearing the way for the Act to proceed through the Senate. The President asked Killian to serve as NASC executive secretary and exercise restraint on its civil service strength; it was Eisenhower's way of holding Johnson back.

Eisenhower knew he would have little time for the Council, leaving Killian to chair the meetings in his absence. He rarely attended NASC gatherings and had confidence in Killian's loyalty to the policies of the administration. The Council was kept in the place Eisenhower reserved for it while he remained in office but Johnson would climb in as chairman himself when Kennedy became President in 1961. Exactly three weeks after Johnson went to the White House to fight for a Space Council the Space Act became law. Instead of an Agency, civilian space operations would be handled by a National Aeronautics and Space Administration. Elevated to an Administration, NASA would be a more powerful arm of the government than anyone in the White House had intended. Killian was brought in to help select an administrator for NASA; the obvious choice of NACA director Dryden was made difficult by congressional doubt over his qualities for leadership and management. After discussing the matter with Doolittle, the NACA chairman, Killian settled on two candidates, one of whom was Dr. T. Keith Glennan, president of the Case Institute of Technology. Eisenhower selected Glennan and nominated him NASA Administrator on 8 August. Together with Dryden as Deputy Administrator, Glennan was sworn in eleven days later. NACA would cease to exist from close of business on 30 September and from 1 October would re-emerge as the National Aeronautics and Space Administration.

In 1958, NACA comprised a headquarters in Washington, employing 170 persons, the Langley Aeronautical Laboratory, with 3,200, the Pilotless Aircraft Research Station close to Langley, with 80 people, the Ames Aeronautical Laboratory, with 1,450, the Lewis Flight Propulsion Laboratory, with 2,700, and the High-Speed Flight Station, employing 300 people. In all, 7,900 scientists, engineers, technicians, clerks, assistants and sundry civil servants. Largely because of the efforts of the Johnson space machine, NASA came into existence with a powerful legislative back-up and a large measure of autonomy responsive to the dictates of the National Aeronautics and Space Council. A year had gone by since Sputnik I sent shivers through the bones of patriotic Americans, and during that period the Air Force had bequeathed its nascent manned space objectives to the new civilian authority. When NASA opened for business on 1 October, 1958, it had as its most stimulating objective, responsibility for placing the first American in space – perhaps the first human being. No one knew all the answers on how to get that job done but T. Keith Glennan set the NASA mood when he simply said on 7 October: 'Let's get on with it.'

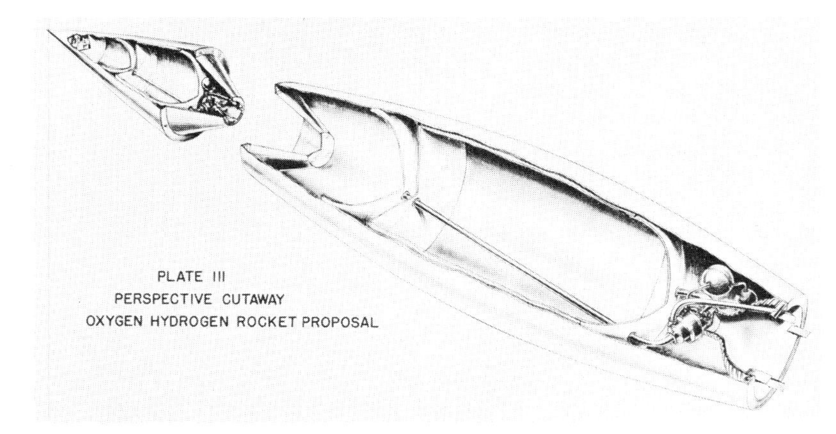

This early post-war proposal for a satellite launcher used high-energy propellants which would be adopted for the upper stages of Saturn launch vehicles.

CHAPTER THREE

Seven to get Ready

In the furore of post-Sputnik fever, diluted temporarily by the successful launch of America's Explorer I satellite on 31 January, 1958, the US armed services gave up the hopes they had for sending men into space. During much of that year, Congressional action on the National Aeronautics and Space Act made it increasingly clear that when NACA became the National Aeronautics and Space Administration there would be little chance of getting the money to carry on planning a manned military presence in orbit. Yet throughout 1958 attempts were made to get a toe-hold in space by all three services, and secure a contribution that would not dissipate when the new civilian administration set up shop. In the event, they failed to make it. The Army had very little to bargain with in the technological stakes; von Braun was working up a clustered booster that could deliver a thrust of more than 600 tonnes, and Rocketdyne was confident it could develop a single rocket motor of that power. But the real stimulus came from the launch of Explorer I. In the wake of development trouble with the Navy Vanguard project, von Braun's Army Ballistic Missile Agency team got approval to send up a satellite and when it successfully accomplished the feat less than four months after Sputnik I the mood was euphoria tinged with pride.

For its part, the Air Force had more to lose. At a time when 'experts' predicted the end of manned combat aircraft within a decade, superseded by missiles and rockets, a viable high speed rocket aircraft program was essential to the survival of a complete functional area of Air Force responsibility. As it turned out the Air Force need not have worried: manned combat aircraft are unlikely ever to be replaced by automated strike systems. During January, 1958, William M. Holaday, Guided Missiles Director at the Department of Defense, received an Air Force five-year plan for space applications. In it, manned and unmanned programs would proceed together through an increasingly sophisticated pattern of missions designed to provide Air Force domination of the space near Earth.

From simple manned capsule flights in Earth orbit would come the knowledge and the technology to build Earth-orbiting space stations, followed by circumlunar flights and colonies on the surface. Concurrently, reconnaissance satellites, weather satellites and communications satellites would administer the needs of Earth-based military forces. In all, the five-year effort would require an estimated $1,700 million. No mention was made of the boost-glide vehicle envisaged during studies performed in 1956 and 1957 but that concept was an integral part of future Air Force plans for a post-X-15 vehicle applicable to operational needs. Had Sputnik I not ascended in October, 1957, it is possible the Air Force would have concentrated on the boost-glider, intended even before the Russian flight to reach Earth orbit at the end of a lengthy development period during which it would perform suborbital skip flights high in the upper reaches of the atmosphere.

With the boost-glide vehicle rightly seen as a longer term commitment to Air Force needs, Sputnik I and the visible move toward a civilian space authority accelerated plans for manned space flight and set the Air Force's sights on a more immediate mode of access to the space environment. The January plan was certainly more ambitious than the Army's 'National Integrated Missile and Space Vehicle Development Program,' in which clustered rockets would power a large booster for manned and unmanned flights to Earth orbit. In that scheme, $850 million would be spent over five years embracing 30 flights leading to full operational status for the 680 tonne thrust launcher by 1963.

In a three-day period beginning 29 January, 1958, the Air Research and Development Command held a conference at Wright-Patterson Air Force Base, Ohio, to examine alternative design concepts for a manned space vehicle–one that could achieve quick success without too much consideration to long-term growth potential. Eleven companies attended, presenting their respective findings submitted formally to the ARDC a month before. AVCO chose a 680 kg sphere placed in orbit by Titan rocket, with a stainless-steel cloth parachute for decelerating the capsule to re-entry and for stability on the descent; Aeroneutronics chose a conical shape containing a sphere which would rotate to line the pilot up with the g forces induced by acceleration, launched by two-stage rocket; Bell designed a spherical vehicle weighing less than 1.4 tonnes but preferred some form of boost-glide concept for longer term application; Goodyear proposed a sphere carrying a tail to which flying control surfaces could be attached for drag modulation during re-entry, launched by Atlas or Titan with an upper stage; Convair spoke of a large manned space station but said a 450 kg spherical capsule could carry a man on top of an Atlas rocket; Northrop preferred a lifting vehicle from the beginning, suggesting a modified Sänger plan whereby a glide vehicle is boosted to orbital speed by conventional rocket; Republic had the most innovative concept: a triangular-shaped vehicle with a 0.7 meter diameter tube from the leading edge around to the rear with propellant for solid rockets in the wings and a manned capsule on top from which the pilot would eject on the way down; North American Aviation put forward a modified X-15 launched by three-stage rocket on a single circuit of the Earth followed by pilot ejection and loss of the spacecraft.

The most significant proposals, however, came from Martin, Lockheed and McDonnell. The Martin plan envisaged a non-lifting ballistic vehicle sent into orbit by Titan rocket in 1961. Lockheed pursued a cone-shaped capsule with the pilot facing rearward and sent aloft on an Atlas-Hustler launch vehicle. The Hustler grew from the 'Manned Ballistic Rocket Research System' study begun in February, 1956, to provide the Atlas with a second stage for boosting manned capsules into orbit; it was later renamed Agena. McDonnell presented a ballistic vehicle in the form of a truncated cone weighing nearly 1.1 tonnes launched by Atlas with a suggested 1959 flight date.

The last configuration closely matched studies then beginning at the Langley NACA facility, primarily at the instigation of Maxime A. Faget, head of the Performance Aerodynamics Branch at the Pilotless Aircraft Research Division. Faget graduated from Louisiana State University with a bachelor of science degree in mechanical engineering during 1943. After service with the US Navy, Faget joined the Langley Memorial Aeronautical Laboratory and provided valuable work on ramjet inlets and supersonic

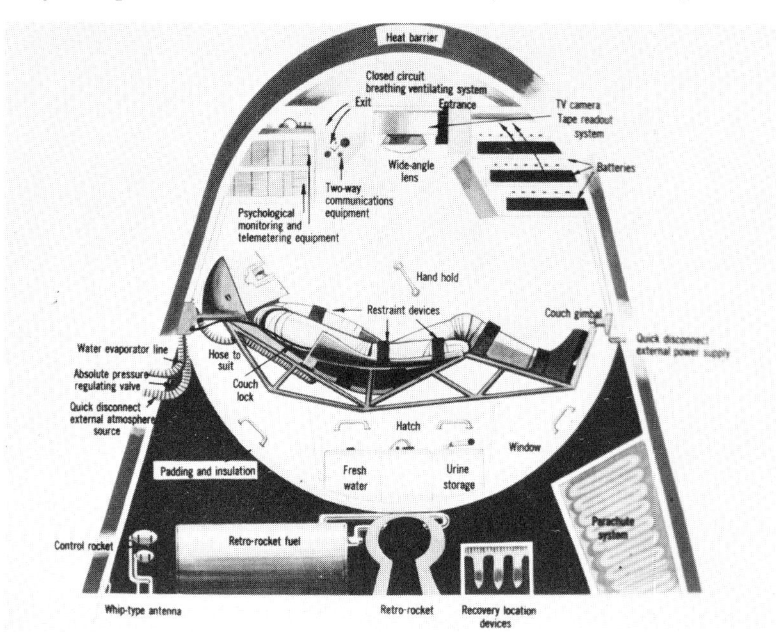

Restrained on a contoured couch, the pilot in this April 1958 suggestion for a manned orbital capsule would rotate 180° for re-entry, presenting his back to the direction of descent through earth's atmosphere.

aerodynamics. His contribution to the design of the first US manned space capsule was as significant as the contribution by Allen and Eggers to ballistic missile nose cone development. In January, 1958, however, Faget's attention was only just turning toward the problems of safe flight up and down from an orbit about the Earth.

In the weeks following the Wright-Patterson conference, Air Force steps to secure a role in manned space flight quickened. On the day it ended, and the day Explorer I reached orbit, the ARDC ordered the Wright Air Development Center to liaise with the Ballistic Missile Division on studies aimed at defining the quickest way to get a man spaceborne. As mentioned earlier, the Advanced Research Projects Agency was effectively in operation from 7 February on orders from Defense Secretary Neil H. McElroy. Roy W. Johnson was brought in from General Electric to direct the office and by the end of the month had determined that the Air Force had 'a long term development responsibility for manned space flight capability with the primary objective of accomplishing satellite flight as soon as technology permits.' Clearly, Johnson had no idea of relinquishing to a civilian agency the tasks he thought more fitting to Air Force application.

In anticipation of the need for a more powerful delivery system, the Defense Department approved full development of the Hustler for use on Atlas or Thor launchers, a project coded 117L. On 27 February, Air Force Vice Chief of Staff, Gen. Curtis E. LeMay, received briefings on ARDC concepts. They centered on three basic approaches: an upgraded X-15 launched by rocket like the North American proposal of the previous month; an improved and accelerated boost-glide program; a simple ballistic capsule like Martin, Lockheed and McDonnell proposals. ARDC went away with a clear mandate to choose a concept and come back with firm plans, which it did just nine days later. Time was running out if the Air Force wanted to establish its presence in space.

Just six weeks earlier, Lt. Gen. Donald L. Putt, Air Force Deputy Chief of Staff, Development, sent NACA a letter inviting their participation on a long term boost-glide vehicle and on a ballistic, non-lifting, capsule for early flight. Hugh Dryden responded as NACA's Director by agreeing to assist with the former but deferred a decision on the capsule due to existing NACA studies which would, he said, have to be concluded first; Dryden knew that events were moving toward transforming NACA into a civilian space agency and was reluctant to concede this important area to the Air Force, which he thought he might do by inference if NACA responded to the call for assistance. On 8 March, a proposed 'Manned Space Flight to the Moon and Return' was laid out by the Ballistic Missile Division embracing a series of interlocking projects leading toward a manned landing on the lunar surface. On 10 March a three-day conference began at the BMD facilities, Los Angeles, at the instigation of the Advanced Research Projects Agency.

ARPA chief Johnson was pressing for a formal plan and in addressing the participants at the opening, Gen. Bernard Schriever imparted a sense of urgency. In attendance were nearly 100 representatives of industry, service, and NACA interests to hear the Air Force outline its tentative plan for placing a man in orbit by ballistic capsule weighing about 1.3 tonnes, a conical structure approximately 1.8 meters in diameter and 2.4 meters tall supporting life for two days. Human factors experts from the Air Force had already decided that a load of 12 g would be the highest recommended design value for acceleration during ascent and descent and because the Thor could induce a force of up to 20 g under certain, albeit unlikely, circumstances, it was decided to adopt a two-stage configuration by incorporating a second stage on the Thor launcher.

Thor was unanimously approved as the most suitable boost system and no-one suggested using Atlas; in early 1958 there were mixed feelings about the big ICBM and with good reason since by the time the BMD conference began Atlas had successfully flown on only two of six attempts. In all, the objective was expected to consume 30 Thor rockets, 20 upper stages using a fluorine-hydrazine propellant combination and about 10 upper stages from the Vanguard launcher. Combinations of the smaller rockets would be used to lift biophysical research experiments designed to test the reaction of living things to orbital flight prior to sending a man. Other considerations anticipated a fully automated flight control system in which the 'pilot' would be little more than a passenger; in defining the stresses and forces imposed upon a spaceman biomedical personnel continually erred on the cautious side, believing that the man might be rendered emotionally or psychologically imbalanced! The pilot would be seated on a couch that could be rotated so as to place him in such a position that the g forces would be imposed from chest to back.

Tests had shown the extreme danger of placing a man in a high g environment with the acceleration from head to feet or vice versa. The blood supply and internal organs would tend to move to the extreme end of the body and cause severe draining of the head or lower torso, depending on the line of force. Much higher g loads could be tolerated if the force moved against the rib cage, avoiding any significant change in the blood flow to the brain. By swivelling the couch, the pilot could place his back against the force of acceleration so that on the ascent and on re-entry, he would be pressed into his seat and so minimize the stress on his body. The projected mission profile required the capsule to ascend on top of the booster and point in the same forward-facing direction for re-entry, hence the need for the rotating couch. Rockets in the nose of the capsule would slow it for descent; during deceleration in the atmosphere the pilot could be expected to experience a load of 9 g and internal capsule temperatures of about 65°c.

After the conference, ARDC prepared its proposal and sent it on 14 March to Air Force Headquarters for approval. On 19 March the Advanced Research Projects Agency received a request from Under Secretary Marvin A. MacIntyre for $133 million to get started on the manned space project; Gen. White, Air Force Chief of Staff, got the Joint Chiefs to approve the plan on 2 April. During the month, Ballistic Missile Division personnel worked out the details of a four-stage program. First 'Man-in-Space-Soonest' (MISS) would provide an early opportunity for a ballistic space flight, following precursor tests with apes. This would be followed by 'Man-in-Space-Sophisticated,' envisaging a larger capsule based on engineering developments with the MISS capsule, capable of sustained orbital flights of 14 days' duration. Third, 'Lunar Reconnaissance' would perform a survey of the Moon's surface with unmanned spacecraft. The final 'Manned Lunar Landing and Return' phase would require circumlunar flights by apes, then by men, followed with manned landing on the surface.

All this was to be accomplished by 1965 at a total cost of $1,500 million. The launchers would be of three types: Thor-Vanguard, Thor plus new upper stage, and a modified version of the new Titan ICBM with additional upper stages. The first segment, MISS, was expected to deposit a man in orbit on the tenth flight of the 'Thor plus fluorine upper stage' configuration in October, 1960. A slight hiccup to ARDC plans for the four-stage program came during May, 1958, when a joint AVCO-Convair proposal envisaged a modified version of the AVCO design presented at the January Wright-Patterson Air Force Base contractor conference. This would adopt a steel-mesh parachute system for re-entry as a drag brake designed to slow the capsule, essentially a sphere with a man inside. The device would be launched on an Atlas missile, without upper stage, would shave up to four months off the Air Force MISS schedule and probably save considerable sums of money.

ARDC were unhappy about the basic design approach but Gen. LeMay was intrigued by the idea, and with the possibility of saving some money asked ARDC to examine it. LeMay was persuaded that the technical problems could lead to massive escalation of the supposedly reduced development cost and that the Air Force would be better off with the currently approved concept. Upstream from ARDC, the Advanced Research Projects Agency balked at the thought of just how much the space plan would cost, even though ARPA had received tacit approval from the National Security Council Planning Board for an early start. Careful examination of the requirements indicated that most of the money would go into developing the new rocket stages or making hybrid combinations of existing designs.

During May, MacIntyre and his assistant, Richard E. Horner, passed down to ARDC a suggestion that if the Ballistic Missile Division could examine the possibility of using an Atlas missile and find a solution to high-g accelerations on the pilot, adoption of the ICBM as a manned satellite launcher might put costs below $100 million. BMD decided that a high-g abort was unlikely and that it need not necessarily be prohibitive. Consequently, on 16 June, the ARPA were able to approve the new MISS plan with its reduced cost estimate of $99.3 million. In the same month, North

In response to a request for industry proposals on Project 7969 (Manned Ballistic Rocket Research System), issued February, 1956, the Air Force received in January, 1958, these eleven concepts which formed the basis for a review later that month of manned orbital vehicles.

	LOCK.	MARTIN	AERO-NEUT.	McDON.	AVCO	GOOD-YEAR	CONV'R	BELL	NAA	REPUB.	NORTH'P.
MIN. MANNED SATELLITE	9' 14'	8' 14'	30° 7' 7'	7'	7'	7'	5'	L/D=5	X-15B STRIPPED	2d	L/D=6
WGT. LB.	3000	3500	2545	2400	1500	2000	~1000	18,000	~10,000	4000	11,000
BOOSTER	ATLAS + HUST.	TITAN	ATLAS + HUST.	ATLAS + POLARIS	TITAN	A or T + 3RD ST.	ATLAS + HUST.	"DYNA SOAR" APPROACH. "WINGLESS WOULD BE ONLY A STUNT"	4 G-26 + XLR-99	ATLAS + POLARIS	"DYNA SOAR" APPROACH
ORBIT	150-300 3 REV	150 m. ~1 DAY	—	100 m. 1 REV.	120 m. 1 WEEK	200-400 5 DAYS	170 mi —		—	100-150 [illegible]	
* TRACKING	—			"MINITRACK" SYST.							→
* ORBIT CONTROL	RETRO. ΔV=200 FT/SEC	RETRO. ΔV=500 '/s	RETRO.	RETRO	VAR. DRAG.	RETRO. ΔV=800 '/s	RETRO. ΔV		VARY L/D	ΔV=65 + VAR. L/D	
ATTITUDE CONTROL	ROCKET + EL. MOTOR	ROCKET	—	ROCKET	'CHUTE	—	—		ROCKET + PILOT	ROCKET	
PILOT FUNCTIONS	NONE	NONE	NONE	NONE	NONE	NONE	?		FLY AIRP.	MONITOR	
MAX. DECEL.	~8g	8-15g	—	8.5g	7-9g	—	—			"low"	
STRUCT. TYPE	ABLATION OR [illegible]	ABLAT.	GRAPHITE HT. SHIELD	HT. SINK (BERYL)	RADIATION (MOLY OR [illegible])	ABLAT.	—		BeO + RENE 41	INCONEL	
SAFETY	EJECT CAPSULE AT LAUNCH	EJECT AT LAUNCH	EJECT AT LAUNCH	POLARIS + CHUTE	EJECT TO 3000' CHUTE	EJECT AT LAUNCH	—		?	—	
W/C_DA #/SQ FT	100	100	61	60	1.5	50	50		W/β=50	W/β=26	
LANDING AREA, MI.	400 × 20	100 × 100	100 × 50	400 × 400	"KANSAS" 400 × 200	800 DIA	—		—	—	
TIME TO MANNED FLT.	2 YR.	2½ YR	6 YR	2 YR	2½ YR	2 YR	1 YR	5 YR	2½ YR	1¾ YR	—
COST, MILLIONS	10-100	—	—	—	40	100		889	120 (8 [illegible])		

*All non-winged designs use zero-C_L reentry with no flight path control. Winged designs have complex ground controlled systems monitored by pilot.

American Aviaticn and General Electric got to work on contracts let by the Wright Air Development Center for design of the capsule, its interior layout, and a system for maintaining a pressurized environment. By July the mood was changing. The National Aeronautics and Space Act was moving through Congress; the Advanced Research Projects Agency's Man in Space Panel under Sam B. Batdorf was unable to secure a major commitment from the Defense Department; the Air Force was pressing ARPA for a funded start.

The Defense Department's position was understandable. With NASA emerging from the old NACA structure within the next few months, with the attitude of the Eisenhower administration firmly set on allocating civilian, scientific and exploratory space projects to the nascent Space Administration, there would be controversy over a military manned satellite project. The DoD was reluctant to allocate funds for what might well turn out to be a NASA project. But there was still just a chance, especially if it could be shown that the Air Force had already gone a long way along the technological road to actually achieving the objective. Roy Johnson, ARPA boss, made a last effort by imploring his subordinates to find ways of cutting further the projected cost, which by now had grown back over the $100 million mark. For his part, Schriever sent a letter to Samuel E. Anderson, the head of ARDC, with confirmation that the Ballistic Missile Division was wound up and ready to go with a release of specifications to potential contractors if he could give the final go-ahead.

As a result of this effort, ARDC received a briefing from the Air Force Ballistic Missile Division in a two-day session beginning 24 July. Johnson would not move. The ARPA, he said, could not give their approval even though he wanted the Air Force to have the job of placing man in space 'soonest' because the Eisenhower camp was immovably committed to a civilian run space program; only those projects which carried an unquestionably military content would get passed to the Defense Department. Just four days later the Space Act was signed by the President and Roy Johnson met with NACA Director Dryden to consider the future of US manned space flight. Deferred until the White House had proclaimed on the matter, the official coup de grace came on 18 August when President Eisenhower decided that NASA should have the job of placing an American in orbit. It was decided that money already allocated for early work on the Air Force MISS plan should be part of the allocation transferred from the Department of Defense to NASA.

This did not affect the separate agreement to cooperate on the Air Force's boost-glide concept, which proceeded more or less independently of the orbiting capsule. Known as Dyna-Soar I, for the 'dynamic soaring' capability it was expected to possess, the NACA signed on 20 May, 1958, an agreement relating to design and development. In this it was assumed that the project would be carried out as a joint NACA-Air Force venture, allocate technical financial and management control to the Air Force, require a separate contractor to design and construct the system, and set up a joint NACA-Air Force test program. The following month the Air Force selected Boeing and Martin from seven potential contractors to perform limited feasibility studies on a design concept that envisaged a hypersonic glider boosted by a modified Titan II missile; Titan was much more powerful than Atlas and was capable of carrying a heavier vehicle.

Other, less viable, proposals for a manned space project emerged during the months of decision in 1958: one from the Army and one from the Navy. Spurred by Explorer I and the application of Redstone in its Jupiter C configuration, the von Braun team at ABMA proposed early in 1958 that a similar missile be used to boost a 'Man Very High' on a ballistic flight to the edge of space. In every respect it was a re-worked adaptation of the British Interplanetary Society's Megaroc concept for a V-2 launched ballistic capsule ascent. Man Very High would reach an altitude of 240 km before dropping back into the denser regions of the atmosphere. Falling toward the Atlantic from a Cape Canaveral launch site, the capsule, about 1.2 meters in diameter and 1.8 meters tall, would deploy drag flaps to stabilize the system and slow it for deployment of parachutes.

An abort procedure was suggested whereby if the Redstone experienced a malfunction on the launch pad, the pilot would be lobbed into a tank full of water at a conveniently safe distance. The thought of Army personnel being thrown like high-divers into a circular tank of water sent ripples of humor through the service. The Army were unsuccessful in their approach to the Air Force for cooperation on the project which, in the following April, was renamed Project Adam. It was supposed that Adam would restore a measure of confidence in the US missile system following the Sputnik episode and contribute toward a vague idea that future troop transporters would consist of rocket-launched vehicles. Not one high ranking official in the Pentagon approved of the Adam proposal, although the Central Intelligence Agency supported the project for the good it would do in international prestige. The idea was killed on 11 July by ARPA chief Roy Johnson when he notified the Army Secretary Wilbur M. Brucker that it was no longer considered a viable project. As it turned out, within three years a Redstone would send a manned capsule on a ballistic flight to a height of 188 km and back to an Atlantic Ocean splashdown.

The suggested Navy design was more exotic and certainly unrelated to anything that actually flew. Called Mer I, for Manned Earth Reconnaissance, it envisaged a cylindrical capsule with hemispherical ends, the latter expanding to form two telescopic wings, inflated internally and securely attached to the rigid body of the vehicle. Launched in the stowed configuration by a two-stage rocket, Mer I would deploy its wings and use them as lifting surfaces when flying back down through the atmosphere. Proposed initially at the beginning of April by the Bureau of Aeronautics, the Mer concept was not received warmly by the ARPA as the Navy submission for a manned space vehicle. It was revised toward the end of the year and reappeared with minor modifications. But nothing could do very much to make the bizarre concept even feasible and it died with the Army project when NASA officially got the manned satellite work.

A joint NASA-ARPA Manned Satellite Panel was set up in September comprising Robert Gilruth, Maxime Faget, Alfred Eggers, George Low, Warren North and Walt Williams from the then still official National Advisory Committee for Aeronautics, and Robertson Yougquist and Samuel Batdorf from the Advanced Research Projects Agency. In a letter to NASA historian Charles C. Alexander in 1963, the Administrator for NASA, T. Keith Glennan, said that he was not sure precisely who authorized NASA to take responsibility for the manned satellite plan but that, 'It seemed the natural thing for me to accept the recommendations of the only people who knew very much about the matter and initiate the program as soon as NASA became an operating agency. In short, I do not recall that President Eisenhower actually assigned the manned space flight program to NASA – I guess I just accepted the tasks which we would have to undertake.'

In fact, NACA had for all that year 'accepted the tasks' associated with manned satellite design. From the start of the first meeting convened by the Special Committee on Space Technology on 13 February, 1958, NACA assumed it would play at least a leading role in the development of a manned space capsule. The Special Committee had been set up on 21 November, 1957, and at the first gathering seven working groups were assigned responsibility for: basic objectives; the vehicle program; re-entry; range, launch and tracking facilities; instrumentation; space surveillance; and human factors. Just how serious NACA was in committing itself to missile and satellite research began to filter through during the opening months of 1958. The Committee on Aerodynamics was re-named the Committee on Aircraft, Missile and Spacecraft Aerodynamics on 20 February. More and more work had moved across the broad field of aeronautics research to embrace research tasks directly applicable to space activities.

The most detached facility was the Ames laboratory, traditionally the home of scientists who preferred the non-political, academic atmosphere of this remote California base. There the work was detached from industrial in-fighting, services pressure, or bureaucratic interference; a haven for the research scientist engaged in exploring the far corners of an aerodynamic frontier. At the High Speed Flight Station the mood was different, one in which research rubbed shoulders with practical trials and tests that constantly brought a flow of Air Force and Navy chiefs from desks all over the United States. It was the home too of the test pilot, that rare breed that never seems to have a specific niche yet fills the corners of offices and hangars in any but the expected places, men who brought a cynical nonchalance to over-confident predictions about the future of manned flight. It was also the place where work conducted at Ames one year would emerge five years later as 'hardware,' a practical, three-dimensional realization of slide rule and calculation.

Because of that it had an air of operational bustle; Walt Williams' group were closer to the front line of service flying than Ames would ever want to be. Lewis had a different relationship, performing work on propulsion units and soon to bequeath its Director, Abe Silverstein, to NASA Headquarters to set up a space flight program. Langley was different again. Here, research was married to technology to provide a visible and practical end product. Aerodynamicists built scale models of tomorrow's aerofoils, flew rocket-boosted models of new aircraft designs and recorded numerous data on performance and flow characteristics. It was also the place of wind tunnels and laboratories where the marriage between theory and practicality was consummated from Walt Williams' desert runways.

It was from Langley that NASA's manned space flight program emerged. Or, to be precise, in the Pilotless Aircraft Research Division (PARD) where Maxime Faget headed the Performance Aerodynamics Branch. Faget was in the small group that gathered one day in March, 1958, to discuss with Langley's Deputy Director Robert Gilruth the future direction NACA should take in designing a manned satellite. The direction was clear: a ballistic capsule with little or no lift. This was the Langley view presented at the NACA Conference on High-Speed Aerodynamics held at Ames in California between March 18 and 20. It was intended to familiarize military, industrial and contractor personnel with the NACA research activity as it related to manned space flight. Some 46 papers were read, including a very significant contribution from Faget, Benjamin Garland and James J. Buglia titled, 'Preliminary Studies of Manned Satellites, Wingless configurations, Non-Lifting.' It was a description of the ballistic concept and strong argument in favor of adopting such a design for the initial orbital flights.

The three authors said that a ballistic shape would simplify the design of stabilization and control equipment, could be brought back from orbit simply by firing small rockets – 'retrograde' rockets firing forward along the direction of travel, rather than 'posigrade' rockets firing backwards – and that attitude control could be maintained in the vacuum of space with small

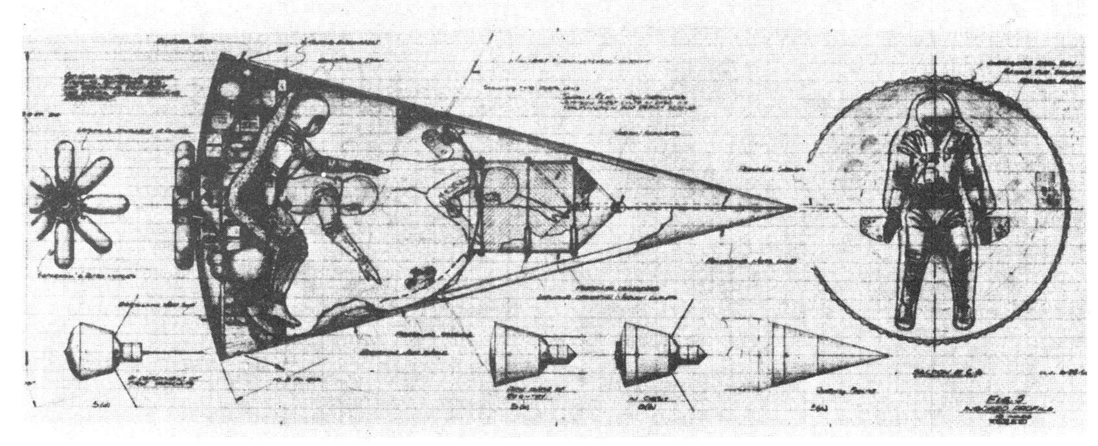

By summer, 1958, the first tentative layout for a manned capsule had been completed.

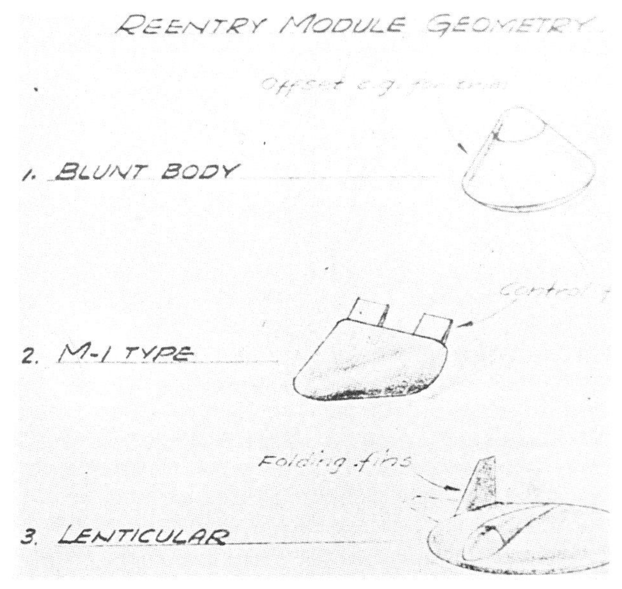

Three contenders for the concept of manned orbital flight, a competition in principles won by the → blunt body shape.

The success of Mercury hinged on a competent management team, represented here by (left to right) Charles J. Donlan, Associate Director; Robert R. Gilruth, Director; Maxime A. Faget, Flight Systems Chief; and → Robert O. Piland, Assistant Chief for Advanced Projects.

rocket motors or 'thrusters', devices that could be made to expel at high velocity a gas or liquid stored in propellant containers; this latter concept had been successfully adopted for the X-1B, the X-2 and the X-15 rocket research aircraft. In the proposed configuration, the capsule would be about 2.13 meters in diameter and comprise a cone with 15° sides and a length of 3.35 meters, in which a single occupant would lay with his back against the flat end of the vehicle. It was suggested that the thrusters used for attitude control could turn the capsule around so that it would face backward for re-entry. In this way, the pilot would absorb the g loads of ascent and re-entry along the same axis and in the same direction. In this way the Faget, Garland and Buglia concept differed from other designs where the seat, or couch, was made to swivel inside a capsule that would maintain a fixed attitude.

But if Langley was wedded to a ballistic concept, one that picked up experience from work by Allen and Eggers on re-entry nose cones, Ames could not relinquish its hold on some form of lifting body. In another paper presented at the NACA conference, John Becker from Langley joined the call from Ames and proposed a conical capsule with a modest lifting surface which could be used during a steep re-entry to serve as a heat shield. But the line from Ames itself centered on a concept called M-1, devised late the previous year by Eggers as the best of all possible designs for a re-entry vehicle. M-1 was designed to establish a lifting profile between the hypersonic boost-glider (Dyna-Soar being a good example) and the purely ballistic re-entry capsule from Langley by presenting a forward pointing cone shape flattened on top and well rounded underneath.

It was estimated that M-1 would be capable of selectively controlling its touchdown area 1,300 km along the ground-track or up to 320 km in a lateral, or 'cross-range,' direction. The primary competition from Ames came in the form of a paper presented at the conference by Thomas J. Wong, Charles A. Hermach, John O. Reller and Bruce E. Tinling, titled, 'Preliminary Studies of Manned Satellites, Wingless Configuration, Lifting Body.' It was essentially a smoothed version of Eggers' M-1, producing a lift/drag ratio of 0.5 and a modicum of controllability with flying surfaces on the trailing edge of the flat-topped vehicle. It is doubtful if Ames could have got anywhere with this approach; by March, 1958, it was clear that speed was considered to be of the essence and that the ballistic shape would require less development and place less responsibility on the pilot.

But apart from that, Ames was not exactly pushing for a part in the manned satellite saga; having done its job by gently pointing out just how efficient, energy-wise, the lifting-body could be, they were only too prepared to bow out and let Langley get on with the 'Buck Rogers' job. A significant contribution was made, however, by Ames' Dean Chapman, during 1958, in a paper on 'An Approximate Analytical Method for Studying Entry into Planetary Atmospheres.' In this, Chapman explored for the first time aspects on the re-entry characteristics of vehicles flying different approach trajectories, and came to some surprising conclusions. He found during the study that the heat load accepted by a spacecraft is less, and the deceleration greater, as the angle of entry into the atmosphere increases. Above an entry inclination of 3° from the horizontal, a ballistic capsule would accept decelerations that could prove to lie beyond the threshold of human tolerance. Also, a vehicle with moderate lift in its design would tend to skip out of the atmosphere from shallow entry angles and while the total heat load would be less than for a ballistic re-entry, heat 'pulses' at the shallow end of each skip would considerably exceed that experienced by a non-skip trajectory.

Chapman also discovered that at the speed of a spacecraft returning from orbit a lift/drag ratio of only 0.5 could be considerably effective, generating a lower g load and experiencing only half the heat rate of a purely ballistic design. From this came appreciation of the critical nature of entry dynamics: minor changes in the basic shape of the capsule would produce a very different aerodynamic response; small errors in the angle of atmospheric entry could prove devastating, unless the capsule had an excessive amount of heat protection – a most unlikely situation.

So, by the end of the March conference, NACA was clearly heading toward a ballistic cone for its manned satellite. At first the capsule's shape took the form of a short dome with a flat heat shield at the base but with a narrow lip extending round the circumference to deflect the airflow away from the sides. This was seen to be dynamically unstable at subsonic speeds, so an elongated cone came next but this had inherent thermal problems for the forward end – actually the rear of the capsule when it reentered base end first – where the recovery equipment and parachutes would be installed. The shape of the recovery enclosure went first to a conical profile, then to a round-top cylinder and finally a truncated cone with a smaller diameter than the cylindrical narrow end of the capsule. A rounded heat shield was seen in tests to transfer too much heat to the side walls; a flat shield would focus too much heat into the shield itself. The final compromise eliminated the extended lip and produced a 203 cm diameter shield with a 304.8 cm radius of curvature. The bottom of the shield would be 101.5 cm below the sides in cross-section.

William E. Stoney, Jr., of the Pilotless Aircraft Research Division, was primarily responsible for the heat shield design, a structure that in the first half of 1958 was considered to require heat-sink materials for protecting the structural spaceframe. Ablative materials – ceramics that 'ablate' or char in a controlled manner to remove materials heated to their thermal capacity – were too far from an operational condition; most missile nose cones were heat-sink in concept. At the suggestion of Maxime

Faget, PARD engineers drew up a preliminary design for the launch escape system that would propel the pilot, and his capsule, to safety in the event of a problem with the booster rocket on the pad or during the early part of ascent. Not for NACA the water-filled tank suggested by Project Adam!

The Faget proposal was for a tractor rocket motor attached to a tower mounted on the top of the capsule which, upon ignition, would fire to pull the capsule from off the launcher. An earlier Air Force proposal envisaged a pusher rocket situated at the base of the capsule. By placing the tower on top, the whole assembly could be jettisoned when it could no longer serve as an abort measure. This had an added advantage of saving weight and getting rid of unnecessary propellant and pyrotechnic devices that could in some highly unlikely situation provide a danger to the capsule it was designed to save. Willard Blanchard and Sherwood Hoffman, PARD engineers, put finishing touches to the basic escape rocket design during late summer.

Air Force reluctance to adopt the Atlas, essentially a single stage rocket with jettisonable booster engines, centred on the g forces the missile would impose in the event of a malfunction. Some predicted trajectories envisaged to simulate an abort from Atlas indicated a 20 g load on the pilot. Normally, the Atlas would impose no more than 6 g on the ascent, while the capsule would incur no more than a 9 g load during re-entry. Faget had an idea that if the seat on which the pilot reclined could be made so that it fitted the pilot's physical form he could withstand higher g loads. With Faget, PARD engineers William Bland, Jr., and Jack C. Heberlig designed a glass-fiber form-fitting couch and sent it to a centrifuge at the Navy Aviation Medical Acceleration Laboratory, Johnsville. Several tests showed that a pilot could easily withstand 20 g for a few seconds, a condition that would be imposed with a capsule entering the atmosphere at a steep 7.5°; the ballistic capsule was expected to enter at an angle of 1.5° and produce a 9 g load.

Meanwhile, Faget and Paul E. Purser, again from PARD, developed the idea of assembling a cluster of small solid propellant rockets into a single makeshift booster for low cost tests of dummy capsules. Lobbed to a height of up to 160 km, the 'High Ride' flights would provide a means for repeated qualification trials for heat shield, structural layout, escape system, etc. This idea was temporarily, and expediently, dropped while the Army laughed at Project Adam; the idea of sending manned spacecraft on ballistic flights found little favor among the non-engineering fraternity that held the purse strings. However, Faget, Bland and Ronald Kolenkiewicz returned to the concept in August, 1958, recognizing the need for the multipurpose test rocket.

When NASA officially began business that October, Faget, Caldwell Johnson, and Charles Mathews from the Flight Research Division had completed the capsule's basic design, a ballistic vehicle built to fly on top of an Atlas ICBM. Scale models were already complete and had been used in water drop tests at Langley. A regular two-way street of commuting engineers was moving between Lewis and Langley, with the former group concentrating on subsystems development. One of the first tasks as a NASA facility sent Lewis representatives to the ABMA shopping for Redstone and Jupiter missiles. Later in October, the Air Force received a delegation at the Ballistic Missile Division looking for Atlases. One of Abe Silverstein's first tasks as Director of Space Flight Development was to set up the administrative machinery for manned space flight.

NASA Administrator Glennan and Deputy Administrator Dryden would come to rely on Silverstein for distributing the technical and managerial assignments to various centers – from inception the NASA facilities were re-named: Langley Memorial Aeronautical Laboratory became Langley Research Center; Lewis Flight Propulsion Laboratory became the Lewis Research Center; the Ames Aeronautical Laboratory became the Ames Research Center. Silverstein wished Robert (Bob) Gilruth to organize a manned program and to become project manager for the emerging Space Task Group (STG) which would command and coordinate the operation. Charles J. Donlan, Gilruth's Technical Assistant when he was Assistant Director of Langley, would be his deputy at the STG. STG began business on 5 November, with 33 Langley personnel. It was time to complete preliminary specifications and to put out tenders for a contractor to build the capsule.

More than 40 prospective bidders received invitations by the end of October to attend a briefing at Langley on 7 November. Representatives of the interested companies heard NASA engineers, among them Maxime Faget, Jack Heberlig and Alan Kehlet, discuss in detail the basic objectives of the program and the configuration preferred by Langley. Faget made it clear that NASA would welcome any alternative configuration to the ballistic capsule so long as it incorporated three fundamental concepts: a retro-rocket for returning to Earth, a non-lifting re-entry body, and a heat-sink thermal shield. About half the number of attendees expressed continued interest and were mailed copies of the historic 50-page 'Specifications for Manned Space Capsule' document on 14 November, 1958.

But what to call the project? It was clearly a venture of substantial magnitude which would surely receive wide public acclaim. That, and the potential overtones – military and political – the enemies of America might seek to invoke, was uppermost in the minds of top NASA management as they sought a suitable title. Bob Gilruth wanted Project Astronaut, but that seemed too deliberate a reference to one or two select individuals. In the end it was Abe Silverstein who fell back on Greek mythology and put forward the messenger of gods, a swift, fleet footed son of Zeus, grandson of Atlas: Mercury. Administrators Glennan and Dryden agreed with this name on 26 November and it was officially announced on 17 December, 1958, that America's first manned space effort would be called Project Mercury. By 11 December, just one week before the public announcement, eleven contractors submitted bids for the Mercury capsule.

AVCO, Convair, Lockheed, Martin, McDonnell, North American, Northrop and Republic had all been part of the gestation period earlier that year when they submitted preliminary feasibility studies for defining configurations. Douglas, Gruman and Chance-Vought were participating for the first time – latecomers to a game where experience, even at this early stage, meant a lot. To assess the different submissions, the Space Task Group set up 11 component assessment teams under a Technical Assessment Committee chaired by Donlan. The component teams would look individually at systems integration; load, structure and heat shield; escape system; retrograde and landing systems; attitude control systems; pilot displays and navigation aids; pilot support and restraint system; environmental control systems; communications; instrumentation, recorders and telemetry; and power supply equipment. The separate areas were evaluated on a scale of one to five covering a range from unacceptable to excellent and totals were averaged to provide a single numerical value displaying technical competence.

Over the Christmas period the Langley teams worked at the proposals, activity made all the more difficult by a move from facilities on one side of Langley Field to the other! By 29 December, STG had completed its work and sent it to NASA Headquarters in Washington, at that time installed in the Dolley Madison House previously occupied by President James Madison's wife between 1837 and 1849. There, a Source Selection Board under Silverstein, but including Charles H. Zimmerman from Langley, would scrutinize the proposals and make a recommendation to the Administrator. A Management, Cost and Production Assessment Committee under Carl Schreiber examined eight of the eleven contenders that got through the Technical Assessment Committee and on 6 January just half that number were passed up to the Source Selection Board.

Staffed by Silverstein, Zimmerman, Ralph Cushman, George Low, Walter Schier and De Marquis Wyatt, the Board met on 9 January to give their final recommendation to Glennan. It had, in the end, been a contest between McDonnell and Gruman. The latter was deemed to be fully committed with Navy contracts and it was felt wise not to load work on to a company which was expected to complete aircraft production orders as rapidly as possible. Glennan approved the decision to give McDonnell the Mercury contract, and telephoned James S. McDonnell, Jr., later that day to give him the good news; a formal announcement was issued to the press three days later.

McDonnell had gone for the contract from the beginning. In fact, as early as October, 1957, the company assigned 20 personnel to a space team that swelled to 40 during preliminary work on the Air Force Dyna-Soar boost-glider. By June, 1958, the company had 70 people at work on what would become the Mercury capsule and in October, Raymond A. Pepping, Lawrence M. Weeks, John F. Yardley and Albert Utsch, headed a team that prepared the 427 page report on company proposals. Langley worked closely with McDonnell during that summer but withdrew when the formal proposals went out, leaving McDonnell to re-work its

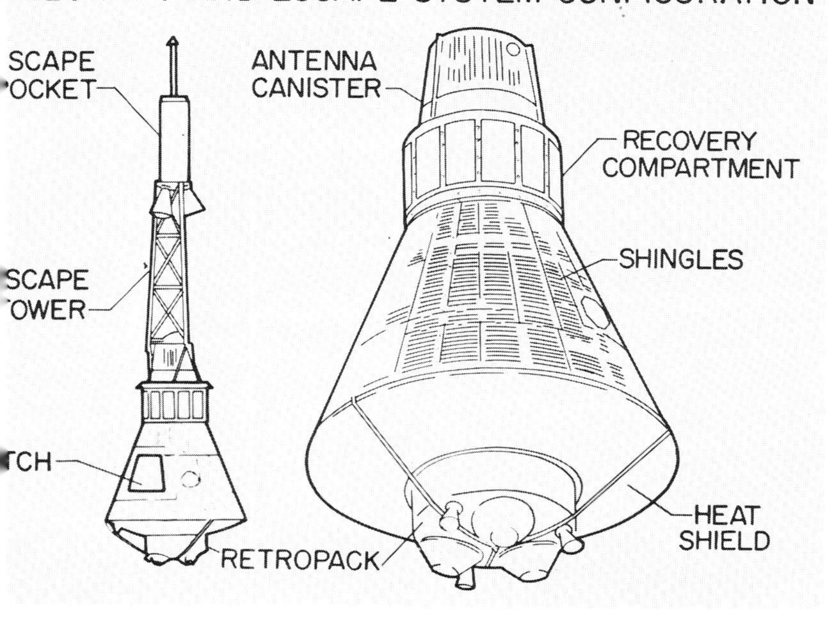

presentation. There was little doubt in the minds of many that McDonnell would get the contract; the company was visibly out in front and had every genuine qualification for the job – experience, even at this early stage, meant a lot!

The formal contract was signed on 6 February, 1959, covering procurement of 12 identical capsules for an estimated cost of $18.3 million and a company fee of $1.5 million. By this time, STG had defined the Mercury requirement and done most of McDonnell's work for them by committing to a rigid set of criteria based on study, test and evaluation carried out by Langley personnel. As outlined earlier, the Mercury capsule would have a truncated-cone shaped structure topped by a smaller truncated cone for the parachute and recovery section. Heat shield selection was still open, but the consensus favoured a heat-sink material. Launch escape would be provided by a tractor motor in the form of a solid propellant rocket attached to a lattice tower installed on top of the capsule; McDonnell preferred pusher rockets at three points around the base of the capsule but the STG decision stuck.

The capsule would be equipped with small attitude control thrusters so that it could be turned around from its forward facing orientation for retro-fire – ignition of small, solid propellant rockets at the base which would decelerate the capsule and start it down toward the atmosphere. Held at the correct attitude by the thrusters, Mercury would protect its pilot from the extreme heat of re-entry by couching him in a form-fitting supporting seat and

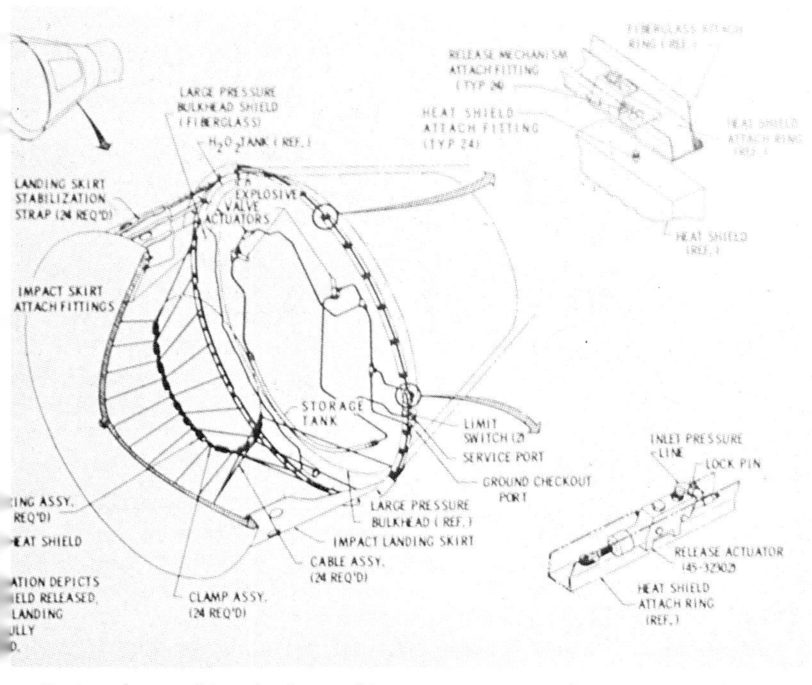

Designed to cushion the force of impact on water, and to prevent serious injury during an emergency land touchdown, the air-filled bag was folded up inside the base heat shield until the latter was released during the final minutes of descent.

bring him to a safe landing by deploying parachutes for the final descent.

Mercury was required to descend on water because land covered only a small fraction of the globe over which the spacecraft would fly. No one knew for sure just how reliable a manned spacecraft would be; it might have to be brought out of orbit at short notice, causing it to descend in an unplanned recovery area. Odds were that in any kind of abort it would come down over water. Another reason for a planned water landing arose because the Mercury capsule had little or no lift. Falling on a ballistic path, water would serve to cushion the impact and descending to a splashdown permitted engineers to use conventional parachutes which, although slowing the spacecraft to a descent rate of 33 km/hr, would have been unacceptably high if landing on a hard surface. In addition, the capsule was designed to deploy a landing bag stowed between the heat shield and the aft bulkhead. After re-entry, the bag would be deployed, pushing the heat shield away from the blunt end of the capsule so it could hang 1.22 meters below the capsule on straps slightly shorter than the bag, or 'skirt,' which was perforated with holes. Air trapped inside the landing bag was expelled through the holes when the heat shield hit the water and the weight of the capsule forced it to collapse, cushioning the impact from about 40 g for a few milliseconds without the bag to no more than 16 g under the worst sea conditions. The system was satisfactorily tested in winds up to 30 km/hr.

Four boosters would be employed for the Mercury program: Little Joe, Redstone, Jupiter and Atlas. Little Joe was the name given by Faget to his proposed ballistic launcher for sending dummy capsules on test flights to high altitude. Originally called High Ride, the idea evolved to acquire the name Little Joe from engineering drawings that showed four hole locations for the modified Sergeant solid propellant rockets like 'the crap game throw of a double deuce on the dice.' These rockets were called Castor or Pollux according to the specific modifications incorporated and the quartet later acquired four supplemental solids in the form of Recruit rockets.

Little Joe was highly adaptable, its eight possible solids could be arranged in any firing order desirable, and some rockets could even be removed for specific test requirements. At maximum, Little Joe could propel Mercury skyward on a thrust of 104 tonnes, enough to accelerate the capsule to a speed of 6,600 km/hr and a height in excess of 160 km. But if capable of reaching orbital *height*, Little Joe was far below the required performance level for taking a capsule to the 28,000 km/hr of orbital *speed*. Nevertheless, it could provide valuable opportunity for qualifying systems critical to the safety of the pilot. North American Aviation's Missile Division got the contract on 29 December to build seven Little Joe boosters for the Mercury program.

The second Mercury booster – Redstone – was expected to carry Mercury test capsules on ballistic 'up and down' flights prior to manned rides on a similar trajectory. This would expose the pilot to several minutes' weightlessness and prove the capsule's ability to provide a habitable environment. With a thrust of 34 tonnes, Redstone would burn longer than the solid propellant Little Joe to push the capsule to a speed of more than 8,000 km/hr. On 16 January, 1959, NASA ordered eight Chrysler Redstone missiles from the Army Ordnance Missile Command at Huntsville, plus two Jupiter rockets.

Not to be confused with the Jupiter C missile developed for re-entry tests and adapted as the Juno I space launcher, which was in effect a modified Redstone, the Jupiter A Army missile projected for application to Mercury would have taken a test capsule on one high-speed heat shield qualification flight. With a thrust twice that of Redstone, the Jupiter would have carried Mercury to a speed of 17,600 km/hr; two were ordered to ensure a back-up in the event of failure. Atlas, with its thrust of 140 tonnes, would place test, unmanned and manned capsules in orbit. NASA's Space Task Group placed the first Atlas order with the Air Force Missile Division at Inglewood, California, on 24 November, 1958.

As the new year came in everyone at STG and McDonnell was confident of quick success. The plan called for the first operational launch in July, 1959, when a Little Joe would perform a test of the escape system. Also that month, an Atlas would send a Mercury capsule on a heat shield development flight. The first Jupiter launch was set for the last quarter of the year with a Redstone launch in December. The first Mercury-Atlas unmanned orbital mission would fly in January, as would the first manned ballistic Redstone attempt. Manned orbital flight was to have been achieved in April, 1960, with five more such attempts up to the September of that year. In short, 19 months were deemed sufficient to move from project go-ahead to the first US manned orbital space flight. If there was a surfeit of anything that spring, it was confidence.

Within the first three months, however, it looked as though finances would get completely out of hand. NASA signed a contractual agreement with McDonnell for 12 capsules totalling $18.3 million – $800,000 higher than the company had originally estimated. Believing McDonnell to have been allocated a price loaded on the generous side, Abe Silverstein was unprepared for the statement, a month after the definitive contract had been signed, that the project would, in McDonnell's estimation, cost at least $41 million with spares, test equipment and ground support items. To place this in context, NASA's first nine months of operation covered the last three-quarters of Fiscal Year 1959, which ended on 30 June, 1959. For that period it had a budget authority of $353 million. Because the financial year was already running when NASA came into being, the funding profile for that first period is complex but essentially provided the Administration with a sum comprised of 50% from the Defense Department, in projects transferred to the new civilian space effort, 25% transferred as unspent money from the FY1959 NACA budget, and 25% as new money.

For FY1960, covering the period from July, 1959, to June, 1960, NASA had requested a budget totalling $485.3 million. Of that, $333 million was for research and development from which $70 million would be allocated to manned space flight. When Silverstein heard the alarming news that McDonnell had increased project estimates by more than 100% in a single month, he knew it would, if left unchallenged, play havoc with total NASA estimates and budget predictions for the coming financial year. But more than that, given that a new agency or administration would have great difficulty in accurately planning the detailed cost structure of untried development projects, it would, nevertheless, be a bad start to have need of Congressional requests for supplemental awards. Silverstein was dismayed to find that cost escalation seemed to be hitting not only the private contractor segment but intra-governmental negotiations as well. For instance, in January, 1959, the Air Force BMD informed NASA that the cost of an Atlas had suddenly grown to $3.3 million, a 32% increase. Moreover, the ABMA had also just put up the cost of Redstone and Jupiter missiles. In fact the cost of a Jupiter now equalled the cost of the more powerful Atlas and this forced a reassessment of the test program. Jupiter was cancelled on 1 July, its mission objectives being presumed accessible from Atlas flights.

Another test segment also got moved out because of cost increases. It had earlier been suggested that a large balloon could carry a capsule to an altitude of approximately 25 km from where it would measure the near-space environment over extended periods, determine the amounts of radiation that could be expected, and 'soak' the spacecraft in the temperatures it could be expected to reach. Instead, a large high altitude wind tunnel at the Lewis Research Center would be used to simulate the suspended test. Despite good progress, the realities of development planning and mission schedules, not to mention a myriad minor problems that loomed over the horizon, began to push out the timeline for accomplishing significant milestones. Within two months of the original schedule, outlined earlier, STG had re-drawn its objectives in the first of many adjustments that signified a learning curve more appropriately applied to following projects.

In March, 1959, Jupiter and balloon flights were still in the schedule, and NASA anticipated the first manned ballistic Redstone flight no earlier than the end of April, 1960. and the first manned orbital Atlas launch in September, 1960. Already the NASA work force was building up. From 35 Langley personnel and 10 from the Lewis Center, STG grew to embrace 150 by January, 1959, and 350 by mid-year. At the beginning of the second quarter, the Task Group got an unexpected influx of predominantly British engineers; more than 100 moved out from A. V. Roe in Toronto, Canada, when a decision was made to cancel the CF-105 Arrow all-weather interceptor. Arrow had first rolled from its assembly plant in October, 1957, but the British Government felt its role could be filled by the US Bomarc ground-to-air missile. Mercury got 25 ex-Avro aero-engineers.

But sheer manpower was not enough to keep hold on the optimistic schedule and as the lines of development and flow charts reflecting milestones grew on walls in offices at Langley, management became aware of the enormous task they had been given. On 17 April, Bob Gilruth announced during a speech before the World Congress of Flight in Los Angeles that the first manned orbital Atlas flight would not take place 'within the next two years.' In January it had been set for April, 1960; in March it was pushed back to September, 1960; now, a month later, no one expected it before May, 1961. In reality, it would be nine months beyond that date. But NASA was not exclusively concerned with manned space flight, indeed Administrator Glennan was concerned that Mercury should not consume the other, albeit less spectacular, segments of the emerging space program.

In May he paid his first visit to STG at Langley and returned to Washington impressed by the pace of effort, the drive exhibited by enthusiastic personnel, and the balanced autonomy that gave it momentum. STG could not for long remain at Langley, the size of the operation precluded it from staying at Langley Field. In anticipation of this, the National Aeronautics and Space Act, passed through Congress in the summer of 1958, incorporated legislation approving construction of a Space Projects Center 'on land already owned by the federal government' in Greenbelt, Maryland. The new facility would be the home of the Naval Research Laboratory's Vanguard group, the Project Mercury personnel, scientists and engineers working on space science and applications, and specialists planning new tracking and data acquisition sites around the world. Known as the Beltsville Space Center, it came into formal existence on 15 January, 1959; the first construction contracts were let in April.

On 1 May, NASA publicly announced that the facility would be re-named the Goddard Space Flight Center, in memory of Dr. Robert H. Goddard who flew the first liquid propellant rocket in 1926. Bob Gilruth would become GSFC's Assistant Director for Manned Satellites, divisional head of one of six departments. Had the Space Task Group been fully integrated into the Goddard Center, a major division of interests that would permeate NASA for more than a decade might never have gained hold. As it turned out, manned space flight grew in size and complexity, consuming large portions of the NASA budget, and from a facility far removed from other NASA Centers, came to be seen as a dangerously hungry element in the organization. Glennan gave STG a level of perpetuation, allowing it to spawn apart from the rest of NASA, providing a climate in which manned space flight could become a separate arm of the Administration. In coming years, many scientists and engineers would curse against the apparent success manned flight had in securing money from Congress; while other projects were cut and trimmed, manned flight became the bright star in international affairs.

It may not have been so had Glennan not allowed STG to be apart from the beginning and had the Langley team moved into the Goddard Center as planned. They did not actually set up shop in Maryland and would construct their own Manned Spacecraft Center in a later year. In 1959, however, the divisions were clear, and growing fast. It was inevitable that the Mercury pilots would be a breed apart from their peers in the adopted attitude of the press and public; in reality, they were just like any other service test crew and a lot more amiable than some. Nevertheless, the seven men that could call themselves the first American astronauts would be a special team assigned a unique job.

In 1959 no one really knew for sure that NASA would have a manned mission in space beyond Mercury. Plans certainly existed for a follow-on program, but Congress had yet to pass judgement and even the scientists and engineers putting together the new space program were unsure of the goals and objectives a future project should have. Many simply wanted to wait and see how well a man reacted to the space environment before committing time and money to future plans. So in the year that NASA chose its first seven astronauts there was no clear mandate for the future and the select group of pilots was in every sense a unique group.

The selection process began as a set of criteria laid down by STG in November, 1958. The plan at the time was to gather 150 candidates from open applications; select 36, from which 12 would go forward for a nine-month period of intensive training; at the end of which 6 would be formally invited to the post of astronaut. With a measure of chauvinism impossible today, applicants were to be exclusively male, between 25 and 40 years of age, less than 180 cm tall, in excellent physical condition, hold a degree in science or engineering, have at least three years experience in either physical, mathematical, biological or psychological sciences, three years experience in a field of engineering, or three years experience of aircraft, balloons or submarines, alternatively a Ph.D or 6 months medical work. Each applicant would have to submit a sponsor to ensure the bona-fide nature of the request.

Before the end of the year the plan had been abandoned at the behest of President Eisenhower, who preferred to see NASA select a group from the armed services' test pilot schools. Early in January, 1959, leading NASA management personnel laid down the new guidelines. Age, size and educational requirements stayed the same, with the applicant now required to be the product of a test pilot school with 1,500 hours flying time as a qualified jet pilot. From the beginning it was believed the pilot would have a large measure of control over his situation; ostensibly relegated to flying around the Earth like 'spam in a can,' STG was convinced that 'success of the mission may well depend upon the actions of the pilot.'

The Defense Department provided 508 records, from which Stanley C. White, Robert B. Voas and William S. Augerson chose 110 men: 5 Marines, 47 Navy personnel, and 58 from the Air Force. STG's Assistant Director Charles Donlan arranged to have three groups come to Washinton for further screening. Beginning on 2 February, White, Voas and Augerson interviewed the men in the first and second groups, a total of 69 applicants so well qualified for the job that it was readily apparent the third group would be an unnecessary burden on the selection process. Of the 69, six were eliminated on height grounds, 56 were passed along for written examinations and psychiatric tests, with 36 qualifying. The final selection process would first require volunteers to undergo rigorous and demanding medical tests at the Lovelace Clinic in Albuquerque.

Thirty-two accepted and the first group of 6 men arrived at Lovelace on 7 February, there to undergo 7½ days and 3 evenings of exhaustive physical examinations. The other 26 candidates were returned to their home bases to wait until their turn came for the same period of tests at the Clinic. Four more groups of 6 arrived at Lovelace on succeeding Saturdays, with a final party of 2 a week after that. Complete medical histories were examined with emphasis on a study of blood and tissue samples, x-rays, eyes, ears, nose and throat, heart and circulation, nerves and muscles, and general medicine. For assessing physical competence, candidates were required to pedal a static bicycle while breathing into a mask. Heartbeat and oxygen consumption were measured at progressively more demanding pedal pressures, with final measurements determining the load that could be pedalled at a heart rate of 180 beats per minute. The Atomic Energy Commission's Los Alamos Laboratory moved in with a total body radiation count to determine the level of potassium in each applicant.

Specific gravity was determined by submerging the subject in water, and weighing him in air. Blood volume was obtained by giving each candidate a small quantity of carbon monoxide and observing the rate of dilution. The presence of heart chamber openings were determined by measuring the amount of blood oxygen after a Valsalva exercise, where the nose is blocked and the lungs exhaled through a tube. At the end of the rigorous analyses, only 1 applicant failed to pass through. The remaining 31 were subjected to stress tests at the Wright Air Development Center, with five groups of 5 and one party of 6 each spending a week at the Center from 15 February; the first group moved to the WADC immediately after their examinations at Lovelace to maintain a constant flow.

Physical fitness was measured by having each subject step up 51 cm to a platform and down every 2 seconds for 5 minutes. A treadmill was also employed comprising a moving platform which elevated 1° every minute. Seeming to remain still by walking in the opposite direction to the moving belt, the subject's workload was assessed by determining the angle of the belt when his heart rate reached 180/min. Pulse and blood pressure were taken as the subject plunged his feet into a tub of ice water. Heart condition was measured as the body compensated for a tilting table, each applicant laying flat on his back for 25 minutes at an acute angle. Adaptability and orientation were determined by placing the applicant in a dark, soundproof room for 3 hours. Centrifuge tests spun each subject to high g loads. Chambers were heated to 54°c, in which subjects were placed for 2 hours. Equilibrium tests required each candidate to stabilize a chair on which he sat, which moved in two axes simultaneously, using a small control stick, with and without a blindfold. Long periods of high frequency noise were a means of measuring susceptibility to different sounds. Psychological tests were the most punishing; each applicant knowing himself to be required to match criteria unknown to any but the medical personnel. Motivation tests, peer ratings, spatial visualization. In all, 13 separate psychological analyzes.

The final selection procedure took the records of 18 recommended candidates to Langley where Donlan, North and White passed names to Bob Gilruth. It was found impossible to get the final number down to the magic 6 stipulated originally. Gilruth approved 7 names and passed these to Abe Silverstein; final concurrence was obtained from Administrator Glennan. Seven men for space: Lt. Col. John Herschel Glenn, Jr. (Marine Corp); Lt. Cdr. Walter Marty Schirra, Jr., Lt. Cdr. Alan Bartlett Shepard, Jr., Lt. Malcolm Scott Carpenter (Navy); Leroy Gordon Cooper, Virgil Ivan Grissom, Donald Kent Slayton (Air Force). The seven astronauts were presented to the world on 9 April at a press conference in Washington, D.C. Most of that year would be spent in thoroughly familiarizing them with every aspect of the Mercury program and its capsule, with academic studies in space and 'astronautics.'

But from the beginning there was a vigorous debate over the type of environment the spacecraft should encapsulate; removed from any semblance of an exterior atmosphere, the interior of the spacecraft would be a miniature planet for the pilot. It is difficult to appreciate the reassurance an astronaut would receive from knowledge that his spacecraft was a warm, friendly, comforting place to be in amidst the cold, radiation-filled vacuum of orbital flight. Despite cramped quarters, the spacecraft's hospitable environment would substantially counteract the psychological fear of being so close to an uninhabitable environment; its confined volume seemed literally to grow in magnitude as the capsule reached orbit. There is literally no psychological compensation for this effect which is completely lost to a casual observer peering inside a spacecraft which excites claustrophobic tensions. But none of this was known before a man actually flew into space and the environmental options were based on scientific and engineering considerations rather than on the psychology of space flight.

The scientific requirement was quite clear, and demonstrated that if compromises were an acceptable part of any detailed and complex engineering venture they could play no part in the physiology of human needs. Atmospherically, this meant that the pilot must receive about the same partial pressure of oxygen as he does at the Earth's surface. On Earth, the atmosphere consists of a 760 mmHg mixture of 21% oxygen and 79% nitrogen, plus several trace gases which can be ignored for temporary and artificial environments. Because the quantity of oxygen in the lungs regulates the amount entering the blood, any appreciable increase in a partial oxygen pressure of 160 mmHg will cause irritation of mucous membranes and could disturb the function of critical enzymes. However, even a moderate fall below the normal partial pressure value quickly causes brain damage and severe side-effects.

The options available to Mercury engineers were that the capsule could contain either a two-gas mixture of oxygen and nitrogen at sea-level pressure (760 mmHg) or a single-gas atmosphere of pure oxygen at about 160 mmHg. Tests revealed that the pure oxygen pressure limits were actually between 150 mmHg and 345 mmHg. Oxygen pressure for a single-gas atmosphere outside these limits would cause severe, if not permanent, damage. Elimination of a second gas from the capsule's environmental control system eased weight problems, made the equipment less complicated and therefore more reliable, and because of the dramatically reduced pressure level reduced the likelihood of a cabin leak. It was inevitable that whatever atmosphere was chosen it would have to be contained within the thin walls and the welded joints of a lightweight capsule. With a pressure less than one-third that at sea-level the casule would be less prone to bleed its atmosphere to the vacuum of the space through the inevitable cracks and porous joints. In any event, as with high performance aircraft, the astronaut would wear a pressure suit into which he could withdraw for all essential life-support functions.

This would be the design philosophy throughout the next two decades: a welded internal pressure vessel as prime

The select seven, America's first astronaut team, from right, clockwise: L. Gordon Cooper, USAF; M. Scott Carpenter, USN; Alan B. Shepard, USN; Virgil I. Grissom, USAF; Donald K. Slayton, USAF; John H. Glenn, USMC; Walter M. Schirra, USN.

environmental container, backed up by a suit capacity for keeping the crew alive until the capsule could be returned to Earth. But selection of a 100% oxygen environment had one potential drawback: materials soaked in the single-gas atmosphere combusted more easily and burned ferociously. In fact an oxygen fire was almost uncontrollable, and certainly would be fought with an almost impossible chance of extinguishing it within the cramped confines of a Mercury capsule. Nevertheless, Stanley C. White, one of several engineers from STG, working with environmental control system contractor AiResearch Manufacturing, pressed hard for the pure oxygen approach. John F. Yardley and John R. Barton of McDonnell agreed with Faget and Richard S. Johnston from STG, that a pure oxygen atmosphere at about 258 mmHg was the better choice because of its inherent 'reliability of operation,' unlike a two-gas system which could cause a major 'increase in complexity' of the 'control system and in monitoring and display instrumentation.' In short it would alleviate potential problems with the cabin structure by allowing a much reduced internal pressure, and simplify the environmental equipment.

The pilot's physiological requirements led AiResearch engineers to propose an environmental control system, or ECS for short, capable of handling over a period of 28 hours adequate heat rejection capacity at a pressure of 258 mmHg. Faget, White, Voas and the astronauts' personal physician Dr. William K. Douglas, late of the Office of the Surgeon General, USAF, agreed that the planned rejection capacity should be increased and that the assumed water production rate of 2.72 kg/day should be increased by at least 15%. Oxygen consumption was estimated at 500 cc/min at standard temperature and pressure, a cabin leak rate of 300 cc/min was accepted and carbon dioxide production was assessed at 400 cc/min, or 1.18 kg per day.

Odors were to be removed from the cabin by activated charcoal and carbon dioxide would be absorbed in lithium hydroxide. Excess heat would be removed by a water-evaporator acting as a heat exchanger. To operate the system, oxygen, cooling water, lithium hydroxide and an electrical source were required. The ECS equipment would be situated below and behind the astronaut's form-fitting couch, located for the most part between the pilot and his heat shield. Although primarily designed for automatic operation, controls placed within reach of the astronaut permitted manual override for critical functions and selectable conditions. The latter included cooling water flow rate, temperature, etc. In preparing the ECS equipment it was apparent that high standards of cleanliness would be essential for the satisfactory operation of the many small components and the successful performance of the complete system: with several tiny orifices, miniature pipes and joints, small specks of dirt or grit could make all the difference between a working ECS and a malfunctioning death trap.

In the loneliness of space, the most important function of the spacecraft was to ensure the survival of the human passenger in as fit and responsive a condition as possible. Accordingly, AiResearch built a surgically clean work environment in which the ECS could be assembled and tested. These 'white rooms' would come to be a familiar feature of manned space operations. Astronaut 'Wally' Schirra was assigned to monitor the development of the Mercury space suit. Each astronaut would assume responsibility for the development and production of a key element of Mercury and report to the rest at meetings designed to familiarize the group with the pace of the program and any knotty problems that looked like needing special attention. Maxime Faget and Stanley White were the first to suggest adoption of an existing pressure suit used by the Navy and the Air Force, suitably modified for the unique requirement of orbital flight. Because it was to serve as a back-up to the capsule's environmental control system, only coming into operation to support the astronaut's metabolic requirements if the cabin failed for some reason, it was possible to base its design on currently available models.

During the first half of 1959, the David Clark Company, the International Latex Corporation, and the B. F. Goodrich Company, vied with each other for the prime space suit contract. By 1 June, each competitor had submitted a representative product and the Air Force Aeromedical Laboratory at the Wright Air Development Center had the job of testing them, with support from the Navy Air Crew Equipment Laboratory, the former favouring its principal contractor – Clark – and the latter selecting its own pressure suit manufacturer – Goodrich. A final decision was left up to Langley but STG wanted more time, deferring until 22 July selection of the Goodrich Company. Their Navy Mk IV suit would serve as the base-line from which to develop the Mercury garb. Because the suit was an integral part of the environmental control system, in that it was required to operate at short notice from a malfunctioning capsule, the specification was changed continually during the next two and a half years. A major procedural change was necessitated by a manned test of the complete life support system in April, 1960.

As designed, the suit would start to provide primary life support functions if the capsule's pressure dropped below 237 mmHg. On the launch pad, however, the capsule was expected to contain a mixture of oxygen and nitrogen at sea-level pressure. As the capsule ascended, its internal pressure would fall in accordance with the reduction in atmospheric pressure until, at a height of about 8.2 km, internal pressure had bled down to 258 mmHg. A cabin relief valve would seal at that pressure, leaving the capsule to continue on into space with a cabin mixture of oxygen and nitrogen. However, from a height of only 3 km, a 0.45 kg supply of pure oxygen would flow into the cabin so that by the time the spacecraft reached the height at which the cabin relief value was activated, it contained 66% oxygen and 33% nitrogen. This would provide a partial oxygen pressure of 171 mmHg, just above the minimum recommended level of 150 mmHg. As the mission progressed, oxygen from the ECS would replace nitrogen lost by normal leakage. But pure oxygen woald be used in the suit circuit and the April, 1960, test was designed to check that element of the system in a simulated run of the aforementioned launch policy.

A McDonnell engineer had been under test for about one hour when he collapsed into unconsciousness due to hypoxia – a decrease in the amount of oxygen reaching the brain. It was found that leaks in the instrumentation lines reduced the partial pressure oxygen level because of a negative difference in pressure between the suit and the cabin. Consistent failure in the circuit led to a decision to change the procedure and to purge the cabin with 100% oxygen for several hours so that there would be insufficient nitrogen to leak up into the suit loop. Although it got rid of a potentially serious deficiency in the ECS, it put in operation a procedural change that compromised the safety of future ground tests.

The reason for carrying a mixed-gas cabin into orbit had been to dilute the pure oxygen environment and so reduce the risk of uncontrollable fire on the launch pad. For normal operation, the suit would deliver oxygen through a waist connector at a temperature of 35.5°C and a relative humidity of 58%. It would circulate up to the face area, with discharged gases exhaled by the astronaut passed through the ECS lithium hydroxide scrubbers, odor absorber and heat exchanger from where it would be delivered back to the suit topped up with pure oxygen. Pressure in the suit was maintained by a demand regulator, which also sensed a loss of cabin pressure and put the suit circuit into operation.

The suit would also be called upon to support the astronaut if a fire began in the capsule while the spacecraft was outside the atmosphere. Plans here were to de-pressurize the interior immediately and so dump the oxygen gas that would feed the conflagration. Without oxygen it would immediately go out. Initially, 13 space suits were ordered to fit astronauts Schirra and Glenn and other personnel at McDonnell and NASA before the final order for 8 operational suits. Dr. 'Bill' Douglas had a suit. But then, Bill had one of everything, for it was his lot to duplicate every test, every simulation, every physical stress any or all of the select seven were exposed to. In every sense, he was the eighth astronaut – an unsung hero of those early days.

Schirra chased every little problem with the suit design, and there were many. Basically transferring technology and operating concepts from Navy and Air Force experience, the dictates of the space age were making their presence felt. Most changes came from the need to reach seemingly obscure regions of the cabin, discomfort brought about by a less amenable attitude from service technicians. By May, 1960, basic ground rules had been established for the suit and ECS, requiring the latter to deliver to the cabin no more than twice the pressure applied to the suit, limiting to 2% or less the volume of carbon dioxide in the system at sea level, and with further limitations on carbon monoxide and potentially toxic fumes. The anticipated requirement for 1.8 kg of oxygen was matched with a supply of 3.6 kg contained in two spherical bottles at 527 kg/cm^2, reduced to 7 kg/cm^2 upstream of a demand regulator, and delivered to the cabin and suit circuits at about 0.35 kg/cm^2.

But there were grave warnings from psychiatrists who cautioned about the optimism reflected in ebullient anticipation of space flight. D. G. Starkey warned that, 'Isolation produces an intense desire for extrinsic sensory stimuli and bodily motion, increased suggestibility, impairment of organized thinking, oppression and depression, and in extreme cases, hallucinations, delusions, and confusion.' From this and other warnings, but more probably from the basic inability of any of the seven astronauts being capable of relinquishing control to a set of automatic controls, considerable attention was given to setting up a global communications network. Doctors were concerned that the extensive medical analyses planned for the post-flight debriefings should be adequately prefaced by continuous transcripts of air-to-ground communication during the mission.

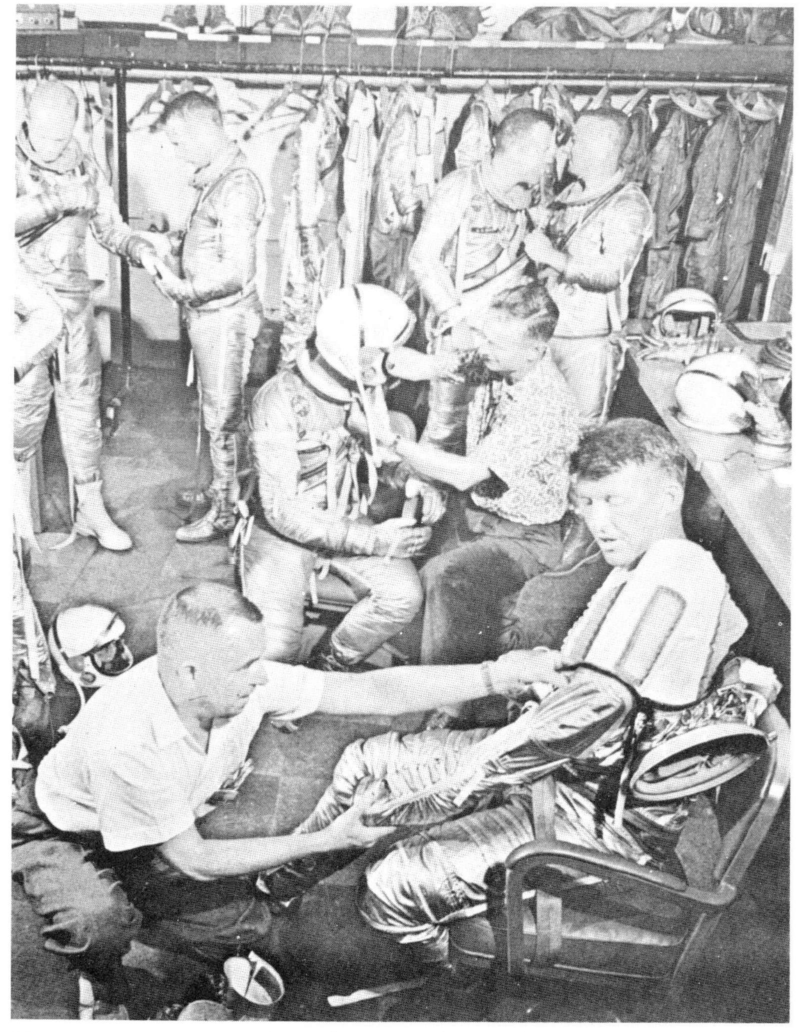

Mercury astronauts try suits on for size in the world's most expensive tailors.

It would be virtually impossible to cover every minute of orbital flight with a two-way communications and tracking capability. For one reason, the comparatively low altitude of the spacecraft's path across the Earth would carry it from horizon to horizon in short order; for another, the Earth would slowly spin beneath the capsule so that its orbit would appear to 'walk' westward on successive passes. Had it been conceivable to plan and build a network of ground stations covering every segment of a single orbital track it would have been impossible to do so for every possible groundtrack during a complete mission. A compromise was obtained whereby voice and telemetry from the spacecraft would not be interrupted for a period exceeding 10 minutes, with the additional proviso that each period of communication would last for not less than 4 minutes.

Two segments were involved: the link established between the station on the ground and the spacecraft in orbit, and the inter-station communication lines leading back to the Mercury control centre. In addition, functional requirements included provision of adequate tracking and computing facilities to determine the capsule's orbit accurately for both normal and aborted conditions, a command capability for the control centre to initiate re-entry over the planned or contingency recovery areas, and a capacity to command an abort in the event the astronaut was disabled for some reason. The basic communications network was established on the premise that a typical Mercury flight would last 3 orbits, or nearly 5 hours. This meant that there would be an infringement of the rule stipulating a 10 minute maximum between stations, another necessary compromise. The communications would originate from the Mercury Control Center located at Cape Canaveral and would coordinate global operations. Control of the network, switching operations, and routing distribution of returned data, would be handled by the Computing and Communications Center, at the Goddard Space Flight Center in Maryland.

By arrangement with the Department of Defense, NASA was able to use the Atlantic, Pacific and White Sands Missile Ranges, and the Eglin Gulf Coast Test Range, which, with sophisticated tracking and command facilities, saved considerable cost and development time. Australia provided some of its existing stations and loaned its Department of Supply to assist with setting up new ones with the Australian Weapons Research Establishment. At this time Australia had a test range at Woomera which was primarily concerned with building up a test programme with Britain's Blue Streak missile; the project was cancelled before Mercury reached the operational phase, however. The United Kingdom gave its permission for sites to be built on Canton Island and Bermuda, Nigeria leased land for stations at Tungu and Chawaka, and Spain provided land for a Canary Island station.

Network requirements covered several important criteria: ground tracking was essential for position data which would be immediately sent to Goddard; computations during the mission about events still to come were essential for predicting re-entry displays at Mercury Control; real-time telemetry display was needed at the network sites (real-time meant a direct display of information received without having it taped before examination); provision of a voice, teletype and radar communications network between stations. Radar was essential during the early phases of the flight when the spacecraft would be riding an Atlas missile into orbit, and during the final portion of re-entry when the capsule's position would be computed for the recovery forces waiting to pick the pilot up. It would also be required, to a lesser extent, along the orbital path for periodically updating the known parameters: information on height, speed, predicted orbit, etc.

Standard FPS-16 radar units were improved to expand their capability from a range of 463 km to 926 km. One unit, located at Bermuda, was improved to a capacity range of 1,852 km so that it could track for a longer period the critical orbit insertion phase when the Atlas sustainer engine was shut down. Verlort radars filled the requirement for S-band tracking and little modification was needed to bring them in with the rest of the network. The Mercury spacecraft would have both C- and S-band beacons: the former was accommodated by the FPS-16 radar, the latter by Verlort. (C-band is the designation, applied during World War Two, to frequencies between 3.9 and 6.2 gigahertz; S-band covers the range of frequencies between 1,550 and 5,200 megahertz.) But it was realized that these radar units would have great difficulty acquiring the tiny capsule speeding across the sky, so an analog computer was suggested which could be supplied with predicted spacecraft arrival times and position date. This was no good because the reliability would be lower than acceptable. A new acquisition aid was finally conceived which would pick up the spacecraft's telemetry signal and automatically track the capsule.

But the area that concerned many more people than engineers actively engaged on tracking the spacecraft and controlling the networks was that which carried the astronaut's voice from the spacecraft to the ground, and words from the stations to the orbiting capsule. Voice transmission and reception was to be on both HF (high frequency) and UHF (ultra-high frequency) channels; bands in the 3,000–30,000 kilohertz and 300–3,000 megahertz range, respectively. The former would serve as back-up to the latter. Command functions between the ground and the spacecraft were accommodated with dual 500-watt transmitters which had a range of 1,300 km.

On the spacecraft, communications equipment included two primary and two secondary voice radio links, a high frequency and a low frequency telemetry transmitter, two command receivers and two (C-band and S-band) radar beacons. Two prime and one back-up recovery beacons were provided. It was also planned to have a seven-track tape recorder on board to record telemetry and voice transmissions. The layout of Cape Canaveral's Mercury Control Center (MCC) was based on the need to acquire pertinent data on a regular and accessible basis, to talk with the pilot from a single console position, to monitor the pilot's physical condition as far as possible from telemetry data, and to co-ordinate support functions such as networks, recovery teams and support personnel. Overall, a flight director would be responsible for the conduct of the mission and in a very real sense serve as conductor to the select orchestration of specialists and experts in the Operations Room. His word was law, although the effort was a cooperative one which could only serve effectively through a team. Discipline was tight and no one was allowed to step beyond his authority. When it came to sending manned capsules into orbit the success of the mission would depend as much on the cool and controlled response of the flight team as on the technical perfection of the equipment.

The Operations Room was a moderately large area covered along one complete wall by a large map of the Earth in Mercator projection, a display with alphanumeric indicators that told controllers exactly where the spacecraft was at any moment along the orbit lines that snaked like sine-waves across the illuminated board. Facing the display, at a respectful distance, were two rows of consoles. The front row, from left to right, was for the support control coordinator, the flight surgeon, the environmental control system monitor, the capsule communicator, and the systems monitor. On the second row, elevated from the first to allow personnel a clear view of the wall display, was the central platform of the flight director, suitably raised on his commanding position of authority, flanked by the recovery status monitor and the range safety observer to his left, and the network controller and the missile telemetry monitor to his right.

Along the right-hand wall, between the back of the room and the global display, sat the retro-fire controller and the flight dynamics officer. These positions were supplementary to the main mission operation in that they were concerned with specific events that occurred only once during a flight or produced data that would support other activities. The flight dynamics job was required to generate orbital data and to predict spacecraft lifetime – not of the systems but of the orbit itself which, comparatively close to the Earth, would feel the effect of the tenuous outer layers of the atmosphere pulling at the capsule and slowly changing its path around the globe.

The retro-fire officer was in charge of computing just where and when the solid propellant retro-rockets would have to be fired so that the capsule could be brought down as close to recovery forces as possible. These two positions would grow into significant importance during later manned programs. Facing them, on *their* wall, were TV monitors and recorders. The illuminated world map was flanked by two large panels, each angled to partially face the double row of consoles, on which were carried 16 trend charts. Operations from the MCC would commence nearly two weeks before a flight and continue on through launch, orbital flight, re-entry, recovery and the return of the astronaut to a surface ship.

In addition to the Canaveral facility, NASA set up a secondary control center at Bermuda. Situated downrange, it would ensure sufficient time following orbital insertion to assess the viability of the vehicle – a tightly timed event from Canaveral – and serve as a back-up control center in the event of communications failure from the prime site. About half the size of Canaveral's MCC, the Bermuda center was arranged in a similar way to the main center, with a smaller global wall display and five consoles in a single row: flight surgeon, environmental control monitor, capsule communicator, systems monitor, and the flight dynamics officer. A flight supervisor behind the row of five would double for the flight director at Canaveral.

In addition to the fixed sites serving communications, tracking and telemetry needs, two ships – *Rose Knot Victor* and *Coastal Sentry Quebec* – would be deployed as and where needed to fill in gaps along the orbital track. Between March and October, 1960, the Mercury network stations were developed. Completion of the last station, at Kano in Nigeria, was effected by March, 1961. Most of the work had been performed by the Western Electric Company, with the National Academy of Sciences' Arnold W. Frutkin giving valuable advice resulting from his experience organizing the interaction of foreign participation in the International Geophysical Year. In all, the world-wide system was unlike anything that had been attempted before, going far beyond the comparatively limited capabilities of the Vanguard effort, filling in several gaps discovered to exist in normal channels of global communication. When complete, the Mercury network consisted of 97,000 km of telephone line, 165,000 km of teletype wire, and 24,000 km of data circuits; sufficient cable to girdle the globe more than seven times.

At the Space Task Group, management was being refined and honed for the arduous flight tests. Under Gilruth, Charles W. Mathews controlled the Mercury Operations division, Maxime Faget organized the Flight Systems division and James A. Chamberlin the Engineering and Contract Administration. By the end of 1959, Gilruth had assumed the new title Director of Project Mercury and Chamberlin took on the role of Capsule Co-ordination Committee head. Packed tight in an old building on the eastern side of Langley Field, STG acquired Walter C. Williams from the High Speed Flight Station, moved at the behest of NASA Deputy Administrator Dryden to liaise between Gilruth and Maj. Gen. Donald N. Yates, Commander, Air Force Missile Test Center. Williams was brought in to serve as an associate director for spacecraft operations, reflecting the need for an administrator used to shakedown trials with jet aircraft and rocket research projects.

He brought a new mood to Langley Field, one far removed from the slightly introverted nature of many STG engineers from the Langley and Lewis Research Centers. Kenneth S. Kleinknecht and Martin A. Byrnes came with Williams to assist in transforming a research project into an operational product. By the middle of 1960, more than 700 NASA people were on the Project Mercury staff. Over at St. Louis, McDonnell were making good progress with the capsules. Although only 400 people were actually working on the capsule production contract, the company recognized the prestigious position it would acquire as the builder of America's first 'can with a man.' But from the outset, as soon as they were selected, the astronauts played a vital part in preparing the design for satisfactory operation. As pilots they saw the capsule differently to STG or McDonnell engineers who were looking for optimum systems layout and performance criteria. Recognizing troublesome areas of equipment installation, window arrangement, controls location, or accessibility, they added a new dimension to the already daunting problems associated with building a manned spacecraft.

In the end, their contribution was certainly greater than any other group of the same size, for they helped pre-empt awareness of poor design or inadvisable layouts. One early contribution was a request for deletion of two poorly sited circular windows, a change that would have been inevitable but certainly costly and time-consuming if left until flight tests began. As it was, a considerable number of changes were to be made as a result of flight

experience and it emerged that no two capsules were exactly the same – a state of affairs that led to some contractual problems with McDonnell.

If Maxime Faget was the star of STG to whom everybody looked for unsolicited guidance on the basic specification for a manned capsule, McDonnell's John F. Yardley was the undisputed head of spacecraft builders. Although second rankers in actual administrative placing, Faget and Yardley established a rapport that led to a smoother dialogue between government and industry than could have been expected in their absence. Yardley had been with McDonnell since 1946 as a stress engineer and project manager.

Yardley's boss, McDonnell Mercury project manager Logan T. MacMillan, sought the services of tooling superintendent William Dubusker to design and prepare a white room area where dust and grit would be expelled by air flowing into a surgically clean room and where engineers and technicians looked for all the world like aliens from another planet, clad in long white coveralls, white hats, protective boots and face-masks behind transparent walls that gave tantalizing views of gleaming components and a spacecraft in the making – a new womb for a technological child of the space age. Dubusker was also called upon to perfect new welding techniques for the titanium skins that covered the structure. In fact, Mercury was almost the perfect advertising venture for the Titanium Metals Corporation of America, selected as the prime supplier by McDonnell.

Titanium had been in commerical use since 1948, in which year some 3 tonnes had been delivered. In 1960, almost 1,000 tonnes was available for the US aerospace industry, most of it going to the missiles emerging in assembly plants across the nation. McDonnell chose titanium because it is 44% lighter than steel and has excellent strength and temperature qualities. Mercury was to have two concentric structures: one to serve as an inner pressure vessel and the second as an external heat shield – not to be confused, however, with the main ablative or heat-sink shield attached to the base of the capsule. The inner vessel began life as 0.025 cm trapezoidal sheets formed into skins and fusion-welded to comprise two cones. The inner cone was made from flat-rolled titanium while the second comprised a beaded titanium outer wall; the two were joined together and

Corrugated for stiffness, the exterior skin of Mercury was supported above a flat-rolled inner layer by structural stringers, the whole assembly welded together to form a truncated cone.

Attitude control propellant was contained by the tubular reservoir seen here recessed around the periphery of the spacecraft's base, to which would be attached the impact bag and the exterior heat shield.

spot-welded while the seam welding was performed. Circumferential seam welds joined the cones between rows of beading and longitudinal welds. Hat-section stringers were spot-welded to the pressure vessel with a layer of ceramic-fiber insulation to the top face of the hat-sections.

Two bulkheads, one at the front and one at the rear, were similarly formed from welded titanium. Because of beading on the outer wall of the pressure vessel, the joined sections comprising an 0.058 cm thick structure had a rigidity equal to a conventional 0.13 cm thick structure. In all, each capsule would have 622 meters of seam and butt weld and 521 meters of spot weld. The new welding concept introduced by Dubusker replaced the normal method of performing the work in a gas chamber which surrounds the operation in an inert atmosphere. Instead, the fusion welds were done with tungsten arc equipment in the open, but shielded by inert gas on both sides; temperature control was effected by chilling the welded area. Sceptics were impressed by the standards achieved, each seam weld looking not only like a metal insert but also stronger than the flanking material. But fabrication was not the constraint that tugged at schedules during 1960.

Inevitably, the most contentious areas arose from debate on the balance between automation and manual control. Or, as some would say, between a 'bummy dummy in the can' or a 'pilot in the pot-shot seat!' It meant the same. Pilots will almost always be pilots and have unyielding faith in a system if they can personally slip in to the loop. So it was with the astronauts. They wanted control of the capsule's basic functions, if only as a selectable alternative to the primary methods of operation. It led to a maze of confusing control modes because the selection was made at the hardware level; in later manned spacecraft it would come in the software. Hardware was an engineer's term for mechanical systems, tangible, physically controllable entities; software was originally procedures and design concepts, but later came to embrace computer operations and taped command programs.

In providing several different items of control hardware the design was cluttered with redundant mechanical systems. During the early months of Mercury development, the attitude control system software contractor, Honeywell, preferred an essentially simple configuration using a digital electronic system. A completely separate system was evolved which would provide the pilot with a manual control through mechanical linkages. But the latter were expensive on fuel and STG engineers provided a facility where the pilot could switch electrically to solenoid valves that operated the automatic thruster jets. No one had real confidence that these alternate modes of control were sufficient for every contingency, so a rate orientiation system was devised which would cut in on the 'fly-by-wire' system, providing a final mode as back-up to the prime.

By the beginning of 1960 the system had essentially four control modes in two separate systems. Each reaction control system (RCS) had its own supply of hydrogen peroxide, decomposed into steam and oxygen for the exhaust jet. One system was controlled automatically, and provided 12 small thrust chambers, separated into 4 for each pitch, roll and yaw axis, with thrust levels of 0.45 kg, 2.7 kg, or 10.9 kg. This system, called system A, was operated by the Automatic Stabilization Control System (ASCS) mode or the Fly-By-Wire (FBW) mode. The second, system B, had 2 jets per axis with a variable thrust output of 0 kg to 10.9 kg or 0 kg to 2.7 kg. System B was aligned with the Manual Proportional (MP) or the Rate Stabilization Control System (RSCS) modes. The two systems had completely separate control functions, separate hardware, and separate fuel tanks. Each was completely independent of the other, except in control modes which could in some instances be crossed between the two. In all, there were 18 thrusters around the exterior of the capsule, capable of changing and directing attitude. The thrusters were incapable of moving the spacecraft from one orbit to another and were limited to rotational movement around the center of mass: rolling around the long axis, spinning around the yaw, or lateral, axis, and pitching up or down.

The ASCS mode was automatic in that it could provide all the necessary attitude functions for a complete mission without any action on the part of the pilot and was the mode used during unmanned tests of the spacecraft. The FBW system would be operated by a hand controller operating the thruster solenoids by electrical command. The MP mode worked proportional control valves via mechanical linkages similarly operated by the hand

SPACECRAFT INTERNAL ARRANGEMENT

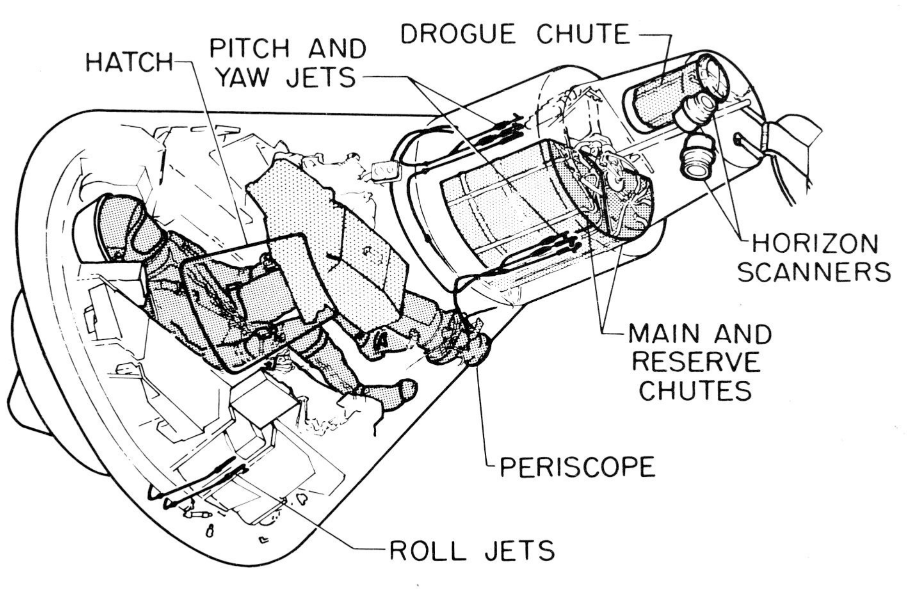

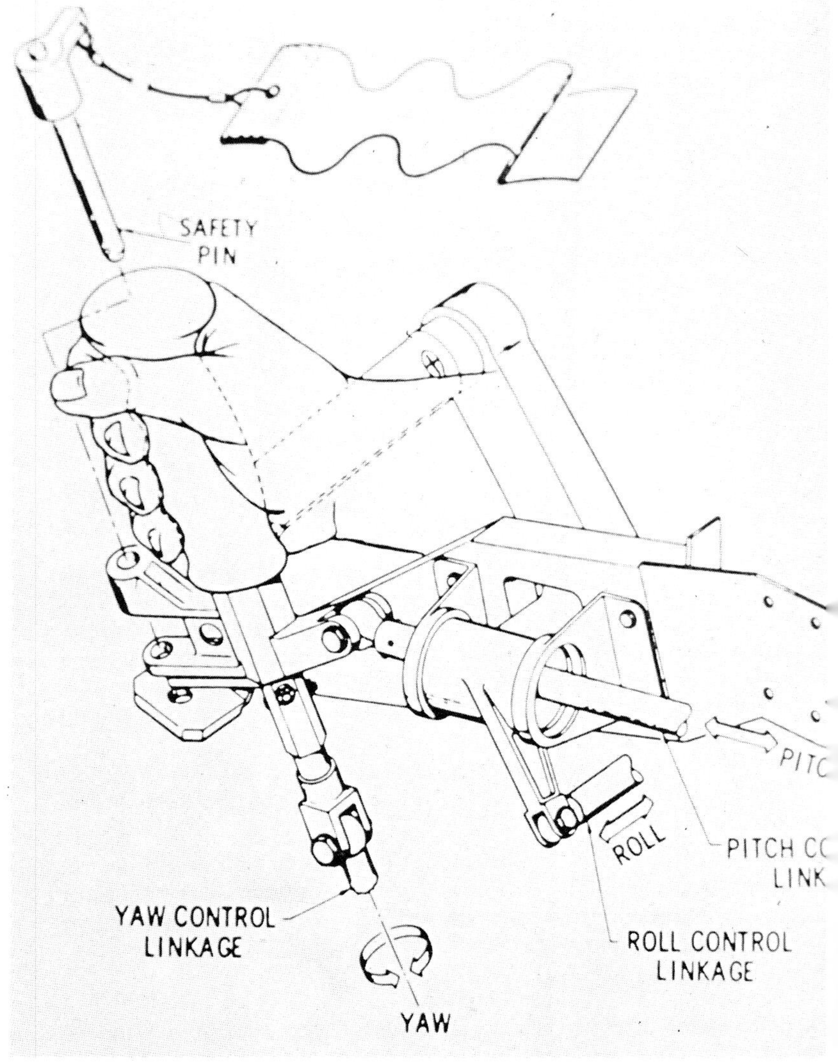

This integrated hand controller was used for rotational movements of the spacecraft about its pitch, roll and yaw axes.

controller. The RSCS mode would use a combination of hand controller positions to provide rate control in conjunction with the automatic system. All modes except the Manual Proportional used electrical current to activate the thrusters, thus in the event of an unexpected power failure the astronaut still had control of the spacecraft's attitude. In addition ASCS and RSCS used DC and AC power, whereas FBW used only DC.

If only partial control was available with any two systems, assuming two had failed completely for some reason, combinations of semi-operational modes would still ensure total attitude authority. For instance, the ASCS could be mixed with the MP, or the FBW and MP modes could be used together. In ASCS mode, the spacecraft would maintain programed attitudes to within 3° of the set orientation. Hand control functions were made easier by positioning the stick so that the astronaut's right hand would naturally fold around it when his arm was laying in the appropriate recess to the form-fitting couch. Thus, he could use the handle under high-g loads, a design concept adopted from the North American X-15. The attitude control thrusters were developed by Bell Aerosystems from earlier work with rocket propelled research aircraft like the Bell X-1A. To use the thrusters, the hand controller was moved from its neutral position at a rate which produced proportional thrust levels from the jets. To make the spacecraft yaw to the left, the controller was twisted in an anti-clockwise direction; to make it turn to the right, the controller was twisted in a clockwise mode. Rolls to left or right would be effected when the hand controller was tilted to the left or right, respectively; pitch-up would be ensured by tilting the controller back toward the wrist, pitch-down would occur when the controller was tilted forward.

Considerable emphasis was placed upon the attitude control equipment because it could be argued to be the most important system on board the capsule. If the capsule could not be properly aligned for firing the retro-rockets, and if it could not be held at a very precise angle during the initial phase of re-entry, the spacecraft would not enter Earth's atmosphere at the correct flight angle, or be positioned for thermal protection as heat built up. Three gyroscopes were fitted to the stabilization system to sense any tendency of the capsule to tumble and to send signals instantly to the thrusters to correct the movement, and two displacement gyroscopes sensing pitch, roll and yaw.

But just when so much additional capability was being built in to the capsule, Mercury engineers were given the word to start taking anything out that looked as though it could be eliminated. At the start, weight assessments indicated a spacecraft mass of 1,225 kg; by the end of the first NASA summer, weight was above this target by 6% and by the end of the year indications were that it would be up 11%. But weight growth is a fact of life for engineers, and always appears more critical when the system is unforgiving. Atlas was unforgiving and would accept little additional mass without seriously re-shaping the available trajectories. As it was, almost no weight saving was possible and the performance became a compromise between a booster weight reduction program and moderate improvements in engine capacity.

Less than one year after NASA formally accepted responsibility for launching America's first man into space, Mercury flight tests got off to an inauspicious start. In a test of the launch escape system from a spacecraft model situated on top of LJ-1 – Little Joe 1 – engineers and technicians were preparing, while pressmen watched, to complete evacuation of the Wallops Island launch site on 21 August, 1959. Suddenly, 35 minutes before the planned test time, a thunderous roar preceded billowing smoke as the escape tower and capsule lifted cleanly from the rocket and shot into the sky to a height of 600 meters, leaving Little Joe bathed in an acrid haze. Near the high point of its premature trajectory, the tower separated from the capsule clamp ring and the jettison motor ignited, carrying the tower away from the capsule.

As expected, the spacecraft's drogue parachute deployed but the main 'chute failed to operate due to insufficient electrical power. The test was planned to stress the launch escape system under conditions of severe dynamic loading. Instead, it ably demonstrated a pad-abort sequence although the spacecraft was lost on impact with the water. The sequence had been correct: the 23.6 tonne thrust solid propellant escape rocket fired for 1 second, and the single 390 kg thrust tower jettison motor burned for $1\frac{1}{2}$ seconds as planned. It was discovered that a stray electrical signal had inadvertently crept into the escape rocket sequencer.

Little Joe rockets were specially developed for tests with the Mercury launch escape system and its tower-mounted rockets.

Failures of this kind were more annoying than potentially hazardous for the program; Little Joe was a tailor made booster especially built for the Mercury program. It would be employed for dynamic tests with unmanned 'boiler-plate' capsules – full-size vehicles but without the interior systems of a flight rated spacecraft – and with prototypes of the production model. In no circumstance would any Little Joe Mercury capsule be manned during a test, unlike Redstone and Atlas boosters which were ultimately destined to carry astronauts on suborbital and orbital missions respectively. Redstone was the first booster scheduled to carry a manned capsule to the edge of space and pioneered the concept of 'man-rating' which would permeate the development program of every booster assigned to the manned space flight program.

Man-rating was a process where the basic rocket was taken to a higher state of reliability, given an abort destruct system if one had not been present, and provided with safety features not usually found on a military missile. In the case of Redstone, NASA had a reliable system, proven in several critical test flights already. But it still needed surgery. To match the performance requirement, Redstone's propellant tanks were lengthened to provide a longer burn time during ascent and to reduce complexity the standard guidance platform was replaced by an LEV-3 autopilot. Also, the Redstone's engine was changed in anticipation of improvements that were already planned for later tactical missile versions of the Redstone, and engine prevalves, located between the propellant tanks and the main valves, were deleted. A total of 800 changes were effected, although the vast majority were minor.

An important addition was 312 kg of ballast added to the forward instrument compartment to compensate for instability during the supersonic region 88 seconds after lift-off, induced as a result of lengthening the tanks. A stringent set of abort criteria was essential to the safety of the pilot, critical limits being an attitude deviation of ±5° in pitch or yaw and ±10° in roll; an abort would also be triggered if acceleration rates in the pitch or yaw axis exceeded ±5°/sec. Redstone was required to fly a precise and unyielding trajectory with very specific limits on performance, but if an abort situation developed a signal would immediately cut off the engine, wait 3 seconds while the Mercury launch escape system wrenched the capsule from the booster and away to safety, and detonate the destruct system to prevent the rocket running amok over populated territory.

In a typical suborbital trajectory, the pilot would be fed into the abort loop at 3 seconds in the countdown (from which point he could initiate the escape system), lift from the pad and for 15 seconds fly straight up, tilt over on a pitch maneuver and reach the period of maximum dynamic pressure at a height of 11 km. Known as max q, this event exposed the ascending configuration to the maximum air pressure, a combination of ascent speed and atmospheric density which, for a Mercury-Redstone mission, imposed a force of 395 N/cm^2 (39 times atmospheric sea-level pressure) on the front of the vehicle. At a maximum acceleration of 6.3 g, the rocket engine would be shut down 143 seconds into the flight, followed by separation of the capsule's launch escape rocket tower, deactivation of the abort system, and separation of the spacecraft 9.5 seconds later. Maximum speed would be about 8,000 km/hr, the spent Redstone falling back to Earth on a similar, but different, path to the spacecraft.

NASA's Atlas requirement centered on manned orbital flight. Only this big tri-engined liquid propellant missile was sufficiently far into development to warrant selection for the Mercury mission. The only other ICBMs then in development were the Titan, still in the design stage, and the Minuteman, which, being a solid propellant missile of limited performance, was quite incapable of lifting anything into orbit about the Earth. The original idea was to have the Air Force Ballistic Missile Division supply one 'C' series Atlas followed by 9 'D' series models, the latter being the then most advanced and potentially reliable missile of the family. The order was subsequently changed to ten 'D' series models, with four more added later. The final meeting of a series between NASA and the Air Force came in April, 1959, when a memorandum of understanding emerged as protocol for liaison between the two groups.

The most important modification accommodated an Abort Sensing and Implementation System (ASIS) specially designed to connect what was essentially a ballistic missile to a manned spacecraft operational requirement. From careful examination of the Atlas flight performance record it was clear that the majority of failures occurred with detectable warning signs in advance of a catastrophe. A few critical parameters could be wired to sensors which in turn were fed to the ASIS unit, thereby capturing the most likely areas of potential disaster. Dual sensors were subsequently attached to observe the pressure in the liquid oxygen tank, the difference in pressure across the common bulkhead that separated the oxygen and kerosene tanks, the missile's attitude rates as it ascended, the rocket engine injector manifold pressure, the hydraulic pressure in the sustainer engine and the launch vehicle's AC power supply. Anomalies in any of these sensors would cause the ASIS to cut power throughout the vehicle, immediately activating the abort system with a 3 second time delay between ignition of the Mercury escape tower motor and the self-destruct command.

Several system changes were made to the basic structure and to engine operation sequences to ensure a smoother ride and a more reliable operation. One of these related to the post-boost phase. In a normal mission, where Atlas was required to send a warhead off on a suborbital trajectory to a destination nearly half-way round the globe, two small vernier engines would continue to burn after the main sustainer engine was shut down; the big booster engines would have been jettisoned on the ascent. To save weight and reduce complexity the verniers were deleted, their job of fine-tuning the precise speed imparted to the payload unnecessary for the Mercury orbital mission where a few meters per second either side of the ideal velocity would not be all that serious.

One attempt to save precious kilograms of weight had a reverse effect, underscoring the delicate balance between too much emphasis on off-loading and insufficient attention to the consequences. In an effort to save 27 kg, a new 'wet start' procedure was adopted whereby water injected in the combustion chamber would cushion the ignition phase, allowing the already thin walls of the tank structure to re reduced in gauge. Subsequent tests showed the need for a stiffer skin and back when the weight so carefully taken off. But in a number of respects the man-rating attention given to Atlas was a vindication of the procedures and modifications believed to lie at the heart of turning the missile into a reliable workhorse for manned space flight. Confidence in the missile was never particularly high, changing for the better when test flights of the configuration brought hopes for a worthy respect in the launcher. In the first 50 flights of the basic Atlas ICBM, between June, 1957, and May, 1960, the missile failed on 21 occasions; a success rate of only 58%. Mercury-Atlas was to have a better record.

As with other systems being readied for the first missions, astronauts were assigned to monitor booster development and to report to the group on progress, deficiencies, adaptations, or anomalies that could speed or defer the planned flight dates. Gordon ('Gordo') Cooper was assigned the Redstone; Donald ('Deke') Slayton stayed with Atlas. Concurrently, John Glenn spent much of his spare time – expressed in a comparative sense meaning hours not spent on punishing trainers – working with Mercury engineers preparing cabin layout and spacecraft simulators. Walter ('Wally') Schirra looked closely after the environmental control and space suit systems, Malcolm Scott Carpenter and Alan Shepard followed preparations at Langley and the Goddard Center for a global network before moving to spacecraft navigation equipment and recovery operations at sea respectively. Virgil ('Gus') Grissom watched over Mercury's complex electrical systems.

But it was not all work for the select seven who thought themselves destined to fly the tiny spacecraft into orbit; in reality, only four would ever sit on top of an ascending Mercury-Atlas. Publicity was a major concern to planners and the planned alike. Simulator engineers who bravely attempted to duplicate an experience their subjects were destined to obtain for the first time during an actual flight, wrestled with the ever present media – electronic and press – who insisted on peering through this goldfish bowl the astronaut training program had become. Suddenly, everybody wanted to see how an astronaut prepared for flying into space. From the time the seven astronauts collectively sold their private lives to Time-Life Incorporated for $500,000, not one member of the team had a minute to themselves. Or so it seemed.

From the outset, NASA and the Defense Department made it clear that astronauts may be special people to the public but to the respective departments to which they owed allegiance, no one was going to get rich on tax-payer's money. Basic government wages were paid for their appropriate ranks, plus small supplements for expenses. But the private sector – the vast public that would soon swell to become a world-wide audience – dictated that the very special spacemen would reap the rich rewards of daring and candour. Exclusively committed to Time-Life, they were prevented from other prizes in the press marketplace but in several different ways they each amassed a fortune. It was, said the government, only a form of insurance after all.

The astronauts' public voice, the official NASA view, was cited by John A. 'Shorty' Powers, recruited by the space agency to handle the difficult job of telling the press the things they were not really interested in, the things NASA felt they should know about technical and program details. Circumventing this carefully orchestrated voice, the media surged upon the men of space in ever increasing numbers. There was a modicum of protection during 1959. From the moment they checked in on 27 April, the seven astronauts spent most of the year in academic work and spacecraft familiarization. The following year, physical adaptation became an agonisingly integral part of class-room lectures and visits to contractors. Public interest welled up to fever pitch. Speculation abounded on the effort required to get one man in orbit; many felt it to be only a matter of time before a Russian spaceman flew overhead, sending more bleeping messages like the electronic signals from Sputnik I two and one-half years before.

But if the technical road to man in space was a grudgingly difficult track hacked out with sweat and over-worked hours, the physical response of the human body was tested in equally crushing terms. From the beginning it was believed that man would face little danger from weightless flight, but the peripheral effects were completely unexplored. With the threat of stress and fatigue should the capsule malfunction, astronauts were required to train far beyond the point where they could safely fly in space. They were expected to reach states of endurance beyond that asked of any other pilot before them, for contingencies they hoped never to experience. Robert B. Voas, a Navy Ph.D. in psychology, was primarily responsible for putting the astronaut training program on track when, late in 1959, it became obvious that blossoming responsibilities prevented them from achieving carefully organized schedules laid out before anybody really knew how to train spacemen.

If there was one man who ably demonstrated in articulate fashion the fiendish depths to which the human mind can sink it was George C. Guthrie, responsible for astronaut training aids that resembled machines for torture and simulators with a morbid sense of the macabre. Delegated to fitting human beings into what must rank as the most bizarre form of masochism known to man, Robert R. Miller was project head of MASTIF: Multiple Axis Space Test Inertia Facility. Some day, Guthrie and Miller may get an award for work they were unprepared to sample but for which a generation of astronauts were grateful. MASTIF was a strange looking contraption comprised of tubular steel cages, a semi-circular spine, gear wheels, ball joints and a Mercury-type seat in the center. Looking for all the world like some medieval form of rack, a place upon which to stretch the limbs of giant Gulliver while Lilliputians looked on with relish, the cage was 6.4 meters in diameter and supported at the center a replica of the inside of the spacecraft the seven men would ride. Conceived by Miller, James W. Useller was responsible for adapting it to their training program.

The basic idea that drove men to build MASTIF grew from a need to test the attitude control system of the Atlas booster. It was capable of moving in three axes of rotation – roll, pitch and yaw – and two degrees of linear freedom at the same time. Encapsulated like some trapped insect about to be dismembered, an astronaut at the center would be accelerated to a complete revolution every two seconds in all three axes at the same time. In theory, using a hand controller identical to that he would employ for attitude control in space, the astronaut would reach this multiple tumble in 15 seconds and spend the next half minute damping the rates to zero, so bringing the cage to a halt; motor brakes were activated by the controller to simulate thrusters restoring the balance of a spinning capsule in orbit. The entire operation was made worse by the knowledge that MASTIF could move at twice the rate normally applied for human test! An over-enthusiastic operator could literally have driven its occupant to despair.

Designed to simulate in the most realistic way imaginable the tumbling motion of a spacecraft caught in wild gyration, and more importantly to give the occupant experience in bringing it under control, MASTIF was ready by early 1960 to receive the astronauts. Al Shepard was the first astronaut to ride the monstrous device but all it seemed to be from a visual inspection loomed worse when the machine started to rotate. Totally surprised by the vicious sensations MASTIF imparted, Shepard quickly aborted the run by punching the panic button. On the second attempt he knew what to expect and rode it out as few would ever do, the efficiency with which the contraption was brought to a halt enhanced by a deep desire to escape its clutches. In all the many and varied training sessions on a myriad different devices, nobody who ever sat in MASTIF would forget its screeching metallic grinding sound – the epitome of a nauseating whirligig trapped in Hades!

The value of MASTIF lay more in demonstrating the adaptability of the astronaut than in the experience it provided. Basic responses each pilot possessed from years of finely tuned reactions to situations common in test flying gave them the necessary ability to stablize a spinning capsule from instinct. It did show, however, that few untrained men could have retained consciousness seated in the device that at the same time pitched its occupant head over heels, hurled him over and over on his side and spun him round and round, each motion simultaneously accelerated to 30 revolutions per minute. Ironically, the two astronauts who would experience conditions close to this in orbit never flew the MASTIF, a product for Mercury alone developed and installed at the Lewis Research Center's altitude wind tunnel.

The astronauts were spared one devilish suggestion, that they each sit on a Mercury seat attached to the top of a 78 tonne thrust rocket motor as it thundered away on static test. It was thought that by exposing him to the sound level of 140 decibels, valuable information would be gained on a pilot's ability to remain coherent under extreme noise and vibration. Physician Bill Douglas drew the line at this and assigned Carpenter to a less dangerous form of test seated near wind tunnels. Another type of dynamic training device similar in one respect to MASTIF was the spinning ride at the Revolving Room installed at Pensacola, Florida. Designed to provide an effect many engineers felt would be necessary to counteract long duration weightlessness, the

The MASTIF trainer was used to give astronauts experience with accelerations about all three axes, simulating a tumbling spacecraft.

room spun at 10 rpm to create an impression of artifical gravity on the outer surface. This too was found to be a somewhat nauseating sensation. At the other end of the scale, multiple forces of gravity were experienced in a centrifuge set up at Johnsville.

In the words of Deke Slayton: 'A gondola is mounted on the end of a large revolving arm. Within the gondola we installed a mockup of our total instrument panel with active flight instruments, driven by the centrifuge computer and our Mercury hand controller, and also a complete environmental control system from the Mercury spacecraft. The gondola was then sealed so that we could depressurize that gondola to the actual flight pressure of (258 mmHg) . . . with a 100% oxygen atmosphere, and we could note the effects, if any, of applying high g under reduced pressure.' None were found and some astronauts were exposed to levels as high as 18 g, protected by the form-fitting contour couches that would be particularly helpful for riding the comparatively high accelerations of Redstone and Atlas boosters.

Tests were also performed on the astronaut's reaction to the sudden loss of high g loads such as would occur when the rocket shuts down, plunging him to a state of weightlessness in less than a second. Weightlessness itself was difficult to simulate, although it was possible to fly a parabolic curve in a conventional aircraft and experience the sensation for several seconds. As John Glenn said in a letter to a friend early in the program, 'I think I have finally found the element in which I belong. We have done a little previous work floating around in the cabin of the C-131 they used at Wright Field. That is even more fun yet, because you are not strapped down and can float around in the cabin doing flips, walk on the ceiling or just come floating the full length of the cabin.'

No one knew for sure just how hot it could get inside a Mercury capsule during a bad re-entry, one in which angle of descent was outside the anticipated 'corridor' or spacecraft attitude offset from the desired angles. Garbed in vented pressure suits the astronauts were put into closed steel boxes, heated to 121°c by quartz lamps on the walls. In a very particular sense, the astronauts were reassured by tests like this. It demonstrated that an environment that could seem impossible to tolerate would in fact be endurable with controlled response of the mind and the body. 'We no longer have any qualms . . .,' said Deke Slayton. A situation others would find difficult to accept but one which highlights the evaporation of fear that comes with knowledge and experience.

Potentially damaging to the body, excessive quantities of carbon dioxide were inhaled by the astronauts in a special chamber which raised the atmospheric CO_2 level from 0.05% to 4% in three hours, a simulation of conditions that could prevail if a lithium hydroxide filter packed up in orbit. Difficulty with providing a reliable monitor for excessive CO_2 levels made it imperative that each pilot knew the sensations associated with a carbon dioxide build-up in the cabin.

To balance the large amount of time spent in test rigs and simulators that could, theoretically, be switched off if the environment became unbearable, astronauts were expected to maintain their high level of proficiency flying high performance jet aircraft. It compensated for the build-up of an artificiality that surrounded the ground simulations, putting them back in a real world where mistakes invariably mean disaster. Mostly, they flew F-106 Delta Darts, all-weather interceptors capable of twice the speed of sound. At a more down to Earth level, physical fitness was a prerequisite for efficiency and mental agility. Stimulated by a rapid flow of oxygenating blood, the brain peaks under conditions of controlled physical stress and the general condition of the body was tuned to an exceptionally high standard by continuous exercise and vigorous activity. Seeking an activity that combines all the physical toning required of the body, astronauts spent some time with the Underwater Demolition Team, eventually getting so proficient they could swim for nearly 2 km underwater.

A primary concern of mission planners was that, having endured the rigors of high-speed, high-g, space flight, astronauts should not be lost through drowning in a sinking capsule. Mercury was designed with re-entry dynamics in mind, not flotation, and as one astronaut would say of a later capsule, 'it was a lousy boat!' Bobbing on the surface, an astronaut would be centimeters from the ocean swell and the side hatch affording access could become a rapid means of flooding the capsule when the time came to get out; astronauts would normally remain in their spacecraft but if the cabin sprang a leak, or if the capsule was far off target with recovery several hours away, it would be desirable to get out and in to a raft. Egress training was essential to guarantee the safety of the pilot. First, an egress trainer was used in a water tank at Langley, then the astronauts took it down to the Gulf of Mexico, dropped it into the sea and practiced getting in and out during rough conditions.

Several methods of egress were devised in this way. If required to exit unaided, an astronaut could inflate his raft by pushing it out the top of the spacecraft, or he could move into the water first and then inflate the raft; swimming practice gave them all an ability to remain afloat for some time while wearing a full pressure suit. Normal egress would be attempted by ducking under the forward instrument panel and working up behind the displays, past the parachute container and out the top. It was a difficult route and one best worked out according to individual choice. Its most appealing virtue was that it brought the astronaut into fresh air at the highest possible level above the waves, albeit only 1.85 meters from the sea.

If close to recovery forces, a helicopter could be used to lift the capsule partially from the water while the astronaut fired explosive bolts to release the side hatch before climbing over the sill and into a 'horse collar' winched down from the same helicopter. This was the cleanest and driest way back to terra firma. An alternate method would be to have remained inside the capsule while a helicopter hooked up, lifted it free of the water and carried it to a waiting aircraft carrier. That idea sounded fine, but no-one really got used to it after a helicopter suddenly dropped a dummy capsule during an unmanned rehearsal! The sight of the capsule sailing down to a mighty splash determined the future options more assuredly than any desk decision. The final method of exit would be used if the capsule shipped water through split hulls or bad leaks. In this event, the side hatch would be blown immediately which would give the astronaut about 10 seconds before the spacecraft began to sink due to water slopping in over the sill. There would be little time for panic or indecisive action.

But all this assumed the pilot would be within comparatively easy reach of recovery forces. What if the capsule came down several hundred kilometers from the nearest ship, or deep within arid desert? There was just a possibility that if the Atlas sustainer engine shut down a little early, putting the spacecraft on a long arching suborbital trajectory, Mercury and its occupant could descend into the central Sahara. To prepare for this

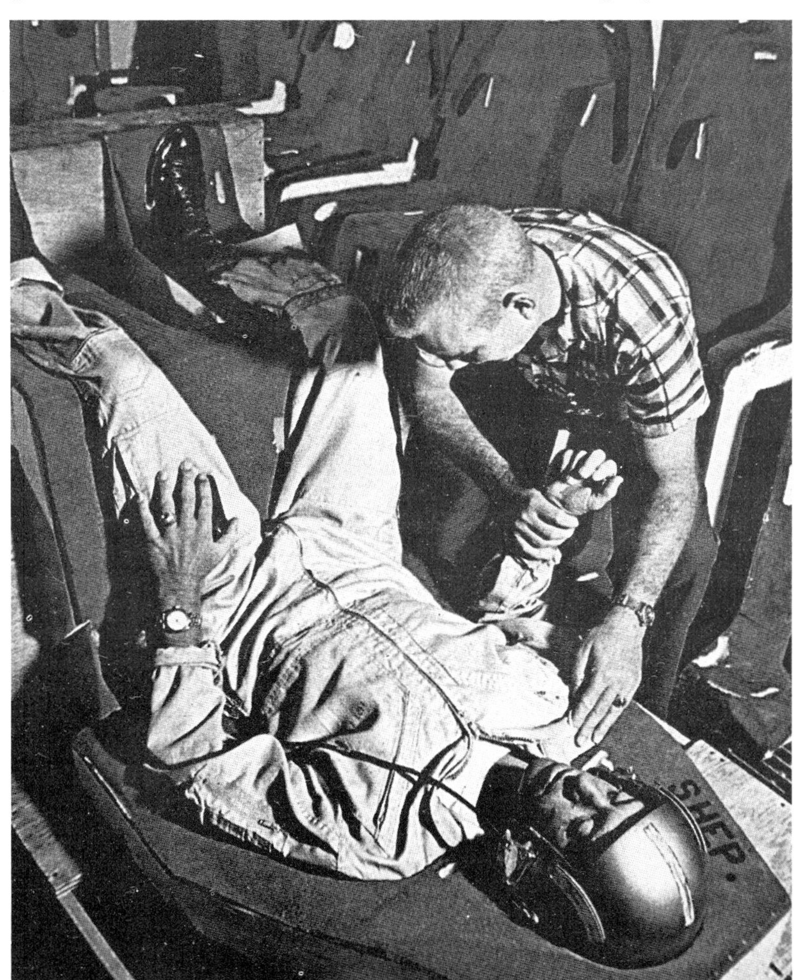

Unique to Mercury among all manned spacecraft, the contour couch would absorb the stresses of re-entry, calculated to reach 16 g for some missions.

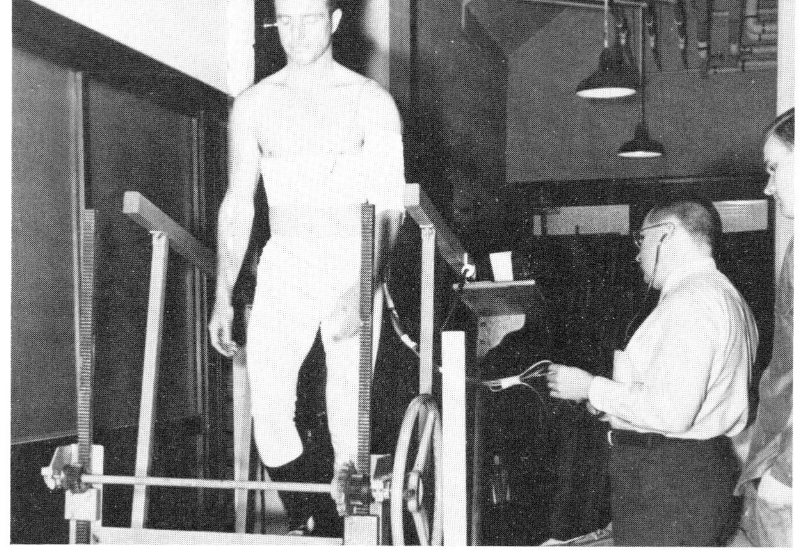

Seen here at the Wright Air Development Center, Ohio, astronaut Malcolm S. Carpenter walks a treadmill during physiological stress checks. In this, the platform is elevated in one degree increments each minute until the subject's heart rate records 180 beats/min.

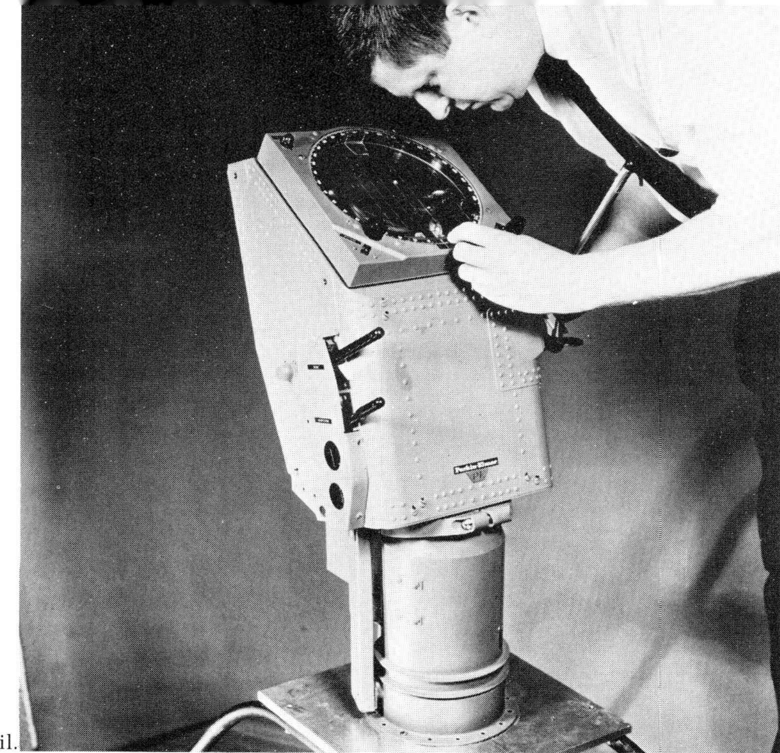

A periscope designed and built by Perkin–Elmer for Mercury's prime contractor → McDonnell Aircraft Corporation (now McDonnell Douglas) would allow the astronaut to determine altitude and attitude from space should other instruments fail.

eventuality the seven went to Nevada and learned to survive in a hostile environment with nothing but the contents of a typical capsule. Fashioning protective tents from torn parachute silk, using a survival kit to kill, gut and cook small animals, they lived for days in the arduous conditions of desert heat. As for being lost at sea, each astronaut spent a full half day alone in a small raft, learning to distill water from the salty sea, protect his head from a blazing sun, and signal ships or aircraft in the vicinity.

But if most of the training sessions revolved around things that could conceivably go wrong, rehearsals for the unexpected contingency action following an anomaly, considerable time was spent learning how to pilot the capsule. Unlike an aircraft, Mercury had no flying control surfaces through which a pilot could change the course of the vehicle, but rather an attitude control system that could operate independently of the flight path. This was perhaps the most radical departure for men trained to fly conventional aircraft. Captured by the gravity of planet Earth, apparently weightless because inverse centripetal force counteracted the downward pull, the spacecraft could be spun, rolled or tilted at any angle required to perform a particular function.

For instance, upon reaching orbit, the capsule would be pointing in its direction of travel with the astronaut sitting up facing forward. Immediately after separating from the Atlas launcher, by firing three small solid propellant 'posigrade' rockets to move the capsule forward, the capsule would start to turn around so that it was facing back along the path it was flying, with the nose pitched down 34° below the local horizontal, essentially flying backwards. This was the attitude which would be required for firing the retro-rockets at the end of the mission and the basic flight plan put the capsule in the necessary attitude from the start.

Theoretically the capsule could be pitched, rolled, or yawed from side to side, around all 360° available in each axis. It had no effect on the orbital path, the Earth being more interested in the capsule's centre of mass than in the direction the physical shape was pointing. Later spacecraft would be able to fire more powerful rocket motors than carried by Mercury, changing the geometry of their orbits to rendezvous with other vehicles. But that luxury was denied to Mercury, limited to attitude control maneuvers. Trainers to put the astronauts through all the peculiarities of spinning, rolling, and yawing while still flying in a precise and unchanging direction were called fixed-base simulators.

Procedures trainers were used to give the astronauts practice of firing the retro-rockets and performing re-entry maneuvers; for instance, although pitched down and flying backwards at an angle of 34° for retro-fire, the capsule was required to pitch at only 1½° during re-entry, and to slowly spin as it moved through the dense layers of the atmosphere. Attitude control thrusters are the only means of moving anything in space, and so it was with Mercury: the crew had to learn to start and stop each critical maneuver. Installed at Cape Canaveral and Langley Field, the procedures trainers evolved to a sophistication that enabled instructors, monitoring from outside every movement on displays and controls seen and performed by the astronaut inside, to set up failures and faults that could develop at critical times. In this way, the seven learned how best to compensate for malfunctioning systems, faulty instruments, situations over which they had no control but around which they had to find an alternative to survive.

It was a test of reaction, response, and mental agility. The procedures trainers kept pace with changes made to the spacecraft, copying in every detail conditions and reactions that the flight hardware would have during an actual flight. In this way, the astronauts developed a feel for the amount of time each action would require, the times they could devote to unscheduled activity, and the way their own actions were influenced by the capsule's response. In many cases, while testing an abort procedure that put stress on systems not usually operated in such a manner, failures occurred that would have invited disaster during an actual emergency. Different procedures would then be adopted and tested until a safe way could be found around a failure.

To simulate the movement of a capsule turning in response to thrusters fired by a hand controller, the ALFA (Air Lubricated Free Attitude Trainer) device was regularly used by the astronauts. Comprising a contoured couch mounted on an air bearing, ALFA moved in roll, pitch, and yaw, in response to inputs from the hand controller wired to compressed air jets doubling for the thrusters. Initial tests were performed with ALFA completely open but later refinements included a mock capsule incorporating a periscope similar to equipment installed aboard the Mercury spacecraft. Because Mercury would fly facing backwards and pitched down 34°, only the horizon would be visible from inside the capsule. To obtain a view directly below the spacecraft, a 19.5 kg periscope manufactured by Parkin-Elmer was made to extend through the hull on the opposite side to the forward viewing window.

The periscope normally showed a view spanning 130°, a portion of the Earth 3,150 km in diameter projected up inside the device to a screen in front of the astronaut 20.3 cm in diameter. If required, a lens could be swung into the center of the image, magnifying an area 20° across but surrounded with a dark ring to provide distinct separation of the two powers. The periscope was designed to back up a failed attitude control monitor by providing visual references from which the pilot could manually align the spacecraft; training sessions at the Moorehead Planetarium gave astronauts experience in setting attitudes by star alignments viewed on the night side of the Earth through the forward window. Various indices on the periscope window supplied alignments for determining the capsule's relative bearing to the Sun and the Moon. ALFA showed an astronaut could track the Earth via periscope and perform all the necessary attitude and retro-fire procedures to get his capsule back to Earth.

But for most activities performed in trainers and simulators an important item of equipment was the contour couch, making possible safe exposure to high g forces induced in real flight by the use of military rockets. Each astronaut was fitted for a couch that remained his throughout the program, every contour and line on his suited profile accommodated in the glass-fiber molding.

Great effort was made to ensure no pressure points chafed the suit, or uncomfortable protrusions threatened to break bones during high g loads. Later capsules would obtain a reprieve from the punishing accelerations of the Mercury ride and this spacecraft was the only one to need form-fitting couches.

Early in 1959 McDonnell engineers tested the consequence of a capsule descending to a touch-down on land to see how effective crushable aluminum honeycomb cells would be in attenuating the shock; pads placed between the aft bulkhead and the heat shield would not, however, prevent accelerations from reaching as high as 50 g. Live subjects were needed as final proof of Mercury's ability to impact safely on land and Yorkshire pigs were placed individually on a support structure that trussed their legs upward and made them look for all the world like overturned turtles. The pigs successfully came through the thudding deceleration that got to between 38 g and 58 g, while technicians watched in amazement as the swine got up and walked away!

As the months rolled by, everybody associated with the project knew that time was running out. Preoccupied with the day to day events of the busy work schedules, pulling together a myriad elements of a completely new endeavour, few could sit back and contemplate its value, the direction America was seeming to go in space flight, or ponder on the real value such a venture could have for the nation. Time was running out because every day could bring news of a Soviet man in space. Very few in those days at the beginning of this century's seventh decade really thought of a space race; most people involved with the project saw it as a very exciting extension of their individual careers. Politicians were certainly aware of the value foreign governments would put on success in high technology, but the sheer magnitude of the effort was daunting.

Very few people in NASA understood just how much work would be involved in getting a man in orbit and as the project accelerated through the months of 1959 and 1960 it came to be just one element of an expanding space program. In the first and second years of operation NASA had drawn together the elements it had failed to obtain at inception, principally the Army Ballistic Missile Agency's Development Operations Division where von Braun had his clustered booster in final design stages. Originally called Juno V, von Braun's suggestion that the project be named Saturn was approved by the Advanced Research Projects Agency in February, 1959. Juno I had been the modified Redstone used to launch Explorer and adapted from the Jupiter C nose-cone re-entry test vehicle. Juno II was a modified Jupiter A missile employing the same solid propellant upper stages as Juno I. Juno III was proposed for use between 1959 and 1962 carrying 300 kg payloads with powerful upper stages launched atop a modified Juno II first stage; Juno IV would have improved further upon Juno III and lifted 450 kg into orbit.

Following NASA's failure to secure the services of von Braun's Huntsville team, a cooperative agreement was signed between the Space Administration and the Department of Defense in December, 1958, two months after NASA officially came into being. For most of 1959 the DoD looked with increasing dissatisfaction at the large space projects still floundering, without a mission, in Army and Air Force establishments; Congress didn't like the distribution of programs and prodded the DoD to look at the matter. As a result, the Defense Department decided in September, 1959, to reduce the ARPA and transfer its projects to the appropriate services. They even thought of moving the Saturn team, lock, stock and barrel, to the Air Force; the Army were left high and dry by the National Aeronautics and Space Act of 1958 that re-grouped most projects into NASA tenure, and the Air Force was the only logical user of a large booster. But since NASA was thinking of developing a launcher called Nova, very much larger than Saturn (which was itself 5 times more powerful than Atlas), the civilian space agency was the logical home for the von Braun group.

Their contributions to the American missile market ended with the Pershing, although Redstone and Jupiter were the only two military weapons wholly developed by the Development Operations Division. For a time it looked as though NASA would get the entire ABMA. In the end it was agreed that the Development Operations Division and the Saturn project would go over to NASA management. Secretary of Defense McElroy approached NASA Administrator Glennan about the possible transfer, a meeting was held at the White House on 7 October, 1959, and two weeks later to the day, President Eisenhower approved an agreement transferring the ABMA Development Operations Division. The plan went before Congress on 14 January, 1960, and the transfer took effect from 15 March, the plan residing in Congress for the stipulated period of 60 days.

The agreement signed by the President in October 1959, called upon NASA to take charge of the nation's super-booster program and for the Development Operations Division to become a NASA facility. Accordingly, the President signed on 15 March, 1960, an order naming the installation the George C. Marshall Space Flight Center, or MSFC for short, in memory of America's only Nobel Peace Prize winning soldier, George Catlett Marshall, who died in 1959 at the age of 79. Established on 1 February, 1956, the Army Ballistic Missile Agency ceased to exist on 22 July, 1960. The Army's dream of supervising a massive exodus to space would be realized in the civilian space authority it had for long contested. The addition of nearly 5,000 personnel from the formal transfer of the ABMA Development Operations Division on 1 July, added a 50% increase to NASA's previous manpower level of less than 11,000 employees.

But although the formal move came half way through 1960, concern over the responsibilities NASA was being asked to take on forced a change in relations with the White House. Up to the beginning of that year, in effect for the first 15 months of life, NASA was left in no doubt that it was there but for the grace of the Administration – Eisenhower's White House staff were not about to place America on a crusading journey to the planets and the space agency was given no encouragement to plan expansive endeavors. To get events in perspective it is advisable to remember that NASA got $353 million for the 9 months remaining in Fiscal Year 1959, and $485.3 million for the full twelve month period of Fiscal Year 1960.

For FY1961, the space agency was allowed to request a sum of $802 million, which included $176 million for the Huntsville facility and the Saturn work. This was about 20% less than NASA originally told the Budget Bureau it would need to run all the many projects that were emerging as part of its responsibility. The cut was in line with White House attitudes. But three weeks after the original budget presentation on 18 January, 1960, the White House followed a letter request for Glennan to report on how the Saturn program could be speeded up by placing an immediate amendment of $113 million to provide a grand total of $915 million. Congress went even further and boosted the FY1961 budget to $970 million. This action focused two very different lines of communication into a single, coordinated effort that would rapidly weld a structure from which the manned space program could be dramatically improved one year later.

Those two lines were moves by Senate majority leader Lyndon B. Johnson, and by NASA estimates on Saturn funding requirements. Johnson maintained a constant watch on the way space funds were going and noticed, by carefully searching the public correspondence, that when asked by the White House how much more it would need for Saturn a figure of $125 million had been mentioned. Johnson found that the Budget Bureau had argued over this and cut the request by $12 million, giving Johnson the ammunition he needed to go to the Senate and lay grounds for suspicion that the Eisenhower administration was penny pinching over issues of national importance. Believing that the Budget Bureau had placed unfair constraint on NASA, the Senate Space Committee put its weight behind a motion to give NASA the financial freedom it needed to get the big booster project off the ground.

This set a mood in Congress whereby it found itself playing God's advocate for NASA pleas requesting additional funds. Saturn was vital to future manned space projects and Johnson's influence on the Appropriations Committee ensured a smooth ride on the floor when it came through the door from the Space Committee. It was this action that cleared a precedent for major and highly dramatic events that would sweep aside the last vestige of apathy inherent in the conservative Eisenhower administration. As 1960 drew to a close dark clouds were gathering on the horizon. A new President would soon be in office, Soviet leaders would press hard challenges on the new administration, and NASA would get the most spectacular goal it would ever receive. The world was changing, in a very real sense leaving the fifties far behind, and the dream-like aura that surrounded the Mercury project would soon be gone for ever. Manned space flight would be a major political issue – and an international challenge.

CHAPTER FOUR

Planning for Apollo

If 1959 and 1960 were years in which the Mercury project gathered momentum toward its first major goal of sending an astronaut on a suborbital space flight, they were also years of gestation for NASA's long range plan. Representing the last two years of the Eisenhower administration it was a time for cautious optimism about the future of manned flight. No great goals – just a steady progression toward a more promising capability. Pressure for a more vigorous approach to space operations came from the Army, whose Ballistic Missile Agency was smarting under the loss of mission for its Saturn booster when NASA clearly got the national mandate for space activity. Initial attempts to have the ABMA transferred to civilian authority failed when the Army stiffened its resolve to retain control of the von Braun team. As head of the Development Operations Division, Wernher von Braun was sure where America was going, stating in a staff report to the House Select Committee on Astronautics and Space Exploration in January, 1959, that he believed circumlunar flights would be performed within ten years and that manned Moon landings would follow after that.

The Army rocket engineers knew that dissenters within its ranks chafed at the prospect of funding what seemed to be a massive development exercise. But Saturn was so much larger than anything yet proposed that control of such a massive booster could place its owners at the hub of a future space program. Or so thought Maj. Gen. John B. Medaris of the Army Ordnance Missile Command when in March, 1959, he ordered a task force study of a manned lunar outpost and a schedule for its operation. It was more a job of defining the maximum mission capability of Saturn than fitting the big launcher to a nationally advantageous space project. But the task force's real job was to demonstrate the importance of the ABMA work and to bolster support for a distinctly flagging program. Medaris was right in thinking that the Department of Defense was disinterested in putting more money into Saturn. A contest arose between the ABMA and the DoD in the closing months before Eisenhower decided to transfer the ABMA Development Operations Division to NASA, one based largely on the project Horizon lunar outpost study authorized by Medaris.

The report from the task force was completed on 8 May and presented to the Secretary of the Army and the Chief of Staff on 28 July. It was decided that an American lunar post was in the national interest, that the plan as outlined by Project Horizon was feasible, that the Army had access to essential capabilities for achieving the objective, and that Army presentation of the proposal should not unduly prejudice a balance between the military or civilian control of such a venture. Because of this latter factor, it was decided to re-write the report so as to delete implications that a Horizon lunar outpost had military application – it was to be sold as a pertinent activity in the national interest, not one integral to America's defense strategy. This was, of course, a play on words, a subtle bow to Congressional preference for an ostensibly civilian space program. The Army knew only too well that few would accept military sponsorship of such a major undertaking when NASA had been organized so recently as a civilian caretaker of such objectives.

It also knew that only by coming up with a desirable mission objective could the ABMA press ahead with Saturn acitivity; a lack of adequate roles had prevented earlier support and the Army's inability to present a clear task for man in space had bled it dry of funds requested to support post-Saturn booster studies. The 'Project Horizon Phase 1' report comprised a four-volume proposal in which the establishment of a manned lunar outpost would, 'demonstrate American scientific leadership in space; serve as a communications relay station, as a laboratory for space research and development, and as a stable, low-gravity launch site for deep space operations; and provide an emergency staging area, rescue capability, or navigational aid for other space activity.' It was all rather vague and unspecific, the 'scientific leadership' argument being central to a sense of peaceful colonization rather than military domination.

In the plan, a vigorous Saturn launcher development program would precede cargo deliveries to the surface of the Moon beginning in January, 1965, with a manned Moon landing in April of the same year. Through a series of launches in the 'build-up and construction phase,' a 12-man colony would be established; extra cargo would be delivered in the year following its inauguration, planned for November, 1966. The initial build-up phase would require 149 Saturn launches in little more than two years, and a further 64 flights in the first year of colonization. It was an incredibly ambitious proposal, and one which would seem in the light of hindsight to have been dramatically under-priced at $6,000 million. Despite this attempt to give the Army a manned role in space, Herbert F. York, the DoD Director of Defense Research and Engineering told ARPA Director Roy Johnson that he had 'decided to cancel the Saturn program on the grounds that there is no military justification' for it.

The Advanced Research Projects Agency had recently given the Army Ordnance Missile Command authority to negotiate with the Martin Company for Titan rocket stages to form the second stage of the basic Saturn and five days later, on 29 July, came York's deliberation. But the Saturn got a reprieve while additional studies were carried out. In a three-day meeting of the ARPA-NASA Booster Evaluation Committee which began 16 September, York presided over debate on the relative merits of the Saturn and the so-called Titan-C, a modified Titan II. Titan-C was to be the launch vehicle for the Air Force Dyna-Soar hypersonic glider and was believed by York to be suitable as a heavy launcher for the space program. The meeting determined that Saturn had a substantially greater payload capability than the Titan and that it could be available a full year ahead of its Air Force rival. Accordingly, York decided not to cancel the Saturn but immediately thereafter began talks with Administrator Glennan about NASA taking over the Army Ballistic Missile Agency and its Saturn programme. At that point the President picked up the sequence of events culminating in the transfer of von Braun's Huntsville team by July, 1960.

With the Air Force effectively tied up with Dyna-Soar, allowing its space protagonists to consider it as a foot in the orbital door that could, later in the '60s, lead to a fully fledged Air Force space program, all military competition to civilian space designs was effectively halted. For a while. The contest would flare up again when John F. Kennedy stirred a hornets' nest in 1961 by juggling with space priorities and agency objectives. But if Pentagon support was lacking for projected missions on a grand scale, the civilian space agency was only too aware that any developments beyond the Mercury concept would be its responsibility. When NASA took over proposals for a manned space mission and turned a part of the Langley Research Center into a base for Space Task Group personnel, it knew the project to be an effort aimed at getting a man into space as a proof-of-concept exercise rather than an operational program. With a planned life of less than 5 hours, the tiny one tonne capsule was strictly limited and future developments would need a more vigorous approach.

NASA had been formed in October, 1958, to be responsive to the national objectives in space by first demonstrating capability and then proposing avenues of activity which would, or would not, be adopted by Congressional decision. No one really knew how space would turn out after experimental research projects opened new opportunities. In establishing a demonstration of capability, NASA resolved to move beyond the limited nature of the one-man Mercury capsule. But there were no accepted patterns of space activity as precedent for the decisions that would have to be made about future developments. Serious evaluation of a myriad potential paths had to be made, adopted projects being the result of a filtered analysis regarding financial requirements and launch vehicle capability.

Set up on 12 January, 1958, before NASA was a formal proposal in Congress, the NACA's Stever Committee submitted its report in the month NASA began life. In part, it urged that 'a vigorous, coordinated attack should be made upon the problems of maintaining the performance capabilities of man in the space environment as a prerequisite to sophisticated space exploration.' Three months later, in January, 1959, discussions between NASA and the Department of Defense produced a formula for booster policy that envisaged four prime categories of space launcher to be developed. The report, 'A National Space Vehicle Program,' called for Vega, Centaur, Saturn and Nova launchers to support specific mission objectives in projects yet to be determined. It was an attempt to pick up existing hardware and paper designs and weld together a series of launchers spanning the potential payload requirements spectrum.

Vega and Centaur both adopted the Atlas as first stage to sequentially more powerful upper stage combinations. Saturn was the clustered 680 tonne thrust brainchild of the von Braun team. Nova was a NASA concept for putting together at the base of one launcher five F-1 engines, each of which could deliver a thrust of 680 tonnes; F-1 had been one of several ABMA projects transferred to NASA when it was formed in 1958. Nova's second stage would use a single F-1 engine burning liquid oxygen and kerosene while a third stage would employ high-energy liquid hydrogen/liquid oxygen propellants. Although difficult to handle, these 'cryogenic' liquids would substantially improve the payload capability of this massive launch vehicle which, with a first stage thrust of more than 2,700 tonnes, could have lifted 100 tonnes into Earth orbit.

But boosters were only a part of the decision-making process that sought to define the geometry of a viable space program. Accordingly, in April, 1959, NASA notified its Ames, Lewis, and Langley Centers, the Jet Propulsion Laboratory, the Office of Space Flight Development and the High Speed Flight Station, that a Research Committee on Manned Space Flight was to be set up under Chairman Harry J. Goett, from Ames. Early that month, a three-day Staff Conference was held at Williamsburg, Virginia, to discuss the type of manned space program that should follow Mercury as a, 'preliminary step to development of spacecraft for manned interplanetary exploration.' Nobody actually sat down and decided that manned exploration of Mars, the Moon, or any other body in the solar system was a logical phase to introduce the new age of space operations. It seemed a natural goal from the outset and in the month Goett's Committee began work, NASA Administrator Glennan went before Congress to explain the content of NASA's Fiscal Year 1960 budget request, which included a sum of $3 million for seeking solutions to problems surrounding orbital rendezvous and docking operations with a manned space station.

It was only part of what top NASA officials believed to be a steady progression toward ambitious goals that culminated on the surfaces of other planets. Clearly, the joining together of spacecraft launched at different times was a fundamental part of any competent space program, but the problems were daunting. At a time when rocket flights were performed with little regard to schedules, the precise launch time essential to an orbital rendezvous seemed impossible to achieve. Large rockets like Atlas were hardly ever launched on time, a factor critical to the use of such systems for national defense but one which was conveniently avoided in building public confidence. Yet a delay of one second in launching a spacecraft to an orbit from which it was to find a satellite already in space, would mean a distance error of nearly 8 km! Furthermore, accurately guiding a spacecraft to a precise matching orbit was an exercise that seemed in 1959 beyond the capacity of existing systems, calling for special vehicles, new tracking capability and unwritten computer programs.

Mercury was certainly incapable in existing form of performing rendezvous and docking activities such as would be required of the next generation of manned spacecraft. It had propulsion systems for attitude control but was unable to change its orbit at any other time except when required to bring itself back to Earth at the end of the flight. Definition of future space program goals began with the first meeting of the Goett Committee late in May, 1959. It was time to examine current systems, appraise future requirements, and set down guidelines for further work. Prime topics were Mercury, the ballistic orbital capsule, and Dyna-Soar, the Air Force boost-glide vehicle strongly backed up with NASA assistance. Participants included Alfred Eggers, Maxime Faget and George Low, the latter from NASA Headquarters. It was Low who probably did more that month to get the Committee thinking big: NASA should, he said, pursue a goal culminating in the manned exploration of the planets, adopt interim objectives as stages on the road to that accomplishment, investigate advanced forms of propulsion like nuclear rockets, and study the possibility of using Saturn launchers to land men on the Moon as a precursor step to Mars flights.

Faget backed Low up, calling for lunar exploration as the next step beyond Mercury. But the general consensus was one that opted for the more cautious approach. In presenting its summary findings, the Goett Committee suggested Mercury should be followed by a program embracing manned Earth orbital operations with maneuverable space vehicles, space stations in Earth orbit, manned circumlunar and lunar orbit (the former meant flying around the Moon and back to Earth, the latter implied a vehicle placed in orbit around the Moon) reconnaissance, lunar landing, planetary reconnaissance, and landings on Mars and Venus. But, for the immediate future, it felt NASA should limit its technical plans to the needs of a lunar landing mission and define the further requirements when progress exposed critical needs for the grander objectives.

On 27 May, Mercury Director Bob Gilruth took his STG confidants Paul Purser, Charles Donlan, James Chamberlin, Ray Zavasky, W. Kemble Johnson, Charles Mathews, Maxime Faget and Charles Zimmerman, to Washington and a meeting with George Low where the team began formal discussion on the manned spacecraft to follow Mercury. It was an historic session and one given impetus by the second Goett Committee meeting in late June. The first meeting had heard STG proponents argue for an expanded Mercury program. H. Kurt Strass proposed a two-man Mercury sufficient for three-day flights, a similar vehicle but with an added cylinder to support 14-day missions, and a Mercury plus cylinder attached by cables to the second stage of the launch rocket so that the two together, spun around their common center of mass, would set up artificial gravity. It was not as lone an idea as it sounded: the NASA FY1960 budget then before Congress requested money for studies on advanced Mercury vehicles that could be used as small two-man laboratories.

At the second Goett Committee meeting, however, recommendations on future policy led to agreement that a manned landing on the Moon would be appropriate as the next manned space objective after Earth orbital flights in Mercury capsules. Space stations were proposed, and indeed given support for the lessons they could provide in running large manned structures, but the Moon mission seemed to have it all: a sufficiently advanced objective to stretch existing technology and justify the effort; one that had relevance to future goals for manned planetary flight; and an operation that could be self-sustaining, an endeavor of singular achievement rather than a mere stepping-stone to some more distant and less well-defined goal. The first meeting had brought a call for each NASA Center to prepare its own studies and recommendations on manned lunar landing operations; the second heard divergent views on the road to a Moon landing.

Bruce T. Lundin from the Lewis Research Center resurrected an argument that had been around for a long time: a lunar-orbit-rendezvous technique whereby a manned spacecraft is first put into orbit around the Moon from where it can descend to the surface, employing rockets to slow it down for a gentle landing on the surface. It was not surprising that Lundin's work should present an argument based on propulsion; that was, after all, the work his center had been assigned. Lundin's argument used comparative figures developed for an Earth-orbit-rendezvous model to show how economical the lunar-orbit mode would be. Whereas the lunar-orbit method could employ a single rocket weighing 4,000 tonnes, returning to Earth a spacecraft weighing 4 tonnes, the assembly in Earth orbit of a structure sufficiently large to fly from there direct to a Moon landing and return would require 14 Saturn launchers with Centaur upper stages. Centaur was a cryogenic (liquid hydrogen/liquid oxygen) upper stage then in development at Lewis. Coupled with all the problems of rendezvous and co-planar orbits that the Earth-orbit-rendezvous plan required, the Lundin method was seen as a more economically viable mode.

The lunar-orbit method should not be confused with a

lunar-orbit-rendezvous plan proposed by Langley personnel later in 1959. It was, in fact, closer in concept to the Direct-Ascent method where a single launcher sent a space vehicle to a direct landing on the Moon without first going into orbit about either Earth or the Moon. However, Lundin suggested that the Committee should address the problem of the boosters because Saturn's would be needed for the Earth-orbit-rendezvous plan, whereas his method of going first into lunar orbit would need a single Nova class launcher. Since launchers were likely to require several years to develop, especially launchers as large as the big Nova vehicle, it was felt advisable to recommend which booster to back as the more likely contender for post-Mercury mission duties. At the meeting, representatives from the Army Ballistic Missile Agency discussed with John H. Disher from NASA the position vis-à-vis large Saturn type booster studies at von Braun's Development Operations Division. They said that Huntsville was proposing a Saturn C-2 vehicle, considerably more powerful than the 680 tonne thrust C-1 formed from the clustered assemblies of existing Jupiter engines.

The C-2 would have a cluster of improved engines delivering a total first stage thrust of 907 tonnes and upper stages providing an enhanced payload carrying capability. In fact, said the ABMA, von Braun's team had been studying post-Saturn booster configurations and were ready to go into detailed design on a launcher delivering up to 5,400 tonnes thrust from the first stage alone. Also, limited studies had been carried out on the best way to fly a Moon mission – work principally conducted so as to give the rocket engineers an idea of the size of rocket required – and had come up with either an Earth orbit refuelling mode or a so-called lunar-surface-rendezvous plan whereby several separate packages would be landed on the moon in advance of the manned landing so the crew would have equipment on hand to supplement existing systems for the return trip; the refuelling concept required several Saturn type launchers to deliver to Earth orbit tanks necessary for supporting a direct Moon flight.

In supporting the concept of a manned space station, Langley Research Center's Laurence K. Loftin, Jr., suggested an Earth orbiting facility as an essential element in manned Moon landing plans. George Low cautioned the Committee that intermediate steps should be planned with great care since the cost of such a multi-disciplinary range of mission types would lead to inflated expenses; money would be tight and each step should be made to accomplish significant goals. The Ames contribution came from Alfred Eggers, who suggested placing manned vehicles into highly elliptical orbits, first by using Vega or Centaur stages to boost the capsule 80,000 km from Earth, then by flying a spacecraft by Saturn out to a distance equal to that of the Moon (400,000 km) prior to circumlunar missions. Eggers' plan envisaged also the use of common spacecraft design for all post-Mercury objectives, a commonality of systems providing economic development of a single vehicle for multifarious roles.

Senate space committee chairman Sen. Robert S. Kerr confers with NASA Marshall Space Flight Center Director Dr. Wernher von Braun before a model of the Saturn launch vehicle.

At the end of the two-day meeting, the Goett Committee's second session concluded that manned flight to and from the Moon should be the next objective for NASA and that toward this end the Lewis Center should study ABMA's Nova concept and Earth refuelling modes, and that the High Speed Flight Station should examine Earth-orbit-rendezvous concepts. At STG, H. Kurt Strass suggested setting up a New Projects Panel which would take the recommended objective of the Research Steering Committee on Manned Space Flight – the Goett Committee – and develop a rationale for the design of post-Mercury spacecraft, rather than concentrate, as some at Langley wanted to do, on advanced versions of the basic Mercury. Pro-Mercury supporters were gathering ranks for presentation of ostensibly cheap and reliable follow-on missions with a manned vehicle which would soon be proven in flight. McDonnell was already working on an improved Mercury capsule which could carry a heat shield to protect it from re-entry after a mission simulating the dynamic conditions of a lunar flight.

The Strass Panel met on 12 August and decided manned Moon flights would require precursor missions designed to answer fundamental questions regarding, for instance, the precise nature of Earth's radiation belts, the susceptibility of organisms to solar radiation, the nature of the lunar environment, conditions on the Moon's surface, and the ability of astronauts to remain in space for comparatively long periods of time. It decided that a new type of capsule would be essential not only for the Moon flight but also for addressing fundamental issues, a capsule that could maneuver during descent through the atmosphere in addition to changing orbits in flight. It was also considered desirable to have the capsule descend to a land landing rather than use the splashdown technique adopted by Mercury. The second New Projects Panel meeting, six days after the first, placed as first priority the design of the capsule and displayed a schedule culminating in a Moon landing during 1970. But Strass had steered his colleagues on an ambitious planning exercise, one that was seen to be increasingly optimistic as the summer gave way to fall and by the onset of winter the mood was one of realistic awareness that none of the proposed launchers would be capable of actual landing missions.

Slowly, just as in the first half of 1959 talk had moved from manned flights to Mars to manned landings on the Moon, the second half of that year was seen to project circumlunar flight as a creditable goal rather than an actual touchdown on the surface. In fact, not everybody fully supported the Goett Committee mandate for an ultimate mission to the Moon, or the STG New Projects Panel in designing the post-Mercury spacecraft around lunar flight requirements. Langley Research Center hosted a meeting during July, 1959, where E. C. Braley and L. K. Loftin, Jr., proposed an Earth-orbiting space station from which physiological responses to long duration flights could be analysed, and from where complex new space vehicle systems could be developed and tested. It was also said that a space station would afford an opportunity to discover new uses for space, both for scientific observation of the Universe and for potential applications that might benefit Man in broader context on the Earth itself. The general strategy would, according to Braley and Loftin, require a space station to develop the expertise from which manned Moon landings could begin about 10 years later.

This was a similar conceptual approach to ideas about space travel put forward earlier this century, a balanced, steady-paced, progression from one space-borne capability to the next logical development. Yet just as space station protagonists were uncoupled from any political or industrial motive for their approach to an evolving space program, so too were the manned Moon landing supporters free from partisan cause that influenced their conclusions. The two projected lines of development reflected the psychology of divergent groups: the entrepreneurial extravaganzas, the spirited followers of spectacle and drama; or the clinically scientific motivation of considered rationales for human development. Both sides were well represented throughout government, industry and in Congress, for they are the opposing sides of any endeavor and as such stimulate technologists with thoughts that go beyond the immediate application of an existing capability.

During this period of decision about America's manned future in space, companies across the United States proposed designs centered mostly on existing hardware; recognizing the ease with which politicians suspect partisan reasons for company proposals, application of existing equipment defers a feeling that corporate gain has stimulated the motivation! Convair's Krafft Ehricke designed a space station adopting the interior of Atlas propellant tanks, utilized as a makeshift work area after the launcher had propelled itself into orbit. The Douglas Aircraft Company were working on a converted liquid hydrogen tank that could, they said, be used to house astronauts in a laboratory-like environment after the rocket stage, launched by a Saturn, reached orbit. It was the basis of a project destined to reach maturity 14 years later, but for the time being the idea of a manned space station seemed to evaporate in the light of planned post-Mercury space vehicles.

There was one other reason for expanding the NASA presence in space, one more perceptible to NASA Headquarters staff than the scientists and engineers at Ames, Langley or Lewis. In the words of NASA's John A. Johnson at the Mayflower Hotel, December, 1959: 'Is there any real alternative? It hardly needs to be said that if we don't pursue with vigor the presently foreseeable goals of space exploration, there are others who will. Is it conceivable that the American people will be content to be onlookers ten and twenty years from now while another nation, eager to seize all the symbols of world leadership, triumphantly parades before the world the Magellans of space?' Few in NASA really thought in terms of political gain, loss of prestige or international pride; but the background to events associated with man's progress in space could not be divorced from the generations of suspicion and mistrust that enveloped great nations and empires.

By November, 1959, the Space Task Group was clearly moving toward definition of the post-Mercury specification. At a meeting on the 2nd of the month, Director Bob Gilruth chaired sessional debate with Paul Purser, Donlan, Faget, Bob Piland, Strass, Mathews, John D. Hodge (an Englishman), James Chamberlin and Caldwell C. Johnson, where it was decided that the post-Mercury spacecraft should have a three-man crew for circumlunar flight, be adaptable to Earth-orbit operations, and capable of doubling as a crew cabin for the more advanced spacecraft that would be required for Moon landings. Piland, Strass, Johnson and Hodge would comprise a panel assigned analysis of critical systems elements set up by arriving at preliminary decisions regarding vehicle weights, flight corridors, propulsion, etc. Gilruth protected his growing Mercury team by stipulating progress on a strictly non-interference basis, fearing that in future months as much as 10% of STG personnel could be working on the post-Mercury capsule specification. Just 17 days after the meeting, Harry Goett sent his Research Steering Committee on Manned Space Flight a memorandum stressing the importance of generating a weight figure for the spacecraft that would be required to fly the circumlunar mission.

This was especially important, as NASA Director of Space Flight Development Abe Silverstein had inferred to Goett, because his (Silverstein's) Saturn Vehicle Evaluation Committee were soon to decide on the upper stage configuration for the launcher. On 8–9 December, 1959, the Goett Committee met for the third time and discussed post-Mercury capsule weight criteria. It was believed the spacecraft would comprise a man-carrying capsule weighing about 3 tonnes to which would be attached a small cylindrical work laboratory weighing about 2 tonnes. The launcher would, therefore, require a lifting capacity of at least 5 tonnes to a trans-lunar trajectory. It was about in scale with the Saturn C-2 payload capability and on 15 December Silverstein's Saturn Vehicle team decided to recommend development of a high-energy liquid hydrogen/liquid oxygen upper stage based on a 91 tonne thrust J-2 engine under development at Rocketdyne.

This ended a long struggle to get high-energy stages funded for Saturn, the combination of sheer brute force in the liquid oxygen/kerosene based first stage married to cryogenic upper stages providing a payload capability greatly in excess of a multi-stage launcher limited to low-energy fuels. With ABMA's Development Operations Division going over to NASA in the new year, decisions could be made with hope for future Saturn developments; the program had limped along for more than a year because the Army, loathe to spend money on a project for which it had no use, resisted plans for funded schedules. However, the Goett Committee went on to define further the requirements of the new three-man capsule.

The number of crewmembers was chosen as the most effective balance between an ideal engineering value and the preferred size of the crew for physical and psychological reasons. Mercury had forced decisions about the degree of control its single occupant would have during his planned three-orbit mission. STG had resisted from the outset attempts to limit his capacity for decision making and control effectiveness, believing that situations developed in the space environment should be met and controlled by the man at the helm and not some fresh technician sitting at a desk. This philosophy carried on to the post-Mercury capsule but with increased emphasis in the degree of on-board control. If the spacecraft was designed to fly around the Moon and back it would need a measure of autonomous control unthinkable for the smaller one-man vehicle.

As a result, it was agreed that one man should have responsibility for monitoring the spacecraft systems, such as the environmental control equipment, propulsion units, electrical power supply, communications equipment, etc., and that a second should be responsible solely for guidance and navigation which, unlike an Earth orbit vehicle, would be an important element on flights far from the home planet; if communications broke down it would be up to the crew to calculate their own maneuvers. A third crewmember would serve as the mission commander and assume responsibility for major decisions, critical piloting tasks and overall authority on the spacecraft.

From H. H. Koelle of the ABMA, the Committee heard a report that Saturn C-1 development envisaged an operational flight date of 1967 after a succession of unmanned test flights to fully evaluate the performance and operating characteristics of this mighty, 680 tonne thrust, launcher. From Lewis, Seymour C. Himmel told the Committee that Direct Ascent would require a six-stage launcher: stages one to three would be employed to send the spacecraft to Earth orbit, stage four would accelerate it from 28,000 km/hr to the 40,200 km/hr of escape speed from Earth's gravity, stage five for landing on the Moon and stage six to accelerate the spacecraft off the surface of the Moon and back to Earth, sending the 4 tonne return vehicle into the atmosphere. Koelle reflected feelings in the von Braun camp when he argued for consideration of an Earth orbit refuelling technique which would, although requiring many more flights than the single launch of the Direct Ascent mode, be performed with rockets more closely aligned in performance to existing Saturn concepts, reducing development time and keeping the launcher within acceptable design limits.

Lundin reaffirmed his belief that Earth-orbit refuelling would demand the successful operation of many highly dangerous and untried activities such as the transfer of super-cold propellants from one rocket stage to another, multiple docking operations, and control of several manned space vehicles in orbit at the same time. An important contribution made by the Research Steering Committee on Manned Space Flight was in providing a recommended program to Homer J. Stewart's Office of Program Planning and Evaluation when the latter drew up the official blueprint for NASA activity during the decade beginning 1960. The OPPE had been one of Administrator Glennan's first decisions. NASA simply had no structured program when it came into being, in fact it could be said that in setting up a National Aeronautics and Space Administration, the Congress was in effect condoning the implementation of a space program for the first time. Military activities in space are not a program but merely a functional adaptation of existing options – technological and administrative – by which national defense could be obtained; space, for the military, was a place in which to operate rather than a program. That is an important distinction and one which will be seen to have provoked debate throughout the life of NASA.

President Dwight D. Eisenhower jokes with von Braun during a top level inspection of the nascent space program.

Set up to evaluate the type and nature of objectives it was expected to design, the Office of Program Planning and Evaluation produced a working draft of long-range planning tasks by June, 1959, mostly composed of recommendations from Abe Silverstein's Space Flight Development Office at Headquarters. Between September and November, the OPPE modified its plan during consultation with the President's Science Advisory Committee, the Space Council, the Department of Defense, and internal NASA elements. When it appeared, the 'Long Range Plan of the National Aeronautics and Space Administration' was a classified document marked as a secret publication but with a less detailed version marked as confidential. The original would remain secret for 10 years before relegation to the confidential category. As an indicator of technical possibilities, reflecting predictions from experts and specialists on the nature of future developments, it was unable to present a structured policy for NASA implementation.

The space agency had been required to match initiatives from the Soviet Union and to establish a breeding ground for national technology; the plan was not drawn up with any consideration of these requirements in view and represented a shopping list for agency administration and management to take before Congress. Nevertheless, it indicated the projected levels of maximum development and set up a broad chronology upon which objectives could be hung. It was certainly a timely document. With NASA's Fiscal Year 1961 budget request coming before Congress early the following year, it would be useful to have a set of long-range objectives to show how financial investment would foster technological growth and, perhaps more important for Congress, to point out ways in which the space program could be made to enhance national prestige.

When the OPPE document appeared in December, 1959, it projected for the following year flights with the first meteorological satellite, the first passive reflector communications satellite, the first suborbital flight by an astronaut, and launches of three new rockets: Scout, a solid propellant launcher for very small payloads; Thor-Delta, a liquid propellant launcher for medium size payloads; and Atlas-Agena B, a two-stage version of the Atlas ICBM with a capacity for lifting payloads weighing $2\frac{1}{2}$ tonnes. Other target dates envisaged the first flight of an Atlas-Centaur, with a 4 tonne lifting capacity, in 1961, followed by a lunar impact vehicle, and the first Mercury manned orbital flight. In 1962, probes to Mars and Venus could be launched. A year later, the first two-stage Saturn could be launched. In 1964, the OPPE plan projected soft-landing robot vehicles to the Moon, and the initial flight of the post-Mercury manned space vehicle with circumlunar capability.

Between 1965 and 1967, the 'first launching in a program leading to manned circumlunar flight and to a permanent near-Earth space station,' would be followed sometime after 1970 by a 'manned lunar landing and return.' Launch vehicle development for the big Nova foresaw a first flight in 1968 and it was with the emergence of that class of lifting capacity that lunar landing flights were aligned. To accomplish this schedule, NASA anticipated an increase in its annual budget request from the $802 million it expected for FY1961 to more than $1,000 million in 1962, nearly $1,200 million in 1963, and $1,350 million for 1964. In FY1965 the budget would start to level out at $1,455 million, going to $1,505 million in 1966, $1,550 million in 1967 and to a constant $1,600 million from 1968. When the OPPE forecast went to the House Committee on Science and Astronautics it included an added item: the soft-landing of a mobile vehicle on the lunar surface, precursor to the manned explorations to follow in the 1970s. NASA's unmanned exploration of the Moon at this time anticipated a soft-landing vehicle called Surveyor and the mobile lander called Prospector; the latter was eventually deleted but Surveyor reached fruition in 1966.

But just as 1959 had seen the emergence and somewhat rapid demise of McDonnell's proposed modifications to the basic Mercury capsule, not to arise from the ashes of its own defeat for another two years, 1960 was to witness the near defeat of post-Mercury manned space flight plans. The year opened with a suggestion from Abe Silverstein at a Washington luncheon that the new manned space project should be called Apollo and by mid-year that was to be officially recognized as the most appropriate title. In February 1960, President Eisenhower moved to change the language of the National Aeronautics and Space Act, in which control of civilian space policy was vested in the National Aeronautics and Space Council headed by the President with dialogue between NASA and DoD lodged with the Civilian-Military Liaison Committee. Abolishing the NASC and the CMLC would give Eisenhower greater control since it would move direct responsibility for major program goals and the allocation of booster development from a body on which the White House was only represented to a position from which it could effect real authority. Eisenhower ordered NASA to prepare legislation and to submit it to Congress but after getting a smooth ride through the House in June it was held unattended in the Senate from August, blocked by Johnson who had been instrumental in setting up the very condition that the President now sought to repeal.

LBJ had no desire to see it go under just a few months before a new administration came in to the White House. Eisenhower was unable to stand a third time for the office of the Presidency and whether Republican or Democrat the new President would have different views to his predecessor. As it turned out, LBJ was to get more than passing interest from the new President. Nevertheless, the attempt by Eisenhower to abolish the NASC represents an interesting observation on how concerned the President was about a mushrooming space program. By moving quickly he could pre-empt decisions concerning long range objectives without having to wait for formal presentation to the Executive Office. But Eisenhower was right to be aware of strong moves within NASA for a post-Mercury manned flight program, and as if to counter attempts to block ambitious goals the agency itself contrived to reach an unchallengeable posture before the new administration came to office.

During April, the Space Task Group set out its requirements for the next manned vehicle in a series of hearings presided over by Bob Gilruth. The spacecraft was to be capable of 14 days'

Selected in October, 1960, to prepare feasibility studies for the Apollo manned spacecraft, then perceived as an earth orbit/circumlunar research laboratory, General Electric, Martin and Convair prepared these separate proposals, submitted to NASA in January, 1961.

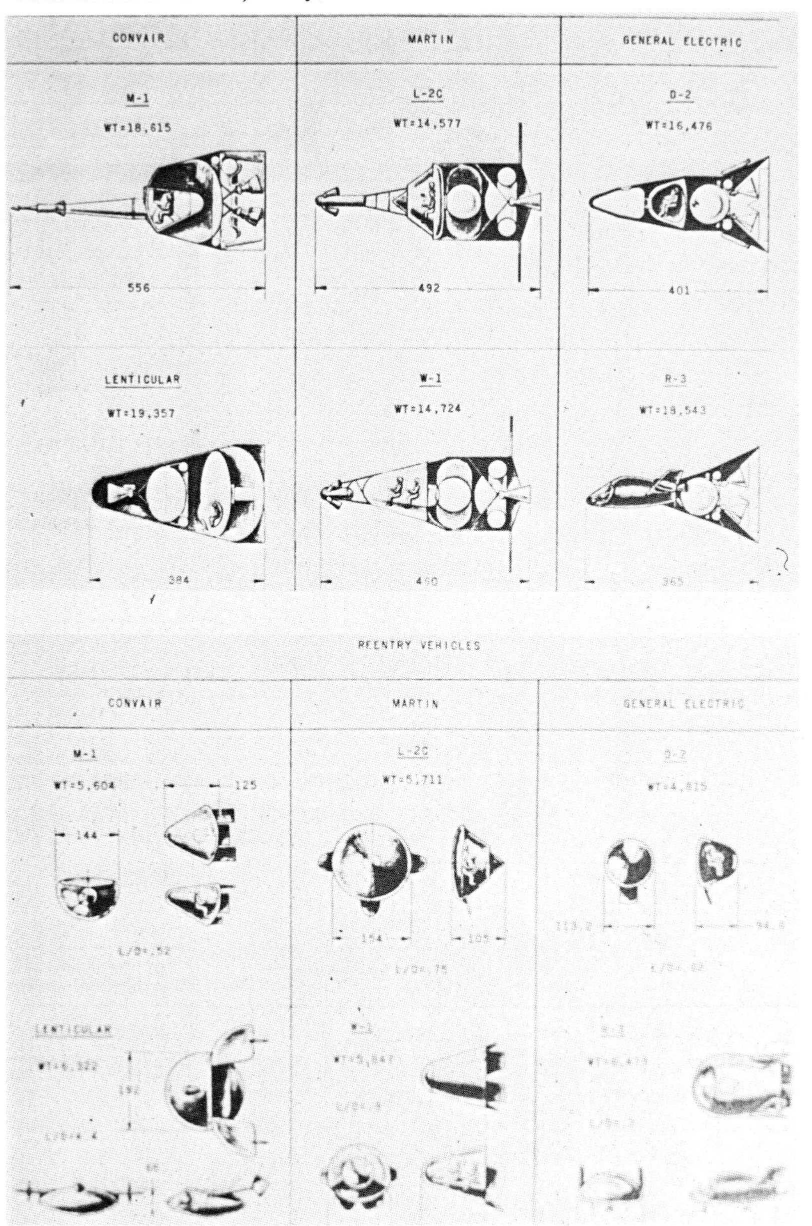

operation in space, supporting a three-man crew on flights to circumlunar objectives and Earth-orbiting laboratories. Lunar reconnaissance would be achieved with a lunar mission module assembly weighing no more than 6.8 tonnes and launched by either Saturn C-1 or C-2 rocket. Targeting would probably require the lunar mission to approach within 80 km of the Moon's surface, presumably on the far side as it looped around the lunar sphere on its way back to Earth. Before this could be achieved, development and qualification flights would take place in Earth orbit with a lunar mission module and a space laboratory weighing no more than 11.3 tonnes. Designed primarily to land at a ground location, the spacecraft would also be required to descend to a splashdown in emergency and to provide safety for the crew for at least three days after landing. It was also expected that the new spacecraft should afford its crew a shirtsleeve environment; that is, one which permits the crew to live in lightweight garments and to wear pressure suits only during emergencies or when going outside the vehicle. It was hoped that development could lead to a planned Earth orbit qualification flight during 1966.

For most of that spring and early summer, STG consolidated the requirements, planning to present basic specifications to industry by mid-year. On 25 July, 1960, Abe Silverstein informed Harry Goett that NASA Administrator Glennan had formally approved the suggested name of America's second generation manned space vehicle. It was to be called Apollo and referred to as such at the forthcoming presentations to industry. That event took place over two days beginning 28 July, an event that brought the civilian space agency into the open with public discussion of future planning schedules culled from the OPPE report. For the first time since its formation, NASA was explaining its policy for a viable space program that centered on exploration of the near solar system (that is, planets with orbits adjacent to the Earth's path around the Sun) by way of an Apollo spacecraft capable of flying round the Moon or performing Earth orbit laboratory functions.

The NASA-Industry Program Plans Conference, held in Washington, was to be the first of a regular series of meetings at which private industry could be made aware of NASA's objectives; most of the space budget would go to outside contractors rather than remaining in-house at government facilities. In summary, George Low presented a chart that showed Mercury leading to Apollo, capable of manned circumlunar flight or Earth orbital laboratory operations, before 1970. After that date, lunar landing flights and large space station operations would proceed simultaneously, followed by interplanetary reconnaissance flights as an outgrowth of the technology developed for Moon landings. Beyond that, the planets would receive landing flights as the final leg of scientific exploration of the solar system. The third quarter of 1960 was occupied with re-shuffles to the management structure – Faget was upgraded to Chief of the STG Flight systems Division, Piland became Assistant Chief for Advanced Projects, and at NASA Headquarters an Apollo Project Office was set up also under Piland – and with awarding feasibility contracts to successful bidders at an Apollo briefing for industry held at Langley.

A surprisingly large number of companies – 64 in all – expressed interest in bidding for Apollo studies and 14 responded to the 'request for proposals' (RFP) packages mailed by STG, 2 later withdrawing. On 25 October, 1960, three contractors were selected to conduct Apollo feasibility studies: Convair's Astronautics Division from General Dynamics; the General Electric Company; and The Martin Company. The contracts would be effective from 15 November. Before the end of the year, Space Task Group had proposed a structure whereby all NASA Centres would be kept in an integrated loop of information and program awareness through Technical Liaison Groups assigned specific areas of Apollo requirements: Trajectory Analysis; Configurations and Aerodynamics ; Guidance and Control; Heating; Structures and Materials; Instrumentation and Communication; Human Factors; Mechanical Systems; and Onboard Propulsion. These would form the backbone via which the different field Centers would feed program needs like a nervous system energizing an organic structure. In December, all three feasibility study contractors presented their first interim reports to NASA personnel. As 1960 ended, the configuration of Apollo was still undecided but the program elements were falling into place with due clarity. Earth orbit and lunar roles were to be met with a modular design incorporating three structures: a command center module, a propulsion module, and a mission module.

The command center module would comprise the crew compartment, a habitable area where the three-man crew would be situated during launch and re-entry, doubling as a flight control area during the mission. The propulsion module would provide a maneuverability denied to Mercury, in effect being a rocket engine with propellant tanks and a capacity to stop and start many times. It would be used for aborts during ascent, if required, for changing from one orbit to another around Earth, and for returning to Earth at the end of the flight. For a circumlunar role, it would provide mid-course corrections – propulsive maneuvers necessary to remove small errors imparted by the departure from Earth orbit – and align the trajectory with the required miss-distance from the lunar surface. It was even proposed that this module might be able to place Apollo in Moon orbit by decelerating it as the assembly coasted toward the lunar sphere, pulled at ever increasing speed by the Moon's gravity. Fired again on the far side of the Moon, Apollo would climb up out of the Moon's gravitational field and coast back to Earth.

But that was just the inevitable stretch in the basic design inherent in any engineering plan. Nobody was then ready to incorporate such an ambitious undertaking as a mission objective, just as an actual lunar landing played no part in the preliminary design specification. Indeed that role was presumed to be the mandate for a post-Apollo development and not Apollo itself. The mission module would be used as and when required; for some flight objectives, the first two modules would be used alone. But Earth orbit tasks would benefit from the pressurized interior afforded by the mission module, equipped as a small laboratory for a multitude of tasks tailored to specific objectives. It would provide a more comfortable habitation for the crew and serve as a research facility according to the needs of a particular flight. Only the command center module would be capable of returning to the Earth, propulsion and mission modules left in Earth orbit or separated en route back from the Moon to be destroyed in the atmosphere.

There were some, particularly at Headquarters, who felt things were moving away from the recommendations of the Goett Committee or the STG's New Projects Panel where a Moon landing was cited as the basic goal, circumlunar and Earth orbit laboratory flights being mere precursors in a lengthy development chain. George Low, Chief of Manned Space Flight in Abe Silverstein's Office of Space Flight Programs, was one. In a memorandum to his boss on 17 October, Low said, 'It has become increasingly apparent that a preliminary program for manned lunar landings should be formulated. This is necessary in order to provide a proper justification for Apollo, and to place Apollo schedules and technical plans on a firmer foundation.' He went on to tell Silverstein of a four-man working group he had set up to monitor Apollo design criteria, establish basic policies for manned Moon requirements, and to lay down a schedule tied to a realistic funding profile. It was a timely reminder that if the technical discussion of future NASA objectives was moving confidently forward, the political obstacles remaining could prove decisive to the future of manned space flight. For in only the previous month it had become apparent that NASA would be starved of Apollo money for Fiscal Year 1962, a budget year that would begin the following July.

As early as February, 1960, Administrator Glennan issued guidelines to departments anxious to submit costing requirements for FY1962 advising of a $1,100 million ceiling. The following month a Staff Conference of senior NASA personnel discussed the budget needs and found justification for raising this target. In May, the Budget Bureau heard of NASA's intention to ask the administration if it could request a sum of $1,376 million, money for Apollo and high-energy upper stages for Saturn as the main inflationary factor. In the general back and forth of budget tussles, the Bureau got NASA to reduce this to $1,250 million for formal presentation. Then, in December, a further reduction was imposed to $1,139.5 million. NASA was deeply concerned. If money was not to be available for advanced manned space vehicles and if Saturn was divested of its potentially most beneficial adjunct, the space agency would be stymied in the face of more Sputnik shocks believed by many to be just around the corner.

It was during the formal presentation of the $1,250 million during August that NASA's plan to move ahead with Apollo came out. This was far from being an approved project; a venture of such magnitude would need executive decisions at the White House, and for this very reason NASA had wanted to maneuver into the most advantageous position before presenting the plan to Congress. Eisenhower was dismayed to see his efforts to constrain NASA seemingly going awry and in December the President's Science Advisor George Kistiakowsky (Killian had retired from the appointment earlier that year) was asked to prepare a report on the 'goals, the missions and the costs,' of manned space flight. Brown University's Donald Hornig chaired the committee that issued the report to the President. By this time Eisenhower was aware that Republican candidate Richard Nixon had been defeated by the calculated plans of the John F. Kennedy camp, and that a Democratic administration would assume office in January. He was unwilling to leave loose ends lying about for the new chief executive.

The 'Ad Hoc Panel on Man in Space' reported a gloomy picture of Mercury and cast doubt on the whole future for manned vehicles. It approved of the optimism inherent in NASA's long range plan for technical solutions to Moon landing goals, but seriously questioned the need for such goals in the first instance. It concurred with the NASA view that, 'none of the boosters now planned . . . are capable' of sending spacecraft to the surface of the Moon, and that such a project would require either many Saturn C-2 class vehicles assembling in Earth orbit the elements of a Moon mission or a single Nova class booster of at least 5,400 tonnes thrust. The PSAC report agreed that development of massive super-boosters was mandatory for implementation of such goals but cautioned that, 'among the major reasons for attempting the manned exploration of space are emotional compulsions and national aspirations.' Furthermore, in the view of the PSAC, 'man-in-space cannot be justified on purely scientific grounds.'

To show Eisenhower the ramifications of an expanding manned space program, the committee prepared what in the light of later events would transpire to be a fairly accurate assessment of the financial price to be paid for such aspirations. It said that the Mercury program, although being a 'somewhat marginal effort' incapable of providing 'adequate safety for the Astronaut,' would cost $350 million to completion of the basic objective. It did also caution that 'a difficult decision will soon be necessary as to when or *whether* a manned flight should be launched' (author's italics) in the light of its finding that the project would not have 'a high probability of a successful flight.' But the report was critical too of the initial Apollo goal of circumlunar flight, a venture which was assessed at $8,000 million, including various unmanned projects essential to the ultimate success of a manned mission around the Moon and back.

Implicit in the PSAC's criticism was the surmise that

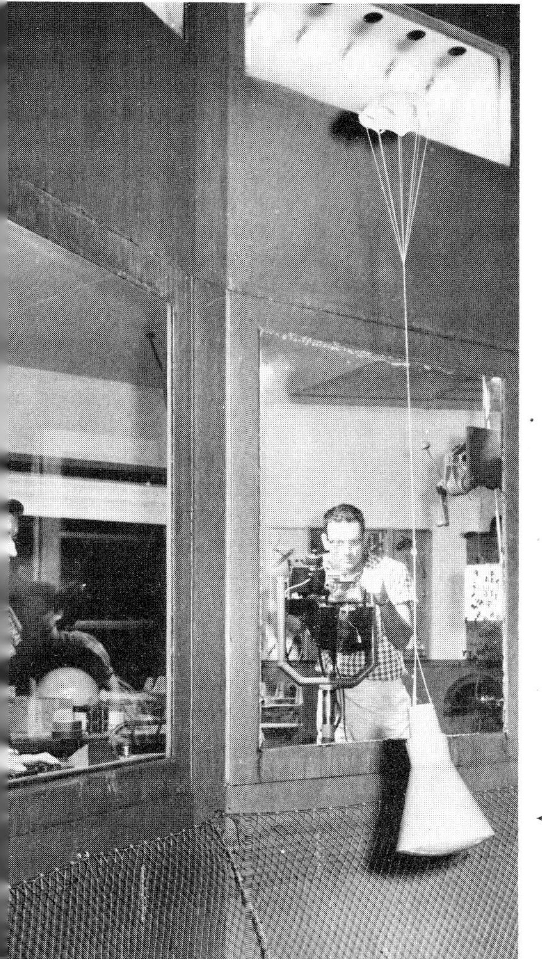

NASA's Langley Research Center played host to Project Mercury when wind tunnel facilities were used to check stability and drag coefficients.

← A scale model of Mercury is released into the airstream of a NASA Langley Research Center wind tunnel to investigate stability of descent by drogue parachute before the model falls into a net.

circumlunar flight would be pointless without moving to the next logical step, perhaps the raison d'être of the entire post-Mercury undertaking: a manned Moon landing, which for the purpose of the study was assumed to be in 1975, a fairly accurate date consistent with NASA's plans at that time. That objective was said to require between $26,000 million and $38,000 million on top of the $8,350 million spent on Mercury and Apollo. With little difficulty it would be seen that if these estimates were true, the average annual NASA budget would have to increase by more than 100% and hold that posture for 14 years! When Eisenhower got the report he was greatly disturbed and called NASA to the White House to explain the runaway escalation implied by current plans. It has been said by a confidant of one who was privy to the meeting, that Eisenhower on hearing quotes about Columbus and the discovery of America affirmed that he was not about to 'hock his jewels' to get a man on the Moon, referring to the efforts of explorer Columbus to get money from Spain's Queen Isabella for his expedition.

This was the first time anybody at the White House had been seriously expected to consider the possibility of landing a man on the Moon. NASA's internal policy reviews had been no concern of anybody outside the agency, indeed the executive branch of the government was as disinterested in the technicalities of space projects as they prove consistently to be about any form of technological application to unprecedented needs. It has been said that there was more laughter at the Moon landing prospect than serious consideration, when NASA put its case before the President, with comments suggesting the utter absurdity of the ultimate request: manned landings on the planets. The pitiful response by politicans to new technical capability has been said by many to have been more responsible for upsetting the balance of an ordered planet than all the creations of scientists and engineers; it is reactions similar to those received from the Eisenhower administration that endorses such conclusions.

The ridicule infesting the December White House session was given more serious ammunition when ex-PSAC head James Killian unsheathed an assault on ambitious space plans. In the same month that Eisenhower heard abortive pleas for a workable budget, Killian said at a speech in New York that he was sure 'we must never be content to be second best, but I do not believe this requires us to engage in a prestige race with the Soviets. We should pursue our own objectives in space science and not let the Soviets choose them for us by our copying what they do.' Moreover, he went on, 'Decisions must soon be made as to how far we go with our man-in-space program. Unless decisions result in containing our development of man-in-space programs and the big rocket boosters, we will soon have committed ourselves to a multi-billion dollar space program.' There was clearly a mandate from advisers and ex-PSAC heads to limit the initiatives expressed by the space agency. Coming just a few days before the important meeting at the White House it reminded Eisenhower of an earlier battle to ignore public demand for a new posture toward space technology when Sputnik I bleeped overhead.

Throughout the years of his administration he was seen to be a 'status quo' man, never allowing himself to be persuaded away from a conservative line on national or international affairs. Eisenhower saw that Apollo plans were inconsistent with his mandate that NASA only partake in programs of notable scientific merit. For its part, the space agency believed it would have to reach the Moon before presenting a base from which to assess future space goals. Eisenhower firmly believed that commitment to the kind of program being sought by NASA was an irresponsible waste of human and technological resources; the budget had already grown more than 100% in two years and that was as far as the President was prepared to see it go. Accordingly, further reductions were authorized and the Budget Bureau informed NASA that its already limited fiscal plan to spend $1,139.5 million in 1962 should be cut by $29.87 million. The final FY1962 budget request, at $1,109.63 million, represented a cut of 20% on the $1,376 million originally sought.

Most affected were Apollo and continued development of the cryogenic, high-energy, upper stages for Saturn – key elements in a growing space program. Minimal funds were available for studies on a post-Mercury vehicle and Administrator Glennan, with his Deputy Hugh Dryden, had to intervene to prevent the President from including in his State of the Union address an executive commitment to abandon manned space flight after the Mercury three-orbit flight. Nevertheless, the President urged that

Three of these small solid propellant rockets were used as retro-rockets to slow the Mercury capsule by firing against the direction of flight, allowing the astronaut to return to earth.

it would be necessary to determine 'if there are any valid scientific reasons for extending manned space flight beyond the Mercury program.' Had Eisenhower not been moving from the White House, handing over as chief executive to a young Democrat, he would undoubtedly have cut further studies and erased the Goett Committee recomendation for a manned Moon landing 'sometime in the 1970s.' NASA had all the technical reasons for pressing ahead with a modest program maturing in concept and capability; the incumbent administration was totally unconvinced that any good would come from the space program as then being planned and deferred the coup de grace only because it would have been political suicide to freeze options available to the new occupants at the White House.

No one at NASA knew which way Kennedy would move on decisions relating to the space program. Contradictory statements from the President-elect left little ground for opinion on Kennedy's view. It was a time of indecision, but one which would very soon lead to an even bigger commitment than anyone at NASA had dreamt of so far. But if Apollo had its troubles, Mercury too looked none too healthy in that winter of 1960/61. It seemed as though the technical milestones on the road to manned orbital flight were coming with increasing difficulty and at a slower and slower pace.

Flight tests began in 1959. Following the abortive Little Joe flight of 21 August, where the launch escape system wrenched the capsule to freedom before its scheduled time, a major heat shield test was timed for September. Called Big Joe, as a natural extrapolation of the Little Joe designation, because it involved a ballistic test from an Atlas missile, the shot was expected to evaluate and qualify the ablative heat shield designed to comprise the broad base of the truncated cone-shaped spacecraft. It was a proof-of-concept flight if ever there was one. No one knew if Faget's blunt faced design for Mercury really would stabilize itself in free fall during re-entry, and the ablative shielding was considered essential as the only alternative to the originally proposed heat-sink shield. The latter would absorb heat and slowly radiate it back to the atmosphere after splashdown or an emergency land landing.

It was considered too hot a structure to hold on to once the capsule was down so it was suggested that the shield should be jettisoned while the capsule descended on parachute. That idea was unanimously dropped when on test one day in 1959 a heavy heat shield sliced through the air like a boomerang and returned to the capsule from whence it had been jettisoned, slicing into it with great force; the thought of a pilot being hacked to pieces by his own heat shield turned engineers to an ablative concept where material from the phenolic fiberglass structure would burn away in a controlled manner, leaving the capsule with a much thinner, but comparatively cool, shield which would in fact be released to pull down the landing bag impact skirt described in the previous chapter.

Big Joe would prove that the ablative shield would withstand the harsh environment. The capsule itself was a boilerplate, tailor-made for the gruelling test a real production capsule would experience later in the program. Supporting more than 100 thermocouples to measure temperatures inside the structure, the boilerplate carried a custom-built nitrogen attitude control system which would turn the capsule around after separating from the Atlas and point it down, blunt end first, to the atmosphere on a long ballistic path expected to end about 3,220 km from the Cape Canaveral launch pad. Precisely at 3:19 in the morning of 9 September, Atlas 10-D, an early 'D' series production model, thundered to life at the start of a long, arching trajectory out over the Atlantic Ocean. The missile was programed to reach a height of 160 km and pitch over below the horizontal to achieve the speed of more than 27,000 km/hr necessary to simulate the heat of re-entry. On thundered the missile past the point where the two outboard booster engines should have been jettisoned; instead of falling away as planned they rode on up with the main structure, decreasing the final speed to 23,910 km/hr and the impact point by 805 km. But more than that; out of contact with ground tracking stations, the capsule was more than 2 minutes late separating from the booster, during which time the small nitrogen thrusters tried in vain to turn the capsule through 180° for re-entry. When finally it separated the capsule was an inert body falling down through the atmosphere but Faget's design got more of a workout than expected and the boilerplate turned itself around, finding its own centre of gravity as it neared the upper atmosphere.

On down through the hot phase of re-entry the capsule experienced a larger heat pulse than expected, although the heat load was less than planned due to the slower entry speed. Far from its planned impact point the recovery fleet hurried to find the capsule while NASA representatives stressed the boilerplate nature of the spacecraft to press men who just might have envisaged a similar malfunction to a production Mercury. The capsule was finally recovered about 10:00 am and late that night it was delivered to Hangar S at Cape Canaveral where Bob Gilruth stood with Maxime Faget waiting to give their protege cursory inspection. It came through with flying colors, paint hardly having been singed on the conical afterbody, and some of the waiting STG representatives separated two halves of a makeshift interior pressure vessel and handed Bob Gilruth a note, which read: 'This note comes to you after being transported into space during the successful flight of the "Big Joe" capsule, the first full-scale flight operation associated with Project Mercury. The people who have worked on this project hereby send you greetings and congratulations.' It was signed by 53 members of the Space Task Group.

Over the next few weeks inspection of the heat shield revealed it to be in the expected condition, to have ablated only 30% of the total thickness applied and, although having experienced a higher heat pulse, to be capable of a normal re-entry trajectory. But a cautionary clause crept in to written recommendations that Mercury engineers could plan on using an ablative shield for orbital flights: in the event the capsule's retro-rockets failed to fire, and the capsule gradually decayed back into the atmosphere over several grazing entries in a period of up to 24 hours, the total heat load would be excessive, as it would also if only one of the three retro-rockets fired and caused the capsule to dip into the atmosphere at a very shallow angle.

The retro-rockets were solid propellant assemblies contained in a flat, circular, drum-like structure attached to the base of the heat shield with three straps that would be jettisoned after retro-fire. The rockets would fire for 10 seconds each with ignition timed at 5 second intervals; each motor delivered a thrust of 450 kg and collectively the three would slow the capsule by about 560 km/hr. The reporting engineer from Faget's theoretical heat transfer section stressed that 'under no circumstances should the weight of the heat shield itself be shaved.' For the next six months flight test personnel busied themselves with a flurry of Little Joe launches and a pad abort test with a production capsule.

Little Joe 6 proved itself in a booster test on 4 October, 1959, when four Pollux and six Castor solid propellant rockets were sent to a height of 64 km. Little Joe 1A was intended to test the Mercury launch escape system under conditions of max-q (maximum dynamic pressure) when the capsule mounted on an Atlas would experience the greatest build-up of aerodynamic pressure. Following a successful 4 November launch the booster and spacecraft passed through max-q with the launch escape rocket wrenching the capsule free 10 seconds late when dynamic pressure was down to only 10% of the planned value.

No one ever found out why the escape sequence had been delayed but a month later Little Joe 2 carried Rhesus monkey Sam on a flight to more than 30 km in a successful test of a high altitude abort. Sam withstood the short period of weightlessness and a tiresome wait of six hours bobbing around in the Atlantic. But still the elusive test to check escape system response to a critical abort at max-q had evaded engineers: Little Joe's 1 and 1A evading the stressful test for separate reasons. On 21 January, 1960, Rhesus monkey Miss Sam flew out on Little Joe 1B with responsibility for pulling a lever when a light flashed during a simulated max-q abort. From a height of 14 km and a speed of 3,220 km/hr the escape motor pulled the boilerplate capsule free of the ascending booster and hurled the passenger to one side and away from what was simulated to be an Atlas running amok. Temporarily disoriented, Miss Sam suffered from mild nystagmus although the high g loads were withstood as expected. Monkey and mission were deemed to be performing well as the grinning pioneer splashed down on time.

But things with the program were, in general terms, not going well. Delays were predicted for production capsules essential to unmanned flights expected to provide information vital to programing the manned missions. Mercury-Atlas flights were seen as an important segment of the pre-operational phase, tests where the capsule would be fledged in orbit and brought back down through the atmosphere exactly as it would with a man aboard. McDonnell had already received a supplemental award for an additional 6 production capsules to add to the 12 contracted in January, 1959. The first spacecraft was expected to be delivered by the end of that November, with the second, ready for a Mercury-Redstone flight, completed by year's end. The first Mercury-Atlas capsule – No. 6 – was expected in February. But the schedule was slipping and McDonnell managers knew only too well that the design of the spacecraft was likely to cause an even worse bottleneck in 1960. The capsule was just too small for more than one or two technicians to work inside the pressure hull at one time, final checkout tasks were made more difficult because systems had been designed into the vehicle without regard for accessibility, and quality control was not all it should be.

Early in the new year an inspector was moved to comment on his log that he could not proceed with his tasks until 'the filthy condition of the capsule was cleaned up.' By January 1960 it had been agreed in consultation involving McDonnell's Yardley and

Mercury-Atlas 1 prepares for a July, 1960, lift-off. Planned as the first ballistic test of a production Mercury capsule the flight ended in disaster one minute after launch when the Atlas broke up.

NASA's Faget and Gilruth, that capsule 6 would be replaced by the less complete capsule 4 for Mercury-Atlas 1. Concern ranked high on the list of managerial priorities during meetings and the regular status reviews held by STG, resulting in 'a new capsule delivery schedule . . . to reflect a delay in delivery of over 3 months in the early capsules. Although various proposals for improving the situation have been considered, there does not seem to be any practical avenue open at this time for effecting any worthwhile change.'

It was at this time NASA decided not to move Space Task Group personnel and their rapidly expanding volume of paperwork to the planned location for manned space flight at Goddard until Mercury was over. There were enough problems between government and its industrial contractors without exacerbating the tension with internal moves. When capsule 4 arrived at Langley there were several points in need of attention but as Paul Purser reported, 'Although there are evidences of careless workmanship, I don't think it is too much worse than standard aircraft practice.' Nevertheless, 'these can only be avoided by inspiring in some way, better workmanship,' a procedure that would probably require, 'inspection by STG people of the flight capsules now on the line. This could be repeated in 6 to 8 weeks to catch the next batch and probably would cure the troubles.' It was a case of the uninitiated leading the novices and nobody wanted to start directing criticism during this sensitive learning phase.

Nevertheless, some kind of re-think was essential to step up deliveries; many tests were being performed without due regard for schedules that could accommodate similar actions while other tasks were being carried out. As Abe Silverstein commented in a letter to von Braun at the Marshall Center, 'A detailed study of the checkout program at McDonnell, Huntsville, and Cape Canaveral has revealed that there exists a great deal of duplication; in particular all the booster capsule compatibility checks are performed both at Huntsville and at the Cape.' But here as elsewhere, there was a mild difference of opinion epitomized by Purser who believed that 'One of the major problems facing Mercury management is the conflict between a real desire to meet schedules and the feeling of need for extensive ground tests.' But as the months rolled by, and spring gave way to summer, the mountain of system and component testing grew, militating against a smooth progression to Mercury-Atlas flights.

Management changes at NASA were being urged in a report submitted to Administrator Glennan although the general infrastructure was considered adequate for the job. Capsule No. 1 was ready for the first Mercury test involving a production spacecraft by May and in a 76 second flight demonstrated its ability to come through a simulated pad abort safely when the launch escape system carried the capsule away from the Wallops Island beach. By this time it was clear that the first manned suborbital Redstone flight would not get off the launch pad before January, 1961. Far short of the orbital flights promised by Mercury-Atlas, it would nevertheless place an American in space if only for a few minutes. And even that achievement was nothing to chafe at as piece by piece evidence built up predicting an imminent Soviet spectacular. On 15 May, a $4\frac{1}{2}$ tonne spacecraft called Korabl Sputnik 1 (Cosmic Ship 1) ascended to orbit from the launch site at Baykonur. It remained passively in space for four days before a failure in the propulsion system sent it to higher orbit instead of back to Earth; it would stay in orbit for more than 2 years. Intelligence from an Air Force Brigadier General, recently back from Russia, led to a conclusion that the Soviet Union would put a two-man laboratory in orbit before summer's end. Public discontent with the pace of America's manned space program was beginning to build up, a mood that would plunge further from the existing level of confidence by the end of the year.

In what was planned to be a repeat of the Big Joe flight 10 months earlier, Mercury-Atlas 1 was expected to demonstrate that a production capsule could live up to its design expectations in a 16 minute flight that would carry the spacecraft to a height of 181 km, a speed of nearly 21,000 km/hr, and bring it down to a splashdown 2,400 km from the launch pad. Minor delays held back the launch until, on 29 July, 1960, the spacecraft and its Atlas 50-D launcher stood ready on Launch Complex 14 at Cape Canaveral. Problems with the capsule were resolved but heavy rain driving across the Cape area kept back the final countdown preparations. In the blockhouse some distance away from the pad, Walt Williams conducted operations while Defense Department officials coordinated the recovery forces that would, very shortly after launch, stand by to recover the spacecraft. It was an important showing, one that the press were keenly watching in this year that would see a new President elected to office, require the incumbent chief executive to approve plans for a follow-on manned space flight program, and hopefully move Mercury to success before the Soviet Union sent a man into orbit. It was the first time an Atlas, carrying a production space capsule, could demonstate the systems later combinations would lift toward orbital flight.

At 9:13 on that Friday, July morning, the three powerful Atlas engines thundered into life and carried the assembly at ever increasing speed toward a spot in the sky where capsule and booster would separate. Less than a minute into the flight, at a height of 9,750 meters, all telemetry from the missile ceased as it broke up under the force of maximum dynamic pressure. Nobody could tell what had happened, the test was a complete write-off and the spacecraft itself remained attached to falling debris, breaking up as it tumbled back to the Atlantic 12 km from the pad; the escape system was not to be armed until three minutes into the mission, fatally in excess of the time available. STG was struck with gloom; NASA was deeply concerned about the effect on budget moves then under way; politicians asked pointed questions. And so did the press.

Two weeks later the weekly publication *Missile and Rockets* said that 'NASA's Mercury manned-satellite program appears to be plummeting the United States toward a new humiliating disaster in the East–West space race. The program today is more than a year behind its original schedule and is expected to slip to two. It no longer offers any realistic hope of beating Russia in launching the first man into orbit around the Earth – much less serve as an early stepping stone for reaching the Moon.' This was a cynical reference to emerging plans, for Apollo had been officially named only four days before the MA-1 disaster. Responding to this, Bob Gilruth sent his team a memorandum acknowledging that 'members of the STG have expressed concern about these articles.' It contained a modest pat on the back and urged everybody to 'put on your thickest hide, to continue your concerted efforts to make Project Mercury the kind of program it was designed to be, and to reflect with me upon our past accomplishments.'

All things considered the latter were not considerable, but the message was a genuine statement that 'there are a number of people around our country who do understand how much work and how much blood and sweat go into an undertaking of this kind.' It was certainly a bad time for failure. Political campaigns were hotting up in anticipation of the coming elections; John F. Kennedy dealt the current administration a sideways blow by snapping that if a man was to have chance of orbiting the Earth that year (1960) 'his name will be Ivan.' Delays were inevitable after the MA-1 accident and it would be nearly seven months before a similar configuration stood ready to repeat the test. Meanwhile, as engineers studied data, results of simulations, and coded telemetry tapes in vain efforts to obtain the cause of MA-1's failure, Little Joe 5 added insult to injury by balking attempts to get an almost fully equipped production spacecraft off an ascending booster in simulation of a max q abort from an Atlas.

Spacecraft 3 was identical in almost every respect to later capsules destined for operational flights. Attached to Little Joe 5 in preparation for flight on 8 November, the day America went to the polls to elect a new President, the capsule was originally to have carried a primate passenger but Gilruth deferred the live test for a mechanical evaluation of this critical abort phase. It was to be the first test with a production capsule for an abort at that critical portion of the flight where excessive g forces were once thought to be a prohibitive factor in selecting Atlas for manned flight. Just 16 seconds after lift-off, both escape and tower jettison motors ignited together prematurely. On an actual abort, the launch escape motor would fire first to remove the capsule from its booster, with the tower jettison motor firing after the main rocket shut down to remove the tower from the front of the capsule. On LJ-5, both fired together and the spacecraft remained attached to Little Joe. It was a ridiculous situation and one which ended by the entangled assemblage falling to the Atlantic. Only 40% of the spacecraft was ever recovered and nobody could tell for sure what had happened, except that there must have been a failure in the sequencer.

But worse still, the first flight of a Mercury-Redstone less than two weeks later bruised injury almost to the point of toleration. It began when Little Joe was being prepared for its flight. On 7 November, 1960, the MR-1 combination was ready and waiting for the order to fire when helium pressure in the attitude control system's pressurization circuit fell to a fraction of its planned value. As an excellent example of the problem engineers had with servicing vital systems, due to the sandwich packing method, the spacecraft – capsule No. 2 – had to be removed from the Redstone booster, the heat shield was then taken off, and the toroidal propellant tank for the attitude control system disconnected so that a single helium relief valve could be replaced. Assembled in the reverse process, engineers performed a minor modification to the sequencer in hopes of avoiding a repeat of the Little Joe fiasco. On 21 November all was ready and the countdown had only a single delay for minor repairs as the configuration neared the time for its crucial test, designed to hurl the capsule on a suborbital lob 105 km high and 370 km downrange.

At the planned time the main Redstone engine fired into life, thundered for a second or two, then promptly shut down. Keen-eyed technicians watching from a safe distance observed the assembly to rise a few centimeters then settle back on to exactly the same points that previously supported it on the launch pedestal. No sooner had the engine stopped, the thunder claps of aborted sound echoing across the marshy Cape, than another sound superimposed a roar across the hazy panorama and Mercury's escape tower rocketed into the air, climbing out straight and true to a height of more than 1 km before falling slowly down to the ground 400 meters distant. Just three seconds after the escape tower sped away, the capsule's forward cylindrical antenna housing popped away from the conical afterbody, its own drogue parachute bringing it down as superficial debris before a trailing line wrenched free the main parachute which promptly hauled out the reserve chute!

It was as if Pandora's box had suddenly opened and colored scarves were unfolding themselves from the inside. Draped in the limp silk of unnecessary canopies, the cylindrical white rocket held the tiny stunted capsule while smoke and paraphernalia fell to ground across the vacant concrete pad. With the booster destruct system still armed, nobody could approach the Redstone until the batteries that powered the system ran down next morning, the only worry being that a wind might get up and balloon the parachutes, pulling the rocket and its ridiculous cargo over in the process. Everything had gone wrong and the humorous sight of a Mercury shedding and unfolding its interior piece by piece on the launch pad was an unfunny thing to concerned engineers, frustrated technicians and depressed managers.

The reason was soon found for MR-1's premature sequence,

In this series of eight views, Mercury-Redstone 1 fails to lift off as planned and promptly activates the entire launch escape sequence during an aborted mission attempt in November, 1960.

for, as George Low wrote to NASA boss Glennan, 'The MR-1 failure is now believed to have been caused by a booster tail plug which is pulled out about one inch (2.54 cm) after liftoff. . . . This two-prong plug is designed so that one prong disconnects about one-half inch before the second one does. This time interval between disconnect of the first and second prongs for MR-1 was 21 milliseconds. The booster circuitry is such that if one of these prongs is disconnected prior to the other and while the booster is not grounded, a relay will close giving a normal engine cutoff signal. The time interval between successive disconnects was apparently just sufficient to allow the relay to close.'

Low went on to say that the heavier Mercury-Redstone would accelarate slower than a military Redstone which would have 'shorter time intervals between disconnects between the two prongs' and so 'allow the relay to close, thus having avoided this type of failure in the past.' It was now painfully obvious that the scheduled flight of a manned Mercury-Redstone could not take place before the following spring. A hail of ridicule and cynical criticism filled news pages in papers across the nation, the uninitiated moved to more voluble emotion than knowledgeable participants. But it was a clear breach of the expected and NASA was determined to get a good shot before year's end, proving not for the first time, but early enough in NASA history for it to be unexpected, that the space agency would perform best under pressure and with the heat turned full on. And full on it was from Capitol Hill down. Immediately, preparations were made for a repeat attempt at what should have been with MR-1 the first suborbital space shot with a production capsule, albeit unmanned.

It was decided to use the Redstone scheduled for MR-3, the original spacecraft (No. 2) from MR-1, the escape tower from capsule 8 and the forward antenna fairing from capsule 10. From Huntsville three weeks after the aborted flight, a message of reassurance: 'The November 21 type event will be avoided in the future by the addition of a ground cable sufficiently long to maintain a good ground connection until all umbilical plugs are pulled.' Everything went well on 19 December when project managers, project engineers and technicians gathered in the blockhouse from where the re-designated MR-1A would be controlled and monitored. Through periscopes that showed clearly the view a short distance away from where the seemingly tiny propulsion system would fire into life and, this time, sustain the rocket as it flung the 1 tonne capsule on course for a suborbital ride, eyes were strained for signs of some indicator that all was not well. But all was very well as test conductor Paul Donnelly brought every sequence in the countdown to the required point at the assigned time.

Shortly before mid-day, the Mercury-Redstone sped away from its pedestal on a flight that was to be every bit as successful as the plans had predicted. At 2 min 23 sec the single engine shut down with the assembly at a speed of 7,812 km/hr. Coasting on up to a height of 210 km the capsule, separated now from the adaptor that held it firmly to the top of the booster, turned around and prepared itself for the punishing deceleration during re-entry. From a slight over-shoot, the capsule was recovered by the carrier *Valley Forge* and returned to Cape Canaveral for checks and examination. It was a good shot, one that restored flagging confidence among men whose job it was to marshall and shepherd the myriad components that went into each production capsule. It was the week before Christmas, the first time an American production line manned spacecraft had been into space, a present indeed befitting the season.

Yet for all the internal back-patting that was allowed for a brief – and all too short – period after MR-1A, it was becoming apparent that Soviet Russia had more chance than America of sending a man into space within the next few months. Few had expected to get through 1960 without a Russian voice calling from space and there were signs aplenty that plans were afoot for a Soviet man in orbit. Following the launch of Korabl Sputnik 1 in May, a second spacecraft was launched on 20 August. Instead of being an unmanned test of a manned space vehicle design, Korabl Sputnik 2 carried two dogs called Strelka and Belka on a flight lasting one day before successful recovery and return of the canine cargo; Russia had launched animal capsules before but this was the first time a spacecraft designed to carry men had been tested with a live payload.

Korabl Sputnik 3 followed on 1 December, after the MR-1 fiasco and before the successful MR-1A mission, with two more dogs called Pchelka and Mushka. It seemed to be another test of the life support systems aboard the capsule but things did not go as well as they apparently had on the previous flight. Although placed into an orbit 166 km high at the low point and 232 km out at the high point, from where the spacecraft would naturally decay to re-entry within the life of the onboard support systems, an attempt was made on the second day of the flight to conduct a normal retro-fire maneuver and return the spacecraft to Earth in simulation of a manned flight. Something went very wrong with the attitude control system, the angle of entry was too steep and Pchelka and Mushka burned to death in the plummeting capsule that was unable to protect its passengers from the incinerating heat. It was an object lesson to all systems engineers everywhere, and a setback for Soviet plans.

Yet, compared with the American effort, Russia's space program was sparse in the extreme. After the launch of Sputniks 1 and 2 in October and November, 1957, respectively, the following year had seen only one Soviet success: Sputnik 3 on 15 May, 1958, designed to study the Earth from space. In 1959 Russia launched just three vehicles, all of them supporting limited study of the Moon: Luna 1, in January, failed to achieve its primary objective of striking the Moon, flying instead more than 5,000 km from the lunar surface as it went on to become an artificial planet of the Sun. Luna 2, in September, hit the Moon as planned – the first man-made object to do so. The third, Luna 3, accomplished a unique feat in that during a flight past the Moon in October, 1959, it obtained the first pictures of the far side hemisphere ever seen by human eyes. Locked in synchronous rotation with Earth, where one complete side of the sphere faces the planet around which it rotates, only the near side can be observed. The pictures from Luna 3 were fuzzy in the extreme, showing dark blotches on a light background, but it was Man's first view of the far side and as such achieved worldwide recognition for the Russian spectacular.

In 1960, the Soviets launched the three prototype manned spacecraft on tests in which the second and third missions carried two dogs apiece, and two unsuccessful attempts to send spacecraft to the planet Mars. But more significant than the fact that in the same period America launched 34 satellites and probes to Russia's 9, was that the Soviet space machine had shown modest growth far beyond the capability of the nascent NASA. The basic Sapwood launcher, adapted from Russia's first ICBM, was capable of lifting about 2 tonnes into Earth orbit; more than Atlas. With Luna 1 in January, 1959, an upper stage was introduced comprising a comparatively low-energy propulsion system and propellant tanks which, mounted to the basic Sapwood by a truss structure, provided payload growth to approximately 4.7 tonnes. When applied to escape missions – spacecraft accelerated to a speed in excess of 40,000 km/hr so that they 'escaped' the Earth's gravitational pull – it was capable of lifting the Luna series to the Moon, the heaviest of which weighed nearly 1 tonne. As an Earth orbit combination, Sapwood plus upper stage could easily accommodate the manned spacecraft developed during 1959 and 1960.

By the turn of the year things looked decidedly brighter as activity hummed around three focal points: STG at Langley, the Marshall Space Flight Center at Huntsville, and Kurt H. Debus' Launch Operations Directorate at Cape Canaveral. Basic launch escape tests were nearly complete, the system all but qualified; Mercury-Atlas flights had not gone as expected but nothing incipiently grotesque had been seen in malfunctions that seemed for the most part to be a result of inadequate knowledge rather than bad design; Mercury-Redstone had proven capable of short suborbital hops of the type scheduled for America's first manned trip into space, albeit far less demanding than the orbital flights to follow on Mercury-Atlas. If all went well on the next Mercury-Redstone flight, the following mission could carry an astronaut and that prospect spurred Task Group personnel to husband their charge to a better state of honed perfection. But flight preparations were necessarily protracted and what seemed to the public a simple matter of raising a rocket and lighting the engine was, in reality, merely the end product of a long and painstaking process.

In readying Mercury-Redstone 2 for a mission on which it would fly a chimpanzee called Ham down across the Atlantic, spacecraft 5 was completed by McDonnell in May, 1960, and checked out during comprehensive tests by Navy Bureau of Weapons personnel at St. Louis, Missouri. In September, capsule 5 was flown by air to Huntsville where von Braun's group at the Marshall Space Flight Center were preparing Redstone rockets.

Dr. Kurt Debus (left) confers with von Braun in a Cape Canaveral blockhouse.

At MSFC the spacecraft was mated to the Redstone's adaptor ring for checkout and compatibility tests designed to confirm that capsule and booster would fit as planned when finally mated for launch. Spacecraft 5 arrived at the Cape during October for acceptance checks and more tests taking 50 days to perform and 60 days to put right discrepancies found along the way; by now more than 100 tasks had been necessary merely to put the spacecraft in the advertised condition. Meanwhile, Redstone checkout at Huntsville put the booster through its paces in facility buildings at MSFC and at static firing stands not far away. It was air-lifted to the Cape just before Christmas where modifications were made to the systems so the booster would fly a flatter trajectory than the MR-1A mission, which had gone further downrange than planned, inducing a load in excess of the forces calculated for the flight.

But for all the attention of putting right minor errors in construction and checkout, engineers were still climbing a learning curve that left them embarrassed on occasions where normal hardwork and attention to detail was visibly seen to be insufficient for the rigorous demands of manned space flight; post-flight examination of film from inside the MR-1A capsule showed a storm of wire clippings, nuts, bolts, washers, springs, etc., suddenly emerging from hidden niches all over the spacecraft, floating in the minutes of weightlessness as the Mercury descended to splashdown. In several respects, MR-2 was a dress rehearsal of equipment and support teams required for a manned shot. It was the first Mercury flight from the Cape with a live payload and recovery forces were required to give just as much consideration to retrieving the capsule as they would later for an astronaut.

Chimpanzee Ham was one of six primates brought from Holloman Air Force Base in readiness for the selection process to assign one for MR-2. Chimpanzees were well suited as stand-ins for the human astronauts in that they had reaction times almost identical to *homosapiens* and supported a skeletal frame similar in several respects. In training for space flight, they each learned to respond to flashing lights by pulling levers, a task rewarded with banana pellets. But if primarily a qualification of Mercury's

Ham, the first chimpanzee to fly in space, gets ready for the MR-2 flight in January, 1961.

Provided to ensure a ready means of escape, this 'cherry-picker' remained alongside the Mercury capsule until a short time prior to launch.

life support system, the chimp itself was a proud advertisement for Lt. Col. Hamilton Blackshear's Holloman Aeromedical Laboratory. As preparations neared the critical point where Ham would have to go aboard spacecraft 5, Walt Williams coordinated operations from the Mercury Control Center at Cape Canaveral.

Just before 8:00 in the morning of 31 January, 1961, the chimp was placed in the capsule and the pad cleared. Flight Director Christopher Kraft monitored a minor systems glitch as an inverter read too hot, causing first one delay, then another, until after holds collectively totalling nearly four hours all seemed about ready for the final countdown. Then the gantry affording access to the Redstone and its charge suffered a stuck elevator, fresh problems with the environmental control system showed up, and booster cover flaps refused to work properly. Finally, after nearly 4½ hours in the supine position aboard spacecraft 5, Ham rode the Redstone toward space, targeted for a high point of 185 km and a downrange splashdown 487 km from the launch pad. But as the booster climbed the cabin inflow valve opened and suddenly reduced pressure from 284 mmHg to a potentially disastrous 52 mmHg! Exactly as designed, the chimp's suit circuit cut in and maintained a comfortable environment. Prior to this, however, radar data acquired by tracking stations indicated the booster was climbing too steep and at 2 min 17.5 sec the liquid oxygen supply was depleted just about the time the engine should have shut down as scheduled.

Designed to observe such phenomena, the Redstone's abort system was triggered, sending the capsule further up the climbing trajectory imparted by the main boost phase. This provided spacecraft 5 with a maximum speed of 9,426 km/hr and an apogee of 253 km–1,180 km/hr too fast and 68 km too high. Because the spacecraft had aborted, the retro-rocket pack was jettisoned prematurely which caused the spacecraft to re-enter at a faster speed than planned, taking the capsule 212 km further downrange from where the recovery forces were waiting. Down through the atmosphere came the capsule, giving its occupant a load of nearly 15 g compared with the planned 12 g, to a hefty thump on an Atlantic swell. That caused the beryllium heat shield, hanging beneath the capsule on the skirt designed to cushion the effects of splashdown, to skim the water on landing and strike a bundle of wires which pushed a bolt, punching two small holes in the titanium bulkhead. Wave action quickly tore the heat shield free, leaving it to sink while the capsule slowly shipped water.

Very soon the capsule's center of gravity was upset, causing it to capsize and ship more water through the cabin pressure relief valve port, a device designed to bring air into the capsule during descent. Unbeknown to recovery personnel moving in on the sinking capsule, Ham was in serious danger of drowning. First sighting had been made from a Navy aircraft nearly one-half hour after splashdown but by the time helicopters arrived an hour later the capsule was on its side and going under. Secured by a line, spacecraft 5 was slowly raised from the Atlantic as water poured from the hull like the cascade from a punctured water-can. Ham was safe, but emotionally disturbed by his experience. Back at Cape Canaveral the surge of eager newsmen upset the fraught chimp, who became aggressive and threatening at the noisy, unruly crowd. Calmed by trainers who could reassure him in quiet he again displayed agitation when taken back to spacecraft 5. It was clearly a one-time event for chimpanzee Ham, who was blissfully unaware of cosmic trail-blazing or the need to suffer on behalf of pioneering humans!

So far so good. If not exactly as planned, the Mercury-Redstone system had shown itself capable of adapting to changing operational situations and the mission was considered to have proven the worth of back-up environmental systems and the abort equipment. Examination of the tattered residue of landing bag skirt straps moved STG to re-design the system substantially and add retention straps designed to survive for 10 hours in a heavy sea. Also, an additional glass fiber bulkhead was to be inserted between the heat shield and the titanium pressure bulkhead to prevent recurrence of the MR-2 accident. Before the Mercury-Redstone flight got off the launch pad, however, program officials convened a meeting at which the fate of Mercury-Atlas 2 was settled.

Long overdue because of catastrophic failure in the MA-1 launcher the previous July, the meeting was to review the findings of an investigation committee set up to find out why the MA-1 had broken up at max q. It was believed that structural failure brought on by excess loads on the thin walls at the forward end of the booster had been primarily responsible for this and other Atlas failures, nearly all of which were found to occur when a particularly heavy payload was carried on the nose. Accordingly, a proposal was put forward whereby NASA would begin to receive 'thick-skinned' Atlases, suitably strengthened at the nose area, from spring, 1961, but that MA-2 could fly with a steel band around the nose to prevent a similar MA-1 accident.

As George Low reported to the NASA Administrator: 'Atlas 67-D will be the launch vehicle for this test, . . . the last of the "thin skinned" Atlasses to be used in the Mercury program. It differs from the booster used in the MA-1 test in that the upper part of the Atlas has been strengthened by the addition of an 8-inch (20.3 cm) wide stainless steel band. This band will markedly decrease the stresses of the weld located just below the adaptor ring on top of the Atlas; the high stress region is shifted by about 8 inches, to a point where the allowable stresses are considerably higher.' But there was debate as to whether STG was wise in flying a makeshift mission when delay of only a few weeks would result by waiting for the thicker skinned missiles. Imminent Soviet successes, a new administration in the White House that might not look favourably on a stymied space flight program, and the urgent need to qualify a production capsule of the type designed to carry men, argued strongly for backing a hunch and flying with the trussed-up Atlas.

It paid off, and again ringmasters Williams and Kraft coaxed the Mercury-Atlas through the countdown and up to ignition. It was as much a test of the booster as of the spacecraft, and the mood was tense in Launch Complex 14's blockhouse as eyes and ears trained on the view through the periscope and status checks from consoles around the walls. Only two Atlas boosters had lifted off in this configuration, the first Mercury-Atlas to carry a launch escape system: Big Joe almost 18 months earlier had carried a boilerplate Mercury on a suborbital flight; MA-1, seven months before had broken up at max q. At 9:12 am, the three Atlas engines thundered into life, seconds later the assembly slipped its mooring, and the big bird was on its way into space. On through max q it swept a minute after launch, proving the brave optimism of Debus' launch crew, accelerating to the point where Big Joe had not gone according to plan – separation of the two big booster engines straddling the single sustainer. That event came off as planned and MA-2 moved on and up, the smiles and relief breaking out all over Complex 14 with each passing second.

On time, the sustainer engine shut down, the escape tower having already jettisoned, the capsule separated, turned itself around, the retro-rockets fired and the re-entry began. Passing over the USS *Greene*, a Naval destroyer stationed in the Atlantic, sailors caught sight of a streaming smoke plume as the capsule heated in the atmosphere, a brilliantly glowing Atlas tumbling behind the descending spacecraft kilometers away. Less than half an hour after splashdown at the conclusion of an 18 minute flight, spacecraft 6 was being air-lifted back by helicopter. It had been to a height of 183.5 km and a speed of 21,290 km/hr.

For a month now, astronauts Glenn, Grissom and Shepard had been training for the first manned Mercury shot, a ballistic suborbital flight on a Mercury-Redstone which would be flown by one of the select three on a last minute choice according to who seemed most ready at the time. Everybody was ready to go in principle, the astronauts were eager to ride the MR combination, NASA and the politicians had an eye on the fortunes of their Soviet counterparts, and the public were being whipped up by enthusiastic newsmen ever ready to create a milestone from an event. Although prima facie everything had seemed to go well with both recent Redstone and Atlas shots, von Braun's team at Huntsville were worried about the off-nominal trajectories of Redstone MR-1A and -2 flights. Space Task Group wanted to move ahead with immediate plans for a manned Redstone flight, and that had been their concern when narrowing the choice for astronauts from seven to three in January. The Marshall Center pointed to work still needed on Redstone and the basic ground rule written in to Mercury test operations that at least one full scale rehearsal covering every aspect of the anticipated flight envelope should be demonstrated prior to committing a man to the Mercury couch.

Williams, Kraft, Yardley, Mathews, Gilruth, all agreed that they had absolute faith in the Redstone booster based on a critical and searching examination of the MR-1A and MR-2 flight results – the word from STG was 'go'. But von Braun had a deputation from booster specialists, Eberhard Rees his Deputy Director for Research and Development, and Debus from the Cape to testify that one more Redstone flight was necessary before they could feel confident about a manned shot. So, MR-3 would wait until MR-2A had proved the validity of modifications still to be made. The first manned shot would not take place until at least 25 April, one month after the newly appointed Redstone shot. STG was not happy about this deliberate delay and had no more production capsules to spare so Marshall employed an old boilerplate capsule previously used for the Little Joe 1B flight in January, 1960. As if to completely divest themselves of association with the sudden emergence of this Mercury Redstone-Booster Development flight, the missions MR-2A designation was scrubbed in favor of MR-BD; only the booster would be fully equipped with operational systems, STG would not attempt to recover the capsule, in fact it would not even be separated from the adaptor.

Meanwhile, STG took the opportunity caused by the requirement for the Marshall Center to launch a Redstone to test, in a repeat of the Little Joe 5 attempt that failed the previous November, Mercury's launch escape system under conditions 'that represent the most severe . . . that can be anticipated during an orbital launch on an Atlas booster.' Dubbed Little Joe 5A, the test scheduled for 18 March would require spacecraft 14 to abort itself 34 seconds after lift-off where the dynamic conditions of mach number, altitude, dynamic pressure, and flight path angle would closely approximate an ascending Mercury-Atlas configuration one minute after lift-off. Again, as with LJ-5, the launch escape motor fired too soon, 14 seconds before the scheduled time. A single retro-rocket motor was fixed to the base of the capsule and this fired off 23 seconds after tower motor ignition, sending the capsule tumbling away from the Little Joe.

The tower and the retro-package were torn off with the force of the spacecraft's gyration, at which point both the main and the reserve parachutes popped out simultaneously, one trailing the other as they slowly lowered the space craft with a descent rate of only 19 km/hr to the sea. But again, as with so many aborted tests, spacecraft systems had been stressed far beyond their design limits and in that regard proved in fine order the enormous punishment Mercury could take; in the LJ-5A flight, parachute loads were six times those expected during a normal flight. STG was running out of hardware. A simulated Mercury-Atlas abort under 'worst case probability' conditions was essential but only one Little Joe booster remained from the initial batch constructed by North American, and there were seemingly no more production capsules to spare from planned man-carrying missions on suborbital and orbital flights. Amost immediately it was decided to commit the final Little Joe launcher to sending the refurbished spacecraft 14 (dubbed 14A) on LJ-5B in April.

But little over a week before the unsuccessful LJ-5A flight, a three month hiatus in tests with their own prototype manned spacecraft ended when the Russians launched Korabl Sputnik 4 carrying a single dog named Chernushka. Sent into space on a single orbit flight, the capsule returned to Earth safely 100 minutes after launch. Whatever had plagued the previous attempt, where two dogs perished in a misaligned re-entry, had evidently been corrected. It was becoming obvious that Russia stood close to being able to achieve manned orbital flight, although at this time it was not known in the west that the Korabl Sputniks were in fact prototype manned spacecraft. What was so impressive was their $4\frac{1}{2}$ tonne weight, provision of live TV from the capsule, the presence of dogs as test subjects, and the supply of food necessary to keep the animal alive for 10 days if the retro-rocket failed and the capsule was left to decay naturally; it was doubtful, however, that the spacecraft would have survived the considerably increased heat load occasioned by the shallow re-entry. Development of a massive booster rocket, and the application of an upper stage, had given the Soviets a head start in the race that carried bonus for the side with the biggest lifting capacity. With a weight of less than 1.4 tonnes in orbit, the tiny Mercury was limited by the meager capacity of Atlas, the largest American rocket anywhere near operational status. Resumption of Soviet flight tests with the heavy spacecraft came only three weeks after the STG agreed to fly another Redstone; there were many who quietly questioned after 9 March the wisdom of a move that delayed the first manned Mercury shot.

Nevertheless, preparations moved quickly ahead and at 12:30 on the afternoon of 24 March the MR-BD mission got under way, blasting the eardrums of a pad rescue crew situated only 300 meters from the ascending Redstone. They were there to prove a similar crew could survive without injury while another Redstone sent a man into space; in the event of a major problem with the booster a 'cherry-picker' would be used to pluck the astronaut from his capsule and lower him to a converted M-113 armored personnel carrier nearby. Seven engineering changes to the Redstone worked as expected, placing the booster's trajectory plot on top of the planned curve. The uncomfortably wide performance margin of MR-1A and MR-2 had been successfully narrowed to more nearly approximate the planned path, increasing confidence that a Mercury capsule would not be thrown to intolerable limits. There was a very real consequence of not being precisely sure that Redstone would perform as planned; chimpanzee Ham had experienced g loads only slightly below a level that could have disorientated a man for several crucial seconds. MR-BD flew to a height of 183 km and a speed of 8,244 km/hr before hitting the Atlantic 495 km from the Cape.

Everybody was happy with the flight: it had qualified the Redstone booster for lifting a manned Mercury capsule on a suborbital mission. But on 25 March, a day after the elated launch crew saw Redstone prove itself, the Russians launched Korabl Sputnik 5 with the dog Zvesdochka on a repeat flight to the previous one-orbit circumnavigation of the globe on 9 March. The last two missions had gone well. How much longer would the Russians hold off from sending a man into space? How many more test flights would they need? Space Task Group was unanimous in their choice of the qualifier. The next Mercury flight would be manned, a mission that would have been second to the MR-BD launch slot had that flight not been relegated to just one last test. It was a test that lost the race to Russia, and gave America the Moon. For if NASA had beaten their Soviet counterparts the events that were to unfold might never have happened, and US manned space flight may have ended right there, on the sands at Cape Canaveral.

CHAPTER FIVE

Challenge

For days the atmosphere had been tense. In and around Moscow, foreign journalists, international correspondents, and radio reporters prepared themselves for official word on a major space spectacular rumored to be imminent. During the early months of 1961, while in America NASA prepared the final qualifying trials with STG's tiny 1½ tonne Mercury capsule, stories spread around the world that Russia had already launched a man into space and lost him in a catastrophe similar to that which killed dogs Pchelka and Mushka in December, 1960. Korabl Sputniks 1, 2 and 3 were precursor test vehicles for the fully equipped manned space vehicle prototypes launched on 9 and 25 March, 1961, respectively. Korabl Sputniks 4 and 5 were in every essential respect identical to the vehicles then being prepared for manned flight. Each contained a dog and selected biological specimens to demonstrate on a single orbit the full operation of flight and ground systems prior to committing the design to the task it was built to perform.

Alexei Ivanov writing years later recalls that careful selection of the tapes carried by each spacecraft was made so as to prevent a misunderstanding that the capsules actually contained men. A taped voice reading numbers was rejected, as was a recording of a singing voice – for fear, it is said, that western reporters would pick up the transmission and spread the tale that 'Ivan' had gone mad – so a choir was included to defer suspicion. The recording was required to qualify the radio communication system and purportedly emanated from a dummy in a white smock sitting in the seat designed for a cosmonaut! It has been said that never before or since has the Piatnitsky Russian choir been heard by potentially so many listeners. Although tracking stations around the world were able to confirm the absence of space vehicles not known to fill one or other of the well identified categories, rumor still abounded that Russia had lost men in space accidents. The nature of east-west relations in the early 1960s would have prevented anything but propaganda being made out of each scrap of information, true or false, and space flight was rapidly becoming the most successful source of scurrilous comment.

As April came the rumors increased and seasoned observers of the Russian scene knew it was only a matter of time. On 12 April, at one minute to ten in the morning, while factory workers toiled at the start of another day, shopworkers prepared for the mid-morning queues, and children sank into the lessons of the morning, teleprinters from the Tass news agency began to chatter with the following announcement: 'The world's first spaceship, *Vostok* (East), with a man on board was launched into orbit from the Soviet Union on 12 April, 1961. The pilot space-navigator of the satellite spaceship *Vostok* is a citizen of the USSR, Flight Major Yuri Alekseyevich Gargarin. The launching of the multi-stage space rocket was successful and, after attaining the first escape velocity and the separation of the last stage of the carrier rocket, the spaceship went into free flight on a round-the-earth orbit.'

In London, it was approaching 8:00 am; in Washington it was nearly 4:00 am. Within minutes reporters jammed telephone lines, security offices were alerted, the foreign diplomats received messages for ambassadors to see first thing that day, and someone somewhere telephoned John 'Shorty' Powers, head of NASA's publicity machine for Mercury, to receive the faux pas of the day when a tired voice groggily replied, 'We're all asleep down here.' And to the press that Thursday morning it seemed an apt quote for why America had, once again, and almost exactly 3½ years after Sputnik 1, been beaten in the space race.

Before breakfast that morning the new NASA Administrator – James E. Webb, appointed by the new President John F. Kennedy – appeared before millions of Americans to congratulate the Soviets and to cushion the blow by expressing calm and unruffled confidence in the Mercury program. But this was no tortoise versus hare race in which an ultimate goal would rescind the hostile criticism from fellow Americans. In being for 2½ years, NASA had been set up by a Congress who believed it could legislate to prevent this very event from being staged. The fact that it had happened at all set firm faces against the space agency, angrily demanding an answer as to why America had not been first with man in space. From the television studio, Webb drove back to headquarters, picked up Dryden, had a hurried conference with top management, and moved on to Capitol Hill, knowing that this morning of all mornings would be the worst NASA would probably have to face. He was not correct in that belief, but never again would the agency face such anger from Congress.

Representatives were appalled that a Russian would enter history's immortal annals as the first human being to leave the Earth, confused as to why the United States should have allowed such a thing to happen, concerned that America had divided itself among too many space programs instead of focusing effort and attention on singular objectives, and ready for assurance that Webb was unable to give. Instead, he turned the issue over and in a brilliant piece of statesmanship held America responsible for the lack of foresight and delayed initiative in not giving NASA the proper tools to do the job. But not at once. Calmly answering with reasoned self control, he was even then planning to turn the situation to NASA's advantage and would only deploy his strategic options over the following months, gathering unparalleled support as the General heading a new army of patriotic Americans; it was to be a marshalling of resources beyond the ability of his predecessor.

That April morning saw a new resolution that if Russia had not really wanted a space race – and there is very little evidence to suggest that it did prior to this – it was certainly going to get one now. The event that caused such a furore really began on 14 March, 1960, thirteen months before Vostok 1 rose from its Baykonur launch site, when 20 Soviet cosmonauts were secretly chosen to train for a manned orbital flight. Among them were Belyayev, Bykovsky, Gagarin, Gorbatko, Komarov, Khrunov, Leonov, Nikolayev, Popovitch, Shonin, Titov, and Volynov; these would go into space, the others falling out for a variety of reasons. Even before Gagarin's epic flight, a man called Valentin had been discharged for medical reasons; when diving into a lake at Zvyozdnii Gorodok, the village near Shelkovo not far from Moscow, Valentin damaged a vertebra which required a month in hospital and vacation of a potential cosmonaut's seat. Zvyozdnii Gorodok (Stellar Village) was commanded by General Nikolai P. Kamanin as a facility where trainee cosmonauts could limber up, study, receive academic tuition and prepare mind and body for the rigors of space travel.

About 1,500 km to the south, Zvyezdograd (Star City) was growing into a small town, a place where workers, engineers and technicians from nearby Baykonur could live with their families remote from intrusion. These were the two primary centers of activity: a place where the astronauts would be finally selected, trained, and prepared for space flight and a community alongside the sprawling launch sites in Kazakhstan where technical preparations of the launch vehicles and spacecraft could be completed. Baykonur was not the oldest Russian launch site, that distinction belonging to Kapustin Yar situated nearly 1,000 km to the north-west, a place from where early post-war rocket flights were carried out, from where sounding probes were hurled several tens of kilometers toward space and from which the first biological experiments were planned. If Baykonur was Russia's Cape Canaveral, Kapustin Yar was the Wallops Island where small, scientific, often suborbital payloads were handled; not for another year would it be used to send satellites into orbit.

Baykonur had risen from the flat wastes in 1955 during the design phase for Russia's first ICBM, merely a year after a new Soviet administration heard from Sergei P. Korolev of the big booster rockets that could be built. Korolev was born in January,

1906, to parents who were schoolteachers in the Ukranian town of Zhitomir. His father died soon after Korolev's birth but a step-father would inculcate a respectful interest in technology. A year old, Sergei was taken to Kiev where the family suffered from the rigors of poverty until the boy was placed with his grandparents in Nezhin, his mother later marrying an engineer called Grigori Balanin. Back in the family environment he had missed for several years, Sergei busied himself with aeronautical pursuits when, after the October, 1917, revolution, small clubs and organizations emerged in towns and rural communities.

Prevented from receiving a good education as a child, Korolev went on to study aeronautical engineering at a technical school in Moscow and worked under Andrei Tupolev, the designer. At the age of 21, Korolev was introduced to the works of Konstantin Tsiolkovsky, the designs of Friedrich Tsander and other international pioneers in rocketry and theoretical space travel. It had a profound effect on his attitude toward the future and set him on a course that would play an instrumental part in getting a man into space.

Korolev involved himself with the efforts of these GIRD groups and the first primitive attempts to fly liquid-propelled rockets, until the Stalin purge of 1937 sent him to a labor camp. Lucky to escape alive, Korolev was allowed to work on aeronautical problems in the confines of a life sentence, but at night he busied himself with calculations about rocketry and space flight; the Soviet secret police had denied Korolev the right to pursue such work, believing that because rocket pioneers were among those 'purged' from the Red Army the technology itself was anti-Soviet. Korolev left prison for a reason that still remains obscure, but probably because Tupolev gathered around him the men he needed to continue his aeronautical design work – an activity that spared the lives of several engineers – until in 1941 he was taken to work with Valentin Glushko; Stalin respected the skills of aeronauts and protected their potential value to the state by assigning them a special category that spared their lives but little else.

When the German armies marched into Russia, rocketry had a respectful place in the research and development laboratories where pressures of conflict stimulated novel applications. But the purge had done its worst and the country suffered from rocket engineers dying in concentration camps, and still more lying in graves throughout the Soviet Union. Korolev's professional contacts enabled him to live and work on designs for aeronautical projects but the association with Glushko ended more than 5 years of hiatus in Russian rocketry. At the end of the war, Korolev accompanied Glushko to Germany in search of V-2 engineers and redundant equipment left behind by the Americans when they retrieved vast stocks of rocket components. It has been said that Korolev, in the uniform of a Lieutenant-Colonel, was made to stand outside a fenced area where British troops fired a V-2 rocket. Prohibited from entering the secure compound where technicians worked on the world's first long-range rocket, frustration must have been deep indeed, as Glushko and other approved personnel walked right in and handled the machined components that gave Germany such a marked lead in ballistic missilry.

Korolev was made responsible to General Gaidukov organizing a resumption of operations at the German rocket factories and at the assembly lines deep in the Harz mountains. In 1946 he supervised the firing of German V-2 rockets from the Kapustin Yar test range and worked with surviving theorists on the problems to be solved before space flight could become a reality. In several respects he was Russia's von Braun, although perhaps more versatile: von Braun was a great organizer, a champion in debate and at meetings of concepts he knew others would be required to perfect, a man who quickly won supporters and drew approval and sincere accolades; Korolev was probably more of an engineer than von Braun and, although gifted with an ability to recruit support for capable ideas, was less of an orator and more of a pragmatist.

President Kennedy gets a briefing in space hardware from astronauts Cooper (center) and Grissom.

Proposals were made for manned flights to the fringe of space on ballistic shots using V-2 missiles, reminiscent of the British Megaroc concept and one which would have undoubtedly been practical. At first employed controlling the ex-Peenemunde rocket engineers brought to the Soviet Union in October, 1946, Korolev spent the latter half of the decade designing missiles and working with Glushko on new liquid propellant motors. His ideas were far beyond anything that had at that time been approved. Intent on re-arming the Soviet armed forces with a strategic capability, Stalin was, however, concerned with any project that could provide a unique weapon system. But the great Soviet leader was as disinterested in what he considered to be scientific playthings as his German counterpart of the previous decade was in the V-2. Korolev's star seemed again to dim in the wake of disapproval and for several years, from the end of the 1940s to the death of Stalin in 1953, the rocket designer was under close arrest.

A lot changed in Russia with the demise of the most powerful Soviet leader since 1917: a more liberal attitude was pursued from the Politburo down to the factory floor; people were made to appreciate the virtues of willing support for the state; and a new leader – Kruschev – courted support on personal appeal and international flamboyance. Korolev joined the Communist Party and was elected to the position of corresponding member of the Soviet Academy of Sciences. At about this time, V. A. Malyshev was appointed to head a new government department responsible for the design and construction of missiles and support facilities. Liquid propellant rocket motors were being developed which held great promise for application in medium range missiles, albeit a class that, compared with the V-2, would have been very long range but in the 1950s was relevant to new standards in weapon performance. A short range rocket of 1953 was a very long range rocket of 1943!

Glushko's engines were reliably producing more than 70 tonne thrust levels and Korolev recognized the technical strides that still prevented consideration of a truly intercontinental rocket. Missiles with ranges of nearly 1,000 km were on the drawing boards, rockets similar in design concept and lifting potential to the Redstone in America, but nobody had any idea how to build an engine with several hundred tonnes of thrust. Yet this was the power required for space flight and, realizing that many comparatively small engines clustered together could provide the same kind of performance that a single, very large, rocket motor would possess, Korolev followed a road travelled by von Braun 20 years earlier. In finding a technically unprecedented military use for his idea, he presented a viable reason for political approval. Korolev knew, as von Braun had discovered, that money would not be provided for a scientific stunt but could be made available for a completely new weapon system. On that basis, he took Glushko's engines and married them to a booster design called the R-7 which at that time was only an idea backed up with blueprints and technical drawings but one nevertheless that caught the imagination of Nikita Kruschev.

Introduced to the concept of an Inter Continental Ballistic Missile at a Politburo meeting in the Kremlin, Soviet leaders were dumbfounded at the suggested performance of such a device. Devoid of any technical knowledge, and lacking an awareness of rocket potential, the political leaders found a tool that seemed to have risen before them as an omen of a new age. In truth, it personified the dramatic changes then being planned to get the Soviet economy moving and to make Russia a mighty world power, a product of Russian engineering principles that appealed in every dimension to the new occupants of the Kremlin. Work progressed quickly on the rocket, potentially capable of sending an atomic bomb to the United States, until the first tests were held in 1957. Unsuccessful at first, the rocket finally flew on 3 August, the ground reverberating to the thunder of 20 rocket motors each with a thrust of around 25 tonnes. An unparalleled sight anywhere on earth, it inspired the designer to think beyond the limitations of the basic R-7 to a modified version suited to the task of launching an artificial Earth satellite.

All the while, Korolev quietly planned for the day when large rockets would bring his dreams to reality, recognizing the need to inspire political leaders with tantalizing promises of global power and prestige. When NATO code namers categorized the R-7 it was called the SS-6 Sapwood but the rocket itself had little military value and was made the focal point of a program planned around the political objectives of the Kruschev camp – one that would use the big rocket for prestige purposes inspired by an American announcement about the International Geophysical Year, in which the US would launch artificial Earth satellites. Because popular opinion in the west questioned the existence of a big Soviet booster, and because America had announced a modest satellite program in which small spheres would be placed in orbit, Korolev was able to convince Kruschev that a similar feat, adopted by the Soviet Union and using the R-7 rocket, could in one stroke prove the rocket's existence and beat the US into space. It didn't really matter that the R-7 was unsuited to military duties for which it had been designed, the rocket could wave the hammer and sickle with threatening realism before the nation that assumed it had the biggest and best of everything. Russia was internally upset by American attitudes, continually harassed by statements that it was incapable of applying modern technology to national challenges. The catalyst, a genuine belief that the world still thought of Russia as a third rate nation, moulded the segments of a space program.

R-7 had a strategic range of about 6,800 km, with an atomic bomb in the nose. With a small satellite at that location, the rocket was capable of reaching orbital speed and that was the task for which it was found to be most valuable. Exciting America into a race they never wanted, Kruschev's Sputnik team was the brainchild of a man who, among the few who can, succeeded in fulfilling a personal dream. Korolev was wedded to the idea of space flight and worked long and hard throughout his life to consummate that marriage. When it was possible, he became the possession of a powerful man who saw in him a means for political growth. Kruschev demanded a spectacle to back up his foreign visits and to impress world leaders, but Korolev was to be kept from public view and common knowledge. The space program planned during 1958 and 1959 was made to seem a product of Soviet workers striving against international capitalism and not the brainchild of a single engineer. Korolev was not concerned for the praise or recognition he richly deserved. It was reward enough to be a key element in forging a new age in which Man would reach for the planets. He was certainly an inspired dreamer, a man remote from the mass. But he also recognized the incremental steps that would have to be achieved if the R-7 and Sputnik 1 were not to be a single event isolated from further development.

Korolev showed Kruschev that manned flight was possible and used the recently established State Commission for Organization and Execution of Space Flight to prepare a strategy based on the availability of improved, multi-stage versions of the R-7. At a time when Kruschev was being criticized for depleting the armed forces of manpower in favor of rockets and nuclear weapons, the public could be made to feel pride in the political success of international respect. And that respect was already being seen to belong to nations demonstrating a progressive technology. Space programs were just what the party leader wanted to gather public support and to show positive results for revolutionary changes in Soviet administration. By late 1959 the elements of a manned space program were falling into place. The R-7 would be modified and receive an upper stage in which configuration the complete assembly could be fitted with a variety of payloads including Moon vehicles and manned spacecraft for Earth orbit flights. By 1960 the design of the Vostok was nearly complete, leading to the unmanned test flights in that year and the following, with or without dogs, on board to check the life support system apparatus.

Preparations for the flight of Vostok I required the spacecraft to be assembled and then mated to the launcher while the latter was in a horizontal position on a rail-mounted bogie. Contained inside a flat-roofed assembly building, the Rakyeta Nosityel Vostok was moved on its combined transporter-erector out through large doors and along the rail tracks to a launch base on which it would be placed for ascent. Rotated 90° to a vertical position, the rocket and its payload was placed in the cradled grasp of four large arm assemblies that gripped the rocket at a point immediately forward of the four large boosters that lay alongside the single main sustainer section. The four arms and the complete launch structure was mounted on a giant swivel table which rotated to the desired azimuth so the rocket could be launched toward the point in space its orbit was expected to reach without having to yaw to left or right and steer itself like an automobile; it was merely required to change pitch during the climb toward space.

After erection was accomplished, two long access arms, hitherto lying to left and right of the rocket, were also rotated through 90° to lie along the tall side of the rocket, affording platforms and ladders by which maintenance technicians and service engineers could reach critical parts of the assembly. The service platform arms were moved up and down by hydraulic rams operated from large motors; the four girder-like support arms were counter-balanced so that when the rocket's engines ignited and lifted it free of the launch base, loads that hitherto kept the arms in their sockets on the rocket were removed to allow the arms to swing back so the launcher could ascend. At the back of the launcher, on the far side to the rails that brought the transporter-erector from the assembly building, was a large propellant service mast, similarly operated by hydraulic ram, carrying lines that delivered the liquid oxygen and kerosene used in the central sustainer and the four large boosters. In all, there were 10 large propellant tanks that had to be cooled, filled and pressur-

Seen here grooming a Vostok spacecraft, Soviet engineers check the shroud that will protect the payload from atmospheric pressure during ascent.

Placed on a wheeled cradle, Vostok's conical equipment section supports the → spherical re-entry capsule carrying Soviet cosmonauts on earth orbit missions between 1961 and 1963.

ized before flight. While on the launch pad, combustible chemicals would be placed aboard the rocket, with several potentially dangerous liquids required to support the operation of the launch vehicle.

An example of how dangerous the business of launching rockets can be came dramatically to the fore during October, 1960, when Kruschev visited the United Nations in New York, arriving in the ship *Baltika*. Anxious to summon up a space spectacular and so emphasize the might of the Soviet Union, Kruschev expected his space engineers to launch two spacecraft to the planet Mars, the alignment of the two worlds falling coincidentally with the Russian leader's foreign trip. *Baltika* carried models and display materials to commemorate the achievement, the first flight of a spacecraft to another planet. Back at Baykonur consternation gave way to dismay as a rocket similar to the one which would launch Gagarin into space, failed to ignite at the end of the lengthy launch preparations.

In this situation, standard procedure should have been adopted, requiring the launch teams to perform a very specific set of operations which, although time consuming, would make the fully fuelled vehicle safe to approach. Aware that Kruschev expected to sail into New York and personally announce the successful launch of at least one probe to Mars, Korolev's team abandoned accepted safety procedures and sped to the base of the rocket to see why it had not fired. Suddenly, a tremendous explosion ripped across the concrete apron. Within seconds the tanks of booster and sustainer sections had burst open, fuelling the fire that raged uncontrollably, incinerating everything and everybody in its path. The people who died included leading figures at Baykonur, although Korolev was not among those who rushed to attend the rocket. Accidents like that were not common, but they left their mark on the morale of all who ever went to work for the Baykonur engineers.

Coming only months before the final preparations for Vostok I the accident heightened tension among the workers and the technicians readying the first manned spaceship to carry a human payload. Kruschev never did get his Mars flights in 1960, a thorny reminder that technology cannot be made an expedient messenger for political prestige without adequate attention to its every need. Yet, despite the need for careful application to safety procedures, the space program in Russia had been forged as a political tool and intervention was the price Korolev paid for selling himself to the taskmasters from the Kremlin.

A lot depended on the successful completion of the Vostok I mission as dawn broke across the flat Kazakhstan landscape that morning in April, 1961, when Yuri Gagarin would take his place among the great names of history. It had been a disappointing year for Soviet space activity, and Kruschev was not pleased with the massive American response to Sputnik – a completely unexpected reaction where Soviet leaders had thought the US would smart under technological failure – nor was he enamored with intelligence reports about NASA plans for manned circumlunar flight. Sputnik had been sold to him as a one-off demonstration of Soviet superiority, not the spur to a vastly superior endeavor from the Americans. Korolev probably knew that his fate hung on reassuring the Soviet leader that whatever the Americans promised to do tomorrow, Russians could do today. It was neck and neck, and Korolev would have breathed a sigh of relief had he known that Mercury engineers had called for more booster tests, effectively delaying the first manned suborbital flight until at least the end of April.

Yuri Gagarin was just 27 years old when he made his space flight. Born in 1934, he spent the early years of his life attending primary school in the town of Gzatsk until Hitler's troops disturbed his education. Furthering his education at Technical College to obtain a school certificate, he moved to Saratov to learn metalworking before learning to fly at the Pilot Training College, Orenburg. He graduated as a fighter pilot in 1957, married Valentina Ivanova, and was selected as a candidate cosmonaut by early 1960.

It was shortly before 7:00 on that bright 12 April morning when Gagarin walked in an orange colored space suit and white helmet toward the rocket that would carry him into space. Standing in the light of an early Sun, the gleaming white rocket carried an ice encrusted frost around the flared boosters as liquid oxygen froze air on the outer skin. The spacecraft this young cosmonaut would ride was considerably larger than Mercury and adopted very different design criteria. The payload capability of the launcher was, at about 4.7 tonnes, considerably greater than that of the single stage Atlas which was hard pressed to reach orbital speed with a $1\frac{1}{3}$ tonne capsule. Because of this, the critical weight constraints that beseiged Mercury designers were not the most severe problems that gripped their Soviet counterparts. With weight to spare, Vostok was built as a sphere with an offset center of mass protected on the outside by an ablative coating, attached to an instrument section which also carried the retro-rocket necessary to bring the capsule out of orbit.

The spherical capsule was 2.3 meters in diameter, weighed 2.4 tonnes, and contained the pressurized cabin for the pilot. This was essentially a padded, circular interior pressure vessel with the cosmonaut lying on an ejection seat behind a large circular hatch. Simple instruments were attached to the interior, and views outside the capsule were obtained via three portholes, one of which contained a novel device for visually determining the orientation of the spacecraft if the automated equipment failed. Although pressurized with a sea-level atmosphere of oxygen and nitrogen, the pilot would wear a pressure suit as insurance against rapid decompression due to a systems failure of the environmental control equipment or because of a rupture in the cabin hull. This redundant philosophy was used in Mercury also.

Attached to the capsule throughout its flight in space but separated from it immediately after retro-fire, the instrument section weighed 2.3 tonnes and comprised a cylindrical upper section with inward sloping sides and a conical lower section truncated by a retro-rocket nozzle at the center. The unit was slightly more than 2.4 meters in diameter and about 2.2 meters tall;

To reach orbital speed Vostok employed a second stage rocket motor seen here structurally attached to the spacecraft.

because the spherical capsule lay within a concave part of the instrument section the complete assembly was 4.3 meters tall including a small pack of electronic equipment on the top of the sphere. Four large straps held the two sections together until separation and a systems 'umbilical' carried electrical wires and gases from the instrument section to the crew capsule; straps and systems umbilical would be jettisoned at separation.

Radio equipment aboard the Vostok comprised a signal system operating at 19.999 MHz for tracking and telemetry, two short-wave channels at 9.019 and 20.006 MHz for radio and telephone communication, and a VHF channel at 143.645 MHz for voice contact when distance to the ground station was within a range of 1,500–2,000 km. Vostok would veritably bristle with antennae when the capsule reached orbit and extended its appendages: four u-shaped antennae were attached to the top of the capsule for control commands, two 3.3 meter whip antennae came out the top inclined 90° to each other, and four 3.8 meter whip antennae splayed out from a single location on the side of the instrument section. Four more antennae were mounted on the capsule, lying flush with each restraint strap.

The top portion of the instrument section carried 16 spherical containers for oxygen and nitrogen. It also supported the tiny thrusters which would be used to position the spacecraft in attitude when it separated from the upper stage of the launch vehicle. The orientation unit could be commanded by Sun sensors or by hand and included gyroscopes and the optical device in the cabin. When the instrument section was jettisoned at completion of retro-fire, the spherical capsule had no means of orientating itself other than by aerodynamic effects on the offset center of mass. This was very different to Mercury, where several redundant attitude control systems were available not only for orbital operations but also during the descent. Mercury showed itself in at least one test to be capable of finding its own attitude much as Vostok would seek a stable posture, but the possibility of a capsule turning turtle was a very real threat without control thrusters.

Attached to the second stage of the launch vehicle, the configuration was 7.35 meters tall when it got into orbit and weighed 6.17 tonnes before ejecting from the inert upper stage. The entire assembly forward of the upper stage was protected during ascent through the atmosphere by a shroud which would be released and blasted free by solid propellant rockets at the point where the shroud was attached to the stage. The protective cover had a circular cut-out adjacent to the location of the spherical capsule's ejector seat cover. The shroud would be released when the launcher passed out of the denser layers in the atmosphere.

Breakfast for Gagarin consisted of meat paste, marmalade and coffee, followed by medical checks, and the suit-up phase: the orange over-garment covered a blue pressure suit modified from a similar garment worn by pilots of high speed aircraft. A tight fitting cap held earphones in place and the white helmet went on last as the first operational cosmonaut rode the short distance to the launch pad. Accompanying him, from the wake-up welcome to the foot of the gantry, was Gherman Titov, the back-up cosmonaut ready to stand in at any point in the preparation if Gagarin was for some reason unable to keep his appointment. Korolev was there when Gagarin paused in his walk to the gantry elevator, turned, and waved with both hands high above his head to the small crowd gathered at the rocket's base. At the top, the platform on which Gagarin and bare-headed technicians stood ready to assist was open with artificial lights playing on the spacecraft hatch.

Feet first, Gagarin slid down the length of the ejector seat, his feet coming to rest on steps at the foot of the rails, and sank back on to the soft backing. Strapped to the ejector seat he could use on the way down from orbit, the hatch was sealed by 30 threaded sockets around its circumference. He was on his own for a ride into space. To his immediate right, along the sloping side of the curved, inner wall, was situated the control handle which would enable him to align the spacecraft's attitude manually should anything go wrong with the automatic system. This was a signal indicator of the fundamental difference between Mercury and Vostok: the former was designed to carry a pilot who would be very much in control of his own ship; the latter would be more of a passenger than any ordinary aircraft pilot could possibly be, the product of a design philosophy deeply rooted in the brand of communism adopted by the Soviet Union where individual achievement is a threat to the system and a group effort more acceptable.

To the right of the control handle was the radio equipment with a few simple controls, and to the side of that, near the cosmonaut's shoulder, was a small food container. To Gagarin's left, on a level with his chest, was a square, flat, black control panel with three rows of switches for basic functions, to be operated by his right hand reaching to the left. The cramped conditions inside the capsule required each control to be in a position from which it could be operated by the opposing hand. In front of the cosmonaut's head, slightly above the level of his eyes, a round porthole was set into the padded walls that surrounded the ejector seat. In front of the pilot's chest, on the curved 'ceiling,' was the square instrument panel containing all the essential information he would need to know.

Dominating the panel was a single globe, set in to the round frame that held it in position so that it could rotate in accordance with the spacecraft's path around the Earth, continually displaying the position over the surface below. Above the globe was an indicator marking the globe's function and to the right, a knob for re-setting it. Below that was a push-button to reset an orbit counter that showed the number of times the spacecraft had circumnavigated the Earth. Across the bottom of the square instrument panel were four dial housings. The one to the extreme left contained a set of chronometers telling the exact time from the start of the flight. Next was a dial containing meters for cabin pressure, cabin humidity and atmospheric temperature. Then, to its right, another dial with information on carbon dioxide levels, oxygen content (a vital ingredient with narrow percentage limits) and radiation levels. On the extreme right, a dial providing details of pressure in the nitrogen, oxygen and pressurization containers. Above this right-hand dial, up to the top of the panel, were indicator lights telling the status of on-board systems.

Directly below this panel, toward the cosmonaut's feet, was a TV camera looking at the pilot's face, and below that the Vzor orientation device. This comprised two annular mirror-reflectors with a light filter and lattice glass. As light from the Earth's horizon struck the first reflector and passed to the second it would be directed up through the lattice glass to the cosmonaut's eyes. The correct attitude could be obtained by slowly moving the spacecraft in pitch using the manual controller to obtain a view of the horizon which appeared as a true circle in the lattice glass. When this was obtained the retro-rocket could be fired with confidence that the spacecraft was within acceptable limits of the preferred attitude.

The atmosphere of the cabin was maintained at a sea-level environment with a permitted carbon dioxide level of 1%, a temperature of between 15°c and 22°c, and a humidity level of between 30% and 70%. All functions were controlled automatically and the cooling system adopted a radiator attached to the exterior through which fluid passed; shutters were used to control the temperature of the fluid according to the requirements for heating or cooling that prevailed inside the cabin. Vostok's internal systems were designed to give it a 10 day capability for one man, orbits being selected which would cause a natural decay to the atmosphere at the end of that period irrespective of retro-fire or not.

While Gagarin waited in the spacecraft the countdown to ignition was delayed briefly when engineers returned to the capsule to replace a valve in the orientation system. Time passed and after two hours in the capsule the final minutes of the countdown proceeded as planned: automatic equipment switched in to control the terminal events, nitrogen flowed through the propellant lines and the combustion chambers to purge hazardous chemicals, the pressurization system was isolated, and the gyroscopes were checked out. With seconds to go before lift-off, the big cable mast at the rear of the launch pad disconnected from the rocket as the fuel valves opened and the turbines whirred into action, sending tonnes of propellant gushing into the combustion chambers.

To Gagarin, seated more than 30 meters above the engines, the launch began with a gathering thunder and a rippled vibration similar to that he had felt in Mig trainer aircraft with the engine on straining against the brakes. The real difference came when a smooth lift-off, hardly perceptible in the cushioned cabin, gave way to an increasing sense of acceleration. Pressed further and further down into the ejector seat the body itself felt as though an invisible hand was pressing it deeper into the metallic structure; there was a sense of pain that could not be identified as such and always the unbroken thunder until, hardly a half-minute after

Protected from the searing heat of re-entry by an ablative coating, the spherical Vostok capsule is seen here shortly after landing.

launch as the rocket went supersonic, all sound stopped, undercut by a rhythmic oscillation as though the capsule was on the end of a gently moving spring. Suddenly, the acceleration too ceased, accompanied by distant echoing sounds of metal latches, hinges uncoupling and strange, sideways motion. The boosters had shut down and separated.

Shortly thereafter the crushing acceleration picked up again and crept over the body, pressing it again down into the seat. Before long a sudden, surprising roar, very close and rapidly dying away, signalled the ejection of the shroud covering the spacecraft on top of the rocket. After that, again a sudden cessation of the g force as the main sustainer stage shut down, hurling the cosmonaut forward against straps that held back the unsprung energy that sought to fling him into the roof of the capsule. The sudden loss of acceleration had pitched him forward like a car suddenly braking from a steady speed. No more than a second or two later, more sounds like metal doors slamming shut only this time very much closer, a sudden total quiet, and then a rapid build-up of acceleration and a louder roar as the second stage engine ignited and the sound traveled through the metal structure. It seemed to Gagarin seated for his ride into history as though the entire process was one of being pressed harder and harder into the seat only to be flung forward at some unexpected moment – but the ride on this second stage was brief, and in a way more sporty as though the massive rocket had shed great weight to perform like the fighter aircraft he knew so well. Suddenly, this time with more warning, the engine was shut down and Gagarin pitched forward to stay in suspension as slowly, item by item, obscure artifacts sailed out from nowhere and spaced themselves around the cabin.

He was in orbit, 181 km above the Earth. Gagarin would later recall: 'When weightlessness appeared I felt excellent. Everything was easier to perform. This is understandable. Legs and arms weigh nothing. Objects were swimming in the cabin, and I did not sit in the seat as before, but was suspended in mid-air.' Vostok 1 had been launched at 9:07 am local time toward a dawn that gave Gagarin the quickest day of his life. In an orbit inclined nearly 65° to the equator, his spacecraft coasted on across the Siberian wastes, over the northern territory of the USSR. Later he would say that 'The sunlit side of the Earth is visible quite well, and one can easily distinguish the shores of continents, islands and great rivers, large areas and water and folds of the land. Over Russia I saw distinctly the big squares of collective-farm fields, and it was possible to distinguish which was ploughed land and which was meadow. During the flight I saw for the first time, with my own eyes, the Earth's spherical shape.'

Passing south of the Kamchatka Penninsula, Vostok I drifted at more than 28,000 km/hr out across the Pacific where it passed into darkness. Of this he would recall: 'It is possible to see the remarkably colorful change from the light surface of the Earth to the completely black sky in which one can see the stars. This dividing line is very thin, just like a belt of film surrounding the Earth's sphere. It is of a delicate blue color. And this transition from the blue to the dark is very gradual and lovely. It is difficult to put it into words.' During the drift from day, through night, and on into day again, Yuri Gagarin practiced eating and drinking from the contents of the small food container close by his right shoulder. Room inside the cabin was limited and weightlessness gave a feeling of increased volume, as though corners of the interior hitherto ignored were suddenly available for human occupation, accessible to a drifting body allowed to move at will.

'I was working in that state, noting my observations. Handwriting did not change, though the hand was weightless. But it was necessary to hold the writing block, as otherwise it would float from the hands.' Just 45 minutes after leaving the Baykonur launch pad, Vostok I was passing south of Cape Horn with the automatic orientation system switched on in preparation for firing the retro-rocket. Gagarin reported he was feeling well. He was dismayed, admitting only to his own thoughts the displeasure he felt at having to deliberately bring to an end an experience no other human being had gained. Approaching the coast of Africa, having drifted up over the South Atlantic Ocean, the on-board timers set in motion the programed sequence of events culminating in retro-fire.

Over Africa, 1 hour 18 minutes into the flight, the spacecraft was flying backwards, pointing down, the rear of the instrument section up above the horizon. For a while, as the engine fired, Gagarin felt as though he was back on Earth. With his back to the engine, the decelerating influence produced a deceptive force of 1 g, returning his body momentarily to the feeling it last had back on Earth. For several minutes after the engine shut down Gagarin was almost too busy to hear the thumping sounds of linked mechanisms separating as the instrument section divorced the spherical capsule. Just ten minutes after retro-fire the spacecraft entered the outer layers of the atmosphere, beginning a long ride back across Asia Minor to a landing 8,000 km from the point over the Earth where the engine ignited. As it struck the upper regions of Earth's canopy, Gagarin felt a sudden twitch in the capsule's attitude, a comforting reassurance that the center of mass was bringing the sphere base end on to the direction of travel.

Through the portholes, a rich orange glow appeared, replacing the dazzling brilliance of the Sun with a burning fire that paled the light of day almost to insignificance. For an instant Gargarin gazed in wonder at the changing intensity of color before his eyes returned again to environmental instruments on the square-shaped instrument console. At this stage of the mission anything could go wrong. With the g forces building again it was an unpleasant reminder of the sensation during ascent but, more important, it indicated the magnitude of stress the capsule was experiencing alone from its instrument section for the first time. A sudden systems failure at this point could still bring disaster to the mission and Gagarin knew too well the fate that befell two small dogs in an earlier prototype of this capsule. But all was well and nothing marred the safe return of Man's first excursion to the black void of space. At a height of 4 km, timers aboard the spacecraft jettisoned a hatch on the opposite side of the capsule to that used for the ejector seat, releasing a single drogue parachute employed for braking the attitude excursions that rocked the capsule from side to side.

From the time when the instrument section had been jettisoned shortly before the capsule hit the atmosphere, the spherical vessel had been stabilized only by the offset center of gravity; it was necessary to achieve a stable condition at subsonic speeds and prevent the capsule turning over and over. It had been a rough ride back down through the atmosphere and the drogue parachute was a welcome sight. At a height of 2.5 km this parachute was suddenly released, a much larger main parachute unfurling to jerk the capsule to a slower rate of descent. At 10:55 am, 1 hour and 48 minutes after lift-off, on a day in which cosmonaut Yuri Gagarin had seen the Sun rise twice, Vostok I came to Earth in the Saratov district on the banks of the Volga.

For most Americans that April, Yuri Gagarin's single orbit of the Earth was another demonstration of Soviet achievement. It was believed to portend a major commitment on the part of the Soviet government to invade the peaceful domain beyond planet Earth and lay claim, or territorial right, to the Moon and the planets. In reality, the Soviet space programme had less direction in 1961 than had NASA in the same period. In Russia too, space protagonists had worked long and hard to convince a reluctant administration that exploitation of this new technology could harvest rich crops: few politicians were able to see what might lay in store for the nation that commanded the near space environment.

In responding so vigorously to what was intended as a single demonstration of Soviet capability, American politicians provided the backcloth against which a major contest would be performed. And in a very real sense, the events that were to unfold in the remaining years of that decade were more a reaction to US ambition than they were an intrinsic Soviet desire for supremacy. Excluding any advantage that might accrue from a vigorous space program, Eisenhower had been right to let the Russians have their day; in consideration of technical feedback, he had been wrong to underestimate the value of an effective space policy. But others now held the reins of power and Eisenhower's approach was already outmoded.

At Space Task Group, Mercury personnel had very few steps to complete prior to sending an American astronaut on a ballistic flight. Although only suborbital, a 15 minute flight on an arching trajectory 185 km from Earth carrying the capsule nearly 500 km down across the Atlantic, it would be a valuable mission to man-rate the vehicle design that would, several months later, fly in orbit. While preparations for the Redstone flight went ahead as scheduled, engineers at Cape Canaveral made ready the fourth flight of a spacecraft on top of Atlas: the first (Big Joe) had used a boilerplate capsule to prove the integrity of the heat shield, the second (MA-1) resulted in an unexpected structural failure, and the third, (MA-2) qualified makeshift repairs designed to prevent a recurrence of the same accident, simultaneously qualifying the heat shield with a thermal environment in excess of the design limit.

Atlas 100-D, first of the 'thick-skinned' rockets, would be used to demonstrate orbital capability with a crewman simulator installed in the Mercury cabin sending telemetered data to stations on the ground as though the spacecraft was occupied by a human cargo. Before the Gagarin flight, MA-3 had been intended as a suborbital abort system test but a meeting of senior STG personnel on 14 April, two days after Russia eclipsed NASA plans, changed the flight to one which would effectively demonstrate the capacity of the team to place a manned (unmanned) capsule in orbit. Three days later a recommendation on the next (MA-4) flight envisaged a chimpanzee flight to Earth orbit during July. The MA-3 launch came on 25 April, 1961, at 11:15 am. After only 40 seconds of flight the range safety officer at the Cape sent the command to detonate the rocket when telemetry on the ground showed it going straight up, ignoring a programed autopilot designed to lay it on course in roll and pitch. As expected, spacecraft 8 was successfully aborted when Atlas received the destruct command, the launch escape system lifting it clean and true away from the renegade missile to deposit it in the Atlantic for recovery.

If nothing else, it proved an early abort from an ascending Atlas was likely to perform as expected. Just three days later, the Little Joe technicians at Wallops Island attempted the LJ-5B flight in which for the third time engineers would attempt to prove that a Mercury capsule could effectively separate from an aborted Atlas at maximum dynamic pressure. But even this did not go exactly as planned. One of Little Joe's solid propellant motors failed to fire, causing the trajectory to flatten from an anticipated apogee of 13.7 km to one which carried the assembly only 4.5 km high. As planned, the abort was triggered 33 seconds after lift-off, but under conditions which doubled the dynamic pressure anticipated for the worst Atlas abort. The capsule, spacecraft 14A refurbished from the previous Little Joe attempt, experienced three times the expected g load, returning successfully to a splashdown from where it was recovered.

Since 22 February, 1961, Shepard, Grissom and Glenn had been short-listed to prepare together for the first ballistic flight, one of whom would be selected to actually fly the mission. The other four astronauts – Slayton, Carpenter, Cooper and Schirra – would each get the chance to fly a later mission, or so was the intention, with the unlucky two from the short list probably getting second and third flights. Most of the time the three candidates lived with the capsule – spacecraft No. 7 – designated for the first manned suborbital mission. From the beginning it had been

Vostok's escape system employed a single ejector seat on which the cosmonaut reclined during flight.

honed for this event, one of McDonnell's best, assigned to the task since October, 1960.

Delivered to STG's Cape Canaveral checkout teams in Hangar S on 9 December spacecraft 7 was originally set for a 6 March, 1961, launch. Delays in the preparation schedule put this back first to late March and then to early May. Redstone booster No. 7 was delivered to the Cape in late March. During this time the three potential crewmembers watched humorously the frantic attempts of pressmen and reporters to guess the identity of the one man who would fly the first mission. In fact, during the third week in January, about a month before the three candidate astronauts were shortlisted, Bob Gilruth had called upon the seven men at their Langley office building to explain that he had chosen Alan Shepard as the best choice for first flight. From that date a month later, Grissom would be the back-up crewman ready at any time to stand in should a change be necessary.

Training became intense as simulators at Langley and the Cape whirred ceaselessly through many different mission situations day in, day out. Before his flight, Shepard was to go through 120 simulations of the profile he hoped to fly on MR-3. During March and April, altitude chamber runs were made more frequent, and the pilots were required to participate in rehearsing the journey from Hangar S to the transfer van, along to the pad 5/6 area, up to the spacecraft and in to the pressure cabin for full scale countdowns. The Mercury-Redstone pad 5/6 site was nestled between Complex 30 for Pershing missiles to the west and Complex 26, an old Jupiter range, to the north-east. To the north lay Lighthouse road serving a row of launch pads along this curving strip of beach south of the Air Force Station where most of the big missiles had their pads. Of the two designated pads at the 5/6 site, MR-3 would use No. 5. Spacecraft 7, launched by Redstone 7, opening a series of flights in which 7 astronauts would ride Mercury, leading Shepard to call his particular capsule Freedom 7, a cynical reflection on the politics of another country that pre-empted the historical niche each astronaut had wanted for himself.

The capsule was one containing several features unique to this suborbital mission. With two circular portholes, one at the astronaut's right side below the hatch sill and one to the upper left of his forward facing position, it represented the last of a production batch to feature early design approaches rectified in later spacecraft at the astronauts' request. None of the seven liked the porthole arrangement but spacecraft 7 was the best for the first flight, everyone had the most confidence in it, so the very first manned mission would use this design. It also had a mechanical hatch weighing a hefty 31.3 kg, released by a handle behind the astronaut's right shoulder, and would be replaced in subsequent spacecraft by a pyrotechnically releasable cover of the same size but weighing only 10.4 kg. Also, No. 7 was fitted with only three attitude control modes: the Automatic Stabilization Control System, the Fly-By-Wire system, and the Manual Proportional System; the fourth (Rate Stabilization Control System) would be employed for the second and subsequent manned missions.

As the weeks before the flight saw increasing public interest in the mission, with press coverage visibly growing, several key officials speculated on the wisdom of exposing a highly technical operation to the pressures of such wide examination. Several Senators raised their voices in question over this issue but the overriding consensus was for as full and as public a show as possible. International suspicion about the validity of Gagarin's flight made the NASA exposure refreshingly open, enabling the US program to eclipse for a brief period the Russian success, reportedly to the enraged annoyance of Premier Kruschev. It would have been exceedingly difficult to hide the events of this Mercury mission. Soon after launch, the capsule would descend to open water where a host of recovery vessels lay waiting to pluck it from the sea.

Launched from a more northern latitude, Vostok I flew in orbit across large areas of land and recovery from water would have taken place outside Soviet territory, the very same type of exposed situation that gave the comparatively meager achievement of a suborbital Mercury flight temporary precedence over news about the Gagarin flight. For this reason, the Russians would continue to recover their spacecraft on land, while America found it simpler to fish its astronauts from the sea. Recovery forces from the Department of Defense would include for MR-3, an aircraft carrier and 8 surface ships, 7 aircraft, and 7 helicopters dispersed to an elliptical pattern more than 800 km down the range. Under the command of veteran Mercury recovery coordinator Rear Admiral F. V. H. Hilles, the DesFlot-Four group would stand by for any conceivable contingency from an off the pad abort to an over-shoot due to a flat, extended trajectory. On experience from the MR-2 flight, an amphibious aircraft was stationed downrange of the predicted impact point ready to drop a pararescue team should the capsule land far beyond its expected location. Tests with the recovery operations displayed problems with a number of proposed methods, until a preferred concept where a helicopter would hold the capsule free of the water as the pilot climbed into a horse-collar, was selected by unanimous vote.

Launch day neared and widespread concern for a successful flight percolated all the way to the White House. NASA should only launch MR-3 when everything was absolutely ready and the President himself continually expressed worry that the mission should provide an unqualified space success for the United States. On 1 May, a day before the planned flight, Kennedy's personal secretary Evelyn Lincoln called Paul Haney, NASA Headquarters public affairs officer, to question the reliability of the Mercury launch escape system. The President, she said, was worried lest it prove a fatal addition to equipment otherwise provided for the safety of the astronaut. After a short delay, Press Secretary Pierre Salinger handled the call and reassured the President that NASA had every confidence in the device.

By the first week of May, more than 300 newsmen had registered their intention to cover the flight and radio and television crews arrived like the vanguard of some colossal circus to relay the event live all across the United States. But if it was a carnival, the mood at circus maximus was optimism personified as Shepard, still an unknown appointment to the public, prepared for his historic day. On 2 May, the astronaut sat fully dressed in his silver space suit for 3 hours while rain and squall advanced upon the Cape. Preparations involved a full countdown to the point, 2 hours 20 minutes prior to launch, where astronaut participation was essential. Only then was the mission 'scrubbed' for that day, a term that would become a familiar word in the techno-pop jargon of NASA talk. Following the announcement that the mission would be abandoned for 2 May, the public officially learned the identity of the chosen figure who had sat for so long in Hangar S. Speculation was over.

Another attempt was scheduled for 4 May and again it had to be cancelled. However, at 8:30 am that day the first phase of a two-stage countdown began with an anticipated launch the following morning. The intensive medical evaluation of Alan Shepard began the day prior to MR-3's original 2 May launch date. The pilot was observed by the psychiatrist to be 'relaxed and cheerful. He was alert and had abundant energy and enthusiasm. Effect was appropriate. He discussed potential energy hazards of the flight realistically and expressed slight apprehensions concerning them. However, he dealt with such feelings by repetitive consideration of how each possible eventuality could be managed. Thinking was almost totally directed to the flight. No disturbances in thought or intellectual functions were observed.'

Physiological inspection of the entire surface area of Shepard's body was performed, followed by an opthalmological survey of the eyes. To quote from the official pre-flight medical survey for an appreciation of the detail required by the NASA medical team for even this short 15 minute flight: 'Examination of the oral cavity, mucous membranes, teeth, and tongue disclosed slight reddening of the mucosa at the medial margines of the posterior tonsillar pillars. In the neck, the thyroid was found to be just barely palpable, smooth, and symmetrical. The thorax was symmetrical; movement was full and equal bilaterally. Over the lung fields, percussion and auscultation revealed no abnormality. Palpation of the anterior thorax disclosed the point of maximal

Vostok formed the base design for a generation of unmanned spy satellites and production runs like this served dual functions of military and manned civilian research programs.

This amphibious US Army lark stood by on the Cape Canaveral beach ready to retrieve the Mercury capsule should the astronaut, waiting for launch, need to trigger an abort with the launch escape system.

cardiac impulse to be in the left intercostal space 11 cm from the midline.' Information of this kind, a definition of Shepard's physical status, set a precedent among medical personnel that would stimulate the same thorough approach for the duration of manned space operations.

After the examination, most of the medical team moved to Grand Bahama Island where they would set up camp to receive the pilot at the end of his suborbital lob, ready to go through the entire examination again for physiologists concerned with the performance of the human body in space. Medical activities at NASA had recently entered a new phase with the Director of Life Sciences at NASA Headquarters planning a vigorous research program involving men, primates, and biological subjects down to insects and plants. Drs. White and Douglas were responsible for STG medical operations and the personal health and physiology of the astronauts, respectively. Shortly after the first phase of the countdown began on 4 May, a launch readiness briefing was held at which the astronaut participated, key personnel reporting in on the status of their particular concerns, receiving word on the status of other elements that impacted their own areas of work. The first phase of the dual countdown proceeded to the T-6 hr 30 min level at which point a 15 hour 'hold,' another unique NASA phrase meaning suspension of existing activity for subsequent resumption – began. The clocks started moving again at 11:30 that night, long after astronaut Shepard had gone to bed in his room at Hangar S.

With lift-off planned for 7:00 the next morning, Shepard was geared to a schedule that required him to get as much sleep as possible up to the early hours of the following day. He slept well, without dreaming, and set a precedent that non-astronauts would find hard to understand: with concentrated training sessions for months before a flight, simulations of situations the astronaut would be most unlikely to experience, realization that the strenuous repetition of endless rehearsal was over put pilots into a sense of relief capped only by the exhilaration of finally achieving what had been to that time only a second best approximation; totally aware of every conceivable facet of the spacecraft's design, operation and performance, fear was an unknown dimension and, had it appeared, would have eliminated that particular pilot as a candidate for space flight, although many times an astronaut would be scared of situations he was called upon to accept willingly – a very different emotion to fear, which is usually a product of ignorance or shocked surprise.

Bill Douglas entered Shepard's room in Hangar S at 1:10 am, tapped the astronaut on the shoulder and woke him on the day Mercury would carry a man into space for the first time. For several hours, technicians had been preparing the spacecraft, Freedom 7, as it stood on top of the Redstone rocket groomed for the all important ascent. John Glenn was to check out the capsule, spending several hours inside the tiny pressure cabin going over the switch positions, running combined spacecraft tests with a sequence monitored from the Cape's Mercury Control Center. Before vacating the spacecraft shortly before Shepard was due to arrive, Glenn would place all the switches back to the positions the prime pilot had arranged to find them in when he took over for a personal systems check. From his bedroom, Shepard went to the shower and then to breakfast with John Glenn, Bill Douglas, and a few privileged operations personnel, eating enthusiastically a fillet mignon wrapped in bacon, scrambled eggs and orange juice. For the past three days, Shepard had been on a low residue diet, and the pre-launch meal, high in protein, conformed to a schedule compatible with the stringent medical requirements for a concise record of pre- and post-flight analysis of his physiological response.

The schedule was relaxed, meticulously precise in its execution, and carried Shepard from a conversational breakfast during which he discussed topics only loosely associated with the flight to a brief physical examination. Dr. Douglas would report afterwards that he found no reason to suspect Shepard's condition and found it unnecessary to do anything other than take the astronaut's blood pressure, performed, he said, because he felt sure someone would ask him what he had actually done to satisfy himself that Shepard was fit to fly. It was 2:40 am. After the cursory physical, Shepard went to receive bio-sensors that would telemeter to the ground information on his condition at every point in the mission. Development of a suitable sensor package had been the concern of STG for some time.

Based on experiments conducted from the Man-High Balloon Gondola series, where human subjects volunteered to ascend to great altitude suspended from a balloon, the temperature of the body was considered to be one of the most important parameters to monitor during the flight. Consequently, a rectal thermistor-tipped probe was designed whose shape, in the test of the descriptive report, 'took cognizance of the anatomy of the rectal sphincter, and whose rigidity was sufficient to permit easy introduction.' About 3.5 mm in diameter, the bulbous device was taped in place.

Respiration was measured by a tiny thermistor attached to the communications microphone that would be positioned close to the astronaut's mouth, inside his helmet. The tiny thermistor carried a funnel for catching air breathed through the nostrils while air from 'the mouth passed directly across the instrument.' By measuring the movement of air as the pilot breathed in and out, a rudimentary value would be obtained from which medical personnel would determine respiration. The third parameter, heart rate measurements, was obtained from two electrodes on an axillary position and two at a sternal location. Glued to the skin, the cup contained an electrode paste from which potential was to be picked up via a shielded wire. The package of sensors conveying data on temperature, respiration and heart rate was developed by McDonnell and tested in centrifuge runs and on primates in the test flights with Redstone.

Concern was expressed about the need for a blood pressure reading and hardware was under development for use on the manned orbital flights. The precise location of various sensors on Shepard's body had been determined in advance, the places marked by small tattoos. Harnessed with the wires that would transmit his physiological condition, Shepard went next to be suited up by STG technician Joe W. Schmitt, a veteran who would service the majority of astronauts through three manned space flight programs. Pressure checks were made to see that, if called upon to provide a miniaturised environment, it would not fail; it was in every sense a back-up to the capsule's environmental control system, a personally tailored spacecraft with a flexible shell.

By this time, John Glenn was at launch pad 5, checking Freedom 7. Countdown preparations had been underway late the previous evening, Gordon Cooper had been monitoring events in the blockhouse, actually an old structure originally used for monitoring tests with the Bomarc missile, and word came that everything was in a 'go' condition. The precise sequence of events culminating in ignition of the Redstone's single engine was designed to halt at T-2hr 20 min for a quick resume of launch vehicle and spacecraft status before passing word to Hangar S that the astronaut could be brought to the pad. It was 3:55 am when the silver-suited astronaut entered the transfer van for the slow journey to pad No. 5. Aboard were Joe Schmitt making final adjustments and a couple of technicians. At the pad 40 minutes later, Cooper brought Shepard up to date with events of the preceding hours, giving him a resume of countdown anomalies and other items that might impact his preparation. Fully suited now, gloves going on last just before leaving the transfer van, Shepard walked from his conveyance toward the illuminated rocket as white liquid oxygen vapor swirled around hoses that led to the base of the Redstone, more vapor seemingly reluctant to detach from the frosted cylindrical sides.

It was 5:15 am when Shepard ascended the gantry that carried him eight stories to the top of the converted missile.

Groomed for its responsible event, Freedom 7 was ready to receive the man. Five minutes later he put his legs across the sill of Mercury's side hatch, winced as his right foot slipped off the couch's right elbow support, but continued to lower himself down and across easing his body into the contours of a form-fitting profile designed especially for his specific dimensions; this couch had been his support for 17 simulated Redstone launch runs in the Johnsville centrifuge, now it was in the flight capsule atop a very real Redstone ready to carry him into space. Shepard's knees were just below the lower edge of the instrument panel and the closely spaced equipment all around seemed tailored to the space between capsule wall and the extremities of the pressure suit. On the instrument panel was a note, left by checkout man John Glenn: 'No handball playing here.'

Schmitt was there to help Shepard into the restraint harness designed to constrain his movement during periods of extreme stress. To a basic shoulder strap and lap configuration similar to conventional harness from a military aeroplane, the Mercury pilot had a chest strap to support the upper torso, an inverted V-shaped strap to keep the lap strap over the abdomen, and two knee straps, which with the lap strap were expected to keep the pelvis aligned with the couch and so reduce the probability of injury to the lumbar region. Strapped firmly but with a freedom that allowed Shepard to shift position, the pilot was connected to the suit system in the spacecraft's environmental control unit before technicians gave a hearty 'Happy Landings, Commander' and secured the hatch. A quickened heart rate at this point was seen to settle down into the 80s as Shepard began to check through the switch positions he faced on the instrument panel.

Comprised of three flat sections, the left panel provided controls for the attitude and stabilization systems and the retro-rockets. Three handles were provided by which the pilot could manually select attitude control thrusters in pitch, roll or yaw. To the right, the second panel carried a display of sequencing indicators, rectangular lights showing green to confirm that a function occurred and red if an automatic system should fail. From the top down they indicated, tower jettison, capsule separation, retro sequence start, retro attitude, retro-pack jettison, 0.05 g on re-entry, drogue parachute deploy, main parachute deploy, and reserve parachute deploy. To the left of each indicator, a handle or switch allowed the pilot to command the relevant system manually should a red light warn of failure. At the bottom was a switch for rapidly decompressing the cabin and another for re-pressurization; emergency decompression would be used above the atmosphere to extinguish a fire.

The main control panel directly in front of the pilot was dominated by a circular screen displaying the view through the periscope, a retractable device affording visual observation of the view out from the underside of the capsule during orbital flight; as the astronaut lay in his couch the view would be directly through the instrument panel and out the side of the spacecraft adjacent to his feet. During an orbital mission the spacecraft could be manually positioned by following the view through the periscope. At the top of the panel were a cluster of dials showing spacecraft attitude and the attitude rate in all three axes; that is, the speed with which the spacecraft moved to a different axial alignment. Because the pilot had availability of either automatic or manual control modes, each with a completely separate set of propellant tanks and thrusters, and because he could align the spacecraft by either one of two completely separate references, the capsule had a very effective layout of two completely redundant attitude control systems. For automatic or semi-automated attitude control, the ASCS control mode would be selected, with attitude alignment and excursion rates monitored on the cluster of dials. For manual authority, the hand controller would be used to command the FBW system with visual alignment from the view through the periscope.

An instrument just above the periscope screen would tell the pilot the time of day and the elapsed time from launch in seconds. Later spacecraft, required to fly for several hours in space, would be equipped with a suitably modified clock to indicate hours, minutes and seconds. Three other dials to the upper left of the central console indicated acceleration during ascent and descent measured in g forces, the altitude of the capsule, and the descent rate. At the top to right of center, alongside the attitude and attitude-rate dials, a ready-switch which the astronaut would punch to confirm a preparedness for flight to the test conductor in the blockhouse, and a light which would glow in the event of an imminent abort. Down the extreme right side of the central panel were a cluster of dials showing cabin pressure, temperature, humidity, coolant quantity and oxygen quantity. Below those, dials to show DC amps, DC volts, AC volts, and communications switches for operating the different channels. On a separate box to the pilot's left, the emergency oxygen and cabin temperature controls.

There were no foot controls, the pilot's boots being firmly retained by the contour of the couch pan and the leg straps. In his right hand, the three-axis attitude controller; in his left hand, a single twist grip device that would instantly command an abort sequence by way of the launch escape system. In terms of available space, the pilot had as much room as would be available to a man with two overcoats on, in the driving seat of a compact car, totally constrained to left and right with a capacity to drift as far forward as the dashboard. In reality, Mercury astronauts would feel very little sensation from weightlessness due to their belts and restraint straps.

It was light outside Freedom 7 as the countdown moved forward; shortly after 6:00 am the hatch was bolted. Minutes later Shepard started breathing pure oxygen as part of the process to purge his body of nitrogen. At 6:37 the gantry, a modified oil derrick, was moved back from the white rocket in a standby mode ready to be returned quickly to the spacecraft if anything should

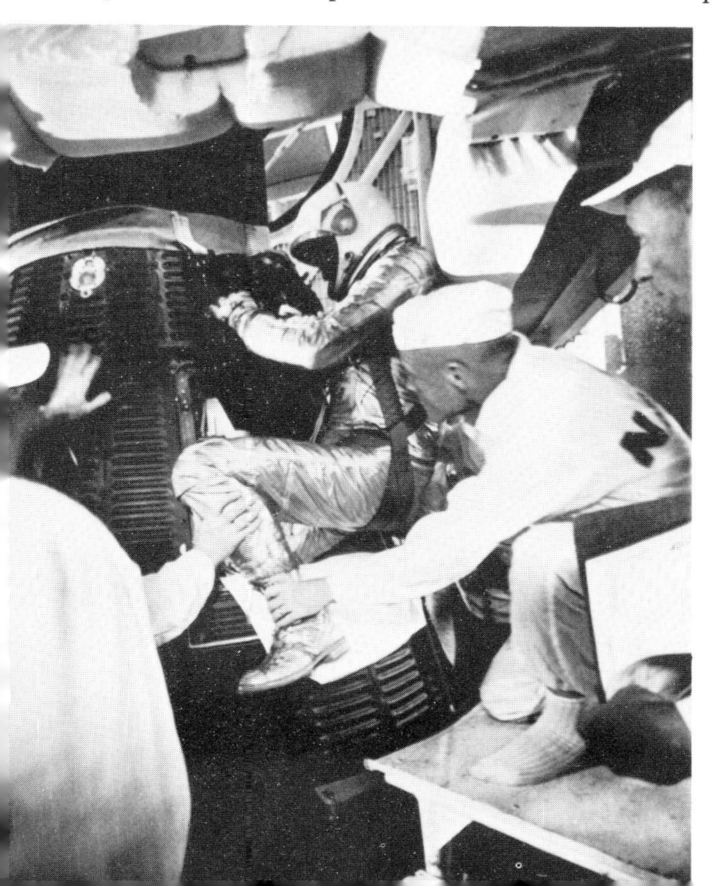

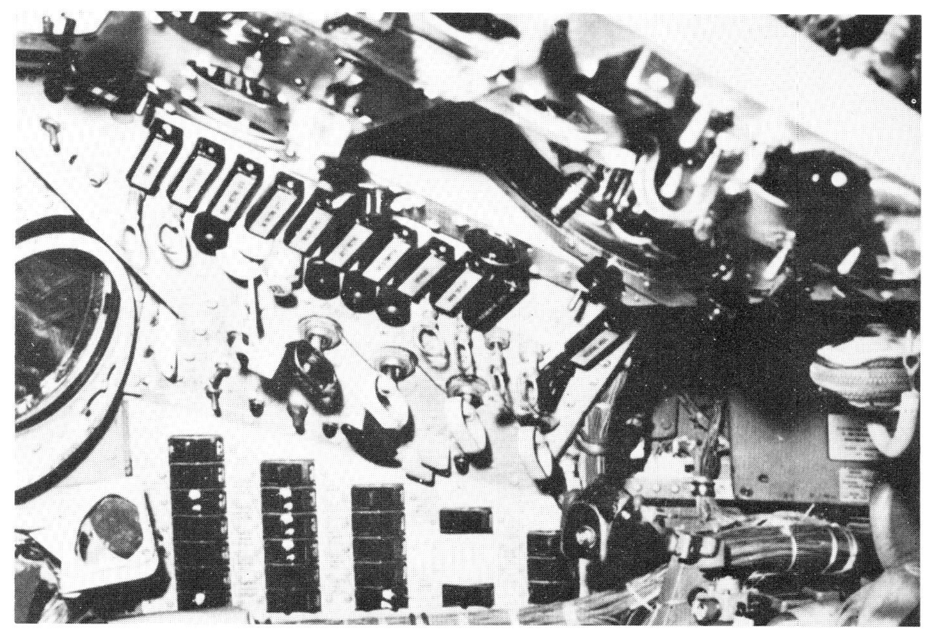

Unique to the MR-3 mission, Shepard's cockpit for the 15 minute ride through space reveals the circular window (left), the main display console (top) and the four ring handles for manual attitude control below the sequencer lights set in rectangular black boxes (center).

Technicians help Alan Shepard ease his way across the sill to Mercury spacecraft Freedom-7.

require the astronaut to leave his capsule in a hurry. It was a poetic bow to history. From the same launch complex that more than three years before had been used by the Army's von Braun team to launch America's first Earth satellite, with a rocket as closely related as any to the Juno 1 launcher, using support equipment like the mobile gantry visually identical to that employed for Explorer 1, an American astronaut was to make a flight into space.

The countdown went smoothly from the time it resumed at the T-2hr 20 min hold to a point just 15 minutes before the planned lift-off. Low cloud in the area would preclude the possibility of photographing the ascending Redstone and meteorologists favored a wait of 35–45 minutes after which the conditions were forecast to be clear. During the 86 minutes that actually elapsed before the countdown resumed, Shepard viewed the Cape through his periscope and discussed events as engineers returned to the launch vehicle to replace an inverter ordered by the test conductor. Re-cycled back to the T-35 min mark, countdown operations resumed until again at T-15 min it was necessary to order a hold for checks on the computer to be used for giving rapid information during ascent on trajectory and possible impact points.

For about 4 hours now, Shepard had been lying on his back in the confined cabin of Freedom 7, a period almost twice as long as expected. It was of modest concern to the medical staff on hand at Canaveral; a thwarted astronaut is more susceptible to continuous delay than the tension of an actual launch. But this time the resumption was for real as clocks ticked down to the fateful zero and ignition of the Redstone motor. Chris Kraft and Walt Williams were in the pad 5 blockhouse monitoring each event in the final preparation of the combined launch vehicle and spacecraft. Gordo Cooper was there too, talking to Shepard and keeping his morale high during the long wait, flanked by Gus Grissom and Carpenter in a passive role – the worst at times like that.

Across the Cape at the Mercury Control Centre, Deke Slayton prepared to 'feed' Shepard instructions or comment from the nerve center where consoles stood poised to receive the telemetry from Freedom 7. John Glenn had joined Slayton by this time, watching the needles and dials that read the condition of spacecraft and launcher to engineers and technicians around the room. Wally Schirra was airborne in an F-106 fighter, circling ready to pursue the rocket as far and as fast as he could, photographing all the while. Not far away, on the Canaveral beach that had this day attracted an unusual number of spectators, stood Dolores O'Hara, the Air Force nurse assigned to the seven astronauts, tensed by events of the preceding days.

The sun was high now as people across the nation stayed at home past the time they would normally have been arriving at work to listen and watch the final stages of this historic countdown. In the capsule, Shepard's heart rate was quickening from a rate that averaged 80/min during the preceding hour, to nearly 100/min in the final few minutes. His left hand gently folded around the abort handle, his right hand resting loosely on the attitude controller. Two minutes prior to zero, the cherry picker moved back to a standby position, ready to close back on the tiny capsule if anything went wrong in the final seconds. Shorty Powers in the Mercury Control Center provided the link between hundreds of expectant technicians and the wide world outside, detached now from the combined attentions of personnel across the Cape.

'T-30 sec,' there was an inevitability about the voice. 'Roger – periscope has retracted,' came the astronaut's confirmation as, on the side of the capsule's conical wall, the barrel-shaped device slid back to lie flush with the external shingles that covered the capsule. Shepard's heart rate was 108/min now, and quickening. 'T-15 sec,' the voice from Mercury control was firm and expectant. Communications to the spacecraft had switched now from Cooper to Slayton as Shepard kept his eyes on the abort light at upper right on the center console. Newsmen and reporters rubbed shoulders with jostling TV cameras, film cameras on tripods, and spectators with binoculars. The mood was tense and alive with static exuberance, constrained now but awaiting the anticipated eruption when Redstone roared into life. As if in telepathic contact with the pilot, hundreds of men whose job it was to tell America the second by second events established a link that crossed the flat Canaveral waste: 'This is it Alan Shepard, there's no turning back now, good luck from all of us here at the Cape,' said an excited voice to millions of listeners on radio sets in homes, cars, offices, factories.

With the access gantry wheeled away, the 'cherry-picker' stands ready to afford ready access in the event of a major problem.

From Shorty Powers, the voice of Mercury Control, a steady countdown that held with bated breath the response from people along the sandy shore: '10–9–8–7–6–5,' Shepard reached up to the console and firmly punched the 'ready' button that lit a telling light on the flight director's console in Mercury Control, '4–3–2–1–zero.' A roar, blinding light, white smoke, and Shepard's heart rate shot to 126/min, his respiration at a heavy 40 breaths/min. Redstone's engine was alight and running. Lift-off. 'Ah roger, lift-off and the clock has started,' came the shaky voice from Freedom 7, unsettled by reverberations moving upon the tiny capsule from a source deep inside the Redstone. Shepard had started the clock inside the capsule that ticked away the seconds from launch and would keep voice communication at least every 30 seconds during the ascent. From outside, a moment of disbelief as the crowds stood, hands shielding the morning glare from the Sun rising in the east, as the crackle of Redstone's engine clapped like thunder across the flat landscape, before shouts ripped forth 'Go baby, go baby, go baby!'

Like unleashed springs, emotion tore from the vocal chords of hundreds, tears rolling down the cheeks of others, as the rocket rose higher and higher, its metallic roar increasing as the sound waves came straight down, unshielded now by concrete or earth. But there was no time for reflection inside the capsule or at the consoles in Mercury Control as needles quivered and dials bounced from one reading to another. At the instant of lift-off Shepard had seen through a cabin porthole the thin umbilical cable fall away from the side of MR-3 and about one-half minute into the flight he called with a status report: 'This is Freedom 7, the fuel is go, 1.2 g, cabin 14 psi, oxygen is go.' The '14 psi' referred to internal pressure in pounds per square inch, equivalent to 724 mmHg. His voice was steady with the clarity of a test pilot beginning a lengthy process of flight evaluation and to his prepared mental state he allowed himself no greater emotion.

About 45 seconds off the pad, vibrations began to set in at a more severe rate than before as the combination spacecraft and launcher accelerated toward the speed of sound, moving through the period of buffeting associated with the transonic regime. Outside, the dynamic pressure on the vehicle increased, tearing at the adaptor that held Freedom 7 to its launcher, until peaking at 1 min 24 sec just after Shepard called to confirm that 'Cabin pressure is holding at 5.5, cabin holding at 5.5 (psi).' This was the first critical stage of the flight and simultaneous events were occurring: max q, and the point where the cabin valves sealed and held pressure at 258 mmHg instead of bleeding down to the vacuum of space as the assembly accelerated higher through the atmosphere. The environmental control system worked and the cabin held pressure, Shepard watching the dial to ensure its integrity. If the cabin had failed to hold pressure at this point Shepard would have aborted, limiting his time spent above 15 km and reducing his abort altitude to no more than 21 km. He had just a few seconds to decide whether to twist the abort handle before passing through a point, at 1 min 29 sec into the flight, where his exposure above 15 km would be considerably beyond the 60–70 seconds considered safe. But those fleeting seconds told Shepard all he needed to know: the cabin was 'holding at 5.5.'

During max q the vibrations reached their worst, sound levels increasing but not to the point where it blocked the friendly voice of Deke Slayton. Vibration caused by the air ripping at the spacecraft shook Shepard's head with a high oscillation rate, partially blurring his eyes and making instruments temporarily difficult to read. But only for seconds as the vehicle passed firmly into supersonic flight. The sound level changed here, falling away to a point where vibrations conducted through the structure of the rocket were the only indication of a running engine, albeit an engine that delivered almost exactly three times the afterburner thrust of the F-106 fighter Shepard was so used to flying. At two minutes into the flight, a call from Freedom 7: 'Fuel is go, 4 g, 5.5 cabin, oxygen go, all systems go.' Those last three words would form the headline for newspapers across the nation as the exuberant success call from the edge of space: 'All systems go!'

Shepard's anticipation reached a peak as the Redstone's engine shut down at 2 min 22 sec after lift-off, revealed by a heart rate of 138/min. At that point the acceleration gave Shepard a 6.2 g load but the sudden release of thrust brought on a sensation of weightlessness not dissimilar to that experienced by the sudden deceleration of a high-speed elevator on the way up. Instead of his body pressed hard and firm into the cushioning effect of the contour couch, Shepard was now aware of his straps as the torso sought to float from the stressed position it had occupied just seconds before. At the point of cutoff, Freedom 7 was moving at 8,340 km/hr, an accuracy to within just 1.1% of the planned value. Immediately after the Redstone ceased to thrust, ground computers predicted an impact point just 5 km from the exact spot the spacecraft would actually descend upon nearly 13 minutes later. Launch escape tower jettison occurred exactly at the point of Redstone shutdown, an event unobserved by Shepard despite an attempt to see smoke from the solid propellant motor through the side porthole.

Ten seconds later, exactly as planned, the Marman clamp ring attaching the spacecraft to the Redstone adapter was severed; at the same time three posigrade rockets on the retro-pack fired for 1 second producing a total thrust of 544 kg to shunt the capsule forward from the booster at a speed of 33 km/hr. Shepard watched the green light for tower jettison followed by a similar green light below the former indicating Freedom 7 was indeed free from the booster. All it needed now was for Shepard to throw the switch to disarm the retro-pack jettison mechanism and tests could begin. He had little more than two minutes before setting up the retro-fire sequence.

At 2 min 37 sec, 5 seconds after spacecraft separation, the Automatic Stabilization Control System (ASCS) began a turn-around sequence designed to yaw Freedom 7 from a heat shield-aft to a heat shield-forward position, having already damped out oscillations set up by the posigrade rockets. Turned round 180° so that Shepard now faced backward to the direction of travel, Freedom 7 pitched down so that the nose of the capsule was 14½° below the local horizontal; that is, 14½°, with the heat shield facing up, to a line level with the Earth's surface below and not the flight path trajectory which still had the capsule climbing toward the high point of its ballistic curve. By this time the periscope had extended and Shepard could use visual sightings to confirm spacecraft attitude.

For 38 seconds after spacecraft separation the ASCS had control of Freedom 7 while Shepard closely monitored the attitude and attitude-rate needles until, at 3 min 10 sec, he switched to the Manual Proportional (MP) system to demonstrate that a pilot in the weightless vacuum of space can control and orientate his capsule; it was as much a vindication of the simulators on the ground as it was of the flight hardware because the result of this phase would be a direct product of the many hours spent rehearsing attitude control. First, Shepard designated the pitch axis to manual operation, while the ASCS held yaw and roll steady on automatic. It took Shepard 10 seconds to pitch the spacecraft down to –34° and another 10 seconds to bring it back to –14½°. Then, with pitch still under manual control, he switched yaw to the MP system at 3 min 20 sec and slewed the capsule 20° in yaw before returning it to the normal attitude, retaining the –14½° pitch. Next, to get full manual authority, Shepard switched the roll control to Manual Proportional at 3 min 45 sec and rolled the capsule 20°, again holding pitch at –14½° and yaw at 0°. The rate used for MP operation moved the spacecraft at about 4°/sec in each axis.

At 3 min 50 sec, having spent only 40 sec proving a man could control his spacecraft, Shepard began Earth observations through the periscope, reporting 'Cloud cover over Florida; 3 to 4 tenths near the eastern coast observed up through Hatteras; can see Okeechobee; identify Andrews Island; identify the reef.' To get this view, Shepard had to look through a gray filter placed across the periscope before launch to filter glare from the rising sun. He had attempted to pull out the filter during the automatic turnaround after spacecraft separation, having forgotten to do so at the appointed time when he retracted the periscope seconds prior to lift-off, but his hand banged into the abort handle; he resolved not to bother with removing the filter and continued his

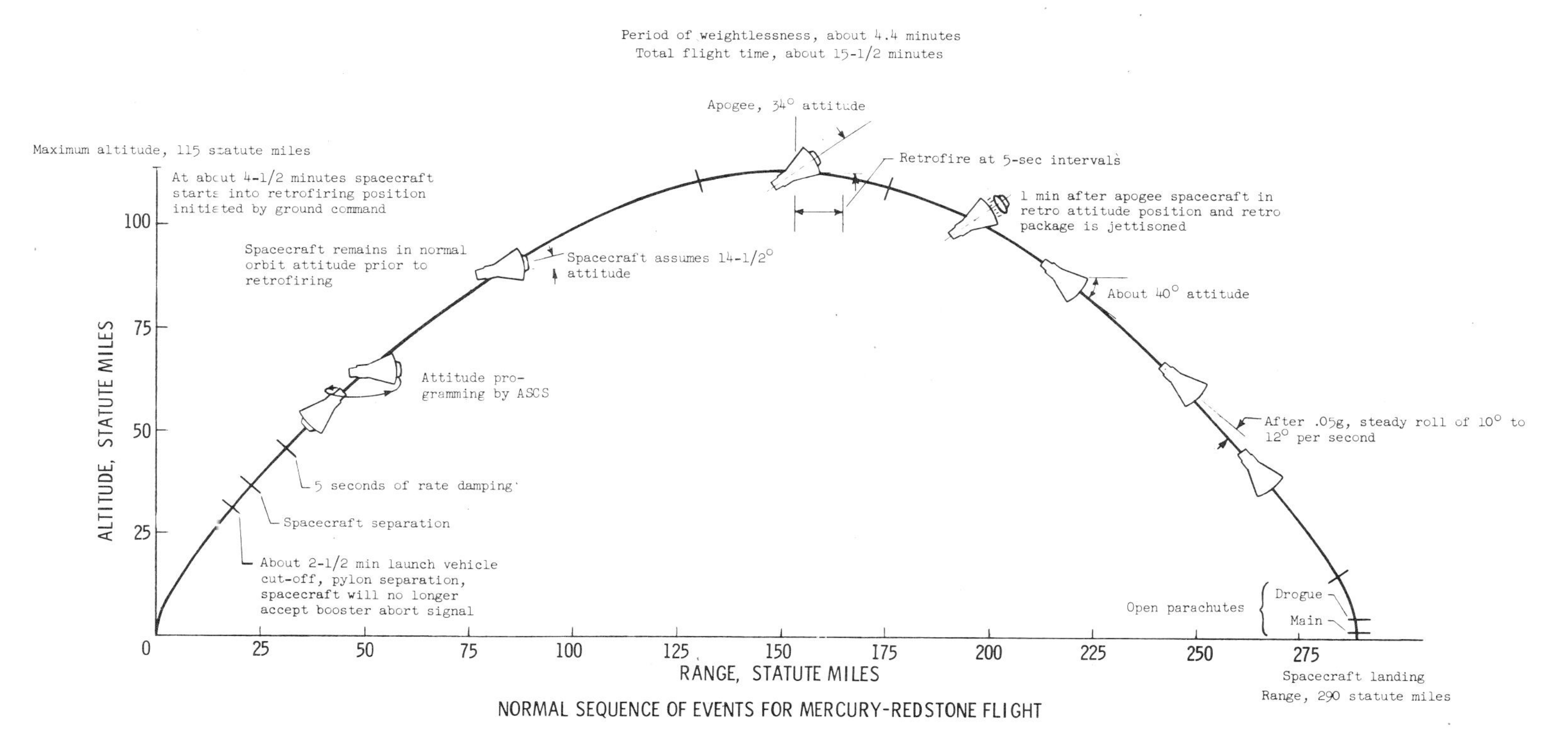

NORMAL SEQUENCE OF EVENTS FOR MERCURY-REDSTONE FLIGHT

view on a darkened screen! At 4 min 44 sec the retro-fire sequence was introduced by onboard timer at which point the spacecraft pitched forward from $-14\frac{1}{2}°$ to $-34°$; that is, with the heat shield and retro-package pointing up above the horizon in the direction of travel, with the opposite end of the capsule (the forward end to the pilot) inclined toward the Earth.

Several seconds before, Shepard had noticed an unusual maneuver where the capsule yawed 20° but this was satisfactorily negated before retro-fire. Switched now to the Fly-By-Wire (FBW) mode, Shepard watched the dials for ignition of the three solid propellant retro-rockets. At 5 min 11 sec, Freedom 7 reached the high point of its arching trajectory at 187.5 km and began the long fall back toward the atmosphere. Three seconds later the first solid fired, giving Shepard a sharp punch in the back, followed 5 seconds later by the second and 5 seconds after that by the third. Each lasted 10 seconds and built up a 1 g deceleration that momentarily felt as though he was back on the ground. Spacecraft attitude was held on manual control during retro-fire and Shepard noticed the only cockpit malfunction of the flight when the retro-pack jettison light failed to come on. Jettison occurred unmistakably 40 seconds after the third retro-rocket burnt out and Shepard punched the manual override button to the left of the indicator, which then lit up as expected.

With attitude control in FBW, Shepard experimented with manual control in this second and alternate mode of operation, moving yaw and roll rates by up to 20°. By 6 min 20 sec, just six seconds after retro-pack jettison, Freedom 7 went to re-entry attitude on ASCS by pitching up to +40°. The actual pitch excursion here was 74° because, although still flying backwards, the capsule had changed attitude from one in which the nose was pointing down 34° *below* the horizon, to one in which it was pitched *above* the horizontal. This put the heat shield in the proper attitude with respect to the atmosphere, taking 15 seconds to reach +40°. At 6 min 44 sec into the flight, the periscope retracted and Shepard prepared for the re-entry. At this point he became temporarily disorientated; the view through the two small portholes incapable of showing anything he could use as a reference. This soon changed when he brought his attention inside the capsule to instruments that fed his senses the information they needed for positioning himself once again in the frame of coordinates he knew.

Little more than a minute after periscope retraction, exactly 1 min 28 sec after going to re-entry attitude, Freedom 7 was down at a height of 61 km and picking up 0.05 g indicated by a light on the console. This was the beginning of deceleration, a force that built up rapidly on the capsule, reaching the peak of 11 g in just 32 seconds. Pressed to the contour couch with a force almost twice that experienced during ascent, Shepard maintained attitude on Fly-By-Wire until, a mere 20 seconds later, the load was down to just 1 g. Switching to ASCS, the pilot missed the altitude calls he was scheduled to make between 24 km and 27 km and sensing a faster descent rate than expected he prepared himself for the drogue parachute deployment, reporting an altitude of 9.1 km as the altimeter unwound on the instrument panel.

During the high-g phase of re-entry Shepard continually reported all the way down to indicate his conscious state to Mercury Control at Canaveral. With a simple, repeated, 'O.K.' the controllers knew he was withstanding the punishing forces that increased his weight from 76.7 kg to 844.5 kg. But now, dropping low on the horizon, his voice was becoming less clear as communications were routed via Grand Bahama Island. At 6.4 km the drogue deployed, 9 min 38 sec after lift-off. Shepard's heart rate had peaked at 132/min during the g load on descent but now it was sinking comfortably back toward 100. By this time the cabin relief valve was open, ingesting air as outside the capsule the atmospheric pressure increased. At a height of 3 km a pyrotechnic charge blew off the antenna canister situated on the nose of Freedom 7, pulling out the 19.2 meter diameter ringsail parachute painted red and white for recognition. For a brief second or two Shepard was jerked up with a force of nearly 4 g as the canopy expanded and arrested the downward motion of the spacecraft to a comfortable 38 km/hr.

For five minutes Freedom 7 would fall toward the Atlantic, providing the pilot with the most relaxed period of the flight. Ironically, for all the fervent concentration on accomplishing all the many tasks and activities scheduled for the mission, fully one-third of the flight duration was spent falling to Earth on the main parachute. During this period Shepard opened the faceplate of his helmet, its services like that of the pressure suit happily not required in the absence of a failure to the environmental control system, disconnected the hose that supplied oxygen, removed the chest and knee straps, but retained the lap belt and shoulder harness; he knew that splashdown could induce some sickening shocks and wanted to retain a modicum of restraint against the whiplash effect of a capsule slicing into heavy seas.

On the console, as he punched the switch to dump excess hydrogen peroxide attitude control propellant, Shepard noticed the glowing light to indicate landing bag deploy. Pushing the heat shield in front of it, the canvas bag had fallen 1.2 meters below the aft bulkhead, ready to attenuate the shock of impact. Communications were getting rough now as the capsule sank low over the horizon. At Canaveral, everyone was jumping up and down, making use of this five minute hiatus during the slow descent to pat each other on the back, shout their approval, or merely stand and listen to the periodic reports from the cool, calm, articulate voice of Shorty Powers – the voice of Mercury Control. At sea, the carrier *Lake Champlain* was steaming toward the predicted impact point as sailors on the flight deck watched the tiny capsule just a few kilometers away.

At one point even the precisely factual Shorty Powers let his enthusiasm get the better of him – and instantly corrected himself without a wince – when he reported that 'The Mercury recovery forces downrange report a visual sighting of the spacecraft descending in the parachute – or hanging from the parachute.' *Lake Champlain* had watched the capsule during re-entry, observing its vapour trail streaking across the sky, and now helicopters were swarming toward the spacecraft. Five were in the area, with one right over the parachute, as Freedom 7, swaying lightly in the wind, hit the ocean with a great splash. It was 9:49 am on the morning of Friday, 5 May, 1961. Shepard's flight, lasting 15 min 22 sec, was over.

Immediately after splashdown, recovery aids, superfluous as it turned out, were deployed with main parachute jettison to prevent it filling with wind or water and pulling the capsule over. At first the capsule listed heavily to Shepard's right, but within about a minute it was upright. Slowly, the dye marker began to creep across the water while a Sarah recovery beacon was switched on. A HF recovery antenna had deployed from the top of the capsule, a long thin whip antenna designed to send signals to patrolling aircraft in the event a capsule landed far from the expected area. Recovery operations for Freedom 7, however, were under way within 2 minutes of splashdown. Inside, Shepard checked for leaks and removed his helmet and remaining straps. Helicopter 44, with pilot Wayne E. Koons and co-pilot George F. Cox, was maneuvered into position above the swaying capsule so that a line could be let down to hook a loop visible out of the top of the parachute container. Sudenly, the long HF antenna struck the bottom of the helicopter and broke off, but no matter, the capsule was snared and very slowly being lifted from the water.

Koons asked if Shepard was ready to undo the hatch and get out: he was not, water still splashed against the portholes. Seconds later, the spacecraft a little higher this time, Shepard notified the helicopter that all was ready. The hatch was unlocked, Shepard emerged over the sill, grasping for the horse-collar sent down from a separate winch aboard the helicopter. Slipping the sling over his head and under his arms he was quickly winched up. Despite problems with communication during descent and recovery – Latin American radio stations were interfering with the transmissions – the whole operation had gone exactly as planned. A few minutes after getting inside the helicopter, the capsule was being gently lowered to the flight deck of USS *Lake Champlain*.

As the landing bag gently collapsed over the heat shield, positioned on a special stand, attendants on the carrier placed wooden chocks between it and the aft pressure bulkhead so that the full weight of the spacecraft, lowered to the chocks by the helicopter, would not damage the prospects of a full and extensive post-mortem on Freedom 7. When the helicopter had gently placed the spacecraft on deck it moved aside and touched down. Just 11 minutes after splashdown, Shepard stepped from its wide access door as cheers erupted from sailors and Navy personnel standing by. Medical attendants moved toward Shepard, arms partially raised as though expecting him to collapse into their supporting embrace, but the sprightly astronaut moved confidently to the back of the helicopter and across to where the capsule stood on its wooden blocks. Reaching inside, he looked around, retrieved his helmet, jumped down off the dias and walked briskly to the port side of the flight deck. He was supposed to have gone directly from the helicopter to the admiral's cabin but that sequence was now resumed as he ran down the steps and disappeared inside. At that instant, 486 km away at Cape Canaveral, Wally Schirra thundered across the flats in his F-106, the afterburner howling as he corkscrewed up rolling vertically into the sky. A fitting tribute from one pilot to another; from one brother astronaut to the first American in space. Shepard was taken to Grand Bahama Island that afternoon for extensive medical examination where attendants learned he had lost nearly 1.4 kg in weight.

But the psychiatrist seemed to sum it all up when he wrote that the 'subject felt calm and self-possessed. Some degree of excitement and exhilaration was noted. He was unusually cheerful and expressed delight that his performance during the flight had actually been better than he had expected. It became apparent that he looked upon the flight as a difficult task about which he was confident, but could not be sure, of success. He was more concerned about performing effectively than about external dangers. He reported moderate apprehension during the preflight period, which was consciously controlled by focusing his thoughts on technical details of his job. As a result, he felt very little anxiety during the immediate prelaunch period. After launch, he was preoccupied with his duties and felt concern only when he fell behind on one of his tasks. There were no unusual sensations regarding weightlessness, isolation, or separation from the earth. Again, no abnormalities of thought or impairment of intellectual functions were noted.'

Post-flight examination of the pilot and his capsule revealed little that had gone wrong on the 15 minute lob into space. One or two spacecraft sequences had minor problems associated with their operation – the launch escape tower jettison had been performed with a manual back-up signal; the retro-pack jettison light failed to come on when it should – and a few mechanical problems were discovered – a microscopic particle of dirt found in the attitude control system plumbing may have caused a slight leak of propellant from out of the thrusters; the landing bag was found to have a few rips – but in general confidence ran high in the ability of the team to launch a man into space for orbital flight rather than a simple up-down ballistic lob. One or two procedures were modified as a result of MR-3. Shepard found his wrist pressed hard against a personal parachute pack when the hand controller was put to an extreme yaw position; the pack, a questionable method of final escape should the main and reserve parachutes fail, was carried on future flights only at the pilot's specific request.

But accolades came thick and fast for the man from space, beginning on the carrier shortly after starting a de-briefing session when President Kennedy telephoned the astronaut to extend his congratulations. It was to be a regular occurrence for Mercury astronauts. On 8 May, Alan Shepard was awarded the NASA Distinguished Service Medal in a ceremony attended by all seven astronauts and their wives at the Rose Garden of the White House. Later that day they were driven up Pennsylvania Avenue to the cheering warmth of 250,000 people before being greeted by Congress, a downtown press conference and a NASA dinner that evening. The recognition America achieved internationally for the open nature of its space program and the way the whole world had been invited to share this unique experience did much to anger the Soviet Premier, who was reportedly furious at the publicity achieved for what, technically, seemed to be such a poor second to Russia's single Earth orbit the month before.

But the world was much impressed with the NASA achievement and questions of cost, delays, dangers, and sluggish development, evaporated in the wake of global acclaim. Riding on the success of this suborbital mission, NASA arranged for the Air Force to open sections of the Canaveral launch area on 19 and 20 May, and sent the Freedom 7 capsule to the Paris Air Show where it was on public exhibition from 26 May to 4 June. Before the flight, media representatives were non-committal in their attitude to what seemed at times to be a pitifully small-scale project compared to sustained successes from the Soviet Union. Some were faithfully in support of America's attempt at restoring international prestige through a manned space program; others were cynically hostile to the use of public money for what they saw as a technological game so few could participate in. But suddenly it seemed that all those assessments had been wrong as the completely open nature of pre-flight, flight, and post-flight events indicated a new chapter for media relations with the purveyors of great technical events.

Never before had the media been allowed so close to the unfolding drama of history, never so welcome by men who frequently avoided the press in other situations. It was the perfect formula for success because it recruited the media – and promptly converted them. Great words and flowing rhetoric gushed from the typewriters at Cape Canaveral. Reporters even adopted the expression 'A-O.K.' used by Shorty Powers to feed the media an impression of minute by minute events. An old telegraphers expression, it had been used by technicians during earlier Mercury test flights although it was never once used by Shepard. Leaving Canaveral at the end of that historic day, newsmen were shouting across to each other in boyish exuberance: 'A-O.K! – A-O.K!'

But there were more profoundly emotive utterances that 5 May, as evidenced by a radio reporter who even as Freedom 7 was being recovered from the ocean said that, 'Alan Shepard stands on the shoulders of many men, many men who have made this day come true. This is a beginning of a fantastic new age where two nations are marching into space in a competition with one another; America has finally forged back tremendously in this particular shot. This brings closer, very much closer, that time when we will attempt to send a man orbiting the Earth at seventeen and one-half thousand miles an hour.'

And again, at the close of a long session broadcasting live the events of the day, an anchor man for another station: 'I can't help but think of how things will be 10, 20, or even 30 years from today when we'll look back on the events of the past few hours and think to ourselves how crude the first (manned) rocket flight was. But then I suppose regular trips to the Moon will be undertaken by more refined spacecraft and I can't help but think that this wonderful, strange, and almost other-wordly event that took place today is only the beginning for this country and that so many greater things are to come and that many other brave men like Alan Shepard will follow him into space.'

Floating gently to earth, the drogue parachute and forward canister is photographed by Shepard from his Mercury capsule as it descends on the single main parachute.

CHAPTER SIX

Decision and Commitment

With a chillingly accurate telepathy, the media had judged the mood of the nation. Congress saw it, politicians sensed it, the public confirmed it. What had seemed an idle boast was vindicated in a trice: America *could* come back from a defeat with sure hope of winning the war; what had seemed an uncrossable gulf when Russia launched Sputnik 1 and, so recently, Yuri Gagarin into space aboard Vostok I, was now a chasm of technical competence being rapidly bridged by an industrial commitment capable of much more than this meager effort if backed by political will. But more than all of this, in a way in which few Presidents could copy, John Fitzgerald Kennedy had – secretly – made the awesome decision that made Mercury pale in the light of new objectives so monumental in kind that they were to stun the nation within days of Shepard's ballistic flight. The story of how President Kennedy came to *pre*-judge the post-MR-3 mood begins late the previous year when, toward the end of 1960, JFK began to look closely at the NASA operation in general and the Mercury project in particular.

John Kennedy began life in a middle class family in Boston, Massachusetts, and was to have five sisters and three brothers. Born in May, 1917, he was the second child of Joseph P. Kennedy and the former Rose Fitzgerald. The family sent him to local private schools in keeping with a domestic environment in which servants, maids and nurses were part of the large household. Father was climbing fast with a prosperous career in banking that would eventually lead to the motion picture business, industry and an ambassadorial post in Great Britain.

John Fitzgerald entered a Catholic school at the age of 13 – the family were strong Catholic Democrats, descended from an Irish immigrant couple that arrived in America in the mid-19th century – and then on to a non-sectarian school in Connecticut. From there he went to Princeton, at the age of 18, and on to Harvard the following year. John Fitzgerald was a keen sportsman, not so enthusiastic for academic study, and was sent to London in 1939 where his father had been ambassador for two years. In the hope of stimulating a sense of history, tutors and family stressed the opportunity for personal research and observation by using the time in London to examine the gathering war clouds over Europe and reach conclusions about England's apparently pacifistic attitude. This formed the basis for a thesis which later appeared as a book titled *Why England Slept*, and it immediately became a best seller.

In 1940, JFK went to South America after graduating from Harvard and attending a short business course at Stanford, before joining the Navy in 1941 to command a motor torpedo boat in the Solomon Islands. In one incident in August, 1943, Kennedy had to swim 5 km to the safety of a deserted island, after a Japanese destroyer rammed and sank his PT boat. Friendly natives rescued Kennedy and other survivors but he was invalided home after receiving the Purple Heart. Kennedy's elder brother Joseph had always been marked down for politics but when he was killed in Europe during 1944, John Fitzgerald decided to step into his shoes. Retired from the Navy in April, 1945, for medical reasons, JFK entered the House of Representatives after a three-year effort campaigning as a Democrat.

Kennedy was a staunch supporter of the New Deal social policies of the Truman administration and spent considerable time, expending enormous energy, fighting for better housing, higher wages, and more attention to the social welfare schemes proposed in post-war America. He was not so staunch a supporter of the incumbent President, however, when it came to foreign policy. Although agreeing with Truman over the principles about European aid and the Marshall Plan, he felt the Europeans should be made more responsible for their own economic and social recovery, the need to block communist gains in the poverty stricken nations of a war-torn Europe very substantially second to other considerations.

Constrained by the limitations of the lower House, Kennedy followed re-election successes in 1948 and 1950 by running for the Senate in opposition to Henry Cabot Lodge for the Massachusetts seat. He won, and entered the upper House at the same time the Republican President Dwight D. Eisenhower went to the White House. Kennedy quickly showed himself to be a 'home-state' senator, preferring the causes of fishermen and longshore workers to the lofty heights of federal committees and affairs only remotely connected with Massachusetts. JFK married in September, 1953, taking as his bride Jacqueline Lee Bouvier, daughter of a very wealthy Rhode Island family. Much of 1954 and 1955 was taken up with a hospital operation on a spine deformity aggravated by war wounds, and the resulting period of convalescence. During this period Kennedy wrote another book about leading statesmen and politicians who had taken unfavorable decisions which were inevitably in the national good; it reflected his own view that a constituency voted for a man of strength to work for the good of all, not merely the local group responsible for his election.

After his illness Kennedy became very concerned with international events, studying foreign affairs to a greater extent than had been his habit before 1954. He transferred much of his senatorial time to wider issues than those with which his earlier term had been concerned. In 1956 Kennedy tried to get on the Democratic ticket as Vice-Presidential candidate to Adlai Stevenson but Estes Kefauver beat him in a marginal contest and both eventually lost the election to the Eisenhower-Nixon duo. It was at this point that Kennedy decided to run for the highest office in the land at the next Presidential election. For much of this period Kennedy spent time writing magazine articles and stories for the national press while the media in turn hailed him for his youthful approach to politics and his vigorous attitude to national issues. He had the press very firmly on his side, and that was a major attribution.

Aware that his record over the next four years could make or break his chances at the Presidency, JFK devoted even more time to international issues, taking up the challenge on behalf of oppression and depressed minorities, and championing the cause of free societies before Third World countries. He was particularly keen to see aid for Asia and Africa, believing that a strong economy in these developing states would stave off the threat of communist takeover; this was a very different view to the one expressed about Europe after the war but one that set him on high at home and abroad as a man of peace intent on preserving freedom. Kennedy also supported moves to court the friendship of east European states (Poland, East Germany, etc.) and to impress upon them the free choice all nations must have in reconciling national ambition with political rule.

But the overriding aspect of his foreign policy that was to cause great change in America's manned space program was a belief he held that negotiated agreements with the Soviet Union would only arise from a position of military, economic and industrial strength. He felt sure that Russia would not respect the position of the United States unless America achieved world recognition for accomplishing great technical steps, although he believed in this merely as an instrument for political gain that would, in the long run, benefit the people of the free world. He was not enamored with the pursuit of technical projects for their own sake. In presenting the American image abroad, the incumbent administration failed miserably to impress the west with US capabilities, failing further to secure respect from Third World states. In polls taken after Sputnik 1, an alarming number of western countries believed the Soviet Union might be on the brink of world supremacy, reflecting a view that America had fallen from its zenith at the end of the war.

In his Presidential campaign during 1960, Kennedy fought this view and strongly supported an attack on Eisenhower and the Republicans for having vacillated while America fell from its dominating posture. Before election day, Kennedy used the flagging space program as a pointer to his contention: 'Because we failed to recognize the impact that being first in outer space would have, the impression began to move around the world that the Soviet Union was on the march, that it had definite goals, that it knew how to accomplish them, that it was moving and we were standing still. That is what we have to overcome, that psychological feeling in the world that the United States has reached maturity, that maybe our high noon has passed and that now we are going into the long, slow afternoon.'

Kennedy became the most vocal supporter of a major commitment to space projects that would do most to secure world respect, believing that 'Control of space will be decided in the next decade. If the Soviets control space they can control Earth, as in past centuries the nation that controlled the seas has dominated the continents.' To Kennedy it was a physical manifestation of the gulf that divided the two political camps: Soviet Russia, militarily committed to pushing into every conceivable niche of the free world for purposes of possession and extortion; the US, bastion of freedom and the preservation of human rights. It was a very black-white situation to the campaigning Democrat who really seemed to believe that an American presence in orbit 'does not mean that the United States desires more rights in space than any other nation.' Whether that was meant to imply that America was just as eager for 'territory' in space as any other country, or as disinterested as most were presumed to be, was a lack of clarity Kennedy would not have expected as an interpretation of his words.

But the paradox was there in reality. To strive for free use of space implied a dominating presence by American astronauts that would be unacceptable to the Soviet leaders. Kennedy was perhaps the first to see it as a real contest believing that, 'we cannot run second in this vital race. To insure peace and freedom, we must be first.' In a peripheral sense, Kennedy epitomized an America that many thought died with the massive industrialization of the mid-war years, and capitalized on that thought by asserting that 'This is the new age of exploration; space is our great New Frontier.' But for all the bold assertions, that America must don the cloak of challenge and face the future with a vigorous dedication to space technology, Kennedy was never convinced of the need to pursue this course. It stemmed from a deep regard he had for America's image worldwide and the pre-election rhetoric would evaporate for a while after he came to office in January, 1961.

President John F. Kennedy; the man who signed the historic commitment for fulfillment of a centuries old dream that human feet would tread the Moon's dusty surface.

That Kennedy had little real idea of exactly where the United States stood in the space technology stakes, or an appreciation of the ultimate objectives, emerged in the period between election and inauguration when he asked Jerome Wiesner to form a task force – one of 29 such groups assigned to examine domestic and foreign issues – to study current programs and report on the existing NASA structure. Called the 'Ad Hoc Committee on Space,' it drew eight participants, including Donald Hornig from the PSAC's Ad Hoc Panel on Manned Space Flight set up by Eisenhower to examine the state of NASA policy regarding Apollo, etc., Trevor Gardner, once the Air Force Assistant Secretary for Research and Development, Kenneth BeLieu and Max Lehr from the Senate Committee on Aeronautical and Space Sciences, Edwin Purcell, a Harvard physics professor, Bruno Rossi from the Massachusetts Institute of Technology, Edwin Land, President of Polaroid, and his assistant Harry Watters; Purcell, Land, Hornig and Wiesner had been or were PSAC members with the Eisenhower administration.

In the interim period between Presidents, NASA was concerned about its own future. During the closing months of 1960, the White House had acted in a manner that drew criticism from Administrator Glennan, starving it of funds needed for feasibility studies on the Apollo concept and for further work on developing high-energy upper stages for the big Saturn launch vehicle. And in spite of Kennedy's vociferous defense of a viable space program, nobody knew what his specific policies were or if he would favor military control over the wholly civilian NASA structure. During the afternoon of 10 January, 1961, the Ad Hoc Committee on Space reported to President-elect Kennedy in Lyndon Johnson's office at the Capitol. Johnson had been Kennedy's running mate during the election and favored with relish the opportunity to dramatically influence the nation's space program.

Influential with decisions concerning the National Aeronautics and Space Act of 1958, the architect of the National Aeronautics and Space Council bought from Eisenhower on the promise of a smooth ride for the NASA plan, Johnson also hosted that day Senator Kerr and Representative Brooks. They heard from the Wiesner Committee a review of the nation's space program which recommended sweeping reorganization. The Defense Department would benefit, thought the Committee, from combining its fractionated program into one assigned authority, a single 'agency or military service.' NASA, it said, was too concerned with building a massive government research and development machine, that it was devoid of a 'vigorous, imaginative, and technically competent top management' and that its performance had simply not been impressive enough in the face of Soviet achievements. This latter emphasis was in line with the Committee's view that motivation for a space program was first and foremost a matter of national prestige, followed by national security, science, military applications and international co-operation, in that order.

The Committee was mindful of the need to develop both the Saturn booster and the giant F-1 engine, capable of delivering on its own a thrust of 680 tonnes achieved by the combined thrust of all eight Saturn first stage engines, acknowledging the enormous lead held by the Soviets in heavy booster operations. Generally, manned space flight came out worst of all in the Committee's report, calling the Mercury program a 'marginal' effort. NASA had, it said, exaggerated 'the value of that aspect of space activity where we are less likely to achieve success, and discounts those aspects in which we have already achieved great success and will probably reap further successes in the future.' Moreover, it urged the President not to allow 'the present Mercury program to continue unchanged for more than a very few months,' advising instead an attempt to 'diminish the significance of this program to its proper proportion before the public, both at home and abroad.'

There was little validity in the prognostications, the Committee itself had met only a few times and talked to nobody in key positions within the military or civilian space establishments. Nevertheless, it did heighten the President's awareness of potential ills with the essentially Eisenhower-orientated space policy. But more than that, it firmly endorsed a plan evolved through meetings at Palm Beach late in December, where Kennedy,

Johnson and Kerr reversed a decision by Ted Sorenson to go along with Eisenhower's plan to drop the NASC. The National Aeronautics and Space Council had nearly gone under the previous August, blocked effectively by Johnson, who was now seen by Kennedy as his right hand man on space affairs.

Johnson was a man used to climbing higher up the political ladder and Kennedy knew that without a cause to champion, the Texan Vice President would become a potential thorn in the President's side. It was a safe niche to secure his limited power, so Johnson was to become chairman of the NASC in legislation that would change the original mandate. Whereas before, the President had first call on the NASC recommendations, now Johnson would be totally responsible for national space policy. It was a role he carved out three years earlier but was only now able to fill; yet it was but a prelude to the most dramatic recommendation he would ever make from that position, a decision at that time only weeks away.

Johnson's successor as chairman of the Senate Committee on Aeronautical and Space Sciences would be decided from between two prime candidates: Senators Kerr and Anderson. Clinton Anderson was divided between chairmanship of space or interior committees but a telephone call from Kerr telling him that Kennedy and Johnson wanted *his* presence as space committee chairman persuaded Anderson to back down. Under its new head, the Senate space committee was to report direct to Johnson, a procedure quite acceptable in White House administration but one which presented the Vice President as author and architect of plans and policies embracing both civilian and military space operations. As head of the NASC, Johnson would administer national space policy; as the Executive Office link-man with the Senate committee he would receive reports and reviews on the way those plans and policies were being orchestrated.

In fact, Kerr imposed a set of limitations on Johnson's intrusion; for instance when the latter proposed sitting on committee hearings Kerr made it clear that he would do so only as a guest divested of Senatorial power. But Johnson was pushing to increase his influence as NASC chairman with propositions that delayed legislation to re-structure the Space Council when Kennedy came to office in late January, 1961. Johnson proposed, in a draft order presented for Kennedy's signature, that he be given authority to supervise NASA and that all space documents and reports which would normally go to the President be routed to the Vice President's office instead. Kennedy received the draft and for a while put the whole issue into abeyance until he had dealt with other matters and come to considered decisions about not only the NASC but the space program in general.

For a while, Johnson's ploy was spiked, but only for two months. In that period, NASA got a new Administrator. T. Keith Glennan resigned his office as NASA's top man on 28 December, 1960, and vacated his office 19 January having prepared in the interval copious notes and materials for his successor. Deputy Administrator Dryden also handed in his resignation but that was not accepted and he stayed on as unappointed, unpaid, acting administrator. Kennedy had asked Johnson to select a new head for NASA but differences of opinion held back a decision until, on 25 January, Dryden prodded the White House by officially informing Wiesner, Kennedy's Presidential Science Adviser, that NASA expected to launch the Mercury-Redstone suborbital flight with chimpanzee Ham within the coming week and that the administration should be aware of this because of the publicity associated with the attempt. That day, Kennedy told newsmen that a new boss for NASA would be announced within the next few days.

Johnson's dilemma had been caused by a preference for a man who held politicial expertise, a view shared by Senator Kerr but one at variance with Jerome Wiesner, who desired a man already schooled in engineering administration. An alternative view, one openly expressed over the preceding weeks during the search in Washington for a suitable candidate, was that the scientific community with which NASA would increasingly have to deal with, would feel more at home with a scientist or an ex-university administrator. Johnson interviewed several candidates, probably as many as 26 although there is no written record now to confirm the precise number, before Kennedy personally intervened with a suggestion that the job be offered to General James Gavin, a prominent figure from the Army's research and development effort following Killian's dramatic warning about Soviet arms production, delivered in 1955.

Johnson was concerned that the public should identify this military giant with the ostensibly peaceful pursuits of the civilian agency and thereby assume that the new administration was erasing the lines of demarcation between NASA and Defense Department projects, successfully convincing the President of the need to think again. Several candidates approached for the job had rejected the offer because they felt they would be no more than caretakers responsible to Johnson himself, or because they felt the reductions imposed by the Eisenhower cutback in Fiscal Year 1962 money heralded a decline in NASA fortunes. Among the most notable candidates were Lloyd V. Berkner, chairman of the Space Science Board of the National Academy of Sciences, William Pickering, of the Jet Propulsion Laboratory, Trevor Gardner, Lee DuBridge, and General Draper.

Spurred to action by Kennedy's personal intervention, Senator Kerr persuaded Lyndon Johnson to put forward the name of James E. Webb, a friend of Kerr who failed to get a job with the Treasury in the new administration. Webb, born in 1906 in North Carolina, came to Washington to serve congressman Edward Pou from the House Rules Committee. For three years Webb had been director of the Budget Bureau under Truman and for a further three served as Undersecretary of State before heading a subsidiary company of the Kerr-McGee Corporation with heavy financial investments in oil. Webb was on McDonnell's board of directors and had a reputation for running a tight ship. From 1959 he had been heavily involved with the Municipal Manpower Commission, educational programs to promote high-school physics, an advisory center for foreign visitors, and several smaller public service programs.

As a candidate he was well placed, a personal friend of Berkner and a well known acquaintance of the President's science adviser Wiesner. Wiesner was approached by the Kerr-Johnson duo and asked to nominate Webb to the President. Wiesner did so, praising Webb's credentials, and was asked by Kennedy to invite Webb to Washington for talks. It was the end of that week, 27 January, when Webb received a message during a lunch in honour of Robert Kerr to go to the seat of federal government. He agreed to go to Washington that evening for a meeting with the President and the Vice President the following Monday, 30 January. After a weekend spent discussing the whole issue of space, NASA's likely role in the coming years, and the probable function of the Administrator under the new administration, Webb decided not to take the job and resolved to find a gracious way out.

Webb met Deputy Dryden, and then talked with Lyndon Johnson, who was completely unsympathetic with his views: Webb was the man recommended and Webb was the man he wanted. Webb left Johnson's office frustrated over his attempts to bow from the scene, and fell upon Clark Clifford whom he was advised by publisher Philip Graham to seek out. Clifford had been with Truman and was helping the administrative changeover from Republican to Democratic administrations. Clifford was no help at all, claiming that he had already recommended Webb for the post. Unlike earlier candidates, who felt that Johnson was the man nominated by Kennedy to run the space show while he, the President, got on with more important and prestigious matters, Webb reasoned that although Johnson was prominent in space affairs, the President would ultimately be the man to whom the NASA Administrator would be responsible. This, decided Webb, warranted a personal invitation to head the space agency from the President himself and this was his attitude when, during the afternoon hours of 30 January, he met with Kennedy at the White House.

Kennedy affirmed the need for a strong policy-maker and took Webb somewhat by surprise when he asserted a desire for the new Administrator to put together the bones of a new space program, not a left-over legacy from Eisenhower, also stating that Dryden would be kept on as Deputy Administrator. Webb accepted the post and an official announcement was prepared for release that very afternoon. Webb relinquished all business and professional relationships that could in any way influence his new position, was heard as nominee before Senate space committee hearings 2 February, and received the full approval and confirmation of the Senate exactly one week later. He was sworn to office on 14 February and went straight to NASA Headquarters to start work, his most important task being 'to review all the programs of the agency and to make his recommendations.'

Kennedy charged the new Administrator to establish a level of confidence as an effective counter to rumor, emanating from

NASA Administrator James E. Webb; the man without whose leadership and fire Americans would never have reached the Moon.

discussion of the two Hornig and Wiesner anti-Mercury committees, that manned space flight was being run down, requesting in addition submission of new budget proposals for the President to take before Congress. One of Webb's first tasks was to settle once and for all the continuing competition, that flared up whenever opportunity arose, for lead agency on national space projects. With an eye on the manned space flight potential, the Air Force seized hold of this issue during the months preceding the election in 1960, seeking to influence the candidates, to pry open further the suspicious view held of NASA by the Eisenhower administration, and to cement a relationship with the new administration that would pay dividends when new policy decisions concerning space were made, as everyone knew they would be.

The Air Force was already concerned that although it retained authority for military programs, NASA's space budget outpaced that of the Air Force in Fiscal Year 1961. When Kennedy won the 8 November election, the Air Force prepared a memorandum which the Secretary circulated to commanders and industrial suppliers on 1 December, affirming that in view of the Democratic party's stance on space during the run-up to the election there was now 'a realization at the highest levels that military supremacy in space is as essential to our security as military supremacy at altitudes near Earth.' It was presumptuous to say the least. The document went on to lay claim to prime reasons for being in the vanguard of America's major space commitments and was clearly intended to influence the new faces going to Washington more than being a confirmation of new direction to the Air Force personnel among whom it was officially circulated.

Aerospace journals, backed by service interests and eager courtiers at the palace of industrial trade, lent their support to the Air Force call for a stiffer resolve toward the militarization of space. To many, particularly in industry, the big ideas held by the Air Force gave more promise of large contracts and new exploitation than did the quasi-scientific, seemingly low-key, pursuits of the civilian space agency. *Aviation Week* magazine assured its readers that the Air Force space plans were 'extremely imaginative and broad scope' proposals that if approved would substantially increase the effective value of the 'Air Force space system program.' For their part, the Air Force did little officially to influence the efforts of space-orientated segments of the service, retaining a non-committal approach that excluded bureaucrats and administrators.

Several reasons emerged to seal the fate of Air Force plans for a more vigorous space program of its own. One of these cut away the principle point of contention around which so much discontent had revolved when new Secretary of Defense Robert McNamara gathered all Defense Department space programs into Air Force aegis from 6 March, 1961. A second reason, one that sucked support from the Air Force across to NASA, came from Congressman Overton Brooks, chairman of the House space committee, an enthusiastic body that had for some time been concerned with the lack of impetus behind NASA programs. Brooks sent a three-page letter to President Kennedy expressing deep concern over the 'persistency and strength of implications reaching me to the effect that a radical change in our national space policy is contemplated within some areas of the executive branch. In essence it is implied that United States policy should be revised to accentuate the military uses of space at the expense of civilian and peaceful uses.'

Brooks went on to say that he could not 'ignore the suggestion, implicit in the unabridged version of the Wiesner report, that the National Aeronautics and Space Administration role in space is purely one of scientific research and that the military role in the development of space systems will be predominant.' In pressing home an implied contravention of the 1958 Act, Brooks reminded the President that any such change of emphasis would 'disregard the spirit of the law.' Furthermore, he stated, 'if NASA's role is in any way diminished . . . it seems unlikely to me that we shall ever overtake our Soviet competition.' Mindful of the contention from Air Force circles that a civilian space program would inevitably lack strength and positive effect, Rep. Brooks believed the NASA program had been 'peculiarly effective because of its public emphasis on scientific and peaceful uses.'

In answer to Brooks' letter, President Kennedy assured the Congressman that 'It is not now, nor has it ever been my intention to subordinate the activities in space of the National Aeronautics and Space Administration to those of the Department of Defense.' But how much that 'intention' had been nullified by a successful attempt from James Webb to weld, rather than separate, the two sides is unclear and must remain so. On 23 February, Deputy Defense Secretary Roswell Gilpatric signed an agreement with NASA about national launch vehicle development programs, resolving to have joint discussions prior to embarking on the development of a new launch system. A day later, Webb had Dryden with him at a meeting attended by McNamara, Gilpatric and Herbert York, Director of Defense Research and Engineering. Approval was stamped all over the agreement and this high level dialogue effectively sealed the fate of Air Force pressure to give more emphasis to the military uses of space.

Four days after this agreement was signed, on 27 February, Webb was made aware of the findings of the Space Science Board chaired by Berkner that had met on the 10th and 11th to consider its posture on manned space flight. The Board would carry a great deal of influence with the new administration. Formed in 1958 several months before NASA officially started business, the arm of the National Academy of Sciences had some of the most influential and respected scientists among its number. It would be a valued aid to Kennedy, and Webb knew this. Berkner and Webb were very good friends and the new NASA Administrator knew he would stand a better chance of getting approval for an expanded budget, the details of which he had already been instructed to work out, if an official body like the Space Science Board presented a favorable report; there had, after all, been two exceptionally gloomy committee reports already. There was no subversive attempt here at deliberately favoring a specific view. Both men knew their roles, and where their loyalties lay.

Prime architects of NASA's budget amendment were Webb, Dryden and Associate Administrator Robert C. Seamans, who met with Budget Bureau chief David Bell two days after Webb came to office and learned that the BoB favored a cautious approach to changes in the Fiscal Year 1962 request, a financial period due to begin 1 July, 1961. Webb received a briefing on NASA's program efforts on 24 February, when he discussed with Dryden and Seamans progress on the Saturn launch vehicle studies, conceptual designs for the massive Nova rocket, proposed schedules and development paths for Apollo – the circumlunar/Earth orbit laboratory successor to Mercury – and integration of existing projects with future goals, in particular the expectation of manned lunar landings after 1970. During a two-day session beginning 9 March, Webb met and discussed future projects with heads from all the NASA space centers and on 15 March convened a meeting with senior management at Headquarters.

Webb not only endorsed the planning objectives of the previous year, where NASA had requested $1,376 million during negotiations with the Budget Bureau for Fiscal Year 1962 – subsequently reduced to $1,109.63 million before presentation as the official Eisenhower submission – but actually approved an accelerated program beyond that envisaged by NASA when drawing up its earlier requirements. Webb served notice on the nation 16 and 17 March that he was pursuing a radical reassessment of space goals and, in speeches designed to announce publicly the mood of the new look at NASA, reminded the administration that space projects could no longer be considered for the limited value of their technological worth but rather within the framework of 'broad national and international goals'. Excited to action by their bold leader, senior management testified before Congress that the program actually approved by Eisenhower was not one in which they concurred and that they too would personally wish to see a more vigorous approach to substantive goals.

On 17 March, the Budget Bureau received Webb's Fiscal Year 1962 amendments totalling $308.191 million, an increase of nearly 28% on the final sum allowed by the Bureau under the previous administration and a total value that was now nearly $42 million above the figure NASA had originally asked for permission to request. It was too high, even for the 'new look' Bureau, which promptly recommended cutting the amendment to a mere $50 million, still giving the agency 4½% more than it would have got under the Eisenhower administration. Large increases had been requested for development work on the Saturn C-2, the Apollo design configuration, and the mighty F-1 engine; the cut to just a $50 million increase precluded major work on any of these projects, Budget Bureau's excuse being that adoption of a post-Mercury spacecraft project implied a national commitment that the President had given no clear indication he was willing to fund.

Webb recognized in this an attempt to do a similar cutting job that the agency had received under Glennan and pressed Budget Bureau Director Bell for a hearing before the President. Bell responded that he did not think the President had the time to bother with new policy moves at that time, but Webb retorted that Kennedy would have to review the situation quick 'whether he likes it or not!' Bell reasoned that NASA could always come back in the Fiscal Year 1963 budget for additional increases to advanced projects; Webb realized only too well that if he lost momentum that year, the agency would stand little chance of a smooth transition from Mercury-type endeavors to the substantially more advanced concept of circumlunar flight, with all that implied regarding new boosters and spacecraft. Consequently, Webb arranged with the White House to meet the President on 22 March.

Bell's reference to the current schedule regarding the President's time had a certain credence in that at this period Kennedy was facing the first major international crisis of his term. Pro American forces in Laos were close to defeat at the hands of the Communist Pathet Lao and on 20 March Kennedy had the first of several National Security Council meetings to decide what to do. The President knew that escalation was inevitable once ground and air forces were committed and received advice from the Joint Chiefs of Staff that implied a military recommendation for major displays of force. Kennedy was advised, as his predecessor had been over Korea, that military success, and a rapid solution, could only be obtained with maximum assault operations backed up with tactical nuclear weapons. On the day before the scheduled meeting with Bell and Webb, Kennedy put American forces in Asia on alert, sent the Seventh Fleet in to the Gulf of Siam, and arranged to make a public announcement 23 March that would warn the Russians to arrange a cease-fire or watch as US troops went in. Reluctant to go as far as his Joint Chiefs recommended, Kennedy was, nevertheless, mindful of the nature of this confrontation: he knew the Soviets were only too pleased to see the youthful American President under stress and international pressure.

On the evening that Kennedy first met the National Security Council over the Laotian problem, Jim Webb sat in his Washington office preparing a justification of the full budget amendment increase. It would be Webb's first showing before the executive, Kennedy's first decision-making session on space, and Johnson's first role as potential mediator between probable contestants. By this time Kennedy had been alerted to pending legislation on the National Aeronautics and Space Council, held in limbo from the day in January when Johnson prepared a memo Kennedy chose to ignore which, if signed by the President, would have delegated authority for national space policy to the Vice President. On 21 March Edward Welsh was appointed executive secretary of the Space Council and provided the NASC with a useful pool of experience, tuned to the President's thoughts during pre-election campaign statements he had prepared for Kennedy. Kennedy was keen to see the NASC become more than just the 'box on the organizational chart as it has been heretofore'. Welsh would be primarily responsible for drafting the legislation designed to restructure the NASC with the Vice President at the head. By introducing Welsh at this juncture, Kennedy was sure that he could safely leave the legal language to a source that had more than average allegiance to the Presidential seat.

On 22 March, the NASA triumvirate of Webb, Dryden and Seamans went with David Bell from the Budget Bureau, his deputy Elmer Staats, and Willis Shapley, the man at BoB responsible for NASA negotiations, to brief Johnson and Edward Welsh at the White House on the new amendment. The Vice President was much impressed with the potential planning opportunities available to the government. After some time, the group was joined by Kennedy, McGeorge Bundy, Wiesner and Atomic Energy Commission chairman Glenn Seaborg. Two principal discussion points were central to the main issue of just how much the new administration would consider allowing NASA to inflate beyond the Eisenhower mandate.

One concerned the pace at which America could eliminate the enormous Soviet lead in booster lifting capacity; the second related to the most desirable post-Mercury manned space flight policy. At this time, of course, Mercury was recovering from tests that plunged STG hopes during the final months of 1960 and had yet to make the MR-3 suborbital flight. NASA Associate Administrator Seamans took center-stage and briefed the President on the program envisaged when it drew up a recommended budget amendment of more than $308 million, the figure Budget Bureau Director Bell had tentatively cut to $50 million causing contention that resulted in the meeting. Seamans outlined a moderately paced development in both propulsion and manned space flight capability. By moving the development of Saturn C-2 up a notch, a first flight could be expected in 1967, two years earlier than feasible under existing plans. Soviet boosters were already capable of lifting weights exceeding 4 tonnes, the Saturn C-1 was even then being groomed for a first flight later that year but under the Eisenhower profile it would be a further eight years before NASA had a booster capable of lifting an Apollo circumlunar spacecraft beyond Earth orbit. By that time, it was assumed, Soviet lifting capacity would be enormous and a program that reduced the delay in C-2 development from eight to six years seemed a conservative balance between priorities.

Moreover, said Seamans, NASA wanted to press ahead with F-1 engine development, accelerating the availability of a Nova-class booster – one capable, if called upon, of landing a manned spacecraft on the Moon – to 1970 instead of the existing projection of 1973, fully twelve years ahead. For post-Nova colonization and manned flight to the planets, a more efficient propulsion system could be employed and for this task Seamans told the President about conceptual studies on nuclear propulsion as an upper stage to the chemically-propelled Nova lower stages. This would require considerable 'lead time' in which to bring to simultaneous maturity the technology of nuclear rocket motors and the development of boosters with large lifting capacity sufficient to warrant use of such a device. For the time being, however, nuclear engine development would be a low key operation.

In the operations sector, Seamans said the NASA proposals envisaged a modest acceleration in the preparation of the Apollo spacecraft concept expected to follow Mercury manned orbital flights. Already the subject of six-month industrial feasibility contracts with three aerospace companies, Apollo should be ready by 1965 instead of 1967. This reasoning, said Seamans, was justified by the need to flight test the hardware in Earth orbit by first having it launched on Saturn C-1 boosters before a full operational capability emerged with the Saturn C-2 in 1967, a launcher capable of sending Apollo round the Moon and back to Earth if necessary. Dryden followed Seamans in presenting a rationale for Apollo development. It could he said, be groomed for two-week, Earth-orbiting, laboratory flights, or be made an integral part of a grander Moon landing objective, a goal which he said would be possible *by* 1970 rather than *after* 1970.

Webb backed up the confident assertions from his colleagues that the type of program they were recommending would 'permit a steady closing of the gap caused by Russian successes.' He called to mind the 'procrastination for a number of years' that had been 'based on a very real skepticism by President Eisenhower personally' of the goals and objectives in space programs of this sort. Pulling no punches he stiffened his criticism of the days under Eisenhower where 'His decision emasculated the ten year plan, before it was one year old, and, unless reversed, guarantees that the Russians will, for the next five to ten years, beat us to every spectacular exploratory flight.' Going on to list a depressing catalogue of Soviet achievements he said that 'We have already felt the effects of the fact that they were the first to place a satellite in orbit, have intercepted the moon, photographed the back side of the moon, and have sent a large space craft to Venus. They can now orbit $7\frac{1}{2}$ tonne vehicles about the Earth, compared to our $2\frac{1}{2}$ tonnes, and they have successfully recovered animals from orbital flights lasting as much as 24 hours. Their present position is one from which further substantial accomplishments can be expected, and our best information points to a steadily increasing pace of successful effort, on a realistic timetable.'

Of the implications for the future funding profiles an expanded space program would incur, Webb said that 'The future effect of our recommendations will be to increase expenditures to an annual rate of $2,000 million by 1965 or 1966,' from the $1,417 million it wanted for Fiscal Year 1962. Appealing to Kennedy's sense of internationalism and his concept of 'New Frontier' politics, Webb reminded the President that 'The United States space program has already become a positive force in bringing together scientists and engineers of many countries in a wide variety of cooperative endeavors. Ten nations . . . all have in one way or another taken action or expressed their will to become part of this imaginative effort. We feel there is no better means to reinforce our old alliances and build new ones . . . the extent to which we, as a nation, pioneering on a new frontier, will be in a position to develop this emerging world force and make it the basis for new concepts and applications . . . for nations willing to work with us in the years ahead.'

The argument was strong and convincing, falling directly into line with Kennedy's personal concept of ensuring freedom of the west through political, economic, and technological strength, cementing existing friendships in the process. The President was a vocal participant at that 22 March meeting in the Cabinet Room, plying NASA with questions and obtaining in the process the first full-scale briefing he had received on the nation's space program. The representatives from the Budget Bureau were vocal too in expressing their opposition to the proposals. Next day, Kennedy again called Johnson, Welsh, Wiesner and BoB Director Bell to discuss, this time in NASA's absence, the possible state of the administration's support. Kennedy was unsure of his posture on manned space flight simply because it was to him a very new technology and a potential money spender he was not yet willing to equate with acceptable returns. He needed more time in which to evaluate the context of such an undertaking within a framework of national policy decisions, a formula he was only just beginning to put together. In any event, the two reports, one by Hornig and one by Wiesner, were highly critical of NASA's Mercury program and the President was keen to see a few manned flights before going all out on an expensive follow-on program.

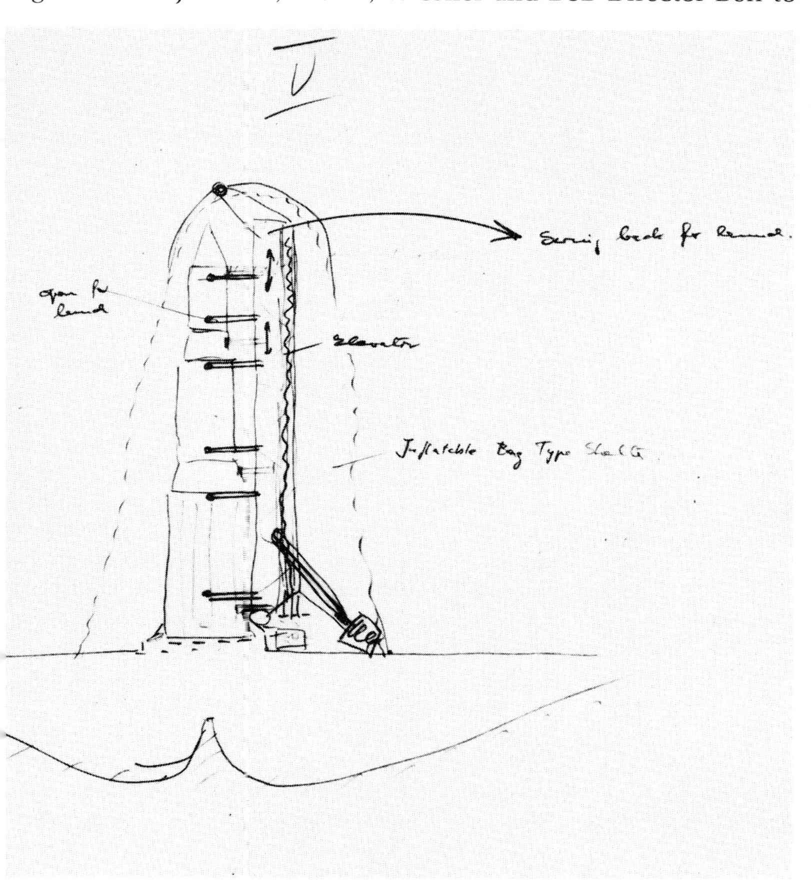

Two months before the public Moon commitment, engineers at Cape Canaveral sketched ideas on how to erect and service massive space boosters.

Of the booster situation he was more sure. The argument in favor of an accelerated Saturn schedule was undeniably watertight and Kennedy decided at this second meeting to approve an amended allocation of $56 million for Saturn, $25.6 million for the Centaur high-energy upper stage, $9.32 million for advanced liquid propulsion systems (essentially the F-1 with a thrust of 680 tonnes), $4 million for the nuclear propulsion concept, and $17 million for new facilities at the Cape Canaveral Atlantic Missile Range. In all, the Fiscal Year 1962 budget amendment added $125.67 million to the $1,109.63 million proposed by the outgoing administration, for a total of $1,235.3 million. With regard to the manned space flight proposal, Kennedy decided that he would await the suborbital and orbital flights of Mercury, both of which were at that time expected before the end of the year, prior to considering future plans for a follow-on project. Consequently, no funds were allowed for research and development except money already being spent for in-house studies and the limited contractor activity already approved.

In this way, the agency would be in an appropriate posture to request again in Fiscal Year 1963 sums that could lead to a re-evaluation of the Apollo project. The 1963 budget would be negotiated with the Budget Bureau during the closing months of 1961 and the President would formally present to Congress the respective sums he was approving for consideration in first the House and then the Senate, reviews that would get under way in the spring of 1962 prior to the commencement of Fiscal Year 1963 in July, 1962. In certain respects the President was dealing with unknown quantities: Webb was new to Washington's Democratic assemblage; Kennedy had yet to define the man's frame of reference; the President was still feeling his way into matters of space and high technology. By agreeing to accelerate booster development, the President secured loose ends that threatened to need the longest lead time and preserved his options on other space matters for consideration at a later time. In total the March amendment was an increase of 11% on Eisenhower's January package and would have represented the definitive posture on space policy as applied by the Kennedy administration in its dealings with the Fiscal Year 1962 budget.

As President, Kennedy had responsibilities that moderated his view expressed during election debates. While recognizing the need for a strong and independent United States, he remained aloof from suggestions that America should be overtly isolationist in its thinking. The balanced view he held that March when deciding which elements of the space program needed more money and which should wait for a later evaluation is best summed up in two sentences from his State of the Union address when taking office in January: 'Today this country is ahead in science and technology of space, while the Soviet Union is ahead in the capacity to lift large vehicles into orbit. Both nations would help themselves as well as other nations by removing these endeavors from the bitter and wasteful competition of the cold war.' Kennedy, therefore, saw the entire space effort as a tool for political success in achieving parity with Soviet endeavors aimed at recruiting support from the uncommitted nations of the world. Because of this he was incapable of evaluating the true value of space flight, unable to address technical decision factors as reflected in budget appropriations, and unwilling to invest human and financial resources in research and development that would take several years to show a return.

NASA was content for a while to see Fiscal Year 1962 as a period of preparation, a time to marshal its resources for a convincing assault on Congress the following year to justify the future proposals for manned flight. The amended budget went before Congress on 28 March as only one small part of an extra $3,200 million for civilian and domestic programs over and above the existing recommendations from the Eisenhower submission.

The events of the next two months were to dramatically change the decision on national space programs that seemed, by month's end, to have settled the questions raised about the new administration's policy but which in fact were merely a prelude to new commitments. To appreciate fully the delicate web of circumstance that so purposefully influenced the President, it is necessary to re-visit the events associated with two activities already discussed in this chapter and the last, and to add a third, political, crisis that certainly had a peripheral effect: the surprise orbital flight of Yuri Gagarin aboard Vostok I, the successful suborbital flight of Alan Shepard in Mercury-Redstone 3, and an incident generally known as the Bay of Pigs fiasco.

Three days after Congress got Kennedy's amended Fiscal Year 1962 budget proposal, preparatory to a lengthy process of hearings and debate that characterize the spring and early summer of each year's activity on Capitol Hill, Lloyd Berkner officially released the recommendations of the Space Science Board committee set up early February to consider manned space flight. Berkner had sent Webb a summary of the Board's conclusions on 27 February, but not before 31 March did the NASA Administrator receive a copy of the report, as did Herbert York, Director of Defence Research and Engineering, Wiesner, and Alan Waterman, the National Science Foundation Director. In what seemed a complete reversal of existing scientific thought on issues connected with manned versus unmanned space flight, the Space Science Board determined that 'planning for the scientific exploration of the Moon and planets must at once be developed on the premise that man will be included. Failure to adopt and develop our national program on this premise will inevitably prevent man's inclusion.'

Moreover, the Board judged that for maximum return of scientific information, 'man's participation in the exploration of the Moon and the planets will be essential.' Board members included physicists Harold Urey and James Van Allen in addition to Donald Hornig who had prepared findings from his own 'Ad Hoc Panel on Man-In-Space.' It was a somewhat unexpected response from the august National Academy of Sciences, that held distinction as a body responsive to the most moderate and conservative evaluation of science and the technology that science would use to achieve its objectives. The Space Science Board also appealed to the grander and more broadly based concepts of destiny and celestial responsibility when it affirmed that 'members of the Board as individuals regard man's exploration of the Moon and planets as potentially the greatest inspirational venture of this century and one in which the whole world can share; inherent here are great and fundamental philosophical and spiritual values which find a response in man's questing spirit and his intellectual self-realization.'

It was all stirring stuff, and somewhat out of character for a Board that carried distinguished members unused to emotional campaigning and flowery rhetoric. But it did what Lloyd Berkner and Jim Webb wanted in that it played a major role in determining the attitude of the President's Science Advisory Committee; PSAC looked with respect at anything the National Academy of Sciences had to say and at a time when no one was sure how he *should* or was *expected* to feel, positive direction could, and did, carry the day with the background of advisers and analysts the President would be duty-bound to call upon when next considering major decisions on space. That effectively set the scene for the three outstanding events of April and May outlined earlier. The first came when Gagarin orbited the Earth in the morning hours of 12 April.

Few in Washington doubted the imminence of this first manned space flight; Wiesner informed the President on the evening of the 11th that he could expect to hear that night that a Russian was orbiting the Earth, and Press Secretary Pierre Salinger had already drafted a statement for the President to issue when the event occurred. Many in America felt the impact of Vostok I more than they had that of Sputnik 1 three and one-half years before. Reconciled to having been beaten at a game in which they were unwitting participants, the public diluted total shock and horror by accepting that it may happen once but surely not twice, and if it did, certainly not with a break of several years between during which period America would be aware of the stakes. Throughout the world it seemed to indicate that US technology was running to catch up with a giant 20th century machine ambling on to greater and more impressive achievement at modest pace but with greater stride.

NASA officials were cautious to point out that it was none of their doing that the Russians were in front, indicating to Congress that displeasure over the slow pace of Eisenhower's response was not totally compensated by the partial commitment of the incumbent administration. Gagarin's mission came when the House was discussing NASA's 1962 budget and so found an easy vehicle within which committees and hearings were primed for space talk. Representative Daddario urged that America acquire a new 'sense of dedication to the goal of being first in space,' and Rep. King spoke of a 'lump of lead in my heart' that morning and the thought that no one remembered 'the name of the second man to fly non-stop from New York to Paris' meant that America would be excluded from a major chapter in the annals of history. Rep. Randall said he felt that the vast majority of Americans were 'not simply shocked and awed, they were literally stunned,' by the suddenness of the flight (*sic*) and were 'downright hurt,' while Rep. Dodd believed Soviet space achievements to be a series of 'devastating humiliations' for the United States.

In an urgent appeal for a champion to rise and lead America in renewed technological vigor, Rep. Fulton implored Webb to 'Tell us how much money you need and we on this committee will authorize all you need. I am tired of being second to the Soviet Union. I want to be first.' For his part, House space committee chairman Overton Brooks believed it imperative 'to make a determination that we are going to be first in the future, if we continue our space program.' The mood in Congress was infectious and conducive to rash decisions. Rep. Fulton became alarmingly concerned and frustrated asserting that 'I would work the scientists around the clock and stop some of this . . . scientific business.' Next day, in a continuation of hearings held in a state of fever that gripped Washington for days, Rep. Anfuso gave warning that 'I want to see our country mobilized to a wartime basis because we are at war. I want to see our schedules cut in half. I want to see what NASA says it is going to do in 10 years done in 5,' because to get America back in the race 'I want to see some first coming out of NASA, such as the landing on the moon . . .'

Amid this frenzied hysteria, President Kennedy maintained a similar posture to that which had characterized his earlier attitude, reflecting a concerned but unflustered response. In a statement he made on the day Gagarin orbited the Earth, Kennedy came so close to an earlier statement – one that this author has used already in this chapter to typify his innermost feeling about space ('Both nations would help themselves as well as other nations by removing these endeavors from the bitter and wasteful competition of the cold war') – that it stands as evidence regarding the President's unruffled approach to this most severe national challenge: 'It is my sincere desire that in the continuing quest for knowledge of outer space our nations can work together.' Also that day, Kennedy frankly admitted that 'no one is more tired than I am' of seeing America come second but 'it is a fact that it's going to take some time' to build up an efficient counter program and 'As I've said in the State of the Union message, the news will be worse before it is better, and it will be some time before we catch up.'

As if to consolidate the interpretation that Kennedy felt unchanged about his basic attitudes toward space he also said that 'We are, I hope, going to go in other areas where we can be first and which will bring more long-range benefits to mankind.' Again, this is in line with his more considered statement made 2 months before when addressing Congress on the proposed legislation anticipated by his administration. Then the press had their say and in a lead article the *Washington Post* said that Gagarin's flight was 'a psychological victory of the first magnitude for the Soviet Union,' and that although as a single event it proved nothing about Russia's technical developments in wider field 'what people believe is as important as the actual facts, and many persons will of course take this event as new evidence of Soviet superiority.'

Moreover, said the paper, 'The general excitement from Europe to Asia, Africa and the Americas will not be diminished by the recognition that no immediate military, commercial or other actual advantage accrues to the Soviet Union.'

As if teasing Kennedy into action, the *New York Times* retorted that 'the President's image has been beset by the difficulties he has had with Congress, by his failure to spell out the promised "sacrifices" to be required of the American people and by the continued recession,' and that it had, perhaps, been premature for the President to 'present himself as an image of a young, active, and vigorous leader of a strong and advancing nation.' The *Times* added dimension to the *Washington Post*'s recognition of public opinion being somewhat distorted on a mass scale by single events of great magnitude when it said that 'Even though the United States is still the strongest military power and leads in many aspects of the space race, the world – impressed by the spectacular Soviet firsts – believes we lag militarily and technologically.' And as if in rebuttal of Webb's confident assessment of the existing NASA program when called to the White House for the 22 March budget meeting, the *Times* reminded its readers that 'The neutral nations may come to believe the wave of the future is Russian; even our friends and allies could slough away. The deterrent, which after all is only as strong as Premier Kruschev believes it is, could be weakened.'

Two days after Vostok I, NASA's Robert Seamans appeared before the House committee to discuss at a previously scheduled meeting the plans for Fiscal Year 1962. Rep. King was succinct in his summary that 'In our race for the exploration of space there are three major breakthroughs or dramatic successes: the first satellite, the first man in space, and the first man to go to the Moon and back. The score is two to nothing, favour the Russians. We still have the third prize to obtain. I think the third is probably worth more than the first two together. So we are still in the race. But I would like to know specifically, of the plans that we now have programmed . . . are such as will enable us to reach the Moon ahead of the Russians.' To this, Seamans explained that a lunar flight to the surface and back using a Nova rocket might be possible in 1969 or 1970 and that this estimation depended upon authorization and approval by Congress of the amended Fiscal Year 1962 budget put up by Kennedy.

Rep. King was anxious to press on with such an undertaking and asked Seamans if it would be possible to accelerate the pace still further to achieve a landing in 1967. The significance of that particular year was profound for a consideration of Soviet plans; as the 50th anniversary of the October 1917 revolution, it was a logical target for Soviet extravaganzas. Seamans replied that 'This is really a very major undertaking. To compress the program by three years means that greatly increased funding would be required . . . I certainly cannot state that this is an impossible objective. It comes down to a matter of national policy. I would be the first to review it wholeheartedly and see what it would take to do the job. My estimate at this moment is that the goal may very well be achievable.'

Rep. Chenowith then asked Seamans whether or not he thought such a major commitment was rightfully appropriate to consider in line with other national priorities or whether it was not perhaps a little too ambitious or grandiose for the international mood. Seamans said he felt it was a decision that should be made by the people of the United States through their appointed representatives in Congress and there then began a mini-debate on the responsibilities of Congress to take upon themselves awesome decision-making tasks about issues on which the general public could not be expected to decide. In any event, he said, a crash program aimed at getting a manned Moon landing under way by 1967 would require annual NASA budgets as high as $5,000 million versus the anticipated $2,000 million a year envisaged by the plan presented to Kennedy on 22 March. Either way it was a discussion that echoed similar conversation shortly to begin at the White House.

The President had called Webb and Dryden to the Cabinet Room on the evening of the 14th to meet with Ted Sorenson, David Bell and Wiesner before Kennedy joined them to discuss the significance of the Vostok I mission. Kennedy's attitude had changed considerably in the two days since his public statements immediately after Gagarin performed his feat. With an impatience that bordered on irritability, the President cut small talk and got straight to the issue in question: 'Now let's look at this. Is there any place we can catch them? What can we do? Can we go around the Moon before them? Can we put a man on the Moon before them? Can we leapfrog?' Dryden spoke up and reaffirmed a belief in all-out technological confrontation with the Soviets he had held for at least the previous three years, warning that although such a project, similar, he felt, in several respects to the Manhattan project for the first atomic bomb, might cost up to $40,000 million in the end, it would be the only way America could stand a fair chance of beating the Russians but that even this course held no built-in promise of success. Webb recognized that bitterness could set in if Kennedy sensed NASA viewed his earlier decision to defer a commitment to Apollo as cause for the Soviet's continuing lead. In typical diplomatic style, deploying tactics few would ever detect beneath a usually brusque and firm exterior, Webb reminded the President that the space agency was doing what it could and that they had faith in Kennedy's leadership which was, said Webb, already responsible for NASA 'moving ahead now more rapidly than ever.'

But the President was concerned about the cost and turning to Bell he voiced this worry: 'The cost. That's what gets me.' Bell was not so much concerned about the money. In fact, in the face of a depressed economy the injection of public money into a high employment program might be just what the nation needed, but to Kennedy, uncommitted to space for its own sake, the cost was alarmingly high. Wiesner had the cautious approach typical of the man, a character who more than once favored the traditional role of his office in playing devil's advocate in the face of promised glory from so many promises around the President. But Kennedy was in no mood for vacillation and, turning again to the group of men before him, appealed for more information: 'When we know more, I can decide if it's worth it or not. If somebody can just tell me how to catch up.' And as if to leave the participants in no doubt about the seriousness with which he viewed the whole issue: 'There's nothing more important.'

Webb, Dryden, Bell and Wiesner left the room and Kennedy stayed on to discuss with Sorenson the possible repercussions of a step up in space activity. Kennedy was well aware that public opinion around the world in the wake of Gagarin's flight adversely affected the hitherto influential position of the United States. In England, wide acclaim was accompanied by renewed anti-American protests, although feeling in Great Britain was already running high over defense issues and military purchases that many felt should more rightly have gone to internal UK procurement (this would enhance any pro-space decision the President came to, however). In France and Italy the press were favorably aligned with Russia's triumph and on the streets people seemed even more favorably disposed toward Vostok than they had been to Sputnik 1. In 1957, America was temporarily toppled from its prestigious position in Europe by the satellite that left many ordinary people confused as to Soviet intentions. But after more than 3 years during which Soviet reaction to world events seemed, at a public level, no more hostile than before, the man in the street was not as prone to suspect Russia's continued successes.

When Vostok sent a man around the Earth, Europeans in general were willing to accept that Russia was claiming rightful distinction for remarkable industrial progress. These reports must have weighed heavy on Kennedy's mind that evening he sat with Sorenson two days after the first manned flight. They were fate-filled hours in which the President struggled to find a way out of the inevitable path he recognized even at this time as the only one he could follow. As if debating with his other self, he turned to advisers more for a reflected communication with views that wrestled with his own than for totally new and unique solutions he knew would not be forthcoming. Two pivotal points about which he could turn seemed to conspire in presenting a fixed response: whether to react at all to Gagarin's flight, and, if inevitable, what to do in answer that would at best not worsen his international position.

He was unable to find any way of ignoring the flight; it had happened and headlines were acclaiming that fact all across the world. As for a suitable answer, that too was inevitable: a similar endeavor of such consummating enormity that the Russians would be totally eclipsed to the advantage of the United States. And so it came back full circle because that was the dilemma: he didn't approve of a major space program at a time when an equivalent sum of money would, he thought, do more for the population. At this point Kennedy was still acting within the rational framework that typified his reaction to matters of space programs all along. Reluctant to act when further study might reveal some way in which he could avert the inevitable, Kennedy sought opinion from literally hundreds of people. But decision was stymied, Kennedy had all the information he would ever need on 14 April, and other circumstances essential to jolting him to a decision would have inevitably stayed the day had they not happened at all. But they did happen and within a week of the Soviet flight Kennedy had been moved to act.

In the days after 14 April, the meeting that confirmed again to the President the availability of technical options within the framework of an expanded space program, Kennedy raised the issue with House appropriations committee chairman Albert Thomas and obtained a heartening response. If he, the President, executed bold authority over the decision-making process the country had elected him to possess, Congress would not only support this single act of nationalism but look with favor on other measures he was likely to want their support with. This was a peripheral issue, although probably more encouraging than agreeable noises from the Space Science Board. By buying in to Congressional favor, he could secure Congressional support so far lacking. Most of the Kennedy camp followers that came to Washington in January, 1961, were young bloods keen to show the old Washington hands how to run a nation. It had been a dispiriting scene as one by one enthusiastic proposals for radical legislation were divested of their energy on the solid door of traditionalism. If in answer to the Soviet challenge a US space commitment could purchase smooth passage in the Congress, the returns on an investment of that sort would be worth consideration.

From Lyndon Johnson, set to the task of sounding Senate attitudes, came affirmation in the form of Sam Rayburn, House Speaker and a loyal friend of LBJ, that the Johnson clique would similarly look with due objectivity at Kennedy's proposals. Assured of the House and now the Senate, the latter an institution that so recently had Johnson as Majority Leader, Kennedy was gathering a creditable following to ensure that, should he decide to move, any decision he made would not run into unexpected opposition. It really only needed a catalyst and that came in the form of a clandestine invasion of Cuba. The sequence of events and the timed relationship of one to another was very important to the action Kennedy was to take. Gagarin's flight of Wednesday, 12 April, had been followed by raging critiques of the space program in the press the following day. The meeting at the White House came on Friday the 14th at which time the President made it clear he was committed to a reciprocal move.

Saturday, 15 April, just three days after the Vostok flight, American aircraft bombed and strafed airfields of Fidel Castro's left-wing forces. Castro had emerged as Cuba's pro-communist leader in January 1959 when he led a guerilla movement that seized power and quarrelled with the United States. Backed by America's Central Intelligence Agency, Cuban exiles trained to fight modern tactical war, prepared to invade the island with funds and equipment provided by the CIA. With possible air power eliminated by US attacks on the Saturday, the invasion force landed in the Bay of Pigs on Monday, 17 April, running headlong into Castro's defense forces forewarned of the impending assault. It quickly became apparent that the exiles, supported by American 'advisers,' would not last long before the Castro forces overwhelmed them on the beach.

By Tuesday night, Kennedy was in a state of abject depression and sought advice at the White House on what to do. Faced with a situation similar to that he had backed away from in Laos the previous month – also a time when space matters weighed heavily on his mind – the President could see no alternative but to withdraw official support. For much of that night and the early morning hours of Wednesday the 19th, Kennedy pondered on the possible repercussions of backing up the failing assault with a strong Marine presence at the Bay of Pigs supported by massive naval firepower from the sea. The arguments were legion and none of the answers could be bought without price. If dubbed a fiasco, Kennedy knew it would lead to bitter attacks on his administration at home and abroad; if he backed up the clandestine operation by openly declaring support for the exiles, Russian intervention was all but inevitable – and Cuba was only 250 km from the continental United States. He had well and truly slid into a cleft from which there was no return but back the way he had got in. Consequently, the President ordered an evacuation with an attempt to rescue as many men as possible.

For the past several days now, Kennedy had been seeking further comment on plans for a major increase in the space program, seeking a means by which he could exploit America's scientific and technical base in eclipsing Soviet successes. The previous Friday he had talked at length with Sorenson about possible projects and there was nothing in the next four days that led him to a specific project. Suddenly, in the wake of three terrible days during which Kennedy was observed by a White House official to be in a period of 'somber stocktaking,' Lyndon Johnson was called to the President's office on Wednesday the 19th and asked to provide recommendations for an accelerated space programme. In doing this, in bringing in a man who had no executive authority at the White House, the President revealed a fundamental acknowledgement that he was unable to make the necessary decisions alone. Johnson was known to be a supporter of vigorous space policy, a champion for big increases in financial appropriations.

By delegating authority for recommendations, Kennedy conformed to the theory of decision making based on crisis motivation which states that the decision maker will recruit an informed executive to support his own, professedly informed, conclusion; by sharing the decision-making event, the decision maker in person solicits support for actions he may feel basically insecure in implementing. Kennedy was conforming to the formula for executive response and reaction to crisis motivation and the outcome was now an inevitable commitment to a massive extravaganza, but for a while the executive process was required to run its course. The President was blocking recommendations from the scientific community and turned from this point on increasingly toward the pro-space lobby, appealing to their certain confidence as counter (and bolster to a bitter feeling of decreasing self-efficiency) for the supposedly hostile factions among the PSAC. There is evidence that from the previous Friday, the President had already decided to evade dissident opinion by avoiding confrontation; when asked by Jim Webb if the President felt it would be advisable to get the view of opponents, thereby ensuring a balanced judgement on the course of post-Gagarin reactions, Kennedy dismissed the suggestion as irrelevant.

But it can be argued that Kennedy had already set his sights on military, industrial and professional space administrators for opinion that in the Eisenhower administration would have been subservient to scientific advisers and policy makers. This epitomised the fundamental difference between the two governments: Eisenhower believed that space programs could only be justified on the basis of scientific or military value; Kennedy was primed for using space as a route to prestige which would, in turn, purchase power directly proportional to the level of domestic and international respect gained through prestige. Kennedy would have received comment adverse to this tenet had he depended more fully on his science advisory team, the PSAC. However, in calling Johnson in on a decision-making ratification of his own conclusions about the inevitable course of action, Kennedy set in motion the mechanism by which others would choose a suitable response.

The two vital ingredients that sealed off alternative options – Gagarin's flight and the Bay of Pigs – channeled Kennedy toward the only solution available while simultaneously retaining Congressional and public respect. But even at this late hour he still fought the inevitable decision. Although loathe to gather opinion from the PSAC, Kennedy spent considerable time, especially between Friday the 14th and Tuesday the 18th, discussing with Wiesner possible alternatives. But this was verbal parrying of the kind that presaged his decision to call Johnson in to conduct the final moves. In really asking himself if he could escape the commitment to massive space expenditures, Kennedy told Wiesner that 'it's your fault. If you had a scientific spectacular on this earth that would be more useful – say desalting the ocean – or something that is just as dramatic and convincing as space, then we would do *it*.' Wiesner certainly put alternatives before the President, but, as if rebutting his own question, Kennedy provided some excuse for not adopting any of those.

Johnson's recommendations to Kennedy on the 19th centered on Congressional hearings on some project that should be contrived by discussions in the Space Council. Kennedy agreed to provide Johnson with a memorandum authorizing a search for suitable projects. The next day, Thursday, 20 April, the Congress approved revisions to Section 201 of the National Aeronautics and Space Act passed by the Senate the previous day that effectively placed the NASC within the executive branch of the White House and put the Vice President as chairman. But more important, Kennedy sent Johnson the promised memorandum outlining his objective discussed at the Wednesday meeting:

'In accordance with our conversation I would like for you as Chairman of the Space Council to be in charge of making an overall survey of where we stand in space. 1. Do we have a chance of beating the Soviets by putting a laboratory in space, or by a trip around the moon, or by a rocket to land on the moon, or by a rocket to go to the moon and back with a man. Is there any other space program which promises dramatic results in which we could win? 2. How much additional would it cost. 3. Are we working 24 hours a day on existing programs. If not, why not? If not, will you make recommendations to me as to how work can be speeded up. 4. In building large boosters should we put our emphasis on nuclear, chemical, or liquid fuel, or a combination of these three? 5. Are we making maximum effort? Are we achieving necessary results? I have asked Jim Webb, Dr. Wiesner, Secretary McNamara and other responsible officials to cooperate with you fully. I would appreciate a report on this at the earliest possible moment.'

This was action precipitated by the Bay of Pigs fiasco, an opinion endorsed by McGeorge Bundy in that if the attack had gone as planned the political heat would have come away from the immediacy of the need for a public proclamation to bolster flagging spirits at home and in the free world abroad. The following day, Friday, Kennedy leaked news that he was on the verge of a major new commitment when he said at a press conference that 'We are attempting to make a determination as to which program offers the best hope before we embark on it . . . In addition we have to consider whether there is any program now, regardless of its cost, which offers us hope of being pioneers in a project (the President was clearly still subconsciously questioning his decision) . . . If we can get to the moon before the Russians, then we should.'

Saturday, 22 April, and the end of a week that had seen American prestige plummet in the aftermath of a bungled CIA invasion which had in turn followed the previous week in which Soviet supremacy in space technology had been driven home before an impressed world. Despite the weekend, Johnson met with NASA officials in the first of several shakedown debates at which a sense of urgency invaded talk about options for a major increase in space operations. NASA's Director of Program Planning and Evaluation, Abe Hyatt, briefed Johnson on future plans from the NASA field centers and brought him up to speed on the results of feasibility studies then being carried out on an Apollo concept by Convair, General Electric and Martin in six month contracts effective 15 November, 1960. NASA dismissed the idea of being first with a manned laboratory: the Soviet lead in heavy boosters permitted them to move quickly toward such feats and the Saturn rockets would not give America lifting parity for a few years yet. It did suggest, however, that 'There is a chance for the US to be the first to land a man on the moon and return him to earth if a determined national effort is made. It is doubtful that the Russians have a very great head start on the US in the effort required for a manned lunar landing. Because of the distinct superiority of US industrial capacity, engineering, and scientific know-how, we believe that with the necessary national effort the US may be able to overcome the lead that the Russians might have up to now. A possible target date for the earliest attempt for a manned lunar landing is 1967, with an accelerated US effort.'

The idea of mounting a lunar landing mission *before* 1970 had first been seriously proposed eight days earlier when Seamans heard frantic calls for action from the House space committee. Prior to that, approval by President Kennedy to increase Fiscal Year 1962 funds for booster development, agreed at the 22 March White House meeting, made possible the consideration of a manned Moon landing option *by* 1970 rather than *after* 1970. In the period of little more than a month, consideration of a Moon landing goal had gone from the vague idea that such a feat might be possible by, say, 1973, to a serious proposal for such an undertaking by 1967, cutting the lead-time for development from 12 years to 6 years. No considered technical evaluation had taken place; in the previous two years Apollo had gradually matured from a follow-on with ill-defined objective to a dual purpose orbital laboratory and circumlunar spacecraft. The belief that NASA could dramatically upgrade the mission for Mercury's successor, and halve the development time in the process, was a bold statement coming as it did from senior management without technical portfolios to back up their confidence.

Nevertheless, Johnson heard that April Saturday about spacecraft that could return Moon samples by 1964, spacecraft for unmanned planetary exploration and satellites for weather forecasting and intercontinental communications. Cost estimates revealed the high price for prestige. A Moon landing in 1970 would cost $22,300 million; that was, after all, the plan made feasible by Kennedy's agreement to accelerate booster development. A scheme to put American astronauts on the lunar surface by 1967 would cost an additional $11,400 million. NASA had been prepared for the Johnson talks, submitting on 22 April a full answer to the President's five-point memorandum with the above programs. And Johnson was the last man to urge restraint, seeing as he now did with unparalleled clarity, opportunities for expanding America's space capability. The following day was one of considered evaluation, a brief period of stock taking.

Monday, 24 April, and a major meeting chaired by Johnson at which Webb, Dryden, John Rubel from Defense Research and Engineering, Wiesner, and Kenneth Hansen from the Budget Bureau, met with three men picked for their perspicacity within the industrial and public sectors of business management and administration: Frank Stanton, President of the Columbia Broadcasting System, George Brown, from Brown and Root, and Donald Cook, Vice President of the American Electric Power Corporation. Also present were General Bernard Schriever from the Air Force Systems Command set up by McNamara to conduct military space projects, Vice Admiral John T. Hayward, Deputy Chief of Naval Operations, and Wernher von Braun, now Director of the Marshall Space Flight Center.

Air Force interest in space operations had been firmly isolated from the civilian (NASA) operation by the McNamara memorandum of the previous month and by Webb's February agreement to conduct cooperative arrangements for liaison regarding booster development. Nevertheless, Schriever saw an expanded NASA program as possibly providing the impetus for discarded Air Force proposals and reasoned that a significant NASA commitment to manned space flight could provide the stimuli for Air Force involvement, at that time limited to the slowly evolving Dyna-Soar manned hypersonic glider. He agreed that 'If we had this sort of (Moon landing) objective, there were so many things that would be required that you couldn't avoid having a major space program.' A subtle argument based on suspicion that most opposition would polarize around the cost factor, led Schriever to point out that within a few years the heavy funding profiles on large-scale missile programs would decline, opening a funding wedge which could be adequately, and beneficially, exploited by a major Moon landing project. Schriever was not arguing for Air Force control of such a venture but clearly anticipated fallout to military proposals.

NASA's von Braun argued for a definite commitment to 'sending a 3-man crew around the moon ahead of the Soviets,' for which the agency had a 'sporting chance,' but also a major goal to 'the first landing of a crew on the Moon.' Here, he thought NASA stood an 'excellent chance of beating the Soviets.' Von Braun argued that the Russians still had to demonstrate a booster capacity several orders of magnitude greater than anything they had yet flown if Moon landing flights with manned space vehicles were to be a reality of their program, something of the order of the Nova rocket being essential. In summary, he believed in an 'all-out crash program' which could see America on the Moon in 1967 or 1968. Again, these promises were being made with very little technical information to back up assertions that Nova *and* a Moon landing spacecraft could be designed, built, tested, and flown operationally within six years. Saturn C-1 had yet to make its first flight and Nova could be up to ten times the size of that rocket.

Opposing von Braun in this view was Vice Admiral Hayward who felt that while a manned Moon landing could effectively serve as the central driving theme for a broadly based space program, total dedication to that singular objective could cloud consideration of other more directly beneficial projects. These, he envisaged, were applications satellites for weather forecasting, communications, navigation, reconnaissance, etc. Several days after this 24 April meeting, von Braun pressed home his arguments in a private memo to Johnson in which he called for 'a few – the fewer the better – goals in our space program as objectives of highest national priority,' so that 'the most effective steps to improve our national stature' could center on a 'man on the moon in 1967 or 1968.' Donald Cook also put thoughts to paper and sent Johnson a letter that stressed the 'fundamental premise that achievements in space are equated by other nations of the world with technical proficiency and industrial strength.' Reminding Johnson of the importance of world power and prestige, Cook prescribed a remedy for the national ill that would seek 'achievement of leadership in space – leadership which is both clear-cut and acknowledged. Our objective must be, therefore, not merely to overtake, but substantially to outdistance Russia.'

From every quarter the Vice President, now firmly established by Senate and Congressional action as the NASC Chairman, was receiving the kind of advice the President wanted to hear. But an early manned Moon landing, while setting up a date that could parallel Soviet attempts to get a man to the lunar surface for the 50th anniversary of the revolution and at the same time precede the 1968 US Presidential election, was a suggestion from the politicians devoid of technical appraisal. With hindsight it is quite remarkable that so many NASA officials went along with the call for a manned flight to the Moon's surface by 1967. But concur they did in discussions with Johnson and Congressional leaders. Sensing the evolving mood, however, NASA had responded to House pressure on 14 April, when Seamans was first asked whether a 1967 flight would be possible, by setting up an internal study of the best way to accommodate this early attempt.

In a circular released just five days later and titled 'Manned Lunar Landing via Rendezvous,' John C. Houbolt of the Langley Research Center, together with John D. Bird, Max C. Kurbjun, and Arthur W. Vogley, outlined a plan embracing three principle steps: MORAD (Manned Orbital Rendezvous and Docking) in which a manned Mercury capsule would practice docking in Earth orbit with a small payload launched by a small solid propellant Scout rocket; ARP (Apollo Rendezvous Phase) where Apollo Earth-orbit spacecraft would practice complex operations like docking with small laboratories, crew transfer, in-orbit refuelling, etc; and MALLIR (Manned Lunar Landing Including Rendezvous) in which two Saturn C-2 rockets would launch a combined command module/lunar module/propulsion module configuration and a separate booster rocket respectively into Earth orbit. The booster would dock with the three modules and accelerate them to the Moon before separating, leaving the assembly to fire itself into Moon orbit from where the lunar lander would descend to the surface. One man would remain orbiting the Moon while his two companions were on the surface until they re-joined him for the flight back to Earth. But the Langley paper was not a summary of general opinion at that time, the Direct Ascent and Earth Orbit Rendezvous modes being the preferred options.

George Low, a keen proponent of the lunar landing objective long before anyone in Congress became interested in the issue, kept alive the basic tenet of the Goett Committee's recommendation by chairing a Manned Lunar Landing Task Group, also known as the Low Committee. Its first job had been to establish a working paper for Fiscal Year 1962 budget presentations and it met first on 9 January, 1961, when it was established that NASA would be advised to aim for lunar landing, lunar colonization, and the establishment of a scientific base as fundamental goals underpinning the entire space program. It would, they said, require a launcher capable of sending between 25 tonnes and 35 tonnes to the Moon, implying a booster with a thrust of at least 3,000 tonnes compared with 680 tonnes for the Saturn C-1 then under development. The Low Committee concluded its work on 7 February by submitting to Robert Seamans a final report confirming that in its view a manned landing on the Moon could be achieved with either a Direct Ascent Nova rocket or via Earth Orbit Rendezvous of several payloads from the multiple launch of Saturn C-2 rockets.

It had been an exercise in theoretical possibilities assuming a maximum-paced effort. At the time it performed its work NASA was still under the impression it would not get funds for the Saturn C-2 or the Apollo manned spacecraft, only getting approval for accelerated booster development late in March; Apollo was still a deferred project, shelved for possible consideration during Fiscal Year 1963 budget discussions late in 1961. But if achieving nothing else, the 24 April meeting told Johnson that the technical capabilities were probably in existence already to mount an operation like Moon landing flights by 1967, whereas in fact nobody could state with certainty that such a feat was possible. For the rest of that week the Vice President tied up loose ends prior to preparing a 5½ page reply to the memorandum from Kennedy. But it was his decision to recommend the response to Gagarin and he made sure that a minimum of turbulence ruffled the conclusion.

In several National Security Council meetings over this period he kept silent on the talks with NASA officials and respected associates. He did telephone Secretary of State Rusk to discuss the possible repercussions and ramifications on foreign policy of a Presidential commitment to a major space spectacular and received positive response. Rusk could see only benefits accruing from such a decison and a later statement to the Senate space committee confirmed this. Also considered at this time was growing concern on the part of Jim Webb about the proposed pace of NASA's future program. While many of his colleagues, all experienced NASA men originating from the early days of the agency, were enthusiastic for bold commitment, Webb had lingering doubts on the validity of verbal agreements to match political goals hung on specific dates. In this he was in opposition to his own deputy; for long, Dryden had been a fervent advocate of manned lunar flight goals. But to Webb it seemed that exuberance breaking out within management was enthusiasm for a unique capability to convince politicians that a major technical and scientific undertaking was not only worthwhile but needed with urgency.

Rarely had these men, all from engineering or scientific backgrounds, encountered such apparently unsolicited support, and Webb felt their euphoria – albeit controlled and kept within the internal structure of NASA – might be due primarily to a recognition that at last their wildest dreams could actually come true, rather than considered approval for suggested policies from outside the agency. Webb was not convinced that NASA had the technical resources, or as yet know-how, to mount such an extravagant operation and quizzed the President's science adviser on this. Wiesner himself had grave doubts about the wisdom of pursuing this single-minded objective. But neither Webb nor Wiesner objected to the memorandum sent by Johnson as summary reply to the call from Kennedy for quick answers to searching questions.

What Johnson said in those 5½ pages influenced Kennedy to the extent that it provided positive patterns for action: a manned lunar landing goal as the central objective for a vigorous space program embracing lunar and planetary unmanned exploration in addition to applications satellites; renewed acceleration in Saturn and Nova class boosters then considered essential to Moon landing flights via Earth Orbit Rendezvous or Direct Ascent, respectively; concurrent development of liquid, solid and nuclear propulsion systems, the technology for solids providing insurance against failure with very large Nova class liquid propellant boosters. The memo reached Kennedy on Saturday, 29 April, and two complete weeks had now gone by since Gagarin's flight; nearly two weeks since the American public learned of the Bay of Pigs fiasco.

Webb made an official request to Wiesner for his considered opinion on the decision that seemed to be inexorably focused toward manned Moon flights in a letter written and dispatched on Tuesday of the following week. It was 2 May. Webb considered the development of a strong US space program to be of prime importance but felt that it must be one 'from which solid additions to knowledge can be made, even if every one of the specific so-called "spectacular" flights or events are done after they have been accomplished by the Russians.' He reminded Wiesner that as head of the President's Science Advisory Committee, and as head of the space agency, they both had responsibility 'to make sure that this component of solid, and yet imaginative total scientific and technological value is built in.'

Wiesner too was divided on the issue. On the one hand he felt the traditionally conservative and solid foundations of the American scientific community had no part in this impetuous race for national prestige and openly expressed dissatisfaction with the use of ostensibly scientific institutions for political gain by partisan groups. On the other hand, he felt the security of those institutions, and others unrelated to science and technology, could perhaps be most appropriately served by the application of a portion of that science for political policy-making strategy. It was the age-old dilemma: whether to use political power to ensure scientific objectivity, or to use scientific and technological capacity for carving political muscle. Wiesner refused to bring the issue before PSAC meetings because he felt the issue was unrelated to scientific or technical justification. It was more, he felt, a political decision divorced from arguments that could call to question the very foundation of the executive decision-making process. Nevertheless, Jim Webb was concerned lest critics at a later date should call upon the decision that seemed about to be made as argument for accusation; Webb was not about to conduct the agency in a manner that would bring it into disrepute as a tool for battering US prestige into the uncommitted nations of the Third World.

On the same day he wrote to Wiesner, Webb set up, through NASA's general manager Robert Seamans, an Ad Hoc Task Group for a Manned Lunar Landing Study, chaired by William A. Fleming from Headquarters. Its purpose was to provide answers to all the many technical, managerial and financial questions that hovered over the proposed Moon commitment: what tasks were critical to the success of the goal? What were the important decision points in the plan? How much money would be needed and in which years? What were the NASA resource requirements for such a plan? Which portions of the program would be best carried out internally and which by outside contractors? They were questions essential to a considered appraisal for and against the proposed mission. If there is any one point at which it became obvious that the commitment was totally political, made outside the normal and accepted conventions of technical decision-making criteria, it was in the setting up of the Fleming task force, for even before the group had time to call their first meeting the decision had been made.

Events were now moving faster than either NASA or the White House expected, and largely at the instigation of Lyndon Johnson. In any controlled decision response to a crisis motivation, a curve of expectant euphoria rises to a peak and rapidly declines, a period during which conclusions are drawn and decisions made which would, at another time and in different circumstances, be considered irrational and an over-reaction to acceptable events. Such was the case with the impetus applied to the question of how the space program could be used as a response to Gagarin. Had it not been for the Bay of Pigs, the peak of expectancy (that a major commitment was pending) would have probably dissipated prior to the technical organization of the response. But Bay of Pigs added energy to the crisis motivation and sustained the impetus long enough for a major commitment to be made.

By 2 May participants had drawn their lines of demarcation: Johnson was going for an all-out space commitment, riding the President's need for a spectacular show to get the broad-based policy he had wanted for several years; Webb was behind the idea of a manned lunar landing, but wanted to wait for the Fleming Report to assess and balance the technical and financial arguments; Wiesner never did make up his mind which side to commit his affiliation to; Dryden, Low, von Braun, and other NASA managers didn't anticipate major technical problems and were willing to agree to anything to get the program going – Low and von Braun wanted a single-purpose commitment to Moon flights while Webb and Dryden campaigned for a broadly based program with permanent objectives.

Wednesday, 3 May, and a critical time for action as Johnson had another large meeting. The purpose here was to determine just how agreeable Congress would be to an acceleration of the type he now envisaged recommending to the President for approval. Accordingly, Senators Kerr and Bridges were with the Vice President when Webb, Dryden, Welsh, Stanton, Brown and Cook met to hear Johnson press for a technical concurrence so that Congressional moves could get under way. Webb still pressed for a delay until Fleming had the pertinent answers but Johnson was not prepared to wait any longer. Bridges was the senior Republican on the Senate space committee headed by Kerr and both were influential friends of Lyndon Johnson. Both agreed to support fully the impending action to dramatically accelerate the nation's civilian space effort. Webb left the meeting knowing that the pace of events was moving faster than considered evaluation could accommodate, and advised his subordinates that NASA was about to get a major injection of funds.

Thursday, 4 May, and more than three weeks had elapsed since Vostok I. Preparations for the launch of Alan Shepard aboard Mercury-Redstone 3 began in anticipation of a flight the following day; the mission had been waiting all that week for weather to clear at the Cape. It was incredible that plans embracing spacecraft weighing tens of tonnes launched on rockets more than 100 times as powerful as Redstone carrying men to the Moon should be maturing even before America's first suborbital hop. Such thoughts formed the nucleus of Webb's unease. Having assured himself of Senate support the previous day, Johnson telephoned James Fulton and Overton Brooks and without detailing the specific nature of the plan, advised them of an imminent commitment to new programs. Fulton rang back and assured Johnson of Republican support on the House space committee; Brooks went even further and vouched for the Democratic team on the committee by presenting Johnson with a lengthy memorandum of suggested projects – a typical Brooks response from this giant among peers who lived in an aura of self induced over-work and a supreme feeling of insecurity.

The House, it seemed to Johnson, was keen to see many branches of this space program stepped up, not just manned flight or exploration, and this jelled with his own idea which would confuse Kennedy in a later year as to just what he had been asked to approve. While final countdown preparations were getting into strident expectation at the Cape that Thursday evening, Johnson learned that he was to go to South East Asia the following Tuesday, 9 May. He would have to move quickly to catch the President with definite proposals, for it would be the end of the month before he returned to Washington. By that time the sense of urgency would have evaporated and Johnson knew only too well that few White House personnel were keen on the idea of an expanded space program.

Friday, 5 May, and Johnson telephoned all the relevant agencies involved in talks over preceding weeks to urge the preparation of formal plans and proposals. He wanted them on his desk by the following Monday for review and presentation to the President before he would have to leave for Asia. NASA and Defense Department officials were surprised by this sudden request. It was to be a long, busy, and fateful weekend. But other things that day temporarily took thoughts away from future goals to more mundane, but infinitely more pleasurable, preoccupations: Alan Shepard successfully flew Freedom 7 and fell back to the Atlantic in his 15 minute ride to the edge of space. At least everyone now knew that American manned spacecraft designers could produce a workable product and that plans for the first manned orbital Mercury missions, still confidently expected by the end of the year, could proceed along the designated schedule. Shepard's flight was a great boost to morale, and a vital ingredient for successfully manipulating Johnson's Moon landing policy first through the White House and then through Congress. A failure that Friday would probably have dashed for ever all hope of rallying support for the 'big one.' If nothing else, MR-3 would dispel fears raised by committee reports that twice in recent months cast doubt on the program's validity – and even applicability – in the light of national priorities.

That afternoon, several hours after the successful splashdown, Kennedy made a statement to the press in which he reaffirmed the nation's need for a 'substantially larger effort in space.' asserting his belief that 'today's flight should provide incentive to everyone . . . to redouble their efforts in this vital field.' But preoccupation was short lived. Following hard on the heels of Johnson's urgent calls, Defense Department, Atomic Energy Commission, and NASA officials agreed to convene a meeting the following day at the Pentagon to pool, for the first time, their combined study efforts on issues relating to possible manned Moon flight goals. It was to be a high-powered session from which a single primary objective would emerge by day's end.

Saturday morning, 6 May, and as senior government officials sped to the Pentagon in cars from all over Washington, newspapers gloated over the Mercury success; unaware that even then their delegates were about to plan a dramatic spectacle, the American public lounged on that first day of the weekend break in sure knowledge that success in space was not an exclusive preserve of the Soviet Union. At the Pentagon, Secretary of Defense McNamara, his deputy Roswell Gilpatric, the new Director of Defense Research and Engineering Harold Brown and his deputy John Rubel, hosted NASA Administrator Webb, his deputy Hugh Dryden, Associate Administrator Seamans and Abe Silverstein, head of the Office of Space Flight Programs. Willis Shapley was there from the Bureau of the Budget, and Glen Seaborg dropped in later to put views for the Atomic Energy Commission.

The NASA view, based on data gathered without priority or haste before the sudden emergence of the new proposals, generally supported the view that a manned Moon flight in 1967 was possible, given the availability of very large boosters, although Webb persisted in reminding the attendants that he would have preferred to see the Fleming task group report before committing his agency to a definite objective of this magnitude. Nevertheless, it was conceded that the Moon landing goal would serve to focus attention on the technical capabilities of the United States and would present a very real opportunity of beating the Russians. Everyone in the meeting knew that Soviet success had been achieved with limited technology and that the vast and broadly based expertise in American industry could outspace the Russian industrial machine if the project was big enough.

NASA presented its argument for a Moon landing commitment on the basis of studies favoring the Direct Ascent method, where a single, very large, booster would be launched from Earth on a 'direct ascent' to escape velocity, accelerating a spacecraft toward the Moon where it would turn around, slow down by firing braking rockets, and land. Return to Earth would be by a 'direct ascent' from the lunar surface. To accomplish this NASA proposed to continue an existing, albeit low paced, study of a booster using liquid propulsion and with a first stage thrust of up to 5,000 tonnes in the expectation that a launcher of this size and capacity would eventually be selected for the mission. For its part, the Defense Department felt that alternative booster concepts should be pursued, and the group agreed to support studies and initial development of a solid booster with similar capability; concurrent development of liquid and solid propulsion systems had long been an element of NASA studies, although the agency was not as enthusiastic as the Air Force about the desirability of the latter. In adopting a recommended plan to propose the two launcher concepts, the group accepted that NASA would only decide on the precise method by which lunar landing operation would be conducted when further development exposed unknown aspects of large rocket performance and operation; several years could elapse before final commitment.

On through lunch, taken in and eaten during the talks, the group considered the value of several levels of program effort. NASA presented planning scenarios focused on several levels of program pace, separating the options into an essentially application, Earth-orbit orientated, space program, one in which a Moon landing would be the focal objective, and a more advanced and extensive preparation for manned flight to the planets. The Defense Department questioned the wisdom of targeting for the Moon when planetary missions were a possibility, raising an idea that perhaps the Russians even now were putting together their own Moon landing project. The general consensus approved of the NASA view that such a technically distant goal might well prove a stumbling block to US capabilities and that its implementation would certainly project well into the 1970s. In seeking to define the best compromise, 'national prestige' was to be the yardstick against which to measure 'the will of the nation to harness its technological, economic, and managerial resources for a common goal.' At the end of the meeting it had been unanimously agreed that the formal policy proposal would embrace a multitude of Earth application projects, in addition to high technology ventures for the third phase of the space program after initial Moon landing operations.

Before dispersing, McNamara suggested that sections of a space policy study Defense Research and Engineering had been preparing for his review could form the basis for a memorandum drawing together the recommendations of the group. NASA agreed to this, and it was decided the NASA's Robert Seamans, the Defense Department's John Rubel, and Budget Bureau's Willis Shapley should get together over the remainder of the weekend and assemble the formal presentation for Johnson. Webb was still concerned that events were moving faster than NASA could adequately handle, and said that he would look in on the three whose job it was to write the recommendation for a manned Moon landing. On through the rest of Saturday, the three worked at the memorandum, adjourning about midnight with considerable work still before them. Across Washington, Saturday night fever received its own mini-boost in the wake of Mercury. The astronauts were coming to town and it was a bright light on the administration's horizon, one of the few occasions when John Kennedy's warming smile was mirrored in the faces of associates. It had been a difficult time these 3½ months in which 'New Frontier' campaign promises were seen to be making little progress through the entangled corridors of Congress. But that night it was different, a hope surged up and Washington was going to make the most of its new hero.

Sunday, 7 May, and Alan Shepard's family was in town for dinner with Jim Webb and social events before decorations on the White House lawn next day. All through the day, Seamans, Rubel and Shapley worked on the draft memorandum which sought to give the President a reason rather than a cause for making the momentous decision he had asked for justification to present. 'It is man, not merely machines, in space that captures the imagination of the world. All large-scale projects require the mobilization of resources on a national scale. They require the development and successful application of the most advanced technologies. Dramatic achievements in space therefore symbolize the technological power and organizing capacity of a nation. It is for reasons such as these that major achievements in space contribute to national prestige.'

The memorandum continued to argue for justification: 'Major successes, such as orbiting a man as the Soviets have just done, lend national prestige even though the scientific, commercial or military value of the undertaking may by ordinary standards be marginal or economically unjustified. Our attainments are a major element in the international competition between the Soviet system and our projects such as lunar and planetary exploration are, in this sense, part of the battle along the fluid front of the cold war.' During the evening of that May Sunday, Jim Webb took Alan Shepard and his family to dinner before leaving them to join Seamans, Rubel and Shapley for the promised visit. It was about 10:00 pm and for a further two hours the four men worked on closing passages in the memorandum before Webb signed off an introductory letter as a new day was coming in.

Monday, 8 May, the day the President would receive the recommendations. During the morning, Webb took the draft proposals to McNamara, the two examined the document in detail, and both agreed to present it to Johnson at mid-day. With Welsh from the Space Council, the Vice President quickly looked over its recommendations and joined Kennedy out on the White House lawn where the President was about to present Shepard with NASA's Distinguished Service Medal. After this Shepard drove to address Congress and then went on to a luncheon at the State Department. Johnson had time to snatch a further look at the document before going to the luncheon, stealing himself away as soon as respectably possible to get to the White House where Kennedy was waiting to receive him. It was late afternoon when Johnson gave Kennedy the 'Recommendations for our National Space Program: Changes, Policies, Goals' document. The die was cast. Just 26 days after Vostok I the Vice President had manipulated the national reaction and married it to projected capabilities for the most dramatic expedition ever given government sanction.

On Monday, 9 May, the day before Johnson flew out of Washington sure in the knowledge that America was on the move, Senator Kerr got the green light to leak reports to the press about a massive new space goal. On Tuesday, Kennedy brought Webb, Dryden, Wiesner, McNamara, Welsh, Bell, Staats, Sorenson, Bundy and McNamara together for final discussion, a sort of proposers question-time in which the President could get answers to matters that still confused his interpretation of the Webb-McNamara memorandum. Kennedy decided to approve the proposals without modification or clause; attendants testify that the President had no ear for proviso or qualification on the stated commitment to land a man on the Moon. He had made up his mind to accept in full the Johnson concept of total growth in the nation's space program.

Basically, the plan called for manned Apollo flights in Earth orbit in 1965, instead of 1967, circumlunar missions in 1967 instead of 1969, and a manned Moon flight in the 1967–1968 period rather than in the early 1970s. To make this possible, the existing Fiscal Year 1962 budget request, then before Congress, would have to increase by a further $549 million. In the past six weeks, Eisenhower's $1,109.63 million space budget had grown by $125.67 million (11%) and now the Congress would be faced with a further 44% increase. But more important for the long range financial projections, it would demand annual outlays totalling more than $5,000 million within the following three or four years. In terms of the national economy, it was only 4% of the total federal budget expected by about Fiscal Year 1964, however, and there were few critics in the Budget Bureau, who saw the financial commitment as only a partial move to re-vitalise the economy. Director Bell was not at all concerned that these sums would be essential, unavoidable outlays for much of the decade.

Following his decision on 10 May to accept in full the recommendations of the Webb-McNamara memorandum, Kennedy sought further financial advice from many sources. All agreed that it was probably too low-key for the definitive

The day of decision: 8 May 1961. Kennedy pins a medal on Alan Shepard's chest and hours later receives recommendations from Vice President Johnson to mount a Moon landing expedition.

recession-avoiding move it purported to be. Many tried to get the President off a technology boosting mood, preferring instead to see a massive public works program where the manpower effect of a given sum would be higher in the arena of low paid workers than with the professional and technical staffs embraced by space money. Advisers and analysts were all agreed, however, that it was a move in the right direction, recognizing that the President had wanted to answer Soviet challenge with Vostok and Communist pressure from Cuba with a program that simultaneously embraced the need for economic stimulus. Only modest changes would be made to the Defense Department's space budget for Fiscal Year 1962: $32 million extra for advanced work on the big solid propellant booster that it was agreed would be developed as insurance against problems with liquid propellant Saturns or the big Nova; military space funds would claim $1,298 million for that year, nearly $500 million less than NASA.

President Kennedy decided to include with his statement to Congress several other proposed moves aimed at stemming the spread of Communism, improving the nation's economy, and speeding toward successful ratification of treaties and agreements with foreign states. It was to be a message on 'Urgent National Need,' appealing to Congress for massive commitments in science and technology that ranged far beyond the space program. Kennedy had originally intended to send Congress this message in a written statement, but was persuaded to present it personally in a speech scheduled for Thursday, 25 May. But although the Webb-McNamara memorandum had not specifically mentioned the expected date for NASA's first Moon landing, the initial draft of Kennedy's speech announced 1967 as the target. NASA management, on seeing this sentence, advised the President that experience with Mercury revealed potential pitfalls in being too positive about a specific date – especially one that deliberately threw down the gauntlet of challenge to the Soviet Union. With due regard for the potential delays that might accrue, recognizing that the whole thing would back-fire if delayed even a few months, Kennedy agreed to delete that reference and give instead a more flexible date based on achieving the event sometime before the 1970s.

Thursday, 25 May, 1961. A drive from the White House to Capitol Hill and an address that would hold the attention of Congress for 47 minutes. Toward the end of an appeal for national resolution, with brief references to new programs, the President came to announce his monumental decision – one that had, in reality, been the product of the Vice President and leading government officials: 'The dramatic achievements in space which occurred in recent weeks should have made clear to us all, as did the sputnik in 1957, the impact of this adventure on the minds of men everywhere who are attempting to make a determination of which road they should take. Now it is time to take longer strides – time for a great new American enterprise – time for this Nation to take a clearly leading role in space achievement which in many ways may hold the key to our future on Earth. I believe we possess all the resources and all the talents necessary. But the facts of the matter are that we have never made the national decisions or marshalled the national resources required for such leadership. We have never specified long-range goals on an urgent time schedule, or managed our resources and our time so as to insure their fulfilment.

'Recognizing the head start obtained by the Soviets with their large rocket engines, which gives them many months of leadtime, and recognizing the likelihood that they will exploit this lead for some time to come in still more impressive successes, we nevertheless are required to make new efforts of our own. For while we cannot guarantee that we shall one day be first, we can guarantee that any failure to make this effort will find us last. We take an additional risk by making it in full view of the world – but as shown by the feat of Astronaut Shepard, this very risk enhances our stature when we are successful. But this is not merely a race. Space is open to us now; and our eagerness to share its meaning is not governed by the efforts of others. We go into space because whatever mankind must undertake, free men must fully share.

'I therefore ask the Congress, above and beyond the increases I have earlier requested for space activities, to provide the funds which are needed to meet the following national goals: First, I believe that this Nation should commit itself to achieving the goal, before this decade is out, of landing a man on the moon

Dr. Kurt H. Debus, Director of NASA's Cape Canaveral launch site, played a vital role in processing launch vehicles and spacecraft, moving them off the earth with superb precision.

and returning him safely to earth. No single space project in this period will be more exciting, or more impressive to mankind, or more important for the long-range exploration of space; and none will be so difficult or expensive to accomplish. In a very real sense, it will not be one man going to the moon – we make this judgement affirmatively – it will be an entire nation. For all of us must work to put him there.

'Let it be clear – and this is a judgement which the Members of the Congress must finally make – let it be clear that I am asking the Congress and the country to accept a firm commitment to a new course of action – a course which will last for many years and carry very heavy costs, $531 million in the fiscal year 1962 and an estimated $7–9 billion additional over the next 5 years. If we are to go only halfway, or reduce our sights in the face of difficulty, in my judgement it would be better not to go at all. This is a choice which this country must make, and I am confident that under the leadership of the space committees of the Congress and the Appropriations Committees you will consider the matter carefully. It is a most important decision that we make as a Nation; but all of you have lived through the last 4 years and have seen the significance of space and the adventures in space, and no one can predict with certainty what the ultimate meaning will be of the mastery of space.

'I believe we should go to the moon. But I think every citizen of this country as well as the Members of Congress should consider the matter carefully in making their judgement, to which we have given attention over many weeks and months, as it is a heavy burden; and there is no sense in agreeing, or desiring, that the United States take an affirmative position in outer space unless we are prepared to do the work and bear the burdens to make it successful. If we are not, we should decide today.'

Congress was moved beyond the point of response: some sat dumbstruck that the one-time Congressman from Massachusetts so concerned with the rights of ordinary, not very wealthy, Americans should now be committing the Nation to a multi-billion dollar 'Moon-doggle'; some shuffled and bit their nails uneasily, nodding or shaking the head in mute agreement; others sat stiff and proud that America was throwing down a challenge at last. But the real reactions would spill out over the next several years, even during the coming months as both Houses debated the issue. Kennedy's Moon landing flight would get the blessing of Congress, yet within a few weeks the President would be making speeches promoting cooperation with the Soviet Union to halt the wasteful drain of financial resources; according to Kennedy, a product of technological competition.

When the President drove back to the White House with Ted Sorenson, he discussed the mediocre response given by Congress at the end of his presentation. The President had revealed his uncertain attitude by continually digressing from the prepared text of his address – a sure indication that he even now questioned the wisdom of such a major challenge to both Soviet technology and American resources. His verbal request for reassurance during the journey from Capitol Hill merely endorsed what he would put into words months later. At almost every turning point it had been a Johnson decision: in setting up the National Aeronautics and Space Council by promising Senate support for Eisenhower's 1958 Space Act; by persuading Kennedy to re-write the terms of the Act placing the Vice President as chairman; by providing informed executive support while Kennedy remained unsure of the Gagarin response; by accelerating the need for managerial plans before technical feasibility studies could begin; by recruiting Senate support for the major new commitment.

But more than that, he organized a major space program under the cloak of a single objective and gathered along with the necessary hardware for the Moon landing a broad inventory of projects only peripherally related to the prime goal. Kennedy had no idea of the value of a space program to technical and economic resources, only adopting the technology for prestige gain to a national cause. He intended the Moon decision to be a single thrust into the side of Communism and never expected it to be a long term commitment, a permanent fixture in the budget process. In this way, Kennedy engineered a built-in stress point which would threaten to fracture when the Moon landing goal had been achieved.

For some reason, the President seemed not to realize the tenacity with which scientists and technologists would hold on to the new capability for space travel and transportation engineered by the Moon goal; when the time came to pack up and change direction, inherent in Kennedy's plan, stiff opposition emerged to the dissipation of what was considered an almost exclusive national resource. Had Johnson not built up a broad based space program, protagonists may well have been beaten. As it was, when the Moon goal had been achieved, there was sufficient strength within unrelated space programs to ensure a momentum through the transition to a permanent manned space flight project. But all that was in the future, and is only of value here to show how different were the views of politicians and scientists in the early days. Suffice it to say that, figuratively, Kennedy wanted to build a giant highway on contractor labor that could be laid off when completed, deferring to a still later date development of traffic for the permanent use of that particular route.

In NASA, the atmosphere was suddenly different. Having already expanded to annual budgets exceeding one and one-half billion dollars, employing 17,000 people 'in-house' and providing work for nearly 60,000 more in industry, the space agency would grow, within five years to command a budget three times as great and a national workforce six times as strong. From now Apollo was to be a Moon landing program committed to that task on manned flights within the next 8 years. Much had yet to be accomplished, not least of which was to get a man in orbit. With such a grand objective, the NASA manned space flight time of 15 minutes 22 seconds seemed pitifully short of the several thousand man-hours of space flight time deemed necessary prior to Moon flight. To some it was a challenge, but to many it was stretching capacity almost to breaking point.

George M. Low, the inspiration behind the bold drive toward lunar exploration

CHAPTER SEVEN

The Race Begins

The Moon landing goal announced by President Kennedy on 25 May, 1961, less than three weeks after the 15 minute sub-orbital flight of Alan Shepard, brought mixed emotions to the still comparatively young NASA organization. Many were clearly amazed at the sheer audacity of the promise; many were openly enthusiastic, but privately cautious, about the opportunity to dramatically expand the national space program; a few were deeply opposed to any such commitment. While manned flight had so recently claimed prominence in Congress, the White House, and even it seemed the Washington Headquarters of NASA, deep-rooted interests in the scientific exploration of space cast scientists and engineering personnel into separate camps. When Mercury became official NASA manned space flight responsibility in 1958, Langley's elite Space Task Group had been granted a level of autonomy essential to moving rapidly with decisions concerning the nascent flight operation. Now, three years later, a deeper gulf emerged, one that would not go away for as long as science and Earth-based space applications were made to pay for massive financial requirements for manned flights – missions that to many a NASA scientist seemed more for prestige than the serious business of reaping rewards befitting the investment.

Soon, the entire NASA organisation would be involved with a dramatic restructuring made necessary by the new mandate: new space centers would be chosen, plans for building up the Cape Canaveral launch site would be gone over, multibillion dollar contracts would be let, and a major increase in funds and personnel would completely transform the agency. It was not a time for vacillation. There was just $8\frac{1}{2}$ years to build a road to the Moon and as yet nobody really knew for sure the best way to do that. So while elements of NASA became wholly concerned with planning for Apollo's grand finale on the decade, Space Task Group at Langley Field entrenched itself in preparing Mercury for its baptism; the ballistic flight of MR-3 was a far cry from orbital flight and much had yet to be done before that step could be taken.

The current plan was for each astronaut to fly a suborbital mission on top a Redstone booster before going into orbit riding the big Atlas missile, an altogether different experience and one thought to benefit from providing the pilot with a taste of rocket flight on a less powerful machine for a flight considerably shorter than the circum-terrestrial mission. Accordingly, Virgil 'Gus' Grissom was already being groomed for his brief 15 minute hop when Shepard performed his own flight to get the capsule space-wet for the first time; John Glenn was back-up as he had been for Shepard and as such assumed responsibility for nursing the technical developments associated with the capsule – spacecraft 11.

Delivered in October, 1960, the capsule was more typical of the orbital model than Shepard's spacecraft 7, incorporating changes recommended by the astronauts and introduced as a result of tests and subsequent evaluation during the long process of design gestation. Like Freedom 7, Grissom's capsule would carry a berylium heat-sink shield at the base instead of the ablative design selected for production spacecraft destined for orbit. Having already been ordered by the time Big Joe qualified the alternative concept in September, 1959, heat-sink shields were carried on all Mercury-Redstone suborbital hops; returning through the atmosphere at less than one-third the speed of a descending capsule from orbit, suborbital spacecraft would heat up to only 530°c.

The most noticeable difference about spacecraft 11 was the replacement of two 21.6 cm diameter circular side windows with a single window of trapezoidal shape situated on the spacecraft centerline directly in front of the astronaut's helmet but, because of the conical sidewall, sloping away from him. The new window was, like the earlier design, made of Vycor glass fabricated by the Corning company and had a length of 53.3 cm, with a maximum width of 28 cm narrowing to 19 cm at the forward end. This window was to be standard for all future Mercury capsules and comprised four separate panes: an outer panel 0.89 cm thick and an inner panel 0.43 cm thick, both in Vycor glass, with two additional inner panes each 0.86 cm thick made from tempered glass. The two outer panels could accept temperatures up to 980°c (temperatures here would never exceed 650°c); the inner panes could withstand the interior pressurization and a magnesium fluoride coating inhibited glare. When seated on his contour couch, an astronaut could look forward and slightly up to obtain a view outside his capsule, framed 30° in the horizontal plane and 33° in the vertical.

Another aspect of Grissom's capsule at variance with Shepard's Freedom 7 was in the nature of the side hatch through which the pilot gained access to his cabin and from which he was required to emerge at the end of the flight. The original design adopted 70 bolts around the sides, the hatch for chimpanzee Ham and astronaut Shepard was a weighty latch-operated device, while that for Gus's capsule was the first of an explosive type designed to facilitate speedy egress should the spacecraft seem to be in danger of sinking after splashdown; the original inventory of 70 bolts was reinstalled but with a 0.15 cm hole drilled through each one so they would fracture when stressed by detonation of a mild fuse wrapped round a small groove in each bolt. Thus liberated, the hatch would pop away from the capsule, exposing the exit route in seconds. Hatch activation responded to signals from a knobbed plunger close to the astronaut's right hand, requiring a depressed force of 3.3 kg after removal of a safety pin. If activated from outside, a lanyard would be found beneath a panel on the side of the capsule which, with a force of 18 kg, would shear the pin and operate the detonation sequence. The new design saved 21 kg over the latch-operated cover.

Also new for MR-4 was a more streamlined form of shroud covering wires – the umbilical – linking capsule to booster. A measured amount of buffeting had been induced by the previous design and the new cover would flip away from the connections at spacecraft separation, unlike the sliding action of MR-3's cover which, it had been suspected, caused minor damage to the wires. With regard to spacecraft systems, Grissom would have a chance to use the new Rate Stabilization Control System (RSCS), a method of controlling the spacecraft whereby the rate at which the capsule changed attitude would be sensed and controlled. Shepard's only manual option was with the Manual Proportional system, a means of directly authorizing capsule attitude changes through the hand controller; RSCS would allow the system to damp to the commanded rate to within ±3°/sec or, without manual control, to zero within the same tolerance. Grissom would use RSCS only after retro-fire and remain on this system for the remainder of the mission.

MR-4's instrument panel was different to Freedom 7 in that an earth-path indicator provided visual determination of the capsule's position over the planet; it would not be needed for the ballistic shot but was an integral part of cockpit displays for the orbital model. The indicator, a square shaped display containing a rolling sphere, had not been carried in Shepard's capsule and its incorporation for spacecraft 11 required re-location of several other instruments. In fact the displays were grouped according to experience with ground test runs and served to emphasize the evolving nature of spacecraft development. The life raft too was an improved model, saving 45% in weight over the MR-3 type, and Grissom's pressure suit incorporated nylon-sealed ball bearing rings in the glove wrist connectors to improve hand dexterity in an inflated condition.

As a direct result of the first ballistic Mercury flight, additional foam insulation was provided in the head restraint of Grissom's contour couch; vibration during the ascent phase had produced uncomfortable movement of Shepard's head, temporarily blurring his view of the display console. Also for Mercury-

Redstone 4, there were two procedural changes that would influence the mission. The first concerned the ascent trajectory, the second re-configured the sequence of test objectives during the brief 5 minutes of weightlessness. Careful analysis of Freedom 7's flight trajectory, and that of the Redstone booster after separation, showed that the two elements would be 1,220 meters apart when the capsule initiated retro-fire and that the booster would land 26 km beyond the capsule's planned splashdown point.

Theoretically, there seemed adequate separation from the time of retro-fire on down through re-entry, a distance enhanced by the posigrade rockets on the capsule firing into the closed adaptor between spacecraft and booster and achieving a 78% increase in effective thrust compared to similar thrusters firing in the open. But lingering doubts about the adequacy of spacecraft-booster separation during the fall back to Earth inspired changes to the ascent trajectory where the Redstone would propel the capsule about 215 km/hr faster than Shepard's booster on the first hop, albeit an increase of less than 3% on the total velocity. Instead of firing the retro-rockets at the apex of the ballistic trajectory, MR-4 would ignite the solids 6 seconds *before* reaching apogee rather than 3 seconds *after* like MR-3. This would have the effect of removing the additional velocity imparted by the booster, allowing the capsule to fly only as high and as long as Shepard's mission, but move the capsule's trajectory away from the booster in space and time.

The second procedural change came as a result of the inadequate time for Shepard to accomplish the planned number of attitude control tests; the duration of weightlessness between separation and re-entry would be the same but Grissom was briefed to go on full manual control for his checks rather than proceed on an axis-by-axis routine, which was very time consuming and far too conservative a sequence.

By 8 June, Kurt Debus had received the Redstone missile at Cape Canaveral and within a few days Merritt Preston's capsule team had installed the spacecraft on its adaptor. Numerous checkout operations followed but events moved well and by 13 July a full readiness review gave the green light for final flight preparations. On 15 July, Walt Williams heard from booster systems engineers and capsule checkout teams that all was as it should be before pronouncing that Mercury was go for a second suborbital hop. Bob Gilruth publicly announced that STG was anticipating a launch three days hence, and re-affirmed Gus Grissom as the pilot; for his part America's second operational astronaut announced that he had decided to call the spacecraft Liberty Bell 7 – because it looked similar to *the* liberty bell and denoted freedom, with a numeral befitting the select seven whose job it was to ride Mercury into space. Almost at once the recovery teams slipped port to be on station by 18 July but a day later, on 16 July, weather reports began to look bad.

Early on the morning of the planned launch day the flight was scrubbed and re-cycled for the 19th in the hope that better conditions would prevail. Grissom had been at the crew quarters in Cape Canaveral's Hangar S since the 15th, preparing for a launch three days later. Having flown 100 simulated missions in the procedures trainer and run 36 simulated flights on the ALFA trainer, he was as ready as any pilot could be for the actual event. Bill Douglas woke Grissom early in the morning hours of 19 July, expecting to prepare him for a flight into space. Instead, it was an elaborate dry run when the mission was cancelled just 10 min 30 sec prior to lift-off, planned to occur at 7:00 am. The following day, at 6:00 am, the first half of the two phase countdown swung into action in anticipation of a launch 21 July; a two-day postponement had been forced upon the launch team due to Redstone being full of liquid oxygen prior to cancellation. Grissom got to bed in Hanger S at 9:00 pm on the 20th for a fitful sleep lasting little more than four hours. He was unlikely to sleep as soundly as Alan Shepard for, although the trauma of bated anticipation was a very personal thing to Gus Grissom yet considerably tempered by awareness that he was not the first to fly Mercury into space, the very nature of the man made him more likely to depress to his subconscious the emotional response that Shepard would simply have ignored.

Gus was a little man, stocky by frame, and at 68.5 kg one of the lightest astronauts in the team. He was by nature less buoyant than Shepard, or several other astronauts, and kept within an aura of professionalism that seemed to some an insulation from boyish enthusiasm. Gus felt things deeply, not least the responsibilities he carried for team and country, and was the astronauts' astronaut who never hurried a test, dodged a check or evaded analysis. It was to be his lot to fly once more in space, then die in the cramped confines of another spacecraft on the ground. Grissom woke to physician Bill Douglas's call at 1:10 am, Friday, 21 July. His day for spaceflight had dawned before the Sun rolled up across the Atlantic horizon on launch pad 5 and Liberty Bell 7, waiting atop the last Redstone to carry a man into space.

The launch team were confident of a 6:00 am lift-off and the late wake time for Gus Grissom followed a re-structured preparation schedule; the astronaut had shaved before retiring and a smoother status report arrangement prior to getting in the spacecraft compressed the sequence of pre-flight activities. After a low-residue breakfast in accordance with a three-day menu set by Beatrice Finklestein from the Aerospace Medical Laboratory of US Air Force Systems Command's Aeronautical Systems Division, Grissom moved through for his medical check. It was a little before 2:00 am. The control on food intake was irrelevant for the 15 minute flight but it was a useful rehearsal of procedures essential to the comparatively long orbital flights to come. Three meals were prepared for each eating period in the three days prior to flight: one to be consumed by the prime pilot, one eaten by the back-up crewmember, and one placed in a freezer for a full day so that any physical reaction by either astronaut could be checked against a preserved duplicate of the intake.

In the medical room, Grissom was observed to be slightly tired, fully aware of the dangers of his mission (a vital and necessary product of a stable mentality), and to have a mild sore throat. Next came the suit-up preparation, with bio-sensors similar to those employed for Shepard's flight, attached to Grissom's body. At this point a technician at Hangar S checked with the pad 5 blockhouse to confirm the status of the flight preparation; if holds were under way the astronaut could be relieved of uncomfortable time in the pressure suit. Weather checks had shown conditions to be acceptable before the second half of the countdown began at 2:30 am. By shortly after 3:00, Grissom was ready for pressure checks, wearing for the first time a urine collection device put together the day before; the tensions of MR-3 pre-flight activity had stimulated Shepard and caused moderate discomfort until he was able to remove his suit. Before donning his own suit, Grissom was assured of its integrity by the results of a pressure check which had it inflated to 258 mmHg. Nevertheless, he would be an integral part of further checks aimed at final qualification in the event it was required due to cabin de-pressurization.

With the suit on, the pilot was placed in a couch while the suit was again inflated to 258 mmHg, but this time the ventilation flow was turned off so that careful monitoring of the quantity of oxygen required to maintain pressure would demonstrate the dynamic leak rate: 400 cc/min for Shepard's suit but only 175 cc/min for Grissom's. The 'leak' rate of course included oxygen consumed by the pilot. From this point on Grissom's suit would not be disturbed until he returned to Earth, although the faceplate could be opened as required.

From the suit donning room Grissom made his way back to the examination room where a final check of biosensor connections on an oscilloscope confirmed the validity of medical preparations. Next, a portable ventilation pack was plugged in to the suit while Grissom walked out to the transfer van, allowing him to connect to the van's ventilation system for the slow ride to pad 5. An 8-channel recorder continued to monitor his biophysical condition. It took just 25 minutes to ride across the Cape. A final reading was taken from the recorder and hurriedly delivered to the blockhouse for calibration before word was sent to the van for Grissom to plug in the portable ventilation pack and walk to the elevator waiting to carry him to Liberty Bell 7 six stories above the pad.

It was just 3:58 am as Grissom slid over the open hatch sill and on to the contoured couch, situated in the centre of what seemed like a giant watch containing a myriad intricate dials, switches, wires and connectors. It was stark contrast to the massive ironwork of the booster gantry, and it had a very different smell. Like the cockpit of a fighter plane it was a custom built office for a potentially hazardous occupation. The suit technician reached across and attached the main ventilation hose, a communication line, the bio-sensor connector and the helmet visor seal hose. Suitably plugged in, Grissom was now to be strapped up. Loosely at first as buckles slid over straps. Minutes later a pure oxygen purge began, quantities of the gas flowing into and through environmental and suit control systems until the technicians were sure the loop carried at least a 95% pure 0_2 content. Now the restraint straps could be fully tightened before a final look around the capsule. At 4:50, just 90 minutes before the planed launch time, technicians secured the hatch's 70 bolts while Grissom's pulse rate rose to 96 beats/min then fell back to its usual 80; this man was more physiologically responsive than Shepard had been indenting his bio-data with each minor event associated with capsule preparation.

Soon after the hatch was placed in position a cabin purge began similar to that performed earlier on the spacecraft's environmental system but at 4:58 a technician on the gantry reported that a single hatch bolt was misaligned; Grissom's pulse shot to 98. At 5:15 a hold was authorized while engineers examined the bolts, judged the hatch to be both safe for the flight and capable of separating at fuse detonation, before resuming the countdown at 5:45 am. Delayed now by thirty minutes, a further postponement was called at 6:00 am while pad lights were turned off. That delay lasted only 9 minutes but at 6:24 am a third hold was ordered for better cloud conditions to accompany the flight. The count resumed at 7:05 am from the T-15 minute mark, Grissom's heart rate pushing up and frequently exceeding the 100 beats/min level.

The pilot had consistently presented a coupled response to external events, tracking the rise and fall of tensions with reciprocal elevation and reduction in pulse and respiration. About 1½ hours before launch the gantry moved back from the capsule and viewing this event Grissom felt as though he was falling over backwards. Clearly no two pilots would sense the same response, exhibit the same reflex or contrive the same reaction. It was 7:20 am when the countdown reached zero and Redstone's single engine thrust into life. Shepard was communicating from the Mercury mission control center and the confident, calmly knowing voice of 'Shorty' Powers fed information to the press who more often than not connected his monotones direct to the listening public rather than attempt a more satisfying commentary; Mercury was becoming a self-contained package, forced by the need to keep a distance between the operations in the control center and the eager press anxious to report every heart beat and hiccup.

'Lift off,' came the voice from the blockhouse, one second after the minute. 'Ah Roger, this is Liberty Bell 7. The clock is operating.' Came the voice of Alan Shepard: 'Loud and clear, José, don't cry too much.' This last plea referred to a maudlin character created by humorist Bill Dana, a Mexican 'astronaut' adopted for a comedy routine where, upon being asked how he expected to spend the long, lonely, depressing hours of space flight, concluded that he planned to 'cry a lot!' Adopted by the seven Mercury pilots the call became a watchword, a symbol of friendly banter, whenever one was in a compromising situation. 'O.K. it's a nice ride up to now,' replied Gus Grissom as the rocket sped on up, visibly faster than Shepard's Redstone.

On through max q, the period of maximum aerodynamic pressure, the pilot reported his condition as one by one the scheduled events cycled off exactly as planned. Maximum g occurred just at the point of booster shut down where Grissom weighed 6.3 times his Earth value. At 2 min 23 sec the Redstone had completed its task and propelled the capsule to a speed of 8,317 km/hr. Just 10 seconds later the spacecraft separated, damped oscillations for 5 sec and then turned around on ASCS control to face backward and pitch down to the −34° retro-fire attitude. At this point in the MR-3 mission, Shepard had been required to move first to Manual Proportional in pitch, then yaw, and finally roll. Grissom switched all three axes to the MP mode and assumed full control from the start. For a minute or so Grissom was expected to retain control by flying on instruments but the satisfying view out the new trapezoidal window, with the Earth occupying the bottom two-thirds and the blackness of space filling the top third, momentarily snatched his attention; with the spacecraft pitched down it was a picture-window on the home planet, unlike Shepard's limited view.

After trying basic maneuvers in the three axes on dials indicating vectors on the display console, Grissom attempted to control attitude by viewing the Earth, a period in the flight plan where Shepard had resorted to the periscope, his only means of external reference. Grissom was glad to have a full minute of time in which to observe the Earth, using it as a reference for attitude control, a determination as to whether a visual reference would be acceptable for orbiting astronauts if the instruments should fail. He found it easier to control pitch and roll than yaw but ran out of time as retro-fire approached. Just 1 min 41 sec after assuming manual control of the capsule, little more than 2 min after separating from the booster, Grissom monitored the retro-fire sequence and held attitude control on the MP system. Manually punching the retro-rocket ignition button at 5 min 10 sec into the flight he switched to the RSCS mode shortly before the retro-rocket pack was jettisoned. This period was covered by Fly-By-Wire, ASCS and Manual Proportional systems on Shepard's flight.

Falling down toward the Atlantic now, Grissom's heart rate reached its highest point, peaking at 175/min; at lift-off it had reached 140 and steadily climed from there. Nevertheless, Grissom found control on the new RSCS mode delightfully responsive although propellant consumption was extremely high.

MR-4 pilot Virgil I. Grissom chats with Astronaut Flight Surgeon Dr. William K. Douglas before the flight. Note the portable ventilator carried by Grissom and the bio-sensor test box over Douglas's shoulder.

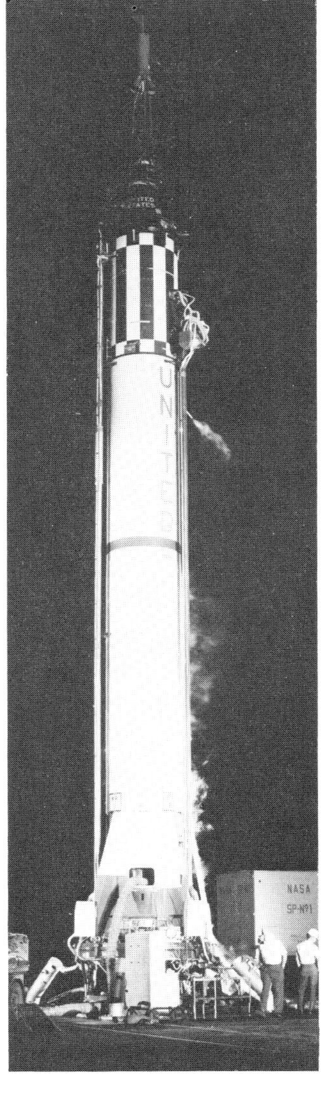

Astronauts (left to right) Glenn, Schirra and Shepard wait tensed for the launch of MR-4 as the latter communicates with Grissom aboard the spacecraft.

← Mercury-Redstone 4 stands ready for flight in the pre-dawn hours of 19 July, 1961, as technicians wait for astronaut Grissom at Pad 5.

It's wheels awash, helicopter 32 struggles to retrieve the Liberty Bell 7 capsule.

Shortly after the periscope retracted at 6 min 40 sec, Liberty Bell 7 automatically moved to the pitched up position, the nose of Mercury pointing 40° above the horizontal. Safe now in re-entry attitude, Grissom checked HF (High Frequency) radio communications and then moved to the UHF (Ultra HF) system. Re-entry began at 7 min 46 sec and deceleration built up rapidly, reaching a maximum 11.1 g just 34 seconds later, rapidly falling off again as the spacecraft decended through the atmosphere. On the way down Grissom noticed a mild buffeting sensation as the capsule oscillated slightly during its slow spin, but more evident was the roaring sound as the berylium heat shield took the brunt of deceleration.

At 9 min 41 sec into the flight, the drogue parachute popped out and partially arrested the descent rate as Liberty Bell bounced slightly under the spreading canopy. Seconds later the drogue was released as planned, the capsule accelerating now in free fall, but another and more comforting jolt came when the big main parachute pulled free to the reefed condition, and then, fully deployed itself, with a much milder secondary shock. Grissom could see most of the parachute as the spacecraft swayed slightly in the air currents and two small splits gave momentary concern but they got no worse and the pleasing view of an unfurled canopy brought comfort. His heart rate now hovering around the 150/min level, Grissom opened his faceplate and disconnected the visor seal hose in preparation for splashdown. The final significant event was deployment of the landing bag when with a loud clunk the heat shield fell from the aft bulkhead to hang on the restraint straps holding it to the capsule. Just 15 min 37 sec after launch the spacecraft sliced into the Atlantic, going under as water covered the window and the forward parachute section, righting itself seconds later after Grissom ejected the reserve parachute.

Now it was time to uncouple himself from the capsule, the helmet oxygen outlet hose coming off first, followed by release of chest, lap, shoulder and knee straps. The suit carried a rubber diaphragm, or neck dam, rolled out from under the helmet ring to hold firm against the pilot's throat, so preventing water from pouring into the suit should it be necessary to leave the capsule and swim to safety. With the oxygen inlet hose still attached for cooling, and the helmet communications lead still plugged in to the capsule's receiver, Grissom set about the task of recording instrument data while the recovery helicopters moved in for the pick-up. Less than three minutes after splashdown the first recovery helicopter was orbiting the capsule, calling Grissom to request his status and enquiring as to his readiness for proceeding with egress.

From inside the bobbing capsule came the response: 'Roger, give me about another 5 minutes here to mark these switch positions, before I give you a call to come in and hook on. Are you ready to come in and hook on anytime?' There was an affirmation. 'O.K., give me about another 3 or 4 minutes here to take these switch positions, then I'll be ready for you.' Grissom completed his checks and removed the oxygen inlet supply, coupled now only by the communications lead. Ten minutes after splashdown the astronaut called in the recovery helicopters: 'O.K., latch on, then give me a call and I'll power down and blow the hatch, O.K?' It was, and the recovery team moved in on Liberty Bell. The reference to 'blow the hatch' concerned the fuse linking the 70 bolts that, upon activation, would jettison the cover. Grissom had removed the cover from the safety pin, and pulled the pin from its socket. He would fire the door when the helicopter had the capsule on winch.

Grissom lay back on the formed contour couch and mused over the possibility of smuggling out a knife he had removed from the edge of the hatch cover and just then placed in the survival pack, retrieved as a souvenir from his first space flight, when suddenly with a loud bang the bolts detonated and the hatch cover flew away from its mounting! Blue sky spread across the open hatch as water spilled into the capsule. Instinctively, Grissom pulled off his helmet, dropped it in the capsule, and plunged through the hatch into the water. Swimming free he felt a tethered line connecting a dye marker can restrain his shoulder but with a shrug he pulled away and scanned the situation. Lt. James L. Lewis was in the prime helicopter, with co-pilot Lt. John Reinhard, when the hatch prematurely blew. At that time, Reinhard was about to cut the 4.2 meter long whip antenna from off the top of the capsule and both men sprang to a reflexed routine.

Realizing that Grissom was safely away from the listing capsule, he could be seen some distance away floating with his shoulders out of water, Lewis lowered the helicopter while Reinhard grabbed the recovery pole after quickly snipping off the long antenna. The spacecraft was sinking fast and Lewis's helicopter was well on its way to follow as all three wheels trod water. Suddenly, the line snapped taught; the capsule was snared. Slowly the helicopter rose from the heaving sea and Reinhard started to lower the horsecollar for Grissom, who by this time had swum back in to the vicinity of the spacecraft to help if needed. He was not, but the comforting sight of the descending collar was aborted as it suddenly started to rise back up! Fighting to lift the submerged capsule back above the surface, Lewis got a flashing red light on his instrument console, indicating an overheated engine, and immediately advised Reinhard not to recover the pilot but to call up the second helicopter co-piloted by George Cox who had so recently snatched Alan Shepard from a spot not very far away.

It seemed to be a battle the prime helicopter was slowly winning as Liberty Bell almost cleared the water at one point. But the flashing red warning light threatened engine failure at any time; the water-filled capsule hanging below the chopper had a weight of $2\frac{1}{4}$ tonnes, just above the helicopter's lifting capacity. Seeing the struggle going on, and observing the second helicopter moving in, Grissom tried with difficulty to swim away from the scene as he fought the downwash from two sets of whirling blades, the water pounded by throbbing engines above his head. Grissom knew it was getting increasingly difficult for him to keep above water as air gradually leaked from his suit through an inlet valve he had forgotten to close. By the time the second helicopter was anywhere near a position from which it could lower its own horsecollar, the Mercury pilot was going under, swallowing water and fighting hard to remain afloat.

Meanwhile, across at the first helicopter, Lewis decided to heed the red light, cutting Liberty Bell free. Slowly the capsule sank down into the Atlantic at the start of a long drifting descent to the ocean floor 5,120 meters below. Grissom had been in the sea for only five minutes but it already seemed much longer as the swell broke continuously over his head. Grasping for the horsecollar he slid into it backwards on and lay there in its restraint as slowly he ascended to the open door of the helicopter. Inside, his first instinct was to don a life preserver – just in case the chopper failed and he ended up back in the water! It had been an ordeal, and that was very evident at the cursory medical examination on the carrier *Randolph* where medical attendants who had seen to Shepard tended a very weary astronaut from the sea.

It transpired that Liberty Bell could probably have been recovered after all The red light was a false alarm. There was no hope of getting the capsule from the ocean floor, however, and Grissom's careful note taking was just so much wasted effort. Apart from the problems with the premature hatch jettison, MR-4 was a good flight, its pilot performed well and as expected, very little difference was observed in his physical state compared to that of his predecessor, and the spacecraft stood up well to its albeit limited flight test. Wally Schirra was assigned to head a review committee set up to look into the hatch problem. It was a crucial part of the system. Had the bolts blown in space the pilot would have been effectively protected by his pressure suit but once started down toward the Earth there was little possibility of him surviving the searing heat of re-entry. From orbital flight temperatures would reach about 650°c on the hatch cover. No one ever did find out why the Liberty Bell lost its cover and tests conducted both prior to and as a result of the MR-4 incident confirmed its integrity as a design concept. There would be continual agreement among engineers that a combination of insidious circumstances conspired to breach convention; it would happen again to Gus Grissom $5\frac{1}{2}$ years later when other insidious combinations took his life.

Lt. John Reinhard watches Liberty Bell 7 as the Marine helicopter begins to lift it free of the water. Seconds later the capsule sank.

In early June, about six weeks before the second suborbital flight, Mercury management decided to tentatively cancel MR-6, and to decide the fate of MR-5 after analysis of results from Grissom's shot. The original plan had been for each astronaut to ride a suborbital mission but that seemed too expensive on both time and hardware in the light of Soviet feats and Presidential commitments. Abe Silverstein convinced Bob Gilruth that MR-4 data would be necessary to confirm the wisdom of transferring attention to Atlas orbital flights. Space Task Group was quietly smarting under the single orbit flight of Yuri Gagarin and nursed a hope that a Mercury-Atlas three orbit mission would steal thunder from the Soviet space machine and raise American space prestige not only across the globe but, perhaps more important, in Congress where questions came thick and fast whenever Russian success pre-empted US achievements. And then it happened again. On 6 August, 1961, cosmonaut Gherman S. Titov reached orbit in his Vostok II space capsule for a 17 orbit flight that would last more than a day!

Because of the time delay, most Americans were fast asleep when Titov was launched at 9:00 am Moscow time. Yuri Gagarin was in Canada on a goodwill tour and knew the flight was about to take place but kept the news to himself until moved ebulliently with the drama of the event. The flight of Vostok 2 was something very different to the pioneering single-orbit mission of the first Soviet manned space flight. It would carry the spacecraft over almost every inhabited part of the globe and it would require the cosmonaut to go to sleep for at least part of the mission. Because the Earth spins from west to east as it moves round the Sun the path of the spacecraft would seem to migrate west on each successive orbit until, a day later, it once again came back over the USSR, the planet in fact having spun round once on its axis while the actual orbit of the spacecraft remained fixed with the Earth's center of mass. Tass correspondent Alexander Romanov was there when Titov left for space and sets the scene by describing first the events of the preceding day as they appeared to him when he arrived at Baykonur.

'It is hot. Our car devours kilometer after kilometer of the smooth macadamized road. Villages, towns, high-tension pylons flash by. Suddenly, at a turn, in the distance there looms up a kind of filigree construction. Coming nearer we begin to discern through the maze of steel the slender cigar-shaped body of a multi-stage rocket. There are men all over the construction surrounding the spaceship. Time and again terse commands come over the public-address system. The last preparations for the flight are under way. An elevator cabin runs down from the top of the rocket and on to the launching pad there steps out a middle-sized fellow in a checked, dark blue shirt tucked into light grey trousers. Quickly running down a ladder, he came up to the group surrounding the chief designer, the man in whose hands converge all the threads in preparing the celestial journey. We would probably have never paid any attention to the boy in the dark blue shirt, who looked like any other worker, had not the man next to me whispered: "That is the cosmonaut – Gherman Stepanovich Titov." By evening all the work had been finished, and we saw the spaceman again. Wearing a flying suit, he was surrounded by comrades who were getting the spaceship ready for its long journey. . . .

'Gherman spoke warmly and sincerely. He thanked his comrades for their good wishes and said he was proud of the trust placed in him and promised to come to the cosmodrome and tell them about the flight on his return to Earth. On the morning of 6 August, the Sun burning hot over the steppe, lit up the silver spaceship, the cosmodrome, and the flowers growing beside the pavement. Members of the Government commission, distinguished Soviet scientists, designers, engine testers and the staff of the launching command assembled at the command post. The chairman of the commission opens the meeting, the last before the spaceship's flight. The reports are terse to the extreme. Their meaning could be expressed in one or two words: "Everything is ready for the flight." The State commission names the moment for launching. The launching area is cleared, only essential staff remain.

'Here comes the light-blue bus carrying the spaceman. Titov alights, crosses to the launching area and the lift takes him up to his cabin. He is wearing an orange-colored flying suit with

"USSR" written on it, and walks rather clumsily. The flying suit, of course, is not for strolling on Earth. Under the suit are various transmitters, which will send back information about the pilot's physical condition. Having said goodbye to his friends – future cosmonauts – Gherman Titov walks to the lift. . . . The chairman of the commission, scientists and the chief designer shake the pilot firmly by the hand, embrace him and wish him a successful flight. Gherman ascends the stairway to the platform where the lift is situated and addresses those assembled below, and all the Soviet people. He says that he will fulfil with honour the task which has been entrusted to him by the party and the Government.

'All those on the cosmodrome warmly applauded him. We see Titov enter the cabin. The command "Take off" will be given shortly . . . Titov was then brought a small book, bearing on its cover the words "Log Book of the Spaceship Vostok II." A pencil is attached to the cover and the cosmonaut checks that it is firmly fixed, so that it will not be lost during the state of weightlessness. Yuri Gagarin did not attach his pencil firmly and lost it. Yuri found it afterwards, Titov laughs . . . To observe the spaceship take off we go to a special square about two kilometers from the launching place, but we are not cut off from what is happening on the spaceship. The radio brings the final commands. Ten minutes to go . . . the chairman of the State Commission asks the cosmonaut how he feels. "I feel wonderful, wonderful," he says. "Thank you for your attention." Five minutes to go . . . and at last, in the silence, rings out the last command: "Take off."

'From a distance we can see the silver rocket, already completely free from the supporting gantry. Another second, exactly 09:00, and the rocket – propelled by some unbelievable, miraculous force – slowly, it seems very slowly, leaves the Earth. Gathering force it streaks more and more quickly upwards like a roaring fiery globe. At a comparatively low height the spaceship inclines to the side and flies on its set course into orbit. . . .'

Romanov's report was the first such account of a rocket flight to emerge from the Soviet Union. Titov's spaceship was very similar to Gagarin's, weighing 4.7 tonnes and comprising the same elements as Vostok I. Shortly after completing the first orbit – each path around the planet would take nearly 89 minutes – the pilot began to feel nauseated, a sensation of vertigo welling in his head. Moving down across the Pacific and past Cape Horn the sickness grew worse until Titov felt it could not be relieved unless the flight was terminated. At this period he was in a communications blackout and resolved to fight the permeating sensation. Much of the time during the third and fourth orbits the cosmonaut was required to radio 'greetings' to the people of Africa and then Europe as the orbital migration took him on a world circling trip. During the fourth orbit he rested for an hour, then performed physical exercises and work 'as set by the flight task.' Then more greetings, this time to the citizens of South America. They were political gestures for only large communication stations could even receive the broadcasts and the vast majority of the public had no knowledge of the transmissions. Unlike Gagarin, Major Titov was required to manually test and evaluate the attitude control and orientation system, which he undertook to perform on the fifth orbit. As Tass, the Soviet news agency, would later report: 'During the flight over Soviet territory the radio-television system transmitted pictures showing the calm and smiling face of the Soviet cosmonaut. Through the multi-channel radio-telemetric system extensive information of a scientific nature continued to come in, as well as detailed data about the functioning of the spaceship Vostok II. While flying over Kwangchow Maj. Titov sent greetings to the people of Asia, and while flying over Melbourne he transmitted greetings to the people of Australia.' But this was Titov's most nauseating time, the sickness he felt earlier had reached a peak, and the cosmonaut was far from the alert test pilot he had trained to be for more than a year.

But what the cosmonaut experienced was not to be exclusively the preserve of Soviet spacemen, although it would be expressed in a slightly different form by American astronauts. Space sickness was not altogether unexpected and several physicians had projected hypotheses concerning the effect of weightlessness on the human body. It may be that Soviet pilots did not receive quite the same degree of training in areas specifically associated with this malady but whatever the cause, America's astronaut corps found it more difficult to induce a similar feeling. It all amounted to a simple lack of temporal adjustment to spatial conditions. And that simply meant that without gravity the otolithic functions of the inner ear were upset. Information about the body's orientation is passed to the brain by nerves which respond to small variations in the position of tiny bones, each of which reacts to the physical changes in direction and motion of the body of which they form a part.

In effect, the otolithic function provides the information the brain needs to balance the whole structure and is extremely sensitive to the force of gravity, and small changes in gravity vectors as the body tilts or changes direction of stance. In space the otolithic function is so disorientated that it upsets the delicate balance between the tiny bones in the inner ear and the nerve responses the brain is conditioned to receive through millions of years of evolution and acclimatisation to a 1 g environment. As a result, peripheral side effects emerge which induce the nauseating feeling of motion sickness, an effect not unlike that caused by jolting bus journeys or roller coaster rides at the fairground. It was a condition made worse by the availability of space; the cramped confines of the Mercury capsule, while not all that much smaller than Vostok, may have been, nevertheless, sufficiently sized to inhibit the full effects of this phenomena which would only become a real concern with larger manned space vehicles several years later.

Experience with later manned space flights would show that the level and magnitude of conditioning each astronaut received determined his susceptability to space sickness; the more training a man received the less prone he seemed to be to the effects of weightlessness, at least in the otolithic functions. However, in Titov's case the condition seemed to stabilize around the end of the fifth orbit, a path around the world which carried the Vostok II spacecraft up the continent of South America, across the North Atlantic, past the northern tip of Scotland, down across the Soviet Union, past the coast of Vietnam, across Australia and the Pacific to ascend again up South America and past Cuba to start the sixth orbit.

During that pass Titov received greetings from Gagarin and once more tested the manual attitude control system. On the seventh pass the spacecraft passed up the coast of Florida and, had the pilot wished, he could have gazed down on the sprawling, and expanding, launch facilities at Cape Canaveral to view the competition as NASA planned Moon flights and other ambitious ventures in space. Rest time was approaching, the sleep period scheduled to begin at 6:30 pm Moscow time as the spacecraft coasted over Africa. Titov settled back down into the ejection seat and made a special point of securing his arms to prevent them floating in front of his face in the weightless environment. As he would later testify, 'I slept the sleep of the just and spent 35 min longer than envisaged by the program. My sleep was good, without dreams. In contrast to Earth conditions, I didn't feel the necessity of turning from side to side.'

Titov woke at the start of the 13th orbit with the spacecraft passing over Indonesia. Shortly after reporting his continued success, Tass chose that time to announce 'The spaceship also carries scientific apparatus for obtaining supplementary data on the influence of cosmic radiation on living organisms. On board are biological objects. During the flight observation of Earth and sky is carried out through three portholes. The cosmonaut may also, if he wishes, use an optical instrument of 3–5 magnification.' Refreshed by his 8 hr rest, Titov felt better for not having performed rapid head movements and quickly learned to move as slowly as possible when transferring his attention from one point to another. But it was time for a short meal and then preparations for re-entry, the food consumed from tubes and containers designed to constrain the free movement of particles that could break free and endanger the safe operation of critical systems.

At the beginning of the 17th orbit Titov monitored the automatic orientation of the spacecraft to a retro-fire attitude, followed shortly after by ignition of the retro-motor and separation of the spherical capsule from the rest of the assembly. His re-entry path carried him across the Sahara and over the Nile Delta to plunge toward the Saratove region in the Soviet Union. 'I was very interested to observe the brilliant luminescence of the air around the spaceship as it entered the denser layers of the atmosphere, and the changes in color of this luminescence as my speed and altitude changed,' he recalled later. 'When the deceleration forces began to act, the weightless condition came to an end, but I observed no sharp or abrupt transition – I felt simply that I had returned to a normal state.'

During the descent Titov elected to eject from the capsule,

and fired himself free at a height of 6.5 km to float slowly down to his recovery zone, more than 700 km south-east of Moscow. His flight had lasted 25 hr 18 min during which time he had travelled a distance of 600,000 km between 183 km and 244 km above the Earth's surface. Titov had flown his spaceship for a period exceeding that for which the Mercury capsule had been designed and the day-long flight of Vostok II pressed home the large amount of work still required to get an American in orbit. It seemed for a while that the limited success of two ballistic Mercury shots had won support from American pressmen for after Vostok II's 25 hr 18 min flight media representatives questioned the validity of Soviet claims where once they would have lambasted NASA.

But the message was clear: if NASA couldn't pull off a manned orbital flight within the next few months the program would find trouble in Congress, for Congressmen and Senators were critical of the further slip in race positions. Just one week after Titov returned to Earth, Bob Gilruth sent Silverstein a memo recommending termination of the Mercury-Redstone program in favor of full scale maturation of the Mercury-Atlas flights. Some of the astronauts wanted to fly just one more Redstone mission, the spacecraft and booster were, after all, ready for use. But the pressing necessity to move ahead with orbital plans denied MR-5 a place in history. The coup de grace was delivered in a formal announcement from NASA headquarters on 18 August. Just two days later the first stage of NASA's adopted ABMA Saturn launch vehicle – the S-1 stage brought from the Marshall Space Flight Center by a barge called *Compromise* – was erected by Henry Crunk's team on the newly constructed Launch Complex 34. It was to be launched within a couple of months with a thrust equal to more than four Atlas missiles, a fitting tribute to a new era for the National Aeronautics and Space Administration. It was a time when NASA was looking for a site on which to build its new Manned Spacecraft Center, a home for Langley's Space Task Group befitting the magnitude of their responsibility.

The first Mercury capsule to demonstrate orbital flight, spacecraft 8A is seen here atop the MA-4 booster prior to launch in September, 1961.

But for most Mercury technicians, engineers and managers, it was the home stretch toward the first American in orbit. At the beginning of that year, the Mercury-Atlas 4 mission was scheduled as the first three-orbit flight with a crewman simulator on board; it was to have followed MA-3, the suborbital test. Unfortunately, MA-3 was a disaster and MA-4 was reduced in consequence to a single-orbit flight to gather information on the performance of a Mercury capsule in orbit, a feat that had yet to be accomplished. The nominal Mercury mission required the capsule to support a man in space for three orbits of the Earth and a test flight of that duration was put back to MA-5 at the earliest. The schedule, it seemed, was slipping again and, despite good progress during that year with Redstone ballistic flights, both manned and unmanned Atlas missions were proving difficult to accomplish. It began to look as though an American astronaut would have to wait until the turn of the year to fly the capsule on the task for which it had been designed. But there was no time for recrimination and engineers worked around the clock preparing MA-4 at Cape Canaveral.

The spacecraft for this fifth Atlas shot in support of Mercury was capsule 8, dubbed 8A, originally flown on the aborted MA-3 flight and retrieved from the Atlantic by the Navy recovery force. A substantial amount of work was necessary to ready it for an operational test and several minor modifications kept the capsule at McDonnell's St. Louis plant from late April to early August. Atlas booster 88-D had been ready since mid July at Cape Canaveral's pad 14 and on 3 August the capsule joined it in preparation for a launch less than three weeks later. That plan was thwarted, however, by reports that transistors similar to those fitted to both booster and spacecraft were found to contain balls of solder that could cause the equipment to fail and endanger the mission.

Back to Hangar S went capsule 8 while engineers crawled over booster number 88, changing the defective transistors. On the first day of September the spacecraft was once again mated to the Atlas adaptor as personnel from STG, McDonnell and Convair fussily accompanied the Cape's launch team. A lot would ride on MA-4. NASA desperately needed to get the orbital operations with Mercury under way to clear the decks for manned flight; STG wanted final confirmation of their confidence in the capsule's ability to support human life in orbit; McDonnell keenly awaited flight performance data to see how well their protege stood up as America's first manned spacecraft; Convair had buried a host of technical changes and modifications to existing systems within the smooth exterior of the Atlas booster as what was essentially a military ballistic missile rode the learning curve of new technology. It was to be the first time a Mercury spacecraft went into orbit, albeit for a single trip around the globe before re-entry. Yet in the wake of what the Russians had accomplished already that year it was eagerly awaited to bolster brave confidence and strong words that had already, several months earlier, spoken of an American on the Moon by 1970.

Despite threatening hurricanes that lashed the Corpus Christi tracking station and ships in the prime recovery area, MA-4 counted down to lift-off at 9:04 am on Wednesday, 13 September. Because the Atlas booster was considerably more powerful than the comparatively small Redstone employed for ballistic flight, contingency plans embraced a more copious repertoire of recovery ships and aircraft on standby to retrieve the capsule. Area A supported the possibility of an abort during the first 4 min 58 sec of the ascent phase and provided recovery teams in an area extending 3,540 km down-range beginning 21 km from the launch pad. It was divided into two sections: the first 885 km covered the first 1 min 12 sec of flight, and was 96 km across; the second occupied the rest of the elongated ellipse across a width of 48 km. Area B was a small ellipse further down-range to cover an abort at 4 min 58 sec. Area C was a similar zone for contingencies at 5 min 1 sec. Area D spread across 32 km and extended a length of 196 km down-range and would cover the point where the spacecraft would be committed for a go/no-go decision to orbit; Area E was an ellipse 39 km by 372 km for aborts up to 5 min 4 sec after the decision to commit the spacecraft.

Detailed and precise abort contingency areas like those arranged for MA-4 were typical of the substantial support functions required of the Navy and were to be duplicated in one form or another for every orbital mission. During the ascent, engineers watched closely the performance of Atlas 88-D; it was, after all, only the second flight of a 'thick-skinned' booster modified as a result of the MA-1 accident more than a year before where vibrations had caused the whole assembly to break up in the air. But MA-4 rumbled on toward space, thrashing the Cape with thunder claps of exhaust noise superimposed over the shattering boom of supersonic flight. A lot of vibration was observed in data that streamed back to the blockhouse and the two booster engines shut down a little early before sliding free of the missile's aft end to leave the single sustainer engine still thrusting on. That engine too cut down early but the speed of the bird at booster separation had been high so on balance the system intelligently sorted itself out.

Seconds later the posigrade rockets on Mercury's retro-package popped the capsule away from the cylindrical adaptor, attitude thrusters damped out oscillations, and 5 seconds later the spacecraft started to swing round from nose-forward to nose-aft, pitched down 34° toward the Earth. It took 50 seconds to accomplish this, longer than expected, and the thrusters burned more than four times the anticipated 1 kg of hydrogen peroxide propellant in getting there. At a speed of 28,205 km/hr, spacecraft 8A was in orbit, the first Mercury to achieve that feat. The complex, elongated box structure that reclined where on another flight an astronaut would lay on a contour couch, was now to show whether man could safely fly the mechanical can. High oxygen consumption rates seen on telemetry coming from Grissom's Liberty Bell capsule were also observed from capsule 8A. First the cabin sealed off at the required one-third atmospheric pressure level, then fell quickly before restoring itself to a higher value than expected.

Spacecraft control responded well to commands to keep the vehicle pitched down 34°, although it did drift from this commanded pitch value because two 0.45 kg thrusters failed. As the spacecraft swept around the world it mimicked the manned vehicles yet to fly, providing ground tacking station personnel valuable experience with acquiring, tracking, and communicating with a capsule programed to respond to calls from the transmitters. The astronauts too were part of the 'dry-run' and participated as though the capsule was a manned spacecraft: Carpenter was stationed at the Muchea tracking station in Australia; Cooper was at Pt. Arguello in California; Wally Schirra sat it out at the retro-fire station near Guaymas, New Mexico; Deke Slayton was at Bermuda for critical ascent and recovery operations; Al Shepard, Gus Grissom, and John Glenn were on hand at the Cape. It was the first full test of a global tracking network, and the Goddard Space Flight Center received vast quantities of data pouring in from all around the world. As communication and tracking operations hopped from one station to the next, keeping pace with the rapidly moving spacecraft overhead, engineers passed across information aimed at predicting coordinates where the spacecraft could be found as it rose above the horizon.

The main stations linked in the primary net connected to Cape Canaveral's control center were situated at: Grand Bahama Island; Grand Turk Island; Bermuda; Grand Canary Island; a tracking ship in mid-Atlantic; stations at Kano, Nigeria and Zanzibar in Africa; an Indian Ocean ship; Muchea and Woomera in Australia; Canton Island and Kauai Island in the Pacific; and at Point Arguello, Guaymas, White Sands, Corpus Christi and Eglin on the American continent. In addition, 20 domestic communication agencies around the world provided telephone, teletype, land, radio and cable facilities to plug the various stations into one integrated operation which at times passed data to Goddard at the rate of 1,000 bits per second. A 'bit' is a specific piece of information, in effect the smallest parcel necessary to differentiate between alternatives.

At Cape Canaveral's Mercury Control Center, 15 flight controllers monitored consoles and wall displays as the room lit up like a Christmas tree, flashing information almost faster than it could be humanly absorbed. In every sense it was a machine set in motion to monitor another machine, patronisingly spitting out information to humans intent on controlling the panoply of events; yet in a very real way the entire structure was coordinated by engineers and technicians without whom the operation would be meaningless. The most important job for MA-4 trackers was to demonstrate a 'real-time' capability for digesting information pertinent to understanding the condition of the spacecraft. Earlier tracking requirements for unmanned space projects could automate many of the systems that had to be dealt with at once in a manned project. The life of the astronaut was of paramount importance and there were so many unknowns related to the performance of both man and machine in space that it was at one stroke a test of each separate element and of the coordinated whole.

Real-time data relay was to be the system of nerves that laced the world on every manned space mission and the MA-4 network would prove that the concept was up to the task of its demanding specification. As the spacecraft neared Hawaii for the first time the attitude control system held the capsule steady for retro-fire as first one, then the second, and finally the third solid propellant motor lit off to decelerate the vehicle for re-entry. On autopilot, the capsule swung into re-entry attitude and sliced into the atmosphere at a shallow 2° to the ground below; Redstone ballistic flights had the spacecraft entering the atmosphere at 6–7°. This was the first time a Mercury capsule would experience the kind of re-entry for which it had been designed. The two earlier Atlas boosted re-entry missions had been from ballistic trajectories where peak heating was experienced for less than a minute – capsule 8A would sustain temperatures above 1,100°c for nearly seven minutes!

Although travelling almost level with the ground below, severe friction with the atmosphere would dramatically slow the capsule until, just a few kilometres from the surface, it would fall vertically to its Atlantic Ocean splashdown. At a height of about 88 km the ablative heat shield at the base of the capsule began to take the full force of the enormous heat pulse building up. Standing free from the blunt end, the entry shock wave had a temperature of more than 5,250°c, far beyond the level any material could possibly survive, and, but for the unique design of Al Eggers and Max Faget, also the temperature the capsule would have heated to during re-entry. At a height of 74 km the spacecraft began to rapidly decelerate from its re-entry speed of 27,400 km/hr and, in the course of flying a slant range of 740 km, dropped to a height of 19 km and a speed of 2,150 km/hr. It took just three minutes to reach the lower level, the most severe period of deceleration producing a force of 8 g which although severe to an untrained participant was only two-thirds the g force experienced during descent from a Redstone boosted ballistic shot.

At a height of 12.8 km the drogue parachute popped out and at 3 km the antenna fairing was jettisoned, pulling free the main ringsail parachute, first to the reefed and then to a fully deployed position. The capsule splashed down just 1 hr 49 min 20 sec after lift-off to be recovered by the USS *Decatur*, a destroyer situated more than 50 km away which arrived to collect the spacecraft 1 hr 22 min later. Capsule 8A was one of the old type, very similar to spacecraft 7 employed for Shepard's flight, with two circular windows badly situated for optimum visibility, and an old style hatch cover. But it had served its purpose, the first Mercury to demonstrate orbital flight, and it was a good omen for MA-5. That flight could now proceed as the three-orbit mission demonstrating the specified design potential, but with a chimpanzee as precursor qualification to a manned flight projected for MA-6.

As plans moved ahead to speed the preparation of Atlas and Mercury for the 'chimponaut' mission, NASA management personnel were busy with selection of a new site for Space Task Group personnel. Cognizant of the large demands placed on the Langley Research Center by the squatting STG camp, NASA Administrator Glennan had favored moving the manned space flight personnel to facilities he believed should be built at the Ames Research Center. When Webb replaced Glennan in January, 1961, the new Administrator pressed for money to build a laboratory specially configured for the unique requirements of the man in space program but failed to convince the Budget Bureau. In the new and dramatically upgraded posture following Kennedy's 25 May Moon landing speech a manned space flight facility was not only inevitable, but specifically budgeted in the amended request for funds.

The Fiscal Year 1962 amendment of May 1961 (FY 1962 began 1 July 1961) added $549 million to the March budget of $1,235.3 million, which was itself more than $125 million up on Eisenhower's January proposed budget of $1,109.63 million. The new FY1962 total ($1,784.3 million) included $60 million for a new manned flight laboratory in addition to $160 million for the Apollo project, an increase for Apollo of $130.5 million over the sum allowed for by the preceding administration in the White House. Clearly manned flight would require major new commitments in facilities construction and NASA calculated a need to anticipate a work force exceeding 3,100 secretaries, technicians, engineers and managers at the new manned flight center.

NASA's manned space flight director Brainerd Holmes (left) shows a model of the Saturn V launch complex to Florida Representative Edward T. Gurney.

John F. Parsons, Associate Director of the Ames Research Center, spent most of August, 1961, heading a site selection team on visits to potential areas for the new center. They visited 20 cities and judged each on ten critical points embracing accessibility by water, air and land, proximity to industrial resources, availability of power, academic and educational institutions, placement of recreational facilities and shipping areas, and capability for performing advanced research and development. In Florida they visited sites at Tampa and Jacksonsville; in Louisiana they saw New Orleans, Baton Rouge, Bogalusa and Shreveport; at Texas sites they examined Houston, Beaumont, Corpus Christi, Victoria, Liberty and Harlingen; in California they went to Los Angeles, Berkeley, San Francisco, San Diego, Richmond and Moffett Field; in Massachusetts they went to Boston; in Missouri they visited St. Louis. By early September the site selection team had considered the candidate locations, made their recommendation, passed the decision along to Administrator Webb and President Kennedy, and received a confirmation of acceptance by letter.

On 1 September, the Army Corps of Engineers were requested to manage construction of a new NASA facility to be located outside the city of Houston in Texas. Just 18 days later NASA formally announced the decision, naming the laboratory the Manned Spacecraft Center (MSC). But the selection brought accusation and criticism from pressmen who thought the influential position held by Lyndon Johnson, as Vice President and Chairman of the National Aeronautics and Space Council and Albert Thomas, Chairman of the House Independent Offices Appropriations Subcommittee, and a Houston Congressman to boot, had swung the NASA decision. There is very little written record of the entire selection process but there is no doubt as to the benefits Houston derived from the Manned Spacecraft Center.

The city had been founded in 1836, a non-existent location advertised by two brothers, land speculators Augustus and John Allen, as the 'great interior commercial emporium of Texas.' Having convinced the Congress of the Republic of Texas that they had better move to this dusty plot of land on the bayou, they promptly set about building the city; the first plots were sold in January, 1837, and by June the city was complete and the Congress ensconced! Tremendous industrial growth characterized Houston's development as German immigrants poured into the area to add European accents to what had hitherto been a predominantly negroid population. In 1850 the population was little more than 2,000; a hundred years later it was nearly a million. Rapid growth between 1940 and 1960 made Houston the cotton city of the south, the resource capital of America, the chemical hub of synthetic manufacturing processes. In 1961 oil dominated Houston's empire, and Houston dominated Texas. It was a city in a hurry on the road to an uncertain future, but the manned space program was just what it needed for the technical showcase extravaganzas it would festoon itself with in the coming decade.

The site for NASA's Manned Spacecraft Center was situated 32 km south of downtown Houston. The Humble Oil and Refining Company cut out two specific sites from what had been the Clear Lake Ranch, donating each to Rice University, Houston's proud center of academic and intellectual stimulation. Founded in 1912, the William Marsh Rice University could accomodate 3,500 students and had an eye on the assets to accrue from the presence of a major government research and development center. The first tract of land was sold to NASA for $1.4 million, the second, almost twice as large as the first, was donated by Rice and the two were transacted simultaneously. It would take some time for the Army to build NASA's new facility and in the interim Space Task Group's personnel, numbering nearly 1,000 by this time, would move into temporary accommodation at various rented properties and sites in and around the city of Houston and at nearby Ellington Air Force Base.

Within a year the manned flight personnel would be spread across eleven separate locations until the MSC facilities were ready for occupation late in 1963. But for the moment, STG personnel were migrants with only adopted homes as during the closing months of 1961 they moved lock, stock and barrel from Langley, Virginia, to Houston. It was a contrast for personnel who approved of the rambling scenery in Virginia, but a more fitting monument to work in Houston, the city that prided itself on development and future goals. On 1 November, NASA headquarters ordered that the Space Task Group would henceforth be known as the Manned Spacecraft Center.

Major changes were also being introduced as a result of the new manned space flight objectives, and because the agency was required to accelerate plans for several other, unmanned, projects in support of the lunar landing goal. Announced 24 September, 1961, and effective from 1 November, the headquarter's structure was completely re-vamped. James Webb had considered the reorganization for some time, discussing possibilities for smoothing the administrative structure at a meeting in Virginia during March, 1961. But the Apollo Moon landing goal was clearly the incentive Webb needed as Administrator, and his subordinates needed as lieutenants, to do away with the old, somewhat clumsy and time-consuming machinery for a more streamlined look capable of accepting greater stress without fracturing the infrastructure.

Webb, Dryden and Seamans offered D. Brainerd Holmes, then with the Radio Corporation of America, a position as organizational head of NASA's manned flight projects. Silverstein, coaxed from the Lewis Research Center by Hugh Dryden in 1958, was not interested in joining Webb's lofty administrative set-up, electing to return to Lewis as Director of that facility. In the re-organization, out would go the former headquarters program offices: Office of Advanced Research Programs; Office of Space Flight Programs; Office of Launch Vehicle Programs; and Office of Life Science Programs. Instead, five new offices were set up to replace them: Office of Advanced Research and Technology; Office of Applications; Office of Space Science; Office of Manned Space Flight (OMSF); and the Office of Tracking and Data Acquisition. Silverstein contested the credibility of the new set-up and believed it could not support the requirements of a dramatically upgraded space program; in the new structure, NASA field centers would report to the general manager – the Associate Administrator – so raising the status of field center Directors and concentrating requests for funds and personnel to a general management structure, relieving Headquarters to align its own program offices with the day to day running of project evolution. In this way it did away with the old system whereby field centers communicated with their respective office at Headquarters for administration of all elements essential to the running of that respective facility *and* the conduct of space flight projects.

Although the flow of Mercury development programs migrated in emphasis from Redstone ballistic flights to Atlas orbital missions during the summer of 1961, a peripheral venture came and went without success. Some technicians felt it important to place a satellite in orbit with a capacity for broadcasting signals to Mercury tracking stations, in effect providing a dynamic checkout of the global network. George Low and Harry J. Goett saw Silverstein in May, 1961, and convinced him, reluctantly, of the need for a small satellite, one that could be launched by a four stage solid propellant Scout rocket. Scout was a new launcher designed by NASA for placing small satellites in low orbit. Delays ensued when several minor technical problems plagued preparations, postponing a planned launch date in July until 1 November.

The Mercury-Scout 1 launch was intended to bolster confidence in the ability of the 18 primary tracking stations around the world to successfully find and track the unmanned capsule groomed for MA-4. When that mission came and went it left the M-S1 team apparently without an objective but work went ahead just the same; it would provide valuable experience before the chimponaut flight on MA-5. Unfortunately the project was doomed to the end. A technician had incorrectly wired pitch and yaw gyros before launch and the rocket flung itself all across the sky just seconds into the flight, confused by signals that worked the wrong muscles! The range safety officer provided the coup de grace 43 seconds after lift-off. Mercury-Scout was a fiasco; no phoenix for *its* ashes.

Slowly, preparations for MA-5 gathered momentum; if all went well it was to be the final test before manned orbital flight, little hope for which remained that the feat could be accomplished before the new year. Atlas 93-D arrived at the Cape on 9 October, delayed for minor modifications resulting from the MA-4 transistor problem. Spacecraft 9, selected for this mission, had been at Hangar S since 24 February, its objective switched first from ballistic flight with instruments to a ballistic flight with chimpanzee, then to a three-orbit flight with crewman simulator and finally the three-orbit chimpanzee flight selected as the final full dress rehearsal for manned operations. Launch date was originally set for 7 November, then the 14th when engineers found several problems during checkout, and finally 29 November.

Holloman Air Force Base delivered five chimpanzees to the Cape on 29 October – Enos, Duane, Jim, Rocky, and the veteran Ham from MR-2 – all of whom were fully trained to perform specified tasks during the flight. Enos was selected on the basis of which primate seemed the most ready at the particular time, having spent 1,263 hours in training over 16 months of which 343 hours had been performed in a restraint harness of the type he would be contained by in spacecraft 9. The chimpanzees had exhibited tremendous capacity for learning set operations. In one activity, they were required to pull a lever exactly 50 times to receive a reward of banana pellets. Each chimp would rattle away the first 45 pulls, slowly count off 46, 47, 48 and 49, before determinedly hauling the lever once more with hand cupped under the pellet dispenser! In another case, one chimpanzee operated levers 7,000 times with only 28 mistakes, a success factor of 99.6%.

Trainers and attendants at Holloman were convinced the chimpanzees generated delight over showing off their skills and eagerly demonstrated their capacity for learning. There is some indication too that they recognized temporarily unpleasant activity such as acceleration conditioning on a centrifuge as an important part of the more welcome tasks. None of them were ever placed in situations for which they displayed apprehension or dislike and several technicians whose job it was to remain with the chimpanzees for long periods expressed a belief that criticism of their use in the rocket tests was more a product of human misunderstanding about primate intelligence than a genuine expression of concern.

Again, as with MA-4, the flight would serve to put key teams through their paces. All the tracking stations were staffed, 17 ships and 13 aircraft supported recovery operations, and additional contingency landing areas were provided in case the capsule should have to return through the atmosphere on any other but the scheduled third orbit. The primary landing area for orbits 1, 2, and 3 would migrate slowly from one position to another as the capsule circled the globe and prime recovery teams would move to keep pace with the changing ellipses configured as splashdown areas. The complex pattern of first orbit contingency sites, strung out from the Cape Canaveral area, designated for the MA-4 mission were again provided for on MA-5, with the addition of other contingency sites for a possible abort during the second orbit. These included areas in the Atlantic near the coast of Africa, in the Indian Ocean near the east coast of Africa and across near the west coast of Australia, and in the Pacific Ocean near Hawaii and another near San Diego, California.

At the Cape, Carpenter was stationed in the blockhouse, Grissom and Glenn served as prime and back-up capcoms in the Mercury Control Center, while Shepard was down at the Bermuda tracking station, Schirra was in Australia, Slayton was at Guaymas and Cooper manned Point Arguello. Newsmen began to gather several days before the planned launch date, setting up cameras, sound broadcasting vans, TV sites and tripods. It was becoming a regular feature of the flat, marshy terrain, and there was an increasing number of events to witness as the space program gathered momentum. Everybody recognized the importance of MA-5. If successful, it would lead directly to the first American to orbit the Earth. Not that year perhaps, but certainly sometime early in 1962.

The countdown began on 28 November and Enos was brought to pad 14 for his ride into space. Strapped to a couch inside a container that would be positioned in the place a human would ride, the chimpanzee was in fine condition, temperature, heart rate and respiration flowing into the control center. An annoyingly troublesome countdown brought numerous holds for various minor problems to be corrected, until Walt Williams left the Control Center and drove to pad 14. His presence had the desired effect; as mission director he was concerned that the probability of success for the chimponaut would decrease as launch time moved back. Several minor problems with telemetry coming from the booster plagued the final minutes until, at 1:08 am, the big Atlas thundered into the air and slowly rose on its way into space. Flight director Christopher C. Kraft committed the spacecraft to a full three-orbit mission shortly after the sustainer engine shut down and the capsule was safely on station. Turned around by the ASCS control mode and pitched down 34° to the re-entry attitude, spacecraft 9 sailed on through its first full orbit.

During the planned three orbits, Enos was required to perform a series of tests designed to prove a primate could not only survive the rigors of space travel but also conduct useful and coordinated work. In fact, this activity began two minutes before the launch vehicle rose from pad 14 and was designed to demonstrate psychomotor response to the maximum 7.6 g experienced during ascent; this was the first intelligent animal flight aboard an Atlas missile and the geometry of ascent and descent trajectories determined that the acceleration on the way into orbit would be about 1½ g more than on a Redstone and about 30% less on the way down than that experienced at the end of a ballistic hop. Enos was watched by telemetry that informed flight surgeons and medical monitors on the ground of body conditioning similar to that required for a human occupant.

The tests Enos would perform were considerably more complex than those given to Ham on MR-2. In the first of four different types, the chimpanzee would operate hand levers to turn off lights that came on in random sequence. In the second, Enos would be required to wait 20 seconds after a green-light came on to depress a lever and receive a drink of water as reward. The third test required the chimponaut to pull yet another lever exactly 50 times to get a banana pellet; neither this nor the preceding test would carry a penalty or punishment. The first and fourth did, the latter being a complex series of shapes in two rows. Odd symbols in adjacent pairs were observed by Enos and singled out by depression of a lever. Between tests, built in rest periods were to determine how controlled the chimp really was. If agitated, Enos would be seen to panic and rapidly pursue the sequence of tests as though completion would terminate the experience.

It was the first time medical staff assigned to Mercury had checked out their global operation with a live subject. Enos' reactions to the flight, and his response to having to perform the lever pulling tests, would be examined after the flight by careful analysis of film from a 16-mm movie camera attached to the instrument panel and set to run until splashdown. This would occur about 4½ hours after lift-off. A second 16-mm camera faced the instrument panel for a record of the instrument readings, and a third continually watched the view through the periscope, a valuable indicator of the spacecraft's position should telemetry during the mission or analysis of instrument readings after splashdown generate a suspicious contradiction. A fourth camera, 70 mm, was positioned to record the view through the large spacecraft window – large being here a comparative value!

In addition to the visual evidence, four radiation packs were placed in the capsule at the positions normally occupied by a human pilot's head and feet; two more were placed at remote positions to measure various types of radiation. As Enos worked on through his flight, 78 temperature sensors continuously beamed to Earth data on the cabin atmosphere, and through the communication channels to ground tracking stations went recorded astronaut voices of the type that could be sent from a manned capsule: 'Capcom, this is Astro. Am on the window and the view is great. I can see all the colours and can make out coast lines. Environment is O.K. I feel great.' And so on, from two tape recorders sequenced to allow the ground station to answer and for that transmission to be recorded for analysis after the mission.

It was activity of this type that brought a degree of levity to the morning's Presidential press conference at the White House. Pausing to receive a typed note handed to him from an aide, John Kennedy smiled and read from the paper: 'The Chimpanzee took off at eight minutes past ten. He reported that everything is perfect and working well!' Yet everything was not perfect, although the mission was going well. Ground controllers observed the spacecraft's on-board clock to be 18 seconds fast as the capsule neared the end of its first pass around the Earth, a seemingly trivial defect but one which would bring the spacecraft to Earth 150 km off target if left uncorrected. It was demonstrating the essential need for real-time data reduction and response. Then the spacecraft inverter, a device designed to convert direct current to alternating current, was sending an indication of overheating. That had been observed on previous flights and it was decided not to do anything unless the problem got worse; the spacecraft had a second unit if needed.

In the environmental control system, engineers were trying out for the first time, equipment designed to support manned operational flights. The two oxygen bottles were pressurized at 527 kg/cm^2 instead of the 246 kg/cm^2 of earlier missions, and the system carried an active water separator. On the second orbit, however, as the spacecraft was passing over Africa for the second time, telemetry from spacecraft 9 indicated that the heat exchanger had frozen up, as Enos' temperature rose from 36.7°c to 38°c and then stabilized. Physically, the chimp was in good health and his spirits seemed high. At lift-off $2\frac{1}{2}$ hours before his heart rate had risen from 94 beats/min to 122 as physical tasks and the stress of the mission brought moderate tension. But his activity had been frustrated by malfunctioning equipment that gave him mild electric shocks at times when he performed correctly when it should only have operated when he was wrong. Although confused, and not a little puzzled, he kept on pulling the levers in the sequence he knew to be correct in the hope that it would register a satisfactory test.

But even though his physical condition was none the worse for the malfunctioning test device and the frozen evaporator, a problem with the spacecraft's attitude control system threatened the full three-orbit mission. Because the spacecraft was being flown in an unmanned mode, where the chimponaut was a passive test subject denied opportunity for altering in any way the control functions of the capsule, the attitude and orientation mode selection was not as generous as when a pilot was aboard. In fact, it admirably demonstrated that a space vehicle designed from the outset to respond and to be controlled by human hand and brain cannot readily be equipped to perform a completely automated mission; the desirability of testing and proving out every possible element of a manned vehicle before committing the lives of crewmembers quickly reaches a level of diminishing returns where the concept turns back upon itself and actually exhibits less proficient tendencies. From capsule separation after orbital insertion until the spacecraft was on the point of reentering the atmosphere at the end of orbital flight, spacecraft 9 would be on ASCS and only go on to RSCS mode when the acceleration sensor indicated 0.05 g on the way down, at which point the capsule would be commanded to roll slowly during re-entry.

However, as MA-5 was picked up by the Muchea tracking station in Australia telemetry indicated that the spacecraft was drifting off its −34° attitude. This could only happen if some physical force was turning on its axis, a thruster leaking propellant for instance At Woomera, technicians were cautioned to look out for this but they were unable to confirm the drift. From Canton Island came confirmation of Muchea's suspicion, however, and it was realized that the capsule was swinging round to one side before being heaved back into the correct attitude by the ASCS logic. Something was seriously wrong with one of the thrusters, although it did not really seem like a stuck orifice bleeding fuel. No matter, the capsule was being forced to use unacceptably high quantities of propellant and something had to be done.

At the Cape, Christopher Kraft, sitting in on the mission as flight director, had to make a decision as to whether to let the spacecraft go around the Earth again or bring it back at the end of the second orbit. Early retro-fire would mean a manual command going out from the California tracking station so he sent a message back to Hawaii asking them to stand by in case Cape-flight wanted them to re-configure the on-board clock. Coasting on at a speed of 28,200 km/hr, Enos was aware only of the repeated tasks he was required to perform. On the ground, Chris Kraft discussed with controllers the probable result of committing the capsule to yet another pass around the globe. Of course, the retro-rockets could have been fired at any time, bringing the spacecraft safely down to Earth. But mission rule books written for just this type of situation were only credible if obeyed and one very important criterion for continuing a flight required the normal termination to be a logical product of fulfilling the mission objective. In this case, re-entry at the end of the third orbit might be too late. For if the excessive use of propellant continued at the same rate as that observed on the second pass, the capsule would have only marginal supplies with which to perform attitude control functions for retro-fire and re-entry.

At launch, the attitude control tanks had been loaded with 27.9 kg of hydrogen peroxide; 2.7 kg had been used turning the capsule around after booster separation, 0.68 kg had been used keeping attitude by occasionally blipping one of the several thrusters during the first full orbit, and about 4.3 kg had been consumed so far during the second orbit correcting the capsule as it continually drifted off alignment by bringing it back into the assigned attitude. Kraft talked with Hawaii, then Point Arguello in California, and finally conferred with his flight controllers at the Cape. Gordo Cooper was out in California and immediately received word that the station should stand by to command retro-fire manually. In the Mercury Control Center, Walt Williams sat through Kraft's masterly orchestration of men and machines, bringing the correct sequence of events on line as required to achieve a recovery in the second orbit prime splashdown zone.

Suddenly, with critical communications buzzing between North American coasts, the vital telephone link between Point Arguello and the Cape went dead: a farmer in Arizona had trundled his tractor and plough across a field and ripped up the telephone cable! But no matter; seconds later the link was re-established via another circuit and the process of bringing Enos safely back to Earth continued, Arizona farmers notwithstanding! From the trajectory controller came the word: 'You got 12 seconds.' Kraft had to make a spot decision. Within seconds the capsule would be past the point where the retro-rockets had to fire if capsule No. 9 was to make it to the second orbit recovery zone. A calm voice commanded the Point Arguello tracking station: 'All right, go ahead with retro-fire,' and to the accompaniment of Gordo Cooper verbally counting out the last five seconds, the command went up and the capsule started home.

Enos had no idea his ordeal was to be cut short by one-third and he satisfactorily rode out the re-entry phase where g loads were not as severe from this orbital flight as they had been for Shepard and Grissom – in fact never again would astronauts experience acceleration forces as great as those endured by the suborbital astronauts. Little more than an hour after splashdown, the USS *Stormes* had picked the capsule from the water and set free the first orbital chimponaut, liberated when, after placing the spacecraft on deck, the explosive hatch was freed by a sharp tug on the exterior lanyard. Careful analysis of a capsule from space would be a post-flight necessity following every mission, but capsule 9 would be the last test before committing astronauts to orbital flight and its condition was of especial concern to engineers from MSC.

In general, it came through well, although there were some deformations in the exterior skin, and the heat shield had punctured the false bulkhead as it hung on its landing bag. Some time after the mission, careful examination of the attitude control system revealed a tiny sliver of metal that had blocked a fuel line, so preventing one of the thrusters from firing to keep the spacecraft in position. Without a means of preventing slight drift, the capsule had inevitably accepted the large attitude excursions that took the capsule into the axial range of other thrusters, at which point the spacecraft had been swung into the correct attitude from where, once again, it would slowly drift off. But this was not of major concern to NASA. The capsule had performed well and most if not all the problems could have been circumvented had an astronaut been aboard.

The chimpanzee too had done splendid service, ably following in the spaceprints of his predecessor Ham when he had opened the way for manned ballistic flight. Now, having been shown the way by Enos, US manned space flight would mature to orbital operations, almost eight months after Yuri Gagarin first circumnavigated the globe in space and nearly three months after Titov's 25 hour journey on 17 orbits. One newspaper humorist paid tribute to the part played by Enos in reaching this position by showing in a cartoon a chimpanzee wearing dungarees walking away from the Mercury capsule with another kitted in a space suit, saying to his companion: 'We're a little behind the Russians and a little ahead of the Americans!' It was a cynical reflection of the apparent caution that characterized the public view of America's first manned space project.

Enos lived on at Holloman Air Force Base for ten months after his two orbits around the Earth. In September, 1962, he contracted shigellosis and dysentry. Day and night his keepers observed him, but nothing could be done; resistant to antibiotics, he died on 4 November. Nobody ever suggested, nor could any evidence be found, that Enos had in any way suffered from his space flight. There was no connection between MA-5 and his premature death. By contrast, squirrel monkey Baker, who rode with a colleague called Able on top of a Jupiter rocket in 1959, was alive and well 20 years after his flight, a respected resident of the Alabama Space and Rocket Center.

Before MA-5, the anticipated flight of a man in orbit around the Earth had been set for 19 December, using Atlas 109-D and Mercury spacecraft number 13 on the MA-6 flight planned to duplicate the expected three-orbit mission of the chimponaut. But it would be a tight schedule. Only one launch pad was available for Mercury-Atlas flights and the preparation time required to clean the pad for another mission threatened to push MA-6 over the Christmas period. Apart from that, several minor corrections were thought necessary and the fact that several thousand recovery personnel would have to cancel their holiday leave also played a part in moving the flight into 1962. Bob Gilruth, Mercury's project manager and now director of the Manned Spacecraft Center, announced that the first manned orbital attempt was expected to take place in January. The 16th had been set as the earliest available date. NASA had learned a lot in the previous few years. Early in 1959 Space Task Group had written up a schedule that envisaged the first manned orbital mission in April, 1960, with completion of all six anticipated manned Mercury Atlas missions six months later.

STG had been unable to give America pride of place as the host country to the first manned space flight and smarted considerably under the realization that it would not achieve manned orbital flight in even the same year that Russia accomplished similar feats. But STG was now MSC, with responsibility for continuing to fruition the effort put into the first nascent manned flight project but also for bringing together a project designed ultimately to put men on the Moon. On the very day Bob Gilruth spoke to the Houston Chamber of Commerce, and used the event to inform the assembly casually of his decision to launch MA-6 in 1962, he fired their enthusiasm and expectation for the coming decades of prosperity by announcing that the Manned Spacecraft Center would also have responsibility for an interim project called Mercury Mk. II, a two-man follow-on to the one-man Mercury.

Mercury Mk. II really began more than two years before and for a while languished in the net linking NASA and industrial contractors until it gradually became necessary as an essential element of a developing manned space flight capability. Concern about the future of manned flight was voiced at the Goett committee meeting held 25–26 May 1959; John W. Crowley, NASA's Director of Aeronautical and Space Research had appointed Harry J. Goett to head the Research Steering Committee on Manned Space Flight charged with mapping new capabilities and recommending future policy. At sessions chaired by Goett, the committee agreed with proposals for maneuverable spacecraft developed from the basic Mercury design concept. During the early months of 1959, Bob Gilruth listened to a presentation from an engineer at the Langley Research Center on the possible application of an inflatable lifting surface that could be used to bring manned space vehicles to a controlled landing on land, instead of having them fall into water. The engineer's name was Francis M. Rogallo.

Soon to be known as the Rogallo wing, his device comprised an inflatable fabric frame joined at the forward point with tubular members trailing aft like a 'V' on its side, with another member down the center so that fabric draped across all three arms would form a load-bearing structure actually capable of generating more lift than drag. Unlike conventional parachutes it could be made to support quite large loads and to bring them to assigned spots on the ground by alternately generating or dumping lift from the wing. Because it was a semi-rigid structure, the 'paraglider' as it became known could be contained in a comparatively small space attached to a space vehicle, and be deployed like a parachute in the denser layers of the atmosphere, at which point the ballistic re-entry vehicle would acquire a substantial lifting surface to control its descent precisely.

Immediate applications seemed legion. Large and expensive rocket booster stages, like those then being developed for the Saturn 1 program, could conceivably be lowered to a land landing if protected with thermal insulation for a descent through the atmosphere. Manned space vehicles after Mercury could benefit from a Rogallo wing, deployed so as to bring the capsule to a safe and controlled descent on land. Rogallo had been inspired by an illustrated article in a 1952 edition of *Collier's Magazine*, the same journal that published details from America's first space symposium, where he saw delta-winged re-entry vehicles launched by enormous booster rockets. If the wings could be made inflatable, he mused, the simpler ballistic shape would suffice for all phases of a flight into space and back, only adopting a deployed lifting surface when they had descended to the safer and more dense levels of the atmosphere.

The Rogallo wing would soon play a leading role in plans for a Mercury successor, but for most of 1959 NASA had its eye on grander objectives leading toward early design proposals for the Earth-orbiting laboratory and the circumlunar Apollo vehicle outlined in an earlier chapter. But that did not stop McDonnell from moving ahead with proposals of its own. In September, the company issued a report suggesting modifications to the basic Mercury capsule where it could be made to fly around the Moon and return to Earth. McDonnell's inability to convince STG successfully of the need for a completely new Mercury type spacecraft, when the space agency was moving toward agreement for a much grander objective, created a limited posture where talk of controlled re-entry – either by Rogallo wing or some form of lifting surface during descent – was deemed more appropriate for Apollo, a radical departure from Mercury.

By early 1961 NASA began to look again at the possibility of stretching the basic Mercury design to accommodate the long duration missions and lifting re-entry concept, suggested by McDonnell when NASA turned down STG's request in May, 1960, for money to pursue further studies. It had become clear that rendezvous between spacecraft in Earth orbit was to be a key feature of future space programs, an act implicit in the concept of manned Moon flight then being formulated. From the beginning in 1958, post-Mercury manned flight plans polarised around two centers of administration: Space Task Group, ostensibly set up just for managing Mercury, and NASA Headquarters, where future policy was rightfully determined. STG naturally wanted to maximize its usefulness, and did this by proposing development of spacecraft conducive to the operational requirements it believed Headquarters would need in planning national strategy in space. Headquarters set up plans for major space projects linked only by the common thread of progress, a visible improvement technologically and programatically. Hence, Headquarters spawned Apollo and STG fathered post-Mercury proposals; the latter would have to convince the former of the viable assets inherent in moving the proposition from project stage to program level before any post-Mercury design could receive funds.

Bob Gilruth and Maxime Faget were early proponents of a follow-on Mercury design and centered their studies around two primary objectives: long duration and rendezvous operations. Serious consideration was already being given to producing an 18-orbit Mercury capsule. McDonnell and STG had already proved that by improving interior layouts, adding more consumables, fitting additional modules, and generally upgrading the spacecraft's capability, a basic Mercury capsule could be made to support manned flight for more than 27 hours versus the $4\frac{1}{2}$ hours it was designed from the outset to accommodate. But rendezvous was a very different matter. That implied a considerably larger propulsion system capable of moving the spacecraft from one orbit to another and, in turn, a large spacecraft to accommodate the additional propellant. Doubts too about the ability of an Atlas launch vehicle to lift the weight of a spacecraft capable of flying for up to two weeks in orbit – the amount of time Moon bound astronauts would have to spend in space before returning to Earth – indicated a need for improved Mercury capabilities.

Toward this end, Maxime Faget turned to James A. Chamberlin for assistance with preliminary engineering analysis of the requirements. When Chamberlin arrived at McDonnell's St. Louis plant to set up shop for a virtual re-design of Mercury, his existing position as head of STG's Engineering Division was retained but Andre J. Meyer, Jr., and William M. Bland, Jr., assumed greater charge of the operations at Langley. Chamberlin probably went beyond the original idea Gilruth had in mind when dispatching him on his new task. But nobody would accuse Chamberlin of undercutting his responsibility and soon he was busy completely re-designing the spacecraft.

On 13 February, NASA met with McDonnell to discuss preliminary conclusions about just what re-designed Mercury might be capable of. Within two weeks, Maxime Faget heard that studies extended to the suggestion that the pilot could actually leave his capsule for a walk in space – an activity known as 'extra-vehicular activity', or EVA – at which point Faget proposed that the modified spacecraft should be designed to contain two pilots. Faget's counterpart at McDonnell – John F. Yardley – dicussed this with McDonnell's vice president Walter F. Burke who ordered engineering drawings to be prepared showing the feasibility or otherwise of stretching Mercury and not merely modifying it. By mid-March, 1961, STG briefed Abe Silverstein on work under way at McDonnell and two weeks later James T. Rose joined Chamberlin at St. Louis to work on trajectory planning and mission operations.

Chamberlin's idea, that the Mercury spacecraft could form the base on which a successor should be shaped, focused attention on a major problem with the existing one-man design. Mercury had been limited in weight and size by the availability of the Atlas launcher and the constraints that vehicle imposed on the payload it could lift. The capsule, in turn, had been designed with weight saving criteria uppermost and represented a compromise between operational ease and optimum systems layout. With the possibility of making substantial changes to the way the spacecraft was fitted out, Chamberlin showed Silverstein the way a two-man vehicle could be made more efficient. Instead of laying systems around the pressure module's interior facilities in conformity with available space, spacecraft equipment should be placed outside the pressure module and beneath the outer skin. It will be remembered that Mercury comprised an internal pressure vessel surmounted by an exterior arrangement of skin panels and shingles designed to offer thermal protection along the flat face of the conical sidewalls.

Chamberlin's idea meant that the interior pressure vessel should be very much smaller than the outer skin, the space between occupied by systems that could be readily removed by undoing panels on the outside. In this way, astronaut activity would center on work *inside* the cabin while engineers and technicians could perform complicated work in the uncluttered area *outside* the capsule. This would avoid the time-consuming operations associated with removing systems from Mercury that invariably required several other components to be taken out first; technicians would work from the outside instead of from the inside out. On learning of the methodology behind Chamberlin's work, and as a result of talking over the issues involved with Bob Gilruth, Silverstein gave STG the green light to enter detailed discussions with McDonnell over a possible Mk. II version.

On 14 April, Space Task Group signed a contract with McDonnell whereby the contractor was required to procure long lead time items for an additional six Mercury capsules, on the understanding that they would be used in a Mk. II version. This, only two days after McDonnell had suggested a further study effort to reduce checkout time for future capsules by seriously studying the Chamberlin concept of modularised packages for spacecraft systems, components that would be placed outside the pressurized crew compartment. On hand at the contractor's plant, Chamberlin meanwhile gathered around him experts from the Engineering Division so that lessons they had learned in the hard won fight to get Mercury space-borne could be applied to the post-Mercury vehicle: Fred J. Sanders' staff from Langley; William J. Blatz and Winston D. Nold from McDonnell.

It was about this time that, albeit for a very different reason, STG personnel started to examine the Titan II launch vehicle seriously. Titan had been conceived as an Air Force ICBM to be developed concurrently with Atlas, ahead by several years on development and production. Atlas carried many novel and innovative features that many Air Force people thought too radical for the role it was designed to fulfil: thin-wall tanks that could buckle without adequate pressurization when empty; a stage-and-one-half concept whereby the booster engines were thrown away leaving the tanks integral with the single remaining sustainer engine. Moreover, Atlas took many hours to prepare for launch whereas a rapid-fire requirement would need a missile that could be ready in minutes. That was the principle behind the Titan concept, a missile that adopted conventional two stage design with storable propellants.

Manufactured by the Martin Company, Titan was theoretically capable of lifting at least $3\frac{1}{2}$ tonnes into orbit but when

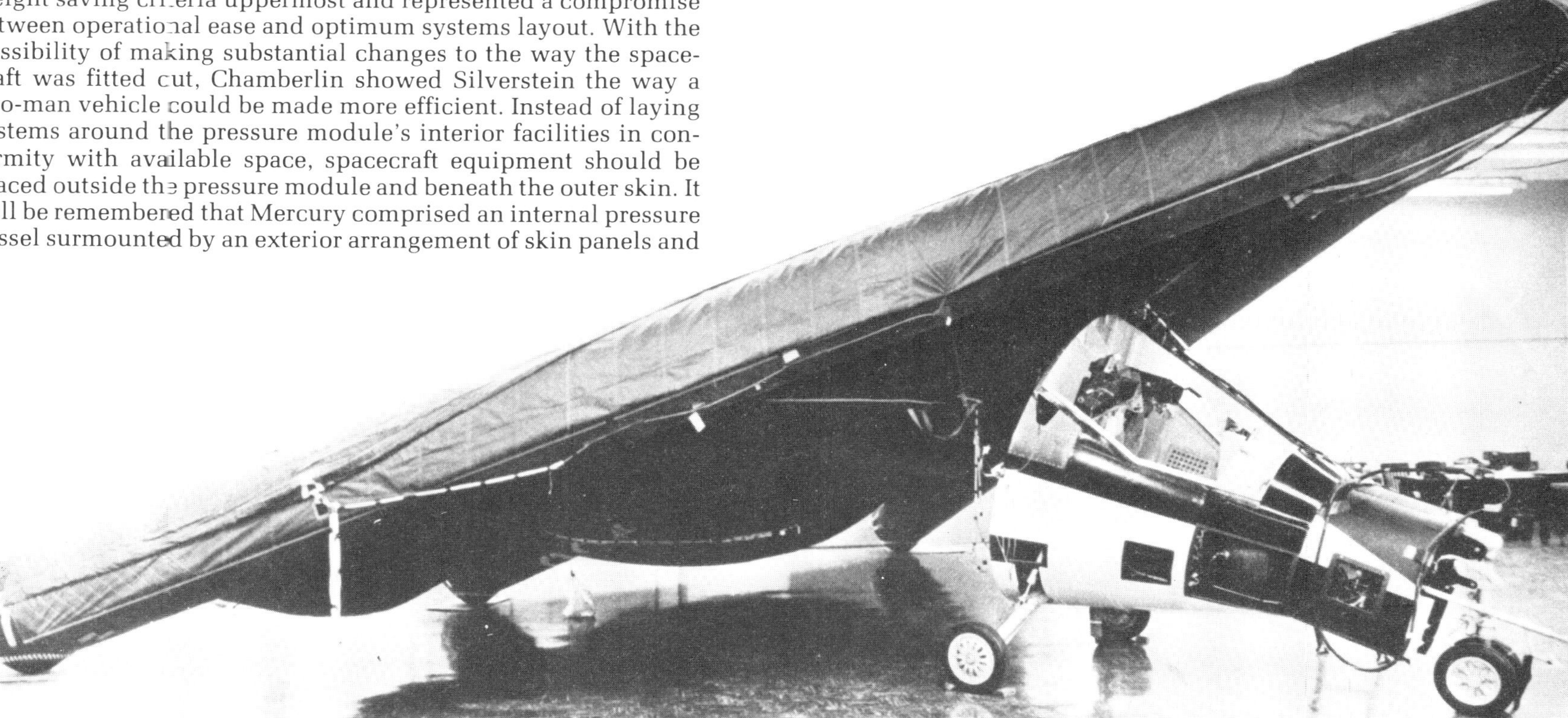

Developed from an idea by Francis M. Rogallo, paraglider tests at the then North American Aviation were conducted in the hope of providing future manned spacecraft with a land landing capability.

Martin representatives briefed NASA in Washington on 7 April it was to persuade Associate Administrator Robert C. Seamans that the launch vehicle could be applied to lunar landing objectives. Abe Silverstein was at the 8 May meeting called by Seamans to further examine the rocket and, although unconvinced about the missile's capacity for this ambitious undertaking, he advised Bob Gilruth that his Space Task Group should seriously examine the Titan II capability for serving the NASA manned space program in some capacity or other. Gilruth was interested in studying the possibility of using Titan to launch a Mercury Mk. II and told Silverstein who reported that fact to Seamans. By this time the increase in space activity presaged by Kennedy's Moon decision was common knowledge and it seemed a logical step that just as Mercury had flown on the most powerful rocket available at the time, an interim venture between Mercury and Apollo should fly on the launcher that possessed a lifting capacity greater than Atlas but less than Saturn.

Meanwhile, Chamberlin, aided by Blatz, worked to prepare a resume of their activity for presentation to top NASA management in early June. Alan Shepard had flown a short suborbital hop, Kennedy had set the wheels rolling on a massive new industrial commitment; the time seemed right to press on with a spacecraft that would mature from the proven record of Mercury to test, long before Apollo could be ready, the many advanced activities the larger spacecraft would be expected to perform. But what STG personnel defined as a Mercury Mk. II was not quite the same as that envisaged by the Chamberlin team.

On Friday, 9 June, a Capsule Review Board meeting was held at which Chamberlin and Blatz briefed Bob Gilruth, Walt Williams, Paul Purser, Maxime Faget, Charles Mathews, Bob Piland, Wesley L. Hjornevik, George Low and John Disher. What they learned was that approval given the previous March for Chamberlin to continue work on the proposal was a ticket for a totally new spacecraft, one that seemed to inject new engineering to a greater extent than it retained basic Mercury design details. Moreover, Chamberlin had transformed the essentially automated systems control functions to one largely controlled by the pilot, relieving the old Mercury design of one tedious feature blessed by technicians and engineers alike. A classic, almost extreme, example of this was Chamberlin's solution to the problem of over-sensitive abort equipment in the original Mercury capsule. By using a heavy lattice tower to support a solid propellant escape rocket, Mercury engineers frequently experienced the result of a trigger-happy capsule wrenching itself free from a booster that gave very little indication of exploding.

The super-sensitive switches and relays designed to command an abort at the slightest provocation removed a level of pilot-induced control that could have prevented premature release of the spacecraft on several occasions. No astronaut actually had cause to fire his escape rocket, nor did any experience the result of a trigger-happy abort system, but the system was recognised to be too responsive. Moreover, the complete escape tower weighed a hefty 580 kg, fully 30% of the capsule's launch weight, and this unnecessary load was only applicable to the first minutes of the flight. To escape these compounded problems, Chamberlin designed his Mercury Mk. II to use ejector seats. But that would only be possible in reality if the capsule rode atop a Titan launcher.

Atlas missiles, with their volatile combination of liquid oxygen and kerosene, would explode with a reaction that could not be exceeded by the speed of any ejector seat; Titan used blended hydrazine and unsymmetrical dimethyl hydrazine as fuel and nitrogen tetroxide as the oxidizer, a less volatile combination that should allow crewmembers to escape before being overtaken by the fireball. Chamberlin had a ready dislike for the escape tower concept and was convinced of the better escape mode promised by the ejector seat. But such seats could only be used in the lower regions of the atmosphere and satisfactory abort procedures beyond an altitude from which a man could safely descend by parachute required the use of some form of capsule separation device that would bring the entire spacecraft back as though it was returning from orbital flight.

But such things were not really of concern at this stage. The Capsule Review Board was somewhat taken aback by the major re-work suggested by Chamberlin and Blatz. Although Faget and one or two other STG engineers had suggested a two-man crew for the Mk. II, Chamberlin had retained the original Mercury concept of a single crewmember. In fact, with hindsight, the proposal now being presented to STG managers was conservative, in every respect a re-worked Mercury with a capacity for 18 orbits. But Gilruth and Williams had assumed that Chamberlin would merely rearrange Mercury's internal components to smooth preparation, checkout and countdown procedures, all of which had been seen as potential pitfalls in a matrix of schedules leading to manned orbital flight. The STG managers were unhappy about the radical nature of Chamberlin's proposal and adjourned for the weekend. Blatz returned to St. Louis, leaving Chamberlin to field the inevitable questions when the Board re-convened the following Monday, and Disher, Low and Hjornevik went back to Washington.

Employing a two-man Mercury capsule, engineers at the then Aeronutronic Division of Ford Motor Company suggested this scheme for rendezvous tests with a separate structure.

On Monday, 12 June, the Board concluded that while Chamberlin's plan was certainly attractive, a major new effort implicit in the presentation was inconsistent with STG's wish to keep Mercury modifications to below $10 million for the 1962 budget year. This was, in effect, a retraction to the position STG held two years earlier when it called for minor changes to the capsule for the 18 orbits seen as the basic capsule's maximum potential. McDonnell had a willing supporter in Chamberlin and when it appeared that the Capsule Review Board shunned plans for a substantial follow-on project, the dissident STG engineer worked all the harder to get some sort of compromise.

On 7 July, McDonnell's Walter Burke presented STG with company proposals, a selection from which Space Task Group could choose the more attractive design. There were three configurations: a basic Mercury capsule modified by hatches fixed to the exterior which would allow technicians access to interior systems, a proposal that could reach flight status within 11 months of a go-ahead and cost $79.3 million total; a similar re-work of the basic Mercury to that envisaged by Chamberlin at the June Review Board hearing, ready within 20 months for a complete cost of $91.3 million; or a two-man version of the second proposal, ready to fly within 22 months for an overall cost of $103.5 million.

The prospect of doubling up the crew cabin kept on appearing! Maxime Faget, the force behind Mercury, was quietly pushing for logical growth. Having first planted the seeds of a two-man pilot idea with Chamberlin the previous March, and reiterated the idea at the June Capsule Review Board meeting, he was quietly gathering support from other sources. And in presenting their three proposals, McDonnell showed how pointless the minimum modifications would be in the light of new capabilities the spacecraft could acquire. During July, while Gus Grissom was getting ready for his suborbital flight aboard Liberty Bell 7, George Low reported to Robert Seamans at NASA Headquarters that although McDonnell had seemingly gone beyond their original mandate in presenting design studies for a substantially new capsule, he had advised them to 'proceed on the basis of minimal changes to the existing hardware and to approach design modifications on this basis.'

But Low need not have played the concept down, for Seamans was a convert to a greatly expanded program to fill the gap between basic Mercury flights and Apollo missions. It seemed that Mercury three-orbit flights could be accomplished in 1962 and Headquarters was already planning to fly four 18-orbit Mercury missions in 1963 followed by eight Mercury Mk. II missions between October 1963 and the end of 1964. These latter flights would pursue problems inherent in rendezvous and docking, seeking and finding solutions to feed Apollo with valuable experience on how to bring two separate spacecraft together and connect them in space. Just six days after Grissom flew his mission, Gilruth joined Abe Silverstein from Headquarters in touring the McDonnell works at St. Louis. They were there to view models of the proposed McDonnell spacecraft, including an 18-orbit version of the basic Mk. I (a retrospective designation), a modified Mk. I with panels cut in the sides, called Mk. II, a Chamberlin-type Mk. II similar to that presented the previous month, and a two-man Mk. II pushed by Faget.

In addition, a full crew station mock-up of the two-man design enabled astronaut Wally Schirra to gain comfort that engineers had 'finally found a place for a left-handed astronaut!' It was an important viewing. Space Task Group had pushed the Mk. II concept as far as possible, and were now appealing to Headquarters to come up with the money and officially approve the project. Silverstein was impressed, and there was little doubt in his mind that NASA needed a manned space project between the basic Mercury and the three-man Apollo. Without a Mk. II program, Mercury would end in 1962 and begin a three year hiatus until Apollo Earth orbit flights could start in 1965 at the earliest. Only two years would then remain before NASA had the Moon landing scheduled and the period between 1963 and 1965 could serve very well to rehearse the type of operations Apollo would be expected to perform.

On the second day of their visit to McDonnell, NASA officials confirmed optimism expressed during the previous day's tour and inspection by authorizing the company to pursue a two-man Mercury Mk. II from that point on. In characteristic vein, McDonnell pressed right ahead with detailed design studies on the chosen concept without waiting for formal approval through legal channels or the allocation of funds, spending its own money to push on with the work rather than fall into pre-set work schedules. Less than two weeks after Silverstein approved the Mk. II, Chamberlin met with Lockheed Missiles & Space company representatives to talk over the possibility of using an Agena rocket stage as the basic vehicle to which the Mk. II could dock in orbit. Agena had been developed for the Air Force as an upper stage to Thor missiles, a booster system for the Discoverer satellites. The 'B' version was capable of firing twice if needed for burn times totalling 4 minutes, delivering a thrust of more than 7 tonnes from a structure 8 meters long and 1.5 meters in diameter.

Clearly, Mercury Mk. II was to be more than an upgraded version of the three-orbit capsule, almost a completely new program in which rendezvous and docking, long duration flight, and two-man operations would build confidence in manned space operations prior to operational three-man Apollo missions. During August the various elements of the new program came together for the first time, inciting an optimistic enthusiasm that advanced the first flight date to March 1963 rather than October of that year proposed by Headquarters the previous month.

Conceived as a two-man version of Mercury, Gemini was to grow into a completely different vehicle, retaining only superficial resemblance to its generic ancestor.

The first flight would be an unmanned 18-orbit qualification run, followed by a manned mission, also for 18 orbits, and two manned missions of up to 7 days' duration. Flight 5, the last in 1963, would be the first to attempt rendezvous and docking with an Agena B launched separately by Atlas. Flight 6, in January, 1964, would carry a chimpanzee on a 14 day mission, followed by another rendezvous and docking and a second chimpanzee flight of long duration. Flights 9 and 10 would be the third and fourth docking tests, completing the basic Mercury Mk. II program in September, 1964. Seven days was considered the minimum required for sending men to the Moon and bringing them back; constraining manned Mk. II flights to a duration of one week would suffice to balance the need for information about physical response to long flights with the engineering requirements for a two-week mission.

The inventory of 10 flights would require eight new capsules – two were to be re-used – ten Titan II launchers, four Agena B rocket stages and four Atlas boosters. The total cost was estimated to be $347.8 million. But there was an appendix to the basic ten-flight plan that presented two alternate proposals for a dramatically grander objective: one that would seize some of Apollo's thunder and, in a very real sense, pave the way for Moon landings. Inspired by Chamberlin, these alternatives envisaged availability of the Centaur high-energy upper stage exploiting liquid hydrogen and liquid oxygen. At this time, Centaur was under development for Atlas missions and promised to significantly increase the payload capability of that booster.

The first appended proposal required four more flights after the ten indicated in the main body of the plan. These would all incorporate rendezvous and docking operations in each orbit, but with Centaur upper stages instead of the Agena B. After docking to a Centaur, Mercury Mk. II would employ the stage's propulsion unit to boost the combination 80,000 km from Earth on a looping trajectory that would carry the assembly far beyond Earth's radiation belts before it returned to the atmosphere and a safe splashdown; the Centaur would be separated after firing the combination to an elliptical path, and burn up in the atmosphere behind the capsule. Two elliptical missions would be flown followed by a further two missions where the Centaur would boost the spacecraft to the Moon. Coasting around the far side of the lunar sphere, Mercury Mk. II would return to Earth and re-enter the atmosphere. The two elliptical-orbit missions were proposed for November, 1964, and January, 1965, with the two circumlunar flights in March and May, 1965, respectively. Apollo would then proceed with its own flight test schedule later that year, Mercury Mk. II having pioneered the way. McDonnell calculated the four additional flights would cost an extra $60 million.

The second proposal incorporated the circumlunar objective, but in a compressed schedule aimed from the outset at achieving a flight around the Moon as soon as possible. Instead of the 10 Earth orbit missions, 2 high-orbit flights and 2 circumlunar attempts, the first three profiles outlined for the basic 10-flight program would be followed by two rendezvous and docking missions with Atlas-Agena B. Flights 6 and 7 would use Atlas-Centaur for high-orbit elliptical rehearsals followed by the two flights around the Moon and back. In this way, the first circumlunar flight could be run in May, 1964, and the 9-flight program would cost $356.3 million. The logic was clear: NASA could save money by agreeing from the outset to aim quickly for circumlunar missions in 1964. But the grand proposals were too radical, and seemed by implication to bypass timely study over several missions of rendezvous and docking problems and the reaction of the human body to long duration flight. There was a danger here that Mercury Mk. II might eclipse the very objective for which Apollo was being groomed and senior management recognized the need to limit the new spacecraft's mission envelope for an expanded potential within less glamorous objectives.

Before the month of August ended the two appended extensions of the basic 10-flight program had disappeared but James Chamberlin tenaciously stuck to the notion that the basic capsule design was applicable to broad examination of new mission roles and on the penultimate day of August he presented 'A Lunar Landing Proposal Using Rendezvous' plan whereby a Saturn C-3 would be drawn in to a program embracing 16 flights. The object would be to land on the Moon, rather than just fly around it and return to Earth, and Chamberlin suggested that a small lunar landing craft weighing only 4 tonnes fuelled and barely 680 kg empty could make this possible. Under this scheme, which Chamberlin expected to begin in March, 1964, an unmanned test flight would be followed by a manned mission in Earth orbit. Two long duration flights would prepare the project for Moon missions, and three rendezvous and docking flights would prove the suitability of theoretical link-up procedures.

Next, two Centaur boosted missions would carry their respective spacecraft to an 80,000 km elliptical orbit, with three more flights to prove the Mercury Mk. II could meet and connect with the unmanned lunar lander placed in Earth orbit by separate launcher. Next, two Centaur boosted missions would demonstrate circumlunar capability, followed by a flight into Moon orbit from a Saturn C-3 launcher. The last flight would place Mercury Mk. II in lunar orbit, allowing the tiny lunar lander to descend to the surface. Saturn C-3 was a proposal then before NASA from Wernher von Braun's Marshall Space Flight Center. Stimulated by the need for a very large and capable booster to send the then nascent Apollo on its way, von Braun's Center provided a steady flow of booster configurations, C-3 being the latest. As proposed, it would generate 1,360 tonnes of thrust at lift-off (compared with more than 180 tonnes for the Titan II selected for the upgraded Mercury and more than 600 tonnes for the Saturn C-1), and comprise three stages, the upper two powered by high-energy liquid oxygen/liquid hydrogen engines, with a capacity for sending 45 tonne payloads into orbit.

As conceived by Chamberlin, with not a little prodding from McDonnell, the Mk. II Moon plan would reach fruition in January, 1966. But this was not much more than a year before Apollo was expected to accomplish the same feat and such conclusions were uppermost in the minds of STG staff members when the Langley engineer presented them with the whole idea. Apollo itself was taking up much of NASA's time, a massive project that threatened to sweep everything else aside and one that seemed at times almost to consume other, less ambitious, projects. As will be discovered in Chapter 9, things were not, however, going well with selection of a mission mode for Apollo. It would still be some time before the contest over several ways to reach the Moon was won by a single, selected, method, and decisions about Mercury Mk. II concerning Moon objectives seemed only to cloud the muddy waters even more. Nevertheless, it was the first feasible Moon flight proposal NASA received and as such deserves credibility. In fact, STG was right behind the idea to expand the Mk. II role, and even suggested in September that Headquarters should adopt an 'Integrated Apollo Program' whereby the whole Chamberlin Moon plan would be married to Apollo so that the mini-lander concept could pave the way for the bigger and more capable landing operations that would follow.

Silverstein reviewed the proposal in September and early October, approving the onset of contractual negotiations for spacecraft, Titan boosters and Atlas-Agena target vehicles. But still the question of Moon flights for Mk. II remained open. Much of October was spent re-writing the operational Mk. II plan, for in the light of increasing Headquarter's hostility to the idea of a stretched Mercury flying to Moon orbit even Chamberlin concluded that it was a theoretical cul-de-sac. In drawing up the final draft of a Project Development Plan, work on which had begun 14 August, when the McDonnell-STG team laid down the 10-flight profile, Chamberlin doffed his hat of vision and acquired another hat more properly tuned to the requirements of NASA. Mk. II would form the basis for the manned flight program beginning where the basic Mercury left off, a program designed to explore the unique requirements of Apollo rather than pre-empt or anticipate it. Mk. II would pursue studies, in Earth orbit only, of the physiological response to long duration manned flight. It would provide ample opportunity to rehearse and test procedures and equipment for rendezvous and docking, and evaluate several methods by which the essentially ballistic re-entry capsule could be brought to a controlled landing at a pre-determined spot.

STG was aided by George Low and his assistant, Warren J. North, in drawing up the schedule for Mk. II operations, preparation of which was completed 27 October, the very day NASA's first Saturn C-1 made its inaugural test flight from Cape Canaveral. Under the revised Mk. II plan, an unmanned qualification flight would be launched in May, 1963, the two month delay between the date cited in the final Project Development Plan reflecting the time that had already passed since the August proposal for a March, 1963, start to flight operations. An 18-orbit manned flight would then be performed, followed by the first of two long duration manned flights lasting up to 14 days. Gone were the two primate flights, manned mission duration doubled now from the originally projected 7 day operations. The rest of the Mk. II program would be devoted to rendezvous and docking: first, two flights (in January and March, 1964, respectively) to demonstrate satisfactorily the ability of the manned vehicle to find and move up to an unmanned Agena B target vehicle, then six flights exploiting the Agena B for various types of docking operation, new forms of rendezvous, and tests of rendezvous activity aborted during various phases in anticipation of similar problems that could befall Apollo; only the rendezvous task was to simulate abort, the mission itself continuing to demonstrate a work-around procedure.

In all, 12 flights, 11 of which would be manned, would be flown to the end of March 1964. Hardware required for this program considerably exceeded the inventory from the August proposals: 12 spacecraft would be needed, 15 Titan boosters would have to be purchased, and 11 Atlas-Agena combinations would be required. Three spacecraft were to be refurbished, but only for use as spares, and 3 Titans and 3 Atlas-Agenas were to be marked up as spares also: 12 capsules and 26 launchers versus the 8 capsules and 14 launchers necessary for the August plan. Consequently, costs went up considerably. The entire program was now estimated to require a total $529.45 million, a sum which included money for paraglider development, application from research work already by then under way for its possible use with Apollo. The Rogallo wing would be a good way of achieving a major project objective: demonstration of precise crew control over the spacecraft, its trajectory and its landing activity.

By the end of October, the Mercury Mk. II program was adequately defined to the satisfaction of STG and Headquarter's personnel alike. And so it was that when Space Task Group became the Manned Spacecraft Center on 1 November, 1961, the arm of NASA responsible for manned space operations had three projects to administer: Mercury, Mercury Mk. II, and Apollo. For a while it looked as though the booster for Mercury Mk. II might cause problems leading to substantial delay. Under the terms of the agreement worked out between NASA and the Department of Defense, the latter organization would develop new boosters for space activity only if they served the same requirement being sought by the civilian agency. John H. Rubel, Deputy Director of Defense Research and Engineering, informed Robert Seamans that DoD intended to cancel further developments of the Titan II and concentrate instead on a much improved model logically called the Titan III.

At first look that would seem to further improve NASA's posture vis-à-vis the weight of the Mercury Mk. II. Not a bit of it. Instead of merely upgrading the two-stage Titan, DoD planned to attach two enormous solid propellant boosters to the side of the first stage so that they could lift the assembly to altitude before the first stage was ignited, in effect acting as a new first stage. To accomplish this the first stage would have to be strengthened considerably, increasing the weight of that stage in the process. NASA had no mind to fly a manned space vehicle on solid propellant boosters, considered at this time to be too unreliable for the more demanding safety requirements, and so could only use the two-stage liquid propellant core of the new Titan III which, because of its beefier construction, would be capable of carrying less payload into orbit without its solids than the basic lightweight Titan II.

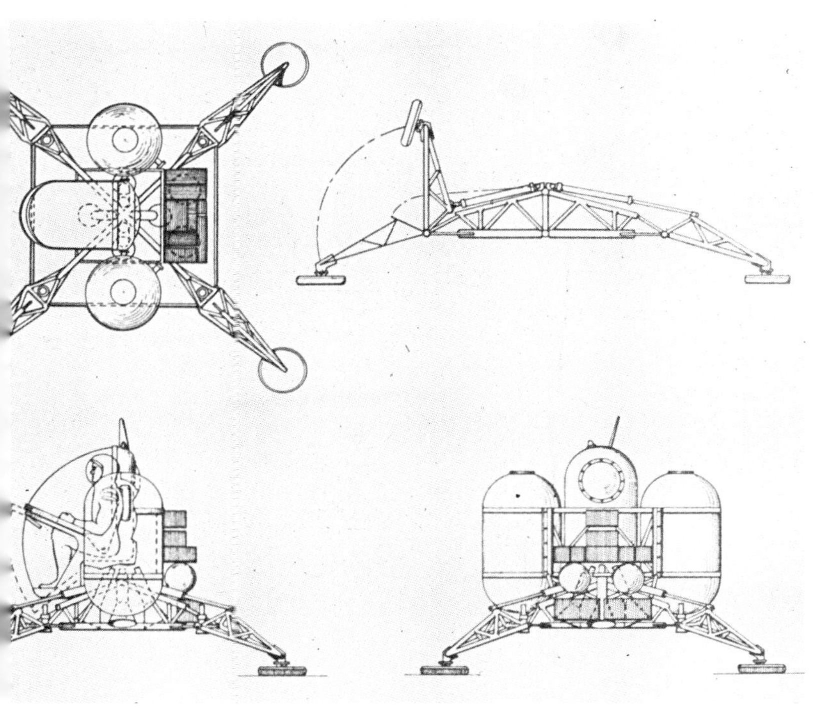

Drawn by Harry C. Shoaf from the Space Task Group Engineering Division, this proposed lunar lander would have been used with manned Gemini spacecraft to achieve a Moon mission purportedly for less cost and at an earlier date than the Apollo concept.

To accomplish the planned Mercury Mk. II objectives, NASA wanted DoD to give a production Titan II larger propellant tanks and several new systems that would substantially improve its safety performance; DoD fielded NASA displeasure at the announcement about Titan III by reminding the space agency that this too would mean a new development program likely to cause delays. But the two were hardly comparable. Titan III was a very different launcher to the II and was being set up for the Air Force Dyna-Soar boost-glider. During November NASA tried to extricate itself from what it defined as a potential brake on plans for a rapid follow-on to the three-orbit Mercury missions. The three boosters in question (Titan II, Titan II-½ – the name given to the long-tank version – and Titan III) were considered at a meeting of the Large Launch Vehicle Planning Group on 20 November and based on conclusions drawn at the event Robert Seamans of NASA and Rubel from the DoD issued a joint recommendation to Secretary of Defense Robert McNamara that the space agency be allowed to procure 15 Titan II missile weapon systems for use as Mercury Mk. II launchers.

They had lost the Titan II-½, but they had secured a reprieve from the consequences of being saddled with a lengthy development run for the new Titan III core, delivery of which would have left NASA with less lifting power for its new spacecraft. Silverstein's imminent move from heading the Office of Space Flight Programs to being in charge of the Lewis Research Center caused a temporary slow-down in the pace at which the program evolved, but on 6 December, 1961, Bob Gilruth formally asked the new Director of Manned Space Flight, D. Brainerd Holmes, to approve the project. The following day, Robert Seamans wrote his formal sanction on a document detailing Mk. II aims and objectives identical to those presented on 27 October, and at the Houston Chamber of Commerce Bob Gilruth officially announced the new project to loud cheers and wide acclaim.

But what to call the substantially modified Mercury successor? NASA Space Task Group Contracting Officer Glenn F. Bailey, and McDonnell's John Y. Brown, had dubbed the project 'Mercury Mk. II' from vague references to an 'Advanced Mercury' early in 1961. A more catching name was essential, one that the public could identify with and one that expressed the true function of the spacecraft. Personnel at NASA headquarters were encouraged to submit suggestions, and at a speech before the Industrial College of the Armed Forces, Robert Seamans offered a prize for the successful postulate. Among suggestions that included 'Diana,' 'Valiant,' and 'Orpheus,' the final name was chosen, having been submitted by a member of the Armed Forces College audience and Alex P. Nagy from the NASA Office of Manned Space Flight. Both received a bottle of scotch whisky from Bob Seamans and on 3 January, 1962, the Mercury Mk. II became Gemini. From that point on, Mercury and Gemini would be two separate programs, with their own project offices.

It was an apt choice for astronomers, denoting a two-man spacecraft to astronauticians, even the symbol for the zodiacal sign was similar to the 'II' designation applied throughout 1961. As the year came to an end it pulled the curtain on a momentous period for NASA, when Apollo was boosted far beyond its original objective, when American astronauts first rode into space – albeit on short, suborbital flights, and when a completely new tool for long duration flight and rendezvous and docking emerged from the basic design of the first US manned spacecraft. Ahead lay hard tasks and strenuous work: Apollo was still without a mission profile, its staff unsure of just how best to fly men to the Moon, Mercury had still to sustain manned orbital flight, and at least 2,000 hours of space flight time had to be accumulated before the Kennedy commitment could be fulfilled. But there was a deep sense of having begun a long road to regular space operations. Decisions had been made, it was time to accelerate the pace of flight activity.

CHAPTER EIGHT

Wings of Mercury

Seen here familiarizing himself with the MR-3 capsule configuration, John Glenn's training for the first US manned orbital flight involved active participation in the preceding missions.

For more than a year before his actual flight, John Herschel Glenn had known he was in for an early Mercury mission. At the beginning of 1961, the precise number of manned ballistic Redstone launches was an uncertain quantity known to rest on the actual performance of at least the first two attempts. Shepard, Grissom and Glenn were selected for the short-list from which a single astronaut would be chosen to fly the first manned suborbital hop. That had been flown by Shepard, the second Redstone flight by Gus Grissom. As back-up pilot to the first two, Glenn was prime pilot for what would otherwise have been Mercury-Redstone 5 but actually turned out to be Mercury-Atlas 6 – the first manned orbital flight. Scott Carpenter would be John's back-up, and that placed him as prime pilot for an early three-orbit mission. Deke Slayton had been chosen for MA-7 in November, 1961, with Schirra his back-up. The spacecraft, meanwhile, had a longer and more checkered gestation to flight readiness.

Spacecraft 18 had originally been slated for the MA-6 flight, then spacecraft 13 when the former was re-cycled back for modification. No. 13 went into fabrication during May, 1960, and its history is typical of many production capsules during the two years of change and re-distribution of mission roles, flight assignments, and spacecraft requirements. By the end of the year, component parts for 13 were held up due to teething troubles with the environmental control system but early in 1961 work was back on the original schedule. Late in April, however, a re-assignment of objectives carried word to McDonnell that 13 would carry the first man into orbit and should be groomed accordingly. The capsule was a late production model and carried all the modifications suggested by astronauts and test engineers. When delivered to the Cape on 27 August, 1961, it was in fine fettle for the mission.

There was little room inside a Mercury capsule for major change in spacecraft layout but Glenn's No. 13 was a little different to Grissom's capsule 11. From a fuse panel at his left, Glenn would survey the main control panel in front carrying flight control displays on a center segment with its periscope screen at the bottom. Between the main panel and the fuse panel were the primary event indicators, the lights and appropriate manual activation switches described for Shepard's capsule; also, the attitude control levers for manual operation, and cabin pressure control knobs. To the extreme right of the center panel, the environmental control, electrical systems, and communications equipment. Between the bottom of the center panel and the place occupied by the pilot's knees, a map and chart kit hung beneath the periscope screen. Above, and again to the left, two floodlights for lighting the cabin when the capsule moved round the dark side of the Earth and a small flashlight for illuminating specific areas. At launch, the spacecraft and launch escape tower would weigh a combined 1,935 kg, reduced to an orbital 'weight' – more appropriately expressed as mass in that weightless environment – of 1,355 kg. On the water at mission's end, spacecraft 13 was calculated to weigh 1,131 kg, having lost parachutes and excess consumables – propellant for the thrusters, water and gases, etc., – and to weigh 1,099 kg at recovery.

Despite the excellent condition of the spacecraft upon arrival at the Cape, some 225 design changes were made to a host of minor components and systems, alterations arising from recent Project Mercury experience with other production capsules during post-flight teardown. For instance, fuse holders were taken out and reinforced because engineers had noticed a susceptibility to fracture, plastic flare seals were taken out of the automatic reaction control system inlet and outlet connections and replaced with soft aluminum washers, and wiring to the thruster solenoids in the electrical system was brought through a common connector so that the solenoids could be disabled and so prevent overheating on test. A myriad minor changes like these reflecting the operational stance of Cape engineering facilities, applied to standard spacecraft produced at St. Louis all the many lessons learned from each individual mission.

Spacecraft 13 entered a strenuous checkout test in Hangar S and began a simulated flight test on 25 November, 1961. Completed on 12 December, a second run was deemed necessary when NASA management scrubbed the launch for December, another series of simulated tests beginning 19 December. The capsule was on a schedule that anticipated it being at the pad on top of Atlas 109-D for 13 days prior to launch but on the very day it was taken to its booster – 3 January, 1962 – lift-off was re-cycled from 16 January to 23 January at the earliest. Activities to be accomplished before the spacecraft could fly included systems tests following the mechanical mating of capsule and booster, electrical interface and abort system tests, the Flight Acceptance Composite Test (FACT), simulated launch, flight configuration sequence and abort checks, simulated flight, general spacecraft servicing, pre-countdown preparations including cleaning up the capsule interior and placing all the switches in the right position, and the final countdown sequences; it was a lengthy, time-consuming operation repeated consistently throughout the Mercury-Atlas program.

Unlike the manned Redstone missions, the big Atlas booster provided pad support facilities greatly in excess of equivalent services at the smaller Redstone pad. Taking the standard Atlas launch pad, NASA modified the site according to Mercury requirements, which included a collapsible 'white-room' inside the service tower that enclosed the combined spacecraft and booster during pre-flight preparations; only at T-55 minutes would the tower roll back to expose the vehicle for launch. This white-room comprised floor, roof and walls that could fold away prior to moving the tower but which would encase the spacecraft in an environmentally controlled atmosphere to minimize the effects of dust flying in the atmosphere, humidity, etc., during preparation for launch. The astronaut could escape via the white room if indications of an imminent explosion were received. After the service tower had been rolled back at T-55 minutes, an emergency egress tower attached to the umbilical tower, permanently fixed to the concrete pad for continuous topping up operations with the Atlas launcher, could be used as a means of escape. This egress tower was little more than an open cage on a pivoted platform, which would move vertically just 30 seconds before launch to clear the way for the booster to rise.

Several rehearsals of a possible escape from a capsule on the pad demonstrated that an astronaut could be out of his spacecraft and into a safe position within 2½ minutes of an alert! From the egress tower, a fire-proof vehicle would rush the pilot to a safety zone. Also, four remotely controlled water or foam nozzles could be directed at any place within range of the jets from a console in the pad blockhouse. In the time between T-30 seconds and 10 seconds into the flight, a radio command system could fire the launch escape system and lift the spacecraft free of a malfunctioning booster. After that, escape procedures would be controlled from the ascending vehicle itself.

Compared with the Redstone flights, an astronaut would experience the booster operation for much longer and to considerably higher speeds: Atlas would burn for 5 minutes versus Redstone's 2¼ minutes and accelerate the spacecraft to 28,000 km/hr versus 8,300 km/hr. About 2 seconds after lift-off, the Atlas roll program would turn the booster round 30° to align its 105° launch azimuth with the 75° required for Mercury orbit insertion. This would be completed 13 seconds later and the pitch program started, tilting the booster over in incremental steps so that it would be travelling horizontal to the ground outside Earth's atmosphere. At 2 min 10 sec a ground guidance station would issue a radio command to shut down the two booster engines – until this point the launcher had a thrust of more than 160 tonnes

Unlike Redstone, Mercury's Atlas booster afforded direct access from a catwalk alongside the spacecraft. Here, Friendship 7 waits for launch.

–, lock up the single sustainer engine, release the boosters, unlock sustainer engine guidance, and continue to ascend on the thrust of its engine until at 5 minutes into the flight a further radio command would be issued to shut that engine down.

At that point the assembly would be in orbit and about 2 seconds after sustainer engine shutdown the capsule would separate from the booster adaptor by firing the pyrotechnic charge and operating the rear facing solid propellant posigrade rockets for 1 second. Immediately thereafter, the capsule would turn itself around from a forward to a rearward-facing position, pitching its nose down (and consequently its heat shield up) 34°. This would be the attitude Glenn would retain, except for brief periods of attitude control tests, during the remainder of the three-orbit mission until, about 4½ hours after orbital insertion, the retro-rockets would be fired. Thereafter, it merely remained for the retro-pack to be jettisoned and for the spacecraft to pitch up 1½° above the local horizontal so that the capsule would be in the proper attitude for re-entry. The landing bag would deploy, pushing ahead of it the ablative heat shield, about 12 seconds after main parachute release to a furled then unfurled position.

Orbital insertion requirements were demanding on both Atlas launcher and the ground tracking teams whose job it was to compute rapidly the actual dimensions of the trajectory in real-time. Critical go/no-go decision gates would be opened and shut as rapidly as possible after sustainer engine shut-down so that astronaut and mission operations personnel could be satisfied the orbit was a safe one. To achieve a safe orbit the launcher had to perform within tight constraints, narrow corridors of speed and altitude outside which the spacecraft would either return to Earth prematurely or enter an elliptical path from which it could not be safely recovered. For instance, for every 2.2 km/hr on top of the actual speed the booster was planned to achieve, the spacecraft's orbital high point would increase by 1.6 km; if the Atlas burned too long and imparted a speed 44 km/hr in excess of the planned value, the orbit would be 32 km too high at apogee. From the optimum value of 28,200 km/hr the booster would generate an abort situation if it deviated by more than 220 km/hr; too fast by 220 km/hr and the spacecraft would re-enter at too steep an angle, and too slow by the same amount and the capsule would not make it around the Earth before sliding back into the atmosphere, the unsatisfactorily shallow path heating the spacecraft far too long for it to survive.

With consideration of the desired trajectory in a nominal orbit, plus contingency operations that would be necessary if the mission should be aborted, flight controllers recommended a path around the Earth about 161 km high at the low point (perigee) and 261 km high at the high point (apogee). But the angle of the orbit with respect to the Earth's equator also had to rely on specific factors. For instance, the disposition of suitable tracking facilities and the path of the spacecraft during various abort situations. It was determined that an orbit inclination of 32.5° was the most suitable, all factors considered. As for the time of day, that influenced the lighting conditions at various points in the flight but all in all it was appropriate to aim for as early an hour as possible, thus preserving the greatest number of daylight hours at the end of even a normal flight, where the astronaut might have to ride it out several hundred kilometers from the nearest recovery ship if the spacecraft came down off target. But that was most unlikely, for increasingly effective and broad-based recovery procedures had tracked the expanding capacity of the program itself, until, for MA-6, the inventory of Naval support embraced 24 ships (including 3 aircraft carriers), 14 helicopters, and 49 aircraft.

Recovery zones were arranged similar to those employed for MA-5 chimponaut flight support: areas A to E strung out in an arching path between the Cape and the Canary Islands ready to retrieve the spacecraft from an early abort during ascent or from a near orbit insertion point, with areas F, G, and H comprising ellipses strung in a line from east of Bermuda down to Grand Turk Island on a south-west pointing line. Areas F, G, and H would form prime recovery zones for the first, second and third orbits. But preparation was not exclusively the concern of ground support teams, the astronaut too was required to plug in to a mission-orientated machine that delivered him to the pad at the correct time.

Largely the concern of Robert B. Voas, the astronaut training program began where basic conditioning left off. For two years after selection, the seven Mercury astronauts went first through academic and classroom study where they learned how spacecraft fly, why they are designed the way they are, how they are expected to perform, and what to do under a variety of unpleasant circumstances; they also acquired information on basic astronomy, the recognition of surface features from a great height, and celestial navigation. Then they were required to sample the various training devices designed to take a physically well conditioned human being and subject him to horrendous and dispiriting contraptions in the interests of pioneering new paths to space travel. Not all the astronauts were convinced of the need for such extreme measures, believing that abort situations and emergency plans could be well accomodated by the repertoire of physical resilience they possessed already. But in the days of unknown response to an uncertain environment, it was wise to take precautions.

But academic and physical conditioning produced a pilot capable of flying a mission; to learn that mission, and to acquire detailed familiarity with the spacecraft he was detailed to fly, required a third phase of training, one that could only begin when he was assigned a specific mission. From selection, an astronaut would adopt a very different life style until the end of the flight when he would ordinarily assume responsibilities similar to those he had left. Voas was concerned to include in the astronaut's spacecraft time – the period inside the capsule in orbit – significant contributions not only to the engineering needs and requirements of the program but also to matters of science associated only loosely with the particular flight in question. The pilot should, thought Voas, become something of a research scientist, experimenting when possible to contribute real-time information that might prove valuable for later projects. In effect, Voas wanted to increase the background knowledge of space flight rather than have the astronaut perform like a pilot unconcerned with how his flying may or may not influence future aircraft design!

It was a switch to a research-orientated role that few astronauts relished. Bred as high-performance aircraft pilots, they were averse to cluttering the flight plan with unnecessary experiments. A good example occurred when Voas suggested that Glenn should experiment with spatial orientation under conditions of both light and dark. With his eyes open, Glenn was to move his arms and try to reach certain parts of the instrument panel, a control experiment for the important part of the test when he would close his eyes and repeat the procedure. Cooper expressed the sentiments of the astronaut corps when he reminded Voas that the cramped confines of a Mercury capsule, packed with switches and levers designed to activate critical systems, was no place to 'be reaching over on this panel with your eyes shut!'

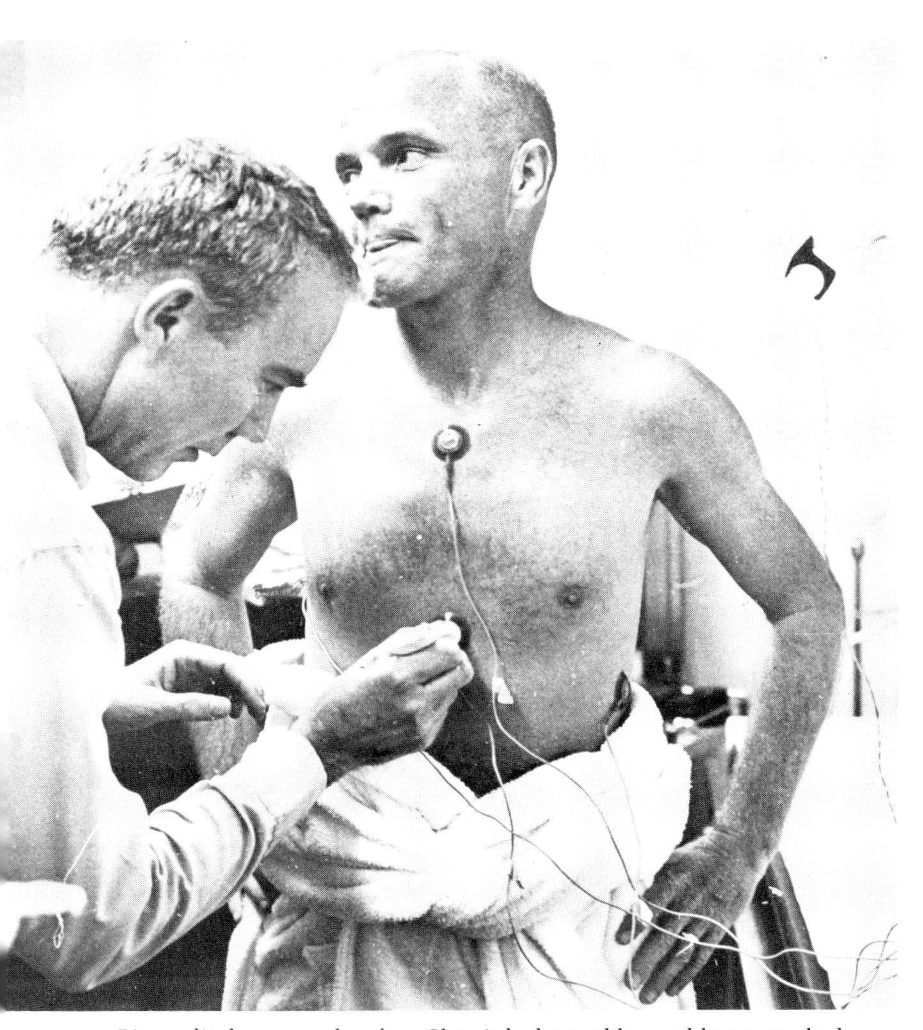

Bio-medical sensors placed on Glenn's body would record heart rate, body temperature, respiration and blood pressure. Here physician William Douglas applies paste to an electrode.

Nevertheless, the introduction of scientific tasks to the astronaut's flight plan was an inevitable consequence of moving into new frontiers, although bitter controversy was to break out over this very issue, causing some astronauts to resign and others to hurl abuse at scientists brought in for training programs. Early in 1962, however, the few scientific tasks assigned to John Glenn were hardly a threat to the safety of the mission, as many would feel their introduction to be just a few years hence. For the most part, the MA-6 pilot would limit his scientific study to reports of astronomical phenomena on the dark side of the Earth and to meterological observations during daylight; in a typical orbit lasting barely 90 minutes, the capsule would be in darkness for about 40 minutes.

But as the pilot and his back-up made final preparations for launch, with engineers and technicians readying spacecraft and booster at pad 14, newsmen from home and abroad converged on the Cape in anticipation of a flight on 23 January, postponed from the anticipated date one week earlier because of trouble with the Atlas propellant tanks. It was a fitting date, the day that Alfred Eggers from NASA's Ames Research Centre received the Institute of Aerospace Sciences' award for contributions to solving spacecraft re-entry problems and for leading the way in designing the blunt shaped re-entry body. But it was not to be. Everybody was ready, the flight operations people and technicians at pad 14 were confident of success, but the meteorologists pointed to the cloudy sky and gave a thumbsdown sign. Around the world, 19,300 people were ready and waiting, 600 newsmen crammed the parkways and caravan sites, and astronauts sat pensively waiting for the command to go. Days came and went, Tuesday the 23rd passed, and Wednesday, Thursday, and Friday produced more cloud and a lot of frustration, among the press at the Cape and among political circles in Washington. But Saturday, 27 January, looked promising and the flight preparations got under way. The astronaut was made ready for the flight but after 5 hours in the capsule he climbed out, the mission postponed at T-29 minutes. The gods over Mercury belied the spacecraft's name: Friendship 7 – selected by Glenn's family with an emblem painted on capsule 13 by artist Cecilia Bibby.

Walt Williams was glad the flight had had to be postponed because of weather. A seasoned veteran of Mercury operations, he recalled afterwards that 'nothing was wrong – but nothing was just right either.' The assembled combination of several hundred thousand components in a conical container so small it could easily stand in the corner of an average living room produced a vehicle that had a character of its own. Each time the system was powered up it reiterated its own unique way of responding, no two capsules behaving exactly the same. So much so that engineers and checkout crews quickly learned to read the 'personality' of each spacecraft, and to sense when it was behaving as expected but with a certain undefined reluctance to give of its absolute best. It would happen that flights would be deliberately scrubbed because the launch operations manager or the spacecraft manager felt uneasy about their mechanical proteges. So it was on Saturday, 27 January.

Because the booster had been fully fuelled with kerosene and oxidizer, the launch had to be re-cycled to no earlier than 1 February leaving time to drain and purge the tanks. During a kerosene fuelling operation on Tuesday the 30th technicians noticed what they thought was propellant seeping through the common insulated bulkhead that separated the fuel tank from the oxidiser tank. The engineers now wanted at least 10 days to fix the problem, having to physically remove and then re-install the insulation essential to keeping a thermal lid on the super-cold oxygen. Consultation with the recovery force commander, Rear Admiral John L. Chew, suggested a new projected launch date of Tuesday, 13 February. Almost as soon as they got word of the near two-week delay, newsmen began to pack up and make their way back from the Cape to cities and suburbs of national and home town newspapers. Glenn too left the Cape, returning home for a few days' rest to Arlington, Virginia.

Two days later, glum words from Representative James G.

Fitted with the new trapezoidal window, Glenn's Friendship 7 spacecraft gets a checkout in one of several tests and simulations performed with a suited pilot.

Fulton, the senior Republican on the House space committee, when he said that 'There's no doubt our overall space program is slipping despite the high words and fine praise coming from the White House. . . . If it continues to slip, we'll be lucky to get a man on the Moon before 1980.' The next day pressmen caught up with Glenn, flooding over his carefully trimmed lawn to get a few words from the astronaut or to click away for yet another batch of photographs, moving him to comment that 'it looks like Hangar S was not such a bad place after all.' On Monday the 5th of February, Glenn called on President Kennedy at the White House and personally briefed him on the status of the Mercury project, the condition of MA-6 flight preparations and his own feelings about the upcoming mission. Just two days later the press enticed comment from the President to the effect that 'I have said from the beginning we've been behind. And we are running into difficulties which come from starting late.'

At the Cape only 200 newsmen were on hand to monitor flight preparations. On the 14th the succession of delays began all over again, with another cancellation on the 16th due to bad weather. By the 19th all looked set for a flight the following day, and the split-level countdown began. During the afternoon, Walt Williams heard from downrange recovery ships that the weather prospect looked good for a flight on the 20th, and pessimistic weather men at the Cape were out-voted as Williams sensed momentum had reached just the right level. He ordered the second phase of the countdown to proceed, called a mission review and advised the astronaut to prepare himself for a launch early the next day. Glenn went to bed and read a systems manual on the spacecraft's attitude control equipment.

Came the Tuesday and long before the final preparations began, Glenn awoke. It was 2:20 am. Along the beaches past the growing inventory of launch towers, tens of thousands of people shuffled about, some still sleeping, others – veterans of countdown delays – had been there for a month or more and knowledgeably kept the late arrivals informed on what was happening minute by minute. Several thousand more scurried about between rows of trailers, pick-ups, convertibles and caravans, the paraphernalia of a shanty town that grew up on the marshy flats of Cape Canaveral in the preceding weeks, 'elected' spokesmen running up and down the rows of temporary abodes banging on the windows and doors to tell occupants of rumors that this was it, the great day had arrived at last. Some preferred to sleep before waking with the Sun for events still several hours away.

At Hangar S there was a mood of expectancy, the feeling that indeed this was the day, a hum of confident activity pervading the preparations. At the blockhouse, Atlas launch vehicle technicians groomed their charge as over at the Mercury Control Center flight director Chris Kraft fed his team into the stream of events moving down the long activity-filled countdown. It was 3:40 am. Spacecraft checks began, network controllers got status reports from around the planet, medical monitors squared away for the bio-physical hook-up, and booster systems engineers linked up with the blockhouse to serve as interface between capsule and launcher. Within 40 minutes Chris Kraft got the word that Earth checked out – the tracking and communications stations around the globe were all in a 'go' condition.

By now Glenn, having had a brief breakfast meal, was in the suit-up room where technician Joe Schmitt eased on the pressure garment over a biological harness strapped to Glenn's body by physician Bill Douglas. Measurements of the astronaut's physical condition during orbital flight would include telemetry conveying his respiration rate, from a thermistor anemometer set to detect the flow of expired air, a set of ECG sensors for heart rate, body temperature from a thermistor mounted in a rectal catheter, and blood pressure from an inflatable occluding cuff on the astronaut's left arm. The Blood Pressure Monitoring System (BPMS) was new and arose from a determined effort to develop a flight-rated device from a project that began as late as June, 1961.

By 5:01 am Glenn was on his way to pad 14, arriving in the transfer van 16 minutes later but a little behind the planned schedule. Minutes later launch vehicle technicians observed a malfunction in the Atlas guidance system but coming in a scheduled 90 minute 'hold' designed for just this sort of anomaly its impact on the resulting countdown was minimized to 45 minutes. The astronaut remained in the van while engineers changed the guidance equipment module responsible for the problem, then stepped out, walked to the service tower elevator and ascended to Friendship 7. At 6:03 am, white clouds, that had up to this point filled the Florida sky, began to move away, John Glenn slid across the sill and slowly settled down into the spacecraft's contour couch. All around, in the cramped quarters of the white room, technicians quietly worked away, waiting for Glenn to settle in before starting the closeout procedure where lines and leads would be plugged to Glenn's suit and straps and buckles tightened down after leak checks and bio-physical sensor calibration.

By 7:00 am concern about the weather reached a peak but Harlan G. Higgins, one of the Cape Canaveral meteorologists responsible for predicting impending trends, observed the clouds scudding away and the air warming up. At Mercury Control the telephone rang. Ernest A. Amman, the Center's contact with Cape weather men, picked it up. Chances for a launch were good and Higgins was backing a clear sky for later that morning. This was it, the big day really had arrived. Newsmen gathering now to begin their work for the day thought otherwise, but casual preparations soon gave way to haste as it became apparent that the countdown would not stop until, expended at zero, the three big engines on Atlas 109-D thundered into life.

At 7:10 am spacecraft technicians noticed that, just like the hatch on Grissom's capsule, a single bolt among the 70 that would hold it tight against the frame was broken. Walt Williams had the final word and ordered it changed, causing another delay, this time of 40 minutes. When the count resumed at 8:05 am other technicians outside the enclosed security areas guarding eager spectators from the anxious Mercury men began to put together cables and equipment for the breakfast show to end all breakfast shows. Across the United States, more than 100 million people would soon be irretrievably committed to their television sets in 40 million homes. Gradually, as the Sun moved up above the Atlantic sending bright rays from behind diminishing clouds, it seemed as though the entire world had walked to Cape Canaveral to witness a very historic event.

Suddenly, it seemed, tens upon tens of thousands of people had risen from the marshy flats, crept in like a human tide from the dark coloured beaches, to press in surprising order against secure fences and preserved sanctuaries. They were several miles from where the silver rocket stood, the rust-coloured capsule distinct beneath a bright red lattice tower, but that mattered little. It was important to a lot of people that February morning that a representative of theirs was about to fly in space. They were not to be denied a part of this great adventure. They were there because they could say they were as time dulled the memory of a morning when the world seemed to be moving to new horizons.

In the capsule high above the concrete and steel that made up pad 14, John Glenn was in the final phase of closeout. Soon the technicians and engineers had left the white room as it was folded up like a house of cards on hinges that moved floors and walls aside in preparation for the great steel tower to slide back a safe distance. Glenn felt a strange sensation as the rocket sat exposed, the view through the periscope showing much of the Cape area. As he moved, carefully distributing his body within the couch, he felt the missile quiver and sway, an apprehensive feeling subdued by the realization that what he sensed was movement imperceptible to an outside observer.

At one point it was necessary to top up the liquid oxygen and Glenn was aware of distant sounds like some subterranean gong booming through a stone dungeon. Then a more positive indication of fluids moving inside the great belly of Atlas as the thin metal skin shuddered and flexed, adjusting to the surging oxidizer. Outside, viewed through the trapezoidal window in front, vapor swept across the field of view, a comforting sight for without the vented ullage – a brewer's term meaning excess gas on the top of a barrel of liquid – the liquid tank would build up pressure and rupture like the outside face of a fractured dam. For a while the dark green landscape of Cape Canaveral paled as white plumes crossed the tiny cylindrical periscope. Then, at precisely 8:58 am, another hold, this time for a problem to be sorted out involving the fuel pump outlet valve that delayed the countdown for 25 minutes. At 9:25 am, T-22 minutes, the count resumed, halted again 15½ minutes later while checks were made on a network computer at the Bermuda tracking station. Two minutes came and went and the countdown clocks moved on. Across the Cape area the tenor voice of Shorty Powers announced each event, breaks in transmission from the operations room occurring at increasingly less frequent intervals, always the commentaries preceded by the now familiar words: 'This is Mercury Control. . . .' Finally, the seconds ticked away as events cycled up on engine ignition. In the blockhouse and at the Mercury Control Center, technicians paced their preparations toward the point of lift-off. From a test sequencer, an unbelieving confirmation: 'Looks good old man. Boy! Can you imagine, here we go.' The test conductor read out console data; the launch director ran around the room getting status reports in the last few seconds. 'All recorders to fast, T-18 seconds and counting.' Over the communication lines, last sentiments, almost subconscious appeals for perfection: 'May the way winds be with you.' 'Good Lord ride all the way.' 'God Speed John Glenn.' And from another controller speaking to several hundred million people around the world: '10–9–8–7–6–5–4–3–2–1; ignition, lift-off.'

At thirty-nine seconds past 9:47 am the three engines thundered into life, noticeable movement now in the tall structure as Glenn felt the point of ignition send shock waves through the big rocket. With a sudden upward surge, Glenn was pressed deep into his couch. 'Roger. The clock is operating. We're underway.' It was a little incongruous. For the first time a man was riding inside a tiny capsule on top of America's biggest operational Inter-Continental Ballistic Missile, a place universally accepted as the rightful location for a nuclear tipped warhead. 'Hear loud and clear,' came the voice from capcom down at the Control Center. Glenn felt the rocket suddenly twist to one side. He glanced at a small mirror positioned by his own hand during the long countdown to provide a view of the ground below as a reflected image from the forward window. The launch pad was clearly visible, flame and steam fanning out like snaking whirls of smoke. But the view *was* turning as the rocket rolled in its correct azimuth. Soon the pitch program would start, thought Glenn, unprepared to feel so strongly the twitching movement of the big bird's guidance commands.

'Roger. We're programing in roll okay,' said Glenn, his voice belying the surprise at the Atlas's motion. Just as he radioed a 'Little bumpy along about here,' the vibrations he had felt the moment Atlas left its restraint arms suppressed themselves beneath the larger rumble of booster and sustainer engines. But the vibration never did completely go away before increasing in volume again as, one minute into the flight, the booster neared max q, that point on the ascent where dynamic forces reach maximum. A louder roar took over to accompany the vibrations, compressed air tearing at the capsule and its escape tower, almost drowning out the muffled roar from the three main engines. Just 10 seconds later the confusion subsided, the Atlas through max q without a problem, accelerating now – so much more pleasing going in a straight line than round and round the central support structure of a centrifuge. There was a difference, and it *was* quite a pleasant experience, the force of acceleration going through 7 g at most rather than the 11 g of a Redstone flight.

Then, at 2 min 9.6 sec, booster engine cut-off (BECO) and a sudden reduction in the force of acceleration as thrust dropped appreciably. A loud thud and a sound like metal runners gathering speed as the booster skirt assembly slid back away from the single sustainer engine firmly attached to the base of the rocket. Lighter now, the accelerations began to pick up but seconds after BECO Glenn saw a swirl of smoke cloud his window. 'The tower fired: could not see the tower go. I saw the smoke go by the window,' reported Glenn. But he was wrong, having mistaken a vortex of booster engine smoke for premature ignition of the launch escape tower jettison motor. A half-minute later he realized what earlier had seemed to be his means of escape flying loose was, in fact, just smoke: 'There the tower went right then! Have the tower in sight way out. Could see the tower go. Jettison tower is green,' indicating the glowing light on the left status panel confirming satisfactory release of the escape system.

Immediately prior to jettisoning the tower the flight programer had pitched the assembly forward, dipping its nose slightly, which Glenn felt as he scanned his instruments. Seconds after the tower sped away from Mercury's antenna housing, the booster pitched back up slightly, but while down had given Glenn a brief view of clouds below a segment of Earth's horizon. The minutes passed, acceleration moving on up toward a peak of 7.7 g at which point the sustainer engine would be shut down on command from the ground. As propellant drained from the two

Glenn's spacecraft became the proud showpiece at exhibitions and museums around the world, as had Shepard's Freedom 7, seen here on display at the Science Museum in London.

Recorded on a film camera, pilot activities were studied after the flight to determine how well he performed. Measurement of eyeball movement showed the areas of instrumentation he observed at specific times.

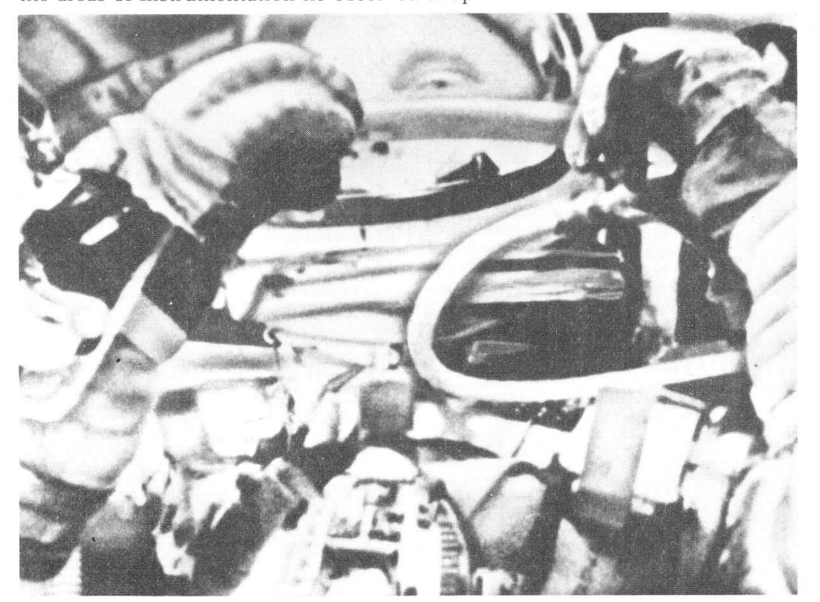

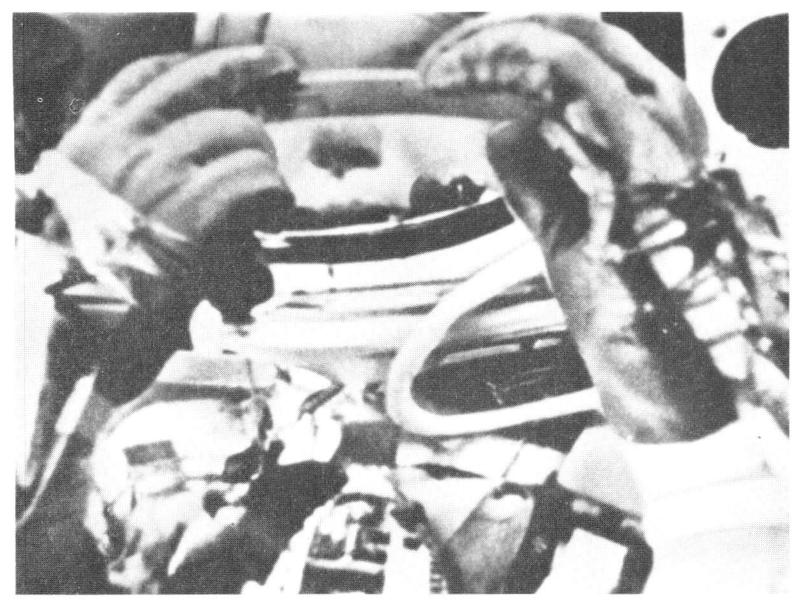

enormous tanks, pressure within the thin aluminum balloon reduced. Glenn felt random movements as the guidance platform sought to keep the single main engine bell thrusting through the rocket's centre of mass. Increasingly, it felt as though he were sitting on the end of a very long springboard, waving, flexing from side to side, as the rocket swung to keep track of the preselected trajectory. But noise too was increasing as depleted propellant replaced liquids with gas to conduct the sound of the engine more efficiently through to the tiny capsule riding on the rocket's nose.

MA-6 was now high above the sensible atmosphere, tearing through the partial vacuum of near-space at a speed greater than any other man but two Russians had ever experienced, flying horizontal to the ground more than 150 km below. 'Roger. Cape is Go and I am Go. Capsule is in good shape. Fuel 103–102, oxygen 78–100, cabin pressure holding steady at 5.8, amps is 26. All systems are Go.' This was Glenn's counter to Shepard's 'A-O.K.': 'All Systems are Go.'

'Roger. 20 seconds to SECO,' came the word from Cape Canaveral as the capcom gave Glenn the word that the sustainer engine would soon cut off. The g forces built up with almost frightening rapidity now, the big rocket getting lighter by the second as tonnes of propellant were consumed in the main engine.

'Indicating 6 g's,' said the Friendship 7 pilot at 4 min 49 sec into the flight. Within 10 seconds that had increased to $7\frac{1}{2}$ g. Noise was now severe, as the intensity of acceleration and the thundering sound from way below reached a peak. Suddenly, with cutting precision, the engine shut down, Glenn sensing a forward motion as the g level sank to zero in milliseconds, his whole body pitching forward, seeming to tumble end over end. But it was not, merely the extremes of sensation as the human autopilot told Glenn's brain to continue feeling forward acceleration denied by the now silent propulsion system. The main engine cut off just 5 min 1.4 sec after lift-off. John Glenn was in orbit. On the ground flight controllers noted that just 5 minutes into the flight, 'We are through the gates.' The booster had performed well, giving Friendship 7 a speed only 8 km/hr below the desired value. From the spacecraft: 'Roger, the capsule is turning around and I can see the booster during turnaround just a couple of hundred (metres) behind me. It was beautiful.'

'Roger, Seven, You have a go, at least 7 orbits,' came the reassuring word from Cape Canaveral.

'Roger. Understand Go for at least 7 orbits. This is Friendship Seven. Can see clear back; a big cloud pattern way back across towards the Cape. Beautiful sight.' Glenn's capsule had successfully separated from the inert Atlas, turned around to face backward, and pitched forward 34°, exactly as planned. Shortly after six minutes into the flight the capsule moved out of Cape tracking and seconds later switched to Bermuda. Around the world, the Mercury tracking and communications network stations at Canaveral, Bermuda, Canary Island, an Atlantic ship, Kano, Zanzibar, an Indian Ocean ship, Muchea, Woomera, Canton Island, Hawaii, California, Guaymas, White Sands, Texas, and Eglin, would support the flight, with astronauts dispersed to selected stations.

The first break in communication with the spacecraft came as Friendship 7 passed from the Bermuda horizon and drifted for nearly a minute before coming up on the Canary Island station, 11 minutes into the flight. With the spacecraft held in the grip of the automatic (ASCS) attitude control mode, major systems checks could get under way as the serious business of orbital flight pushed aside the freedom to comment on Earth below or the sensations in space:

'I am in orbit attitude for your tracking. Status report follows: Fuses all number one except tower Sep numbers two Emergency Retrosequence, Emergency Retrojettison, and Emergency Drogue are in the center-off position. Squib is armed. Auto Retrojettison, is off. ASCS in normal. Cabin lights are on both. Photo Lights are still on. Telemetry Low Frequency is on. Rescue Aids are on automatic. Ah, Jettison Tower, and Sep Capsule lights are out. The pressure regulator is still in the "in" position. Launch control is on. All sequence and panel positions are normal except Landing Bag is off. Are you receiving? Over.' Confirmation followed. 'Attitude: roll 0 (degrees), yaw 2 (degrees) right, pitch –33 (degrees). Rates are still indicating zero. I am on ASCS at present time. The clock is still set for time to ret, for retrograde time of 04 plus 32 plus 28. I have retrograde times okay from Bermuda. Cabin pressure holding steady at 5.7. Cabin air 90 (degrees). Relative humidity, 30 (percent). Coolant quantity is 68 (percent). Suit environment is 65. Suit pressure is indicating 5.8. Steam temperature 60 (degrees) on the suit. I am very comfortable. However, I do not want to turn down just yet. Primary oxygen is 78 (percent); secondary, 102 (percent). Main bus is 24. Number one is 25, 25, 25. Standby one is 26; Standby two is 25; Isolated, 29, and back on main. Ammeter is indicating 23. ASCS is 112. Fans are 112. Over.'

And so it would go on, status checks, switch positions, measurements of every important aspect of the capsule's performance, backed up by telemetry streaming to Earth with more data from deep within the spacecraft's systems. Glenn had launched to the voice of Alan Shepard from the capcom console at Canaveral. Bermuda had brought the deep, growling voice of Gus Grissom. Ahead, Gordon Cooper would pick up communication from the Muchea station, the first of two on the Australian continent. But first, a pass over Africa where he saw dust storms, and a confirmation that he found Fly-By-Wire a good control mode as the spacecraft yawed to the right in a maneuver planned to allow Glenn to check if he could orientate his capsule by visual flight routines. Looking out the window in the normal rearward facing attitude, like sitting in the back seat of a car looking through the rear window, provided a perspective reference whereby it was comparatively easy to judge pitch by observing the Earth's horizon, and even easier to determine roll by checking the same horizon for tilt. But judging yaw, or the pointing angle to left and right of the capsule's forward end, was more difficult.

In the normal orbit attitude position with the spacecraft pitched down 34°, the ground below gave a poor indication of whether the spacecraft was in line with the flight path. But when he pitched his capsule fully 60° down from a level position, the visual movement of the capsule across the Earth enabled Glenn to judge slight yaw angles more effectively. Passing across Africa

and coming within range of the tracking ship stationed in the Indian Ocean, Friendship 7 swept on into its first night: 'At this present time . . . I still have some clouds visible below me, the sunset was beautiful. It went down very rapidly. I still have a brilliant blue band clear across the horizon almost covering my whole window. The redness of the sunset I can still see through some of the clouds way over to the left of my course. . . . The sky above is absolutely black. I can see stars though up above. I do not have any of the constellations identified as yet.' And then, less than two minutes later after an exchange of possible abort times for the first, second, and third orbits: 'This is Friendship Seven. I am having no trouble at all seeing the night horizon. I think the Moon is probably coming up behind me. Yes, I can see it in the (peri)scope back here and it's making a very white light on the clouds below.'

Shortly thereafter, 49 minutes into the flight, his enthusiasm carrying the communication beyond range of the Indian Ocean ship, Glenn's on-board recorder picks up an enthusiastic astronaut: 'Friendship Seven, broadcasting in the blind, making observations on night outside. There seems to be a high layer way up above the horizon; much higher than anything I saw on the daylight side. The stars seem to go through it and then go down toward the real horizon. It would appear to be possibly some 7 or 8 degrees wide. I can see the clouds down below it; then a dark band, then a lighter band that the stars shine right through as they come down toward the horizon. I can identify Aries and Triangulum.' And then, four minutes later while talking to fellow astronaut Gordon Cooper at the Muchea tracking station: 'This is Friendship Seven I have the Pleiades in sight out here, very clear. Picking up some of these star patterns now. Little better than I was just off Africa.'

At Woomera, coming up on one hour since lift-off, the station communicator told Glenn they were reading a blood pressure of 126 over 90, and on the medical consoles his heart rate remained a steady 80–85 beats/min. Reporting to Australia that he could see lights down below the communicator told him that they would probably be the lights of Perth, switched on in full for the passing astronaut. Shortly after coming within range of the Canton tracking station, Glenn made ready the periscope to see the sunrise from behind his capsule: 'In the periscope, I can see the brilliant blue horizon coming up behind me; approaching sunrise. Over. . . . Oh, the sun is coming up behind me in the periscope, a brilliant, brilliant red. . . . It's blinding through the scope on clear. It's started up just as I gave you that mark; I'm going to the dark filter to watch it come on up.'

And then, seconds later, as if suddenly moved by several amazing events at once: 'This is Friendship Seven. I'll try to describe what I'm in here. I am in a big mass of some very small particles, that are brilliantly lit up like they're luminescent. I never saw anything like it! They're round a little; they're coming by the capsule and they look like little stars. A whole shower of them coming by. . . . They're going at the same speed I am approximately. They do have a different motion, though, from me because they swirl around the capsule and then depart back the way I am looking. There are literally thousands of them!' Glenn was not sure he was still in range of the Canton antenna, calling to the ground for acknowledgement. The Guaymas station answered and Glenn reiterated his experience with the tiny particles. In full daylight now, the 'fireflies' disappeared and Glenn got back to his status checks and control tests.

Then exactly 1½ hours after the mission began: 'This is Friendship Seven. Yaw is drifting out of orbit attitude and will bring it back in. Over.' With just seconds to go before loss of signal from Guaymas, Glenn noticed what seemed to be a repeat of the thruster problem that plagued the previous mission and brought chimpanzee Enos back to Earth prematurely. But California took over communication and the spacecraft was held to the correct attitude on manual control, the intervention of a human pilot directly responsible for saving the mission. It would be no real problem so long as Glenn could manually override the ASCS, but in-flight experiments would have to be curtailed and propellant consumption carefully watched. Through the Canaveral station, Glenn summarized:

'This is Friendship Seven. I'm going on fly-by-wire so I can control more accurately. It just started as I got to Guaymas, and appears to be it drifts off in yaw, to the right at about 1° per second. It will go over to an attitude of about 20°, and hold at that and when it hits about a 20° point it then goes into orientation mode and comes back to zero, and it was cycling back and forth in orientation mode. I am on fly-by-wire now and controlling manually. Over.' And just a minute later, he had performed a logical in-flight conclusion: 'What appears to have happened is, I believe, I have no (0.45 kg) thrust in left yaw. So it drifts over out of limits and then hits it with the high thrust. Over.'

From the ground, agreement: 'Roger, Seven, we concur. Recommending you remain fly-by-wire.' At this point the network controller expected to patch a call from the White House through to Friendship 7 for a real-time conversation between President Kennedy and the orbiting astronaut but difficulties delayed and finally cancelled the conversation. There were other problems looming larger than this in the Mercury Control Center, and as Glenn moved on to be picked up at Bermuda, Alan Shepard at his capcom console got a call from William Saunders the 'TM' controller in charge of incoming telemetry from the spacecraft.

Keyed to extract dubious data among the mass of information rolling in through the Goddard Space Flight Center, Saunders noticed an indication of segment-51, a telemetry code, that if correct spelt disaster for John Glenn. Segment-51 was a signal that flashed notification of landing bag deployment, the skirt, packed between the capsule's heat shield and the inner bulkhead, which would pop out before splashdown and push the shield away from the spacecraft to cushion the force of impact. If segment-51 was correct, if the landing bag had deployed in orbit, there would be nothing to hold the heat shield in place at the back of the capsule. Three strong metal straps held the retro-rocket package in place but, once released, the heat shield would drift away from the spacecraft's aft bulkhead so that when the capsule began to re-enter Earth's atmosphere it would be torn away, leaving Glenn helplessly exposed to a temperature of 1,650°c. He would be incinerated to ashes along with the interior of his spacecraft. And there was nothing he could do.

'Capcom, TM,' said Saunders.

'Go ahead TM,' came the reply.

'Roger, you do have a valid impact bag signal up at this time.'

'A valid what?'

'Impact bag signal, segment-51 is indicating a valid signal.'

'O.K. Let me check it out,' said Shepard.

'Roger.' Almost at once the word went out via land lines and radio links for the global tracking stations to watch especially for signs of segment-51 as the capsule swept into view, and to remind Glenn that the landing bag deploy switch should be in the off position. It was conceivable that the switch had been accidentally knocked out of detent, perhaps by the blind touch tests pressed on Glenn by Voas! If so, grist to an astronaut's mill for no more science! If not, it could merely be a faulty relay or a problem with the telemetry switching circuit. There was no way of knowing without fully involving the astronaut and no-one was about to tell John Glenn that his heat shield might drift away at any moment.

While Friendship 7 moved along its second orbit around the Earth, telephone calls and anxious conversation between technicians and specialists on the ground sought answers to the puzzling signal. Almost at once Chris Kraft favoured a contingency plan to leave the retro-rocket package in place during re-entry. No spacecraft had ever been expected to come back with the drum-shaped contraption still attached by its three metal straps. No studies had ever envisaged the necessity for such a need and there were no wind-tunnel tests to show what might happen to a re-entering Mercury capsule with the pack left on. A call to Maxime Faget in rented accommodation at Houston told William Bland that Mercury's chief designer had no objections to a plan like that, if telemetry proved that each of the three retro-rockets fired. If one rocket failed to fire the solid propellant still intact would ignite during the heat of re-entry and possibly burn a hole right through the heat shield.

The main advantage in leaving the pack on, however, would be to provide a means of keeping the shield strapped up until the force of atmospheric friction took over. Providing the capsule could be kept firmly base-end on to the direction of flight the pressure of air would hold the shield against the spacecraft. But would the projecting retro-rocket pack cause the capsule's shock wave to attach itself to the capsule walls, raising temperatures above the level for which the vehicle was designed? Nobody knew for sure, and conversation between Shepard and flight controllers went on.

'What's going to happen when we cut the retro straps loose?'

'Now, we know also that the retro-pack will hold the heat shield up.'

'This is Al Shepard. How much does this bag have to displace before this limit switch is actuated?'

'Ah, I couldn't answer that one Al. I don't know.'

'Well if the straps are going to hold the goddamn thing in there it's not going to move anyway, is it?'

'That's right, it shouldn't move see, I would assume that it would be tight enough to hold it up there. See Al, we might get trouble the other way – by keeping that pack on.'

'Yeah, but you're in a hell of a lot less trouble with the pack on than with the bag off!'

'This is right.'

'Yeah, we want to be damn sure on this one Ed because if that landing bag comes down it's disaster, whereas it's not disastrous if we make a re-entry with the retros on.'

During the hotter part of descent the metal pack would almost certainly burn away, cleaning up the smooth contour of the slightly convex ablative shield, and by that time the straps would have done their job. During the pass across the Atlantic, Glenn reported a surprising switch in his attitude control problem, that to flight controllers on the ground now seemed trivial by comparison but to the astronaut aboard Friendship 7 was a niggling malfunction inhibiting his freedom to perform attitude tests and exterior observations. 'My trouble in yaw has reversed,' he reported over the Zanzibar station. 'When I had trouble over the west coast of the United States, I had a problem with the yaw, with no low thrust to the left; now I have thrust in that direction but do not have low thrust to the right. When the capsule drifts out in that area, it hits high thrust and drops into orientation mode, temporarily. Over.'

Before passing on to the Indian Ocean ship, Glenn again described the view he had of orbital sunset: '. . . I want to make a mark here when the Sun goes down. Sun is on the horizon at the present time, a brilliant blue out from each side of it. And I'll give a mark at the last. The sun is going out of sight. Ready now, MARK (2:09:06 ground elapsed time, or GET). There's a brilliant blue out on each side of the sun, horizon to horizon almost. I can see a thunderstorm down below me somewhere and lightning.' Minutes later over the Indian Ocean Glenn was advised that his tracking ship had released some flares amid turbulent seas and asked Glenn if he could see them through a thickening storm. He could not. And then, at 2 hr 19 min:

'We have a message from MCC for you to keep your Landing Bag switch in off position. Over.'

Glenn replied with a curt, 'Roger.'

On toward Australia now and seven minutes later another query from Gordon Cooper at Muchea: 'Will you confirm the Landing Bag switch is in the off position. Over.'

'That is affirmative. Landing Bag switch is in the center off position,' came the reasuring voice of John Glenn. But Cooper pressed the issue a little further.

'You haven't had any banging noises or anything of this type at higher rates?'

'Negative,' replied Glenn.

Cooper's reference pursued a notion that if the heat shield actually was loose, Glenn might feel or hear slight movement as he swung the capsule round to change attitude. And then again, five minutes after the Muchea contact, Woomera felt obliged to do its part in confirming the position of the switch in the spacecraft, but this time in a more subtle manner as Glenn was asked to run through his checklist and then, a minute later, to repeat it more slowly! If Glenn could possibly have come through the repeated, suspiciously casual, requests for information about this one item, without having realized what was wrong, all was completely lost when, 15 minutes after hearing from Woomera, the Canton capcom boldly assumed that Glenn knew the complete story.

'Friendship Seven. This is Canton. We also have no indication that your landing bag might be deployed. Over.'

From the capsule: 'Roger. Did someone report landing bag could be down. Over.'

'Negative, we had a request to monitor this and to ask you if you heard any flapping, when you had high capsule rates . . .'

'Negative,' replied Glenn, who immediately assumed a very different reason for the surprising comment. 'Well, I think they probably thought these particles I saw might have come from that, but these are, there, there are thousands of these things, and they go out for, it looks like miles in each direction from me and they move by here very slowly. I saw them at the same spot on the first orbit. Over.' There was no amplification, nor indeed a correction to Glenn's assumption. Network tracking capcoms were not constantly in touch with the capsule's communication when it was out of range of their station and the Canton communicator had assumed the Cape had already briefed the astronaut on segment-51, which they had not.

By this time the hydrogen peroxide propellant supply for the ASCS mode was down to 62% and Glenn had been advised to let his capsule drift, bringing it back to the rearward facing, pitched down, attitude only when the attitude excursions got too severe. Nevertheless, it was expensive on fuel and Glenn tried to balance drift with conservatism for his thrusters, control jets which would be essential for retro-fire and the re-entry phases. On, across the United States, sped Friendship 7 for the second time, about to start the final orbit. For much of the time the conversation centered on minor technical issues relating to the attitude control modes, but over Zanzibar for the third time, Glenn again commented on the view below.

'There's quite a big storm under me. It must extend for, I see lightning flashes, as far, way off on the horizon to the right. I also have them almost directly under me here. They show up very brilliantly here on the dark side at night. They're just like firecrackers going off.' At Muchea, Cooper joked with Glenn about a request the astronaut had just made that the three orbits about to be completed should qualify for that month's 4 hours flying time. Cooper wanted to know if that should be flying time or rocket time? 'Lighter than air, buddy,' quipped Glenn. But at the Cape the mood was far from jocular as senior management personnel wrestled with decisions that either way could spell disaster for the mission. Should the heat shield be secured by leaving the retro-package on through re-entry, or should they back the most likely cause of the signal and proceed with a normal sequence in the hope that segment-51 was in fact erroneous? One more time the pilot would be quizzed on the possible malpositioning of the switch.

A call went up through the Hawaii capcom, leaving Glenn in no doubt as to what was happening: 'Friendship Seven, we have been reading an indication on the ground of segment-51, which is Landing Bag Deploy. We suspect this is a erroneous signal. However, Cape would like you to check this by putting the Landing Bag switch in auto position, and see if you get a light. Do you concur with this? Over.' If the light came on, it would imply a deployed bag, if not the bag would in all probability be in its stowed position.

'Okay. If that's what they recommend, we'll go ahead and try it. Are you ready for it now?' There was an affirmation and seconds later Glenn came back on the air: 'Negative, in automatic position did not get a light and I'm back in off position now. Over.' But still the problem niggled Chris Kraft in the Mercury Control Center, and conversing with Walt Williams, the mission's overall director, the two men decided to ask Glenn to leave the retro-rocket package on after retro-fire in the hope that it would not unduly upset the dynamics of the falling spacecraft, or compromise the integrity of the heat shield as it burned off and threatened to cause a local hot spot. Support for the adopted posture was divided, some preferring to have the pack jettisoned and others happy to see it left on. In the closing minutes of the mission now, Wally Schirra in California prepared to talk Glenn through the retro-fire sequence. Status checks came at frequent intervals and a test was made to see if the clock aboard the capsule agreed with the master clocks on the ground. Earlier in the flight it had been learned that Friendship 7's clock was 1 second out and if left uncorrected that would have put the capsule a minimum 8 km off target at splashdown. Calculations about the precise time to fire the retro-rockets were made on the ground, computations that were valid only if the two sources agreed.

Although on ASCS, the automatic mode that used the defective yaw thrusters, Glenn backed it up manually and helped to keep attitude rates within 3° of the desired values: facing backward in a straight line, pitched down 34°. Just before counting down to ignition of the first retro-rocket, Glenn reported reading 39% fuel on his attitude control propellant quantity guage. Then, with less than a half minute to go, Schirra fired a request at Glenn: 'John, leave your retro-pack on through your pass over Texas.' It was the first indication that controllers on the ground retained doubt about the condition of the combined landing bag and heat shield assembly just centimetres behind his back. It suddenly became very apparent to the Friendship 7 pilot that Cape Canaveral harbored serious concern for the astronaut's safety. At Hawaii he had realized the extent of the Cape's worry about the apparently erroneous segment-51, but thought that to be over. Now it was chillingly fresh in his mind again as he sat through the flight controller's verbal countdown to retro-fire, Wally Schirra joining in during the closing seconds.

'Roger, retros are firing,' confirmed Glenn as the first solid propellant motor gave the weightless astronaut a firm kick in the back. 'It feels like I'm going back to Hawaii,' commented Glenn as he temporarily felt the force of 1 g again.

'Don't do that, you want to go to the east coast!' Shirra's humor hid gathering tension across the United States, gathered again when the California capcom reminded Glenn to leave his pack on 'until you pass Texas.' Following a rapid status check read out to Schirra on the ground, Glenn again requested information as to when he should jettison the retro-package, reminding the capsule communicator that the ready light was glowing red and that he was standing by to release the sequence and let the automatic event timer cut the straps holding the pack. 'Texas will give you that message. Over.' Schirra, acting on orders from Chris Kraft, was not committing himself. If the pack was going to be jettisoned after all it would have to be done quickly, before the capsule started entering the upper layers of the atmosphere. If not, there was no hurry.

Back at the Cape, engineers hurriedly examined telemetry to see if the violent shock of the retro-rockets firing had changed the nature of the segment-51 warning. It had not. As soon as Friendship 7 passed across to the Texas tracking station at Corpus Christi, the communicator filled Glenn in on the full sequence the pilot was now required to follow. All the while, the Mercury spacecraft was descending toward the atmosphere.

'This is Texas capcom, Friendship Seven. We are recommending that you leave the retro-package on through the entire re-entry. This means that you will have to override the 0.05 g switch which is expected to occur at 04 43 53 (GET). This also means that you will have to manually retract the scope. Do you read?' To Glenn, it was an increasingly threatening situation, and he wondered for several minutes just what it really was Mercury Control knew that he did not.

'This is Friendship Seven. What is the reason for this? Do you have any reason? Over.'

Came the clipped reply: 'Not at this time; this is the judgement of Cape Flight.'

Glenn asked for the instructions to be repeated. They were, with a reminder that events would start happening less than 4½ minutes hence. From Corpus Christi, reiteration that it was merely a change in procedure: 'Friendship Seven, Cape Flight will give you the reasons for this action when you are in view.' Now, the minutes ticking away, Glenn had to get busy manually overriding sequences that should already have put his capsule in re-entry attitude, a maneuver inhibited by the cancelled retro-package jettison sequence. Swinging the capsule up to its correct attitude, where the nose is placed 1.6° above horizontal, the spacecraft flying backwards, Glenn then turned to the periscope and pumped furiously on a handle as the cylindrical protrusion smoothly slid back into the main body of the capsule. By now, the spacecraft was within range of the Canaveral capcom, who confirmed again the instructions from Corpus Christi:

'. . . we are not sure whether or not your landing bag has deployed. We feel it is possible to re-enter with the retro-package on. We see no difficulty this time in that type of re-entry. Over.' Seconds later, Shepard passed up to Glenn the exact time tracking stations on the ground had computed the capsule would be at the 0.05 g level. From this the capsule would automatically damp oscillations in re-entry attitude as the spacecraft descended, but manual actuation would be essential due to the unique procedure now being adopted. Now, manual attitude control propellant quantity read only 15% remaining on Glenn's console and he elected to switch from ASCS to fly-by-wire for re-entry control. That judgement was approved by Shepard, who continued to talk to Glenn as the capsule fell toward the atmosphere. Seconds before punching the 0.05 g button, Glenn sensed he was entering a very low gravity field, and felt the predicted time was in fact a little late. Seconds after that the deceleration began to build up and an ionization layer around the capsule, a shield effectively blacking out all radio communications built up by the terrific heat of re-entry, cut off Shepard's voice.

Only recently having pieced together why the Cape had been so concerned for the past two orbits, Glenn settled back to his main task of ensuring the spacecraft kept within acceptable limits of attitude control. There was a general rule that a descending Mercury capsule should be kept to within 10° of its attitude line during the fierce minutes of re-entry. Suddenly, the thunderous roar mixed with a distant metallic tearing sound indicated to Glenn that this was it. Not only was Mercury carrying a human cargo back from orbit for the first time, but the most unusual and unplanned sequence of events was a very real part of that tension-filled situation. Through the window, Glenn saw a bright light moving upon the exterior walls of the capsule, a deep orange glow, all the while a scraping sound like giant fingernails clawing at the metal shingles covering his tiny spacecraft.

Soon, pieces of debris began to appear, moving fast past the window. Great chunks of what Glenn really believed were pieces of his heat shield swept past the window as the light outside changed to a white hot brilliance. He was physically aware that the interior of Friendship 7 was noticeably hotter, increasing the air temperature as just a very short distance from his back the ablative heat shield charred and burned at a temperature of more than 1,600°c. Convinced now that incineration was a very real possibility, he decided to concentrate hard on the attitude control task; no astronaut would ever take into space pills or drugs designed to speed what could otherwise be a terrible death, for despite misunderstanding on this score from newsmen and the public alike, such thoughts are consumed by the passion of a trained mind fixed purposely to a set task. And so it was with John Glenn as other problems began to emerge.

No sooner had Friendship 7, still totally cut-off from tracking antennae, even now calling up to the sky in vain hope of penetrating the capsule's enveloping sheath, emerged from the peak g force of deceleration than the attitude control propellant dropped dangerously low. Glenn sensed increasing movement, a massive swinging pendulum-like motion where the capsule flung itself dangerously past the recommended limits. With a rate damping system switched in, the oscillations abated somewhat and Glenn realized that he was actually past the point where the spacecraft would have been consumed had anything in fact gone seriously wrong during the descent. But thoughts of perishing in the bowels of a charred capsule had never been allowed to get deep within his conscious mind, so the more real problem of depleted attitude control fuel caught his firm attention.

With almost two full minutes still to go before the drogue parachute was due to pop out and stabilize the swinging spacecraft, the manual fuel tanks ran dry, followed a minute later by the automatic supply. By this time the heat of re-entry was beginning to get through to the capsule's interior and the warmed oxygen he felt during peak deceleration was but a prelude to an increasingly hot interior. Now, the oscillations were getting worse and Glenn decided to manually release the drogue parachute early, but as he reached up the automated sequence anticipated his action and fired the drogue that unfurled a welcome canopy. The drogue had released itself more than 2 km higher than the programed altitude of 6.4 km. By now communications had been under way with the Cape for several minutes with a relieved pilot reporting that 'My condition is good, but that was a real fireball, boy. I had great chunks of that retro-pack breaking off all the way through!'

In the final drifting minutes of descent, Friendship 7 put out its main parachute and John Glenn released pent emotion: 'Main chute is on green! Chute is out in reef . . . and beautiful chute. Chute looks good . . . and the chute looks very good! The chute looks very good!' And then, recovery teams called to the falling capsule:

'Friendship seven, this is Steelhead. Be advised according to my surface gadget, your range (11 km) from me, on the way. Over . . . Remain in capsule unless you have an overriding reason for getting out. Over.' By this time the landing bag had been manually deployed by Glenn, acting on a reminder from the Cape. The solid clunk it gave as the heat shield fell away to hang on the extended skirt gave proof positive that the segment-51 signal had indeed been erroneous. But Glenn would never forget the feeling inside Friendship 7 as earlier, during descent, the capsule had physically warmed up while conducted thermal energy seeped inside the cabin, spreading thoughts that this indeed might be a day for others to remember.

Glenn had little time to muse over the day's events, his mind fully on the final seconds before impact. When it came, the capsule's impact into the Atlantic was a harder jolt than Glenn had anticipated, but a very welcome one nevertheless. Only seventeen minutes elapsed between splashdown and the destroyer *Noa* arriving alongside Friendship 7; another three minutes and the *Noa* had winched the capsule on board to the cheers of sailors crowded on the ship's upper deck. Since atmospheric descent the spacecraft had become increasingly hot and Glenn's temperature was up as he perspired in the tiny cabin, trying hard to remove equipment from inside to ease his egress through the upper part of the spacecraft. But this seemed tiresome so Glenn asked the attendants to stand clear while he blew the hatch off, flooding the interior with fresh air considerably cooler than that which had pervaded his capsule minutes before. Unfortunately, Glenn sustained the only injury of the entire flight: striking the plunger that detonated the pyrotechnic cord connecting the 70 hatch bolts, it rebounded and cut his knuckles. Glenn was found to be 2.3 kg lighter, most of that having been lost during the period of profuse heating while waiting on the sea for a pick-up.

On board the *Noa*, Glenn's first task was to complete a brief recorded summary of his immediate recollections and feelings, leaving out technical data or other information he knew he would readily recall at the scheduled post-flight de-briefing. Then, by helicopter to the carrier *Randolph* from where, after a physical examination which included x-rays, he was transferred to Grand Turk Island later in the day. Grand Turk would be a sanctuary from where thorough medical examinations could be conducted and at which first conclusions about technical issues related to the flight could be reviewed. There would be extraordinary pressure on the astronaut when he returned to the mainland, little opportunity for casual de-briefings or private conference with spacecraft designers and builders. But it was *their* day, the many men who took and nurtured the responsibilities bequeathed with the approval of the National Aeronautics and Space Act of 1958. Late in the day compared with Soviet achievements, Mercury had finally performed the job it had been conceived to accomplish, less than 40 months from start.

20 February, 1962, had been an unusual day for everybody in America. When the flight started most people should have been on their way to work, just starting a day's labour like the populace of any other industrialized nation. But that day things were different. Children stayed home from school, shops stayed shut, or allowed their staff to cluster around portable television sets, while customers came in to join the celebration. In the Congress there was universal approval to recess the Senate at 2:30 pm for the all important splashdown broadcasts. In New York's Grand Central Station, a huge screen carried pictures throughout the morning and early afternoon, scenes of flight controllers at the Cape, people being interviewed in the street about this unprecedented event, and relatives of the pioneer astronaut in recorded interviews and re-runs seen so many times on network shows preceding the launch.

At the Cape, a hive of activity, a sense of not really knowing the true significance of what many regarded as a monumental landmark in the evolution of the American nation. Abroad, radio channels kept up running commentaries, fed from American network broadcasts, the air-waves filled with American accents and semi-technical jargon. In Britain, millions crowded around radios and caught televised reports as workers filed from factories and families sat down to the evening meal. Only months before, public opinion had been influenced by the flight of Yuri Gagarin, and later by the day long space mission of Gherman Titov. Now, with an almost universal voice, the balance had been adjusted, the massive public relations job had paid off. In the months that followed, the interest shown by foreign nations in the exploits of America's astronaut corps far outstripped the expectations of many in the administration who, when asked to approve the allocation of money to the space program, had never been able to reconcile the technical nature of the program with claims about its massive potential for propaganda.

But the public could identify with this activity as though, from the comfort of their homes, they could participate in events akin to wagon-train treks across the mid-West a century before. No Hollywood fiction this, but the real thing with real people risking their lives as ambassadors of planet Earth. For a while this is precisely how the man in the street saw the exploits of both America *and* Russia. It was as though some giant public relations machine was caught up in a race akin to that performed by Scott and Amundsen when the two great polar explorers reached out for the South Pole in 1911. And the tension between East and West sustained by military and political initiatives by first one side and then the other could be given full vent in the cosmic void. Somehow it was a jousting confrontation where two great nations entered a neutral zone to do battle, not with each other but rather with the common dangers that faced both sides.

Responding to the call for more information, the United States Information Agency made a film called 'John Glenn Orbits the World' and released this within seven days of Friendship 7's splashdown to 106 countries in 32 languages. In all, 1,300 prints of the film went out, including 400 in 35-mm for cinema audiences and the rest in 16-mm for informal gatherings and mobile units set up by the USIA. It has been estimated that more than 200 million people around the world saw this film in the four months after MA-6. Following it, a much longer color film called simply 'Friendship Seven,' made by NASA and distributed by the USIA, went to 71 countries in 11 languages – including Swahili! From all over the world, congratulatory messages poured in to the White House and NASA facilities at the Cape. President Kennedy responded to a telegram from Premier Kruschev by asserting America's interest in being able to 'work together in the exploration of space.' Moreover, he said, 'I am instructing the appropriate officers of this Government to prepare new and concrete proposals for immediate projects of common action, and I hope at a very early date our representatives may meet to discuss our ideas and yours in a spirit of practical cooperation.' How different was the public voice to sentiments expressed behind closed doors in the White House.

Just three days after the flight, Kennedy flew down to the Cape and bestowed upon Bob Gilruth the NASA Distinguished Service Medal, awarding Glenn the DSM also. By this time, the greatly expanded role of the Manned Spacecraft Center – still in a host of different, rented, buildings spread over Houston – had created new roles for the men from Langley who just three years before had occupied a shed at the edge of the flying field. Gilruth, his administrative responsibilities now embracing Mercury, Gemini, and Apollo projects, assumed the title of MSC Director, and set up a Gemini Project Office under James Chamberlin with Kenneth S. Kleinknecht head of the Mercury Project Office; staff for these two offices were drawn from the old Engineering Division, now dissolved. On 26 February, a motorcade made its way through rain-soaked streets in Washington as a quarter million citizens roared and cheered the diminutive figure of John Glenn, on his way to address Congress.

On 1 March the astronaut was in New York when more than four million people lined the streets to hurl confetti as Glenn made his way toward Mayor Robert Wagner's celebration, jointly offered to Glenn and Gilruth. The next day, a visit to the United Nations where Glenn gave a formal address, and on the 3rd a visit to his home town of New Concord where the population of 2,300 were swamped by more than 70,000 visitors. But the technical inspection of spacecraft 13 proceeded apace; another Mercury flight was even then in an advanced stage of preparation and questions raised by the ASCS problem and the phenomena of segment-51 had to be resolved.

Careful inspection of the thruster orifices revealed metal fragments similar to other pieces found in the MA-5 capsule. This constituted a problem in that missions of longer duration could be threatened not so much by a physical breakdown in the attitude control system's ability to work but rather in the increased fuel consumption made necessary by switching to a less conservative control mode. Margins were already tight and further constraints could inhibit mission length. Accordingly, a substantial effort was made to prevent thruster problems occurring on later flights: dutch-weave screens designed to control the flow of propellant through the thrusters were replaced with a stainless steel fuel distribution plate and platinum screens; the bore and size of the heat barrier was changed; and the fuel metering orifice was relocated. These were subtle changes but they paid off and no more problems were experienced.

As for the apparently erroneous segment-51 signal, that was put down to a loose rotary switch connected to a stem which would signal landing bag deploy when rolled open. Nevertheless, two lessons emerged from the in-flight problem: keep the astronaut informed at all times of observations on the ground that appear to threaten the scheduled conduct of the mission; and never go to a contingency plan on suspicion that something *may* be wrong, if the alternative plan puts the pilot in greater danger. John Glenn had some significant things to say about the protracted route taken to give him information about the segment-51 signal, learning from post-flight conferences that the first indication of an anomaly had taken place 2 hr 40 min before he was purposely informed.

Glenn gave good reason for a change in this reluctance to tell the pilot news that might cause anxiety, citing the applicability of on-sight evaluation where, had he known about the signal, he could have observed the reaction of his spacecraft and interpreted on-board sounds with a broader base from which to work. However, the build-up in Mercury operations was as much a means of giving controllers on the ground valuable experience in the operational set-up as they were a method for performing orbital tasks. Without the lessons and tears of early manned operations, the complex and sophisticated activity of later flights in the Gemini and Apollo programs would have been impossible to achieve.

The period between May, 1961, when Alan Shepard first flew a Mercury capsule into space, and February, 1962, when John Glenn became the first American to orbit the Earth, produced a unique set of circumstances that first inspired the address before Congress to accelerate the pace of the US space program and then deflated that ideal in the attitudes of the one man who had initiated the response: John F. Kennedy. While pressed into a corner over the Bay of Pigs fiasco, memory of the Laotian debacle, and the very real challenge of Yuri Gagarin's space flight, Kennedy formed different judgements on the very same issues when, later in 1961, the pressure came off and the reaction curves on the Presidential performance graph resumed normality.

Congress had shown an almost unanimous approval of the 25 May decision to head for the Moon when hearings quickly revealed the essence of the national mood. But throughout the rest of the year, and by early 1962, there were critical noises rumbling from Congressional dissidents, opponents of the Kennedy administration, that would erupt in 1963 to oppose the commitment openly. Moreover, Kennedy too became diffident about his own conclusions and increasingly sought means by which the unilateral decision could be made to accomodate cooperation with the Soviet Union. But by the end of 1962 the enormous industrial build-up would be too far along the road to absorb a political change, and Kennedy had no recourse but to continue with his original decision. Had Apollo nursed a more protracted gestation, Moon flights may have been diverted to some other less expensive goal. That would not have been necessarily good for the space program, but it would have conformed more fully with Kennedy's own beliefs regarding priorities; he never, ever, understood the need to develop a major space program, seeing only the propaganda value of extravagant space spectaculars. But to the newly named Manned Spacecraft Center, there was a job to be done, and that meant pursuing Mercury flights to their inevitable conclusion.

Thoughts were beginning to emerge on the desirability, or otherwise, of 'stretching' Mercury to its logical maximum. An 18-orbit Mercury had been the original intent of the then STG management concurrence with Chamberlin's plan for studies on an improved capsule. When those studies led, by the end of 1961, to approval for a completely new, two-man, spacecraft based on the design principles of Mercury but with a greatly expanded

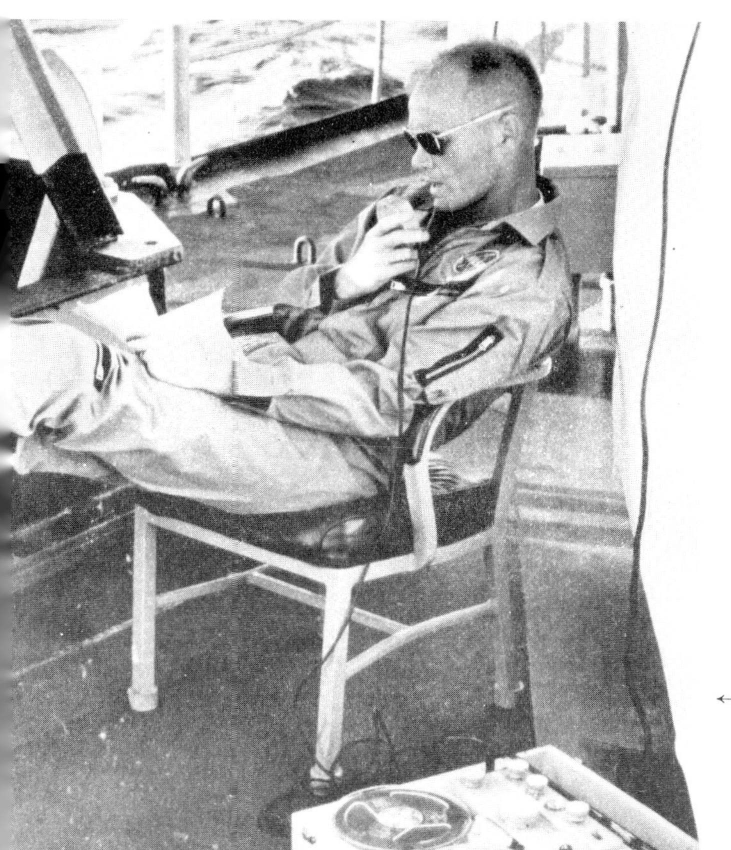

John Glenn receives a Presidential award from John F. Kennedy to recognize his unique position as America's first orbital astronaut.

← Voicing his immediate post-flight reactions to a portable tape recorder, Glenn relaxes on board the USS *Noa* before returning to the Cape and a comprehensive de-briefing session.

Struck from the flight list by a minor heart condition, Donald K. Slayton tries out the centrifuge at Johnsville, Pa.

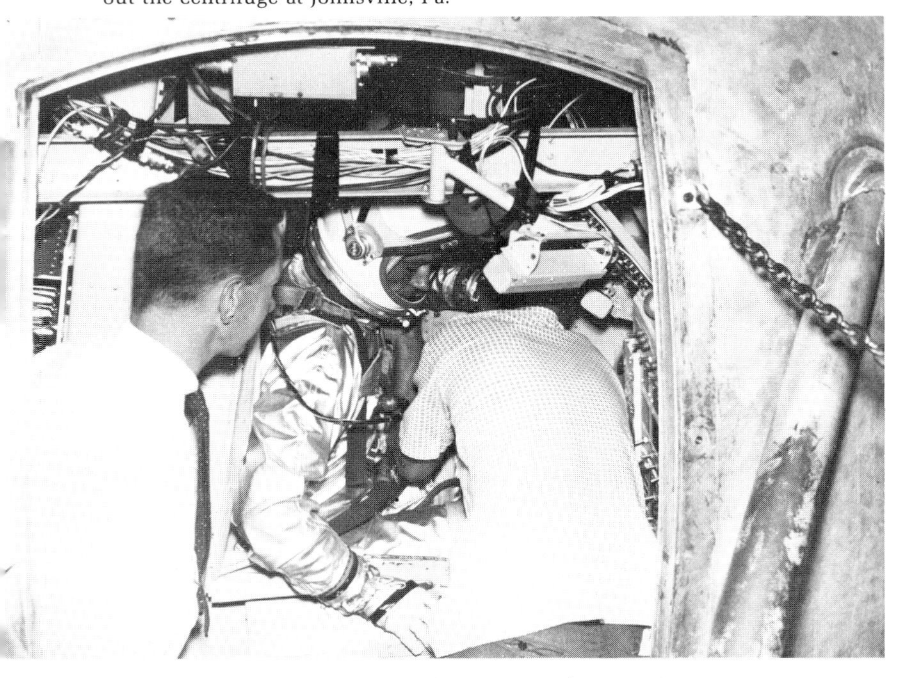

capability embracing orbit changes and 14-day flights, the Manned Spacecraft Center recognized that that still left the 18-orbit objective unfulfilled. So, after Glenn's successful three-orbit mission, the seeds of a long-duration Mercury flight were sown, with the conclusion that by pushing Mercury to a day long endurance capability the program would be in better position to hand over to Gemini in 1963.

Since November, 1961, Deke Slayton had been training as the MA-7 pilot, with Wally Schirra as his back-up. Suddenly, or so it seemed to an unsuspecting public, Deke Slayton was declared unfit for operational duty and was removed from flight status. On the day that announcement was officially made, 15 March, 1962, NASA named Malcolm Scott Carpenter, originally the back-up to John Glenn, as the pilot for America's second manned orbital flight attempt. It was a bitter blow to Slayton, who would carry on for more than a decade a determined fight back to operational status. But in 1962 it seemed a cruel twist of fate that many thought totally unjustified. In fact, Deke Slayton had been under medical examination on and off since a day in August, 1959, when the Johnsville centrifuge showed him to have a mild idiopathic atrial fibrillation – a slight irregularity in the heart's response to stress and fatigue caused by unusual activity in a muscle at the top of that organ.

Bill Douglas, for several years the astronaut's chief physician, decided to get opinion about Slayton's condition from a cardiology chief at the Philadelphia Naval Hospital, who agreed with Douglas that the man was perfectly fit for duty. However, not wanting to put at risk the life of a man who could find himself in crippling difficulties far from help, Douglas took Slayton to the School of Aviation Medicine, a facility of the Air Force, where he received a similar response. For much of 1960 the issue abated, although Slayton did attend several more clinics for further tests, and in the November he was named as the prime MA-7 pilot. Shortly thereafter, recalling a report from Douglas to NASA management in Washington, James Webb ordered a fuller investigation of Slayton's condition which led Douglas to debate the issue intensively with physicians White, Augerson and Henry. They all agreed that Slayton was fit to assume his astronaut role on MA-7. Doctors Roadman and Knauf at NASA Headquarters agreed with Douglas' recommendation but the matter was brought to the attention of the Air Force Surgeon General who, with Slayton only on loan from the Air Force, had a very special interest in the case.

It also prodded a file that had been on the Surgeon General's desk for some months. Douglas and Slayton appeared before an eight-man board convened by the Air Force to review the astronaut's mild heart condition and they too concluded that Slayton was fit and well up to the stresses he was expected to experience. Then Jim Webb requested assistance from three famous cardiologists: Proctor Harvey, a professor at the Georgetown University; Thomas Mattingley, from the Washington Hospital Center; and Eugene Braunwell, of the National Institute of Health. As scientists, they were unable to state categorically that Slayton would *not* suffer because of this atrial fibrillation, a condition about which there was almost universal ignorance among the medical experts in cardiological treatment. Erring on the safe side, they came to a conclusion that did nothing to reduce suspicion about Slayton's abilities, but in the absence of a positive assurance that nothing should be expected to go wrong, they recommended that if NASA had an astronaut that did not have this condition then it would be unreasonable to select a man who *did* fibrillate.

Webb could do nothing but order Slayton removed from the list of flight ready astronauts, although his demotion was more a product of ignorance about the real and potential stresses of space flight than it was an affirmation of the severity of his condition; quite simply, nobody knew in 1962 just how debilitating, or serious, idiopathic atrial fibrillation could be. Medical examinations continued and in June, 1962, Paul Dudley White – Eisenhower's personal physician – gave Slayton a rigorous inspection before concluding that nothing was evident that should permanently remove him from the goals and objectives for which he had been selected in April, 1959. It had been nearly three years, during which period Slayton had lived, trained, and worked with six other astronauts, for the privilege of flying a capsule in space. The bitter news changed Slayton for the rest of his career, instilling in him an uncommon tenacity and purposeful drive so that many who came within range of his aggressive manner were not easily to escape his lashing tongue or his decisive attitude. Yet, at the conclusion of a long struggle to fight back from crippling defeat, Slayton's was the inspiration that drove good astronauts to be even better – and unsuitable candidates to resignation. For Slayton, there were to be many years during which he would train others to fly in space until finally he got the chance himself – 13 years later.

The elimination of Deke Slayton for consideration as a flight rated astronaut left vacant the position he held as pilot for MA-7. Schirra was his back-up but MSC decided that when Slayton's couch was removed from spacecraft 18, then being prepared for the second orbital mission, the one inserted in its place should be the couch of Scott Carpenter because, as back-up to John Glenn for more than a year, he had more experience in the pre-flight preparations and in mission simulations. In this way, Schirra would get to fly the third mission and, in preparation for that event, retain his back-up position for MA-7. From this point on, the preferred routine would put the back-up pilot of a given flight into the position of prime pilot for the following mission, thus ensuring an increasing commitment to training and simulated rehearsals for an actual flight before allowing an astronaut to obtain a flight seat.

MA-7 was to be different to Glenn's flight in that having already proven the ability to fly in space and satisfactorily bring a human occupant back to Earth, the capsule could be made to support substantial scientific and engineering activity during the periods in orbit when the astronaut had very little to do. Science, in fact, was to be the downfall of Carpenter's flight – at least from an operational point of view. Mercury project manager Kleinknecht set up a Mercury Scientific Experiments Panel under Lewis R. Fisher so that all the many suggestions and proposals for scientific experiments, coming from both inside and outside NASA, could be properly filtered and selected.

During April, 1962, the panel decided that among all contenders the most promising were an inflated balloon proposal to be released from the capsule but connected to it by tether, an experiment to observe the behavior of liquids in zero-gravity (weightlessness), observations of flares set off on the ground as the pilot flew overhead, weather photography with several different cameras and film types, and a study of the airglow layer, the luminous band seen around the Earth's horizon. The tethered balloon was to be a 76.2 cm diameter sphere inflated by a gas bottle on command from the capsule, weighing 0.9 kg in all, and observed at the end of a 30-metre long nylon line to measure air-drag effects and to study the reflectivity of sunlight on the different coated surfaces it would carry.

As a peripheral experiment, Carpenter was to see if he could provide information that might influence controversy over Glenn's 'fireflies' which the MA-6 pilot reported he saw at sunrise flying around the outside of his spacecraft. At the de-briefing held shortly after that mission, Glenn had been cruelly teased over reports on these luminous particles, and never more so than when a psychiatrist had responded by asking, 'What did they say John!?' Much of the photographic work was to determine further the value of cloud views. NASA's Tiros weather satellite program was getting into its stride and the Office of Applications was intent on determining how certain films were suited to the task of enhancing meteorological picture analysis.

The flight was scheduled for late May. Mercury-Atlas preparations had been paced at maximum several months before Glenn made his successful flight. It had been by no means certain that the first manned orbital attempt would come off as hoped and a tight schedule augured well for timely follow-up should that first flight go awry. So it was that Carpenter had barely two months to rehearse the flight operations for MA-7, to learn the requirements of the five assigned science tasks and to gain expertise in handling the comparatively crowded flight plan. Because the flight experiments for MA-7 had been inserted only after Glenn's successful flight, the minor changes and modifications that naturally attend any scientific task pressed hard on Carpenter's ability to absorb constant changes to the flight plan. In fact, amendments and deletions were common right up to the week of launch and it was a very busy astronaut who tried to keep pace with so many inputs from so many people, all of whom prided perfection as their work's hallmark.

By 23 May, 1962, all was ready for the second three-orbit flight, and the first phase of the two-part countdown began at launch complex 14, the now familiar site for Mercury-Atlas launches. Spacecraft 18 had been originally assigned the MA-6 slot, but spacecraft 13 was finally selected for that flight, and Carpenter followed tradition by choosing the name for his own particular vehicle: Aurora 7, because, he said, 'I think of Project Mercury and the open manner in which we are conducting it for the benefit of all as a light in the sky. Aurora also means dawn – in this case the dawn of a new age,' with a number indicating the astronaut corps traditionally established by Alan Shepard. In light of experience with three manned flights, MSC deleted several recovery aids, deemed unnecessary due to good recovery procedures having been established, and removed the knee and chest straps which tended to restrict an astronaut beyond the point where he was adequately protected from unexpected movement or gyration. Also, the Earth-path indicator was taken out along with the camera viewing the instrument panel and the red window filter.

Aurora 7 had been at the Cape since 15 November, 1961, and numerous modifications and replacement items were worked on the capsule before it was mated to Atlas 107-D, which arrived 8 March, 1962, and was immediately erected on pad 14. The original plan called for a launch on Tuesday, 15 May, but booster systems problems delayed the flight first to 19 May, because of a problem with Aurora 7's attitude control system, and then to 22 May, because of a modification deemed necessary to the parachute deployment system's altitude-sensing equipment. Finally, the launch was put back to a 24 May attempt when engineers found irregularities in a temperature control device in the Atlas flight control system. Thursday, 24 May, was good to MA-7 and the nearly flawless countdown moved toward a lift-off at 16 seconds past 7:45 am.

Scott Carpenter examines the honeycomb protective material on the main pressure bulkhead of his spacecraft located in Hanger S at Cape Canaveral.

It was one of the cleanest countdowns so far in the Mercury program. Carpenter had been up since 1:15 that morning to follow a routine similar to his predecessor by first having breakfast, then being suited up after receiving bio-instrumentation that on this mission gave only moderately satisfactory performance. At 3:45 the astronaut had begun his move from Hangar S to pad 14 in the transfer van that also carried suit technician Joe Schmitt, a familiar figure for all Mercury launches. Again, Chris Kraft was in the Mercury Control Center to conduct operations at the nerve center, while booster technicians in the pad 14 blockhouse groomed the combined assembly for lift-off. Moderate ground fog hugged the Florida shore line that morning but a burning Sun moved it away in good time for the weather men to give their approval for flight clearance.

Several built-in holds kept the launch crews in a casual schedule during the last few hours; Carpenter even chatted with his wife over the communications line that linked his capsule to any one of several selectable sites around the Cape area. Not so many people watched TV this time as the second manned Mercury-Atlas lifted cleanly into the sky on a column of fire and smoke that punched a hole in the early morning haze, only apparent when the bright light of the rocket's exhaust became momentarily dimmed by water vapor in the atmosphere. Theirs was an aura of relaxed confidence as flight controllers monitored Aurora 7 and its big booster on the way into space. Carpenter was cryptic in his comments back to the Cape, the most voluble statement coming 1½ minutes after lift-off when he noted that 'The sky is getting quite black at 01 30 – elapsed. Fuel and oxygen is steady, cabin pressure is levelling off at 6.2, 22 amps and the power is still good, one cps sway in yaw.'

Always, the technical jargon of a new language, the sophisticated preserve of a few privileged participants. On the way up, Carpenter noticed that the sky changed quickly from light blue, to deep blue, to black, and that the transition through max q was not very noticeable, the vibrations not at all severe. In truth even boosters, like spacecraft, differed from one to the other; uncomparable too were the reactions and responses of astronauts, each the summed product of a lifetime crammed with experience of mechanical high-speed contraptions. During the ascent, Gus Grissom was capcom calling out checks and transferring information to the Aurora 7 pilot, but there was little time for casual comment or impressions from the capsule. At staging, where the booster engines slid free from the ascending sustainer, Carpenter felt hardly a judder, but the release of the launch escape tower was more severe than he expected. Smooth too was the separation when Atlas' single sustainer shut down 5 minutes after lift-off.

As the capsule turned itself around, to face rearward and pitch down 34°, Carpenter caught sight of the booster, flashing silver in the sunlight from space and noted that there was a 'steady stream of gas, white gas, out of the sustainer engine.' Turnaround had been good, expelling a conservative 0.72 kg of hydrogen peroxide versus the 2.27 kg used up by Glenn's capsule at this point. But the apparent ease with which the flight was progressing lulled both pilot and the ground control teams into false security as, shortly after turnaround, the horizon sensor proved to be off in pitch by 20°. Also, he became increasingly hot during the first part of his orbit and made the first of 13 adjustments to the suit temperature control knob.

Carpenter was surprized to find that he had absolutely no sensation of speed as he moved round the Earth at more than 28,000 km/hr, and that he had positively no feeling of disorientation when the capsule moved out of alignment with the flight path; slewing round out of attitude hold, he found it a comparatively simple matter to let the spacecraft roll over to an inverted position so that he could observe the surface of the Earth through the trapezoidal window and use ground cues to re-orientate his attitude. In all respects, he felt at home in the new medium and adopted from the outset a familiarity with his environment.

Passing over Africa he was intrigued to find recognition of surface features an easy task, and to see roads, even dust tracks, on the ground that he had thought would be impossible to identify from such a great height. But this was to be a common experience of both Soviet and American astronauts and for several years many scientists disbelieved the claims by pilots of things they said they could see. Over Canton Island the biomedical telemetry gave the ground an indication that Carpenter was over-heating, and the capcom asked him if he felt well; the instrumentation was apparently playing up and Carpenter reassured controllers that he was far from the state they assumed him to be in. So far, the pilot's performance had been good, observing flares over Australia he was hampered by cloud and failed to catch sight of the one flare sent up during the first orbit; the exercise was discontinued, abandoned because of the weather. Over Hawaii, the astronaut caught sight of Glenn's fireflies: 'I have the particles. I was facing away from the Sun at Sunrise – and I did not see the particles – just – just yawing about – 180 degrees, I was able to pick up – at this. Stand by, I think I see more. Yes, there was one, random motions – some even appeared to be going ahead. There's one outside. Almost like a light snowflake caught in an eddy. They are not glowing with their own light at this time . . . It could be frost from a thruster.'

Soon after that, a cryptic comment on the state of zero-g: 'The weightless condition is a blessing, nothing more, nothing less.' Working hard on his crowded schedule, Carpenter was moving faster through the planned activity list than any previous astronaut had been called upon to do. Several times he accidentally knocked the spacecraft control systems from their isolated position, unnecessarily cutting in the manual system when the spacecraft was on automatic. This was expensive on fuel and from an initial load of 11.3 kg in the manual tank and 15.8 kg in the automatic supply, Carpenter was down to quantities of 64% and 56% respectively by the end of the first orbit. This concerned Cape flight operations personnel who saw the emerging specter of depleted propellant tanks before the all-important retro-fire attitude maneuvers. Taxed by the high work load, Aurora 7's pilot was wasting too much propellant, operating the high-thrust jets too frequently and inadvertently activating several modes simultaneously. But he was discovering the penalty for too many operational tasks and found difficulty in remaining within planned margins of fuel consumption when called upon to place the capsule in an unusual attitude for some engineering task or scientific experiment such as ground photography or air-glow observation.

But Carpenter was prone to come up with the unexpected, and passing over the Guaymas tracking station, situated in New Mexico, at the end of the first orbit, radio stations carrying live commentary for the spacecraft picked up the following message with an intended destination impossible to mistake: 'Hola, amigos, felicitaciones a Mexico y especialmente a mis amigos de Guaymas. Desde el espacio exterior, su país está cubierto con nubes – and – es – also – es muy bello. Aquí el tiempo está muy bueno. Buena suerte desde Auror Siete,' which translated means, 'Hello, friends, greetings to Mexico and especially to my friends of Guaymas. From outer space, your country is covered with clouds and is very beautiful. Here the weather is very good. Good luck from Aurora Seven.'

About 1 hr 38 min into the flight, the balloon experiment began with release of the sphere from the forward end of the capsule, but it failed to inflate to its anticipated volume and the balloon lazily drifted away from the capsule at the end of a snaking line. Nevertheless, scientific observation of the erratic motion was attempted: 'The balloon is partially inflated. It's not tight. I've lost it at this moment. . . . The balloon not only oscillates in cones in pitch and yaw, it also seems to oscillate in and out toward the capsule; and sometimes the line will be taut, other times it's quite loose. . . . The balloon is oscillating through an arc of about 100 degrees. It gets out of view frequently. At this moment, it's nearly vertical, mark a coastal passage at this time . . . what I'm trying to tell you is that it oscillates 180 degrees, above and below.'

As the capsule passed over the Kano tracking station a conversation developed over the increasingly rapid consumption of propellant in both automatic and manual control systems: 'The only thing of – to report regarding the flight plan is that fuel levels are lower than expected. My control mode now is ASCS. I expended my extra fuel in trying to orient after the night side. I think this is due to conflicting requirements of the flight plan. I should have taken time to orient and then work with other items. I think that by remaining in automatic, I can keep – stop this excessive fuel consumption. . . . I think I can cut down the fuel consumption considerably on the second and third orbits.' Still busy with scientific tasks, it was up to Carpenter to balance the remaining quantities of propellant so that at the end of the mission he would not run out of fuel for attitude control during retro-fire.

Most of the second orbit was spent reducing the attitude changes made necessary by conflicting tasks the pilot had been called upon to perform by the crowded flight plan. Automatic fuel reserve was now down to 34% so Carpenter elected to remain on manual authority, although between Woomera and the second pass across America quantity in the manual supply fell from 64% to 40%. Throughout the second orbit the pilot was hot and uncomfortable, continually trying to cool his capsule down while bio-telemetry indicated a body temperature of 38.8°c. Over Hawaii, toward the end of the second orbit, the flight surgeon suggested that Carpenter should take a drink of water, concern was that high over his condition which, as would be learned after the flight, gave undue cause for alarm because of the malfunctioning biological sensor harness.

Passing across the Cape area at the start of the third orbit, Carpenter came to a similar conclusion as that already being worked out on the ground: that for the remainder of the mission, he should remain in drifting flight as frequently and for as long as possible, thus conserving a dangerously low fuel state. For much of that final pass around the world, the spacecraft slowly drifted, Carpenter performing selected experiments and others at will. Over Canary, he commented on the re-appearance of the 'fireflies': 'They no doubt appear to be way, way far away. There are two that look like they might be 100 (meters) away. I haven't operated the thruster for some time. Here are two in closer. Now a densitometer reading on these two that are in close. Extinct at 5.5, the elapsed time is 3:27:39 (GET). I am unable to see any stars in the black sky at this time.' Reference to the lack of thruster activity referred to a suggestion that the 'fireflies' were frozen hydrogen peroxide particles drifting with the capsule; and reference to the densitometry study indicated yet another scientific task pressed into Carpenter's flight plan.

And then, during the handover from Canary to Kano, an unsolicited comment on the sensation of looking at the Earth from different attitude positions: 'I could very easily come in from another planet, and feel that I am on my – on my back, and that Earth is up above me.' Then, over the Indian Ocean ship, a comment on the fact that the pilot had been unable to jettison the balloon as scheduled because the nylon line had entwined itself around the cylindrical forward antenna canister. With less than half an orbit to go before re-entry, casual remarks on Carpenter's physical impressions of vigorous head movement designed to reproduce the sensations cosmonaut Titov had experienced: 'I'm shaking my head violently from all sides, with eyes closed, up and down, pitch, roll, yaw. Nothing in my stomach; nothing anywhere. I was a little disoriented as to exactly where things are, not sure exactly what you want to accomplish by this but there is no problem of orienting . . . you just adapt to this environment. It's a great, great freedom.' At Woomera, Carpenter tracked the star Phecda Ursae Majoris, a member of the Great Bear constellation, as it rose through the haze layer hugging Earth's horizon. Soon, he would be in daylight again.

Suddenly, 'Ahhhhh! Beautiful lighted fireflies that time. It was luminous that time. But it's only . . .' and then, as if remembering that he was about to pass Woomera to the Hawaii station, 'okay, they – all right – if anybody reads, I have the fireflies. They are very bright. They are capsule emanating. I can rap the hatch and stir off hundreds of them. Rap the side of the capsule; huge streams come out. They – some appear to glow. Let me yaw around the other way.' And that was the start of an ill-advised use of additional propellant to move the capsule around and literally chase the darting particles that now seemed to have been resolved. In the rigid, highly disciplined, nature of test flying, the pilot should never depart from a pre-set plan to pursue phenomena unanticipated at the beginning of a mission. Many people associated with the astronaut training program thought this philosophy was a sound spring-board to space, others believed the astronaut, unlike the pilot, should be responsive to new experiences and use the equipment at his disposal to investigate unknown situations. The former group would have had their day when evaluating the wasteful use of propellant that now ensued aboard Aurora 7.

The activities to be performed over the Hawaii station required Carpenter to place the capsule in retro-fire attitude and then to perform, and complete before loss of signal, a checklist that would, theoretically, hand him over in a clean condition to the California capcom, where he would receive the final updates for monitoring retro-fire less than two minutes after acquisition of signal. It was a tight schedule, one that required the Aurora 7 pilot to have completed all orbital activity by the time capsule communications opened up with Hawaii. Unfortunately, Carpenter was thrown out of sequence by the euphoria at apparently resolving the question of the 'fireflies,' and rather than attending to essential pre-retro activities he intruded on the time that should have been occupied with preparations for coming back to Earth by moving the capsule around in a search for more particles.

To this point the mission had gone well. Carpenter had done a good job with what would later be seen as an unacceptably crowded flight plan. Unfortunately, Carpenter ran into further trouble with his attitude control system, discovering that the ASCS would not hold the capsule steady. Switching back to Fly-By-Wire, he forgot to turn off the manual mode and depleted further the already sparse quantity of fuel for the thrusters. Several times Carpenter had to be asked to prepare certain critical switch positions essential to successful retro-fire, and was required to manually punch the retro-rocket ignition button because of the particular control mode he was in due to the ASCS problem. It appears that Carpenter became confused at this point as to his actions, although he would be particularly obtuse about this point during later de-briefing sessions. However, he waited for the retro-rockets to fire and when they did not he backed up the command with the manual operation he should have adopted in the first place, wasting 3 seconds of time.

The events associated with the period between acquisition of signal at Hawaii and the manual retro-fire command over California, a crowded 11 minute period on that third and final orbit, is best told by Carpenter himself in the following frank description: 'I think that one reason I got behind at retro-fire was because, just at dawn during the third orbit, I discovered the source of the space particles. I felt that I had time to get that taken care of and still prepare properly for retro-fire, but time slipped away. The Hawaii capcom was trying very hard to get me to do the pre-retrograde checklist. After observing the particles, I was busy trying to get aligned in orbit attitude. Then I had to evaluate the problem in the automatic control system. I got behind and had to stow things haphazardly. Just prior to retro-fire, I had a problem in pitch attitude, and lost all confidence in the automatic control system. By this time, I had gone through the part of the pre-retro checklist which called for the manual fuel handle to be out as a back-up for the automatic control system. When I selected the Fly-By-Wire mode, I did not shut off the manual system. As a result, attitude control during retro-fire was accomplished on both the Fly-By-Wire and the manual control modes.

'At the time, I felt that my control of spacecraft attitude during retro-fire was good. My reference was divided between the periscope, the window and the attitude indicators. When the retro-attitude of $-34°$ was properly indicated by the window and the periscope, the pitch attitude indicator read $-10°$. I tried to hold this attitude on the instruments throughout retro-fire, but I cross-checked attitude in the window and the periscope. I have commented many times that on the trainer you cannot divide your attention between one attitude reference system and another and still do a good job in retro-fire. But that was the way I controlled attitude during retro-fire on this flight. Although retro-sequence came on time, the initiation of retro-fire was slightly late. After receiving a countdown to retro-fire from the California capcom, I waited 2 seconds and then punched the manual retro-fire button. About 1 second after that I felt the first retro-rocket fire. If the California capcom had not mentioned the retro-attitude bypass switch, I would have forgotten it, and retro-fire would have been delayed considerably longer. Later, he also mentioned an auxiliary damping re-entry which I think I would have chosen in any case, but it was a good suggestion to have.'

As the three retro-rockets fired off, Carpenter noticed smoke whirling up in the capsule, but he was not to know this was from two fuses that had blown under the influence of the sudden shock caused by the solids firing. The retro-rockets had fired at an elapsed time of 4 hrs 33 min 10.3 sec (04:33:10.3 GET), during conversation with Alan Shepard through the Point Arguello station in California. There was about 32% fuel remaining in the manual system when Carpenter made the mistake of leaving that system on during Fly-By-Wire control prior to retro-fire. Within minutes the entire system had drained to depletion. The 0.05 g level indicating entry, when the capsule, penetrating the outer layer of the atmosphere, would be dynamically stable was still about 9 minutes away. A fateful comment came down to Point Arguello: 'I am out of manual fuel, Al.' Shepard advised Carpenter to use sparingly the propellant for Fly-By-Wire which he would extract from the ASCS tanks to get the capsule in to re-entry attitude. From a pitch down below the horizontal of 34° the capsule would have to point up 1.6°.

But already a tracking station on the ground recognized that Aurora 7 would be long on splashdown: 324 km further downrange because at retro-fire the capsule was inadvertently slewed 25° to the right; 28 km further downrange because of the 3 second delay in firing the retro-rockets; and a further 111 km downrange because the rockets actually produced 3% less thrust than expected. The spacecraft would come down 463 km off target. Recognizing that he now had very little automatic propellant left, and realizing that this was in fact the only attitude control fuel remaining, Carpenter wisely elected to let the capsule drift until the last few seconds before re-entry, when he would commit whatever was in the fuel system in an attempt to bring the spacecraft's attitude into reasonable alignment with the desired values. A minute or two later, Al Shepard reminded Carpenter to retract the periscope, an action he had forgotten to take. In contact with Gus Grissom through the Cape station, Carpenter remarked that the balloon had finally gone, but had to be reminded to close his faceplate which would provide a better re-entry environment and preserve the integrity of the suit should the capsule rupture during the descent.

Sparingly swinging Aurora 7 into an approximate re-entry attitude – his confidence about the attitude control needles had been badly shaken by the problem with orientation just prior to retro-fire – Carpenter watched oscillations begin to build, but a switch to the auxiliary damping mode solved that. With the capsule in communications black-out brought on by the ionization sheath, Carpenter continued to talk into his voice recorder, reading out values on g forces and descent rate. As he watched for visible signs of the spacecraft heating up an orange glow spread over the exterior, pieces of retro-package strap were torn off – fragments left intact when the retro-package had been jettisoned only minutes earlier. At maximum, Carpenter felt a force of 7.5 g, about 0.3 g less than accelerations experienced during ascent more than $4\frac{1}{2}$ hours earlier. Coming out of black-out, Gus Grissom told Aurora 7 that the spacecraft was long and that Carpenter would have to wait 'about 1 hour' for recovery force aircraft to reach his splashdown point, just north-east of the Virgin Islands.

Preparations for Carpenter's recovery had actually been underway since early in the third orbit, when an Air Rescue Service SC-54 manned by a pararescue team had been dispatched from Roosevelt Roads in Puerto Rico. Just 11 minutes before the capsule hit water, an Air Rescue Service SA–16 amphibious aircraft was also sent out from the same base with instructions to head for the calculated landing position at 19.4°N latitude by 63.88°W longitude. The calculated impact point for Aurora 7 was far from the planned ellipse where the carrier USS *Intrepid* controlled the 2 ships and the aircraft and helicopters assigned to prime recovery, but the *Intrepid*'s commander assumed responsibility for launching a contingency rescue plan according to approved plans established by the Recovery Control Center.

In the descending capsule, Carpenter was concerned about excessive oscillations as Aurora 7 swung crazily from side to side without the benefit of thruster fuel to damp the wild motion. Deciding, like Glenn before him, that the capsule was in danger of turning over completely if left to its own resources, Carpenter reached up and punched out the drogue parachute at a height of 7.6 km, deploying the main canopy less than a minute later. The flight had lasted nearly five hours when it struck the Atlantic swell, alone and far from the comforting sight of recovery vessels, which were even then closing fast on the splashdown point far to the south-east of the prime H zone. The astronaut had no idea just how far off target he was, but the absence of Gus Grissom's friendly voice indicated he was well out of communication range with the Cape.

To Carpenter, the spacecraft seemed to be listing rather more than it should and he prepared to exit the forward end rather than blow the hatch and suffer an intentional fate similar to that brought upon Grissom by accident. He had forgotten to lock up the suit inlet hose and when he slid down into the water he felt the inside of his boots begin to get wet and immediately rectified the procedural lapse. Within a few minutes Carpenter had scrambled into a raft deployed from the top of the spacecraft on his way out, there to wait until recovery helicopters came to pick him up. As planned, for contingency operations like this, the *Intrepid*'s recovery commander took an inventory of ships in the area to see if any merchant or civilian shipping could be called upon for assistance.

The Coast Guard advised the *Intrepid* that they had a cutter at the Virgin Islands, and the Navy passed along the information that a merchant ship was only 57 km north of Aurora 7. However, the destroyer USS *Farragut* was 167 km to the south-west and would get there first so its commander was ordered to make full speed for the new recovery coordinates. Aurora 7 had landed at 12:41 pm local time and by 1:34 pm the Air Rescue Service's SC-54 was overhead and getting ready to drop a pararescue team, which it did just 6 minutes later. Just as the swimmers hit the water, out of sight of Carpenter who was looking around the sky for more aircraft, *Intrepid* launched two HSS-2 helicopters with a doctor in the lead chopper. It would take them more than 1½ hours to reach the downed spacecraft.

Meanwhile, at 1:50, only minutes after pararescue swimmers popped up over the side of Carpenter's raft, the Air Force SA-16 amphibious aircraft began circling overhead. Back at the Cape, the recovery coordinator ordered the SA-16 not to land on the water unless the helicopters, at that time only just having left *Intrepid*, were ultimately unable to secure a retrieval. In for a long wait, Carpenter opened his survival rations and promptly set about satisfying a moderate hunger while offering his fare to the bobbing swimmers; they declined to eat his packaged rations. But already, an argument had broken out among personnel at the Cape unfamilar with the strict lines of demarcation between Air Force and Navy protocol. Having already arrived at the scene, the SA-16 was forbidden to pre-empt Navy retrieval. To have flouted tradition would be to have courted reprisal from the Navy! Carpenter had to remain in the water until the Navy – officially recognized as the responsible agent for sea operations – arrived on the scene. It was to provoke vigorous debate in press, NASA and Congressional circles for some time after the flight.

As it was, Carpenter had to wait until the first Navy HSS-2 arrived over his spacecraft at 3:30 pm – 1 hr 40 min after the Air Force SA-16 began circling. Within 10 minutes the astronaut was in the helicopter, having been dunked in the sea first, and attendants found him in good spirits. Sitting with his leg hanging over the open door sill he cut a hole in his suit sock to drain the water out! Carpenter was on *Intrepid*'s deck at 4:52 pm, and Aurora 7 was retrieved by the USS *Pierce* two hours later. The following day, the inert hulk of Atlas 107-D fell to Earth over the South Atlantic, fragments falling near Barkly East in South Africa. Glenn's Atlas 109-D had come down within a few hours of the spacecraft and it too had showered fragments which were later recovered from the ground – also in Africa. Because, in placing their payloads in safe orbit, the boosters too followed a similar path, their return through the atmosphere presented minimal hazard, but one thought too small to bother about because the rocket was essentially an inert titanium cylinder, the only structure of any substance likely to survive the heat of re-entry for which it was not protected being the comparatively bulky sustainer engine at the base end. Each redundant Atlas, 20 meters long and 3 meters in diameter, weighed approximately 3.4 tonnes after expending its propellant.

At post-flight celebrations it was Walt Williams' turn to accompany the astronaut at a ceremony where Kennedy awarded both men the NASA Distinguished Service Medal. And just as his predecessor had a John Glenn Day, so too did Scotty Carpenter – on 29 May; it was a time for heaping accolades and praise: on the day Carpenter roared into space, John Glenn had been named National Father of the year following the title of World Mother of the Year bestowed on his parent earlier that month. It was certainly a time of national rejoicing. For much of 1961 the American news media had saturated their readers with plans and preparations for Mercury flights, the ballistic shots of Shepard and Grissom, announcements from NASA about bold new initiatives, messages before Congress by an energetic and youthful President calling upon the nation to back a Moon landing goal. Now, half way through 1962, space flight was visibly seen as an exercise in which Americans were going to compete strongly with Soviet counterparts, a challenge fit to stir the hearts of patriotic citizens. Or so it all seemed, and so indeed it was projected by the press and the electronic media across the United States. But the honeymoon was over and by year's end the critics would be firm behind their entrenched barriers, calculated arguments conspiring to demolish the very essence of Kennedy's sudden commitment made in May, 1961.

After two successful Mercury flights in Earth orbit, during which NASA had demonstrated the spacecraft's ability to perform the standard three-orbit mission, discussion within the agency's own ranks questioned the next moves. Not for MA-8, the third manned flight, for its objective had already been established, but rather for flights beyond MA-8 if indeed, as some proposed, there could be justification for a fourth mission. Gemini was by this time not considered capable of providing a manned flight before 1964 and the need to expand upon the very limited duration put up by Mercury provided stimulus for the 18-orbit flight first seriously considered in early 1961. But Gemini was in deep trouble by the second half of 1962, as will be seen in chapter 10, and Apollo too was running a gauntlet of technical indecision.

Inside NASA there were deep concerns over the ultimate magnitude of budget requirements, and the agency's ability to deploy so rapidly a major industrial effort involving thousands of contractors and several hundred thousand workers. By this time, NASA's in-house work force had grown from little more than 16,000 personnel when Kennedy came to office to more than 23,000 by mid-1962. Moreover, the contractor work force had increased from about 40,000 to more than 115,500 in the same period as a result of the expanding investment in a host of new space programs, not all of them associated with manned flight. Nevertheless, the NASA budget was not only growing but allocating increasingly significant sums to manned flight.

For instance, the final Fiscal Year 1962 budget – a period extending for 12 months from 1 July 1961 – totalled $1,855.3 million. Of this, nearly 51% would go to manned space programs. Prepared late in 1961, the Fiscal Year 1963 budget, effective from a few weeks after Carpenter's three-orbit flight, requested $3,787.3 million. Of this, about 61% was for manned flight activity, mostly research and development money for Gemini and Apollo, the big Saturn rockets, and construction needs of the rapidly expanding space port at Cape Canaveral. During the second half of 1962, while technicians were preparing for the third Mercury flight, NASA officials were putting figures together for the Fiscal Year 1964 request. That would be announced before Congress by President Kennedy during the annual State of the Union message in January, 1963. For that budget, the space agency would request $5,712 million of which nearly 66% would be earmarked for manned projects.

It was growth on a staggering scale, although the massive Fiscal Year 1964 request would be only 4.8% of the sum spent on all government programs that year. Nevertheless, it was in marked contrast to Eisenhower's Fiscal Year 1962 budget which would have left NASA with $1,109 million. In just two years of office, Kennedy had increased the civilian space budget by more than 400%. The 18-month period betweeen mid-1962 and the end of 1963 were to transform the space agency radically, and to inject a degree of reality in manned space projects that had hitherto been excluded by the euphoric wonder of first success.

By June, 1962, the Manned Spacecraft Center was deeply concerned with the need to exploit more fully the inherent capacity of the basic Mercury capsule, and to demonstrate that fact operationally with an expanded mission for MA-8 by the end of the summer. It would go some way toward matching the 25 hour flight of Gherman Titov aboard Vostok II the previous year, and it would serve as a fitting prelude to the 18-orbit mission tentatively planned for some time in 1963. If accomplished on schedule, it would allow Mercury to exceed the 17 orbits of Titov's flight, albeit two years after the Soviet mission! But that presumed the Russians would stand still and not expand further their own manned flight duration record. It had been nearly a year since Vostok II's 17 orbit mission; would another year go by without a Soviet manned flight, enabling the United States to snatch the record?

Vanguard blows up during a planned launch, 6 December, 1957.

The seven Mercury astronauts complete survival training at Stead Air Force Base, Nevada. Left to right: Cooper, Carpenter, Glenn, Shepard, Grissom, Schirra and Slayton.

MR-3 carries Shepard cleanly into the air on NASA's first manned space flight.

Alan Shepard strides toward MR-3 for his suborbital hop through space, 5 May, 1961.

Awash and heavy with water, Liberty Bell 7 strains the capacity of the recovery helicopter and seconds later begins to sink.

– Spacecraft Aurora 7 carries Carpenter aloft on MA-7, May, 1962.

Faith 7 is recovered from the Pacific by the USS Kearsarge, ending the Mercury Program after two suborbital and four orbital space flights.

Mission control, Cape Canaveral, Florida, before operations shifted to Houston.

The Manned Spacecraft Center, Houston. Purpose built for expeditions to space.
Charles Conrad looks toward Gordon Cooper as the two astronauts orbit the earth for eight days, Gemini V, August, 1965.

Ed White dances at the end of his gold coated umbilical during NASA's first space walk, June, 1965.

Earth views like this from Gemini V, August, 1965, helped show how much information could be gathered by manned space vehicles in orbit.

The clean lines of Gemini are revealed in this rendezvous picture, December, 1965.

Gemini VIII brings spacecraft 8 to its Agena target vehicle; less than one hour later, Armstrong and Scott will be fighting for their lives.

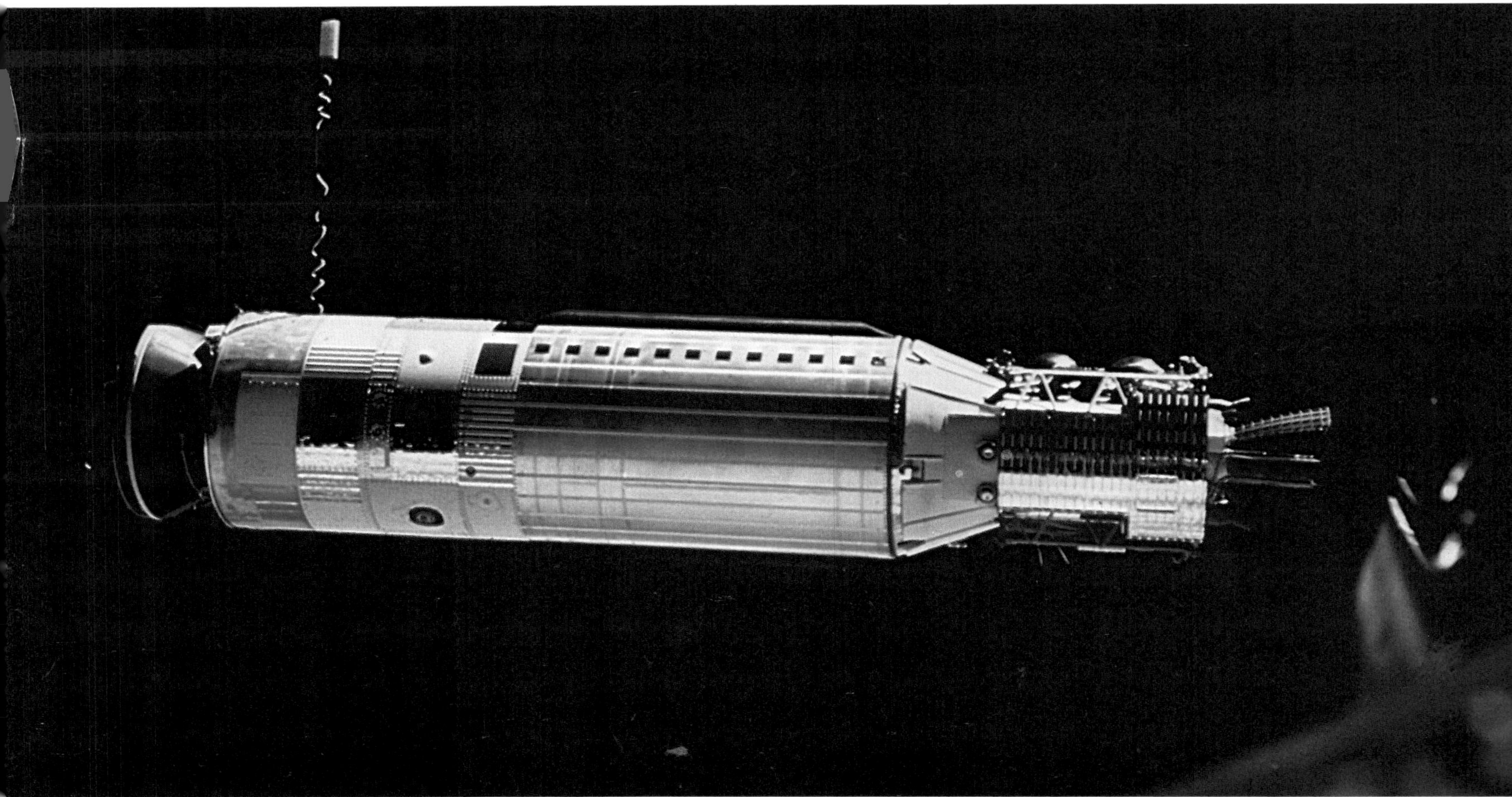

CHAPTER NINE

Paths to the Moon

The lessons from Scott Carpenter's MA-7 flight were clear and unmistakably distinct. The flight plan had been excessively loaded with a myriad activities, each the product of a scientist or an experimenter anxious to have the astronaut obtain the best possible data from his project. The real trouble was that flight plan coordination had not really been applied, and that mistake was more a product of poor integration than any deliberate disregard for the pilot's need to allocate time for systems management – housekeeping, as it became known. One particular experiment would have the astronaut position his capsule in a certain attitude, the next task would call for the spacecraft to be in a totally different attitude, while a third activity would have Mercury return to angles only marginally different than those for the first operation. The need to place the first and third activities in a time sequence, followed by the second after that, was only brought home in the post-flight de-briefing sessions. Intelligent management of the spacecraft's attitude, where experiments would be placed in an attitude-dependent sequence so that the capsule could sweep across successive pitch or roll arcs rather than move continuously back and forth, became a prime consideration for all future missions.

Nobody really understood just how critical attitude propellant margins were until both first and second manned orbital Mercury capsules came home with dry tanks; it was almost a legacy from the aeronautical experience most MSC engineers had received before the Mercury days, where attitude control was as free as the air against which control surfaces reacted. In space, it was very different for if a pilot ran out of attitude control propellant there was no physical means of pointing the spacecraft in the appropriate direction. From an awareness of the critical nature of spacecraft systems, and the need to plan scientific tasks around the operational trajectory, came an understanding of the balance between spacecraft systems management – housekeeping – and peripheral activities irrelevant to the safe conduct of the flight.

Glenn's mission had been a blend of both engineering *and* science; Carpenter's flight had been overcrowded with science experiments, most of which demanded considerable attention from the pilot. So, in addition to sequencing science operations in a logical train that placed minimum impact on the technical conduct of the spacecraft's housekeeping, flight planners continually sought to get experimenters thinking about automated equipment. The more functions removed from the pilot's activity list, the more attention he could give to the satisfactory performance of the spacecraft. The net effect of this would be to increase, rather than decrease, the time and quantity of equipment devoted to science operations, for, by relieving the pilot, the overall security of the mission was enhanced, improving the likelihood of any given flight going for the full planned duration.

So it was that MA-8 was assigned only minimal experimentation which required pilot intervention, the main responsibility of flight planners being to provide an essentially engineering flight in which the pilot could demonstrate the spacecraft's ability at flying for a longer period than that for which the capsule had originally been designed. As early as the post-flight debriefing for Glenn's mission, in February, 1962, Kleinknecht had put some MSC people to work with Yardley's men at McDonnell on the possibility of flying a seven-orbit mission after MA-7 before the 18-orbit flight then under consideration. In effect, a seven-orbit MA-8 flight would pave the way for the day-long mission while generating astronaut experience and testing mission operations on the ground. But the spacecraft had been designed as a three-orbit capsule, with systems sized for a mission lasting only 5 hours and quantities of consumables balanced by the need to provide wide margins.

The critical systems were environmental control, with the need to provide an oxygen environment for the full duration of the flight and to remove excess carbon dioxide exhaled by the astronaut, the attitude control system, with adequate propellant in the manual and automatic systems for a full flight, and the electrical system which provided the batteries for power to all the equipment in the spacecraft. The spacecraft had been designed to an acceptable leak rate not exceeding 1,000 cc/min in a pressurized condition, and to contain two separate oxygen tanks each containing 1.8 kg of life-giving gas. Adopting the same mission rules written down for the three-orbit flights, systems engineers found that a total supply of 3.9 kg would be necessary, but if the cabin leak rate could be reduced to a maximum 600 cc/min the full 3.6 kg quantity from the two existing tanks would be well within limits, enabling the spacecraft to support a seven-orbit mission. As for the increase in carbon dioxide exhaled by the pilot, it was found that if the lithium hydroxide cleanser was filled to its maximum 2.45 kg capacity, instead of the 2 kg level normally used for three-orbit flights, that too could accept the requirements and demands of the longer mission.

Attitude control propellant was extremely sensitive to use, as had been ably demonstrated by the first two flights, and if the spacecraft was to be controlled for seven orbits in the automatic mode the mission would require a fuel quantity of 12.7 kg. As designed, the automatic system, with ASCS and FBW control modes, carried 15.8 kg; the manual system, with MP and RSCS control modes, contained 11.3 kg. This had hardly been sufficient for the three-orbit flights, and the problem of using the same quantity for a mission of more than twice that duration seemed insoluble. But the seriously depleted flight of Scott Carpenter had provided ample engineering data from which to plan a more expeditious use of hydrogen peroxide. MSC engineers formulated a control plan whereby a subtle combination of automatic and manual attitude control tasks would expend 10.4 kg and 8.2 kg respectively, thus ensuring ample reserves at the end of the flight. The pilot was to rigidly conserve his attitude propellant throughout the mission, a schedule which if adhered to would ensure sufficient quantity for every operational task.

As for the electrical needs of a seven-orbit flight, here again study showed that the spacecraft would consume an anticipated 11,190 watt-hrs of the 13,500 watt-hrs available – far too high a fraction of the loaded supply to leave adequate safety margins; a standard production Mercury capsule of the type used for all preceding missions carried three batteries of 3,000 watt-hrs each and three of 1,500 watt-hrs each. Each battery was of the silver-zinc type with a discharge rate of 4.5 amperes. The main 24-volt direct-current bus was hooked to the three larger batteries; two of the smaller capacity batteries were slaved to a stand-by circuit of 24, 18, 12, 8 or 6 volts dc, with the third on an isolated circuit of the same selectable range of voltage levels. MSC engineers wanted a greater margin between the power consumed and the amount of energy remaining. After all, if the spacecraft was required to go around the Earth again for some reason it would still have to supply its own electrical needs. By carefully examining the power consumption level of all the systems on board, it was decided that if the spacecraft shut down systems not in use at any given time the capsule could fly a seven-orbit mission with the existing electrical equipment.

By the time MA-7 was being readied at launch complex 14, approval had been granted for a long duration flight attempt with MA-8. Capsule 16 would be modified by engineers from the Preflight Operations Division at the Cape facility, with assistance from McDonnell people from St. Louis. But flying a seven-orbit mission would be very different from the standard three-orbit flights. With the Earth spinning eastward the effect of seven orbits would be that the spinning planet shifted recovery zones further west: orbits 5 and 6 would bring the spacecraft down at points successively further away from the continental United States in the Pacific Ocean north-east of Midway Island; a return on orbit 7 would bring the capsule's groundtrack directly over the Hawaiian Island chain but, more important, would necessitate a major increase in the number of ships covering recovery operations. Because of this, NASA adopted a six-orbit mission and by the end of July the pilot assigned to fly the MA-8 mission had his flight plan.

Carpenter was convinced that MA-8 would not need the periscope that had been an integral part of the spacecraft's equipment from the beginning and flight analysts agreed with mission engineers that it did seem to be a weighty piece of superfluous paraphernalia. Useless for attitude alignment on the night side of the Earth, it could be replaced by window references on the day side. But MSC technicians noted that contributing to Carpenter's overshoot was the yaw attitude error compounded by using marks on the window which in event did not seemingly prove as accurate a method of orientation as the old periscope. So back it went, but only for MA-8; any other Mercury mission would probably dispense with the equipment.

To alleviate problems further with the attitude control system, minor modifications were made to spacecraft 16 in the hope that neither intentional switch modes nor accidental duplication of control function would again cause rapid and thirsty consumption of vast quantities of hydrogen peroxide propellant. Changes to the automatic ASCS mode increased the deadband range – the limit of tolerance – from ±3° to ±5.5°; during the actual flight it was found that the thrusters for some reason exhibited a deadband operating cycle within ±8°. Also, an attitude selection switch was provided to enable the pilot to put the ASCS orbit mode in either retro- or re-entry-attitude control, with another switch to disable the high thrust (10.9 kg) jets in the Fly-By-Wire mode. In addition, a special cover was placed over the pitch horizon scanner to protect it during ascent from heat that would build up on the forward section of the spacecraft, thought by some to be the reason for its off nominal alignment during Carpenter's flight.

All these ostensibly confusing modifications and additions were necessary to adapt the attitude control system to lessons learned in the realistic environment of space operations. As it turned out, they were wise changes. Experimental equipment attached to capsule 16 included exotic ablative compounds attached to 9 of the 12 berylium shingles around the forward, cylindrical part of the spacecraft. Each 38 cm by 12.7 cm sample would be removed after the flight to see just how well they had endured the test designed to qualify their use in later programs. Also, nuclear emulsion packs were placed inside the capsule in an experiment from the Goddard Space Flight Center designed to measure galactic radiation. Two radiation-sensitive films were provided by the Naval School of Aviation Medicine for measurements of near-Earth radiation budgets. Photographic equipment consisted of a 70 mm Hasselblad with film magazines and filters for the astronaut to take pictures of Earth, surface features, and cloud patterns. Finally, a light flash experiment was provided whereby the astronaut would look for flares sent up from Australia and South Africa.

Since late 1961, Wally Schirra had been down for an orbital flight. Serving first as back-up to Slayton, then to Carpenter when the former was laid off through heart trouble, he was well suited for the role of prime pilot on MA-8. As back-up astronaut, Gordo Cooper could expect to fly the fourth flight, then being prepared as a much longer mission. By the time Carpenter was back on Earth preparations for the six-orbit mission were well and truly under way. The spacecraft had checked out well and engineers had managed to get the leak rate down to 460 cc/min, well below the 600 cc/min deadline set for the long flight. By late July, G. Merritt Preston was confident that he could have the spacecraft ready for a 18 September launch, but it was 8 August before the Atlas booster arrived and then the Air Force wanted to static-fire the propulsion system because this particular bird – 113-D – incorporated a baffled injector modification which eliminated the normal two-second hold-down before lift-off. Recent failures with this particular piece of hardware on other Atlas development models eliminated confidence in going straight from manufacture to igniton for the mission. That would delay the flight until at least 25 September.

And then, before Americans fully woke to start the week-end on Saturday, 11 August, an announcement from Radio Moscow: 'On 11th August 1962, at 11.30 a.m. (Moscow time), the spaceship Vostok III was launched into the orbit of an Earth satellite. The spaceship Vostok III is piloted by the Soviet citizen and Cosmonaut-Pilot Comrade Maj. Andrian Grigoryevich Nikolayev.' Moscow went on to say that Vostok III was circling the Earth in an orbit with perigee and apogee of 183 km and 251 km, respectively, and that it made a complete orbit of the planet once every 88.5 min. Almost exactly a year after Titov's 17 orbit flight, Russia's third manned space flight was under way. As usual, official news was sparse with little information about the technical developments leading to this mission. In fact the capsule was almost identical to the earlier Vostoks used for the first two Russian manned space flights, weighing 4.7 tonnes and with the usual stock of food and supplies sufficient for a flight lasting 10 days, just in case the spacecraft's retro-rocket failed to fire; a Vostok orbit was planned so that atmospheric drag would bring it to re-entry within that time, theoretically ensuring that the pilot had sustenance for the necessary period.

Throughout that day reports flowed from Moscow at infrequent intervals: 'Cosmonaut Maj. Andrian Nikolayev, at 11.45 (Moscow time) reported from spaceship Vostok III while flying over the territory of the Soviet Union "I feel well. Everything is normal on board. The Earth is visible through the porthole." The flight of the spaceship Vostok III is proceeding normally. All the instruments and systems on board the ship are functioning accurately and without fail. The systems of regulation and air conditioning keep up the necessary temperatures, pressures and humidity . . . The television image of the cosmonaut and the reports received on the Earth show that he feels well . . . Nikolayev sends greetings to the peoples of the Soviet Union from his ship at 13.08 (Moscow time). The state of the cosmonaut is good, his pulse varies from 78 to 92, respiration from 12 to 20. In conformity with the program of the flight, Cosmonaut Nikolayev unharnessed himself, got up from the chair and moved freely in the cabin. No upsetting of the vestibular apparatus was noted. At 19.00 hrs on August 11th the elements of the ship's orbit were: period of revolution 88.32 minutes; apogee and perigee, 234 and 180 kilometers . . .'

Although described as a 'chair' the cosmonaut's ejector seat was of similar design to that employed for the Gagarin and Titov flights, and special reference to the pilot unharnessing himself and moving about indicated attention to the spatial awareness reported by Titov. Physicians were concerned about the long-term effects of space flight and the vestibular disorientation sensed by the cosmonaut triggered a set of experiments on Vostok III aimed at providing additional information on the report: was it a natural phenomena from comparatively long periods of weighlessness, or did it stem from Titov's own susceptibility to the environment? After more than 9 hours in orbit, Nikolayev transmitted live pictures to Earth and the Moscow Television Center broadcast the signal to privileged citizens who were fortunate enough to have a receiver. An hour later, the cosmonaut ate supper, having had breakfast about 50 minutes after launch, and 'dinner' at 3:00 pm Moscow time.

Nikolayev's rest period began soon after he finished the third space meal of the day, sleep being assigned between 10:00 pm and 5:00 am on the Sunday morning. The Soviet news agency Tass dutifully reported that 'In accordance wtih the flight program, Cosmonaut Nikolayev went to sleep . . .' The following morning it was revealed that during the night, the cosmonaut's pulse varied between 60 and 65 beats per minute, and that his breathing was even 'in frequency and depth.' Soviet physicians were meticulous in their concern to retrieve all the important medical information they could, going to great pains to miniaturize everything and to place on board the spacecraft a recorder which would continue to provide a permanent trace of the cosmonaut's reactions during the communication blackout on re-entry. Also, the pilot wore electrodes in the outer corner of each eye so that eye movement to right or left could be monitored as well as eye 'blinking' in the hope that this information would enable post-flight analysis of muscular reactions to be enhanced. Respiratory function was monitored by a thin rubber tube filled with carbon powder; the resistance of an electric current applied to the pick-up changed according to the expansion of the pilot's chest and this enabled information on breathing rate to be supplied. A microswitch would operate at chest movements exceeding 2 mm. Heart rate was monitored via a harness supporting electrodes pasted to the chest. Also, skin galvanic reflexes were measured with electrodes on the foot and lower third of the cosmonaut's right shin. Whenever a change occurred in the electrical potential of the skin the instrument attached to the electrodes registered the magnitude of the change and telemetered the data to Earth, or recorded it for later transmission.

With Vostok III, Soviet space science began to get into its manned space flight stride. The first two flights had proven the feasibility of multi-orbit missions, validated the technical and engineering design, and demonstrated the capacity of human passengers to not merely survive but intelligently contribute toward the objectives of the flight. Soviet engineers had provided a comparatively crude vehicle, certainly not as responsive to traditional 'pilot' roles as the Mercury capsule, weighing nearly four times the mass of its American counterpart, but with the basic capacity for sustained orbital flight. It could be commanded to selected attitudes – indeed that function was essential for accurate pointing during retro-fire – but it was designed and built with a high degree of automation where the pilot was more of a passenger than US designers preferred for their spacecraft. Inside there was considerably more room than in the diminutive Mercury, but still insufficient space literally to move about the cabin. That Tass agency reference to Nikolayev removing his harness and moving freely implied greater internal volume than was actually available, the pilot in fact being capable of drifting forward to the opposite wall and turning about his long axis but with little freedom to change position within the capsule.

During the Saturday of Vostok III's launch, Soviet officials sent a message to the US government imploring them not to let off any nuclear tests at high altitude. A nuclear detonation in the far reaches of the upper atmosphere the previous month had upset the Russians and they reminded the Americans that 'The common interest of all countries in the exploration of outer space for peaceful purposes unquestionably imposes definite obligations on all States. This means in particular that States must refrain from any measures that could in any way hinder the exploration of outer space for peaceful purposes or endanger the cosmonaut's life. It is known, however, that the Government of the United States of America plans to hold new high-altitude nuclear explosions.'

The message was handed to the temporary charge d'affaires in the Soviet Union by G. M. Pushkin, Russia's Deputy Foreign Minister and the US responded with confirmation that it did not intend to detonate any nuclear device nor would 'take any action which could harm the Soviet cosmonaut in any way.'

Pavel Popovich, framed by his pressure helmet and communications soft-hat, became Russia's fourth space man when he ascended in August, 1962, to join Nikolayev in orbit.

After Nikolayev awoke on the Sunday morning observers throughout the world wondered just how long the cosmonaut would remain in space. Almost exactly 24 hours after announcing the launch of Vostok III, the Soviet news agency Tass again warned teleprinter watchers that an important announcement was imminent. Three minutes later came a statement, quite different from the one many had supposed would tell of Nikilayev's safe return to Earth: 'The spaceship Vostok IV piloted by Pilot-Cosmonaut Pavel Popovich was put into the orbit of an Earth satellite at 11:02 hours (Moscow time) on 12th August. In accordance with the tasks set, the Vostok IV spaceship was launched while the Vostok III spaceship . . . was on its (own) orbit (of the Earth). Now two Soviet spaceships . . . are simultaneously flying in outer space.' There had been rumor of a major Soviet space spectacular for some time, and Titov several months before had spoken of an achievement which would be seen to eclipse his own 17-orbit mission completely.

Tass announced also that the Vostok IV spacecraft was circling the Earth at the same inclination as that of Vostok III. In fact, the latter had been sent into an orbit inclined 64.98° to the equator while the second ship slipped into a 64.93° orbit. It was hailed by the Russians as the first rendezvous mission, the close orbiting of two spacecraft – in this case, both manned – with communication both with the ground and each other. Popovich had earlier that year replaced the Soviet cosmonaut Anatoli, one of the original group selected early in 1960, because the latter was observed to have a mild heart irregularity during centrifuge tests where he was exposed to a force of 12 g. Soon after the flight was over, and he had returned to a foreign press conference, Popovich would say that 'Knowing where Vostok III would be in relation to me, once I was in orbit, I looked for it immediately and saw it at once. It was something like a very small moon in the distance.'

But Soviet enthusiasm for their new found capability was an over-stated case, and really represented an ability to plan and launch the simultaneous operation of a dual mission rather than perform a rendezvous flight of the type they claimed to have demonstrated; neither spacecraft could maneuver and each Vostok was only capable of changing attitude. Once injected in to the orbit given to the spacecraft by the launch vehicle, that orbit could only be changed when the retro-rocket fired to bring the spaceship home and in that regard it was no more a performance of rendezvous than would be demonstrated by flying two Mercury vehicles at the same time. But the popular view prevailed and once again the world was duly impressed by the Soviet achievement. As it was, the dual flight adequately proved the increasing competence of Soviet space engineers. It needed real precision to track Vostok III's orbit, determine the precise position of the vehicle at specific times, and to prepare for launch a second vehicle that would have to get off the launch pad at just the correct second of time to arrive in orbit close to its companion.

Soon after arriving in space, only a few kilometers from Vostok III, Popovich congratulated Nikolayev on breaking the 25 hour record flight of Ghermann Titov. Throughout that Sunday, television transmissions came down to earth from Vostok IV and routine communications sessions were held with the few tracking stations operated by the Soviet Union. Periodically, conditions aboard the two spacecraft were monitored. Pressure and humidity were normal; the temperature aboard Vostok III was seen to be a comfortable 17°c while that on Vostok IV was said to be 24°c. Trajectory controllers watched carefully as the assymetrical gravitational mass of Earth pulled and tugged at the orbital parameters, and as tenuous outer envelopes of the atmosphere helped gradually modify the elliptical paths. By midnight, Vostok III was observed to be in 176.7 km by 227.6 km orbit while Vostok IV was in a path 177.9 km at the low point and 234.8 km at apogee.

Both men seemed well suited to their respective assignments: Nikolayev's pulse rate remained within 64–72 beats/min while Popovich's pulse was between 60 and 72; respiration was measured at between 15 and 19 breaths each minute for both. Nikolayev had begun his sleep at 9:00 pm, followed by Popovich a half-hour later but both awoke at 4:30 am on Monday 13 August. Vostok IV was cooler now, having radiated heat built up in the capsule during launch to the vacuum of space and by midday both ships were between 15°c and 18°c. By now, both cosmonauts had exceeded Titov's 17 orbits of the Earth, with Nikolayev on his 33rd. Throughout that day the dual mission continued, but all the while the elliptical paths of respective spacecraft were taking the two vehicles further apart. Without the capacity to change their orbits they could do nothing but drift along in their respective paths. During that Monday, Soviet Premier Nikita Kruschev sent personal greetings and the two cosmonauts pursued their schedules of eating, observing, communicating, and building up a valuable inventory of bio-physical reactions for doctors and physicians to probe when they returned to Earth.

Gradually, and almost imperceptibly, their respective orbits were changing. By 10:00 pm, Vostok III was in a 173 km by 221 km orbit, while Vostok IV was in a 176 km by 229 km path about the Earth, now inclined 64.83° and 64.95°, respectively. Careful comparison with figures presented for earlier periods of the flight will reveal the subtle changes introduced by the uneven distribution of mass within the planet, causing any spacecraft in orbit about the Earth to migrate continually from one path to another. This would be a source of major concern during rendezvous flights performed by both Russia and America, beginning several years after the Mercury and Vostok missions, but the dual flight of Vostok III and Vostok IV serves to illuminate the changing co-ordinates of targets in space.

Nikolayev and Popovich entered a scheduled 7 hour rest period at 9:00 pm on the 13th, and when they awoke on the morning of the 14th both men were found to be in good health. Each reported to Earth the condition they felt themselves to be in, and the atmospheric parameters of their respective spacecraft: temperatures were between 15°c and 20°c; humidity was an even 75%. Within the next few hours they pursued normal flight activities and Nikolayev completed three full days in space. Soviet scientists had been unsure of the ability of humans to eat normally, or to consume normal food, in the unique weightless environment. Lessons from Vostoks I and II showed ready adaptability and Nikolayev and Popovich enjoyed simple food prepared with special attention to the low-residue consideration for orbital requirements. Several drinks were provided, including water, coffee, and fruit juices.

For much of the flight, Soviet news sources had kept up a slightly more informative narrative than had been their custom for the earlier flights or for the early hours of the Vostok III mission. But Tuesday the 14th saw a gradual slide to political ideology, propaganda exercises, and general criticism of the competitive Western beaurocracy. In fact, said *Pravda*, Nikolayev and Popovich would welcome US space vehicles to follow them in the exploration of space so that 'the peoples of the world would applaud America just as warmly as they are today applauding the Soviet Union.' Pravda also criticized 'military strategists' for planning 'battles in space' and said that if the United States would lay down its arms there would be no need for the 'secrecy' attending Soviet space launches and orbital operations, reminding the world that 'we, the peoples of the socialist countries, find to our regret that in the USA . . . businessmen who worship the golden calf are beset by dreams of the immeasurable riches they could grab by exploring for instance, the treasures of the Moon. The atomic maniacs, for their part, are ready to transform the Moon into a military base. What monstrous plans!'

Furthermore, echoed Tass, '(The) twin launching of Soviet spaceships has caused unofficial dejection in Washington over the technical and propaganda superiority of Moscow in space exploration . . . dejection also in NASA headquarters.' There followed a bitter assault on the attitude of Congress and a condemnation of 'denizens of the Washington Capitol.' A Moscow correspondent even resurrected a statement from von Braun several years before who, upon being asked what US astronauts can expect to find on the Moon, replied, 'Russians.' The cynical reference implied Soviet initiatives already made that prospect more likely than cautionary. Fully exploiting the advantage, the same correspondent confidently predicted that 'bourgeois observers fully realise that millions of people throughout the world are today drawing the inevitable conclusion concerning the superiority of the socialist system over capitalism.' In exceptionally bad taste, it was even said that a little girl called Caroline might ask her father in the United States why Russia was technically superior and whether she should be frightened for the future – John F. Kennedy had a daughter named Caroline.

On through the 14th the two flights continued, the cosmonauts going to their rest period a little earlier than on preceding days. By 9:00 pm the orbits of Vostok III and IV were calculated to be 170 km by 214 km and 173 km by 224 km, at inclinations of 64.98° and 64.95°, respectively – again, noticeable changes to parameters measured the previous evening.

On Wednesday, 15 August, the two spaceships came home. Vostok III landed at 9:55 am, while Vostok IV touched down six minutes later but 300 km away. At closest approach, the two spaceships had been just 6.5 km apart, shortly after the second arrived in orbit, but at the same time of retro-fire they were separated by 2,850 km. Both cosmonauts elected to eject from their respective capsules at a comparatively low altitude, decisions taken, it was reported, because both were enthusiastic on parachuting as a sport; a more likely reason is that a gentle float to Earth was infinitely preferable to the bone-shaking thud of the Vostok cabin hitting solid ground. Nikolayev had been in space for only 1 hr 38 min less than four full days, his companion in Vostok IV space-borne for 1 hr 3 min less than three days. Soviet manned space flight now totalled more than eight man-days in orbit; America's, less than 10 hours. Simply stated, Vostok was merely exploiting built-in capabilities. From the outset it had been designed to support life for several days and the comparatively lengthy missions were fulfillment of that objective.

But the effect of flights lasting three and four days was stunning. Not for a further two years could America hope to match flights of that duration, while work then actively leading toward the longest Mercury flight, envisaged a day-long mission sometime in 1963. But then, Vostok was incapable of changing the orbit it received from the final stage of the launch vehicle, while Gemini, hopefully ready for manned flight by 1964, would move freely from one orbit to another, rendezvousing and eventually connecting to another spacecraft already in orbit. Nevertheless, public acclaim was won on performance, not promise, and American space officials were anxious to press ahead with plans for Gemini and Apollo. For a while, concern about the apparent gap opened up by Soviet achievements rallied military support for a more active Defense Department role in the nation's accelerated space program. Several Air Force chiefs made stirring speeches calling for greater emphasis on the military role in space, purportedly claiming that Air Force jurisdiction would get things done quicker and with increased relevance to Russian initiatives.

There was no justifiable reason to think military administration of the US manned space flight program would have got things done any quicker. The plain truth was that nobody in American political circles was at all interested in advancing the technology of manned flight until it became politically expedient to rally support for partisan causes by supposedly polarizing public desires. The man in the street had little idea in 1957 of the need for any kind of space program, but Sputnik 1 changed all that and gave politics another opportunity to proclaim a public rally-call to super technology. Far from advancing at a quicker pace, military control of the national manned space program would probably have restricted the national advantage in paying for projects like Gemini and Apollo by concealing them behind a net of security conscious restrictions. In full view of the world, public openness was the one ace America had up its sleeve during the days of manned flight when US technology seemed far behind concealed Soviet initiative. Public following of a program held entirely in the open, with announced launches pursued by cohorts of pressmen and women, held a certain magic appeal compared to the secret schedules of a closed society.

But the 64 Earth orbits of cosmonaut Nikolayev, and the 48 orbits of Popovich, were to be widely reported in the world's newspapers. This fact alone seemed to rally support temporarily for what many felt was a race with the bottom suddenly dropped out. It had been a year since the Titov flight and several US politicians wondered about the wisdom of setting Moon goals in the face of such apparent apathy. With hindsight, it was but the beginning of a lengthy contest where public announcements about NASA plans would stir Soviet reciprocation designed to purposely eclipse American events. Behind a cloak of total secrecy about their own space program it would seem for many years that the Russians were always first while the Americans followed behind, seeming to race on unable to ever catch up.

Yet in quickly putting together by 1962 what in lay terms seemed to be a rendezvous mission of the type NASA spoke openly of flying in 1964, Soviet scientists were made to bend their projects according to political expediency, and Kruschev's ideological propaganda machine could make much out of such advantageous short cuts to success. But as with most applications in those early days of being first, once applied the Soviet initiative was never exploited. Where observers may have expected a new

Walter Schirra discusses with flight director Christopher Kraft aspects of his forthcoming space flight aboard spacecraft 16.

application of space capability to be more fully explored in repeated and increasingly useful follow-ups, the Soviets endured on the instant success of a one-shot venture. It was an apt demonstration of the shallow value of a politically-orientated technology project. While the Russians were busy impressing their communist bloc neighbours with record-breaking space flights, Manned Spacecraft Center engineers at Cape Canaveral prepared Mercury capsule 16 for the six-orbit flight then planned for 25 September.

Apart from minor modifications to the spacecraft, fitting it out for a mission twice the length of flights for which it had been designed, the most noticeable change to pre-flight preparations was in the planning of tracking and communication coverage and in the deployment of Defense Department recovery forces. It would be the first time a manned flight would wander appreciable distances from the primary tracking sites laid out several years previously for three-orbit missions. But in six orbits the Earth would spin 135° on its axis, displacing the path of the capsule so that it would move away from the now familiar stations. To fill in gaps that would otherwise prevent the pilot from talking to the ground, the former Atlantic Ocean relay ship was fitted with a command capability, where it could send instructions to the capsule, and moved to a position south of Japan. Also, three ships – the *Huntsville*, the *Watertown*, and the *American Mariner* – were introduced to the Mercury network and stationed in the Pacific near Midway. Finally, a station at Quito in Ecuador was brought on line for spacecraft communications during the sixth orbit, a path around the Earth that would carry capsule 16 too far south to appear over the Cape Canaveral radio horizon.

As it was, with all the shuffling around for optimum coverage, there would be periods of nearly 30 minutes when nobody could talk to Wally Schirra in orbit. It was to be the beginning of an acceptable drift away from the original need for continuous communication with stations set up to ensure gaps no greater than 10 minutes for any orbit of the Earth, a stringent mission rule during the early days when Mercury was still in the planning stage. Recovery operations, moved to the Pacific for planned sixth orbit retrieval because spinning Earth placed the American continent in the way of a normal Atlantic splashdown, required 83 aircraft and helicopters versus the 63 for Carpenter's flight, with 26 ships in all, 6 more than for MA-7. Around the world, Department of Defense support would call on 17,000 personnel, 4,000 more than on the previous Mercury mission. The increases were due, primarily, to the need for contingency recovery areas in both Atlantic and Pacific Ocean areas.

On three-orbit flights, the mission rule stipulating a prime splashdown contingency zone for each orbit could be accommodated by placing all recovery forces within one large area in the central Atlantic, spread through areas A to H and discussed in the previous chapter. For MA-8, recovery areas A to F were provided as usual for emergency aborts during the first few minutes of the mission, with specific zones laid out between the Cape and the ocean south of the Canary Islands. Emergency recovery areas for the second and third orbits were designated 2-1 and 3-1 zones, replacing the G and H areas of the first two Mercury manned orbital missions. But there was also a fourth orbit contingency recovery zone, called 4-1, immediately north of Grand Turk Island, and a second, called 4-2, in the Pacific south-east of Midway Island. The emergency recovery zone for the fifth orbit, and the planned splashdown site for the sixth orbit, occupied virtually the same ellipses north-east of Midway. Accordingly, zones 5-1 and 6-1 employed the same recovery force comprising the carrier USS *Kearsarge*, three destroyers, four search aircraft and three helicopters.

In effect, capsule 16 would make $5\frac{1}{2}$ orbits of the Earth, but nobody made the subtle differentiation between a partial or a full orbit and the mission remained earmarked as a six-orbit flight. At 1,962 kg including the heavy launch escape tower, it was the heaviest manned Mercury so far – Glenn's capsule 13 weighed 1,936 kg and Carpenter's capsule 18 grossed 1,925 kg – and Schirra chose Sigma 7 as the name to be painted on the side of spacecraft 16; sigma for the engineering symbol denoting summation, and 7 for the seven astronauts including Slayton whom nobody in the corps was about to write off even if medical academics *did* say he had a fibrillating heart. Almost a month after the dual Russian success, Mercury's Atlas booster sprang a leak through a seam in its fuel tank and that put paid to hopes of a September launch, deferring the first flight attempt from the 25th of that month to 3 October at the earliest. Pre-flight preparations were re-cycled to be compatible with that schedule and Schirra began the final period of intensive training, having already gained advantage from the early flight plan delivery date recommended by astronaut Carpenter, who blamed his own flight problems on the hurried pace of pre-mission activities.

Schirra was well trained for MA-8, having spent 45 hours in the Aurora 7 spacecraft in his previous role as back-up to astronaut Carpenter. Now, with an added 31 hrs 27 min inside Sigma 7, checking it, testing it, familiarizing himself with its idiosyncracies, he was well prepared for the six orbit flight. At the Cape he added 29 hrs 15 min in the Mercury trainer to the 28 hrs already spent in the same device prior to MA-7 and the 8 hrs in the Langley procedures trainer, flying 37 simulated missions, 40 capsule turnaround maneuvers and 53 retro-fire operations. Schirra's preparation schedule benefited fully from the lessons of earlier flights and the astronaut himself was keenly enthusiastic for the rigorous and demanding pre-flight work load. As a man, Schirra was outspoken, more than once causing moderate embarrassment over his frank and pointed comments. Shortly before MA-8 he challenged the wisdom of loading John Glenn with so many public appointments and talks when the entire astronaut corps was supposedly under pressure to work at maximum pace. But he was a tough opponent of temperamental spacecraft systems and had a reputation for digging in and pulling through, a character trait that would elevate him among his peers when the time came to fly Apollo on shakedown tests. Schirra was not a man who necessarily went out to be liked but he certainly earned the respect of managers and administrators at MSC.

The pace of manned space flight was taking an upward turn, the first tangible evidence of that being the momentous decisions of 1961 that placed upon NASA responsibility for getting men to the Moon by 1970 and provided the space agency with its second manned spacecraft to follow Mercury: Gemini. Now, the tools for the job were flocking in as contractors were signed up to build machines for space the like of which nobody had rationally conceived in all their wild imaginings. And so were the astronauts. On 17 September NASA announced the names of 9 lucky candidates from among a pool of 253 applicants who were to join the original seven Mercury pilots in training for seats aboard Gemini and Apollo, and it also announced that Deke Slayton was to be the new Coordinator of Astronaut Activities. Four of the new men were from the Air Force: Maj. Frank Borman; Capt. James A. McDivitt; Capt. Thomas P. Stafford; and Capt. Edward H. White II. Three were from the Navy: Lt. Charles Pete Conrad, Jr.; Lt. Cdr. James A. Lovel, Jr.; and Lt. Cdr. John W. Young. One, Elliot M. See, Jr., was a test pilot for the General Electric Co., and one was a civilian test pilot already with NASA: Neil A. Armstrong.

Bob Gilruth announced their names, and told the public that 'Assignment to flight crews will depend upon the continuing

Sigma 7 thunders into orbit atop MA-8 on Project Mercury's penultimate mission, October, 1962.

physical and technical status of the individuals concerned and upon the future flight schedule requirements.' More immediately, flight schedule requirements centered upon Mercury-Atlas 8, the third manned NASA flight, chalked up for a 3 October lift-off. But the announcement that more astronauts were to join the 'select seven' which for 3½ years had sole access to American space only opened again the old arguments about a fitting wage for the job. Large sums of money had changed hands only on paper when Time Life Inc. purchased all rights to their personal stories, and the majority voice from public opinion deplored the fact that these men received only the pay due to their respective service ranks plus expenses. The day after Gilruth's announcement, the *New York Times* began anew its campaign for what it believed to be only rightful and moral justice:

'Since the National Aeronautics and Space Administration is a civilian agency, it might be wiser to make all the astronauts civilians so that no questions of inequality or discrimination arise among them and also so that they might receive more adequate pay than is provided by the low military pay scales. . . . In permitting the astronauts to cash in on their exploits, the Kennedy Administration is following an unwise precedent set by the Eisenhower Administration. While the practice of profiting from memoirs of Government service is an old one, such memoirs are normally written by persons who have already left Federal employment. . . . The Government would be far wiser if it paid its astronauts a sufficiently generous salary so that it could in good conscience ask them to observe the same practices of discretion and modesty which have hitherto been considered normal for all other Government employees.'

In the years ahead, many astronauts would indeed reap the rich rewards of momentary fame, cashing in on business ventures offered in the first flush of success – not all of which left them as profitable as when they went in – but a lot of the astronaut pool would get little sight of grand offers and invitations to high places. It would result in bitterness more frequently seen in the communities that grew up around Houston, more readily understood within the social clubs and classes that spawned from a feverish activity leaving wives, mistresses and mates enmeshed within a conflicting drama of frustration and insecurity. The astronauts would find affinity with each other away from the bickering and backchat that very often pervaded the bungalow campus residence of families and friends. It was a logical price for an uncertain future, one in which always the goals and ambitions of 'tomorrow' took precedence over the happiness and achievements of 'today.' But much of that was still to come when, in October, 1962, Wally Schirra made ready to fly six orbits in space.

The by now familiar routines of pre-flight preparation – up early, shower, breakfast with selected friends, Joe Shmitt's suit-up procedure, check with the blockhouse, drive to the pad – all went like clockwork in the morning hours of Wednesday 3 October. Upon sliding into the form-fitting couch, now devoid of its lower section but with new and more comfortable toe, heel and knee restraints, Schirra found a small ignition key dangling from the safety pin attached to his right-hand attitude controller. And in the compartment reserved for personal items, a steak sandwich wrapped in a plastic bag. Nobody knew except a few pad technicians privy to the conspiracy that would cause Schirra so much pleasure, and NASA so much fury! It was the hallmark of a happy team, the calling card of good-luck gestures.

The countdown had gone very well that morning as, once again, Walt Williams looked on from above while Christ Kraft pulled his flight controllers into harness; like the conductor of a gigantic orchestra, he would mould and hone the console technicians into a single organic body dedicated to the success of MA-8. And it was a process that would be repeated through the manned space flight program. From the blockhouse, conducting its own orchestra of events centered on Atlas 113-D, a hold in the countdown at T–45 min: Canary Island reported a problem with one of its radar tracking units. It was 6:15 am and just 15 minutes later the clocks started running again and events moved smoothly to a fast ignition in the three rocket motor chambers that would send Schirra to space. First, the two small vernier motors ignited and then the sustainer and booster engines, followed rapidly by a sudden upward thrust as the big bird shook itself free and leapt into the air. Schirra was momentarily surprised by the impetus of lift-off but Deke Slayton's cheerful voice brought him rapidly back to the task at hand as the roll program twisted the big rocket round its central axis, turning it to face the trajectory it would need to put Sigma 7 in the desired orbit.

'Ah, she's riding beautiful, Deke,' came the voice from capsule 16, at the beginning of what was to be one of the more success filled missions in Mercury's checkered career. During the powered ascent Schirra was very aware of the dynamic events associated with booster cut-off, escape tower jettison and sustainer cut-off; more so than Carpenter reported after his flight. There was one event that nearly brought the one and only abort to Mercury: during the roll maneuver just 10 seconds off the pad, the booster programed round to within 20% of triggering an abort, an activity watched with horror by technicians in the blockhouse. Also during ascent, the noise associated with max q sent the radio microphone into an inconvenient mode whereby Schirra had to push a 'talk' button every time he transmitted to the ground. This required him to frequently take his left hand from the abort handle, something he was reluctant to do since that was the means by which he would escape from a rogue booster.

Thrust on the single sustainer engine seemed lower than expected; Schirra noticed that the build-up in acceleration after booster staging seemed slower than he had been trained to expect. This was borne out when the rocket continued to fire several seconds past the predicted cut-off point, finally shutting down 5 min 15.7 sec after lift-off. However, at 28,233 km/hr, speed was about 16.5 km/hr faster than desired and this resulted in a 161 km by 283 km orbit – Schirra was the fastest and highest flying Mercury astronaut of them all, but only by a tad. During turn-around his manually operated Fly-By-Wire mode expended just 0.14 kg of propellant, significantly less than that used by Carpenter, and much less than Glenn's performance. But this was what Schirra was especially charged with, responsibility for flying an engineer's mission, one in which the spacecraft was to be made to really show its colors without consideration to scientific tasks that could carry the pilot dangerously close to propellant depletion.

On the first orbit Schirra was well pleased with the modifications made to capsule 16's attitude control thrusters, remarking on how crisp were the jets of hydrogen peroxide used in slowly turning the spacecraft to keep track visually of Atlas 113-D's spent carcass as it tumbled away, although by this time he was in ASCS before again switching to Fly-By-Wire; as he would recall later: 'it was a thrill to realize the delicate touch that it is possible to have with Fly-By-Wire, low. This touch is an art that a pilot hopes to acquire in air-to-air gunnery for getting hits. In this case the control system was so effective that it just amounted to a light touch and maybe a few pulses in either axis to get the response I wanted.'

During the initial part of the first orbit telemetry on the ground indicated a rapidly rising suit temperature, going from an indicated 23.3°c at lift-off to 32.2°c by the time Schirra came across to Muchea, Australia. But Schirra had given problems like this consideration in ground simulators where he learned to very carefully adjust environmental control system equipment in discrete steps, rather than immediately turning to a much cooler setting; in fact, indiscriminate switching between settings designed to cool the suit and others intended to heat it up had prevented Carpenter's suit temperature from adequately settling down on MA-7. Realizing that rapidly increasing suit temperature would cause concern on the ground, Schirra gradually advanced the control knob regulating temperature from the number 4 setting it had at launch, through to position 8, each increment being retained for 10 minutes to prevent the system over-reacting and upsetting the balance.

At Mercury Control, Chris Kraft debated whether to order the capsule down at the end of the first orbit, or back the indications of stability in the increasing suit circuit temperature, leaving the capsule for one more orbit to see if Schirra really had it under control. The flight surgeon expressed real concern that Schirra might be suffering from the very hot suit environment and was disturbed at not being able to monitor the astronaut's body temperature, the sensing instrumentation having failed for that parameter. Between Muchea and the Canton Island station, ground readouts revealed an indication that the suit inlet temperature might actually have peaked out and that if left for a while could actually start to fall. Weighing all these factors, Chris Kraft decided to let Sigma 7 go round once more and as he passed over the Guaymas station Schirra indicated that he was beginning to cool down. The gamble paid off. Over Australia on the second orbit, temperature inside the suit was down to 22.2°c and before the end of the mission it would occasionally read 19.4°c. But the cabin's atmosphere remained warmer than expected, staying within a range of 40.5°–43.3°c.

Much of the second orbit was taken up with tests to see if an astronaut could adequately align the spacecraft yaw axis without looking at the attitude indicator on the display panel. Using terrain features viewed through the trapezoidal window, Schirra set about the first of these as the capsule sped across the Bermuda tracking station, 1 hr 41 min after lift-off. With the attitude indicator covered up, Schirra positioned the capsule in the retro-fire attitude – roll zero, pitch −34°, yaw zero. He then yawed the spacecraft 8° to 10° from neutral and brought it back using observation of the ground below to provide visual alignment. The second test, carried out in daylight like the first, came 9 minutes later and for this Schirra yawed Sigma 7 23° to the right and attempted to bring it back to retro-fire attitude. At completion, the spacecraft was out only 2° in yaw and 1° in pitch, proof that visual references were quite adequate for alignment and that a capsule could be safely placed in the retro-fire attitude if the instruments failed. He also found the window view the most favourable means of determining the yaw errors, considering the periscope a useful back-up but nothing more; Carpenter's earlier conclusion was vindicated.

The second two tests were performed over Australia in the dark using the window only. They came within two minutes of each other, with the first beginning 2 hr 26 min after lift-off, and again Schirra brought the capsule back but with little increase in error. Considering the lack of adequate lighting, observing star fields and horizon alignments, the maximum 3° drift from zero yaw was remarkably proficient. Schirra observed that the window was too small for good stellar reference, although his performance was well within the anticipated limits. Starting in on his third orbit, Schirra finally got around to performing the test Robert Voas had wanted nearly a year before. With eyes closed he reached out for the touch test to selected points on the instrument panel. He found no disorientation compared with control tests on the ground, reaching all points to within 5 cm of the actual location.

Unlike Glenn and Carpenter, Schirra voiced prolific commentary on to his vox (voice actuated) record system. While out of contact with successive ground stations he would continue talking, backing up his actions with verbal descriptions and observations. Several times he was heard to come grumbling up over the radio horizon, talking away to himself about this sensation or that activity. At 4 hr 22 min Sigma 7 got within range of Gus Grissom at the Hawaii station for a status report and a message passed along from Christ Kraft at the Cape: 'Okay. Fine. Cape feels you are in good shape, Wally, and so I have good news. They give you a go for 6 orbits.' In fact, Sigma 7 was in arguably better condition after nearly three orbits than Friendship 7 or Aurora 7 had been at the end of their respective first orbits.

At Cape Canaveral's Mercury Control Center, technicians monitor Schirra's progress in space. On the plotboard, his six orbital tracks span the mission with circles depicting the coverage from 14 communications stations.

Suit circuit temperature was well under control now and the cabin temperature was dropping to a more normal level. In the attitude systems, automatic fuel was down to 82% full and manual was pegged at 92% remaining. Fuel quantities for attitude control were actually higher than expected for this half-way point in the flight. For a period on the third orbit, Sigma 7 had gone into a drifting flight mode and now, passing across the United States to start the fourth orbit the spacecraft was powered down and once again put in the passive attitude control mode. For a complete orbit of the Earth, spacecraft 16 just drifted slowly in attitude while Schirra made numerous observations and checks on his physiological condition. Passing across Califronia toward the start of his fifth orbit, John Glenn talked Schirra into a live broadcast relayed to much of America and Europe by radio networks covering the flight. Powered up again, Sigma 7 was slowly moved back to retro-fire attitude, automatic and manual fuel quantities reading a healthy 74% and 80% respectively.

Suit temperature was now down around 16.7°c and the complete inventory of spacecraft systems were in the most acceptable condition so far observed on a manned space flight. At the beginning of the sixth orbit, Sigma 7 was far to the south of normal Mercury groundtracks and Schirra called through the Quito station, specially configured for this long duration flight, reporting acquisition of good weather pictures and to dutifully inform capcom that he had momentarily opened the faceplate on his helmet to 'wipe my nose!' Now it was time to pack away the several items floating loose around the weightless capsule, and to prepare the spacecraft for retro-fire over the Pacific. On across the Indian Ocean tracking ship, Schirra confirmed that he was going through his checklist and then, passing out of range for several minutes, kept up a verbal discussion with his recorder about his sensations inside Sigma 7: 'One gets the illusion that you're on a train or some other vehicle, due to the humming, and you feel that you should be on a track of some kind and you're driving down. Much like the sound of the ship when you're under way at sea. The blower noise I assume, and the inverters give you the same illusion.'

Just 1½ minutes before retro-fire Schirra passed into range of Al Shepard on the Pacific Command Ship. As the three solids fired the spacecraft held rigidly to its pre-set attitude. Immediately after the third motor burned out, Al Shepard passed along the remarkable news that the capsule was observed on telemetry to have 68% automatic and 78% manual propellant remaining in the tanks. On down through re-entry the events Schirra experienced were recorded on his onboard tape machine, the colours coming from the forward shingles as materials heated in the fiery descent adding realism to his descriptions of high g forces, a seemingly unbroken commentary all the way down to the Pacific. During much of the descent, Schirra controlled spacecraft attitude with the fuel expensive RSCS mode but the large quantities still remaining made the selection comfortably acceptable. When dumped deliberately during the final minutes prior to splashdown there was still more than 50% hydrogen peroxide in the manual propellant tank while the automatic supply ran to nearly 60% remaining. It would have been inadvisable to land on the water with quantities of propellant still on board; the shock of splashdown would undoubtedly cause minor leaks and hydrogen peroxide is a toxic chemical under the best conditions.

Sigma 7 came down only 8 km from the planned coordinates, sailors on the USS *Kearsarge* seeing the contrail in the sky as capsule 16 descended, hearing the sonic booms as it decelerated through the speed of sound, watching as it slowly fell under the spreading canopy of the main parachute. Just a few minutes after splashdown, Schirra notified the recovery communicator that he would prefer to remain in the spacecraft and have a boat crew come across from the *Kearsarge* and tow him to the carrier's hoist. A few minutes later a five-man whaleboat came chugging across the waves, attached by line to the parent ship. Pararescue men had already installed a flotation collar to the sides of the capsule, steadying it in addition to ensuring bouyancy. But no water appeared inside and Schirra steadily went through his shut-down procedures. Just 41 minutes after hitting the water, Sigma 7 was being carefully hoisted up to a retrieval deck on the side of the carrier. It was gently lowered and Schirra punched the door jettison knob receiving like Glenn a hand injury as the plunger sprang back.

Everyone back at the Cape was in jubilant mood, gratified that their optimism had been vindicated and more sure than ever that plans for a day-long flight could proceed as scheduled. For Schirra there were three days aboard the *Kearsarge* for medical examination and initial mission de-briefings. Physically, he appeared in good condition, although there were notable effects not seen on the three-orbit missions. For instance, when placed on the tilt table, a horizontal support structure designed to pitch the occupant at various pre-set angles, his heart rate was a comfortable 70 beats – min. When he got off the table it shot up to 100 beats/min. Changes were observed too in the blood pressure readings, but none of these expected effects were lasting and the pilot quickly adapted to the familiar gravity of Earth. As for the capsule, it too came through its ordeal with flying colors – in fact, better than earlier capsules on preceding missions. All previous heat shields exhibited a loose plug at the centre of the circular structure; some had come right out during splashdown. Sigma 7's shield plug was firm and tight with no evidence of it having ever worked loose.

As the last flight of the basic Mercury capsule it represented the peak of design performance. The next flight – MA-9 – would use a much modified spacecraft equipped with larger quantities of consumables for the day-long flight. In the light of glittering success with Sigma 7, Bob Gilruth decided to aim the final flight for an April 1963, launch. NASA Headquarters had already decided that MA-9 would be the last Mercury operation and that following the 18 orbit flight, manned space flight operations would move as quickly as possible to Gemini missions, hopefully before the end of 1964. There was still a long way to go before Moon landing Apollo flights, in fact 18 months after the public announcement by John F. Kennedy that NASA would fly its astronauts to the lunar surface in that decade the strategy for action was only just becoming clear.

It had been a long and painful process finding out how to fulfill Kennedy's promise and while, through the second half of 1961 and the first six months of 1962, NASA dispatched astronauts on suborbital and orbital Mercury flights, and while Chamberlin was pulling together the elements of Gemini that promised to provide long duration and rendezvous missions, the space agency was wrestling with many different proposals for Apollo design configurations. Central to all its problems was the tantalizing uncertainty of how best to build a Moon ship. When Kennedy publicly asked Congress for the financial support to go to the Moon, NASA was unsure of the technical steps essential to satisfying the requirements. It will be remembered that upon realizing the Kennedy administration leaned toward a major space spectacular, Associate Administrator Robert Seamans had set up the Ad Hoc Task Group for a Manned Lunar Landing Study, chaired by William A. Fleming from NASA Headquarters; it was to be more generally known as the Fleming Committee.

That was on 2 May, 1961, and purported to construct an appraisal of the technical viability of such a mammoth undertaking. It's mandate was to consider the technical feasibility, or otherwise, of NASA mounting a lunar landing operation by 1967 – the date then favored by Johnson – using large Saturn boosters of the type being designed by von Braun's engineers at the Marshall Space Flight Center. Then, before the Fleming Committee had formulated basic conclusions, Seamans set up a second body with a broader mandate and formed from senior staff members at the Office of Launch Vehicle Programs, the Office of Advanced Research Programs, and NASA Headquarters. Authorized on the day Kennedy made his famous speech before Congress (25 May, 1961), the study group was chaired by Bruce T. Lundin, and it was thereafter known as the Lundin Committee; it was to examine all the many possible ways men could be carried to the Moon's surface, generate information on the recommended size of the launch vehicle best suited to the job, evaluate the advantages and disadvantages of different mission modes, and generate cost comparisons for each. In effect, the Fleming Committee would report on the general feasibility of a lunar landing program taking into account all the many technical steps that would be essential for success, while the Lundin Committee sought definition of the way it would be actually accomplished. Both groups were expected to report their findings to Seamans within the following few weeks.

Several choices were open to students of lunar flight modes, but one, which involved a novel form of rendezvous operation, made persistent appearances whenever officials got together for discussions. To appreciate its genesis it will be necessary to return briefly to a period when plans for spacecraft beyond Mercury

What the well dressed astronaut wears on a desert survival course. Front row, left to right: Frank Borman, James A. Lovell, John W. Young, Charles Conrad, James A. McDivitt, Edward H. White. Back row, left to right: Ray Zedehar (astronaut training officer), Thomas P. Stafford, Donald K. Slayton, Neil A. Armstrong, Elliot M. See.

embraced Earth-circling space stations adapted from vehicles capable of flying around the Moon: to 1959 and work being carried out in the Aerospace Mechanics Division of the Langley Research Center. John M. Eggleston from the AMD worked closely with John D. Bird of Langley's Theoretical Mechanics Division throughout 1959 on problems arising from rendezvous operations and from that work, then only loosely related to future manned space projects, the Chance Vought Corporation learned of advantages inherent in using a rendezvous mode with modular vehicles to accomplish at less cost and with less complexity major space objectives initially thought to require large space-borne systems.

Chance Vought worked on a *M*anned *L*unar *L*anding *A*nd *R*eturn (MALLAR) concept, a privately funded study aimed at seizing a major initiative for the company by presenting NASA with a mission opportunity only vaguely considered by the agency itself. Thomas E. Dolan was in charge of the study, which occupied the company between March, 1959 and January, 1960. When the Eggleston-Bird equations were studied it seemed to Chance Vought that here was the nucleus of an idea that could dramatically reduce the weight of a Moon landing vehicle. By first placing into orbit around the Moon a spaceship designed to house crew and supplies for a two-way trip from Earth, the addition of a second, comparatively light, vehicle to fly down to the surface and back up to the mother-ship would provide in total a realistically-sized capability for Moon landings. The study, handed to NASA in January, 1960, envisaged a 3 tonne crew carrying mother-ship, a 4 tonne mission module with supplies and propellant to put all elements into Moon orbit and accelerate the mother-ship back to Earth, and a 12.2 tonne lunar landing module which would descend to the Moon's surface.

Dolan remains unsure where he first heard mention of a lunar-orbit rendezvous concept for Moon landing but believes it to have been at a meeting of Langley's Pilotless Aircraft Research Division (PARD), re-named the Applied Materials and Physics Division late in 1959, where H. Kurt Strass, then with Space Task Group's Flight Systems Division, raised the possibility of parking orbits around the Moon for equipment not essential to the actual landing operation. In fact, many men before Dolan, or even Strass, had recognized the advantage of such a concept, and numerous instances can be found in the earlier chapters of this book; most notable were Kondratyuk and Ross. Dolan made a further presentation of his Manned Modular Multipurpose Space Vehicle on 4 February, 1960, during a meeting at which John Bird urged the proposer to apply further study to this promising concept.

At that time, nobody was interested in moving rapidly toward such exotic capabilities and little stimulus remained for further work. Nevertheless, while Chance Vought had to put their money into more profitable channels, Langley retained a coveted hold on lunar-orbit rendezvous, not necessarily as a Moon landing concept but as a technique applicable to many potential space operations around Earth too. When Langley began serious consideration of a space station as the next best goal beyond the limitations of Project Mercury, John C. Houbolt from the Dynamic Loads Division pressed home the advantages of rendezvous; when the Goett Committee came to Langley in December, 1959, Houbolt was already a vociferous supporter of the concept.

The emphasis given by Langley to rendezvous operations in Earth orbit spilled across to Space Task Group studies of a post-Mercury spacecraft, injecting future mission studies with a basic need for expertise in the art of matching the orbits of two vehicles so that one could close in and dock with the other. It was eventually to inspire Chamberlin's fight for the spacecraft ultimately named Gemini, the development story of which has already been told. But Space Task Group as a whole was averse to the idea of rendezvous, and the Goddard Space Flight Center too was concerned about the value of such a seemingly exotic activity. So Langley sought support from the Marshall Space Flight Center, the home of the heavy-weight boosters then being proudly conceived on von Braun's drawing boards. Because Marshall favored the 'grand' objectives linked to dramatic Moon landings, and because von Braun's concept of a space program stopped at nothing short of manned planetary exploration, the study application for rendezvous activity centered on the use of this technique for Moon flights.

Caught up with the way-out nature of the topic, Langley too began studies of rendezvous in conjunction with lunar landings.

In John Bird's Theoretical Mechanics Division, William H. Michael, Jr., presented Houbolt's ad hoc rendezvous committee with a paper laying out in clear detail the way manned Moon flights could be accommodated within the framework of lunar-orbit rendezvous. That was in May 1960, at the beginning of a 12-month period during which Houbolt became the self-appointed advocate for lunar-orbit rendezvous – LOR. But still rendezvous was only a reluctant visitor among a variety of post-Mercury techniques even then deemed necessary for a successor to the one-man spacecraft. But that year did see recognition that if NASA was to move significantly beyond the simple Earth orbital tasks assigned to Mercury, rendezvous would be an important part of any future manned space program.

And the point was not lost to Headquarters personnel, for on 17 October, 1960, George Low, as chief of manned flight programs, opined the view to Space Flight Programs Director Abe Silverstein that it had 'become increasingly apparent that a preliminary plan for manned lunar landings should be formulated'. As a result, Low set up a working group comprised of himself as chairman with Oran Nicks and John Disher from the office of Space Flight Programs and Eldon Hall from the Office of Launch Vehicle Programs. They were to spend a couple of months pulling together earlier work at NASA and Air Force facilities for a set of 'ground rules for manned lunar landing missions,' and 'to prepare an integrated development plan.'

By the end of the year Bob Seamans had been brought from RCA to serve as NASA's Associate Administrator, carrying with him a wealth of experience from the Missile Electronics and Controls Division where RCA had been putting together the bones of a military project called Saint which aimed to produce a satellite interceptor.

Rendezvous was clearly an essential part of any project that aimed to move one satellite up close to another and Seamans was in tune with the forecasts of orbital rendezvous coming from Langley and Marshall. In fact, during his tour of NASA facilities late in 1960, Seamans spoke at length with Houbolt on the concept of lunar parking orbits and was much impressed with the work already performed at Langley to clear the way for formal evaluation at a higher level. Much of the work performed during 1960 was done at the behest of John Houbolt who, although not the first at Langley to seize upon the LOR concept, certainly did more than anybody else to see it survive through to 1961. That fact was probably more significant than any other associated with the technical direction of America's Moon landing plans. The fact that nobody seemed really interested in LOR at a Headquarter's level is more a product of the lack of mission than a lack of foresight: NASA was actively studying Moon flights during 1960, but only as circumular or lunar orbit flights; techniques associated with actually descending to the surface seemed largely irrelevant.

But what the agency was not to know was that within six months events around the world would conspire to give NASA an awesome responsibility, one in which Houbolt's LOR was like a guiding light at the end of a long dark tunnel. Seamans' grounding in LOR was cemented on 14 December, 1960, at a meeting postponed from the previous day because the Associate Administrator was ill on the assigned date. Clinton E. Brown and Ralph W. Stone, Jr., briefed Seamans on the Langley plan for lunar-orbit rendezvous for a Moon landing objective. The real reason given here for adopting the concept was that it enabled the estimated weight placed on the Moon to be significantly higher than a similar spacecraft could land by direct means. A direct flight to the Moon would require a much larger rocket to accomplish the same task but the analysis presented 14 December was based on the premise that with identical launchers available for use, a larger payload could be set down on the surface. Eventually, the captivating logic of LOR would center on the much smaller rocket that would be required to land a similar weight, but for the time being the payload advantage was good enough reason to justify the work.

John Houbolt spoke at this meeting on the general advantages of rendezvous in consideration of any ambitious space objective, John Bird addressed the benefits to be had from assembling, either in Earth or lunar orbit, modular elements of a system brought up in stages from the Earth, and Max Kurbjun spoke on the ease with which he felt pilots could be made to bring vehicles together in a docking operation. Nobody proposed a specific launch vehicle at this meeting, assuming for the purposes of argument that the modules would be launched separately into Earth orbit first. By early 1961, rendezvous was gathering influential advocates in Headquarters: in addition to Seamans, George Low, a passionate supporter of manned landing plans – which at that time were not formally in existence – embraced the concept when he set up the Manned Lunar Landing Task Group on 6 January, making sure the LOR and Earth-orbit rendezvous modes were well aired.

One day before, George Low had presented the studies which evolved from his working group established in October, 1960, to members of the Space Exploration Program Council (SEPC). Throughout that day, 5 January, the SEPC heard presentations from several NASA facilities on divergent lunar landing modes. The Space Task Group favored a direct ascent approach using the Nova booster for a direct flight to the surface and back. Von Braun's Marshall men preferred the Earth-orbit rendezvous method employing several Saturn rockets to assemble in Earth-orbit the various elements which, when docked together, would make the lunar trip. Langley favored the Lunar-orbit rendezvous mode.

Administrater Glennan was concerned that the studies should not migrate to a formal NASA project, offering caution that such a grand undertaking must await higher decision. President Eisenhower was about to leave office and it was inappropriate at that time to consider such a boldly unparalleled initiative. Nevertheless, the Low Committee, established by consensus at the SEPC briefing, set about the task of defining just what NASA thoughts were on the technical issues involved – just in case somebody from the White House began to think big!

Bob Seamans got word of the Low recommendations later that month and on 7 February he received a formal report to the effect that 'no invention or breakthrough is believed to be required to insure the over-all feasibility of safe manned lunar flight.' The Low Committee was uncommitted on the mode NASA should adopt, stating only its belief that rendezvous of some sort 'could allow us to develop a capability . . . in less time than by any other means.'

Low felt a Moon landing could be made in 1968 or 1969 employing Saturn C-2 rockets. A maximum seven such rockets would be needed to send a 4-tonne spacecraft to the Moon, the minimum weight cited by the Space Task Group. Alternatively, only one large Nova would be needed. But the Nova would not be ready before 1970 at the earliest. The Low Committee projected an estimated cost of $6,997 million to achieve a manned Moon landing by 1969 but said that 'the program objectives might be met earlier with higher initial funding. . . .'

In April, Houbolt circulated the MORAD plan (see Chapter 6) which was a product of the Langley Research Center subcommittee on rendezvous, set up originally to investigate the role of rendezvous in relation to space stations close to Earth, and from this came the proposed method of incorporating both Earth and lunar orbit rendezvous. When the Fleming and Lundin Committees were set up on May 2 and 25 respectively they had to consider essentially seven different ways in which astronauts could land on the Moon, finding from the list of candidates a single mode to recommend for further analysis. The Fleming Committee had been purposely set up to suggest a mission mode, but, given only a few weeks in which to do the job, they chose the easy way out and while concurring with the NASA view that a manned Moon flight was feasible claimed Direct Ascent as the best of all possible concepts.

The reasons they gave all had one thing in common: it was the best way of getting to the Moon with the least fuss and the lowest risk. Basically, Direct Ascent required a rocket to be built so large that it would ascend from Earth and accelerate to escape velocity, a speed of 40,200 km/hr, by dropping off two or three stages like several smaller rockets then being introduced. Coasting to the Moon, the spacecraft would turn around and fire rockets to decelerate and land. From that position, perhaps several weeks later, an upper stage engine would fire the spacecraft off the Moon, leaving the bottom section on the surface, and return the crew to Earth, accelerating first to lunar escape speed of 8,600 km/hr before discarding the propulsion module and coasting toward the atmosphere for a normal re-entry. Preliminary studies of a Direct Ascent spacecraft showed that it would be probably 20 meters tall, 6.5 meters in diameter, and weigh about 70 tonnes.

To send such a large vehicle to the Moon by this method would require a rocket producing at least 5,450 tonnes of thrust.

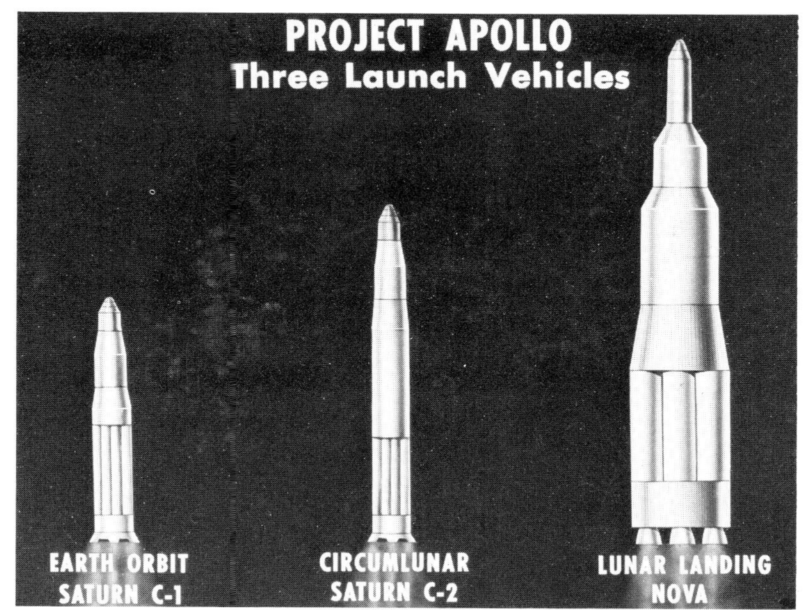

← Compared to the Saturn launchers conceived for earth-orbit and circumlunar roles, Nova represented a major step up in the size of a manned space launch system.

NASA had such a behemoth in the conceptual stage. Called Nova, it was the latest in a line of proposed super-boosters. The Nova design concept of 1959 had five F-1 engines at the base delivering a total thrust of 3,400 tonnes with second and third stages using a single F-1 and a single J-2 engine, respectively. Growing in increments to keep pace with the increasingly ambitious mission profiles seriously considered between 1959 and 1961, Nova was now expected to have eight F-1 engines in the first stage, four M-1 engines delivering a total thrust of 2,177 tonnes in the second stage, and a single 91 tonne thrust J-2 in the third stage. This enormous rocket would just be capable of throwing 70 tonnes to the Moon. But it was a monster, and should be compared with the existing Atlas (160 tonnes thrust) and Titan (181 tonnes thrust) for scale.

Guidelines set up for the Fleming Committee required it to find a 'feasible and complete approach' which it took to mean one which offered 'the greatest assurance of success based on existing experience, technology, and knowledge.' Rendezvous was an unknown area of space operations while the big Nova chosen for Direct Ascent was thought to be so closely based on the smaller Saturn rockets that it would be less of a gamble to go that route than via some as yet un-tried operation in orbit. Nevertheless, the Fleming Committee was the first to use the Sequenced Milestone System for integrating essential requirements with steps necessary to achieve the goal.

SMS was an adaptation of the Program Evaluation Review Technique (PERT) developed by the US Navy for the Polaris fleet ballistic missile. The result of several computer runs demonstrated that a manned Moon landing would be possible by Direct Ascent in July, 1967, using Nova. The basic Saturn C-1 would be employed for manned Earth orbiting shakedown flights beginning in late 1964, followed one year later by flights around the Moon with a larger Saturn called the C-3 before the first lunar landing with Nova. Embracing a program of 167 launches to support all the manned and unmanned missions necessary for developing operational expertise, the Fleming Committee recommendation cited a total cost of $11,683 million.

Direct Ascent had been the traditionally accepted method for carrying men to the lunar surface, and was the favored method at Headquarters. But the size of Nova remained outside the inventory of designs then being formulated at the Marshall Space Flight Center, NASA's only big booster facility. A one time supporter of Direct Ascent himself, von Braun was leaning now toward some form of rendezvous where modules sent first to Earth orbit on smaller rockets could be assembled for the Moon flight. Houbolt wrote to Seamans just 17 days after the Fleming Committee was set up, citing the lack of adequate booster power as good reason for concerted evaluation of various rendezvous techniques, all of which promised to save weight and reduce the necessary power of the rockets employed for such a task. Von Braun's engineers were working almost around the clock trying to juggle big engines and large boosters. The first super-booster was Saturn C-1, then being groomed for a launch later that year, which combined eight engines in the first stage to generate 680 tonnes of thrust. The C-2 would generate 1,360 tonnes of thrust from two F-1 engines in the first stage, but even this was being seen as too conservative a design for the projected needs of the space agency, so larger designs were prepared.

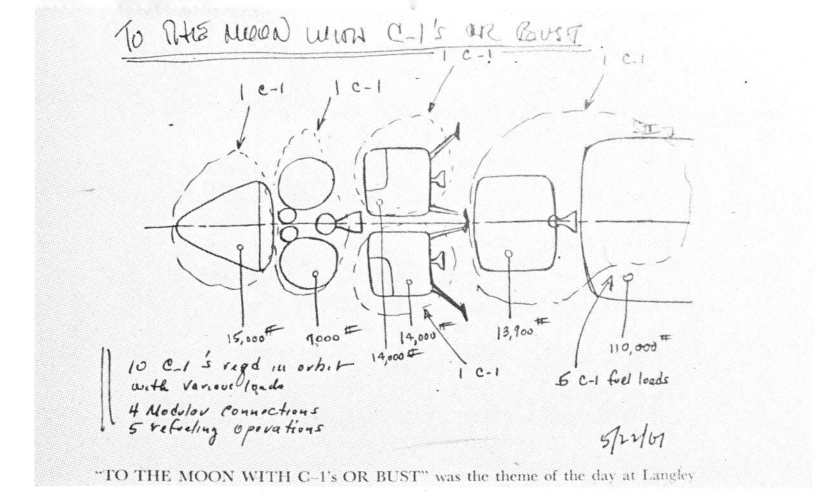

"TO THE MOON WITH C-1's OR BUST" was the theme of the day at Langley

↑ Proposed as a means of reaching the Moon via C-1 rockets, this method would require 10 separate launchers to assemble the vehicle in earth orbit.

↓ Incorporating earth- and lunar-orbit rendezvous, Saturn C-2 could support the lunar landing objective with only two launches: one for the hardware, the other for propellant.

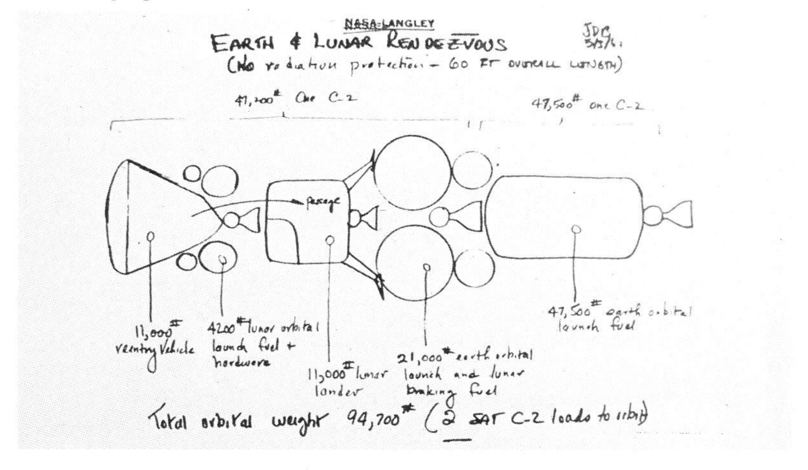

Before the C-2 disappeared, however, Houbolt's letter to Seamans caused ripples of concern: Seamans knew that Houbolt was right and that Direct Ascent, then being proposed by the Fleming Committee, would require many years in which to develop the Nova booster. So it was that on 25 May, the Lundin Committee was formed to go one better than Fleming and examine *all* the proposed concepts for Moon landing flights in the hope that a way could be found of cutting booster development time, seen at this period as the pacing item for the whole program. Direct Ascent was one of seven possible options. The others were in several ways even more ambitious in that, while demanding less booster power, they all required more sophisticated techniques and operational procedures, activities that looked prohibitively complicated in May 1961 when Alan Shepard had only just performed his ballistic flight into the Atlantic.

An unlikely contender because of the extreme nature of its requirements was the lunar-surface rendezvous mode; here, many Saturn C-1 launchers would land cargo pallets on the Moon so that a manned capsule descending to the surface would 'rendezvous' with equipment already sent up for the return trip, a risky process in the extreme and one that never really gathered serious support. A third concept was the so-called Earth Orbit Rendezvous (EOR) mode employing ten Saturn C-1 launchers to place in orbit about the Earth equipment which, upon being assembled into a single multi-modular spaceship, could be accelerated to the vicinity of the Moon. Houbolt's preferred application of the EOR concept was at variance with that of the von Braun group; the former believed it would be possible to put the Moon modules into lunar orbit first, from where a lunar lander could descend to the surface, while the latter wanted to assemble in Earth orbit a spacecraft identical to that which would be employed for Direct Ascent, merely launching it on several rockets rather than the one big Nova.

John Bird drew up a flight configuration for an alternate EOR mode, one which would employ two Saturn C-2 rockets to first place in Earth orbit the equipment necessary to fly to the Moon's surface. These two distinctive categories were soon conceptually separated into EOR and LOR: Earth Orbit Rendezvous and Lunar Orbit Rendezvous. The EOR concept would employ a single

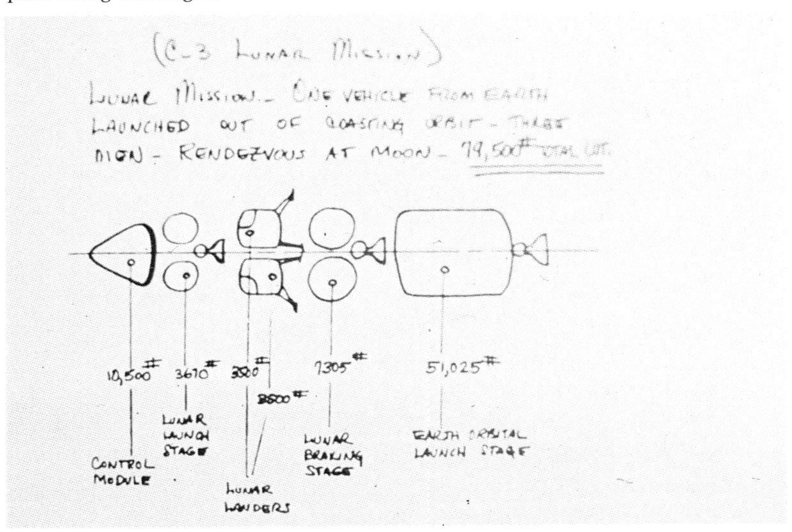

Adopting a proposed Saturn C-3 launcher, engineers sketched a Moon landing plan using one flight.

spaceship to fly to the Moon after having been assembled by the payload from two or more launchers in Earth orbit; the LOR concept would embrace those schemes that envisaged a lunar lander coasting to the surface from Moon orbit. Now, the two schemes from John Bird were firmly placed in the LOR category and formed the fourth and fifth options. One employing ten C-1 rockets, and another employing two C-2 boosters. The sixth option embraced a rocket not even designed at this stage: a large vehicle capable of lifting by itself the LOR payload of crew module, mission module, and lunar module. The final option was a variation on the basic EOR method whereby, instead of assembling the separate elements of a Moon landing spaceship in Earth orbit, the main spacecraft would be sent up first followed by a tanker to fuel the Moon-bound configuration.

In truth there seemed almost a surfeit of modes and methods by which astronauts could fly to the lunar surface. Yet in many respects the C-2 was just too small and von Braun withdrew it as a viable option for mission mode basing on 27 May, 1961, replacing it during the coming month with the C-3. Carrying two F-1 engines like the C-2, it would possess more powerful upper stages and increased payload lifting capacity. The Lundin Committee was the first to submit its findings, sending Seamans on 10 June their conclusions that endorsed some form of Earth Orbit Rendezvous using two or more Saturn C-3 rockets to send a spacecraft straight to the lunar surface. This was, after all, the most promising compromise between the Direct Ascent mode which required a prohibitively large rocket and the Lunar Orbit Rendezvous method which, with the currently available launchers, would require complex rendezvous operations to be performed both in Earth *and* lunar orbit; C-3, with its comparatively large capacity, could provide a multiple launch facility for a direct flight to the Moon from Earth orbit.

The Fleming Committee reported just six days later, and endorsed lunar landing flights as technically feasible objectives for the 1960s. Four days after that, on 20 June, 1961, Seamans set up an Ad Hoc Task Group to define the parameters of a plan for flying to the Moon in 1967 using the EOR concept with Saturn C-3 launchers. Chaired by Donald H. Heaton from Don R. Ostrander's Office of Launch Vehicle Programs, it comprised members from the OLVP, the Office of Space Flight Programs, the Office of Advanced Research Programs and the Office of Life Sciences Programs. Meanwhile, acting on the recommendations of the Fleming Committee, in which doubt had been expressed over the confident assumption that massive boosters could be effectively handled, Seamans asked Kurt Debus, Director of the NASA Launch Operations Directorate, to perform a joint analysis with Maj. Gen. Leighton I. Davis, Commander of the Air Force Missile Test Center at Cape Canaveral, on the feasibility of big booster operations.

Known as the Debus-Davis study it made great progress over the following weeks in preparing ground rules for launching large rockets, paved the way for pad design, and supported the projected feasibility of erecting, preparing, and launching, rockets of the C-3 and Nova class, boosters for which the Debus-Davis study had been commissioned. Yet throughout these early months of the Moon commitment, it was the availability of launch vehicles that weighed heavily in the decisions on which mode to select. By mid-year, it was a fight between Direct Ascent and EOR, with the latter just a neck ahead, and NASA realized it would need more positive information on the advisability or otherwise of planning massive boosters. Accordingly, on 7 July, NASA and the Department of Defense set up a DoD-NASA Large Launch Vehicle Planning Group which would report to the NASA Associate Administrator and to the Assistant Secretary of Defense. The Group's Director was Nicholas E. Golovin from NASA, with a deputy, Lawrence L. Kavanaugh, from the Office of the Director of Defense Research and Engineering.

Charged with responsibility for selecting optimum launch vehicle configurations for a wide range of NASA and DoD requirements over the coming years, it soon became the forcing house for even larger designs from the Marshall Center. For at least a year, von Braun's engineers had been working toward increasingly powerful boosters and in accordance with the sudden emergence of the Moon landing commitment the Center was turning out increasingly ambitious concepts. When the Golovin Committee, informally named after its Director, pursued studies of these ambitious designs, it was an enthusiastic customer for even larger concepts. By the end of August the Committee informed various NASA centers that it wanted supporting analyses for several different Moon flight modes. Marshall was to study the EOR concept, Langley was to examine LOR, Headquarters would put its Office of Launch Vehicle Programs on the Direct Ascent mode, and the Jet Propulsion Laboratory would report on lunar surface rendezvous.

Meanwhile, also in August, the Heaton Committee reported its conclusions, supporting a Saturn C-4 for Earth Orbit Rendezvous operations; the C-4 would have four F-1 engines for a total first stage thrust of 2,720 tonnes and a second stage thrust of 363 tonnes. Two C-4 rockets would be needed to reach the Moon, one carrying the modules for habitation and the other propellant for getting off the surface. Using the C-4, the Heaton Committee said a Moon mission would be possible by mid-1966 for a total program cost of $9,883 million. If a Direct Ascent mode was developed alongside the EOR concept, as a hedge against problems, that would add a further $3,034 million. Direct Ascent alone was believed to cost $11,683 million.

Rendezvous was clearly the favorite, but a joining together in Earth orbit of separate sections of the main spacecraft rather than the use of a small lunar landing vehicle from Moon orbit. Lunar Orbit Rendezvous was just too risky a venture for the conservative minds who pereferred to keep the complex operations close to Earth. It had too many unknowns, and few had confidence in the ability of astronauts to perform elaborate manoeuvres around another body in space far removed from Earth. Nobody had yet shown how rendezvous could be performed. Was there perhaps some insidious drawback to the whole idea that separate spacecraft could be brought together in space? While every step of a Moon landing operation could be fully tested in Earth orbit first, it would be catastrophic to the program for NASA to pursue an operational cul-de-sac. So the urge to keep rendezvous away from lunar landing operations persisted, and when the Golovin Committee reported, it too endorsed an Earth Orbit Rendezvous plan where separate sections could be brought together around the Earth – but with Saturn C-3 launchers rather than the larger C-4.

But developments elsewhere in the agency were priming NASA facilities for the inevitable re-organization essential to a successful Moon project. On 22 June, 1961, Administrator Webb met with Hugh Dryden and other senior officials to formulate policy planning profiles to get Apollo off the ground. From the several committees that had already reported back it was obvious the project would be immense in scope and requirements. On 6 July, Bob Seamans set up a Manned Lunar Landing Steering Committee to begin immediately the job of organizing the space agency for the new role prior to formal adjustments then being planned.

In three meetings held during July, the committee streamlined development of the Saturn C-3 by advising von Braun to stop work on the detailed design of Nova, developed mission planning needs, and finalized the configuration preferred for C-3. But for all that, the Golovin Committee's close examination of the broadly based need for large rockets added the last remaining perspective on mode selection criteria: Fleming had studied Direct Ascent and costed a project on that basis; Lundin offered choices of rendezvous and costed those; Heaton looked at the broad spectrum of both categories and mixed the options to generate a third set of figures. Now, Golovin ignored operational preference and presented the first realistic evaluation of the program from a launch vehicle consideration.

The committee found that dates projected by the Fleming and Heaton studies were optimistic, based on spacecraft development, and ignored the time required to develop the big boosters. Golovin found the Lunar-Orbit Rendezvous mode would absorb 28 Saturn C-1 flights and 38 missions with the C-4 for a landing in October, 1967, at a total cost of $7,330 million. The Earth Orbit Rendezvous mode, however, would require 32 Saturn C-1 flights and 53 C-4 missions. A landing could be made in July, 1968, at a cost of $8,160 million. The Direct Ascent mode was the cheapest, at $6,390 million, but could not guarantee a landing before October, 1968, and would absorb 22 C-1 missions and 38 Nova flights using the configuration with eight F-1 engines in the first stage. In all these estimates, the number of launch vehicles referred not to the total used per landing mission but the quantity expended for qualifying the rockets and rehearsing the operations.

By October, 1961, Harvey Hall, Chief of Advanced Development in the NASA Office of Launch Vehicle Programs, completed the study he had been asked to perform for the Golovin Committee conclusions from the separate field analyses. Hall's subcommittee favored the rendezvous concept, and even suggested a re-examination of Lunar Orbit Rendezvous, when it reported back to Golovin on 10 October. But for most NASA officials the thought of such a radical concept was not commensurate with the President's mandate to get to the Moon as fast as possible. Nevertheless, in continued studies of launch vehicle developments, the von Braun group came up with yet another concept, the C-5. This would be essentially a developed version of the C-4, with five F-1 engines in the first stage, five J-2 engines in the second stage and a single J-2 in the third stage. The larger C-5 had been prompted by the increasing weight predictions for the Moon spacecraft, and the three-stage launcher promised to send more than 40 tonnes to the Moon. It was still far short of the 70 tonnes required for the Direct Ascent mode, but Langley could hardly believe their eyes. In the process of increasing the size of projected boosters, to match the Earth Orbit Rendezvous requirements they favored, the Marshall Center had given John Bird's group the biggest prize of all.

Only a few quick calculations were needed to show that Saturn C-5 was capable of sending a spacecraft to the Moon on its own! Until now, Lunar Orbit Rendezvous had required the use of at least two smaller Saturns, first to put into Earth orbit the separate vehicles that would go into Moon orbit, and then to the surface. Now, a launcher was proposed that could do the entire operation in a single flight. It did, in fact, get the Marshall engineers out of a spot of trouble with the C-4 design. By placing four F-1 engines in a square pattern – the only logical arrangements for such a big rocket – a disturbing backflow from each engine would produce excessive heat in the space at the center. With a fifth F-1 now placed at that location, it would produce additional thermal energy to prevent the backflow from creeping up into the base of the rocket. It was also suggested that structural mounting of the fifth F-1 would cause little problem if a cross-beam arrangement provided to support the existing four engines was made to support the additional motor.

In anticipation of the need to integrate more fully the increasing array of launcher proposals with the evolving requirements of the Office of Manned Space Flight, Milton W. Rosen from the OMSF Launch Vehicles and Propulsion section, informed D. Brainerd Holmes that he had set up a working group to sift the bewildering inventory of launchers. Rosen went all out for the C-5 and visited the Marshall Center in November during a hectic two weeks in which he put together all the wise counsel proffered by the Fleming, Lundin, Heaton, Debus-Davis, and Golovin Committees. On 20 November, Rosen informed the new Manned Space Flight boss that rendezvous was a prime mover in getting to the Moon and that the Saturn C-5 should be developed with all speed so that it could be used for lunar landing missions by rendezvous, but here Rosen was including the Earth Orbit rather than Lunar Orbit Rendezvous mode. Rosen also said that the primary mode should be the Direct Ascent method and that towards this end the agency should start work on a Nova class launcher comprising a first stage with eight F-1 engines, a second stage with four M-1 engines, and a third stage with a single J-2. But not everyone agreed with Rosen and Seamans reflected the changing view at Headquarters when he leaned more toward the conclusion of the Golovin Committee that preferred some form of rendezvous operation using rockets less powerful than Nova.

Although unwritten in reports and submissions from the many committees and study groups set up in 1961 to examine lunar flight modes, an attractive feature of the Direct Ascent method was undoubtedly the enormous size required of the launcher built for such a mission. Clearly, some form of rendezvous operation would dramatically increase the payload of a specific rocket so if NASA developed a Nova class booster for Direct Ascent missions to the Moon, it would be well on the way toward possessing a launcher for more ambitious operations necessitating greater lifting capacity later on. It would be more desirable to invest in a booster that had very real possibilities for growth in mission potential than one tailored to a single objective. Few thought of this at the time; many would wish they had later when post-Apollo planning left a void in the availability of heavy launchers. However, the Rosen recommendations of 20 November did include reference to a three-engine version of the Saturn, a configuration more closely associated with a broader range of NASA and Defense Department missions than the five-engined C-5. But that was peripheral to the central issue of lunar landing flight modes.

On 15 November, 1961, Houbolt wrote again to Seamans, this time in a private letter addressed direct to the Associate Administrator in person, in which he poured forth his dissatisfaction with the way LOR was pushed to the bottom of the pile in discussion and debate about the method to select. He appealed to Seamans' logic by passionately defining the way in which LOR could be made to accelerate the pace of Moon landing preparations dramatically; he lambasted his colleagues for their hostile attitude toward the concept and strongly protested the treatment he had received in committee sessions aimed at resolving the issue. It was in every respect a desperate plea for a fair hearing, an appeal for ears to listen with because until now, he said, he had been just a 'voice crying in the wilderness.' Seamans was impressed by the sincerity of Houbolt's epistle and passed the letter to Holmes with instructions to see that better evaluation of *all* the proposed landing modes preceded commitments on launchers or spacecraft design.

The first half of 1962 was spent turning around from Direct Ascent and Earth Orbit Rendezvous to Lunar Orbit Rendezvous and the advantages it promised. Before the old year was out, on 21 December, Holmes formally announced a new Manned Space Flight Management Council, including Bob Gilruth and Walt Williams from MSC, Wernher von Braun and Eberhard Rees from the Marshall Center, George Low and Milton Rosen from Headquarters, and Joseph F. Shea. Shea was a new boy from Space Technology Laboratories, Inc., and was Deputy Director for Systems Engineering.

By now, NASA had decided that the C-5 would be developed and on 15 December, 1961, the Boeing Company got the contract to build the first stage; six days later at the first meeting of the Manned Space Flight Management Council it was decided that C-5 would have five F-1 engines in the first stage. Less than a month later Seamans received from Holmes a preliminary C-5 development plan for the vehicle that would, 'by two launches, inject enough payload into Earth orbit to accomplish the desired mission with one rendezvous.' The final configuration was announced 25 January and although Nova studies would continue for several months the big booster was already dead. Holmes was impressed, like Seamans, with the letter from Houbolt to the extent that he ordered Shea west to visit the Langley Research Center and discuss the LOR concept further with its adopted author.

On 13 February, 1962, Headquarters began a three-day meeting at which the whole question of rendezvous was integrated with the Gemini program, now an official successor to Mercury. A month later the Marshall Space Flight Center completed an initial program survey for the Saturn C-5, optimistically asserting its confidence in a scheduled first launch late in 1965 and the first manned flight with an Apollo spacecraft on top during the last quarter of 1967, but this assumed the massive launcher could be man-rated on the eighth flight, a process that required full systems clearance for manned space flight. Contractors were now working on the final design details and launch facility design was nearing completion. Clearly, other elements of the program, mission mode selection, spacecraft design concept, etc., would have to appear soon to keep pace with rapid developments in the launch vehicle schedule. But not only the booster awaited decisions essential to its development, for the Apollo spacecraft too was moving ahead.

Before Kennedy turned Apollo into the Moon landing program, Mercury's one-time successor was seen as a spacecraft concept defined in only the vaguest terms. It was generally agreed within NASA that the Apollo objective should be a two-fold capability: Earth orbital operations with a command center module supporting a three-man crew, a propulsion module for changing orbits, and a mission module in which scientific equipment could serve to provide the functions of a small orbital laboratory; and the lunar operation where command center module and propulsion module would have the capacity to carry three men around the Moon and back or, in some undefined role, place them in Moon orbit prior to accelerating crew and propulsion modules back to Earth.

It will be recalled that NASA had been unable to gain firm funding profiles from the Eisenhower administration for what was seen as an engineering exercise to keep the space agency in the arena of manned flight. The fact that, after 25 May, 1961, Apollo became the Moon landing program was the consequence of it being the next project in line to Mercury. It would have to be substantially changed in concept to achieve that objective. The original Apollo objective had been publicly announced in July, 1960, at a combined government-industry conference and in October of that year, General Electric, The Martin Company, and General Dynamics had been selected for six-month feasibility studies of the advanced manned spacecraft. That work officially began 15 November and was concluded by mid-May, 1961, just 10 days before Kennedy publicly announced the Moon landing goal.

Two companies – General Electric and General Dynamics – preferred a crew command center module shaped more like an aeroplane than a spacecraft, a configuration known as a 'lifting body' in which a combined wing and fuselage blended into a shape capable of generating lift. The Martin study produced a ballistic cone-shaped structure like Mercury, but with flaps to modulate the angle of the capsule and so influence the re-entry trajectory. But these studies were suddenly made redundant by the grander and more ambitious objective and no time could be lost in choosing a contractor to build the definitive Apollo, which at that time was seen as a space vehicle of three elements albeit each somewhat different to the center command/propulsion/mission module arrangement. The new Apollo concept would still need a crew command module, and a three-man arrangement seemed appropriate too, but the propulsion section would now have to be a landing stage, separated from the crew compartment by an equipment module carrying the supplies. This was the Direct Ascent concept where the entire spacecraft would lower itself to the Moon's surface.

Ironically, all thoughts of introducing a more sophisticated means of descent – one based on some form of lifting surface rather than a ballistic cone – would have to wait for the next manned spacecraft requirement. Back in 1958, NASA had been unable to adopt the technically preferable lifting-body shape, or glider, because of the need to outpace Soviet developments with a project that promised to get a man in space at the earliest opportunity. Now, Apollo too was on a time schedule that prevented NASA taking the lifting-body as the selected shape. Lifting-bodies needed considerable design and development time unavailable within the Apollo mandate, so it seemed even in May 1961 that the crew module for the Moon landing vehicle would look more like the Mercury capsule than any more revolutionary form of re-entry design. If boosters and mission modes were still up in the air, spacecraft design had to be based very firmly on the ground from the beginning. It would be more than a year yet before NASA finally chose the way it would fly to the Moon but a contractor would have to get busy on the drawing boards long before that if the spacecraft was to be ready on time for the great event.

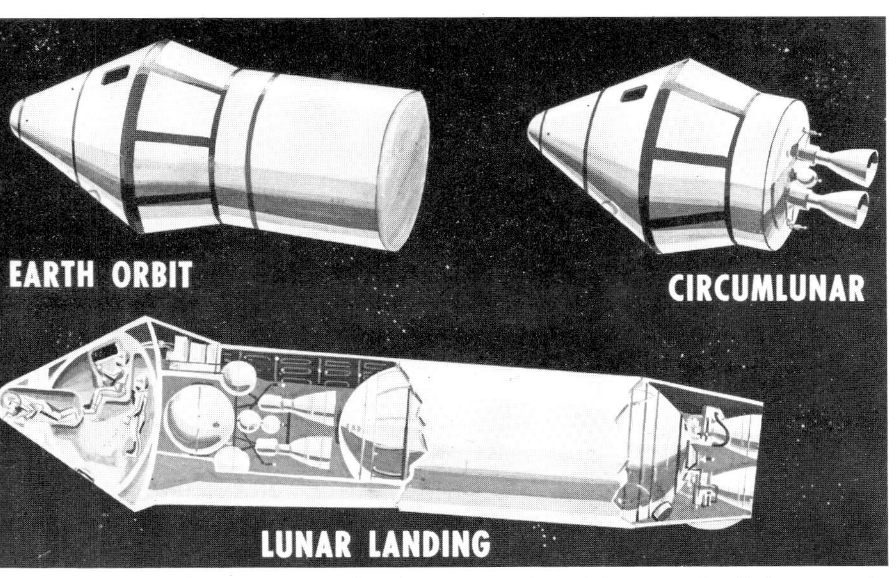

Seen originally as an earth-orbit or circumlunar laboratory, Apollo's new objective gave it a lunar landing role satisfied, in theory, by the addition of a separate descent section.

Within a week of President Kennedy's statement, a major gathering at NASA Headquarters defined the requirements of the new program, embracing several specific projects, not least of which was the actual spacecraft. In a three-day meeting beginning 18 July, 1961, the three feasibility contractors who had so recently completed studies on the older, less demanding, Apollo requirement, presented at an Apollo Technical Conference in Washington, D.C., results and recommendations to 300 potential contractors keen to get in on the new manned project. At the meeting, attendees were told that NASA expected to define the Apollo spacecraft within the next few weeks and that it would publish a final specification. No time could be lost in choosing a prime contractor because although the precise details of the mission mode to be adopted would be essential to the final configuration of Apollo, considerable work could begin on elements of the spacecraft that would be unchanged by any preference for the way the vehicle would perform its landing activity.

The crew compartment could be designed, maneuvering engines could be developed, environmental control systems could be defined, thermal protection could be selected. During July 1961, NASA Headquarters approved the Statement of Work and the Procurement Plan laid out for distribution to companies invited to submit proposals. The Statement of Work explained that submissions would be required in three parts: a technical proposal, a business management proposal, and a cost estimate and cost control proposal. On 28 July, twelve companies were invited to respond to a request for proposal on Apollo, and were required to have their documents with NASA by 9 October. They were, the Boeing Airplane Company, Chance Vought Corporation, Douglas Aircraft Company, General Dynamics/Convair, the General Electric Company, Goodyear Aircraft Corporation, Grumman Aircraft Engineering Corporation, Lockheed Aircraft Corporation, McDonnell Aircraft Corporation, the Martin Company, North American Aviation, Inc., and Republic Aviation Corporation.

In the Statement of Work, three separate phases were outlined: Phase A would embrace manned Earth orbit flights, launched by Saturn C-1, of up to two weeks' duration to effectively qualify all the spacecraft systems, and re-entry from high elliptical orbits to test the ability of the heat shield to withstand severe heating effects; Phase B, where Apollo would fly circumlunar, lunar orbit and parabolic re-entry tests employing Saturn C-3 rockets; and Phase C, the manned lunar landing operation performed by either Saturn C-3, Earth Orbit Rendezvous, or Nova, Direct Ascent, modes – at this time the Lunar Orbit Rendezvous method was still an outsider. All Statement of Work documentation was completed by 2 August, and appropriate specifications were sent to the 12 potential contractors invited to bid; four additional companies were sent Request for Proposal documents upon request: Radio Corporation of America, Space General Corporation, Space Technology Laboratories, and Bell Aerospace Systems. As now defined, Apollo would comprise a command module, supporting in a pressurized environment the three crew members that would fly to the Moon, a service module, being an unpressurized cylindrical structure containing all the life support, electrical, propulsion, and cooling systems, and a lunar landing module, comprising propulsion systems for descending to and lifting from the Moon's surface and landing legs upon which service and command modules above would rest.

In the light of later events it is interesting to note that at this time the command module was considered to need a two-gas atmosphere pressurized at roughly half that of the sea-level envi-

Viewed from left to right, Apollo backs down to a horizontal landing in this artist's concept based on the modified pre-1961 design.

ronment: 362 mmHg versus the 760 mmHg on Earth. Mercury adopted a pure oxygen, single-gas, atmosphere pressurized at 258 mmHg for space but purged with pure oxygen at 775 mmHg on the launch pad to prevent nitrogen from further reducing the oxygen content of the pressurized cabin (for further details see Chapter 3). The choice for Apollo of a 50-50 mixture of oxygen and nitrogen, where oxygen would comprise 180 mmHg of the total pressure, was based on the suspicion that a two-week exposure to pure oxygen could induce pulmonary atelectasis, or collapse of the lung tissue. Ideally, the spacecraft design would benefit from a single-gas atmosphere pressurized to the lowest level permissible physiologically. The lowest acceptable pressure level was known to be 150 mmHg, but the effects of lengthy exposure to the single gas were unknown.

Systems engineers disliked the complex two-gas atmosphere and concerted efforts were made to determine the effect of a two-week exposure to pure oxygen. The project to determine these effects was carried out by the Republic Aviation Corporation at Farmingdale, Long Island, New York, by the US Air Force School of Aviation Medicine, and the US Navy Air Crew Equipment Laboratory. The National Academy of Sciences established a select committee to examine the tests and evaluate the results, finding that in the several evaluations made with six volunteers exposed to pure oxygen at varying pressures for 14 and 30 days, a 258 mmHg pure oxygen atmosphere was acceptable. No deleterious effects were observed. Accordingly, because it was physiologically acceptable to fly three men for at least 14 days with a single-gas atmosphere, and because a preliminary study showed a weight saving of 16 kg from deleting a second gas in the Apollo command module atmosphere, NASA formally adopted a pure oxygen environment pressurized at 258 mmHg for space operations, retaining the high-pressure purge adopted for Mercury and now planned for Gemini pad operations. At no time was it considered necessary to dilute the oxygen atmosphere for fear of combustion within the capsule. It was well known that fires burning in a pure oxygen gas are almost impossible to extinguish, but the dangers of such a conflagration starting in the first place were not accepted as tenable postulates. As will be seen in a later chapter, a minimal effort was made to reduce the quantity of combustible items carried inside the cabin. But all this was still

Used before the House space committee in March, 1962, this illustration shows proposed configurations for the Apollo mission.

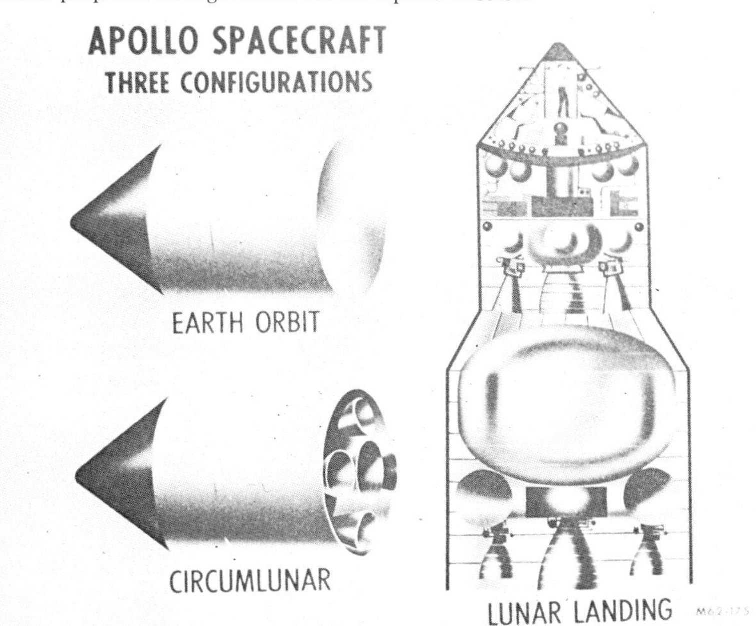

ahead when between the beginning of August and the end of the first week in October, 1961, five companies of the 12 invited responded to NASA's request for Apollo contract proposals.

On Monday, 9 October, NASA formally accepted the submissions as time ran out for prospective proposals on how to build America's Moonship. On the Wednesday of that week executives from 12 companies met NASA officials in the Virginia Room of the Chamberlain Hotel at Old Point Comfort, Virginia. Submissions were grouped into three team efforts: General Dynamics with Avco; General Electric, Douglas, Grumman, and STL; McDonnell, Lockheed, Hughes, and Chance Vought. Two were based on the prime subcontractor approach where only one company controlled the project: Martin, and North American Aviation. After the formal reports, eleven separate panels responsible to Business and Technical Sub-committees began an intensive evaluation of the documentation associated with each of the five submissions. The Technical Assessment Panels had completed evaluation of engineering criteria by 20 October and within five days had submitted their recommendations to the Technical Subcommittee.

On 1 November the Source Evaluation Board received the final reports. Members were to examine the recommendations and findings of the Technical and Business Subcommittees, appointed by the Board, which in turn had called upon panels of specialists. In all there were about 190 people from NASA and the Defense Department. The Source Evaluation Board was formed from a group of 12 persons, chaired by Maxime Faget and including Bob Piland, George Low, Kenneth Kleinknecht, Charles Mathews, James Chamberlin, Wesley L. Hjornevik, A. A. Clagett, Dr. Oswald Lange, D. W. Lang, James Kippenhaver, and ex-officio member Bob Gilruth. Their job was to award a system of points to each submission in each of three separate categories: Technical Qualifications, Technical Approach, and Business Management and Cost. The first referred to the experience offered by the respective candidate or team in the technical performance displayed previously by the contractor or companies. The second pertained to the Apollo submission and evaluated the technical competence of the proposed design. The third would rate the management structure of each company, examine and mark its business efficiency, and analyse the proposed funding structure with a view to determining the integrity of each cost projection. From the combined totals of all three categories, the Board would submit summary findings with a recommendation for a specific contractor passed up to NASA management.

In its final evaluation, the Board concluded that 'The Martin Company is considered the outstanding source for the Apollo prime contractor. Martin not only rated first in Technical Approach, a very close second in Technical Qualifications, and second in Business Management (to General Dynamics), but also stood up well under the further scrutiny of the board. The Martin Company appears to be well prepared to undertake the Apollo effort. This was evidenced by a Technical Proposal that was complete, well integrated with balanced emphasis in all areas, and of high quality overall with a minimum amount of superfluous material. Martin's proposal was first in five of the eleven major Technical Approach areas . . . Martin, therefore, scored high in planning, design, manufacturing and operations, reflecting the quality across the complete scope of the job . . . Their inhouse experience in many of the required technical areas results in a high confidence as to their capability of a systems integrator. The individual key technical personnel Martin proposed to assign to the project were evaluated as excellent both in competence and experience. Martin's proposed management arrangement of a prime contractor with subcontractors appears technically to be the most sound both as far as reaching technical decisions quickly and properly and also for implementing these decisions . . . Martin proposed a strong project organization for Apollo . . . Martin's cost proposal compared well with the others. Their cost estimate was considered to be both realistic and reasonable.'

In fact, Martin had been third highest out of the five submissions, projecting a total cost of $563 million; North American Aviation had been lowest, at $351 million. The Board also noted that 'North American Aviation, Inc., is considered the desirable alternate source for the Apollo spacecraft development,' but said that 'Their project organization, however, did not enjoy quite as strong a position within corporate structure as did Martin's. The high Technical Qualifications rating resulting from these features of the proposal was therefore high enough to give North American a rating of second in the total Technical Evaluation although its detailed Technical Approach was assessed as the weakest submitted . . . North American submitted a low cost estimate which, however, contained a number of discrepancies.' The Source Evaluation Board decided that although there were indeed attractive features contained within the proposals from the three team submissions, the spread of capability throughout two or more separate companies necessary to achieve equality with the single submissions of Martin and North American Aviation should, in the light of good results from analysis of the other two, result in their elimination. Accordingly, the teams led by General Dynamics, General Electric and McDonnell were not recommended.

It is interesting to note that in the category of Technical Approach, that which, it could be argued, carried more relevance than any other, Martin came first, followed by McDonnell, General Dynamics, General Electric, and finally North American Aviation. In terms of Business Management and Cost, General Dynamics came first, followed by Martin, General Electric, McDonnell, and North American Aviation. Only in Technical Qualifications, the category based on company experience, did North American Aviation not come last. But the Board was adamant, it seems, in eliminating team submissions and put forward The Martin Company as outstanding first choice, with the only other lone submission, from North American, pulled in as second.

North American had been after the Apollo contract for several months. J. Lee Atwood, NAA's Chairman was keen for the Space and Information Systems Division to have a go and SISD's head, Harrison A. 'Stormy' Storms, agreed with Atwood on the general direction the company was heading. In fact, Storms had seen to it that the Space Division had the right people on hand when NASA went over the corporate executive sheet. As Storms' number two, Dale D. Myers, ex-program manager for the Hound Dog missile, was a comparative newcomer, having been appointed to the position only a year before. But his presence so close to the top was a key factor in structuring the NAA management list. Storms had a dynamic approach to project management and had ridden through trial and tribulation on many government contracts. He was not at all sure that North American was ready to take on something as demanding as Apollo but, assured of a place at the helm, he backed Atwood when the NAA Board gave the venture their blessing.

During November, 1961, Jim Webb, Hugh Dryden, and Bob Seamans heard the Source Evaluation Board put forward their findings. The three then retired to another room, taking with them the head of NASA's procurement section, the senior attorney to the space agency, and Bob Gilruth who, as head of the Manned Spacecraft Center, would be responsible for working with the selected contractor. Board appraisals were received on the 24th and just four days later NASA formally announced that North American Aviation had won. NAA would receive a prime contract immediately for the command and service modules, and a further contract for the landing module in due course; by this time, it will be recalled, the agency was beginning to listen hard to Houbolt and his Lunar Orbit Rendezvous plan which would, if adopted, require another prime contractor to build a separate lunar module.

Jim Webb had made the final choice and in overturning the recommendation of the Source Evaluation Board put greater emphasis on the company's technical experience, choosing to rewrite the score-sheet by allocating points for NAA's corporate record rather than that of the Space Division alone. In selecting the root reasons why he decided to award the contract to North American, Webb says they were based on the management list where 'specific persons proposed to do the work were judged to be better qualified' and that 'North American submitted the lower cost proposal . . .' During their preliminary assessments of the five submissions, the Board had been concerned that existing government work might interfere with Apollo activity, requesting explanations from several companies about the management assignments should they be rewarded with a contract. In the case of NAA, the Board questioned the recent contract for a large rocket stage designed to fit on top of a large Saturn booster. Called the S-II, the stage was to be built by the Space Division at Seal Beach. It was a major development effort in that it was to contain cryogenic propellants (liquid hydrogen and liquid oxygen) for consumption in five J-2 engines for a total stage thrust of 454 tonnes. Nobody had built such a large stage to carry such difficult liquids and it would be the most powerful high-energy rocket system anywhere in the world. NAA assured the Board that there would be no problem in manpower allocation and management chains as had Martin when similarly asked about their existing commitments with the growing family of Titan boosters.

Rapidly now, the major contracts were being let for NASA's big assignment: NAA would build Apollo, McDonnell Douglas would continue with their Mercury Mk. II work in the form of the two-man Gemini capsule, Chrysler was to build the first stages for Saturn C-1, NAA and Douglas would build the Saturn C-5 second and third stages respectively, and, in a contract awarded 15 December, 1961, Boeing got the job of putting together the massive Saturn C-5 first stage with a thrust of more than 3,400 tonnes.

Apollo evolved through conceptual changes made to support re-directed mission goals, ultimately embracing a separate lunar landing module.

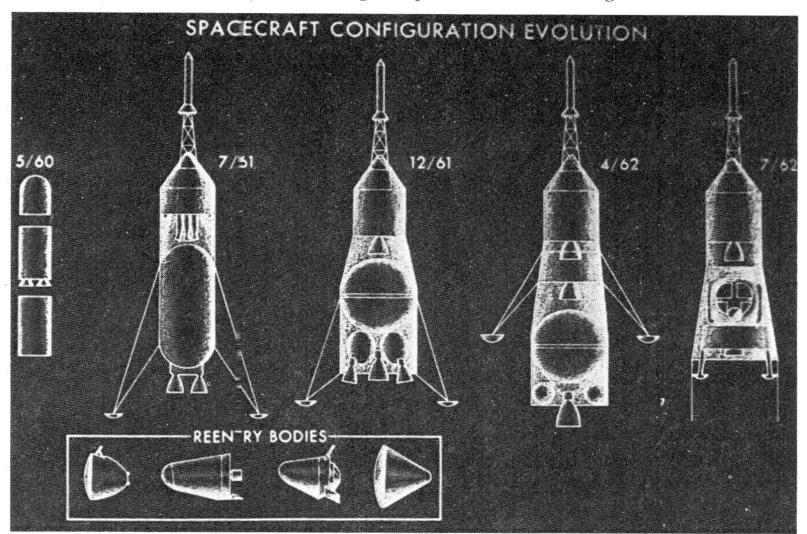

Rocketdyne was to supply all the engines for the entire family of Saturn launchers.

So it was that when Joseph Shea returned from visiting John Houbolt at the Langley Center early in 1962, and reported favorably to his boss, Brianerd Holmes, on the LOR concept. North American cast a wary eye toward Seamans and his bosses Webb and Dryden. North American had come in on Apollo to do the whole job, not to give up the Moon landing contract to another company brought in to build the lander. If NASA moved across to LOR, the main Apollo spacecraft would go no further than lunar orbit. During the early months of 1962 NASA passed Houbolt's figures on LOR to several leading analysts, many respected laboratories, and not a few respected engineers, requesting opinion and comment. It all seemed to favor the Lunar Orbit Rendezvous approach. Availability of the big Saturn C-5 was the real advantage in a concept that had, up to late the previous year, required the assembly of elements in Earth orbit first from the payload of several smaller Saturns. The entire operation seemed complex and risky, until it was shown that the 40 tonne lunar payload capacity of the C-5 could simplify the whole operation by launching Apollo and a lunar lander in one go. There would still be the delicate sequence of separation and landing for the smaller spacecraft in Moon orbit, and then the ascent and rendezvous with the orbiting mother-ship, but the whole operation began to look very much better in 1962 than it had the previous year.

By April Bob Gilruth's Manned Spacecraft Center was convinced of the advantages in LOR, seeking a meeting with von Braun's Marshall men to put the case and try to get a consolidated front. Seamans, Holmes and Shea were already sympathetic to Houbolt's idea, but Gilruth would have to run the mission and von Braun would have to supply the boosters. Marshall was still firmly following the Earth Orbit Rendezvous line, preferring to use Saturn C-4 or C-5 vehicles for sending into Earth orbit the spacecraft and propellant on separate missions, keeping all the complicated rendezvous activity close to Earth. From there the spacecraft would fly to the Moon and land on the surface without any part first going into orbit.

Wiesner too was in favor of EOR, although he even pleaded the case for Direct Ascent where no rendezvous of any kind would be necessary, the whole assembly launched by one monstrous rocket from the surface of the Earth. Von Braun's engineers even came up with a suitable booster. Called the Saturn C-8 it only existed for a few months, and then only on paper. But it was a behemoth in every dimension. The first stage would have carried eight F-1 engines – and each F-1 was as powerful as the eight-engined Saturn C-1! Total thrust would have been a phenomenal 5,443 tonnes, about 60% more power than the C-5. In every design aspect it was equivalent to the Nova quoted so frequently when talk about Moon flight wandered to the Direct Ascent mode.

But the only sensible contenders were EOR and LOR, the Direct Ascent mode demanding too much rocket power to make it a wise choice where time was of the essence. When summarizing the Apollo contract proposals in November the previous year, the Source Evaluation Board had concluded that whereas the first manned Apollo flight was scheduled for mid-1964 on a Saturn C-1 (the eleventh flight), availability of the C-4 or C-5 could not be expected before 1966 at the earliest and that the Nova class launcher would not appear before 1967/68. Any slight delay would put the Nova-class launcher beyond the end of the decade already marked out for Moon flights.

A lot hinged on what the von Braun team thought and the matter reached a peak of expectancy on 7 June, 1962, when Joseph Shea visited Marshall and heard the conclusions of the rocket men. In a series of presentations that lasted well into the afternoon, von Braun summarized the view by endorsing the Lunar Orbit Rendezvous mode because it 'offers the highest confidence factor of successful accomplishment within this decade,' and because 'it offers adequate performance margin.'

Von Braun also hit a critical nerve in the whole equation when he offered the view that 'The designs of a maneuverable hyperbolic re-entry vehicle and of a lunar landing vehicle constitute the two most critical tasks in producing a lunar spacecraft. A drastic separation of these two functions into two separate elements is bound to greatly simplify the development of the spacecraft system.' But the Marshall engineers did want to see a logistics vehicle sent to the surface ahead of the Moon landing mission to insulate the flight more effectively against possible failure. Von Braun was adamant about the eventual need for bigger rockets, however, stressing that: 'Our recommendation against the Nova or C-8 mode at this time refers solely to its use as a launch vehicle for the implementation of the President's commitment to put a man on the Moon in this decade. We at Marshall feel very strongly that the Advanced Saturn C-5 is not the end of the line as far as major launch vehicles are concerned! Undoubtedly, as we shall be going about setting up a base on the Moon and beginning with the manned exploration of the planets, there will be great need for launch vehicles more powerful than the C-5.

'But for these purposes such a new vehicle could be conceived and developed on a more relaxed time schedule. It would be a true follow-on launch vehicle. All of our studies aimed at NASA's needs for a true manned interplanetary capability indicate that a launch vehicle substantially more powerful than one powered by eight F-1 engines would be required. Our recommendation, therefore, should be formulated as follows: "Let us take Nova or C-8 out of the race of putting an American on the Moon in this decade, but let us develop a sound concept for a follow-on 'Supernova' launch vehicle".'

Only the previous month, MSC had successfully won support from Seamans and Holmes who, being in elevated positions ostensibly removed from decisions more properly made by the field centres, were reluctant to openly commit the agency to one particular concept against another. Soon after von Braun gave his approval to LOR, Wiesner began a move inside his Science Advisory Committee to generate hostility toward a concept that would remove from North American Aviation the pride of building America's Moonship.

But NASA was getting advice from several outside contractors asked to provide independent analyses of the mode selection criteria. In May, Chance-Vought reported to Headquarters and personnel at Marshall and Langley. Two days before Shea met with the von Braun team, a Grumman-RCA team informed Headquarters on its own conclusions of the LOR mode, and on 15 June Douglas Aircraft told NASA about their own results. On 16 June, NASA's Office of Systems Engineering prepared a Manned Lunar Landing Mode Comparison study, a sort of balance sheet with arguments for and against each candidate method of reaching the Moon.

It came out in favor of Lunar Orbit Rendezvous because, from evidence gathered by previous committees and its own studies, the cost of getting to the Moon by Earth Orbit Rendezvous or Direct Ascent would be at least $10,600 million versus an expected $9,200 million for LOR, and because it could be accomplished by July, 1968, whereas the other two modes could not guarantee a landing before early 1969. The fact that 'the Directors of both the Manned Spacecraft Center and the Marshall Space Flight Center have expressed strong preferences for the LOR mode,' weighed heavily in favor of that choice.

As defined, LOR would, said the study, require 15 flights with the Saturn C-1, 13 with a two-stage version called the C-1B, and 20 flights with the C-5, the first manned flight with this big booster projected for January 1968, just six months before the first manned landing attempt. Nine manned C-5 flights were planned, at two month intervals.

Joseph Shea presented the Manned Lunar Landing Mode Comparison Study to the Management Council on 22 June. After a

Manned flight boss Brainerd Holmes talks about the Apollo program at the 11 July, 1962, press conference which revealed the plan to public view. At his right, Robert Seamans (spectacles) and James Webb; at his left, Joseph Shea. In front, the first lunar lander concept.

long discussion, taking in all the operational requirements of each mode, the Council finally recommended Lunar Orbit Rendezvous by C-5 as the best of all possible methods to be adopted for reaching the Moon. There was not time to develop a back-up, or contingency, mode, said the Council, and reiterated its support for the two-stage C-1B as an early Earth-orbit launcher for flying Apollo development models.

During the next few days Headquarters buzzed with discussion and deliberation on the recommendations of the Management Council. It was time to put away the options and choose a final and lasting method that promised to get the job done in the shortest time at the least cost, providing a fair balance of safety and with the highest overall probability of success. Six days after the Management Council met to decide its preferred mode, Webb and Seamans met the Council in full to hear their recommendation and its defence; Hugh Dryden was ill at the time and received a separate briefing.

Webb felt the decision to be sufficiently important that only the top NASA leaders should make the final selection; Brainerd Holmes believed that since he would have to implement the decision, he should give the final word on choice. For nearly two weeks the NASA triumvirate discussed and pondered the implications and ramifications of each candidate mode, finally dispatching a memorandum to Holmes that he should 'Adopt Lunar Orbit Rendezvous (LOR) as the prime mission mode for our first manned lunar exploration,' and that development 'of the lunar excursion vehicle should be initiated immediately.'

Webb, Dryden and Seamans thought development of the logistics vehicle proposed by von Braun was a good idea but that work on that concept was to be 'withheld pending further study.' Next day, 11 July, the press were formally told of the decision. But Webb was reluctant to admit the choice was final because the PSAC, led by Wiesner, was still critical of LOR. Holmes, urging momentum, said that 'There comes a point in time, and I think the point in time is now, when one must make a decision as how to proceed, at least as the prime mode.' Fourteen days later, seizing the initiative in a manner not unlike the effort applied to getting control of Mercury in 1958, and promoting Gemini in 1961, the Manned Spacecraft Center invited 11 companies to submit proposals for a Lunar Excursion Module (LEM). They were, The Boeing Airplane Company, Northrop Corporation, Lockheed, Ling-Temco-Vought, Inc., Grumman Aircraft Engineering Corporation, Douglas Aircraft Company, General Dynamics Corporation, Republic Aviation Corporation, Martin-Marietta Company, North American Aviation, Inc., and McDonnell Aircraft Corporation. In the Statement of Work distributed to these companies, MSC outlined the mission operations and provided a preliminary specification.

The LEM would be carried within an adaptor placed between the command and service modules (CSM) of Apollo and the third stage, the S-IVB, of the Saturn C-5. After first being placed in Earth orbit for a checkout, the third stage with its payload would ignite its single J-2 engine to reach escape speed of about 40,200 km/hr. Coasting freely toward the Moon, the CSM would separate from the adaptor, ejecting panels that covered the LEM on top of the S-IVB. Turning through 180°, Apollo would move back in on the LEM and dock with a special fitting on top, extracting it from its position atop the redundant third stage. A rocket motor at the base of the CSM would be used to decelerate the complete assembly into an orbit about the Moon 185 km above the surface. After separating from the CSM, having taken on two astronauts, leaving a third in the CSM, the LEM would transfer to an elliptical path which carried the vehicle to within a few kilometers of the surface. From the low point of this equi-period orbit, designed to bring the spacecraft back to the CSM should anything go wrong during descent, the LEM would fire its engine and lower itself to the surface like a helicopter.

There would be no need for the LEM to return to Earth, so its design could avoid the necessity for aerodynamic re-entry. The entire configuration could be biased around lunar landing requirements. When the time came to leave the Moon, the LEM would fire itself away from a lower, landing leg, section, and rendezvous with the orbiting CSM. The crew would transfer to the Apollo mother ship and use its powerful engine to accelerate to lunar escape velocity, arriving back in the vicinity of Earth two or three days later where the crew would re-enter and land on either water or the ground according to design requirements which were, at that time, still to be determined.

On 30 July, 1962, the Office of Systems at the Office of Manned Space Flight in NASA Headquarters outlined its findings from careful study of reports and recommendations made during the past 12 months. It found that the least complicated method was certainly Direct Ascent, needing none of the orbital rendezvous and docking required by Earth Orbit or Lunar Orbit

Apollo mission plan.

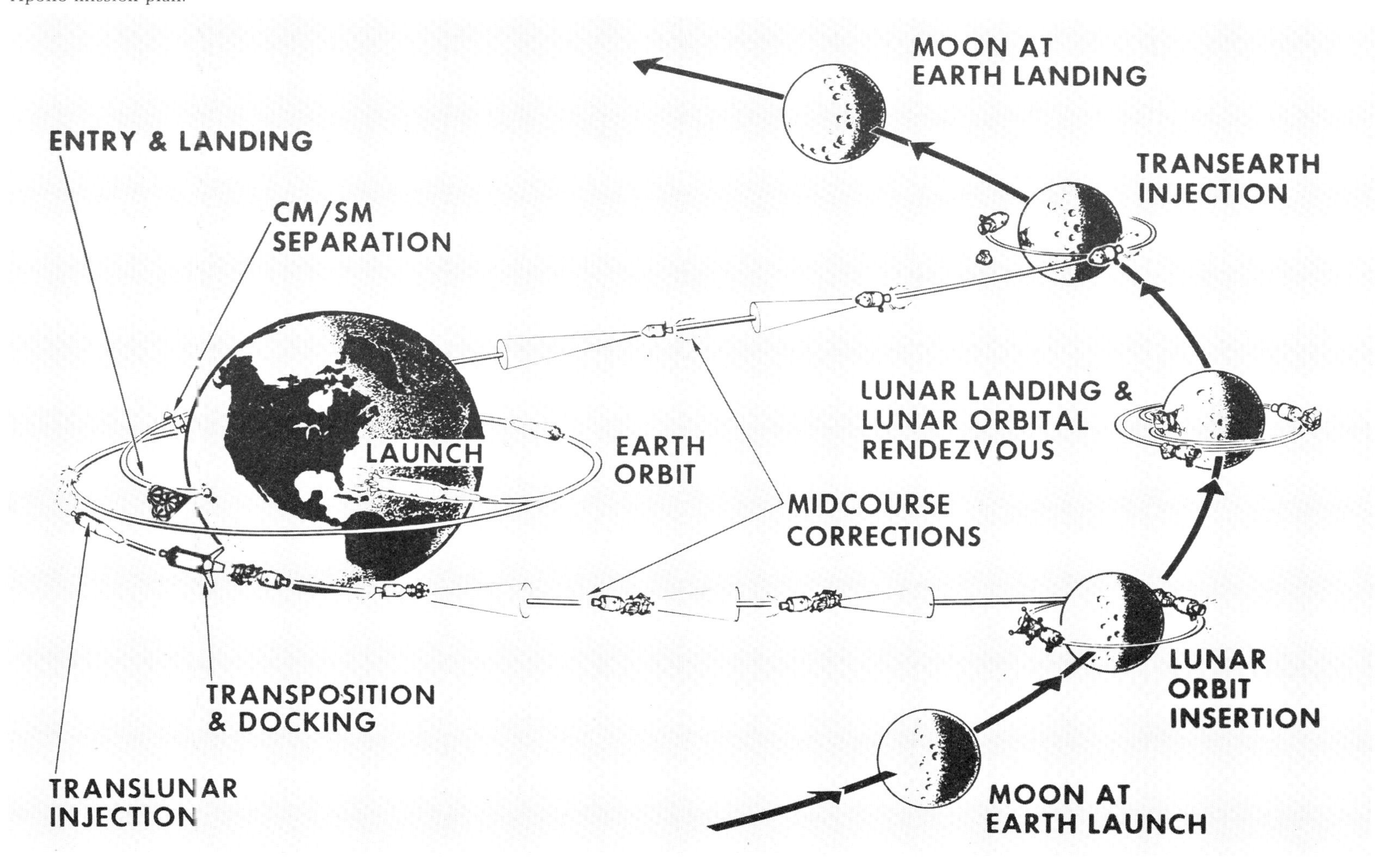

Rendezvous. But this mode was prohibitively expensive on time and money because of the enormous size of rocket required to complete the task. It found also that EOR was in fact least likely to succeed of all alternatives, since it required the launch of at least two Saturn C-5 launchers for rendezvous and docking operations that if for some reason were found impossible for a particular attempt would completely destroy that launch opportunity. Since only EOR and LOR modes were economically feasible under the situations in existence at that time, LOR was again deemed to be the better method.

Clearly, Manned Spacecraft Center interests centered on getting the LEM work in as advanced a state of preparation as feasible and on 1 August J. Thomas Markley was appointed Project Officer for the Apollo CSM contract, with William F. Rector performing a similar function for the LEM. By 11 August three companies, NAA, Northrop, and McDonnell, had made it clear they would not bid for the LEM contract, just a few days before the first Apollo command module boilerplate model was delivered to MSC. It would be used for water recovery and flotation stability tests and was known as BP-25, the first real piece of Apollo hardware. Wiesner had been unsuccessful in last ditch efforts to get NASA to change its mind. North American Aviation were deeply involved with learning just how demanding, and in some ways thankless, a task it was going to be to build the big mothership. They had no heart for the LEM contract as well, recognizing that they would probably not have got it anyway. The company was already in for some heavy government funding and NASA had a code of contractual conduct regarding too much to too few.

The story of a parallel confrontation between NASA and the PSAC extending from early 1962 until the irrevocable decision to use Lunar Orbit Rendezvous late that year contains many issues dimmed now by the passage of time. But a cursory glance at the main issues shows again how numerous were the interests and causes vying for supremacy. Dissent had its origin in the transfer of Nicholas Golovin from NASA to the President's Science Advisory Committee. Golovin had carried out the 1961 launch vehicle study and was already well known inside NASA for his unconventional views on the decision-making process. The PSAC's Space Vehicle Panel developed a sense of responsibility about the many decisions then being taken by NASA which it felt should be reviewed by the White House monitoring group. With Golovin now among its ranks, that seemed a natural thing to do.

In late April the Panel met with NASA officials in Washington to hear about future manned space flight plans embracing Mercury, Gemini and Apollo. Concerned about the way NASA was proposing to organize the technical and managerial interfaces, the PSAC attempted to get into the lunar landing mode selection study, an attempt contested by the space agency who felt that internal resolution of divergent views should precede any open discussion with an outside body. At the Management Council meeting 29 May, 1962, Holmes proposed to present to the Panel a coordinated and confident picture of broad NASA policy rather than get the PSAC Panel involved in the dispute. But Golovin and Wiesner intervened and had NASA present a full discussion of EOR, LOR and Direct Ascent modes at the Council meeting held 5 and 6 June.

Before the end of the month, the Panel had also heard about technical aspects of EOR and LOR at the Manned Spacecraft Center but since by that time informal agreement had been reached within NASA to choose the Lunar Orbit Rendezvous mode it was largely an academic presentation. Back in Washington, the PSAC requested a briefing from Headquarters. Shea sent Wiesner all the contractor reports recommending LOR and the Manned Lunar Landing Mode Comparison study. In all, fourteen documents were sent across on 2 July. Next day Wiesner and Golovin began to go through the books and came upon a discrepancy about safety values for the three modes.

A single table in the NASA mode comparison book carried two calculations that showed LOR to be much more dangerous than the other two. Wiesner called Webb who told him he would send Shea across to explain the problem. Meeting with Wiesner and Golovin, Joseph Shea pointed out that the numbers came from different sources and that a Marshall Space Flight Center figure was arithmetically incorrect and should not have been included.

The second error, said Shea, derived from an over-pessimistic view projected by the launch vehicle committee chaired by none other than Nicholas Golovin himself! The blatantly undiplomatic assertion that the defendant was protesting innocence by citing the accusor for uninformed assault did nothing for the NASA case! Wiesner and Golovin were convinced somebody was trying to make LOR look too attractive and decided to do their own calculations. Using the books already supplied by Shea's office, the two PSAC men failed to come up with the figures projected by NASA.

A few days later Shea supplied additional figures even more favorably disposed to LOR, but Wiesner had even greater difficulty in agreeing their totals. Now Wiesner and Golovin were absolutely sure that Headquarters was developing figures suited to their case and on 6 July Wiesner and some PSAC Panel members told Webb, Seamans, Holmes and Shea about their concern. It was because of this protest that Webb deferred the memorandum authorizing Shea to adopt the LOR mode. The Administrator wanted time to complete a review of one other mode pushed by North American Aviation and Wiesner. This projected the use of a two-man spacecraft on a Direct Ascent using a single Saturn C-5.

It would just be possible if cryogenic propellants were used for decelerating to the lunar surface and lifting off again, since these high-energy fluids would be more efficient than storable propellants and so do more work for less weight. But it was an inherently dangerous approach and the handling of liquid hydrogen and liquid oxygen was difficult enough on Earth to preclude ready acceptability of it for deep-space operations. Nevertheless, Space Technology Laboratories were asked to study this mode and for a while it opened again prospects for competing firms.

Not only did North American propose a two-man Apollo, for McDonnell studied a direct C-5 mission with a modified Gemini spacecraft.

By early August, Wiesner met Webb in a mood where the former distrusted the reasons NASA gave for selecting LOR. But Webb pointed out that documentation used by Wiesner and the PSAC was now somewhat dated and that the technical pace of mode evolution outstripped the paperwork. But Wiesner felt his deliberations, and the conclusions of the PSAC, should be brought to the President while Webb asserted that NASA 'would have to find some method of review that did not prevent the initiative of moving ahead . . .'

On through August the differences boiled while yet another study was completed, this one from the Bureau of the Budget on projections for the President. In early September, Kennedy decided to make a two-day tour of space facilities at the Cape, Huntsville, Houston and St. Louis. At the Marshall Center on 11 September von Braun used a Saturn C-5 model to explain Lunar Orbit Rendezvous but Wiesner, as the President's science adviser, stepped in and contested the endorsement of LOR. James Webb cut him off with strong approval for what von Braun was saying and a short but volatile debate broke out in front of newsmen reporting the visit.

For a few minutes the President listened, then stopped the debate with a request that a final report should be sent to him in due course. A month later NASA had the STL and McDonnell studies to hand proving that the original decision had been the best way after all. But Wiesner was unconvinced, believing still that NASA was chopping up the big Moon mission among several contractors so that threat of contract cancellation because of poor work would not impact the entire program. It could only do that by going to LOR, for only one Apollo spacecraft contractor would be required to build a direct landing vehicle thereby placing considerable reliance on a single firm. Moreover, Wiesner believed NASA was concerned to limit the work given to North American.

On 16 October, Webb reported to Kennedy aide Kenneth O'Donnell that the two-man capsule idea was not feasible and that LOR was still the only way to go. Eight days later the NASA Administrator wrote to Wiesner justifying his decision. But for the involvement of the White House in a major international crisis, Jerome Wiesner might have taken the issue all the way to the Oval Office. As it was, the President's preoccupation with the Cuban missile build-up inhibited a move of this kind. In the light of a potential war with Soviet Russia, technical decisions about a space program were of little concern to Kennedy. And Wiesner knew it, so the issue officially died right there. Unofficially, Wiesner and the PSAC pulled out their heavy guns for one last assault on NASA's LOR mode but it was a veiled contest of little concern to anyone. Internally, it caused a rift between the space agency and the President's Science Advisory Committee and Jim Webb felt the need to inform Kennedy personally of his decision.

In a letter dated 10 December, the NASA Administrator said that 'despite very extensive study efforts . . . we are dealing with a matter that cannot be conclusively proved before the fact, and in the final analysis the decision has been based upon the judgement of our most competent engineers and scientists who evaluated the studies and are experienced in this field.'

Wiesner reluctantly accepted the arguments denying Direct Ascent or Earth Orbit Rendezvous the validity he had wished they could carry. But it was to remain a thorny issue and Jerome Wiesner would never forget losing the battle for North American Aviation.

As it was, there was another contractor about to come on line – the contractor selected to build the Lunar Excursion Module. Nine companies responded to MSC's request for proposals by the deadline date of 4 September, 1962. Just nine days later a two-day session began at Ellington Air Force Base where industrial proposals were examined, special visits to company sites being made across a three-day period starting 17 September. NASA was keen to get the LEM contractor on board and promised a quick deliberation. But while NASA was moving toward selection, another administrative shake-up transformed the agency into an even more efficient machine, although the alignment moves were not as dramatic as those made late in 1961. Under that reorganization, the Office of the Associate Administrator had authority over one staff office, one staff support office, an agency wide support office, four program offices and nine field centers, giving Seamans a control span of 16. There were inevitable problems with feeding quickly the varied needs of support and user facilities through one Associate Administrator. A temporary improvement had been effected in March, 1962, when the new Launch Operations Center was placed under the authority of the Director of Manned Space Flight, reducing Seamans' load to 15.

By October more drastic change was necessary and from 30 October a new Deputy Associate Administrator was appointed. Called the Deputy Associate Administrator for Manned Space flight Centers he would be responsible for Marshall Space Flight, Manned Spacecraft, and Launch Operations Centres. The original Deputy Associate Administrator, Dixon, would have authority over the other seven centers and would eventually acquire the impossible title of Deputy Associate Administrator for Other Than Manned Space Flight Centers! Brainerd Holmes would be the Deputy Associate Administrator for Manned Space Flight Centers, in addition to his existing role as Director of Manned Space Flight. In this way, Dixon's superior, Seamans, had his authority comfortably cut from 16 positions under the November 1961 reorganization to a more welcome 8. But the real value gave Holmes effective control of both manned space flight institutions and program operations, a tidier infrastructure than previously provided. It also placed the Launch Operations Center (Cape Canaveral) on an administrative level with the other NASA centers and generally oiled the wheels of what could easily have become a very sluggish machine.

NASA kept to its word in rapidly choosing a contractor to build the LEM. On 7 November James Webb personally made the announcement that told the public Grumman Aircraft Engineering had won: 'We are affirming our tentative decision of last July . . . The results of these studies added up to the conclusion that Lunar Orbit Rendezvous is the preferable mode to take.' Formal negotiations would now have to begin but NASA expected the contract to be worth at least $350 million, almost as much as the CSM contract to North American Aviation. Grumman had worked long and hard for the LEM contract, and had an impressive plan for both technical development and management control. It was in many respects more difficult a job than building the Apollo mother-ship. Without parallel, divorced from aeronautical considerations, LEM engineers would be the first real spaceship builders, producing as they would a vehicle incapable of flying in the atmosphere, incapable of bringing crew members back in a conventional re-entry, unable even to stand on the planet from which it came; designed for the one-sixth gravity of the Moon, it would have no need of the additional strength, and weight, necessitated by an Earth-standing capability.

As Grumman's vice president for space programs, James G. Gavin, Jr., was well aware of his company's responsibility. It was up to Grumman to build the vehicle assigned to carry two astronauts from the protected environment of Apollo to the hostile lunar landscape, set them safely down on its dusty surface, and send them back up to the only spacecraft capable of returning them through Earth's atmosphere. The eyes of the world would be on the LEM, it was to be the ugly duckling of the entire project, but few vehicles would again have such responsibility. For LEM would convey the hopes and ambitions of an entire generation committed to making space travel a reality, and it would be the only means by which the two astronauts could survive during the several hours away from Apollo.

As 1962 came to a close the space agency was charged with new ideals, unique opportunities, and a growing inventory of young scientists and engineers flocking to join the ranks of America's new breed. It had been a fantastic two years, during which time the agency had flown America's first astronauts in space, set up unmanned planetary exploration programs for pioneering flights to Mars and Venus, responded to the challenge put down by President Kennedy, and paved the way for massive new commitments. At the end of 1961, NASA began to recruit scientists from colleges and campuses all across the nation, dispatching 15-member teams to all the major cities and towns in America. In all, 181 connurbations had been visited by these scout groups, making contact with 14,000 people in which 5,000 interviews were held. Just seven months later NASA had taken on an extra 3,000 scientists, young men charged with zeal and determination to contribute toward this great adventure. During the 12 months beginning 1 July, 1962, a more consolidated drive, where NASA settled recruitment offices in large cities rather than move from place to place, produced a further 3,500 scientists and engineers. But they did not come for money. About 49% received approximately the same wage that they were receiving already, and a further 12% took a cut in salary.

As news about space programs – unmanned Earth and planetary satellites as well as the planned Mercury, Gemini, and Apollo projects – spread across the nation, people volunteered their services and brushed up academic score ratings to get a chance to work with NASA. Inside the agency, there were mixed feelings about the enormous effort being given to the manned space flight program. Planetary exploration projects were being cut to open financial coffers for other programs which could be made to support in some way the long term goals of Apollo. Between late 1958 and early 1961, a 2½ year period during which science carried great weight in intra-departmental planning sessions, objectives were laid down for major initiatives to learn the

NASA Flight Research Center director Paul F. Bickle displays a model of the lunar landing research vehicle designed to develop techniques for Moon landing and eventually to be used for astronaut training.

origin of the solar system, probe the mysteries of the Universe, and deploy useful applications satellites to Earth orbit. All that was in keeping with a balanced space program in which manned flight projects like Mercury were contained in proportion to their usefulness to the broad mandate detailed by the Space Act of 1958.

Only a year before Kennedy issued the Moon landing objective, the Jet Propulsion Laboratory had begun detailed work on Ranger, a vehicle designed to 'acquire and transmit a number of images of the lunar surface.' The unmanned spacecraft would crash to the Moon's surface after sending back TV pictures, and later versions were expected to carry scientific instruments that could be 'hard landed' on the surface, that is they would release shock-protected spheres designed to roll to a stop before operation – 'soft-landing' implies a controlled descent to a gentle touchdown. Also in 1960, JPL pursued development of Surveyor, another Moon-bound unmanned project but one designed to land on the surface instruments and equipment that could provide lengthy surveys from a fixed location. As part of an initial attempt at learning more about the Moon they were but the vanguard of several spacecraft then being considered for development. When Apollo received its broader mandate, Ranger and Surveyor were immediately swept up into a major effort to support the manned program. Unmanned Moon missions would now be single-mindedly committed to paving the way for LEM landings.

On the very day that President Kennedy made his famous congressional speech, NASA's Director of Program Planning and Evaluation, Abraham Hyatt, issued a new strategy for Moon missions whereby existing programs would abandon the integrated, timely, schedule for remotely exploring Earth's natural satellite to be re-orientated at the feet of Apollo. There is no doubt that in the minds of several top NASA executives, this pinnacle of manned space flight endeavor was in truth becoming a god to which other less captivating endeavors would be sacrificed. Throughout the agency, plans emerged for scientific payloads that could most effectively provide information deemed necessary for the early design phase of Apollo and the LEM. Information about the Moon had been acquired for centuries, and no other world beyond Earth was as well mapped. But so little was really known about the surface conditions that designers could only guess at the real requirements of a spacecraft designed to set foot on its surface.

Although prevented by fate and mathematical principles from designing the vehicle that would actually land, North American Aviation were compensated in part by not having to actually construct the LEM. It was to give Grumman enough headaches, and the full resources of NASA were marshaled toward the goal of finding out what the surface was really like. Only then could final design details be applied to landing legs for the LEM, to footpads for the legs, and to mass distribution of the entire vehicle. In the final analysis, NASA was to put too much emphasis on obtaining information in this manner. In projects like Ranger, Surveyor and another called Lunar Orbiter still several years away from realization when the LEM contractor was announced, the space agency spent large sums and tied up many people in a desperate attempt to give the engineers solid data about the lunar surface from which to work in finally designing the LEM. As it turned out, not one scrap of information provided by the 13 unmanned spacecraft that orbited, crashed into, and landed on, the Moon proved useful to the design of Apollo or its LEM, euphemistically nicknamed the 'bug' from the start.

True, the orbiters did provide helpful shots of the surface from which landing site selection committees could recommend touchdown coordinates, but that was about the extent of the application. In NASA circles, the rift between manned and unmanned programs was growing deeper by the month. Scientists really believed, in those early days before the first Moon landings, that Apollo would open a new era in which Earth exploration would spill over to the solar system. To major NASA centers, it was an engineering job first and foremost. If, after the operation was shown to be feasible, scientists wanted to provide the stimulus for continued exploration, that was all right by them. But the two factions – engineering on the one hand, science on the other – were distinct arenas from which protagonists would plead respective causes. It was to be a long time before the two began to understand each other. In late 1962 there seemed nothing to stop the machine from delivering up the promised achievement. There were just seven years left. And a lot of work ahead.

CHAPTER TEN

New Themes

On 14 November, 1962, less than six weeks after the successful six-orbit flight of MA-8, NASA formally appointed Leroy Gordon Cooper as prime pilot for the day-long mission of MA-9, with Alan Shepard as back-up. Cooper, youngest of the original seven astronauts, was the only Mercury pilot still to experience the exhilaration of space flight, excepting Slayton who had been off the flight list for a year now with a fibrillating heart. For most of 1962, Manned Spacecraft Center engineers worked long and hard with McDonnell to adapt what was essentially a three-orbit capsule designed to survive the rigors of space flight for about 4½ hours into a fully fledged one-man spacecraft capable of sustaining life for more than a day. Conservatism had been not only the key word but more the rule book during early design stages of this first US manned spacecraft. Now, several flights later, those rules could be relaxed in the light of experience and learning processes that served to instill confidence in the capsule's ability to survive. Based on the steps taken to assure success for Schirra on MA-8, technicians now tinkered with capsule 20, the last of McDonnell's proud batch, for a mission nearly four times longer.

There had been something magic about an 18-orbit flight that set that target as one in which Mercury men could gather confidence. It was a full day by implication for the time required to go 18 times around the world, and it was a full orbit longer than Russia's second manned space flight when Titov flew Vostok II. But that record had been greatly exceeded by the 94 hr and 71 hr flights respectively of Vostok III and Vostok IV, and other considerations pressed upon the space agency to change the plan. Whereas MA-8 had been originally intended as a seven-orbit mission, reduced to a six-orbit objective by consideration of the splashdown point, MA-9 was found to benefit from an increased flight duration; re-entry on the 18th pass would cause the capsule to descend on the United States if the retro-rockets fired too late, and optimization of recovery forces revealed a preference for recovery on the 22nd orbit. In fact, of course, the spacecraft would make 21½ orbits of the Earth. The extra four orbits would extend the mission by some six hours, an increase that was itself in excess of Mercury's original performance requirement! In total, MA-9 would last for about 34 hours. Could it be made to do so?

Engineers now exploited the over-conservative design trends when, in 1959 and 1960 during final blueprint stages, nobody had been really sure what margins would be necessary for space flight. As though to signal its new image, Gilruth's Manned Spacecraft Center, spread incongruously among buildings and sheds around Houston while the Army Corps of Engineers built on, decided to call MA-9 the Manned One-Day Mission (MODM), informing Headquarters in a scheduled report that MA-8 was to be considered the last Mercury mission. It was an unwieldy change, and one that never did catch on. But preparations for the flight entailed substantial modification to Mercury systems and to flight procedures arranged to fit the consumables (water, propellant, oxygen, electrical power, etc.), around the mission requirements.

Weight saving was an important counter to added supplies essential to the mission's safety, and everything inside the spacecraft was carefully examined in the hope that it could be eliminated. But things did nearly get out of hand. One suggestion questioned by implication the absolute necessity of a form-fitting contour couch. It had, after all, been substantially reduced to just a torso and head support for Schirra's flight. When it was suggested that the couch be replaced with a net designed to fully support the pilot in an elasticated grip, astronauts were horrified. The thought of Gordon Cooper being battered to death on his own instrument panel as he bobbed up and down brought humor to the serious business of life support. The couch remained where it was! But other items could be removed from capsule 20, including the 34 kg periscope, a 1.4 kg back-up UHF transmitter, a 0.9 kg back-up telemetry transmitter, and a low-level commutator. Also taken out was the Rate Stabilization and Control System (RSCS), one of four attitude control modes provided on all Mercury flights except the first suborbital mission flown by Shepard. It saved a further 5.4 kg and reduced automatic control functions to just the one ASCS mode; RSCS had been expensive on fuel and deemed largely redundant by most crewmembers.

Wally Schirra had shown it was feasible to power down the Mercury spacecraft and to conserve not only electrical energy by switching off systems and equipment unnecessary for large portions of a mission but also attitude control propellant: hydrogen peroxide. Nevertheless, even the frugal consumption of power and propellant by Sigma 7 revealed a need for more of both on the 34 hour flight of MA-9. Accordingly, two of the three 1,500 watt-hr batteries aboard the spacecraft were removed and replaced with two of 3,000 watt-hr, increasing the inventory of large capacity batteries from three to five, thus ensuring a mission capacity of 16,500 watt-hrs. That would be sufficient, if the powered-down phases could be implemented. Regarding hydrogen peroxide, more of that was provided in the form of a supplementary tank capable of holding 6.8 kg, with an interconnect valve linking the manual and automatic supplies should one run dry and control authority be preferred for the drained system.

The basic propellant supplies had been carried in two semitoroidal containers which, placed together, constructed a complete torus around the circular base of the capsule between the aft bulkhead and the heat shield. The auxiliary tank, also of semitoroidal shape, was fitted adjacent to the existing manual tank and increased the total available supply to 34 kg. Demands that could not be so easily alleviated embraced the atmospheric control systems, where increased consumables were a necessity that even the figure-juggling systems engineers at MSC could not avoid loading. So, more oxygen had to be provided, and more water was needed in the cooling system; for the former, a single 1.8 kg capacity tank in addition to the existing two which provided earlier capsules with a total oxygen supply of 3.6 kg, and for the latter, an extra 4.1 kg of cooling water in addition to the existing total of 17.7 kg. For drinking purposes, Cooper would have use of a 4.5 kg water supply versus 2.5 kg available to previous pilots. Lithium hydroxide, carried as a carbon dioxide remover, also had to be supplemented, a further 0.36 kg being added to the original 2 kg. An over-conservative quantity in the design of the basic capsule, understandable for the serious effects an increase in CO_2 would produce on the ability of the pilot to concentrate, was made to absorb the greater demands of the 34 hour flight. In the odor absorber, the quantity of activated charcoal could actually be reduced, from 0.45 kg to 0.09 kg.

In addition to supplementing the consumables, modifications were made in continuing attempts at refining the spacecraft systems, hopefully to clear up problems like those that afflicted Wally Schirra when both suit circuit and cabin temperatures rose above the desirable limits. To smooth the integration of both suit and cabin cooling circuits more effectively, engineers provided an improved condensate trap which promised to remove water more efficiently from the suit circuit, and they added a parallel coolant control valve in the form of a suit bypass device serving as a redundant unit to the prime valve; Schirra's high suit temperature had been attributed to a blocked coolant control valve in the one and only system that spacecraft carried. In a continuing attempt at finding a better design for the sensitive low-thrust attitude control jets, MSC came up with yet another configuration for the 0.45 kg and 2.7 kg thrusters. In all, there were 183 changes to spacecraft 20 compared with the configuration of capsule 16 used for MA-8.

But the balance between equipment taken out of the spacecraft and new items placed on board for the greater demands of the 22 orbit mission was a masterpiece of systems engineering. Capsule 20 would weigh only 1.8 kg more in orbit than Schirra's Sigma 7! In fact the difference in orbital weight between Glenn's capsule and the last was only 21 kg, a remarkable feat made possible not only by stringent and careful inventory of the on-board systems but also by the dedicated team-work within the McDonnell facility and between the contractor and NASA. What weight growth there had been in the program occurred between initial feasibility studies in 1958, where a projected weight of 1.1 tonnes had been adopted, and the first manned orbital flight in 1962, where the capsule weighed 1.35 tonnes, a growth factor of nearly 23% across 3½ years of development.

Although the increasing tendency to load the spacecraft with scientific equipment and tasks, which reached a peak with Carpenter's MA-7 mission, had been largely cut back from Schirra's flight plan, the substantial increase in MA-9's flight duration allowed a modest re-introduction of tasks and activities deemed pertinent to the mission. In fact, the allocation of scientific equipment to the Mercury capsule grew from an assigned 5 kg for Glenn's flight, to 8.2 kg for Carpenter's mission, and 10 kg in Schirra's capsule. Cooper was to carry a science load weighing 28 kg, comprising the optimum selection of the Mercury Scientific Experiment Panel set up in April, 1962, to choose and integrate candidate tasks for orbiting astronauts.

For MA-9, there would be the tethered balloon experiment previously flown on Carpenter's capsule, a flashing light experiment where a xenon-gas discharge lamp would be jettisoned from the spacecraft and visually tracked by Cooper, a 70 mm Hasselblad hand-held camera for terrain and cloud photographs, and a radiation measuring experiment using various emulsion packs placed around the capsule. As with earlier missions, microscopic maps would be constructed, before the flight, of the entire exterior surface of the single vycor window and compared with another set of maps drawn up after the mission. By carefully comparing the two, a history of the tiny pits caused by collision with microscopic particles could be performed, information useful to an increasing need for reliable data on the density and distribution of small particles near the Earth. Although not considered a scientific experiment, engineering data on a TV camera would also be gathered by a device designed to transmit to ground monitors during flight images of the MA-9 pilot every two seconds. Three different lenses were provided and ground controllers were permitted real-time views of Cooper via a scan converter that took the unprocessed image and passed it to a commercial monitor set up in a Mission Control Center at Cape Canaveral.

John Glenn examines gloves with a suited Gordon Cooper before the latter flies the last Mercury mission.

But for all the paraphernalia of engineering and science equipment placed aboard capsule 20, Cooper would be more responsible for his own mission than any US astronaut before. Schirra's flight had lasted long enough to allow the Earth to drift a significant way round a full revolution; Cooper's mission would see the Earth spin almost 1½ times on its axis. That meant that capsule 20 would roam far from the established band of tracking stations, positioned in the optimum locations for a short 5 hr flight at an orbital inclination of 33°. To help fill in areas devoid of any tracking or communication coverage, the network was expanded by the addition of two ships: the *Twin Falls Victory*, positioned between Florida and Bermuda, and the *Range Tracker*, stationed near the Gilbert Islands in the Pacific Ocean. The existing ships *Coastal Sentry Quebec* and *Rose Knot Victor* were positioned north-east of Okinawa and north-east of Easter Island, respectively, in the Pacific.

Despite efforts to capture as much coverage as possible, Cooper would drift for long periods of the flight out of communication range with any station. It would be particularly quiet during orbits 10 to 14 but after that the spacecraft would once again fall increasingly into contact with ground and ship stations. In fact, ground tracks for six orbits – 16 to 21 inclusive – would almost precisely duplicate the six orbit ground tracks of Wally Schirra's MA-8 mission; orbit 16 would follow almost the same path as orbit 1, the Earth having spun once on its axis, while orbit 22 would closely follow orbit 7. Because of this repetition, contingency recovery procedures were made a little easier. Nevertheless, it was a major task to assign recovery forces to contingency zones where there was little chance of the spacecraft actually coming down, so compromise necessarily followed on the prediction that full mission support functions would be beyond the capacity of the Navy. Accordingly, although an emergency recovery zone was plotted for each of the 22 orbits, minimal support was provided for the less probable areas. Despite the compromise, 26 naval vessels were assigned recovery tasks, with 110 aeroplanes and 14 helicopters. USS *Kearsarge* would be the prime recovery carrier in the Pacific south-east of Midway Island. Emergency zones were strung out across the North Atlantic between America and Africa, around Midway Island in the Pacific, and south of Japan around the Mariana Islands north-east of the Philipines. In total, 18,000 men from the Department of Defense deployed globally. As many, in fact, as were concerned with the first US manned orbital flight more than a year before.

Yet for all the apparent attention given to what was even then being taken for granted as the last Mercury mission, the amount of NASA resources dedicated to MA-9 was far short of public impact both inside and outside the United States. Manned space flight was getting into its stride, regular trips by commercial network broadcasting personnel to the Cape saw that the public at large were made an integrated part of the entire momentum, and popular interest in space and anything associated with extraterrestrial travel reached an all time high. But that upward trend showed no sign of reaching peak levels, and the press serviced the public with all the show stoppers they wanted. Nevertheless, Mercury was on the way out, and projects like Gemini and Apollo gathered increasing interest. And NASA too was growing, commensurate with the national will to follow the progress of astronauts and space men.

By the early months of 1963, as final preparations for MA-9 once again brought confusion to the Cape with choked beaches and crammed roads, Project Mercury boasted fewer than 500 assigned NASA personnel, down by 50% from the force just a year before when John Glenn went into space. An equal number of personnel were busy working on the two-man Gemini program, grooming that out-growth from Mercury for manned flights then expected the following year, but more than 7,000 people in the space agency were already assigned to Apollo which greedily consumed manpower and effort in almost equal proportions. But even this was a mere prelude. As Mercury, and eventually Gemini too, made way for Apollo, the Moon landing project would claim nearly 10,000 NASA personnel within four years. But by then the space program would be providing work for more than 20,000 companies and more than a half million people in private industry. As it was, the early months of 1963 demonstrably clarified the scale of America's ascending space program. The Manned

Spacecraft Center was considerably bigger than when Langley Field played host to the Space Task Group, having expanded from a peak of about 800 people prior to the Kennedy Moon commitment to a complement of more than 3,300 persons by spring, 1963, just two years later.

Preparations for MA-9 got off to an inauspicious start when, in January, 1963, Atlas 130-D was returned to its General Dynamics/Astronautics factory for errors in wiring to be rectified. When next the big booster appeared through the hangar doors, it was in a better state of preparedness than any previous Atlas booster had been in support of the Mercury program. By 22 April the Atlas and its spacecraft were sitting together for the first time on pad 14 – the last time a live Mercury-Atlas combination would be seen. Astronaut Cooper was perhaps a trifle more philosophical than most space pilots about their mission and the program to which it belonged. A few astronauts would consistently exhibit tendencies that endeared them to public images of spacemen, proferring a poetic cliche, recounting an emotive connotation, repeating a dramatic simile. For Gordon Cooper, the transitory nature of his particular role had deeper meaning than the slimline history books would allocate to a flight that years hence would seem but a very meager step toward interplanetary travel.

So, recognizing that he was part of something bigger than the sum total of its parts, he called his spaceship Faith 7, 'as being symbolic of my firm belief in the entire Mercury team, in the spacecraft which had performed so well before, and in God.' It was the latter that caused a ripple to move through NASA, and the press to muse that should the capsule sink like Grissom's Liberty Bell, it would not look good to see reported that 'The United States today lost Faith. . . !' But the name endured, much to the chagrin of certain elements in the United States then campaigning to disassociate the country from higher authority.

There had been speculation that MA-9 might in fact not be the last one-man space mission and that a flight of even longer duration might be considered. NASA Associate Administrator Bob Seamans confirmed that 'in planning flights of that importance we always have a back-up possibility. We do have two back-up capsules as well as the two Atlas boosters that could be used in the event that we don't obtain all the information that we anticipate obtaining in the Cooper flight . . .' It was certainly the time to incorporate any further Mercury missions; with several months required to prepare pilot and hardware adequately for a specific flight the preparation of an MA-10 flight would necessarily have to begin prior to the launch of MA-9. Walt Williams, Deputy Director for the Manned Spacecraft Center, named also as Deputy Director for Mission Requirements and Flight Operations in January, 1963, supported the idea of a fifth manned orbital flight with this vehicle but Seamans and Webb felt it was about the right time to call a halt on the one-man missions. Accordingly, despite Seamans' cautious comment about MA-10, there was virtually no chance of any such flight being approved. Interestingly, not everybody was as keen as NASA management in writing off the fate of the fifth Mercury. Martin Caidin, writer and broadcaster, wrote a dramatic novel describing what might happen if retro-rockets on a hypothetical Mercury flight failed to fire, using a fictitious MA-10 mission as the scene for his plot. With the uncompromising title 'Marooned,' it was eventually made into a film, although the scene shifted then to a much more advanced space mission.

NASA had originally planned to launch Cooper on his day long flight during April, but a succession of delays put this date back to 14 May. Early in the morning hours, Cooper suited up and rode the short distance from Hangar S to pad 14, arriving at the capsule shortly after 6:30 am. Alan Shepard had given Cooper a plumber's suction cup and handle with an instruction attached to 'Remove before launch.' Unfortunately the appeal extended to the pilot too, for after four hours in the capsule, Cooper heard that the mission had been cancelled for that day, due first to problems with radar units at Bermuda and then to a stuck gantry at pad 14, before more problems with the Bermuda station made it impossible to launch. Preparations were re-cycled for a planned lift-off at 8:00 am local time the following day, the second part of the split level countdown beginning precisely six hours earlier.

Again, Cooper went through the traditional pre-flight activity list, attaching the bio-sensors before suiting up. Because MA-9 was to remain in space for more than a day, body temperature would be obtained with an oral probe instead of the rectal thermometer used on previous flights, but the blood pressure monitoring system and heart rate measuring equipment were more or less the same. Countdown operations on the morning of May 15 went exceedingly well, with only a 4 minute hold for minor problems with some ground equipment. Just thirteen seconds past 8:04 am, MA-9 was on its way, the three Atlas engines sending thunder rolling across the warm Florida morning as Faith 7 justified its name. 'Sigma Seven, Faith Seven on the way,' said Cooper to the listening ear of capcom Wally Schirra in Mission Control. 'Feels good buddy.'

Atlas 130-D was a lady and gave Cooper little of the oscillations and movement reported by Glenn and Carpenter. Flying straight and true right down the middle of a plot limiting extremes of attitude permitted before abort, Faith 7 accelerated into space and put Cooper in a good orbit five minutes after lift-off. And it was a good turnaround too, as the spacecraft popped away from the cylindrical adaptor on top of Atlas and turned 180° to pitch down facing backwards as every Mercury had since the first reached orbit, albeit unmanned, 20 months before. Cooper took his time turning, carefully nudging the capsule through its attitude changes, taking 100 seconds to reach the desired position but consuming only 0.09 kg of propellant versus Schirra's 0.14 kg. As the capsule sped across the Atlantic and Indian Oceans, heat acquired during ascent (temperatures during the climb normally reached 705°c on the exterior walls) began to permeate the interior atmosphere – a normal occurrence on all Mercury flights and one that on MA-9 again caused temperatures to rise. Over Muchea the cabin read 47.8°c but thereafter declined as the cooling system won out over the finite heat pulse imparted to the capsule after launch.

Down below, the lights of Perth, Australia, were clearly visible, and at sunrise Glenn's 'fireflies' made their customary appearance – frozen droplets clinging to the exterior surface. Gus Grissom's deep and friendly voice told Cooper he was good for at least seven orbits as the capsule swept across Mexico, and Al Shepard talked to MA-9 when TV pictures came down every two seconds during the mission's first space-borne pass across Canaveral. Minutes later Cooper began a power down sequence designed to conserve electrical power and attitude control propellant as Faith 7 moved into the second orbit. On across the Indian Ocean again, Cooper dozed for a few minutes and then prepared to release the flashing light experiment scheduled for activation early on the third orbit. The sound of the small device breaking free was evident to Cooper inside the capsule but not for a full pass around the Earth would he unintentionally catch sight of the object he had strained to observe shortly after jettison. From that point on it came increasingly to view and Cooper was able to report its flashing beacon, although within four hours of release the device had drifted more than 20 km from Faith 7, increasing its separation continually.

As Faith 7 coasted through to orbit number four, passing over the Cape as most Americans were sitting down to lunch, capcom Wally Schirra passed along good words from the ground: 'Faith Seven, this is Cape Capcom. We are very impressed with the work you're doing. . . . We lay a pat on the back from Walt Williams.' It was indeed a smooth flight, less eventful than Wally Schirra's own mission. Electrical conservation was better than expected, as was the meager use of attitude control propellant. Suit temperature was down to a comfortable range never exceeding 21°c, and cabin temperature was satisfactory at around 35°c. During the next two orbits Cooper ate some of the prepared food placed on board and voided into the urine collection device for the all important analysis post-flight that would tell medical teams so much about his chemical reactions during weightlessness. Medical considerations were uppermost in a flight that promised to all but double the total accumulated time acquired in space by America's five previous space flights. Record breaking indeed, but only for American astronauts still dramatically short of the four-day flight of Russia's Vostok III.

On the sixth orbit Cooper prepared to release the inflatable sphere designed to demonstrate the reaction of two objects in close proximity during weightlessness. Nothing happened, nobody ever found out why, and the entire experiment was a non-starter. On the seventh orbit, passing over Hawaii on one of the last sessions with that station before a gap of about 15 hours as Earth revolved on its axis, Cooper mumbled comments into the microphone about a recently attempted meal: 'I'm eating a pot roast of beef. I've had considerable difficulty getting the water in it from this water device on the McDonnell water tank. I spilled

Cooper flies a simulated 22-orbit mission in the Mercury simulator at Cape Canaveral. →

Cooper checks over the instrument panel for his Faith-7 spacecraft with Mercury engineer Robert Graham at Cape Canaveral.

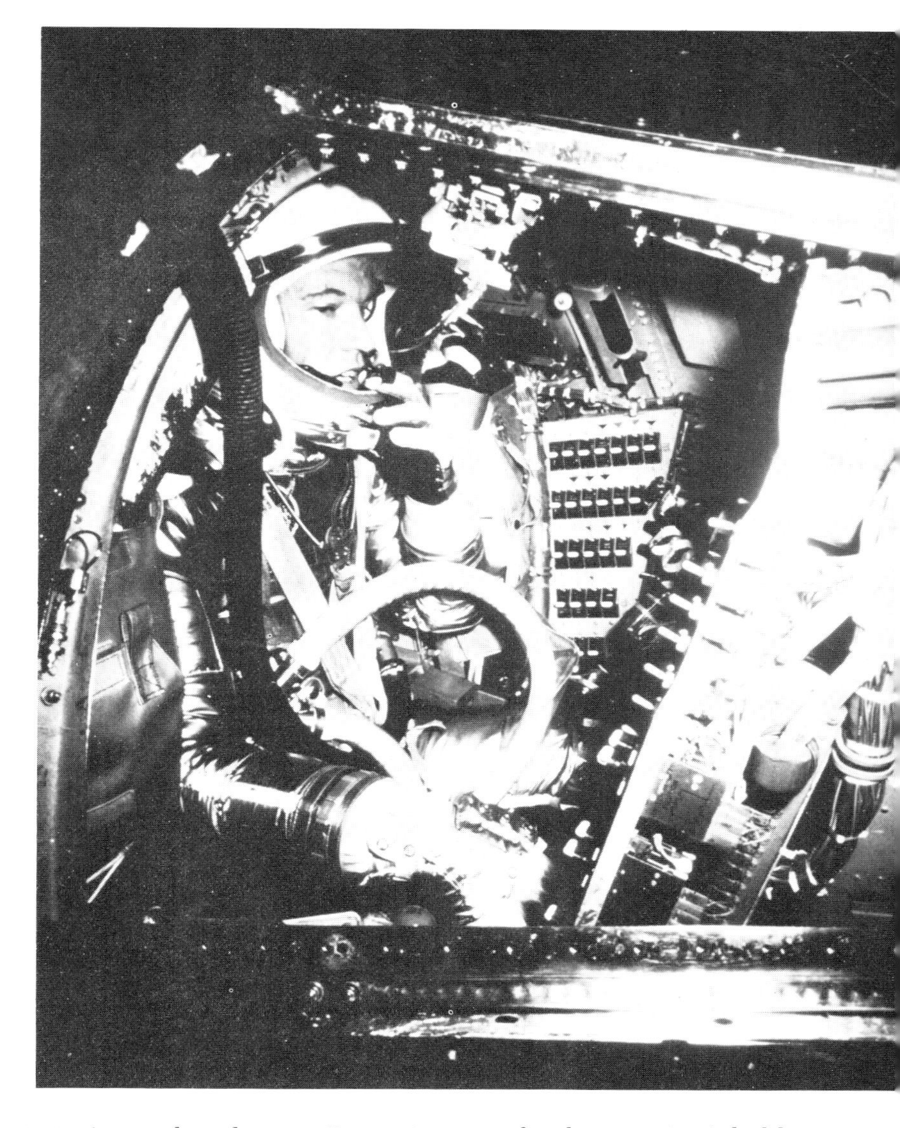

water all over my hands and all over the cockpit here trying to get some in it. I have succeeded in getting about half of it dampened and am proceeding to eat. . . . I am washing my face with a damp cloth now. Certainly feels good. . . . This is ridiculous. Come out of that damned ditty bag – Pandora's locker.'

Communications with Faith 7, increasingly intermittent as the capsule moved out of range of the established Mercury network, were sparse and brief. No lengthy commentaries like that heard from Wally Schirra's capsule, nor ebullient banter like that from John Glenn. Yet for all the brevity and attention to detail, a light-hearted conversation typical of the casual, confident, personality of Faith 7's pilot. On orbit 8, new records for Mercury as capsule 20 flew on past Sigma 7's duration. But for Gordon Cooper, checks on the spacecraft systems prior to beginning a rest period while the spacecraft drifted far from the fixed ground stations.

Whoever had planned that Gordon Cooper should rest while his flight path carried him across the visual splendor of South America, the Himalayas, Tibet, China and South-East Asia, must by nature have been a highly unimaginative individual. And so found Gordon Cooper as during orbits, 9, 10, 11 and 12, the Faith 7 pilot snapped away busier than a tourist, impressed by the host of colours, the visible undulations as mountains rose toward him more than 6 kilometers from the Earth's mean surface. Cooper was flying across some of the most dramatic territory on Earth, crossing continents and lands never before seen from space by an American astronaut. Peering through the clean, unpolluted air of rural lands, Cooper surveyed the surface below and produced the greatest contribution MA-9 would offer up to manned space flight: a description of things he believed he could see that for several years would bring disbelief and an inability to comprehend the view from space.

All the Mercury pilots before had been too busy, too concerned with the necessary demands of comparatively short flights, to sit and study the ground beneath the speeding spacecraft, passing by comparatively slowly as though from a high flying aircraft yet sufficiently fast to change one scene for another with exciting frequency. It was to be the greatest contribution of the flight, a statement that would bring scientists, military strategists and analysts to Cooper's report for the promise it held in surveying the planet from orbit. 'I could detect individual houses and streets in the low humidity and cloudless areas such as the Himalayan mountain area, the Tibetan plain, and the south-western desert area of the U.S. I saw several individual houses with smoke coming from the chimneys in the high country around the Himalayas. The wind was apparently quite brisk and out of the south. I could see fields, roads, streams, lakes. I saw what I took to be a vehicle along a road in the Himalaya area and in the Arizona-West Texas area. I could see the dust blowing off the road, then could see the road clearly, and when the light was right, an object that was probably a vehicle. I saw a steam locomotive by seeing the smoke first; then I noted the object moving along what was apparently a track. This was in northern India. I also saw the wake of a boat in a large river in the Burma-India area.'

Nobody had reported with such detail exactly what the view from space revealed, and no US astronaut had until then the time to take it all in with considered judgement for what exactly he was actually looking at. Cooper had no means of enhancing his view from orbit, it was all eye-ball description, and because of that many people who would discuss with Cooper long after the mission things he believed he could see felt that the astronaut was suffering some mild form of visual hallucination. No-one had believed it would be possible to make out tracks and roads, wisps of smoke drifting in the wind, and trains moving along rail tracks. The implications were significant, both for the future of Earth observation from space and for the value of manned vehicles in particular. At this time, just 5½ years after Sputnik 1, space photography was an undeveloped science, having grown largely from existing techniques and applications in high-altitude airborne photo-reconnaissance work. The pictures sent back by early weather satellites, of the type the first development vehicles were by now transmitting, were not at all clear, although sufficient for simple and crude observation of major meteorological phenomena. Now, Cooper's claim to define with certainty objects and artifacts on the ground below would prejudice his peers to send him back into space two years later for a longer, assisted, observation of objects on the ground. But for the time being

Cooper was unaware of the debate his claims would inspire, and on he coasted across the most beautiful scenery on planet Earth, observed for the requirements of science but indulged upon because of its sheer aestheticism.

But the events of the day were already taking their toll and Cooper felt compelled to rest. The last communication before sleep came at 13:35 GET while passing across the western Pacific and not for nearly eight hours did anybody talk to Gordon Cooper. But he slept fitfully and continuously awoke to voice comments into the onboard recorder before dozing off again. After the flight, he would remark that 'One indication of my adjustment to the surroundings was that I encountered no difficulty in being able to sleep. When you are completely powered down and drifting, it is a relaxed, calm, floating feeling. In fact, you have difficulty not sleeping. I found that I was catnapping and dozing off frequently. Sleep seems to be very sound. I woke up one time from about an hour's nap with no idea where I was and it took me several seconds to orient myself to where I was and what I was doing. I noticed this again after one other fairly long period of sleep. You sleep completely relaxed and very, very soundly to the point that you have trouble regrouping yourself for a second or two when you come out of it.'

On the fourteenth orbit, about 22 hours into the flight, Cooper ran through a systems check. Consumables were in very good condition. There was still more electrical power available than pre-flight systems evaluation had predicted there would be for this part of the mission and attitude control propellant was way above the expected quantities: about 65% remained in the automatic supply, and about 90% in the manual. Drifting, powered-down flight was paying dividends for the operational capability of the Mercury spacecraft. Across the Muchea tracking station Cooper exchanged humor with the capcom on duty and joked about receiving a cup of tea or coffee, having just emerged from the planned rest period. He was feeling in need of refreshment and responded by affirming that 'I will have tea, thank you. . . . In fact hot black tea would go very well right now!'

Passing over the Pacific, 21 hr 49 min into the flight, Cooper recorded the first declaration of faith transmitted from space in a prayer offered spontaneously to the silent tape: 'I would like to take this time to say a little prayer for all the people, including myself, involved in this launch and this operation. Father, thank you, for the success we have had in flying this flight. Thank You for the privilege of being able to be in this position, to be up in this wondrous place, seeing all these many startling, wondrous things that You've created. Help guide and direct all of us that we may shape our lives to be good, that we may be much better Christians, learn to help one another, to work with one another, rather than to fight. Help us to complete this mission successfully. Help us in our future space endeavors that we may show the world that a democracy really can compete, and still is able to do things in a big way, is able to do research, development, and can conduct various scientific, very technical programs in a completely peaceful environment. Be with all our families. Give them guidance and encouragement, and let them know that everything will be okay. We ask in Thy name. Amen.'

By now, Faith 7 was drifting back into range of ground tracking stations, repeating from orbit 16 the first orbits he had flown a day before. Project Mercury had broken through that invisible barrier – the day-long goal that had always been an objective, albeit little more than a dream-like hope – and come out on the other side of a full Earth spin. Cooper had girdled the globe in two dimensions: by orbiting in a path that carried him around the planet in 1½ hours, and by remaining in that fixed orbit as the planet spun on its axis beneath him. Physically, he was in good shape, adapting well to what was already a stressful experience. The cramped confines of a hot Mercury capsule are not conducive to a tranquil state of being.

Activities aboard capsule 20 increased in pace now, the assigned, but largely ignored, sleep period well behind Cooper as he sped on for the last leg before returning to the Pacific waters. Systems checks, alignment on the onboard clock that had drifted off by 16 seconds, events that hastened the passage of time for the orbiting astronaut. Over Africa and across the Indian Ocean on orbit 16, Cooper took more photos, and coming across Australia he counted off the exposures, methodically confirming to a precise photographic schedule carefully worked out on the ground before the flight. Orbit 17, more status checks, and a request for a new blood pressure reading from the Muchea station.

The flight was going very well, but systems problems were even then building up to give Cooper a taste of spacecraft malfunction that after the mission would be seen to have given him an uninvited, but nevertheless satisfactory, opportunity to display a cool attitude and a coordinated response. Good performance at times like that would increase his chances for another flight. On through orbits 17 and 18 Faith 7 kept up the sustained activity of scientific photography of Earth, the far distant horizon – or limb – of the planet, and infrared pictures of the clouds. And all the while the oxygen pressure in the capsule was slowly falling, almost imperceptibly. From the start of the mission pressure had been going down from the 258 mmHg, at which the environmental control system was set to operate, to about 181 mmHg. That was not dangerously low, but the trend needed watching in case it dropped lower toward the physiological threshold where Cooper would suffer permanent damage.

Also, partial pressure of carbon dioxide was going up slightly higher than expected, and now read 3.5 mmHg, the danger level being different under varying situations but no higher than 5–7 mmHg. Carbon dioxide would dull Cooper's mind, slow his reactions, and lull him into a state of apathy before quietly putting him to sleep. But there were other sensations associated with high CO_2 levels, and all astronauts were trained to detect in advance an increase in build-up of this potentially lethal gas that Cooper himself was producing. Like the falling pressure level, it was something to watch, and if it got too high Cooper could have switched to his emergency oxygen flow in the suit circuit; capsule within a capsule, the system was designed to provide this back-up.

But more ominous and threatening problems emerged when, on the 19th orbit, Cooper noticed the spacecraft was decelerating, an indicator that a re-entry sequence was triggered and as such certainly not the kind of information it was comforting to see with three orbits still to go! Cooper had been switching the display panel lights off to become dark-adapted for scientific tests and the 0.05 g light came on when he returned the switch to

The view from MA-9.

the 'dim' position. Responding to the malfunction, Cooper immediately turned off the switched fuse that also controlled the ASCS gyroscopes and cut the emergency fuse, inhibiting the relay functions.

On passing across the Hawaii station, Faith 7 reported the condition and told the ground about the fused switched settings. There was great concern now to define the extent of the peripheral effects, and across Guaymas the pilot was asked to apply power to the ASCS bus. This was necessary to determine the state of the amplifier-calibrator, the automatic pilot, and to probe the system further. On across the United States, benefiting from the advantage of continuous communication that would, only a few hours before, have been impossible, the pilot kept up the test sequence, slaving the gyroscopes to the horizon scanners and waiting to see if any drift was detected. It was not and that confirmed the suspicion that both gyroscopes and scanner were without power, as had been assumed from the reported switch positions, and that the spacecraft would have no automatic control through retro-fire and re-entry. It would be a manual operation all the way.

Suddenly, from several hours in monitoring, where flight controllers watched entranced by the flight's perfection, the Mercury Control Center erupted to a frenzy of activity. Flight controllers at the global tracking stations were warned to stand by for updates they would be required to pass to Cooper, and, across the *Coastal Sentry Quebec*, Faith 7 was asked to switch on the telemetry and the C-band tracking beacon. Radar data would be of prime importance because, operating the retro-fire sequence manually, the pilot would need to have precise information on ignition times. There was a possibility that the ASCS could provide automatic attitude control, and set up the necessary roll rate, during the descent phase after the 0.05 g event time.

A test carried out over Hawaii on orbit 20, where Cooper switched back on the prime and back-up 0.05 g fuses and placed ASCS in automatic mode, confirmed that indeed the logic had sensed the sequence initiation and that the capsule had begun to slowly roll as though it was about to re-enter. Fooled into stepping ahead of itself, the system proved it would take over past the normal 0.05 g point. But for the rest of the flight up to that point, the gyroscopes would be inoperable. Over *Coastal Sentry Quebec*, stationed north of the Mariana Island, instructions were relayed to Cooper on the manual retro-fire sequence and a status check for the preceding events; ignition of the solids would come 90 minutes later over that same tracking ship. The geometry of the final Earth circuit carried Faith 7 out across toward the Pacific and then to the Zanzibar station, with no communication over the Americas as the capsule swept down across South America and on toward Africa.

Over Hawaii on that 21st orbit, however, came another unexpected problem. With just two minutes remaining in the communication pass before going 'over the hill' and on toward South America, the ASCS inverter, a device designed to convert alternating current to direct current, blew a fuse. This was the main 250 v-amp inverter without which all automatic control would be completely lost, removing the hope that re-entry could be made with normal roll and attitude control. Switching to the back-up inverter, Cooper noticed that this too was inoperative, so he switched off AC power and decided there was nothing for it but to fly himself home on manual control throughout the remaining portions of the mission. In an almost casual way, with literally seconds to go before losing contact with Hawaii, Cooper informed the ground of his condition: 'Well, things are beginning to stack up a little. ASCS inverter is acting up, and my CO_2 is building up in the suit. Partial pressure of O_2 is decreasing in the cabin. Standby inverter won't come on the line. Other than that, things are fine.'

A short while before, Cooper had been told to take a dexedrine tablet to boost his awareness of the situation; the Cape flight surgeon was concerned that lack of adequate sleep during the rest period may have dulled his ability to remain alert. For about one-half hour the spacecraft was out of communication as it moved across to the Zanzibar station in Africa. Coming over the radio horizon more status reports came from the capsule, and final procedures from the ground about what to do during the manually controlled retro-fire and re-entry. It would be the first Mercury to come home entirely on the Manual Proportional mode, and it would be up to Cooper to fire his retro-rockets at the precise time required; moving at a speed of nearly 8 km each second a microsecond hesitation could make all the difference between a quick retrieval or a long wait for the pick-up.

Across Zanzibar, the station questioned Cooper on the condition of his increasing carbon dioxide level and advised him that he could go to emergency oxygen flow if the situation got too dangerous. Cooper informed capcom that the situation seemed to be steady and that he would monitor the rate. Only 18 minutes separated Zanzibar and the *Coastal Sentry Quebec*. John Glenn was the communicator that would talk Cooper down from his 22 orbit flight. Providing time calibrations by counting down from the ground, Glenn gave Cooper the information he required in a timely manner, more efficiently than had been possible during the rushed final orbit of Scott Carpenter's MA-7 mission when he too faced major systems problems. Cooper certainly displayed greater awareness and concentration than had his predecessor, but in Carpenter's case there had in truth been so little time to generate good work-around procedures and transmit them to the spacecraft. However, Cooper held spacecraft attitude in the retro-fire position and punched off the three solids exactly at the prescribed time.

His manual control of the spacecraft's pointing angle was superb, and the following events cycled off with deft precision. Only four minutes separated the tracking ship from the Hawaii radio horizon but Cooper had firm control of the situation and when the capsule landed it was an estimated 1.8 km off target! – the best yet, from a retro-fire and re-entry on manual authority. Recovery operations were equally as smooth: a helicopter dropped pararescue teams who placed a flotation collar around the spacecraft, while other choppers moved in, one of them to pick up the spacecraft and carry it to the *Kearsarge*, which had been only 7 km away at splashdown. Cooper blew the explosive hatch 40 minutes after his spacecraft hit water, and lay there in the couch while a cursory medical inspection ensured his capacity for standing erect in the 1 g environment of planet Earth. Cooper felt dizzy as he stood and slowly adjusted to the force of gravity. But then he was away for a more concentrated medical examination and de-briefings aboard the carrier.

Ahead, celebrations, a telephone call from President Kennedy, and tickertape parades – for many, the worst, the most arduous, part of the mission. Faith 7 had flown for 34 hr 20 min. It was a magnificent climax to a manned space flight project that had gone some way toward redressing the balance, an inventory of successful flights on which to base the future endeavors so recently defined. In special ceremonies at the White House, where all seven Mercury astronauts were present, senior NASA officials were awarded the medals of exceptional government service. In New York, it was reported that 2,500 tonnes of tickertape rained down on the 4½ million people estimated to have lined the streets when Cooper was paraded through America's gateway to the continent.

Public interest had been intense. In New York's Grand Central Station, an estimated 8,000 people stood and watched televised sequences of the launch. In Europe, aborted attempts on the first day at transmitting TV pictures from the United States via the Relay 1 communication satellite were followed on the second day of the mission by good coverage for the excited European audience. In the other direction, Americans saw their efforts praised and applauded in a special program beamed to domestic US television via Relay 1. From Radio Moscow: 'The Soviet people, who were pioneers in space, are hoping that this scientific experiment works out successfully. We sincerely hope that its results will serve the development of peaceful cooperation and above all cooperation between the Soviet Union and the United States in the study of the laws of the universe.' But not everybody on the home front was as convinced as NASA that, in the words of Brainerd Holmes, 'Although manned flight is more expensive than unmanned flight in the development phase, the increased reliability of a manned vehicle will ultimately enable us to carry out complex tasks in space more effectively and, very possibly, at less expense than with automatic equipment.'

President Kennedy was publicly applauding the achievements of NASA: 'I know there are lots of people who say, "Why go any further in space?" When Columbus was halfway through his voyage, the same people said, "Why go on any further?" And they want to stop now. I believe the United States of America is committed in this decade to be first in space, and the only way we are going to be first in space is to work as hard as we can here and all across the country, and support not only Major Cooper, but all those who come after him.'

Dwight D. Eisenhower, late of the Presidential office, reflected in the light of recent developments the stance he had always taken: 'Now clearly the strong competitive spirit of the American people has been aroused by the so-called space race. But let us step back for a moment from emotion and be objective. . . . We didn't and don't want to be a second-best nation, not in any important field, and certainly not in total accomplishment. But can we best maintain our overall leadership by launching wildly into crash programs on many fronts? This is where we seem to have got out of focus. Let me make it perfectly clear that we should have an aggressive program of research and exploration, so broadly based that in the long run there will be no question of our space leadership; but we should pursue it in an orderly, step-by-step way to enlarge systematically our knowledge of the scientific, military, and industrial potentials in space. . . . The annual cost would not include money for stunts and unnecessary contests. . . . Proud as we may be of our astronauts . . . this racing to the Moon, unavoidably wasting vast sums and deepening our debt, is the wrong way to go about it . . . the average citizen may be loath to question the huge sums now being requested. But he should. . . .'

To Vice-President Lyndon B. Johnson, chief architect of NASA and now Apollo, it was very clear and rather simple: 'I do not believe that this generation of Americans is willing to resign itself to going to bed each night by the light of a Communist moon. . . .' And further, on the question of Cooper's flight in particular: 'In 1942, President Roosevelt called together our wartime leaders for the final decision on continuing or abandoning the Manhattan project which produced the atomic bomb. One of the most eminent leaders present heard presentations from all sides. Then, he solemnly gave President Roosevelt his verdict: "The bomb will never go off – I speak, of course, Mr. President, as an expert on explosives." In that first, uncertain spring of the space age 5 years ago, some conscientious experts took the same attitude toward Project Mercury. History has proved them grossly wrong . . . Hitler once predicted the Nazis would wring England's neck like a chicken . . . Winston Churchill said to the Commons, "Some chicken. Some neck." We have heard some say recently that the civilian space program is only "leaf-raking." Considering . . . the vast technological cooperation which made his mission a success – I would say today, "Some leaf. Some rake." '

But Cooper's flight alone was not the reason for what seemed then to be a veritable outpouring of opinion to question or debate the central issue about priorities, nor indeed was it caused by the end of Mercury flights. It was perhaps the last vestige of the response to national and international failures that the Kennedy administration tried to reverse by claiming major stakes for American fortune. From first announcing a dramatically accelerated space program in May, 1961, which included the expanded role of Apollo for landing on the Moon, Kennedy carried the nation with him for no more than a year. Early in 1962 there were ominous rumblings from both public and Congressional quarters that sought to question the seemingly vast expenditures on space technology. Because the decision to go to the Moon had been made more for the impact it would have on international affairs than for any domestic reward it could secure, danger to America on an international level allayed for a while the criticism about Apollo that was inevitably rising.

By early 1962 the bubble had seemingly burst, but the coup de grace on NASA's honeymoon period was postponed when Soviet missile bases on the island of Cuba appeared to threaten the security of the United States. Because a vigorous and successful space program *was* seen as an effective bulwark against communism, the presence of an overt threat stiffened resolve at a domestic level to preserve and be grateful for the major technology venture already in being. For a while the whole NASA machine was off the hook, because it was credited with being partly responsible for the Russians backing down when Kennedy blockaded Cuba and threatened to go to war. The situation was so explosive that nuclear armed bombers were actually in the air ready to fly to their targets when Kruschev pulled back from Cuban waters the supply ships that carried missile components.

The placing of missiles on Cuba in the first place had been a Soviet concession to Castro's feeling that American plans to invade the island might not stop at the Bay of Pigs fiasco. Honouring Castro's request, Russian missiles began to arrive and only when American spy photographs revealed the launch sites did the Kennedy administration reach the most dramatic peak of its tenure. But after the confrontation subsided, questions were again raised about the relevance of a vast space program to the domestic needs of the American people.

Many people, especially in the scientific community, had thought a vigorous space program would add measurable value to the pool of knowledge about the world and its place in space, claiming justification for such an effort on the grounds that it would establish a precedent by which Man would measure human progress and that it would bring untold benefits via applied science. Politicians thought otherwise, as reflected by the Kennedy decision, and hailed space as the greatest flag-waving bonanza for national propaganda ever devised. But there was a third reason, one that an increasing number were attracted to, and one that came from the very depths of the scientific community so recently disillusioned by the replacement of true science by a vast engineering commitment such as Apollo.

It said that the prime reason why circumstances had conspired to give the nation such a bold objective was that there is, within each free society, a nucleus of inspiration seeking to do the seemingly impossible and that rather than pursuing an irreversible path, where once challenged there was no alternative but to strike for dramatic options, the Moon decision had been taken because of some undefined urge to excel. Dr. Harold Urey, a world famous Nobel Laureate in chemistry, framed the analogy to ancient Greece: 'Athens built the Acropolis. Corinth was a commercial city, interested in purely materialistic things. Today we admire Athens, visit it, preserve the old temples, yet we hardly ever set foot in Corinth.' It would, said Urey, be possible for America 'to be a fat cat, merely be wealthy, rich, comfortable, and not take a dare,' and in the process lose all identity with progress and invention. Dr. Urey was convinced that the fraction of people in society concerned most with internal domestic issues such as more education for deprived children, better health care for the aged and the poor, greater attention to the rights of minority groups, was about equal to those who were 'interested in doing these daring things' like space exploration or some other risky endeavor.

But he felt that, while it was logically more acceptable to commit large resources to solving social and domestic ills, more would actually be done in those areas if the nation stood proud in the shadow of great accomplishments and meritorious achievement. It was a view gathering increasingly favorable support and was encompassed by a single quote from Dr. Urey: 'There is nothing so deadly as not to hold up to people the opportunity to do great and wonderful things, if we wish to stimulate them in an active way.' In other words, proper attention to Earthly needs of the poor, the depressed and the downtrodden, would naturally evolve from dynamic, articulate, spirited awareness of the great goals for Man and the society he conspired to erect.

Dr. Colin S. Pittendrigh, Professor of Biology at Princeton, saw Apollo as a nationally advantageous project 'Largely because it is so tangible and exciting a program and as such will serve to keep alive the interest and enthusiasm of the whole spectrum of society. . . . It is justified because . . . the program can give a sense of shared adventure and achievement to the society at large.'

Britain's noted astronomer Sir Bernard Lovell: 'The challenge of space exploration and particularly of landing men on the moon represents the greatest challenge which has ever faced the human race. Even if there were no clear scientific or other arguments for proceeding with this task, the whole history of our civilization would still impel men toward the goal. In fact, the assembly of the scientific and military with these human arguments creates such an overwhelming case that it can be ignored only by those who are blind to the teachings of history, or who wish to suspend the development of civilization at its moment of greatest opportunity and drama.'

Martin Schwarzschild, Professor of Astronomy at Princeton, the man who would give his name to the core of a black hole: 'The idea of man leaving this earth and flying to another celestial body and landing there and stepping out and walking over that body has a fascination and a driving force that can get the country to a level of energy, ambition, and will that I do not see in any other undertaking. I think if we are honest with ourselves, we must admit that we needed that impetus extremely strongly. I sincerely believe that the space program, with its manned landing on the moon, if wisely executed, will become the spearhead for a broad front of courageous and energetic activities in all the fields of endeavour of the human mind – activities which could not be carried out except in a mental climate of ambition and confidence which such a spearhead can give.'

Dr. Lloyd V. Berkner: 'Human society – man in a group – rises out of its lethargy to new levels of productivity only under the stimulus of deeply inspiring and commonly appreciated goals. A lethargic world serves no cause well; a spirited world working diligently toward earnestly desired goals provides the means and the strength toward which many ends can be satisfied . . . to unparalleled social accomplishment.'

Such were the views and deliberations of men averse to cluttered rhetoric, great scientists without sociological portfolios, but men who nevertheless claimed the betterment of human beings as the all-embracing justification for a bold space initiative. Kennedy, indeed the entire political machine, had with but few exceptions seen the space program as a national asset for partisan reasons or professional cause. It was left to the scientists and the thinkers, even the writers of that year when Gordon Cooper closed the book on Mercury, to liberate the project from lesser levels that tried to reason for its continued existence. In 1963 Alfred North Whitehead wrote that 'The vigor of civilized societies is preserved by the widespread sense that high aims are worth-while. Vigorous societies harbor a certain extravagance of objectives, so that men wander beyond the safe provision of personal gratifications. All strong interests easily become impersonal, the love of a good job well done. There is a sense of harmony about such an accomplishment, the Peace brought by something worth-while.'

Whitehead's observation would not do for the cost conscious sitters on Capitol Hill, and the Senators and Representatives elected to pay the nation's bills were not prepared to put high price on nebulous values or peripheral side effects. By the end of 1962, and certainly when Cooper stepped ashore from his record flight, the honeymoon was over. Had the decision not been made with the haste in which it emerged it is questionable whether Man would ever have gone to the Moon, or indeed whether any great manned space adventure would have begun. This much was readily apparent by 1963, the year in which deep and searching questions were asked about the goals that had, less than two years before, seemed unavoidable in their construction. Kennedy continually doubted the value of the space race for domestic policies he knew he would have to rely on for any future popularity or election potential – foreign affairs never draw faint hearts to a President. Russia set in motion an assemblage of spectaculars far beyond the value of such events, and in so doing misjudged the American mind, which is never likely to be bludgeoned into subservience like the will of some petty satellite state under the crushing treads of tank tracks. For its part, the United States misjudged this initial surge forward as but the vanguard of a long and lasting parade of technical successes, drawing all the while the international support it courted. It is ironic that Kennedy's Moon landing decision came at a time when this momentum was all but spent, when the American national mood was synthesized for action, and when technical studies laid a basis for promised accomplishment.

But Russia was made to go further than she planned, and to counter increasingly sophisticated space endeavors with reciprocal initiative. It is ironic too that Kennedy invited Kruschev to a cooperative space endeavor imploring the Soviet leader to join America and 'go to the Moon together,' just one month after rallying Congress to the Moon goal – further evidence here that the President had acted out of haste in appealing to Johnson for some means of beating the Soviets, a haste he very quickly regretted, a posture from which he rapidly sought to withdraw. Within six months of the Moon commitment, the Department of Commerce pressed hard for international cooperation by setting up a workshop for foreign and domestic participants; unfortunately, the Russians ignored the event and failed to turn up. Again too the following year, Kennedy's administration tried desperately hard to excite the Soviet mind and for a while seemed to succeed. In an exchange of letters early in 1962, Kennedy and Kruschev moved closer to cooperation on some aspects of space activity than at any other time. There was little hint in early correspondence of any manned space flight venture between the two countries, but as the contact between heads of state deepened there were certain subtle suggestions that any preliminary set of talks that might be arranged for the near future should not eliminate the possibility of joint manned flight operations.

In March, 1962, Deputy Administrator Hugh Dryden met with Academician Anatoli A. Blagonravov in New York to sound out the possibility of formal talks. Dryden was impressed by the largely apolitical nature of Blagonravov's conversation, detecting in the man a genuine desire to lay the ground for the start of serious discussions. Back in Moscow a month later, Cosmonautics Day, arranged to celebrate the anniversary of Yuri Gagarin's space flight, provided a platform from which Gagarin and Titov praised the concept of cooperation, and even the President of the Soviet Academy of Sciences, Mstislav V. Keldysh, verbally supported the concept. At the end of May Dryden and Blagonravov met a second time, this time for nearly two weeks in Geneva, during which period they secured three projects of mutual interest to be considered as candidates for cooperation; the projects embraced scientific goals related to a study of the Earth and its physical environment and were ratified by an exchange between Webb and Keldysh in October.

Debate in Congress over the intentions of the Soviet government ranged long and hard from positive approval of cooperative initiatives to outright denial of any validity to claims of Soviet benevolence. But soon the issue was moving to new horizons and while Congress generally approved of moves to placate the international scene, Kennedy's assistant McGeorge Bundy sent the President a memorandum in which he drew attention to the need for consolidation down one track or the other: cooperation or competition. Recognizing the frayed state of domestic opinion regarding an intensive and costly race with the Russians, Kennedy was aware that decisions made at that time would have significance far down the political road. A day before Bundy sent the memorandum, on 17 September, 1963, Bob Gilruth openly expressed his concern at talk about Soviet-American cooperation when he doubted the ability of any technical development that could satisfactorily unite the two technologies. The following morning, after receiving Bundy's memo, Kennedy called Webb to the White House and informed him of the general pattern of thinking then permeating the Executive Office: cooperation would get the administration off the hook as far as criticism of space budgets was concerned, provide a very real justification for the initial Moon landing goal (it would be seen to have pressured the Soviets into agreements), and present a stable platform for international relations, so easing foreign pressure on the domestic economy.

Kennedy also wanted to know if Webb was 'sufficiently in control' to prevent any of his administrators or managers making impromptu statements that could prejudice Soviet responses. Webb assured the President that if that was what the administration wanted, he would keep tight rein on his own people. On 20 September, Kennedy went before the United Nations and called upon the Soviet Union to join the United States in a cooperative flight to the Moon, proposing that the two countries link up in this technological feat as a symbol of the ability of two nations to live together on Earth in peace and harmony. It was, of course, motivated by a different reason, but, in the response, that was not to matter. The Soviets virtually ignored the appeal and the US Congress was visibly hostile to any such suggestion. Having got his administration into an irreversible commitment, the Washington Fathers were not about to help Kennedy find his way back to national acceptance. In fact, in language attached to the bill authorizing NASA funds for 1964, Congress stipulated that 'No part of any appropriation made available to the National Aeronautics and Space Administration by this Act shall be used for expenses of participating in a manned lunar landing to be carried out jointly by the United States and any other country without consent of the Congress.'

Less than two months before Kennedy's appeal, Sir Bernard Lovell sent Hugh Dryden a report in consequence of a visit he made to the Soviet Union during which he was shown extensive radio and astronomy equipment. Keldysh had, said Sir Bernard, told him that the Soviet Union had abandoned an original idea to send men to the Moon because they could neither justify the expense nor satisfactorily find a way around the technical problems. Apart from that, they suspected the solar radiation translunar cosmonauts would be exposed to could permanently damage human tissue and body organs. Consensus of opinion on the matter had directed the Soviet lunar research program to concentrate on automated exploration using unmanned vehicles and space probes. Between the issuance of the Lovell report to the press and the announcement of September 20, Kennedy supported the Apollo project with the following statement: 'The point of the matter always has been not only of our excitement of interest in being on the moon, but the capacity to dominate space, which would be demonstrated by a moon flight, I believe is essential to the United States as a leading free world power. That is why I am interested in it and that is why I think we should continue, and I would not be diverted by a newspaper story.'

Whether enjoined by the newspaper story or not, the President was to make his appeal to the Soviet leader just a few weeks later. The momentum of Project Apollo had gone just too far down the road for a change of national policy, even though Kennedy nursed to the end a bitter regret that he had ever had to make the commitment. It was too late to stop. On 22 November, 1963, John Fitzgerald Kennedy succumbed to an assassin's bullet at the parade in Dallas, Texas, he had been urged to avoid. It was nearly 2½ years to the day since he had stood before Congress, shoulders slightly hunched, to rally Americans forward on one of their greatest national commitments. Now the space program's chief architect, Lyndon Baines Johnson, was to hold the reins of Presidential power for more than five years. During that time, he would also be the architect of its downfall.

During the six months in which there flared up for a moment the specter of an about-face for NASA's Apollo project, America's space activity began the transformation from an essentially nascent organization into what would soon become a competent operational platform from which to build a Moon mission. No sooner had the flight of MA-9 concluded than speculation rose anew on the possibility of the agency flying another long duration flight. On the day after Cooper's splashdown, Chris Kraft from MSC's Flight Operations Division claimed a greater capability for subsequent capsules by carefully totting up Faith 7's performance. Kraft found that, even by retaining respected mission rules and safety margins, NASA could fly a Mercury mission for 92 orbits of the Earth. A flight lasting more than 5½ days would indeed have captured public interest in the wake of Soviet missions that had already extended to all but four days. But plans tentatively laid down the previous September, even before the six-orbit flight of Wally Schirra, had cancelled the four day-long flights scheduled to commence with MA-9. Instead, the three deleted missions would carry their objectives forward to early flights of the Gemini vehicle.

Beginning 7 June, 1963, Holmes, Gilruth, Williams and Kleinknecht spent two days deliberating with Webb, Dryden and Seamans on the immediate future course of manned space operations. It was to be Webb's decision in the final analysis, but his verdict was agreed by Dryden and Seamans. There would be no more Mercury flights. And as things turned out it was just as well, for money was tight and schedules were moving slower than hoped for with the project designed to rehearse techniques and operations essential to Apollo. When NASA finally decided late in 1962 that it would fly Apollo to the Moon on top of a single Saturn V – the designation had changed from C-5 to V three months before the last Mercury flight – it bequeathed Gemini even greater responsibility. With highly complex rendezvous and docking operations coming at several key stages in the flight, NASA astronauts would need the experience provided by two-man flights to build confidence in their ability to meet and connect with another vehicle in space. In June, 1963, it seemed likely that Gemini would not carry men into space before 1965; although the first astronaut mission was scheduled for October, 1964, the increasing pace of delays and schedule slips denied confidence to the managers honing what had become a very sophisticated vehicle.

And then, nearly a month after Cooper returned to Earth, the Russians launched Vostok V with Valeri Bykovsky on board. His 4.7 tonne spacecraft got off the ground at 3:00 pm on 14 June and moved quickly into the familar orbit inclined approximately 65° to the equator. There was little doubt about the significance of the mission. It had been ten months since the dual flight of Nikolayev and Popovich and both rendezvous and long duration objectives were suspected. At that time nobody outside the Soviet Union knew what Vostok really looked like, nor what it was capable of doing. Its range and operational flexibility were a complete mystery and even the rocket used to launch the spaceship was an unknown development of some vague reference to ballistic missiles; the relationship between space rockets and military missiles was the persistent excuse made by the Russians for the great secrecy that shrouded their endeavors. Because of the lack of information surrounding Soviet manned space flight, speculation was, and remains, a high factor in determining future patterns of development. But in 1963 it was little more than a wild guessing game, based on logical patterns of activity laid down by American spacemen.

The purpose of the Vostok V flight was reportedly never guessed, although rumor leaked from Moscow on the days preceding the flight that something special was about to happen. Launch day came and went, the cosmonaut still circling the globe, and the next day was the same. The usual flow of Tass news reports kept up a steady commentary on the routine operation of tests, experiments, eating, and sleeping. 15 June saw Bykovsky complete a full day in space and on through another night for Earthlings he swept around the globe. Then, on 16 June, after two days in space, when it seemed that a dual flight like that performed the previous year was not the reason for Vostok V, a second spacecraft was launched from the Baykonur complex. On board, the world's first cosmonette: Valentina Vladimirnova Tereshkova, launched shortly after noon local time to arrive in orbit minutes later only 5 km from the spacecraft carrying Valeri Bykovsky. Both could see each other's ship although as time went by the slight difference in their respective orbits, and perturbations in the Earth's gravity field, would move them increasingly apart. The flight was basically similar to that of the preceding mission but the added interest in medical studies of space effects on a woman was to be heightened still further by subsequent events.

Tereshkova had been in training for a year, assisted and encouraged by other cosmonettes, with specific preparation for Vostok VI under way since August, 1962. She did not adapt to the weightless environment as well as her male colleagues, although this may have been due to her inexperience with high-g flight and rapid accelerations. Tereshkova was a parachutist, not a pilot, and came from a humble factory job to train for space. There is circumstantial evidence that the Soviet scientists were intent on carrying out a unique experiment made possible by the genuine relationship she had with cosmonaut Nikolayev, the pilot of Vostok III. Concern about the effects of cosmic radiation on the human body was greater in Russia than elsewhere and strenuous tests were conducted on male and female cells in an attempt to study the response of living things to unnatural doses.

It has been reported that Professor Alexander Neyfach discovered the ability of some cells to continue growing long after the nucleus is destroyed and for embryonic development to be moderated accordingly. Tereshkova and Nikolayev were to marry in November, 1963, and to have their first child the following June. Considerable care was taken to study every phase of the child's development but the girl – called Yelena – grew into a healthy young woman without apparent side effects from the unusual dose of primary cosmic radiation her parents had voluntarily received. But all this was still ahead for Tereshkova as she fought to overcome extreme nausea and bad effects of disorientation. Much has been made of the problems experienced by this first woman in space, most of it for political affect, but there were undoubtedly times on her flight when serious concern overcame the medical specialists watching her condition on the ground. There seemed little compensation for not having had a lengthy acclimatization to stresses and strains of high-powered aircraft, and the ability of ordinary people to step aboard a manned space vehicle was a dream fast receding into the future. Conditioning was a weak point in Tereshkova's preparation for Vostok VI and the effects she experienced were a valuable indication of the need for careful cosmonaut selection.

Tereshkova's spacecraft came down to Earth after a flight lasting three days, the capsule thudding down empty while the cosmonette landed by parachute, having ejected during the descent. It was 11:20 am, 19 June. Bykovsky went round on two more orbits before he also returned, similarly ejecting, to land at 2:06 pm. The two spacecraft were 800 km apart and the latter – Vostok V – had set a record that would stand for more than two years: five days in space, minus 54 minutes! Recovery forces were quickly on hand, holding back several inquisitive spectators who milled around the charred sphere. Less than a week later the two pilots were received at Moscow's Vnukovo airport by Premier Kruschev in person. The entire event had gripped Muscovites, transforming them into a cheering throng lined all along the road that led to city center. All Russia's cosmonauts who had flown in space were there, following the celebrities in black limousines while the people shouted, waved flags and scarves and literally jumped up and down in the streets. Kruschev's post-Stalinist era was apparently working; Russians had not laughed and cheered so much for almost a generation, and it was all very good for the portly little man with the rough peasant accent who had worked so long and hard to control the Kremlin.

The flight of Vostoks V and VI caused a stir in America, not for the duration of the flight so much as for the sex of Vostok VI's pilot. Senator Ernest Gruening from Alaska was livid with an article by Clare Boothe Luce in *Life* magazine which said that 'The astronaut of today is the world's most prestigious idol. Once launched into space he holds in his hands something far more costly and precious than the millions of dollars' worth of equipment in his capsule; he holds the prestige and the honour of his country. . . . But the astronaut is also something else: he is the symbol of the way of life of his nation. In entrusting a 26-year-old girl with a cosmonaut mission, the Soviet Union has given its women unmistakable proof that it believes them to possess these same virtues. The flight of Valentina Tereshkova is, consequently, symbolic of the emancipation of the Communist woman. It symbolizes to Russian women that they actively share (not passively bask, like American women) in the glory of conquering space.' Sen. Gruening expressed concern that NASA had not envisaged just this sort of situation and criticized the agency for having seemed to discriminate between male and female astronauts. There was no end to the political responsibilities the space agency was retrospectively plagued with; no limit to what it was supposedly to provide Congress with.

Vostok had been stretched to its capacity. With two dual flights now behind them, Soviet space engineers were already being made to think in terms of greater achievements. But the spacecraft itself was already at the limit of its potential. In several regards, the impatient Kruschev had shown his hand too soon, been challenged too readily by the United States, and now had to embark on a desperate scheme to fulfil the pretence of being ahead in the technology race. Brashly displaying the might of his ace tool – the big, powerful, launcher used for Sputnik, Vostok, and other prestige flights – before the less easily bought experience of space technology had been acquired, he would now pay the price for having professed a capacity far beyond that possessed by the Soviet Union at the time. Well aware of the great strides now taking place in the United States, a feeling for pace, accelerated momentum, invaded Russian plans and in the wake of news about NASA's two-man Gemini, Kruschev took a gamble and ordered a daring variation of the established Vostok theme. Meanwhile, Gemini was becoming a thorn in NASA's side.

For two years now the project had been a major endeavor, tying up personnel and valuable engineering talent in a program

Soviet cosmonette Valentina Tereshkova checks Vostok equipment with a colleague.

far removed from the grand objectives of Apollo. In every dimension, Gemini was to be a stop-gap measure, a filler between the limited capacity of the one-man Mercury capsule and the broad capability of the three-man Apollo CSM. Yet even as the project got under way during the first half of 1962, it was becoming increasingly apparent that it could be more than just a training device for later missions. It could be as useful in a limited sense as Apollo was expected to become, providing a full operational instrument for manned Earth orbital activity. What had started out as an exercise for a better Mercury, had very quickly become a vital component of the entire Moon machine and, as it would turn out, an essential element without which Man would have been unable to reach the lunar surface when he did.

The Manned Spacecraft Center issued McDonnell a letter contract to build the 'Two-Man Spacecraft' on 15 December, 1961. There was no need here for requests for proposals, no bidders conferences. As a progression from Mercury, the intent of the project would have evaporated under the necessity to design a completely new, two-man, vehicle had another contractor been required to do the work. With all the Mercury experience behind them, using Mercury as the design base, McDonnell engineers were the only choice. By mid-January, 1962, Bob Gilruth had set up a Gemini Project Office under Ken Kleinknecht, Chamberlin was Project Manager, and André Meyer had joined Chamberlin as his deputy. Spacecraft systems would be managed by Duncan R. Collins, brought from General Dynamics by Chamberlin, while launch vehicle management would be the responsibility of Willis B. Mitchell, Jr., also from General Dynamics. Chamberlin was instrumental in setting up a group of six coordination panels, aimed at easing the development of critical Gemini elements into a homogeneous whole. Three panels covered spacecraft activity (mechanical, electrical, and flight operations) and three covered support elements (paraglider, Atlas-Agena, and Titan). Each panel would meet once a week and submit their decisions up to Chamberlin, effectively tying the Project Manager in with day-to-day developments at a technical level, but liberating the executive control of specific developments.

Gemini was unique in one respect. It was the first and only American manned space project not to go before Congress for formal sanction before approval as an official program. That was made possible by the fact that it was an outgrowth of Mercury and as such could be sustained financially by funds requested at the usual time. The formal presentation to Congress came as part of the NASA Fiscal Year 1963 budget plan brought before the two Houses in the spring of 1962. By then the project had taken on several sub-contractors and there was little concern by Congressmen and Senators for the comparatively meager sums it required. Mercury had come out at about $400 million, Apollo was known then to require several billions of dollars by the end of the decade, and Gemini was said to need about $350 million.

In designing the Mercury Mk. II, a paper predecessor to Gemini, Chamberlin had sought to replace the layered system concept with a modular approach that placed all serviceable equipment between an inner pressure vessel, containing the atmospheric environment for the two astronauts, and an outer structural shell to which would be attached shingles similar to those attached to Mercury. Detachable panels would afford access to the modular components for rapid servicing and trouble-free countdowns. Speed was of the essence in Gemini, not haste for that was an unnecessary stop-gap for bad planning, and success with key operations would hinge on the project's ability to respond quickly to real-time situations. Considerable effort was to be placed on the spacecraft's ability to get off the launch pad on time. Rendezvous plans being formulated in 1962 envisaged an Agena target vehicle placed in orbit by an atlas rocket similar to the boosters used for Mercury. Some time later the manned Gemini vehicle would ascend to an orbit from where it could begin the rendezvous maneuvers necessary for bringing the chase vehicle to the target. The two could then dock together and form a single rigid structure.

Launch 'windows' were the key to success with rendezvous, and a single window would appear for each orbit of the target vehicle. With only a limited capacity for maneuvering, the Gemini would have to be launched within very tight constraints, or wait another orbit for the target to appear. Rapid solutions to unexpected problems on the launch pad during countdown were an essential part of the quick-fire plan for rendezvous. The lengthy holds experienced during Mercury countdowns could be devastating for Gemini rendezvous flights. With launch windows coming only once every 90 minutes – the time taken for the Agena

With a cylindrical nose designed to dock in space with a modified Agena, and a base diameter compatible with the Titan II launcher, Gemini carried two astronauts beneath hatches hinged for space walking.

to go once round the Earth – a succession of unexpected holds could keep the flight in tension for several days, an unacceptable situation. Fortunately, in using the only launch vehicle available for sending Gemini into space, flight planners would have the benefit of storable propellants; Titan II used hypergolic fuel and oxidizer which would greatly ease the problems encountered with the cryogenic oxygen in Mercury's Atlas.

But improvements to the operating schedule of Atlas was itself a necessary part of streamlining flight operations. Responsible for launching the Agena target vehicle, Atlas could keep a manned spacecraft waiting uncomfortably long if it was unable to send the rendezvous vehicle into orbit on time. For Gemini rendezvous flights, the Cape would be preparing, fuelling, checking and counting down two separate launch vehicles, one of them manned, for a staggered launch but with the second unavoidably paced by the first. Chamberlin knew well the problems with Mercury, where replacement of a failed system could entail the removal of several other systems totally unrelated to the offending component. It was, after all, promise of easier accessibility that gave Chamberlin the excuse for designing the Mk. II and now that this endeavor had evolved into Gemini, the two-man spacecraft would be built around the concept of modularized packages separately installed behind removable panels.

Much of 1962 was taken up with selecting contractors for specific system designs firmed up the previous year. Since Mercury was to form the basis for Gemini systems concepts, it was only logical that wherever possible responsibility for fabricating Gemini components would go to contractors familiar with the one-man spacecraft. Such a case was AiResearch Manufacturing Company, who were directed in the February to proceed with development and fabrication of an environmental control system. Demands would be high for performance and reliability, but consumables quantities too would be considerably in excess of Mercury demands. With planned flights of 14 days duration, the optimum system would put the emphasis on trouble-free operation over long periods. Nevertheless, as manufacturer for Mercury's environmental unit, AiResearch was familar with the exacting requirements of manned space vehicles and took basically the same system, married two together, and produced the Gemini ECS.

Again, as with Mercury and the planned Apollo command module, Gemini would have a pure oxygen environment pressurized at a nominal 3.45 N/cm^2, employing lithium hydroxide for removing the pilots' exhaled carbon dioxide, and a charcoal cartridge for taking away odors from the fouled gas. In addition, an active coolant circuit would also be required, and demands here too would be greater than for Mercury; Gemini would move through different thermal environments by having to point in fixed directions for comparatively lengthy periods, would be in space considerably longer than any manned vehicle had then experienced, and would have greater heat generation to handle from systems inside the spacecraft designed to draw heavy electrical loads. In fact, the spacecraft and crew would generate three times the heat generated inside the one-man Mercury, and that system's cooling ability had been drawn into serious question on several flights where the astronaut was overheated.

Designed from the outset to handle considerably greater thermal loads, the Gemini cooling system adopted the principle where coolant fluid flows at controlled rates through cold-plates and cold-rails to which would be mounted the equipment primarily responsible for generating heat. In this way, a plate or rail would absorb the heat by conduction and pass it by the same process to fluid on the inside. When the fluid absorbed its capacity it would move through tubes and pipes to radiators located in a large adapter section at the rear; this adapter would serve to contain all the weighty systems and equipment necessary for the flight, but would be jettisoned before the spacecraft re-entered Earth's atmosphere, leaving the re-entry module to support the lives of its occupants with limited consumables designed for the short time it took to bring the capsule to a splashdown. The entire outer skin of the adapter served as a radiator, accepting the silicate ester coolant fluid through hollow stringers which would absorb heat and radiate it to space. The fluid would itself be protected through a range of different temperatures by an additive that gave it a freezing point of $-73°c$.

Power for the Gemini spacecraft threatened to be a major problem. Operating for two weeks, equipped with complex electronic equipment drawing heavy loads, and with many critical functions needing comparatively high power levels, conventional spacecraft batteries were not first choice. In fact, three

GEMINI EQUIPMENT ARRANGEMENT

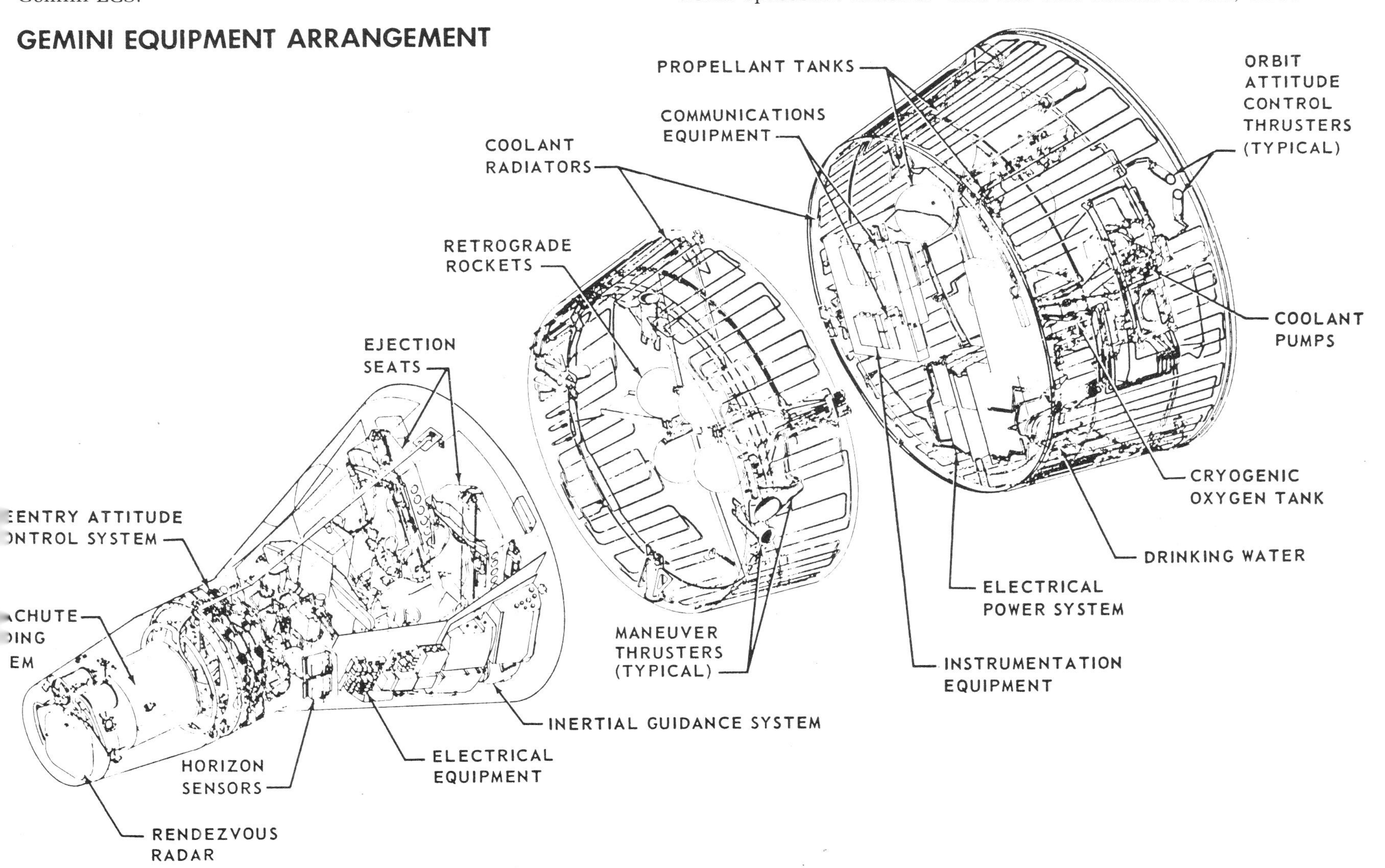

Rocket systems for attitude and orbit changes were carried on the adapter and in the forward area of the re-entry module.

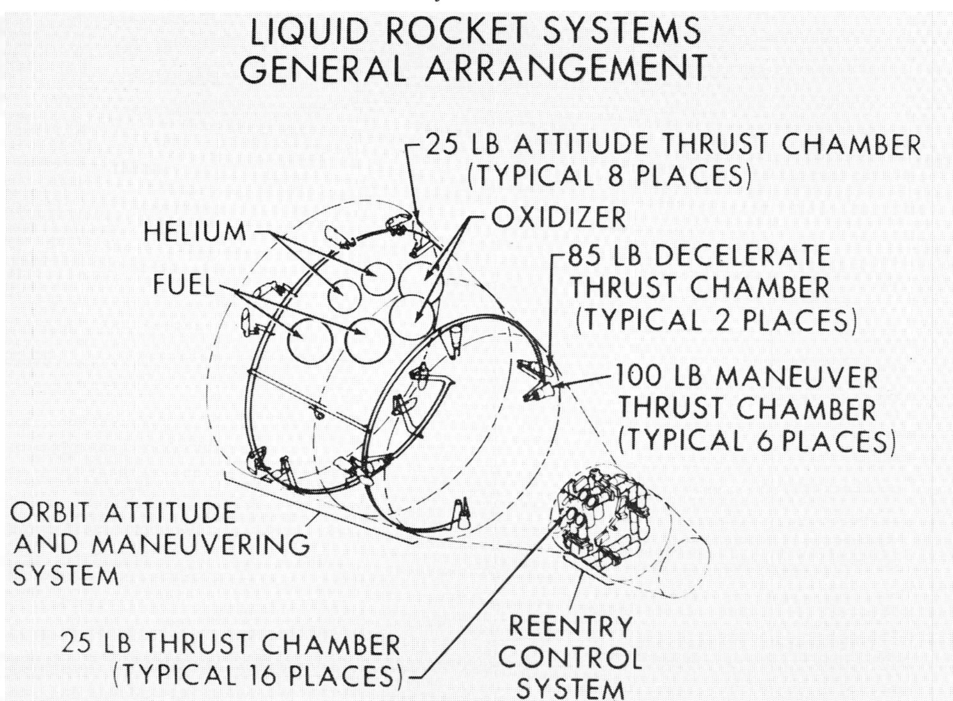

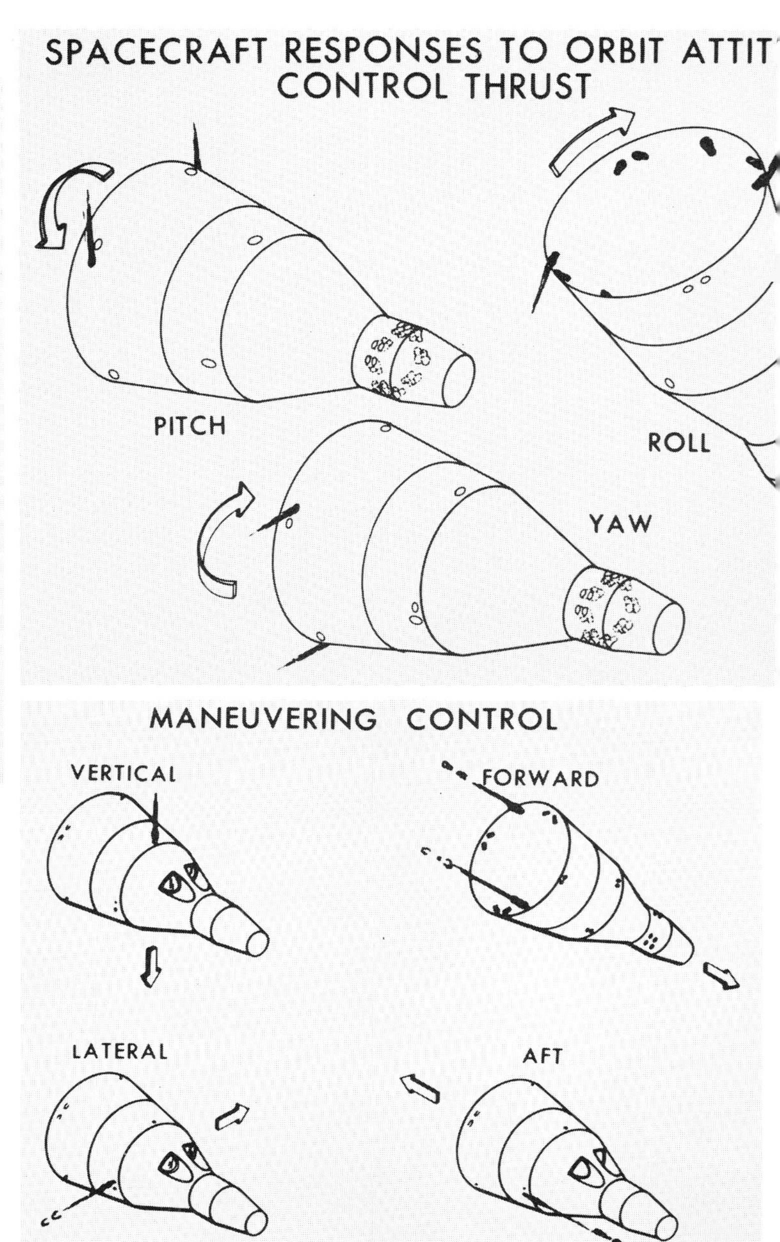

possible power sources were available for selection: silver-zinc batteries similar to but better than the types used in Mercury; solar cells laid out on deployable panels; or fuel-cell units. The latter gave great promise for the high power output versus size and weight they provided. Batteries would have been prohibitively heavy, and solar cells were too vulnerable for a spacecraft that would be maneuvering from one orbit to another, shunting up against rocket stages in space, and performing propulsive bursts from rocket motors attached to the spaceframe.

Fuel-cells work on the principle of reverse electrolysis whereby instead of applying an electric current to water, and so breaking it down into its constituent elements (hydrogen and oxygen), energy is liberated in a chemical reaction that takes place when hydrogen is brought into contact with a catalyst bonded to a polymer plastic sheet. Electrons are released from the hydrogen when the gas contacts an anode and the electrolyte frees the resulting ion to move to the cathode, combine with oxygen, and, with ions from a separate source, produce water. But that is a by-product, since the electrical balance is maintained by the migrating electrons and the movement of ions through the electrolyte. To simplify, hydrogen and oxygen stored in separate containers will, when brought together at the fuel cell, combine to generate electricity and produce water. Some of the reaction energy comes off as heat, and that was acceptably removed by the coolant system.

Fuel cells had never been used in a spacecraft, although the technology for their application to several unique requirements was emerging when Gemini looked round for a power source. Weight for weight they were the best system, although as a concept not as highly developed as silver-zinc batteries or solar cells. Gemini would need silver-zinc batteries as well, however, but only for the re-entry module to power its systems during descent and for activating pyrotechnic devices such as the squibs designed to fire apart the separate modules of the spacecraft before re-entry. Seven batteries would be carried; four main 45 ampere/hr units for the re-entry module, and three 15 ampere/hr squib batteries. The Manned Spacecraft Center decided in January, 1962, that Gemini would use fuel cell power for its long duration flights and within two months had selected the General Electric Company to build two units for each spacecraft; Eagle Picher was to supply the batteries.

The hydrogen and oxygen, called reactants, would be stored at cryogenic temperatures in separate spherical containers, the complete assembly and supporting equipment finding a place in the spacious adapter section. Reactants would have to be stored as liquids to conserve volume and weight otherwise taken up with a structure required to contain the fluids in a gaseous state. Oxygen would be loaded at −183°c, but hydrogen would enter the tank at −253°c. Tiny heaters placed within the tanks would be activated to cause a modest expansion within the contained volume, raising the pressures to the desired operating level: 64 kg/cm^2 for the oxygen; 17 kg/cm^2 for the hydrogen. The heaters would be operated automatically, or with a manual override, but careful operation was essential. Too high a pressure and the tanks would burst, liberating the contents and eliminating the source of sustained electrical power.

But more than that, for the oxygen carried as fuel cell reactant would also feed the environmental control system, a portion of the total quantity carried in the single tank. A secondary oxygen supply would be contained within two separate tanks in the re-entry module, sufficient to supply the crew with oxygen during re-entry after separation of the adapter. It was a self-contained system serving two needs: electrical and atmospheric. An efficient means of reducing weight and a tested way of improving reliability by eliminating unnecessary components. Drinking water would be stored in a 19.1 kg capacity tank in the adapter, and in a 6.6 kg tank within the re-entry module, with the contents of both supplied by Canada Dry to a specification that reduced bacteria to zero and dissolved solids to less than 5 parts per million.

Two tanks would be required to store water produced as a by-product from fuel cells, one tank pressurized with nitrogen gas before launch, the other filled with drinking water within a bladder. As drinking water was used it would be replaced with water from the fuel cells, serving to keep a pressure in the supply for consumption which was made to pass through the cabin water tank in the re-entry module, thereby ensuring that this would be the last supply to drain. A pressure relief valve would be provided in case fuel cell water production got unexpectedly high. In electrical and environmental control systems there was a concept of duality in both series and parallel. The two systems were inter-dependent in that oxygen for the cabin was drawn from the supply primarily installed as a reactant for the fuel cells, water for drinking was pressurized by waste water from the fuel cells, cooling fluid was supplemented by water from the same source, and heat generated by systems powered from the fuel cells was removed by coolant diluted with fuel cell by-product.

But duality also arose from the expanded role of Gemini. Unlike Mercury, designed from the outset for a three-orbit mission in a fixed orbit, Gemini had to remain operational for a maximum 14 days, carry two men, provide facilities for rendezvous and docking, support an active propulsion system designed to move the spacecraft from one orbit to another, and carry out a set of experiments greater in capacity and potential than anything carried aboard the one-man predecessor. It was inconceivable that all the equipment necessary for a full operational role could be contained within the truncated cone that formed the re-entry module, a crew compartment for two men down through the atmosphere. So, the concept of the adapter section, akin to Apollo's service module, evolved so that all the operational equipment necessary for supporting the spacecraft's activity in space could be carried within a section of the vehicle designed to be jettisoned prior to re-entry. That required additional – dual – systems in all essential areas: electrical, environmental, and propulsion. So, like the separate orbital and re-entry electrical supplies, and the two separate environmental control systems, Gemini's propulsion would also demand a duality – one system for operation in space, keeping the attitude aligned with required pointing angles and the vehicle's orbit on the required track, and one for re-entry, where attitude control alone would be necessary.

From the project's outset, a basic mission rule stipulated that the re-entry attitude control thrusters would be switched on only during descent, and that if they were needed in orbit, the spacecraft must be returned to Earth at the next available opportunity. Mercury had escaped accident by fortunate circumstances; depletion of attitude propellant several times threatened to flip the capsule and incinerate the contents. Designed from the outset to take the better ideas from Mercury and build in a new methodology where life-critical systems were sacrosanct, mission engineers were adamant about the need for an effective re-entry control system. Gemini would be less tolerant than its predecessor, threatening to burn up if its attitude wandered too far from predetermined angles, because the spacecraft was designed with an off-set center of gravity, giving it a moderate degree of lift denied to the purely ballistic Mercury. Although it was dynamically stable during all anticipated re-entry attitudes, the elongated cone could very easily flip over if it oscillated from a set value.

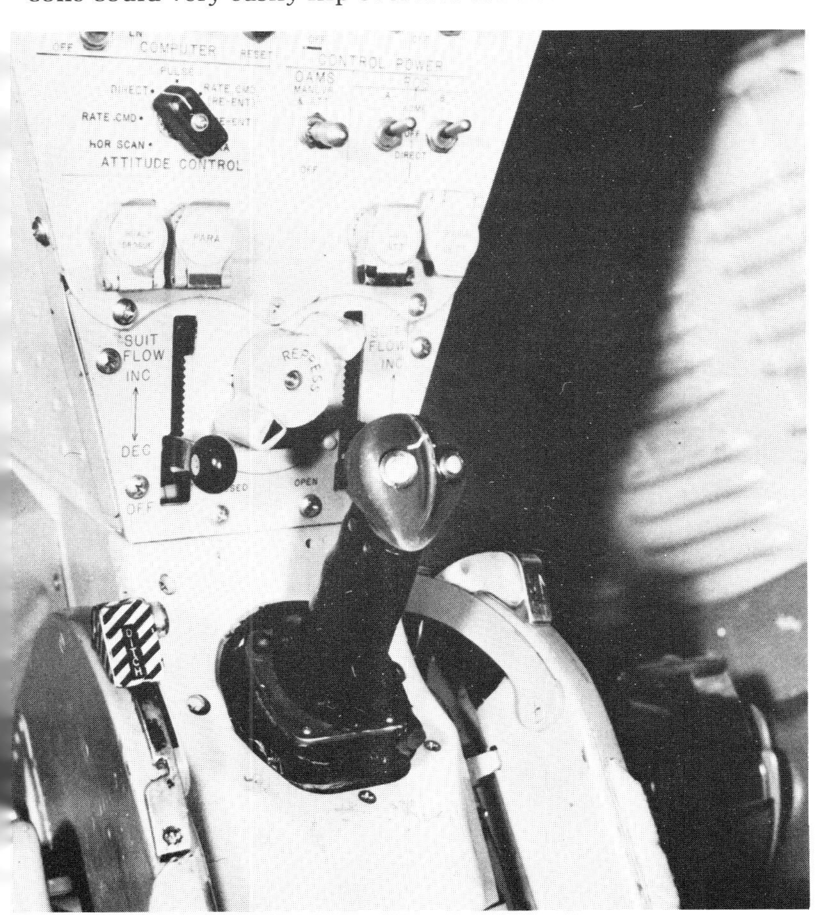

Gemini attitude changes were made through the hand controller located between two seats.

Propulsion for Gemini in orbit was contained with the Orbit Attitude and Maneuvering System, or OAMS, a system that provided 16 thrusters for all attitude and orbit translation maneuvers. Unlike the Mercury thrusters which were more like gas jets than active propulsion units, Gemini's OAMS equipment would comprise a major systems layout based on hypergolic rocket motors. The complete system would be installed in and around the adapter section to the rear of the re-entry module, with fuel, oxidizer, and gas pressurization tanks inside and thruster motors outside. The largest attitude control jets on Mercury had a thrust of 10.9 kg but for Gemini control rockets would average between 11.3 kg and 45.4 kg thrust according to type. Eight thrusters, each producing 11.3 kg thrust, were to be located in pairs at four points around the base of the adapter, situated and aligned so they could be used for attitude control in pitch, roll, and yaw.

Four other thrusters, each producing 45.4 kg thrust, were to be situated 90° apart around the adapter's exterior, pointing radially outward from the centre of the spacecraft. They would be used for up, down, left, or right translation maneuvers while the other eight kept a constant attitude. Two more 45.4 kg thrusters would be installed in the back of the adapter, one at the top and one at the bottom, for moving the spacecraft forward; another two thrusters, each of 38.6 kg thrust, would point forward for shunting the spacecraft backward. Eight thrusters would therefore be employed for attitude control; eight would be used for changing the geometry of the spacecraft's orbit or, in small bursts, for moving it sideways, vertically up or down and forward or back. All three functions – attitude control, orbit translation, relative movement – would be necessary for rendezvous and docking operations of the type envisaged for Gemini. Originally, the two forward firing thrusters were to be of 45.4 kg thrust but that was reduced when careful study showed that if one failed it would throw a torque on the spacecraft outside the capacity of the attitude thrusters to return the vehicle to a stable condition; the forward firing thrusters were off-set from the spacecraft's center of gravity while the others were aligned close to the spacecraft's centerline.

All OAMS engines would use nitrogen tetroxide oxidizer with monomethyl-hydrazine fuel, expelled from respective tanks by helium gas contained in a separate sphere. Gemini would carry single fuel and oxidizer tanks on initial flights, and additional tanks when required later in the program. With just two propellant tanks the spacecraft carried about 146 kg of OAMS propellant. A completely separate system was provided for control of the spacecraft after separating from the adapter, and, logically called the Re-entry Control System (RCS), it comprised two separate loops of eight 11.3 kg thrusters. Propellants similar to those used in the OAMS were adopted for the RCS but with a nitrogen pressure gas rather than helium, and all the fluids were contained in tanks carried in the nose of the re-entry module. Operational control of both RCS and the attitude thrusters in the OAMS would be provided via a hand control stick situated between the two seats. Translation control of the spacecraft was to be effected by a separate stick known as a maneuver hand controller, one for each pilot.

To return the re-entry module to Earth, the back half of the adapter would be jettisoned, exposing four solid propellant retro-rockets supported on a cruciform cross-beam assembly behind the convex heat shield. In 1961 an early design concept envisaged five standard solid propellant rockets but the final design that emerged by mid-1962 possessed four rockets each with a capacity of 1,130 kg thrust wired to fire sequentially at 5.5 second intervals. Like the OAMS and RCS, the retro-rockets could be fired automatically or manually. The remaining half of the spacecraft adapter would be jettisoned 45 seconds after the first solid fired, less than a half minute after retro-rocket burnout. Each rocket comprised a 32.2 cm spherical motor case with a pyrogen igniter and a nozzle. Both sections of the adapter, and their profuse contents, would fly separate trajectories, the rear section remaining in orbit before decaying naturally, while the retro-rocket section would follow the re-entry module down through the atmosphere to burn up through friction.

Gemini's guidance system, the nerve center linking computer programs with spacecraft systems and the heart of the vehicle's intelligence network, was to be supplied by International Business Machines' Federal Systems Division's Space Guidance Center. It was to be an altogether different concept from anything flown in space up to that time, far more sophisticated than any Mercury guidance or control unit and a direct precursor to the sort of complex software control options to be available to Apollo pilots. Software was the name of the game in the early 1960s, when extravagant needs met complex electronics head on, avoiding a collison only by selectively streamlining specific functions, inter-leaving computer programs with spacecraft systems.

Heart of the system would be the Inertial Guidance System, controlled by the inertial measurement unit, or IMU, including a 'platform' comprising a four-gimbal structure and a stable element with three rate integrating gyroscopes and three accelerometers. The accelerometers were there to measure motion about any of the spacecraft's prime axes, and to generate information on the rate at which that motion was being applied – the acceleration. The gyroscopes were there to show which way the spacecraft was pointing with reference to a pre-set attitude. Complementing the IMU was a Digital Computer capable of storing 159,744 bits in 4,096 computer words, able to add, subtract or conduct a transfer operation in 140 microseconds, multiply in 420 microseconds, perform division in 840 microseconds, or conduct three operations simultaneously. Used in conjunction with the Auxiliary Tape Memory, the third element in the triad of Gemini guidance equipment, it was a formidable autopilot and storage facility. The Tape Memory provided ten operational modes, individually selectable on board the spacecraft:

prelaunch, for diagnostic checkout of computer routines and for preparing to load the computer with a specific set of commands;

ascent, for closely monitoring in the spacecraft the condition of the trajectory as the launch vehicle propels the vehicle to orbit, and for providing the crew with immediate information for IVAR (Insertion Velocity Adjustment Routine) where the crew have to fire thrusters in the OAMS to correct an error in speed at launch vehicle separation;

catch-up, where information is displayed to the crew, and corrections automatically implemented, for correct velocity changes and pointing angles during thrust operations for rendezvous;

rendezvous, where the computer accepts target information from the radar and resolves the measurements into the spacecraft coordinate system by the IMU, and where the crew obtain details of the velocity necessary to put the Gemini on target for the Agena;

relative motion, where the crew insert into the computer information on the spacecraft's position and velocity relative to a target and obtain in exchange the velocity information necessary to perform a desired maneuver;

orbit navigation, where the computer is allowed to navigate during periods of spacecraft thrusting;

orbit prediction, for position and velocity calculations of the spacecraft relative to the target up to one orbit in the past and up to three orbits in the future;

orbit determination, for accepting star alignment settings performed by the crew in six measurements to predict future orbit paths;

touchdown prediction, where the computer calculates and displays to the crew a predicted touchdown latitude and longitude;

re-entry, for control of the module's lift vector by changing its angle of attack.

Of all the computer guidance modes, the last was to be the means by which manned space vehicles could demonstrate a precise landing, and Gemini was the project at this time charged with developing a capability for land touchdown rather than water recovery like its predecessor. For essentially no lift during descent, Gemini would roll at about four revolutions per minute, abandoning the advantage of the off-set centre of gravity purposely built into the re-entry module to shorten, extend, or provide crossrange capability to the flight path. No-lift ballistic descent would be targeted for a point about halfway along the downrange extension available to the spacecraft, but control of the touchdown point by using the lift vector would be obtained by rolling the spacecraft to a set attitude. With the modest lift available, Gemini could fly up to 565 km downrange, or up to 50 km either side of the ground track. Careful monitoring of the spacecraft's roll rate and position, plus continuous updates through the guidance system, ensured a better opportunity for placing the re-entry module at the planned location on Earth. It was to be a precursor technique to more sophisticated routines employed by Apollo.

In addition to specific computer programs for storing, implementing, or presenting information to the crew, Gemini pilots would have the use of up to seven control modes for orbit or re-entry: rate command; direct; pulse; re-entry rate command; platform; horizon scan; and re-entry. They were selectable options available through the attitude control and maneuver electronics, enabling the pilots to select at will fully automatic or fully manual attitude control authority, and a variety of switched alternatives between the two extremes. Horizon sensors would be employed for aligning the inertial reference platform and for controlling pitch and roll axes automatically. The sensor observed the infrared horizon, discriminating between the strong radiation from Earth and the almost complete absence of infrared in space.

Although many of the functions and roles performed by spacecraft equipment and systems were designed to relieve the crew of mundane, time-consuming activity, during complex and difficult operations, a considerable amount of information would have to be presented to the computer by one or other of the two crewmembers on board. To do this, a Manual Data Insertion Unit (MDIU) would be provided on the main display console. The MDIU would comprise a keyboard of nine digits and a zero entry key, the ten pushbutton switches providing a means by which the crew could insert data into the computer. A display capable of handling seven digits would be placed alongside the keyboard, with the first two digits indicating the address and the last five digits the message to be inserted to compliment the relevant program. Errors would be erased by a button marked CLEAR, after which the correct keying process would be inserted by a button marked ENTER; data display on the seven-digit panel would be called up keying the appropriate two-figure address via the keyboard and depressing the READOUT button.

Commands placed in the computer by ground transmission would be routed via the digital command subsystem, built by Motorola and housed in the adaptor. Telemetry would be transmitted to the ground stations on various critical functions aboard the spacecraft, not least of which would include the bio-physical condition of the crew. For tracking the vehicle, two C-band radar beacons – one in the re-entry module and one in the adapter – would provide responses to signals from the ground for orbital flight and recovery operations. Voice communications with the ground were to be effected via one HF and two UHF transmitter-receivers, built by Collins Radio. Communications would be lost during the anticipated black-out on descent, and for about 30 seconds prior to recovery antenna deployment. The single HF and two UHF whip antennae were to be located in the adapter section, with the adapter-mounted C-band beacon.

Gemini was a radical departure from Mercury in its mission role and operational capability, a long-duration spacecraft designed to move around in space, pursue target vehicles, maneuver its nose into a docking collar, and provide for one crewmember to open his hatch in space and stand up; with more generous provision for oxygen it became technically feasible to envisage de-pressurizing the cabin and having an astronaut egress the spacecraft although no definite plans were made in this regard until quite late in the program and then amid debate and disagreement. But if substantially different from its one-man predecessor in objectives, Gemini was made to resemble as close as possible the design configuration of the hull and the pressure vessel built for Mercury. It was to benefit as far as possible from learning curves built up through hard won experience at McDonnell's St. Louis plant. Seen in this light, Gemini was in every dimension a logical outgrowth of the experimental precursor that carried American's first astronauts into space, an operational successor honed for demanding roles.

The re-entry module was essentially a Mercury capsule increased in size to provide 50% more cabin volume for the two-man crew. It was elongated in the forward section, equivalent to Mercury's recovery compartment and antenna housing, to accommodate parachutes and other aids but also, and as equally important, the rendezvous radar that would provide the information with which the spacecraft would approach the Agena target vehicle. The nose was also required to provide the docking interface with the collar on the target. Because the optimum size of the re-entry module dictated a maximum diameter of 2.29 meters, and because the Titan launch vehicle's second stage had a dia-

← McDonnell employed a new welding device capable of joining internally the many separate sections of the pressure module.

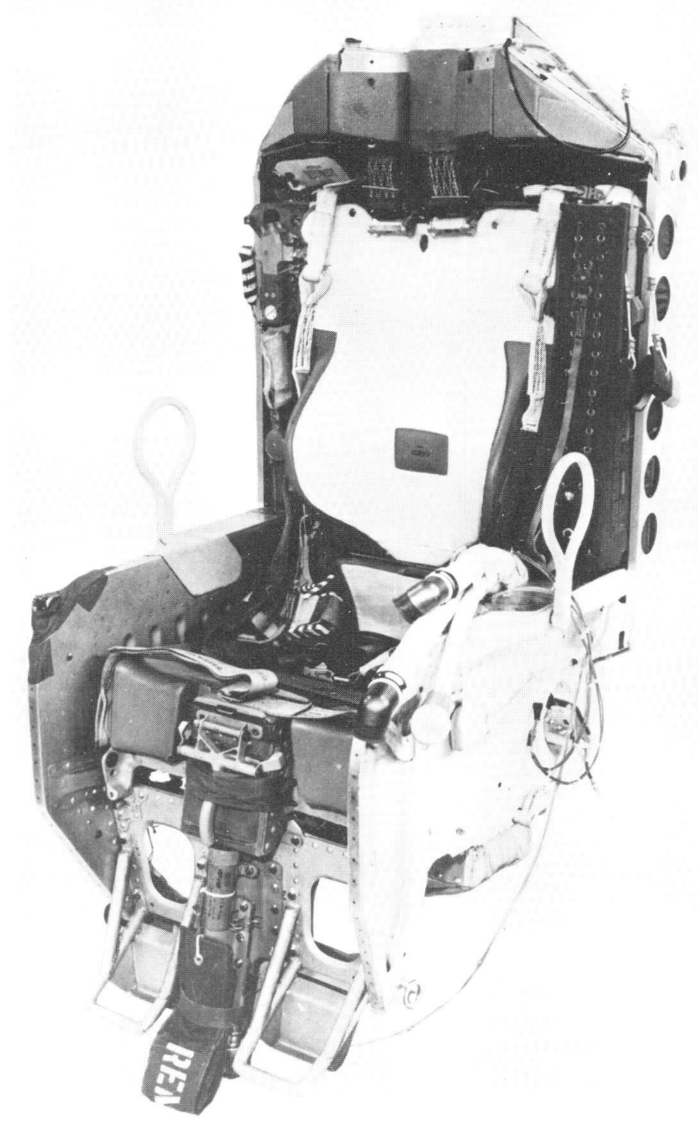

The Gemini ejection seat provided opportunity for new technology developments which helped push the state-of-the-art.

meter of 3.05 meters, the adapter section attached to the re-entry module was a tapered cylinder, effectively 'adapting' the diameter of the spacecraft to the size of the launcher. The re-entry module comprised a truncated straight-sided cone 1.79 meters long, surmounted by a re-entry control system section 69.85 cm long and a rendezvous and recovery section 96.03 cm long. From the base of the module to the forward end of the rendezvous and recovery section, the spacecraft was 3.45 meters in length, tapering in diameter from 2.29 meters at the heat shield end to 98.2 cm at the re-entry control system section.

Gemini's crew cabin contained the pressure vessel comprising a fusion-welded titanium frame attached to side panels and fore and aft bulkheads, the latter of double thickness, thin-sheet titanium (0.025 cm) with the outer sheet beaded for stiffness and the inner sheet smooth skinned. The pressure vessel was built to contain the atmospheric oxygen, a sealed environmental compartment designed to support two pilots for up to a fortnight in space. Because the cabin would be pressurized with pure oxygen at 258 mmHg, it was necessary to select materials with no pyrophoric reaction – metals that would create a spark when struck violently – for fear of causing an uncontrollable fire on the interior. Titanium was tested in a variety of situations, immersed in pure oxygen at various pressures, and it was satisfactorily determined that at the Gemini cabin pressures there would be no danger from using this alloy. It is interesting to note, however, that the oxygen vessels containing gas under extremely high pressure were fabricated from Inconel because of the suspicion that titanium did in fact carry a pyrophoric reaction under this circumstance.

The cabin section containing the pressure vessel took the form of a truncated cone, with space between the outer wall and the inner pressure vessel designed to accommodate numerous equipment bays, access to which would be gained through several panels and easily removed hatches. The basic load carrying shell was to be fabricated from titanium because of the thermal exposure these members would receive from heat coming through the skins. Aluminum was to be used on interior sections adjacent to the outer surface of the pressure vessel, with Rene 41 and L-605 alloys used adjacent to the exterior. The large pressure bulkhead at the rear carried two ejector seat beams, with strengthened beads on the outside and a smooth skin on the inside, both skins being of 0.025 cm thickness. Two outward opening crew access hatches were provided, each with semi-circular scoops to permit the incorporation of flat faced windows looking forward along the nose of the spacecraft.

The triple-paned windows were supplied by the Corning Glass Company, specially designed with an air space between each pane, the left window containing two outer panes of 96% silica and the innermost pane of toughened alumino-silicate glass; the right window was similar to the left on early spacecraft but later models would replace the innermost alumino-silicate glass with a silica pane increased in thickness by 42% to provide equal strength. Later models too carried an extra transparent pane designed to catch smoke and smuts blown on to the window at staging on the ascent, jettisoned by turning a thumbscrew on the inside when the spacecraft reached orbit.

Particular care was necessary when fabricating the hatch sill, since the doors were designed to have the capability of opening in space and as near a perfect seal as possible was essential when the time came to close up. The sill comprised a groove, 1.9 cm wide and 1.27 cm deep, around each door frame, a door hinge located on the outside having a capacity for being activated from either inside or out. The unique concept of using crew ejector seats was made possible by the use of a launch vehicle containing propellants that would burn during an explosion with less violence than other launchers. It was an improvement over Mercury in that the total weight assigned to emergency escape was a fraction of that required by the large escape tower carried by Gemini's predecessor. However, the fact that Gemini astronauts were able to adopt a standard seat on which to lie for ascent and re-entry, arose from the reduced forces of acceleration compared to Mercury launch and re-entry phases.

During the ascent, astronauts would experience a maximum 7 g just prior to second stage engine cut off. On descent through the atmosphere it would be less than 4.5 g. Careful observation of the physical response of the human body during Mercury flights provided confidence in leaving behind the concept of a contoured form-fitting couch. Weber Aircraft would build the ejector seat for Gemini pilots, but the system was to move through several troublesome phases prior to qualification. Nevertheless, it was an integral part of the crew's cabin, an assembly which forced McDonnell to prepare special equipment for the welding operations. The complete cabin assembly required 76.2 meters of automatic and hand welding, 472 metres of seam and stitch welding, and 6,117 spot welds. Mating the thirteen pieces of each hatch would require 7.24 metres of hand fusion welding, and each complete double-hatch sill would be made up of 85 welds requiring over 178 metres of automatic or hand fusion welding.

The main heat shield, attached to the aft bulkhead to protect the spacecraft from friction with the atmosphere, was developed in McDonnell laboratories with the aid of the General Engineering Division, the Materials and Process Division and the Research Division, to a specification developed by the company. The ablative material was a substance marketed by Dow-Corning called DC-325, a paste-like material which hardens when exposed to the atmosphere. The basic structure comprised a fiberglass honeycomb sandwich structure consisting of two 5-ply faceplates of resin-impregnated cloth separated by a 16.5 mm thick fiberglass honeycomb core, with an additional fiberglass honeycomb bonded to the convex side of the sandwich. A fiberite ring encircled the shield. In all, the honeycomb core provided about 250,000 cells to be filled with the organic compound. But internal integrity was essential if the shield was to protect the spacecraft and crew satisfactorily. To check each individual cell, the completed shield was to be radiographically inspected, an x-ray film taken of sectioned areas of the complete shield. A 5 mm thick fabric cover was tailored to the precise contour of the shield and aligned with index markers on the shield proper. Grid lines in horizontal and vertical section were marked on the cover to facilitate precise reference to specific cells. Only three or four exposures were necessary to determine voids, lack of bonding, or unwanted inclusions, and the offending cell would be marked with a mapping pin. The new ablative compounds developed by McDonnell, and the new method of fabricating the heat shield as a completed unit, saved 56.7 kg over the Mercury shield.

The sides of the conical re-entry module forward to the small front bulkhead were covered with overlapping shingles for aerodynamic and heat protection, a direct successor to similar fitments on Mercury. With a thickness of 0.405 mm, each Rene-41 beaded shingle was identical in composition and manufacturing technique to the Mercury shingles. Un-beaded shingles of cross-rolled beryllium were to be attached to the cylindrical re-entry control section and rendezvous and recovery section areas. Ranging in thickness from 0.71 cm to 0.76 cm, the shingles would be attached to the spacecraft by beryllium retainers that allowed them to expand. For Mercury, sections were fabricated from hot-pressed beryllium block but the new cross-rolling procedure would ensure higher strength and greater resistance to shock, essential qualities for the maneuverable Gemini. Beryllium produces toxic vapor and tiny oxide particles at temperatures above approximately 900°c, a prime reason why Mercury engineers opted in favor of the lighter ablative shield, and the reason for limiting beryllium shingles to surface positions on both Mercury and Gemini where temperatures were kept below that level; during ascent, the forward section of the spacecraft would heat to only 540°c at maximum, and during re-entry the backward-facing re-entry module would protect the nose section from high thermal loads.

The re-entry control system section was to be fabricated as a thin-skin cylinder with nine external radial webs supporting longitudinal stiffeners enclosed by machined titanium rings. The titanium skin was submerged from the outer face of the stringers and rings, separated from the beryllium shingles by a gap of about 15 cm. The rendezvous and recovery section, forming the extreme forward nose section of the re-entry module, would comprise a semimonocoque construction of titanium skins, rings and channels used as stringers. Beryllium shingles were attached direct to the rings and channels and a nose fairing designed to protect sensitive radar equipment on the flat face was to be fabricated from fiberglass honeycomb and jettisoned after ascent. Thermal blankets were to be attached to the skin interior, except in the re-entry control section where they were held in place by mesh screens supported from the stringers.

Gemini's adapter module was functionally divided into two sections: the retrograde section immediately behind the re-entry

The Gemini adapter section contained pressure vessels for attitude control propellant, water, hydrogen, oxygen and pressurizing helium.

← The forward section of the re-entry module carried thrusters for attitude control during descent and a parachute recovery system. On top, the nose section used for docking would be jettisoned prior to releasing the parachutes.

module heat shield, and the equipment section that carried systems and consumables for orbital flight between the retrograde section and the launch vehicle. Until shortly before retro-fire, the adapter would function as an integrated whole, only separating when the equipment section was jettisoned, exposing the retro-rockets. The retrograde section was to be jettisoned in turn when the solid motors had fired to decelerate the re-entry module.

The adapter was constructed as a stiffened skin-stringer structure consisting of circumferential aluminum rings, extruded alloy stringers and magnesium skin. The T-shaped stringers were designed to have a hollow bulbous portion through which coolant fluid could flow, transferring heat from cold rails and cold plates to the skin of the adapter, serving as a radiator. Three titanium tension straps kept the adapter rigidly secured to the re-entry module and a shaped pyrotechnic charge was employed to jettison the equipment section prior to retro-fire. The retrograde section was to be about 76 cm deep and contain crossed aluminum 'I' section beams on which would be mounted the four solid propellant retro-rockets discussed earlier in this chapter.

The equipment section was about 1.53 meters long with a flat aluminum blast shield across the entire top area; under certain critical abort conditions, the re-entry module and retrograde section would be required to fire away from the equipment section on top of the launcher, and the protection of systems in the back of the lower adapter and propellant in the upper stage of the Titan from solid motor exhaust improved the chances of the spacecraft satisfactorily separating to safety. The blast plate was made up from a 2.54 cm thick honeycomb consisting of a forward 0.23 cm fiberglass faceplate and an aft 0.05 cm aluminum faceplate with an aluminum core separating the two. The entire outer surface of the adapter was to be coated with white ceramic paint and the inner surface would carry aluminum foil to distribute heat. During launch the adapter would reach a temperature of about 315°c, and both sections would burn up in the atmosphere at the end of the flight.

The adapter was attached to the Titan launch vehicle by a forged aluminum ring, 3.05 meters in diameter, with 20 lugs through which bolts would be fastened to secure the two structures. When the spacecraft was separated from the upper stage, a pyrotechnic charge would instantly sever the adapter 38 mm above the mating point in much the same way a metal shear would cut through sheet metal. To move the spacecraft away from the launch vehicle's upper stage, the two aft-facing rocket motors on the rear of the equipment section would be fired to nudge the vehicle forward in a similar fashion to the action of posigrade rockets on the retro-package of the Mercury spacecraft. In a flight configuration, the Gemini spacecraft would be 5.8 meters long, a mazimum 3.05 meters in diameter, and weigh about 2.2 tonnes, although there would be program growth during later missions to a maximum in-orbit weight of about 3.7 tonnes.

Compared with Mercury's orbit weight of 1.35 tonnes, Gemini was heavy, and the additional space available for extra systems made it firm choice as the first spacecraft to be designed for a land landing. At least, that was the intention. The Rogallo wing, or paraglider as it became known during 1961, was not the unanimous choice of all concerned with Gemini, however. James Chamberlin favored it but Chris Kraft's Flight Operations Division was unconvinced of its value, or the desirability of incorporating it into what many considered was already an almost overworked design. Nevertheless, largely at the instigation, certainly the pressure, of Chamberlin, North American got a contract in November, 1961, to begin work on the paraglider. At this time there was a vague idea that paraglider should be a non-aligned concept applicable to any manned spacecraft NASA cared to commission, but early troubles visible within the management schedules soon put paid to thoughts that a great new era of land landing was about to dawn. Paraglider was quickly written up as a Gemini-unique concept – for the time being at least.

Norbert F. Witte was North American's project manager for paraglider and he immediately placed schedules and plans for a detailed test phase at a level consistent with the tight pace of Gemini's projected development. Paraglider was expected to fly on Gemini missions from the second launch, the first manned flight in the projected twelve-launch program. But just in case of trouble with the schedule, it was decided to plan for a possible fall-back posture by preparing to have the second flight carry conventional parachutes. From the outset, paraglider's development program was doomed. At El Centro, California, on 24 May,

Inflated on the ground and released by helicopter, this paraglider wing was conceived for bringing Gemini spacecraft back to earth and is seen here descending under the control of test pilot E. P. Hetzel.

1962, North American ran into problems following a good first drop test with a parachute recovery system, an emergency system for a small scale model in case the real article should run foul of a dilemma. On what should have been the final drop to qualify the back-up system, an electrical short-circuit sent the device back to the workshop where a basic design flaw was uncovered. More tests – and more failure – followed slowly in the weeks ahead. Things got even further behind when, during tests of the full-size paraglider, back-up recovery system parachutes failed, weather postponed drops, and the test vehicle was finally destroyed when none of the emergency devices worked. But hopelessness expressed by work teams trying hard to get a successful test run with the emergency system was paled by the near catastrophe of trials with the paraglider itself. Towed into the air by helicopters, half-scale models broke loose, their wings fell off and incorrect radio commands sent via a remote control unit brought disaster to seeming chaos.

By September, 1962, Chamberlin ordered a halt for time in which North American could re-write the test sequence and regroup their development schedules. Already, it was clear that paraglider would not make an operational flight before the third Gemini mission; North American's Space and Information Systems Division was strained to the very bone by the NASA contract to build Apollo, awarded the previous November. Its attention first and foremost was on the Moon ship and in the wake of another major contract, for the second stage of the mighty Saturn V (then called C-5) launch vehicle, the company all but abandoned its paraglider responsibilities. But Chamberlin was late in pulling North American to order, for only four weeks earlier Norbert Witte had been replaced by George Jeffs as paraglider project manager.

Jeffs had been with North American since 1947 and spent several years in the Aerophysics Laboratory soothing troubles with the Navaho missile before being appointed Chief of Advanced Design and Chief of Systems Engineering at the Missile Division, moving up as Manager of Corporate Technical Development and Planning in 1959. Jeffs had a good reputation among the engineers working on paraglider. They knew him to be a tenacious troubleshooter who would surely seek out and find the bugs in the system.

By the end of October tests at Edwards Air Force Base in California had shown paraglider – at least a half-scale model – was stable in free flight. As conceived earlier in the year, the paraglider landing system would permit the Gemini spacecraft to land on skids housed within the space beneath the crew pressure vessel, behind the outer skin. Following a convenient re-entry, a drogue parachute would be released at a height of 18 km, withdrawing in the process the furled paraglider. The wing would then be inflated so that the spacecraft could remain suspended horizontal to the ground, lift generated by the paraglider being used by the crew literally to fly the vehicle down to the ground, extending the skids before touchdown. It would be a somewhat precarious mode of operation at best, completely unsuited to contingency landings away from prime recovery zones which, for paraglider flights, would be on land; for water recovery, it could prove dangerous at best, or lethal, if in gliding to a horizontal splashdown the re-entry module were to be thrown across the sea, tumbling over waves.

With a successful test flight on the half-scale model providing modest confidence, the next step was to attempt a wing deployment sequence in flight. On the first attempt, everything became fouled in the tumbling panoply of lines, lanyards and canopy, until the paraglider hardware was jettisoned from the test model so that it could deploy its emergency recovery system. If anyone doubted that paraglider was doomed he or she would have had their faith tested when, early in 1963, a small squib switch designed to deploy the reserve parachute failed, causing the test model to crash; the switch was one of a type that had never, ever, been known to fail, a standard item used frequently in the space program. During the early months of 1963 repeated failures did little to ensure paraglider a place in Gemini, especially in the light of financial troubles afflicting the entire project.

By April, North American received instructions from NASA on how to extricate themselves from what was becoming a momentum of disorder following an improbability of success. Considerable time, effort, and money was being wasted on trying to get test vehicles working properly, models adequately fabricated, and reserve parachute systems satisfactorily operating. Final design, fabrication and test of the definitive hardware was receding further from Gemini's projected launch schedule; by this time there was little hope of getting paraglider on a space flight before the seventh Gemini mission! A new contract was signed in May as a last-ditch attempt to get the system aboard at least one or two spacecraft, and NASA closed out the earlier agreements for a sum of $7.8 million and agreed to a negotiated work contract of $20 million for the new effort. Trimmed to the essentials, paraglider's test program achieved modest progress during the summer of 1963, but questions about instability, and debate concerning an alternate concept called the parasail – a flexible non-rigid 'flying' sail with less lift than paraglider but simpler and hence more acceptable – proposed by engineers at the Manned Spacecraft Center, threatened the pace of development.

MSC's Flight Operations Division and the Engineering and Development Division were hotly opposed to the paraglider, but the astronauts, and MSC's Flight Crew Operations Division, favored the concept. Headquarters was briefed on the parasail as a possible contender for land landing techniques, an argument based on the historic trend of cost increases and development delays afflicting the inflatable rigid-wing paraglider. Parasail was presented as a simpler and less costly way of getting a Gemini spacecraft down on land, with a schedule that proposed its first use on the seventh mission; by this time the paraglider was not expected to fly before the tenth mission. But stopping paraglider when it was so far along a road that many thought would only be repeated by parasail was deemed too risky for the entire land-landing concept. In talks between MSC and Headquarters, it was decided to restrict North American to work associated only with proving that the paraglider wing could be inflated and deployed for a stable flight through the lower atmosphere, and that all work at McDonnell on fitting paraglider components and skids into Gemini re-entry modules should cease forthwith. McDonnell was to retain the option of putting that capability back into the last three capsules in the unlikely event that paraglider tests took a dramatic turn for the best.

It was a losing battle trying to keep the land landing system on Gemini. With a weight penalty of 360 kg over an equivalent parachute water landing system, engineers worrying over inevitable weight growth in the spacecraft were only too eager to seize the initiative and use that newly found balance. Over the following months, on into 1964, the paraglider died an inglorious death. Starved of funds by an agency already deeply committed to more pressing, and costly, activities, the concept was just too far from realization and while tests limped along for much of that year it was made clear that Gemini would adopt a conventional parachute landing system for all the scheduled twelve missions. In fact, before the end of 1963 work was well under way toward completing a parachute recovery system for the re-entry module. It transpired that Gemini would descend to a water landing using a concept left over from the paraglider but also made necessary by the shape and design of the module.

To control and fly to the planned land landing a spacecraft suspended beneath a paraglider, the spacecraft was made to rotate 55° from a vertical, base down, attitude adopted for re-entry, to one in which it was nearly horizontal with the pilots in the normal sitting position. This was very different to Mercury, where the parachute came out the top and the capsule hit the water base downward, a landing bag absorbing the shock of impact on the heat shield. Because Gemini was less stable in the vertical position than Mercury, and because the crew access doors were large, covering a wide surface area of the exterior skin, it was necessary to put the spacecraft down in a near horizontal posture so that it could float more like a boat than the bobbing cork Mercury had seemed to be at splashdown.

During descent, a 3.27 meter diameter drogue parachute would deploy in reefed condition from a container at the top of the rendezvous and recovery section, unreefed 16 seconds later to bring the spacecraft down from a height of 15.2 km to an altitude of 3.26 km. At the lower point, having stabilized the re-entry module, the drogue would be cut free by a guillotine, causing the parachute to pull out a pilot canopy, 5.58 meters in diameter. This action would be initiated by the crew and within 2½ seconds of deployment to the reefed condition, the pilot parachute draws the 106 cm long rendezvous and recovery section away from the remainder of the re-entry module, separated from the re-entry control section by a mild detonating fuse activated at the same time. The rendezvous and recovery section, essentially a structural canister, would float down separately; about 3½ seconds after liberating itself the pilot canopy would unreef. Meanwhile, from a fiberglass cylinder attached to the top of the re-entry control section, a 25.6 meter diameter ring-sail main parachute would be deployed to a reefed condition. Put out at a height of 2.96 km, the parachute would unreef 10 seconds later but at this time the spacecraft would still be suspended in a vertical position.

To place it in the proper attitude for splashdown, a pilot would activate the two-point suspension system. Within a fiberglass tray in the support structure between the two hatches, a forward bridle leg would tear free through frangible insulating material that would have kept it covered throughout the mission. Suspended by one leg from the nose of the re-entry module, the aft leg would pull back to suspend the spacecraft also from a position just forward of the heat shield, simultaneously pitching the spacecraft down (by bringing the back up) to within 35° of horizontal. By transferring from a single to a two-point suspension, the re-entry module would be in the correct position to impact the water, with the crew descending at about 32.9 km/hr in a more comfortable sitting position than the supine posture afforded by Mercury and to be adopted for Apollo. After splashdown, the main parachute would be jettisoned, and the crew could call upon several recovery aids. This equipment would include a flashing recovery light, dye marker, survival kits, HF and UHF rescue communications and beacon devices, splash curtains that could be pulled up across open hatches to prevent water spilling in over the side, a hoist loop, flotation material, and emergency electrical supplies. A life raft would also be on board, just in case the anticipated pin-point landings were not quite so accurate.

Back-up equipment for the spacecraft recovery devices met and adopted the selected mode of escape from an ascending booster running amok: the unique ejection concept where the pilots would individually eject on rocket-propelled seats in which they would be supported for a normal mission. If anything went wrong on the way down, the ejection seats could be employed to get the crew away from the re-entry module. With this escape function available, engineers could do away with a reserve main parachute and rely on the ejection capability. Although made possible by the calculated expansion rate of the fireball that would result if the Titan blew up, Gemini's ejection seat capability pressed hard on existing technology. No-one had actually designed and built an ejection seat to *guarantee* the safe return of the occupant, such escape devices being constructed to give the pilot a better chance of survival at best. Such loose criteria were too nebulous for manned space flight, each separate spacecraft system being required to fail without endangering life. Consequently. new design approaches were essential for the preparation of an adequate Gemini escape procedure.

The rocket catapult (rocat) for the escape seats was to be provided by Rocket Power Inc., of Mesa, Arizona, and the seat would be manufactured by Weber Aircraft. Tests of the system aimed at simulating an escape from a Titan launcher on the ground – a so called off-the-pad abort – would use the 45 meter tall tower built at the Naval Ordnance Test Station, China Lake, designed originally for missile development firings. The specification was demanding. If Titan blew up the resulting fireball would expand to a radius of 23 meters in 0.5 seconds, to 61 meters in 3 seconds, and to 70 meters in a further 3 seconds. The calculated trajectory required to *guarantee* the pilot's safety would carry him out a distance of 167 meters in 3 seconds and 305 meters in 6 seconds. At this pace, acceleration would momentarily hit 24 g, but for less than ¼ sec while the rocket propelled the seat from rails on the aft bulkhead. Other requirements placed on the system demanded precise discrimination between a potentially disastrous situation and one in which the crew could afford some discretion about the need to eject. Riding on an ejection seat all the way into orbit, remaining in space on the device for several days, and returning at the end of the flight to rely on it again for escape, would place unusually stringent demands on the design.

Simulated off-the-pad ejection (sope) testing began in July, 1962, under the competent management of Kenneth F. Hecht from the Naval Ordnance Test Station. The ejection system was as new a concept for manned spacecraft development as the unique paraglider, but the former met and passed progressively less restricting obstacles by separating for as long as possible development trials with the rocket system and the seat itself. Hecht was concerned with the dynamic envelope involving the ability of the rocket to propel the seat an adequate distance in the appropriate time while retaining a constant center of mass; if the latter got out of control the seat would tumble under high-g loads and literally destroy itself in flight. By August the sope tests had gone as far as they could in satisfactorily demonstrating the rocat design. In September McDonnell completed the integrated rocat and seat design, clearing the way for combined systems testing. But problems arose this time with the rocket, and still the seat could not be made to remain in a constant attitude throughout the ejection sequence.

Hecht had insisted from the beginning on using dummies to simulate the mass distribution of a human body, duplicating as far as possible the changing center of mass as the rocket consumed its solid propellant and the line of thrust moved across the seat centerline. Before the end of the year a ballute had been added to the ejection system to prevent an astronaut from spinning excessively if aborting from a height greater than 2 km. A ballute was a combination balloon and parachute, providing the functions of both. Constructed of inflatable rubberized fabric it would be a maximum 122 cm in diameter and 137 cm long, deployed at heights of up to 22.5 km after the pilot had separated from his ejection seat; a personal parachute strapped to the pilot's back would then open at a height of 1,737 meters and lower him gently to the ground. Without the stabilising effect of the ballute a man would be unable to release his parachute properly.

By April, 1963, separate elements of the integrated design seemed ready once again to assume sope tests and progress was made in a series of trials where several minor defects were cleared from the basic design. The next most critical series of tests involved the now familiar process of simulating high dynamic pressure by ejecting the seat from a fast moving sled. For Gemini qualification, two ejection seats were placed in the boilerplate spacecraft mounted on the sled at the Naval Ordnance Test Station. Rapid forward acceleration would simulate the atmospheric pressure of a booster ascending to space, simulating in turn ejection from the re-entry module of astronauts triggering a high altitude abort. Early attempts to get sled ejection data were compromised when solid propellant rockets designed to accelerate the boilerplate along the track tore loose and set fire to the structure. While repairs were made, sope tests during May inspired confidence that the entire system was in a good condition and by June, 1963, sled tests proved successful. By the fall, more design alterations were brought to the ejection system, this time to accommodate sequence changes. By early 1964 a concentrated schedule of ballute and sled tests fully qualified the design changes and a series of integrated system trials was all that remained before certification.

But then more modifications were deemed necessary when a shortage of pyrotechnic charges held up the trials, and NASA inserted another test phase to prove, as sope and sled tests had been unable to, that the complete system would work from high altitude. The agency wanted to show that the complete sequence of events would perform as planned. Accordingly, an F-106 aircraft was modified to carry a production Gemini escape system, tested with a live ejection on the ground first just to prove it was possible to fire off a Gemini ejection seat from the new mount. Again, after a good ground proof firing, trials had to wait for the delivery of pyrotechnic charges, delays that held up certification trials for four months and F-106 firings for three months. It was almost a case of overcooking the turkey, for on 5 November, 1964, a sled test ended in failure when part of the seat collapsed during ejection, bringing the device plummeting to the ground and completely demolishing the dummy.

It is pertinent to note here that on another occasion, when a selection of astronauts was on hand for a formal demonstration before NASA and Air Force officials, the re-entry module hatches failed to open as they should about 0.3 seconds before the ejection rockets fired. Out popped the seats through respective hatches, cleanly punching two neat seat-shaped holes in the metal, effectively decapitating the dummies. An astronaut close by took one look at the result – and walked away! On another occasion, a Gemini astronaut would literally have a fraction of a second to decide whether to remain with the spacecraft or eject. Events like these would not instill a desire to choose the latter. Even as 1964 drew to a close, the ejection system was still in a state of suspended preparation, waiting for pyrotechnics. And neither was the ejection system the only device marking time while tiresome minor discrepancies fouled the schedule.

Although selected early in the program, Gemini's planned use of fuel cells for generating electricity, and water as a by-product, forced a major research and development effort that nobody anticipated at the outset. The idea of using hydrogen and oxygen reactants was certainly not new, but neither had anybody successfully produced a fully operational system. From the beginning, General Electric applied consistent efforts to the fuel cell contract, completing in 1962 new facilities for its Direct Energy Conversion Operation at West Lynn, Massachusetts. During the rest of that year, and well into 1963, GE encountered numerous problems: tests failed, equipment leaked (not for any predictable reason), schedules were seen to be too inflexible for breakdowns frequently encountered, and technical re-design was a constant process. The latter was made more difficult as the company switched its effort from an almost wholly research and development orientation to an increasing emphasis on production-line work, designs and configuration being modified and changed on a continual basis.

By the fall of 1963 NASA was concerned about the slack pace of fuel cell preparation. George E. Mueller, brought in as NASA's Deputy Associate Administrator for Manned Space Flight, commissioned three engineers from the Bell Telephone Laboratories to enter GE's facility and report on the state of the project. The agency was already looking at contingency plans, flying the first few missions on batteries only, limiting the seven planned rendezvous and docking flights to two days so they could use batteries also, etc., and it was concerned that the other prime function of Gemini – long duration flight – just might not be possible at all if nobody came up with a viable fuel cell. GE tests showed that although theoretically good for more than 600 hours continuous operation, the units were consistently performing for only one-third of that time.

When the Bell Telephone Laboratories' engineers reported to NASA the conclusions of their quick inspection GE was endorsed as still the best bet for coming up with an operational system but they cautioned that time-consuming delays were almost inevitable. By November, 1963, General Electric halted all production activity on the fuel cells and set up tiger teams to attack key problems facing engineering clearance of the definitive design. For its part, NASA instructed Walter Burke, head of McDonnell's Gemini effort, to design the first spacecraft to operate on batteries. Late in the day, major obstacles loomed on the horizon threatening to significantly cut the major objectives set for Gemini, an essential precursor to Apollo. And time was running out.

CHAPTER ELEVEN

Doubts and Delays

From the beginning Gemini project management was beset with unparalleled cost increases that threatened to substantially modify the entire program. The original Project Development Plan envisaged an unmanned qualification flight in May, 1963, followed by the first manned mission two months later. But that had been drawn up toward the end of 1961, before John Glenn made his historic three-orbit flight. Inherent in the Plan was a fundamental flaw that escaped obvious notice, a product of the lack of experience in building and flying manned space vehicles that would not be repeated for succeeding spacecraft, and that inevitably led to major delays. It was the insistence on the part of George Low and Warren North to program the second and subsequent flights on a time interval identical to that which separated the first from the second. The first flight of any new manned spacecraft would be a time for trial and evaluation demanding considerable post-flight inspection of components and associated hardware. The need for an extended inspection period between the first two flights was not so apparent in the early 1960s because the only vehicle that could have provided experience with this policy – Mercury – was itself in the vanguard of spacecraft operations; Mercury was enmeshed within a multi-flight test program designed more as a proof-of-concept vehicle than a true manned spacecraft. So it was left to Gemini to provide valuable lessons on the most efficient procedures to adopt when introducing a new vehicle. And it was to include in those lessons a cost penalty and a visible indication of the very different abilities of engineers and administrators.

No sooner had concerted efforts led to the first contract awards in early 1962 than the costs began to grow. The Project Development Plan quoted a program estimate of $529.5 million that included $240.5 million for McDonnell to supply the Gemini space vehicles, $113 million for modified Titan II launch vehicles, and $88 million for development of the Atlas-Agena combination to be used for lifting the target vehicles into orbit. Technical meetings to define the contractor's costing more effectively got under way no earlier that April, 1962, and by the following month the Gemini Spacecraft Cost and Delivery Proposal cited a minimum figure of $391.6 million. Several months after formal Headquarter's approval, the project was only just getting down to the studies necessary for a truly accurate assessment of what the program was likely to cost. But the 63% increase in prime contractor estimate was a deep shock. In fact, McDonnell had not been at all sure just what the program entailed when it spent most of 1961 building up a good case for the two-man spacecraft with NASA's James Chamberlin.

For one thing, several test articles would be needed, vehicles on hand for the numerous tests and trials that came along as the project progressed but which could not be accurately predicted in advance. Boilerplates would be needed for dynamic tests and structural analysis, and new simulators would have to teach old and new astronauts how to fly the spacecraft. And then NASA learned the boosters would cost a lot more than anticipated. By March, 1962, the estimated Titan II development sum had risen by 45% to $164 million and included a more comprehensive plan for adaptation than had been thought necessary several months before. For reasons which even at the time were quite hard to define, Atlas-Agena prices also went up, by 20% to $106 million. Although approved in late 1961, it was March the following year before NASA and the contractors had come up with the first reliable cost estimate, predicting a program bill totalling $744.3 million, up more than 40% on the Project Development Plan.

But worse than that, NASA had presented its Fiscal Year 1963 requirements to the President for official announcement that January, sums based in part on the estimated cost of the Gemini project that was now seen to be far below actual predictions. The total program cost would be spread over several years, but any predicted total growth would have an impact on the first financial year of the project, a time which must as countless aeronautical projects prove get the best fiscal attention and fully adequate resources. Penny-pinching in the initial years of a project can seal its fate for ever, whereas modest fund-juggling during later years has less of an impact on schedule. But worse was to come. Despite persistent requests for finite cost goals, McDonnell refused to come up with figures to which it could tie itself, claiming that contractors and subcontractors were still arranging purchase orders and that the fluid state of system and subsystem development funding profiles prevented the highly accurate estimates NASA wanted so badly.

Ere long it got those important estimates, and in August, 1962, when the space agency sat down to draw up a definitive contract with McDonnell, the contractor quoted a price of $498.8 million – just for its own work to produce the spacecraft. This represented an increase of 27% over the figure quoted as a best estimate just three months earlier, fully 107% up on the figure put forward by McDonnell when the Manned Spacecraft Center pushed Headquarters to approve the project less than a year before. By October, the Air Force Space Systems Division informed NASA that the boosters would now cost $172.6 million, and that, despite a reduction of from 11 to 8 in the number of Atlas-Agena targets, those launchers would now cost $118.3 million. In other terms, whereas the 11 Atlas-Agena vehicles had cost a breakdown average of about $9.6 million apiece in May 1962, the July figure showed that 8 would each cost an average $14.8 million. If Atlas-Agena production depended on Gemini program needs it would have been understandable to see a unit cost increase, but it did not. The Agena program was backed up with production needs for many military and civilian space projects and the few taken by NASA for the Gemini project had little effect on production unit costs.

As Fiscal Year 1963 drifted in on 1 July, 1962, NASA was left

McDonnell engineers assemble the Gemini pressure vessel within its outer shell

McDonnell's Gemini production line reflects the expanding US commitment to manned space flight in the early 1960s.

waiting for late action by Congress to approve sums that were badly needed for that financial period. The Gemini program had not been funded at all in the Fiscal Year 1962 budget and 1963 money was essential to picking up momentum toward anticipated launch dates; what had been spent on Gemini up to now was money gleaned from other projects but with formal contracts under way the full 1963 sum was in demand. It would certainly be tight going even with the full 1963 sum requested for the two-man project; in the light of what suppliers and contractors had been telling NASA during the preceding weeks its original 1963 estimates were far too low. In mid-August the Congress finally passed a bill that authorized the fiscal sum requested for 1963, but it was the end of September before the appropriation bill was passed. Contractors were all but knocking on NASA's door for payment, urgently sending telegrams and messages threatening dire consequences if NASA failed to come up with sums essential to the project. It affected other programs too and would remain a consistent thorn in NASA's side for many years; sometimes Christmas would arrive before contractors got word on the money they could expect to receive for projects officially funded from 1 July in any given year. The problem affected many federal agencies and administrations and would finally be solved by streamlining the annual budget hearings and by shifting the start of fiscal year periods from 1 July to 1 October. But not before 1977, and in 1962 that was a benefit denied to government administrators.

As part of an overall budget plan, the Manned Spacecraft Center's financial needs for FY1963 topped $687 million under tight constraints applied by Associate Administrator Bob Seamans; the FY1963 budget approved by Congress for all NASA projects was $113 million below the figure requested. From the total available to MSC, Gemini needed $299 million, but to stay within ceilings imposed by Seamans only $234.1 million would be available. Brainerd Holmes was deeply concerned about the impact of these cost cuts on the pace of the entire space program and prepared schedules to reveal long delays in Gemini launches if the lower sum was to be the limit. NASA tried hard to get President Kennedy to present a supplemental request to Congress but after hearing arguments from administrators he decided early in October not to go to Congress. NASA would have to live with what it had already, and MSC would have only $660 million for FY1963.

Houston wanted to divide the $27 million cut equally between Gemini and Apollo, but Headquarters stipulated that Gemini absorb all the sum to prevent the Moon landing program from getting any further behind; at this time the space agency was only just pulling itself out of the lengthy decision making process of choosing how to send Apollo to the Moon. MSC was horrified at the constraints placed upon them by the financial limits, believing that it could mean the end of paraglider and possibly abandonment of all rendezvous plans for Gemini. By the time NASA learned it would have no help from the White House initial negotiations had been conducted with several contractors over the question of reprograming the entire operation. NASA had earlier battered McDonnell's dramatic cost submission down to $464.1 million, and now sought further reductions in the way the program was to be handled in efforts to drive the costs down further still.

Meetings between McDonnell's Walter Burke, NASA's Bob Gilruth and MSC's James Chamberlin successfully got the message across: McDonnell could not hope to progressively inflate the project and still see it through to completion; cosmetic support systems had to go in the interests of retaining at least the semblance of original objectives. Tests were to be reduced to an absolute minimum – at subcontractor level as well as with the prime contractor – and one of four proposed static test models was to be deleted. The end result of reprograming the spacecraft contract cut the estimate to $438.2 million, $60.6 million below the figure McDonnell had put to NASA during August. Now it was the turn of the launch vehicle.

Richard Dineen was Space Systems Division's chief of Launch Vehicle Development, and received word from Gemini program manager James Chamberlin that $59.28 million would be all the money available during FY1963 for development activity essential to qualifying the Titan for manned flight, compared with $77.5 million Dineen said he needed. Back and forth went the requests and the denials for money that simply was not available. To economize on propulsion development, the more than 170 engine test firings planned originally were cut to just 34, with rigorous development and qualification standards placed upon the project. SSD was concerned not only for its own work but for that of contractors associated with Titan: Martin built the launch vehicle and Aerojet produced the engines. In one sense, SSD was in the middle, a branch of the defense administration arranging and bargaining for and with civilian contractors on the one hand and the government civilian space agency on the other.

But if beset with a troubled development chief at SSD, Chamberlin was even more depressed by having to contact the Agena systems manager at the Marshall Space Flight Center, Friedrich Duerr, and inform him of the need to trim anticipated FY1963 expenditure on the Agena program from $27 million to $10.3 million. As with the Titan procurement, Agena economies were to entail drastically restructured test programs, combining several desirable test objectives wherever possible and re-defining the magnitude of reliability required of the program. At best it would mean lengthy delays in readying Agena for its rendezvous role; at worst it would result in deletion of this entire role for the complete Gemini program. Just a few months before the financial rot set in, the Air Force had successfully persuaded NASA to plan for using the Agena D, an improved version of the Agena B originally believed to be the best target vehicle. Air Force requirements dictated development of the D model independent of the NASA manned space flight program, but it was in the space agency's interest to use the more refined version.

But while Duerr was moving back and forth between NASA and the Air Force, trying to find ways of re-grouping schedules and assignments on the Gemini-Agena development board, the space agency got itself in an internal debate on the actual value versus cost of the entire Gemini rendezvous objective. Joseph Shea claimed that having now settled how American astronauts would fly to the Moon, Agena was not really necessary for Gemini and that the practice of rendezvous could be readily attained without the Lockheed-built target vehicle. Apollo itself was expected to embark on a protracted series of rendezvous and docking tests following initial flights in Earth orbit. But Bob Gilruth stuck to ground he had for long held about the phasing of Gemini with Apollo, believing that the two-man vehicle should take as much slack out of the manned space program as possible; if Gemini could significantly advance the state of the rendezvous and docking art to the point where it could reduce the number of Apollo rehearsals, any unforeseen problems with the Moon ship could be accommodated within a time frame that originally dictated lengthy rendezvous practice. It was fortunate that Gilruth had this concept of making good use as soon as possible of as much as possible. It was to be one of two key program concepts primarily responsible for putting men on the Moon by 1970.

However, Shea convinced Holmes that MSC should look again at the role of Agena in the Gemini program, and the latter ordered Gilruth to return to Houston and review options. By early November, 1963, Chamberlin heard of Shea's criticism of the Agena concept and strongly protested against the move to rid Gemini of an important element in the drive to the Moon. But Shea recognized the need for Gemini to practice rendezvous; he just did not think a fully fledged Agena D fitted with a specially designed docking cone into which the nose of Gemini could be inserted, was in the best financial interests of the total manned program. So he proposed an alternative. Studies showed that Gemini could carry in the back of its adapter for launch a package which would be separated in orbit for a stable rendezvous platform. Maneuvering some considerable distance from the package, equipped with radar transponders similar to units being considered for Agena, the Gemini spacecraft could then compute re-rendezvous operations as though it were approaching an Agena or, more to the point, as though it were an Apollo mother ship pulling alongside a lunar landing vehicle.

It sounded fine in theory, but André Meyer from the Gemini Project Office believed the device would weigh twice the 90 kg suggested by Headquarters. If it did, that would put paid to the paraglider, and at this time the land landing concept was every bit as sacred as, for instance, the long duration flight capability. Nevertheless, the Manned Spacecraft Center designed a device of its own called the Radar Evaluation Pod (REP) weighing a scant 30 kg and possessing a small battery powered radar transponder, beacon and light. McDonnell was told to proceed with final design and development for planned flights on the second and third missions, precursor tests to Agena flights MSC hoped to have for subsequent missions but for which it realized there would be a delay due to the drastic cuts. Dedicated efforts in the MSC, plus helpful dialogue between ardent Agena supporters and the Lockheed company, scaled down dramatically the sophisticated research, development and test program for an operational Gemini Agena Target Vehicle (GATV).

Threatened by total shut-down if they barked too loud for money that seemed suddenly very scarce throughout the space agency, supporters of an independent Gemini rendezvous plan honed the program down to an estimated total cost of $44.1 million versus $76.8 million it had been estimated to cost at last accounting. But gradually the decision whether to keep or eject the Agena rendezvous objective in Gemini's operational envelope slipped from Bob Gilruth and his stalwart MSC men as Shea grew louder in his criticism. At a Management Council meeting on 27 November, 1962, Gilruth impressed upon the administrators how successful had been combined MSC-SSD-Lockheed efforts at trimming the program, and how essential a part of the run up to Apollo the whole concept of docking exercises would be seen to be. A final decision would have to be made soon if budget plans for the next financial year were to be effectively compiled for presentation to the President. Holmes agreed to give the matter urgent attention, but Shea continued to talk aloud about doubts permeating NASA on the desirability of proceeding with Gemini-Agena.

During December a detailed analysis of the operation was given by an investigative committee charged with deciding the issue. James Rose came across from the Gemini Project Office and swung the day in favour of the MSC viewpoint. Four days before Christmas Day, Headquarters wired the Manned Spacecraft Center with a message to go ahead with Agena plans and spend the $10.3 million originally sanctioned. The Manned Center was to manage the project from then on, but only the Gemini segment. All other Agena project work – it was fast becoming the standard upper stage workhorse for many NASA projects – would be taken from the Marshall Center and given to Lewis. MSC could now deal direct with the Space Systems Division and with Lockheed on matters concerning project management and administration, a more effective control concept where both money and technical decision emanated from the same location. For a while it looked as though the Atlas program too would come under the budget axe, and a contingency plan was formulated whereby NASA would use three Atlas launchers left over from the Mercury program. In the event, that was not necessary, however, and money was found for the NASA share of a Defense Department Atlas improvement program aimed at giving everybody a better product.

By the turn of the year the first Gemini flight had been put back to December, 1963, an unmanned suborbital flight originally proposed by McDonnell to test the heat shield and structurally qualify the spacecraft for manned flight. That would come on the second launch, planned for March, 1964, with succeeding missions every two months until the last flight in November, 1965. It was an ambitious plan, as events would indicate later in 1963. In several respects that year marked a turning point for the space affairs of not only NASA but the Defense Department also. In what was to prove the last year anybody – Americans or Russians – sent one-man spacecraft into orbit, directions changed in the affairs of the US Air Force, for several years now actively working on their own space vehicle: Dyna-Soar.

Concern about the continued applicability of this aerospace lifting vehicle reached a high-water mark in 1962 as studies of the next decade's needs in military space technology disgorged the Manned Orbital Development System, MODS for short, in which a series of interlocking objectives would be met with common hardware. The Air Force looked at Gemini and recognized a true production line vehicle that could be used for ferrying supplies between Earth and an orbital space station. Although military interests in space, particularly *manned* space flight, had taken a downward turn shortly after the major manned objectives were sought, between 1958 and 1961, Air Force interest in an operational space role persisted and in the wake of a dramatically increased civilian space budget the Defense Department hoped to creep in unquestioned by suspicious Congressmen and get the foothold in orbit they had coveted for a decade.

Dyna-Soar had gone through several changes since Boeing and Martin were selected in June, 1958, to perform limited feasibility studies. By April, 1959, these two companies presented detailed proposals for consideration by the Air Force and on 9 November were selected for contracts: Boeing was to manufacture the Dyna-Soar glide vehicle, integrate the systems and assemble and test the configuration; Martin was to produce the Titan I launcher selected for the program. By March, 1960, Boeing had submitted a final design configuration to the Air Force and the Aerospace Vehicles Panel of the Air Force Scientific Advisory Board met to review both this and the findings of a preliminary Phase A study embracing all Dyna-Soar technology up to that time. On 25 April, the Air Force gave its permission for full scale development and in the September held a bidders' conference as the first step in embracing private industry.

On 13 January, 1961, the Air Force decided to substitute the improved Titan II for the Titan I launch vehicle, a better choice since payload capability was higher and the booster used storable propellants; experience with Mercury showed the wisdom of choosing a launch vehicle with rapid turnaround. Before the end of the year, however, the Air Force had changed its mind again. Dyna-Soar weight was going up all the time and even the Titan II was seen as inadequate for orbital flight. So the Titan III came along in a plan announced on 28 December, with solid propellant strap-on boosters and a lift-off thrust exceeding that of NASA's Saturn C-I. On 26 June, 1962, a designation change was formally announced. From then on Dyna-Soar would be known simply as the X-20.

On 19 September, the six Air Force pilots selected to fly X-20 were named: Maj. James W. Wood, Maj. Russell L. Rogers, Maj. Henry C. Gordon, Capt. William J. Knight, Capt. Albert H. Crews, Jr., and NASA's civilian test pilot Milton O. Thompson. As an aerospace glider, the X-20 was to carry a delta planform wing with a single vertical stabilizer at each tip, aerodynamic control surfaces for maneuvering during descent, and a small crew compartment. In fact it was a small vehicle in dimension, about the size of the Hawker Siddeley Gnat used by Britain's Red Arrows Royal Force aerobatic team for many years or, by comparison, shorter than the Hawker Hurricane of World War Two and only half the wingspan; 10.7 meters long, 6.09 meters in span, and 2.4 meters high standing on the wire-brush skids it would use when landing. With a weight of about 4½ tonnes, X-20 would be launched into orbit by a Titan III, descending through the atmosphere protected by heat-sink materials which could be used again without refurbishment.

Capable of maneuvering on the descent to landing sites selected in advance, the vehicle would have been used as a research and development tool, an orbital platform for the Air Force to evaluate the benefits of manned space activity, but a vehicle without portfolio in every sense. Ardently supported by personnel assigned to its development, X-20 found less favor with Pentagon planners and strategists putting together the aerospace development plans for a decade hence. Congress liked the project, allocating embarrassingly inflated sums which the Air Force had to refuse. In Fiscal Year 1961 the Air Force was given $58 million, a sum it spent as feasibility gave way to program development. For 1962 the outgoing Eisenhower administration proposed expenditure totalling $70 million, but the Kennedy camp raised this to $100 million and Congress made it clear it would authorize more if requested. On the recommendation of the appropriations committee of the House of Representative, Congress appropriated an additional $85.8 million but the new Secretary of Defense, Robert McNamara, was against speeding up the project and the money was not spent. For Fiscal Year 1963, Kennedy asked for $115 million on the behalf of the Air Fcrce, and Congress pressed again for an accelerated program, allocating a further $42 million on their own initiative.

By this time, three years after the formal go-ahead, plans embraced a series of air drops from B-52 aircraft of models designed to show stability and control at low speeds; optimized around aerodynamic requirements in the hypersonic (Mach 6 to Mach 25) region, slow flying at low altitude would require attentive piloting. Those test activities would be performed in 1965, with unmanned and manned flights of the full size X-20 beginning in 1966. But X-20 was still very much an experimental precursor to an operational vehicle with undefined mandate; X-20 was designed to fly between Earth and near-space and prove that pilots could safely move in and out of the new environment, seeking a role for military man in space. With Gemini coming along in the meantime, the payload lifting capability of that civilian successor to Mercury was an infinitely more attractive proposition as a ferry vehicle to the manned orbital laboratories then being planned for the late 1960s. X-20 would simply be incapable of lifting cargo and crews, its existence demanding a successor in turn to justify in an operational setting the very experimental nature of its own role.

Ideally, the lifting re-entry flown by the X-20 would be an essential element of any permanent ferry vehicle; X-20 type vehicles could be re-used many times over, unlike the semi-ballistic Gemini type spacecraft. What the Air Force wanted was a reusable, winged space transporter but, in the wake of X-20, funds would be eliminated for any operational role the vehicle would be built for in the first place. So, late in 1962, the Air Force began to consider the possibility of using Gemini spacecraft in rendezvous and docking exercises. A need for a permanent orbital laboratory was emerging as a research facility for military space activities and for a while it looked as though the Air Force would assume control of the entire Gemini program. Some NASA personnel, notable among them being Bob Gilruth, the MSC Director,

Five pilots who might have flown a Dyna-Soar to space. Left to right: Albert Crews, Henry Gordon, William Knight, Russell Rogers, James Wood.

← Lt. Gen. Roscoe C. Wilson unveils an early model of Dyna-Soar, seen here atop an early Titan booster.

The USAF Manned Orbiting Laboratory followed Dyna-Soar as the high hope of a military man in space, depicted here with a Gemini capsule in front.

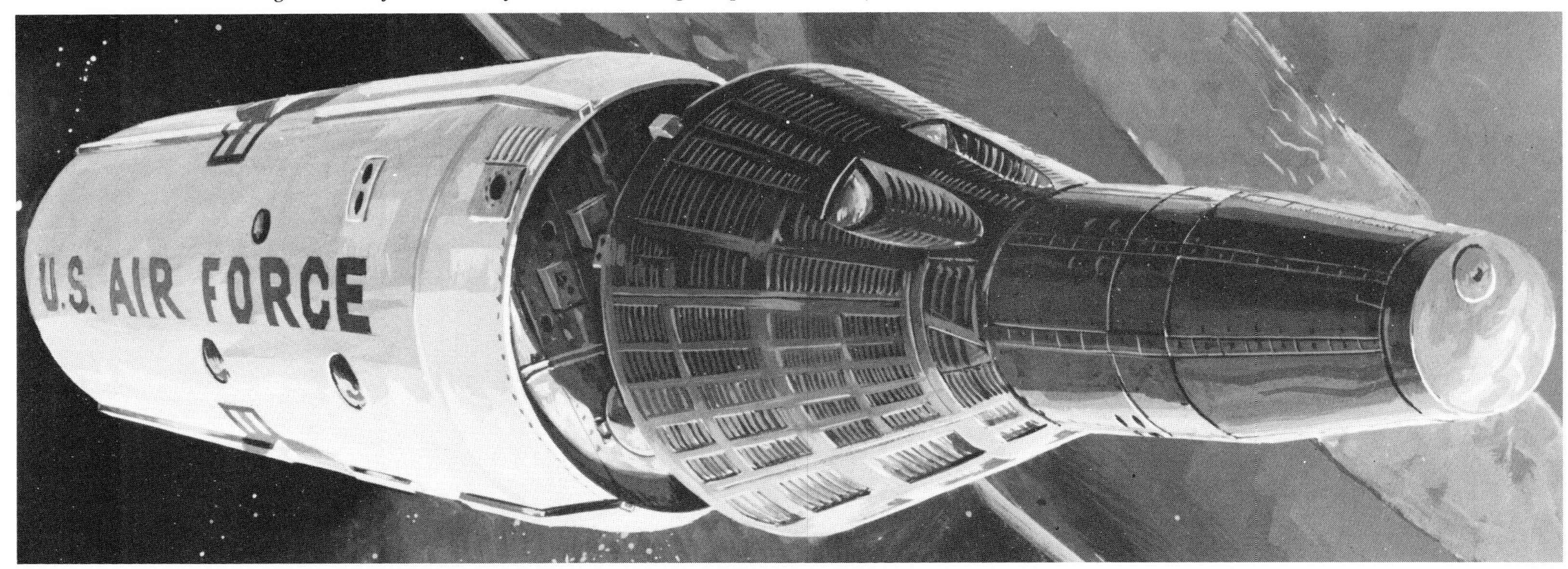

approved of wider service participation and saw nothing wrong with providing the Air Force with what was essentially a production line spacecraft. But the argument in favour of retaining a clear and distinct mandate for the civilian control of what was after all a step on the road to civilian Moon flights was strong.

For its part, the Air Force was hostile to a Gemini takeover. The whole idea had been dumped in their lap by McNamara and it was 1963 before the matter was sorted out. The Defense Department would contribute services and facilities to the NASA Gemini program, and would fly DoD experiments on some missions but it would not participate in program management, nor would it take over any of the project's elements. What did emerge was the Gemini Program Planning Board, co-chaired by Bob Seamans for NASA and Brockway McMillan for the DoD, with responsibility for phasing military experiments in with NASA mission schedules and flight opportunities.

Between the spring and the fall, 1963, the Air Force restructured its military manned space plans, deciding in favour of a Gemini B procurement program; the 'B' referred to Blue Gemini, the unofficial tag applied to plans the previous year for an Air Force takeover. The X-20 would have to go and in its place the Air Force proposed the Manned Orbiting Laboratory (MOL) program where a modified two-man Gemini would rendezvous and dock with a cylindrical laboratory launched separately on a Titan III. McNamara liked the idea and officially announced the planned extinction of the old Dyna-Soar design at a press conference on 10 December, 1963. By this time, half way into Fiscal Year 1964, the Air Force had already spent a further $52.2 million. In all, the X-20 had cost nearly as much as the complete NASA Mercury project.

Against this background of re-grouping that greatly influenced the Air Force's plans for future manned missions, Gemini men spent most of 1963 battling with nagging systems problems and a burgeoning cost tree. In fact late in calendar year 1962 program manager James Chamberlin knew the project was in deep financial trouble again, with technical solutions to first paraglider, then Titan launch vehicle, problems eating up the finances at an unprecedented rate. But Chamberlin thought he could integrate the increases with economies made elsewhere. The trouble with that idea was that events overtook the ex-Langley engineer when, in February, 1963, the projected total Gemini development cost of $834.1 million was exceeded a month later by an improved MSC prediction that went above $1,000 million. Bob Gilruth received the full force of displeasures emanating from Headquarters in Washington and on the 19th he stripped Chamberlin of his position, re-assigning him as Senior Engineering Adviser to the Director.

To the post of Gemini Program Manager came Charles Mathews of MSC's Engineering and Development Directorate. In fact, Mathews had been appointed Deputy Assistant Director in addition to his existing position as Chief of the Spacecraft Technology Division, a member of the original team from Langley's Pilotless Aircraft Research Division and head of the Space Task Group's Spacecraft Research Division. Chamberlin had been with Gemini since he brought it to life more than two years past, from a concept that emerged two years before that. He was an engineer's engineer, probably one of the most brilliant men ever to work for NASA, pruning and honing the design to pack within an incredibly small space a whole generation of new technology, transforming Mercury into a two-man operational spaceship from a tiny one-man capsule. He was the single most important link in the chain that gave NASA the tool it needed to pave Apollo's way, but in March 1963 he was seen as an inappropriate choice for senior management, a technical man with a personality very different to that which was required for project management. Chamberlin had been in charge too long, in fact there were some that said he should never have been afflicted with the post, but only because events had outpaced his ability to keep them under control. In another type of program he may have performed an excellent role in project management, but Gemini was a beast of its own from inception to fruition and needed a tough ringmaster to keep it tame.

In almost every dimension, the spacecraft was a very real product of the times, conceived as an afterthought, prepared in haste, and with responsibilities piled high upon its trembling shoulders. What had once been presented as an interim vehicle between Mercury and Apollo was rapidly becoming more a precursor to Apollo than a successor for Mercury. The first job facing Mathews was to put the program on a realistic schedule. A new inventory of flights was ready within three weeks, adding a second unmanned flight to the suborbital mission settled a few months earlier. There would now be ten manned flights out of 12 planned missions. That first flight was still tentatively planned for December, 1963, but the second mission, an unmanned ballistic flight, was put back from March to July, 1964, deferring the first manned mission by five months to October, 1964. General consensus felt the first manned mission should be targeted for a three-orbit flight rather than the 18 orbits originally planned, and by June it was agreed.

The following month Gemini's second manned flight was tentatively programed for January, 1965, as a seven-day mission carrying the Radar Evaluation Pod for practice rendezvous. The third and fourth manned flights were to be rendezvous and 14-day missions respectively, with six three-day rendezvous and docking flights closing out the series by January, 1967. By now it was apparent that a two-month separation between flights was optimistically conservative regarding flight team turnaround and mission preparation, so missions would be scheduled around three-month centers. By now too it was becoming increasingly clear that Gemini represented a major new effort in manned space flight, acquisition of a new capability, using new technology where once it had been thought possible simply to transfer from Mercury the systems and subsystems originally designed for substantially less demanding objectives.

The shock at seeing in MSC's Fiscal Year 1964 estimates the major financial increases that led to Chamberlin's replacement woke Headquarters to a more sensitive dialogue with Houston over the dimensions of the two-man program. During May, 1963, Holmes spelled out to Congress just how valuable a tool the Gemini spacecraft could be made to be for ferry duties with space stations or-Earth orbiting laboratories. Taking a leaf out of the Air Force book, it was support the rising cost columns could do with before the money-sorters in Washington. But in truth, Gemini was a boon for Apollo during the difficult years when design details were chasing production requirements for North American Aviation's three-man Moon ship. Systems packaging and test programs researched and applied by Gemini management was watched with more than casual interest by the managers of Apollo, and much that was to speed the larger program when it too had a major re-shuffle came directly from the work done by Mathews in 1963.

By the end of that year the project had drifted further away from the flight schedules confidently laid out when Mathews took over that spring, the product of tiresome troubles with Titan II development, echoing the Atlas delays prior to Mercury orbital flights. Nothing escaped the scourge of failure and delay that year and at times it looked as though the entire program would have to be re-written. Clearly the first Gemini mission would not fly in 1963 as planned and doubts were expressed about the ability of the program to secure a flight within the first quarter of 1964. In the meantime, further administrative changes were made necessary by the colossal growth in NASA as a whole.

The re-grouped infrastructure of November, 1961, and the adjustments enacted from October, 1962, were incomplete for the ascending magnitude of the space agency's responsibilities. The two new Deputy Associate Administrator positions set up in 1962 were ill-defined and in need of re-grouping. By mid-1963, Brainerd Homes had announced he would be returning to industry in the autumn having fulfilled the obligation to serve the government for two years required of him when he came as the first Director of Manned Space Flight to NASA Headquarters in 1961. From 1 November, 1963, when the re-organization would take effect, Dr. George E. Mueller (pronounced Miller) was to take his place, a recruit from Space Technology Laboratories. The reason for the changes was inspired by drastic increases in the size of the space agency, but the timing was a product of Mercury's end and Apollo's commitment to an upward development curve. The new look would combine program and center management by putting the field centers under control of the respective Headquarters' program directors rather than general management. The Marshall Space Flight, Manned Spacecraft, and Launch Operations Centers would be under Mueller; the Goddard Space Flight Center, Wallops Station, Pacific Launch Operations Office, and the Jet Propulsion Laboratory, would be under Homer E. Newell, the head of the Office of Space Sciences and Applications; and the Langley, Lewis, Ames, and Flight Research Centers, the old NACA facilities, were placed back under Bisplinghoff's Office of Advanced Research and Technology. Mueller, Newell and Bisplinghoff were given Associate Administrator positions.

(On 29 November, 1963, President Johnson signed an Executive Order honoring the memory of the late John F. Kennedy, and changing the name of the NASA Launch Operations Center: 'Whereas President John F. Kennedy lighted the imagination of our people when he set the Moon as our target and man as the means to reach it; and whereas the installations now to be renamed are a center and a symbol of our country's peaceful assault on space; and whereas it is in the nature of this assault that it should test the limits of our youth and grace, our strength and wit, our vigor and perseverance – qualities fitting to the memory of John F. Kennedy: Now, therefore, by virtue of the authority vested in me as President of the United States, I hereby designate the facilities of the Launch Operations Center of the National Aeronautics and Space Administration and the facilities of Station No. 1 of the Atlantic Missile Range, in the State of Florida, as the John F. Kennedy Space Center; and such facilities shall be hereafter known and referred to by that name.' It has remained the Kennedy Space Center to this day. Also on 29 November, President Johnson said that 'I have also acted today with the understanding and the support of my friend, the Governor of Florida, Farris Bryant, to change the name of Cape Canaveral. It shall be known hereafter as Cape Kennedy.' On 12 December, Rep. Edward J. Gurney (R.-Fla.) said that the renaming 'will not be official unless it is acted on by the Florida Legislature,' quoting Arthur Baker, Chairman of the Board of Geographic Names, as saying that the name will only apply to federal maps and charts and is not compulsory for the people of Florida. The Committee on Domestic Geographic Names said that citizens of Cape Canaveral town, numbering 4,000, did not want the name change and that they need not have it. Publication of the name change in the official decision list of the US Board on Geographic Names completed the legal procedure on 2 April, 1964. Following continued campaigning by residents, and the presentation of several resolutions before Congress in 1973, Cape Kennedy officially reverted to Cape Canaveral on 9 October of that year by action of the Dept. of the Interior. NASA's Kennedy Space Center, at Cape Canaveral, retains its assigned name.)

Thus was established with greater clarity the chains of responsibility under Seamans. Much further down the management chain, Ken Kleinknecht left his post as Mercury program

Four key officials responsible for much of Gemini's success. Left to right: Charles Mathews, Gemini Program Manager; George Low, Deputy Director Manned Spacecraft Center; Flight Director John Hodge; Flight Director Christopher Kraft.

manager and went to work as Mathews' deputy – the responsibilities of his former office now defunct – Walt Williams went to NASA Headquarters as Deputy Associate Administrator for Manned Space Flight Operations, to leave James C. Elms in the post of Deputy Director of the Manned Spacecraft Center, a post he vacated two months later to return to industry. In March, 1964, George Low filled Elms' position. But other plans too were being formulated for the day when Gemini would perform unique functions in space. Not least among this increasingly long list was the science – some would say art – of extra-vehicular activity (EVA), which the world would come to know as space-walking.

The first serious mention of an astronaut leaving his spacecraft in orbit came during talks in March, 1961, between Maxime Faget from the then STG and McDonnell's John Yardley; responding to Faget's suggestion that any prospect of EVA would demand a two-man vehicle for safety, Yardley had a McDonnell Vice President, Walter Burke, start work on a two-man Mercury. It could be said that the prospect for EVA was the single cause of Gemini emerging in the form it did. For a long while that prospect seemed almost lost among long duration flights, rendezvous and docking, guided re-entry, paraglider developments, etc. During the latter half of 1962, however, NASA's Life Systems Division reported on work it had conducted concerning pressure suits, ventilation, thermal protection, maneuvering units, and insulation for a Gemini astronaut leaving his re-entry module. Likewise, in February, 1963, the MSC Crew Systems Division laid down guidelines for EVA and asked McDonnell to report back on requirements for a crewmember to open a hatch and stand up – stand-up EVA, or SEVA – and a full exit, or full EVA.

A month later a meeting at the Manned Spacecraft Center approved suggested operational requirements, proposed a 30 minute period of EVA activity, determined that the astronaut should remain attached, or tethered, to the spacecraft at all times, and agreed that McDonnell be requested to accomodate provision for a crewmember to leave his cabin on every capsule from spacecraft number 4 up. By the end of the year MSC had received and evaluated proposals for an EVA life support package to be worn by the astronaut for autonomous environmental provision, and had selected the Garrett Corporation to build their submitted design. In January, 1964, the program plan for Gemini EVA was complete.

At this time Mathews had re-worked his earlier launch schedule to accommodate continuing delays, and envisaged a ballistic flight, GT-1, in March, 1964, an unmanned orbital flight, GT-2, in the August, and the first manned orbital flight, GT-3, in the November. GT-4, the first flight with an EVA capability, would have the pilot open his hatch and stand up for a brief period. That would be scheduled for February, 1965. GT-5, in May, would provide practice in full egress and ingress procedures followed by a translation to the adapter section on GT-6 in August, 1965. The next two flights would practice maneuvering along hand holds and tethers on the outside of the spacecraft, with evaluation of an astronaut maneuvering unit on GT-8. Other more advanced maneuvering equipment could be tested on the last three flights. Thus, by the beginning of 1964 Gemini had a firm purpose in providing a platform for EVA. It was a fully integrated element in the program and was to provide one of three prime project objectives, the others being long duration flight, and rendezvous and docking. All other objectives were secondary to those requirements, and each was a potential stumbling block to a smooth Apollo schedule.

But it was budget time and mini-crisis emerged again when, in March, 1964, NASA estimated that the projected spacecraft cost submitted at the final contract agreement with McDonnell a year before would rise from $456.6 million to $667.3 million before the program was over. Moreover, the Space Systems Division's estimate for Atlas-Agena rose from $103 million in September to $137 million by March, 1964. In the last financial crisis during March, 1963, the Air Force estimated a $240 million price tag for Titan II work. By March, 1964, that figure stood at $324 million. In all, the project now carried an estimate of $1,220.3 million from start to finish – and it was still a long way from being finished. Again the scrutiny set in, but this time Mathews was at the helm. Everything was looked at with a view to economizing, even plans for dropping a flight were examined. Rendezvous got another appraisal, gaining little help from a continuing round of technical problems that instilled little confidence in the Atlas-Agena combination being ready on time for flight. Yet again, re-programing got the project out of the woods without too many injuries, but it was another reason for dropping ailing elements like the paraglider.

Work on spacecraft 1 for the first unmanned test flight came to an early halt when several component failures were discovered during checkout, July 1963. It was finally accepted by the space agency on 30 September following exhaustive systems tests and altitude tests in a vacuum chamber. At the Cape on October 4, s/c 1 was devoid of full systems layouts of the type necessary for manned missions but because of delays to preparation of the Titan launch vehicle it was 3 March, 1964, before the vehicle was delivered to launch complex 19. Gemini Launch Vehicle-1 (GLV-1) had gone through several tests itself. First, following delivery of propellant tanks from Martin's Denver division to the Baltimore facility in October, 1962, the second stage oxidizer tank was found to be cracked and had to be replaced. Then, in May, 1963, the fully assembled launcher was found to have a short circuit where a defective clamp had cut electrical wire insulation, necessitating inspection of all 1,500 clamps throughout the vehicles. Several minor problems held up functional verification tests: pipes were found to be dirty, solder balls came loose and fouled up the gyroscope, electrical problems continued.

By 11 September, the Air Force was presented with the launcher but in acceptance inspections several electrical connectors were found to be contaminated, and a broken component was discovered in a turbopump. GLV-1 was rejected by the Air Force inspectors and when Martin engineers scrutinized the connectors they found it necessary to replace 180 of the 350 carried by the vehicle. When the launcher finally did arrive at the Cape, on 26 October, it was the start of yet another time-consuming round of preparation and test. Dragging further behind schedule with each passing week, Mathews took steps to speed up the checkout process, re-grouping coordination panels into a single authority. Nevertheless, problems arose with the combined systems test designed to check the integrity of the two stages, then the wet-mock simulated flight was delayed; this was a full countdown up to and including filling the tanks with propellant. Errors in processing the sequence of events in that test were responsible for further delays, and in the sequenced compatibility firing, where countdown and ignition would be performed in a test designed to demonstrate a full launch sequence, success came on only the third attempt. Up to this point the launcher was still separated into its component stages but on 31 January, 1964, GLV-1 came together on launch complex 19 for the first time. By March, the spacecraft had arrived for installation on top of GLV-1's second stage.

Launch complex 19 was very different from Atlas sites. The Titan II launch area was up the coast from the Atlas complex used in Mercury, the same complex 14 that would be employed for launching Atlas-Agena targets for Gemini. In fact complex 19 was past Titan complex 15 and complex 16, the latter scheduled for static firing tests with Apollo's big service module engine, and could be found at the intersection of ICBM Road and Heavy Launch Road across from the Air Force Station. Titan's launch stand was a three-storey concrete and steel structure measuring 8.53 meters tall, 19.8 meters wide, and 137.2 meters in length, supporting the umbilical tower, erectors, thrust mounts and flame buckets. Beneath the launch deck was an equipment room to control the erector which would raise the separate stages, a power distribution substation, a transfer room where cabling from the blockhouse nearby joined umbilical tower lines, a guidance equipment room, winch pit and motors for the erectors, and an actuater room for the giant steel arms that would raise the launcher. The stand also contained 80 water nozzles and a cooling system from which would pour 94,625 liters per minute through the flame bucket designed to control the direction of the Titan's spent exhaust. The blockhouse was designed to receive and survive a direct impact from a Titan running amok, its walls 6.4 meters thick. The building comprised a concrete, sand and steel structure 47.4 meters in diameter and 15.24 meters tall, a self contained two-storey center from where consoles and flight control equipment would monitor and control every step of the ascent.

The launch stand's umbilical tower stood about 31.1 meters tall and carried electrical cables, fuel lines, hydraulic pipes and cryogenic supplies to the launch vehicle and spacecraft, a fixed structure permanently attached to the ground. The erector was designed to lift the two stages of Titan II into a launch position on

Technicians groom a Gemini spacecraft (No. 3) atop the service tower, an area known as the 'white room' because of its extreme cleanliness.

McDonnell technician George Baldwin examines equipment in Gemini spacecraft 1 prior to the first flight in April, 1964.

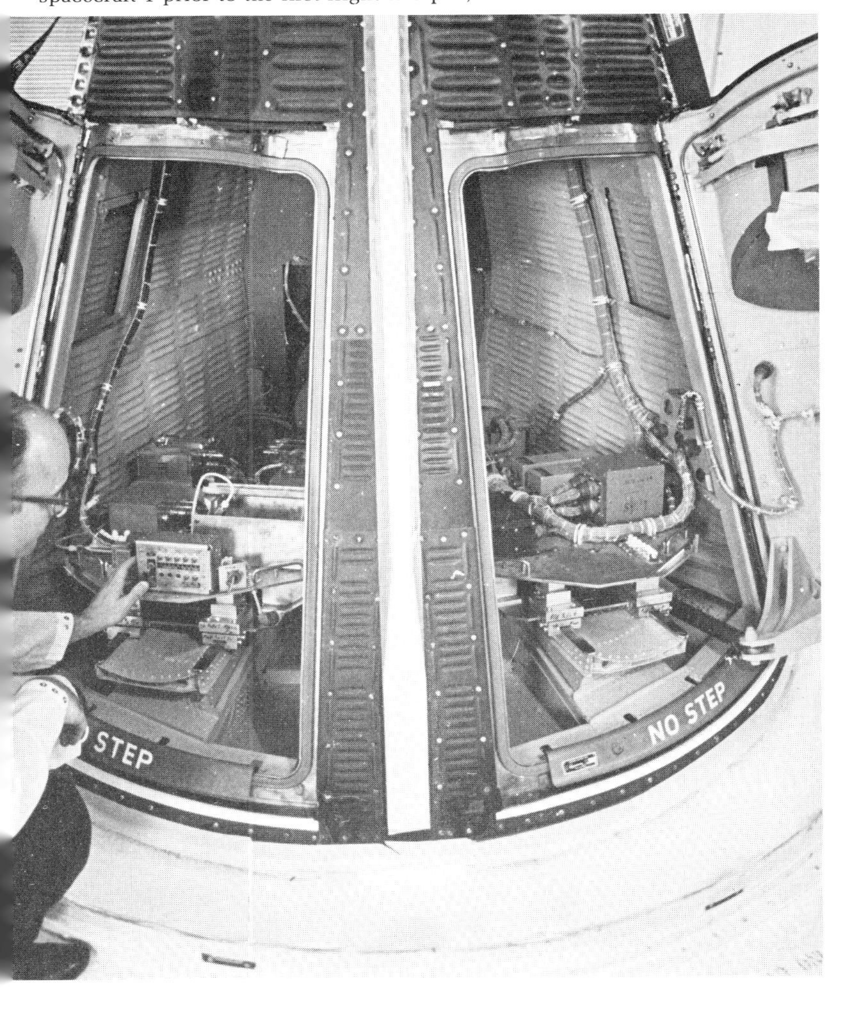

the thrust mount, with eight movable platforms extending up the inside affording access to areas on the launch vehicle. The erector weighed 127 tonnes and comprised a long box structure, 7.62 meters square and 42.06 meters long. In the absence of a service gantry, such as was provided for the Mercury-Atlas configuration, access to the spacecraft was obtained via a specially built white room fabricated from aluminum, 7.62 meters square and 15.24 meters tall, located at the top of the erector. At the top was a large crane capable of lifting 4.5 tonnes, and around the sides closed windows to permit the passage of radio waves for communication. The temperature inside the white room, specially protected against dust entering the enclosed area by maintaining a higher pressure inside than out, would be kept at a constant 22.2°c with a relative humidity of 50%.

A day after spacecraft 1 was placed in the white room adjacent to GLV-1's second stage, engineers conducted a pre-mate systems test to clear the way for mating the two structures. During the test an engineer accidentally dropped a wrench on the top of the second stage oxidizer dome, causing a scratch 0.0038 cm deep. The scratch was burnished, tests confirmed the validity of the tank dome, and the preparations continued. Interference checks came next where the configuration was tested for systems compatibility, an essential precursor to confirmation that the entire assembly would operate as expected and without the performance of one system impacting another. On the third day circuit amplifiers in the first stage propulsion system gimbal actuaters showed noisy outputs until test engineers recognized the problem to be in the ground equipment.

By the end of March the pace of testing picked up momentum: on the 27th, a combined systems test and simulated flight went off well; Pogo (longitudinal vibration) suppression equipment arrived on the 31st. During preparations for a late night wet mock simulated launch, personnel noticed smoke pouring from a transformer on the launch stand. More delay while a replacement was found and brought to the pad. Then the elements took a hand. Because the wet mock test required the movement of toxic chemicals from supply sources to the interior of the launch vehicle, it was necessary to have a good atmosphere above the launch complex for efficient dissipation of the fumes. When everything was technically ready for the wet mock test the Cape was covered by a thermal inversion, where cooler air was trapped by warmer air above, and engineers had no option but to sit it out until the inversion itself slowly dissipated. Fortunately the test, on 2 April, 1964, went well, although a minor hold in the simulated countdown was called.

The next day, Walt Williams convened his Spacecraft Flight Readiness Review Board, including Mathews, Kraft, Slayton, F. John Bailey, Jr., from Reliability & Flight Safety, and Merritt Preston, from MSC Florida Operations. All seemed ready for the GT-1 mission, a fact that would be confirmed by one last test: the simulated flight planned for two days hence. Sunday, 5 April, came and the final test sheets were signed off. Williams called the Mission Review Board together and everybody there, representing anything that was a part of the planned flight, passed the word that they were happy to commit the vehicle to flight. By mid-day a launch time had been set for 11:00 am on Wednesday, 8 April. Toward that end the split countdown started the next day.

Spacecraft 1 was not at all representative of the vehicles yet to carry men into space. It carried dummy equipment in the adaptor, with only the racks in which systems would be installed similar to those on a manned vehicle. The retro-rockets were dummies too; for some time now GT-1 had been an orbital flight where only the first three orbits would constitute the mission, leaving the spacecraft to circle the globe and decay back down through the atmosphere naturally after three days or so, and recovery was not planned. In place of crew couches the re-entry module carried instrumentation pallets with pressure transducers, temperature sensors and accelerometers installed at various locations around the cabin, wired to telemetry transmitters so that engineers on the ground could get a good idea of the internal environment. In all, there were 104 measurements concerning temperature, acceleration, and pressure. The spacecraft's environmental control system would function only to maintain a differential pressure between the interior of the cabin and the atmosphere, and the electrical system comprised a single battery. Around the world, NASA tracking stations at the Cape, Bermuda, Australia, Point Arguello, White Sands and Eglin Air Force Base would monitor the progress of the spacecraft, which was not designed for this flight to separate from the second stage of the launch vehicle.

The final countdown began promptly at 6:00 am on the morning of Wednesday, 8 April. Walt Williams was in the blockhouse. It was to be his last launch, all but the last official role he played in NASA's affairs. Three weeks before the mission he had resigned from the space agency and accepted a post as vice president and general manager of Aerospace's Manned Systems Division. Chris Kraft would move across as Gemini Operations Director, who had also been appointed Manned Spacecraft Center Assistant Director for Flight Operations the previous November. Nothing marred the progression of events that closed in on 11:00 am when the two Aerojet-General propulsion systems would fire into life. At just 1 second after that time the two engines did ignite, an acceptably late start that moved Williams jokingly to question the accuracy of the blockhouse clocks, and for 1.8 seconds the propulsion system built up to 77% thrust at which point a two second timer was started. Just 3.8 seconds after ignition the four bolts holding the booster down on the thrust mounting were severed and GLV-1 slid gracefully away from complex 19, the transparent flame of the hypergolic propellants a stark contrast to the billowing smoke of an Atlas launch, the yellow flame stabbing viciously into open air.

About 7 seconds into the flight the roll program began and the Titan turned on its axis to an azimuth of 72°, ceasing 10 seconds later for the pitch program to start 20 seconds off the pad. About 2 min 29 sec after lift-off, with GTV-1 at a height of 64 km, the first stage shut down and was followed one second later by ignition of the single second stage engine in a 'fire-in-the-hole' staging sequence technically known as FITH. Unlike Atlas, or many other launchers, Titan II was designed to ignite the second stage prior to separation of the first, although the bolts holding the two separate stages together were blown apart just milliseconds later; the interstage structure had several cut-outs which allowed heat from the second stage engine to escape through the sides. The second stage continued to burn for a full 3 min 3 sec before shutting down. The combined second stage and spacecraft was 160 km above the Earth, 1,000 km from the Kennedy Space Center.

It had been a near perfect performance, the launcher imparting a speed only 25.2 km/hr greater than the planned value of 28,372 km/hr, giving the configuration an orbit that at apogee would be 21 km higher than expected. The only impact that would have would be to keep the assembly in orbit just that little bit longer. By 3:50 pm the GT-1 mission was all but over. Three orbits provided all the telemetry engineers needed to test the ability of the spacecraft to conform to predicted responses. The launcher had done its job admirably, raising hearts and hopes that the two years of hard work had paid off and that Titan II was up to the task for which it had been groomed. Just four days later, on Sunday, 12 April, the second stage and spacecraft slid back into the atmosphere on its 64th orbit, burning up in a flaming shower of sparks over the South Atlantic. The message had been clear several days before: Gemini flies.

But could it carry men? Whether it was ready for that was GT-2's responsibility, an unmanned suborbital flight into the Atlantic from where the spacecraft would be recovered for careful examination. The men who would ultimately have to prove the case on GT-3 were presented to the public audience, even then gearing up to receive more space spectaculars, on Monday, 13 April: Virgil I. Grissom, and John W. Young; it would be Young's first flight but Gus had been there before in the sinkable Liberty Bell 7. Back-ups to the two prime crewmembers were Walter M. Schirra from the six-orbit Mercury flight, and rookie Thomas P. Stafford. Grissom had unwittingly perpetrated a caricature of his diminutive stature on the complete interior of the Gemini spacecraft. Astronaut assignment during early 1963 had given him responsibility for the two-man vehicle itself, requiring Grissom to spend considerable time at the McDonnell plant and to be the 'model' for design layouts, crew station fit checks and relative positioning of ejection seats, control displays, stowage lockers, etc. So much so that the spacecraft soon became know as the 'Gusmobile'.

The problem was that when other astronauts came to try the vehicle their greater average body volume was not as comfortable in the tiny crew cabin as design teams had expected it to be. Tom Stafford was 12.7 cm taller than Grissom and his substantial frame threatened to squeeze apart the seat arm rests while his knees nudged the control console and his head pressed into the hatch. In the end, when it was found that none of the astronauts were comfortable, substantial modifications had to be made to the positioning of equipment adjacent to the crewmembers' positions. But before any thought of a manned mission could replace concerns still uppermost about the spacecraft's ability to survive the hostile environment, a second unmanned test flight was necessary.

GT-2 would be flown in August, with the first manned attempt in the November. Spacecraft 2 for the unmanned ballistic shot would be a fully configured production vehicle equipped with two crewman simulator packages in the absence of live pilots. The simulators consisted of sequencers, batteries, cameras, lights, a timer, instrumentation, and a tape recorder to perform as a normal crew and to record vibrations and temperatures during launch and re-entry. It would be the first time everything was planned to work like it should on an orbital mission: the spacecraft would separate from the launcher, the equipment section would jettison, retro-rockets would fire, then the retrograde section itself would be released, the re-entry module would orient itself for entry, and the recovery system would safely lower it to the Atlantic about 3,460 km from pad 19. In effect, the crewman simulators, strapped to ejection seats that were, for this mission, bolted securely to their guide rails, would provide the inputs for spacecraft sequences and operations normally requiring human intervention. It was the simplest and most satisfactory way of flying unmanned a vehicle designed for full manual control if necessary.

Preparation of spacecraft 2 proved to be a lengthy and pro-

Pad 19, Cape Canaveral, and technicians prepare the GT-1 space vehicle from this concrete blockhouse.

tracted process; systems tests had not started before January, 1964 and even the design engineering inspection was late. This would normally come at the end of the manufacturing process and prior to the systems tests which would show whether each spacecraft element functioned according to plan when mated with the structure adjacent to other systems. The design inspection board, comprised of nine members including personnel from the Gemini Program Office in addition to the program manager and the head of reliability and flight safety, convened a two-day meeting at which about 100 experts and observers would scrutinize every scrap of paper associated with the manufacture and assembly of spacecraft 2, determining whether specification changes had been adopted, component histories were acceptable, or manufacturing details adequately expressed. Systems testing held things up for a while as several minor snags crept in to the spacecraft. In fact it was 3 July before all spacecraft modules were mated for final tests involving the fully assembled configuration. Hopes for an August launch were slipping quickly away.

By this time, spacecraft 3, honed for the Grissom/Young flight, was moving out of its own detailed engineering inspection having been completed on the production line in December, 1963, and brought to the McDonnell white room for six months of equipment installation and the minor changes that permeated the system. To smooth the flow of changes wrought by re-written operations procedures or simply a change in component test sequences, Charles Mathews introduced a Gemini Configuration Control Board so that whatever was altered from then on would have to pass through a single route; its first meeting on 27 July, signified that Gemini was theoretically locked up for production and manned operations, and that any further changes were to be considered only in the light of their non-interference with other elements. The McDonnell white room used for equipment and subsystem installation was similar in concept to that employed for Mercury. Wire harnesses, delicate equipment of all kinds, would be placed in the spacecraft within the environmentally controlled area covering a floor space of 5,016 square meters. Temperature was kept at a constant 23.88°c, humidity at 55%, with pressure marginally higher inside the white room to prevent fine particles flowing into the protected area on air streams. In fact, the filtered air re-entering the room was scrubbed of 99.9% particles greater than 0.001 mm in size and 90% of all particles between 0.0003 mm and 0.001 mm. No hospital operating theater was ever as clean as this, but the statistical probability for keeping the crew alive was theoretically higher than that accepted by any medical team.

Another sign that the program was out of the development phase came when the SSD and NASA began awarding incentive fee contracts, first to Martin for work on Titan, then to Aerojet for the launch vehicle engines and to Lockheed for the Agena, still moving along in the development stage but considerably ahead of earlier fears. McDonnell's was the biggest, and potentially the richest, incentive agreement. The final figure for Gemini topped $712.3 million for spacecraft development and fabrication, plus a sum between $28 million and $55.7 million for company performance. Spacecraft 2, meanwhile, was held up at McDonnell with nagging systems problems. In the event that mattered little for Gemini Launch Vehicle teams at the Cape were running into problems themselves.

GLV-2 arrived at the Kennedy Space Center on 10 July and by the 14th the launcher was standing erect on complex 19. Unfortunately, on 17 August, a thunderstorm moved across the Cape area, pad 19 was struck by lightning, and tests later showed several items of equipment on the ground had suffered damage. Mathews flew into the Cape four days later and a full examination by a special committee endorsed an earlier decision to perform a complete re-test of all booster systems equipment as though the launcher had just arrived at the pad. That would take two weeks, but the spacecraft itself was now well behind time, so no problem. However, just ten days after the storm, Hurricane Cleo passed perilously close to the Cape only hours after Martin personnel had removed the upper stage, lowered the erector and lashed down the first stage as torrential rain and high winds swept through. The second stage was put back on the launcher during the night of 31 August–1 September. Just two days later the second set of functional verification tests got under way. At last all seemed to be going as hoped for.

Then, suddenly, Hurricane Dora was seen heading straight for the Cape. On Tuesday, 8 September, Martin technicians hurriedly pulled the second stage down, then raised the erector to lower the first stage as the storm raced straight for launch site 19. The following day everybody sat it out while the two Titan stages lay protected in a hangar. The storm changed course and managers and engineers planned on getting back to work on the Thursday. But that was not to be as Hurricane Ethel sped in from the Atlantic, threatening to batter the Florida coast by week's end. The Titan stayed where it was and homes in the area prepared for another round of deluge and storm. The following Monday, 14 September, all seemed clear and the stages were taken back to pad 19 and erected again. Once more, engineers attempted to resume the second verification test sequence, beginning that work on the 21st as spacecraft 2 finally arrived from McDonnell's St. Louis plant. But it would be a further four weeks before subsystem and combined system tests cleared the vehicle for mating to the launcher.

Charles Mathews knew by now that all hope of a manned flight before the end of the year had disappeared with the subsiding winds of Cleo, Dora, and Ethel. The second flight was by now re-scheduled for mid-November and GT-3 could not possibly fly before the end of January, 1965. At the Cape, engineers managed to get GLV-2 out of the combined system test phase by 6 October, ready to receive the spacecraft that was again lagging behind schedule. In the three weeks since arriving at the Cape, spacecraft 2 was fully eight days adrift. But America was not alone in preparing an unmanned test flight of a successor to the one-man spacecraft era. In remarkable parallel to NASA efforts at reworking the basic Mercury design, Soviet Russia, probably under the personal instigation of Nikita Kruschev, was about to introduce what at best could be considered a makeshift vehicle with limited capability.

As the genius behind Russia's space program, Sergei Korolev was already working on a second generation spacecraft that would not appear for three or four years at best. This was simply not good enough. With full knowledge about not only Apollo Moon landing plans but also the intermediate Gemini spacecraft, ostensibly the first multi-man space vehicle, Kruschev is believed to have pressed Korolev for an interim plan aimed at eclipsing what he considered to be a major initiative for the

The first two-man crew named for space: Grissom (right) and Young.

United States. On 6 October, 1964, while engineers and technicians were preparing hardware for the unmanned Gemini flight, Cosmos 147 was put into an orbit between 177 km and 413 km. A day later it was returned to Earth. On 12 October, at 7:30 am, Moscow time, a spaceship called Voskhod 1 was launched from the Soviet Union with three cosmonauts on board. The vehicle weighed 5.3 tonnes, and one like it had clearly been sent on a test flight six days before.

Very little is known to this day about the project, the Russians have maintained a strict silence on its very existence and no photographs, drawings, or references have been officially released. It is with great interest therefore that the origins of the venture are sought by intelligence specialists and analysts throughout the world. Considerable evidence exists to support the hypothesis that Voskhod was merely a 'stretched' version of the one-man Vostok spacecraft, a design that would, nevertheless, continue to be used for biological and reconnaissance test flights up to the late 1970s. But in the early 1960s, when America's fortunes in space took a dramatic turn for the better, Kruschev was concerned about the continued emphasis placed on the implied race by people everywhere. Vostok was so limited in capability that it soon expired its potential for greater and more spectacular feats. The five days of Vostok V had taken the vehicle to just about the limit of its endurance, assuming that an adequate safety margin was an essential feature of a manned capability. What was needed for political and prestigious continuity was a demonstration of multi-man flight to presage the inauguration of that capability by the United States with the Gemini spacecraft.

So, it would seem, Kruschev pressed hard for Korolev's design team to come up with some way in which the political and propaganda needs of the Soviet premier could be met with existing hardware. There was only one way that could be done, and it seems that for the sake of at least his continued activities at Baykonur, Korolev was forced to bring compromise to what had been a perfectly satisfactory engineering solution to the problems of manned flight; he was required to make the basic Vostok less safe and place greater credence on reliability so that Kruschev would get his spectacular flight and the world would once again believe the bold claims to pre-eminence in technology and space flight operations. Kruschev had once before interfered with the aims and objectives of space flight development, being primarily responsible for the dual flights that characterized the latter Vostok missions and for the selection from a totally unconditioned background of the only woman to go into space. Intelligence about the Soviet premier's activity, and that of the Baykonur space engineers, indicates that late in 1963 the flurry of concern about potential American strides in space so worried the Soviet leader that this was the time when work on the so-called Voskhod spacecraft began. It may too have been the time when serious design work began on what would emerge as the true successor to Vostok, a spacecraft called Soyuz that would not appear before 1967, but for the interim, with safety given only cursory attention in so far as it affected the ability of the team to pull off a spectacular flight, a re-modeled one-man Vostok was to be rushed off the assembly line.

Gemini had two prime objectives that worried Kruschev, and in as much as they too enjoyed competing with their American counterparts, the Baykonur team as well: multi-man capability and the expectation for EVA – spacewalking. These two potentially rewarding feats were prized by the Soviets as good propaganda material, the long duration flight capability of Gemini being not so well defined as a show-stealer before the world. Apart from which, both multi-man missions and EVA operations could be accomplished from a common design change to the existing Vostok vehicle. During the first nine months of 1964 the teams worked furiously to get Voskhod flying before Gemini, a launch date envisaged for the latter having been set down for November. But there would be problems associated with stretching Vostok. The one-man capsule had only just kept below the maximum payload capability of the modified ICBM launcher and its adopted upper stage; with a weight of 4.7 tonnes the spacecraft was only a few kilograms below the configuration's maximum lifting capacity. Any modifications to the basic spacecraft would further increase the weight, to a point certainly beyond the capacity of the original launcher.

At this point it is beneficial to adopt the nomenclature proposed and followed by Dr. Charles S. Sheldon II, at the Science Policy Research Division, US Library of Congress, where the SS-6 ICBM plus Vostok-type upper stage was known as the A-1. But there was, in 1963, another upper stage which could be used on top of the original SS-6 booster. In fact the more powerful stage could have appeared as early as 1960 when Russia abortively attempted to send two spacecraft to the planet Mars, unmanned robots designed to fly past the planet and return data from scientific instruments they were designed to carry. The more powerful upper stage was about twice the size, in length and weight, of the type used to push Vostok into orbit, but with the same diameter of about 2.6 meters. With a length of nearly 7.5 meters, the new stage could place up to 7 tonnes in Earth orbit, send 1.6 tonnes to the Moon, or dispatch 1.18 tonnes to the nearer planets. It had already been flight tested several times and was capable of being used with the modified Vostok, allowing the design team a weight growth of more than 50% over the design weight of the one-man spacecraft.

But all that weight increase was not necessary, the essential feature of the quick re-work being to keep modifications to an absolute minimum; Gemini proves the dictum that sophistication and broadened capability leads to protracted development and complex preparations tests. Korolev quickly resolved the opportunity for modifying Vostok and presented an engineering solution to the need for a dramatic propaganda tool: the re-worked Vostok could carry three men for a comparatively short flight, or it could carry two men and an inflatable airlock so that a cosmonaut would be able to leave the spacecraft and drift outside. Two options for dramatically improving upon the by now hollow feat of merely putting cosmonauts into orbit and returning them to Earth. But there was a price, as inevitably there would be with a product engineered by politically motivated need for world acclaim and favour. The cosmonauts would have to forego the back-up protection of space suits, except of course for the space walk mission, and the escape system, employed by Vostok for both ascent and descent, would not fit within the rearranged interior now designed for two or three men.

It was a matter of volume, and confidence in the new system would need to be high for if anything went wrong with the booster during the early part of ascent there would be no means of getting away to safety. The three-man version of Vostok – Voskhod I – would have a weight of 5.32 tonnes, too high in fact for the tried and tested parachute system employed for the 4.5 tonne Vostok. That parachute design had only just been satisfactory for the one-man spacecraft; with a descent rate of almost 44 km/hr on to solid ground, cosmonauts elected to eject on the way down (except for Gagarin who bravely stayed on board). But with ejection seats gone there would be no way to get out of the Voskhod vehicle, descending, due to its heavier weight, at a greater speed than Vostok. So a new idea was adopted, one already designed, developed and tested for the Army: a retro-rocket housed on top of the spherical capsule which would fire close to the surface and quickly arrest the rapid descent rate to one which would gently set the capsule down on the ground.

Everything else about the Voskhod was essentially the same as Vostok. With three lightweight couches replacing the single ejection seat carried hitherto, three men could squeeze inside the cramped sphere if the centre seat was raised slightly above the level of the other two. No space suits to protect the crew if pressure was lost inside, nor if a meteorite unexpectedly punctured the pressure vessel; no launch escape system and no escape on descent should the recovery system fail. With no problem of overloading the lifting capacity of the new launcher configuration, extra weight incurred by the retro-rocket system to be employed at touchdown was a satisfactory solution to the increased descent rate; if it failed to work bones would certainly be broken but lives would in all probability be saved.

The crew for Voskhod I, selected as the trial run of this re-worked Vostok, was chosen to include a medical doctor, an engineer, and an existing cosmonaut from the original team that began training in 1960: Yegorov, Feoktistov, and Komarov, respectively. The engineer and the doctor were new to space flight, having been selected only four months prior to the anticipated launch date. Feoktistov had been around manned space flight activity for some considerable time, a leading personality associated with the development of the Vostok spacecraft, and seen in a Soviet movie film accompanying Yuri Gagarin. Yegorov too had a history of association with space operations. The son of a leading cancer scientist, he had been granted a position with early experiments leading toward manned flight and was seen

This view of the Voskhod control panel reveals little new or different from the Vostok predecessor.

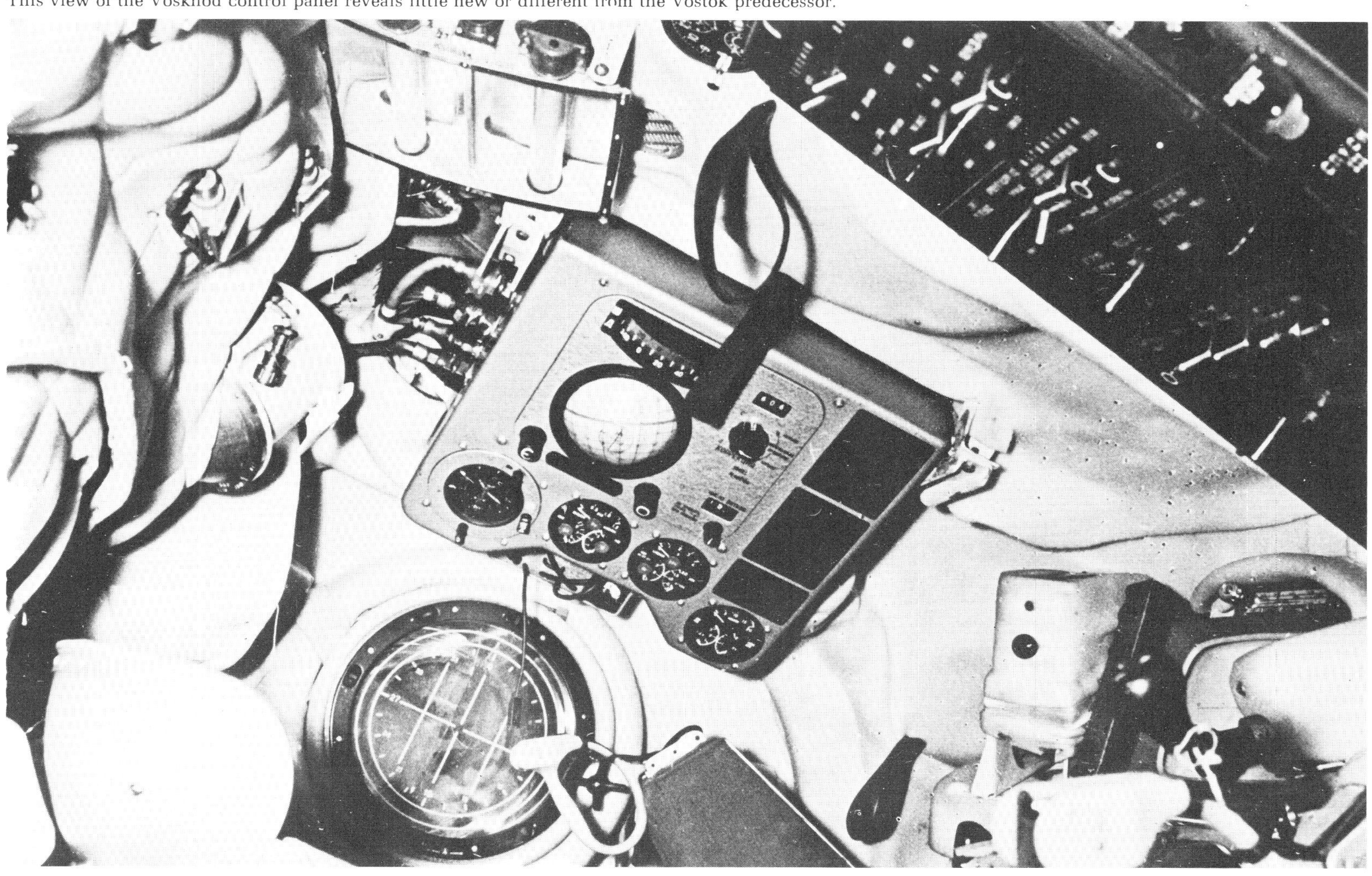

attending a dog trained for test flights preceding the Gagarin mission. For his part, Komarov was lucky, a lot more fortunate than his American counterpart, for this test-pilot cosmonaut had been rejected for space flight due to an abnormal heart response, but later tests confirmed his suitability for an operational role.

So it was that three publicly unknown cosmonauts, two so recently plucked from research duties they had been pursuing for some time, came to fly the re-worked Vostok called Voskhod 1. It must indeed have been with a certain feeling of trepidation that the three men walked to their spacecraft early on the morning of Monday, 12 October, 1964. It had been a year since Cooper's 22 orbit Mercury flight, fully 15 months since the dual flight of Bykovsky and Tereshkova. Now, bridging the gap between Vostok and the still nascent Soyuz program, was a new model of an old chassis. Although launched nearly a week before, the only A-2 launcher configuration to carry a Voskhod-type spacecraft did little to instil confidence and without a means of escape during the ascent any major malfunction would probably have brought disaster to the attempt. There is considerable evidence that not everybody in the Kremlin, or the Politburo, approved of this crash attempt at snatching favor from a spectator world watching US–Soviet space jousting events. In fact, the attitude that drove scientists and engineers to do the bidding of Russia's pompous premier was already moving forces against the great dictator which, before the end of the very flight he had pushed so hard to have, would depose him from his powerful seat. Kruschev was not to know, but his days were strictly numbered.

Preparations for the flight went well, the new A-2 launcher configuration behaving well during the lengthy countdown. Only minor holds were picked up during the closing hours before lift-off. Promptly at 10:30 am Moscow time the four boosters and single sustainer stage of the redundant ICBM thundered into life, smoke and flame pouring from the twenty main rocket engine nozzles. Only at the beginning of ascent were the crew devoid of a means by which they could escape, but the slow acceleration of the ascending trajectory soon quickened as the configuration slid effortlessly through the lower atmosphere. Faster than most A-2 launches – the configuration was light for the capability it possessed – Voskhod I and its launch vehicle performed as planned. Soon, the four boosters were spent, then the sustainer. Now it was the turn of the new second stage, more powerful than the smaller propulsion unit used to propel the Vostoks into orbit, and again the acceleration picked up rapidly to exceed that of any early Russian manned flight.

Finally, the assembly was in orbit and the spacecraft popped off the top of the long second stage to begin a flight with an apogee more than 400 km above Earth. With the exception of Gagarin's single circumnavigation of the planet, no Soviet cosmonaut had been higher than about 244 km; Vostok I had the unusually high apogee of 327 km because it was returning to Earth at the next perigee point. In fact Voskhod I was higher than any men had been before, for the highest Mercury mission took Wally Schirra 283 km above Earth. It is uncertain just why the Soviet launch team chose precisely this type of trajectory. Perhaps it was because the new upper stage, substantially under-loaded, would have accelerated the payload to an excessive 'g' force in reaching the speed necessary to get into orbit. By aiming for altitude rather than speed, the necessary velocity would have been achieved, for moderately less 'g' force, while the launcher sought to raise apogee; it would be one or the other: high speed or higher altitude, and seemingly the latter was chosen.

From the outset, Voskhod I was a publicity exercise prepared in advance and completely expected. Strong rumour of the impending flight surrounded press men in Moscow and everybody was in a mood of anticipation. It was like Yuri Gagarin's Vostok I mission all over again. In Tokyo, Soviet participants at the Olympic Games would join the celebration and during the short flight the three orbiting cosmonauts sent their greetings to the participating athletes. Conversation via radio was held between the cosmonauts and Kruschev and then between Voskhod I and Mikoyan. Each crewmember had a set of tasks to perform, and each fell back upon his particular speciality to learn for himself the practical reality of phenomena until now just a theoretical study. Yegorov in particular was concerned to sense the physiological stimuli, the body sensations in this unfamiliar weightlessness; Feoktistov, the scientist, was also associated with the development of new space operations, and the needs of future vehicles for performing new tasks; Komarov was ever the commander, the calm engineer who sought to orchestrate the spacecraft systems and to husband the finite resources of the microcosmic environment. At one point Feoktistov saw Glenn's 'fireflies' accompanying the spacecraft, and all three tried to estimate size and distance from the spherical pressure vessel.

Voskhod I returned to Earth early the next morning, having spent 24 hr 17 min orbiting the Earth 16 times. They landed 312 km north-east of Kustani in Kazakhstan, the shattering thud of the spherical capsule striking ground prevented by satisfactory operation of the braking rocket just a few meters off the surface. When the cosmonauts appeared before the press eight days later, 2,000 journalists from all over the world were in Moscow to question the three men. Throughout, Voskhod was hailed as a new spacecraft, one that could in fact lead to many more spectacular achievements, and one in which the future goals of Soviet space activity would be enhanced. Nobody was prepared to tell the newsmen that the spacecraft was just a re-worked Vostok, hurriedly put together at the insistence of a showman. But that really did not matter any more, for during the closing stages of the flight Nikita Kruschev was quietly replaced in a coup that took place in Moscow while the premier was in his Black Sea *dacha*. A new force had entered the Kremlin, and Brezhnev was now the power behind 220 million Russians.

It was a fortunate move, at least as far as Russia's space program was concerned. Under Kruschev the popular appeal of steel-nosed rockets blasting spaceward from the Kazakhstan steppes was a firework parade of pyrotechnic splendour for the post-Stalinist regime he inaugurated; a visible demonstration of Soviet might, Russian perseverance, communist superiority. But the fortune Russia's premier poured into the development of space spectaculars between 1955 and 1964 far outweighed the political value viewed by Kruschev's opponents. Others saw what America was never fully to ever realize, that Kruschev had teased from the United States a response aimed at getting Americans to play their technological cards and be found wanting; in the event, it spurred Americans on to greater goals than the Soviet technico-industrial machine was then capable of matching, stretching ever more thinly the veneer of technical success so ably spread by Korolev's brilliant ingenuity. Kruschev had in truth pulled out the rafters – and brought the whole house crashing down around his ears.

The administrative changes that now came with the new rule were to bode well for the space program, although it would be nearly a decade before the Russian manned space flight program was fully back on course, having been diverted in the interim by an attempt to meet, and match, American challenges. But just a matter of days after the successful day long flight of cosmonauts Komarov, Feoktistov, and Yegorov, the world acclaimed Soviet prowess. In the wake of NASA plans to launch a two-man Gemini vehicle early in the new year, and the first of three-man Apollo flights probably by 1966, Soviet engineers had already, in October, 1964, sent three men around the world in what they claimed was a new spaceship. To the public, it all smacked of a major Soviet response to Kennedy's Moon landing challenge.

Developed to generate electrical power for long duration flights, this Gemini fuel cell would make possible missions longer than the four days matched by conventional batteries.

Apollo was a three-man vehicle and so was Voskhod. It mattered litte to them that the spacecraft weighed only a half-tonne more than the one-man Vostok, even though that information came from western analysis of tracking information and from Russian announcements after the mission. It was the sudden impact of a multi-man flight that got the most attention, the very response Kruschev knew it would.

On the day of the launch, Professor Leonid I. Sedov claimed that Voskhod I 'opens up new horizons in space' and that it would speed the development of an 'orbiting space platform, a flying space institute' that would serve to provide a 'springboard for further interplanetary expeditions.' Even Pavel Popovich, the pilot of Vostok IV, assessed the flight as having been particularly significant 'because our aim is to send space stations to distant planets.' Press representatives at the post-flight Moscow conference actually asked the cosmonauts if Voskhod 1 could land on the Moon, and of course they received a non-committal reply to the effect that this particular vehicle was not designed to but others then being built would be capable of doing so. The launch day too brought political argument from Britain's Labour Party leader Harold Wilson when he said, then just three days from a general election, that the development of space technology would inevitably lead to the emergence of missile-launching satellites which 'would mean that all-American deterrent on which Sir Alec Douglas-Home bases his defense argument will soon be made absolete by space missile development.' It was a twisted argument, but one that a left-wing politician could have been expected to make.

Yet whichever way the statistics were turned, it revealed a significant lead for Soviet engineers. Russian cosmonauts had by now assembled a staggering 455.2 man-hours in space; America's total was a scant 53.9 man-hours. Russia had launched manned spacecraft on seven orbital flights; NASA had sent four manned vehicles into orbit. Russia had flown 9 cosmonauts into orbit; America had just 4 orbital astronauts. The Russians knew, more so the departing Kruschev, that when the Gemini program got under way American space successes could multiply like mushrooms in a well-watered meadow; the all-out Voskhod bid was to ensure that Soviet meadows were at least damp. And even as NASA's Manned Spacecraft Center moved toward the last test flight before two-man flights could begin, Soviet engineers were planning another space spectacular using again the re-worked Vostok that had served their propaganda purposes so well already. The pattern had been set. Russian space efforts would seek to eclipse as fully as possible the announced objectives and intentions of America's civilian space program, and where it could not achieve pre-eminence or pre-emptive activity, it would turn to some other facet of the space options available to both sides. It was a rigid conduct of operation that would prevent the Russians from flying to the Moon, but in late 1964 the end result of this technico-political philosophy was unknown. All that mattered was to get another spectacular flight in before NASA ran its first Gemini mission.

Meanwhile, things were not going precisely as planned at the Kennedy Space Center where, as Gemini Titan 2 Mission Operations Director, Chris Kraft was bringing together elements of a reluctant mission. Held up already by late hardware, three hurricanes, and a lightning strike that forced new checks on the Titan II launch vehicle, it was 5 November before spacecraft 2 was finally mated to the launcher's second stage. From then on preparations went tolerably well, toward the planned launch date of 9 December. The meticulous set of pre-flight tests and checks were performed almost as expected: tanking exercises to calibrate the propellant quantity indicators; electrical interface integrated validation, where electrical signals were sent across the launcher/spacecraft interface to test all possible modes and combinations of power; the joint guidance and control test, where tests were carried out to check switching operations during an ascent at the point where primary guidance gives way to secondary guidance; the joint combined systems test, where fully suited astronauts hooked up to the environmental control unit fly every conceivable part of the planned mission, duplicating electrically and systems-wise every operation of the spacecraft, but where for GT-2 there was no crew, just a spacecraft; the wet mock simulated launch, dropped after the first few Gemini missions, but necessary for GT-2 as a test of the entire launch operation involving full servicing of the launch vehicle and spacecraft up to T-1 min before lift-off.

Crewman simulators fill vacant seats in spacecraft 2, ready for the second Gemini flight in January, 1965.

On 28 November, final spacecraft systems tests were under way, a full duplication of checks made at the pre-mate verification stage conducted at the beginning of preparations and before the spacecraft was mated to the launcher. This confirmed that after the considerable power-up/power-down sequences, strenuous operation of specific elements of the spacecraft, and continuous plugging and unplugging of wires and lines leading to complex ground test equipment, the electrical and mechanical condition of the spacecraft was as near perfect as it had been prior to installation on the launcher. Then, five days later, came the simulated flight test. Here, spacecraft 2 was put through the electrical sequences it would be required to support during a normal mission and with any combination of failures or contingency modes. On 7 December, the launch team carried out the scheduled precount, an event fixed for F-3 ('F' signified firing day; 'T' denoted a specific time on that day), where stray voltage checks were made with the spacecraft powered up.

Final servicing operations got under way in readiness for the midcount, the first part of the formal countdown where the spacecraft was again powered up, this time to check that, when engineers went in to place the various pyrotechnic charges on board, the electrical and switch configuration would not cause premature detonation. Propellant loading finally began during the night before the planned launch, the day the midcount procedures were satisfactorily completed. Planned launch time was 11:00 am on Wednesday, 9 December, 1964. During the previous hours a total of 41 minutes hold was accumulated but right on 11:41 am the Titan's engines roared to life – only to shut down precisely one second later! Hydraulic pressure, lost when a housing on a servovalve burst draining hydraulic fluid, transferred to a back-up unit which notified the Malfunction Detection System that something was wrong and that since the launcher was still on the ground and building up thrust it would be expedient to shut the propulsion system down and prevent bolt detonation.

It happened far faster than a human eye could read this explanation, and saved the hardware for another day. Man-rating was a fine art and Titan had been man-rated to a tighter set of constraints than had Atlas for Mercury. MDS was an intelligent watch-dog for events that happened quicker than the human response, a valuable additon that sometimes displayed a trigger-happy reaction pathetically pessimistic about the Titan's ability to fly. It would happen yet again, and next time with men aboard. But for GT-2, it was no-go for that day. It also posed a problem in that nobody had established procedures for 'safing' a spacecraft wired and armed with dangerous pyrotechnic charges. Although comparatively small, designed to trigger one-time events like stage separation, adapter jettison, or retrograde section release, the little squibs were lethal for engineers or technicians close by. It took time to invent a method of extricating pyrotechnic isolation valves from propellant lines that could easily loose their contents over the workmen, but eventually a procedure of chilling the lines down was formulated.

Lessons here certainly contributed toward a production line modification whereby later spacecraft carried a special shut-off valve. It turned out, however, that new servovalves for the Martin booster were long in arriving back at the Cape. Special tests had to be carried out at the subcontractor's plant to re-design that part which had failed on the first launch attempt, meticulous activity that denied all possibility of getting GT-2 off the ground in 1964. The new valves finally arrived on 6 January and a launch schedule was written up with lift-off 13 days hence. This time everything went well, although hope for a flight test of the troublesome fuel cell electrical production unit went awry.

Most of 1964 had been taken up with careful troubleshooting by General Electric, but the development pace for these new, and revolutionary, forms of power was all but stalled. At the end of 1963, when it looked as though NASA would probably have to put the whole fuel cell program into another contractor, McDonnell was keyed to prepare the second spacecraft – the one groomed for GT-2 – to fly with a combined fuel cell and battery combination. During January, 1964, the space agency informed the Gemini builder to start work on a battery version of spacecraft 4, and on the 27th of that month a meeting was held to evaluate the situation. It was agreed that the existing design, called the PB2, would be discontinued forthwith, and that effort would be made to prepare for flight a more promising variation on the same design, called the P3. However, as an exercise, the PB2 type would be flown on spacecraft 2, more to qualify the reactant supply of hydrogen and oxygen, a completely separate contractual effort with little in the way of technical problems, than to test the fuel cells themselves. General Electric recognized that although it could not reasonably be held to blame for much of the delay in getting an operational fuel cell product for Gemini, further development work would be assisted by changes to the company's management structure. This gave the entire program an optimistic momentum as tests with the new model began by mid-year.

However, flight schedules seemed likely to compromise the planned fuel cell development plan and NASA Headquarters decided to have MSC develop a new electrical production pack for Gemini V; GT-2 would use the old model for the unmanned ballistic shot, GT-3 and GT-4 would use batteries. The third manned mission was now deemed a fit candidate for adopting supplementary battery power to accomodate peak loads, with fuel cells flown just for normal electrical demands. It would mean taking out some of the scientific experiments that could otherwise be carried on this flight, planned to last seven days, but McDonnell were confident they could come up with a net change of only 30 kg additional weight in the completed spacecraft by flying the combination pack. McDonnell was also notified of an impending decision to use batteries for all two-day rendezvous missions. By November, spacecraft 6, to be used on a rendezvous flight, was also to have batteries. The use of Gemini as a precursor test-bed for the electrical system adopted for the Apollo spacecraft was a failing bid to maximize the usefulness of this two-man vehicle. It began to look as though fuel cell for Gemini would follow paraglider into the waste project bin. Considerable progress had been made by GE in extending the guaranteed life of fuel cells, a capability discovered when tests revealed the sensitive response the system had to power demand and reactant temperature, but still development was flagging.

As designed, each fuel cell would carry three stacks in a cylinder 61 cm long and 30 cm in diameter. Together, the two cells would produce up to 2 KW at peak power. Each stack consisted of 32 series-connected cells, for a total of 96 per fuel cell unit. Preparation of the PB2 design aboard spacecraft 2 fell into disarray when erratic behaviour accompanied an attempt to get them operating for the 9 December launch schedule. One of the six stacks was found to be operating when tested several days later but flight preparations for GT-2 on 19 January, 1965, led to further problems where, to keep the single stack operational, engineers would have had to isolate the spacecraft's entire electrical system. That ended all hope of getting any of the fuel cell components working for GT-2, and so the dead equipment merely went along for the ride.

This latest attempt to get Gemini II spaceborne, delayed by one of the most tiring and ill-fated sequences experienced at the Cape, went well. It was a mission long overdue and one on which hung the fortunes of astronaut teams even then preparing for a rapid-fire run of successes, where spacecraft would be sent aloft every three months to rendezvous and dock, stay for two weeks circling the Earth, or provide opportunity for astronauts to walk in space. If there was some combination of design errors that required attention, it would set the program back many months at best. A good ballistic flight, followed by design clearance from post-flight scrutiny of the returned spacecraft, was an essential precursor to the first manned flight, now confidently set for April, 1965. The first flight, GT-1, had proven the soundness of launch vehicle/spacecraft interface and the structural integrity of the assembled combination; GT-2 would qualify the spacecraft for manned flight.

The countdown to launch of spacecraft 2 went well that January morning when NASA moved a step closer to operational tests with Apollo's precursor. At precisely 139 milliseconds before 9:04 am on the 19th, Gemini II was on its way. Just 2 min 31 sec later the first stage engines shut down, a fraction of a second later the second stage ignited and milliseconds after that the pyrotechnic bolts holding the two stages together shattered and freed the two slithering cylinders. Guidance commands noticed the slight tremor as the second stage trimmed out the minor shock of staging and for a further 3 min 4 sec the assembly continued to ascend until SECO (second stage engine cut-off) brought an end to the propulsive phase. Now, for the first time, a new sequence of spacecraft functions would be tried out in the environment for which they were designed.

About 22 seconds after SECO, when the single engine had tailed off, Gemini II separated from its second stage and fired the two rear facing OAMS thrusters, adding a speed of 3 metres/sec to the speed of the spacecraft at separation. Within 28 seconds of separation, the spacecraft had begun a turnaround maneuver designed to place the forward part of the vehicle in a rearward facing attitude and the adapter section in the direction of travel. This was accomplished 15 seconds later, some 6 min 40 sec into the mission. Seventeen seconds after that the equipment section of Gemini II's white painted adapter was pyrotechnically jettisoned and an automatic sequence initiated prior to retro-fire, not an essential activity for this suborbital flight but one aimed at proving that the system worked in space. After the retro-rockets fired, the four solid propellant rockets burning in salvo, the retrograde section was jettisoned and a roll rate of 15° per sec was initiated in the re-entry module via the reaction control system thrusters in the nose.

Falling back down through the atmosphere, the spacecraft experienced a higher thermal load than it would normally encounter during a return from orbit, reaching a temperature on the heat shield bondline at least 33°c higher than the design limit. At separation, spacecraft 2 weighed about 3,130 kg; devoid of the two adapter sections and parachutes employed to lower it gently to the sea, it weighed only 2,130 kg. On the descent everything went according to plan: at a height of 6.4 km the cabin vent valve inlet snorkel opened to admit atmospheric air; at 3.23 km the pilot parachute popped out, followed by release of the rendezvous and re-entry section at 2.93 km; and at 2.74 km altitude the main parachute opened, with the spacecraft dropping into its two-point suspension position just 2.04 km above the waves. From there it took little more than four minutes to splash down, nearly 63 km off target. In all the flight had lasted a little more than 18 minutes, and the spacecraft had been to a height of 171.1 km above the Earth, and to a speed of about 28,240 km/hr. At splash-down, spacecraft 2 was 3,422 km from the Kennedy Space Center, and it would wait in the water for 90 minutes more while the aircraft carrier USS *Lake Champlain* cruised to the spot.

Everything had worked well during the flight, and examination of the exterior, interior, systems and subsystems, after recovery revealed the re-entry module to be in good condition and for the jettisoned adapter sections to have behaved almost as expected. Most gratifying was the performance of the OAMS thrusters on the adapter sections, and the RCS thrusters in the nose of the re-entry module. Rocketdyne had been given the contract to build these motors in February, 1962, but pressure of work for a myriad other small-engine space contracts, technical difficulties with the Gemini propulsion units, and a poorly managed program kept development hanging far behind the desired schedule. A major report on Rocketdyne's performance was made, and an audit taken of their financial structure, during the latter half of 1964. It revealed, when completed during the spring of 1965, gross mismanagement, frequent turnover of key personnel in high positions, poor cost estimates, and a badly structured development schedule.

Serious technical problems had kept Rocketdyne moving from one concept to another for several months during 1963 and 1964 but in the end the company recognized the imminence of warning messages from its customer and re-structured the program in a modestly satisfactory way during 1965. It had been close, Bob Gilruth would have changed contractors had his personal view prevailed. Nevertheless, all seemed cleared at last for the first manned Gemini flight and as the crew accelerated their strenuous training and preparation schedules, engineers and technicians at the Cape were preparing the hardware that would open the second phase of NASA manned space flight operations.

It had been 20 months already since Gordon Cooper climaxed Mercury by his 22 orbit flight, and there was a nagging suspicion that the Russians were not finished yet. During 1964, and the early part of 1965, there grew within the Manned Spacecraft Center a cadre of flight personnel and mission controllers dedicated to pushing American technology, a band of Robin Hoods sworn to pushing forward at maximum pace toward the great Moon prize that loomed ahead. Mercury had been managed, operated, and dissected, by ex-Langley men, engineers who cared more for the intricacies of a system, or the application of a new component, than the political ideology that drove NASA forward with money allocated as a form of national technical security. But the new breed of crew-cut flight controllers recruited from universities and campuses all across America, saw in the manned space flight program a means by which they could give expression to their own particular way of doing things.

Increasingly, a degree of lethargy had been creeping in to America's established academic fraternity, tutors preached social awareness, deans hailed the freedom of societies divorced from materialism. For some students who lounged and lazed through the hot summers of the early 1960s, it was food for thought – and little else; a fitting excuse for detachment and an abrogation of responsibility. Many then entering college would appear three years later on the university campus smoking pot or sticking flowers in rifles carried by militia brought to the streets by threatened anarchy. But to many in the early 1960s, there was a need to find old principles, to have done with the preoccupations of an idyllic life out of tune with the modern world.

It was from that breed that NASA got most of its young technical brains to man the mission control room, young men in their 20s burning to equal the deeds of fathers and relatives recently returned from World War or Korean War; it was to them, conflict without bloodshed, but a battle nevertheless. Change was also reflected in the new teams of astronauts selected to fly Gemini and Apollo missions. Mention has already been made of the second selection in 1962, but by the time the two-man flights were about to begin, NASA had chosen a further 14, bringing the total to 30 astronauts in training. The third selection had been completed with the formal announcement on 18 October, 1963, of the New Gemini and Apollo pilots.

From the US Air Force: Maj. Edwin E. Aldrin, Jr., Capt. William A. Anders, Capt. Charles A. Bassett II, Capt. Michael Collins, Capt. Donn F. Eisele, Capt. Theodore C. Freeman, and Capt. David R. Scott; from the US Navy: Lt. Alan L. Bean, Lt. Eugene A. Cernan, Lt. Roger B. Chaffee, Lt. Cdr. Richard F. Gordon, Jr.; from the US Marine Corps: Capt. Clifton C. Williams, Jr.; and from civilian backgrounds: R. Walter Cunningham, from the Rand Corporation, and Russell L. Schweickart, a researcher at the Massachusetts Institute of Technology.

The second group, chosen in 1962, had been exposed to less stringent familiarization trials than the pioneer group of seven who, in 1959, opened a new chapter for medical science as much as for the technology they heralded. The first astronaut to leave the space program announced his resignation in January, 1964, after the last Mercury flight but prior to the first Gemini unmanned test flight. John Glenn said he was retiring from the Marine Corps and leaving NASA to stand for Democratic nomination as Ohio candidate for the Senate. At the end of February, however, he suffered a bad fall and the resultant injury put him in hospital and removed the opportunity to enter politics. But only for a while, for he was eventually to take up the seat he coveted. Another astronaut was almost forcibly removed from the program. In January, 1964, Alan Shepard had a small benign tumor removed from his thyroid gland by surgery at a local hospital. Another astronaut was not so lucky. On 31 October, Theodore Freeman crashed his T-38 jet at Houston's Ellington Air Force Base, losing his life in the accident. He was the first astronaut to be killed.

The third group that came to the Manned Spacecraft Center in 1963 were a part of the new school of trainees inclined toward scientific and engineering studies, academics who played an operational role in the armed forces. The first group of seven had been test pilots pure and simple; the second group were more like the third. To the original Mercury men it was important to train for a disciplined response. To the second and the third groups it was important to temper conditioned response with initiative and a degree of autonomy the original team thought too loose a mandate for the tight constraints of space flight. Nevertheless, the second and third groups made good commanders, while the first group were the best test pilots by far. The degree of intervention levelled by the new astronauts at program schedules and plans configured and arranged by management personnel infected the attitude of the ex-Mercury pilots.

No better example can be had than the continuous campaigns run for more demanding missions by astronauts eager to match the true pioneering efforts of men like Shepard, Grissom, Glenn, Carpenter, Schirra and Cooper. In January 1965, just 6 days prior to the Gemini II flight, NASA Administrator James Webb had to personally rule out the possibility of flying the first manned mission as an open-ended flight. To the astronaut corps it was their lives on the line and if they wanted to go for 30 orbits that should be acceptable to the space agency. It was only a part of the continuing difficulty most astronauts had in equating their own

Sailors and technicians aboard the aircraft carrier Lake Champlain retrieve Gemini II from the sea after its suborbital flight through space.

popularity and importance with an ostensibly subservient posture within the organization. The public clamored increasingly for astronaut pictures, autographs, and space photographs showing them dangling from helicopters or climbing out of spacecraft. It was a difficult role for any normal human to play: star of TV and space spectacular on the one hand, disciplined component of a bigger machine on the other.

To many it would be a role only just possible, but to a few it would be too much, to an even lesser minority it was no problem at all. There was certainly room for the astronaut viewpoint. As test pilots, most of them were quite used to flying missions with less likelihood of pulling through than the carefully controlled operations of a manned space flight. Living day after day with a spacecraft built to provide the pilots with every conceivable means of escape from the most unlikely combination of errors imaginable, the element of danger and stress that made these men test pilots in the first place was all but eliminated. Compared to the development of a new aircraft, the risk to life was infinitely less, but the consequences were appallingly more dramatic. One area that consistently proved troublesome, right up to the final preparations for the first manned Gemini flight, was final qualification of the ejection system carried by the spacecraft as a means of escape during ascent, or from a spacecraft disabled on the way down from orbit and unable to deploy its parachute.

Final tests had ground to a halt at the end of 1964 when pyrotechnics necessary for simulated-off-the-pad escape (sope) runs failed to turn up as planned. However, by 16 January, 1965, sope qualification runs were under way again, preceded five days earlier by a final stage of parachute testing. Resumed too was the plan to simulate a high-altitude abort, and prove the consecutive sequence of events essential to saving life, from an F-106 fighter aircraft. The first ejection from an altitude of 4.7 km on 28 January was a success but in mid-February the parachute failed to deploy when an aneroid trigger refused to work. Minor failures kept the test program staggering on through February, until the final jump on 13 March put the final piece of test data into the program, fully qualifying the entire system. There were just ten days to go before Grissom and Young flew Gemini III into orbit!

In operation the device would be a fearsome beast to ride. A D-handle between the pilot's legs would, when pulled, activate a sequence designed to eject both crewmembers rapidly from the re-entry module. Both hatches of the spacecraft would be flung wide open followed, milliseconds later, by ignition of the solid propellant escape rockets powering the seats upward on rails built in to the aft bulkhead. Just 1.1 seconds after ejection, about 45 meters away from the spacecraft they had just left, the seats would separate from their occupants and fire, 2.3 seconds later and some 140 meters from the spacecraft, drogue guns to deploy pilot parachutes from each astronauts backpack. At 6.5 seconds after ejection the main 8.53 meter diameter main parachute would open followed by release from the backpack of survival equipment the astronaut could need on the water. About 10 seconds after an off-the-pad abort the crewmembers could be on the water, 200 meters from the booster. In aborts from altitude above 2,953 meters the sequence would be similar, except for a ballute deployed to stabilize the descending pilot before release of the main personal parachute automatically at a height of 1,737 meters; Gemini pilots could use the ballute up to 22.5 km.

But ejection from the re-entry module was not the only means of escape from a booster threatening to run amok or literally blow itself apart. Direct ejection (Mode I) was the first means of escape while the vehicle was still on the launch pad, and for the first 50 seconds of flight. From an altitude of 4,570 meters, however, the Mode II procedure would perform a delayed retro-abort. This means that the spacecraft would wait about 5 seconds until aerodynamic pressure on the ascending vehicle was reduced before simultaneously separating the retrograde section and re-entry module from the equipment section and launcher and firing the retro-rockets to propel the upper sections forward and away. The retrograde section would then separate from the re-entry module so that the crew could perform a normal descent on parachutes in a sequence similar to that employed when returning from orbit. Above 21,335 meters and up to a speed of 22,700 km/hr the Mode III abort procedure would duplicate the Mode II sequence, except there would be no delay since aerodynamic pressure would already be low. For velocities exceeding 22,700 km/hr, up to within 330 km/hr of orbital speed, normal abort procedures would be followed where the spacecraft would separate and immediately go through turnaround, retro-fire, and re-entry. Aborts in the final seconds of powered flight would probably not be critical and would permit the spacecraft to push off early from the second stage of the launch vehicle and use additional on-board OAMS propellant to make up the deficiency in speed necessary to reach orbit. The future conduct of the flight would then be considered.

Although Gemini's ejection escape system was the last element of spacecraft hardware to be qualified prior to the first manned mission, other program elements adopting a supporting role were still far from complete. The most important of these was the world-wide data retrieval and tracking facility essential to complete and thorough monitoring of each mission. Very soon after Cooper's 22 orbit flight ended the operational phase of Project Mercury in October, 1963, engineers and technicians got to work with managers and administrators to develop a new concept for manned flight support. Long after Mercury appeared on the scene manned space flight had been injected with new projects and demanding requirements that augured well for a complete re-think on the needs of future missions. Where once it had been thought appropriate to control each flight from the center at Cape Canaveral, it was now expedient to place flight control functions at the new Manned Spacecraft Center near Houston, Texas. Accordingly, plans were made to transfer to that facility all responsibility for flight operations immediately the launch vehicle and spacecraft left the Earth; until then the launch operations teams would groom and cosset the hardware for flight.

The modifications to the global communications facility set up to support Mercury were performed in a $56 million program aimed at transforming the makeshift sites into elements of the Manned Space Flight Network – affectionately known as 'misfin' for the acronym derived out of key letters in the new title: MSFN was to be the backbone for Gemini, Apollo, and future manned operations against which would be laid the living organs of flight control and mission management. For Gemini alone, the network would be asked to absorb and distribute forty times the amount of information generated by Mercury. From the spacecraft in orbit, telemetry would stream down to Earth from 275 data points, compared with about 90 for the one-man precursor. In a nationwide search for contractors involved in modern telecommunications and tracking development, NASA pulled together an industrial team comprised of ITT, Canoga, Bendix, Electro-Mechanical Research, RCA, IBM, AT & T, Collins Radio, Radiation Inc., and Univac.

By early 1964 the net was ready and three scheduled launches of existing programs were used to exercise the system: SA-6, a Saturn I development flight; GT-1 the first flight of Gemini-Titan hardware; and AC-3, a test flight of the new Atlas-Centaur launch vehicle. Technicians around the world used these flights to track, as though it was a manned operation, respective vehicles passing overhead. The most important change brought about by the MSFN work was a change from analog to digital formats, speeding data handling capability and improving the transmission load. Central to the network would be the Mission Control Center-Houston, MCC-H for short, designed to replace what would be called MCC-K at the Kennedy Space Center until Houston became operational with the fourth Gemini mission. Two stations were lost from the old Mercury network. Muchea in Australia was replaced by a new and improved station at Carnarvon, and the Zanzibar facility was abandoned in favor of a station at Tananarive on the island of Madagascar to cover the large communications gap that opens up when spacecraft pass across Africa to the Indian Ocean.

Speed would be important for sending to the spacecraft information essential in rendezvous and docking operations, and extended coverage would be beneficial to medical monitors watching the response of crewmembers during EVA. To link all the many stations around the world NASA introduced NASCOM (NASA Communications Network) under the control and supervision of the Goddard Space Flight Center. Goddard had been responsible for filtering and routing information to the Cape control center from global tracking stations. Now, that role was to be expanded and placed on a more formal basis. NASCOM would link 89 stations, including 34 overseas points, with messages, voice and data communications in circuits and terminals spanning more than 160,000 route km and more than 800,000 circuit km. Also, because the likelihood of specialists and technicians needing to talk to each other over conference lines would be dramatically expanded compared to Mercury, a complex switchboard system was introduced so that up to 100 lines could be channeled to the same circuit, a capability that even as preparations to launch Gemini 3 were under way was being expanded to handle 220 lines. Known as SCAMA II, it had ten times the capacity of the Mercury conference line.

At the Goddard Center also, two IBM 7094 computers were set to operate in parallel and accept position information from the MSFN stations so they could compute separate phases of a flight and present plot-boards and position charts on spacecraft orbital parameters. In all, 1,500 people would be employed in the Gemini communications net, at Goddard and at the remote stations. By the time Gemini 3 was being groomed to fly the global tracking network was up and ready, but administration and control of the flight was retained at the MCC-K facility at the Cape.

One set of supporting elements essential to early preparations was the inventory of training aids provided for Gemini pilots. There were to be flight simulators at Houston and at the Cape, replicas of actual spacecraft wired to control consoles and to the Mission Control Center for authentic and complete simulation of any mission and any contingency operation the controllers chose to select. Familiarization with the dynamic environment of space flight embraced centrifuge runs at the Naval Air Development Centre, Johnsville, where a gondola was fitted out with a duplicate re-entry module interior so that astronauts could experience the forces of acceleration during ascent and re-entry. Originally, the Houston Gemini mission simulator was installed at the McDonnell plant in St. Louis, where it provided engineering familiarization of crew systems layout until moved from there to the Manned Spacecraft Center in July, but it was October before work was completed to convert the Cape simulator for Gemini 3 layouts.

Useful too was the dynamic crew procedures simulator built by Ling-Temco-Vought in Dallas, Texas, where crewmembers could experience with vivid reality the noise and vibration of various flight events. Escape from a sinking capsule was an experience not to be had without prior conditioning and astronauts would practice procedures in a Gemini mock-up at the McDonnell plant, go through the same operation in a giant water tank at Ellington Air Force Base, and finally perform a realistic simulation from a dummy spacecraft in the Gulf of Mexico. As an example of the intensive preparation for the first, 5 hour, Gemini flight, the prime crew spent 188 hours in the mission simulators and 18 hours on egress and parachute training, a further 17 hours being spent in the launch vehicle simulator. In addition to the formal simulator and training sessions, astronauts would spend time experiencing for a few seconds weightlessness induced by a parabolic flight in a KC-135 aircraft, backing up the simulated space flight with continuous flying duties to sharpen their responses to real situations; too much time on simulators would develop an immunity to full anticipation of the dangers a real event could bring.

As preparations for Gemini missions began to pick up through 1964 and early 1965 a flight crew training schedule evolved where astronauts would embark on a 24 week period of intensive preparation and familiarization. Up to that point all astronauts on flight status were kept at a peak of physical and engineering capability, ready to be plugged in to the flight schedule when appointed to a specific mission. The pace of the Gemini program was beginning to accelerate now and it was apparent that flights could perhaps get off the pad at intervals more frequent than the anticipated three-months. To have a ready and capable supply of Gemini pilots was an essential feature of good mission results. Hurried preparation and incomplete training would waste expensive hardware and costly flights. It was a gruelling schedule that kept many astronauts rotating between prime and back-up positions on an almost continual basis, stressing their physical, mental, and social resilience to the full.

Each Gemini flight would call for two men, one designated the command-pilot occupying the left seat and the other, the pilot, assigned to the right seat, living and working together for at least six months. In reality, the 24 week flight crew training schedule would apply to few assignments, most being named well in advance of that date. For the first time the psychological ability of two men to get on with each other was a crucial factor in flight crew selection. In some respects it was worse than Apollo requirements. Selecting three men for a single mission meant that one would probably be a loner, one would be a commander, and one would be an integrating type. For Gemini, it would be all too

Employing manikins for high-speed ejection tests, engineers prepare a test run at China Lake, California.

easy to challenge the appointed command-pilot's decision making criteria, for these men were highly trained specialists and, where groups of three naturally recognized the need for a single leader, a duo would stimulate contest for the last word. It happened more than once.

Because Gemini astronauts would spend considerably more time in their spacecraft than Mercury pilots had ever been expected to, the comfort of the space suit (pressure-garment assembly, or PGA) was of paramount importance. Goodrich had done good work on Mercury and the proud possession of a NASA contract to build the first American space suit was fitting for the company that had as one of its employees the man who produced a crude altitude suit for aeronaut Wiley Post in 1934. But stringent new requirements for the Gemini era put Goodrich back as bidder in a competition to find a suitable contractor for space flight that could include EVA. Up to now, the pressure-garment assembly had been a mere back-up for failure in the spacecraft's environmental control system. Now, a man would deliberately expose himself to the almost total vacuum of space, and all the thermal and radiation hazards that included. However, for the early development period there would be consideration of the special requirements of EVA.

Goodrich began work on an unspecified suit during March, 1962, on contract to NASA for prototype garments developed for advanced space operations. On 19 September, NASA amended the contract specifically to include the requirements for Gemini. The test suits were delivered for examination on 6 October to MSC's Life Systems Division, a partial-wear design to enable astronauts to remove elements of the suit designed for doffing inside a spacecraft on long duration flights. However, the David Clark Company, an unsuccessful bidder for a Mercury space suit contract in 1959, submitted proposals for design and fabrication of Gemini garments, finally getting the prime contract during negotiations concluded during May, 1963. By the end of the year, the proposed G2C training suit was examined by MSC's Crew Systems Division and a production version, called the G3C, was approved for production. It was late August, 1964, before the first four Gemini flight suits, to be used for Gemini III, arrived from the contractor and preliminary fitting trials had already been carried out due to the pressing needs of the flight schedule. By this time the log-jam of astronaut flight assignments was beginning to build up, precursor to the time rapidly approaching when crews would have to conform to a strict schedule for time on the simulators, visits to contractor plants and a myriad minor tasks such as seat fitting for spacecraft, reach tests for control sticks, suiting-up checks for pressure garments, sizing for experiment equipment designed for a particular crewmember. Following the award of the two Gemini III seats, Grissom and Young as Prime and Schirra and Stafford as back-up, came the Gemini IV crew assignments on 27 July, 1964. James McDivitt would fly command-pilot to Ed White, with Frank Borman and Jim Lovell as back-up. Shortly after the suborbital flight of Gemini II, the crew for the third manned flight was announced. On 8 February, 1965, MSC publicly proclaimed Cooper and Conrad as Gemini V command-pilot respectively with Neil Armstrong and Elliot See as back-up.

The first mission was to be a three-orbit flight, the second was to go for four days and probably include EVA, the third was to try for a week in orbit. The fourth manned flight, Gemini V, was to be the first official rendezvous and docking attempt and since its success depended on good crew familiarization with all the difficult procedures associated with that event, Deke Slayton, as coordinator of astronaut activity, gave Schirra and Stafford unofficial word that they would fly that mission with Grissom and Young as their back-up. It would be April, 1965, before an official announcement to that affect was made. If all went well with the first four manned flights, the Gemini IV back-ups would fly the two-week long Gemini VII mission, and the Gemini V back-up team would rotate to Gemini VIII, the second rendezvous and docking mission. But those last assignments were only part of a broad plan and in the first quarter of 1965 were of little concern since all four were busy with existing mission duties. For a time there were twelve astronauts filling prime and back-up slots, with at least four pressing for simulator time.

The Gemini IV crew was anxious to press forward with preparations for what could turn out to be the first EVA from an orbiting spacecraft. Gemini EVA plans had been prepared in a program document in January, 1964, where it was assumed the GT-4 pilot could open his hatch in space and stand up on the seat, his head and shoulders above the hatch line. McDivitt and White lost no time from the date of their crew selection announcement in pushing for this historic first on their mission. So much so that they discussed with the David Clark Company potential development problems that could be encountered if NASA was to authorize work on a special version of the standard G3C suit designed for full EVA. Largely at their instigation, NASA did authorize development of a G4C garment for McDivitt and White, retaining the option of adopting EVA for that mission while not committing the agency to a program objective as bold as this at such an early stage.

Ken Kleinknecht had casually told the press in July, 1964, when they met the Gemini IV crew, that McDivitt and White might actually participate in EVA, although at this time the Manned Spacecraft Center had not received permission to plan actively for it. MSC departments and divisions were working away at a comparatively leisurely pace looking at a variety of life support equipments that would be needed for a full EVA away from the re-entry module. In fact only days earlier a full design review of an extravehicular life support system (later called the ELSS) chest pack had been carried out at MSC with AiResearch personnel. One addition to a potential Gemini experiment load that helped EVA gain a respectable niche in Gemini flight programing was the contract AiResearch received that August from the Department of Defense to integrate the ELSS into a Modular Maneuvering Unit (MMU). Scheduled to fly in the adaptor section of Gemini IX, the device would permit an astronaut to move in a controlled manner around the outside of his spacecraft. As experiment D-12 it was one of several Defense Department payloads to be carried by Gemini.

By November, 1964, the Gemini III crew was actively contributing toward the move for an EVA on the second manned flight, assisting the case presented by their colleagues with an offer to use scheduled vacuum chamber tests with their own spacecraft for showing that an astronaut could open the hatch and stand up in a space environment. Spacecraft 3 was at that time in the altitude chamber for tests controlled and monitored by McDonnell, when Grissom suggested opening the hatch at a simulated height of 46 km. Neither Grissom nor Young was able to convince the company that the test should be carried out, but Kleinknecht saw the logic of qualifying the concept. If left until spacecraft 4 entered the altitude chamber it would be too late to include EVA on that mission. So McDonnell got a memo from the Gemini Project Office authorizing Grissom and Young to see if they could safely go through a stand-up EVA sequence in the chamber at a simulated altitude of 12 km, a compromise so that if anything went wrong the delay time before they could be removed would be lower.

Grissom and Young went through their cycle, simulating a stand-up EVA, but then had difficulty closing the hatch. The concept was proven feasible, however, and although impossible for Gemini III due mainly to its brief orbital duration, the prospect of EVA on the Gemini IV flight moved Bob Gilruth, the MSC Director, to approve formal altitude chamber runs for spacecraft 4 and the McDivitt/White crew. Management was now fully convinced of the need for an early EVA and the strenuous efforts of both Gemini III and IV pilots to get it in the flight program provides a good example of the different approaches to Gemini operations.

To the astronauts, space flight operations were a continuing dialogue with natural dangers, threats in space not totally understood. To managers, administrators, engineers and technicians, the Gemini spacecraft was a completely new vehicle which required at least a partial return to the cycle where the product is first checked out under conditions of minimum stress with few loads placed on the system. Mercury had first flown in space with a man for three orbits, so Gemini too, they thought, should be eased into operation with a brief shakedown mission. But the astronauts saw Gemini as a continuation of Mercury, not a return to first principles, and were frustrated by the overtly cautious approach that to them bespoke a lack of confidence in the product. To the pilots actively training for Gemini flights, the 3, 3, 6, and 22 orbits of Mercury's four missions logically pointed to a 30 orbit flight for Gemini III, and this philosophy of approach resulted in the official clampdown by Administrator Webb mentioned earlier. But their fortitude paid off with EVA for Gemini IV, a capability that would certainly not have existed had it not been for the crew themselves.

But others elsewhere were looking with interest at NASA plans for EVA, and had been for some time. And although as instigator of show-case flights Nikita Kurschev had been removed, the presiding authority over Soviet Russia permitted one more flight of a re-worked Vostok spacecraft, this time called Voskhod II. The launch came as Gemini III was on the last day of the simulated flight that preceded countdown preparations; NASA's two-man spacecraft would fly just five days later. Development of Voskhod II was centered around a capacity to allow one man to leave the confines of the spherical cabin and drift in the vacuum of space for a few minutes. It started life as the basic Vostok with a substantially rearranged interior and the necessary landing rocket designed to brake the high descent rate. However, because the life support system used in Voskhod was the tried and tested equipment previously developed for Vostok flights, engineers preparing the Voskhod II mission provided an inflatable airlock. For Voskhod II was to secure a space 'first' by having one of two crewmembers crawl from the pressurized cabin into the lock so that it alone needed to be depressurized for EVA.

The precise reason for this is officially unknown, although analysts have suggested that because Soviet manned vehicles adopt a mixed-gas, sea-level, atmosphere the large quantities of both nitrogen and oxygen needed to re-pressurize the cabin after an EVA would have been too great for the storage capacity of the existing system. By confining de-pressurization to just the small volume of an inflatable airlock, the existing expendables originally designed for one man would be made to support the two men of Voskhod II. Why not three as with Vostok I? There was simply no need, having already exceeded by one the number of crewmembers NASA would fly into space within the near future. Also, the equipment necessary to support the inflatable airlock for Voskhod II intruded upon the interior volume of the cabin and combined with the need for pressure suits, unlike the Vostok I crew that flew without any additional protection, there was simply insufficient room.

The exterior shape of the spherical capsule was modified by the bulbous housing that contained the collapsed airlock section, and the fairing that covered the spacecraft on top of the A-2 launcher can be clearly seen in photographs of pre-flight preparations. In the flight configuration, Voskhod II weighed 5.68 tonnes, some 360 kg more than its predecessor. Pilots chosen for the flight were cosmonauts Pavel Belyaev and Alexei Leonov, the latter a long standing friend of Yuri Gagarin. The interior of Voskhod II carried two couches similar to the three that had been used for the earlier mission, with two instrument panels overhead containing switches for the airlock chamber. On the extreme right of the two couches, adjacent to the position occupied by Belyaev, was the main control panel. In aligning the seats within the re-worked Vostok the dual positions had been rotated 90° from the alignment of the single ejector seat carried by the one-man design. In this configuration, with Leonov to the left of Belyaev, the airlock chamber was placed in the position occupied by the circular ejection hatch on Vostok flights, and was to Leonov's immediate left. Thus, the man who would remain continually inside the pressurized cabin was situated against the far wall where he would not have to move to allow the space walking cosmonaut to maneuver himself into the airlock. From Belyaev's position the spacecraft would be controlled in attitude by a long handle and further down the spherical wall from the main control box, a smaller unit with protected switches and toggles for manually firing the retro-rocket.

To Belyaev's left, across the space occupied by Leonov, was a panel of indicators and instruments for displaying the position of the spacecraft over the Earth on a rotating globe similar to one carried aboard Vostok. Because of the significance of this flight, live TV was especially catered for, additional antennae being an essential feature of continuous transmission from a slowly rotating spacecraft. The system would provide pictures for Belyaev to monitor activity outside the cabin, and for observers on the ground to watch Leonov as he moved about outside. Transmissions were to be on 625 lines at 25 frames/sec. Movie cameras also were carried inside the cabin, with another TV camera on the interior wall.

The two cosmonauts began training for Voskhod II in April 1964, with studies of the vehicle design and simulated excursions through mock-ups mounted in Tu-104 aircraft performing parabolic curves for momentary weightlessness. Other tests were carried out in a large vacuum chamber where activity similar to that performed by Grissom and Young at McDonnell's altitude chamber cleared crew and equipment for flight. Leonov was even committed to a soundproof room for a month, after which he was suddenly liberated, taken to a Mig-15 jet aircraft and accompanied for a rigorous session of aerobatics to evaluate his ability at adjusting to the new environment. But it was a lucky crewman that trained in 1964 to be the first man literally 'in space'; only a year before he had been traveling in a car that plunged off a road into water, Leonov's wife and their driver nearly drowning before he swam out of the car and rescued them.

The flight of Voskhod II began at 9:00 am Moscow time on Thursday, 18 March, 1965. Within minutes the spacecraft was in orbit, 173 km at perigee, 498 km at apogee. Almost at once the preparations for EVA began with Leonov starting to breath pure oxygen so that nitrogen would be purged from his respiratory system. Toward the end of the first orbit the space suit and the airlock chamber were checked out, the latter extended now to provide a mechanical chrysalis through which the space-walking Leonov would emerge. Leonov's suit was colored white and had a special thermal insulation, but inside the pressurized cabin the 205 mmHg suit pressure made movement considerably more comfortable than when pressure was raised for the EVA. As the spacecraft moved to complete the first orbit, Leonov slipped through the open hatch that led directly into the cramped confines of the airlock, about 2 meters long and 1 meter in diameter. Inside, the hatch separating Leonov from the Voskhod II cabin was closed, and Belyaev watched his colleague on the TV monitor as pressure was evacuated from the airlock.

Pressurized now to about 300 mmHg, Leonov's suit was awkwardly rigid, the ballooning affect of oxygen stiffening the joints. But everything was working well: the life-supporting oxygen pack on his back, the moisture removal system in the suit, the communications line that snaked like a very real umbilical as slowly the outer hatch was opened. From inside the airlock, Leonov drifted straight out, grasping for a few minutes the lip of the circular outer hatch, until he pushed free and felt the spacecraft give a modest shudder. A space man had been truly born; another step forward to space colonization accomplished. Through his communication circuit he could hear a radio announcer telling listeners about the flight and inside the capsule his companion watched as the white mass drifted into view, an indefinite haze surrounding his suit, now brilliantly lit in the Sun.

Later, Leonov would recall, 'I looked through the light filter and through the glass of the spacesuit. The stars were bright and unblinking. I could distinguish clearly the Black Sea with its very black water and the Caucasian coastline. One could even see what the weather was like there – I saw the mountains with their snow tops looking through the cloud blanket covering the Caucasian range. The Volga appeared and disappeared; the Urals floated under us; I saw the Ob and the Yenisey. When we were over the

Yenisey the order was given for me to return to the ship. I complied. It was somewhat more difficult to get into the ship because I had to dismantle the cine camera and climb into the airlock with it. I did all this within the set time-table.'

Leonov had been outside for about 10 minutes but the effort to get back inside the airlock was made more difficult when the movie camera hitherto fixed to the exterior refused to stay inside the inflatable device. Every time the cosmonaut pushed it inside the cylindrical lock it came floating back up again as he moved to place his legs inside. Finally, perspiring profusely under the effort, he put his foot on the object and moved down the airlock tunnel until his feet touched the inner hatch. Now, the outer hatch had to be closed and locked so that oxygen and nitrogen could pressurize the interior. Leonov's space suit had a regulator for controlling the pressure and now he placed the knob in a setting designed to reduce the level to about 1.65 mmHg. Within a matter of minutes the airlock had been pressurized and the inner hatch opened, allowing the cosmonaut to slide feet first into the cabin where he met Belyaev.

For the next several hours the crew kept up a semi-continuous dialogue with Earth, speaking to Soviet leaders and generally expressing their personal euphoria. Leonov had drifted about 5 meters from the spacecraft, restrained only by the tether that carried a communications line. As a human being, he was totally dependent on the autonomous backpack for life-giving oxygen and environmental control, albeit in a basic and limited form. Soon, it was time to rest, and to prepare for the second day in space when the two cosmonauts would return to Earth.

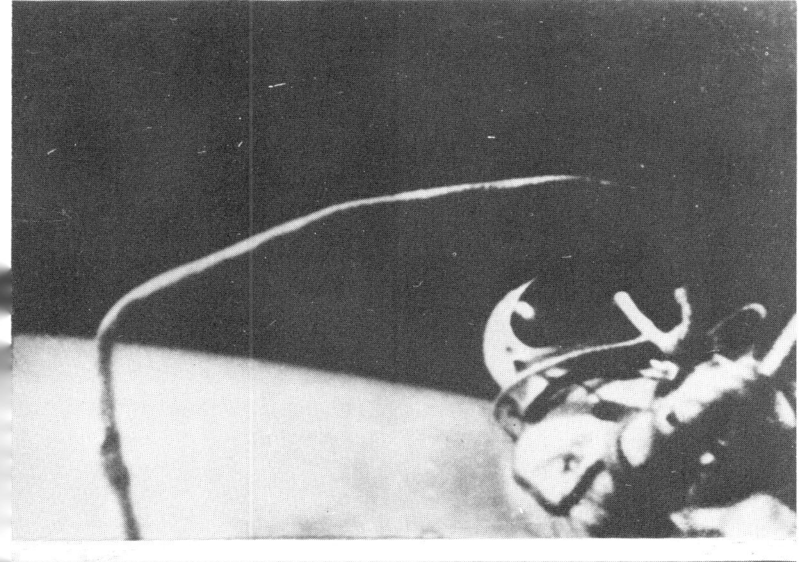

Cosmonaut Leonov becomes the first man to leave his spacecraft in orbit.

Re-entry was planned to take place on the 17th orbit and preparations toward that end went ahead as scheduled. Suddenly, Belyaev reported that the solar orientation system was malfunctioning and that the automatic sequence designed first to lock the spacecraft's attitude in alignment with the Sun was not responding to inputs from the solar sensor. It was clear that without a correct attitude trim the spacecraft would return through Earth's atmosphere at an incorrect flight path angle, and that the capsule and its contents would burn up. Belyaev reported that it took 18 seconds for the ground to advise the crew that a completely manual landing operation was the only option and that they should go round once more for an attempted re-entry on the 18th orbit. They did, but the Earth continued to spin on its axis and their orbit was far to the west of the prime zone when, for the first time in a Soviet manned space mission, the crew actually performed a manual operation during a major flight event.

For a while flight controllers on the ground were tense with apprehension: it was not in the nature of Soviet space engineers to relinquish such delicate activity to the probability of human error. But everything went well, the spacecraft was manually orientated, held at the required alignment, and the retro-rocket was fired followed by jettison of the redundant equipment section. But re-entry was not as accurate as had been hoped and the capsule came down to Earth in a densely wooded area about 20 km from the nearest clearing. In fact the spacecraft was in the Urals near Perm, 1,200 km north-east of Moscow. Snow was on the ground and there were real fears of the men suffering frost-bite – or worse. Within 5 minutes the two cosmonauts had released themselves from their capsule but more than $2\frac{1}{2}$ hours went by before the first helicopter arrived. In rough terrain, with tall pines blocking potential landing areas, the helicopter could do no more than drop supplies and food for the men. All that day, Voskhod II having landed at 11:02 am, 19 March, Belyaev and Leonov made ready their nocturnal camp, realizing they would not be rescued before the next morning.

It was cold in the forest that night as the blue frosting glazed across snow laden slopes flanking hills and mountains, but on and off the two men slept, trying to maintain a schedule where one kept awake. Next day, 20 March, helicopters landed in a clearing some distance away. All the cosmonauts could do was stay exactly where the helicopter that dropped supplies had marked their position. Soon, skiers appeared moving up and down across endless dune-like hillocks crowded with slender brown trunks and winter foliage. Joined now by their rescuers a plan evolved whereby the group would be supplied with more food and warm clothing so that they could rest up for another night and head for the clearing 20 km away on the following day. It was past noon now and there was little chance of the men making it to the clearing before dark. It was best to stay where they were known to be rather than have to stop over at a night location unknown to the airborne rescuers.

Came the 21st and the long trek to safety, hours of climbing and clambering across the bleak and chilling terrain until they finally arrived where the helicopters stood waiting. It was time to get back to civilization. Voskhod II had been a great success, despite the problem with the orientation sensor. The mere fact that Belyaev and Leonov were capable of quickly recovering from potential disaster and restoring the flight to normality by manual intervention displayed a growing maturity within the Soviet space program undetected hitherto. Time and again it would be shown that whenever the rigid authoritarian control of mission events are relaxed and relegated to pilot control the success of the flight increases proportional to the degree of human intervention. This was one very difficult lesson for the Soviet mind to accept. But it was a basic Russian capacity for survival that on more than one occasion would pull success from imminent catastrophe. Wide public acclaim rightfully flowed to the creditable achievement of Voskhod II. And in reporting on the use of the new suit, Vladimir Krichagin writing for Tass said that 'This spacesuit may be used . . . for landing on the lunar surface.' Time was running out.

CHAPTER TWELVE

Gemini Flies

On the day before cosmonauts Pavel Belyaev and Alexei Leonov rose into space, Gus Grissom and John Young told the world the name of Gemini III. After turning down 'Titanic' because of its cynical reference to the sinkable quality of Grissom's Liberty Bell 7, NASA finally condescended to approve the name 'Molly Brown' after that *un*sinkable lady of Broadway. But it was a very unofficial nod that ended a tradition of naming spacecraft. Unofficial names would surface from time to time but not for some years would the practice resume, and then only because with two spacecraft in space at one time it was decidedly confusing not to refer to either one by name.

Grissom was a quiet man, a calm, unflustered command-pilot by nature. Young was very similar, and the two together presented a distinctly unexciting duo to pressmen and newscasters intent on resurrecting the flight-fever that accompanied Mercury. Grissom rarely condoned publicity in the attitudes of fellow astronauts who eagerly sought the limelight, but there were times when he shared his boyish enthusiasm with those around. It was a good team to fly a new spacecraft; there would be no antics on this mission, and certainly no unnecessary conversation. It had been proposed that Gemini III carry in its adapter the Radar Evaluation Pod commissioned by NASA when flagging progress with the Agena target vehicle raised the possibility of delay. Designed to be pursued following release by a spacecraft simulating a target rendezvous, the pod was finally struck from Gemini III's flight plan on 4 January.

Despite the brief duration, Gemini III would carry three science experiments, two of which had been left over when prospects for a Mercury Atlas-10 flight disappeared: cell growth in weightlessness, and radiation and low gravity effects on human blood samples. The third experiment, more a technology research exercise, was designed to attempt a process whereby the plasma sheath that enveloped descending spacecraft would not cut communications with the ground. The device to be carried was installed behind a 'door' in the re-entry module originally put on the spacecraft to cover a landing skid when planned use of a paraglider incorporated land landing attempts. Water, sprayed into the enveloping plasma at timed intervals, would hopefully permit UHF radio signals to reach antennae listening on the ground.

Spacecraft 3's three modular sections were first mated in September, 1964, and the assembled vehicle arrived at the Kennedy Space Center on 4 January, 1965. Recognizing that the sustained test procedures adopted for GT-2 would be unnecessarily time-consuming, Cape personnel devised an abbreviated preparation sequence where they assumed the vehicle to be in a flight ready condition, installing only the equipment assigned to Cape teams. However, the spacecraft was taken to the Merritt Island Launch Area radar range where antennae and communications equipment were tested in simulated checks of an actual flight, and to a static test area where the propulsion systems were thoroughly evaluated in live firings. Gemini Launch Vehicle stages arrived in separate flight loads aboard a converted Boeing Stratocruiser called the Pregnant Guppy; both were at the Cape by 23 January, GLV-3 was erected on the 25th, and the spacecraft arrived in the erector white room on 5 February to be mechanically mated 12 days later.

Originally assigned an April flight date, work leading toward the final preparations of GT-3 inspired confidence that the mission could be moved up to March. During the early days of that month it was decided to aim for a first launch attempt on Tuesday, 23 March. It was certainly a trimmed spacecraft. Only a single tank for each propellant chemical was carried, limiting the load to about 140 kg of oxidizer and fuel, plus a single tank for helium gas to pressurize the system; a small quantity of oxygen too, only 6.94 kg in the primary system, plus the standard 5.9 kg in the secondary supply for re-entry. The last major physical examination on the prime crew was performed on 21 March and on the following evening the two men had dinner in the Manned Spacecraft Operations building, the MSOB, or Gemini's equivalent to Mercury's Hangar S. Following a brisk workout in the gymnasium they lounged in front of a television and retired at 9:00 pm.

Deke Slayton, now Assistant Director for Flight Crew Operations, woke Grissom and Young at 4:40 am on the morning of the 23rd. A short medical check and then breakfast with twelve invited guests, during which talk centered around Mercury flights and the coming mission. They ate the traditional 1 kg steak with eggs, half a cantaloupe, toast and jelly, tomato juice and a glass of milk for Grissom. It was just before 6:00 am that the two astronauts took a short ride to pad 16 and donned their suits; noticeably different from the silver Mercury garb, the soft white material of David Clark's G3C development befitting the new era of two-man flights. Pad 16 was the site for the suit-up trailer, while back-up crewmembers Schirra and Stafford spent closing minutes in the spacecraft, readying the vehicle for its assigned charges; they had been in spacecraft 3 since 3:00 am, one hour after the final countdown began. Grissom and Young wore a new and improved biomedical harness equipped with sensors for monitoring heart rate, respiration, pulse and blood pressure, a much more convenient package than the troublesome equipment first developed for Mercury.

It took just four minutes for the transfer van to carry the two men to pad 19 where the cylindrical black and white rocket stood straight and true in the morning dawn. It was 7:09 am and within three minutes they had walked determinedly to the erector elevator and ridden it to the white room, the plain flat-faced steel doors opening upon a scene of white gowns bustling in calm anticipation, the spread of heads turning to greet the two astronauts.

Chris Kraft was the Gemini III Mission and Flight Director, and Maj. Gen. Vincent G. Huston was in charge of Defense Department forces responsible for picking up the two men at the end of their three planned orbits. Medical affairs were now the concern of Dr. Charles ('Chuck') Berry, Chief of Medical Programs at the Manned Spacecraft Center. But responsibility for having got the hardware to the readiness it exhibited that March morning reflected the responsive skills of Gemini Program Manager Charles W. Mathews, the one man above all others to equal James Chamberlin's technical dexterity with managerial expertise; without either one, each in his respective position, the program would not have matured in the time it did. There had been just three full years to pull it all together and here it was, at last, ready for the checkout flight that would hopefully, fire the start gun to a rapid sequence of breathtaking spectaculars. To see if they could integrate the many elements of a Gemini mission as efficiently as expected, flight controllers in Houston would shadow the activities of their colleagues at the Cape which for this mission would be controlled from the Kennedy Space Center.

Gemini was certainly a very different spacecraft to enter than Mercury, the two massive ejection seats standing proud of their respective mountings on the aft bulkhead, the two large doors open each side. Compared with the one-man vehicle, Gemini presented an almost luxurious interior to the two men as they quickly and quietly slid down on to the supine seats. Soon, Guenter Wendt's McDonnell closeout crew would button up the hatches and depart the white room but for a few minutes there were checks to perform, straps to fasten, switches to secure. The crew station displays were a far cry from the tiny Mercury panel, positioned on that capsule almost as an afterthought in front of the pilot with wires and plugs protruding from behind the instruments.

Gemini had three main display positions. One, in front of the command-pilot – Grissom, now in the left seat – carried a large *attitude director indicator* (ADI), a rotating ball capable of moving through 360° in any axis, slaved to the inertial guidance platform to show the attitude of the spacecraft at all times in pitch, roll, and yaw. It also carried needles to display control movements required to place the spacecraft in a commanded attitude, or the rate at which the vehicle was moving to the desired attitude. Below this, at the bottom of the panel, the *incremental velocity indicator*, with three sets of digital displays showing changes in velocity with reference to pre-set values. For instance, it would show the difference in speed at orbital insertion between desired and actual values, displaying to the crew the dispersion in all three axes for subsequent correction (Insertion Velocity Adjustment Routine, or IVAR). It could also be set to indicate rendezvous rates, and the change in speed during retro-fire. Each display had three digits and two lights, the latter to show in which direction the thrusters should be fired to correct the velocity dispersion.

To the left of this display, a *propellant quantity indicator*, to tell the command-pilot the amount of OAMS propellant remaining, and above that, a *rate-of-descent indicator* for reference during re-entry. Above, to the right of the attitude indicator, an Accutron 24 hour clock, emergency parachute release switch and controls for switching modes on the ADI, affectionately known as

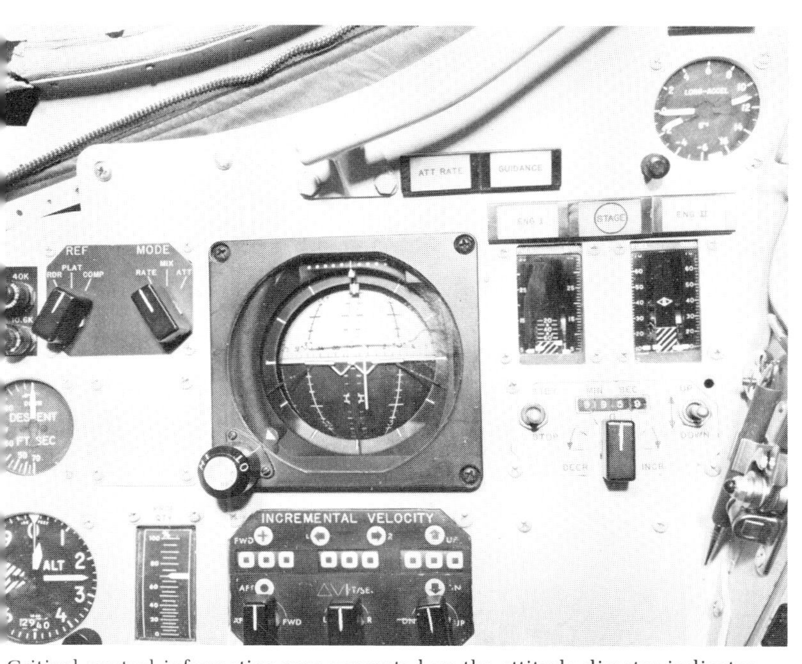

Critical control information was presented on the attitude director indicator ball and the incremental velocity indicator immediately below. Note the zip by which a curtain could be retained against the window in the hatch.

the 'eight-ball'. Other displays on the command-pilot panel included a *range and range-rate indicator*, for use in conjunction with a target vehicle in orbit, propellant quantity gauges reading from respective tanks in both stages of the Titan launch vehicle, an *event timer*, for setting up timed operations in the spacecraft, a *longitudinal accelrometer*, to tell the pilot forces of acceleration in the spacecraft's long axis. At the top, to the right, was an abort light that nobody wanted to see illuminated.

Across the other side of the cabin was the pilot's panel which carried an *attitude director indicator* identical to the one provided on the command-pilot panel, and to the right of it a 24 hour clock, below which was situated the computer *manual data insertion unit* (MDIU) and the associated readout unit. In the left lower quadrant, the fuel cell indicators displayed critical parameters necessary to maintain a close watch on electrical power production, and above those a prolific set of fuel cell controls for purging, section switching, and electrical bus switching between the two cells. Again, an abort light was at the top of the panel.

The largest group of displays was situated on the center panel separating the two pilot displays, with information provided on the state of critical elements in the environmental control system, cabin temperature, cabin pressure, suit pressure, carbon dioxide partial pressure, communications switching equipment, transmitter and receiver signal strength displays, telemetry controls, tape recorder status displays, and communication mode selectors. Between the two pilots, on a console known as the pedestal, angled so as to be available to either astronaut, were the thruster controls and switches, the computer controls, and the rendezvous radar instruments. Down the left side of the center panel was a vertical column of status lights designed to illuminate and confirm functions related to one-time spacecraft events. At the top of the center panel; a row of five ring-handled controls for environmental functions such as opening the cabin vent, oxygen circulation, etc.; at the bottom of the angled pedestal, cabin repressurization controls and suit oxygen flow switches.

The sides of the crew cabin and footwell walls had a corrugated appearance, thin panels stiffened for rigidity, but either side of the seats, just below respective hatch hinges, were left and right circuit breaker panels. Overhead, between the two seat positions and on the structural support between hatches, was another circuit breaker panel for the thrusters and environmental equipment. The abort handle was on the left, adjacent to the command-pilot, and D-rings between the astronaut's legs when pulled would initiate the ejection sequence. In orbit, spacecraft attitude would be manually controlled by a handle on the center console below the pedestal panel, and manual control of orbit changes or movement back, forth, up, down, and from side to side

Although a dramatic improvement on Mercury, the cramped cabin on Gemini called for close coordination of crew functions at all times. Here, two McDonnell engineers check out the controls.

would be governed by use of a knobbed handle below the command-pilot's panel. In all there were 384 levers, dials, strip displays, toggles, switches and buttons. Yellow lights beamed like stars into the pale grey interior, rectangular yellow status indicators shone forth like tiny searchlights picking out the console's myriad recessed cavities, glinting off the metal rods guarding circuit breaker panels from accidental knocks.

The total pressurized volume inside Gemini's comparatively spacious re-entry module grossed 2.26 cubic meters, reduced by one-half when crew and equipment was installed. Floodlights with continuously variable rheostat control provided either red or white light as required. Stowage lockers were positioned wherever standard spacecraft equipment presented a vacant recess, the favorite place for designers, the least accessible for the crew, being between the two seats on a line with the pilots' shoulders. Although each spacecraft was built to a standard specification each vehicle was individual in that it carried equipment, instrument displays, and fitments specifically assigned to the objectives of a particular mission. In that regard, nomenclature assisted continuity: spacecraft 3 was to fly GT-3; spacecraft 4, GT-4, etc.

First in the series destined to lift a human load, spacecraft 3 was well groomed for its duties that March morning as Grissom and Young moved in to the 'office' to do their day's work; test conductor George F. Page, responsible for the spacecraft from the time the first components came together, had done a good job. Just 200 meters from the concrete hardstand supporting the 154 tonne Gemini-Titan assembly, launch controllers sat pensive in the blockhouse ablaze with light and the hum of anticipation. It was 7:34 am when the hatches were shut, locked by a barrel-key on the outer face, and NASA engineers accompanied McDonnell technicians away from the white room. Between launcher and spacecraft and the dome-shaped control center, 1,300 km of wiring carried unheard messages back and forth through computers, consoles and space vehicle. It was a dialogue essential to near-perfection, the only acceptable state for a sophisticated manned space vehicle assembled from 1,230,000 parts on top of a volatile booster with more power than 150 jet trainers of the type used commonly by the astronaut corps.

Nearby, at the Mission Control Center, waiting for the blockhouse to launch GT-3 into their custody, the mark of Gemini had changed the place from Mercury days. Four new consoles had been added and to supplement the four existing plot-boards a fifth had been put up. The now famous world display map, 2.4 metres high and 7.6 meters long, carried new signs, of tracking stations newly assigned around the globe to more important tasks than those which kept the same room tense more than two years before with the sound from space of a single human voice in a shingled metal can.

Consoles were added, and now there were new functions to perform: a display coordinator would channel the flow of information put up on screens for flight controllers; a support control coordinator was also there; a flight surgeon would look at heart beat, pulse, respiration, and blood pressure; an electrical, environmental, and communications systems engineer – 'eecom' – would similarly take the pulse of spacecraft 3; the booster systems engineer would have a short day, but an important one nevertheless; guidance, navigation and control monitor – 'guido' – would shape the path of Gemini III to duplicate the planned values; a network controller would balance 'misfin' stations around the world; the flight director and assistant flight director would orchestrate the complete assembly into one harmonic accord.

And then there were the supernumaries: the public affairs officer telling the outside world what was happening, deciphering the language of another age for ears in the 20th century; the Department of Defense representative, ready for strategic decisions should spacecraft 3 land unexpectedly out of range; the operations and procedures officer, a valued middle-man. One man would look to Gemini III's return – 'retro' – the retro-fire officer in charge of placing ignition of the four powerful solids at the right place in space and time. And finally, the link across the vacuum of airless space: 'capcom' – the capsule communicator who would talk the language of machines to men accustomed to the words few would know; the man to keep crews sane.

On board the spacecraft all was set for lift-off, but in an otherwise flawless countdown there was a minor problem when at T-35 min an oxidizer line in the Titan's first stage leaked nitrogen tetroxide. The count was stopped, an engineer tightened a nut, the count resumed. At 9:22 am a deluge of water cascaded across the pad thrust mounts and flame bucket, cooling the metal and building a barrier against fierce heat. Precisely 640 milliseconds past 9:24 am, GT-3 rose from pad 19 – Grissom and Young were on their way. Little more than 2½ minutes later the launch vehicle shed its first stage and thrusted toward space on the single engine of stage two. At 5 min 34 sec the booster shut down and Gemini III was in orbit. Gordon Cooper was capcom at the Cape and warned the spaceborne duo of a slight overspeed, Titan having burned a little too long put the spacecraft into a slightly different orbit to the programed value.

There was no mistaking shut-down as both men pitched forward into their secure restraint straps. Just 20 seconds later the pyrotechnics fired to sever the launch vehicle's second stage from the adaptor module and Grissom kicked the spacecraft forward with the two aft-facing thrusters. With a perigee of 161 km, Gemini III was on target, but the 224 km apogee was 17 km below the planned high-point. Good enough for first attempt. Shortly after passing across the Atlantic, while out of range of tracking stations, Young observed a sudden drop in the oxygen pressure gauge but correctly concluded the source of the problem and switched to a secondary electrical converter. All was well. A major go/no-go decision on the future of the mission was scheduled for the pass across Australia and tracking stations provided the confidence for the Mission Control to approve at least two orbits. Across the Pacific, the crew observed thruster firings and noted the extent and appearance of exhaust plumes coming out from the adaptor orifices.

Then, across the United States at the end of the first orbit, Grissom and Young got ready to make a historic 'first' that seemingly was of little consequence but in reality opened a door on future possibilities: the crew would use the translation thrusters to actually change the spacecraft's orbit, proving that spacecraft could almost literally fly from one path to another. It was the secret of rendezvous and docking. The actual event began at 10:57 am, just 1 hr 33 min after lift-off, when Gemini III fired two thrusters for 74 seconds, slowing the spacecraft by 15.2 metres/sec across the future home of Gemini controllers: the Manned Spacecraft Center. The new orbit was 157.7 km at the low point and169 km at the high point. In effect, apogee had been lowered by 55 km and perigee had dropped a little too.

A second translation maneuver was performed over the Indian Ocean on the second orbit, where Grissom fired both forward and aft thrusters in a sequence lasting 15 seconds designed to alter the speed of the spacecraft by 3 meters/sec, changing the inclination of the orbit by one-fiftieth of a degree. The third and final translation maneuver began at 4 hr 21 min over the Pacific. In a burn lasting 109 seconds, Gemini III was slowed by 29.26 meters/sec so that the orbit would be lowered to a vacuum perigee of just 83.7 km. Vacuum perigee is an expression implying the theoretical low point where actual perigee would lie were it not for Earth's atmosphere slowing the vehicle before it reached that point. This maneuver ensured that, if the retro-rockets failed, the spacecraft would still come home. Already, there was a touch of conservatism compared to Mercury. No back-up existed on the earlier capsule, and no preserved supply of propellant in a separate reaction control system designed to hold the spacecraft at the prescribed attitude during descent.

Preparing for ignition of the solids that would put the spacecraft on course for splashdown, the adapter punched off on schedule startling the crew momentarily who had not quite expected the sharp crack of the pyrotechnics and the jolt of separation. Exposed now to space, the four retro-rockets barked into life and again the crew were surprised at the g force building up to decelerate the re-entry module. Seconds later, the pyrotechnics fired on the retrograde section and that too cut loose and spun away. Molly Brown was on her way home and on the way down John Young started the plasma sheath experiment as UHF signals sought, and found, ground antennae. But indications showed that Gemini III was too short of the prime recovery zone and that the spacecraft would splash down far from the waiting ships. The difference could be accounted for by Molly Brown not possessing the precise degree of lift wind tunnel simulations had told the engineers it would have.

During the descent, Grissom modulated the spacecraft's bank angle in an attempt at lengthening the descent path, but still the flight path was too short. Everything else about the descent

After their three-orbit flight, Young (left) and Grissom arrive back on the recovery ship *Intrepid*.

went well, the sequence of operations cycling off as planned. When it came time to transition the spacecraft from the vertical to the pitch-forward position – the two-point suspension attitude for splashdown – the unexpectedly violent motion hurled the two astronauts into their respective windows, smashing Grissom's faceplate and badly scarring Young's. When Gemini III splashed down 4 hr and 53 min after launch it was more than 80 km off target. However, the US Coast Guard cutter *Diligence* was only a few kilometers away, 80 km uprange of the prime recovery carrier, USS *Intrepid*. A helicopter was sent aloft from the *Diligence* and arrived over the spacecraft just 17 minutes after splashdown.

By this time an Air Rescue Service C-54 had dropped pararescue personnel into the sea, followed by swimmers from the first Navy helicopter to arrive. Securing a flotation collar to stabilize the rocking motion of Molly Brown, the men stood by in the water while Grissom opened his hatch. It was hot inside the spacecraft, and stuffy. But more than that it was a heaving, pitching, tossing nausea machine, that caused the command-pilot to lose his breakfast. For some time they waited in the 1.7 meter swell for helicopters from *Intrepid* to pick them up. About $1\frac{1}{4}$ hours after splashdown they were on *Intrepid*'s deck, joined by the spacecraft 93 minutes later. For the next few hours the crew chatted on the telephone to President Johnson, talked to Vice President Humphrey, had a meal, voiced into a recorder the initial impressions of their flight, and received a minor physical check. The next day they had a more strenuous examination followed by a lengthy de-briefing session and an address to the ship's crew assembled on the hangar deck. The day after that they were flown to the Cape for more medical checks, more de-briefings, and a motorcade procession from the Kennedy Space Center to Cocoa Beach and the first post-flight press conference.

There were few problems to worry senior management about the future course of manned space flight. Gemini had performed very well. The lower lift/drag ratio proven by the short re-entry profile indicated that Gemini was more of a ballistic vehicle than had been hoped, but no matter, the controlled descent where Grissom rotated the re-entry module to modulate what lift there was augured well for future landings. It was in truth the first fully manual landing where the pilot had a degree of control over the angle of the flight path. There was one worrying event during the flight: the spacecraft continually tried to yaw to the left, like a car with its front wheels out of alignment. But that was soon resolved when it was discovered that the water boiler was venting liquid to space and that the jet behaved like an uncommissioned thruster continually trying to spin the spacecraft. It was a relief to discover that the water boiler was responsible; worst fears supposed it was a flaw in the reaction control thrusters.

Chris Kraft was especially pleased with this first flight of a new spacecraft: 'The entire flight was completely satisfactory from the standpoint of the control of the astronauts. . . . The environmental control system, I think, outdid itself in that for the first flight it could not have performed any better. It was just about perfect. The cabin and suit temperatures were what we had experienced in the vacuum chamber runs and, certainly, in orbit the system performed extremely well.' Great credit for that was due to Garrett AiResearch Manufacturing for having done so much to transform what began as a marriage between two Mercury units into what became a completely different and unique Gemini ECS layout. Many had thought the simple coupling of the old one-man units would suffice, but how naive that thought had been seen to become!

There was one other point that greatly amused the astronauts, but gave managers straight faces. Wally Schirra had bought a corned beef sandwich at an eat-by on North Atlantic Avenue, Cocoa Beach, before the flight and given it to John Young. Half way through the flight the pilot pulled out the smuggled sandwich and handed it to a beaming Gus Grissom, who ate a ceremonial bite or two and then put it away for fear loose crumbs found their way into equipment in the cabin. It had happened once before – but on that occasion the roles had been reversed!

The day after they arrived at the Cape from *Intrepid*, Grissom and Young flew to Washington with their families to be received by Lyndon Johnson. Ceremonies scheduled for the Rose Garden at the White House were moved to the East Room when rain poured down but it could not dampen the pride Gus Grissom felt at receiving a cluster to the NASA Distinguished Service Medal awarded nearly four years before by a President who set the space agency on course for the Moon. Grissom was the first man to fly twice in space, even though his first trip was only a 15 minute suborbital hop. Both men received the NASA Exceptional Service Award, and Bob Seamans, NASA's 'general manager' got the DSM. In the hours remaining of that Friday the two astronauts were paraded to the Capitol for a luncheon, and then were given a tour of the Smithsonian Institution before a Congressional reception.

After the weekend spent resting, there was a flight to New York for more parades, the keys of the city from Mayor Robert F. Wagner and a visit to the United Nations where Secretary General U Thant presented medals before the astronauts and their wives were hosted at an official reception at the Waldorf-Astoria. Next day, Tuesday – Chicago and a parade from O'Hare Field to City Hall, a route flanked by a million people shouting and clapping all the way. Then, an official luncheon at Sherman House and a question and answer session with children from Chicago schools participating. Wednesday, and a flight to Houston where 10,000 students flocked to the airport, surging upon the home-coming heroes. On Thursday the initial round of celebrations were brought to a halt by the presentation of a flag made from parachute silk by members of the MSC Technical Services Division. The flag had been carried aboard Molly Brown and was to be flown at MSC during all future Gemini missions.

But the parties, the celebrations, the cheering, could only last for a little while. It was after all, only a beginning for the ten Gemini missions scheduled to carry astronauts into space. And only recently, Voskhod II had pressed home what looked increasingly like a challenge to engineers and technicians preparing hardware for a possible space walk on Gemini IV. But it was the astronauts themselves who were the driving force behind plans being formulated at the Manned Spacecraft Center. It was a reflection of the new breed, matched equally by rookie astronauts and rookie technicians, with one single purpose threading all their efforts: the burning desire to win, for to run the race was not enough, but to run it as a team was everything; there were to be no places for the prima donnas.

Just days after Alexei Leonov squeezed through the inflatable airlock attached to Voskhod II, George Low met Bob Gilruth to examine a maneuvering unit that an astronaut could carry in his hand, to look at a chest pack that would serve as a back-up to the spacecraft's oxygen system, and to inspect the new suit designed for EVA. But although ready, the equipment essential to space walking was not yet tested and qualified. Permission had been granted for White to open the hatch of spacecraft 4 and stand up in the altitude chamber at McDonnell on 24 March. For this test, performed the day before the spacecraft was qualified and packed up for delivery to the Cape, air was pumped out to a simulated altitude of 45.7 km. But it was a very different activity to that which would be performed in orbit. Then, on Monday, 29 March, just six days after Gemini III, Bob Gilruth hosted Deputy MSC Director George Low, Richard Johnston from the Crew Systems Division, and Warren North of the Flight Crew Operations Division, at an informal meeting where the possibility of greater things was discussed.

Fresh from their bench tests and statistical graphs, the engineers convinced Gilruth and Low that Ed White could safely

open his hatch and, connected to the spacecraft by a tether providing oxygen, electrical power, and communication, go all the way out to a distance of more than 5 meters. Up to this point nobody had seriously thought of anything more for Gemini IV than having the pilot stand up in an open hatch. By the end of the week personnel from the Crew Systems Division were on their way to Washington; on Saturday, 3 April, they formally presented the plan to George Mueller, head of manned space flight at Headquarters. But the hardware had not been qualified and, although interested in the idea, Mueller was not totally won over to an EVA on Gemini IV. It all seemed to smack of a rushed response to Voskhod II. Why, if EVA had been planned for Gemini VI, should an already intense schedule present the busy Gemini IV flight crew with further training for complex activity. Nevertheless, Mueller was not about to call a halt on MSC's initiative and the engineers returned to Houston resolved to press ahead with qualification tests.

For the remainder of that month the Crew Systems Division worked long and hard in conjunction with personnel from the Flight Crew Operations Division to get the EVA equipment fully tested and qualified. At the Cape, signs of an imminent launch were gathering: spacecraft 4 was hoisted to the erector white room on the 14th; pre-mate systems tests got under way on the 19th; a pre-mate flight test began on the 22nd; the spacecraft and launch vehicle were mechanically mated on the 23rd. The joint combined systems test was over by the end of the month. On 14 May, Bob Gilruth set up a demonstration of EVA equipment at the Manned Spacecraft Center for Bob Seamans and by the end of the day MSC had won a staunch supporter to the cause; Seamans returned to Washington determined to discuss the matter with Webb and Dryden. Qualification was now complete on elements essential to the plan formulated for approval by Headquarters.

Basically, this resulted in converting the standard Gemini suit, the G3C, into an extravehicular version, called the G4C, and providing the astronaut with a Ventilation Control Module (VCM), a tether and a maneuvering unit. The G3C comprised an inner layer of nylon, to minimize pressure points that could build up on the body from equipment attached to the exterior, covered by a neoprene-coated nylon pressure layer. Next came a link-net material as a restraint layer, ensuring that the suit kept its tailored shape and did not balloon. The outer layer was fabricated from HT-1, a high-temperature resistant nylon to which was attached the external fittings: two disconnects for oxygen in and out, and an electrical connector also carrying bio-sensor wires from beneath the inner suit layer. The suit ventilation system provided flow to the entire body, with 60% of the flow ducted by a manifold system to the boots and gloves and the remainder passing through an integral duct in the helmet neck ring to be directed across the visor to prevent fogging and to provide fresh oxygen to the face area.

Modifications to transform this garment into the G4C EVA suit included the addition of an extravehicular cover layer, two additional faceplates on the visor, and thermal gloves designed to minimize heat transfer from parts of the spacecraft the astronaut might touch while outside the vehicle. The cover layer comprised, from the inside out, two HT-1 nylon micrometeoroid protection layers, seven layers of aluminized Mylar for insulation, a single HT-1 micrometeoroid absorber, and an outer protective covering of HT-1 nylon. The cover layer was 0.5 cm thick and formed an integral part of the original G3C suit structure. The G4C helmet was equipped with a detachable EVA visor assembly comprising two over-visors, the outer one of which afforded protection against the Sun and was made from grey-tinted Plexiglass covered on the exterior surface with a thin gold film to reduce light transmittance by 88%. The inner over-visor was made from a polycarbonate material to form an impact protection for the fixed faceplate. It afforded protection against ultra-violet energy by containing an inhibitor within the polycarbonate. It also inhibited thermal flow to the fixed faceplate by supporting a low-emittance coating. The inner visor was fixed to the support frame and served as a pressure restraint; it would remain closed all the time the suit was pressurized, unlike the two components of the over-visor which could be moved separately as desired.

The Ventilation Control Module was provided to control suit pressure at about 202 mmHg, to provide ventilation within the suit, and to contain a 0.15 kg supply of oxygen pressurized at 281 kg/cm^2, sufficient to keep the astronaut alive for 9 minutes should his prime supply of oxygen from the spacecraft fail to come through for some reason. A small pipe connecting the VCM to a port on the neck ring, easily seen on photographs of this equipment, would carry oxygen to the astronaut's oro-nasal region. In effect, the umbilical line that connected him did more than extend the normal hose connection for environmental oxygen and supply electrical power and telemetry lines; it also restrained him from drifting away in space, considered a very real danger during the early days of EVA. It was officially called the '7.62 meter umbilical' but in fact that oxygen line within the umbilical was 8.28 meters in length, whereas the tether designed to take all strain between the pilot and the spacecraft was only 7.23 meters long to prevent undue stress being put upon the line that carried life-giving oxygen. Designed to take a load of up to 450 kg, the tether was covered with a thin layer of gold film to distribute the heat it would receive from the Sun.

From early in the evolution of EVA concepts, engineers recognized the need for some method of attitude control outside the confines of a spacecraft. Once free of the vehicle, an astronaut would have no means of changing his body position or, more important, of moving from one place to another. It could only be a matter of time, thought the engineers, before astronauts would need to move from one space structure to another, so a form of hand gun was necessary, expelling gas at will, to change position and move locations. Tests revealed a preference for a means of incremental thrust; the mass of the human body was so low that a fixed-thrust emission of gas from a hand-held gun would lead to a fixed speed between any two points, and that was no good because short distances required slow movement while greater distances called for more rapid movement. So a variable thrust device was designed where a trigger applied thrust proportional to the pressure on the switch.

Because the gun, called the Hand Held Maneuvering Unit (HHMU), was to be carried inside the cabin, it was required to use a gas that would not contaminate the spacecraft atmosphere in the event of a leak. So, oxygen was the logical choice and development time was saved by 'borrowing' two 0.16 kg capacity tanks from the ejection seat supply provided for the astronaut obliged to eject from a spacecraft at high altitude. By fitting two together in a

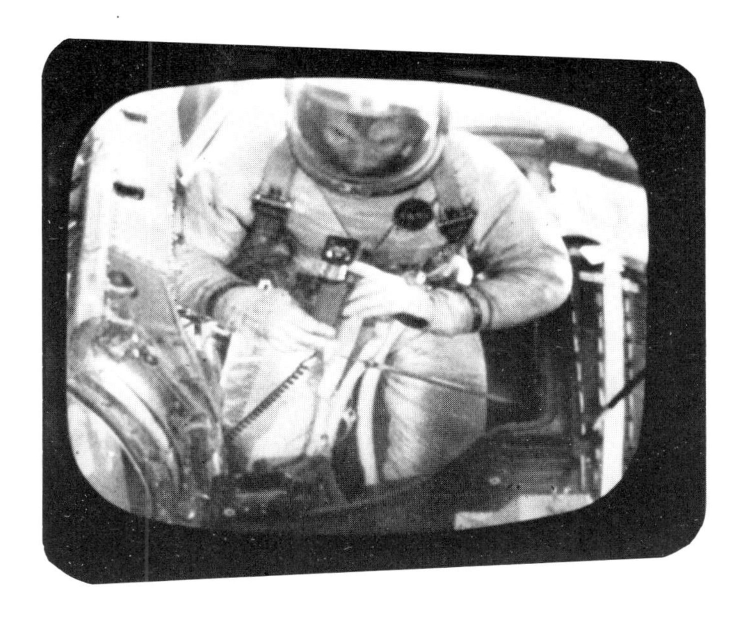

Astronaut Ed White practices extravehicular activity (EVA) in the high altitude pressure chamber at McDonnell, St. Louis, 24 March, 1965.

hand unit the HHMU would be capable of a total impulse of 18.14 kg-sec. A single pusher nozzle was placed at the rear of the hand section, with two tractor nozzles at the extreme ends of a split boom across the front of the device. Thrust up to 0.9 kg could be applied by a maximum hand pressure of 9 kg, proportionately reduced to zero as hand pressure fell past 6.8 kg – the pressure required to operate the valve poppet. In this way, a maximum 0.9 kg thrust could be maintained for 20 seconds, but in reality the HHMU would be used sparingly and at lower thrust levels. The device weighed 3.4 kg and came in two sections for assembly in the spacecraft.

In all, the EVA pilot would have a mass of 97.5 kg equipped for a limited excursion outside his vehicle. Russia's cosmonaut on Voskhod II had carried an autonomous life support system, NASA's astronaut would get his oxygen from the spacecraft. But Leonov had no means of controlling himself other than by pulling on the tether designed to prevent him floating away. If White was allowed outside his spacecraft on Gemini IV he would have limited maneuverability with the HHMU. This was the equipment inspected by Bob Seamans on 14 May, just a few weeks before the flight that could give an American astronaut a chance to step outside his spacecraft. Next day, qualification certificates were signed for the G4C suit and the David Clark Company confirmed that MSC would have six such garments by the end of the month. It was time for a decision.

On the 15th, Charles Mathews was in Washington talking to Mueller about the EVA prospect but the apparent haste with which the plan had been formulated still troubled the manned space flight administrator. Four days later Mathews telephoned Mueller again to confirm that final tests had qualified every item of hardware under simulated Gemini IV conditions and that everything was completely ready for the go-ahead. Meanwhile, Seamans had talked to Webb and Dryden, but the latter disagreed with his boss in that he thought the plan was too soon after Leonov's flight to give the public confidence in moving ahead progressively and independent of Soviet achievement. That had been the one mistake NASA made about the whole EVA issue. For two years the plans had been slowly evolving at the Manned Spacecraft Center, sufficiently so for the Russians to have known about the space-walking plans for some time, but insufficient public awareness had been generated to prevent possible criticism of risky chances taken in haste. NASA had played down the whole idea of EVA on Gemini flights, and now it would have to announce it suddenly as a major part of what was otherwise intended as a four-day flight, in itself a considerable achievement compared to Cooper's 34 hour flight in Faith 7.

Nevertheless, Webb asked Seamans to prepare a memo for him detailing why he thought EVA should be included in Gemini IV's flight plan, and Webb then handed it to Dryden, who was against the proposed activity from the beginning. On Tuesday, 25 May, Dryden called Seamans to his office and silently handed him back the memo he had compiled with an approval note by James Webb written over one corner. That day NASA announced that Ed White would open the hatch of spacecraft 3 and leave his cabin early in the flight, then scheduled for launch on 3 June. But there was concern about the flight from medical men assigned to prepare and monitor McDivitt and White during their planned 98 hour mission. It was significantly longer than any American

Test subject models the Gemini EVA suit complete with chest mounted ventilation control module and gold-coated umbilical line.

astronaut had flown, and impinged upon the flight duration experienced by the Russians in the last four Vostok missions.

Since 1963 the Russian medical personnel had warned of significant effects from long periods in space, presenting papers at conferences telling of disorientation, faintness, depleted cardiac resistance for several days after a flight, and possible loss of bone mineral. It all added up to grave caution and since nobody knew whether the Soviets were over or under-estimating their case, the newly appointed Chuck Berry as physician in charge of astronaut health – in and out of the spacecraft – was not about to throw caution to the wind. Pessimists pointed to the effects already observed in the latter Mercury missions, flights where cardiovascular response had given cause for some concern about the ability of the human body to adapt to the strange new environment. But all that could really be done was to provide some form of exercise device, which in the limited confines of a Gemini spacecraft was no mean design feat, with which crewmembers could tone their muscles and keep fit. Beyond that, it was a matter of keeping careful observation on their condition through the bio-sensors attached to their bodies, and providing for every conceivable contingency.

The exercise device finally chosen was a bungee cord attached to a handle and a foot strap that required a force of 300 N to extend it 30 cm. It would be used at daily intervals until preparations for re-entry, when it would be used every few hours to pre-condition the heart. The only undesirable aspect of the Gemini spacecraft's operational function would come at the end of the flight when the spacecraft dropped into the two-point suspension mode, suddenly pitching the crew forward into an upright position where blood would be required to act against the force of gravity in moving around the body; until then the crew would have experienced g loads in a safer, supine, posture.

Now that the space agency was planning increasingly longer flights into space, the physiological response of the human body was only a part of mission support requirements. Food too played a considerable role in levelling the condition of the astronauts. On Gemini IV, the crew would eat four meals a day, corresponding roughly to breakfast, lunch, dinner, and supper. The meals were to be stored in 18 packages, 14 two-man meals and four one-man meals, in a compartment above the command-pilot's left shoulder. They were marked by day and meal and were placed in the compartment in order of use, so that the first day meal would be on top. All packages were connected to a nylon lanyard to prevent them getting out of order. The food itself was chosen with due consideration for individual preference, consisting of freeze-dried items, dehydrated items in powder form, and compressed bite-size lumps. A special water gun was provided for reconstituting the freeze-dried and dehydrated items. After consumption, freeze-dried food packages would be biologically neutralized by a tablet placed in the food pouch by the astronaut.

A typical day's meals began with breakfast consisting of apricot cereal bars, ham and applesauce as a reconstituted item, cinnamon toast, and reconstituted coffee. Lunch would comprise beef bites, reconstituted potato salad, pineapple fruitcake bites, and reconstituted orange juice. For dinner, banana pudding, chicken salad, peaches and beef sandwiches, and for supper, potato soup, chicken and gravy, toast, peanut cubes and a drink of reconstituted tea. Each day's meals provided an average intake of 2,500 calories, the actual value of the calories consumed changing with the level of work performed for that day; this fact is conveniently missed by authors of standard diet sheets, where no two people can ever have the same response to a given calorie intake. In reality, the Gemini IV crew would consume slightly less food than the total provided, McDivitt averaging an intake of 2,066 cal/day, White about 2,230 cal/day. The difference can be accounted for because the pilot was required to perform several physically stressful feats, such as the EVA, while the command-pilot had a comparatively low physical work load on average.

There was, however, one consequence of eating four meals a day on a four-day flight, and that was taken care of by the first real NASA attempt to come to terms with the most unpleasant aspect of a long duration mission inside a vehicle not much larger than two telephone kiosks. While a urine collection device was a comparatively simple engineering problem, defecation was a technically difficult task to both perform and accommodate. For Gemini flights fecal bags were provided and the crews went to a low-residue diet several days before a mission. The bags carried an adhesive surface designed to stick to the surface of the astronaut's buttocks. The process of defecating into a tiny bag would be hindered by the absence of gravity and the unpleasantness was enhanced by a requirement to physically squeeze the contents into the bottom of the bag, which then had to be liberally kneaded so that they would be mixed with a biological neutralizing agent. Urine was to be flushed overboard on early flights from a line containing a circular orifice with a small tap on the end to prevent the contents floating back into the cabin.

It was all very different to the Mercury days and despite the recent flight of spacecraft 3 it was Gemini IV that really seemed to carry the mood of operational expectancy, heralding a new era of manned flight closer by far to the day when three men would set sail for the Moon. As usual, final major medical examinations were completed two days before the flight. The night before the launch, McDivitt and White retired to bed at 8:30 pm for a 7 hr 40 min rest before the day's events began on the morning for which they trained so long. After breakfast, suit technicians Joe Schmitt and Clyde Teague helped ease the two crewmen into their new suits, White distinguished now by the bright gold visor cocked back on top of his gleaming white helmet. It was exactly 7:08 am when the two pilots reached pad 19, Borman and Lovell having checked their switches hours before and trimmed the spacecraft's interior for their arrival. Four minutes later they were inside the re-entry module, checking for themselves the instruments, dials, and switches.

Spacecraft 4 was a little different to its predecessor, although it still did not carry the fuel cells designed to generate Gemini's electrical power. In fact that had been an instrumental factor in limiting the second manned flight to a duration of four days. Chuck Berry had not wanted to go from the first three-orbit flight to a planned seven-day mission for fear of running into medical problems that could threaten later missions. When General Electric continued to have problems with getting the definitive fuel cells ready, it sealed the fate of Gemini IV; four days was the limit with batteries. But spacecraft 4 had more oxygen than its

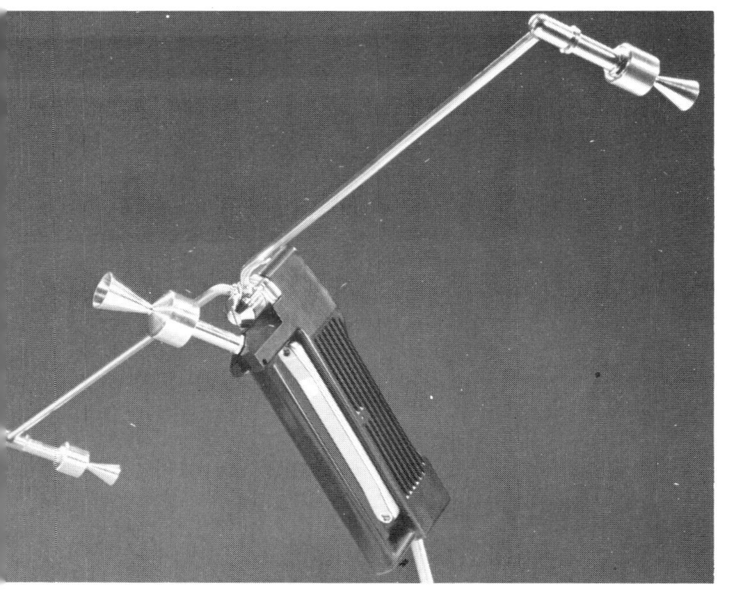

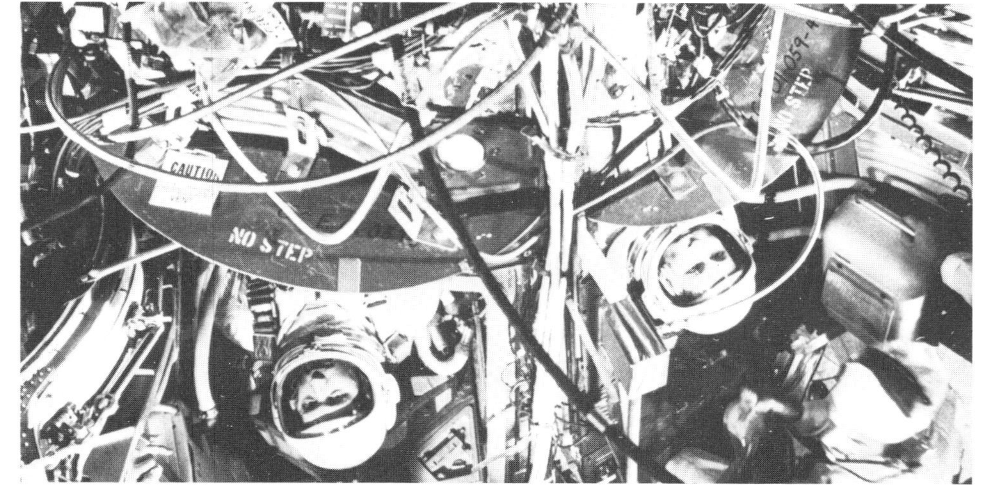

In last minute checks of their spacecraft, McDivitt and White prepare for their flight, only days hence.

← The hand-held maneuvering unit was used by White to move around outside Gemini.

Training continues for Gemini IV as McDivitt and White go through checkout procedures in the NASA Mission Control Center simulator at Houston.

predecessor: 23.6 kg versus 6.94 kg, and, with propellant totalling 300 kg, more OAMS maneuvering capability too.

The supreme feeling of having begun true space flight operations permeated the preparations. Across at Houston there was every reason to be thrilled with this particular flight. It was the first time a mission would be managed from the Manned Spacecraft Center. In the Mission Operations Control Room (MOCR), pronounced 'moker', Christ Kraft was plugged in to the events at Kennedy Space Center. The layout was a considerable improvement on the old Mercury control room at the Cape. At the back, against a transparent screen separating flight controllers from public viewing stands, was the Mission Director console, raised above the rest and flanked by the Public Affairs Officer to his left and the Department of Defense representative to his right. In front of them, from far left to far right, would sit the Operations and Procedures Officer handling and integrating events throughout the MOCR and its interface with the ground operational support system, the Assistant Flight Director ready to stand in for the Flight Director at his right, responsible for handling the detailed events of the flight from launch through splashdown, and the Network Controller, keeping alive the flow of tracking and communications information streaming to Houston from the Goddard Space Flight Center.

On the next row forward, again from left to right, came the Flight Surgeon's console, then Spacecraft Communicator – capcom – , with a Vehicle Systems Engineer monitoring electrical and mechanical support functions aboard the spacecraft, the Guidance Officer monitoring Titan performance during ascent and spacecraft inertial platform activity during the flight, and another Vehicle Systems Engineer looking after the environmental equipment. On the front row, from the extreme left, the Booster Systems Engineer monitoring propellant and tank pressures, the Retro-fire Officer, the Flight Dynamics Officer, watching the orbital tracks and the go/no-go gates, and his assistant. On a separate console to the right, but down on a level with the front 'trench,' the Maintenance and Operations Supervisor looking after the MOCR equipment ready to call upon technicians or engineers to work problems with consoles or equipment.

With Gemini moving to long duration flights for this mission and the next, a posse of flight controllers would constitute a separate team under one of three Flight Directors: Christ Kraft would take the first shift, looking after orbital insertion, the trims necessary to give the orbit a long life, and the EVA. Eugene F. Kranz would take the second shift, followed by Englishman John D. Hodge. Kranz would be primarily responsible for systems monitoring during what would turn out to be the latter part of the crew's day, following the active part worked by Kraft's controllers. Hodge would spend the night cranking new flight plan instructions into the next day's routines, ready to give the crew updates to work with that day. Kraft, Kranz and Hodge had taken turns practising as Flight Director during a week in October 1964, joined at that time also by Glynn S. Lunney when all four tested the Manned Spacecraft Center's interface with the global network routed through Goddard. It was just a small part of the intensive training program devised for mission support personnel, for the safety of astronaut's lives would depend as strongly on the competence of people on the ground as on their own orbital performance.

It was 7:32 am when technicians in the erector white room sealed spacecraft 4, White now wearing a fabric protector over his gold-coated visor, and both men calmly proceeding with their respective checks. In the blockhouse close by, astronauts Borman and Williams sat in with capcom Al Shepard, all three anxiously scanning the status boards and console screens. Then, at 8:25 am, as the erector was being slowly lowered to the concrete hardstand, a badly positioned connector brought the massive steel structure to a halt, 12° from vertical. It took more than an hour to find the offending component and after a hold lasting 76 minutes the countdown resumed.

There was something unique about the centers of attraction that June morning. For long the hub of launch day activity, the scene of major celebration, the Cape had given up a large number of its usual supporters to the new three-storey building at the Manned Spacecraft Center where events just a few hours after launch would claim more attention than the lift-off. White was scheduled to open the right hatch of spacecraft 4 early in the second orbit; there would simply be no time to get from Florida's Kennedy Space Center to Houston's base of operations. It was also the first flight where attention would switch from the Cape just minutes into the flight, and Houston would be the place where, for the first time, the deep grinding voice of Paul Haney would announce 'This is Gemini Control.' suspense kept high at the Cape during previous flights by the continuing events of a prolonged space mission would now shift, like a phantom departing in the night, to Houston's proud center. More than 1,000 newsmen were installed at the Manned Spacecraft Center, far more than had been expected, and NASA had no recourse but to hire facilities intended as office space.

But all that was of no concern to McDivitt and White as alone inside their spacecraft they monitored the tiny points of light on display consoles and status panels and listened to the comforting and cheery voices coming through the communications line. More people were watching this flight live than had seen any other space mission, as an Atlantic communications satellite called Early Bird relayed TV pictures and sound from the Cape to 12 European countries. It was 10:16 am local time when the two first stage engines roared into life, gently lifting Gemini IV on its way. It was a smooth ride, far more comfortable than Atlas, and hardly gave an indication of lift-off, just a gently accelerating push – many would say they could only feel the sensation intuitively while looking at dials that confirmed the big bird was flying. Little more than five minutes elapsed before the spacecraft was in orbit, pushed free of the booster's second stage by a 12 second burst of the aft-firing thrusters, adding 3 meters/sec to their orbital speed of 28,250 km/hr. Flight control had now shifted to the MOCR at Houston, and clustered around the capcom console were John Young, Gus Grissom, Wally Schirra and Deke Slayton, all listening to the chatter from space.

One of the first duties to be performed was an attempted rendezvous with the second stage of the launch vehicle, a cylindrical shell now, 5.8 meters long, 3 meters in diameter, with a large engine bell on the aft end, in all not much larger than the Gemini spacecraft itself. This exercise was prompted by deletion of the Radar Evaluation Pod being developed as an early test for rendezvous while the project waited for Agena target vehicles, and by a humorous remark from capcom Gordon Cooper during the previous flight to the effect that Gus Grissom might like to try and rendezvous with the booster. MSC chiefs Gilruth and Low heard the commentary and took the idea seriously, suggesting its adoption for Gemini IV. In reality, Jim McDivitt and Ed White found it harder than expected. When they turned the spacecraft around after imparting the separating thrust the inert hull seemed to be about 200 meters away and off to one side, but developing a tumbling motion; it would be dangerous to get too near a spinning hulk, which, although weightless, still had plenty of mass and inertia.

As the spacecraft and its attendant booster swept into the night side of planet Earth, the two were drifting apart. Maneuvers performed by Jim McDivitt seemed useless for the task of reducing that distance and although aided by a pair of flashing lights placed on the second stage for just this very task, every move made by Gemini IV's pilot seemed to place the spacecraft further away. It was paradoxical: in apparently thrusting toward the steel shell, the spacecraft seemed to move to a greater distance. It was the first time anyone had tried to approach another object in space and it seemed to carry a strange mirror-like reversal to the instinctive skills exercised by McDivitt. From the spacecraft: 'We still have the booster, we're off quite a ways from it now, er, its taking a little more fuel than we had anticipated. Do you want to make a major effort to close with this thing or to save the fuel?'

From his position in the MOCR, an immediate decision from Mission Director Chris Kraft to the capcom console: 'You might tell him, er, as far as we're concerned we want to save the fuel; we're concerned about the lifetime more than we are matching that booster.' With finite reserves of thruster propellant, spacecraft 4 could quickly run out of maneuvering capability, and with four days to go there was little point in pursuing a negative exercise. So the rendezvous was abandoned, but it was hardly that anyway since the two structures were only a few hundred meters apart. Soon, however, the respective motion of Earth's two neighbor satellites sent them far apart and it would only be a couple of days before the Titan's second stage would burn up over the Atlantic. The more immediate area of concern centered on White's preparations to open the hatch and go outside, but the first orbit had been completed before that activity began so McDivitt elected to delay the planned EVA from the second to the third orbit.

It certainly took far longer to complete the check lists, fasten hoses and tethers, and get all the equipment ready for this seemingly easy task. In reality, it was a major advance on space operations thus far. To de-pressurize a space cabin and open a hatch that, if not closed properly, could bring disaster during re-entry, reflected a level of confidence and expertise many believed the program had but few were willing to test. It was as significant a step on the road to the Moon as long duration missions or the mechanical process of rendezvous and docking, for a man would have to do this very act when leaving his spacecraft for a walk on the lunar surface.

By the end of the second orbit preparations on board the spacecraft were nearly complete. The first test was to evacuate the cabin to a pressure of 103 mmHg. With the suits designed to hold a pressure of between 187 mmHg and 202 mmHg, the reduced cabin atmosphere would soon start to rise if oxygen leaked from the pressure garments. It did not and the command-pilot got a 'GO' to start complete depressurization as the spacecraft moved across the Carnarvon tracking station in Australia on the third orbit. Coming up on Hawaii the crew prepared to crack the hatch, and as the spacecraft came above the radio horizon the hatch was open and White was standing by for the word to move out. From the Hawaii capcom came the welcome voice of approval: 'Gemini IV, Hawaii capcom. We just had word from Houston, we're ready to have you get out whenever you're ready.'

MCDIVITT 'O.K. We've got our "go" now, is that right?'
CAPCOM 'Affirmative.'
MCDIVITT 'O.K. We've still got a little work right here.'
CAPCOM 'Roger, understand.'

The hatch had opened at an elapsed time of 4 hr 18 min, about 2:34 pm Cape time, lunchtime in Houston. White's first task was to install a 16-mm movie camera and to place an umbilical guard on the hatch sill. Seven minutes later he was standing on the seat fitting together the two halves of the HHMU – or 'gun' – that would propel him around the outside of the spacecraft, and placing on the front of it a small camera through which he would take shots of the vehicle. Twelve minutes after the hatch opened, White started out:

'O.K. I'm separating from the spacecraft. O.K. my feet are out. I think I'm dragging a little bit but I don't want to fire the gun yet. O.K. I'm out.'

Slowly drifting up through the rectangular cut-out, the big hatch door angled across to his right, White gently squeezed the trigger on his maneuvering unit, literally pulling himself out in a tractor mode but slightly higher than he wanted to go. Within seconds he was about 2½ meters from the spacecraft, out past the nose of Gemini IV. 'O.K. I put a little roll there – took me right out.' A gentle movement with the HHMU put White into a slow tumble. 'O.K. I rolled off and I'm rolling to the right now. Under my own influence – here goes. Looks like a thermal glove Jim,' as a small object drifted up through the hatch and spun away into space, the right hand over-glove that White had elected not to wear in the hope of improving his hold on the maneuvering gun. 'All right. Now I'm coming above the spacecraft. I'm coming back down now. I'm under my own control. O.K. I'm coming over. Must be worth a million dollars! I'm coming back to you. The gun works real good Jim.' White was in a very slow tumble around the nose of the spacecraft when he tugged gently on the tether. It carried him back across the top of the vehicle and put him in the area of the adapter, far behind the re-entry module. Just where he did not want to be.

Stable again in attitude, White decided he would give the maneuvering unit a good workout and, with his left hand out, he put the HHMU into a line with where he thought his center of gravity was located and fired a long burst. It carried him straight down the center between the two hatches, along the extended rendezvous and recovery section, and out past the nose. Stopping himself with another burst on the pusher nozzle, White performed a few yaw tests and then squeezed the trigger to move back across the spacecraft. No luck. The small supply of oxygen gas was totally expended:

'O.K. I think I've exhausted my air now. I have very good control with it, I just needed more air . . . Capcom, it was very easy to maneuver with the gun, the only problem I have is I haven't got enough fuel. This is the greatest experience, it's just tremendous . . . Right now I'm standing on my head and I'm looking right down and looks like we're coming up on the coast of California . . . as I go on a slow rotation to the right. There is absolutely no disorientation associated with it.'

MCDIVITT 'One thing about it. When Ed gets out there and starts whippin' around it sure makes the spacecraft hard to control.'
WHITE 'O.K. I'm drifting down underneath the spacecraft. There's no difficulty with recontacting the spacecraft . . . particularly as long as you move nice and slow . . . Feel very grateful to be having the experience to be doing this. I'll bring myself in and put myself out of your view.'
CAPCOM 'Are you taking pictures?'
MCDIVITT 'O.K. You want me to maneuver for you now, Ed?'
WHITE 'No, I think we're doing fine. What I'd like to do is to get all the way out Jim, and get a picture of the whole spacecraft. I don't seem to be doing that.'
MCDIVITT 'Yeah, I noticed that, you can't seem to get far enough away.'
WHITE 'I'm coming back down on the spacecraft. Listen, it's all

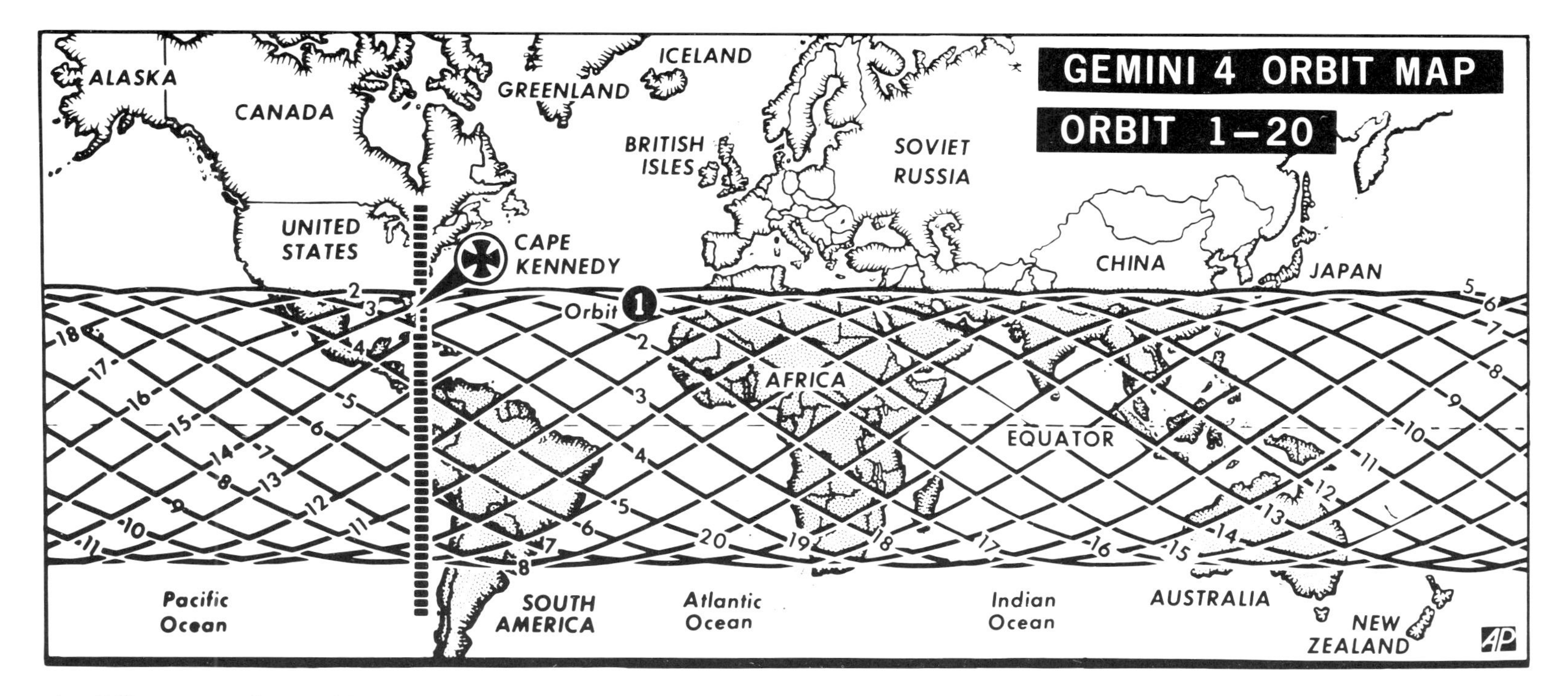

the difference in the world with this gun. When that gun was working I was maneuvering all around.'
CAPCOM 'You got about five minutes.'
WHITE 'O.K. I'm going to let myself go out now.'

White ran out of compressed oxygen for his maneuvering gun about four minutes after pulling himself out of the hatch and spent time using the camera and learning the difficulties associated with tether dynamics. With no other means of attitude control, he had little alternative but to use the long, snaking line to pull himself around. In the weightless void it was an almost impossible task; like some bob-weight dancing around on the end of an elastic cord, White was going everywhere but in the actual direction he wanted to. From the front of the spacecraft, he again found himself at the back near the white adapter, then he was under the spacecraft and gently bumping the skin with his helmet. But it was an exhilarating experience and the capsule communicator had been trying for a few minutes to get the attention of two very pre-occupied pilots. Moving at 7.8 meters a second, Gemini IV would soon be out of range of tracking stations as the vehicle moved across the continental United States heading for an Atlantic pass.

McDivitt goes through a weight and balance check during a wet mock simulated test at the Cape.

CAPCOM 'Gemini IV, Houston.'
CAPCOM 'Gemini IV, Houston capcom.'
MCDIVITT 'Gus, this is Jim, got any message for us?'
CAPCOM 'The Flight Director says, get back in!'
MCDIVITT 'O.K.'
WHITE 'Where are we over now Jim?'
MCDIVITT 'I don't know . . . we're coming over the . . . and they want you to come back in now.'
WHITE 'Back in!?'
MCDIVITT 'Back in.'
CAPCOM 'Roger, we've been trying to talk to you for a while here.'
WHITE 'Coming in.'
CAPCOM 'You've got about four minutes to LOS.'

Loss of signal would occur before White returned to the spacecraft unless the Gemini pilot made a determined effort to terminate his EVA and close up the vehicle. Just 22 minutes after opening the hatch, White was back on the seat preparing to slide down into the spacecraft. But there were problems. Reluctant to pull to, the hatch was difficult to bring down from its open position, and when it was back in place the locking mechanism refused to function. Each time White tried to apply torque to the latch, he succeeded only in lifting himself back up out of the seat. McDivitt had to hold on to White's legs to prevent him drifting up, and then White felt a little movement in the handle. Within seconds it was locked tight.

The simple action of closing the hatch was the most strenuous part of the entire EVA. During most of his space walk, the Gemini pilot had a heart rate of between 145 and 155 beats per minute; closing the hatch down raised his heart to 180. At an elapsed time of 4 hr 54 min, the EVA was officially over, counted for this flight as the period between hatch opening and closing. On that criterion the event lasted 36 minutes, with White physically outside the spacecraft for about 21 minutes. Inside, White was wet with perspiration, his face streaming with beads of moisture. It had been too brief a period to fully evaluate the hand maneuvering gun, too new an experience to consider the value of using the tether to move about. In fact, it had done little more than prove a man could go through the physical process of EVA from a Gemini spacecraft. It would be left to later missions to prove just what an astronaut could do in the way of work tasks and assigned activity. It certainly seemed that White became exhausted earlier than expected, only enthusiasm preventing him from recognizing the self-induced symptoms.

Because the spacecraft had been on an internal communications line for much of the EVA, Houston was unable to call up the command-pilot when it was time to have White get back inside. For that reason it appeared to several newsmen that White was reluctant to end his walk in space and papers next day would carry questions about the ability of an astronaut to remain in full control during these dangerous and exhilarating experiences! As he later admitted, White was reluctant to get back inside for a very down to earth reason – it was too good a thing to cut short! But would an astronaut tire quickly outside his spacecraft? Was it a stressful activity more complicated than would be thought in theory? Answers would have to wait. But good training had already paid dividends: when the hatch stuck it took knowledge to get it latched acquired when McDivitt and White completely disassembled a specimen hatch at McDonnell's St. Louis plant before the mission; without that experience neither would have known the precise sequence essential to getting it locked, and that operation took three hands simultaneously working the ratchet.

The pre-mission flight plan required each crewmember to sleep at a different time to his partner, but in practice that was made difficult by the activities of the active astronaut. The communications circuit also was noisy, and the banging noise of thrusters firing from time to time, stabilizing attitude, was a disturbing sound. Gemini IV carried many more experiments than any previous manned space vehicle and while McDivitt slept on the sixth and seventh orbits, White performed scientific tasks, photography, and an in-flight exercise routine. Then it was White's turn to sleep while McDivitt performed more experiments, and so on throughout much of the flight. But it did not turn out that way in practice, considerable sleep being lost because of the interruptions; Gemini was too small a vehicle for two men to conduct very different activities adequately without mutual interference.

On the second day Gemini IV got a message of congratulation on passing Cooper's 22 orbit record. Houston received a curt but truthful: 'Roger, we've got quite a few more to go.' Considerable periods were spent with the spacecraft powered down, without fuel cells electrical conservation was a necessity, in a drifting mode between functional tasks. Considerable photographic activity had been scheduled and the crew did good work in operating the various cameras on board, hand-held devices pointing out the forward windows. Each man was settling down to a group of responses and reactions totally his own. White found his physical condition flagged before meal time, but after an intake of food he felt thoroughly charged and ready for work. McDivitt, being a little taller than his companion, was becoming increasingly frustrated by the top of his helmet rubbing on the hatch lining causing it to twist slightly every time he turned his head. After two days McDivitt removed his helmet for a while, as he would again just before re-entry. Irritating little nuisances that loomed larger than the major events on board the cramped spacecraft.

On into the third day swept the spacecraft, systems performing remarkably well and everybody clearly amazed at the smoothness of the flight. On board the spacecraft it was getting harder to find a spare cavity for rubbish that seemed to grow out of all proportion to the duration of the mission. Carefully packed before launch by experts trained to get the most amount of equipment into the smallest container, it was not an activity conducive to prolonged flight. Moreover, it was getting increasingly uncomfortable inside the suits where the pure oxygen environment had a strange reaction on the body's sensitivity to perspiration and very tiny particles that eventually found their way out from behind panels and from under containers. In many respects the tasks were repetitive. But the purpose of the flight was to prove men could remain in space for a period half as long as that required to get to the Moon and back. It was a medical exercise first and foremost, and would lead directly to two more long-duration flights culminating in a two-week mission around the end of the year. If, that is, the fuel cells were ready in time, for without them the spacecraft would have insufficient electrical power for such a long period.

Periodically throughout the flight the crew made orbital maneuvers to prolong the spacecraft's lifetime and to test further the thrusters first used by Gemini III. The only technical problem that gave cause for real concern came on the third day when McDivitt could not turn the computer off following an update from the ground. Several troubleshooting tests were contrived, and instructions passed up to the crew. But no amount of switching or workaround procedure could bring it back and from that point the crew had to perform several functions manually. It was no real problem, just an annoying thing to have happen. It did mean, however, that instead of flying an automatic re-entry profile the crew would have to go to a rolling descent mode where the spacecraft would fly a ballistic path instead of a modulated lift the trajectory was intended to use.

As the final hours of the four-day flight ticked away, the crew made final preparations for re-entry. Shortly before firing off the adapter's equipment section, the crew fired the OAMS thrusters to slow the spacecraft down, lowering perigee in a maneuver similar to one carried out on the previous flight, an insurance in case the retro-rockets failed to fire; the logic of this operation took cognizance of the fact that only when the OAMS capability had been jettisoned would the astronauts know if their solids were going to fire. As it turned out, they did, slamming McDivitt and White into the backs of their respective ejection seats with thudding deceleration as each motor burned with a thrust of 1,134 kg. Then it was the retrograde section itself that had to go, and finally the spacecraft was turned around to prepare for re-entry.

Russian cosmonauts had come back to Earth, surviving the punishing g forces of re-entry, after longer flights than Gemini IV, but still there was the nagging knowledge that no American astronaut had yet performed the feat. Chuck Berry was keen to see their condition, and gratified to hear that more time had been spent on the bungee exerciser than planned. Safely descending through the atmosphere, both men kept up a continuous conversation, first with capcom Gus Grissom and then with each other as the enveloping plasma sheath cut communications with the ground. McDivitt stopped the spacecraft's slow roll at a height of 27 km, and shortly thereafter the drogue parachute popped out followed by the main canopy. They jerked into a two-point suspension position just 1.5 km off the water, and neither crewmember felt faint. In fact, despite a hefty bump as they landed unusually heavy in the Atlantic swell, both men felt better than they had prepared themselves to feel.

The spacecraft was 80 km off target when it landed, but swimmers were soon on hand with a flotation collar to calm the vicious pitching and rolling motion. Just 57 minutes after splashdown the pilots had been retrieved by helicopter and were touching down on the carrier USS *Wasp* to an uproarious welcome and a red carpet reception. Over the next few days strenuous medical examinations were performed. The word from the *Wasp* could re-write the next flight, Gemini V, intended to go for twice as long. Chuck Berry was well pleased with the crew's performance and gave an optimistic report about the ability of astronauts to remain in space for extended periods. There was a little concern at the loss of bone mass, and apprehensive surprise at the depleted blood plasma volume, but nothing that could stop the next mission. Gemini IV had captured the interest of the space-watching world. The usual round of celebrations and accolades were inevitable.

But the momentum of the program was now picking up speed, and there was more to be achieved before year's end. For a few days though, the two Gemini IV pilots were at the center of a happy present-giving session. Newsmen at Houston presented them with an abacus – just in case the computer broke down again – McDivitt and White presented President Johnson with a photograph album of the flight when he visited the Manned Spacecraft Center, and with a flag carried aboard the spacecraft when he hosted the crew at the White House. But the most important thing to come out of GT-4 was the operational perfection with which the controllers had handled the mission: Glynn Lunney's team at the Cape for launch, and MSC's red, white, and blue teams headed by Kraft, Kranz and Hodge, respectively. Devoid of a tight and responsive flight team, a space mission would be like a ship without a captain. Gemini IV had proven the captains to be both competent and well primed for the task at hand. A test of that ability would come with the very next flight.

Gemini V was in several respects a further step toward operational confidence. For two flights now, crews had been required to save considerable quantities of precious OAMS propellant for the pre-retro-fire de-orbit burn, a contingency against failure in the solids. McDivitt and White had been annoyed over having to fly this fail-safe routine, and campaigned to have the next mission fly with a full utilization of OAMS propellant for operational maneuvers in support of scientific or technical experiments. Houston was behind the move, and Mueller agreed; the propellant-expensive fail safe burn would not be budgeted in the consumables plan. Moreover, the mission would fly with fuel cells as the primary source of electrical production, a flight made possible only by their availability. Gemini IV employed six 400 ampere-hr silver-zinc batteries to supply the electrical needs that totalled about 2,073 ampere-hrs; Gemini V would consume an estimated 4,200 ampere-hrs and for that fuel cells were essential to the spacecraft designed for such a system. In general, requirements exceeding 800 ampere-hrs are better and more efficiently accommodated by the use of cryogenic reactants, winning out over batteries in weight and volume.

But at least one questionable effect of the particular design adopted for Gemini's environmental control system had been answered in good time for the long duration mission coming up. Tests revealed a possibility that condensation would form on interior surfaces and McDonnell took urgent steps to line the cabin walls with a moisture-absorbent material in time for Gemini IV. Careful inspection of the re-entry module after the four-day mission showed no signs of any moisture.

With the coming of extended and long duration flights the means by which the flight could be measured had to change. For comparatively short trips, the constant spin of the Earth around its polar axis mattered little to the defined value of the orbit: one 'orbit' of the Earth meant, quite understandably, the circumnavigation of the globe from landing to splashdown. But where the Earth would go round several times on its axis, the term would be confusing. Instead, 'revolutions' were considered the most appropriate yardstick for measuring the parameters of a particular flight. One revolution equals the time it takes the spacecraft to pass 80°W longitude on successive passes, about 96 minutes. One orbit equals one 360° circuit of the Earth, which takes about 90 minutes. Therefore, a spacecraft will make 16 orbits of the Earth in a full day, but only 15 revolutions.

The reason it takes the spacecraft longer to make a revolution of the planet is because the Earth is spinning counterclockwise when viewed from above the North Pole. In going one full 360° circuit of the planet, the spacecraft will not arrive back over the longitude of launch, but will have to continue to fly for an additional 6 minutes until it has caught up with the spinning planet and is once again over the Cape. And that is all the difference really is: an orbit is one 360° pass around the planet, whereas a revolution would be a complete flight around the world as seen from the surface. Gemini IV made 62 revolutions of the Earth. Gemini V would be scheduled to make 121.

NASA Administrator James Webb chose the 23rd annual meeting of the Hampton County Watermelon Festival on 26 June to publicly announce that the third manned Gemini mission would be an eight-day flight. Previously billed as a seven-day mission, confidence generated by the superb performance of crew, controllers and equipment on the Gemini IV flight made it more logical to fly the next mission for the full duration required to get to the Moon, stay a day on the surface, and return home. That would clinch the medical question still hanging over Apollo plans. Not even a Russian had been in space for that long.

One day later NASA announced the names of six scientist-astronauts picked from a group of 400 names submitted by the National Academy of Sciences to train for Apollo flights. They were: Owen K. Garriott, associate professor of physics, Stanford University; Edward G. Gibson, senior research scientist, Applied Research Laboratories, Aeroneutronic Division, Philco Corporation; Duane E. Graveline, flight surgeon at the Manned Spacecraft Centre; Lt. Cdr. Joseph P. Kerwin (USN), staff flight surgeon, Air Wing 4, Cecil Field Naval Air Station; Frank Curtis Michel, assistant professor of space sciences, Rice University, Texas; and Harrison Schmitt, astrogeologist, US Geological Survey. The recruits would report to the Manned Spacecraft Center and receive their initial orders from Deke Slayton, a man who strongly disapproved of the new influx. He balked at the prospect of having a bunch of what he considered to be undisciplined scientists trampling all over the Manned Spacecraft Center, getting in and out of spacecraft without, in some cases, even having a pilot's licence. But it was a move engineered by NASA to placate the scientific community that seemed to think the Apollo Moon landing operation was beneficial to science and should therefore be considered predominantly in that light.

Many at NASA strongly disagreed with this dictum, and with good reason. Nobody had ever tried to sell manned space flight on the premise that it was a scientific exercise. But, inevitably, if the scientific community blessed a venture with its attention, the motivations of that endeavor would necessarily be transformed and re-orientated. It was but the beginning of a great debate that NASA would sustain for many years: science, or engineering as first priority in manned space operations? Each side had its own view, and those not strong enough to stand up to Deke Slayton's verbal milling machine would find some cause more worthy of them elsewhere. Not a few were to fall by the wayside because of this man, a personality that groomed the best and rejected the flaws. But in June, 1965, it was Gemini V that concerned the Manned Spacecraft Center, and preparations for the missions to come.

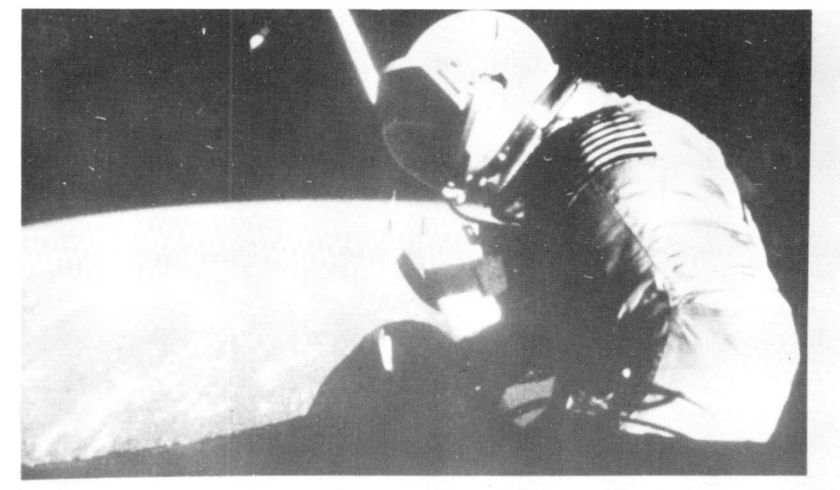
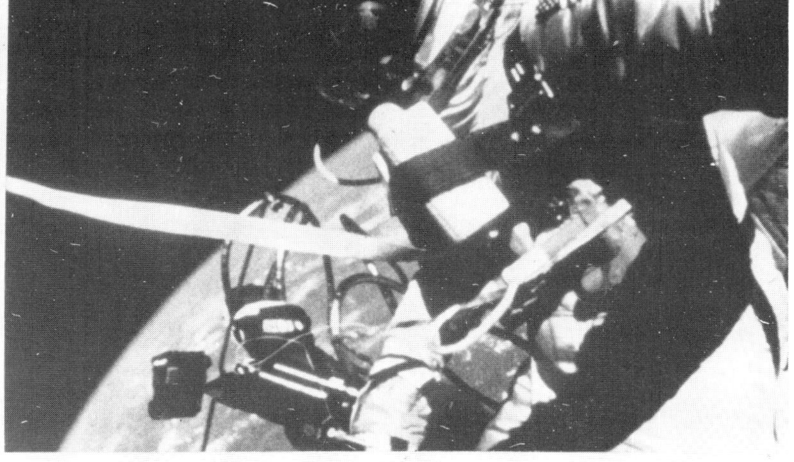

Ed White drifts lazily away from the right hatch on spacecraft 4 during NASA's first EVA, 3 June, 1965.

Experience with hauling Titan launch vehicles from the Martin works, flying Gemini spacecraft from St. Louis, and integrating the two at the Kennedy Space Center developed an accelerated schedule that pilot preparation activities could not match. With an accelerated technical preparation of hardware necessary for each mission, flight crews were getting less and less time for practice and training. The Gemini V crew had been working since February for a mission only six months away; half the time given to Grissom and Young for a three-orbit mission. It was a crippling routine, and one that caused Deke Slayton to fly to Washington and knock on George Mueller's door to argue for a postponement in the assigned launch date. His men were not yet ready and nobody was going to tell him when to clear them until he was sure they had all the training they required. It needed just a couple of weeks delay, that would make all the difference. Deke got his way, Mueller agreed, and Gemini V was re-cycled back to late August, giving the launch teams a slight respite.

On 1 July, the training schedule was further compromised by the selection of Gemini VII pilots Frank Borman and Jim Lovell, with Ed White and Michael Collins as back-up; the flight was scheduled for late that year or early 1966 and would fly the full 14 days for which the spacecraft had been designed. Stafford and Cernan were already putting in overtime hustling along for the Gemini VI rendezvous and docking mission, and the Gemini VII crew would accelerate their training when Cooper and Conrad lifted off for Gemini V and vacated the simulators. And there was much to simulate, for although the training program for flight longevity was not as demanding as that for operational tasks, the Gemini VI crews were busy putting new touches to rendezvous concepts that demanded many simulated hours in the trainers. It would be difficult enough without inadequate preparation.

For its part, Gemini V was reduced in scope by Ed White's successful EVA. Pledged originally as the first flight to try space-walking, Gemini V was now seen very much as a long duration flight only, the additional day made easier to achieve by eliminating procedures that would call for cabin re-pressurization. Nevertheless, and despite a vigorous capaign to get permission to remove their flight suits in orbit, Cooper and Conrad were told they must fly their eight-day mission wearing the G4C garment developed for Gemini IV; having already ordered them in anticipation of performing an EVA, they had little alternative but to use them. Postponed from Gemini IV, Cooper and Conrad would practise rendezvous with the Radar Evaluation Pod, or REP, designed to be secured in the spacecraft's adapter and released by a switch in the cabin. A considerable amount of time was spent simulating the exercise in trainers on the ground; it would provide an interesting comparison with the impossible operation planned for Gemini IV where the crew were to approach and keep station on the spent booster's orbiting second stage.

But for Gemini V, rendezvous would be more an electro-mathematical problem than an eyeball exercise. With a transponder similar to equipment planned for the Agena target vehicle, the REP would produce a signal response to radar in Gemini V's nose. Rendezvous charts and computations made on board the spacecraft would complete the information theoretically needed to pull alongside from a great distance. Before leaving Earth, Gemini V and its REP would go through compatibility tests to ensure the electrical dialogue necessary for success. And in that regard too, Gemini V was the first completely operational spacecraft of its type. For in addition to fuel cells, and elimination of the fail-safe re-entry process, the spacecraft would be equipped with a rendezvous radar for the first time.

Gemini's rendezvous radar was built by Westinghouse Defense and Space Center to a specification constructed on the premise that the two man Gemini spacecraft would be required to search out and find an Agena target vehicle equipped with a radar transponder that would provide the crew basic details about the target's orbit, facilitating information for an eventual approach and, ultimately, docking. Information essential to rendezvous would include range and range-rate details. The chosen design provided a radar beam projected from the spacecraft through a 60° cone. The transponder on the target, upon receiving an interrogating pulse, would transmit a signal delayed in time and shifted in frequency. This would permit the receiver to discriminate between the transponder on the target and other radar signals either from the ground or from the skin of the target. This time delay would improve near-target tracking since, unlike conventional pulse radar, it would not confuse reflected signals and direct transmissions.

When the interrogator radar received a transponder reply it would measure the total trip time and generate analog and digital readouts of target range. The radar system would employ four antennae located beneath a fiberglass cover in the nose of the spacecraft. One antenna would transmit pulses to find the target, two more would receive signals from the target, while a fourth would act as a phase reference for the other two which would rotate, thus enabling the radar to generate directional information. Three antennae would be employed on the target: a cylindrical shaped dipole antenna 12.7 cm in diameter, mounted to a 2.44 meter long boom, and two spiral antennae on the target facing up and down to provide complete coverage for the stable target wherever the spacecraft happened to be in the sky. The actual radar interrogator would measure 33 cm by 48 cm by 79 cm and weigh less than 33 kg. The transponder on the target would measure 23 cm by 25 cm by 48 cm and weigh 14.2 kg. Gemini's radar would seek and find the transponder target to an estimated range of about 402 km.

It was a sophisticated piece of Gemini hardware, and extremely important to every remaining mission except the 14-day flight of Gemini VII. In the flight plan for Gemini V, the REP would receive the radar's coded signal, modify it and return the altered pulses to the Gemini spacecraft which, upon receipt, would inform the crew of acquisition by a light on the main display console. The spacecraft computer would be used to process range, range-rate and bearing information necessary to calculate the maneuvering tasks prior to rendezvous.

Yet another important way in which Gemini as a program was adding new dimension to space operations came in the form of expanded experiment capability available within the larger volume of the spacecraft itself, and within the expanded format of each mission's flight plan. Gemini IV served to introduce this enhanced role. By carrying eleven experiments it opened the possible application of manned space activity to theoretical research and practical tests in a way impossible with smaller vehicles of limited duration. Compared with Mercury, the experiment loads that were to be carried by spacecraft starting with Gemini IV outclassed in every dimension the Mercury program it replaced. On Gemini IV, apart from the medical experiments, such as the in-flight phonocardiogram for measuring the electrical activation of the heart muscle, the bone demineralization study, and the in-flight exerciser, experiments were devoted to synoptic terrain and synoptic weather photography. Terrain views were commissioned by the Goddard Center to determine the value or otherwise to geologists and related scientists of high-quality Earth pictures from space; the weather shots, performed for the US Weather Bureau, had a similar purpose, but also to serve as quality references for TIROS meteorological satellites coming into regular use.

In addition, scientific experiments included a study of radiation measured inside the spacecraft, a determination of the electrostatic charge on the exterior – a valuable measurement for future docking operations –, the measurement of protons and electrons around the Earth, a determination of the magnitude and direction of the Earth's magnetic field, and two-colour limb photographs in an attempt to determine the separate colour components of the observed horizon. Supporting future navigation requirements for Apollo, an eleventh experiment had the crew gathering observable information on celestial references and evaluation of a sextant device. These experiments were of general or scientific interest; the selected experiments for Gemini V encompassed 17 separate studies divided into medical, Defense Department, and scientific categories.

In addition to the three medical experiments flown by its predecessor, Gemini V would also study the otolithic function, changes to sensors in the inner ear, by having a crewmember don a pair of goggles with one eye piece containing a light source in the form of a movable white line; the astronaut would position the line in what he regarded as the pitch axis of the spacecraft while the other pilot recorded the result. A fifth medical experiment had the astronauts wearing an inflatable cuff around each thigh, automatically inflated for two minutes out of every six in the first four days of the mission to measure the effectiveness of the cuffs in preventing deterioration of the blood distribution system. The Defense Department experiments included a measurement of radiation from the exhaust nozzles of rockets during launch,

Under test at the Westinghouse Defense and Space Center, Baltimore, the radar evaluation pod is tested for its role in Gemini V.

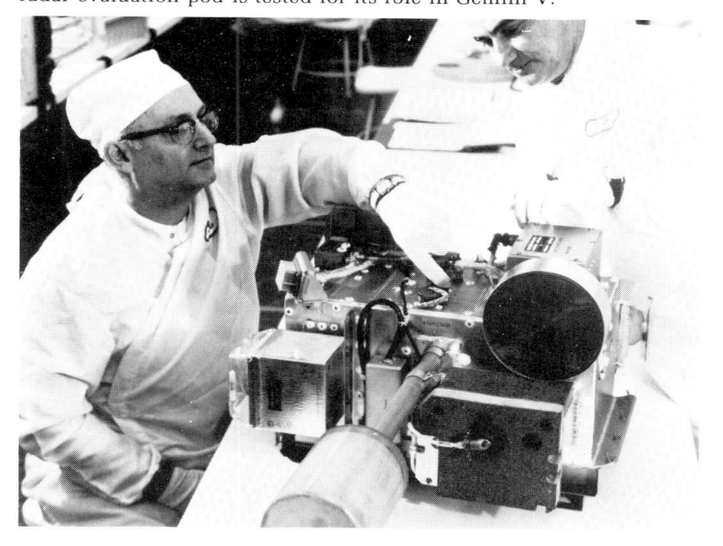

photography of objects close to the spacecraft in orbit, a measurement of radiant energy from ultraviolet to infrared of the Earth, stars, planets and the Sun, surface object photography, photographs of moving objects, and a visual acuity test similar to one of the science experiments.

The six science tasks would seek to photograph the zodiacal light, a hazy illumination seen in the west after twilight, take spectrometer readings of clouds, measure electrostatic charge on the surface of the spacecraft, obtain terrain photographs, provide weather pictures, and test the visual acuity from space of astronauts observing specific patterns laid out on the ground. It was no mistake that this last experiment was to be performed on the very flight commanded by Gordon Cooper, the man who for two years now had puzzled scientists and geologists with his claims about what he saw on MA-9 during May 1963. The visual acuity test actually consisted of several component experiments, most prominent of which was the pattern of marks laid out near Laredo, Texas, and near Carnarvon, Australia. Much of this was of interest to the Department of Defense, which sought to use the operational Gemini flights to determine levels of potential usefulness for orbiting astronauts and to define further significant areas that could be exploited in the military space program.

Interest was heightened by the imminent announcement of official Presidential sanction for the Defense Department's Manned Orbiting Laboratory program, itself a recipient of redundant Gemini hardware. But that would await the end of the Gemini V flight. For whichever application, the experiment load was significant in itself, and a true reflection of the growing concern for new applications. On flights like Gemini V, where the crew would spend long periods drifting in the weightless environment in an ostensibly benign mode, experiments and tests to determine the future usefulness of man in space were both applicable and appropriate. Before committing itself to flight, the hardware for Gemini V was used in one unique training session. On 22 July, less than a month before the planned launch, GT-5 on pad 19 and an Atlas on pad 14 were simultaneously counted down in a test of the procedure which would follow on the next Gemini flight. The crew participated in this rehearsal to the point of simulated launch, but had to be retrieved by a cherry-picker reminiscent of the old Mercury-Redstone days when the pad 19 erector could not be raised back up; appropriately, it was Gordon Cooper who had pressed so hard to have the cherry-picker brought to pad 19 as a stand-by!

Preparations for the launch of spacecraft 5 went well up to the day before the scheduled lift-off. On 18 August computer predictions of the response onboard fuel cells might produce to certain anomalous situations seemed to indicate a fault in one of the sections installed in the adapter module. Further study seemed to confirm the unlikely nature of this pessimistic viewpoint and the preparations for launch the following day at 9:00 am local time went ahead. At 4:00 am on the 19th, however, the countdown stopped at T-5 hrs when monitoring equipment hooked up to the spacecraft's fuel cells showed an unusual quantity of hydrogen boiling off the cryogenic tank. Eventually the problem was rectified by re-filling the cryogenic supply with cooler hydrogen and tests showed it to be a satisfactory solution. The crew, meanwhile, had slept in due to the hold and were given a leisurely pace at which to prepare for their eight-day flight. It was 10:35 am, Cape time, when they slipped into their seats, the fuel cell problem having already shifted the launch time to noon.

From that point on the countdown seemed to go well, until, at T-10 min, a thunderstorm crept up on the launch area and a telemetry programer wavered. At the same time a power surge caused by lightning in the vicinity produced some suspicious response from instruments wired to the affected circuit. With a laconic 'Oh, gee, you promised us a launch today and not a wet mock-up,' Cooper prepared to climb down from the spacecraft and take a ride to the ground in the erector elevator. Within minutes the thunderstorm vented its fury over the Cape, underscoring the inevitability of a cancellation for the day. It had been one of those times when nothing seemed to go right. A sudden security scare sent guards and FBI agents racing toward the launch pad area, only to find two youths had somehow got through the fencing which sealed off the prohibited area. They were arrested on the spot and questioned long into that Thursday night.

At 5:00 pm visiting pressmen were sitting down to a conference with NASA management when word suddenly arrived that a fire had broken out in an underground cable shaft linking the pad with the blockhouse. It was a day better contemplated at its passing! The problems of 19 August required a mock countdown to be carried out before an intent to launch, so that a full engineering check-up could be carried out to see why power surges had apparently occurred during Thursday's count. Checks were held throughout the next day and a launch attempt for Saturday, 21 August, seemed possible. The computer checked out well, new techniques had been developed for loading cryogenic reactants, and the telemetry programer had been replaced.

Cooper and Conrad were woken at 4:30 am on the Saturday with the good news that GT-5's countdown was going well and on schedule. Less than an hour later they sat down with Drs. Howard Minners and Eugene Tubbs, and astronauts Schirra, Stafford and Slayton to a breakfast of filet mignon, scrambled eggs, toast, coffee, grape jelly, and orange juice. It was a jocular affair, as inevitably it would be with Pete Conrad around. It was a perfect day for launch, and technical preparations echoed the mood of the crews preparing launch vehicle and spacecraft for their big event. Right on the planned launch time of 9:00 am the slender assembly swept away from pad 19 as nearby on the beach, watching with tens of thousands of spectators, Cooper's wife, Trudy, stood with daughters Camala and Janita, as a father and a husband rocketed spaceward; Jane Conrad and four sons watched it all from their home in Texas, leaping up and down, hands clapping, as the large color TV in their lounge brought events so far away closer to reality.

The ride into space was not as smooth as earlier flight crews reported, Titan oscillating along the long axis to a maximum 0.38 g at one point; earlier development troubles with this 'pogo' stick effect, at one time a major technical issue, brought a stipulated limit of 0.25 g to methods proposed by engineers for rectifying the problem. Now pogo was back, but only for this flight since ground procedures adopted during launch vehicle preparation were found to be responsible. Nevertheless, it was a disconcerting experience. Five and one-half minutes after lift-off, spacecraft 5 pushed itself into orbit with a burst from the aft firing thrusters, kicking free from the spent stage to enter an orbit of 162 km at the low point by 350 km at the high point. It was a good orbit, little different to the one planned. It was a good start. But not for long.

As the spacecraft moved out of range of the Kano tracking station, 25 min 51 sec into the flight, unbeknown to anybody on the ground or on the spacecraft, the heater circuit in the fuel cell's oxygen tank failed, causing pressure to start a long, slow decline. Experience had shown in tests carried out on the ground that fuel cells tended to survive longer if operated at as low a reactant pressure as possible. Cooper was determined to operate the cells at a minimum level but Conrad noticed that pressure was falling below the level even his command-pilot wanted to see so, on a recommendation from Houston, he flicked the cryogenic oxygen heater switch on. It did nothing to halt the gradual decline in pressure. However, there was no immediate concern, the fall was very shallow and there were other tasks to perform.

Over Carnarvon, half way round the world, Gemini V fired its thrusters to raise perigee from the 162 km imparted at launch to a safer 171 km, preserving the life of the orbit, and setting up a path from which the spacecraft could chase the Radar Evaluation Pod, due for release at the beginning of the second orbit. But still the fuel cell oxygen problem persisted. In principle, the oxygen and hydrogen supply for fuel cell operation was a delicate balance between the leak rate of the spherical containers and the maximum pressure the system could be allowed to accept. With a boiling temperature of $-183°c$, the liquid oxygen was contained within its spherical tank at a nominal pressure of about 57.3 kg/cm^2. On Gemini V, 81.5 kg of oxygen was loaded in the reactant tank. However, in operation a small amount of 'ullage' is taken up with gas which forms at the top of the tank. Tiny heaters – a comparative word, the actual increase in temperature is extremely low – cause the liquid to expand, raising the pressure within the enclosed volume of the tank, driving the gas molecules together into association with the liquid. In this way a two-phase condition is restored to a single-phase supply, an acceptable condition for transferring oxygen to the fuel cells where it will mix with hydrogen to produce electricity.

Gemini V carried 10.5 kg of hydrogen at a temperature of $-253°c$, pressurized at a normal 16.8 kg/cm^2. Correct pressure in the oxygen tank, therefore, required the tiny heaters to remain on until the normal leak rate of the spherical container allowed heat from outside to enter the liquid and perform the function for which the heaters had been designed. Without a source of heat in the tank, pressure would be lost as oxygen was consumed. Experience with fuel cells on the ground, and in the General Electric test rigs, revealed that it would not be wise to depend on fuel cells where reactant pressure dropped below about 14.1 kg/cm^2 at the gas regulators. It was, therefore, with not a little concern that at the start of the second revolution, flight controllers at Houston noted the fall in oxygen tank pressure from 57 kg/cm^2 to a mere 31.6 kg/cm^2 across the space of about one hour.

The pressure was already well on its way down toward the minimum text-book recommendation. Nevertheless, the spacecraft was working well in other respects and the chirpy reactions of the crew helped bolster flagging confidence at Houston in the ability of the electrical system to last much longer. For the time being, normal operations could go ahead. At 2 hr 13 min into the flight, with Gemini V passing across the Indian Ocean on the second revolution, Cooper yawed the spacecraft 90° out of plane to the south and let go the Radar Evaluation Pod. It moved away from them at about 8 km/hr, and the rendezvous radar successfully found the transponder. Fifteen minutes later over Carnarvon there were other things to worry about. Telemetry showed the oxygen tank pressure was down to 23.2 kg/cm^2 and falling rapidly. Even the jaunty spirits of Pete Conrad were tested, and Gordon Cooper's heart sank as more and more it looked as though the flight would have to be called back.

Gemini V's eight-day mission had been important to Cooper, the first long duration trial run for flights of Moon-mission length. He had even tried coaxing NASA management from their somber officialdom in requesting acceptance of a mission badge showing a Conestoga wagon like those used to move west a century before, with the motto 'eight days or bust!' written underneath. NASA had frowned on this unwarranted levity, and Cooper never did get to stitch his patch to the flight suit. Now it looked as though there had been good cause to reject the proposed caption. Coming across toward acquisition with the Canton and Hawaii tracking stations, the situation deteriorated. At the mid-Pacific point oxygen pressure was down to 9 kg/cm^2, well below the recommended minimum.

Flight controllers at Houston were not worried for the men's safety: the re-entry module batteries would provide electrical power should the fuel cells fail completely – but only for a short while until they could perform an emergency re-entry. It looked bad for the mission, however, and while Chris Kraft was pulling together a consumables analysis team to work out just how long the spacecraft could be made to fly on the four batteries alone he told the crew to power down the spacecraft to just 13 amperes. If possible, he wanted to reach the contingency recovery zone assigned to sixth-orbit re-entry paths and one way of using whatever life remained in the cells was to reduce the load and get more orbit time from their faltering performance. All hope of tracking the 'Little Rascal' – Pete Conrad's name for the ejected Radar Evaluation Pod – had gone. There was no power for the large electrical demands made by Gemini's rendezvous radar. It was now simply a matter of shutting down to a minimum posture and waiting for the orbits to drift by until a good recovery zone hove in sight.

It would have been possible for Gemini to come home on the third revolution, just beginning as the power-down decision came into effect. But the situation was not so critical that it required a crash reaction and revolutions four and five would take the descending path of the spacecraft's orbit across the United States, so sixth-orbit recovery was the best bet at that time; fuel cell power had not completely gone yet and Kraft was assured that the re-entry module batteries, although designed only to provide electrical power for the descent after ejecting the equipment section, would provide energy for a good 13 hours. Good enough not to panic.

On board Gemini V, the mood was one of supreme disappointment. Cooper reflected the annoyance tinged with a precipitous feeling of uncertainty that pervaded the crew compartment by reaching to the locker separating the two ejection seats and scrawling upon its surface a little drawing of a Conestoga wagon – tottering on the brink of a cliff! Would it make it? Or would it not? That question was still very much in the balance as pressure continued to spiral downwards. Passing around the globe, now on their third revolution, tracking stations kept watch on the dying oxygen tank. Down went the pressure with little sign of halting, until it reached a low of 4.2 kg/cm^2 – a far lower pressure than the bale-out level of 14.1 kg/cm^2! That was it. Gemini V would have to come home on the sixth orbit. Cooper and Conrad started packing away the myriad things they had unpacked within the first hour of launch, expecting to remain where they were for more than a week. They had removed their helmets and gloves, donned a lightweight microphone and headset, and rolled cuffs and neck dams out to keep the suit oxygen flow going. Then passing across Australia on that third revolution the situation suddenly looked very different. Pressure had stabilized.

By the time the spacecraft came up on the Hawaii tracking station, 4 hr 22 min into the mission, oxygen tank pressure was steady at 5 kg/cm^2. Instead of working toward an assumed recovery on the sixth pass, revolution four was taken up with some very careful tests and checks of the complete electrical system. A test rig at St. Louis, set up a couple of hours before to simulate the problem on board Gemini V and to keep track of what could be

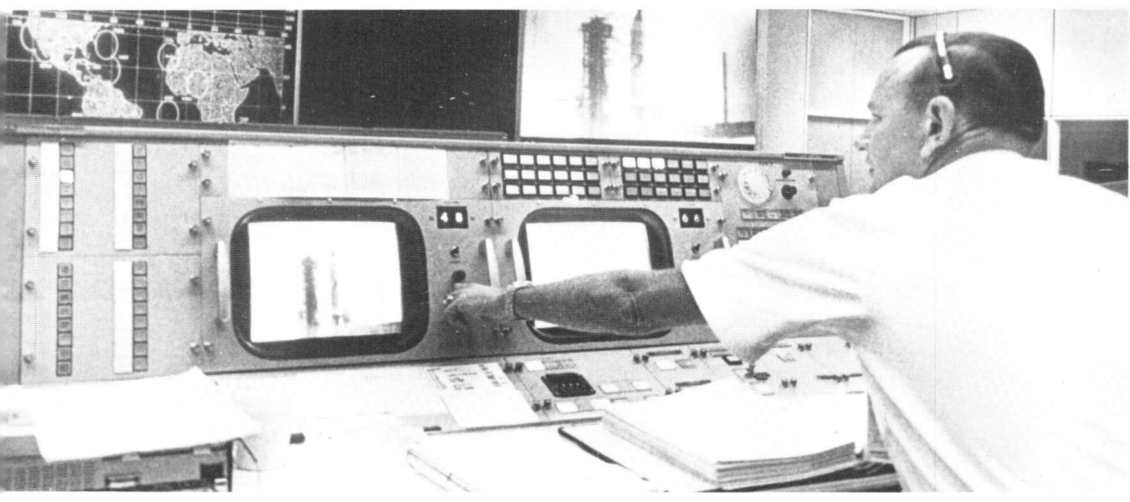

Flight director Chris Kraft waits expectantly for the launch of Gemini V.

US Navy destroyer *Dupont* retrieves the hull of Gemini's Titan II first stage from the sea off Bermuda shortly after the launch of spacecraft 5.

happening in space, seemed to show that the gas regulator upstream of the fuel cell inlet was working, just holding its own, at the extremely unusual pressure displayed by spacecraft 5. Revolution 4 provided the hope that perhaps after all the flight could be saved. The fuel cells seemed to be working correctly; the gas regulator was correctly feeding oxygen through to the sections at the required 1.56 kg/cm^2, despite the fact that on the other side of that valve, pressure was 5 kg/cm^2 instead of the normal 57.3 kg/cm^2.

But it was holding its own and the cells themselves seemed oblivious to the very abnormal condition of the plumbing on the supply side of the regulator. On through revolution five the spacecraft continued to shape up well in every other regard, the stable condition of the reactant supply a source of relief. As every tracking station routed telemetry through to Houston, controllers sat nervously watching the displays, almost unable to believe their good luck. Pressure was holding. At the end of the fifth revolution Christ Kraft passed up the unanimous decision that spacecraft 5 should continue for the next few orbits, giving the mission a 'go' for at least one full day.

With 13 hours electrical energy in the re-entry batteries, with pressure holding now, even rising a little, in the troublesome oxygen supply, and with electrical production in the fuel cells going as planned, it was a safe decision. But one, nevertheless, made only because the global tracking network could so efficiently monitor the status of the spacecraft. It had been a close haul. Tests, and post-flight inspection of records and telemetry, would show that Gemini V had reached, at 5 kg/cm^2, the saturation point for the liquid line, with the contents of the oxygen tank separated into a two-phase condition. Fortunately, most of the fluid taken from the tank was low-energy liquid rather than high-energy vapor which would have caused the cell sections to fail. Nothing could have been closer to writing *finis* on the mission destined not only to prove survivability for humans, but for the whole fuel cell concept as well; had fuel cell problems caused an early return of Gemini V it could have postponed, or delayed at best, the planned two-week flight of Gemini VII, injecting lost confidence in the fuel cells being built for Apollo. As it was, Gemini fuel cells worked for spacecraft 5 and saved not only the mission but future plans as well.

When Gene Kranz replaced Kraft at the Flight Director console, the spacecraft's electrical health looked better by the minute. Ever so slowly, pressure climbing back up, the result of heat leaking in from outside and doing on a small scale the job of the faulty heaters, provided confidence in powering up the spacecraft on a limited basis. And to save the lost rendezvous opportunity with the REP, astronaut 'Buzz' Aldrin was working a procedure with MSC's Mission Planning and Analysis Division to have the crew perform a re-written series of rendezvous maneuvers, should power climb back to the level approaching normality. John Hodge was in the MOCR room when Kranz told him and Kraft about plans to increase the power level. All three agreed, and that was good news for the crew. In a powered down condition, temperatures had fallen very low in the spacecraft, and at one point breath froze on the window. Outside, Little Rascal kept coming into view, albeit on a slow drift getting further away from the spacecraft.

The transformation of judgement about the fate of the mission both on the ground and in the spacecraft was nothing short of remarkable. Everybody recognized they had been given a lesson in systems management and within a few hours confidence in the future of the mission was rising rapidly. On the second day, Chris Kraft, a man of chosen words and considered comment, said that as of that time he did not 'see anything that would stop us from going eight days . . .' In a reaction to the widespread reporting of problems with Gemini V's fuel cell system, John Hodge strongly rejected the Soviet accusation that the flight had been prepared in haste merely to score points in the space race.

Another problem emerged at the end of the 27th revolution about 43 hours into the mission. This time it was the hydrogen supply to the fuel cells automatically venting overboard excess gas from the spherical container. The tank's heat leak rate was a little higher than expected and as temperatures rose in the hydrogen supply the pressure increased to the point where, at 24.6 kg/cm^2, the vent valve opened, expelling the excess to space. This boiling off process caused the spacecraft to roll slightly, compromising the spacecraft's attitude control.

Working with Aldrin's rendezvous simulation, flight controllers were sufficiently confident about the condition of the spacecraft to plan for the crew to rescue at least an element of the originally scheduled operation with Little Rascal. Buzz Aldrin had studied rendezvous operations for some time, having completed a thesis on guidance in manned orbital rendezvous for his doctorate from the Massachusetts Institute of Technology. He was a valued addition to the Gemini program in that as an operational pilot he could provide meaningful rendezvous plans from a position of knowledge about crew capabilities. Aldrin's plan was that the Flight Director should feed the spacecraft a set of hypothetical coordinates related to a phantom target. The Gemini V crew could then proceed to rendezvous with this point in space, the computer believing it to be an actual vehicle. For the purposes of the simulation it would suffice to practice maneuvers planned for real with the next flight along.

Chris Kraft was wholeheartedly behind the plan, and its announcement to the crew aboard the spacecraft lifted spirits to a higher level than they had been for two days. At last there was real work, and workaround procedures to act with in salvaging as much as possible for the threatened flight. Kraft told the spacecraft's computer that a target was in an orbit 227 km at perigee and 338 km at apogee, providing the position so that the maneuvers necessary to rendezvous with this spot could be worked out. In fact the plan was a compromise between the ideal of a full rendezvous simulation and the need to conserve both electrical power and OAMS propellant. So Gemini V would arrive at a point where, on an operational rendezvous mission, the spacecraft would begin the terminal events to final orbit matching.

The exercise began on revolution 32 and for two orbits of the Earth the spacecraft made four maneuvers and arrived at the assigned position in space to within 320 meters in perigee and 965 meters in apogee of the desired location. The computer, the computations, and the men all checked out. The radar had received a work out on the second day; when fuel cell problems looked like cancelling a good many tests, engineers set up an 'L'-band transponder on a tower at the Cape and read the performance of the spacecraft's radar as it swept overhead and locked on. It too behaved like a lady.

On the day following the rendezvous test, the fourth day of the mission, Cooper and Conrad observed huge checkerboard marks laid out on the ground as part of their visual acuity test, and watched a Minuteman missile climb a smoke-filled staircase to space from the Vandenberg Air Force Base in California. On the next day, 25 August, they again watched a Minuteman launched from Vandenberg and everybody on the ground heard an ebullient Pete Conrad acknowledge the fact: 'There it goes. I see it. I see it.' Shortly thereafter, they watched the smoke of a rocket sled at Holloman Air Force Base and sought the eye of tropical storm Doreen south of Hawaii. Then , sweeping across the United States at the end of the 75th revolution during the breakfast hours at Kennedy Space Center on 26 August, Gemini V coasted into a new record as they exceeded the 4 days 23 hrs of Vostok V.

It was more than just the passing of a single record. From this point on the manned space flight programs of the United States would pull further away from Soviet achievements until, more than a decade hence, Russia would develop new initiatives so very different from American interests. It was the beginning of a new age of technical expertise in a medium that until this point in time had been a purely experimental preserve. But all was not perfect aboard the record-breaking spacecraft. Powered down again to conserve electrical power, the crew were cold and showing signs of mild irritation, not unexpectedly after their testing preoccupation with fuel cell problems.

They criticized ground procedures personnel for 'poor planning' in setting out exercise periods, chaffed at the almost freezing conditions in the cabin, and now had to contend with two malfunctioning OAMS thrusters. Cooper was snappy when controllers on the ground inquired into his physical condition, and reported a generally sluggish feel to the rest of the OAMS thruster assemblies: 'We saw great blobs of liquid coming out of them drifting by us when we were firing them in pulse mode.' But other records were being broken too, and thoughts of a bet Cooper had with his wife brought a modicum of light relief when he asked the Houston capcom to tell her she owed him a dollar – she too had thought he would never make five days, but for a totally different reason!

As the days rolled by the condition of spacecraft 5 got progressively less like an efficient spaceship and more reminiscent of a rundown restaurant in a Chinatown backstreet. It became the wardroom, the waste room, the bedroom and the kitchen for two men loaded down with cameras, lenses, film magazines, crowded lockers, crammed containers, and parts of space suits. Neither Cooper nor Conrad were particularly tidy, and the continual process of packing, then unpacking, large quantities of specialized equipment grew tedious and nerve racking. Sleep became a major problem early in the flight – first because of problems with the fuel cells, and then because of the same reasons first encountered by McDivitt and White. Finally, Cooper and Conrad tried going through sleep, eat and work cycles together, but that was not a great success. Nevetheless, Pete Conrad provided the usual high points by entertaining flight controllers on the ground to the most appalling prose ever to drift from space to Earth. On the sixth day, a tortuous rendition far removed from anything akin to the poetic verse it purported to be:

'We were drifting along, by the CSQ,
When the radio said, "Here's word for you,
Your controls are dead, but you're not through,"
So here we are for three days more, with the end quite far.'

Signs of agony were visible throughout the MOCR! Then, from the ground came a barrage of dixieland jazz, excused by George Mueller as 'an extended test of the high frequency transmission system!' Yet for all their spirit and determination to see the mission through, conditions aboard the tiny spacecraft were not pleasant. On the seventh day, Friday, 27 August, the fuel cells were observed to be producing an unexpectedly high proportion of water. The reserve tank used to accumulate this waste was also the drinking water tank, with the polluted contents separated by a bladder. There was no way of dumping any of the water, so flight controllers recommended switching one fuel cell section off so that the water production rate would drop. Calculations showed that the spacecraft could get through the eight days without the tank reaching capacity. That only perpetuated the chilling, clammy atmosphere inside the spacecraft, by now a cabin filled with a complexity of odours. The heady feeling brought on by breathing pure oxygen eased the problem in one respect: it was not so easy to differentiate the various smells.

Cooper gave Chuck Berry a cynical view of life aboard Gemini V when the physician asked how they were exercising: 'I hold Pete's hand once in a while. I use a cleansing towel. Then a couple of days we chewed gum.' Toward the end of that day, McDivitt got a contemptuous view of life on a long duration flight when he asked Gordon Cooper if they could sell the experience as a ride at a carnival: 'I don't think you could sell this day-to-day drifting flight as a ride anywhere.' The next day, Saturday, was uneventful, except for a further burst from Pete Conrad:

'Over the ocean, over the blue;
From Gemini V, here's thanks to you.'

Conrad's wife watched Gemini V in the morning dawn from her home in downtown Houston, and composed a message for capcom to read up to the spacecraft:

'Twinkle, twinkle, Gemini V,
How I want you back alive,
Up above the world so high,
I saw you today as you went by.
Twinkle, twinkle, Gemini V,
Tomorrow you take a great big dive,
Zinging toward the ocean blue,
And I send my love to you.'

Recovery day, 29 August, the two astronauts performed an interesting communication test with Malcolm Scott Carpenter, the second man to fly a Mercury spacecraft in orbit, while he rested on the sea floor in Sealab II, a US Navy experiment 62.5 meters below the surface off La Jolla, California. Carpenter and four other aquanauts were to remain in their 3.6 meters by 17.7 meters accommodation of 45 days and the space to seabed hook-up was made out of respect for the unique pioneering role of a man still technically on active status with NASA.

During the next few hours weather men grew anxious about the approach of Hurricane Betsy from the coast of South America. Gemini V was told of the situation and all agreed that the flight should terminate one revolution early to avoid the crew landing in storm filled seas. Again, all the old questions were raised about the capacity of the crew to suddenly experience the force of deceleration after what was a world record-breaking flight outlasting anything yet put up by the Russians. Gemini V had not given its crew an easy time, and they had not exercised with their bungee cord as much as Chuck Berry would have liked. But there was no way round it, and it was, after all, the primary purpose of the eight day flight to find out if man could withstand a Moon-length mission.

Three Gemini flight controllers light up in the traditional splashdown congratulations following a successful manned flight. Left to right, Christopher Kraft, John Hodge, and Eugene Kranz.

Just before punching off the adapter section equipped with the fuel cells and their reactants, ground telemetry displayed 0.68 kg of hydrogen remaining, with 33.1 kg of oxygen unused. Oxygen tank pressures had been rising for seven days, reaching 18.3 kg/cm^2 at the end of the mission. Close by the island of Hawaii, seven days, twenty-two hours after launch, the four retro-rockets fired and Gemini V started home. However, hopes for a landing close by the assigned location were again dashed, as they had been on the first two manned Gemini flights. The reason was infuriating.

Gemini V was to be the first to fly a constant bank-angle re-entry whereby guidance logic in the computer compares the lift vector of the spacecraft with the navigation updates performed prior to retro-fire. There had been no time to prepare new computer programs accommodating the revised lift/drag value of 0.19 determined from careful study of re-entry profiles for the first few flights, so spacecraft 5 carried erroneous data no longer valid, values that could be interpreted by the crew in the light of preflight training that enabled them to incorporate the correction. However, when the guidance update was sent to the spacecraft, human error provided a range angle out by 7.9°, causing a navigation error in the computer equal to 878 km. Throughout the re-entry the computer presented incorrect information to the crew and by the time they realized the error the spacecraft did not have the maneuver capability to do more than get to within 168 km of the planned splashdown coordinates.

Spacecraft 5 splashed down at 7:56 am and within 43 minutes swimmers from a helicopter launched from the prime recovery carrier, the USS *Lake Champlain*, were in the water attaching a flotation collar. About one and one-half hours after landing, Cooper and Conrad stepped from a helicopter and walked on the carrier's deck. They had proved the point, and were glad to be back, a jesting, jocular duo, proud that spacecraft 5 had seen them through. It had been a long haul, but one worth pursuing.

Gemini V was a turning point in several ways. The three two-man missions culminating in the splashdown of Cooper and Conrad increased by more than tenfold the man-hours spent in space by American astronauts. On its own, Gemini V lasted longer than all America's previous manned space flights, and proved men could survive the weightlessness of orbital flight for a period at least as long as that required to get to the Moon and return. But it also carried the accumulated man-hour record in space beyond that accumulated by the Soviet Union for the first time. Not since the beginning of space activity in 1957 had America been so clearly in the lead, and not since Yuri Gagarin could the American public gain pride from having so distinctly excelled in the endeavor where once they lagged so far behind. Up to and including the space-walking flight of Voskhod II, Soviet cosmonauts accumulated 507 man-hours on eight missions. When Gemini V splashed down after 120 revolutions of the Earth, NASA astronauts had tagged up more than 641 man-hours on nine space missions. From August, 1965, the Soviet Union was to fall increasingly behind records claimed by the United States for manned space flights until, more than a decade later, they once again excelled in the science of manned orbital flight.

It was a significant mission, this eight-day flight around the world, and President Johnson captured the mood of the public when he hailed the venture as a 'moment of great achievement, not only for astronauts Gordon Cooper and Charles Conrad, but for those whose hopes have ridden with Gemini V. I am so happy that Mr. Webb and Mr. Seamans, who had so much to do directing this very successful venture, are here to share with us the pride we all feel today. And I deeply regret that our late, beloved President Kennedy, under whose leadership all of this work was so carefully planned and thought out, can't be here to enjoy the fruits and success of his planning and his forethought. . . . Only seven years ago we were neither first nor second in space – we were not in space at all. Today the capacity of this country for leadership in this realm is no longer in valid question or dispute. . . . This flight of Gemini V was a journey of peace by men of peace. Its successful conclusion is a noble moment for mankind – and a fitting opportunity for us to renew our pledge to continue our search for a world in which peace reigns and justice prevails. To demonstrate the earnestness of that pledge, and to express our commitment to the peaceful uses of space exploration, I intend to ask as many of our astronauts as possible – when their schedule and program permit – to visit various capitals of the world. Some, I hope, will be able to journey abroad soon.'

Johnson clearly recognized the momentum of record-breaking achievements gripping the entire space program. The public were getting used to the tri-monthly spectaculars, without much thought for the crushing pressure it brought to families involved in the program, and now NASA manned space flight activity had exceeded Russian initiatives it was the right time to maximize the asset. The President's idea of turning astronauts into ambassadors was not altogether welcomed by the Manned Spacecraft Center as a whole, and certainly not by Deke Slayton in particular. The Gemini program was only one-third over, in terms of manned flights at least, and there was the upcoming Apollo program with its own special requirements and flight assignments. To have members of the astronaut corps flying all around the world at such a time was not conducive to a peaceful aura round Slayton's office that summer of 1965. As it turned out, the Cooper-Conrad mission was brief and limited. But their responsibilities were almost beyond endurance.

Behind the casual flight by flight routine of launch and recovery, most public on-lookers were unaware of the price even then being paid for the pace of the schedule. So much was being asked of the man who flew the spacecraft that Gemini V marked the beginning of a much tougher regime. Demanded without compromise by Deke Slayton, the limits on personal contact were far beyond a level any man had been asked to endure before. The flight of Gemini V is a good example of how strenuous were the efforts to keep the men at peak performance. Because the atmosphere on board the spacecraft was tuned to an operational situation, and because the necessity to wring every last drop of value out of each minute in orbit was uppermost, Deke Slayton forbad the astronauts' wives from visiting the Manned Spacecraft Center, or from talking direct to the crew in space.

It had been more casual in the past, but now the heat was on and Deke for one was not going to let up one jot on the demanding requirements for perfect conditioning. His men were honed, spring-loaded, to act like super-men in the face of any adverse or critical situation they met. But super-men or not, in reality it was the beginning of a sorry trail of wrecked marriages, wrecked minds, and total disillusionment.

After splashdown, the Gemini V crew were hurried into concerted medical examinations, technical de-briefings, and scientific reviews lasting eleven hectic days: four at sea on the carrier steaming to the Cape, and at the Kennedy Space Center itself, and seven days at the Manned Spacecraft Center. When the crew touched down at Ellington Air Force Base from the Cape, their wives were allowed to be on hand for a reception, then the astronauts were confined by the non-stop pace of post-flight activity.

On 9 September they were allowed to hold a press conference to tell the world about their adventure, and five days later they were flown to Washington for the now familiar medal awards at the White House: Cooper, Conrad, and Chuck Berry received NASA Exceptional Service Medals from the hand of the President. Then, on to the Capitol where Vice President Hubert Humphrey presented them to the Senate and then to the House of Representatives. Next day, they left for a six-nation tour, taking in Greece, Turkey, Ethiopia, Madagascar, Kenya, Nigeria, and the Canary Islands. Just five days after leaving by aircraft from Washington, Pete Conrad was assigned to the back-up position on Gemini VIII. And so the cycle would begin again, only just as intensively as for the prime crew, announced 20 September as Neil Armstrong and David Scott; Richard Gordon would back-up the pilot while Conrad backed Armstrong. Just five days after Gemini VIII flew, in March, 1966, Conrad would begin training as the prime command-pilot for Gemini XI.

His is but one example of the almost continuous pressure placed upon astronauts during this period, a time when the public were openly responsive to the glories and the spectacle, events that from the inside of a Gemini spacecraft never did look quite so glamorous. But if President Johnson was concerned about the propaganda value of space successes, and the benefit to US prestige foreign goodwill tours could bring, he was similarly reluctant to pass up the chance to push for a more permanent role in space. Convinced that the Air Force should be given every opportunity to exploit the new medium, he announced mid-way through the flight of Gemini V that he was giving the official go-ahead to the Manned Orbiting Laboratory (MOL) program under analysis for some time. In this, redundant Gemini capsules would be used to fly Air Force crews to a cylindrical laboratory launched by Titan rocket from the Kennedy Space Center. The program was to cost $1,500 million, about half as much again as the entire Gemini program, enabling airmen to spend up to 30 days in orbit performing reconnaissance experiments and research into possible defense roles for man in space. It was to infuriate the Russians, and to open a new hostility between the countries concerned.

For some time, Soviet news agencies had been denigrating the US space program, and now official mention had been made of a military space station planned for later that decade it was open house on every form of scurrilous comment imaginable, however irrelevant it was in reality. In fact before the President announced a formal start to MOL plans, the Russian military newspaper Red Star, accused the United States of conducting clandestine military operations with Gemini V: 'Pentagon bosses do not conceal the fact that manned spaceships may provide a clue to strategic domination in outer space. American propaganda praises the scientific aims of the Gemini program and tries to conceal its military character. But these attempts are futile. The real purpose of the program is obvious.'

The paper went on to say that the orbital path of Gemini V had been calculated to carry it over China 40 times, across North Vietnam 16 times, and across Cuba 11 times, with special cameras designed to spy on activities down below. In fact, what the paper had done was to count up the orbital passes made across pro-communist states within the limited constraints of Gemini's 32.6° orbital inclination. It failed to mention that this orbit would not once carry it across the Soviet Union. But that did not really matter to the argument in question; it was a case of disappointment tinged with anger over the apparent ease with which NASA successes were accumulating popular support not only within the United States but around the world too.

And it certainly did seem that nothing could halt the break-neck progression from experimental flights to operational missions as one by one the objectives of the Gemini program were being fulfilled. Although it had not gone the full 14 days promised for Gemini VII, the Cooper-Conrad mission had indeed given medical personnel a sample of long duration flight. And although Gemini IV provided only minimal opportunity for Ed White to leave his spacecraft and perform an EVA, it had shown that space activity outside a pressurized cabin is indeed possible. It was now up to Gemini VI to show that manned vehicles could rendezvous and dock with target vehicles placed in orbit at an earlier time, a step essential to the manned Moon landings to come. But the intent was to prove more promising than the deed, and for a while at least the string of successes would be more difficult to sustain than anyone could have thought in the first half of 1965. It was going to be up-hill from now on – all the way.

'. . . death or the ejection seat'

The futile attempt made by Jim McDivitt and Ed White to pull alongside the spent second stage of Gemini IV's booster proved how difficult that operation was going to be. Simple in principle, the mechanics of orbital rendezvous were far from convenient in reality. What McDivitt sensed in trying to approach the rocket stage was an all too real demonstration of the mirror-like qualities of space flight. For in the ever-circling world of orbital flight, to approach a vehicle from behind requires a thrust in the opposite direction, and to slow down means the spacecraft must speed up. These weird characteristics, far removed from the eyeball flying so many astronauts were used to, were to generate a new method of flying – one ruled by computer language, maneuver charts, and navigation marks. For this was real *space flight*, not a projected course set by the final stage of a launch vehicle, and unlike aeronautics the 'logic' of the human brain could be the misleading influence between safety or disaster. It was vitally important to demonstrate that orbital rendezvous not only worked, and could be performed by any astronaut crew, but that it could be conducted with sure confidence of success. For the price to be paid for failure around the Moon would condemn Apollo Moon landing crews to eternal desolation.

It was the one single point around which the entire lunar mission would pivot, and after so much debate on the method to choose, NASA had selected the one which seemed, *prima facie*, to be the most perilous: complex orbital rendezvous maneuvers around another world in space, with only one half that orbital path even visible from Earth. Rendezvous had to be shown as a feasible proposition, for it was the key not only to Apollo but future manned space operations anywhere. To demonstrate the routine nature of what was considered a difficult feat to achieve in theory, early Mercury Mk. II plans quickly settled on the Lockheed Agena as the only possible candidate for a target vehicle. In January, 1962, NASA began discussion with the Air Force Space Systems Division regarding potential acquisition of Agena vehicles, and was briefed by Lockheed Missile and Space Company executives on the applicability of the vehicle to space agency needs.

Agena was a proven upper stage design, far more reliable than any other upper stage then in existence, and was matched to the Atlas launch vehicle as a booster. Agena, developed as a stage designed to carry payloads into orbit, thereby improving the lifting capacity of the rocket used to send it on its way, was itself to be the payload for Gemini missions. Placed in orbit by Atlas, the modified Agena would carry a docking cone on the opposite end to its single liquid propellant rocket motor in to which the nose of the manned Gemini spacecraft could be inserted. But the task of adapting a standard rocket stage to such a demanding task was not as easy as at first thought: Agena would need a radar system, tracking aids, a restartable propulsion system, an improved attitude stabilization system, and a reliable docking cone on the front end.

In June, 1962, it was agreed that NASA would use the Agena D model, a more powerful version of the 'B' series. Modifications steered by the Gemini program were to labor the program through to the first operational flights. Because the Gemini-Agena-Target-Vehicle, or GATV, would need to have a capability for firing its main engine up to five times, correcting errors and changing its own orbit on command, the existing model on which the engine was based had to be modified. The old design used solid propellant starter cartridges to provide a flow of gas with which to spin the turbine prior to ignition, the number of times the engine could be started limited by the number of cartridges the propulsion unit could safely accommodate. In the end it was decided to dispense with the solid starters and employ liquids contained in a tank which, when fed to a gas generator, would serve the same purpose but on a repeatable basis.

The engine unit itself was contracted to Bell Aerosystems, and based on their successful Model 8247 engine, a 7.25 tonne thrust motor burning unsymmetrical dimethyl hydrazine fuel and inhibited red fuming nitric acid as oxidizer. The new motor would possess a Primary Propulsion System (PPS) comprising the same main engine as the old 8247 and a Secondary Propulsion (SPS) placed as two packages, one each side of the big engine, like saddle-tanks. Each Model 8250 package was to be a self-contained propulsion unit with two thrusters: one 90.1 kg unit and one with 7.25 kg thrust. Fired in unison, the two larger thrusters would be used for up to 20 minor orbital corrections while the latter would be used for fine trim maneuvers or for settling the Agena in the correct attitude for main engine ignition.

Throughout 1963, Bell worked long and hard on the PPS engine, testing the system at the Arnold Engineering Development Center and eventually re-designing components in the gas generator to give the engine a reliable start sequence. Financial troubles that rocked the entire Gemini program took their toll of Agena development too, and in the fall of 1963 the whole issue of rendezvous with Agena vehicles entered active debate. By the end of the year technical problems were piling up, on both PPS and SPS hardware, and Charles Mathews made concerted efforts to pull the program together, bringing in Aerospace Corporation as a mediator for technical directions and general systems analyses. On through the following year, the GATV program was tossed in the stormy seas of flagging development, cost problems, and the juggled threat of outright cancellation along with paraglider. When major systems testing resumed in 1964, after hoped for solutions to development problems, defective start tanks produced more trouble and further delay. Trouble too with the SPS engines led to flagging spirits and doubt about the ability of the program to match the requirements.

In April 1964, the Air Force suggested flying a model of the GATV to demonstrate an operational capability, but NASA baulked at the extra money this would cost and suggested instead allocation of a target vehicle to development test and trouble-shooting activity. Consequently, AD-71, the first fully configured GATV delivered, was modified as the GATV-5001, a fully fledged test vehicle; GATV-5002 would fly the first rendezvous attempt. By using the first model as a ground test vehicle, NASA saved money by cancelling one Atlas booster, otherwise needed for launching it into orbit. But the first model was to be put through a full preparation and pre-flight grooming session to de-bug the systems and to speed experience with launch scheduling. It was late 1964 before the vehicle emerged from preliminary testing, and January, 1965, before McDonnell's docking cone had been installed for a full systems firing of both propulsion modules. It was a good test, although minor discrepancies put the schedule a month behind.

Meanwhile, Atlas SLV-5301, one of the new standard space boosters, arrived at the Cape from General Dynamics at San Diego and went into storage during February until the first flight-rated Agena target was ready. But the date at which that event could take place seemed further away still when, on 27 May, the ground test Agena was found unacceptable. But GATV-5002, by now at the Cape, was in a much better condition and that target was assigned the Gemini VI mission, planned for October, when inspected during the month of August.

For many months the prime crew, Schirra and Stafford, had been pushing for a chance to fire the PPS engine in the docked configuration, using Agena's main engine to push the assembled configuration to high orbit. It had been an issue debated for more than two years: how far to extend the technical capacity of the Gemini hardware? Conceivably, the spacecraft itself could have been made to fly around the Moon, but that had been thought too foolhardy and, in any event, that was Apollo's purpose during the run-up to Moon landings. Now there was the question of firing up the Agena's 7 tonne engine, but was that really wise considering that the propulsion system had not been designed for manned operations? Nobody was prepared to give a final decision at this early stage, certainly not without seeing how the target behaved in space, so Gemini VI would be limited to rendezvous, docking and a test of the docked configuration for attitude control, etc. In fact the mission was to be quite brief.

Schirra was reluctant to include EVA on a flight where so much crew time would be taken up with the complicated rendezvous maneuvers, cluttering the cabin with charts and manuals, so the mission was to last two days, completing 29 revolutions. Spacecraft 6 was the last of the battery-powered ships, carrying, like spacecraft 3, three 400 ampere-hr silver-zinc batteries in addition to the four smaller batteries for the re-entry module and the three squib batteries for firing the pyrotechnics. Gemini VI would absorb about 1,100 ampere-hrs during orbital flight. Later spacecraft would all carry fuel cells for electrical energy production.

The sequence of events necessary to prepare a Gemini-Agena for launch were carefully laid down during preparation for this first rendezvous operation. Launch Pad Pre-mating Tests comprised inspection and leak checks before filling the propellant tanks with fluids on the launch pad ramp. A Weight Determination and Vehicle Mating operation followed to determine the GATV's mass before lifting it to the Atlas launcher. Blockhouse Compatibility Checks were next and here all the electrical and instrumentation lines were monitored to establish the electrical health of the assembled configuration. Then the F-1 Day Electrical Functional Checks, where some of the preceding test sequences were re-run, but where the complete vehicle was brought to a condition for plugging in the Aerospace Ground Equipment (AGE) and the ground telemetry lines.

Next came J-FACT, the Joint Flight Acceptance Composite Test, an integrated check of all contractors, the range personnel, and the vehicle and support equipment systems in a simulated countdown and flight, concentrating on electrical rather than mechanical aspects. Concurrent with J-FACT, other engineers tested the L-Band radar transponder on the Agena and the docking cone support equipment. Then, a Radio Frequency Interference Test was to be conducted with the vehicle in a launch condition to ensure that no adverse combinations would occur among electromagnetic radiations at the Cape. Next was a Launch Simulation Test, or full-dress rehearsal of the countdown stopping just short of Atlas engine ignition, concluding with a command to extend the Agena boom antenna to test L-band frequency links with the Gemini spacecraft on pad 19. The spacecraft, meanwhile, would be cycling through its own complex and demanding preparation sequence, and the crew would be running final tests and simulations in the trainers and in the actual spacecraft atop its Titan II.

Like the human cargo in Gemini itself, everybody breathed a sigh of relief when launch day came and the flight was on; to the public suddenly exposed to launch day operations it all looked very cool and calm, but to everybody involved it was time to sit back and cross fingers – apart from the flight control teams at Houston, who were winding up for the non-stop tensions of orbital operations. Now that flight activity and mission control operations switched to Houston just minutes after launch, there was a vacuum that usually swept across the Cape facilities, expended only when the beach parties got under way that evening and personnel who for several weeks had worked night and day to groom the big birds for space could break out pent emotions and soak in a relaxed and carefree mood. Until the next one. So it was in October 1965, when Wally Schirra and Tom Stafford got ready for their planned rendezvous in space with the first operational Gemini Agena Target Vehicle.

Gemini VI crewmembers Walter Schirra (foreground) and Tom Stafford face the press.

The development of rendezvous operations had languished awhile during the early part of that year, but Schirra and Stafford threw themselves energetically into developing flight techniques almost guaranteed to bring success. Several methods had been evaluated in the preceding two years until three concepts stood out from the rest: tangential, coelliptic, and first-apogee. All three were based on the use of different orbital altitudes to make the chase vehicle move at a different speed to the target, and on the principle of the Hohmann transfer ellipse. From the former it was recognized that a chase vehicle in a lower orbit would gain on the target and eventually catch it up. Adopting the latter, maneuvers made at perigee would affect apogee, 180° around on the other side of the orbital path, and thruster firings at apogee would similary affect perigee. In this way, orbits could be made to do many things essential to rendezvous.

The tangential approach assumed the target to be in a circular orbit and required the chase vehicle to be placed in an elliptical path behind the target with apogee at the same altitude as the target. This would cause the approaching spacecraft to move closer and closer to the target, dropping into a low ellipse once on each orbit and so speeding up, until the two were together. Of course, this required the chase vehicle to rise to its final apogee at a point in space and time where the target was to be found. To achieve this a number of course corrections would have to be made during the final climb from perigee to apogee and rendezvous with the target, maneuvers particularly critical to the success sought.

The coelliptic method was similar to the tangential in that it assumed the chase vehicle to be first placed in an elliptical path below the circular orbit of the target vehicle. But here, apogee would be below the target's orbit and provide an opportunity for a small number of trim burns to be performed before circularizing the chase vehicle's orbit at a constant height below the target. On an inner orbit, the chase vehicle would gain on the target until, at the correct moment, it fired thrusters to raise apogee. During the final intercept climb to the target, mid-course maneuvers would be made to bring the chase vehicle up from the lower orbit below and in front of the target. In the final phase of rendezvous, the spacecraft would have to fire its thrusters as if increasing speed in the forward direction, raising the orbit's perigee and slowing it down relative to the target.

The third alternative, or first-apogee method, would place the spacecraft in an orbit where it would intercept the target at the high point of its elliptical path. Of course in this method the required mid-course maneuvers would be conducted rapidly and with little tolerance for inaccuracy but in one respect it was similar to the tangential method. It was not considered good for the first rendezvous operation because it could easily get the spacecraft into a condition where large quantities of propellant would be required to achieve rendezvous. While valid as a rapid rendezvous operation, useful to have in case a quick link-up was ever necessary, it was best to go by a simpler and more forgiving route than first-apogee promised to provide. The only logical method was the coelliptic routine and this was selected for pre-flight training.

From June, 1964, flight controllers worked to perfect the geometry of each maneuver, with Dean F. Grimm from MSC's Flight Crew Support Division working with astronaut Buzz Aldrin to get the concept refined and tidied up in time for Gemini VI. It was with flights like this that solid engineering training was useful and where a mathematically agile mind was essential. The final flight plan was to go something like this. Agena would get into a circular orbit 298 km above Earth, inclined 28.87° to the planet's equator, and move in such a path that after one full revolution it would pass directly across the Cape. Exactly 1 hr 40 min 52 sec after the Atlas-Agena launch, Gemini VI would lift off pad 19 in pursuit, getting into orbit 5½ minutes later, about 1,925 km

The instrument panel of spacecraft 7, home for two on a two-week sojourn in space.

behind the target. With the spacecraft in an orbit varying between 161 km at perigee and 270 km at apogee the spacecraft would be catching up slowly. At the end of the first orbit, Gemini would have reduced the distance separating the two vehicles to 1,185 km. At that point a very small trim burn would be performed to raise apogee about 0.9 km, essentially a burn to null the dispersions incurred during orbital insertion.

The first major rendezvous maneuver would be carried out about 2 hr 19 min into the flight, when the spacecraft reached second apogee, to fire the thrusters and increase speed by 16.3 meters/sec. This Phase Adjustment would result in the spacecraft's perigee being raised from 161 km to 215 km which, with apogee still 28 km below the circular orbit of the Agena, would reduce the catch-up rate. At third apogee, one full revolution later, about 3 hr 48 min into the flight, Gemini VI would execute the Coelliptic Maneuver designed to circularize the spacecraft's path at 270 km, placing it a constant 28 km below the Agena target. This maneuver would speed the vehicle up by 16 meters/sec and would take place with the two vehicles 259 km apart. Gemini would be trailing Agena, at a constant 28 km below the target.

The next sequence would be the most critical and required the spacecraft to pitch up 27° from local horizontal so that it would point at the Agena target precisely at the correct time to exercise Terminal Phase Initiation. This would place the Gemini spacecraft on course for an interception with the target, 63 km away. Normally, TPI would require a 9.75 meters/sec forward burn pitched up 27°. It would take Gemini VI about 30 minutes to raise its orbit 28 km and traverse the distance separating chase and target vehicles. But the precision with which the whole operation had to be conducted becomes evident when terminal phase requirements are examined.

TPI had to place the chase vehicle 915 meters from the target three minutes before sunrise, when the target vehicle ahead first became illuminated by the solar rays. This was to provide a switch in visual references from the more acceptable Agena rendezvous lights, up to a distance of 915 meters, to purely visual tracking of the target vehicle's shape in sunlight. The relative distance between the two would be difficult to judge if the Agena was in darkness close up; on the other hand, Agena would be too far away to see if terminal phase maneuvers were made in the full light of day. Hence the need to switch to the day side of the Earth at the point where visual cues would be better served by sunlight. To get close to Agena just as the two vehicles were breaking daylight, Gemini VI had to perform the Terminal Phase Initiation maneuver 130° around the Earth from the point where it would be required to be 30 minutes later. That point was one minute after sunset, the point where TPI would be performed.

Between TPI, 63 km from the Agena, and rendezvous, Gemini VI would perform two mid-course corrections, tracking the target vehicle's lights against the stars and performing inertial angular measurements to calculate the precise magnitude of the thruster burns necessary for rendezvous. The first would occur twelve minutes after TPI, based on information from the on-board computer, with the second twelve minutes after that. Range would be about $7\frac{1}{4}$ km at this point, from where the crew would commence to shift from visual observation of the Agena lights to eyeball views of the sunlit target vehicle. Final braking maneuvers would be performed by thrusting toward the Agena, further raising the Gemini orbit and so slowing the spacecraft down with respect to the target; a paradox in reality for pilots used to moving in the observed direction. If the spacecraft fired its rocket thrusters forward in the hope of slowing down, as logic would dictate, the spacecraft's orbit would dip to a lower perigee, Gemini VI would accelerate, and move below and past the target.

Constraints placed upon the time of rendezvous, by the precise number of revolutions required to perform the maneuvers and by the need to pull within 915 metres just at sunrise on Gemini's fourth orbit, severely restricted the time each day when the flight could be scheduled. The launch window for Gemini VI would be about 2¼ hours, separated into panes through which the spacecraft would have to pass for each of several alternative missions. Obviously, with the Agena target moving so fast, and with a requirement for the spacecraft to get into orbit at a precise distance behind the Agena, rendezvous on the fourth revolution would only be possible if the spacecraft lifted off within the first 100 second pane of the launch window.

The geometry of both the Earth's orbit and the lighting conditions in space would shift the rendezvous to a progressively later orbit for incremental delays beyond that point: a launch during the next 100 seconds would shift rendezvous to the fifth orbit; lift-off at the 200 second window pane would shift it to the sixth revolution; a launch beyond 200 seconds from the scheduled time would move rendezvous out to revolution 16. If the Gemini was to be delayed beyond the 25 minute point Agena maneuvers would be required to effect a rendezvous at all. Delays beyond 2¼ hours would demand a complete re-cycle for the next day, at which point the window panes would again open in sequence. Agena could last about five days on the batteries it carried. Plane changes would be incorporated in the rendezvous maneuvers during ascent to the target, the Gemini spacecraft being capable of shifting orbital inclination by as much as 0.5°. Any further dispersion between the Agena and Gemini orbits would call for the Agena to use its own propulsion system, considerably more powerful than any of the propulsion units aboard Gemini, capable of orbit changes of up to 10°.

The flight plan for Gemini VI called for the pilot to reduce the speed of the spacecraft as it approached the Agena, reducing the relative velocity to less than 1.5 meters/sec at a distance of only 600 meters, then to less than 0.3 meters/sec at docking, where the long nose of the Gemini rendezvous and recovery section would slip gently into the conical docking collar of the target vehicle. Docking could come as early as 5 hr 33 min into the flight, in daylight over the Indian Ocean. Following this, both Schirra and Stafford were to separate and re-dock the spacecraft in daylight and in darkness, before beginning a well earned 7-hr rest period. During the 12th revolution, 18 hours into the flight, the spacecraft was to pull away from Agena with a 2.1 meters/sec thrust, enter darkness over the Carnarvon tracking station five minutes later, and use the target for tracking tests with a sextant developed for Apollo.

For the remainder of the flight the spacecraft would pull ahead of the Agena by nearly 34 km per revolution until, at ignition of the four solid propellant retro-rockets 46 hr 10 min after lift-off, the two vehicles would be 570 km apart. There would be an option to re-enter early, on the 15th revolution, after one day in space, but that was not expected. There was also a plan formulated by the Manned Spacecraft Center which would power up the Agena after spacecraft 6 returned, moving it to a higher orbit, 445 km above Earth. In four months, this orbit would decay to about 270 km, where the Agena, although inert, could be used for a re-rendezvous exercise with Gemini VIII and its own Agena. Sophistication was creeping in to confidence brought on by successful flight operations.

Although brief in duration, Gemini VI was to be one of the most ambitious space missions so far, the complexity of its operation clearly distinct from the comparatively passive role played by earlier crews. Countdown preparations went well, the Atlas-Agena cycling down to zero just as assuredly as Gemini-Titan moved progressively closer to its own lift-off time, staggered nearly 101 minutes beyond the target vehicle's launch hour. At 9:45 am Schirra and Stafford slid on to their respective ejection seats in spacecraft 6 as, a short distance away on pad 14, blockhouse teams passed through T-15 minutes with the Atlas-Agena. It was to this very pad, nine years earlier to the month, that General Dynamics delivered the first Atlas launch vehicle ever to stand erect. Just four seconds past the hour of 10:00 am on Monday, 25 October, the first Atlas-Agena assigned a role in Gemini lifted cleanly into the sky at the behest of launch chief Thomas J. O'Malley.

Thunder rolled across the Florida marshes once again, the yellow flame, the billowing smoke, and all the cacophony of an Atlas launch. Just 2 min 11 sec later the booster engines slid away from the sustainer section, and at 4 min 41 sec the single main engine too shut down. At 5 min 8 sec into the mission, Agena slipped away from the adapter that held it to the top of Atlas and coasted on up while readying itself for ignition prior to injecting itself into a safe orbit. Six minutes twenty seconds off the steaming pad, the Primary Propulsion System lit up ready to thrust for more than three minutes and add speed for global flight. Suddenly, all telemetry ceased, and the tracking beacon too was lost. Gemini Project Office's Agena man in blockhouse 14, Jerome B. Hammack, sensed something was dreadfully wrong, and sought the opinion of SSD's Atlas man, Colonel L. E. Allen. Consensus was that it could have exploded.

Fourteen minutes into the flight, Canary reported no Agena climbing over its horizon, no telemetry streaming into the tracking terminals. Across at pad 19 a hold in the Gemini VI countdown was called, while communications engineers advised the Carnarvon station to be particularly alert; there was just a slim chance that something had happened to cut telemetry over Canary and that Agena 5002 was alive and well. Forty-eight minutes into the phantom flight, Carnarvon kept reporting 'No joy – no joy.' Nothing had appeared and the only logical conclusion was that Agena had blown itself apart at PPS ignition and dropped to the Atlantic depths, deficient of the speed necessary for orbital flight. At 10:54 am, the mission was scrubbed and two very sad-faced astronauts climbed out of their ejection seats.

Gemini Mission Director William Schneider was already on his way from Houston to the Kennedy Space Center. He was to attend a meeting set for the following day at the command of Colonel John B. Hudson, SSD Deputy Commander for Launch Vehicles. Initial thoughts were that perhaps 5001 could be made ready for a quick launch, thereby saving the mission and preserving the sequence of flight operations. Two days after the aborted rendezvous attempt, NASA would set up an Agena Review Board under Bob Gilruth from MSC and Major General O. J. Ritland, Deputy Commander for Space, Air Force Systems Command.

Back at complex 19, astronauts Frank Borman and Jim Lovell, assigned the next Gemini mission, were with McDonnell's Walter Burke and his deputy John Yardley, when Burke casually suggested using a Gemini as target. Yardley was interested in this and asked Raymond Hill, McDonnell's chief of spacecraft preparation at the Cape, to recall the proposal made several months before by the Martin Company along the same lines. Soon, all five were enthusing over the simplicity of the scheme. Until, that is, Borman heard of the idea to place an inflatable docking cone in the back of the spacecraft adapter. Any scheme to use Gemini as a target would probably involve his own mission and he was not about to let another spacecraft poke round

Head sensors attached to the naked scalp of Frank Borman would record sleep patterns during the two weeks of Gemini VII.

among all the delicate systems necessary for operating the vehicle.

The Martin plan would require two spacecraft to be launched within the orbital lifetime of the first, and where once that idea seemed totally impossible, rapid launch schedules had shown the plausibility of getting another Gemini-Titan on the same pad within a couple of weeks. At least on paper. Burke and Yardley were fired with the possibilities opened up by such a scheme, however, and sought out George Mueller and Charles Mathews to raise the issue with them. The two NASA men dismissed such a rapid turnaround as being impossible in reality. It was one thing to launch two Gemini vehicles in two months. But within two weeks? That was very different. Martin's chief of Titan operations at the Kennedy Space Center, Joseph Verlander, and the Aerospace Test Wing's chief of the Gemini Launch Vehicle Division, Colonel John Albert, had put together a plan some months before whereby a fully configured Titan II would be physically removed from pad 19 to the redundant pad 20 by a Sikorsky Skycrane helicopter. Aerojet stipulated that once main stage engines had been installed and the vehicle checked out, the stack must remain vertical; the Albert-Verlander proposal would give pad crews time to prepare a Titan II fully, park it over on the neighbouring pad while another Titan was prepared, fitted with a Gemini and launched, rush back with the first Titan II, fit *it* with a spacecraft, and complete only those checks necessary to launch. That idea got nowhere and the suggestion from Burke and Yardley that Mueller and Mathews should consider using a Gemini as a target seemed to the NASA executives to recall the Martin plan.

However, not to be done out of their idea altogether, the two McDonnell managers walked from the Launch Control Center to the Manned Spacecraft Operations Building to join an informal post-mortem then going on among the engineers and technicians. They got no better reception there either, the consensus of opinion being that spacecraft 7 should be switched with spacecraft 6 so that Borman's 14-day flight could proceed on the GLV-6 launcher. There was no question of having spacecraft 6 fly Gemini VII's mission, since the rendezvous vehicle was fully equipped with systems and crew provisions for the two-day mission; spacecraft 7 was the only vehicle with fuel cells and cockpit layout capable of supporting the 14-day flight. At the meeting in the MSOB, Albert and Verlander backed away from their proposal of earlier that year and even in the face of this disastrous end to GT-6 refused to back Burke and Yardley pushing for a dual launch within a 10 day period so the latter could rendezvous with the former. Burke and Yardley sustained the discussion about switching spacecraft, but called for GLV-6 to be de-erected and replaced by GLV-7.

Webb got to hear about the proposed switch from George Mueller at the Cape and the following morning the NASA Administrator called in his deputy, Hugh Dryden, and Associate Administrator Bob Seamans, and his deputy Willis Shapley to meet Mueller, back overnight from Florida, for discussions. The alternate options were that if GLV-6 then on the pad could launch spacecraft 7 on its two-week mission, a launch would be possible by 3 December; if not, the Borman flight would be five days after that. Rendezvous would then begin in late February or early March 1966, with spacecraft 6 and, presumably, another Agena.

Meanwhile, as Mueller flew to Washington to debate the options calmly, Burke and Yardley headed for Houston and on the day Headquarters talked of switching spacecraft on the GLV-6 Titan, MSC Director Bob Gilruth heard first hand from the two McDonnell engineers the most astounding suggestion he felt he would be asked to support: that two manned Gemini vehicles should be launched less than two weeks apart. Gilruth was at a loss to point up any tangible obstacles, and called in his deputy, George Low, to stir up the debate. Low liked the plan, he had an ear for a daring scheme well constructed as had been evidenced by his role in settling NASA on the right road to the Moon. But there was the question of flight control and for that Chris Kraft was invited to the 'conference.'

By this time Charles Mathews was in Gilruth's office, the man who was really responsible for the whole program. He too thought the idea sound and logical but Kraft was amazed when he heard the plan. It simply could not work, or so he thought. But then he felt that, after consultations, there might be a way round the most daunting problem of all: how to communicate, track and receive telemetry from two spacecraft in orbit at the same time? Kraft called his deputy, Sigurd A. Sjoberg, and told him to convene a meeting of flight operations personnel directly after lunch, and then called Deke Slayton and asked him to get a response from the astronauts who would have to fly the mission.

Meanwhile, back at the Cape the day after the abortive attempt from pad 14, John Williams, Director for Spacecraft Operations, told a meeting of minds in his office that the weight of spacecraft 7 was too great for the Gemini 6 Titan, and that if the 14-day mission was to go next, GLV-6 would have to be pulled down and replaced with GLV-7; spacecraft 7 was just 117 kg heavier than spacecraft 6. That was that. A switch was out of the question. So, would the Burke-Yardley proposal work after all? Deputy Mission Director for Launch Operations Merritt Preston felt that it might just be possible to take down GLV-6, erect and check out Gemini VII, launch that mission, and quickly put back on pad 19 the Gemini VI hardware already in an advanced state of checkout. There might just be time to launch GT-6 with Gemini VII still in orbit.

Over in Houston, Chris Kraft had been to lunch with George Low and when they returned to the meetings called a few hours earlier for 1:30 pm, John Hodge's team were beside themselves with enthusiasm. This was just the sort of challenge they relished. It was one of Hodge's men that came up with a brilliant solution to the problem of data flow and dual tracking and communications procedures. If the passive, target vehicle (Gemini VII), was handled like a Mercury spacecraft, where tracking stations summarized data coming in by telemetry and sent it to the Mission Control Center by teletype, the active, chase vehicle (Gemini VI), could continue to use the normal Gemini data return format where computer processors interrogated the telemetry and sent it direct to Houston. Gemini VII could proceed as a normal mission up to the point where Gemini VI ascended in pursuit, the former vehicle going in to the passive mode while all the resources of the Gemini tracking and data relay network would concentrate on the second spacecraft. Gemini VI need last no longer than a single day, and the long duration mission would then resume its hold on the prime network. It had one more thing in its favour. Within a couple of years, NASA would be operating two spacecraft in space in support of the Apollo program and it would be an unprecedented opportunity to rehearse the simultaneous control of two manned vehicles in Earth orbit.

It was all beginning to come together. It took Kraft just 90 minutes to pull broad enthusiasm from the meeting with flight controllers. At 3:00 pm he rang Bob Gilruth and told him that he was prepared to discuss the matter again. At 4:00 pm Chris Kraft arrived at Gilruth's office for a meeting with the MSC Director, George Low, Charles Mathews, Deke Slayton, Walter Burke, and John Yardley. Kraft laid out the Hodge plan for doubling up the communications and tracking capabilities and broadly discussed the minor modifications that would have to be carried out at tracking stations so that two spacecraft could be on the communication loop with Houston. Everybody was pleased with the plan; nobody had any negative comment.

To Bob Gilruth, it provided an opportunity to show off the paces of his Manned Spacecraft Center – organizationally and managerially; to George Low, it was valuable experience within the entire span of manned space flight, and it was an unexpected bonus in rehearsing the kind of operations Apollo would introduce; to Charles Mathews, it was a further expansion of the Gemini concept, and a demanding role for the entire program team; to Deke Slayton, it provided valuable crew experience, got his rendezvous men back in the business for which they had trained, and justified the many hours Schirra and Stafford had already spent preparing for just this sort of routine; to Burke, it was a proud gesture for his company's product; and to Yardley, well, he was behind anything that flexed the muscles of manned space flight and carried forward the two-man spacecraft he and James Chamberlin had dreamed up 4½ years before.

The events of that day, the first complete day since Atlas shed its load into the Atlantic, had seen a complete turnabout for projected use of the two upcoming flights. That evening, the press got wind of something and Mueller's deputy, James C. Elms, rang Houston, spoke to George Low and asked him what all the fuss was about. Low briefed Elms on the gathering enthusiasm at the Manned Spacecraft Center that would put two manned Gemini spacecraft into orbit at the same time, and Elms, recognizing that support was growing at a rate unexpected in Washington, decided to telephone his boss and tell him about the plan. Only that morning, Webb, Dryden, Seamans, Shapley and Mueller had

To survive aboard Gemini VII for two weeks, engineers developed a new soft-hat suit worn over this 'bone dome.' Note the hat zipper connecting it to the pressure suit torso.

sat around talking spacecraft switches, and not one person had told the Administrator about the Burke-Yardley proposal. Mueller, on hearing from Elms the news passed to him by Low, decided it was too hot to leave until office hours, and telephoned Bob Seamans at his home. Seamans said he would get with Mueller the very next morning and go over the proposal.

Wednesday, 27 October saw Mueller cautiously telling Seamans all about the plan; Seamans was very interested in the idea but told Mueller to keep quiet until he had passed the idea up to Dryden and Webb, which he did that very afternoon accompanied by Willis Shapley. Seamans withdrew from debate as to the desirability or otherwise of sending four men into space at the same time – although Webb, ever aware of the public role he played, recognized it was a nice touch by way of eclipsing the Soviet Voskhod I flight with its three-man crew (!) – and the Administrator telephoned Mueller immediately to ask his manned space flight man if the idea was tenable. Put so obviously on the spot, Mueller asked Webb to hang on while he checked with Bob Gilruth at the Manned Spacecraft Center. On hearing that Webb was warming to the proposal, and that Dryden too was agreeable, Gilruth nevertheless recognized that he too was now on the line. So he asked Mueller to call back in thirty minutes after he in turn had called key MSC managers.

George Low came across to Gilruth's office and together they rang through to Chris Kraft, Deke Slayton, Charles Mathews, and Merritt Preston down at the Cape. Everybody agreed to back the dual mission, and that was just as well because Mueller rang Gilruth fifteen minutes later for an answer. Webb and Dryden were notified and from that point on it was official: Gemini VII would be launched first, followed by Gemini VI-A, renamed now to differentiate it from the Agena mission it had been only two days before.

Jim Webb telephoned Presidential assistant Joseph Laitin and the latter told the NASA Administrator to forward the proposal to the White House from where it could be sent to the President at his Texas ranch. The formal notification was at the White House by the evening of Wednesday and next day, 28 October, Lyndon Johnson formally announced to the world that Gemini VI-A would chase after Gemini VII some time during January. Across the state in Houston, Bob Gilruth was backing a December launch date! At the Kennedy Space Center everyone erupted into a hive of industrious activity, Air Force teams got together to plan how best they should integrate the two launch vehicle checkout schedules, technicians took down GLV-6 on Friday 29 October, and pad leader Guenter Wendt nearly had a fit. Believing NASA to have gone beyond the point of tolerance, he promptly laid down a dozen reasons why he could not possibly turn around a manned space vehicle in nine days, and then proceeded to set about learning how he could!

If the idea had been passed around for general comment, there is little doubt that the plan would have died a well trodden death on and in the hangars and workshops at the Kennedy Space Center. But having been given the job, they were fired with intent to see it through. In fact, from an operations viewpoint almost nothing would change, the only real flight plan addition to Gemini VII's mission schedule being the circularization maneuver to put the spacecraft in a path identical to that Agena 5002 should have had 25 October. That would put Gemini VI-A into a similar flight situation to the one it had presumed to have in chasing the Agena, spacecraft 7 being the hardware switch and nothing more. The only real deficiency would be that neither spacecraft could touch; electrostatic tests with Gemini flights so far indicated there was a limited danger from the electrical discharge of one vehicle to another. For that reason, three electrostatic vanes were designed and fitted to the Agena target to alleviate the Gemini spacecraft of its charge before actual docking took place. But first, the hardware had to be prepared.

With the launch vehicle de-stacked, spacecraft 6 was placed in a hangar and attention focused on preparing the Gemini VII mission for a December launch. Spacecraft 7 was an all-up systems vehicle in every dimension. Built from the beginning to last 14 days in space, it carried two fuel cell sections to accomodate the 5,600 ampere-hr usage planned for the mission. It would have been impossible without them. Since the trouble with the fuel cells aboard spacecraft 5, strenuous efforts applied both by General Electric and NASA had placed high confidence on the selected sections, seven 14-day profiles having been flown in simulation of the Gemini VII flight. To inhibit the thermal leak that caused the hydrogen vessel to vent off so quickly, Gemini VII's hydrogen tank had additional insulation to defer the transfer of higher temperatures through to the cryogenic contents.

If any one area of spacecraft 7 got more attention from the astronauts it was probably the stowage space. Demand for free space on board the 14-day vehicle was high. Cooper and Conrad had been almost inundated with garbage floating around the spacecraft's interior and from the outset Borman and Lovell were determined not to get in an untidy situation. They felt that this would be a major factor in their ability to remain aboard the spacecraft, floating forward and backward in the sitting position with, as Borman preferred to compare it, no more room than 'the front seat of a Volkswagen.' Ken Kleinknecht accompanied the two astronauts to St. Louis to search for more room in the spacecraft and they did find several spaces usefully applied as garbage disposal crannies. But the main problem was in the ability of the men to endure the flight wearing cumbersome space suits.

Back in June, Gordon Cooper and Elliot See operated without suits in a spacecraft in McDonnell's altitude chamber to a simulated height of 36 km, trying to prove that they need not wear suits for the full eight days of Gemini V. McDonnell was not too happy about the environmental impact, however, because the cabin could become excessively hot if the crew doffed their space suits; each suit carried an oxygen system which played a major role in keeping the astronauts cool and balancing the thermo-atmospheric environment of the crew compartment. However, further tests with the environmental controls specially set up for the doffed suit mode showed that the Gemini ECS would perform better with suits off than with them on. But the impact on safety was high. If a fire broke out in the pure oxygen atmosphere it would spread quickly. The only way to put it out would be to dump the cabin oxygen as rapidly as possible, certainly in less time than it would take to put a space suit on.

James V. Correale from MSC's Crew Systems Division came up with an idea that looked promising. By reducing the size and bulk of the simple G3C suit, transforming it into a soft suit, the crew would have sufficient protection to return to Earth within a depressurized cabin, yet not be hampered by the bulk of the stiffer G4C style. That proposal looked good and the David Clark Company immediately got to work on what would emerge as the G5C suit. The standard suit weighed 10.7 kg, but the new lightweight unit weighed only 7.3 kg and was built to accommodate a soft hood connected to the torso by a pressure-sealing zipper, unlike all standard suits that comprised a hard shell pressure helmet connected by a circular steel neck ring. The hood could be unzipped and folded back behind the astronaut's head, with communications retained via a headset within a standard aviator's hard hat. The suit could be completely removed inside the spacecraft but its limp condition when doffed enabled it to be placed at a point of convenience rather than stowed away because of bulk like a standard suit. It would take about 17 minutes to don the suit, or considerably less in an emergency. The idea was that the crew would remove their suits during the period following orbit insertion and leave them off until retro-fire, but opinion would vary as to the intent of the flight planners on this activity.

Regarding the flight plan itself, experience with Gemini IV and V showed the futility of pre-planning every minute operation aboard the spacecraft. Gemini VII would fly with only a general schedule of experiments, tasks and requested activity, beyond the orbit insertion point. Moreover, the crew would not sleep separately, but would perform work, eating and sleep routines approximately on the same schedule as they would at the Cape, preserving their individual 'day' clocks amidst the rapid day-night cycles of orbital flight. This would also smooth the flight control functions at MSC: Chris Kraft would get the early shift to implement the flight plan operations, Gene Kranz would do the housekeeping and take stock, and John Hodge would get the graveyard shift to work up plans and procedures for Kraft to review and transmit the following day. There would certainly be much to integrate aboard spacecraft 7; with 20 experiments it almost qualified as the world's first orbiting laboratory. Medical, technical, and engineering experiments were to be performed and Gemini VII was an ideal vehicle for that activity; designed as a long-duration spacecraft, its prime function was to stay in space and give the vehicle, its crew, and the flight controllers on the ground, a 14-day soak.

By early November, plans were firming up and it was agreed that Gemini VII should be launched during an early afternoon hour on 4 December and that an attempt should be made to send Gemini VI-A aloft on the 13th. The reason for Gemini VII's late launch originated from the desire to have the crew start 'work' in accordance with their normal daily routines so that the conventional cycle would be retained throughout the mission. And it was to be a valuable one. Nobody knew when physicians would get the chance to improve upon the two-week record; Apollo was not quite capable of matching a flight of that length and plans for manned projects beyond the Moon mission were fragmented at best. Consequently, medical preparations were a priority and while engineers and physicians wanted to see how well their respective charges survived the mission, it was with a great deal of experimental interest too that the medical scientists prepared Borman and Lovell for their flight.

Careful logs would be kept by each crewmember of their daily food intake and body waste cycles for two weeks before the mission and for nine days after. Nutritional studies were to be conducted by technicians from the National Institute of Health, carefully weighing and recording every morsel of food consumed in that period. During the flight they would carry all the medical experiments performed on preceding missions, and in addition a study of Borman's brain waves during sleep, monitored via electrodes pasted to his scalp. Data for the in-flight sleep analysis could only be obtained during the first two days, until hair grew to upset the sensors.

The delay in getting Gemini VI spaceborne was proving a nuisance for simulator planners, but Schirra and Stafford were well trained for their rendezvous operations, letting Borman and Lovell get their due slice of rehearsals, only cutting in every so often to keep their proficiency high on techniques with which they were already familiar. Nevertheless, the rapid pace of program schedules required another crew to get their flight assignments, and on 8 November the Gemini IX crew was announced: Elliot See and Charles Bassett, with Stafford and Cernan as back-up. Their mission was to last two or three days, fly in the third quarter of 1966, and include in addition to rendezvous and docking with an Agena target, EVA from the re-entry module to the Air Force's Astronaut Maneuvering Unit (AMU), a cumbersome back-pack designed to provide full attitude control for a space-walking pilot.

EVA had fallen behind other space priorities during the previous months, although Armstrong and Scott were scheduled to perform EVA on the Gemini VIII mission, but only to try out new life support packs developed at the Manned Spacecraft Center. Although one-time campaigners to have EVA dropped from their Gemini VI mission, Schirra and Stafford felt they could embellish the new flight plan by opening their hatch and perhaps swapping pilots with Gemini VII in space. Spacecraft 6 had, after all, been originally built specifically as an EVA vehicle. Frank Borman was not too happy with the idea, and in any event it would upset medical research into the reactions of two different human beings to a flight lasting two full weeks. George Low was recruited to the 'bring back EVA' campaign now waged by Wally Schirra but, although they reviewed the suggestion, NASA Headquarters though it too much of a Buck Rogers stunt to warrant inclusion. So it was that Gemini VI-A settled back as a one-day flight, the expanded role of its original two-day profile jettisoned along with Agena 5002, which would attempt to fly to Gemini VII nine days into the long duration mission.

One last piece of the jigsaw had to fall into place. Martin had stipulated that once erected and checked out, a Titan II must remain vertical or back up further down the checkout schedule, a cautionary requirement stressed by Aerojet. Further analysis showed that this was not absolutely necessary and that it would be possible to de-stack the vehicle without going all the way back to the beginning when checkout resumed; Martin acquiesced and NASA got the fast re-fire procedure it wanted. By mid-November everything looked good for the Gemini launch on 4 December, but Chris Kraft nearly failed to make it for his own vital slot in the MSC machine. On 17 November, while flying from Florida to New Orleans with public affairs officer Paul Haney, a youth pointed a gun at Kraft and squeezed the trigger. It failed to go off and he was restrained by a passenger and the two NASA men. The young man had wanted to fly to Cuba, thinking that by killing a passenger he would instill the fear deemed necessary to get control of the aircraft.

There was just one problem still lurking that could conceivably threaten the Gemini VII launch. Fuel cells intended for the spacecraft were flooded with hydrogen during an unexpected increase in pressure. Nothing could be seen to damage the cells, but they were replaced on the 17th. On 22 November spacecraft 7 was mechanically mated to GLV-7 and the last major hurdle was crossed.

Just two days before the launch, the space agency was denied the wise counsel of its Deputy Administrator, Dr. Hugh Latimer Dryden. On 2 December, he died of cancer after suffering for some time. Born in 1898, Hugh Dryden once said that he and the aeroplane grew up together. He was named Director of the National Advisory Committee for Aeronautics on 2 August, 1947, replacing Dr. George W. Lewis, who had to resign through illness. Dryden's appointment marked the transition for NACA from an essentially engineering organization to one more appropriate to the science of aeronautics. A former Associate Director of the National Bureau of Standards, and for 45 years an ordained minister in the Methodist Church, Hugh Dryden became the Deputy Administrator of NASA when it was formed in 1958, staying on as acting administrator during the transition from the Eisenhower to the Kennedy administrations in early 1961, and eventually accepting an invitation to continue as Deputy Administrator under the leadership of James Webb.

Hugh Dryden was a respected scientist, and an inspiration to the young space agency, a very real link with the defunct NACA and all it stood for. At his passing, the space agency began a drift away from the essentially research orientated role it had been set up to perform to the operational organization it was inevitably to become. More than a decade later NASA's aeronautical research facility, the Flight Research Center at Edwards Air Force Base, would be re-named as the Hugh L. Dryden Flight Research Center. At Dryden's funeral, two days into the mission of Gemini VII, the first NASA Administrator, T. Keith Glennan, and astronauts Carpenter, Cooper, McDivitt and Glenn, served as pallbearers with former NACA Chairman Gen. James H. Doolittle. Four days later Robert C. Seamans was named the new Deputy Administrator.

It was 7:00 am Cape time on Saturday, 4 December, when Frank Borman and Jim Lovell were woken for their long sojourn in space. For several hours the final countdown had been under way at pad 19, where Ed White and Mike Collins had been checking out spacecraft 7. They soon finished that and briefed the prime crew on how things stood in preparation for a 2:30 pm launch. It was a leisurely schedule. At 10:00 am they had breakfast: tenderloin steak, eggs, toast, jelly, orange juice and coffee, in the company of John Young, Pete Conrad, Dick Gordon, Deke Slayton, Dave Scott, Neil Armstrong, Gus Grissom, Al Shepard, Wally Schirra and Tom Stafford. Soon they were on their way to the suit-up trailer alongside pad 16 where they put on the new lightweight suits, looking for all the world like some alien beings as they walked down the wooden steps to the transfer van, each man holding a ventilator.

Nothing could have gone better as they arrived at the Titan pad and leisurely moved up the erector, entered the white room, and paused to be received by the spacecraft technicians. Within minutes they were hooked up and the closeout crew busied themselves with the thick manual detailing every minute process required to seal the spacecraft. Just three seconds past 2:30 pm, Gemini VII lifted off into clear skies and within minutes had threaded its nose through that hole in the sky ensuring a safe orbit. In fact it was so good it was almost a record. Spacecraft 7 was in an orbit with a 161.2 km perigee and 327 km apogee inclined 28.89°, only 10.9 km off on apogee, and 0.02° out of the planned orbital inclination. Borman blipped the aft firing thrusters for just two seconds to shift his spacecraft away from the inert Titan second stage and then turned the spacecraft through 180° to face the cylindrical structure.

First priority was to try a station-keeping exercise on the stage, a feat McDivitt had been unable to achieve on Gemini IV. Because Gemini had launched into an afternoon sun, the bright ball in the sky washed out good visual identification of the stage, but after a half-minute adjusting to the conditions, Borman positioned the spacecraft close by, moving to within about twice the length of the stage; hours of practice in the simulators paid off and Gemini VII showed how easy it was to keep close beside a vehicle

Borman and Lovell walk to Pad 19 for their historic ride into space.

in space, the secret being to get turned around before it drifted too far away. Toward the end of the day, the crew saw what they took to be unidentified satellites drifting some distance away, then it was time for a meal and some housekeeping chores before settling down to sleep. A trim maneuver designed to raise the spacecraft's perigee had been performed some hours earlier where the crew fired the OAMS thrusters for 75 seconds, raising the low point from 161 km to 222 km. And as if reminiscent of the Gemini V saga, a fuel cell warning light blinked on to tell the crew and the flight controllers that pressure was dropping. It was not, and a cursory look at other measurements showed that it was probably a faulty light.

Down on pad 19 it was a very different post-launch atmosphere. Nobody was wandering around in relaxed mood, few thought of any parties that night. Within minutes of the GT-7 mission blasting off the concrete hardstand, engineers, welders, plumbers, and inspectors, converged on the complex. Pad damage was not excessive, many had seen worse from a Titan launch, and while Borman and Lovell got ready to hit the sack, searchlights were already on and a full shift of workmen were banging and clattering away under bright blue welding torches and overhead arc lights; after any launch there was minimal damage to support structures, umbilical arms, electric cables, cooling lines, etc. Pad leaders reported that GLV-7 had been a clean bird and left its nest in tidy condition, confidently predicting that by dawn they would be ready for the Martin people to move in the GLV-6 launcher.

5 December, and furious activity as first the erector lifted into a vertical position the 19.8 metre long first stage, lowered itself, then attached the second stage and cranked up into a vertical position to place that load on top of the first. Next, it was the turn of the spacecraft, and by the time dusk crept across the flat Florida horizon it was mechanically mated to the launcher. Late that night, when it was apparent that pad preparations were going better than anybody had dared to hope, a planned phasing maneuver scheduled for Gemini VII's sixth day was cancelled and the crew informed that they would be required to perform a partial phasing burn two days earlier with another performed later so that Gemini VI-A could still rendezvous if launched on the eighth or the ninth days. That way, if GT-6 was ready on the 12th instead of the planned 13th, it could still go a day early and find Gemini VII in the correct orbit. During that day in space, 5 December, the crew attempted to sight markings laid out on the ground in Laredo, Texas, but found difficulty with obtaining a clear view.

Next morning, Jim Lovell took off his space suit and that seemed to improve the temperature inside the cabin; Borman was hot and both men had stuffy noses, a condition brought on by the pure oxygen. Lovell decided to sleep that night with his suit off, but Borman was disgruntled over a mission rule stipulated by George Mueller that only one astronaut should go suitless at a time, and that both should be suited up during launch, rendezvous, and re-entry. But whatever discomfort Frank Borman felt over having to keep his lightweight pressure garment on, Jim Lovell never quite got over the humorous sight of himself floating up and down between ejection seat and instrument display wearing nothing but a pair of regulation Air Force issue longjohns! Before ending the day, their third in space, the astronauts observed a Polaris A-3 missile fired from the USS *Benjamin Franklin* submerged off the Cape.

That night, Lovell slept in his underwear while Borman sweltered in his suit amid grumbles and complaints to Houston about the temperature. A good reason for Lovell to keep his suit off was that being the larger of the two it was a more troublesome activity for him to have to pull and tug at the separate components. But Borman failed to see it that way, although he blamed ground controllers for not allowing him to doff his own suit rather than Lovell for not changing turn and turn about.

During Tuesday, 7 December, Borman fired up the thrusters for an orbital adjustment, placing the spacecraft in an orbit with perigee of 188.3 km and apogee of 301.7 km ready for circularizing the orbit two days later so that Gemini VI-A could find it in a path almost identical to that which would be provided by an Agena target vehicle. Later that day, he agreed to let Lovell keep his suit off for the second night while he tried to get some sleep with his own suit on, albeit without gloves and with the hood unzipped and folded back.

Wednesday, 8 December, and the only notable event was

cancellation of a planned laser experiment due to cloud over New Mexico; they were to have shone a laser beam at a selected site as part of an optical communication test. But on the ground there was a unique 'first' as the government brought a temporary injunction against Safety Steel Services Inc., of Corpus Christi, Texas, preventing them from running internal combustion engines as Gemini VII passed overhead. Engines operated by this steel fabricating firm were apparently interfering with spacecraft communications. At the Kennedy Space Center, things were working out well for preparation of GT-6, although a slight ruffle pulsed through the system when the spacecraft computer had to be replaced. It was still looking good for an attempted launch the following Sunday. During the evening hours, Frank Borman asked capcom Eugene Cernan to discuss with Chris Kraft the possibility of him removing his suit, but to go through Deke Slayton first. But controllers were trying to persuade the crew to follow a plan of their own: for Jim Lovell to put his suit on, and for Borman to take his off, so that medical monitors could obtain data on the two of them in similar and comparable conditions. Again, for the third night in a row, Lovell lounged in his longjohns while Frank Borman put his trust in Deke Slayton – the strategic guardian of astronaut interests.

Thursday, 9 December brought with it confidence that Gemini VI-A would rise to its occasion on the 12th, a day earlier than originally planned. A quick conference between Charles Mathews and SSD Commander Ben Funk confirmed that schedule and the MOCR at Houston was given a 'go' for final orbit adjustment maneuvers to be transmitted to spacecraft 7. Houston had been busy, and was now preparing to hook up a spare control room to the simulator for Schirra and Stafford to go through a complete flight, rehearsing the procedures they would follow just three days hence. In space, Frank Borman fired the OAMS thrusters for 78 seconds, raising perigee to 299 km, and then one half-orbit later, with the spacecraft at its new low point, Borman fired the thrusters again for 15 seconds, lowering apogee to 303 km. In effect, the hitherto elliptical path had been made into a circular orbit by raising perigee by more than 100 km and lowering apogee by a mere 2 km, hence the difference in the thruster burn times. Although not a true circle, the orbit was considered adequate for Gemini VI-A to rendezvous. Everything was set up for the launch of spacecraft 6.

Next day, Friday, 10 December, temperature in the spacecraft had risen to an uncomfortable 29.4°c and once again the issue of the suits raised itself. Borman wanted to wait until after the rendezvous activity to fulfill a request from Chris Kraft for the men to change suit modes, but the Flight Director insisted that it was now essential for Lovell to put his suit on and for Borman to take his off. Late in the afternoon, with the flight of Gemini VII now more than six days old, the suit swap took place 300 km above the Earth; Frank Borman was now in his longjohns, and getting cooler by the minute.

Saturday, 11 December, the day before the planned rendezvous, spacecraft 7 successfully beamed a laser down to Hawaii but the contact was not strong enough to test a communication circuit. On the ground, Cape technicians put the finishing touches to GT-6, and in Houston, flight controllers got ready to implement their makeshift plan for switching data handling modes. At orbit insertion, spacecraft 6 would trail spacecraft 7 by nearly four minutes flying time at orbital speed, but as the former closed in on the latter the separation in talk times over respective tracking stations would get progressively shorter as Schirra and Stafford met Borman and Lovell.

Sunday, 12 December, and the two grounded astronauts arose before most Americans to lay the phantom of mission VI. Gus Grissom and John Young had checked the cabin and Gordon Cooper was with the Gemini VI-A flight crew for breakfast. Then it was off to the suit-up trailer at complex 16 followed by a drive the short distance to GT-6. With time to spare, the crew moved across the hatch sills, the two big doors were closed, and the pad closeout crew performed their duty on spacecraft 6 for the second time. Lift-off was planned for six seconds after 9:54 am and just 26 seconds before that time spacecraft 7 swept overhead, out of sight in the morning light.

With the familiar good wishes, the capsule communicator committed GT-6 to launch: 'You are cleared for takeoff . . . adios.' Then, across the silence that suddenly stilled the chattering headphones, a single voice from the blockhouse: '10–9–8–7–6–5–4–3–2–1–ignition.' A wailing banshee whine signified ignition of the hypergolic propellants, followed a split second later by the thunder of the rocket motors. And then – silence. 'We have a shutdown, Gemini VI.' The voice sounded disbelieving, but urgent, anticipating the inevitable. By every written law in the mission rule book, Wally Schirra should now reach down to his D-ring and pull sharply up toward his chest, triggering the escape system and ejecting both men from what milliseconds later could be a gathering fireball of explosive energy. But nothing.

Inside the spacecraft the mission clock started rolling, sure indication that the big Titan booster had slipped its shackles: 'My clock has started.' Then, cutting in over the open communication circuit, willing the crew to eject, Chris Kraft, steady as a rock, voiced confirmation, endorsing the first warning: 'No lift-off, no lift-off.' Titan's two engines had burned for 1.2 seconds, long enough to shake loose an electrical plug attached to the base of the rocket, setting off the on-board programer, but because the vehicle had not sensed forward motion it automatically shut itself down.

Another comment from Schirra: 'Fuel pressure is lowering.' Sitting atop 136 tonnes of hypergolic chemistry, Wally Schirra knew the price for error. But he had been this way before, on an Atlas admittedly, and he instinctively realized in that microsecond between judging the event and making a commitment to stay put or eject that Titan was still firm on its thrust mount and that it was probably safe. But a sequence had been started in the belly of the ticking booster that moved those dangerous propellants closer to each other than before, and with the sure knowledge that if they came into contact they would automatically ignite, Wally Schirra decided to call the Titan's bluff and to sit it out. That he did so saved the hardware for a possible rendezvous after all, for if he had ejected that would have put paid to any hope of getting spacecraft 6 spaceborne to meet Gemini VII. None of the astronauts were particularly keen on the idea of using the ejection seats but Schirra was firm in acknowledging that if the Titan was about to blow up, 'It's . . . death or the ejection seat.'

Minutes later, after blockhouse 19 confirmed that the launcher was indeed safe, Guenter Wendt prepared to go back again to the reluctant spacecraft. The erector clattered its way to a vertical position and McDonnell personnel moved into the white room. It was 11:33 am, 99 minutes after the shutdown, when the two astronauts were liberated from their supine supports, bitter disappointment again etched across their solemn features. But it was not the catastrophe it could have been, and the hardware was intact and capable of supporting another attempt. In the belief that the apparently loose plug had caused the entire incident, launch teams immediately began re-cycling the launch vehicle and spacecraft for another attempt on Thursday, 16 December. Gemini VII would not return before the Saturday, which meant there was time for a quick 24-hr mission before recovering the two vehicles on successive days. But was it that simple? Or was there something more insidious about the way GT-6 had shut itself off?

Cursory examination of data traces from several places in the Titan propulsion system showed suspicious indications that thrust tail-off began fractionally before the plug pulled loose. Colonel Albert received a telephone call from Aerospace Corporation who, upon looking at the records of telemetry, thought the Gemini Launch Vehicle chief should see the anomaly. Albert took the graphs along to a meeting where turnaround plans were being worked and more calls, this time to the engine's manufacturer, put Aerojet engineers at the Cape on to a detailed search in the area of the gas generator. By 7:00 pm the examination was under way and for many hours searchlights again sustained the nocturnal activity. On through the night they worked, searching in vain for the source of the thrust decay.

Shortly after 9:00 am on Monday morning, while Charles Mathews began to think the worst, the cause was discovered. An engineer came running from the Titan's boattail with the news that a dust cap placed over an inlet port on the gas generator had been left on during assembly of the launcher and that due to the extremely remote location of the generator nobody could possibly have seen it. Once missed, it would never be found by any process of inspection GLV-6 went through in the time between leaving Martin, Baltimore, and ignition on the pad. Within a few hours of beginning the re-cycle of GLV-6, engineers were confident they could have the bird ready to fly within three days, buying back an extra day from the earlier projection of an attempted launch the following Thursday; Wednesday, 15 December, was the new target.

The day that Gemini VI-A back-fired on the launch pad went

Shutdown: Gemini VI-A stands poised atop a potential fireball as the Titan rocket ignites then cuts out on the pad.

Schirra and Stafford walk away from their failed booster nearly two hours after it fired for 1.2 sec before prematurely shutting down.

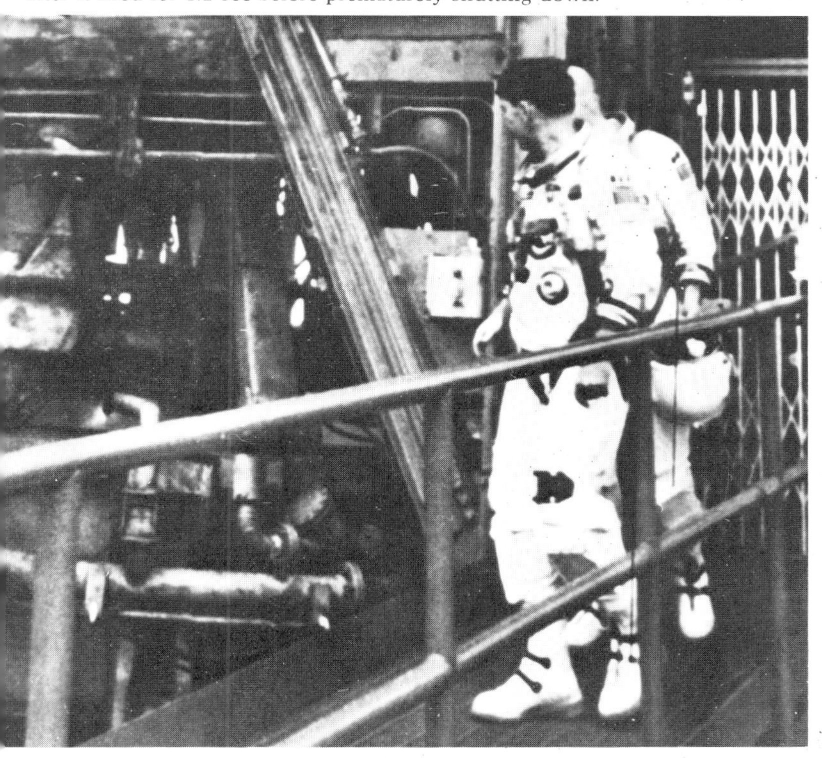

smooth in space for Borman and Lovell, and during that afternoon they exceeded the eight-day mission of Gemini V. But it was all uphill. On Monday, 13 December, the VII crew ran up against another fuel cell problem when a warning light flashed telling them that excess water had built up in the waste tank. It was flushed from the system by forcing in extra oxygen from the crew compartment's reserve supply.

On Tuesday, 14 December, Borman and Lovell observed a Minuteman missile launched from Vandenberg as its nose cone re-entered the atmosphere over Eniwetok, the first time a missile re-entry had been observed from space. But the strain of the flight was taking its toll and neither astronaut was in the best of spirits this day. 'Jim and I are beginning to notice the days seem to be lengthening a little,' came a doleful comment from Frank Borman. 'We're getting a little crummy!' The two astronauts were certainly showing different reactions to the environment, and different responses to their respective roles. Aware that ultimate responsibility rested with his own command of the mission, Borman slept fitfully, woke frequently to scan the instruments before dozing back again, and generally got less sleep than Lovell; and less each night as the flight progressed. During the first few days, Borman slept for periods averaging nearly 6 hours; by the end of the first week this average was down to between 4½ hours and 5 hours. Lovell got an average 6 hours continually.

Borman had a pronounced physiological reaction to the mission too, and not until the fifth day did the command pilot defecate while his partner was regular from the second day in space – although this was only what had been demonstrated by the eight-day mission. On Gemini V Command-pilot Cooper first discharged fecal waste on the sixth day while his accomplice on that flight defecated on day three. Combined with the generally warm conditions aboard the spacecraft, the abundance of equipment for experiments, and the lack of adequate toilet facilities, Gemini VII was a pungent assembly of various smells and sensations far removed from the idyllic life an astronaut supposedly led while romantically voyaging among the stars. It was a reality few people were keen to sample, and there was nothing very pleasurable about conditions in spacecraft 7.

Dawned the day when Gemini VI-A got another try; Wednesday the 15th. Launch this time would be at twenty-seven seconds after 8:37 am and to meet that deadline Wally Schirra and Tom Stafford rose at 4:00, showered, breakfasted with Al Shepard, and for the third time heard Grissom and Young report that a healthy spacecraft awaited them, ready for birth but a little reluctant. Into the spacecraft they slipped at one minute past seven o'clock and again they counted down the lengthy sequence of checks and communication tests. On arriving at the white room to enter spacecraft 6, Schirra and Stafford were welcomed by a white card telling them 'Good luck from 2nd shift.' Now that was gone, as were the technicians that had spent more time with this particular spacecraft than any other. As the final seconds ticked down to zero, a determined voice, almost straining into the microphone, implored the assembly to leave the pad and get space-wet: 'Gemini VI, you are GO!'

And from the crew compartment atop the black striped Titan, a further appeal from Wally Schirra: 'Go! d'you hear the man, GO!' And 'go' she did as the booster lifted from the thrust mounts on pad 19, quivered slightly as if a little nervous to be finally on her way, and slipped quickly along the pre-planned trajectory toward Frank Borman and Jim Lovell. GT-6 was on its way into space. At staging a tremendous flash of fire swept up along the sides of the second stage, past the windows of spacecraft 6, telling the encapsulated crew that the second stage was alight and firing. Shortly before six minutes after lift-off, the second stage shut down and Gemini VI-A was in orbit, 1,070 km from the Kennedy Space Center, 1,992 km behind Gemini VII. For the first time, there were four men in orbit at the same time. The first orbit would be spent preparing for the rendezvous maneuvers and now that their chase vehicle was in space, Borman and Lovell put on their suits and settled down to cooperate with the steadily approaching vehicle.

But spacecraft 6 was 3.6 meters/sec slow at insertion which resulted in an orbital apogee lower than the required value; instead of being in a 161 km × 270 km orbit, with the high point 28 km below the orbit of Gemini VII, the spacecraft was in a 161 km × 259 km orbit, 11 km too low on apogee. A contingency option had been built into the flight for just this sort of eventuality, a programed opportunity provided in the rendezvous schedule worked up for the Gemini VI/Agena mission scrubbed the previous October. This height adjustment maneuver was performed 1 hr 35 min into the flight as the spacecraft crossed North America at the end of the first orbit and comprised a 4.27 meters/sec OAMS burn to raise apogee by 11 km. The next maneuver, the first that would start the Gemini VI-A spacecraft up toward its target, was made at an elapsed time of 2hr 18 min with the two vehicles 713 km apart. Called the phase adjustment maneuver it called for an addition of 19 meters/sec so that perigee would be raised to 224 km, reducing the rate at which the spacecraft was gaining on Gemini VII and bringing the Gemini VI-A orbit closer to that of the target.

The first one and one-half revolutions of the Earth had been a catch-up period, with Gemini VI-A narrowing the distance on VII by more than 1,200 km. Now, with the phase adjustment burn having molded the shape of the chase vehicle's orbit, it was time to start the gradual process of rendezvous on the fourth orbit. But one more trim had to be performed. A maneuver not planned in advance but one that could be carried out if the plane of the chase vehicle's orbit was not exactly that of the target's orbit. Accordingly, about one-half hour after the phase adjustment burn, Schirra fired the thrusters in a 10.4 meters/sec burn lasting 40 seconds which, with the spacecraft turned 90° to the flight path, changed the orbital inclination by the necessary 0.007°. The precise position at which this could occur, called the node, represented the place where the two orbital planes intersected, each orbit presenting two such opportunities for a plane-change burn. Coming across the United States at the end of the second revolution, with Gemini VI-A at its second perigee point, the final height adjustment burn was conducted, a one second tweak firing to change speed by 0.24 meters/sec. The first height adjustment maneuver performed at the end of the first orbit had carried Gemini VI-A a little higher than planned, so it was necessary to slow it down and in so doing trim the apogee height by 740 meters, putting the spacecraft precisely 27.8 km below Gemini VII at the high point of its elliptical orbit.

Now, the elliptical path could be made more circular and as Gemini VI-A moved on across the African continent on its third revolution, the two vehicles were closing to the point where radar contact should soon be possible. Up to then, it was a matter of aligning one set of orbital trajectories with another. But slight differences would arise between where one spacecraft thought the other was, and vice versa, until radar would seek, find and precisely define the position of the transponder on the target. Elliot See, the capcom on duty at Houston, told Schirra that contact should be made possible very soon. Suddenly, a flicker, and then a solid lock-on. Distance between the two vehicles: 434 km. Gemini VI-A now had access to the precise details of Gemini VII's position relative to its own for the coelliptic sequence maneuver to be performed at third apogee, two and one-half revolutions into the mission. Up to this point the rendezvous charts considered each spacecraft as a separate vehicle referenced to the Earth, now the reference used for calculating maneuvers would be the target, assuming that to be a fixed point in space toward which the chase vehicle was moving.

At 3 hr 47 min, over the Carnarvon tracking station, Schirra fired again the aft facing thrusters that would push Gemini VI-A faster by 13 meters/sec. At the end of the 54 second burn, Gemini VI-A was in a 270 km × 274 km orbit, the burn having raised perigee by nearly 50 km. The two spacecraft were now in nearly circular orbits, Gemini VI-A, on the inner track, about 27.8 km below the path of Gemini VII, trailing it now by 318.5 km but closing all the time. In fact, the coelliptic maneuver had been performed 3.7 km further uprange from the desired position due to a slight mis-calculation which resulted in too great a separation at the time of the burn by precisely that distance. That error was erased by starting Terminal Phase Initiation two minutes late. Now, having used the radar to refine the position for the coelliptic burn, it was time to hand over to the computer.

When the coelliptic maneuver was performed over Carnarvon Tom Stafford reached across to his clock and started the stopwatch facility; it would be used for all rendezvous maneuvers from that point on. The event timer on Wally Schirra's panel, slaved to the pilot's stopwatch, would be the command-pilot's cue for specific events. Precisely four minutes after the burn, the event timer was synchronized and the computer switched to the rendezvous mode. From now on, both men would busy themselves with taking readings on elevation angle and range, the former obtained by recording the pitch angle of the spacecraft as its nose slowly rose above the horizontal to track the target above, the latter by direct radar readouts. Inside Gemini VI-A the lights were now turned down ready for the crew, their eyes adjusted to the light outside, to acquire the target visually. Coming across South Africa, 5 hr 16 min into their flight, the Gemini VI-A astronauts performed the TPI burn that would carry them up to Gemini VII.

The target spacecraft carried lights similar to those fitted to Agena vehicles, a hasty addition when the re-written flight plan included a rendezvous with Schirra's spacecraft, with a single 'docking' light and two acquisition lights flashing once a second. The crew expected to see the acquisition lights at a distance of 55 km, but not until they were within 47.6 km could they faintly discern the docking light, the other two not visible until a range of 26.8 km. During the terminal phase, spacecraft 6 made two course corrections, the first at 5 hr 32 min, the second twelve minutes later, although neither had the advantage of a visual cue. During the long climb up to Gemini VII the vehicles were in darkness coasting together across the Indian Ocean.

At a range of 1.4 km Gemini VII broke into the ascending dawn and minutes later, at a distance of 900 metres, about 5 hr 50 min into the flight, Schirra began the final braking maneuvers. At a distance of 450 meters the relative speed between the two vehicles had been reduced to 2 meters/sec and Schirra let spacecraft 6 coast across to its final parked position about 36.5 meters from Gemini VII. It was 2.33 pm at Houston, 5 hr 56 min into the flight of Gemini VI-A, and the first space rendezvous had been achieved.

In the MOCR at Mission Control, everybody erupted, clapping, shouting, waving flags, lighting cigars, and congratulating each other. In space, high above the Mariana Islands, four very happy astronauts joked with each other across the short distance that now separated them. 'You guys sure have big beards,'' quipped Schirra as he gazed across through the windows at Borman and Lovell, their smiling faces up against the panes. And from Frank Borman: 'Howdy guys . . . for once we're in style!' For several hours the two spacecraft would remain together, drifting in and out of day and night, watching the Sun flashing across the white adapter casings of each other's vehicle, experiencing the variation of colour that comprised the Gemini spacecraft, looking for signs of heat during ascent, or for damage unsuspected but perhaps inflicted on remote parts.

There were surprises. 'You've got a lot of stuff all around the back end of you,' Schirra told Borman as Spacecraft 6 drifted slowly round the back of Gemini VII's adapter. Long streamers and tails hung motionless from the periphery of the adapter, remnants of the tape that sealed and kept clean the line of pyrotechnic separation between booster and spacecraft. Such things could not be seen from the forward facing windows in the re-entry module. At another time, Borman and Lovell watched fascinated by the tongues of flame that stabbed the crystal vacuum of space as thrusters aboard spacecraft 6 fired up from time to time. Each astronaut took turns controlling respective vehicles; one would remain stationary while the second cavorted round the first, then the former would perform its pirouette around the latter. At one time the two approached to within 30 centimeters of each other, flying away as far as 90 meters at another.

The night passes were the most intriguing. Submerged by the eclipsing mass of planet Earth, the docking and acquisition lights were quite bright close up, and cabin lights could be seen through the two eyelid-like windows, bathing the interior with an orange glow; when dimmed, smaller penlights cast dancing shadows across the bristled faces that moved like dolls heads behind the thin transparent screens. Below, the ghostly presence of Earth lent a deep feeling of quiet, an aura of remoteness, a gathering of men from a planet in the awesome hostility of black night. But soon – too soon – it was time to resume the planned schedule, and final positioning for the sleep that all four astronauts were now glad to accept.

That day had been the highlight of the mission for Borman

Gemini meets Gemini in a rendezvous high above earth.

and Lovell, kept up late by the antics of Gemini VI-A. Schirra fired his OAMS thrusters one more time, and the two vehicles slipped further and further apart, eventually drifting more than 30 km from each other. It was good to have visitors, a distinct comfort on that eleventh day in the now all too small Gemini VII cabin. As the sleep period set in, a pleasing comment addressed to no one in particular, drifted down to a tracking antenna: 'We have company tonight.' Next morning, 16 December, Schirra and Stafford were up fresh and perky as usual, packing for the homecoming they elected to perform. It would have been possible to keep Gemini VI-A in orbit another full day, but there was little point, especially since the tracking and communication modes were operating in the makeshift manner especially developed for the VII/VI-A mission.

But suddenly, a report from Tom Stafford that sent Chris Kraft's headphones ringing: 'This is Gemini VI. We have an object, looks looks like a satellite, going from north to south, up in a polar orbit. He's in a very low trajectory . . . looks like he may be going to re-enter pretty soon. Stand by . . . it looks like he's trying to signal us.' And then, down from Gemini VI-A, specifically for children all across the planet below: 'Jingle bells, jingle bells, jingle all the way . . .,' from Wally Schirra's harmonica, and bells tinkled by Tom Stafford on that tenth day before Christmas. Santa Claus was coming back, and new warmth filled many homes that December morning.

With a final 'Really a good job, Frank and Jim. We'll see you on the beach,' Wally Schirra put spacecraft 6 into re-entry attitude. One by one the four retro-rockets fired, slowing Gemini VI-A by 100 meters/sec as the solid propellant motors burned into the night sky. On down through the atmosphere, spacecraft 6 tracked a computer guidance curve from a height of 85 km, banking to trim the descent path. From a height of 24 km the spacecraft came straight on down, the parachutes coming out in sequence from 14 km. Gemini VI-A splashed down 13 km from the recovery carrier USS *Wasp* and the crew remained aboard until it had been retrieved, chatting to frogmen through the open hatches, the spacecraft stabilized by a flotation collar secured to the water-line. Just 62 minutes after landing, Wally Schirra and Tom Stafford were winched to the carrier deck.

Meanwhile, high above the Earth, spacecraft 7 was deficient of two thrusters used for yaw control. About 9:30 am that morning, Borman and Lovell reported thrusters 3 and 4 were emitting unburnt propellant and that they could not be used. But no problem, just a nuisance; other jets could be used instead. A similar problem had affected Gemini V, where two yaw thrusters had gone out on the fifth day, and it had been assumed that because electrical conservation had turned off heaters used to prevent OAMS propellant freezing the thrusters, oxidizer had built up an ice barrier. But thrusters were not the only problem facing Gemini VII. Shortly after Gemini VI-A arrived close to spacecraft 7, Borman and Lovell removed their suits and reported water draining from the inlet hoses, indicating a flooded suit heat exchanger. If so, the lithium hydroxide canister used for scrubbing carbon dioxide from the spacecraft atmosphere might also be waterlogged, inhibiting its valued role aboard the spacecraft. Accordingly, the flight controllers had the crew dump excess water overboard by boiling the water boiler, a piece of equipment designed to remove excess heat during ascent until the radiators could be used. This seemed to work and within two hours the crew reported the cabin warm and dry. Fuel cell problems had emerged on the mission's sixth day, when several rapid performance drops began with one of three stacks in section 2. Before Gemini VI-A arrived on station, at about 9:30 am on the 15th, the section resumed its decline which resulted in two stacks being removed from the line.

When Schirra and Stafford departed spacecraft 7 on the 16th, spirits were lost once more and the two Gemini VII pilots noticed themselves getting less tolerant of each other. On the 17th, an almost continuous fuel cell light raised question on the duration of the flight but the cells settled down once more and the mission was given a final 'go' for the full term. But it was two happy astronauts who entered their last period during the afternoon hours of that December Friday. Sleep periods had been stepped back 30 minutes each day so that they would gradually adjust to the final rest session timed to have the crew awake and refreshed for Saturday morning re-entry. The last two days of the flight provided opportunity for Borman and Lovell to get some reading done; the Gemini V crew had regretted leaving books behind and VII's crew took note. Frank Borman, a lay preacher at his local church, took *Mark Twain* along, while Jim Lovell read *Drums Along the Mohawk* by Walter D. Edmonds.

One of the final chores to perform was to stow all the equipment they had used over the preceding 13½ days. It was easy to leave some item floating under a lip of the display console ready for the g forces of re-entry to pluck it from weightlessness and hurl it at a crewmember. Neatness was not just for the sake of being tidy. Getting the suits on was a less pleasant experience than sitting back for retro-fire. After two weeks without as much as a wash or change of the now familiar longjohns, the flexible suits were clammy, soiled, garments that transformed the two returning astronauts into bedraggled, whiskered, replicas of their former smartness.

Right on time that Saturday morning in the blackness of a Pacific night, spacecraft 7 lost its equipment section on schedule. Only 3.9 kg of hydrogen remained of the 10.7 kg loaded at launch, and the oxygen tank retained only 27.6 kg of its orginal 82.5 kg supply. When the retro-rockets fired the deceleration gave Borman and Lovell a momentary force of 1 g before returning them to the familiar weightless state until the effects of the atmosphere squeezed a 3.9 g load on during descent. Neither crewmember was particularly troubled by the experience, most of their attention centered now on getting the spacecraft back at the appropriate spot in the Atlantic.

Spacecraft 7 splashed down 11.8 km off target but within easy reach of the USS *Wasp*, out to repeat its fishing trip for astronauts following a successful 'bag' two days before. Thirteen days and eighteen hours after lift-off, Frank Borman and Jim Lovell were back. They landed at 9:05 am and by 9:37 were being put down on the carrier's deck, followed 31 minutes later by spacecraft 7. Apart from a general feeling of fatigue the crew were in remarkable good health, quickly restoring themselves to a pre-flight condition. There were effects of the two-week mission, but most of those could be circumvented on another mission by adequate exercise or conditioning prescribed on the knowledge gained from Gemini VII. As a mission designed to prove that man could survive for the duration of an Apollo Moon flight it had been a great success, and as target vehicle for spacecraft 6 it had adequately supported the objectives of Gemini VI-A.

One week before Christmas it seemed that what a year before had been the latter end of an age of experiment and preparation had been truly transformed into an era of space operations packed with adventure and achievement. But warning signs were there to caution over-optimism and to wag a figure at lethargy and poor quality control. Increasingly, minor problems were creeping into Gemini missions: the fuel cells never ever worked really well; thrusters balked at lengthy use; lights fused in spacecraft cabins; radar systems went on the blink. It was nothing very threatening but with hindsight would be seen as the product of an increasingly blasé attitude. The feeling was rife that Chris Kraft's controllers could rectify anything, work around any malfunction, put right any systems failure, and press on regardless. Gradually, insidiously, a barrier was building up between the ability of engineers and technicians to fabricate the components of a spacecraft and the very possible lethality an incorrect assembly of potentially faulty components could set up. Nobody really believed space flight was as dangerous as Space Task Group engineers considered it to be just seven years before when deciding how to put a man in space.

But Gemini was only half way through its manned flight schedule and there was more work ahead, polarized principally by three specific types of activity: EVA, with the new generation of support equipment then being worked at MSC; docking practice with Agena targets; and several different forms of rendezvous techniques that could be adopted by Apollo. Whereas the first half of the operational Gemini program had been a time for breaking records, setting new ones, and demonstrating the practical feasibility of EVA, long duration flight, and rendezvous on the fourth revolution, the second half would be characterized by less dramatic consolidation of the former and the latter. It was probably just as well no more long duration missions were being staged, for the fuel cells were working fine in short durations and they would be the prime source of electrical power for the remainder of the program. Each Gemini mission from this point on would probably last three days, sufficient time to fly new and unique types of rendezvous, and test the EVA procedures being developed at MSC.

The Air Force was coming along with its Astronaut Maneuvering Unit under the aegis of Colonel Daniel McKee, head of the Systems Command Field Office at MSC, and planned, with NASA, to evaluate its performance on Gemini IX, when Gene Cernan would perform an EVA. Yet as early as mid-1965, while Cooper and Conrad were flying their record-breaking eight-day mission in space, MSC's Flight Crew Support Division and the Crew Systems Division were completing initial tests with two new items scheduled to get a workout on Gemini VIII. Recognizing the limited supply of emergency oxygen and the poor ventilation capabilities of Ed White's VCM chest-pack, engineers at the Manned Spacecraft Center designed a completely new pack meant to extend the availability of reserve oxygen, improve the thermal control characteristics, and raise the efficiency of suit flow and pressurization.

Like the Gemini IV Ventilation Control Module, the Extra Vehicular Life Support System, or ELSS, was to be connected to the spacecraft suit system via two multiple gas connectors and the umbilical between the crew compartment and the astronaut, and was to be fitted to the pilot's chest with velcro. Weighing 19 kg, the ELSS was to be a rectangular box 45.7 cm high, 25.4 cm wide, and 15.2 cm deep, containing an ejector pump for circulation, a heat-exchanger for cooling air, and a 30 minute reserve oxygen supply. Controls and a warning system for the emergency oxygen supply were mounted on top. The ELSS, more simply referred to as just the chest pack, would be carried by every space-walking Gemini pilot.

Recognizing that Ed White needed more maneuvering control with his oxygen hand gun, the Crew Systems Division designed a larger capacity tank into a back-pack specifically built for Gemini VIII as a trial run of procedures to be followed on the following flight when the Astronaut Maneuvering Unit would be tried. The ELSS, or chest-pack, could only pressurize an astronaut's suit with oxygen supplied from the spacecraft, and there were limits to the length an umbilical could be with its oxygen line inside; too long and it could get snagged, or kink, cutting off vital supplies. What was needed was a separate life-support system capable of liberating the pilot from the spacecraft's supply.

The back-pack designed at MSC for Gemini VIII, called the Extra Vehicular Support Package, or ESP, was a 41.7 kg structure 66 cm high, 53.3 cm wide, and 43.2 cm deep. The ESP was to be mounted in the rear of the adapter's equipment section within a cradle. Upon reaching that location, the Gemini VIII pilot would seat himself at the cradle and back on to the ESP, securing it to his suit by a nylon strap which would come across his shoulders and fasten to the front of the chest-pack. Major components in the ESP included two high pressure gas storage bottles, the one on the astronaut's left holding a 3.17 kg oxygen supply, and the one on the right side containing an 8.2 kg supply of Freon-14, the fuel supply for a Hand Held Maneuvering Unit (HHMU). It also carried a 28-volt battery for instrumentation and communication via a UHF transceiver and wedge antenna mounted on top for radio communication with the spacecraft.

Additional equipment prepared for Gemini VIII included two tethers. One, a 7.6 meter umbilical similar to that used by White on Gemini IV, carried an oxygen line, a nylon tether, and an electrical line for communication and bio-instrumentation. The oxygen line would be protected from excess temperature by layers of aluminized mylar and the whole unit would be encased in a white nylon sleeve. A second umbilical was provided for Gemini VIII, contained within a bag attached to the top of the back pack in the adapter. Of 22.8 meter length it comprised a nylon tether and an electrical hardwire with 13 connections for communications and bio-instruments from the astronaut's body. The bag containing the tether allowed the line to be pulled out from either end: one end would be attached to the parachute harness, the other to the 7.6 meter umbilical.

Borman (right) and Lovell feel the breeze aboard the recovery ship USS *Wasp* at the end of fourteen days in space.

The new Hand Held Maneuvering Unit utilized a 351.6 kg/cm^2 store of Freon-14 from a 7,194 cubic centimeter bottle in the back pack. The 1.4 kg hand gun was connected to the ESP back pack by a flexible line that delivered gas to the pusher and tractor jets on depression of a trigger. Maneuvers totalling more than 16 meters/sec would be possible with the new HHMU, compared with 1.8 meters/sec available to White.

It was all a complex and confusing assemblage of tethers, bags, jump cables, back pack, chest pack, zip-gun, and plugs. Neither Armstrong nor Scott were ever completely satisfied with the demanding work load their simulations had trained them to accept for Gemini VIII. After a rendezvous similar to that performed by the Gemini VII/VI-A mission, Armstrong was to push Gemini's nose into the docking collar of the Agena target and perform some tests, powering down the spacecraft for a sleep period. Donning the chest pack and fitting up the short umbilical, Dave Scott was to open his right hatch about 20 hr 30 min into the flight, emerging from the spacecraft at sunrise eleven minutes later on the 13th revolution. Covering a Stateside pass, Scott was to mount a camera, retrieve a package from the side of the spacecraft set up to study cosmic radiation, move across to the Agena and retrieve a micrometeoroid detection pack, return to the spacecraft, change film in the camera, point it in an aft facing direction and move to the rear of the adapter. Checking that the back-pack and tether bag were still in their affixed location, Scott was then to return to a plate on the side of the adapter's retrograde section and demonstrate the use of a special power tool designed to be used in weightlessness; in that environment a normal power tool would turn the astronaut rather than the chuck and the Defense Department had developed a non-torque power tool for evaluation.

Returning to the adapter for the following pass around the dark side of Earth, Scott was to don the ESP back pack and at second sunrise wait while Armstrong undocked from the Agena and moved 18 meters out of plane to station-keep on the target. Scott was then to move out from behind the adapter, and position himself in increments to the full extent of the 22.8 meter tether. Across at the Agena, Scott would wait as Armstrong maneuvered the spacecraft up to him, and then follow as the command-pilot gently pulled back from the Agena to a distance of 76 meters, hauling Scott along in the process. The pilot was to get back in the Gemini spacecraft after an EVA of about 2 hr 10 min.

Following that activity, several tests would be performed with the target vehicle, including a separation and re-rendezvous from a maximum distance of 21 km using a hand-held sextant and the on-board computer. The purpose of this was to test procedures envisaged for a later Gemini rendezvous where the target would be completely passive, demonstrating and rehearsing a technique made essential by a disabled vehicle. It was a contingency operation and as such provided valuable experience in approaching a spacecraft, for instance, around the Moon but without operable radar systems. The EVA tests were contingency training too: in gently nudging the spacecraft up to the astronaut on the Agena, emergency rescue operations should an astronaut find himself physically unable to make his own way back to a motionless spacecraft. After final undocking the Gemini VIII crew were to back away and remotely command the Agena engine to fire, photographing the resulting exhaust plume. Retro-fire was to take place during the spacecraft's 44th revolution, bringing Gemini VIII back to Earth nearly 71 hrs after lift-off.

It was certainly an ambitious mission and during the second half of 1965 prime and back-up crews trained long and hard to perfect the many different techniques of its crowded flight plan. Primary activity excepting the space walk centered on the availability of an Agena target vehicle and the 25 October flop when GATV-5002 fell into the sea sent doubt through program schedules that another Lockheed stage would be up and ready for the new year. Initial trouble-shooting activity at Lockheed's Sunnyvale plant seemed to reveal the possibility that 5002 had suffered a 'hard start' where too much fuel had been injected to the engine's combustion chamber too soon.

At the Cape, Wolfgang C. Noeggerath from Lockheed teamed with MSC's Horace E. Whiteacre to hunt out the cause similarly sought by Colonel John Hudson's Agena Flight Safety Review Board. Noeggerath and Whiteacre believed an electrical failure cut engine power and that pressure built up in the tanks causing the assembly to blow itself up. They also pointed out that Agena engine tests had not been carried out at simulated altitudes greater than 34 km. By mid-November, recommendations had been made to test Agena's engine at simulated altitudes exceeding 76 km, and to re-design the flow profile so that oxidizer entered the combustion chamber first. During that fall too, McDonnell's John Yardley hosted NASA managers at his hotel in Cocoa Beach, Florida, and introduced them to an idea that would take McDonnell hardware and assemble an Agena stand-in which could be used if the Lockheed product again showed itself unwilling to fly.

Wearing an ELSS chest-pack and an ESP back-pack, an engineer tries out the Gemini VIII EVA hardware with its integral hand-held maneuvering unit.

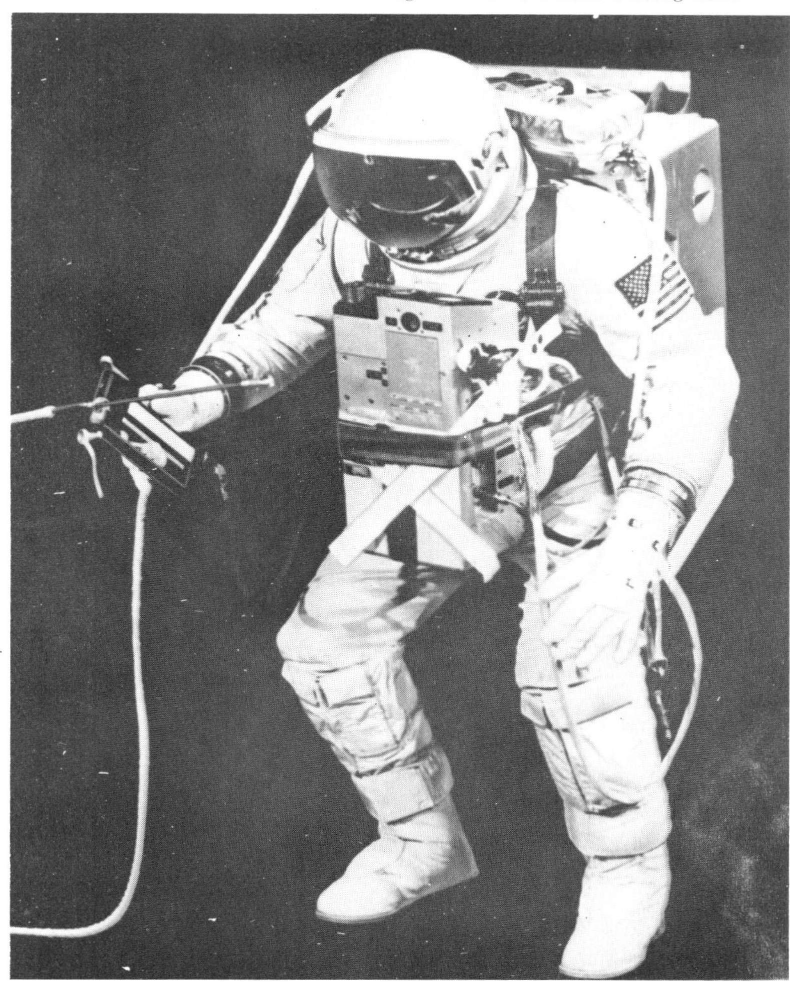

Called the Augmented Target Docking Adapter, or ATDA, it would comprise the rendezvous and recovery section of a Gemini spacecraft, bolted to an Agena docking cone; everything on the docking end would be identical to equipment on the front of an Agena. McDonnell manufactured the cone used by Agena, so the entire venture could be put together under the existing Gemini contract without bringing in subcontractors and causing attention among the legislators. George Mueller liked the idea, especially after the experience with Agena 5002, and told Mathews to get with Yardley and prepare a watertight case for going ahead with it. He would then take the proposal to Seamans for sanction. First, a telephone call to General Dynamics confirmed that a basic Atlas could launch ATDA; being a simple assembly of non-propulsive elements, it had no power of its own to get into orbit and would necessarily rely on the Atlas for lift all the way, unlike the heavier Agena which propelled itself on the final leg.

On 5 December, plans were complete and Mueller explained the idea to Bob Seamans, now the acting Deputy NASA Administrator at Dryden's death until officially appointed a few days later. Seamans gave the green light and four days later a formal work statement was ready. McDonnell began work on the ATDA immediately and by 2 February NASA had formally inspected and accepted it as a stand-by target just in case the Agena failed to iron out its own weighty problems. In one superb psychological move, NASA's Jerome Hammack sent Lockheed's Gemini Manager, Lawrence A. Smith, a photograph of the makeshift target! But it would take more than threats to get Agena on course, and Project Surefire, set up to run altitude tests on Agena propulsion systems, ran into trouble.

George Mueller wanted Lockheed to static-fire Agena 5003 slated for the Gemini VIII flight but Bernhard A. Hohmann from Aerospace Corporation disagreed and pointed out the delays in time and the cost in money. Between the beginning of February and the end of the first week in March, superhuman efforts were applied to get all the tests recommended by review bodies set up after the 5002 disaster completed for qualification. At any point during that period it was made quite clear that NASA could decide to adopt the ATDA as spacecraft 8's target. Finally, Mueller backed down on his stance about static firing – some said he never really wanted to achieve anything other than put the fear of God into Lockheed – and two weeks before Gemini VIII was due to fly, Agena 5003 was signed out for the mission. The ATDA was to be kept available should it be needed in the future.

Meanwhile, Dave Scott spent considerable time training for his EVA, logging more than 20 hours on the air-bearing trainer. This device consisted of an air-cushion floating 0.002 cm off the ground moving across a surface flat to within 0.001 cm. Standing on the floating platform, Scott practised maneuvering with the hand gun, sliding with ease across the flat surface. He was to spend more time with the air-bearing trainer than any other Gemini pilot. After the hectic nine months of 1965, when ten astronauts flew five Gemini missions, the brief respite in crew preparation was a welcome repercussion from Agena problems. But the pace was quickening again and exactly one month after Christmas Day, John Young and Michael Collins were named to fly the Gemini X mission; Alan Bean and Clifton Williams would be their respective back-ups. Now there were three mission crews again training for their flights. Simulators were in much demand and schedules were tight.

The Gemini IX crews were already booked in for a two week session on the docking simulator at McDonnell's St. Louis plant and set out from Houston's Ellington Air Force Base in two dual seat T-38 aircraft on the morning of 28 February, 1966. It was bright that Monday morning across the flats of Texas but warnings of bad weather over St. Louis brought caution to the routine flight. Weathermen briefed prime pilots Elliot See and Charles Bassett, and back-ups Tom Stafford and Eugene Cernan, that cloud base was at 180 meters and visibility only 3 km, with rain and fog. It would be done by instruments and See contacted St. Louis flight control to advise them of their Instrument Flight Routine (IFR) plan. See was in the front of one T-38, Stafford in the front of a second. It was just 7:35 am as the two single-engined aircraft rolled down the Ellington runway and thundered into the clear sky.

One hour 15 minutes later they were approaching St. Louis. Down below them, unseen in the fog, was Building 101 where McDonnell engineers were putting final touches to spacecraft 9, their own personal charge. By now the overcast had lifted to 240 meters, but visibility was worse, at 2.4 km. Fog hung low across the grey jumble of flat-roofed buildings, and rain dropped heavy water spots on the windows of the control building, snow swirling in the misty morning. Coming low out of the bottom of the cloud, its blinking navigation and landing lights brightly visible through the repelling weather, See and Bassett's T-38 came on through a little low but on course, perhaps a bit to the left of the runway centerline. But the runway was not exactly where they thought it was, the T-38 too far down the approach and going too fast. No problem: wave-off and try again. But instead of climbing straight ahead and turning at altitude to re-align with the runway, See and Bassett curved round in a low sweeping left turn, hugging the field.

Suddenly, it was apparent the aircraft was losing altitude, its sink rate too high. With a crack and a roar like thunder on slow playback tape, the T-38's afterburner cut in and howled, desperately trying to push the little two-seater high above the buildings. Too late. The aircraft's underbelly came heavily upon the roof of Building 101, tearing support structures and sending beams crashing down inside, the T-38 bouncing to the courtyard beyond and exploding in a ball of fire. Both men were dead. Inside Building 101, technicians were injured, but not seriously, and spacecraft 9 was undamaged. Two days later a low-loader moved slowly past an American flag at half-mast on the entrance to McDonnell's plant, on its back the vehicle they would have ridden into space; next day Tom Stafford and Eugene Cernan put their colleagues to rest at Arlington National Cemetery.

In time few would remember the day Elliot See and Charlie Bassett crashed into Building 101; in the light of later events there would be others to capture flagging memory. For all of that their contribution was just as great, perhaps more, for they never got the chance to be better than their peers. Within three weeks, Stafford and Cernan were named as the new prime crew, the first time a back-up team had moved into the hot seats, and Jim Lovell and Buzz Aldrin were to be the new back-up crew. Dark days indeed; but darker days ahead.

Preparations for the Gemini VIII mission proceeded on schedule, the tensions of the previous months broken at last when it really looked as though Agena would fly. The launch was scheduled for 15 March, the second time an Atlas-Agena had tried to deliver a target to space, but problems with both spacecraft and Atlas launch vehicle postponed the attempt for one day.

At 7:00 am on the 16th, Armstrong and Scott were woken to the now familiar routine. But behind the regular procession of events that accompanied each flight, the publicized schedule with key personalities, a host of support personnel busied themselves as usual for unsung efforts. Agena was to go first, sent aloft by Atlas from pad 14. Gemini VIII would follow about 101 minutes later from pad 19. But the visual separation of the two points was not easy for tourists, sightseers and spectators who, for the most part, were far removed from the perilous scene of operations. Down on the beaches, up along the parkways, lining the makeshift concrete roads, people were everywhere as there had been since Al Shepard roared into space atop a tiny Redstone booster nearly five years before. Not everybody was too familiar with exactly where he or she was supposed to go, and security personnel had the thankless task of ushering, cajoling, appealing, and persuading.

At 8:17 am, the two Gemini pilots left their Merritt Island quarters and moved to the suit-up trailer at launch complex 16. For this flight they wore a new G4C model with a thinner micrometeoroid layer to reduce bulk, an integrated thermal skin on the main part of the glove, and a few minor improvements. Across at pad 14 blockhouse personnel reported a good countdown toward lift-off at 10:00 am. During countdown operations for spacecraft 8, however, problems were noticed with heaters in the OAMS thruster circuit. A quick check by pad technicians revealed a fairing that had severed an electrical wire causing a short circuit that failed the heater. It was repaired and back in operation without any delay in the overall preparation.

Across at Houston a new chief controller, John Hodge, moved up from the blue team shift to replace Chris Kraft who was now occupied full time with running active preparations for Apollo. The first manned flight of this three-man vehicle was now expected within 12 months and Hodge would open the mission by running the activation shift. From his position as Flight Dynamics Officer, Clifford E. Charlesworth was training to become a Flight Director. Hodge decided it would be best to split the 24-hr day into two 12 hr periods divided between his own shift and Eugene Kranz's. Back at the Cape, Armstrong and Scott slipped into their seats. It was 9:46 am, fourteen minutes before Atlas-Agena lift-off.

Across the Cape people jumped up and down, shouting, cheering, some sending verbal encouragement to the ascending astronauts, as Atlas-Agena lifted cleanly into the morning sky precisely on time; only when they were driving back to Florida motels and coastal towns further north did they hear on car radios that Armstrong and Scott were about to head for space, and that it was the target vehicle they had mistaken for GT-8. It happened to a few every time, and several came back for the next one – just to get the manned launch they had missed the first time!

Five minutes' flying time from pad 14, Atlas shut down and Agena slipped free of the adapter. The moment of truth; the point in flight where GATV-5002 failed. Good telemetry streamed back to the blockhouse as GATV-5003 lit up – and continued to burn on the $7\frac{1}{4}$ tonne thrust of its single rocket motor. Three minutes later the Agena was in orbit 298 km above the Earth. At seventeen seconds past 11:32 am, as Agena was coming back over the United States at the end of the first revolution, John Hodge ordered a hold in the countdown. But this was a planned hold scheduled to last 5 min 45 sec, a cushion against some activity that could have needed attention. When it resumed at T-3 min the condition of both spacecraft and Gemini's Titan launcher looked good. At two seconds after 11:41 am, Armstrong and Scott were on their way.

Into orbit six minutes later, they punched off the second stage and slipped into a 161 km × 272 km path, 1,945 km behind the Agena. In darkness, half way round the planet, Carnarvon capcom told them not to worry about streaming water from the radiators; the pilots were worried by flaming tongues pulsing from the attitude thrusters. The first orbit was an easy adjustment to their new environment but coming back across the United States they fired their thrusters for 5 seconds to lower a slightly high apogee. For the next half revolution Armstrong and Scott broke out the rations, stored in the right aft food box behind Scott's right shoulder. While they tucked in to chicken and gravy followed by apricot pudding, bread cubes, brownies, and fruit drinks, the two astronauts aligned the computer for the upcoming perigee adjustment burn, a 15 metres/sec firing to raise the low point of their elliptical orbit 50 km; then 2 hr 46 min into the flight, a plane change to push spacecraft 8 into the same orbital inclination as the target vehicle.

Radar lock-on was obtained at a distance of 322 km and over Madagascar the coelliptic maneuver pushed perigee up a further 50 km and circularized the spacecraft's orbit about 28 km below that of the Agena. It was all going very well, just like the VI-A flight. Pulling up to within 140 km of the target, 4 hr 40 min after lift-off, the crew thought they saw what looked like artificial lights on the Agena and sure enough, when the range had narrowed to 102 km, it was unmistakable. The TPI burn came a few seconds before an elapsed time of 5 hr 15 min and Armstrong went over to the automatic rendezvous mode, checking alignments all the way up through darkness toward the blinking lights.

Across the Indian Ocean swept Gemini VIII and after two course corrections the Agena came closer and spacecraft 8 thrusted to a stop, 46 metres away. At six hours into the flight, Hawaii heard the news that two man-made objects were sitting over the Pacific, very close together. Cautiously, Armstrong approached the target vehicle and held station just 60 centimetres in front of the docking cone. The two spacecraft were now in line, flying perpendicular to the orbital path so that the Sun would pass from left to right across the nose of Gemini, the manned vehicle facing south. For nearly a full half-hour the two would sit in that position while the vehicles drifted across to within range of the tracking ship *Rose Knot Victor* stationed in the South Atlantic. It was advisable to get good telemetry feed-back during this first ever docking in orbit.

Agena's Target Docking Adaptor, or TDA, comprised the forward section designed and developed by McDonnell accommodating the docking cone itself, latches to secure the spacecraft to the target, a drive system to retract the cone, and a display panel visible to the crew of a Gemini spacecraft moving in to dock. The 147.3 cm diameter docking cone was supported on the forward end of the 1.5 meter diameter TDA by seven dampers filled with hydraulic fluid to absorb the energy of impact. Small mooring latches on the inner face of the cone would engage receptacles on the nose of Gemini on contact. When engaged, an electric motor would drive three gear boxes to pull the cone back into firm contact with the main structure of the TDA, rigidizing the complete Gemini-Agena configuration. A notch on the top of the cone allowed the crew to view the status panel indicating the condition of Agena systems; it was, in this regard, like an adopted addition to the internal crew displays within the Gemini spacecraft itself.

The Agena could be programed from Gemini by way of an encoder control circuit – a device comprising two concentric wheels and a lever down on the right side of the crew cabin. By using the programer, and observing the condition of the Agena from displays on the external panel, pilots could consider the docked configuration as a single structure from which to perform larger maneuvers than would otherwise be feasible with the undocked Gemini. Nobody was anxious to have Armstrong and Scott fire up the Agena propulsion systems, even the smaller SPS packages, until technical data about the behavior of the vehicle under docked conditions in space had been fully analyzed. But there lingered a remote hope that later flights could use the big 7¼ tonne thrust engine to push Gemini to new heights of achievement.

As Gemini VIII came across the *Rose Knot Victor*, capcom Keith Kundel gave the final approval to go ahead and dock. Telemetry was coming in solid and everything was ready. By this time the Sun was going down in the west: Neil Armstrong's position in darkness; Dave Scott's window still admitting light from the horizon off to his right. At a very cautious pace, moving at less than 20 cm/sec, Armstrong moved Gemini toward the Agena. It was quite dark when the mooring latches engaged and with a little movement detected in the spacecraft the electric motor whirred into life and pulled the manned spacecraft in to a 'hard dock' condition. The glowing status lights showed up well in the night sky as the two docked vehicles drifted across the southern countries of Africa. In a simple statement issued shortly after 6:15 pm, Cape time, Neil Armstrong uttered the words that said it all: 'Flight, we are docked!'

The first order of business was to send commands to the Agena, 16 in all, so that its attitude control system could operate cold-gas attitude thrusters and yaw the assembly round 90°. Across the Madagascar station, Jim Lovell voiced his support for the crew efforts and, prophetically, warned them to turn the Agena control system off if it malfunctioned and assume attitude control with the Gemini thrusters. It took just 55 seconds to yaw the configuration around and passing on over the Indian Ocean the crew were feeling well pleased with their day's work. But all was not well.

About 27 minutes after docking, Dave Scott started the Agena's tape recorder and noticed on his instrument display that the docked combination had yawed out of alignment by about 30°. In darkness still, there were no visual cues to indicate a change of attitude, and the motion had been so slight that neither crewmember felt the yaw maneuver. In any event, the interior lights were full on and nothing except the dim shape of the TDA stuck fast on Gemini's nose was visible. A short while before, the crew had experienced difficulty in programing commands in to the Agena and, thinking that some anomalous bit had invaded the attitude control logic, decided to switch the Agena off and assume attitude authority with the OAMS thrusters. Using his encoder, Scott turned off Agena's command system and deactivated the horizon sensors and the attitude rate system.

Over the next three minutes the attitude excursions in both roll and yaw accelerated, however, and the crew busied themselves repeatedly trying to turn the Agena attitude control system off; they had succeeded in doing that with the first command but the continuing motion led them to believe it was still on and thrusting. Then, just as Armstrong and Scott were debating what to do, the rates dropped and for four minutes the system seemed to partially restore itself. Eight minutes had gone by since Dave Scott first noticed the yaw. Then suddenly the momentum of roll and yaw started up again, accelerating faster than before. Very soon the attitude excursions were so violent that there was a very real danger of Gemini breaking away from the nose of the Agena, the structural integrity of the two assemblies likely to surrender before the gyrating configuration.

Going through every conceivable spacecraft control mode, realigning every possible thruster valve and feed system, the crew were completely unable to find the source of the wild motion. In the course of wrestling with the attitude logic, however, they restored the system to a more normal state at which point they decided to undock while the rates were comparatively low. The moment Gemini's nose pulled free, the manned spacecraft took off like a bucking bronco, turning fast in combined roll and yaw maneuvers. Obviously something was very wrong with their own spacecraft and not the Agena after all. It seemed now that the hand controller on the plinth between the two ejection seats was completely useless. Nothing Armstrong could do would null the accelerating rates. Soon, the structure of Gemini itself would be called into question. How long before pieces of the adapter, the life-giving shroud concealing valuable systems, began to break off? Three minutes went by while the roll and yaw motion picked up rates to a horrifying level of 360° every second. The spacecraft was swinging wildly in roll and yaw at sixty revolutions a minute.

By now the crew were more than dizzy: eyes glazed as vision blurred and instruments became difficult to read. Was there nothing that could be done? It was a seemingly hopeless situation. Everything they could think of had been tried – and nothing worked. In going through procedures they did not really think would work, but had to try, they completely and totally isolated the OAMS thrusters, starving them not only of propellant but electrical power as well, simultaneously switching on the re-entry control system thrusters around the nose of the spacecraft. The RCS rings were only to be used after separation from the adapter, the only means of control for the re-entry module descending through the atmosphere. In a state of desperation, the crew turned to anything that might help.

Gradually, the rates began to drop off, the spacecraft started to slow down. Furiously pumping with the hand controller that seemed at last to have authority, they regained control and within 30 seconds Gemini VIII was stable again. It had been a crazy roller-coaster ride; the phantom of the MASTIF machine had stalked them into orbit! At the height of gyrations there was real danger of the crew physically blacking out – losing consciousness from which there would have been no escape. For without the complete electrical disconnection of the OAMS thrusters the spacecraft would have accelerated further and probably broken up in orbit. Soon after spacecraft 8 had pulled free of the Agena, the crew came up on the tracking ship *Coastal Sentry Quebec* in the Pacific. Capcom James R. Fucci was puzzled by the indications of separation – Gemini VIII should have been in a docked mode at this time – and by his inability to get a fix on the rolling spacecraft. It was certainly desperation that turned them to the re-entry control system but switching the small thrusters on during orbital flight brought an inevitable price: the flight would have to be aborted. At last, an eventuality that even Houston could not excuse.

Over at the Mission Operations Control Room, anxious astronauts crowded in as John Hodge made the decision to bring Gemini back on the seventh revolution. That revolution was chosen because it provided the earliest opportunity to get the spacecraft on the water at a comparatively safe contingency point – after eleven successful manned missions, an in-flight abort and use of those contingency sites that nobody hoped ever to have to use. Gene Kranz was in the MOCR by now and Hodge accepted a suggestion that his team should relieve the tired red shift; his own men would have covered a normal Gemini VIII recovery anyway. At the end of five orbits, more than 7½ hours into the flight, the fate of spacecraft 8 had been sealed. Now that a modicum of control had been restored, Armstrong and Scott set to work finding out what was wrong with the OAMS thrusters. One by one they brought the thrusters on line, methodically examining the state of the attitude control system.

When they switched on thruster number 8 the spacecraft again started to drift off its fixed attitude at the start of what would have been another mad circus ride. Quickly disconnecting it from the line, they realized what had happened. The valve system worked correctly. Only electrical power seemed to upset the thruster. It must be another electrical short-circuit, like the one that caused a problem on the pad several hours before, like the one that cut off the thrusters on Gemini VII, like several suspected shorts for minor and irritating problems during the last few missions. But post-mortems would have to wait until after splashdown. On the ground there had been deep concern.

Lockheed personnel at the Cape feared the worst when they heard over radio circuits that Gemini VIII was in some sort of control problem. George Mueller was in the air flying to a dinner at the National Space Club in Washington when the pilot of his aircraft heard the reports over the communication line, told Mueller, and received instructions to turn around and head back for the Cape. In Washington, at the very dinner Mueller was scheduled to attend, Bob Seamans was called to a telephone. He made an announcement but disliked having to go public without adequate knowledge of the situation. Also in the air were top McDonnell engineers flying to Houston from the Cape, ready to be on hand at the MOCR should anything go wrong with the spacecraft. They could do nothing until they arrived at Ellington Air Force Base, but the lesson learned here re-wrote the schedules for top Gemini men; never again would all the contractor specialists be airborne at the same time.

At sea, the USS *Leonard F. Mason* steamed at full speed for the contingency recovery zone; naval ships had long since stopped covering *every* conceivable landing location and the sites were now mere points on a map where, in the unlikely event of abort, a spacecraft could be targeted. Now it was the real thing. There was just one full revolution to get things squared away on board spacecraft 8, many items that Armstrong and Scott had planned on dumping into space: TV monitor, film canisters, EVA visor, etc. There was no time to open the hatch and throw out the redundant items. They had to come back and places had to be found where they would be safe during the deceleration of re-entry. Because the spacecraft was beyond range of land tracking stations except Hawaii, the ships *Rose Knot Victor* and *Coastal Sentry Quebec* bore the brunt of communications dialogue, passing along to the crew instructions from Houston for retro-fire.

Almost exactly three hours after the spacecraft began its unplanned gyrations, the four solid propellant retro-rockets each burned their prescribed 5.5 seconds at the end of a countdown from the Kano capcom. It was dark when the motors burned, consecutively the third spacecraft to fire its retro-rockets on the night side of the planet. Before the crew splashed down in the eastern Pacific, 1,000 km south of Yokosuka, Japan, a C-54 had been dispatched from the Tachikawa Air Force Base in Japan and another aircraft of the same type took off from the Naha Air Base, Okinawa. Within just 45 minutes of landing a flotation collar had been placed around spacecraft 8 by pararescue men from the Naha C-54. Armstrong and Scott rode the re-entry module until the *Leonard F. Mason* drew alongside. Exactly three hours and six minutes after hitting the water, the two crewmen were on deck, joined by their spacecraft nine minutes later. At the Cape it was 1:28 am. In the Pacific it was early afternoon. The flight had lasted barely 10 hr 41 min.

Wally Schirra was at Okinawa when the ship docked 18 hours later; along with Dr. Duane Catterson and other NASA officials, he was on a goodwill tour. Just before separating from the Agena in orbit, the two astronauts had passed control of the guidance back down to the ground, so it was possible to command the on-board propulsion system to burn and push Agena to a higher orbit, a parking orbit so that it could be used again for a rendezvous later in the program. A complex set of maneuvers were carried out where the Agena's initial orbit of approximately 298 km × 298 km was changed to an elliptical path of 298 × 407 km, and then to a circular 407 km; the first occurring 21 hr 42 min after launch, the second 5 hr 21 min later. A third maneuver was planned whereby the plane of the orbit was to be adjusted. During that maneuver the stage was observed to drift in yaw and to raise apogee to 622 km as well as changing the orbital inclination. Jerome Hammack analyzed the problem and found it to be an off-set centre of gravity causing the Agena to turn slowly about its own mass.

Over two days, seven additional burns were made until Agena was in a 407 km × 411 km orbit, having expended propellant that would have constituted a danger for crews of later Gemini missions should they wish to rendezvous with the inert stage. The high orbit took cognizance of the tenuous atmosphere at that altitude which would cause the path to decay down to about 300 km within a few months; it was at that height that the next crews could visit the stage. Agena had thoroughly vindicated itself, firing its main engine eleven times in all, and receiving 5,439 commands versus the 1,000 stipulated by the specification given to Lockheed. But what of Gemini VIII? As a mission it had totally failed to perform any useful task except rendezvous and docking with an Agena. All the EVA planning was lost, along with tests of the Extra Vehicular Support Package, the back-pack serving as a practice device for the big Air Force maneuvering unit scheduled for Gemini IX. It was the one and only built and consequently was the only experiment ever lost from Gemini; others got the chance to fly more than once.

As for the spacecraft itself, the equipment section carrying thruster number 8 was jettisoned before retro-fire, burning up on the way down. It was more than a month before Scott Simpkinson's team were through with their meticulous examination in a preserved area at McDonnell's St. Louis facility of the only element that did come back: the re-entry module. But when the final report appeared it had ominous tones between the lines of fact. In closely stripping apart the components and systems in the re-entry module, they found that the thruster malfunction was 'probably caused by an electrical short,' and that in their careful search through almost every centimeter of wiring in the recovered section of spacecraft 8, 'there were *several locations* . . . at which the fault could have occurred (author's italics).'

Was the program moving too fast? This thought was already haunting key officials in government and industry. During the post-flight examination of documents and data, it was suggested that a special check should be made of the adapter section to spacecraft 9, then at the Cape in preparation for a May launch attempt, another Agena plus EVA mission. In going over wiring and installed equipment associated with the OAMS thrusters, technicians found scores of places where electrical shorts were likely to occur, modifying each point as they came to them. At St. Louis, a special circuit was designed into future spacecraft that could, at the flick of a switch, electrically isolate the thrusters collectively and at the same time. A harness was rushed to the Cape for spacecraft 9 and it was installed before the flight. Deep within the program the hectic pace of quick-fire launch schedules was taking its toll. Nobody questioned the significant change in attitude reflected by installation of a safety circuit for future shorts. There would be some still to come that could not be so easily turned off.

CHAPTER FOURTEEN

Twins Arising

The year of 1966 was expected to draw the curtains on Project Gemini. That was increasingly apparent as the year moved on and as preparations for the launch of the first Apollo spacecraft accelerated at the Kennedy Space Center. On 5 May, Joseph F. Shea, NASA's Apollo Program Office Manager at the Manned Spacecraft Center, said that he believed the agency could 'make the first (Apollo) manned flight this year, although it is not scheduled until next year.' There was certain to be a sense of urgency about activity on Apollo in light of the program's objective. The tremendous success of the past six Gemini missions – even the aborted Gemini VIII flight was seen as a success for having saved the lives of the crew – spurred the engineers, technicians, managers, and directors of the three-man vehicle to greater hopes for an early flight test. And increasingly, during the first half of 1966, Apollo leaned toward Gemini for checking operational procedures the Moon missions might adopt. Rendezvous was not one capability acquired and then possessed; it was a variety of different ways by which one spacecraft could link up with another and very many different types of rendezvous would have to be tried and tested before Apollo could fly to the Moon with confidence.

In part, that was what Gemini IX was all about. The standard fourth-orbit rendezvous mode used by Geminis VI-A and VIII bore little resemblance to several different forms of lunar-orbit rendezvous being worked on by the Apollo program offices. Personnel closely associated with planning the three-man missions thought it would be a good idea to follow more closely the timeline to be adopted for rendezvous around the Moon when the top of the lunar landing vehicle flew up to meet the mother-ship. Performed in Earth orbit that would call for a rendezvous on the third revolution, an M=3 type rendezvous versus the M=4 type flown twice before. Discussion of Gemini flying an M=3 rendezvous began before the VI-A mission of December 1965, and MSC tentatively planned to put it in the flight plans of Gemini IX and Gemini X.

In effect, an M=3 rendezvous was very similar to the fourth-orbit mode, except that the first revolution would be taken up with maneuvers. But from the end of the first revolution the standard M=4 rendezvous was little different to the faster method. By the early new year, Gemini IX had adopted three different types of rendezvous operation, all to be performed on the one mission. First, an M=3 type rendezvous, then an equi-period orbit rendezvous where the spacecraft would deliberately move away to a distance of about 21 km and perform optical and computer based re-rendezvous with the Agena, so called because if left to itself the spacecraft would still return to the vicinity of the Agena. This re-rendezvous operation would demonstrate the ability of a manned spacecraft to search out and find another, possibly disabled, spacecraft; it was a carry-over from the aborted Gemini VIII flight.

The third rendezvous was to simulate an aborted Moon landing where, having descended to a height of 15 km from the lunar surface, the lander would decide to abort, a maneuver which would place it in front and ahead of the Apollo spacecraft in Moon orbit. On Gemini IX, the Agena was to fire its SPS motors to move below and behind the manned spacecraft so that it could play the passive role of an Apollo while Stafford and Cernan became the lunar lander. It was, in effect, a re-rendezvous from above and in front of the target rather than from below and behind, placing unique demands upon the optical abilities of the crew, observing their target against an Earth background. It was certainly a full inventory of activity for the three-day mission. Spacecraft 9 was the last to carry a 316 kg load of OAMS propellant. The last three spacecraft would have a 50% greater capacity, and since maneuvers required propellant to make them happen the ambitious flight plan for Gemini IX called for frugal use of the thrusters. The three rendezvous operations alone would demand 16 orbital changes using OAMS propellant; a total twelve Agena burns were called for, including four maneuvers scheduled after the crew returned to Earth. But there was one visible sign of growing confidence in Agena. For the first time the crew were to be allowed to fire up the target vehicle's Primary Propulsion System in the docked configuration. The multiplicity of firing tests performed with GATV-5003 had paid off and NASA management finally approved of the plan to use Agena's big engine with a manned vehicle on the other end. If successful, it was planned to use Gemini X's Agena to propel the docked vehicles to a rendezvous with one of the inert Agenas from either Gemini VIII or Gemini IX.

The program was growing out of all proportion to the original concept of what Gemini should be and do. But it was worthy expansion, a very real product of success. The only other major activity planned for Gemini IX was the EVA, a scheduled 2 hr 25 min excursion to don the Astronaut Maneuvering Unit and demonstrate its capability for attitude control of a space-walking pilot. The Air Force hoped to fly the AMU on Gemini XII as well, and were anxious to see how it performed on IX so they could plan

Agena docking target with Gemini spacecraft, at the Cape Canaveral boresight range, displays respective size of both elements.

further test operations for later that year or early in 1967. It is a reflection of just how efficient the spacecraft and launch vehicle preparation phases were going that all the hardware for Gemini IX was at the Cape before Armstrong and Scott took off for Gemini VIII: spacecraft 9 arrived 2 March, the GLV-9 launcher's first and second stages arrived 8 and 10 March respectively; Atlas 5303 had been at KSC since 13 February and Agena GATV-5004 booked in on 12 March.

Yet the build-up of hardware was matched by an equally embracing flurry of astronaut assignments and selections. Just five days after Gemini VIII returned, NASA formally announced the selection of the crew to fly spacecraft XI: Charles Conrad and Richard Gordon as prime, and Neil Armstrong and William Anders as back-up. Also announced were the prime and back-up crews for the first Apollo mission: Virgil Grissom, Edward White and Roger Chaffee, and James McDivitt, David Scott and Russell Schwickart, respectively. It was to be a busy year for NASA all round. Atlas-Centaur, the new variant of the old stager first developed as an ICBM, had just about scraped through its protracted development trials and was about to perform an important flight carrying NASA's first soft-landing Moon vehicle called Surveyor to the lunarscape. Development of the Saturn I launch vehicle into a variant called the Saturn IB was hurried along to support manned Earth orbiting Apollo flights, and three qualifying missions with unmanned payloads were scheduled for the year. And a new contingent of astronauts was about to clock on at the Manned Spacecraft Center.

On 4 April, 19 pilots were announced. From the Air Force: Capt. Charles M. Duke, Jr.; Capt. Joe H. Engle; Maj. Edward G. Givens, Jr.; Maj. James B. Irwin; Maj. William R. Pogue; Capt. Stuart A. Roosa; and Capt. Alfred M. Worden. From the Navy: Lt. John S. Bull; Lt. Cdr. Ronald E. Evans; Lt. Thomas K. Mattingley; Lt. Bruce McCandless II; Lt. Cdr. Edgar D. Mitchell; and Lt. Cdr. Paul J. Weitz. From the US Marine Corps: Maj. Gerald P. Carr; and Capt. Jack R. Lousma. And four civilians: Vance D. Brand; Fred W. Haise; Dr. Don R. Lind; and John L. Swigert. Recruitment had been under way since September the previous year and NASA chose 159 that met basic requirements from 351 applicants. But the business of operational flying was going on apace and while the neophytes still had to experience the traditions of the astronaut corps Tom Stafford and Gene Cernan made ready for their three-day flight aboard spacecraft 9.

Tuesday, 17 May, dawned bright for the launch of an Atlas-Agena from pad 14. Only minor trouble was encountered during the countdown and while Stafford and Cernan moved to their spacecraft on GT-9, the final events ticked smoothly to conclusion. At 10:15 am, the three Atlas engines roared into life and again the spectators and the listeners turned eyes and ears to the by now familiar sound. But it was short lived. Two minutes into the flight the rocket was approaching the point where it would lose its two booster engines. But a mere ten seconds before separation, in the words of a flight safety review board report some time later, 'The Atlas No. 2 booster engine swivelled to an extreme hard-over position.' Thrown completely off track, the launch vehicle received commands from the Cape to shut down the single sustainer engine and to inhibit the ignition of the Agena.

At the proper time the Agena slipped away from the Atlas adapter as the entire assembly dropped limply toward the Atlantic. Signals continued to come from a perfectly healthy Agena target vehicle until 7 min 36 sec, at which point they suddenly ceased, 198 km from the Kennedy Space Center. Mission Director Bill Schneider had no option but to cancel the mission and to re-group wtih the Augmented Target Docking Adaptor, already prepared by McDonnell for just this eventuality. Ironically, ATDA had been designed, authorized, and built to back up an ailing Agena program. Through no fault of its own, Agena 5004 had been ditched and because the next Agena in line was still in preparation for Gemini X, the stand-in was the only possible candidate for a Gemini IX target vehicle. Next day, Colonel Hudson got a call from Charles Mathews requesting that the next Gemini Atlas be moved up for an ATDA flight on 31 May.

Why had Atlas 5303 pitched itself into the Atlantic? That was a question wrestled by Richard W. Keehn, General Dynamics Program Manager for the Atlas used in the Gemini missions. Only a few days elapsed before the answer arrived on a NASA desk. In assembling and checking out the launcher's auto-pilot, a technician somewhere, somehow, had pinched an electric wire causing a short circuit. With all the information now in, the re-worked, and re-numbered, Gemini IX-A mission was officially scheduled for 1 June. But now, with another three flights left, what to do if another Agena ditched itself? With the one and only ATDA gone there would be no back-up, and only the actual number of target vehicles needed to complete the program were on order. The broad solution to that could be left to the event itself – if it ever happened – but for Gemini IX-A, the flight would still go ahead and concentrate on the EVA objectives. Agena 5003 left over from Gemini VIII was still too high above the Earth to make it safe for Gemini IX-A to chase after it.

On 1 June the simultaneous countdown of Atlas 5304 and GLV-9 progressed satisfactorily to the point where Stafford and Cernan once again ascended the east elevator at pad 19, stepped into the white room and occupied the two ejection seats in spacecraft 9. Right on time at four seconds past 10:00, the Atlas engines thundered into life and once again the mission was under way. This time the launcher passed safely through the critical events leading to orbital insertion of the ATDA. Capable only of controlling its attitude, the 1,008 kg payload separated from Atlas and aligned itself in space. Down on pad 19 Stafford and Cernan went through final checks and word was passed along that the ATDA was in a safe orbit. Orbiting between 294.5 km and 298 km, the only anomaly seemed to be the lack of confirmation on telemetry sent minutes before from the launch vehicle that the ATDA's 296 kg shroud had been jettisoned as planned. If it had not, the docking cone would be covered and Gemini IX-A would be unable to link up. It was a point that could only be resolved by a personal inspection from close up, and that was just what Tom Stafford and Gene Cernan planned on doing.

An innovation for countdowns on rendezvous and docking missions provided a built-in hold at the T=3 minute mark. This was nominally for a period of four minutes, and allowed final and very precise guidance updates from target vehicle tracking to permit a small variation in the specific second of launch. Ground equipment continually fed information to the spacecraft's computer based on the continual refinements in the known positon of the target. At the prescribed time, Gemini IX-A entered its four minute hold, and the countdown resumed as planned. At T=1 min 40 sec it stopped – unintentionally. Final guidance commands could not get through to spacecraft 9's onboard computer and the launch sequence was inhibited. Cycled back to T=3 min almost immediately, the ground feed tried again to give Gemini's computer the latest target vehicle coordinates. And again it failed. A third try was similarly unsuccessful. Mission Director Bill Schneider had no option but to scrub the flight, and re-schedule it for two days later. The target vehicle was in a safe orbit and could be reached by daily windows. But it was a sorrowful crew that descended yet again from the erector white room on pad 19.

As the only man to fly two space flight missions within a period of seven months, Tom Stafford was a lucky man. But others saw it differently. On Gemini VI he had been through the countdown before the flight was cancelled, on Gemini VI-A he sat through Wally Schirra's big moment of decision, he went through a full countdown for Gemini IX, and on the latest attempt Gemini IX-A seemed reluctant to go. Tom Stafford's philosophy was simple: 'You know, I think Gene was starting to think that I was jinxed. A long time ago, Wally Schirra and I went up and down the elevator a few times and then after I had been up and down the elevator with Gene, I finally decided what it was – it was Wally that was jinxed and Gene that was jinxed and I wasn't the jinx at all!'

At the third try, Gemini IX-A back-ups Lovell and Aldrin were in the spacecraft shortly after 2:00 am on Friday, 3 June, preparing the vehicle for the prime crew who were woken little more than an hour later for the familiar routine. To prevent another recurrence of the 1 June trouble causing further holds, Houston would update the spacecraft's computer with ATDA target information at 1 hr, 30 min and 15 min prior to the planned launch time in addition to the T=3 min slot still planned. If it failed to get through, Stafford and Cernan would lift-off with what they had from the earlier inserts. Again the crew had a cursory physical, and again they suited up.

It was a different suit for Gene Cernan to that anybody had worn before. Hot gases from the Astronaut Maneuvering Unit control thrusters Gene Cernan was to strap himself into on the adapter mounting would impinge upon the legs of the astronaut and extra thermal protection was essential. The basic suit for Cernan started life as the G4C variant, a type used by Dave Scott on the preceding mission. This differed from White's Gemini IV

← Sitting atop its Atlas booster, Agena waits for launch in support of Gemini IX.

Stafford and Cernan (left) relax as they hear word of Gemini IX's target which failed to reach orbit, canceling their flight for that day.

suit in that the protective outer covering comprised two layers of neoprene-coated nylon in lieu of the high-temperature resistant HT-1 nylon and nylon felt micrometeoroid layers. But temperatures as high as 704°c were possible around the astronaut's leg area, far higher than the neoprene-coated nylon would tolerate, far higher too than the aluminized Mylar insulation, layers which separated the neoprene-coated outer covering from the micrometeoroid stopper layers beneath. HT-1 was not suitable for temperatures above 260°c so the Crew Systems Division prepared a stainless steel fabric, a woven metal cloth, called Chromel-R applied as an exterior covering to replace, in the leg area alone, the HT-1 nylon outer layer of the earlier design.

Tests also showed that the aluminized Mylar would melt beneath the temperature of the AMU thrusters so the layers of this insulating material used in the arm and torso areas of the G4C were replaced about the legs with layers of aluminized H-film inter-leaved with layers of fiberglass cloth. The new suit weighed 15.9 kg versus 10.4 kg for the standard G3C suit worn by Tom Stafford. Cernan's Chromel-R suit was tested at the General Electric Valley Forge vacuum facility and a total of eight models were constructed, three for qualification and five for training and flight use by Gene Cernan and his back-up Buzz Aldrin.

The new EVA suit for Gemini IX-A adopted improvements in helmet technology: Cernan would wear a new pressure sealing visor made from polycarbonate materials, about ten times as strong as the Plexiglas used before. This eliminated the need for an inner impact visor on the pivoting sun-screen attachment, leaving just a single gold coated plexiglas visor. It was an uncomfortable garb. Even unpressurized the Chromel-R metal reinforced leg area felt stiff and the whole apparel was bulky and difficult to work with. Cernan certainly had his own thoughts about its application; it was one thing to transform an astronaut virtually into a separate spacecraft, now they were making the suits out of the same material!

The two astronauts were in their spacecraft and buttoned up more than 100 minutes prior to the scheduled launch time, Tom Stafford having presented the closeout crew with a 1 meter long imitation match to get the rocket off the pad! This time there was no accompanying target vehicle counting down: ATDA was already in orbit. As on 1 June, the computer on board spacecraft 9 refused to accept the final updates but everybody agreed to go with what had been accepted nearly 15 minutes earlier. At 8:39 am local time, Gemini IX-A lifted away from pad 19, the reluctant launch vehicle leaping cleanly from the hot steelwork. Six minutes later the spacecraft was in orbit for an M=3 rendezvous. There was no time to look at the view on this flight. First order of business was an IVAR maneuver – Insertion Velocity Adjustment Routine – whereby the OAMS thrusters were used to trim the parameters of the orbit to the planned value. The spacecraft was in a 160 km × 267 km path, 1,060 km behind and below the Augmented Target Docking Adaptor.

Speeding across toward the African coast, Stafford and Cernan got a report on initial tracking solutions from ground radar, providing data to feed the computer ready for the phase adjustment maneuver planned for 43 minutes after orbital insertion. When it came, in the dark approaching Australia, the additional 22.7 meters/sec imparted by the thrusters rammed Gemini IX-A into a more nearly circular orbit by raising perigee 70 km. One of the more difficult burns was still due, a maneuver intended to accomplish height adjustment, phasing correction, and plane correction at the same time. Not surprisingly, it was called the corrective combination maneuver and grouped into a single operation the separate functions of three specific maneuvers performed with the fourth orbit rendezvous.

Coming across the Atlantic on the second revolution the crew got ready for the burn, feeling more comfortable now with helmets and gloves off. Minutes later, at an elapsed time of four minutes short of two hours, Tom Stafford fired the thrusters for a corrective combination maneuver, adjusting the catch-up rate, phasing the orbit, and eliminating a slight plane error. Gemini IX-A was now little more than 300 km behind the target, and gaining fast. Too fast for the final climb so the standard coelliptical maneuver came next, at an elapsed time of 2 hr 25 min just one-half hour after the triple loaded corrective combination burn. Performed at second apogee, again close to Australia, the 16.2 meter/sec speed-up burn lifted Gemini's perigee and put it on a par with apogee. The spacecraft was now in the required circular orbit, 22 km below the ATDA, trailing it by little more than 200 km.

By this time, in fact just before the coelliptic burn, Stafford got a good radar lock-on and was tracking the target with his on-board range-finder. Terminal Phase Initiation was similar to TPI with an M=4 rendezvous, only the spacecraft should theoretically be 53 km behind the target versus 74 km for a fourth orbit match. Stafford and Cernan first caught sight of the ATDA from a distance of 93 km, before TPI was to begin, the Sun glinting off its silver structure. Shortly after beginning the transfer maneuver to lift up to the ATDA's orbit, both crewmembers recognized the blue flashing light, the target now swathed in the black of a space night. That was a good sight, for whereas the reflected sunlight told them the target was actually there, the flashing light told them the shroud protecting the docking cone at launch had in fact broken free as scheduled; the light was mounted beneath the shroud and allayed fears generated by the telemetry signal shortly after launch two days earlier.

Gemini was over the Atlantic north of Brazil when the TPI

maneuver began with a single burst from the aft firing thrusters, carrying the spacecraft in a long arching loop up and toward the ATDA. As with all terminal phase maneuvers, having raised apogee to exceed the height of the target slightly, the approach and braking phase would be performed by thrusting toward the target, increasing the speed of the Gemini vehicle so that perigee would increase in height on the opposite side of the world and the chase vehicle slow down with respect to the target. Gemini IX-A swept into darkness soon after the TPI burn and from then on would remain on the night side of the Earth until after the braking maneuvers. Cruising 130° around the planet, Stafford and Cernan would slowly pull up to and match the circular orbit of the Augmented Target Docking Adaptor, visible now in moonlight that at times threatened to blind the view from spacecraft 9.

Within less than a kilometer of the small, cylindrical target, the astronauts caught sight of the docking light, sure indication they thought that the split shroud had in fact jettisoned since the docking light was situated beneath the clam shell-like doors. But then, in the strong light of a full Moon, from a distance of no more than 300 meters, the ATDA took on a strange and unexpected form. This was not the blunt faced target vehicle they were expecting to see. An extended snout seemed to be yawning from one end of the vehicle. Closer still, and it was obvious what had happened. The shroud *was* still on the front end, but half open like the jaws of an 'angry alligator' according to Tom Stafford. Without removing it from its apparently fixed position, spacecraft 9 would not dock with the cone. In contact now with the tracking station at Hawaii, Stafford gave a depressing report: 'We have a weird looking machine here . . . both the clam shells of the nose cone are still on but they are wide open. The front release has let go and the back explosive bolts attached to the ATDA have both fired . . . the jaws are like an alligator's jaw that's open at about 25 to 30 degrees and both the piston springs look like they are fully extended. . . . It looks like an angry alligator out here rotating around.'

Almost at once Mission Control looked for ways round the dilemma, discouraging the crew from moving in with Gemini's nose to pry loose the steel band that still held the two halves of the shroud together. Across the Pacific in full daylight, Tom Stafford maneuvered close in the ATDA, inspecting closely the tangled lines that bound the doors, clearly noticing, and photographing, two lanyards taped neatly around the half open shroud. Bill Schneider had astronauts McDivitt and Scott visit the Douglas plant to see if the shroud could be cut free by an astronaut outside spacecraft 9. It could, but the high tensile strength of the steel tapes could lacerate a space suit or damage the front of the Gemini spacecraft. While the crew stood by, and after they backed Gemini IX-A away from the offending target, ground commands sent to the ATDA moved the docking cone in and out in attempts to free the shroud. It nearly worked, but the two halves closed up slightly and then stopped. There was nothing for it but to re-group and save what could be saved from a mission already late and now stunted in its capacity.

Although planned for the 28th hour of the flight, it was agreed to press ahead immediately with the first of two re-rendezvous exercises to test procedures that could be adopted by the Apollo program. In the first, begun about an hour after the initial rendezvous, Gemini IX-A fired two thrusters directly away from the Earth resulting in an orbit that carried the spacecraft about 4.6 km above the orbit of the ATDA in a loop extending back down to the target's altitude but 20 km behind it. Cernan was doing all the calculations, with the radar on only for comparison between the purely optical mode they were now performing and an automatic rendezvous that would have employed radar. The idea was to demonstrate an ability to optically track and match the orbit of a target seemingly dead without transponder feed-back. It was at maximum separation, when the crew began to move into daylight again ready for the terminal phase maneuver, that the ATDA's shroud was a useful appendage. It reflected sunlight with about four times the brilliance of the target's main structure, greatly aiding in optical tracking for the manual rendezvous. Gene Cernan's calculations and alignment plots were near perfect as the spacecraft moved back up to rendezvous for the second time, little more than 6½ hours into the flight.

But no sooner had they arrived back at the ATDA than they prepared for another separation burn with the OAMS thrusters, this time to place them in a lower orbit so that during the coming sleep period they would gradually pull ahead of the target vehicle. At 7¼ hours into the mission, Stafford fired his thrusters once more, placing spacecraft 9 in a 289 km × 296 km orbit, leaving the target in an almost circular orbit at 296.5 km. Thus, on each revolution of the Earth, Gemini would rise and fall below the orbit of the ATDA, increasing its distance ahead of the target. The original flight plan called for the Agena to make the initial maneuvers, Gemini IX-A only firing up to perform the terminal phase thrusting for rendezvous. Without an Agena propulsion system on hand, the contingency operation had to be implemented whereby time was allowed to carry the spacecraft a comparatively great distance from the target to accomplish over a longer time frame the activity originally planned for a brief set of maneuvers.

After a fitful sleep period, during which noises in the spacecraft kept both crewmembers awake and prevented continuous sleep periods exceeding 40 minutes, spacecraft 9 was 111 km ahead of the ATDA. Because the rendezvous was to be attempted from above and in front of the target, Stafford fired thrusters to put Gemini IX-A up above the ATDA's orbit, slowing it down relative to the target's faster orbit and so allowing it to catch up. This height adjustment burn came at 18 hr 23 min into the flight and threw spacecraft 9 into an elliptical path that carried it about 13 km above the ATDA orbit. A second burn of the OAMS thrusters about one-half orbit later circularized Gemini IX-A at the higher altitude and so allowed the ATDA to gain space on the manned vehicle. In the course of lifting above the target's path, the spacecraft had moved to a maximum separation distance of 155 km but now the two began to move toward each other, although in reality it was Gemini slowing down for the ATDA.

Rendezvous operations involving eyeball observation of the target, lit now against the North African continent, were made difficult; as the tiny pinpoint of reflected light moved rapidly against the backgrond of sand dunes and desert patterns the crew completely lost sight of the target and had it not been for the radar they would probably have been unable to compute the maneuvers to a re-rendezvous. From its position in a near circular orbit 13 km above the path to the ATDA, Stafford began the Terminal Phase Initiation burn, an almost mirror-image reversal of the normal approach made from below and behind. Dropping down, in effect speeding up to match the faster target in the lower orbit, the crew finally came to rest alongside the ATDA about 3 hr 20 min after the morning's events began with the height adjustment burn. It had been difficult, but successful nonetheless, proving that the rendezvous from above and in front of the target would be tricky over the moving surface of a planetary body.

Not having the Agena, and not being able to dock with the ATDA, had thrown Gemini IX-A's original flight plan right out the window. The crew had performed three rendezvous operations within the first 24 hrs of the flight, however, and were to have gone into the major EVA on that second day also. Standing close by the ATDA, Stafford felt he should postpone the spacewalk until the third day and in any event he wanted to burn away from the target vehicle so that in the intervening period the two would drift apart, completely removing the risk to Gene Cernan should one of the tensile steel straps suddenly spring loose from the captured shroud. Through capcom Neil Armstrong, Flight Director Cliff Charlesworth agreed, and the two astronauts rested for the remainder of their 'day,' caught up with some of the experiments on board and prepared for the EVA.

Before separating from the ATDA for the last time, Stafford moved very close to the moose-like structure, nudging to within 8 cm of the slowly rolling stand-in. This was a good opportunity to try out the Defense Department experiment that called for evaluation of pilot ability at closely inspecting and flying formation on a non-stable satellite. ATDA was rolling about very very slowly and Stafford played a mirror-image attitude excursion, rolling around underneath the silver cylinder as it moved to lay itself across Gemini's nose. The precise and finely tuned control Stafford had over his spacecraft was a perfection of man-machine relations that would increasingly characterize the art of space-piloting. It made a lot of difference to have such delicate maneuverability.

On the third day, Stafford and Cernan made ready for the pilot to leave his right seat. The single most important task for the planned EVA was to have Gene Cernan strap himself into the Astronaut Maneuvering Unit, disconnect his life support supply from the spacecraft and fly around the spacecraft with the thrusters on his back-pack, breathing oxygen from the same source. With an autonomous life support system, Cernan would be a completely separate satellite of the Earth. The plan was to have the pilot exit from his hatch as the spacecraft moved into daylight about 49½ hours into the flight. Preparations began some 4½ hours earlier but half way through the activity Stafford reported a failure in one thruster to hold the spacecraft in the correct attitude and a malfunction in the attitude indicator. It took McDonnell's James Walker just minutes to suggest that the crew had inadvertently knocked the horizon scanner heater circuit breaker. They had, and when corrected everything was back to normal.

Flight Director Eugene Kranz was in the 'hot seat' for this EVA, the second performed within the Gemini program, and at their pass over the Carnarvon tracking station the crew were given a go-ahead for cabin depressurization. It was completed at 49 hr 19 min as the spacecraft swept across Hawaii and just three minutes later Cernan was standing in the hatch. Attached by the 7.62 meter umbilical, and wearing the ELSS chest-pack originally carried on the aborted Gemini VIII flight, Cernan had no maneuvering gun with which to position himself, but during the first daylight pass he found little need for one and worked well using the dynamic response of the umbilical to move from place to place. First tasks were associated with setting up mirrors, cameras, jettisoning redundant equipment, deploying handrails and taking pictures. Passing across the United States, Gene Cernan thought he saw Los Angeles, then wondered if he could see the Flight Research Center at Edwards Air Force Base. Cernan felt at times that the snaking umbilical was a handicap, and Stafford too was well aware of the dynamic reaction on the spacecraft. Thruster firings were inhibited by the presence of the astronaut outside his spacecraft and at one point spacecraft 9 had been turned round 150° in yaw, rolled to an inverted position, and pitched down 40° as Cernan pulled and tugged at the tether, turning the spacecraft more than himself at times.

Looking like the 'angry alligator' it was dubbed by Stafford, the ATDA target docking cone is obscured by the snagged shroud.

The first part of the EVA was going well as, just 30 minutes after opening the hatch, Cernan was taking a planned rest back at the hatch. Then, before going to the back of the adapter to don the AMU, he carefully closed the hatch until it rested on the 3 cm thick umbilical hooked up to the spacecraft's environmental control system. This was to preserve the thermal design load on the soft hatch sill seal and to prevent the interior of the spacecraft heating up from direct exposure to the Sun's rays. Apart from that, Stafford was wearing the standard G3C suit without special micrometeoroid protection and he was partially exposed to the particles that could penetrate his suit, entering the spacecraft at a slant angle through an otherwise open right hatch.

Moving to the back of the adapter, Gene Cernan now began to realise how poor had been their understanding of the necessary restraints and footholds an astronaut was expected to require in the weightlessness of space. Gemini IX-A had been fitted with hand bars, foot rests, and hand holds comprising strips of Velcro, a hook-and-eye material developed as a ready means of 'fastening' objects to interior surfaces of a weightless spacecraft. But even these were inadequate as Gene Cernan discovered when, in the back of the adapter, he wrestled with the arm rests on the AMU.

By now, 58 min after the hatch opened, the spacecraft was moving into the night side of the Earth and Cernan was up to schedule with his planned activity. But thermal insulation between layers of his suit pulled apart and he began to warm up as he stood and floated in the adapter with his back to the descending Sun. Now the Sun was down and the temperature of his pressure garment was decreasing, he felt he could better tackle the difficult tasks that in simulation had seemed so easy. But it failed to improve significantly problems of inadequate body restraints and positioning aids. Nobody had satisfactorily estimated the complexity of movement a weightless astronaut induced when trying to remain at a fixed work station. Every movement of an arm or a hand against the solid surface of the spacecraft would send him turning around his own center of mass. Considerable physical effort was required just to remain in the vicinity of the task; fulfilment of that task was virtually impossible unless straps or tethers were holding him firm.

As Cernan wrestled to unpack the AMU arm rests and to keep himself fixed in one attitude to ease problems with torque and inertia, his physical condition began to outstrip the capabilities of his chest-pack. Up to Sunset time, Cernan was stressing his heart to between 140 beats/min and 160 beats/min, occasionally going as high as 180 beats/min. It was building up an environmental condition that the ELSS chest-pack was incapable of accepting. Switching the pack's oxygen control to the high flow position at sunset, Cernan continued to have difficulty with the foot rests placed so as to enable him to secure his position, and with the two hand rails, one either side of the AMU, mounted for hand-hold supports while sitting down into the back-pack.

The AMU itself comprised a rectangular aluminium back-pack 81.3 cm high, 55.9 cm wide, and 48.3 cm deep, with a form-fitting cradle on the inside where the astronaut would seat himself during flight. The unit had 12 small thrusters mounted on the corners of the pack, four firing forward, four firing aft, two up and two down. About 10.9 kg of hydrogen peroxide fuel was stored in the AMU, with firing controlled by two sidearm supports attached to the back-pack structure. The left hand assembly would be used by the pilot for translation control in four directions, a switch for selecting manual or automatic stabilization and volume control of the communications. The right hand arm contained controls for positioning the crewmember in pitch, roll and yaw. The pack carried an internal oxygen supply totalling 3.4 kg, with a battery powered UHF transceiver on top of the pack for communication with the spacecraft. Green running lights on the pack, one above either shoulder, one on the bottom, and one on top behind the UHF wedge shaped antenna allowed the movements of the pilot to be observed from the spacecraft. The back-pack had a total velocity impulse of 83.8 meters/sec, each thruster producing about 1 kg of thrust. Its development had been the responsibility of Major Edward G. Givens of the Air Force Systems Command, selected as a NASA astronaut with the group named a few weeks before the Gemini IX-A flight.

The scheduled procedure required Cernan to turn around and back on to the AMU, with his feet in rests at the bottom of the spacecraft adapter. He was to have hooked up his 7.6 meter umbilical to a 38.1 meter tether folded in a bag open at each end and carried on top of the AMU; the other end was to be clipped to the parachute harness on Cernan's suit. The next procedure would have the pilot released from the adapter by a switch operated by the command-pilot from inside the spacecraft which would unlatch restraint toggles gripping the AMU. Using the thrusters, Cernan would have moved out when satisfied that the back-pack could also pressurize and supply oxygen to his suit, would have disconnected the 7.6 meter umbilical, affording himself maneuverability some 45 meters from the spacecraft. The long tether was merely a restraint, a means by which he could be returned to the spacecraft in the event of trouble; oxygen and communication equipment was integral with the AMU.

It was the sheer multitude of preparation tasks that proved the downfall of Cernan's attempt to play the Buck Rogers role. After several minutes of darkness, Cernan was able to deploy first one arm assembly and then the other, but that activity alone caused much physical effort combining the task itself and muscle power trying to stay in one place: any movement down, trying to pull free the stowed arm rest, only succeeded in sending his own legs floating over the top of the adapter; the foot rests were next to useless. Finally, Cernan made some progress and hooked up the AMU communication lines to his suit. But quality was poor as he tried in vain to get a good signal to the spacecraft's antenna from the back-pack unit. Within minutes of Sunset, Cernan's visor began to fog over, drops of moisture gathering quickly on the interior. Half way through the night side pass, Cernan sat on the form-fitting seat and talked with Stafford over the umbilical line.

When fogging occurred it was either a short-term misting up, or it was a complete layer of condensed moisture impossible to clear. As the minutes went by the faceplate became almost totally obscured, the pilot able to see the one light fixed for night-side illumination as though through a haze. Although nobody knew it at the time, the ELSS evaporator had dried up, virtually eliminating the heat exchanger for correcting the high humidity inside Cernan's helmet brought on by a high respiration rate. Shortly after sunrise, about 1 hr 34 min after opening the hatch, both Cernan and Stafford decided it would be pointless, perhaps even dangerous, to continue with the maneuvering tasks ahead while Cernan's visor was almost totally opaque. There was no sign that it was about to clear and there was nothing to do but terminate the EVA and climb back in. Resting again for a few minutes, Cernan regained about one-quarter vision but this was too little to warrant going ahead with the fly-around.

Without even releasing the AMU from the adapter, Cernan re-coupled himself back from the maneuvering unit and returned to Tom Stafford in the spacecraft. By the time he stood in the hatch the visor was almost one-half cleared but as he performed a picture-taking operation it again fogged over. The hatch was finally closed about 2 hr 7 min after it was first opened on what many would consider to have been an unsuccessful EVA. In at least one respect the failure to prepare adequately for conditions that would be met in attempting to fly the AMU was a vindication of what Gemini was all about. Without a test vehicle from which experience could be gathered in advance, the upcoming Apollo missions would be cluttered with experimental tests of techniques and equipment. It was Gemini's job to seek out pitfalls in seemingly straightforward routines and had the program been completed with perfect success on every flight it would not have been justified as a shakedown project designed to smooth operationally rough edges.

Nevertheless, it was a bitter disappointment for Defense Department technicians anxious to see how their maneuvering unit stood up under operational conditions. But it did reveal in dramatic fashion just how important it would be to anchor the pilot at his required work station. Almost from the moment Gene Cernan pulled the lid down on his EVA, technicians and engineers at the Manned Spacecraft Center were back at the drawing boards planning a new family of restraints and positioning devices designed for better body control on the space-walks planned to give three more Gemini pilots experience in working outside their spacecraft. Inside spacecraft 9, Stafford and Cernan prepared to come home. Little more than 1½ hours later Mission Control passed up figures for a small thruster firing recommended to place the vehicle in a better orbital position for retro-fire.

After another sleep period the crew were brought to action by a call from the Carnarvon capcom and less than four hours later the spacecraft in turn reported to the Canton Island tracking station that the four retro-rockets had fired and they were on their way to the Atlantic. Again, USS *Wasp* was on standby as the prime recovery vessel. In superb style, spacecraft 9 splashed down barely 0.7 km off target, and within four minutes the flotation collar was attached. For the second time in the program, a Gemini crew requested to be lifted aboard the *Wasp* in their spacecraft, cheekily waving to the carrier like two hitch-hikers thumbing a lift. Within 53 minutes of hitting water, spacecraft 9 was on deck. Three rendezvous operations, no docking, a stalled EVA, and a superb re-entry to a pin-point landing. Success was mixed with frustration; elation with failure.

All in all it had taught engineers a lot about the dynamics of a human body in EVA, and about how unforgiving space hardware can be when left to its own resources. The reason for the target's gaping jaw was simple human error. Douglas was responsible for building the shroud that covered the ATDA, as well as all the Agena shrouds which were identical, and Lockheed was responsible for attaching it to the forward end of the target vehicle. The Douglas engineer at Kennedy Space Center responsible for wiring up the pyrotechnics and for fitting the shroud left the leads free for a test before the flight aimed at proving the device would work. When he departed to attend to a pregnant wife, McDonnell technicians were left on launch day to connect everything up. The instructions written by Lockheed made reference to a drawing which had not been included because the Douglas engineer knew exactly what to do. But he was not there. In fixing up the ATDA ready for flight, the technicians were not sure what to do with the loose lanyards. So they taped them neatly to the sides of the shroud! When Gemini IX-A's cameras were unloaded, and when film magazines had been developed, the tidy work was there for all to see.

The second year of Gemini operations was proving less successful than the first only in regard to the degree of sophistication spacecraft and crews were now called upon to possess. Where the early phase of Gemini was characterized by single-point objectives, the latter sought to integrate several different

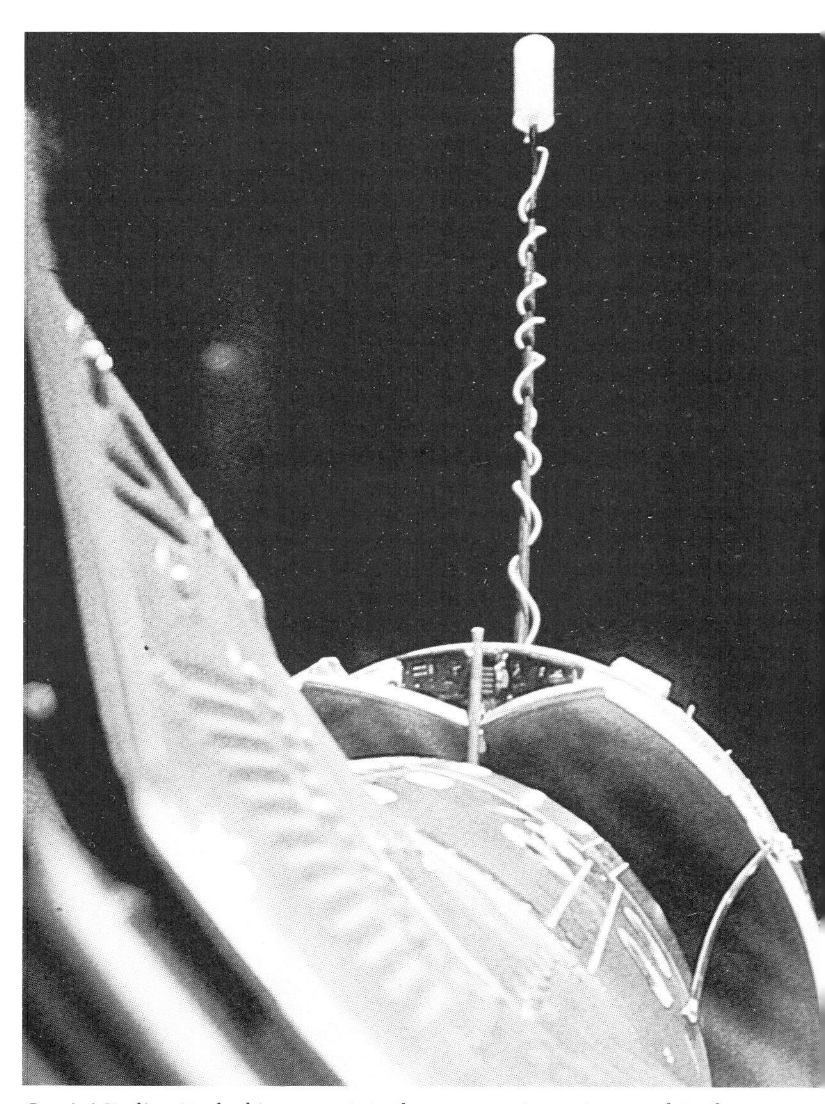

Gemini X slips its docking nose into the cone on Agena to complete the second US docking operation.

Gemini X drifts along with its Agena target in convoy, sunlight glinting off the window.

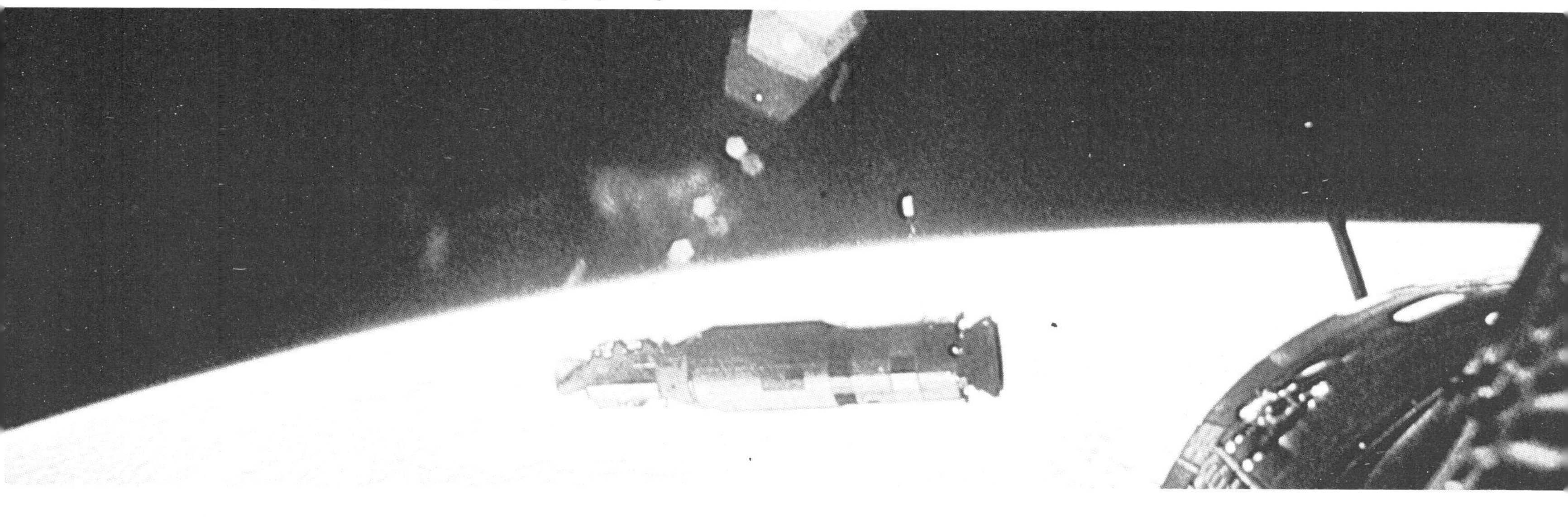

requirements within a single mission. No better example of that can be had than the highly successful flight of Gemini X. Astronauts Young and Collins had been training for their flight since January, nursing hope that theirs would be the mission granted opportunity to fire up the Agena's main engine to propel them higher than any other man had been. It had been in Gemini IX's flight plan until GATV-5004 plummeted to the Atlantic. Now Young and Collins were to get a chance.

But experience with EVA on Gemini IX-A confirmed suspicion that the increasing number of tasks assigned to space walking activity would be better served by several shorter periods outside the spacecraft rather than one long excursion. So Mike Collins began training for a stand-up EVA – SEVA – in addition to a full space walk where he was to collect micrometeoroid packages from the exterior surface of Gemini VIII's Agena, orbiting the Earth since March. The first rendezvous, accomplished like that of VI-A and VIII on the manned vehicle's fourth orbit, would carry the spacecraft to its own Agena launched earlier that day. In the docked configuration, Gemini X would use Agena's engine to place it in a correct phasing position for rendezvous with the Gemini VIII Agena at higher altitude. Separating from the first Agena, Gemini would perform terminal phase maneuvers to pull alongside the VIII Agena and it was from that vehicle Collins would collect the sampler package. Multiple rendezvous planned to accommodate two targets; multiple EVA tasks separated into two specific periods of activity: SEVA at the beginning of the second day, full EVA at the beginning of the third.

In the light of failure already experienced with two out of three Gemini Agena launch attempts, it was decided to plan a contingency whereby in the event of a failure to GATV-5005, the mission would be re-designated X-A and fly to a rendezvous with the Gemini VIII Agena on the 16th revolution. Preparations for the flight went superbly well and on the day before the mission John Young and Mike Collins rested while thunderstorms threatened the Cape and tropical storm Celia lumbered toward Florida. But nothing could deter the activities on Monday, 18 July, as pad technicians prepared two launchers and two spacecraft, radar tracking antenna being boresighted on a fifth piece of hardware in space destined for a visit two days hence. Young and Collins were to launch at the respectable hour of 5:20 pm, an unusual time but one dictated by the orbital path of GATV-5003.

Their own Agena, GATV-5005, was sent aloft at 3:39 pm. The Gemini VIII target was in an almost circular orbit 395 km above the Earth as Agena 5005 slipped into its own path almost 100 km closer to the planet. Spacecraft 10 was similar in many respects to its predecessor, except for the additional propellant provided by two spherical tanks which raised on-board quantity to 426 kg. Mission planners budgeted 118 kg for the initial rendezvous operation. The crew for Gemini X were in their spacecraft with the hatches down when Atlas 5305 thundered into space from pad 14. They had gone to bed at 2:00 am that morning, risen shortly after noon and brunched with fellow astronauts Cernan Lovell, and Slayton shortly before 2:00 pm. Now it was their turn and spacecraft 10 gave them a smooth ride all the way into orbit. Launched at 5:20 pm, they were to be docked with the Agena target vehicle by midnight Cape time.

Gemini X's initial orbit was a 161 km × 270 km ellipse, the spacecraft 1,800 km behind the Agena. It was another Gemini VIII-type operation and 5 hr 21 min after launch, the crew reported through the Tananarive tracking station that they were 12 metres away from the target, looking straight at the stars and stripes, carried for the first time by this Agena, having performed the phase adjustment and coelliptic maneuvers as planned. Terminal Phase Initiation was performed with the spacecraft slightly out of alignment, demanding costly corrective maneuvers during the coast up to the target. This burned considerably more propellant than budgeted and required a re-think about docking practice, an operation which in itself was expected to consume comparatively large quantities of propellant. The early phases of rendezvous had not gone too well either. Collins was unable to obtain good navigation sightings with his Kollsman sextant and the complete rendezvous operation was a combined effort, the crew working with information fed up from Glynn Lunney's flight team.

Nevertheless, cooperation being the keyword that day, Young and Collins slipped the cylindrical nose of the Gemini spacecraft into the docking cone on GATV-5005 about 5 hr 52 min into the flight. It had only happened once before – and on that occasion for barely 30 minutes. The flight plan called for several maneuvers following the initial docking involving each crew-member trying out the link-up sequence – backing away from the cone, then driving forward to re-contact – and flying around the Agena for scientific experiments designed to measure the target vehicle's ion wake. All of that would be cancelled said Lunney, with the next major event being the docked Agena PPS burn. Facing the target vehicle, the crew would see the engine burn at the opposite end of the Agena, and feel the acceleration push them with a force of at least 1 g in to their straps.

Ignition of the big Agena engine came at an elapsed time of 7 hr 38 min 34 sec and for 13 seconds 7¼ tonnes of thrust pushed the assembled combination, weighing about 6.8 tonnes docked, faster by 464 km/hr. It was an impressive sight to the crew aboard spacecraft 10. Looking forward along the nose of their spacecraft, the view blocked considerably by the circular docking cone with its V-notch cutout, and the circular cross section of the Agena stage beyond that, their eyes were almost blinded by the brilliant white light at ignition. John Young recalled that 'At first, the sensation I got was that there was a pop, then there was a big explosion and a clang. We were thrown forward in the seats. We had our shoulder harnesses fastened. Fire and sparks started coming out of the back of that rascal. The light was something fierce, and the acceleration was pretty good.'

Both astronauts felt the whole combination yaw off to one side, a phenomenon predicted in advance and partially compensated by ballast attached to the interior of Agena following the severe turning moment observed during ground-commanded firings of the Gemini VIII Agena engine. But Young was much impressed: 'The shutdown on the PPS was just unbelievable. It was a quick jolt . . . and the tailoff . . . I never saw anything like that before, sparks and fire and smoke and lights.' At completion of the burn, the docked configuration was in a 294 km x 763 km orbit, almost twice as high at apogee than any other human had ever been. From that altitude the curvature of the planet was pronounced, the contrast between the land and sea areas more revealing. But it had been the PPS firing that really impressed the two astronauts, and not surprisingly for it was the first time anybody had sat backward facing a comparatively large rocket motor firing in the vacuum of space.

Little more than an hour after the burn Young and Collins settled down for a nine-hour rest period, drifting in their elliptical path around the world. Shortly after waking up they were given the figures for another scheduled PPS burn, this one to lower the apogee of their orbit to an altitude about 17 km below the orbit of the Agena left in space since Gemini VIII. Shortly after 20 hr 20 min the engine fired up for 10 seconds, cutting their speed by 380 km/hr and lowering the high point of their orbit to 381 km. Again a comment from John Young: 'It may be only 1 g, but it's the biggest 1 g we ever saw! That thing really lights into you.' The next maneuver was to fire the PPS for the third and last time in the docked configuration to circularize the orbit about 17 km below the path of Gemini VIII's Agena. It came at 22 hr 37 min in a brief 2 second burst that added 90 km/hr and raised perigee to 378 km. About 8½ min later the Secondary Propulsion System fired for 10 seconds to refine the orbit, now 379.7 km × 388.7 km. The docked configuration was now in a near circular orbit and would slowly overtake the other target vehicle at which time the crew would separate from their own Agena and fire up the thrusters for a terminal phasing burn to rendezvous.

Between burns with the PPS on this second day of Gemini X's mission in space, Mike Collins prepared to carry out his stand-up EVA where he was to open the hatch and remain attached by a short tether to prevent strain on the oxygen inlet hose, supplemented for the SEVA by a 45.7 cm extension, and the outlet hose, with an additional 61 cm extension. Collins' suit was very similar to that worn by Dave Scott on Gemini VIII except it adopted the polycarbonate visor used by Gene Cernan on Gemini IX-A, with a single lens visor modified to be attached with Velcro rather than metal pivots, red-colored lenses on the fingertip lights to avoid damage to photographic film, and underwear arms and legs removed at the torso seams. The helmet was to be removed before the EVA in the pressurized cabin and coated on the interior surface with an antifogging solution to prevent a similar occurrence to that which inhibited Gene Cernan's activity. While John Young set up the third PPS burn, Mike Collins pushed ahead with readying equipment, umbilicals and tethers for the planned EVA. It was to take place as the spacecraft entered darkness to accommodate the S-13 Ultra Violet Astronomical Camera experiment, a device designed to explore the ultraviolet spectra of stars and some of the planets.

Sunset closed around spacecraft 10 at an elapsed time of 23 hr 23 min, little more than 35 minutes after the last brief Agena burn. One minute into night, the hatch came open and Mike Collins floated head and shoulders out into space. Over the next 33 minutes he set up the UV camera and proceeded to shoot off 22 exposures of the Milky Way from Beta Crucis to Gamma Velorum. Sunrise came 34 minutes after the hatch was cracked and Mike Collins handed John Young the ultraviolet camera and prepared to take 70 mm photos with a Maurer camera of color patches to see if National Bureau of Standards color slates registered the same hues when exposed in space. About 40 minutes after hatch opening, Collins began to feel a strong irritation in his eyes, a sensation similarly felt by Young. Soon, tears were rolling down the cheeks of both men as the spacecraft's two compressors fed lithium hydroxide traces back into the suit loops. There was nothing to do but terminate the EVA, which they did promptly, closing the hatch at 24 hr 12 min. Inside the spacecraft the crew turned off one of the two suit compressor fans and the eye irritation cleared itself up; clearly there would be no more simultaneous use of both loop compressors!

After a meal it was time to rest, both men settling down to a well-earned sleep. Little more than 39 hours into the flight the crew were awake and actively preparing for the rendezvous with Gemini VIII's Agena, only a passively inert target now but the scene of near disaster four months before. First, a plane change burn with the Agena's two SPS engines, fired for 18 seconds at 41 hr 4 min into the mission. Then, 31½ min later, came the last docked burn, again with the SPS, for a 4 second burst to lower apogee by 1.8 km. This phasing burn set up the correct path as Gemini X closed on the other Agena in a higher path. It was time to discard their own Agena and go hunting for Gemini VIII's target. At 44 hr 40 min John Young uncoupled spacecraft 10, having been docked securely for more than 38½ hours.

All the maneuvers computed and commanded so far were based upon tracking information from the North American Air Defense Command (NORAD) antennae and not until the two were within visual range could Gemini X's crew perform onboard calculations for the terminal braking maneuvers. At one point John Young reported the crew could see the Agena, but at a range of 176 km that would be impossible. It turned out they were really looking at their own Agena still only 5½ km distant. Without radar transponder, and devoid of any lights, GATV-5003 was a dead target. Nevertheless, final maneuvers began at 47 hr 26 min and soon the Agena was in sight. The terminal phase operation was designed to take place in daylight, since the target had no means of illuminating itself or communicating its presence. Consequently, from the first TPI burn, through the two planned mid-course corrections, to the final series of braking burns, the spacecraft would travel only 80° around the globe versus a 130° transfer employed for earlier rendezvous operations.

During the time from separation with their own Agena to rendezvous with the second target, Collins busied himself getting ready for the full EVA planned to begin shortly after John Young drew alongside the Gemini VIII Agena. When mission clocks at Houston and on spacecraft 10 clocked up a clear two days elapsed time in space, Gemini X was sitting about 3 meters away from the second Agena it had met during its travels. The EVA was scheduled to last a full daylight pass around the Earth and at an elapsed time of 48 hr 41 min, five minutes after Sunrise, the hatch opened and Mike Collins stood up. He was restrained by a tether approximately 15.2 meters in length carrying oxygen to the ELSS chest-pack in addition to providing a tension-pull inhibiting stress on the life support line.

But unlike other umbilicals, Collins' line also carried a 0.95 cm diameter nitrogen hose which would be connected to a Hand Held Maneuvering Unit at the astronaut end and to a nitrogen supply port on the side of the spacecraft adapter at the other end. The nitrogen hose joined the umbilical 2.9 meters from the supply port connector, and at the other end provided the astronaut with 1.26 meters of free line with which to maneuver the hand gun. In addition, the umbilical also carried the mandatory electrical line for communications and power. The system had been devised to prevent an astronaut having to don the cumbersome Extra Vehicular Support Package carried on Gemini VIII but never used. A hard-line from the nitrogen port led back inside the adapter to two 7,194 cc pressure vessels containing the HHMU's gas. About 4.9 kg of nitrogen was available for use, offering a total velocity increment of 25.6 meters/sec. The hand maneuvering unit itself was an outgrowth of the Gemini VIII device. Nitrogen was used instead of Freon 14 because of the temperature problems associated with the colder gas.

Gemini X had moved along quickly with preparations for the EVA and as Mission Control passed up approval for the space walk to begin John Young threw back a confident comment: 'Glad you said that, because Mike's going out right now.' The first order of business was for Collins to float out the hatch, move back along the adapter, retrieve a micrometeoroid package and place it back in the cabin, then connect up the nitrogen line from the other end of his umbilical to the port on the side of the spacecraft. Momentarily drifting back down into the open hatch, he paused for a moment, fired a few squirts from the hand gun to assure himself it worked, and prepared to move back out and across to the target vehicle. Young had maneuvered spacecraft 10 close up to the docking end of the redundant Agena so that it was hanging above the hatches, about 1½ meters distant, sloping away at a 45° angle behind the adapter. Collins gave Young final instructions to position the spacecraft and while the command-pilot held Gemini at a fixed distance from the Agena, pushed off and up, floating to the rounded lip of the docking cone under the impetus of his own mass.

Working his way round the cone, Collins found great difficulty in using the blunt lip for control, several times nearly slipping off as he moved counter-clockwise to where the micrometeoroid pack had been sitting exposed to space for four months. Unable to stop when he wanted to, the inertia of his body carried him head over heels as his legs just kept on going. Drifting off the Agena he could make his way back either with the hand gun or by pulling on the umbilical. Pulling up on the nitrogen line until the floating maneuvering unit came into his heavily gloved hand, Collins grasped it firmly and squeezed the trigger, sending bursts of nitrogen spurting into space from the small nozzles.

More by luck than judgement, Collins found himself flying straight down upon the spacecraft and feet first in the hatch, promptly coming to a halt on the seat! His movement had pulled the spacecraft out of attitude with the Agena and Young gently blipped the hand controller to nudge spacecraft 10 back into position with the target vehicle hanging overhead.

For the second time Collins drifted up to the docking cone, preferring to use the HHMU this time, but in trying to correct a pitching moment he translated too high and narrowly missed grabbing hold of the docking cone. Instead of trying to do it the planned way, Mike Collins fumbled for bunches of wires in the back of the cone, firm hand grips with which to pull himself up and over to the micrometeoroid package. During this activity Collins' heart rate never exceeded 120 beats/min and his respiration was a steady 20 breaths/min. But now, having retrieved the plate he had been working for ten minutes to get, it was time to return to the spacecraft. He was supposed to emplace another package for possible retrieval on a later flight but he abandoned that idea when it became clear that in using his hands to clamber around the cone he would probably lose the one he had just removed.

Using the umbilical to pull himself back in, Collins was prevented from giving the maneuvering gun a good workout because OAMS propellant was getting low and Young was cautioned about station-keeping on the Agena. So that was that. The EVA would have to be terminated so that spacecraft 10 could pull back from its target and drift, far away from the danger of a collision with the Agena. But the process of getting back in caused Collins great difficulty. First he became wound up in the umbilical, then he had difficulty moving arms and legs within the pressurized suit. Finally, a combined operation freed the pilot and the hatch was closed at a mission time of 49 hr 20 min. Denied the opportunity to test the HHMU fully, Collins had, nevertheless, retrieved a valuable package that would tell scientists about particles moving in the vicinity of Earth and contribute toward an understanding of near-space environmental conditions. One hour and twelve minutes after the hatch was shut on the second EVA of the mission, it was opened for a third time to eject EVA equipment no longer required.

Having discarded their original target vehicle, and having now retrieved from the Gemini VIII Agena the micrometeroid package they had come to collect, it was time for an orbit-shaping burn to lower perigee and set up the path for retro-fire. The OAMS burn came off on time at 51 hr 38 min and after taking pictures of the terrain below and the weather between, the crew aboard spacecraft 10 settled down for their last sleep period in space. Awakened about 63 hours into the mission, John Young and Mike Collins packed away their scientific equipment and loose items floating in the cabin. At 70 hr 10 min the four retro-rockets punched off, decelerating the spacecraft for a splashdown 26 minutes later only 5.4 km off target.

The USS *Guadalcanal* had charge of recovery operations that bright July afternoon in the western Atlantic and swimmers dropped from a circling helicopter had the flotation collar around the spacecraft within minutes. Only 27 minutes after splashdown, Young and Collins were standing on the ship's deck, followed a further 27 minutes later by their re-entry module. In the twelve hours following spacecraft 10's return, ground controllers commanded Agena to perform three maneuvers. The first, performed with the big PPS engine, changed its orbit to a path with an apogee of 1,390 km and a perigee of 385 km. For nearly seven hours telemetry revealed very little temperature difference at the high altitude compared to the lower orbit and the second PPS burn brought perigee down to 352 km. The final, SPS, burn put GATV-5005 into a 352 km × 347 km orbit.

With just two flights to go there was an increasing shortage of hardware and spare parts. For some months skeptics had questioned the need for pressing ahead with all twelve flights while pessimists felt they just knew that NASA management would cut the tail end of the program like many had felt they had done with Mercury three years before. Yet while NASA still believed it could get to the Moon by the end of 1969, the date for Apollo's first flight had slipped into 1967. Some wanted to combine the last Gemini with the first US three-man mission but the complexity involved in a dual manned launch operation was regarded as unnecessary. Most astronauts were training hard in Apollo simulators, learning every possible detail about the big spacecraft, and rehearsing a variety of complex and demanding flight routines.

During the second half of 1966 there were just two crews training for the last Gemini missions. Pete Conrad and Dick Gordon had been preparing for Gemini XI since March. Just eleven days after spacecraft 9 splashed down in June, Jim Lovell and Buzz Aldrin were named as the prime crew to close out the Gemini program with spacecraft 12. Gordon Cooper and Gene Cernan were to be their back-ups. After Gemini X came home it was time for stock-taking, time to call a quick inventory of objectives and accomplishments.

Two specific tasks still remained to be achieved: a quick rendezvous on the first orbit, demonstrating the type of lunar orbit rendezvous urged upon Gemini officials by the Apollo Program Office; and a satisfactory demonstration of good EVA preparation, the key to successful space walking. In addition, engineers wanted to tether Gemini to an Agena with a long rope, have the two vehicles separate to the extent of the tether, and experiment with gravity-gradient stabilization. In a similar test, it was also felt desirable to experiment with tethered vehicles made to rotate around their common center of mass so inducing a modicum of centrifugal force.

As for the first objective, gravity stabilization merely implied the pendulum effect that keeps clocks of this type ticking on the surface of the Earth. By aligning the two masses with the center of the planet, Earth's gravity acting along the tether should keep the two vehicles stabilized with respect to each other, although in reality they would be slowly rotating as they orbited the Earth continually pointing down at the surface of the spherical body. In any event it was an interesting exercise and program managers accepted the operational challenge and planned for both flights to demonstrate the experiments. So, while personnel began to trickle away from Project Gemini to other activities within the NASA organization, preparations moved ahead with the last remaining launch vehicles and spacecraft.

During the previous year, production of spacecraft and launch vehicles had accelerated to outpace the flight schedule and most of the hardware was in storage at the Kennedy Space Center in readiness for the penultimate and final missions. The day after Young and Collins came back to Earth, GLV-11 was taken from storage and erected on pad 19. Power was applied on 27 July and a day later Atlas 5306 was set up on pad 14, electrical power flowing to it on the following day. Spacecraft 11 arrived atop Titan II on 28 July and the day after that even Gemini XII's launch vehicle completed acceptance tests. On through August preparations moved toward a 9 September launch date and combined tests of hardware and crew moved ahead on schedule. There was little room for delays now in a program officially given until 31 January to complete all space flight operations.

Pete Conrad was never willing to forego a spectacular feat and upon receiving assignment for mission XI, he pressed for revival of an earlier plan whereby the Agena's main engine would be employed to push Gemini into a highly elliptical orbit of the Earth. Only the previous year support had mounted for a Large Earth Orbit (LEO) mission where Gemini would have been propelled around the Moon and back. That idea gathered a posse of supporters but Webb and Seamans remained firm in their conviction that Gemini should pursue troublesome operational tasks and shake down activities that could otherwise hamper the pace of Apollo's progress when that vehicle began flight operations. Conrad tried everything to revive the LEO plan, whipping up support by tramping a bogus excuse that weathermen would gain immeasurably by a very high orbit flight. It really had little chance of working, and it didn't. Instead, Pete Conrad would have to be content with a similar Agena boost to that performed by Gemini X, although he would be propelled to higher altitude than Young and Collins.

During the early stages of the 9 September countdown a small leak was discovered in the first stage oxidizer tank of GLV-11, causing a postponement to the following day. On the 10th, Conrad and Gordon got up, had breakfast, suited up and transferred to pad 19. In the time it took them to move from the ready-room to the spacecraft at the top of the erector, launch engineers found cause to hold the countdown and to have the crew stand by in the white room until the problem, a malfunction in the target vehicle's launcher, had been resolved. It was not and at 7:19 am, about 56 minutes after the hold took effect, Mission Director Bill Schneider called the crew down and scrubbed the flight. It was re-cycled to Monday, 12 September.

America's first astronaut, Alan B. Shepard Jr., was the breakfast guest of Conrad and Gordon in the early morning hours of launch day. Breaking what had become a tradition, the crew substituted sirloin strip steaks for filets with their scrambled eggs, toast, coffee and fruit juice, before moving to the suit-up area at pad 16, then to the erector white room surrounding their spacecraft. Arriving at 7:25 am, Armstrong and Anders brought Gemini XI's prime crew up to date with a countdown status report, then checked switch positions before leaving. When the crew arrived at the white room they had been greeted by a notice stuck to Dick Gordon's hatch: 'This is ABSOLUTELY your last chance!' During the final minutes of the Atlas-Agena countdown, proceeding within the pad 14 blockhouse nearly 2 km away from spacecraft 11, Pete Conrad called the closeout crew in on a suspected oxygen leak through his hatch. It was re-opened checked, found indeed to have been poorly secured, re-locked and closed out a second time.

That delay cost 16 minutes of time but at precisely 8:05:01.725 am, Atlas 5306 roared away from the Kennedy Space Center. Minutes later it was in a 284.6 km × 302 km orbit, circling the Earth in 90.55 minutes. Precision from here on was the watchword to success. Gemini XI would go in a hurry for this M=1 rendezvous, docking before starting the second revolution to simulate the ascent of a lunar landing vehicle from the surface of the Moon to the orbiting mother ship, Apollo. Gemini's T-3 minute hold would be critically tuned to phase the launch of spacecraft 11 with the precise orbital parameters of the Agena target vehicle. Pete Conrad asked the control center for updates on the precise launch time and was told that it would be plotted for 9:42:26 am, with ignition of the two Titan engines three seconds earlier. There would be no time on this flight for compensating launch errors, the window itself was only two seconds in duration. If delayed longer than that, the flight would be pointless.

At exactly 9:37:05 am, Gemini XI paused for a two minute and twenty-one second hold, phasing the spacecraft's lift-off to the precise state of the target's orbit. When the bolts holding GT-11 shattered, the sound of pyrotechnics firing was concealed beneath the thundering roar of the thrusting Titan, lifting smoothly into clear skies for a rendezvous within the first orbit. Coming up on six minutes into the flight, the second stage shut down and spacecraft 11 pushed off. Through his window, temptingly distracting him from important guidance tracking, Dick Gordon stole a view of the vast amount of material that blew away around the spacecraft when the pyrotechnic charge separating adapter module from rocket stage fired. It was a colorful and unexpected sight, although he had been told about the view by pilots on missions less critical than this one.

From now on it would be all up to the spacecraft itself. Unaided, Conrad and Gordon were to perform the computations for rendezvous. First would be an IVAR maneuver to trim dispersions imparted by the launcher and place them exactly at the correct position in space with respect to the target, now 430 km distant. Spacecraft 11 was in a 161 km × 279.6 km orbit, a path that would place the chase vehicle about 18 km below and 28 km behind the Agena ninety seconds before Gemini reached first apogee on the other side of the world. That was the keyhole in space through which Gemini XI would thread itself for Terminal Phase Initiation. Calculations on board the spacecraft were cycling off at the assigned times, producing confidence that everyone had done their sums correctly and would shortly receive the reward.

But first, a short burn of the thrusters to push spacecraft 11 into precisely the same orbital plane as the target: 28.85° to the equator. That maneuver came off over Africa 23 minutes after obital insertion. So far so good. The only anomaly occurred when Houston was late in passing up to the crew information that actual lift-off had been 0.5 seconds late, putting the spacecraft into orbit $3\frac{1}{2}$ km from where the crew thought they were at the IVAR burn. But no matter, that dispersion could be nulled out during TPI. When the time came to point the nose of Gemini XI up toward the target, still ahead in orbit, the radar light flashed a strong signal to the watching duo. Switched now to the rendezvous mode, the computer would relieve Conrad and Gordon of much time-consuming alignment navigation.

Over Tananarive, John Young called the crew from Houston and fed to them verbal information on their trajectory obtained via the tracking antenna over which they had flown since getting into orbit. The determinations as to position, and the recommendation on the TPI burn, were so close to their own onboard calculations that they decided to go autonomously on to the bitter end with spacecraft data. The terminal phase section of the rendezvous was to encompass a 120° radius of travel around the globe, harder to perform than the normal 130° but not as critical as the 80° traverse adopted for the Gemini IX-A equiperiod re-rendezvous and the Gemini X/Agena VIII rendezvous. Just 49 min 58 sec into the flight, moving toward Australia, Pete Conrad fired up the thrusters for Terminal Phase Initiation. Spacecraft 11 was 18 km below and 28 km behind the Agena target vehicle. It would now coast on up to the Agena's orbit via a sequence of two mid-course corrections.

Gemini XI views the Gulf of Aden.

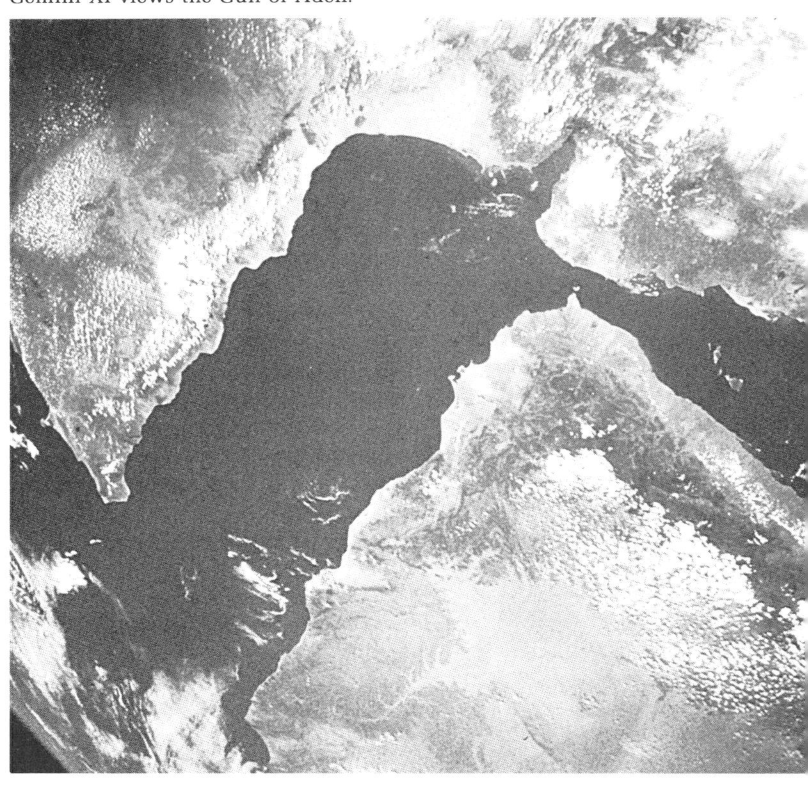

Watching with curiosity the blinking lights on the target, Conrad and Gordon tracked the object ahead and trimmed the path of their ascending spacecraft. Suddenly breaking into the dawn of a new space day, the target flashed confusingly bright reflections to the eyes of the Gemini crew. Pete Conrad fumbled for his sunglasses, then dropped them, found them again and quickly put them on. Launching 16 minutes late (due to the countdown hold to check a suspected leak), the Earth had spun further into the Sun and because the rendezvous was fixed with respect to the center of the planet, both vehicles emerged from the night-side terminal phase earlier than expected. It was no time to lose focus on instruments or target, and no time to be blinded by the bright Sun outside.

At an elapsed time of 1 hr 18 min, moving across the Pacific toward North America, spacecraft 11 began the braking thrusts that would halt the relative movement between chase vehicle and target. Just seven minutes later the two spacecraft were stopped. Rendezvous had been achieved, and from the spacecraft: 'Mr. Kraft – would (you) believe M equals 1?' It had been Chris Kraft's show from the beginning. Even Mission Director Bill Schneider had doubted the ability of the crew to perform the necessary operations to pull alongside an Agena within one orbit of the Earth. Now he was sure it was possible. In the MOCR capcom John Young passed up the word that Conrad could go ahead and dock with GATV-5006. And at 1 hr 34 min 16 sec, that was what he did, the command-pilot casually but determinedly confirming what everybody on the ground saw on telemetry: 'We are docked!' Down below were the southern States of the Union. It was the fastest rendezvous and docking from a ground launch to be achieved in the history of manned space flight.

But it had cost propellant: 131 kg, more than one quarter of the total tanked supply. Paradoxically, that was better than the Gemini X performance where Young and Collins consumed 163 kg during their fourth orbit rendezvous. On board spacecraft 11 it was experiment time, each crewmember practicing the docking maneuver in daylight and in darkness. Scientific activities kept Conrad and Gordon busy for a while, activating a nuclear emulsion experiment, photographing the gegenschein and two comets. Less than three hours after the first docking, Conrad commanded the main Agena engine to fire for 3 seconds in a plane change maneuver to check out re-ignition satisfactorily before going for a high altitude burn on the second day. It had been a tiring day and both crewmembers were glad of the opportunity for a rest, beginning about 6:00 pm Cape time.

On the mission's second day, Gordon was to go space walking to attach a 30 meter long tether contained within a bag on top of the Agena docking cone to the forward index bar on the nose of the Gemini spacecraft so that tethered gravity-gradient stabilization tests could be performed. Also, he was to try out a maneuvering gun similar to the HHMU carried by Gemini X but contained by a fitting within the adapter to alleviate the pilot of tricky operations in the confines of the cabin. Nitrogen gas was carried from tanks in the adapter identical to the Gemini X fitting, but through an umbilical which allowed the pilot to move 9.1 meters from the spacecraft rather than 15.2 meters.

The flight plan gave Conrad and Gordon four hours to get the pilot ready for this activity – far too long, as the crew found to their discomfort. Suited up and ready to go within 50 minutes, Conrad called a halt to the proceedings, waited for an hour while Gordon sat cluttered up with bulky equipment, then connected the pilot's suit to the ELSS chest-pack for an oxygen flow check. That had the effect of releasing gas to the pressurized spacecraft cabin which had to remove the excess by dumping it through relief vents to the vacuum of space, wasting valuable supplies. Then Gordon's heat exchanger, not designed to work in a pressurized cabin environment, caused the pilot to overheat, creating further discomfort as both men idly waited for time to pass. As command-pilot, Conrad even considered requesting an early start to the space walk, but finally decided against that. To have changed the sequence at this point could bring repercussions further down the flight plan.

By and by, the time came to finally button up and proceed. But fumbling with a catalogue of bulky fittings, Gordon found it nearly impossible to fasten his sun visor. Suddenly, the pace had to quicken, for what had been a depressingly protracted schedule now seemed inadequate for the difficult job of fastening up the helmet. Gordon soon began to perspire, the heat exchanger in the ELSS again stressed beyond its performance. Conrad tried to help, but the confined space in Gemini XI made assistance a casual thing. Wrestling with his visor, Gordon inadvertently cracked it, although it was a post-flight examination that revealed the fact; serving merely to shade the pilot's eyes, the Sun visor had no critical function. Finally, panting heavily, Gordon got his hatch open at an elapsed time of one day and two minutes, nine minutes after Sunrise.

Suddenly, everything loose in the cabin drifted straight up and out the open hatch. Debris, garbage, bags – even Dick Gordon went sailing straight up until a 'Hey, grab me, I'm leaving you,' brought Pete Conrad's hand to the pilot's leg strap for a helping restraint. A 23 cm long strap folded inside a Velcro pocket had been sewn to Gordon's left calf area for just this purpose. Finally stabilized, the pilot checked that the adapter hand rail had deployed properly, manually put out the forward rail, retrieved a nuclear emulsion package from the exterior, and set up the 16-mm Maurer EVA camera. The first major task was to go to the docking interface between the spacecraft and the Agena and attach the tether, without which the combined gravity-gradient tests could not be performed, to the docking index bar. Gordon pushed off and moved under his own impetus, then grabbed the open orifices of the Re-entry Control System thrusters as hand holds to pull himself forward. Like Cernan and Collins before him, Dick Gordon re-discovered the unexpected difficulty in accomplishing even the simplest task outside a weightless spacecraft and overshot the docking adapter before Pete Conrad reached across from inside the re-entry module and pulled him back with the long umbilical tether. Again Gordon tried the task, and this time he managed to halt at the top of the docking cone.

Intending to straddle the space between the nose of the spacecraft and the Agena cone with his legs jammed around the cylindrical structure, Gordon found it more difficult than in the simulations. Attempting to use his legs and feet to hold himself in position while his hands were free to fasten the 30 meter tether to the spacecraft, Gordon looked for all the world like a showground bronco-buster! 'Ride 'em, cowboy!' yelled Pete Conrad as Gordon slowly sank and rose from the top of the spacecraft's nose. Struggling hard to attach the line he finally had it snagged on the Gemini index bar. But it had nearly exhausted him in the process. So much energy just to stay in one place. Even the very act of breathing seemed to involve its own motion as the pilot panted and rested for a minute.

Back at the hatch now, Gordon was faced with a decision that Conrad relieved him of making. The command-pilot could see how tired he was, how over-taxed his chest-pack seemed to be, how risky it would be to have him go further with more physical tasks. Gordon's faceplate was free of the fogging that afflicted Gene Cernan because of the fluid wiped on before the space walk began. But perspiration had run into his right eye and it was almost impossible to blink. It stung badly, and he felt wet around his neck and face. Standing in the hatch, Gordon exchanged conversation with Conrad. The next part of the EVA would have the pilot move hand over hand along the adapter rail to the back of the spacecraft where, up inside the equipment section, he would find the hand maneuvering gun he was to evaluate by moving all around the spacecraft. Conrad made up his mind that Gordon had had enough, and the EVA was brought to a close, 33 minutes after it began.

Again, body positioning aids were found to be essential, again ground preparation had fallen short of the necessary equipment. The two adapter hand rails were the best of all help but so much still was needed to keep an astronaut satisfactorily fixed to the area of his work. Considerable physical effort was called for in staying at one place, and nobody on the ground really seemed to understand fully how serious this requirement was. Apart from that the ELSS chest-pack, although much more capable of environmentally conditioning the wearer than the Gemini IV Ventilation Control Module, was too inefficient for the strenuous exertions space walking now involved. But it was too late to develop EVA equipment unique to Gemini requirements, and the Apollo back-packs were a different thing altogether. Made by a separate contractor, they were not yet ready and in any event would not have been compatible with the available space within Gemini.

Cut short by nearly 1¼ hours, Gemini XI's aborted EVA failed again to test the power tool first carried aboard Gemini VIII, and neither could Dick Gordon evaluate the hand maneuvering gun. It was the last time an EVA pilot would have such an opportunity outside his spacecraft. In one regard, the failure to pursue EVA tests with equipment not directly applicable to the Moon landing goal reflected the limitations wilfully placed upon what could otherwise have been the development of a broad manned space capability. There were clearly serious problems with space walking, but it was researched only as long as Gemini flights lasted. Without a purpose on Apollo, there were deemed to be more pressing activities demanding attention. Nevertheless, it was clear from Gordon's first EVA that Gemini XII would be the last chance to finally get it right.

Shortly after the space walk ended, the Gemini XI crew got ready to jettison excess equipment. At an elapsed time of 25 hr 37 min, they opened the hatch again and threw out the redundant baggage. The rest of the 'day' before their second sleep period was

Gemini XI astronaut Dick Gordon floats between the spacecraft and its docked target vehicle.

spent reporting medical details and taking more photographs, this time of the airglow phenomena on the horizon. Conrad and Gordon began their 7½ hour rest session at an elapsed time of 31 hr 30 min – 5:12 pm, 13 September.

Awake and ready to fire up the Agena's PPS by midnight, the crew commanded ignition at 1:12 am in the morning hours of the 14th. 'Whoop-de-doo!' yelled Conrad, 'the biggest thrill of my life.' It was indeed a pyrotechnic delight as a bright ball of fire and flame filled the rear of the Agena, right down along the spacecraft's nose. Within a second or two the light was almost invisible again, then, when tailoff signalled shut-down, there was once more a shower of sparks and fire. Agena's engine had burned 25 seconds, adding 1,009 km/hr to their orbital speed, and propelling the docked configuration to a new record height of 1,373 km above Australia. The burn came on revolution 26, the assembly coasting downhill to reach perigee at 289 km above the Atlantic before climbing again to the incredibly high apogee and returning once again to the low point in their elliptical path.

The two orbits spent climbing high above the planet enabled Conrad and Gordon to take 300 photographs, terrain views and weather views, as well as more airglow photography. The view was breathtaking. When the Carnarvon capcom asked Conrad what it was like up there he replied in typically jocular fashion: 'I'll tell you, it's GO up here, and the world's round . . . you can't believe it . . . I can see all the way from the end, around the top . . . about 150 degrees . . . the water really stands out and everything looks blue . . . the curvature of the Earth stands out a lot . . . a lot of clouds over the ocean . . . Africa, India, and Australia clear . . . looking straight down, you can see just as clearly . . . there's no loss of color and details are extremely good. . . .'

The elliptical path of spacecraft 11 and the docked Agena target vehicle had been carefully selected to place apogee over the southern hemisphere at a longitude approximately over Australia so as to avoid exposing the crew to the higher radiation levels in Earth's Van Allen belts found elsewhere around the globe. After two orbits it was time to bring apogee down and, over the United States nearly 3 hrs 23 min after the big kick, Agena's PPS burned again, this time for 22.5 seconds lowering the high point to a less dramatic 304 km. The configuration was once again in an almost circular path. Next on the list of day two's activity was a stand-up EVA from the pilot's hatch for a lengthy picture-taking session. Timed better this time on the experience with the first EVA, but in any event uncluttered because of the absence of the chest-pack, the two astronauts worked quietly and smoothly toward the hatch opening time clocked at 46 hr 07 min.

Restrained by a short tether identical to the one used by Mike Collins two months earlier, Gordon had no difficulty this time slowly performing the assigned tasks. With the ultraviolet astronomy camera installed Gordon coaxed Conrad in positioning the spacecraft at the correct angle for good targeting; the command-pilot's vision was obscured by the mass of the Agena sitting on the spacecraft's nose. Sunset came down 19 minutes after the hatch opened and for 36 minutes Gordon snapped away at Shaula, Antares, and Orion. Moving now toward the United States, the crew wondered whether to have the pilot slip down on to his seat and close the hatch until the next night side pass began 49 minutes after Sunrise. The Hawaii capcom said they were well within the limits on oxygen and that they might want to take some pictures of the United States since it was clear of cloud. The hatch stayed open and the two astronauts got a magnificent view of the continent stretched out below them.

At an elapsed time of 47 hours, spacecraft 11 was drifting toward Florida and the Atlantic Ocean pass and not for another 40 minutes would the crew have anything to do. On what was all too rare an opportunity, the crew could just sit back, stand back, or lie where they were, taking in the view. The Atlantic was not conducive to eyeball concentration of the Earth and so each man remained quietly drifting with his own thoughts. To Pete Conrad it was a claustrophobic feeling of being blocked in from all sides. With the pilot standing in his hatch there was nothing to look across at.

But to Dick Gordon it was a tranquil time. With the full glare of the Sun dimmed by the visor he wore, the steady and continuous hissing sound of pure oxygen flowing through umbilicals to the suit outlet points in his helmet area, and the utter silence of an encapsulated environment, he drifted far away in thought and peaceful rest. It was one of the most pleasing experiences he had encountered. Like floating in a micro-world submerged in a swimming pool, the sound of living things blocked, almost pleasantly discarded. There was no tendency to float free of the hatch now, the short tether gently tugging at his harness from time to time, his whole body very gently swaying back and forth. Quietly, he drifted into a shallow sleep. And so had Pete Conrad, the exertions, pressures and strain of two days taking their toll. It was a unique and deeply restful tranquillity. The only time an extravehicular astronaut simply went to sleep.

Minutes later, Pete Conrad woke with a start and calling to Dick Gordon on the communication loop, woke the pilot from his slumber. Five minutes to go before Sunset. It was time to shake awake and start aiming the ultraviolet camera again. Passing across Australia, they saw fires burning in the bush, but kept on snapping views of Shaula and Orion as planned. With just a few minutes to go before Sunrise again, and having captured photographs along two consecutive night side passes, Gordon slipped back on to his seat and closed the hatch. It was an easy operation now, not as difficult as Ed White had found that apparently simple task to be fifteen months before. McDonnell had done a good job on re-working the locks. Gordon's stand-up EVA had lasted more than two hours, and now it was time to prepare for the gravity-gradient exercise.

Conrad unlatched the Agena target vehicle at 49 hr 55 min, the spacecraft pointing directly down at the Earth. The idea was to back away from the Agena and in so doing pull free the full extent of the 30 meter long tether connecting the two. Over the Pacific the operation began but when just a few minutes out from the Agena's cone the Gemini moved to one side. Conrad had no difficulty bringing it back but in doing so the tether got stuck in the deployment bag. A sharp backward jerk with the hand controller and it was free, but that put excessive rates into the system and the tether stuck itself to some Velcro patches strategically placed to prevent the line pulling free too soon. Both vehicles were oscillating, two inertia-filled masses torquing each other at the end of the long tether. It seemed to be taking far too long to settle the configuration down into a stable gravity-gradient posture so Mission Control advised the crew to try the spin-up maneuver, where both spacecraft would cartwheel around their common center of mass.

In attempting to fire up a rotational movement, Conrad noticed something nobody had predicted. Like a skipping rope, the tether was looped out between the two vehicles, rotating around. 'Man! Have we got a weird phenomena going on here. This will take somebody time to figure out,' reported Conrad. But the skipping rope also appeared to have tension, for it was almost impossible to pull straight. It was indeed a strange situation. Whatever they did, the tether seemed to retain a memory of its looped configuration, the two vehicles oscillating at either end as the energy moved through the rope. Finally, for a seemingly inexplicable reason, the tether straightened itself out and the two spacecraft were aligned. With a light blip on the thrusters the configuration started turning, but the loop reappeared and Conrad wanted to try once more to straighten it. Houston advised the crew to sit tight and leave it alone. Within minutes it had absorbed the energy and was in line again.

Rotation was now at a modest 38° per minute, the two vehicles slowly cartwheeling around the Earth. When urged to speed up the rate by the Hawaii capcom, Conrad had his doubts about the stability of the configuration. And doubt proved right when it broke free and whipped back and forth, the Gemini spacecraft wobbling back and forth like the whip-end of a rope. Thruster fire soon reduced the whiplash effect, stabilizing the Agena also, and the rotation rate settled down at a faster 55° per minute. Neither crewmember could feel any 'gravity' effect induced by the centrifugal force at work in the rotating configuration, but when a camera was taken from a mounting and released in front of the spacecraft it moved straight back to the rear of the cabin. At an elapsed time of 53 hr 58 min Conrad jettisoned the docking index bar on the nose of the Gemini spacecraft, releasing the tethered Agena.

There was one final task to perform before the crew could return to Earth – a stable orbit re-rendezvous exercise with the Agena. The first, equiperiod, re-rendezvous of Gemini IX-A had been used to study the lighting conditions for the dual rendezvous performed by Gemini X. The second Gemini IX-A rerendezvous simulated an aborted Apollo Moon landing and the type of rendezvous from above and in front of the target that would be mandatory for bringing the two vehicles together. Gemini XI's re-rendezvous was to be different again, and pursue

The last of the batch: spacecraft 12, packaged for the final manned Gemini mission, like its predecessors was delivered by a modified Stratocruiser.

in the course of its execution an intercept course beginning with the chase vehicle in exactly the same orbit as the target but trailing it by about 46 km. All other types of rendezvous provided phasing maneuvers where the chase vehicle was either above or below the target. So, to set up the relative positions of the two vehicles, the Gemini XI crew fired the thrusters putting their spacecraft in a higher path to the Agena.

Because the target was now in a lower orbit it was moving faster and so began to move ahead of the Gemini spacecraft. Another maneuver reduced the altitude of the manned vehicle so that it was in the same orbit as the Agena but separated from it by 30 km. For about 12 hours the two would very slowly drift further apart. Having set up the required conditions for the next day's performance, Conrad and Gordon went to sleep. Both spacecraft were in the same path and because of that the situation was called a stable orbit rendezvous sequence. Having achieved the stable orbit, the rendezvous operation commenced at an elapsed time of 65 hr 27 min when Agena was 46 km ahead of the manned spacecraft. The ground computed intercept maneuver was targeted for a rendezvous with the Agena after travelling 292° around the Earth. A second and final maneuver was made when 34° of orbital travel remained.

The first burn had placed the Gemini spacecraft in an elliptical path so that it swooped 9.3 km lower than the Agena, catching up with it in the process; the second maneuver set up the alignment of that path for a terminal phase burn prior to rendezvous. The operation from intercept burn to braking maneuver lasted about 1 hr 13 min, completed in little more than three-quarters of a full Earth orbit. Just minutes later, Conrad fired up a 1 meter/sec separation burn from his OAMS thrusters and the two spacecraft moved apart for the last time, leaving 'the best friend we ever had,' according to Pete Conrad. For the next few minutes the crew swapped banter with Mission Control, suggesting that Flight Director Glynn Lunney might like to send up a tanker so they could start activities with the Agena all over again. If only it had been possible, Pete Conrad would have been the first to hook up hoses! But it was time to come home, and at 70 hr 41 min the retro-rockets fired and spacecraft 11 headed Earthward all set for yet another 'first' to add to the score.

Unlike earlier flights, Gemini XI would fly home to a 'hands-off' re-entry and splashdown. The hand controller was to be locked up and the crew were to sit out a fully automated, computer-controlled, descent. Conrad kept a wary eye on the needles, and on the response of the autopilot to navigation inputs from the inertial platform. When they hit the western Atlantic the spacecraft was only 4.9 km off target. It had worked beautifully, bringing Conrad and Gordon comfortably close to the USS *Guam*, a helicopter platform carrier. Retrieved by helicopter, the crew were standing on the *Guam*'s deck only 24 minutes after landing, and the spacecraft that had behaved so well during their 71 hr 17 min flight joined them 35 minutes later.

There was little time to waste. What had seemed after Gemini X to be a reasonable fix for troubles experienced by Cernan during his EVA, now re-appeared with depressing similarity to the Gemini IX-A problems. Gordon had been unable to continue with the full EVA primarily because of body fatigue, made unnecessarily worse by the lengthy period spent with his equipment on inside the spacecraft. Nevertheless, whatever the excuse, Gemini XI exhibited the same problems as Gemini IX-A; only Collins on Gemini X had not had his space walk terminated by physiological stress. One thing was clear. Gemini XII, the last manned flight in the series, would have to dedicate itself to probing in a controlled and responsive manner the apparent stumbling blocks to a successful long duration spacewalk. So what were the real problems?

EVA pilots to a man were surprised at the sheer workload required to perform what seemed in theory to be a simple task. Cernan listened to White, and thought he appreciated what EVA was all about; Collins took note of Cernan's surprise reaction to the work he was expected to perform; Gordon listened and talked with Collins about how difficult it is to fix the body in one place. None of them fully understood what it was like until they each in turn opened the hatch and moved away from their spacecraft. They had convinced each other because they had now all gone through the same experience. And that had the effect of presenting a consolidated front to engineers and technicians at the Crew Systems Division and the Flight Crew Support Division. Where once the ground-based designers could have thought a particular astronaut or two had perhaps bungled an activity, they now realized the men who flew the vehicles were trying to tell them something of which they had no adequate knowledge.

So there was increasing pressure, not least from the flight crews themselves, for a more intelligent placement of hand holds, tethers and body positioning devices. And Gemini XII was to be the mission to carry it all into space. But there was a contest between running the last flight as a true experiment in EVA energy management or using part of the space walk to fly the Air Force AMU, held over from Gemini IX-A because of a fogged faceplate. It was to be the responsibility of the Gemini Mission Review Board, comprising Charles Mathews, James C. Elms, Edgar M. Cortright, and Major General Vincent C. Huston, to decide the fate of the Astronaut Maneuvering Unit. At the first

Gemini XII pre-mission meeting, held 14 September at the Manned Spacecraft Center, the Board decided to recommend deletion of the AMU. That decision was ratified at the second meeting nine days later and the deliberation was passed up to George Mueller who decided to go along with that on 30 September.

In drafting a memorandum to the Air Force explaining why Gemini XII would not carry the AMU, Mueller adequately summed up the whole problem: '. . . it becomes increasingly apparent that techniques and procedures devised for EVA have evolved from analyses, theories, and experimental concepts that in certain critical instances, and for reasons currently beyond our grasp, are not entirely accurate. Consequently, I feel that we must devote the last EVA period in the Gemini Program to a basic investigation of EVA fundamentals . . .' And that was that. The Air Force would wait several more years to get their maneuvering unit back-pack tested in the weightless vacuum of space. The next three years were surely to be filled with more pressing manned space flight objectives. The Gemini Mission Review Board was proving a valuable precursor to the inevitable contest between competing factions for specific experiments and tests to fly on a particular mission. Set up by Elms' boss, George Mueller, after Gemini IX-A ran into EVA trouble, the Board met to plan and integrate candidate objectives for the last three Gemini flights.

But there very nearly was a flight without a rendezvous objective. When Atlas flew with a defective Agena in October 1965, the loss helped stress the need for additional spare launchers but when Atlas 5303 threw away a good Agena the following May it made such a need critical for full employment of manned vehicles awaiting launch. Early in 1966, the Gemini Project Office acquired an Atlas originally intended for the Lunar Orbiter program run by the Office of Space Science. When that program ran late it enabled the Gemini target to hitch a ride into space on an Atlas not too different from models used earlier for Agena flights. A new Atlas contracted from the Air Force would require purchase of a more sophisticated variant untested for the kind of work it would have to do in supporting the Gemini program. So the new Atlas went to the Lunar Orbiter program instead, lessening the risk to the last Gemini flight.

Also, the Agena itself was a reworked model. The old 5001, previously condemned as unfit for flight operations, was returned to Lockheed on 23 November, 1965, for refurbishing. When it emerged as Gemini Agena Target Vehicle 5001R on 21 July, 1966, it was assigned to GT-12 and proceeded through the usual inventory of test schedules in the months before the flight. For some months, Jim Lovell and Buzz Aldrin, whose role in rendezvous planning had been more than mere inspiration, had been training for the last flight since mid-June and in nearly five months up to launch day saw the erosion of several candidate activities, not least of which had been the trials with the AMU. It soon became very clear that Gemini XII was going to sweep up all the loose ends left hanging from earlier flights and because so much effort was to be placed on EVA evaluation the flight was allowed to grow into a four-day mission planned for launch on 9 November.

For the first time, but wisely, EVA itself would be the experiment and not tasks performed within the profile of a space walk. Measurable tasks were to be performed so that engineers could obtain tangible values for determining the amount of energy expended on a given task in weightless space versus 1 g Earth conditions. There would be three EVA periods, two of which would be stand-up activities involving scientific experiments. Two supporting facilities were to go far in making Gemini XII a candidate for success with EVA tests: a water tank for zero-g rehearsals, and numerous tie-downs and hand holds every EVA astronaut since Ed White had campaigned for. The water tank tests were a boon for practicing body movement around a mocked spacecraft and target vehicle and use of 'neutral bouyancy' facilities would characterize future training for weightless EVA. But what ground engineers had erred in leaving off earlier spacecraft they made up for on Gemini XII with its numerous hand holds and restraints.

First, there were the two rectangular hand rails standard since spacecraft 9. These were in two sections fixed on a line running back along the adapter module from a point between the center of the two hatches. Flush with the adapter at launch, the hand rail to the rear automatically popped up 3.8 cm at launch vehicle separation but the forward rail, along the retrograde section, was manually deployed by the astronaut during EVA. They were rectangular in shape to give an astronaut better grip and to prevent him rotating over the top of the rail. Because of the ELSS chest-pack, the EVA pilot was required to move along the side of the rail, gripping it with both hands, rather than pull himself hand over hand. A pair of large cylindrical handrails identical to those carried by spacecraft 9 were also provided in the back of the adapter, and two small hand rails were fixed to the cylindrical forward portion of the Agena, just to the rear of the docking cone. Recognizing how well received had been the two adapter section hand rails on earlier flights, engineers provided Buzz Aldrin with a telescopic rail that he would manually place in a socket between the hatches at the spacecraft end and in a special receptacle on the target docking cone at the other. This would provide an almost continuous hand rail capability from the back of the adapter to the nose of the spacecraft. Two short, fixed, hand holds were attached to the back of the docking cone proper and rigid Velcro-backed portable hand holds were provided at the work stations around the two vehicles. They could be used as tether attachment points, each carrying a pip-pin device through which the restraint strap could be secured.

Waist tethers, adjustable through a range of lengths from 53 cm to 81 cm, would hold Aldrin at his work area and pip-pin hand holds and tether attach devices would provide tie-down points or simply serve as handles by which to move from one position to another. Pip-pin anti-rotation devices were attached to the Agena, consisting of depressed receptacles into which a pip-pin could be placed at a fixed angle rather than retain freedom to rotate. Nine U-bolt hand hold/tether-attach devices were fixed at the Gemini XII work stations, again serving as waist tether points or hand grips. Molded foot restraints were fixed to the inside edge of the adapter section so Aldrin could slip his feet into a positioning aid designed to prevent his legs drifting around. In all, there were 44 devices to aid his EVA positioning. Two work stations were provided. The one in the rear of the adapter was 76 cm × 76 cm square, comprising a panel on which was mounted hardware such as electrical and fluid connectors, hook-and-ring combinations, Velcro strips, a fixed bolt, and a removable bolt. A similar, but smaller, work station was attached to the area behind the Agena docking cone.

As for extravehicular equipment, Aldrin would use the ELSS chest-pack worn by all EVA astronauts since Gene Cernan and employ a 7.6 meter umbilical identical to the one used by Cernan on Gemini IX-A. Aldrin would have no use of a maneuvering gun, neither would he liberate himself, environmentally, from the hose that carried oxygen to his suit. As for the suit itself, that was a modified Gemini IX-A type, with the stainless steel fabric on the legs replaced with high-temperature nylon, and four layers of aluminized H-film and superinsulation deleted. The coverlayer thermal layup was quilted to the first micrometeoroid layer, with a rectangular pattern quilted over the torso, to prevent the aluminized H-film or Mylar tearing as it had on Gene Cernan's mission.

It was a motley collection of makeshift hardware that arraigned itself across pads 14 and 19 for the last Gemini launches. On the former, the newly acquired Atlas purloined from an unmanned Moon mission supported a refurbished Agena that had never been intended as a fully operational target; on the latter, the last Titan II purchased for the program supported a spacecraft that had more than its fair share of spares obtained from here and there at the Cape. And it was a reluctant mission that prepared for flight on 9 November. A full day prior to that event, Mission Director Schneider called a halt and postponed it for 24 hours when the secondary autopilot in the Gemini launch vehicle showed erratic behavior. Then, on the 9th, another delay, this time for two days. It did provide Lovell and Aldrin, and back-up crew Cooper and Cernan with additional practice time in the Cape simulator, however.

Gemini XII was assigned a late launch and it was 10:30 am Cape time on Friday, 11 November, when the prime crew woke. After a cursory physical examination they had breakfast with ten guest astronauts, returning to the preferred filet mignon abandoned by the previous crew in addition to the usual, by now almost mandatory, eggs, toast, coffee and fruit juice. Two hours after waking, Lovell and Aldrin made their way to pad 16 and the white suit-up trailer that had seen nine Gemini crews walk out into history. Here, too, a minor problem as Aldrin's suit seemed to have blocked air tubes in the left sleeve necessitating removal and another try at donning the bulky garment. This time it was fine and the team moved out to pad 19, arriving as Atlas 5307 made

Edwin Aldrin practices EVA procedures for his Gemini mission.

ready to fly. Around the astronauts, the trappings of a last close-out, Lovell and Aldrin wore square placards around their necks, 'The' on one and 'End' on the other. And at the pad, a poster: 'Last chance. No re-launch. Show will close after this performance.'

For their Veterans Day flight, Lovell and Aldrin were in peak condition and despite the holdup at pad 16, they were inside spacecraft 12 when the Atlas-Agena lifted off 1.3 seconds before 2:08 pm. For the last time, the crew received updates on their launch time based upon the actual position of the Agena orbiting in a 294.5 km × 303.5 km path. Not quite a circular orbit, but near enough for rendezvous capabilities of the Gemini spacecraft. At thirty-three seconds past 3:46 pm, GT-12 slipped its berth and rose on a transparent flame toward a clear sky. Across the Cape, hardware for the upcoming Apollo shakedown flight was being prepared for Grissom, White and Chaffee to ride on a Saturn launcher into space. Ascending straight and true on the last converted missile ever to carry a US astronaut, Gemini XII made it look easy as the first stage separated and the second thrust on up toward space.

Rendezvous would occur on the third revolution, an M=3 type operation like that performed by Tom Stafford and Gene Cernan the previous June. It went well until the two vehicles were separated by a mere 120 km after the coelliptical sequence burn that caused spacecraft 12 to slip into an almost circular path 18 km below the Agena. From that point on it was a manual operation because the radar failed for the first time on a Gemini flight. There could have been no better pilot to handle the clipboard chores than Buzz 'Dr. Rendezvous' Aldrin. With sextant in one hand, navigation charts in the other, he gave Jim Lovell good information to make a near perfect approach through the terminal phase events. Shortly after an elapsed time of 4 hours, the two vehicles docked but during the planned undocking and redocking exercises the spacecraft's nose hung up on one of the three latches snagging the docking cone. Short bursts of first the aft and then the forward facing thrusters made Gemini XII rock like a rowing boat in the wake of an ocean liner. Suddenly, it came free and another docking proved it had been crew error; too far off-center and the device would foul up.

Then another gremlin crept out and revealed itself. During the latter stages of the ascent phase engineers watching telemetry coming to Earth from the Agena, observed a drop in chamber pressure, only by about 6% but enough to indicate a possible malfunction. Flight Director Glynn Lunney made the final decision. Gemini XII would not fire up that engine a second time just to go to high altitude. There were to be no docked burns with the PPS. But that helped re-instate a failed experiment lost when the mission was cancelled from the originally planned 9 November flight date.

A solar eclipse was to have been the target for Gemini XII's cameras but cancellation of the high-altitude ride retained the spacecraft in an orbit from where, with a small burn from Agena's two smaller SPS engines, spacecraft 12 could migrate to a brief rendezvous with the eclipse as it passed across the South American states. It was a long shot, and a heavy addition to the existing flight plan, but Lovell fired the engines at 7 hr 5 min into the mission and just under nine hours later the last Gemini became the first manned vehicle to synchronize itself with a celestial event of such aesthetic majesty. This was orbital mechanics at its best. In carefully plotting a trajectory that would cause the flight path to intersect the moving line of the eclipse, ground controllers gave Lovell and Aldrin a seven second period in which to snatch two photographs with the Moon directly between their spacecraft and the Sun. It was done on the product of a broadening capability for rendezvous operations of a different kind but with similar finesse.

Resuming the flight plan, Buzz Aldrin got ready for his first EVA, a stand-up activity designed to catch two dark side passes for ultraviolet and star photography with the 70-mm Maurer camera. By this time a nagging and persistent light told the crew of potential trouble in the two fuel cell stacks back in the adapter. Not for several missions had anything been wrong with the unique electrical production plants but there were strong indications that too much water was being produced. That seemed in fact to be the case when the crew drew off some drinking water, allowing the bladder to deflate slightly and accommodate more contaminated water on the other side. It was not critical and the EVA preparations continued. At 19 hr 29 min the hatch opened and Buzz Aldrin stood up. 'Man! Look at that!' he exclaimed, as his eyes caught sight of the magnificence below and the awesome wonder of space above. Soon it was Sunset and the spacecraft was bathed in the cloak of night. For 33 minutes he stood and photographed Creation, an intruder on the velvet doorstep of infinity.

At daybreak – installation of an EVA camera, hand rail deployment ready for the next day's space walk, micrometeoroid package retrieval, installation of the new hand rail from the hatch to the docking cone. Then some general photography of the Earth and its gossamer shrouds of cloud before getting ready for the next night side pass and more ultraviolet photography with the Maurer. Aldrin terminated his EVA eleven minutes after the next Sunrise, ending a stand-up sequence that lasted an incredible 2 hr 29 min. What wonder. What things to be seen. Aldrin was much impressed with the view from space. It had been a full day and the crew were determined to get a good night's rest before the full space walk planned for day 2. Nevertheless, having risen quite late for an afternoon flight, it was midnight Cape time before their rest period began.

Next morning, problems again from the fuel cells. At about 4:00 am Cape time, the power from fuel cell stack 2B, one of three in each of two sections, fell dramatically low and, as dawn broke across the Atlantic, Mission Control in Houston decided to remove it from the line. Then, about 2 hr 20 min before the EVA began, Gemini XII reported a problem with the OAMS thrusters. A pitch and a yaw thruster were not apparently working as they should and Aldrin was determined to give them a look when he got outside. Timing his preparation so as to leave spacecraft 12 at the scheduled hour, the hatch was opened at 42 hr 48 min: mid-morning at Houston, nearly lunch time at the Cape. This was the big test, the one that would de-jinx the entire concept of space walking. If it worked! First, the EVA camera was set up to record his activity, then, moving perpendicular to the long hand rail, he worked his way to the Agena, attached a pair of waist tethers and rested. These were the key to keeping a good environmental balance between the level of his metabolic output and the necessary work load. In previous EVAs of this type, each specific task had been well below the limits of the chest-pack but the mistake came when another activity began before body energy generated by the first had been removed. Built-in rest periods, only about two minutes at a time, were valuable in keeping a low metabolic profile.

Coming up on 20 minutes into the EVA, Aldrin had the Agena tether hooked on the spacecraft's index bar so that gravity gradient stabilization tests could repeat the activity first performed by Conrad and Gordon. Next, the Gemini pilot prepared the Agena work station, one of two 'busy-boxes' at which he would evaluate different mechanical tasks in the next hour or so. Then, he moved back up to the nose of the spacecraft, handed Lovell the EVA camera and received in return an adapter camera, and worked his way to the back of the white equipment section. Restrained from drifting too far by the comparatively short tether, movement was easier without the mass dynamics of a long umbilical trying to twist him off his perch. That was another secret of good body stability in a weightless environment. It seemed to be working.

At the adapter busy-box just 40 minutes after hatch opening he felt cool and well adjusted to the task at hand. And there were other indications that this time round it was a well balanced EVA. Monitoring his heart and respiration rate in the MOCR at Houston, the Flight Surgeon reported average cardiac rates of 100 beats/min and a pulmonary function of less than one breath every three seconds. With his feet in the molded shoes, Aldrin evaluated the ability of the restraints to hold him securely. He found he could lean over 45° to either side, and lay his back over up to 90°. Now it was time to unstow some penlights and to start work at the adapter station. Within minutes the Sun set and he was in darkness, assisted by the increased ability of the ELSS chest-pack to remove body heat. The EVA had been timed for potentially strenuous tasks to be performed in darkness where no Sun would beat down on the suited astronaut.

A torque bolt was worked with ease and Aldrin performed the task repeatedly with fewer restraints each time so ground observation of metabolic activity would demonstrate the energy involved. Next came connection and disconnection of electrical plugs and with cutters removed from a special pouch an exercise to show how strands and hoses could be severed in weightlessness. Aldrin had been outside his spacecraft for more than one hour now and halfway through his 33 minute night side pass began work on a Saturn bolt of the type that may be required for EVA attention should the space agency proceed with plans then emerging for a space station based on the big launcher. If Saturn stages could be modified in orbit there was hope for orbital engineering on a grand scale and this particular task demonstrated that type of bolt loosening and tightening. Finally, there were more electrical connectors to play with, then an evaluation of the Velcro-backed tethers before Sunrise.

Moving back from the adapter, Aldrin re-installed the EVA camera and worked his way to the Agena for more tasks on the smaller busy-box. Thus would the pilot demonstrate similar tasks at two different work stations, one in Sunlight, one at night. Coming back toward the hatch, Aldrin wiped Lovell's window to collect a sample of contaminated material smeared across the outer pane after launch. Since spacecraft 9, a protective shield had been carried, jettisoned in orbit, to protect the windows from the chemistry of launch and staging. But it had done little good and it was a last opportunity to really determine the composition of the film. Standing in the open hatch before completing his space walk, Aldrin observed the thrusters on the back of the adapter as Lovell blipped the attitude hand controller. The command-pilot had been right. There was something wrong with at least a pitch thruster, issuing streams of vapor when commanded to fire.

About the only activity that had gone askew during the entire EVA was a problem with installation of the adapter camera designed to record for later study his activity at the larger busy-box. It did not seem to fit its holder and the pilot returned it to the cabin intending to bring it back for analysis. After 2 hr 6 min, spacecraft 12's right hatch was shut as effortlessly as it had on the two previous flights and the most successful EVA to emerge from the Gemini program was over.

There had been one other event, performed early in the spacewalk, specifically planned as a private commemoration of the dead and dying from two World Wars. Had Gemini XII launched on the planned date, the space walk would have been carried out on the day America mourned her war victims. Pulling from his pocket a small blue-bordered pennant about 24 cm long and 15 cm wide, Aldrin announced that 'To commemorate our launch on Nov. 11 I have an emblem here I would like to leave in orbit . . . I'd like to extend the meaning to include all the people of the world who stand for now and continue to strive for peace and freedom in our world.'

But now, with the EVA behind them it was time to start the tether exercise and Lovell commanded the Agena to pitch the two vehicles directly down at the Earth. At thirty-seven minutes prior to completing two full mission days, Gemini XII began the gravity gradient exercise by slipping free from the Agena's docking cone. For a while Jim Lovell was unable to get the tether taut but in the end a stable situation was set up as the combination floated through day and night passes. Four hours twenty-seven minutes after separating from the cone, spacecraft 12 jettisoned its docking index bar and the tethered operation was over. Twenty-three minutes after that a short burst from Gemini's thrusters put spacecraft 12 in a slightly different track to the target vehicle. Combined operations were over, 52 hr 14 min into the four-day flight. From that point on it seemed almost routine to support personnel on the ground. Even the press were casual now that space walking had been shown to have no subtle in-built obstacle. With release of the Agena all the spectacle had gone from the Gemini program and it only remained for a second, and last, stand-up EVA on the following day to complete the flight's major objectives. But first, another sleep period for Lovell and Aldrin.

Early into their third day, the crew prepared to open the pilot's hatch for the last time. This stand-up EVA had been put in the flight plan to catch the solar eclipse which, had the spacecraft got off the ground on schedule, would have occurred at the right place in space and time. There was no need now to rush and so it was 66 hr 6 min when the lid came open and Buzz Aldrin floated up to the maximum length of his harness tether. First, equipment was thrown free of the spacecraft that would otherwise clutter the cabin; 'It looks like after this one you can call us the litter-bug flight,' exclaimed Jim Lovell. Then, as Gemini XII flew into night, Aldrin started snapping pictures of the heavens with the ultraviolet camera. As dawn approached, additional shots were taken of the Sunrise and that terminated the activity. Nothing further was planned and the hatch came shut after 55 min.

For the rest of the day, experiments performed from inside the cabin included an attempt to photograph a sodium cloud released by a French rocket from Hammaguir, Algeria. They were unable to get a visual identification of any such cloud but the experiment marked the first cooperative international effort during a manned space flight. During that day, two more yaw thrusters went out and continued trouble with the fuel cells required the crew to consume as much drinking water as possible and to purge the system at frequent intervals. After another sleep period, completed shortly after Gemini XII clocked up 80 hours, more experiments and preparation for retro-fire.

Soon, it was apparent that the adapter drinking water supply was exhausted. Then, coming up on 89 hours, the crew performed a test of the OAMS thrusters, confirming that two yaw right, a yaw left, and a pitch down jet were inoperable. With less than four hours to retro-fire, scheduled to occur over the Canton Island tracking station, there was more trouble with the fuel cells. Two stacks in section 2 were dropping rapidly and the four main re-entry module batteries were brought on line to back up the power requirement from that point on. Gemini was limping along to a sad conclusion, only just escaping an early re-entry due to depleted systems capability. Nevertheless, there was plenty of power in the silver-zinc batteries and retro-fire came at the assigned location, 93 hr 59 min 58 sec after lift-off.

Again, it was to be an automated ride down through the atmosphere. But the spacecraft had one more surprise. During the onset of peak g forces, a pouch containing weighty flight documents and lightweight equipment broke loose from a position on the cabin wall and shot toward Lovell's lap. Not wishing to suffer the full effect of its impact, nor wishing to inadvertently grab the ejection seat D-ring between his legs, Lovell snapped his knee-caps together and successfully caught the package, fighting its attempt to ram home with a force of 4 g by gripping his knees until they almost hurt! When the spacecraft dropped into its pitch forward angle, he could relax – and the package fell harmlessly to the floor. It was a humorous end to a mission that had packed surprises all the way. Gemini XII was only 4.8 km off target as it bobbed in the calm Atlantic swell, the carrier USS *Wasp* standing close by.

Within eight minutes swimmers had a flotation collar around the pitching spacecraft and just 22 minutes later the crew were on the *Wasp*'s deck. It was over. The operational phase of Project Gemini had drawn to a close. And what a host of records it had seized since inception five and one-half years before. No longer was America second runner in a race where once they had been last. In the entire period of operational two-man flights, not a single Soviet cosmonaut had been into space. Having accumulated a total 507 man-hours in space by March 1965, the Russians were thoroughly beaten within just 20 months by America's new record which now stood almost four times as great: 1,993 man-hours. The momentum of government/industry teamwork was just too great to exceed and not for a decade would the Soviets regain a measure of superiority in manned space operations.

But Gemini was more than a record breaker. Designed from the outset to answer deeply disturbing and fundamental questions about complex and demanding operational tasks, Gemini had burst the bubble on pessimistic voices intent on castigating the bold initiatives inherent in such activities as rendezvous and docking, space walking, and long duration flight. It had shown that all these things were possible, proved that man need not be a mere passenger on the technological wizardry of design teams and laboratories. Integrated from the outset, manned participation in each Gemini task had carried the program through to success in every fundamental objective assigned from the beginning.

Perhaps one of the most valuable assets to accrue from the two-man flights was the broad generality of tasks involved in 'turnaround' – turnaround from near disaster; turnaround from operational failure; turnaround from collapsed hardware; turnaround from flight plans suddenly thrown away by some procedural error. At the beginning, flight objectives had to be arranged around the availability, or otherwise, of the all important fuel cell. Long duration missions were put back to the second half of 1965 because of initial development problems with the electrical production system. Then when Gemini V ran into trouble with its own fuel cells, McDonnell had just 40 days before spacecraft 7 had to be delivered to the Cape in which it was to install additional cryogenic storage tanks and a crossover feed valve between the reactant supply and the environmental system.

Proving it could be done. Aldrin shows how to work effectively in space.

Changes of this magnitude would have taken months in the days of Mercury.

Again, on the very next flight, mission personnel had 38 days between the aborted attempt to launch Gemini VI and the long duration flight of Gemini VII in which to conceive, fabricate, construct and implement, a plan where spacecraft 6 would rendezvous with the 14-day vehicle already in space. Nobody had seriously planned for a dual manned mission, although that prospect had been looked at from time to time, but, again, it was a procedure unthinkable in days of caution and discretion that opened the manned space flight era. In 1966, within 85 days, an Augmented Target Docking Adaptor was conceived, approved, built, and fully tested, before sitting in a Cape hangar against the day another Agena would fail to reach orbit. Seven weeks later it was used as the target for Gemini IX-A, a major piece of program hardware literally fashioned from left over items scrounged around the facility.

And then, in the last few missions, there was real-time flight planning whereby only the basic skeleton of a mission plan was formulated, the precise detail and sequence of events left to the Flight Director and the Mission Director according to how the mission progressed and what was encountered en-route to completion. From the accelerated pace of EVA equipment development, to the record launch pad turnaround to get Gemini VI chasing after its predecessor only a few days after the launch of Gemini VII, program managers and personnel worked in a way difficult to have envisaged only a couple of years earlier. It was the biggest single cause of success in later manned projects, and paved the way for dramatic events of greater magnitude. In one sense, Gemini did much to speed Apollo on its way, but in areas of program management and operational sequence testing only.

The design and construction of the Gemini spacecraft was very different from that of its successors. Within the giant Apollo machinery there were two very different spacecraft, neither of which bore much resemblance to Gemini. The Apollo ship itself was designed around a very different set of criteria, and the lunar landing vehicle was a true space-ship not designed to return to Earth. Project control of both was as different from each other as it was from Gemini. Yet, for all that, the Apollo Project Office sought advice from Gemini during the latter's busy operational phase, and procedural changes promising to speed Apollo were adopted.

Gemini ran out for a total cost of $1,281 million, about three times the cost of its predecessor. But the scale and scope of the program had increased beyond the earliest expectations. Twelve spacecraft had been fabricated, ten of them man-rated, and twelve Titan II launchers had dispatched themselves from pad 19. As for the targets, seven Atlas launchers carried six Agenas and one augmented target vehicle: 19 payloads; 19 rockets. Around the United States, 3,500 contractors, subcontractors and vendors contributed to the Gemini success, and from Canada and England young men came to grow older on the edge of a new frontier. In truth, Gemini had seen the peak of the space years. NASA's budget was in the first phase of decline, its all-time high already a thing of the past.

To the public, Gemini saturated for a while their need for a preoccupation from foreign wars and internal unrest. The high summer of Gemini's success had seen race riots in Alabama, student violence on University campuses, and an increasing commitment in Vietnam. During 1965 and 1966, the operating years for the two-man spacecraft, President Johnson walked along an irreversible one-way street to disaster by indulging in covert policies concealing very different priorities. Fearing for the safety of his Great Society program of social reform, he denied the American public knowledge about massive financial commitments to Vietnam, clouding over the escalating arms build-up and diverting attention from what was inevitably to become one of the bloodiest confrontations since World War Two.

Amid civil unrest, an increasingly disparate Washington clique, and a burgeoning US foreign involvement, NASA made ready to go to the Moon. Compared with Apollo, Gemini was a dream. And indeed they had been halcyon days as month after month teams of astronauts had rocketed away to success after success. For many, those days were to remain a memory, for when a new year dawned, a year in which Apollo was expected to prove it could fly in space, euphoria was plunged to depression amid a succession of disasters unlike anything the space program had ever known. Shock waves would be felt for a long time. The fun days were now a thing of the past.

CHAPTER FIFTEEN

Machines Made Ready

Just eight days before Lovell and Aldrin flew the last manned Gemini mission, engineers at the Cape sent a Titan launch vehicle into space from a pad several kilometres north of the facility where spacecraft 12 waited to fly. But it was a very different Titan to that being groomed to launch Gemini XII from pad 19. Yet in their respective roles, both launch vehicle and spacecraft were outgrowths of existing programs pointing toward a broader capability in future objectives. The launcher was a Titan IIIC; the payload, in part, was Gemini spacecraft 2 previously flown in January 1965. Both were an integral part of the Air Force's Manned Orbiting Laboratory program designed to accommodate two-man crews for up to a month in space. The Air Force used the recovered Gemini 2 spacecraft to qualify in a second suborbital flight modifications made to the re-entry module.

MOL envisaged development of a cylindrical laboratory based on the redundant second stage of a Titan II, affording up to 34 cubic metres in habitable area divided between pressurized work and living compartments. Access to the laboratory would be gained via a forward hatch leading to a Gemini spacecraft carried on the forward end. The complete assembly would be launched by a Titan IIIC, essentially a Titan II of the type used to launch the NASA Gemini flights but with two large solid propellant strap-on boosters for lift-off. In the configuration adopted for the November, 1966, test flight, a Transtage third stage was placed above the Titan II mock laboratory. Transtage, the modified Titan II tank and the Gemini B adapter would go into orbit and release three small engineering and technology satellites placed there by the Department of Defense.

But the real purpose of the configuration was to prove that a circular cut-out offset from the center of the Gemini re-entry module heat shield, plugged with a hatch carrying the same heat shield material on the outside, could survive the fiery return to Earth. The hatch, placed between the two ejection seats, would afford access to the laboratory via a pressurized tunnel through the adapter section. In this way, MOL would carry its crew into orbit in an attached Gemini spacecraft, provide facilities for up to a month in space, and permit the crew to return to the re-entry module, close the hatch and re-enter the Earth's atmosphere.

After the first two Titan III stages and the two strap-on boosters were jettisoned during the unmanned test flight on 3 November, 1966, the Transtage fired to place the configuration 167 km above the Atlantic at just below orbital speed. At this point the Gemini B re-entry module separated from its adapter and fell in a long arching trajectory toward the sea and a splashdown close to Ascension Island. In space, meanwhile, the Transtage fired again to place itself and the forward payloads into an elliptical path from which the three piggy-back satellites were sequentially released. But the object of the exercise was to prove that a 'long shape' Titan IIIC could fly safely and that a modified Gemini heat shield containing a circular hatch could also return to the surface of the Earth intact and undamaged. It was the first, and as time would tell, the only, MOL test flight.

At this time, manned flights were scheduled for late 1968 and considerable progress had been made in selecting systems and design layouts for an orbital facility operated by and for the Defense Department. In no way was MOL considered suitable for any hostile military action, nor did it ever have an active defense function. It was to be purely a passive research facility, a laboratory from where technology directly applicable to US defense requirements was to be conducted and from where astronauts could evaluate different types of surveillance equipment. From time to time the possibility of using orbital flight as a means of threatening foreign states was mooted by military and civilian proponents alike. With no point on Earth more than 40 minutes away from a ground-launched missile, however, the 90 minutes it would take any given weapon system to travel back to a location it had just passed in orbital flight precluded the use of that environment for storing hostile weapons.

As conceived in 1963 when it became apparent that the X-20 Dyna-Soar program was limited in its ability to support complex manned operations in space, MOL would transfer emphasis from a sophisticated controlled re-entry technique to a broadly based operational facility designed for EVA, laboratory operations, and long-term stay in the weightless state. As such it was directly inspired by the emergence of the civilian Gemini program between 1961 and 1963 in a similar way that the Apollo Moon landing objective had inspired NASA to look beyond the lunar expedition at other, perhaps more ambitious missions, that could utilize Apollo hardware. As early as October, 1961, Langley Research Center's Emanuel Schnitzer suggested the use of Apollo vehicles in the orbital assembly of a space laboratory which he called 'Apollo X'. But it was the design and development of basic Apollo equipment that kept NASA and the contractors busy in the period from contract award in late 1961 to the planned manned mission of early 1967.

Compared to Gemini, the development of Apollo was more reminiscent of the Mercury days. Comprising a completely new spacecraft design, Apollo would be qualified in a series of suborbital and orbital flights of boilerplates and production spacecraft employing a variety of different launch vehicles, all of which were an order of magnitude more powerful than the rockets used for either Mercury or Gemini. For suborbital tests of the launch escape system and for qualifying the established abort procedures to be employed on operational flights, a new launcher was conceived by the Manned Spacecraft Center in June, 1961. Like Mercury's Little Joe, it was to comprise a collection of solid propellant rockets, the specific configuration and the precise firing sequence of which would be tailored to the needs of a particular mission. Dubbed Little Joe Senior in preliminary design studies aimed at selecting an optimum configuration, the concept was finally called Little Joe II by the time bidders were requested to submit proposals in April the following year.

At this time NASA anticipated five flights: two max q (maximum dynamic pressure) abort tests and one high altitude atmospheric abort in 1963, and one very high altitude and one max q abort in 1964. In May, 1962, General Dynamics/Convair had been selected as the prime contractor. Little Joe II was to employ combinations of two different types of rocket: the 46.8 tonne thrust Algol and the 15.1 tonne thrust Recruit. A typical combination would group four Algols and five Recruits. Two Algols and all Recruits would be ignited on the pad, with two Algols firing later as a second stage. Thrust at lift-off could average 169.4 tonnes, more in fact than the three engines of the Atlas launch vehicle. Total lift-off weight for this assembly would be nearly 65 tonnes and the Little Joe II/Apollo would stand 27.2 meters tall.

Pad abort and dynamic suborbital hardware tests with unmanned production spacecraft and boilerplates would prepare equipment for orbital evaluation. For this, the Saturn C-1B would lift Apollo into Earth orbit, or send an unmanned lunar landing vehicle on a similar path; as an outgrowth of the earlier Saturn C-1, it was incapable of lifting the more than 40 tonnes of both Apollo and lunar lander modules. Saturn C-1 first flew in October, 1961, in a Block I configuration. Only the first stage was live, the upper stages and fairings being dummy items incorporated to smooth airflow and approximate the mass and balance charateristic of a fully staged launcher. A second Saturn C-1 flew in April, 1962, and a third and fourth followed in the following November and March, respectively. All four were suborbital, where only the first stage was live. A Block II configuration appeared for flights in 1964 and 1965 employing an S-IV liquid hydrogen/liquid oxygen upper stage. Several of these two-stage Saturns carried boilerplate Apollo models and pioneered the use of high-energy cryogenic propellants for large launch vehicles.

Early in 1962, however, the Marshall Space Flight Center proposed development of a C-1B variant. This would have a

Beginning a new era of large launch vehicles, the first Saturn I lifts off at 10:06 am, 27 October, 1961, at the start of a ten-vehicle test series.

lightweight first stage based largely on the C-1, but with improved and more powerful engines and a much larger and more powerful S-IVB upper stage. Improvements and additions would enable the C-1B to lift 18 tonnes into Earth orbit, twice that of its predecessor. By the middle of 1962, the Manned Spacecraft Center had formally adopted the C-1B for Apollo Earth orbit flights, proposing to begin manned missions on the fifth launch following four development flights. Saturn C-1B would stand 69 meters tall and deliver a lift-off thrust of nearly 726 tonnes, nearly four times that of the Titan II rocket assigned at this time to launch two-man Geminis into orbit.

By October 1963 George Mueller got his way in a suggested plan to speed up Apollo development by eliminating unnecessary steps in hardware tests. Headquarters buzzed with a concept called 'all-up systems testing' where several new elements of a particular launcher would be tested simultaneously rather than in a series of several different flights. This was Mueller's approach and one that found favor where cost conscious administrators detected a means of cutting large numbers of flights with big and expensive boosters. As part of the philosophy, Mueller suggested cancelling an original plan to use the last four basic Saturn C-1 launchers for lifting manned Apollo spacecraft into Earth orbit on shakedown trials. By this time the 'C' prefix had been dropped, Saturn C-1 becoming I and the C1B becoming IB.

To accomplish the full mission objectives of the Apollo program, the Saturn V was the only rocket capable of lifting fully fuelled Apollo and lunar lander modules and throwing them to the Moon. It was altogether different from the Saturn I and Saturn IB launchers and represented the ultimate creation of the Marshall Space Flight Center. Von Braun was justifiably proud of the Center's role in developing this vehicle which, as events would dictate, remained the world's largest rocket ever developed to operational status. It was to be capable of lifting more than 100 tonnes into Earth orbit, and of sending more than 40 tonnes to escape velocity. In comparative weight terms, the giant booster could have delivered in one flight to Earth orbit all the manned space vehicles flown by America and the Soviet Union in Mercury, Gemini, Vostok and Voskhod programs; 24 first generation manned spacecraft. Much of the design and development for the smaller Saturns had been done by the Marshall Center, Chrysler being brought in to manufacture later stages for the I and all the IB stages with Douglas contracted to build the S-IV and S-IVB upper stages. Boeing, North American and Douglas would do much of the design work, as well as build, the first (S-IC), second (S-II), and third (S-IVB) stages of the Saturn V.

Commonalty crept in to reduce the additional cost of a completely new trio of Saturn V stages: the third stage would use the second stage of the Saturn IB, suitably modified for multiple starting and a longer in-orbit lifetime. A major departure in the Saturn V's first stage was the quintet of F-1 rocket motors at the base. These engines were completely different to the H-1 used in the smaller Saturns which were developments of the propulsion systems designed and built for military ballistic missiles. Each F-1 produced a thrust equal to that of all eight H-1 engines in the Saturn I or IB and each would receive propellant from just two oxidizer and fuel tanks. The Saturn I and IB used eight cylindrical tanks clustered round a ninth. But the main advantage in the enormous lifting power of the first stage was the selection of cryogenic high-energy upper stages burning hydrogen and oxygen propellants. The enhanced efficiency further increased the theoretical lifting capacity of the completed assembly. When the Manned Spacecraft Center outlined Apollo flight schedules in September, 1962, the seventh Saturn V flight was tentatively assigned as the first to carry a manned Apollo vehicle. After that, test flights of both Apollo and the lunar lander would precede the first manned Moon missions.

The three basic types of launcher to be used in Apollo – Little Joe II, Saturn IB and Saturn V – would support four phases of spacecraft design development and qualification: mock-ups, boilerplates, Block I manned and Block II manned spacecraft. The mock-ups were to be used for final design configuration changes, fitment checks with other hardware elements, and for testing such things as crew panel layouts, ingress/egress ability, systems installation, etc. When a basic configuration had been established inside and out, the boilerplates would demonstrate structural integrity and fly simulated missions of extreme situations, such as aborts from off the pad, aborts from high altitude, recovery characteristics, etc. Block I manned spacecraft were the first fully operational man-carrying vehicles incorporating sufficient systems hardware for a full orbital shakedown evaluation of the complete assembly. They would sometimes carry equipment installed as an interim measure just to get the spacecraft flying while a more sophisticated, definitive, component completed its development cycle. Block II would be the final design configuration incorporating all the equipment assigned to support of the Moon landing operations, although of course the Apollo spaceship was not to actually land on the lunar surface, only to go into orbit about the Moon with the lunar lander docked to its nose. In some respects the Block I vehicles would be the prototypes; all Apollo spacecraft destined for manned Saturn V flights would be Block II types.

Where the launch vehicle contractors were spread among the most experienced and renowned aviation companies in the United States, Apollo's architect lived at Downey in California and assumed responsibility for all spacecraft hardware. Grumman Aviation in New York was just getting into development with the lunar landing vehicle but it was North American that had responsibility for the manned mother ship. From the outset, Bob Gilruth had confidence in Harrison 'Stormy' Storms and his deputy Dale Myers at North American when final contracts were signed late in 1961; Myers would be busy on the Hound Dog project for a couple more years but from 1964 was almost full time on Apollo work. As chief engineer on Apollo it would be Charlie Feltz's responsibility to take a design set broadly by the dictates of MSC's Maxime Faget and fashion it into an operational spacecraft. But management went through a critical shift both at Headquarters in Washington and at the Manned Spacecraft Center in Houston in six months beginning late 1963.

As Associate Administrator for Manned Space Flight, George Mueller replaced Brainerd Holmes, and Walt Williams, having groomed himself for management at the Cape in administrative experience to add to his operations work at the Flight Research Center, moved in as Mueller's deputy. Bob Gilruth was to stay as Director of the Manned Spacecraft Center but Elms, his deputy, was replaced from March, 1964, by George Low. Thus, the team that took Apollo through its conceptual and early contract stages in 1961 and 1962 were a different group to managers and administrators responsible for hardware development from 1964. One final personality dropped into place in October, 1963. From a position as Deputy Director for Systems at Headquarters, Joseph F. Shea moved in as MSC Apollo Spacecraft Program Office Manager. As such he was directly opposite Harrison Storms.

As the pace and magnitude of Apollo activity accelerated during late 1963, Mueller saw the need for a deputy in charge of the program office at Headquarters, a right-hand man to the role of director he personally played in addition to his position as head of manned space flight. On the last day of that year, Mueller announced selection of Brig. Gen. Samuel C. Phillips as Deputy Director of the Apollo Program Office at Washington and within the next few months Phillips was to take up the full Directorship, relieving Mueller of specific program administration.

The differences that would have normally evolved in the Block II vehicle were accentuated in Apollo by the completion of Block I design prior to the selection by NASA of the Lunar Orbit Rendezvous mode in late 1962. Consequently, because a lunar landing vehicle would be carried on the forward end, North American had to design into the existing cabin a means by which the crew could move from one spacecraft to the other within a pressurized environment. Devoid of a role whereby Apollo itself would land on the Moon, the spacecraft was divided into two separate elements: a pressurized Command Module, or CM, designed to carry the crew and return them safely to Earth, and a Service Module, or SM, comprising an unpressurized structure containing the big liquid propellant engine for placing the vehicle in lunar orbit and returning it to the vicinity of the Earth and systems necessary for supporting a flight of up to two weeks in duration; together, they would be known as the CSM, the Command and Service Modules.

From the beginning of detailed design analysis in 1962, the CSM configuration was limited to a weight-lifting capacity dictated by the Saturn V launcher, minus the weight of the lunar landing vehicle. At that time, the lunar lander was allowed a target weight of 11.5 tonnes which gave the CSM an allowable weight of nearly 27.9 tonnes fully loaded; the adapter covering the lunar landing vehicle at launch was to weigh 1.4 tonnes, making up the Saturn V's estimated 40.8 tonne capacity. Compared to the 3.7 tonne Gemini, Apollo was big and heavy. But in several other ways it was more a return to Mercury design concepts than a follower of innovative trends set by the two-man vehicle. Because Apollo would be designed to survive a re-entry from the vicinity of the Moon, its heat shield would experience much higher temperatures. The 39,500 km/hr re-entry speed was much greater than the 27,000 km/hr Earth orbit re-entry velocity and temperatures would reach more than 2,800°c. To navigate a path safely through the atmosphere, whereby the lifting characteristics of the Command Module could be made to change the landing site or eliminate slight errors in the alignment of the trajectory, the spacecraft would have need of a heat shield all

AVCO workers inject phenolic resin into the honeycomb matrix of an Apollo heat shield. Portion shown here is the base of the Command Module.

around its exterior surface, not just on the base end of the blunt-faced conical shaped body.

Because the vehicle would pitch and roll as it moved quickly through Earth's atmospheric envelope, heat would be experienced on the conical sidewalls and because the temperatures were so much greater than anything felt by earlier manned vehicles the thickness of the shield would be more than that needed by Mercury or Gemini's base-end protection. So whereas the first two manned spacecraft designs incorporated a heat shield on the aft area, Apollo had to incorporate a complete shield all across its exterior surface. Because of this, heat shield development posed several unique problems. For instance, reaction control system thruster orifices would have to be accommodated within the heat shield structure, as would vent valves and attachment points between the Command Module and the Service Module. The AVCO Corporation's Research and Advanced Development division was selected by North American in March, 1962, to fabricate heat shields for the CM. Ablative compounds were the only plausible contenders for shield application; heat-sink materials would have been too heavy.

The company had a history of research into possible heat shield materials dating back to the days when Space Task Group looked around for Mercury protection. For Apollo, AVCO selected a phenolic epoxy resin compound, a sort of reinforced plastic, which it proposed to fill in 370,000 fiberglass honeycomb cells. The total shield was separated for manufacturing and installation into three sections: the aft section, covering the base of the Command Module; the crew compartment section, forming the conical sidewalls of the spacecraft; and the forward section, covering the access tunnel and the parachutes. The main structure of the shield served to support the ablative compound and its fiberglass matrix. It comprised a stainless steel brazed honeycomb core bonded between steel alloy sheets. The fiberglass honeycomb was attached to the outer alloy sheet and filled with the resin.

The main structure of the Command Module itself comprised aluminum honeycomb bonded between aluminum sheets, the complete section varying in thickness from 3.8 cm at the base to about 0.63 cm at the forward section. It was to this inner structure, designed to contain the pressurized atmosphere, that the heat shield structure was to be attached, separated from it by a fibrous insulation. The heat shield varied in thickness from 6.35 cm at the base to 1.27 cm at the forward section. It weighed about 1,360 kg, or roughly one quarter the total mass of the Command Module. CM structures were delivered to AVCO's Lowell plant in Massachusetts for precise weight and centre of gravity checks. Then it was precisely measured – each would be fractionally different in size to its predecessor – before the honeycomb panels were fitted and the fiberglass bonded and cured at a temperature of 176.7°c. Ingredients for the ablator material were then prepared, mixed as a composite and injected with catalysts and hardeners. Temperature at this point was critical for the necessary density.

Stored in a freezer, batches of the ablator were then heated when technicians applied the material to each individual fibreglass cell with guns designed with special heaters for controlling the thermal state of the resin. The final process involved machining the heat shield on a computer-controlled lathe to the precise thickness required on the various heat shield sections. X-ray inspection verified correct density within each cell, a sealer was added to the surface and the Command Module then painted white. At several points around the shield, access doors and ports had to be built in during manufacture. These accommodated, for example, the legs of the launch escape system tower, the umbilical connection between the Command and Service Modules, and the attachment points to the SM.

But there was little opportunity for building in to Apollo the panels and access doors provided on Gemini. The more stringent operating environment drove engineers back to a Mercury concept where everything would have to be installed and inspected from the inside. This compromised the work of several different engineering groups and posed a major problem for systems technicians and wiring harness designers. The habitable volume of the pressurized interior was a comparatively spacious 5.95 cubic meters in which would be accommodated couches for three astronauts, displays and controls, storage bays and stowage lockers, provision for navigating in deep space via optical instruments built in to the main structure, a main access hatch and four windows. In the original designs, only the Block I Command Module had a window in the hatch. Two, square-shaped, windows were provided in the conical walls of the structure for outer crewmembers to look sideways into space. Two other windows were recessed within scooped sections rather like Gemini, for forward vision along the nose of the conical spacecraft. As each outer crewmember sat in his couch, the forward looking windows would present a view of the lunar landing vehicle during docking operations, or any other space vehicle that might be approached.

The windows comprised inner and outer panes, the former consisting of two 0.63 cm thick tempered silica glass sheets 0.25 cm apart, the latter being an amorphous-fused silicon pane 1.78 cm thick. Softening temperature of the outer pane was 1,538°c. The Block I Command Module did not incorporate provision for docking and had a rounded nose cap giving the spacecraft a height of 3.66 meters. The Block II Command Module was 3.22 meters tall to the tip of the mandatory docking probe designed to latch up on the lunar landing vehicle. All Command Modules were to be 3.3 meters in diameter.

Apollo's Service Module was a direct outgrowth of early studies on the use of three-man space vehicles for supporting Earth orbit space stations or circumlunar flights, a genesis dating to the 1959–1961 period extensively covered in Chapter Four. In supporting lunar mission operations the Service Module was to contain the engine used for propelling Apollo and the lunar lander into Moon orbit, and for accelerating the CSM back toward Earth at the end of a normal flight. In addition, the engine was to be responsible for trajectory corrections and course changes above a velocity of about 1.5 meters/sec; any corrections below that speed would be made with the attitude thrusters firing together.

The Command Module base (foreground) before mating to the pressure module seen here encased by a support trolley.

This early Block I Command Module, although much bigger than the diminutive Mercury and Gemini vehicles, represented the only part of a near-3,000 tonne structure designed to return to Earth.

The SM's Service Propulsion System, or SPS, was therefore required to fire velocity changes totalling about 1,800 meters/sec (6,480 km/hr). This comprised a deceleration of 3,180 km/hr to drop into lunar orbit, and an acceleration of 3,290 km/hr to leave the Moon's orbit and return to Earth. These figures are average values, and would change from mission to mission according to the precise position of the Moon with respect to the Earth and with the specific type of trajectory selected. There would also be additional velocity increments trimming the Moon orbit and performing the course corrections.

The optimum engine was to have a thrust of about 9.3 tonnes, fire for as long as 12½ minutes or as short as 0.4 second, be re-startable 50 times and burn a combination of 50% hydrazine and 50% unsymmetrical dimethylhydrazine as a fuel and nitrogen tetroxide as the oxidizer. But more important than any of that, it had to work – every time. For if the engine failed to fire when called upon to accelerate out of lunar orbit, the three astronauts would have no means of returning to Earth. It was one of two propulsion systems that simply had to work, the other being the engine designed to lift the top of the lunar lander back into Moon orbit from the surface.

When the Service Propulsion System was designed, orbital re-starts were not always successful. Even as late as November, 1965, Gemini astronauts were forbidden the use of Agena's main engine, smaller than the Apollo SPS, because of suspected trouble. Such a situation would be intolerable around the Moon. The propellants selected for the engine were a compromise and at the same time an insurance. Being hypergolic they would ignite on contact and as such have no need of complex and potentially troublesome ignition sources. It was merely necessary to bring the two chemicals into contact within the combustion chamber, and they would fire.

To accommodate the engine and its large load of propellant – about 18 tonnes would be necessary for the scheduled speed changes – the Service Module was built as a cylindrical structure divided into six pie-shaped sectors. The basic structure was built on six solid aluminum alloy radial beams machined and chemically milled to thicknesses varying between 5 cm and 0.046 cm. Forward and aft circular bulkheads covered the opposite ends of the Service Module with six sector doors 2.5 cm thick built from aluminum honeycomb core between aluminum facing sheets. Taking the form of a cylindrical drum 3.56 meters tall and 3.3 meters in diameter, internally divided into six vertical pie-shaped segments, the SM carried a cylindrical tunnel up the center, 1.12 meters in diameter. The bottom of the tunnel was taken up with the Service Propulsion System engine while the top accommodated two spherical helium tanks, one above the other, for pressurizing the propellant tanks and forcing fuel and oxidizer into the engine.

Four cylindrical propellant tanks were carried, one each in four Service Module sectors. Two of the tanks, 3.91 meters tall and 1.29 meters in diameter, served as sump tanks while the other two, 3.92 meters tall and 1.14 meters in diameter, were storage tanks. Feed lines connected the storage tank to the sump tank so that propellant would flow from the former to the latter before going from the sump tank to the engine. In that way, the sump supply was to be the last to run dry. In all, 18 tonnes of propellant were arranged so as to ensure a balanced vehicle under all levels of depletion. The cylindrical tanks were longer than the basic Service Module structure and the hemispherical end-domes would protrude above and below the fore and aft bulkheads. A special thermal cover was applied to the lower ends of the four tanks both as protection from damage and from thermal backflow during long burns with the big SPS engine.

At the forward end, six radial beam trusses would extend above the bulkhead support and secure the Command Module on top. Three of the beams carried shear compression pads and tension ties, the other three carrying only compression pads. Each tie carried an explosive charge at its center to sever the structural connection between the two prior to re-entry. Only the Command

Module was required to come home intact, the Service Module discarded like Mercury's retro-package and Gemini's adapter sections. The hemispherical forward domes of the four propellant tanks were contained within the area beneath the six attachment pads. The space between the convex lens-shaped Command Module aft heat shield and the flat Service Module bulkhead was to be enclosed by a fairing 55.9 cm high and 1.3 cm thick. The Block II Service Module would carry radiators on the outer face of the fairing, which from the exterior looked like a continuation of the flat side walls.

Both generations of Apollo were to carry two radiator panels on the main walls of the Module, each 2.79 square meters in size, but at different locations. In addition, the arrangement of systems within the six pie-shaped sectors differed between Block I and Block II, the latter carrying a large high-gain antenna for deep space communications on an arm extending outward from the aft bulkhead. With four dish antennae, each 78.7 cm in diameter, the 2 GHz system was essential for transmitting and receiving large quantities of data over great distances.

Aerojet-General was under contract to build the Service Propulsion System by the end of April, 1962, and made good progress in preparing the hardware. In bringing propellants to the combustion chamber, it was decided that redundancy should be the corner-stone of reliability. So, the bipropellant valve assembly comprised two separate pneumatic control systems, either of which could effectively fire and control main engine combustion. Nitrogen gas was to be used to move the valves open or closed, stored in two separate pressure vessels. But there could be only one injector, and only one combustion chamber. And here the very essence of simplification, where there would be no need for an ignition source, paid off. Nothing could be simpler downstream of the injector valves. Upstream, there was to be plenty of redundancy.

Each pneumatic system had two solenoid control valves and four propellant ball valves, two each for fuel and oxidizer providing eight for the two systems. It would be very difficult indeed to see how the engine could malfunction. Control of the thrust vector, essential if Apollo was to remain on course during engine burns that at times could exceed six minutes, would be effected by gimbal actuators capable of moving the engine ±4.5° in yaw and pitch axes. Each actuator had its own redundancy, two DC motors controlled from four electromagnetic clutches operating through a jack-screw. In operation, the SPS engine was simplicity itself. When nitrogen pressure was applied to the solenoid control valve actuators, the propellant ball valves would rotate allowing fuel and oxidizer to flow to the injector. Shut-down would result from removing power to the solenoid control valve, closing them and thereby eliminating nitrogen gas from the actuator piston. Spring

The Service Propulsion System would be the only means of getting back to Earth from Moon orbit. Here, the cylindrical combustion chamber is flanked at right by the square-shaped pitch actuator and at left by the main fuel delivery pipe. The oxidizer pipe runs from left to right around the base of the chamber.

presure would force the propellant ball valves to close and shut off the flow of fuel and oxidizer. Each cylindrical titanium propellant sump tank carried a retention screen at the outlet end so that in the weightlessness of space there would always be a quantity of propellant for starting the engine. Once ignited, the force of acceleration would push the propellant to the rear of the tanks and ensure continued operation.

But like Mercury and Gemini, Apollo would need attitude control thrusters. In March, 1962, North American selected the Marquardt Corporation to develop and build Command and Service Module reaction control systems. Five months later, Marquardt shifted Command Module RCS design to Rocketdyne with NASA's consent. But it was the Service Module thrusters that were to bear the brunt of all positioning responsibility during a normal mission, the Command Module thrusters only coming into use after separation of the two elements prior to atmospheric re-entry. Four clusters of thruster systems were to be carried on the side walls of the Service Module, each cluster supporting four rocket motors of 45 kg thrust apiece. Each cluster was 90° from its neighbour and all thrusters would serve for attitude and stability control as well as minor course corrections en-route to the Moon or the Earth or for moving (translating) from one location to another with reference to a secondary object close to Apollo.

For instance, quite early in the program, the Manned Spacecraft Center decided that the Apollo CSM would conduct a free fly-around from its position atop the launch vehicle adapter shroud back in toward the adapter where the lunar landing vehicle would be housed. Turning through 180°, Apollo would dock with the cone on the lunar lander, latching securely on to it before pulling the vehicle free of its stored position. In that docked configuration the two elements would coast to the Moon, having separated from the third stage of the Saturn V that had propelled them out of Earth orbit to escape speed. So the Service Module RCS thrusters were to play an integral part in moving the CSM around in space, and were to be employed for firing in salvo to nudge and jostle the docked configuration into the correct trajectory or orbit. Consequently, they would fire many times and would need to be exceptionally reliable through several operating cycles.

The design specification, placed by NASA and met by Marquardt, required each thruster to fire for as long as 500 seconds or a period of only 12 milliseconds, to guarantee a service life of nearly 17 minutes accumulated firing time, and to have a capacity for firing 10,000 times. Propellants used by the Service Module RCS thrusters were the same as those employed by the Service Propulsion System, hypergolic for reliability and simplicity in that it eliminated an ignition system essential for chemicals that will not ignite on contact. Each RCS quad was to be attached to a panel forming part of a Service Module sector wall, installed on a removable plate for ease of servicing. On the interior face of the plate, inside the respective Service Module sector, Block I spacecraft would carry a single fuel tank, a single oxidizer tank, and a spherical helium tank used for pushing propellants into the thrusters. Block II spacecraft would allow each thruster quad the use of propellant from two fuel and two oxidizer tanks, although each propellant would have two tanks of slightly different size.

The method of expelling propellant from a tank was one carefully researched by Bell Aerosystems, contracted to build the RCS propellant tanks for Apollo Command *and* Service Modules. Inside each tank, a Teflon bladder was provided containing either fuel or oxidizer. A central diffuser, essentially a tube with a series of holes around the outside, would allow the fluid to pass through to a feed port at one end. When helium gas was introduced to the space between the metal wall of the titanium tank and the flexible bladder, the fluid would be pushed through the diffuser. In this way, a constant pressure within the bladder could be maintained by admitting increasing quantities of helium to the volume between the tank and the bladder. This 'positive expulsion' concept was quickly developed for application to several other space projects and has consistently worked well. Block II Service Modules would carry about 608 kg of RCS propellant in the 16 fuel and oxidizer tanks distributed around the four quads.

The RCS thrusters built by Rocketdyne for the Command Module provided a nominal thrust of 42 kg and would use the same propellants as the Service Module thrusters. The system comprised two separate sets of 6 thrusters, with two groups of propellant, pressurization, dump and purge equipment. The 12 thrusters were all grouped around the Command Module, 10

An evaluation engineer here checks two flight director attitude indicators that would show Apollo astronauts the attitude of their spacecraft. He is touching one of two attitude control handles. The translation controller is to the left. To the right of the indicators is the panel displaying engine fuel pressure and the panel carrying thumbwheels for dialling up roll, pitch and yaw angles for attitude instructions.

along the edge of the broad base and two in the forward section near the apex of the conical spacecraft. The propellant tanks were contained between the heat shield structure and the Command Module's inner pressure vessel, each tank being of identical design to the smaller of the two different sizes employed for each Service Module RCS quad.

Each of the two redundant systems employed single fuel and oxidizer tanks and a single helium supply. In all, a 122.5 kg propellant load. Because the propellant could be hazardous after splashdown, residual fuel and oxidizer would be burned through the orifices during descent on the parachutes, the propellant feed lines then being purged with helium gas to remove trace deposits. The Command Module's thrusters would be a vital part of the spacecraft's control system during descent. The spacecraft, designed with a lift/drag ratio of 0.35, would fly a very precise re-entry controlled only by the angle of attack along the flight path.

Operation of the Service Module thruster quads during the flight up to separation prior to re-entry could be manually obtained via a rotation hand controller, two of which were provided inside the Command Module. They could be used as right-hand controllers on either of the two outside couches, or one could be removed and fixed to a socket on the wall of the spacecraft furthest from the astronaut's heads. This is the navigation station where precise angular rotation of the spacecraft about pitch, roll, or yaw axes would be necessary so as to align the optics on selected stars. Standing in the space between the couch foot-pans and the extreme wall, an astronaut would command full attitude pointing for this vital activity. Translation, moving the spacecraft up to or away from another vehicle, changing orbit around Earth or Moon, or performing course corrections, was to be manually controlled, if required, by a translation T-handle mounted to a box secured to the end of the left armrest on the left couch.

Both rotation and translation hand controllers were to be slaved to the three prime axes and mounted in such a way that their operation followed normal conventions. For instance, the rotation controller would be moved forward to pitch the spacecraft's nose down, moved back to pitch it up, tilted to the left to roll left, and tilted to the right to roll right. Yaw right or left would be performed by twisting the controller clockwise or counter-clockwise about its central axis. The translation controller, used with the left hand, would accelerate the complete spacecraft in each of those axes rather than turn it about its center of mass: with respect to a target vehicle in front, tilting the handle up would cause the down-facing thrusters on opposite sides of the Service Module to fire allowing the spacecraft to rise; tilting the handle down would cause the spacecraft to move down; moving the handle to the left or the right would cause the spacecraft to move in respective directions; pushing the handle in or pulling it out would move the spacecraft forward, or cause it to back away. Both attitude control and translation maneuvering could be delegated to manual or automatic modes of operation. An ultimate sophistication allowed manual control via the rotation controller of gimbal angles for the Service Propulsion System during firing.

More than any other manned vehicle, Apollo comprised two very different modules mated for most of the flight but designed to go their separate ways just before re-entry. Because so many of the systems used by the Command Module were contained within pie-sectors in the Service Module, the physical mating of the two elements required a substantial structure for carrying electrical and environmental leads from one to the other. The umbilical designed for the purpose was to be covered by an aluminum fairing 1 meter long and about 46 cm wide. On the Block II vehicle it was positioned between the Command and Service Modules around the spacecraft on the opposite side to the crew entry hatch. On the earlier, Block I design, it was at the front, to the immediate left of the hatch. When the time came to separate the two structures physically it required several events to occur almost simultaneously.

The three 6.3 cm wide stainless steel straps, or tension ties, holding the two together were to be severed milliseconds after isolation of the umbilical. At the point of separation the electrical wires would be deadfaced – power removed from both sides of the interface – with valves closing off the fluid connections. A guillotine would then slice through electrical wires and plumbing inside the umbilical shroud, allowing the latter to rotate back away from the Command Module as small explosive charges severed the three tension ties underneath the base heat shield one-tenth of a second later. From that point on the Command Module would rely on its own internal systems for all operational functions down through re-entry and on to the sea. At the instant of separation, the Service Module's thrusters would fire to back it away from the Command Module for 5.5 seconds; just 2 seconds into the burn, however, the roll engines would also fire to impart a slow spin and, like a rifle bullet spin-stabilized in flight, the Service Module would move continuously further away from the flight path of the manned vehicle maintaining a fixed course.

Descent through the atmosphere inevitably led to dependance on Northrop-Ventura's parachute recovery system which, compared to problems posed by Mercury and Gemini was a record breaking headache maker. The Command Module would weigh nearly 5½ tonnes falling to Earth, not far short of three times that of a Gemini re-entry module. Nobody had any very good idea at first how to build a reliable recovery system capable of safely lowering three men, with valuable Moon rocks, gently to a splashdown. Wesley A. Steyer was Northrop's Apollo parachute recovery system manager, with William Freeman and Theodore W. Knacke providing the answers. Under contract to NAA's Space and Information Systems Division, Steyer had to pack his entire recovery system into space nobody else wanted around the docking tunnel beneath the forward heat shield; moreover, tucked in under the spacecraft's apex cover, Northrop would have to find a system that weighed no more than 220 kg.

The company got an early start on its problem, having received in late December, 1961, the contract for recovery design. The two constraints (volume and weight) militated against the use of conventional techniques where redundancy ensured safe recovery in the event of partial failure. There would be no room for additional back-up canopies and what the spacecraft carried it would have to use in full for bringing its crew down intact. The only way Northrop could increase the reliability factor, dismally absent from parachuting in general until space flight brought stiffer demands, was to perfect a technique of packing and to adopt a method of manufacture far in advance of anything then developed. The wizardry summoned by Wesley Steyer is a tale in itself only eclipsed by the 100% success record his product was to achieve.

Calculations quickly showed that Apollo would need three canopies, each 26 meters in diameter, to lower the CM at a descent rate of 34 km/hr required by NASA. In fact two alone could do the job, with a modest increase in the descent rate, but a third provided the only modicum of back-up allowed within the entire system. One day it would be needed. But the big orange and white ringsail 'mains' could only be deployed safely at speeds below about 350 km/hr. They would also have to pop out and deploy to

← Master parachute riggers Norma Cretal and Stan Enger offer up an Apollo parachute to a packing fixture which will hydraulically ram the envelope into a suitcase-sized space.

Situated beneath the apex cover, an Apollo main parachute is stowed ready for use.

the reefed condition within fractions of a second of each other. If one came out appreciably sooner than the other it would be torn apart by the falling spacecraft; $5\frac{1}{2}$ tonnes could only be braked and supported in mid-air if all three canopies took the strain at the same time. But worse than that, the blunt-faced cone carrying men and rocks set up such a turbulent wake that Steyer thought it would preclude safe deployment even if they did all come out together. A stable condition could only be obtained if the risers were long enough to extend well beyond the zone of turbulence. That sounded good in theory, until longer and longer lines seemed to suffer increasingly from wake turbulence.

Steyer confronted Maxime Faget with the problem and the two were soon in disagreement over the cause. Faget believed the basic design was wrong; Steyer knew from long experience in the parachute business that it was right. The solution was simple. Despite theoretically being considered a rigid line, the risers were responding to the dangling weight of the Command Module and bouncing up and down, the effect magnifying with increased length. The answer was eventually found in a commercially available fishing line, procured from a local hardware store. When used to fabricate a suspension line of sufficient strength to hold the load it had the desired effect on tests that for some months during development had everybody hung up on a potentially insoluble problem.

Sixty-eight risers were employed for each main parachute, the spacecraft hung by 249 fishing line cords each 36.6 meters in length. The problem of a simultaneous deployment was solved by two sequence controllers, each 7.6 × 7.6 × 15.2 cm in size weighing 1.8 kg, containing four barometric pressure switches, four time delays, and four relays for parachute release. The Command Module would be descending through the lower regions of the atmosphere at speeds approaching 500 km/hr when the mains should ideally be deployed. The only way to slow the vehicle down before putting out the mains was to carry two drogue parachutes to cut the descent rate down to about 282 km/hr. These were to be 4.17 meters in diameter and serve to stabilize as well as slow the spacecraft. They would be suspended 19.8 meters above the Command Module. Smooth operation of the mains would benefit from three 2.2 meter diameter pilot parachutes deployed first and suspended from a height of 17.7 meters.

When it came to packing all eight parachutes in the forward section of the Command Module, ingenuity was no longer the only means by which the deed could be accomplished. Sheer brute force was essential. Because nylon possessed excellent 'memory' characteristics it would readily unfold from a densely packed condition. But even those qualities were tested when Northrop went to the extent of squeezing each main parachute under a 21 kg/cm² press, in a vacuum so that all the air molecules would escape and so provide just that extra bit of compressibility that made all the difference between putting it on board or leaving it off. Folded and pressed to the density of wood, each 57.6 kg main ringsail parachute was a packing machinist's pride. But would it work in practice?

The entire sequence of events would begin when the sequencer barometers detected an altitude of 7,315 meters. About $1\frac{1}{2}$ seconds after jettisoning the forward heat shield the two drogues would be deployed, fired from two barrel-shaped mortars. At a height of 3,050 meters the drogues would be jettisoned and the pilots fired from small mortars around the side of the recovery compartment. These would pull free the main parachutes packed around the forward access tunnel, to first an initial reefed condition, then to a second reefed condition, and finally the fully disreefed condition. To prevent the full deceleration of a sudden impact, the spacecraft would hang with a pitch of 27.5° during its descent on the three mains. At splashdown the spacecraft would slice into the water with a glancing blow. Nevertheless, the crew could receive a substantial jolt if the wave motion slapped the base end of the heat shield.

To prevent this, an impact attenuation system was provided comprising four crushable ribs installed in the aft compartment between the aft heat shield and the inner pressure vessel. Under its downward momentum the pressure vessel would continue to move, the energy absorbed by the bonded laminations of corrugated aluminum collapsing inward. In addition, the three crew couches were to be suspended from the spaceframe on eight struts, each about 6.4 cm in diameter, designed with internal steel wire rings between inner and outer pistons that would crush when subject to forces that would otherwise be detrimental to the crew.

Because Apollo was considerably bigger than either Mercury or Gemini, and because the large amount of equipment the crew would carry made it important to use every available space, the couches were literally to be suspended in the center of the inner pressure vessel. The Weber Aircraft Division of the Walter Kidde company got the contract to fabricate the couches, made from hollow steel tubing and covered with a heavy fiberglass cloth called Armalon. The decision to use a separate escape rocket for lifting the entire Command Module away from its launcher in the event of an abort, unlike the Gemini which employed ejection seats, was based in part on the proven escape system concept pioneered by Mercury. Apollo was basically designed long before Gemini and much of its conceptual allegiance owed its origin to Mercury philosophy. Gemini was more the product of James Chamberlin's useful ideas for adapting Mercury. Apollo was a product of Maxime Faget's Manned Spacecraft Center engineering divisions.

But apart from related hereditary, Apollo's mission requirement precluded the use of large hinged hatches spring loaded to fly open at a second's notice. As with the internal systems layout, heat shield design placed heavy demands on the number of cutouts that were allowed to populate the exterior surface. For that reason alone, the couches were merely required to support the crew internally under all possible mission conditions. Although the three couches were in fixed installations, the center one could be folded down and pivoted back from a head beam that ran across the entire breadth of all three head supports.

The Apollo main display console presented instruments across the full width of the Command Module. Note access hatch at left, two couch support struts and the docking tunnel behind the display console.

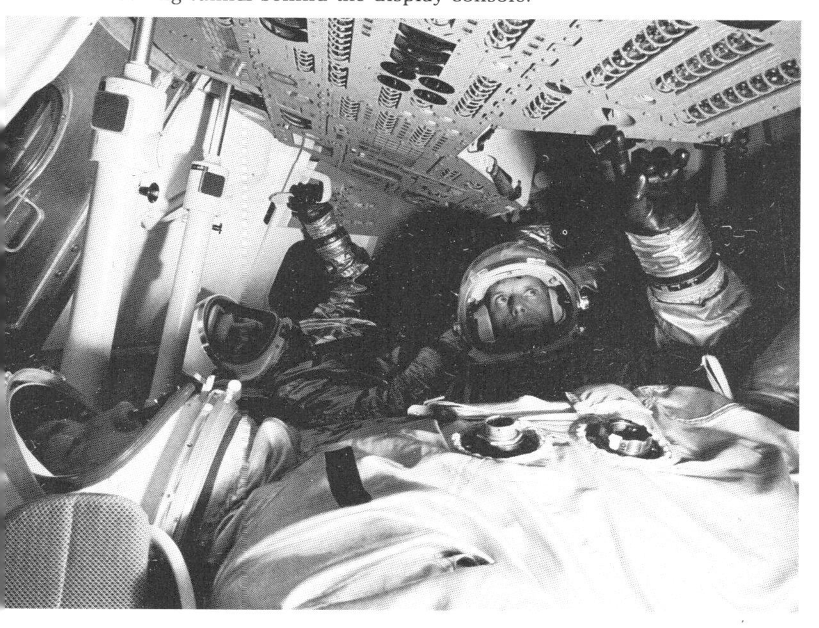

Folded up and laid back down on the aft bulkhead – the floor – of the internal pressure vessel, it allowed the crew additional space to move and work

Each couch comprised back, seat, leg and foot pans, with a head rest recessed below the level of the body support to accommodate the bulbous configuration of the astronaut's helmet. Four vertical struts attached to the forward structure of the pressure vessel would support the weight of the couches and their occupants on the launch pad, during ascent and on re-entry. Two were attached at the head beam and two were attached at points adjacent to the knee positions. In addition, two side struts were fixed to left and right walls of the cabin, with a further two carried from the aft bulkhead diagonally across to a position underneath the couches. In this way, with the eight struts, motion was to be attenuated in the vertical, lateral, and horizontal axes.

The interior layout of the crew compartment, and the selective distribution of instruments on the main display console, was arranged and grouped around the displacement of the three astronauts. The left seat would be assigned to the Commander (CDR) in charge of the mission, ultimately responsible for events in both the Apollo CSM and the lunar landing vehicle. The center couch would be occupied by the Command Module Pilot (CMP), specifically responsible for flying the Apollo mother ship, and the crewmember whose lot it would be to remain aboard while his colleagues explored the surface of the Moon from the lunar landing vehicle. Control of that spacecraft was to be the responsibility of the Lunar Module Pilot (LMP), assigned to the right couch. The two outer couches were designed to carry left and right arm rests installed on the back pan; the center couch had none. All four rests could accept the rotation and translation hand controller boxes with their handles.

The multiplicity of controls and indicators positioned around the spacecraft far exceeded anything conceived for Mercury or Gemini. So large was the range of systems switching functions that each crewmember had responsibility for a specific range of equipment, only the Commander having a general monitoring capability. The main display console comprised a flat panel extending across the entire pressure vessel with controls and displays accessible to respective crewmembers in their flight couches. Circuit breaker panels were located to right and left of the console, angled inward for ease of reach; smaller panels were arranged around accessible areas. When free of the couches, astronauts would move to other control stations distributed around the spacecraft. Beyond the foot pans, on the pressure vessel wall opposite to that containing the side hatch, an area known as the lower equipment bay, were the guidance and navigation instruments. Along the right wall of the spacecraft, down past the Lunar Module Pilot's couch, were to be located the waste management controls, equipment and switches for urine and fecal collection. Along the opposite wall, down past the Commander's couch, were the environmental control systems equipment, canisters containing lithium hydroxide for cleaning the atmosphere, oxygen regulation controls, etc.

The philosophy adopted for spacecraft crew arrangements provided status and control capability while the astronauts were in their respective couches, with equipment that required periodic attention distributed in areas that made the most out of the limited space available and for which the crewmember would have to leave his couch. Thus, any switch or indicator that an astronaut could possibly want during a critical situation or an emergency would be within reach from the place he would most likely be found during dynamic events: the couch. Astronaut categories during the early part of the program were more an indication of role: command pilot, senior pilot, and pilot. Not before manned flights began was NASA to adopt the CDR, CMP and LMP designations. However, the arrangement of panel instruments and controls changed considerably from Block I to Block II spacecraft, reflecting the additional responsibilities of the latter. In all, the Command Module contained nearly 600 switches, 40 mechanical event indicators, 71 lights, and 24 instruments. They represented the nerve ends from more than 24 km of electrical wiring packed into the Command Module.

But astronaut safety required that a means of escape be provided for all phases of the ascent trajectory. The big SPS engine at the bottom of the Service Module could be used to re-orientate the complete Apollo spacecraft, but only during later stages of the long climb spaceward. From the outset, it was clear that Apollo, like Mercury, would require an escape tower supporting a solid propellant rocket that could wrench free the Command Module alone during the period of first-stage thrusting on the giant Saturn V. After that, the complete CSM configuration could lift free using the spacecraft thrusters.

Contractors got in early on the launch escape system project. By February 1962, North American Aviation had selected the Lockheed Propulsion Company at Redlands, California to design and build the equipment. And from the beginning, Irwin Spitzer was project manager, a post he enjoyed with relish and enthusiasm. It was to be one of the more pleasant, incidental, jobs created *by* Apollo and *for* Apollo. Designed to stand by in case it was needed for only the first three minutes of flight, the launch escape system weighed, at 3.7 tonnes, about the same as a manned Gemini spacecraft, yet it would be thrown away one-half minute after the Saturn V's second stage ignited.

Initially, NASA decided to specify a solid propellant rocket motor generating 91 tonnes of thrust, with an active thrust-vector control system which would transfer the line of thrust away from the centre of mass causing the assembly to move to one side and so haul the Command Module to a safe area. Soon, however, preliminary studies revealed a better, and much simpler, means of accomplishing that objective. With the single solid rocket motor firing through four nozzles, the flight path of the ascending escape system would automatically move to one side if the diameters of the nozzle exit throats were slightly different. In other words, instead of designing the rocket to fly a straight course, deliberately design it not to stay on course so that in firing upward it would also drift to one side, a phenomenon not usually associated with deliberate design practice! Also, if a pitch motor was employed to kick the whole assembly to one side as it lifted the Command Module away, it would increase the separation rate from what could be the exploding fireball of a runaway booster.

A third rocket motor could be carried within the same structure to separate and jettison the escape system during a normal flight. But if the escape system *was* used, something would have to pitch the assembly over so that the base of the Command Module heat shield faced down toward the atmosphere; although the escape system was designed to free the astronauts within the atmosphere, it would also be called upon to serve vacuum aborts up to 90 km. That could be accomplished by two canards, small lifting surfaces designed to pop out from their stowed location flush with the sides of the escape cylinder's forward end about 11 seconds after abort ignition. That would throw the whole assembly over and pitch up the spacecraft's base end. The tower carrying separation, pitch and jettison motors would be released at this point, leaving the Command Module free to deploy its parachutes at the appropriate time.

Protection for the exterior surface of the Command Module from the exhaust products of separation or jettison motors was to be ensured by a white conical cover completely shrouding the Command Module. Contoured to the CM's exterior profile, it was to be attached to the legs of the escape tower and pull free when the escape system was jettisoned, exposing the manned vehicle for the first time. Called the boost protective cover, it solved in one stroke the problem Gemini had with smears and stains across the windows. Made of layers of impregnated fiberglass and cork, it had ports through which the reaction control system thrusters could fire, and a 20.3 cm diameter window aligned with the Commander's forward viewing window; all other windows were covered until the protective cover was jettisoned along with the escape tower at a height of 90 km. Not for Apollo astronauts the thrill of an outside glance!

In all, the escape tower and its torpedo-shaped structure was 10 meters tall. The cylindrical structure that contained the escape, pitch and jettison motors was 66 cm in diameter. In abort conditions, the escape motor, delivering a thrust of 66.7 tonnes – greater than the thrust generated by Alan Shepard's Mercury-Redstone rocket –, would lift the Command Module away in a 3 second burst initiated by a malfunction detection system or by manual command. If aborted from off the pad, it would lift the Command Module to an altitude of 1,220 meters for immediate activation of the spacecraft's recovery equipment. The pitch motor would fire in an off-the-pad or low altitude abort for 0.5 second with a thrust of 1 tonne. In the case of an abort at altitude, the escape system would remain with the Command Module until the assembly had fallen to an altitude of 7,315 meters, where normal recovery operations would begin with tower separation. Here, and in the case of tower jettison on a normal flight, the 14.3 tonne thrust jettison motor would fire radially downward through two exhaust ports, one either side of the forward section. The jettison motor was carried forward of the escape motor in a tandem configuration.

The extreme nose of the cylindrical structure housed the two canards and a Q-ball, a device equipped to sense different dynamic pressures on the nose of the ascending vehicle and to monitor the space vehicle's angle of attack during flight. The system was the first to be fully qualified. By 1964 it had passed through its design, development, and qualification phases and although it was never called upon to operate on a manned mission the availability of an escape system was an essential feature of flight safety. Aborts above 90 km would use the Service Propulsion System Engine and require the complete CSM to lift free of the launch vehicle adapter. Not all aborts would use the SPS, however, but if employed it could be used to decelerate the spacecraft prior to separating so that the Command Module could descend, or it could be used to kick the spacecraft into Earth orbit if the launcher malfunctioned at sufficient speed and altitude to bring orbital flight within the range of the SPS performance.

Early in the Apollo design phase, the Manned Spacecraft Center was convinced of the advantage in fuel cells for electrical energy aboard the manned vehicle. But the cells would have to operate with greater reliability than the fuel cells on Gemini and with higher energy outputs too. Long before subcontractors were busy with the two-man spacecraft, NASA had chosen Pratt and Whitney Aircraft Division of the United Aircraft Corporation to develop a fuel cell for the Lewis Research Center suitable for use in Apollo. That was in April 1961, before President Kennedy gave Apollo a re-written objective to go for a lunar landing. At that early date, it was believed the spacecraft would probably need a power supply of between two and three kilowatts at a nominal 27.5 volts. At the upper limit that would provide a 109-amp supply. Ten months later, Pratt and Whitney got the definitive contract to build hydrogen/oxygen fuel cells for the CSM. Developed at the company's attractive rural facility near Hartford, Connecticut, the fuel cells for Apollo were to operate at between 27 and 31 volts and comprise 31 separate cells in each stack.

Individual cells would comprise a hydrogen and an oxygen electrode, a hydrogen and an oxygen gas compartment, and an electrolyte of potassium hydroxide and water. Thermally controlled by a flow of water-glycol, the non-regenerative cells would normally operate at temperatures between 196°c and 260°c. Theoretically capable in the definitive model of producing 2.3 kilowatts direct current at 20.5 volts, the cells normally generated between 563 watts and 1,420 watts at the required voltage range specified by NASA. Three cells would be carried by each spacecraft, attached to an equipment shelf in one of the two Service Module pie-shaped sectors not containing an SPS propellant tank. Maximum power capacity at 30 volts would be a total 4.2 kW, or little more than a three-bar domestic electric fire. Two cells would be sufficient to handle all primary electrical loads, and one alone could supply emergency power. By comparison, the maximum capability of the General Electric fuel cell developed for use aboard the Gemini spacecraft provided a total power production from the two units of about 2 kW at a nominal 25 volts DC.

But Apollo's fuel cells produced pure water as a by-product of the reaction between hydrogen and oxygen, suitable for drinking purposes and for use in the cooling circuit. In addition to the fuel cells, the Command Module would carry three silver oxide-zinc storage batteries supplied by the Eagle Picher company. Each battery supplied 40 ampere-hr current at 37.2 volts and all three would supplement the fuel cells during flight. Their primary function was to provide electrical power for the CM after separation from the Service Module. Also, two 0.75 ampere-hr batteries similarly housed within the CM were to be used to power the pyrotechnic devices employed for separation from the SM, parachute deployment, separation from the launch vehicle, etc. The storage batteries were coupled to a charger that would ensure they remained at full capacity. This would be important for if they were required in an emergency there would be no time to charge them up. That provision would be called upon on one historic flight.

The spacecraft's electrical distribution system provided two redundant buses for DC and AC power. The two main DC buses (Main A and Main B) were coupled to the three fuel cells or the three storage batteries, occasionally both. Two battery bus circuits (A and B) were each connected to individual storage batteries with the third battery capable of providing power to either or both circuits in the event of a failure. Three solid state inverters contained within the Command Module would be used for converting 28 volt DC power to 115 volt alternating current with a 3-phase, 400-cycle output.

Cryogenic reactants for the three fuel cells, hydrogen and oxygen, were contained within spherical pressurized vessels constructed by Beech Aircraft. The two oxygen vessels were fabricated from a nickel-steel alloy called Inconel to prevent inadvertent combustion through impact. The two spherical hydrogen dewars were fabricated from titanium and contained a total capacity of 25.4 kg. Because of the exceptionally low boiling temperature of these two liquids, the insulation required to maintain the fluids in the semi-liquid state posed major problems. In affect, the pressure vessels were extraordinarily efficient vacuum bottles, comprising inner and outer spheres. The oxygen tank was insulated with fiberglass, paper and aluminum foil. The hydrogen tank contained a vapor-cooled shield suspended in a vacuum to serve as a heat barrier. Both sets of tanks carried internal heaters to increase pressure and fans to stir up the contents.

The cryogenic tanks were so isolated from the ambient environment that Beech Aircraft was proud of the boast that if ice cubes were placed inside they would take 12½ years to melt and reach room temperature. Furthermore, said the manufacturer, the pressure loss was so slow that if an automobile tyre leaked at the same rate it would take more than 32 million years to go flat! Together, both oxygen tanks carried 290 kg of liquid, used as a gas in the nitrogen pressurized fuel cells. But only 190 kg of the tanked supply would be available for electrical production. The balance was made available to the environmental control system – the ECS – for cabin pressurization.

The inextricable involvement between electrical and environmental systems pioneered operationally by the Gemini spacecraft was an integral part of Apollo from the outset. The selection of a pure oxygen atmosphere for the three-man vehicle has been discussed in Chapter Nine and, to summarize the choice, it can be said that an important consideration from the beginning was the complexity inherent in a two-gas system. Moreover, with oxygen pressure fixed at a minimum level, the total atmospheric pressure of a two-gas environment would threaten the structural integrity of the pressure vessel. Leak rates would be higher and in attempting to provide a less flammable spacecraft the product could well have picked up potentially dangerous characteristics from the structural re-design that would have been necessary, apart from the added complexity of the two-gas system. A basic tenet of spacecraft design, as in every electrical or mechanical device down to household and domestic level, is that simplicity leads to reliability; complex products with dials, switches and knobs in profusion may look advanced, but they are commercial gimmickry designed to attract buyers influenced by sophistication.

The fundamental design principle for Apollo retained simplicity in preference to complex and ingenious equipment.

Nevertheless, the danger inherent in a pure oxygen environment was brought rudely home in two separate incidents. On 9 September, 1962, a fire broke out on the thirteenth day of a two-week test in a pure oxygen environment at the US Air Force School of Aerospace Medicine. Pressurized to the 258 mmHg planned for Apollo, and used in Mercury and Gemini, the atmosphere rapidly increased the severity of the fire which had broken out in a test panel used to evaluate the psychomotor abilities of two pilots. Both men suffered smoke inhalation and were rushed to a local hospital. During the first night after the accident, one subject's vital signs dropped alarmingly but his pulmonary edema quickly cleared the next day. Both men were eventually returned to operational duty.

On the second occasion, 17 November the same year, a fire broke out in an identical environment on day sixteen of a twenty-day test at the US Navy Air Crew Equipment Laboratory. Three subjects and a flight surgeon were inside the test chamber when one man attempted to replace a light bulb in a standard 24 volt DC fixture of the type commonly used in aircraft. A wire in the connection must have been disconnected, and the resulting arc caused a flame to emerge from the socket. Asking through the communications line for water to dowse the flame, technicians outside the chamber told one of the subjects to use a towel. When he placed it over the socket it immediately burst into flames and the fire spread to the man's own clothing. An asbestos blanket nearby was grabbed by another man but when placed over the flames this too caught fire, spreading the conflagration to the other three men in the chamber merely by contact. All four were eventually rescued from the chamber but two men were very seriously burned and most combustible items within the chamber had been consumed by the time the men were liberated.

Because materials soaked for long periods in a pure oxygen environment can substantially change their susceptibility to flame propagation, materials selected for use in the Apollo Command Module were, theoretically, screened for possible rejection. In reality, many items were creeping in to the interior layout without full regard for the safety factor involved. But in principle the decision to proceed with a pure oxygen environment, taken early in the program's development phase, was wise and appropriate. Weight was certainly saved by adopting a one-gas system, and further equipment savings were made by integrating the supply with the reactant storage capacity of the fuel cell system.

Pratt and Whitney fuel cells built to produce electrical energy are seen here on test.

Oxygen was to be supplied through the suit circuit until the spacecraft reached orbit, when the pilots would remove their helmets and suits and breathe the cabin atmosphere. During reentry, when the Service Module could no longer provide gas from the reactant tank, the crew would use a small oxygen surge tank carried in the Command Module. Two water tanks were carried in the CM, one for accepting water from the fuel cells, called the potable water tank, and one, called the waste water tank, equipped with a bladder to accept water from the suit heat exchanger. This tank was to also supply water to the glycol evaporator designed to supplement the main cooling system and on Block I vehicles it was also to supply an evaporator section in the suit circuit. Excess water was to be converted into steam and vented overboard.

Thermal control was provided by a series of coldplates and coldrails on which electronic and electrical equipment was mounted similar to the method adopted for the Gemini spacecraft. Conducted to the water glycol solution flowing through the pipes and tubes, excess heat would be radiated to space from tubes on the exterior surface of two Service Module sector panels. An aluminum reservoir was provided for water glycol storage. As in earlier manned space vehicles, Apollo chose to use the cleaning effect of lithium hydroxide for removing carbon dioxide from the cabin atmosphere and the type of canister selected for use on the three-man spacecraft was also designed to remove odor by the same process. Two canisters were provided in aluminum housings down the left wall of the vehicle, their respective elements being changed alternately at 12 hour intervals. A quantity of elements would be stored in a nearby locker. The environmental control unit was designed and built by Garrett Corporation's AiResearch Division at Los Angeles. Like Northrop-Ventura, Garrett had supplied for Mercury and Gemini the equipment it was contracted to build for Apollo. Like Northrop-Ventura, Garrett was experienced at the art of getting their products to work.

But Apollo was a complex vehicle, and each system, while seeking simplicity, became increasingly sophisticated. It was no easy task building the Command Module, even worse to integrate all the many elements of Apollo into a single-minded machine. What Lee Atwood accepted on behalf of North American Aviation became a nightmare of increased demands from the customer, flagging schedules from the subcontractors, and a major head ache at integration and test phases. It was a pace undeniably motivated by crass ignorance among politicians and officials concerned only with summoning a technical miracle from the over-stretched facilities of American industry. There were some who saw the writing on the wall. As early as January, 1964, Republican Representative Thomas M. Pelly introduced a resolution to change the goal of landing a man on the Moon within the decade of the 1960s to a period before 1975, 'to allow our scientists greater flexibility in meeting the challenging technological problems of the Apollo project. Especially, I have in mind the desirability of eliminating the pressures of meeting time schedules, of minimizing perils . . . , and frankly I would stretch out the fantastic cost of the manned space exploration program over a longer period of time. . . .'

John Kennedy had been dead only two and one-half months and already the critics, although present for counting fully two years earlier, were now demanding a re-think on national priorities. It was proof positive that had the incumbent President not made the questionable decision he did in May, 1961, the United States would never have embarked upon a lunar landing mission. At least not in the 20th century. As the decade of the 1960s unfolded, bringing with it civil unrest at home, deepening and bloody commitments abroad, and the realization of finite global resources, the picture of national ambition was repainting itself away from the pursuit of adventure and challenge to one in which the debilitating encumbrance of a stagnant society smothered vision and ideals. Representative Pelly failed to gain support for his resolution, but more because at such a time of internal stress surely the lawmakers could not be seen to waver. And the Apollo machine moved quickly along its well-lubricated tracks.

Speed was of the essence and from cursory inspection of the many hardware tests conducted between 1963 and 1966 an observer would see evidence of progress and achievement. But the difference between the Block I spacecraft and the Block II, designed for the Moon flights, widened until the former became a complexity of wire bundles and stretched leads, re-worked pipes and bent conduits. Block I was rapidly becoming a waste ground of disorder and chaos. The need to have a Block I vehicle at all was a requirement dictated by the schedule. If manned lunar landing operations were to get underway by 1969 at the latest, a precursor vehicle would be essential. There was still a complete generation of launch vehicles and spacecraft to assemble, check, test, and fly, not to mention rehearsal and qualification of extremely complex and demanding flight techniques in Earth orbit first.

Saturn I launchers were still working methodically through a protracted development schedule that took each potential pitfall in turn and examined it with a unique flight. In all, ten Saturn I missions flew between 1961 and 1965, and at the end of that period the rocket was recognized to have proven it could work and to have orbited a few payloads thought up as a good way of using the ostensibly research-orientated flights. If Saturn I's operational successor, the IB, or the massive Saturn V, were to be ready within a few years, some radical transformation in test philosophy was essential. Wernher von Braun had performed a miracle few believed could happen in bringing together the separate elements of the Saturn I. But his test operation was lengthy and unwieldy. It was at this point, late in 1963, that Mueller's 'all-up' systems testing philosophy came appropriately to the fore.

It is best expressed in the minutes of the NASA Management Council Meeting held at Headquarters on 29 October, 1963, where, in part, it records that, 'Doctor Mueller stressed the importance of a philosophical approach to meeting schedules which minimizes "dead-end" testing, and maximizes "all-up" systems flight tests. He also said the philosophy should include obtaining *complete* systems at the Cape (thus minimizing "re-building" at the Cape), and scheduling both delivery and launch dates. In explaining "dead-end" testing he referred to tests involving components or systems that will not fly operationally without major modification.' In essence, this meant launching a rocket with all the stages it would ultimately carry, moving from one set of tests to the next as each was seen to be successful in turn. In that concept, the ability of the test to extend beyond predicted levels was up to the capability of the hardware and not set by the limits of the available systems. It was an approach – a philosophy – that did much to put man on the Moon by the end of the decade.

Mueller too was responsible for setting up the Apollo Executives Group where Center Directors concerned with supporting the manned program would periodically get together for a review of program elements pertinent to the common goal. It also ensured a direct management liaison between government and industry heads, where corporate executives could meet with their state customers to learn of progress, or otherwise, at a lower level of the management chain. Employment on manned space programs reached a peak during 1965, where more than a quarter million persons in government and private industry were employed directly as a result of the Apollo and Gemini programmes. On Apollo alone there were more than 240,000 persons of which only 8,700 were in government service. A creditable achievement where less than 4% of the workforce retained civil service pay cheques. And throughout the United States, more than 20,000 companies, businesses large and small, in addition to the big corporations building the hardware for NASA, were building equipment for Apollo. As President Kennedy had said four years before: 'In a very real sense, it will not be one man going to the moon – it will be an entire nation. . . .'

The official cancellation of four proposed manned Earth orbit flights with the Saturn I came on 30 October, 1963, two days after Mueller received word that Webb approved the plan he had presented to the Administrator for 'all up' systems testing. The flagging Saturn IB program was to be accelerated by direct intervention of Headquarters. Or so it seemed to the von Braun cadre at Huntsville's Marshall Space Flight Center. The German engineers who migrated west with their good looking leader at the end of the War, and others drawn to the fold in later years, resented interference as much as they were themselves opposed by the polarization of traditionalists in Washington. Nevertheless, winds of change were essential to the expedient availability of the improved Saturn. From the outset, said Mueller, Saturn IB launchers should carry fully operational Block I Apollo vehicles equipped to test their systems extensively in Earth orbit shakedown missions. On 1 November, 1963, Mueller informed Bob Gilruth at the Manned Spacecraft, Wernher von Braun at Marshall, and Kurt Debus at the then Launch Operations Center, Cape Canaveral, that 'all up' testing would henceforth permeate every spacecraft, launch vehicle and program milestone.

Less than three weeks later he set out the tentative schedule for introducing Saturn IB and Saturn V rockets into the Apollo flight schedule. The tenth and last Saturn I mission would be flown in June, 1965, he said, followed by the first Saturn IB mission in the first quarter of 1966. The first manned Saturn IB Apollo mission was to take place within a twelve month period starting in the third quarter of that year. The first Saturn V mission was to take place in the first quarter of 1967, with the first manned Saturn V Apollo flight between the third quarter of that year and the second quarter of 1968. Mueller put the initial manned flights of each launcher on the third operational mission of each respective vehicle, compressing the schedule that would have pertained under the sequential philosophy of the original concept evolved at Huntsville.

It should perhaps be said that in the face of so many remarkable developments achieved by the Marshall Center, it was beyond the scope of von Braun's men to change conceptual modes of test and qualification completely. That would perhaps have been expecting too much, the achievement of which may have prevented other more important successes. However, nearly two years later, the tenth Saturn I flight was only a month beyond Mueller's projection and by the end of 1965 General Phillips, the Apollo Program Manager, forecast a first flight of the IB in January 1966, with the first flight of the V a year later. As it turned out, the first Saturn IB mission flew on 26 February, 1966, with the first Saturn V in November, 1967. Not bad shooting for a Headquarter's man!

From time to time the precise nature of the launch designations has caused confusion to historians and observers of the space program. Some flights were designated Apollo-Saturn, while at other times being referred to as Saturn-Apollo, abbreviated to letters AS or SA. Saturn I missions received numerals 1 to 10, so that the first flight was designated SA-1 and the last SA-10. Flights with the IB adopted a three-number system beginning SA-201, with SA-202 the second flight, SA-203 the third, et al. Saturn V flights adopted numbers in the 500 range, the first being designated SA-501. When Mueller came to NASA in late 1963 the most visible piece of hardware was the Saturn launch vehicle and von Braun's billowing personality stamped the program with the SA prefix from Headquarters to the Texas manned flight facility; according to von Braun, it was a Saturn launcher flying Apollo spacecraft, hence SA.

When Mueller moved in and settled down, representing the iconoclastic Washington view, the head of manned space flight saw it as primarily the Apollo program giving work to Saturn launch vehicles, hence AS. From then on the formal designation would be to label the first Saturn IB flight the AS-201 mission, and the first Saturn V flight as AS-501. Marshall had to concede, but nearly a decade later the SA prefix crept back in to some internal documents generated by that Center. At least until von Braun left.

Increasingly, during 1964 and 1965, NASA was concerned about the slipping schedules in hardware from contractors across the nation. More particularly, Block I spacecraft troubles were multiplying along with the number of changes and modifications introduced by the space administration. As requirements became more clearly defined, systems that were assumed to be almost ready for installation had suddenly to be dismantled and completely re-worked. The magnitude of the program was becoming apparent, more so as each month went by. Worst of all was the interior design and layout of the Command Module. It was such a small vessel to carry so many integrating systems, wired up with sufficient electrical cable for 50 residential homes. There were inevitable problems, and not only with the process of system layout and fabrication.

North American Aviation had 8,000 people working on the CSM contract at its Downey plant in California, and as unmanned development flights picked up, a further 1,000 at the Kennedy Space Center. More than 12,000 suppliers were feeding the Space Division with equipment large and small for some element or other of the CSM and the company's investment alone rested on a contract worth at that time nearly $2,500 million. Not everybody came up with the goods. Small firms sometimes went bankrupt before the technical excellence of a unique bolt or fixture could materialize at Downey. Always there were problems, and always it was 'Stormy' Storms who had to weather the turbulence. NAA's

Apollo man, the Space Division's program manager Dale Myers, had a good deputy in Charles H. Feltz, the true engineering interface between design and development and the company administrators; in a functional role, he was equivalent to McDonnell's John Yardley during the Mercury program.

Sam Phillips moved in as Apollo Program Director at NASA's Washington Headquarters during October, 1964, relieving Mueller as manned space flight boss so that he could orchestrate the symphonic harmony of Gemini, Apollo, and post-Apollo plans. Shortly after assuming the full office, General Phillips and Dr. Mueller performed an evaluation of the work under way at Rocketdyne on the J-2 engine, the prime propulsion system for the two upper Saturn V stages. Serious problems loomed on the horizon, and NASA examined the possibility of moving to another contractor. Months later, however, improvements were made and the pace of the program gathered momentum. But the performance of North American Aviation in pursuit of the Apollo CSM contract was giving greater cause for worry. Much of the period between May and December, 1964, was spent by selected teams in developing the Block II spacecraft requirements. It was the first opportunity to finalize the specification based upon what NASA wanted to do with Apollo during the lunar landing operation.

Most of 1965 was spent preparing drawings and design layouts for Block II hardware and because the definitive article was more clearly understood at this time than hitherto, engineers found more and more uses for Block I systems, more and more things the precursor vehicles could test, and more and more ways of measuring performance with the Block I before completing Block II hardware. Consequently, Block I spacecraft of the type to be used in initial manned shakedown flights, were increasingly modified in critical areas such as electrical wiring and environmental system plumbing. Long after completing manufacture and assembly schedules, new devices were continually brought for inclusion on vehicles that should have been sealed from further intervention. Wires were spliced, re-spliced, others rolled over and taped back, until the complexity of harness, conduit and pipe evolved into a mesmerizing nightmare of confusion. At times, the accumulation of electrical wiring became so great in confined areas that engineers had no alternative short of physically squeezing it in behind access panels that burst open when unscrewed.

The effect of an increasing torrent of requirements and improved specifications from NASA and its field centers was to slow the programme dramatically down and to severely challenge the ability of the contractor to hold the project together. It was a time when flight tests were accelerating at the White Sands facility and when late Saturn I research flights were carrying boilerplate vehicles on missions from the Kennedy Space Center. Moreover, NAA's Space Division was beginning major vacuum chamber trials at the Manned Spacecraft Center with NASA engineers and technicians. Manpower was stretched to breaking point; pace had caught up with capability.

In a period of little more than twenty months beginning in November, 1963, the space administration launched 10 major test operations with Apollo equipment. Activity began on 7 November when a boilerplate fabricated by NAA successfully demonstrated the ability of the launch escape system to conduct an off-the-pad abort. Just 15 seconds after ignition from a special stand at White Sands Missile Range in New Mexico, the escape tower jettisoned and from a height of 1,494 meters the dummy Command Module returned on parachutes deployed seconds later. Six months later on 13 May, 1964, White Sands set up another Apollo test, the first employing a Little Joe II rocket in which another boilerplate Command Module escape system was tested, this time from a low altitude abort. At a height of 5,182 meters the solid propellant booster was intentionally blown up and the escape motor fired immediately, lifting the CM to an altitude of 7,315 meters. The escape tower jettisoned as planned, but during mains deployment one of the parachutes broke free, the boilerplate returning to Earth safely on two canopies, 7½ minutes after lift-off.

One week and one day later, from the Kennedy Space Center, the sixth Saturn I launcher carried a boilerplate CSM into orbit in an almost perfect flight marred only by premature shutdown of one of the eight first stage engines. It was a valuable development flight leading toward the Saturn IB launcher, and showed that the front-end configuration of a Saturn plus Apollo was aerodynamically acceptable. On 18 September, another Saturn put another CSM boilerplate in orbit, a mission on which the configuration was measured during exit from Earth's atmosphere. On 8 December, from White Sands, the third Little Joe II simulated abort mission began with a flight to max q, the region of maximum aerodynamic pressure where the booster and payload could break up if dynamically unstable. The rocket was intentionally destroyed at a height of 8,840 meters and the escape system successfully worked a third time when it lowered the CM to the desert floor. It was the first test with the boost protective cover representing the configuration of a flight rated spacecraft.

Little more than three months later, on 16 February, 1965, Apollo cooperated in a combined exercise designed to place in orbit a structure comprising a wing-like panel 29 meters in length, folded inside the shell of a Service Module. Launched on the eighth Saturn to fly, the device, called Pegasus 1, was to unfold itself in orbit after separation of the boilerplate CSM and send information back to tracking stations on the ground about meteoroid population in Earth orbit. The flight was a great success, Apollo separated on time and went into its own orbit while the Pegasus structure remained with the S-IV stage as planned.

On 19 May, a third Little Joe II was prepared for launch on a mission intended to demonstrate the ability of Apollo's escape system to perform a high altitude abort, using the two forward canards for stability. Little Joe was to burn for 89 seconds and carry the boilerplate CSM to a height of 35 km for separation via the escape system, boost to 56 km, followed by a normal recovery procedure. However, the Little Joe II ran amok within seconds of leaving its White Sands test stand and following a series of excessive rolls the escape system automatically took over and removed Apollo from the offending projectile, safely returning it to the ground as it would be called upon to do under similar operational conditions. Although the test objectives were not achieved, it was good for morale; the escape system was intelligent enough to know when it should cut in!

Seven days later the Kennedy Space Center loosed a Saturn I with the second Pegasus on board in a repeat of the earlier flight. Again, another CSM boilerplate went in to orbit, and the meteroid wing unfolded as planned. A month later, 29 June, the second off-the-pad escape system test was carried out from White Sands, using the same CM boilerplate employed for the max q Little Joe II flight on 8 December, 1964. All went well and the system proved itself for emegency use while still attached to the launch pad. From the Kennedy Space Center on 30 July, a CSM boilerplate previously used for static dynamic load tests was sent into orbit by the last Saturn I, covering the third Pegasus to fly. It had been a gruelling test schedule and no more boilerplates would fly. One more Little Joe abort mission remained, but that would use a Block I spacecraft. And two Saturn IB test flights would also carry Block I CSMs before the first manned Apollo, again with a Block I, perhaps by the end of 1966.

In the period occupied with the ten Apollo hardware tests, NASA's Office of Advanced Research and Technology performed tests through the Langley Research Center on the thermodynamic environment of a high speed projectile entering Earth's atmosphere. It was research directly supporting qualification of the Apollo Command Module configuration and required a small conical body, 67 cm in diameter and 52.6 cm in length, to be boosted back into the atmosphere from a ballistic trajectory imparted by an Atlas rocket. The heavily instrumented cone weighed 90 kg and would reach a speed of more than 41,000 km/hr, simulating the speed of a returning Command Module entering Earth's atmosphere from the vicinity of the Moon. Nobody had tested the performance of a conical body at this speed, in fact nothing had been observed and monitored during descent from a speed higher than Earth orbit velocity. The blunt-faced body would generate unique chemical and environmental reaction in atmospheric molecules, pushing ahead of it a bow shock wave heated to nearly 11,000°c. The bow wave temperature in front of a spacecraft returning from Earth orbit was less than half this value.

Called Project Fire, the first launch was successfully carried out on 14 April, 1964, when an Atlas D rocket propelled the conical test body on an arching path to a height of about 800 km. As the projectile fell back on its ballistic path an Antares solid propellant rocket accelerated the cone to a speed of 41,520 km/hr, driving it down toward the atmosphere along a flat trajectory, slicing into the atmosphere and sending back telemetered results at the rate of 2,000 each second. Splashdown came in the Atlantic more than 8,400 km from the Cape. A second Project Fire flight was carried out under similar conditions on 22 May, 1965.

Meanwhile, Bob Seamans and Hugh Dryden, the two top NASA men under Webb, were watching the sluggish performance of North American Aviation with increasing concern for the future pace of the program. And as Apollo Program Director, Sam Phillips was asked to put together a task force and go in to the Downey plant for constructive analysis of the company's entire operation. George Mueller, Phillips' immediate superior, and as head of NASA's manned space flight endeavour directly responsible to Seamans and Dryden, had discussed the flagging productivity for some time. Lee Atwood, North American Aviation's President, explains: 'It was a period of rapid engineering action. There were changes coming from every major subsystem supplier, there were changes coming from our own fit and function activity which we follow very carefully with our engineering detailed drawings. And during that late fall, about October, we began to notice, of course, the difficulty in getting our milestone on schedules accomplished as we had hoped to do. A little later on, Dr. Mueller phoned me and said he noticed that we were having indications of milestone slippages.'

Mueller suggested to Atwood that General Phillips bring his teams in and examine the activity under way at Downey. He made it quite clear that NASA was deeply worried about the repercussions of what seemed to be a badly run program and that if North American wanted to help put the program back on track it had little alternative but to cooperate to the full. There was even a suggestion from Jim Webb that NASA would shop elsewhere, at this late stage, for alternative contractors if North American failed to put right things that were soon to be reported in the pages of a secret document sent by General Phillips to Lee Atwood. Sam Phillips began to scrutinize the Downey facility on 22 November, 1965, examining, with his review team, engineering operations, manufacture of hardware, quality control, program control and the practice of subcontracting. General Phillips said he would return in April of the following year and review changes recommended by the task force. He reported to George Mueller and Lee Atwood on 19 December, and supplied the NAA President with a letter, dated that day, which left him in no doubt as to the severity of the review team findings:

Dear Lee:

I believe that I and the team that worked with me were able to examine the Apollo Spacecraft and S-II stage programs at your Space and Information Systems Division in sufficient detail during our recent visits to formulate a reasonably accurate assessment of the current situation concerning these two programs.

I am definitely not satisfied with the progress and outlook of either program and am convinced that the right actions can now result in substantial improvement of position in both programs in the relatively near future.

Inclosed are ten copies of the notes which we compiled on the basis of our visits. They include details not discussed in our briefing and are provided for your consideration and use.

The conclusions expressed in our briefing and notes are critical. Even with due consideration of hopeful signs, I could not find a substantive basis for confidence in future performance. I believe that a task group drawn from North American Aviation at large could rather quickly verify the substance of our conclusions, and might be useful to you in setting the course for improvements.

The gravity of the situation compels me to ask that you let me know, by the end of January if possible, the actions you propose to take. If I can assist in any way, please let me know.

Sincerely,
(signed)
Samuel C. Phillips,
Major General, USAF,
Apollo Program Director.

In explaining the content of the notes he submitted to Atwood, Sam Phillips says that he 'felt that the top management of both the corporation and the division were not giving sufficient attention to the details of the direction and execution of these contracts and recommended more attention from that level of management to the details of their problems and progress.' On the issue of program planning and control, 'I felt here that they had not gone nearly far enough to make a work structure breakdown or what we have come more recently to call work packages. . . . There were planning groups in several places but I was critical of the manner in which they were bringing all the planning together so that the total job could be properly understood and directed and I was critical also of what I call the visibility that program management had.'

On the question of the Block II spacecraft falling behind, '. . . I felt there was no reason why it should be behind schedule if it was properly addressed by the management of the engineering organization.' Moreover, 'We were critical of their test operations with respect to the timely development of the procedures which tell the test engineers the steps to go through. . . . The late design releases gave manufacturing a problem. . . . I was critical of the way in which manufacturing was divided between the program organization and the central manufacturing activities. . . . We were critical of the behind schedule position of some of the components and subsystems and of the practices that were being followed in managing certain subcontracts and in expediting some of the materials.'

In examining the manufacturing end, Sam Phillips was critical 'of what I considered the effectiveness of supervision. The equipment that we get, of course, is the result of work of individual humans and the skill with which they do their work is where good products start. . . . I was critical of the efficiency of the work force. In other words, the work per man-hour.' But there were other observations that impacted the liaison between separate contractors and between the prime contractor and the government: 'I was critical also of the protracted negotiations that had become more or less the way of business between our respective contracting organizations and recommended that actions be taken to improve our ability together to negotiate contracts and changes to contracts in a more timely fashion.' But in a more sinister conclusion aimed at improving the standard of workmanship on Apollo, Sam Phillips was concerned that 'our Government inspectors were finding what I considered to be too many discrepancies over and above those that the company inspectors identified.'

Throughout the searching examination of the company's practices and procedures, North American cooperated fully, although they did not agree with every criticism and conclusion passed along. Atwood's reaction was immediate. He brought in Harvard W. Powell and C. Wesley Scott to head a team responsive to the Phillips review and to construct a reply requested by the Program Director in his letter of 19 December. From the top down, NAA flushed its plumbing with prompt and active measures to get the program under control. Storms pressed hard on Dale Myers and Charlie Feltz and they in turn lost no time in reworking several procedures, tightening quality control all the way.

George Jeffs came in from a position as corporate executive director, Engineering, to serve as Dale Myers' deputy and as chief program engineer. George B. Merrick was brought to Apollo as director of CSM systems engineering from the corporate research and engineering section. Before the formal presentation of the Phillips report essentially concerned with the CSM contract, NAA had already shifted its top management on the S-II rocket stage program under way at Seal Beach, California. This was in response to soundings from an investigation earlier that year on progress with both the S-II and Rocketdyne's J-2, a subsidiary company of North American Aviation. In a situation not unlike the one that confronted James Chamberlin during problems with Gemini management, the existing S-II manager, William F. Parker, was replaced by Robert E. Greer, fresh from a career in the Air Force on satellites and guided missile programs, and employed by NAA as an assistant to Harrison Storms. Atwood saw the need to drive a more prominent wedge between the CSM and the S-II programs, allocating to each a degree of autonomous responsibility, not control because that still rested with Storms as head of the Space and Information Systems Division.

By mid-April, 1966, Sam Phillips re-convened his survey team to examine in detail the progress made by NAA in implementing the recommendations. What he saw was encouraging. In fact he was assured that the program was 'in tremendously better shape at that point in time than it had been several months earlier.' On the 22nd of that month Phillips gave verbal reports to Mueller, Atwood, and their respective staffs, to the effect that 'progress that had been made on the problems I had identified was proper. The company had been cooperative and responsive. The

Astronaut Neil Armstrong (left) and Charles Feltz, Apollo's chief designer.

Called in as chief program engineer, George Jeffs brought experience to the Apollo engineering team headed by Dale Myers. ↑

progress that had been made was such as to give me reasonable confidence that we were on the way to having the problems that I had identified earlier solved properly.' As for Dr. Mueller, he was as much 'satisfied with the work of North American as we were with the average contractor we have. In other words, they weren't better than or worse as a prospect.' The head of manned space flight continued to have reservations.

During this time preparations for the first space test with a production Apollo vehicle went ahead at the Kennedy Space Center. Before that test (a ballistic flight on top of the first Saturn IB) spacecraft 2 was made ready at White Sands for the final abort flight involving a Little Joe II. On 20 January, 1966, the solid propellant booster fired Apollo to a height of 24 km and the launch escape system did its job in removing the Command Module for recovery by parachute. It was a repeat of the aborted abort test which failed to fire to high altitude in May the previous year, and was the last qualification flight for the Lockheed launch escape system – one of the very few elements that never failed when called upon to work.

But it was the launch of NASA's first Saturn IB and its payload, spacecraft 9, that everybody waited for. On that mission, the Service Propulsion System would ignite following separation of the Service Module from the booster's stage to drive the spacecraft toward Earth's atmosphere in a test of the Command Module's ability to withstand temperature and pressure of re-entry. This was the type of launch vehicle designated to fly the first manned Apollo mission, then seen as probable during 1966, and Block I was the CSM design configuration that would carry astronauts into Earth orbit. It was all very significant for the future Apollo schedule. Success here was vital, especially in light of the Phillips report.

Saturn IB had been suggested in 1962 as a means by which the space administration could fly early development models of the Apollo vehicle, and it was now about to perform that significant role. It was also to be the first flight since 1963 from Launch Complex 34. Although planned to accomodate later Saturn I flights, it was assigned to Saturn IB missions when George Mueller's 'all-up' testing concept broke surface in late 1963. The Saturn IB first stage, logically called the S-IB, was erected on the pad 18 August, 1965. The S-IVB second stage followed in October, an important visitor to the Cape since this would be the standard upper stage for all Saturn IB launchers and the all important third stage for Saturn V charged with responsibility for blasting Apollo from Earth orbit to the Moon. For the more moderate Saturn IB flights, it would propel the Apollo CSM, or the lunar lander, into Earth orbit for shakedown trials.

Spacecraft 9 checked in at the Kennedy Space Center on 25 and 27 October; the Command Module first, followed by the Service Module on the second date. The Service Module went to pad 16, where days before Wally Schirra and Tom Stafford had stopped off to suit up for the reluctant Gemini VI, to be electrically tested. The Command Module went to the hypergolic building for servicing and checks of the environmental system. By early November it was clear to G. Merritt Preston's launch operations staff that too much remained to be accomplished and that Hans Gruene's rocket team would have to put off a scheduled launch date planned for late January. Minor problems continued to appear and it was late February before the AS-201 mission was ready to fly.

Standing 68.27 meters tall, the eight first stage engines would deliver a thrust of more than 725 tonnes and at lift-off the stack would weigh 597.8 tonnes. The Apollo CSM was attached to the second stage of the launcher by an adapter identical to the type that would enclose the lunar landing vehicle when boosted by a Saturn V; the Saturn IB's second stage was to be dimensionally identical to the Saturn V's third stage. The adapter weighed 1,724 kg and comprised four aluminum honeycomb panels 4.3 cm thick and 6.4 meters long. They were attached to a cylinder 2.1 meters tall and 6.6 meters in diameter at the base. The cylinder rested on top of the launcher's upper stage, and the four panels provided a tapered structure on top of which the Service Module was fixed during ascent. The big SPS engine bell protruded down into the adapter and the lower area would house the lunar lander. But only on the Saturn V flights; Saturn IB did not have the lifting capacity to carry both Apollo *and* lander so the adapter was empty when the CSM was carried. In orbit, the CSM would pyrotechnically fire a separation cutter, following which it would move forward while the four doors simultaneously splayed open to an angle of 45°. On a Saturn V flight the CSM would turn around, fly back in, dock to the lunar lander, and pull it free. But for this first flight on a Saturn IB, the Service Module was to be only partially filled with propellant for the big SPS engine.

Designed to carry a maximum of more than 18 tonnes of propellant, the normal weight of the Service Module would be about 25 tonnes. Spacecraft 9's Service Module weighed only 10.3 tonnes which, with the 5 tonne Command Module, provided a total Apollo weight of 15.3 tonnes for this mission. Even with minimal propellant in the SM tanks, it was more than four times the weight of a manned Gemini vehicle. The ballistic flight of AS-201 began one second past 11:12 am on Saturday, 26 February, 1966, as the Saturn IB slipped its moorings under the baton of test conductor Paul Donnelly. At 2 min 21 sec the four inboard H-1 engines were shut down followed 5½ seconds later by the remaining four outboard engines. Small motors on the S-IVB second stage fired in an ullage maneuver designed to settle propellants properly for main stage ignition, the first stage separated, and the single J-2 on the S-IVB ignited. It fired for 7 min 33 sec and shut itself down. About 22 seconds into that long burn the launch escape system jettisoned as programmed. It was a typical Saturn IB/Apollo ascent trajectory, one that would be repeated many times.

For this flight the assembly had been accelerated to sub-orbital velocity, but to simulate the excessive heat from a re-entry steeper than would be expected normally – a crucial objective for this unmanned test shot – the SPS engine was to fire twice and speed the spacecraft up to about 29,770 km/hr, driving it hard into the atmosphere in a demanding test of the AVCO heat shield. A high heat rate, high temperatures over a comparatively short period, was the aim. Coasting up to an apogee nearly 500 km above the Atlantic, the spacecraft separated from the petal-like doors of the adapter and after beginning a shallow curve back toward the atmosphere the reaction control system thruster quads on the sides of the Service Module fired for 30 sec to settle propellants in the main tanks.

At the end of that period the SPS engine lit up, about 20 minutes into the flight, to accelerate the spacecraft. At 2 min 20 sec into the burn thrust chamber pressure began to fall and by 3 min 4 sec, at shut-down, had decayed to only 70% of the planned value. The second SPS burn came almost immediately, again preceded by an ullage burn from the thrusters, and throughout the 10 sec period chamber pressures oscillated from 70% down to 12%. Because of this the speed of the Command Module when it re-entered the atmosphere was only 29,057 km/hr. Shortly after, the CM experienced a force of 14.3 g as it decelerated, considerably higher than any load planned for manned missions but a good test of the structure and of the ability of the AVCO shield to tolerate high heating rates.

Spacecraft 9 splashed down in the Atlantic nearly 38 minutes after lift-off, about 9,000 km from the Cape – a little short due to the reduced thrust from the SPS, but in good shape. It had re-entered at a speed faster than that experienced during a return from orbit, heating the base of the shield to about 2,000°c. Almost exactly 2½ hours after launch the USS *Boxer* retrieved the Command Module for an extensive post-flight inspection. Spacecraft 9 had not carried equipment necessary for manned occupation of the Command Module, and the electrical supply originated from batteries and not fuel cells, but it was a significant milestone in the preparation of Apollo hardware. And it was a good start to a welcome year. Far to the north, preparations were under way on the massive Launch Complex 39 site for roll-out of a Saturn V facilities check vehicle.

LC-39 had been built for handling these larger-than-life monsters designed by the von Braun engineers to blast a 40 tonne-plus load to a speed 30 times that of a supersonic jet. Dubbed 'Moonport,' an autonomous attraction on its own, the Saturn V facilities were awesome, candidates in their own right for superlatives beyond access to human speech. With a huge Vehicle Assembly Building inshore, from where Saturns would be assembled and arraigned, a large Launch Control Center to one side, in which more than 400 technicians would monitor machines launching the machines, and a massive Crawler-Transporter capable of walking along at 1.6 km/hr with nearly 6,000 tonnes on its back, the total impression of a nation on its way to the Moon suddenly came together.

For several years now, engineers and construction workers had been transforming the Florida swamp into a gargantuan space port from whose berths would slip the ships of discovery bound for another world in the solar system. Nothing, up to this point, had quite synthesized the feeling that now welled up from deep inside the machine that moved inexorably toward that day when Man would become a cosmic being.

Launch Operations Director Rocco Petrone began pulling together the elements of 500-F early in the year. That was the designation applied to the facilities vehicle, identical in every external aspect to a flight-rated Saturn V with provision for propellant loading, but without live engines and incapable of actually lifting off the pad. It was to be the ceremonial precursor to operational Saturns yet in the making, a full size mock-up with which all the electrical wires, cables, conduits, pipes, tubes, fastenings, fixtures, and connectors installed on LC-39's pad could be checked out. Nobody could get a realistic impression of its size enclosed as it was by girder and steelwork inside the Assembly Building. But when it slowly crept out through the 45-storey doors in the morning hours of 25 May, it was a sight nobody present would ever forget for the remainder of their lives.

As tall as a Gothic cathedral, slender and white with black patches and tiny US flags, it epitomized Apollo. Nothing like it had ever walked at the Cape before. Even the Saturn I and IB vehicles were erected within shrouding covers actually on their launch pads, and the much smaller Titans had to be wheeled to their launch stands lying down. It was incredible to see such a behemoth clanking its way through massive open doors. Its first stage alone was capable of holding twenty times the propellant load of John Glenn's Atlas launcher; the total thrust of all three stages of an active Saturn V would exceed 115 Redstone rockets of the type that sent America's first astronaut into space. Watched by several hundred guests, Apollo officials and their families, Saturn 500-F was a welcome sight, a visible indication of intent. And it was five years to the day that the late John F. Kennedy gave America the Moon by announcing before Congress that men would walk its dusty plains that decade.

25 May Kennedy Space Center provided a generous sample of hardware for guests and visitors to see, and anyone taking a trip down the beach south of Launch Complex 39 that day would, in addition to the Saturn V facilities vehicle, have witnessed busy preparations for a flurry of important flights. First, down at complex 37, the Saturn IB AS-203 vehicle was in final preparation for a unique test to destruction where the second stage would demonstrate its ability to restart in space, a vital operation for future Moon flights. Next, at complex 34, another Saturn IB stood, with Apollo spacecraft 11, ready to fly AS-202 as a repeat of the first Saturn IB flight three months before. Further south still, a Gemini-Titan combination was being carefully groomed for the Gemini IX-A flight, and down the coast that day engineers were mating an Atlas launcher on complex 14 to the Augmented Target Docking Adapter, destined to support the same Gemini mission. And right at the southern tip of ICBM Road, on pad A at launch complex 36, an Atlas-Centaur made ready to launch the first Surveyor Moon soft-lander. It was to set itself gently down on the lunar surface to photograph the surrounding desolation, a project long in the making and one originally expected to provide cursory data about the surface of Apollo's destination.

As it turned out, AS-203 flew before AS-202, allowing more time for hardware preparation with what was considered to be the last qualification flight before the first manned Apollo missions. AS-203 lifted into clear skies on 5 July, 1966, devoid of an Apollo spacecraft, its top end dominated by an aerodynamic shroud. The second stage itself was the test objective on a flight that would put the S-IVB through every kind of stress to prove it was up to the important tasks to come. Mueller had pressed hard for an 'all-up' test in every sense of the word for the first unmanned Saturn V launch, then scheduled for early 1967. It was essential to see if the S-IVB could satisfactorily stop and start several times in space because engineers planned to fire it up twice on the initial Saturn V mission and AS-203 would need to prove it could do just that.

The S-IVB successfully put itself in orbit and then broke up during revolution 4 in a planned destruction test to see how well it withstood planned pressures. Re-ignition had been simulated and the engineers passed it with flying colors. Now it was the turn of AS-202 from complex 34. Spacecraft 11 was more like a manned Block I Apollo than its predecessor on the first Saturn IB mission, with fuel cells for electrical production and fully operational instrumentation in the 5.4 tonne Command Module. The Service Module had 10.4 tonnes of propellant for the SPS engine, and together the CSM combination weighed nearly 20.3 tonnes. It was a 25 August mid-day launch for AS-202, a mission designed to lob spacecraft 11 along a flat suborbital trajectory from the Cape to a point south of Wake Island in the Pacific.

Within seconds of separating from the second stage, Apollo ignited its SPS engine for a long 3½ minute burn to boost itself to a maximum height of more than 1,140 km over Africa. On the descending leg minutes later, coasting through space over the Indian Ocean, spacecraft 11 lit up the SPS for a 90 second burn, and followed this with two more burns of three seconds each to demonstrate a rapid re-fire capability. That accomplished, the spacecraft separated into Command and Service Modules, the former flying a long shallow trajectory designed to expose the heat shield to low heat rates but high heat loads – in short, lower temperatures but over considerably longer periods. Combined with data from the AS-201 mission, it was to qualify the shield for manned orbital flights and go a long way toward proving the design was sound for lunar missions too.

Spacecraft 11 splashed down more than 300 km too short of its target because predictions on the precise lift/drag ratio of the blunt-shaped cone were not sufficiently accurate. It was a learning process already experienced with Gemini. The flight had lasted 93 minutes and the Command Module was safely picked up by the USS *Hornet*. George Mueller was confident for the future: 'the results of today's flight – once examined – will provide us with the information necessary' to make a final decision on whether the next flight with a Saturn IB could be manned. When it came it told astronauts Grissom, White and Chaffee that the next ride would be theirs. The way was clear for the first manned Apollo mission.

Five weeks later, on 29 September, the three-man crew for the second flight officially received word of their assignment: Wally Schirra, Donn Eisele, Walter Cunningham, with Borman, Stafford and Collins as back-up. Like the first, it too was expected to be an Earth orbit shakedown flight lasting up to two weeks. Everything seemed ready at last to test the Moonship. Only *Old Moore's Almanac* cast a cynical eye at future events; the new edition just released predicted that 1967 would see the first space disaster.

This time-lapse picture shows erection and launch of Gemini X, July, 1966.

Gemini XI brings the separately launched elements of its mission together in space. Here, the Agena is only meters away from spacecraft 11, September, 1966.

To check its ability to withstand extremes of temperature in space, an Apollo command module structures simulator roasts one side to 315°c while dowsing the other side with liquid nitrogen at −178°c.

25 May, 1966. The Saturn/Apollo facilities check vehicle trundles out the 45-storey high doors of the Vehicle Assembly Building at Cape Canaveral. Two launch umbilical towers can be seen parked and the turning basin where rocket stages arrive by barge is in the foreground. The launch control center where nearly 500 engineers will monitor lift-off blocks the base of the emerging vehicle.

Hard down at the pad, complex 39A, the Saturn V facilities vehicle proves it → all fits together, 25 May, 1966. Note the blast trench beneath the launch umbilical tower and mobile launch platform.

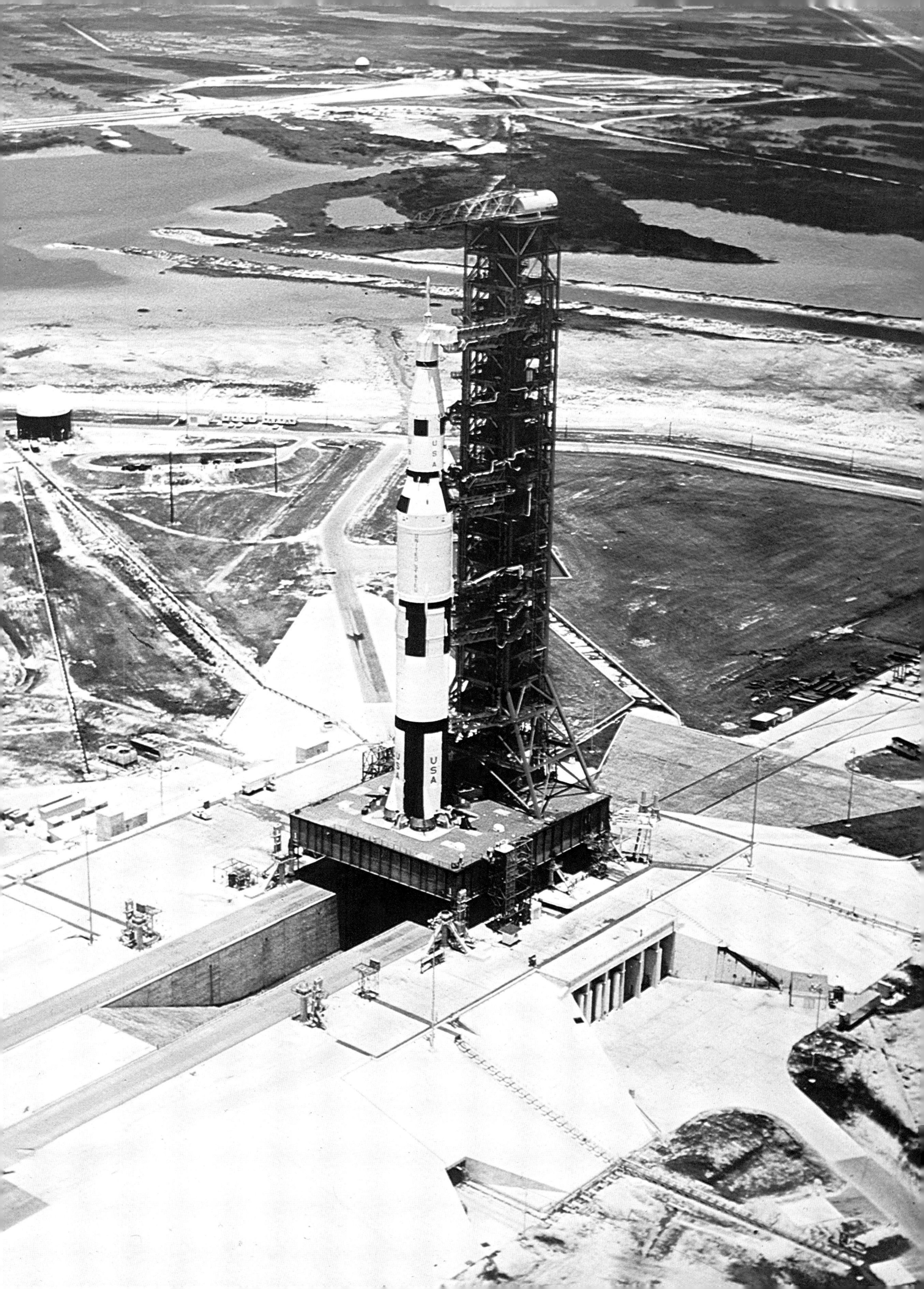
USA
USA

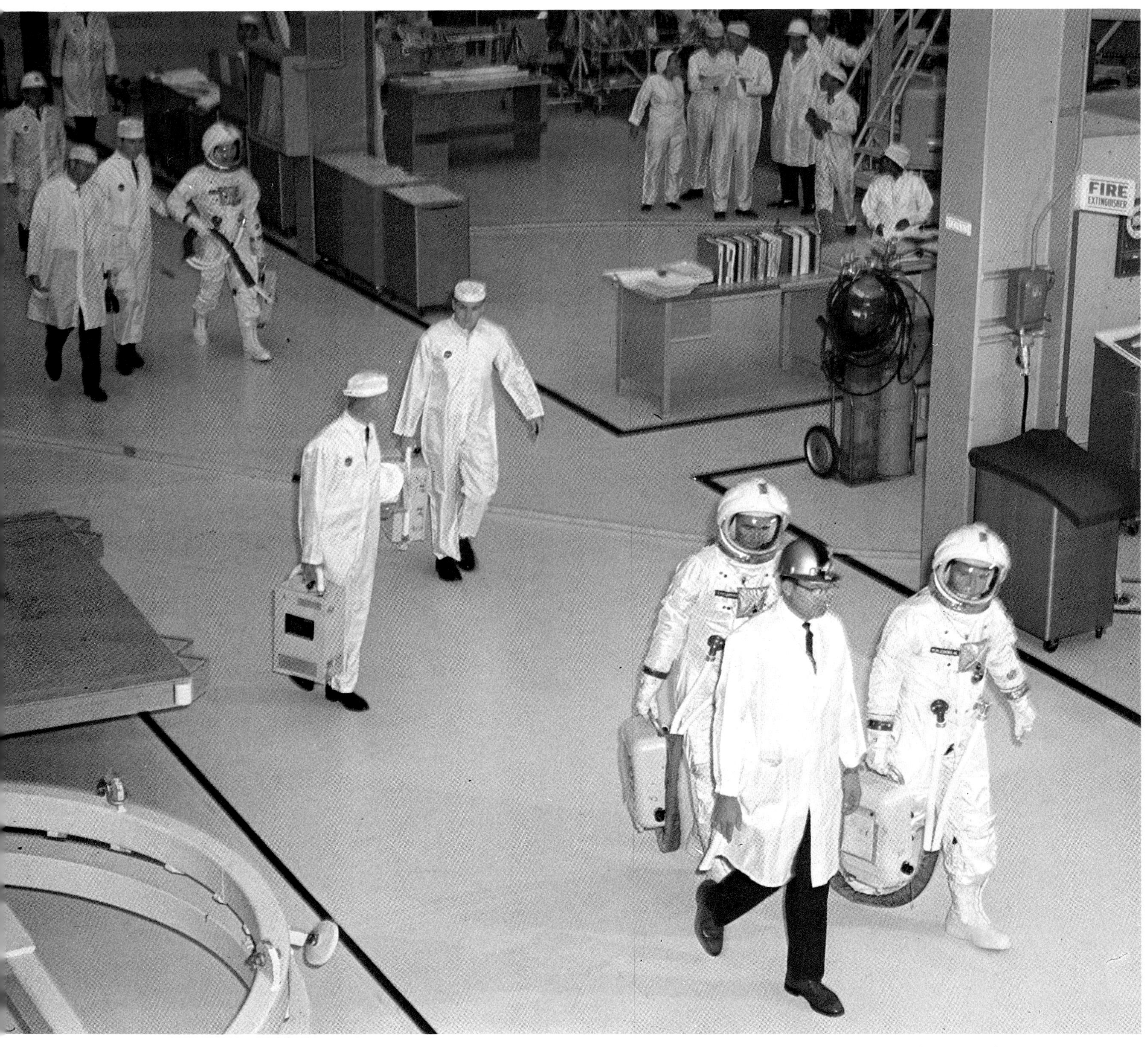

Grissom, White and Chaffee prepare for the first manned Apollo flight, scheduled at that time for February, 1967.

In this charred interior of spacecraft 12, Grissom, White and Chaffee died on the evening of 27 January, 1967.

Two-dimensional wire harness jigs were suspected of causing many electrical arcs throughout the Apollo spacecraft by chafing when bent round conduits and access panels. Three dimensional jigs were to be standard for all operational Apollo spacecraft.

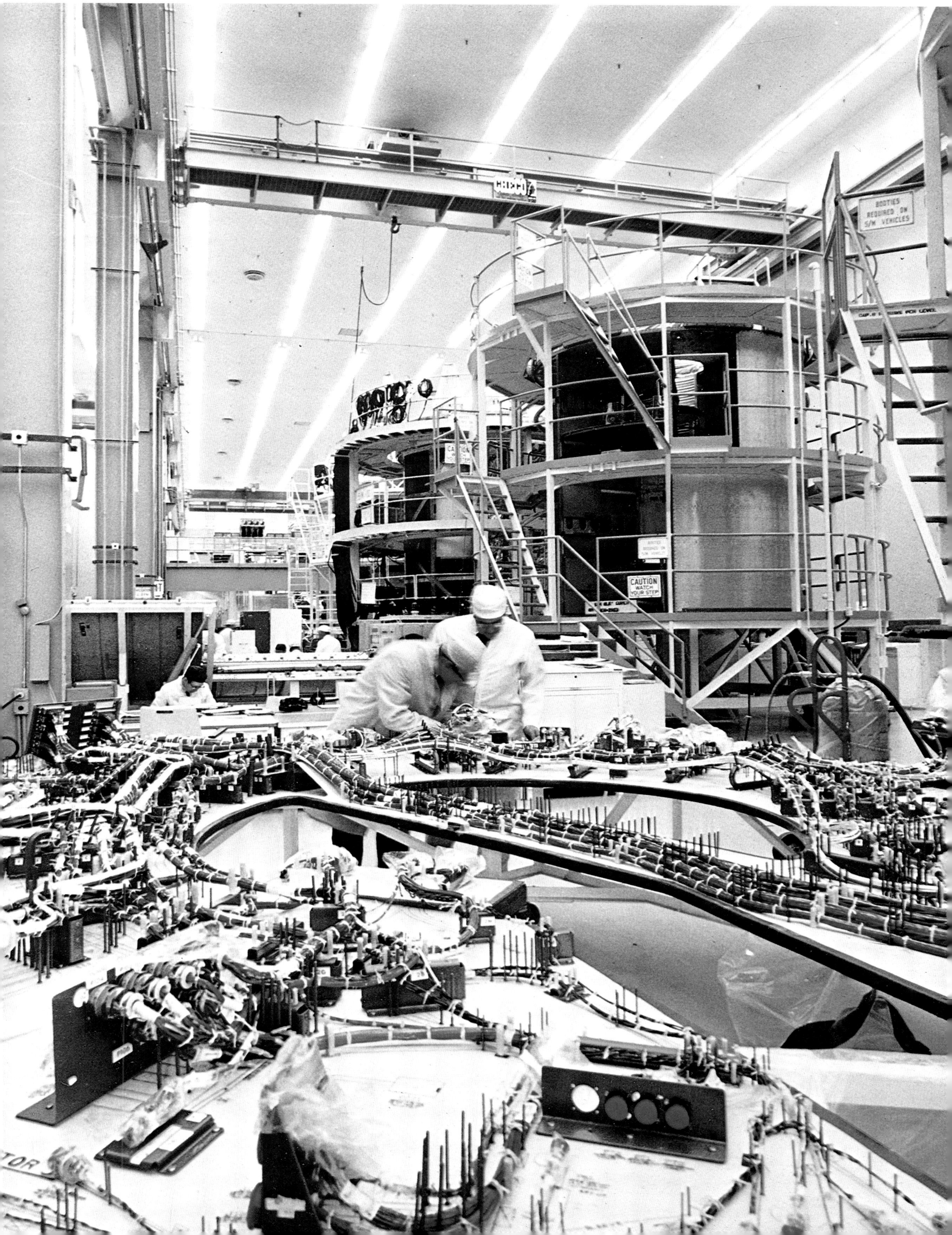

The Block I Apollo spacecraft almost completely re-designed into a new and much modified Block II after the fire of January, 1967.

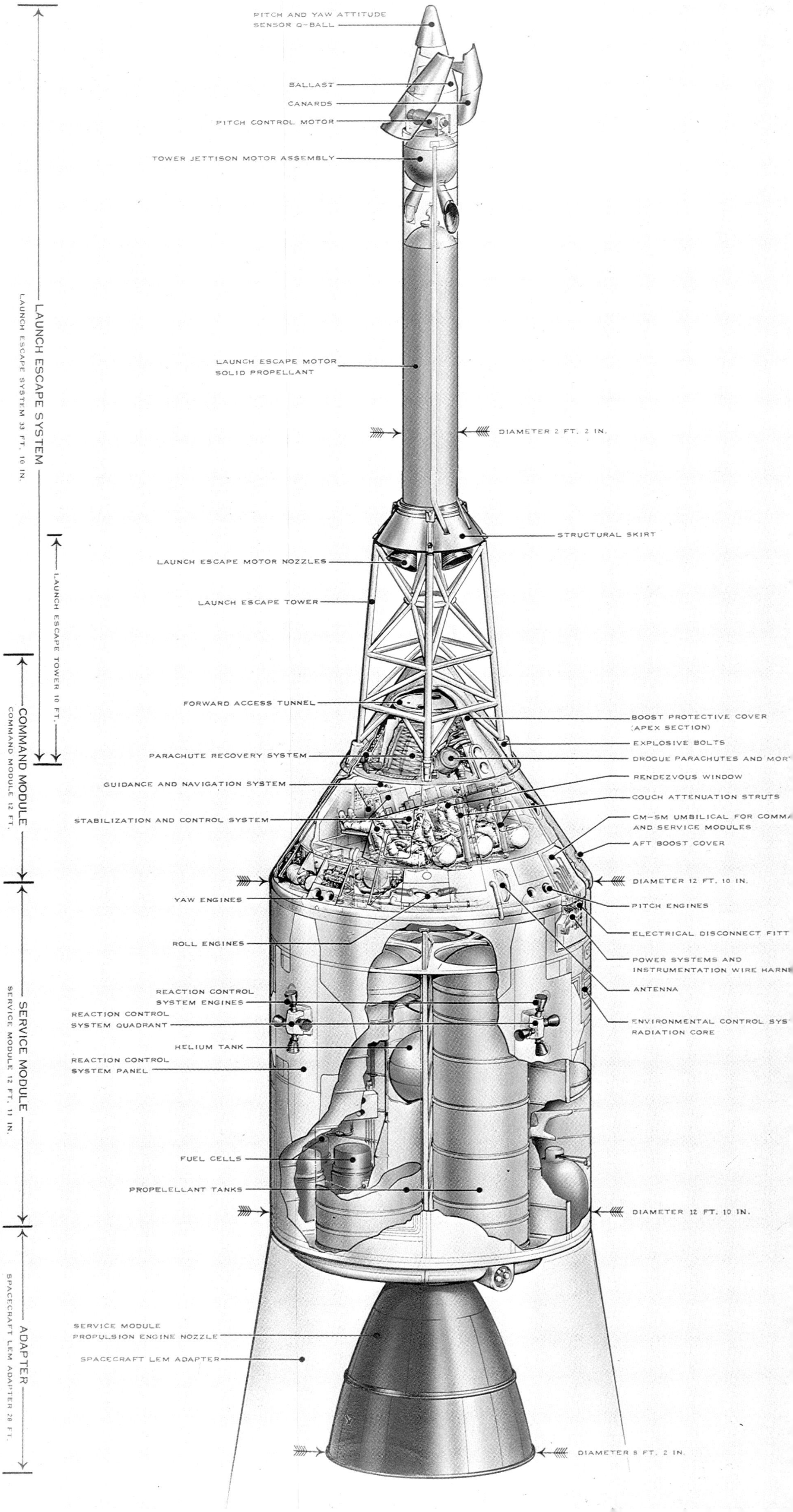

LAUNCH ESCAPE SYSTEM
LAUNCH ESCAPE SYSTEM 33 FT. 10 IN.
LAUNCH ESCAPE TOWER 10 FT.
COMMAND MODULE
COMMAND MODULE 12 FT.
SERVICE MODULE
SERVICE MODULE 12 FT. 11 IN.
ADAPTER
SPACECRAFT LEM ADAPTER 28 FT.

PITCH AND YAW ATTITUDE SENSOR Q-BALL
BALLAST
CANARDS
PITCH CONTROL MOTOR
TOWER JETTISON MOTOR ASSEMBLY
LAUNCH ESCAPE MOTOR SOLID PROPELLANT
DIAMETER 2 FT. 2 IN.
STRUCTURAL SKIRT
LAUNCH ESCAPE MOTOR NOZZLES
LAUNCH ESCAPE TOWER
FORWARD ACCESS TUNNEL
BOOST PROTECTIVE COVER (APEX SECTION)
EXPLOSIVE BOLTS
PARACHUTE RECOVERY SYSTEM
DROGUE PARACHUTES AND MOR
GUIDANCE AND NAVIGATION SYSTEM
RENDEZVOUS WINDOW
COUCH ATTENUATION STRUTS
STABILIZATION AND CONTROL SYSTEM
CM-SM UMBILICAL FOR COMMA AND SERVICE MODULES
AFT BOOST COVER
DIAMETER 12 FT. 10 IN.
YAW ENGINES
PITCH ENGINES
ELECTRICAL DISCONNECT FITT
ROLL ENGINES
POWER SYSTEMS AND INSTRUMENTATION WIRE HARNE
ANTENNA
REACTION CONTROL SYSTEM ENGINES
REACTION CONTROL SYSTEM QUADRANT
ENVIRONMENTAL CONTROL SYS RADIATION CORE
HELIUM TANK
REACTION CONTROL SYSTEM PANEL
FUEL CELLS
PROPELELLANT TANKS
DIAMETER 12 FT. 10 IN.
SERVICE MODULE PROPULSION ENGINE NOZZLE
SPACECRAFT LEM ADAPTER
DIAMETER 8 FT. 2 IN.

CHAPTER SIXTEEN

Black Days

Despite nagging problems with hardware, equipment that seemed in need of considerable attention, and differences with the prime contractor over quality of management and product control, the Apollo program was in moderately good shape during the latter half of 1966. Considering the magnitude of the endeavour, progress had been rapid, and flight tests so far had gone well. Project Mercury needed 21 spacecraft development flights over a period of 27 months, including abort tests, suborbital flights and orbit missions with boilerplates and production capsules, before the first manned orbital flight. Apollo required 10 development flights over a period of 33 months. It was time to plan for Grissom, White, and Chaffee to take spacecraft 12 into orbit for a shakedown test of all the Block I systems and subsystems in a mission left in duration to the decision of the command pilot but probably destined for 14 days.

AS-201 and AS-202 had proved the validity of the heat shield design and qualified the structural integrity of the spacecraft and Saturn IB launcher. AS-203 showed that the S-IVB stage was capable of firing several times in space and that it was up to the job given it when that stage was selected as the third stage of Saturn V. Block II Moon vehicles would fly on the big Saturns when initial flights of Block I on the smaller Saturn IB proved Apollo could satisfactorily perform its duties in earth orbit. AS-204 would be the first manned Apollo mission, employing spacecraft 12, a mission designed to follow closely on the heels of the last Gemini flight.

Preparation of the hardware ran into customary problems surrounding any radical prototype. Although production spacecraft had flown on unmanned missions, none had carried the full inventory of equipment and layouts required for a long duration manned flight. North American Aviation began to build the Command and Service Modules in August, 1964, at their Space and Information Systems Division facility in Downey, California. Thirteen months later the structures were essentially complete, a period in which all the many separate components and subsystems were fabricated, inspected, tested, and qualified. The Preliminary Design Review was held in a two month series of meetings beginning November, 1964, and by September of the following year NAA began installing all the systems and subsystems, a job that took six months.

In February and March, 1966, Critical Design Review meetings were held at the contractors' plant after which the many subsystems were checked out and tested in an integrated environment. This means that systems normally qualified on benches totally isolated from other systems against which they would be located in the spacecraft were finally run in the flight location to see that they neither received nor incurred damage. NASA moved in force to NAA during July and August for the Customer Acceptance Readiness Review following which a Certificate of Flight Worthiness was awarded, clearing the CSM-012 configuration and assuring themselves that everything was acceptable for manned flight. By the end of August, however, the Command and Service Modules were at the Cape for a mission expected to fly in December.

Only two manned Gemini missions remained in the two-man program and Gemini Project officials were hurried along to get their program completed. Several major tests were to be carried out, however, before Grissom, White and Chaffee could take their first ride on a Saturn. Following tests in the Kennedy Space Center altitude chamber, where the complete CSM would demonstrate its ability to operate in the simulated environment of outer space, the two elements would be mated with the Saturn IB launcher prior to installation of the spacecraft's launch escape system. This represented the integration, henceforth, of hardware and systems tests, checks that until now would have been performed separately on specific equipment.

Finally, there were to be four major pre-launch tests designed to prove in every appropriate way possible, that nothing would fail when called upon to support the lives of the crew and the success of the mission: an Overall Space Vehicle Plugs-In Test, where the full launch and ascent sequence would be simulated on the launch pad with the spacecraft connected to the ground via umbilicals carrying electrical power; an Overall Space Vehicle Plugs-Out Test, where the CSM would simulate a countdown and launch up to and including the point where electrical cables and umbilicals would be ejected to allow the CSM to operate on its own internal power; a Flight Readiness Test, to test the quality of all the ground equipment but without ejecting power umbilicals; and a Count Down Demonstration Test (CDDT), designed to rehearse the accurate time sequencing of every event associated with a planned launch.

The terminal countdown would begin with its own series of periodic tests and checks of hardware and systems equipment leading to an actual lift-off on the start of the mission. Spacecraft 12 was taken to the KSC altitude chamber early in September where many discrepancies and malfunctions were observed and corrected. By the end of the month, plans to conduct the first vacuum tests were running two weeks behind schedule and it was 11 October before the first simulated test was performed. Four days later the spacecraft successfully survived an unmanned altitude test, despite continuing problems with equipment, and the prime crew began the first manned run on the 18th. It was soon stopped, however, when a transistor in one of the inverters failed at a simulated altitude of 3,960 meters. It was continued the next day. A second run, this time with the back-up crew, came two days later, but that was discontinued when the environmental control system's oxygen regulator packed up. When it was removed later engineers discovered a basic design flaw and the complete unit was taken out and returned to the Garrett Corporation after being removed on 27 October when further problems emerged. Spacecraft 12 would need a new environmental control unit.

In the meantime, a propellant tank failure in the spacecraft 17 Service Module at Downey prompted engineers to remove spacecraft 12's Command Module from the Cape altitude chamber and then take out the Service Module for checks to make sure it could not happen to this vehicle. But no indications were found that spacecraft 12's SPS tanks would fail, and the two elements were back in the altitude chamber by early November. On the 8th, a new environmental control unit was installed in Command Module 12 and two days later the equipment was given its first test run. But two weeks after that more trouble showed up as water/glycol used as coolant fluid in the thermal control system leaked from the unit. In six specific instances recorded during the tests, 2.1 liters of fluid leaked out, on one occasion flowing on to and across a bundle of electrical wiring. On 5 December, this second unit was taken out and sent back to the manufacturer; nine days later it was ready and on 16 December the spacecraft received its repaired environmental unit.

Apollo's environmental equipment had had an unfortunate history. Earlier that year, on 28 April, an Apollo ECS unit burst into flames during a 500 hour test of operation in a pure oxygen cabin like the Command Module. At an elapsed test time of 480 hr 37 min an electrical arc apparently ignited heater tape; enclosed in a special test chamber, there was no fear of injury to personnel nearby. A review of the incident seemed to support the view that equipment pertinent only to test monitoring contributed toward the fire and that there would not have been a fire path of that type had the incident occurred in a manned spacecraft.

On each of four consecutive days beginning 27 December, spacecraft 12 was put through its altitude paces in the vacuum chamber, occupied on the last two days by the back-up crew. Problems were observed, and corrected, before the tests completed the initial phase of preparation. Removed from the altitude chamber on 3 January, the spacecraft was mated to launch vehicle

AS-204 three days later. The trouble with environmental control equipment had been responsible for shifting the flight into 1967 but all seemed to augur well for a late February lift-off. Only four specific checkout phases now prevented Apollo from flying.

On the 18th, the electrical mating and emergency detection system tests were performed and just two days later the Plugs-In test was carried out with only a few minor discrepancies. But they were in sufficient abundance to warrant a second Plugs-In run on the 25th, a test that went well and according to schedule. The Plugs-Out test was chalked up for 27 January, a simulated flight run where the spacecraft would go on to internal power. Most, if not all, of the 113 engineering tasks still left to be done when spacecraft 12 arrived from Downey had been carried out, although since delivery North American had plagued its Cape work force with a further 623 changes to the spacecraft. It had been a thankless task, preparing and checking out a spacecraft considered only an interim vehicle leading to the definitive Block II ship. Yet it was slowly coming together and but for the unexpected environmental control unit trouble, may have stood a chance of flying in 1966. As it was, when engineers prepared equipment for the 27 January Plugs-Out test, Apollo mission 204 seemed well set for a late February flight.

A month before, Apollo management changed the flight assignments and switched crews. Both moves were based upon a premise that the first manned Saturn V flight could come on the third launch of that assembly, allowing only two unmanned test missions to qualify the booster for manned operation. Early in November Schirra, Eisele and Cunningham switched roles with the Apollo 204 back-up team of McDivitt, Scott and Schweickart, giving the latter the second manned Apollo mission, AS-205. Since March, this trio had been training along with Grissom, White and Chaffee and it was believed they were in a better condition to fly the second mission where docked operations with a lunar landing vehicle were planned. Assigned in September to the second mission, Schirra, Eisele and Cunningham were now expected to get seats on a later Saturn V flight. But the first mission with the big rocket would be the AS-503 flight and for this Borman, Collins and Anders would get the prime role, with Conrad, Gordon and Williams in back-up positions.

The flight schedule went like this. The first manned flight (Grissom, White, Chaffee), would be followed within a few months by the launch of the second, AS-205, mission (McDivitt, Scott, Schweickart). A day later, AS-208 would send an unmanned lunar landing vehicle into space with which the 205 crew would dock to perform preliminary tests in Earth orbit, flying the lander into different orbits and returning to the Apollo mother ship. In this period NASA would fly two unmanned Saturn V missions, clearing the way for AS-503 (Borman, Collins, Anders) with both CSM and lunar lander modules on the same launcher before the end of 1967. There would then be two years in which to build toward the first lunar landing mission.

In anticipation of this NASA had planned to order 15 Saturn V rockets, 15 Block II Apollo CSMs, and 15 lunar landers from Grumman. Six Block I CSMs were to qualify the basic Apollo systems before handing over to the Saturn V flights with Block II. It was hoped to fly at least 4 Apollo Saturn V missions in 1968 and a similar number the following year, by which time all the tricky operations necessary for a lunar landing would have led to the first successful touch-down. Two years before, NASA believed it would need 15 Saturn V flights to get the first duo to the Moon but budget cuts pushed two or three Saturn V missions beyond the end of 1969. Several careful studies seemed to show that the space administration could accomplish the Apollo objective with less than the 15 Saturn Vs and administrators planned to target the landing on the ninth or tenth Saturn V launch, thereby getting to the Moon in 1969 as proposed by Kennedy. Even so it would be tight, but just possible. How prophetic had been the cautionary amendment to the President's 1961 speech that aimed the Moon goal for the end of the decade rather than 1967!

The pressing need to get Apollo in orbit to see just how well it performed was an essential prerequisite for firm schedules, and there was considerable pressure on the launch vehicle and spacecraft teams at the Cape that January in 1967. The spacecraft had already consumed 47 weeks of test and checkout procedures, compared with an average 9 weeks for Gemini and 14.5 weeks for Mercury. But 11 weeks of that total had been concerned with problems to the environmental control unit so progress was not as bad as it looked. What was significant emerged in the figure showing 21 weeks at the Kennedy Space Center for modification and test. This put too much manufacturing and assembly work on the Cape's already tight schedules and North American Aviation still had considerable engineering orders to fulfil after delivery. Yet by January the pace of preparation had quickened and most of the checkout schedule was being met on time.

But not everyone felt confident about flying Apollo mission 204. Thomas R. Baron, until January, 1967, a quality control inspector and receiving inspection clerk, was convinced Apollo was a lethal machine unsafe for men to fly in space. Based largely on his observations while working for North American Aviation, Baron was deeply worried about what he believed was a frightening lack of adequate quality control, personnel supervision, documentation, and safety in tests involving toxic chemicals and potentially inflammable materials. He expressed these views to Michael Mogilevsky early in December, while a patient at the Jeff Parrish Hospital, Titusville, Florida. Mogilevsky was a technical writer with NASA and thought that Baron's story, backed up as it purportedly was with documentary evidence, warranted official investigation. So he told NASA quality control officer Frank Childers on the 16th and repeated the story when Childers called in an engineer from the Office of the Director of Quality Assurance.

The story about Baron's allegations moved fast. That same day Rocco Petrone, now a launch operations man at the Kennedy Space Center, told John M. Brooks, the chief of the Regional Inspections Office, to seek Baron out and get the full story. Which he did, reporting not only to Petrone but direct to the KSC Director, Kurt Debus. What Baron had to say reflected strongly upon the industrial integrity of North American and his claim to have strong evidence of malpractice was tinged with intrigue when he told Brooks that he had been prevented from presenting his report to higher authority by his immediate superior. He said he wanted NASA to know what was going on, that he had no case against any of the Kennedy Space Center personnel, and that he wanted the allegations dealt with by John Hansel, Chief of Quality Control at Downey.

Baron trusted Hansel and believed this man would give him a fair hearing and appreciate what he was trying to expose. Among the examples of poor control cited by Baron was a claim that one of spacecraft 9's rendezvous windows had been improperly installed, that spacecraft 12's heat shield had been incorrectly fitted to the pressure module, and that poor standards of electrical wiring in the RCS thruster quads threatened the safe operation of the equipment. He also said that North American Aviation failed to conform to standard procedures for documenting modifications, that test personnel were permitted to smoke immediately after hazardous activity, and that emergency elevators designed to provide a means of escape were frequently found not to be operating. But worst of all, Baron said that poor workmanship and faulty quality control procedures led in some instances to incorrect installation or assembly going uncorrected when he drew these things to the attention of supervisors.

Quality control inspectors were not necessarily required to be competent engineers, a practice well supported by many aerospace operations. They inspected equipment and installations with a view of applying defined criteria in judging the standard of workmanship or the quality of the assembly. Using precise measuring standards, they would check spaces between wire clips, distances separating several different pipes, security of bolts on panels, etc. Consequently, checks were not arbitrarily the judgement of individuals, nor a reflection of personal opinion. They were the product of careful measurements and inspection. Because of this, Baron's allegations were taken seriously. But the man himself was in fastidious pursuit of a perfection very few expected to see. He had a reputation among his peers for enthusiastic scrutiny beyond the normal standards of workmanship inherent in programs of this type. So much so that his colleagues dubbed him D. R. (Discrepancy Report) Baron!

Six days after he became privy to the allegations, Petrone met with Wiley E. Williams from the Test and Operations Management Office at the Directorate for Spacecraft Operations and the two agreed that the matter was one primarily of concern to North American Aviation; meeting with Hansel and two other NAA officials the same day, they agreed among themselves that Baron should meet the man he had asked to see. Next day, the day before Christmas Eve, 1966, Hansel heard Baron report gross deficiency in several areas and received a copy of a report the man had prepared itemizing all his complaints and warnings. Hansel was concerned that Baron appeared to him to have insubstantial evidence, and was reluctant to believe everything the man said since it was noted he did not have authority in critical areas of the Downey plant.

On 4 January Petrone got a verbal report from NAA and next day it was publicly announced that Baron's services were being terminated. Without access to the Kennedy Space Center, due to his status having been removed by the contractor, Baron asked Brooks to meet him at his home so they could continue to discuss the issue. Baron gave Brooks a report that the NASA quality man duly had duplicated before returning the original to its author. A copy was delivered to Rocco Petrone's office on 26 January but by this time the press also were interested and John Wasik of the *Titusville Star Advocate* sought an interview with Petrone after speaking at length with a NAA public affairs spokesman. Considerable effort was expended trying to get at the truth behind the so-called 'Baron Report,' but it was NASA and NAA that did most of the probing.

In light of later events there is little doubt that what Baron saw was true in part, but there is equal certainty about the over-enthusiasm of the man who cried 'Wolf!' once too often. Both NASA and the contractor were aware of Baron, the former being recipient of endless discrepancy reports when he worked for NAA. The entire issue seems to have become an obsession with him, for only a few months later he was to make wild accusations against both NASA and his former employer; it is important to note that some of his statements offended congressional committees and that he was severely reprimanded on at least one occasion by a body of politicians investigating procedures at North American Aviation. The issue has been dealt with here in full due to the suggestion frequently made about grim warnings unheeded and timely information supposedly disregarded. Thomas R. Baron was a man caught up by events. What he believed he could see as the only inevitable result of what he thought was shoddy work clouded both his judgement and the opinion of others as to the validity of his accusations.

The day after Petrone got the duplicated report detailing inefficiency, poor quality control and devious malpractice, astronauts Grissom, White and Chaffee walked in white space suits to Launch Complex 34 and Apollo Saturn-204. It was Friday 27 January, the day of the Plugs-Out test, a further milestone on the road to first flight for NASA's Apollo. If successful, a Flight Readiness Test and a Count Down Demonstration Test would clear the hardware for launch, perhaps within the coming month. Because Apollo was a radical new concept in manned space operations, the procedures had been laid out the previous July, the first draft had been completed 26 September, and formal acceptance by NASA of the way the test would be carried out was signed off on 13 December. Only the previous evening, at about 5:30 pm on the 26th, the last major revision had been prepared and even as the crew prepared to suit up at 10:00 am on the 27th, four late pages arrived at pad 34.

The test was not considered hazardous, there were no propellants involved, fire crews were on standby rather than at alert in the pad vicinity (as they would be for a test classified hazardous), and medical attendants were not at a state of readiness for emergency. Quick escape from the spacecraft was a matter of vital concern to the crew and at Grissom's suggestion the procedures included as a final objective a simulated emergency. At the end of the Plugs-Out test the crew would exit the spacecraft as though it were a danger to their lives, proof they could escape, if called upon, unaided. The route they would take in quickly getting away from the spacecraft brought them from the interior of the Command Module out on to the white-room area from where they would run along a gantry to elevators ready to convey them rapidly to surface level.

Nearly twice as far from the ground as pilots of Mercury and Gemini spacecraft, the Apollo crew were keen advocates of well rehearsed safety precautions and the need for rapid escape from a malfunctioning Saturn IB launcher. It was a dedication to the unexpected that led Grissom to make one of his rare comments to a newsman: 'Everything's got to be exactly right in there or somebody's going to get it.' But Grissom was always aware that one day someone would not walk out of a manned spacecraft and that fatal accidents were an inevitable part of the price for human progress.

When the three suited Apollo astronauts arrived in the white room the mobile service structure was around the complete stack. The white room itself, and the gantry arm that connected it to the fixed umbilical tower, were not a part of this giant skeletal assembly designed to afford workmen, engineers and technicians access to various parts of the launch vehicle and spacecraft. Unlike white room areas designed for Mercury and Gemini, Apollo's environmentally controlled access area was on the extreme end of an arm that would, on an actual mission, rotate back from its position against the Command Module shortly before launch. The big service structure would have been moved away 5½ hours prior to lift-off. The white room enclosed the side of the spacecraft supporting the ingress hatch, rather than surrounding it completely as with earlier structures, and the crew would move into their spacecraft by passing through an opening in the white room wall designed to fit the size of the hatch opening in the side of the spacecraft. From outside, the white room looked like a concertina-box structure fixed to the conical Command Module.

It was shortly before 1:00 pm when Grissom, White and Chaffee arrived in the white room resting tight against spacecraft 12 nearly 56 metres above the base of the eight first stage engines. Because the Plugs-Out test was designed to simulate every electrical condition of the planned launch, then scheduled tentatively for 21 February, and because the spacecraft was to be pressurized with pure oxygen as it would be on the day of lift-off, the main access hatches would have to be sealed and locked. It was important that nitrogen should not leak into the spacecraft interior, so in further simulation of the procedures for launch day the pressure inside the Command Module would be raised to 860 mmHg, about 100 mmHg above sea-level atmospheric pressure. This would ensure that if any leak occurred it would be from the inside out rather than from the outside in.

Power had been applied to the spacecraft 5 minutes before the crew slipped across the hatch sill and onto their respective couches: Grissom, as command pilot, was on the left; White, as senior pilot, was in the center (the last to enter); and Chaffee, as pilot, occupied the right couch. Their respective roles were well rehearsed. Grissom would monitor flight control displays directly in front of his couch. White, assigned navigation tasks for the forthcoming flight had a monitoring role during launch but remained ready to operate key spacecraft elements. Chaffee had the communications switching equipment near his couch and he would serve as the ship's radio officer during powered ascent. But for this full-dress rehearsal it would all be simulated.

Grissom was first in and almost immediately noticed an odor coming up through the environmental control system suit oxygen loop. At 1:20 pm the simulated countdown was brought to a halt for samples of the gas to be collected and analyzed. It proved to be nothing serious and Test Conductor Clarence 'Skip' Chauvin agreed that the count should proceed and that the spacecraft should be sealed. At 2:42 pm the clocks started ticking down to zero and three minutes later the crew on pad 34 began the close-out procedure. With the three astronauts on their respective couches, hooked up to the suit loop and with faceplates closed, the white room crew closed the hatches.

First, directly behind Ed White's head, the inner pressure hatch was installed. This was designed to provide a seal for the internal atmosphere and could only be opened from the inside by the pilot in the center couch reaching for a special ratchet lever, inserting it in a total of six sockets, cranking open in turn the six bolts that held the hatch in place, and then placing it on the floor of the Command Module beneath the head rest of his own couch. The next hatch to go on was the thermal shield, essentially a square plug of ablative heat shield material designed to present a smooth surface to the exterior conical profile of the spacecraft's shape. That could be opened quickly from the inside by a handle device. The third hatch was not really a hatch at all, merely a square section of the white boost-protective-cover installed in such a way that it would swing open with the outer hatch should the crew need to remove themselves from the spacecraft before launch. It was made of the same fiberglass and cork from which the complete BPC was fabricated and was designed to come away with the launch escape system, having protected the Command Module windows from soot and smoke of the escape rocket motor.

All three hatches – pressure, thermal, and BPC cover – could be removed within about one minute and all three astronauts could be out and on their way to safety within a total of 90 seconds. Nobody had any cause to expect trouble with the launcher on the pad, but the combination of a pure oxygen spacecraft environment pressurized above atmospheric sea-level value was a combination considerably more lethal than anyone had anticipated. The original hatch design was a single door designed to retain internal pressure and incorporate thermal protection, explosively released like the hatch first used on Mercury flights. But when Gus Grissom's door unexpectedly came off, causing his capsule to flood with water and eventually sink, NASA changed the design to a more conservative approach. None of the astronauts liked the double door design, however, and pointed out that it would be almost impossible to conduct an EVA through such a cumbersome exit.

Space walking was not essential to the Command Module's basic mission of carrying men and supplies to an orbit about the Moon. But if the Apollo spacecraft failed for some reason to dock with the lunar lander after it came up from the lunar surface, a space walk from one spacecraft to the other would be the only way the crew could re-join their companion in the Command Module. NASA agreed and started design of an integrated hatch designed to swing open on a hinge and perform both pressure and thermal protection roles. But it would not be ready before Block II flights began on Saturn V launchers, so for Block I flights only the double door system was installed. Another example of the way Block I was made to fly with equipment on evolutionary branch lines toward definitive hardware.

When the pressure and heat shield hatches were installed and locked, pad technicians closed up the BPC hatch cover. It could not be locked because wire bundles passing under the edge of the cover to the space between the cover and the heat shield hatch distorted its shape. At 5:40 pm, after nearly three hours in the sealed Command Module, a further hold was called due to poor communication between the spacecraft and ground technicians. At times it was impossible to interpret what the crew said, and a problem with communication between facilities at the Cape added nagging problems to what had been a comparatively normal countdown test. Then the crew had a problem with a microphone that refused to turn itself off and between 5:45 and 5:53, Chaffee busied himself with various switch positions in an attempt to isolate the fault. The simulated count was about to come upon a busy period for systems checks and switching operations, a place in the timely preparation for flight where last minute checks are made to give the Test Conductor a final 'go' for launch.

Several electrical circuits would be powered up as the spacecraft received energy from its own internal sources, divorced from the ground supply. This was, after all, the object of the Plugs-Out test. There was a vague suspicion that the environmental control system had perhaps sprung another leak and that the problem with communication between the spacecraft and the ground was caused by water-glycol saturating wires. Grissom was concerned about the multitude of tiny problems that had arisen that afternoon and as darkness gathered across the Cape he felt annoyed about the apparent condition of the spacecraft. Several weeks before he had openly expressed a need for more testing and improved systems performance. But for some time he had been adamant about the need for immediate and spontaneous corrections when systems failed or performance fell below specification. Some had thought Grissom too critical, none were prepared to say so!

During the hold for communications troubles to be put right, the blockhouse continued to check off countdown items as best it could through the poor audio circuits. At 6:20 pm the checklist reached the point where, at T−10 minutes, the spacecraft would go to simulated fuel cell power. Pending clearance of the audio circuits for this critical event, the blockhouse went into a hold, waiting for the technicians and engineers to improve the quality and reliability of the ground circuits. On the service structure surrounding spacecraft 12 and in the white room area attached to the Command Module, there were 27 personnel watching and waiting for completion of the long test. It was dark now and lights blazed all across the Cape, twinkling beads of illumination that marked the route Grissom, White and Chaffee would take as they rapidly freed themselves from the spacecraft and sprinted across the gantry arm in a dress rehearsal of evacuation procedures they hoped never to need.

Minutes ticked by. The simulation should be over soon, the astronauts had been in their couches nearly 5½ hours already, and wives and families at homes all across the Cape area were getting ready for the evening meal. At Houston, playing an active part in preparations for the first Apollo flight, it was only just getting dark and wives and families there too would soon begin the evening activity. Wives and families of astronauts intent in spacecraft 12 on clearing the final bugs from the system. Gus Grissom was a tenacious pilot, dedicated to wringing the problems out of Apollo, very concerned that the first flight should go well. Ed White was a well chosen companion for Grissom, a perfectionist but one capable of initiative. Roger Chaffee had a modicum of ebullience that comes from long anticipation of a meaningful event.

In the spacecraft, lights illuminated the dials and knobs, the switches and the levers, the handles and the toggles. Three television cameras grouped around the white room and above it showed a view through one of the spacecraft windows and activity on the vehicle. Engineers stood waiting, expecting the simulated countdown to resume any minute. In the blockhouse 300 meters away William H. Schnick, the assistant test supervisor, sat at his desk dutifully noting in the log every event associated with the test. Nearby, Rocco Petrone watched the TV monitors. There was a mood of quiet deliberation, the feeling of tiredness from a long and demanding activity. Telemetry channels that, on an actual flight, would send down to Earth information on the status of the spacecraft were wired to recorders at the Cape. Engineers watched the electrical loads, the power supply, environmental pressure level in the Command Module, oxygen flow rate. Even movement inside the spacecraft would be seen as the gimbals in the inertial platform responded to every minute shift in the position of the astronaut's bodies, each man harnessed and buckled to his own couch.

It was 6:30 pm. Exactly. Twenty-one seconds later, biomedical data coming to the blockhouse from Ed White showed a momentary rise in pulse rate; nine seconds after that, muscles were at work as the electrocardiogram picked up increased heart activity. But these were almost imperceptably small responses – a measure of the monitoring capability designed to scan an astronaut's condition in space. Thirty seconds after the half-hour, Gus Grissom shifted in his couch, the familiar brushing sounds of breath on an open microphone clearly audible. For several seconds now the gimbals in the spacecraft's guidance and navigation section showed movement inside the cabin. Nothing too violent, but significant motion nevertheless. An increase in the oxygen flow rate confirmed this.

Then, suddenly, at the very precise moment of 5.15 seconds before 6:31 pm, telemetry coming from the spacecraft showed a momentary power surge in AC Bus 2, the second of two alternating current buses at this time powered from inverter no. 1 still connected to the main DC ground source. It was hardly noticeable on trace graphs of the event, but indicative nonetheless of a short, an electrical arc, somewhere in the more than 20 km of wiring threaded through the spacecraft. But significant also was the interruption in normal response from equipment connected to this bus, sure indication of a major short.

Less than nine seconds later, 4.7 seconds after 6:31 pm, a voice called from the spacecraft with tones of horror: 'Fire! We've got a fire in the cockpit!' It was Gus Grissom, knowingly aware that this was it, that nothing could possibly be done to halt the inevitable progression of events. Once started, a fire in pure oxygen was lethal. At above normal sea-level pressure, as it was in spacecraft 12, every appalling condition was magnified. From a position below Grissom's left couch, down past his foot pan and along to his left side, flames licked quickly across to the top of the cabin, moving up behind the wide instrument panel laid out before all three astronauts.

Netting placed within the spacecraft dripped flames down to boxes, lockers, wire conduits and bags on the floor below the couch foot pans. It was so bright. Grissom was nearest the tongues, so long and white in the pure oxygen, but White had already started frantic motion to find the hatch handle and undo the six bolts. Six bolts! Flame and pure oxygen would not wait. No sooner had Grissom completed his first call than the display panel lit up with warning indicators and a howling siren, activated by the sudden increase in oxygen flow 15 seconds before – the delay was built in to prevent a momentary surge from setting off alarms. Twelve seconds past the minute, two seconds after Grissom completed his call, and temperature and pressure began to rise alarmingly.

Flames were now white heat sheets wrapping the interior of the Command Module in searing panels that hugged the aluminum sidewalls. For the first few seconds the intense heat

Exterior effects of the flash fire in Command Module 12 can be seen here in the white room. Note the lattice tower supporting the launch escape rocket that could have ignited had the fire spread upward.

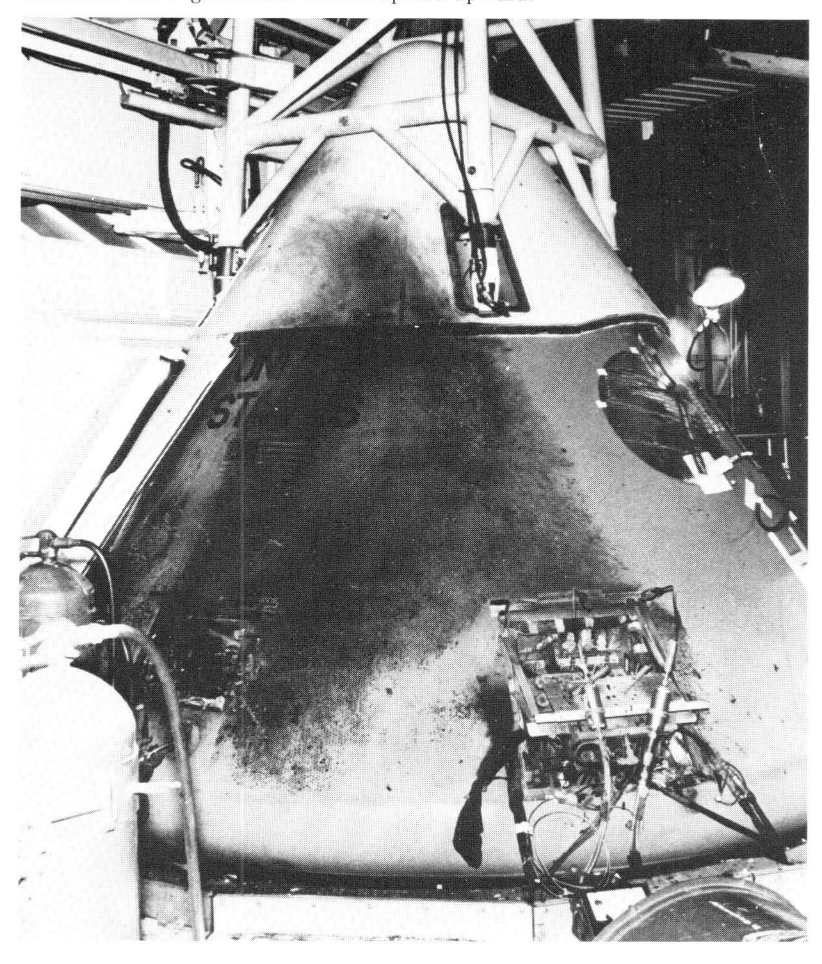

had been absorbed by the metal. Now, with increasing quantities of combustible materials within reach of the flames, the temperature was almost beyond toleration. The fire had broken free and was running amok along anything that carried it. But still the fierce intensity was concentrated down past Grissom's left side, beyond and below the foot pans. A place where the lithium hydroxide canisters would be carried. Quickly now, the hot flames set up a wall of fire between the command pilot and an oxygen vent valve control on the left wall. Several times Grissom tried to reach it, but it would have put his whole arm under burning fire for several seconds. In any event, the vents could not have evacuated the oxygen quickly enough to put out the fire, or to relieve the pressure.

Now it was 16.8 seconds after the minute, and another call came through the communication lines: 'We've got a bad fire – let's get out . . . We're burning up!' The horrifying reality of this appeal, probably from Roger Chaffee, was all too apparent even before the voice completed its call. Before it finished, pressure built up to a point where the spacecraft could no longer take the stress. No one can ever know for sure at just what precise pressure the cabin split itself apart; certainly more than 1,500 mmHg but no greater than 2,070 mmHg, the predicted strength of the cabin. But when it did occur, 19.5 seconds after the minute, it sealed the fate of Grissom, White and Chaffee. For up to this time the fire had been concentrated on the left lower side of the cabin.

Under a load of twice normal atmospheric sea-level pressure, the hull split around the bottom right section causing a terrific rush of gas, smoke and fire diagonally across the couches, up over the instrument panel and through the positions of the three astronauts. When Roger Chaffee gave his last agonizing statement, 'We're burning up,' the fire was sucked over their bodies with a terrible roaring sound that sent the conflagration billowing out into the white room area and around the back of the spacecraft where gantry platforms from the service structure hugged the spacecraft. There was one last, chilling cry of agonized pain from the communications circuit. And then nothing. Flames and smoke gushed through the ruptured Command Module and, within 6 seconds of splitting, pressure inside was down to atmospheric level and the fire was moving into its third phase.

The second phase, beginning when the hull split at the join between the floor and the wall below and to the right of Chaffee's position, had been the most intense, with predominantly flame and heat present in the Command Module. Now virtually devoid of oxygen, the fire released billowing, choking clouds of smoke and soot. Fractured oxygen lines and tubes carrying water/glycol coolant fluid fed the point of origin and a small white fire continued to struggle against the suffocating black soot. Ten seconds after the rupture, the atmosphere inside the spacecraft was lethal to life. Just 25 seconds had elapsed between the first call from Grissom and the end of combustion.

The spacecraft contained the products of the flame and the fire well, although considerable quantities of smoke were leaking to the white room. But it had happened so fast, nobody had time to do a thing that significantly altered the fate of the three men trapped inside. One full half minute had elapsed since the probable start of the fire and the density of smoke and soot in spacecraft 12 was so intense that nothing could survive. The flames had gone now, their worst done, and within a few seconds all three astronauts were unconscious. Their burns were indescribable, but they were still alive, in a state of unconsciousness brought on by asphyxiation. Within a matter of three or four minutes they suffered the inevitable result of a heart attack caused by oxygen starvation, indirectly preventing oxygen from reaching the brain.

To personnel outside the spacecraft the call of fire instantly brought about a desire to evacuate the area as quickly as possible. On television monitors in the blockhouse, Deke Slayton and Stuart Roosa watched horrified as a bright, blinding, light filled the interior of the Command Module. The terrible sound of metal physically ripping itself apart led many to believe the spacecraft was blowing up. Lower down the service structure men ducked when the pressure vessel ruptured and billowing clouds of smoke and flame shot through the space between the cabin floor and the heat shield. Fearing the worst, technicians fled the white room and ran out on the gantry attached to the umbilical tower. Seizing fire extinguishers, and realizing that the anticipated explosion had not happened, they turned and ran back into the smoke-filled area. Over the communication line, capcom Stu Roosa called again and again to the burning spacecraft and Deke Slayton shouted for medical assistance. Along with Joseph Shea, Slayton had considered joining Grissom, White and Chaffee inside the Command Module for the duration of the test, squatting in the space between the couch footpans and the lower equipment bay; the decision not to saved their lives.

Small fires had broken out all around the white room area, repeating the scene seconds earlier in the spacecraft itself as fire brands sailed through the air; it had been the showering fire brands inside spacecraft 12 that set several patches of material alight before the pressure vessel burst. In the blockhouse, attention switched to the television cameras in the white room, filling quickly now with black sooty smoke. Outside that area, Jessie Owens, a North American engineer, ran from the pad leader's desk toward the confined room and was pressed back by flames gushing forth 'like an acetylene torch.' Donald Babbitt, another North American engineer serving as pad leader, was the first to re-enter the white room; smoke now replacing flame as an effective deterrent, Babbitt ran back out gasping for breath, then in again frantically concerned to get to the hatches on spacecraft 12.

But it was no use. No lights could penetrate the thick soot and he ran out to find his communication headset. Equipped now with small fire extinguishers found on the gantry, technicians soon put out the fire that burned around the base of the spacecraft near the boost-protective-cover hatch. Five North American engineers now embarked upon a potentially dangerous mission to get the three hatches off the spacecraft as quickly as possible, totally disregarding the fact that the large escape motor just above their heads could by now have heated to the point of combustion. They could, like other technicians on the gantry that night, have run for their lives. But they did not. They stayed, returning in sequence to the dense smog of the white room, clawing their way to the Command Module, groping for the single small tool that would enable them to remove the doors. Babitt was in charge of these men and James D. Gleaves, Jerry W. Hawkins, Steven B. Clemmons, and L. D. Reece took it in turns with NASA's Henry H. Rogers to move back in to the confined area.

Visibility was down to a few centimetres, the special removal tool was eventually found and little more than a minute after the spacecraft burst, Babbitt reported to the blockhouse that his team were trying to open the hatches. Ignoring the inevitable, the men clawed at the outer boost-protective-cover hatch which, although unlocked, still needed the special tool to open it. That soon came off, followed by the first of the two spacecraft hatches. The inner, pressure, hatch took longest to undo, work significantly hampered by the swirling smoke and soot. About five minutes after the first report of fire from inside the spacecraft, the square-shaped inner hatch was lifted free, an unsuccessful attempt being made to place it on the floor of the cabin below the line of couch headrests. Something seemed to be in the way so it was pushed to one side.

The fierce heat that met the technicians when the last hatch came off was accompanied by a sudden expulsion of soot and smoke to add to the generally choking conditions in the white room area. Don Babbitt was there when the hatch came off. Through the all too small opening he peered for signs of life, anything that moved. He could see nothing, not even the crew. Lights still shone dimly through the grey-black smog, an unreal scene of charred chaos confusing interpretation of what lay inside. It seemed at first as if nobody was there at all, and the positions of the crew were discovered only by sight and feel.

The left couch, Grissom's, was in the '170°' position where it was essentially flat out from head rest to foot pan. The center couch, White's, was in the '96°' position where the back was flat and the lower portion was in a raised (knees upward) setting. The right couch, Chaffee's, was in the '264°' position with the back horizontal but with the leg sections dropped down to the floor, or aft bulkhead, area. Chaffee was supine on his couch, his harness undone but his oxygen hoses connected and his faceplate closed. White, instead of being on the center couch, was found lying transversely across the aft bulkhead, below the line of headrests and just below the hatch sill. Grissom, worst affected by the flames and heat, still had his feet on his own couch but his body was underneath the center couch, the space for his legs made possible by the raised knee supports on the couch used by White. Both Grissom and White had closed faceplates.

It was clear what had happened and positions bore testimony to the last seconds of action performed by all three. Grissom, nearest the hottest part of the fire, close to the environmental control system where ignition came only seconds before the flames, had disconnected his harness and oxygen hoses to seek refuge under White's couch; the fire had spread up and over the couches during the instant prior to rupture. White, frantically trying to open one of the bolts that held the inner hatch shut, had abandoned the attempt to crawl between the couch headrests and the hatch sill. But his bravery had been extreme, for the harness was still closed and locked, the webbing having burned through before he groped for safety. He had been right to abandon the escape attempt for increased pressure would have made it physically impossible to pull free the inward opening hatch against the choking smoke. Chaffee had undone his harness but he remained on the couch to maintain communications with the blockhouse. His was the least burned suit, but his had been the cry that ended all communications from spacecraft 12. The other two had no communication lines when the cabin split apart sending a rush of fire across the couches.

What Don Babbitt saw was impossible to absorb in the instant he took to dive back on the gantry catwalk and grab a headset. Less than 5½ minutes after the start of the fire, the pad leader reported that all hatches were open and that he could not describe what he saw. This was to prevent other personnel hooked up to the same communication circuit from hearing that the crew were dead before the Test Conductor could be informed. The firemen had been summoned at the instant of a reported breakout but it took them about eight minutes to reach the white room, getting to the vicinity of the spacecraft about three minutes after the last hatch came off. They immediately went into a systematic search of the area to seek concealed fires, although the technicians and engineers had done a good job already in getting the situation under control. Once free of the spacecraft there was not a lot that would burn, and in any event the dense choking smoke starved the flames of oxygen. A couple of them tried to get the crew out but found that impossible.

Within three or four minutes of the firemen arriving the three doctors reached the white room and confirmed that the crew were dead. By this time Don Babbitt had been relieved, along with several others, to attend the clinic for medical treatment and to receive superficial examination. They had all inhaled considerable quantities of smoke in their desperate attempts to get at the trapped astronauts. The doctors had not brought oxygen equipment of any kind and so could not remain for long in the cramped area still shrouded in smoke. No oxygen equipment of any kind was anywhere near pad 34 which could have effectively handled dense smoke, only small face masks being available to combat the effects of fumes. A short time later the doctors returned when the spacecraft had been purged of smoke to attempt removal of the three bodies. But what they found when the confusion had cleared away prevented them from doing so.

All three suits had been extensively burned, Grissom's suit the worst. Only about one-quarter remained unconsumed by the fire. But worse than this, the material had melted and welded its occupant to the cabin by fusing with liquid nylon, now solidified, at the peak of the fire. It was quite apparent that the crew could not be removed without extensive effort. That took place over a period of 1½ hours beginning at 12:30 am on the morning hour of 28 January, six hours after the start of the fire.

Shock was deep, and felt everywhere very intensely – nowhere more so than in the homes of Gus Grissom, Ed White and Roger Chaffee. Deke Slayton's wife went with Charles 'Chuck' Berry, the astronaut's doctor, to Betty Grissom's home in Houston. Pete Conrad broke the news to Pat White. Mike Collins called on Martha Chaffee and was soon joined by Gene Cernan. The sense of personal loss was nationwide. At a Cocoa Beach restaurant often visited by the astronauts, the owner draped a black scarf across a board citing Grissom's achievements. The Cape's Saturn Hotel illuminated a flashing sign: 'Gus, Ed and Roger – here and on the Moon you will be remembered.' Further down Route A1A, another sign: 'We mourn the loss of three fine Americans.' Everywhere, certainly across the United States, there was a bitter and deeply felt sorrow for the men and their families. Messages of condolence poured in from overseas, sentiments of regret for the disaster, offers of hope for the future. Only from the Soviet Union came a bitter note of recrimination, accusation that the crew were 'victims of the space race created by American space program chiefs,' and that the pace had been set by 'hate.'

The most immediate concern of managers and administrators concerned with the Apollo program was to impound everything associated with the fire for an official board of inquiry. That came on the day after the fire and at mid-day on the 28th, only 17½ hours after the fire, the chairman and several members assembled at the Kennedy Space Center to meet Bob Seamans, Sam Phillips and a host of other NASA officials. After discussing procedure in the Mission Briefing Room, they left for pad 34 and a personal inspection. The Board was set up formally that day by a memorandum from Bob Seamans acting on management instructions laid down in April, 1966 for just such an eventuality. Its chairman, Dr. Floyd L. Thompson, was the Director of NASA's Langley Research Center. Other members included astronaut Frank Borman, Maxime Faget, E. Barton Geer from Langley, George Jeffs from North American Aviation, Dr. Frank Long from Cornell University, Col. Charles F. Strang the Air Force Inspector General, George C. White of NASA Headquarters quality and reliability office, John Williams from the Kennedy Space Center, and George Malley from Langley.

Within the first few days of the investigation, NAA's George Jeffs was told he could no longer consider himself part of the Board and he was eliminated from its list along with Dr. Frank Long who resigned due to pressure of work. He was replaced by Dr. Robert W. Van Dolah from the Explosives Research Center, Bureau of Mines. A replacement memorandum from Seamans on 3 February set up the new list of Board members. All but two were from NASA, a fact which led to speculation in the press about the level of ambiguity likely to bias the inevitable judgements.

There was much to be done. For the immediate future there could be no hope of moving ahead with other tests until some answer had been obtained as to what caused the 204 fire. Suddenly, or so it seemed, hope of reaching the Moon by the end of the decade, now less than three years away, was significantly lower than it had ever been before. It was imperative to the future course of manned space flight that technical examination of the spacecraft get under way as quickly as possible. On Sunday, 29 January, special memorial services were held across the United States for the three dead astronauts, and in Houston a service was held for Roger Chaffee; another service for Grissom and White was held in Houston the following day. But on that Sunday activity at the Cape concentrated on allowing the press a limited access to the spacecraft.

Dr. Thompson permitted three men from the press to record

the scene through the hatch of the Command Module. A motion picture cameraman for the electronic media, a still photographer for the newspaper and magazines, and a writer for the words that would tell the world what it looked like inside the seemingly small crew compartment. The following day procedures were set up for the inauguration of several review panels that could examine in great detail specific sections of the spacecraft and relevant segments of the operation. It would result in 21 such panels, each monitored by a member of the Apollo 204 Review Board. On Tuesday, the last day of January, the Board reviewed procedures for disassembling the spacecraft by accepting a plan to remove the three couches and then suspend a transparent floor from the four vertical strut attachment points; until then a plywood floor would be fixed just above the couches to prevent activity inadvertently destroying evidence that could contribute to an understanding of the fire's origin.

But elsewhere that day there was renewed sorrow for relatives of the three men. All three were buried with full military honours: Grissom and Chaffee at Arlington National Cemetery and White at West Point. But sorrow too overtook the families of Aircraftman 2/C William F. Bartley, Jr., and Airman 3/C Richard G. Harmon, when they died of severe burns several hours after a fire in a pure oxygen pressure chamber at Brooks Air Force Base. Entering the chamber at 9:33 am to take samples from rabbits under observation for analysis of the effect on blood of long periods in a pure oxygen environment, the two men were trapped when, twelve minutes later, a fire broke out from an apparent malfunction in some piece of equipment. Within 13 seconds the chamber was brought down to atmospheric pressure but it was not possible to save them from inevitable death.

The pure oxygen environment was taking its toll in human life. A full report of the accident was made to the Apollo 204 Review Board that same day and Dr. Thompson asked Col. Strang to report further on this accident when additional information became available. Two days later, on 2 February spacecraft 14 arrived at the Kennedy Space Center from Downey so that technicians preparing to remove item by item the several thousand pieces they would obtain from spacecraft 12 could familiarize themselves with every action needed to get at respective systems by performing that activity first on an identical Command Module. Both this vehicle and the items removed from spacecraft 12 would be kept in the Pyrotechnic Installation Building at the Cape. Progress was similarly achieved with 'safing' the AS-204 assembly when technicians removed the launch escape tower so that members of the Fire Propagation Panel could gain entry on the 3rd before any equipment was removed.

In hunting down the cause of the accident it was necessary first to ascertain the direction and intensity of flame and fire patterns within the assembled structure before removing each piece of equipment for individual analysis. On the 4th, the three couches in spacecraft 12 were taken out and replaced by a false floor the following day. By 7 February the Board began to receive evidence from experts in materials combustion and fire propagation, learning that the amount of oxygen in the Command Module at the time of the conflagration would permit combustion of only 6.8 kg of material at most. These included plastics, nylon, polyurethane, ethyl glycol coolant fluid, and rubber. From the beginning of manned space flight intentions had been good, but over the years increasing quantities of potentially inflammable substances had crept in to spacecraft interiors until Apollo represented a considerable hazard. Spacecraft 12 contained nearly 32 kg of combustible materials.

Members of the Apollo 204 Review Board Fire Panel examine a component from spacecraft 12 in the Bonded Storage Area at Cape Canaveral. Left to right: Al Krunpnick, Manned Spacecraft Center; Dr. Robert Van Dolah, US Bureau of Mines; Dr. Homer Carhart, Naval Research Laboratory; Irving Pinkel, Lewis Research Center.

On the 8th, engineers removed the batteries from spacecraft 12 and determined that they played no part in the ignition of the fire. Over the next few days the Board was shown several short films compiled by a variety of industrial and government sources in connection with research on fire propagation and flame characteristics, on the ability to discern and adequately interpret activity viewed through a typical Command Module window, and on the flammability of various materials in a pure oxygen environment. By the 17th, removal of critical items from the spacecraft reached a point where the Command Module could be removed from the launcher and on that day it was raised from the Service Module and taken to the Pyrotechnic Installation Building. Tables had been laid out to support the many items taken from the spacecraft, each providing inspection opportunities with as little necessity as possible to actually touch the hardware. Over the next few weeks inspection of relevant items reached a peak, with visual and x-ray techniques being employed to examine equipment situated close to the fire's point of origin.

By 6 March the Board heard evidence that inspections revealed the existence of many arcs and short-circuits, each of which could have caused the fire, and on the 7th the Command Module heat shield base was removed, revealing for the first time the rupture in the floor extending two-thirds around the circumference of the pressure vessel. Until it became possible to inspect the cabin from underneath, nobody suspected the split to have been so extensive. But by this time the Board had also seen film of flame propagation tests in an Apollo boilerplate at the Manned Spacecraft Center, patterns of flame motion identical to those mapped for spacecraft 12.

It was now six weeks after the fire and the careful process of photographing, removing, examining and documenting each piece of equipment was paying off. By the middle of March the rest of the heat shield, those sections covering the crew compartment sides and forward cone, had been removed and nearly 1,000 individual items catalogued. In anticipation of the impending removal of what was left of the extensive wiring system, Dr. Faget recommended use of special jigs on which wires and harnesses removed from spacecraft 12 could be placed for sizing and examination in the configuration similar to that it had in the Command Module. Particular attention was to be paid to the electrical distribution system since it was that area specifically suspected of having caused the spark that ignited combustible materials.

As each wire was removed it was examined under 7 power magnification. Laboratories at the Cape, and at special facilities to which certain items were distributed for examination, provided answers that should, under ideal circumstances, have been known before the fire. Teflon wire insulation, while serving as an excellent insulator with good fire protection characteristics, was found to have poor cold flow resistance and to be especially vulnerable to bad handling or poor installation. It could, for instance, be easily stripped bare by scuffing or pressure behind access panels or doors. Tests performed by the board showed that a piece of teflon-coated electrical wiring could be directly exposed by opening and shutting a panel pressing against it, and that an arc in such an exposed wire could readily ignite combustible material up to 10 cm away.

The most valuable test conducted for this analysis comprised the full-scale simulation of the spacecraft 12 event in a boilerplate at MSC. With combustibles arranged in the same order as those in the AS-204 vehicle, flame patterns, propagation rates and duration of fire was identical to that which occurred on pad 34 – even to the build up of pressure which ruptured the vessel. Careful examination of the plumbing aboard spacecraft 12 revealed potential flame propagation sources from coolant fluid liberated when soldered joints or plumbing itself came apart under the influence of intense heat at close quarters. Although composed of 35.7% water and 1.8% corrosion inhibitor, the glycol solution was rendered dangerously inflammable by rapid evaporation of the two additives, leaving in the liquid state a solution of ethyl glycol electrically conductive and a ready fire path. Moreover, the solution was found to be corrosive when exposed to electrical wiring. The Board believed that all the solder joints it examined were inadequate for unplanned loads that could, theoretically, be applied to the plumbing. Laboratory tests confirmed that although neither water glycol nor the poor state of the plumbing contributed toward ignition, it certainly played a part in conducting the fire around the interior of the Command Module.

By the end of March disassembly was complete. Since 1 February, hearings before committees of both Houses of Congress brought NASA officials into close contact with politicians intent on finding a cause for the fire that took Grissom, White and Chaffee. They heard, for the first time, about the task force implemented by Sam Phillips in December, 1965, about the criticism from Jim Webb of the way North American was performing under contract, and about the 20,000 equipment failures including 220 in the environmental control system alone recorded during the entire program. Failure levels of this magnitude seem alarmingly high, but are no greater than with similarly based technology programs. The real disturbing news for Congress was in the amount of inside doubt it heard in testimony by NASA officials. Up to now the space agency had presented a successful series of spectacular technological feats and the feeling that nothing reasonable lay beyond the capabilities of the administration injected every Congressional attitude.

It was as though the space agency was thought of as a political propaganda tool funded for sophisticated feats of international prestige and that for it to fail, or fall short of its tasks, was to challenge the very idea that America was technologically supreme. It revealed, in one awful event, that NASA had been labored by pressure from every political side to achieve within an alarmingly brief period a feat so great that disaster was almost inevitable. Among the technicians and engineers close to the Apollo program, the fire brought a sense of relief. The fact that it had happened at all could not be reversed, and to an aerospace industry well used to crashes and deaths among prototype high performance aircraft and their crews, the agony of this one incident cut deeper scars than could be expected. But there was an increasing feeling as the space program evolved that the odds on continued success were shortening. These men knew that behind the public face of brave, heroic astronauts blazing new tracks of discovery across the celestial frontier were a succession of incidents and events close to causing total failure.

On occasion, light bulbs were accidentally broken in space, switches broke off due to unexpected shocks, or wires were pinched creating arcs and short circuits. Now, in a ground test for a mission to fly the first manned Apollo, three men had died, and set the odds right once again. There is a strong sense of fate and superstition among aeronauts and test pilots that framed the base of the manned space program and nobody doubted that sooner or later a terrible accident would happen. Now it had, and in a way only aviators will fully understand, there was once again that unmistakable smell of fresh air after a terrible storm. It was not in the nature of men behind Apollo to doubt their efforts in the face of tragedy. To do so would have been to deny the inescapable victims their rightful place in the development of a new space machine, and when the men had been laid to rest thoughts turned more assuredly than ever before to pressing on and reaching the goal set six years before.

But to the politicians and the public an answer was sought as to why it had to happen, and that was the responsibility of the Board. In closely examining all the evidence it found the fire to have moved quickly through three separate stages. The first, lasting only 15 seconds from the initial verbal report, 'was not intense until about (8 seconds into the event). The slow rate of build-up of the fire during the early portion of the first stage is consistent with the view that ignition occurred in a zone containing little combustible material. The original flames rose vertically and then spread out across the cabin ceiling. The debris traps not only provided combustible material and a path for the spread of the flames but also firebrands of burning molten nylon.'

The second stage of the fire began 15 seconds after the first verbal report and lasted until an elapsed time of 21 seconds, during which it was 'characterized by the period of greatest conflagration due to the forced convection that resulted from the outrush of gases through the rupture in the pressure vessel. . . . Evidence that the fire spread from the left hand side of the Command Module toward the rupture area was found on subsequent examination of the Module. For example, the leg rest control handle on the left side of the left hand couch is fabricated from aluminum tubing. Tongues of flame pouring over the control handle melted its left side. However, a nylon button at the base of the handle was unconsumed and only slightly deformed. Similarly, flames spreading across the floor beneath the couches caused more burning on the left side of three nylon helmet covers than on the right. . . . Fire across the floor of the spacecraft lasted but a few seconds and spread from left to right. . . . Evidence of the intensity of the fire includes burst and burned aluminum tubes in the oxygen and coolant systems at floor level.'

The third stage began about 21 seconds after the first verbal call and 'was characterized by rapid production of high concentrations of carbon monoxide. Following the loss of pressure in the Command Module and with fire now throughout the crew compartment, the remaining atmosphere quickly became deficient in oxygen so that it could not support continued combustion. Unlike the earlier stages where the flame was relatively smokeless, heavy smoke now formed and large amounts of soot were deposited on most spacecraft interior surfaces as they cooled. The third stage of the fire could not have lasted more than a few seconds because of the rapid depletion of oxygen. It is estimated that the atmosphere was lethal five seconds after the start of the third stage.'

However, a local fire continued in the area adjacent to the point of origin due to open oxygen lines and dripping water/glycol solution, burning a substantial hole in the aft bulkhead, or floor, of the Command Module in the left corner opposite the hatch. This was where the fire began, down past Grissom and below the level of the suspended couches. If the main component of the fire's third stage did in fact last a brief 5 seconds as the Board says, and if the temporary power surge five seconds before 6:31 pm caused ignition (a mere 9 seconds before the verbal report), the complete sequence was over within 35 seconds. And as has been related earlier, the crew lost consciousness within 15 seconds of the end of the third stage. To have opened the three hatches and escaped would have taken at least 90 seconds. No one knows if Ed White actually managed to undo any of the six bolts holding the inner (pressure) hatch; in undoing the hatch from the outside, the same set of bolts are turned. But the engineer that did throw all six believes one to have been partially turned when he came to give the expected movement he thought it would need for complete release. The tool was found near the hatch area inside the spacecraft.

But another object was also found inside the spacecraft: a 0.6635 cm wrench socket pressed into a bundle of wires. In procedures established by North American to maintain control of the movement of tools, all such instruments and equipment carried from a special box alongside the spacecraft into the pressure module proper were to be logged out to a specific recipient at a recorded time and day. In testimony from Dale Myers, 'A review of the logs shows that the procedures and instructions were not being fully implemented in practice. Specifically, the records do not show the date on which the . . . wrench socket was left in the spacecraft or the identity of the person who left it.' Although indicative of an obvious breakdown in procedures, the errant wrench socket was eliminated as a possible cause of the short that ignited cabin materials.

During the course of its investigation, the Board heard summarized reports from the relevant panels set up to look in great detail at specific areas of hardware design, fabrication and operation on the probable causes of ignition. It was possible to eliminate, for instance, spontaneous ignition, or ignition from electrostatic discharge, or from chemical or mechanical activity. In every test carried out, including that in which a spacecraft boilerplate was deliberately burned at the Manned Spacecraft Center, electrical ignition was the only source that satisfactorily met the facts. The next job was to find what sort of electrical source caused it.

At first it was thought that perhaps some electrically powered item within the cabin had malfunctioned in such a way that it caused the fire. But scrutiny of more than 1,000 individual items removed this possibility. Then the probability of overloaded conductors was examined, and similarly eliminated. The third possibility, that an electrical arc caused ignition, was the most likely candidate, especially in the light of several arc points having been found in areas remote from the point of origin liberally distributed throughout the wiring harness taken from spacecraft 12. From flame and heat flow patterns the fire obviously

← Wrench socket found between two wire bundles in spacecraft 12 was eliminated as a likely cause of the tragic fire.

This view of severely damaged equipment near the floor in the lower forward section of the left hand equipment bay, below the Environmental Control Unit, shows the area where the fire probably started in a power cable. It is believed the cable crossed over the horizontal tubing in the center and under the lithium hydroxide door, removed when this picture was taken.

started in the vicinity of the lithium hydroxide access door, the panel covering environmental control equipment which would be used in flight to remove carbon dioxide and odour from exhaled gases.

In a region at the base of the door, where it meets the floor (or aft bulkhead), all wiring and adjacent structures were completely destroyed, consuming in the process all evidence of an arc, or indeed a wire, in that region responsible for ignition. Moreover, it was the only area where all the consumable items on and around the equipment were totally destroyed. In other areas there was partial consumption of combustibles but, due to the brief period of flame propagation, not total destruction. Clearly, the environmental control system lithium hydroxide panel was the point source of origin. Across the floor from the panel covering wire bundles, a large Raschel net, or debris trap, served as an effective propagator for the fire that resulted when an electrical arc occurred. Bearing in mind that the first verbal report of fire came 9.7 seconds after the electrical glitch observed in the telemetry from spacecraft 12, the Board found it relevant that in a simulated fire from exactly this location flames became visible 8 seconds later – sufficiently close in time to further support the theory.

In testimony from George Mueller, the head of manned space flight, the Board heard that NASA recognized 'that there were design deficiencies, engineering design deficiencies, in the wiring of the Block I spacecraft and we had taken steps in the Block II design to improve the manufacturing process, to improve the cable layout, to improve the ability to build a harness and use it without having as many changes as were necessarily incorporated in the Block I wiring harness.' Mueller stated that the major problem with Block I harnesses arose from the use of two-dimensional jigs, from the continued and increasing number of measurement points calling for extensive splicing after installation, and from the introduction of new wires after the original harness had been built in. There were plans to standardize a more stringent code of wiring practices for the Block II vehicles and Mueller testified that the examination of spacecraft 12, a Block I product, by engineers at the Cape revealed deficiences in the harness that would not be tolerated for more advanced mission profiles.

It was the same problem that haunted continued operations with Block I spacecraft: interim procedures and early development concepts being used for operational space flights. Block II would employ three dimensional harness jigs, where twists and turns through three axes would be used when laying out wiring for sizing and installation. Block I had employed two-dimensional jigs, in essence flat tables, for sizing the lengths of wire prior to actual installation. So determination on the exact cause of the fire centered not on the source of ignition, nor the location since that was certainly known, but rather on the probability of an arc having occurred in wiring now completely destroyed and missing from the site.

In its conclusions about the state of wiring aboard spacecraft 12, the Board was clear in language that left little to the imagination: 'Some areas of wiring exhibited what would be referred to as 'rats nests' because of the dense, disordered array of wiring. In some instances excessive lengths of wires were looped back and forth to take up the slack. Also, there were instances where wires appeared to have been threaded through bundles which added to the disorder. . . . A circuit breaker panel was pressed so close to a wire harness that wiring indentions were left in the circuit-breaker potting. . . . The floor wiring and some connectors in the LEB (Lower Equipment Bay) were not completely protected from damage by test personnel and the astronauts. This is evidenced by mashed 22-gauge wires found in some of the wire harnesses. . . . Wire colour coding practices were not always adhered to. . . . Some wiring in these was found with damage to the sleeve which covers the shielded wire.'

For its part, North American Aviation contributed valuable testimony to the fact that when originally designed the wiring was of an excellent standard but that continued insistence on the part of NASA to add, re-distribute, or change equipment connected or serviced by the electrical distribution system led to the general chaos: 'The original installation of the wiring to these feed-through connectors was orderly but due to changes which were ordered after the original installation, disarray did occur in some areas. In some cases excessive lengths of wire had to be stored or looped back into the bundle because they were to calibrate resistances for the instrumentation functions, and the instrumentation would be affected if these wires were not to the calibrated lengths. In other cases due to changes ordered, equipment was relocated, thereby leaving lengths which could either have been cut and spliced or looped. It was considered that looping was as fully acceptable a practice as cutting and splicing.'

In the face of extensive criticism from the Board, both NASA and North American were forced to a defensive posture rather than one aimed at eliminating as erroneous the published conclusions of the various panels. But it was not easy to see how wiring could be so damaged that it was directly exposed to combustible materials for ignition from an arc affecting that particular wire. Elsewhere in the spacecraft, the Board found evidence that although regulations stipulated doors and panels were not to be closed if in doing so they pressed hard against wire bundles, there were in fact several panels not completely destroyed that revealed scuffing on wire harnesses. Moreover, the Board criticized the practice of routing wires 'across and along oxygen and water/glycol lines.' It was pointed out to the Board that the regulations allow this if the electrical wire is no less than 1.3 cm from the piping. As tests carried out during the investigation revealed, combustible materials can be ignited up to 10 cm away from an arc in a section of electrical wire. And as documents showed, there had been extensive leak problems within the environmental control system coolant loop.

Because the evaporation of water from the coolant solution leaves a highly inflammable residue of glycol when leaked to an exterior surface, the close juxtaposition of potential ignition source and probable fire path was a significant factor in spreading the fire once it was under way. Aluminum tubing had been selected over stainless steel in order to save weight and while the soldered joints were found to be of a high quality, the 'creep' caused by stress from remote sources could quickly produce leaks. And did. The Board's chairman, Dr. Thompson, summed up the situation he believed his team had found: 'Very well fabricated solder joints, not subject to anything but the loads which they were really designed to withstand, or the pressures in the line in the protected area, could very well stand up. The facts of life are that in putting these things in and having them exposed to the problems of installation, other activities around the area, the movement of people, and subject to the vibration of the spacecraft, that the loads on those joints, the stresses on those joints, even though they may be very well made, would fail because they just do not have the tolerance for abuse that some of them almost certainly get.'

And on the question of the electrical distribution system, undoubtedly the cause of ignition, Dr. Thompson had this to say: 'The wires – in order to avoid these problems of having wires go over sharp edges or get in front of doors that have to be opened and then have to go around elements of the vehicle in such a way as to avoid any abrasion or sharp bends – have to be engineered in a very careful way and should use three-dimensional forming to do that. The more wires were added, the conflicts were added, and then the wires were wedded up without just an engineering analysis of just where they should go and how they should be channeled around to avoid trouble or abrasion, how they should be channeled to avoid the danger of people stepping on them or misusing them after installing. Fundamentally, I think this is what is back of what we have seen there, too much building without the real intensive use of engineering to formulate the design before allowing people to put wiring in.'

When the Board issued its formal report the public was stunned by the lethal nature of the implications. Never before had a single official body cast such doubt on the engineering practices of a government-industry team dedicated to one of Man's most ambitious endeavors. But inherent throughout the critique was the common thread of ignorance rather than deliberate malpractice, lack of awareness rather than devious cost-saving measures. It all seemed to show how much there was still to be learned about systems management and integration with complex and sophisticated space vehicles. NASA, for its part, had showered the contractor with a host of changes and additions until the very concept of separating Block I (precursor) and Block II (definitive) designs became submerged within a flood of modifications and amendments. As for North American Aviation, they were visibly climbing higher up the learning curve from the days when Sam Phillips took his task force to Downey, but there was still much to acquire in the form of knowledge about possible repercussions from new engineering practices.

There was one other concern of the Board. The relationship between NASA and the contractor had not been good enough to sense in advance minor problems that arose in respective camps. From the government side there was a feeling that the contractor was hostile to repeated additions in hardware needs and specifications. From the industry side came suspicion that the customer wanted to move in on project management at a direct level. Dialogue was stifled and much was left to be desired in relations between the two. At a technical level, there were four significant areas of primary concern to the future operation of manned Apollo vehicles: the state of the wiring, the condition of the plumbing, the use of many combustible items inside the cabin, and the lack of adequate escape procedures or escape routes from an internal fire.

Upon reviewing the findings of the Board, NASA agreed that there were significant wiring deficiencies even in the Block II design and planned to cover all exposed wires with additional protection. Metal covers would be used over conduits across the floor, or aft bulkhead, with wire or harness wraps in areas where this was not possible. Three-dimensional jigs would become standard, as indeed they had for almost a year before the fire at North American Aviation's Downey plant, and special attention would be made to sharp corners or 90° turns. On the question of repeated failures in the environmental control system, NASA planned to adopt immediately the Block II ECS design rather than the Block I hardware installed in spacecraft 12. This improved ECS had features significantly safer than the earlier model, and was to be further modified so that potential leak sources were eliminated. Also, aluminum oxygen lines would be replaced by stainless steel lines, and coolant pipes carrying water/glycol would be armored at the solder joints. Also, vibration tests, a process never performed on a Block I vehicle, would check the integrity under stress of plumbing carrying the potentially combustible liquid.

As for the use of combustibles, major and drastic revision would be made in the use of such items. In spacecraft 12, the Board found 1,412 items fabricated from non-metallic materials. NASA agreed to give this area sufficient attention to render the Apollo Command Module virtually non-inflammable, replacing an estimated 2,500 combustible sources in the spacecraft with other less flammable materials. Considerable effort would be expended on this area, even to the point of providing space suits that could resist flame impingement for several seconds, manufactured from material known as Beta cloth. The Nomex used in the suits worn by Grissom, White and Chaffee, was tested for flammability characteristics in the course of the post-fire investigation and found to propagate flame at a rate of more than 8 cm/sec under the conditions inside spacecraft 12. Moreover, Nomex would ignite at a temperature of 480°c in a pure oxygen sea-level pressure atmosphere. Beta cloth would start to melt without ignition at a temperature of 840°c.

As an ultimate test of the re-designed interior, an Apollo boilerplate was to be intentionally ignited for analysis of the response from new and safer materials, better plumbing, and improved protection for wiring. As for the lack of adequate escape procedure, that included reference to the three cumbersome hatches needing 90 seconds of time to afford egress for a trapped crew as well as to the lack of adequate fire-fighting equipment in the white room or on the gantry arm. NASA failed to classify the Plugs-Out test as a hazardous activity because propellants or igniters were not involved. In future, adequate equipment would be on hand as well as oxygen supplies to administer in the event of fire or the release of toxic fumes. Moreover, three emergency oxygen masks would be provided independent of the normal environmental control supply, with hardened and protected plumbing to insure against asphyxiation from smoke and soot entering the suit loop.

The big question that arose from the fire concerned the established use of a pure oxygen environment. If a change to a two-gas system at higher pressure was deemed necessary it would mean complete re-design of major elements in the combined Command and Service Modules. There were arguments that dissipated the implied threat of a raging fire in space. Tests seemed to show that a fire in weightlessness could smother itself; devoid of convection in the same way that a fire moves on Earth, some people said it would virtually put itself out. Only a few were convinced. Yet by eliminating virtually all possible sources of ignition, by providing the crew with suits affording protection against heat, and by designing out of the Command Module interior all conceivable sources of flame propagation, it was deemed satisfactory to operate the spacecraft at 100% pure oxygen under 258 mmHg pressure as originally planned.

During ground tests, and while the crew were in the spacecraft on the launch pad, the cabin could be pressurized to a 760 mmHg sea-level environment with a mixed atmosphere of 40% nitrogen and 60% oxygen. Inside their pressurized suits, the crew would breath pure oxygen, but the diluted environment of the pressure vessel would ensure that a fire started prior to lift-off

The new unified hatch could be opened in seconds by a pumping action on the handle, activating 12 latches to swing the 102 kg structure back on its two hinges.

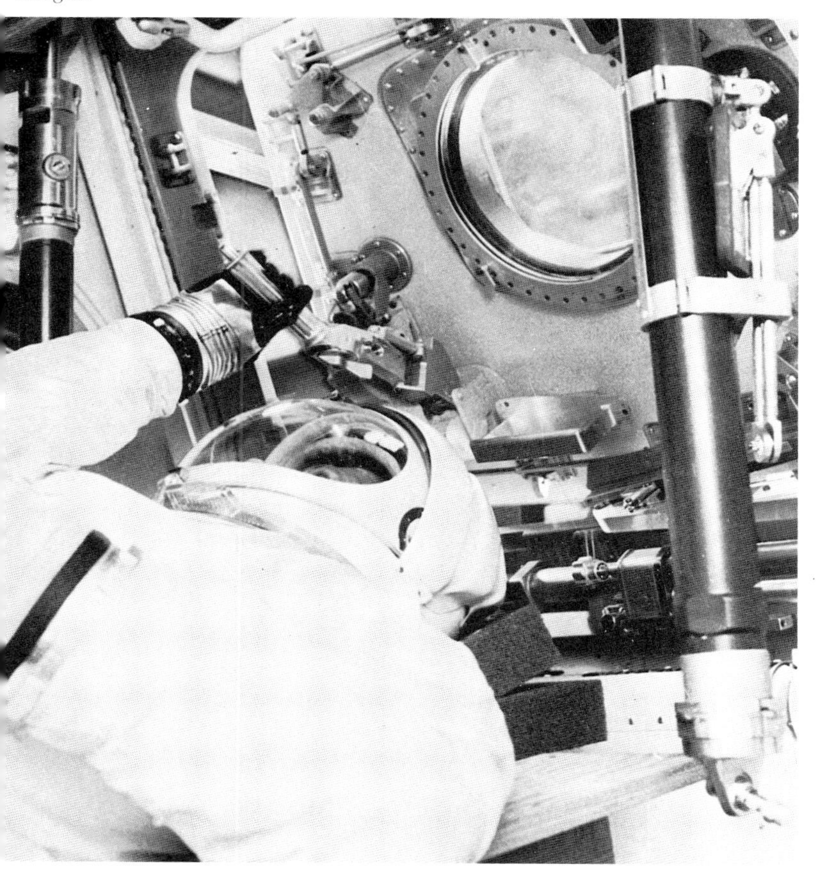

would propagate no faster than under normal Earth conditions. As the spacecraft ascended to orbit the environmental control system would quickly replace the nitrogen with pure oxygen as normal. In this way, under the worst conditions, the cabin would be pressurized to no more than one-third sea-level pressure with pure oxygen.

Finally, the new quick opening hatch was to be developed for use on the first manned Apollo flight, an integrated device capable of being opened within about 5 seconds. The inner hatch on spacecraft 12 could not be opened if interior pressure was more than 17 g/cm^2 greater than the exterior pressure; even before the fire quickly raised the internal pressure, spacecraft 12's atmosphere was 140 g/cm^2 above ambient pressure. It would have been physically impossible to remove the hatch until the interior and exterior pressures had been equalized. By adopting a sea-level, mixed-gas, atmosphere on the pad, future Apollo missions would not be hampered by the need to raise internal pressure above ambient level. The new integrated hatch, providing both pressure and heat shield protection, would be opened by the pilot on the right couch pumping an attached handle to activate a pressurized nitrogen cylinder employed to swing it open. Unlike the earlier hatch, where the inner component was designed to be pulled inward, the new device would swing out on hinges.

The astronaut on the center couch would have removed his straps by the time the hatch was open, ready to swing up and out to be followed by his two colleagues in turn. In effect, the new hatch could be opened in less time than it took to gain freedom from the couch. If a fire should develop in orbit, despite the improvements and almost non-inflammability of the interior, a vent valve in the door could be rapidly wound open to dump the cabin to a vacuum within 60 seconds.

But if a fire should start, and evacuation of the cabin be prevented by either a stuck vent valve or a jammed hatch, special aluminum panels would isolate the conflagration to a limited area behind doors and covers strategically designed for this purpose. And if, despite the absence of combustible items, fire should persist, the crew would use an extinguisher to expel an aqueous gel into dispensing ports located at various places across the main display console. With individual oxygen masks and fireproof space suits, the probability of another disaster like that in spacecraft 12 was reduced to an almost incalculably low value.

By the end of March much of the Board's work had been completed. By the beginning of April the Congressional debate was under way and by May the space agency had a good idea of what was required for it to get Apollo back on course. No more Block I spacecraft would fly after the unmanned Saturn V test launches required to man-rate the big booster for space operations. Flight schedules had been changed substantially by the fire and NASA had only a general idea of when it could resume manned operations on Block II. It all depended on when the re-configured design could be made ready. There was no criticism in the Review Board report of basic spacecraft systems, but considerable quantities of material had to be taken out and others put in to fireproof the Command Module.

It would be wrong to believe that changes were limited to modifications outlined in preceding paragraphs. In looking at the spacecraft considerable effort was expended to make sure nothing like this could possibly happen ever again and much equipment was modified or altered in minor ways to significantly improve the operability of the complete system. There is no doubt that what came out of North American's Downey plant a year after the fire was a very different product to that delivered some time earlier to pad 34 for a Block I flight with Grissom, White and Chaffee. Everybody learned a lot, and nobody ignored the obvious lessons. But the politicians and the public demanded heads for the guillotine and they got two: one from North American Aviation, and one from NASA. They were inevitable replacements, but little more than figureheads to represent society's disgust at what had happened down at the Cape that dark January evening.

At the contractor plant, Lee Atwood and Harrison Storms were obvious candidates for removal; Atwood made it quite plain to Storms that he was not about to suffer martyrdom and the latter departed from his position as head of the Space and Information Systems Division. In his place came a man with a distinguished record considerably higher up the management and administration tree than the S and I D represented. William B. Bergen left his job as President of the Martin Co., in 1966 because of different views held by the chairman of the board George M. Bunker. Bergen was shopping around for a comfortable position with a predominantly aeronautical company when Lee Atwood seized him up as North American Aviation Corporate Vice-President for the Space and Propulsion Group. He would have Rocketdyne and the then Space and Information Systems Division directly under him.

When Storms left, Bergen suggested to Atwood that he should be allowed to move in as President of the S and I D. It would mean going down a notch, but the respect Bergen held among his peers could only do good for the hurt pride of the space engineers. In April, Bergen moved in to his new job, and from the following month his section was to be known simply as the Space Division. On 22 September, North American Aviation merged with the Rockwell-Standard corporation to become North American Rockwell. Willard F. Rockwell, Jr., former President of Rockwell-Standard, replaced Atwood as the chairman of the board. Transition was complete. The names had changed, the figureheads had been replaced. Only men like the old S and I D Vice President, Dale Myers, and his subordinates stayed on to see through the modifications put into effect by the Apollo 204 Review Board report.

In 1967 also, Ralph R. Ruud moved from Vice President, Manufacturing, to executive Vice President, Space Division, to bolster top management on both Apollo and Saturn S-II stage contracts. Finally, Bastian (Buz) Hello became Vice President of Space Division for Launch Operations. It would be his job to accept and condition the Apollo hardware for flight at the Cape. In this regard, the company stiffened its corporate control on activity down at the Kennedy Space Center. But it was to a very important triumvirate that the job fell of putting the spacecraft program back on the road: Dale Myers and under him George W. Jeffs as assistant program manager and chief engineer, and George B. Merrick as director of Apollo CSM systems engineering. Both Jeffs and Merrick had moved in the year before the fire to improve corporate structure after the Phillips report of December 1965, and both knew how to pull the program into shape. With Myers approving decisions and Bergen at the helm, the re-configured Space Division now had one of the strongest management structures around.

The second executive to go because of the fire was a token sacrifice in the form of MSC's Apollo Program Manager Joseph Shea. He was replaced by George M. Low in the April. Second to Gilruth for more than three years, Low was replaced as MSC Deputy Directory by George S. Trimble several months later. To Shea the Apollo program was the lifeblood of his own existence. It was the one thing that held him together, for he was one of the very few who really believed in what Apollo was and what it stood for. As the most prominent government representative facing the contractor, Shea's job was to see that the company did a good job, completed a task for which it was commissioned, and that as the MSC manager for Apollo the spacecraft was in an acceptable condition.

The fire all but destroyed him. In August 1966 during the formal acceptance handover of spacecraft 12, Grissom, White and Chaffee were photographed, their heads bowed, during a prayer of dedication. The three astronauts later sent Shea a copy of this view, with an ironic statement scrawled across the bottom: 'It isn't that we don't trust you, Joe, but this time we've decided to go over your head.' Shea would keep that photograph in a prominent place at his Massachusetts home for several years. He never, ever forgave himself for what others felt had been his responsibility. When he left NASA it was to join D. Brainerd Holmes at an electronics firm near Cambridge, Mass.

By May 1967, little more than three months after the fire, Mueller had laid the general schematic for a new and improved flight schedule. It was hoped that two unmanned Saturn V flights could be launched from the Kennedy Space Center before the end of the year with one unmanned and three manned in 1968 and five manned missions in 1969. The first manned Saturn IB flight would have taken place toward the middle of 1968 with a total of nine manned missions culminating in a Moon landing on the eleventh Saturn V flight at the end of 1969. Prior to the accident Mueller had planned for 13 Saturn V flights before the end of the decade. In effect, the accident had delayed the program by two Saturn V missions. It would be tight going but if the space agency could get the Moon mission launched with only nine manned rehearsals of all the various techniques for a lunar landing, Kennedy's goal was still within reach.

When Bergen set up shop at Downey he scoured the plant for personnel deemed a drag on the new pace of the program. Charlie Feltz was clearly in a leading role within the engineering directorate, so too was Norm Ryker. But Scott Crossfield, the head of Quality and Reliability, was out, replaced by T. C. McDermott from the Autonetics division. This was one of the most important positions at the company. Quality control and the reliability of the product was an important front for what happened in the rows of machine shops and fabrication plants buried deep within the bowels of the facility. Quality control was the one job nobody wanted, but the one that could bring warm praise for a job well done. If somebody else failed in a trusted task, the position was

Changes to the Apollo suit were mandatory in recommendations made by the Review Board. Here, a Hamilton Standard engineer models a development version of the type used aboard Apollo 204.

not worth a dime. And so it was with Scott Crossfield, a test pilot who weaned North American's X-15 rocket 'plane. When the fire burned holes in NAA's quality assurance program, the head was irrevocably severed. Scott Crossfield was a valuable man to have around, however, a good man in bad situations, and his head was one that should not have had to go.

The only decision about exactly when the space agency could again fly manned vehicles into space came when Jim Webb agreed that modifications to the basic Apollo Block II would precede final fabrication of spacecraft 101, the ship now designated for the first manned Apollo. Batch numbers assigned to production spacecraft did not follow the numerical sequence used by McDonnell for Mercury and Gemini. Because Apollo generated two specific Apollo Block designs, the first employed sequential numbers from 001, while Block II spacecraft would begin from 101. Command and Service Modules 12, the hardware involved in the fire, would not be used again for manned operations; the Command Module was to be impounded at the Langley Research Center in an environmentally protected cocoon for 50 years from the date of the fire, just in case it could serve as a research tool when more advanced equipment became available.

On 9 May, Webb, Seamans and Mueller publicly announced the selection of AS-204's back-up crew to the mission spacecraft 12 should have flown. CSM-101 was expected at the Cape some time before the end of 1967 and Schirra, Eisele and Cunningham would fly it into space three months later on top of another Saturn IB launcher. The delay in manned Apollo operations would allow the development of Saturn V to catch up with that of the spacecraft, permitting early migration to the big Moon booster at an earlier point in the flight schedule than had been originally anticipated. Immediately after the fire all existing crew assignments were cancelled, but with Schirra, Eisele and Cunningham flying the initial shakedown mission it was hoped that the dual launch of two Saturn IB's – one with a manned CSM, the second a day later with the lunar lander – could be dropped in favor of going immediately to a combined CSM/lunar lander test flight with the first manned Saturn V. Borman, Lovell and Anders had spent considerable time up to the fire practising for a Saturn V mission with the two manned spacecraft necessary for a Moon landing. Their experience would be retained if early movement to Saturn V became possible. McDivitt, Scott and Schweickart, originally assigned to the dual rendezvous mission on the second manned Apollo flight, could then use their rendezvous training for the third manned mission in the new schedule. But that was speculative thinking and what was needed first was a successful 10-day flight in Earth orbit flying the basic Apollo Block II CSM.

America was not the only country to introduce a new manned space vehicle. For several years Korolev had pressed ahead with development of the Soyuz spacecraft, a true second-generation vehicle for Russian cosmonauts. When he died in 1966 from a failed heart the Soviet space program lost an important advocate among a large group of politicians with mixed emotions about what had ostensibly begun life as a Kruschev project. It is unclear even to this day exactly why the Soyuz program began. Its applicability to manned space stations in Earth orbit is, and always was, an obvious feature of its design. Coupled from the outset with rendezvous and docking, it would form the backbone of Soviet manned flight operations for more than a decade. But there were other activities for which it could have been used, events closely associated with the race for global acclaim.

Although Russia has never publicly decreed its commitment to a manned Moon landing attempt, evidence is clear that the Soviet Union intended to pre-empt in every conceivable manner possible the activities of the United States in this new field of technology. Having been set upon that road by an over-eager Kruschev, nothing could be done by turning back the clock. So while Soyuz may have always had primary roles associated with manned space stations, its design was sufficiently flexible to permit it to carry cosmonauts at least around the Moon and back, and probably into Moon orbit also. None of this was fully understood in 1965 when strong rumors about a new Soviet spacecraft leaked to the West. At a time when NASA's Gemini program was accelerating through successive manned operations, when Voskhod had already flown three cosmonauts into orbit and given Leonov a space walk, most observers expected big things from Russia that year or the next.

But nothing came and it was late 1966, immediately after the last Gemini flight, before tracking stations in the West picked up

Russia's Soyuz spacecraft appeared in 1967 as the first truly operational manned spacecraft developed by that country. In capability it was about as good as Gemini but would remain in use for at least fifteen years.

the anticipated signs of unmanned test operations with a radically new manned spacecraft. On 28 November, Cosmos 133 was detected in space, flying a low orbit with telemetry channels reminiscent of frequencies usually reserved for manned vehicles. It came down two days later, too short a period in completely the wrong orbit to be a reconnaissance satellite; apart from which initial estimates on its size and mass indicated it to be too large for the usual class of short-period satellites. It was clear that a manned flight was in the offing when a second spacecraft of this type flew a similar mission beginning 7 February, 1967. And there were other indicators of renewed Soviet activity.

For nearly two years the Russians had been flying a launch vehicle equal in lifting capacity to that of NASA's Saturn IB, although early flights during this period failed to cater adequately for the booster's potential. Also, according to NASA Administrator James Webb, construction of very large launch facilities began in 1965 and by the following year photographs revealed an engineering model of a Saturn V class launcher. Nobody was ever shown pictures claimed to have been taken of this vehicle by military reconnaissance satellites but it was an effective instrument for pressing the urgency of space funds upon a reluctant Congress. Was this the super-booster Russia needed to get cosmonauts to the Moon? American intelligence experts thought so and the project was soon dubbed 'Webb's Giant' because the NASA boss frequently brought it up at meetings with politicians.

In the nomenclature of space launchers, the big booster was called the G-class rocket, while the Saturn IB type launcher was called the D booster. The latter could lift more than 20 tonnes in its developed version and it was with this that Western experts expected to see the next series of manned vehicles launched. It came as something of a surprise, therefore, when Soyuz 1, flown by Vladimir Komarov, ascended on the same combination employed for the earlier Voskhod, itself an adaptation of the Vostok manned vehicle. Rumor had been widespread that a major Russian flight was about to begin and attention centered on the possibility of a rendezvous and docking mission. In the period since Russia's last manned flight the US had flown ten manned Gemini missions accomplishing docking, EVA, and long duration flight. Surely the Soviets would try at least to equal this record of achievements? Rendezvous and docking had been anticipated for some weeks before the launch of Soyuz 1 and rumor had it that a Soyuz 2 would ascend to link up with the first so that crew exchanges could take place via space walking. That feat would certainly accomplish much to gather public acclaim, even if it did have little significance operationally.

The flight of Soyuz 1 got under way during the early morning hours of 23 April, 1967. It was 1:35 am in Moscow, two hours later at the Baykonur launch site, when Soyuz 1 rose from the nested steelwork of support beams and umbilical tower. The spacecraft rested on top of the launcher's second stage, an A-2 vehicle capable of lifting $7\frac{1}{2}$ tonnes, within an aerodynamic shroud designed to separate in two halves during ascent and expose the manned vehicle. During the early part of ascent, while the shroud was still attached, an escape rocket was carried on the forward end of the assembly, ready to whisk the manned module from off the launcher in the event of a malfunction.

The spacecraft comprised three main sections. An Equipment Module about 2.8 meters long, essentially a cylindrical structure designed to carry propulsion units at the rear, attitude control thrusters on the exterior and solar cell arrays on opposing sides, provided volume for systems and control equipment. On the forward end was mounted the Crew Module, about 2 meters in length, comprising a bell-shaped structure with flat sides and a hemispherical heat shield across the base. It would carry three cosmonauts in a shirt-sleeve environment of nitrogen and oxygen at sea-level pressure, similar to earlier manned vehicles but affording accommodation without pressure suits.

The Soviet view was that an accident serious enough to split the hull of the Crew Module was one in which the cosmonauts could not be expected to survive so the use of bulky space suits was irrelevant now it had been proved comparatively safe to fly in orbit. Suits would be needed for EVA, or whenever the cabin was to be depressurized, but with the new Soyuz vehicle confidence was so high that shirt sleeves would be quite safe. To the forward end of the Crew Module, Soyuz carried an Orbital Module, a barrel-like structure with domed ends into which cosmonauts could pass via a hatch separating the two pressurized areas. For launch the crew would lie on couches in the Crew Module, but once in space they could open the forward hatch and drift up into the Orbital Module which was designed to support a docking unit on the forward exterior surface.

The total habitable volume was almost exactly that of the combined Apollo CSM with the lunar lander docked to the forward part of the Command Module. But Soyuz was limited in that the launcher had a maximum lifting capacity considerably below the Saturn V designed to carry the 40 tonne Apollo. The Russian spacecraft had a maximum weight of about 6.8 tonnes and could accommodate little propellant for major maneuvers, at least in its basic form configured for space station ferry duties. The introduction of the Soyuz ferry several years before the ultimate availability of a space station provides strong evidence that the pace of the program's development was matched by the secondary objective: that of supporting manned flight around the Moon and back just as quickly as possible.

It is not entirely clear if the Orbital Module was attached to the forward end of Soyuz 1 or if it was left off for this inaugural flight. Tracking data seems to confirm that it was in fact attached. With a total length of about 7.5 meters and a diameter of 2.2 meters across the habitable modules and 2.7 meters across the Equipment Module. Soyuz had a span of 8.37 meters. In capability it was similar to NASA's Gemini spacecraft. Two 400 kg thrust rocket motors, one prime and one back-up, were installed in the rear of the Equipment Module, and twelve attitude control thrusters – eight thrusting at 1 kg and four at 10 kg – were carried for orientation. Soyuz was the first Soviet manned space vehicle capable of moving from one orbital path to another by firing onboard propulsion systems. Rendezvous was to be a key factor in future Russian plans. The choice of solar cell arrays as the means of generating electricity was one based on the power levels necessary to run the ship. Stored in the folded configuration against the side of the Equipment Module during launch, they were unstowed and extended in orbit after separation from the launcher.

For possible Moon flights, the Orbital Module would be removed, the Crew Module heat shield would be improved and additional communications equipment would be installed. Apart from the obvious need for additional supplies, Soyuz could readily accommodate a lunar role, launched from a D-class booster proven capable of sending 6 tonne payloads to escape velocity by two test flights into Earth orbit during the two months preceding Komarov's mission. In addition to the two A-2 precursor flights, the two D-class missions strengthened interpretation of Soyuz as an essentially circumlunar spacecraft. However, the initial manned shakedown flight was probably only expected to last two days at most since both unmanned A-2 test missions had been for periods of that duration.

For the first few hours of Komarov's flight everything seemed to go as expected but as the mission progressed toward the end of its first full day there was increasing awareness that major systems problems had developed in the stabilization and control equipment. The Russians have retained their customary silence about the performance of the spacecraft in orbit but several reports indicate that one of the solar cell arrays failed to deploy as it should and that systems were going awry almost from the outset. There is little to indicate a major malfunction, however, before the end of the mission. What is significant, is that Komarov began his sleep period only 9 hours after lift-off and completed it at a mission elapsed time of 17 hrs 45 min. Had his mission really been intended to last only 17 orbits, as official Soviet reports would claim, the sleep period would probably have been shifted closer to his re-entry time, allowing the activities of the second day to begin with a well rested cosmonaut. As it was, there elapsed a period of some nine hours between the end of his rest period and the actual re-entry.

There was good reason to put his sleep period at the early time in that this was the period when Soyuz 1 wandered from communication with ground stations for many of the orbital passes. But it was also a schedule closely followed by later flights involving rendezvous and docking. Whatever the intent, Soyuz 1 was in grave difficulty by the 15th orbit and Komarov attempted to bring the spacecraft to Earth at the end of the 17th. Unable to align the spacecraft correctly for retro-fire, Soyuz went round again and fired its engines to return on the 18th orbit. It is possible that at this time the pilot placed the spacecraft in a high spin to execute a ballistic re-entry rather than the controlled descent for which the Crew Module was configured. A systems failure in attitude control and stabilization forced the cosmonaut to maintain a base-end first attitude by rifling the flight path in a spin-stabilized condition.

Immediately after this the Equipment Module was jettisoned from the rear of the cabin and the Orbital Module was released from the front, leaving the Crew Module to descend on its own along a ballistic path generating up to 10 g deceleration versus the maximum 5 g from the lifting re-entry originally intended. With a problem in the stabilization system there was no way to control a lifting vector; ballistic descent was the only chance for survival. On the way down the single main parachute deployed at an altitude of 7 km but the slowly spinning spacecraft induced a curling motion into the parachute risers and the canopy twisted around itself. Devoid of any means of reducing speed, the Crew Module slammed into the ground at a shattering speed of 450 km/hr, reportedly bursting into flames and incinerating everything inside. The structure was essentially sound but nothing on the interior was left intact. Komarov was dead.

It is highly probable that Soyuz 1 was to have been the target for a second vehicle scheduled for launch on the day Komarov returned. It is even possible to speculate that the prime crew comprised Bykovsky, Yeliseyev and Khrunov. What is not so certain, however, is the precise nature of Soyuz 1's malfunction, or the precise reason for aborting the flight. Clearly it was that which prevented launch of the second spacecraft leading also, but indirectly, to the fatal crash.

Russia was dazed and deeply shocked by the tragic incident, ironically the victim of accusations it had made only weeks before about American capitalism and the deaths of Grissom, White and Chaffee. Now the Soviet Union was mourning its own space dead, and probably for the same reason. Komarov was cremated and his remains interred in the wall of the Kremlin on 26 April. The Soviet news agency had waited half a day before officially announcing Komarov's death, and the US astronaut corps sent a message of sympathy. Gordon Cooper and Frank Borman were to have attended Komarov's funeral as a mark of respect from America's astronauts, but the Soviets refused to allow them in to the country claiming that the event was 'an internal matter.'

It clearly hurt Soviet pride and dashed hopes of pulling back the expansive lead set up by the American manned space flight program. For some months after the disaster, leading Russian scientists and engineers expected to see propaganda made out of the event by politicians and partisan causes in the West. When they failed to materialize it surprised the Soviet technocrats. They had been well doctrinated in the interpretations of extreme political viewpoints and fully expected a tirade of abuse.

The year was proving to be a sad twelve months for many people everywhere. In the background to technological invention and the forward marching pace of space engineering, increasing military involvement in South East Asia served as but the portent of violent conflagrations to come. But it was within the ranks of pilots and aeronautical engineering that the personal sadness reach a desperately low level. The toll was getting almost beyond parallel: Grissom, White and Chaffee burned to death on 27 January; Bartley and Harmon burned to death in an oxygen fire on 31 January; Komarov killed by concussion on 24 April.

On 10 May, lifting body pilot Bruce Peterson crashed the M2-F2 while attempting to land at Edwards Air Force Base. Distracted by a helicopter his flying machine rolled over several times, bouncing along from wing tip to wing tip, incurring terrible facial injuries on the pilot whose torso was battered by fragmenting sections of the aircraft's nose. The lifting bodies were a family of small research craft similar to wingless re-entry vehicles thought appropriate for flying down through the atmosphere from space. Designed to be launched from under the wing of a B-52 carrier aircraft, they provided aerodynamicists with data on low speed handling characteristics. Peterson was subject to considerable surgery of a revolutionary type and the film of his crash was used by producers of the TV film series 'Six Million Dollar Man.'

Then, on 6 June, astronaut Edward G. Givens, the man who did so much to develop the Astronaut Maneuvering Unit that never did get the planned workout on a Gemini mission, died when his car careered off the road and crashed. Two other persons in the car were badly injured. Givens was buried at Quanah, Texas, three days later, following a memorial service at Houston the previous day. It was attended by many astronauts who knew only too well the price for an overworked mind and body.

On 5 October, astronaut Clifton Williams, the Marine Corps Major who stood with Alan Bean as part of the Gemini X back-up crew, crashed his T-38 aircraft when a failure in the oxygen supply drew him from consciousness. He was buried with full military honors at Arlington National Cemetery on 9 October. Williams had been working on the lunar landing vehicle and would probably have had a Moon mission.

And then yet again on 15 November, Maj. Michael J. Adams was killed when X-15 No. 3 crashed while out of control over the Mojave Desert close by the Edwards Air Force Base. Built by the then North American Aviation, the X-15 had been America's most successful and dramatic rocket powered research aircraft venture. Flown first by NAA's test pilot Scott Crossfield in 1959, it had been delivered to NASA for high speed and high altitude tests. Joined eventually by two more of the same type, it raised the speed record from Mach 3 to beyond Mach 6, and reached altitudes exceeding 100 km. It was the one and only fatality during the entire program. Adams had been assigned to the Air Force Manned Orbital Laboratory program but requested and got a transfer to the X-15. His death came on the 191st flight, ironically only eight flights before the end of X-15 operations in October 1968.

But if 1967 was the year of human tragedy, it was also the time when an equally enduring debility struck plans for future manned activity. It was to bring home to many NASA people the reality of Kennedy's one-shot approval for a manned Moon landing as being a single goal rather than a precedent for continuing operations. The Apollo Program had the seeds of its own extinction sown within its design, a vehicle inappropriate for sustained space activity. Selected as a quick answer to Russian prowess and daring, the spacecraft was the shortest route to heady fame and was neither the engineering concept preferred nor the economic transport system it needed to be for open-ended exploration.

Put simply, it was the costliest alternative of several ways to reach the Moon, the remainder of which would have pushed the first landing beyond the 1960s. Each Moon landing would cost almost half as much as the entire Mercury effort and it would gobble up funds that could have been applied to the development of more economic, and enduring, space transport systems. But this was not fully understood at the beginning of the decade when there seemed no limit to the money Congress would allocate for space spectaculars. Most planning sessions at the Washington Headquarters centered on the continued operation of deep space exploration. Between 1958 and 1963 momentum built rapidly toward proposed Moon exploration, colonization, and use. But it also spurred plans for manned expeditions to the planet Mars – the Red Planet beloved by science fiction writers and authors of fact alike.

At this time no spacecraft had penetrated the black void between Earth and Mars to see close up what caused the apparent waves of vegetative growth during warmer seasons on the planet, and coloration that waxed and waned like verdant greenery on Earth. It was all a very tantalizing view, this limited awareness of what the Red Planet was really like. Earth-based observation seemed to indicate a comparatively dense atmosphere, probably nitrogen and oxygen, with simple life forms populating its primitive surface. In places, pools of life-bearing water were thought to break the vast expanse of desert plains while in others thick fields of lichen, spawning ferns and strange plants were suggested. It seemed a logical place to visit, and a suitable goal for post-Apollo plans.

But as the imaginative elements of space gave way to the hard reality of practical operations, the level of naivety concerning the supposed ease with which Mars landing flights could be performed dissipated with each succeeding year. By 1964 the assumption that NASA would move toward such operations by the mid-1970s had been relinquished in the face of engineering problems that seemed to loom increasingly large. In examining favorable launch opportunities it was agreed that the mid-1980s appeared more logical for manned Mars flights; for the time being all thought of post-Apollo plans developing such ambitious objectives were put to one side. Although not forgotten, more rewarding application of Apollo hardware could be proposed.

The origin of post-Apollo planning can be fairly accurately fixed as having originated on 30 January, 1964, when President Johnson sent Jim Webb a letter expressing his interest in receiving an appraisal of future mission opportunities, a shopping list from which the Administration could select suitable goals. Coming as it did only a few weeks after the death of President Kennedy, the request is not unconnected with Johnson's own, personal ambition to see his mark on the space frontier. Chief architect of the National Aeronautics and Space Administration at its inception nearly six years before, prime instrument in pressing Kennedy to a Moon landing commitment, Johnson wanted a more strident pillar upon which to hang his own far reaching achievements. The Moon goal was a Kennedy hallmark, something else would have to be constructed for the Johnson monument. It was merely another manifestation of the competition that continually dogged the Kennedy-Johnson dialogue.

That the President was thinking big is seen in a portion of the January letter where he suggests to Webb that 'we be in a better position to make future decisions involving . . . *large space programs* (author's italics).' Within four months Webb had responded; Johnson gave him six so that decisions could be factored into the Fiscal Year 1966 budget to be prepared in the fall of 1964. Web's reply was low key and reiterated the inventory of hardware that would be available by the turn of the decade, suggesting it would be useful to fly astronauts in space first for a period of one month, then for three months, then for a year. A large laboratory housing 24 men would be useful. Manned exploration of the lunar surface could get under way with Apollo hardware by keeping men there for up to two weeks, using the lunar lander as a shelter and modified versions as unmanned supply vehicles. Permanent lunar bases could appear simultaneously with the 24-man laboratory in Earth orbit. Manned fly-by missions to Mars and Venus would usefully reconnoiter the two nearest planets to Earth in anticipation of manned landings later on.

There were plenty of ideas, but little direction in the long letter from Jim Webb. Not that the space agency had no plans for the post-Apollo era. As early as October, 1961, just five months after Kennedy's commitment, Emanuel Schnitzer of the Langley Research Center suggested an adaptation of Apollo hardware into an 'Apollo X' vehicle similar to conceptual designs in the period up to the May lunar landing decision. Readers will recall the basic Apollo design comprising a command centre module, for the crew, a propulsion module, for making orbital changes, and a mission module, or small laboratory for scientific activity in Earth orbit (see Chapter Four). Even Schnitzer's suggestion in October, 1961, to incorporate a small laboratory beneath the main Apollo vehicle, was long before the suggested use of Apollo as a mothership to Grumman's lunar lander. But it did restore the purpose of Apollo originally set out in 1959 as a general purpose successor to

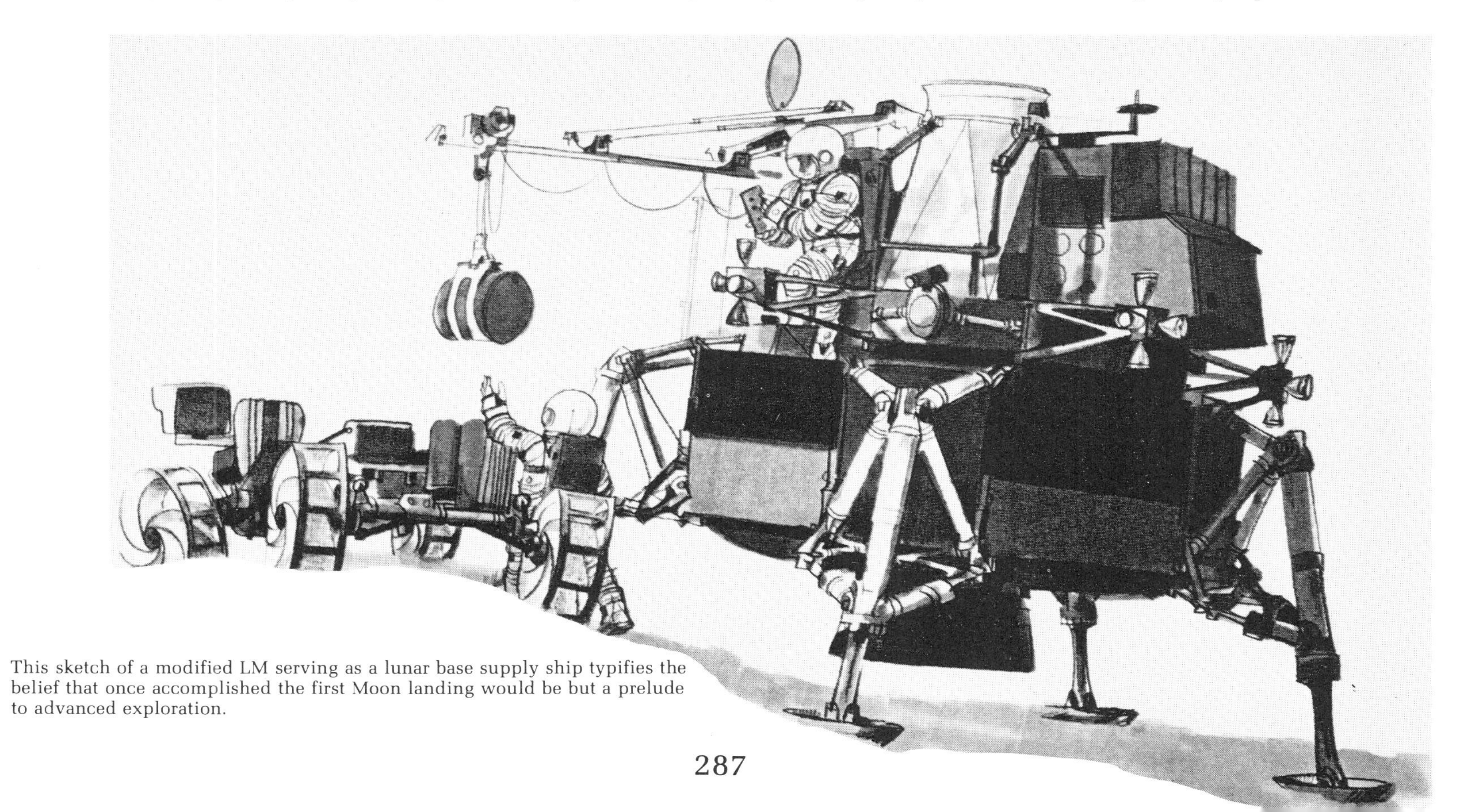

This sketch of a modified LM serving as a lunar base supply ship typifies the belief that once accomplished the first Moon landing would be but a prelude to advanced exploration.

Mercury for expanded Earth orbit and, possibly, circumlunar roles.

Throughout 1962 and early 1963, however, attention centered on the development of a completely new space capability, a manned facility in orbit about the Earth carrying up to 24 people which would phase in after the initial Moon landing. In June, 1963, the Manned Spacecraft Center let two contracts, one to Lockheed and one to Douglas, for studies on a space station designed for a useful life of five years with supply missions and crew changes at periodic intervals. Three weeks later, the Langley Research Center let contracts to Boeing and Douglas for detailed studies of their proposed Manned Orbital Research Laboratory (MORL), a four-man space station capable of supporting at least one continuous period of habitation for a full year. Elsewhere, NASA was bristling with ideas on how to capitalize on the Apollo investment by using elements of that program for extended operations in the 1970s.

Wernher von Braun suggested to Joseph Shea in December, 1963, that two Saturn Vs could be used to deliver an integrated taxi/shelter spacecraft to the surface of the Moon. Called by von Braun the Integrated Lunar Exploration System, it was never fully supported but did lead on to some intriguing adaptations of lunar hardware. Throughout NASA there was a feeling of having only just begun; unlike the intentions of the President nearly three years before, nobody was about to give up on the Moon after fighting for so long to get an active space program. While recognizing the limitations of basic Apollo hardware it was appreciated, nevertheless, that much could be done with the equipment. In fact there was a growing awareness of just how valuable Apollo could be for, as Joseph Shea commented in January, 1964, if the unmanned precursor missions like Lunar Orbiter and Surveyor failed to get adequate information about the surface of the Moon a CSM flight could effectively reconnoiter the lunar sphere to map its own targets. 'It might set our landing back six months to a year, but if the other programs don't work, we aren't dead,' he commented.

Yet it was to initiatives *beyond* the date of the first landing that future missions analysts concentrated during 1964. Prodded by the President, NASA was not about to pass up an opportunity to construct a viable program for the future. During that year, Apollo X was revitalized and made to form the projected basis for extended Earth orbit activity tentatively programmed for the 1969–71 period. By removing unnecessary equipment otherwise incorporated for Moon flights, by shaving off unnecessary heat shield material carried originally for return from lunar distance, and by eliminating one of the three crew members, the basic CSM could, if docked to a similarly trimmed lunar lander, perform around Earth valuable scientific tasks as a precursor space station.

But the degree of importance still placed in 1964 on the competitiveness vis-à-vis the Soviet Union can be seen in a comment from a leading US aerospace weekly of the day: 'The Russian achievement of three-man orbital flight (Voskhod I) robs two-man missions of much of their propaganda charm and may make it impossible to judge this question on a purely technical basis.' Apollo X was deemed applicable to both Earth and lunar orbit missions of about one month in duration and parallel development of an expanded capability of the lunar lander envisaged extended exploration of the Moon in 1971. Several variants of the basic Grumman lander were possible and the company spent considerable time and effort on contriving a list of viable options including landers carrying supplies, landers ferrying mobile vehicles, landers adapted as lunar cranes, etc. The true post-Apollo objectives, as seen in 1964, were to lead toward a manned Earth orbit space station by 1972–73 embracing year-long physiological experiments. As an add-on, the plan also anticipated manned Mars flights by the mid-1980s, but planning for that was in a very tentative stage.

The main thrust of Apollo X thinking, by now called Apollo Extension System studies, was the concurrent operation of a mini Earth-orbiting laboratory based on the CSM and a lunar orbit operation lasting up to one month, with surface exploration for two weeks at a time. This was the magnitude of post-Apollo studies when Jim Webb wrote President Johnson his uncommitted response on 20 May, 1964. By early 1965, however, the NASA boss had received back a Future Programs Task Group report put together by Francis R. Smith of the Langley Research Center. Compiled at Webb's behest to constitute a more definitive recommendation on NASA needs, the report was sent to the President on 16 February, 1965, in time for the Congressional hearing on the Fiscal Year 1966 budget. NASA had $48 million earmarked that year for Apollo Extension System studies. It would be necessary to keep the President fully aware of NASA thinking on future goals. There was certainly a great deal of concern within NASA that the hardware it had been provided with money to develop was a far cry from the scientific apparatus some members of the national community thought it to be.

In 1965, Dr. Donald F. Hornig, Director, Office of Science and Technology at NASA, said that 'although the return to Earth and subsequent scientific analyses of lunar materials will be an enormous step forward, the first Apollo will not provide an adequate opportunity either for exploration or for scientific study of the Moon. In view of the very large investment we will have made in the development of this system, it is clearly important that its capabilities be extended so that after the first expedition or two, astronauts can stay on the Moon a week or 10 days.' Late in 1964, the Space Science Board of the National Academy of Sciences responded to a request from Jim Webb for an evaluation of its view regarding the future pace of America's space program, and within that effort the magnitude of attention manned space flight should secure.

For several years the scientific community had been critical of too much emphasis on manned activity at the expense of generally less costly unmanned programs and while NASA enjoyed general approval for what it was trying to achieve, most scientists believed the agency to be too orientated toward engineering ventures or purely operational spectaculars. In presenting its recommendations for the post-Apollo era, the Board singled out the unmanned exploration of Mars by robot vehicles as the 'major goal past 1970.' Second, it felt that 'The Lunar program should be continued but subordinated to the Mars effort.' The third priority was scientific study of the Earth's environment from space and of astronomical phenomena from orbit. Fourth, but as 'a secondary – not a primary – goal,' was the manned Earth orbit program. On the question of manned Moon missions after the basic Apollo goal had been achieved, the Board was quite clear.

Believing that 'a post-Apollo manned lunar follow-up has much value,' it nevertheless considered that 'results from a high priority, unmanned planetary program (with emphasis on Mars) will contribute to expansion of knowledge over a broader field.' Dr. Gordon J. F. MacDonald from the Space Science Board put the scientific consensus with clarity and distinction when he testified before the Senate space committee that 'the bulk of the information that we will secure about the planets . . . can be obtained and should be obtained by unmanned vehicles . . . but I don't believe, and I want to make this quite clear, that we require man to answer the scientific questions that we are now able to raise as far as the planets are concerned.'

And from Senator Clinton P. Anderson, Chairman of the Senate space committee, a timely warning about further expensive manned ventures: 'We are involved in a great many things that we decided long ago with very little discussion. I mean the goal of President Kennedy to get a man on the Moon and return to Earth in this decade was not discussed with the congress or the committees very much before it was announced. I don't worry about that. I think it was a fine goal. I am glad we have gone ahead with it. But the expenditure of billions of dollars, a great many of them might frighten some people who didn't understand how it was to be done and for what purpose.'

As a final note of caution to politicians eager for national commitments, Dr. Lloyd V. Berkner, noted scientist and one time chairman of a panel set up by the President's Science Advisory Committee, said that 'It is quite clear that none of us would today recommend a manned expedition to Mars.' The final crunch on grand planetary expeditions came that same year, for in 1965 the first spacecraft to fly past Mars sent back 21 photographs by radio which revealed a very different world to the one projected by popular opinion. Although the resolution was low, and taken across a very small area of the planet, Mariner IV photographs showed craters and dust bowls strikingly reminiscent of Moon views. Although inconclusive from a scientific standpoint, it was the public image that counted most when considering massive multi-billion dollar ventures. And the public impression of Mars was a barren, hostile world with little atmosphere and a dried-up surface. By the end of 1965, all hope of moving beyond the Moon was dashed, although a Mars expedition would gain brief support

M&SS for Earth and lunar orbit work was dashed under the weight of funding crises. By October, following on the heels of one postponement that year already, the first orbiting workshop was put back to a 1970 launch. That month also, plans to develop an improved version of the Saturn IB by fitting four solid propellant Minuteman strap-on boosters to the first stage were cancelled. If developed, it would have raised lifting capacity from 18 to 22 tonnes, supporting improved Apollo CSMs carrying additional supplies in the Service Module. By early 1968 all hope of accommodating a lunar role for AAP had evaporated and when NASA officials came before Congress to plead for funds to accommodate Fiscal Year 1969, a period beginning 1 July, 1968, it was with a very much reduced Applications Program schedule.

Assuming an initial manned Moon landing in 1969, NASA said it hoped to send an S-IVB workshop aloft in 1970 as AAP-2. The first manned 28-day habitation would be launched as AAP-1. At its conclusion, AAP-3A would send another crew for 56 days aboard the workshop. A manned AAP-3 Apollo would rendezvous with AAP-4, the Apollo Telescope Mount launched by another Saturn IB, and carry it to the S-IVB for a third stay, also of 56 days. Next year, 1971, a further three manned CSM flights, AAP-5 to -7, would each provide 56-day stays in orbit, overlapping so as to maintain continuous operation of the Apollo Telescope Mount. Funds that would otherwise have gone on expanded lunar operations with long duration exploration trips, and on the second Earth orbiting workshop, were now to concentrate on the most ambitious AAP element: the Saturn V launched workshop, fitted out on the ground for habitation in space.

NASA believed it would still be possible to get the 'dry' workshop operational by 1972 and to design it for advanced operations involving many complex scientific activities. It was to support continuous operations for at least a year, leading directly by the middle of the decade to a fully fledged manned space station developed separately. Essentially, the budget cuts reduced AAP to six manned visits with the 'wet' workshop converted in orbit for habitable functions, and several, as yet tentative, flights to a dry workshop launched by Saturn V. In one statement before the Senate space committee during February, 1968, George Mueller passed cynical reflections on the prevailing situation: 'We recognize the Nation's need for imposing funding limitations on programs not directly involved with Vietnam. We also believe in the necessity to maintain our space capability on a sound basis at the minimum cost during this critical funding period. We cannot afford to cast aside all of our gains in space science and technology but neither can we afford to progress at the rate allowed by the capabilities developed.'

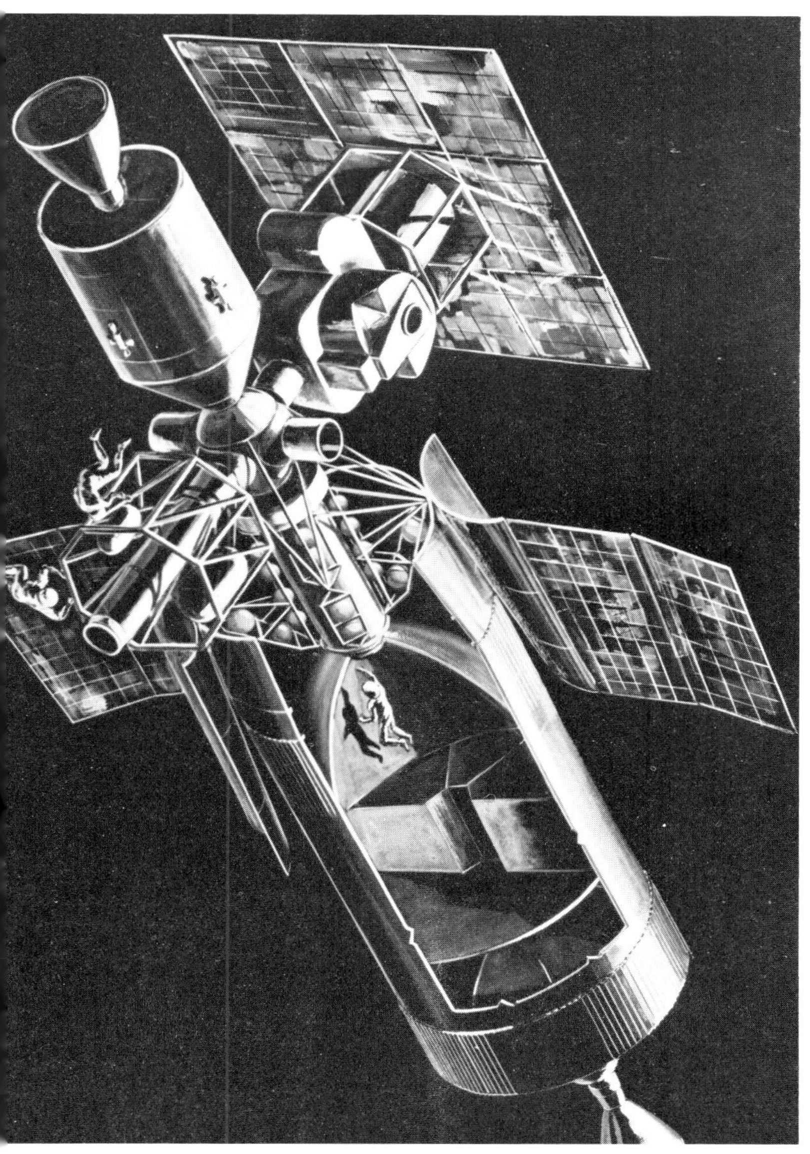

Seen here in artist's concept, the entire AAP cluster includes workshop, Airlock Module, Multiple Docking Adapter, Mapping & Survey System, Apollo Telescope Mount and Apollo spacecraft.

The considerably reduced AAP plan cut the anticipated annual launch rate of four Saturn IB and four Saturn V vehicles in half. Jim Webb and George Mueller had testified a year before, when defending the most ambitious AAP plan outlined earlier, that competence in managing and operating a safe and reliable space program depended on launch rates no less than four of each vehicle per year. Below that figure the experienced teams would be fragmented and reduced pace would lead to lethargy and disinterest among engineers and technicians. Accidents would be more likely to occur, and unit costs would rise out of proportion. But it cut no ice with Congress and was merely the final step in a series of cuts affecting the Apollo Applications Program. The Fiscal Year 1969 budget could support no more than the much reduced effort, and that was that.

Back in Fiscal Years 1965 and 1966 it had been no problem. NASA requested, and got, $14 million and $51.2 million respectively for Apollo X, Apollo Extension, and Apollo Applications planning. Desiring a start on AAP in FY 1967, NASA was unable to get the $264 million it wanted for that year, having to make do instead with a meager $80 million. For FY 1968, NASA asked the Budget Bureau for permission to request $626 million, hoping that it could finally get under way with actual AAP development. The Bureau cut the figure to $454.7 million and Congress slashed it further to a final authorized value of just $347.7 million. It was that double stroke that eliminated all hope of running the very ambitious schedule outlined earlier.

But the revised program detailed above was tailored to a funding profile which the agency had modest confidence in getting approval for, despite the fact that the Budget Bureau again slashed the permission for request – this time from a proposed $525.6 million to $439.6 million. When that latter figure was presented for Congressional sanction it was duly examined, priorities reviewed, and discussions held. At the end of the day, FY 1969 produced only $253.2 million for AAP, a mere 48% of the sum originally proposed to the Budget Bureau. Even the limited six-flight program for the wet workshop concept was threatened.

By mid-1968, when Congressional action on the FY 1969 budget came through, the future for manned space operations looked decidedly gloomy. It had been nearly two years since the last manned Gemini mission, three astronauts had lost their lives in a pad fire, and the first manned Apollo vehicle had yet to lift free from the Kennedy Space Center. It was a frustration mirrored by the general reduction in NASA budgets. In FY 1965 the space agency received funds totalling more than $5,200 million. By FY 1969 NASA's overall budget was down to $4,000 million. In real spending terms it was an even worse catastrophe, because numerical money values take no account of inflation. When expressed as a function of the Federal budget in those years, taking cognizance of inflation and the increased wealth of the nation, the 1969 budget was only 50% that of 1965.

It was very clear what was happening, and nobody could do a thing about it. With peak funding levels for big engineering projects like Apollo now out of the way, the reduced financial burden was not being absorbed by investment in future projects. The downward trend of the development curve was going unexploited, because a war in Vietnam and a disastrously inept management of domestic policies once lauded by Congress were starving technology of the money it needed to keep America in front. Devoid of the will to retain the tremendous advantage in space capability held by NASA, Congress apathetically turned its back on grand objectives for the future. Not because of the fire. Not because of Vietnam. Not because of campus riots. But because of all three.

CHAPTER SEVENTEEN

Brighter Horizons

No sooner had NASA received the recommendations of the Apollo Review Board than it put together plans for flight operations scheduled to prepare and test all the many elements of the program. For much of 1967 the possibility of landing two men on the Moon by the end of the decade waxed and waned. A considerable amount of work had to be done to fireproof the Command Module, and to ensure that safety procedures and contingency plans for many other activities were similarly protected from the sort of accident that took the lives of Grissom, White and Chaffee. Nevertheless, by year's end it was apparent that, if everything went well from this point on, NASA might just make it to the Moon within the assigned period after all. There were just two short years to get everything ready. And one of the most questionable pieces of hardware still left untested was the gargantuan Saturn V, a rocket so big it could throw the equivalent of 50 limousines to the Moon in a single shot.

But delays in the preparation of the first launcher, AS-501, put back the flight by several months and the flight assignments set by Mueller in May, 1967, were re-written to accommodate fewer flights with the massive booster. The new schedule accommodated only one Saturn V flight in 1967, an unmanned 'all-up system' test involving three live rocket stages, a dummy lunar lander in the adapter and a Block I CSM fitted with the new unified hatch, the development of which had been accelerated after the fire. This was Mueller's radical philosophy stretched to its logical, but frightening, end. Where von Braun's building block approach to rocket flight had required ten Saturn I development flights, each more ambitious than its predecessor, Mueller's revolutionary approach to qualification assembled the entire structure and tested succeeding elements in turn on the same mission.

Flight operations for 1968 envisaged, under the new schedule, an unmanned Saturn IB flight (AS-204) to place in Earth orbit the first production model of the Grumman lunar lander. Without heat shield and designed only to operate in space, it could not be returned to Earth but would provide opportunity for engineers to put it through its paces, firing the engines, checking the guidance platform, etc. Then, two more unmanned Saturn V flights (AS-502, -503) would qualify the stack for manned operations. A second unmanned lunar lander flight test (AS-206) would follow the big Saturns, with the first manned Apollo mission (AS-205) launched during the third quarter of 1968. All hope of getting Apollo off the pad by the end of 1967 evaporated quickly when it became clear just how much work remained to be done to improve safety. The last flight planned for 1968 would be the first manned Saturn V mission (AS-504), where Apollo CSM *and* lunar lander would be placed in Earth orbit for a rehearsal of operations involving both spacecraft.

Mueller planned five Saturn V flights for 1969 (AS-505 to 509) to practice first in Earth orbit and then around the Moon every essential activity without actually descending to the surface. If all went well, AS-509, the last flight in 1969, might descend to the Moon's grey dust. At the beginning of serious Apollo planning, 15 Saturn Vs had been considered the minimum necessary to get two men on to the Moon. Budget cuts up to the end of 1967 deferred two Saturn Vs to 1970. A further two were deferred by delays resulting from the fire in the first schedule change, and now two more were put back. In all, six of the big Saturns would fly after 1969, but Mueller still thought he could pull off a landing with the available hardware. It really all hinged on how well the big Saturn performed during the three unmanned test flights. It was a considerable step up in size and power from anything launched anywhere on Earth before, and both second and third stages employed the high-energy propellants liquid hydrogen and liquid oxygen that, although very desirable for the improvement in performance they afforded, could prove temperamental in use.

Because the fire dramatically changed schedules and flight assignments, the inevitable hiatus in operations was used by Headquarters to change the system of nomenclature for Apollo. Whereas prior to the fire flights would be recognized by the serial number of their respective Saturn launch vehicles (the first manned mission being known as AS-204 because it was the fourth Saturn IB flight), flights after the fire were retrospectively designated in honor of the Grissom, White and Chaffee mission. Under this system, AS-204 became Apollo 1, the Saturn IB flight of 26 February, 1966, became Apollo 2, while that of August 25 became Apollo 3. Thus it was that the numerical start to the Apollo sequence was based on the memory of the three dead astronauts. It also had more logical significance. Three production Apollo spacecraft had been used in flight tests (spacecraft 2 on 20 January, spacecraft 9 on 26 February, and spacecraft 11 on 25 August 1966), retaining the significance of three Apollo test operations. Whichever way it was determined – emotionally or logically – the first unmanned Saturn V flight would be Apollo 4, with spacecraft 17 as the payload, although the Service Module was from spacecraft 20 due to the original hardware being damaged by an explosion. It was an ambitious mission in every dimension but equally as appropriate was the operating experience it would provide for the significantly improved global tracking facilities known as the Manned Space Flight Network – 'misfin' in name but very different from the communications and tracking operation set up for Gemini. Even as preparations at the Kennedy Space Center moved toward the first Saturn V mission, technicians and engineers were building up the MSFN facilities around the world. At completion, eleven Earth stations would support 9.14 meter diameter dish antennae, three would provide 25.9 meter diameter dish antennae, three tracking ships would each provide a 9.14 meter antenna, two ships would contribute 3.66 meter diameter dishes, and eight ARIA (Apollo Range Instrumentation Aircraft) would move around the world as required, filling in vital gaps for telemetry coverage and voice communications.

Although much of Apollo's operation would be performed in Earth orbit, during tests, rehearsals, and checkout, the network was required to support for the first time, deep space tracking and communication. S-band frequencies were selected early in the Apollo program for telemetry and voice relay, with VHF for Earth orbit use or as a back-up. But there were many operational modes to confuse the comparatively simple methods developed in the Mercury and Gemini programs. Two manned vehicles would be talking to each other for short periods – one in Moon orbit, the other on the surface – each would wish to talk to Earth stations, two astronauts moving across the Moon's surface would be in communication with each other, and would also have an open line to Houston. Tying together the many Earth stations, the NASCOM facility based at the Goddard Space Flight Center would have a far more demanding role than it performed for Gemini.

Through circuit lines equal in length to four times the Earth-Moon distance, network controllers would keep the data flowing from respective ground stations to the Univac 642 and more powerful 494 computers employed to switch and sort the millions of information packages streaming to Earth from the Apollo space vehicles. In the completed system, 48 Sperry Rand Univac 1230 computers situated at the remote tracking sites would relay information and data to and from the spacecraft. In addition, 33 Univac 1218 computers would be employed pointing and positioning the radar dish antennae on instructions from the Manned Spacecraft Center. Six massive Univac 494 switching systems were located at Goddard and at Houston, waiting to read messages sent from the 1230s and the 1218s distributed around the world for checking, filtering, switching, and distributing. Each could store 2 million bits of information in a core memory, or expand this capacity to 46 million bits with a FH-880 fast storage drum.

Capable of recalling any one of the 46 million bits in 17 thousandths of a second, the 494 would ingest data at the rate of 40,800 bits per second from remote sites around the world. Seven Univac 1108 computers would be used to tackle complex calculations to produce detailed engineering data or scientific information, and seven 418 computers at Goddard and Houston would provide routine switching functions or be employed for training. The amount of information to be handled by the Apollo network was in sharp contrast to that for Mercury. In equivalent terms the primitive one-man capsule generated information each second equal to the contents of a single printed page; Apollo would generate each second information equal to that contained in a novel. In real terms, network traffic for a typical Apollo mission would equal the words contained within 500 sets of Encyclopædia Britannica.

No longer could routing rely on ground lines. Satellites would be employed to leap-frog oceans and connect tracking sites to Goddard within a fraction of a second. In every dimension, Apollo 4 was switching on the lights. It would be the first real test for 'misfin' – Apollo style . Although primary attention focused on the flight hardware, the preparation of ground facilities and support equipment was in every way as essential as the rockets and spacecraft central to the Moon effort. In truth, without complex software systems the entire operation would be impossible. And it was Apollo 4's function to prove not only that the hardware worked but that the tracking stations, computers and ground technicians could knit together as an integrated whole.

But Saturn V was an eye-catcher at the Cape during 1967; only once before had workers and visitors witnessed the impressive sight of a Moon rocket roll-out, and then only with the facilities vehicle the previous year. Delays in getting the stack assembled for roll-out from the Vehicle Assembly Building to pad 39A was due in part to the searching checks made of spacecraft 17 after the Apollo fire of 27 January. All the Saturn hardware had been delivered, and Debus hoped to get AS-501 into space sometime during May, but the inspection of spacecraft 17 on top of the three stages revealed many wiring errors similar to those suspected of having caused the fire. With time hanging due to work still being performed on the Saturn's S-II second stage, the Command and Service Modules were removed and taken to the operations and checkout building.

Saturn V, the world's largest launch vehicle stands ready on the Florida beach for its first flight in November, 1967.

By March the wiring deficiencies were found to be so numerous that the repair-as-find process had to be stopped and a full review conducted of the problem. More than 1,400 errors had been located. Over the next few weeks it became painfully apparent how bad the electrical distribution system was on spacecraft 17 and hopes for an early launch were dashed. By late June the spacecraft was back on top of its launcher and preparation of the S-II second stage was well along. But checking out a vehicle of the dimension and complexity of a Saturn V/Apollo stack was never easy; on the first launcher it was an excruciating nightmare. By late August all was ready for roll-out and on the 26th the stack moved ponderously through the open doors of the VAB into Sunlight for the first time. It was as appealing a sight as its predecessor. But this one was for real – a live Saturn V supporting a production line spacecraft for a flight further from the planet than any previous vehicle designed for manned flight had journeyed.

It was an incredibly important mission, one that could restore or dash the hope for a manned Moon landing by the end of the decade. One complete Saturn V failure, particularly with the first, would almost inevitably put the entire program back into the 1970s. There was no more slack in the schedule and a re-design of even one key Apollo element was out of the question. But more than Apollo program considerations were the implications of success and failure for the political confidence in NASA and the entire space program. It was the first major launch since the 27 January fire. A failure now, albeit with an unmanned system, could sacrifice ground the space administration would never get back. Rain, wind, minor technical problems and a determination not to fly until everything was as ready as it would ever be, delayed the launch to early November.

Preparation of AS-501 showed it would be a difficult and lengthy exercise to set up a smooth checkout and readiness schedule with the massive configuration. This was no Redstone, Atlas or Titan. With several million parts, each Saturn V/Apollo was a spaceship in its own right, a machine rather than a projectile. The countdown operation took six days, with a final intensive sequence beginning 24 hours prior to the planned lift-off time. From all across the nation and several countries around the world, people flocked to the Cape, anxious to get a glimpse of the black and white rocket before it thundered into space. Turnpikes and highways were crammed with vehicles of every shape and size converging on the Florida peninsula as they had for Al Shepard, for John Glenn, and for ten Gemini missions. It was an event more to some than others, and the Kennedy Space Center guest list read like an inventory of great names in the race for space.

Von Braun was there, witnessing 30 years after the days he first pondered problems of Moon flight the final hours to the launch of his most ambitious creation. George Mueller looked on, his thoughts turned to the future uses of this giant launcher. Kurt Debus, the one man who more than any other had been responsible for turning marshy flats into spaceports for the planets, was there along with Deke Slayton, the grounded astronaut whose lot it was to train and condition other men for the 'big ride.' General Phillips, concerned to bring together all the many industrial segments of the massive Apollo contract, was earnestly wishing success for this maiden flight.

As the hours ticked away in the dark morning skies of Thursday, 9 November, managers and administrators gathered in Firing Room 1 of the Launch Control Center. Situated alongside the Vehicle Assembly Building it afforded visual access to the concrete launch pad nearly 5 km away. Launch was scheduled for 7:00 am and as the Sun rose beyond the man-made complex, casting silver strands across the Atlantic, it illuminated the beach-side scene. More than 700 invited guests moved, chatting excitedly, around the confined areas to which they had been assigned. On the sand tens of thousands stood, uninvited but welcome nonetheless, willingly exposed to the power of Saturn V. It was now a matter of machines talking to machines, for the volume and rapidity of questions and answers that now moved between computers in the pad and launch equipment was far beyond a level at which the human brain could react.

From throughout the 36-storey stack came information on

more than 4,000 items, many of them read several times a second, computers recording in the flick of an eyelid several ten thousand statements about the condition of this gargantuan structure. Weighing now in the fully fuelled condition more than 2,700 tonnes, AS-501 creaked and groaned, the sound of metals adjusting and settling under the enormous weight of propellant in the three rocket stages. The first two stages, accounting for 60% the height and 58% the weight of the complete stack, had never flown before and even the third stage, the S-IVB used for Saturn IB, was a different version intended to re-ignite in Earth orbit.

At T-3 min 10 sec the sequence switched to a fully automatic mode. Nobody would press buttons, and there would be no levers to remotely control the flow of propellants down into the five first-stage engines. At lift-off, nine huge swing-arms bridging the Saturn to the umbilical tower alongside would release the rocket and pivot back to free the ascent. Restrained to the mobile launch platform by supports at the base of the first stage, the Saturn V would only leave the pad if computers agreed it was safe to do so. The ignition command was sent to AS-501 at T-8.9 sec, and little more than a second later the five engines burst into life.

As viewed from the VIP stand or the press site holding 500 newsmen, the base of Saturn boiled with fire. Billowing clouds of red and yellow burst out from under the first stage, rapidly followed by two elongated troughs of flame panning out along the pad, channeled away from the engines by a flame trench designed for this purpose. There was no sound. Saturn V was too far away for the sound waves to reach the nearest humans before the massive machine lifted from the pad. That event came six seconds after the engines spewed forth their combusted products, an interminably long period when everybody stood and watched, many literally holding their breath. Then, suddenly, with mushroom clouds of smoke building up in two elongated colums stretching at an angle either side of the rocket, the swing arms could be seen moving back and Saturn lifted slowly, very slowly it seemed, into the air, accelerating on a pillar of flame that went down to the launch pad exciting shock waves of fire as the energy bounced off concrete and steel.

Thundering on to the pad itself, a veritable deluge of water, cooling the mass of steel and pipe heated by the rocket's hot blast. About ten seconds after lift-off the base of the Saturn cleared the top of the umbilical tower 137 meters in height. The brightness was unbelievable, greater than the morning Sun, and the column of fire upon which the rocket stood seemed incalculable in length. Tongues of flame licked down into the bowels of the concrete pad, pouring out the energy needed to sever the gravitational bonds of Earth. And then the sound came – first in tremors through the ground that built up like a million Vikings thundering toward Valhalla, and then in a physically moving vibration magnified by the concrete stands. Through the air between man and machine the shock waves of unimpeded noise broke upon the unsuspecting crowds. It arrived like thunder and built rapidly to a pitch akin to the expected sound of a nuclear detonation – in fact, in all the world, the noise was second only to a Hydrogen Bomb.

Compressed air battered on the roofs of broadcasting vans near the press stand, reporters clapped their hands to deafened ears, some bowed over close to the ground, trying to escape the rising volume of noise – noise so loud that it pressed against the human rib cage and seemed to move right inside the chest. From the CBS news van, the usual calm, controlled, dialogue of Walter Cronkite broke down as the very structure of the vehicle threatened to collapse under the pressure of the thunder, pounding like fists on the roof: 'The building is shaking. . . . Boy it's terrific, the building's shaking. This big glass window is shaking as we're holding it with out hands . . . LOOK AT THAT ROCKET GO!'

Engineers and technicians put their hands upon the thin walls that for several seconds seemed helpless under the barrage of moving air waves assaulting everything with a physically present force. The sight and sound was an indescribable experience that left some men weeping under the sheer magnitude of so much power. Others screamed, their voices unheard by the metallic crackle of the big Saturn engines. For a decade von Braun coveted the notion of F-1 engines in his Saturn boosters, and now it was happening. The whole Earth seemed to shake and tremble at the pulverized debris of shattered calm. To the spectators who had watched earlier manned vehicles, this was something beyond anticipation. It was in truth a Wagnerian overture to a symphonic exodus.

In 150 seconds of flight the first stage completed its task, accelerating the booster to a height of 63 km and a speed of 9,695 km/hr. The center engine had been shut down 15 seconds before the remaining four to lessen the shock of thrust cut-off. In the $2\frac{1}{2}$ minutes since leaving complex 39, Saturn V consumed more propellant than would be used by an average motorist in more than 1,000 years of normal driving! Less than a second after shutdown, the first stage was separated from the second as eight forward-pointing retro-rockets burned briefly from the four engine fairings at the base to reduce forward momentum and prevent the first stage shunting into the second. A second after that the five J-2 engines ignited on the second stage to the delight of technicians and pressmen alike as cheers and clapping followed the report of that event from public affairs officer Paul Haney.

At separation eight ullage rockets attached to the cylindrical adapter connecting first and second stages fired to impart acceleration to the S-II and settle propellants in the tanks. They burned for four seconds, by which time the S-II stage was alight and firing. Burning hydrogen and oxygen propellants with a thrust equal to three Atlas rockets, the stage continued to fire for 6 min 8 sec during which time it accelerated the payload to a height of nearly 190 km and a speed of more than 24,600 km/hr. Just 29 seconds into the long burn, the cylindrical adapter was jettisoned, sliding down past the five J-2 engines, and $5\frac{1}{2}$ seconds after that the Apollo launch escape tower fired itself away from spacecraft 17. But now the S-II had shut down, 8 min 40 sec into the flight. Less than a second later the second stage separated from the third, braked for $1\frac{1}{2}$ sec by four forward facing retro-rockets mounted in the conical adapter which came away with the stage. Almost immediately the third, S-IVB, stage ignited, its J-2 preceded by a four second burst from two ullage motors. The S-IVB continued to burn for 145 seconds, the ullage motor cases having been jettisoned 12 seconds into the burn, accelerating the combination to a speed of more than 28,000 km/hr. Apollo 4 was in orbit, a nearly circular path 190 km above the Earth. From this point on, the flight would simulate two important phases of a lunar landing mission: acceleration from Earth orbit to an elliptical path close to escape speed, and re-entry of the Command Module at lunar trajectory return speed.

The first phase went well as, at an elapsed time of 3 hr 11 min 27 sec, the S-IVB lit up for the second time and fired for 5 min, raising the speed by 5,630 km/hr. This placed the third stage and spacecraft in an elliptical path with a high point of 17,209 km. Had the third stage burned a little longer it would have sent Apollo 4 to escape velocity, but for this mission flight planners wanted the spacecraft to fall back to Earth during daylight in the Pacific so the burn was short of the duration required to reach escape speed. Ten minutes after stage shut-down, the Apollo spacecraft separated from the stage adapter and orientated itself in space for a short burn with the Service Propulsion System engine. That burn, which lasted 16 seconds, boosted apogee to 18,092 km. Immediately thereafter, with the spacecraft climbing toward its new apogee, Apollo turned to place its hatch in full Sunlight so that valuable thermal data could be obtained on the new unified design.

Apollo 4 reached the high point of its trajectory less than six hours into the mission and then began the long plunge back toward the atmosphere. On the course set by Apollo's first SPS burn the flight path would intersect the atmosphere at an angle of 8.75°, incurring a force of 16 g during deceleration. To approximate more closely the angle and re-entry speed of a spacecraft returning from the Moon, that had to be modified by a second SPS burn to speed up the vehicle and flatten the path. The burn came at an elapsed time of about 8 hr 10 min, nine minutes prior to re-entry, and lasted $4\frac{1}{2}$ minutes. It was slightly longer than planned, resulting in a speed of 40,100 km/hr versus 39,867 km/hr, reducing deceleration from an anticipated 8.3 g to 7.3 g, but increasing the thermal load by 6%.

During re-entry, the Command Module, separated now from the Service Module, flew a 'roller-coaster' path similar to that of the Project Fire cone used for tests in 1964 and 1965. Dipping first into the atmosphere to reduce speed, the spacecraft used its lift vector to rise again, dissipate thermal energy, and sink back down into the lower regions of the atmosphere for the final deceleration. On the second entry spike, deceleration was 4 g versus a planned 4.5 g. This was the type of re-entry manned vehicles would fly when returning from the Moon and greatly alleviated the heat

Standing before a full-size mock-up of an early LM configuration, President Kennedy holds a model of the conical Command Module.

shield from the severest environment. With an external temperature approaching 2,760°c, the Command Module interior climbed only 6°, and throughout the mission never once exceeded 21.1°c.

As the spacecraft fell effortlessly under the open canopies of three Ventura parachutes, it was obvious Apollo 4 had been a resounding success and, even before splashdown in the mid-Pacific, program managers were talking freely about eliminating a third unmanned Saturn V flight if the second, planned for early 1968, was equally successful. That would accelerate the pace at which manned operations with the big Moon booster could begin. It was a welcome triumph, a fitting note on which to end a year of tragedy. But Saturn V, having proved it worked, now gave way to concern about the availability of the all-important lunar lander. The Moon rocket was the propulsion necessary to leave Earth, Apollo Command and Service Modules were the taxis to lunar orbit, but descent to the surface would be made in the weird looking contraption developed by Grumman. And as good progress emerged on fire-proofing the CSM, the lander was increasingly seen as the pacing item. It had to be ready for manned Saturn V flights which, on the performance of AS-501, were probably nearer than originally thought. It would be uneconomic to use the big boosters for CSM flights only, and in any event there were many operations that would have to be tried and tested prior to the first Moon landing.

The lander had been an unusual subject for Grumman to pursue. With a rich history of naval aircraft behind them, the company had little experience in space hardware. Corporate operations began in humble fashion shortly after the economic depression. In 1930, Grumman Aircraft Engineering opened for business with 15 employees but soon found itself developing, designing and building amphibious aeroplanes, hydrofoil boats, and sea-planes. During World War II, Grumman produced 17,000 combat aircraft including the famous Hellcat, Wildcat and Avenger series, before moving to commercial aircraft post-war. In 1960, Grumman successful competed for a NASA contract to build the Orbiting Astronomy Observatory, a series of satellites designed to put astronomical equipment in Earth orbit.

Grumman was keen to expand but unsuccessfully attempted to get the Mercury contract in 1958. Two years later the company presented NASA with results of a lengthy study compiled with ITT cooperation on a proposed Apollo concept. Again they were unsuccessful. A year later, Grumman responded to an invitation for bidders on the prime Apollo contract, when the spacecraft was seen as the landing vehicle itself. It again failed to win first place. Sensing the shift during 1962 toward a Lunar Orbit Rendezvous mode of operation, where Apollo would only go to Moon orbit so that a separate vehicle could descend to the surface, Grumman put considerable finance into in-house studies aimed at completing a winning proposal for the lander it believed the space agency would shortly request.

In early November, as recorded in Chapter Nine, Grumman won the contract to build the lunar lander, a device that nobody really knew how to design. Used only in the vacuum of space, it would be the first spacecraft equipped to carry a crew but not designed for re-entry. Its function was to operate at lunar distance as a taxi from Apollo to the lunar surface and back to Moon orbit. During launch it would be carried in the location Apollo's landing section would have been attached to had that vehicle been developed for a Direct Ascent or Earth Orbit Rendezvous mode. Protected from the atmosphere on the way up, it would have little need of the aerodynamic shapes so necessary for re-entry. It would be a totally new vehicle for an unparalleled role, and as such would look literally like nothing on Earth.

During conceptual evolution in 1961 and early 1962, Manned Spacecraft Center engineers referred to the lander as the Lunar Excursion Vehicle (LEV), but in April of 1962 the vehicle was re-designated the Lunar Excursion Module (LEM), bringing it into the line with the Apollo designation. Affectionately called the 'lem,' it was to lose the Excursion part of its name in June, 1966, because NASA thought that implied a mobility which it did not possess. The latter year was, as referenced in the previous chapter, a period when lunar mobility was being requested as an Apollo Applications Program. The agency wanted to make sure Congress knew the limitations of the re-named Lunar Module, or LM; despite the change, it retained its 'lem' pronunciation.

One of the first areas of concern for Grumman's vice-president of space programs, Joseph G. Gavin, Jr., was to identify the major elements of the LM. Under the guidance of design engineer Thomas J. Kelly, Grumman recognized quite early in its in-house study period an inherent advantage in building for one environment. Unlike McDonnell who built Mercury to fly in space and the atmosphere, Grumman could concentrate on the perfect layout for a true space vehicle – one that would never have to fly through the atmosphere. Grumman had been looking at LM requirements since shortly after they lost the Mercury bid and were ready in late 1962 with a baseline design anticipated as a starting point from which the definitive article would evolve. It was just as well that the company had this attitude from the beginning; the final LM configuration would be a very different 'bug' from the one proposed in 1962.

An early conundrum concerned the nature of the LM's two stages, called Ascent and Descent Stages respectively. The lower, or Descent, Stage would support the legs necessary for standing on the lunar surface in a vertical attitude; the Ascent Stage would contain the pressurized crew compartment where two astronauts would live during descent and ascent. The problem centered on whether to use one main engine for landing and lift-off, or two separate engines, one in each Stage. With only a single engine employed, it would be attached to the Ascent Stage and protrude through an opening in the Descent Stage. If two were carried, the Descent engine would fire to lower the LM to the Moon while the Ascent engine would be used to fire off the Descent Stage and carry the crew back into orbit.

Surprisingly, weight studies revealed little difference in either concept. There was great advantage, however, in having a separate Ascent Stage engine: if the Descent engine failed as the LM was lowering itself to the surface the crew could immediately fire up the Ascent engine and lift off the Descent Stage. Without a second engine they would be doomed to destruction on impact with the surface.

The initial LM design was known as the M-1 configuration, set out in mock-up form early in 1963, 6.1 meters tall and a mere 3 meters in diameter. The Descent Stage supported six propellant

This 1962 model of the Grumman LM had five legs and a bulbous front section

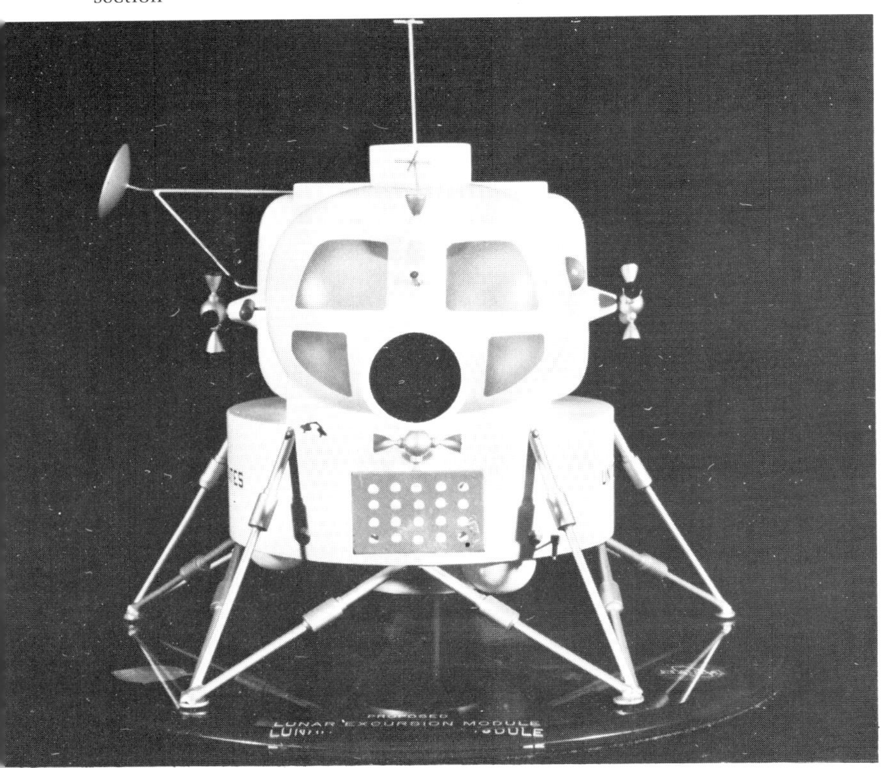

By 1963 the LM was defined as a four-legged vehicle with aluminium front cabin face and two recessed windows.

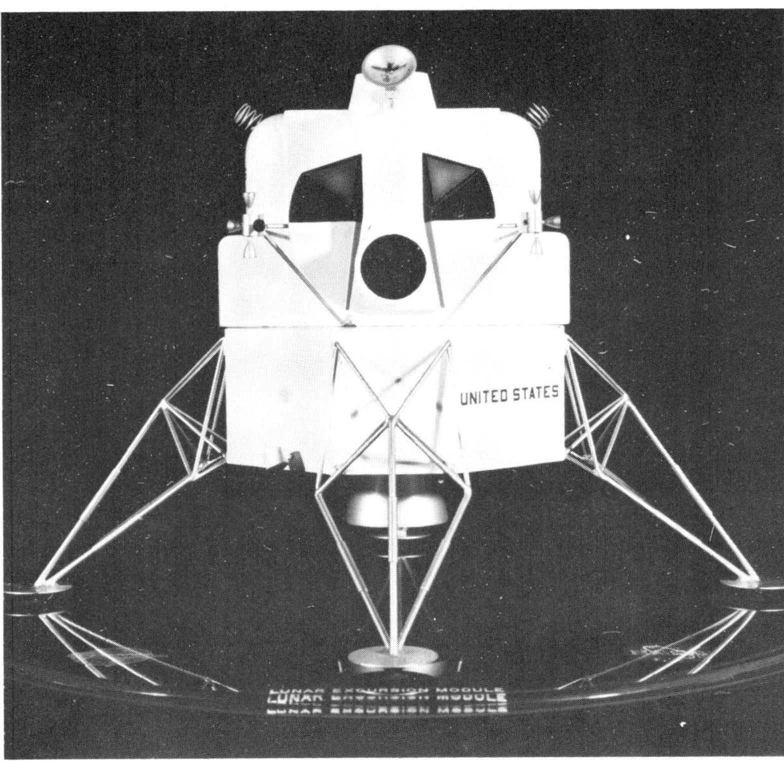

tanks feeding a single 4½ tonne thrust Descent engine, set in the middle of the tanks. On the exterior were five fixed landing legs each terminating in a circular pad. The Ascent Stage was essentially a glass bubble with a circular docking port on top and a second on the front. Four sets of four attitude control thrusters were positioned at front, rear, and either side. In the center was a throttleable Ascent engine generating a maximum thrust of 1.58 tonnes. The top port was to be used during initial docking, when the Apollo spacecraft would pull it from the adapter on top of the Saturn V third stage; the forward port was to be used for docking with the Command Module when it returned to Moon orbit from the lunar surface. In all, the M-1 configuration weighed less than 10.9 tonnes.

But it was unequal to the tasks demanded. Unprotected from solar radiation, the glass bubble would discolor, the interior would heat up to excessive levels, and the weight of the structure was too much for allowed limits. As early as February, 1963, Grumman recommended modification to the landing gear, pointing out the inherent instability in a five-leg design, By 17 April the Manned Spacecraft Center accepted Grumman's argument and agreed to a four-legged landing gear with a capacity for folding under its support beams, thus enabling the size of the Descent Stage to grow within the allowable volume of the Saturn V adapter. Also, the number of propellant tanks in the Descent Stage was reduced from six to four, and the Ascent Stage crew cabin was changed in favour of a cylindrical structure about 2.34 meters in diameter.

By July, 1963, Grumman sought and obtained permission to change the frontal design to one incorporating two triangular windows set in a steel face, and to delete seats in favour of a restraint system which would support the crew in a standing position. As much as 27 kg could be saved in this way and it brought the eyes of the crew closer to the triangular windows, enhancing their downward visibility. Designed basically to support 24 hr operations on the lunar surface, necessity to stand for the duration of the excursion would be little hardship in a 1/6th gravity environment.

Before the end of the year, Grumman suggested reducing the number of Ascent Stage main propellant tanks from four to two, a move designed to save an additional 45 kg. Whereas before the left and right sides of the Ascent Stage had a symmetrical appearance, adoption of single oxidizer and fuel tanks introduced a distinctly asymmetric configuration. Because of the different masses of oxidizer and fuel, the latter was positioned further out than the former due to the greater mass of the oxidizer. Only in this way could designers preserve a satisfactory center of mass. The modified configuration also re-located the four attitude control thruster quads from positions at the sides, fore and aft, to locations either side at the front and either side at the rear. This had the advantage of removing the thrusters from positions close to the forward docking port, and from locations directly above the landing gear struts.

By March, 1964, Grumman had completed its TM-1 mock-up incorporating all the recent changes, including the new landing gear deployment design, and among those present at the Review Board meeting were Maxime Faget from MSC, Deke Slayton on behalf of the astronaut corps, and Grumman's Tom Kelly and M. Carbee. From this meeting came an idea aimed at eliminating the forward docking port – the one which was to have been used to dock with the Command Module after reaching lunar orbit from the surface. Astronauts Ed White and Elliot See were among those who felt it was a cumbersome and unnecessary practice to have two docking ports on one vehicle.

The forward port would, by necessity, serve as an egress hatch from the Ascent Stage to the lunar surface. If it also had to incorporate a docking interface, optimum design would be compromised between the two requirements. Instead, a rectangular window placed in the roof of the crew cabin would afford a view upward so that the top docking port could be used to link up with Apollo. Moreover, it would reduce the number of docking aids required and eliminate a docking cone for the forward port. MSC formally recommended this change on 24 April and on 22 May, 1964, the Apollo Spacecraft Program Office authorized Grumman to eliminate the forward port and use that location for a hatch optimized around the requirements for getting on to the lunar surface. The only unavoidable penalty for this innovative change required the pilot conducting the docking to look directly upwards through the rectangular window. That was considered a minor inconvenience.

The last change required to transform the Lunar Module into the shape it eventually came to have was the addition, at a recommendation laid in June, 1964, to place a short platform between the forward egress hatch and the top of the forward landing gear. This would enable an astronaut to back out along the 'front porch' (as it became known) and down a ladder fixed to the forward leg, on to the surface of the Moon. Several other changes were in the offing but those were primarily confined to internal systems design. It had taken just 18 months to reach the definitive external shape. Inspection of the new M-5 configuration got under way at the Grumman Bethpage facility on 5 October, 1964, where modifications reflecting all the latest amendments to specification and design were reviewed by government and industry officials.

On the 8th, the Review Board heard requests for changes resulting from the detailed examination of the vehicle and NASA formally requested Grumman to implement those approved

By 1965 the LM had given up its forward docking port for a square egress hatch through which astronauts would leave the vehicle on the Moon.

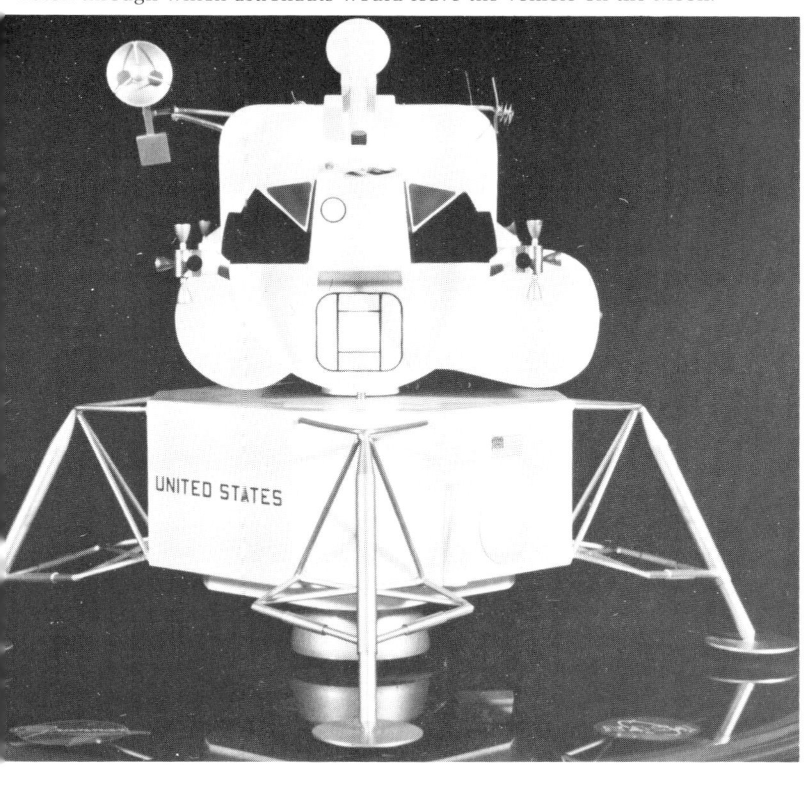

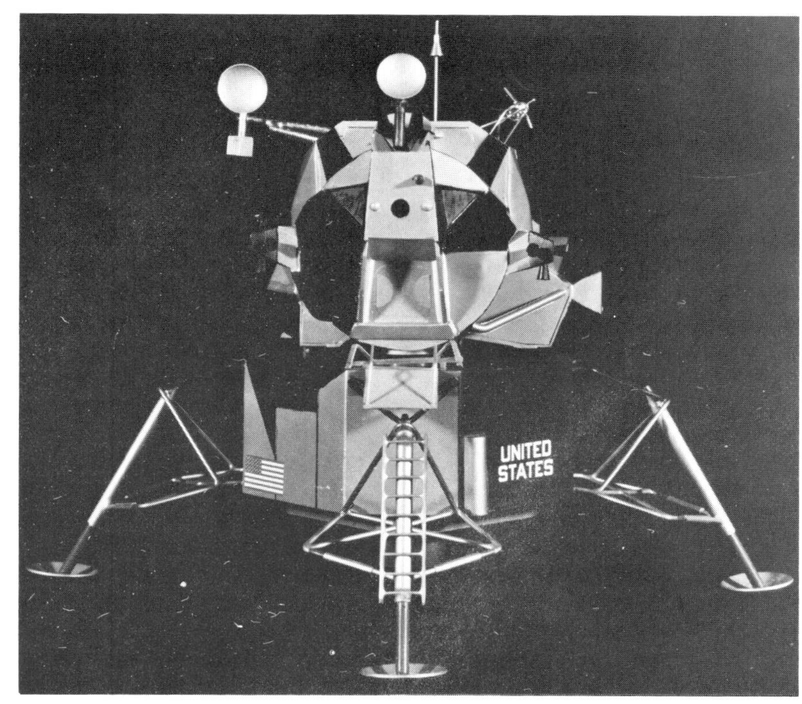

The definitive LM.

Lunar Module Ascent Stage and Descent Stage structure.

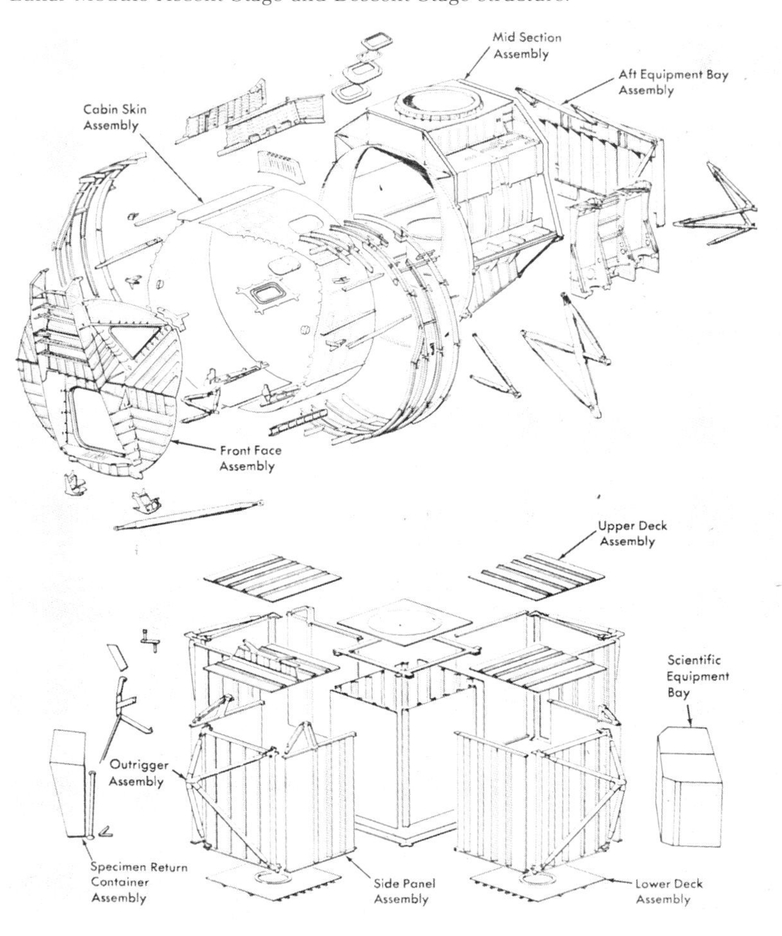

alterations arising from inspection of the mock-up. The new configuration was cleared for tooling and fabrication, marking the end of preliminary design and the beginning of development. But if success seemed easy for Grumman engineers seeking a match for NASA requirements, the pursuit of a reliable and safe lander went hand in hand with escalating weight problems. In December, 1962, the LM control weight had been 11.6 tonnes. By December, 1964, that value had risen by 15% to 13.3 tones; within twelve months the control weight would have risen to 14.5 tonnes.

The dramatic increase in Lunar Module weight was a very necessary product of improved design concepts at a systems level, but one acceptable at vehicle level only because of improvements in the projected performance of the Saturn V launcher. As the big Moon rocket progressed from initial design to major development tight constraints set by Huntsville were relaxed, permitting an increase in the total payload prediction from 40.8 tonnes in December, 1962, to 43.1 tonnes three years later. Moreover, because Apollo CSM design was essentially frozen a full 18 months before the Lunar Module, the latter was beneficiary to the relaxed margins from the Marshall Center. In this way, the Lunar Module accounted for a slightly greater percentage of the total CSM/LM payload than had been anticipated.

By the end of 1965 the weight problem was critical, however. Continual inputs from NASA, program orientation from Headquarters and technical changes from Houston, pushed the projected weight higher and higher. In August, Grumman issued a report on the effort it was making to cut excess mass and received approval from the MSC to implement a Super Weight Improvement Program (SWIP) similar to one employed on the F-111 aircraft project. Within three weeks 70 kg had been shaved from both stages with more promised before the major design reviews got under. In preparing LM-1, the first production Lunar Module, the company set up Operation Scrape aiming to cut 57 kg from the all-up weight. In September it was agreed to employ a single-wire electrical distribution system in non-critical areas, a move designed to remove 36 kg from the system. By the end of the month, 10 kg was taken out from the landing gear in a re-design aimed at making the legs more reliable as well as lighter.

Throughout the year, growth in weight ran concurrent with expanding budget requirements. When Grumman negotiated the contract in late 1962, NASA estimated the LM to cost $400 million – about the same as the CSM contract put up by the then North American Aviation. Lunar Module would eventually top $2,000 million, and the CSM cost would exceed $3,500 million. A lot of learning went into the early years of manned space flight evolution. The original Grumman contract required 11 production Lunar Modules, each fitted ready for manned flight, in addition to several boilerplates called Lunar Test Articles, or LTAs, and mock-ups. By July, 1965, Apollo Spacecraft Program Manager Joseph Shea requested Grumman to have LM-1 delivered by November, 1966, with five each in 1967 and 1968. On 15 December, MSC requested from Grumman a proposal for an additional four LMs to be delivered between December, 1968, and June, 1969. Three of these would eventually be cancelled and subsystems on hand relegated as spares.

Throughout 1966 the Lunar Module fell further behind, due primarily to the constant effort applied to reduce weight, but also because major subsystems proved troublesome to develop. Had the entire Apollo program not been delayed because of the fire in spacecraft 12 it is doubtful that the schedule extended in late 1966 would have been possible. As it was, the first manned Lunar Module would not be ready before early 1969. Several times, beginning in 1965, Grumman suggested flying unmanned an

← Looking back from the astronaut's position, the Lunar Module's aft section contained the Ascent Stage engine cover (bottom) and the docking tunnel hatch (top) through which the crew would enter the vehicle from Apollo.

Astronaut Charles Conrad checks out a LM mock-up, restrained by the straps that would hold crew members in place at their respective stations.

early Lunar Module devoid of those systems still giving trouble. Apollo management was averse to this, intruding as it would upon Mueller's 'all-up' systems concept; many feared that once concessions were made to fly a LM without several systems it would open the door for reduced effort. Nevertheless, by late 1966 it was apparent that the full subsystems installation on LM-1 would not match the need for an early unmanned flight test to qualify the propulsion systems in Earth orbit before manned operations began with subsequent models.

Accordingly, LM-1 was delivered to Kennedy Space Center in June, 1967, devoid of critical environmental control and communications equipment; it was not absolutely necessary for the first, unmanned, evaluation to carry life support items, nor the full complement of tracking and relay hardware. Neither did LM-1 have the four landing legs fitted to subsequent models, for it was not to rehearse the descent operations of later missions. But in almost every other dimension, it was representative of man-rated variants.

The Descent Stage had been designed to accommodate five requirements: support the entire Ascent Stage; support the landing gear; support the complete LM in the Saturn adapter on top of the S-IVB; support scientific equipment for use on the lunar surface; serve as a stable platform from where the Ascent Stage would ascend to lunar orbit. But its task was far more demanding in reality. Rocket launches on Earth were traditionally reminiscent of a high-school chemistry test: nobody really knew what chemical quirk lay within the seething mass of propellants. Only with the prolific launch rate brought on by the accelerated pace of the space program could guesswork give way to calculated precision. Nevertheless, the paraphernalia of a launch complex required much hardware and considerable numbers of personnel. From the surface of another world, 386,000 km from Earth, two men alone would countdown their own personal rocket and perform the launch, remaining with their charge during the long ascent back to the only spacecraft built to bring them home: the Apollo CSM.

The Lunar Module's Descent Stage would become the Moon's Kennedy Space Center. To accomplish its job, the Descent Stage was built around an engine, four spherical propellant tanks, and four spider-like landing legs. The frame to contain all this comprised two pairs of parallel beams 1.65 meters apart, arranged in a cruciform with a deck on the upper and lower surface. The ends of the beams, 4.1 meters apart, were closed off, as were the internal compartments framed by the crossed beams, thus forming five box structures – one in the center and four on its outer surfaces – with four triangular quadrants between the boxes. The center box carried the Descent Stage engine, while the remaining four carried propellant tanks. The quadrants housed environmental control, electrical, and other systems equipment.

The complete Descent Stage was enveloped in a thermal and micrometeoroid shield, additionally covered on the top and side panels with a nickel inconel mesh; a teflon coated titanium blast shield would deflect engine exhaust from the Ascent Stage when it took off from the Descent Stage. The Descent Stage engine itself would radiate temperatures of 1,030°c into the engine compartment so that too had a titanium shield to protect the structure and four outboard boxes. Elsewhere, Mylar and H-film blankets were fitted to distribute heat from the Sun and absorb the energy of tiny micrometeorites.

The landing gear was designed around a requirement allowing the LM to come to rest on a maximum incline of 6° with depressions 61 cm deep, total vehicle inclination being 15°. Beyond those limits there was no guarantee that the Ascent Stage would correct itself after lift-off before slipping sideways into the surface. Landing gear constraints would accommodate a vertical descent rate of 3 meters/sec, a vertical descent rate of 2.1 meters/sec where the horizontal motion would be less than 1.2 meters/sec, or slow descent to a 6° incline where attitude rates would be less than 2°/sec – tight constraints for an undulating Moonscape.

The primary landing gear strut was attached at the upper end to the outboard outrigger; the lower end terminated in a ball joint to which was attached a 94 cm diameter aluminium honeycomb footpad. Essentially a telescoping strut, the primary landing leg would absorb loads by compressing a crushable cartridge of aluminum honeycomb located in the upper sleeve. With a maximum compression stroke of 81.3 cm the leg could absorb the impact of several bounces. Two secondary struts were attached to the primary strut at one end and separate side frames at the other. Cross members linked the secondary struts and down lock mechanisms were positioned at each side frame. Stored energy in springs within the deployment mechanism would, when released, drive the side frame out, moving the primary and secondary struts to the downlock position. In this regard it was a one-way operation; once opened the lock was impossible to close.

Under lunar conditions of 1/6th Earth gravity, the Lunar Module would weigh less than 1,300 kg, having expended almost all the Descent Stage propellant. Dry weight on Earth would be 3,900 kg (Earth weight) so even when devoid of propellants or liquids of any kind, the landing gear was incapable of supporting the complete vehicle on any surface other than that of the Moon. Even an empty LM would collapse on Earth. The ladder fixed on the primary strut of the forward leg provided nine rungs, the bottom one of which was about 76.2 cm above the ground so that if the lower section of the strut compressed heavily the ladder would remain extended. As it turned out, LM landings were comparatively light and the gap between the bottom rung and the Moon's surface required a noticeable leap from descending astronauts. The platform, or front porch, was placed across the forward landing gear outrigger, an aluminum sheet 81.3 cm wide and 114.3 cm in length. Earlier plans to have the descending astronaut climb down a knotted rope were abandoned when tests showed how difficult this would be in a full pressure suit.

The Ascent Stage was altogether different, designed to serve three primary functions: provide a controlled and protected environment for two men; provide propulsion and control for ascending from the lunar surface and visibility for rendezvous and docking; provide a base from which astronauts could embark on lunar surface exploration and to which they could return with Moon samples. Like the Descent Stage, propulsion was one feature central to its design, the second being a crew cabin with shirt-sleeve environment of pure oxygen at 248 mmHg. Unlike the Descent Stage, it presented an appearance configured by a thin steel frame across which was placed layers of thermal insulation and micrometeoroid protection. With the foil covering removed, the Ascent Stage looked very different, comprising an aluminum alloy cabin using conventional aircraft-type construction supporting clusters of spherical and cylindrical propellant tanks, quads of attitude control thrusters and associated plumbing.

Skin and web panels were employed to make up the cylindrical cabin section, chemically milled for minimum weight, with structural members fusion welded where possible to minimize leaks. Separate assemblies were mechanically joined with fasteners and epoxy as a sealant. Grumman intended to weld all cabin sections when it laid out program requirements, but by early 1964 had changed to a partially riveted cabin. For several months Grumman engineers examined riveting operations on the big Saturn stages, learning quickly that a pure oxygen environment was not conducive to good stress or shock resistance in the thin gauge aluminum.

The front face of the cylindrical cabin was fabricated from ten separate welded and machined sections, fitted with stringers and brackets for rigidity. The mid-section comprised the structural core of the Ascent Stage, providing a secure mount for the Ascent engine and for the forward and aft crew compartment assemblies. Inside, the mid-section provided a habitable area with outward curving walls, about 1.42 meters wide, 1.52 meters in height. The drum-shaped top of the Ascent engine protruded through the floor, a ready made seat for tired astronauts. In back, the aft bulkhead closed out the pressurized midsection with supports for coldrails and coldplates carrying electronic equipment. The floor was 1.37 meters from the aft bulkhead to the step that formed the front section of the crew compartment. This circular area was closed out by the front face assembly with its two triangular windows each 0.18 square meters in area and the single square-shaped hatch through which the astronauts would pass to reach the lunar surface.

The front section was 2.34 meters at maximum width, 1.98 meters high and 1.1 meters from the hatch to the step that led up to the mid-section. From the front face, the most forward part of the crew compartment, to the aft bulkhead, was a distance of 2.47 meters, a total habitable volume of 4.5 cubic meters, considerably less than that provided for the three astronauts in the Command Module. Above the Ascent engine cover was the upper hatch through which two crewmembers would pass from or to the Command Module. It was fitted with a 83.8 cm diameter cover designed to open inward by rotating a latch handle 90°. Beyond the hatch, a short 40.6 cm long tunnel section would lead to the Command Module. The tunnel section also supported the drogue, a conical structure to which the Command Module docking probe would be secured until the two vehicles were mechanically joined and pressurized. The probe and drogue would then be removed for movement between the two spacecraft.

The peculiar shape of the Ascent Stage was occasioned by the micrometeoroid and thermal insulation materials, and by the offset placement of Ascent engine fuel and oxidizer tanks, discussed earlier. Brackets, or 'stand-offs,' supported thermal blankets built up from at least 25 layers of aluminized Mylar or H-film away from the structural shell. Aluminum frames contoured the insulation around the propellant tanks and plumbing. On the outside, forming the exterior skin, but separated from the thermal blankets by a continuation of the same stand-offs, was a micrometeoroid shield of sheet aluminum varying in thickness from 0.01 cm to 0.02 cm. It was this flexible skin, capable of being kicked in by the average astronaut, that led to humorous reference about the frail nature of the 'lem.' It was, of course, merely the protective covering of a very strong structure.

Four attachment points secured the Ascent Stage to the Descent Stage, pyrotechnically separated milliseconds after electrical isolation of the two elements by explosive bolts that freed the former to ascend from the latter. At the same time, explosive guillotines would sever electrical and environmental connections between the two.

Propulsion was arguably more important for the Lunar Module than for any other manned vehicle; without an efficient Descent Propulsion System (DPS), or 'dips,' the LM could crash to the Moon, its ability to control its rate of descent or to hover entirely dependent on the downward thrust of the engine; devoid of an Ascent Propulsion System (APS), or 'aps,' it could not return to the Command Module, nor carry the crew away to safety should the DPS fail during descent; sudden loss of the Reaction Control System (RCS) would deny attitude control and orbital maneuvering essential to pointing accuracy for DPS or APS burns. Both main engines simply had to work.

Because the Lunar Module would consume more than half its own weight by burning Descent Stage propellant during the descent, the DPS had to be a throttleable system capable of matching thrust output to the slowly decreasing weight of the vehicle – sometimes greater, sometimes about the same. The Ascent Propulsion System, on the other hand, was to be a fixed thrust system designed for simplicity of operation and maximum reliability.

Built by TRW, the Lunar Module Descent Engine (LMDE) emphasized reliable throttle operations to safely put the LM on the Moon.

The Rocketdyne division of North American Aviation was selected to design the Descent Propulsion System engine on 30 January 1963; Bell Aerosystems was awarded a contract for the APS engine on the same day. Almost immediately, Grumman began formal discussions with Rocketdyne over the development of a helium-injected engine with a throttleable range of 10:1, designed to a specification laid down by the Manned Spacecraft Center. But because the project was so important, and because it was considered as a potential pacing item within the entire Apollo program, Grumman invited four companies on 14 March to bid for an alternate design, one employing mechanical throttle linkages rather than chemical thrust reduction.

By 1 May, Grumman had authorized Rocketdyne to go ahead with the helium injection concept in response to preliminary proposals from that company. In the same month Space Technology Laboratories (STL), later to become TRW Inc., were selected to proceed with the mechanical engine. STL had been working on stable throttleable engines for the previous three years and when asked to work on the DPS concept had already developed a small 227 kg thrust unit of that type. It gave the company a head start and illuminated their efforts when throttleable engines were difficult to fabricate. By January, 1964, STL had successfully carried out the first full-throttle firing test. Just three months later the company set up a permanent Lunar Module DPS facility at its new San Juan Capistrano site in California.

Most of the development work would be performed here. Inevitably, initial success soon led to minor problems, and in May a hot gas leak in the injector (where propellants are brought together for combustion) set off a design hunt for causes that threatened the safety of the design. Modest success was achieved, however, with the ablative thrust chamber tested later that month. In July, 1964, STL began an exhaustive series of altitude tests at a simulated height of 39 km from the Reno facility in Nevada. A heavyweight engine skirt was used for this test. But the concentrated effort put into the LMDE (Lunar Module Descent Engine) program by STL paid off in January, 1965, when Rocketdyne were ordered to stop work on their competitive concept; STL was to be the sole contractor and the advantages of the mechanical throttle were too good to ignore.

A nagging problem for some months now had been the threatened explosion caused by backflow when the LM landed on the Moon and exhaust products from the Descent engine were trapped by the long nozzle. One meter long probes extending from the landing gear were chosen as a means of automatically cutting off the engine while it was still just above the surface; with only 1/6th gravity the LM would not be damaged in the resulting free fall. But tests showed that in most cases the engine was either still running at touchdown or shut off at too low an altitude. By March, 1965, MSC had come up with a pilot activated cut-off switch and this promised to do the job. But reaction time from the first flashing light would have to be 0.4 seconds or less to cut the engine before touchdown.

Inevitably, as total LM weight grew, so too would the velocity requirement from the LMDE. In March, Grumman advised STL that the engine would have to burn longer and that simple request put back the engine development program by almost one full year. But 1965 was as bad for the Lunar Module as it proved to be for the CSM builders. By fall that year STL was in deep trouble with the engine program: apart from the inevitable overweight problem, throttle mechanisms went awry, combustion instability grew steadily worse, chamber erosion became critical, and valves proved troublesome. But progress with the shutdown indicators resulted in an increase in sensor probe length to an eventual 1.52 meters.

On through 1967 the engine progressed, accumulating tens of thousands of seconds in repeated firing trials aimed at wringing the problems out. Finally, in August, almost exactly four years after receiving the official contract, STL completed qualification firings. The LMDE was ready for flight test aboard LM-1. It was not a big engine. Thrust, at a maximum 4.5 tonnes, was less than half that of Apollo's Service Propulsion System. But it was reliable, rugged, and precise, from maximum down to the minimum throttle setting of approximately 476 kg thrust output. Weighing 174 kg, the LMDE had a titanium encased ablative thrust chamber designed to an operating life of more than 17 minutes; in mission operations it would be required to fire twice: once for 28 seconds to change orbit about the Moon; once for a maximum 12 minutes to decelerate the Lunar Module for a smooth touchdown.

Designed to accept an exhaust temperature of 1,482°c, the engine nozzle was made from columbium alloy in four welded segments reducing in thickness from the top down so that it would compress 71 cm in 0.5 sec should the primary landing struts telescope on impact, bringing the engine bell into contact with the lunar surface. Engine thrust was to be controlled by flow valves and adjustable orifice sleeves in the injector operated either by a manual hand controller in the crew cabin or by the guidance computer.

Meanwhile, the Ascent Propulsion System engine had moved through a checkered history. From initial contract award in 1963, Bell Aerosystems moved ahead with design of a fixed thrust motor, making an early decision on Grumman's recommendation to employ an ablative thrust chamber. This retained the option of an ablative or radiation cooled exhaust nozzle. Within a matter of months Grumman selected the ablation nozzle with a reservation that if a radiation bell became feasible it would achieve desirable weight savings. Much of 1964 was taken up with tests on different nozzle throats and by August of that year a satisfactory design emerged, actually increasing performance over the evolutionary concepts. A potential problem challenged Bell engineers when doubts prevailed about the engine's ability to withstand the shock waves that would inevitably bounce back from the Descent Stage at ignition. 'Fire-in-the-hole' tests carried out at the Arnold Engineering Development Center relaxed suspicion and good response was measured with the existing nozzle.

By year's end Bell faced similar problems to those encountered by STL. With a rapid growth in Lunar Module weight the specified 1.59 tonne thrust would have to increase to 1.81 tonnes. The only alternative required the APS engine to burn longer, a solution chosen by the LMDE contractor and one adopted by Bell. But problems with the injector plate haunted engineers seeking a level of stability denied so far. The ablative thrust chamber wall eroded too fast and performance dropped below specification at mixture ratios that alleviated the ablation problem. By January, 1965, the engine's weight had grown by 15 kg, estimated development cost had doubled, and the combustion instability was so severe that Grumman and MSC entered the development program to see if they could help with engineering reviews.

A new baffled injector emerged by the spring and in April Bell tested a one-piece ablative chamber designed to replace a molded-throat design which cracked during firings. In late summer the new injector was tested at the Arnold Engineering Development Center but on 1 September an engine blew up on a third run, due probably to propellants leaking into the injector before the test was scheduled to begin. Instability was a nagging brake on APS engine qualification all the way through 1966 and into the following year. By early 1967 NASA was concerned about the impact this would have on the Lunar Module program and in late July offered Rocketdyne a contract to design, develop and qualify a new injector. Instability and excessive erosion were the main pitfalls with the Bell Aerosystems engine and if a suitable injector could be produced the remainder of the APS engine hardware was considered excellent for the job.

By early 1968 Rocketdyne showed good progress with the new injector, earning itself further contracts to assemble the Bell engine with their own design. Unlike the Descent engine which could be gimballed 6° in each axis, it was a fixed thrust motor securely bolted to the mid-section of the Ascent Stage but with a 1.5° tilt toward the forward landing leg. Descent engine propellants would be forced to the engine by a supercritical helium supply, an ambient helium supply being used from a separate source to first pressurize the volume between the propellant and the tank wall. Ascent engine propellants would be delivered to the combustion chamber by helium pressurization also: stored in two separate tanks, either supply of which would be sufficient to pressurize the propellants, the system evolved from a basic need for simplicity and efficiency. Both engines employed nitrogen tetroxide as the oxidizer and a blend of hydrazine and unsymmetrical dimethylhydrazine known as Aerozene-50 for fuel in separate tanks for each stage. The oxidizer and fuel was identical to that tanked for the Reaction Control System thrusters.

The Marquardt Corporation was selected in January, 1963, to design and fabricate thrusters for the Ascent Stage. Known as the R-4D engine, it was identical to the RCS thrusters attached to the Apollo Service Module quads and obtained propellant from either one of two systems, or both, or from the Ascent Propulsion System engine tanks via a crossfeed connection. Systems A and B each supported a fuel and an oxidizer tank and eight of the 16 thrusters. A normally closed crossfeed connection could transfer propellant from either System to all 16 thrusters. Although preferably constrained to periods when the Ascent engine was firing, the RCS thrusters could obtain propellant from the two spherical APS tanks also supplying the main engine. In this way, the System A and B supply could be reserved for attitude and translation manoeuvres during free flight, the APS supply being assigned to both main engine and RCS support between lunar surface lift-off and Moon orbit insertion.

The Lunar Module's electrical system, like the environmental equipment also, could use the larger volume of the Descent Stage for subsystems hardware, only carrying aboard the Ascent Stage sufficient supply for the trip back into Moon orbit and rendezvous with the Command Module. The Manned Spacecraft Center provided Grumman with estimated mission requirements for electrical power and in January 1963 the company selected a fuel cell system supplemented with batteries: three fuel cells would be carried in the Descent Stage with batteries in the Ascent Stage. Based upon the reference mission at that time, MSC estimated a total electrical requirement of 61.3 kilowatt-hours.

By mid-year, Grumman had selected Pratt and Whitney to build the fuel cells and the Manned Spacecraft Center concurred. But as the LM mission was more accurately defined, the anticipated electrical requirement went up to 76.53 kilowatt-hours by August and to 121 kilowatt-hours in March, 1964. Based on this increased need Grumman suggested moving to a two-cell system with a Descent Stage peaking battery, and an Ascent Stage survival battery. The Apollo Spacecraft Program Office disagreed, however, and ordered Grumman to proceed with the 900-watt model, three of which would be carried in the Descent Stage. This decision was based on the premise that if one failed the mission could still continue and if two failed it could be safely aborted on the power of the third. Within a few months, however, the Program Office revised its thinking and advised Grumman that in configurating the battery needs it should be assumed that the mission would be aborted if one cell failed.

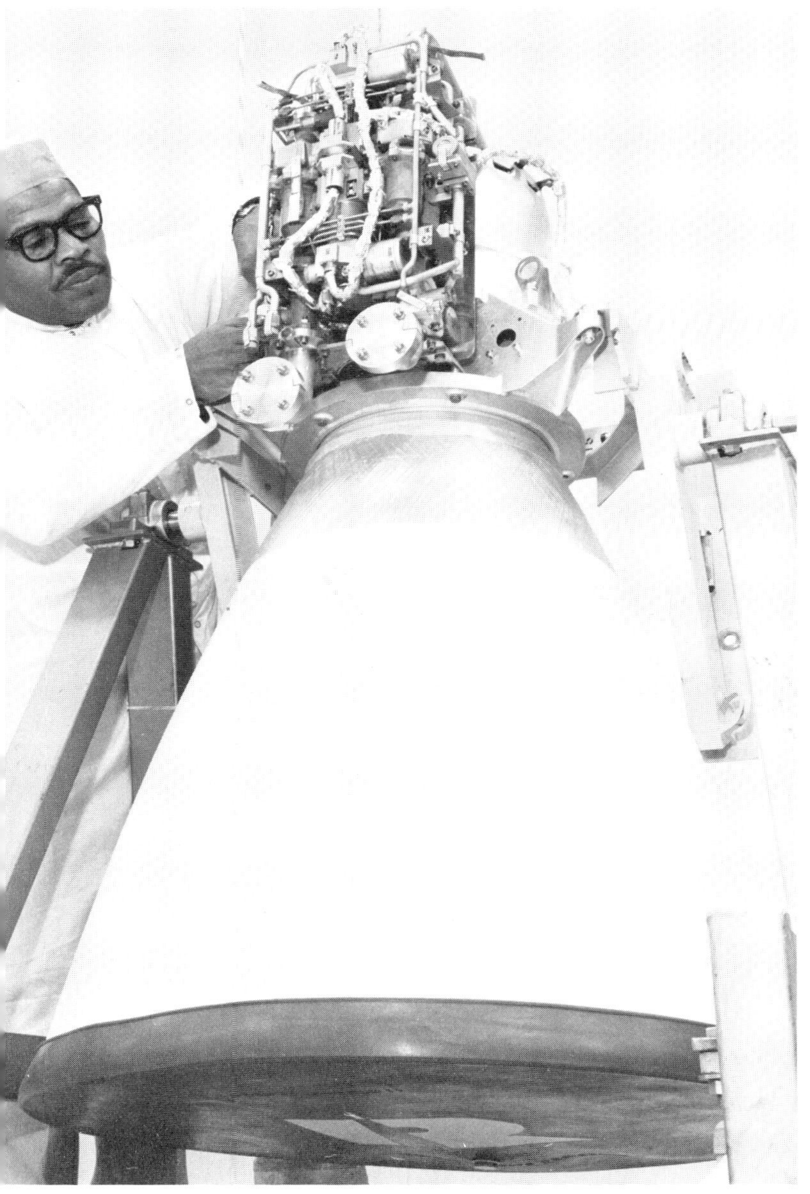

Designed initially by Bell, the Lunar Module Ascent engine was assembled by Rocketdyne to criteria stipulating simplicity of operation; it simply had to work to get the men off the Moon.

During the first year of development Pratt and Whitney made steady progress on design, solving a potential problem with excess potassium hydroxide deposits and demonstrating continuous operation of up to 400 hours. Awareness that mission success could be threatened by even minor failure in fuel cell operation, evidenced by the cautious move to anticipated aborts should only one cell go off line, began a gradual move toward large batteries for supplementary or emergency power. In October, 1964, Grumman requested permission to incorporate a large 1,800 watt-hour battery for auxiliary power. MSC concurred. By the end of the year, the fuel cell design proved a 980 watt-hour capacity and regularly ran for more than 600 hours. Each cell had 15 sections but Grumman had been denied opportunity to increase this by two additional section plates.

Yet still it was a losing game. Weight growth was now critical for not only the fuel cells proper but also the cryogenic storage system required to house and deliver super-cold hydrogen and oxygen. In November, 1964, the Apollo Spacecraft Program Office reported back on a preliminary survey of the trade-off between cells and batteries. Recent advances in the technology and potential of silver-zinc batteries made their choice a more attractive proposition than a year before. Also, Grumman completed its own study on the precise power needs of a Lunar Module during worst-case abort situations and found it would have to increase further the size of the auxiliary battery recently added to the electrical system. Within a few weeks Grumman was deep into a Lunar Module re-sizing program and among the candidates for consideration was the fuel cell system in the Descent Stage.

But Pratt and Whitney were keen to report optimistically that they had apparently solved minor problems with deposits and that they could guarantee 400 hour lifetimes even with the development cells. By February, 1965, the Apollo Spacecraft Program Office began extensive reviews of Grumman's proposal to delete the fuel cells completely and adopt an all-battery LM. The arguments were unquestionably persuasive and on 1 March the Manned Spacecraft Center agreed. A month later the contract with Pratt and Whitney was terminated and two weeks after that MSC had approved the suggested layout and size of conventional batteries.

That decision did have one good bonus. By eliminating the need for cryogenic plumbing between Descent and Ascent Stages, the complex disconnect fitting could be eliminated, saving weight there too. Also at this time, it was agreed that power for the initial Lunar Module checkout, while still docked to Apollo, should come from the LM's own electrical distribution system rather than from the Command Module supply as originally intended. By mid-May, Eagle-Picher had been selected as the battery contractor. Many were happier with battery power than electrical energy supplied by fuel cells; this was a period of deep trouble for Gemini fuel cell engineers and the concept was retarded by the bad name it had from General Electric's worrisome teething troubles.

Final evaluation of a typical lunar landing mission showed a nominal requirement for a total 2,192 ampere-hr capacity, of which about 1,600 ampere-hr would be needed in the four Descent Stage batteries and 592 ampere-hr from the two Ascent Stage batteries. All six batteries would operate in parallel during separation, descent to the surface and lunar operations. Only when the time came to lift the Ascent Stage off the Descent Stage would electrical power from the landing stage be cut; in reality, the Descent Stage would provide power for all phases of the mission up to lift-off preparations, the Ascent Stage batteries powering that stage during ascent and rendezvous. All six batteries weighed a total 358 kg with a nominal direct current of 30 volts and each incorporated silver-zinc plates with a potassium hydroxide electrolyte. The Descent Stage batteries and the electrical distribution system were stored in the two forward quadrants, immediately behind and either side of the forward landing leg structure, while the Ascent batteries were housed behind the aft bulkhead at the back of the Stage.

But if Pratt and Whitney were denied the LM fuel cell contract another United Aircraft division – Hamilton Standard – successfully negotiated the contract for an environmental control system. Awarded in January, 1963, the job stipulated a requirement for 24–36 hr stays on the lunar surface, requiring a unit capable of operating 48 hours, providing a pure oxygen environment with odor and carbon dioxide removed from the cabin, ports provided for re-charging the astronaut back-packs, a supply of coolant fluid for removing heat, and drinking water for the two-man crew. All the ECS equipment had to be contained in the Ascent Stage, since that was the element designed to support the crew and the only structure required to carry them back to Apollo. But tanks containing the gaseous oxygen and the cooling water for use during the descent and surface stay could be housed in the Descent Stage. Smaller quantities would be carried in the Ascent Stage for use after lift-off from the Moon.

The atmosphere revitalization section maintained the crew compartment at 248 mmHg by circulating pure oxygen first through the suits then to the cabin. The suit and cabin circuits were connected but would be isolated when the latter was depressurized for EVA. Passing from the suit circuit in normal operation, oxygen would be filtered through a debris-trap before passing along a lithium hydroxide cartridge to scrub out additives picked up in the suit or cabin environment. Two lithium hydroxide cartridges would be carried, each capable of operating for 20 hours of normal cabin use, with a small canister of activated charcoal in each cartridge; a smaller third cartridge would serve as back-up when the prime units were being changed.

The liquid cooling section was installed to supplement the main heat transport section and comprised a 0.14 kg supply of water in a package in the floor to which could be connected hoses attached to liquid cooling garments – underwear comprising stitched tubes to remove body heat. The heat transport section proper employed primary and secondary coolant loops through which a water-glycol solution would pass to porous plate sublimators. These were designed to freeze the heat-carrying water into ice and sublimate the ice directly into vapor which would then be discharged into space from the back of the Ascent Stage. Equipment generating excessive heat was attached to a system of coldplates and coldrails through which the coolant flowed.

The water management section supplied the sublimators and drinking ports in the cabin for consumption or food preparation. Three separate water tanks were provided: one large tank in the Descent Stage and two small tanks in the Ascent Stage. Nitrogen gas was employed to expel liquid from the aluminum spheres by compressing a rubber bladder containing the water. The oxygen and cabin pressure control section was responsible for maintaining a supply to the atmosphere revitalization section and for providing extra oxygen to re-pressurize the crew compartment a maximum four times after it had been de-pressurized for surface exploration or to discard unnecessary equipment. Life-support back-packs could also be filled up to three times from this source. The oxygen was to be stored in three tanks supplied by Grumman: one large Descent Stage tank in the aft right quadrant and two small spherical vessels in the aft compartment on the Ascent Stage.

Development of the Lunar Module ECS went well compared with other systems but management problems in 1965 threatened to slow the program. An inspection of the company's procedures revealed excellent engineering practice but a poor management structure; with help that too soon paralleled the quality of Hamilton Standard's reliable little product. At one time Grumman favoured use of a super-critical oxygen supply but weight savings here were minimal so in favor of an increased reliability MSC recommended an all-gaseous oxygen design. In fact because of delays in deciding whether one astronaut or two should be allowed to roam the lunar surface simultaneously, engineers were forced to build an increased flexibility into the system, one that allowed one man or two man operation.

Preparations for the first Saturn V flight were well under way when Lunar Module-1 arrived at the Kennedy Space Center in June 1967. It was three months late but flight preparations chief John J. Williams pitched in with gusto as several hundred engineers and technicians began to prepare Apollo 5 for a Saturn IB launch. Launch vehicle 204, the two-stage Saturn carrying the spacecraft that killed astronauts Grissom, White and Chaffee, had been removed from Complex 34 in March and set up on Complex 37A the following month after a careful inspection. It would be used to convey the unmanned LM into orbit for a shakedown flight aimed at testing the two main propulsion systems, the electrical system, and guidance equipment.

Unlike Apollo Saturn IB flights, the LM mission would generate a unique appearance for the rocket, usually topped by a conical Command Module and the familiar launch escape system. Unable to carry both Lunar Module *and* an Apollo CSM, this launcher would be topped by a shroud similar to that employed as an adaptor on the Saturn V, with no escape system necessary since the LM was not stressed to withstand the rigors of emergency release, had no parachutes with which to descend, and was unable to survive the extreme thermal environment of a high altitude abort.

By late November Mission Director William Schneider had approved the documentation for Apollo 5, NASA managers had successfully reviewed the hardware for flight, and Jim Webb received notice that the Cape was aiming for a 18 January launch. LM-1 had been encapsulated on top of the Saturn IB's second stage since erection on 19 November, devoid of landing legs and without water in the Descent Stage tank. But both Ascent and Descent Stage propellant tanks were full, 2.3 tonnes in the former and very nearly 8 tonnes in the latter for a total Lunar Module weight of 14.4 tonnes. One additional piece of hardware was carried inside: a LM Mission Programer to perform control functions normally assigned to a two-man crew and to store 64 programed contingency procedures just in case something went wrong to interrupt the planned sequence of events. The interior was devoid of any crew support equipment, and aluminum panels covered the triangular forward windows and the rectangular overhead window.

Minor problems hit Rocco Petrone's launch schedule and delays plagued the smooth countdown, interspersed with lengthy holds for analysis. Kranz and Hodge were assigned mission control when the launcher cleared the pad, effectively transferring operations from the Cape to Houston just seconds after lift-off. But it was 22 January before AS-204 thundered spaceward with the first production Lunar Module; the opening flight for 1968, a year that would hopefully see the first manned Apollo mission sail into space.

During the preceding weeks several key activities converged to set out the definitive flight assignment schedule leading to a manned lunar landing. Activity at the Manned Spacecraft Center during late 1967 centered on getting procedures ready for flight operations with Apollo and the Lunar Module. George Low was flat-out working to perfect the series of steps essential to success. Owen E. Maynard, at work with MSC since the Space Task Group days when he studied the way Apollo systems should fit together fully twelve months before the Kennedy commitment, came up with a seven-step program aimed at minimizing flight categories but maximizing the rehearsal and experience factor in procedures culminating in the manned landing:

Mission A would test Saturn V and bring a Command Module back through the atmosphere at lunar-return speeds (already carried out by Apollo 4);

Mission B was to use Saturn IB for Earth orbit Lunar Module tests (Apollo 5);

Mission C was to be the first manned Apollo CSM mission in Earth orbit from a Saturn IB;

Mission D was to place both CSM and LM elements in Earth orbit for full manned systems checks, using either one Saturn V or two Saturn IBs;

Mission E was to repeat D but in a high Earth orbit, a further check of docking, rendezvous, and combined operations;

Mission F was to be a deep-space repeat of E, possibly carried out in lunar orbit;

Mission G was to be the manned Moon landing profile.

Of all the integrated elements, only the Lunar Module remained to be tested. Already, by January 1968, NASA management was prepared to cancel a planned second test of the LM if Apollo 5 answered questions on performance and capability, moving instead to a manned Lunar Module flight on the next attempt. Any flight that could possibly be deleted was time saved for the complexity of manned operations preceding the lunar landing. And in several regards the feeling of commitment to a manned Moon mission before the end of the decade was stronger now than it had ever been; with three lives taken by inadequate attention to detail, everybody was concerned to see their efforts justified by pulling off the task at hand.

Apollo 5 lifted from Complex 37A in the dusk of evening on Monday, 22 January. Ascending on the 725 tonnes of thrust from eight H-1 engines, AS-204 burned its first stage for 2 min 22 sec

Developed and put together by Hamilton Standard, a division of United Aircraft (now United Technologies Corporation), the Lunar Module's environmental control system had an unusually successful gestation.

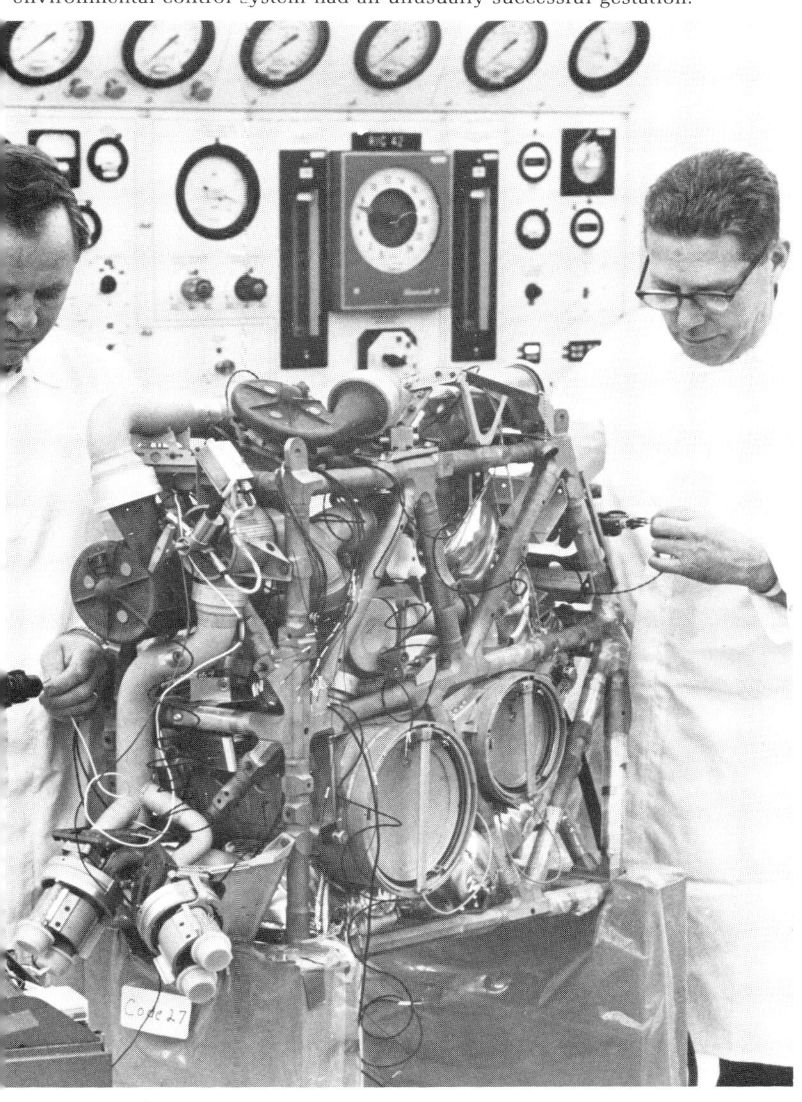

and then fired the single J-2 on the second stage for 7½ minutes, placing the stage and its payload in a 163 × 222 km orbit inclined 31.6° to the equator. Coming up on 54 minutes into the flight, half way round the world, LM-1 pushed free from the adapter on top of the S-IVB stage, the energy thus imparted raising perigee by 3.7 km. First order of business was for the Descent Propulsion System, or DPS, to fire for 38 seconds; the first 26 seconds would be performed at 10% thrust followed by a build-up to 92.5% at the time of shutdown This would simulate the type of maneuver to be carried out around the Moon where, after separating from the nose of the Apollo CSM, the LM would fire itself into an elliptical path with a low point above the Moon of only 15 km.

Ignition of the throttleable DPS engine came shortly before an elapsed time of 4 hours with the legless spider-like space vehicle coming up on Australia in the third revolution. But a conservative guidance computer sensed insufficient thrust build-up and automatically shut the engine down just four seconds later. It was human error: incorrect software programming had placed an impossibly constrained margin on the guidance. At least it proved the computer was doing its thing! But the premature shutdown changed the complete situation and Mission Control had no option but to call up the mission programmer and proceed with a contingency sequence built in for just this sort of eventuality. The second DPS burn was to have simulated a full 12 minute burn of the type that would lower a Lunar Module from 15 km to the surface of the Moon on a long curving flight path, with increments of throttle percentage assigned specific periods of the burn. That had to be abandoned and instead the DPS would fire two shorter burns followed immediately by ignition of the Ascent Propulsion System, or APS, engine to simulate an abort during lunar descent where the Ascent Stage would fire off the Descent Stage and propel itself back into orbit.

Ignition came with LM-1 moving across the United States to complete four revolutions. The DPS engine burned for 26 seconds at 10% thrust, ramped to full thrust for 7 seconds, shut down for a 32 second drift, fired again for 26 seconds at 10% and terminated with a 2 second burst at full throttle. Immediately at the point of shutdown the APS ignited and the two sections were pyrotechnically severed as the Ascent engine lifted the tiny crew section away from the Descent Stage. That engine fired for 60 seconds at fixed thrust to prove it was possible to abort a Lunar Module safely during descent. Following this activity, Houston put LM-1 back on to its primary guidance computer but the information inadvertently assumed the Descent Stage was still attached and commanded the attitude control thrusters to fire as though the Lunar Module still had the greater mass of its two sections.

This confused the computer and quickly depleted propellant from the Reaction Control System tanks through prolific thruster activity. The second firing of the APS engine was, consequently, made without attitude control capability. Sustained to APS propellant depletion, the engine burned for 5 min 47 sec. Controlled in attitude at first by open crossfeed lines delivering propellant to the small thrusters from the big APS tanks, the computer closed the interconnection at an assigned time denying all attitude control to the Ascent Stage, which proceeded to tumble uncontrollably. Nevertheless, the contingency sequence loaded in the mission programer had successfully retrieved the minimum test operations to satisfy the primary flight objectives and the Lunar Module came through with flying colours.

By early March, while preparations were nearing completion for the second unmanned Saturn V test, George Mueller decided that another LM test was unnecessary and that LM-2 need not be made ready for flight. Although NASA considered each spacecraft as a baseline vehicle suitable for adaptation to varied mission requirements, in reality most production vehicles, whether LM or CSM, were closely tailored to projected flight requirements for a peculiar mission objective. So it was with LM-2, a spacecraft incapable of carrying a human crew, which now had to give way to LM-3, the first production model equipped for habitation. But there was one shadow on the horizon: stress corrosion cracks were discovered in the aluminum structure of some vehicles. George Mueller was concerned about this, revealing his annoyance to subordinates and contractor representatives. It was just the sort of thing that could dash all hopes of a manned Moon mission, a nagging time-consuming problem.

Within a few weeks of Apollo 5 opening the way for manned LM activity, engineers at Grumman's Bethpage facility had inspected the first five man-rated vehicles and in areas accessible to inspection had found none of the suspected cracks. Major decisions had to be taken when certain areas of LM vehicles already built could not be reached without disassembly. Inspection would certainly delay flight operations so Grumman assumed that, because it had been unable to find evidence of corrosion in the parts it could get to, the chances of corrosion in the few hidden component structures was sufficiently low to assume there were none. The company switched to a different alloy for all subsequent fabrication.

But it was inspection of LM-2 at the Kennedy Space Center that led to a suspicion about the condition of electrical wiring in all Lunar Modules. Brig. Gen. Carroll H. Bolender, the MSC Lunar Module manager, urged inspection of LM-3. As suspected, several wire segments were found to have broken during reconnection following systems testing. This could not be tolerated and Grumman was urged to take greater care. Nevertheless, time was running out for the preparation of Lunar Module hardware and it seemed unlikely that equipment for the LM-3 flight would be ready to support a manned mission in 1968. But there was at least one more unmanned Saturn V and one manned Saturn IB mission before thoughts could turn to LM operations again. So while Grumman got stuck in to concentrated preparations for the first manned LM mission to be flown probably no earlier than early 1969, engineers and technicians at the Cape prepared hardware for the preceding events.

April, 1968, brought punctuated reminders of what the decade had been all about. In a week beginning 31 March, President Johnson announced his intention not to stand for re-election during the coming campaign, Civil Rights leader Martin Luther King was brutally gunned down in Memphis, and von Braun's technological wonder shook the Florida beach once more on a none too perfect attempt to repeat the success of Apollo 4.

Johnson had been unwittingly propelled into office at John F. Kennedy's assassination in November, 1963. Now, four and one-half years later, the great architect of NASA's civilian space program, the one man above all others responsible for the great

American space spectacle of the decade, was quitting because, he said, he felt it would improve the chances for peace in a war he had for long kept hidden. A full decade after the debate from which NASA emerged as a civilian space agency in response to the challenge from Soviet technology epitomized by the spirit of Sputnik 1, Johnson was walking away from the Great Society goal it had been his lot to propose but never construct.

Hindered by forces of depraved and virulent aggression flooding south from North Vietnam, the Great Society plan and the NASA dream walked side by side down the slope of diminishing priority; the war in South East Asia was accelerating to horrendous proportions and there seemed so much to put right within American domestic affairs that few Congressmen gave much priority, or indeed thought, to the wider aspirations of the nation. The death of Martin Luther King was the last insane finality on a decade of violence and street aggression the like of which America had not seen for decades, another sad mirror of the disillusionment that came with the 1960s.

As 1968 moved to its spring, few people in America, or any foreign state, were as concerned about the Apollo program as they had been about Alan Shepard or John Glenn. It was as if the light had gone from the dream once held so reverently high, as if the real point had been missed. So many now felt that having stood up in challenge and defiance to the threatened pre-eminence of Soviet technology, the end was an unnecessary step, the actual moon landing an untimely event out of context and not a little irrelevant. It was as if the side-shows had stolen a march on the set-piece event at the carnival.

On the very day King died, AS-502 thundered into life and clapped shock waves of deafening sound across the Cape. It was to be a repeat of AS-501: into Earth orbit, re-ignition of the S-IVB stage on the second revolution, separation of the Apollo CSM, ignition of the Service Propulsion System to drive the spacecraft to a lunar return speed, and re-entry followed by splashdown in the Pacific.

Hardware for Apollo 6 had been at the Cape more than a year, Boeing's S-IC stage checking in during March, 1967, but erection awaited delivery of the S-II. It was July before AS-502 stood tall in the Vehicle Assembly Building; December before Service Module 14 and Command Module 20 were mechanically connected to the adapter. Electrical mating, emergency detection system tests and an overall test were conducted in January, 1968, prior to roll-out on 6 February amid squally showers and wind. On through driving rain the massive configuration lumbered to Complex 39A, temporarily halted when the storm disrupted communications with the Launch Control Center. As the mighty Saturn crawled up the ramp darkness fell across clearing skies but high winds lashed the Cape and for two days the big Mobile Service Structure was unable to move from its parked position to a location alongside AS-502.

Nevertheless, a flight readiness test was completed the following month but the launch was delayed beyond its originally assigned 28 March slot by a succession of minor problems. Lift-off was finally set for 4 April and as propellant surged through the large diameter pipes carrying appropriate quantities of kerosene, liquid hydrogen and liquid oxygen to the six large tanks the vehicle settled down, shorter by more than 20 cm due to the 2,600 tonnes of liquid for combustion in the eleven main engines compressing the 36-storey rocket.

When ignition came six seconds before launch it set hearts racing among the several thousand people standing in the humid morning air, a repeat performance from the first Saturn V five months before. Very soon after lift-off things began to go wrong. First, toward the end of first stage thrusting, about 2 min 13 sec into flight, portions of the spacecraft adapter supporting the Apollo CSM on top of the third stage, broke away under the influence of massive longitudinal oscillations causing the entire rocket to pulse up and down as it ascended. This was a product of resonant frequencies from the first stage engines that nobody really believed would plague a Saturn V. It had been a big problem with Gemini's Titan II booster, but that rocket employed hypergolic propellants while Saturn used a cryogenic oxidizer and kerosene.

Measurements of this 'pogo' effect on Apollo 4 had lulled engineers to a false security probably because only on Apollo 6 did the simulated mass of the missing Lunar Module closely approximate the real shape and distribution of the latter's volume; Apollo 4 carried a mass simulator that apparently served to damp the pogo effect. However, despite this, the first stage completed its performance with the center engine shutting down $3\frac{1}{2}$ seconds before the four outer engines; 1.4 seconds later the first stage had separated and the five J-2 engines on the second stage were up and running. But not for long.

Instead of burning for a planned 6 min 9 sec, engine 2 shut itself down after giving warning of a malfunction by two separate fluctuations. Just 1.3 seconds after that, engine 3 also cut off. Aware of its own malfunction, the Saturn V's Instrument Unit commanded the remaining three engines to burn for as long as there was propellant flowing through the lines, fuel and oxidizer which would have been burned by engines 2 and 3 now being consumed by engines 1, 4 and 5. Thrust was down by 40% and the S-II was accelerating much slower now than it should to deliver the third stage effectively to its assigned speed and altitude for ignition. The S-II burned 59 seconds longer than planned and then ran out of propellant.

Within precisely one second, the second stage had separated and the S-IVB ignited to put the payload in orbit. But it had to fire 29 seconds longer than planned to compensate for the lower performance from the S-II, finally putting itself and the Apollo payload into a 178 × 366 km orbit versus the planned circular path at 185 km. While engineers in non-critical areas began an immediate analysis of what went wrong during ascent, flight controllers at the Manned Spacecraft Center planned the next major maneuver: re-start of the S-IVB to simulate Trans-Lunar Injection, or TLI. For this mission it would place Apollo in a highly elliptical Earth orbit rather than send it to the vicinity of the Moon.

When the command went out from the programed sequencer to start the S-IVB a second time, preparatory events cycled off as planned but ignition never came. Controllers had no alternative but to fall back on a contingency plan, reducing the operation to one considerably short of the planned objectives. Already programed to separate from the S-IVB, the Apollo spacecraft pyrotechnically severed itself from the adapter while pulsing forward with the attitude control thrusters. Less than two minutes later the Service Propulsion System fired up for a long 7 min 22 sec burn, driving the CSM into an elliptical orbit with apogee 22,242 km above the Earth. The big Apollo engine had done what the S-IVB refused to accomplish but a second SPS firing planned originally to increase the speed of the spacecraft for a lunar-type re-entry, was not possible since propellant for only 23 seconds remained in the main tanks.

Six hours twenty-nine minutes into the flight, Apollo 6 reached the high point of its day and began the long accelerating fall toward the Pacific. Slamming into the atmosphere at a speed of 36,024 km/hr, Command Module 20 successfully survived the punishing heat and deployed its parachutes for a perfect descent, culminating in a less than perfect splashdown when the spacecraft turned turtle, apex down, in the heaving sea. But this eventuality had been provided for in a recovery righting system attached to the forward section of the Command Module. Known as Stable II, inverted position, it required activation of three inflatable spheres pressurized by two air compressors. A few minutes later the Command Module righted itself to the Stable I, apex up, position.

Immediate conclusions on the failure during ascent with two engines in the S-II and on orbit with the unfulfilled re-ignition of the S-IVB dashed hopes of eliminating a possible third unmanned Saturn V flight. But within a matter of days Mueller sought opinion from Sam Phillips, Bob Gilruth and senior managers on the desirability of scheduling the next flight as a manned mission with a reservation that if workaround procedures which it was believed could solve the anomalies failed to satisfy the engineers, AS-503 would revert to a flight test role. The pogo problem in the first stage could be solved by adding helium to the propellant feed system, effectively breaking the natural resonance of the rocket. As for the premature shutdown from two J-2 engines, tests revealed an incorrect wiring procedure – human error – that need not have caused a problem. The S-IVB, it seemed, failed to start because a propellant line ruptured, delivering insufficient fluid in time for the start procedure.

Kennedy Space Center launch teams recognized the need for an accelerated pace to preparations for the launch of Saturn V rockets. AS-501 and -502 required lengthy periods from stacking in the VAB to the date of actual launch. This was too protracted a schedule for flights that would ripple off at the rate of 4 or 5 a year.

The Cape's Apollo Program Manager, Admiral R. O. Middleton, was determined to demonstrate a rapid-fire sequence to instill confidence in flight assignment planning for 1969. Everybody realized by now that 1968 too would be a year for pulling it all together. From hardware changes driven hard into design blueprints the previous year, 1968 would be a time for accelerating the pace of flight preparation and launch readiness, for 1969 would be like none other in manned space flight. It was either going to go that year, or it was not going to go, and if the manned Moon mission was to be accomplished by the end of the decade, 1969 would see a more frenzied pace of interlocking events than anybody cared to anticipate too far in advance.

To clear the way for a running start to 1969, Middleton wanted to have the third Saturn V up and ready in record time, evidence to prove to Headquarters management that several Saturn V missions could be accommodated in 1969. In fact Middleton boasted a confident ability to have 503 rolling to the pad within two weeks of 502. As it turned out, the unexpected problems with 502 stayed the start flag for 503, and Middleton had to wait to show off his newly honed procedures. By 23 April George Mueller was openly confident of going manned on 503 rather than have it repeat the 501 and 502 objectives. After all, everything that happened to 502 showed it to be a forgiving design, safe to fly through a gremlin-filled sky, and capable of avoiding total failure in a flurry of malfunctions. And that was the most beautiful thing about Saturn V. More than any other launcher before or since, the big bird imbued a feeling of indestructability, correcting and apologizing for its own malfunctions. Good enough reason, thought Mueller, to let it have its head.

Within three days Jim Webb agreed with Mueller's decision but signed off only those preparations necessary to preserve an option to man the next Saturn V; a firm decision would be made later as to whether it should be flown manned or not. KSC could have an unmanned 503 ready for flight by mid-July but if it was manned, with CSM-103 and LM-3, it could not depart the spaceport before late November. And Lunar Module 3 was in no shape to fly before early 1969. Charles Mathews, late of the Gemini program and now with Mueller as a troubleshooter, expressed concern in mid-1968 about the Cape's ability to process all the necessary hardware for missions set at three or four month intervals. What worked for Gemini might not be so accessible with the added complexities of Apollo hardware.

Grumman was finding it hard going to come across with the necessary assurances about delivery dates; North American Rockwell was still hampered by sub-contractors more concerned about their image in the light of Apollo work than hard-nosed management schedules; Saturn V contractors were still wrestling with a dozen or more fixes made necessary by the 502 failure. Moreover, Kurt Debus was now being asked to accommodate a Saturn V launch schedule that projected three big birds in flow at the same time, rather than the two required by earlier schedules. Matching Owen Maynard's Mission A-G sequence to the available time to the end of the decade demanded a launch rate that tied Kennedy Space Center to processing nine big rocket stages for three Saturn vehicles equipped with six separate spacecraft simultaneously. It was a mammoth undertaking, one made all the worse by a constraint on the launch operations team; it all had to be accomplished with the same number of personnel because no more money was available from an already depleted budget. Apollo was going to make it – but only just, and on a shoestring.

Lunar Module 3 arrived at the Cape during June 1968, and very soon it was discovered that considerable work would be necessary to get the vehicle ready for a manned flight. Leaks sprang from Ascent Stage tanks; wiring repairs were so numerous, the quality control inspectors were unable to log them all; major problems developed from tests on the guidance and navigation equipment; the communications units were appallingly deficient in performance; the rendezvous radar needed replacement parts from Bethpage; systems interference kept engineers tied up shunting coaxial cables around to minimize crossfeed; parts sent from New York were sometimes the wrong type. Frustration with LM-3 would endure until the vehicle eventually flew. It was the first of its type and as such gave Grumman headaches over schedule dates and delivery plans. The company was only now moving through a learning gate passed by North American Rockwell three years before when the first Block I CSM vehicles were delivered for test.

By early August it was clear that LM-3 really would not be ready to fly before early 1969, a fact suspected for some months

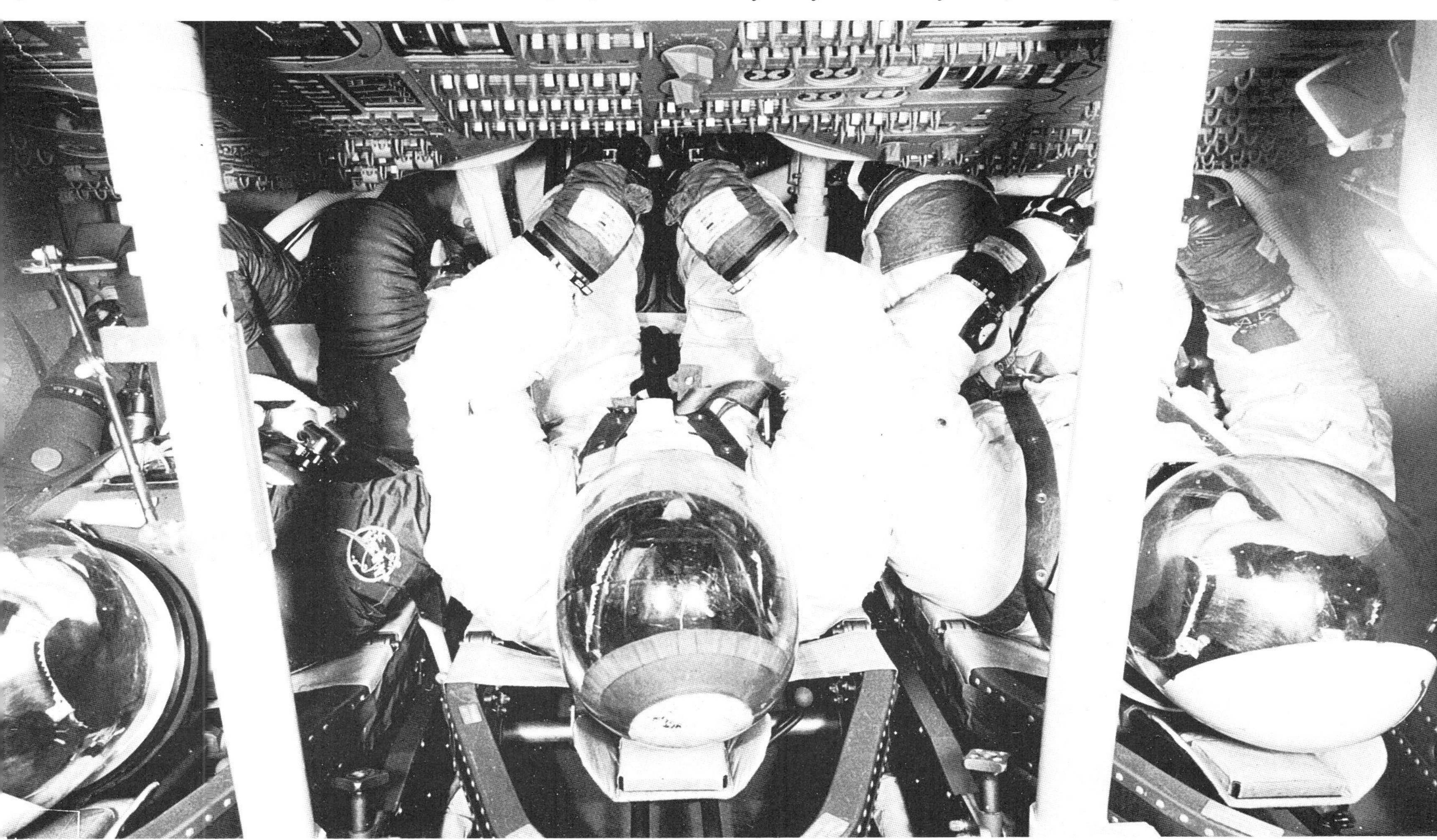

Modifications to the Command Module during 1967 and 1968 made it the safest spacecraft yet developed, retaining the general layout but significantly improving reliability.

but now, in the wake of a decision to fly AS-503 with astronauts on board, an even worse drawback to plans for an early LM shakedown run with men in the vehicle. And it was at that point that one of the most daring and audacious schemes surfaced from desks that had for several months already hatched discrete plots to accommodate a meaningful package aboard flight 503: if the first manned Saturn V was denied the ability to fly a Lunar Module, why not use it to push the second manned Apollo flight to Moon orbit? Qualification of the Apollo CSM would come on Apollo 7, the first manned CSM Block II flight planned for a Saturn IB launch early in October. If that went well there was nothing to stop Apollo 8 becoming a Moon orbit flight with CSM only but launched by Saturn V to inaugurate manned operations with the big booster.

The concept had been with Apollo for many years. Some people thought a precursor step to the Moon landing proper would be satisfactorily met only by a flight either around the Moon or into Moon orbit. When Maynard laid out the A-G plan it did not expressly stipulate a Moon orbit mission, but it certainly made good sense to slide one in to the schedule. After all, operating a spacecraft in Earth orbit would be very different to manipulating systems during a translunar coast: in Earth orbit the vehicle would spend almost equal periods of a 90 minute pass in light and dark, using the night side pass to reject heat acquired during the day side flight; between Earth and Moon the spacecraft would be continually exposed to heat from the Sun and require a delicate balance between hot and cold surfaces.

Moreover, celestial navigation would take on an entirely new slant during deep space flight, and communications would require unique equipment both on the spacecraft and at the ground receiver sites. All this equipment had been built in to the program as an essential part of Apollo, but because it would form merely the background against which the major operational techniques were to be rehearsed, a precursor flight to the Moon and back – without the added complexity of a Lunar Module mission – had much going for it.

Called E-prime, the plan that emerged during the opening months of 1968 suggested using Maynard's Mission E (CSM and LM operations in high Earth orbit) around the Moon; now that the Lunar Module was unlikely to fly on the first manned Saturn V, it was logical to stretch the Mission C plan (CSM only operations in Earth orbit) to a C-prime and Moon orbit on Apollo 8. George Low was the first to see the opportunity, and lost no time in pushing the idea; instigator of many a daring manned space mission in the past, Low was again the man behind a revolutionary new idea. But for the moment, all attention focused on preparations for the first manned Apollo flight – Apollo 7 on Saturn IB AS-205. If successful, it would release all subsequent Saturn IBs from Moon mission support and make them available to whatever the Apollo Applications Program ended up with from Congress.

Schirra, Eisele and Cunningham had known they were in for the first shakedown flight ever since the fire that killed Grissom, White and Chaffee vacated respective positions. It would be their lot to step in and prove Apollo was a safe vehicle for orbital operations. The first Block II vehicle, spacecraft 101, would be their mount. By May, 1967, Schirra's crew were at Downey to live with 101 all the way through to the pad at Kennedy Space Center. They were to get to know their spacecraft like no other crew before them, searching deep within its systems for signs of potential problem areas that nobody expected them to tolerate. That three astronauts had already lost their lives to failure elsewhere in the program served only to spur North American Rockwell to an almost super-human effort that was already visibly apparent. The dedication that permeated the Downey facility augured well for the future reliability of the CSM production batch. Hard won, the lessons were not ignored.

Flammability tests inside a mock-up of the re-designed Command Module proved that once begun a fire would actually extinguish itself in pure oxygen. Searching and deliberate survey of materials and structure incapable of propagating a fire had lined the new Block II vehicle with the safest inventory of equipment ever placed inside a spacecraft. In almost every conceivable dimension, it was a vehicle in which the manufacturer could have justifiable pride. By the end of 1967 a valuable precursor to the manned operation of a Block II spacecraft neared completion at the Downey facility. Called 2TV-1 it was identical in every important aspect to spacecraft 101 and was to prove the concept in thermal and vacuum tests at the Manned Spacecraft Center's Space Environment Simulation Laboratory. Placed inside a spherical vacuum bell, the mock spacecraft was put through rigorous trials aimed at revealing potential problem areas.

Astronauts Joseph Kerwin, Vance Brand and Joseph Engle lived in 2TV-1 for a week, troubleshooting systems and monitoring the performance of the environmental control unit, a very different piece of hardware to that employed in the earlier, Block I, vehicles. Capable of producing thermal environments from −143°c to 93°c, the interior of the vacuum chamber was 36 metres high and 18 metres in diameter; a complete CSM could easily fit inside. Tests with 2TV-1 built confidence in the re-worked Apollo ship and endorsed the opinion of Schirra's crew that the spacecraft was now a very different vehicle to that which took the lives of Apollo 1's brave crew.

By year's end major crew assignments had been announced, and a new category had been appended to the usual prime and back-up nominations: a three-man support crew provided for all the many time-consuming duties that were not the real concern of the flight assigned crews. Now there would be nine astronauts working for each Apollo mission. Although the second Saturn V had not flown when the assignments were made public–on 20 November, 1967, the personnel for the second and third manned Apollo missions expected to fly a launcher of this type; even in late 1967 NASA was confident of moving from Saturn IB to Saturn V flights on the second manned operation.

After the Schirra, Eisele and Cunningham flight on the IB, McDivitt, Scott and Schweickart would get to fly both CSM and LM vehicles in Earth orbit on the Saturn V; Stafford, Young and Cernan would be the back-up team, with Swigert, Evans and Pogue as the support crew. The third manned Apollo would be flown by Borman, Collins and Anders; back-ups were Armstrong, Lovell and Aldrin, with Mattingly, Carr and Bull as the support team. If all went well, the back-ups would rotate in turn for the fourth and fifth manned Apollo flights. By spring, 1968, when it looked as though the next Saturn V shot would be manned, the second Apollo flight was assigned the task of checking in Earth orbit all the many complex operations associated with two vehicles: the CSM and the LM. After McDivitt, Scott and Schweickart had proven the Lunar Module, the Borman, Collins and Anders team would fly a CSM to high orbit, moving 7,400 km from Earth in a test of major Apollo systems in deep space.

The two Saturn V missions (Apollo 8 and 9) were then scheduled for November, 1968, and early 1969, respectively. The next flight (Apollo 10), probably manned by Stafford, Young and Cernan, would take the CSM and LM to Moon orbit and rehearse all the essential steps to a landing without going down to the surface. That mission could take place in May and be followed, perhaps as early as mid-July, 1969, by Apollo 11, manned conceivably by Armstrong, Lovell and Aldrin, as the first Moon landing attempt. Thus, the objective could be achieved as early as the fifth manned Apollo flight on the sixth Saturn V. It was the most ambitious planning schedule yet proposed; where once NASA had felt it imperative to fly 15 big Saturns before the actual landing attempt, the agency was now capitalizing on the certain success of Apollo hardware.

But one failure now could upset the entire schedule. And while astronauts received their assignments to fly complex routines with Saturn V missions, Schirra, Eisele and Cunningham made ready for the inaugural manned Apollo test. It was a rapid-fire sequence nobody really expected to pull off, an ideal target to aim for in the absence of anything going wrong on any flight. Mueller was enthusiastic for the ambitious nature of the schedule, recognizing the spur to personnel and morale from aiming high, but he never once expected to see it follow through, believing at least one flight would have to be re-done. Two more flights could be squeezed in to 1969, available slots in case anything went wrong on precursor attempts, and Mueller was determined to keep firing Saturn V missions at $2\frac{1}{2}$ month intervals until the first manned landing was achieved.

Spacecraft 101 was late getting to the Kennedy Space Center. Erected in mid-April, 1968, launch vehicle AS-205 was sitting on pad 34 six weeks before the CSM arrived. Groomed to perfection under the strict eye of North American Rockwell's John Healey, the new Block II design incorporated more than 1,800 changes driven in by Bill Bergen's insistence that 101 should break all records for quality; working hand-in-hand with George Low, Shea's replacement at Houston, the new head of the Space Division lost countless hours of sleep watch-dogging the new

The crew who gave Apollo a reputation for success. Left to right, Cunningham, Schirra, Eisele.

hardware through fabrication. The astronauts were pleased with John Healey's workman-like approach, a man who held respect for the views of the Schirra team, but an engineer nevertheless who insisted the spacecraft was *his* responsibility until it arrived on dock at KSC.

When it did finally get to Florida, spacecraft 101 was as clean as anything yet delivered from a contractor; in the final acceptance reviews, North American Rockwell cleared the vehicle with only 13 discrepancies, and all those were put right before it left Downey. At the Cape, Rocco Petrone pressed hard for a September launch but the sheer determination to get things right postponed several critical tests. Unmanned altitude runs got under way by the end of July, followed within days by a manned run where Schirra, Cunningham and Eisele spent several hours at a simulated altitude of 68.9 km. Then the back-up crew went through the same process while other personnel made ready a new safety precaution at LC-34, one recommended by the post-fire review boards.

To enable astronauts to flee the vicinity of an impending explosion on the pad, slide-wires had been attached to the service structure. About 360 meters long, the wires were tested in early August with dummies riding the device to the ground. Brake settings on the mechanism were not quite right and two dummies sailed through the embankment put there to halt the downward motion! But adjustments took care of that and the system was declared operational for Apollo 7; a similar device was set up on the Saturn V pads at LC-39.

By 10 August, the spacecraft was mechanically mated to the launcher and by early September the countdown demonstration test had been successfully run. A Flight Readiness Test followed before the end of the month with a final countdown getting under way early in the afternoon of Thursday, 10 October. All across the nation, and in several foreign countries, Apollo watchers settled down for a long night as technicians prepared the Saturn IB for its responsible job the following day.

Thoughts were varied that evening: in a restaurant in downtown Houston, several personnel gathered together for a celebratory feast, some from the Manned Spacecraft Center, others from contractor support groups; in Washington, squawk-boxes kept Headquarters personnel informed of progress at the Cape, as lights burned long after midnight in offices along Maryland Avenue off the Mall and within sight of the Capitol; at Huntsville, Saturn engineers broke temporarily from detailed support activity on future projects to gather round radio sets and other squawk-boxes. It was the dawn of a new era, a time of professionalism where there was little or no room for the Buck Rogers feeling that permeated the old Mercury days. How far away they seemed that night, all across the nation at select sites where the full meaning of what was about to happen fired the mind and quickened the pulse.

The last two years had been months filled with disillusionment, fears about the future of manned space activity, concern at toppled budgets. For a while, perhaps for 10 months or so, there would be spectacular relief from the creeping reality of a nation that had suddenly lost interest. For a brief period, there would be a magic ride toward the immortality of historical recognition. NASA was on the home stretch toward a goal placed $7\frac{1}{2}$ years before. But for one man professional involvement with the Moon had already passed. In September Jim Webb announced he would resign as Administrator, a position he held with outstanding success since early 1961 when President Kennedy sought new men for bold initiatives.

He said it would leave him free to talk openly and truthfully about the space program, not that he had any secrets to divulge, just a concern that it should not be made the whipping boy of budget planners. He also said it was to smooth the way for his successor, who would surely be appointed by the new President elected to succeed Lyndon Johnson. Webb stepped down four days before the countdown began for Apollo 7, his position temporarily taken by Thomas O. Paine. Deputy Administrator Bob Seamans resigned from NASA earlier in the year, his departure effective from 5 January, and Tom Paine was nominated by President Johnson to replace him. Paine came from General Electric's technical Military Planning Operation. Webb's departure marked a point in NASA's history where the 'old guard' no longer considered themselves the backbone of America's civilian space program.

Many of the original seven Mercury astronauts were out of the program too. Deke Slayton was still off the flight list due to his irregular heart condition; John Glenn was now a Senator; Gus Grissom had been killed the year before; and Scott Carpenter was so busy with his aquanaut program that he could no longer be considered an active astronaut. Before final preparations for Apollo 7, Wally Schirra announced this would be his last mission, leaving only Shepard and Cooper in the program.

Friday, 11 October, brought confidence that all would be well for the planned launch: early fears about the weather were allayed; procedures during the night countdown had gone exceptionally well; tracking and recovery teams around the world were keyed up and ready to go. From the Manned Spacecraft Operations Building, the three crewmembers emerged for the short ride to pad 34. The suits they wore, the A7L model, were 10 kg heavier than the 14.5 kg suits originally designed for Apollo but infinitely more effective in protecting the wearer from flame. More flexible than their predecessors, the suits were fabricated from layers of aluminized Kapton, neoprene-coated nylon, Beta cloth, and sections of Chromel-R at the knees, elbows and shoulders. The pressure helmet consisted of a transparent polycarbonate shell attached to an aluminum ring designed to fit into and lock with a matching ring on the torso limb suit. The helmet contained a feed port at the front and a vent pad at the back through which oxygen would flow to the face area.

Tight on the head, each astronaut wore a communications hat fitted with two microphones, two earphones, three laced straps for attachment, an open pocket for a radiation dosimeter, and a 21-pin electrical connector. Dubbed the 'snoopy hat,' it would be worn even when the crew removed their suits. Unlike Gemini days, astronauts would doff their weighty pressure garments a few hours after lift-off and put them on again only when it was time to come home. For the initial period after launch, the suits would provide the crew with a pressurized pure oxygen environment within the mixed-gas cabin. Only when the cabin vented to one-third sea-level pressure and replaced the 40% nitrogen content with oxygen could the crew breath cabin gas.

But wherever the new suit was worn it would be considerably more comfortable, for unlike the Mercury and Gemini suits it contained convoluted bellows at the elbows and knees to ease the problem of limb movement in a pressurized garment. More bulky than the earlier designs, the A7L was a very different suit to the pre-fire model. Finished in a white cloth, it had the appearance of a pressure suit for the Moon. However, two versions had been developed by ILC Industries: intravehicular and extravehicular models; the latter was strengthened with additional protection from micrometeoroids and contained an Integrated Thermal Meteoroid Garment as an overlayer. But for Apollo 7, the intravehicular version would suffice in the absence of EVA.

By the time the crew arrived at the pad, AS-205 was fully loaded with propellants in both stages. John Young and Gene Cernan had been in the spacecraft several hours before, checking the systems, making sure the fuel cells were operating correctly, and setting up the appropriate switch positions. Moving into their spacecraft about 2½ hours before launch the crew were served by the friendly Guenter Wendt, a pleasant reminder of the Mercury and Gemini days. As a member of McDonnell's closeout crew he had served all three generations in the manned program and was now with the Apollo team. When the hatches finally closed on the three Apollo astronauts, Wally Schirra advised listeners on his communications line that 'The next face you will see on your television screen is that of Guenter Wendt.' And from another circuit outside the spacecraft: 'The next face you fellows better see is that of a frogman – or you're in trouble!' Guenter was a charged shot of humor; a veritable white room mascot.

Inside the Command Module the crew were connected to the environmental control system by twin pipes that ran to chest connectors on each suit – a blue fixture for the inlet; a red one for the outflow. Elsewhere, an electrical connector carried biomedical data, and from the thigh a special pocket supplied a thin tube to the waste management system from a urine device inside the suit. In the two hours that separated the crew from closeout to ignition, there were checks of the abort advisory system, the emergency detection system, range safety tracking equipment, and radio frequency systems in the launch vehicle.

At T-40 minutes the pad area was cleared, final command checks were made from Houston, and minutes later the Apollo access arm moved to its standby position. Exposed since the big service structure moved away during the night, spacecraft 101 was now isolated on top of its launcher. At T–28 min the final range safety command checks were performed, and the crew pressurized the spacecraft thrusters. At T–20 min they blipped the tiny jets to satisfy themselves they worked and 5 min later the spacecraft went to internal power. Around the blockhouse Launch Operations Manager Paul Donnelly got affirmative responses from his 'go/no-go' call. One minute later the Apollo access arm fell back to the fully retracted position. From this point an abort would trigger the spacecraft's launch escape rocket. At T–2 min 43 sec. the Saturn went to an automatic launch sequence as computers intensively scrutinized every facet of the stack on pad 34.

At 28 sec in the count the configuration switched to internal power, electrically autonomous for free flight under its own control. Just three seconds prior to the planned lift-off time the eight H-1 engines roared into life sending billowing steam and fire belching from the rocket's blunt base as slowly, ever so slowly, it lifted its 579 tonne load off the concrete nest. And for the fifth time – a Saturn IB bird flew; the first time manned. To Wally Schirra it was a familiar experience, but to his rookie companions it was the thrill of a lifetime.

At the point of ignition they heard the gurgling sound from deep within the big bird's belly as the spinning turbines drank 2.8 tonnes of propellant each second. The sudden explosion of power shook the Saturn IB as 720 tonnes of energy burst upon the channeled flame deflector, a surprising sway on the vehicle as it balanced itself on the tower of fire thundering back down to Earth. Acceleration was slow, but building quickly as the Saturn moved gracefully spaceward. Down below, more than 600 newsmen leaped about and shouted encouragement to the eight powerful motors on whose success the fortune of Apollo 7 ascended. All around the world, the view transmitted by communications satellites, people watched in awe as the mighty Saturn – nearly four times more powerful than Gemini-Titan – sped cleanly through the fleecy clouds into a deep blue sky. After 23 months during which NASA had ridden through its most traumatic incident, manned space flight had resumed with the inaugural flight of a spaceship destined one day to carry men to the vicinity of another world.

Considerable interest had accompanied the preparation of Apollo 7, with all eyes on the vehicle that could vindicate Apollo or dash its hopes for ever. Within 11 minutes the two stages of the Saturn launcher had performed as planned, placing the second stage and CSM into a 285 × 228 km orbit versus the planned 281 × 228 km path selected before launch. In space, CSM-101 performed impeccably. On the second revolution the crew took over control of the S-IVB stage as they could be called upon to do in an emergency, checking the attitude command capability.

One of the first operational tasks called for the CSM to separate from the S-IVB adapter, which on a Saturn V mission would house the Lunar Module, to turn around after moving some distance away and to fly back in toward the adapter in simulation of a move to be performed on a later flight where the LM would be extracted from its 'hangar'. Actual separation came 2 hr 55 min into the flight, a noticeable thump from the Command Module as the pyrotechnics fired and Schirra pulsed the aft-firing thrusters to push the CSM forward. Firing the thrusters again to halt the forward motion, and again to turn the spacecraft 180°, the CSM moved back toward the yawning cavern of the S-IVB adapter. Designed to swing outward to an angle of 45°, one of the four panel sections had stuck at a partially deployed position. Had a Lunar Module been aboard it could have proven difficult to extract the vehicle. Later flights would jettison the panel sections of the adapter so this situation would not occur again.

But gently keeping station on the big stage was a thrill, especially to Schirra who had the experience of a Gemini rendezvous to contrast with the big cylinder he now had hanging off in front. Down below, at the end of two revolutions, the Cape could be clearly seen to one side of the S-IVB. But new tasks were now at hand. Apollo 7 was to perform a maneuver with the RCS thrusters that would carry it into an orbit slightly below that of the S-IVB. On an inner track around the world, spacecraft 101 would gradually move ahead of the stage so that approximately a day later it would be about 139 km in front. From that position, the Apollo 7 crew were to use combinations of the Service Propulsion System and the RCS thrusters to effect a rendezvous, looping first up and above the target to move behind and below it, then to conduct a rendezvous from that position similar to conventional Gemini operations.

First was the phasing maneuver to set up the separation rate. That came about 25 minutes after physically separating from the S-IVB, a 17 sec burn from the RCS thrusters. Over the next six revolutions, however, the orbit of the big S-IVB decayed more than expected and another burn, also for 17 sec, was performed at 15 hr 52 min into the flight. The next maneuver was a corrective combination burn to bring the spacecraft into an elliptical loop 75 km above the S-IVB. Being much higher than the stage, the spacecraft naturally fell behind to a position where it dropped down below the S-IVB. The burn to accomplish the corrective combination maneuver came at 26 hr 25 min, a 10 sec burst from the big SPS engine to increase speed by 64 meters/sec. It was the first SPS firing performed with men aboard and slapped all three back into their couches with the sharp punch at ignition. Schirra was very pleased with the engine's performance, enjoying the dynamic experience of thrusting acceleration with a cry of 'Yabadabadoo!' over the communication circuit.

For 1 hr 35 min the spacecraft orbit rode high above the S-IVB, reaching apogee at 360 km before dropping back down, below and behind the S-IVB. Unimpeded, the spacecraft would have continued to fly its elliptical path, moving further away from the stage with each revolution. But on the first drop down behind the target the crew again fired the SPS engine, this time for 8 sec to achieve a coelliptic orbit. Apollo 7 was now 148 km behind the S-IVB, in a path 14.5 km below it. But what should have been a burn to place Schirra's spacecraft in an orbit with a precise differential altitude was slightly off from the desired value, bringing the Terminal Phase Initiation burn 4½ min earlier than planned. But that was of little consequence. In a 46 second burst from the RCS quads, Apollo 7 increased speed by 5.4 meters/sec and started the gradual ascent to rendezvous with the target.

TPI was carried out 1 hr 16 min after the coelliptic burn and lasted 28 minutes with two minor course corrections before the first braking maneuver, again with the RCS quads. The Terminal Phase sequence had been set up on computer solutions from inside the spacecraft on data from a sextant tracking the S-IVB, and by 29 hr 56 min the spacecraft completed rendezvous alongside the inert rocket stage. In the 26½ hours since firing the first phasing burn, the S-IVB had decayed from 232 × 309 km to an orbit of 226 × 298 km, tangible evidence for the traces of atmosphere present at that height pulling and tugging at the big structure. For about 20 minutes the spacecraft remained alongside S-IVB-205 before firing the thruster quads for 5 sec in a separation burn. Over the next 24 hr period the crew would perform sextant calibration tests, a rendezvous navigation test, an attitude control test, and a primary evaporator test. The crew discovered they were able to track the S-IVB up to 593 km away by using the sextant.

Coming up on the third day in space, Apollo 7 had already completed a major demonstration of guidance, navigation, and propulsion systems performance, overlaying a continually prosperous performance from the electrical and environmental control systems. For the first two days managers and engineers hugged the MOCR consoles at the Manned Spacecraft Center, aware that if there was some underlying anomaly in the Apollo design it could emerge with frightening speed. But as the hours wore on into days the tension relaxed and technicians developed confidence in a spacecraft that really did seem to be as good as the handbook said. For Wally Schirra and his colleagues, the spacecraft itself was a joy to fly. Packed with sophisticated electronics and multiple switching capabilities, an astronaut could enter manual authority to any level of participation he required. Hands off, the ship could virtually fly itself.

Compared with Gemini, it was space travel first class style. With considerably more room inside than the two-man vehicle, Schirra welcomed the space to move around, floating in the weightless state to nooks and crannies unavailable to ground trainers. The center couch could be removed easily, folded down upon itself to expose a center 'aisle' along which the crew could move, with left and right couches supporting hammocks slung underneath for sleep or rest. Nobody liked the hammocks and Walt Cunningham strapped himself down on to his couch but that did not really work because the broad expanse of the main display console always had some switch that had to be operated, always by the man awake disturbing the Cunningham asleep! Flight plan routines required at least one astronaut to be awake at all times but, just like on Gemini, it was a bad plan that gave nobody much rest.

Apollo could be a quieter vehicle than Gemini if nobody operated anything or switched equipment on and off. Occasional thruster firings sounded like gongs booming in a distant cylinder, and then there was the hiss of the suit loop, or the flow control valves opening to admit more oxygen to the cabin. But the lights on the display console were fascinating to watch, especially with the main interior lights out and the windows covered. And there too there was an improvement over Gemini: the windows were much cleaner, free from the oily grease that always seemed to smear up the outer panes. Down past the foot pans on the outer couches was the lower equipment bay, or the navigation station where sextant and telescope were firmly installed in the cabin wall, their optics protruding through the heat shield to lie flush with the conical exterior on the opposite side to the hatch. On the ground the lower equipment bay was the only place an astronaut could stand up; in the weightless state it did not really matter because rarely was a person completely straight and the floor was no constraint. It did provide opportunity for privacy however.

By floating up behind the enclosed display console into the docking tunnel it was easy to find seclusion, looking back down past the eyepieces on the lower equipment bay wall. But the waste management section was on the right wall down past Cunningham's couch so there was not much chance to obtain privacy when it was really needed; nevertheless, that tunnel really was a place to get some peace and be alone with one's thoughts – if only for a minute or two.

As the mission hours passed ground tracking facilities kept constant watch on the drifting spacecraft. At Mission Control, the flight dynamics officer concurred with a plan to move up the third SPS burn by 15 hours to improve the back-up facility whereby if the Service Propulsion System failed to work the RCS thruster quads would be required to retro-fire the spacecraft down from orbit. Altitude was too high for the RCS quads to make it before the propellant ran out so at 75 hr 48 min Schirra fired the engine for 9 sec, reducing speed and cutting perigee by about 130 km. With a new perigee of 166 km the thrusters could de-orbit the spacecraft if the SPS malfunctioned. It was a contingency previously adopted for early Gemini flights. The velocity required to cut perigee had been insufficient to obtain good test data on the performance of the stabilization and control system so controllers built in an out-of-plane component that kept the engine burning longer than strictly necessary but still reduced forward speed by the required value. It was just the first of several changes and modifications to a flight plan that Wally Schirra fought to preserve intact.

Toward the end of the first day, first Schirra and then all three suffered from head colds that were particularly uncomfortable. In the weightless state their nasal congestion was difficult to clear, and the pure oxygen environment made that sensation worse. With puffed cheeks pressing into the sinuses because of the lack of gravity, mucus in the nostrils found its way from the point of exit and deeper into the sacks. Just when the decongestant tablets were beginning to gain ground on the tenacious germs, repeated changes to the schedule brought Wally Schirra close to verbal assault.

His annoyance at the pressure he felt was being applied from the ground first appeared on the second day when he was scheduled to stage a live TV show from orbit. Denied this sophisticated facility in Gemini, the NASA public affairs people had pressed very hard to have the first manned Apollo convey a good image to the watching world. But the TV slot clashed with Schirra's head cold, as Mission Control was made to understand: 'The show is off! The television is delayed without further discussion. We've not eaten. I've got a cold, and I refuse to foul up my time.' Developed by RCA, the tiny, but highly effective, TV camera would wait another day. Finally, on the third day of the mission, Schirra complied with NASA wishes and sent down an entertaining program of interior scenes.

For the first time, Americans were able to appreciate what it was like to float weightless in a vehicle big enough to allow full movement. Shortly after starting the transmission, Schirra held up a card on which was written: 'Hello from the lovely Apollo room high atop everything.' And on a second card: 'Keep those cards and letters coming in folks.' Likening the coming Apollo flights to a TV series which employed the same phrase, Apollo 7 was inviting its audience to keep tuned in! All across the United States, people in their homes could watch three astronauts casually dressed in light blue coveralls gently drifting before a portable camera in orbit. It achieved the desired result and brought home to many Americans the reality of history in the making.

The TV show seemed to ease increasing tensions between the crew and ground controllers, even minor communications interference causing a humorous quip from Cunningham: 'I'm getting a hot tip on some hospital-insurance plan from some guy.' From the ground: 'Maybe they're trying to tell you something.' Then, fourteen orbits after the first show, another burst of TV on revolution 60. This time, one of the cards that appeared on the TV screen had a question for a colleague on the ground: 'Deke Slayton, are you a turtle?' In well established bar room tradition the addressed would be obliged to return 'You bet your sweet ass I am' or pay for a round of drinks. Placed deliberately in a compromising position where millions of listeners would hear what NASA considered an indelicate reference, Deke Slayton cancelled the microphone for a second or two, recorded his answer, and resumed communications with Wally Schirra! Not for Deke the price of a bar round. But the cheerful faces were a mask, for Schirra woke in decidedly grumpy mood that morning: 'We three have colds. I asked for an hour and a half extra sleep for each last night and that was apparently ignored.' Bill Pogue was on capcom duty at the MOCR and tried to quieten the tension: 'We acknowledge the error here. We are setting up for you to get a 10-hour sleep cycle tonight.' Schirra was in no mood to condescend: 'We can't do that! Let's get nearer an average eight. That'll be plenty.'

Nevertheless, when the TV began, Schirra rose to the occasion, introducing 'The one and only original Apollo roadshow starring the great acrobats of outer space.' But systems tests were still the reason for Apollo 7 remaining in space, and Schirra was determined to go the full eleven days planned from the beginning. During revolution 48 the crew had tested the rendezvous radar transponder by locking on to a radar unit at White Sands Missile Range. Then, for 4½ hours, a special test of the environmental control system radiator to qualify it for a lunar flight. At 120 hr 43 min another burn with the SPS, just for 0.5 sec this time to test the engine for a minimum firing resulting in a velocity change of only 3.9 meters/sec.

A point of major concern to Schirra was the biomedical harness each man wore. They were proving none too effective under the improved mobility in the comparatively spacious Command Module, their tiny wires frequently breaking. Schirra more than once referred to the danger of sparks from these broken wires and reminded ground controllers about the consequences of that type of problem. In reality there was little danger, the wires carrying an exceptionally low current. But it was just one more nagging intrusion on what Schirra considered would have been a text-book flight without flight plan changes and continual modifications to the schedule. In fact, personnel in the MOCR were exercising the spacecraft in a way designed to eliminate all possible doubt about the vehicle's ability to function effectively under a variety of differing circumstances. But the irritation spread and when asked to perform a particularly tedious operation, Cunningham responded firmly: 'I would like to go on record here saying that people who dream up procedures like this after you lift off have somehow or other been dropping the ball the last three years.'

And then there were more SPS firing tests. At precisely 165 hr elapsed time the crew fired the main engine for a 67 sec burn, both to further test the propulsion system and to phase the ellipse of the spacecraft's orbit so that during retro-fire at the end of the flight the Hawaii tracking station would have at least 2 minutes of continuous coverage for data acquisition. This and the preceding burn, in addition to those still to come, were arranged so as not to appreciably alter the low perigee point, set up on the third burn to permit an RCS de-orbit if necessary. A second minimum firing burn was carried out at 210 hr 8 min, the ninth day of the flight, for 0.5 sec. And then Schirra was adviced of new figures for a penultimate SPS burn on the following day which would be greater than originally planned to test the stabilization and control system more effectively.

It began a new hostility between Apollo 7 and the ground when Schirra responded: 'I've had it up here today. We have a feeling that you down there believe some of these experiments are holier than God. We are a heck of a lot closer to Him right now.' And later: 'We are not going to accept any new games like adding to the delta-V for a burn, or doing some crazy tests we never heard of before.' Even Donn Eisele put in a quip: 'I want to talk to the man, or whoever it was, that thought up that little gem – That one really got us.' Nevertheless, the questioned burn came off at 239 hr 6 min for 8 sec and satisfied the ground controllers. It was the seventh performed during the flight, exactly the number written in to the flight plan. But Schirra was persistently opposed to continual changes and reminded Mission Control that 'from now on I am going to be onboard flight director for the updates.'

And as if to mock criticism from the ground, where even Schirra's colleagues had nicknamed him 'The biggest Bolshie in space', the penultimate TV show was a demonstration of close order drill. 'It is known in spacecraft talk that we have a crew commander,' Schirra said, 'but what is not known is that we run a taut ship, and to maintain moral discipline we carry on a close order drill instruction period.' Most people watching the show thought it was a sarcastic reference to what Schirra would have considered sloppy flight control; only a very few were ever to know that life aboard Apollo 7 was not too far from the reality of that strict code. Schirra was a man from the Mercury days, in every dimension a test pilot as well honed as Deke Slayton or Al Shepard. But Apollo Commanders were expected to be more like airline captains than fighter pilots and Schirra was one of the last from the early years who still held firm to 'old breed' ideas.

Now it was time to come home, a lighter relief from the pressures of the previous few days. At conclusion of the last telecast from space, Schirra signed off: 'As the Sun sinks slowly in the West, this is Apollo 7 cutting out now.' But there was one final confrontation. Still suffering from bad nasal congestion, all three crewmen worried about their inability to blow their noses during descent to free pressure in their ears; with helmets on, relief would be impossible. Schirra told controllers he was bringing his men home with the helmets off, Deke Slayton tried to convince him that this was unwise since only unmanned Command Modules had ever come back from space until now and if anything went wrong they may need the pressure afforded by the suit loop, but the Commander was adamant. So the crew came back with their helmets stowed.

Coming up on retro-fire, Public Affairs Officer Paul Haney, spokesman for the Manned Spacecraft Center since 1964 when Julian Scheer in Washington replaced 'Shorty' Powers, talked his global audience through the final seconds: 'Mark. One minute from the de-orbit burn. Within two minutes after that de-orbit burn the spacecraft and the Service Module should separate. Spacecraft will be in a pitch down, 48°, attitude at the time of the de-orbit burn, about a (91 meter/sec) burn. Some ten seconds duration.'Flying backwards and pointing down at the Earth, controllers on the ground counted down to retro-fire as the crew monitored the systems and stood by to intervene should anything go awry.

Precisely at the expected second, the bit SPS engine lit up for the last time on Apollo 7, as Schirra acknowledged ignition: 'We're burning right on the mark.' In the MOCR at Houston, mission clocks began counting up from the time of retro-fire, signalling the sequence of events that would bring spacecraft 101 back through the atmosphere. Four minutes later the Service Module was jettisoned giving the crew a 'slap in the face when we separate', freeing the Command Module for its journey home. Entry interface, the point where the spacecraft sensed 0.05 g deceleration, came ten minutes after that followed by a five minute communications blackout due to an enveloping plasma sheath around the vehicle. All Apollo vehicles carried identical heat shields. Designed for re-entry at lunar return velocities, spacecraft 101 would be comfortably inside its performance capability.

At an elapsed time of 10 days 20 hours 9 minutes 3 seconds, the 'Wally, Walt and Donn Show' ended with a resounding splash in the Atlantic, a mission acclaimed as one of the best so far in the manned space program. But the crew were not to get away with an uneventful return. Hauled over to an inverted position by parachute lines, Schirra had to right the Command Module by inflating the three bags on top of the spacecraft. For a few minutes recovery ships lost all trace of the bobbing spacecraft, until it popped back into communication at the Stable I position. It was drizzling when three bearded astronauts arrived on the deck of the aircraft carrier *Essex*. Unlike the Gemini days, astronauts were obliged to exit their Command Modules on the water due to the weight of each conical spacecraft; ships' derricks were the only lifting gear capable of retrieving them.

Everybody was well pleased with the performance of Apollo 7, but Wally Schirra had revealed to the world a volatile side to his nature. In every respect he was a Navy man, blunt and professional when working, entertaining and jaunty in play. It was the same Wally Schirra that complained at the heavy public relations load bequeathed to John Glenn back in 1962, but a side to his personality few would recognise until Apollo 7. Increasingly, astronauts were expected to be set-piece ambassadors for what the space agency considered its most responsible role so far: the final run up to and the accomplishment of Man's first landing on another world.

Following the initial selection of NASA's first seven Mercury astronauts in 1959, the agency had recruited additional teams: 9 in 1962, 14 in 1963, 6 scientist astronauts in 1965, and 19 in 1966. Partly to accommodate the expanding need for crew positions envisaged by the Apollo Applications Program, then seen as a very ambitious attempt to sustain lunar exploration and set up Earth-orbiting space stations, and partly to offset attrition, the agency had selected 11 more scientist astronauts in August, 1967. These men reflected in the most extreme way the changes of a decade. They were to give Deke Slayton more than one headache.

The list comprised: Joseph P. Allen; Philip K. Chapman; Anthony W. England; Karl G. Henize; Donald L. Holmquest; Brian T. O'Leary; William B. Lenoir; John A. Llewellyn; Story Musgrave; Robert A. Parker; William E. Thornton. All had doctorates in either science or medicine – one in both – and they were all typical of the incumbent academic fraternity that sought to bring science to Apollo. As such their motives were impeccable, but their fortunes were varied.

Welshman John Llewellyn resigned two months before Apollo 7 because he was unable to accept the rigors of high speed flight and because the pressure and pace of a military-style existence took its toll. Brian O'Leary had left within eight months of selection because he could not accept the style of a program already ten years into its genesis; unable to accept that space flight was still essentially an engineering venture he went on to use his intimate knowledge in a published condemnation of the entire Apollo program, which he continues to do today. Holmquest took leave of absence in 1971. Chapman resigned a year later. In 1975, after serving effectively as a well loved support scientist, Joe Allen took up an administrative position. None of the others ever flew in space.

CHAPTER EIGHTEEN

To the Moon

With the successful return to Earth of the Apollo 7 Command Module, NASA had regained confidence lost in the Moon mission hardware by spectators and observers of the manned flight program. From what Sam Phillips considered to be a mission with '101 per cent success' came the green light for more ambitious activity. And there were plenty of suggestions about what the next steps should comprise based not only on open knowledge of the pace of Apollo's schedule but the discreet sources watching Soviet space plans. When Jim Webb resigned from NASA in October he was bitterly disappointed with the lack of support from the Johnson administration. In the light of sweeping budget cuts affecting every federal agency, NASA was forced to cancel Saturn V launchers beyond the 15th vehicle, and Saturn IBs beyond the 12th. Already, as seen in Chapter Sixteen, Apollo Applications flights had been cut drastically. Webb was deeply concerned that the Soviet Union was still ahead in the space race and that they would continue to pull in front.

For several years Webb had warned of the development in Russia of a massive booster to equal the lifting capacity of Saturn V. But it seemed unbelievable in the light of so many Gemini flights, and now the success of Apollo 7, that America was actually *behind* Russia. What Jim Webb knew, and the public did not, was that even as he spoke plans were being finalized in the Kremlin to fly men around the Moon before year's end, or at least in January, 1969. Intelligence sources laid a dispiriting carpet of information about concerted efforts by Russia to upstage American Moon landing plans and if details were lost to an increasingly apathetic public, anybody associated in any way with the space program watched the final pieces of the Soviet jigsaw fall into place. Toppled by the death of Komarov in April, 1967, Soviet plans now embraced an all-out assault on two separate operational procedures. The first concerned rendezvous and docking, a feat still to be achieved by Soviet spacecraft.

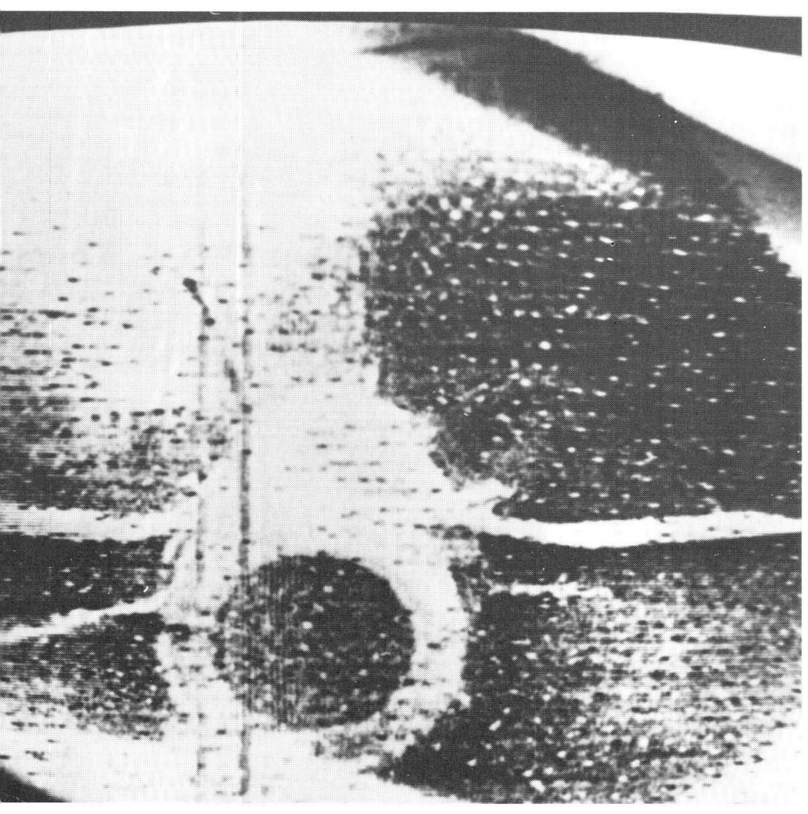

Pictures from Moscow Television show Cosmos 213 pulling away from Cosmos 212 following a successful docking operation in April, 1968, activity essential to Soviet space plans.

On 27 October, 1967, Cosmos 186 was placed in Earth orbit until its sister ship, Cosmos 188, was launched on 30 October. Ascending from the launch pad to a rendezvous on the first revolution, the chase vehicle closed to within 24 km at first apogee. From this point a completely automated rendezvous and docking sequence began, resulting in Cosmos 186 assuming the active role for the actual docking operation. The two spacecraft were stripped-down versions of the Soyuz spacecraft first flown in the manned configuration by Komarov six months before. After $2\frac{1}{2}$ orbits of the Earth the two vehicles undocked over Soviet territory, Cosmos 186 making a re-entry and soft landing one day later followed by Cosmos 188 two days after that. It was a fitting tribute to Russia's celebration of the November 1917 revolution, a 50th birthday many suspected would see a Soviet man on the Moon.

Then again, six months later on 14 April, 1968, Cosmos 212 ascended to a low Earth orbit, followed a day later by Cosmos 213. Again, each vehicle was in fact an unmanned Soyuz test model. Launched to within 5 km of the target, Cosmos 213 became the passive partner as 212 locked on and completed an automatic rendezvous operation. Following the docking, TV cameras were switched on to present a view from outside looking across from one vehicle to the other. The ships separated after nearly 4 hours, each taking up a separate set of flight routines, with 212 returning on 19 April, followed by 213 a day later. Clearly Soviet engineers were using the time required by Soyuz technicians to correct the faults responsible for Komarov's death of flying unmanned vehicles of a similar type designed for rendezvous and docking.

On 28 August, 1968, as final preparations were underway at the Kennedy Space Center for the launch of Apollo 7, the Russians launched Cosmos 238. Another Soyuz like the previous four, it carried equipment specifically intended to support a manned test activity. Four days later it returned to Earth, having performed a complete evaluation of the several systems it contained. It was the final qualifying flight for changes wrought by the dramatic failure of Soyuz 1 sixteen months before. Apollo 7 counted down, flew its eleven day mission, and splashed into the Atlantic north-east of Puerto Rico on 22 October. Just three days later, emerging now from the clandestine Cosmos designation, Soviet launch teams sent Soyuz 2 into low orbit, an unmanned vehicle clearly related to a manned mission yet to come.

Observers waited just one day before, on 26 October, an announcement that Colonel Georgiy Beregovoy had been placed in orbit aboard Soyuz 3. In a flight profile similar to the two earlier rendezvous missions, Beregovoy rode out an automated sequence that brought his spacecraft to within 200 meters of the unmanned target, Soyuz 2. TV cameras on the outside showed a Soviet audience the view of the opposing vehicle. Several times the cosmonaut approached Soyuz 2, but never achieved a docking. He was undeniably thwarted in what had been intended as the first Soviet manned docking operation; the Russians would insist that this had never been a part of Beregovoy's flight plan. Following a scheduled sleep period, the pilot again brought his spacecraft alongside the target, this time from a distance of several hundred kilometers due to the drifting separation of the intervening period.

With operations controlled from Earth on information processed from rendezvous antennae on each vehicle, the engines on Soyuz 3 fired to alter the orbit, Beregovoy again taking over manually at a separation of just 200 meters. Close up, the side-facing windows of the inhabited compartment prevented the pilot from seeing the target. A periscope was provided for this purpose, an essential addition for close-in maneuvering. It protruded from the cabin wall to align with a target marker on the unmanned vehicle. Two days after arriving in space, Soyuz 3's pilot watched as Soyuz 2 moved away. During the morning hours of 28 October, Moscow time, the retro-fire process began and the target vehicle's re-entry module returned to Earth. Beregovoy came back on the 30th after a flight lasting four days during which time he had carried out several scientific and engineering tests of

Much modified from the Gemini days, Mission Control Room (MOCR) facilities at the Manned Spacecraft Center were shaped around Apollo Moon mission needs.

the spacecraft in addition to televising frequent shows to the ground tracking stations.

Rendezvous and docking was clearly an important element in the Soviet manned space flight program and the dual flight of Soyuz 2 and 3 served to qualify the manned vehicle following what must have been extensive modifications to prevent an accident of the type that took Komarov's life. But this was only one of two new operational procedures quickly developed by Soviet engineers in 1968. The second concerned an ability to control a space vehicle in circumlunar flight and to return it through Earth's dense atmosphere at lunar return velocities.

Under the guise of the Zond designation, a Soyuz spacecraft was propelled to deep space by a D-class launcher on 2 March, 1968. Devoid of an orbital module on the front, the spacecraft was propelled by a rocket capable of delivering 6 tonnes to the vicinity of the Moon but for this flight the spacecraft, called Zond 4, weighed about 4.9 tonnes and was boosted to a deep space elliptical orbit in a direction opposite to that of the lunar sphere. This was probably an engineering flight to evaluate the performance of a fourth 'escape' stage added to the basic D-1 launcher and to check the Soyuz heat shield. The presence of the Moon would have complicated the trajectory unnecessarily.

An attempt to launch this mission profile was made as early as 22 November, 1967, but the launcher failed on that occasion. It seems that the Zond 4 mission was not the success expected either and another attempt failed on 20 April, 1968. However, Zond 5 was put on to a trans-lunar trajectory following launch on 14 September, less than a month before Apollo 7. Three days later it swung around the Moon only 1,950 km from the surface and headed back toward Earth, aiming for a narrow corridor of space between 35 km and 48 km above the planet; 10 km too low and it would burn up, 24 km too high and it would skip out of the atmosphere and never return. Skimming into the air at 39,240 km/hr, the descent vehicle of Zond 5 was aimed for entry over the south pole, splashing down in the Indian Ocean as it flew north.

On board were biological payloads and film cameras, turtles, meal worms, plants, wheat, pine and barley seeds, wine flies, bacteria, and chlorella. Zond 5, a modified Soyuz without the forward orbital module, took seven days to fly to the Moon and back. It had not gone into orbit but that had not been intended. Clearly the concept worked and manned flights of a similar type were feasible. But the precise technique still needed refinement. In re-entering over the south pole, Zond 5 imposed between 10 and 16 gs on its payload since it used the simplest and safest form of trajectory. One more flight would be needed to demonstrate a lifting, or 'skip,' technique where the descent vehicle would use its offset centre of gravity to fly a roller-coaster re-entry rather like Apollo would when returning from the Moon. This would reduce maximum g loads to less than 7, an acceptable value for manned flight.

So far, nothing had been sent into space by the Russians to indicate a Soviet preparedness for manned *landings*, but the Soyuz missions of 1968, while seeking to hide their purpose under the different Zond designation, pointed strongly toward a Soviet interest in manned *circumlunar* flight. But because Soyuz had also sought a viable means of rendezvous and docking, was the basic manned vehicle merely one component of two elements designed to dock together in support of a Moon landing operation? Was Soyuz perhaps like the Apollo CSM, merely the manned mother-ship to serve another vehicle? The D-class launcher was certainly incapable of launching more equipment than one modified Soyuz (Zond) to the vicinity of the Moon. But was the D-class rocket perhaps the equivalent to NASA's Saturn IB, responsible only for launching the mother-ship on checkout tests? If so, where was the Soviet equivalent of Saturn V?

The answer was already known to US intelligence sources when Jim Webb retired from NASA. For more than a year preparations had been moving confidently toward the first flight of the so-called G-class rocket – 'Webb's Giant' as it became known. Tests with mock-ups, load test vehicles, and dynamic test articles all preceded the planned assembly of a flight rated launcher. Early in the spring of 1968 a facilities test mock-up was sitting on the Soviet pad but major problems still attended the flight hardware. The facilities vehicle was returned to shelter, only to re-emerge in the July for more checks and tests. As for the first flight-rated booster, that was getting further and further behind schedule and by the time Zond 5 proved the spacecraft was nearly ready for manned circumlunar operations it was apparent to everybody concerned with the G rocket that a flight could not be expected before 1969.

In the wake of the Soyuz 1 disaster, and as a consequence of delays to the G-class launcher, Soviet hopes for a manned Moon landing slipped away. But the D-class rocket was available, and that could support a manned circumlunar flight at least with just the modified Soyuz components. One more test and it would be ready for the cosmonauts. But one essential element of the entire operation required a politically clear run for the engineers and technicians working on the mission. Soviet Russia was unwilling to be visibly seen dragging its feet behind American contemporaries. If the United States successfully flew a team of astronauts around the Moon before they could launch a manned Soyuz on such a mission, the whole idea would have to be abandoned. For if the Americans did that, it was odds on that they would also land on the lunar surface within a few months. And that would leave the Russians second to a circumlunar or Moon orbit flight, and nowhere at all in the race to reach the surface. Politically, it was a doomed plan if America flew that type of mission first. If left unattempted, Russia could always claim it had never been in a Moon race to begin with.

The scheme that snagged the Soviet plan emerged, ironically, as a completely separate issue among top level NASA personnel. As related in the previous chapter, by mid-1968 it was painfully apparent that LM-3, the first manned Lunar Module being prepared for launch, would not be available until early 1969. That seriously threatened the continuity of the A-G sequence of increasingly complex mission profiles set up by Owen Maynard. Under Mueller's most recent plan to this time, Apollo 8 would employ a Saturn V to launch McDivitt, Scott and Schweickart on an Earth orbit flight testing both CSM and LM vehicles before year's end. Early in 1969, another Saturn V would launch a deep-space CSM flight manned by Borman, Collins and Aldrin. It was the impending absence of LM-3 for Apollo 8 that led to George Low's plan to upgrade the mission to a CSM-only Moon orbit flight.

But by this time Mike Collins had been grounded temporarily to have an arthritic bone spur removed from his spine, an operation said to require six months recuperation. So his place aboard Borman's crew was taken by Jim Lovell, the back-up in Collins' seat. This shuffle was to give Collins a ride on the greatest flight of all, and forever prevent Jim Lovell from landing on the Moon. Apollo 9 was now teamed by Borman, Lovell and Anders.

George Low began his move to get a more ambitious mandate for Apollo 8 on Wednesday, 7 August, when he called on Chris Kraft and asked him to start work on a suitable flight plan. Next day, Low took Carroll Bolender, Owen Morris and Scott

Simpkinson to the Kennedy Space Center for a discreet chat with General Phillips, KSC Director Kurt Debus, Rocco Petrone and Roderick Middleton. His task was to seek the status of the next Saturn V launcher, AS-503. Middleton had been spring-loaded to wheel 503 right out after 502 (Apollo 6), but trouble with the second and third stages on that flight stayed his command to start wheeling. Nevertheless, everybody thought January 1969 was attainable and, satisfied with what he heard, Low flew back to Houston.

Friday, 9 August. Early in the morning hours, Low sought Bob Gilruth, MSC's Director, and told him the whole plan. Instead of flying a LM type mission in Earth orbit, Apollo 8, the first manned Saturn V flight, should operate the complete CSM leg of a typical Moon flight, going through all the routines into and out of Moon orbit before returning to Earth. Gilruth was fired by the idea of such a dramatic role for Apollo 8 and immediately sat down with Low to call up Debus and Petrone at the Cape and Phillips and George Hage from Washington. Within a very short while it was arranged that with Kraft and Slayton, Gilruth and Low would fly to the Marshall Space Flight Center that afternoon and meet the four executives from the Cape and NASA Headquarters.

At Huntsville, von Braun, Eberhard Rees, Ludie Richard and Lee James hosted the visiting party. Throughout that afternoon, the twelve men discussed the availability of hardware, the opportunity for processing the necessary launch equipment, and the status of technical changes induced by the problems with Apollo 6. The Marshall men were confident of giving the spacecraft people a good bird, and the KSC managers agreed they could get 503 ready by early December. If Apollo 8 was to have any meaning for the future objectives of the program it would have to fly a mission very similar to that projected for the first landing flight. That meant the lighting angles had to be the same which in turn demanded a launch on or about the third week in December.

Low had been particularly concerned to press for a December launch rather than one in January because that would allow five flights in 1969, preserving a maximum number of missions in case something went wrong before the landing attempt, necessitating a repeat mission. It was also in the interests of the agency to snatch such a dramatic flight just as early as possible, for any delay could give the Soviets valuable time to mount their own spectacular. Sam Phillips was concerned to keep the plan secret until they had all had a chance to discuss finer details with systems specialists. Jim Webb and George Mueller were not to be approached until everybody had all the answers they might need. At the end of the day the conspiring twelve agreed to meet in Washington five days later to decide finally on the plan of action. If everyone still agreed to proceed with a recommendation, Sam Phillips would fly out to Vienna where Webb and Mueller would be attending the United Nations Conference on the Exploration and Peaceful Uses of Outer Space.

Back at Houston that Friday evening, George Low talked with Ken Kleinknecht and Carroll Bolender, George Abbey and Dale Myers, about the suitability of spacecraft 103. Third in the production batch of Apollo CSMs, 103 was next in line to 101; 102 had been used to check out pad 34 and was not considered suitable for flight. Dale Myers left for Downey that evening not only to hurry North American Rockwell along with 103, but to make sure it carried any changes Kleinknecht might wish to make in light of its impending switch of roles. Bolender, meanwhile, was on his way to the opposite coast, to Grumman's New York facility in search of a LM-3 stand in. Because launch vehicle 502 had exposed vibrations created by the presence of a lunar test article simulating the mass distribution of a Lunar Module, 503 would have to carry a similarly authentic stand-in to see if the corrective changes worked.

By Monday morning, Chris Kraft had put together a preliminary flight plan, recommending 20 December as the launch date. Selected so as to fly with the same lighting angles around the Moon as a landing flight would have, it also was the only available slot for satisfying all the abort and contingency constraints. Deke Slayton too had satisfying news. Over the weekend he talked with Jim McDivitt about the proposed change of role and got a very disinterested response, not surprisingly since his crew had spent long hours in the simulators training fervently for a Lunar Module flight. So Slayton then approached Frank Borman and offered him the mission. Since the Borman team had been training already for a CSM-only flight, the high altitude component of their original mission needed stretching only a few hundred thousand kilometers to keep most of it intact!

So that was that. If approved, Frank Borman, Jim Lovell and Bill Anders would fly Apollo 8 to Moon orbit, while Jim McDivitt, Dave Scott and Rusty Schweickart continued to train for the CSM-LM Earth orbit flight of Apollo 9. It did mean a switch of Command Module assignments however. As Command Module Pilot, Dave Scott would have to change 103 for 104 and Jim Lovell would get the spacecraft Scott had groomed from the assembly line. But since Lovell had only just slotted in from the back-up position, that bothered him less than it affected Scott.

On Wednesday, 14 August, Low, Gilruth, Debus, Petrone, Phillips, Hage, Kraft, Slayton, von Braun, Rees, Richard, and

The altitude chamber at the Manned Spacecraft Center was built to contain Apollo spacecraft on systems tests.

The size of the altitude chamber is scaled here by people standing close by the main door, concave in profile because the greater pressure would be on the outer surface when the interior of the chamber was reduced to a vacuum.

James met at NASA Headquarters in Washington. They had asked to have a session with Tom Paine, Deputy Administrator since March. Also present were William Schneider and Julian Bowman. Between them they presented a coherent picture of a team ready and waiting to go: spacecraft were available, launchers were being prepared, astronauts were primed, and flight operations personnel supported the initiative. Paine was not so easily convinced and prodded the group with questions. He could find no technical or programatic reason why the proposal should not go through, but it was not up to him to give the final word. As head of manned space flight since 1963, George Mueller had to agree, and as the long and respected head of NASA, Jim Webb's concurrence was the final requirement.

Before the meeting broke up Mueller was on the line from Vienna and during a preliminary conversation about the proposal he expressed his displeasure at this sudden turnover in the well prepared schedule. Mueller implored Phillips not to come to Vienna. When he heard from Mueller that his subordinates had been hatching plots in his absence, Webb was furious. But he gradually came around to the idea and telephoned Sam Phillips to discuss details, and then talked with Tom Paine on the matter. Tome Paine sensed that Mueller and Webb merely needed adequate details at their disposal to see how correct the decision was. LM-3 would not be ready before the early months of 1969 and the next Saturn V would fly merely a repeat of Apollo 7 if it had nowhere to go but Earth orbit. Yet Apollo 7 had not flown and nobody knew it would be the success it eventually turned out to become.

On Thursday the 15th, Paine got Willis Shapley, Julian Scheer and Sam Phillips to draft a lengthy memorandum to Webb, sending it as a cable for the Administrator's perusal. It was a highly detailed explanation of the sequential steps necessary to accomplish a Moon orbit flight and suggested a draft statement that Webb might care to make from Vienna in announcing the plan publicly. But Jim Webb was not about to go public on the strength of a plot hatched by executives in the US. He had a very detailed discussion with George Mueller before cabling Washington on the Friday, 16 August, authorizing Paine to go ahead with preliminary preparations for the advanced mission but without making an official announcement that the orginial sequence had been changed.

Sam Phillips telephoned George Low to set up a meeting for the following day in Houston, where the two met with Hage, Gilruth, Kraft and Slayton. Because of the sensitive nature of a mission from which the agency would have to back away should Apollo 7 not go precisely as planned, the managers were to prepare officially for an expanded Earth orbit flight with Borman's crew on an Apollo 8 mission for 6 December. If, in the meantime, Apollo 7 went well, they could go public on the Moon orbit plan and re-cycle launch preparations to 20 December, the appropriate launch window for the expanded flight. Until that time, everybody was to treat flight and crew preparations as though Apollo 8 was a conservative step beyond Apollo 7.

Sam Phillips wrote the directives for Apollo 8 Monday, 19 August, and on that day NASA officially announced to the press that it was switching the McDivitt team for the Borman crew, flying spacecraft 103 as planned on Apollo 8, and deferring manned Lunar Module operations until the next flight – subject to the anticipated success of Apollo 7. And that was that. Borman, Lovell and Anders moved on with their training at an accelerated rate, having now moved up from the third to the second manned Apollo mission, and embraced critical simulations relevant to a trans-lunar flight objective. As for the McDivitt team, they had an additional two and one-half months in which to get ready for the first Lunar Module manned shakedown operation, hopefully ready for flight at the end of February.

But by this time CSM-103 had arrived at the Cape, all the launch vehicle stages were on hand and a final design certification review imbued confidence in the changes made as a result of the 502 anomaly. About the only operational change would occur when the center F-1 engine in the base of the first stage was shut down earlier than before, leaving the remaining five engines to burn about six seconds longer. This had the effect of easing acceleration as the vehicle became much lighter toward the end of first stage thrust. Nothing that had happened to 502 was now seen as a potential peril to 503 and Apollo 8. Apollo 6 had been an unfortunate set of circumstances that conspired to multiply a set of errors. Small price to pay for the by now creditable 'all-up' systems test philosophy. And as an extreme example, Apollo 8 itself was in pursuit of that concept.

A decade before, probably ten flights would have preceded the ambitious performance now asked of spacecraft 103. As it was, many people on the edge of the program had grave doubts about the wisdom of flying only the second manned Apollo all the way out to the Moon. But immediate concern focused around the activities of Apollo 7: preparations throughout September, launch 11 October, recovery 22 October, post-flight analysis in the weeks that followed. A Certification Board meeting at the Marshall Center on 19 September cleared the Saturn V in its modified configuration, giving the administrators an assurance the engineers possessed a month before, and final weight lifting potential was calculated. Decision time emerged early in November. Detailed inspection of Apollo 7's performance record, meticulous analysis of telemetry records and data sheets, proved that nothing now stood in the way of a technical 'go' for the Moon orbit flight. On 7 November, a full hardware status review was made at the Certification Board meeting where officials learned that spacecraft and launch vehicle were ready.

On Sunday the 10th, Sam Phillips, George Low, Rocco Petrone, Deke Slayton and Chris Kraft met with George Mueller and key officials from the industrial contractor force responsible for prime systems in Apollo: Thomas Morrow from Chrysler; Joseph Gavin from Grumman; Lee Atwood and William Bergen from North American Rockwell; T. A. Wilson and G. H. Stoner from Boeing; Hilliard Page and Gerald Smiley from General Electric; Walter Burke from McDonnell Douglas; Stark Draper from MIT; B. O. Evans and R. W. Hubner from IBM; B. P. Blasingame from AC Electronics; George M. Bunker from Martin Marietta; Robert E. Hunter from Philco-Ford; William Gwinn from United Aircraft. The 16 contractor representatives were each asked in turn for their opinion of the plan presented by NASA management. All unamimously agreed that Apollo 8 should head for the Moon; only one, Walter Burke from McDonnell Douglas the Saturn's third stage builder, thought the spacecraft should simply go around the Moon rather than burn itself into Moon orbit from where the Service Propulsion System would have to work again to return the crew to Earth.

Final flight scheduling had stipulated launch in the morning hours of 21 December, a refinement from the earlier date. Analysis of the electrical needs showed that even if two of the three fuel cells failed, the spacecraft could still get back on the power from one. Studies of the environmental system showed that even if one oxygen tank failed the crew could get back on the supply from the other. Understandably there was a degree of searching appraisal about the risks involved, but those present soon realized that any negative conclusions they came to were not a critique of Apollo 8 per se, but of the complete Apollo operation, for all lunar landing missions would be required to fly the profile selected for the second manned flight. If it was not now appropriate for Apollo 8, it might never be appropriate for *any* spacecraft. And then there were the arguments that it was, of course, only the second time a manned Apollo would have been sent into space, only the first time men had ridden a Saturn V, and only the first time astronauts had been sent beyond the immediate vicinity of planet Earth.

Support of the mission in light of these questions fell back upon a sophisticated variant of Mueller's 'all-up' systems testing where each sequential step in a complex flight is attempted only on the success of the preceding event. Mueller had set out nine 'plateaux' levels each of which would have to go well nigh perfect before the lunar landing mission proceeded to the next. Apollo 8 would fly along similar lines, moving step by step to increasingly ambitious routines. At the end of the day everybody at the meeting approved of NASA's proposal. It was time to take it to Tom Paine, Acting Administrator now at the resignation of Jim Webb.

The following morning, Monday, 11 November, Sam Phillips, Lee James, George Low, Chris Kraft and Rocco Petrone presented Paine with the results of their determinations. After hearing their case, the Acting Administrator satisfied himself that key elements not represented at the meeting were in full accord with the proposal. He spoke to Gerald Truszynski about the readiness of the MSFN stations to track and communicate with the lunar-bound spacecraft, to Vincent G. Huston about the availability of recovery forces to meet a December splashdown in the mid-Pacific, to personnel at MSC, KSC and the Marshall Center, and to Frank Borman – the man who could command Man's first exodus

MANNED SPACE FLIGHT NETWORK (APOLLO-8)

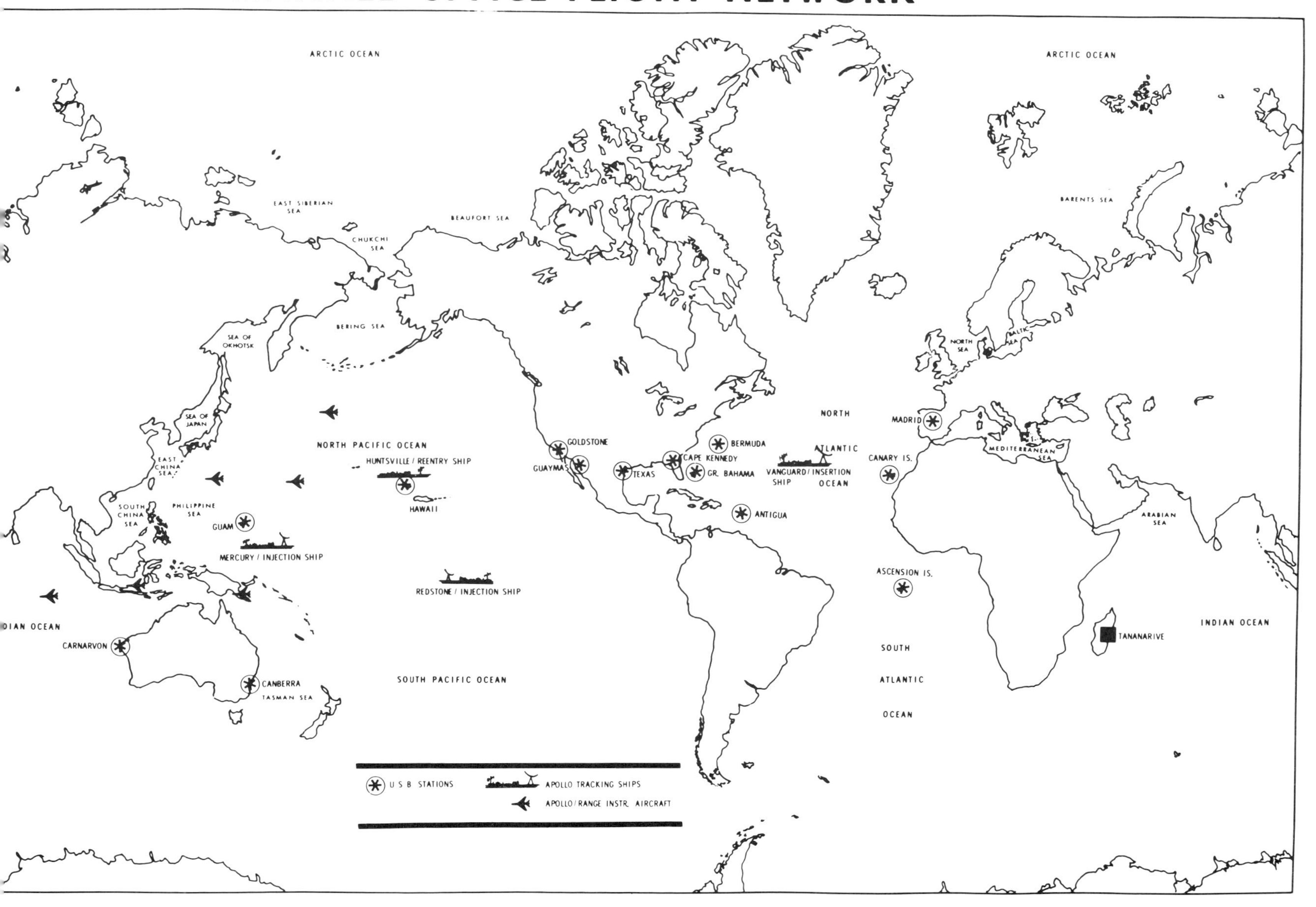

to a new world. At the end of Monday's work, Paine passed final judgement – go for the Moon. His decision was inevitable in the wake of so much success with earlier flights but the final commitment was certainly made in the sure knowledge that the Russians were probably not far behind.

Only the day before, Zond 6 left Earth on a flight to the Moon. Passing around the far side, 2,250 km above the surface, the Soyuz instrument and descent modules looped back toward Earth for an approach over the south polar region on 17 November. But instead of flying a ballistic re-entry like its predecessor two months earlier, Zond 6's descent module skipped in the atmosphere, reducing speed from 39,600 km/hr to 27,360 km/hr before starting the final descent. This roller-coaster re-entry cut g forces in half and carried the vehicle on to a land recovery in Soviet territory. It was clearly an advance on Zond 5 and brought the degree of sophistication essential for manned operations. Following the flight, Soviet news sources openly spoke of a development program to send men around the Moon and cited the flights of Zond 4, 5 and 6 as precursor steps leading toward that goal. But, they said, more tests would still be required.

The day after Paine's formal decision, while Zond 6 was en-route to the Moon, the Acting Administrator met the press along with Sam Phillips to announce the upgraded responsibilities of Apollo 8: 'After a careful and thorough examination of all the systems and risks involved, we have concluded that we are now ready to fly the most advanced mission for our Apollo 8 launch in December, the orbit around the Moon. . . .' One day later Tom Stafford, John Young and Gene Cernan were formally announced as the crew for Apollo 10, back-ups comprising Gordon Cooper, Donn Eisele and Edgar Mitchell. After McDivitt's Apollo 9 test with CSM and LM in Earth orbit, the Stafford crew would probably fly a Mission F profile where CSM and LM would be taken all the way to Moon orbit for full dress rehearsal of the landing attempt. Or, if McDivitt's flight failed to produce good results, Apollo 10 could be limited to an Earth orbit mission. At the earliest, Apollo 11 would make the touch-down attempt, probably in July, 1969. And if crew rotations held good, Neil Armstrong and Buzz Aldrin would make the historic descent.

Hardware for Apollo 8 was ready in the Vehicle Assembly Building by 7 October, the day CSM-103 was mated to the Saturn V. Two days later, AS-503 rumbled through the big doors at the start of a journey destined to end around the Moon just before Christmas. Celestial mechanics dictated the time line, but it certainly was a nice touch. A fitting Christmas present for the world, and a timely reminder to the incoming President that the space program was an important tool for international goodwill. By early November the stack had been electrically mated on the pad before the Overall Test Plugs. It verified compatibility among all the many systems in the rocket, in the spacecraft, on the pad, and in the surrounding support equipment. A Flight Readiness Test followed by the middle of the month with both prime and back-up crews participating at various times.

Preparation of 503 was unique in that it brought men into the loop. Being the first manned Saturn V operation it was to set a precedent for later missions. The Count Down Demonstration Test (CDDT) was to be divided into 'wet' and 'dry' portions, the former being performed without the crew and made to represent a full rehearsal of an actual countdown with propellants brought aboard the vehicle at appropriate times, the latter being a repeat with the crew brought into the spacecraft at the appropriate time but without propellants loaded; only on the day of launch would the potentially hazardous fuelling operation involve the crew.

The first wet CDDT attempt began on 5 December but problems dragged it out beyond the expected schedule. Three days later a second attempt revealed software problems causing a postponement to the 9th, until additional delays moved it out to the 10th. Late in the day the wet CDDT was completed, followed on 11 December by the dry run involving the crew. But holds like this on launch day would be unacceptable. The Saturn had a $4\frac{1}{2}$ hour launch window on successive days from 21 to 27 December, the earliest beginning at 7:51 am on the 21st. Yet the extent of this tight window period was only possible due to a variable azimuth concept, where changes in the trajectory during ascent would permit the launcher to rendezvous with the appropriate position in space for the boost to the Moon. The 4½ hour window represented, in fact, the limits on the azimuthal change made possible by launch rules at the Cape.

As final preparations moved ahead at the Kennedy Space Center, public opinion was making itself felt, at home and abroad. In the United States, press coverage began to pick up momentum, the rendezvous of one singularly historic event with Christmas not going unnoticed. Said the *Washington Star*: '. . . this promises to be one Christmas when the thoughts of all . . . will contain more than visions of sugarplums, of laden stockings, of gifts about to be received and bills about to come due. It is, in fact, just possible that NASA will succeed in putting more missing ingredients back into the yule season, and that more prayers will be offered this Christmas than at any time in the past 2,000 years.'

And from the *New York Times*: 'Space contains more than enough opportunity for fruitful application of the energies that all mankind can devote to its exploration, development and eventual settlement. There is no need here for wasteful rivalry deriving from earthbound nationalistic and political ambitions. In the face of the most breathtaking challenge humanity has ever faced, the only rational response is cooperation to make space an arena of unity and international brotherhood. Man's hopes and prayers ride with the pre-Christmas voyagers.' But it was Frank Borman that summed the views of all three crewmembers: 'When you're finally up at the Moon, looking back at the Earth, all these differences and nationalistic traits are pretty well going to blend and you're going to get a concept that maybe this is really one world and why the hell can't we learn to live together like decent people.'

But not everyone agreed that it was right to fly Apollo 8 to Moon orbit, that the risks were worthy of the goal, or indeed that Apollo was a creditable venture at all. In an interview for British television, Sir Bernard Lovell, Director of the Jodrell Bank radio telescope station near Manchester, England, was sure that 'On a scientific basis. . . . We've reached the stage with automatic landings when it's not necessary to risk human life to get information about the Moon. Within a few years this information could be obtained by automatic, unmanned instruments,' and that as far as the Moon orbit flight was concerned there was 'a dangerous element of deadline beating in it.' Yet Sir Bernard merely reflected the popular misconception that the Apollo 8 flight, even the entire program, represented an attempt to acquire scientific information. Later, he would rally to Apollo as one of its most ardent supporters.

Dr. Ralph Lapp, prominent figure in the development of the world's first atomic bomb, went further by criticizing the very reason that inspired Apollo 8. 'We are pushing our luck, gambling that everything will work perfectly. . . . The basic factor is not really technical. We are racing the Russians to the Moon. A lot of people in NASA and in industry are hoping that a successful Apollo 8 orbiting of the Moon – or even circumnavigation – will build up public support for an invigorated manned space program.'

At this the *Washington Daily News* found it wise to counsel readers on the real motives. 'There are perhaps sound reasons involving national prestige for trying to be the first nation to send men into a Moon orbit. But surely no such reasons are compelling enough to cut corners on safety. The technical arguments advanced by Dr. Lapp are far too complex to be resolved by laymen. But after the tragic fire that took the lives of three of our Apollo spacemen two years ago it should not be necessary to urge that (NASA) exercise all due prudence – even at the risk of losing the race around the Moon.' What the *Washington Daily News* failed to grasp was that the operations performed by Apollo 8 would be an essential ingredient of every manned Moon landing and that in the light of a Lunar Module touch-down flight the orbital objective of this second Apollo mission would pale into insignificance.

But Britain's *New Scientist* magazine, traditional critic of America's manned space program, thought that it would be 'prudent' to 'test one or two more of the dozen Saturn V vehicles before venturing into lunar flight.' In the final analysis though, said the magazine, 'we are forced to the conclusion, time and time again, that the whole manned space business, if its relationship to reality is not of a more ominous nature, is mere prestigious prancing; that, despite offended denials that the USA could be concerned in anything so trite, it is a plain technological set-to between the Americans and Russians in which the only outcome is the right of the winner to strut before lesser nations cocking a snook at the loser. The technological spin-off, though it exists to some extent, is no more than the poor apology of somewhat guilt-troubled NASA men who feel at heart some pangs about the billions of dollars that pour through their hands in this empty, obsessional quest.'

Unbalanced and emotive comment such as this was not common, however, the basic reference of the argument being false in content when it implied a large proportion of the nation's wealth going on space activity; Americans spent more on looking after national parks. Nevertheless, Apollo 8 catalysed all the expressed thoughts of a new generation with time to examine priorities. If nothing else, Apollo 8 served to open again the old debate as to the value of Kennedy's commitment. But if nothing to some, the success of a decade in space was meaningful to many.

On 9 December a White House dinner was held for astronauts, space administrators, contractor representatives and Government leaders. Hosted by Lyndon B. Johnson it was a farewell to comrades and friends: 'I asked you to come here tonight in the twilight of this administration, so I could pay the respect and the honor and the affection that I felt for the man who has directed your efforts and directed them so well, and so that I could express my personal admiration and respect for you.' Responding to this, Jim Webb said that 'The challenge of space is large and so is NASA. In all such human endeavors, organized institutional efforts are essential, and we know, in the words of Emerson, that they are "the lengthened shadow of one man." We in NASA know, Mr. President, that you are the man of which our civilian space effort, conducted for the benefit of mankind, is the lengthened shadow.'

Gentle emotion swept through the White House that Monday evening as the architects of a nation's space program reminisced about the early days, months and years that the swift veil of time had screened. Within six weeks a new President would occupy the White House. Richard Milhous Nixon was the Vice President to Dwight D. Eisenhower when NASA came into existence in 1958 – and now the President elect. Lyndon Johnson had come a long way since tramping the road to California as a young boy seeking his fortune. Leader of politicians and Congressmen, LBJ had been the single most important catalyst for events and personalities that contrived the space program of the 1960s. And he had watched as a bitter war and civil unrest brought low the ambition he sought to put his nation among the stars.

The countdown for Apollo 8 began after dusk on the evening of Sunday, 15 December. Already, the very first visitors had arrived, setting up camp on the beaches north of Merritt Island. By launch day, they would be joined by a multitude in the tens of thousands. In the Manned Spacecraft Operations Building, final flight preparations for Frank Borman, Jim Lovell and Bill Anders. From time to time, runs in the Cape simulator, briefings from officials, and jogging trips to keep in trim. On Wednesday the 18th, they watched a Delta rocket launch a communications satellite from pad 17A. On Friday, the day before their own launch, the crew rested in the MSOB isolated from the prying press and the eager crowds. Much of this area was restricted to officials but elsewhere the gathering throng grew in intensity. Around the globe, 10,000 men of the armed services were on station ready to rescue the crew from contingency zones in Pacific and Atlantic Oceans. Across from the Cape, stretching in a 6,200 km ellipse toward Africa, was the abort contingency area where ships and helicopters stood ready to retrieve the astronauts. From the Goddard Space Flight Center there were checks and double checks on the 14 ground stations, 4 ships, and 6 aircraft assigned to support communications and tracking of the Apollo 8 spacecraft.

During the afternoon of that Friday before the flight, a crawler-transporter moved in beneath the massive Mobile Service Structure to walk it away from pad 39A, a monstrous moving giant 45 storeys tall with a combined weight of nearly 7,200 tonnes. From the MSS, several hundred engineers and technicians had prepared 503 and its spacecraft payload for the dramatic Moon flight. Now the rocket stood alone, flanked only by the launch umbilical tower feeding the Saturn V via eight massive arms with the electrical power and propellant lines injecting life to the big bird.

The first men to ride a Saturn, leave Earth's gravity, or fly into Moon orbit: (left to right) Lovell, Anders, Borman.

All across the launch complex, 5,000 personnel went about their ordered business while in Houston, 1,600 sat waiting for the moment of flight. As night fell across the Cape area, searchlights focused on pad 39A, lighting up the sky for kilometers around. It was like a giant Christmas tree, this white and black needle pointing skyward, lit up by the candle-power of dozens of illuminating beams. Roads across the Kennedy Space Center were lined by the rows of street lighting laid out across the marshy flats. But far to the south where the early days of Cape activity bequeathed a host of rusting, redundant, launch pads to the new tomorrow etched by the magnificence of pad 39, the searchlights could be seen brightening the black sky.

It was a star-filled sky, that night before men left for the Moon. The air was a chilled reminder of another night nearly seven years before when John Glenn waited to ride into orbit with the coming dawn. Now, three men were getting ready to circle another world. Nothing it seemed could stop the momentum of Apollo. Earlier that month, Russian tracking ships had sailed for home from dispersed locations all round the globe, even the *Komarov* left its Caribbean position to return to Russia. And on the wall of one NASA office was chalked a prediction: 'Next Russian unmanned attempt – March 30!'

The flight crew were woken at precisely 2:36 am and 15 minutes later were receiving a final medical check. Special care had been taken to isolate them from likely sources of colds and 'flu in the wake of Apollo 7's epidemic of nasal germs and it seemed to have paid off. By 3:30 am they were breakfasting with a select group of guests and donned their suits shortly before 4:00 am. It took 45 minutes to garb the men for space and just after 5:00 the pad transfer van arrived at complex 39A. Ten minutes later they were in the white room and preparing to enter spacecraft 103. For this trip, Apollo had a full load of propellant in the Service Module tanks. Apollo 7 had carried a load totalling only 4,417 kg; Apollo 8 would fly with 18,253 kg of fuel and oxidizer.

As dawn crept across the Atlantic and dowsed the brilliance of artificial light, roads and highways leading to the Cape were jammed with cars and trucks. The first day of a weekend brought more visitors than usual to see the departure of Man's liberation from the home planet. The sight of Apollo 8 proud and erect several kilometres distant was an awesome prelude to the shock at launch when thunder would loose itself upon the watching crowds. The press stand had been full for several hours, but more journalists arrived with each passing minute, hundreds of representatives of opinion-makers from across the globe. From helicopters ferrying dignatories and officials, the noted and the notable walked with smiling anticipation to be met by NASA personnel sporting identity badges flapping in the light breeze wafted by whirling rotor blades. Occasionally, the distant whine of a security car could be heard moving fast along some preserved access road, or the slow movement of ambulances deploying to pre-set locations.

Jim Lovell was last to enter spacecraft 103, his ingress monitored on one of 15 television screens in the Launch Control Center wired to 60 cameras all over pad 39A. During the checkout and countdown engineers had access to 112 separate communication channels, but only the most essential voices now chattered with the necessary talk of the final countdown as 450 consoles dazzled with readouts and feedback; in the Launch Control Center alone, more than 400 personnel were monitoring the machines. Without a fully automated countdown sequence, an innovation for Apollo, the big bird would never have been prepared in time for launch. Just 15 minutes before lift-off the vehicle transferred to internal power, a critical time when pad electricity is isolated and the Saturn V draws energy from its own batteries. On its own now, the stack delivered final messages informing computers of its status in a myriad systems and subsystems.

At T–3 min 7 sec the automatic sequence began. Nothing could now halt the inevitable progression toward lift-off; only a failure somewhere would stop the clocks. At 17.2 sec the guidance reference release command slid in, locking up the launcher with target coordinates already set down. Seven seconds after that, more than 100 meters below the crew's perch atop the mighty Saturn, a multitude of nozzles dumped a deluge down on to the flame deflector positioned beneath the five F-1 engines; more than 500 liters of water cascaded down each second, cooling the volcanic rock placed on the deflector to channel and divert the inferno from above. Scanning the more than 2,600 measurable points on the complete stack, computers confirmed that all was ready for ascent, checking quickly on the 20 telemetry channels aboard 503. From the first stage itself, two TV cameras and four recoverable film cameras wound into action.

At precisely 8.9 seconds before the scheduled lift-off time of 7:51 am, the ignition command was sent to the rocket. Less than three seconds later all five F-1 engines burst into life, smoke and flame boiling away from the bell-shaped exhaust nozzles. Channeled to left and right by the inverted-V flame deflector the appearance of fire and bright light brought a gasp from the watching multitude. It took just 500 milliseconds for all five engines to burst into life, a staggered start to lessen the shock from a thrust load exceeding 3,400 tonnes. For six seconds 503 sat squat upon the concrete launch pad while the engines smoothed their combustion to steady thrust. In the spacecraft, vibration moved quickly through the shimmering stages, attenuated in the crew cabin by couch struts. From deep down in the rocket a thunder moved toward the three men, unable to hear the dramatic sound above the squaking communication hats strapped tight upon their heads.

But to the watching horde it was starburst, birth of a new experience for those who imagined they could visualize the power of a Saturn V but now stood open mouthed at the fire and the fury. No sound yet, the echoing shock waves still moving inevitably toward the crowds who now gave vent to emotion: leaping up and down, some shouting, some standing on car roofs, others wafting giant American flags too big to handle, more still just awestruck by the drama. Vibrations took hold of the giant behemoth and shook the Saturn V as though it were a cardboard cut-out. Several hundred kilogrammes of ice, great chunks of flaking white shapes, rained down the sides of the gleaming white rocket from the chilled cryogenic tanks that held the sub-zero liquids to feed the Saturn's power. More than 550 kg in weight was added to that of the rocket itself by frost clinging to the stages, freed now, shaken loose by the shimmering movement.

At T–0 the four restraint points shattered and 503 slipped its shackles. Bormann, Lovell and Anders were ascending on a rocket eighteen times more powerful than the Titan they had ridden on another December day three years before. But to veterans Borman and Lovell it all felt very familiar; to Bill Anders it was a totally new experience. As the large swing-arms moved back to release the rocket twenty-nine nozzles burst into life, pouring down upon the launch deck, at the rate of 3,150 litres each second, water to cool the exposed steel. At the same instant, a spray system drenched the separate arms on the umbilical tower. In several respects Saturn V was so big that to ride the bird aloft was to take a first-class trip on a sophisticated transport system. But the first stage was a little rough, twitching and twisting to keep the vehicle on course, gimbal motion with the F-1s whipping the forward end of the rocket back and forth perceptibly, the guidance keeping the line of thrust aligned with the center of mass.

In little more than 2 minutes, long after thunder claps battered the watching Earthlings, the first stage had done its required job. First the center engine shut down, then the outer four, and at the instant of separation when the retro-rockets briefly fired to shunt the stage backward, the entire vehicle was enveloped in a sheath of flame, visible from the ground as a burst of light that brought gasps of 'ooh!' and 'ah!' from crowds across the Cape. Then the clear, transparent flame of the five second stage J-2 engines lit up followed, seconds later, by separation of the cylindrical adapter shrouding the base of the second stage. Shortly after that the launch escape tower jettisoned, taking with it the boost protective cover. Now the crew could use their five windows although attention focused on the instruments at the display console. Aborts from this point on would require separation of the complete CSM.

It was a difficult ride, the 6 min 9 sec spent on S-II power was accompanied by slight pogo oscillation, disturbing enough to make shut-down a pleasant anticipation. When it came, the second stage shunted away and milliseconds later the single S-IVB engine lit up. A brief burn this, lasting only 2 min 40 sec, to apply the final speed required to stay in orbit; desired altitude had been achieved with the S-II. At an elapsed time of 11 min 35 sec, Apollo 8 reached the first plateau: Earth orbit insertion. It would remain at this point for more than 2½ hours, circling the globe while final guidance checks were made and targeting information compared with the pre-flight predictions.

As the assembly passed from one tracking station to the next, each in turn confirmed the readiness of equipment and technicians to track the vehicle to the Moon. Finally, on the second revolution, the crew heard the historic message from Houston, word from capcom Mike Collins, recuperating from his operation, that the mission could proceed with trans-lunar injection: 'Apollo 8, you are go for TLI, over.' And from Jim Lovell: 'Roger, understand. We are go for TLI.' And it was as simple as that. For the first time, men would break free from the gravitational bonds of planet Earth. By firing up the S-IVB's J-2 engine for the second time the assembly would gain an extra 10,994 km/hr, accelerating to a trajectory that would carry Apollo 8 to the Moon's sphere of gravity. At that point the Moon would draw the spacecraft toward itself, accelerating it faster all the while, until, on the far side, Apollo's SPS engine would brake the spacecraft into lunar orbit. But first, the S-IVB re-ignition that had failed on the last, albeit unmanned, Saturn V test.

Heading north-east across the Pacific the crew of spacecraft 103 monitored final preparations and, when it came, ignition of the J-2 was smooth and steady. Passing across Hawaii, radio and TV audiences dashed out into the pre-dawn morning to watch the bright star trail a point of light toward the horizon. Minutes later the spacecraft burst into Sunlight and 5 min 18 sec after it began the single engine at the base of Saturn's third stage shut down. Three men were travelling faster now than any other human before: 38,988 km/hr. Within mere minutes of time they had passed and exceeded the altitude record set by Gemini XI and just 25 minutes after trans-lunar injection the CSM separated from the S-IVB and its forward adapter. In free flight now, the 28.9 tonne Apollo turned around to face the S-IVB for a brief station-keeping exercise. On later flights the CSM would be required to move in and pull a Lunar Module free. Then, a small burn with the RCS quads to impart a separation maneuver; in drifting flight the two could nudge into each other unless both were placed on marginally separate paths.

But soon it appeared that the separation burn had been insufficient and at an elapsed time of 4 hr 45 min the crew fired the thrusters again for a satisfactory maneuver away from the S-IVB. Residual propellants from the S-IVB had been used to displace the path of the inert stage from that of the CSM so that it would loop around the Moon 1,225 km from its surface. Using the gravity of the Moon to perform a 'slingshot' boost, the S-IVB would be accelerated out of Earth and Moon gravity fields to an independent path around the Sun. Like Earth and Moon it would continue to orbit the Sun for ever. The Saturn V had performed well, almost flawlessly, confounding the suspicions of critics who thought NASA should have flown at least one more rocket of this type before committing it to manned activity.

With launch, TLI, and separation behind them, the crew got down to housekeeping duties. But first, the suits were removed to get into a more comfortable situation. Almost immediately the crew left their couches and started moving around they all sensed a nauseating motion sickness probably brought on by the same disorientation that afflicted Soviet cosmonauts several years before. In addition, Frank Borman seemed to have contracted gastro-intestinal 'flu which, despite all the pre-flight precautions, had invaded the Commander. First order of business in the spacecraft was for Jim Lovell to go to the lower equipment bay – the opposite wall of the crew compartment to that containing the access hatch – and to perform the first of several trans-lunar navigation sightings.

Guidance and navigation (G&N) aboard a Moon-bound spacecraft had for long been of deep concern to those whose job it was to plan the trajectory of a deep space vehicle. The Massachusetts Institute of Technology got an early start on basic problems as early as the late 1950s when they received study contracts from the Air Force for Mars flight path analysis. Director of MIT's Instrumentation Laboratory, Dr. Charles Stark Draper was a national asset for automatic pilot equipment and sophisticated gyroscope mechanisms. It naturally fell to him to point NASA on the right road for an Apollo G&N system. By mid-1961 MIT was involved with the manned Moon plan, under contract to develop the concept of trans-lunar navigation and then to hand over fabrication of hardware to civilian contractors at the appropriate time.

It was not an easy road. Many times the space agency was concerned about the ability of Draper's laboratory to come up with a fail-safe way of guiding a manned spacecraft to the Moon. At one point, cynically referring to suspicious prods from Headquarters, Draper offered to 'volunteer for service as a crewmember on the Apollo mission to the Moon . . . let me know what applications blanks I shall fill out. . . .' NASA backed off and Draper eventually produced a remarkably efficient system. In 1962 the AC Spark Plug Division of General Motors was chosen to build the complete G&N system, the Raytheon Company was selected to build the guidance computer, and the Kollsman Instrument Company received a contract to build the optics. These were the three principle elements about to be used for the first time enroute to the Moon by Apollo 8's Command Module Pilot.

Containing more than 40,000 parts, the G&N system carried as its most important element an inertial guidance platform from which the computer would obtain information to calculate the position of the spacecraft and to compute necessary course corrections. The inertial measurement unit, or IMU, provided a stable reference against which all motion of the spacecraft could be measured. It comprised a platform mounted within three gimbals, or pivots, set at right angles to each other, and stabilized by three gyroscopes. Every rotation or movement of the spacecraft around the stable platform would be monitored by instruments within the IMU, determining the attitude in roll, pitch and yaw. Spacecraft acceleration would be measured by three pulse-integrating pendulous accelerometers (pipa's) mounted to the stable platform. Situated at right angles to each other, any translational force applied to the spacecraft would cause an acceleration or deceleration sensed by one or more of the three accelerometers. A pulse train signal delivered to the computer continually updated information held on the velocity of the spacecraft. The complete IMU weighed 19.3 kg, comprised a sphere 32 cm in diameter, and was rigidly mounted to a navigation base precisely aligned with the primary axes of the spacecraft.

The digital computer would store and use signals from the guidance platform and sighting from the optical assembly to calculate course corrections. With a fixed memory accommodating 38,864 words, and an erasable memory containing up to 2,048 words, the computer provided all information necessary to fly to

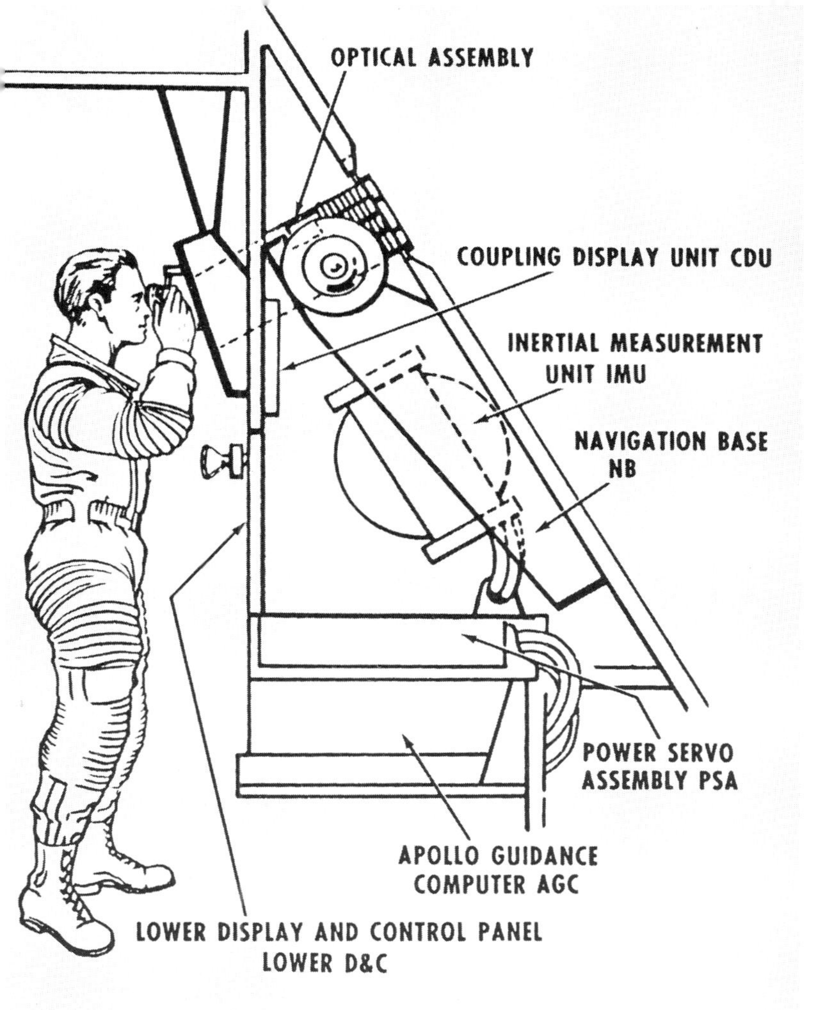

CATION OF GUIDANCE AND NAVIGATION EQUIPMENT IN SPACECRAFT

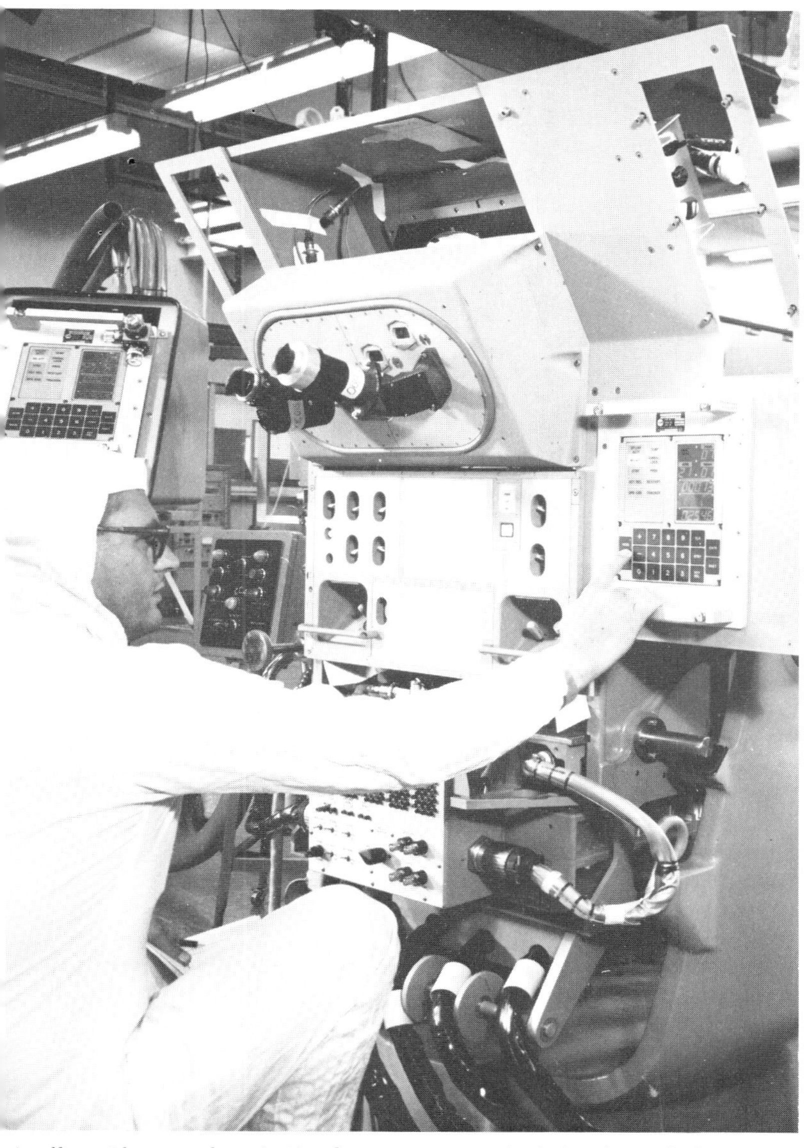

Apollo guidance and navigation here seen on test includes the optical assembly (top), keyboard (right), and control equipment (center).

the Moon and back. It was linked to two display keyboards: one in the lower equipment bay and one on the main display console. Each keyboard contained ten numerical keys labelled 0 to 9, two key signs (+ and −), and seven keys for inserting instructions into the computer. Verb, Noun, Clear, Enter, Proceed, Key Release, and Reset. Numerical data would be fed to the computer via the Verb key, designating the action to be taken, and the Noun key, informing it of what the action was to be applied to. For instance, two digits would be keyed in as separate numbers after depressing the Noun key, and two numbers would be depressed after pushing the Verb key. The + and − signs merely instructed the computer as to decimal values. The Clear key removed data from the displays, successive depressions continuing to clear the other addresses. The Proceed key would instruct the computer to hold on standby, or if already on standby to go ahead with normal operations. The Key Release function would wipe away illuminated displays of numbers just entered and bring up information from the computer program. The Enter key told the computer to accept the numerical instructions. Reset would clear the address.

As an example of the way the computer would be set up for a burn with the SPS engine, an astronaut would first enter the program, which for this event would be a P30, and then start to load the erasable memory with pertinent information. Noun 47 would require a + key insert, followed by five digits. That would tell the computer the weight (or mass) of the CSM. Next, two noun 48 inserts of three digits each would tell the computer the pitch and yaw SPS gimbal angles desired. Then, separate insertions of three, two and four digits, each sequence preceded by a + insert, would give the computer a noun 33: the ground elapsed time in hours, minutes, seconds and hundredths of a second when the burn was to take place. Several other inserts would be necessary to load in the exact velocity to be applied, the duration of the burn, etc., prior to completing the P30 pad. In all, a long list of numbers usually fed up from the ground to the crew on the communications circuit from which the computer would take over the automatic operation of the programed sequence. Many different master programs were prepared by MIT. The one used by Apollo 8 was called Colossus; that used for Earth orbit flight by Apollo 7 had been called Sundisk.

The third element of the G&N system was the optical assembly, which, like the inertial platform, was securely mounted to the fixed navigation base. It comprised a dual line-of-sight, electro-optical, sextant with a 28x magnification and a 1.8° field of view. Capable of measuring the included angle between any two targets to an accuracy of 10 arc seconds, one line of sight would be established by changing the axis of the spacecraft while the other would be aligned on the shaft and trunnion axes. The second optical instrument comprised a single line-of-sight refracting telescope similar to a theodolite in that it could measure elevation and azimuth to a single target from an established reference. Again, shaft and trunnion angles would be used for reference to the spacecraft attitude.

It was with the optical assembly and the computer keyboard that Jim Lovell busied himself for a cislunar navigation update using program 23. It was to prove the validity of Stark Draper's calculations that Apollo 8 was flying to the Moon at all and careful monitoring of the G&N system was essential for confidence in later missions where such fundamental issues would be merely the backdrop against which more dramatic activity could take place. It took Lovell almost an hour to make his alignment sightings on the three stars for this purpose. And by that time the tracking stations on Earth had carefully observed the path of Apollo 8, compared it with the pre-planned trajectory and found need for a small course correction.

It would take the spacecraft about 69 hours to reach the far side of the Moon. During the trip between Earth orbit and the lunar sphere, flight controllers had planned four mid-course correction, or MCC, maneuvers, opportunities to fire the big SPS or the smaller RCS quads to trim the trajectory for accurate flight around the Moon. MCC_1 occurred about 6 hours after trans-lunar injection; MCC_2 would come on the second day at an elapsed time of 26 hr 30 min; MCC_3 at 47 hours; MCC_4 at 61 hours as a final trim before going into Moon orbit. Nobody expected that all four opportunities would be needed; they were there as the pre-designated spots when corrections would be most conveniently applied.

As it turned out the first mid-course correction burn was a little later than the flight plan dictated. At 11 hours elapsed time the big engine fired for exactly 2.4 seconds, slowing the CSM by 4.6 meters/sec. Since leaving Earth orbit, gravity had tried to pull back the spacecraft and succeeded in slowing it to a speed of just 8,983 km/hr, a mere fraction of the speed imparted to Apollo 8 by the Saturn V third stage. But the trajectory took account of this continual slowing effect, arrested only when the spacecraft would enter the Moon's sphere of influence and start to speed up.

In Houston it was evening; on the spacecraft it was time for Frank Borman to get some sleep. Then, after a scheduled 7 hr rest period, Lovell and Anders could sleep. After that, more P23 navigation sightings and preparation for the first public relations exercise of the flight, a short telecast from Apollo looking back at the Earth. For the first time, hundreds of millions of people would see themselves as only three others saw them.

It was mid-afternoon at the Cape, just about lunchtime in Houston and evening in London on Sunday, 22 December, when Apollo 8 beamed its first TV pictures at the home planet. Thirty-one hours twenty minutes into the mission, spacecraft 103 showed viewers the whole Earth for the first time on live TV: 'This transmission is coming to you approximately halfway between the Moon and the Earth . . . we have about a little less than 40 hours left to go to the Moon. Show you the Earth. It's a beautiful, beautiful view, with predominantly blue background and just huge covers of white cloud, and particularly one very strong vortex up near the terminator. Very, very beautiful.' The terminator represented the fuzzy line between night and day, a word that would soon be applied to the same line dividing the two hemispheres of the Moon. Earlier that Sunday the attendant minister at Frank Borman's Episcopalian church in League City, Texas, offered a prayer for their lay reader, now, this Fourth Sunday in Advent, half way to another world: 'O eternal God in whose dominion are all the planets, stars and galaxies, and all the reaches of time and space from infinity to infinity, watch over and protect, we pray thee, the astronauts of our country and all those who with their learning and skill support them.' Several thousand kilometers away, in Rome, Italy, the Pope offered his blessing: 'We open the window and instinctively the eye, the thought, the heart, go to the heavens. We pray to the Lord for them, and for the world, which is dazed at the conquest of science and of human endeavor.'

By this time Borman was feeling considerably better than the day before and the spacecraft was in good shape. But two windows were fogging over, leaving only the forward facing rendezvous windows clear; the circular hatch window had started to mist over shortly after TLI and this followed a similar occurrence noted on Apollo 7 where, although the windows remained clear for some time they eventually became almost opaque. Pictures of the Earth provided by the first telecast had not been very clear but the second screening brought improved results. Coming up on an elapsed time of 55 hours, with the spacecraft 326,000 km from Earth, another TV session began with distinctly recognizable features covering the Western Hemisphere: North and South America, the North Pole, Antarctica and the Pacific Ocean.

But it also brought a philosophical comment from Jim Lovell: 'Frank, what I keep imagining is if I am some lonely traveler from another planet what I would think about the Earth at this altitude, whether I think it would be inhabited or not . . . I was just curious if I would land on the blue or the brown part of the Earth.' Bill Anders passed a humorous quip in response: 'You better hope that we land on the blue part.'

With the TV off, just 38 minutes after the elapsed time of 55 hours, Apollo 8 made the historic transfer from the gravity field of the Earth to that of the Moon. It was 3:29 pm Cape time, and three men were quite literally out of this world. Earth had done its worst; in pulling hard on spacecraft 103 it slowed the cruising ship to a mere 4,385 km/hr. But from this point on, the Moon would tug it faster and faster, drawing it ever closer across the 62,600 km that still separated the two objects destined for a celestial rendezvous. The course imparted by the third stage of 503 more than two days before had been a free-return ride that would, if nothing was done to slow Apollo on the far side of the Moon, bring the spacecraft back on a giant figure-8 loop to the vicinity of Earth. It was conceivable that the course back would need some correction, but nothing that could not be accommodated by the small RCS thruster quads.

It was insurance, that if anything was found to be wrong with the SPS engine it would not have to be used to get the crew back safely, that sold the Moon orbit plan to doubting officials at Headquarters. But once used to put Apollo 8 in orbit, it would have to fire again to bring them back. It was for this reason that flight controllers wanted to see a slight error in the course to the Moon so they would have cause to fire the SPS and prove it out before the all-important lunar-orbit insertion, LOI, burn around the far side.

Yet the first burn of the big Service Module engine had gone so well that the next two opportunities for mid-course correction were not needed, only one last 'tweek' burn with the small RCS quads being necessary at MCC_4. It was an evening hour in Houston, Monday, 23 December, when Borman, Lovell and Anders fired their thrusters for a brief 13 second burn to trim the path that would carry them around the far side of the Moon; 61 hours into the flight, 8 hours to go before lunar-orbit insertion. One more sleep for Lovell and Anders before the big event. And then, during the final hours before LOI, ground controllers and technicians checked every critical system aboard the spacecraft. Only if CSM-103 was in a well nigh perfect condition would the spacecraft get a 'go' for lunar orbit insertion.

Word came to Apollo 8 via the capcom console manned by Gerry Carr just minutes before the three men passed out of sight around the left side of the Moon as viewed from Earth. And one minute before loss-of-signal (LOS) Jim Lovell responded to a good luck call from Carr: 'Thank's a lot troops. We'll see you on the other side.' For 33 minutes nobody on the planet would know if Apollo 8's SPS engine had fired for the correct duration. Moving quickly across the lunarscape at more than 9,200 km/hr, the spacecraft would cut its speed by some 3,220 km/hr in a 4 min 7 sec burn designed to drop the CSM into an elliptical path. If it burned too long the spacecraft would brake fast and crash into the Moon; if it burned too short the spacecraft would enter a path from which it could only escape by another SPS firing. It had to go just right, and the margins were very tight. If the SPS failed for some reason to fire, spacecraft 103 would reappear around the right side of the Moon ten minutes earlier than expected.

That time came and went and almost exactly 33 minutes after disappearing from view, Apollo 8 swung into view – safely in an elliptical path 111 km by 312 km. It would spend four hours – two revolutions of the Moon – in this orbit before burning into a more nearly circular path using the Service Propulsion System again. But the event was greater than the moment, and while Public Affairs Officer Paul Haney fumbled for words to tell the world it was a success, cheers went up in the support rooms at the Manned Spacecraft Center, and at many places around the globe. Three men were in orbit around another object in the solar system. Not since Creation had the evolution of planet Earth relinquished its hold on the life to which it had been a raft of survival for 4,000 million years.

The moment was profound, and would be seen increasingly so in retrospect. Before Apollo 8 no human being had travelled further than 1,373 km from the Earth. Now, three astronauts were more than three hundred times that distance from the world that gave them life. Man had truly broken an umbilical with Mother Earth. On the planet itself, humanity was touched for a moment of time by the deeper significance of this Christmas pilgrimage. The *New York Times* felt that '. . . the drama and interest of yesterday's view of Earth from space transcended any prosaic considerations of practical utility. Rather the excitement these pictures aroused among millions of stay-at-homes flowed from the visual evidence they provided of Man's successful entrance into a completely new realm, one which poses challenges, opportunities and dangers, such as the human species has never before faced.'

In America and around the world, it was Christmas Eve; people in Europe went about their lunchtime business, shops stayed open for the last hours of the customer rush, churches kept open doors for the thoughtful gatherings that knelt that day in awesome respect of an event nearly two millenia distant in time. In the United States it was still dark, few people were yet up from the night's slumber, but many to whom this day would be a lifelong reminder of a Christmas never to be forgotten listened in as the voices from the Moon etched the event with perception and clarity:

CAPCOM 'Apollo 8, Houston. What does the old Moon look like from (113 km), over.'

LOVELL 'O.K., Houston. The Moon is essentially gray, no color; looks like plaster of Paris or sort of a grayish beach sand. We can see quite a bit of detail. The Sea of Fertility doesn't stand

Christmas Eve, 1968. Apollo 8 reaches lunar orbit.

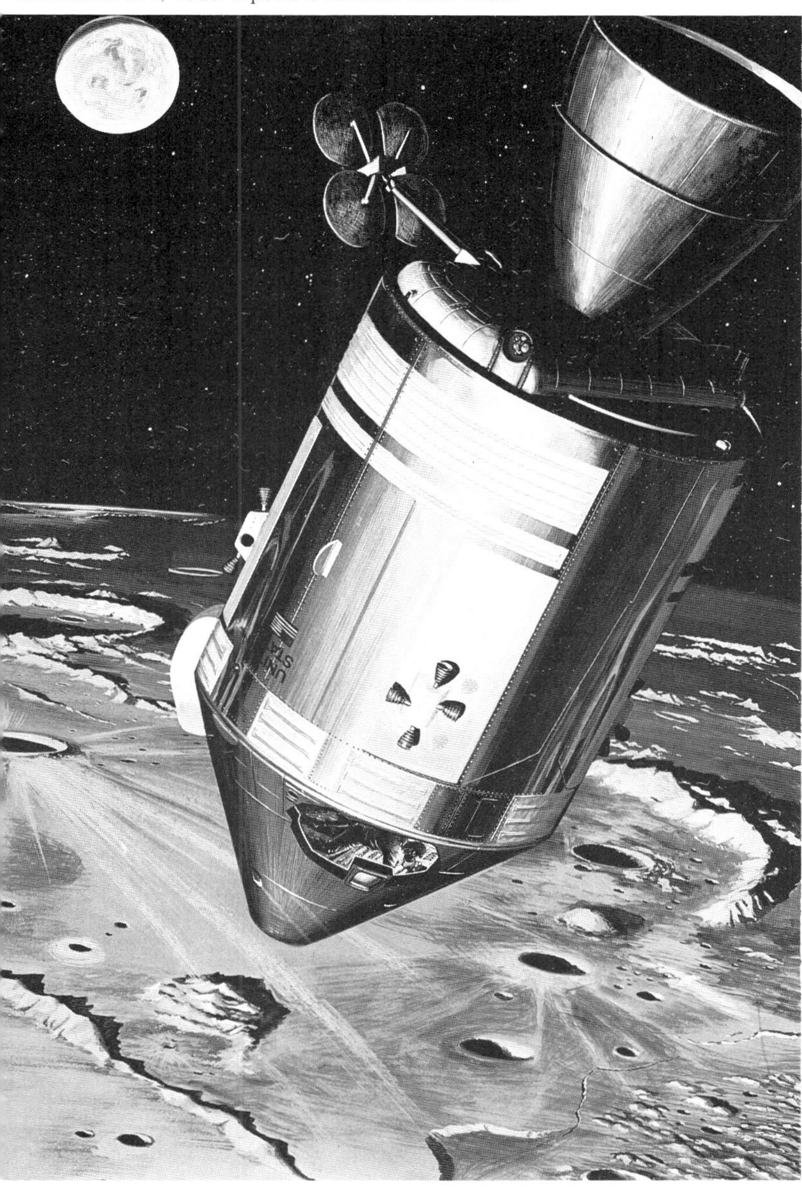

out as well here as it does back on Earth. There's not much contrast between that and the surrounding craters. The craters are all rounded off. There's quite a few of them; some of them are newer. Many of them . . . – especially the round ones – look like hits by meteorites or projectiles of some sort. Langrenus is quite a huge crater. It's got a central cone to it. The walls of the crater are terraced, about six or seven different terraces all the way down.'

CAPCOM 'Roger. Understand.'

LOVELL 'And coming up now on the Sea of Fertility are our old friends Messier and Pickering that I learned about so much on Earth.'

CAPCOM 'Roger.'

LOVELL 'And can see the rays coming out of, I believe, Pickering. We're coming up now to our P-1 initial site, which I'm going to try and see. Be advised, the round window – the hatch window – is completely iced over and we can't use it. Bill and I are sharing a rendezvous window.'

CAPCOM 'Apollo 8, Houston. Roger, got any more information on those rays, over?'

LOVELL 'Roger, the rays out of Pickering are quite faint from here there's, er, two different ribs going to the left. They don't appear to have depth to them at all, just rays going out.'

CAPCOM 'Roger.'

LOVELL 'They look like just changes in the color of the maria. O.K. Over to my right are the Pyrenee Mountains coming up. And we're just about over Messier and Pickering right now.'

Pointing down at the lunar surface the view was one of total lifelessness. Views came across the spacecraft window revealing P-1, a candidate Apollo landing site, the waste expanse of the volcanic floor called Sea of Fertility, Pickering, Messier, and Langrenus, the craters they had learned to recognize from Earth. Within 32 minutes of coming around the limb of the Moon, Apollo 8 swept into night, the Sun illuminating a little of the near and most of the far side; moving round Earth, and on its axis, in one Earth month, the Moon's daylight would last more than two weeks at any particular spot. But just like around Earth a spacecraft would spend almost one half each revolution in darkness around the far side from the Sun, so too would this Moon-orbiting Apollo move continually through day and night hemispheres.

At the end of the second revolution the SPS engine was fired again, just for 9 seconds this time, to lower the spacecraft to a safe 110.6 × 112.4 km orbit. Much of the time had been taken up with control point sightings through the optics, navigation marks on set targets to help the computer establish an accurate value on where the spacecraft was. Revolution three was taken up with preparations for rest, more photography sessions, landmark tracking, and IMU alignments. On the spacecraft swept, circling the Moon in an almost circular path, at a speed of about 5,860 km/hr.

Most of Christmas Eve was spent recording on motion and still film the shifting scene of the lunar surface and gathering engineering information that would ease the flights yet to come, missions that would bring a Lunar Module for descent to the dusty lurrain. Much of the photography the crew had used to learn the visual appearance of features over which they would fly had been provided by the Boeing Lunar Orbiter, a five out of five success program where spacecraft flew to Moon orbit between August 1966 and August 1967, building up in the five missions a valuable record of the topography; to this very day the Lunar Orbiter Photographic Atlas is an impeccable reference for lunar scientists. But to Borman, Lovell and Anders, Lunar Orbiter pictures had been an invaluable training aid without which they would have had a considerably harder time differentiating the various features.

The sleep routines were spasmodic to say the least. First Lovell was to rest for a couple of hours, then Borman was to doze for three, followed by Anders for two before Lovell again slept for two hours. In practice it became a matter of 'real-time planning' and in the absence of critical activity for the 15 hours following the second insertion trim, all three got what rest they could. At one point Frank Borman asked to have a message passed to Rod Rose and the congregation at St. Christopher's Episcopalian church. Informing capcom that he was to have read the Christmas Eve prayer but that he 'couldn't quite make it', the Commander read from a sheet of flight paper the words he hoped would be conveyed to the congregation:

'Give us, O God, the vision which can see thy love in the world in spite of human failure.

Give us the faith to trust thy goodness in spite of our ignorance and weakness.

Give us the knowledge that we may continue to pray with understanding hearts.

And show us what each one of us can do to set forward the coming of the day of universal peace.

Amen.'

On Earth, as Christmas Eve moved through its busy, bustling hours, everywhere people went they saw television sets bringing shoppers and carol singers alike a view such as Man had not seen before at any other time in history. From around the Moon, for the night before the Christ child's birthday, came a new awareness of what the next day would mean in countries all across the world. And from Moon orbit too, a new reverence, a sudden feeling of awe, that after many years hurtling toward a distant goal there was suddenly something very inspiring about the great event now being staged.

To the accompaniment of carols old and new, a billion people on planet Earth prepared to celebrate the coming day with three ambassadors of all mankind quietly observing the rugged features of a world little changed since a billion years before Man walked. It touched the hearts of people everywhere and moved the normally unemotive and objective commentary from broadcasters and radio reporters to words of feeling and rhetoric, an openness usually veiled from the vast audiences they addressed.

As a final bow to the great interest and concern expressed by people around the world, Apollo 8 prepared to send one last telecast from around the Moon on the penultimate revolution before the SPS would once again be fired, this time to bring the crew back to Earth. It was early evening in Houston, time for the office party in Washington, but in London and other European cities it was already three and one-half hours into Christmas Day; on both continents night had fallen. From a makeshift studio at the Manned Spacecraft Center, correspondent Ed Hickey waited with his listeners for the last live views of the Moon's dusty surface and words of description from the astronauts that would surely accompany the indescribable scene:

'On the earlier transmission today, the Apollo 8 astronauts said the old Moon looks sort of grayish, sandy, very little color, that there appeared to be craters which had been impacted by meteorites from other planets. They are putting in a call down here from Mission Control, we're standing by waiting to hear from them, and we will let you hear what we here, around the world, looking at the Moon from a spacecraft just above it. And it seems impossible as we sit here and await this, and the earlier pictures we have seen, how anyone who can sit and watch their television monitor, as they are perhaps right now around the world, and not wonder, cease to wonder, at what Man has wrought on this Christmas day. . . .'

And then the picture came and with it the fluent voices of Frank Borman, Jim Lovell and Bill Anders. It was a moving experience, one that seemed to be an unfamiliar accompaniment to space spectacle but nevertheless a mood set firmly by the solemn reverence of the crew and one that befitted the occasion. The dialogue began just minutes after the Manned Space Flight Network acquired Apollo 8 on its ninth lunar revolution:

BORMAN 'This is Apollo 8 coming to you live from the Moon. We've had to switch the TV camera now, we showed you first a view of Earth as we've been watching it for the past sixteen hours. Now we're switching so that we can show you the Moon that we've been flying over at (111 km) altitude for the last sixteen hours. Bill Anders, Jim Lovell and myself have spent the day before Christmas up here doing experiments, taking pictures, and firing our spacecraft engine to maneuver around. What we'll do now is follow the trail that we've been following all day and take you on through to a lunar Sunset. The Moon is a different thing to each one of us, I think that each one carried his own impression of what he's seen today. I know my own impression is that it's a vast, lonely, forbidding type existence, or expanse of nothing. It looks rather like clouds and clouds of pumice stone and it certainly would not appear to be a very inviting place to live or work. Jim, what have you thought most about?'

LOVELL 'Well Frank, my thoughts were very similar; the vast loneliness up here at the Moon is awe-inspiring and it makes you realize what you have back there on Earth. The Earth from here is a grand oasis in the big vastness of space.'

BORMAN 'Bill, what d'you think?'

ANDERS 'I think the thing that impressed me the most was the lunar Sunrises and Sunsets. These in particular bring out the stark nature of the terrain and the long shadows really bring out the relief, that is here but is hard to see on this very bright surface that we're going over right now. The horizon here is very, very stark, the sky is pitch black and the Earth, or the Moon rather, excuse me, is quite light, and the contrast between the sky and the Moon is a vivid dark line.'

LOVELL 'Actually, I think the best way to describe this area is a vastness of black and white – absolutely no color.'

For some minutes the view to Earth embraced the many different types of structure the old Moon's face presented to these newcomers from a neighbor planet. And again, for listeners around the world, broadcaster Ed Hickey described the view on TV monitors in the Houston control center:

'From this height, 109 km above the Moon, the lunar surface isn't streaking by, just moving slowly much like if you were travelling in a jet aeroplane, just to give some comparison of what we're seeing here now, looking out the window of Apollo 8 as the men inside look down on their Moon, the first to see the Moon with the naked eye from so close a distance . . . all of us had hoped we could see the spacecraft as it goes around the Moon. It has been sighted in orbit around the Moon by an observatory in the mid-Western United States. It showed up as a dot. Of course we couldn't see it with our naked eye but the powerful telescope from this observatory has been used since it has been in orbit some 18 hours now and the description from those who glanced through the telescope said that it looked like a tiny, minute dot as it whizzed around the surface of the Moon.'

By now the spacecraft had reached that point where Sunlight gave way to darkness and the lengthening shadows of surface features heralded the coming night, hastened by the speed of Apollo 8 as it flew from east to west. It fell again to Ed Hickey to describe to the listeners the impressions he received from this strange, un-Earthly, sight:

'. . . these are the craters, the mountains, the rilles, formed aeons – thousands of millions of years – ago, untouched by Man. The purpose of this flight, to find out more about it. What it hides inside its core. What wealth is stored. Could it be used to go further out into interplanetary space. . . . The Moon has always been a glamorous place for Man, from what the astronauts have told us on this historic day – Christmas Day. Something never before seen by any other men, to the best of our knowledge. And clearly visible, carved out of the lunar surface, ridges forming sort of a mountain as they go down into the valley like the cone of a volcano. Whether there is volcanic activity or what it is, we don't know yet. . . . Truly amazing pictures of the Moon.'

But the crew had a very precise and determined impression they wished to impart in the closing minutes of the final telecast from Moon orbit, a special and carefully considered message to all people of many different faiths, beliefs and philosophies all across the world on this Christmas night:

ANDERS 'We are now approaching lunar Sun(set) and for all the people back on Earth, the crew of Apollo 8 has a message that we would like to send to you. "In the beginning, God created the heaven and the Earth. And the Earth was without form, and void; and darkness was upon the face of the deep. And the spirit of God moved upon the face of the waters. And God said, Let there be light: and there was light. And God saw the light, that it was good: and God divided the light from the darkness."

LOVELL '"And God called the light Day, and the darkness he called Night. And the evening and the morning were the first day. And God said, Let there be a firmament in the midst of the waters, and let it divide the waters from the waters. And God made the firmament, and divided the waters which were under the firmament from the waters which were above the firmament: and it was so. And God called the firmament Heaven. And the evening and the morning were the second day."

BORMAN '"And God said, Let the waters under the Heaven be gathered together unto one place, and let the dry land appear: and it was so. And God called the dry land Earth; and the gathering together of the waters called Seas: and God saw that it was good." And from the crew of Apollo 8, we close with goodnight, good luck, a Merry Christmas, and God Bless all of you – all of you on the good Earth.'

PAUL HANEY (Public Affairs Officer) 'This is Apollo control, Houston. The speakers in the order that they read from what we believe to be chapters from Genesis were Bill Anders, then Jim Lovell, and to close out with, Frank Borman. That was both a Biblical and a geological lesson that none of us will forget.'

So far the mission had gone surprisingly well. After the '101 per cent' success of its predecessor, flight planners for this C' mission profile anticipated minor systems problems reminiscent of the Gemini days. But again, an Apollo spacecraft was performing with outstanding regularity, although there were irritating minor anomalies which hardly credited the 'failed' category. Passing around on the far side of the Moon, spacecraft 103 made its last full circumnavigation of the lunar sphere, orbital speed for Earth's only natural satellite being a mere 21% that of a satellite orbiting the Earth itself. There was just time to get a bite to eat before preparing to burn the SPS, the all-important firing that simply had to work. Trans-Earth Injection – TEI – would add 3,840 km/hr to Apollo 8's Moon orbit speed of 5,860 km/hr, boosting it Earthward on a journey that would take $57\frac{1}{2}$ hours. Normal far side passes kept Apollo out of radio communication for some 46 minutes but the spacecraft would appear early if the SPS had burned as planned.

Coursing its way round the left limb of the lunar sphere on revolution 10, communications were cut and from the duty Public Affairs Officer Terry White, came a comment on the coming event: 'And at 88 hr 51 min, we show loss of signal with the spacecraft. Our next communication with Apollo 8 should come in about 37 minutes. We are now about 28 minutes prior to our Trans-Earth Injection maneuver. As the spacecraft went over the horizon, capsule communicator Ken Mattingly passed along for the second time the word that "all systems are go" and we got a very terse "Roger" back from the spacecraft.'

Again, the atmosphere was tense in Mission Control at Houston, and like the LOI burn a day before where an early appearance round the right limb would have indicated no ignition so would a late appearance now signal a similar anomaly. The engine simply had to work, for there was no other propulsion system that could boost them out of Moon orbit. But it was a reliable motor and right on the anticipated mark the following exchange bounced its way from radio station to radio station all across the world:

LOVELL 'Houston, Apollo 8, over.'

CAPCOM 'Hello Apollo 8 loud and clear.'

LOVELL 'Roger. Please be informed, there *is* a Santa Claus.'

Immediately thereafter, an exchange of technical information on the status of the engine burn which had lasted 3 min 23 sec. It was so accurate, and the trajectory so well matched with the desired path, that the SPS engine would not be needed for any of the three trans-Earth mid-course corrections the spacecraft would be allowed to make if necessary; only a squirt from the RCS quads would be made 14 hours later. But the fact that the engine had accelerated the crew out of Moon orbit was a source of much prayer and thanksgiving that Christmas morning.

In Europe it was already long past the time when millions of children had woken early to search for evidence of at least one night time trip, but in America it was still too early for excited youngsters to have fallen, exhausted, fast asleep. In the few hours after firing out of Moon orbit the crew turned their TV camera on and brought viewers a picture of the Moon as the spacecraft climbed up from its gravity well back toward Earth. Then they turned it in upon themselves, something they had been somewhat reluctant to do, revealing scenes of Jim Lovell exercising with a bungee cord device similar to isometric equipment carried on Gemini flights, Bill Anders busily preparing a meal, and Frank Borman playing cameraman. It was a lighthearted crew, still serious about their duties, somewhat more somber than any other crew had ever been seen to be, as though by edging ever closer to the Moon's very surface they were touching an experience none had prepared themselves for. But there was certain relief in their smiling faces.

Down on Earth, two-year-old Jeffrey Lovell could no longer resist the big parcel at the foot of the large family Christmas tree. Opening it, he found the only sensible present a boy could receive from the man in the Moon: a toy space helmet. Later that morning, Marilyn Lovell and Susan Borman took their respective families to the local Episcopalian church. Frank Borman once said that 'We live on a sort of an island, and our ferries to the mainland of the community are the church and school.' That morning as the families and friends of many astronauts went to give thanks for the safe departure from Moon orbit of their respected colleague, Marilyn Lovell wore a new mink coat delivered that very morning by a delivery man from Neiman-Marcus. Valerie Anders likewise attended her own place of worship, the local Roman Catholic church. But she would have to wait for her Christmas present; even then en route to her with husband Bill, it consisted of a figure-8 gold pin with a moonstone crest, tucked away in Anders' ditty-bag of personal effects, the only non-flight items allowed on board.

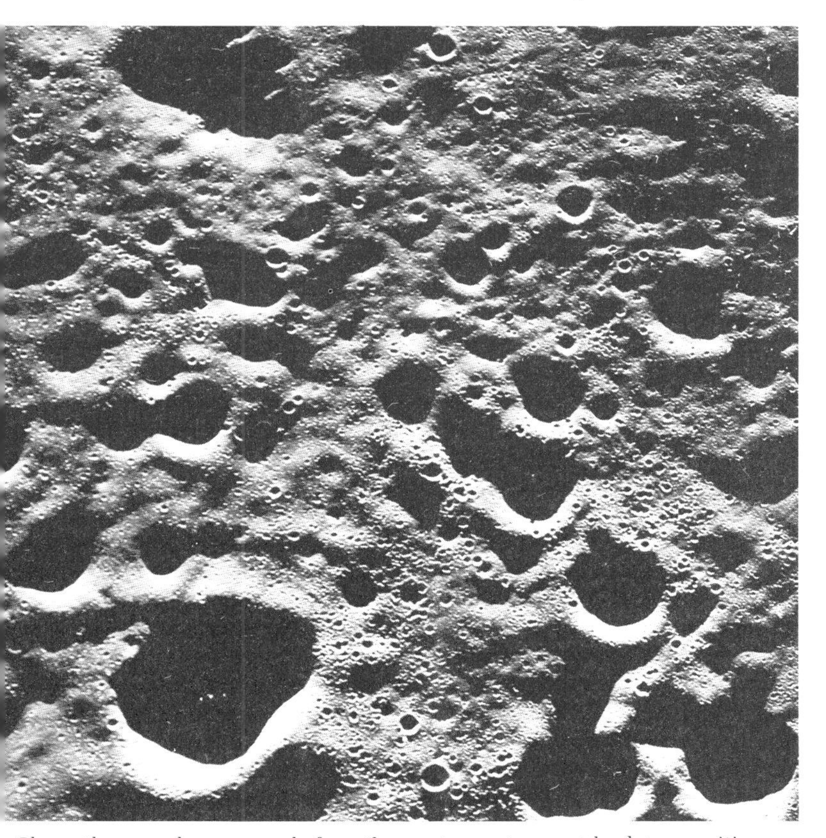

Closer than any human eye before, three astronauts report back to a waiting world from their lunar orbit what they see below on the primordial Moon.

When Apollo 8 sailed back into communication, the chart of the Moon used for the last day in plotting the progress of spacecraft 103 suddenly switched to a map of the Earth and a moving indicator showing the spot, deep in space, over which the vehicle was passing – the sub-spacecraft location. And to make the tired and overworked flight control teams just that little bit more responsive to the day it was around the world, somebody had wheeled in a big Christmas Tree and set it up, lights twinkling from the moment Apollo 8 re-appeared. It was indeed a great day for celebration, a time to re-kindle lost thoughts about other gifts long ago, a pausing point for thoughts dimmed by the pressure of daily routines. For Borman, Lovell and Anders – the sudden awareness of how tense they had been, how hard they had worked during the 20 hours spent orbiting the Moon. More navigation sightings, snatched sleep, meals, and engineering tests filled the balance of Christmas day.

Next day – another TV show from the returning spacecraft, more homespun philosophy from a crew that had been truly to the edge of human experience, and increasing anticipation of splashdown. But to Bill Anders the return home was a blessing: 'I think I must have the feeling that the travellers in the old sailing ships used to have, going on a very long voyage away from home and now we're heading back and I have this feeling of being proud of the trip but still happy to be going back home and back to our home port. And that's what you're seeing right here.'

Jim Lovell said that he could notice the increase in Earth's apparent size from the previous day and the last telecast showed a bright jewel alone in the solar system, 179,000 km from Apollo 8.

During the cruise home the spacecraft had first slowed to its minimum trans-Earth speed, at which point it crossed from the Moon to the Earth's gravitational field and then began to speed up; at the time of the last television transmission it was travelling at 6,675 km/hr, and increasing in velocity all the time as spacecraft 103 was pulled inexorably toward that invisible funnel in space down which Apollo 8 would thread itself for safe entry.

Very early in the morning hours of Friday, 27 December, preparations were under way for the re-entry and subsequent recovery of the Command Module and its three-man crew. Events would happen quickly. Thirty five minutes prior to entry interface, electrical power would switch to the entry batteries in the Command Module, followed, fifteen minutes after that, by separation of the Service Module. There was no critical time slot within which these events were required to happen, so long as the batteries came on line sufficiently near to entry for adequate electrical power throughout the descent and the Service Module was separated with sufficient time for the Command Module to orientate itself. It was the first time an American spacecraft had been returned through Earth's atmosphere from the vicinity of the Moon and it was the first time astronauts would experience a re-entry at lunar return speed.

During the automated sequences controlling spacecraft events, the guidance computer would move through four programs (P61 – P64) before splashdown. At any point the crew could take over manual control, but this would not be like an Earth orbit re-entry. Adopting the optimum trajectory for dissipating heat from the heat shield and keeping g forces as low as possible, Apollo 8 would fly an initial descent from entry interface at 122 km down to 55 km before lifting to an altitude of 64 km. From there it would again begin a long descent path terminating in splashdown, for this mission a location south-east of Hawaii in the Pacific Ocean. In all, from entry to splash, spacecraft 103 would fly 2,500 km around the curvature of the planet, a trajectory typical of all Apollo lunar re-entry paths.

In the Pacific it was dark when Apollo 8 approached its rendezvous with the recovery teams and as the spacecraft neared entry interface it passed into the shadow of the Earth's mass. The Service Module was separated at an elapsed time of 146 hr 28 min, followed by orientation for descent. Just 24 seconds after sliding into the atmosphere on a path inclined only $5\frac{1}{2}°$ from horizontal, the Command Module entered blackout where communications would be cut for more than five minutes. Nobody would know for sure that Borman, Lovell and Anders had successfully survived the blistering heat, temperatures approaching 2,800°c, until the plasma cleared from around the descending cone and radio signals could get through again.

Safely back on Earth, Borman, Anders and Lovell are picked up by the USS *Yorktown*.

But like everything on Apollo 8 it was almost perfect to the letter and 14½ minutes after hitting the atmosphere, three Moon men were bobbing in the Pacific. A second after splashdown the Command Module flipped over like its predecessor but that was soon corrected when the three inflation bags brought it back to a Stable I position. Apollo 8 splashed down in the pre-dawn darkness, the first time ever that a manned spacecraft had not landed in daylight, and it was ninety minutes before the carrier *Yorktown* had the crew safely on board. Freshly shaven with an electric razor carried aboard the helicopter at Borman's request, the ex-West Point man led his triumphant crew to the deck. Next day they would fly to Hawaii and on Sunday meet briefly with their families at Ellington Air Force Base en route to an intensive week of de-briefings at the Manned Spacecraft Center.

What followed would determine if the tentative plan set forth months before could indeed signal a lunar landing attempt for mid-1969. Only a year earlier the Mission A-G plan required at least three flights with both CSM and Lunar Module combinations. Now, in the wake of a successful Moon orbit flight, only two might actually precede the first manned landing attempt: Apollo 9 for an Earth orbit shakedown mission, and Apollo 10 as a full Moon orbit CSM-LM rehearsal barring touchdown. If careful examination of the telemetry records revealed nothing unduly troublesome in spacecraft systems, and if full technical interrogation of the crew confirmed the belief that all had gone as expected on Apollo 8, the Christmas Moon flight would have done its job and qualified the main Apollo spacecraft for more demanding roles to come. As a basic spacecraft it had been pressed to its full capacity, but the potential was greater still and it was that extra '10%' that was to be probed during the Lunar Module flights.

Recognition of Apollo 8's success was both prolific and abundant. Words flowed like paper at a ticker-tape parade, in praise of the mission and in adulation of the crew. Messages arrived from countries all over the world. Her Majesty the Queen sent a special note from London to the effect that 'We have all followed with greatest admiration the thrilling and historic journey ...' Kenneth Gatland, Vice President of the British Interplanetary Society, thought that 'The world now stands on the brink of entirely new experiences in interplanetary exploration.' From President Podgorny of the Soviet Union a request to 'Please accept, Mr. President, our congratulations on the occasion of the flight of the Apollo 8 spacecraft, which is a new achievement in man's quest to conquer space.'

Official Soviet comment was laudatory and in high praise of the three men 'who have accomplished this outstanding scientific and technical experiment.' Nevertheless, while recognizing Apollo 8 as 'a great scientific conquest,' Soviet scientist Leonid I. Sedov quickly pointed out that 'There does not exist at present a similar project in our program. In the near future we will not send a man around the Moon.' Sedov then went on to say that Russia preferred to investigate other worlds with 'automated soundings.'

And from Sir Bernard Lovell, a reflection on the flight: 'The success of Apollo 8 has been absolute in the sense that the vast engineering and organizational task has been accomplished with precision. We should not seek for isolated reasons which might justify the expense and hazards involved. Apollo has the deeper meaning that man is still willing and able to struggle with the almost impossible. The fringe benefits to science, industry and national aggrandisement are an acceptable bonus for an age which so often submerges the vision of man in the search for cost effectiveness.'

But of all the opinion offered up by politicians and scientists around the world, the most important views to NASA were those expressed by President-elect Richard Nixon. Nobody knew quite how the new President would respond to calls by the space agency for decision on future goals and objectives. Clearly, the increasing financial commitment to a war in Vietnam had stayed the hand of Lyndon Johnson's administration since major escalation in that conflict during 1967. Would the new President's firm stand on ending the war permit a less constricting attitude toward the NASA budget, freeing it to develop new initiatives in manned space flight? Existing cuts had effectively limited Saturn V and Saturn IB production lines. If Apollo 11 performed the first Moon landing, NASA would have only nine large Saturn vehicles left for the Earth orbit and lunar landing Apollo Applications Program. Beyond that the agency would require new projects and restructured goals.

Clearly, even before Apollo 8 left its Cape Canaveral launch pad, the Moon was seen as only a technological forcing-house for other, more directly beneficial, programs; the mood was clear and distinct: Congress was not about to fund a permanent lunar exploration project. Whatever NASA wanted to do in the future would have to be tempered with the calculating logic of budget books and accounting sheets. If ever there was a time when the space agency stood imperiled by the success of its own achievements it was now. Nixon was quiet on opinion about Apollo 8; it was, after all, a product of another great political and personal rival. But his gathering administration approved of a strong and viable U.S. space program, although his new science adviser, Lee DuBridge from the California Institute of Technology, was skeptical about the emphasis on manned activity. Nixon had commissioned a study on future space goals from the Nobel physicist Dr. Charles Townes.

Now, several top NASA men had left for other posts, Presidential parties had come and gone, administrations had orchestrated their own unique set of priorities, America had moved on. There were just twelve short months remaining to the end of the decade. If ever a decision was needed on the direction for future space activity, it was now. But for a while discussion would pale into insignificance before the great events that would unfold during the coming year. And as scientists, technologists, engineers, and administrators, looked with increasing awe and wonder at the prospect of a man on the Moon in 1969, the magnitude of Apollo 8's achievement seemed to dim in the light of anticipation. If not forgotten, it would soon be seen as a mere prelude to the great voyages of lunar exploration. The year had arrived; the year of the Moon.

CHAPTER NINETEEN

Pathfinders

The exhausting pace of Apollo 8's pre-flight schedule resumed when Frank Borman, Jim Lovell and Bill Anders returned to Earth on 27 December, 1968. Following an extensive de-briefing in Houston, the crew was sent to Washington on 9 January for an appointment with President Johnson at the White House, one of the last official LBJ functions before handing over office to Richard Nixon, where they were awarded NASA's Distinguished Service Medal. Following that, a joint session of Congress heard Frank Borman say that 'The one overwhelming emotion that we carried with us is the fact that we really do all exist on the small globe . . . If Apollo 8 was a triumph at all it was a triumph for all mankind.' The following day the crew were hosted by the city of New York, with the mandatory ticker-tape parade, with a visit to Newark, New Jersey, on the 11th, receptions at Miami, Florida, on the 12th, and official homecoming welcome in Houston on the 13th.

It was purportedley the biggest parade in the city's history, with a quarter-million people lining the processional route. Next day – a parade in Chicago where 1½ million persons saw Borman, Lovell and Anders made honorary citizens of the city. Before the end of the month they were to begin a goodwill tour of England, France, Belgium, the Netherlands, West German, Italy, Spain, and Portugal. In a tightly packed schedule lasting three weeks the Apollo 8 crew were to witness ebullient European reaction to the great Christmas flight, but to learn to their surprise, that 'they would hesitate to ask us questions, because they assumed . . . information about the flight . . . would be classified.' But if, until the end of 1968, intelligence about the direction of Soviet manned space operations had been somewhat vague, the events of January 1969 added clarity to an increasing suspicion that having been eclipsed by the Apollo 8 Moon orbit flight cosmonauts would now be directed toward the one application where America still lagged behind, and threatened to remain behind for a few years at least: Earth orbiting space stations.

For on 16 January, two manned spacecraft linked up in space for what the Soviet Tass news agency claimed was the 'first experimental space station', which 'makes it possible to carry out operations in space such as replacing crew members of permanent space stations and rescuing cosmonauts in emergencies.' This clearly alluded to criticism voiced at America's Apollo program where, once launched, a spacecraft would be unable to seek rescue from another. Until now, Russia had succeeded in docking only unmanned vehicles, that event coming as recent as October, 1967. But on 14 January, 1969, at 10:39 am Moscow time, Col. Vladimir Shatalov ascended on the plume of a Soyuz launcher, reaching orbit minutes later.

At the launch site, snow lay thick on the ground, air temperature had been below freezing, and rumor filled the press as correspondents speculated on the nature of this Soyuz 4 mission. Russia had traditionally avoided winter launch windows for their manned vehicles. With the docking flight of Apollo 9's CSM and Lunar Module spacecraft only six weeks away, the purpose became apparent when Soyuz 5 went aloft one day later. On board were three cosmonauts: Lt. Col. Boris Volynov, Master of Technical Sciences Aleksey Yeliseyev, and research engineer Yevgeniy Khrunov. During the first lone day in space, Soyuz 4's pilot had used the spacecraft's propulsion systems to circularize his orbit more nearly at about 207 × 237 km, ready for the second spacecraft launched at 10:14 am on the 15th. Through a series of maneuvers lasting 17 revolutions of the Earth, Soyuz 4 became the active participant in the game of celestial snooker as the manned vehicles closed in on one another to a distance of 100 meters. Taking over manual control from the automated rendezvous sequence, Shatalov steered Soyuz 4's docking probe into a receptacle on Soyuz 5 at 11:20 am on the 16th, accomplishing the first docking operation between two manned vehicles in space.

On the next orbit, Soyuz 5's 19th, Yeliseyev and Khrunov moved into the forward orbital module to which was attached the docking mechanism linking it to the orbital module on Soyuz 4. Donning space suits with great difficulty, due to the weightless state, the two men sealed themselves in by closing the hatch through which they had moved from the re-entry module, depressurizing the orbital module, and moved outside. Khrunov came out first, connected as was his partner by tethers, and the two spent one half hour outside before moving into Soyuz 4's orbital module via a hatch similar to that through which they had passed from Soyuz 5. Joining Shatalov, first Khrunov then Yeliseyev moved into Soyuz 4, greeted by a 'Welcome' sign placed in the depressurized module by its pilot. Hatches were closed, the two visitors from Soyuz 5 moved into the re-entry module, and preparations were made for a return to Earth early the following morning.

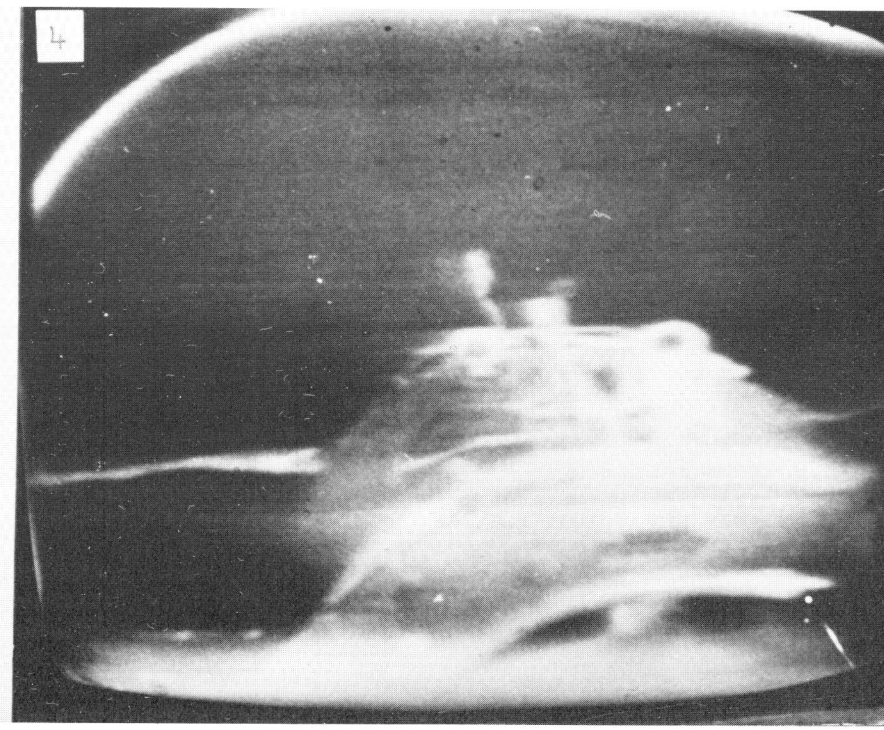

Soyuz 4 and 5 about to link up, as seen on Moscow TV.

The docked configuration of two Soyuz spacecraft.

Soyuz 4 landed at 9:53 am on the 17th, followed a day later by Volynov in Soyuz 5 at 11:00 am. Russia was clearly thinking of permanent orbital space stations, but it was significant that Soyuz 4 had a water-landing capability like the Zond versions of the basic Soyuz used to test during the previous year, unmanned, operations involved in a circumlunar return mission. Reaction to the dual rendezvous and docking flight was bland. The type of operation represented by this flight had been conducted many times in the Gemini program more than two years before; only the transfer of crews was a novel and innovative attempt to score a 'first.' Extra-vehicular transfer would never be a part of Soviet missions in the future, except as a back-up. Not everybody in the Soviet Union either agreed with the political leaders about the desirability of such activity. In a parade through Moscow to commemorate the flight, cosmonaut Beregovoy along with a driver and a security guard were injured when bullets from what the Soviets' described as a 'schizophrenic' gunman tore into a limousine passing the Kremlin's Borovitsky Gate.

With onset of the new year, preparations for Apollo 9 at Kennedy Space Center and in Houston's Manned Spacecraft Center reached new heights. If successful the mission would open the door on just one final Moon orbit rehearsal before the landing attempt with Apollo 11. On 9 January the crew for that mission were formally announced and against all speculation that the Borman crew would be given that flight, Neil Armstrong, Michael Collins and Edwin (Buzz) Aldrin were assigned to prime seats on the first Moon landing attempt; Armstrong and Aldrin would descend to the surface leaving Collins in the orbiting CSM. The back-up team would comprise Lovell, Anders and Haise. Stafford, Young and Cernan were already busy with preparations for Apollo 10, and McDivitt, Scott and Scweickart pressed ahead with readying Apollo 9, a Mission D flight scheduled to begin 28 February.

Hardware for Apollo 9 arrived late at the Cape, but first in was the second (S-II) stage for Saturn V AS-504 during May, 1968, followed by the S-IC and S-IVB stages in September. Lunar Module 3 had been at the Cape since June, and CSM-104 booked in during October. It was the Lunar Module that caused Borman to get the chance of flying the ambitious Apollo 8 mission but troubles that plagued early availability of this spider-like vehicle evaporated under the sheer weight of attention placed upon it by technicians and engineers. Once the wiring defects were cleared up LM-3 began a preparation schedule aimed at qualifying systems and equipment on board both stages prior to launch commitments signed off on all flight elements. Altitude chamber tests were performed on the Ascent Stage to check its environmental and thermal control qualities, but, when examining the main propulsion system, cracks were discovered in the engine and it called for a replacement unit. Installed in good time to proceed with mating, Ascent and Descent Stages were mechanically joined before combined testing began.

At year's end the hardware was ready and three days into 1969 a dawn roll-out saw 504 chug its way to pad 39A in a day-long trip before a flight certification review cleared the assembly on 7 January. For the first time, KSC personnel were preparing a full-up Saturn V/Apollo: S-IC; S-II; S-IVB; Instrument Unit; Lunar Module; CSM; launch escape system. Every element of a Moon mission would be tested on Apollo 9, even down to the Portable Life Support System (PLSS, pronounced 'pliss') backpack designed to provide life support, cooling and communications needs for a Moon walking astronaut. McDivitt, Scott and Schweickart had been preparing for this type of mission since December 1966 when they were given the second manned Apollo mission after the then scheduled Grissom, White, Chaffee flight. Delays to LM-3 had held their flight back from second to third manned Apollo slot but now all was ready for an intensive Earth orbit shakedown mission.

In several respects it was considered more dangerous than the Moon orbit flight of Apollo 8: for the first time in the history of manned space flight, astronauts would enter and fly away with a vehicle incapable of returning them safely back through the atmosphere unless they successfully re-joined their mother-ship; it would call for the first operation of the docked CSM-LM configuration, with propulsion burns from all three main engines on the two vehicles; and it would call for an EVA – a space walk – where Rusty Schweickart would leave the Lunar Module via the front hatch, make his way along hand rails to the Command Module, place his torso in the open hatch to demonstrate an ability to transfer this way from one vehicle to another in time of emergency, and then return to the Lunar Module back along the same route. Although performed in orbit about the Earth, almost every Apollo 9 operation was a potential danger point since in proving the ability of spacecraft systems to operate in emergency situations the spacecraft would be taxed beyond normal mission requirements – the extra 10% of performance essential to confidence in operating a Moon landing flight.

In all, nearly 1,700 hours were spent training for a flight that in all probability would fade from memory behind glory yet to come for Moon landing astronauts. Yet it was an essential precursor, one of two pathfinder missions before the big one. In preparing flight plans and procedures for the complex and demanding flight objectives, the crew resurrected an old astronaut practice of naming their respective space vehicles. The practice had been discontinued when Gus Grissom dubbed his Gemini 3 vehicle 'The Unsinkable Molly Brown.' But with two spacecraft in operation simultaneously, Apollo 9 would have to use identifiable code names for each. The team chose Spider for the Lunar Module, and Gumdrop for the CSM; the former for obvious association with arachnids and the latter because when the crew saw the Command Module arrive in its blue colored tape in looked for all the world like a hard-boiled sweet! Public Affairs chief Julian Scheer winced at the irreverent appellation, but the crew got their way.

Public interest waned a bit for Apollo 9. It was, in their view, not as spectacular or as enticing as the idea of three men flying around another body in space. But the newsmen were there in force again, counting down the missions to the flight that would fully release all the emotion pent up through a decade of preparation. Unfortunately, the crew proved reluctant spacemen when, after more than two years getting ready for this one flight, they went down with head colds just one day before the scheduled lift-off. As Dr. Mueller said, 'We certainly don't want to send the astronauts into space if they are ill. There will probably be a three to five day delay unless their health changes drastically.' Viral symptoms detected by the doctors could be masking something more serious.

As it was, the Cape-recycled lift-off for 11:00 am, Monday, 3 March. A germ-free crew was almost imperative for, following two consecutive flights where cold viruses struck, Apollo 9's team would spend considerable time in their space suits flying the Lunar Module, performing EVA, and testing out both manned vehicles; stuffy noses would be a decided hazard, encapsulated by a polycarbonate bowl! Was there some insidious reason why three Apollo crews in succession had contracted some form of nasal or gastro-intestinal upset? Physicians thought so and cautioned NASA management that unless extremely strenuous training sessions relaxed, more astronauts would succumb to mild germs; working long days, week in, week out, the men were physiologically weakened to the point where they could no longer resist infection. It was a punishing schedule, this final sprint to the Moon, but once accomplished the pace would relax; Headquarters already planned to cut Moon missions to one every four months for as long as the hardware lasted, but that for budgetary reasons rather than health consideration.

Came Monday, 3 March, and final preparations went ahead as scheduled following medical examinations the previous day. Again, the pre-flight countdown, the early wake-up, followed by breakfast in the MSOB, suit-up, transfer to pad 39A, and entry to CSM-104 with little more than two hours to go before lift-off. At launch the stack weighed more than any other rocket before it, marginally heavier than any of the three previous Saturn V's: 2,902 tonnes. The Command Module weighed 5,627 kg, but the Service Module to which it was attached was not loaded to capacity with SPS engine propellant. Instead of full tanks totalling 18,250 kg, as they would be for a normal Moon flight, SM-104 would carry fuel and oxidizer weighing 11,400 kg. Apollo 9 was not going to the Moon so SPS use could be relaxed, an advantage because the Lunar Module would have to carry full tanks and the Saturn V was scheduled to perform engineering tests necessitating two re-starts of the third stage from orbit. LM-3 weighed 14,525 kg, including propellant totalling 10,290 kg; CSM-104 weighed a combined 22,028 kg.

Again, McDivitt, Scott and Schweickart got a trouble-free ride on the Saturn's first stage, but the S-II revealed some oscillation around the seven minute mark in the flight. AS-504 was a little late getting into orbit, about 16 seconds, but the ride on the third stage had been smooth and when the J-2 finally shut down the crew knew they were 'go' for a circumnavigation of the globe: 504 put them in a 189.5 $\times$ 191.3 km orbit versus the planned 190.7 km. Good enough.

The first new task for McDivitt's crew was to check out the docking probe, carried by the Command Module for the first time. The docking mechanism worked on the probe and drogue concept where an extendible device would seek and latch on to a circular hole in the middle of a cone on the target vehicle. Retracted by pressurized nitrogen gas, the probe would pull both vehicles together so that latches on the docking ring would engage the back surface of a flange on the target. Silicon seals compressed when the two vehicles were pulled together would provide a pressure-tight interface across the two surfaces.

Apollo's probe comprised aluminum cylinders, one inside the other, with 25.4 cm of free telescopic travel. Carrying a cone-shaped head with three small capture latches, the probe was secured to the circular docking ring by three support beams. The drogue located in the Lunar Module's docking tunnel was 79.9 cm in diameter and 33 cm deep to the circular hole. It was secured by six bolts arranged in pairs at 120° intervals. At initial capture with the three small probe head latches, an operation called 'soft docking', the pilot would fire the nitrogen bottle to retract the piston, pulling the two vehicles into contact. Although 12 docking latches were provided for final locking, 'hard dock' could be achieved with only 3 in the sprung position. The others would be locked up when the crew entered the pressurized tunnel linking the two spacecraft.

First order of business for Apollo 9 was to extend the probe, which the crew duly accomplished shortly after reaching orbit. At 2 hr 45 min, Dave Scott fired the CSM free from the Saturn's adapter, jettisoning the four panels shrouding the Lunar Module. Firing forward, stopping, turning around and moving back in, Apollo 9 approached LM-3. To assist in lining up the two structures, the pilot used a Crewman Optical Alignment Sight (or 'coas') comprising an optical device providing an illusion that the image is the same distance away as the target. With the crossed lines on the lens aligned with a docking target on the Lunar Module, the pilot was required to thrust forward while using the RCS jets also to maintain attitude. An indication of range to the target was provided by the relative size of the image to the target. Fixed in the left rendezvous window, the coas was designed to be used by the pilot's right eye.

The LM's docking target comprised a black circle marked on the outer skin of the vehicle, with a tee-cross fixed on a short post at the center. Docking was accomplished at the first attempt, coming seventeen minutes after separation, and the crew pumped oxygen into the tunnel separating two opposing hatches on respective vehicles. Still attached to the S-IVB, the LM would be connected with electrical umbilicals prior to releasing it from the third stage. First, after checking that the tunnel was in fact

Astronauts (left to right) Schweickart, McDivitt and Scott prepare for Apollo 9 at the Cape Canaveral Apollo Mission Simulator.

With a second spacecraft in tow, Apollo 9 was at least twice as complex as previous Apollo missions. Here, LM-3 is fit-checked with CSM-104 in the Cape Canaveral altitude chamber.

pressurized, the forward Command Module hatch was opened. This was a circular hatch, 76 cm in diameter covered on the outer face with a thick wad of insulation and a layer of aluminum foil. Secured to the inner face of the lower docking tunnel by six latches, it could be removed quickly by pumping once a handle offset from the center of the hatch. A small pressure equalization valve in the center ensured compatability of the two environments before the hatch was opened affording access from the Command Module to the tunnel.

With the hatch now off, McDivitt and Schweickart checked that all 12 docking latches were in fact secure, connected up the umbilical lines that would feed electrical power from the CSM to the LM during checkout, and put the hatch back in place. At an elapsed time of 4 hr 18 min the combined CSM-LM assembly separated from the S-IVB. At pyrotechnic release, four spring-loaded latches released the Lunar Module from the Saturn's third stage as the CSM thrusters pulled the docked configuration free. It worked. So far so good. Much criticism had been levelled at the docking mechanism. Installed in the tunnel through which astronauts would have to pass to enter or return from the Lunar Module, the bulky probe and drogue assembly was a potential source of trouble; if it could not be removed an attempted Moon landing would be aborted, or if it could not be re-installed a docking would be impossible. Virtually anything that unexpectedly fouled the docking mechanism could cause abandonment of a Moon landing operation; many questioned the wisdom of building a mechanism so easily rendered inoperable.

Shortly after pulling free of the S-IVB, the third stage fired a second time. In a 70 sec burst from the powerful J-2, the stage propelled itself to a 207 × 3,087 km orbit, a highly elliptical path. Reaching the high point of this path 1 hr 20 min later, the engine was fired again, this time for 4 min 2 sec, boosting the stage out of Earth orbit. It was destined to circle the Sun for ever like its predecessor.

Meanwhile, back down at lower altitude, McDivitt got his spacecraft ready for the first SPS burn. With a combined mass of 36.5 tonnes acceleration would be slow to pick up but engineers wanted to test the docked configuration's bending moment and to test the ability of the Lunar Module to successfully ride out a punch from the pusher-rocket on the opposite end. Just short of 6 hours into the flight, the SPS lit up for a 5 second burn, adding 10.4 meters/sec to orbital speed and raising apogee to 234 km. Vibration levels in the LM were within design limits and the short test demonstrated a stable condition between the two vehicles. With a combined length of about 17 meters, the SPS gimbal actuators were required to steer a very different mass balance to that of the CSM on its own.

Apollo 9 returned to the more civilized practice of allowing all three crewmembers to sleep simultaneously. With so much activity planned for both vehicles it would need all three astronauts to be awake at the same time. But it was a welcome change from the practice that demanded at least one crewman to stay awake at all times. McDivitt, Scott and Schweickart ended their first day shortly after nine hours into the flight, receiving a wake-up call from Houston nine and one-half hours later. Their second day would comprise three docked SPS burns for further tests with the Service Module engine, and some systems checks aboard the Command Module.

The first SPS burn that day began about $3\frac{1}{2}$ hours after their sleep period ended, a 1 min 50 sec firing performed out of plane with the orbital path to set up the orbit for rendezvous operations later in the mission. It also provided an opportunity to check the ability of the SPS gimbals literally to steer the docked configuration in and out of a potential thrust alignment problem, and to test the digital autopilot. The second burn on day two came three hours later, lasted 4 min 41 sec, and raised apogee to 503 km. Its main purpose, however, was to consume propellant and lighten the CSM, and to shift the orbital ground track 10° east for better lighting angles during the rendezvous on day five. By lightening the Service Module the configuration's center of mass was shifted further forward, providing additional test data on the next SPS burn.

That came three hours after the second, a brief 28 sec firing designed to shift the orbit a further 1° east. Another engine gimbal test was also carried out, more severe than the first. The system came through with flying colours.

Apollo 9's third day was devoted almost entirely to the first manned activity with LM-3. It began with all three astronauts donning their space suits, an activity that caused Schweickart to feel severely nauseated. A poor traveller at best, he had taken an anti-motion-sickness pill before launch, but its effect had noticeably worn off! Suddenly, overcome with a queasy sensation, he vomited but managed to retain it in his mouth until he reached a bag. Released into the weightless cabin, the products of his breakfast would have been impossible to contain – a further unpleasant side effect of orbital flight. McDivitt too felt nauseated, but not as bad as Schweickart.

Dave Scott opened the forward hatch after checking pressure in the tunnel, and set about removing the probe, then the drogue. This required the three small capture latches to be set free by depressing upward a plunger set in the base of the probe body. Both probe and drogue were passed back into the Command Module where they would be secured until re-installation. Schweickart was first up the tunnel to open the Lunar Module hatch which, unlike its counterpart in the Command Module, was hinged to the side of the LM docking ring. Having already adjusted to 'floor' and 'ceiling' in the Command Module, the crew would find it strange to move down through the roof hatch into the Lunar Module orientated in exactly the opposite dimension; with the two vehicles docked end to end, the floors of respective vehicles faced opposite directions.

McDivitt and Schweickart would spend a full working day in the Lunar Module, checking out its systems and firing the big Descent Propulsion System, or 'dips,' engine. But they would not undock from the CSM; that activity was planned for two mission days hence. When Schweickart floated into the LM his first task was to activate and check the electrical system. Powered up, he could then bring up the separate elements of the environmental control system; until then, oxygen in the Lunar Module had been provided from the Command Module, first by bleeding through the appropriate valve and then by free passage through the tunnel.

About an hour into the checkout session, McDivitt joined Schweickart, leaving Command Module Pilot Dave Scott alone in Gumdrop. The drogue was put back in the tunnel, followed by the probe and then the forward Command Module hatch. The Lunar Module's overhead hatch was closed up and the two vehicles provided their own pressurized environments. Spider was checking out well, responding as expected to the intricate complexity of many different switch settings and instrument checks. Communications were checked out via the several different modes by which the two spacecraft could talk to each other and through which Spider's crew could talk with the ground.

Then it was time to look over the guidance and navigation equipment – the G&N system – which for the Lunar Module comprised primary and back-up systems: the Primary Guidance and Navigation Section (PGNCS, or 'pings'), and the Abort Guidance Section (AGS, or 'ags'). The PGNCS would act rather like a Lunar Module autopilot, a system into which all the many different modes of control and operation could be programed, receiving data from the navigation equipment in return. Like its Command Module counterpart, the PGNCS accommodated inertial, computer, and optical subsystems.

The inertial platform was similar in function and operation to that of the CSM equipment: three gimbal rings with three integrating gyroscopes and three pulse integrating pendulous accelerometers. The LM guidance computer comprised a 38,868 word memory, a 2,048 word erasable memory, provided a 745.65 hour clock for displaying ground elapsed time, and a facility for entering information via a display and keyboard assembly – called 'diskey' for short – identical to that described in the previous chapter for the Command Module. The optical subsystem comprised an Alignment Optical Telescope (AOT), a device installed in the roof of the forward crew cabin. It was with this instrument that the LM crew would obtain star or landmark navigation sightings for position information fed to the computer.

The Abort Guidance Section was developed as a back-up to the PGNCS and had a capacity to determine trajectory data for return to the CSM from any point in a typical lunar mission. It incorporated a 'strap-down' inertial system, where the gyroscopes and accelerometers were fixed rigidly to the structure rather than the gimballed concept adopted for the PGNCS and the Command Module G&N equipment. Less accurate than a free inertial platform, the AGS was sufficient to rescue the crew from a disabled guidance section. It carried a data entry and display assembly (or 'deeda') serving a similar function to that of the display and keyboard assembly described earlier.

Shortly after closing up the Lunar Module, isolating it from the CSM, an appropriate button was pressed and the four landing legs sprang into their deployed positions, secured by the downlock mechanism. But Rusty Schweickart was still suffering from a bout of motion sickness, at which point Jim McDivitt requested a private communication loop with the ground controllers so that he could discuss the future conduct of the mission frankly. But it was not as severe as McDivitt had thought and after vomiting a second time, Schweickart felt much better. Now it was time to hot-fire the RCS thrusters on the LM's four outrigger structures to check that they operated as expected and that they would respond to inputs from the hand controller.

Manual attitude control would be accomplished via one of two hand grips carried in the cabin. It would operate in precisely the same way Mercury, Gemini and Apollo CSM controllers were designed to work: push forward to pitch down; pull back to pitch up; twist left to turn left; twist right to turn right, etc. Each time the controller was moved from its central, detent, position the appropriate thrusters would fire to change the LM's attitude. Translation of the LM from one orbit to another, or for docking where the vehicle would be the active participant, was accomplished by a thrust/translation controller assembly. This comprised a stub-handle – two were carried in the crew cabin – by which the thrusters would be commanded to fire and physically move the Lunar Module to left or right, up or down, and forward or backward.

With a lever on the side positioned to a secondary notch, the controller would also allow the pilot to throttle the main Descent Stage engine by preserving the original thruster's translation functions except for the up and down movement, where it was replaced by thrust command to the 'dips' motor. With the attitude controller in the right hand, and the translation controller in the left hand, a pilot could fly his Lunar Module in a way similar to that of a conventional helicopter: balancing its attitude while simultaneously controlling where it went and at what speed. This was an essential feature of a vehicle designed to lower itself gently to the surface of the Moon, allowing the pilot to hover at will and move horizontally in picking out a precise spot to touch down.

To check out the LM thrusters, Command Module Pilot Dave Scott put the CSM in a free drifting mode while McDivitt and Schweickart across in Spider blipped the RCS quads. Now it was time to check out the Descent Stage engine with a comparatively lengthy burn across a wide range of different throttle settings. Scheduled to last 6 min 10 sec, the engine would be controlled by the PGNCS autopilot, iginition at 10% thrust, increasing to 40% after five seconds, and then to maximum after a further twenty-

Separated from the S-IVB stage, Apollo turns around and moves back in upon the Lunar Module high above Earth.

one seconds. For the last full minute of operation the engine would be manually controlled, ramped between 10% and 40% in several cycles until shut-down. In all, the burn would shift Apollo 9's orbit 6.7° east without significantly altering apogee or perigee. When the engine burned into life McDivitt was amazed to observe the precise degree of control he had hoped for, and when he took over manual authority the 'dips' motor responded exactly as commanded. Using the right-hand controller, he was able to bring the engine to the precise throttle settings simulated so many times on the ground. It was a satisfying first try at manually controlling the only throttleable engine on Apollo.

Now it was time to open up the tunnel between the two spacecraft and to rejoin Dave Scott in Gumdrop. McDivitt and Schweickart had been in Spider for about eight hours already. But they had satisfactorily checked out the Lunar Module ready for the active rendezvous two days hence. There was just one more operational activity – another burn with the SPS engine on Apollo to re-shape the orbit for better rendezvous conditions. Six hours after firing the Lunar Module's Descent Stage engine, the SPS lit up for the fifth time on Apollo 9; burning for 43 seconds, it placed the docked configuration in a 229 × 239 km orbit. Helmets, gloves, and space suits could now be removed. It had been a long and exhausting day, meticulously checking off several hundred itemized switch positons, sending telemetry down to flight controllers on the ground.

The following day was to have provided Schweickart with a space walk hand-over-hand from Spider to Gumdrop so that he could demonstrate an EVA route from one spacecraft to another. McDivitt was concerned that Schweickart's sickness posed a potential hazard and suggested that the exercise be limited to one in which both spacecraft hatches were open with Schweickart wearing the portable life support system back-pack. Under this plan he was not to go outside but merely demonstrate that all the systems worked. But when the crew awoke on day four, Schweickart felt considerably improved and after scurrying to open up the LM tunnel again McDivitt decided to let him go out on to the LM's front porch. For his part, Dave Scott would open the Command Module side hatch after depressurizing the spacecraft and float up with head and shoulders outside. While McDivitt remained inside the Lunar Module, Rusty Schweickart prepared to move through the hatch below the main display panel, the door designed for Moon walking astronauts:

MCDIVITT 'Mr Schweickart. Proceed out the door.'
SCHWEICKART 'O.K. Did you get the camera on there?'
SCOTT 'It's running.'
SCHWEICKART 'O.K. Proceeding on out. O.K. In the golden slippers. Hello there!'
SCOTT 'Hello there! That looks comfortable.'
SCHWEICKART 'Boy oh boy. What a view.'
SCOTT 'Isn't that spectacular?'
SCHWEICKART 'It really is. There's the Moon, right over there.'
SCOTT 'Why don't you say hello to the camera?'
SCHWEICKART 'Hello there camera. Oh! There's Baja California.'
MCDIVITT 'Things are still falling out up there. What you doing? Throwing everything overboard?'
MCDIVITT 'O.K. Dave, you ought to start getting your hatch closed.'
SCOTT 'Say again.'
MCDIVITT 'Better start getting your hatch closed if you're not already doing it.'
SCOTT 'It's closed, just not locked. Ooh! It's closed; locked.'

The EVA lasted just 37 minutes and instead of moving from one spacecraft to the other before returning to the LM, Schweickart rested on the porch, his feet in the 'golden slipper' foot restraints attached to the spacecraft. It was another step along the road toward qualifying Moon equipment; for the first time an American astronaut was detached from the life giving support functions of his spacecraft by breathing oxygen wholly contained within the back-pack. Gemini IX came close to achieving that when Gene Cernan almost got to try the Astronaut Maneuvering Unit. Now an autonomous life-support capability had been demonstrated by Apollo 9. Considered almost a spacecraft in his own rite, Schweickart was nicknamed Red Rover during the brief space walk, a reference to his hair color. Despite the abbreviated schedule, equipment had been satisfactorily tested and a collection of photographs obtained in addition to thermal samples retrieved from the Command Module exterior by Dave Scott.

In preparing for the EVA, Schweickart doned the integrated thermal micrometeoroid garment worn over the space suit and comprising a one-piece, form-fitting, multilayered unit made from several layers of aluminized Mylar and laminated Kapton and Beta fabric. Special overgloves were worn to protect the astronaut from excessive changes in temperature and an EVA visor was attached over the transparent helmet. Special boots too provided extra protection. The back-pack developed for lunar use was designed to operate for 4 hours and carried a primary oxygen supply of 0.47 kg. Contaminated gases returned from the suit were cleaned by a canister containing deactivated charcoal and lithium hydroxide. The water cooling system flowed through the liquid-cooling garment containing polyvinyl chloride tubing in a nylon Spandex material, worn next to the skin. Body heat would be removed by the water and lost to space via a sublimator in the back-pack.

The communications system provided a VHF link carrying voice and biomedical data to an EVA antenna on the back of the Lunar Module. Two 16.8 volt batteries provided electrical power, and an emergency oxygen pack on top of the back-pack would provide a 30 min supply if the primary unit failed. With a weight of 32 kg, the Portable Life Support System (PLSS) would support limited exploration of the lunar surface. But on Apollo 9 it had shown itself capable of supporting human life. It would be the only try-out in space before use on the Moon. Backed up by several thousand hours of simulated space environment testing on Earth the unit was qualified for operational duty.

Television pictures were sent to Earth as McDivitt and Schweickart closed up the Lunar Module prior to re-joining Dave Scott. The most demanding activity was to come on day five of Apollo 9's flight. The separation of the spacecraft, followed by rendezvous and docking, would be the first chance to really fly the Lunar Module and put it through its paces. McDivitt and Schweickart had found difficulty in keeping up with the flight plan when moving into the LM on the two previous mornings so this time they got up early and hustled along, receiving permission from Houston to move through the tunnel unencumbered by space helmets or pressure hoses; only after the LM had been checked out would they don the helmets and connect up to the suit circuit. Then the probe and drogue were re-installed and the capture latches snubbed on the inner flange of the receptacle before all 12 docking latches were manually cocked.

With both hatches closed, one from each side of the tunnel, the spacecraft were ready for undocking. That came at an elapsed time of 92 hr 39 min and was accomplished by Scott activating a

Astronaut Scott stands in the open hatch of Gumdrop as Schweickart, outside Spider, takes this embracing view.

As seen in this representation, Schweickart was to have crawled from Spider to Gumdrop, proving the access route should the docking tunnel become unpassable.

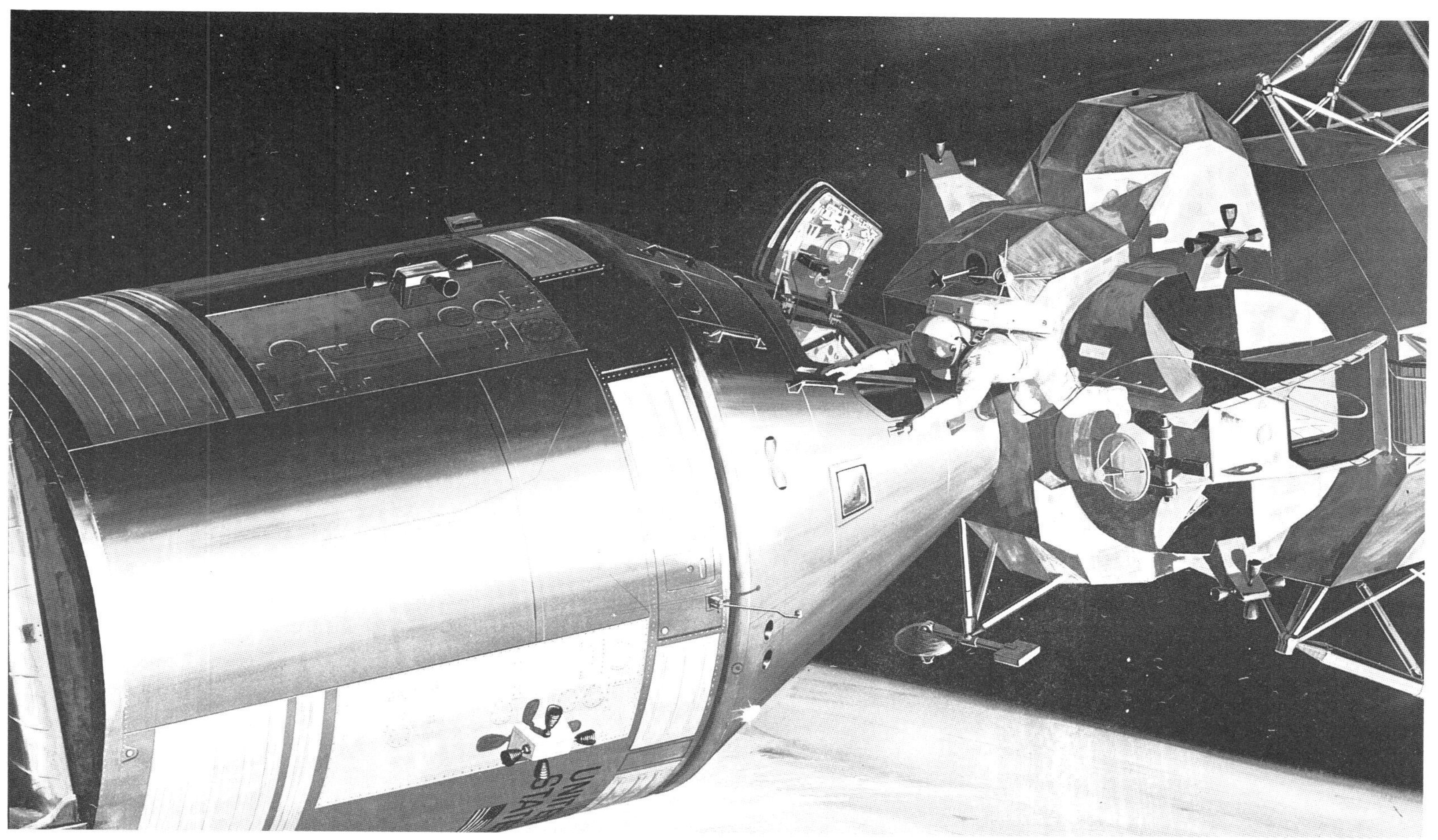

capture-latch release switch which electrically retracted the three small latches holding on to the drogue. At first it refused to work, then it retracted. Mechanically unlatched, the CSM pulled back away from the Lunar Module, free for the first time to go its own way in space. But first it had a fly-around inspection where Dave Scott visually inspected the bug-like vehicle. Then the CSM fired its RCS thrusters to change orbit slightly so that within 45 minutes it would be about 4 km away from the LM. During that time, known as the 'mini-football' maneuver because the ellipse drawn out on the trajectory charts looked like a baseball, systems aboard the LM would be checked out in readiness for the first major separation burn.

That came at 93 hr 47 min with a 19 sec burn of the Descent Stage engine, carrying Spider into an elliptical path 22 km above and below the circular path of Gumdrop, which would now assume a passive role on standby in case it was needed to go off in pursuit of the LM. Spider climbed toward the high point of its elliptical path, slowing down and pulling behind Gumdrop. At apogee the LM was 22 km above the CSM and 50 km behind it so rather than allowing Spider to coast on down to perigee, where it would speed up and close back in toward the CSM, the next maneuver was carried out to circularize the orbit 22 km above that of Gumdrop. To accomplish this the main engine was fired at 10% for 22 sec. Spider would now continue to fall further behind Gumdrop unless another burn was performed to lower the orbit and allow it to catch up.

The first burn with the main engine had given McDivitt a suspicion that all was not well. When he ramped the engine on throttle it groaned noisily so McDivitt backed off. The noise stopped, he ramped forward again, made up the velocity required and shut down. The second burn was quiet, but at only 10% thrust that may have been misleading. Drifting back further behind Gumdrop, Spider prepared for the next major operation where the Ascent Stage would be pyrotechnically separated from the Descent Stage immediately followed by a 30 sec burst from the RCS thrusters in a concentric-sequence burn designed to place the vehicle in an elliptical orbit with perigee 18 km below the path of Gumdrop. This would halt the continual separation and set it up for another maneuver that would cause the Lunar Module to slowly gain on Apollo.

When staging occurred there was a noticeable 'thump' in the cabin and bits of foil insulation flew out from the top of the Descent section. Now there was just the 4.4 tonne Ascent Stage alone with its two-man crew, 139 km behind Gumdrop and pulling away to a maximum 185 km before dropping down below the CSM's orbit. At the low point of the new ellipse, 144 km from Gumdrop, the Ascent engine was fired in a 3 second burst to circularize the orbit 18 km below the CSM. In just the same way that the LM had drifted further behind Gumdrop by remaining on an outer track, so now would Spider gradually close that distance from its inner path around the globe.

About one hour later, when Spider had closed to within 40 km of Gumdrop, the Lunar Module thrusted into its terminal-phase-initiation burn with the RCS quads. The closing maneuver, lifting the Ascent Stage orbit the 18 km altitude that vertically separated the two vehicles, lasted 33 minutes. Spider was back alongside Gumdrop having proved that all propulsion, guidance, and navigation systems worked as expected. Now there was another inspection from Dave Scott in the Command Module as the Ascent Stage danced around on its axes.

But docking the two 'craft proved more difficult than expected. Glare from the sun shone on the docking target prescribed around the right rendezvous window as McDivitt struggled to get a clear view through his coas fixed to the rectangular overhead window in the roof of the Ascent Stage cabin. All the controls were arranged so as to be used in the forward facing direction and it was difficult aligning the two spacecraft with the head looking directly upward. Nevertheless, the drogue was snagged on the probe's three little capture latches, Scott punched the retract button which drew the two together, and hard dock was completed at 98 hr 59 min. The Lunar Module had flown on its own for 6 hr 20 min and it was a very happy crew that got ready for return to Dave Scott in the Command Module. It had been a good rendezvous, but not without its moments. Particularly at the end when the two vehicles were docked for, as Jim McDivitt said, 'That wasn't a docking, that was an eye test.'

After McDivitt and Schweickart had crawled back into the Command Module, all three crewmembers prepared to jettison the LM Ascent Stage for a remotely commanded firing of the main Ascent engine in further engineering tests monitored through telemetry sent back to the Manned Space Flight Network. Separated from Apollo, the Ascent Stage fired up at 101 hr 53 min for 5 min 50 sec by which time the oxidizer was depleted and the engine safely shut down as required. The little vehicle was propelled into a 230 × 6,940 km orbit as a result of this burn. It would no longer play a role in the Apollo 9 flight.

The three days of activity completed by LM jettison had taken a lot out of McDivitt and Schweickart, the men who gave a green light to all LM crews after them by proving the vehicle spaceworthy. But they were still less than half-way through the ten-day mission which, in addition to qualifying the Lunar Module, was also required to demonstrate spacecraft survivability for a duration equal to that required for a Moon landing mission. But all the tiring and strenuous activity had been concentrated early in the flight and the crew could now take their time with a list of more engineering tests, housekeeping chores, photography, and navigation alignment checks.

Next day, the mission's sixth, the crew performed another SPS firing, this time for 1.3 seconds, so as to provide the CSM with an orbit from which the RCS thrusters could de-orbit the spacecraft should the main engine fail at retro-fire. And then they unpacked the main experiment of the flight: an earth resources multi-spectral camera system with which they would obtain pictures of the southern United States, Brazil, Mexico, and Africa. Consisting of four 500-EL Hasselblad cameras operated by electric motor and installed in a common mount, the experiment provided each camera with between 160 and 200 frames, a standard 80-mm focal length lens and separate film filter combinations. It was a major effort to expand on the ability of space-borne cameras to reveal geologically useful results from targets on the ground.

On day seven there were landmark tracking sightings, navigation platform updates, and more spacecraft housekeeping. Day eight brought a planned SPS burn, the last before retro-fire, to shape the orbit, and more earth resources photography. On day nine, the crew worked on inertial platform alignment, shot some more film off, and performed more mundane chores like purging the fuel cells and changing the lithium hydroxide canisters – for the sixteenth time. Day ten was much the same as the one before, except for some stowing tasks to prepare for return to Earth. Then, just before starting into their final rest period, the crew caught sight of the little Ascent Stage by squinting through the coas. At this time it was more than 1,200 km away. Only the previous day they caught a view of the Pegasus II micrometeoroid research satellite launched by one of the early Saturn I rockets in 1965. On that occasion the object was 1,800 km away.

Retro-fire came one revolution later than planned so as to move the splashdown point from a location south-west of Bermuda to a position 1,111 km east of Cuba, avoiding bad seas and high wind. The SPS ignited over Hawaii at 240 hr 31 min and Apollo 9's Command Module splashed down only 5.5 km from the USS *Guadalcanal* 30 minutes later. Weather was perfect, the spacecraft remained in Stable I, and within forty-nine minutes all three astronauts were safely on the deck of the recovery ship. In George Mueller's opinion, it was 'as successful a flight as . . . any of us have ever seen,' having 'fully achieved all of its primary objectives and in numerical count, we accomplished more than the planned number of detailed test objectives.'

In the sequential series of flight assignments, Mission F would be the next to fly, leap-frogging Mission E since its call for a repeat of Apollo 9's Mission D in high Earth orbit rather than low was too conservative. So Apollo 10 was all set for a lunar orbit flight during which astronauts Stafford and Cernan would dive down toward the Moon on a practice run of the path Neil Armstrong and Buzz Aldrin would take on Apollo 11, while Young rode alone in the Command Module and awaited their return. George Low had thrashed out the basic mission plan with Owen Maynard at MSC during late 1967, but John Mayer, Carl Huss and Howard W. Tindall engineered the precise details of the flight a year later. By the end of 1968 Tindall was an eager supporter of a daring profile for Mission F, believing that the two pilots should actually begin the powered descent to the Moon before aborting with a 'fire-in-the-hole' LM staging to demonstrate the ability of the system to abort successfully and return to the Apollo mothership.

This was indeed a daring plan – too daring as it turned out. There was only a low probability that something catastrophic would ever terminate a landing attempt once the actual descent began; to run a lunar orbit mission deliberately just to demonstrate an abort was akin to launching a Saturn only to blow it up during ascent! So Apollo 10 reverted to the next best scheme for getting maximum return from the mission. In an eight-day flight

Spider drifts ahead of Gumdrop. Note the four sensing probes which would be used on a Moon mission to tell the crew when to cut the engine; actual Moon missions carried only three probes.

during which the spacecraft would spend $61\frac{1}{2}$ hours making 31 revolutions of the Moon, Apollo's CSM would follow a timeline similar in every essential aspect to that which would be followed by the Moon landing missions to follow. Lunar Module operations would perform every scheduled Moon landing maneuver with the exception of a powered descent to the surface and a lift-off into an intermediate lunar orbit. But preliminary separation burns, and all the rendezvous sequences, would be carried out as though the spacecraft was going down to the surface and then coming back up to rejoin the CSM.

The only major engineering re-work needed as a result of Apollo 9 was to the S-II Saturn V second stage, which for AS-505 would fly with a phased engine shutdown sequence in a last attempt to remove the disturbing pogo effect. On 8 April, 1969, AS-508's second stage was test fired at the Mississippi Test Facility for 6 min 25 sec, with only the four outer engines burning for the last 1 min 26 sec. In a similar sequence to that employed for the first stage, the center engine would be shut down early, leaving the remaining four engines to burn on. Apollo 10's crew should hopefully get a smoother ride than previous Saturn V fliers.

Although nobody yet knew just how President Nixon would react to America's space activity, preliminary signs revealed that his administration was behind the idea of a solid base upon which to build future goals and objectives. The Townes study was still being prepared and science adviser Lee DuBridge was non-committal in his opinion about future priorities from the White House. But NASA felt that it could use remaining Saturn V launch vehicles to continue a modest program of lunar exploration. Production had been pegged at the 15th vehicle by past budget cuts; there was no hope of re-opening the production lines once they closed, and as the big rocket was the only launcher capable of sending Apollo to the Moon it was considered essential to use them all to the maximum value possible.

Accordingly, the new Administrator Tom Paine – President Nixon had Paine sworn in on 3 April – agreed to approve three manned Moon missions a year. If Apollo 11 successfully landed on the lunar surface, nine more attempts would follow at about four month intervals, each one going to a uniquely interesting site in an attempt to inject meaningful science into what had begun life as a technological goal. There was much interest from the scientific community in visiting different locations on the Moon and NASA was concerned that it would never get stability in funding or program status without widespread support from the scientific community.

On 10 April, less than a month after Apollo 9 splashed down in the Atlantic, Conrad, Gordon and Bean were announced as the crew to fly Apollo 12, four to six months after Apollo 11 if that initial landing attempt was a success. The crushing pace of activity at the Kennedy Space Center, where for the past six months the Merritt Island facilities had been processing hardware for three missions simultaneously, would ease after the Apollo 11 flight.

The hardware for Apollo 10 began arriving in October, 1968, when LM-4 was delivered from Bethpage into the receiving care of inspector Joseph M. Bobik, who found the vehicle in considerably better condition than its predecessor. Grumman had caught up with the learning curve and joined performance with the CSM from North American Rockwell. CSM-106 checked in during November, as did the launcher's massive first stage; second and third stages arrived the following month. Two days before Apollo 9 came home, AS-505 rumbled and snorted its way to pad 39B, the first use of the more distant twin to pad 39A that had been used exclusively to this time for all Saturn V flights. A third pad for Launch Complex 39 had been planned, but that went out when increasing costs met descending predictions on the anticipated launch rate.

Less than a week later Sam Phillips moved the lift-off date back a day to 18 May so that the shadow line across the Moon would have advanced further west, casting light upon sites P-2 and P-3. By mid-April the two spacecraft had checked out on their flight readiness reviews and everything was well up to schedule. Tom Stafford agreed with his crew that the respective names of their two vehicles should immortalize the comic strip 'Peanuts' feature from the pen of Charles L. Schulz. Accordingly, adopting these characters as their mascots, CSM-106 became Charlie Brown and LM-4 became Snoopy. Again, Julian Scheer was horrified at the flippant names; it would be the last time the astronauts were allowed such levity, for beginning with the first lunar landing attempt the names of respective spacecraft would be required to take on a more refined application.

On Sunday, 18 May, Rocco Petrone's launch team counted down to the planned launch of Apollo 10 and at 12:49 pm thunder again broke loose across the Cape. Thousands were awed by the blinding column of fire stretching down 400 metres to the launch pad as the cool voice of Tom Stafford called out events indicated aboard the spacecraft. But the tone of the flight was set by Gene Cernan's jaunty comment: 'What a ride babe, was a ride,' as supersonic shock waves bounced off the dense layers of the atmosphere. At Houston, flight director Glynn Lunney monitored the succession of events that cycled off as planned.

It was a heavy Saturn V, weighing in at 2,908 tonnes in the instant of lift-off, with an Apollo CSM that grossed 28.87 tonnes and a LM that weighed almost exactly 14 tonnes; LM-4 carried Ascent Stage main engine tanks with only 63% propellant: 1,188 kg versus a nominal 1,880 kg.

Pressed back into their respective couches with a force of 4 g, the deceleration of first stage shut-down slammed all three crew-members into their harnesses as the sudden cut-off pitched them forward toward the instrument displays. Then the S-II fired up and once again the pogo effect could be felt. For nearly five minutes the five J-2 engines propelled Apollo 10 toward space; bouncing and shaking its way out of the atmosphere the spacecraft gave Tom Stafford's crew a rough ride. Then the center engine shut down and, for another 1½ minutes, thrust continued on the outer four. They finally shut off and within one second the S-II punched off the third stage while the single J-2 on the S-IVB fired into life. But that too was a rocky ride as the stage groaned and vibrated for the 2½ minutes it took to propel the payload into a safe Earth orbit. Clattering and bouncing its way into space, the big Saturn excited John Young to comment: 'Charlie, are you sure we didn't lose Snoopy on that staging?' They had not, but it felt at times as though everything must be falling off.

William R. Frye would write of this event in the *Philadelphia Evening Bulletin*: 'TV cameras do not do it justice. It is like 100 claps of thunder, each following the other with machine-gun speed. The flame that leaps from behind the rocket could have come straight from Dante's inferno. It is too bright to be seen with comfort by the naked eye. The earth trembles beneath the feet, (4 km) away. Then the towering rocket, nearly twice as high as Niagara Falls, two-thirds the height of the Washington Monument, creeps with an agonizing slowness the first few (metres) off the ground, enveloped by a white cloud. Then it is gone – and man is left to wonder and to pray.'

About 32 minutes after lift-off, Apollo 10 swept into its first Sunset. One hour later, passing across the United States at the end of the first revolution, the docking probe was extended. From the ground tracking stations, new state vectors for the Command Module computer came up on the communications line. Around the globe, checks on spacecraft telemetry as controllers searched for some reason to cancel the upcoming TLI burn. They could not find one and over Australia on the second orbit the S-IVB thundered into life, shaking and rocking all three astronauts for the 5 min 45 sec it took to boost the combination out of Earth orbit and on to a free-return trajectory of the type first flown by Apollo 8. It had once again been a rough ride, 505 was certainly no lady, and Gene Cernan thought positively about the possibility of aborting the burn. But as with several points on this pathfinder flight, the mission would benefit from having a vintage crew on board, men who had flown in space before and would not be easily panicked by what rookie crews would consider to be a threatening situation. Between them, Stafford, Young and Cernan notched up five orbiting Gemini flights.

Soon after receiving confirmation that the S-IVB was stable, John Young got ready to separate from the adapter, move forward, turn around, and pull Snoopy from its kennel. Stafford and Young had swapped couches for this, placing Tom in the center and John on the left for a good view out that rendezvous window. Minutes after punching free from the four panels that shrouded the Lunar Module, TV viewers on Earth got their first view of the weightless event. The camera would continue to transmit the view until the Snoopy-bug had been extracted. It took just 14 minutes to move back in upon the circular docking cone, the four panels having safely spun away on springs that flipped them end-over-end to the blackness of space. At 3 hr 56 min, fifty-three minutes after pulling away from the S-IVB, Charlie Brown dragged Snoopy by

Apollo 9 provided the first opportunity for an exterior view of Apollo in space. Gumdrop is photographed from Spider.

the ears to let the inert rocket stage go its own way. Docked together for the long trip to the Moon, the two spacecraft weighed 42 tonnes – more than all ten manned Gemini vehicles combined! Such progress, so evident to these ex-Gemini pilots.

The crew had unexpectedly turned on the TV camera to show viewers the scene as the Lunar Module came away from the S-IVB, setting the tone for a mission that would break all records for the concern the crew had for sending pictures of their operation; it was a public relations dream, very different to the reluctant 'Wally, Walt and Donn' show three flights before. Now it was time to settle down into the routine of trans-lunar flight, but not before an SPS evasive maneuver to place the docked assembly on a slightly different path to that of the S-IVB. At 4 hr 39 min the Service Module engine fired for 2½ seconds, satisfactorily starting Apollo 10 along a path that would cause it to move increasingly away from the inert stage which, soon after, vented residual propellants through the J-2 engine bell sending it to a slingshot maneuver where it would pass the Moon and head for solar orbit. It would pass within 3,245 km of the Moon three hours after Apollo burned into lunar orbit.

Before settling down for their first night in trans-lunar space, the spacecraft was put into the 'barbecue' mode to prevent the Sun heating one side more than the other; the radiator on the solar side could easily be saturated with increased temperatures on the skin of the spacecraft, while radiators on the opposite side could freeze over. So Apollo 10 was aligned at right-angles, pointing vertically upward, and set in a slow roll at about two revolutions per minute, evenly distributing the thermal effects of the Sun. The spacecraft could certainly accomodate asymmetric heat loads at times when it would have to stop rolling and point in a fixed direction – for navigation marks or mid-course correction burns – but the rolling mode had proven useful on Apollo 8 and was to be standard procedure for all Moon-bound Apollo vehicles.

But housekeeping aboard a lunar vehicle en-route to its destination was a progression of systems chores designed to keep the spacecraft in peak condition. The cryogenic oxygen system supporting the fuel cells used to produce electrical power had to be purged every 12 hours; the hydrogen supply, every 48 hours. Potable water had to be chlorinated before each sleep period. Waste water dumps had to be timed away from navigation sightings to prevent a flurry of icy particles filling the view through the optics. Two batteries would be charged up at specific intervals. A lithium hydroxide canister would be changed every 12 hours, or sooner if the carbon dioxide partial pressure in the crew compartment reached 7.5 mmHg. There were 18 replaceable LiOH filters plus two in the environmental control system at launch. Redundant components would be checked over every 24 hours, and there were always the maneuver pads voiced up from capcom in Mission Control.

These were mainly data sheets of information that the computer would need in the event of a sudden need to abort the flight, which at lunar distance would require many hours of continued flight time to get back to Earth. Aborts during trans-lunar coast could take one of two forms: immediate return to Earth or a loop around the Moon. If the abort came early enough the big SPS engine, or the Descent Stage engine on the LM, would be used to fire in the direction of flight, braking the forward speed and in so doing transforming what would have been a circumlunar trajectory to a highly elliptical orbit of the Earth, causing the planet's gravity to pull the spacecraft back before it crossed into the Moon's sphere of influence. Nevertheless, depending on the point on the trajectory where the abort came, it would take anywhere between 18 hours and 65 hours to get back home. If the abort call came close to the point where the spacecraft would pass into the Moon's gravity field it would be quicker to go around the Moon and come straight back. But aborts from lunar orbit would take at least 40 hours and probably much longer. So maneuver pads were an important part of looking after possible contingencies, and for that reason the crew were kept advised of the various abort 'gates' so that the computer could be updated with the necessary information.

The second day on their way to the Moon the Apollo 10 crew fired the SPS engine for 7 seconds to trim the path at the MCC_2 opportunity; so good was the burn that no other mid-course corrections were needed to carry the vehicles behind the Moon 113 km from the surface. Before starting their first rest period, the crew sent back two additional TV transmissions showing the Earth and portions of the spacecraft interior. On this second day they sent another telecast to Earth, this time for nearly 28 minutes. Apollo 10 carried a 5.4 kg Westinghouse color camera producing a 525-line picture at 30 frames per second. Equipped with a small 7.6 cm monitor, a zoom lens for distant shots and a wide angle lens for interior views, the crew were a floating television studio, eagerly seizing unscripted opportunities to beam pictures to the home planet.

Communications with the Moon-bound Command Module came via the steerable S-band antenna attached to the base of the Service Module. Mounted to a 99 cm long boom the antenna comprised four 78 cm diameter parabolic dishes and was capable of transmitting at wide-, medium-, and narrow-beam settings. The wide-beam horn would be used near Earth, out to a distance of about 48,150 km to ensure coverage of half the planet's diameter. As the spacecraft moved away, and the Earth got smaller, one of the four parabolas would be used to transmit a medium-beam signal. From a distance of about 198,000 km, almost half-way to the Moon, all four parabolas would be employed to transmit a narrow-beam.

The four reflectors could move in each of two axes and be positioned by one of the crew to lock on to the signal transmitted from Earth. The antenna would then automatically track the signal to the limits of its maneuverability. Four omnidirectional S-band antennae were spaced equally around the base of the Command Module for communications at low altitude or in Earth orbit; two omnidirectional VHF antennae were attached to opposite sides of the Service Module for communication between the two manned vehicles in Moon orbit, or from the CSM to Earth at low altitude.

As with Apollo 8, and all succeeding Moon flights, the Earth's gravity pulled down the speed of the travelling trio but just before starting into their second rest period:

CAPCOM 'Apollo 10, this is Houston. Mark. You are half way, over.'
STAFFORD 'Roger. Thank you.'
CAPCOM 'And based on present trajectory analysis, looks like no more mid-course corrections will be needed prior to LOI, over.'
STAFFORD 'That sounds beautiful.'
CAPCOM 'Right on the money.'
YOUNG 'Looks like it'll be cheaper to keep going than turn back, huh?'
STAFFORD 'I'll tell you, it looks beautiful going away and it's going to look even better coming back.'

The third day was spent leisurely in space: the inevitable housekeeping, star alignment checks, and four more telecasts. The crew were a little perturbed by chlorinated water and by gas building up in the liquid used for drinking. It produced some near crippling stomach aches, but nothing more unpleasant. Frank Borman's Apollo 8 crew got no chance to see the Moon until they arrived in its vicinity, suddenly aware of its dark looming presence hoving into view. Stafford's crew, however, watched it from early in the mission, although they were unable to get any landmark plots on surface features.

Just before starting their third rest period, the Apollo 10 crew crossed the equigravisphere – 352,870 km from Earth, and 62,635 km from the Moon. At this point the speed of the Apollo 10 vehicles had dropped to its lowest level (3,423 km/hr) and from now on they would speed up as the Moon pulled them ever closer.

Day four came and with it a busy schedule during which they would burn into lunar orbit and open up Snoopy to check the LM over ready for the all-important separation and fly-away exercise the day after. Shortly before passing around the western limb of the Moon, Apollo 10 spent 90 minutes in the shadow of the lunar sphere. And then, 'Apollo 10, Houston. Two minutes to LOS. Everybody here says God Speed.' Loss of signal came about seven minutes before the SPS engine would begin its long burn to decelerate the docked assembly for lunar orbit. Apollo 8's lunar-orbit-insertion burn had lasted 4 min 7 sec but, with the LM attached, Apollo 10's burn would last 5 min 50 sec to compensate for the heavier load – 42 tonnes versus 28.8 tonnes. As the docked vehicles swept to the point of engine start, speed was 9,022 km/hr and at the end of the long firing the docked vehicles dropped into lunar orbit at a speed of 6,004 km/hr; pointing in the direction of travel, the big engine had successfully cut their speed by more than 3,000 km/hr and Apollo 10 was now in a 110.4 × 315.6 km path. The burn was sufficiently accurate to produce an error of only 740 meters compared to the planned orbit. Thirty-four minutes after going 'over the hill,' Apollo 10 appeared around the eastern limb:

CAPCOM 'Hello Apollo 10, Houston, over.'
STAFFORD 'Er, roger Houston, Apollo 10. You can tell the world that we have arrived.'
CAPCOM 'Roger 10, it's good to hear from you.'
STAFFORD 'The guidance was absolutely fantastic and we'll give you the burns right now.'
YOUNG 'This engine is just beautiful.'
CERNAN 'I'll take my hat off to the guys in the trench – I love 'em!'
YOUNG 'Yeah, kiss that man that runs MSFN.'
CAPCOM 'I don't know whether I can do that but I'll say thank you.'
CERNAN 'Yeah. Say thank you, big.'

Like tourists at a popular stop-over, all three astronauts were fascinated by the view out the windows. Observation from Apollo 8 revealed areas that seemed distinctly gray in color, while at other times the surface had a definite brown tint; now Apollo 10 saw the same phenomena and it would stimulate considerable debate until careful analysis of photographs revealed a tendency for the surface coloration to apparently change with different Sun angles. But as Gene Cernan told capcom Charlie Duke: 'It might sound corny, but the view is really out of this world.'

Although there had been disagreement about the precise use of terminology by the inevitable hawks of English language use, the high point of Apollo's orbit about the Moon was called apocynthion while the low point was referred to as pericynthion. At the end of the second revolution the SPS engine was fired again, for just 14 seconds, to slow the spacecraft by 152 km/hr and lower apocynthion to the same value as the low point, effectively circularizing the orbit. When the docked vehicles came around the limb, tracking showed Apollo 10 to be in a 109 × 114 km path. Very close to the planned value. Minutes later the crew took out the TV camera and entertained viewers to a 29 minute show from the Moon.

Passing across the lunar surface at a noticeably slower speed than that at which they had orbited the Earth earlier this mission and on the previous Gemini flights, the crew came up on landing sites further west than had been illuminated by the Sun during the Apollo 8 survey. Originally, ten sites had been selected as candidate locations for LM landings, spread out across a broad band centered on the equator: sites 1 and 2 were on the eastern side; site 3 was at the center of the visible hemisphere; sites 4 and 5 were far to the west. Based on photographs from the Boeing Lunar Orbiters, site selection took cognizance of the Lunar Module's design criteria and was based upon smooth, flat, crater-free, areas with slopes of less than 2° in the approach path and landing areas. Apollo 10's job was to perform a reconnaissance of sites 2 and 3; site 1 had been examined and photographed by Apollo 8.

For several reasons, engineers in charge of planning lunar landing missions preferred the eastern sites. With the daylight creeping along from east to west, any delay in a particular launch would merely move the prime landing site to the next one west of the original; if a western site was selected as the prime, launch technicians would have to wait until the Moon came round again for early Sunrise on an easterly site. This was because ideal landing conditions would be met only with the Sun at a comparatively low angle behind the descending Lunar Module.

For the first time this mission, the crew opened up the tunnel connecting the two vehicles to remove the probe and drogue. When that was done, Lunar Module Pilot Gene Cernan opened Snoopy's hatch to go on through when a flurry of Mylar insulation fragments blew out into the Command Module. Scurrying around retrieving the bits and pieces, Cernan continued to activate the Lunar Module, checking all the equipment first before powering it up and switching on the communication system. Tests completed, neither Cernan nor the ground controllers could find any reason to abandon the planned rendezvous activity planned for the following day, so the LMP returned to the Command Module and all three crewmembers got ready for sleep. Instead of keeping the tunnel open as planned, Tom Stafford elected to put the probe and drogue back in place and clear space in the crew compartment. Not all the tasks planned for Cernan's inspection had been completed, some of the anticipated activity was cancelled due to the time involved, but everybody was confident of a good performance from Snoopy.

The crew were all up early after a long rest, already well into their 'morning' activities when capcom gave them a call. It was late morning in Houston, nearing lunchtime at the Cape, and it was late afternoon in Europe when Tom Stafford and Gene Cernan moved into Snoopy for power-up and systems activation. Soon, John Young would be alone in the only spacecraft that could return all three to Earth, busily zipping around the suddenly very spacious Command Module. But first there would be 3hr 20 min of detailed and meticulous inspection and checks that would prepare the Lunar Module for its exacting task ahead, activation of the primary and secondary guidance systems, radar tests, thruster pressurization, etc., etc.

From the moment communication was switched on from the LM, ground control adopted the 'Peanuts' call signs. It came in handy when a slight problem emerged prior to passing around the western limb of the Moon on revolution 11. In checking the physical alignment of the two spacecraft, it was observed that Charlie Brown and Snoopy had slipped out of alignment in yaw as though, at the time of docking more than three days before, the CSM had been rolling slightly and barrelled into the Lunar Module's docking ring. To separate the two vehicles now might shear any or all of the 12 docking latches and prevent a satisfactory redocking later. Flight director Glynn Lunney immediately conferred with his men, but George Low, in the MOCR at the time, advised Lunney that so long as the vehicles were not twisted out of alignment by more than 6° it would be acceptable to go ahead and undock. A decision was critical, for only minutes remained before the two vehicles moved out of sight around the Moon's west limb prior to undocking on the far side. The alignment was in error by only 3.5°, so all was well:

CAPCOM 'Charlie Brown, Houston. We're concerned about this yaw bias in the LM and apparent slippage of the docking ring. We'd like you to disable, and keep disabled, all roll jets until after undocking, over.'
YOUNG 'Roger. O.K. fine.'
CAPCOM 'And, er, Snoop. We got 3 min 50 sec to LOS, over.'
STAFFORD 'Roger, three-fifty to LOS.'
CAPCOM 'And we'll see both Snoop and Charlie Brown at 98:25.'
CAPCOM 'O.K. Charlie Brown and Snoop. Three minutes, going over the hill, you are go for undocking, we'll see you round the other side.'

Shortly prior to appearing round the east limb of the Moon on revolution 12 at an elapsed time of 98 hr 25 min, the two spacecraft would slip apart ready for John Young in Charlie Brown to fire his thrusters briefly, placing the CSM in a mini-football separation maneuver similar to that performed at the start of Apollo 9's rendezvous operation. This would carry the CSM away from Snoopy so that the Lunar Module could check out the rendezvous radar and other systems. When the communications resumed, Snoopy and Charlie Brown were apart and standing off 9 meters from each other. John Young turned the TV on and gave Earth viewers a look at the quadruped that looked like something from another world; with landing gear now deployed, the Descent engine pressurized, and guidance systems turned on, Snoopy was ready to prove itself out.

Undocking occurred three minutes before the vehicles came round the east limb. Twenty-five minutes later Charlie Brown blipped its thrusters and slowly moved away. It would separate to a maximum distance of 3.7 km half way round the Moon, just after passing out of sight one hour later. In the intervening period, more systems checks and preparations for the next maneuver where the Lunar Module would briefly fire the 'dips' engine to lower itself toward the Moon. Called descent-orbit-insertion, or DOI, it would be standard procedure for Moon flights. It was a critical operation, one that would slow Snoopy's speed by 78 km/hr causing it to descend to within 15.5 km of the lunar surface.

Decelerating into a an elliptical path with pericynthion close to the Moon, there would be almost no margin for error: if the Descent engine burned for just a few seconds too long it would slow the vehicle too much, causing it to crash to the surface. Continuous tracking by Charlie Brown, and later by Earth stations, would confirm the orbit to be a safe one. But if not, if the Descent engine *had* burned too long, Snoopy would have to fire a 'bail-out' burn to speed itself up and raise pericynthion back up to a safe altitude.

DOI came at an elapsed time of 99 hr 46 min, about 40 minutes prior to appearing round the Moon's east limb. Snoopy's main engine burned for precisely 27.4 seconds, dropping the spacecraft into a 113 × 15.5 km path. But there was a temporary problem in the communications circuit where John Young in Charlie Brown could not hear the Houston capcom, although Earth could hear Charlie Brown. So Young called up Stafford in the Lunar Module to query the position:

YOUNG 'I'm all locked up on 'em Tom but I just don't read 'em.'
CAPCOM 'Charlie Brown, Houston, over.'
YOUNG 'Roger, read'n you loud and clear. Snoopy was "go" for DOI.'
CAPCOM 'Rog, great, sounds great, we copy.'

Young then proceeded to pass down the ranging data to confirm that Snoopy was descending at about the expected rate and that the burn had been good. Because of the separation between Charlie Brown and Snoopy, the Lunar Module was a few minutes from resuming communication and John Young commented on their jocular spirits: 'They're down there among the rocks, mumbling about the boulders and things right now.' To John Young, tracking Snoopy as it curved down so close to the lunar surface it really did seem as though Stafford and Cernan were in fact leap-frogging mountains and rilles. The relay from John Young continued: 'They just saw Earthrise. They say they're looking up at the horizon now.' And then, from speedy little Snoopy racing low across the lunar landscape:

CERNAM 'Hello Houston Houston, this is Snoopy.'
CAPCOM 'Rog Snoop. Go ahead.'
CERNAN 'We's goin', we's down among 'em Charlie.'
CAPCOM 'Rog. I hear you're weaving your way up the freeway, can you give me a post-burn report, over.'
CERNAN 'Yeah. Soon as I get my breath! Okay. Our burn was on time, our residuals were minus point one, minus point three, and minus point five, and that's the residuals with the dips burn, we did not null anything out. We're in a fifty-one point two by nine point two and the AGS has us in an eight point six.'
CAPCOM 'Roger Snoopy, we copy all the residuals and looks like we are all "go." Your dips is looking good; it's go. Over.'
CERNAN 'Ah Charlie, we just saw Earthrise and it's got to be magnificent.'
STAFFORD 'There's enough boulders around here to fill up Galveston Bay, too.'
CAPCOM 'We copy it Tom.'

Cernan's verbal report on the engine burn included residuals – the difference between the actual versus the planned speed change in each of the spacecraft's three axes – and a comparison between the orbit calculated by the primary (PGNCS) versus the backup (AGS) guidance systems. It all looked very good as Snoopy descended to the low point just 15,000 meters above the mean lunar surface. The actual point of closest approach was 15° uprange of landing site 2, the preferred spot for Apollo 11, and on a future mission the Descent engine would be switched on at that point for a long powered-descent to the lunar surface. On this flight, the pericynthion point provided a brief opportunity for landing radar tracking tests and photography of the approach path. Pitched down pointing directly at the surface below, Stafford and Cernan saw the increasing speed of the lunar surface as Snoopy swept along toward pericynthion. At 100 hr 43 min Snoopy reached that point and capcom Charlie Duke asked Cernan to tweak up the high-gain antenna. Communications with the Lunar Module came via a steerable S-Band antenna attached to the top of the Ascent Stage by an outrigger assembly. Minutes later all communication was temporarily lost and Duke called John Young with a request for him to contact Snoopy and relay the message for new antenna angles. That worked, the rushing crackle stopped, and Snoopy was back on line.

Suddenly the excited voice of Gene Cernan: 'We're low babe, man we're low!!' And from Tom Stafford, ever the calm Commander, information on crater rims and rilles. Five minutes after passing pericynthion Snoopy swept across landing site 2 and toward the Moon's terminator – the line separating day from night. And, again, communication worsened to the point where Snoopy was no longer accessible so when the word came to give a 'go' for the next maneuver Charlie Duke asked John Young to inform Tom Stafford. Nine minutes later Snoopy's crew lit up the Descent engine for the second time, a 40 second burn to speed LM-4 by 194 km/hr and raise apocynthion from 113 km to 352 km.

This would cause the Lunar Module to sweep on a wide arc more than 230 km above Charlie Brown before dropping down to the low pericynthion for a second pass across the landing site. At the time of the burn – a phasing maneuver – Snoopy was about 280 km below and *ahead* of Charlie Brown and, for the rendezvous, had to get back about 490 km below and *behind* the CSM. It could only do that by slowing down so that the Lunar Module could be overtaken by Apollo, hence the phasing maneuver sweeping out on a slower arc.

Stafford had momentary trouble getting into the correct attitude, and the shock of ignition activated several warning lights and alarms in Snoopy's cabin: 'It was real good, the burn was ready, we had a Descent quantity light on, we had an engine gimbal light on, the master warning, and all those good things, but we just pressed right on, over,' reported the Commander. The LM phasing burn came at 100 hr 58 min and not for nearly two hours would another major event occur. In that period Charlie Brown would continue to orbit the Moon in a nearly circular path while Snoopy looped out and dropped down below and behind the CSM. Passing around to the far side of the Moon at about 101 hr 37 min, it was time for Tom Stafford and Gene Cernan to get a bite to eat and rest for thirty minutes or so.

At 102 hr 22 min, John Young swept back around the east limb, more than 600 km ahead of Snoopy and almost beyond the range of the VHF communications system between the CSM and the LM: 'I am no longer in voice contact with Snoopy. I think we're just flat out of range.' Nearly four minutes later, the voices of Tom Stafford and Gene Cernan, down-linked to Earth on the S-Band frequencies, could be heard trying to contact Charlie Brown. Not for the first time, Earth served as relay between the two vehicles as Snoopy dropped down once more to the low point that had been so exhilarating first time around. But there were more important matters at hand. Within minutes, Stafford and Cernan would jettison Snoopy's Descent Stage and then, ten minutes after, fire the Ascent engine to drop the LM into a path with apocynthion 28 km below the CSM's orbit.

Mission control quickly scanned telemetry and passed along the word that all looked good for the staging sequence which would simulate the position a Lunar Module would be in with respect to the CSM after lifting off the lunar surface into Moon orbit. It was for that reason Snoopy carried only partially full Ascent Stage propellant tanks; arriving at the correct spot below and behind the CSM, Snoopy could also provide a weight simulation which assumed propellant had been used up lifting off the Moon. Stafford was early getting Snoopy into the correct attitude from where the Descent Stage would be pyrotechnically jettisoned. In the closing minutes to staging, Gene Cernan read out aloud the detailed checklist for Tom Stafford to check every switch configuration relevant to the coming event. Listening in on the ground, Houston monitored the activity, checking on Cernan's readout.

At the very instant of pyrotechnic release, the Ascent Stage would thrust forward on a blip from the RCS jets to prevent re-contact with the Descent Stage. Unknown to the crew, a switch had been inadvertently left in the wrong position. The Abort Guidance System – the back-up navigation equipment – had been left connected to the 'automatic' mode rather than 'attitude hold'. In the former the Ascent Stage, upon sensing Descent Stage jettison, would immediately cause the vehicle to hunt around the sky in search of the Command and Service Modules so that the radar could lock on for rendezvous. In the latter mode, the Ascent Stage would remain stable. Notice of the sudden and unexpected maneuver brought a curt 'Son of a bitch!' from Cernan and it took two minutes to get Snoopy back under control. 'Okay, something went wild there on that staging, but we're all set, we didn't lock it, we're going ahead with the "auto" maneuver,' said Stafford. Still out of communication with the Lunar Module, John Young in Charlie Brown asked capcom Charlie Duke in Houston how the situation looked:

YOUNG 'Charlie, how was the staging?'
CAPCOM 'Charlie Brown, Houston, they got staging, they had a wild gyration though but they got it under control, over.'
YOUNG 'Roger.'

Charlie Brown and Snoopy rehearse Moon mission operations without actually landing on the surface.

Now came the Ascent engine burn to lower apocynthion and to drop Snoopy into position below and behind the CSM:

CERNAN 'Got a lot of time. Seven minutes. I don't know what the hell that was babe.'
CAPCOM 'Snoop, Houston, you're looking okay for the insertion burn.'
STAFFORD 'Roger Charlie, that was something we've never seen before, it was real good, we went to AGS, then er . . .
CERNAN 'The computer's yours Tom.'
STAFFORD '. . . let me tell you what happened there real quick as we come around to this insertion burn. Went to attitude deadband, started thrusting aft, the thing just took off on us.'
CAPCOM 'Roger, we copy.'
CAPCOM 'Snoop, Houston, stand by for a mark, five minutes to the burn.'
CERNAN 'Okay, Charlie, we're with you, I think we got all our marbles. It's sure coming down to that ground, I'll tell you. Okay babe, I got good AGS and everything's looking good, I got the attitude set so if we have to switch we'll be all right. Okay, four zero seven, I'll monitor it 'till the burn, we're at four minutes, okay, four minutes, boy that's hard to do with helmet and gloves on. Give me a monitor Ascent pressure one and two, let's take another look at it, that's looking good. Engine-stop pushbutton reset and abort-abort stage reset. AGS translation, I mean AGS translation four jets, okay Tom.'
YOUNG 'Houston, I'm not reading them, so if they don't make it you got to tell me, huh?'
CAPCOM 'Roger, they're counting down now, looking good Charlie Brown.'

There was audible tension in Gene Cernan's voice as a torrent of technical data flowed from the alert Lunar Module Pilot feeding Tom Stafford with information for the coming insertion burn: 'Burn time is fifteen seconds, so it's going to go in a hurry.' In fact the Ascent engine fired for precisely 15.5 seconds at an elapsed time of 102 hr 55 min. Snoopy was now in a path that would carry it back up to apocynthion of only 84 km instead of the long, looping trajectory of the previous revolution. The Ascent Propulsion System burn came at the low pass across the lunar surface and provided once more an opportunity to rehearse a critical mission event similar to one which would be performed on a landing flight: rendezvous and docking with the CSM.

Snoopy spent 50 minutes coasting up toward the high point of its new orbit, closing the distance to the CSM from 460 km to 280 km. At an elapsed time of 103 hr 45 min the small RCS quads were used to more nearly circularize the orbit in a 77 × 87 km path provided by the concentric-sequence-initiation maneuver; 58 minutes later yet another RCS burn put Snoopy into a more nearly circular orbit from the constant-delta-height maneuver. The Lunar Module had now made one full revolution since staging and trailed Charlie Brown by about 140 km in an orbit approximately 25 km below the CSM.

Slowly catching up, Snoopy made the terminal-phase-initiation burn shortly before going round the west limb of the Moon on the 15th revolution. Ascending now to Charlie Brown, Tom Stafford and Gene Cernan made the final computations to close in and circularize their orbit alongside John Young in the CSM. Most of the terminal-phase trajectory would be made in the darkness of a lunar night, rendezvous being completed shortly before the two spacecraft moved back into communication with Houston around the Moon's east limb, 106 hr 19 min into the flight. Three minutes later a relieved Tom Stafford called Houston: 'Hey. Snoopy and Charlie Brown are hugging each other.' Apollo 10 was a united duo once more. And again, to a comment for more work tasks on spacecraft systems: 'You're right Jose, it's been a long day.'

In fact, Snoopy had loafed around the Moon, firing its engines, kicking its Descent Stage off and sporting around the lunar sphere for exactly eight hours, but the crew had been inside for a total twelve hours already and a further hour would pass before they could get back in the Command Module and button up the transfer hatch. The crew had proved the routines worked in practice: descent to a point where landing maneuvers could begin; rendezvous from a position an Ascent Stage would arrive at after lifting off from the Moon. Minor problems had caught the crew unawares, and some of the cameras failed at the critical low pass across the No. 2 landing site, favorite for Apollo 11. But all in all it had been a rewarding day and throughout the long sequence of activity, models of Snoopy and Charlie Brown sat on consoles in the Mission Operations Control Room. Peanuts came through. The pathfinders had illuminated the way.

Next order of business was to transfer equipment back into the CSM, although considerable quantities would be left in Snoopy, and close up the two hatches. Two items did get an unexpected ride back – the Maurer sequence camera that failed earlier and a lithium hydroxide canister that also gave trouble. One full revolution later, again around on the Moon's near side, Snoopy was jettisoned and the CSM thrusted into a slightly different path so that it would move increasingly away from the Ascent Stage. Then, at 108 hr 51 min, eight minutes after separating, Earth tracking stations sent a command to Snoopy to fire the Ascent engine to propellant depletion. The burn lasted 3 min 33 sec and carried Snoopy out of lunar orbit and into a path about the Sun. Ground stations would continue to track it for 12 hours until the Ascent Stage batteries ran down.

But it had been a very long day and by this time the crew were heading for their well earned rest. Next day, most of the time was spent tracking landmarks on the surface of the Moon, building a comprehensive navigation index to the lunar environment and the changing orbit of the lone CSM. Fatigued by their busy day, two television transmissions were cancelled and the extra day around the Moon was a welcome rest amid continuing activities within the Command Module. The CSM remained an extra day in orbit so as to give the crew time to rest from their exhausting rendezvous activity before squaring away for TEI – Trans-Earth Injection. Following another rest period, Apollo 10 sent a twenty-four minute TV show to Earth with shots of the lunar surface and then the interior of Charlie Brown on the 29th revolution. At the end of revolution 31 the SPS engine was fired for 2 min 44 sec and Mission Control eagerly awaited resumption of communication when the CSM appeared round the Moon's east limb to see how the burn had gone:

STAFFORD 'Hello Houston, Apollo 10.'
CAPCOM 'Hello Apollo 10, this is Houston, how'd the burn go?'
STAFFORD 'Roger Houston, we are returning to the Earth, over.'
CAPCOM 'Glad to have you on the way back home ten.'
STAFFORD 'Ah, roger, the burn was absolutely beautiful and Geno has the report and we got a fantastic view of the Moon now, over.'
CAPCOM 'Mighty fine Tom, standing by for that report.'

Before sending back technical information on the performance of the engine, Cernan advised Mission Control that the TV was coming on, and that they had a little something they would like to send down. Almost immediately, from the encapsulated environment of Apollo 10, came the recorded voice of Dean

Martin crooning his way through a very familiar song: 'Going Back to Houston.' The onboard tape-recorder had varied applications! For forty-three minutes the crew transmitted remarkably clear views of the Moon as Charlie Brown climbed rapidly out of its gravity well, slowing down continually until it crossed the equigravisphere where it would once again be predominantly attracted by the Earth.

Two hours later, another TV session was transmitted, and then the spacecraft was put in a slow roll – passive thermal control – similar to the attitude mode used en-route to the Moon three days before. The journey home was uneventful and yet just as important for the navigation practice it afforded and the time it allowed for systems checks; engineers were still concerned to explore every facet of Apollo's operating potential and to divine every nuance of its response.

It would take less than 55 hours to return from the Moon, little more than two days for a journey that had once seemed impossible. On the day before entry, Sunday, 25 May, Tom Stafford asked capcom Joe Engle to apologise on his behalf to Pastor Bob Parrott and the congregation at Seabrook United Methodist Church for being 'out of town for church today.' He requested the minister to read a special quotation from the Bible in church that morning, verses from Psalms and Isaiah:

'Oh Lord our Lord, how excellent is thy name in all the Earth! who hast set thy glory above the heavens . . . I was glad when they said unto me, Let us go into the house of the Lord . . . for my brethren and companions' sake, I will now say, Peace be within thee . . . I will praise thee with my whole heart: before the gods will I sing praise unto thee . . . In the day when I cried thou answeredst me, strengthened me . . . And he shall judge among the nations, and shall rebuke many people: and they shall beat their swords into ploughshares, and their spears into pruning hooks: nation shall not lift up sword against nation, neither shall they learn war any more.'

The performance of the Service Module engine had again confounded the skeptics: Apollo 10's homeward path was so accurate that the first two mid-course corrections were cancelled but MCC_7 was performed just three hours prior to entering Earth's atmosphere, a small tweak burn with the RCS thrusters to change speed by 1.7 km/hr. Like Apollo 8 before it, the Apollo 10 Command Module had to enter the atmosphere at a flight path angle of 6.5° with a 1° margin either side. The actual entry path was only 0.03° off the planned value. Although the margins were almost alarmingly tight, mission operation teams responsible for tracking and for performing guidance and navigation, demonstrated remarkable accuracy that belied the severity, or consequences, of error in precisely targeting the spacecraft for its tiny slot in the sky.

During the trip back, a waste water dump was timed so as to minimise the need for a mid-course correction. By aligning the spacecraft so that the expelled water would produce a modest thrust simulating the effect of a Reaction Control System jet the trajectory was pressed closer to the precise value required. Again, the Command Module flew a roller-coaster descent, dipping into the atmosphere with a deceleration of about 7 g, lifting out to higher altitude, then dropping down again for a 4 g deceleration culminating in parachute deployment. Splashdown came close by the USS *Princeton*, standing in the prime Pacific recovery zone east of Pago Pago in American Samoa, and the Command Module remained in Stable I. In just thirty-nine minutes the crew had been recovered by helicopter and placed on the deck of the ship.

Apollo 10 was over. The last hurdle had been crossed. Just fifty-one days remained to the launch of Apollo 11 and Man's first descent to the surface of another world in space. Tom Stafford, Gene Cernan, and John Young had not only lit the path that led to the Moon's surface , they had re-kindled public interest with a flurry of TV shows aimed inside and out – at remarkable views of Snoopy, at the fantastic scenes of lunar ruggedness, and at the fragile raft of life called Earth. It took just two weeks for key program officials to sift the already filtered data on Apollo 10 and come up with a final decision for the anticipated Moon landing attempt – Apollo 11 *would* be launched on 16 July with a plan for lunar descent. On 12 June, Sam Phillips cautioned that if anything developed to prevent the flight, NASA would not launch Neil Armstrong's crew until all was 'ready in every way.' Moreover, he said they would not, once launched, 'hesitate to bring the crew home immediately if we encounter problems.'

But only discretion reminded a global audience that any one of several anomalies could strike Apollo 11 from imminent launch. Very few believed the big bird would not fly on time and from across the planet pressmen and others began a migration that would end on the marshy ground of Merritt Island, home of launch vehicle 506 and Apollo 11's hardware. The public affairs operation began in earnest one calendar month before the planned launch day: 16 June, 1969. It was a massive operation designed to absorb and cater for the requirements of 3,500 newsmen and photographers eager to get new angles on the Moon flight, unique shots of expensive hardware, and 'exclusive' interviews with astronauts and key personnel.

Florida would take the brunt of the spectator sport, only the genuine and the sincere making it across to Houston and the Manned Spacecraft Center; many who came arrived for the fun, some rode in on press tickets like party-crashers. Nevertheless, 500 telephones were wired up at the Kennedy Space Center in anticipation of the deadline needs from key correspondents. In addition, the White House invited 18,000 guests to the launch and 183 buses were assigned the task of ferrying people to the viewing site. Nearly 40,000 people would be within the restricted zones, several hundred thousand more were expected to arrive outside. Yet behind the official facade there was still much to be done before Armstrong, Collins and Aldrin could climb the fire-stairs into space and while people all over the United States, and many in foreign lands, made plans for a trip to the Cape in July, engineers wrestled with last minute details of the all important flight plan.

This was a central event for each mission; without the meticulous care each document received, Apollo flight plans would be incapable of providing an essentially flaw-proof list of activities. From the flight plan came many documents to tell the crew how to perform this maneuver or that, checklists to go through before specific operations, cue-cards to prompt pilots of spaceships. It was an interlocking series of pre-planned and newly determined schemes, each phase of the flight detailed with excrutiating precision. Apollo 11 would fly to the Moon carrying a 9 kg flight data file containing flight plans, landmark charts, star maps, photograph logs, checklists, procedures documents, crew logs, orbit maps, contingency plans, abort plots, computer cards, and navigation charts. There were twenty volumes, stowed in the Command Module on the right side of the cabin in three separate containers.

In a very real sense the mission would be performed on the ground – and only carried out in practice during the actual flight. While the crew flew the profile carefully written up in the preceding years, engineers, managers, specialists of every kind, would follow the progression of the men in space, checking off every step they made so that if anything untoward should arrest the steady progression of events there would be someone somewhere with a ready reason for what was happening. One example of how meticulous pre-flight planning was can be had from the sequence of decisions that ultimately assigned Neil Armstrong the role of first man on the Moon.

From the first tentative Moon landing plan back in the early 1960's, NASA assumed that one man would remain aboard the Lunar Module while the second roamed the lunar surface. As technical decisions were made about the life-support back-packs, equipment was included which would permit each PLSS to be recharged after separate three-hour excursions. From an early date, the space agency considered it desirable to have only one man out on the surface at any specific time. Then, in August, 1965, Bendix Systems Division, TRW Systems Group, and Space-General Corporation were each awarded low-cost contracts to design preliminary Apollo Lunar Surface Experiment Package (ALSEP) equipment which could be carried on the Descent Stage of the Lunar Module for deployment by a Moonwalking astronaut.

Bendix eventually received a definitive agreement to build the ALSEP experiments, separate packages each comprising six to nine instruments powered by a small nuclear generator which could be left on the Moon for continued monitoring after the crew came home. But retrieval from the Descent Stage, manual transporting to a suitable site close by, and the physical laying out of the various items was a time-consuming activity. Late in August, 1968, astronauts Don Lind and Joe Schmitt performed a simulated ALSEP deployment which quickly revealed the difficulty experienced with handling the equipment. At this time, two periods of lunar surface EVA were planned for the first Moon landing, each of up to three hours in duration.

George Low suggested limiting exploration on the first landing to one excursion only, in which a set of priorities would structure the number and type of tasks involved: a quick 'grab' sample as contingency against not being able to retrieve any soil at all should an emergency develop; photography of the LM and the terrain; an assorted collection of lunar samples; deployment of an abbreviated ALSEP. Sam Phillips had seen the demonstration by Lind and Schmitt and immediately asked Bob Gilruth to gather a consensus of Manned Spacecraft Center opinion as to the preferred sequence of lunar surface activity and forward conclusions to George Mueller's Manned Space Flight Management Council.

Opinion quickly polarized around two opposing views: Gilruth wanted two men out on the surface to spread the work involved setting up ALSEP; Phillips preferred one man outside and no ALSEP deployment at all. All of this referred to the initial Moon landing. Everybody agreed that once established as a safe and plausible procedure, Moon flights should accommodate just as much science as possible. But the first landing crew would have considerable pressure over them pioneering a completely new set of operational procedures, and it was considered foolhardy to unneccessarily risk the men's lives for posterity; there would be more landings to follow, after all.

Phillips was concerned that valuable time spent training to set up scientific experiments would detract from the need for rehearsals on possible contingency plans and pointed this out to Mueller's council. But the Manned Spacecraft Center was concerned that even one flight should not be a scientific waste and pressed hard to convince the council that two men should get out and that an 'Early' ALSEP pack comprising just a few small experiments would be an acceptable compromise. Accordingly, Early Apollo Scientific Experiment Package (EASAP) was born and Bendix received immediate notice to put together an abbreviated ALSEP, not to run on nuclear generators but rather to use solar cells on the exterior.

Considerable opinion has been expressed about the amount of science incorporated into Apollo plans. It should be especially noted that both MSC and Mueller's management council made strenuous efforts to broaden Apollo's scientific base and that the sheer magnitude of the engineering problems involved could not reasonably have been expected to allow more science than was incorporated. The program's original intent would have rested on the routine operation of at least four manned Moon flights a year, such was the plan embraced by the lunar segment of the Apollo Applications Program discussed in Chapter Sixteen. Complex scientific activity would have been a hallmark of the expanded program.

But even that activity would have been impossible without several missions almost wholly dedicated to proving it was safe for Apollo to make repeated landings. As it turned out, the number of landings that were actually accomplished only just exceeded those that were for almost purely engineering purposes, and so the scientific community was denied the expanding scientific lunar base it coveted. But not because management or engineering factions were averse to scientific exploration; to a man, they knew that the more justification they generated for each flight the more chance they stood of receiving funds for extra missions.

The prime crew for the first Moon landing: (left to right) Neil Armstrong, Michael Collins, Edwin 'Buzz' Aldrin.

On 1 November, 1968, the management council decided to allow two men to leave the Lunar Module at the same time; weighing up the arguments both for and against simultaneous EVA, it was recognized that should one astronaut suddenly become ill it would be a decided advantage to have a second man out on the surface. As for the LM systems, ground stations could monitor telemetry and call the astronauts back in should a major problem develop.

Now there was the question of who should get out first. And almost as an afterthought the realization dawned that whomsoever did descend the ladder first would be forever immortalized in the history books. Whenever that point had been raised before the Lunar Module Pilot was the natural selection. A Commander always stayed with his ship: the second crewmember always performed EVA in Gemini. It was a logical decision. The crew for Apollo knew their selections in late 1968 and even at that time it looked as though this would be the flight to perform the historic landing. By late December work charts designed to show the allocation of tasks assigned Neil Armstrong the job of going outside first followed by Buzz Aldrin, the Commander followed by the Lunar Module Pilot.

Armstrong certainly felt strongly about his prime role as Commander but Deke Slayton too thought the procedures would work better if the CDR left first followed by the LMP. But these were internal procedures unknown to the press and not before early April did the public get to hear that Neil Armstrong was to leave first; earlier press reports had continued the original story that Aldrin would be first out. Armstrong *had* exerted his right as mission Commander but the decision would undoubtedly have been the same anyway. Nevertheless, it was late in the day when mission planners could finalize the lunar surface procedures document laying out in detail the functions and responsibilities of both crewmembers.

The very selection of Armstrong, Collins and Aldrin had been a chance affair based on unpredictable events. When originally replaced by Jim Lovell as Command Module Pilot aboard Apollo 8, Mike Collins was promised an early assignment just as soon as he recovered from the bone spur operation. If he recovered in time he could replace Haise as CMP for Apollo 11. He did – on both counts – relegating Fred Haise to the back-up LMP slot, with Jim Lovell as the CDR and Bill Anders as the CMP.

In a number of significant ways the Apollo 11 crew epitomized the 'in-house' dream of what every good astronaut should be – not the public view, because that had been well typified by several swashbuckling crews on earlier flights. Neil Armstrong was a firm commander of men and machines, more the latter than the former, and a professed loner. Never very easy in company, he kept counsel with himself more than others and attracted very few friends. Moreover, he was an introverted recluse when it came to personal achievement. His assignment to lead what could emerge as Man's first expedition to the lunar surface was a matter of reflection, not a matter for conversation, and even Armstrong's father heard the news from press sources. But his greatest quality was that which allowed him to quickly scan a variety of different situations, pick out the important components and reject the rest – a quality known as perspicacity and one that would stand him in good stead.

Buzz Aldrin was one of those men who inevitably makes good at his assigned profession and seems to reflect an aptitude that would often be mistaken for sheer hard work. He had been rejected as an astronaut at the 1962 selection on the grounds that he was not a graduate from test pilot school. The following year that condition was relaxed and Aldrin was accepted. But the man was a brilliant mathematician and for several years became the source of ideas and theories about space rendezvous, as related already in connection with the Gemini program. Yet he too was something of an introverted individual, accepting a 'team' attitude only so far as it was necessary to impress selection boards with his ability to get on with colleagues. More than once he expressed a belief in the need for individual expression, and would frequently go his own way if to do so would allow him to express himself to a higher level.

As for Mike Collins, the Command Module Pilot whose lot it would be to go no closer to the Moon than lunar orbit, everybody wondered why he was in the space program at all. Dedicated above all else to the events of the moment, Collins represented another segment of astronaut personality, the one that refused to accept any task or duty as exceptional. What some would label indifference, others a blasé attitude out of tune with the dimensions of the task, astronaut selection boards would judge to be the perfect personality trait for group activity in a dangerous situation across a hostile frontier.

Nothing would raise adrenalin in either Armstrong, Aldrin or Collins that they would be likely to encounter on the flight. And probably for that reason, out of the three, one would sense a never ending frustration with mundane activity and a second would move from job to job never escaping from the haunting realization that his historic step for ever precluded him from standing equal with other men. For there is nothing so lonely as to reach a hitherto unassailable height and never be allowed to return. All three were unemotive individualists, disciplinarians in their own ways, but all three were fired with an ambition to climb peaks.

The six months leading up to the launch of Apollo 11 was an intensive period during which the crew had to take their turn at the simulators, field a barrage of questions from press and public, participate in the myriad details concerning spacecraft and associated hardware, and prepare not only to duplicate the activity of Apollo 10 but also to perform a fundamentally new set of procedures. For the first time an astronaut would have to be a very good helicopter pilot, for in the last few thousand meters to the Moon, the Lunar Module would gradually translate from a horizontal flight path descending like an aircraft approaching a runway to a vertical descent ending in a hover 50 meters or so above the Moon. From that point the final touchdown site would be selected. But quickly. For there would be only a few seconds of propellant remaining in the tanks.

That had been one of the hardest decisions to make: how much propellant to allow the pilots for hovering over the surface. Critical kilograms of weight were taken out of LM systems in several programs aimed at reducing the total mass. But enough propellant would have to be carried for at least a minimal hover time. To aid the LM crew in balancing hover time against the need to translate across the surface, Mission Control would call out the seconds of engine propellant remaining. Nevertheless, there would be little more than a minute to spare from a planned trajectory that had the LM coming straight in for a touchdown.

The hardware arrived at KSC in good time. Launch teams were now well into their processing schedules cycling three vehicles simultaneously. First in during January were Ascent and Descent Stages of LM-5, moved initially to the altitude chamber at the Manned Spacecraft Operations Building. Unmanned, then manned, altitude tests were carried out followed by reception of CSM-107 and a further set of tests with that vehicle. The Lunar Module for Apollo 11 weighed slightly over 15 tonnes and the Command Service Modules a hefty 28.8 tonnes. During the first two weeks of April, while technicians prepared AS-505 for Apollo 10, the various elements came together: the LM landing gear was installed, the CSM's SPS engine bell was attached, the LM was encapsulated in the adapter, and the CSM was stacked on top. On the 14th, AS-506 received its charge in the Vehicle Assembly Building.

The main difference between Apollo 11's Saturn V and the five predecessors was that 506 had less test instrumentation. In several respects it was a fully operational Saturn versus the essentially developmental orientation of the former vehicles. AS-506 would generate only 1,348 measurements during ascent versus the 2,342 of the Apollo 10 launcher. Moreover, a successful weight improvement program vigorously applied to all three stages considerably reduced the dry weight of respective elements so although the combined weight of the spacecrafts was greater than that of its predecessor, Apollo 11 was a slightly lighter stack because of dramatic weight reductions in the rocket. At lift-off, 506 would weigh an estimated 2,902 tonnes.

The stages had been delivered to KSC during January and February and on 5 March the vehicle stood erect on its mobile launcher in the Assembly building. Following mechanical mating of the spacecraft adapter with the top of the rocket, electrical connections were made and a plugs-in test performed to check the compatibility of the systems. Three days after Apollo 10 thundered spaceward, Apollo 11 began its journey to the Moon. Early in the morning on 21 May, AS-506 moved ponderously through the vertical open doors of the VAB and along the crawler-way to pad 39A. On 4 June a Flight Readiness Test began with the astronauts participating. Then there was a wet Count Down Demonstration Test and a dry CDDT where the crew entered the simulated countdown procedure after cryogenic propellants had been removed.

But this time not only were there launch and checkout teams at work to ready Apollo equipment; at Houston and on the recovery ships, hardware of a unique kind was awaiting the return of men from the Moon, astronauts who could conceivably return biologically contaminated. It was a remote possibility, but one which could not be ignored. A special Lunar Receiving Laboratory had been set up at MSC to accept the containers of Moon rock and the bags that contained lunar soil as well as astronauts and close attendants. Medical evidence was diverse as to the period of incubation for potentially dangerous germs. A consensus revealed satisfaction with plans to quarantine the men for 21 days from the time they blasted away from the Moon's surface. But getting them from the spacecraft to the LRL posed further problems until a special trailer was devised that could be carried on the prime recovery ship, enabling the crew to remain environmentally isolated from other human beings.

The tremendous heat of re-entry would effectively sterilize the exterior of the Command Module, and when the crew emerged from the spacecraft they would quickly don biological isolation garments for the transfer, via helicopter, from raft to trailer. Crew and samples would be delivered to Houston in the quarantine trailer on the third day after splashdown after first being flown from Hawaii to Ellington Air Force Base. Inside the LRL, a complex series of physical, chemical, and biological tests would hopefully detect evidence of life, however primitive, and only when those tests proved negative would the crew be allowed out. Until then, de-briefings would take place through a glass screen and all contact would be strictly limited to personnel on the inside.

It was the need to isolate the crew both before and after the flight that gave Apollo 11 an aura of almost reverent dedication. At the final set of press conferences arranged in Houston, Armstrong, Collins and Aldrin appeared before newsmen from a transparent box designed to insulate them from cold germs. Even managers and engineers wore masks when close to the chosen three.

But the tempo of preparation for Apollo 11 infected even the teams preparing to transmit the event to the world. Increasingly, through Mercury, Gemini and Apollo flights, the public affairs office at MSC was associated directly with events in the various control rooms. First 'Shorty' Powers, then Paul Haney, assumed the role of interpreter for the listening public. It was NASA's intention at first to provide a public affairs commentator for the exclusive use of the press, but the dramatic tones of 'Mercury Control,' then 'Gemini Control' and finally 'Apollo Control' eclipsed the rhetoric of newsmen to the point where the press eagerly routed the public affairs commentary to the air-waves, leaving the arbitrator to tell it in his own way.

This led to empire-building in the exlusive world of MSC public affairs activity and Headquarters public relations head Julian Scheer was intrumental in facing Haney with one of two alternatives: either he operated as the commentator in the MOCR or left that function to someone else and concentrated on organizing the press coverage. Haney wanted both, Scheer ordered him to a Washington job, but he left NASA rather than vacate to a less visible niche. For Apollo 11, Brian Duff would organize the press coverage, and public commentary would be shared by John McLeaish, Terry White, John (Jack) Riley, and Douglas Ward.

As for live coverage of events, the crew would operate two TV cameras: the Westinghouse interior color camera carried first by Apollo 10, and a 3.3 kg lunar surface camera with 320 lines per frame at 10 frames per second also built by Westinghouse. The black-and-white surface TV camera would be housed in the Modularized Equipment Stowage Assembly – or MESA – attached to the forward right quadrant of the Lunar Module Descent Stage. Retained in the stowed position, the MESA would be deployed to its open condition by a lanyard which the descending astronaut would pull on his way down the ladder. This would also swing the TV camera into position from which it could transmit views of the forward landing leg. The MESA would open along a horizontal hinge at the bottom, permitting it to fall down like a wall-mounted table and support a variety of equipment used on the lunar surface. The EASAP array of scientific experiments would be contained in the left rear quadrant diametrically opposite the MESA.

Two TV transmissions were scheduled going to the Moon, a third was to show lunar orbit scenes, one of the Lunar Module flying free alongside the CSM, one of the surface of the Moon from orbit, a long telecast of surface activity, and two during the long coast back to Earth. All but the surface scenes would be in color. The quality of surface scenes would be technically inferior to theoretically possible standards. It was only possible to use a portion of the communications signal for TV, however, and Apollo 11 would have to use the spacecraft antenna for transmitting this view unlike later missions which would use a separate antenna deployed by the crew. This umbrella-like antenna was carried aboard the Lunar Module but in view of the complexity involved in setting it up it would only have been used if the main antenna proved unacceptable.

The final Apollo 11 flight plan emerged only weeks before the planned lift-off. If launched on time, LM-5 would land on revolution 14, remain on the surface for nearly 22 hours, include a 2 hr 40 min walk, and ascend to rendezvous with Apollo on the latter's 25th orbit, docking on revolution 27; the CSM would burn out of Moon orbit after 30 revolutions and return through the atmosphere for a total flight duration of 8 days 3 hours. But not everybody rejoiced over the timeline, set primarily by the position of the Moon and the need for crew sleep periods, for if adhered to it would have Neil Armstrong descending the ladder at 2:15 am, Eastern Daylight Time, with the Moon walk ending shortly before Sunrise. Broadcasting networks were concerned that this could lower audience levels – considerably!

No one was to know that the crew had already discussed the possibility of going out early, officially rejected the idea, then adopted it when far from the disapproving comments of missions planners. As it was, touchdown would occur during the mid-afternoon hours of Monday, 21 July, and few employers expected normal attendance on this day the nation had moved toward for more than eight years.

The week before the flight saw the entire world rise in anticipation. From his abode at Castel Gondolfo, Pope Paul VI appealed to Christians everywhere to pray for the success of the flight. On the last Sunday before the launch, the Rev. Paul H. A. Noren from Mount Olivet Church, Minneapolis, led 300 people in prayer at the White House. From southern States across the Union, mules and wagons converged on the Cape as the Rev. Hosea Williams led a host of the nation's poor in an attempt to get 'as close as possible' to the launch, saying that they were 'not against things like the space shot.'

Rev. Ralph Abernathy, spiritual successor to Dr. Martin Luther King, led 25 poor families from the south to a gate at the Kennedy Space Center. NASA Administrator Tom Paine went down to the gate and heard their plea for more attention to the nation's poverty-stricken, asserting that 'it will be a lot harder to solve the problems of hunger and poverty than it is to send men to the moon, but that if it were possible for us not to push that button and solve the problems you are talking about, we would not push that button.' Paine let the families into the VIP stand.

From everywhere they came, flying in by helicopter and fixed-wing aircraft, driving in with plush limousines, walking from endless traffic jams. On the day before the launch, the intensity of preparations reached a crescendo matched only by the flood of humanity that arrived outside the Kennedy Space Center. It was as if they had come to see a new world born, a new age for Man in which hope could replace frustration, plenty replace poverty, food replace the famines. Almost all were ordered and disciplined, many visibly moved as though the event they were about to witness was worth the endless privations of the dense multitude.

In the air, hundreds of aircraft flew over Brevard County, and at sea, thousands of tiny boats pitched and tossed on the Atlantic swell as cutters and Coast Guard vessels tried almost in vain to keep them from the abort recovery areas. It was hot that July – very hot. The air hung like wet velvet across the dense congregations and by the evening before launch more than one-half million people were on the beaches, along the sandy strips, and by the roadsides. Nobody had ever seen anything like it before and the intense enthusiasm evident among the spectators belied the polls that said more Americans approved of a halt to space activity than thought it should continue. Across the nation, retailers reported massive sales of color TV sets, radios, toys related to space themes, and models of Snoopy and Charlie Brown, the beloved mascots of Apollo 11's pathfinder. But more significant than anything else was the statistic that in the months preceding this day, the death rate dropped dramatically as a euphoria set in, culminating in this one flight. Within weeks of the mission's end, the rate would be back to normal.

With just hours to go before the launch, hundreds of thousands of people arrived at the Cape. Now almost a million people had arrived and airports, police, emergency services, and security teams were stretched almost beyond organized capacity. Two hundred Congressmen were on their way to Merritt Island, 60 foreign ambassadors would soon arrive, 19 governors were expected, and 40 mayors booked in. On the beaches, row upon row of tented humanity paying homage to the great Moon mission. In all his wild imagining, Jules Verne had never contemplated a scene such as now unfolded at the Cape. But in a very real way, he was remembered, for Neil Armstrong chose Columbia, Verne's mythical spaceship, as the code name for Apollo. Julian Scheer approved. As he did for the name of the Lunar Module: Eagle.

Now at last, everything was ready. Amid the hustle and bustle of activity, small groups of dignatories and officials met in hotel rooms, administrative blocks and operations buildings. George Low was there, the man who nursed the idea of a manned Moon landing long before Johnson gave Kennedy the target. Wernher von Braun was there, the engineer and publicist who designed the rocket that put America's first satellite into space and now took credit for the biggest rocket ever launched into space. Jim Webb, the administrator who moved NASA along the road to this special day from humble beginnings and a faltering start. George Mueller, the organizer who contributed ways and means to speed up development flights. Each had his own special thoughts that night before Man left for the Moon, but many were deeply saddened by the prospect of cutbacks, although few really thought the flights would ever be cancelled.

Many would have liked to have been there but responsibilities kept them elsewhere. At the Manned Spacecraft Center, four Flight Directors who would command four separate shifts as Apollo made history: Eugene F. Kranz with the white team; Glynn S. Lunney with the black team; Clifford E. Charlesworth on the green team; and Milton L. Windler with the maroon team. And then there were the astronauts who would serve at the capcom console: Duke, Evans, McCandless, Lovell, Anders, Mattingly, Haise, Lind, Garriott, and Schmitt. And in overall charge of operations, was Mission Director George Hage. In support rooms and back-up facilities, hundreds of specialists stood by, representatives of companies that had built just a little piece of the grand machinery that now stood ready for launch on the Florida Peninsula. They would be needed to give expert opinion should anything go wrong after lift-off, stood down only when relevant segments completed their assigned tasks and were jettisoned or discarded.

To each and every participant in this lifetime's experience the event had a different meaning. But to everyone who had worked many years to put men upon the Moon it was the culmination of a dream inherited from generations gone, the forefathers of engineers and scientists into whose custody the great task had been bequeathed.

From the White House, President Nixon sent a last message to the crew: 'On the eve of your epic mission, I want you to know that my hopes and my prayers – and those of all Americans – go with you. Years of study and planning and experiment and hard work on the part of thousands have led to this unique moment in the story of mankind; it is now your moment and from the depths of your minds and hearts and spirits will come the triumph all men will share. I look forward to greeting you on your return. Until then, know that all that is best in the spirit of mankind will be with you during your mission and when you return to Earth.'

And in humble comment upon the indescribable impact of Man's Earthly liberation, the Los Angeles Herald-Examiner: 'It is with an almost breathless sense of awe that we await tomorrow's blast-off from Cape Kennedy – the launching of three space explorers on the most ambitious and fearsome adventure in all human history. Mere words cannot capture the immensity of the flight of Apollo 11. Quite literally, man will be attempting a final break of the chains which have bound him to this earth.'

CHAPTER TWENTY

'The Eagle Has Landed'

21 July, 1969, had arrived and Apollo 11 would begin its journey to the lunar surface from the tropical temperatures of Moonport, USA. It was very hot, hardly a breath of air moved upon the gathering throng that morning as nearly a million people sat, stood, or lay out in the early dawn, waiting for boiling fire to propel the Moonship spaceward. Humid and unyielding, the temperature was well up already and by launch time would register 29°c. But to the men who had spent long night hours preparing 506 for the historic voyage, weather was only of concern when it came to forecast predictions for launch. More than 450 people sat through the long countdown at 14 rows of consoles in firing room 1 at the Launch Control Center, the flat topped building alongside the Vehicle Assembly Building. While propellants were ushered aboard the Saturn V, engineers and technicians stood back until, just before it was time for the crew to arrive, the closeout team returned to spacecraft 107.

The astronauts had been up for some time. Woken at 4:15 am they breakfasted on orange juice, steak, scrambled eggs, toast and coffee before moving to the suit-up room for Joe Schmitt to fit them with the pressure garments; veteran space flight attendee Dolores O'Hara had taken physical particulars from all three before physicians performed a brief examination. In the MSOB, final procedures were conducted before transferring the crew to pad 39A eight kilometers away. They left in the transfer van at 6:27 and arrived at the Saturn V twenty-four minutes later. In the white room adjacent to spacecraft Columbia, Fred Haise was in the Command Module making final preparations. Alongside, Guenter Wendt stood by to install the suited astronauts and spacecraft test conductor Skip Chauvin made ready to perform checkout operations once the crew settled into their couches.

Neil Armstrong was the first man in, sliding over the hatch sill and across to the left couch. Astronauts did not always adhere to the strict intent of the rule-book writers in assigning couch positions and Michael Collins went in next to occupy the right couch. Then it was Aldrin's turn, who for several minutes now had been standing to one side in the white room contemplating a variety of subjects to prevent himself from thinking too intently about the events ahead. With little more than ninety minutes to go before lift-off, the closeout crew shut the hatch, closed the boost protective cover and pressurized the spacecraft to check for leaks. None were detected and thirty minutes later the team left the white room and rode an elevator to the pad; apart from a rescue crew standing 1 km away with protective clothing and converted army vehicles, nobody was within blast range of 39A. From the spacecraft test conductor, earlier that morning, came the opening invitation to go for the Moon:

STC 'CDR, STC, how do you read.'
ARMSTRONG 'STC, loud and clear.'
STC 'Good morning Neil.'
ARMSTRONG 'Good morning.'
STC 'Welcome aboard . . . Neil, let me know when you can verify some switch positions.'

Encapsulated by their Moonship, not for another eight days would the astronauts speak openly to anybody but each other. Across the Atlantic, morning sun broke bands of light through traces of wispy cloud to spread a panorama of beauty along an otherwise vacant horizon. But now, at five minutes before launch, the Sun was gathering elevation on this blistering summer day and from the side of the gleaming spacecraft, the Apollo access arm moved the white room back to the fully retracted position: for thirty-eight minutes it had stood by only 1.5 meters from Columbia in case the crew had to make a rapid exit down the elevators or the slidewire. Now an emergency would be met by ignition of the launch escape motor.

But that was not to occur and in the final minutes to lift-off the care and attention given 506 by Hans Gruene in the months preceding this day paid off as launch vehicle test conductor Norman Carlson assumed command of the countdown. Launch operations director Rocco Petrone was eyefully concerned that his responsibility should be adequately met; just twelve seconds after leaving the pad, control would go to Houston and flight director Cliff Charlesworth, standing by now to receive command. And in the background to technical preparation, Kurt Debus, the man responsible for shaping Cape Canaveral into the Kennedy Space Center, now its Director, stood by with bated anticipation. Launch operations manager Paul Donnelly passed along a final message before Skip Chauvin called in:

DONNELLY 'This is the launch operations manager, the launch team wishes you good luck and God speed.'
ARMSTRONG 'Ah, thank you very much, know it'll be a good one.'
STC 'And CDR, STC, how do you read?'
ARMSTRONG 'STC, loud and clear.'
STC 'Okay Neil, have a good one.'
ARMSTRONG 'You bet.'

To several hundred thousand people within earshot at the Cape, and to several hundred million people around the world watching on television sets, the voice of public affairs officer Gordon Harris carried news of a spaceport exodus that would captivate a large portion of all humanity:

'Eighty second mark has now been passed. We'll go on full internal power at the 50 second mark in the countdown. Guidance system goes on internal at 17 seconds leading up to the guidance ignition sequence at 8.9 seconds. We're approaching the 60 second mark on the Apollo 11 mission. T-60 seconds and counting. We have passed T-60. 55 seconds and counting. Neil Armstrong just reported back: "It's been a real smooth countdown." We have passed the 50 second mark. Our transfer is complete on to internal power with the launch vehicle at this time. 40 seconds away from the Apollo 11 liftoff. All the second stage tanks now pressurized. 35 seconds and counting. We are still go with Apollo 11. 30 seconds and counting. Astronauts reported, "feels good." T-25 seconds. 20 seconds and counting. T-15 seconds, guidance is internal, 12, 11, 10, 9, ignition sequence starts, 6, 5, 4, 3, 2, 1, zero, all engines running, LIFTOFF. We have a liftoff, 32 minutes past the hour. Liftoff on Apollo 11.'

It was a beautiful launch, with Saturn V ascending into a clear sky populated by fleecy lines of white cloud at 5,000 meters. The bright star-like beam that shone from the booster's broad base topped a column of boiling white smoke capped under the five big engines by glowing orange and red flames. For many, it was their first Saturn V launch and for all it was a uniquely spectacular sight – one that could never be repeated, for this was the flight to carry men to the surface of the Moon for the first time.

The crew was quiet during ascent, monitoring control functions now taken over by the green team of flight controllers. The more experienced teams, led by veteran flight directors, were held in reserve for handling the really tricky parts of the mission; launches were now a routine affair. But the ride was smooth for Armstrong, Collins and Aldrin, with no detectable pogo effect from the efficient second stage. Shut down at an elapsed time of 2 min 41 sec, the first stage had tilted Apollo to an acute angle from vertical to begin the gradual translation to horizontal flight. It was the job of the S-II to gain maximum altitude and, at the time of cutoff, to be traveling almost horizontal to the surface below. The S-IVB ignited 9 min 9 sec into the mission and just 2½ minutes later Apollo 11 was in Earth orbit. From a weight of 2,902 tonnes at lift-off, the mass of the assembly had been reduced to an orbital weight of just over 135 tonnes.

Capcom Bruce McCandless: 'Apollo 11, this is Houston. You are confirmed GO.' The combined third stage and payload was in an almost perfect path 191 km above the Earth. Within minutes, flight dynamics officer Dave Reed busied himself with radar data from ground tracking stations to refine the precise value of the orbit: only when the exact position of the assembly was known could definitive figures be calculated for trans-lunar injection. Apollo 11 moved out of communications range with the continental US stations less than three minutes after orbital insertion, but little more than two minutes later the Canary antenna picked up the vehicles passing over the Atlantic. At 24 min, that station too was left far behind. At 32 min, Apollo 11 came within range of Tannanarive for a 5½ minute period and then Carnarvon was acquired at an elapsed time of 52 min. With a drop-out lasting mere seconds, the Australian pass lasted nine minutes. Apollo 11 was now more than half way through its first orbit.

It had been the crew's intention to test the on-board TV camera on the first pass across Goldstone, California, but when the spacecraft appeared above the horizon at 1 hr 29 min problems with signal switching prevented the transmission from reaching Earth. Now it was time to hot-fire the RCS thrusters for checkout, and to extend the forward docking probe. Passing out of range with Canary at 1 hr 56 min, Apollo 11 moved to the Tananarive station at 2 hr 9 min on the second orbit. Loaded with computer pads – instructions read up from the ground for manual insertion via the 'disky' –, the crew awaited a final 'go' from Houston for the all-important TLI burn that would set them on course for the Moon. Word came via the Carnarvon station at 2 hr 29 min. From this point on communications would be almost continuous through TLI, routed via Apollo Range Instrumentation Aircraft cruising the Pacific airlanes and the tracking ship Mercury at sea.

Ignition came at 2 hr 44 min and with a firm push in the back all three astronauts felt the exhilaration of a speed boost that within 5 min 47 sec accelerated Apollo 11 to Earth escape velocity: 39,030 km/hr. At cut-off the assembly was already 328 km high and within ten minutes had climbed to more than 2,300 km. But gravity continued to slow the vehicles down and just fifteen minutes after TLI speed was down to 31,000 km/hr and decreasing. Neil Armstrong had congratulatory words for the big Saturn: 'Hey Houston, Apollo 11. This Saturn gave us a magnificent ride.' Depleted of S-IVB propellant, the assembly now weighed only 63 tonnes.

Separation from the S-IVB came at 3 hr 17 min with Apollo 11 more than 9,000 km from Earth. Twelve minutes later Columbia docked with the LM and then, with Lunar Module Eagle firmly attached to the CSM's forward docking ring, Apollo broke free from the S-IVB and its adapter. The combined weight was now 43.9 tonnes; it was the first time both vehicles had flown together with full propellant tanks in all sections. Mike Collins had a little difficulty with control switches during transposition and docking but the operation was satisfactory. Final separation from the S-IVB occurred at 4 hr 17 min and twenty-three minutes later the SPS engine fired for 3.4 seconds to shift the trajectory away from the third stage.

Without that burn the docked vehicles would have passed within 1,300 km of the Moon's far side; now, Apollo 11 would come within 333 km of the surface. Clearly, a mid-course correction would be necessary to lower pericynthion to a desired miss distance of about 111 km. For its part, the S-IVB would miss the Moon by 4,334 km and head off into deep space. Just before settling down to a P.23 navigation plot, the Apollo 11 crew sent back a commentary on their view of the Earth: 'Well, we didn't have much time, Houston, to talk to you about our view out the window, so when we were prepared for lunar injection, but up to that time, we had the entire northern part of the lighted hemisphere visible including North America, North Atlantic, and Europe and Northern Africa. We could see that the weather was good just about everywhere. There was one cyclonic depression in Northern Canada . . . Greenland was clear and it appeared to be we were just seeing the ice cap.'

In space, exactly three hours after reaching escape velocity on the S-IVB, Apollo 11 had already succumbed to the planet's gravitational grasp. Although 47,500 km from Earth, speed was now down to 13,100 km/hr – exactly one-third the value achieved at TLI. In Houston, Cliff Charlesworth's team handed over to Gene Kranz and the white shift. Before leaving the MOCR, Bruce McCandless received from the crew comment on an important event:

APOLLO 'Down in the control center you might want to join us in wishing Dr. George Mueller a happy birthday.'

CAPCOM 'Roger. We are standing by for your birthday greetings.'

The night before the launch, astronaut Mike Collins relaxes at supper with astronaut Mattingly.

APOLLO 'I think today is also the birthday of California and I believe they are 200 years old and we send them a happy birthday. It's Dr. Mueller's birthday also but I don't think he is that old.'
CAPCOM 'Roger. We copy. And looking back in the viewing room, I don't see him.'
APOLLO 'He may not be back from the Cape yet.'
CAPCOM 'Roger. I believe Dr. Mueller is on his way back from the Cape. We will relay his greetings for you.'

On the ground, Kranz agreed with Charlesworth, that the first mid-course correction maneuver was too small to bother about and that a period of further tracking would be desirable to get a better idea of the precise trajectory. Then it was time to put Apollo 11 in the 'barbecue' rolling mode to distribute heat from the Sun evenly around all sides of the configuration. Capcom Charlie Duke was now on shift and at 9 hr 58 min Buzz Aldrin had a word for him on the Earth view, a planet now 90,000 km away:

'Hey Charlie, I can see the snow on the mountains out in California, and it looks like LA doesn't have much of a smog problem today.'
CAPCOM 'How's the Baja California, look, Buzz?'
ALDRIN 'Well, it's got some clouds up and down it, and it looks pretty good – circulation system . . . off the west coast of California.'

Minutes later the crew sent a TV show down to Earth for recording from the Goldstone tracking station, a seventeen minutes session essentially for calibration. The flight plan called the crew to sleep at an elapsed time of 13 hr 30 min but the MOCR flight surgeon reported that Mike Collins was asleep thirty minutes early and that Armstrong and Aldrin were only an hour behind him. The busy and fragmented activities of the day had put Apollo 11 well on course for the Moon, although TLI had given the configuration a predicted lunar miss-distance of 1,300 km versus the planned 642 km. But that was of little consequence since the evasive burn was designed to move it closer and there were three opportunities remaining for mid-course corrections.

Soon after the crew began their rest period flight director Glynn Lunney took over from Gene Kranz as the black team moved in to the MOCR. But when the ground put in the first call on day 2, capcom Bruce McCandless was taking over from Ron Evans with flight director Charlesworth controlling events. The crew had been up for some time, according to the flight surgeon, and were well on with their first meal, navigation alignments, and the news from around the world read up by McCandless. At an elapsed time of 25 hr 53 sec, Apollo 11 passed the halfway point in distance: 193,256 km from both Earth and Moon, and speed was now down to just 9,700 km/hr.

Less than one hour later the crew fired the SPS engine for a second time to trim the course and depress the trajectory so the vehicles would pass closer to the Moon. The burn lasted 2.9 seconds and as a result of the 23 km/hr change in speed Apollo would approach to within 116 km of the Moon's far side instead of 333 km. That was good enough and no more mid-course corrections would be necessary. The only systems problem to emerge with significant consequence was a repeated master alarm indicating high oxygen flow to the No. 3 fuel cell. But no emergency was likely and it was a nagging problem only. Jim Lovell was talking to Apollo 11 when the commentary from space remarked on the view from 204,000 km:

'Hey Jim, I'm looking through the monocular now and to coin an expression the view is just beautiful. It's out of this world. I can see all the islands in the Mediterranean. Some larger and smaller islands of Majorca, Sardinia and Corsica. A little haze out over Greece. The Sun is setting on the eastern Mediterranean now. The British Isles are definitely greener in color than the brownish green we have in the islands near the peninsula of Spain.'

And commenting on the available space inside Apollo, remembering his EVA on Gemini XII, Buzz Aldrin had a word on the weightless sensation: 'I'll tell you, I've been having a ball floating around inside here, back and forth up to one place and back to another. It's just like being outside, except more comfortable.' Even astronauts forgot how roomy a spacecraft becomes when weightlessness opens every nook and cranny to a floating body. At 30 hr 28 min the crew sent an unscheduled TV picture to Earth and shortly thereafter a new shift took over in the MOCR:

APOLLO 'Houston, Apollo 11.'
CAPCOM 'Go ahead, 11.'
APOLLO 'Oh Charlie, that you?'
CAPCOM 'That's me, how are you there?'
APOLLO 'Oh, just fine. How's the old white team today?'
CAPCOM 'Oh, the old white team's bright eyed and bushy tailed. We're ever alert down here.'
APOLLO 'Ever alert and ready. Hey, you got any medics down there watching high grade? I'm trying to do some running in place down here, I wondered just out of curiosity whether it makes my heart rate up?'
CAPCOM 'Well, they will spring into action here momentarily. Stand by.'
CAPCOM 'Hello 11, we see your heart beating (!).'

Now Gene Kranz would hold the Houston reins while the spaceborne crew got ready for a scheduled evening telecast to be distributed among the domestic networks world wide. But then, news from capcom Charlie Duke about a White House announcement: 'President Nixon has reported – declared a day of participation on Monday for all Federal employees to enable everybody to follow your activities on the surface. Many state and city governments and businesses throughout the country are also giving their employees the day off, so it looks like you're going to have a pretty large audience for this EVA.' At 33 hr 59 min a TV transmission began with Apollo 11 more than 240,000 km from Earth. Lasting thirty-six minutes, it included a view of the home planet and then Buzz Aldrin brought the camera inside and showed interior scenes with Mike Collins commentating:

'We are very comfortable up here. . . . We do have a happy home. There's plenty of room for the three of us and I think we're all willing to find our favorite little corner to sit in. Zero g's very comfortable but after a while, you get to the point where you sort of get tired of rattling around and banging off the ceiling and the floor and the side, so you tend to find a little corner somewhere and put your knees up, or something like that to wedge yourself in, and that seems more at home.'

And then Mike Collins proceeded to show his global audience the food preparation equipment:

'Well it looks like it's probably almost your dinner time down there on Earth. We'll show you our food cabinet here in a second.'

CAPCOM 'Eleven, roger.'
CAPCOM 'Eleven, Houston. We see a box full of goodies there, over.'
COLLINS 'We really have them, Charlie. We've got all kinds of good stuff. We've got coffee up here in the upper left and the breakfast items and bacon in little small bites, beverages like fruit drink and over in the center part we have, oh all kinds of things. Let me pull one out here and see what it is.'
CAPCOM 'Rog.'
COLLINS 'Would you believe you're looking at chicken stew here? All you have to do is . . . hot water for 5 or 10 minutes. Now, we get our hot water out of a little spigot here with the filter on it that filters any gasses that may be in the drinking water out, and we just stick the end of this little tube in the end of the spigot and pull the trigger three times and then mush it up and slice the end of it and there you go. Beautiful chicken stew.'
CAPCOM 'Sounds delicious.'
COLLINS 'The food so far has been very good. We couldn't be happier with it.'

After the TV, time to prepare for the second night in space, a scheduled 10 hr rest session; the trans-lunar coast, although littered with the many housekeeping chores pioneered by the two previous lunar voyages, provided opportunity for building up a good reservoir of energy for the demanding activities and the tight timelines of coming events. Shortly after 38 hr elapsed time, flight surgeon John Ziegleschmidt reported that Armstrong and Collins were already asleep but that Buzz Aldrin was still awake. In the MOCR, during that late evening hour at Houston, Glynn Lunney moved his team on to the consoles, replacing Gene Kranz. They would not communicate with the crew until Cliff Charlesworth took over nine hours later.

Flight surgeon Willard Hawkins kept a watchful eye on the sleeping astronauts as the flickering lights on his console indicated their condition. The flight director decided to let the crew sleep on when Hawkins told Charlesworth there was no sign of them waking even as the scheduled time of 47 hr came and went. One hour later the surgeon said they were stirring so the team sent a wake-up call and received a status report on the sleep duration, radiation readings, etc.

For the first part of the day, systems checks, a fuel cell purge, and then a preliminary look inside the Lunar Module, carried on a TV transmission to Earth. At about 55 hr 20 min Neil Armstrong removed the probe and drogue, having already taken out the forward hatch which, unlike Apollo 10, did not have the dense Mylar insulation that came loose in fragments like a 'snow storm.' But just like Apollo 10, the two vehicles were observed to be slightly out of alignment. At a mere 2.05°, however, that was nothing to worry about. At 55 hr 38 min Aldrin moved up the tunnel into the Lunar Module, followed seventeen minutes later by Armstrong. When the LM hatch came open it automatically turned on the interior lights, causing capcom Charlie Duke to quip: 'There is that same guy when you opened up the door, why, he is waiting there for you and he turns the light on.'

With television pictures beaming to Earth, now nearly 326,000 km away, viewers received scenes of what the flight plan called 'LM familiarization.' There were several items to check off, equipment given quick examination to save time for the busy landing day. The emergency oxygen bottles on top of the backpacks had to be checked, and items stowed after handling in positions which would not cause them to fly around the LM's interior during the long lunar-orbit-insertion burn. After a ninety-six minute telecast, pictures ceased at 56 hr 44 min and the crew were back in Columbia about an hour later setting up another passive thermal control roll for 'barbecue.' During the meal which followed, Armstrong, Collins and Aldrin put on some of the tapes carried on the spacecraft for storing personal recordings of their observations and comments. Rather than carry empty cassettes, they took along pre-recorded music tapes which would erase as comments were recorded but which would provide entertainment and light relief in the meantime. Mike Collins said that it was the third anniversary of Gemini X and that he 'just had the urging to' play some music. Collins had been pilot on that mission to John Young.

At 60 hr 27 min Charlie Duke said a 'goodnight from the white team' at the start of what should have been a nine hour sleep period. But minutes later the spacecraft called back with a request for information on the distance the S-IVB was calculated to be from the spacecraft. The answer came back quickly: 'Apollo 11, Houston, the S-IVB's about (11,112 km) from you now, over.' What caused this request originated from observation of a strange object which appeared to be quite close to Columbia. Mike Collins thought he felt a 'thump' on the spacecraft moments before what looked like an L-shaped structure hove into view. At times it seemed cylindrical, then it would drift into a different angle and its L-shaped profile became clear. It kept the crew up for some time, talking about the strange sight, but it was a phenomenon they would reserve for discussion with ground controllers after the mission. The observation came very close to the point where the Moon's gravity field exceeded that of the Earth – the point of equigravisphere. It was an area in which debris could become 'trapped' by the summed gravitational forces between the two massive bodies.

The precise time when the spacecraft crossed this point, 61 hr 40 min, Columbia was 345,281 km from Earth, 62,638 km from the Moon, and travelling at a speed of only 3,281 km/hr. From now on it would speed up until passing around the west limb of the Moon almost exactly 14 hours later. But it was at an elapsed time of 62 hr that the flight surgeon reported all three crewmembers to be asleep, just as flight director Gene Kranz was handing over to Glynn Lunney. In Houston it was late evening, the day before the Moon. Five hours later Lunney conferred with flight dynamics officer Jay Green and both men agreed that the final mid-course correction opportunity should be passed up – Apollo 11 was on a good course and minor dispersions could be accommodated by the LOI burn on the far side of the Moon.

Very soon after the appointed wake-up time of 69 hr, the spacecraft called down to the ground to find out if MCC_4 was to be performed. On hearing it was not, the caller promptly turned over and went back to sleep with a 'Okay, I'll see you at 71 hrs.' When the next call came Cliff Charlesworth was at the helm with capcom Bruce McCandless. Apollo 11 was now within 22,000 km of the Moon. Over the next few hours the crew commented on their view of the Moon ('It's quite an eerie sight.') and received instructions from the ground, ever ready with exactly the right information, on the type of camera, film magazine, and shutter settings should they wish to take some pictures.

Then it was time to do a little housekeeping on the fuel cells ('We'd like to do a little cryo(genic) balancing, so if you could position the oxygen tank number 1 heater switch OFF and hydrogen tank 2 heater switch to OFF leaving all the rest of the cryo switches the same, we'll let it run that way for a few hours.') before trimming up the radiators ('On the secondary loop check when we went to flow on secondary radiators, the quantity dropped from 40 per cent down to 36 in the first 10 seconds and then stabilized at 36 for the remainder of the 30 seconds'). All the time, systems checks and advice from the MOCR on just what to do to preserve the balanced integrity of spacecraft performance. Although it would have been possible to return to Earth in an emergency without any communication between the spacecraft and the ground, success in the expanded mission roles now sought depended largely on the ability and experience of flight control teams in the MOCR. Theirs was the glory, albeit unsung.

Then there were the 'pads,' lists of technical information to be written down on computer cards in the spacecraft so that the crew could enter in the computer's erasable memory detailed information on what to do in emergency situations. For instance, if the LOI burn failed to come off, or if the engine burned too short, or too long. And if it went into orbit, what to do if that orbit was unsafe and likely to bring the spacecraft dangerously close to the surface. Then there was the LOI pad proper, the one to instruct the computer how to respond to programs it already carried about SPS start sequences, shut-down operations, etc. When all that activity was tucked away, light relief in the form of a capcom duo – Fred Haise and Bruce McCandless – taking it in turn to read up the mornings news which in part informed Armstrong that 'Even Pravda is headlining the mission and calls Neil, "The Czar of the ship".' Shortly after, Mike Collins informed McCandless that 'the Czar is brushing his teeth, so I'm filling in for him.'

At 75 hr 25 min flight director Charlesworth went around the room polling for go/no-go decisions on individual assignments. Everybody agreed that the spacecraft was in a fit and ready condition so McCandless got the word to pass the crew the news they wanted to hear: 'Eleven, this is Houston. You are GO for LOI.' Little more than 10 minutes remained to LOS – loss of signal – as Apollo 11 went around the Moon, passing out of view on a course that would carry it to within 116 km of the surface. In the MOCR, astronauts began to file in and lounge across the capcom console, sliding in as close as possible to events that were now beginning to pick up momentum. Bill Anders and Jim Lovell joined McCandless, then Fred Haise slipped in, with Deke Slayton already at the console. In the viewing room at the back, John Glenn, Tom Stafford, Gene Cernan, Dave Scott, Al Worden and Jack Swigert could be seen in the front row.

Time would tell the story: if the spacecraft failed to burn its engine for lunar-orbit-insertion, the crew would appear round the east limb at 75 hr 5 min, ten minutes later if the SPS had again responded on command. For twenty-three minutes nobody would know if it had worked, and then for each minute without communication an increasing hope that the burn had gone as planned. Lunar-orbit-insertion would need a 6 min 2 sec burn of the big SPS engine to brake the 43.5 tonne configuration – 400 kg lighter than at launch due to venting of excess fluids and gases – by 3,200 km/hr. But it worked, and the burn was precisely as planned.

Nobody in the MOCR knew that and the mood was one of quiet anticipation. Hardly a sound came from the control consoles, only a few people standing chatting in low tones but from the very second the vehicles would have appeared had something gone wrong, the background hiss of the noise from space underscored the tension. If voices broke through, the MOCR would spring into a contingency plan. But nothing came until the hoped-for time at 75 hr 15 min when the Madrid antenna captured the good news and capcom McCandless asked the crew how the burn figures looked. 'They were like perfect,' said Neil Armstrong. The assembly, now weighing less than 32.7 tonnes due to the consumption of SPS propellant, was in an elliptical orbit of 113.5 × 312.6 km. Not bad when targeting for a 113.2 × 313.2 km path. Descriptions of the surface followed within minutes.

ARMSTRONG 'Apollo 11. We're getting this first view of the landing approach. This time we are going over the Taruntius crater and the pictures and maps brought back by Apollos 8 and 10 give us a very good preview of what to look at here. It looks very much like the pictures, but like the difference between watching a real football game and watching it on TV – no substitute for actually being here.'

CAPCOM 'Roger. We concur and we surely wish we could see it first hand, also.'

ARMSTRONG 'We're going over the Messier series of craters right at the time, looking vertically down on them and Messier A we can see a good size block in the bottom of the crater. I don't know what the altitude is now but that indicates that those are pretty good size blocks.'

Apollo 11 was 235 km from the surface, climbing up toward apocynthion. But ever the watching dog of gremlins and unexpected anomalies, flight controllers quickly scrutinized sample traces of telemetry before sending up final instructions for the LOI-2 maneuver, the circularization burn at the end of revolution 2: 'Eleven, Houston. During your SPS burn as played back on tape down here, we've observed the nitrogen tank Bravo pressure in the SPS system dropping a little bit more than we anticipated. It's holding steady right now. We'll continue to watch it and keep you posted if anything comes up.'

Nothing did, but it was comforting to have such a lot of caring nannies! But just to be on the safe side, the crew was advised to perform the circularization burn on bank A rather than with both operational. Used to pressurize the SPS propellant ball valve drivers, nitrogen had apparently leaked from bank B during the LOI-1 burn: bank A was currently showing 158 kg/cm^2; bank B showed 138 kg/cm^2; the red-line was based at 28 kg/cm^2, a long way below the levels shown.

Docked together, Columbia and Eagle went round the Moon's west limb and out of sight at 77 hr 40 min, and in the MOCR the absence of communication was used to change over flight control teams. On revolution 2, the spacecraft went into inertial attitude, rather than track the Moon's surface, for a sextant star check and preparation for the next SPS burn to circularize the orbit. But TV pictures were also sent down, spectacular views of the lunar surface on the approach path to where Eagle would descend a day later. Then the view got progressively darker as the spacecraft headed toward the terminator and darkness.

APOLLO 'And as the Moon sinks slowly in the west, Apollo 11 bids good day to you.'

CAPCOM 'Roger, we sort of thought it was the Sun setting in the east.'

With nothing to show the TV came off and then the crew copied the maneuver pads read out on the air-ground loop for the circularization burn coming up on the next far side pass. The LOI-2 burn lasted 17 seconds and dropped Apollo 11 into a 99.6 × 121.7 km path, not a true circle because the Moon's gravity would pull and tug at the trajectory, distorting the orbit to a near circular path when the time came for Eagle to go down to the surface. The benefit of a circular orbit would come during rendezvous operations so it was not necessary to have that orbital shape this early in the mission. By knowing how the Moon would change the path, controllers could calculate a deliberate ellipse which would become a circle at the required time.

When the crew came back into communication range, at 80 hr 33 min, they began preparations for checkout in Eagle, making sure everything was ready and where it should be for the planned descent and landing. Mike Collins would, meanwhile, perform some landmark tracking using the optical equipment in the lower equipment bay. The LM was pressurized with oxygen at 80 hr 50 min and twenty minutes later the probe and drogue had been removed. At 81 hr 25 min, shortly before passing out of sight again, Armstrong and Aldrin reported they were in Eagle and starting the checkout. On the next front side pass, the ground tracking and communication stations tested relay modes into Eagle, performing voice checks and telemetry link checks. Before again moving behind the Moon, at 83 hr 44 min, the final systems checks were voiced down and Armstrong and Aldrin prepared to move back to Columbia. Instead of installing the probe and drogue, the crew agreed to leave it out in the Command Module that 'night' to ease the work load for the following day.

When next the crew appeared, 84 hr 30 min into the mission, they were buttoning up Eagle ready for the next rest period; the Lunar Module would remain pressurized during their sleep session. But the flurry of activity put them behind time and it was two and one-half hours past the scheduled start time of 85 hr when the flight surgeon reported all three asleep. They rested, but fitfully. None of them slept as they had the previous nights; it would have been an almost impossible feat. Nevertheless, with flight director Glynn Lunney manning the black team, flight surgeon Ken Beers reported all three were sound away at 91 hr 40 min, just before the spacecraft went around the far side on revolution 8; toward the end of the next front side pass capcom Ron Evans would give them a call. It came at 93 hr 35 min.

CAPCOM 'Apollo 11. Apollo 11. Good morning from the black team.'

COLLINS 'Oh my, you guys wake up early.'

CAPCOM 'Yes, you're about 2 minutes early on the awake up. Looks like you were really sawing them away.'

CAPCOM 'Eleven, Houston. Looks like the Command Module's in good shape. Black team has been watching it real closely for you.'

COLLINS 'We sure appreciate that because I sure haven't.'

So far, mission events had been comfortably set up for sleep periods closely aligned with the crew's diurnal rhythms: like travellers abroad, they would suffer 'space lag' if moved away from the biological clock. But this would be a busy day, and they would not get back to sleep for more than 22 hours, effectively throwing them off the comfortable cycle. In Houston it was early morning, only a couple of hours past dawn, and for many thousands this too would be a long and tiring day – the day men would walk on the Moon. On the ground in the MOCR, while Armstrong, Collins and Aldrin had their last breakfast together for two days, displays and plotboards gradually assumed a more operational configuration than heretofore. Flight director Lunney was getting ready to hand over to the white team led by veteran Gene Kranz. At acquisition of signal on the next pass – revolution 10 – Houston read the news up to the crew.

CAPCOM 'The Black Bugle just arrived with some morning news briefs if you're ready.'

APOLLO 'Go ahead.'

CAPCOM 'Roger. Okay. Church services around the world today are mentioning Apollo 11 in their prayers. President Nixon's worship service at the White House is also dedicated to the mission, and our fellow astronaut Frank Borman is still in there pitching and will read the passage from Genesis which was read on Apollo 8 last Christmas. The cabinet and members of congress with emphasis on the Senate and House space committees have been invited along with a number of other guests. Buzz, your son Andy got a tour of MSC yesterday. Your Uncle Bob Moon accompanied him on the visit which included the LRL. Among the –.'

APOLLO 'Thank you.'

CAPCOM 'Roger. Among the large headlines concerning Apollo this morning there's one asking that you watch for a lovely girl with a big rabbit. An ancient legend says a beautiful Chinese girl called Chango has been living there for 4,000 years. It seems she was banished to the Moon because she stole the pill for immortality from her husband. You might also look for her companion, a large Chinese rabbit, who is easy to spot since he is only standing on his hind feet in the shade of a cinnamon tree. The name of the rabbit is not recorded.'

APOLLO 'Okay, we'll keep a close eye for the bunny girl.'

While the LM crew donned their liquid cooled underwear, Collins put on his pressure suit and changed a lithium hydroxide filter; the CSM would be standing by for the next nine hours, supporting LM activation or getting ready at any time to perform a rescue manoeuvre should the Lunar Module get into trouble after separation. It was wise to get housekeeping chores out the way in good time. Going around toward the start of revolution 11, the black team bid farewell to the three astronauts: 'we'll be going off here shortly and we'll pick you up in the morning for sure . . . three minutes to LOS. AOS at 96 plus 20.' Three minutes later the ground got loss of signal; acquisition of signal would occur at 96 hr 20 min.

In the intervening forty-six minutes Gene Kranz's team slid in to operation ready to support the activity they had been training for nearly a year to control: descent and landing, the only operational flight role not yet carried out by an Apollo support group. Around on the far side of the Moon, Aldrin moved into Eagle and switched on the Descent Stage batteries at 95 hr 54 min.

When the spacecraft reappeared Armstrong was not yet in the LM but Aldrin was proceeding as planned with activation. For the duration of the front side pass on revolution 11 a brisk technical dialogue filled the communication circuits between Columbia and Eagle and between each vehicle and the ground. It was time to dispense with the humorous anecdotes and settle down to serious business. Neil Armstrong was soon in Eagle helping Aldrin prepare the systems, and Mike Collins in Columbia set up the CSM for mission support functions. A portion of the exchange testifies to the new pace of preparedness:

CAPCOM 'Okay, now you want to give me CSM verb 05, noun 01, ENTER?'
APOLLO 'Okay I give verb 05, noun 01, ENTER. We're on 17 06 ENTER.'
CAPCOM 'Roger.'
APOLLO 'Get ready to copy.'
CAPCOM 'Go ahead.'
APOLLO 'Roger. Register 1, 5 balls; register 2, 20017; register 3, 20616. Over.'
CAPCOM 'Understand R1, 50s; R2, 20017; R3, 20616.'
CAPCOM 'That's correct. I'm standing by configured to record your PCM data. And I'm ready to start on a . . . time . . . course align when you are, and when you're ready go min deadband and hold.'

And so it went on, for the seventy-two minutes Columbia and Eagle moved from east to west across the visible face of the Moon. During that time, the electrical systems were brought fully on line, the primary guidance and navigation section was turned on, S-band and telemetry was switched on, environmental control equipment was brought up to operation, the clock in Eagle was synchronized with the clock in Columbia, high-gain and steerable antennae were activated, and inertial measurement gimbal angles were aligned. Armstrong had gone back into Columbia to don his pressure suit, without helmet and gloves, and now it was Aldrin's turn to do the same. Progress was conducted at a stiff pace. By the time the docked vehicles disappeared around the Moon's west limb the probe and drogue were in the tunnel and the hatches were about to be closed; starting thirty minutes early, the crew had maintained that pace and were still one-half hour ahead of the flight plan at LOS.

Coming round on revolution 12 they would briefly don helmets and gloves, check out the integrity of the environmental control system, check the cabin regulators and then doff those items for a further two hours of test activity prior to undocking from Columbia. In Windsor Locks, Connecticut, a red telephone sat on the desk of Lunar Module environmental control systems chief Kenneth L. Hower. Connected direct to the Manned Spacecraft Center, it would be used should Eagle develop trouble in that system during the independent flight, a hot-line direct to Hamilton Standard, the people who built the unit. It was but one of many links with systems specialists backing up the incredible scale of activity now unfolding.

At acquisition of signal, 98 hr 18 min into the mission, the LM crew were well ahead with their preparations. Houston uplinked a state vector to the Lunar Module computer, telling it where it was and what platform reference it should use for guidance alignment. Meanwhile, Mike Collins was turning and pitching the docked assembly getting navigation sightings on pre-designated Moon targets so that tracking stations would have a better knowledge of exactly where the vehicles were. Then Eagle's landing gear was deployed and the RCS thrusters pressurized for action. Final updates went into the Abort Guidance Section which would haul them away from trouble in the event of a problem during descent. Mike Collins copied down a maneuver pad for separation, a small burn to be performed by Columbia just after undocking to carry the CSM a short distance from Eagle so that ranging data could instill confidence in the ability of the Lunar Module to generate rendezvous data.

At 99 hr 20 min flight director Gene Kranz called up his team one by one to get status checks for undocking. Everybody came up 'green' and capcom Charlie Duke gave Eagle the good word: 'Apollo 11, Houston. We're GO for undocking. Over.' The flight plan called for television coverage from the Command Module of the upcoming undocking but Mike Collins told the ground that 'There will be no television . . . I have all double windows either bullet heads or cameras, and I'm busy with other things.' The coded message was understood; the MOCR concurred. At 100 hr 14 min, two minutes before coming around into view again on the 13th revolution, the two spacecraft undocked and capcom Charlie Duke quizzed the crew at acquisition of signal.

CAPCOM 'Eagle, Houston. We're standing by. Over.'
CAPCOM 'Eagle, Houston. We see you on the steerable. Over.'
EAGLE 'Roger. Eagle. Stand by.'
CAPCOM 'Roger. How does it look?'
EAGLE 'The Eagle has wings.'

Almost at once Charlie Duke passed up the pad for descent-orbit-insertion, or DOI, the maneuver that would carry Eagle down to a low point of 16 km from where powered descent to the lunar surface would begin; Apollo 10 had been down to low altitude after a DOI burn, but powered descent was the final maneuver that would bring Eagle to the Moon's surface. And then the PDI – powered-descent-initiation – pad was voiced up as Buzz Aldrin busily copied them down on computer cards. This information could not be sent earlier because the magnitude and characteristics of the burn would change according to minor irregularities in the precise position of the vehicle versus the required coordinates. Only by tracking up to the last minute could the most accurate data be fed to the computer.

About the time the two vehicles, only a few meters apart, swept across the area of the landing site, Columbia fired its thrusters for 8 seconds to begin a drift that would carry it about 3.5 km away one-half revolution later on the far side of the Moon. If everything still looked good in both spacecraft at that time, Eagle would burn descent-orbit-insertion and begin a long coast down to the low point of 16 km. There was no conversation with the ground when Collins monitored the separation burn at 100 hr 40 min but gradually Columbia drew away leaving Eagle a diminishing star in the black void. But a star with shape, for the two would not be so far apart.

On through the rest of the front side pass, technical information bounced along the communication links as flight controllers in the MOCR prepared for the hectic flurry of activity now less than one orbit away. And minutes before passing out of sight, the three astronauts heard Charlie Duke pass up a 'GO for DOI' shortly after Gene Kranz received quick-fire status reports from his controllers in the trenches. At 101 hr 28 min, the two spacecraft passed around the far side. Nobody would know if the descent-orbit-insertion burn had come off as planned and on time until Eagle and Columbia reappeared, forty-eight minutes later.

Meanwhile, Armstrong and Aldrin set up the spacecraft for ignition of the Descent engine – the first time it had been fired on this flight – and positioned it with the propulsion system firing against the direction of travel in a 'heads up' attitude; this essentially put the crew on their backs with respect to the lunar surface 110 km below. When the engine ignited, less than eight minutes after passing around the west limb as viewed from Earth, the crew in Eagle could feel no deceleration, nor could they hear the thunder of its combustion. But 15 seconds later when thrust increased from 10% to 40% they could sense the decrease in speed; after a total burn of nearly 30 seconds the engine shut itself down, having slowed the LM by 84 km/hr. Because Eagle had slowed down, Columbia would overtake it and appear first around the Moon's east limb at a predicted time of 102 hr 14 min. The Lunar Module would come into sight two minutes later.

In the Mission Operations Control Room, Gene Kranz had by now accepted the command of his men, a tight ship on this shift and one that had simulated every possible emergency countered by many conceivable contingencies that could occur to Eagle as it descended toward the point where powered descent would commence. Only if everything looked good up to the time when the big Descent engine was required to fire would Kranz give a 'GO' for ignition; if anything at all threatened the operation, Eagle would simply remain in its safe elliptical path with a pericynthion of about 16 km. But once powered descent began, the only way to extricate the crew safely from inevitable impact with the Moon would be to have them abort by stage-firing the Ascent Stage to use its engine for an ascent back to Columbia.

Among officials crowding into the rear viewing room, separated from the trenches in the MOCR by a large glass screen, were NASA Administrator Tom Paine, Marshall Space Flight Center Director Wernher von Braun, Manned Spacecraft Center Director Bob Gilruth, Lewis Research Center Director Abe Silverstein, Electronic Research Center Director Jim Elms, Rocco Petrone from KSC, Eberhard Rees from MSFC, Kennedy Space Center Director

Kurt Debus, Langley Research Center Director Edgar Cortright, and Dr. Draper from MIT's Instrumentation Laboratory. Astronauts too began to move in: Pete Conrad, veteran of Gemini's V and XI; Fred Haise, back-up on this very flight; Tom Stafford, from Gemini IX-A and Apollo 10; Gene Cernan, also from Gemini IX-A and Apollo 10; Jim McDivitt, from Gemini IV and Apollo 9; Jim Lovell from Gemini VII and Apollo 8; Bill Anders, from Apollo 8; and John Glenn, the first American to orbit the Earth less than 7½ years before.

Chris Kraft was there, his flight controllers carrying the shifts that conducted the missions. Sam Phillips was there, the Apollo Program Director for nearly five years now, the one man who had carried the manned project through fire and tragedy. Bob Seamans was there, now Secretary of the Air Force, the 'general manager' that helped build NASA into the machine it now was. George Low was there, the man more than any other responsible for the outstanding success of manned space operations. And Deke Slayton, the man whose job it was to train others for space where once he had hoped to fly himself.

With just a few minutes to go before the spacecraft would appear from around the Moon, Gene Kranz called up his controllers via the head sets they wore at all times in the MOCR: plugged into whichever console they happened to be at, they were all in immediate range of the flight director. It was time to settle down and to get back to respective positions. The mood had been tense, everyone felt the awesome responsibility they had, stomachs turned with apprehension, and palms were wet with perspiration. Now it was time to suppress all that and become just for a while impersonal, unemotive, beings tuned fully to the task at hand. A final word from Kranz before the chatter subsided and excess conversation was inhibited: 'Good luck to all of you.'

On the plotboards up front, charts to indicate guidance and navigation tracking, information essential to rendezvous with the lunar surface. It was like a docking operation with an inert mass of incalculable magnitude, for setting Eagle upon its perch was above all else a guidance problem with little or no margin for error. It was 102 hr 14 min. Columbia should sweep into view any second, followed by Eagle two minutes later, hopefully on its way to a 16 km pericynthion. Network controllers tossed information to Kranz that Columbia was back in view; telemetry would be streaming to Earth within seconds.

CAPCOM 'Columbia, Houston, Over.'
COLLINS 'Houston, Columbia. Reading you loud and clear. How me?'
CAPCOM 'Roger, five by, Mike. How did it go? Over.'
COLLINS 'Listen babe, everything's going just swimmingly. Beautiful.'
CAPCOM 'Great. We're standing by for Eagle.'
COLLINS 'Okay, he's coming around.'
CAPCOM 'We copy. Out.'

A minute later, Eagle appeared with a call to Houston. There were just seventeen minutes to powered-descent-initiation when the Descent engine would come on for the twelve-minute drop to the Moon. For a few minutes Houston had a problem locking up on Columbia's high-gain steerable antenna, then data from Eagle suddenly dropped out. The communication link between Earth and Eagle was the most vital element of the operation. Without continuous tracking the LM's position could be estimated only to within unacceptable tolerances. And without telemetry the condition of the vehicle would be guesswork at best. Gene Kranz threw his controllers into the problem and from that electrifying point the MOCR ran up to speed for the remainder of the operation. It was a fortunate spur to activity and set the pace for the next twenty minutes or so.

Kranz snapped requests to first this controller then that, punching verbal orders to his men like a traffic director. Within a minute or two the problem was under control, communication restored. Five minutes to powered descent, Gene Kranz called up his men for status checks. Then the communications link dropped out again, the trouble being essentially the inability of the steerable antenna to hold lock on Earth stations. When it came, the 'Go for powered descent' call went via Mike Collins in Columbia, who had his own communications link with Eagle. And that was it. Eagle could head down toward its assigned spot on the Moon's Sea of Tranquillity. Seconds later, Eagle broke through the crackling static and flight controllers could hear Buzz Aldrin reading off a checklist as Neil Armstrong configured the switches. Standing in front of the two triangular windows, firmly restrained by lines and harnesses that connected each man's waist and torso to fixed points on the floor and the ceiling, the crew made final preparations for ignition.

By this time the Lunar Module had descended very close to the 16 km pericynthion point and in speeding up along the low point of its trajectory had moved ahead of Columbia orbiting above. At the time of powered-descent-initiation (PDI) Eagle would be 240 km ahead of and 95 km below the orbit of the CSM. For most of the descent the Lunar Module would be flying legs forward in a heads-up attitude until, beginning at about 3 min 50 sec into the burn, Eagle would gradually rotate to a vertical position from where the crew could see the landing site at 8 min into the descent. Four minutes later Eagle should be on the surface, having flown a downrange distance of 480 km to descend from the 16 km pericynthion point. Only in the last few seconds would Eagle slowly fall vertically to the Moon.

The precise time of ignition for PDI was 102 hr 33 min 4 sec and before the Descent engine ignited the RCS thrusters were fired for 7 seconds to settle propellants in the main tanks. Just at the time of Descent engine ignition Houston momentarily lost communication but events began well. Burning for 26 seconds at 10% throttle, Armstrong and Aldrin again failed to sense the thrust, until the engine ramped up to full power and deceleration pressed their feet into the floor. The main braking phase would last about 6½ minutes, reducing speed all the while, until throttle-down to about 55%

The precise altitude at PDI was 15.2 km with the crew flying heads-down toward the surface, but 3 min 52 sec into the long burn, Eagle yawed 174° to a heads-up position so the landing radar fixed to the base of the Descent Stage could 'see' the surface and provide altitude and velocity updates; lights in the spacecraft would go out when the beams scanned and acquired data for the computer. Altitude was now 12.9 km. The excited voice of Charlie Duke kept up constant dialogue with the crew:

'Eagle, Houston. You are go. Take it all at 4 minutes. Roger, you are go – you are go to continue powered descent. You are go to continue powered descent.'
EAGLE 'Roger.'
CAPCOM 'And Eagle, Houston. We've got data dropout. You're still looking good.'
EAGLE 'PGNCS we got good lock on. Altitude lights out. Delta H is minus 2900.'

It was 5 min 38 sec into powered descent. Altitude, 9,620 meters; speed, about 2,200 km/hr. But nobody knew that the landing computer program was loaded with erroneous information about the relative speed of the Lunar Module with respect to the Moon. And when the landing radar updates flooded the program the lunar guidance computer rejected sets of data, causing alarms to ring in the spacecraft and on the ground. Quite simply, it was getting too much information and if something was not removed from the range of tasks it was required to perform all program information would wipe out and the LM would have to abort.

But to make tasks even more critical, the retro-fire officer was just now coming up with a final time on the throttle down prediction which capcom would pass to the crew. It was a mixture of cross-talk in the Mission Operations Control Room and conversation between Charlie Duke and Neil Armstrong. Steve Bales was the LM Guidance and navigation officer.

EAGLE '1202, 1202.'
RETRO 'Flight, retro.'
KRANZ 'Go, retro.'
RETRO 'Throttle down 6 plus 25.'
EAGLE 'Give us the reading on the 1202 program alarm.'

The 1202 computer program alarm was an executive overflow condition and Gene Kranz called Steve Bales up on the intercom to get a status; decision here was critical because mission rules would call for a rapid abort-stage fire within seconds if the alarm recurred.

BALES 'We're go on that, flight.'
KRANZ 'We're go on that alarm?'
CAPCOM 'Roger. We got – we're go on that alarm.'
BALES 'If it doesn't reoccur we're go, he's taken out the delta H now.'

KRANZ 'Rog.'
EAGLE 'Roger. P30.'
CAPCOM '6 plus 25 throttle down.'

What Houston would now do was to monitor Eagle's 'delta H', the difference in height between the actual value read by the landing radar and the value the primary guidance system expected to see when the radar locked on. When the program alarm sounded, the two had been in variance by 853 meters. For a few minutes, the crisis was over, but Houston would have to monitor that segment of the update format extracted so as not to overload the computer.

CAPCOM 'Eagle, Houston. We'll monitor your delta H.'
EAGLE 'Delta H is looking good now.'
CAPCOM 'Delta-H is looking good to us. Right on time.'
EAGLE 'Throttle down's better than in the simulator.'
CAPCOM 'Rog.'
EAGLE 'AGS and PGNCS look real close.'
CAPCOM 'You're looking great to us, Eagle.'
EAGLE 'Okay, I'm still on slew so we may tend to lose as we gradually pitch over. Let me try auto again now and see what happens.'
CAPCOM 'Roger.'
EAGLE 'Okay, looks like it's holding.'
CAPCOM 'Roger, we got good data.'

For the seven minutes since ignition, Eagle had been gradually pitching forward into an upright posture. But very gently. At this point the Lunar Module was still pitched backward 65°. But as the vehicle slowly continued to right itself concern had risen over the ability of the high-gain antenna to maintain lock on the Earth, for as the spacecraft pitched up so the antenna would have to move in the opposite direction to sustain communication. Earlier, contact had been lost because of this, and lost telemetry at this point was information that could prove vital to understanding the course of any accident that might occur.

At throttle down, 6 min 25 sec into the burn, Eagle was 7,130 meters above the Moon and had slowed to 1,540 km/hr. At 7 min 30 sec with the throttle setting around 55% thrust, Eagle was down to 4,970 meters and pitched over 60°. One minute later – the end of the braking phase and the start of the approach phase. At this point 8 min 26 sec into the burn, Eagle switched to program 64. Horizontal speed was 550 km/hr, descent rate was 140 km/hr, Eagle was 2,300 meters above the Moon and noticeably pitching forward now, translating horizontal flight into a vertical descent.

Now the crew could see the landing site, still 7.6 km ahead; amazingly they had travelled about 472 km in the past $8\frac{1}{2}$ minutes, descending from 16 km to 2.3 km. The purpose of the approach phase – initiated by 'high gate,' an old aeronautical term from the barnstorming days indicating an aircraft landing approach position – was to allow manual command of the LM so that the crew could migrate from computer information to visual perception about the precise area for the final, landing, phase.

CAPCOM 'It's the P64.'
EAGLE 'Good. Roger.'
CAPCOM '8 30 you're looking great.'
CAPCOM 'Eagle you're looking great, coming up 9 minutes.'
EAGLE 'Manual auto attitude control is good.'

Eagle was now pitched back from vertical only 35–38° and forward speed was down to 230 km/hr with a 70km/hr descent rate. It was time for Gene Kranz to go around his controllers once more on a final 'go/no-go' for landing:

KRANZ 'Okay, have we still got landing radar, guidance?'
BALES 'Affirm.'
KRANZ 'Okay. Is it converged?'
BALES 'Beautiful.'
KRANZ 'Has it converged?'
BALES 'Yes!'
KRANZ 'Okay. Okay all flight controllers, go/no-go for landing. Retro?'

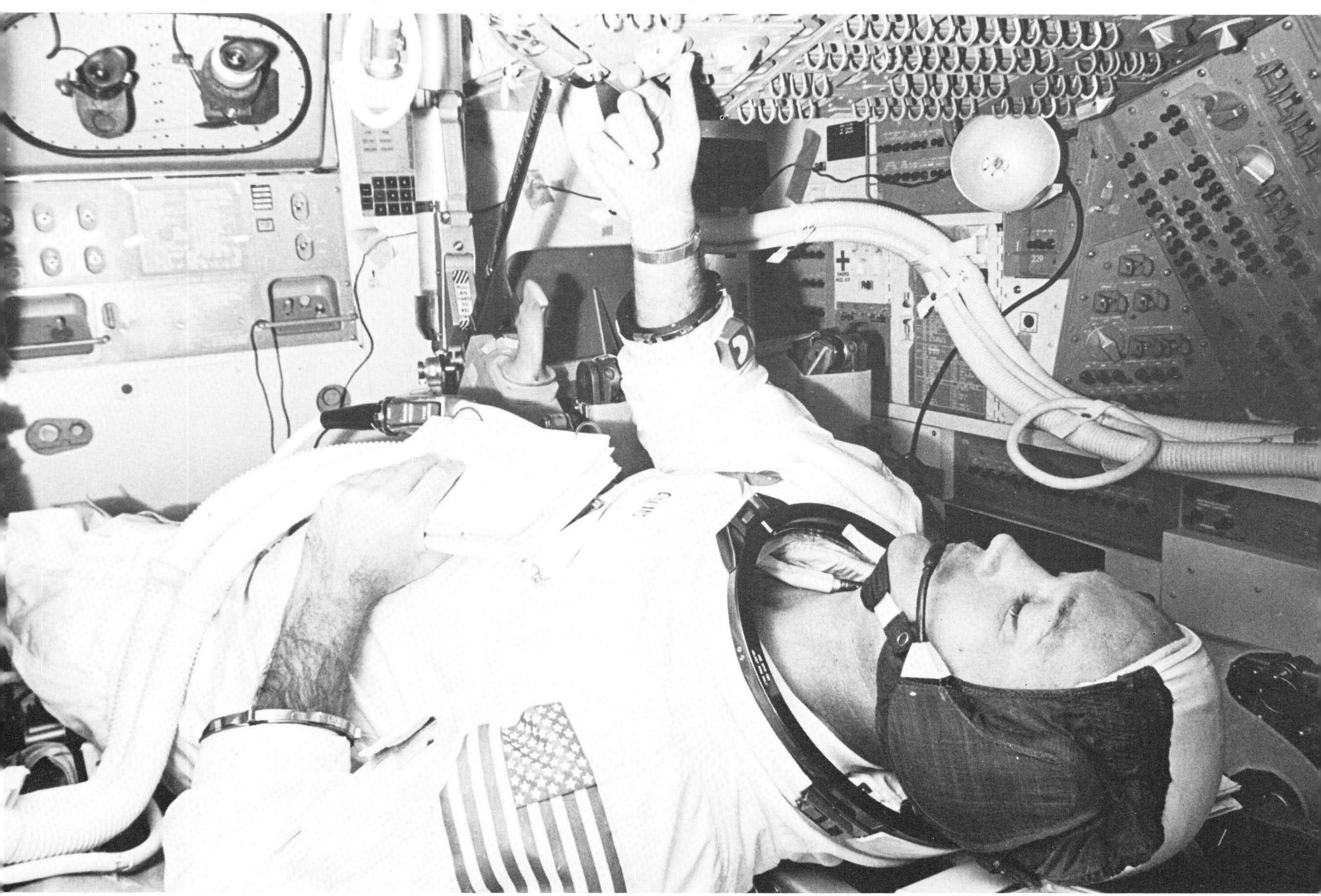

During Eagle's excursion to the lunar surface, Collins remained alone in the Apollo Command Module, an activity practiced here in ground tests.

RETRO 'Go.'
KRANZ 'Fido?'
FIDO 'Go.'
KRANZ 'Guidance?'
GUIDANCE 'Go.'
KRANZ 'Control?'
CONTROL 'Go.'
KRANZ 'G & C?'
G & C 'Go.'
KRANZ 'Surgeon?'
SURGEON 'Go.'
KRANZ 'Capcom, we're go for landing.'
CAPCOM 'Eagle, Houston, You're go for landing, over.'
EAGLE 'Roger, understand. Go for landing.'
And then it happened again:
EAGLE '1201 alarm . . . 1201.'
CAPCOM 'Roger, 1201 alarm.'
CAPCOM 'We're go, PGNCS high, we're go.'

This alarm was a different segment of the same executive overflow that occurred earlier. Again, the flight was dangerously close to losing valuable data on displays essential for accurate guidance. But now Eagle was passing into the landing phase where Armstrong would soon switch from program 64 to P66, where rate of descent would be automatic, allowing him to translate manually the LM to a suitable touchdown spot. Armstrong could see the landing site clearly now, and off to the left of the groundtrack a 200 meter diameter crater – West Crater as it would later be named – caused him to translate the Lunar Module manually further downrange, extending the targeted touchdown spot to a smooth region beyond.

Eagle was about 125 meters above the lunar surface when P66 was entered, with the spacecraft pitched back only 11° now. If left to its own resources the LM would gradually descend to a hover point 500 meters downrange and then gently lower itself the remaining 35 meters. In the MOCR, capcom Charlie Duke was tense with anticipation, the last remaining minute or so hanging with excrutiating agony. Continually interjecting the commentary from Eagle during this final critical maneuver, Charlie Duke received a sharp crack on the arm from Deke Slayton and a crisp 'Shut up!' The words from space were telling their own story:

EAGLE '35 degrees. 35 degrees. 750, coming down at 23. 700 feet, 21 down, 33 degrees. 600 feet, down at 19. 540 feet, down at 30 – down at 15. 400 feet, down at 9. 8 forward. 350, down at 4. 330, 3½ down. We're pegged on horizontal velocity. 300 feet, down 3½. 47 forward. Down 1 a minute. 1½ down. 70. Got the shadow out there. 50, down at 2½. 19 forward. Altitude-velocity lights. 3½ down, 220 feet. 13 forward. 11 forward, coming down nicely. 200 feet, 4½ down. 5½ down. 160, 6½ down, 5½ down, 9 forward. 5%. Quantity light. 75 feet, things looking good. down a half. 6 forward.'

Eagle was now very close to the Moon and increasingly dependent upon the visual ability of Neil Armstrong to see the surface. Radar was not very much use here, except for information on altitude and rate-of-descent. The selection of a boulder-free zone rested entirely upon the piloting skills of the astronaut. At a height of 35 meters radial fans of dust and fine soil spread out all around the LM from a point directly beneath the main engine. Standing on a column of thrust, Eagle was kicking up a sheet of material through which the surface could only be seen with difficulty. The radial motion made it difficult to judge the position of the LM and at 25 meters the sheet of flying particles intensified.

Now, Eagle was little more than 20 meters above the surface and almost at a hover as Armstrong flew the vehicle downrange at a slow pace, searching for a flat area. But propellant was getting low. Margins were already tight; in electing to fly beyond the 200 meter diameter crater the crew had slowed down the landing procedure and were already cutting in to reserve time. Charlie Duke came in with the agreed warning as to how much flying time the crew had left:

CAPCOM '60 seconds.'
EAGLE 'Lights on. Down 2½. Forward. Forward. Good. 40 feet, down 2½. Picking up some dust. 30 feet. 2½ down.'

Armstrong and Aldrin had less than a minute flying time left; beyond that point the engine would stop and they would crash to the surface. Eagle was now 10 meters above the Moon and particles were flying up toward the vehicle. In trying to halt the sideways movement, Armstrong sensed the Lunar Module begin a backward drift. Eagle must be essentially hovering over one spot before starting down the last few meters; any appreciable backward motion could bring it on to a boulder, or cause the rear landing leg to dig in and tip the spacecraft.

EAGLE 'Faint shadow. 4 forward. 4 forward, drifting to the right a little. 6 . . . down a half.'
CAPCOM '30 seconds.'

The thirty second call was the final statement that this was it: if Eagle was not put immediately on the surface, the engine could cut at any moment. The quantity gauge read out on telemetry sent to Houston had an accuracy of 2%. Eagle was now into that error margin. An abort would be made at the end of the 30 seconds now remaining unless Eagle got to the surface first. But there could be no hurry; the descent rate must be as low as possible.

EAGLE 'Forward. Drifting right . . . contact light. Okay, engine stop. ACA out of detent. Modes control both auto, descent engine command override off. Engine arm off. 413 is in.'
CAPCOM 'We copy you down, Eagle.'
ARMSTRONG 'Houston, Tranquillity base here. The Eagle has landed.'
CAPCOM 'Roger, Tranquillity, we copy you on the ground. You've got a bunch of guys about to turn blue. We're breathing again. Thanks a lot.'
EAGLE 'Thank you.'

On the Moon, a light haze obscured the view as Armstrong and Aldrin peered between busy tasks at the view outside. Dust blown up by the scouring effect of Eagle's engine hung for several seconds in the one-sixth gravity. Gradually, through the clearing fog of grayish-brown sand, blocks and boulders appeared, a scene emerging as though unveiled in solemnity for these first creatures from another world. For in all of time since Creation, the beings that evolved on planet Earth had been condemned to their own backyard. Now, 4,700 million years after it formed, the Moon received its visitors, ambassadors from a world that for all that time had been so close across the sea of space. Together they had grown; suddenly, in an instant, they were joined. It would never be the same again. Man was on the Moon.

To the flight controllers in the Mission Operations Control Room it was an unbelievable moment. Gene Kranz momentarily froze, while beyond the large glass screen separating the trenches from the viewing area, people were clapping and cheering. But no respite. Eagle had settled upon the Moon and could be in a dangerous condition. There would not be much Descent Stage propellant remaining in the tanks but what there was could be leaking to the surface; the shock of landing could have ruptured propellant lines or feed points. Mission plans provided two possible emergency lift-off times: T1 and T2. Gene Kranz called his controllers up and got a status check for stay/no-stay at the T1 opportunity, passing the word to Charlie Duke who then called Eagle: 'And you are stay for T1. Over. Eagle, you are stay for T1.'

The two possible lift-off times were based upon the position of Columbia up ahead in orbit. Minutes later, Charlie Duke spoke to Mike Collins:

COLUMBIA 'Houston, do you read Columbia on the high gain?'
CAPCOM 'Roger, Columbia. He has landed, Tranquillity base. Eagle is at Tranquillity, over.'
COLUMBIA 'Yeah, I heard the whole thing.'
CAPCOM 'Rog, good show.'
COLUMBIA 'Fantastic.'

Then it was time for a T2 stay/no-stay status. That lift-off option came less than nine minutes after touchdown. Eagle was in good condition, nothing prevented it remaining in a safe condition on the surface and that too was passed along to the Lunar Module. On board the spacecraft, a flurry of activity accompanied post-landing elation. Vents were opened, systems checks were run, computer loads set up ready for imminent lift-off should the need arise. Then, gradually, the pace lessened, and everybody realized – on the Moon and in Houston – that Eagle was on its perch for at least the next two hours. With Columbia swinging around the Moon at two-hour intervals, possible lunar surface lift-off times were aligned with the mother-ship's orbit. Then the

Aldrin descends the ladder to become the second man on the Moon.

comments, first on landing and then on what the Moon looked like from the surface, the first descriptive words from Tranquillity base:

ARMSTRONG 'Houston, that may have seemed like a very long final phase. The auto targeting was taking us right into a football field sized crater, with a large number of big boulders and rocks for about 1 or 2 crater diameters around us, and it required a . . . on the P66 and flying manually over the rock field to find a reasonably good area.'
CAPCOM 'Roger, we copy. It was beautiful from here, Tranquillity, over.'
ALDRIN 'We'll get to the details of what's around here, but it looks like a collection of just about every variety of shapes, angularities, granularities, every variety of rock you could find. The colors are pretty much depending on how you're looking relative to the zero phase point. There doesn't appear to be too much of a general color at all, however, it looks as though some of the rocks and boulders, of which there are quite a few in the near area, it looks as though they're going to have some interesting color to them, over.'
CAPCOM 'Rog, Tranquillity, be advised there's lots of smiling faces in this room, and all over the world, over.'
EAGLE 'There is two of them up here.'
CAPCOM 'Rog, that was a beautiful job, you guys.'
COLUMBIA 'And don't forget one in the Command Module.'
CAPCOM 'Rog.'

It had been touch and go. At landing, Eagle had about 20 seconds flying time left and Neil Armstrong had put the LM down 300 meters downrange and 100 meters to the left of the targeted spot, the point computer programs 64 or 65 would have set the Lunar Module down. By going to P66, Armstrong had secured manual authority to translate across the surface and in doing so very nearly run out of propellant. Armstrong would later apologize for what he considered sloppy piloting; a lesser pilot would never have reached the surface. But shortly after settling on the dusty surface of Tranquillity, it became apparent that the entire descent operation had been shifted farther west than intended.

Final guidance updates had been passed to the LM on revolution 13 based on ground station tracking on revolutions 11 and 12. Irregularities in the Moon's gravity field and an unintentional error in position data built up a dispersion so that when the final calculations were sent to Eagle for PDI the spacecraft was 6,800 meters further downrange and 1,400 meters further south than ground controllers believed. As a result, at touchdown, the spacecraft was about 7 km from the planned site.

About two hours after landing, capcom Charlie Duke bid the crew farewell and prepared to hand over to Owen Garriott as Gene Kranz pulled his flight controllers from the MOCR. Milton Windler would handle the upcoming events. The flight plan had the crew starting a four hour rest period just three hours after touchdown but Armstrong and Aldrin told Houston they would now like to get ready for EVA and the historic walk. Houston agreed. The TV networks would get to cover the Moon walk during prime evening time after all.

It was late afternoon when they landed; in Europe, it was mid-evening. On the mission, an elapsed time of 102 hr 45 min 43 sec. The crew were to have had a 4 hr rest, performed the Moon walk, then rested for about 4½ hr before lift-off. Now, the first sleep period would be slipped and joined to the second. Cliff Charlesworth's team had trained to handle the walk, just as Gene Kranz was set up for the landing operations, so Windler was replaced by Charlesworth at 106 hr 15 min into the flight. Just before that event, at about 105 hr 25 min, Buzz Aldrin asked people all over the world to stop what they were doing.

EAGLE 'Houston, Tranquillity. Over.'
CAPCOM 'Tranquillity, Houston. Go ahead.'
ALDRIN 'Roger. This is the LM Pilot. I'd like to take this opportunity to ask every person listening in, whoever and wherever they may be, to pause for a moment and contemplate the events of the past few hours, and to give thanks in his or her own way. Over.'
CAPCOM 'Roger, Tranquillity base.'

In the silent moments that followed Buzz Aldrin bowed his head in prayer, a thoughtful gratitude for the events that had brought them to this moment. Then it was time for a meal, the first meal on the Moon, and Houston heard that the crew of Eagle would begin preparing for EVA in about a half-hour. It would take at least two hours to get everything ready, and to double-check that everything was in order. Bruce McCandless kept up a dialogue with Mike Collins in Columbia on each near side pass and with Armstrong and Aldrin, donning their cumbersome extravehicular-mobility-units.

It took longer than expected to prepare for the EVA, and when the two Moon walkers put their helmets and gloves on, switching to the Portable Life Support System ('pliss') communication equipment, the antennae on top of the back-packs scratched against the roof and hampered the verbal dialogue. At last all was ready and Cliff Charlesworth gave a 'go' for cabin depressurization. This would be the last real check on the integrity of the suits, before Neil Armstrong would open the hatch and back out on the porch. It was 108 hr 55 min, Eagle had been on the Moon for more than six hours. Just seven minutes later the doorway to the Moon was unlatched and pulled back.

It was a small hatch, square in shape, only 81 cm along each side. But it was big enough. Neil Armstrong kneeled down on his hands and knees and slowly backed out on the porch, carefully feeling his way, moving, edging, crawling very slowly backwards on instructions from Buzz Aldrin who verbally guided him out the door. Hinged along the right side, the hatch seemed to want to close at first, but Aldrin held it back until Armstrong was well into the open space. At 109 hr 19 min 16 sec: 'Okay, Houston, I'm on the porch.' Before starting down the ladder, Armstrong fixed up a clothes line device designed to aid the crew in lifting equipment, primarily samples, back into the Lunar Module. Then he began the backward climb down the nine rungs of the forward landing gear ladder, pulling the lanyard that deployed the MESA carrying the television camera and surface equipment.

CAPCOM 'Man, we're getting a picture on the TV.'
ALDRIN 'Oh, you got a good picture. Huh?'
CAPCOM 'There's a great deal of contrast in it, and currently

it's upside-down on our monitor, but we can make out a fair amount of detail.'
ALDRIN 'Okay, will you verify the position, the opening I ought to have on the camera?'
CAPCOM 'Standby.'
CAPCOM 'Okay Neil, we can see you coming down the ladder now.'
ARMSTRONG 'Okay, I just checked – getting back up to that first step, Buzz, it's not even collapsed too far, but it's adequate to get back up.'
CAPCOM 'Roger, we copy.'
ARMSTRONG 'It takes a pretty good little jump.'

The view that came down to Earth filled the flat screen at the right of the MOCR projection wall and flooded into the homes of several hundred million people all across the world. In the United States it was evening, in Europe it was several hours past midnight and in the far east it was morning. On the Moon, the strong rays of the Sun shone from almost directly behind Eagle, casting a long shadow out in front of the forward leg. The footpad itself was in the shadow, and how difficult it was to see exactly where the rim of the pad was; without an atmosphere to scatter light, shadows were pools of blackness. Backing down the ladder made it even more difficult to see the surface, the entire underneath of the Lunar Module shaded from the Sun. On TV, a ghostly apparition moved down the ladder and jumped down from the lower rung into the circular pad, a single leg feeling for height before hopping down.

ARMSTRONG 'I'm at the foot of the ladder. The LM footpads are only depressed in the surface about 1 or 2 inches. Although the surface appears to be very, very fine grained, as you get close to it. It's almost like a powder. Now and then, it's very fine.'

Armstrong was standing, one arm on the ladder, with his left leg raised out across the surface.

ARMSTRONG 'I'm going to step off the LM now. That's one small step for man. One giant leap for mankind.'

The historic step came at an elapsed time of 109 hr 24 min 15 sec. The goal set by President Kennedy had been accomplished. A man was standing on the surface of the Moon.

ARMSTRONG 'As the – the surface is fine and powdery. I can – I can pick it up loosely with my toe. It does adhere in fine layers like powdered charcoal to the sole and sides of my boots. I only go in a small fraction of an inch. Maybe an eighth of an inch, but I can see the footprints of my boots and the treads in the fine sandy particles.'
CAPCOM 'Neil, this is Houston. We're copying.'
ARMSTRONG 'There seems to be no difficulty in moving around as we suspected. It's even perhaps easier than the simulations at one-sixth g that we performed in the simulations on the ground. It's actually no trouble to walk around. The Descent engine did not leave a crater of any size. There's about 1 foot clearance on the ground. We're essentially on a very level place here. I can see evidence of rays emanating from the Descent engine, but very insignificant amount. Okay, Buzz, we're ready to bring down the camera.'

Now it was time for Aldrin to descend the nine steps, unaided by his colleague who was to photograph the event.

ALDRIN 'Now I want to back up and partially close the hatch. Making sure not to lock it on my way out.'
ARMSTRONG 'A good thought.'
ALDRIN 'That's our home for the next couple of hours and I want to take good care of it. Okay, I'm on the top step and I can look down over the RCU, landing gear pads. That's a very simple matter to hop down from one step to the next.'
ARMSTRONG 'Yes, I found it to be very comfortable and walking is also very comfortable. You've got three more steps and then a long one.'

When he reached the surface, Aldrin had two words that summed up the view: 'Magnificent desolation.' But Armstrong was now trying various movements, bouncing up and down, using what spring there was in his suited legs to leap up and down. When he felt a tendency to pitch over backwards, the exercise was quickly stopped: 'The mass of the back-pack does have some effect on inertia.' With the soft soil under his boots, it was difficult to find a solid surface. To show viewers on Earth what the surface looked like, Armstrong went to the MESA and prepared to take the TV camera out some distance from Eagle to scan the horizon before setting it upon a tripod about 25 meters away. Buzz Aldrin checked his balance and stability, learning that much of the surface seemed slippery, giving him an insecure footing on the glassy granules.

Prior to moving the camera, Armstrong described a plaque fixed to the LM forward landing gear strut, clearly visible between two rungs on the ladder: 'For those who haven't read the plaque, we'll read the plaque that's on the front landing gear of this LM. First there's two hemispheres, one showing each of the two hemispheres of the Earth. Underneath it says "Here men from the planet Earth first set foot upon the Moon, July 1969 AD. We came in peace for all mankind." It has the crew members' signatures and the signature of the President of the United States.' Then it was time for the TV audience to receive a conducted tour around the lunar surface while Buzz Aldrin removed a solar-wind-composition experiment, essentially a flat aluminum sheet hung vertically from a pole; it would remain exposed to radiation from the Sun for approximately 77 minutes before retrieval for return to Earth.

With the camera installed on the tripod it was easy to see how difficult normal walking activities became under one-sixth Earth gravity conditions. Encumbered by the bulky suits, the astronauts found 'it takes two or three paces to make sure you've got your feet underneath you,' and that it would need about 'two to three, or maybe four, easy paces,' to come to a stop. Loping along, they found it easier to adopt a kangaroo hop, but that too carried them further along than they wanted to go. Gravity might be less, but inertia was exactly the same!

As Columbia got a call from Houston about the activities going on below, Armstrong and Aldrin set up the flag. But the ground was hard and the pole only went in a short distance. In any event, the telescoping sections stuck so it was only partially extended. It was almost certain to fall over when the Ascent Stage blasted away so this was only a token gesture. Coming up on one hour out on the surface, the two Moon walkers were called in front of the camera while Houston relayed a telephone call from the President:

NIXON 'Neil and Buzz, I am talking to you by telephone from the Oval Room at the White House. And this certainly has to be the most historic telephone call ever made. I just can't tell you how proud we all are of what you (are doing). For every American, this has to be the proudest day of our lives. And for people all over the world, I am sure they too, join with Americans in recognizing what a feat this is. Because of what you have done, the heavens have become a part of man's world. And as you talk to us from the Sea of Tranquillity, it inspires us to double our efforts to bring peace and tranquillity to Earth. For one priceless moment, in the whole history of man, all the people on this Earth are truly one. One in their pride in what you have done, and one in our prayers that you will return safely to Earth.'
ARMSTRONG 'Thank you Mr. President. It's a great honor and privilege for us to be here representing not only the United States but men of peace of all nations. And with interest and a curiosity and a vision for the future. It's an honor for us to be able to participate here today.'

The two men then went about separate tasks. Armstrong to collect rock samples within about 30 meters of Eagle; Aldrin to photograph all sides of the Lunar Module. Their back-packs were holding up well under the extreme conditions, and the men themselves were in good spirits and excellent condition. Even the excitement coupled with sometimes strenuous work could not raise their heart beats above 100 for long. Periodically, both Moon walkers reported back to Houston on the indicated suit pressure and the amount of oxygen remaining in their tanks. Houston could monitor their condition but wanted to know if both sources agreed.

Aldrin reported that the lunar surface beneath the Descent Stage engine bell was hardly marked and that it had not scoured out a crater as some had thought it might. Armstrong had not been able to see the initial touchdown, nor to feel it, and engine shutdown was commanded with the spacecraft virtually on the surface. It was a 'walk round' inspection of a very unique kind, an activity pilots the world over were used to accomplishing before getting in to fly away. Buzz Aldrin was now at the rear of the

Lunar Module, extracting the experiments to deploy them a short distance away. Contained in the left rear quadrant of the Descent Stage, they comprised a laser ranging retro-reflector and a passive seismic detector. While Aldrin set up the seismometer, Armstrong carried off the retro-reflector. It had its own levelling device, a small bubble in a tube, but it refused to adjust to the required position, until Armstrong walked away and came back to find it perfectly aligned!

The pressure suits were a definite hindrance now it was time to manipulate equipment and special tools had been provided so that the crew could avoid having to bend down. Not that they could have had they wished to. Modifications would be needed before Moon walkers could become lunar gymnasts. Based on the status of their consumables, capcom Bruce McCandless passed along a MOCR decision to extend their EVA by up to 30 minutes. 'Okay, that sounds fine,' said Armstrong. Now it was time to get a documented sample. That implied several activities including rock collection, core sampling and procurement of a gas sample within a container. Unlike the earlier, bulk sample, collection, where the astronaut ran around grabbing whatever he could, the documented samples would be the first retrieved scientifically. As such, they were relagated to the end of the mission since priority was low for this initial manned landing.

Buzz Aldrin rolled up the solar-wind-composition experiment and as the first Moon walk drew to a close the two very dirty astronauts prepared to get back inside the LM. Aldrin was back inside at 111 hr 29 min, followed by his Commander ten minutes later and with an 'Okay, the hatch is closed and locked,' the two men began the long process of pressurizing the cabin, removing their PLSS backpacks after re-connecting their suits to the Lunar Module environmental control system, and gathering together items that would be left on the Moon. Two and one-half hours later, the LM was depressurized a second time, the forward hatch was opened and equipment was pushed outside. This included the back-packs, the used lithium hydroxide canisters, the arm rests, and the other incidentals. Within minutes the hatch was closed and the LM pressurized again.

At 114 hr 25 min Buzz Aldrin threw the switch on the TV and the screens went blank. For the past five hours it had shown scenes of indescribable achievement. The wispy view seemed almost beyond acceptance to many who had worked long years to put a man on the Moon. The imperfection of the picture, the transparent forms of Armstrong and Aldrin loping along, their crackly voices pitched appropriately for the mood, it all seemed so 'other worldly' that many would feel the event had been enhanced by the feeling of remoteness it gave. Now, Armstrong and Aldrin were to get some rest. For there was less than ten hours to lift-off, the first launch from the surface of another world. But there was no break, no halt to the rapacious enthusiasm for answers from the two Moon men:

Even during the first Moon flight, scientific experiments were carried to the lunar surface. Here, Aldrin carries the passive seismometer (left) and the laser ranging retro-reflector.

CAPCOM 'Tranquillity, this is Houston. We also have a set of about 10 questions relating to observations you made, things you may have seen during EVA. We can either discuss a little later on this evening or sometime later in the mission. It's your option. How do you feel? Over.'
EAGLE 'I guess we can pick them up now.'

For the next twenty minutes Owen Garriott discussed surface conditions, rock types, the apparent angle of the Lunar Module – which engineers calculated was leaning only 4.5° – and general impressions of surface conditions. At 114 hr 50 min, the ground said 'Thank you and I hope this will be a final goodnight.' It was. But not for the crew, who found it almost impossible to sleep. How could they? Neil Armstrong made a foot sling from tethers and lay across the Ascent engine cover that protruded through the cabin's mid-section floor. Buzz Aldrin curled up on the floor in front of the forward hatch.

Both men wore their helmets to breathe clean air; the inside of the LM was like a dust bowl from all the lunar soil they brought in on their suits. They pulled window shades down to shut out the glare from the lunar surface, increasingly bright as the Sun moved up into the sky behind Eagle, but they were little use. Even features on the outside could be seen with the shades down. Then Armstrong sensed increasing light levels in the cabin and discovered the Sun pouring through the overhead guidance telescope. Fitfully, and restlessly, they tried to doze but the Lunar Module was a noisy spacecraft at best and ever so often they heard hissing sounds, cracks as material expanded, and the occasional gurgling noise. Unlike the Command Module, Eagle wrapped its many systems around the crew cabin and it was like trying to get to sleep inside a packing factory!

Lunar lift-off was scheduled for Columbia's 25th revolution, calculated for an elapsed time of 124 hr 22 min. On the ground, flight surgeon Kenneth Beers watched Neil Armstrong's physiological response; Buzz Aldrin was not wearing a telemetry set. Neither man was seen to fall into deep sleep. Six hours fifty minutes after the last call, Houston resumed communications with Eagle on the Sea of Tranquillity. There were just 2½ hours remaining to lift-off, and much to do in preparation for that event. But during the dramatic activities at the surface, the orbiting CSM had been performing a limited set of tasks.

It was a lonely role for, as Apollo Control said, 'Not since Adam has any human known such solitude as Mike Collins is experiencing during this 47 minutes of each lunar revolution when he's behind the Moon with no one to talk to except his tape recorder aboard Columbia.' It was indeed a very special kind of solitude; one man, totally divorced from every other human being alive on each far side pass around the Moon. But that would soon change, for Armstrong and Aldrin were now well along with their launch preparations. For nearly two hours the exchange between Houston and the two space vehicles maintained a technical dialogue aimed at checking equipment that would carry them to rendezvous and docking. Shortly after passing around the Moon's west limb at the start of the 24th orbit, Mike Collins heard a message from capcom Jim Lovell, as also did Eagle on the surface.

CAPCOM 'Eagle and Columbia, this is the back-up crew. Our congratulations to yesterday's performance, and our prayers are with you for the rendezvous. Over.'
EAGLE 'Thank you, Jim.'
COLUMBIA 'Thank you, Jim.'
EAGLE 'Glad to have all you beautiful people looking over his shoulder. We had a lot of help down there, Jim.'

Then there was time for Eagle to convey further thoughts about the landing site when, with little more than an hour to go, Houston received a description of the craters and the small rocks. Flight director Glynn Lunney would conduct operations in the MOCR during the critical ascent phase when Eagle would put itself into an elliptical orbit of 17 × 84 km, there to begin three and one-half hours of rendezvous maneuvers. Just as it was vital for ground controllers to know the exact position of the Lunar

Module prior to powered descent, so it was essential for the Lunar Module to know where it was with respect to Columbia's orbit. By the time of lift-off, controllers thought they had a fair knowledge of the LM's position, and proved it by loading a good state vector into the lunar guidance computer.

Slowly, the countdown to lift-off cycled the Ascent Stage on. Its batteries were switched over and seen to be producing the necessary current and other systems appeared satisfactory. From an original weight of 15 tonnes, Eagle had an Earth weight of 7.2 tonnes on the Moon – 1.2 tonnes at one-sixth gravity – and would lift away with an Ascent Stage mass of 4,908 kg. About twenty minutes before ignition, the network controller told Lunney that the tracking stations around the globe were equipped with their 'battle shorts' waiting to relay telemetry to Goddard. At times such as this it was imperative that every last bit of telemetry should be obtained from the Ascent Stage and to prevent a normal short cutting in a circuit breaker during critical periods, engineers fitted mechanical links in communications equipment at the network stations that would inhibit loss of capability even to the point of equipment burning out before failing completely.

In the closing minutes, controllers in the MOCR checked primary and abort guidance systems, found them to be in agreement, and waited for ignition. At lift-off, Columbia would be up ahead in orbit, 152 km slant range distance in front of the ascending Lunar Module; at orbit insertion, Eagle would trail the CSM by about 490 km. The communications line to Tranquillity was calm in the closing seconds. It was an event that *had* to happen if Armstrong and Aldrin were to get off safely. Nobody had ever launched anything from the Moon into orbit and the Ascent Stage had never performed this operation in exactly the sequence it was now programed for. After 21 hr 36 min 13 sec, Eagle was taking flight.

EAGLE 'Forward 8, 7, 6, 5, abort stage, engine arm, ascent, proceed. That was beautiful. 26, 36 feet per second up. Be advised of the pitch over. Very smooth.'

It worked. With a noticeable crack from pyrotechnics, and a spray of dust and debris shooting out from all around the Ascent Stage, Eagle gently accelerated off the Descent Stage. Eight seconds later, at a height of 52 meters, the lunar guidance computer responded to commands from program 12 and pitched Eagle over to begin the translation from vertical to horizontal flight. In another eight seconds it had pitched to 52°, the fixed thrust from its engine accelerating the seemingly flimsy Ascent Stage along a clear-cut path into space. Now Armstrong and Aldrin could see the lunar surface, the spacecraft's attitude giving them a magnificent view of the rough terrain below. This was a better view than on descent, and far more enjoyable.

The crew easily recognized craters they had been trained to observe: 'That's Sabine off to the right now. . . . There's Ritter out there . . . there it is right there (!) . . . Man, that's impressive.' Eagle's engine would shut down after a burn lasting 7 min 20 sec, having accelerated the Ascent Stage to a speed of 6,660 km/hr. On through the final seconds of burn, Eagle kept up a running commentary on the speed yet to be achieved prior to shutdown, the spacecraft displays providing their own unique countdown to cutoff:

EAGLE 'About 800 to go. 700 to go. Okay, I'm opening up the main shut-offs. Ascent feed closed, pressure's holding good, crossfeed on, 350 to go. Stand by on the engine arm. 90, okay, off, 50, shutdown.'
CAPCOM 'Eagle, roger. We copy. It's great. Go.'

The orbit was almost precisely as planned: 17 × 84 km. Inserted on to that path at the low point, Eagle would ascend to the 84 km apocynthion around on the far side of the Moon. At that point it would fire the Reaction Control System thrusters in a coelliptic-sequence maneuver designed to place the vehicle in an almost circular path 27.8 km below the orbit of Columbia. In Mission Control, a comment from the public affairs officer: 'Flight Operations Director Chris Kraft commented that he felt like some 500 million people around the world are helping push Eagle off the Moon and back into orbit.' And so they were, for the event carried as much drama as the landing. Without the engine burn, Eagle would have remained on its nest.

There was now only a short period before both spacecraft would disappear round the Moon's west limb and tracking by ground stations would refine navigation marks being taken on board. Lift-off occurred at 124 hr 22 min and Eagle went out of sight at 125 hr 9 min, preceded two minutes earlier by Columbia up ahead in higher orbit. At CSI – coelliptic-sequence-initiate – distance between the two vehicles was less than 260 km. The burn came at 125 hr 19 min and fired Eagle into a 84 × 90 km path from where it could perform a second maneuver on the Moon's near side, the constant-delta-height burn, designed to set up Eagle at a constant altitude below Columbia.

Eagle lifted off the Moon with a weight of 4.9 tonnes but consumption of Ascent Stage propellant on the seven minute burn into orbit reduced this to a mere 2.7 tonnes. Columbia appeared first on its 26th revolution, forty-six minutes after loss of signal, followed by Eagle three minutes later. Flight controllers at Houston quickly learned that the CSI burn had gone very well and that a possible plane-change would not be needed. This would be used if the plane of the Lunar Module's orbit was found to be at variance with the plane of the CSM's orbit about the Moon.

For much of the front side pass Eagle and Columbia performed the ranging information necessary to compute the rendezvous maneuvers. This was where Gemini had done good work, providing valuable experience in seeking a target in space, and the crew, all of them veterans from rendezvous flights, included 'Dr. Rendezvous' himself. For this operation, flight controllers on the ground were looking over the crew's shoulder while the two space vehicles performed their own computations. Calculations flowed like pennies from a slot machine as the two spacecraft narrowed in distance from each other.

Columbia was waiting in a 105 × 116 km path and, at acquisition of signal, range between the two was down to 190 km. About twenty minutes later Eagle burned its thrusters for 18 seconds to trim the altitude difference in a CDH maneuver aimed at setting up the orbit for final rendezvous. The LM was now in a 76 × 87 km path, 170 km behind Columbia. Forty minutes after the constant-delta-height burn, Eagle fired its thrusters yet again for the terminal-phase-initiation maneuver. Thrusting along the line of sight to the CSM, Eagle began the ascent that would bring it up to Columbia. But minutes after starting into the final lap, both spacecraft swept around the west limb of the Moon; distance between the two: 71 km.

Very little communication went up from Houston during that front side pass which included two major rendezvous maneuvers. Communication between the two spacecraft was all-important. Without the experience of Gemini this operation would have been an incredibly difficult task, and without the adequate timelines paved by Apollo 10 the risk would have been greater. But Eagle was now on the final leg of its rendezvous activity and on the far side of the Moon pulled up close to Columbia, finalizing the match at 127 hr 45 min, forty-two minutes after TPI.

When Columbia and Eagle came 'over the hill' and back into contact, controllers eaves-dropped on busy talk between the two vehicles only meters apart and preparing to dock. At 128 hr 3 min, twelve minutes after reappearing, the two spacecraft linked up, but only after a confusing gyration seconds after the three small capture latches on Columbia's probe snagged the receptacle in Eagle's drogue.

COLUMBIA 'That was a funny one. You know, I didn't feel it strike and then I thought things were pretty steady. I went to retract there, and that's when all hell broke loose. For you guys, did it appear to you to be that you were jerking around quite a bit during the retract cycle?'
EAGLE 'Yeah. It seemed to happen at the time I put the contact thrust to it, and apparently it wasn't centered because somehow or other I accidentally got off in attitude and then the attitude hold system started firing.'
COLUMBIA 'Yeah. I was sure busy there for a couple of seconds.'

What happened was that Columbia moved slowly up to Eagle and latched in the 'soft-dock' condition for Eagle to drive forward into the drogue. But the autopilot in Eagle cut in with fierce attitude maneuvers that threatened to twist the two vehicles out of alignment. All was well when Mike Collins took over, stabilized the systems and retracted the probe; the vehicles were pulled together into a 'hard-dock' and the main docking latches fired, rigidly clamping hold of Eagle. All three crewmembers were back together and little more than ten minutes after docking Mike Collins had Columbia's hatch open ready for Armstrong

Rendezvous.

and Aldrin to remove the probe and drogue and stow them in Eagle.

But procedures here were a little different to the many other occasions when tunnel hatches between Lunar Module and Command Module were opened. This time the Ascent Stage contained a considerable quantity of loose dust, and material that could otherwise clog or contaminate the Command Module environmental control system was to be prevented from escaping out the LM by two special procedures. First, after securing a good pressure in the tunnel between the two vehicles, Collins would stand by while Armstrong and Aldrin vacuumed the space suits, the lunar sample containers and other items scheduled for transfer.

Meanwhile, Collins would have pressurized the Command Module by closing the vent valves and pumping up the cabin at least 26 mmHg above the pressure inside the Lunar Module Ascent Stage. With the pressure valve open in the LM's forward hatch, and oxygen bleeding from the Command Module into the Ascent Stage, a positive flow of gas into the LM would create a draft and prevent material flowing back into the Command Module.

With the Command Module's direct oxygen supply valve set to deliver 0.36 kg/hr the oxygen supply from Apollo would exceed the normal leak rate in the LM and keep up a constant flow, bled from the Ascent Stage via the forward hatch dump valve. The docked configuration was around on the far side of the Moon when Armstrong and Aldrin opened the LM hatch and began their movement back into Columbia. When the vehicles appeared again, they were united with Collins and ready to jettison the Ascent Stage.

COLUMBIA 'Houston, this is Columbia, reading you loud and clear. We're all three back inside, the hatch is installed. We're running a pressure leak check. Everything's going well.'
CAPCOM 'Roger, Eagle, correction, roger Columbia, we copy. You guys are speedy.'

Apollo 11 was on its 29th lunar revolution and final preparations were made to eject the Ascent Stage from the docking ring. That would be accomplished by firing a pyrotechnic charge around the base of the docking ring immediately forward of the circular Command Module tunnel hatch, jettisoning Eagle along with the docking ring and the twelve latches. The crew scurried along with that activity and were ready about 1 hr 30 min earlier than the flight plan stipulated. At 130 hr 10 min the pyrotechnic charge severed the tunnel and Eagle sprang away. 'There she goes, it was a good one,' came a voice from Columbia.

Now there was to be a separation burn to pull Columbia away from the vicinity of the Ascent Stage so that when Apollo burned out of lunar orbit the remaining section of the lander would not threaten a collision. Just twenty-one minutes after jettisoning Eagle, Columbia fired up its thrusters for 7 seconds, kicking the CSM into a slightly different orbit so it would drift away from the LM. At trans-earth-injection, the Ascent Stage would be 37 km behind Columbia and 1.8 km below. Minutes after the separation burn, a call from Columbia to capcom Charlie Duke:

COLUMBIA 'Imagine that place has cleared out a little bit after that rendezvous. You can find a place to sit down almost, huh?'
CAPCOM 'Rog. Our MOCR's about empty right now. We're taking it a little easy. How does it feel up there to have some company.'
COLUMBIA 'Damn good. I'll tell you.'
CAPCOM 'I'll bet. I think you'd almost be talking to yourself up there after 10 revs or so.'
COLUMBIA 'No, no. It's a happy home here. It'd be nice to have company. As a matter of fact, it'd be nice to have a couple of hundred million Americans up here.'
CAPCOM 'Roger. Well, they were with you in spirit.'
COLUMBIA 'Let them see what they're getting for their money.'
CAPCOM 'Rog. Well, they were with you in spirit, anyway. At

least that many. We heard on the news today, 11, that last night – yesterday when you made your landing, *New York Times* came out with a – headlines, the largest headlines they've ever used in the history of the newspaper.'
COLUMBIA 'Copied. I'm glad to hear it was fit to print.'

At 131 hr 1 min Columbia swept round to the far side of the Moon and reappeared on schedule at 131 hr 48 min. Busy with a meal, virtually no conversation emerged and the air-waves were almost totally quiet for the entire front side pass – except for a news read-up with timely comment on how the world was reacting to the momentous event; in Houston it was mid-evening, a full day after the landing which had occurred the previous mid-afternoon:

CAPCOM 'Starting off, congratulatory messages on the Apollo 11 mission have been pouring into the White House from world leaders in a steady stream all day. Among the latest are telegrams from Prime Minister Harold Wilson of Great Britain and the King of Belgium. The world's press has been dominated by news of Apollo 11. Some newsmen estimate that more than sixty percent of the news used across the country today concerned your mission. *The New York Times* which we mentioned before has had such a demand for its edition of the paper today, even though it ran 950,000 copies, that it will reprint the whole thing on Thursday as a souvenir. And Premier Alexei Kosygin has sent congratulations to you and President Nixon through former Vice President Humphrey who is visiting Russia. The cosmonauts have also issued a statement of congratulations. Humphrey quoted Kosygin as saying "I want to tell the President and the American people that the Soviet Union desires to work with the United States in the cause for peace." And Mrs. Goddard said today that her husband would have been so happy. "He wouldn't have shouted or anything. He would just have glowed." She added, "That was his dream, sending a rocket to the Moon." People around the world had many reasons to be happy about the Apollo 11 mission. The Italian police reported that Sunday night was the most crime-free night of the year. And in London a boy who had the faith to bet $5 with a bookie that a man would reach the Moon before 1970 collected $24,000. That's pretty good odds. You're probably interested in the comments your wives have made. Neil's Ann had said about yesterday's activities, "The evening was unbelievably perfect. It is an honor and a privilege to share with my husband, the crew, the Manned Spacecraft Center, the American public and all mankind, the magnificent experience of the beginning of lunar exploration." She was then asked if she considered the Moon landing the greatest moment in her life. She said, "No, that was the day we were married." And Mike, Pat said simply, "It was fantastically marvellous." Buzz, Joan said, "Apparently couldn't quite believe the EVA on the Moon." She said, "It was hard to think it was real until the men actually moved, after the Moon touchdown I wept because I was so happy." But, she added, "The best part of the mission will be the splashdown."'

Columbia wrapped itself around the Moon's west limb at 133 hr. Just 46 minutes later it was back in communication on the other side. Charlie Duke put in a call as the crew began preparations for the burn home. Maneuver pads were read up, the Command Module computer was updated with a new state vector and final systems checks were performed. Then it was time to align the platform in the inertial guidance system, ready to guide and control the trajectory of the CSM during the scheduled 2 min 28 sec it would burn its SPS engine on the far side of the Moon. Gene Kranz was in the hot seat again, and once more called up his controllers on a 'go/no-go' for TEI. Everyone was happy with what they saw and Charlie Duke passed word to the crew: 'Apollo 11, Houston. You are GO for TEI, over.'

Around on the far side of the Moon, spacecraft 107 fired its engine for 2 min 30 sec, speeding up by nearly 3,600 km/hr to blast free of the Moon's gravity. At ignition, Columbia weighed 16.6 tonnes but when the spacecraft appeared round the limb of the lunar sphere at an elapsed time of 135 hr 34 min it had a mass of only 12 tonnes, the balance having been propellant consumed during the burn.

CAPCOM 'Hello Apollo 11, Houston. How did it go? Over.'
COLUMBIA 'Time to open up the LRL doors, Charlie.'

The burn had been good. So good that the SPS would not be used again, even for the one mid-course correction en-route home. Shortly after conferring with Charlie Duke on the status of the burn, and a technical exchange on the performance of the SPS, Apollo 11 received a call from Deke Slayton:

'Rog, 11, Houston. This is the regional capcom. Congratulations on an outstanding job. You guys have really put on a great show up there. I think it's about time you power down and got a little rest though, you've had a mighty long day here. Hope you're all going to get a good sleep on the way back. I look forward to seeing you when you get back here. Don't fraternize with any of those bugs en route except for the *Hornet*.'

COLUMBIA 'Okay. Thank you boss. We're looking forward to a little rest and a restful trip back and see you when we get there.'

Columbia was on course for a Pacific splashdown and a rendezvous with the aircraft carrier *Hornet*. It had been an exhausting series of events. Armstrong and Aldrin woke from their last full sleep period at an elapsed time of 93 hours; forty-five hours later they settled down for the next full rest period. In the intervening period they had slept only fitfully, dozing lightly on the floor of the Lunar Module at Tranquillity base. But Mike Collins too had received only a nominal rest session. Sleeping when the LM crew slept, he had received only 5 hours' rest while Armstrong and Aldrin dozed on the Moon. Now they could catch up with a light day ahead and a possible mid-course correction.

Flight director Cliff Charlesworth and capcom Bruce McCandless were in the MOCR when the next call went up at 149 hr 45 min. The crew had been given a $9\frac{1}{2}$ hr rest session but Buzz Aldrin rose first followed by Armstrong and Collins before McCandless broke silence: 'Good morning Apollo 11, this is Houston, over.' The crew responded and a normal day's activity aboard a home-bound spacecraft began. At 148 hr 7 min Apollo 11 crossed back into the Earth's gravitational environment and began to speed up. Little more than two hours later the Service Module thrusters burned for 11 seconds, aligning the re-entry trajectory with the planned corridor. Without that burn the spacecraft would not have entered the atmosphere but with it Apollo 11 was within 0.04° of the targeted 6.5° flight path angle.

Removed from the rolling passive thermal control mode for the mid-course-correction, Columbia resumed its barbecue attitude, gently turning beneath the spit-roasting Sun. While controllers on the ground waited to see if the PTC roll attitude was stable, a dialogue developed over the precise point on the Moon's surface where Eagle touched down. Then there were more questions about surface rocks and features. At 154 hr Gene Kranz relieved the Charlesworth shift and a call from the ground broke into what sounded like garbled conversation coming down to Earth.

CAPCOM 'Hello Apollo 11, Houston. Your white team is now on. We're standing by for an exciting evening of TV and a presleep report, over.'
CAPCOM 'Apollo 11, Houston. Are you sure you don't have anybody else in there with you?'
COLUMBIA 'Houston, Apollo 11. Say again, please.'
CAPCOM 'We had some strange noises coming down on the downlink and it sounded like you had some friends up there.'
COLUMBIA 'Where – where do the white team go during their off hours anyway?'

Apparently intruding upon the communication link between Earth stations and the spacecraft, strange noises sounded in the earphones as network controllers searched for a cause. Soon it was time to get the lines configured for a live TV transmission from the spacecraft, but, while controllers busied themselves with preparations, a vase of long stemmed red roses was brought into the MOCR. Attached was a note: 'To one and all concerned. Job superbly done. From a Moonstruck Canadian.' The picture came in at 155 hr 35 min and after showing a view of the Moon, Neil Armstrong introduced his audience to an interesting payload:

ARMSTRONG 'We've got a lot of scientists from a number of countries standing by to see the lunar samples and we thought you'd be interested to see them as they really are here. These two boxes are the sample return containers. They're vacuum packed containers that were closed in a vacuum on the lunar surface, sealed and then brought inside the LM and then put inside these fiberglass bags zippered and resealed around the outside and placed in these receptacles in the side of the

Command Module. These are the two boxes and as soon as we get onto the ship I'm sure these boxes will immediately be transferred and delivery started to the lunar receiving laboratory. These boxes include the samples of the various types of rock. The ground mass is the soil, the sand and silt and the particle collector for the solar wind experiment and the core tubes that took depth samples of the lunar surface.'

Buzz Aldrin then proceeded to demonstrate 'for the kids at home' how globules of water could be made to remain on the bowl of a spoon and how, in the weightlessness, it would remain attached to the spoon even in the inverted position. After fifteen minutes the transmission ended and the crew made ready to start their next sleep period. Then Charlie Duke had a family note:

COLUMBIA 'I was just wondering how everything is going at the homefront. Are all our kids in one piece?'
CAPCOM 'Roger. Sure are. Everything doing fine. All the gals are having a little party tonight as far as I know.'
COLUMBIA 'Roger. Glad to hear it.'

The penultimate rest session aboard the speeding Apollo 11 began at 160 hr 30 min with the spacecraft 262,000 km from Earth and travelling at a speed of 5,000 km/hr. With the crew asleep it was time to check on station configurations around the world for status reports. The network officer heard that the Guam tracking antenna was back at peak performance. Earlier, during an important phase of the mission, a 10 year old boy named Greg Force had been recruited to work his arm through a 6.3 cm hole and pack a new bearing without which the station would have been unable to track the spacecraft. But everything seemed in order, and Columbia too looked to be in very good shape.

After eight hours the crew were reported awake by the flight surgeon and at 171 hr 10 min capcom Owen Garriott put in a call. There was a report that weather looked good for splashdown and that a tropical storm that had earlier threatened to shift the planned landing point was, after all, moving away. Three hours after wake-up, Apollo 11 passed the halfway point in distance – 190,549 km – moving at a speed of nearly 6,200 km/hr. P23 navigation sets were made on three pre-selected stars and the crew began breakfast. During the meal Mike Collins called Houston with a comment about the quantity of food consumed by the infamous Al Bean, assigned to fly on Apollo 12.

COLLINS 'We're trying to calculate how much spaghetti and meatballs we can get aboard for Al Bean.'
CAPCOM 'I'm not sure the spacecraft will take that much extra weight. Have you made any estimates.'
COLLINS 'It'll be close.'
CAPCOM '11, Houston. The medics at the next console report that the shrew is one animal that can eat six times its own body weight every 24 hours. This may be a satisfactory baseline for your spaghetti calculations for Al Bean, over.'
COLLINS 'Okay, I guess there's been worse.'

This 'day' would be short. No more course corrections were planned and only one television transmission remained. One that would provide a last opportunity for the crew to express their reflective thoughts on the flight. It began at 177 hr 32 min with Mike Collins showing his audience the interior of the spacecraft, remarking on the need for almost perfect performance from all the many items that made up the Saturn–Apollo Configuration, and ending with a message for the men who had worked long years to put man on the Moon: 'This operation is somewhat like the periscope of a submarine. All you see is the three of us, but beneath the surface, are thousands of others, and to all those, I would like to say thank you very much.'

Then Buzz Aldrin addressed all the people on Earth: 'Good evening. I'd like to discuss with you a few of the more symbolic aspects of the flight of our mission, Apollo 11. But we've been discussing the events that have taken place in the past two or three days here on board our spacecraft. We've come to the conclusion that this has been far more than three men on a voyage to the Moon. More still than the efforts of one nation. We feel that this stands as a symbol of the insatiable curiosity of all mankind to explore the unknown. Neil's statement the other day upon first setting foot on the surface of the Moon, 'this is a small step for a man, but a great leap for mankind,' I believe sums up those feelings very nicely. We accepted the challenge of going to the Moon. The acceptance of this challenge was inevitable. The relative ease with which we carried out our mission, I believe, is a tribute to the timeliness of that acceptance.

'Today, I feel we're fully capable of accepting expanded roles in the exploration of space. In retrospect, we have all been particularly pleased with the call signs that we very laboriously chose for our spacecraft, Columbia and Eagle. We've been particularly pleased with the emblem of our flight. Depicting the US Eagle, bringing the universal symbol of peace from the Earth, from the planet Earth to the Moon, that symbol being the olive branch. It was our overall crew choice to deposit a replica of this symbol on the Moon. Personally, in reflecting the events of the past several days, a verse from Psalms comes to mind to me. "When I considered the Heavens, the work of Thy fingers, the Moon and the stars which Thou hast ordained, what is man that Thou art mindful of him." '

To Neil Armstrong, who spoke next, 'responsibility for the flight lies first with history and with the giants of science who have preceded this effort.' Then, in Armstrong's view, it lay with the American people, to four Presidents, to Congress and to the industrial team that built the hardware. 'To those people tonight we give a special thank you, and to all the other people that are listening and watching tonight, God Bless you. Good night from Apollo 11.' And from a distance of almost 170,000 km, a final view of the Earth before the picture went down.

Shortly after, Gene Kranz took over from Cliff Charlesworth for Columbia's period of sleep and rest. But in the preceding hours weather predictions for the re-entry worsened and a decision was made to use the Command Module's lift vector to fly a trajectory that would carry the spacecraft 398 km further downrange, shifting it away from a region of heavy seas and thunderstorms. And then, with a comment to Mike Collins that he would get his 'chance at landing tomorrow,' the white team signed off as the crew began their sleep period. It was 182 hr 5 min.

The splashdown controllers headed by Milton Windler came into the MOCR six hours later, taking over from Gene Kranz who had had his last shift for this historic flight. A final midcourse correction opportunity would not be needed so the crew could lie in a little longer than required by the flight plan. But the crew called in at 189 hr 29 min and were soon heading into their last Apollo breakfast. Columbia was 75,000 km from Earth moving at a speed of 10,600 km/hr. Entry pads were read up to the crew with details of the modified re-entry profile; spacecraft 107 would fly 2,598 km from the point where it hit the atmosphere to splashdown nearly 14 minutes later. And a knowing comment from the Houston capcom to Mike Collins: 'Our crew is still standing by. I just want to remind you that the most difficult part of your mission is going to be after your recovery.'

At the console in the MOCR, capcom Ron Evans was joined by Jim Lovell, Bill Anders and Fred Haise. Deke Slayton was there also, as he always was at critical mission times. An Apollo had come back from the vicinity of the Moon twice before, but never with such a rich cargo, and re-entry represented the last hurdle before the mission could be labeled a complete success. Failure at this point was all too easy, and the tension built up as spacecraft 107 accelerated at terrifying speed toward the narrow slot in space.

At two hours before entry, range was 33,000 km at 15,000 km/hr; one hour prior to entry, the spacecraft was 17,000 km away and speed was 21,000 km/hr. In the Pacific the prime recovery force stood by, everything ready for the 'big one,' with TV cameras and newsmen standing close by the mobile quarantine facility in which the crew would spend their time before delivery to the Lunar Receiving Laboratory in Houston. Now it was time to power up the Command Module and sever the Service Module, jettisoning it to its own fiery entry from which it would reduce to ashes all the intricate systems and subsystems that had kept the astronauts alive all the way to Moon orbit and back.

Aligned at right angles to the flight path, explosive connectors severed SM-107 and the crew in the Command Module saw it from a window as the RCS thrusters fired, depleting maneuvering propellant and shunting the Service Module away from a possible re-contact with the descending manned spacecraft. With just fourteen minutes to go before re-entry, Apollo 11 was 1,500 km away from Earth traveling at 36,000 km/hr. It was early morning in the Pacific, noon in America, and late afternoon in Europe on Thursday, 24 July.

At the point of entry, Apollo was 122 km above Earth, slicing into the atmosphere at what would visually have been perceived

Lunar Receiving Laboratory, Building 37, at the Manned Spacecraft Center where samples and crew would temporarily reside.

to be an almost horizontal flight path, at a speed of 39,800 km/hr. Less than one-half minute later all communications were cut off as an intense thermal plasma built up around the Command Module, preceded by 'And 11, Houston, you're going over the hill there shortly, you're looking mighty fine to us,' from capcom Ron Evans. Less than four minutes later scratchy communications began to get through and seconds after that recovery aircraft reported a visual contact as the spacecraft streaked through the air.

Five minutes after re-entry, the USS *Hornet* picked the spacecraft up on radar and it was readily apparent that everything had gone just as planned. Apollo 11 was coming home. Mild oscillations with the spacecraft's attitude damped quickly when the drogue parachutes deployed nine minutes into the descent and when the mains popped out less than a minute later the crew felt a reassuring jerk. Sonic booms had been heard at sea as the Command Module ripped through the atmosphere but now it was lowering gently to the Pacific. Fourteen minutes after re-entry, Command Module 107 splashed down – and promptly flipped over into the apex down position. Minutes later the three inflated landing bags successfully righted the spacecraft into Stable I as the recovery forces were converging upon the three Moon men.

Deployed from recovery helicopter number 66, John McLachlan, Terry A. Meuhlenbach and Mitchell L. Bucklew were first in the water close by the gently pitching Command Module to fit a flotation collar and plug into the spacecraft's communication circuit via a small connector on the outside. Then they moved away and left another swimmer to don a biological isolation garment and to toss three such suits into the Command Module, opened for sufficient time to allow access. The three astronauts then donned the garments, equipped with soft helmets and a breathing filter, opened the hatch once more and climbed out into a raft tied to the flotation collar. Neil Armstrong was first out; Aldrin was last. They had been on the water just thirty-nine minutes. Twenty-four minutes after that they were on the deck of the *Hornet* and five minutes later they were in the mobile quarantine facility.

The hatch area of the bobbing spacecraft had been sprayed with decontaminant and then carefully closed up. Later, the lunar sample containers would be retrieved from the spacecraft and sent to Houston via separate routes on different aircraft – just in case of accident or planned sabotage. In Houston, the big plotboards that kept controllers glued to figures and trajectory data for more than eight days lit up with a TV picture from the recovery ship, a large plaque of the Apollo 11 badge, and a message across the center-spread which read: 'I believe that this nation should commit itself to achieving the goal before this decade is out of landing a man on the Moon and returning him safely to Earth . . . John F. Kennedy to Congress, May, 1961.' And on the top of the mission emblem: 'Task Accomplished . . . July, 1969.'

What the public never saw: Neil Armstrong strums a guitar and leaves his colleagues to accept an ovation from the Mobile Quarantine Facility.

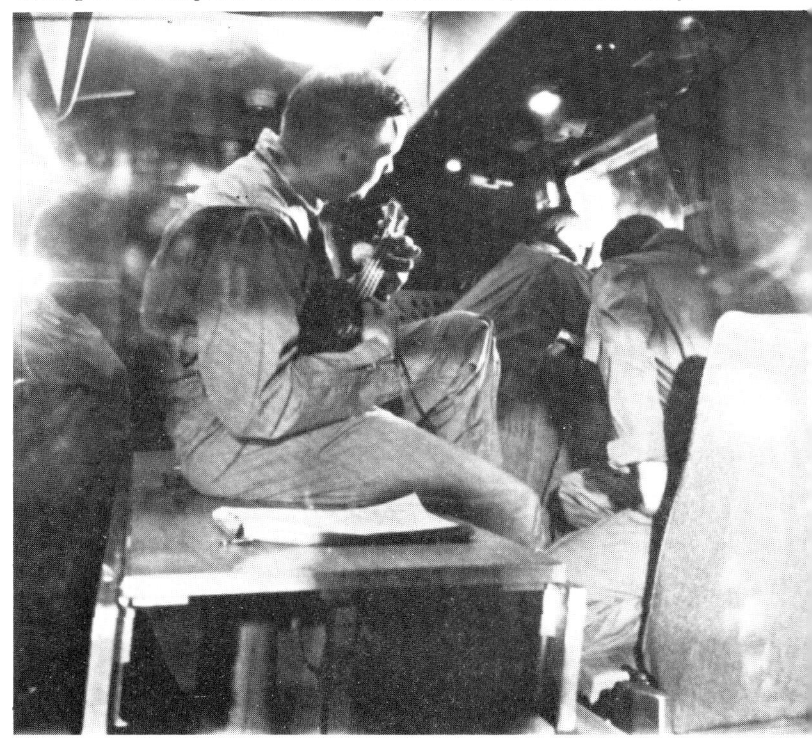

The Mission Operations Control Room was awash with surging technicians and controllers, jostling gently in transfixed euphoria at the achievement. Cigars were lit up all round when the crew stepped on to the carrier, flags hung from little stands on every console. Soon, the astronauts appeared at the large window on one end of the mobile quarantine facility. Mike Collins had grown a mustache but the others were clean shaven. All three pressed against the glass, and beamed very wide smiles.

President Nixon was there and stepped forward to talk to the men via a microphone. His personal enthusiasm shone through and for once the President of the United States seemed lost for words. Perhaps he was. It was an incredible end to a remarkable decade. And then it was time for Chaplain Pierto to say a prayer of thanksgiving for the safe return of Armstrong, Collins and Aldrin. But that too reflected the message of the times, calling on the same spirit that fired men to head for the Moon to 'inspire our lives to move similarly in other areas of need.'

Very early in the morning hours of Monday, 28 July, while it was still very dark in Houston, an aeroplane carrying the mobile quarantine facility landed at Ellington Air Force Base. Within hours the crew were in the Lunar Receiving Laboratory, there to spend a further thirteen days isolated from the rest of the world with 20 other persons. They emerged early. On 10 August, early in the morning, the men were released by Dr. Berry. It was the beginning of endless discussions, de-briefings, lectures, conversations, processions, ceremonies, appearances, and presentations that would fill their lives for many months. During the first full day out of quarantine, 11 August, all three crewmembers were given a day off, and none of them revealed where they were going to spend those hours.

But next day it began, the endless parade of formal events. First, a press briefing at the Manned Spacecraft Center narrating a 45 minute film of their mission. 13 August: all three, with their families, flew with Tom Paine to New York for a three-hour ceremony involving a greeting at City Hall by Mayor John Lindsay, a motorcade to the United Nations where they were met by Secretary General U Thant, a ticker-tape parade to John F. Kennedy International Airport, and a flight to Chicago. There, they were greeted by an estimated $3\frac{1}{2}$ million cheering people represented by Mayor Richard Daley who entertained them to a party at the Civic Center before the astronauts were required to address 15,000 young people in Grant Park.

Then a helicopter rushed them off to O'Hare International Airport for a flight to Los Angeles, where they were met by Mayor Samuel Yorty. After a brief address, they were rushed to the Century Plaza Hotel for a state dinner where Mr. and Mrs. Nixon dined with them along with other astronauts. Also there were the widowed wives of Robert Goddard, Gus Grissom, Ed White and Roger Chaffee.

16 August: a quarter-million persons watched the astronauts drive through Houston to a gala at the Astrodome where paper and confetti was hurled at the motorcade to a depth of one meter in places.

17 August: all three astronauts met the electronic media on the CBS program 'Face the Nation,' where they were asked to comment on the possibility of landing men on Mars. All three were reluctant to agree that such a commitment was a logical next step. Then there were appearances at their respective home towns, speeches before a joint session of Congress, and ceremonies at the Smithsonian Institution where a piece of Moon rock was presented to the Secretary, Dr. S. Dillon Riply.

In September they began a crippling world tour with an itinerary that included Mexico, Colombia, Argentina, Brazil, Canary Islands, Spain, France, Holland, Belgium, Norway, West Germany, Berlin, England, Italy, Yugoslavia, Turkey, Congo, Iran, India, Pakistan, Thailand, Australia, Guam, Seoul, Japan, Honolulu, and Canada.

Meanwhile, lunar samples were being distributed to several hundred scientists from all over the globe eager to begin a careful analysis of the material brought back from the Moon's Sea of Tranquillity. Nearly 150 principal investigators, each supported by at least one co-investigator, represented teams numbering at least 1,000 people directly involved in handling the 21 kg of samples returned by Apollo 11. It was not a large load, but not much would be needed to start the scientific ball rolling, and there would be flights to follow with grander objectives and a more scientific orientation. As it was, Apollo 11 proved two men could safely land on the lunar surface, walk around outside in the airless vacuum, be supported by their PLSS back-packs, and successfully return first to Moon orbit and then the Earth.

For the first time, all the many elements of Apollo had been tested in their design environment. The Lunar Module behaved exceptionally well and proved itself a generous vehicle by accommodating the products of human error. The back-packs, frequently considered by many at the time as an afterthought to success, were put through a full checkout, clearing the way for more ambitious lunar surface EVA operations. And once again, the CSM drove all the way to the Moon and back with hardly a rustle.

During and after the flight the media were full of comment, not all of it laudatory, and for several weeks stories continued to flow from the world's presses. Rhetoric ran higher than at any other time this century, and poets and painters gathered where appropriate to offer their contributions. At the Kennedy Space Center, artists were commissioned by NASA to sketch and illustrate the records of Apollo 11 with their own visual impressions. And throughout the nation reporters and journalists turned to the age-old precursors of their trade and broke into prose for descriptions of this consuming event.

Some samples from the first Moon landing in their return container.

Apollo 11 stands separate from the story of man in space, for it epitomizes what Man will inevitably aspire to in the midst of human tragedy and suffering. No better epithet can be found than that contained in the words of Wernher von Braun, at a celebration dinner in Huntsville two days after Neil Armstrong, Mike Collins and Buzz Aldrin returned to Earth:

'We worked together and together we accomplished our part of the mission. The Moon is now accessible, and someday, because of the beginning that we have made here, the planets and the stars may belong to mankind. This reach toward the heavens, toward the stars, can eventually loose the human race from the confines of this Earth and maybe even this solar system and give it immortality in the immense and never-ending reaches of space . . . life has left its planetary cradle and the ultimate destiny of mankind is no longer confined. When the Mayflower landed on American shores the pilgrims did not envision the nation that would eventually evolve. Neither can we truly say what will eventually spring from the footprints around Tranquillity base.'

CHAPTER TWENTY-ONE

New Horizons

The success of Apollo 11 did much to discolor public attitudes toward the space program in general, and manned space flight in particular. Most people, especially Americans, saw in President Kennedy's Moon goal what engineers and scientists failed to appreciate: that the Apollo objective was conceived as a limited, one-off, technology boost aimed to procur global prestige for flagging nationalism. To the man-in-the-street, space was useful as a diversion from more mundane, certainly less savory, events but once taken to its peak the gains outlived their usefulness and could be put away.

Not so to the several hundred thousand aerospace workers, or to the gathering ranks of scientific opinion that leaned more favorably than ever before to the notion of broad science on the Moon. Even before Apollo 11 put the first human footprints on the dusty Moonscape, plans were building up toward a further nine manned landings. But to the public, the very success was a good enough reason to call a halt and come home to more Earthly problems. It was the beginning of the end for grand dreams of manned Mars flights and massive space stations. But the end was still a few years off.

Several key NASA officials were disturbed that Apollo had a built-in momentum tuned to early technological success. They were concerned that timely, and adequate, attention to scientific exploration would be forgotten in the race to the lunar surface. Bob Gilruth was one of these men and fifteen months before the first successful landing, six months before the first manned Apollo flight, the MSC Director pressed George Mueller for attention to further activities on the Moon after the first landing. At that time satisfaction of Kennedy's goal by mid-1969 seemed a remote prospect and Gilruth felt confident that modifications to basic Apollo hardware could wait at least a year.

If there is anything an engineer likes better than creating a new machine it is to make that machine better. Lee R. Scherer, late of the Lunar Orbiter project as program manager at the Langley Research Center, was given the job in 1967 of heading the newly formed Lunar Exploration Office at Headquarters in Washington. His task was to come up with improvements and changes designed to extend the basic Apollo capability and expand the opportunities for lunar surface exploration. In the wake of Gilruth's expressed belief in a future expansion for Apollo Moon flights, Mueller approved formation of a Lunar Exploration Working Group in October, 1968.

Headed by John Hodge, late of the Manned Spacecraft Center, it was to develop new items of hardware – nothing as ambitious as the Apollo Extension System developed two or three years earlier, where Grumman's lander would have been adapted as shelter, taxi, lunar crane, or survey vehicle, but simple equipment geared to significantly improve mobility and stay time. There were a finite number of 15 Saturn V launchers, and if Apollo 11 could pull off the first Moon landing, nine more flights would follow, mused NASA heads early in 1969.

By this time, Hodge's men had come up with basic modifications to the LM that would extend its stay on the lunar surface. Also, studies of scientifically interesting sites revealed that almost all the places scientists wanted to go lay in dangerous areas or zones inaccessible to Apollo. From the outset, conservative mission planning required that initial Moon landing sites be located within 10° latitude of the Moon's equator, and that specific places be flat areas free of mountainous topography. The equatorial region could be reached without any plane changes after trans-lunar injection; but excessive increments of latitude would reduce propellant margins built in for safety.

As confidence with Apollo equipment built up during early 1969 several constraints were relaxed and NASA accepted the logic of flying the second landing to a site more scientifically interesting than the first, an area selected purely on safety criteria. If Apollo was to send its lander to precise locations, as would be mandatory for rough sites, some specific marker should be found by which to judge unequivocally the accuracy of touch-down. So, a Surveyor spacecraft site was suggested and approved as the second target. Surveyor had returned valuable information between 1966 and 1968 on the bearing strength of the lunar soil and one of the five successful unmanned soft landers was to be visited by the second Lunar Module flight to the surface late in 1969.

All three lunar research projects (Ranger, Lunar Orbiter and Surveyor) came too late to affect design of the Lunar Module significantly but after the first manned Moon landing the results returned by these valued precursors would play an increasingly useful part in choosing successive sites. If unused for engineering design, they were certainly vindicated by mission operations planning. George Low pressed Sam Phillips in April for a decision soon on when MSC could begin work on the modified Apollo vehicles. Phillips opened NASA coffers for initial studies and by the end of May, 1969, Administrator Paine received word that the Manned Spacecraft Center had been authorized to proceed with changes to production spacecraft from CSM-112 and LM-10 onward. These would support three days on the Moon, three periods of surface EVA, expanded science from orbit, and a mobile roving vehicle or lunar-flyer. They were expected to be ready by late 1970, but that date was too optimistic as proven later.

Until these expanded vehicles could fly, as Apollos 16 to 20, the second, third, fourth and fifth missions would comprise Mission H profiles. This was a new addition to Owen Maynard's Mission A-G sequence, where the latter represented the first lunar landing capability. Mission H was conceived as a maximum stretch of existing hardware before the three-day (Mission J) surface stay time equipment was ready.

In President Johnson's outgoing Fiscal Year 1970 budget, a recommendation made public in late January, 1969, for the period beginning 1 July that year, there was no money for anything beyond Apollo 14. Four manned Moon flights were funded, just in case the first scheduled attempt – Apollo 11 – failed to make it. Within two months of assuming office, Richard Nixon reviewed NASA's overall FY1970 budget submission and, while reducing the total amount added $79 million to the meager $11 million assigned by Johnson to future Moon exploration within the capacity of hardware already built or under contract. It was this single gesture that gave NASA the ability to transform Apollo into a modest lunar science program.

The H-series flights would remain on the Moon for a maximum 36 hours, provide two astronauts with two surface EVA sessions of up to six hours each, and return to Earth with up to 40 kg of lunar material – twice that brought back by Apollo 11. And it would all be accomplished with maximum utilization of the existing CSM and LM designs. The J-series missions would permit lunar surface stays of up to 78 hours, support three two-man EVAs of six hours each, and bring back up to 100 kg in samples and soil.

But if Nixon favored maximum exploitation of existing equipment, he was not so keen on developing a vigorous Apollo Applications Program, AAP flights designed to pioneer the development of Earth-orbiting space stations. Under the proposed Fiscal Year 1969 plan NASA would begin AAP operations in 1970 with a 'wet' workshop converted from a Saturn IB second stage propelled into orbit by its own propulsion. Teams of astronauts would occupy the assembly on three missions of up to 56 days in duration, followed by a further three visits in 1971. A 'dry' workshop would be launched by Saturn V in 1972, fitted out on the ground and propelled into space by the first two stages of the launch vehicle. It, too, was to be visited by teams of astronauts for one-year manned operations.

As detailed earlier, actual AAP funds for FY1969 dropped by 52%, so out went the comprehensive schedule, already a dim reflection of former plans. By early 1969, when FY1970 was being

Boeing's Molab was proposed during the early days of the Gemini Program → and although replaced in the Apollo era by more suitable configurations, nevertheless provided useful research information on Moon vehicles.

Headed by Lee R. Scherer, the Apollo Lunar Exploration Office paved the way for later developments of basic hardware and the design of mobile vehicles.

discussed in Congress, NASA had abandoned all hope of the Saturn V launched dry workshop, and cut in half the wet launch concept. While Apollo was to support two Moon flights in 1969, three each in 1970 and 1971, and two in 1972, AAP was to proceed toward a wet workshop launch on a Saturn IB late in 1971 to support three manned periods of habitation lasting 28 days, 56 days, and 56 days, respectively. And that was that. Both Apollo lunar exploration and Earth orbit AAP flights would end about the middle of 1972. But even as budget hearings got under way, officials expressed doubt that even this depleted schedule could produce initial AAP flights in late 1971, believing 1972 to be a more realistic expectation.

One casualty of broadening plans for Earth orbit NASA missions was the Air Force MOL program, itself a Phoenix from the ashes of Dyna-Soar. MOL had been dragging its feet for two years. In 1967 the first manned operation was put back two years from the original flight date of 1968 due to technical difficulties. At the end of 1968 inevitable pressure on other military programs, plus an unwillingness on the part of the outgoing Johnson administration to compromise the policies of the Nixon team led to a decrease in the development budget. By early new year, 1971 was considered the earliest MOL would fly.

On 10 June, 1969, the Defense Department announced it was cancelling the entire program, at a saving of $1,500 million over the next four years and in August NASA received seven astronauts who had been in training for the manned missions: Lt. Col. Karol J. Bobko, Lt. Col. Charles G. Fullerton, Col. Henry W. Hartsfield, and Col. Donald H. Peterson, from the Air Force; Cdr. Robert L. Crippen, and Cdr. Richard H. Truly, from the Navy; and Lt. Col. Robert F. Overmyer, from the Marine Corps. They were to train for space station operations of the future: possibly AAP but certainly the big manned facility planned for launch in about 1975, and as pilots for the reusable space shuttle.

It was inevitable that considerable interest should be generated in Congressional hearings by the obvious conclusion of Apollo operations in 1972. For more than a year NASA too had been giving considerable thought to how it should conduct the post-Apollo era. Moon landing flights were a suitable focus for popular opinion but they could not go on for ever and the agency found widespread agreement in its ranks for an Earth orbit capability similar in concept to that proposed back in the late 1950s when a small group of Space Task Group personnel put together a post-Mercury plan for space stations and Apollo ferry vehicles.

As it turned out, and as it was viewed in the closing months of the next decade, Apollo and the Moon landing goal had been a digression. Now that the objective was within sight, it was time to return to former strategy and build a permanent manned flight capability in tune with current needs. One that would build upon the vast storehouse of technical knowledge stimulated by Apollo, but one which was likely to bring greater benefits to a broader group of humans.

But not everybody agreed with altruistic attitudes toward the application of a space-faring capability. Plucked from comparative obscurity by Richard Nixon, Vice President Spiro T. Agnew was charged with the new-found capacities of high office and following the launch of Apollo 9 pledged Kennedy Space Center workers that he would 'lend whatever thrust I can to nudge the President into an awareness of what I consider of overriding importance.' When Apollo 11 thundered from the launch pad it gripped Agnew and propelled him to new heights of expectation, saying that 'It is my individual feeling that we should articulate a simple, ambitious, optimistic goal of a manned flight to Mars by the end of this century.'

As if in reciprocation, Edward Kennedy said that same day 'I think . . . the space program ought to fit into our other national priorities.' The prospect of heading for Mars with manned spaceships certainly seemed absurd, even in the unreal wake of Man's first Moon landing. Few were willing to support such a goal openly, but several officials in NASA believed the venture would serve to catalyze remote areas of the space program into a formidable tool for more beneficial use on Earth. George Mueller in particular was concerned that momentum would be lost if the only ambitions and aspirations were centered upon purely altruistic sentiments. Holding to the belief that benefits would increase if additional attention was paid to the sharp end of technology, Mueller proposed an integrated space program for the post-Apollo era, the period after Apollo Moon and AAP Earth orbit flights were completed in 1972.

His plan was based in part on a space task group report published in September 1969. The group comprised Bob

Mueller's integrated space program envisaged development of four primary elements from which would grow colonies on the Moon and Mars, orbiting bases and large space stations around Earth.

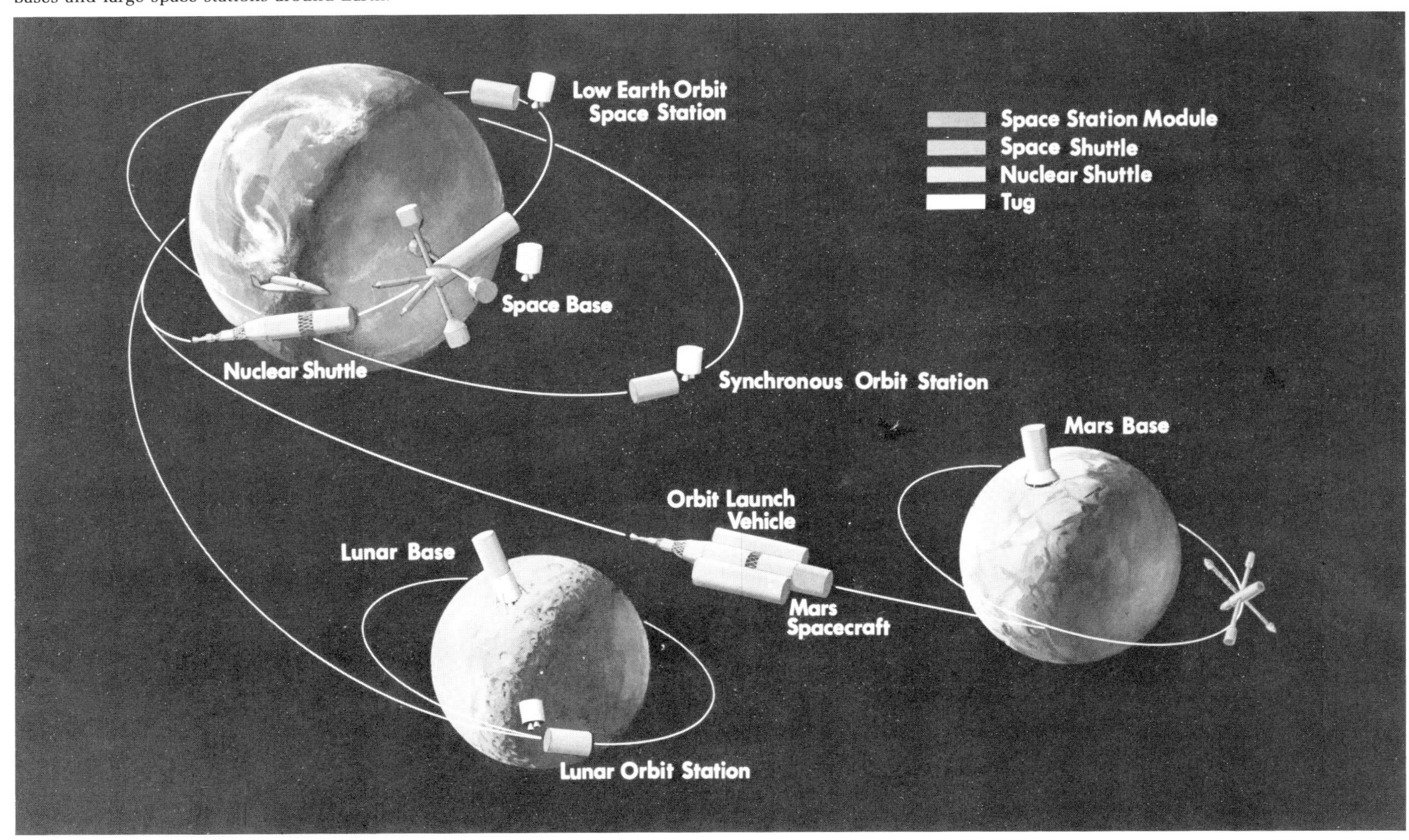

Seamans, now Secretary of the Air Force, Tom Paine as NASA Administrator, and the President's science adviser, Lee A. DuBridge; observers included the Under-Secretary of State for Political Affairs, Alexis Johnson, the Chairman of the Atomic Energy Commission, Glenn Seaborg, and the Budget Bureau Director, Robert Mayo. Vice President Agnew was chairman. The task group was set up in a memorandum from President Nixon dated 13 February, 1969, in which he requested 'definitive recommendation on the direction which the US space program should take in the post-Apollo period.'

Instrumental in restoring NASA plans for Apollo Moon exploration, Nixon had severed virtually all funds for follow-on projects until the task group made its recommendation. When it did it spoke of current 'achievements as only a beginning to the long-term exploration and use of space by man. We see a major role for this Nation in proceeding from the initial opening of this frontier to its exploitation for the benefit of mankind, and ultimately to the opening of new regions of space to access by men.'

Most of the recommendations came from existing strategy formulated in the preceding year, principally by Bob Gilruth and George Mueller. Gilruth in particular was concerned about the extended use of existing hardware, believing NASA should invest and exploit new avenues based on extant technology but capable of pushing into hitherto unexplored avenues. In December 1968, the Nasa Science and Technology Advisory Committee for Manned Space Flight had met at La Jolla, California, to discuss space policy in the period 1975–1985, the post-Apollo decade. The integrated space program that emerged from La Jolla, and the work it inspired across the following months, pivoted on three prime requisites: reusability; commonality; and cross-application.

Reusability called for an end to wasteful expendables in the space program, citing the availability of new technology that could generate space shuttle vehicles capable of flying regularly between Earth and space many times, landing on conventional runways and returning to Earth with a nominal payload. Reusability also implied deep-space propulsion systems stored in Earth orbit for refuelling and re-use. This embraced the nuclear shuttle, a comparatively low-thrust rocket motor using a nuclear reactor for the heat necessary to vaporize a fuel; without combustion there would be no need for an oxidizer, and with hydrogen as propellant the nuclear shuttle would significantly increase the payloads transferred from Earth to Moon, or Earth to the planets. Plying between Earth orbit and Moon orbit, for instance, it could transport massive cargoes assigned to support space stations and surface bases.

Commonality addressed the problem that so far specific vehicles had been developed for unique roles. That would have to stop in the space program of the future, replaced instead with a family of spacecraft embracing a variety of applications. For instance, a large habitable structure designed to support astronauts in Earth orbit could form the core of a lunar orbit station, or indeed a lunar surface base; as a self-contained environmental support system it was applicable to habitability almost anywhere in space. Commonality was also a feature of the proposed nuclear shuttle, for it too could serve as propulsion for almost any deep-space requirement of the future.

The third component – cross application – sought to integrate manned and unmanned programs so that vehicles developed for one category could be easily adapted to the needs of the other. For instance, the space station core would conceivably support unmanned activity on the Moon by containing a large inventory of scientific instruments; common satellite 'buses' could be built on production lines for a variety of applications: common objectives could be assailed by manned and unmanned systems alike rather than find separate applications for each.

That was the philosophy of the post-Apollo program. But what of the hardware? Here again, Gilruth and Mueller worked on issues of transportation and habitation, with von Braun's Marshall Center instrumental in structuring the large engineering projects envisaged. First, NASA would proceed with development of a reusable shuttle to replace existing expendable launchers in the medium-size payload class. Second, the agency should continue development of the nuclear rocket motor, work on which had been under way for a decade already. Third, a basic 12-man space station should be developed for re-supply and servicing by shuttle. The shuttle and the station would be developed concurrently. The station would then form the basis for a lunar orbit station and a lunar surface base, core sections being transported to the Moon by nuclear shuttle (essentially a rocket stage and not, like the shuttle, equipped with wings to fly back down through the atmosphere).

From this would emerge the first Moon colony, a base with probably six personnel equipped for lengthy scientific study of the lunar surface and substructure. By this time a space tug would have emerged, a small propulsion unit capable of supporting either an unmanned control station for automatic use with computer controls, or a manned cabin supporting a two-man crew. The tug would move around in Earth orbit, shunting experiment modules to the stations, moving cargo from shuttle to station, or propelling payloads into high orbit. These developments would carry NASA through the decade of the 1970s and place it in an operational posture for more ambitious developments in the 1980s. The lunar base would be expanded to a 24-man facility, space station core modules would be placed in geosynchronous orbit (like an old AAP dry workshop proposal), and the manned Mars landings would begin.

It was this latter activity that provided Gilruth with his grand objective as a forcing house for continued technological development. The integrated space program would conspire to provide all the equipment necessary for a Mars flight by seeking to build up a viable Earth orbit and lunar science and applications program first: the space station core module would serve as a habitable base for the Mars crew en-route; the nuclear shuttle would propel the assembly out of Earth orbit; the conventional shuttle would fly back and forth to Earth's surface, feeding supplies to the Mars hardware before departure. In this way, when the decision came to fly to Mars, NASA would have to develop only the landing vehicle – the Mars Excursion Module (MEM) – to carry the crew from Mars orbit down to the surface and back. Additional nuclear shuttles would be taken along to brake the assembly into Mars orbit and return it to Earth. In this way, there would be no need for a crash program with peak funding levels many times the normal agency budget. All the hardware, except the Excursion Module, would be in use and need minimal modification to support the Mars mission.

Launch windows appeared in 1981, 1983 and 1986. If accommodated at the beginning of the decade initial landings would build toward a Mars surface base with 48 people by the end of the 1980s. Meanwhile, expansion of the Earth-orbiting space station into a space base supporting up to 100 persons would evolve from additional elements lifted to the station by Saturn V type launchers or modules sent up by shuttle. Plugged onto the station core, additional modules could be tailored to the needs of the planet: unique manufacturing processes could be exploited; land use management could be advanced; new weather observation systems set up; ocean surveillance made practical. The potential list of benefits was endless.

Following publication of the space task group report, manned space flight officials set about a vigorous program of public education in the practical benefits of an expanded space program. How expanded? That would depend upon budget allocations, but all these projects could be developed with funding levels no higher than the peak NASA budget years in the mid-1960s. To a certain extent various proposed money levels would alter the timetable rather than the variety of hardware elements. Below a certain level, however, almost all the manned operations would be uneconomic since their utilization would not amortize the investment. What the task group did was to present the President with the 'definitive recommendation' requested by the White House. Nixon would consider the proposed space program of the future and make a decision within a few months.

For its part, NASA had already moved along with feasibility studies of a 12-man space station, that work having emerged from several industrial study contracts in the preceding years. But feasibility studies on the Earth-to-orbit shuttle still had to be made and in February, 1969, NASA awarded contracts to North American Rockwell, McDonnell Douglas, General Dynamics, and Lockheed. They were each to examine a range of different configurations and report back to NASA in September with preliminary conclusions about the optimum configuration for the reusable transporter. In July, meanwhile, NASA chose North American Rockwell and McDonnell Douglas to perform 11-month definition studies – a step beyond feasibility work – on the 12-man space station core module, a structure probably supporting four to six separate floors and launched by a Saturn V type vehicle.

In an uncertain climate of Presidential indecision, while NASA waited to receive word from the White House on the new administration's space policy, tentative plans envisaged a start to space station operations by about 1975, with the shuttle becoming operational the following year. The nuclear shuttle was well on its way to an operational role by about 1977 and the tug was being studied at the Marshall Space Flight Center.

For its part, the Manned Spacecraft Center was fully occupied with preparation for the second Moon landing flight with hardware in the final stage of readiness at the Kennedy Space Center. Anticipating an early need for AS-507, possibly in September if Apollo 11 failed to put two men on the surface, all three stages of the giant Saturn V were erected in the Vehicle Assembly Building while 506 was being rolled to the pad and

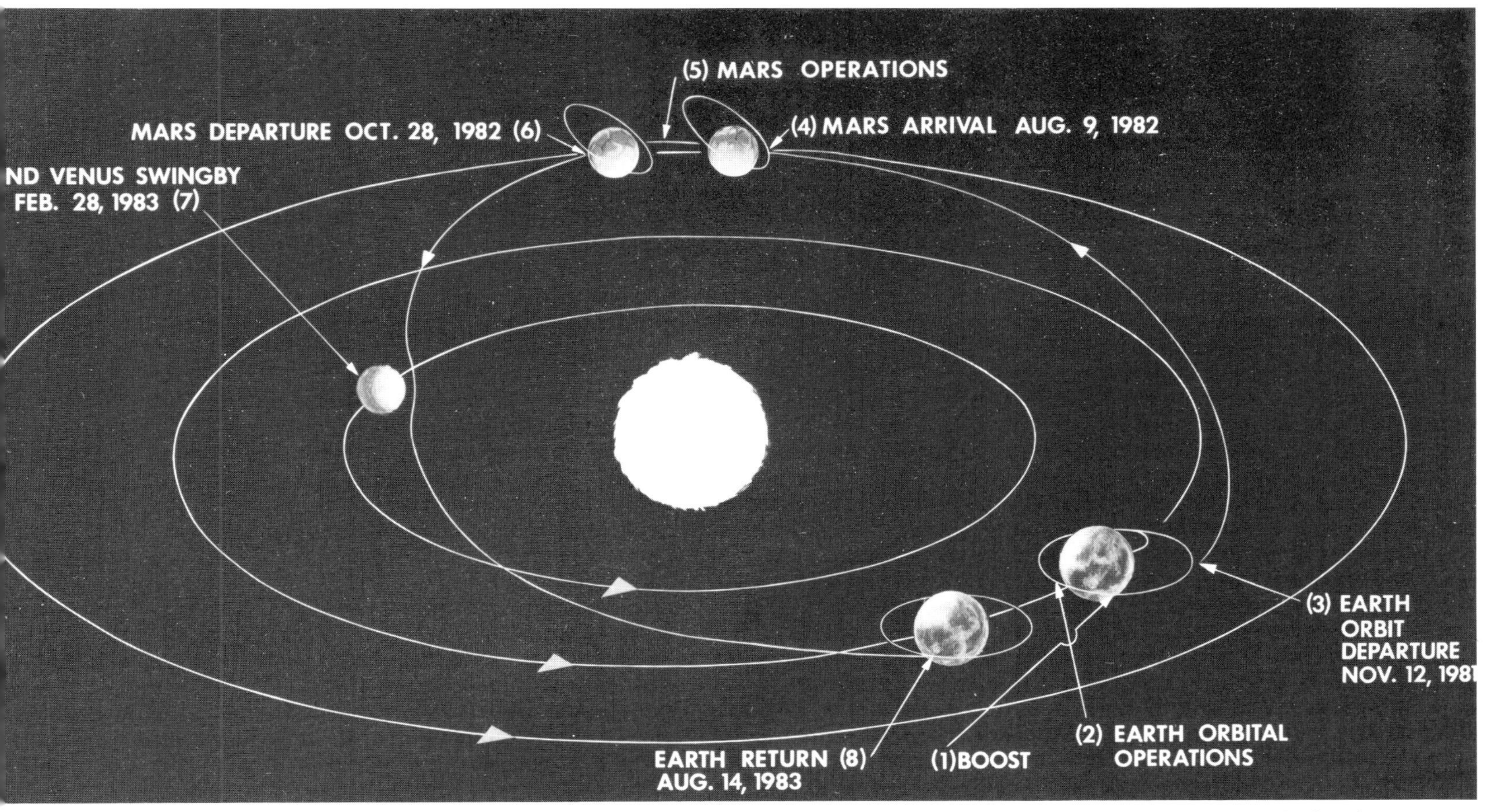

This schedule for initial Mars exploration displays the earliest launch date (12 Nov. 1981) from Earth orbit proposed by the Gilruth-Mueller think-tank.

Apollo 10 was performing its pathfinder role around the Moon. By early July, the payload was assembled in the VAB but after Armstrong and Aldrin successfully splashed down in the Pacific the pace slackened; never again would the Kennedy Space Center work to such tight launch schedules. Instead of flying in September, Apollo 12 would make its way to a landing site on the Moon's western quadrant beginning with a launch on 14 November.

It was September before 507 rolled to pad 39A and by late October the Count Down Demonstration Tests had been satisfactorily carried out. CSM-108 was almost identical to the previous combination, and LM-6 was not very different to the spacecraft called Eagle. It did, however, carry a full ALSEP array of scientific instruments, complete with a small nuclear generator attached to the side of the Descent Stage in a graphite cask to inhibit thermal flow; the case was shock-tested and capable of withstanding re-entry should the Lunar Module be part of an aborted configuration during ascent, thus preventing radioactive contamination of the atmosphere. Names chosen for the spacecraft were modest: the CSM was to be called Yankee Clipper, and the LM would answer to Intrepid.

Apollo 12 would head for a touchdown spot 340 meters from the Surveyor III spacecraft; the unmanned vehicle had bounced and slid to a stop on the inner sloping wall of a 200 meter diameter crater in the Moon's Oceanus Procellarum, some 1,537 km west of the Apollo 11 landing site on 20 April, 1967. But the first manned landing had been precisely 6.87 km off target and the prime objective of the second was to demonstrate a capacity for pinpoint landing. Astronauts Pete Conrad and Al Bean were to clip off parts of Surveyor for return to Earth in an attempt to unravel secrets of lengthy exposure to the hostile space environment; Dick Gordon would circle the Moon during the 32 hr period Intrepid remained on the surface.

Mission planners thought they knew why Apollo 11 landed so far off course: the descent had gone well, apart from the overloaded computer, but the entire profile had been displaced 7 km west by inaccurate data on the exact position of the LM. So Intrepid would receive a final guidance update from ground computers on the 14th revolution, a state vector and a radius-to-landing-site figure precisely aligning the longitude at which to begin powered descent. By this time the LM would be on its way down to the pericynthion height of 15 km but there would just be time to uplink final trajectory measurements based on tracking performed after the spacecraft came round the Moon's east limb.

Apollo 11's LM had only 17 minutes between appearing on the communications line to igniting the engine for powered descent. Because the landing site was further west, Apollo 12's Intrepid would be visible for 37 minutes before powered descent and the Mission Control Center would hopefully get a state vector up within twenty minutes of acquisition. That should pin down the precise point at which to begin the descent and give Intrepid a good chance of landing at the appointed location, north east of Surveyor III. But landing site navigation was not the only innovation for Apollo 12.

Unable to reach that site on a free-return trajectory, one that would automatically loop the spacecraft around the Moon for a safe return to Earth if nothing was done to brake the vehicles into Moon orbit, Apollo 12 would fly a non-return trajectory in the final hours of trans-lunar coast. But it would be sent on its way along a free-return trajectory until the LM had been extracted from its shroud atop the S-IVB, when Apollo's SPS engine would change its path to a non-return type. In the case of Apollo 12's trajectory, the spacecraft would miss Earth by more than 91,000 km after looping round the Moon. This 'hybrid' trajectory was insurance. If the SPS failed, Apollo 12 would still come back; if it failed after extracting the LM the Descent engine could be used to restore a free-return capability. Only when a second engine was known to be available would Apollo 12 switch tracks to a non-return path. This form of hybrid trajectory would be used throughout the remaining Moon missions.

The main science inputs to Apollo 12 centered on the deployment of five ALSEP instruments and a limited start on geologic activity. ALSEP would measure things like the local magnetic field levels and direction, protons and electrons streaming from the Sun, and any trace gases drifting across the surface. A solar wind collector would also be set up, as it had been on Apollo 11, for return to Earth along with Moon samples gathered during the first EVA. Both periods outside Intrepid were to last 3½ hours, the second providing time for the crew to make their way to Surveyor and retrieve sections of it during a geologic traverse on which they would describe and gather interesting rock samples.

With several scientist-astronauts in the program there was pressure for one of these men getting an early flight but, as the Lunar Module Pilot also had to be able to fly the LM, Deke Slayton and Al Shepard, whose job it was to select the crews, were reluctant to bring in men who were comparative newcomers to flying anything, let alone manned lunar spacecraft. Their view, quite rightly so, preferred lost science to a crashed LM.

Final preparations went well for Apollo 12 as a new Director of Launch Operations, Walter J. Kapryan, took over from Rocco Petrone. Sam Phillips resigned from his position as Apollo Program Director in August to join the Defense Department's space program and Petrone moved into his shoes. In September, Roderick Middleton resigned as Kennedy Space Center's Apollo Program Office Manager, and Edward R. Mathews moved in, a one-time head of the Saturn IB Systems Office. That month too, George Low vacated his position as MSC Apollo Program manager, first to stand in a special advisory capacity to Director Gilruth, then to receive nomination for the post of NASA Deputy Administrator under Tom Paine. He would be sworn in by the end of the year. Apollo 9 Commander Jim McDivitt took over Low's former position.

Launch window for Apollo 12 was a brief 3 hr 5 min. If delayed beyond that time a two-day postponement was inevitable, when the mission would head for a touchdown at another site further west. Consequently, tensions rose just two days before the planned 14 November launch date when technicians discovered one of the fuel cells' cryogenic hydrogen dewars had sprung a leak during a filling operation. Fortunately, CSM-109 was on hand to support Apollo 13 and it was replaced with a hydrogen tank from that spacecraft. Several operations were needed to get the old tank out and the new one in, including removing electrical wires and plumbing, but the job was done on time for a predicted 14 November launch.

Weather was poor for the launch of NASA's second Moon landing attempt. Rain showered down between thunderstorms on the day before the flight, and cold fronts pushed up across the Cape. Nevertheless, during the dark night preceding lift-off, the weather cleared and Mission Director Chester M. Lee gave a 'go' for the terminal countdown. At 6:05 am on the 14th, Conrad, Gordon and Bean were woken by Tom Stafford to be examined by two doctors in the MSOB. Pronounced fit, they received an update on the weather and moved to the pre-flight breakfast appointment with Stafford, Jim Irwin, Jim McDivitt, Paul Weitz and Chuck Tringali, head of the support training group for Apollo 12. Conrad had been given a mascot – a life-size stuffed gorilla – which sat in on the breakfast of steak, eggs, orange juice, coffee and toast.

Several celebrities were on hand for the launch, not least of whom was the President of the United States Richard Nixon, the only Chief Executive to witness an Apollo launch. Only a day before, Conrad, Gordon and Bean had howled across the Cape in their T-38 jets, acknowledging the effort made to get the mission under way and saluting Walt Kapryan's men. But now two aircraft were patrolling quietly over the Cape, measuring the approaching weather systems which threatened to dump showers on the rocket at the assigned launch time of 11:22 am.

The crew were on their way to complex 39 by 8:35 am and at T−2 hr 50 min entered spacecraft 108: Conrad on the left couch, Bean on the right, Gordon in the center. It was raining when the five giant F-1 engines thundered into life. Conrad saw water trickling down across three Command Module windows inside the boost protective cover and knew it was to be a wet lift-off. Spectators – more than 3,000 guests had been specially invited, stood beneath umbrellas or simply let the rain do its worst as the Saturn V lifted toward the clouds only 240 meters above the pad.

Launch Operations Manager Paul Donnelly conferred several times with Kapryan about the state of the weather but no mission rules were actually infringed by what was reported and they agreed to let the big bird go. There was ebullient comment from Pete Conrad in the seconds before near disaster struck Apollo 12:

CONRAD 'A pitch and roll program, and this baby is really going.'
CAPCOM 'Right Pete.'

CONRAD 'Roll's complete.'
CAPCOM 'Roger Pete.'
CAPCOM 'Mark 1 Bravo.'
CONRAD 'Roger; we got you on that.'

Just 36 seconds into the flight, at a speed of 380 km/hr and a height of 1.8 km, the crew saw a bright white light flash around them, and Conrad saw the spacecraft's caution and warning panel light up like a Christmas tree. On the ground, a few observers saw lightning snake down to the pad steelwork along the Saturn's steaming exhaust plume. In the Command Module and on the ground the warning tones sounded and from an ascending spacecraft gathering speed on top the world's largest rocket, the most important navigation device – the guidance platform – then went out, as did several other critical systems:

CONRAD 'Okay, we just lost the platform, gang; I don't know what happened here; we had everything in the world drop out.'
CAPCOM 'Roger.'
CONRAD 'Fuel cell lights, and AC bus light, fuel cell disconnect, AC bus overload, 1 and 2 out, main bus A and B out.'

The electrical strikes occurred until 52 seconds, when another fork of lightning snaked to the launch pad. Had Saturn's guidance platform gone out the mission would have been aborted immediately; without guidance signals the giant rocket would have careened across the sky uncontrollably. In Houston, flight director Gerry Griffin went around the room polling his men for staging. Everything appeared good from the ground, although the crew would have to re-set the fuel cells and get the platform realigned when they reached orbit. The first stage came away as planned while Conrad brought the fuel cells back on.

'We are weeding out our problems here; I don't know what happened; I'm not sure we didn't get hit by lightning,' said Conrad as the S-II stage lit up. 'Okay, I have a good GDC and Al has got the fuel cells back on, and we'll be working on our AC buses,' confirmed the Commander. 'I think we need to do a little more all-weather testing,' said Conrad. Seconds after releasing the launch escape tower, another report from Pete Conrad to the effect that 'We've got an ISS light on and we have a cycling CO_2 partial pressure high, which don't bother me particularly, and we have reset all the fuel cells, we have all the buses back on the line, and we'll just square up the platform when we get into orbit.'

When offered the view that 'that's one of the better sims (simulations), believe me!' the Houston capcom came back with 'We've had a couple of cardiac arrests down here too, Pete.' Thrust on the S-II was noticeably rougher than on the first stage and Conrad commented that 'she's chugging along here, minding her own business.' The center engine shut down as planned and one and one-half minutes later the remaining four J-2s also quit on schedule. The S-IVB lit up and Apollo 12 was on the final leg into orbit.

Conrad was a little concerned over the rolling platform,

Apollo 12 astronaut Alan Bean adjusts the portable life support system for Charles Conrad during rehearsals at Cape Canaveral. Note the portable hand tool carrier.

seeking confirmation that controllers on the ground were thinking 'about how we're going to get that thing, cause it's just drifting; just floating.' Now that spacecraft 108 was riding on the rocket's terminal stage time could be spared to cage up the gyroscopes and reset the platform power. The undervolt had momentarily taken energy from several systems but the inertial platform was unable to return to its alignment without a deliberate sequence put in by the crew. But it was good news from Gerry Carr that 'You've got a GO orbit; you're looking good,' when the S-IVB shut down and put them in to a 185 × 190 km path. Much of the first orbit was taken up with realigning the guidance platform and checking systems to prepare for trans-lunar injection.

Shortly before re-igniting the S-IVB on the second revolution, controllers noticed a state vector error in the Saturn's guidance computer but decided to ignore it. The only consequence would be that the initial, free return, trajectory would swing the spacecraft closer to the Moon than originally planned but since a mid-course correction was to be made after Intrepid had been pulled from the S-IVB that was of little consequence. On the second pass over the United States Conrad was concerned that his thrusters were apparently not operating. When set to minimum-impulse he could neither see any plumes out the window nor hear any combustion up through the spacecraft's structure. But Neil Armstrong was in the MOCR and Gerry Carr told Conrad that he too had not heard the thrusters at low levels of activity. Engineering data indicated they *were* working and that all was well.

The burn out of Earth orbit was monitored by two ARIA aircraft over the Pacific, relaying voice communication and recording telemetry for playback later; ARIAs could not relay engineering information real-time. Hawaii would pick up the assembly shortly before S-IVB cut-off and send telemetry to Houston. Voice contact was scratchy through the ARIAs and solid communication began only when the spacecraft appeared over the Hawaii horizon, climbing high away from the Earth. At cut-off, Apollo 12 was more than 350 km out and speeding along at 38,860 km/hr. But from the second the engine stopped, Earth's gravity would drag that speed down all the way to the equigravisphere.

Nevertheless, when the time came to separate from the S-IVB just 10 minutes later, Apollo 12 was already more than 7,000 km above the planet. And from Al Bean: 'We can see the whole United States.' But water on the windows during ascent had frozen and partially obscured vision. Nevertheless, Pete Conrad swung spacecraft 108 round to face Intrepid, still within its shroud on the S-IVB, and with the television camera running to show Earth viewers the docking operation, moved in upon the LM's conical drogue. It had taken just nine minutes to turn Apollo round and link up with the LM. Now, a few quick checks before releasing both vehicles from the Saturn stage.

First, the forward Command Module hatch was opened after oxygen had been pumped into the tunnel, then a check on the 12 docking latches; the dynamic force of physically separating from the S-IVB would cause minor shock or vibration that could damage a partially snagged coupling. Three latches were found only half-cocked so Pete Conrad closed them up and then plugged in electrical umbilicals that would transfer power to the LM when necessary. With the LM pressurized the two docked vehicles were in an acceptable state for separation, that event occurring at 4 hr 13 min. The S-IVB was now on its own with a set of tasks planned to give it a slingshot impulse around the Moon's west limb and on into solar orbit like its predecessors.

First, residual quantities of oxygen would be dumped through the J-2 engine bell, then the attitude control thrusters on the stage would fire in unison to move the trajectory physically still further. But a multiplicity of errors set in. With a slightly incorrect state vector prior to TLI the path on which both S-IVB and spacecraft had been placed was one which would carry both structures to within 846 km of the Moon rather than the planned 3,428 km. The consequence of the same state vector error during slingshot activities propelled the S-IVB to a path which would carry it 5,724 km past the Moon more than two hours after the manned vehicles got into lunar orbit. But instead of accelerating past the Moon to solar orbit, the S-IVB would have an orbit about the Earth approximately 165,000 km at the low point and 850,000 km at the high point: an ellipse, with apogee about twice the distance of the Moon from Earth.

Apollo 12 too was on a path that would carry it within 846 km of the Moon, but a course correction would be made nearly 31 hours into the mission, transferring the free return trajectory into a non-return path. Pete Conrad was overcome with fascination at the spectacular sight of liquids and gases venting from the S-IVB, still within visible range of the CSM: 'Boy, is that thing venting, what's it keep venting anyhow, Houston? Keeps throwing out big clouds of something . . . it's throwing stuff off the sides and in the back like crazy . . . looks like it's got a halo around it now. . . . It's really weird Houston, there is something that is venting radially, and then there is something that is venting along the axial axis . . . right now it reminds me of some guy standing back there with a water hose just spraying it in any old direction.'

Television views were still up and the crew brought the camera in upon themselves while minor activity kept them occupied: 'They say this thing shakes, rattles and rolls when you fire the thrusters. It's like being on a jerking train.' And on the question of the electrical transients during ascent, Pete Conrad mused over the sensibilities of his LM Pilot: 'I'll tell you, it's a terrible way to break Al Bean into space flight.' Shortly before the S-IVB began its evasive maneuver Carr called up with a suggestion from the flight controllers:

CAPCOM 'We've been thinking here a little bit, and we'd like for you to consider a proposal here. It's the idea of getting into the LM tonight before bedtime, and going through the housekeeping portion of your checklist short of the communications, and powering up the CMC and giving us a check and an E mark dump.'

GORDON 'Yes, that sounds like a good idea. We've been up here talking about what the launch may have done for the LM, and I think we can do that between two P23. What do you think of that?'

CAPCOM 'We're not going to do MCC_1, Dick. It looks like you won't need it, so you can do that during the time you would normally be doing an MCC_1.'

GORDON 'Okay. It sounds good. We really don't have any place to go tonight so we don't mind working late.'

It made good sense to open up Intrepid and give it a quick check, and the interval between two scheduled P23 navigation sighting periods would be suitable in the absence of a need for the first mid-course correction opportunity. Now that Intrepid had been extracted, and its Descent engine was available if needed, the SPS engine could fire Apollo 12 on to a non-return trajectory at the MCC_2 opportunity.

Six hours into the mission, flight director Pete Frank took over the MOCR and Ed Gibson relieved Gerry Carr at the capcom console. Telemetry from the two spacecraft seemed to indicate that the abort sensor assembly – part of the Abort Guidance Section – had a failed heater circuit breaker which could have popped out during ascent and because the thermal limit on that component was 8 hours, Houston suggested that Conrad should hustle along for a LM checkout before that time. But first, a lengthy list of abbreviated checkout procedures was read up by Gibson and copied down in the spacecraft.

Shortly after an elapsed time of 7 hours the crew busied themselves with opening up the tunnel between Yankee Clipper and Intrepid, before Pete Conrad slipped through into the LM at 7 hr 20 min quickly followed by Al Bean. LM power came on sixteen minutes later, battery current was observed to be good at 34.2 volts on all six and the VHF and S-band communications equipment looked satisfactory. Within one hour of entering Intrepid the crew were on their way back to the Command Module. Everything checked out as hoped for; there was nothing to cause concern for the full mission activity.

At an elapsed time of 8 hr 30 min Apollo 12 was being set up for the passive thermal control 'barbecue' mode, set to roll slowly under the spit-roasting Sun, after the probe and drogue had been re-installed. But technicians scanning telemetry from the LM detected an increase of just 1 amp in the power drawn from the CSM. Suspecting that the crew had left a circuit breaker in the wrong position during checkout, Gibson asked Conrad to go across to Intrepid a second time and check the switches. Conrad suggested that the current increase might be due to the interior light still being on.

Fitted with a microswitch operated by the Lunar Module overhead hatch, Conrad said he tried to close it gently and sneak a glimpse of it going out as the door closed. He could not see it extinguish and wondered if it was still on. There was only one way to find out. At 10 hr 35 min Conrad and Bean moved back to

Intrepid, confirmed that although the light switch worked when depressed, the hatch did not cancel the light, pulled the appropriate circuit breaker and returned to the Command Module. 'That did it, we're down where we were,' said flight director Pete Frank.

It was a relaxed spacecraft as Apollo 12 cruised to the Moon. Pete Conrad was a cooperative Commander who ran his ship by agreement rather than order and always got the confidence of his men. As they looked through the windows at a diminishing Earth, country and western music filled the Command Module and Ed Gibson asked for a downlink transmission into the MOCR. But the crew in Yankee Clipper were later getting to their first sleep period than the crew of the previous mission. This was to accommodate the lengthier coast; arriving at the Moon eight hours later than Apollo 11 the crew would have to phase their rest periods with the demands of later activity. Flight director Cliff Charlesworth and capcom Don Lind were in the MOCR when the Conrad crew finally put to bed at 18 hr elapsed time.

Nearly nine hours later flight director Gerry Griffin was on shift with communications handled by Paul Weitz as Pete Conrad called in at the start of another day aboard Apollo 12. Five hours after that, at 29 hr 17 min, the docked assembly passed the halfway point between Earth and Moon – 209,089 km from both bodies – as the crew prepared for another TV transmission and the important MCC_2 burn which would take them from their free return trajectory to a non-return path. Pete Frank handled those events in the MOCR and viewers on Earth could see for the first time scenes inside the Command Module as the computer-controlled engine burn came off exactly as planned. The SPS fired for nearly 9 seconds and cut Apollo 12's speed by 68 km/hr, trimming the approach path to within 111 km of the Moon's surface. There was no noticeable vibration as the engine ignited and the entire operation went smoothly. Shortly after the TV went off the crew got down to some personal care.

APOLLO 'We are trying all these things we didn't have in Gemini, like toothpaste and shaving – we are really having a ball up here.'
CAPCOM 'Roger. All dressed up and no place to go.'
APOLLO 'Oh, we're going someplace. We can see it getting bigger and bigger all the time.'

And then, from Dick Gordon, a reflective comment on the Earth: 'Hey, did you hear, somebody's probably said this before, but that place looks like an oasis down there.' Most of this second day was a relaxed time with only a few housekeeping chores and no navigation sightings. But just as Apollo 11 had been accompanied by unexplained phenomena, so too was Apollo 12 apparently the attraction for some celestial porpoise: 'We have had an object which is in the same place all the time and appears to be tumbling. We have had it ever since yesterday and it just seems to be tagging along with us.' Chances were that it was the S-IVB, uncontrollably drifting on its own path past the Moon. But with Pete Conrad around it was bound to be a point of joculation:

APOLLO 'Okay. We'll assume it's friendly anyway, okay?'
CAPCOM 'Roger. If it makes any noises it's probably just wind in the rigging.'

What Houston knew, that may or may not have been comforting to the crew, was that when they spoke of an object in their immediate vicinity, the S-IVB was more than 4,600 km away.

Conrad, Gordon and Bean awoke at 52 hr 30 min after a nine hour rest period to begin a day in which they would send more TV down, check out the LM as scheduled in the flight plan, possibly perform a mid-course correction maneuver, and get some rest for the lunar orbit days ahead. Apollo 12 would make 45 revolutions of the Moon, almost half as many again as Apollo 10 and 11, supporting nearly 32 hours of lunar surface activity. Tracking information showed the docked configuration to be on a good course for the Moon and no more corrections would be needed. The crew decided to move up the TV transmission to avoid losing live relay to the MOCR in Houston, and to TV audiences around the world. Soon after the scheduled start of transmission, communications would switch to the Honeysuckle station, thereby preventing relay. Pete Conrad chose to move it forward and preserve real-time transmission through Goldstone.

Apollo 12 would approach the Moon on the dark hemisphere and as the day progressed the crew commented on the diminishing crescent as the spacecraft cruised further toward the west limb as viewed from Earth. The TV came on at 62 hr 52 min and Pete Conrad moved up the tunnel into the LM after removing the probe and drogue: 'We're putting the CSM hoses down inside the LM since there's no ventilation in there now. We just lay them around down there where we're going to be working with the LM and it makes it real nice for cooling and gives us clean air down in there. Without it gets a little stale after a while.'

And then the crew participated in a description for the global audience of significant parts of the Lunar Module: 'Okay, what we've got on this side is what we call a left hand stowage compartment and it's pretty unique in that it's made out of beta cloth and with these straps you can see we can remove it. So what we do after the first EVA, we put a lot of our things that we don't need anymore inside these bags and then we can put them outside on the lunar surface and after the second EVA we can put some other things in this . . . and then put it out on the lunar surface also, so we end up being very tidily put some of our gear on the outside . . . so we'll be a lot lighter when we get ready to leave the lunar surface which is handy since we get more rocks and what have you.'

Pete Conrad also remarked that the LM guidance telescope points directly in one of the Command Module rendezvous windows when the two vehicles are docked and 'I was looking out to see what I could see and I saw this face looking back at me and it was Dick in the other window!' In less than one hour, Conrad and Bean checked off the appropriate items and were back in the Command Module. Apollo 12 crossed the equigravisphere at 68 hr 30 min and began to accelerate as the Moon drew it closer. Then, with a 'Goodnight Dick,' from capcom Gerry Carr, the crew settled down for their third night in space. It was extended by two hours because MCC_1 was not necessary and the crew were advised to get as much rest as possible.

At 78 hrs elapsed time, Houston played up 'Sweepers Man Your Brooms' on the boatswain's pipe and Dick Gordon responded with a crisp 'All persons accounted for, sir.' All three crewmembers were Navy men. Activities for the day would be two lunar orbit burns and a brief check on the Lunar Module's VHF and S-band communication systems. Flight director Glynn Lunney and the black team was on shift with capcom Paul Weitz when Apollo 12 passed into the Moon's shadow at 82 hr 53 min. In the closing hour of Apollo 12's trans-lunar coast astronauts and managers wandered into the MOCR and stood around at the capcom console, willing the spacecraft on to a successful SPS burn for lunar-orbit insertion. At 83 hr 11 min Apollo 12 swept out of view. Just fourteen minutes later the SPS lit up as planned and fired for 5 min 52 sec, dropping the two vehicles into a 116 × 313 km path.

Thirty-three minutes after loss-of-signal, Apollo 12 appeared round the Moon's east limb: 'Hello, Houston. Yankee Clipper with Intrepid in tow has arrived on time. . . . I guess like everybody else that just arrived, we are all three of us plastered to the windows looking . . . but for the Navy troops, it doesn't look like a very good place to pull liberty, though.' Minutes later the TV came on and viewers were given a picture of the Moon's surface accompanied with liberal commentary from a much impressed trio. But this time there would be more of the front side pass visible for cameras because, with a landing site far to the west, the correct Sun angle had the light falling across a wider band of front side longitude.

The surface was illuminated for about one hour of the eighty-five minutes it took Apollo 12 to pass across the Moon's Earth-side face, and then the spacecraft was out of sight again. The second revolution was taken up with preparations for the LOI-2 burn, a brief 17 second burst from the SPS engine to circularize the orbit. Successfully carried out on the far side of the Moon, the docked assembly slipped into a 100 × 122 km path designed with gravity anomalies in mind so the orbit would be at the required value when Intrepid headed down toward the surface. Passing out of sight again, on revolution 3, Conrad and Bean were clearing the tunnel for a check on Intrepid.

Less than two hours later their tasks were complete and the crew moved back to Yankee Clipper as the vehicles were disappearing around the Moon on the fourth orbit. Now it was time to get another 'night's' sleep before the busy day of landing operations and the first surface EVA. Unlike Apollo 11, Conrad and Bean planned from the start to go out on their first Moon walk immediately after touchdown; programed for two periods of surface activity it would have been impossible to budget sleep periods any other way than between the two EVAs. The traditional wake-up call was put through about eight minutes before Apollo 12 passed out of sight on revolution 9.

When the spacecraft reappeared on orbit 10, Al Bean continued a conversation he began shortly before starting his sleep about a nasal congestion that concerned him. Medical advice flowed from the MOCR for Bean to use decongestant tablets every six hours and to use a nasal spray; ever since getting into Earth orbit five days earlier, the LM Pilot had the symptoms of a head cold but that was probably due to weightlessness (moving mucus up into the head) and the pure oxygen atmosphere.

While Conrad remained in the Command Module to don his bulky space suit, Bean moved back into Intrepid to switch on the Descent Stage batteries, to check the caution and warning systems and to switch on LM communications. Then Pete Conrad moved to the LM. Power initially went on at 104 hrs and time checks were made to calibrate the master clocks in each spacecraft. Before disappearing round the Moon on revolution 11 the crew had Intrepid buttoned up ready to undock: the probe and drogue were back in, the hatches were sealed, and the Lunar Module's legs had been deployed. Conrad and Bean were almost an hour ahead of schedule.

The front side pass on revolution 12 was taken up with some thruster checks and antenna tests, before the docked vehicles swept round toward the thirteenth orbit. When they reappeared, at 107 hr 45 min, Conrad and Bean were ready to undock Intrepid from the Command Module. But here too, a change of procedure aimed at enhancing their chances of a pinpoint landing. Instead of undocking in-plane and with the probe retracted, the vehicles turned 90° to the flight path before Dick Gordon extended the probe so that Intrepid was hanging on the end, snagged by only the three small capture latches. This 'soft' undocking prevented any impulses from nudging Intrepid's orbit away from the predicted parameters.

Nine minutes after acquisition-of-signal, the two vehicles undocked in full view of a TV audience watching through the lens of a camera in Yankee Clipper. Then, maneuver pads were read up to both vehicles: descent-orbit-insertion (DOI); powered-descent-initiation (PDI); and abort contingency pads in case of emergency. At 108 hr 25 min Yankee Clipper fired its thrusters to move away from Intrepid for the communications and radar ranging checks that would qualify operations for the coming descent. But this was where efficient tracking from ground stations began to pay off. A message went up from Houston: 'The way things are looking right now you are going to be starting PDI about (9.3 km) north of track and during the descent you are going to be steering south.'

Precision was the order of the day. Intrepid would gently steer out the 9.3 km discrepancy between the line of the flight path during descent and the precise location of the landing site. At 108 hr 57 min the two spacecraft went out of sight. Descent-orbit-insertion came in a 29 sec burn on the far side of the Moon; Intrepid was now in an orbit that would carry it to a low point, or pericynthion, of 15 km ready for powered descent. Upon acquisition of signal at 109 hr 43 min ground stations began a concentrated effort to resolve the precise orbital paramaters in readiness for a late guidance update. Doppler tracking put the LM 1.28 km further along the groundtrack than the on-board computer thought so this error was prepared for uplink during the first 90 seconds of powered descent: Intrepid would now steer out a 9.3 km crosstrack error, and shorten the descent range by 1.28 km.

The crew selected the braking phase program (P63) shortly after acquisition of signal to give the system a quick look, and then switched it off before receiving state vector and radius-to-landing site updates from the MSFN ground stations. Intrepid would begin powered descent in a heads-up attitude and remain so until the LM pitched forward gradually to acquire visual sight of the landing point. At 110 hr 20 min 31 sec four RCS thrusters fired in an ullage maneuver to settle propellants, followed seven seconds later by the Descent Stage engine for powered descent; it would continue to thrust all the way to the lunar surface.

Intrepid's engine throttled up to full thrust 27 seconds after ignition and for 5 min 56 sec it continued at full bore, a braking phase shorter than planned because of the range correction fed to the Lunar Guidance Computer 1½ minutes into the burn. But 3 min 22 sec after ignition, the landing radar locked on for altitude data at the phenomenal height of 12,630 meters, followed four seconds later by velocity data from the same radar at 12,222 meters; considerably higher than expected. Intrepid was flying heads-up to give the rear-mounted radar good chance for early acquisition of the surface, and it seemingly responded well.

Lying on their backs during this main braking phase, the force of deceleration gave the crew a feeling that they were back on Earth ('Feels good to be standing in a g field again.') It was a rocky ride through the full thrust period, however, with Pete Conrad crying out that it was 'really banging around' as the thrusters fired almost continuously to keep Intrepid pinned down to the precise guidance track carried by the computer. Now the descent phase was coming up on the end of P63 and the start of P64, the approach phase, where a gradual pitch-over would bring the LM up so that the crew could see first the Moon's horizon and then the approaching touchdown site.

At 8 min 33 sec into the powered descent, the crew switched to P64 and the computer immediately gave figures for the landing point designator, a set of numbers inscribed on the window telling the crew where on the surface the automatic guidance program was taking them. Intrepid was at high-gate, 2,130 meters above the surface and nearly 7,500 meters from the target point. But how accurate was Intrepid going to be in settling toward the appointed spot? A lot rested on the precise location of the LM; without close positioning the crew would be unable to reach the Surveyor spacecraft. Without a demonstration of pin-point accuracy more ambitious mission profiles for later flights would have to wait.

As Intrepid pitched forward, the computer gave Conrad the numbers he needed to look through the window, and what he saw was almost unbelievably similar to what he had seen many times in the trainer on Earth. 'Hey, there it is! There it is! Son of a gun, right down the middle of the road!' The landing point designator produced a number that had Intrepid heading for a point slap in the middle of the Surveyor crater! Not only was the spacecraft on target to within a few tens of meters, it was actually closer to the unmanned spacecraft sitting on the surface than it was supposed to be.

Conrad decided he would have to take over and foreshorten the trajectory a bit and move it to the right of the 200 meter diameter Surveyor crater. But it was an incredible sight, to come all this way and suddenly find the spacecraft pitching over right on the desired course: 'Hey, it started right for the center of the crater. Look out there! I just can't believe it. Amazing! Fantastic!'

The approach phase lasted 1 min 39 sec, during which time Intrepid swept up to the landing area while Pete Conrad redesignated the automatic touchdown spot; this 'stepped' the guidance program out to extend the auto-pilot in range and carry the LM beyond Surveyor crater. But Conrad's intention was to take over manually and pull short of the crater to set Intrepid down to its right, as planned. From a height of 250 meters, however, Conrad did not believe this to be the best area, so he continued on down the landing phase and put in program 66 for manual override. Intrepid was 10 min 12 sec into powered descent; 112 meters above the Moon.

Moving on down, further along the ground track, Conrad put Intrepid on a path that would carry it beyond the Surveyor crater but still to its right. Pulling a steep descent path now, the crew detected dust coming up at a height of 54 meters and from then on conducted an almost vertical drop to the surface. But very gently. At a height of 15 meters the ground was totally obscured by not only the seemingly transparent sheet picked up by Neil Armstrong but a thickening cloud of material lifting away from the surface. While Conrad piloted, Bean read out the displays and Houston cut in from time to time with relevant information:

> BEAN '40 coming down at 2. Looking good, watch the dust. 31, 32, 30 feet. Coming down at 2. Pete, you got plenty of gas, plenty of gas babe. Stay in there.'
> CAPCOM '30 seconds.'
> BEAN '18 feet coming down at 2. He's got it made. Come on in there . . . Contact light.'
> CAPCOM 'Roger, copy contact.'

From a height of about two meters the Lunar Module dropped to the surface, touching down with a vertical descent rate of less than 4 km/hr. There were no solemn words this time – no snappy response from Houston – only the methodical, but jaunty, Pete Conrad shut-down procedure as Intrepid's crew made the vehicle ready for immediate lift-off should anything go awry. But like Eagle before it, Intrepid was firmly on its lunar perch. Descent

Stage propellant tanks were vented as planned and the Lunar Module seemed in good shape, having landed only 163 meters north-west of Surveyor on the flat surface just beyond the crater rim opposite the unmanned spacecraft. But Conrad recognized the wisdom of having aimed for a high descent profile: had he come in at an angle to the surface, excessive dust would have totally prevented good site selection. 'Man oh man, Houston. I'll tell you, I think we're in a place a lot dustier than Neil's,' said the ebullient Commander.

One of the first activities was to perform inertial measurement unit alignment sightings, and to determine precisely where the spacecraft was with respect to the surrounding features, all of which had looked very familiar to the LM crew. Circling the Moon in Yankee Clipper, Dick Gordon made an attempt to sight Intrepid through his 28× telescope and at 114 hr 20 min, about 3 hr 50 min after touchdown, Houston got an excited call from the CSM:

GORDON 'I have Intrepid! I have Intrepid!'
CAPCOM 'Well done Clipper. One crater diameter to the north. Is that affirm?'
GORDON 'He's on the Surveyor crater. About a fourth of a Surveyor crater in diameter to the north-west.'

By this time Conrad and Bean were well along with their preparations for the first Moon walk, although a little behind the flight plan schedule. Pictures of Pete Conrad descending the ladder were sent back from the same location on the Lunar Module as used with Eagle, but this time color embellished the scene; the camera was from Apollo 10's Command Module, refurbished for Apollo 12. After deploying the MESA, and coupling up the lunar equipment conveyor, Conrad moved down to the forward footpad:

CONRAD 'Whoopie! Man, that may have been a small one for Neil, but that's a long one for me! I'm going to step off the pad. Right. Up. Oh, is that soft. Hey, that's neat. I don't sink in too far. I'll try a little – boy, that Sun's bright. That's just like somebody shining a spotlight on your hands. I can walk pretty well, Al, but I've got to take it easy and watch what I'm doing. Boy, you'll never believe it. Guess what I see sitting on the side of the crater. The old Surveyor.'
BEAN 'The old Surveyor, yes sir.'
CONRAD 'Does that look neat. It can't be more than (200 meters) from here. How about that?'

Grabbing a contingency sample, Conrad moved quickly along with his tasks, observing that he had no tendency to slip on the granular materials, before Al Bean joined him on the surface. The S-band erectable antenna, carried by Apollo 11 but never deployed, was to be set up to give viewers a better picture from the improved color camera. The umbrella-like antenna was set up a few metres from the MESA table and, about thirty-five minutes after reaching the surface, Pete Conrad went to re-locate the camera on a tripod some distance from Intrepid.

Within two minutes the public relations exercise collapsed when Conrad inadvertently pointed the camera at the Sun, burning out the vidicon tube in the process. There would be no TV from that point on, a fact which did considerable damage to NASA's attempt at sustaining media interest in Moon mission coverage. Now, only sound would carry notice of Apollo 12's Moon walking events. While Bean put out the solar wind collector, Conrad put up the flag on its telescopic pole, suitably held out by a bar along the top, before both men set about removing the ALSEP instruments, carried in the LM Descent Stage in the left rear quadrant.

One and one-half hours into their first EVA the crew had the first of two science packages off-loaded and on the surface. Next, the fuel cylinder containing plutonium-238 dioxide was to be pulled from its graphite cask alongside the ALSEP stowage bay. But that proved difficult and it took Conrad ten minutes to loosen it from the container. Gentle hammer taps freed the element as it radiated thermal energy through gloves but it came out eventually and was placed immediately in the finned thermal generator. The two packages of instruments were connected together by a bar which would form the antenna designed to transmit information to Earth.

Bean walked a distance of 130 meters from Intrepid, grasping the bar but finding it difficult to keep his hands tightly gripped. It took the two astronauts more than one hour to set up the five instruments, and to connect them to the radioisotope thermal generator, or RTG, to which the fuel element had been delivered. Then the central station was erected, a square-shaped container with the bar antenna fixed on top, and all the instruments attached.

Approached from the south-east, Surveyor 3 reflects sunlight 200 meters south-east of the Lunar Module.

Apollo 11's Early Apollo Scientific Experiment Package – EASAP – carried only one active instrument: the passive seismometer protected from the cold of lunar night by tiny heaters and powered during the lunar day by two panels of solar cells. Unfortunately, the experiment operated for just three weeks after Armstrong and Aldrin left Tranquillity, and on 27 August the prospect of getting it working again was abandoned. Lunar days and nights each lasted nearly two Earth weeks. ALSEP-1, deployed by Conrad and Bean, was designed to continue operating for a full year. But Bendix had done good work. Nearly eight years later it, along with others, would be shut down intentionally, their lifetimes having outlived their continued usefulness.

In retrospect that would make this hour spent by Conrad and Bean setting it up a worthwhile activity, but to the two Moonwalkers it was a task well out of the way; there were many interesting rocks around to investigate. Now, three hours into the EVA, they could move up to Shelf Crater, named by the astronauts for the 300 meter depression just north-west of their position. They had already been given a nominal extension of 30 minutes on their planned 3½ hour excursion outside the LM, but there was time only to get a quick look at Shelf before heading back to Intrepid. Much still to do. Bean got a core sample at the vicinity of the LM and then packed samples in a return container. First up the ladder, the LM Pilot was followed by Pete Conrad and the first Apollo 12 EVA ended after 3 hr 56 min, a creditable increase on the 2 hr 40 min set by Apollo 11. And there was still another EVA to go before lift-off.

While the LM crew were preparing to re-charge their PLSS backpacks from the environmental control system in Intrepid, Dick Gordon fired Yankee Clipper's SPS engine for 18 seconds to shift the orbital plane 4°, placing the mother-ship in an orbit ready for ascent and rendezvous. Margins built in to the EVA plan for consumables carried in the back-packs were adequate. Conrad had 42% oxygen, 44% cooling water, and 34% battery power remaining in his PLSS at the end of the Moon walk; Bean's pack had similar reserves. Clearly, the two men could have gone on a lot longer, probably extending their EVA by another hour or so. But emergency reserves were an essential feature of safety. Nevertheless, metabolic rates for both men were lower than predicted, a welcome change from the situation when NASA last had men indulging in suited EVA.

While Conrad and Bean prepared for their coming rest period, preceded by a welcome meal, Houston chatted to them about the day's events, de-briefed them on a highly successful EVA, and discussed plans for the next walk based upon refined information as to their whereabouts. Clearly, there would be no difficulty reaching Surveyor, but what of the remaining period outside Intrepid? Mission planners agreed a plan that would take Conrad and Bean on a circuitous route from the LM, almost due west to Head Crater, a 90 meter diameter crater next to Surveyor Crater, down to a still smaller Bench Crater depression, west again to tiny Sharp Crater, then back 600 meters due east to the unmanned lander waiting for them to snip off suitable parts for return to Earth. After that, back along the rim of Surveyor Crater to Intrepid.

The crew were to have got to sleep about ten hours after touchdown but two hours late they finally said goodnight to Houston for a scheduled 8½ hr rest period. But more than one hour early, at 129 hr elapsed time (18½ hr after landing) the MOCR flight surgeon reported the crew were up and busy in the LM so a call went up from the ground. Pete Conrad was not about to be late for his second Moon walk.

INTREPID 'Hello, Houston, Intrepid. How are you this morning?'
CAPCOM 'Good morning, Intrepid. How did you sleep?'
INTREPID 'Short, but sweet. We're hustling right now, and we're going to eat breakfast, have a little talk with you, and get about our business.'

Already, it seemed, Apollo operations were swinging into a new phase. Astronauts were switching hats, assuming the role of explorers as very worthy stand-ins for professional geologists, and doing so with relish and verve. How different this second landing was turning out to be. Devoid of the rhetorical acclaim fashioned for a great historical event, Apollo 12 was setting a brisk pace.

EVA-2 officially began at an elapsed time of 131 hr 32 min, some two hours ahead of schedule. Mission Control agreed with a timely suggestion from Pete Conrad that they should move on out just as soon as possible: 'Okay, Intrepid, Houston . . . Pete, whenever you're ready, at your own pace, you can go over the sill.'

At 131 hr 39 min, Conrad jumped down from the ladder's bottom rung: 'Whoops, long step. Okay, Houston, mark. I'm on the lunar surface.' Bean joined him ten minutes later. Moving past the ALSEP instruments, the seismometer picked up their loping walk as vibrations through the surface. 'Pete, we're watching you down here on the seismic data. Looks as though you're really thundering right by it,' radioed Houston. 'Oh, okay. Meet you at Head Crater pal,' said Conrad as Bean stopped to pick up a rock: 'Here is a dandy extra grapefruit size type goody.'

At Head Crater, Conrad rolled a rock down the slope with his foot and then photographed the track it made – valuable pictures for scientists to study. Minutes later, with both men stopped, Conrad rolled a much larger rock down the same slope, telling Houston when it started and when it stopped; engineers watching seismic traces from the ALSEP would know when tremors on the charts were due to this activity. But geology was very much the order of business, and the crew quickly settled in to their exploratory activity.

From Pete Conrad: 'I've been concentrating, Houston, as I came walking over here to Head Crater, to see if there is any possible changes in either texture, slope, color, anything you can think of, that would say to me that I was walking on a different surface than I was when I started. And I can't identify a thing yet, it all looks the same.' With a shovel attached to an extension handle, the crew then dug a small trench to obtain samples from a few centimeters beneath the surface. Mission Control was watching the timeline, however, and told them to move on up to Bench Crater; at 1 hr into the EVA, they did just that.

On the way, Bean noticed some interesting samples: 'These rocks obviously came out of the crater, because they are scattered more uniformly around it. There's a bunch of them on the rim and there's not many far away. We probably ought to grab a big one of them.' But all the while, time was against them, and after surveying the area between Head and Bench, Al Bean quickly summarized: 'I noticed when I was looking at that rock back there up real close that it had been hit by meteorites so much I guess it had given it a rounded appearance something like those in the hole except there's a couple over there like you say that don't look that way.' Then it was time to head across to Sharp Crater for a 10 min stop with core tubes and sample bags. Conrad noted how soft the rim of this comparatively small, 10 meter diameter, crater seemed to be as he helped Al Bean with the site tasks.

The two astronauts were now as far away from Intrepid as they would get on this mission: 400 meters. At 1 hr 50 min into the EVA they packed up and headed east, passing Bench Crater, to begin the traverse to Surveyor Crater. On their way they stopped at another small depression called Halo Crater, and then they walked across to the 200 meter bowl that contained the unmanned lander which had been sitting there for thirty-one months.

CAPCOM 'Okay, Pete. You're 2hr and 7 min into the EVA. And we show you leaving Halo at around 2:15. And now that's for a 4 hr EVA. We extended you 30 min for total EVA of 4 hr. We'd like before you go on, to figure out a plan of attack on the Surveyor. Make sure . . . you remain away from directly below the Surveyor as you move up to it.'
CONRAD 'Okay. We concurred with that.'

The path to Surveyor was trod from one slope across to the other and both astronauts tested the stability of the loose surface material to see that they could get back up out of the crater the same way they came – just in case they walked into trouble, perhaps a section of deep soil. Then they rested for a few minutes while looking across the crater's panorama and the precise position of the Surveyor spacecraft. When they got to the tripedal vehicle a number of tasks confronted them: photograph certain structural components; remove and collect glass from the thermal boxes; photograph scoop marks on the lunar surface; cut and retrieve cable samples for return; and retrieve the TV camera. The latter item took the crew ten minutes to remove and to stow in Conrad's sample bag. Then it was time to return to Intrepid and button up.

Back at the LM 3 hr 10 min into the EVA, the crew stowed the solar wind collector, which had been out on the lunar surface for nearly 19 hours, hauled a sample return container aboard the

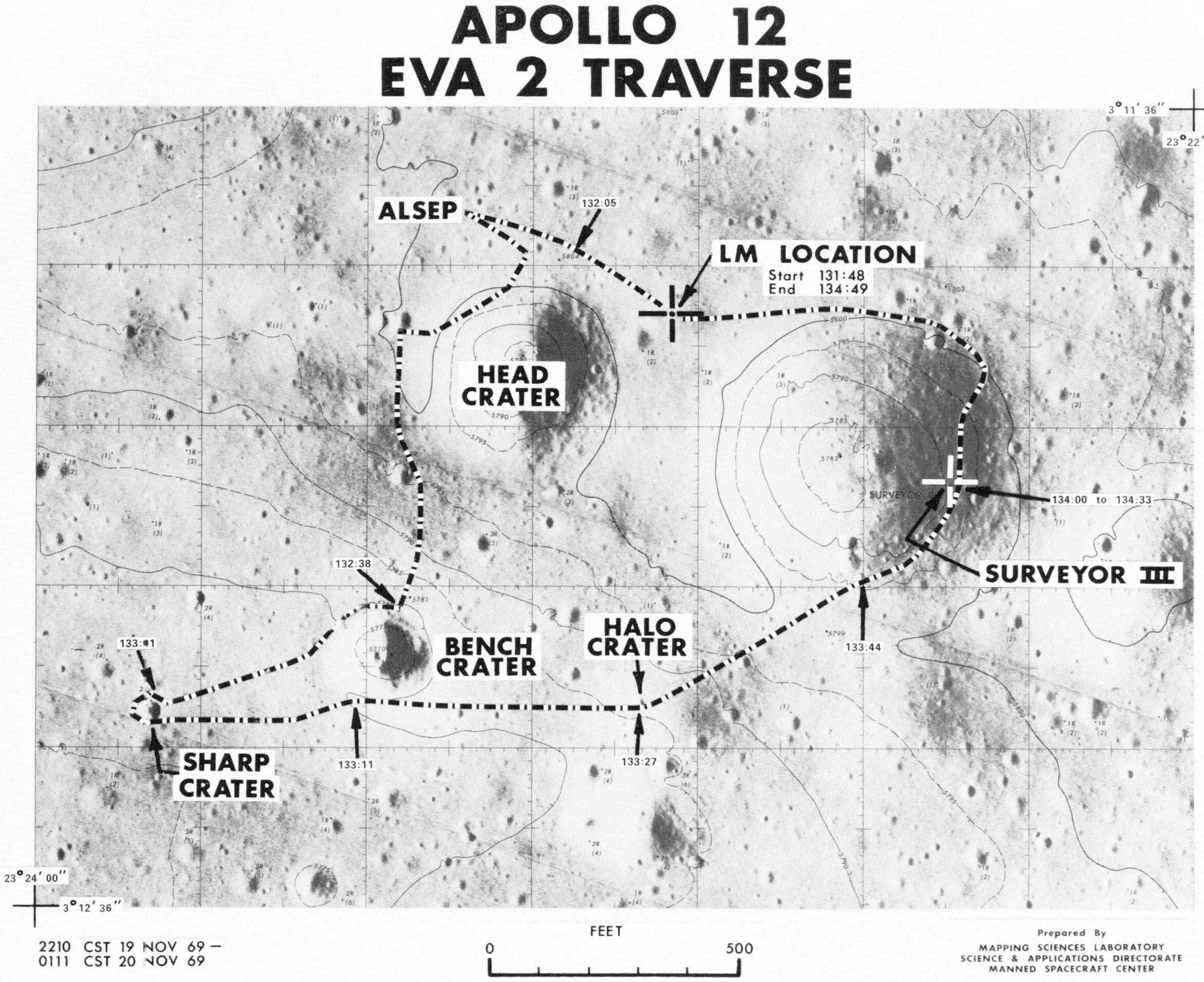

APOLLO 12 EVA 2 TRAVERSE

ALSEP
132:05
LM LOCATION
Start 131:48
End 134:49
HEAD CRATER
134:00 to 134:33
SURVEYOR III
132:38
133:44
133:01
BENCH CRATER
HALO CRATER
SHARP CRATER
133:11
133:27
3°11′ 36″
23°22′ 36″
23°24′ 00″
3°12′ 36″

2210 CST 19 NOV 69 –
0111 CST 20 NOV 69

FEET
0 500

Prepared By
MAPPING SCIENCES LABORATORY
SCIENCE & APPLICATIONS DIRECTORATE
MANNED SPACECRAFT CENTER

spacecraft, and climbed back up the ladder. By the time the hatch had been closed and the EVA terminated the two Moonwalkers had accumulated a total time of 7 hr 45 min on the two excursions. And they had 34 kg of rock and soil samples. The activity had been a creditable start to lunar surface geology and although leaving much to be desired from a scientific standpoint, provided valuable experience for more sophisticated missions to come.

Conrad and Bean had roamed 2 km across the surface, carrying basic geologic tools and a hand-tool carrier with extension handles, brushes, scoops, tongs, and a gnomon – a weighted staff suspended from a tripod and equipped with a color chart for comparison with lunar materials – which was put down on the surface within the frame of each general site photograph. Each crewmember had approximately the same reserves as measured for the first EVA and metabolic levels were a little lower, due to the reduced physical nature of the activity; it had been hard work putting up the ALSEP station.

Even with their 30 min extension, Conrad and Bean were nearly a full hour ahead of the flight plan. Time to prepare for lift-off without scurrying. Just as the crew got under way with their post-EVA checklist, Dick Gordon in Yankee Clipper was informed by Mission Control that observers on Earth had reported a transient event in the crater Alphonsus. Transient Lunar Phenomena, or TLP, had been reported at various times for several decades, characterized as brightening of certain crater floors as seen in the telescope. Nobody knew what they precisely were but Apollo was responsive to these timely reports and as the crater in question was under the CSM groundtrack it would be good opportunity for the astronaut to get some photographs.

Gordon failed to see anything different about Alphonsus, but took some pictures nevertheless. Then Conrad and Bean, hooked up to the Lunar Module's suit circuit, opened the forward hatch a third time and dumped excess equipment on to the porch and down the ladder before closing up and re-pressurizing the cabin. But efficiency was the password with Apollo 12, and within two hours of crawling back in, the two astronauts had the ship in shape:

CONRAD 'We have everything stowed, geared properly, and we are ready to start the launch countdown at the proper time, and if you'll give us about 15 or 20 minutes to chow down here, we'll come back with you and a little chitty chat about EVA.'
CAPCOM 'Roger Pete; that sounds like a good plan.'

But lift-off was still four hours away; time to get some food before the six hours of ascent, rendezvous and docking. Both crewmembers actively described their EVA sessions to listening ground controllers, and discussed surface conditions and rock samples. But Gordon too had been busy; equipped with a multi-spectral camera assembly he shot reels of film covering large portions of the lunar surface, and obtained Hasselblad views of future candidate landing sites in Fra Mauro and Descartes regions.

Fra Mauro was the tentative site for Apollo 13, a difficult area to get to, but one which would explore vast sheets of ejected material thrown out close to the beginning of the Moon's history when a massive crater 900 km across – called a basin – was excavated by a large object slamming into the surface. Called Mare Imbrium, the basin materials extended far to the south and Fra Mauro was the type name for this Imbrian ejecta Blanket. So began a tradition of using one flight to reconnoitre surface conditions for another.

After 31 hr 31 min on the surface, Intrepid lifted away to rendezvous with Yankee Clipper. Ten seconds before engine ignition Pete Conrad pushed the abort stage button which mechanically separated the Ascent and Descent Stages; five seconds later the Ascent Stage engine was armed and the PROCEED button was depressed telling the computer's Luminary 116 program to go ahead with ascent activity. At ignition time, Conrad pushed the ENGINE START button to back up the automatic sequence and the motor roared to life. 'Liftoff, and away we go,' was the word from Intrepid as the Ascent Stage lifted cleanly away from the site in Oceanus Procellarum.

Four minutes later the LM yawed 20° right, on commands from Conrad, to keep the high-gain antenna pointed at Earth, while on the displays from the computer Al Bean monitored the decreasing numbers telling him the velocity still to be gained before shutdown and orbit insertion. More than seven minutes after ignition the engine was shut down, but manually because an incorrect switching sequence prevented the computer from cutting the burn. A resulting 1.2 seconds overburn propelled Intrepid to a 17 × 115 km path about the Moon. Quickly recognizing the discrepancy, the crew computed a corrective burn and fired the thrusters to drop Intrepid back to the planned 16 × 86 km orbit. The normal rendezvous sequence could now begin.

When Intrepid and Yankee Clipper passed out of sight round the limb of the Moon the two spacecraft were 350 km apart but on the far side Conrad fired the thrusters again for a concentric-sequence maneuver and when they reappeared the LM and the CSM were separated by only 200 km. During the front side pass Pete Conrad suggested they use the Abort Guidance Section to fire the constant-delta-height burn since Bean had scurried around updating its computer memory; normally, rendezvous maneuvers would only be made with the PGNCS computer, the AGS being there in case of emergency. The CDH burn was to nearly circularize Intrepid below the orbit of Yankee Clipper. That maneuver was carried out on the front side of the Moon, one-half revolution after the concentric-sequence burn. One-half orbit later, round on the far side again, Intrepid started up toward Clipper with the terminal phase firing. When the two vehicles reappeared, Intrepid was closing at 42 km/hr on Clipper, still 3 km ahead. Five minutes later the rates were essentially zero and the two spacecraft were sitting together in orbit waiting to dock.

Several minutes later the two were linked together with a 'Super job . . . that was cool, wasn't even a ripple,' from Intrepid. Viewers on Earth got good TV pictures of the docking from the color camera carried by Clipper, the refurbished model used by Apollo 11's Columbia. Small consolation for no TV on the Moon walks. In the preparations for transfer that now followed, a brief exchange ensued on a piece of metal seen by Intrepid dangling from the rear of Yankee Clipper, just like 'some of the stuff we used to get back . . . on Gemini.' Two hours after docking, Intrepid's crew were moving back to join Dick Gordon. The same vent procedures used on 11 were followed for 12 but there was more equipment to move through the tunnel, and maneuver pads to load Intrepid's computer with for burns not performed hitherto.

Back in Clipper, Conrad and Bean helped Gordon prepare for jettisoning the LM. At 147 hr 59 min Intrepid was cast free and five minutes later the CSM fired its thrusters for 5.4 seconds to move away. Passing round the far side of the Moon the two vehicles slowly drifted apart in readiness for a long burn from Intrepid's thrusters which would cause it to curve down on an impact trajectory with the Moon. Apart from removing potential debris hazards for later flights, crashing Intrepid on the Moon would produce a sizable vibration for the seismometers, allowing modest information to be obtained on the substructure and good calibration for the detectors. Intrepid was back on the Moon's near side when the thrusters fired for 82 seconds, carrying the Ascent Stage to impact twenty seven minutes later at a speed of 6,043 km/hr. The resulting crater was 11 meters long, 6 meters wide, and about 5 meters deep.

By this time the crew were busy stowing everything in Clipper but that kept them awake past the planned sleep period, scheduled to begin at 150 hr into the flight. Three hours later they were reported getting down by the MOCR flight surgeon but only four hours after that they were up and scurrying, getting ready for a planned plane change to set up Clipper's orbit for a pass across Fra Mauro and Descartes, the latter a crater area east and to the south of the Moon's equator. The 19 sec firing of the spacecraft's SPS engine came at 159 hr 5 min on the 39th revolution.

Bootstrap photography was the order of the day on the final orbits of the Moon, so called because the crew would obtain stereo photographs of important future landing sites by aiming the camera through the spacecraft sextant. By taking one shot approaching a target on the surface and one shot going away, the overlay would provide three-dimensional views of the surface – important information for refining contour profiles along projected descent paths. Then there were more questions from the MOCR about surface activity, about all the dirt the crew complained of in the spacecraft, and about conditions on the surface. Al Bean had a comment regarding the gross features: 'The backside is a lot more worked over, a lot more worn and smooth and where the frontside's got all these mare areas and lots more contrast and a lot more sharp features, I personally like to look at the front side, the hills and the higher mountains, the contrasts and the mares.'

The trans-Earth injection burn came as planned at the end of the 45th lunar revolution and with a 'Hello, Houston. Apollo 12's moving home,' the spacecraft reappeared on course for a Pacific splashdown. TV pictures flowed for 38 minutes, shots of the receding Moon, before once more the crew prepared for a sleep period. That began three hours after burning TEI and not for a further twelve hours did Houston try to rouse the weary explorers. By that time Clipper had passed through the equigravisphere and was back in Earth's gravity grip, building up speed all the time. The mid-course correction maneuver, a small thruster burn, was postponed to allow them to rest as long as possible. More geological de-briefings followed, with another sleep session lasting more than nine hours.

Another TV transmission was broadcast beginning at 224 hr 7 min at which members of the press passed questions to the MOCR capcom for response by the crew. A final sleep session preceded preparations for entry and 3½ hours before separating from the Service Module, Al Bean had a splendid view of the planet: 'This has got to be the most spectacular sight of the whole flight. We can see now that the Sun's behind the Earth. We can see clouds sort of on the dark part of the Earth and of course the Earth's still discernible by this thin narrow or thin blue and red segmented band . . . The clouds appear sort of pinkish grey and they're sort of scattered all the way around the Earth.'

Eleven minutes after re-entry the spacecraft was visually sighted and landing occurred 14 minutes after entry interface at an elapsed time of 10 days 4 hr 36 min. In the Stable II position for five minutes after splashdown, Command Module 108 was only 6.5 km from the recovery ship USS *Hornet*. Again, a Moon crew was subjected to the biological isolation of first an overgarment, then the mobile quarantine facility, and finally the Lunar Receiving Laboratory. But scientists were well pleased with not only the return of information, data, and samples, but with the performance of ALSEP-1, still sitting on the surface. From switch-on, the equipment performed as required and very soon the RTG was up to design power levels of 73.59 watts. But even as Apollo 12 returned to Earth storm clouds of dissatisfaction were gathering over NASA headquarters.

During the two or three months preceding the November flight, agency officials had been deeply involved with discussions on the Fiscal Year 1971 budget, the period which would begin mid-1970 and for which President Nixon would make his annual request in the January. Where even the abbreviated schedule set up some months previously envisaged a total ten Moon landing attempts and three AAP flights, both categories terminating late 1972, or early 1973 as pessimists believed, flights now envisaged were dramatically changed in content and number.

Plans for hardware to support the J-series missions (Apollo 16–20), three day stays on the surface, expanded science role for orbiting CSMs, etc., crystallized during 1969 with a decision by George Mueller on 23 May to develop a lunar roving vehicle for enhanced mobility. The Marshall Space Flight Center would direct the project and industrial proposals were requested 11 July, shortly before Armstrong, Collins and Aldrin set off for the first Moon landing. For a time after the Apollo 11 mission debate continued on the relative value of committing a potential payload of 180 kg to four wheels and two seats; Armstrong believed traverses on foot would be no problem and contested suggestions that a rover would assist with trafficating the surface. Ex-Gemini man André Meyer was not convinced by Armstrong's argument, and won the day when MSFC awarded a contract to Boeing in October for four lunar rovers scheduled to fly with Apollo 17, 18, 19 and 20.

By early fall, the flight schedule for these Moon landings had slipped to include a final mission in 1973. There would be three H-series flights in 1970, generating maximum advantage from basic Apollo hardware, followed by two J-series flights in 1971, two in 1972 and the last in early 1973; the first rover would be carried on Apollo 17, the second J-series flight, in July or August, 1971. Mueller was keen to maintain an active launch schedule and baulked at the prospect of six-month intervals between launches. His scientific colleagues were concerned that as the flights evolved into more productive expeditions the maximum value could only be fed to the next flight in line by careful and timely preparation. So Mueller got H-series launches at the rate of three per year while the scientists got the more complex J-series flights at two per year.

But by year's end there was little doubt that the depleted NASA budget would hardly support anything else, especially as the AAP flights would be coming along in 1972. And that was where another change came in. As early as 21 May, 1969, George Mueller asked NASA officials for their views on the possible alternative options for AAP. Two days later von Braun responded with an unqualified recommendation for using a dry rather than a wet workshop. Under the existing Apollo Applications Program plan, the agency would launch a Saturn IB so that its S-IVB second stage could be vented of residual propellants in orbit and converted to a habitable station for three stays of 28, 56 and 56 days, respectively.

In von Braun's plan, NASA would resurrect the old Saturn V launched dry workshop, using just the shell of an S-IVB to fit it out on the ground as a fully habitable space station, the wet workshop – so called because it would be launched full of propellant – giving way to a dry station suitably fitted with all the necessary equipment, consumables and supplies all three crews would need. This would, said von Braun, give NASA a firmer hold of space station technology and serve as a very real precursor to the planned 12-man space station for which North American Rockwell and McDonnell Douglas were performing definition studies.

A day later, Bob Gilruth too recommended switching to a dry workshop concept, essentially the Saturn IB station launched on the first two stages of a Saturn V. On 18 July Tom Paine officially approved the change from Saturn IB to Saturn V launched station, requiring contract termination with Grumman on Lunar Modules that would have been used for Apollo Telescope Mount (ATM) solar telescopes to be attached at the station's forward end. Launched by Saturn V, a separate telescope mount more appropriately fashioned around the specific requirements of such a module could be built by the Marshall Center.

It was at this time also that McDonnell completed and submitted to NASA a study on the use of a Gemini spacecraft design for supplying men and equipment to orbiting space stations; not the AAP workshop but the more sophisticated post-Apollo designs then being generated for the mid- to the late-1970s. It called for extension backward of the same conical configuration used by the two-man Gemini in either of two versions: Min-Mod Big G, and Advanced Big G. The former would accommodate nine men in an extension to the conical re-entry module presenting a maximum diameter of 4.2 meters; the latter would hold 12 men (clearly with the 12-man space station concept in mind) and provide more sophisticated guidance and tracking systems. Both would carry a cylindrical cargo propulsion module and land by parawing on skids. In the light of NASA plans for developing a reusable, winged, shuttle craft for space station ferry duties it was not likely to get anywhere – and did not.

However, when it became apparent during the late fall and early winter budget negotiations that NASA would not get the necessary funds to re-open Saturn V production beyond the fifteenth vehicle, the earlier decision to use a launcher of this type to carry the AAP workshop into orbit necessitated cancellation of one Moon landing attempt. Consequently, one of the planned five J-series missions was taken out of the schedule in a decision officially made 7 January, 1970, but one implicit in NASA plans for several months. More than a year earlier NASA chose the name Skylab for the AAP workshop suggested by Donald L. Steelman from the Air Force in response to a request from Headquarters from a name other than Apollo Applications Program. Divested of all the grand objectives presented for planning purposes only three years earlier, AAP would from 17 February, 1970, be officially known as Skylab – a one-shot workshop serving as precursor to a fully fledged space station still in the design and definition stage.

But Fiscal Year 1971 budget prospects looked bleak and in the wake of severe cuts from what NASA wanted for that period it was decided to uncouple simultaneous operation of both Skylab and J-series Moon landing flights. It was a further move at cutting lunar missions and presented a plan whereby two H-series flights would be flown in 1970, one H- and one J-series flight in 1971, and a second J-series flight in early 1972. Skylab would be launched later that year complete with the first 28 day habitation. The second and third crews would occupy the workshop during the first half of 1973. The last two Apollo flights, both J-series, would fly six months apart in 1974 and close out all operations with Apollo type hardware. Basically, it reduced to two the maximum number of Moon landing flights per year and halted lunar activity during the 12 month Skylab operation, extending the sequence out to the middle of the decade.

The long-awaited announcement from President Nixon on future US space policy appeared early March, 1970, in response to the recommendations of the task group chaired by Spiro Agnew during 1969. After reiterating a belief that 'there were no clear, comprehensive plans for our space program after the first Apollo landing,' he went on to claim three 'general purposes' to guide the future selection of goals: exploration; science; and practical applications. 'We must realize that space activities will be a part of our lives for the rest of time. We must think of them as part of a continuing process – one which will go on day in and day out, year in and year out – and not as a series of separate leaps, each requiring a massive concentration of energy and will and accomplished on a crash timetable.'

Still trying to shape his attitudes around a long-held hatred of Kennedy policies, Richard Nixon clearly criticized the attitude of his predecessors and felt that by not structuring deliberate goals or responsibilities for the space program he would be seen as a more suitable decision-maker. He did, however, say that America 'should continue to explore the Moon,' and that 'our decisions about manned and unmanned lunar voyages beyond the Apollo Program will be based on the results of these missions.' While emphasizing the value of unmanned planetary probes for exploring the solar system, Nixon said that there 'is one major but longer-range goal we should keep in mind as we proceed with our exploration of the planets. As a part of this program we will eventually send men to explore the planet Mars.' This bowed to the superior logic of the Mueller-Gilruth integrated space program discussed at the beginning of this chapter although it proposed nothing that would speed such a commitment.

Nixon also nodded approval in his statement for 'reusable space shuttles as one way' of achieving low-cost transportation to orbit, and for the 'orbiting workshop' fitted out from a redundant Saturn V third stage. But on the question of a large, 12-man station, he felt a decision should be deferred until 'on the basis of our experience' with Skylab it is learned 'how to develop longer-lived space stations.' At most it was a summary without commitment and set the pace for the next few years where in the absence of dynamic direction from the White House it was left to NASA to plead before Congress for funds to continue work on the reusable shuttle and the space station.

Four days before Apollo 12 flew, George Mueller announced his resignation as Associate Administrator for Manned Space Flight. His had been the daunting task of putting Apollo on the road, of grooming Gemini for successful precursor tests of operational Apollo procedures, and of putting his reputation on the line by introducing a completely new mode of vehicle qualification: 'all-up' systems testing. Said Tom Paine, of Mueller's achievements: 'It is due to Dr. Mueller's creative leadership of the magnificent manned space flight organization that the flight of Apollo 11 . . . achieved the national goal set in May, 1961 . . . We regret that Dr. Mueller has made the decision to return to private life, but recognize that decision comes at a time when the task he accepted is complete and a sound foundation for our future national space program has been established.'

George Mueller's resignation became effective 10 December and he moved to General Dynamics as Vice President immediately thereafter. On 8 January, 1970, NASA formally announced the appointment of Dale Myers to replace Mueller. Myers had managed Rockwell's CSM contract and was now joining the customer as head of manned space flight. His appointment was effective on 12 January.

On the same day Mueller retired from government service, NASA announced it would send Apollo 13 to Fra Mauro on 12 March, a vast highland area east of Apollo 12's landing site, and one directly in support of scientific exploration of the Moon. The crew was announced 6 August, 1969: Lovell, Mattingly and Haise, with Young, Swigert and Duke as back-ups. It was to be Jim Lovell's fourth space flight and after flying a 14-day flight in Gemini VII, still the space flight record holder, a rendezvous and docking mission in Gemini XII, and the first Moon orbit flight on Apollo 8, he was now to command a landing mission designed around a search for material excavated from deep inside the Moon. Mattingly and Haise were getting their first rides into space.

Hardware for the flight had been at KSC since June, 1969, when a maximum paced effort was underway in case Apollo 11 failed to make it to the surface. After the successful landing a possible launch in November of that year was cancelled and the 508 launcher was not erected until 1 August, with CSM-109 and LM-7 going on top in December. The flight was to be a little different from Apollo 12 in that a descent-orbit-insertion burn was to be performed on the second lunar revolution by the big SPS engine, placing the docked configuration into a elliptical 13 × 106 km path where it would remain until the Lunar Module separated for powered descent on revolution fourteen.

On earlier flights, the DOI burn had been performed by the Lunar Module only after pulling apart from Apollo on revolution 13. Now, both vehicles would spend approximately 22 hours in this elliptical orbit with low pericynthion uprange of the planned landing site. Also, the S-IVB stage would be targeted for lunar impact rather than a slingshot to solar orbit, this to provide a measurable impact for the Apollo 12 seismometer left in Oceanus Procellarum. Based on consumables used during the two previous landings, Lovell and Haise were to spend 4 hours on each of two EVAs, with possible extensions. Surface stay time would be increased to 33.5 hours, and an extra orbit added to Moon activity. Bootstrap photography would be obtained of future landing sites in areas called Censorinus, Descartes, and Davy Rille.

By using the SPS to lower the LM for powered descent, considerable propellant would be saved for landing operations, the precise geometry of the initial orbit would permit even better navigation during descent, and less information would have to be pushed into the computer after acquisition of signal on revolution fourteen, 34 minutes prior to powered-descent-initiation. Before commencing the landing run, the Apollo CSM would burn itself into a circular orbit from where it could await rendezvous if the Lunar Module aborted. Lovell and Haise would put out another ALSEP collection of scientific instruments, and use a lunar drill to obtain a core sample from 3 meters below the surface. But its most important task would be to drill two holes for a heat-flow experiment which would tell scientists on Earth the temperature of the outer layers and the rate at which thermal energy was escaping from the Moon.

By January, 1970, preparations for launch were eased when Headquarters decided to defer the flight until the next opportunity on 11 April. Due partly to the relaxed schedule demanded by reduced budget levels, it allowed more time to prepare crew and hardware, and also phased Apollo 14 out to October, 1970, rather than the earlier launch date of July. Apollo 13 moved to pad 39A on 15 December, before the deferred launch, and satisfactorily passed two systems tests before countdown preparations gave 508 a wet CDDT beginning 18 March. Seven days later an unusual event occurred at the Kennedy Space Center. To pre-cool the plumbing used for pumping liquid oxygen, technicians emptied 39,000 liters of oxygen into a ditch to one side of the launch complex. Reduced to a gas, the oxygen accumulated across banking near the ditch and when security guards stopped their cars near the perimeter fence the oxygen-rich atmosphere caused ignition sparks from the engines to set the vehicles on fire. Spontaneous ignition and oil across engine components added a bizarre footnote to Apollo 13's flight preparation. It was the first of many. Several persons warned of danger to this flight, all warnings generated by the mission's number, and it seemed indeed to be jinxed as preparations moved ahead.

During the CDDT activity, the two cryogenic dewars containing liquid oxygen were filled for checks that the spacecraft fuel cells were working properly. Each oxygen tank was 67.4 cm in diameter, designed to contain 145 kg of oxygen with approximately two-thirds going to the fuel cell electrical production units and the balance for cabin atmosphere. Each tank contained a temperature sensor with a small heater and a fan. When the time came in the CDDT to empty both tanks to 50%, only No. 1 tank bled down through the vent line; No. 2 tank remained at 92% capacity. Of comparatively minor significance to the Count Down Demonstration Test, the problem was put aside while the main activity on the stack went ahead as scheduled.

With CDDT over, further attempts were made on 27 March to de-tank the contents. Engineers attempted to remove the oxygen through the fill line but that resulted in a reduction to only 65%. It was suggested that oxygen was leaking from the vent line back up the fill line without appreciable loss from the tank. Then, a normal de-tanking sequence was performed and again nothing happened. Finally, a pressure was applied to the vent but here too the oxygen refused to move. The only solution was to boil off the oxygen by switching on the internal heaters and KSC personnel activated the tank heaters with a 65 volt current from standard ground equipment for 1½ hours before turning the fans on.

Four and one-half hours later the quantity still read 35% and it was agreed to perform a pressure-cycling operation. It took five such cycles to empty the tank completely after which the fan and heater were turned off, having completed 8 hours' continuous operation. A decision had to be made as to the tank's ability to safely support the coming mission and in support of this a test was run on 30 March where tank No. 2 was filled with gaseous oxygen. Then it was 20% filled with liquid oxygen and although operating satisfactorily, it could only be emptied through repeated pressure-cycle operations with heaters on. It was determined that tank 2 probably had a loose or ill-fitting fill tube and that there was nothing known that could cause trouble. Tank No. 2 would stay where it was. There was no back-up site for Apollo 13, the importance of Fra Mauro being sufficient to prefer a month's delay should the launch be postponed beyond a three hour twenty-four minute launch window beginning 2:13 pm, 11 April. To have changed the oxygen tank shelf in Service Module 109 would have put at risk the first opportunity.

The formal countdown for Apollo 13 began on Sunday, 5 April, but persistent trouble with the Lunar Module's Descent Stage super-critical helium tank, used to pressurize the Descent engine propellant tanks, caused doubt about the ability of the system to support a lunar landing attempt. When loaded with helium, the tank appeared to warm the super-cold liquid too quickly, threatening to increase pressure to the level of the burst-disc safety valve. If that occurred the valve would open, venting helium. Without helium, no landing could be attempted. The following day more tests were conducted and a decision was reached to go ahead with the countdown; if it got no worse the helium pressure would not reach a critical level before the supply was used to pressurize the propellant tanks for landing.

But the same day the countdown began, Lovell, Mattingly and Haise were suspected of having been contaminated by Charlie Duke. Suddenly taken with German measles, Duke was known to have been in the presence of the 13 crew during the critical incubation period. Lovell's son too had measles – the rubella variety. On 6 April Dr. Berry decided to recommend removing Mattingly from the flight list when tests carried out on all three showed him to be the only one without immunity to rubella. However, Duke was back-up to Fred Haise so a straight-forward crew switch was impossible. Nobody liked breaking up a crew once appointed, and within five days of the launch it was an almost unthinkable travesty of accepted procedure.

The only alternative, however, was to move Mattingly's back-up, Jack Swigert, in as the Command Module Pilot. But was Swigert ready, and would Lovell and Haise readily accept him? Much crew integration relied on the intimate knowledge each acquired about the other's way of doing specific operations, second-guessing at times an awareness of trouble or problems. The only way to see would be to have Lovell and Haise go through simulator runs with Swigert to find out if they could weld together as a team. On Thursday, 9 April, two days before the flight, the three men entered the Command Module simulator for a rough shakedown ride through every conceivable contingency, emergency, and failure Flight Crew Operations Branch chief Riley McCafferty could dream up.

By the following afternoon the results were to hand, McCafferty conferred with Deke Slayton and both agreed that if Lovell and Haise were happy about it, Swigert would fly in Mattingly's seat. They were, and the most unusual switch in crew assignments ever to smite a NASA manned flight took place. But Jim Lovell too was there because Mike Collins had an operation nearly two years before. Lovell would have been with Armstrong and Aldrin as back-up to Apollo 8, and been along for the ride on the first Moon landing, had not Collins developed a bone spur that pushed Lovell, as back-up, into his seat. He would not be sitting in as Apollo 13's Commander, had he not flown Apollo 8. Nothing was right about 13: the hardware was troublesome; the crew were there because of some physical ailment afflicting a colleague. Was 13 a duped flight?

The Reluctant Hills

The third lunar landing attempt was an apathetic affair for newsmen, TV stations, and media representatives around the world. Houston TV decided not to disturb their planned transmission with live pictures of launch, less than 700 newsmen gathered at the Cape to describe lift-off, and only 300 planned to move to the Manned Spacecraft Center for the flight itself. Among the public at large, it was not the attraction Apollo 11 had been, less than one-tenth the number flocking to the Kennedy Space Center at the hour of launch. The astronauts too felt the disinterest. Very few pressmen attended briefings at which the crew described what they would do on their flight, and Ken Mattingly felt decidedly redundant after being dropped from the team. He never did contract rubella, and found joy in reminding the doctors so.

But the main point of discussion in the Nation's press concerned the marked lack of awareness that Apollo 13 was building a creditable pillar of scientific achievement. Many had criticized NASA for not injecting more science earlier in manned operations; many now ignored the fact that as each mission went by it leaned more heavily toward the acquisition of scientific data and information.

Apollo 13 was going at a good hour for crew sleep schedules. Launched in the early afternoon, Cape time, it would have the crew begin their rest period a lot earlier than the Conrad crew on Apollo 12. Nothing marred the smooth countdown that only served to underscore the wide acceptance that one more Moon mission was slipping berth at Spaceport, USA. But to Jim Lovell, Jack Swigert and Fred Haise it was to be the journey of a lifetime.

The crew were in the spacecraft shortly before noon, having breakfasted and suited up in the MSOB. This was the heaviest Saturn to fly. With each mission engineers incorporated modifications and minor alterations, changes designed to wring every last drop of performance from the giant rocket. The CSM was called Odyssey, after having the name Auriga rejected because it sounded too much like Aquarius, the name of the LM. In the MOCR at Houston's Manned Spacecraft Center, Ken Mattingly reluctantly walked into the room and up to the capcom console. Flight director Milton Windler tossed him a wry glance. 'Sorry to see you here, Ken ' he said, and Mattingly knew he meant it.

John Young and Joe Kerwin were at the console, listening in to the communication between spacecraft 109 and the Launch Control Center at KSC in Florida. Outside, in the open, Vice President Agnew hosted West German Chancellor Willy Brandt; perhaps it was for this reason that West German correspondents accounted for 15% of the press contingent. But despite the reduced influx of humanity to Cape Canaveral, Brevard County's Sherriff had a busy day, patrolling the lanes and roads clogged with 25,000 cars, keeping an eye on the Banana and Indian Rivers, virtually filled to capacity with small boats and launches, looking after the 500 aircraft that had flown into Brevard's private airstrip.

It was warm, reminiscent of that day in July when Man first left for the Moon. But it was easier to get around; much more pleasant to watch. Shortly before the three-minute mark in Apollo 13's countdown, launch operations manager Paul Donnelly signed off with a 'Good luck, head for the hills.' Fra Mauro awaited them, exactly 397,011 km away. In the Launch Control Center, capcom Paul White kept up a countdown sequence to Commander Jim Lovell as the final events ticked to the inevitable conclusion. Launch rules had been tightened since Apollo 12 put its big needle nose into charged clouds five months before, but nothing stood between fire in the Saturn and a space flight for the Apollo trio.

At exactly 2:13 pm, the Saturn V lifted away and up through good skies toward orbit. The first stage shut down first its center engine, then the outer four, and at 2 min 44 sec shed its load from the ascending stack. Less than two seconds later all five J-2 engines in the S-11 second stage were up and running, but vibrations, by now a common feature of Saturn V rides, increased in amplitude. Oscillations in the liquid oxygen feed system caused pressure to drop below a minimum level and the liquid oxygen pump cavitated, gulping a gasfilled bubble through the pump. This in turn momentarily chopped engine thrust below an acceptable value and, detecting loss of pressure in the center engine's thrust chamber, limit switches sent a signal to shut that engine down. Cut off 2 min 13 sec early, the other four engines continued to burn.

LOVELL 'Houston, what's the story on engine 5?'
CAPCOM 'Jim, Houston, we don't have the story on why the inboard out was early, but the other engines are GO and you are GO.'
LOVELL 'Roger.'

On the ground, everybody prayed the other four motors would keep on firing, which they did for an extra 44 seconds. But it was not quite enough to make up for the reduced performance and the third, S-IVB, stage had to fire 9 seconds longer than planned. Was there sufficient propellant for the important TLI burn out of Earth orbit to a free-return path for the Moon, a trajectory that, like Apollo 12, would be changed to a non-return type when the LM had been safely extracted? Engineers were concerned to do their sums and come up with an answer. As were the crew. Arriving in orbit 44 seconds later than planned their path was, nevertheless, quite close to the planned value.

All was well, there would be enough S-IVB propellant for trans-lunar-injection, and the crew sent the first of several scheduled TV transmissions down as the assembly passed over the Cape area 98 minutes after launch. Toward the end of the second revolution, the worthy S-IVB ignited a second time and boosted Apollo 13 out of Earth orbit. 'We see the booster doing all the right things, and FIDO says your trajectory looks good, and it looks like we'll stick with a pretty close to nominal mid-course 2; we'll have some numbers for you later,' said a calmly confident Joe Kerwin.

The TLI burn put the vehicles on course for a 770 km pass around the Moon at an elapsed time of 77 hr 51 min; about 30 hours, however, into the mission Odyssey's SPS engine would burn Apollo 13 to a non-return trajectory, carrying it to within 120 km of the lunar surface 23 minutes earlier at 77 hr 28 min. But first came transposition and docking with Aquarius before extracting it from atop the S-IVB. 'On with the TV now,' said Fred Haise as he began a live transmission of the event. Swigert had moved to Jim Lovell's left seat to conduct the docking and Haise pointed the camera through a window at Aquarius. Odyssey had already separated and was turned back upon the target.

Within minutes Swigert had successfully snagged Aquarius' drogue on the three small capture latches before retracting the probe, pulling the two vehicles together. After opening the tunnel, checking the docking latches, connecting the LM-CSM umbilical and putting back the hatch, Odyssey extracted the LM from the S-IVB. But Jack Swigert was one ahead of capcom Joe Kerwin:

CAPCOM 'And we have word that the propellant usage for T and D (transposition and docking) was nominal.'
SWIGERT 'What is nominal, please?'
CAPCOM 'Well I didn't ask that yet Jack!'

At 4 hr 18 min, seventeen minutes after extraction, the ullage thrusters on the S-IVB fired for 8 seconds to move the stage away from the docked Apollo vehicles; 21 min after that the excess liquid oxygen propellant was vented through the engine bell and at 5 hr 48 min the attitude engines fired for 217 seconds to move it on course for lunar impact close by the Apollo 12 seismometer. Devoid of attitude control, the stage began tumbling but its course was already well placed for a good impact, although not precisely at the exact spot intended. Some 13½ hours later the

S-IVB would mysteriously change speed, moving the target point almost exactly to the planned coordinates. Nobody ever knew why.

Early tracking of Apollo 13's trajectory showed no need for a course correction before the planned hybrid transfer burn at MCC_2, so the crew set up the scheduled passive thermal control mode with the space vehicles set to roll slowly at 3 revolutions per hour, evenly distributing the Sun's heat. Then it was time to charge a re-entry battery, and then to get ready for a sleep period. But Fred Haise had a comment.

HAISE 'I guess the world really does turn. I can see some of my land masses now. It must be Australia down there in the bottom and I guess we haven't really figured out what's over the – to the left. It must be some part of Asia. China probably.'
CAPCOM 'Hey, maybe the fact that you verified that the Earth really turns, we can call this the Haise theory, huh?'

In the MOCR, Gene Kranz was bringing his white team on shift to replace Gerry Griffin's men from the gold team, and in space three tired astronauts settled down for a 10 hour sleep period. Before they woke, Glynn Lunney had his men on board and soon after an elapsed time of 23 hours the crew were back talking to the ground. Lunney's capcom console buzzed with information back and forth on the Saturn V performance and on the vibrations felt in the spacecraft during the severe oscillations that took out engine 5. After a meal, preparations began for the hybrid transfer burn to place Apollo 13 on a non-return path, the only one that would put it into the correct orbit for the Fra Mauro site.

At 27 hr 20 min elapsed time the spacecraft passed the halfway point to the Moon: 207,553 km from Earth and its natural satellite. Less than three hours later the crew televised the SPS burn that slipped the trajectory away from its free-return slot. Ignition came at 30 hr 41 min and lasted three seconds. Odyssey and Aquarius would pass within 120 km of the Moon's far side. Then the TV camera was used to scan activity inside the spacecraft as the screens on Earth showed Jack Swigert drifting in the weightless environment: 'And Vance, thought we'd get a picture of Jack so that all the girls will know that he's still here.' Among all the astronauts, Jack Swigert was the *Don Juan* of the Manned Spacecraft Center; Houston would never be quite the same without him!

When it finished, the 50 minute TV transmission brought broad smiles from four people in the MOCR viewing room, the large area behind a glass screen through which visitors could watch the flight controllers; Mary Haise and children Mary, Frederick and Stephen had sat through it all. Jack Swigert busied himself with a P23 navigation sighting and before long the spacecraft were being placed in the slow role that characterized trans-Lunar coast.

About two hours before starting their second rest period, the crew received a suggestion from the MOCR that they might like to move up the planned checkout of Aquarius from 58 hr to 55 hr elapsed time and read off the pressure in the Descent Stage supercritical helium tank. Although it had seemed erratic earlier in the countdown preparations, when loaded for launch the pressure rate had actually been less than would normally have been expected. Nevertheless, if it was rising the system would need attention before it reached the burst-disc safety valve trip level of about 132.7 kg/cm^2.

Controllers agreed that if it read between 54 and 56 kg/cm^2 all would be well. If it read above 56 kg/cm^2 Houston would watch the system for a couple of hours on telemetry and if it threatened to exceed 70 kg/cm^2 the only way pressure could be relieved and still preserve the integrity of the helium supply would be to briefly fire the Descent Stage engine, opening a squib valve which in turn would cause some helium to flow into, and pressurize, the four main propellant tanks. This would be only a short, 5 second, firing, changing Apollo 13's speed by no more than 1 km/hr – insufficient to alter the trajectory significantly but enough to keep the pressurization system operational for Aquarius' powered descent to the Moon's surface.

As the vehicles cruised toward the Moon, still slowing under Earth's influence, the crew played some music from the on-board tapes and settled down for a full 10 hr rest beginning at 37 hours elapsed time. Within a few hours most of Houston was up and awake on the morning of 13 April and while Lovell, Swigert and Haise slept on, newsmen on Earth attended briefings on the big space station, the proposed shuttle transportation system, and the so-called space tug, projects for the future but ones for which publicity was a prerequisite to congressional sanction.

Awake early, Jim Lovell called down to the MOCR capcom shortly before 47 hours but within seconds there was a moan from spacecraft 109: 'It might be interesting that just after we went to sleep last night we had a master alarm and it really scared us. And we were all over the cockpit like a wet noodle.' But it had been nothing to worry about – only low pressure in hydrogen tank No. 1 which tripped the alarms before the heaters saw it and cycled on to increase the internal activity. About 30 minutes after the crew began their housekeeping chores it was observed that oxygen tank No. 2 was reading off-scale high, an indication of more than 100% capacity when only a short while before it had read an acceptable 82%, about right for that time into the flight. Houston came up with a theory that the sensor broke when the fans were activated to stir up the tanks, 46 hr 40 min into the flight; more likely, an electrical short or arc between wires in the quantity gauge.

Shortly before an elapsed time of 49 hours flight director Gene Kranz took over from Gerry Griffin. It was a fortuitous move, as later events would denote. In space, the fans in oxygen tank No. 2 were turned on again at 47 hr 55 min and 51 hr 8 min, with no adverse effects but still the off-scale high reading on the display console and the telemetry. Then it was time for a meal before getting the TV set up ready to send down pictures of LM checkout. Throughout this period, more caution and warning tones rang when the hydrogen quantity dropped to the trip level only to be cancelled by the crew when they turned on the fans and heaters.

Jim Lovell suggested opening Aquarius early so that when the TV transmission began as scheduled at an elapsed time of 55 hours the crew would have completed preliminary checkout including reading the supercritical helium pressure. Vance Brand passed along concurrence from Kranz's MOCR men and the crew got ready to open the tunnel. By 53 hr 51 min cabin pressures in Odyssey and Aquarius had been equalized and fifteen minutes later Fred Haise drifted through to the LM. Aquarius fed its own power from 54 hr 46 min and five minutes later the crew had a reading on the supercritical helium pressure: at a comfortable maximum of 50.6 kg/cm^2 it was within expected limits and would give no cause for concern.

Minutes later, Jack Lousma slipped in to the capcom seat to replace Vance Brand. Kranz would soon be taking his men off shift. At 55 hr 12 min the TV picture appeared on the large screen at the right in the Mission Operations Control Room and controllers settled down with LM systems analysis, or simply snatched glimpses of the interior, opened for the first time this mission. The transmission was not the eye-catching scene-stealer earlier Apollo flights had presented. It was all very familiar and few networks carried pictures live of yet another weightless trio drifting about in a pressurized bubble. Even many of the contractor personnel put homely duties before space spectacle.

But in Windsor Locks, Connecticut, Hamilton Standard engineers Ted Jansen and Charles Wigmore were on hand in the special control room set up to support Mission Control should anything go wrong with the Environmental Control System during checkout or lunar operation. With Aquarius opened up, they stood by – just in case. Telephones linked support rooms at the Mission Control Center; a red telephone waiting for emergency calls. In the halls and corridors outside the small room where Jansen and Wigmore sat comfortably watching TV between checking strip charts and diagrams, janitors and maintenance personnel walked quietly about their duties. It was mid-evening, and nobody could forsee the surge of activity that would soon transform that room into a hive of bustling work.

In space, Jim Lovell showed his diminished audience the Ascent Stage engine cover protruding through the floor like a hat box. Then Fred Haise went around the interior of Aquarius showing viewers the drinking water bag and tube fitted to Apollo 13's EVA suits when Conrad and Bean complained about thirst outside on the Moon. At 55 hr 38 min, Lovell decided they had shown all they wanted to: 'Okay, Houston. For the benefit of the television viewers, we've just about completed our little inspection of Aquarius and now we're proceeding through the hatch gap into the tunnel and going back up to Odyssey.' It had been a brief checkout, LM batteries had been on for only $12\frac{1}{2}$ minutes prior to the TV show, and a sleep period awaited them back in the Command Module.

At 55 hr 47 min Jim Lovell signed off for his viewers: 'This is the crew of Apollo 13 wishing everyone there a nice evening and we're just about ready to close out our inspection of Aquarius and get back for a pleasant evening in Odyssey.' Seconds later the TV image went dead and the crew began to button up the tunnel. At 55 hr 52 min 30 sec, the master alarm sounded. Again, it was low pressure in hydrogen tank No. 2, and 28 seconds later Houston passed up a request: '13, we've got one more item for you when you get a chance. We'd like you to stir up your cryo tanks.' Jack Swigert responded with a simple 'Okay,' and moved across to the left couch from where he could reach the center of the main display console to flick the four fan switches situated directly beneath the four meters that showed cryogenic tank pressure and quantity.

Telemetry showed power going to the cryogenic fans at 55 hr 53 min 20 sec but unknown to the crew just 2.7 seconds later an electrical short circuit produced an 11.1 ampere spike in the current from fuel cell 3 which began a fire in Teflon materials inside the No. 2 oxygen tank. Just 16 seconds after power was applied, the pressure began to rise in oxygen tank No. 2. Two seconds later voltage decreased on AC bus 2 and three seconds after that a massive 22.9 ampere spike affected fuel cell 3 current. Pressure was now building rapidly in tank No. 2 and fluctuating electrical activity restored the quantity reading that had shorted out more than nine hours before. Then temperatures in the oxygen tank rose quickly as combusted Teflon led a fire path inside the dewar to bunched Teflon on the tank's dome.

When that happened the fire ruptured the oxygen tank before all data ceased from the No. 2 vessel at 55 hr 54 min 52.7 sec. The entire contents of the tank, about 125 kg of by now gaseous oxygen, was dumped through either a split or a burned hole in the dome to the shelf area containing the two dewars. Sealed off above and below by separate shelves, and on the outside by the Service Module panel, the confined area was incapable of containing the gas. About 0.4 second after the No. 2 tank ruptured, oxygen now probably feeding several small fires in the Service Module's bay 4 mid-shelf area blew the complete panel out, exposing the pie-shaped bay section to the vacuum of space. At that instant accelerometers in the spacecraft picked up violent motion in all three axes as the shock reverberated through the vehicle.

Dressed for drilling. An engineer displays equipment carried aboard Apollo 13. Left, the lunar surface drill; right, the core tubes and carrier.

In the Command Module, Swigert was still in the left couch, Lovell was down in the lower equipment bay putting the TV camera away and Haise was in the tunnel between Aquarius and Odyssey returning from the Lunar Module. A loud bang accompanied the shock and 0.4 seconds after the bay 4 panel blew out all data was lost for nearly two seconds as the spinning aluminum sheet clipped the S-band high-gain antenna on the base of the Service Module, knocking it off alignment with Earth tracking stations. But the crew knew nothing of events that had so nearly led to worse disaster: extinguished by the vacuum of space, the fire that spread through the shelf area could have penetrated other systems, other equipment leading to the Command Module.

All they heard was the loud bang as the spacecraft shook violently and the caution and warning clanged in the spacecraft, telling them of a voltage drop on DC main bus B. They did not know that oxygen tank No. 2 had ruptured, nor that one complete Service Module door had blown off. Already at the displays when the master alarm light came on, also illuminating the appropriate warning indicator at the top of the panel, Jack Swigert sped across to the right couch from where he could watch the condition of the electrical systems and monitor voltage that he now knew had dropped; Jim Lovell sped up from the lower equipment bay.

SWIGERT 'Okay Houston, we've had a problem here.'
CAPCOM 'This is Houston, say again please.'
LOVELL 'Houston, we've had a problem. We've had a main B bus undervolt.'
CAPCOM 'Roger. Main B undervolt. Okay, stand by 13 we're looking at it.'
HAISE 'Okay, right now, Houston, the voltage is looking good. And we had a pretty large bang associated with the caution and warning there. And if I recall, main B was the one that had an amp spike on it once before.'
CAPCOM 'Roger Fred.'

When tank No. 2 blew out, the shock closed oxygen supply line valves to fuel cells 1 and 3. Within about three minutes oxygen already in the supply lines was exhausted and the two fuel cells ceased to function, tripping more warnings and illuminating more lights on the display panel. Apollo 13 could not complete its mission and land on the Moon with only one fuel cell working in the Command Module and realization that the flight would probably have to be aborted sent a wave of disappointment through the spacecraft. But soon they would be more concerned than ever. In checking through the systems, Jim Lovell moved across to the center of the display panel and was dismayed to see the oxygen tank 2 quantity indicator read zero; it had stuck in the off-scale high position earlier but the new reading was ominous. Only two oxygen tanks were available not only for reactant by which the single remaining fuel cell could produce electrical power but also for life-giving oxygen in the spacecraft. Dismay turned to horror when Jim Lovell looked to the No. 1 tank indicator and saw that too falling, noticeably dropping as he watched it.

What had happened was that the exploding No. 2 tank, or perhaps the fire that permeated a shelf area in bay 4, ruptured a line or part of the No. 1 tank dome, causing the contents to bleed slowly away. Lovell was now deeply concerned for the lives of his crew and as he looked through one of the windows more evidence of a very unusual situation appeared: 'and it looks to me looking out the hatch that we are venting something. We are venting something out into space.' Whatever it was it must be a prolific supply, for not only was the bay 4 area down below the Command Module but it was also round on the opposite side of the spacecraft. Unknown to any of the crew, oxygen from tank No. 1 was leaking to space. Down in the MOCR, Gene Kranz knew well the implications of what he saw coming through on telemetry.

KRANZ 'Okay, let's everybody think of the kind of things we'd be venting. G & C, you got anything that looks that normal in your system?'
G & C 'Negative, Flight.'
KRANZ 'Okay, now let's everybody keep cool, we got LM still attached, let's make sure we don't blow the whole mission.'

With only one fuel cell on line, and only main bus A operational, the first order of business was to power down the spacecraft by 10 amps to conserve electrical energy and ease the load.

Kranz ordered the spacecraft powered down with the standard procedure in an onboard checklist. About 25 minutes after the collapse of main B power and the loss of oxygen tank No. 2, the crew informed Houston that whatever was venting from the base of the spacecraft was causing a combined pitch and roll drift that could only be corrected by direct manual thruster firings.

CAPCOM 'Okay 13, we've got lots and lots of people working on this, we'll get you some dope as soon as we have it, and you'll be the first one to know.'
APOLLO 'Oh, thank you.'
APOLLO 'Okay Jack, and the weird configuration we're sitting in now, is we have the hatch installed, we still have the probe and drogue inside the Command Module, and we're going to stay in this situation until you kind of give us an okay to reinstall the probe and drogue. Or if necessary to use the LM consumables.'

It was the first time anybody had mentioned on the air-to-ground loop a possible need to power up Aquarius and use its systems for life support. But such a conclusion was inevitable. Where only minutes before it had seemed disappointing to lose the landing operation it now began to look as though the crew would be lucky to return to Earth alive. Oxygen tank No. 1 was the only bridge between a life-giving environment and an inert spacecraft.

About 45 minutes into the event, the crew reported the venting to have apparently eased although the spacecraft was still rolling round as though propelled by escaping gas. Ten minutes later Fred Haise went through a systems check from the main display panel and reported the two hydrogen tanks to be at normal pressure but that tank No. 1 was now down to 21 kg/cm^2. The lower limit at which a warning alarm would normally sound was 56 kg/cm^2. Nothing could seemingly save it now. By this time the Command Module's oxygen surge tank had been isolated, preserving the 1.7 kg of gas it was designed to provide for atmospheric entry after Service Module separation. Now it would be the only oxygen supply in the Apollo spacecraft.

The flight controllers tried one last effort to isolate the leak from tank No. 1, thinking that it might be associated with the loss of fuel cells 1 and 3 and about one hour after losing tank No. 2 the reactant control valves were isolated. That effectively terminated their usefulness, for having once been shut down through the reactant valves there would be no possibility of restoring power to the line. But dead anyway, they were useless for anything else. The leak did not stop and at that point Gene Kranz made the irrevocable decision to move back into Aquarius and power it up. The Command Module was operating as it should, but the Service Module was in an unknown state, and unable to supply fuel cell power or cabin atmosphere. About 1 hr 25 min after the problem revealed itself, the crew reported the pressure in tank No. 1 down to 14.2 kg/cm^2 and still falling:

CAPCOM 'It's still going to zero, and we're starting to think about the LM lifeboat.'
APOLLO 'Yes, that's something we're thinking about too.'

But first, the Command Module batteries must be left in top condition while the last remaining power from the drained Service Module was still available. So the crew started charging battery A and then received a special procedure for powering the Lunar Module:

CAPCOM 'And we have a procedure for getting power from the LM, we'd like you to copy down.'
HAISE 'Okay, Jack. About how long is it?'
CAPCOM 'It's not a very long procedure, Fred. We figure we've got about 15 minutes worth of power left in the Command Module. So we want you to start getting over in the LM and getting some power on that.'

One hour forty-five minutes after the alarms rang, Jim Lovell and Fred Haise opened up the tunnel and moved back into Aquarius while Jack Swigert copied down the procedure. Minutes later Aquarius' electrical system was powered up ready to bring on the lander's environmental control equipment. By now, pressure in oxygen tank No. 1 was down around 7.1 kg/cm^2. In the MOCR, Glynn Lunney was relieving Gene Kranz who would pull his men out for detailed studies of recovery procedure. In the docked assembly of Odyssey and Aquarius, now opened up all the way through, Jack Swigert continued to power down the Command Module, making sure to leave on the inertial platform until Aquarius performed a course alignment with its own IMU. Without power in the Command Module the Lunar Module would have to know which way round it was in space so as not to lose guidance information. Once Aquarius' platform had been aligned, Jack Swigert turned off the Command Module computer. Never before had an Apollo been so inert en route to the Moon.

There had been little time aboard Apollo 13 to worry about getting back. Jim Lovell's immediate concern had been to sustain life by bringing up the Lunar Module and safely preserving surge tank oxygen and battery power in Odyssey. But flight controllers knew full well that it was touch and go. They suspected that other systems in the Service Module might be affected by whatever ailed the oxygen tanks and soon realized that it would be unwise to operate other equipment. Clearly, an engine burn would have to be carried out because the docked assembly was on a non-return path that after looping around the Moon would bring it to within only 74,000 km of Earth. Aquarius' Descent Stage engine, the one designed to land the LM on the Moon, would have to be used; the Service Module's SPS might be damaged so that engine was not available.

Apollo 13 was more than 329,000 km from Earth when the explosion ripped through bay 4 and the abort opportunity loaded for use would have required an SPS engine burn to slow the spacecraft by 6,700 km/hr. Performed at an elapsed time of 60 hours, it would have decelerated the spacecraft from crossing the equigravisphere, looping it back without going around the Moon, for a splashdown less than 60 hours later. But that was out if only because Aquarius' Descent Stage engine was incapable of slowing the nearly 44 tonnes of docked spacecraft by that amount; its total velocity change capability was designed around its own mass without the CSM attached. So that left the DPS engine on Aquarius the task of burning before the spacecraft crossed the equigravisphere to place Apollo 13 back on a free return trajectory.

As for consumables aboard the LM, that too would depend on the duration of the flight. If left to the looping path following the DPS burn, Apollo 13 would not reach Earth before an elapsed time of 155 hours. If another burn was performed shortly after coming round the Moon to speed up the return path, the crew could target for either an Atlantic splashdown at 133 hours, or a Pacific recovery at 142 hours. Was there enough water for the cooling system, oxygen to breathe, battery power for electrical energy, and lithium hydroxide for removing carbon dioxide in Aquarius? The Lunar Module was designed to support two men for 33–35 hours on the lunar surface. Could it be stretched to allow three men to survive for 85–100 hours?

Almost as soon as it became apparent that Aquarius would have to play a more demanding role than any other LM before or since, engineers at the Manned Spacecraft Center began a series of frantic calls to team specialists and contractor plants across the nation. As builder of the LM Environmental Control System, Hamilton Standard were in the front line of battle for life aboard Apollo 13. At best, it would be a race against time; at worst, a pitiful incapacity to bring the crew home.

In Windsor Locks, Connecticut, Ted Jansen and Charles Wigmore sat through the panoply of events from the first report of a major problem, the realization that the crew would have to use Aquarius as a lifeboat dawning on them with slow reluctance. From Houston, Hamilton Standard's PLSS back-pack manager Cal Beggs received a call. He was in bed asleep at the time, as was Andy Hoffman, program manager for the Environmental Control System, when Beggs called him. Hoffman was at the Windsor Locks control room within ten minutes, having driven 18 km in the interim. Chief engineer on Aquarius' lift-support systems, Warren Pinter, was stirred by Hoffman, who then proceeded to bring in all the systems engineers who could possibly be of help.

By the early morning hour after midnight, Tuesday, 14 April, Hamilton Standard's control room hosted a motley collection of bleary-eyed technicians, all sporting garments hastily donned. Across in New York, Grumman pulled its people in from all over the city and beyond. One employee speeding to the Bethpage facility was pulled up by a police car, only to receive a welcome escort, sirens wailing, when he quickly explained what had happened. While guidance and propulsion engineers in Houston got down to working out a maneuver pad for the free return burn coming up, Hamilton engineers with their Houston colleagues calculated the status of consumables.

Oxygen would probably be adequate. With 21.8 kg in the

Descent Stage tank and a further 2.2 kg in the two Ascent Stage tanks, the supply would theoretically last much longer than needed by Apollo 13. Designed to support two re-pressurizations plus back-pack filling operations, it was more than enough for a constant cabin atmosphere without EVA; each re-pressurization after a Moon walk used about 20% of the total quantity. As for water, that would be very tight. Operating through to a sublimator designed to draw as much as needed, it would be up to electrical systems engineers to keep heat sources down as low as possible and conserve water. With 157 kg in the Descent and Ascent Stages, supplies would be at a premium.

When Lovell and Haise powered Aquarius up the system was drawing nearly 2.9 kg of cooling water per hour. At that rate the tanks would be dry at an elapsed time of 110 hours! It was imperative to get that rate down as soon as possible. But how low could the temperatures be allowed to safely get before equipment froze up? Critical cooling areas included gyroscopes in the guidance platform but every piece of systems equipment was a candidate for being switched off to reduce heat, and consequently cooling water consumption, levels. Battery power was probably the most critical problem area. With four Descent Stage and two Ascent Stage batteries, the spacecraft had a total 2,192 ampere-hr supply, of which less than 2,000 ampere-hr was usable due to earlier operations in Aquarius. Under normal conditions, Lunar Module systems would use about 50 amps when powered up, or about 20 amps completely powered down to life-support levels. Apollo 13 would have to draw an average of less than 15 amps to get back with power to spare.

Within one hour of powering up Aquarius, the systems in Odyssey were switched off. Fuel cell 2 was finally shut down at 58 hr 38 min, irrevocably cutting all electrical power from the Service Module. Then it was time to power down Aquarius to as low a level as possible. Communications would have to go through the omni-directional antennae rather than the power-consuming high-gain dish on top of the Ascent Stage. From the spacecraft, Jim Lovell suggested rigging up the urine hose to the dump valve on the side hatch to prevent heater power being used, and to prevent the urine dump equipment freezing up. On the ground, maneuver pads were being worked up to put the trajectory back on a free return path, and also for a second burn two hours after passing close across the far side of the Moon – a speed-up maneuver called Pericynthion plus 2.

In the simulators, astronauts Cernan, Evans and Engle were already running through procedures for the docked burns and for star alignment checks from Aquarius. It was the middle of the night in Houston and already people were working on a proposed schedule for crew sleep periods. But first, the hybrid burn to restore a return path and for that the LM landing gear had to be deployed. Coffee flowed strong and hot on the rows of consule positions as the final details were generated. The burn came at 61 hr 30 min. The DPS engine was ignited at 10%, increased to 40% at five seconds and continued to thrust at that level for a total 30.4 seconds. Based on the results of the burn, Apollo 13 was predicted to pass within 252 km of the Moon's far side. Now Aquarius was to be powered down as low as possible before the Pericynthion plus 2 maneuver nineteen hours later.

But the next problem was what to do about the thermal environment. Normally, Odyssey's RCS thrusters would have been used for the passive thermal control roll but now Aquarius would have to perform that function and the torque generated by spinning two vehicles from the top would be very different to the dynamics of rotation imparted from the central plane of the docked mass. So new procedures would be necessary here too. Mission Control came up with the idea that the computer aboard Aquarius should be powered up just to point the vehicles in the correct attitude before powering it down again and using the hand controller to impart a slow roll. From the ground came a suggested schedule designed to provide maximum possible rest between activity. Haise would begin a sleep session lasting 6 hours at an elapsed time of 63 hr, all three would eat a meal at 69 hr, and Lovell and Swigert would sleep for 6 hours from an elapsed time of 70 hr.

Already, debates were raging in Mission Control over the type of Pericynthion plus 2 speed-up burn to perform. Flight directors Glynn Lunney and Gerry Griffin talked with Deke Slayton, Chester Lee and Rocco Petrone about a possible super-fast return which would put the crew back on the Pacific at 118 hr. But to achieve that, the Service Module would have to be jettisoned, cutting weight by about 50%, leaving the Command Module still attached to the Lunar Module until the time came to move back into Odyssey and separate from Aquarius. But that in turn would require the Command Module's base heat shield area to be exposed to the thermal environment of space for 40 hours. Nobody knew how it would react if soaked in unfamiliar temperatures for that duration.

A preliminary maneuver pad had already been voiced up to Fred Haise containing figures for a burn that would bring them back for a recovery at 142 hr. Could the consumables last that long? If they could it would be a preferable way to go. There seemed reasonable chance, if electrical consumption could be chopped to a mean of just 17 amps, which in turn would draw upon 1.21 kg of cooling water per hour. That should get the crew home – but only just. At an elapsed time of 66 hours, Glynn Lunney's black team came off shift as Gerry Griffin's gold men clocked in. Up ahead was the pass around the Moon and whatever Pericynthion plus 2 burn was decided.

As a possible means of supplementing the water supply, the crew suggested a procedure whereby the back-packs could be used to fill the LM tanks, reversing the role for which the system had been designed. That was a problem for Hamilton Standard to wrestle, but they were already working on that one, as well as a procedure where if things went from worse to worst the crew could feed urine into the supply lines. Then there was a problem with carbon dioxide possibly building up in the Command Module so Mission Control suggested 'that you take the Commander's hoses in the LM and put a cap over the red return hose so they'll blow up in the CSM by extending as far as possible . . . we'll get some flow off the blue side circulating up and around the Command Module to keep the CO_2 level down.' As for the passive thermal control roll, the immediate answer was to have the crew manually fire Aquarius' thrusters every eight hours to turn the vehicles physically 90° at a time.

The news of Apollo 13's trouble broke like waves of disbelief at home and abroad. By now the United States was waking up. It was Tuesday, 14 April, and what they heard on the radio and TV, and what they read in late issues of the newspaper, stunned many people. Nobody quite expected anything like this and it shook the apathy that had once been 13's trademark. At 6:40 am, Tom Paine flew in to Ellington Air Force Base and was driven swiftly to Mission Control. Little more than one hour later, Richard Nixon telephoned Paine from the White House: the President was deeply concerned and asked to be kept in constant touch with the NASA facility at Houston. Already, President Pompidou of France had offered the full resources of his Navy should Apollo 13 have to land in unassigned water. It was but the first of many offers of assistance from several nations around the world.

At an elapsed time of 69 hours, Fred Haise was woken from his cool station in the darkened Command Module. The spacecraft were still some hours away from passing behind the Moon and after a snack Lovell and Swigert would try to get some rest. Communications with Earth were scratchy and indistinct; powered down to minimum, the configuration was not conducive to good links. Mission Control advised the crew to get the lithium hydroxide canisters out of the Command Module before they swelled up and proved difficult to extract from storage containers. Equipped with only two primary lithium canisters, Aquarius would be unable to clean carbon dioxide from cabin air throughout the long coast home, and the Command Module canisters were a different size and shape so they were unable to fit the LM Environmental Control System. Nevertheless, a procedure had been worked up on the ground where adhesive tape and cardboard would provide a makeshift oxygen cleanser from Odyssey's canisters.

But not yet. Where lithium canisters would normally be changed when the partial pressure carbon dioxide level reached 7.5 mmHg, controllers agreed, after consultation with the flight surgeon, that Aquarius could wait until the CO_2 level reached twice that value. Then the primary canister would be changed for the secondary and only when that was saturated would the contingency procedure be implemented. Until then, tests could proceed in ground simulators, just to prove there were no adverse effects. Simulators were over-worked already. Dave Scott joined fellow astronaut Gene Cernan at Houston to practice procedures worked up by staff support rooms, and at the Kennedy Space Center Dick Gordon busied himself in the Cape simulator with other procedures.

Shifts of flight controllers elsewhere worked new methods of guiding, controlling, and navigating the unusual configuration, and specialists across the United States came up with unique ways of operating equipment designed for a different purpose. While Hamilton Standard's James Morancey worked out a way to manage the water supply, Carl Beggs proposed a means by which Command Module water could be fed to the PLSS back-packs, encapsulating it for transfer to the Lunar Module's life-support system. Not only could Aquarius get water from the tanked back-pack supply, it was perhaps possible now to actually transfer it from one spacecraft to another.

The tide had turned. With Apollo 13 now on a free return path to Earth, with still a few hours to go before it looped round the Moon, controllers had the measure of consumables budgeting – so much so that it actually began to look as though the LM's electrical supply just might support a long, slow, charge-up on the Command Module's entry batteries, topping them with a good reserve of power for the final descent. It was a weird atmosphere in Aquarius. Only darkness met a gaze up the tunnel to Odyssey, and only minimum lighting was on in the Lunar Module. The spacecraft was not too cool at this point, but there was a strange feeling of eerie detachment. Few lights shone in the grey spacecraft and the normal sounds were strangely muted. But Jim Lovell got only a short nap before drifting back into Aquarius and resuming conversation with Charlie Duke in the MOCR about star alignment checks prior to the Pericynthion plus 2 burn.

Most of the procedures were for manual operations that in the absence of computerized programs for this sort of activity were the only means by which the docked vehicles could be run. Long lists of checkout activity and systems operations were read up to the crew as one by one the LM systems were brought back on line in readiness for the speed-up burn. Final tracking indicated the spacecraft would pass within 254 km of the lunar surface and at an elapsed time of 76 hr 32 min, Apollo 13 went into the shadow of the Moon. Lovell was early getting the vehicles into a correct attitude for the burn and capcom Vance Brand told the crew they could sit in that attitude until making the maneuver before resuming passive thermal control. The vehicles went behind the Moon's west limb at 77 hr 8 min and for twenty-five minutes nobody could communicate with them.

Eight minutes after disappearing the spacecraft broke into Sunlight and Fred Haise and Jack Swigert busied themselves with cameras, snapping away at the barren lunar surface it was their fortune to see so close but never visit. Fate and mission 13 had seen to that. Jim Lovell was still deeply concerned about their prospects. In Houston, it all looked good on paper. But out there in the black cold of deep space it was an uncomfortable thought that everything would have to keep operating at least as efficiently as it was then to last a further three days. When Apollo 13 came back into view at 77 hr 33 min it was on its way home. For the first time since the explosion, the distance was narrowing with each passing second. That was a welcome, albeit psychological, comfort to both crew and ground personnel.

An event of almost inconsequential interest occurred shortly after the spacecraft reappeared. At 77 hr 56 min, the Saturn V's third stage hit the Moon 137 km from the spot where Pete Conrad and Al Bean had deployed the ALSEP seismometer. It was the first time an S-IVB had been dumped on the lunar surface.

Vance Brand was at the capcom console as flight director. Eugene Kranz got status checks from his controllers about the readiness of Apollo 13 to perform the Pericynthion plus 2 burn. A decision had already been made that the 4 min 23 sec DPS firing would speed the return by nearly 10 hours and drop the Command Module into the Pacific at an elapsed time of about 142 hr 53 min. A lot of people were in the MOCR. It was an important firing. Without it, no one was too sure the consumables aboard Aquarius would last. Tom Paine was there, with George Low, his deputy, astronauts Frank Borman, Lovell's Commander the last time he flew that route, Ken Mattingly, Al Shepard, Ed Mitchell, and Stu Roosa.

Ignition came at 79 hr 27 min and Swigert was in Odyssey with Lovell and Haise manually controlling the burn from Aquarius. For 5 seconds it fired at 10% thrust, then increased to 40% for 21 seconds, and continued at full thrust for nearly four minutes. At shutdown, Apollo 13 had speeded up by 945 km/hr. Without that burn the spacecraft would have arrived back on Earth at an elapsed time of 152 hr to an Indian Ocean splashdown. Now it was on course for a Pacific recovery nine hours earlier.

Astronaut Swigert holds the 'mailbox' built to use the Command Module lithium hydroxide canisters to purge carbon dioxide from the Lunar Module on Apollo 13.

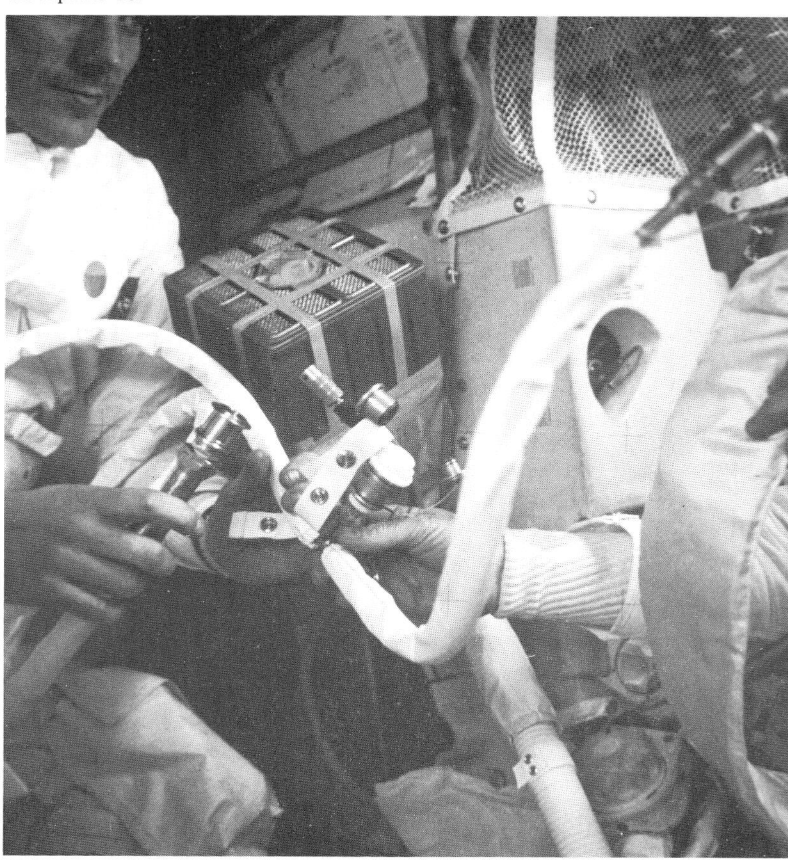

Power for the burn ran at 50 amps so it was important to get back down to a low consumption level as quickly as possible. It took about 45 minutes to set up for the passive thermal control roll, and then a further 30 minutes to stabilize the rates, but the new procedural method adopted for this mission worked well eventually and by 81 hr 20 min flight director Gene Kranz had trajectory reports back from the tracking stations that told him the Pericynthion plus 2 burn had been good. Only a small course correction would be needed at an elapsed time of about 105 hours; both PGNCS and AGS had been powered up for the speed burn and it looked as though the primary system had done the job intended.

By 82 hr 30 min the spacecraft was powered right down to a consumption rate of only 12-14 amps and a quick check on the consumables showed they were just within tolerable limits, although water was still the leading contender for extinction, with lithium hydroxide a close second. One hour later it looked as though the first lithium hydroxide canister would reach the new toleration level of 15 mmHg by 83 hr 30 min. After that, the secondary canister carried as a cleanser for when the primaries were changed over, would last a further six or seven hours after which the second primary could be used. By this time Fred Haise was waking up after a sleep as Jim Lovell and Jack Swigert tried to get some rest. There were 2½ days to go before splashdown, and still a long way to coast.

Shortly after the two men went to the Command Module, Fred Haise reported having seen some additional venting from the direction of the Service Module and a small piece of metal drifting past the window. It gave brief concern that what he saw was something new, but nothing untoward showed up on the systems so it was a matter of academic concern. Alone in Aquarius, Fred Haise had time for a few casual words:

HAISE 'From the sounds of all the work that is going on and is still going on, this flight is probably a lot bigger test for the system on the ground than up here.'
CAPCOM 'Yeah, we've been working it out a little bit.'
HAISE 'You guys have really got a tough job right now. . . .'
CAPCOM 'Well everybody down here is 100 percent optimistic. Looks like we're on the top side of the whole thing now.'

For the next several hours capcom Jack Lousma kept up a steady dialogue with Fred Haise, conversing with him through the long lonely hours while his colleagues slept in Odyssey. At 85 hr 24 min, Haise switched from the first primary lithium hydroxide canister to the secondary as CO_2 partial pressure reached 15 mmHg – twice the level ever seen before in a manned spacecraft, but acceptable for this emergency; the second primary would be switched on when the second canister reached maximum soak. The first had lasted 27 hours. If the secondary lasted 6 hours, and the last primary operated 27 hrs also, there would be oxygen cleanser up to an elapsed time of 118 hr. After that, a jury-rigged system would suffice on canisters from Odyssey. Then, at 85 hr 50 min, a call from Jim Lovell.

LOVELL 'Tell me how do you read now? Fred went back to get some rest. This is Lovell here.'
CAPCOM 'Gee whiz. You got up kind of early didn't you?'
LOVELL 'It's cold back there in the Command Module.'
CAPCOM 'Well, what we were really thinking about doing is letting you sleep a little bit longer because we figure you're pretty worn out. . . . Is Jack up there with you?'
LOVELL 'No Jack's still sacked out.'
CAPCOM 'Okay, Jim. . . . And we just went on to the secondary CO_2 canister. Fred swapped out the primary but we want to stay on the secondary until it is all used up.'

As the hours drifted by, Lousma talked with Lovell about products still apparently venting from the Service Module, and about the mid-course correction burn coming up. At 90 hr elapsed time, Jack Swigert joined Lovell in Aquarius, leaving Fred Haise still resting. Gradually, spirits were building up in the face of this adversity, and on the spacecraft Jim Lovell played some music on tape.

CAPCOM 'You got a Chinese band going up there?'
LOVELL 'Oh, sorry. I forgot I was on mike.'
CAPCOM 'Sounds pretty good.'

In Houston, it was the start of another day as Jack Lousma read up a recommended plan for everybody to get a meal at 95 hr, followed by a six-hour rest period for Lovell and Swigert starting one hour later while Fred Haise kept watch. It was the second full day since the explosion and on Earth as Glynn Lunney's team replaced Milton Windler, Americans were getting up.

Wednesday, 15 April. The first full day had been spent putting Apollo 13 back on course and in a fit condition to survive the long coast home. This day would be spent improving systems operation, jury-rigging a new CO_2 cleanser, and working up a procedure for partially powering up main bus B in the Command Module prior to transferring a charge into the entry batteries. It took Houston a full hour to read up the procedure for that last activity and by 92 hr elapsed time all three astronauts were in Aquarius getting a meal. Jim Lovell had received the least sleep of all three, but Fred Haise had just completed a good eight hour period. Medical monitoring of the crew's condition was made more difficult by the absence of the biomedical telemetry harness normally worn by at least one or two astronauts during sleep periods. With the communication configuration optimized around good voice data, biomedical monitoring had been abandoned when the data came in too scratchy to use. But even voice communication was difficult to achieve. With only two sets of equipment in Aquarius one crewmember was always off line, and for the most part only one astronaut wore a headset.

By 93 hr 25 min Houston advised the crew to hook up the contingency lithium hydroxide canisters. During the hours that preceded this, Lovell and Swigert had taped a small Command Module canister to each of the two suit outlet hoses carried in Aquarius. Normally attached to the suits, they were taped to the hydroxide cans so that oxygen from the spacecraft would pass through them before going along the plumbing and back into the cabin. Sucked through by the pumps, atmospheric oxygen would be cleansed before re-distribution. Mission Control had earlier decided not to switch to the last LM primary canister, but to start operating the jury-rigged system first so that, if it failed to work, controllers would have a buffer from which to derive new procedures. There were 16 unused Command Module canisters in all, each of which would last a nominal 12 hours. There were also two PLSS canisters that could be similarly attached to the outlet hoses for a further 14 hours of cleaning. Carbon dioxide levels no longer seemed the problem they threatened to be a day before.

Now it was time to go into Odyssey and configure the switches ready for a check on main bus B before Lovell and Swigert moved in for a rest. Mission Control knew the next 36 hours would in some ways be the worst of all: devoid of power for many hours, things were going to get a lot colder and nobody wanted the entry batteries in a condition where switches would not activate the flow of current; it would be a terrible thing to have come all the way back only to find the Command Module systems frozen up.

In operating the mission control procedures, flight directors Windler, Lunney, and Griffin agreed to free Gene Kranz and his men to work up procedures for re-entry. It would be a difficult and hitherto untried operation. Not only was the CSM coming back with an almost full load of SPS propellant, but the Lunar Module was still attached as well, and would remain so right up to the latest time possible. Three teams would do 8-hr shifts, and Kranz would come on for the final tense activities: getting back into the Command Module, powering up minimum systems with the entry batteries, and pushing free from the Service Module *and* the Lunar Module.

In Aquarius, the powered down configuration was limping along toward Earth with a drain of only 10 amps on the batteries. And from the spacecraft itself, down through the communications loop, came a very appropriate song: 'The Age of Aquarius.' Nobody questioned that, except the female voice did bring a jibe from capcom Joe Kerwin: 'Hey, have you guys got a woman on board?' Now, Apollo 13 was back inside Earth's gravisphere, and noticeably accelerating. That too provided a lift, a morale booster that seemed to make the preceding hours more worth while than ever.

But again Jim Lovell had difficulty sleeping and at 98 hr 30 min he was back on line to capcom Vance Brand, boasting about the '4 or 5 hours' he had managed to snatch in Odyssey. Now there was additional venting from the Service Module area, in fact it seemed hardly to have stopped since rounding the Moon, but Houston passed along word of a very positive vent that was building up for an inevitable release: the supercritical helium was building to the level where it was expected to exceed the burst-disc pressure, at which point it would suddenly pop open and vent the helium to space – but not for another eighteen hours or so. Before that there would be a course correction burn using the Descent Stage engine again.

At an elapsed time of 100 hours, Apollo 13 was at the halfway point between the time Aquarius began its lifeboat role and the time it would be cut free from Odyssey prior to re-entry. The consumables looked quite good, not as marginal as before, and satisfactory for a return at 142 hours elapsed time. There was about 55% water remaining, about 55% electrical energy in the batteries, and about 75% oxygen remaining. Water and electrical

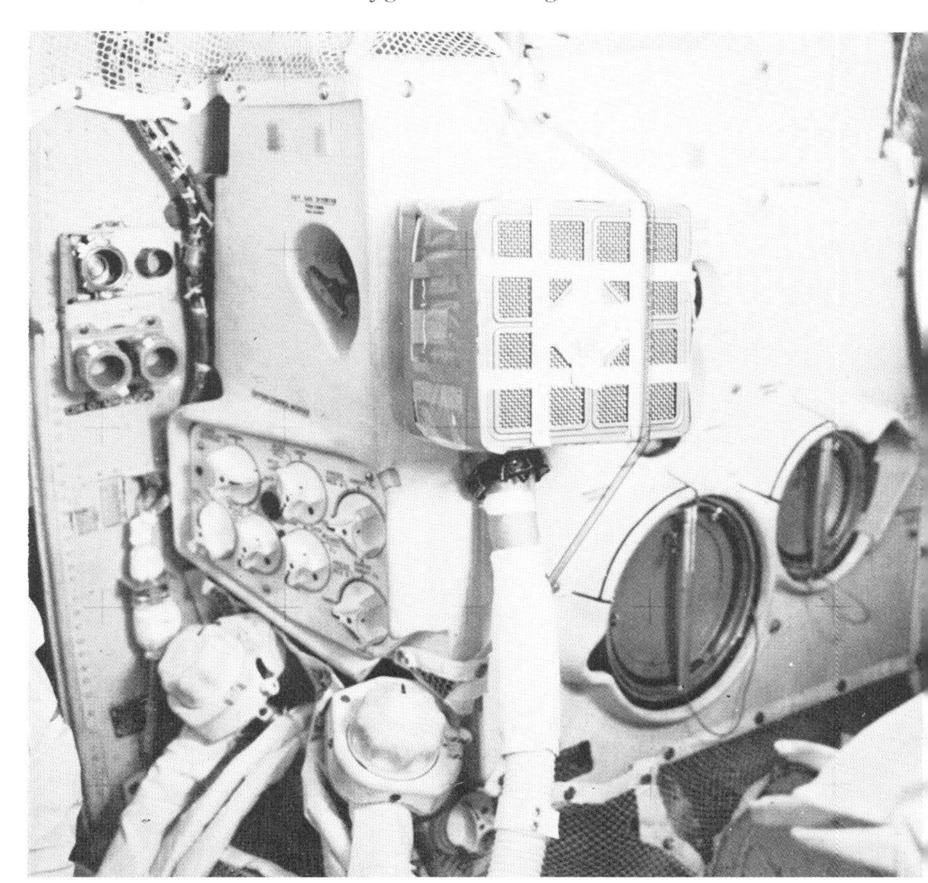

In this jury-rigged configuration, Command Module odor canisters were adapted for use aboard the LM.

power were the two to watch, with as much time still to go as had been taken already to consume the balance.

Timing for the mid-course correction burn turned on the rising pressure in the helium tank; it was necessary to get that maneuver out of the way before the burst-disc valve popped. New to Apollo operations, it was a predictable consequence of having fired the engine twice without consuming larger quantities of propellant and did not signify a new problem or an unexpected event. The two earlier burns had been made with a small degree of error in Aquarius' guidance platform; there had been little time to get a good alignment. Consequently, the course was not quite on the desired re-entry path and it was essential to move it across for, without any further engine burns, Apollo 13 would just miss the Earth's atmosphere.

Also, there could be no chance of trimming any slight overburn because the appropriate LM thrusters would impinge directly upon the docked Command Module. So the crew were told to shut down the engine one second before theoretically reaching the correct velocity change, then if any more burns were necessary they could be made for a minor course correction using thrusters other than those facing the Command Module.

Before the burn that would place Apollo 13 on the correct course for home, ground controllers took a sneak look at systems in the Command Module as Jack Swigert and Jim Lovell powered up some equipment on bus B for 7 min 37 sec of telemetry that told engineers the precise state of the re-entry vehicle. That information was to assist Gene Kranz work up entry procedures and confirmed the condition of CSM systems. But Jack Swigert was concerned about the chilly atmosphere in Odyssey: 'I'll tell you Deke, it's cold up in there, I don't know whether we'll be able to sleep up there tonight, it must be about 35 or 40 degrees (1.7–4.4°c).' But capcom Vance Brand questioned this because readings on the ground indicated that temperatures in the two spacecraft were about the same. Jim Lovell responded: 'Well, we really don't know. There's two people in the LM cabin and it seems to be a lot more compact so we don't notice the coldness down here as we do in the Command Module.'

Now there was a lot of work to do preparing the controls for the next engine burn and it took the crew about one hour to carefully place all the switch positions in locations dictated by procedures voiced up from the capcom console. When it came, at 105 hr 18 min, the 15 second burn was performed at a 10% throttle setting and shaped the trajectory around the planet's predicted position at entry. It was time to power down the vehicle again and to settle back for a long cruise home. So far everything was going well.

In Houston, Wednesday was drawing to a close as Swigert and Haise prepared to get some rest, leaving Lovell on watch. By 106 hr elapsed time, Aquarius was coasting along drawing only 10-12 amps of power. Systems were holding up, and the makeshift carbon dioxide cleanser was working well; the CO_2 level was only 1.2 mmHg. So long as power was kept down to an average of below 14 amps, there would be just enough power to get back *and* top up the batteries aboard Odyssey. In Houston, Gene Kranz and his capable team had just about completed writing the new procedures for events prior to re-entry and there was renewed hope of applying a little power to the Command Module both to warm it up slightly and to push some charge into entry battery A, the one with the least current in it.

There had been just one potentially significant problem: ground controllers received a malfunction signal from one of the four batteries in the LM Descent Stage; without it there would be insufficient power to get back. Checks revealed nothing untoward, the indicator probably being faulty. But now Aquarius was being asked to perform long past the normal duration for which the vehicle had been designed. To support a normal $1\frac{1}{2}$ days on the Moon, the Lunar Module would operate for nearly 48 hours. Aquarius had already lasted more than 50 hours, and there were still $1\frac{1}{2}$ days to go before the crew could switch back to Odyssey's systems.

Thursday, 16 April. Scheduled to get some sleep beginning at an elapsed time of 106 hr, Swigert and Haise remained up as the supercritical helium tank increased in pressure toward the burst-disc rupture level, predicted for the early morning hours. When it blew it would prohibit further use of the Descent Stage engine but the last course correction had Apollo 13 back on an acceptable entry path so the trajectory was well within the capacity of the small thrusters; and there was always the Ascent Stage engine, although that would be a risky process facilitated only by jettisoning the Descent Stage. Capcom Jack Lousma was on duty when the helium tank blew its safety valve, at a pressure of 136 kg/cm², and emptied the tank through two horizontally opposed, supposedly non-propulsive vents.

CAPCOM 'See anything?'
LOVELL 'Yeah, Jack. I was just about ready to call you. Underneath quad 4, I noticed a lot of sparklies going out.'
CAPCOM 'Can you hear or feel anything?'
LOVELL 'I sure did, but I think it changed our passive thermal control. Let me check it.'

The venting gas had knocked Apollo 13 slightly off alignment for the barbecue roll necessary to prevent one side of the configuration heating more than the other. In fact it had stopped the vehicle and set it moving in the opposite direction. That took some time to correct. The burst-disc valve blew at 108 hr 54 min but nineteen minutes later the crew received a master alarm coupled to the LM battery that gave an earlier indication of failure. Again, it was just a sensor problem, but yet another reminder that a single major failure in any system at this point could be disastrous.

It was now a matter of survival. With the supercritical helium tank in a safe condition once again, Swigert and Haise got their heads down, but in the Lunar Module because Odyssey was now very cold and uncomfortable. Jim Lovell stayed on watch. In Houston, it was still dark, and flight director Milton Windler was about to hand over to Glynn Lunney. On the communication loop, Jack Lousma chatted with Lovell about the procedure for transferring a charge to Odyssey's entry battery A, the one that had been depleted of 20 amps when the Service Module fuel cells collapsed. That would take a 120 ampere-hr load from Aquarius but frugal use of the six LM batteries provided a margin sufficient to give Odyssey a full electrical load for the critical entry phase coming up.

On Earth, reaction to the plight of Apollo 13 had been enormous, quite the reverse of apathetic trends displayed by a media that seemed to have discarded the events that once stirred the minds of men everywhere. As soon as the terrible realization of possible tragedy dawned with alarming clarity, the space program had its public audience back. Newsmen flew quickly to Houston, the prospect of fatal consequences exciting the press to new heights of attention. In the Senate on the first morning of the accident, a resolution gained unanimous support calling for 'all businesses, commercial operations, communications media, and others, who wish to and can comply to pause at 9 pm today, 14 April 1970, in order that all persons who so desire may join in asking the help of Almighty God in assuring the safe return of the astronauts.'

In St. Peter's Basilica, Rome, Pope Paul VI told an audience of 10,000 that 'We hope that at least their lives can be saved.' In France, radio stations interrupted regular programs with news of the unfolding sequence, while *Le Monde* said that 'The whole human race is participating with them in the agony of their return.' Ironically, Australian television carried the film 'Lost in Space' but superimposed captions informing viewers of Apollo 13's status. From Moscow there were brief announcements on domestic broadcasting stations but official comment from the Soviet delegation to the Committee on Peaceful Uses of Outer Space expressed hope for the astronauts' safe return.

In Houston, Marilyn Lovell and Mary Haise kept their older children away from school and remained at home, listening to the NASA 'squawk' boxes installed to relay every word from the spacecraft. Next day, 15 April, Premier Kosygin sent President Nixon a message informing the Chief Executive that 'the Soviet government has given orders to all citizens and members of the armed forces to use all necessary means to render assistance in rescue of the American astronauts.' From Britain's seat of government, Prime Minister Harold Wilson offered Royal Navy services should they be of use, as did the Italian Defense Minister for his maritime force. From the World Council of Churches, Dr. Eugene Carson Blake implored people the world over to pray for the safe return of Lovell, Swigert and Haise, and in earthquake-devastated Santa Ninfa, Italy, Mayor Vito Bellafore said that 'all our worry is for those three lonely men.'

Next day, 16 April, President Nixon told Kosygin that 'I will let you know fast if we need your government's help.' It seemed almost an omen of increased cooperation between these two powerful states: the first Strategic Arms Limitation Talks (SALT)

Carefully monitoring conditions aboard the crippled Apollo 13, personnel at the Manned Spacecraft Center show tense expectancy. Seated (right to left): Alan Shepard, Edgar Mitchell, Raymond Teague, guidance officers. Standing (right to left): flight director Pete Frank, astronauts Evans, Cernan, Engle and England.

discussions between the United States and the Soviet Union were just opening in Vienna. By now, thirteen nations had pledged their forces to be held at the ready for possible assistance should Apollo 13 have to return to unfamiliar territory and more than 70 countries said they would not use radio frequencies close to those assigned the spacecraft communication channels. It was as if the world was uniting under the enormous drama now unfolding, growing each minute as the emotions of a global population poured out in support of the three brave men then returning to a Pacific splashdown.

Shortly after 5:00 am on Thursday, 16 April, Swigert successfully completed procedures that began the transfer of electrical power from Aquarius to battery A in Odyssey. It was quite a switch. For sixty hours the Lunar Module had been the only means of electrical and environmental support. So effective had been the power-down exercise that there was energy in reserve to bring the Command Module back to peak capacity. It would take more than 15 hours to put 20 amps in to the battery, feeding current to Odyssey's battery charger through main bus B.

All three crewmembers were now up and working, the pre-planned schedule drifting off the assigned times as the astronauts worked out their own nap times. There was neither the atmosphere nor the environment to get a complete rest, just sufficient drain on the nerves to send one or other into a doze whenever possible. It was getting very cold in Odyssey, and Aquarius was cooling fast as well. There was a danger here in bringing power into the system in the belief that most of the problems were over, but over-confidence now could undo all the previous effort. Jim Lovell tried to get some rest while Swigert and Haise stayed awake. New procedures were developed for changing the lithium hydroxide canisters from their jury-rigged positions; the Command Module canisters lasted only 12 hours and had to be regularly changed.

After only three hours, Lovell woke up and sent the other two into a rest period. The cold was now getting to them, wearing away their resistance to the unpleasantness. Boots were a help, and gloves too kept the cold out for a while, but the men were contained in such a comparatively small space that bulky clothing hindered mobility. Apollo 13 was now about half way back, equal in distance between the Moon and the Earth. But the docked vehicles weighed more than any other combination at this point on preceding flights. In fact it was the first time a Lunar Module had flown back from the Moon's vicinity. After the three Descent Stage engine burns, Apollo 13 still weighed 39.8 tonnes, of which 67% was propellant, mostly in the Service Module SPS tanks.

Yet it still seemed strange to feel the uncomfortable atmosphere inside the Command Module. The walls were perspiring and damp with moisture, the windows were covered with water, and the small food lockers felt like refrigerators. By now, the temperature was down to 11°c, and getting colder by the minute. Now it really was a matter of survival. Where earlier there had been plenty to do for safe return there now came the period of waiting. Waiting for time to pass as the Earth pulled Apollo 13 closer; waiting for the distance to narrow between the spacecraft and the home planet. Morale was high, in space and on the ground, but the pressures of the past two and one-half days were taking their toll. At the Mission Control Center, a message arrived from Henry H. Wilson, Jr., President of the Chicago Board of Trade: 'The Chicago Board of Trade will suspend trading at 11 am today for a moment of tribute to the courage and gallantry of America's astronauts and a prayer for their safe return to Earth.' But still the cold increased, playing a dangerously prominent part in the events aboard Apollo 13.

CAPCOM 'Jim, Houston. You guys put on any extra clothes to try and ward off the nip of Jack Frost? Over.'
LOVELL 'Well, the lunar men are in two pairs of underwear. . . .'
CAPCOM 'Okay . . . incidentally, you've less than twenty-four hours to go.'
LOVELL 'Another note of interest to the crew systems people. Tell them they don't have to bother putting a refrigerator on board. I just went out for some hot dogs and they might be freezing.'

Now, with all the crew awake, it was time to discuss the complex procedures beginning sixteen hours hence when re-entry operations would begin with power introduced to Odyssey's systems. Then a long list of stowage positions for equipment that would fly back with the crew. In Houston it was late afternoon, in space Apollo 13 was 123 hr into its flight, and in Odyssey Fred Haise talked with Vance Brand as telemetry was sent briefly down to the ground for a quick check on Command Module systems. It all looked good; battery A was coming up as expected. But still the temperatures were dropping, reading a cool 7°c.

CAPCOM 'It's kind of a cold winter day up there isn't it? Is it snowing in the Command Module yet?'
HAISE 'No – no, not quite, the windows are in pretty bad shape . . . every window in the Command Module is covered with water droplets. It's going to take a lot of scrubbing to get that cleared off.'

As for the consumables, it was going to be tight but within acceptable margins. The Descent engine burn to speed up the return home had been a wise move. It was calculated that the electrical power from Aquarius was good to an elapsed time of 147 hours with water sufficient to 155 hours; Apollo 13 was scheduled to separate from Aquarius at 141 hr 30 min. Deke Slayton talked to Jack Swigert about the possibility of taking some pictures of the Service Module when it was jettisoned. Photographs of the Module could prove useful in seeking a cause for Monday's explosion. But if temperatures got much lower frost would cloud the windows: 'All the windows in the Command Module are heavily coated with water right now. So I don't know what kind of pictures we'll get out of them. . . .'

At 126 hr, Swigert switched off the charge on battery A but on the advice of capcom Vance Brand proceeded to begin a top-up charge on battery B, which had been used to look at Command Module telemetry some time before charging activity began. With Fred Haise resting, Lovell and Swigert began to copy down long instructions for entry activity, so different from anything they had trained to perform with different switch configurations and completely new sequences. But the Descent Stage systems were getting close to depletion levels and Mission Control recognized that the Ascent Stage water tanks would have to be used from some time late on Thursday evening. So throughout that day engineers at Hamilton Standard worked to perfect a test rig of the water system.

Hamilton's James Morancey completed the schematic by 9:00 am, laboratory workers had constructed the rig by 3:00 pm, and four hours later the tests were completed. Houston wanted as much use out of the Descent Stage water supply as possible, leaving Ascent Stage switch-over until the last safe minute. Morancey's rig provided that time, sufficiently in advance of total depletion to prevent the sublimators freezing over. Grumman's James Sheehan, manager of the LM environmental equipment, explained later that 'We were actively considering letting the . . . tank go down until the regulator said "Hey, I'm out of water," but we . . . decided to switch over ahead of that. The Hamilton Standard test was completed not long before the actual switch-over, so we were beginning to sweat it.' Word was rushed to Houston recommending the time tanks should be switched, and at 128 hr 30 min – 9:43 pm Houston time – the crew cut the feed from the Descent Stage supply and began to use water from the Ascent Stage. Only 31 minutes earlier the battery charging activity was completed, all three Command Module packs now in top condition for entry. It was time for Lovell and Swigert to get some rest, leaving Fred Haise on watch.

In the MOCR, the flight dynamics officer confirmed that Apollo 13 was on a path that would intersect the atmosphere with a slope of 6.03°, close to the centre of the permissible band between 5.5° and 7.5° to local horizontal. Nevertheless, because slight perturbations might be induced when the modules were separated, Mission Control wanted it closer than that so a course correction would be fired off using Aquarius' thrusters about five hours prior to entry interface, the precise point at which the Command Module would strike the atmosphere.

Friday, 17 April. Across the midnight hour, Jim Lovell and Jack Swigert rested while Fred Haise remained on watch, although he too dozed while Mission Control kept off the communication loop. It was the last opportunity to get some sleep before the meticulous procedures involved in getting the men home. Every step in the long, and new, procedure had to go just right and there was an element of danger that one astronaut would fail to place the correct switch in the right position. They had received so little rest for the past three and one-half days. It was 1:45 am in Houston when the activity began; 132 hr 30 min into the flight of Apollo 13. Jack Swigert was asked how he slept. 'Oh, I guess maybe two or three hours. It was awful cold and it wasn't a very good sleep.' Temperatures in the Command Module were now close to freezing.

Mission Control told Swigert he could get some more rest, but the astronaut was concerned about the temperature. 'Well, if I get everything done I'll try, but I'll tell you, it's almost impossible to sleep. All of us have that same problem. It's just too cold to sleep . . . it's just awful cold.' This was the danger period. Reduced to a low condition by the worsening environment, minor irritations assumed paramount importance. It was the point physicians hoped to avoid, the point where usually reliable people falter and slow down. In the MOCR, Deke Slayton talked to the crew.

CAPCOM 'I know that none of you are sleeping worth a damn, because it's so cold, and you might want to dig out the medical kit and pull out a couple of dexedrine's apiece.'
LOVELL 'Fred brought that up. We might consider it.'
CAPCOM 'Wish we could figure out a way to get a hot cup of coffee up to you; it would taste pretty good now, wouldn't it?'
LOVELL 'Yes, it sure would. You don't realize how cold this thing becomes in a passive thermal control mode that's slowing down . . . the Sun is simply turning on the engine of the Service Module. It's not getting down to the spacecraft at all.'
CAPCOM 'Hang in there. It won't be long.'

Deke Slayton had been at the capcom console for most of the previous evening. It was now 2:30 am, Houston time. With sufficient power in hand to complete the remaining hours prior to powering up the Command Module, ground controllers agreed with Slayton that it would be advisable to bring Aquarius to normal operating levels a bit early.

CAPCOM 'Okay skipper, we figured out a way for you to keep warm. We decided to start powering you up now.'
LOVELL 'Sounds good. And you're sure we have plenty of electrical power to do this?'
CAPCOM 'That's affirmative. We've got plenty of power to do it.'

That last statement was a relative comment. Gradually working their way through the revised checklist, the crew slowly brought Aquarius back up to modest consumption levels, bringing up the window heaters when atmospheric temperatures had taken the chill from them. In $1\frac{1}{2}$ hours, the temperature was up to 13°c; one hour after that it was 16°c in Aquarius. Now, electrical consumption was running at an expensive 40 amps, far higher than the 10-12 amp power-down consumption seen most of the way back from the Moon.

In Houston, it was the hour before the dawn on the day Apollo 13 was scheduled to drop its crew to the Pacific waves. In Mission Control, Gene Kranz swung his controllers into their respective console positions. Final computations were being made for the last course correction burn. It would be small, performed for just a few seconds with the RCS thrusters on Aquarius. But it would put Apollo 13 right down the funnel. The schedule called for Aquarius to be powered up six hours prior to entry. That it had happened three hours earlier would reduce the pace in those closing hours. The docked spacecraft were still more than 80,000 km away but each passing minute seemed to bring them closer.

When Jack Swigert drifted up into Odyssey he was pleasantly surprised: 'Hey, it's warmed up here now. It's almost comfortable . . . I'm looking out the window now and that Earth is whistling in like a high speed freight train.' At 136 hr 10 min Swigert went to the Command Module again, this time to power up the systems and get the vehicle ready for the important role ahead. First, a short mid-course burn. That came at 137 hr 40 min for 22 seconds, placing Apollo 13 on a 6.49° entry path. Now it was time to get into a proper attitude for the first procedural departure. With Aquarius still attached, Odyssey would jettison its Service Module about $4\frac{1}{2}$ hours prior to entry, leaving just the Command Module attached to the Lunar Module's docking collar. It was a very unusual configuration, one that would never be repeated for any other Apollo operation.

Seen by Apollo 13's crew in the Command Module, the blasted Service Module sector reveals extensive damage seconds after separating en-route to Earth.

It had never been tried before in space, and there were certain risks in conducting such a unique sequence for the first time. But there was no other way. The Service Module contained almost a full load of propellant totalling nearly 19 tonnes. It would be necessary to have the weighty structure far removed from the Command Module at entry interface, just in case debris caught up with the manned vehicle. Moreover, separation would be effected with the aid of Aquarius' thrusters and there would be no separation burn with the Service Module jets for fear of causing further trouble: nobody knew the condition of that large structure. Five minutes before separating, Jack Swigert fired the Command Module thrusters to check they were still working. Jim Lovell and Fred Haise were in Aquarius ready to help with Service Module jettison. But that too had a unique aspect.

Unlike conventional modes of separation, the ground would be unable to read the telemetered condition of pyrotechnics armed to sever the Command Module from the Service Module. So the crew would have to power up the pyrotechnic circuit and hope all was well, for there was no means of checking it before use. At 138 hr 2 min, the crew in Aquarius fired the thrusters to push the spacecraft at 15 cm/sec. Then Swigert, now in the Command Module, fired the pyrotechnics to sever the Service Module as Lovell and Haise reversed the direction of thrust and fired the forward Lunar Module jets to back away at 15 cm/sec, thereby providing a separation rate of 30 cm/sec between the two structures. At entry, the Service Module would be about 5 km from the Command Module.

Calculations on the ground predicted that the Service Module would appear, drifting away, through window No. 5 – the one adjacent to the right couch – and Swigert moved across and searched in vain for a visual sighting, hoping to get some good photographs. But over in Aquarius, Jim Lovell realized that the shock of jettisoning the Service Module had imparted an unexpected pitch to the docked combination so he pulsed the LM thrusters to bring it back on the specified alignment. Suddenly, through the overhead window in Aquarius, Lovell caught sight of the structure, slowly drifting away. And what he saw brought a chill to the MOCR.

LOVELL 'And there's one whole side of that spacecraft missing!'
KERWIN 'Is that right?'
LOVELL 'Right by the high-gain antenna, the whole panel is blown out, almost from the base to the engine.'

What they saw now for the first time was a severely damaged Service Module, torn out from the inside of bay 4. Wires hung like bundles of spaghetti from the area of the cryogenic shelf, the high-gain antenna was damaged, and even the SPS engine bell was scarred and dented. Had it been used, there would have been serious pressure problems. On hearing the call from Lovell, Swigert sailed up from Odyssey and soon all three had taken pictures: Lovell through the overhead window, Haise and Swigert through the LM's two triangular forward-facing windows.

The computer in Aquarius had a good alignment and now that had to be transferred into Odyssey, so that Jack Swigert could get an even better alignment for entry. But first the Command Module had to be fully powered up. Fred Haise went to assist Jack Swigert in Odyssey and Lovell remained in Aquarius to move switches and support the other two in getting the Command Module ready. There was no time to analyze the explosion; all that could wait until they got back.

In the Mission Operations Control Room, across the console positions and from the viewing area, positions began to fill up. Astronauts Deke Slayton, Dave Scott, Russell Schweickart, Tom Stafford, Charlie Duke, Ken Mattingly, Gene Cernan, Joe Kerwin, Ron Evans, Tony England, and Jim McDivitt were there. From management, Tom Paine, George Low, Kurt Debus, Eberhard Rees, Dale Myers, Walter Kapryan. Also present by this time were Rep. George Miller from California, Rep. Olin Teague from Texas, George Mueller, once head of NASA manned space flight, and Lew Evans, President of Grumman. His product it was that saved the lives of Lovell, Swigert and Haise.

In Odyssey, activity now briskly performed set up the dormant Command Module for entry. But Swigert had a little difficulty pointing the optics at selected stars for a course align because of particles still following the spacecraft. At 140 hr 10 min the crew applied power from batteries A, B and C through main bus B while Swigert monitored the voltage just in case they had difficulty handling the load due to the cool temperatures. But in Aquarius Jim Lovell watched the Earth, seemingly drawing closer all the time, and asked Swigert how he was coming along with the checkout; as he would later testify, it was an uncomfortable feeling floating about in the only piece of hardware still attached that could not possibly return safely through the atmosphere!

At about 141 hr elapsed time, all three crewmen were in Odyssey, and minutes later the hatches between the two vehicles were closed up for the last time. Aquarius had done a magnificent job, having supported them for nearly 84 hours and provided the propulsion to get back home on a safe course. The batteries had just 189 ampere-hrs remaining, sufficient for just half a day at minimum power levels. Of the original water supply, 12.8 kg remained, enough to last a further eight hours or so. Oxygen quantities were very good: 12.9 kg remained, more than 50% of the total loaded at launch. As for carbon dioxide removal, there was plenty of life remaining in the jury-rigged system using canisters from Odyssey for a further five days. But lack of electrical power would have quickly eliminated life from the spacecraft, putting the crew to sleep through lack of oxygen within a few hours of depletion.

With hatches closed, the plan now required the crew to vent the tunnel to about half the pressure in the two spacecraft. This was to establish that there were no leaks by allowing the astronauts to watch the stability of the pressure. When the crew were satisfied that all was well and that the hatches had a good integrity, the normal jettison procedure was followed, except that the 150 mmHg pressure in the tunnel caused Aquarius to pop off like a cork from a bottle and to drift away from the conical Command Module at a rate of about 3.3 km/hr. With a 'Farewell Aquarius; and we thank you,' the crew triggered the jettison at 141 hr 30 min. Totally dependent now on the life-support systems aboard the Command Module, Apollo 13 was in a more normal configuration. But procedures were very different from those on other Moon flights, with still a full hour to go before re-entry.

At jettison, Apollo 13 was 18,020 km from Earth, accelerating all the time, and in the period up to entry interface recovery personnel took up station in the Pacific. Three C-130 tracking aircraft were stationed along Apollo's descent path ready to update the predicted splashdown point. It was only the second time in US manned space flight history that a mission had been aborted, but unlike Gemini VIII there had been time to re-deploy the prime forces. With just 35 minutes to go, ground controllers gave Odyssey's computer a final update and advised the crew that they had a good load of consumables for entry. Then Jack Swigert checked out the entry monitor system on board the spacecraft, a scroll type device designed to allow the crew to track and monitor important parameters of their descent path through the atmosphere, before passing his thanks to the MOCR: 'I know all of us here want to thank all you guys down there for the very fine job you did.' One time physician, astronaut Joe Kerwin was nursing his charges back through the terminal minutes of the mission.

SWIGERT 'You have a good bedside manner, Joe.'
KERWIN 'That's the nicest thing anybody has ever said.'
HAISE 'Sure wish I could go to the Fido (flight dynamics officers') party tonight.'
KERWIN 'Yes, it's going to be a wild one!. . . . We'll cover for you guys. And if Jack's got any 'phone numbers he wants us to call, we'll pass them down.'

Six minutes later, Odyssey slid into Earth's atmosphere and after a few seconds the crew experienced a maximum force of 5.2 g before the spacecraft pulled up and then dipped again into the atmosphere. In the Mission Control Center, scenes from the recovery ship *Iwo Jima* flashed on the MOCR screen and minutes later the spacecraft could be seen descending on three good parachutes: 'Odyssey, Houston. We show you on the mains. It really looks great.'

Suddenly, clapping broke out all across the rows of consoles, cheering could be heard behind the rear glass viewing screen, and across the United States people applauded while others just stood transfixed before television screens. In New York's Grand Central Station, thousands cheered when Apollo 13 appeared on the wide screen that more than eight years before had carried scenes of a triumphant John Glenn returning from America's first manned orbits of the Earth. Now, three men had limped back from the vicinity of the Moon in a blasted spaceship supported by the consumables of a lunar landing vehicle.

At the Manned Spacecraft Center, people were overcome with the emotion of it all. Apollo 13 had already taken a heavy toll: from the events of that week had come at least one heart attack, two marital separations and probably two broken homes. The tension had been incredible. Now that the conical spacecraft was floating to a splashdown, men wept, secretaries cried, and veteran managers and administrators just stood quietly and looked at the TV monitors. Apollo 13 splashed down at 12:08 pm, Houston time, about 6.5 km from the *Iwo Jima*, where sailors stood cheering and shouting their joy at the tiny speck. Odyssey remained in an upright position and within 45 minutes the crew were put down on the recovery ship; no need here for biological quarantine.

As television cameras brought happy scenes into the Mission Control Center, Deke Slayton reached across the consoles and firmly shook Chester Lee's hand as Gene Kranz sucked a very large cigar. The impact of Apollo 13 was enormous, and in some respects more widespread than Neil Armstrong's first lunar bootprint. In New York, ticker tape and confetti showered down from windows high above the busy city streets. At the Manned Spacecraft Center a sign that had read 'Our hearts are with the Apollo 13 astronauts,' now said 'Sigh of relief party here tonight.' Right across the United States, church bells peeled out, linking a fragmented canopy of sounds preserved hitherto for great events in the nation's history.

In Indianapolis, all the traffic came to a complete standstill. In Las Vegas, for the first time in more than a century, dice stopped rolling momentarily. In Los Angeles, for a brief period, there was no crime reported anywhere. And in a Richmond, Virginia, department store, people watched television sets, according to the manager, with a 'quiet smile, but it was curious

Flight director Glynn S. Lunney salutes personnel on duty in the MOCR following successful recovery of the Apollo 13 crew.

the way they stood looking at all that good news – silently and almost as if they were still praying.' From the White House, President Nixon telephoned Lovell, Swigert and Haise and then said that he planned to go 'first to Houston, where we will pick up the two wives, and then go to Hawaii,' to present the astronauts with the Presidential Medal of Freedom, the very next day.

On the 19th said Nixon, a proclamation would assign Sunday a day of national prayer and thanksgiving for the safe return of Apollo 13 and its crew. It was almost an official rebuff to atheist Madalyn Murray O'Hare, who for several years had campaigned for a ban on prayers from space, taking her case to the Supreme Court, appealing a decision by the US District Court that rejected her plea.

In South Africa, a prominent witch doctor, Magomezulu, said that 'these men have interfered with God and he has turned them back. This is a warning that the Americans should stop landing on the Moon.' Across the world, reports poured in telling of wide public concern for the astronauts. The European Broadcasting Union said it believed the live coverage of splashdown may have been the most widely watched TV program ever. During the closing hours prior to entry, the British Broadcasting Corporation allocated one of its two television channels to a continuous transmission of commentary on the air-to-ground loop, presenting a caption telling viewers that the voices were from Apollo 13. The US Embassy in London was saturated with telephone calls. 'People were sobbing with obvious relief and happiness,' said one operator. 'I just didn't know what to say to them.'

Messages poured in to the United States from more than 100 countries, expressing their pleasure at the safe return. And from NBC, a statistical claim that more than 40 million people in the United States watched Apollo 13 splash down in the Pacific. Added to the millions that watched via satellite, the 'global total that joined in the final moment of relief probably can never be measured,' said NBC. Even the press turned in upon itself and criticized the lack of awareness that clouded coverage of the flight's first two days. Broadcast networks 'deserved the criticism they got for slow reaction to the unexpected danger', said the *Washington Post*, but 'deserve praise today for the crisp, professional reporting and for some unusual restraint,' during the closing hours of the mission.

On Sunday, 19 April, Governor Nelson Rockefeller dedicated the 'day of prayer and thanksgiving' while in Honolulu President and Mrs. Nixon attended church services. 'There are very deep ideological differences that divide the world today,' said the President, 'But when it was learned that these men were in danger, there poured into the White House from all over the world messages from the Communist countries, from people of various religions, saying that they wished their best, offering their assistance. When they learned they were back, there was an outpouring of relief and rejoicing from people, regardless of their political or religious differences. . . . If only we could think in that way about every individual on this Earth, we could truly have world peace.'

Across the United States, people attended services held for this special day of thanksgiving and in Ohio a petition opposing attempts, primarily by Madalyn O'Hare, to ban prayers from space got 50,000 signatures. 'The manner in which great numbers of Americans attended special prayer services . . . has been inspirational,' said Sen. Margaret Chase of Maine. 'We feel that this points up the fact that despite all of Man's knowledge and ingenuity there is still a supreme being in control of things, and that we have to trust him,' responded J. F. Meredith of the Seventh Day Adventists' Central State Conference in Kansas City.

Vice President Agnew was charged with admiration for the way the flight had been handled although he believed that 'social levellers of the New Left' would again try to get space funds 'soaked down into the nearest slum' in the same kind of 'shortsighted tragic blunder' that had characterized preceding years. But perhaps the most truthful comment can be found in the editorial column of the *National Observer* which, in its 20 April issue, said that 'At a time when Americans were becoming bored with successful flights to the Moon, the ordeal of Apollo 13 reminded the nation of the dangers and difficulties of space. But because of courage, care, and uncommon resourcefulness – because, in fact, of history's most dramatic field expedient – the astronauts made their way safely back to Earth. The first walk on the Moon was a wonder of wonders, but the return of Apollo 13 was the gladdest moment of all.'

Astronauts Haise (left), Lovell (centre) and Swigert show little sign of stress during recovery operations.

Jim Lovell, Jack Swigert, and Fred Haise, arrived at Houston's Ellington Air Force Base on Sunday, 19 April, to a welcome from a crowd numbering 5,000. It was the beginning of a most important post mortem. All future Apollo flights were postponed pending the conclusions and recommendations of a review board set up by Tom Paine two days before. Edgar M. Cortright, Director of NASA's Langley Research Center, was to be its chairman. Members of the board included Robert F. Allnutt, Vincent L. Johnson, and Milton Klein from NASA Headquarters, Neil Armstrong representing the astronaut corps, John F. Clark, Director of the Goddard Space Flight Center, Walter R. Hedrick, Jr., from USAF Headquarters, and Hans M. Mark, Director of the Ames Research Center.

George T. Malley from the Langley Center would provide counsel to the board, and Charles W. Mathews from NASA Headquarters would ensure technical support. When the review board convened at the Manned Spacecraft Center on Tuesday, 21 April, four panels were formed, each chaired by an official experienced in the specific function of that group's responsibility. In addition to the board's activity, James McDivitt set up an investigation team at MSC, effective in liaison with the board via Mathews as the technical support coordinator.

By early June the review board had completed its work and on the 15th the chairman and members met with Tom Paine and George Low to transmit the final report. The board discovered that tank No. 2, factory number 10024XTA0008, was manufactured in 1966 by the prime cryogenic tank contractor, Beech Aircraft. It was the eighth Block II tank produced. Several minor flaws were found to exist in tank 2 during tests following manufacture. Some of the welding had to be redone, incorrect welding wire had been used and the fan motor was excessively noisy and used more current than predicted. The tank was taken apart, elements replaced, the shells re-welded, and a vacuum pulled between the two walls over the specified 28 day period. The board found that an 'Acceptance test indicated that the rate of heat leak into the tank was higher than permitted by the specifications. After some re-working, the rate improved, but was still somewhat higher than specified. The tank was accepted with a formal waiver of this condition. Several other minor discrepancies were also accepted.'

Shipped to North American Rockwell on 3 May, 1967, tank No. 2 was assembled on its oxygen shelf ten months later for installation in spacecraft 106, then slated for the Apollo 10 mission. That task was accomplished on 4 June, 1968, only for the complete oxygen shelf to be taken out on 21 October for standard modifications, replaced by one already subject to those changes. During this removal, the small crane used to lift it out the Service Module caused it to fall 5 centimeters. It was not thought any damage had occurred to the tank and visual inspection showed no scarring. Tests were completed by 22 November and the shelf installed in spacecraft 109, following which the vehicle was delivered to KSC in June the following year. At the Cape, other problems afflicted tank No. 2, as detailed in the previous chapter. But the board uncovered a major error in administration also.

It found that 'The original 1962 specifications from North American Rockwell to Beech Aircraft Corporation for the tank and heater assembly specified the use of 28 volt dc power, which is used in the spacecraft. In 1965, North American Rockwell issued a revised specification which stated that the heaters should use a 65 volt dc power supply for tank pressurization; this was the power supply used at KSC to reduce pressurization time. Beech ordered switches for the Block II tanks but did not change the switch specifications to be compatible with 65 volt dc.' Quite simply, an amendment noting the higher voltage to which each tank would be exposed at the Cape had not been passed by Beech to the subcontractor supplying the switches. And, as the board found out, 'The thermostatic switch discrepancy was not detected by NASA, North American Rockwell, or Beech in their review of documentation. . . . It was a serious oversight in which all parties shared.'

The board discovered that the switches could accept the 65 volt DC load during pressurization because 'they normally remained cool and closed.' However, when they were opened during special procedures used to remove oxygen from the tank, operations not performed on other tanks up to that time, 'they were welded closed' just as power came through. In fact, they never did open because of the electrical arc that occurred at that point 'and were rendered inoperable as protective thermostats.' Because the tank heaters were powered for eight hours at the Kennedy Space Center in attempts to evacuate the tank, temperatures in the tube assembly probably got as high as 538°c which would have severely damaged Teflon wire insulation inside the tank head.

From that point on the tank was in a lethal condition and at some random point the wiring could have fractured, short circuited and ignited the insulation. Immersed in a high pressure oxygen bath, combustion was inevitable. And that is what presumably happened on Apollo 13. When the fans stirred the tank contents they probably caused the wire to short which ignited the Teflon. The board said that 'The accident is judged to have been nearly catastrophic,' and that 'the accident was not the result of a chance malfunction in a statistical sense, but rather resulted from an unusual combination of mistakes, coupled with a somewhat deficient and unforgiving design.'

But if deep concern was felt over the explosion to Apollo 13's Service Module there should have been gratitude about foresight over the possible contamination of the atmosphere from release of radioactive products: Aquarius came plummeting back to Earth with its plutonium nuclear fuel source strapped to the side of the Descent Stage; designed to survive such an anomaly, it fell to the floor of the Pacific where it will remain. But the message from Apollo 13 had been clear. Despite sophisticated and attentive measures aimed at preventing an incident like the fire that took three lives in 1967, procedural error allowed the wrong equipment to be installed in a contractor product. North American Rockwell specifically instructed Beech that the Block II tanks would be required to withstand 65 volts DC from ground equipment at the Kennedy Space Center but the contractor continued to install switches rated at 28 volts.

As the board found, if nothing untoward happened to a tank it could probably have remained safe. But Apollo 13's No. 2 tank received unusual de-tanking operations and that sealed the mission's fate, 56 hours after lift-off. Had the tank failed 56 hours *later*, all three men would have died, for by then Aquarius would have been at the Fra Mauro landing site. Only because it had all its consumables intact could the Lunar Module play the role of lifeboat. If this accident had happened on Apollo 8, when there was no Lunar Module, it would have had a similar fatal result.

The impact of Apollo 13's explosion on the future pace of lunar landing activity pivoted on the recommendations of the review board and the hardware changes it suggested. Before the board completed its work, Tom Paine made a decision based on the recommendation of the Apollo Program Site Selection Committee at its 7 May meeting at MSC that Apollo 14 could not possibly fly before 3 December, 1970. That date accommodated suitable launch windows for the Fra Mauro site, still the first choice from selenologists, and changes by then considered necessary to future spacecraft. On 27 June, Rocco Petrone and Dale Myers passed a memorandum up to Administrator Paine detailing the proposed actions to get Apollo flightworthy once more.

The oxygen tanks would be modified to minimize combustion through heated wire by replacing all Teflon insulation with stainless-steel sheaths. The quantity probe too would be made of stainless-steel, the two 75 watt heaters in each tank would be replaced with three 50 watt units, and manufacturing methods would change to prevent possible damage during assembly. The spacecraft's caution and warning system was also to accommodate a more specific indication of problems with the fuel cells. A third oxygen tank would be added to the Service Module, placed at the top of bay 1, a hitherto vacant sector, raising total oxygen quantity by 50%. This tank could be isolated from the other tanks in sector 4, and because it was remote from the standard location for cryogenic equipment, any catastrophe in sector 4 would not necessarily inflict damage on the tank in sector 1, which was on the opposite side of the Service Module.

Paine reviewed the recommended changes, as well as a proposed postponement of Apollo 14 to a launch on 31 January, 1971, and announced his concurrence on 30 June. Over the next few months, acting on the review board recommendation to improve where possible the ability of an Apollo CSM to provide reserve consumables, the Manned Spacecraft Center authorized a 400 ampere-hr silver oxide/zinc battery to be placed in sector 4. Weighing 61 kg, it was identical to each of four Lunar Module batteries housed in the Descent Stage, and would permit a reserve in case the fuel cells failed. Five plastic containers, each capable of holding 3.8 liters of water, were carried in a bag stowed in a locker. In emergency power-down situations, water would be drawn from the Command Module storage tank before it froze and dumped through the waste management system prior to re-entry. The summed changes brought an additional weight of 439 kg to spacecraft 110, Apollo 14's CSM.

When Lovell, Swigert, and Haise began their flight the schedule envisaged Apollo's 13 and 14 in 1970, 15 and 16 in 1971, and Apollo 17 in early 1972 before Skylab flights later that year and 1973. The last two Moon flights, Apollos 18 and 19, would fly in 1974. But financial preparations for Fiscal Year 1972, executed in the fall of 1970 for publication by the President in January 1971, made it quite apparent that NASA would not get the money it needed to sustain even this depleted schedule. It was simply a matter of balancing use of existing Apollo systems, in more Moon landings and Skylab space station activity, against the need to begin development of the permanent space station and the resusable shuttle.

In negotiating with the Office of Management and Budget, the re-named Budget Bureau, OMB allowed NASA to propose a sum 10% below the value it originally requested. A plan already in hand to delete two more missions was brought out. Apollo's 18 and 19 would not now fly to the Moon, their hardware being held in reserve. Only one more H-series mission would fly: Apollo 14 in late January, 1971. And only three J-series flights would get off: Apollo 15 in July or August, 1971; Apollo 16 in March, 1972; Apollo 17 in December, 1972. Planned to fly in July, 1972, Skylab could not now be launched before March or April, 1973, the balance of that year being taken with three manned periods of habitation for one month, two months, and two months, respectively. Out of an originally trimmed schedule for 10 manned Moon landings, there would now be chances for only 6: two had already flown and there were four to go.

As for development of the big space station, contractor studies carried out during 1970 showed an economic advantage in a modular concept which, instead of being launched as a single structure by a two-stage Saturn V as originally planned, could be built up in orbit from the assembled payloads of several shuttle vehicles. Built from cylindrical sections fitted out for specific tasks the station could be as small or as large as required to match changing needs. And as a hedge against further budget cuts as year by year the space agency saw its financial status plummet to new depths, NASA began to solicit support from abroad, principally Europe, in the hope of wooing foreign governments into a cooperative venture where cuts and cancellations would bring undesirable repercussions and therefore be less likely to get through the White House.

In its preliminary schedule, NASA now saw little chance of getting the shuttle spaceborne before March, 1978. There was no money in the FY 1972 budget to start on detailed design and initial development, only to keep the contractors going with more definition studies in a protracted attempt to find a cheaper design concept. By mid-1970, North American Rockwell and McDonnell

Fuel cells designed to provide electrical energy and water as a by-product were crippled when two cryogenic oxygen tanks on the shelf below evacuated their contents to space.

This cryogenic oxygen tank is identical to the one which blew up on Apollo 13 and shows the tight packaging that left little room to dissipate energy when the tank over-pressurized and blew up.

Douglas had been chosen to conduct competitive definition studies on the reusable shuttle but the development costs were still too high to convince the OMB that shuttle was a viable project for the future. To gain more weight for the shuttle's case, NASA seized eagerly on the prospect of using it to launch modular elements of a future space station. By the fall, other contractors had been brought in to seek cheaper shuttle designs.

Recognizing a four-year gap between the Skylab and shuttle operations, manned space flight boss Dale Myers was concerned about the hiatus brought on through low funding priorities. Despite statements to the contrary, President Nixon's administration seemed reluctant to give America new leadership in space, failing continually to give NASA assurance of even a minimum level budget. In the absence of any guarantee for specific sums, it was becoming an annual fight for money, using Congress to pressurize the White House. Technically, NASA was an arm of the incumbent administration and top administrators at the space agency were unable to protest openly at the lack of support. Before committee hearings in the House and the Senate, however, they did their best to drop hints about the serious decay in America's manned space flight potential, while lesser heads put themselves on the chopping block by openly lambasting the White House caretakers.

But if discontent reigned over the confrontation between government policy and technical capability, there were rumblings too within the NASA structure. The past year had been one of sadness and disillusionment as project after project was deferred, postponed or cancelled. All the way from 1967, it seemed, cuts and schedule shuffles had reduced once grand plans to meaningless documentation, gathering dust. It was as if the space program was turning back the clock, as if the prospects for space convoys pulsing their way to the planet Mars were now more remote than they had been when John F. Kennedy set such bold objectives before the American people less than ten years before. And in the course of that evaporation, bitterness smoldered over the differences between NASA Centers, and even the men who ran them.

For long the home of teutonic authority, Wernher von Braun's Marshall Space Flight Center was populated by managers and department heads selected and groomed through the ranks of ex-Peenemunde men and those who came from Germany to find work. Where Headquarters reflected the 'American' way of exploiting space, Huntsville sought to satisfy more substantial goals. But it was even deeper than that, for resentment ran high over the place these Germans had in the NASA infrastructure.

The faithful believed von Braun was being prepared for the top job at NASA, but that could never have been. Early in the Nixon administration's tenure, Tom Paine was led to understand that the President was seeking a bold initiative with which to write his name across the dark space lanes of the solar system; indeed, his space task group emboldened that idea. NASA had not strengthened its future planning groups to the level now deemed necessary and Tom Paine was convinced the rhetorical von Braun was just the man to rally support for grand objectives. So he was asked by Paine to move to Washington, leaving his post of MSFC Director to become Deputy Associate Administrator for Planning. Von Braun was reluctant to leave the Center that had been his home for 20 years, unhappy with the decision to relegate him to a Headquarter's post. It was almost a demotion, for he would no longer be directly responsible for a specific package of hardware.

Von Braun left the Marshall Center and went to Washington in February, 1970, his place at MSFC taken by Dr. Eberhard Rees. But Nixon's space dreaming was already over, the overwhelming pressure of foreign and domestic events pushing ever further away the prospect of an expanding manned flight program. Von Braun pitched actively into the fray, arguing for a viable post-Apollo initiative, one in tune with demanding social needs, but one also looking with a vision to the future. Nobody wanted to know, and even at Headquarters a sense of acceptance began to set in. Von Braun was visibly distressed by the apathy among White House tenants, by the mood of acquiescence from those who once blazed a path to the Moon.

Nothing so assuredly transformed national pride and hope into mothballs than the turning of a new decade and the inauguration of a new President. For within eighteen months of assuming office in January, 1969, it was clear that NASA would be hard put to preserve any form of manned space operation, certainly not the visionary conceptions of the German immigrants. On 15 September, 1970, Tom Paine walked away from NASA Headquarters for the last time. With him went the dream still fostered by von Braun that one day men would roam the red dusty plains of Mars or seek a destiny beyond the planets. On his own now, in a world very different from that in which he led America's rocket men to success, Wernher von Braun was visibly depressed. He walked alone where once a nation held him high.

As for Tom Paine, he moved across to General Electric. George Low would hold the reins until a successor could be found. From a scientific standpoint the greatest days of Apollo were yet to come. Boeing was working hard to develop the lunar roving vehicle, Houston was busy modifying the Apollo CSM design to accommodate a battery of instruments for lunar orbital science, Hamilton Standard was developing a new back-pack with improved capability, and ILC was preparing a new and more flexible space suit. It hardly seemed an appropriate time to put up the shutters. And so thought Tom Paine long before he left NASA.

From the time he took over from Jim Webb, Paine nursed hope that the United States and the Soviet Union could get together in joint participation through space science and exploration. Hugh Dryden had similar thoughts when, in 1964 and 1965, he tried to develop a line of communication through Anatoliy Blagonravov. That came to nothing because of entrenched attitudes within the administrations of both countries. Paine had learned to speak Russian at the end of World War II when he considered communication with that country to be an important prerequisite for future cooperation. When he came to NASA as its boss, Paine felt that it was 'time to stop waving the Russian flag and to begin to justify our programs on a more fundamental basis than competition with the Soviets.'

As part of a proposed new level of cooperation, Blagonravov received an invitation from Paine to attend the launching of Apollo 11. Paine stressed the genuine nature of his invitation and assured the esteemed Soviet scientist that 'steps could be taken to avoid publicity.' Blagonravov declined the offer but official comment on the first manned Moon landing was warm and conciliatory, a far cry from the blustering commentary issued frequently by Kruschev. Paine replied with further invitations, addressed now to the President of the Soviet Academy of Sciences, Mstislav Keldysh, the organization within which Blagonravov operated as head of space exploration and application.

Keldysh was interested in what Paine had to offer and when the latter sent a copy of Nixon's space task group report on future programs – the integrated space program discussed in the previous chapter – Moscow warmed to the idea of talks. By this time President Nixon had been made aware of the Paine-Keldysh dialogue and formed an inter-agency committee to examine various ways in which the two countries could get together in a cooperative venture. When Blagonravov visited New York in spring 1970 Paine hosted the Soviet delegate at a private dinner during which he introduced the possibility of Soviet cooperation in unmanned planetary exploration and possible space rescue for marooned astronauts. Keldysh said he would pass these ideas along to appropriate parties in Moscow.

In Leningrad, Neil Armstrong presented a paper at an international science conference while George Low talked at length with Russian delegates. With the US group was a Dr. Philip Handler from the US National Academy of Sciences who recalled having seen a film called 'Marooned' in which an American astronaut was stranded in orbit. Handler felt that this was a very real possibility in fact and talked to Paine about possible space rescue involving the Soviets. Paine urged Handler to go ahead and try opening doors, with the proviso that discussions would naturally evolve back upon the NASA establishment once talks got under way. When Handler met Soviet officials in May, 1970, he pressed hard for cooperation, explaining to them the story of 'Marooned' where a Soviet cosmonaut came to the aid of a stranded American. The Russians were astounded at this, schooled as they were in tales of negative Western attitudes.

Handler spoke to Keldysh and several other influential people and in return the Soviets promised to propose new initiatives to Tom Paine in the form of increased exchange of information, more weather satellite data, and unified communication between spacecraft and ground stations. But Handler wanted more, suggesting that discussions start immediately on a common docking system which could be employed by both countries on future space vehicles, allowing spacecraft from one country to visit the orbital facilities of another. Back in the United States, Ambassador Anatoliy Dobrynin called Handler and asked him to receive a message from Keldysh. What the message said came close to suggesting US-Soviet cooperation on a common docking system.

Handler was getting better success than NASA, but that was because the Russians assumed the National Academy of Sciences to be in charge of the space agency, a structure adopted in the USSR. Handler wrote to the Soviets, laying out the lines of communication that must now ensue, transferring responsibility to the space agency. Paine sent two letters, in July and September, suggesting that Soviet representatives visit the Manned Spacecraft Center for formal engineering discussions. Paine also proposed a Soviet docking cone fitted to Skylab so that a Soyuz spacecraft could connect with the US workshop during periods of manned habitation in 1973. This was an attempt to inject mobility into the proposals – nobody actually expected the Russians to agree to this.

At this time the Russians were working toward the launch of their own space station, capable, as it turned out, of sustained activity over very long periods, unlike Skylab which matured from redundant Apollo hardware originally intended as the forerunner of large, permanent, orbiting facilities. Called Salyut, the Soviet hardware was then in final stages of development and manned activity would get under way in 1971. Earlier, in October, 1969, three Soyuz spacecraft participated in a major flight test.

Soyuz 6 was launched into orbit on the 11th carrying cosmonauts Shonin and Kubasov. Intended to further qualify Soyuz systems developed since the Komarov accident, Soyuz 6 carried equipment for gathering Earth resources data, and special remote-controlled welding instruments to test the feasibility of space assembly; similar tests were then being planned by NASA for Skylab experiments. The Russian equipment, called Vulkan, provided only modest success, although good results were obtained with electron beam welding. Next day, 12 October, Soyuz 7 ascended manned by cosmonauts Filipchenko, Volkov and Gorbatko. It too carried Earth resource equipment. One day later, Soyuz 8 carried Shatalov and Yeliseyev into space, cosmonauts from the previous Soyuz operation and directly involved in rendezvous and docking operations.

Maneuvers carried out on 14 October brought all three spacecraft to within a few hundred meters of each other on the 15th. A flurry of engineering activity put space and ground systems through a rigorous checkout intended as a final test of all the equipment involved for later Salyut missions. Considerable comment surfaced in the West about the intentions of the triple mission. Were Soyuz 7 and 8 intended to dock together? It seemed strange to bring them so close only to move away again. Because docking had been demonstrated earlier that year, it is possible that Soviet engineers were concentrating on tests associated with rendezvous and radar systems; docking would have been superfluous.

In all probability the two vehicles were not programed to link up. Within the framework of a program where so much is kept secret, intentions are difficult to define. Had an uninformed audience heard of Apollo 10's close pass over the Moon without landing, they would doubtless have concluded that Stafford and Cernan aborted the operation when in fact a landing was not part of the test.

Soyuz 6 returned to Earth on 16 October, followed by Soyuz 7 on the 17th and Soyuz 8 a day later, each flight lasting nearly five days. Then, while preliminary discussion was under way on possible East-West cooperation, Soyuz 9 reached orbit on 1 June, 1970, carrying Nikolayev and Sevastyanov. It was to be a long duration flight during which the crew could conduct medical tests, perform biological experiments, obtain Earth resources data, and practice navigation by stellar alignment. Television was provided for controllers to watch the crew at work, new equipment was tried out of the type considered for use in Salyut stations, and many periods of photographic activity were assigned.

From an initial low orbit, Soyuz raised its path first on the fifth revolution, then on the 17th, before lowering its orbit on the 14th day to reduce the orbital lifetime – just in case the retro-fire sequence failed to work. This was Russia's longest manned flight, exceeding by nearly four days the two-week flight of Borman and Lovell aboard Gemini VII $4\frac{1}{2}$ years earlier. Physiological, psychological, and biomedical reactions were studied, during the flight and after. When the crew returned on 19 June they were noticeably affected by the mission, taking several days to get back

Packaged and ready for flight, Soyuz 9 in the hardware preparation building.

into shape, longer to regain normal fitness. Clearly now, Russian cosmonauts were moving toward manned space station work and that made it more imperative than ever to get some form of cooperation on common docking equipment.

Soviet response to Paine's suggestion that a Soyuz type docking cone be fitted to Skylab came in the form of sustained interest in getting together for talks rather than a direct reply on that specific suggestion. Keldysh proposed a meeting in Moscow between Dr. Paine and the Soviet Academy of Sciences, suggesting an agenda which included a possible rendezvous and docking flight to test a compatible docking system. Encouraged by this, George Low reiterated NASA's interest in pursuing this line when he responded as Acting Administrator after Paine's September resignation. A date was set and people selected. MSC Director Bob Gilruth would go with chief flight director Glynn Lunney, Caldwell C. Johnson from the spacecraft design division, and George B. Hardy from MSFC's Skylab office. NASA's Arnold W. Frutkin would be accompanied by William Krimer from the State Department.

By this time, stirred to action by increasing interest from the White House, NASA began to study compatible docking systems that could be attached to space vehicles from both nations. But even incomplete awareness of the full Soyuz potential exposed its inability to make the large orbital maneuvers necessary for rescue purposes; moreover, a three-man Apollo crew would not fit into a Soyuz, even if it carried only one cosmonaut pilot. Because Apollo had been built to accommodate large propulsive maneuvers far from Earth, it alone, among extant hardware, was even remotely capable of chasing after a doomed spacecraft in Earth orbit. That would reveal publicly an apparent discrepancy between the manned vehicles of America and Russia. The Soviets would not approve such an obvious asymmetry, and NASA knew this. Even the Skylab visit seemed on closer examination to be a non-starter. A tiny Soyuz nuzzling up to the giant American space station was again a demonstration of unequal capability. It began to look more and more as though the only way out of the dilemma was to perform a rendezvous and docking exercise between Apollo and Soyuz as a demonstration and engineering test of compatible hardware.

In Moscow, Low's party was given a conducted tour of a Soyuz simulator while specialists discussed technical details of the manned spacecraft. In the United States, there had been little opportunity to study in depth the various accomplishments of the Soviet space program but this direct inspection of hardware brought a degree of reality unseen hitherto. Cosmonaut Shatalov provided the delegation with a tour of several facilities and the team was surprised at the sparse coverings on spacecraft walls, the minimal amount of equipment in the cabin. Discussions were held on the way Soviet rendezvous and docking operations were conducted, hearing that most of the engine burns needed to match orbits came on commands generated at ground tracking stations. Then an on-board guidance system would take over, bringing the two together within 500 meters. The final phase would gently move the active vehicle toward the target by firing small thrusters, although this could be controlled manually – the only element of rendezvous and docking with a dual capability.

Then the Americans described the Gemini and Apollo systems, surprising the Russians with the amount of manual authority built in the control mechanisms. But whereas existing hardware designs employed male and female sides of the docking interface, future, compatible, mechanisms would need to be androgynous in concept – that is they would have to be identical in design, allowing either vehicle to be the active participant. This is where Caldwell Johnson came in, for he had wrestled docking designs in the early 1960s when Gemini and Apollo evolved primarily from docking requirements. But no one was really happy with the drogue-and-cone concept employed for Apollo and Johnson persistently tried to get approval for a design change, seeking in 1967 to have the AAP workshop (later called Skylab) carry a 'double interrupted ring and cone' working as an androgynous device; that biological term implied the characteristics of a neuter.

NASA decided to retain the probe and drogue system but Caldwell briefed the Russians on his ideas, talking over the design with his Soviet counterpart, Vladimir Sergeyevich Syromyatnikov. The main drawback with the Russian docking mechanism was that once fitted to the forward section of Soyuz's orbital module, it could not be removed for intravehicular transfer. Any exchange between spacecraft necessitated a space walk outside. This is uncharacteristic of Soviet practice in other areas of their program and certainly stands as circumstantial evidence that perhaps the main engineering design details of the spacecraft were completed before it was given a comprehensive docking role in association with Salyut space stations. Added to the fact that direct visual observation of the line of docking was possible only via a periscope which could have been an addition to the design, this supports the contention that Soyuz was primarily a circumlunar vehicle. Because human access was prohibited, the docking mechanism also had to carry all electrical connectors linking the docked vehicles, another drawback uncharacteristic of Soviet practice.

Discussions in Moscow also embraced the Skylab plans, George Hardy providing technical details for the Soviet hosts. Later, at a party held in the home of the US Embassy's science attache, Konstantin Feoktistiv claimed responsibility for the majority of design work on Vostok, Voskhod, and Soyuz spacecraft. It provided a unique opportunity for Bob Gilruth to meet the man who for years kept US manned flight protagonists on their toes. From the October meeting, three working groups were set up to monitor design and engineering trends in the search for a compatible docking system. It was only a preliminary discussion, but the seeds had been laid. President Nixon watched these events with more than a little interest. In the opening stages of SALT negotiations, any progress outside the Vienna discussions would contribute to the general climate of goodwill emerging from the discourse. On 28 October, the two parties assembled to sign the agreement leading to further talks six months later. At the end of the day there was general recognition of the need for a test flight specifically aimed at proving whatever docking system emerged from the design effort.

But nobody spoke openly of the need for a special flight, although this was very much in mind when George Low flew at Keldysh's invitation to Moscow in January, 1971. Before going to Russia, Low had discussions with foreign adviser Henry Kissinger and got White House clearance to go as far as possible in setting up cooperative agreements with Moscow. Kissinger was averse, however, to any implied suggestion that reconciliation at a technical level on scientific projects implied similar resolution to political issues.

On 20 January, George Low and Arnold Frutkin met with Keldysh and Feoktistov to formally propose development of a compatible docking system for Apollo and Soyuz spacecraft. The Russians were sincerely interested and promised to refer back to their political masters for an early decision. The scheme was based on several proposals worked out by MSC's Clarke Covington the previous month, each mission profile accommodating a flight in the 1972–1975 period. This new move, imparted by Low personally, moved the negotiations up a notch. Up to now, the discussions concerned compatible docking systems for future space vehicles; now, the Americans were proposing an immediate start on the androgynous docking equipment proper with a test flight within five years.

One of the proposals, a profile incorporating crew transfer between the two vehicles, required a special docking module with a US docking mechanism on one end and a Soviet system on the other. An adaptation of this put an androgynous docking system on the end of the module, with a similar system on Soyuz. In that way, crewmembers could use the docking module between as a sort of airlock; Apollo used pure oxygen at one-third sea-level pressure; Soyuz had an oxygen-nitrogen, sea-level atmosphere. A separate module would improve the options available, permitting astronauts to visit each others' vehicle in space.

When George Low returned home on the 21st he was hopeful that Soviet reaction would indeed favor some form of space demonstration of the compatible docking system. But for the more immediate future, Apollo 14 was only ten days away from setting sail for Fra Mauro. After two engineering flights to the lunar surface, a near-disaster with Apollo 13, and the beginnings of a shift to Earth orbit space stations and reusable shuttles, it was time to resume the manned Moon program with the first of four increasingly complex missions dedicated to extracting as much scientific knowledge as possible from the crusty face of the Moon.

Soyuz 9 comes home and personnel check the spacecraft. Note reporter at left documenting the event for posterity.

Three-quarter rear view of Grumman's Lunar Module which served well beyond specification when called upon to return three Apollo 13 astronauts to earth in April, 1971.

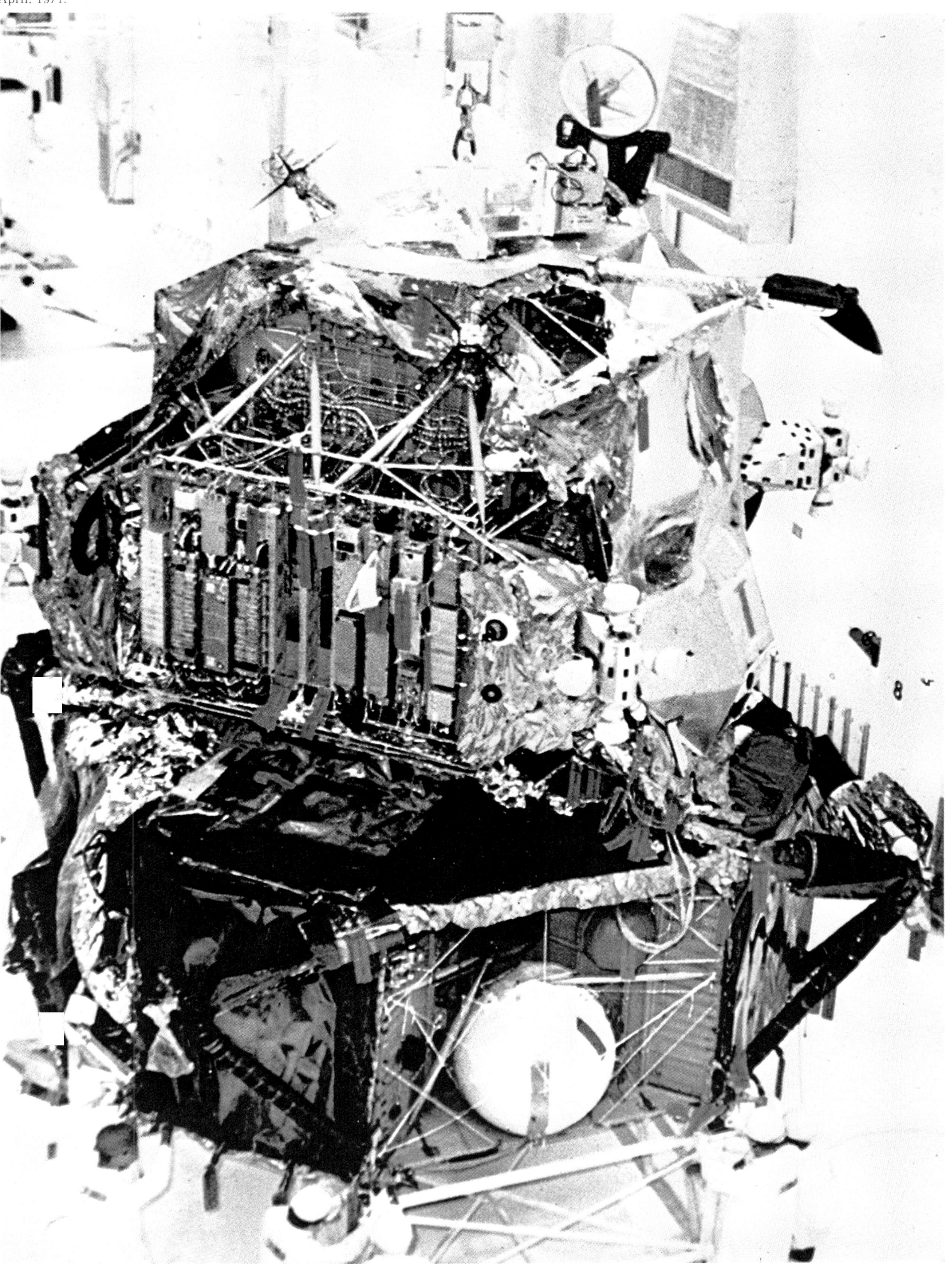

Employed to propel Apollo 7 into earth orbit, October, 1968, this S-IVB rocket stage still carries the petal-doors which when closed supported the Apollo spacecraft on top.

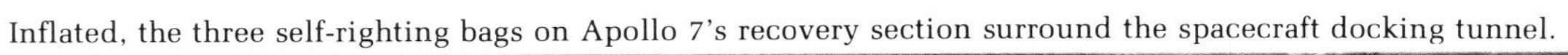

Inflated, the three self-righting bags on Apollo 7's recovery section surround the spacecraft docking tunnel.

Viewed for the first time by human eye, earth hangs like a jewel in the blackness of space. Apollo 8, en-route to the Moon, December, 1968.

The largest launcher brought to operational status, Saturn V launched all but the first of eleven Apollo spacecraft. It was retired in May, 1973, after launching the Skylab space station in a two-stage configuration, $5\frac{1}{2}$ years after the first flight in November, 1967.

The first flight to carry both Apollo spacecraft and Lunar Module in the configuration later missions would employ to land on the Moon. Apollo 9 astronauts McDivitt (left), Scott (center) and Schweickart.

With its back to the Sun, the Lunar Module sits square upon the Moon's dusty surface. Note the crinkled thermal layers and the bent height sensing probe attached to the leg (foreground); antenna on top was used to carry communication during EVA.

Bringing international flavor to the first Moon flight, this Swiss experiment being deployed on the surface would collect solar particles for analysis back on earth.

Lunar surface experiments on the first three missions successfully sent to the Moon were contained in a quad at the rear of the LM.

Apollo 12 lifts off into a thundery sky and seconds later is struck by lightning.

Apollo missions 12 and 14 improved communications between EVA astronauts and earth by using an umbrella-like antenna seen here at right. Note the small hand tool carrier in front of the astronaut, removing equipment from the modularised equipment stowage assembly table.

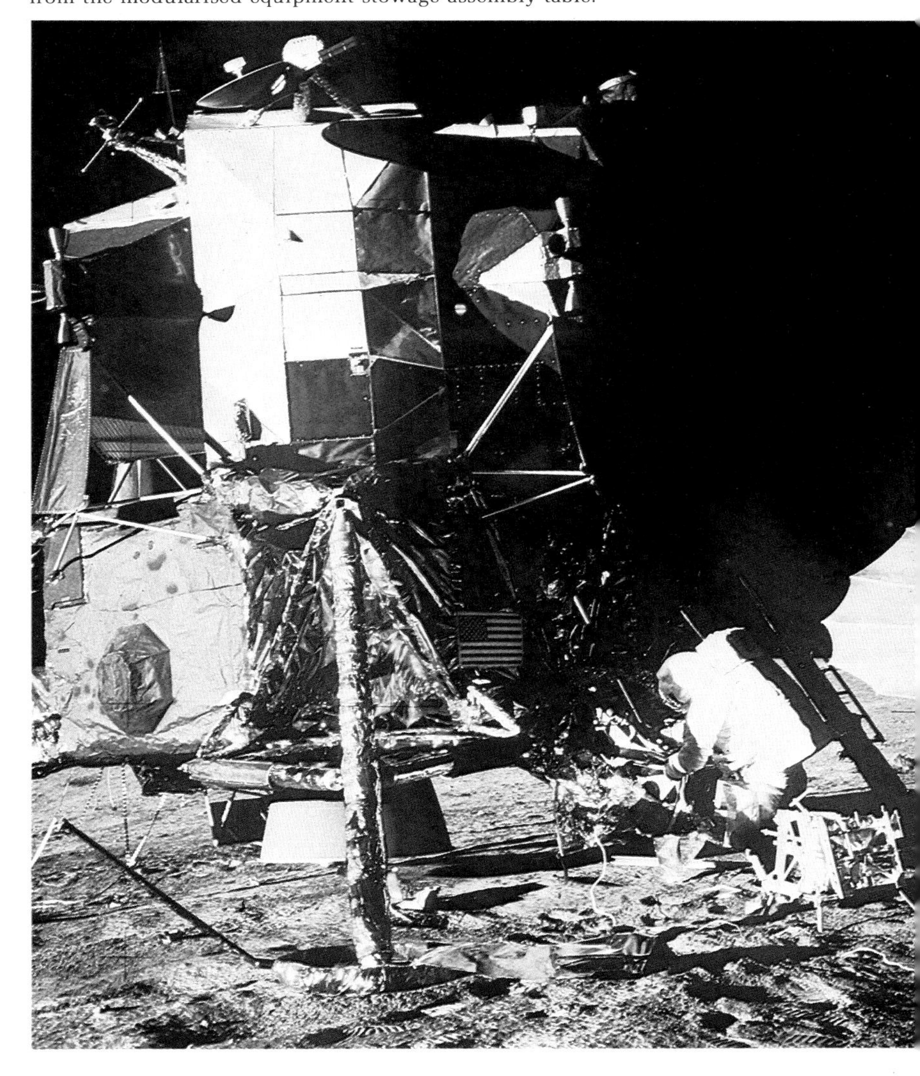

Tracks from Apollo 14's LM, evidence that wheels were first used on the Moon in February, 1971, supporting a small handcart with tools – perhaps the first use of wheels on earth several thousand years earlier.

CHAPTER TWENTY-THREE

Science on the Moon

Although NASA concentrated efforts to expand Apollo on the J-series flights beginning mid-1971 with Apollo 15, the last of the H-series missions was to fully capitalize on everything learned by previous flights. Apollo 14 would be commanded by America's first operational astronaut, Alan Bartlett Shepard, veteran of the first manned Mercury flight, albeit a suborbital shot, on 5 May, 1961. Now almost ten years later, Shepard was to walk on the Moon. Grounded in 1963 because of an ear infection, not even allowed to fly an aircraft unless accompanied by a second pilot, Shepard underwent an operation and was restored to flight status in May, 1969. He, along with companions Stuart (Stu) Roosa, and Edgar Mitchell, was assigned the flight of Apollo 14 simultaneously with the 13 crew.

Between 1963 and 1969, Shepard held the post of Chief of the Astronaut Office. Another grounded Mercury-era astronaut, Deke Slayton, was Coordinator of Astronaut Activities. Some would have said Apollo 14's launch date, exactly 13 years after the launch of America's first satellite, was ill-timed. Determined to break the spell cast on Apollo 13, the agency shrugged off suggestions that flight teams wait a month until the next window opened. Fra Mauro was the target, and 31 January, 1971, the launch date.

Apollo 14 had been originally scheduled to fly in February, 1970, when maximum paced activity plans had Moon missions rippling off every three months. Delayed first by a decision to span out the missions, then by modifications resulting from 13's review board, hardware was at KSC more than a year before the actual launch: spacecraft 110 and LM-8 (Kitty Hawk and Antares, respectively) in November, 1969; S-IC, S-II, and S-IVB stages of Saturn V 509 in January, 1970. Even the astronauts concentrated on their mission in record times. Stu Roosa logged more than 1,000 hours in the Command Module simulator.

Apollo 14 was originally planned for the Littrow area of the Moon, far north of the lunar equator and as such the first site to move appreciably away from the 'safe' zone close to the preferred orbit plane. But when Apollo 13 failed to reach the first science site at Fra Mauro plans changed and the crew began to study in earnest the geological tasks they would perform at that site.

A flurry of minor problems surfaced during preparation of 509's hardware. In April, 1970, a technician accidentally punched a hole in a new inertial measurement unit being installed to replace an older design, pouring water-glycol over the lower equipment bay floor. As workers removed panels and replaced saturated wiring the extent of the leak was seen to be worse than feared. It took one month plus a special drying out procedure to get the Command Module back on schedule. Then changes to the S-II stage were made in response to an investigation of why Apollo 13's second stage shut its center engine down early, with installation of a helium gas accumulator in the center engine's liquid oxygen line, a special cut-off instrument in case the accumulator failed to damp out pogo oscillations, and improved flow control valves.

Then Grumman successfully traced noisy communications in Antares to a faulty signal processor. Altitude chamber tests were delayed while engineers made special checks on ball valves in the Lunar Module. Saturn V stages had been stacked in May but engineers noticed that paint had bonded the first and second stages together and the vehicle was de-stacked, cleaned, then re-assembled on 2 November. The two space vehicles were placed atop the stack and 509 rolled to complex 39A on the 9th. Nothing marred the final preparations for launch.

During the final three weeks, Shepard, Roosa and Mitchell were carefully screened from contact with personnel outside a special group physically examined for traces of contagions. Bells and klaxons warned of the imminent approach of the 14 flight crew as unscreened persons scurried out of the way. Because it came after the nearly disastrous Apollo 13 mission, and possibly also because its commander was America's first astronaut, a record crowd for post-Apollo 11 visits flocked to Cape Canaveral: in all, nearly a million people. Press coverage was down though, to about one-half the contingent that covered Neil Armstrong's flight. It was a pleasant contrast to Shepard's first launch, however. On that occasion, only 440 newsmen turned up, albeit a large number for 1961.

Apollo 14's launch window lasted 3 hr 49 min beginning at 3:23 pm local time. If postponed beyond 31 January, the mission would be re-cycled to the next lunar month, with opportunities on 1, 2 and 3 March. As the countdown progressed it began to look as though the flight might well have to wait. Clouds gathered across the Cape and the overcast sky gave Walter Kapryan second thoughts about launching the stack into suspect weather. More than 30,000 invited guests, including Vice-President Spiro Agnew, Episcopalian priest John B. Medaris – once closely connected with von Braun's entry to the United States – and royalty from Spain were waiting to see Apollo 14 fly. But aircraft up aloft across the Florida skyline detected electrical fields in the gathering cloud and Kapryan, following strict launch rules modified by the Apollo 12 event, called a hold at T−8 min 2 sec.

From firing room 2 the decision was made to re-start the countdown clocks after a 40 min 3 sec hold; if left more than one hour the countdown would have to be re-cycled further back. When 509 lifted off it was the heaviest launcher yet, just a few tonnes heavier than anything flown to that time. Within 36 seconds, soon after thunder claps shook the Cape, Apollo 14 vanished into cloud but all the way to orbit the vehicle conformed to its programed event sequence, injecting third stage and payload into a safe Earth parking orbit. At the Cape, Spiro Agnew had a few words for the launch controllers before Prince Juan Carlos of Spain added his own congratulations.

By this time operations had switched to Houston as capcom Gordon Fullerton kept up a dialogue with the three men in space. There had been some concern about Shepard's physiological

The second team dispatched to the Fra Mauro hills, Apollo 14 astronauts (left to right) Roosa, Shepard and Mitchell.

reaction to the Saturn V flight; cynics questioned the wisdom of sending a 47 year old man to the Moon. For some reason, Shepard's biomedical sensors failed to work during ascent but Stu Roosa's heart shot up to 132 beats per min, while Mitchell's remained in the 80s. Alan Shepard earlier gave the close-out crew a quizzical look when they offered him a walking stick before swinging onto the couches aboard Kitty Hawk. He had his own view about those who thought him too old.

Because Apollo 14 had been delayed 40 min 3 sec by weather at the Cape, mission controllers decided to exercise an option built in to procedures for just this eventuality. The TLI burn on revolution 2 would send the vehicles on course for lunar-orbit-insertion 40 minutes earlier than originally planned, shortening the transit time by that amount. That would put Apollo 14 in Moon orbit at the predicted Earth time calculated to give the mission full coverage from the appropriate tracking stations.

But doing so would shorten by the same amount of time all events written out in the flight plan, since the spacecraft's clocks would start at lift-off and, consequently, be shortened by some 40 min. So it was decided to perform an update by adding 40 min 3 sec to clocks in Mission Control and aboard the spacecraft telling elapsed time. That update would be carried out on the third day but readers should note that to avoid confusion the elapsed times quoted here refer to the actual hours and minutes from lift-off, so corresponding to the real time rather than the false time set up to conform to the flight plan. Apollo 14 would still get to the Moon at the scheduled time of day but on a path some 40 minutes faster than originally planned.

Trans-lunar-injection began with the spacecraft within range of the Carnarvon tracking station and concluded with ARIA coverage over the Pacific. With the S-IVB shut down, it was time to separate Kitty Hawk, turn around, and move in to dock with Antares. But that was where trouble began, for when Stu Roosa guided the extended probe into the conical drogue on top of the LM, the three small capture latches that should have snagged the circular cut-out in the center of the drogue failed to work. It was 3 hr 14 min since lift-off; Apollo 14 was only two or three minutes behind the sequence of events laid out in the flight plan. But failure to dock was a puzzle.

Four minutes later, after backing off a few meters, Roosa drove the Hawk forward a second time, holding the aft-firing thrusters on for four seconds. Still no luck. It was as if the latches were rigidly fixed to the probe's head. In the MOCR, capcom Gordon Fullerton talked with the crew while Pete Frank's flight controllers watched the status of systems aboard the spacecraft. With the TV camera aboard Apollo 14 transmitting a picture to Earth, the abortive attempts were seen with clarity. Moreover, tiny scratches in the drogue could be observed by the crew, as if the capture latches were scoring marks in the alloy as Kitty Hawk pushed hard into the receptacle.

Unable to latch on Antares, the Moon mission would have to be called off. Mission Control had little more than six hours to resolve the problem, for after that the S-IVB's batteries would have run down, preventing the stage from holding attitude. After two failures, Roosa tried twice more, each time driving Kitty Hawk faster toward Antares, and each time it failed to snag. In the MOCR, astronauts Cernan and Young examined a probe and drogue assembly, discussing with Mission Director Chester Lee possible causes of the problem. Perhaps a pin had stuck in the probe's head; connected to the boost protective cover before the launch escape system jettisoned, perhaps the pin that should have pulled free somehow snagged up. Perhaps water – it had been drizzling across the Cape at launch – froze inside the probe head, fixing the three small capture latches to the probe.

Cernan took over from Fullerton and talked with Roosa and Shepard about the problem. The consensus was for Roosa to try a fifth time and retract the probe as the CSM moved in, driving forward for a hard-dock where the 12 main latches would fire, by-passing the normal sequence of soft-dock followed by probe retraction. This time it worked, at 4 hr 57 min into the mission. 'We had a hard dock, Houston,' reported Al Shepard as flight controllers sent up a loud cheer from the MOCR. The operation had been more successful than expected because the capture latches snagged the drogue just before the probe telescoped back in, suggesting that whatever prevented movement in the latches had now worked free. Kitty Hawk had Antares in its talons.

The rest of the trans-lunar coast went well. After separating the docked assembly from the launch vehicle's third stage, Apollo 14's crew prepared for a televised inspection of the probe and drogue mechanism by pressurizing the LM, opening the tunnel, removing the hardware, and showing it to the camera while performing some rudimentary tests to demonstrate its efficiency. The probe seemed to work perfectly now. It would have to once more, when Antares' Ascent Stage came up from the Moon, and nobody wanted to run the risk of an emergency crew transfer along hand rails to Kitty Hawk.

Meanwhile, the S-IVB was placed on course for lunar impact shortly after Apollo 14 got into Moon orbit. Like its predecessor, it was expected to set the Moon 'ringing like a bell' from the reflected waves transmitted through the lunar substructure; crashing redundant stages on to the surface provided valuable seismic data from which to determine the Moon's layered interior. The same task was to be accomplished by crashing empty Ascent Stages to the surface, an operation pioneered on Apollo 12.

The crew's first rest period began at an elapsed time of 16 hours and ten hours later, shortly before reaching the halfway point in distance, Pete Frank's orange team got a call from the spacecraft. It was time to hold a scheduled debriefing on the performance of the Saturn launch vehicle but that conversation came on Milton Windler's shift three hours later, where the crew testified to the smooth nature of the beast with a 'slight pogo . . . nothing of any great magnitude' on the second stage. It seemed at last that all the anomalies had been wrung from the giant Saturn , and that no more premature shut-downs or disturbing periods of vibration would afflict Apollo crews from now on.

The MCC_2 opportunity was taken to transfer Apollo 14 from a free return path to a non-return trajectory, guiding the point of closest approach down from a TLI target of 3,745 km to a planned 111 km for lunar-orbit-insertion. But when it came, at 30 hr 36 min, the SPS engine fired Apollo 14 on course for a pericynthion of 124 km. Another course correction maneuver would probably be needed. The crew began their second rest period at 41 hours, waking nearly ten hours later. Apollo 14's clock, and those in the Mission Control Center that phased specific event sequences, was given a 40 min 3 sec update at 54 hr 53 min, the numerals suddenly leaping on to an elapsed time of 55 hr 33 min, artificially slipping subsequent events to the correct times in the flight plan.

Several hours later the crew took out the probe and drogue, opened Antares' hatch and inspected the interior of the Lunar Module while sending a 42 minute telecast to Earth. Antares was devoid of Mylar fragments or pieces of debris – a washer once floated out from behind a panel – that sometimes met astronauts inspecting their lunar lander for the first time after launch. Then it was time to close up the hatch and begin the third 'night' in space. Minutes after formally beginning their scheduled rest period, capcom Gordon Fullerton woke the crew with a warning that the oxygen flow rate was unusually high, as though the environmental control system was seeking to maintain cabin pressure in the presence of an unexpected leak. No problem. Ed Mitchell cycled the waste management valve and the rate fell; apparently excessive use of the facility had caused a slight vent!

Flight director Gerry Griffin had his gold men on shift when capcom Fred Haise talked Apollo 14 through the mid-course correction burn performed at 76 hr 58 min after lift-off. Ignited for a brief 0.6 sec, the SPS engine changed Apollo's speed by a mere 3.8 km/hr, sufficient nevertheless to tuck pericynthion down to 112 km. Then Mission Control had Ed Mitchell once again transfer to Antares to perform a complex set of switching operations to check the validity of battery No. 5. Earlier, telemetry during the checkout showed a 0.3 volt drop from the predicted 37 volt reading. Mission Control was concerned why that should be and with Apollo 14 only hours from going behind the Moon it was imperative to check LM systems. Antares checked out well and capcom Joe Engle, who had voiced to Ed Mitchell the procedure for testing the battery, handed back to Fred Haise.

But still there was concern about the docking probe anomaly and further word was passed to the crew on the close examination of spacecraft telemetry completed during the preceding hours; nobody saw anything horrendous with the final docking phase, so Apollo 14 was given a 'go' to proceed with a normal mission. Around the Moon swept Antares and Kitty Hawk to fire the SPS in a long 6 min 12 sec burn just like four other spacecraft before. But from this point procedures were a little different.

Instead of burning into a circular path about 112 km around the Moon the SPS firing at the end of revolution 2 was designed, like the plan for Apollo 13, to place the docked configuration into an elliptical orbit of about 109 × 17 km. It would remain in this path until the two vehicles separated on revolution 12, swooping low across the Moon's surface uprange of the planned landing site for more than seventeen hours. Performed by Kitty Hawk, this early descent-orbit-insertion burn would conserve propellant in Antares, for it had always been the Lunar Module that performed this manoeuvre in the past. From Moon orbit, Alan Shepard told Fred Haise what it was like to see the stark contrast of a lunar day.

SHEPARD 'Well, this really is a wild place up here. It has all the grays, browns, whites, dark craters that everybody's talked about before. It's really quite a sight . . . No atmosphere at all. Everything is clear up here. Really fantastic.'
MITCHELL 'Fredo – I think the best description that comes to mind, we mentioned this when we first looked at this thing, is that it looks like a plaster mold that somebody has dusted with grays and browns but it looks like it's been molded out of plaster of paris.'

With the descent-orbit-insertion burn completed, the view down now across the Moon's dry ground was a spectacle to behold. It was the first time an Apollo CSM had been down to that height, and certainly the first time anybody had had time at that altitude to study the scene and take pictures. The DOI burn was potentially dangerous. Performed on the far side of the Moon, ground stations were spring-loaded to acquire the spacecraft as it came round the limb and get some good tracking on the vehicles. If the SPS engine had burned just a few seconds too long it would have lowered pericynthion to a point where it intersected the surface. Ground controllers had just twelve minutes to determine whether Apollo 14 was in a safe elliptical path or not. If needed, a bail-out burn could be performed at that time to lift pericynthion and prevent the docked vehicles crashing into the surface. But it was not needed: the big engine fired almost exactly as required.

Shepard (right) and Mitchell rehearse Moon walk activity at Kennedy Space Center. Note the arm bands worn by the mission commander and the new visors on each helmet.

MITCHELL 'Looks like we're getting mighty low here. It's a very different sight from the higher altitudes.'
CAPCOM 'In about four minutes you'll be at your minimum altitude, which should be about (12,200 meters) above the terrain. We were wondering how things looked down there.'
MITCHELL 'Well, I'm glad to hear you say we're that high. It looks like we're quite a bit lower – as a matter of fact we're below some of the peaks on the horizon, but that's only an illusion.'
CAPCOM 'Roger.'
MITCHELL 'The surface appears to be a lot smoother down here where we can see closer to the detail and particularly at this higher Sun angle it appears to be a softer surface, but it certainly is an unusual sensation flying this low.'

Photographic activity centered on pictures of the Descartes region, a future candidate landing site, nestled in the craggy lunar highlands where old eroded craters and pockets of accumulated debris made pin-point landing a necessity. During an eight-hour rest period taken by the crew prior to commencing the landing preparations, Apollo 14's orbit changed gradually as the asymmetric mass of the Moon pulled and tugged at the spacecraft's path.

The large lunar basins, filled now with dark basaltic lava and called maria, contained more massive materials than the surrounding crust and these areas of mass concentration – called mascons for short – were a potential hazard. When Apollo 14 burned DOI its path was precisely 109 × 17.8 km. When the two spacecraft separated nearly eighteen hours later, pericynthion had decayed to just 14 km. Shepard and Mitchell moved into Antares on revolution 10 to check out the LM prior to undocking, an event that occurred as planned 103 hr 48 min after lift-off. With a 'Boy, you look mighty pretty out there,' Stu Roosa took several pictures of the spider-like spacecraft as Shepard and Mitchell turned it through several attitude changes.

Nearly one full revolution later, on the Moon's far side, Kitty Hawk fired its SPS engine and raised pericynthion, settling into an almost circular path about 110 km above the lunar surface. Antares was still in its elliptical path from where it would fire up the Descent Stage engine for powered descent, but the CSM was in the right type of orbit from which to effect a rendezvous should anything go wrong.

Shortly after coming into view on the 13th revolution, telemetry on the ground, and the display keyboard on Antares, showed a spurious bit – an unexpected electronic signal – flashing up on program 52. Had the electronic spook shown up during descent operations it would have unintentionally aborted the entire landing operation. Because the LM guidance computer was then in a navigation alignment program, however, the bit did no damage. A few quick tests showed it to be the result of minute specks of dirt in the abort switch. Mission Control conferred with MIT, the design team for Antares' computer, and developed a procedure whereby several verbs and nouns punched into the computer between ignition and throttle-up just 26 seconds later would effectively lock out the abort instructions should it reappear.

With the computer slaved to programs 64, 65 or 66, the mission would be suddenly aborted if the random signal crept in. Valid aborts would be handled through the AGS, however. Several taps on the abort switch caused the bit to disappear each time it appeared during checkout. But when Antares came round on revolution 14, thirty-five minutes before beginning powered descent, Mission Control had a new procedure – one that would allow the crew to set up the lockout process before ignition.

Down again to a height of just 14 km above the Moon, the Lunar Module was over 490 km uprange of the landing site when the main engine came alive, almost imperceptible until, throttled up to a maximum thrust, the braking phase began. With a cry of 'Right on the money, right on the money,' Ed Mitchell read out display data as Alan Shepard steered Antares down to Fra Mauro. The only real worry that developed during the 11 min 33 sec powered descent phase was the absence of radar data before descending to 6,900 meters. 'Whew, that was close,' sighed Ed Mitchell, for without comparative data the descent could not continue for long. But when it finally landed, Antares was only 50 meters from the precise spot chosen, a deliberate foreshortening which Shepard intended to attempt; their major activity on the second space walk would be to explore a region east of the LM so any bias in that direction would help.

'We're on the surface,' called Mitchell. 'That was a beautiful one.' Antares was standing at an inclination of 8° but that was no problem and the two astronauts began preparations for their first EVA, scheduled to last 4¼ hours, or longer if the back-pack consumables held up well to the demanding requirements. Only a minor communication switching problem marred the preparation but Shepard was a little late getting out on the surface.

CAPCOM 'Okay Al, beautiful, we can see you coming down the ladder right now, it looks like you're about on the bottom step. And on the surface. Not bad for an old man.'
SHEPARD 'Okay, you're right. Al is on the surface, and it's been a long way, but we're here.'

It had indeed been a long trip for Alan Shepard, one that began twelve years earlier when he applied for the first group of astronauts. Only two were still in the program – but he had made it to the Moon.

The first EVA would be taken up almost completely with setting out the array of scientific instruments carried by Antares. First, the erectable S-band antenna to improve communications, then the solar-wind-collection experiment and a laser reflector. Ed Mitchell joined Shepard on the surface and while the former put up the flag (this activity by now rated no more than a five-minute task) the latter took photographs of the Lunar Module to show how the footpads affected the surface and how the dusty regolith, or outer covering, had been affected by the LM. The ALSEP instruments for Apollo 14 included a passive seismometer like the one left at the Apollo 12 site, 172 km west, two detectors to measure particles coming from the Sun, another to detect solar protons and electrons that might reach the lunar surface, and an active seismic experiment.

The latter comprised a string of three geophones, tiny detectors set up to sense vibrations coming through the surface from detonations close by, laid out between 3 meters and 94 meters from the ALSEP central data station which would transmit to Earth information obtained from all five experiments. The explosive detonations were to be provided by a mortar box which, upon command from Earth long after the crew left, would fire four projectiles to distances of 152, 304, 914, and 1,524 meters. By studying the seismic waves created by the detonation of the high explosives, varying between 0.045 kg and 0.45 kg, scientists would gather useful information about the site's substructure to a depth of about 150 meters.

Another element of the active seismic experiment included a 'thumper' comprising a drum of 21 pyrotechnic charges of the type used to activate squib valves on spacecraft, attached to a short staff on which was mounted a handle and a firing switch. Being essentially miniature seismometers, the three geophones would record wave data as Mitchell walked along the line firing off the tiny charges at 4½ meter intervals. Like all ALSEPs, the instruments would be powered by a radioisotope generator left connected to the central data station.

But before unpacking the scientific equipment, Shepard and Mitchell unstowed the first wheeled trolley to be used on the Moon. Called the MET, or modularized equipment transporter, it consisted of a light alloy frame, a long handle, and two wheels supporting low pressure tyres developed by Goodyear. When installed aboard the MESA on Antares' Descent Stage, the tyres carried a pressure significantly less than that of the surrounding atmosphere, giving them the appearance of collapsed balloons; on the airless moon they were correctly inflated to 105 g/cm². The MET would be used primarily on the second EVA for carrying the geological equipment necessary for exploring the various sites and as such would be a useful improvement on the standard hand tool carrier.

After returning to the LM to switch communications to the erectable S-band antenna, Mitchell helped Shepard deploy the transporter and then set about carrying the ALSEP packages a short distance from the LM. It took nearly two hours to remove the packages and deploy all the equipment before Mitchell was ready to start walking down the geophone line firing off the thumper charges. It took more than 30 minutes to try all 21, but only 13 detonated, of which nine were observed on the geophones; the first four were unusable because one geophone was found to be on its side and until depressed was incapable of picking up vibrations. The plan was for Mitchell to stop moving 20 seconds before and up to 5 seconds after ignition of each charge, but during this time Shepard was busy collecting samples.

Following well-documented procedures, Shepard used various geologic tools to collect soil and small rocks, ending up with a football-sized rock as Mitchell joined him to complete the first EVA. After 4 hr 49 min the LM cabin was repressurized, 10 hr 15 min after touchdown. Shepard and Mitchell were more than one and one-half hours behind their flight plan but everything had been accomplished as required and there was time enough to catch up.

With an 'Okay, we're up and running this morning,' Al Shepard broke the radio silence that prevailed while the two Moon men slept at Fra Mauro; intending to abbreviate the scheduled 10 hr rest period to about 7½ hrs, the crew were in fact awake after less than 5 hrs sleep. The second EVA would give them an opportunity to obtain samples from the Fra Mauro eject a blanket and Shepard was concerned that they should have sufficient time for that activity. Scurrying along in fine fashion, Shepard and Mitchell were ready in almost record time, beginning their second EVA about two and one-half hours earlier than scheduled in the flight plan. The first Moon walk took place during the late morning and lunch hours in Houston; the second was to take place during the pre-dawn hours of the following day, Friday, 6 February.

Exploration of Fra Mauro was essential for access to material excavated from deep within the Moon's interior for in the absence of deep drilling equipment a natural event was the only way rocks from below the surface could be retrieved for analysis. In several ways, the key to an understanding of the Moon's evolution lay within the close examination of materials from the lower regions of the crust. Nearly 4,000 million years ago a large basin was formed by impact which threw out material from several kilometers below the surface. At Fra Mauro that ejecta blanket was about 70 meters deep, overlain with a 9 meter covering of pulverized ejecta from smaller craters formed since the main event. It was on this over-layer that Antares landed.

To gain access to the true Fra Mauro ejecta blanket, mission planners chose a crater 340 meters in diameter approximately 1.1 km east of the Lunar Module. Called Cone, the crater had itself thrown out material from a depth of up to 30 meters, in effect excavating the true Fra Mauro formation and bringing it to the surface. Shepard and Mitchell would now walk east to Cone and collect samples from the rim; crater ballistics determine that material from the deepest regions affected by the impact will be laid out on the rim proper and that materials closer to the surface will be distributed over a wide area. This was the philosophy that took the crew on their search for remnants from one of the most cataclysmic episodes to affect the Moon's surface.

But the road to Cone was uphill most of the way, although the early part of the traverse would take them through a modest depression. Shortly after reaching the surface, Mitchell unpacked a portable magnetometer with which they would measure the local magnetic field at various points on their traverse. Unlike Earth, the Moon has no dipolar field (with north and south polarities) but rocks froze into their memory the magnetic fields present when they cooled and so presented a variety of different magnetic strengths. The hand cart, or MET, would be invaluable on this excursion; they would never make it to the rim of Cone without some form of transport for their tools and equipment.

Before leaving the area of Antares, Shepard pointed the camera up toward Cone. It would not have the range to follow them far but would at least provide an interesting view for Earth-bound watchers. Geological traverses were frustrating for those left behind. It was the last flight where the camera would remain at the landing site. With the roving vehicles available from Apollo 15, the TV would travel with the roaming astronauts.

The first stop, called Station A, was only a few minutes' walk away. A site remote from the landing area but one which was presumed to be still on the overlayer. Here, core tubes were sunk into the surface for a vertical sample. Shepard noted that 'the surface here is textured. It is . . . a very fine grain dust. About the same as we have in the vicinity of the LM. But there seems to be small pebbles – more small pebbles here on the surface than we had back around the LM area. And the population of larger rocks perhaps small boulder size is more prevalent here. The point where we are sampling is just about in the center of three craters of almost equal size. I would say perhaps 20 meters in diameter. I'm pretty sure we're just about where point A is on the map.'

Here, a magnetometer reading was taken. Carried on the hand cart, the magnetometer was attached to a line and a sensor.

Walking slowly along a line of geophones laid out on the surface, Mitchell detonates small explosive squibs at the end of his 'thumper' to induce minor seismic shocks.

The sensor was unreeled 10 meters from the cart for an uncontaminated reading. After thirty minutes at the site, Shepard and Mitchell packed up and moved on. Ten minutes later they were at Station B, another sample site en route to Cone. 'The area here . . . (has) considerably more boulders,' said Mitchell and Shepard collected rocks. 'A large boulder field (with) more numerous boulders than we've seen in the past.' Minutes later they were off again, moving noticeably up hill now.

The going was beginning to get a little tougher. Station B was only 300 meters from Antares; Cone crater was still 800 meters away. When Mitchell turned up the cooling flow on his suit, Mission Control advised the two men to rest a few minutes. 'Get the map and see if we can find out exactly where we are,' said Mitchell, 'The old LM looks like it's got a flat over there, the way it's leaning.' The uphill traverse resumed and heart rates climbed to 120 beats per min. It was not a great slope, but in bulky suits and pulling the MET it was a strenuous walk. The dusty regolith too was soft and felt like loose snow impeding their mobility. Then it got firmer and the going improved a little.

They were out of the valley-like depression and trudging up the slopes of Cone. 'Al's got the back of the MET now and we're carrying it up,' said Mitchell as the hand cart dragged its wheels through the fine granular material. Time to rest again. Although equipped with maps of the site, visual navigation among the undulations, the shallow craters, the block fields, and the pillowing regolith made it very difficult to stay on course for long. As they approached the area where the big crater should be they became increasingly confused. First one rim would appear and, thinking that to be Cone, the crew moved up only to find another rim beyond. Back on their traverse the going was getting strenuous. Heart rates now read 150/min for Shepard and 128/min for Mitchell.

MITCHELL 'The rocks and boulders are getting more numerous toward the top here.'
SHEPARD 'You know, we haven't reached the rim yet.'
MITCHELL 'Oh boy, we got fooled on that one.'

It was time to rest yet again, and capcom Fred Haise talked to them about the location of Cone. Unable to see exactly where they were, Houston had to follow their progress on maps in the science support room and at the consoles.

SHEPARD 'Really got a pretty steep slope here.'
CAPCOM 'Yeah, we kind of figured that from listening to you.'
SHEPARD 'Well, that's apparently the rim of Cone over there. That's at least 30 minutes up there.'
MITCHELL 'Yep. It could take longer than we expected.'
SHEPARD 'Our positions are all in doubt now, Fredo. We've got a ways to go yet. Perhaps you can say what it is you want, I'd say that the rim is at least 30 minutes away. We're approaching the edge of the boulder field here from the south Flank.'

Flank was a crater they were trained to recognize but the problem was an inability to detect easily the precise rim of the crater they had come so far to explore.

'Okay, Al and Ed. In view of your estimate of where your location is and how long it's going to take to get to Cone, the word from the backroom is they'd like you to consider where you are the edge of Cone crater.' Shepard felt they could get a little closer to Cone and advised Fred Haise that they would 'press on a little further.' Houston had just given them a 30 minute extension to their EVA based on telemetry of the state of back-pack consumables. In moving east toward Cone, Shepard and Mitchell wandered too far south and were in fact on the very rim flank of Cone when they decided a few minutes later to stop and begin their sampling activity. They believed themselves to be near the west rim, but were puzzled over not being able to see a noticeable rim edge, surprised even more that they could not see anything that looked like the crater's interior. Nevertheless, the traverse had been exhausting and time was passing. They were already 2 hr 10 min into their EVA, and more than one kilometer from the LM.

Engineers had given due consideration to the problems that could arise should one back-pack fail. With oxygen supplied from the emergency OPS unit on top of the pack, cooling water would, however, be unavailable. This was acceptable within a short distance of the LM, since the OPS oxygen flow would also remove body heat. But far from the LM, oxygen demand would be so high that it would not last until the astronaut could return to the pressurized cabin. So Apollo 14 carried a BSLSS (Buddy Secondary Life Support System) to alleviate this problem. A long connecting hose would link the water supply in one back-pack to the suit of the other astronaut, sufficient water being available for both men to move quickly back to the protection of the LM. If Shepard or Mitchell ran into trouble now, the 'Buddy-pack' would save one life that would otherwise be lost. This conservatism constrained all lunar surface EVA operations. Sufficient consumables had to be available to return in safety from any point on the traverse.

Unknown to either astronaut, the crew had arrived to within about 30 meters of the actual rim of Cone when they unloaded the equipment, took magnetometer readings, and began a sampling sequence among the blocks and the boulders. 'This area that we're in right now, is a pretty darn rugged boulder field area,' said Ed Mitchell as he set about deploying the portable magnetometer sensor. Shepard took more core samples: 'The area is apparently very rocky but I did get down into the second layer of the underlying layer of the regolith which was white as opposed to being dark brown.'

Now, at the selected site, heart rates were down with Shepard reading 108 and Mitchell 86. Sampling the Moon was less strenuous than walking on it. At an EVA time of 2 hr 35 min the two explorers were ready to leave for their return traverse to the LM, stopping off at Stations F and G; two other stops had been cancelled because of time constraints. Moving back down the gently sloping incline the going was much easier and within twenty minutes they were at the first Station stop, about 350 meters east of Antares. Core tube samples, a small trench and sample collection occupied Shepard and Mitchell for 30 minutes at Station G, a site just short of three nested depressions called Triplet crater.

Then, because of high oxygen consumption Fred Haise passed word that the EVA extension had been cut to 15 minutes beyond the planned $4\frac{1}{4}$ hours. 'This country is so rolling and undulating, Fred, with rises and dips everywhere, that you can be going by a fairly good size crater and not even recognize it,' said Shepard. Back at Antares the crew re-positioned the TV camera and loped off to the ALSEP site to realign the antenna on the central station; telemetry indicated it was slightly off the desired angle for good communication with Earth. Then it was time to collect the solar wind sheet exposed on the lunar surface for a total 21 hours since deployed on the first EVA, and to pack the equipment transfer bag which, attached to the clothesline on the side of the LM, would allow the crew to load Antares with valueable Moon rock. Then Al Shepard sprang a surprise.

SHEPARD 'Houston, while you're looking that up you might recognize what I have in my hand as the handle for the contingency sample return and just so happens I have a genuine six iron on the bottom of it. In my left hand I have a little white pellet that's familiar to millions of Americans. I drop it down. Unfortunately the suit is so stiff I can't do this with two hands but I'm going to try a little sand trap shot here.'
MITCHELL 'Hey, you got more dirt than ball that time.'
SHEPARD 'I got more dirt than ball. Here we go again.'
CAPCOM 'That looked like a slice to me, Al.'
SHEPARD 'Here we go. Straight as a die – one more. Miles and miles and miles.'

And so Alan Shepard became the first man to play a shot from the rough on the Moon's dusty surface. He would receive awards later for this unique golfing achievement!

While Shepard and Mitchell worked on the surface, up above in Kitty Hawk Stu Roosa performed important photographic tasks with a Hycon lunar topographic camera. Essentially a modified aerial reconnaissance camera, it weighed 29 kg and carried a 45.7 cm focal length lens providing a resolution of 6 metres from orbital altitude. The camera was fixed in the center hatch window for use. At other times, Roosa used a 700 mm Hasselblad with a 500 mm lens and when the topographic camera failed the hand-held Hasselblad was used instead. During the period Antares rested on the surface, communications with the CSM went via a separate channel, alleviating earlier problems where both spacecraft were on the same loop.

But now, with the second $4\frac{1}{2}$ hour EVA completed, it was time to dust off and prepare for ascent. With less than one hour to go before ignition, Shepard reported that the umbrella-like antenna standing on the lunar surface had blown over when the crew test fired the RCS thrusters. Kitty Hawk had performed a plane change ten hours after Antares landed and was now in a good orbit for rendezvous. At lift-off, the LM ascended for 10 seconds, then pitched over to begin the seven-minute haul toward an elliptical path below the CSM. At insertion, Antares was 250 km behind Kitty Hawk, about half the separation distance on earlier flights. This was because Apollo 14 pioneered a faster lunar-orbit rendezvous, compressing the usual three hours required to pull alongside the CSM to just half that time.

Instead of taking one and one-half orbits to catch up, Antares would rendezvous and dock within one revolution. Here at last was the M=1 rendezvous first tried by Gemini XI four and one-half years earlier. Eliminating the concentric-sequence and constant-delta-height maneuvers, Antares steered to a co-planar orbit from where it could begin terminal-phase-initiation 45 min after lift-off. A small 'tweak' burn was performed three minutes after getting into orbit, and there was provision for a bailout burn that would restore the normal coelliptic-type rendezvous should the spacecraft be outside the required constraints. But Antares was right on course and when the two vehicles swept out of view thirty-five minutes after lift-off they were closing fast and only 95 km apart.

Eleven minutes later Antares fired its Ascent Stage engine for the second time, lifting the spacecraft on course for rendezvous with Kitty Hawk. It was the first time the APS had been fired twice during ascent maneuvers and the 10 sec burn all but consumed the remaining propellant ; the extra thrust was necessary to reduce the high closing rate. Soon after coming around on its 32nd orbit of the Moon, Kitty Hawk transmitted a TV picture of the final rendezvous as Antares came up toward Apollo against the grey lunar surface. During the final braking phase Shepard and Mitchell were unaware that telemetry indicated a failure in the Abort Guidance Section, but there were no caution and warning tones so the crew were asked to cycle the circuit breakers. But to no avail.

'Oh you look clean and nice Stu. Want to come in a little closer. It'll save you some gas?' said Antares, as the LM maneuvered to the docking position. Trouble with the probe shortly after heading out for the Moon made the next operation a well-watched event in the MOCR. But the docking went according to plan.

CAPCOM 'Apollo 14 this is Houston. You're go for the docking.'
ANTARES 'Roger. We got you.'
ANTARES 'Yeah. How about that.'
ANTARES 'Say again.'
ANTARES 'Okay, we capture.'
CAPCOM 'Beautiful. Normal docking.'
ANTARES 'Okay. And we have hard dock.'
CAPCOM 'Beautiful. There's a big sigh of relief being breathed around here.'
KITTY HAWK 'All over the world there is.'
ANTARES 'You want to try it from up here.'
CAPCOM 'This world and out of this world too.'

Fifteen minutes later the crew were preparing to open the tunnel and within two hours everything scheduled for return to Earth had been placed in Kitty Hawk and the tunnel closed up. On this flight the probe would be brought back for examination. Just in case a mechanical fault threatened later flights.

The LM was jettisoned on revolution 33 and five minutes later Kitty Hawk moved away by firing its thrusters for 7 seconds. On the next revolution, minutes after coming into view round the west limb, the empty Lunar Module began a long burn with its RCS thrusters that resulted in the vehicle crashing to the lunar surface 28 minutes later. Seismometers at the Apollo 12 and 14 sites picked up the reverberations. At the end of that revolution the CSM burned its main engine to bring Kitty Hawk back toward Earth.

During the long coast home several tests were performed with future missions in mind. On the three J-series flights beginning with Apollo 15, an astronaut would leave the Command Module during trans-Earth coast to retrieve cassettes and film magazines from instruments located in sector 1 of the Service Module, equipment that would have been used to obtain scientific data from Moon orbit. That would require the cabin to be depressurized, incurring high oxygen flow rates.

To check the ability of the oxygen storage system to accommodate these high rates the crew opened a dump valve in the

Command Module, increasing the flow from about 0.045 kg/hr to 2.7 kg/hr. Oxygen quantity in the new No. 3 tank was down to 21% and this provided a check of the system at low level, another requirement of the test. It worked well and proved that the oxygen manifold system could meet the deep-space EVA requirements, but when pressures fluctuated in the manifold the crew were advised to end the test, 1 hr 10 min after it began. The problem was found when engineers discovered a urine dump had been made, increasing the flow through the main regulator and so exceeding the system's capacity.

A few hours later, during a scheduled telecast, the astronauts demonstrated tasks conducted in weightlessness that might lead to industrial processes performed aboard orbiting space stations. An electrophoretic separation test hoped to show how the lack of convection would assist the movement of different molecules passing through an alkaline solution within an electric field. It was thought this capacity might prove useful in producing new vaccines and other biological preparations. Then the crew performed a liquid transfer test to show the advantage of baffles in preventing liquid sloshing around the inside of a container. Engineers could study the movement of fluids with and without baffles to aid in the design of future orbital refuelling systems.

Then a heat flow and convection demonstration was performed to show the movement of heat transfer patterns in weightless liquids and gases where electric heaters immersed in separate samples of water, a sugar solution, and carbon dioxide gas, showed thermal contours. A fourth test was designed to show flow around a sample of oil containing aluminum flakes. This latter experiment was not as successful as the other three, however. Finally, a composite casting demonstration was to be activated in space for analysis on Earth. Designed to show how improved metallurgical products could be produced in weightlessness, a crewmember would place up to 18 samples in a special heater box, each containing various composite materials. In all, the four demonstrations weighed 17 kg and were housed in separate containers in stowage lockers or on the aft bulkhead below the couches.

About 21 hours prior to entry, Shepard, Mitchell and Roosa answered questions from pressmen covering Apollo 14, passed up through the capcom console during final television transmission. In the Mission Operations Control Room at Houston, a dozen red roses arrived from Cindy Diane of Montreal, Canada. Ever since Apollo 8 the young lady had sent flowers for the flight controllers, each bouquet accompanied by a brief message. Gerry Griffin accepted them and they were placed, in a vase, on the table below the US flag.

Effective April, 1971, James B. Fletcher began his tenure as NASA boss. At left, Kurt Debus, and at right, George Low.

The flight home had been uneventful. Only one small correction had been necessary. The weighty probe was carefully packed and restrained, and all loose items stowed. Fifteen minutes prior to entry interface the Service Module was jettisoned and during the descent communications blacked out for the expected 3½ minutes. At splashdown the Command Module remained in an upright position, but one of the three parachute lines failed to sever. Although they still had to be qarantined, the crew were spared the need for biological isolation garments, donning instead clean coveralls and a filter mask before moving to the mobile facility on board the recovery ship *New Orleans*.

Apollo 14 splashed down less than 2 km from the planned location on what could have been a mission of restoration, placing confidence shaken by the near disaster of its predecessor back in the public mind. But if Apollo 14 had a good audience when it left, it lost its support en route, for even as the flight progressed interest waned and the media all but abandoned it. Even the President passed comment by proxy and unlike attention lavished upon Apollo 11, this third Moon landing excited only a telephone call from the White House to Deke Slayton asking for good wishes to be passed along to everybody.

In Europe, it no longer caught prime viewing time, and throughout Britain the media were more concerned with a bunch of outrageous terrorists in Northern Ireland than with the resumption of lunar exploration. In Italy, an earthquake claimed the headlines, while in France the television broadcasts waned. From east Europe, rancor at the thought of a privileged few 'shut out of science by the expense,' boiled to the surface, and from behind the Soviet border statements that manned vehicles were unnecessarily complex and more prone to failure than the Russian's automated exploration vehicles. Even in the United States, New York's blackout vied with stories about the flight. Where once the world acclaimed the forward march of mankind, the cynics had come from their lairs.

There was one other experiment on Apollo 14 in addition to the published list. Ed Mitchell's personal interest led him to set up with four friends an experiment in extra-sensory-perception – ESP – far from Earth. In a complex sequence of tests later analyzed with the aid of Dr. J. B. Rhine from the Foundation for Research on the Nature of Man, in Durham, and Dr. Karliss Osis of the American Society for Psychical Research, in New York, Mitchell completed a demonstration that gives strong evidence for the ability of one mind to respond to another. It seemed to show, moreover, that time and distance are irrelevant parameters. Mitchell remained silent about the test; carried out as a personal exercise, one that would stimulate him to pursue ESP after leaving NASA; it was unique to manned space flight and provided an unprecedented opportunity for research into unexplored avenues.

Three months after the flight, Alan Shepard, accompanied by his wife and his mother, attended a special ceremony at the Kennedy Space Center to unveil a commemorative plaque marking the tenth anniversary of his first flight into space. It said: 'The first American to penetrate outer space began his flight from this launch complex in Freedom 7 on Mercury Redstone No. 3 at 9:34 A.M. May 5, 1961. From this beginning, man reached the Moon.' By that time, NASA had its new Administrator. Nominated by President Nixon on 27 February, James B. Fletcher was presented to the Senate on 1 March and sworn in on 27 April after ratification by Congress. Fletcher was a scientist-administrator serving as President of the University of Utah and College of Eastern Utah when called to head America's space program. He had organized the Space General Corporation, a subsidiary of Aerojet-General, served as its President and then Chairman of the Board, and had contributed to the Air Force Science Advisory Board.

In several respects he was a caretaker, a safe man to have in the seat for a President unmoved by calls for new objectives in space technology – a very different man to his predecessors in that he took up reins cast aside by others who could not operate within a holding posture. Great decisions lay ahead, and there was much to prepare for. It was not a time for faint hearts. One project upon which increasing attention was being focused, albeit behind the scenes, was the proposed joint docking flight with a Russian Soyuz spacecraft. As Apollo 14 flew to the Moon and back, and as Jim Fletcher slid in to his new position, engineers at the Manned Spacecraft Center worked feverishly on the technical proposal first made by George Low and Arnold Frutkin to Keldysh in Moscow during January.

The following month two documents were sent to Russia. One of them, called 'A Concept for a Union of Soviet Socialist Republics/United States of America Rendezvous and Docking Mission,' had been drawn up under the supervision of René Berglund. This backed up the administrative proposal and received Gilruth's blessing in a covering letter to Boris Petrov. The Russian responded by confirming that the proposal required 'further study which our specialists are now engaged in.' NASA wanted to hurry things up and schedule an interim meeting ahead of the planned reunion for May. So far the two sides had only got together for the one meeting held in Moscow during October, 1970, when Gilruth brought a delegation from MSC. But the Russians preferred to leave the schedule intact.

Then the Soviets sent word that their party could not visit the US as planned but that they would let NASA know when their specialists could be released. In fact, the Russians had been engaged in two difficult tasks: convincing Soviet opponents of detente that the idea was in the best interests of the Soviet Union, and making final preparations for manned activity with the first true space station. Within a few weeks of NASA's third Moon landing, the Russians began to speak openly about the Salyut station, claiming that the record breaking flight of Soyuz 9 would soon be exceeded.

Salyut 1 ascended to a 222 × 200 km orbit on Russia's D-1 rocket, the most powerful space launcher so far used by the Soviet Union and equivalent to NASA's Saturn IB. It was 19 April, 1971, the beginning of a decade of achievement for manned Earth-orbiting laboratories. About 21 meters long, the station weighed nearly 19 tonnes and provided 100 cubic meters living space in a cylindrical structure 4 meters in maximum diameter. A small 3 meters long transfer tunnel afforded access from a docking port on the forward end. Only 2 meters in diameter, it was connected to the main habitable section formed from two cylindrical sections joined end to end by a cone section; the smaller section was 2.9 meters in diameter.

Two sets of solar cell arrays were attached to the exterior and a propulsion system at the rear provided attitude control and limited maneuverability for changing orbit. Television was provided from both without and within the station and several small circular windows provided a view of the Earth and facilities for scientific instruments. The interior comprised a single open compartment with wall panels, lockers, storage bays, and controls. Large chairs were provided to which cosmonauts could strap themselves.

Three days after launch, Soyuz 10 ascended for rendezvous and dock with the Salyut station. Unlike earlier rendezvous operations tried by the Soviets, cosmonauts Shatalov, Yeliseyev and Rukavishnikov took a complete day making the three orbital changes needed to bring the two together; Salyut made four maneuvers by remote command. The two vehicles remained docked for 5½ hours while TV cameras showed views of the opposing spacecraft. But nobody opened the hatch and moved across to Salyut 1. This was the first time a manned Soyuz had flown with the new docking equipment that permitted intravehicular access and several analysts speculated on the true intent of Soyuz 10.

It had been a tricky business docking with the comparatively large structure, although it was only a fraction of the size of an S-IVB stage to which Apollo docked during each Moon flight. But it was a new experience for the cosmonauts and all three crew-members played a part in successfully linking up. When they separated, Soyuz 10 flew around Salyut 1, inspecting it and photographing the exterior for signs of possible damage during launch. They found none and Soyuz 10 returned to Earth after only two days in space. Many observers believed the mission had been intended as a long duration flight but there are several features of the flight which indicate it was intended as an operational pathfinder, among which were the several new items of equipment which always inspired manned test activity from the Soviets – the need to inspect visually what was considered a revolutionary new manned vehicle, and final qualification of the definitive docking hardware.

Six weeks went by, during which Salyut was maneuvered to a higher orbit for prolonged life, until Soyuz 11 was launched on 6 June, 1971, with cosmonauts Dobrovolskiy, Volkov, and Patsayev. The day long rendezvous of Soyuz 10 proved to be the new standard, for its successor too took 24 hours to arrive alongside Salyut. From a distance of 100 meters Dobrovolskiy manually controlled the docking which was finally completed about 3½ hours after rendezvous. First there was a soft dock, then the two vehicles were rigidly linked for hard dock, after which electrical leads were connected and the pressure integrity of the docking seal checked. When pressure between the two vehicles had been equalized, the hatches were opened and Patsayev moved into Salyut 1, followed by Volkov and, later, Dobrovolskiy.

For three days the crew maintained a rigid check on Salyut's systems, tested the life support unit in a variety of operating modes, unpacked equipment for scientific research, and tried out simple exercise devices designed to combat the affects of weightlessness. It was mid-morning Moscow time on 7 June when the crew of Soyuz 11 became the first men to inhabit the first space station. A day later they performed an engine burn that raised the orbit, and then again on the 9th, another maneuver to put Salyut 1/Soyuz 11 into a safe path high above the atmosphere. On the fourth day the crew completed activating the station and fully checking it for regular operation. Now it was time to begin a concerted program of research during which the cosmonauts would observe the Earth for cloud, weather, and geological information, perform astronomical observations, and conduct biological experiments inside.

Each day in space became a regularly planned sequence of exercise, meals, rest, and work. Television programs were broadcast to Earth and pictures relayed to Soviet citizens and released as film to the world's press. On 19 June, Patsayev celebrated his 38th birthday and a few days later all three began an intensive period of Earth observation, taking pictures and recording measurements of the ionosphere. On the 28th they packed away loose equipment and prepared to return to Earth.

Eight days before, the Soviet delegation had arrived in Houston for the second joint meeting, a return visit to the one made by Bob Gilruth to Moscow. They were noticeably proud of the Salyut 1 achievement and keen to get on with technical discussions about cooperative missions. During the course of several sessions spent exchanging views on the way to develop the next steps, Petrov expressed the preference of the Soviet Academy of Sciences for a joint docking flight employing the androgynous docking system originally discussed. It was gratifying to the MSC men to hear the Russians approve of the joint flight idea and also to propose development of the common docking unit for that event.

Petrov said he felt that a joint flight could involve either a Soyuz spacecraft docking with a Skylab/Apollo configuration, or an Apollo CSM docking with a Salyut/Soyuz combination. Either way, the docking equipment would be of the androgynous kind. Various working groups were set up to study the plan and for once it seemed international rivalry was evaporating under a new mutual understanding. Not yet endorsed politically by either side, Russian engineers were enthusing with their American counterparts about the possible flight of joint manned vehicles.

During their stay, the Russians visited shops in Houston, ate in restaurants, and were suitably confused by the tax system. Exposed to the Texas way of life, the Russians seemed surprised by the general level of affluence where they had been told so often of poverty and urban decay, and by the jovial nature of the American people. And then more meetings, this time to discuss technical details relating to which mission profile should be adopted.

It was generally agreed that the baseline planning should centre on an Apollo flight to a Salyut space station, although this was only a reference mission and would probably change. When the meetings ended on 25 June, it was in a spirit of accord and approval for futher discussion in November. Glynn Lunney and Konstantin Bushuyev were to be directors for the project in their respective countries.

But the mood in which the Russians departed for Moscow was soon to change. On 28 June, NASA held a press conference at which the discussions about possible cooperative flights were revealed for open debate. A day later the cosmonauts that spent 22 days on Salyut 1 were dead. Back inside Soyuz 11, Dobrovolskiy, Patsayev, and Volkov had prepared for return to Earth by first sealing Salyut's forward hatch and then a similar cover on the forward section of the orbital module. Moving back down into the re-entry module they next secured the hatch separating Soyuz's two pressurized compartments and pumped the atmosphere out of the forward compartment.

Separated from Salyut they readied the instrument section

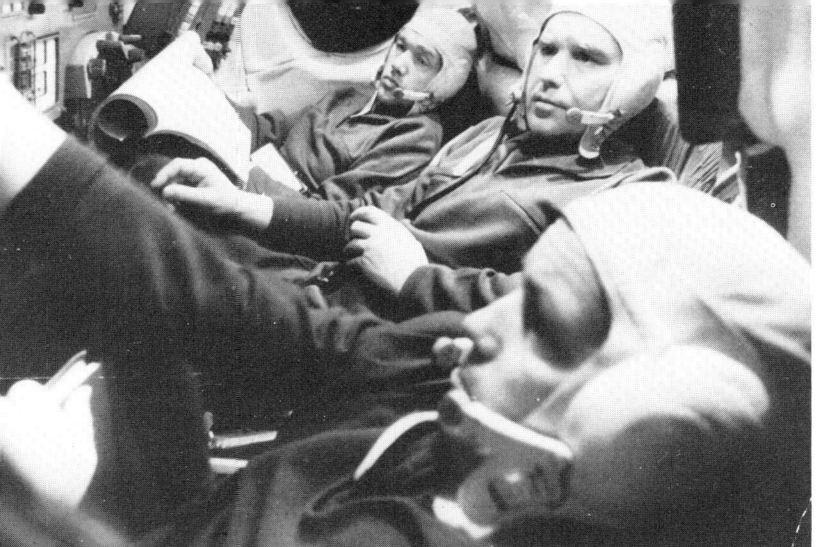

Crewmembers of the Soyuz 10 spacecraft.
Cosmonauts (left to right) Alexei Yeliseyev, Vladimir Shatalov, and Nikolai Rukavishnikov.

Soyuz 10 is moved to its launch site.

for retro-fire, an event which took place as planned at 1:35 am on the morning of 30 June. All three cosmonauts were seen on telemetry to have low heart rates and ground controllers watched and monitored the various events. When the three modules separated, however, leaving the re-entry module exposed for descent, one of two pressure equalization valves popped open, evacuating the oxygen-nitrogen atmosphere inside. Within 50 seconds the interior was a virtual vacuum unable to support life. The crew tried to crank closed the valve that had opened prematurely, an opening in the cabin wall designed to allow Earth's atmosphere to flow in during the final stages of descent. But there was insufficient time and the three men rapidly lost consciousness.

All the way down to a landing ground personnel tried to call up the crew – to no avail. Fears rose that the flight had uncovered some hideous repercussion of long periods of weightlessness and recovery forces sped to the landing site. Two helicopters touched down and doctors ran to the spacecraft, only to find the three cosmonauts still strapped to their respective couches. They died from pulmonary embolisms. It was a bitter tragedy, one that cut deep within the ranks of space-faring pioneers and one that hurt the nation and its people. Next day the three bodies lay in state in Moscow and Tom Stafford flew in from America to pay respect.

At first the Russians leaked a story that the cosmonauts had failed to secure the hatch separating re-entry and orbital modules. Then the press speculated that the more than 23 days of Soyuz 11 had been too much; evidence from NASA's Gemini program seemed to agree that limitations might appear to restrict man's free exposure to weightlessness. Only gradually did the Russians allow the truth to emerge, as technical studies revealed the complete story. Coming only fourteen months after the near disaster to Apollo 13, three years after the death of cosmonaut Komarov aboard Soyuz 1, it was a blow to protagonists of manned space flight. The Russians affirmed their continued interest in manned operations but it gave opponents an argument with which to fight their case for more attention to unmanned projects. The three dead cosmonauts were cremated and their remains placed in the wall of the Kremlin. Russia now had four victims of the space program to mourn.

For his part, the new NASA Administrator was a staunch supporter of international cooperation and stressed quite early in his tenure the importance of uniting with the Soviets in a joint project. But Jim Fletcher knew that a hiatus would exist between the last manned Skylab occupation in late 1973 and the start of shuttle flights, probably no earlier than 1977 or 1978, and that a joint rendezvous mission with the Russians would serve to sustain a manned US presence in space until the reusable launcher and a permanent space station came along later in the decade.

Several projects had already been examined by Dale Myers in an attempt to find a comparatively cheap mission that would use up some of the now redundant hardware originally built for Moon landings. There would be four spare Apollo CSMs and one scheme suggested a single Earth resources survey flight each year from 1975 to 1977 inclusive, each mission lasting up to one month in duration, with a docking flight to the Russian Salyut in addition. As an alternative, back-up Skylab hardware could be utilized for a second workshop mission known for planning purposes as Skylab B. But that would invite large budget requirements and only put the whole program back into a situation where all NASA's financial resources were going into finite applications; the shuttle and the large orbiting space station were still seen as the real targets for which to aim.

Competitive definition studies begun in 1970 on shuttle designs were completed by June, 1971, when North American Rockwell and McDonnell Douglas presented documentation to NASA. The message was clear, distinct, and very unpalatable. A fully reusable shuttle, where both booster and winged orbiter would be flown back to runway landings, would cost at least $10,000 million; the Office of Management and Budget wanted it half that value and reiterated warnings projected in late 1970 that until NASA was able to prove it could fully develop the shuttle for only $5,000 million the administration was loathe to authorize a start.

Clearly a compromise was essential. And doubt, sown earlier by cynics, now took on a dismal reality that NASA would be unable to simultaneously fund both shuttle *and* space station. Even the modular station, built up in orbit from the collective payloads of several shuttle flights, was seen as too costly for the likely budget levels envisaged for the coming decade. So by mid-1971, NASA contracted extension agreements with the prime shuttle competitors and asked other companies to study a 'research and application module' concept, RAM for short, whereby single modules would be carried in the shuttle's cargo bay for use either as single elements in orbit or as mini-laboratories retained within the shuttle.

It was the beginning of the end for all hope of getting a permanently manned facility in orbit. So by the middle of the year, plans for post-Apollo hardware were slowing down under increasing threat from an apathetic administration at the White House. It became increasingly urgent that NASA get agreement for a joint US-Soviet docking flight both in the interests of international cooperation and for keeping the full technical resources of several space centers at work while future programs evolved.

After the Houston visit from the Russians, MSC got down to detailed examination of a docking module that seemed essential for the joint flight. René Berglund swung his team into action on the necessary hardware for such a mission, and Glynn Lunney laid out the necessary working groups for integrating activity between Russia and America. By the end of June attention centerd on just one cooperative mission for the period between Skylab and shuttle and all mention of the proposed annual Earth resources flights disappeared.

But by that time, J-series hardware was complete for the first Apollo Lunar Exploration Mission where basic equipment had been re-worked to provide a significant improvement in performance and sophistication. For the past two years engineers at contractor plants and at the Houston and Huntsville NASA facilities developed equipment for use on the last three Moon landings, the true science of Apollo at last given full exposure. Changes to the CSM and the LM would make the payload for Apollo 15 some 2.4 tonnes heavier than its predecessor and push the combined Apollo-Lunar Module weight to nearly 46.8 tonnes by far – the heaviest yet.

To what had been a competent and remarkably successful range of hardware was brought a clutch of new requirements far beyond the needs envisaged a decade before when engineers at North American and Grumman wrestled bold goals against practical technology. Hitherto, Moon flights had lasted a total 8–10 days; the J-series flights would operate for more than 12 days. The CSM would make detailed scientific observations of the lunar surface from orbit, while the LM doubled the time spent on the surface by Apollo 14. Where the record for lunar exploration provided two periods of EVA totalling little more than 9 hours, Apollo 15 would support three excursions totalling at least 20 hours – and give the crew a ride to distant areas and interesting geological features. It was like the birth of a new age in scientific discovery, both of new frontiers and of new ways to explore.

To support the extended missions, changes were made to CSM-112 whereby a third cryogenic hydrogen tank was added to the upper section of sector 1 in the Service Module. A third oxygen tank had been added for Apollo 14, and now the standard Block II supported a full 50% increase in cryogenic fluids for the fuel cells and the environmental control system. The two tanks occupied the top 1.37 meters of that sector, leaving the rest to accommodate the Scientific Instrument Module – or SIM – bay containing specialized instruments for photographing and scanning the Moon at a distance. A 2.54 cm thick panel, 2.8 meters long and 1.5 meters wide, covered the SIM bay and would be pyrotechnically jettisoned to expose the equipment. Internal volume was, at 8.5 cubic meters, slightly larger than the habitable volume within the Command Module! Hand holds were attached to the exterior of the Service Module, and an EVA panel added inside the Command Module below the hatch from where a 7.6 meter long umbilical would provide oxygen for an astronaut going outside to retrieve canisters from the SIM bay. With added consumables and the new scientific equipment, the CSM was transformed into a 14-day orbiting laboratory – very close to the original concept of Apollo back in 1960.

But it was to the Lunar Module that the most significant changes came for the J-series flights, for that vehicle had a 100% increase in capacity. The four quads formed by the cruciform box-structure in the LM's Descent Stage carried substantial quantities of support systems and some of this had to be rearranged to vacate one quad for the lunar rover. Quad 1, to the right of the ladder when facing the LM, was assigned to carry the Moon explorers' electric car so two 415 ampere-hr batteries were taken out, as were the two from Quad 4, to the left of the ladder, and, with a fifth added for the J-series flights, were re-located around on the back of the Descent Stage inboard of the rear landing leg support structure. The added battery raised total available energy to 2,075 ampere-hrs.

The two quads at the rear remained unchanged: Quad 2 carried the ALSEP packages, improved and expanded versions of the equipment carried by Apollo 12-14, and Quad 3 now supported two small pallets that could be unlatched and let down to reveal geological equipment, a brush, and similar items to be packed aboard the rover. Where Quad 4 previously contained the MESA, carrying the TV camera, sample containers, antenna cables and basic tools, that package was now hung on the exterior of the quad, making room inside for a 50 kg water tank to supplement the 120 kg tank carried in Quad 2.

Quad 4 also now carried an extra waste management container, an additional oxygen tank, a gaseous oxygen module, and a facility for the S-band erectable antenna (Apollo 15 did not carry this item). The new MESA suspended on the exterior of Quad 4 was made larger to carry new batteries for the PLSS backpacks and a cosmic-ray detector. To accommodate the additional mass during descent, an extra 6.3% of Descent Stage propellant was provided, increasing by 8.6 cm the height of the four main tanks within the box sections.

But because the Descent Stage engine would burn longer than on previous missions, changes were made to the combustion chamber by replacing the silica lining with a quartz substitute, reducing erosion rates. The engine skirt was also extended in length by 25.4 cm with a new 'astroquartz' liner imported from France. The skirt was designed to crush on impact with the surface and actually saved 6 kg in weight over the original design. The net effect of all these changes was to increase the LM weight to more than 16.4 tonnes. But it was the lunar rover that would give the J-series missions radical new potential for real scientific work.

Studies had been under way for a decade on various mobile devices capable of traversing the Moon with astronauts and samples. Boeing pitched in early in the 1960s and came up during 1964 with MOLAB (Mobile Laboratory) and LSSM (Lunar Scientific Survey Module) designs. But these were all too large, or too heavy, for the limited capability of the Lunar Module. The Lunar Roving Vehicle, or LRV, contracted in 1969 had to weigh less than 200 kg, fold up within the Descent Stage quadrant in which it would be stowed, survive temperatures from −38°c to 120°c, carry 450 kg on slopes of up to 25°, and provide a navigation system for returning to the Lunar Module from a distance of several kilometers.

NASA's Marshall Space Flight Center was to be in charge of the project and less than three months after Apollo 11 returned home, the Boeing engineers at the Seattle, Washington, facility were hard at work on detailed design. By the end of the year a static mock-up had been completed and at MSFC engineers finalized crew station layout and subsystems design. From the beginning, astronauts Engle and Carr worked long hours to perfect the LRV. And from the outset, Delco Electronics (formerly AC Electronics) of the General Motors Corporation was responsible for the drive system.

Mobility in the form of a four-wheeled electric car posed problems in that low surface gravity would compromise stability: whenever the car went over a bump pneumatic tyres would give the vehicle a pronounced bounce. Some form of wheel was necessary which would absorb more energy than normal. An open mesh wire 'tyre' was finally chosen, held in place by a titanium 'bump-stop' inside. Titanium chevrons fitted round the outside would provide additional grip and the flexible wire tyre was to be 22.9 cm in width and 81.8 cm in diameter. Each wheel was to be powered by a separate electric motor with an 80 to 1 reduction gear and mechanical brake shoes.

The rover's suspension comprised a double, horizontal wishbone, giving a 43 cm ground clearance unladen, or 35 cm laden. The steering system worked on the Ackerman principle where a coupling vane moved the front wheels in an opposite direction to the rear wheels to turn left or right, movement initiated by a hand controller. Speed would be controlled by pushing the handle forward, braking activated by pulling it back, and when inclined to left or right the vehicle would turn in response. The LRV had a wheelbase of 2.3 meters and a length of 3.1 meters, built up from three chassis sections where front and rear would fold over the center for stowage. Two fully folding seats were attached to the center section, with a foldable console providing instruments for navigation and control. Two 121 ampere-hr batteries ensured a 36 volt current for the electric motors, with silver-zinc plates in a potassium hydroxide electrolyte. Battery temperatures were to be maintained between 4.4°c and 52°c by a passive heat sink covered by removable blankets to balance thermal retention against rejection.

All in all it was a unique vehicle, the first powered car designed to drive the Moon's dusty road. Shaped wheel guards would inhibit spray from fine particles, seat belts were provided to prevent the crew falling out, and plenty of space for tools was provided at the back. Deployed to the surface by two lanyards pulled in sequence, the LRV would unfold itself with only a little attention and require less than fifteen minutes to set up and check out.

When placed in October 1969, the contract called for seven test units and four flight articles. But that was when NASA envisaged four LRV flights on Apollo 16-20; budget cuts reduced that to three for Apollo 15-17. Boeing had just 18 months to get the first flight article delivered. A 'LM-LRV Glob' was built to evaluate the stress imposed upon the LM's Descent Stage structure, a strange angular device built to simulate the car's mass when folded. In June 1970 a critical design review was carried out where final plans were approved and two months later Delco completed the 1g trainer; like the Lunar Module, the LRV was incapable of standing in the full Earth gravity field and driving tests required the use of a specially stressed trainer using inflatable tires.

A special mobility test unit had been used in preceding months but now flight crews could get the 'feel' of a real LRV for themselves. By this time too a vibration unit proved the design could withstand stresses and strains of launch and propulsive boost, and a special 1/6th gravity device worked final problems with the deployment mechanism. Finally, the qualification unit was completed in late November, a unit identical to the first flight vehicle. After a hectic two months in assembly, LRV-1 was completed in February, 1971. On 10 March, Marshall Space Flight Center Director Dr. Eberhardt Rees attended a formal hand-over ceremony and the vehicle arrived at the Huntsville facility five days later, little more than sixteen months after Boeing commenced work on the project. During this time, and while MSFC conducted exhaustive checks on LRV-1, a separate chassis unit mapped performance limits and stress margins.

Yet the lunar rover was but one element of a greatly expanded EVA capability embracing a new suit and a new backpack. Astronauts had used the A7L suit since Apollo 7, but it was ill equipped for the broader limb movements necessary for Moon exploration. So ILC came up with a new, A7LB, design accommodating flexible convolutes instead of a block and tackle mechanism built in to physically extend or foreshorten the limb sections. Convoluted joints fitted at the knees, wrist, elbows, ankles and thighs made the suit less tiring when worn over long periods, allowing a wearer literally to bend over and touch his toes or crouch down and touch the ground. A pressure bladder was incorporated to prevent ballooning, and zippers were moved to accommodate the new convolutes. The suit also provided an increased supply of drinking water in a flexible container inside the neck ring, a facility provided first on Apollo 13 but not used until the following flight.

The new -7 Personal Life Support System stretched Hamilton Standard's basic design to the limit, and gave EVA crewmen a potential seven hours on the surface at each excursion. The

Although a l-g training unit, this view of a Lunar Roving Vehicle clearly identifies the prime design points: Ackerman steering, double wishbone suspension, batteries for power, two-seat operation.

With the new A7LB suit, astronaut Worden is able to bend to a crouch.

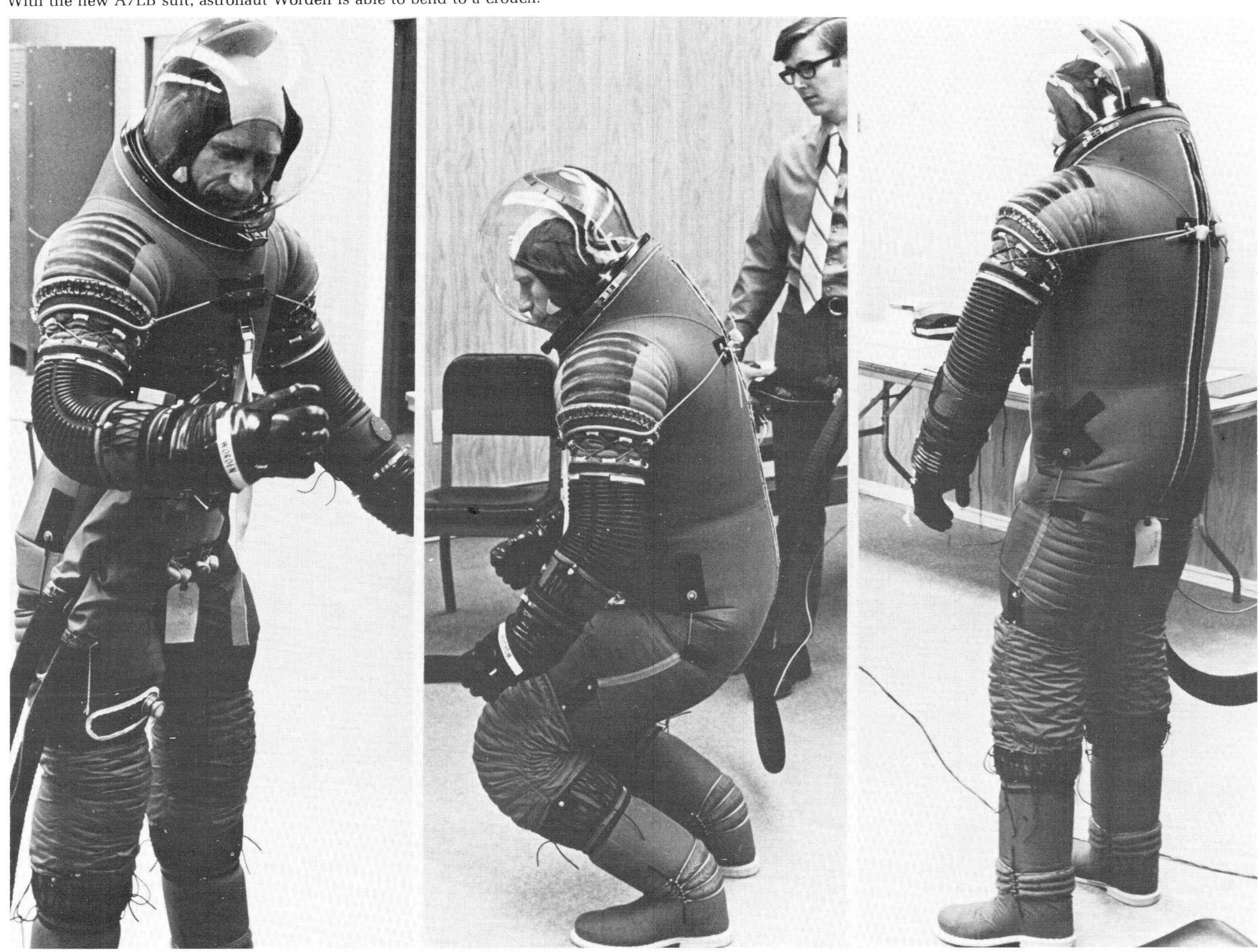

oxygen tanks were thickened to withstand an increase in bottle pressure from 71.7 kg/cm² to 100 kg/cm², providing an extra 50% gas, and the water supply was increased by 39% carried within a second, elongated, tank in the back-pack. The single battery employed by each PLSS was increased in capacity by 30% and an additional quantity of lithium hydroxide ensured longer life. The emergency pack, or OPS, carried on top of the main back-pack in case the PLSS failed, had a limited supply of oxygen and NASA wanted to develop a Secondary Life Support System to greatly increase this capacity; roaming far from the Lunar Module astronauts would take longer to get back to safety. But budget cuts prevented this so engineers placed a new emergency feed port at the back of the helmet and by reducing the oxygen flow rate emergency OPS duration was extended to 75 minutes.

This was to place an effective limit on the distance Moon walkers could roam, for if the vehicle broke down they would have to walk back, and if a back-pack failed that would leave only the OPS to provide life giving oxygen. In the final analysis Apollo astronauts would be denied the full range of the LRV because of constraints placed upon the stretched emergency pack. The radius of travel was, therefore, restricted to a ride-back duration of 75 minutes assuming an average speed of 8 km/hr.

Apollo 15's crew had been in training a long time. Named to fly this mission on 26 March, 1970, Dave Scott, Al Worden and Jim Irwin were delayed by modifications made necessary after the Apollo 13 accident, and by budget cuts that stretched Apollo flights out beyond the schedule in existence when they were selected. But during the first half of 1971 preparations for the fourth Moon landing accelerated as the mission came together. In February a site was chosen high in lunar latitude, nested within a range of mountains on a valley cut at one end by a deep incision in the Moon's crust; to north, south, and east lay high mountains, and to the west lay a long sinuous rille.

It was the most dangerous site selected for a manned landing, and one only chosen after pin-point landings had been achieved by Apollo's 12 and 14. In April, after careful study of the lunar samples from earlier flights and of the biological condition of the crews, it was decided to abandon the stringent quarantine regulations in force since Apollo 11. That was very welcome news to the crew of Apollo 15 for it was a frustrating experience to be physically separated from friends and family. But it was the site itself that provided stimulus for much of the training that now ensued.

Called Hadley-Apennine, from the range of mountains nearby and the chain to which the local peaks belong, the valley was flanked by massifs towering several kilometers above the floor. Clearing 4 km mountains by only 2,700 meters, the Lunar Module would come up to the landing site on a steeper approach path than flown hitherto. After an essentially horizontal braking phase, the approach path was inclined 16° for Apollo 14; Scott and Irwin would descend on a 25° slope before entering the landing phase and dropping vertically to the surface. Scientists wanted Apollo 15 to land at this site because it promised access to several features covering different periods of the Moon's history.

The massive Apennine mountains were in fact old eroded remnants of the rim formed when an asteroid-type body slammed into the Moon close to the beginning of its history. Unfolded to form walls of a basin more than 150 km across, the mountain flanks formed the margins for blankets of basaltic lava that oozed from the Moon nearly a billion years later, filling up the basin – now called Mare Imbrium – and forming the floor upon which Apollo 15 would land. The sinuous rille was probably carved in more recent times by a flow of hot lava running just below the surface which, allowing the roof of the rille to collapse, left an incision more than 300 meters deep and up to 1 km wide.

By sampling the floor of the region, and visiting in the LRV Apennine mountain flanks and the sinuous rille, Scott and Irwin would collect material from a wide range of geological events. And compared to their predecessors the crew were almost professional geologists for considerable time had been spent training them to recognize certain rock types and to make qualitative observations. But the extra hardware packed aboard the two vehicles raised total payload weight far beyond the Saturn V's design capability, so engineers set to work removing excess weight from the big rocket and improving its performance.

To begin with, four of the eight retro-rockets carried in the base of the first stage were removed; fitted so as to apply a braking effect after staging, studies showed the job could be effectively carried out with only half the design number. Also, the four ullage rockets fitted to the adapter separating first and second stages were removed; attached so as to settle second stage propellants by applying a modest acceleration prior to main engine ignition, observation of performance on earlier flights showed them to be superfluous. Then the F-1 engines were modified to accept a slightly higher flow rate; AS-510 would consume first stage propellant at 13.4 tonnes/sec versus 13 tonnes/sec for earlier launches.

It all added up to a significant increase in performance, but other changes were necessary. By altering the permitted launch azimuth from the Cape, and by shaping the trajectory to achieve orbit at a slightly lower altitude than before, engineers bought an extra 300 kg in payload capacity. Added to the major alterations in vehicle systems, it was sufficient to give AS-510 clearance for launch.

Hardware began to arrive at the Cape in 1970 with delivery in November of LM-10 followed by the CSM two months later. But even after delivery, hardware still kept arriving as pallets and packages were placed on and in the various modules. With so much equipment installed for use on or around the Moon, Earth's gravity environment prevented full and adequate test of deployment mechanisms and systems checks. The Service Module's SIM bay, for instance, carried three spectrometers, two of which would extend on booms 7.5 meters from the side of the spacecraft. Unable to support their appendages in 1 g, the booms never did get a satisfactory workout before launch.

When the LRV arrived in March 1971, Cape engineers began an intensive series of checks to complete tests begun at the contractor plant. On the 26th the flight crew spent many hours doing a crew fit and function test where all the equipment they would pack on the rover was removed from stowage and assembled in simulation of activity carried out on the Moon. Then there were tests with the LRV systems, the navigation equipment and the motor drives. By the middle of the following month, Boeing engineers were ready for the final test and after packing it into Quad 1 on the Lunar Module they checked a full deployment sequence by using a wooden board to brake its fall and so simulate the shock of unfolding in 1/6th gravity. Then it was time to perform a second set of crew fit and function tests concluding with LRV-1 being folded up and re-packed in the Descent Stage on 25 April; nearly 100 days later it would unfold on the Moon.

Saturn V had been erected on 17 September, 1970, and on 8 May, 1971, the spacecraft were placed on top prior to roll-out three days later. There was a new mood at KSC as 510 walked slowly to complex 39A. The astronauts chatted with spectators, signed autographs on request, and talked about the science they would conduct at Hadley-Apennine. Even the names they chose for CSM-112 and LM-10 reflected the desire to unfold frontiers: Endeavour, after the ship that carried Captain James Cook on the first scientific expedition to the South Pacific in 1768, and Falcon, for the bird that soars beyond visible horizons, respectively. Even managers and administrators seemed to join in the general acclaim for Apollo 15 as the first true scientific mission to the Moon. For NASA the prospect of continuing lunar exploration was a campaign that could be easily lost to the man-in-the-street. Where once space spectaculars filled the public with admiration and lay spectators cheered the select contingent of astronauts, attitudes now questioned the wisdom of doing it over again. It would be difficult convincing the public that each new flight would gather rich harvests of scientific knowledge.

But everybody joined in the attempt. NASA public affairs personnel wrote fatter press releases, staged more news conferences, and pressed for more media time. Gene Simmons, chief scientist at the Manned Spacecraft Center, wrote on his own initiative the first of three publications detailing for public consumption the scientific goals of the J-series missions. The 46-page brochure for Apollo 15 was widely distributed in the United States. But the public found it very difficult to think in terms of spectrometers, calorimeters, and gegenschein where once they thrilled to the Buck Rogers bonanza of less complex technology. It was a losing game, for having proven they could land men on the Moon the engineers and technicians received increasing comment that their services were needed down on Earth. Clearly, the last three Moon missions were going to be for the specialists; even scientists found it hard to keep pace with the dramatically enhanced complexity of the J flights.

Not everything was going well down at the Cape in those final weeks before launch. Several problems plagued preparation of LM-10: trouble with the rendezvous radar; then the antenna; communications were poor; the environmental control system played up. In all, LM-10 received four rendezvous radar units before the vehicle was fully primed for the flight readiness test, carried out on 14-15 June 1971. Jim Irwin was niggled by all the many seemingly minor technical troubles, outwardly asserting his displeasure at a press conference which prompted Dave Scott to rebut the comment. Then there was talk that because the schedule had been stretched out much of the hardware had been around on shelves for a few years and that faults and failures were an inevitable consequence. There was debate too on the depleted morale among a work force that knew large numbers of its contingent would get the sack when Apollo 17 sped spaceward eighteen months later.

This was over-stated. But there was a measurable loss of confidence in the intentions of the government regarding future prospects. Skylab was a limited asset, to be completed within a year of first flight operations, and nobody knew just how valid the rumors were about a future docking flight with the Russians. Congress was stalling on the funds needed to continue shuttle definition studies, and the White House had bucked the decision as to when NASA should go ahead with development. But it was true too that talk about incompetence creeping in to KSC operations only served to spur engineers and technicians to new heights of attention and dedication, although the weather again tried to spike preparations.

During the flight readiness test lightning struck the mobile service structure around 510, hitting it again a day later. Tests were run to check the equipment and, although the energy travelled along the attached mast, power units suffered minor damage. Ten days later, the pad was struck yet again, necessitating replacement of more power equipment. Then, on 2 July, while technicians were loading hypergolic fluids, another lightning strike was measured. Checks were incorporated in the Count

An oblique view of the Apollo 15 landing site as viewed from orbit. Hadley Rille meanders through the lower center of the picture. The Apennine Mountains (actually the rim of a giant basin or crater) are at right, the Caucasus Mountains are at upper right. The Marsh of Decay is at lower left. The northernmost crater is the 55 km diameter Aristillus, while that to the south of it is the 40 km diameter Autolycus. The landing site was just to the east of the 'chicken beak' of Hadley Rille in the small bay flanked to the north by Mount Hadley and to the south by Hadley Delta.

Down Demonstration Test beginning 7 July. The wet CDDT was completed on the 13th and on the following day Scott, Worden and Irwin participated in a dry run. Launch Director Walter Kapryan showed Administrator Jim Fletcher the final activity associated with Apollo 15 during the latter's first formal visit to the Cape since his appointment. But more lightning strikes at the pad after the formal countdown began on 20 July, persuaded Kapryan to delay moving back the protective mobile service structure until the evening of the day prior to launch.

In the final hours before ascent, scheduled for 9:35 am Cape time on Monday, 26 July, the flight crew followed the procedure so many had conformed to and were at pad 39A on time. In the firing room, Jim Fletcher watched more than 400 launch personnel check flickering screens as the countdown progressed. In Houston, Gerry Griffin had his flight control team in position. Talking to the crew, astronaut Vance Brand carried word on how the final minutes were moving in the launch control center. Endeavour's test conductor 'Skip' Chauvin passed along a 'go' for flight, as did Falcon's watchdog Fritz Widick. Paul Donnelly, the launch operations manager, and Walt Kapryan received in turn approval from mission director 'Chet' Lee to proceed on.

Broad coverage of Apollo 15's uniquely scientific role brought large crowds to the Cape. Some 30,000 people were inside the KSC boundary, including 5,570 specially invited guests, and nearly a million flocked around Brevard County confident of seeing and feeling the thunder of the Saturn bird flying aloft. When it came, ignition was no less spectacular than it had been nine times before. But this time the trajectory was a little different, although imperceptible to the eye. The point of maximum dynamic pressure, or max q, came 80 seconds after lift-off, but this time experiencing a force greatly in excess of earlier Saturn V flights. On sailed the Saturn, however, and shut down the outer four F-1 engines 24 seconds after cutting out the centre engine. Thrust on the second stage was good and the vehicle ploughed on under a greater load than ever before. 'Okay, Gordo. Thank you. Looks good up here,' came the call from Endeavour to capcom Fullerton when word was passed that the Saturn flew well. Cut-off times were closely matching scheduled events although the ride on the S-IVB third stage was a little short due to a hot engine. At insertion the spacecraft and S-IVB were very close to the planned 166 km target orbit.

During the nearly two orbits prior to TLI, Houston generated a guidance update for the third stage and the crew performed their pre-ignition tasks. Trans-lunar-injection for this mission would be a little different in that to reach Hadley-Apennine at the correct orbital inclination the spacecraft would have to get into a path about the Moon inclined 26° to the equator. That required a non-return trajectory but instead of flying TLI to a return path until the LM was extracted from the S-IVB, when a Service Module main engine burn could transfer the path to the less safe non-return type, Apollo 15 would incorporate the hybrid transfer burn in the TLI maneuver when boosting out of Earth orbit.

That would put the spacecraft on a non-return path from the beginning and have the advantage of using S-IVB propellant to do what the SPS engine would otherwise have been called upon to perform with a consequent consumption of Service Module propellant. With a heavy load to put in Moon orbit, and several orbital plane changes planned for this flight, SPS propellant was at a premium. Immediately after the burn to put Apollo 15 on course for the Moon, Scott, Worden and Irwin got ready the color TV camera that would show viewers on Earth the transposition and docking maneuver.

Unlike its predecessor, Apollo 15 performed a near perfect docking with Falcon and after pressurizing the LM interior, Endeavour towed it free from the redundant third stage. But now engineers in Houston were puzzling over a strange indication in the spacecraft that said SPS engine valves were in the wrong position. Just to be on the safe side the flight director ordered the crew to pull the appropriate circuit breakers in case an electrical signal got through. It was soon apparent that the solenoid valve drivers were receiving an electrical short, or that the switch itself was shorting. But there was time enough to diagnose the problem.

By now, the S-IVB had performed a separation maneuver to move away from the docked space vehicles so that it could safely fire up its thrusters to place it on an impact trajectory with the lunar surface. At TLI Apollo 15 had been put on a course that would carry the configuration to within 257 km of the Moon's far side versus a planned pericynthion of 126 km. A course correction would be necessary, but not before the MCC_2 opportunity. Just as the crew were beginning their first eat period, Houston passed up a series of simple tests to clarify the situation on the apparent short. By wiggling the translation hand controller and throwing a few switches, it was hoped that the SPS thrust light would go out. It did not.

Shortly after an elapsed time of seven hours, with systems engineers hard at work processing the fault, Griffin's gold team vacated the MOCR for Milton Windler's maroon men and capcom Karl Henize. Thirty minutes later, Henize told the crew that Mission Control believed the fault to be either a short to ground which lights up the indicator, a short that could ignite the SPS early when activated, or a short upstream of the pilot valve which would disable Bank A when activated. Over the next two hours the crew worked at procedures to define the problem and eliminate contenders. At an elapsed time of ten hours, Mission Control was reasonably sure it was a short on the downstream side of the switch itself, as extensive tests at North American's Downey plant indicated, but that a course correction burn performed first on Bank B and then on A would prove this one way or the other.

If one bank was affected it would be sufficient to cancel the lunar landing, for then, with only a single bank left, the main engine on Falcon would be required as a back-up to Endeavour's SPS engine. But during the sleep session beginning at 14 hrs, a new procedure for firing the SPS required the engine to operate on Bank A alone, and astronaut Dick Gordon tested the procedure in the Command Module simulator at Houston. A flurry of minor problems kept the crew alert throughout the trans-lunar coast phase, but the first course correction burn, combined with an SPS test, blipped the main engine for a brief 0.72 sec to lower pericynthion.

Tracking carried out in the hours following TLI showed that the force of separating Falcon from the S-IVB changed the trajectory, lowering the point of closest approach to 233 km. But the course correction itself was slightly adrift, chopping pericynthion to within 117 km of the Moon; another minor adjustment would be required. But at least the firing proved to everybody's satisfaction that the short was in fact on the downstream side of the wiper switch and that Bank A could be used for manual firings with Bank B reserved for computer-controlled burns.

At 33 hr 47 min, one of the tracking antennae at the Goldstone station suffered a failure in its power amplifier but that was soon corrected by switching to another dish. It was just about the time the crew were getting ready to inspect Falcon and came almost simultaneously with what looked ominously like the problem that afflicted Apollo 13. With the master alarm sounding a tone throughout the spacecraft, the crew reported a main bus B undervolt and an AC bus 2 caution light. But checks on the system seemed to indicate a faulty signal for power was back to normal. 'Karl, Jim just reset bus and everything seems to be okay now. He checked the voltages and they were fine,' came word from Endeavour. Then the TV picture came on and while systems engineers began a close scrutiny of telemetry data to troubleshoot the electrical glitch, the crew aboard Apollo 15 moved into Falcon. It was 33 hr 56 min when they went in, about fifty minutes ahead of schedule, and good systems checks were carried out. About thirty minutes later the voice communication circuit was given a workout, but there was one casualty from all the vibration and stress the vehicle endured.

SCOTT 'One little problem we ought to discuss with you before we go on. It seems that somewhere along the way, the outer pane of glass on the tape meter has been shattered, I don't know whether you can get a picture of it on the TV or not. . . .'
CAPCOM 'Roger Dave, we are reading you loud and clear.'

It was the range/range-rate exterior cover glass that was broken, removing a helium barrier between it and an inner pane. Tests were immediately run on the ground to confirm the meter's function in that condition. The dual purpose instrument would tell the crew altitude and altitude-rate during descent, and range and range-rate for rendezvous. But Houston was still concerned about the voltage glitch and quizzed the crew on what they were doing when it occurred, after which, with the anomaly still unresolved, the spacecraft was placed in a slow roll before the crew got their heads down for another rest.

Throughout the hours Scott, Worden, and Irwin slept aboard Apollo 15, flight director Glynn Lunney manned his black team while capcom Robert Parker sat back with nothing to say. But

Engineers check deployment of a Lunar Roving Vehicle from its stowed location on the side of the LM Descent Stage.

Gerry Griffin's gold men were on shift when capcom Joe Allen rang the bugle call at 49 hr 5 min.

> WORDEN 'Okay, Joe we certainly did have a nice sleep and we think your tracking data must be right; the Moon is getting bigger out the window.'
> CAPCOM 'Roger, Al. At least our direction is right.'
> WORDEN 'Appears that way.'

Apollo 15 was now close to passing from Earth's gravity field into that of the Moon. After a breakfast meal, the crew began an experiment into visual light phenomena reported by all trans-lunar crews of earlier flights. At first, astronauts had been reluctant to speak of what seemed like flashes of light across their line of sight when eyes were closed, fearing an extended session with the local psychiatrist. But slowly the knowledge surfaced and scientists took them seriously when everybody reported the strange phenomena. One theory was that cosmic rays traveling through the astronauts' skulls impinged upon the optic nerve, causing a message to travel to the brain. With their eyes shaded by an opaque cover, and all three facing the same direction, they spent fifty minutes calling out whenever they saw a flash. In that period the crew reported 54 sightings.

At 56 hr 26 min Scott and Irwin re-entered Falcon for another sequence of checks and housekeeping activity, but that struck a literal note when they were required to use the suction pipe and vacuum the broken glass from the tape meter! About 90 minutes later they were finished and back in Endeavour, having replaced the probe and drogue. Then a leak developed in the chlorine port and it took several minutes for Mission Control to come up with a procedure to stop it. Dave Scott was getting worried about the amount of water forming as a large bubble on the exterior, a familiar effect of weightlessness. It was finally attributed to a loose gland, tightened back up with a spanner from the onboard tool kit.

> SCOTT 'And our trusty LMP came up with an interesting analogy relative to the last event. He wondered if the original Endeavour had ever sprung a leak like that.'
> CAPCOM 'Okay, that's a good question. We'll put our historians out to check that one.'
> CAPCOM 'Hey, what did you do with all that extra water. Stick it overboard or drink it, or what?'
> SCOTT 'Oh no we've got a bunch of towels hanging up in the tunnel, right now. It looks like somebody's laundry.'

After another rest period during which all three got a good $7\frac{1}{2}$ hours' sleep, it was time to prepare for the final course correction burn. Performed on Bank B, the SPS fired for 0.92 sec and shaped the trajectory to a pericynthion of 126 km, exactly the planned value. Little more than one-half hour later the SIM bay door was pyrotechnically jettisoned, ejected to go spinning off far from the docked vehicles. 'I felt a little shudder, but not too much,' said Dave Scott. When the CSM reached Moon orbit the instruments could be employed to scan the surface.

Joe Allen was at the capcom console when Apollo 15 disappeared round the Moon's west limb prior to burning its SPS engine: 'Gentlemen, everything looks perfect down here and all we can say is "Have a good burn." ' It was becoming almost routine, the seventh vehicle to pass around the far side of the Moon. 'Roger,' was the only reply. Programed to decelerate the vehicles by some 3,292 km/hr, the 6 min 40 sec firing was the longest ever performed by an Apollo CSM burning into Moon orbit. With a heavy vehicle coming around slightly faster than any other, the SPS consumed 12 tonnes of propellant before dropping into an elliptical 107 × 315 km orbit. Bank A had been shut down 32 seconds before the scheduled cut-off time to obtain performance data on Bank B.

At acquisition of signal, Dave Scott broke radio silence: 'Hello Houston, the Endeavour's on station with cargo, and what a fantastic sight.' Comment flowed freely from the trio orbiting the Moon.

SCOTT 'And you know as we look at all this after the many months we've been studying the Moon, and learning all the technical features and names and everything, why when you get it all at once it's just absolutely overwhelming. There are so many different things down there, and such a great variety of land forms and stratigraphy and albedo that it's hard for the mental computer to sort it all out and give it back to you – I hope over the next few days we can sort of get our minds organized, and get a little more precise on what we're seeing. But I'll tell you, this is absolutely mind boggling up here.'
CAPCOM 'Gentlemen, I can well imagine that a foreign planet must be a weird thing to see.'

Less than one hour after Apollo 15 centered Moon orbit, the S-IVB slammed into the surface about 184 km east-north-east of the Apollo 14 ALSEP. Seismometers at both that and the Apollo 12 site picked up reverberations, and measured the profile of the substructure to a record depth of 80 km. All across the front side pass the Apollo 15 crew kept up a constant description of the surface. Passing along a different orbital plane than any previous Apollo vehicle, they saw craters and features unseen hitherto. But their enthusiasm was due also to the great part they were made to feel they had in unfolding new frontiers for science.

All through their training the emphasis had been on really getting to grips with true Moon exploration; all three were enthusiastic participants. And every once in a while, the crew passed messages for Farouk El Baz, the noted Egyptian geologist actively involved with science training sessions for Apollo astronauts at the Manned Spacecraft Center. Back on the front side pass during revolution 2, there hardly seemed time to copy down a maneuver pad for descent-orbit-insertion, to be performed at the end of that orbit in a sequence first adopted for Apollo 14. If impressed at the high altitude, the crew were to become ecstatic down low. On the far side once more the SPS engine burned a fourth time and dropped Apollo 15 into a 17 × 108 km path. When the spacecraft reappeared spirits were high.

SCOTT 'Hello, Houston, Apollo 15. The Falcon is on its perch.'
CAPCOM 'Good to hear you coming around that corner. How do things look.'
SCOTT 'I'll tell you, it's really spectacular when you can see the central peak (of a crater) coming up over the horizon before you see the rim!'
CAPCOM 'Hey, that's an interesting astrophysical observation.'

The first SIM bay operations began on revolution 3 as scheduled. Equipped with three spectrometers (gamma-ray, x-ray, and alpha-particle), the SIM also carried two cameras: a panoramic camera for high resolution views of the surface and a mapping camera for high quality imaging to assist with cartographic studies of the Moon. There was, in addition, a mass spectrometer designed to detect trace gases present in the Moon's vicinity. Mass and gamma-ray spectrometers were the two instruments mounted on retractable booms.

Shortly before passing out of sight on revolution 4, Mission Control said goodnight to Apollo 15 for the last rest session the crew would get before Scott and Irwin landed on the Moon. Yet even as they slept tracking stations on Earth were watching the spacecraft's orbit as the Moon's asymmetric gravity pulled and tugged at the elliptical path. Milton Windler conferred with Gene Kranz, then getting ready to come on shift, and both agreed the crew would be woken just a little early: Apollo 15 was now down around 14 km and the new ellipse was moving the point of pericynthion further away from where it would need to be for powered descent. The crew would fire the SPS again and trim the orbit to one which would raise pericynthion and put the low point back where it should be.

Awake by 93 hr 35 min, Scott, Worden and Irwin ate breakfast and then, coming around on revolution 9, capcom Robert Parker passed word that the trim maneuver would, after all, be performed with the small RCS thrusters. If left unattended, Apollo 15 would be in a path down as low as 10.9 km at powered-descent-initiation on revolution 14. The trim firing lasted 21 seconds and shifted pericynthion back up to 17.8 km. On the preceding orbit the crew sent down television pictures of the projected landing site, giving watchers an unprecedented view of the Hadley region from so low an altitude.

In Houston's Mission Control Center, flight director Glynn Lunney was relieving Kranz for the important descent phase. Scott and Irwin entered Falcon a little earlier than scheduled, turning on the LM electrical power at 97 hr 35 min, and received a 'go' for undocking shortly before passing out of sight on revolution 11 nearly two hours later. Among the final guidance computations was knowledge of a consistent downrange position error of 4.5 km which would be uplinked to Falcon's computer for correction during the main braking phase. But there were other problems to worry about. 'Okay Houston, we didn't get a sep, and Al's been checking the umbilicals now on the probe,' was the message from Dave Scott when the vehicles came round on revolution 12.

Connected electrically, the two vehicles had refused to separate when Worden activated the appropriate sequence. But the Command Module Pilot found a loose umbilical in the tunnel and when pushed home, with the hatch sealed up once more, Endeavour slipped its shackles – only 25 minutes late. One hour later, with Falcon left in the descent orbit, a 4 second SPS burn pushed the CSM back into a circular orbit about 112 km above the Moon from where it could respond to an abort should Falcon falter. Performed on the far side immediately prior to the start of revolution 13, the circularization burn preceded landmark tracking activity to locate the CSM more precisely above Falcon, information that would be essential during a quick rendezvous.

All through the next front side pass, maneuver pads were voiced up to both spacecraft, and Falcon checked its landing radar while the crew donned helmets and gloves. At acquisition of signal on revolution 14 Houston told Falcon that precise landmark tracking by Endeavour indicated the LM to be 6 km to the south of the desired groundtrack at PDI. That error would be steered out by a slight rolling maneuver during powered descent as the spacecraft translated north, shifting slightly to the right of its flight path. In the MOCR, Glynn Lunney took a final status check and then passed Falcon the word to proceed with powered descent.

At ignition the engine was hardly noticeable, but ramped to full thrust deceleration was now very apparent as the Lunar Module rattled and shook its way down the flight line. Throttled back down again seven minutes later, program 64 took Falcon into the approach phase. At pitch-over the landing site ahead lay visibly similar to the simulators they had practiced in for hours but this time the effects of engine blast on the fine dusty particles would raise an obscuring cloud. Descending almost vertically in the landing phase from a height of about 60 meters, Scott and Irwin saw dust snaking along the surface seconds later and from a height of 18 meters the surface was impossible to see; final touchdown would be on instruments like those of Apollo 12 but with worse conditions than ever before. Unlike previous landing operations, Dave Scott brought Falcon down an almost vertical shaft which helped because the final landing spot was chosen early enough to get a safe location.

'Okay Houston. The Falcon is on the plain at Hadley,' said David Scott amid technical readouts and immediate preparation for ascent should anything have gone wrong on touchdown. For the first time the LM's engine bell had contacted the surface, crushed slightly beneath the Descent Stage due to its greater length. Falcon was tilted back approximately 8°, but that was safe enough both for stability and for deployment angles when the time came to unfold the rover. The landing took place during late evening in Europe, giving media networks opportunity for prime viewing time, and in Houston it was early evening when people arriving home from work could snatch a televised view of scenes at Mission Control. It had taken Falcon nearly 12½ minutes to descend, and Scott had dropped the Lunar Module more slowly during the landing phase than earlier crews, but the spacecraft performed well with a modest reserve of propellant.

It was late in the day, Friday, 30 July. The first of three days on the Moon, three periods of surface exploration each preceded by a sleep period and a meal. First order of business was to prepare for a stand-up EVA, or SEVA, where Dave Scott would connect to a long umbilical, depressurize the cabin, stand on the Ascent Stage engine cover and with his upper torso out the top hatch describe surface conditions. From this information mission planners hoped to get a better and more immediate fix on the precise location of the spacecraft among the myriad of small craters and depressions across the Hadley Apennine floor.

The crew was about 30 minutes late beginning the SEVA activity but when Dave Scott emerged through the upper hatch,

Exposed by a panel jettisoned earlier, Apollo 15's Scientific Instrument Module bay reveals scientific equipment for six days of research.

his view was spectacular. Perched 7 meters above the lunar plain, he could see around the planned traverse routes and identify features near and far.

SCOTT 'And all of the features around here are very smooth. The top of the mountains are rounded off. There are no sharp jagged peaks or no large boulders apparent anywhere. The whole surface of the area appears to be smooth with the largest fragments . . . in the walls of Pluton. There are no boulders at all on St. George, Hill 305, Bennet, or as far as I can tell looking back up at Hadley. Hadley's sort of in the shadow. It's a gently rolling terrain completely around 360 degrees, hummocky much like you saw on 14 . . . I can see, as I mentioned before, Chain . . . and Pluton, very rounded subdued craters. It looks like the southern rim of Pluton is on the same level as our location here.'

In the science support room, geologists listened to the verbal description of craters and features arbitrarily named for this flight; many of the places and depressions called out were too small to have been officially mapped. From the capcom console, Joe Allen enthused with the crew over what they could see. It all augured very well for a significant increase in scientific activity.

SCOTT 'I can see the Sun from the other side of the north rim of Pluton. All of it very flat, smooth, and gently rolling. Inside walls of Pluton are fairly well covered with debris. Fragments up to, I'd estimate, maybe 2 to 3 meters . . . just sort of scattered around. I look on around, and our Mt. Hadley itself is in shadow, although I can't see that the ridge line on the top . . . it too is smooth. I see no jagged peaks of any sort. The hill I would call number 22 on your map, far distance, also looks smooth and rounded, no prominent features. I'll skip the distant field to my 6 o'clock because it's all in the shadow and looking into the Sun, of course obliterates almost everything.'

The Falcon's Commander was moving round in a clockwise fashion, describing things directly in front, the 12 o'clock position, round to the rear, or 6 o'clock, area. But all the while, personnel in the Mission Operations Control Room at Houston were poring over lunar surface EVA maps, trying to fix the LM's precise location.

SCOTT 'As I look down to my 7 o'clock I guess I see Index crater here in the near field. But back up on Hadley to the east of Hadley Delta, again I can see smooth surface; however, I can see lineaments. I'll take a picture for you, there are some very interesting. . . . But there appear to be lineaments running – dipping – through to the northeast parallel and they appear to be maybe 3 percent to 4 percent of the total elevation of the mountain. Almost uniform. I can't tell whether it's structure or internal stratigraphy or what, but there are definite linear features there dipping to the northeast, at about – oh, I'd say 30 degrees. As I look up to Hadley Delta itself, I can see what appears to be a sweep of linear features that curve around from the western side of Hadley Delta on down to Spur down there. And they seem to be dipping to the east at about 20 degrees. These are much thinner lineations on the mountain than I saw before, these probably are less than 1 percent of the total elevation of the mountain. The craters on the side of Hadley Delta are rather few.'

Falcon was sitting on the flat floor of Hadley plain about 1½ km due east of the rille. To the north-east lay the dominant mass of Mt. Hadley, with Hadley Delta, a little smaller, to the south. St. George crater was on a bend on the rille where it curved to avoid the flanks of Hadley Delta, while Index, Pluton, and Scarp, were craters around the Lunar Module. Hill 305 and Bennet were features far to the west beyond Hadley rille. The first EVA would take Scott and Irwin south-west to the area of St. George and the rim of the rille, while the second and third would explore the flanks of Hadley Delta and the rille wall respectively. Houston was delighted with the result of this first SEVA ever carried out prior to surface exploration.

CAPCOM 'Superb description, Dave. Got every single word, beautiful. And we'll ask you to hustle on around and give us something on the near field, plus a comment on the ALSEP deployment possibility. Superb communication though, beautiful.'

SCOTT 'Okay, coming on around to St. George, which again is a very subtle old crater, but in this case I can see some lineaments running, dipping to the west at about 20 degrees parallel to the rim of the crater. The rim of the crater is very subdued and smooth. Coming around I'll just take a quick look at the near field for you here. It's about generally the same. The crater density is I'd say quite higher, somewhat higher, than I expected. Sizes are mostly less than about 15 meters. The only large crater that I can see is what I believe to be Index back here about the 8 o'clock, and it has a very subtle rim, almost no shadow at the bottom of it. I think that's one of the things that was deceiving on the descent. There are very few deep dark craters in this area. . . . Trafficability looks pretty good. It's hummocky, I think we'll have to keep track of our position, but I think we can manipulate the rover fairly well in a straight line and I can see the base of the Front. . . . Looks like we'll be able to get around pretty good.'

CAPCOM 'Roger Dave. We copy.'

SCOTT 'The continuity of the surface that I see in our general position, I don't think we'll have any trouble taking the ALSEP out 300 (meters) or so and placing it.'

CAPCOM 'Roger, Dave . . . sounds like we are in business, old friend.'

SCOTT 'Yeah, but we are indeed in business, and I think once we get through here and I hop back down, why we can talk over more of what I've been seeing up here.'

CAPCOM 'Roger, Dave. You're coming up on 30 minutes into this SEVA and we don't have any more questions. You've answered everyone beautifully, outstanding.'

SCOTT 'Okay Joe, I'll take another quick look around and see if anything looks unique. There's just so much out there, I could talk to you for hours. Do you have any specific questions before we call it quits?'

CAPCOM 'Dave we're hoping you will be talking to us for hours about it. We don't have any specific questions right now, we'll think about it and talk to you again once you button up.'

Scott used a 500 mm telephoto lens on the Hasselblad camera to shoot panoramic pictures all around the area, and to get views of specific features. Nobody had ever been quite so enthusiastic to get down to science as Dave Scott now seemed to be, throwing his team into a brisk pace of lunar activity from the moment Falcon landed. It was to set the scene for a busy three days. After five hours sleep, commencing 5 hr after touchdown, the two explorers got ready to go outside. But first, flight director Pete Frank had them address a possible leak in the oxygen system which showed ominous signs on telemetry; checks revealed an open urine dump valve leaking oxygen to the surface which, when closed, solved the problem.

The first EVA formally began fifteen hours after touchdown. In Houston it was 8:13 am, Saturday, 31 July. It would be mid-afternoon, mid-west time, before the crew returned to Falcon. Dave Scott came down the Falcon's ladder just fourteen minutes into the EVA, deploying on his way the externally mounted MESA with its color TV camera pointing at his shadowy frame.

SCOTT 'Okay. Okay, Houston. As I stand out here (in) the wonders of the unknown at Hadley, I sort of realize there's a fundamental truth to our nature. Man must explore. And this is exploration at its greatest.'

CAPCOM 'Roger, Dave.'

SCOTT 'Rather interesting sight, Houston. I can look straight up and see our good Earth back there.'

First, the TV was removed from the MESA and set up on a tripod so as to give Houston a view of the rover deployment sequence, that event taking a little longer to complete than expected. Jim Irwin had followed Scott to the surface by this time and the two astronauts carefully unpacked their transport, removed from Falcon's bay after 95 days in storage. Then it was time for a quick test drive before discovering a minor problem. 'I don't have any front steering, Joe,' said Dave Scott as he pulsed the hand controller. Different switch positions failed to restore full Ackerman effect but that would be no real problem since the rear wheels could do all the turning.

Then it was time to install the lunar communications relay unit, or LCRU (pronounced 'lacru'), through which all information from the roaming astronauts would be sent to earth. Traversing several kilometers from the LM, rover travellers would have to carry their own means of communication since the antenna carried by the spacecraft would be out of range. Weighing only 25 kg, the LCRU was a self-contained package attached to the front of the rover with three separate antennae: a low-gain helical antenna for voice and telemetry from the back-packs during motion; a 1 meter diameter high-gain dish antenna for TV; and a VHF omni-antenna for receiving crew date and voice communication. The omni-antenna would pick up transmissions from the astronauts' packs while the helical antenna sent it to Earth.

For the first time, viewers would move from site to site with the lunar explorers. But TV could only be sent to Earth when the rover stopped, because of the need to align the rib-mesh parabolic dish precisely. A 16 mm movie camera would shoot views from the rover in motion for playback when the film was developed back on Earth. The relay unit worked in conjunction with a remote control system where an engineer in the support room at Mission Control could move the TV, scanning the surface while the crew worked, or following their activity on the surface. It was altogether different from anything ever developed for Moon use before.

When the TV had been attached to the remote control equipment on the front of the rover, tools were loaded on the geology pallet behind the seats. Presenting a rack for easy access to many different types of hand equipment, the pallet was a valuable carrier. It was time to move out on the first geological traverse, for on this mission the first order of business was to conduct a scientific survey and then return to deploy the ALSEP experiments close by the Lunar Module. But navigation on the Moon would be a vital necessity. The wandering crew of Apollo 14 showed how easy it was to get lost when looking for a specific crater among many, a time-wasting exercise with lost science as a by-product. So the rover was given a dead reckoning navigation system both to aid in finding specific features and in quickly returning to the LM by the shortest route should anything go wrong during the traverse.

Three items of equipment comprised the navigation unit: a directional gyroscope; a single odometer in each wheel traction drive for distance information; and a small computer for calculating bearing and range to the Lunar Module, the distance travelled, and the speed of the car. The gyroscope was set up by using a special shadow indicator to measure precisely the orientation of the rover with respect to the Sun. Mission Control would calculate a heading from the Sun angle measured by the crew, and tell the astronauts what that value was believed to be so they could adjust the gyroscope until the heading indicator read the same.

Scott and Irwin were packed up and strapped in the rover about two hours into the EVA. They would head south west to the rille rim, follow it due south to a sharp bend, and stop off at their first geological station, a crater called Elbow just before St. George and approximately 2.2 km from Falcon. They would move out on a 208° heading (that is, 208° clockwise from due north), passing craters Rhysling and Earthlight en route.

SCOTT 'Okay, we're doing 10 kilometers (per hour) now. Now we're heading up a hill and when we (do) it drops down to about 8. No dust, Joe. No dust at all.'

CAPCOM 'Yes sir, sound great.'

SCOTT 'Okay, could this be Rhysling right here, Jim.'

IRWIN 'Probably is – this depression off here to our left.'

SCOTT 'Yes. Man, I can see I'm going to have to keep my eye on the road!'

SCOTT 'Boy, it's really rolling hills, Joe. Just like 14. Up and down we go. Oh, this must be Earthlight, huh? Could that be? Boy look at that.'

IRWIN 'There's a long depression here before you get to Rhysling. I don't think we're to Rhysling yet – Rhysling ought to be about 1.4 (km) and we've only gone, see, 0.4.'

CAPCOM 'Roger, Jim. We think you're short of Rhysling now.'

On across the rough lunar surface, the rover bounced and rolled, through trough after trough, multiple depressions, small craters and bumpy ridges. Moving at only 10 km/hr, the very nature of the surface made it the experience of a lifetime.

SCOTT 'Okay Joe, the rover handles quite well. We're moving at, I guess, an average of about 8 km (per hr). It's got very low damping compared to the 1 g rover, but the stability is about the same. It negotiates small craters quite well although there's a lot of roll. It feels like we need the seat belts, doesn't it Jim?

IRWIN 'Yes, really do.'

SCOTT 'The steering is quite responsive even with only the rear steering. It does quite well. There doesn't seem to be too much slip. I can maneuver pretty well with the thing. If I need to make a turn, sharply – why it responds quite well. There's no accumulation of dirt in the wire wheels.'

CAPCOM 'Just like in the owners manual, Dave.'

IRWIN 'Boy, it really bounces doesn't it?'

SCOTT 'Well, I think it sort of – the rear end breaks out at about 10 to 12 clicks (km/hr).'

CAPCOM 'Roger, Dave. It sounds like steering a boat with the rear steering and the rolling motion.'

SCOTT 'Yes, that's right. It sure is.'

About ten minutes into their motorized traverse the crew came upon the edge of the rille, the wide and very deep incision that cut a vertical gorge through the outer layers of dried magma that flowed from Mare Imbrium into the Apennine foothills. It was a very spectacular sight, one not seen before by Moon explorers from Earth. A view altogether unique to the lunar panorama.

SCOTT 'Hey, you can see the rille! There's the rille!'

IRWIN 'There's the rille!'

SCOTT 'Yea. We're looking down on it. Down and across the rilles we can see craters on the far side of the rille.'

SCOTT 'Yea, now we're getting into the rough stuff.'

IRWIN 'We're right at the edge of the rille, I bet you.'

SCOTT 'Yes sir, we're on the edge of the rille. You better believe it.'

Unlike views from above, scenes observed from lunar orbit, features on the surface with very little height above the mean level were difficult to see until they were only a few meters away. So it was with the rille. Suddenly, although expected, the appearance of this deep gorge falling away in front of them caused adrenalin to flow faster, the heart to race a little quicker. Across in the far wall, more than a kilometer away, parallel blocks could be seen, large boulders stuck like stone in a quarry adhering to the shallow slope. Turning south to follow the rim, Scott and Irwin pressed on, bouncing and rocking to Elbow crater and their first geological stop.

IRWIN 'And again looking at the, looking to the south along the edge of the rille that faces to the northwest I can see several large blocks that have rolled downslope. Very large blocks that are about three quarters of the way down the slope into the rille.'

CAPCOM 'Roger Jim, copy.'

IRWIN 'I can see the bottom of the rille. It's very smooth. I see two very large boulders that are right on the surface there. On the top of the very smooth portion of the bottom of the rille. And the one to the southeast I can see the track of where it's rolled down slope.'

CAPCOM 'Roger, Jim, copy. And is the bottom V shaped or fairly flat?'

IRWIN 'I'd say it's flat . . . I'd estimate maybe, oh, 200 meters wide on the flat area of the bottom. Oh, and I can see what we thought was Bridge crater. And it definitely would not have been a place to cross Hadley rille. It's just a depression in the west wall of the rille.'

There had been talk during EVA planning sessions long before the flight of crossing the rille via what seemed from space photographs to be a bridge where slopes were at a minimum. That suggestion was now seen to have been premature for the slopes were in fact quite steep at that location.

CAPCOM 'Dave, are the front wheels wandering off of straight ahead as you drive along there?'

SCOTT 'No, they're okay Joe. It's just there are a lot of craters and it's just sporty driving. I've just got to keep my eye on the road every second.'

CAPCOM 'Roger. We understand that, just trying to get some engineering information here. Apparently your front wheels are tracking straight ahead. Is that correct.'

SCOTT 'That's correct and of course when we turn they dig in and it makes the rear end break out but it's okay. We can handle it.'

IRWIN 'Oh, this really a sporty driving course. Man, oh man, what a grand prix this is!'

SCOTT 'How are we doing on time, there, Houston?'

CAPCOM 'Like gangbusters, Dave and Jim. Continue on and we'll give you the exact number in a minute.'

In meandering across the Hadley plain, rover 1 had traversed a distance in 25 minutes greater than that accumulated by the crew of Apollo 14 on their two EVA periods. Elbow crater was a good place to stop and sample the material, with a small trench dug and several rocks retrieved. After a stop lasting 14 minutes they climbed back on the rover and prepared to move out a short distance to the rim of St. George crater.

SCOTT 'Okay, Joe. The time consumer here is the seatbelt operation. In 1/6th g we don't compress the suits enough to be able to squish down and get the seatbelts locked without a certain amount of effort.'

CAPCOM 'Roger. We understand.'

It took just eleven minutes to reach station 2, where the crew would remain for nearly one hour, sampling, describing the surface materials, and performing geological tasks. Dave Scott revealed the depth of his training while searching for valuable specimens:

'Very angular; very rough surface texture; looks like it's partially – well, it's got glass on one side of it with lots of bubbles and they're about a centimeter across and one quarter of it has got all this glass covering on it; seems like there's a linear frature through one side; it almost looks like that might be a contact; it is within the rock. It looks like we have a – maybe a breccia on top of a crystalline rock. It's sort of covered with glass, I can't really tell, but I can see it. A definite linear feature through one side of it which is about a fifth, and the glass covers both sides of what I guess I'm calling a contact, and there's also – parallel to that contact one surface, which is quite flat; only for about (20 cm) or so. Looks like it's been chipped off. The boulder itself is on the order of about a meter across and maybe a – gee it looks like a half meter thick or so. It's got a fillet up one side and the other side is in a shadow; I can't really tell whether – it doesn't look like it's filled. It's got a fillet on the downslope side and the upslope side is open and free, as a matter of fact, it really looks like it's almost excavated beneath it.'

CAPCOM 'Roger, Dave and Jim. You look crystal clear and we've got a beautiful pan of both you and the boulder on the TV. And it probably is fresh; probably not older than $3\frac{1}{2}$ billion years.'

SCOTT 'Can you imagine that, Joe? Here sits this rock and it's been here since before creatures were in the sea on our little Earth.'

CAPCOM 'Well said, Dave.'

Higher than at their previous station stop, the crew settled down to the assigned geological tasks for this location but the wandering eyes of the television camera, moved by a controller in the support room at Houston, came upon the sinuous rille and presented a view to millions on Earth unlike anything seen from the lunar surface before. The sheer depth and size of the gorge was immensely impressive. A breathtaking scene that re-kindled interest among many; but sadly, most people were untouched by the importance of the science now being played out on Earth's neighboring world. Scott and Irwin were nearly 4 km from Falcon, having travelled a winding route that carried them $5\frac{1}{2}$ km across the lunar plain. And then, when the Houston controller had difficulty elevating the TV camera by remote command, a request was passed up to the astronauts.

CAPCOM 'Could you see if we have a TV cable hung up on the "lacru" some-place? We are having trouble commanding the direction.'

IRWIN 'Yea, you do. I think the wire from the high-gain antenna has got your cable to the TV.'

SCOTT 'I'll get it.'

CAPCOM 'Roger, could you give that unmanned vehicle a little help, please?'

SCOTT 'Okay. Done.'

Nobody was about to miss a point demonstrating the importance of having a man at the site! Core samples were obtained in the loose wall material of a fresh 10 meter diameter crater while stereo pictures were obtained of interesting rock samples. It was all done according to the geological text book, using procedures developed over decades of similar work on Earth. The crew had absorbed 3 hr 47 min of their EVA when the return traverse began. They were to drive almost due north back to Falcon from where they would unpack and deploy the packages of ALSEP equipment.

But ever the adventurous explorer, Dave Scott had a word for

The first operational Lunar Roving Vehicle stands on the Moon, July, 1971. Note the distant flanks of Hadley Delta and St. George crater, 4 km to the south.

capcom Joe Allen: 'If anybody come back here, Joe, and wants to go down into the rille. Have them come talk to us 'cause there's a good place to do it here.' From this position the crew could not see the Lunar Module, relying entirely on their navigation system which told them the bearing back to the LM; that piece of information was continually available, just in case they had to scurry for home. Then they came upon some familiar tracks.

SCOTT 'There are some rover tracks. How about that. Yeah, here we go.'
IRWIN 'Somebody else has been here.'

All the way back the two explorers kept up a constant description of the scene before them, and of rocks along the way. They stopped about 1.7 km from Falcon to get more samples, before pressing on once more until they caught sight of the LM; rover 1 had brought them straight back from St. George. The display console told them they had traveled a total 10.3 km. The EVA was nearly 4½ hours old, and both men still had more than 40% oxygen remaining in their new back-packs. One of the reasons for putting ALSEP deployment last was to keep the astronauts within close range of the LM in the closing hours of this first walk, for nobody had worked on the lunar surface longer than 4½ hours and it would have been unwise to place the crew far from safety when breaking in the new back-packs.

It would be a more strenuous activity, unpacking the ALSEP packages and deploying them in sequence, than the comfortable tasks conducted so far on this first Moon walk. Scott drove the rover almost due west while Irwin loped behind, until they were at the best site for putting up the instruments, about 300 meters from Falcon. Irwin would put out the equipment while Scott unpacked the lunar surface drill, a Black and Decker tool similar to that carried by the unsuccessful Apollo 13 mission. ALSEP incorporated a passive seismometer like those left at the Apollo 12 and 14 sites, a solar wind experiment, a lunar surface magnetometer, and two small packages designed to monitor and measure the effect of the Sun on any trace lunar gases that might be present.

Put down too was an improved laser retro-reflector, with 300 corner reflector segments versus the 100 carried by similar equipment left at the Apollo 11 and 14 locations. They would be used to accurately measure the relative position of the Earth and the Moon with unparalleled precision, and to carefully monitor the drift of continents across the Earth's mantle. The third seismometer would provide a triangulation capability essential for pinpointing the area of casual impacts from meteorites or rocky debris.

Scott's task during this time was part of a sequence of heat flow experiments to be set up by all three J-series Apollo missions. He was to emplace two heat-flow probes 3 meters deep and 10 meters apart which would be connected to the ALSEP central data station to send information on the rate at which heat was escaping from the Moon's interior, a valuable indicator as to the level of geophysical activity in the core and mantle. But first there was a problem with the drill chuck.

Two probes were to be inserted in the surface, one above the other, supporting a string of thermocouples designed to measure heat flow. But the drill stems would not penetrate the stiff lunar substructure beyond a depth of 1.62 meters at the first hole, so Scott was told to put the probes in as far as possible and leave the top thermocouples out on the surface. But even the drill was so tight in the chuck holding the core stems that he had great trouble getting it free and only after a strenuous heaving to and fro with a small vice did the drill come away. Then, while attempting to drill out the second hole, the chuck stuck again, then the vice hiterto used to get it free stuck on the drill! So much energy was being expended through the back-pack that Joe Allen passed along a recommendation to abandon the activity and move back to the LM. They could try to drill a deeper hole at the end of the second EVA; the first was already more than 5½ hours old.

Less than one hour later the crew had put the rover in a parked location, placed the thermal covers at the appropriate position, set up the familiar solar wind collector, and transferred the 25 kg of Moon samples up into Falcon. Their first EVA had lasted 6 hr 34 min, cut short by 30 minutes due to excessive oxygen consumption while deploying all the ALSEP equipment. The resistance met by Dave Scott when attempting to drill a couple of 3 meter deep holes was unexpected, calling for more thought on how to get the heat flow experiment working. But there were words of comfort for the tired duo, exhausted by the long day spent working the Hadley Base site as Joe Allen passed up information on Endeavour's orbital activity.

CAPCOM While you are working around there you might be interested in a little conversation from down here. The SIM bay's chewing up data like it's going out of style. We're working beautifully and as far as we can determine the ALSEP is working as advertised, getting all kinds of data from it and I'll get a good accurate reading on that for you later on. And I think that your traverse goes without comment; it was beautiful, and we're just trying to digest some of the data from that right now.'
SCOTT 'Okay, I'll tell you one thing Joe. Time sure goes fast out there.'

During the post-EVA activity, Scott reported that a bacterial filter on a water spigot inside Falcon was broken and that liquid was escaping in a steady flow. A small puddle had accumulated on the cabin floor but the leak stopped when the system was recycled and turned off. After a clean-up and a meal, the crew answered a comprehensive set of questions about the EVA activity, discussing the problem with the heat flow probes. 'I had the impression we were drilling through rock,' said Scott, adding that there was considerable torque on the drill stem which caused the chuck to seize up. It was mid-evening in Houston, about 26 hours after touchdown for Apollo 15; time for the crew to get some sleep.

After a solid seven hours' rest, Houston called up to the crew at the start of another day. In Texas it was 1:35 am on Sunday, 1 August. First order of business sent Scott and Irwin searching for pools of water. The systems engineers detected a loss of more than 11 kg which, because the LM was tilting backward 8°, they believed could have moved behind the circular Ascent Stage engine cover. It had, and the crew lost valuable time removing equipment netting and other items to mop it up before filling two used lithium hydroxide containers which could be partially dumped through the urine collection system, the residual contents being jettisoned overboard with redundant equipment. Careful examination of Falcon's wiring diagrams revealed little to support the contention from some that seeping water could short the electrical equipment.

Based on the consumption of back-pack supplies on the first EVA, Mission Control aimed to accomplish all the tasks on this second EVA in 6½ hours versus the planned 7 hours. Then Houston asked the crew to re-fill one of the back-packs. Filled out of vertical, experience showed that bubbles would form in the water tank unless held straight up during re-fill. Prior to depressurizing the LM, Scott reported that Irwin's back-pack antenna was broken. That was taped across the top of the emergency OPS pack, and the second EVA began – almost exactly one hour later than the flight plan. It was 6:38 am, Houston time, Sunday 1 August Nearly one hour was spent packing up the rover, but after being requested to cycle the circuit breakers on the console, the car's front steering action worked and for the first time the crew would get the feel of the LRV with the double-Ackerman moving as designed.

SCOTT 'It's working my friend.'
IRWIN 'Beautiful. You know what I bet you did last night, Joe. You let some of those Marshall guys come up here and fix it didn't you?'
CAPCOM 'They've been working, that's for sure.'
IRWIN 'It works Dave, yes sir. It works my friend.'
SCOTT 'Beautiful.'
CAPCOM 'Lots of smiles on that one, Dave. We might well use it today.'
SCOTT 'Well, Boeing has a secret booster somewhere to take care of their rover.'

They now began a traverse that would take them far to the south, across 5 km of the Hadley plain to the 'front,' a region high on the flank of Hadley Delta from where they could look back down upon the Falcon's perch. Down past Index they drove, the turning ability enhanced with both sets of wheels steering. On past Earthlight, and a group called South Cluster which they skirted to the right, avoiding little Dune crater on their long journey south. Now they were starting up slopes that followed the gently curving lines of the Apennine front, their speed considerably reduced. Then a short rest before pressing on again. Their first sampling stop was at station 6, a location just beyond Spur crater, which they reached 43 minutes after leaving Falcon.

IRWIN 'As we drive up Sun here, I'm looking to the left and I can see the Mount Hadley and the linear patterns in it are really remarkable. Dipping to the north west. And the pattern runs from the very top of a whole mountain has the same pattern – linear pattern.'

What Jim Irwin saw on the mountain on the opposite side of the plain was linear features caught in relief by Sun angles that were slightly higher this day than they were on the first EVA. For more than an hour they sampled the site, allowing TV viewers a picture of their activity, but giving geologists in the science support room good scenes from which to gather information in planning further episodes in this EVA and the next.

CAPCOM 'We've got a beautiful picture. We're trying to look into the Sun at the moment, somewhat unsuccessfully. But the TV's working beautifully.'
IRWIN 'You ought to look up toward Mt. Hadley. You can see the linear pattern.'
CAPCOM 'Roger, we'll take a look. And thanks for the recommendation.'

All the while the crew worked at their assigned tasks, collecting rocks, gathering dust samples, raking small pebbles out, digging sample trenches, the Houston controller moved rover's TV to survey the scene around Hadley Base. Sometimes fixing on one astronaut, then the other, then both together helping to accomplish a single task, sometimes just panning around the spectacular mountain scenery. Color was good in the picture on Earth, a remarkable improvement over the first fluffy pictures sent back two years before from Apollo 11. But still it could not really capture the serene beauty of this virgin world upon which no living thing had stood in nearly five billion years since the Creation.

It was time to move on, just a short distance, for much of this EVA would be committed to sampling the front where plain floor met mountain flanks. It took just three minutes to move 400 meters west to station 6A. They would remain there twenty-one minutes, collecting more samples, filling more bags. When they left 6A it was to ease the rover gently down a steepening slope to station 7 – Spur crater itself. They spent 49 minutes at that location, describing more rocks and small surface features. The second EVA was more than 4 hours old when they departed Spur to

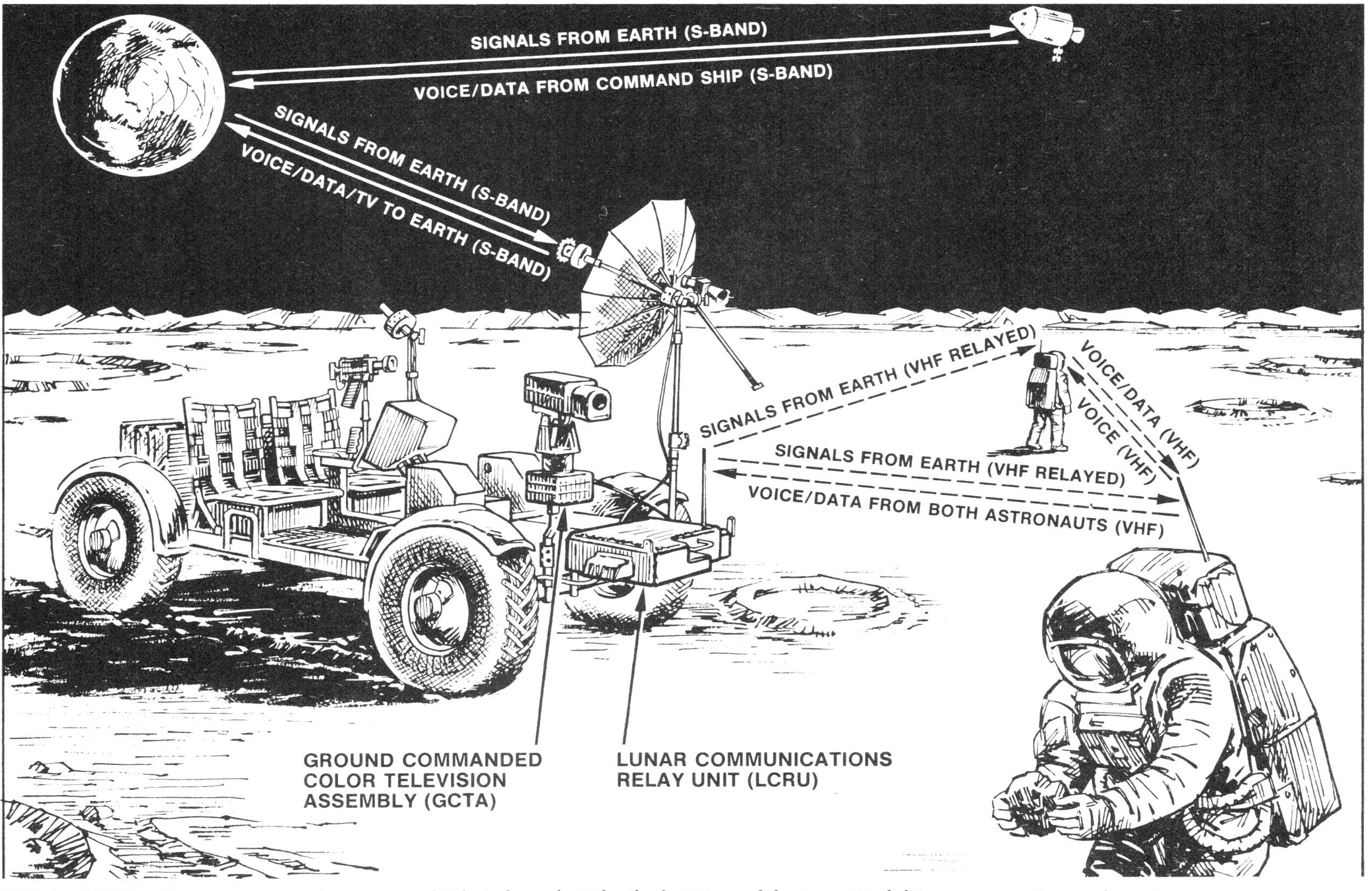

With the LCRU on the rover, astronauts were completely independent, for the first time, of the Lunar Module's communication equipment.

drive back across the plain, pausing on the south west rim of Dune crater, 3.4 km from Falcon, to sample the soft material at what was called station 4. As on the first EVA, Dave Scott was consuming more oxygen than Irwin, but both men had more than 40% remaining at an EVA time of 4½ hours.

Within visual range now, Falcon lay dead ahead and the roaming duo skirted Earthlight, re-joining their familiar tracks past Index. The LRV navigation system had again provided a precise and ready access to a specified location across the floor of Hadley plain. For nearly twenty minutes, Dave Scott tried to drill a deeper second hole for the heat flow sensors. But it refused to penetrate below about 1.6 meters and the experiment was only partially successful; Joe Allen advised him to leave it as it was and to obtain a core sample by using the drill to extract a vertical segment of the regolith. But that too proved more difficult than planned, for the core stems remained rigidly fixed in the surface while Dave Scott slithered and tumbled all over the area, heaving and pulling on the drill by hooking his elbows underneath and pushing down with his feet and legs.

Joe Allen recommended Scott to abandon the task, leaving the stem in the surface with the drill still attached for attention on the third and final EVA. Sampling activities not far from the LM at what was considered to be station 8 occupied the crew for the final 30 minutes before driving back to the Lunar Module. After quickly setting up the American flag, the crew moved back inside the Ascent Stage to complete their second day on the Moon, an EVA that actually lasted more than the seven hours originally scheduled. Consumables had been used more extensively than expected, but experience with the new back-packs afforded a more generous commitment in what would have been considered reserve time on an earlier mission.

The crew were more than two hours late getting into their second surface rest session, and that fact decided Mission Control to shorten the last EVA to between four and five hours; they must not be late back inside Falcon for the important lift-off time at the end of nearly three days. Because of this, tasks would be limited to a visit to the rille, stations to the north at a cluster of large craters abandoned in favor of pictures and samples of the spectacular gorge. The crew awoke after six hours sleep and began preparations for their last Moon drive. It was 3:52 am in Houston, Monday, 2 August.

First operations centered on retrieving the long core stem from the station 8 activity. Fully rested now, both men set about the activity with verve and determination, finally removing the full 2.2 meter long core stem, but leaving it unbroken until they returned from the exploration of Hadley rille. They started their drive west at 1 hr 3 min into the EVA, pausing momentarily to take pictures on the way, before approaching the side of the rim known as the terrace. Dave Scott was persistent about adventuring further, however.

SCOTT 'Pretty good slope, we could probably drive down there.'
IRWIN 'I think we can drive over – straight ahead and stay on a fairly level contour.'
CAPCOM 'Dave, when you climb off could you dust off our TV lens please?'
SCOTT 'Certainly We're off and stopped and going to get on with the task here.'

It took them just fourteen minutes to travel the 1.6 km separating Falcon from station 9, the first stop alongside the rille terrace. The crew moved on from that location at 1 hr 59 min into their EVA, arriving three minutes later a short distance closer to the rim. Called station 9A, it would occupy them for nearly one hour. Standing on the edge of the rille, Scott gave capcom Joe Allen, and all the many scientists crowding the science support room, an unexpurgated description of the deep gorge.

SCOTT 'From the top of the rille down. There's debris all the way and it looks like some outcrops directly at about 11 o'clock to the Sun line, it looks like a layer about 5% of the rille wall with a vertical face on it and within the vertical face I can see other small lineations, horizontal about maybe 10% of that unit. And that unit outcrops all the rille; its about 10% from the top and it's somewhat irregular, but it looks to be a continuous layer.'

And so it went on. For many minutes Houston listened to Dave Scott's detailed discussion of what he could see in the opposite wall, and watched the TV reveal a scene only moderately less precise than the commentary. It was very impressive to hear these men of science discussing their reports from the Moon, an age of discovery for Apollo like nothing received before. While there, Scott and Irwin obtained a rake sample, drove core tubes into the soft rille flanks, and collected small hand-sized blocks reminiscent of material that may once have formed the roof of this collapsed lava tube. They left station 9A at 2 hr 56 min, driving, again for three minutes, to a point further north along the wall of the rille. But all too soon it was time to pack up the rover and head back for the Falcon, new procedures now available for breaking down the four united core stems from the drill site.

CAPCOM 'Dave, we're standing by for a mark when you're rolling and we'd like for you to press on back towards the drill site, we've got a procedure for you to separate two sections of the deep stem from the other two sections and we're going to carry the two halves into the LM that way.'
SCOTT 'Okay.'

At an EVA time of 3 hr 11 min the two Moon explorers were driving back toward Falcon. The Sun, high now above the eastern horizon, facing them as they moved across the undulating plain, beamed glancing blows of golden light at the lineated slopes of Hadley, to the left, and Hadley Delta, to the right.

SCOTT 'Oh, look at the mountains today, Jim, when they're all Sunlit, isn't that beautiful?'
IRWIN 'It really is.'
SCOTT 'By golly that's just super. You know, unreal.'
IRWIN 'Dave, I'm reminded about my favorite biblical passage from Psalms: "I'll look unto the hills from whence cometh my help." But of course we get quite a bit from Houston too.'

It took only 13 minutes to travel 2 km back to Falcon and Scott separated the core stem into a single 3-section string before moving it back toward the LM. Then Scott obtained a special sample of lunar material significantly contaminated with exhaust from the Descent Stage main engine, folded up the solar wind composition sheet which had been exposed for 41 hr 8 min, and franked a stamp envelope with a small franking machine carried along for the purpose.

SCOTT 'Okay. To show that our good postal service delivers any place in the Universe, I have the pleasant task of cancelling here on the Moon the first stamp of a new issue dedicated to commemorate United States achievements in space. And I'm sure a lot of people have seen pictures of the stamp. The first one here on an envelope, at the bottom it says "United States in space, a decade of achievement," and I'm very proud to have the opportunity here to play postman, pull out a cancellation device. Cancel this stamp. Is August the second, 1971, first day of issue. What can be a better place to cancel a stamp than right here, the Hadley Base. My golly, it even works in a vacuum. But not too well. But it's the first time, so I guess they're just learning.'
IRWIN 'Now I'll stick this back in a special mail pouch here and we'll deliver it when we return.'
CAPCOM 'Roger.'
SCOTT 'I think that's pretty good after only ten years. Here we are spending three days on the Moon. That's moving ahead.'
CAPCOM 'Dave, this is Houston. We're wondering if you could use that to mail home an ounce of rocks, please.'
SCOTT 'Well, alright. I'll do that. I bet we could.'

Then the Apollo 15 Commander proceeded to demonstrate a very old scientific principle.

SCOTT 'In my left hand, I have a feather. In my right hand, a hammer. I guess one of the reasons we got here today was because of the gentleman named Galileo a long time ago who made a rather significant discovery about falling objects in gravity fields and we thought where would be a better place to confirm his findings than on the Moon? And so we thought we'd try it here for you. The feather happens to be appropriately a Falcon feather for our Falcon and I'll drop the two of them here and hopefully, they'll hit the ground at the same time.'
SCOTT 'How about that.'
SCOTT 'This proves that Mr. Galileo was correct. And his findings.'
CAPCOM 'Superb.'

With those little ceremonies completed Scott drove the rover 100 meters east of the Lunar Module and left it so the TV camera could view the Ascent Stage at lift-off. But he left something else, too. In a small depression six meters north of the rover, a tiny human figure alongside a white plaque, black edged, bearing the names of 14 Soviet and American space men who died of various causes since joining their respective programs, symbolizing 'the fallen astronaut.' Loping back to the Falcon, Scott joined his colleague to complete the third EVA after less than five hours on the surface. One hour later the LM hatch was opened again to eject redundant equipment in bags carried aboard the Ascent Stage. There was little time to waste; lift-off was scheduled for an elapsed mission time of 171 hr 37 min.

During the third surface EVA, Al Worden up above in Endeavour burned his SPS engine to change the plane of the orbit in readiness for the upcoming rendezvous. In the Mission Operations Control Room, Joe Allen got ready to vacate the capcom console where he had tirelessly tracked the Moon men through three full working days on the surface. Unlike earlier flights the scientist trainer slept when the crew slept to be on hand for communications relating to surface exploration.

CAPCOM 'Ed's coming on the line down here just wanted to say I enjoyed it '
SCOTT 'Oh, well, thank you Joe. You did a super fine job. Appreciate you keeping such good track of us.'
CAPCOM 'Wouldn't have missed it for anything.'
IRWIN 'Thank you Joe.'
GRIFFIN 'This is the flight crew. The whole mission control team wants to take their hats off to you for a fine job. It was a lot of fun.'
SCOTT 'Well thank you Gerry. We'd like to take our hats off to the whole team. By golly you guys are really sharp down there and we sure appreciate it. 'Cause you know as well as we do, we sure couldn't do it without you.'

It was time now to leave Hadley Base and re-join Endeavour.

CAPCOM 'You're go for lift-off, and I assume you've taken your explorer hats off and put on your pilot hats.'
SCOTT 'Yes sir, we sure have. Ready to do some flying.'

When it came, nothing quite like the lift-off of Apollo 15's Lunar Module had been heard before, nor, as it turned out, since.

The Apollo 14 crew left on the Moon this plaque carrying the names of fourteen astronauts and cosmonauts who had lost their lives since selection, plus a small replica of the 'fallen astronaut.'

For at the moment the Ascent Stage engine fired, the downlink signal to Earth broke out with a stirring rendition of 'Off We Go into the Wild Blue Yonder.' It took most people by surprise and received frowns from several officials. But the message was clear and distinct: Apollo lunar landing operations were by now an almost regular ferry route for the NASA manned space flight teams, devoid of the nail-biting tensions characteristic of the first flights. The tape of military band music ran for several seconds until the crew switched to a verbal description of the scene passing below.

Ascents were always a visual spectacle; pitching forward to steer into an elliptical path about the Moon, the astronauts were suitably placed to look directly down upon the stark gray surface. Near to the end of the seven minutes required to achieve Moon orbit, Falcon's back-up guidance system produced a master alarm. But that was no problem and the sporty stage rattled into an elliptical path so precise that a trim was not required. Like Apollo 14, this fourth ascent from the lunar wastes would pull alongside the CSM within one Moon orbit. At orbit insertion, Falcon's Ascent Stage was closing on Endeavour at a relative speed of 472 km/hr but that reduced as the two-man vehicle climbed toward its apocynthion. It was further cut by burning the terminal-phase-initiation maneuver shortly after both spacecraft went around the Moon's west limb; Endeavour was about to start revolution 49.

When they came round on the east limb just 46 minutes later, Falcon was 6 km from Endeavour and closing the distance at 35 km/hr. Little more than one-half hour later the two were linked together and the lengthy process of transferring all the equipment and samples began. Three hours later the crew were back in the Command Module, and the hatches had been closed, when a check on the tunnel pressure indicated a leak from either Falcon or Endeavour; a tunnel pressure integrity check was a good way of discovering any escaping gas, since it would slowly rise if oxygen was bleeding through.

Flight director Glynn Lunney conferred with his flight dynamics officer and systems specialists before deciding to delay the jettison and de-orbit firing until the crew had opened up the tunnel hatches, re-sealed each one, closed them up and checked the pressure integrity once more. By the time that had been accomplished the spacecraft passed round to the Moon's far side and ground controllers resolved to wait until it reappeared to read again the pressure indication on telemetry. But minor concern developed when ground stations could not lock up on the spacecraft antenna, and then when they did there were no voice communications from Endeavour. Seconds later, after seeing a comforting pressure indication on telemetry to confirm the spacecraft was operating as expected, Dave Scott's voice broke through and the communication links were refined.

Five minutes later Houston confirmed that the pressure check looked good and that the crew could prepare for jettisoning the Ascent Stage and firing the separation maneuver. More than two hours later than planned, Falcon was released from Endeavour and five minutes after that the CSM fired its thrusters to move slowly away. Nearly one revolution later, the Ascent Stage fired its own thrusters in a de-orbit burn designed to bring it crashing down to the lunar surface for another seismic test. It came down 93 km west of the recently deployed ALSEP site, about 1,100 km north of Apollo 12 and 14 seismometers.

Aboard Apollo 15 the crew prepared to begin a well earned rest; there were still two days of orbital science to conduct before firing up the SPS to return home. But capcom Joe Allen was concerned to pass along a story regarding Capt. Cook's ship *Endeavour* and the suspicion that she too, like the spaceship, may perhaps have sprung a leak. Technicians at Australia's Honeysuckle tracking station provided evidence that indeed it had.

CAPCOM 'I'm going to go on if you're still listening to read some history that was sent to us by the Honeysuckle people. It was 11:00 pm on 11 June, 1770, a clear moonlit night, when His Majesty's Ship *Endeavour* under the command of Captain James Cook, sailed serenely under full sail within the waters of the Great Barrier reef off Australia's northeast coast. Then disaster struck. The ship had got upon the edge of the reef which lay to the northwest . . . The captain, clad only in drawers, rushed on deck. He summoned all hands to the pumps and ordered all unnecessary stores to be thrown overboard . . .The

original *Endeavour* was finally freed from the reef by means of oakum and wool wrapped in a sail sunk under the ship and plugged into the hole in the hope that it would be sucked into the leak . . . The experiment was entirely successful and I quote from Cook's diary, "In about a quarter of an hour to our great surprise, the ship was pumped dry and upon letting the pumps stand she was found to make very little water," unquote. Subsequently the *Endeavour* arrived at the Australian mainland . . . and after two months the damage had been repaired and the ship returned to England, and that's the end of your history lesson for today. Over.'
SCOTT 'That's quite an analogy, isn't it.'
CAPCOM 'Quite an analogy, Dave. Certainly is.'

Shortly after waking, all three crewmembers participated in another light flash experiment where they reported apparent impact of cosmic rays with the optic nerve, and then set about the extensive orbital science programmed into J-series flights. The lone pilot aboard the CSM for three days of Lunar Module surface activity was a busy man compared to his predecessors, even busier now there were three men to share added tasks. A minor problem ailed the retractable mass spectrometer boom, and telemetry indicated that a height sensor on the panoramic camera was not working correctly. But all in all the operations were a great success. After another sleep period, at the end of which capcom Karl Henize woke the crew with the theme tune from '2001 – A Space Odyssey,' ground controllers scanned the Hadley Base site with the rover's TV camera; geologists wanted to see the area at high Sun angles and there was a limited amount of power from the batteries.

During lift-off the surface camera beamed pictures of the Ascent Stage firing away from the landing segment but the controller had elected not to elevate the mounting because problems during the third EVA seemed to presage total failure. Tests with the camera now showed it to be working properly in the closing hours of its life. An impromptu discussion evolved with senior geology advisers in Houston, Scott and Irwin enthusiastically breaking into a detailed conversation with Leon Silver and Jim Head at the capcom console.

Only one major event lay between the passing hours and trans-Earth-injection. At the rear of the SIM bay, Apollo 15 carried a very special payload – a small subsatellite which would be mation to ground stations long after the spacecraft left. To put Apollo 15 into the correct orbit for releasing the subsatellite, a shaping burn was conducted on revolution 73. Fired for 3 seconds, the SPS engine pushed the spacecraft into a 100 × 141 km path. More than two hours later the small 35.6 kg subsatellite was spring-ejected from the SIM bay to measure nuclear particles and electric fields. Powered by solar cells, the hexagonal structure, 79 cm long and 35.5 cm in diameter, would also measure the Moon's mascons and minicons – areas of gravitational variation within the lunar sphere. Constructed by TRW, the subsatellite carried three booms, one of which supported a magnetometer, and was spin stabilized during ejection.

Released just before starting home, this subsatellite was spring-ejected from the SIM bay to remain in lunar orbit.

One orbit later, at the end of revolution 74, Apollo 15 burned out of lunar orbit and began the long trip home. As the spacecraft moved quickly away from the Moon the crew spent several minutes looking at the spectacular sight and describing the view. For Moon-orbiting spacemen, this was one of those events not to be missed. After circling the Moon for six days the amazing sight of the complete sphere was relief and spectacle combined. Shortly after crossing the equigravisphere, and after another long rest, the crew put on their space suits ready for de-pressurizing the spacecraft; Al Worden was to go outside and work his way to the back of the SIM bay to retrieve cassettes and film magazines from the orbital science experiments.

This deep-space EVA would be a feature of all three J-series missions and required special equipment in the Command Module to accommodate the activity. Tethered by a 7.4 meter long umbilical with a tensile strength of 272 kg, Worden also wore an OPS emergency oxygen unit identical to the type fitted on top of a Moon walker's back-pack. Normal supply would flow through the umbilical but the OPS would be there if the primary umbilical failed. The spacecraft was depressurized at 241 hr 56 min and less than ten minutes later the side hatch was open. Making his way out along hand rails, Worden placed the TV camera on a special mounting, then moved back along the Service Module to retrieve the mapping camera cassette. Working back once more along the hand rail, he handed this to Jim Irwin standing in the open hatch who in turn passed the cylindrical drum back into the Command Module.

Worden's second trip was to collect the panoramic camera cassette and a third visit to the SIM bay was performed to inspect a probable fault with the altitude sensor in the pan camera housing and a sticking mass spectrometer boom. The EVA was over in 38 minutes. During re-pressurization Worden dumped the contents of his OPS into the cabin, assisting the environmental control system bring the Command Module back to 760 mmHg. During the EVA the CSM thruster quads likely to impinge upon the drifting astronaut were inhibited, but now the spacecraft could return to the passive thermal control roll it had been in before, with several operations of the spectrometers in the SIM bay.

Another sleep period began at 253 hrs, almost back on the usual Houston rest times for down on Earth at the Manned Spacecraft Center it was late evening, Thursday, 5 August – less than two days to go before splashdown. Tracking data indicated the level of precision achieved with the TEI burn out of Moon orbit and the first three mid-course correction opportunities were cancelled. After waking, more tests with the light flash experiment, more navigation sightings on stars, and the usual housekeeping activity. The customary press conference followed, where newsmen selected questions to be read up to the crew for a formal answer on the communication loop. The final sleep session began at 277 hr 30 min and ended when the Mission Control Center uplinked the music from a Hawaiian war chant, sending heart beats leaping on the physician's console.

There was just one small course correction to perform before separating for re-entry: a 24 second RCS thrusting maneuver to give the entry path a 6.49° angle. Fifteen minutes before entry interface the Service Module separated and the Command Module swung round to present its broad conical afterbody to the atmosphere. On down through descent the spacecraft performed as expected, putting out the drogue parachutes before three strong mains. Seconds later, however, one of the three parachutes collapsed. 'Apollo 15, this is Okinawa; you have a streamed 'chute. Stand by for a hard impact, Okinawa over,' came word from the recovery ship. It would later be discovered that residual propellant from the RCS engines, vented during the final minutes of descent, probably burned through several riser lines, causing

During trans-Earth coast, astronaut Al Worden (left) worked his way back along the SIM bay to retrieve cassettes of data from the instruments, depicted here by an artist's impression.

the parachute to fail. Slicing down faster than normal, the crew received a noticeable jolt when the Command Module came to rest in the one meter swell, about 10 km from the prime landing spot. The spacecraft remained in Stable I and the crew were retrieved after donning clean flight suits.

After landing on the USS *Okinawa* from a helicopter sent to pick them up, Scott, Worden and Irwin were given an extensive medical examination before flying, next day, to Hickam Air Force Base, Hawaii, and on to Houston's Ellington Air Force Base. Acclaim for the three men was wide and spirited. Much work had been accomplished on their 12 day 7 hr flight, where twice as much time had been spent out exploring the lunar surface than on any previous mission, where they returned with nearly twice as many samples, and where they travelled across the Moon accumulating almost ten times the distance traversed by their immediate predecessors. Despite several irritating problems, the mission had lived up to its expectations, and the crew revealed the measure of long weeks spent learning geology in several places across the United States.

Five days after splashdown, NASA formally announced that Scott, Worden and Irwin would be the back-up crew to the prime Apollo 17 astronauts: Eugene Cernan, Ronald Evans, and Harrison Schmitt. Selected as a scientist-astronaut in June, 1965. Schmitt would be the only professional scientist ever to make a Moon landing. To enable him to fly, Joe Engle, the LM Pilot from the Apollo 14 back-up crew, had to stand down. Schmitt was to have flown with Gordon and Brand on the now cancelled Apollo 18, but politics moved him back up to the 17 seat.

A bitter cloud paled the achievements of Apollo a year after the flight, when it was revealed that all three astronauts had passed, via a third party, 100 unauthorized covers carried to the Moon and franked for sale at exorbitant profit to dealers in Europe. Moreover, NASA's conduct standards document was contravened so as to set up a trust fund for their children. When officials found out about the deal, said to have netted more than $150,000, the astronauts pulled out of the arrangement.

NASA was deeply disturbed at the indiscretions, and promptly moved all three off flight status. Scott was sent to the Flight Research Center after serving a year as 'special assistant on Apollo' before forming his own company in 1977; Irwin retired altogether to set up the High Flight Foundation, a religious organizaton; Worden moved to the Ames Research Center as head of the Systems Studies Division before leaving to serve as President of the Americans for the National Dividend organization. What Congress found particularly despicable, when it investigated the sales, was that all three had profited by a deal to market replicas of the tiny 'fallen astronaut' figure.

But there was another side to post-flight activity. In investigating the 77 kg of samples returned from Hadley Apennine, scientists found what they believed to be the oldest rock yet returned from the Moon. Called the Genesis Rock, it was aged at 4,150 million years. 'This is unquestionably one of the most important rocks ever returned from any lunar mission,' said selenologist Dr. Leon Silver. 'It has a recording in its minerals of many series of events in lunar history – events of high temperature and high pressure.'

It was apparent from the moment the rock boxes were opened in the Lunar Receiving Laboratory that a new age of lunar science had indeed dawned. What was to be gathered from the next two sites would keep scientists busy for more than a decade to follow. It would have to – for as the summer of 1971 gave way to the annual fall budget tussle, it became apparent that NASA would be hard pressed to retain a manned capability at all, let alone a resumption of Moon exploration.

CHAPTER TWENTY-FOUR

On the Shoulders of Giants

While Apollo 15 epitomized newly acquired growth in lunar exploration techniques, prospects for manned flight after Skylab Earth orbit operations dimmed noticeably. The last of three Skylab visits would finish in late 1973 and the proposed space shuttle was unlikely to fly in orbit before 1978. From an original expectation of manned operations commencing in 1975, NASA had been forced to postpone the date because of delay in getting approval to start work on the project. Since early 1970, major US aerospace corporations had been working hard to define a shuttle configuration acceptable both technically and economically, but the constricting financial limits placed on the program by the Office of Management and Budget (OMB) kept engineers busy trimming the design and shaping it into a cheaper proposition. Extensions applied in June, 1971, to definition studies by McDonnell Douglas and North American Rockwell were further extended that October for a period of not less than four months and a possible duration of six months.

During the fall, while NASA heard from the OMB that its projected needs for Fiscal Year 1973 (to begin 1 July, 1972) were unacceptably high, the space agency pitched full tilt into detailed examination of several cost-saving schemes in the shuttle configuration. In January, 1971, in an attempt to woo Defense Department support for the reusable transportation system, NASA met with military planners at Williamsburg, Virginia, to hear suggestions about the necessary payload requirement for this winged vehicle. The civilian space agency believed it would need a shuttle capable of sending 11 tonnes of cargo (payload) to an orbiting space station 500 km above Earth. But the Air Force needed a vehicle capable of carrying 29.5 tonnes to a low Earth orbit. Moreover, military requirements demanded high maneuverability in the atmosphere during descent to allow the shuttle to fly back to its launch site after only one revolution, during which time the planet would have spun several hundred kilometers east.

High payload and major flight path control requirements implied higher development costs. And this was one of NASA's problems during 1971: to accommodate the Defense Department requirements, and so obtain their support when NASA came to present the project at Congressional committees, and to fit the configuration within OMB's financial limits. So the industrial contractors came up with several candidate schemes that looked, late in the year, as though the agency would match the demands. But would the Defense Department support be sufficient to convince cost-conscious lawmakers on Capitol Hill? To further enhance their chances for the coming battle, NASA commissioned a financial assessment from analysts Mathematica Inc., and got the company's head, Oscar Morgenstern, to work with finance expert Klaus P. Heis on an economic evaluation.

NASA was firmly resolved to present the shuttle as a reusable replacement for all the many expendable rockets used to launch satellites and space probes into orbit. Mathematica had done a preliminary study in August, 1970, that showed great cost advantages in developing a reusable launcher. It certainly made good sense, space station or not, to build a launch vehicle capable of many flights back and forth to space. The new study was even more favorable to the shuttle system, and NASA stiffened its resolve not to let another financial year slip by without getting approval for a formal start on the project. While the OMB tried to slam the approval door in NASA's face yet again, Administrator Jim Fletcher went straight to the top, flying out to President Nixon's retreat in San Clemente, California, for a meeting on 5 January, 1972.

It was an historic get together. Shortly before noon the President's response was announced: 'I have decided today that the United States should proceed at once with the development of an entirely new type of space transportation system designed to help transform the space frontier of the 1970s into familiar territory, easily accessible for human endeavour in the 1980s and 90s.'

By the end of the following month NASA had technical studies to hand which enabled it to decide the definitive configuration. On 15 March, 1972, a formal announcement presented the choice everybody recognized as the obvious contender. A reusable orbiter propelled by engines carried in its tail, fed with fuel and oxidizer contained in a separate external tank, assisted for the first leg of its ascent by two solid propellant boosters. Each winged orbiter would make at least 100 flights into space, carry the stipulated 29.5 tonnes to low orbit, make a controlled descent through the atmosphere, and land like a conventional aeroplane. As for the space station which inspired shuttle's origin, that was now relegated to a modular concept for some distant date. In hearings before House and Senate space committees in early 1972, Jim Fletcher would assure congressmen that NASA did not anticipate spending any money on further space station studies or proposals. The idea was officially dead.

Space shuttle would marry manned and unmanned programs by relegating astronauts to the role of ferry pilots; if and when a desirable reason could be found for setting up a manned orbital facility, it could be developed then, but only when financially expedient and not as a concurrent appendage to the shuttle. It was a far cry from George Mueller's 'integrated space program'. Even the nuclear rocket engine was on its way out. Reduced in power from 38 tonnes thrust to 6.8 tonnes during the closing months of 1971, the entire project would be shelved little more than a year later. Without any missions in need of the enhanced performance, there was simply no point in pressing on with development. Yet it was a project as old as NASA itself, embodying grand objectives with bold initiative.

The protracted shuttle schedule was due in part to a need to conserve funds in case a flight was scheduled between the end of Skylab and the first orbital flight with the reusable launch vehicle. The program being worked out would rapidly provide a shuttle orbiter for air-launched tests in the atmosphere – glide tests from a mother-plane – while delaying by a year full development of the model destined for space.

At the Manned Spacecraft Center, Glynn Lunney and René Berglund were busy preparing plans for the proposed Apollo/Salyut flight, aiming, for the purpose of the study, toward a launch in the second half of 1974. Christopher Kraft, promoted to the post of Deputy Center Director in December, 1969, called for a time schedule detailing preparation of flight hardware by September, 1971, directing Caldwell Johnson to mobilize his design department to produce a docking adapter configuration two months later. It was to be ready for the next meeting with the Russians. Bob Gilruth was pushing for as much engineering detail as possible, and Johnson assigned Clarke Covington to coordinate engineering and Bill Creasy to work on the androgynous docking equipment. The airlock module itself would be worked up by James C. Jones.

Dubbed the International Rendezvous and Docking Mission, René Berglund prepared a work statement for North American Rockwell which was delivered in July, 1971, authorizing a four-month study of appropriate hardware to support the flight and requesting recommendations on the specific CSM the company felt it appropriate to modify. It was proposed that the mission be launched by a Saturn IB, that it carry the docking module in the adapter normally used to house the Lunar Module, that the flight should include up to two days linked with a Salyut, and that the total duration of the Apollo segment should be two weeks. Meanwhile, Gilruth set up a MSC task group under Berglund to organize the study.

The death of three Soviet cosmonauts aboard Soyuz 11 certainly cast doubt on the continuing interest the Russians might have in cooperating with a joint flight. But in setting up the proposed November meeting, Glynn Lunney wrote to Bushuyev: 'This sad accident has further strengthened our emphasis on the solution of the common docking problems.' The Russians

Apollo Block II Command and Service Modules.

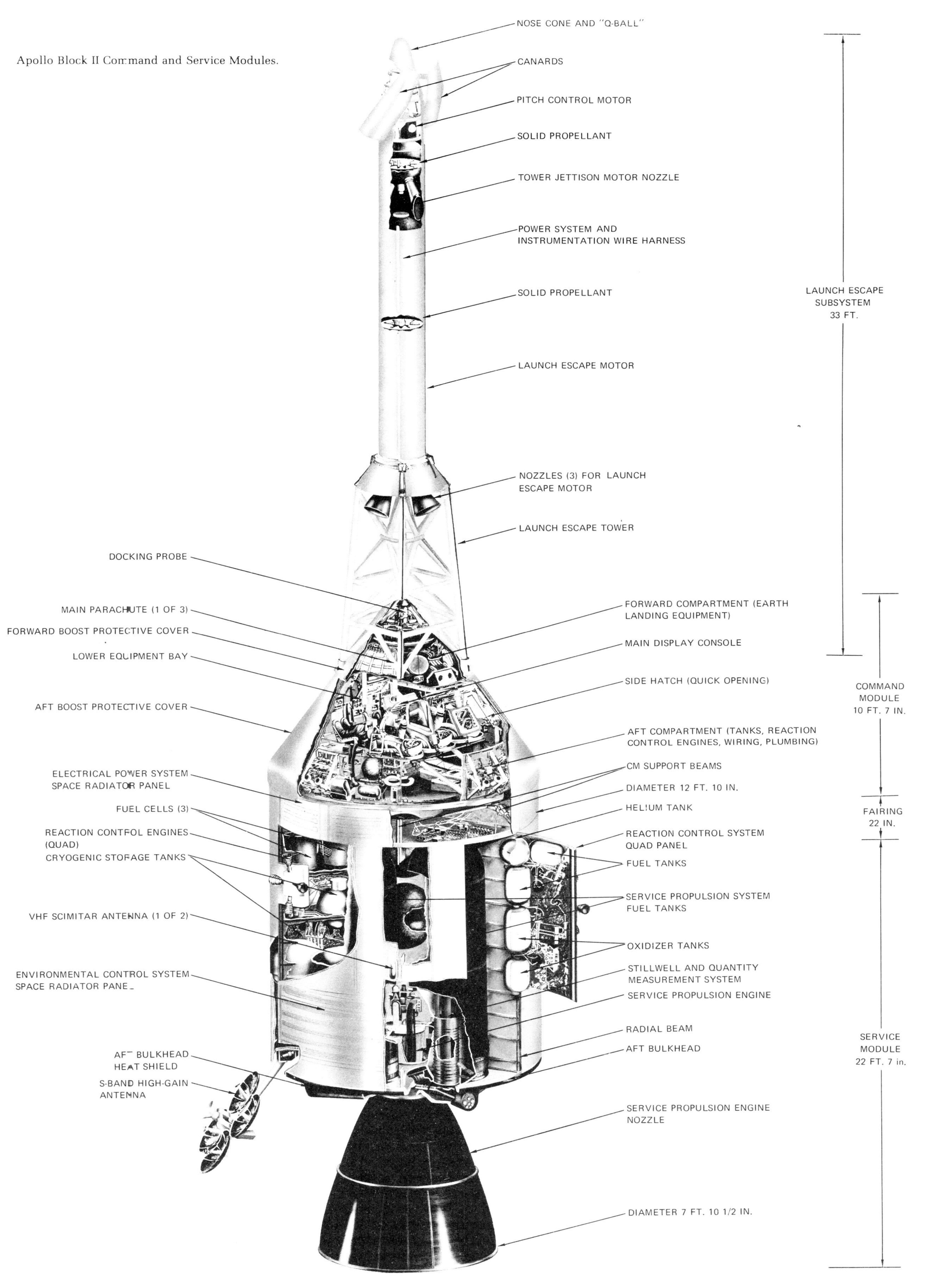

Representative of the configuration finally selected for NASA's Shuttle, this proposal from Grumman Aerospace was unsuccessful in getting the definitive production contract.

responded by agreeing to have the American team visit Moscow between 29 November and 7 December, 1971. Not a few had held their breath, wondering how the Soviets would take their bereavement. As it turned out, they need not have worried, for the Russians had much to gain from a careful scrutiny of NASA operations and were as eager to gain access to mission activity as America was to continue manned flights between Skylab and shuttle.

The need for a separate docking module arose from the different gaseous environments of the two vehicles: returning from the sea-level mixed gas atmosphere of Soyuz to the pure oxygen reduced pressure environment of the Apollo Command Module would necessitate a four-hour period of pre-breathing. Moreover, the docking module would have to withstand the sea-level pressure environment of the Salyut since the two pressures would have to be compatible during movement from the module to the space station. Externally, thought the design team, the docking module should be 2.54 meters long between docking equipments, 1.42 meters in diameter on the interior, with a probe and drogue docking unit on the Apollo end and a new universal (androgynous) docking unit at the opposite end.

In letters from Moscow during September, Bushuyev stressed the importance of looking after 'technical requirements and solutions for long term capability,' rather than being concerned merely with the one proposed Apollo/Salyut mission. Bushuyev knew as well as his American counterparts that approval might be hard to get for continued missions unless they developed a need for further cooperation at an early date.

When Glynn Lunney took his party to Moscow on 27 November, they were warmly received and well presented by the official Soviet news media. With them came detailed plans for the compatible docking system and the proposed docking module. Bob Gilruth was there too, for it was his Center that would develop the US segment. Petrov was reluctant to have his men involved in a NASA suggestion that regular telephone conferences be conducted between the two countries, saying it was too expensive. In reality, Soviet bosses only agreed the idea of a joint flight on the condition that all communication was first passed through them for approval. During the early months of negotiations it had become clear why Russia took so long to move ahead in manned flight activity.

As talks got under way at the November session dates of possible launch opportunities moved into 1975. Nobody knew for sure that even this date was acceptable until official sanction had been sought and obtained, but it was a working figure to go with until then. Bill Creasy's double ring and cone docking concept, incorporating four guide vanes around the periphery, was examined by the Russians but for more than a year now the Soviet docking specialist Vladimir Syromyatnikov had been working on an almost similar design using three guide vanes. In this area too, the men from Houston learned they were there to compromise and not to dictate. The compromise was that the Soviet concept would form the baseline for further engineering studies. There were just seven months before a final design had to be decided, and the universal docking unit was one piece of equipment that had to be the same from both countries for it would be used on NASA's docking module and Russia's Salyut.

On 7 December the American delegation flew back home, travelling via England to visit friends and colleagues. Preliminary costing showed a preference for having North American Rockwell build the docking module; they could do it cheaper, and they would provide a satisfactory interface with the Command Module since they also built the CSM. When it came to choosing a specific model off the production line, CSM-111 was the most appropriate, but that was one of the earlier Block II batch that did not have a SIM bay in sector 1 of the Service Module. NASA wanted to maximize return from the flight by having it perform detailed Earth observation after the docking exercise. CSM-115 and -115A were the obvious choice, since they did have a SIM bay (one serving as the flight article and one as back-up); CSM-116 to -118 were already chosen as Skylab vehicles.

But use of the late production vehicles would cost up to $280 million and at a meeting with George Low, Dale Myers, Willis Shapley and Arnold Frutkin on 24 February, Administrator Fletcher authorized the team to proceed within a ceiling of $250 million. The SIM bay experiments could not be flown because CSM-111 was the only vehicle that could perform the flight for that cost; already built, it would not need additional work demanded by the later models. So all work stopped on CSM-115 and -115A, while CSM-111 and a back-up (CSM-119) went ahead.

By this time, Bob Gilruth had left the Manned Spacecraft Center to take up a position as Director of Key Personnel Development at Headquarters. Chris Kraft moved in as MSC Director when Gilruth left on 14 January, 1972. For more than thirteen years he had looked over and directed the several thousands employed first at the Space Task Group and then the Manned Spacecraft Center. Bob Gilruth was in every important respect a counterpart to von Braun, for where the latter set up America's big booster program the former constructed manned space flight from theoretical beginnings to practical lunar exploration. Now it fell to Chris Kraft to marshall Houston's resources for the first international docking flight.

Vladimir Syromyatnikov headed the delegation from Moscow at MSC between 27 March and 3 April. Engineering drawings were examined and the docking specialists got down to work. It was the final determination of basic concepts to clear the way for a start on fabricating the hardware. But things were moving almost too quickly. On a matter of such international importance, Congress would have to debate and approve authorization for the flight, and assign funds for the work. A start in this direction had been made during debate in the White House; Henry Kissinger asked NASA to prepare a proposal by mid-April.

After conferring with Jim Fletcher, George Low decided to take Glynn Lunney and Arnold Frutkin to Moscow early in April to meet with the Soviet Academy of Sciences where a set of documents could be put together for formal presentation to heads of state. At Houston and in Washington, Lunney and Low made covering excuses for being 'out of town' in the period they would fly to Moscow so as to prevent any discussion or press reports. It was the final reckoning time. If the Russians really were interested in a joint flight, the trip would produce positive results. George Low was quite excited to be caught up in this intrigue and Glynn Lunney remarked on how he felt like a certain very famous British spy!

Petrov and Bushuyev were at Moscow's airport to meet the trio and in the car on the way to the city centre, Petrov informed the group that Prof. Keldysh had been placed in hospital, his role now taken by one Vladimir Alexandrovich Kotelnikov. It was Monday, 3 April an Easter day in Christian countries. At the American Embassy, the trio met science attache Jack Tech and received an invitation to dine with Ambassador Jacob Beam and his guests. George Low was disturbed to learn that Ambassador Kaiser's son Robert would be there. As Moscow correspondent for the *Washington Post*, Robert Kaiser could spill the whole story. Low was assured that the event was a social one and that there would be no need for Kaiser to know anything about the nature of the visit.

Next day, during the afternoon, Low, Lunney and Frutkin met with Kotelnikov, Petrov, Bushuyev, Vereshchetin and I. P. Rumyantsev. Continued at the Club of the Scientists, the meeting went on until the early evening. Next day, Wednesday the 5th, more discussions were held and after a snack at the Embassy, the group of Americans had another session with their Russian hosts before concluding the agreement that afternoon. Frutkin and Vereshchetin were left to edit the text. On Thursday, the two parties signed the agreement in Kotelnikov's office at the Soviet Academy of Sciences before being taken to the room in which Napoleon slept during his last hours in Moscow.

Kotelnikov had been precise in telling the delegation that it proved technically difficult to plan for a Salyut docking and that the Russians wished to plan for a Soyuz docking. The Soviets were firm in their determination to honor the intent of the agreements but equally resolved not to consider use of the Salyut space station. Over the coming months and years it would be apparent that Salyut had a considerable military role to play and that the laboratory was not exclusively preserved for civilian or scientific activities. Low agreed that an Apollo/Soyuz flight would be acceptable but all three found it difficult to get open cooperation from the Russians about lines of communication.

Nevertheless, 17 points of agreement were written up and signed, including issues as diverse as who would be responsible for flight control if an abort was necessary to how much information should be released to the public; the Americans were adamant on that last score: there would be total exposure to the US segment but overall news distribution would conform to the traditions of the respective countries. It was a significant step forward in opening the hitherto closed flight operations mounted by the Soviets, allowing the west to get detailed information on how Russians control their manned missions.

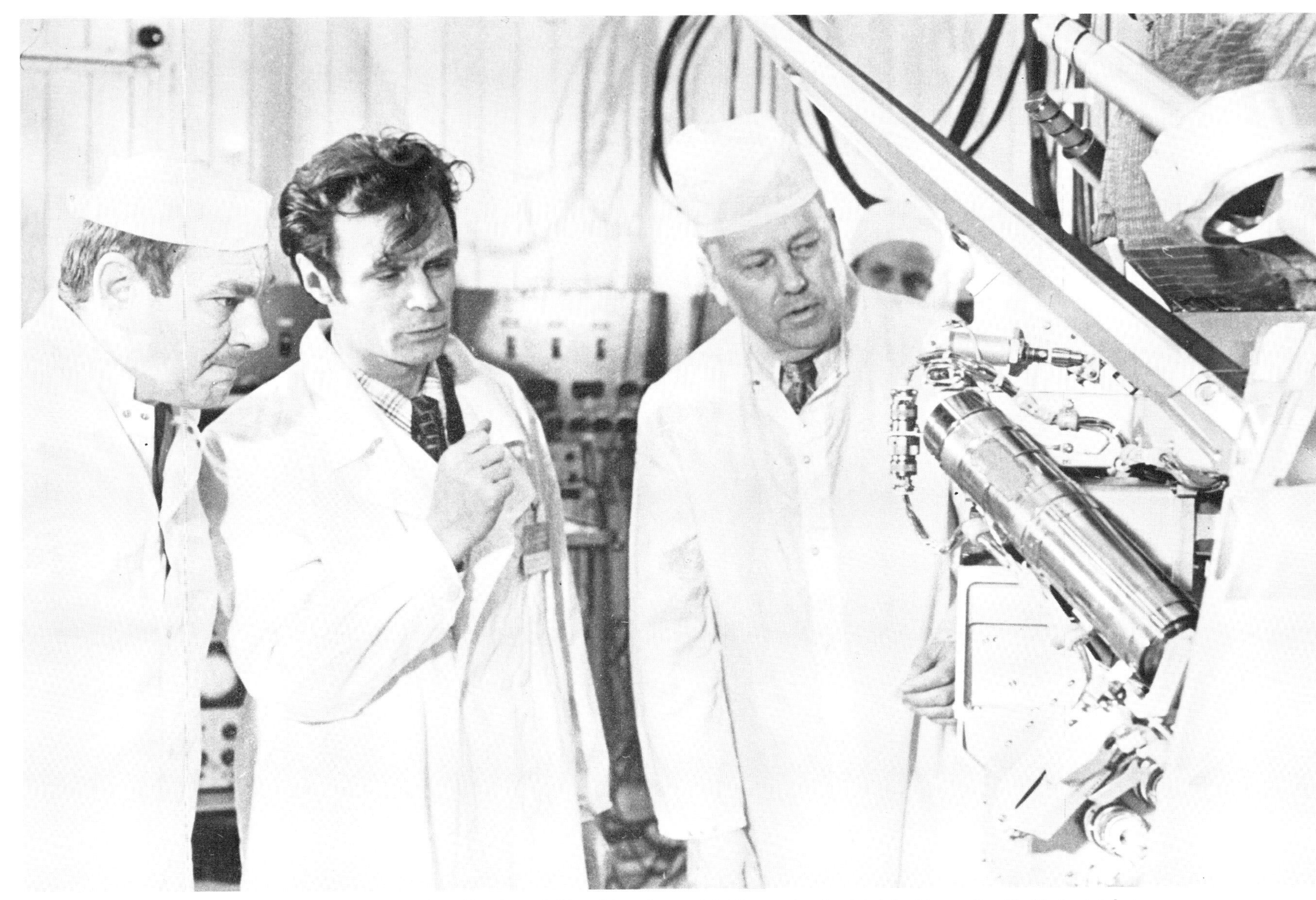

Rockwell International ASTP management representatives Ken Love (left) and Ray Larson (right) examine the joint docking unit with it's Soviet designer Vladimir Syromyatnikov.

Astronaut John Young, commander of Apollo 16, participates in a Kennedy Space Center training session with the Far Ultraviolet Camera designed to map from the lunar surface deep-space sources of hydrogen.

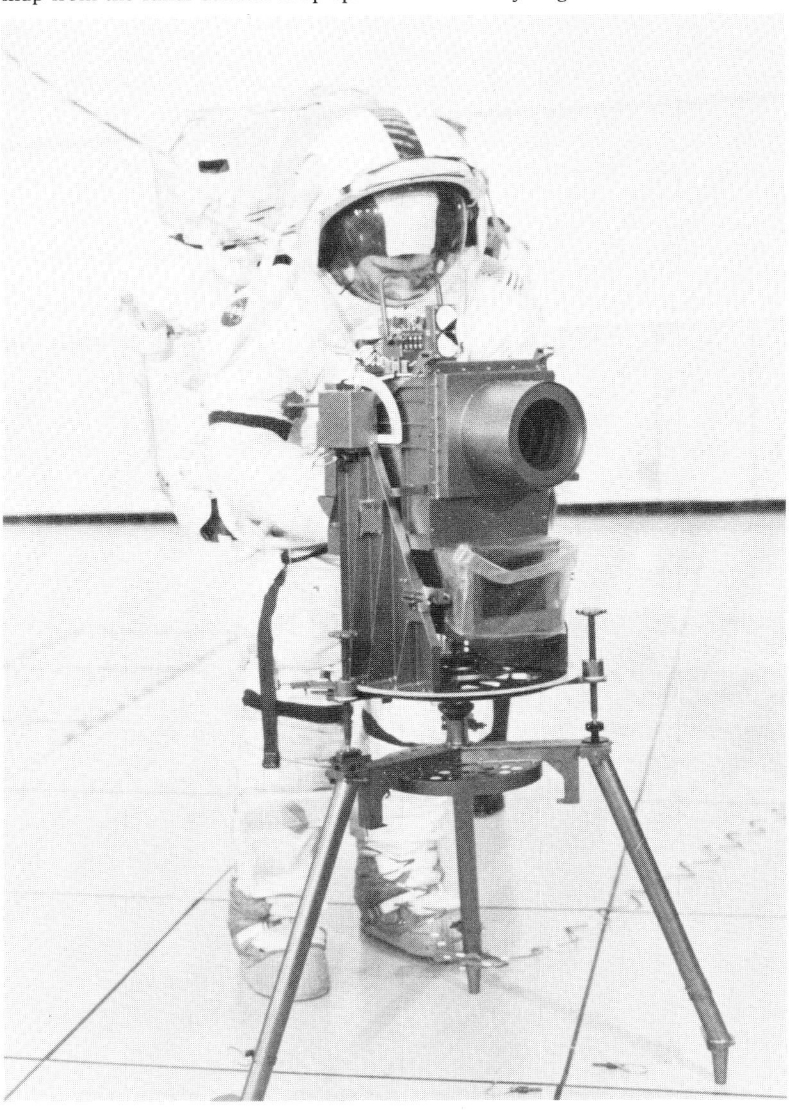

When Low's party returned to the United States confidence was strong that nothing now stood in the way of a joint Apollo/Soyuz flight. Only the State Department could now set up the final links in the chain leading to an actual flight in 1975. Henry Kissinger was briefed on the negotiations, and on the progress made by NASA in resolving lines of communication and technical agreements, with a view to the President incorporating a joint docking flight in documents signed as a result of the successful negotiation of a Strategic Arms Limitation Talks treaty. President Nixon would meet with Premier Kosygin the following month to sign SALT-1. It was the perfect setting against which to organize the cooperative venture, a very real milestone on the road to peaceful coexistence and one so very different from the mood that triggered the several space projects mounted by America and Russia.

But at the Manned Spacecraft Center and at the Cape, preparations were in an advanced stage for the penultimate lunar landing. John Young, Ken Mattingly, and Charlie Duke, had been selected for Apollo 16 on 3 March, 1971, and during the weeks following this announcement the space agency determined that the Lunar Module would descend to a rough area known as Descartes, much photographed on earlier flights. To reach this site the docked vehicles would fly an orbit inclined only 9° to the lunar equator, versus the 26° of Apollo 15, and descend across mountainous terrain located in the lunar highlands. Samples would be sought from two different types of material, both volcanic, which apparently moved out across the surface at different times.

Highlands are characterized by the light, highly reflective, areas of the Moon and are thought to comprise sections of the original crust; it was to study these primordial features that the crew were assigned a site near to the crater Descartes. Originally scheduled for launch on 17 March, Apollo 16 was put back one month, three days after Charlie Duke entered hospital with bacterial pneumonia. But Duke's temporary illness was not the only reason for deferring launch. Several minor problems arose which threatened to crowd timely preparation of personnel and hardware.

One of these brought new demands to the EVA suit, where Young and Duke would bend many times to pick up samples or rocks. A new restraint cable was necessary and re-design of the lunar boot taping layer. Also, in testing batteries designed for the LM, engineers discovered current fluctuations that prompted a re-design in certain battery components with a resulting demand for lengthy qualification tests. Finally, in tests associated with a docking ring destined for a Skylab mission, failure to separate effectively from the mounting, as it would be required to do during jettison, revealed voids in potting used to restrain strands of pyrotechnic charge. That too suggested a need for re-work on 16's equipment, and again more testing.

Changes too were made necessary when one of Apollo 15's three main parachutes collapsed after deployment. It was not possible to determine the exact cause of the failure, although many thought propellant jettisoned during descent was responsible while others pointed to flaws detected in links connecting suspension risers in the one canopy actually recovered. To prevent this happening again, new parachutes were supplied with Inconel links instead of connections made from steel.

Launch vehicle 511 was stacked in the Vehicle Assembly Building during October, 1971, followed two months later by CSM-113 and LM-11. Packing aboard the two spacecraft all the many items and packages necessary for the mission was an arduous task, and one that brought a change of emphasis to experienced KSC personnel. Where once they learned to bring up a Saturn V/Apollo in record time, they now acquired the art of assembling and stowing an unprecedented range and number of intricate items. Boeing had the second rover on dock at KSC just two months after Apollo 15 flew, and that too brought headaches to the stowage chiefs.

The Saturn V was just a little different from its predecessor in that four first stage retro-rockets removed from 510 were put back on 16's launcher because insufficient separation had been observed with only four rockets installed. AS-511 was rolled to pad 39A on 13 December, 1971. But on 25 January, two weeks after the new April launch date had been announced, human error in failing to seat a pressure relief valve properly caused a bladder inside one of the Command Module's RCS propellant tanks to rupture. Installed on the pad the tank was impossible to replace. It would have to be taken back to the VAB where the Command Module could be removed and taken to the Manned Spacecraft Operations Building for a new vessel.

Back went 511 on 27 January and a day later the spacecraft was in the MSOB. Technicians had earlier believed the replacement task would threaten an April launch but Walt Kapryan and Rocco Petrone organized a work schedule that had 511 out again on pad 39A by the evening of 9 February. Just eight days later, human error again caused two pressure discs on oxidizer tanks to rupture but these could be replaced on the pad.

As with earlier crews, the Apollo 16 astronauts were quarantined for three weeks before launch to reduce the risk of illness. But concern too was expressed by Dr. Berry's medical team over fatigue observed in the Apollo 15 crew as a result of the strenuous three days on the lunar surface and cardiac arrhythmias observed at several points in the mission. Because the flight demanded work loads higher than normal over long periods the crew displayed physical signs that they were approaching a dangerous level. To alleviate this kind of body impact, Dr. Berry organized a special diet for Young, Mattingly, and Duke which would ensure adequate potassium levels, depletion of which was judged to be primarily responsible for Apollo 15's experience. But the health and welfare of the Apollo 16 crew was not alone the sole area of medical attention during spring 1971.

Starting back in April, 1967, Deke Slayton was given a twenty-eight month course of Quinadine in an attempt to stop his heart fibrillating, a condition observed since he became an astronaut in 1959 but for which there had been no apparent cause. At the end of the medication Slayton was observed regularly and no further episodes of fibrillation were noted. In December, 1971, he was taken to the Mayo Clinic by Dr. Berry and given extensive medical tests in the laboratories, where he was observed for long periods doing tasks that had earlier caused him to fibrillate. Special care was taken to compare the left and right heart chambers, and to study the blood flow and supply under varying levels of stress. At the end of the day it was not possible to find abnor-

Flight director Eugene F. Kranz seated at his Mission Operations Control Room console on the morning of the launch of Apollo 16. The flight would be controlled from Houston seconds after lift-off.

mality of any kind whatsoever. As a result, Dr. Berry signed him fit for Class II flying status according to rules laid down by the Federal Aviation Agency.

Released from his operational quarantine on 13 March, Deke Slayton was back in the air. But was he fit for space flight? Asked that very question, Dr. Berry said that, 'from a medical point of view, if we are willing to clear him for flying in aircraft with what we know at this time, then he is placed back, as far as I am concerned, into the pool for that type of consideration.' It had been almost exactly ten years since Slayton received word that as far as medical opinion was concerned at the time he would probably never fly in space. Now he was back on the flight list and although the crew for Skylab's three manned visits had been selected in January, if all went as expected there would be three seats available on the joint US/Soviet docking flight. It would probably be the last chance for Deke Slayton; if that flight went in 1975 as tentatively planned, Slayton would be 51 years old.

At the Cape, meanwhile, preparations moved ahead for the launch of Apollo 16. After a successful flight readiness test on 1 March, engineers attended a special meeting where John Young and Charlie Duke explained the need for Apollo 16 and what science hoped to gain from it. CDDT activities kept crews busy until the end of the month and the final countdown began 10 April, six days prior to launch. The flight was to begin at the comfortable time of 12:54 pm and being a Sunday, 16 April, was expected to draw near record crowds. The weather put the final touches to the event and several thousand people flocked in to Brevard County, sending the Sherriff's office into action like it had on so many previous occasions.

More than 7,500 special guests were at the VIP viewing area, a further 38,500 were on the roads and parking lots around the site, and in the Launch Control Center Jim Fletcher and George Low hosted Spiro Agnew. Outside, President Eisenhower's son David and his wife were there with officials and dignatories, including King Hussein of Jordan. The launch window would last 3 hr 49 min but final preparations went well and AS-511 looked good for an on-time flight. Only minor problems plagued an otherwise smooth countdown: a gyroscope fluctuated in the Saturn's instrument unit, and a pressurization tank seemed to be leaking, but no danger was suggested and on the preparations went.

This was a heavier Saturn than its predecessor, grossing 3,185 tonnes at ignition, but additional thrust from the improved F-1 engines helped compensate and when lift-off came at the beginning of the launch window the vehicle lived up to its reputation. But although it was only one of several lunar landing flights, and not even the first J-series mission, Apollo 16 provided a notable 'first' in that Soviet poet Yevgeny Yevtushenko attended the launch. On the evening before the flight began, Dave Scott took Yevtushenko to see the vehicle standing bathed in light from the many searchlights trained upon it.

'It's really a beautiful show, this white tender body of a rocket, supported by the clumsy but sometimes tender hands of the red gantry tower. I absolutely had the feeling of one big brother embracing his sister before a long way, a long road,' said the poet. 'It was wonderful. Silence not people. No press. Nothing. The sky, the ground, the rocket. It was so beautiful.' When he saw the launch, Yevtushenko was visibly moved and greatly impressed. But for one spectator this was to be his last official mission. James McDivitt, Apollo Spacecraft Manager, announced his retirement from NASA and brought heady technology down to Earth when, in explaining why he was not waiting until the flight of Apollo 17, the last Moon mission, he said that it would enable his children to start their new schools from a new home at the beginning rather than the middle of a school term; astronauts, it seemed, were just normal family men after all.

The flight of Apollo 16 was characterized by several minor problems and malfunctions, each one of which could have prevented the mission completing its tasks. Gene Kranz was on shift with his white team in the MOCR, Gordon Fullerton the capcom. Within the first few minutes after getting into orbit ground controllers asked the crew to check the glycol reservoir bypass valve since telemetry indicated a leak in the primary spacecraft coolant loop. Over Australia on the first orbit, problems developed with the S-IVB stage's two attitude control modules. First the No. 2 package experienced a helium regulator failure which caused the gas to vent overboard continuously and then a helium leak was detected in the No. 1 module. Moreover, the instrument unit between the spacecraft adapter and the S-IVB stage leaked gaseous nitrogen from a bottle supporting the temperature control system. The crew were asked to prepare for attitude control with the Apollo thrusters just in case helium was depleted. But that was not necessary and TLI put Apollo 16 on course for the Moon following a 5 min 42 sec burn performed over the Pacific with communication through an ARIA tracking aircraft.

Six minutes later the CSM separated from the adapter, moved away 15 meters, turned round, and headed back to dock with the Lunar Module still gripped by the bottom of the shroud. Within 3 meters the crew could hear the thruster exhaust impinging upon the top of the LM and saw the thermal insulation rippling as the gases breezed across the flexible surface. Slight discoloration could also be detected. Television pictures sent to Earth showed the docking activity between the two vehicles but as the CSM moved slowly in John Young reported a stream of light-colored particles flowing past the window: 'Man, it just looks like a picture book from up here, Gordo. We must have a zillion particles along with us.'

When the probe's three small capture latches snagged the LM's drogue the CSM was moving in at only 1 km/hr. Nobody heard or felt the moment of soft-capture, only the talk-back indicators slowly rolling gave proof that the two were hooked together. When the probe was retracted, however, the thumping sound of main docking latches rigidly clamping the structures in unison reverberated through the spacecraft. Less than one hour later Casper and Orion – the CSM and the LM respectively – pulled free from the ailing S-IVB; although adequate for the TLI burn, helium loss would inhibit accurate targeting for the planned lunar impact.

Then the crew had a chat with the ground about particles seen venting from what appeared to be discrete locations around the LM. It turned out to be paint flakes from a number of places. All J-series Lunar Modules carried 16 panels on the Ascent Stage painted white to reflect heat and reduce the temperature of attitude control propellant stored in tanks beneath the thermal shield material. Tests later revealed that the paint flaked at surface temperatures below −84°c. The flaking particles would be no problem, except when Charlie Duke wanted to use the optics for a set of navigation marks when the flurry of so many reflecting segments would confuse the star field.

At 8 hr into the mission, Apollo 16's crew got the word to open Orion for a quick examination, just in case reports of venting presaged something more ominous. With TV on the astronauts opened the tunnel and checked the LM, finding all well but with considerable flaking visible through the windows. Several hours later the crew began their first sleep period, waking at an elapsed time of 23 hours; for much of that 'day' they would be concerned with more material flaking away, questions to the ground about the thermal validity of exposed surfaces on the LM, and the possible impact it would have on the surface activity. Houston felt there was little to worry about, it was something that they would have to live with.

An unusual experiment was conducted as Apollo 16 passed the halfway point in distance: an electrophoretic test like the one performed aboard Apollo 14. But Apollo 16 was carrying quite a few strange passengers. A rock collected by Bean and Conrad more than two years earlier was being returned to the Moon to see if it acquired a certain type of magnetism thought to have originated elsewhere. If it came back without that component of magnetism the case would be proven. The spacecraft also carried Biostack, a German experiment containing selected biological materials to be exposed to the cosmic rays of space. Most unusual of all was the microbial environment container housing 60 million passengers lodged on five strains of bacteria, fungi, and viruses. Put out on the EVA camera boom during Mattingly's planned trans-Earth space walk it would provide information on the effects of reduced oxygen pressure, weightlessness, and ultraviolet radiation.

Apollo 16's non-return trajectory was close to the predicted pericynthion of 132 km, but to lower the path from an actual predicted pass 252 km from the far side, a course correction burn was made at 30 hr 39 min where the SPS engine burned for 1.8 sec, lowering pericynthion to 133 km. As the flight plan dictated, the crew performed their first scheduled checkout of Orion beginning at 33 hr 15 min with removal of the hatch. Less than two hours later they were back in Casper. No loose particles could be found, although a few nuts and washers did float out from nooks and crannies, left there by an obviously deficient assembly inspection.

Then a master alarm sounded, indicating a problem with the inertial measurement unit. But that was soon recognized as a probable short and it did not impact the mission. The second sleep period ended a little early at 46 hr because another IMU warning flashed on the Houston consoles; again, it was an intermittent short and caused no problem other than the suspicion that other components of the guidance system might suffer a malfunction. It was time to do another light flash experiment, but this time the crew had a sophisticated device called the ALFMED – Apollo Light Flash Moving Emulsion Detector. Resembling two flat boards hinged like a book, the device was placed in front of the face so that permanent records of actual cosmic ray tracks etched in the emulsion on each surface could be correlated with verbal reports of flashes.

The next checkout performed in Orion came at 53 hr 30 min and lasted more than 1½ hours after which Young and Duke returned to Casper. Charlie Duke found his suit very tight and requested permission to slacken laces, not granted because it was felt the suit would fit when pressurized in a vacuum. Crossing into the Moon's sphere of gravitational influence nearly 2½ days into the mission, Apollo 16 cruised to the Moon as the crew began their last trans-lunar night. They got a lie-in because the last course correction opportunity was not needed, finally moving about their duties at 67 hr 15min.

Gerry Griffin's gold team was in the MOCR for lunar-orbit-insertion but with 21,000 km still to go the SIM bay door was jettisoned with considerable material streaming out in accompaniment. Fired free with a noticeable bang the door drifted away at 15 km/hr. Just 4 hr 18 min later Casper and Orion, firmly clamped together, swept out of sight around the Moon's west limb. It was 74 hr 17 min elapsed time; mid-afternoon in Houston, Wednesday, 19 April. Eleven minutes later Casper's big SPS engine lit up and burned for more than 6 minutes, braking the docked configuration into a 107 × 315 km path about the Moon.

'Hello Houston. Sweet 16 has arrived,' was the word from John Young when Casper's high-gain antenna once more caught sight of the Earth, 'super double fantastic burn.' In true circumlunar tradition, Apollo 16's crew spent the first front side pass enthusing over the view, with John Young reminiscing over targets on the surface he had seen the first time round on Apollo 10. Young was one of only two astronauts who would fly twice to Moon orbit; his companion on Apollo 10 was already hard at work training to lead the last lunar landing flight.

At the end of the second revolution Casper fired again the engine that had delivered it to an elliptical path, dropping the docked vehicles into a descent-orbit-insertion ellipse of 20 × 108 km. Apollo 16 would stay in that orbit until the vehicles undocked on revolution 12. But rather than wait a full revolution for Casper to circularize its path before beginning powered descent to the surface, Orion would ignite its descent engine one orbit early, on revolution 13, shortly after the CSM burned into a higher path. It was reorganized from the previous flight to get Orion on the Moon one orbit earlier than usual so as not to significantly affect the budgeted consumables; the LM was to have spent 73 hours on the Moon versus less than 67 hours by Apollo 15's Falcon.

After a good rest the three astronauts were up and busy by 91 hr 30 min, shortly before passing around the Moon's limb on revolution 9. But trouble with the mass spectrometer boom in the SIM bay gave Orion an extra task when it was decided to let the LM crew visually observe the boom after undocking. Electrical power came on in Orion at 93 hr 7 min and as the two vehicles passed round to start revolution 11, Ed Mitchell set up his own capcom console to talk to Mattingly in Casper while Jim Irwin got ready to talk to Young and Duke aboard Orion. With activity compressed into a tighter timeline, separate functional duties called for early separation of communication lines.

The LM crew had entered Orion 40 minutes ahead of schedule, but some of that advantage eroded as checkout continued. Nevertheless, the crew were waiting for acquisition of signal on revolution 11 before performing final tests. But it was not to be the smooth pre-landing preparation carried out on earlier flights. During a test of the steerable S-band antenna shortly after acquisition of signal at 94 hr 22 min trouble developed in the yaw axis, the antenna subsequently working only in pitch. But that could be worked around, with the crew manually inserting data that would normally flow directly to the onboard computer from Earth stations locked on to the high-gain antenna.

While Duke wrestled this problem, Young began to pressurize the RCS thruster systems but system A immediately began to rise uncontrollably. The regulator had clearly developed a leak and at 95 hr 5 min the crew made the first of four propellant transfer operations, moving liquids from the RCS tanks into the Ascent Propulsion System tanks to make room for the leaking helium. With those two problems out the way all seemed set for undocking about 3 minutes prior to coming round on revolution 12. Wit a cry of and we're sailing free,' Orion came round on time just a few meters away from Casper. For the balance of that front-side pass the two spacecraft prepared for Casper to circularize its own orbit on the Moon's far side, and for Orion to get ready for powered descent to the surface twenty-six minutes after acquisition of signal.

But when the two spacecraft came into view again on revolution 13 Ken Mattingly had a sorry tale to tell. In the closing minutes before firing up the SPS engine to circularize the orbit of his spacecraft, unexpected oscillations appeared in the secondary yaw gimbal servo. The motor employed to physically change the angle of the SPS engine, and keep its line of thrust through the spacecraft's center of mass, appeared to oscillate. This could mean that the secondary, or back-up, loop was inoperative and if that was so the Moon landing would have to be called off since Apollo 16 would have no contingency against failure in the primary system. Not that anybody suspected the primary loop; it was one of those ironies of space flight that a back-up *has* to be there – just in case.

Scientists in the Mission Control Center's ALSEP room examine a seismic reading of the Apollo 16 third stage (S-IVB) impact on the lunar surface, 3:00 pm Houston time, 19 April, 1972. Dr. Gary Latham (kneeling) studies a segment of the trace with Mission Director 'Rocco' Petrone looking on (long-sleeve shirt).

But could the engine rely on the juddering control system? Until engineers could come up with an answer the landing would have to be postponed. But it could not delay for long, a limit of 10 hours, or five revolutions, being set by the slow migration of the spacecraft's orbit and the continued use of consumables aboard Orion. If it was determined that the secondary gimbal drive system was inoperable Pete Frank's orange men would move in and control operations where the two vehicles would re-dock and use Orion's Descent Stage engine to fire out of lunar orbit toward Earth; otherwise, Gerry Griffin's gold team would stay on for the landing activity. Before passing out of sight on revolution 13 Ken Mattingly was advised to maneuver Casper back to rendezvous with Orion when they next came within closest range on the far side of the Moon. But two tests of the gimbal actuator were to be completed for engineers to study before deciding 16's fate.

Forty-eight minutes after loss of signal, Casper and Orion appeared on revolution 14 close together and waiting for word from Houston. But Charlie Duke was concerned about the schedule which previously had the LM crew performing an EVA prior to the first surface sleep period: 'Houston, Orion. John and I have been talking about if we get to land this thing, we'd like to probably think about going to sleep first and then we'd get up in a full EVA tomorrow.' Houston agreed that it would be best to do just that.

On around the front-side pass, Casper and Orion slowly narrowed the distance between them. But the Lunar Module's leaking helium regulator was now pressurizing the thruster tanks to a dangerously high level again. The delayed landing had not consumed the propellant that would otherwise have made room for the gas bleeding into the RCS tanks. Before passing again out of sight, Houston told the crew that they would pass up a final decision on revolution 15. In the meantime, exhaustive tests were carried out at North American Rockwell's Downey plant, at the Manned Spacecraft Center, and at contractor plants around the country.

It soon became apparent that vibration caused when the secondary actuator oscillated would not prohibitively affect the structure and at the Mission Control Center Chris Kraft went in to a management meeting to decide between caution and optimistic confidence. Shortly before Casper and Orion appeared on revolution 15, the decision was made to go for a landing attempt on revolution 16. Maneuver pads were passed up to the crew, systems were checked out, and Orion informed that at powered-descent-initiation the spacecraft would be 4.9 km high and 6.1 km south of the original coordinates.

Between the re-rendezvous of Casper and Orion and the ignition of the DPS engine for powered descent on orbit 16, the LM maneuvered several times to deliberately fire off the thrusters and open additional space in the RCS tanks for the leaking helium. Frequent yaw burns were made and several automatic maneuvers. Around on revolution 16 Casper reported it had successfully burned the SPS engine to circularize the orbit and Orion yawed round 20° to allow ground stations to load information to the LM's computer via the omni-directional antenna; Orion's main steerable high-gain antenna was still not working in yaw.

Powered descent would begin from a greater height than ever before: 20.1 km, and, in the final solutions, 4.8 km south of the planned groundtrack. This would mean that Young and Duke would have slightly less propellant for hovering around above the surface since Orion would steer from south to north during the main braking phase and consume extra fuel and oxidizer. Ignition came 5 hr 42 min later than planned. In Houston it was late evening, Thursday, 20 April. Two minutes into the descent the pilots punched in a correction informing the computer that Orion was 244 meters away from the downrange expectation and at the end of the braking phase the spacecraft was right back on the original groundtrack.

At a height of 6 km the landing site became visible when John Young pressed his faceplate against the optical alignment sight and sneaked a preview of the terrain below. Orion was right on target for the Descartes site, placing the touchdown spot between two craters – North Ray and South Ray – to which the crew would travel on their rover. In the approach phase now, the crew saw Orion heading for a spot 600 meters north of the precise target and re-designated the descent by punching corrections in to their computer.

At 150 meters above the Moon a shadow moved across the surface: Orion's presence was darkening the regolith. At a height of 25 meters dust started to move across the surface, increasing as the LM slowly descended almost vertical to the Moon. At touchdown the spacecraft was still moving forward slightly and the crew sensed a pronounced sinking feeling as the engine was cut off when the probe light came on. The comment from Orion was jubilant with glee: 'Wow! Wild man, look at that. Old Orion has finally hit it, Houston. FANTASTIC!' The Lunar Module was only 270 meters north and 60 meters west of the preflight target point.

It was decided to power down the LM to minimum levels for normal operation to make up for the longer time in lunar orbit. Pete Frank came on shift in the MOCR to relieve Gerry Griffin and the crew provided a brief description of the site seen through the two forward facing windows. Then it was time for Young and Duke to get some sleep before beginning their first EVA which, because of the delay and the re-scheduled sleep sequence, would be approximately 17 hours later than the flight plan. But there was an element of danger in this new, powered down, configuration because the mission timer was stopped and that rendered unusable the information on emergency lift-off times. If the communication loop was lost with Earth, the crew would have to use their wrist watches to determine the precise moment to head for a rendezvous.

As the night gave way to day and light fell across the Mission Control Center, managers, engineers, and systems specialists met and talked about the future timeline. Bearing in mind that there would be little reserve left if the mission went its full 12 days it was decided that Apollo 16 should spend one day less around the Moon after completing activities on the surface. That would give the mission time to recover from further problems that could develop with the SPS engine before consumables began to run out. And because it was agreed that the SPS should not be fired more than necessary until it was called upon to return the crew to Earth, elimination of a scheduled plane-change burn would prevent Casper being in the right place for the scheduled surveys of surface targets, further reducing the value of staying in orbit. The third EVA would probably last only 5 hours versus 7 hours originally scheduled for all three to give the crew a better sleep schedule under the new conditions.

It was mid-morning in Houston, Friday, 21 April, when John Young descended Orion's ladder and hopped down to the surface: 'Hey, mysterious and unknown Descartes. Apollo 16 is going to change your image.' The crew was about one-half hour ahead of the revised schedule as they began their first Moon walk but just before Young left Orion, Mission Control performed a clock update by advancing the event timers 11 min 48 sec. Elapsed times here, however, are actual hours and minutes since lift-off.

First order of business after Charlie Duke emerged to join John Young was to deploy the lunar rover, put up the flag, and unpack an ultra-violet camera set upon three legs a short distance from the LM. Essentially an electronographic Schmidt camera and spectroscope, the device would search for hydrogen sources in deep space by pointing at selected star fields, galactic clusters and targets nearer like Earth and regions of the lunar environment. It was carried to the Moon in quad 3. Then it was time to put out the ALSEP experiments: a passive seismometer; a lunar surface magnetometer; an active seismic test; and a second set of heat flow sensors. In addition, a portable magnetometer would be carried on the first and third EVA's for measurement at selected locations, a solar wind composition experiment would be left out, and a set of cosmic ray detectors would be put against one of Orion's landing legs until retrieved, like the SWC, for return at the end of the mission.

While Young shot some exposures with the UV camera, Duke loped off with the two ALSEP packages to select a suitable spot to lay them out; Young was to have driven the packages out with the rover but was too involved with camera work. Then Duke set about drilling the first of two holes for the heat flow experiment in which he was to place a set of sensors similar to those unsuccessfully laid in shallow holes at the Apollo 15 site. The first hole was drilled with little effort, unlike the experience Scott had at the Hadley site, and Duke started on the second hole when the Commander tripped over a data tape lying across the lunar surface.

YOUNG 'Charlie?'
DUKE 'What?'
YOUNG 'Something happened here.'
DUKE 'What happened?'
YOUNG 'I don't know. Here is a line that pulled loose. Uh, oh, what is that? What line is it?'
DUKE 'That's heat flow, you've pulled it off.'

Young had inadvertently torn out the data tape connecting the heat flow electronics package to the main ALSEP data station, or transmitter. The package would feed two further tapes to the separate holes so with the main link between package and data station broken there was no point in continuing to drill the second hole; the complete experiment was lost, tantalizingly so after all the trouble on 16's predecessor. But it was little surprise that the accident happened. During ALSEP deployment there were a large number of wires, lanyards, tapes, and strips of covering littered all across the site; surprise was that such an accident had not happened before.

Now Charlie Duke went about the deep drilling task, using the Black and Decker to obtain a core of the Moon 2.6 meters long. The linked stems were placed on the back of the rover, broken into two sections, and left at the ALSEP site for later retrieval. Drilling here was certainly a lot easier than at Dave Scott's site. Now it was time for John Young to take the rover and deploy three geophones along a 100 meter data line along which he could move, stopping to fire a total of 19 separate charges from a 'thumper' almost identical to that employed at the Apollo 14 site more than a year before. A mortar box with four grenade charges was also set down, but only three of the four legs would deploy.

When laying out the geophone string, Young found the rover's rear steering to be working, unlike initial tests soon after deployment. While Young walked down the line firing off a total of 19 tiny charges for seismic profiling, Charlie Duke inspected the heat flow data tape to see if there was a way to repair it; Mission Control was buzzing with ideas and suggestions but it would be abandoned in the end. It was about 4 hours into the EVA when the two Moon men climbed aboard the rover and traversed to their first geological site, a region almost due west from Orion, about 1.4 km across the surface to a comparatively small crater called Flag. There they sampled half way into its bowl-like depression, obtained a football-sized rock as required, and climbed back on the rover to visit Spook crater half-way back to the LM.

It was time to take a reading from the portable magnetometer and then to collect some samples. It was 6 hr into the EVA when they got back to the ALSEP area for Young to dismount and arm the mortar package, and then for Charlie Duke to put out the solar wind collector back at the LM. Before completing the first EVA Young took the rover on a 'grand prix' test to check its handling and performance, while Duke filmed the whole event.

DUKE 'Man, you're really bouncing it.'
CAPCOM 'Is he on the ground at all?'
DUKE 'He's got about two wheels on the ground. It's a big rooster tail out of all four wheels, and as he turns he skids. The back end breaks loose just like on snow. Come on back, John . . . Man, I'll tell you, Indy's never seen a driver like this. Hey, when he hits the craters it starts bouncing, it's when he gets his rooster tail. He makes sharp turns. Hey, that was a good stop, those wheels just locked!'

Splayed rooster tails of dust and fine particles sprayed out from behind the wheel guards as John Young threw the rover into every conceivable performance trap, all maneuvers carefully worked out beforehand, plus a few thought up on the spot! After taking some shots with the UV camera and passing up some samples into Orion, the crew moved back inside to end their first EVA. Based on examination of consumables expended and energy rates generated, the second EVA was given a 'go' for seven hours, with the proviso that the third excursion would be restricted to five hours.

It was in the final hour approaching midnight in Houston when the crew of Orion began their sleep period, 131 hr 20 min into the flight. When next the flight controllers heard from the Moon men it was 7:00 am, Houston time, Saturday, 22 April. Less than four hours later they were back on the surface at the start of what would be a full geological traverse south of Orion toward Stone Mountain. After loading the rover and checking its systems, the explorers moved out, heading at 164° from north; they had already been out on the surface 44 minutes.

Down past an incline called survey ridge they drove, with large cobbles up to 30 cm in size, then through blocks and boulder fields before rough, undulating, terrain as they gradually climbed up the slopes of Stone. Just 33 min after leaving Orion, they were at their first station stop. They shot 500 mm telephoto views, raked the soil for pebbles and small rocks, jabbed the surface with a penetrometer to measure its hardness, sampled a glass-splattered bead, took core samples with the hammer and scooped up regolith. Then, a panoramic survey for geophysicists back on Earth, as all the while Ed Fendell in Houston operated the TV camera on rover's forward section.

They spent just over an hour at this stop, the so-called station 4, before moving back north more than one-half a kilometer for a further 50 minutes at station 5. Here they obtained a rake sample from a crater wall, obtained a further reading from the portable magnetometer, and picked up a large crystalline rock. Time to move on again, this time north west to station 8, station 7 having been intentionally deleted. They were there 1 hr 10 min, where, on the flank of Wreck crater, they picked up fragments of a shattered rock collected a batch of soil, and completed a documented sample, so called because it required a comparatively extensive series of photographs to fit the samples within pictures of the site. Before leaving for station 9 the crew performed a quick trouble-shooting operation on the rear wheel steering problem that became evident early on the first EVA, discovering a mismatch of switching positions to have been the cause.

It was 4 hr 17 min into the activity when Young and Duke stopped at station 9, only a little further north than the previous station and still 2.6 km from Orion. After overturning a boulder for sub-samples, putting glass crystals in a collection bag, and taking more pictures, they packed for the traverse back to station 10, a position mid-way between Orion and the ALSEP instruments. Penetrometer readings were taken here, and a double core sample, after which they obtained more rock samples and prepared to get back inside the LM.

The second EVA had lasted 7 hr 23 min and it was late evening Houston time when the duo began their well earned rest period. It would be the last they would get prior to ascending back for a rendezvous with Casper, two hours earlier than planned but four hours after the flight plan time. During the nocturnal hours, Houston talked at length with Ken Mattingly about the slight change in rendezvous procedures.

Sunday, 23 April, was the third day on the Moon for Young and Duke. At about 6:30 am Houston time the crew were woken for their last EVA and less than three hours later it was under way. Their first job after loading the rover was to move north 4.5 km to station 11, right on the very rim of North Ray crater, a prominent depression nearly 1 km across. On its rim too, a massive boulder – the largest approached so far by Apollo astronauts – about 20 meters high and 30 meters long. It took the crew 35 minutes to drive across the undulations to station 11, crossing ridges that seemed so subdued when seen from a distance, distinct barriers close up, and across blocky rock fields strewn over the regolith.

North Ray was spectacular, not in the way the massive Hadley peaks or the incised rille had been at 15's site, but because of its sheer size; this enormous hole in the surface was unique in Apollo experience. Rake samples, 500 mm telephoto shots of Smokey Mountain to the north, and sample collection preceded a lope across to the big 'house sized' rock that stood on the rim. The rock impressed the crew with its physical size and with its location right on the edge of this large crater. All too soon it was time to return to the rover, 150 meters away from the big 'house' rock.

A short stop at station 13, 3.8 km from Orion, kept them busy for 30 minutes with more sampling and magnetometer readings, but 3 hr 15 min into the EVA – extended now by 25 min – the crew were en route back to station 10. But it was a re-designated station stop, making a triangle with the old station 10 and the ALSEP site. That kept them busy with a double core and rake samples before resetting the UV camera and parking the rover. Then the solar wind collector was rolled up after 45 hr 5 min out on the surface and the crew got back inside. It was time to prepare for ascent and, although shorter than expected, Apollo 16 had broken new records in the time on the surface, the duration of the three EVA's, and in the weight of samples brought back.

Before firing free from the Descent Stage, the crew heard from Gene Kranz's men that they were to get some sleep after docking with Apollo in orbit, leaving until later the final closeout, separation, and LM de-orbit activity. From the evidence of fatigue on Apollo 15, physicians were concerned that the 16 crew should not be over-stressed. But the product of three tiring days exploring the Moon were taking their toll of equipment and procedures.

On the first EVA, orange juice housed in a bag inside the neck ring leaked out and ran down into the suit and around the ring connector. It became almost impossible to remove the helmet after the first EVA until water had been poured down the neck to dilute the sticky orange juice. Then dirt became a real problem, jamming the wrist connectors on the suits, impregnating the suit outer layers with a fine grime, clogging equipment, and rendering it almost impossible to keep the inside of Orion in a suitable shape for operation.

Before the second EVA, Duke's antenna broke at the end and then master alarms kept sounding because of the helium leak in the RCS regulator. Inside the Ascent Stage, conditions were dirty, cramped, and tiring, ill-fitted to the three full days spent on the Moon. But within such conditions, crew performance was outstanding and testified to the value of careful screening at selection. Through it all, Moon-exploring astronauts inclined to play down the problems. But they were very real, and almost unbearable at times.

With a call of 'What a ride, what a ride,' Orion's Ascent Stage fired off toward Casper in orbit up ahead making its 52nd lunar orbit. A 7 sec burn took place earlier to bring Casper's track in line for optimum rendezvous. It was to be an M=1 event yet again, Orion planning to pull alongside the CSM within one lunar revolution after lift-off. After more than seven minutes thrusting into orbit, Orion had to perform a small velocity change to tweak up the trajectory before firing the terminal-phase-initiation burn shortly after passing out of sight around the Moon.

When next they came into view range was less than 6 km but the planned TV session of the docking operation was impossible due to the inoperable high-gain antenna. The LM crew did describe, however, a spectacular sight of the MESA blankets blowing 200 meters in front of Orion at lift-off, an event not seen in the lens of the rover's TV because the vehicle was behind the Lunar Module. After inspecting Casper's SIM bay to seek answers to problems experienced with the experiment booms, the CSM viewed Orion and saw large strips of thermal blanket virtually hanging off the back of the Ascent Stage. Torn loose at ignition on the Moon, the blankets were probably impacted by reflected exhaust from the Descent Stage.

After successfully docking with Casper, Orion's crew vacuumed the cabin interior until the device failed leaving dust and fine particles floating across into the Command Module with equipment transferred for stowage. Although set up to provide a positive flow of oxygen, crew movement from one spacecraft to the other caused swirl patterns in the gas which threw dust upstream into the Command Module. Mission Control had planned to have the crew get some sleep before jettisoning the LM but in the event this created so many new operational problems that it resulted in a longer 'day' than if the vehicle had been released as originally scheduled. The crew were annoyed that Houston had altered their timelines. Was it the same old familiar problem of interference from the ground when sophisticated procedures became commonplace and controllers developed an over-confident approach? It had happened during Mercury and Gemini to some degree.

It was just after midnight, Houston time, when the three crewmembers finally began their orbital rest session; about 9:15 am, Monday, 24 April, when they woke up. A large number of checklist and timeline book changes were voiced up making these documents into a 'very messy' condition, according to the crew. The alterations were made significantly worse by having delayed jettisoning the Ascent Stage. The basic philosophy for the new flight plan was to follow the existing procedures up to an elapsed time of 192 hrs, go on to procedures starting at 214 hrs and at 216 hr 15 min go back and follow jettison activities marked up at 176 hr 55 min!

It was not surprising that the crew left a switch in the wrong position when they entered Orion for final activities prior to sending the Ascent Stage down toward the Moon. When Orion was jettisoned from the Command Module it began slowly tumbling and never did fire its RCS thrusters for de-orbit; inhibited from operating its maneuvering engines because of the switch position, Orion would remain orbiting the Moon for nearly a year before it finally struck the surface.

A few minutes after releasing its charge, Casper fired thrusters to move away from the Ascent Stage and less than one hour later the subsatellite carried in the SIM bay, like the first deployed by Apollo 15, was sent spinning free. Because the CSM had not performed a planned shaping burn to set up the orbit properly, Apollo 16's subsatellite was not in the desired path and would last for little more than half the planned one year lifetime before it too spiralled to the surface. Concern at the condition of the back-up SPS gimbal control also deleted a plane change originally scheduled for the post-surface orbital science period but that activity was shortened by a full day so the burn was irrelevant.

At the beginning of revolution 65, less than five hours after releasing the subsatellite, Casper successfully fired its SPS engine to return home. Shortly thereafter Mission Control performed a further clock update, advancing the elapsed time recorders by more than 24 hours to put them at the appropriate point in the flight plan for trans-Earth activity. After a lengthy sleep period, and a small course correction with the RCS thrusters, Ken Mattingly went outside for his deep-space EVA to retrieve cassettes from the SIM bay experiments. Much impressed with having the chance to float around outside his spacecraft, Mattingly reported on the condition of paint flaking from the Service Module and on the appearance of equipment inside the bay.

The return home was comparatively uneventful, the crew noticeably disturbed by the change in timelines and procedures brought about by reducing the time spent in Moon orbit; it was a very effective way of disorientating the crew, devoid already of normal day/night cycles. More light flash tests were conducted, the crew talked at length about their lunar surface EVAs, answered questions posed by newsmen, and wrestled minor problems that arose with equipment. Warning lights flashed to indicate trouble with the inertial platform, but a status light went out when the panel to which it was attached was rapped lightly. Just before re-entry, a small course correction was made to move the path of the Service Module away from possible descent on the Pacific island of Penrhyn with 500 inhabitants.

After splashdown, Apollo 16 flipped over into an apex-down position but the effectiveness of the recovery operation resulted in Young, Mattingly and Duke standing on the carrier *Ticonderoga* just 37 minutes after hitting the sea. It had been a successful flight, but one characterized and ultimately shaped by the problems with Casper's SPS engine. It was determined that a wire had probably pulled loose during operation of the gimbal actuator and Apollo 17's engine was suitably modified. Nevertheless, and despite the loss of the heat flow experiment, science on both the surface and from orbit had gone well.

All the other ALSEP equipment worked perfectly and on 23 May, 1972, ground controllers fired three of the four mortar charges contained in the launcher left close to the geophone line: a fourth, the 1.5 km range charge, was not discharged because an attitude indication sent via telemetry seemed to show the mortar package had fallen over.

Yet there was only one more Moon mission with which to tidy up loose ends left unsecured because of some earlier failure. Apollo 17 would be the last chance to get a good heat flow experiment going. But even as Apollo 16 came home, events were edging closer to the possibility of a joint docking flight with the Soviets. Following George Low's clandestine visit to Moscow in April, 1972, the State Department began work with the White House to prepare a document mutually acceptable to both US and Russian negotiators working to resolve final differences over the SALT-1 agreement.

The Russians wanted to embrace wider issues than the docking flight alone and seemed unable to understand why a government could not answer for the commercial interests of communication satellite operators with whom they worked to secure an agreement in its country. But after final wrangling everybody agreed on the proposed text and at 6:00 pm on the evening of Wednesday, 24 May, President Nixon and Premier Kosygin signed in Moscow an 'Agreement Concerning Cooperation in the Exploration and Use of Outer Space for Peaceful Purposes.' It was late afternoon in Washington when Agnew held a press conference at the Executive Office Building at which Jim Fletcher formally told the public that NASA and the Soviet Academy of Sciences would work toward 'the rendezvous and docking of a US spacecraft with a Russian Soyuz spacecraft in 1975.' Two days later Nixon and Kosygin signed the SALT-1 treaty.

At Houston, work accelerated during June on preparations for the flight. Dale Myers dispatched texts to Marshall, Kennedy, and Manned Spacecraft Centers laying out the work plan. Word crept over from Russia that the Soviets would have a back-up vehicle ready and waiting along with the prime hardware, and that they would wish to launch their Soyuz first rather than have Apollo go first. Clark Covington suspected they were afraid of failure and then also learned that the Russians would send up another Soyuz if Apollo was delayed beyond the first day or so; the Russian craft was limited to four or five days in orbit whereas Apollo could last two weeks.

It was generally agreed that Apollo would have to do the maneuvering, since the Soviet spacecraft could only cooperate with equipment carried by the Salyut, or another Soyuz. During examination of the Soyuz environmental control system NASA engineers Ed Smylie and Walt Guy discovered that the spacecraft carried no supplies of oxygen or nitrogen, relying instead on pumping up the cabin before launch and placing their faith in a completely leak-proof vehicle and a potassium superoxide chemical regeneration system. In other agreements, Russia would receive from America VHF transceivers similar to those carried by the Lunar Module and the US would manufacture Soviet VHF/FM radio equipment for Apollo.

A major meeting convened in Moscow during October, 1972, where twenty-seven NASA people from several working groups set up to prepare for the venture thrashed out technical and trajectoral details with their Soviet counterparts. When the Russians discovered, to their embarrassment, that Soyuz would be incapable of reaching a height in excess of 225 km they presented circuitious arguments to seek a rendezvous height of 222 km from the earlier US suggestion of 232 km. Only after considerable debate did the US delegation realize the Soviets were sensitive to limitations on their vehicle. Clark Covington asked them outright if they could not get to the requested altitude, and received a grudging affirmation.

But the Russians were very helpful when it came to seeking a solution to the problems of different pressures in the two spacecraft. To prevent astronauts returning to the Command Module from pre-breathing pure oxygen for several hours in the docking module between Soyuz and Apollo, the Soviets suggested lowering the Soyuz atmospheric pressure to two-thirds the sea-level environment, making it only twice that of Apollo's environmental oxygen pressure. That would speed up transfers between the two spacecraft and smooth out timelines, as well as improve the situation for possible refuge in either spacecraft should one experience a catastrophic failure of some kind.

A month later, chief designer of the Soyuz rendezvous and docking control system, Viktor Pavlovich Legostayev, headed a nine-man delegation to Houston during which time details were finalized on such things as communication frequencies, radar range requirements, docking techniques and procedures, etc. By the end of 1972 it all looked in good shape for the mission two and one-half years later, as contractors in America and establishments in Russia worked to prepare the test and flight hardware.

Improved too was the position regarding the reusable NASA shuttle formally approved by the President three and one-half months before Apollo 16. By April, 1972, the definitive shuttle configuration settled the previous month provided final technical specifications for main engine performance; Rocketdyne received a design award in July, 1971, but not for nearly a year could it finalize the specification. On 17 March NASA requested aerospace companies to submit proposals for design, development and test of the reusable winged orbiter and four responded by the deadline set for 12 May: North American Rockwell, McDonnell Douglas, Grumman, and Lockheed. On 25 July, 1972, NASA formally announced that North American Rockwell had won the contract with McDonnell Douglas coming second.

As for the research and applications module – RAM – which evolved from space station studies now defunct, NASA moved toward a cooperative venture with the European Space Research Organization on what was now called Sortie Module, a pressurized laboratory retained in the shuttle orbiter's cargo bay for research activity.

But it was with the last Moon flight, and the four Skylab launches, that engineers and technicians at the Cape concerned themselves most during the second half of 1972 despite the disappearance of key names in the history of manned space flight. At the Manned Spacecraft Center, Jim McDivitt's place as Apollo spacecraft program manager was taken by Owen G. Morris, and at Headquarters Dr. Wernher von Braun retired from NASA on 1 July.

In announcing on 26 May his intention to resign, von Braun

Apollo 17 landing site was on the dark floor of the Taurus-Littrow area (center) flanked by the South Massif (mountain at left-center) and the North Massif (hilly region at right). Note the liberal distribution of small craters on valley floor where the LM landed and the fingers of light mantling material having slid down the slopes of South Massif to the darker basalt.

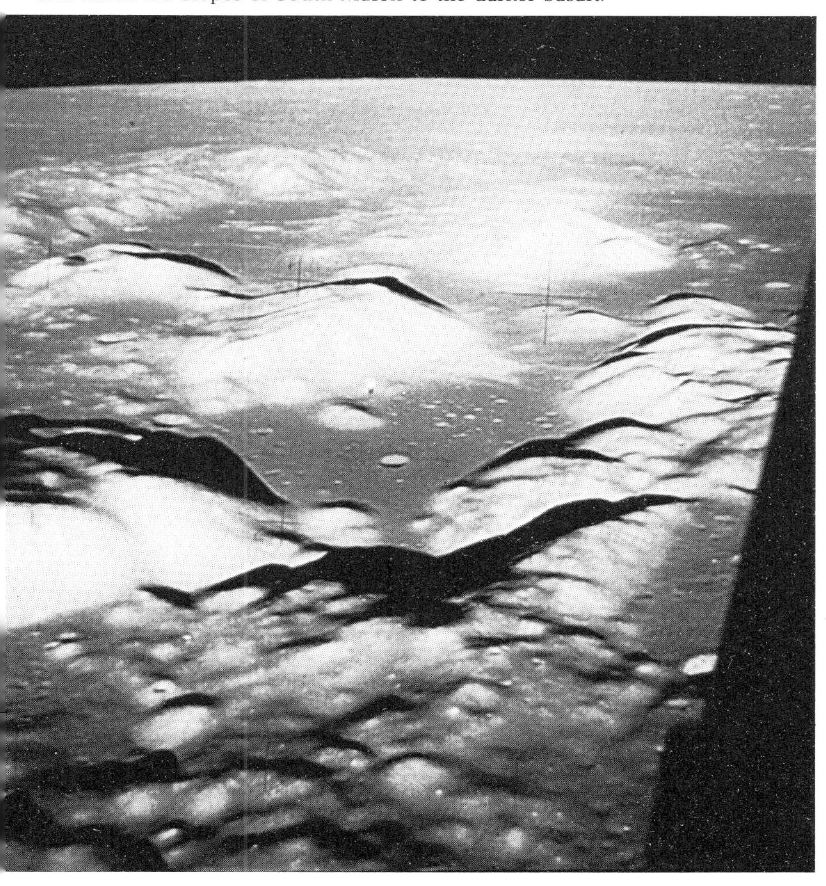

said that he was going 'with a deep feeling of gratitude for the wonderful and unique opportunities the agency has given me during the last 12 years.' In return, the United States had received the launcher that sent America's first satellite into orbit, the free world's first large space booster, the rocket that sent Apollo spacecraft to the Moon – still the biggest ever to fly into space – several first generation battlefield military missiles, and direction for 10 years of the facility that grew from Army auspices to be NASA's largest engineering establishment.

Yet for all that there was bitter irony in Jim Fletcher's parting statement: 'All of us at NASA will miss the daily stimulation of his presence.' For during the nearly two and one-half years in which he headed future missions planning at Headquarters, von Braun was frequently made to feel out of tune with the realities of the day. NASA had become a holding institution unwilling to stand up and fight for what it believed, evolving into a civil service organization where even its Administrator would now argue for only five shuttle vehicles because he thought the White House would not like him to ask for any more!

Von Braun was a tired and disillusioned man when he walked away from Washington to begin a brief career as Corporate Vice President for Engineering and Development at Fairchild Industries. Without ears to hear his words of warning about the price for a flagging technology boost, politicians and administrators of the Washington scene failed to grasp the deep regret von Braun felt at not seeing in his own lifetime men land on Mars or penetrate the depths of the solar system. Denied from ever going into space himself, he was prevented at the last hurdle from seeing the true beginning of celestial man.

At the Manned Spacecraft Center, Deke Slayton, now freed for a flight seat, was concerned about the large numbers of astronauts who stood no chance of a space mission. Urging many of those unassigned to either Apollo 17 or the three Skylab flights to pack their bags and seek success elsewhere, he openly professed a belief that 'our (astronaut) manpower is over our known requirements by a factor of three.' It would be the late 1970s before the shuttle would commence regular manned trips into space again, a hiatus of probably three or four years after the Apollo/Soyuz docking flight in 1975. NASA was learning to live within the tight budgets it would probably get during the rest of the decade, and the old face of the space agency was just a little wrinkled and pale because of that.

Bob Gilruth had now gone from MSC, and of the 118 Germans who came to America in 1945, only 42 were still with NASA. Of the rest, 12 had died, 16 had returned to Europe, and 26 had retired. And then, on 9 July, 1972, Lew Evans, head of Grumman when it worked so hard to perfect the Lunar Module many believed could simply not be done, died at the premature age of 49 years.

Fortunes were certainly mixed that year men last went to the Moon. In congressional hearings prompted by the Apollo 15 crew's misdemeanor, politicians learned that the Franklin Mint had gained financially by melting down and producing 130,000 saleable coins from 200 privately minted silver medals carried in ditty bags aboard Apollo 14. It had been knowledge of this questionable involvement that prompted Deke Slayton, in his capacity as head of crew activities, to warn Scott, Worden and Irwin not to be tempted as well. In the wake of public discomfort with the 15 crew's attempted financial gain, the Apollo 17 crew were cautioned in advance not to try any such deal. But for several reasons, this last Moon flight had a very special crew.

As Commander, Gene Cernan was committed from the start to making this the best flight yet; in the Command Module Pilot seat, Ron Evans worked long and hard to perfect routines for orbital science; assigned as the only professional scientist ever to reach the Moon, Lunar Module Pilot Harrison Schmitt knew a lot hinged on his performance for many had questioned the wisdom of allowing non-career pilots a seat on Apollo. The political need to give at least one scientist a Moon trip ousted Joe Engle from Cernan's crew but nobody showed visible rancor. Several years later, Joe Engle would be named to command one of the first shuttle flights. It was a fair trade. Crew announcements were largely a matter of formality by this time. Since Apollo 12 the back-up crew of one flight skipped two missions and became the prime crew for the third in line.

By now, contributions from the orbital sensors of J-series missions were beginning to produce operational results. Sensors and photographs from Apollo 15 pointed to an area of special interest far from the lunar equator and it was to that area –Taurus Littrow – that Apollo 17 was assigned at the beginning of February, 1972. Surrounded, at the south-east corner of the Mare Serenitatis, by the Taurus mountains and close to the crater Littrow, the site was thought to contain several comparatively fresh volcanic vents, or cinder cones, indicating recent activity within the Moon. Framed by 3 km high mountains called North Massif and South Massif, the valley floor contained many craters and formed when lava flooded the massive basin called Serenitatis excavated when a large asteroid size object struck the Moon billions of years ago.

Formed like Imbrium, the Taurus mountains of which the Massifs were a part threw up from the surface to form the large rim of the basin. By travelling to the mountain slopes the crew would sample material excavated by the impact and by digging the valley floor they would recover deposits of the extruded lava. But by sampling what looked like cinder cones they could possibly provide the best evidence yet that the Moon is still active inside. In addition, orbital sensors carried in the CSM would replace geochemical detectors installed on Apollos 15 and 16. Instead, Apollo 17's SIM bay would support a lunar sounder, an infrared radiometer, and an ultraviolet spectrometer.

The lunar sounder would provide an electromagnetic map of the lunar surface and operate through two antennae carried outside the SIM bay at the base of the Service Module: two high-frequency antenna sections deployed from opposite sides of the SM for a total span of 24.4 meters, and a very-high-frequency yagi antenna 2.7 meters in length. The radiometer and spectrometer measured, respectively, the temperature gradients at the surface and the atomic composition of any atmosphere the Moon might temporarily hold on to as well as solar radiation reflected from the surface.

Apollo 17 was to be a flying laboratory in every respect with another visual light flash experiment, a Biostack experiment from Germany, and more microbial passengers similar to those carried by Apollo 16. The SIM bay would also contain panoramic and mapping cameras for topographical measurements and photo-interpretation of features.

A month before the launch of Apollo 16, Gene Cernan wrote a specification for artist Robert T. McCall in designing 17's mission patch: 'Our desire is that Apollo 17 symbolize not the end of an era, but rather the culmination of the beginning of mankind's greatest achievements in his history – achievements which only have as their bounds the infinity of space and time – symbolization that man's seemingly impossible dreams can become limitless realities. We would like to recognize the historical foundation upon which the thoughts of the future are based– and, so as never to forget, we also hope to pay tribute to the Apollo program and our nation, its people and its heritage, which have made these accomplishments all possible. The symbolism which captures these ideas sounds sophisticated and complicated to create. We hope it is not, because it is our desire to capture our theme with simplicity.'

It was to be a mission like none before or since, for as Ron Evans said: 'We may be the last *crew* going up there, but we're going to be the first *team* . . . that's for sure.' All through the months of preparation, engineers who built and checked the hardware, scientists who designed and prepared the instruments, and personnel across the United States who contributed to crew training, all resolved to give Cernan, Evans, and Schmitt the best and most fruitful ride yet. Rocket stages for Moon flights were by now quite aged. Built quickly in anticipation of an early flight date, and continuous productions runs that never did get ordered, Saturn Vs were placed in storage until needed.

The first stage for AS-512 was put together in a sixteen month period beginning July 1967, although components had been ordered from subcontractors in August 1963. Barged to NASA's Mississippi Test Facility for static firing of the five F-1 engines at its base, the first stage was placed in storage early 1970 back at the Michoud facility where it came together in the first place. Two and one-half years later it was taken out, barged to Canaveral, and put in the Vehicle Assembly Building for Apollo 17.

It was to be the last manned Saturn V, the last rocket of this type to fly with three live stages, and the last to boost a payload out of Earth orbit. It was very definitely the end of an era that began nearly five years before with the launch of Apollo 4, the first Saturn V. CSM-114 and LM-12 arrived in late March and June, 1972, respectively, with the third and final lunar roving vehicle booked in during June also. After unmanned and manned altitude runs in the simulation chamber at the MSOB, Command and Service Modules were transferred to the assembled Lunar Module and adapter, mated to the four petal doors, and trundled up to the Vehicle Assembly Building at complex 39. There the encapsulated LM and the top mounted CSM were lifted on to the Saturn's third stage almost thirty stories above the floor of the massive building.

AS-512's erection had begun 15 May and was completed 27 June, with the spacecraft going up on 24 August. Four days later the complete assembly rolled out to pad 39A. It was an event not to miss. Everybody it seemed wanted to watch the last three-stage Saturn V slowly rumble along the crushed gravel road to the pyramidal shape of the launch pad stand. It had happened so many times before, but never like this. At dawn, Bendix engineers swung up into the turret cabs at front and rear on the big crawler transporter and with crowds of people outside drove slowly through the gaping doors facing east. It was a behemoth to be sure, made all the more impressive by the puffs of smoke as the crawler got up speed close on 1 km/hr.

For brief periods of the long, slow, journey, the 17 flight crew clambered aboard the crawler cabs and chatted with the Bendix men. Along the way, they signed autographs and played to the crowds. It was the beginning of the show – the last in town. Already, back in the VAB, hardware for Skylab was coming together, technicians grooming three Saturns once more as they had in more hectic days three and one-half years earlier. When the Saturn V arrived at 39A and the big mobile service structure had been moved up to the launch vehicle, Grumman engineers hung a sign close by the Lunar Module work level. It simply said: 'THIS MAY BE OUR LAST BUT IT WILL BE OUR BEST.'

Grumman would feel the end of Moon flights at a personal level; 600 technicians would be out of a job when the flight ended. It was a trend mirrored by the fate of so many in this time of transition. Within a few months the space program would have lost a total 300,000 workers from its peak employment level of more than 400,000 achieved in the mid-1960s. Decay had been eating away at the program since Neil Armstrong and Buzz Aldrin returned from the surface of the Moon. In fact, in the year up to Apollo 11, space workers had been laid off at the rate of 1,000 each week. By the end of 1970 the Los Angeles area of California had lost 85,000 aerospace workers, taking with them more than 50,000 other workers in supermarkets, stores, and banks.

Arguments levelled against the space program concerning wasted skill were irrelevant: in 1970 some 93,000 aerospace workers were still claiming unemployment benefit, up 75% on 1968. Major aerospace companies were badly hit by government trends that sought to placate cries of minority groups for more attention to the poor. In one week, Boeing had to lay off 64,000 workers just to remain solvent and three years later nearly 15,000 men and women still queued at street kitchens for free food and a hot drink, facilities organized by benevolent companies still in business. It was a switch unlike anything seen in California before. But the picture was similar at the other coast.

In Florida, the First National Bank closed 4,000 cheque accounts, 2,500 deposit accounts, and made a 75% loss on loans. Kennedy Space Center personnel was down from a peak of 26,000 to a mere 13,000 and the commercial foundation of Brevard County suddenly became very uncertain. House prices sank alarmingly and financial losses multiplied. At the launch base itself, rust overtook the historic launch stands and rocket pedestals that once framed events acclaimed by the entire world. Grass and weed did its worst to metal that eroded in the salty atmosphere of Cape Canaveral; miles of disused concrete and steel disappeared back into the advancing wilderness. Only pockets of attention kept operational pads free of encroaching decay. A nation had turned its back on the great explorations in space; lost its determination to blaze new trails.

In the closing months to the December launch of Apollo 17 attention seemed to increase around the hardware that would make the last trip. Nobody wanted the program to go out on a disastrous note, and the spur to success came in the re-kindled hope that perhaps, after all, this was only a temporary halt. It was to be a spectacular exodus, the first night launch for the world's biggest and most powerful rocket. But if excitement increased for the Cape launch, apathy once more plagued media attention. Very few stations chose to transmit live pictures and the Public Broadcasting Service network was openly critical of requests for television time from NASA.

Yet the lift-off promised to be a unique event and a record 3,503 press representatives booked in for the launch, exceeding even the number that came to view Apollo 11. Two special centers were used to cater for the large number and a fleet of more than 300 buses was laid on to ferry visitors and the press from key assembly points off the base. By the time launch day arrived more than 700,000 people were in the area to witness the nocturnal spectacle. One of the very special guests was Charlie Smith who was 13 years of age when Abraham Lincoln fell dead by assassination. A former slave, he brought his 70-year-old son along for the trip.

As darkness fell across the Cape, engineers took up their final positions for launch. Five hundred personnel were at consoles in the Launch Control Center, a further one hundred were at spacecraft control rooms in the MSOB, and several thousand personnel stood by across the Cape to assist if needed. The launch window for Apollo 17, dictated by the Moon's position relative to the Earth and by the need to enter orbit inclined 20° to the lunar equator, would last 3 hr 38 min from 9:53 pm local time, 6 December. Another opportunity would occur one day later for the same period.

But launch vehicle 512 was a reluctant flyer. With bated anticipation the excrutiating repetition of Apollo 17's countdown was seen by many on illuminated clocks as the public affairs officer at the Cape kept up a running commentary. Suddenly, with just 30 seconds to go, the countdown, now on an entirely automatic sequence, froze, halting all further actions. What happened was that at T−2 min 47 sec the automatic sequencer failed to provide the command to pressurize the Saturn's third stage liquid oxygen tank. Personnel at consoles in the firing room saw this and manually commanded the action. The stage duly pressurized and everything looked satisfactory. But the automatic sequencer, cognizant of its own deficiency, stopped the count.

'It did not take very long to determine that we should bypass the command system, pressurize the tanks manually, and continue on to ignition. We have a preplanned scheme which permits us to employ banana plugs and put in jumpers to bypass any point in the circuitry,' said launch director Walt Kapryan. But first, the countdown had to be backed up to the T−22 min point which itself took nearly 40 minutes to accomplish. But longer than that was the time needed to try out the electrical jumper on a breadboard system at the Marshall Center, home of Saturn V.

It was 10:40 pm when the countdown was ready and waiting at the hold point for a management decision on when to go. At 11:00 pm the countdown clocks started rolling again with a command to halt at T−8 min if no decision had been made by that time. It had not, and the clocks stopped at the new hold time. It was 11:14 pm. Excess hydrogen venting from the second and third stages was now being burned off in a pond far to the north of pad 39A. It was necessary to hold at T−8 min so as to chill down the second and third stage thrust chambers adequately, an activity which had to take place at least 7 min 40 sec before and no later than 20 min after the first stage liftoff time.

For more than an hour the countdown held at T−8 min while engineers convinced launch managers of safety in the jumper procedure. Then a 'go' was given to proceed with the countdown and at 12:25 am, Thursday, 7 December, the clocks started up once more. Eight minutes later light broke across the Cape as intense as day, preceded by the regular voice of launch control: 'T minus 26 seconds . . . Mark, T minus 25. We'll get a final guidance release at the T minus 17 second mark. T minus 17, final guidance release. We'll expect engine ignition at 8.9 seconds . . . 10 . . . 9 . . . 8 . . . 7 . . . ignition sequence started – all engines are started – we have ignition . . . 2 . . . 1 . . . zero – we have a lift-off. We have a lift-off and it's lighting up the area, it's just like daylight here at Kennedy Space Center as the Saturn V is moving off the pad.'

This time there was no hitch, the jumper worked and the countdown progressed to its inevitable conclusion. At ignition, fire and light burst from beneath the gleaming white rocket, blinding eyes adjusted to the dark of night. In the spacecraft, the lights were turned full up to prevent the bright glow temporarily dazzling eyes that would now busily scan the instruments. Boiling fire filled the billowing smoke that gushed from each end of the flame trench and as the huge Saturn V roared toward the sky shock waves of light bounced off the concrete pad sending shells of fire rippling up above complex 39. Nothing like it had ever been seen before.

From 800 km away, places as far from the Cape as Havana, Cuba, Charlotte, North Carolina, and Montgomery, Alabama, people could see the glowing ember thread its way at phenomenal speed toward orbit 169 km above the Earth. At the Cape, the glow from Saturn's five main first stage engines sent a white ball through the sky, illuminating clouds high above the Atlantic veiled by black night mere seconds before. When the sound hit the viewing stands and the launch control center, it seemed a deeper growl than ever it had in daylight, a more threatening spectacle as darkness crept back along the Florida shore. But what an experience, what a thrill!

'Looking great – right on the line,' came a call to Apollo 17 as Gene Cernan's crew bounced and jockeyed their way into space. In Houston, Gene Kranz was on duty with capcom Bob Overmyer. 'Gene, we're going round the room, looks go here. You're looking real good, Gene, right down the line,' was the word from the MOCR. 'Let me tell you, this night launch is something to behold,' responded Cernan.

All the way into orbit events cycled off as planned: F-1 centre engine shutdown; main engine cut-off; first stage separation; second stage ignition; skirt jettison; launch escape tower release; J-2 centre engine shutdown; main second stage shutdown; stage separation; third stage ignition; third stage cut-off. The crew could see the fire of ignition as the mighty second stage thundered into life just seconds after the forward firing retro-rockets on the S-IC temporarily bathed the rocket in light. Then, when the S-II shut down, retro-rockets on that structure brought another glow through the windows. For nearly 12 minutes the big Saturn propelled them on and up until they slipped into Earth parking orbit almost exactly as planned. Minutes later they burst upon a dawning world as speeding over the Atlantic the spacecraft plunged into Sunrise.

During the first orbit the crew received several master alarms aboard the spacecraft, spurious signals as it turned out, which could be induced by bumping part of the main display panel and by operating certain switches. It looked like an electrical short circuit, but nobody would ever find out exactly what it was. Passing across Hawaii on the second revolution, capcom Robert Overmyer passed word to the crew that they were in good condition for trans-lunar-injection.

CAPCOM 'Guys, I've got the word you wanted to hear. You are go for TLI – you're go for the Moon.'
CERNAN 'Okay, Robert. I understand. America and Challenger with their S-IVB are go for TLI.'

For this last lunar flight, Cernan chose names representative of the entire program: America for their Command and Service Modules and Challenger for the Lunar Module. Unlike earlier flights, the TLI burn would come at the beginning of the third revolution and, because of the 2 hr 40 min launch delay, the flight would be a little faster to make up that time by lunar-orbit-insertion more than three days later.

Charlie Smith (with bow tie), a former slave born eighteen years before the start of the American Civil War, prepares to watch the launch of Apollo 17 with his 70-year-old son Chester.

The flight to lunar orbit was uneventful. After completing the 5 min 51 sec TLI burn, Apollo 17's S-IVB held itself steady for the transposition and docking maneuver where the CSM separated, turned around, and moved back in to snag Challenger on the extended docking probe. After checking the docking latches in the tunnel between the two vehicles, and finding three still unconnected, the crew pulled away from the S-IVB which then went about its own activity for lunar impact days later. This was a different trajectory to any that had been employed before, the culmination of a gradual shift toward increasingly less accommodating paths to the Moon.

The two previous manned flights had shifted the hybrid transfer operation back into the TLI burn; that is, although launched to a non-return course, the spacecraft could have put itself on to a free return path by using the RCS thrusters at any time up to 5 hours after leaving Earth orbit. But Apollo 17 was not able to use this trajectory, being launched instead directly to a non-return path that could neither use the RCS thrusters to get to a free return course nor employ the LM's main engine to reach Earth if the CSM's Service Propulsion System failed to fire on the far side of the Moon. The big SPS engine simply had to work for, having been propelled from Earth orbit by the S-IVB, nothing but America's main propulsion unit could get the crew back home whatever the situation.

It was a measure of the confidence in Apollo systems that the program could now gain access to hitherto inaccessible landing sites. But the trajectory actually imparted by the S-IVB's re-burn put Apollo 17 on a collision course for the Moon so a correction would be necessary at the MCC_2 opportunity to steer it back up to the correct miss distance, about 94 km. After a sleep period that ended with the spacecraft more than 125,000 km from Earth, the crew did some navigation work and performed exercises. They also wrote down plans for updating mission clocks so that the indicated elapsed time would read the same as events already in the flight plan. Speeding to the Moon in 2 hr 40 min less time than planned, the crew would be back on schedule for all the lunar activities. Toward the end of the second day in space, Cernan and Evans got their heads down once more, but Jack Schmitt stayed awake, looking out the window at Earth, the most beautiful jewel in Apollo 17's black sky.

SCHMITT 'I think the one philosophical point, if any, that comes out of it is that somebody, probably three and a half billion years ago or so, could have looked at the Earth and describe patterns not too dissimilar. And it was within those patterns that life developed, and now you see what that life has progressed to doing, and I certainly think that all of us feel it has not stopped that progression, and we'll probably see it doing things that even you and I can't imagine them doing.'
CAPCOM 'Roger Jack, we concur.'
SCHMITT 'Bob, you always wish that you had a poet aboard one of these missions so he could describe things that we're seeing and looking at and feeling in terms that might transmit at least a part of that feeling to everybody in the world. Unfortunately, that's not the case, but he certainly couldn't look at that fragile blue globe and not think about the ancient sails of life that are crossing its path and wonder ahead up to the present, to the modern sails of life that are represented by men that developed out of that life that are sitting there next to you and that are in the country in all sorts of different guises and working towards the same end, and that is to put that life farther into the Universe. I certainly hope that some day in the not too distant future, the guy can fly who can express these things.'

Day three began for the crew 33 hours after lift-off; they would get four sleep periods on the way to lunar orbit rather than the normal three since, although compressed now by 2 hr 40 min, the flight path for this mission was unusually long. First order of business was to perform the course correction, a brief 1.7 second SPS engine burn to speed the spacecraft by 11 km/hr. It was sufficient to pull Apollo 17 back out to a lunar miss distance of 99 km. Then the crew got ready for Cernan and Schmitt to inspect Challenger, which they did for about two hours, finding the spacecraft in good condition. One of the 12 docking latches was found to be loose, and a few nuts and washers came drifting out from the LM, but communication checks worked out well.

At the end of the third day in space, Apollo 17 was 142,000 km from the Moon and still slowing down en route to the equigravisphere. It took Mission Control a full hour to wake the crew. They even considered sending a high frequency oscillator signal up the communications line to rouse the sleeping trio, until finally the crew responded.

CAPCOM 'Hello 17, hello 17. How do you read us this morning?'
APOLLO 'We're asleep.'
CAPCOM 'That's the understatement of the year!'
APOLLO 'Never let Evans be on watch.'
CAPCOM 'I think we'll go along with that from here on.'
APOLLO 'Okay, we got you. Our biggest problem this morning is keeping Ron from going back to sleep.'
CAPCOM 'One final thing. Management has informed me that since you've been so late getting to work this morning, we are going to have to dock you all a day's annual leave.'
APOLLO 'All of us! I can understand that for the Commander, since he's always the Commander, but I do not understand why the LMP loses a day!'

There was no real urgency about the scheduled events for the next few hours – a few scientific experiments that could be performed on a comparatively flexible schedule, and another inspection and check inside Challenger. The Lunar Module again came through with flying colours and shortly before starting the final sleep session of the trans-lunar coast phase, Apollo 17 crossed into the Moon's gravitational influence. But this time the crew got an intentional lay-in when controllers decided not to perform another course correction.

Coming up on the Moon in the final hours of the journey, Cernan, Evans, and Schmitt got an unusually spectacular view. Earlier flights put the spacecraft into the shadow of the lunar sphere but because Apollo 17 was flying this unique trajectory the crew were able to watch the Moon as it pulled them ever closer. 'It's an unbelievable view through the monocular now. You can really see down in the depths of some of the larger craters with a great deal of clarity. And you can see some of the higher ridges actually rolling right over the horizon as they go away from you,' called Cernan from his position at a window.

Only a few hours before, the crew jettisoned the SIM bay door, which went spinning away from the spacecraft to reveal the exposed sensors of science equipment that would keep Ron Evans busy for more than six days around the Moon. Passing around the west limb of the lunar disc as viewed from Earth, Apollo 17 was just 730 km above the Moon, curving around and down toward a point less than 100 km from the surface. It was 86 hr 3 min since lift-off more than three days before; just 1:36 pm Houston time, Sunday, 10 December. Nine minutes later the SPS engine lit up for the second time this flight, burning for a long and steady 6 min 33 sec to place the docked vehicles into a 97 × 315 km path around the Moon.

Problems with communication prevented timely conversation between the spacecraft and the ground but within minutes of appearing round the east limb the antennae were locked on solid. 'Houston, this is America, you can breathe easier. America has arrived on station for the challenge ahead,' said Gene Cernan. Shortly thereafter the S-IVB slammed into the surface, but again too far west for the impact to be visible from the spacecraft. The seismometers recorded its vibrations, however, as expected. Jack Schmitt enthused over his view of the surface below, sending professional opinion to the Moon watchers on Earth. But the first few hours of lunar orbit were busy with engineering duties.

At the end of the second revolution the SPS engine was again fired, putting the docked vehicles into a path 27 × 109 km around the Moon with the low point situated 10° west of the landing spot. Instead of reaching pericynthion uprange of the surface target, Apollo 17's first descent-orbit-insertion – DOI-1 – burn was one of two designed, like the single maneuver on earlier flights, to put the LM at a suitable point for powered descent. Higher at pericynthion than normal for DOI, Apollo 17 would stay in this path for ten revolutions while the crew slept with little threat of the orbit decaying dangerously close to the lunar surface.

Schmitt slept the longest during the 'night' that followed, with Evans getting the least rest. It was early light in Houston, Monday, 11 December, as Mission Control transmitted morning music to wake the drowsy trio in Moon orbit. They would land on the Moon this day to begin one of the most fruitful periods of lunar exploration achieved by Apollo. Aboard the spacecraft Gene Cernan's crew had breakfast and as America and Challenger swept out of sight at the end of revolution 9, preparations were moving along for the coming events.

'Okay, Houston, we're with you and we're in the process of getting the tunnel pressurized and moving right towards probe and drogue removal,' came word at acquisition of signal on revolution 10 as the crew began LM activation procedures. At an elapsed time of 105 hr 9 min Challenger's electrical power was switched on and just ten minutes later the docked assembly swept out of sight once again. Forty-seven minutes later they reappeared on revolution 11, deployed the landing gear, checked the guidance platform and after sealing the respective spacecraft Challenger was set free from America's probe on the far side of the Moon shortly after beginning revolution 12. In Houston bookmatches were passed around, specially procured by flight dynamics officers from all the MOCR shifts, which said 'The Trench, Mission Control,' on the reverse of which were the words 'Mercury, Gemini, Apollo, Skylab, Apollo Soyuz, Shuttle.' The first three had ticks against them. Confidence indeed that this final landing operation would go well. And it did.

'Okay Houston, we're floating free out here,' came word from America as the space vehicles moved once more into communication with Earth, 'The Challenger looked real pretty.' It was time for engineering information, maneuver pads, and vital details about the upcoming descent. Mission Control told Challenger the LM would be a little higher than predicted at powered-descent-initiation, and that the T1 stay/no stay point would come 17 minutes after powered descent began, the second point coming $7\frac{1}{2}$ minutes after that. They would have less than five minutes after landing to decide whether or not to remain on the Moon, vital minutes to check readings on instruments and obtain word from Houston about telemetry unseen aboard the frail landing craft. But first both vehicles had to move to different paths before any of this could happen.

On the far side of the Moon, minutes before starting revolution 13 (traditionally the longitude of the lunar-orbit-insertion burn), America fired its SPS for 4 seconds to circularize the orbit. Almost exactly five minutes later, Challenger fired its DPS engine

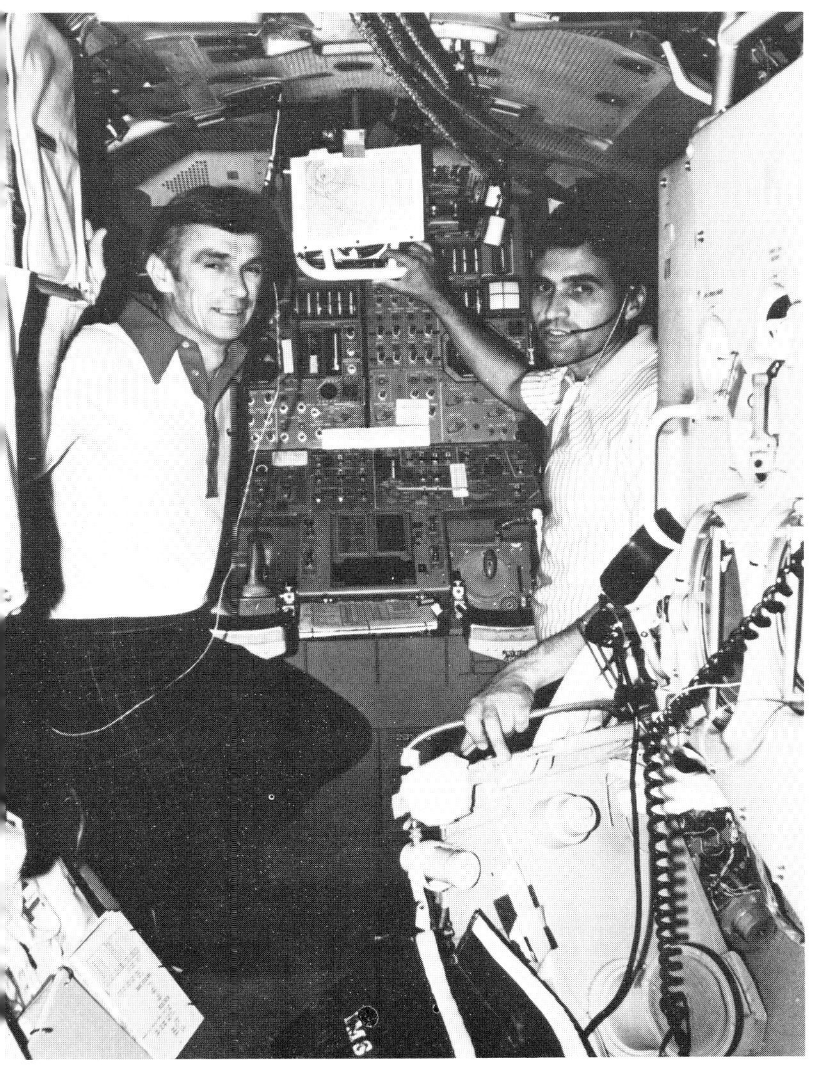

Apollo 17 crewmembers Cernan (left) and Schmitt familiarize themselves with the interior of their Lunar Module. They were to use this as home for nearly four days.

to drop closer to the lunar surface; over the ten preceding revolutions the Lunar Module's orbit had decayed to a pericynthion of just 22 km, but the DOI-2 burn dropped it to a mere 11.4 km. The actual low point of this elliptical path was downrange of the PDI point so the spacecraft would be about 18 km above the Moon when the descent began. It would not matter too much and be better than having Challenger on the ascending leg of its orbit at PDI.

At the MOCR, Gerry Griffin's team were on shift to watchdog Challenger through the twelve minutes of powered descent. Coming round the limb of the Moon the last time for three days, Gene Cernan and Jack Schmitt were able to report a good burn and Ron Evans in America passed along a report about his own circularization maneuver. It was a short run up to powered descent: just 13 minutes between acquiring Challenger's signal and firing up the DPS engine, four fewer than with the first landing which had up to now the briefest coast between acquisition of signal and PDI. Nevertheless, Houston uplinked a state vector to Challenger in record time. 'Man, I'll tell you we're getting close,' said Cernan as the Lunar Module swept down to its new pericynthion. He had been that way before; three and one-half years earlier when it was his lot to reconnoitre the path for Apollo 11, and now he could fire the 'dips' and set Challenger on its perch.

On time, the four down-firing thrusters burst into life followed, eight seconds later, by the main engine at 10% thrust. Then the computer ramped the big engine to full thrust and the power was evident. Their feet pressed down to the floor as the LM decelerated, Cernan and Schmitt monitored the two navigation systems and watched for altitude and velocity lights. At two minutes into the burn, riding on the command from program 63, the crew punched in a landing site correction that told the guidance Challenger had a downrange error of 1,036 meters; it was already smoothing out the extra height, seeking to have the actual parameters right on the predicted lines by the time the LM reached high-gate and the start of program 64. 'The day of reckoning comes in four minutes, Jack,' cried Gene Cernan as they swooped toward high-gate. Radar data looked very good as the differentials nulled to zero.

Nine kilometers above the Moon now, the terrain below looked very familiar, almost identical to the simulator views they saw frequently in training sessions back on Earth. South Massif, off to the left, was visible, as was the far end of the valley floor. Challenger was slowly pitching forward, bringing more and more of the landing site creeping up from the bottom of the two triangular windows. 'Okay, there it is Houston, there's Camelot, right on target. I see it,' called Cernan as he recognized a crater. Challenger was coming down at a steep angle now, less than 2 km above the dusty surface, well below the Massif mountain ranges to left and right of the valley; ahead, a scarp cutting across their site area, and, directly below, the sculptured hills that sealed the valley's eastern flank.

'Challenger, you're GO for landing,' was the word from Houston as Cernan re-designated the landing spot a little to the south. Just 100 meters off the surface, Cernan took over manual control and at 20 meters dust began to obscure the landing site. Watching carefully the movement of Challenger's shadow across the Moon below, Cernan and Schmitt rode the lander down, down to the valley of Taurus Littrow. With a vertical speed of just 4 km/hr, the Lunar Module touched the Moon, settled back on its rear leg into a small 4 meter wide crater, and gave the crew a noticeable backward jolt as it sank a few centimeters into the granular regolith.

But the comments said it all: 'Okay Houston, the Challenger has landed. . . . Boy when you said "shut down", I shut down and we dropped, didn't we. . . . Yes sir, but we is here, man is we here! . . . Houston, you can tell America that Challenger is at Taurus Littrow.' The Lunar Module touched down barely 200 meters from the pre-planned site, tilted back no more than 5°, and in very good condition. 'I'll check everything again. Let's just double-check. Okay. That hasn't changed . . . it looks good . . . the manifold hasn't changed . . . the RCS hasn't changed . . . ascent water hasn't changed . . . the batteries haven't changed. Oh, my golly, only we have changed!' It was true, for they would never be the same again: two of twelve among the billions in history. The last two men to land on the Moon.

By now it was early afternoon in Houston and the crew were scheduled to go out on their first EVA at about 5:30 pm on the first of three surface explorations each lasting seven hours. But Gene

Cernan had a nostalgic quip before they got ready: 'All the way through PDI prior to pitch-over, Jack and I had the real America, or the other America, right out smack out the front window all the way down, which was pretty spectacular.'

Like their predecessors getting ready for the first Moon walk, Gene and Jack were reluctant to skip over the detailed checklist before depressurizing the LM. With only a flexible suit between them and instant death, nobody wanted to assume anything. Consequently, they were about thirty minutes late beginning the first EVA. 'Oh man, that looks like a Santa Claus sack,' said Cernan as he moved the equipment transfer bag. Outside the LM the shadow of the angular spacecraft blocked rays from the Sun that lit the valley's floor beyond: 'Jack, I wouldn't lower your gold visor until after you get on the porch because it's plenty dark out here,' said Cernan as he moved slowly on to the ladder.

SCHMITT 'Okay, Houston. The Commander is about three quarters of the way down.'
CERNAN 'I'm on the footpad. And, Houston, as I step off at the surface at Taurus Littrow, I'd like to dedicate the first step of Apollo 17 to all those who made it possible. . . . Oh, my golly! Unbelievable. Unbelievable, but is it bright in the Sun!'

In Houston, it was 6:05 pm. Now Schmitt was making his way out of the front door, moving backwards on to the porch, feeling with his boots for the top rung on the ladder. 'We gotta go back there,' said Cernan as his colleague pulled the hatch to, 'You lose the key and we're in trouble!' It took both men about one hour to set up the roving vehicle, load it with the equipment they would need and test drive it. 'Can't see the rear ones but I know the front ones turn, and it does move. Houston, Challenger's baby is on the road,' called Cernan as he drove the Boeing Moon-buggy. And then an instruction to his scientific colleague: 'Hey, Jack. Just stop. You owe yourself 30 seconds to look up over the South Massif and look at the Earth.' Then, with the transport set up and ready to move, the crew took a few minutes out to erect the flag and convey a message.

CERNAN 'Houston, I don't know how many of you are aware of this, but this flag has flown in the MOCR since Apollo 11. And we very proudly deploy it on the Moon to stay for as long as it can in honour of all those people who have worked so hard to put us here and to put every other crew here. . . .'
CAPCOM 'Roger, 17. And presuming to speak on behalf of some of those that work in the MOCR, we thank you very much.'

Apollo 17's ALSEP array of scientific instruments was very different from previous assemblages. Commissioned as a special set from Bendix, it incorporated some exciting new equipment. The heat flow was there, as it had been for Apollos 15 and 16, and in the light of important results from 15 scientists eagerly awaited a properly laid out array. New to the Moon, however, was the lunar surface gravimeter which would hopefully report gravitational waves pulsing through the solar system to the lunar sphere. Ever since Dr. Joseph Weber claimed to have detected such waves, physicists and astronomers enthused over the possibility of detecting these signs of violent activity deep in the Universe. Earth is too active to record the minute waves so the Moon was well placed to serve as a unique tuning fork.

Also set up as part of 17's ALSEP was the lunar atmospheric composition experiment, a small but complex box designed to record and report the presence of gas molecules in the vicinity of the surface. It too would be connected to the central data station and powered by a finned cylinder containing a small nuclear power source. A lunar ejecta and micro-meteorites experiment was also deployed. Designed to measure and record the speed, mass, and direction of travel of tiny dust particles striking the Moon's surface, it had a protective cover which would be jettisoned by a small explosive charge after the astronauts left for home. Finally, a lunar seismic profiling experiment, similar in concept to the thumper and geophones carried by Apollos 14 and 16 and to the mortar box designed to lob charges varying distances from a passive seismometer, was attached to the transmitter.

But it was a more sophisticated layout and one which required eight separate charges of between 57 grams and 2,722 grams to be placed at varying distances from four geophones laid out at the ALSEP site, placed in the center and at each corner of a 90 meter equilateral triangle. Left on the surface during all three EVAs, the charges would be precisely located by navigation readings from the rover. With a known distance from the geophones, and a precise knowledge of the detonation times, seismologists could construct a profile of the valley floor by studying seismic waves picked up at the ALSEP site.

Each package contained two delay timers and a plate that had to be physically removed, placing the detonator and the charge in direct contact. Only when three O-rings had been pulled free, and a 1.6 meter antenna extended, would the timers start, at the end of which period the ground could send the signal to detonate the charge; minimum timer delay was 90 hours to ensure the charges would remain locked throughout the period Cernan and Schmitt would be on the surface.

Before deploying the ALSEP equipment, Schmitt put out two cosmic ray detector panels each similar to the one put out at Descartes and returned to Earth. Because of a solar flare, data on that mission had not produced the desired results; the panels would be collected and stowed aboard Challenger at the end of the third EVA. While Schmitt completed deploying the ALSEP intruments, Cernan struggled with the lunar surface drill to get two 2.6 meter holes in which to place the heat flow sensors. Connected to the electronics, and thence to the central data station, the flat ribbon-like cables linking all ALSEP equipment had a special strain relief device to prevent a recurrence of the problem when John Young pulled Apollo 16's heat flow experiment to pieces.

It was hard work drilling the Moon, and the extra pressure exerted sent Cernan's heart rate up to 150 at times. Pressure in the gloves was painful also, and the Commander was surprised at the physical effort required to get the holes drilled. But drilled they finally were and, at the third attempt, an Apollo mission had successfully set up a fully operational heat flow experiment. Now it was time to drill a third, 2.8 meter, hole for the deep-core sample. But this was to be no hole left vacant in the surface. Maximizing the operation's scientific value, Cernan used the hole to deploy a neutron probe experiment comprising a 2.4 meter long section containing boron and plastic sleeves.

Because neutrons readily interact with boron to produce alpha particles the equipment would record alpha tracks etched in the plastic when the two sleeves were rotated into juxtaposition. In this way, neutrons captured by certain elements would be mapped by the probe so that scientists could deduce the mechanics of soil composition within the upper layers of the Moon. The probe would be retrieved on the third EVA, and the boron and plastic sleeves rotated away from each other, effectively turning the detector off and so preserving a valuable score of implanted events. Like the cosmic ray detector, it too would be analyzed back on Earth.

There were few problems putting all the instruments out, but several small delays put the crew behind schedule. First the radioisotope generator had a problem, the dome of the fuel cask being stuck for a few minutes. Then the neutron probe took longer than expected to set up in the deep-core hole. But now the last activity could begin before starting off on the abbreviated geological traverse in what was left of the first EVA. That was to drive off toward station 1 but en route, about 100 meters from the LM, put down the surface electrical properties transmitter for deployment on the way back. The transmitter was part of an attempt to get details of the subsurface structure by observing its electromagnetic properties.

Mission Control told Cernan and Schmitt to cut short the planned geology trip and to go to the flank of a crater called Steno, about 1.1 km from Challenger south-south-east of the landing site. They got there after an eight minute drive, 5 hr 3 min into the EVA. They were about 33 minutes behind schedule. Putting down an explosive package and pulling out its antenna, they then conducted a sampling activity, raking the ground for pebbles, and driving a double-core into the surface with a hammer.

Then they went back toward the LM after 32 minutes at the new station 1, halting at the electrical properties transmitter to set it out. Totally remote from any other power source, the device stood on four short legs and supported a flat solar cell array by which it would obtain electric energy for operation. Two antennae were laid out along the surface at right angles to each other, 70 meters from tip to tip and crossed beneath the transmitter's legs. Sending a continuous signal through the surface, a receiver mounted on the back of the rover would be used periodically to obtain readings during the second and third EVA expeditions,

When the rear fender broke on Apollo 17's Lunar Roving Vehicle, the crew fashioned a makeshift guard from spent card.

contributing information on the electrical nature of the substructure. In this way, large boulders located up to several kilometers below the surface would be 'seen' through the reflected waves picked up some distance from the transmitter as the crew went about their geological tasks. It was similar in concept to the lunar sounder carried by the orbiting America's Service Module.

Before putting out ALSEP, and after beginning the shortened traverse, the crew had taken readings from a traverse gravimeter, also carried for the first time by Apollo 17. That device, mounted to the rear of the rover, would provide data on subtle changes in mass situated within the Moon detected as the crew moved from one location to another; it was a way of recording mascons and minicons known for some time to exist within the Moon and observed in orbital changes to spacecraft about the lunar sphere. It too would be used frequently on the second and third expeditions.

The first EVA lasted 7 hr 12 min. It had been a long and tiring day. The Moon walkers had been up 23½ hours by the time they finally went to sleep on the lunar surface, safely inside Challenger for a full eight hours of rest. 'Well, the Moon's weather is fair and sunny. It's only scattered clouds, and all of those seem to be attached to the Earth,' came comment from the spacecraft. The first lunar 'day' had gone well, with only one minor operational problem still unresolved: broken early in the EVA, a rear fender on the rover came completely away as the crew returned to Challenger, showering the crew with a fine 'rooster-tail' of dust as they bounced and rolled across the surface. Ron Evans had been asleep in the CSM for some time and woke up a couple of hours after Cernan and Schmitt got to sleep.

With musical accompaniment, Houston woke the Challenger men about 1:45 pm, Tuesday, 12 December. In the crew's hours of slumber, John Young had worked up a procedure for mending the rear fender; it was not merely a convenience to prevent dust flying over the vehicle since much delicate equipment was carried and dirty suits were a great inconvenience back in the LM. The recommendation was for lunar surface maps to be taped together and attached to the fender rails with clamps used aboard the spacecraft. It just might work. During preparations for the second EVA, Cernan and Schmitt fell further behind and were about 1 hr 25 min late getting out on the surface. But with a full sleep behind them, both men set about their duties with gusto.

CERNAN 'Okay, I'm going down the ladder . . . God speed the crew of Apollo 17, I think I'll read that every time.'
CERNAN 'Okay, Houston. On this fine Tuesday evening, as I step out on the plains of Taurus Littrow, Apollo 17 is ready to go to work.'
SCHMITT 'Come on hatch. Oh, what a nice day . . . there's not a cloud in the sky – except on the Earth!'

It was time for Cernan to become a lunar rover repair man while Schmitt took telephoto pictures of the North and South Massifs. Then, with the rover loaded up and the traverse gravimeter giving a reading, they set their sights on station 2 at the flanks of the South Massif itself. The mountain was not high, but the slopes were steep, and a light mantle had slid down and partly across the valley floor from some impact millions of years ago. They would sample both the mountain proper and the edge of the light colored mantling. On the way, the first major depression they noted was the crater called Camelot.

CAPCOM 'Okay, one thing I might mention to you guys as you're driving here, Jack, before you start talking again, is that as you go by Camelot you might keep an eye out for blocks along the rim there. A second thing to remind you is if you do stop for a rover sample along the way give us a call and keep us informed because we're timing you on the way out and we're under a 63 minute limit to get you from the LM out to station 2 because of OPS drive back.'
SCHMITT 'Okay, Bob . . . we'll keep you informed.'
SCHMITT 'Okay, the surface is not changing in detail. The surface texture of the fine grained regolith still there is a raindrop pattern. Occasional craters show lighter colored ejectas both all the way down to half a meter in size. Other craters that are just as blocky have no brightness associated with them. Most of the brightest craters have a little central pit in the bottom which is glass lined. The pit is maybe a fifth of the diameter of the crater itself. Okay, we're just south of the rim of Camelot.'

Now more than a kilometer from Challenger, the rover moved on across rough terrain. Soon they came up on Horatio, another crater to their right as they moved almost due west, 2 km from the Lunar Module.

CERNAN 'Yes, that's Horatio. We're right on course, sir. There's a little depression we didn't talk about though between Horatio and Camelot, but it's a depression and not a blocky crater at all. As a matter of fact the total block population has changed – once we get away from the rim of Camelot, block frequency is quite a bit smaller. It's down maybe to less than one percent of the surface.'
CERNAN 'Boy, am I glad we got that fender on.'
SCHMITT 'The scarp looks very smooth from here – no obvious outcrops at this time. Don't seem to be penetrating to any bedrock in the area we're traversing now just to the southeast of Horatio. Horatio has a blocky wall, however, the upper several tens of meters probably of rim looks as if it's either mantle or composed of the light gray regolith material we've been driving on.'

The scarp was not too evident as they headed directly for it; if they kept going they would approach it on the down fault side, appearing to them as a vertical cliff rising up as high as 80 meters from the valley floor. But they were now heading too far south to encounter the scarp and, as planned, would sample the light mantle where it met the valley floor. Several times on their way to station 2 the crew paused to pick up rocks. The next crater they approached was called Bronte.

CERNAN 'That must be Bronte. My, is that big! That's bigger than I expected. I got to go around this thing.'
SCHMITT 'There's not an awful lot of blocks around the rim, just some small ones, compared with what . . . we've seen around Horatio. Nothing at all like we saw yesterday at station 1.'

Then they came upon the light mantle which slid down from the slopes of South Massif when a nearby impact from some meteorite shook the mountain and sent a small landslide down to the valley.

SCHMITT 'That's the light mantle we're coming up on right up here.'
CERNAN 'We're only 100 meters from the light mantle.'
SCHMITT 'Yeah.'
CERNAN 'The craters are much brighter in their walls than we've seen before.'

Then they moved back on to darker material after stopping for another quick grab sample. They were closer to the mountain now, and starting to move up the undulating slopes. 'Yeah, there's no question that there is apparent lineations all over these

Massifs in a variety of directions. Hey, look at how that scarp goes up there,' cried Schmitt as Cernan controlled the rover. 'It looks like the scarp overlays the North Massif doesn't it,' replied the Commander. 'Boy, I tell you. Are those Massifs getting to look big now. Holy Smoley! That scarp looks nice over there too, doesn't it?' said Schmitt.

They drove on past Lara, a crater seen in photographs from orbit, and then by Nansen to stop on its flank at the spot called station 2. 'Boy, you're looking right into Nansen. We're right where we wanted to be for station 2. And it looks like a great place. Big blocks. It looks like quite a bit of variety from here. Different colors, anyway, grays, and lighter colored tans.'

They would spend seventy-two minutes on the flank of this prominent crater and while the TV responded to the commands from a back room at Mission Control, Cernan and Schmitt sampled the regolith, collected rocks, reported readings from the traverse gravimeter, raked the loose material, and retrieved rock chips. The crater itself had been unofficially named after the Norwegian arctic explorer Fridtjof Nansen. 'When you look down into the bottom of Nansen it looks like some of the debris there has rolled off the South Massif and covered up the original material,' commented Schmitt, 'All the boulders that have come down are on the south side of the slope.' The area was such a rich hunting ground for samples that Mission Control told them to spend ten minutes extra at Nansen and delete that time from a later station stop. They were finished and on their way back to Lara crater 3 hr 13 min into the EVA.

At station 2 they were 7.6 km from the Lunar Module but at the next stop only 6 km from their lunar home. The area they sampled was on the rim of a small 30 meter crater about 200 meters from Lara. Again with the TV on they sampled the soil, raked the regolith, and thirty-eight minutes after arriving moved north to a crater called Shorty. What they found there was to make everything else anti-climactic.

Arriving at station 4 about 4 hr 48 min into the EVA, the crew set about sample tasks. Shorty was a dark halo crater 110 meters across and situated close by the light mantle. Thought by some to be a volcanic vent – or fumerole – in the valley floor, Jack Schmitt was concerned to seek evidence for its origin. 'Okay, Houston, Shorty is clearly a darker rimmed crater. The inner wall is quite blocky except for the western portion of it which is less blocky than the others; the floor is hummocky as we thought it was in the photograph. The central mound is very blocky and jagged.' Minutes later, while his Commander attended to the rover, the professional geologist came upon a sight that seemed to show conclusively the volcanic history of Shorty crater.

SCHMITT 'Hey! There is orange soil.'
CERNAN 'Well don't move it 'till I see it. '
SCHMITT 'I stirred it up with my feet.'
CERNAN 'Hey, it is. I can see it from here.'
SCHMITT 'It's orange!'
CERNAN 'Wait a minute – let me put my visor up – it's still orange!'

A typical sample site at which has been placed the gnomon used to show local vertical and the color code of rocks and soils in the vicinity.

SCHMITT 'Sure is, crazy orange. He's not going out of his wits, it really is!'
CERNAN 'Fantastic sports fans. It's trench time.'

What they believed they saw was oxidized material from a vent below the surface but examination back on Earth would tell them it was merely the chemistry that altered the apparent color. But more than half way through their second EVA, it was a jubilant moment for both. They took a reading from the traverse gravimeter, which indicated the possible presence of a buried mass at Shorty. All too soon it was again time to move, this time to the flank of Camelot only 1.4 km from Challenger.

At 6 hours into the EVA they sampled the rim and moved back toward the LM but on the final lap Cernan dropped Schmitt off at the ALSEP array for the geologist to adjust the lunar surface gravimeter at a request from Mission Control. When they unpacked the rover they found abrasive marks on the back-packs from the seats that held them rigid during the long traverses; riding the rover would have been difficult without the firm restraint of the seat sides gripping the square-shaped life support systems.

The second EVA had lasted 7 hr 37 min by the time they re-pressurized Challenger. It had been a good day. After attending to the spacecraft regulator – Mission Control thought there was a slight leak – the crew re-charged their back-packs and had a well deserved meal. Ron Evans in America had been asleep for some time, and now it was time for Cernan and Schmitt to begin their last 'night' on the Moon. Despite a late start, and an extended EVA, they were only one hour behind schedule when Mission Control put them to bed.

Shortly after Evans woke, Mission Control told him that he would probably have to perform a trim maneuver to get the spacecraft in the right path for a scheduled plane change burn with the SPS engine. But that was not an immediate requirement and further tracking would refine the need.

It was early afternoon, Houston time, Wednesday 13 December when Cernan and Schmitt woke to a background of music from the Texas University war hymn. They had slept fitfully, Cernan having managed only three hours' rest. Speed was of the essence now for they would have to lift-off at the scheduled time to reach America on the correct orbit and another sleep period lay ahead. When the third EVA began they were only one hour behind the flight plan.

With each passing 'day' on the Moon the Sun rose higher in the black sky. Taking two weeks to travel from one horizon to the next, the Sun was 15° above the sculptured hills when they went out on their first traverse, 25° at the beginning of the second, but before they completed this day's work it would be 40° high, its dominating presence felt in the higher temperatures. 'Okay, Bob, I'm on the pad and it's about 4:30 Wednesday afternoon as I step out on to the plains of Taurus Littrow – beautiful valley,' said Cernan. Before departing they retrieved the cosmic ray detector which had been outside for $45\frac{1}{2}$ hours and then they drove to the surface electrical properties transmitter site to begin their traverse.

They were to move almost due north to a place called station 6 where, at the edge of the North Massif, they could sample large boulders dislodged millions of years ago by some Moonquake or other seismic event. Moving round the rim of a crater nicknamed Henry, Cernan and Schmitt bounced and rocked along to the base of the mountain; having sampled South Massif the day before, they would now visit the more spectacular scenery at the north end of the valley. From photographs taken by the Lunar Orbiter spacecraft several years earlier, and in pictures returned by the crew of Apollo 15, geologists knew there were interesting boulders to sample at this location. But when the crew approached it across the bumpy valley, it was quite spectacular. 'Oh man, what a slope,' called Cernan as he struggled to find a path. The boulders they wanted to see lay 80 meters up the South Massif on a gentle incline from which snaking tracks led higher up the mountain testifying to the distance the boulders had rolled.

They arrived at station 6, about 1 hr 19 min into the EVA, alongside a massive boulder 18 meters long, 10 meters wide, and 6 meters high, shattered now into five separate segments. It was a geologist's dream. From this vantage point above the lava floor on which their spacecraft stood more than 3 km away, the view was a splendid montage of undulating hills, the bleak face of South

At the foothills of North Massif, Apollo's LRV is dwarfed by this large boulder which had rolled down the mountain.

Massif, and rippled humps across the flat valley. The Sun was noticeably higher today, equal to mid-morning on a Summer day back on Earth. But the lonely, desolate, lunarscape made it unique.

There were brief moments at this high point to reflect mere seconds on what the past $3\frac{1}{2}$ years had done for science. As all the while they busied themselves with samples of the big broken rock, Cernan and Schmitt were never far in thought from the awesome moment of this December day. On an earlier December, two men proved it was possible to remain in space for fourteen days and achieved the first rendezvous between manned spacecraft. On another December, three men left the gravitational bonds of Earth for a flight around the Moon. Now, two explorers were dashing to get the last hand-picked samples before the curtain came down on a decade of exploration.

They stayed at station 6 for $1\frac{1}{4}$ hours and then got ready to drive a few minutes along the North Massif foothills to station 7 just 600 meters away; from there they would move to station 8, a notch like some gulley incised between the Massif to the north and the sculptured hills to the north-east. 'You know Jack, when we finish with station 8, we will have covered this whole valley from corner to corner – I didn't think we'd ever really quite get to that far corner,' mused Cernan. But then there was another 'first' on the Moon.

CERNAN 'Oh, man. Houston, we've got a couple of dented tires.'
CAPCOM 'What's a dented tire.'
CERNAN 'A dented tire is a little golfball size, or smaller, indentation in the mesh. How does that sound to you?'
CAPCOM 'Sounds like a dented tire, that's how it sounds.'

The open wire mesh 'tire' had succumbed to the exertions of the traverse and suffered deformation in the process. As they moved to station 6 the crew looked down into the valley, where, from their position high along the slope, they could see much more of the surrounding terrain.

CERNAN 'We must be about 200 meters up the slope, looking at that little valley down there, Jack. Am I right?'
SCHMITT 'Yes. I think you're right.'

The extra time spent at station 6 abbreviated their stay at station 7 and at 2 hr 58 min into the EVA they moved on to station 8, arriving at that point, 4 km from Challenger, sixteen minutes later. Trench samples, soil samples, soil from beneath a boulder, rake fragments, rock fines, core tube samples, traverse gravimeter readings. They got a lot done in just forty-eight minutes. Now it was back down into the valley for station 9 at a crater called Van Serg almost half-way back to the LM; from the furthest point on the previous Moon walk to the furthest on this day of exploration, Cernan and Schmitt covered nearly 12 km of the valley floor.

They spent almost one hour at Van Serg before heading across to Challenger 2.2 km away, eliminating a planned station 10 because of extra time already spent at previous stops. It was not an important site, only put in so as to get the maximum return should the crew have a few minutes to spare.

The EVA was nearly 6 hours over when the crew went about final tasks around the Lunar Module. The neutron probe experiment was retrieved from the deep-core hole, having been emplaced for 49 hours. The surface gravimeter was adjusted further. Two explosive packages were put down near the electrical properties transmitter; throughout the three EVA sessions, Cernan and Schmitt had deployed other packages at varying distances from the ALSEP. On all three periods outside Challenger, a total 2,200 pictures had been taken and a record 110 kg of lunar samples collected. The crew had traversed a total 35 km including a single traverse of 19.5 km during the second EVA which broke all records. Now it was time to perform the final closeout ceremony, a few minutes of thought for all that preceeded this flight.

CERNAN 'Houston, before we close out our EVA, we understand that there are young people in Houston, today, who have been effectively touring our country. Young people from countries all over the world, respectively touring our country. They had the opportunity to watch the launch of Apollo 17, hopefully had an opportunity to meet some of our young people in our country, and we'd like to say first of all, welcome and we hope you enjoyed your stay. Second of all, I think, probably one of the most significant things we can think about, when we think about Apollo, is that it has opened for us, for us being the world, a challenge for the future. The door is now cracked, but the promise of that future lies in the young people, not just in America, but the young people all over the world. Learning to live and learning to work together.

'In order to remind all the peoples of the world, in so many countries throughout the world, that this is what we all are striving for in the future, Jack has picked up a very significant rock, typical of what we have here in the valley of Taurus Littrow. It's a rock composed of many fragments, of many sizes, and many shapes, probably from all parts of the Moon, perhaps billions of years old. But a rock of all sizes and shapes, fragments of all sizes and shapes, and even colors that have grown together to become a cohesive rock outlasting the nature of space, sort of living together in a very coherent, very peaceful manner. When we return this rock or some of the others like it to Houston, we'd like to share a piece of this rock with so many of the countries throughout the world. We hope that this will be a symbol of what our feelings are, what the feelings of the Apollo Program are, and a symbol to mankind that we can live in peace and harmony in the future.'

SCHMITT 'A portion of a rock will be sent to a representative agency or museum in each of the countries represented by the young people in Houston today, and we hope that they will, that rock and the students themselves, will carry with them our good wishes, not only for the New Year coming up, but also for themselves, their countries, and all mankind in the future. Put that in the big bag, Geno.'

CERNAN 'In the big bag. We salute you, promise of the future.'

CAPCOM 'Roger, Jack and Gene, we thank you for your sentiments and your interests.'

CERNAN 'And now, let me bring this camera around. To commemorate not just Apollo 17's visit to the valley of Taurus Littrow, but as everlasting commemoration of what the real meaning of Apollo is to the world, we'd like to uncover a plaque that has been on the leg of our spacecraft that we have climbed down many times over the last three days. I'll read what the plaque says to you. First of all, it has a picture of the world. Two pictures. One of North America and one of South America. The other covers the other half of the world, including Africa, Asia, Europe, Australia, it covers the North Pole and the South Pole. In between these two hemispheres, we have a pictorial view of the Moon. A pictorial view of where all the Apollo landings have been made. So that when this plaque is seen again by others who come, they will know where it all started.

'The words are: "Here man completed his first exploration of the Moon, December 1972 AD. May the spirit of peace in which we came be reflected in the lives of all mankind." It's signed Eugene A. Cernan, Ronald E. Evans, Harrison H. Schmitt, and most prominently Richard M. Nixon, President of the United States of America. This is our commemoration that will be here until someone like us, until some of you who are out there, who are the promise of the future, come back to read it again, and to further the exploration and the meaning of Apollo.'

CAPCOM 'Roger, Gene. We in Houston copy that and echo your sentiments and Dr. Fletcher is here beside me. Like to say a word to the two of you.'

FLETCHER 'Gene and Jack, I've been in close touch with the White House. And the President has been following closely your absolutely fascinating work up there. He'd like to wish you God's speed as you return to Earth and I'd like to personally second that. Congratulations, we'll see you in a few days. Over.'

CERNAN 'Thank you, Dr. Fletcher. We appreciate your comments and we certainly appreciate those of the President. And whether it be civilian or military, I think Jack and I would like to give our salute to America.'

SCHMITT 'And Dr. Fletcher, if I may, I'd like to remind everybody, I'm sure of something they're aware, but this valley, this valley of history has seen mankind complete its first evolutionary steps into the Universe. Leaving the planet Earth and going forward into the Universe. I think no more significant contribution has Apollo made to history. It's not often that you can foretell history, but I think we can in this case. And I think everybody ought to feel very proud of that fact. Thank you very much.'

It had been a moving moment for in their own simple language Gene Cernan and Jack Schmitt reiterated the sentiments of many on Earth who had worked long and hard to give that dream an undenied reality. Cernan felt a particular responsibility for this last Moon landing and was mindful of the significance in those final few steps. 'Ah, what a nice little machine,' he said before walking back a short distance to the Lunar Module, having parked the rover behind Challenger so that the TV could view lift-off, 'Good old mother Earth is right smack in the center.' Then he had another thought for the several million listeners back on Earth.

CERNAN 'Bob, while we've got a quiet moment here . . . I'd just like to say that any part of Apollo that has been a success thus far is probably, for the most part, due to the thousands of people in the aerospace industry who have given a great deal besides dedication and besides effort and besides professionalism . . . and I would just like to thank them. And I God Bless you and thank you. I guess there might be someone else that has something to do with it too, and I've been reading His signs, maybe not from Him directly, but His in Spirit, as we run up and down that ladder and that's "God speed the crew of Apollo 17," and I'd like to thank Him too.'

The Ascent Stage of Apollo 17's Challenger lifts cleanly away from the Descent Stage in this spectacular liftoff viewed through the camera of 17's Lunar Roving Vehicle parked behind the LM. This scene brought to successful completion a faultless three-mission contribution from the color TV built by RCA.

The last picture taken from the Moon's surface before Apollo 17 returned its crewmembers to earth.

Final last minute adjustments to the ALSEP array followed and Cernan took one last picture of Challenger standing alone at Taurus Littrow. Then it was time to get back up into the Ascent Stage.

CERNAN 'Bob, this is Gene and I'm on the surface and as I take these last steps from the surface, back home for some time to come but we believe not too long into the future, I'd just like to list what I believe history will record, that America's challenge of today has forged man's destiny of tomorrow. And as we leave the Moon at Taurus Littrow, we leave as we came and God willing as we shall return, with peace and hope for all mankind. God speed the crew of Apollo 17.'

That third EVA had lasted 7 hr 15 min, bringing to more than 22 hours the total time spent by Cernan and Schmitt exploring the surface. About one hour later the crew depressurized Challenger a fourth time to jettison excess equipment out the porch. Shortly after that they began their last lunar surface sleep period. While they slept, Ron Evans fired America's thrusters to trim his orbit in preparation for a plane change needed to get the CSM in position for the coming ascent. One-half orbit later the plane change was made exactly as planned using the SPS engine.

After waking up, Cernan and Schmitt collected more items scheduled to be dumped on the surface and at 2:58 pm Houston time, Thursday, 14 December, opened for a fifth time the door through which they had passed to explore Taurus Littrow. 'Here you go, Santa Claus,' came a cry from Challenger as another bag shot through the open space. Two minutes later the hatch was closed and the crew settled back into lift-off preparations.

Ignition of the single APS engine came at 4:55 pm Houston time, ending a record 75 hours on the Moon. At pitch-over, tracking stations lost telemetry from Challenger but that was a ground problem quickly solved. 'Challenger, your trajectory is right on the money, both systems are go,' was the word from the MOCR as

Cernan and Schmitt observed the prime and back-up guidance systems. A small tweak burn with the APS engine was performed after getting into orbit and as the two spacecraft swept across the terminator the Apollo CSM hove into sight about 207 km away. Five minutes after moving around the Moon's west limb and out of contact, Challenger fired again the APS engine that safely brought it up from the surface, pulsing toward America in a terminal-phase-initiation burn designed to bring it alongside the CSM about fifty minutes later.

At acquisition of signal on revolution 52, the two spacecraft were less than 2 km apart. Before docking, Challenger flew around America inspecting the SIM bay and taking pictures. Capture was not achieved at the first try and a second attempt was necessary before the three small latches snagged the Ascent Stage drogue. As Challenger pitched down to get in the correct docking attitude, up across the forward windows came a view of the Taurus Littrow landing site directly below. 'What a super flying machine,' came a voice from Challenger, 'Okay Houston, we have capture.'

Minutes after docking, capcom Gordon Fullerton read a message from President Nixon which, in part, said that 'Few events have ever marked so clearly the passage of history from one epoch to another. If we understand this about the last flight of Apollo, then truly we shall have touched a many splendored thing.' Now there was the familiar routine of pumping up America's cabin pressure, opening the tunnel linking the CSM with Challenger, transferring sample bags and equipment, and closing up the Ascent Stage. Nearly two and one-half hours after docking, those tasks were nearly complete.

CAPCOM 'Challenger, you're go for closeout.'
CHALLENGER Okay, Houston, Challenger is going off the air.'
CAPCOM 'Okay, Challenger. It's been a pleasure talking to you the last few days.'
CHALLENGER 'It seems like an unfitting finish to a super bird, but it's got one more job to do.'

After closing up the tunnel in America, Challenger's electrical power was finally switched off from the CSM; it had been operating for eighty-six hours already, now its own batteries would keep it going down to an impact with the surface. Five minutes after jettisoning the Ascent Stage, America fired its thrusters to move away and at 12:31 am, Houston time, on Friday, 15 December, the thrusters fired on Challenger to send it to the surface. It struck the South Massif only 9 km from the ALSEP array and seismometers at all the other sites recorded the shock waves as they traveled through the Moon.

Now the orbital science part of Apollo 17 could begin in earnest, for with three astronauts on hand Cernan could occupy the left couch monitoring systems and generally running operations while Schmitt and Evans tended the battery of instruments and cameras aboard the CSM. It was the middle of the night in Houston when the three tired astronauts settled down to rest. They were woken shortly after noon that Friday to the tune 'The First Time Ever I Saw Your Face.'

'Good morning Gold Team, this is the Command Module Pilot of the spaceship America, and we're ready to go to work again this morning,' quipped Evans in response. It was to be a fruitful last day in Moon orbit. For five days already, Ron Evans had kept the science of Apollo alive from above the lunar landscape by methodically working with a separate communications channel, monitored and administered by its own flight director and capsule communicator, to operate the SIM bay scanners. By the end of this sixth day around the Moon, the panoramic camera would have obtained a total 1,603 frames, the mapping camera would have secured more than 3,000 frames, and the laser altimeter would have made 3,769 shots to obtain accurate height data.

In addition, the infrared radiometer obtained more than 100 hours of data, the lunar sounder, although plagued by a problem with the antenna extension/retraction mechanism, obtained more than 10 hours data, and the ultraviolet spectrometer worked away above the Moon for 114 hours. Some of these instruments would be used during the trans-Earth coast phase but the impressive achievements of Apollo's orbital science activity served well the objectives of the J-series program. Also conducted that day was a geology de-briefing session where the crew talked at length with Mission Control on issues that could possibly gain advantage from the window view.

In the early morning hours of Saturday, 16 December, the

crew of Apollo 17 settled down for their last night in lunar orbit, but not before Ron Evans sent an indirect message to his family: 'And if my home front's listening, I just want to say good night and sleep tight.' Eight hours later Mission Control piped up a 'Come on Baby Light My Fire,' wake-up call. It was 9:03 am, Saturday, and spacecraft America was just about to complete 71 orbits of the Moon. More science activity lay ahead before the important TEI burn late that afternoon, squeezing the last possible value from the SIM bay equipment.

Spacecraft America went round the west limb of the Moon for the last time at 5:11 pm, Houston time, approaching the end of revolution 75 and the final minutes prior to the trans-Earth-injection; it was the last time Apollo's Service Propulsion System would fire on a Moon flight. But technical information was passed along right to the end.

CAPCOM 'America, Houston, about 2 minutes 'till LOS. One reminder about the DSE, we'd like you to go to low bit rate just prior to LOS as per the flight plan. And then go back to high bit rate at 6 minutes prior to ignition per your burn cue card and you can just leave it in high bit rate from there on through AOS. We just went around the room once more, everything looks good. Have a good burn and we'll see you and the TV picture as you come out on the other side. Over.'
AMERICA 'Okay, Gordie, thank you. We're looking forward to a good burn. And we'll see you coming out the other side.'

And that was it – the last word from lunar orbit almost exactly four years after Frank Borman, Jim Lovell, and Bill Anders first dropped into a circular path about the Moon's sphere. Twenty-four minutes later the trusty SPS fired up on command, accelerating Apollo 17 by more than 3,300 km/hr. Right on time the spacecraft appeared, climbing fast away from the gray lunar surface. 'Houston, America has found some fair winds and some following seas, and we're on our way home.' It was the most welcome message of all. 'I know there's not as many smiling faces up here as there are down there, but we're making up for the difference in numbers.'

For only the second time on an Apollo flight, TV viewers saw a picture of the Moon's far side, or at least a portion of it, for as America climbed out high and fast it sneaked a backward glance partially around the east limb as viewed from Earth. For several minutes the crew enthused over the view they had of features on the Moon below them, and then Gene Cernan reflected again on the true meaning of Apollo:

'We're looking back at some place, I think, we will use as a stepping stone to go beyond some day. . . . It's a faith I truly and dearly have. And I think we will see it in our lifetime not just as a nation, but as a world. . . . I think the Apollo program not only has given us the first step to that sort of impossible dream, but has given us an opportunity to make the first step in bringing a world together as one unit so that we can make that step together. . . . I think it's important that in doing so (we) establish a tradition of peace and freedom within the solar system. From that larger home now, we move to greet the future.'

It was midnight, several hours later, when the crew began their first trans-Earth sleep period, about 7:23 in the morning of Sunday, 17 December, when they awoke to the strains of 'Home for the Holidays,' sung by Jerry Vale.

CERNAN 'Hey, your choice of music is getting better down there. We're going to have to keep you there every morning.'
CAPCOM 'Well if I'm here waking you up on Wednesday morning, fellow you're in trouble!'

The Command Module was due to splash down shortly after noon Tuesday, Houston time. Minutes after waking up, the crew passed from the Moon's gravity to the Earth's sphere of influence; now they would start to accelerate back to the home planet. A significant item read up to the crew on the by now traditional newscast from Mission Control identified the flight of Apollo 17 with that made by Wilbur and Orville Wright exactly 69 years earlier; such progress in man's allotted span.

It was early afternoon in Houston when Ron Evans got to perform his own space walk. Leaving the confined Command Module to drift back along hand rails on the Service Module he made three trips to recover, on separate journeys, the cassettes from panoramic camera, mapping camera, and lunar sounder equipment. 'I can see the Moon right behind me. Beautiful. The Moon is down here to the right. Full Moon. And off to the left just outside the hatch down here is the crescent Earth.' Evans commented on slivers of silver paint flaking from the spacecraft wall, and at the lower end of the SIM bay he reported at length on the condition of instruments and equipment. The EVA lasted more than one hour, nearly 290,000 km from Earth.

During that evening, Mission Control fired the last of the eight explosive packages left at varying distances from the ALSEP array. The first three had been fired while America orbited the Moon, two had been set off within a few hours of leaving orbit, and the last three were detonated between late afternoon and mid-evening that Sunday. 'All eight charges have been detonated and they were all on schedule and produced excellent results,' said capcom Gordon Fullerton as the crew prepared to get another night's sleep. 'These data were used in conjunction with the Ascent Stage lift-off and also its impact data which should give us an excellent picture of the geologic structure of the outer three kilometers of the Moon.' Fullerton was reading from a summary written up by scientist Joe Watkins to keep the crew informed of their experiments deployed so painstakingly at the end of the previous week. It was after midnight when they finally got to sleep.

The morning call to Apollo 17 sounded out with the song 'We've Only Just Begun,' at 7:53 am Monday, quickly followed by suitable music for the coming Christmas season. This final full day en route home was a time of preparation for re-entry on Tuesday morning, for housekeeping duties aboard America, and for a search to locate missing scissors that under the influence of high g loads during re-entry could prove dangerous. Across the noon hour, Houston time, Ron Evans donned the light flash detector but that proved fruitless when after one hour he had seen nothing. Later, during the early afternoon, all three crewmembers answered press questions about their flight.

For nearly one-half hour and with the television on, Cernan, Evans, and Schmitt offered their own peculiar views on aspects of the flight and the experiences they had received. It was left to Cernan to offer the final word of formal acknowledgement: 'I guess I can certainly assume during the flight we've carried many well wishes and very many prayers aboard from people throughout the world. I personally believe that those prayers played no small part in any success that we were able to achieve on this flight. I ask those people however to continue their prayers particularly for some of our friends and some of our comrades . . . who may not have the opportunity to get home and enjoy the Christmas that we're looking forward to. And with that from Apollo 17 spacecraft America on 18 December, 1972, we all wish you a very, very, merry Christmas and a happy holiday season. God speed and God bless you all.'

The crew began their sleep period shortly before the midnight hour, minutes after capcom Bob Parker slipped into his seat for a full eight hour session.

AMERICA 'We bid you hello Bob, and at the same time, goodnight.'
CAPCOM 'What can I say? I'm cryin'.'
AMERICA 'Well, we thought we'd give you about eight hours to think about it.'

Apollo 17 was now 136,000 km from Earth, moving at 7,200 km/hr. When the crew woke to the strains of 'Anchors Aweigh,' America was 81,000 km away travelling Earthward at nearly 10,000 km/hr. It was 6:53 am, Tuesday, 19 December. Little more than three hours later the spacecraft thrusters were fired for nine seconds to nudge Apollo 17 on to a precise course for entry inclined 6.49° to the local horizontal. 'We're burning, Houston,' came the call from America, followed a minute later by readouts from the displays.

Soon it was time for the Moon men to get back into their bulky suits. 'The other guys are crawling around down under the couches – they're trying to get the stuff locked in there,' said Ron Evans. 'I think I must have shrunk; my shoes went on easier now than they did on the fitting.' There was an element of truth in that. Denied the blood pooling that usually accompanies an Earth gravity environment, astronauts in a weightless state experienced puffy cheeks and arms while the legs and feet slimmed down.

But even Mission Control was reluctant to turn off instruments in the Service Module that had done so much for orbital science: 'And it's sad to shut off the SIM bay; it's operated so tremendously on this mission.' It was 12:57 pm when the Service Module was jettisoned from spacecraft America. 'Didn't we get it!?' came a voice from space, 'That thing really bangs, doesn't it?' Moving fast now toward Earth's atmosphere, the Command Module swept across northern Siberia and down across the Sea of Okhotsk north of Japan, returning home at a higher approach latitude than ever before. 'America, you're looking great, we've got a TV picture of the weather in the recovery area, and the ship *Ticonderoga* and it's looking great,' came word from the MOCR.

Neil Hutchinson's controllers were on shift for the final hour of the last Moon mission, the last time for many years a spacecraft would re-enter at such enormous speed. 'That's really moving, isn't it,' said an unidentified voice in the final seconds before a communications blackout enveloped spacecraft America, bringing it home to a Pacific dawn. Banging around in response to guidance commands, the RCS thrusters kept Apollo 17 on course as it fell through the increasingly dense layers of the atmosphere. 'Hey, that's good – beautiful from here,' came another voice minutes later as, out of the blackout now, two drogue parachutes popped away from their respective mortars.

Oscillating wildly on the liners that held the canopies, the crew were surprised at the swaying movement. Then the main parachutes deployed followed by fully open canopies with only one taking longer to disreef than planned. During re-entry, Ron Evans wore a special orthostatic countermeasure garment

Apollo CSM America and the antenna deployed for the lunar-sounder experiment in which HF and VHF signals reflected from the Moon below allowed scientists to determine conditions to a depth of more than 1 km.

designed to apply and maintain pressure to his legs. Medical examination would determine the value of such a device for preventing blood moving into the lower limbs too quickly on re-entry.

When a television picture appeared on the big screen at Mission Control showing spacecraft America descending on the parachutes loud cheers erupted and people clapped. 'Hello, Recovery, it's a beautiful day,' came a voice from the Command Module. Indeed it was. A fittingly triumphant climax to eleven manned missions with North American Rockwell's moonship. It was precisely one second before 1:45 pm, Houston time, when the spacecraft struck the Pacific waters, It remained in the Stable I, apex up, position, about 2.4 km from the planned spot. The recovery ship *Ticonderoga* was 6.4 km away.

In Mission Control, Congressman Olin Teague shook hands with technicians and engineers as excitement moved through the MOCR. As the recovery helicopters moved in, flight director Neil Hutchinson implored his men not to light up their cigars – by now a tradition for Apollo recoveries – until the crew stepped to the deck of the *Ticonderoga*. Perched atop a ladder in the Mission Operations Control Room, network controller Dave Young was ready to hang the final plaque in the Apollo series on the wall with all the rest.

Less than one hour after splashdown, Cernan, Evans, and Schmitt were safely delivered to the Navy recovery ship. The operational phase of Apollo Moon missions was officially over. Later that day the Command Module, like its predecessor still containing residual propellants to prevent chemical decay of parachute lines should the fluid contact the risers while being jettisoned, was secured aboard the *Ticonderoga*. Next day, the three astronauts were flown to Samoa and thence to Ellington Air Force Base and home via Norton Air Force Base in California. The flight itself, the longest Moon mission, had lasted 12 days 13 hours 52 minutes. But the science it bequeathed would last much longer.

Like ALSEP instruments left at four other sites, Apollo 17's battery of scientific equipment would endure far beyond the period for which it was designed, continuing to send back a stream of data for several more years. As for the 379 kg of lunar rock and soil brought back by all six landing missions, that would keep scientists busy for many decades. In one respect the cessation of manned lunar landings was a timely affair, for the sheer magnitude of scientific capability and the volume of data flowing back was now beyond the ability of existing institutions to handle adequately. Only time would allow considered and careful analysis of physical samples and remote sensing data.

The real irony of Apollo's success lay in the lack of public awareness that the J-series flights had been so scientifically rewarding. Where once Apollo attempted to bring political gain, it now stood proud with other great voyages of discovery coupled with scientific exploration. It was finally to parallel the Antarctic explorations of Robert Falcon Scott, and the Pacific journeys of Captain Cook. By the time Apollo had maximized its real potential, however, the public audience had packed up and gone elsewhere.

The last Moon mission marked a significant turning point in the US space program, for where all previous objectives had polarized around a desire to move further away from Earth, the home planet itself now became the focus for attention. In the years that followed the December, 1972, flight of Apollo 17, NASA would be forced to dismantle all the complex technology that carried men beyond Earth's gravity. Within a few years it would not have the vehicles to go anywhere beyond Earth orbit and even if a political need arose once more to strike out for other frontiers the tools to accomplish that job would have to be forged anew. Some would say it was a retrogressive step, this emphasis on Earth applications, but others would see in it a more permanent platform on which to base a truly viable manned space program.

Apollo was a product of the expendable era; the reusable shuttle, although limited to Earth orbit only, was the first of the true aerospace vehicles that would ultimately prove of greater benefit to long-term goals. Yet for all the arguments of economics based on practicalities it was the awesome goal of Apollo that gripped the imagination and for at least a few years this century lit a flame that still burns bright in memory. Among the many words written about the meaning of the Moon flights, those from Richard M. Nixon and Harrison H. Schmitt project what must surely stand as the final, authoritative statements from the head of state and the dedicated scientist:

RICHARD NIXON 'Though our ancestors would have called the deeds of Apollo miraculous, we do not see our age as an age of miracles. Rather, we deal in facts, we deal in scientific realities, we deal in industrial capacity, and technological expertise, and in the belief that men can do whatever they turn their hands to. For all this, however, can we look at the record of 24 men sent to circle the Moon or to stand upon it, and 24 men returned to Earth alive and well, and not see God's hand in it?

'Perhaps, in spite of ourselves, we do still live in an Age of Miracles. So if there is self-congratulation, let it be tempered with awe, and our pride with prayer, and as we enter this special time of spiritual significance, let us reserve a moment to wonder at what human beings have done in space to be grateful.'

HARRISON SCHMITT 'Man's unique character among the living species of nature is manifested in many ways; one such manifestation is that he has the audacity to try to understand his place in the long scheme of things, and the further audacity to try to use that understanding to alter the long scheme of things. The record of Apollo, I believe, is a record of man's audacity to understand his Moon; the record of his use of that understanding is just beginning.

'As to the historical legacy of Apollo, I have found no reason to change my thoughts expressed as we left the Moon, and the valley of the Taurus Littrow. "That valley of history has seen mankind complete its first evolutionary steps into the Universe." With those steps, a tradition of peace and freedom now exists in the solar system. From this larger home, we move to greet the future.'

The last Apollo Moon mission comes home to a safe recovery. Note the inflated self-righting bags.

CHAPTER TWENTY-FIVE

New Directions

Skylab cluster with Apollo CSM at extreme left in this artist's impression of what the space station would look like in orbit around the earth.

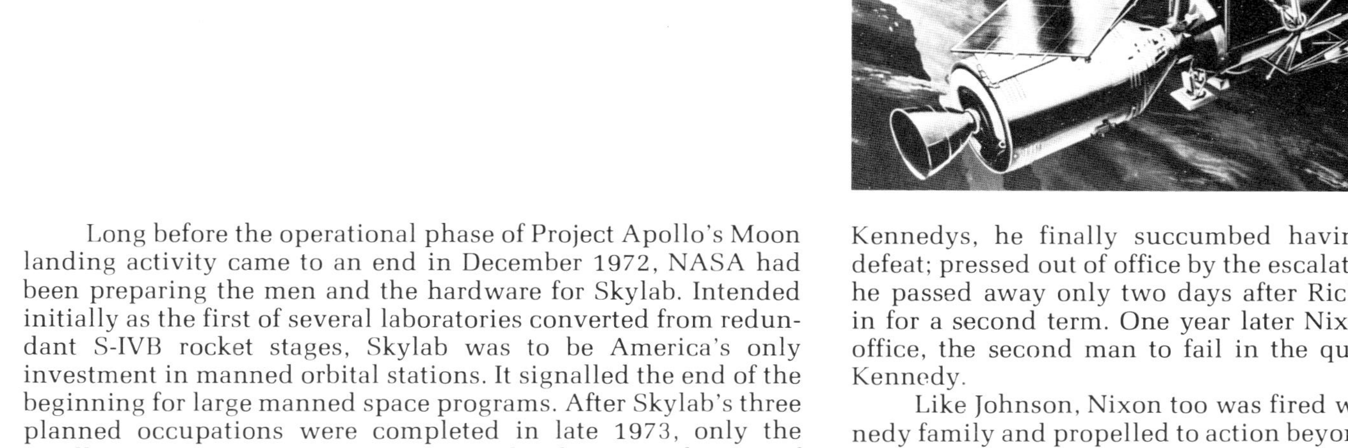

Long before the operational phase of Project Apollo's Moon landing activity came to an end in December 1972, NASA had been preparing the men and the hardware for Skylab. Intended initially as the first of several laboratories converted from redundant S-IVB rocket stages, Skylab was to be America's only investment in manned orbital stations. It signalled the end of the beginning for large manned space programs. After Skylab's three planned occupations were completed in late 1973, only the Apollo-Soyuz Test Project – ASTP – lay between the era of expendable spacecraft, and ocean splashdowns, and the reusable freighter called shuttle (by now referred to as the Space Shuttle) with its land landings and reusability for at least 100 missions to Earth orbit.

Skylab was the last of the first generation hardware that relied for its success on one-shot flights; it was doubly daunting to build not only a winged spacecraft designed to fly back through the atmosphere but also build it for many different missions. So while North American Rockwell put its back into Shuttle development, other teams worked on to demonstrate the last, and varied, uses of Apollo hardware. There was no chance of running even a minimum manned space flight program while the Shuttle topped its own funding curve, so all manned operations would halt from 1975 until the winged transporter became available, perhaps in 1979.

It was a situation forced upon the space agency because the heart had gone out of the fight. With a President in the White House determined to push his own decisions in spite of contrary advice, NASA was fast disappearing down a monetary funnel. From 1966, when the space agency received peak funds, annual budgets continued to fall. Determined in late 1972, the Fiscal Year 1974 request would amount to just 37% of the 1966 sum when discounted for inflation in the intervening years; by the end of the decade NASA's annual budget would be equal, in real terms, to an average 30% of the 1966 value. Within the framework of that total budget, NASA had to apportion sums for manned flight, space science, Earth applications, planetary exploration, research and development, tracking, administration, and new building projects.

It was decline for America's manned operations in space in the face of increasing build-up within the Soviet system. Where once America annually launched nearly 100 satellites for a few years around the mid-1960s, while the Soviets sent up little more than half that number, the United States now launched less than 40 on average each year while the Russians increased beyond the magic 100 figure – and still kept on increasing. The disparity was to prompt verbal conflict with the incumbent administration, fanning flames already growing around the Nixon establishment. Only unprecedented success with foreign affairs kept the President in the White House. It was a story finally closing its own chapters.

More than a decade before, Lyndon Johnson laid the foundation for a National Aeronautics and Space Administration and three years later pressed John Fitzgerald Kennedy into the decision of the decade. One month after an Apollo spacecraft came back from the Moon for the last time, on 22 January, 1973, LBJ died. Like his successor, Johnson had been the victim of a bitter irony: for ever concerned to outdo the brash successes of the Kennedys, he finally succumbed having witnessed his own defeat; pressed out of office by the escalating conflict in Vietnam he passed away only two days after Richard Nixon was sworn in for a second term. One year later Nixon too would fall from office, the second man to fail in the quest teased out by John Kennedy.

Like Johnson, Nixon too was fired with dislike for the Kennedy family and propelled to action beyond his ability to conceal activities that, although far less destructive than those of JFK, would doom him to expulsion. The bitterest curse from the Kennedys worked, it seemed, on those around the proud Massachusetts family and not, perhaps after all, on those of their own kind.

The day after Lyndon Johnson died, Richard Nixon appeared before the nation to tell them the Vietnam war was over. The long and hard struggle to undo what began as a clandestine involvement under John Kennedy had finally borne fruit. Yet Johnson's enthusiasm for a vigorous American space program had pushed the United States to the front of technology. It was a service recognized by all, framed by popular opinion, and shaped by Senate Joint Resolution 37 which called upon the House to ratify a decision that the Houston facility be re-named the Lyndon B. Johnson Space Center. On 17 February, 1973, the President signed that order and from then on the Manned Spacecraft Center was no more.

It was a fitting change presaging the increasing emphasis toward combined programs utilizing both manned and unmanned systems. Skylab was a suitable bridge between the two because it sought an identity for space-faring man, not among the stars where former Presidents had thought, but rather in the dedicated attention given to Earth itself. Ecology and conservation were words heard more and more as the 1970s wore on; Skylab was the technologists' answer to the call for more attention to the home planet. As time would tell, the entire program evolved as probably the most successful manned project of any. For although the endeavor was afflicted with unusually severe problems during the operational phase, Skylab activity came through it all and achieved more in the final analysis than anybody expected.

Success may have been due to consistency throughout the design phase, or to the retention of a solid management team. Whatever the reason, it provided valuable lessons for later projects and when the Shuttle ran into serious problems in the late 1970s, Skylab was seen as a yardstick by which to measure the steps necessary to put it back on the road.

Skylab had a checkered genesis, and that story has been threaded through the evolving histories of Gemini and Apollo, but shortly before Armstrong and Aldrin took 'one small step for man' the then head of manned space flight, George Mueller, responded to Wernher von Braun's suggestion for a 'dry' workshop and agreed to program Skylab around a set of objectives calling for conversion on the ground rather than in orbit; hitherto, the Marshall Center worked to a plan envisaging launch of the S-IVB stage into orbit, expulsion of residual hydrogen fuel, and conversion to a manned workshop. Instead, Skylab would ascend fully kitted out for three periods of habitation (28, 56, and 56 days, respectively) on the propulsion from two live Saturn V stages.

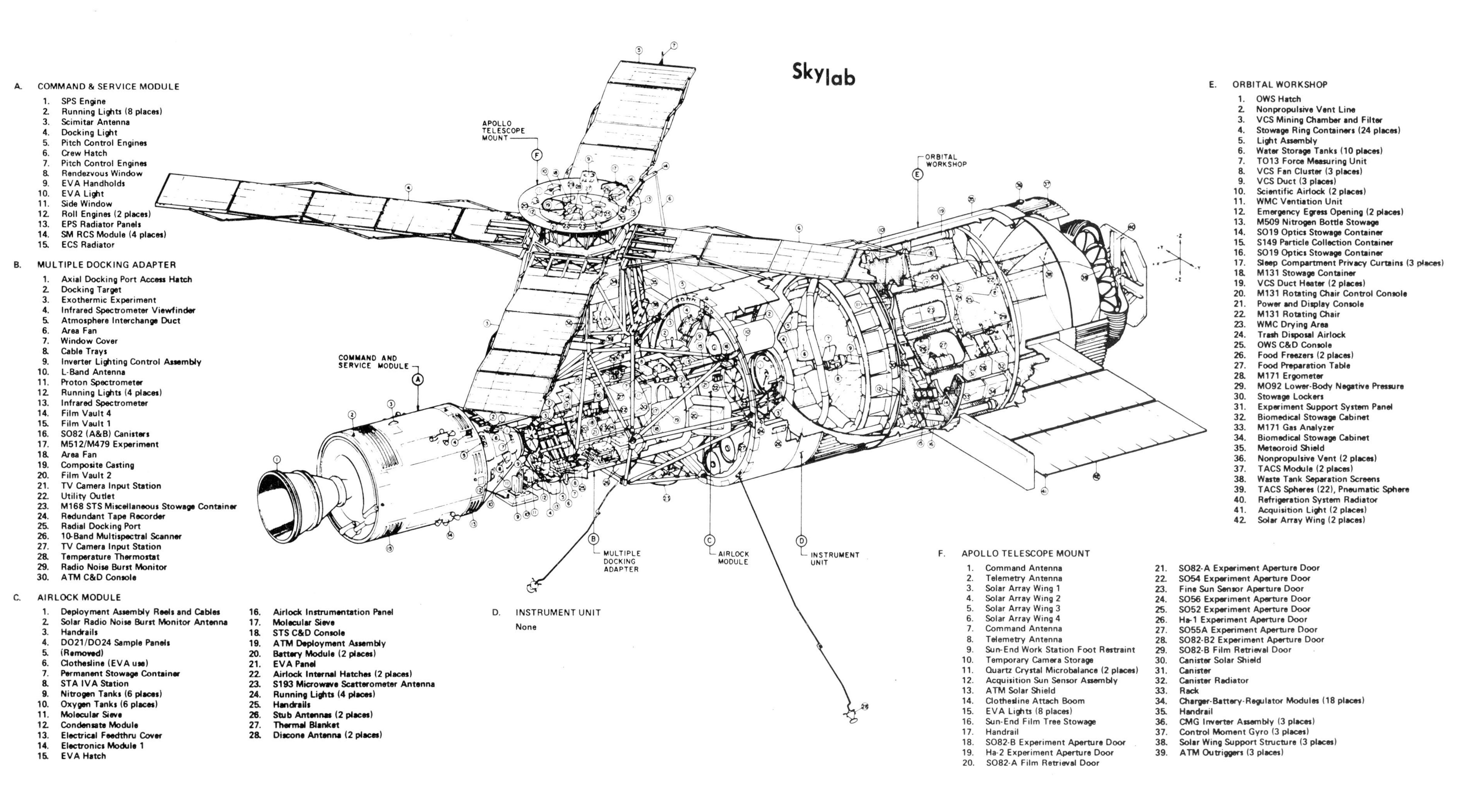
Skylab
APPOLLO TELESCOPE MOUNT
F
ORBITAL WORKSHOP
E
COMMAND AND SERVICE MODULE
A
B
MULTIPLE DOCKING ADAPTER
C
AIRLOCK MODULE
D
INSTRUMENT UNIT
A. COMMAND & SERVICE MODULE
1. SPS Engine
2. Running Lights (8 places)
3. Scimitar Antenna
4. Docking Light
5. Pitch Control Engines
6. Crew Hatch
7. Pitch Control Engines
8. Rendezvous Window
9. EVA Handholds
10. EVA Light
11. Side Window
12. Roll Engines (2 places)
13. EPS Radiator Panels
14. SM RCS Module (4 places)
15. ECS Radiator
B. MULTIPLE DOCKING ADAPTER
1. Axial Docking Port Access Hatch
2. Docking Target
3. Exothermic Experiment
4. Infrared Spectrometer Viewfinder
5. Atmosphere Interchange Duct
6. Area Fan
7. Window Cover
8. Cable Trays
9. Inverter Lighting Control Assembly
10. L-Band Antenna
11. Proton Spectrometer
12. Running Lights (4 places)
13. Infrared Spectrometer
14. Film Vault 4
15. Film Vault 1
16. SO82 (A&B) Canisters
17. M512/M479 Experiment
18. Area Fan
19. Composite Casting
20. Film Vault 2
21. TV Camera Input Station
22. Utility Outlet
23. M168 STS Miscellaneous Stowage Container
24. Redundant Tape Recorder
25. Radial Docking Port
26. 10-Band Multispectral Scanner
27. TV Camera Input Station
28. Temperature Thermostat
29. Radio Noise Burst Monitor
30. ATM C&D Console
C. AIRLOCK MODULE
1. Deployment Assembly Reels and Cables
2. Solar Radio Noise Burst Monitor Antenna
3. Handrails
4. DO21/DO24 Sample Panels
5. (Removed)
6. Clothesline (EVA use)
7. Permanent Stowage Container
8. STA IVA Station
9. Nitrogen Tanks (6 places)
10. Oxygen Tanks (6 places)
11. Molecular Sieve
12. Condensate Module
13. Electrical Feedthru Cover
14. Electronics Module 1
15. EVA Hatch
16. Airlock Instrumentation Panel
17. Molecular Sieve
18. STS C&D Console
19. ATM Deployment Assembly
20. Battery Module (2 places)
21. EVA Panel
22. Airlock Internal Hatches (2 places)
23. S193 Microwave Scatterometer Antenna
24. Running Lights (4 places)
25. Handrails
26. Stub Antennas (2 places)
27. Thermal Blanket
28. Discone Antenna (2 places)
D. INSTRUMENT UNIT
None
E. ORBITAL WORKSHOP
1. OWS Hatch
2. Nonpropulsive Vent Line
3. VCS Mining Chamber and Filter
4. Stowage Ring Containers (24 places)
5. Light Assembly
6. Water Storage Tanks (10 places)
7. TO13 Force Measuring Unit
8. VCS Fan Cluster (3 places)
9. VCS Duct (3 places)
10. Scientific Airlock (2 places)
11. WMC Ventiation Unit
12. Emergency Egress Opening (2 places)
13. M509 Nitrogen Bottle Stowage
14. SO19 Optics Stowage Container
15. S149 Particle Collection Container
16. SO19 Optics Stowage Container
17. Sleep Compartment Privacy Curtains (3 places)
18. M131 Stowage Container
19. VCS Duct Heater (2 places)
20. M131 Rotating Chair Control Console
21. Power and Display Console
22. M131 Rotating Chair
23. WMC Drying Area
24. Trash Disposal Airlock
25. OWS C&D Console
26. Food Freezers (2 places)
27. Food Preparation Table
28. M171 Ergometer
29. MO92 Lower-Body Negative Pressure
30. Stowage Lockers
31. Experiment Support System Panel
32. Biomedical Stowage Cabinet
33. M171 Gas Analyzer
34. Biomedical Stowage Cabinet
35. Meteoroid Shield
36. Nonpropulsive Vent (2 places)
37. TACS Module (2 places)
38. Waste Tank Separation Screens
39. TACS Spheres (22), Pneumatic Sphere
40. Refrigeration System Radiator
41. Acquisition Light (2 places)
42. Solar Array Wing (2 places)
F. APOLLO TELESCOPE MOUNT
1. Command Antenna
2. Telemetry Antenna
3. Solar Array Wing 1
4. Solar Array Wing 2
5. Solar Array Wing 3
6. Solar Array Wing 4
7. Command Antenna
8. Telemetry Antenna
9. Sun-End Work Station Foot Restraint
10. Temporary Camera Storage
11. Quartz Crystal Microbalance (2 places)
12. Acquisition Sun Sensor Assembly
13. ATM Solar Shield
14. Clothesline Attach Boom
15. EVA Lights (8 places)
16. Sun-End Film Tree Stowage
17. Handrail
18. SO82-B Experiment Aperture Door
19. Ha-2 Experiment Aperture Door
20. SO82-A Film Retrieval Door
21. SO82-A Experiment Aperture Door
22. SO54 Experiment Aperture Door
23. Fine Sun Sensor Aperture Door
24. SO56 Experiment Aperture Door
25. SO52 Experiment Aperture Door
26. Ha-1 Experiment Aperture Door
27. SO55A Experiment Aperture Door
28. SO82-B2 Experiment Aperture Door
29. SO82-B Film Retrieval Door
30. Canister Solar Shield
31. Canister
32. Canister Radiator
33. Rack
34. Charger-Battery-Regulator Modules (18 places)
35. Handrail
36. CMG Inverter Assembly (3 places)
37. Control Moment Gyro (3 places)
38. Solar Wing Support Structure (3 places)
39. ATM Outriggers (3 places)

The three members of the prime crew of the first Skylab mission (left to right: Joseph P. Kerwin, Paul J. Weitz, Charles 'Pete' Conrad) go through food preparation activity in the wardroom of the crew quarters of the Skylab OWS trainer at the Johnson Space Center.

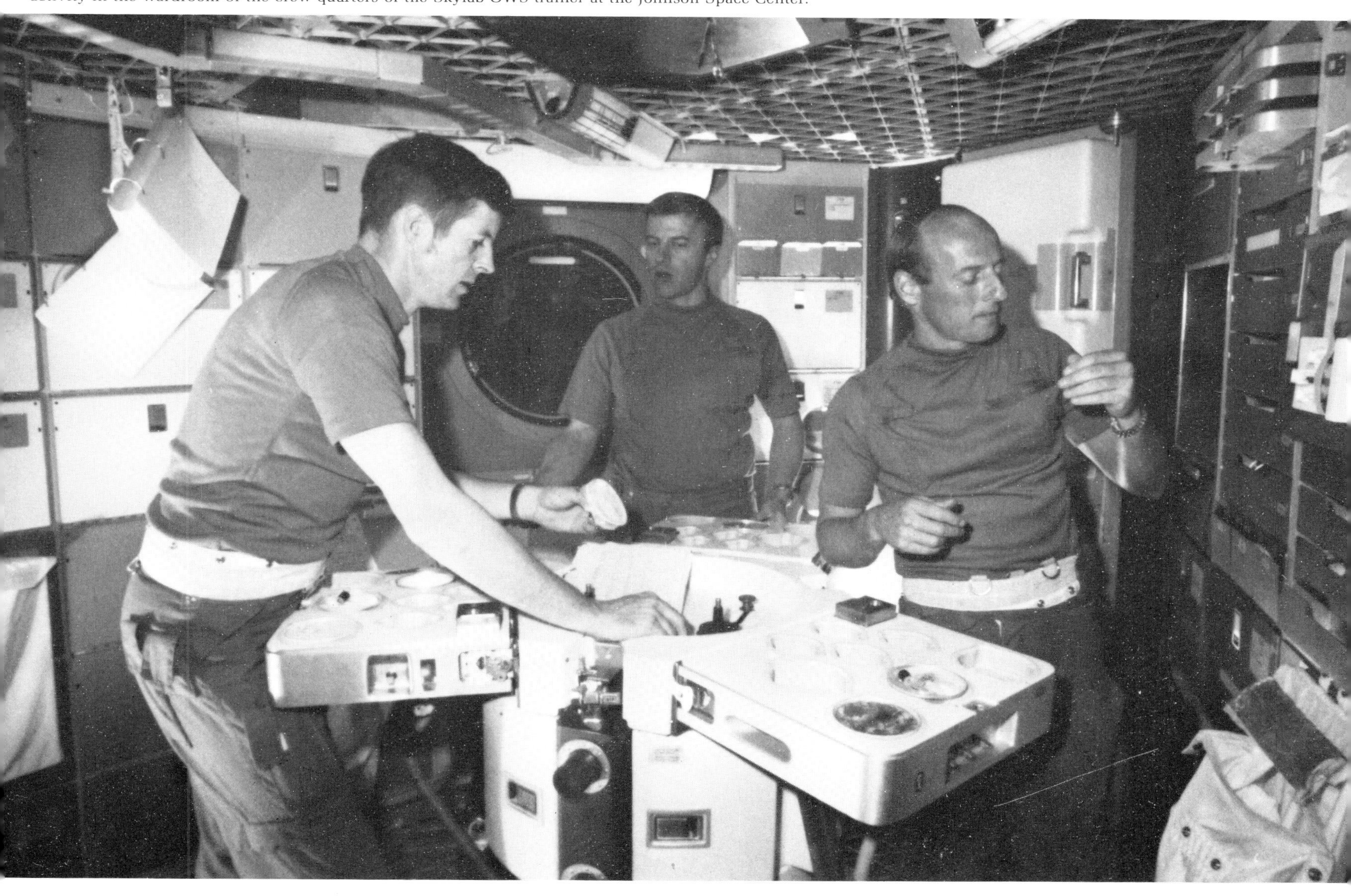

By this time, Leland Belew had been MSFC's program manager for more than three years, laying the engineering groundwork for full go-ahead from late 1969. In 1966 too, Robert F. Thompson was made Belew's counterpart at the Manned Spacecraft Center, and in May 1968 Harold T. Luskin became program director at NASA Headquarters only two months after joining NASA. In November 1968, Luskin died and from 18 December his place was taken by ex-Gemini and Apollo Mission Director William C. Schneider. It would be his task to nurse Skylab through more than five years of development and operation. At the Kennedy Centre, Robert C. Hock was appointed program manager in the same year Belew and Thompson were named to their own respective posts. Like Belew, Hock would see the program through to completion. But MSC's Thompson was replaced on 13 February, 1970, by Ken Kleinknecht when the former moved across to manage the Shuttle program at the Manned Spacecraft Center.

From 1970, no major management shifts occurred, a significant beacon on the road to success. Before Skylab became a dry laboratory in 1969, the space station was already assigned several appendages designed to aid docking operations for an Apollo CSM bringing crew and supplies and to assist with the several periods of EVA that would be necessary to retrieve film cassettes from the Apollo Telescope Mount, or ATM. The ATM was inappropriately named. Although at first conceived as a battery of solar telescopes attached to a converted Grumman Lunar Module, the ATM ultimately emerged as a purpose-built structure specially designed to house the experiments scientists hoped would significantly expand knowledge about the Sun.

But that was only one of four primary objectives. Skylab was also to address biophysical, Earth resources, and engineering objectives, each of which had at least one piece of hardware aboard the station. The main workshop was to be a converted S-IVB stage, essentially a production article from which the single J-2 and all propulsive elements would be removed, leaving it an inert shell to form the structural base. Only the hydrogen tank was to be pressurized, but that was by far the larger of the two cylinders, each separated from the other by a common hemispherical dome. The smaller oxygen tank at the bottom would be used as a waste disposal container to prevent the discarded products of 140 days of occupation contaminating the vacuum of space; many sensitive instruments would peer through the void and experience with earlier missions showed how water and waste products vented to space clouded the view.

The pressurized hydrogen tank, 8.9 meters tall from the top of the liquid oxygen dome, was 6.6 meters in diameter and when fitted out with all the equipment it would need to support the Skylab flights provided a volume of approximately 292 cubic meters with a weight of nearly 35.4 tonnes. It was big – very big – as tall as a house and compared with earlier spacecraft a veritable barn. Properly called the Orbital Work Shop, or OWS, it was divided into two separate sections: a lower living quarters and biomedical area and a forward, experiment, area.

The living quarters were defined by a floor across the top of the convex liquid oxygen tank dome, attached to it at the center with a small airlock through which waste and redundant materials could be passed to a 62.5 cubic meter space in the lower tank; 7.4 cubic meters would be available for liquid waste. The 'roof' on the living quarters separated it from the forward experiment area and carried a hexagonal cut-out in the center through which astronauts could float. Both floor and ceiling were formed from metal grids designed to accept cleats carried on the soles of shoes worn by personnel wishing to 'lock' themselves at one position. In the absence of gravity it would be all too easy to drift away from a work area.

The living area was divided into four sections: wardroom, waste management, sleep, and experiment compartments. The wardroom provided facilities for food storage, preparation and consumption and contained 58 stowage lockers, a food chiller and two freezers. Because Skylab would be sent up with all the supplies three astronauts would need for a total 140 days, careful packaging would be a pre-requisite for ready access to consumables. The wardroom supported a food table at the center to which three food trays could be clipped, each tray accommodating eight receptacles of which three could be heated. Cans would be placed in the tray for use and a separate water spigot on a flexible line would be used to reconstitute the dehydrated items or supply drinking fluid. Most food could be prepared with a consistency that would prevent it floating away from the can, but a plastic membrane was fitted to the top just in case; a spoon or fork could be inserted through a split across the center. In all, Skylab would be loaded with nearly 1 tonne of food stored in 11 lockers and five freezers.

The waste management compartment provided the function of a combined washroom and toilet. It was a rectangular room supporting on one wall a urine/fecal collector unlike anything astronauts had been provided in the past. Similar in appearance to a toilet seat it was lined with a removable bag designed to contain feces. Air sucked down from the lip of the seat would draw fecal material into the receptacle where it could be contained, removed from the collector, dried and its mass measured on a special instrument carried aboard the OWS. The products would be stored on board and returned for analysis. Urine was collected and frozen in a container attached to the same wall unit. A handwasher sprayed water into a contained area, with air flow employed to prevent the droplets from floating into the compartment. Soap attached to magnets was to be carried, with tissues and disinfectant pads. Each astronaut had his own hygiene kit, including nail clippers, hair cream, deodorant, shaving equipment and soap.

The sleep compartment was sub-divided into three cubicles with a total of 22 lockers for personal and private items. Sleeping bags were provided for each crewmember, attached to the floor and ceiling for bat-like rest in the weightless environment. Special attention was paid to the need for privacy and flexible doors could seal each cubicle from the others. Each sleeping bag would be changed every two weeks and the frame to which it was attached allowed the astronaut to assume any sleep posture in which he felt most comfortable – including the foetal crouch. The needs of the crew were accommodated with unusual attention; Skylab was in every way a new deal for man in space, big enough to contain generous living facilities.

Almost one-half of the total available volume within the living area was given over to biomedical experiments using equipment attached to the floor outside the walled wardroom, waste management and sleep compartments. Medical evaluation of the Mercury, Gemini, and Apollo astronauts was conducted in pre-flight and post-flight analyses backed up by lengthy examination of data recorded on tape during the mission. For Skylab all that would have to change because in-flight evaluation of pulmonary, cardiovascular, and musculoskeletal reaction was a vital ingredient of long-duration flight. Without assurance that the daily response of each astronaut was not taking him beyond his physical tolerance the program would be impossible to conduct. Accordingly, the three principal items carried in the experiment area of the crew section were designed to both condition and probe the astronauts' well-being.

This Skylab OWS crew quarters mock-up shows (right) the wardroom, or galley, the waste management section (lower right), the three sleep compartments (bottom left) and the exercise area (top left) containing the lower-body negative pressure device at top, the bicycle ergometry exerciser to the right and the vestibular rotating litter chair near the sleep cubicles. Note the waste disposal hatch in the center of the complete structure.

A static bicycle, called an ergometry exerciser, provided measured stress through which the condition of the heart could be examined. A special face mask obtained exhaled carbon dioxide to measure the response of the lungs and to obtain information on metabolic activity. Another piece of equipment called the lower body negative pressure (LBNP) device comprised a horizontal cylinder with a flexible circular curtain. Inserted to his waist, an astronaut would fix the curtain around his body so that pressure removed from his legs and pelvic region would induce reactions similar to those he would encounter upon return to Earth. His blood pressure could also be monitored during this activity and flight surgeons would obtain valuable indicators to the health of the person concerned.

Finally, a rotating litter chair would measure an astronaut's vestibular reactions, or response to weightless flight, by spinning him at up to 30 rpm in the seated or supine position. Disorientation would be monitored by giving the subject spheres and magnetic pointers and requesting him to place them in specific positions with his eyes covered. Many other medical experiments were to be performed aboard Skylab but the three specific items fixed to the floor of the lower experiment area were to be used consistently throughout all three planned habitations.

Above the crew compartment ceiling, up through the hexagonal cut-out, astronauts would move to the forward compartment, or main experiment area. It was by far the largest and most open area provided aboard Skylab, a circular domed volume within which several experiments would be carried out, including tests with a maneuvering unit of the type previously considered for EVA activity from a Gemini spacecraft. With Skylab, that could now take place in the comfortable environment of a pressurized space station. The forward compartment was 6.6 meters in diameter and about the same in height, from the floor to the top of the tank dome.

Around the circumference, where the dome met the straight sides, ten separate cylindrical tanks were attached containing more than 2.9 tonnes of water for cooling, drinking and reconstituting food, and supporting various experiments. A single, portable, tank with a capacity of 12 kg was also attached to this area. Twenty-five lockers were also fitted containing urine bags, electrical cables, portable lights, hoses, umbilicals for EVA, pressure suits, tape recorders, filters, and lamps. Two small scientific airlocks were fixed over holes in the workshop wall through which instruments could peer; they were diametrically opposed enabling one to view the Earth, the other to scan space, from any fixed attitude.

During preliminary design of the workshop concept in the mid- to late-1960s, engineers thought it would be essential to install a pole for access between the large compartments. In the final concept, this 'fireman's pole' device was provided as an optional extra since the astronauts thought they could move about by pushing off from their local surface and drifting along until they reached the next obstacle. It was like having all the hardships of EVA suddenly turned to the advantage of the crew member, and in a comfortable shirt-sleeve environment too.

But Skylab was more than the Orbital Work Shop. Attached to the forward end of the converted S-IVB, a fixed airlock shroud provided rigid support for two cylindrical sections through which the crew would pass on their way into the OWS and for the Apollo Telescope Mount and the ATM's truss assembly. The Airlock Module, or AM, was fixed to the top of the OWS and structurally supported by the fixed airlock shroud. It was designed to serve a unique function.

In early manned programs access to the vacuum of space was gained by first de-pressurizing the cabin, evacuating the life-supporting atmosphere that kept the crew alive. But de-pressurizing, and then re-pressurizing, the cavernous interior of the OWS would be prohibitively expensive in gas, time, and

Conrad works at the materials processing station in Skylab's Multiple Docking Adapter. The Apollo Telescope Mount console is at the extreme left foreground.

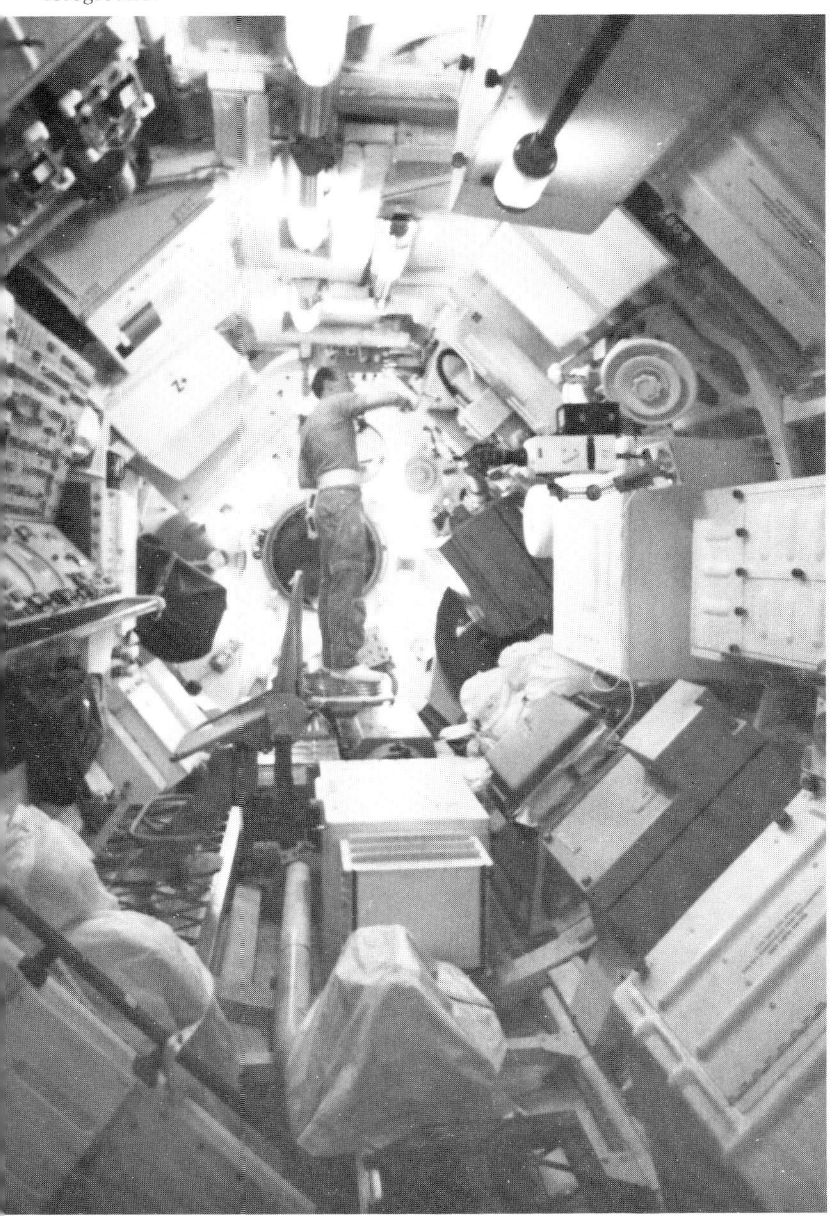

reduced safety levels. EVA would be unavoidable since certain experiments which needed attention were fixed to the exterior, so the Airlock Module was provided with circular hatches at either end to isolate the volumes de-pressurized of atmospheric gas. It also provided a structural link between the converted S-IVB stage – the OWS – and the docking module on front, and provided storage for small items needed in this forward area of the space station. Access to the exterior would be gained via a hatch taken from a redundant Gemini spacecraft and fitted to the circular wall.

The AM was 5.3 meters in length and provided 17.3 cubic meters habitable volume, capable of comfortably accommodating two suited astronauts while a third remained in the docking module on front. In this way it was necessary to de-pressurize only 5% of the total Skylab interior for each EVA, the internal area of the Airlock Module sealed by two 1.2 meter diameter hatches. The main cylindrical section of the Airlock Module had a diameter of only 1.7 meters but the extreme forward section increased to 3 meters where it connected with the Multiple Docking Adapter. The MDA was one of the earliest appendages assigned a role in Skylab, evolving in the mid-1960s as the module that would provide several docking ports to which Apollo spacecraft would connect when bringing crew and supplies to the workshop.

Originally required to provide a single axial and four radial ports, it finally emerged with just one of each, the radial port serving as a back-up to the prime axial port. The MDA weighed nearly 6.3 tonnes, compared with more than 22.2 tonnes for the Airlock Module and truss assembly, with a length of 5.2 meters and a diameter of 3 meters. At the forward end it supported a drogue docking receptor identical to the drogue carried by Lunar Modules, preserving the same docking hardware as that used in the Moon landing program. Inside, astronauts would have 32 cubic meters to move around and operate instruments and equipment.

The most important bank of controls lay across a console for operating the ATM solar telescopes. Crew members would spend many hours in here controlling the observation sessions and recording images of Earth's nearest star, the Sun. Opposite, along the cylindrical wall to the left of the MDA as viewed looking toward the workshop, was a materials processing facility designed to accommodate simple experiments in a furnace or a vacuum chamber. Underneath the MDA, structurally connected to the exterior but operated from the inside floor area, were the Earth resource scanners carried to address a fundamental Skylab role: that of determining the value in advanced study of the planet from space.

Whereas the Airlock Module was the space station's 'engine-room,' supporting equipment vital to environmental and systems control, the Multiple Docking Adapter would become the centre of operations for solar studies, materials science, and Earth observation. It carried a circular hatch leading to the docking ring which, like similar installations on the Lunar Module, received the 12 docking latches attached to the Apollo spacecraft. As with the workshop conversion of an S-IVB stage, McDonnell Douglas built the Airlock Module, while Martin Marietta worked the Multiple Docking Adapter. From the forward hatch on the MDA to the floor of the living quarters in the OWS, astronauts had a continuous open run of 19 meters, equal to the height of a six-storey building. Or, they could close it off into three separate modular elements.

Attached to the fixed airlock shroud was the third modular addition to the workshop, the Apollo Telescope Mount and its battery of eight solar instruments designed to operate in conjunction with the control and display panel in the MDA. In this way, Skylab would provide the world's first manned solar observatory, recording images on film carried in cassettes adjacent to the telescopes. But the main ATM assembly would also serve to support attitude control equipment, and large solar cell panels for generating electrical energy. With the solar panels folded for launch, the ATM weighed 11 tonnes and comprised a box structure 4.4 meters tall and 6 meters across.

From the beginning, it was clear that a converted S-IVB used for manned operations would require more power than could be adequately provided by batteries or fuel cells. Solar cell arrays were the only answer where loads of up to three times that of an Apollo CSM would be called for. But that implied a need for power generation three times the required load so that losses due to power conditioning and cable resistance would not reduce the available supply below that needed for systems and experiments.

The four solar array wings attached to the ATM were each 13.2 meters long and 2.7 meters wide supporting 41,040 individual solar cells. Under ideal circumstances they were theoretically capable of generating a total of more than 11 kilowatts but usable power would be less than 4 kw because of the aforementioned losses. Two other solar cell arrays were attached to the workshop, each wing supporting 73,920 separate cells in a configuration 9.5 meters by 8.3 meters in size. The OWS wings had a theoretical contribution of nearly 13 kw but here too power losses reduced the available supply to one-third. Like the ATM wings, they were to be folded, concertina-like, but its booms were designed to pivot 90° from their launch position alongside the workshop walls. In all, Skylab would carry 312,000 solar cells designed to generate about 8 kw of usable power.

Because Skylab would spend almost as much time in darkness as it would in Sunlight for many of its planned orbits the station carried 26 batteries for storing electrical energy to be used when it was on the dark side of the Earth or when it was turned away from the Sun. The ATM array fed power to 18 charger battery regulator modules, or CBRM units, containing batteries each designed to provide 415 watts at maximum power output. Housed in the ATM structure, they were connected to the four solar array wings. The two large workshop arrays fed energy to 8 power conditioning groups, or PCG units, effectively permitting the workshop wings to have the same autonomy possessed by the ATM array. They were attached to the fixed airlock shroud. Both systems were independent but in reality were to be operated in parallel to reduce the demand on one system and optimize the load-sharing capability.

Multiple Docking Adapter and Airlock Module components attached to the Skylab Orbital Work Shop (out of picture, right) with the Apollo Telescope Mount truss structure (top) and docked CSM (left).

Skylab's electrical system could also receive up to 1.4 kW from the Apollo spacecraft fuel cells but only for about the first two weeks until cryogenic reactants were depleted, following which the station would have to supply power to the CSM and top up its entry batteries for return to Earth. Nevertheless, this would supplement fully one-half of this first Skylab visit and one-quarter of the second and third habitations. Skylab could, therefore, accommodate loads calling for 9 kW electrical power although requirements were expected to call for 7.5 kW even under the most active periods. Throughout the Skylab cluster, 28-volt utility outlets were provided for plug-in equipment in addition to the 'hard wired' instruments built integral with the structure.

But the Apollo Telescope Mount was not only responsible for supporting a major part of the electrical system, for it also carried new attitude equipment in the form of control-moment gyroscopes. To a more precise level than ever before, Skylab would be required to point with great accuracy at specific targets in space or on the Earth, and for that the conventional means of attitude control was no good since it would call for prohibitively large quantities of thruster propellant. Just as electrical requirements fashioned batteries for Mercury, fuel cells for Gemini and Apollo, and solar cell arrays for Skylab, so too would control systems evolve from the monopoly held so far by small rocket motors.

For attitude information, Skylab would employ Sun sensors and star seekers designed to lock on respective targets and provide details of position about all three axes (pitch, roll, and yaw). To keep the massive cluster fixed on one target, however, or to move it from one attitude to another, the control-moment gyroscopes (CMG) would physically torque the station by responding to a well established principle. This dictated that a spinning gyroscope would respond to torques about one axis by precessing around the other.

In theory, very large spinning wheels contained within an inner and an outer gimbal would adequately control the Skylab attitude by moving the entire cluster in reaction to a torque induced by a motor driving the rotor axis out of alignment. This called for three massive (181 kg) rotating wheels, one aligned in each axis, and Skylab's CMG units were placed at three orthogonally displaced locations in the Apollo Telescope Mount. Set to revolve at about 8,950 rpm, they would be employed for all attitude control movements about the cluster's center of mass.

In addition, reverting to the classic mode of attitude control, Skylab carried a thruster attitude control system – or TACS (pronounced 'tax') – at the rear of the converted S-IVB. This comprised a similar system to that employed for a standard S-IVB rocket stage, where two modules, one on each side, supported three thruster jets through which gaseous nitrogen would be expelled to induce a reactive force. But a larger quantity of gas would be needed, even though TACS was considered a secondary mode of control for normal mission phases.

Twenty-two spherical titanium gas bottles were attached to the base, or extreme rear, of the workshop containing a total 647 kg of pressurized nitrogen. The jets could be commanded to deliver 9.1 or 22.7 kg thrust and to fire in 1 second or from 40 to 400 millisecond bursts. The CMG wheels would be employed for attitude orientation when the cluster settled down in orbit, but for the first 7$\frac{1}{2}$ hours, and during selected periods of the mission thereafter, TACS would be used. Yet although the combined CMG/TACS equipment could achieve pointing accuracy of ±2°, the sensitive solar telescopes needed a much finer targeting than that. So, the ATM experiments package incorporated an independent control system designed to stabilize the telescopes' viewing alignment to ±2.5 arc-sec for up to fifteen minutes.

That was accomplished by placing the Apollo Telescope Mount canister inside a roll cage which could be independently steered by electrically controlled actuators, keeping the telescopes 'floating' on target within the coarser alignment set by the CMG units. Information on movement around the cluster's center of mass would be determined by three rate gyroscope processors in each axis; two for control information and a third on standby. Similarly, two of the three 53 cm diameter CMG wheels could be used for attitude control if the third failed.

But Skylab had a third system equally vital to the life of the crew expected to man the cluster: the environmental and atmospheric support system. Unlike earlier manned vehicles operating on a pure oxygen atmosphere at one-third sea-level pressure (258 mmHg), Skylab would need a two-gas system due to the anticipated length of each habitation. Pure oxygen was fine for flights of up to a fortnight, but physicians were concerned about possible damage resulting from prolonged exposure to a single gas system. Yet, as in earlier manned vehicle designs, it was still necessary to keep the total pressure as low as possible for structural considerations. So mission and systems planners were bound on the one hand by a need for a two-gas environment and on the other for a one-third sea-level pressure.

A compromise was reached by providing a 258 mmHg atmosphere of air with 72% oxygen and 28% nitrogen. This effectively reversed the ratio on Earth, where oxygen accounts for only 21% and nitrogen 79%, and preserved a safe margin where the partial pressure of oxygen was 185 mmHg; physicians set a lower oxygen level of 150 mmHg below which severe brain damage would result. For the first time, American astronauts would breathe air in their orbiting spacecraft, albeit at one-third sea-level pressure and in different mixes.

The gases would be contained in twelve separate external vessels: six, for oxygen, attached to the fixed airlock shroud, and six, for nitrogen, mounted on the Airlock Module's support truss. In all, 3.5 tonnes of atmospheric gas for the anticipated 140 manned occupation days. During the launch phase, Skylab would contain an atmosphere of dry nitrogen at 62% above sea-level pressure for structural stability. Vents would allow this to bleed away as the vehicle ascended and to evacuate the interior to the partial pressure of nitrogen the station would provide for manned habitation: 73 mmHg. The balance of oxygen (185 mmHg) would be supplied from the storage tanks and thereafter kept at a constant one-third sea-level pressure by regulators.

This reversal of mixed-gas content, and reduction to a lower pressure, was made possible only because the Skylab cluster was to be sent aloft unmanned. The atmosphere would be at a required level and mix ratio when the first crew docked with the station a day later. In all, Skylab provided a habitable volume of 345 cubic meters – more than fifty times that available in the Apollo Command Module – and good atmospheric circulation was an essential part of the design specification. Pockets of contaminated gas could accumulate unless adequate circulation was provided and the cluster carried fans in all three habitable modules.

Water vapor, odors, and exhaled carbon dioxide were to be removed by passing the atmosphere through one of two molecular sieves. These trapped large particles, cooled the gases through a heat exchanger, passed part of the atmosphere through molecular sieve beds where water and CO^2 were removed, and fed the rest to a charcoal canister where odors were taken out. Ducts and a large mixing chamber situated in the dome of the workshop's forward experiment volume kept the atmosphere circulating through all areas. The environmental system was also equipped to support two space-walking astronauts with pure oxygen and to control venting and re-pressurization of the Airlock Module.

Controlling Skylab's internal temperature was a problem met in concept by passive and active thermal systems, the latter employing a fluid known as Coolanol 15 passed through two separate and redundant loops employing control valves, pumps and heat exchangers. It was a system borrowed from Gemini, using several components designed for that two-man precursor. A radiator situated below the Multiple Docking Adapter was used to reject excess heat to space. The passive elements comprised paint patterns designed to distribute or radiate thermal energy falling on exterior surfaces.

Unlike previous requirements, Skylab would be called upon to operate for nearly one full revolution of the Earth about the Sun and because the plane of the station's orbit around the Earth remained constant the cluster would, at times, be exposed to the Sun's rays throughout complete orbits of the globe for several days on end. This brought unique demands to Skylab and the balance between active and passive control systems was one worked out with this unusual situation in mind. Most of the time Skylab's solar arrays would point directly at the Sun so the side of the station facing the solar disc was painted white. In fact, the workshop carried a micrometeoroid shield designed to protect the thin skin of the station from possible damage or rupture from impacting fragments and this was the surface painted with the thermal cover.

The shield was to be held tight around the workshop during ascent and deployed by torsion bars and spring links to a distance 12.7 cm from the cylindrical hull when the station reached orbit. It would serve to take the initial shock of impact, vaporizing the fragments before they struck the wall. Some said there was no need for this added complexity; others said anything that raised the probability of success was an advantage. There was little or no possibility of impact from an object large enough to penetrate Skylab and as events would have it the shield was to play a leading role in dramatic episodes high above Earth.

Several auxiliary cooling systems were provided for unique operations: thermal control for EVA was met with Apollo style liquid cooling garments using a water loop to remove heat; Earth observation and ATM control equipment in the Multiple Docking Adapter was cooled by a single water coolant loop with three pumps; Coolanol 15 kept food in five freezers in good condition and circulated fluid through the drinking water chiller, the urine freezer, the blood sample freezer, and cooled the pump power supply. The refrigerator loop used an octagonal shaped radiator panel at the extreme rear of Skylab.

The Apollo Telescope Mount was originally designed as an autonomous unit launched separately and docked to the manned cluster. As a result, it evolved with a unique thermal control system using an active coolant loop filled with a water-methanol solution, a special radiator section, and a transport loop operating through cold-plates from which excess heat would be collected. The ATM carried a large, circular, shield at the Sun end designed to protect the systems below from solar rays.

In the definitive configuration, Skylab's ATM was, of course, firmly attached to the three cluster modules throughout launch and operation. But because the launch configuration had to present a smooth profile to the atmosphere during ascent, the bulky 11 tonne assembly was attached to the fixed airlock shroud by a series of struts, cables, and pulleys designed to hold it in front of the Multiple Docking Adapter for launch and then to pull it to one side, moving it through 90°, in orbit. Known as the ATM deployment mechanism it was locked into place in front of the cluster atop the two-stage Saturn V that would place it in space. A release mechanism would be automatically commanded to separate elements of the truss structure, allowing reels to pull cables and swing the massive assembly to its permanently latched position.

But only when Skylab separated from the Saturn's second stage: until then it would be encased by a 11.8 tonne nose shroud comprising four petal-like covers 16.8 meters long. Ordnance systems detonated by command from the onboard computer would sever the joins, causing the four covers to move away from Skylab's forward end at a rate of 22 km/hr. Thus exposed, the Apollo Telescope Mount would be free to rotate to its permanent

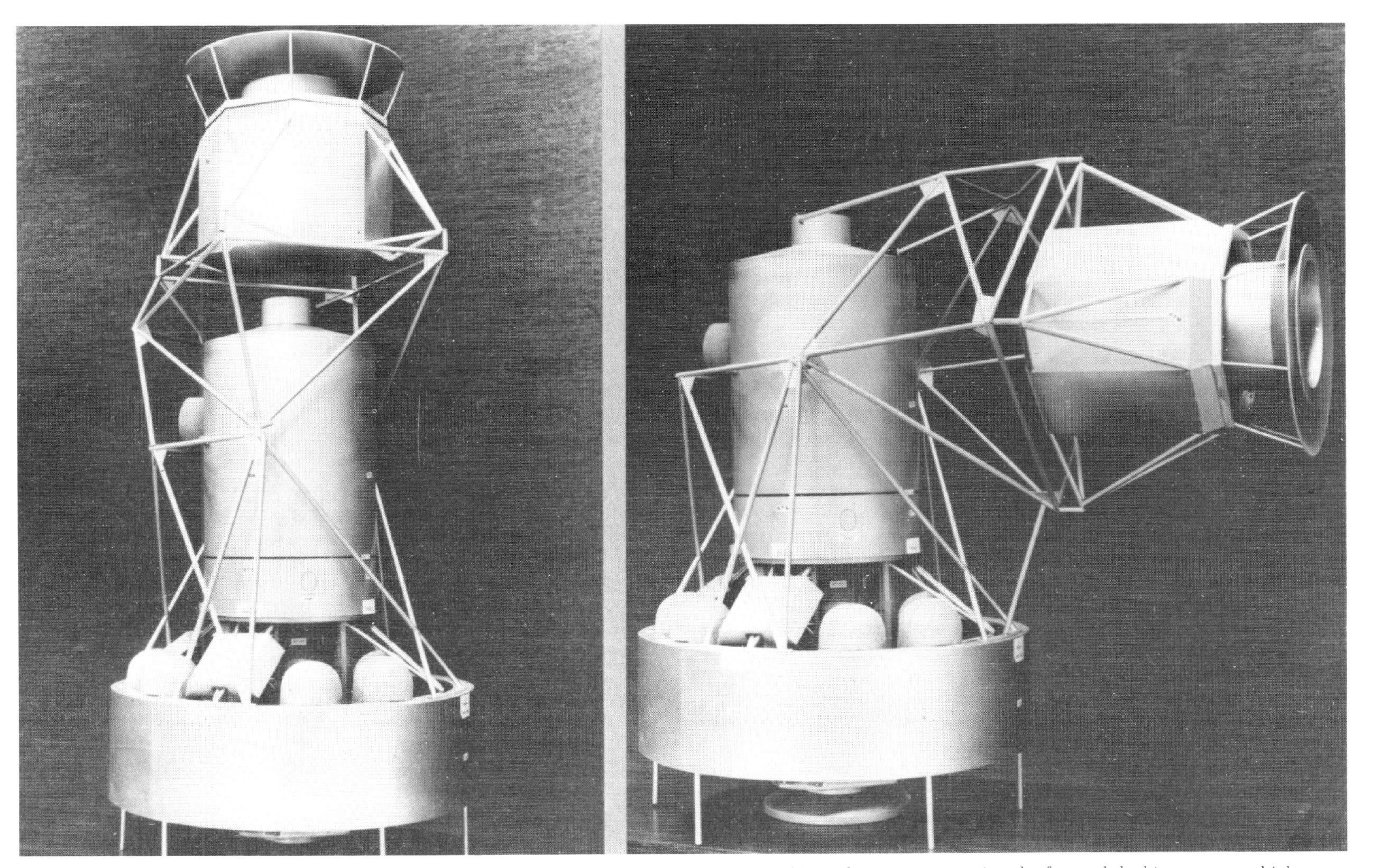

Designed to move the Apollo Telescope Mount structure 90° to one side away from the vertical launch position, vacating the forward docking port to which would be attached the manned Apollo CSM, this deployment mechanism had to work for successful manning of the giant cluster.

position. This shroud was the largest ever used to protect the forward parts of an ascending space vehicle. It simply had to work if Skylab was to be manned since it blocked the only way through to the interior.

Rendezvous and docking would be key events in bringing crews to the space station and Skylab carried several external lights including four strobes attached to the ATM deployment assembly visible more than 1,000 km away through a Command Module sextant. There were eight docking lights on the clusters extremities, two each in red, white, green, and amber, with 20-watt bulbs, arranged so as to present a cross when viewed from the docking position. Four white 0.7-watt lights were also placed at the extreme tip of two long discone antennae, deployable 12.2 meters long telemetry booms splayed 90° apart from attachment points on the fixed airlock shroud. The antennae would be damaged if struck by an orbiting Apollo inspecting the exterior.

Communications would be a vital element in Skylab's success. Data from the more than 200 engineering and science experiments on board would flow fast and continuous, and, because Skylab would be within range of a ground station on average only 33% of the time, high speed data dumps would be essential. Taped information would be sent live to tracking and communication antennae on Earth at a rate of 51,200 bits per second, while recorded information would flow from Skylab at a phenomenal 112,400 bits per second. Each day the number of 'words' flowing to Earth would exceed the word count in 350 volumes of this book. Four times a day, printers would publish a 2,000 page volume containing data sent down from Skylab and at the end of the three periods of habitation those books would, if placed on top of each other, form a pile nearly one-half the height of Mount Everest above sea level.

Communication equipment would be carried in the station's Airlock Module with UHF and VHF antennae but voice contact would go via the docked Command Module's S-Band system. Thirteen speaker boxes were distributed at key points throughout the pressurized interior of the Skylab cluster through which astronauts could talk to the ground or record information. Communication between astronauts in remote parts of the cluster could be effected via the speakers or special headsets worn as necessary. A portable color television camera was also provided for operation at any one of five input stations (three in the workshop; one each in the Airlock Module and the Multiple Docking Adapter), through one of the Earth observation viewing ports, or through one of the two scientific airlocks in the workshop wall.

The ATM also carried a television system via five separate black and white cameras fixed as an integral part of five of the eight solar telescopes. Astronauts operating the ATM console in the Multiple Docking Adapter, or scientists on the ground, could view any one of the images upon which the respective experiment was focused at that time. But if there was one area which threatened to impede the smooth and timely operations aboard Skylab it was the many flight plan instructions which, on previous missions, were meticulously written down while a capsule communicator on the ground slowly read up the changes.

A very different process was evolved for Skylab because the sheer volume of minor changes anticipated for the long flights and the necessity to keep the crew working as productively as possible demanded a semi-automated process for updating schedules and plans. That came in the form of a teleprinter and 156 rolls of spare paper on to which ground controllers would send messages, detailed flight plans, experiment schedules, block data for systems operation, and communication contact times. It was arguably the most important piece of hardware on board, for without it the execution of Skylab experiments would have been a fraction of the achieved results. On all three manned missions, controllers would send a total 3,684 separate teleprinter messages containing up to 50 lines of information, each line supporting 30 characters.

Unlike previous missions, Skylab personnel would man the station with only a skeletal plan from which to work. Because the station was expected to respond to changing situations – sudden storms on the Sun, changes in weather patterns over Earth observation targets, astronomical phenomena observed from the ground, etc. – the daily routine of planning flight operations called for wise manipulation of the crew work schedule to get the most from each day. Communication was the lifeline for this activity.

But the shift from exploratory, deep-space, operations like Apollo Moon flights to Earth-orientated space station activity with Skylab prompted integration of two separate communication groups within the NASA organization: the Manned Space Flight Network ('misfin') and the Satellite Network. The latter embraced Earth-orbiting science and applications satellites and the ultimate integration of manned and unmanned programs argued successfully for a merger that came before the last two Moon flights. The new structure was known simply as Spaceflight Tracking and Data Network STDN, or STADAN, a designation hitherto applied exclusively to unmanned tracking operations. So it was that the old 'misfin' passed away during busy preparations for Apollo 16 early in 1972.

Maintained and operated by the Goddard Space Flight Center, STDN was linked together by NASCOM, the NASA Communications Network employing satellites, submarine cable, land lines, microwave systems, and HF radio across 3.2 million circuit kilometres. The only other tracking and communication facility run by NASA was the Deep Space Network (DSN) similarly linked by NASCOM with large antennae in Goldstone, California, Madrid, Spain, and Canberra, Australia. NASA considered anything at or beyond the distance of the Moon as 'deep-space' and only unmanned robots would patrol that far in future.

Yet, despite the sophisticated support facilities for transferring information rapidly to the Skylab space station, astronauts would need considerable material to which they could refer for basic procedures, operating instructions, and check lists. There would, after all, be more than 20,000 separate stored items aboard. Accordingly, the Skylab flight data file was a considerably inflated version of the document chest carried aboard Apollo, with 46 books weighing more than 42 kg launched in the unmanned cluster. The first manned mission would bring an extra 37 books weighing 23 kg while the second and third visits would account for a further 41 and 38 books, respectively.

As for film carried in a special vault protected from solar radiation, that too would have to be carried aboard at launch. Skylab would ascend with 103 rolls of 16-mm, 70-mm, and 127-mm film capable of storing 88,800 individual frames of the Earth. In addition, Earth resource observation sensors could store 55 km of data on 28-track magnetic tape. Because additional stocks of both film and tape could be brought up in Apollo vehicles, both totals would be exceeded by 50% at the end of the third mission.

Adequate management and control of the many different experiment tasks, plus integration with balanced crew activities, could only be achieved by a unique concept in mission operations. This required in-flight analysis of vehicle performance and astronaut efficiency on a real-time basis, rapidly generating and evaluating a consensus on the mission so as to improve successive events. Flight control would remain at what was by now the Johnson Space Center (JSC), with four teams of flight controllers headed by Neil Hutchinson, Charles Lewis, Donald Puddy, and Milton Windler. Flight director Phil Shaffer would bring a fifth team on line for the second and third missions.

The schedule required each team to work a 5-day-on/2-day-off routine and, because of the need to prepare a detailed flight plan for each mission day, to prepare a preliminary activities' list two days in advance and to finalize the plan before the crew got to rest on the preceding day of activation.

Unlike Apollo missions, Skylab was a routine operation and for that the crew would assume a familiar schedule similar to one they would follow on Earth. That was, after all, an obligatory part of finding out just how well man could work at routine tasks aboard an orbiting space station. The day would begin at 6:00 am, Houston time, and end at 10:00 pm. Eating periods were to be the same for all crewmen (although in reality that would be relaxed) with 1 hr sessions beginning at 7:00 am, 12:00 noon, and 6:15 pm. Two 15-min blocks each day would be allotted to personal hygiene with 30-min per day for exercise. A complete day off was scheduled for each crewmember once a week. There were, therefore, approximately 60 hours to the Skylab astronauts' working week. For flight plans and the daily times schedule, Skylab operations would work to Greenwich Mean Time, the universal clock.

But Skylab was essentially a Marshall Space Flight Center project, unlike Mercury, Gemini, and Apollo, which had been effectively managed by the Space Task Group and its successor the Manned Spacecraft Center. Huntsville would have all the expertise to consult and qualify procedures and changes called for by Houston as the missions progressed. Nobody expected any trouble with Skylab, but with past experience in mind it was inevitable that things would go wrong somewhere along the way and that the engineers who built and fabricated the cluster would be called upon for direction. So MSFC established a Huntsville Operations Support Center (HOSC) for just this sort of activity.

Situated on two floors and a basement of the MSFC Computation Laboratory, the 929 square meters given over to the Support Centre housed an Operations Manager, his deputy, technical teams, administrative personnel, conference rooms, computer rooms, separate rooms for specific Skylab systems, data distribution stations, and simulator facilities. With on average more than 3 million pieces of data streaming into the Support Center each hour, personnel would operate monitoring facilities through the Auto Scan system specially set up for Skylab.

Computers would shunt data for up to twenty television displays where engineers and specialists would examine the cluster's status minute by minute. From the Cape and from Houston, 80 television monitors would provide visual cues and updates for technicians and officials, managers and scientists. Elsewhere at MSFC, the neutral buoyancy facility provided underwater simulation of the weightlessness astronauts would encounter in space, providing test facilities for procedures that might evolve as the missions progressed. Just as Houston would attempt to second-guess the cluster's condition several days ahead, so would Huntsville work along with a changing pattern of operating conditions, standing ready to prove out new procedures as requirements emerged.

But for the men who would fly aboard Skylab's three record-breaking missions, personal care and hygiene assumed unique proportions. Fifteen clothing modules were stored on board in lockers in the wardroom walls. Packed with a 28-day supply of clothing for one man, each module contained various combinations of apparel: in all, 3 jackets, 53 trousers, 58 shirts, 15 pairs of boots, 15 pairs of gloves, 4 union suits, 199 T-shirts, 47 half-union suits, 34 jockey shorts, 102 boxer shorts, 16 knee shorts, 128 pairs of socks, and 7 constant-wear garments; more than 700 individual items.

The waist-length, long-sleeve, jacket was woven from Durette fabric, carried a zipper down the front and three snappers on the waistband to attach it to trousers. The trousers were also woven from Durette and could be converted to shorts by unzipping the lower section at the knees. The shirt was a pullover type in polybenzimidazoic fabric with a single chest pocket. Short sleeve shirts were provided in several sizes. The boots were lined with Durette fabric and a knitted inner sole lining. The gloves were also knitted Durette with palm and front in deerskin.

The short-sleeve union suit was a full-length undergarment with integrated socks in cotton and the half-union was similar but separated at the waist into lower and upper portions. The constant-wear garment was the same type used on Apollo missions. Two contingency clothing modules were packed in case the back-up crewmen flew instead and that contained specially sized items.

When space walking, astronauts would wear the Apollo A7LB suit over a liquid-cooled undergarment. An Astronaut Life Support Assembly worn on the chest would regulate water, oxygen, and electrical power received from the same 18.3 meter long umbilical that connected the astronaut to the space station. An emergency supply called the Secondary Oxygen Pack was carried on the left hip. With the long umbilical, an astronaut could make his way into the workshop from the docked Command Module in the event of a major problem that prevented the station being pressurized as planned.

Late in the preparation of flight hardware, engineers evolved and fabricated an ultimate luxury item: an airflow-assisted water shower. Attached to circular plates fixed to the floor and ceiling in the living area, the circular 'curtain' contained water sprayed by expelled air through a shower head and collected by a suction head located elsewhere inside the unit. A small soap dispenser allowed each crewmember 8 milliliters per shower. The water was provided from a 2.7 kg supply filled from the water heater in the waste management compartment.

Personal hygiene would be important for scientific and mundane reasons. Accordingly, 840 30.5-cm square washcloths enabled each man to use two per day, and 420 towels were carried to be changed daily for personal use. Also, 46 boxes of dry-wipes, wet-wipes, biocide-wipes, and utility-wipes, were provided at several different locations for use in food preparation, cleaning, toilet, personal hygiene, etc.

In addition, Skylab supported 366 trash bags for containing items placed into the waste disposal airlock for dumping in the unused liquid oxygen tank aft of the workshop area, 28 plenum bags containing biologically inactive substances, and stored between the living area and the liquid oxygen tank dome, 140 vacuum cleaner bags, 168 disposal bags for loose items, and 149 urine disposal bags. With an option of three separate tools, an astronaut would regularly use the vacuum cleaner, capable of suction velocity equivalent to 0.3 cubic meters per minute, to clean the air screens, the air mixing chamber, and various filters throughout the cluster.

For the first time, recreation was given serious consideration by spacecraft design teams, Skylab being equipped with a special pack of off-duty items stored in the wardroom. It contained 60 music cassettes, four decks of playing cards, 16 ear pieces for use in four stereo headsets, 43 paperback books selected by crewmen, three balls, three hand exercisers, velcro-tipped darts, and a target board.

Interior lighting in Skylab was required to provide an even illumination in the open spaces and in the nooks and crannies between bulky equipment; it was a designer's nightmare finding the correct location, and the permitted balance, between all the many optional areas for lights. Fifty fluorescent floodlights were provided, each with a switch for selecting 12.5-watt or 9-watt intensity, on or off position, 42 of which were distributed around the two primary workshop sections; eight were located in the Multiple Docking Adapter. The MDA was also lit by fourteen 10-watt incandescent hand rail lights installed in the forward section of the Airlock Module. The AM proper supported an additional six hand rail lights and four 20-watt incandescent fixtures. In all – 74 fixed lights, plus three portable low-power lights similar to the fluorescent floodlight and two portable high-intensity lamps delivering 40-watt or 70-watt light.

Throughout, Skylab was a new experience in manned space flight for design engineers, flight control teams, and the astronaut corps. It was by far the largest satellite to reach operational status. At launch the cluster would weigh nearly 89 tonnes, of which 11.8 comprised the shroud, and stand 38 meters above the combined first and second stages of the launch vehicle, itself 67 meters tall. Payload capability for a two-stage Saturn V at the desired Skylab altitude was about 91 tonnes.

Shortly after orbital insertion, Skylab's weight would drop to about 72.2 tonnes with release of the four-door shroud and then increase to approximately 91 tonnes when an Apollo CSM brought up the first crew. Total length of the docked configuration would be 36 meters. Although conceived in the mid-1960s, Skylab's external configuration went through little change in the decade before flight, a fact due primarily to the existing S-IVB stage used as the workshop's structural spine. But comparatively late in the program, NASA added a major Earth observation role to the cluster's responsibilities and this compromised the planned attitude configuration.

Skylab would point directly at the Sun for energy and so it was logical to place the ATM solar telescopes along that axis, but Earth observation required the cluster to break lock on the Sun and track the curvature of the Earth while using electrical energy stored in the battery modules. These were to be the two primary attitude modes: solar inertial and local vertical. Had Skylab had a major Earth observation mandate from the beginning it would have been possible to provide a scanning platform designed to maintain lock on the ground while the cluster continued to face the Sun. As it was, Skylab would be maneuvered back and forth as the experiment schedule required, and that too placed constraints on the time spent studying the Sun or looking at Earth since the two activities could not be performed at the same time.

Choices open to the orbit planners were the height above Earth and inclination to the equator – how far north and south Skylab would fly. Engineers wanted as low an altitude as possible to get the maximum lifting capacity from the two-stage Saturn V; scientists wanted as high an orbit as possible to reduce the effects of Earth's contaminating atmosphere; mission planners too wanted a high orbit to reduce drag from traces of atmosphere and prolong the station's lifetime; geographers and agriculturalists desired a low altitude for good observation and more frequent passes around the Earth to increase the targets of opportunity. Mission planners wanted to keep the cluster below 518 km since that was the maximum altitude at which the Apollo RCS thrusters could de-orbit the spacecraft should the big Service Propulsion System fail. Solar physicists wanted a height of at least 407 km to study the Sun.

A compromise was sought and found at an altitude of about 432 km on a plane inclined 50° to the equator, close to the limit stipulating ascent over water; a greater inclination would have carried the launch across the eastern States. As it was, Skylab would make more than 14 full Earth revolutions per day and repeat its entire ground track every 71 revolutions. Thus, at approximate four-day 22-hour intervals, the station would return across exactly the same spot. It was a greater inclination than used before by US manned vehicles and would carry Skylab nearly as far south as Cape Horn and as far north as the southernmost tip of England. In all, Skylab would fly over 75% of the Earth's surface, observing areas in which 90% of the world's population lived, scanning 80% of the food producing-land.

Observation of the planet was the responsibility of Skylab's EREP (Earth Resources Experiment Package) instruments attached inside and outside the Multiple Docking Adapter. Multispectral photography would be accomplished via a six-channel 70-mm camera system viewing the ground at different wavelengths and through a single, 45.7 cm focal length, Earth Terrain Camera looking down from one of the workshop's scientific airlocks. An infrared spectrometer would record on magnetic tape information about the ground in near- and far-infrared regions of the spectrum, a multispectral scanner would view areas up to 37 km either side of the groundtrack for observation in visible, near-infrared and thermal infrared regions, and microwave and L-band radiometers would measure the brightness temperature of the Earth's surface as passive sensors receptive to radar backscattering.

Added to Skylab, EREP epitomized the mood of the new decade by applying to existing space operations experiments designed to prove a viable role for space in an era of diminishing global resources, environmental awareness and call for conservation. But it was 1970 before experiments were finalized and contracts let, leading, as discussed earlier, to a conflict between mission priorities. Nevertheless, EREP was to figure large in NASA's public relations on Skylab as precursor to a new generation of Earth applications satellites.

Solar physics experiments, the eight telescopes contained in the big ATM structure, would provide scientists with more information about the Sun than had been gathered throughout history to the time of launch. Unlike Earth observation activity, solar physics research had been an important part of NASA operations for several years, originating with the launch of the first Orbiting Solar Observatory in March 1962.

OSO was a program of comparatively small (275 kg maximum) unmanned satellites carrying a 120 kg load of scientific experiments designed to investigate the solar phenomena; Skylab's solar telescopes would weigh nearly 1 tonne and explore a broad range of events through more than 150,000 exposures.

Two X-ray telescopes would view the solar corona, or outer atmosphere, with a special detector designed to watch for flares on the Sun and warn the astronauts of an impending event. Three ultraviolet telescopes would observe the lower regions of the corona and the Sun's chromosphere, a region just above the actual surface. A white-light coronagraph was also attached to the ATM for providing visible images of the corona by blocking out the bright solar disc, a method known as occulting. Two hydrogen-alpha telescopes were carried for aligning the ATM instruments, for providing continuous observation of solar events and for measuring active regions as they traversed the solar disc. Another solar physics experiment, the X-ray/ultraviolet camera, was operated from one of the two scientific airlocks in the workshop. The eight ATM telescopes were divided into six separate experiments.

Nine astrophysical experiments were carried aboard Skylab. These included a nuclear emulsion experiment to record the relative abundance of primary heavy nuclei outside Earth's atmosphere, an ultraviolet stellar astronomy camera to measure emissions from stars via a 15.2 cm reflecting telescope mounted in a workshop scientific airlock, an ultraviolet airglow experiment to record the behavior of ozone in the outer atmosphere, a gegenschein/zodiacal light experiment for measuring the background illumination above the atmosphere, collection and measurement of interplanetary dust passing the station, X-ray detector mapping of faint sources within the galaxy, a spectrographic analysis of bright stars, a search for trans-uranic nuclei, and measurements of the composition of Earth's magnetosphere.

Sixteen materials processing experiments were scheduled for use with the relevant equipment in Skylab's Multiple Docking Adapter, five with the vacuum work chamber and eleven with the multipurpose electric furnace. Engineering and technology experiments were subjectively divided into separate categories: six were concerned with man and his ability to work effectively in weightlessness, providing an Astronaut Maneuvering Unit (AMU) with which to evaluate the capability of a new mobility aid designed for EVA but tested in the spacious forward experiment area, a foot-controlled maneuvering device, and time and motion studies of specific tasks and work assignments. Six were concerned with the spacecraft environment, measuring radiation in the vehicle, the size, concentration and composition of aerosol particles inside Skylab, and retrieving for examination on Earth coatings and surfaces placed initially outside the space station for prolonged exposure.

Late in the Skylab development phase, NASA and the National Science Teachers Association agreed to promote a student experiment program aimed at stimulating youthful interest in space science and technology. In October, 1971, a competition was announced to find tests or experiments involving a minimum of astronaut time, a low weight or volume requirement, with maximum value to science or technology. From the more than 3,400 proposals received, 301 were selected as regional winners prior to the final selection in March, 1972.

Eleven out of 25 experimenters were invited to prepare their hardware for flight, eight would use existing hardware on Skylab, and four of the remaining six would receive data similar to that which would have been generated by their proposed experiments. Two would be permitted access to scientists and NASA research workers involved in schemes similar to those they had proposed. Experiments accepted ranged from colonies of non-pathogenic bacteria incubated at specific temperatures so that photographs taken at intervals would show development in weightlessness, to microscopic observation of cytoplasmic motion in leaf cells under zero-g conditions.

One student provided equipment to contain and monitor the web building process of the common cross spider (Araneus diadematus), while another studied the root and stem growth of rice seeds and checked their ability to use light as a substitute for gravity when obtaining directional information.

There was little time to prepare and qualify the experiment hardware, decisions about the student experiment program having been taken less than eighteen months before launch. But the determination observed in the participants augured well for future plans to involve students in Shuttle experiments when that reusable vehicle entered service at the end of the decade. It also provided appropriate application of unsolicited funds received by NASA as gifts over the years from admirers and well-wishers. The $5,548 received up to May, 1973, was used to offset student experiment costs. There had never been so many experiments carried by a single vehicle and the major areas of attention were not pursued at the expense of secondary, or corollary, tasks.

But transport to and from Skylab's orbiting cluster was a little different to that employed for earlier Apollo duties. Gone was the need for more than 18 tonnes of propellant in the big Service Module engine. Only enough for rendezvous and de-orbit burns would be needed. The high-gain antenna at the base of the Service Module was an unnecessary appendage since there would be no deep-space communications, and three fuel cells were one too many for the limited needs of the Earth orbit ferry flights.

CSM's 116, 117 and 118 were assigned to Skylab flights, along with a back-up vehicle – CSM-119 – which would serve in a rescue role if needed, about which more later. Structurally, the spacecraft were similar to earlier models used for Moon flights, representative of original batch production assembled for the G mission profile since the Service Modules carried only two oxygen and two hydrogen tanks for the fuel cells; later H and all J series spacecraft carried three of each.

One fuel cell was deleted but original locations for the remaining cryogenic hardware were retained in sector 4. During a normal Skylab mission the cryogenic fluids would be exhausted long before the crew were scheduled to return to Earth. For electrical power aboard the CSM, between power-up and separation

Technicians dressed in white 'clean room' uniforms perform a power-up test of the Skylab's Apollo Telescope Mount at NASA's Marshall Space Flight Center.

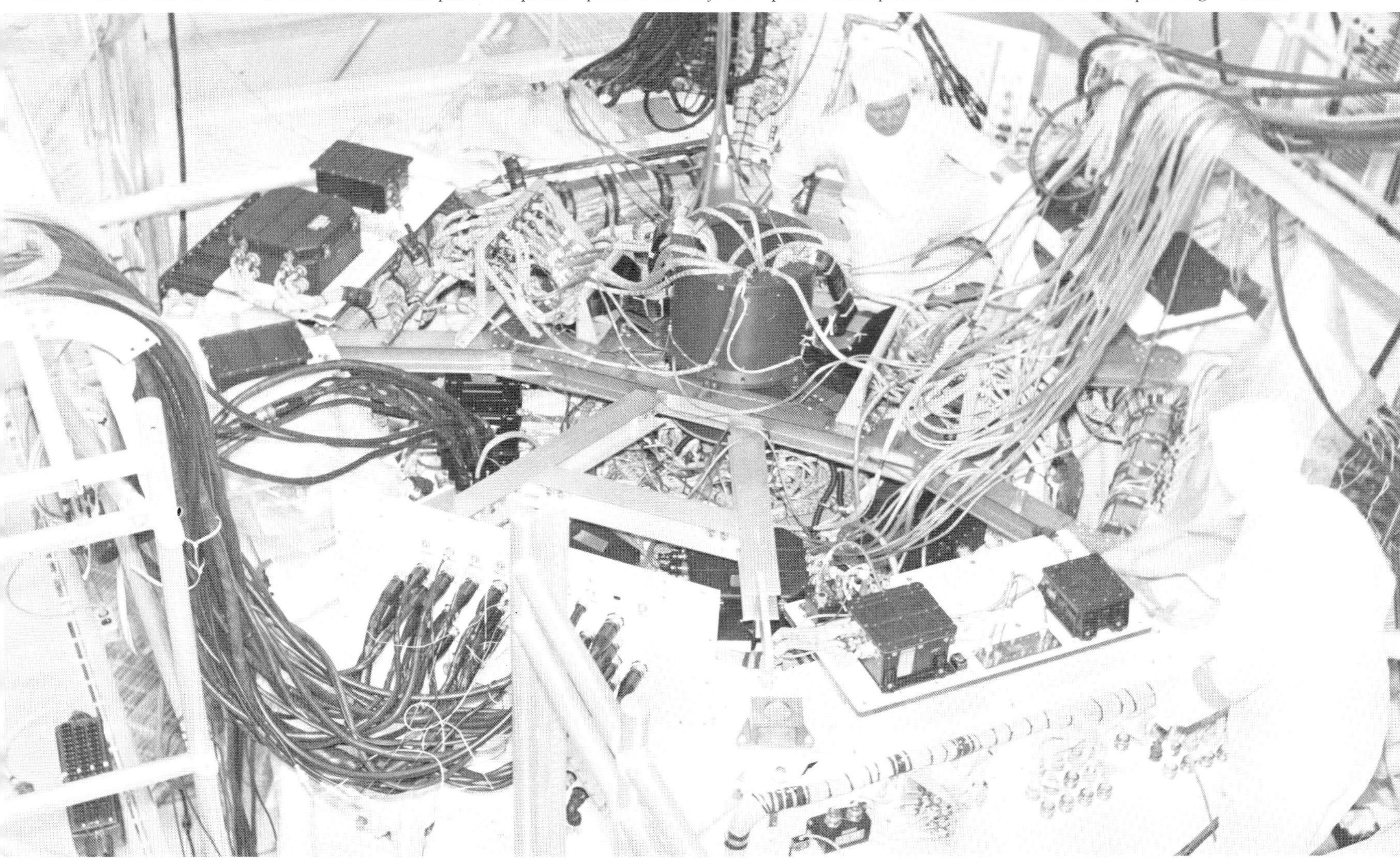

from the Command Module after retro-fire, the Service Module was equipped with a pack of three 500 ampere-hr silver oxide-zinc batteries weighing 111 kg each. Two would be adequate for a normal pre-separation sequence but all three, when used conservatively, could provide electrical energy for up to 18 hours. They were mounted to an upper shelf in sector 1.

The Command Module too had an improved electrical system, the two small pyrotechnic batteries being replaced with two re-entry batteries rated at 40 ampere-hr each, bringing that total to five. Also carried in the CM, a power transfer umbilical allowed Apollo to transfer electrical energy to Skylab and for the cluster to top up the Service Module batteries.

In early 1972 NASA added a new role to the CSM which called for more propellant available to the Service Module RCS thrusters. Recognizing that even 432 km above Earth atmospheric drag would slowly pull Skylab down, engineers designed a supplementary pack carried in the Service Module for extending the thruster burn duration so the small jets could be used to push Skylab back up to the desired orbit at periodic intervals.

Sector 1 carried a propellant storage module comprising five oxidizer tanks, four fuel tanks, and three helium tanks in addition to the plumbing necessary to carry fluids to the 16 RCS engines attached to the four quads; each quad still carried one helium, two fuel, and two oxidizer tanks on the inner face of the respective Service Module door. But the propulsion module's supplementary propellant supply was contained in nine identical tanks 98.8 cm long and 32.1 cm in diameter, much larger than the biggest tanks carried adjacent to each engine quad. The propulsion module's three helium tanks, however, were identical to the existing four carried on all manned Block II vehicles. The reserve supply boosted thruster propellant capacity by 113%.

But Service Propulsion System engine propellant was not needed in the design quantities envisaged for Moon flights so one fuel and one oxidizer storage tank was removed from Service Module sectors 3 and 6 respectively. Furthermore, the two remaining tanks were to be loaded with only 720 kg of fuel and 1,152 kg of oxidizer, or about 8% of tank capacity. Because the two tanks removed from the Service Module carried significantly less fuel and oxidizer than the cylinders left in, the tanked quantity represented about 10% of that carried for a normal Moon flight.

Also taken out was the lower of two spherical helium tanks fitted in the cylindrical center-section, making room for a hat-box shaped tank 101 cm in diameter capable of holding 228 kg of waste water from the fuel cells which would not be needed since the crew would live inside Skylab. It was better to store the water rather than dump it overboard and contaminate the region around Skylab. The only major environmental control change provided a duct for supplying oxygen from the workshop. Small heaters were also added to the RCS thrusters to prevent them freezing and to the SPS plumbing for a similar purpose.

In all, Apollo CSM vehicles for Skylab duty were the same size as earlier models but weighed only 13.8 tonnes, less than one-half their Moon-bound predecessors. However, this was close to the payload lifting capacity of the Saturn IB. Launch vehicles assigned to CSM's 116–119 were AS-206 to -209, four Saturn IB launchers in storage since 1968. Only three were expected to fly, the fourth serving as the rescue vehicle for CSM-119 in the event CSM-118 was disabled in space. It was but one part of contingency plans designed to save the program and its planned objectives in the event of major systems failure.

As related earlier, the S-IVB from which McDonnell Douglas built Skylab's orbital workshop section originally belonged to launch vehicle AS-212, the twelfth Saturn IB off the production line, and was to be launched by the first two stages of AS-513, the thirteenth Saturn V assembled. An identical set of back-up hardware was to be in an advanced state of readiness when the prime hardware lifted off so that NASA had an option to launch the back-up twelve to fifteen months later if the prime equipment became unusable. The back-up workshop originally belonged to AS-515 and the first two stages of that launcher would serve as the propulsion to place it in orbit.

The need to have a full set of hardware in reserve was but a minor incursion on the budget, one that could save the entire program if the prime hardware failed. The existence of the second workshop would tempt several NASA officials to press for a second, Skylab-B, mission following completion of the three visits with the prime workshop, but financial pressures and the need to develop a reusable Shuttle would argue against that option. Nevertheless, the opportunity to provide a limited rescue capability for Skylab emerged in early 1971 in the wake of Apollo 13's near disaster the previous year.

Assuming that astronauts could seek refuge, or remain, in the orbital workshop until a five-man Apollo could be sent up, a major failure in the CSM docked to Skylab need not prevent the crew from getting back to Earth. Ever since Yuri Gagarin's single-orbit flight of 1961, engineers recognized the potential failure of equipment designed to bring spacecraft back through the atmosphere. Embellished by the film 'Marooned' it was a possibility all too real in concept. Careful examination of the Command Module showed that by removing storage boxes on the aft bulkhead (the floor) two additional crew couches could be installed beneath the three already provided but turned 180° from the upper couches in a head-to-toe configuration.

In extreme emergency, two astronauts could fly a modified CSM to the Skylab cluster, dock to the prime or contingency docking port and bring the three men on board for re-entry. The rescue kit could be installed in a standard Skylab Command Module within eight hours and although the concept required Skylab to contain the stranded crew safely in space it did improve their chances of getting back to Earth. Reaction time to a rescue call would depend on the state of readiness in preparing for the next flight since the succeeding CSM scheduled to carry astronauts aloft would be called upon to perform that function.

At the beginning of a Skylab visit, reaction time would be 48 days due to the extensive clean-up at the pad necessary to get it ready for another launch, but that would decrease until at 56 days elapsed time the rescue vehicle could be sent up at 10 days notice. Back-up for the third mission would be provided by CSM-119 and Saturn IB AS-209, a fourth set of hardware processed like the earlier vehicles in case it was called upon for rescue on the last visit.

Skylab was to introduce a significant change in Saturn launch vehicle operations, focusing activity at launch complex 39's two concrete pads originally constructed for Saturn V launchers. When Skylab became a 'dry' workshop in 1969, NASA was already looking to complex 39 for launching both two-stage Saturn V *and* Saturn IB vehicles. Originally built to support the smaller Saturn I and IB launchers, complexes 34 and 37 were considered redundant when Apollo 7 carried aloft the only manned Apollo scheduled to fly on the Saturn IB before Skylab missions began.

By late 1968 it was apparent that keeping operational the single pad at complex 34 and the two pads at complex 37 was an unnecessary drain on resources. A year later, while the Cape prepared to send Apollo 12 to the Moon, Boeing and Chrysler studied with KSC engineers the possibility of launching future Saturn IB's from a complex 39 pad. Pad interface equipment was, after all, the same for Saturn IB's upper stage and payload sections as it was for the Saturn V's third stage and payload. Adaption would be simplified if a launch umbilical tower could be cut down to match the lower height of the smaller rocket, or if the rocket itself could be elevated above the pad base.

NASA had already decided to close out the single pad at launch complex 34 in favor of using the two at complex 37 and now faced good arguments for closing 37 and bringing all Saturn operation to complex 39. But schedules embracing simultaneous preparation of Apollo Moon mission and Skylab Earth orbit hardware seemed to show a probability that one program would get in the way of the other. Only when, in 1971, it was decided to complete all Moon landing flights *before* Skylab began operations was it sensible to follow the more economic plan.

The proposal to use complex 39 suggested remaining Apollo Moon flights (14–17) employ pad A while engineers modify pad B to accept Saturn IB. The two-stage Saturn V lifting the Skylab cluster aloft would similarly use pad A, retaining Saturn V operations at that location (in fact pad A had been continuously used since Apollo 11). And then, while Saturn IB vehicles sent three successive astronaut teams to the Skylab cluster from pad B, modifications could begin on pad A to prepare it for Shuttle launches, then confidently expected before the end of the 1970s.

But that still left undecided the issue concerning raising the rocket to match the upper umbilical tower interface points or lowering the tower to match the launcher. It was more economical to build a pedestal for the Saturn IB and launch it from that. The new structure was mounted to launch platform 1, one of three built in the early 1960s to support Saturn V operations, and comprised a steel tower 39 meters tall, 14.6 meters square at the bottom and 13.4 meters at the top. Resting on four legs fixed outside the mobile launch platform's flame cut-out, the launch table had an inside diameter of 8.5 meters to allow exhaust from the Saturn IB's eight first stage engines to pass down the center. The pedestal weighed more than 226 tonnes and the mobile launcher supported an additional 226 tonnes of support equipment, although three service arms had to be removed from the lower part of the adjacent tower to make room for the pedestal. For access to the engines, a circular work platform was installed across the launch table and removed before ignition.

In all, the modified launch platform and its lightweight Saturn IB weighed about the same as a fully equipped unfuelled Saturn V, minimizing changes to the crawler-transporter employed to carry it from the Vehicle Assembly Building to pad B.

Launch platform 2 was converted for the two-stage Saturn V assigned to place the Skylab cluster in orbit from pad A. Only minor changes were necessary, including modification to the swing-arm configuration and removal of the standard 'white room' swing arm originally fitted to align with the Command Module. Another white room swing arm was fitted lower down at the 73.1 meter level to allow access to Skylab during preparations in the VAB and for emergencies on the pad. The unfuelled two-stage Saturn V/Skylab was about 23 tonnes heavier than an unfuelled three-stage Saturn V/Apollo.

Firing rooms 1, 2, and 3 in the Launch Control Center were used sequentially for the first five manned Saturn V flights but after Apollo 12 firing room 3 was reserved for changes necessary to equip it to support Saturn IB flights from pad B, with firing room 2 last used for Apollo 14 before it was configured for the Skylab launch from pad A; the last three Apollo Moon missions were launched from firing room 1.

In several respects, Skylab was a new dimension for NASA manned space flight managers. For the first time, major flight events would bring into use a facility designed for occupation rather than transportation. For the first time, there could be no flight test of hardware designs; it all simply had to work. For the first time on a manned mission too, solar cells would replace conventional batteries or fuel cells for electrical energy. And for the first time also, manned Earth orbit operations would carry astronauts to higher inclinations than any previous US mission. Many of Skylab's systems were new, both in concept and in design, although the program evolved on the premise that it use as much available hardware as possible.

The prospect of an Apollo Command Module ending up in the water 50° north or south of the equator raised questions about the craft's survivability in the event of a contingency landing. Spacecraft 7A, an old Block I Command Module suitably modified and employed in water tests since 1966, was exposed for long periods to a water temperature of 0.5°c and an air temperature of −4°c in special tests carried out at Eglin Air Force Base, Florida, in late 1971. It came through, having proved that Apollo was a cold-water spacecraft after all.

On 18 January, 1972, NASA formally named the nine men assigned to fly Skylab missions. Pete Conrad would command the first manned visit called SL-2 (SL-1 was the launch of the unmanned Skylab cluster) carrying science-pilot Joseph Kerwin and pilot Paul Weitz. The second manned visit (SL-3) would be commanded by Al Bean, with science-pilot Owen Garriott, and pilot Jack Lousma. The third and final visit (SL-4), would place science-pilot Ed Gibson and pilot William Pogue under the command of Gerald Carr. Only two back-up crews were necessary: Russell Schweickart, Story Musgrave and Bruce McCandless for the first mission, with Vance Brand, William Lenoir and Don Lind for the second and third flights.

Policy throughout was to have one professional scientist on each crew. Kerwin, Garriott, and Gibson were selected as scientist-astronauts in 1965 and the only other contender from that batch still with NASA was Harrison Schmitt, then scheduled for a flight on Apollo 17. Back-ups Musgrave and Lenoir were from the 1967 scientist-astronaut batch, the group from which so many resigned. Skylab was a scientist's program from the outset, lacking the sophisticated pilot requirements inherent in Apollo Moon flights.

The Skylab commander would be in charge of the overall mission and perform systems checks and CSM control functions.

The science-pilot would perform ATM studies of the Sun, or conduct primary medical tests (Kerwin was a physician, Garriott and Gibson were physicists). The pilot would operate the EREP Earth observation equipment and generally run the cluster. As it turned out, Schmitt, Kerwin, Garriott and Gibson were the only science-astronauts out of an original 17 to fly ballistic manned space vehicles.

During 1972, the year of preparation for Skylab, the joint docking flight with a Russian Soyuz spacecraft received approval at the highest level and, as recorded in the previous chapter, mission plans were laid for a rendezvous flight in mid-1975. On 30 January, 1973, three and one-half months before Skylab ascended, the American crew for Apollo-Soyuz was named. Tom Stafford would command the US side of the mission, with Vance Brand and Deke Slayton. Bean, Evans and Lousma would be their back-ups. After sixteen years in the astronaut corps, Slayton, recently put back on the active list, was to fly in space.

Yet while Apollo Moon flights were run out, and preliminary work tasks evolved for the joint docking flight, Skylab hardware reached completion for the year of space operations many thought would presage a new era of space stations and extended Earth observations. But not everybody was completely pleased with the mission's timing. Solar physicists wanted to observe the Sun during the period of maximum Sunspot activity which, coming on average at 11.2 year intervals occurred in 1957 and again in 1968–69. Had the Skylab ATM been sent into orbit on the original Apollo Applications Program plan, it would have observed the solar disc at the last period of maximum activity. As it was, Skylab would view the Sun during a period when observers predicted less than 30 annual spots compared with more than 100 at Sunspot maximum. But the Sun was not to let them down, as scientists would observe when the manned flights got under way.

Another area of more immediate concern rested on the availability of good observation throughout the growing season in Earth's northern hemisphere. It was necessary, said the Earth scientists, to launch before the start of Spring, and to obtain coverage throughout that calendar year. When Skylab was reorientated in 1969, a provisional launch date was set for June, 1972. But when Apollo was extended due to budget cuts and two cancelled Moon landings put Skylab after lunar exploration flights, the date was put back to November. By April, 1971, it was apparent that Apollo 17 would not fly before December, 1972, and Skylab was further delayed to 30 April, 1973. In that intervening period Earth observation came to be an important segment of Skylab operations and scientists pressed for adherence to the 30 April launch slot. Yet in the final months of preparation the schedule *did* slip, but only by two weeks.

On 14 February, 1973, Schneider met with the Manned Space Flight Management Council and the final dates were agreed. The cluster would go up on 14 May, followed a day later by SL-2, the first manned visit. After 28 days, Conrad, Kerwin and Weitz would return to Earth, splashing down on 12 June. Three weeks later, on 8 August, Bean, Garriott and Lousma would ascend for a 56-day stay, returning on 3 October. Less than a month later, on 9 November, Carr, Gibson and Pogue would launch for a second 56-day stay, completing their mission on 4 January 1974.

For a while in 1972 pace at the Cape quickened as it had before during accelerated preparations for the first Moon landing. Hardware for the *last* Moon landings was now being processed as loads arrived in crates for Skylab, rocket stages slipped in on

Picked to fly America's long duration missions, three-man crews for the Skylab missions. The first, 28-day flight, would be manned by the Conrad crew (top) followed by a 56-day mission with the Bean crew (left) and a second 56-day flight by the Carr team (right). In practice, durations would change significantly.

barges, and modules flew in by air. Launch vehicle 513's second stage, the S-II, had been at KSC since April, 1971. Placed in storage for a year, it was brought out for modifications as the big Saturn V first stage, the S-IC, arrived at Port Canaveral. Completed as early as August, 1969, the S-IC was static fired at NASA's Mississippi Test Facility in February 1970 before being refurbished and stored until delivery back to Michoud, its point of origin, in May, 1971. On 16 July, 1972, it arrived at the Kennedy Space Center where it was errected on mobile launcher 2 seventeen days later. The second stage was erected on 20 September.

Two days later the orbital workshop docked at Port Canaveral as an aircraft flew in with the crated Apollo Telescope Mount. A day later, 23 September, a barge carrying the workshop moved up the Banana River and pulled into the turning basin alongside the Vehicle Assembly Building. Five days later the workshop was installed atop the two AS-513 stages and on 6 October the combined Airlock Module/Multiple Docking Adapter arrived by air. They went to the Manned Spacecraft Operations Building for extensive checkout before transfer to a high-bay area two months later. The CSM scheduled to carry the first manned crew was then used in a docking test to check compatibility of all the hardware, but not until Apollo 17 was launched.

Before the end of December the combined airlock and docking modules were attached to the fixed airlock shroud and all elements were taken to the Vehicle Assembly Building where the two-stage Saturn V and orbital workshop was ready to receive them. On 29 and 30 January, technicians fixed the airlock and docking modules to the workshop, attached the ATM to its deployment truss, and fixed the large payload shroud on top.

Meanwhile, hardware for the manned CSM launch had been coming in. First to arrive, on 24 June, 1972, was the second stage followed less than a month later by CSM-116. The Saturn IB first stage arrived at the Cape on 22 August. By 8 September the stages had been mated together in the VAB and while the Apollo CSM elements were tested, checked, and fit-aligned in the MSOB, a boilerplate configuration simulating the CSM was erected on the two-stage Saturn IB. Because AS-206 would fly from a newly configured launch pedestal it was desirable to check out the entire assembly before taking the flight hardware to pad B.

The assembly, complete with boilerplate, was moved out on 9 January, 1973, where it remained while engineers checked cables, plumbing and connections at the pad. After fuelling and systems checks were completed on the Saturn IB during January, the assembly moved back to the VAB where the boilerplate was taken off and replaced with the flight hardware. With CSM-116 on top, it slowly rumbled back to pad B on 26 February just five days after the spacecraft met its launcher for the first time.

One last management shift set up the personnel for Skylab's arduous mission activity. On 26 January, 1973, Eberhard Rees retired as Director of the Marshall Space Flight Center and his post was taken by former Apollo Program Director Rocco Petrone. For more than twenty-two years von Braun and Rees had managed the nation's big rocket development base at Huntsville. Now only Kurt Debus still remained as a high NASA official from the group that came to America in 1945 and within a year he too would announce his retirement as Director of the Kennedy Space Center.

Skylab was a turning point for NASA aspirations that once embraced major expeditions to the Moon and Mars, for now emphasis was firmly on Earth applications. For a decade, men had been simulating the loneliness of long duration space flight in vacuum chambers and isolation rooms. All the leading aerospace companies invested in tests aimed at proving man's ability to perform useful work remote from other human beings: General Electric put four men in a space simulator for one month; Boeing tested the response of five men isolated for the same period; the Air Force had four volunteers live two months in an oxygen/helium atmosphere.

When Skylab became a formal NASA project, studies took on a more urgent profile. It was one thing to design and plan interior configurations for long stays away from Earth, quite another to predict how men would respond to each other after two months in orbit. It was in pursuit of the latter that Marshall Space Flight Center engineer Chester B. May joined the six-man crew of Dr. Jacques Piccard's submersible vessel, Ben Franklin, to see how men reacted after 30 days under water. The research vessel was to drift with the Gulf Stream a distance of nearly 2,700 km, 300 meters underwater.

Two days before the launch of Apollo 11, while the world held its breath in anticipation of man's first Moon landing, the Ben Franklin slipped beneath the waves off West Palm Beach less than 200 km south of Cape Canaveral. A month later, on 14 August, the vessel surfaced south of Nova Scotia and Chester May had the information he wanted. The following year, NASA formally participated in the Tektite II underwater research program where teams of visiting scientists and engineers spent several weeks in a pressurized living area on the floor of the Caribbean.

The lessons were unquestioned for there was little or no parallel to information learned this way. Only submarine crews knew the feelings and sensations associated with isolation. Color was important – psychological tests had shown for some time that emotions, taste, smell, and mental alertness were all shaped around the balance of colors and hues – and a determined set of objectives was important also because nothing was so demoralizing as routine or drudgery. Clear and distinct goals were set early in the astronaut training program, each team set up to achieve the tasks laid down for its particular mission.

On 26 July, 1972, astronauts Robert Crippen, Karol Bobko and William Thornton entered a 6 metre diameter altitude chamber at the then Manned Spacecraft Center specially configured to simulate a portion of the Skylab interior. Called the Skylab Medical Experiments Altitude Test, or SMEAT, activity progressed as though aboard the flight hardware with the crew experiencing the same environmental conditions other astronauts would have in the actual workshop. Fifty-six days later, on 20 September, they were released from this test having satisfactorily shown it was possible to safely live for two months in the design environment.

Elsewhere at the Manned Spacecraft Center, Milton Windler continued grooming flight control personnel in roles for which he had been training them since early that year. In October a single team was pulled from the Apollo operations group then preparing for the last Moon flight and the remaining personnel were assigned to Skylab at completion of that mission. Throughout 1973, flight controllers would man the MOCR in a sustained operation unique to US manned space flight, for even when the first and second crews came home the average three week periods between occupation would be filled with systems checks, partial de-pressurization, automated experiment tasks, etc.

Yet it was the astronaut teams that bore the brunt of training, and the men who groomed the simulators at four locations enabling the crew to become proficient at all phases of planned Skylab activity. Space walks were to take place on all three visits (one at the end of the first; three on the second; two on the third) and they too required special training sessions to develop familiar sequences for retrieving film cassettes from the massive ATM structure. And then there was the scheduled maintenance sessions, periods where housekeeping duties would embrace replacement activity. Skylab would be launched with more than 340 kg of spare parts and tools with which to perform planned or contingency tasks.

Added to many years of professional training as astronauts, each team accumulated more than 6,000 man-hours of training for their respective missions. But there would be no let-up for the second or third crew when the first and second lifted off the launch pad. It was their lot to feed in new methods or procedures developed with the practical experience of the preceding team.

When the Saturn IB and its Apollo payload rumbled to pad 39B on 26 February, 1973, the big Skylab stack was in the final stages of preparation at the Vehicle Assembly Building. Although launched a day after the space station, SL-2 would require more attention at the pad so it went out several weeks before the unmanned cluster. SL-1 was moved to pad A on 16 April, four weeks before the planned 14 May launch date. Next day engineers connected the workshop to an environmental system at the pad and nine days later the formal countdown demonstration test got under way.

Before the end of the month final items had been stowed in the airlock and docking modules and on 1 May the cluster was 'closed out' ready for launch. The countdown began at 2:00 am on the morning of Wednesday 9 May. Just 1,300 meters away at pad B, the countdown for SL-2 began six hours later. It had taken nearly a decade to get the hardware and the experiments into a flightworthy package; Skylab was now ready to fly.

Advanced lunar missions represented by the last three Apollo Moon flights carried sophisticated scientific equipment for scanning the surface from orbit while the LM spent three days hosting the explorers. Here, gamma ray and mass spectrometer booms have been deployed from the Scientific Instrument Module bay while mapping and panoramic cameras photograph the surface.

Apollo 16's Lunar Module ascent stage complete with flapping insulation → blown loose during takeoff from the Moon.

The last – and the best: Apollo 17 astronauts Cernan (seated), Evans (right) and Schmitt pose in front of their Saturn V launcher, the mission badge superimposed top left.

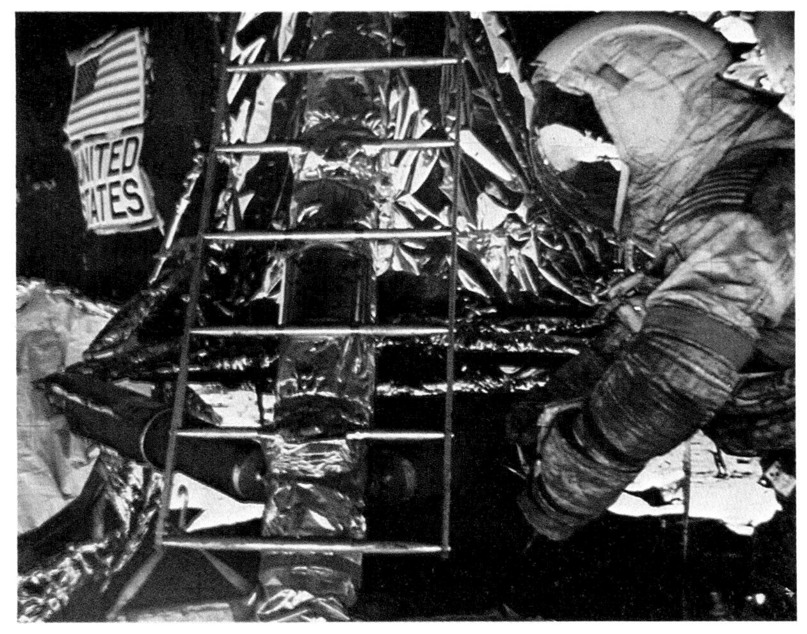

Mission commander Eugene Cernan returns to the LM after the 'day's' work on the dusty lunar surface.

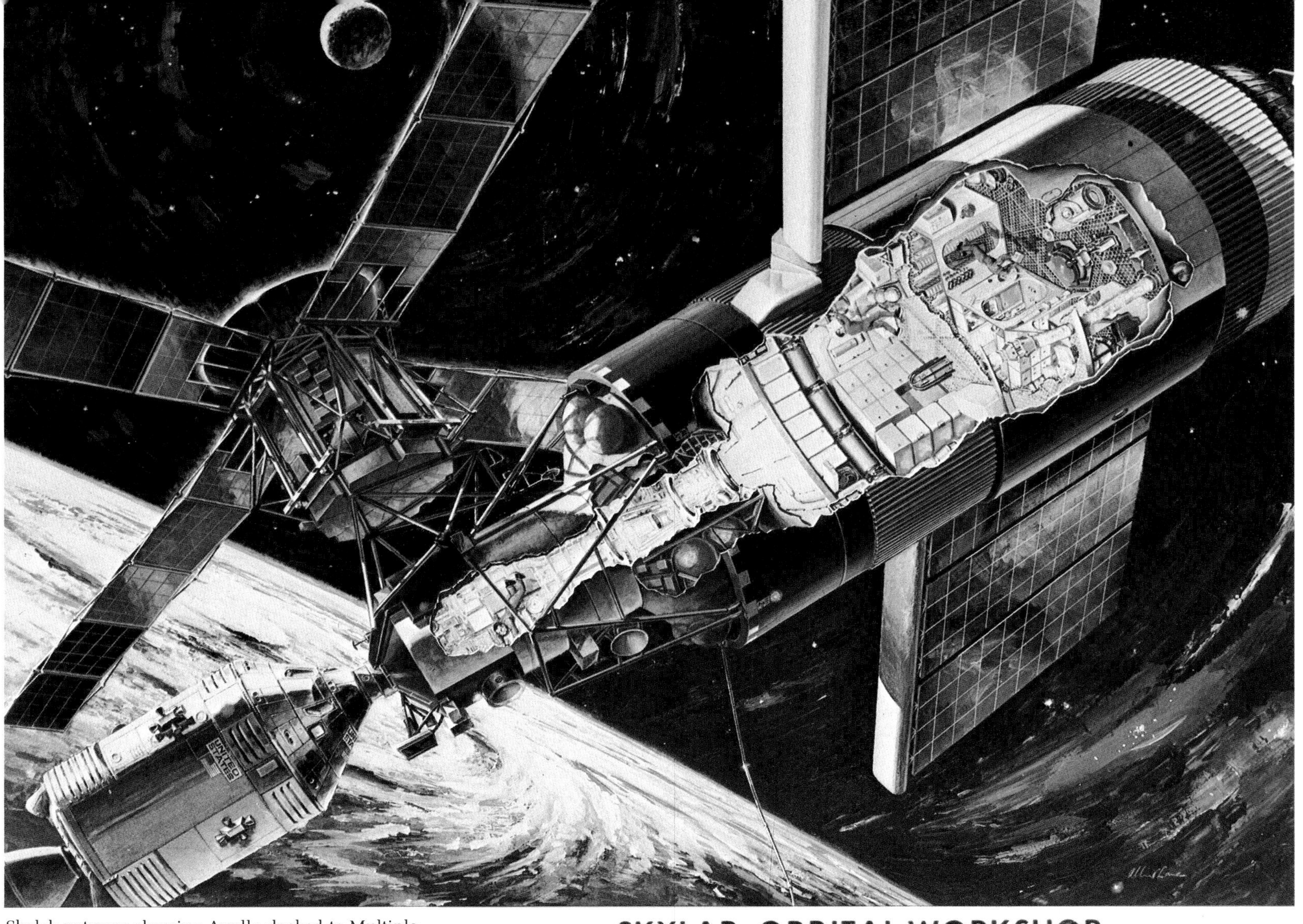

Skylab cutaway showing Apollo docked to Multiple Docking Adaptor which in turn is connected to the Airlock Module in front of the Orbital Work Shop.

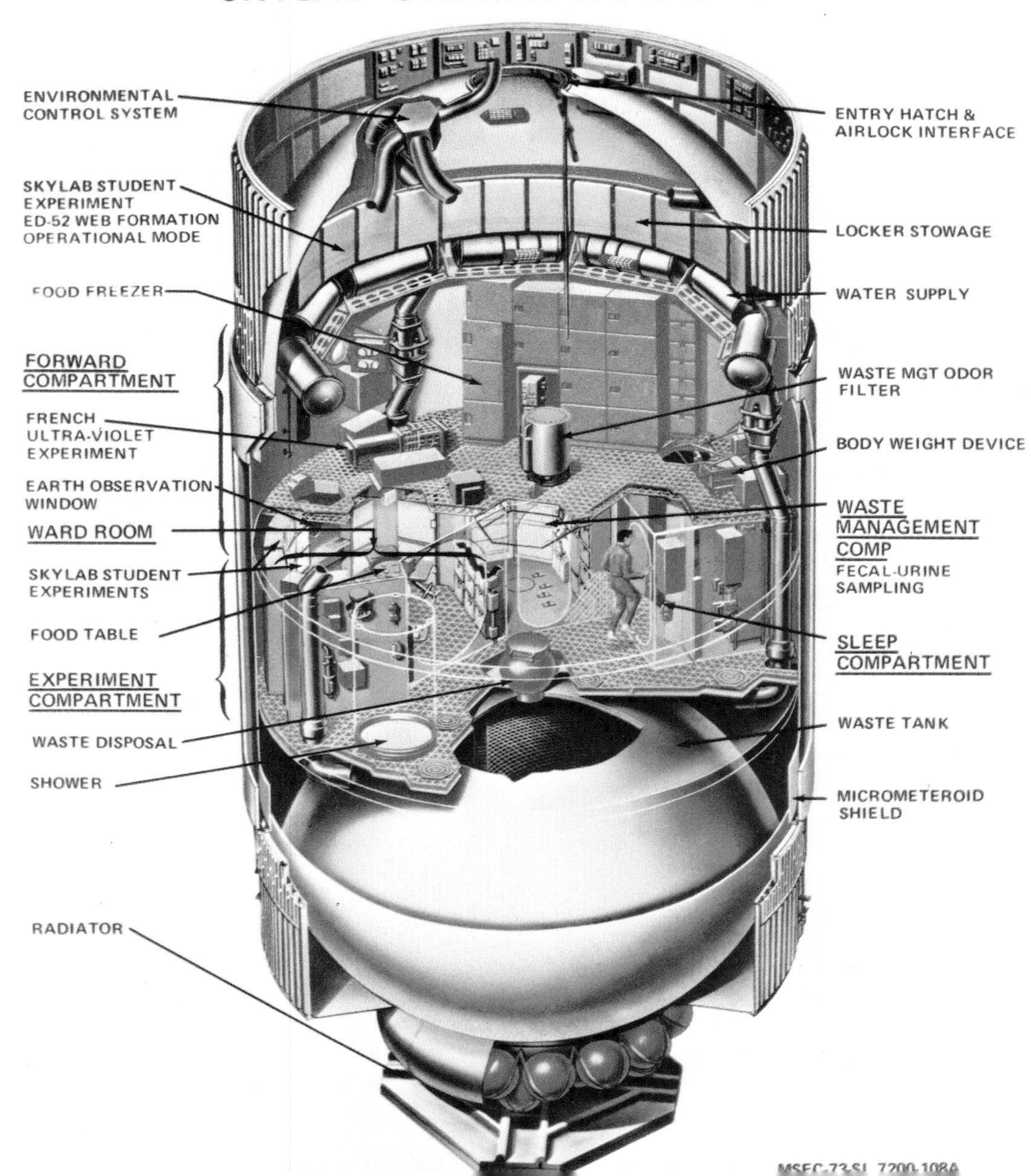

The last Saturn V, a modified two-stage version of the launcher built to send men to the Moon, carries the Skylab space station into orbit, May, 1973.

Skylab astronauts Kerwin and Conrad prepare to deploy the Skylab sunshade in underwater simulation at the neutral buoyancy facility. This A-frame shade was, in fact, deployed by the second crew while the first put out a parasol device.

An overhead view of the Skylab space station during the final 'fly-around' before the last crew return to earth. Note the creased A-frame shade overlaying the parasol deployed by second and first crews, respectively, and the single remaining OWS solar array.

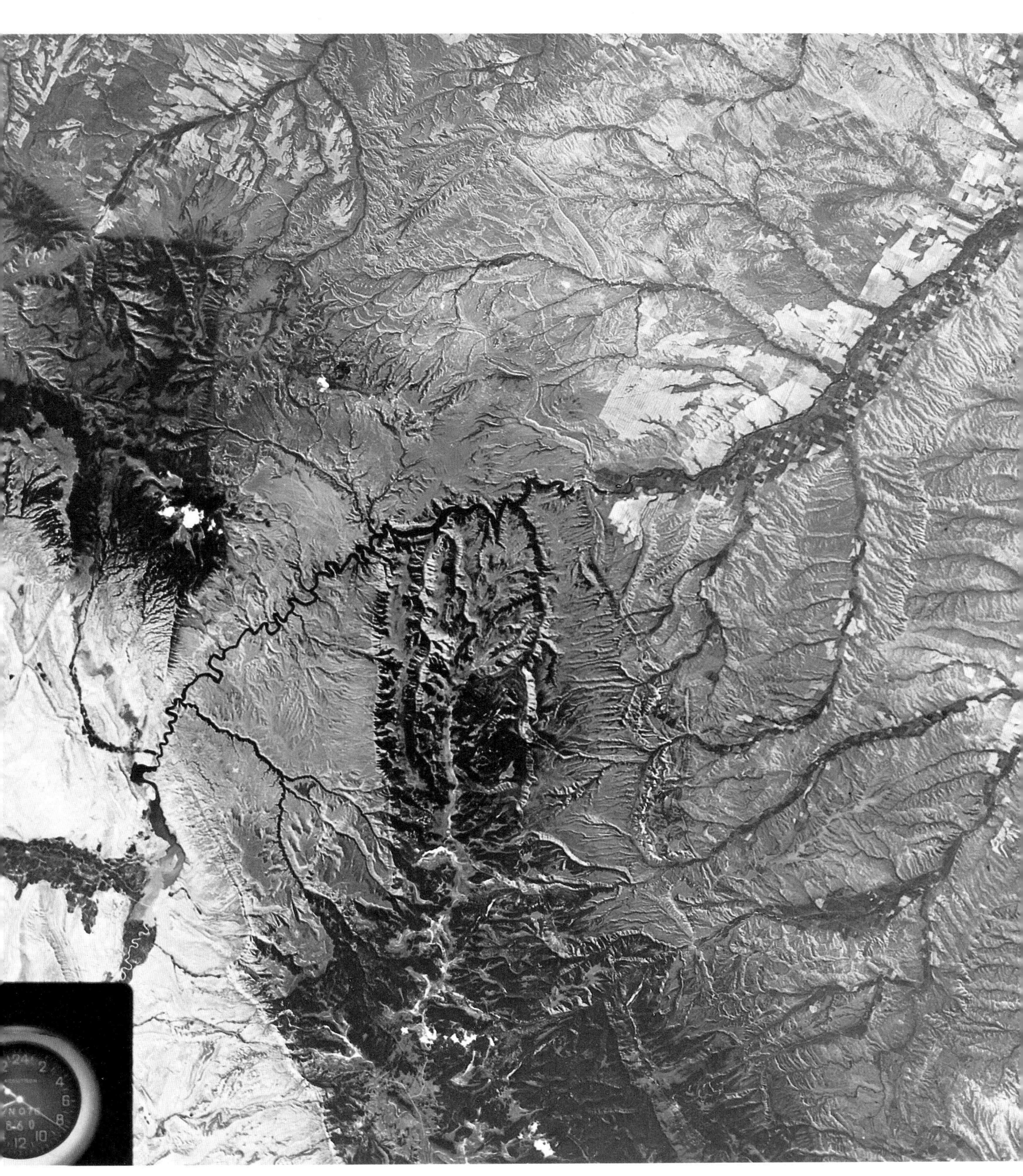

A view of central Wyoming and southern Montana shot by Skylab's earth terrain camera.

A vertical view of the east coast of Sicily with smoke from Mt. Etna (top).

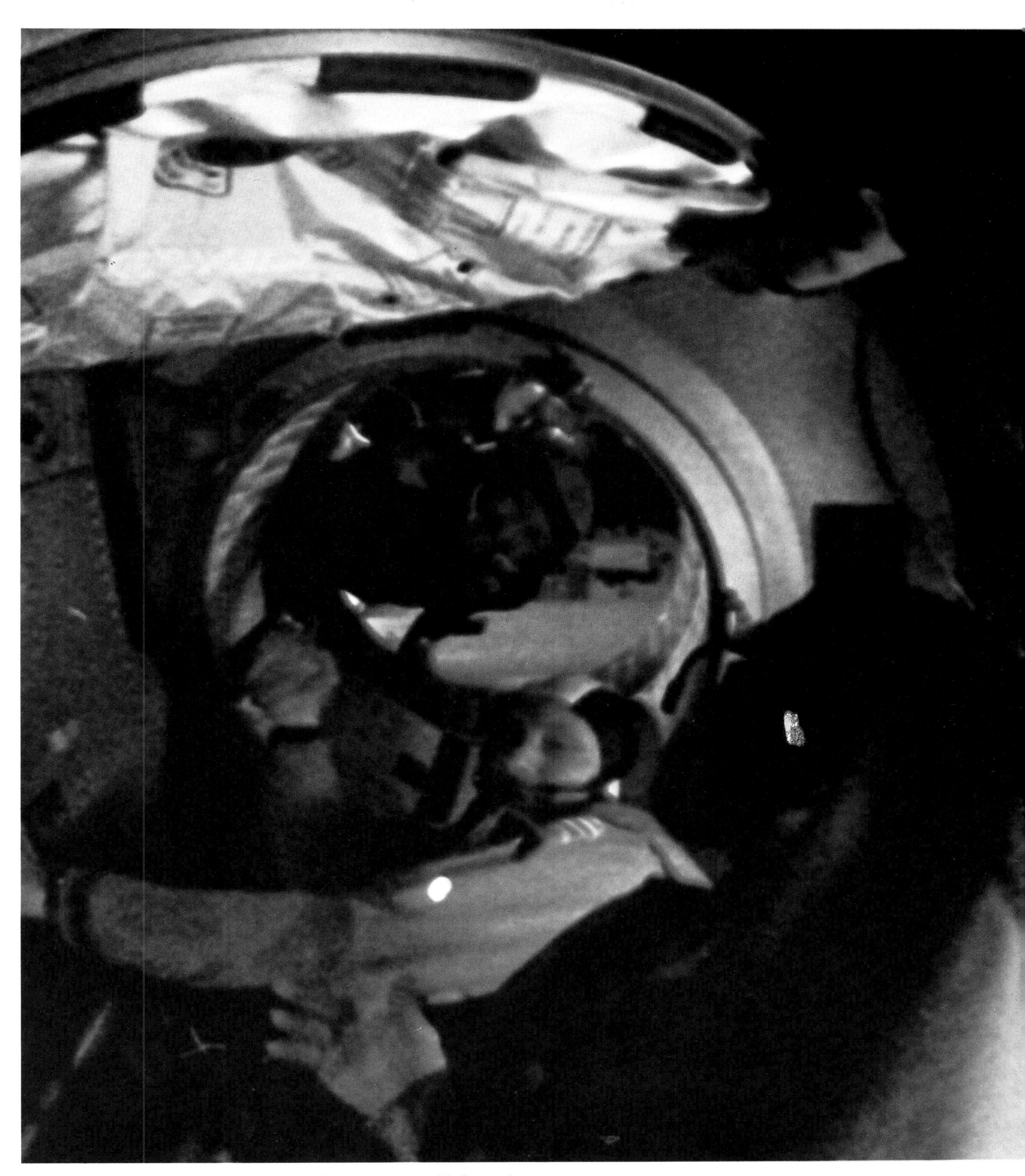

Stafford (foreground) shakes hands with Leonov in the first international link-up, July, 1975.

Apollo (top left), Docking Module (center), and Soyuz's two pressurized modules seen with respective crew positions at the moment of the historic handshake, 17 July, 1975, on the joint docking mission involving the last US ballistic manned space flight.

CHAPTER TWENTY-SIX

Repair and Restoration

Unlike every manned flight before it, Skylab missions would fly with an almost empty flight plan. Maximizing the concept of real-time planning, Skylab astronauts were schooled in sequences, functions, and activities relating to specific operations and duties, but the sequence of those tasks was to be arranged as the flight progressed, thereby ensuring maximum response to changing needs. When the final countdowns began for SL-1, the unmanned cluster, and SL-2, the manned Apollo, flight controllers were aiming for a lift-off at 1:30 pm local time on Monday, 14 May. After getting the cluster settled down in orbit, and the various appendages deployed, the workshop and its attached modules would be pressurized to one-third sea-level pressure ready for the launch of Conrad, Kerwin and Weitz at about 1:00 pm the following day.

Adhering to a strict Earth cycle slaved to the Houston time base, the crew would spend Day 1 flying up to Skylab and docking with it, scheduled to take less than eight hours before the Command and Service Modules were firmly linked with the Multiple Docking Adapter, and the crew were to begin their first night in space at 10:00 pm. Day 2 would be a busy one, with the crew unpacking supplies carried up in the CSM, opening up the station, and re-locating storage boxes placed in the cluster's center of mass for ascent but which could by this time be taken to permanent locations. Activation and experiments checkout would continue on Day 3 but by Friday of launch week the crew were expected to settle down to daily tasks and operations for which they had been launched.

The most critical areas of concern surrounded the several automated sequences the Skylab cluster would have to go through before presenting a suitable configuration for the manned occupation. Everything else had been tried out before – rendezvous, docking, complex operations in a weightless environment – but no one knew for sure that all the sophisticated Skylab apparatus would work precisely as planned.

At launch complex 39, and the Manned Spacecraft Operations Building 11 km south, about 1,300 engineers and technicians began a minutely scheduled sequence of events in the closing hours to launch of the main Skylab cluster. At the Launch Control Center alongside complex 39's Vehicle Assembly Building, 500 personnel manned firing room 2 while another 500 stood by to occupy firing room 3 for the second launch a day later. At the MSOB, 300 technicians watch-dogged the unmanned cluster up north on pad A, while 200 stood by to monitor the condition of the spacecraft on pad B.

Two days before the SL-1 launch, Conrad, Kerwin and Weitz howled into Patrick Air Force Base in their T-38 jets and went to KSC for final medical checks. Only a few days before, on 9 May, lightning struck the launch umbilical tower alongside Skylab but exhaustive tests showed no damage. Now all seemed ready for the dual sequence to begin. The flights, although bringing added attraction of a two-stage Saturn V and a manned Saturn IB launched on successive days, drew fewer spectators than Apollo Moon flights. Barely 500,000 were grouped on and around the Kennedy Space Center when 14 May dawned bright and warm.

Above, clouds moved slowly across the sullen sky as watchers took up their positions, 25,000 guests getting a privileged view of the two launch pads. Pete Conrad, Paul Weitz, their two wives, and Joe Kerwin, watched from one location; Kerwin's wife and his parents watched from another. Test supervisor Chuck Henschel got appropriate 'go' reports from the range safety officer and systems engineers manning the several score of consoles in firing room 2. Launch operations manager Walter Kapryan announced his own 'go' for the launch, and launch vehicle test conductor Norm Carlson pronounced Saturn V AS-513 – the last of its kind – ready for flight. It was almost exactly five and one-half years to the day that the first Saturn V ascended from the same pad; now, thirteen vehicles later, the dying breed was giving birth to a new era in manned space flight.

The trajectory for this two-stage flight was biased for altitude and aimed for a higher orbit than previously targeted for test or manned Moon flights. Separation of the first and second stages would occur about 87 km above the Earth versus 68 km, and shutdown of the S-II second stage would come at a height of 442 km compared with 173 km; the second stage would fly a steeper ascent than ever before.

About the only significant change to AS-513 concerned the programed shut-down sequence of the four F-1 outer engines at the base of the first stage. As usual, the center engine would cut off early, at about 2 min 21 sec, but the four outer engines would be staged in pairs at 0.7 second intervals instead of cutting off together at about 2 min 38 sec. This would reduce the shock of sudden deceleration and possible damage to the ATM structure in the nose.

It was good news to get the 'go' from Norm Carlson for this last Saturn V, albeit unmanned and devoid of a propulsive third stage. With just three minutes to go the automatic sequencer was already controlling a myriad separate events leading to release for flight. At T – 2 minutes the propellant tanks were pressurized in the first and second stages and at 50 seconds the electrical load shifted to an internal battery source, a water deluge having already begun to drench the flame trench below.

At T – 30 seconds the big swing arms moved back from the launcher. The voice of public affairs officer Chuck Hollinshead carried across the Cape from speakers attached at various points and through the air-waves of several commercial and public-service broadcast networks. The world, for the most part, was relatively unconcerned about the launch of this Earth-orbiting space laboratory: in Europe it was early evening, in Asia it was night, and most people had other things to do so the media were unresponsive and only a few channels carried live news of events that Monday that would soon have the world waiting and wondering. In Houston it was thirty minutes past noon and at the Cape it was 1:30 pm.

'T – 20 seconds and the countdown continues to go smoothly. Guidance release; T – 13 . . . 12 . . . 11 . . . 10 . . . 9 . . . 8, we have had ignition.' It took just 500 milliseconds for the five massive first-stage engines to ignite, boiling smoke and fire surging down into the trench, bisected at the flame deflector into two opposing channels roaring aside from the shimmering steelwork. For several seconds computers would ask a thousand separate questions of AS-513, diagnosing its condition all the while as mere mortals stood by and waited for the electronic verdict, leading, on this occasion again, to a perfect lift-off.

'Sequence has started . . . 6 . . . 5 . . . 4 . . . 3 . . . 2 . . . 1 . . . zero, and we have a lift-off. The Skylab lifting off the pad now, moving up. Skylab has cleared the tower.' At that point Houston flight director Don Puddy and his team assumed command of the mission while gasps and cries of awe rippled through the half-million spectators watching the thin white finger of AS-513 write a white pattern across the Florida sky. About one minute after lift-off the launch vehicle accelerated through the speed of sound; in ten seconds it would pass through max q, the period of maximum dynamic pressure where atmospheric forces were at their worst.

Suddenly, almost unnoticed, telemetry from the launcher indicated premature deployment of the workshop meteoroid shield. And worse, release of the No. 2 workshop solar array boom. They were the last appendages anybody wanted to see flapping around at this point! Surely it must be an erroneous signal, quite possible under conditions of extreme stress on the vehicle? Everything else looked normal and nobody raised alarms.

On thundered the first stage, cutting off its center engine on time followed by the outer four in paired succession. Then the five S-II engines ignited, burning a transparent flame of combusted hydrogen and oxygen as the Saturn pushed on, quickening its pace toward orbit. And then, at 3 min 10 sec into the flight, the structural skirt separating the S-IC and S-II stages should have slid away from the second stage but apparently failed to do so when telemetry examined after launch revealed that the shaped charge designed to cut it free had failed to propagate more than half way round the join line. With the 5.2 tonne skirt still attached, the second stage would burn a little longer than planned, albeit only 0.7 seconds.

On and up cruised 513, the five J-2 engines smoothly performing as planned. For 7 min 8 sec the stage fired, shutting down its centre engine 4 min 35 sec before the outer four. The S-II shut down just 9 min 49 sec after lift-off with Skylab 442 km above Earth, 1,800 km out over the Atlantic due east of Newfoundland. Three seconds later the four retro-rockets in the front of the S-II fired forward as a shaped charge severed the connection with Skylab. Less than two seconds after that telemetry indicated unusual attitude changes in the detached cluster but they were soon damped out by the Skylab attitude control system thrusters.

Eight seconds after separation the cluster began pitching over to point down toward Earth prior to blowing apart the four massive clamshell doors of the payload shroud. Kicked into a different orbit by the retro-rockets, Saturn's S-II stage would continue to drift away from Skylab. But the first Skylab event after separation sent a cover over the refrigerator system radiator spinning away from the flat octagonal structure at the exposed rear of the workshop. Seconds later the refrigeration system switched itself on. While the shrouded cluster nosed down to point toward Earth ground controllers received preliminary tracking data which indicated Skylab to be in a very satisfactory orbit. In fact, on average, the orbit was about 1 km higher than planned, height at orbit insertion being only 100 meters from the planned value.

At 15 min 20 sec the four big payload shroud doors jettisoned, exposing the cluster of modules at the forward end; 39 seconds later Skylab was maneuvering itself to a solar-inertial attitude so that when the four ATM solar arrays and the two solar array wings on opposite sides of the workshop were deployed they would be facing the Sun ready to convert its energy into electrical power. So far so good. By now, the big station was approaching the coast of France, almost exactly due south of Lands End, England. At 16 min 39 sec the drive motors responsible for pulling the ATM structure round 90° to its operational position whirred into life and less than four minutes later it was locked in position, having exposed the axial port on the Multiple Docking Adapter.

Skylab was now over the Adriatic, heading for a pass almost directly above Athens. The four solar cell arrays would be commanded to unfold and extend to form a cross spanning 31 meters. It was 24 min 52 sec when they began to deploy and the giant cluster was passing south of Jerusalem. It took less than a minute for them to extend fully and lock in position. Now the workshop's two solar array booms would be instructed to turn out through 90° and then unfurl their concertina-like solar cell panels. Passing through Saudi Arabia on a course parallel with the Red Sea, Skylab moved across the Gulf of Aden into darkness and a flight over the Indian Ocean.

Before crossing the Mediterranean, Skylab went beyond range of the Madrid tracking antenna so two Apollo Range Instrumented Aircraft (ARIA), one flying 160 km off the coast of Greece and a second near the Seychelles, relayed telemetry from the big cluster. Passing north-east of Mauritius, Skylab was to have rotated the workshop solar array booms at an elapsed time of 41 minutes, followed eleven minutes later by the solar cell panels extending from the two fairings. The cluster passed out of range of the second airborne ARIA just as the command to deploy the workshop booms was to have been passed to the deployment mechanism from the onboard sequencer.

Skylab's orbit would take it south of Australia and New Zealand and over the Pacific to intersect the continental United States across lower California. Just before that event, 1 hr 35 min into the flight, it would come within range of the Texas and Goldstone tracking stations for a long 32 minutes pass over the US. Before that, the cluster would be viewed briefly by antennae at Australia's Carnarvon and Honeysuckle stations.

When Carnarvon acquired Skylab at an elapsed time of 54 minutes there was no indication that the two solar array booms had in fact deployed. At 1 hr 5 min Honeysuckle began sending data during a brief 71 second pass as Skylab swept low over the Australian horizon far to the south. Again, no deployment signal. But the environmental systems officer reported the workshop to be doing all the things it should as far as he was concerned. His report would change dramatically when the station moved into Sunlight.

At 1 hr 7 min, Skylab was beyond range of Honeysuckle and starting its 28 minute passage over the Pacific in silence, too far south to come within sight of the antenna at Hawaii. In Houston, the picture was confused. Several questions were already floating around concerning the indication of premature workshop wing and meteoroid shield deployment 63 seconds after lift-off and the unusual attitude excursions after the cluster separated from the Saturn's second stage. When Skylab came within range of the Texas antenna the sequencer was getting ready to command deployment of the meteoroid shield, that event coming one minute after acquisition. It was clear that the signal had not been adequately received for no shield activation was seen on telemetry.

What was puzzling, however, concerned the electrical systems controller at Houston who saw indication of a 25 watt current flowing from the circuit connected to solar array wing No. 1. Timed to extend at an elapsed time of 52 minutes, *both* workshop wing arrays should have been delivering a total in excess of 6 kW by the time Skylab passed over the United States at the end of the first orbit. Yet other systems were coming on line as planned. Shortly after beginning the first full Stateside pass Skylab's attitude control system began to activate the big control-moment gyroscope wheels, spinning them up to a planned 9,000 rpm. Only the workshop solar arrays and the meteoroid shield gave cause for concern. The ATM arrays had deployed, but they contributed at most only 40% of Skylab's total design capacity; the meteoroid shield was not essential to the mission, yet it carried the thermal paint patterns vital for reflecting solar energy and keeping the exterior of the workshop down to an acceptable temperature.

Flight director Puddy asked the booster system engineer to prepare a set of back-up commands and as the cluster passed over the Newfoundland tracking station instructions went up to the ailing Skylab. Minutes later, from the Madrid station, came confirmation that the signals had got through but that they had not changed the situation. Then, passing over the Honeysuckle station a second time, a back-up command was sent to deploy the meteoroid shield 2 hr 45 min after lift-off. Hawaii reported that it too was not functioning.

By that time the environmental systems officer was swamped with information he never expected to see. Temperatures were all wrong, fluctuating wildly, but for the most part going in just one direction – up! There was now no question but that the meteoroid shield had somehow been torn away near max q shortly after leaving pad 39-A at the Kennedy Space Center. Nobody knew the configuration of the solar array booms; at least one was giving indications that it wanted to deploy but could not, while nothing was coming back at all from telemetry on the No. 2 boom.

If Skylab was to get no more electrical power than it had then, the normal sequence of manned missions could not go ahead. Skylab needed at least 8 kW for all the systems and experiment equipment on board running at scheduled levels. As it was, the cluster had little more than 4 kW from the four ATM arrays.

To the Huntsville Operations Support Center the picture was alarmingly clear. From the early hours of that morning George Hopson had been in control of HOSC operations; since midnight he had been watching the relayed signals from Cape Canaveral telling him the internal pressure of the workshop. Hopson wanted to be sure it was not leaking beyond the design limit for that would reduce its usefulness in orbit.

Shortly after lift-off Alvin P. Woosley, the electrical power support group leader, had commented that telemetry seemed to show the meteoroid shield had deployed and then seconds later somebody else said the workshop solar arrays had been released. Now, three hours after launch, Hopson was worrying about the rising temperature. He called on Harold Coldwater, chief of Marshall's analytical mechanics division, to find out what the Skylab cluster could structurally stand and where the thermal limits lay for expansion and contraction.

At the Cape, top management personnel remained in firing room 2 to discuss the situation, hooked up by telephone links to the Johnson Space Center in Houston. Administrator Jim Fletcher was there, as was his deputy George Low, manned space flight head Dale Myers, KSC Director Kurt Debus, the newly appointed head of the Marshall Center Rocco Petrone, Skylab's director Bill Schneider, and the Johnson Center program manager Ken Kleinknecht. At Huntsville, the Operations Support Center fired into action to trouble-shoot the situation, and Marshall's Skylab chief, Leland Belew, had his men scrutinize data that would, under normal circumstances, have waited for a more convenient hour.

As soon as back-up commands had been sent to Skylab on the second revolution it was apparent the profile of activity would have to change drastically. While flight director Don Puddy discussed alternative mission sequences because of the reduced power available to Skylab systems, managers went into conference about the manned SL-2 launch still counting down from firing room 3. Bill Schneider announced he was calling a press conference at the Cape, hooked up live with Marshall and Johnson Centers, for 9:00 pm local time. Then it was postponed one hour. Newsmen sensed the gravity of the situation and tempo increased as more and more personnel were called in on the problem.

From Martin Marietta, experts on the Multiple Docking Adapter were consulted about critical equipment. At McDonnell Douglas, workshop engineers were brought in to discuss and recommend new procedures for operating the systems while other engineers from MDC were consulted about the Airlock Module. At Downey, engineers from the Apollo assembly plant (North American Rockwell had been re-named Rockwell International on 15 February) provided data for the necessary changes. At Houston, astronauts Shepard and Slayton were consulted about the impact on crew schedules, and at Marshall the Deputy Directory of the Astronautics Laboratory, James Kingsbury, was asked to look over candidate options for systems operation.

Toward the end of revolution 4 the Honeysuckle tracking antenna picked up attitude changes which took Skylab away from its solar-inertial mode, causing the four ATM arrays to drift off their lock on the Sun. By the time Hawaii relayed telemetry from Skylab, however, the cluster had corrected itself. But engineers were already working on another problem which first showed up toward the end of the second revolution. The first of several rate gyroscopes to overheat became erratic and threatened to fail. These were not the control-moment wheels designed to torque the spacecraft and *control* attitude but the nine rate gyroscopes, three in each axis, which provided attitude *information*. Without them the cluster would never know which way it was pointing at any one time.

Back at the Cape, Bill Schneider came away from the first intense management session to issue a formal announcement telling the press that SL-2, Pete Conrad's manned launch, was postponed five days to Sunday, 20 May, when the cluster would once again be in a suitable position for rendezvous and docking. Launch would take place at 11:00 am. Then he drove to Cocoa Beach to meet the press and to link newsmen with Belew and Kingsbury at Marshall and Puddy and Gene Kranz, now Flight Control Division chief, at Johnson.

At Houston's control center, flight director Puddy was pulling his men out of the MOCR to leave Milton Windler's team in charge for what promised to be a tension-filled evening. When the NASA men met the press Skylab was barely eight and one-half hours into its mission yet already the re-written plan for manned operation had changed significantly. The Apollo CSM would be able to supply the cluster with about 1.2-1.4 kW of electrical power for a nominal fourteen days but engineers believed there were sufficient cryogenics for between sixteen days and twenty-one days. This would supplement the average 4 kW available from the ATM solar arrays and permit at least a degree of scientific work for the first $2\frac{1}{2}$ to 3 weeks. After that, the CSM would no longer supplement the power supply and there would be little or no margin left for powering scientific equipment.

On the planned four week stay scheduled for the Conrad crew it might be possible to remain aboard Skylab the full 28 days, albeit with the last week spent acquiring medical data alone, before returning to Earth. As for the two 56 day flights, there was an increasing suspicion that they would be reduced to four-week missions at most in repeat of the first 28 day flight. As for the next few hours, Conrad, Kerwin and Weitz would fly back to Houston and work hard re-scheduling the planned activity cycle, while Petrone flew back to Marshall to organize workaround procedures from the Huntsville Operations Support Center.

De-pressurized to 58 mmHg during ascent as planned, the

A group of key Skylab flight controllers cluster around Flight Director Don Puddy at his MOCR console in the Mission Control Center. Johnson Space Center Director Christopher C. Kraft (with jacket) looks over Puddy's shoulder while contemplating the seemingly crippled Skylab mission.

normal Skylab re-pressurization cycle was postponed in light of the delayed habitation and Milton Windler's shift was occupied with busy conversation to and fro concerning relevant systems and how to operate them with limited power available. As the evening progressed, Huntsville recommended a procedure that promised to solve the problem of increasing temperatures. Clearly the atmospheric environment of Skylab was likely to cause concern about food, film, and possible toxic gases released from plastics and other materials in the cluster. With Skylab pointing at the Sun the exposed gold coating on the exterior of the workshop was a good thermal conductor, passing energy through to the interior. But if the station was turned so that the hull was shaded by the large ATM structure, or by the ATM's four solar arrays, electrical power would fall dramatically.

The plan from Marshall was for Houston to command Skylab to go from a solar-inertial attitude to one pitched up 90° so that the Sun shone on the front, or Multiple Docking Adapter, end. Although this would prevent any electrical energy getting into the 18 ATM power conditioning batteries, minimal use of on-board systems would allow the batteries to supply necessary loads. One orbit later, Skylab would pitch back down 45°, allowing some charge to get through, albeit only partially shielding the exposed workshop hull. Shortly after 11:00 pm, Houston time, Marshall engineers completed discussions with MOCR personnel. By 11:30 Milton Windler had decided to accept the recommendation and lost no time in preparing an activation procedure which he planned to have his men pass up to Skylab when it appeared over the Goldstone tracking station toward the end of revolution 7.

It was a brief pass and engineers monitoring communications and telemetry were not convinced they had a secure lock on the cluster so the computer instructions were held over for the next station, tracking ship *Vanguard* located in the Atlantic port of Mar del Plata, Argentina; Skylab was west of the Pacific coast at acquisition. But again the data came in falteringly and Windler held the command uplink for activation over Hawaii nearly one hour later, the next station to acquire. Skylab was now experiencing the protracted coverage that occurred after the first six or seven revolutions, first encountered operationally by Mercury planners more than a decade before.

Officially the mission was now in the second day of operation and at 1:25 am on Tuesday, 15 May, Hawaii gave good data and Windler got the command through for Skylab to slowly pitch up 90°. It took about thirteen minutes for the cluster to get in position, an attitude confirmed when next it came within range of *Vanguard*. Toward the end of the 9th revolution, Skylab pitched back down 45° and one orbit after that resumed solar-inertial pointing. But it had given Marshall good thermal data on just how the cluster responded to different attitudes, a calibration by which to plan a more permanent position balancing the need for electrical power and the necessity to keep temperatures as low as possible.

Neil Hutchinson had been on duty since 1:30 am while Windler pulled his men out to do some systems planning. At the start of revolution 12 attitude control was transferred from the nitrogen jets (TACS) to the control-moment gyroscopes which, having been fully spun up, could now take over and conserve gas. By this time, structural parts of the workshop interior were reading 38°c while the exterior skin indicated 82°c. Temperature problems would become acute this day. At 9:33 am Florida time, Conrad, Kerwin and Weitz howled down the Patrick Air Force Base runway from which they began their flight to Houston. They had trained for more than a year to fly with experiments aimed at getting maximum returns from the planned 28 day habitation. Now it looked as though they would be lucky to stay the full four weeks and certainly have the experiment schedule severely compromised.

But the tenacious resolve of diminutive Pete Conrad was a daunting force for space gremlins and to the astronaut who once rode out a lightning strike on the world's biggest rocket it was just another challenge, one that he knew he could not pass up. In the next few weeks, the indefinable Conrad spirit would make all the difference between mediocrity and outstanding success. Yet even as the three crew members flew to Texas they knew it would be difficult to save much from the stricken mission. Bill Schneider was not even sure the men should be launched on Sunday and decided to hold a management meeting at the Cape Saturday afternoon to decide one way or the other.

Schneider decided to stay at the Kennedy Space Center and oversee things from that end while Rocco Petrone pulled James Kingsbury and his deputy William Horton in to head up a special task force at the Marshall Space Flight Center. Leland Belew would coordinate operations and liaise with the Johnson Center, even then working on some ideas on how to shade the exposed surface of the Skylab workshop. By early afternoon that Tuesday, several proposals had been put forward. It was becoming apparent that perhaps the thermal problem, creeping along with increasing severity as the hours elapsed, would outdo the power loss for the concern it caused.

Fortunately, re-pressurization of the workshop interior had been stopped, for if the station had been raised internally to the 258 mmHg planned for manned habitation, excessive temperatures could quickly increase the level and burst the workshop wall. Nevertheless, if left unprotected, the outside would stabilize at a temperature as high as 163°c. So while the Johnson Center pitched in quickly to study possible shades that could be deployed in orbit, the Marshall Center was restrained in this pursuit by a vital need to find an attitude for Skylab that would restrict interior temperatures yet not starve the solar arrays of energy from the Sun. The tests very early that day had drained the batteries 50% before Skylab assumed solar-inertial attitude once more.

With computers in the HOSC working fast and hard trying to thermally balance the structure at suggested attitudes, physicians and food specialists did their own sums and believed all would be well with the consumables if the temperatures were stabilized soon. For its part, McDonnell Douglas was working mathematical models of the heat profile in support of the HOSC effort. At 1:30 pm, with temperatures reading very different values throughout the structure, flight director Don Puddy commanded the station to pressurize to 225 mmHg with 70% oxygen. This was necessary to check the environmental control system and to provide a thermal soak for extremely hot regions of the interior. It took nearly seven hours to complete the pressurization and by late afternoon good data had been obtained on the gas temperature inside.

On the side facing the Sun, Skylab's exterior workshop temperature was now 146°c, half way round to the back at the shadow line it was 60°c, and on the anti-solar side it was 32°c. On the interior, temperatures varied between 49°c against the wall facing the Sun, and 21°c on the opposite side. If it became much hotter the interior could release carbon monoxide and carbon dioxide gases, or worse, toluene diisocyanate could be released from the polyurethane insulation around the walls.

In the Mission Operations Control Room, Windler was scheduled to relieve Puddy at 7:00 pm. Puddy had been on since 7:00 am that morning and the two men would cycle 12-hr shifts from this point on through launch of SL-2, leaving Shaffer and Hutchinson to work up procedures for the manned habitation. As for the other systems, a rate gyro in one axis was virtually useless (leaving one as back-up to a prime) and additional attitude changes had consumed more than twice the anticipated TACS nitrogen gas. To conserve electrical energy, heaters in the ATM structure had not been activated as scheduled at an elapsed time of 3 hr 11 min; shielded by the large circular shade at the Sun end, ATM's problem was just the reverse of the exposed workshop. But now engineers turned on four heaters to activate and check the telescope mount's separate thermal control loop. It worked well and temperatures followed the anticipated profile. By late afternoon, HOSC came up with a plan to prevent temperatures increasing at rates then seen in telemetry from Skylab. During the night, in the early hours of Wednesday, 16 May, Milton Windler's shift kept the cluster oscillating first to a vertical attitude then to solar-inertial for complete orbits in each mode.

As for protection when the crew finally arrived at the station, there were basically three optional procedures. One required the astronauts to fly around the cluster and from the open hatch of their Command Module deploy a shade which would be fixed to the workshop hull. The second called for the crew to dock with Skylab as planned and then perform a space walk through the Airlock Module to put out some sort of cover. The third, operationally the most simple, would have Conrad, Kerwin, and Weitz extend an umbrella-like parasol through the scientific airlock in the workshop wall facing the Sun, opening it out by push-rods from the interior before pulling it back down against the exterior surface.

As for types of shade, the options were almost endless,

Deployed during tests at the Johnson Space Center, this parasol sun shade was proposed as a means of shielding the outer skin of the Skylab OWS.

ranging from inflated balloons blocking the Sun's rays to shaped covers laid out along booms placed either side of the workshop. But one thing was certain. Whatever Marshall and Houston decided upon would have to be designed, fabricated and packed aboard the Command Module prior to launch, for nothing on board the orbiting space station would suffice for this job.

As Wednesday morning brought new results from the several cycling attitudes performed earlier, engineers decided to change the pattern. Since the night hours, Skylab had been tracking the surface of the Earth in a local vertical mode between full orbits pointing directly at the Sun. Now, instead of returning to a full solar-inertial attitude on alternate orbits, the cluster would pitch to a fixed 45°. This was necessary because the temperature curve was still going up, albeit on a much lower slope than had been seen during the first day. During the early hours, skin temperature on the exterior read in excess of 148°c and the reduction in temperature at local-vertical attitude was just losing out to the heat gained at solar-inertial attitude. By not quite bringing it back to full solar-inertial on alternate orbits, Marshall hoped to bias the drift toward a net heat loss despite the fact that this would severely limit the amount of electrical energy produced by the ATM arrays.

By now, Bill Schneider was back at the Marshall Center pulling daily management meetings at 1:00 pm where status reviews were received from each NASA facility, including Houston and the Cape. It was apparent that in its present condition the station was uninhabitable – at least on a long-term basis – and that deployment of a shade was one of the priorities facing Conrad, Kerwin, and Weitz. The crew were already busy with the Houston simulator trying out various rendezvous, docking and fly-around procedures they might find necessary. For their part, the back-up crews were busy at telephones and computer terminals, collecting reports and engineering information to feed the crew updates on the situation and to shape their rehearsals around the most probable mission profile.

Computer runs of the electrical situation gave reason to believe that if Skylab could be allowed to follow solar-inertial attitude, by having a Sun shade put up, there would be sufficient energy to allocate between 600 and 800 watts to experiments on board the station during the period Apollo supplemented the supply with its fuel cells. This was about half the design supply but one which would save at least the first two weeks of the mission with a modest program of activity. Yet even as this second full day of the Skylab problem progressed, hopes were restored that perhaps the workshop was not permanently doomed to only 40% of its design power supply.

Careful examination of Skylab telemetry by a few personnel not directly involved with working the thermal problem revealed evidence that although it was likely the No. 2 workshop wing was missing, the No. 1 array might in fact still be attached, its deployment restrained by debris from the torn meteoroid shield. If that was true, Conrad's team could possibly remove the debris and free the solar array boom. Something was certainly still there because a few watts of electrical power had trickled through until engineers decided to shut down the workshop charger battery modules and couple circuits to the ATM arrays. Moreover, photographs obtained from very special Department of Defense tracking cameras operated by the Air Force in New Mexico backed up this assertion.

By late afternoon, Wednesday, the results of Skylab holding an acute angle to the Sun had been seen to pay off. In fact, Don Puddy edged it up in two 5° increments so that by the time he came off shift at 7:00 pm it was pitched up 55°. This was the optimum balance between thermal loads and electrical energy from solar rays. The ATM cells were providing about 2.7 kW (versus about 4 kW in solar-inertial attitude) and the internal temperatures were coming down. But it was nothing dramatic. While the solar side of the workshop was cooling slightly, the opposite side was warming a little, leaving a very slight net decrease inside. Before, the average internal temperature peaked at about 51°c; it was now down to about 43°c. However, pitched up 55° the cluster's ATM was heating up in the area of the control-moment gyroscopes.

Thursday, 17 May, came in across the midnight hour, Milton Windler monitoring the thermal condition with Skylab at fixed pitch, reduced to 50° by the time he came off shift. In Huntsville, Joe Kerwin was actively studying the solar shade proposed by the Marshall Center which envisaged poles extending down the sides of the workshop on which could be deployed a large curtain pulled across like a window shade, while at Houston Pete Conrad stayed with the parasol concept pushed out through one of two scientific airlocks. But still proposals came in from engineers at NASA facilities and from the general public all over America.

Some specialists suggested thin Mylar laid taut across the workshop hull by inflatable ribs. Others even proposed the Conrad crew get out their brushes and give Skylab its thermal paint patterns once more by applying new colors! To the Marshall, Johnson and Kennedy Space Centers letters arrived by the sackload with suggestions from young and old alike, some serious, others well intentioned but impossible to adopt. One small boy from Ohio offered his services instead of Pete Conrad as the person best equipped to go up and fix Skylab 'because I'm all for NASA and I'm of no value.'

The earliest idea to evolve, one put forward on the Tuesday morning after launch, suggested a spray method to apply paint and restore thermal balance. Marshall's Robert Schwinghammer, chief of the materials division, flew to Huntsville from the Cape and, arriving at 3:00 pm that afternoon, pressed his team to work on tests to see if paint could be sprayed in a vacuum. Some said it could; others said it could not. Schwinghammer had engineers try it in a specially rigged vacuum chamber. It did work, but the idea was soon out of favor because of the possible contamination it could cause. Fabric materials were the only candidates and the materials division began testing a total of fourteen different types.

There were thermal tests to see how each stood up to rapid temperature changes over long periods. Exposure to ultra-violet light checked how each material would react to harmful solar radiation. Stickiness tests were carried out; it was no use carrying a tightly packed shade into orbit if it refused to unfurl in the vacuum. Elongation tests eliminated several types that threatened, under test, to stretch and deform. Then it was necessary to qualify a radiation resistant rope which would be used for whichever shade concept Skylab managers decided to adopt.

Two prototype shades were fabricated from Tedlar film, sprayed with a coating of special paint designed to balance radiation against absorption. That turned out to be too heavy and by the

17th Marshall had come up with a better and lighter material, suggested by Johnson Space Center and obtained from their stocks. It was produced by the National Metalizing Division of the Standard Packaging Corporation. Essentially aluminized mylar, it was covered with a nylon layer colored international orange.

Donald Fisher had been working Marshall's textile employees hard on the problems of sewing the correct size shade. Now, a plan evolved to fly in specialists and seamstresses with their own sewing machines from International Latex Corporation. Accordingly, Delores Zeroles and Ceal Webb made plans to go to Huntsville and sew up the Skylab shades. But Thursday too was the day thoughts turned positively to deploying what could be the restrained workshop solar array boom, freeing it to extend and unfurl solar cells.

Nobody knew for sure if the No. 1 boom was still on, but there was a fair chance it was and tools would be needed to cut free debris from the torn meteoroid shield. A. P. Warren, chief of Marshall's auxiliary equipment section, set to work finding the necessary tools to do this undefined job. Putting together a picture built by telemetry, and with information from ground-based cameras in New Mexico, it was possible that the boom was restrained by a thin aluminum strap, part of the deployment mechanism for the meteoroid shield, and by small bolts and metal arms. Cable cutters and shear-type metal cutters were needed, thought Warren, on poles 3 meters long so the tools could be operated by an astronaut standing at the nearest point where he could gain restraint. Shears were bought from a local hardware store, a tree-trimming tool with rope and pulley to operate from a distance. It was the nearest commercially available design and Warren set about re-designing it for use aboard Skylab.

Realizing that telephone linesmen used tools similar to those he was seeking, Warren called on the A. B. Chance Company of Centralia, Missouri, specialists in that equipment. An official from the firm was soon en route to the Marshall Center with the company's entire stock of designs for the opportunity of a salesman's lifetime! From building 4619, astronaut Schweickart and Marshall engineers selected two of the tools for modification: cable cutters and a two-pronged tool for prying and pulling. Teledyne Brown Engineering of Huntsville worked with Warren, and his deputy, Bobby Lawson, to get the hardware ready for simulations in the neutral buoyancy facility by week's end. On its way from Houston was a mock-up of the Command Module from which underwater tests would be conducted to see how easy it was to put out a shade or pull free debris snagging a solar array boom.

Meanwhile, Schwinghammer's materials division pulled from stores panels similar to those employed as the workshop interior walls. Placed in a large vacuum chamber and heated to the same temperatures experienced by Skylab, engineers were able to show that aluminum lining would not separate from special insulation. But Schwinghammer's tests did prove that considerable outgassing would result at temperatures of 145°c and considerable study evolved as to whether this would be dangerous. It would result in a decision to purge the internal atmosphere several times before the launch of SL-2.

At Marshall, Schweickart and Kerwin studied EVA procedures for shade deployment and, at Johnson, Conrad continued to rehearse station-keeping and formation flying in the simulator. Also at Houston, Weitz, assisted by Dave Scott and Ron Evans, rehearsed the stand-up EVA which would be necessary for one of the proposed shade deployment techniques. By noon that Thursday, management personnel in consultation with Administrator Jim Fletcher decided to postpone for a further five days the launch of SL-2, allowing more time for hardware to be properly designed and fabricated.

Within the past 24 hours three types of shade had been selected as candidate for the rescue plan: deployment from the Command Module; use of a T-shaped extension device already on board to unfurl a window-blind type shade from the ATM structure down across the workshop; a parasol device extended through an airlock in the hull. The T-shaped device was 9 meters long at maximum extension and comprised a device already designed to convey film cassettes from the Airlock Module up to the Sun end of the ATM telescopes. Three 1.3 meter long extensions would be necessary to provide the required 12.9 meter pole. Carried at the cross-section end of the extended T, the shade would be unfurled by pulling one edge of it back toward the Airlock Module.

The decision to postpone SL-2 until the 9:02 am launch opportunity on 25 May also brought preliminary selection of two shade types which were to be taken aloft in the Command Module: Marshall's T-shaped shade and Johnson's plan to put out a cover from an Apollo spacecraft alongside the workshop. The latter concept would be secured at three points, two at the rear of the workshop and a third at a handhold structure near the ATM. The cover deployed from stand-up EVA in the Command Module would be the first method attempted since it required the Apollo spacecraft to be undocked. If that failed, an EVA from the Airlock Module after docking would allow the crew to put out the T-shaped support for a shade.

By postponing the SL-2 launch from Sunday to the following Friday, Conrad, Kerwin and Weitz would have time to rehearse the deployment procedures in the water tank at Huntsville. Nevertheless, only one week remained in which to get all the hardware prepared and qualified and the astronauts trained to do the job. As the plan stood that Thursday evening, Conrad would fly around Skylab after rendezvous to see if he could pry loose with special tools the debris restraining solar array boom No. 1, after which he would attempt to put out the Houston shade.

It had been a day of turning from possible disaster to one of hope for restoring a usefulness to Skylab nobody thought possible two days before. Pitched permanently now at 50° to the Sun line, internal workshop temperatures inside were stable at an average 40.5°c. By the evening it was agreed to put Chuck Lewis in the MOCR rotation schedule, relieving Puddy and Windler of their gruelling 12 hr-on/12 hr-off routine with all three on 8 hr shifts.

Friday, 18 May. This day, Conrad and Kerwin, back from Huntsville, would practice flying around the workshop in the simulator while Weitz rehearsed procedures for deploying the shade from the Command Module hatch. At the Marshall Center, Schweickart and Musgrave would go through EVA activities suited up in the water tank. But increasing concern about the ability of the crew to put out the shade from an exterior position raised interest in the parasol pushed out like an umbrella through the scientific airlock. Utilizing an existing experiment container, a long elongated box designed to attach at the airlock so that rods could be pushed outside through the workshop wall, the Johnson Center pursued a concept based on spring-loaded ribs designed to open out and spread a shade when fully extended.

McDonnell Douglas were working on a similar approach, one deployed manually at full extension. But there was still a nagging suspicion that debris from the meteoroid shield might have fouled the circular airlock on the outside so the other proposals held favor over the umbrella device.

During the afternoon Don Puddy monitored a partial depressurization of the workshop interior designed to purge possible toxic gases that may have vented from internal materials. Engineers planned to take it down to less than 50 mmHg before re-pressurizing on several cycles, but not before determining the effect on Skylab's attitude since the evacuated gases were vented to space and could upset the precise angle with respect to the Sun. At the front of the cluster, Skylab's Multiple Docking Adapter was running an internal temperature of about 7°c while the Airlock Module behind was a cool 4°c so a decision was made to use the thermal control system to put some heat in the latter and thus prevent the coolant loops freezing. Behind the two forward modules, the workshop was holding steady at 40°c

Yet expenditure of TACS nitrogen gas in the several small manoeuvres necessary to keep Skylab in position was running higher than anybody expected before the flight. About 24% had been consumed and a concerted effort began in the MOCR to find ways of reducing this use level. Studies performed on various foods similar to those packed aboard Skylab revealed the possibility that about 5% would be endangered by temperatures between 46°c and 54°c, the maximum thought to have been reached earlier. Since Skylab carried an excess of nearly 12% over the total necessary for 140 days of habitation, a loss of 5% would cause no problem.

But film contained in lockers and vaults would probably suffer damage. This was not film for the ATM telescopes, rather for the corollary experiments and that was stored in the hot workshop. Kodak was called in to examine problems on two sides: temperature and humidity. Salt packs carried with the film to keep up the humidity were running out of useful life; in four days they would be unable to provide the necessary moisture. Someone suggested the crew should place wet towels in the lockers when they got aboard Skylab, others preferred to have film magazines replaced – just in case the original stock was ruined by the heat. Drugs too might have been damaged, chemically changed because of the environment, and a new pack of those would probably have to fly up in the Command Module.

With the weighty equipment necessary for deploying the solar array boom, and placing some sort of shade over the exposed workshop, many items had to be off-loaded from the spacecraft before launch. It was a compromise between rescue apparatus and the needs of a normal mission. Because of the extra equipment to be carried to Skylab final stowage would be carried out in the morning hours of Thursday, 24 May, just one day prior to lift-off. It all had to be ready by Wednesday and to make that possible training equipment was to be available on Saturday (19 May) for the prime crew to practice with at the Johnson Space Center. The crew would then fly to the Marshall Space Flight Center for simulations in the water tank on Monday and Tuesday.

But that Friday, as plans evolved for the first few days of the planned habitation, news came from the Marshall water tank that Schweickart and Musgrave were getting satisfactory results from a mock-up of a new type of solar shade. Earlier, tests with the T-shaped device showed there could be problems for an astronaut deploying the equipment in space and in the de-briefing that followed 75 Marshall engineers thrashed out a concept significantly better than the original. Instead of using one long pole extension, why not put two poles out, fixed to a single point at the ATM structure but splaying out either side of the workshop hull and positioned so that a shade pulled taut between the two would completely cover the exposed area?

Called the A-frame device (or the Marshall Spinnaker as Schweickart joked), it would be easier to put out *and* be a more rigid structure. Donald Geurkink of Teledyne Brown had built the T-shaped device on the shop floor during the night of the 15th from drawings made by Marshall engineers. It was then taken to the Skylab mock-up in building 4619 and there put through tests that revealed deficiencies. Now there was the A-frame concept and Gustav Kroll organized team leaders to work on separate sections of the design. Richard Dotson and Joe Bryson would work on the shade and its stowage bag; Roy Runkle and Joe White would build the rods and the rod stowage pallet; Richard Beck and Harry Thayer were assigned the lines to deploy the shade; Earl Deuel, Willibald Prasthofer and Watt Bell would fabricate the base plate; Robert Webber and Allious Petty were responsible for a new astronaut foot restraint.

The engineering division now set up two 12 hr shifts to work round the clock on the A-frame shade; by the end of activity, 78 engineers would expend 6,300 man-hours on the concept. Late Friday, Jim Splawn at the water tank facility received preliminary hardware for simulator work but tests showed that one of the rods would probably spread out of tolerance. Bobby Lawson came up with a special sleeve to fit over the rod female ends, and the concept was back in business. But what to use in simulation of the shade material? Jim Splawn wanted to get the A-frame device in the water tank without further ado. Parachute fabric was found to have the correct balance between tendency to float and sink and would suffice to simulate the shade's response to weightlessness.

Into the tank went the equipment and with astronaut Ed Gibson standing by to assist, Schweickart and Musgrave put it through its paces. To deploy the A-frame shade, one astronaut was positioned up on the ATM support structure while another remained near the Airlock Module hatch. Assembling a single 16.8 meter long rod from eleven separate sections, one astronaut would hand the completed pole up to his companion on the ATM who would then secure it to a baseplate attached to the support structure. A second pole would be prepared in the same way and similarly fixed to the single point on the ATM. Thus attached, the twin-pole configuration would have the appearance of a horizontal V. The extreme rod on each pole would have an eyelet threaded with a continuous rope. The folded shade would be drawn out across the workshop hull rather like a flag, but in the horizontal position and secured both sides rather than one. Two more ropes would be employed at the front to tie it down each side. The complete assembly would weigh 51 kg. During the water tests that Friday, Schweickart and Musgrave proved it would work, and that it was probably the best of any concept then developed.

At 10:31 pm that Friday evening, flight controllers began to dump Skylab's atmosphere down to a pressure of less than 30 mmHg in the first of several cycles designed to purge the workshop of toxic fumes. Earlier that afternoon Don Puddy's team

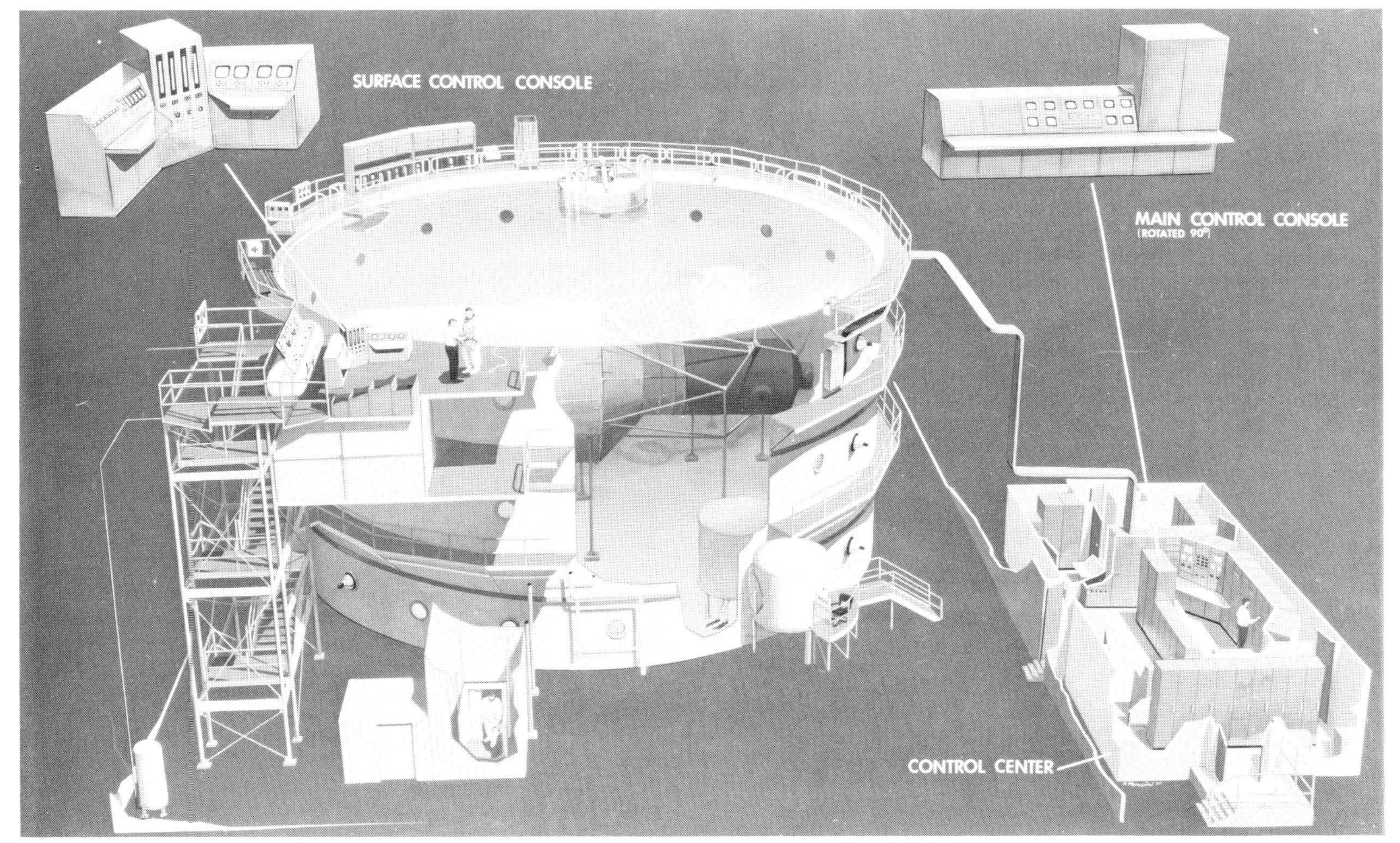

Of vital importance for checking the many potential rescue techniques for getting Skylab operational, this Neutral Buoyancy Simulator at the Marshall Space Flight Center was fitted with Skylab hardware mock-ups enabling astronauts to rehearse procedures in simulated weightlessness before going into space.

had satisfied themselves that venting Skylab's atmosphere to space would not upset the stability of the cluster. However, it was soon apparent that instabilities were now setting in and since it was vitally important to keep the station attitude fixed with respect to the Sun, venting was stopped at 2:19 am Saturday morning when Skylab had de-pressurized to 105 mmHg. Chuck Lewis was on the graveyard shift and decided to halt the activity until a way could be found to prevent it tipping the cluster.

By noon the situation had improved and at 12:06 pm, Houston time, commands were sent for the workshop to resume depressurisation toward a goal of 30 mmHg. Theoretically the venting should not have moved Skylab from its pre-set attitude. Atmosphere was flowing from the workshop to space via two non-propulsive orifices on opposite sides at the rear of the workshop, exhausting the environment at a rate of 1.5 mmHg per hour.

And then, at about 3:00 pm, the Airlock Module primary coolant loop suddenly switched to the back-up loop. It was commanded back to the primary but switched itself to secondary once more. Later Saturday, the workshop venting was terminated at a pressure of 37 mmHg.

For the past day or so, information from the rate gyroscopes on Skylab's precise attitude had become less and less reliable since they had not been updated by the Sun sensors locking on the solar disc and gradual drift carried them out of the precise calibration they had at the start of the mission. So controllers, who by now had developed a very precise knowledge of the effect minute attitude changes, had on the internal and external temperature, mapped the changing profile, observing fractional increase or decrease in temperature, to tell the guidance controllers the precise attitude of Skylab.

By now, one of the suit coolant loops was down to about 2°c so to prevent the water freezing up, Skylab was rolled slightly allowing the Sun to warm that side of the structure and stabilize the temperature.

But Saturday was decision day as far as optional concepts for shading the exposed workshop were concerned. The parasol-type shade became the prime selection since it had the crew safely inside the workshop before beginning deployment. Because it called for the astronauts to dock and then cursorily inspect the cluster it would at least show whether anything else had happened to Skylab which might prevent manned habitation and thus make the shade unnecessary after all. But tests were still proceeding on all types of cover. There was time yet to change the decision after water-tank tests early in the coming week revealed the practical advantages in each concept.

But this day, Jim Splawn's team at the neutral bouyancy simulator were preparing to test tools selected from the A. B. Chance Company for attempting to pry loose the one solar array boom still thought to be intact. John L. Ransburgh of Marshall's process engineering laboratory designed a support mounting for the 900 kg Command Module mock-up flown in from Houston. Placed inside the water tank, it would allow the prime crew to rehearse procedures for deploying the shade designed for use from a stand-up EVA position. Also Saturday, part of the underwater Skylab mock-up was enmeshed in debris of the type thought responsible for snagging the solar boom. Dubbed the 'junk pile' by technicians, it would be used to check suitability of the specially modified tools.

Conrad, Kerwin and Weitz were scheduled to fly to Huntsville the following Tuesday for final tests in the water-tank but Saturday was spent at the Johnson Center in the Multiple Docking Adapter trainer developing timelines and activity lists for an EVA operation to deploy the shade. Late in the day they entered the Command Module simulator to test that end of the operation. Astronaut Bruce McCandless had been working hard on final design of modifications to the deployment tools.

By midnight the first A-frame shade was complete, sprayed with special S-13G thermal paint of the type used on Saturn and Skylab hardware. The shade being developed at Houston in pursuit of parasol and EVA deployment concepts used T-76 material rather than the NMD type adopted at Marshall. It was produced by the G. T. Schjeldahl Company of Northfield, Minn., and Marshall obtained a batch via the Johnson Center to fabricate two additional shades; three prime NMD shades were planned. Without the S-13G paint, however, the material would degrade in the ultra-violet rays from the Sun.

Liaison between the two centers continued throughout the long hours spent developing procedures to rescue the crippled Skylab, thanks in no small degree to the efforts of George B. Hardy, head of engineering development and integration at Marshall. He was the funnel through which all the workaround procedures developed in the HOSC were fed to the flight controllers at Houston. For their part, the Johnson personnel quickly defined the lines of demarcation: Skylab was predominantly a Marshall project and they knew it was no use telling Huntsville what Houston wanted; Marshall engineers knew exactly what was needed and were left to get on with the job while Johnson developed its own rescue concepts.

Morale, depleted by the quickened pace of activity early in the mission, was restored within hours as personnel, government workers, contractors, and consultants, recognized the need for constructive suggestions and productive contributions. The first three days had been the worst. Controllers and engineers who began monitoring Skylab preparations early Monday morning stayed at their posts all that day and the next. Some did not get to sleep in a bed again until late Wednesday. Several had to be ordered away from their consoles in the HOSC or from work stations at the engineering buildings. Nobody wanted to let go before assurance that things were getting better.

Sunday, 20 May, reflected the achievements of the past two days. In the week ahead the crew would want to check equipment and procedures and there was still much to do, tools to complete, documents and plans to compile, test devices to prepare. Marshall's second NMD sail was completed and by 4:00 pm had received its coat of S-13G protective thermal paint, sprayed from a vertical fixture this time for better distribution across the sheet. The first two sails would be reserved for possible use aboard the spacecraft since their effectiveness would improve with time due to better curing. The third A-frame shade made from NMD material would be used for folding and deployment tests.

At Johnson, the prime crew had a day free from formal training or simulation runs spent catching up with flight plan procedures and new EVA activities. In the Command Module simulator, Schweickart, Musgrave and Gibson flew anticipated flight profiles just in case their back-up services were needed.

Monday, 21 May brought minor problems with decreasing temperatures in an EVA suit coolant loop and early that day controllers on Don Puddy's shift at the MOCR juggled specific Skylab attitude alignments to raise thermal flow into that region. Pitching between 40° and 46°, the cluster responded well to what was now a well-mapped response to different inclinations. During the day, controllers rolled the cluster 51° in further attempts to balance cool temperatures at the front with a hot environment in the workshop. But the combined effects of attitude changes over several hours of orbital flight caused the workshop to hot up, averaging 53.5° by the afternoon. Careful study of the cluster's reaction to thermal profiles early in the mission showed that for every hour exposed to excessive heat loads it took 14 hours for the workshop to restore itself to the previous level.

For two days the workshop had been pressurized at 37.5 mmHg, or about one-twentieth Earth's sea-level atmosphere, and Monday saw controllers start the first of four pressurization/depressurization cycles planned to take place before Conrad, Kerwin and Weitz ascended Friday morning. Using nitrogen gas to conserve oxygen, Skylab would be brought up to about 105 mmHg and then dumped back down to 37.5 mmHg in the lengthy operation aimed at purging possible toxic components released from structural and insulation materials.

During the morning the prime crew were again in the Command Module simulator at Houston rehearsing procedures developed for the proposed stand-up EVA activity and in the afternoon they walked through deployment procedures. The two preferred concepts were the parasol device deployed through the workshop wall and the A-frame shield put out during an EVA from the Skylab Airlock Module. The former weighed 43 kg, about 8 kg less than the latter. The stand-up EVA sail put out by having the Apollo spacecraft fly to various attachment points was now firmly relegated as first alternative and equipment to support that operation weighed in at 14 kg. As second alternative, two inflatable structures, one developed by Marshall Center contractors and another at the Langley Research Center, were being considered.

Final decision on the actual hardware to fly – by Monday it had been agreed that SL-2 should carry at least the two preferred concepts – would come only after the crew had tried deployment methods in the water-tank. To do that they were scheduled to leave Ellington Air Force Base later in the day and spend Tuesday working at Marshall with hardware mock-ups. By 5:00 pm Monday, Marshall engineers completed the third A-frame shade in readiness for folding tests and Rocco Petrone told Schwinghammer to coordinate all activities associated with getting it ready for flight. Kennedy Space Center needed the equipment by Wednesday and wall charts were used to track the sequence of preparations on an hourly basis.

Also Monday, Charles Cooper, one of Splawn's men, performed checks on the 'junk pile' laid up in the water tank. By late evening all was ready and Cooper looked everything over for the last time. It was a huge tank, 23 meters in diameter and 12 meters deep, containing more than 4.9 million liters of water. In the course of preparing equipment for the astronauts, Cooper tested the solar array boom repair tools from the Command Module mock-up immersed alongside the workshop mock-up. Ed Gibson and Al Bean assisted with this activity.

Tuesday, 22 May. While flight controllers at Houston juggled Skylab's attitude continuing to wrestle the temperature balance between the front of the cluster and the workshop, engineers commanded a continuation of the pressurization cycles. At the Marshall Center, meanwhile, Conrad, Kerwin and Weitz would be busy giving their approval to final equipment and procedures, while at the Johnson Center engineers would practice rigging and packing the stand-up EVA shade.

During the early morning the prime crew booked in at the Skylab full-scale mock-up residing in building 4619. There, they went through a full deployment sequence, albeit in 1 g, of the Johnson and Marshall alternate concepts but later in the day it was determined that the inflatable shades were not suitable and that the stand-up EVA concept was likely to cause damage through Apollo thruster jets impinging on the workshop exterior during close station-keeping. At 9:00 am the crew arrived at the water-tank for discussion about the 'junk pile' and one hour later they began the underwater simulation, first to rehearse putting up the A-frame shade in a closely matched test that fully qualified the concept.

Splawn was particularly pleased with the crew's reaction. 'Everything went unbelievably well,' he said, 'I was impressed with the attitude of the crew. They really had determination . . . They were definitely going to pull this thing off. We had done our job and they were going to do theirs'.

During the afternoon Weitz went back in the tank and practised releasing debris from the 3.5 meter section of simulated solar array boom, part of the 'junk pile.' By late evening Bill Schneider received reports on the day's events and pronounced his satisfaction before reaffirming the probability of launching less than three days later.

And from the Johnson Space Center came assignments to allocate duties between the three crewmembers: Conrad and Weitz would train to deploy the JSC parasol; Kerwin would stay with the Marshall A-frame shade; and Weitz would familiarize himself with the stand-up EVA concept. Whichever mode was finally chosen there would be one man on board who knew the detailed design of the equipment.

During that Tuesday, Skylab was moved in attitude once more as the troublesome coolant loop hovered above the freezing point of water. Early in the day controllers pitched it 80° for one full revolution to give the workshop temperatures a kick down before restoring a 45° angle to the Sun. By mid-afternoon the interior was at a predicted 52°c.

With good checks completed in the Marshall water-tank, the parasol device was confirmed as the preferred method for shading the workshop, with the A-frame shade a valid second. The parasol from Houston used an existing experiment canister 135 cm long by 21.6 cm square which would contain the telescopic umbrella and shade.

Four ribs, compressed in five sections kept tight around the central post by pressure from the walls of the container, would be pushed into space through the small, square-shaped, airlock on the Sun side of the workshop by seven extension rods fed in through the rear end of the box. Fully extended, the extreme end of the parasol would be 6.4 meters beyond the airlock exit thus freeing the four ribs from the end of the container box. Unrestrained, the ribs would rotate upward 90° under the influence of springs, deploying the 6.1 by 7.3 meter square shade.

The extension rods would then be pulled back in until the shade was close to the exterior surface of the workshop. The central pole to which the four telescopic ribs were attached was offset from the center of the shade so it could be rotated to change the effective cover. Each rib was 6.4 meters long in the fully extended position with a diameter of 2.2 cm at the pole end and 0.95 cm at the opposite end. Each rib carried two springs fabricated from piano wire.

During the evening Conrad, Kerwin and Weitz flew from Huntsville to Patrick Air Force Base, touching down at the Florida field 8:01 pm local time, one hour later than in Houston. The prime crew went straight to their quarters in the MSOB. Less than 61 hours later they were to ascend from complex 39's launch pad B. The parasol's inventor, Jack Kinzler, was as anxious as the crew to prove it would work.

Preparations for the launch of AS-206 on the SL-2 mission had been modified to provide maximum charge in the spacecraft flight batteries. Instead of beginning the countdown at 8:30 pm that Tuesday, engineers deferred operations until 5:30 am Wednesday morning. It was a quiet night and while controllers continued the cycle of atmospheric purges discussion ensued about attitude changes since temperatures in the workshop interior were still a little higher than expected.

During the afternoon of Wednesday, 23 May, flight director Milton Windler planned and executed a major attitude change, placing the cluster in a pitch of 65° for two full Earth revolutions. Commands went up through the Carnarvon tracking station just before Sunrise on revolution 133 and the big cluster responded as ordered. Temperature in the workshop food area was an estimated 53.4°c and in the Airlock Module coolant loop about 1.5°c. The precise maneuver was designed simultaneously to raise the coolant loop and cool the workshop. After two revolutions Skylab pitched back to 45°, increasing the amount of solar energy falling on the ATM arrays and allowing the 18 charger battery modules to receive power from the cells.

Several hours later the temperature in the food area was down to 51.7°C, a small drop by comparison with the high temperature existing in the workshop but one which reflected the considerable effort needed to obtain even a slight change in the situation. During the late evening controllers began another depressurization cycle.

At the Marshall Space Flight Center on Wednesday a major design certification review convened on the various shades and equipment items planned for flight aboard Pete Conrad's SL-2. Senior managers from Headquarters were there, as was MSFC Director Rocco Petrone, Johnson Space Center Director Chris Kraft, Kennedy Space Center Director Kurt Debus, Deputy NASA Administrator George Low, head of manned space flight Dale Myers, Chairman of the Aerospace Safety Advisory Panel General Harold Dunn, and representatives from McDonnell Douglas, Martin Marietta, and Rockwell International.

It was agreed that solar array boom No. 1 was probably still attached and deployed only about 5–10 degrees from the stowed position and that the tools and procedures evolved over the preceding week were adequate for the situation the crew expected to find when they arrived at the Skylab cluster although reservations were expressed over the exact condition of the boom. While still preferring to retain final options until the very last hours prior to launch, managers agreed that the Johnson parasol would be the first choice for shade deployment followed by the Marshall A-frame device put out during EVA after docking. But space was also found aboard the Command Module for a third shade, the stand-up EVA type which, although not ideal, would suffice as an alternative to the back-up.

The three boom deployment tools modified from commercial stock were airlifted from Marshall to the Cape, arriving at about 12:30 pm and later on Wednesday Eugene I. Kirkland was nominated to fly to Florida with the A-frame shade and deployment equipment. Time was getting short and KSC would start packing the spacecraft the following day for launch on Friday. There was really nothing more anybody on the ground could now do to change the probability of success: SL-2 would have to fly with what had already been developed and it was up to flight control personnel to keep the orbiting cluster in as stable a condition as possible for the next several hours. Thoughts could turn at last to why Skylab had run into so much trouble so soon after launch. During Tuesday, 22 May, NASA Administrator Fletcher announced he was setting up a special investigation board under Bruce T. Lundin, Director of the Lewis Research Center.

Activity was now switching to the Cape as hardware arrived

at complex 39 for packing aboard the Command Module. Additional medical supplies were going up, as were extra film cassettes, although the shades were still in final stages of preparation.

Thursday, 24 May; the day before the flight. Launch operations director Walt Kapryan decided to change the activity at pad B and preserve the option of stowing the shades late in the countdown. The Cape had hoped to have everything on hand by this time but extensive tests of materials and equipment delayed delivery. It was agreed to start loading the Saturn with cryogenic oxygen at 9:45 pm, three hours earlier than scheduled, providing a built-in hold of one hour from 1:00 am Friday morning. Personnel would have at least four hours during which time they would install the parasol canister in the Command Module, bolt in the centre couch, and set up the display panel switch positions.

It would be the heaviest Command Module, weighing 6,062 kg, carrying 82 kg of extra equipment. The Service Module too was heavier than it had been on the original launch date, carrying an additional 100 kg of cryogenic fuel cell reactants.

Bill Schneider was keen to have all four candidate shades and their equipment on hand at the Cape for a final decision late that evening: Johnson would send the parasol and the stand-up EVA shades; Marshall had already dispatched the A-frame device; Langley were hustling along with their inflatable concept, deployed like JSC's parasol through the small scientific airlock. Temperature changes on the Skylab cluster again called for a special pitch up maneuver lasting two revolutions where the assembly was turned 68° to the Sun rather than the 50° it had for the past several hours. That activity got under way late afternoon Houston time and by 9:00 pm Skylab was back at 45°.

A few hours later, just minutes into Friday, controllers began to pump up the workshop interior for the last pressurization cycle before launch of SL-2. Yet for all the careful manipulation of thermal load versus electrical needs, the ATM power system already showed signs of distress. With the ever present need to keep Skylab pointing up toward the Sun rather than directly facing the solar rays, ATM cells were producing only 2.7 to 3.2 kW of electrical energy compared with more than 4kW in a solar-inertial attitude. Earlier in the week, depleted levels tripped eight of the ten battery modules off line and when the Skylab incidence angle was reduced, allowing more energy to flow from the cells to the modules, only seven came back on leaving CBRM number 15 unresponsive. It was a further reduction of available power, albeit only 5.5%.

At 5:24 pm, Thursday, the stand-up EVA shade left Ellington Air Force Base, Houston, for Florida's Kennedy Space Center, aboard a T-38. Minutes later a Lear Jet roared down the runway with the parasol and its container, carried by this larger aircraft because the device was just too big to fit a T-38 stowage bay. At the same time, Walt Kapryan's Kennedy Space Center launch team began to roll back the mobile service structure prior to fuelling the Saturn IB. Adjusted to Cape time, it was 7:58 pm when the T-38 arrived with the back-up shade followed thirty-two minutes later by the Lear Jet. Thunderstorms were in the Cape area and the service structure roll-back had already been delayed 54 minutes but the weather reports for Friday morning were optimistic and by the time the shades arrived technicians were getting ready to load liquid oxygen and liquid hydrogen aboard the rocket's two stages.

Time was tight, but with luck everything would be ready. Earlier, SL-2 had been struck by lightning but a quick check showed no damage and the countdown moved on unimpeded. At 9:45 pm fuelling began, with technicians and engineers cleared from all but essential functions. The closeout crew would scurry back to the spacecraft during the early morning hours and pack last equipment items, not least the shades. Time for a break while the rocket men readied the booster, but not for long since the stowage schedule had to be checked against the late arrivals. At 12:45 am the propellants were aboard and the white room crew moved quickly up the umbilical tower.

Shortly after 2:00 am, Friday, 25 May, the two shades were delivered to pad B. High above the concrete pad astronaut Hank Hartsfield worked with the closeout crew as the rescue equipment arrived. Minutes later, across at the Launch Control Center, astronaut Bob Crippen took up his position at the capcom console to cover pad preparation and launch. At 3:45 am Conrad, Kerwin and Weitz were woken from their fitful rest in the Manned Spacecraft Operations Building 11 km south of complex 39.

By this time launch engineers had the countdown in a 1 hr 13 min hold as part of a planned slot designed to accommodate last minute problem-chasing.

After a quick physical examination by Drs. Charles Ross, Jerry Rojenksky and Royce Hawkins, the crew moved down to breakfast with Alan Shepard and Deke Slayton on steak, scrambled eggs and orange juice. At 4:20 am the stowage was complete aboard Command Module 106 and the center hatch was bolted to the attenuator struts. The countdown had resumed. Spacecraft test conductor Bob Reed checked switch positions with Hank Hartsfield in the Command Module and the flight crew moved to the suit-up area prior to leaving the MSOB.

Elated at the prospect of getting off the ground when so recently it seemed they never would, Conrad, Kerwin and Weitz were in jaunty mood as they entered the transfer van waiting to take them on a 20 minute ride to pad B. Plugged in to portable ventilators before they reached the spacecraft and hooked up to its environmental control system, the crew arrived at the base of the pad after 6:15 am. It took several minutes to get Conrad and Weitz installed on the outer couches while Kerwin waited near the elevator that had carried all three to the white room nearly 100 meters above the pad.

Across the calm Atlantic, an unusually beautiful sunrise brought shafts of red light dancing across the rippled surface, visible to Joe Kerwin as he stood waiting. In the Command Module Hank Hartsfield helped the crew plug in their communications leads after removing special boot covers, remove the gas connector plugs and then plug in the space suit umbilicals before removing the ventilator leads. Plastic helmet protectors were then passed out to the closeout crew as the life jackets each man wore were finally adjusted. By 7:00 am the hatch was closed, ready now for a special check before tests with the emergency detection system.

The crowd assembled for Conrad's lift-off were fewer in number than would have seen his team ascend ten days before, yet for all that more people were aware of what was happening that day if only because of Skylab's crippled state. It was the first time, in more than twelve years of manned space flight, that astronauts had been launched toward an objective impossible to achieve without major repair tasks in space; it was their job to make the workshop habitable, by shading the exposed hull, and possibly to extend its capabilities by freeing what was thought to be a snagged solar array boom.

In Houston, two teams of flight controllers were at separate console positions monitoring two space vehicles: Don Puddy's crimson team was nursing the orbiting cluster while Phil Shaffer's purple team, with capcom Dick Truly, waited to assume command of the SL-2 mission when it cleared the pad at Cape Canaveral.

At T-55 minutes the white room closeout crew pulled down a hood in front of the spacecraft so that the access area would be protected from the environment when the arm to which it was fixed moved away from the Command Module. Ten minutes later the arm came back to a 12° standby position, no more than 4 meters from the spacecraft in case an emergency called for a quick exit, and from the Launch Control Center astronaut Bob Crippen continued to talk to the crew and pass along check-out information. Shortly, Joe Kerwin armed the Service Module RCS thrusters, opening hypergolic flow valves by throwing appropriate switches in the spacecraft. With about thirty minutes to go before ignition RP-1 propellant flowed to the four first stage fuel tanks to bring to the desired level quantities that had been placed aboard the Saturn IB prior to the countdown demonstration test. Cryogenic fluids continued to top up the five oxidizer tanks so recently filled.

Above Earth, Skylab was going out of range with the Honeysuckle tracking station prior to continuing across the Pacific on its 156th revolution. Don Puddy gave a 'go' for his side of the operation and Phil Shaffer prepared for the upcoming rendezvous. CSM-106 would perform an M=5 operation, drawing alongside the big station less than eight hours after launch. At T−15 minutes, the countdown for SL-2 was halted for two minutes to allow final adjustments in the precise moment of ignition. Paul Donnelly, the launch operations manager, passed along the by now familiar good wishes traditionally conveyed by whoever held that position and the various test conductors called in with their own 'go' affirmations.

At four minutes in the countdown Paul Weitz reached up and turned on the spacecraft batteries and with less than a minute

This view of the exposed Skylab hull shows the area vacated by the torn micrometeoroid shield where the parasol was later deployed from the interior through a small airlock hatch seen here as a circular patch on a white square. Note the hanging wires the solar cell array ripped off when it prematurely deployed.

to go Pete Conrad inserted the final guidance alignment to the computer. The swing-arm carrying the familiar white room had now swung back to its fully retracted position and the launch sequencer continued on down to ignition of the eight H-1 engines.

It was a curious sight, for this flight gave a unique view of Saturn IB ascending from its pedestal atop the mobile launch platform. With only one-seventh the thrust of a Saturn V the sensations were muted repetition of thirteen earlier flights from Launch complex 39, but it was the first time a Saturn IB had left from this location. To the commentator's cry of 'We have launch commit and we have lift-off,' the three space men rose toward their target which at the very moment of ignition, was 1,445 km ahead at the start of revolution 157.

With an exultant cry from Pete Conrad interspersed with their new assumed role, the space repair men were launched: 'Houston, Skylab 2, "we fix anything," we've got a pitch and roll program.' They were on their way.

Little more than two minutes after release from the launch pedestal, the eight H-1 engines were shut down, in groups of four, three seconds apart. Seconds later, with the two stages separated, the spacecraft rode on up from the thrust of the S-IVB. Then it was time to jettison the launch escape tower. 'Tower jet on time,' came the call from spacecraft 106, and six and one-half minutes later the single J-2 shut down with the assembly in a good orbit of 155 × 350 km. Skylab was in an almost circular, 434 × 444 km, path and Apollo would rapidly gain on the slower target up ahead.

'Skylab, Houston, we confirm that you're in a nominal orbit and you're cleared for a nominal separation sequence,' came word from Dick Truly as flight controllers watched the tracking data speeding through computer terminals to displays in the MOCR. About six minutes after orbit insertion the spacecraft fired the pyrotechnic charge to separate it from the four adapter panels attaching the Service Module to the booster's second stage. With a blip from the thrusters, spacecraft 106 nudged forward as the four panels opened 45°; ejected on Saturn V flights, they were now retained to prevent unnecessary clutter in orbit. Four revolutions later the S-IVB would dump excess propellants through the engine bell in such an attitude that it braked the orbital speed and sent the inert stage plummeting to destruction in the atmosphere north of Hawaii.

In the meantime, Conrad, Kerwin and Weitz would perform the first rendezvous maneuvers. During the first revolution, however, the crew checked spacecraft 106 and performed calculations with the ground on the precise time of the first phasing burn, designed to raise the low point, or perigee, of their orbit. Performed with the big SPS engine, the maneuver changed Apollo's speed by 227 km/hr when the spacecraft was near apogee on the second revolution minutes before coming within range of the Honeysuckle tracking station. An ARIA relay aircraft was cruising in the atmosphere and Dick Truly attempted to contact the spacecraft but communication was scratchy although Pete Conrad did get word through that the burn had been performed as planned.

The spacecraft was now in an almost circular path about 22 km below and several hundred kilometers behind the orbiting Skylab. A second phasing burn of 49 km/hr occurred at an elapsed time of 4 hr 41 min over the Pacific east of Hawaii. Again, it was an SPS burn and put the spacecraft into a 370 × 407 km orbit just 465 km behind Skylab. The spacecraft would continue to gain on the cluster, albeit more slowly.

A so-called corrective-combination burn was fired over Africa just forty-six minutes later with an abortive attempt at communications relayed through another ARIA stationed on the ground at Cape Town. Coming up on the Carnarvon station, the VHF ranging radar picked up Skylab 217 km ahead and Pete Conrad proudly informed Dick Truly that Joe Kerwin was 'able to track the tracking lights, although they are quite dim.' But now, as Wally Schirra had once said, the real job of rendezvous would begin. For the tricky maneuvers were those that brought the vehicle to a complete halt relative to the target.

The next burn occurred at 6 hr 4 min on the manned spacecraft's fourth revolution. Apollo was over the Pacific and through Hawaii's tracking station capcom Truly heard the good news that it too had come off as intended putting the spacecraft in a co-elliptical orbit 18 km below Skylab now less than 170 km ahead. 'All around the world we've been getting it,' quipped Pete Conrad concerning radio interference, 'There is just one whale of a lot of noise on VHF; do you guys have any idea where that's all coming from?' They did not but it was no real handicap.

Passing across the United States to begin its fifth revolution, spacecraft 106 narrowed on the Skylab space station, within 100 km by the time they crossed the Atlantic coast. It was time to get ready for the TPI – terminal phase initiation – burn which would carry the spacecraft up to the Skylab orbit. 'Houston, we had a good TPI burn,' was the word at Carnarvon about activity carried out shortly before passing into tracking view. The trusty SPS had fired five times to bring Apollo on a rendezvous with Skylab. Performed in darkness, the burn raised the orbit of the manned vehicle to one in which it could intersect the cluster, but course corrections, braking burns with the thrusters, would be calculated all the way up. 'Tallyho the Skylab. We got her in daylight,' came the voice of Pete Conrad through the Guam tracking antenna as a television picture flickered into life on MOCR monitors.

The big cluster was at a pitch of 47° and now the crew could see the crippled space station. 'Houston, I can already see the partially deployed solar panel,' said Conrad before, minutes later, affirming that 'the meteoroid shield area is solid gold,' indicating that the painted cover had in fact been torn off exposing the foil tape on the workshop hull. 'Okay, be advised the meteoroid shield is pushed up in under the SAS (solar array system) panel,' said a voice from spacecraft 106 as the details of the damage became clearer with the narrowing distance. A minute later the spacecraft moved beyond range of Guam; fourteen minutes later they were picked up by Goldstone.

'As you suspected, solar wing 2 is gone completely off the bird. Solar wing 1 is, in fact, partially deployed . . . I think that we can take care of that with the SEVA. It looks, at first inspection, like we ought to be able to get it out. The gold foil has turned considerably black in the Sun.' Now Pete Conrad was station-keeping with Skylab and moving slowly around the big cluster inspecting the exterior. Initial assumptions had been correct: the meteoroid shield *had* been torn free, debris *was* restraining solar array boom No. 1, and boom No. 2 *was* missing.

The crew now saw clearly that a metal strap had snagged the boom half way along its length, jamming it about 15° out on a butterfly hinge that was quite visible at the forward end of the boom. Hopes were now high that the crew would deploy the remaining boom by freeing the debris; in its present configuration the ATM arrays would deliver little more than 4 kW, but with the workshop wing out at least a further 3 kW would be available. Spacecraft 106 and the adjacent Skylab moved beyond range of U.S. tracking antennae at the start of the sixth manned orbit, 8 hr 14 min after lifting off from the Kennedy Space Center.

It was 5:14 Cape time on the afternoon of Friday, 25 May. In Houston it was 4:14 pm, the clock that Conrad, Kerwin and Weitz would follow for the next four weeks. They had been launched with hope that a full 28 day stay in orbit would still be possible but with consideration for any change in the situation resulting from what they saw in Skylab. Out of contact with the ground until they swept within range of Carnarvon, the manned Apollo moved round to the front of the Multiple Docking Adapter and, with the probe extended, snagged the capture latches on to the drogue within the docking port to moor the spacecraft while the crew had a meal. But they did not hard-dock the two structures by retracting the probe and clamping the 12 main latches, for the crew would separate from Skylab and fly round to the solar array boom, perform a stand-up EVA and try to pry loose the debris fouling deployment.

When next the spacecraft swept into range of the Carnarvon station the men aboard 106 were eating. It was 4:56 pm. 'Boy, I've had some big things on my nose in space before,' commented Conrad of the view through the forward facing windows, 'but this is by far the biggest. It sure beats the Agena or the LM.' Skylab and its visitor were now passing Guam and the crew were thinking about their plan of action for the upcoming fly-around and stand-up EVA, activity which for the most part would be out of communication range. At 5:34 pm Goldstone picked them up and heard that preparations were under way.

Twenty-five minutes later the tracking ship *Vanguard* relayed a message from Houston: 'Pete, we've talked about a lot of things here on the ground, but I guess about the only thing we feel like passing up is the fact that we probably think if that piece of metal that's bent over the wing is indeed a little piece of angle-iron, that you probably cannot cut it, and so if you want to get it out of the way you'll probably have to bend it, but it'll be strictly your call when you guys get out there.' Weitz thought he saw a jagged piece of angle-iron during the earlier inspection and if that was true the crew had little chance of removing it during the stand-up EVA.

Moving on from the tracking ship, spacecraft 106 prepared to separate from Skylab. 'You're clear for a local flight. Have fun and fly safe,' said capcom Dick Truly as the docked vehicles moved across the South Atlantic. Skylab and Apollo would drift in orbit across the Indian Ocean, up over China and Japan, across the northern waters of the Pacific just out of range of Hawaii and then come within range of the next tracking station – Goldstone in California – more than one hour later at 7:10 pm Houston time. During that time Conrad, Kerwin and Weitz would attempt to free the solar boom.

After unlatching the probe and flying round to the side of the workshop, the cabin was de-pressurized so that Weitz could stand up and use the 3 meter long tool on the snagged boom. Struggling hard to pull loose the metal strap wrapped tightly round the thick structure, Weitz rocked the complete assembly, disturbing the stability and causing oscillations to build up. As he pulled, the spacecraft moved in closer to the workshop causing Conrad to fire the thrusters and back off, but that too put some energy into the cluster which again rocked slowly back and forth. It was no use, the boom was firm and despite strenuous and exhausting effort, Weitz could not move the device. At 7:11 pm, when they came within view of Goldstone, the crew were packing up as Weitz began to stow the long poles, twice nearly hitting Conrad on the head which sent him diving for cover under one of the couches. Drifting into night, they finally got the hatch closed as the spacecraft passed out of range with the Texas antennae, 7:23 pm.

Weitz had been partially outside, restrained in the open hatch of spacecraft 106, for 37 minutes. The attempt had been a failure, but at least the crew now knew what they were up against. It would take an EVA from the Airlock Module to get the boom out – if indeed it was ever to come out. Minutes later Pete Conrad had the spacecraft back round at the Docking Adapter and moved forward to link up with the cluster. It had been a long day; they would rest in the Command Module this night before going into Skylab tomorrow. And then it happened. The three small capture latches refused to snag the drogue and after two attempts the crew heard the squawking voice of Houston's communictor through the *Vanguard* tracking ship.

'I've made two attempts to get a soft dock and now I can't get one and we are just about to start through the emergency procedures and standing by for any of your suggestions,' said Conrad. It was frustrating, and on top of the aborted attempt to free the No. 1 boom brought unprintable language from the Commander, unconcerned at this time that the ears of the world were listening. It was a bad time for trouble. After the few minutes through *Vanguard* it would be one hour before they next came over a tracking station, so Houston quickly passed word that they should try once more and then go through each of three separate procedures to trouble-shoot the probe.

While management personnel talked with controllers in the MOCR about what to do if the docking was not achieved, Conrad, Kerwin and Weitz tried the three back-up methods, cycling switches, pushing forward, backing off, trying again, and always failing. At 8:43 pm they were in sight of Hawaii's radio ears. 'Pete. What's your status,' called the Houston capcom. The dismal news was depressing. As if the boom problem had not been enough, they could now not even dock with the overheated cluster. But the Conrad spirit was certainly not extinguished yet. Houston advised the crew to prepare for one more procedure, which they attempted after a brief interval before coming within range of *Vanguard* once more. It was no good.

There was just one more procedure to go through prior to backing off from Skylab for the night and further deliberations on the ground: de-pressurize the Command Module, take out the forward hatch, remove the docking probe, electrically by-pass the retraction circuit to withdraw the extended section, and re-close the hatch prematurely. Then the spacecraft could be driven forward, without soft-capture, direct to a hard-dock where the 12 latches would hopefully clamp the structures together – just like Apollo 14. Again, there was a long wait before the spacecraft came within communication.

It was 10:19 pm when the crew were heard through Hawaii, about one hour since *Vanguard* with only a few scratchy comments relayed by an ARIA in between. Now, procedures were voiced up on what to do if this last-ditch attempt was a similar failure. The crew were completing the de-pressurization activity and copied down a manoeuvre pad that would take them to a safe distance from Skylab for the night. At 10:32 they were out of range again, waiting for darkness that was thought to make the attempt easier since the docking aids were more visible when lit. Almost exactly twenty minutes later, through the *Vanguard* antenna, Pete Conrad, sang an ebullient comment: 'We got a hard dock out of it!'

It had been done. Three hours and twenty minutes late, after numerous attempts running through every procedure in the book, spacecraft 106 was hard-docked to Skylab. A loud cheer went up in the MOCR followed by ringing applause. What a day! Eight minutes later they were out of *Vanguard*'s window and not for another full Earth revolution would the docked vehicles resume communication. An important day lay ahead, one in which they would deploy the parasol from the workshop to shade the exterior and so get the temperature down, allowing the cluster to happily face the Sun and soak the ATM arrays in solar energy.

Earlier that evening, at 8:50 pm Houston time, controllers began to pressurize the workshop with an oxygen/nitrogen atmosphere. For several days it had been pressurized with nitrogen only to about 105 mmHg during cycles that five times took it down to about 30 mmHg. Now it was to be pumped all the way up to 258 mmHg (one-third sea-level) for the period of manned habitation beginning next day. It would be fully up to the required pressure by 8:01 am.

In the MOCR, mission clocks were being re-set to follow Greenwich Mean Time, five hours later than the Central Daylight Time, or CDT, then prevailing in Houston. But for all practical purposes mission events would follow the Houston day/night cycle, the crew beginning their day at 6:00 am Texas time and commencing eight hours of rest at 10:00 pm. They would be late to sleep this night, remaining in the Command Module without even opening the forward hatch so close now to another in the Multiple Docking Adapter that would provide access to the biggest orbiting laboratory ever sent into space, and likely to remain so for some time.

At twenty-seven minutes past the midnight hour, the familiar horn sounded in the MOCR announcing two minutes to acquisition of signal through *Vanguard*. The crew were busy with their last meal of the day. Minutes later they spoke to Houston through the Ascension Island antenna. The last communication came through Guam with a quizzical statement from Pete Conrad: 'Now that we're docked, I'm not sure how we get undocked.' It was 1:37 am.

The crew's second mission day began at 9:05 am on Saturday, 26 May, after a lie-in beyond the scheduled time. During the night flight director Milton Windler commanded Skylab to reduce its angle to the Sun: the final long pressurization drew considerable power from the seventeen operational ATM batteries and the electrical systems officer wanted more charge from the Sun. It was soon back down to 45°, however. At 7:00 am Neil Hutchinson's team relieved Windler and the day's operation got under way. Skylab systems were good, except for excessive use of the nitrogen gas for the TACS thrusters. Of the total tanked at launch, only 54.5% remained. But the period of high usage was now past and there would be just enough for all three manned missions if used sparingly.

As for atmospheric gases, the supply of nitrogen was a little lower than expected because of the re-pressurization cycles, although oxygen had not been used so there was more of that than planned. Hank Hartsfield was on shift to communicate with the crew when they called down through the Madrid tracking station shortly after 9:00 am. Conrad was informed that Mission Control planned a private conference with the crew at the next pass over Madrid. That was later cancelled because the crew felt well and had not yet entered the spacious workshop.

Routine medical conferences would be held daily on a normal flight to allow the astronauts to converse with their physicians without the world listening in. They would also be allowed private conferences with their families, and if some emergency developed that the crew felt required a lengthy discussion they could similarly request a closed circuit. Statements would be issued to the press after each medical conference, although the precise wording would not be released.

Much of the morning was spent trouble-shooting the problem with the probe, but at 11:30 am the crew opened the forward hatch and inserted a special test tube in the hatch on the Multiple Docking Adapter to see if the atmosphere inside contained toxic products. It did not and five minutes later Weitz opened up and floated through, the first man to enter Skylab, reporting that it was a cool 10°c; it would be a different story in the workshop. Skylab was just passing out of range of Honeysuckle but ten minutes later it came over Hawaii. 'Okay Houston, we're in the MDA and we're

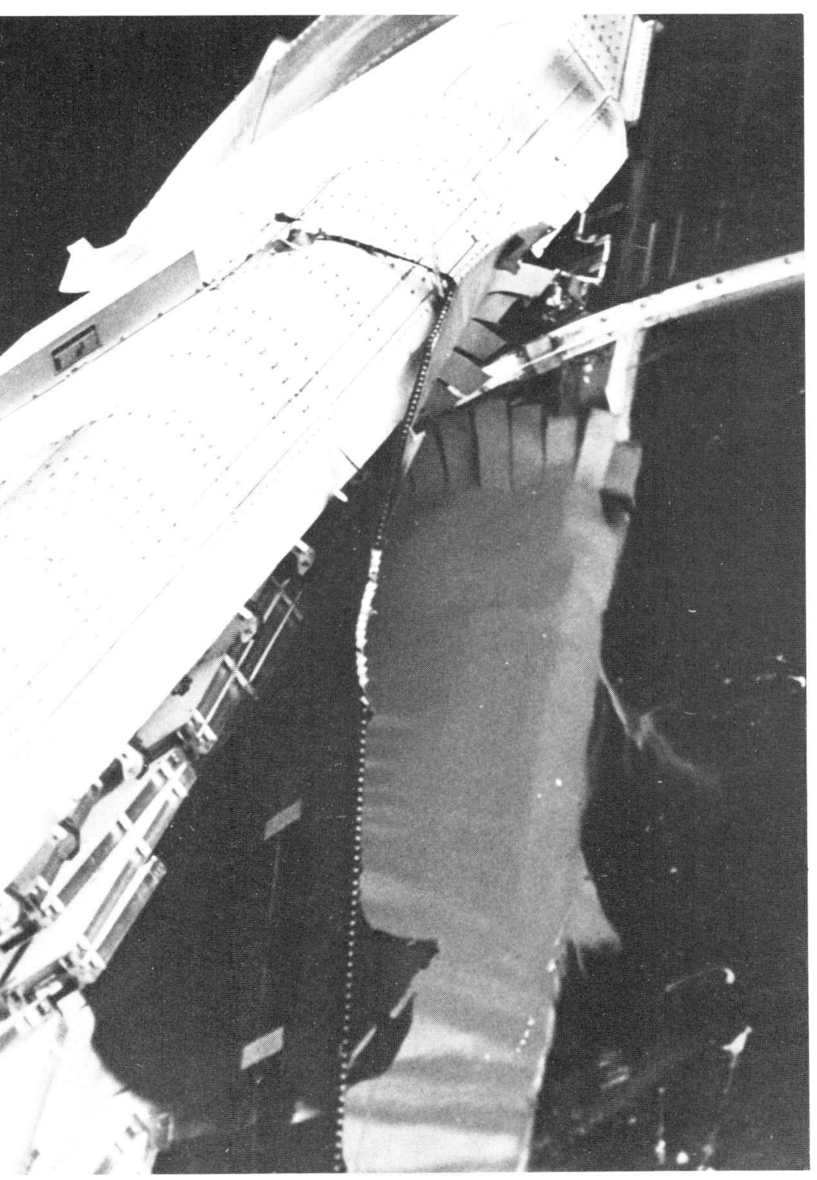

Close-up view of the damaged and partially deployed solar array wing showing aluminum strapping which prevented full deployment until cut free by Conrad and Weitz.

pretty busy . . . So far we've collected one screw, one nut, and one piece of red thread floating around, otherwise it's clean as a whistle,' reported Conrad.

The general procedures now had Conrad back in the Command Module addressing systems, while Kerwin and Weitz activated the Docking Adapter, a comparatively spacious module compared with Apollo, providing more than five times the volume of their transport vehicle. By noon, Houston time, Kerwin was setting up the ATM control panels with appropriate switch positions and one-half hour later the MOCR prepared to test the Skylab teleprinter. Shortly after 1:00 pm, Weitz opened the Airlock Module hatches and moved through into the AM itself and then into the workshop, wearing a special mask in case of carbon monoxide, returning to convey his first impressions via Goldstone at 1:35 pm.

'Okay, on our very quick inspection, the OWS appears to be in good shape. It feels a little bit warm, as you might expect. From the three or five minutes I spent in there, I would say, subjectively, it feels like . . . in the desert. Hank, I could feel heat radiating from all around me, but in the short time I was in there, I never felt uncomfortable. I had the soft shoes and the gloves on, and nothing I touched even felt hot to me.' In fact, the orbital workshop was at about 51°C.

But the big station was a new experience to the drifting trio: the Airlock Module had about half as much volume as the Docking Adapter (three times the volume in Apollo, nevertheless) and the workshop was a veritable cavern, with fifty times the available space in the Command Module. At 1:48 pm, Houston informed the crew that Mission Control was about to command a pitch maneuver, an experience capable of producing a disturbing sensation unless the crew were given warning.

Since 1:30 pm a special sampler tube had been sensing the workshop atmosphere and more than one hour later the crew queried the ground as to when they could open up again and move in. They would have to wait for further tests. As all new astronauts had an irresistable urge to look out the windows, Pete Conrad knew for sure that Kerwin and Weitz would be no exception, and he was right. While they prepared to have a meal, leaving the Docking Adapter and Airlock Module hatches open, but sealed off from the workshop, the two 'rookie' crewmembers found it 'kinda hard to get away from the windows.' There were four windows in the Airlock Module and the view was indescribable.

And then came the word everyone wanted to receive: 'You're GO for entering OWS.' Conrad was happy with body movement in such a large structure and remarked on how easy it was to get about: 'Mobility around here is super. It's turned out to work better than we even hoped for. Nobody has had any problem with any feeling of motion sickness or anything, so we're all squared away on that. Everything that we've been supposed to unfold or move has been easier than we could hope for.'

At Mission Control in Houston, a small crowd was gathering in the MOCR in preparation for deploying the parasol. Most of the testing on Jack Kinzler's shade design had been performed by Don Arabian and he too was on hand in case the astronauts should need advice. They moved into the workshop around 4:00 pm and promptly set about installing the TO-27 experiment canister containing the shade in the scientific airlock attached to the workshop wall. Kerwin remained in the Docking Adapter or the Airlock Module while Weitz assisted Conrad. Network controllers at their consoles were concerned that if delayed too long the astronauts would have lengthy gaps between contact since the space station was migrating to a remote orbital path where there would be little coverage for some time. But it was a job that could not be rushed. Pete Conrad had a comment over the United States: 'Sure gets hot down here on the Stateside pass, and the other thing is that when that TACS goes off, it sounds like somebody's beating on the bottom of the lab with a hammer.'

Minor modifications were necessary in procedures written down for the astronauts but the operation progressed smoothly. In the early evening, about 5:25 pm, flight controllers broke from their consoles and wandered about with bags of sandwiches while others moved off to the snack bar as Skylab drifted for more than one hour out of range of tracking stations. Neil Hutchinson came back on the loop minutes before reacquiring the signal, asking everyone to quieten down. 'We are progressing slow but sure and everything so far is working,' came the call from Skylab. It was 6:30 pm.

In the Command Module now, Joe Kerwin had set up the TV camera against the left rendezvous window in the hope of sighting deployment of the parasol from the workshop wall. By now several extension rods had been attached, moving the folded umbrella-like device further from the cylindrical workshop.

At 6:54 pm Skylab was over the tracking ship *Vanguard* and Weitz told Houston that the last extension rod was about to go on. But controllers wanted Kerwin to observe visually the final deployment sequence where the last rod would push the parasol out, releasing the four ribs which were then to open 90° and spread the shade. That required a pause to wait for Sunrise which would occur at 7:26 pm while Skylab coasted across the island of Sri Lanka and the Bay of Bengal, out of range of a suitable ground station.

The fifty-nine minute hiatus was broken at 8:03 pm with a report from Conrad via Hawaii: 'We had a clean deployment as far as rods, clearing and everything, but it's not laid out the way it's supposed to be.' In fact, tight packing had pressed the shade material limiting it to an effective area 5.5 meters and 6.1 meters on the two long sides by 6 meters and 6.5 meters on what should have been the two short sides; laid flat as designed it would have provided a rectangular shade 7.3 meters on the long sides and 6.7 meters on each short side. 'So, in effect, we have a trapezoid which has the smallest dimension toward the base of the vehicle,' added Conrad.

At this time the crew had not pulled the parasol back down on to the workshop wall, removing the extension rods as they went, waiting for consultation with Houston. It was explained to Conrad that the shade was in fact a rectangle and that the center of the four ribs was off-set from the center of the shade; in the rush to get the crew trained Conrad's team were not unduly concerned with the shade's precise shape. But now they understood. 'That's the way it is,' said Conrad; 'That's the way it should be,' said Houston.

In the MOCR, behind the big glass screen separating visitors from flight controllers, men and women who had worked on the parasol clapped loudly at this news, while discussion immediately broke out as to how best the wrinkles could be evened out. Minutes later, with Houston having advised the crew to leave the shade as it was until next station, communication lapsed for 21 minutes. Then *Vanguard* picked up the signal. Mission Control asked for a description.

KERWIN 'And if you want to know what it looked like when it deployed, I (saw) the thing sticking up, bunched in the middle, billowed a little bit at the top and at the bottom, and when they deployed it, all four legs came up. The front legs, that is the forward ones closest to the Command Module, came up smartly. It looks as if they actually went over center a little bit, then bounced back. The back ones did not come up, it looked like, all the way – didn't come up 90 degrees. They went slowly and they just kind of drifted to a stop.'

CAPCOM 'Okay, we would like for the CDR and the PLT to go back in the workshop and pull her in and we'd like for you to pull as many rods in at one whack (as you can).'

Conrad and Weitz moved quickly into the workshop and hauling on the rods quickly withdrew them through the long box attached to the small, square, airlock in the wall. At 8:35 pm, the parasol shade was close down against the hull and ready to do its job, shielding the exposed gold foil from brutal heat. Nobody expected to see the temperatures start down at once; it would surely take a couple of revolutions to notice a change. At about 9:45 pm controllers sent up the commands to move Skylab face on to the Sun: 'Skylab, Houston. You're on your way to solar inertial now.' This would be the real test of the parasol's effectiveness. But not just yet, for the attitude change was delayed by further trouble with the rate gyroscopes which told the guidance equipment what position the cluster was in with respect to fixed coordinates.

In fact for a few hours the controllers juggled with the cluster's attitude in the hope of finding the solar-inertial position. The rate gyroscopes had been without reference updates from Sun sensors for so long that nobody was really quite sure any more just what precise attitude the station was in. Finally, around midnight, the electrical system controller reported that he was seeing the change he expected and that Skylab was in fact facing the Sun. The gyroscopes could now be calibrated with that position and from then on the cluster would know where it was. The last conversation with the crew came ten minutes after loss of signal through Ascension Island.

Although a vital function for the first manned Skylab mission, repairing the crippled station was not alone the function of the Conrad crew. Here, during ground training, Conrad checks equipment at the materials processing facility rehearsing tasks he would perform in space.

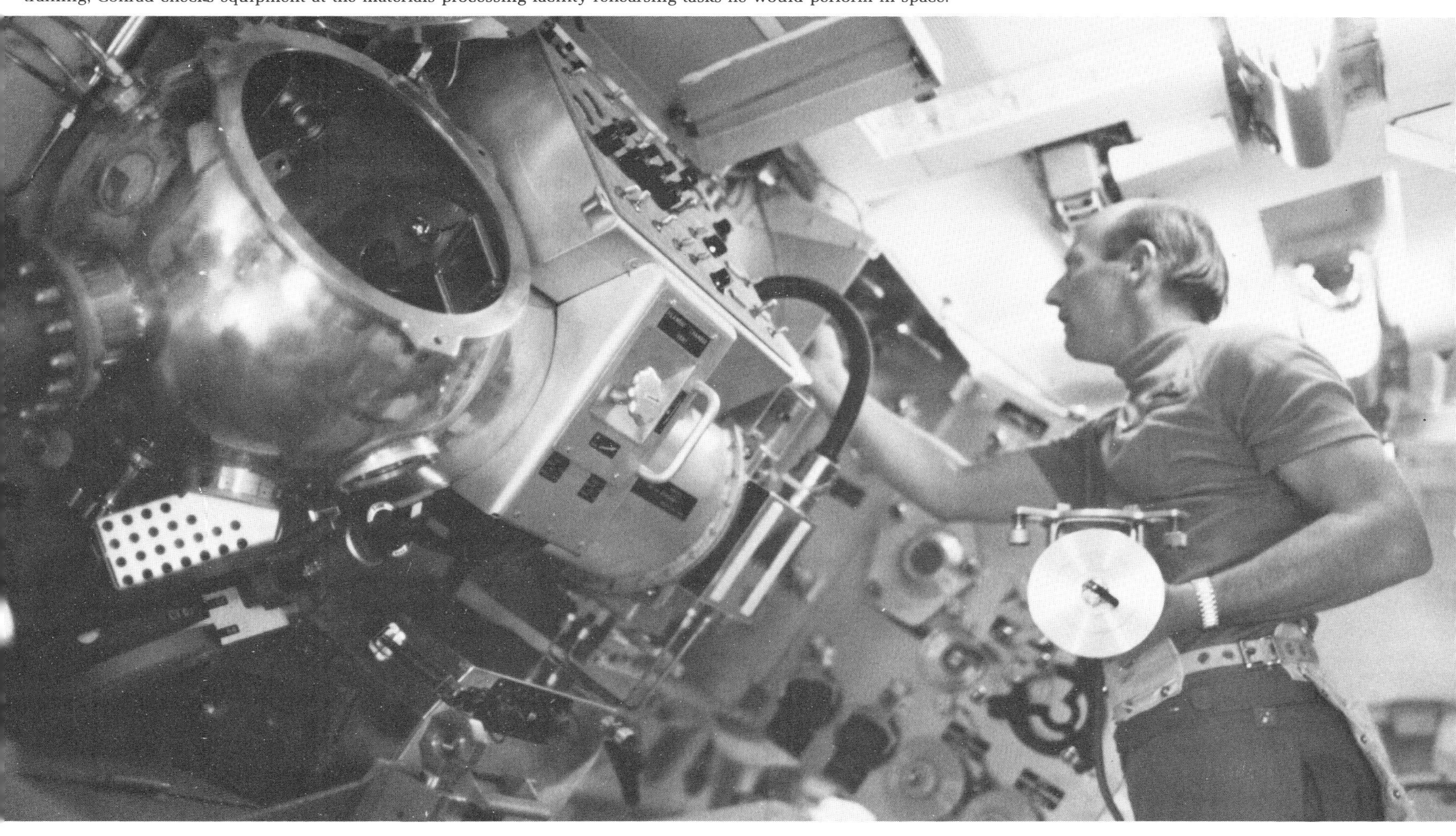

Sunday, 27 May and less than ninety minutes into that day the biomedical officer informed flight director Don Puddy that Joe Kerwin had gone to sleep without his biomedical harness. There had been no medical data on the previous night and the physicians were anxious to have the astronaut woken for him to don the equipment. The Johnson Space Center Skylab manager, Ken Kleinknecht conferred with Puddy and when the latter asked his medical monitor if he thought the astronaut was asleep, receiving affirmation that he probably was, Puddy decided not to waken him. There would be no biomedical data for the second night, and that was that. It was more important to have the man get his deserved rest.

Across the early morning hours in Houston, while Conrad, Kerwin and Weitz slept on aboard Skylab with all the hatches open for good ventilation, temperatures were seen to be dropping. By 2:30 am the workshop was down to around 46°, a drop of 5°c. At 5:00 am the average temperature was around 43°c, falling steadily, and by the time the crew were up and about shortly after 8:00 am it was reducing at a rate of 1°c per hour. There was much unpacking and stowage to do this day and after a breakfast meal from stocks in the Command Module, Joe Kerwin went off to activate the ATM equipment in the Multiple Docking Adapter. The TV camera was turned on over Goldstone showing a view down into the crew quarters area from the dome of the forward experiment compartment.

After lunch, Weitz recorded a report on the precise condition of the snared solar array boom, information that astronaut Rusty Schweickart would work with in developing procedures for going outside later in the mission and attempting once again to free the jammed structure. By late afternoon the air temperature in the workshop was down to around 36°c although one of the food lockers read as high as 47°c. Saturday night had been spent in the Command Module but this evening the crew would sleep in the Multiple Docking Adapter. It was still too warm down in the workshop.

The day went well from a procedures viewpoint, although nobody made as much progress as controllers hoped. So much loose equipment had been left lying around from the previous day's hectic activity, the Command Module was a shambles, and there was a lot of tidying up and putting away in this spacious mansion of a space station. But slowly, throughout the afternoon hours, the last remaining systems were activated in the wardroom, the washroom, the sleep compartments, and the experiment areas.

In the evening Hank Hartsfield told the crew that Houston was going to command some automated checkout sequences on the ATM. Tomorrow the final transfers would be completed before the first medical experiment was run, on Weitz, during the afternoon. Monday also, the press would get the chance to have questions relayed to the orbiting astronauts. By Tuesday, Conrad, Kerwin and Weitz would be setting up a paced schedule of scientific activity, followed by full operations one day later than the original flight plan which called for full activation by day 4 of the manned mission. The last call from the crew came at 9:05 pm that Sunday evening. When they awoke, they would go down to the workshop wardroom for breakfast. Already, the temperature was down to 32°c.

Flight director Neil Hutchinson's silver team was on shift at the beginning of day 4, Monday, 28 May. 'Hi there. Our hands are full of bloody medical equipment,' came a call from Skylab as Dr. Joe Kerwin drew blood samples from his colleagues before spinning them in a small centrifuge and placing the phials in a freezer. Slowly, the thermal balance had been adjusting to the new configuration. Temperatures in the workshop were varied, but averaging 30°c, and in the Multiple Docking Adapter it was now a comfortable 20°c. If all went well this would be the last day of activation prior to full scientific activity for day 5. 'CDR just finished shaving. Breakfast is cooking and I think with a little luck at all we might get on to a good routine,' opined a crewmember through the Texas tracking antenna.

One piece of equipment tried out for the first time was experiment M171, the body mass measuring device. This comprised a seat contained within a spring loaded cage designed to measure inertia and, in the absence of weight, determine the mass of a subject placed inside. Latched while the astronaut settled into the chair, a small handle would be released to set the seat oscillating back and forth at a frequency determined by the 'weight' it carried. It was an integral part of medical experiments to monitor weight changes in the body and could accurately measure up to 100 kg. A much smaller, wall mounted, version was to be used for weighing cans and containers, calibrated to a range between 50 grams and 1 kg.

But electrical energy was still at a premium and Conrad had the crew keep as many lights off as possible, at one point running only 10 throughout the cluster. In solar-inertial attitude the four ATM arrays were drawing maximum solar energy and producing about 4.6 kW. With Skylab systems seen to need about 3.6 kW, that left little more than 800 watts for all experiment activity plus a reserve. Fully powered, the medical devices would draw 450 watts, the Earth resource instruments would take 450 watts also, while the ATM telescopes required 700 watts. It was not an ideal situation for getting the most from Skylab and thoughts turned toward schemes for deploying the jammed workshop solar array boom while in orbit. The crew held a press conference through the Houston capcom and continued to put away storage containers.

The vestibular rotating litter chair was installed and Kerwin examined Weitz's cardiovascular condition in the lower-body negative pressure device (LBNP). Then they tried out the bicycle ergometry exerciser and Kerwin had a few comments for the ground: 'As a lot of us suspected, we've got a significant mechanical efficiency problem in riding the bike, which is going to take us I think a few days to solve. The harness is not efficient enough. He winds up doing a great deal of work with his hands and not being as efficient with his legs and his big muscles and he can't get to the high work loads nearly as well. We terminated that run with a little under three minutes to go – both, for that reason and because of an obvious thermal problem. It's just too damn hot in here. . . .'

At 8:08 pm, two thrusters on the Apollo spacecraft fired aft for sixty-three seconds, changing Skylab's speed by 2.3 km/hr to conduct the first trim burn, moving the cluster 117 km west, back on to the desired ground track. A large quantity of propellant had been used up during the fly-around on the first day, and during the problem with the docking activity. Less than 3 km/hr speed change capability now remained in the Service Module thruster tanks, allowing, of course, a reserve margin for safety in case the big SPS failed at retro-fire. Controllers hoped they would not have to call for a second trim burn on this visit.

So far, almost all the stowed equipment had been placed in its proper location and ATM and medical devices were ready for operation. There would be some trouble-shooting to be sure, for that was what the first manned habitation was all about, but so far things had gone very well considering the slow start and the initial problems. That night, Joe Kerwin donned the biomedical harness and all three astronauts settled down to another night in the MDA. To help bring the temperatures down as low as possible (the average workshop atmosphere was now a comfortable 29.5°c) controllers opened up both primary and secondary coolant loops. One of the three pumps tripped a circuit breaker, however, but the system was switched to a second pump and continued to run normally.

Tuesday, 29 May, sent Kerwin into a full session at the ATM console while Weitz activated and inspected the many Earth resources items carried in the Multiple Docking Adapter. The previous evening Pete Conrad had requested a private conversation and it was arranged for 9:18 am during a pass over the Honeysuckle tracking station. The JSC Director, Chris Kraft, Deke Slayton, and flight director Neil Hutchinson would participate. During the conversation, Conrad expressed his pleasure at the way the flight was going, apologized for having forgotten to get biomedical data two nights back, and talked about the possibility of working on the probe and developing procedures for freeing the solar boom.

The nature of the communication was seen by participants on the ground to have been the type that did not warrant a private loop. Jim Fletcher issued a statement later to the effect that in future the capsule communicator would ask the crewmember if a private loop was really necessary. But nobody rebuked the Commander for having requested a private talk on this occasion for fear of inhibiting his freedom to choose such a course should problems develop later that warranted such a request. It had more public relations impact than anything else. The press were unfamiliar with astronauts requesting private conversations during flight, seeing in it more than actually existed.

By the evening hour Paul Weitz had resolved several minor operating problems with the EREP Earth resources equipment and Kerwin had the ATM set up for a full day of science. That night all three crewmembers slept in their respective compartments down in the workshop's crew quarters, the first time since launch. Temperatures were now averaging 27.5°c and the cluster was balancing itself well after the high heat loads it endured for nearly two weeks.

Day 6, 30 May, began with Kerwin operating the litter chair for vestibular tests on disorientation while Conrad settled into the ATM console. Kerwin relieved him at noon and at 3:35 pm the first full EREP pass began. Skylab was to scan a region between the coast of Oregon and Brazil in a 23 minute run observing twenty-five selected ground sites. But the heavy drain on electrical loads took its toll for when Skylab disappeared into the night side of Earth it had not quite made it back to solar-inertial attitude, having been placed in an Earth pointing orientation for the scan. When the cluster appeared above Hawaii's radio horizon, ground controllers saw five of the seventeen battery modules off line, having disconnected themselves due to low charge.

At Sunrise the voltage surge kicked CBRM-3 off line and at the end of the day the second of eighteen battery modules was inoperative. Each lost CBRM denied a further 200 watts to the cluster and the failure of two such batteries lent impetus to plans for deploying the remaining workshop array. Already, the crew were conserving power by not only reducing the number of lights but also by leaving food out to warm gradually rather than use the heaters provided. Now, Mission Control cancelled all further EREP passes until the situation improved. Conrad asked Houston about the electrical budget.

CONRAD 'How serious does it look on the EREP? Is it just a matter of juggling or are you really going to have to go sèt your heads to work out the power or does it look like we just don't have enough?'

CAPCOM 'Pete, I guess we're still scratching our heads about it. We just really haven't thought it out. We *will* be doing more EREP and we'll plan it so we can get the most out of it.'

CONRAD 'Okay, I understand. It gives us a little more impedence to get that other SAS panel out . . . that thing is sitting there and I think once equipped with the right tools it's just a matter of applying (force) on that strap and that baby would be out and running.'

CAPCOM 'We are very actively working several alternatives that are in people's minds about solving the SAS problem on this mission . . . Rusty (Schweickart) left this evening, he's going over to Marshall to work on a tiger team with some good thoughts along that line. . . . We haven't given up yet.'

During the evening the crew worked at the docking probe, removing its head section in an attempt to find out why the capture latches refused to work during the several docking attempts. Weitz set up the ATM telescopes to run automatically during the night hours and the crew got a comparatively early night.

31 May was the most conservatively planned day of the post-parasol period, for electrical loads would have to be kept down as low as possible. After trouble-shooting the ultra-violet stellar astronomy experiment, and turning off more equipment as they went around the cluster, Conrad, Kerwin and Weitz settled to a day of modest scientific activity. Medical experiments occupied them for most of the time. During the evening Conrad told Mission Control that he preferred not to use the next day as a rest period but rather to press on with budgeting the electrical power with the experiment schedule, and received word that an earlier plan to push a TV camera out through the remaining scientific airlock to inspect the jammed array had been dropped.

Conversation moved back and forth on possible tools that could be used to free the boom but Conrad was sure that a saw carried aboard would not do the job. The edge of the meteoroid shield had whipped round the boom, riveting screws on to the boom itself. Temperatures inside the workshop were now stable at 26.5°c but electrical loads were kept down to an average 3.6 kW because of the tendency for the sixteen operating batteries to switch off at low charge. In Houston, careful examination of the consumables aboard Skylab led managers into a decision concerning the launch of SL-3, the second manned mission originally planned to last 56 days.

It seemed possible to preserve this flight by expeditious use of the available electrical supply although the situation would certainly improve if and when the jammed boom was freed. In any event, as the year progressed Skylab's fixed orbit would cause it to spend increasing time in direct Sunlight which in turn called for less time on battery power. The schedule still called for SL-2 to remain docked with Skylab until re-entry on 22 June at the end of its four-week mission. But instead of waiting until 8 August to launch the second manned mission, it was agreed at a teleconference of the Manned Space Flight Management Council that SL-3 should go up on 27 July.

Moreover, the parasol deployed by the Conrad crew would probably be jettisoned when the second crew arrived on board so they could put out the A-frame shade developed at Marshall; the material used to fabricate the Johnson parasol was thought to lose efficiency under exposure to solar ultra-violet energy. That would call for a special EVA, but probably pay dividends in the longer operational use of the Marshall design. In any case, that shade and the stand-up EVA cover launched aboard Conrad's spacecraft would be kept in Skylab for the second crew. Astronauts Bean, Garriott and Lousma agreed with the early launch and their training schedules were adjusted.

Day 8, 1 June, had the crew attend to several minor problems that arose in preceding days, but, officially a day off, it did provide opportunity to film some sequences of weightlessness. In the early afternoon, Deke Slayton spoke to the crew about efforts under way at the Marshall Center to come up with a means by which they could free the workshop solar array boom.

Next day, Saturday, Schweickart simulated EVA procedures in the neutral buoyancy facility at Huntsville in final tests of equipment already on board Skylab that just might do the job. Managers agreed to make a final decision on the proposed EVA by Tuesday, 5 June, three days hence. But it was Pete Conrad's 47th birthday on the crew's ninth day and during the afternoon he held a private conversation with his wife through the *Vanguard* ship. Earlier, a short EREP scan went off as planned, conservatively shaped around the electrical profile with precise attitude alignment for getting as much power in to the batteries before and after the operation.

Sunday, 3 June, had Paul Weitz pull a long shift on the solar telescope control console, broken only to help Conrad perform another EREP observation, as Kerwin performed more medical tests. The crew noticed discoloration on the parasol now and Houston told them the boom deployment procedures would be sent up the next day by teleprinter. Several methods of supplementing Skylab's meager electrical power had been proposed by government and contractor teams. McDonnell Douglas suggested a roll-up solar array panel that could be carried aboard Apollo for the second manned mission, attaching it to the exterior and connecting it to power lines.

Another suggestion, from Rockwell International, would have used the Docking Module then being developed by that company for the Apollo-Soyuz Test Project rendezvous and docking mission scheduled to fly two years hence. By attaching a spare solar array boom to the Docking Module, the device could be carried to Skylab by the second crew and docked to the radial port provided for emergency. But the preferred method was still to have Conrad and Kerwin go outside and try to free the existing boom.

Monday incorporated more EREP Earth observation and solar studies with the ATM. Over the United States a calibration rocket, a small device launched from a mobile site, had to be destroyed in flight and the crew failed to get the scanners locked on. The electrical situation was improving a little due to familiarity with the system. A trickle charge had now been fed to four of the eight workshop batteries (PCG-5,-6,-7,and-8), from the sixteen ATM batteries, bringing one PCG unit up to 100% charge; it was a lengthy operation though, made possible only because of the interconnecting lines, because the ATM batteries were not to be allowed to fall below a 70% charge due to the possibility of tripping off line as experienced five days earlier. The crew were standing up well to the activity and the mass measurement device revealed body weight loss of less than 2 kg. Conrad requested that the crew delete the next two days off to compensate for the planned EVA to free the boom on Thursday 7 June.

Three astronauts monitor the progress of Skylab's first EVA. (Left to right) Edward Gibson, Russell Schweickart and Robert Crippen.

Tuesday had the crew actively studying the uplinked instructions for EVA, with a proposal that they go through the entire procedure inside the workshop during Wednesday to allow Houston visual monitoring of their procedures on the television link. This day, a second PCG was brought up to full charge, with a third at 95% and a fourth at 50%. Slowly, but very surely, the electrical systems controllers were gaining ground, trickling a charge into increasing numbers of workshop batteries. Then, during the evening, a third charger battery regulator module (CBRM) began to fluctuate, leaving only fifteen good units still on line. The need for supplementary power was becoming an essential part of continued activity. Within a week, Apollo's fuel cells would run dry and the ATM arrays would have to feed that spacecraft as well as Skylab.

On Wednesday, 6 June, Schweickart was on hand in the MOCR to monitor rehearsal of boom deployment tasks and to discuss with the Skylab crew tricky aspects of the job which he noticed and worked in the water tank. But it was also a day for science experiments, with another EREP pass and a special assignment for Kerwin to take photographs of Hurricane Ava along the Pacific coast of the United States. It was a combined operation. Far below, a Hercules transport aircraft equipped with special sensors and a very full load of fuel flew straight through the eye of the hurricane, monitoring conditions all the way.

By the evening, CBRM-17, the ATM battery unit that showed erratic tendencies only the night before, seemed to have re-joined the group although two gyroscopes were not performing as they should. At evening's end the crew was granted a request to cancel exterior television relay of the next day's EVA, thinking the sequence of operations already full without adding tasks. TV would be shown from a window on the spacecraft instead.

Thursday, 7 June, day 14, was devoted entirely to the EVA boom deployment task. Shortly after 6:00 am the crew were busy getting ready for the pre-EVA chores when they received a call through Honeysuckle, Australia. There was a lot of equipment to get ready, most of it having been unpacked and put aside the day before. The stand-up EVA cable cutters would be used to snap the restraining strap; vice grips would be needed; a bone saw was taken from the dental kit; a prying bar was brought up. The EVA would take the crew into an area that nobody thought they would ever have to go, so no hand holds or foot restraints had been provided. It would take special care, and a very sure sequence, to get the task accomplished.

First, Conrad and Kerwin had to get out through the door on the side of the Airlock Module. After donning suits they drifted into the compartment, sealed hatches at both ends, and de-pressurized the interior, leaving Weitz in the Multiple Docking Adapter. This was standard practice to prevent the third astronaut becoming stranded should the Airlock Module become unusable. In Houston, Jim Fletcher, George Low, Dale Myers, Bill Schneider and Ken Kleinknecht sat at management consoles, with capcoms Schweickart, Gibson, and Slayton on the voice loop. At the flight director's console, Milton Windler supervized operations. Like the captain of a ship, it was his day.

At 10:17 am, Houston time, Skylab came within range of the Ascension Island tracking antenna with the Airlock Module slowly de-pressurizing. 'Okay, Houston. The lock's on the way down.' Six minutes later the door was opened and at 10:25 the manned space station was out of communication again. At 10:48 the cluster would come within range of Carnarvon shortly after entering night on the far side of the Earth to the Sun. In the meantime, Conrad and Kerwin drifted outside and began the first tasks, assembling five rods to form a continuous pole 7.6 meters in length, fixed at one end to a metal cutter which would be operated by pulling on a cable laid out along the pole to the opposite end. When they came within sight of the Australian tracking station, that task was almost complete.

CONRAD 'Where the hell's the world, anyway?'

CAPCOM 'Houston, we're right here. We're listening loud and clear.'
CONRAD 'Oh, I didn't mean the world world, I meant the clouds and Earth and sea world underneath.'
CONRAD 'Okay, we have five poles rigged swinging on the hook. And we're just intrepidly peering around out here deciding how far around Joe can get in the dark.'

The next job was for Kerwin to move back to a bag near the Airlock Module and to fully play out the umbilicals connected to each man's suit: Conrad had 16.7 meters, Kerwin had 10.7 meters. Then they would make their way round the fixed airlock shroud to the bottom and fix one end of the long pole to a strut, securing the other end so that the cutter jaws slid over the metal strap jamming the solar array boom.

After a few minutes out of radio contact, Guam picked them up for nine minutes at 11:03 am. Unable to see their way, the crew elected to wait until Sunrise, which came just as the space station drifted over Guam's horizon. Almost exactly thirty minutes later they were within range of the US tracking antennae and Houston heard of difficulty snagging the cutter jaws on the metal strap; without hand holds along the smooth hull of the workshop back toward the array boom, they had to remain at one end of the 7.6 meter pole with the other end flailing around near the debris. Straining with heart rates between 100 and 150 beats per minute, the crew finally had the jaws secured. And then, before passing from *Vanguard*'s antenna, they managed to get it fixed. The most difficult job lay ahead, and as the cluster swept out of radio view at 12:00 noon, Houston time, it was to be a tense period. Not for another 63 minutes would Skylab be within range of another tracking station. Those on the ground could not even watch, only wait.

At 12:10 pm, Skylab moved back around the dark side of the Earth and without artificial light the crew were unable to continue. Completely removed from the listening ears of radio stations, Pete Conrad waited to accomplish what was probably the most daring and risky space-walk activity yet performed. At 12:45 the cluster was back in Sunlight and Conrad made his way, still tethered to the Airlock Module, along the 7.6 meter pole to the jammed boom. Making sure the cutter jaws were tight over the metal strap wrapped round the joint from the meteoroid shield, he checked the rope was in place and held the pole as Kerwin, from his position back at the truss assembly, pulled hard to operate the cutter. Try as he could, the jaws refused to snap the metal.

Conrad moved back up to the cutter and just as he reached it the jaws worked sending the strap flying apart. Suddenly, the mighty boom began to move out sending Conrad cart-wheeling into space. Grappling with the tether he made his way back on to the pole and re-joined Kerwin. But the boom came out only 20° instead of rotating the full 90° so Conrad worked his way back again to perform the second task, already anticipated. A clevis bracket on the actuator had to be physically snapped to free the boom and allow it to rotate.

Having already attached a rope to one of three square-shaped vent modules on the boom, Conrad and Kerwin pulled with all their might, using the rope in a tug-of-war with the boom. But the bracket failed to snap so Conrad moved hand over hand along the rope. Standing on the workshop wall he placed the rope over his shoulder and lifted upward while Kerwin simultaneously pulled from his position back near the ATM. It worked, the bracket snapped and the boom slowly moved out the full 90°.

The array comprised three separate sections which, when fully deployed, would present a flat wing of solar cells. However, because the panels had been in shadow for many days the hydraulic dampers had frozen, inhibiting full extension. When Skylab came in view of the Goldstone tracking station at 1:03 pm, Conrad had a summary for capcom Rusty Schweickart: 'We've got the wing out and locked, the outboard panel and the middle panel are out about the same amount, and the third one is not quite. Now Joe, I think before you come in, you better take a look up there and make sure that third one is clear of all the debris.'

As electrical systems controllers watched the amp rates on their consoles it was deduced that the two outer concertina sections had opened about 40% while the inner section was deployed only 30%. Milton Windler decided to warm the dampers by pitching the cluster 45°, allowing the Sun to fall on the now exposed surface. Within six hours the Sun would have done its job, fully deploying the three solar array sections. Skylab would now have nearly twice the electrical power it had struggled with to date, receiving an additional 3 kW from the single workshop wing in addition to the average 4 kW from the ATM arrays. Meanwhile, mundane tasks still had to be performed before the EVA was complete.

During the first few days of ATM work, one of the solar telescope film magazines developed an erratic feed so Kerwin changed the ultraviolet spectro-heliograph canister. Also, the aperture door covering the X-ray spectrographic telescope had given concern so the Science Pilot pinned it in the open position; both jobs were performed up on top of the ATM. From that position the Skylab cluster looked massive, and a sense of height was felt for the first time during an EVA. Usually, the Earth looked to an astronaut's eyes very much as the surface does to a pilot or an airline passenger; devoid of a structural link the familiar sensation of vertigo was absent. But the sheer size of Skylab seemed to add a feeling of altitude and Kerwin was relieved to climb back down from the big ATM structure and complete the EVA.

Conrad and Kerwin's space walk lasted nearly $3\frac{1}{2}$ hours, but they had accomplished the first major repair tasks in space, not only pulling free the snagged solar array boom, but manually correcting problems that threatened the smooth operation of two solar experiments. But even as Conrad and Kerwin moved back in to re-join Weitz, problems were developing in the coolant loops that had their origin during preparations for the EVA. Hooked up to the primary loop, the two astronauts noticed a decrease in temperature but this was not critical and they carried on, only realizing the potential seriousness of the situation when they got back inside.

For most of the evening the crew worked with ground engineers to stabilize the coolant loops, switching to the secondary system before settling down for the night. But that too began to cool toward the freezing point and they were re-contacted and asked to put the liquid-cooled garments normally worn underneath the space suits on the ends of long umbilicals. Placed down in the workshop near water tanks on the Sun side of the cluster, the LCGs provided a path for heat to reach the cooling system in the Multiple Docking Adapter. Satisfied that the system was stable at least for the night, the crew went back to sleep.

Next day, Friday, 8 June, Conrad, Kerwin and Weitz exceeded the US manned space flight record of 13 days 18 hours 35 minutes, set up seven and one-half years before by Gemini pilots Borman and Lovell. Only the Soyuz 9 flight, in June 1970, and the catastrophic Soyuz 11 flight a year later, exceeded that duration. By the end of the mission both records would be broken.

The second half of this first manned Skylab visit progressed effectively toward the program's mission goals. The cluster was in a better condition than at any time since launch and the crew settled in to routine housekeeping duties, Earth resources scans with the EREP equipment and scientific observation of the Sun.

8 June saw the crew tend the coolant loops that gave trouble the night before and select the secondary system when the primary again seemed ready to freeze up. But life aboard this big station was very different to that on earlier manned vehicles. The crew learned to get around the spacious interior using methods and techniques impossible to plan or rehearse in advance. They quickly grasped the dynamics of weightlessness and 'swam' around the interior from one location to the next in easy movements that belied their inexperience.

It had been thought that a crewmember would become stranded if, after pushing off from one wall, his momentum was spent before contacting another; stranded in the middle of the big workshop, it was believed he would be incapable of moving to a safe handhold. In reality, subtle twists and turns of the body imparted motion and it took very little practice to adapt with agility the constraints of the weightless state to the limitations of the human body. It was a new experience to work in one section of an orbiting vehicle and then float ten or twenty meters to the living area for food or sleep.

While working in the Multiple Docking Adapter, at the ATM console for instance, the conventional 'up' and 'down' axes developed a familiarity and a comforting awareness of direction. When the astronaut left that position and moved up to enter the large dome of the workshop, the brain sensed the enormous length of the station and translated that into height. It was easy to think one was falling toward the grid separating experiment and living quarters seven meters below, a sensation forcibly expelled from the brain but not before hands momentarily grasped for a sure hold! And then there were the fun times, when the sheer joy of liberated movement led to novel exercises, athletic feats impossible to repeat in a 1 g environment.

A common relaxation was to run around the ring of cylindrical water tanks, feet pounding against the outer wall, setting up a centrifugal motion. Body twists, all the contortions and turns effected by a spinning athlete leaping to the ground from a high bar, could be sustained for as long as the astronaut had energy, for here there was no need to snatch precious seconds of freedom while gyrating toward the ground.

Temperatures inside the workshop were still falling, down now at the half-way point in this first visit to about 23°c. On day 16, workaround procedures with the coolant loops paid off, both showing good recovery. The cluster had been put through a lot since launch on 14 May, performing in ways and modes for which it had never been designed. It was a wonder everything still worked as well as it did.

Continued studies of the Earth took the crew to different objectives and sites. On one survey they examined and photographed chlorophyll blooms off the coast of the United States, then they recorded storm fronts near the Great Lakes, studied severe storm conditions in the Mississippi delta or photographed the effects of urban growth in Florida. By mission day 20, 13 June, Skylab was running at peak efficiency, confidence having been fully re-gained in the ability of cluster systems to accommodate high experiment loads.

During an Earth resources run that day the experiments drew a record 5 kW, still leaving a healthy reserve of 2 kW. Next day, shortly after noon, the two fuel cells aboard Apollo were shut down; for the next week, Skylab would feed 1.1 kW to the docked Apollo and with the cluster drawing 4.7 kW that still left 1.2 kW in reserve. But four CBRM units, the battery regulator modules fed from the ATM arrays, were delivering only 10-12 amps versus the designed 20 amps, a situation that would have cut the mission short when Apollo's fuel cells ran dry had not the workshop solar array boom been deployed.

But even in the event of a failure to the CSM, a rescue vehicle was already waiting, for on Monday, 11 June, SL-3 had been rolled to pad B at launch complex 39. The first stage of Saturn IB AS-207 had been stacked on its mobile launcher three days after SL-2 was launched, joined by the second stage a day later and by CSM-117 on 8 June. If everything went as planned, it would carry Bean, Garriott and Lousma to a two-month stay aboard Skylab at the end of July. And all was going very well on the first long-duration flight.

During a management meeting on 15 June, plans for the deployment of a replacement shade were finalized. It had been thought advisable to have Conrad and Weitz go outside on their scheduled 21 June space walk and deploy the A-frame shade since the material used in Johnson's parasol was known to have limited life under solar ultraviolet exposure. However, because the Conrad crew would be at the end of their one month stay, and because they had been loaded with many new tasks on their mission, it was unanimously agreed that Bean, Garriott and Lousma would put up the replacement shade on a special EVA at the beginning of their mission. Development of the supplementary power systems, originated when it seemed unlikely that the workshop boom would be deployed, all but stopped. Only work on the so-called Solar Array Module continued in case it was needed after all.

On 14 June, Conrad, Kerwin and Weitz began a series of clock shifts in which their nightly sleep session would be stepped back ready for the final day's activity. Medical considerations were of paramount importance on a flight greater in duration than any before it and physicians were concerned that the Skylab crew should not have to face the rigors of separation, re-entry, and recovery in a tired or stressed condition. Accordingly, the normal eight hour sleep period that began at 10:00 pm Houston time was changed on day 21 when the crew went to rest at 8:00 pm and got up at 3:00 am on day 22. That evening they began their rest at 6:00 pm and awoke at 2:00 am, day 23. For six nights they would maintain that 6:00 pm/2:00 am rest cycle allowing their bodies to adjust before the final sleep change the day before the mission was due to end.

17 June provided a second opportunity to adjust the Skylab orbit further and refine the path for the second mission in July. At 3:59 pm two RCS thrusters fired for 9 seconds changing Skylab's speed by a mere 0.33 km/hr. During the afternoon Pete Conrad's family were in Mission Control to speak to Skylab's Commander under the supervision of Rusty Schweickart. It was a typical family chat, with some of the children wanting to know when their father was coming home, a junior member calling on dad to come mend his bike, and somebody else wanting to know if Pete's back was still bothering him! Later that day the maneuvering unit was checked out with the nitrogen thrusters fired from the stowed position; originally scheduled for test on the first manned mission with a free 'flight' around the inside of the workshop, an already crowded schedule deferred that activity until the second manned habitation.

Next day, 18 June, the crew swept past the Russian manned flight record and Rusty Schweickart briefed them on the upcoming EVA. During the day, the crew asked for Deke Slayton or Tom Stafford (Commander of the joint US/Soviet flight planned for 1975) to relay their respects to the Russian cosmonauts. Later, on

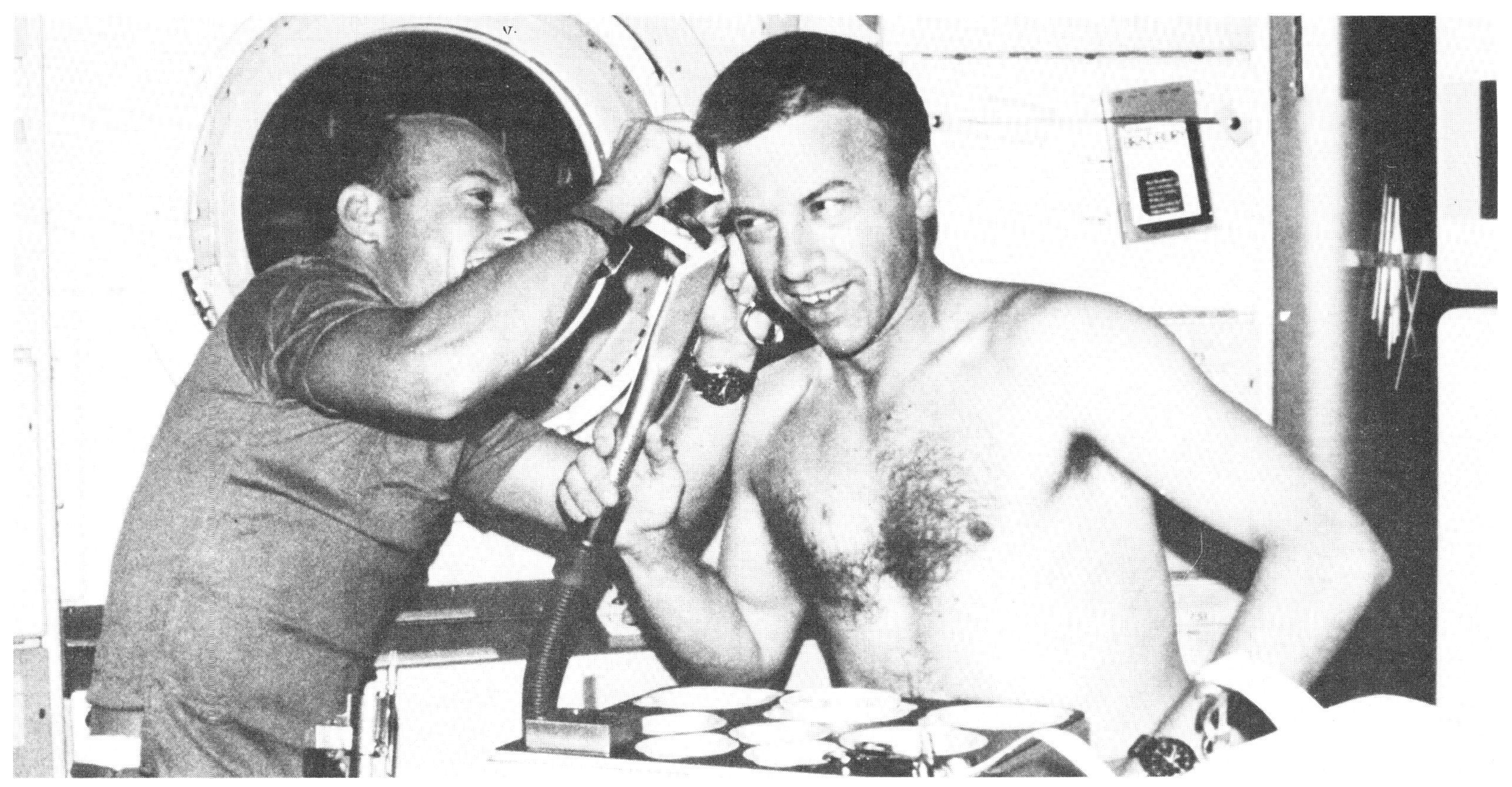

Charles Conrad trims the hair of Paul Weitz during the 28-day Skylab flight.

behalf of his team, the chief cosmonaut, Vladimir Shatalov, sent the following message to Houston:

'To the crew of the Skylab space station, Charles Conrad, Joseph Kerwin, Paul Weitz, we sincerely congratulate the courageous crew of the Skylab astronauts on your achievements in conquering outer space. Wishing you successful completion of your program and safe return to our beautiful blue planet Earth.'

It was the last complete day on which a full workload of science could be accomplished. The first manned mission had already achieved between 81% and 100% of the various task categories originally scheduled, despite the many problems that emerged and the added responsibility for repairing the cluster.

The crew were up in good time for the planned EVA on Tuesday, 19 June, with Conrad and Weitz moving through the Airlock Module door earlier than scheduled. One of the first tasks was for Conrad to move up on to the ATM structure and hit the compartment containing CBRM-15 with a small hammer. It was believed that a stuck relay was all that prevented the ATM battery module from operating and if freed by shock it would provide an additional 200 watts of electrical power, lost since before the Conrad crew were launched. Through the Bermuda tracking station Conrad and Weitz again justified their 'we fix anything' launch cry.

WEITZ 'There it goes. Boy is he hitting it. Holy cats!'
CONRAD 'All right. Did anything happen?'
CAPCOM 'Okay, that's good. It worked, thank you very much gentlemen, you've done it again!'
WEITZ 'How about that.'
CAPCOM 'How hard did you hit it?'
CONRAD 'Pretty hard.'
CAPCOM 'That's what it takes; the old Army technique wins once again.'

Inside the cluster, Kerwin turned on the amps charger for CBRM-15 and it flooded the gauge with power, restoring another failed element of the Skylab space station. Before going about the task of film retrieval for which the EVA had been scheduled, Conrad reported on the condition of the parasol and recommended that when they got back inside they should rotate it about 15° to increase its coverage over the workshop hull. Then the Commander moved up on to the ATM canister and recovered film cassettes from the white light coronagraph, the x-ray spectrographic telescope, the H-alpha instruments, the ultraviolet coronal spectroheliograph, and the ultraviolet spectrograph.

Then Conrad put out a sheet of material of the type used for the A-frame shade. It would be retrieved on the second manned mission for analysis of wear and degradation under solar radiation. The final task was to collect some special thermal coatings fixed to the exterior as part of an experiment to study the effect of the space environment on different materials. During the space walk, both participants observed paint and insulation strips peeling from the surface of the Apollo spacecraft. It had already been in space more than twice as long as any other Apollo and nobody was completely sure that it would not suffer in some way from the prolonged exposure; that was, after all, one of the reasons for setting up a rescue plan using the next CSM in line. Also, shredded particles were seen in the ATM canister, but that was not thought to compromise any of the experiments. The EVA lasted only 1 hr 36 min before Conrad and Weitz moved back inside.

Later in the day they turned the parasol about 15° but had to return it to its original position when temperatures in the living area were seen to rise rapidly; eyeball analysis from the outside was not as accurate as Conrad believed. Nevertheless, by getting CBRM-15 back on line Skylab now had seventeen of the eighteen ATM battery modules working again; CBRM-3 was still inoperative. All eight workshop batteries were functioning as designed. The electrical condition of Skylab was certainly getting better by the day. As the Earth moved in its annual path around the Sun, Skylab spent longer periods in the light of its rays, increasing to 8 kW the available energy from the four ATM arrays and the single workshop array.

Now it was time to deactivate the cluster, time to pack up experiment tapes and film canisters for return to Earth two days hence. Limited science work was carried out on 20 June, the day after the successful space walk, but that mostly comprised medical tasks; it was important to maintain a good profile of each man's physical condition for nobody knew if they would suffer from the stress of re-entry. Final packing and stowage was scheduled for 21 June. Rising by 2:00 am, Conrad had his team scurrying around to complete the tasks scheduled. Again, the teleprinter came in as an indispensable tool. Devoid of an ability to uplink long and complex instructions without the astronauts having to write down each procedure (new activities as requirements dictated) the crew would have had no time for much of the science they had already accomplished. As before, minor adjustments to the checklists could be sent up on the teleprinter.

It was a short day, the crew getting their heads down only a little later than the planned time of 1:30 pm that Thursday afternoon, 21 June. Six hours later they were awake, preparing for the return to Earth. Ahead would be nearly three hours of activating the CSM with the Commander in the spacecraft while Kerwin set up the ATM telescopes for unmanned operations; Weitz busied himself assisting both astronauts. By midnight, however, a problem had been discovered in the refrigeration loop, altogether different from the two Airlock Module coolant loops. Whereas the system, vital for preserving food and essential for the water chiller, the urine freezers and the urine chiller, was normally running below the freezing point of water, temperatures were now rising. Several procedures were tried in attempts to lower the temperature to its normal operating range, but to no avail.

It was quite the wrong time to develop new problems aboard Skylab. If delayed, the re-entry would be off target. Nevertheless, flight director Neil Hutchinson decided to turn the cluster 45° to the Sun so that the refrigerator radiator panel at the base of the workshop would be warmed, hopefully freeing an apparently frozen bypass valve. By 00:45 am, 22 June, all three crewmembers were in the Command Module and one hour later ground commands put Skylab in its temporary attitude, using nitrogen gas from the TACS thrusters. Shortly after 3:30 am the crew were told not to undock because the cluster had not yet stabilized back down at the solar-inertial attitude. Minutes later the controllers gave their approval and capcom Hank Hartsfield told the crew they could probably do a fly-around inspection of the cluster after they undocked.

'We haven't done anything by the flight plan yet, so we'll go by ear again,' quipped Pete Conrad. And then, with a cry of 'Bye-bye, Skylab,' the Commander slipped Apollo from the docking port and slowly backed away. The time was 3:55 am. Forty-five minutes later, passing across the Indian Ocean, the manned spacecraft fired its tiny RCS thrusters in a separation burn designed to place it more than 2 km from the space station at retro-fire. Twenty-five minutes after that, at 5:05 am, Houston time, the big SPS engine lit up for ten seconds, cutting Apollo's speed by 290 km/hr and dropping it into an elliptical path with perigee at 167 km. Nearly three hours later, after completing almost two more revolutions of the Earth, Apollo again fired the SPS engine to bring it back down through the atmosphere.

'Very good, Pete. You're in the groove,' called capcom Dick Truly as the spacecraft communicated via an ARIA aircraft. Minutes later the Command Module splashed down in the Pacific. It was 8:49 am, Houston time, and the USS *Ticonderoga* was waiting to receive three record-breaking space men. Less than forty minutes later the recovery ship was alongside, the crew remaining in their spacecraft to avoid the stress of scrambling around in rafts and helicopters. Lowered gently to its support dolly, Command Module 116 disgorged its human charges to NASA recovery team leader Melvin Richmond waiting with other officials on the deck. For the first time, US astronauts had returned from space wearing only lightweight flight garments instead of pressure suits. Uncertainty about their ability to withstand the rigors of atmospheric descent tipped the scales in favor of an unsuited re-entry. But the only noticeable effect made Joe Kerwin sea-sick while waiting in the Command Module, and nobody seemed much the worse for the experience.

As for effects of the 28 days in space, that needed long term observation. A little weight was lost, but that was inevitable, and all three were slightly disorientated upon return to a normal gravity field. Their hearts too were observed to work a little harder during the day following splashdown, but nothing serious emerged and medical opinion agreed with management plans for two full missions each lasting 56 days. Recovery from serious mechanical problems had rescued the Skylab program and it was the greatest vindication yet that man in space can do useful work, repairing structures and renovating failed systems. Without human intervention, Skylab would have quickly become a stranded hulk in space.

CHAPTER TWENTY-SEVEN

Stations in Space

Although a tentative decision to resume planned Skylab operations came close to the end of the first manned mission, when it was apparent the cluster had been restored to a working condition, final plans were possible only when a full review showed how well the Conrad flight had gone. Before trouble struck the Skylab station one minute after launch, the first mission should have gone up a day later on 15 May. The second mission would have been launched 8 August while the third would have ascended 9 November. The ten day delay that resulted from the need to evolve workaround procedures would, logically, have deferred the second and third flights by equal duration.

But deterioration of the parasol put up by the Conrad crew, and trouble with rate gyroscopes used to measure attitude angles, pressed managers to schedule an early flight for SL-3, the second mission, and to move ahead with the third visit, SL-4, just as soon as possible after that. Preparation of flight hardware could meet a scheduled 27 July launch for the second manned habitation but tracking information showed Skylab to favor an ascent on the following day, Saturday, 28 July, at 7:10 am Cape time. Moreover, examination of groundtracks that would be flown toward the end of the planned two-month mission showed an operational preference for recovery three days later than originally planned. Thus, the second mission was now expected to last 59 days versus 56.

The prime crew for the second Skylab mission: (left to right) Garriott, Bean and Lousma, seen here in a Multiple Docking Adapter mock-up at the Johnson Space Center.

SL-2, the Conrad flight, lasted just twice as long as the previous U.S. manned flight record; SL-3 would be twice as long again, so physicians decided to give the second visit a tentative 'go' for one full month, extended weekly beyond that for the full 59 days based on medical reports and telemetry from the equipment on board. Nobody, not even the Russians, had been in space for two months.

While Bean, Garriott and Lousma were in the final stages of flight training, engineers developed a package of six rate gyroscopes that would be carried to Skylab and installed if needed as replacement for the original gyroscopes on board. There were three rate gyros for each axis, only one being absolutely necessary to inform the attitude control system of the cluster's pitch, roll and yaw from a given reference. But one had already failed and five had overheated at various times since launch on 14 May. So Bean's crew would carry the 66 kg 'six-pack' just in case others failed and a minor repair job became necessary.

The spur that sent engineers reaching for an augmented package occurred 16 July, just twelve days before the launch of SL-3, when the prime up-down rate gyroscope failed completely. Three days later, after tests at Huntsville, one of two secondary gyroscopes was switched on and worked well. Plans were already in hand to fly the 'six-pack' and development was hurried up by construction of three sets of hardware. One went to the contractor plant in St. Louis for fit checks with the back-up Multiple Docking Adapter, a second was sent to Houston for crew training, while a third was retained at the Marshall Center for engineering evaluation. If needed, the pack would be installed in the MDA but connected to an exterior power source by existing conduit. It would require about twenty minutes during a space walk to set up the electrical contacts.

The Board set up two months earlier to seek the cause of failure in the workshop's meteoroid shield and the loss of a solar array boom, reported back mid-July and revealed their determination of the problem. It appeared that 63 seconds after lift-off air rammed down the front of an auxiliary tunnel on the outside of the workshop, exceeded pressures calculated in advance and ripped out fixtures securing the flexible shield. Wrapped tight against the cylindrical workshop wall during ascent, the meteoroid shield comprised 16 curved sheets of aluminum, 0.064 cm thick.

When lifted from the surface of the workshop hull, the pressure of air on the outside of the ascending vehicle took hold of the flapping sheets and tore them from mountings and links designed to gently deploy the shield a few centimeters from the workshop wall in orbit. Ripped from its secure mounting, the shield pulled solar array boom No. 2 from its stowed position. But air pressure prevented it from deploying, holding it firm against the side of the workshop. After the vehicle got into orbit, however, the solar array boom drifted partially open.

When the Saturn second stage came to separate from the workshop it fired its forward-facing retro-rockets to back away and in so doing exhaust plumes struck the partially deployed boom causing it to fly back, snap its hinge at the 90° angle and cartwheel off into space. Fortunately, boom No. 1 had resisted the attempt of the meteoroid shield to pull it loose but became snared instead by debris from the torn sheets, preventing it opening when unlatched by command shortly thereafter. But why had air pressure built up in the auxiliary tunnel to start with?

The Board found that seals, or caps, were missing from two stringers, allowing air to enter, and that there was an inadequate fitting between metal components on the outside, similarly admitting air under pressure. Also, because the auxiliary tunnel was designed to lift up from the surface of the workshop hull and move a few centimeters to one side when the meteoroid shield deployed, it was not possible to design an adequately air-tight join.

Because of those three reasons, the Skylab mission was nearly written off. It all pointed to one basic principle of space systems engineering: never overdesign or reach for more complex operating modes than those absolutely necessary for the safe performance of the mission. In contravening the intent of that dictum, Skylab engineers over-reached predicted safety levels and built in a shield that had questionable value, rendering Skylab less reliable in the process. There is a fine line between safety and over-sophisticated design; it was exceeded in this one Skylab system.

Between visits, Skylab was to follow a series of scientific objectives suited to the unmanned mode in which it then operated. The ATM telescopes were to be operated by command from Earth, and other passive experiments were to be conducted. But trouble with the rate gyroscopes led controllers to go easy on solar observations for fear the system would degrade completely before the second crew could get aboard.

Within hours of Conrad, Kerwin and Weitz pulling away from Skylab on 22 June, engineers evacuated the workshop interior, depressurizing it to less than 105 mmHg. Two days later problems with the cluster's refrigeration cooling system cleared up, largely as a result of last-minute procedures conducted by the first crew before they departed. Final de-pressurization, down to 34 mmHg, began forty hours before the launch of the second manned mission, with the workshop pumped back up to the normal 258 mmHg one day before lift-off. Engineering analysis of the cluster showed it to be holding up well under extremes of temperature experienced earlier. Against a design leak tolerance set at 2 kg per day, the workshop was leaking oxygen/nitrogen atmosphere at only 0.14 kg per day.

As for attitude control gas, sparing use during the latter half of the first manned mission boosted hopes that there would be enough to last for the remaining two flights. Of the total tanked at launch, only 42% remained but most of the balance had been consumed during the extraordinary activities in the first ten days.

For thirty-six days, the orbiting cluster waited for its second set of visitors. On the ground, preparations moved smoothly toward the first full duration stay. At Houston, a fifth set of flight controllers made ready to populate the MOCR when their turn came, upgrading the operations from a four-shift routine set out for Conrad's mission. As events would dictate, even that was inadequate for the task. At the Cape, hardware preparations moved ahead on schedule with the launch vehicle and spacecraft settled on pad 39B, 11 June. A weight limit of 6,124 kg placed on the Command Module by design restrictions on the parachute lines required some equipment to be removed so that the 'six-pack' could fly as two separate packages in lockers originally designed to carry experiments. Total CSM weight was, at 13.86 tonnes, a little heavier than Conrad's ship.

Heightened by trouble early in the previous mission, interest centered on the launch of AS-207 with Bean, Garriott and Lousma. More than 35,000 people were on site at the Kennedy Space Center to see the team ascend for their planned 59 day sojourn in space, despite weather that threatened to delay the flight. Thunderstorms moved across the Florida coastline, cloud obscured the sky and ground fog rolled up during the night before the flight. The astronauts had gone to bed in the Manned Spacecraft Operations Building at 5:30 pm that evening. By the hour of launch ten per cent of the sky was covered with cloud and visibility was about the distance between the launch vehicle and the viewing stands.

In the final hour before ignition, 28 July, a tranquil and apparently unconcerned Jack Lousma drifted off to sleep as physicians watched his heart rate descend to a cool 38 beats/min! A gentle word from the capcom console at launch control brought him back to consciousness. A mere ten seconds before 7:11 am, local time (6:11 am in Houston), AS-207 shook free the gravitational shackles that held it to pad B. It had been a good launch vehicle in checkout and now it was a good bird in flight, carrying the crew into orbit ten minutes later. At Houston, flight director Don Puddy nursed the Skylab cluster while a second team under Phil Shaffer monitored the manned Apollo.

Down below, at the Florida launch site, work continued on SL-4, hardware for the third manned mission; it was to be erected three days later and rolled to pad B on 20 August to support the Carr, Gibson and Pogue flight or to serve as rescue for the Bean crew if needed.

Soon after coming up on the United States at the end of the second revolution, some 45 minutes after the first SPS rendezvous burn, Bean reported seeing a strange sight through his No. 5 window. 'We got some sort of sparklers going by the right window, over by Jack, but we don't have any going by the left,' said the Commander, 'Maybe we've got something spraying out that side.' Almost immediately, Mission Control told the crew to 'secure quad Bravo,' and shut off one set of RCS thrusters by isolating both propellant supply and the helium gas used to pressurize the lines.

There was no immediate danger, but flight controllers saw on telemetry what the crew observed visually: propellant and helium was leaking through the quad to space. With three other quads the mission could progress although maneuvering options would be limited. But whatever was causing the leak could wait until after rendezvous and docking.

On through the pattern of maneuvers similar to those carried out by the preceding crew, Bean, Garriott and Lousma drew alongside the giant Skylab cluster little more than seven hours after lift-off. Observed on television transmitted from the spacecraft via Carnarvon, observers in the MOCR watched the flimsy parasol flapping in the 'breeze' of the RCS thrusters as the CSM maneuvered round the structure. One full revolution later Apollo slipped its docking probe into the drogue on Skylab and the two vehicles were locked together seconds after that.

It had been an early launch, Houston time, and the flight plan called for the crew to move quickly into the stowage and activation tasks, unpacking the Command Module and settling into routine scientific operations aboard the workshop. All the next day, 29 July, would be spent bringing Skylab back into a working condition ready for scientific activity occupying the crew from Day 3. But only on a limited basis. Day 4 was to have Garriott and Lousma go outside and in a 3½ hour EVA put up the A-frame shade developed in May at the Marshall Space Flight Center. When adjusted to normal scientific duty from Day 5, the work period would be shaped around sleep sessions spanning 10:00 pm to 6:00 am on the Houston clock.

Shortly after docking with Skylab, Bean, Garriott and Lousma opened the hatches and moved in to their orbiting home. It was 4:15 pm, just ten hours after launch. By 8:00 pm Bean and Garriott were complaining of dizziness and a feeling of nausea. They slowed down, suspecting that the delicate balance mechanism in their middle-ear region had not yet fully adjusted to rapid movement in the weightless state. Shortly before starting the rendezvous maneuvers, Lousma had taken a tablet for motion sickness, a condition to which he was prone, and now he was unable to eat anything and vomiting at irregular intervals.

Shortly before 9:00 pm the crew were allowed to begin their sleep period; it had been a long day. Communicating with Houston via Carnarvon shortly after 6:00 am on 29 July, the crew were up and about believing their rest to have restored normal sensations. But they were wrong. By noon, the effects were welling up again and Houston advised them to get some more rest. And then word was passed up that they should try some deliberate head movements to orientate the vestibular function and prevent nausea. By evening the crew were up again, slowly activating the workshop. A private medical conversation was granted and a decision made to postpone the EVA to Day 5 at the earliest.

Next day, 30 July, the crew got up early to re-close the door on the waste disposal lock situated in the floor of the crew quarters area. Through a tiny crack, atmosphere was venting from the workshop and Mission Control had to open the oxygen/nitrogen fill ports. Back in their cubicles, the crew slept until 8:00 am and then continued bringing the cluster into operation, accelerating their own activities as head movements and rest familiarized their bodies with the new environment. But another medical conversation with doctors in Houston satisfied the ground that their condition was improving, although the EVA was put back a day once more.

31 July saw Bean, Garriott and Lousma complete most of the outstanding activation procedures and perform a troubleshooting job on the condensate tank that seemed to have malfunctioned. No longer taking medication, and jogging round the ring of water tanks, all three men gained ground on the physical condition they would need for full scientific activity. During the evening ground controllers examined detailed plans for the coming few days and agreed to schedule the space walk tentatively for Day 8 (4 August).

Next morning, Day 5, medical experiments began and the second Skylab mission was scientifically under way. But the switch of pace and schedule occasioned by the mild discomfort and illness brought a testy Al Bean into verbal conflict with the controllers during that evening: 'Say, if you got any friends among the flight planners down there, they keep telling us they're going to give us some little spare time to do this job. It never seems to happen. We've been working from Sunrise to right now, and we're still not finished. We haven't even started the Day 5 transfers and we hustled all the time. Tell them to give us a little more pass if they possibly can, because I've looked out that window five minutes in five days. The rest of the time we've been hustling.'

It was a problem similar to that encountered by any set of complex work procedures performed remote from a central authority: head-office always wanted to interfere and the men in the field could never understand why. Except, in space activity ground controllers could always see more of the total condition – men *and* machine – than could the crew themselves. It was vital for Houston to play lead role in monitoring systems and recommending flight plans and it was necessary for morale that the crew should have an on-board commander and that he should have an unchallenged authority on his own ship. The lines of demarcation were shadowed, frequently intruded. But this was what Skylab was all about, finding out how best to run long duration orbital flight operations on a routine basis. It would emerge with almost alarmingly severe consequences on the last Skylab mission.

Day 6 began like any other for Skylab controllers. Flight director Chuck Lewis was on duty with capcom Hank Hartsfield. At 5:30 am, a controller told Lewis that the temperature in Apollo Service Module quad D was falling and that he might have to bring on the secondary heaters. Sixteen minutes later the crew were woken on board Skylab by a warning tone sounding in their headsets; the system had sent a cautionary message of its own problem. Alongside the RCS quad, opposite the one which had a single failed thruster since shortly after launch, an SPS propellant tank was also cooling rapidly, indicating a super-cold fluid was venting into the area occupied by both thrusters and tank. Moreover, the crew could see crystallized propellant drops streaming past the Command Module window. The Service Module had sprung another leak.

Immediately, Chuck Lewis asked the crew to shut off all propellant valves upstream of the quad manifold, and to isolate the flow of helium. Before that could be effected the RCS propellant supply had been depleted by 5.4 kg, indicating a leaking orifice about 0.023 cm in size. Quad isolation had been requested at 7:07 am through the Goldstone tracking station; by 8:15 am the temperatures were coming back to normal, the leak having been stopped. But with two thruster quads inoperable, what to do? The second mission was less than one week into a planned two-month stay and nearly 50% of the attitude control capability had already gone. A management meeting was called for 8:00 am. The crew were safe, but if they got back into the Command Module and separated Apollo from Skylab, would the attitude control capability be sufficient for a safe return to Earth, and would by the end of the mission all other systems operate as planned?

Bill Schneider decided to play safe. Telephoning Kurt Debus at the Kennedy Space Center he requested full activation of the rescue flow, putting the Saturn launch vehicle preparation teams on a three shift per day, seven day per week, routine until the vehicle was on the pad. As it stood, CSM-118 was about to go into the altitude chamber for vacuum tests. Kennedy told Schneider they could have it stacked to the two stages of AS-208 by 10 August, eight days hence, and roll it to pad B by the 13th. A Flight Readiness Test could be conducted eleven days later and by 27 August the Saturn would be ready for fuelling toward a 5 September launch. But there was no hurry, so Schneider deferred a decision whether to accelerate preparation for a normal flight, just in case a rescue was needed, or proceed with the installation of a rescue kit immediately.

Either way, the Cape was responding to the first rescue call it had ever received. A firm decision one way or the other was necessary before the 27th. At Houston, astronauts Brand and Lind began to rehearse the procedures of a rescue mission. If called upon to fly, CSM-118 would be flown by these two men in a modified Command Module equipped to carry five people. At 1:40 pm, the Skylab crew held a conference with the ground about rescue procedures and the plans then being formulated to get the next CSM up to them if their own vehicle was considered inoperative. Johnson Space Center Director Chris Kraft talked with the men and made them fully aware of the situation. More than a decade before, John Glenn had wished he had the same frank dialogue when problems then seemed to threaten his safe return.

For the time being there was nothing for it but to press ahead with scientific activity while on the ground all speed was made to get the hardware prepared in case it was needed. Earlier that day, Houston informed the crew that they had decided to postpone the EVA yet again, moving it on from the planned activity on Day 8, perhaps to Day 10. Next day, 3 August, Bill Schneider decided to have CSM-118 move to altitude chamber tests in a standard mission configuration, believing that it would probably not be needed for rescue. Analysis over the preceding hours showed the ailing CSM-117 to be safe and flyable if it remained in its existing condition. But the launch vehicle and CSM would still move along the maximum paced schedule and receive the rescue kit on the pad if necessary. The 24 hours involved in that activity would postpone the earliest rescue date from September 5 to 10 because of the five day interval between successive launch opportunities due to Skylab's orbit.

In space, Bean, Garriott and Lousma performed the first EREP Earth resources scan of their mission and busied themselves with increasingly time-consuming science. With a week gone, they were behind the mission's list of daily objectives. The space walk was finally set up for Day 10, preceded by a day of preparation, so that the new A-frame shade could be put out and certain ATM film canisters changed. Full science would begin with runs on the ATM from Day 11. Late in the evening of Day 7, however, a major short affected operation of ATM TV bus No. 2 requiring a checkout of the equipment next morning.

Also that day, the crew had to jettison an experiment stuck in the Earth-facing scientific airlock (readers will recall that the other airlock installed in the opposite side of the workshop, supported the parasol shade). Attached to a telescopic arm, a 16-mm camera and photometer for scanning Earth's horizon were irretrievably lost since other experiments would need the same airlock and it was impossible to free it for retraction inside. But there seemed no end to the succession of minor problems. For now the Airlock Module coolant loops – primary and secondary – were failing.

Careful examination of telemetry showed leaks in both loops, with coolanol bleeding from the primary at about 0.033 kg/day and from the secondary at about 0.018 kg/day. The implication here was that the primary would bleed dry within about three weeks, while the secondary loop would last for the remainder of this mission, throughout the unmanned period between visits, and for the first few weeks of the third mission. Clearly, some form of supplementary supply would be needed so engineers at Huntsville set to in efforts designed to come up with additional supplies of coolanol to be carried aboard CSM-118.

Although Day 9 was important for preparing the EVA equipment, and for discussing final plans with Houston, the crew pitched heartedly into two EREP runs over selected sites. The first began at 9:55 am and lasted 14 minutes as the cluster swept over Montana, the scanners tracking the Earth as Skylab moved around. The second pass began on the next revolution and supported an Earth resources fact-finding survey backed up by two NASA aircraft flying along the groundtrack and 127 fishing boats in the Gulf of Mexico keeping records of fish caught and others sighted. It was the type of survey Skylab could take in its stride, one which could build encouraging evidence of the value accrued from space-based resource scans.

Next day, Garriott and Lousma were up in good time for their planned space walk. Before starting the EVA, the parasol was pulled down from the inside against the exterior surface of the workshop hull so as to allow room outside for the A-frame device to overlie the existing shade, now severely discolored due to prolonged exposure. But it was a lengthy procedure and required the crew to assemble, from 24 poles, two sections each 16.7 meters long which would be attached to a common baseplate and projected flat along the workshop, giving the appearance of a horizontal 'V'. Two ropes serving as halyards would then be used to pull the folded shade out along the poles until it was fully open and covering the exposed hull, including the smaller parasol. When fully deployed the shade was 6.8 meters by 7.4 meters in size.

The EVA began at 12:32 pm over the Goldstone tracking station and one hour later all the equipment was outside with the first of the poles being assembled. At 2:30 pm the crew reported through *Vanguard* that they had the first pole ready; one hour later the second was assembled and fixed to the baseplate. By 4:00 pm the crew reported the deep pleats in the shade seemed to be stuck together and that the sheet looked more like a concertina than a flat cover. Within 30 minutes the deployment tasks were complete and the Marshall A-frame shade was already doing its job as controllers began to observe slight temperature decrease inside the workshop.

From communications through Hawaii came a report by Jack Lousma that he could see discoloration along the entire exposed area of the workshop, even the gold covering of the hull now turning a bronzed and burnished hue. High up on the ATM structure, there to change film cassettes on selected telescopes, he could see down on to the cluster. It was a fascinating view, and he too developed an awareness of height just like Kerwin before him. But another minor repair task was necessary to remove two bolts securing a door on the Sun end of the telescope mount normally used selectively to cover and expose the instrument inside. With constant use it tended to stick so Lousma pinned it permanently open. A simple box-wrench did the job.

Then there was a particle collection experiment, put out during an EVA on the first mission, to retrieve and another small container to place on the ATM structure; prevented from using the Sun-facing scientific airlock, tests on prolonged exposure with various materials required space walking astronauts to leave samples attached to sections of the support truss for later retrieval. Before getting back inside Lousma and Garriott looked for signs of leaking coolanol from the primary and secondary Airlock Module loops. They could see nothing. By 7:03 pm, Houston time, the space walk was over, having lasted 6½ hours, far longer than planned. But an improved thermal shade had been satisfactorily laid out and Skylab was back on routine operations.

The crew had done well and were given a lie-in next morning, the day the real work began with major periods of solar observation. Despite the Sun's quiet period, activity was beginning to increase on the solar disc and in the atmospheric corona surrounding it. Completely unexpected, events would develop on the Sun totally uncharacteristic of this period between solar maxima, bringing useful and rewarding work to physicists on the ground working with astronauts in space to obtain hitherto unrecorded data.

At the end of the second week, Skylab responded to a call from the National Oceanic and Atmospheric Administration that a coronal transient had been observed. Training their instruments on the reported burst, the crew tracked this material as it hurled itself from the surface of the Sun, a mass greater than that of Earth itself ejected at a speed six hundred times that of a rifle bullet. Coronal transients were rare in the period of the quiet Sun and it was a good opportunity to obtain scans with the ATM instruments.

But if Skylab was science at the bold end, it was opportunity also for gathering knowledge about lesser events in nature. Carried along as a student experiment promoted by 17 year old Judith Miles, the web-building device on board the space station was periodically photographed to see if a spider, named Arabella, had spun a web and, if so, how it compared with webs on Earth. The second week of this second mission revealed a fascinating response. Shaken into the transparent box on Day 9, Arabella spun a satisfactory web four days later. But when Owen Garriott partially opened the door and broke one of the web longerons, collapsing the structure in the process, Arabella set about building another web, only this time fixing the longerons to rigid sections of the box so that if disturbed again movement of the door would not destroy the web. Later in the mission, the second spider called Anita similarly spun a good web indicating that whatever reference or stored logic was used to construct the gossamer structures, gravity was not an important element. Only Arabella showed some initial disorientation.

And so it went on throughout the long eight weeks of this second mission: Earth observation scans, sustained investigation of solar phenomena, a host of secondary experiments, and the student investigations supporting youthful interest in science. Major hurricanes were tracked by Skylab and environmentalists got photographs of ecosystems and urban distribution. Hydrologists kept the crew busy photographing and scanning coastal features, water run-off patterns from hills and mountains, and measurement of chlorophyll distribution in the sea. Oceanographers brought new tasks to Skylab even as the mission progressed, asking the crew to study unusual sea states, reported tsunami (large waves), and pollution patterns. Land managers meanwhile were concerned to have the EREP equipment monitor spread, distribution and migration of insect populations as they advanced upon growing crops.

In one remarkable response to a frighteningly unpredictable event, the Bean crew was assigned special scans of the region immediately south of the Sahara desert in central Africa. Mali tribesmen, women and children were dying from malnutrition and thirst as an unusually severe drought carried the arid desert farther south than ever before. With its scanners, Skylab could determine where sub-surface water deposits were likely to be found and where vegetation indicated good soil. Processed in haste immediately after the mission, this information sent aid teams rushing to the area to direct the tribesmen to water and new settlements. Countless thousands were saved in this way.

In other areas, agriculturalists built new crop inventories and studied the results of new land management schemes on food growth. In all, 33 countries outside the United States were studied for one or other of these Earth resource tasks. Studies were made of geographical conditions as different as the lush orchards of California and the tropical jungles of Paraguay. Although Skylab's orbit carried it over much of the Communist world, taking the station as far north as Krakow in Poland and Kiev in the Soviet Union, care was exercised in the selection of suitable ground sites for this experimental work.

Avoiding sensitive areas that could bring accusation of spying from space on an ostensibly civilian mission, the EREP planners worked night and day to process all the requests for data and all the proposed surface sites into coordinated observation scans. It took real energy from the Skylab systems to break lock on the Sun and track the Earth's surface and the maximum use of such maneuvers was of vital concern to the controllers.

By the third week, ground controllers had sorted the optimum schedules for shifts and work hours. With five teams rotating in the MOCR, each week would have one team performing five day shifts between 7:00 am and 6:00 pm to midnight, with three days off. A third team running five consecutive graveyard shifts between midnight and 7:00 am would take four days off before resuming the cycle. But, whenever major problems developed, the five rotating teams were invariably reduced to three, with the other two pulled out of rotation for non-stop troubleshooting. With overlaps and de-briefing periods it often meant a team being on duty for more than 12 hours at a time, day after day.

Inevitably, health and home life suffered and for those hectic months of 1973 a nucleus of personnel at Mission Control resurrected the nightmarish days of the middle and late 60s when divorce rates escalated and breakdowns became an acceptable hazard.

A major systems problem arose on Day 15 when the control-moment gyroscopes were caged up and all attitude control was relegated to the TACS thrusters because of a complex set of situations that evolved during an EREP pass. The rate gyroscopes gave further trouble a day later and it became increasingly apparent that the 'six-pack' would probably be needed. Several times, beginning on Day 17, the crew tested the Automatically Stabilized Maneuvering Unit (ASMU) developed from the Gemini days when Air Force plans to fly their back-pack propulsion device were aborted. Equipped with tanks of nitrogen to prevent contamination of the Skylab atmosphere, the ASMU had 14 thrusters by which to control attitude.

Al Bean was the first to fly it around the spacious workshop dome, effectively proving that such a device would allow space walking astronauts to move from one vehicle to another or station themselves at one location for work or inspection purposes. Such units were being considered for the Shuttle when orbital flight would approach more routine levels of activity.

By this time decisions were in the offing about not only the end of this second mission but also the schedule for the third and last habitation. Bean, Garriott and Lousma were scheduled to return 25 September. Managers earlier intended to fly the third mission as soon as possible after that date, wise policy in the light of systems problems that continually plagued Skylab.

But other options emerged. SL-4 could be sent aloft on 24 September and dock to the radial port, leaving Bean, Garriott and Lousma to return in their own Apollo one day later as planned, or the third crew could be launched before the Bean crew separated from Skylab, station-keep while the second crew departed whereupon they would immediately go in and dock. Or, the third crew could go up after a period of unmanned activity as originally

Built by Martin Marietta, the Astronaut Maneuvering Unit gets a workout in space by Jack Lousma during the second Skylab mission.

planned. The only advantage in an early launch, apart from getting the third visit over before more trouble hit Skylab, was to prevent the lengthy deactivation/activation sequences. In several respects a decision hinged on whether AS-208 would be needed in the rescue role, for, if it was, AS-209 would have to be taken to the pad and that would cause delay.

Bill Schneider was concerned to get this question answered and put concerted effort into finding out what had caused the leaky thrusters. Only one thruster was leaking in quad B, and that was determined to be a partially unseated oxidizer valve. As for quad D, the one that started leaking on 3 August, tests on the ground closely simulating the record of events seen on telemetry, revealed the possibility of human error which, if correct, minimized the impact on other systems and reduced the probability of serious trouble elsewhere. The section of plumbing thought responsible for the leak was joined by a screw fitting which should have been tightened by torque wrench. Tests showed that if the fitting was screwed together by hand alone, the inner face of these dynatube fittings would slowly leak propellant up the metal-to-metal join and erode a rubber ring seal buried in the screw. Records showed that quad D had a bad record and that continual replacement of several parts during test may have left this one join only hand tight. It was a situation that fitted the observations and Schneider decided to accept that explanation for what had happened.

In the dark before dawn on Tuesday, 14 August, Bendix crawler driver K. D. Kelly swung up into the giant cab beneath the mobile launch platform and chugged through the open doors of the Vehicle Assembly Building with AS-208 on top. Before the end of that day it would be hard down on pad B, accelerated by the rescue schedule called less than two weeks before.

During the afternoon, Schneider approved the decision that relegated AS-208 to the normal SL-4 mission role. Preparations would go ahead at maximum pace, but stop on 9 September for a decision about the rescue need. Hypergolic loading on that date would have 208 off the pad six days later on its way with Brand and Lind to rescue the Bean team – but only if things got worse with the docked Apollo in space. As it stood on 14 August, the vehicle would be held for a normal SL-4 launch. So when should that third visit take place?

It was no longer a question based wholly on engineering considerations for the state of Skylab systems or the life of ailing components. For yet again there were strong arguments from both engineering *and* scientific camps. On 7 March, while attempting to photograph an asteroid within the solar system, the Czech astronomer Lubos Kohoutek had discovered what he believed to be a new comet heading for a close pass of the Sun. From his post at the Hamburg Observatory, Dr. Kohoutek sent word to the international astronomical fraternity and several people were able to observe the object up to early May, shortly before the launch of Skylab.

Then the relative positions of the comet and the Earth were such that observation was impossible until it came within view once again and was 're-discovered' by world famous Japanese comet expert Skeya-Seki. From the US Central Bureau for Astronomical Telegrams, Director Dr. Brian Marsden worked long and hard to discover the precise orbit of what was now called Comet Kohoutek, finding it would pass closest to the Sun 28 December that year. With instruments unique to the space sciences ready and available on the Skylab space station, astronomers recruited NASA scientists in the campaign to get SL-4, the third manned visit, operational during the period when the comet could best be seen from space.

So much interference through the atmosphere was present in recordings and observations from the ground that the opportunity to have this complex and sophisticated manned observatory track and observe the comet was invaluable for science – and public interest! But if SL-4 was launched immediately after SL-3, the second manned visit, its duration would bring recovery one month before the comet reached its closest approach to the Sun – perihelion. Because pressure from solar radiation causes condensed gases to stream away from a comet's nucleus, giving it the characteristic tail, observation at this critical period was important for the scientific results.

The third manned visit would have to be spaceborne during Comet Kohoutek's perihelion and that meant putting it back to a launch during November at the earliest. Bill Schneider supported the consensus: place SL-4 on the original 9 November launch date. That would carry it through to 4 January, 1974, and allow Carr, Gibson and Pogue to study the comet with appropriate instruments. Moreover, during the tele-conference held between NASA field Centers and management personnel 16 August, when the final decision was taken to head for a 9 November launch, it was agreed that observation of the comet would benefit from use of a special ultraviolet camera which it was believed could be made ready for flight aboard CSM-118.

Only three days earlier, NASA Administrator Jim Fletcher concurred with a recommendation from manned flight boss Dale Myers that the Skylab workshop and cluster back-up hardware should be stood down; it had been provided in the event the Skylab itself became uninhabitable or was destroyed during launch. Now the back-up was almost irretrievably lost where once enthusiasts at NASA thought it could be employed for a second Skylab operation. Several contractors too had supported what became known as the Skylab B proposal but lack of government work forced companies to lay off key workers earlier than expected and the additional cost of holding this hardware in an advanced state of readiness forced NASA's hand. Whatever happened to the orbiting cluster it was the only station now likely to fly in the Skylab program.

By the fourth week in space, Bean, Garriott and Lousma showed evidence that they were getting into their stride. The problem of disorientation experienced early in the flight was now well behind them, the slow start to experiment activity gave way to accelerated timelines, and the condition of the systems appeared to have stabilized. On 16 August the two Apollo fuel cells were shut down; from now on the single workshop solar array, and the four ATM arrays, would provide all the electrical power and feed CSM-117 too.

Two days later the crew responded to a call from the National Oceanic and Atmospheric Administration, photographing a tropical depression that had now developed into a storm called Brenda moving across the Caribbean toward land. NOAA's hurricane centre reported Brenda would increase in fury, calling up aircraft, weather satellites, and the Skylab instruments to keep track and monitor progress. Next day, Brenda was a priority and brought a quip from Jack Lousma: 'If you know anybody named Brenda now's your chance to tease them about their temperament.' But other Earth scans were called for as well. After taking a look at the tropical storm, the crew photographed the Straits of Magellan. In the afternoon, they hunted for pumice fields below the sea, and then scanned the Laccadive Islands near India before photographing the Antipodes.

A day later, the crew obtained views of strange lines etched in the surface at the Plains of Nazca, an ancient Peruvian feature thought by some to have been laid with the aid of extraterrestrial beings. Formations in the Great Barrier Reef were observed and active volcanoes in a depression on the central part of the southern tip of North Island, New Zealand, while the crew obtained data for rain land measurements over Australia. And then, as Brenda swelled to hurricane force, solar physicists got a bonus in the form of a transient event that took Al Bean by surprise.

The daily log of scientific accomplishments were improving. By now, the scheduled man-hour rates for ATM telescope work were being exceeded and the crew asked for more experiment time each day. The second phase of Skylab orientation was under way.

The first flight had been unable to set formal patterns of familiarization, work orientation, or psychological response, not only because of the many engineering tasks required but because the Commander, Pete Conrad, was a test-pilot's pilot capable of inspiring his men to highly motivated reactions. He was good for morale, an effective troubleshooter more at home with pulling solar array booms out than setting up regular, and lengthy, sessions for scientific observation.

Al Bean was quite the opposite, geared to planning ahead and setting up a motivated team inspired to exceed new goals, set others, beat those and come back with new challenges. The first period of this second mission was, like any other, defined by the crew settling in and establishing their own work rates, their own ways of doing things rehearsed on the ground but refined in orbit, during which they passed through the testy phase of resenting interference from the ground. The second phase built a learning curve through which the crew could individually set their own sights on specific challenges. The third phase would quickly follow, demonstrating maximum and most efficient activity where the ground, instead of shaping the daily routine, would respond to a demonstrated pace observed in work performance and crew morale.

Hurricane Ellen displays its swirling clouds in this view from Skylab taken 20 September, 1973.

Motivation reached a peak on Day 22, the climax of the crew's second phase, when all three worked for sixteen hours. But it was not entirely devoted to science, for it was now imperative to check the leak in the primary Airlock Module coolant loop. Next morning, Al Bean checked the condensate system by pressurizing it with nitrogen gas and listening for leaks with a stethoscope. He found none, but that system too was certainly leaking. The primary coolant loop was now down to 703 g/cm^2; it would fail at 351 g/cm^2. Consequently, the loop was deliberately shut down on Day 27 (23 August). From now on the secondary loop would have to carry the load but plans were already emerging in Huntsville for a supplementary supply of coolanol which engineers thought the third crew might be able to inject into the primary loop via a saddle valve.

Next day, 24 August, Al Bean positioned himself in the Command Module while Garriott and Lousma went outside to retrieve and replace film cassettes up on the ATM structure, fix two more telescope doors that were sticking, and plug in the new gyroscope package installed in the Multiple Docking Adapter; Bean would take over attitude control with Apollo should anything have gone wrong while Skylab was drifting. The EVA began at 11:24 am, Houston time, and lasted $4\frac{1}{2}$ hours, during which time all the scheduled tasks were completed. Lousma was reluctant to get back inside; it was an exhilarating experience floating free more than 400 km above the planet. But now the ailing gyroscopes would have good back-ups from the six-pack, increasing confidence that Skylab would survive.

Next day, the Bean crew passed the record set up by Pete Conrad's men as they moved into their fifth week in orbit. But life aboard Skylab was not totally devoid of its associations with normality on Earth. Early in the morning of Sunday, 26 August, Jack Lousma asked that Mission Control convey his thoughts to brother Don on this his birthday and to the boys at his Sunday School class, Harris County Youth Village, in addition to the Clear Lake Community Church where he worshipped. Each 'evening' before sleep, Lousma spent a brief period reading some part of the Bible, usually Psalms or Proverbs, just as he did at home.

In the evening, Commander Al Bean proved his mission had progressed to the third phase when, during a lengthy conference on the communications loop at refining some procedures, he reaffirmed his desire to have the crew work as hard as the ground could arrange it despite offers from capcom Bob Parker for an easier time in the weeks ahead. Bean had the measure of his men, and they had the measure of the tasks to be performed.

But the Command Module was leaking water-glycol through a small crack and Bean was advised by the ground to stuff used clothing into that area to soak up the liquid. It was such a small amount, not really a threat to the mission. But it was proof positive that no two days aboard the spacious cluster were quite alike! So enthusiastic was Bean's role in the Skylab program that he asked Houston for an extension beyond the scheduled 59 days. It was politely turned down.

The sixth week brought potential threats to the Houston control center while up above the space station had a successful time with science. Hurricanes were advancing upon the Texas coast and flight personnel began to think of moving to the back-up control facility at the Goddard Space Flight Center, Maryland. The storms abated and the MOCR men remained *in situ*. From orbit, Skylab undertook a mapping exercise, previously arranged with the Paraguayan government, whereby in a series of six runs over that country the crew provided sufficient data for a survey to be completed within the next three years; without Skylab, it would have taken 20 years to equal the task.

And then Owen Garriott was put to work flying the maneuvering unit around Skylab's interior. It was the first time an astronaut had ever carried out a scientific exercise without any form of prior training, but it was a good way of seeing just how well an astronaut coped with unfamiliar duties; after all, if used for rescue in the future a crewmember could well have to fly such a device between spacecraft.

August gave way to September with Skylab entering its sixth week in space; for the first time, Earthlings had spent the duration of a single calendar month in orbit. By 9 September, the deadline set by Kennedy Space Center engineers for preparing

Science-Astronaut Owen Garriott in the Lower Body Negative Pressure Device designed to evaluate the cardiovascular conditioning of an astronaut's body in space.

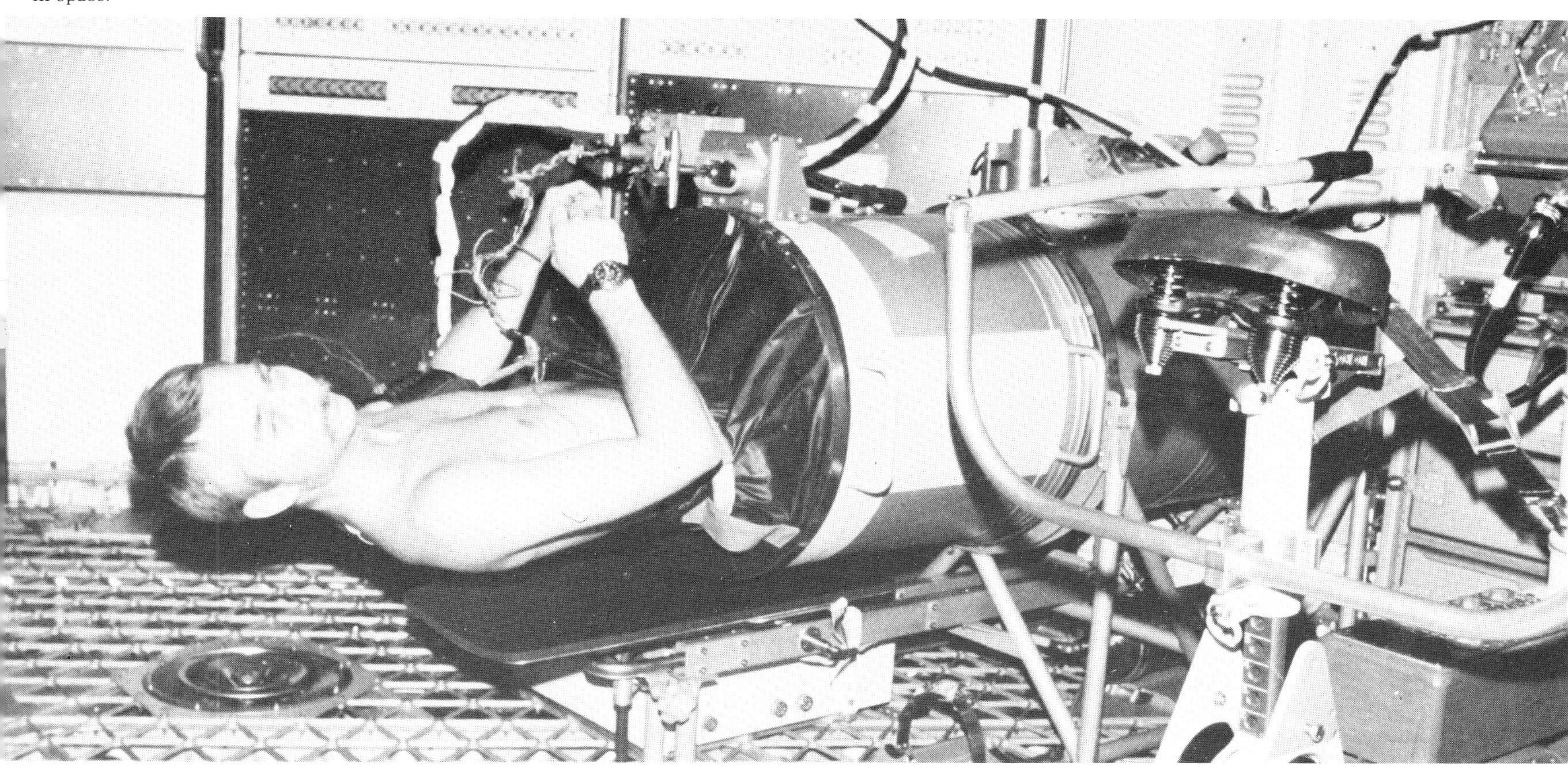

SL-4 as a rescue mission or reverting to a holding posture for the normal 56 day flight, it was apparent that CSM-117 had not deteriorated and that it could be used at the end of their stay to bring Bean, Garriott and Lousma back to Earth. So KSC held AS-207 on the launch pad, ready nevertheless to execute a rescue flight should the need arise.

The next two weeks went very well. All three crewmembers maintained a consistently high performance far above that predicted at the beginning. Despite a late start, total mission allocations for different experimental tasks were already being met and exceeded. What Cernan, Evans and Schmitt did for lunar exploration, Bean, Garriott and Lousma were doing for space station activity. Their performance was impeccable, and even the systems appeared to rally for the second half of this mission; it was remarkably free of the several minor, often irritating, problems that beset other periods.

Re-entry would come during the late afternoon, Houston time, so there was little need for substantial change in the diurnal cycle that regulates all humans. Nevertheless, on Day 51, 16 September, the crew moved to a seven and one-half hour night beginning 8:30 pm, and on the following night scheduled a seven hour sleep from 8:45 pm. For the next six nights they adopted an eight hour session commencing 6:00 pm each evening.

All three men were performing beyond goals set pre-flight. The 295 hours scheduled for ATM observations of the Sun, an aim rather than an expectation, was exceeded on Day 40. The 156 hours targeted for EREP activity was passed on Day 47. And the 276 hours for medical experiments was reached on Day 54. As for corollary experiments, the 21 activities falling into unique and unassigned categories, the 146 hours planned for those was passed on Day 35. By the end of the mission, motivation for the work at hand shone through the statistical records.

Despite an unscheduled 42 hours spent repairing systems or troubleshooting problems aboard the space station, the crew cut their accumulated sleep time by 11%, cut 72% out of the built-in off-duty time, reduced their meal times by 15%, and raised the time schedules for physical exercise. Despite that, ATM, medical, and EREP experiment times were up 54%, 18%, and 43% on the pre-flight goals respectively. By far exceeding, in this very tangible way, every objective set for the second manned Skylab mission, Bean, Garriott, and Lousma proved to be the most productive crew ever flown into space.

In the final week they performed a last space walk, Bean and Garriott moving through the Airlock Module door at 6:18 am on Day 57, 22 September, to remove ATM film cassettes for four telescopes and replace two, to retrieve particle collectors and strips of material put out on the previous EVA, and to clean an occulting disc over the white-light coronograph. The operation lasted more than 2½ hours. Over the next two days the crew packed everything aboard Apollo destined for return to Earth and deactivated the cluster. With the last manned visit scheduled to reach Skylab on 9 November, the station would be unmanned for 45 days.

Up and working by 2:00 am, 25 September, Bean, Garriott and Lousma powered up the CSM and made ready for descent. Careful selection of appropriate switches ensured the crew would safely orientate their spacecraft without using the crippled thrusters. By 9:30 am all the hatches had been secured and the record breaking spacemen were ready for undocking and a small separation burn. At 2:50 pm the Apollo spacecraft slipped away from Skylab's docking cone but because of the inhibited RCS thrusters the crew did not perform the scheduled flyaround and exterior inspection maneuver. Also, the normal process of phasing the re-entry by first putting Apollo into an elliptical shaping orbit was abandoned in favor of firing a single SPS de-orbit burn with the manned spacecraft about 1.8 km ahead of and below the Skylab cluster.

With a call of 'We're looking forward to seeing you guys again, it's been great working with you,' Mission Control gave the crew a 'go' for re-entry. Fired at 4:38 pm, Houston time, the SPS engine burned Apollo out of orbit and back to the Pacific waters south west of San Diego, concluding a flight that had lasted nearly 59½ days – more than twice the length of any other space mission, fully accomplishing the basic objectives originally worked up in the mid-1960s. Probably because of improved conditioning and more prolonged sessions with exercise equipment, Bean, Garriott and Lousma were, if anything, in better condition than their predecessors. There were no outwardly visible long-term effects of their two month stay in the weightless environment and from first examinations on board the recovery ship *New Orleans* it was apparent that man himself was an adaptive creature capable of accepting new environmental challenges of this unique nature.

Very little planned activity had been left out of the second manned visit, although the scheduled trim burns, three separate thruster firings on Days 5, 31 and 53, were cancelled due to the leaks. Because of that, Skylab's orbit had shifted 3.15° east of the groundtrack it would have had with the necessary trim burns. SL-4 would hopefully adjust the path for correct EREP scans over previously targeted sites. As for cluster systems, consumables were in very good order.

Expeditious use of the nitrogen gas left a strong reserve totalling 37% of the quantity available at launch despite high use before the first manned visit. Oxygen and nitrogen were in better supply than predicted for this stage in the program, the cluster itself being less prone to atmospheric leaks than the specification allowed. Water too was in good supply, with more than half the original quantity remaining.

Deployment of the improved shade reduced temperatures from a normal 26°c during the first mission's latter stages to between 20.5°c and 25°c for the second visit. The molecular sieves too were holding up as planned, with carbon dioxide levels kept below a maximum 5.8 mmHg. Although the condensate system had played up, procedures worked out on board restored the function to normal. Based largely on this awareness of Skylab's performance, and a dramatic reduction in the systems failure rate, Bill Schneider conferred with his managers about the prospects for SL-4, the last manned habitation.

Approved on 16 August for launch on 9 November, final computations based on lengthy tracking of the Skylab cluster refined the precise time of the launch window, moving it on a day to 10 November at 11:41 am Cape time (but not before an earlier prediction of 11:04 am, 11 November); minor perturbations in the Skylab orbit would cause amplified changes months later, hence the uncertainty factor reduced to nil only a few weeks prior to launch. As to duration, pressure from scientists wanting to get the maximum observation time on Dr. Kohoutek's comet and from physicians wanting to wring more medical data from the program, influenced Bill Schneider to stretch the program beyond the original 56 days planned for this last visit.

By early October, two weeks after Bean, Garriott and Lousma came home, stocktaking revealed the possibility for food and consumables to support a Carr, Gibson and Pogue flight lasting 70 days – two weeks longer than anybody had seriously considered before. Then, when it was realized that only diminished food reserves called for a return even at the 70 day stage, Schneider tentatively approved plans to carry concentrated foods up in CSM-118 for an additional two weeks on top of the 70-day proposal, stretching this last visit to a maximum 84 days.

A final decision would await detailed examination of the crew's condition after two months in space, with extensions on a weekly basis until they satisfied the full potential of three months continuous operation; physicians were unwilling to go beyond this without interim examinations on the ground, and systems engineers were not sure the cluster could support operations beyond that time. Predicted deterioration and depleted oxygen, nitrogen, and water would follow quickly after mid-February, 1974. And so it was agreed. As launched, SL-4 would aim for a 56 or 59 day stay, extending after that only as the mission progressed. But it gave new objectives for the Carr, Gibson and Pogue team, the first all-rookie crew for seven and one-half years.

Nutritionists went to work on special high-energy food bars to supplement the on-board stocks, an off-shoot of food already developed by scientists from NASA, the Air Force, and the Pillsbury company. Weighing 55 grams each, the bars contained 300 calories and were to be eaten every third day, with a reduced intake of conventional Skylab foods, thus ensuring sufficient food for 85 days plus a 10 day reserve for emergency.

Elsewhere, engineers worked on a coolanol reservicing kit with which to pump additional fluid into the primary Airlock Module coolant loop, bringing it back on line to supplement the secondary system in constant use since 23 August. The kit was delivered to the Kennedy Space Center just five days prior to the scheduled 10 November launch date.

But things were not going smoothly at the Cape. There were two sets of hardware to get ready. Without a follow-on manned mission, AS-208 would call for a special rescue back-up in the form of AS-209, a Saturn IB which was to be prepared for flight, rolled to the pad, and readied for launch by late December. At completion of the third manned visit, AS-209 would be rolled back to the Vehicle Assembly Building and de-stacked for storage. The prime launch vehicle for SL-4 was already at pad B, having been taken there 14 August when it seemed likely to provide a rescue service.

On 15 October KSC management reviewed the status of the launch complex and over 17 and 18 October Schneider hosted a combined flight readiness review. Yet long before that, just one day after the SL-4 stack rolled to pad B, AS-208 was afflicted with the gremlins that surely haunt the historic launch sites at KSC.

On 15 August the mobile launcher was struck by lightning and damaged electronic equipment was replaced in the Apollo spacecraft. Just twelve days later cracks were discovered in an 'E' beam fixture situated at the top of the first stage. There was concern about these load bearing structures and the possible consequence of stress corrosion in other areas of the launcher not easily accessible. The stages had been built in the mid-1960s and were likely to suffer structural fatigue. Metal was removed either side of each crack and additional welds applied to strengthen the beam. Completed by 3 September, the work led to inspection of more than 2,000 other items in the launcher; none were found to have corroded.

By mid-October weather threatened the integrity of the launcher on its exposed pad and director Walter Kapryan prepared plans for quickly moving the rocket back to the Vehicle Assembly Building. The storm, code-named Gilda, blew by and the preparations continued. But flaps placed over exposed relief valves to protect them from heavy rain days later caused a potentially serious problem during fuelling operations on 23 October.

As kerosene flowed into the fuel tanks two domes buckled inward because the flaps, sucked down on to the valve discs, prevented the pressure balance being maintained. The reversed domes were re-formed by pressurizing the interior and gently blowing them out like balloons. Careful checks showed that although visibly dimpled the metal would not fail during the launch and tests with dye penetrant and x-ray proved it was acceptable for flight.

Because of the comparatively tight schedule, AS-208's count-down-demonstration-test was backed up to the actual countdown for ascent. Held during the week ending Friday, 2 November, the CDDT was completed by the weekend whereupon the propellant tanks were drained ready for the final countdown to commence the following Thursday, two days before launch. Carr, Gibson and Pogue were scheduled to arrive at the Cape by Wednesday evening, the day a special Manned Space Flight Management Council meeting would convene at KSC. But routine inspection of mechanical parts revealed suspicious looking cracks near the aft attachment fittings at the base of the first stage.

Breaking in to a management session at the administrative block, a telephone call was made to Pat Young. It was 3:00 pm, Tuesday, 6 November. The news was bad. There was only one decision that had to be made. Launch operations would wait until all eight first-stage fins had been removed and replaced. It was a new level of attention for a Saturn residing on the launch pad. Never before had such an operation been conducted away from engineering facilities. By the early hours of the following day a new launch window had been computed. AS-208 would head for a lift-off five days late at 9:03 am, 15 November.

Meticulous inspection of all eight fins showed two had one crack each and six fins had two cracks apiece. But the longest was only 3.8 cm although other cracks had been found on AS-209 some time earlier. First, special procedures had to be worked out so that no other elements of the launch vehicle would be damaged by the unusual work now required. With the stacked booster resting on seven fin supports, the eighth fin would be replaced, load applied through that leg and the next off-loaded and replaced. Saturn IB was the only launcher then in use that carried pad supports through the fin beams.

Wednesday was spent developing procedures, draining the propellant tanks and bringing lifting equipment up to remove each 220 kg fin assembly. At 2:33 pm on Thursday, work began on the first fin and load tests at the Michoud Assembly Facility proved that replacements were fully up to the work required during launch. On through the weekend the engineers and technicians worked to replace all eight fins. But Monday brought more bad news.

At 11:30 am, engineers scrambling into every conceivable nook and cranny hunting for more cracks had worked their way up to the aft interstage area between the first and second stages. There, they discovered what they hoped not to find in seven of the eight reaction beams. Bill Schneider issued a formal announcement several hours later that in light of the new crack discoveries SL-4 would slip to a projected launch attempt of 16 November. But because the beams were in compression and not tension there seemed little reason for working on them. Stress analysis showed a safety factor of 1.5 versus the mandatory 1.4, in other words the structure could safely withstand one and one-half times the highest calculated load during flight.

Meanwhile, fin replacement had picked up pace. After the first four had been changed the rest were removed in pairs and the last was replaced by 7:04 am, Tuesday, 13 November. The final countdown was scheduled to commence a mere 19½ hours later. The tasks conducted during each fin change required exposed paint on the main body of the stage to be removed by nitric acid and pitting or roughness to be taken out by buffing prior to a special treatment with zinc chromate primer. Reinforcing blocks were placed about the mounting bolts at each location to ensure an adequate load path should cracks reappear after installation.

As for the cracked fins, they were returned to the Marshall Center where extensive tests showed them to be fully capable of withstanding the launch environment without replacement. They were, in fact, accepting a greater load by supporting the vehicle on its stand than they would be asked to endure in flight. For that reason it was technically unnecessary to replace the fins but to have them in place was unthinkable in the light of much public concern; coming as it did in a crucial stage of flight preparations, full exposure by the press forced NASA to deal with the problem to the satisfaction of a lay public. In fact, stress corrosion had always been a problem and Apollo was plagued with many such incidents where exotic alloys were exposed to coastal weather for months on end. By the early 1970s this problem alone prompted special reports analyzing its impact on cost and managerial deficits.

At 2:30 am, 14 November, the countdown for SL-4 began as Carr, Gibson and Pogue attended final briefings, rested, worked out in the gymnasium, or simply re-read flight documents. As with all Skylab crewmembers, the men scheduled to fly this historic mission were subjected to a quarantine from three weeks before the scheduled launch date but on the day before flight they took their wives on a tour of the Cape facilities. Not for nearly three months would they see them again.

Technicians complete installation of a replacement fin from the first stage of the Saturn IB erected to launch the last Skylab crew.

It was a heavy spacecraft because CSM-118 carried several new packages: 72 kg of extra food, including high-energy food bars, a new far ultraviolet electronographic camera for shooting Comet Kohoutek, two additional film magazines, a special Airlock Module coolant loop repair kit, and a unique and fascinating experiment from the Department of Agriculture. Scientists were interested to see if, as theory predicted, 500 gypsy moth eggs in a special container would hibernate quicker than the usual period of 165 days. If so, sterilization of the male insects could lead to population control and a consequent reduction in defoliated areas much used for food growth.

With just a couple of days to go before launch, the Manned Space Flight Management Council met and considered Schneider's proposal that SL-4 should aim for a 59 day stay aboard the orbiting Skylab, extended in weekly reviews to a maximum 84 days but only if everything went well and on the understanding that any deterioration in the condition of the men or the station would result in an immediate return to Earth. The physicians really had the last say, and they agreed that it was well within the capability of all three men to try for three months' continuous flight.

They would do more exercise than earlier crews. The first mission, SL-2, had assigned one half-hour each day to physical conditioning, SL-3 had baselined one full hour, but this last crew would exercise for one and one-half hours each day. As for the orbiting cluster, that had been doing the rounds of planet Earth happily intact for more than seven weeks and the excellent manner in which it was holding up contributed in no small way to Council approval for a three-month try.

The day Bean, Garriott and Lousma came back to Earth, controllers at Houston depressurized the interior to just 103 mmHg and immediately repressurized it with pure nitrogen to 258 mmHg, the most efficient means available for cooling the six-pack of supplementary gyroscopes installed in the Multiple Docking Adapter. With all flow shut off, the interior slowly leaked to space, dropping to 210 mmHg by 24 October on which date it was pressurized back to 232 mmHg. By 14 November it was back down to 195 mmHg and on that day the vent valves were opened and it was evacuated all the way down to 37 mmHg. Several hours later it was re-pressurized with the appropriate mixture of oxygen and nitrogen to 258 mmHg, ready for Carr, Gibson and Pogue.

Two weeks earlier, on 3 November, one of the three control-moment gyroscopes, CMG-1, showed fluctuating wheel speed as the current flowing in the motor winding increased slightly. Within one hour the wheel was back to normal, but it was a prelude to complete failure early in the third manned mission. Apart from a small problem afflicting the ATM pointing control mechanism, everything else was working well and the weeks of unmanned flight were used for automatic solar observations with the ATM telescopes.

In Houston, flight director Phil Shaffer was monitoring Apollo operations while a second team under Don Puddy kept the workshop ready. Lift-off came at the precisely calculated time of twenty-three seconds past 9:01 am, Friday, 16 November. For all their inexperience, Carr, Gibson and Pogue were a cool trio, exhibiting heart rates between 98 and 111 during lift-off. In Houston it was 8:01 am and by 3:19 pm the crew had Skylab in sight. 'She looks pretty as a picture,' called Carr as they performed the final calculations for rendezvous, pulling alongside the big station ten minutes later.

But the docking operation lasted a full 21 minutes beyond the planned time, for it took three goes to snag the capture latches on Skylab's drogue due to slow approach and a sluggish probe. Then, at 4:02 pm, they were hard docked and ready for a long, long stay. Yet it was a bad start for all that. Pogue felt nauseated during the rendezvous and took a drug to settle himself, the physicians telling him to remain quiet and not move about too quickly; careful analysis of the malaise felt by Bean, Garriott and Lousma seemed to reveal a susceptibility for disorientation in large spaces, at least until the vestibular function adjusted to weightlessness, a condition not observed in comparatively small vehicles.

Shortly after 9:00 pm the crew settled down for their first night in space, remaining in the Command Module. Next day they quietly went about their duties, opening up Skylab by 8:30 am and moving inside to find a friendly message on the teleprinter: 'Gerry, Ed and Bill, welcome aboard the space station Skylab. Hope you enjoy your stay. We're looking forward to several months of interesting and productive work. Signed, Flight Control.' And then, as they moved down to inspect the crew quarter area in the workshop they found three stowaways. For there, on the bicycle ergometer, on the lower-body-negative-pressure device, and poised uncomfortably on the toilet seat in the waste management section, were three stuffed space suits looking for all the world like extra crewmembers!

But humor soon paled in the wake of a serious miscalculation on the part of the crew. For during the previous evening Bill Pogue had vomited as a result of the nausea that physicians on the ground understood to be discomforting at most. Instead of reporting this, the crew quietly discussed among themselves the likely repercussions of such an illness and unanimously elected to keep quiet, narrowly rejecting a proposal that they discard the vomit and maintain total silence even after the mission. They reminded themselves that some people on the ground would be happier if they kept quiet and prevented an overcautionary move by physicians.

The real problem from all this came at about 6:20 pm Saturday evening when Skylab came within range of Goldstone. 'And we'll be dumping the data voice tape recorder here,' came a call from the capsule communicator. What nobody knew at the time was that the crew had inadvertently left the recorder running during the entire conversation about covering up the fit of vomitting.

When controllers ran it through there were red faces in plenty. For it told of certain ground personnel who privately opposed the priority given to medical opinion on Skylab operations. Through Hawaii on the next revolution, Al Shepard had a word for the crew. 'We think you made a fairly serious error in judgement here in not letting us know the report of your condition,' said Shepard firmly. 'We're on the ground to try to help you along and we hope that you'll let us know if you have any problems up there again as soon as they happen.'

At least this table was turned right around, for now the *spacecraft* was not letting the *ground* know everything that was going on! 'Okay Al. I agree with you, it was a dumb decision,' said Carr. By this time Carr too was feeling slightly nauseated although Pogue was a little better. Both would be completely recovered within two days and Gibson escaped altogether.

Activation and checkout followed over the next few days, although it would be Day 6 before the crew could address major scientific research work. Many items they found to be in places different to those they had expected, and several times they felt the pressure to move on was pushing them beyond a reasonable pace. From the outset, it was clear the men would have difficulty setting up their duty cycles and would find problems with proposed flight schedules. They were tetchy at times with ground controllers and seemed to resent the quantity of activities loaded on their timelines.

As the crew slid into routine experiments with the ATM, EREP surveys, and materials processing tasks, they encountered frequent duties for which they had little training. In the rush to maximize this last manned visit, engineers and scientists had crowded the crew with an unusually full inventory of work. Noticeably slower than the previous crew in their first week, Carr, Gibson and Pogue avoided the severe bout of sickness experienced by the Bean team.

An early assignment was to partially re-fill the primary Airlock Module coolant loop by stripping insulation from a suitable pipe and clamping a Y-section saddle valve over the exposed area. A screw, tightened down on the piping, penetrated the loop, opening a path for 19 kg of coolanol to flow in through the orifice. Pressurized by a 18 meter long pipe to a nitrogen gas manifold elsewhere, the method proved effective. The secondary loop would continue to be used until it failed, but the primary would be employed for EVA cooling. The job was carried out on Day 4 to the relief of systems controllers at Houston. Again, human hands had restored the cluster to more normal operations.

Postponed three days because of health and systems activation schedules, the first space walk of this third mission began at 11:42 am on Day 7 when astronauts Gibson and Pogue moved through the door of the Airlock Module. During preparations the crew found mildew on the liquid-cooled garments but Houston advised them to wipe them clean and hang them up to dry out thoroughly in a warm part of the workshop.

During the 6½ hour space walk they accomplished a multitude of tasks, repairing an antenna on the radiometer/scatterometer/altimeter instrument underneath the Docking Adapter, installing film in four ATM telescopes, pinning open a sticking aperture door on an H-Alpha telescope, deploying panels for the thermal coatings, particle collection, cosmic ray, and magnetospheric collection experiments, and obtaining photographs of the Earth's horizon at Sunset for a coronagraph contamination experiment.

The most strenuous task, that of repairing the EREP antenna, involved Gibson working his way through the maze of struts in the vicinity of the ATM, round to the underside of the Multiple Docking Adapter, and anchoring himself on foot restraints near the truss assembly so that he could hold on to Pogue while the latter leaned over and pinned the antenna pitch axis at zero. It had given trouble during earlier EREP scans and the procedure had been worked out prior to the launch of SL-4. It was a satisfying day's work and one that cancelled reserve plans for another space walk the following day had the antenna repair not gone as planned.

Next day, trouble struck the control-moment gyroscopes, the large wheels spinning at approximately 9,000 rpm designed to torque Skylab around pitch, roll or yaw axes for pointing control, so alleviating the nitrogen gas jets from sustained operation. CMG-1 had given modest trouble several weeks before, but early in the morning hours of 23 November it began to overheat. At 2:04 am, within minutes of Skylab passing out of sight from the Honeysuckle tracking station, the guidance and navigation controller reported CMG-1 was rapidly over-heating. The big, massive, wheels ran in bearings sunk within oil baths and telemetry was provided to indicate oil temperature, wheel speed and the electrical load drawn by the motors. All telemetry was valid, with only the wheel speed on CMG-3 unknown; that had gone out on 8 October for some unknown reason.

It took Skylab just 38 minutes to go from Honeysuckle to the Bermuda tracking window and in that time CMG-1 spun down to a complete stop from 9,000 rpm, an operation that would require up to 30 hours for a free-running wheel. Clearly, it had seized up and at 2:47 am Phil Shaffer ordered the immediate shut-down of CMG-1. It was a dramatic decision but unavoidable. Now Skylab would have to control itself and perform attitude changes with only two control-moment gyroscopes. Was that possible? And for 76 more days?

With a single failed wheel the cluster was stripped of much of its momentum storage capability and if a second wheel stopped that would require full and almost continuous use of the nitrogen thrusters in the TACS supply, sufficient for about five days at most. Propellant stored in Apollo's Service Module could keep the cluster stabilized for a further 20 days but after that the station would have to be evacuated. Engineers calculated a 0.3% chance that a second CMG would fail before the end of the mission. Within hours of it failing, CMG-1 was considered totally useless and engineers came up with modified control procedures that would, if the other two wheels held up, maintain almost the same degree of control of the orbiting cluster as it had before.

But it was not a good day for the crew either. Woken in the small hours by a warning tone telling them something was wrong with the control gyroscopes, the crew went back to sleep for the scheduled duration after determining the cause of the alarm. But they felt overtaxed by the first full week and next day they spent the period resting, disturbed only by the need to operate a scheduled trim burn and put Skylab back on a five-day repeating ground track; fired for 88 seconds, the CSM thrusters stopped a 28 km/day drift.

By the middle of the second week Gerry Carr was noticeably snappy with ground personnel, believing the crew once more to have received an impossibly demanding work load. Again, there was a measure of truth in this, with the third team having to tidy up some thirty minor repair tasks before getting on with the science of their mission. Frustrated with themselves, the problem was compounded by a lack of real confidence in flight planners at Houston. For their part, the operations people were fired by the remarkable performance of the Bean crew, forgetting that the exceptional performance came as a result of both space and ground operations accelerating together until both found the optimum running speed for space station operations.

Expecting the Carr team to take over where Bean left off, the pace was set by the ground and held as a dominating target for too long into this third visit; only when the ground controllers backed off, thoroughly debated the problem with the men in space, and allowed the astronauts to lead the field, would performance improve. As it was, for several more weeks, friction evolved as a product of disagreement and frustration.

By the third week, control operations had improved, with Rusty Schweickart, the ground-based 'fix it' man, flying simulator runs with new procedures that were uplinked on the teleprinter. Maneuvering Skylab with only two CMG wheels up and running was no mean feat, something akin to crossing the mid-West with a broken wagon axle!

On 4 December, Day 19, the two Apollo fuel cells were shut down and Skylab settled into the electrical profile it would have for the remainder of the flight. But now the retarded performance from all three astronauts was a problem for deliberation and analysis on the ground. Energy expenditure per day was about that of the Bean crew, but accomplishments were considerably less. Also, only half the planned 1½ hours set aside for exercise were actually being taken, and that was beginning to concern the physicians. Ed Gibson was tired, expressing a desire for the full eight hours sleep he was theoretically assigned rather than the 6½ hours he was getting at most.

On Day 20, 5 December, CMG-2 showed signs of fluctuation when the speed decreased by 50 rpm but engineers believed this to be partial binding due to a low lubricant temperature and switched on heaters to bring it back up. Through better understanding of how to operate Skylab in a partially failed CMG mode, engineers now believed that if a second wheel stopped they could keep the cluster stabilized for two weeks on TACS gas alone and for a further 10 days on Apollo propellants through its own RCS thrusters. Fluctuations like those now seen in CMG-2 had affected the failed wheel one month before it stopped. A lot of people were holding their breath that the same problem would not materialize again.

Three days later, another fluctuation was seen in CMG-2 and again on Day 27 the wheel slowed for a brief period. On each occasion, temperatures went up as wheel speed decreased and engineers believed that careful control of the lubricant heaters could reduce this tendency, effectively preventing the problem getting any worse. By this time Skylab had its rescue capability in hand, for on 3 December, Saturn IB AS-209 arrived at pad B to stand ready for any emergency that might call for a second CSM. On 12 December a second trim burn was performed to halt a slow drift and refine the precise overlap of successive groundtracks when CSM-118 fired two aft-facing thrusters for 17 seconds.

By the end of the fourth week it was obvious that Carr, Gibson and Pogue would be hard put to match pre-flight objectives. They were averaging only 25 man-hours per day versus 29 per day for the previous crew at this stage, and all three were keen to accept their rest days rather than work on. The amount of time spent on medical tests was now falling behind the predicted level, ATM work was not yet up to the expected value, but EREP scans were more or less on the anticipated average. Much of the latter was compromised by the failure of CMG-1, which inhibited use of the cluster at different attitudes. However, a preliminary management review of systems, crew performance, and medical opinion gave the green light for a second month in space.

By 15 December the second momentum wheel was making its anomalous presence felt with increasing likelihood of total failure. On that day, wheel speed dropped from 8,912 rpm to 8,870 rpm, again probably because of low oil temperatures, but heaters manually switched on brought the CMG back up to speed. Three days later the crew caught one of the most active solar days in the Skylab diary as the largest prominence seen for twenty years reared up from the solar disc. It was all very uncharacteristic of this period of the quiet Sun, but a welcome departure for physicists concerned at the opportunities for research.

By Christmas, controllers were manually switching CMG-2 oil bath heaters to prevent the temperature dropping below 19.5°c, most of the threatened stalls having been observed when temperatures dropped lower than that. But as the season of goodwill approached systems engineers nursed an ailing Skylab, for now the ATM control console coolant loop was getting very noisy and pumps had to be turned off at night. On the day before Christmas, the crew spent time preparing for the scheduled EVA on 25 December, Day 40 of this second mission. But there was also time for a telecast from the orbiting Skylab.

For much of the year, TV watchers in the United States received regular broadcasts on their sets of Skylab astronauts cavorting in space, performing unique scientific tasks, or simply going through the motions of living. But on this occasion it was time for reflection and personal deliberation, so the view that Christmas Eve was one of three distant travellers in orbit grouped around a mock tree fashioned from old cans and loose items in the workshop. Gerry Carr sent a commentary of opinion that summed the attitudes among so many:

'Words come and go and return again and others never leave us. Words that come from the future like "death," "love," "hope," and "peace." Words of the past from ancient poets conjure up images like "wonderful councillor," "mighty God," "prince of peace," "spirit of wisdom and understanding." Now we would like to add a few of our words from the outside, from Skylab. You know, our Earth seems large to us as we look down upon it, yet those men who have flown Apollo to the Moon say it's small. And as we see it there are vast areas of desolation and great masses of water with man crowded only into the more hospitable zones of the Earth.

'The men from Apollo perceive the Earth as a tiny blue island in the vast sea of space. Well, either way you look at it, the observation is humbling because the tenuousness of our existence is emphasized by the need for man to get into harmony with his environment and with his fellow man. Among Christians the Christmas season serves to heighten our awareness of others and the brotherhood of man. And whether we're Christians or Jews or Mohammadans or Buddhists or Confucionists or atheists and no matter what the season is, I think we all agree that one of man's primary goals for the future should be to learn to live in peace and harmony with one another. So, to that end, I wish for all the world a most fruitful and peaceful day.'

A similar message was echoed by Gibson and Pogue before a host of good wishes was sent up throughout the afternoon from the flight control teams, the families of the astronauts, Johnson and Marshall Space Centers, and even the teleprinter girls. Next day, Christmas Day, Carr and Pogue went outside on a space walk designed to study Comet Kohoutek among other operations scheduled for the EVA.

But space walks were never routine. Having switched pressure control units with Pogue, Gerry Carr noticed a large build up of ice on the chest-pack designed to regulate the flow of water and oxygen. The pressure control unit was an important part of the life support equipment, being the device designed to accept consumables, including electrical energy, from the space station and distribute it to the suit. But switching PCUs forced an O-ring seal on the water connector, causing liquid to leak through the gap and freeze on the outside beneath the Beta cloth cover on his space suit. It could have been disastrous had the frozen mass built up and jammed control valves in the chest pack.

During their near seven hour walk, the crew pinned open the Sunshield door on the ultraviolet spectroheliograph, adjusted the filter wheel on the x-ray spectrographic telescope, replaced ATM film cassettes, retrieved the particle collection sample, used a special x-ray/ultraviolet instrument to photograph the Sun, shot views of Kohoutek's tail with the electronographic camera (actually, the back-up to a similar instrument carried to the Moon by Apollo 16), and used the coronagraph contamination instrument.

The appearance of Kohoutek was something of a disappointment for ground-based observers, its apparent brightness being considerably below predicted levels, but the Skylab instruments got a unique opportunity to view this visitor from the depths of the solar system. Three days later the comet passed the Sun at a distance of just 21 million kilometers, little more than 10% the distance between Earth and Sun, at a speed of some 404,000 km/hr.

Much of Friday, 28 December, was occupied with conversation about the celestial visitor as the astronauts talked with Dr. Kohoutek via the air-to-ground loop. Next day, Saturday, Carr and Gibson performed the third EVA of this mission when they went outside to operate once more the special electronographic camera brought up to survey Comet Kohoutek and to perform a coronagraph contamination measurement. It lasted 3½ hours and after getting back inside Skylab during the late afternoon they sketched their impressions of what the object looked like with only a transparent visor between eyeball and comet. The coronograph experiment was intended for use through the solar facing scientific airlock but the parasol and the A-frame shade prevented that so it became a hand-held EVA task.

Sustained concern at the apparent lack of full performance from Carr, Gibson and Pogue evolved into a major discussion, frank and open in content, with vented pent-up feelings long overdue. During the evening hours on Thursday, 30 December, Day 45 for the second mission, an open conversation began in response to teleprinter messages from the ground and tape recorded comments from Skylab. Carr addressed several poignant questions to flight director Phil Shaffer who answered with a statement on the printer read-out. But the conversation resulting from this earlier exchange cleared the air for the rest of the flight.

Capcom Dick Truly spoke at length about the lack of adequate dialogue between the men in space and controllers on the ground, claiming that these feelings were representative of 'the whole ground team here.' Truly went on to admit 'that we've tried not to be very subtle, so we tried to answer your questions directly.' Agreeing that there had been 'a whole number of changes' that compromised crew time, Truly nevertheless reminded the astronauts that 'you told us the other night, "we'd all kind of hoped before the mission – and everybody had the message – that we did not plan to operate at SL-3 pace." I'd like to answer that one directly, and that is that everybody did *not* have that word.'

But the determination to back off and let the crew have their own pace was, as Truly admitted, 'producing the same output as we were before.' And then Gerry Carr reaffirmed his own resolve to improve the performance levels: 'It's been my feeling that we really do need to work at the fastest and most efficient pace. It looks to me now like we're approaching it from the right direction. I think our problem at the beginning was we started too high. And what we need to do is just ramp up to it until we get to the best level and then maintain that. And like I say, I think we're headed in the right direction now. Also, I think, as I mentioned in my little note to you, that a guy needs time – some quiet time to just unwind if we're going to keep him healthy and alert up here.

'I want also to mention that there's two tonics that really do a lot for us up here for our morale. One of them is just looking out the window and seeing things that you recognize and haven't time to think about, write something down about, or record something about. And the other big morale booster that we have up here is our capcoms, Dick. I want you to pass the word to all the rest of the capcoms that we're greatly appreciative of the attitude you guys take and your cheery words and your occasional bits of music and all that really help make our day.' SL-4 had shifted into the second phase of operations and would quickly determine optimum performance levels and settle down to a routine similar to that of the second mission.

1 January, 1974, began after a shorter night than usual. Although US time would not go to daylight-saving before 6 January, the Skylab crew began their adjustment nearly one week in advance, getting up at 5:00 am for an EREP scan. This third mission was still officially aiming for a 59 day stay and on 4 January the recovery ship *New Orleans* slipped berth at San Diego and made for Pearl Harbor where it would arrive a week later for early use if needed.

But it was not, for on 10 January a full management review gave SL-4 approval for a futher week in space beyond the scheduled return time. The performance of both crew and controllers was improving, although evidence of fatigue was there: despite a total 30 man-hours of science performed on the 8th and the 9th, wrong switch positions on the 10th caused a massive loss of TACS nitrogen gas following an inadvertent loss of film footage during an EREP scan when wrong switch positions fouled the instrument sequence.

On 12 January the crew passed the 58½ day record set by Al Bean, Owen Garriott and Jack Lousma, but Skylab systems were threatening to cave in at any time. Increased temperatures on the interior resulting from the relative position of orbit and Sun required both coolant loops for maintaining equipment at an acceptable level; with the station now exposed to Sunlight throughout each orbit an extra thermal stress was placed on the entire cluster. Control-moment gyroscope number 2 was showing increasing signs of binding, slowing at one point to 8,800 rpm before picking back up to normal speed. More and more time was needed to keep everything running normally. The same hardware had supported five months of manned operation, a man-hours total greater than that accumulated by all US manned space flights since Alan Shepard's suborbital hop aboard Freedom 7 nearly thirteen years before.

But the equipment aboard Skylab was performing beyond expectation, for despite the succession of appalling disasters that befell the orbiting cluster, the original scheme to support three flights of 28, 56, and 56 days respectively reached its total on 6 January. Every day beyond that was a bonus. On 19 January the crew changed an Airlock Module tape recorder after it had given good service for 1,446 hours versus a design limit of 800 hours. But crew error was still creeping in. On the 20th, Bill Pogue accidentally turned off a multi-spectral camera during an EREP pass and lost valuable data.

Next day, the Apollo RCS thrusters were fired for 10 seconds, shifting the groundtrack 6.5 km west. The CMG-2 condition was worsening, however, with the 27th distress condition observed on the 20th. Two days later the familiar symptoms were seen to last nearly six hours before relief and Bill Schneider approved a plan to bring the recovery force quickly back to San Diego for supplies sufficient to last for possible recovery on days 75, 80 or 85; recovery options were optimized around five-day centres because of the characteristics of the Skylab orbit, bringing the cluster back over precisely the same groundtrack every five days. On 23 January the entire crew day was spent with CMG-2 in distress and while they went about their scheduled duties managers on the ground conferred on the available options should the wheel bind and stop.

Everybody agreed CMG-2 was probably in its last hours of life and Mission Control teams worked up a procedure to keep the mission going, with very limited attitude control capability for a further 15 days after total wheel failure. With little spare TACS nitrogen, maneuvers for EREP passes and other science tasks were severely curtailed; the gas would be invaluable should CMG-2 collapse. Opinion differed on the degree of harm incurred by ground tracking maneuvers away from the solar-inertial attitude yet there was little time now for catching up on lost EREP passes: of the total stock available when the third mission began, between 54% and 75% of film loaded for the multispectral camera, the earth-terrain camera, the infrared spectrometer and the multispectral scanner had been used.

Yet stalled schedules upset by the need for timely attention to flagging systems drove workloads far into days assigned for crew rest. By the end of the month CMG-2 wheel speed was averaging 8,840 rpm but when the area of the ATM containing this equipment turned face on to the Sun during maneuvers, speed increased to a more normal 8,900 rpm momentarily, the heating effect of the solar radiation having warmed the oil bath.

Careful study of the Skylab orbit, slowly decaying due to traces of atmosphere at that height, showed it to have a lifetime of about eight years. Before launch the intended orbit was believed to give Skylab a life of nearly 6½ years, but the actual orbit it reached was a little different and management personnel resurrected an old, and frequently discussed, topic: what to do with Skylab at the end of its planned operations.

Nearly four years earlier, officials said the likelihood of damage to life or property was sufficiently low to render an uncontrolled re-entry quite acceptable. In April 1973, just a month before the cluster ascended to orbit, final studies were carried out where de-orbit using the Apollo thrusters or the big SPS engine was considered. However, because this would have significantly increased the potential risk to the astronauts, it was abandoned, as was a third scheme to use the big S-II Saturn stage to bring it back down. The only danger from an uncontrolled re-entry was that such a large structure would spread debris over a very wide area, some of which might just fall on populated zones.

But study of the actual path of Skylab during the last few weeks of the third visit showed it would probably crash to Earth sometime in 1981. Vague ideas circulated on the possibility of re-visiting Skylab at some future date, not for manned habitation but just for a quick visit in space suits from an orbiting Shuttle, and plans were already formulated to leave a special bag in the Multiple Docking Adapter filled with food samples, clothes, a surgical glove, a heat exchanger fan, a fire sensor panel, four film samples, a roll of tele-printer paper, camera filters, communication cords, four flight data files, and two electric cables; samples of materials representing the complete range used in Skylab which could give valuable information on the degradation and decay of such items in the space environment.

So to increase the possibility of one day being able to rendezvous with Skylab in a reusable, winged, Shuttle, plans were drawn up to perform one last trim burn to raise the Skylab orbit and extend its life by about 1½ years.

On 1 February, Day 78, the CMG wheel was going continually in and out of distress, hanging on only by the fanatical attention of controllers on the ground, cycling the wheel heaters and nursing the attitude changes. By this time the shower unit had packed up and science operations were approaching the last full day of activity. On 3 February, Carr and Gibson performed the last Skylab space walk, moving through the Airlock Module door at 10:19 am for more than five hours of activity which included operating the coronagraph contamination instrument for photographs of fine particles in the upper atmosphere, operation of the x-ray/ultraviolet solar photography experiment, retrieval of ATM film cassettes, and retrieval of several exposed modules and particle collection panels for return to Earth. One set of particle collection panels was put out, just in case somebody re-visited the cluster one day. Again there was trouble with a PCU connector on Gibson's suit, the 4 kg water reservoir running completely dry due to a leak at the fitting which necessitated the astronaut going to gas cooling.

When the crew got back inside they completed what history would record as the last EVA performed by American astronauts with a ballistic spacecraft, the final space walk of the 'old' era when men blasted forth from the Florida beach on expendable rockets and one-shot space vehicles. It was the last American space walk of the decade, and it was conducted through a hatch that came from Gemini where the first US astronaut to become a true space man opened new frontiers that led to Moon walks and lunar exploration. It had been more than 8½ years since Ed White spent his famous twenty-one minutes floating above the United States. Gerry Carr and Ed Gibson closed the book on at least that beginning.

Next day, 4 February, the last major medical tests were performed and the materials processing facility was used to study the effect of material combusted in a closed, weightless, environment. It was the last experiment from that facility because of the destructive effects of the exercise. 5 February was stowage day, with the crew gradually packing up all the equipment, items large and small for storage aboard the Command Module. During the evening, flight controllers abandoned their continuing attempts to operate the heaters manually in CMG-2, since friction was now heating the oil bath. It was a miracle the wheel was still going, with detectable binding now evident it could only last a matter of days at most. But it lasted the full four days that still remained in the life of Skylab, seeing the crew through to the full 84 days they hoped to accomplish.

Next day, the CSM thrusters were fired in a 180 second burn which raised the orbit from 431 × 444 km to 433 × 455 km, enough to give it a theoretical life of more than nine years. The cluster was now expected to decay through the atmosphere some time in 1983, sufficiently far ahead to allow a Shuttle the opportunity of re-visiting the first US space station. Or so everybody thought.

Sleep adjustments for return to Earth began with the crew going to bed at 8:00 pm on 6 February. They woke at 4:00 am, 7 February, for a short nine-hour 'day' in which they completed final stowage tasks begun the day before. All the major systems were now turned off, the wardroom had been deactivated, the workshop oxygen/nitrogen replenishment was off, and the CSM was powered up and ready. Trouble with the supplementary propellant module in Apollo's Service Module had shown up during the first trim burn during the second week, indicating a leaky valve to quad B. It was confirmed during the second trim maneuver in December and the crew had turned off that supply to quad B, using the integral tanks instead. There were no further problems now, however, and the spacecraft was in good condition for re-entry. It had survived more than 84 days in the hostile space environment, a design that Rockwell International's predecessor had built for a two-week mission to the vicinity of the Moon.

At 1:00 pm the crew began a seven and one-half hour rest session, receiving a call from the ground at 8:40 pm, Houston time. A few hours later Carr and Pogue entered the Command Module, leaving Ed Gibson to be the last man out of the Skylab space station. With all the hatches open through the entire structure, he climbed across into the Command Module tunnel at 1:15 am, Friday, 8 February, 1974, a camera in hand. Fifteen minutes later the probe was being installed. At Mission Control, a dozen

Skylab, as seen by the Carr crew as they headed for home following a record 84 days in space.

red roses arrived, right on cue. Miss Cindy Diane, former resident of Montreal, Canada, and now living in Pennsylvania, had not forgotten.

At 3:15 am the RCS thrusters were fired to satisfy the crew they worked as they should, and at 5:34 am, Apollo undocked from the drogue. From the ground, a farewell comment: 'Say good-bye for us; she's been a good bird.' Once more, and for the last time, control operations were separated into specific vehicles: Neil Hutchinson monitored the Skylab while Phil Shaffer had responsibility for Apollo. Capcom Robert Crippen was at the communicator's console. 'It's been a good home, Crip,' said Ed Gibson. 'You can tell Al Bean and guys that they did a great job putting that sail up; it's very symmetric. It's been a real useful machine, Crip. Hate to think we're the last guys to use it.'

For some time the crew flew CSM-118 around the vacant cluster, taking photographs of the complete exterior, and generally getting a last long look at what had been home for nine astronauts for much of the previous nine months. At 6:32 am the Service Module fired up its SPS engine and kicked Apollo into an elliptical shaping orbit before the retro-fire burn, also with the Service Propulsion System, at 9:36 am, Houston time. Earlier, during checkout, a leak in the helium pressurization line to one of the two separate re-entry RCS thruster loops was discovered and the crew were told to operate on ring 1, leaving the suspect ring 2 unused. In the event they sensed an unusual smell they were to use the oxygen masks installed aboard the Command Module. At this time it was believed the leak was in a fuel or oxidizer line but subsequent analysis would reveal a 0.2 mm hole in the helium line. Prior to separating from the Service Module, the crew set up incorrect switch positions on a circuit breaker panel which resulted in the need for manual control of pitch and yaw during descent through the atmosphere.

At 10:17 am, Houston time, the spacecraft was bobbing on the Pacific waters, albeit in an ingloriously inverted position until inflated balloons brought it back to Stable I. About forty minutes later, Carr, Gibson and Pogue were on the deck of the *New Orleans*. Physically, the crew were in good condition and over the ensuing months careful analysis of the data, and their return to a completely normal condition, would reveal little that threatened to limit the possibilities for space flights lasting one year.

Within hours of the last crew leaving, Skylab was put through engineering tests too risky to try with men aboard. CMG-1 was powered up, but the wheel stuck fast and refused to spin. CMG-2 was tested under severe conditions, and failed to cave in giving indication that it could have lasted longer than thought possible just a few days before. As for the battery units, the eight power-conditioning-groups serving the single workshop solar array operated efficiently throughout, and of the eighteen charger-battery-regulator-modules only two had failed, although performance from the remainder was significantly less than at the beginning of operations. Throughout this final mission, power requirements averaged 4.8 kW versus a 6.2 kW to 8.7 kW production level.

As for the Airlock Module coolant loops, the secondary system never did need replenishment and kept going until the end, the primary loop being used for supplemental cooling during EVA and in conjunction with the secondary loop for the last seven weeks. Of two molecular sieves built in for maintaining carbon dioxide at an acceptable level only one was ever needed throughout the complete sequence of missions, and all other environmental control equipment performed well. Because Skylab was more leak proof than designers believed it would be in practice, sufficient oxygen and nitrogen remained at the end for a further 212 days of manned operations, a period in itself greater than that accumulated by all three visits. As for water, there was enough left for 90 days at powered levels of use.

But attitude control gas was the one item that threatened even to curtail plans for an 84 day mission with SL-4, careful husbanding of diminishing reserves seeing the flight through to completion with almost no contingency supply left. Had those carefully honed use rates been maintained, however, Skylab could have limped along for a further 30 days before reaching the red line set for emergency. But that would have assumed continued operation of CMG-2, for if that moment-wheel had seized up like its neighbour there would be no attitude control remaining. Despite an extension of more than 31 days beyond the time Skylab was built to survive, it could, theoretically, have sustained a further four weeks of manned habitation.

In the face of unprecedented problems from beginning to end, the space station had proved that manned involvement with mechanical space structures would probably prove more cost effective than redundant robots ostensibly less complex. As conceived, Skylab, in the guise of first AES then AAP objectives, was intended to utilize redundant Apollo hardware to show how effectively man could live and work in the space environment and so point the way toward permanently manned, purpose-built, space stations of the future.

The living part was evidence itself for man's adaptability. As for work, the three manned visits had increased by 33% the time originally assigned for ATM observations, increased by 17% time planned for medical experiments, increased by 145% time allocated to astronomy and astrophysics, and increased by more than 300% time and effort expended on materials processing tasks. Although the last mission did not quite get back up to pre-flight predictions, it too proved a valuable point: that the integration of men and machines within a set of complex objectives attempted far from the planning table depends more on the human relationships and on the levels of communication and integration present than on time and motion studies set up as goals or targets.

In all, Skylab returned more than 128,000 photographs of the Sun, more than 46,000 pictures of the Earth, and more than 72 km of magnetic tape on which data about the planet was recorded from the EREP instruments. The three teams had performed nine space walks from the Airlock Module totalling nearly 42 hours, and they had travelled more than 113 million kilometers on successive orbits around the planet, equivalent to nearly 150 round trips to the Moon. But more than that, they filmed science demonstrations for later use in high-school classes, brought experiments back from space operated on behalf of children recruited to the sciences of Skylab, and opened new libraries of knowledge on the behavior of man in a weightless environment.

Who could have known, for instance, that after adjusting to weightlessness, an astronaut would be less susceptible to motion sickness: on Earth, pilots would experience nausea after rotating at about 15 rpm while performing about 75 head movements, yet in space they could achieve twice the rotation speed and double the number of head movements. And who could have guessed that an astronaut would be totally disorientated with his eyes closed. Neither was it a casual phenomena, for arms extended to grasp an object in total darkness would go off in completely the wrong direction; perhaps coordination of arms and brain required both gravity *and* vision, with emphasis on the former.

Yet greater than all of this, Skylab did more to convert scientists to the technology of manned flight than all the missions that had gone before. More than a decade earlier, the scientific community was suspicious of the emphasis placed on manned space flight, believing it to be a show-business extravaganza out of place in a world full of realistic problems and priorities for research and exploration. For many years the opposing viewpoints polarized in separate camps the different interpretations of where man was, or should be, going with his progressive instincts. That theme runs consistently throughout the history of manned space activity.

From 1974, however, a broadening consensus recognized the need to take man along on most, if not all, the scientific activities he would be called upon to conduct in the future. Leo Goldberg, Director of Kitt Peak National Observatory, as representative of one of science's most entrenched disciplines, reaffirmed that 'Many of us had serious doubts about the scientific usefulness of men in space, especially in a mission which was not designed to take advantage of man's capability to repair and maintain equipment in space.'

Yet for all that he was impressed by what Skylab had demonstrated in that the astronauts 'performed near miracles in transforming the mission from near ruin to total perfection. By their rigorous preparation and training and enthusiastic devotion to the scientific goals of the mission, they have proven the value of men in space as true scientific partners in space science research.' It was a timely conversion, for many scientists would be drawn increasingly toward work with manned missions as the NASA Shuttle emerged from drawing boards and blueprints to take shape in metal and composites at Rockwell International's plant in Downey, California.

The full integration of manned and unmanned programs would follow as a consequence of building the next generation launcher as a piloted vehicle involving human control at every

This overhead, exterior, view of Skylab taken during the fly-around inspection prior to re-entry of the last crew to occupy the station shows the A-frame shade developed at the Marshall Space Flight Center and deployed by the second crew. Note the parasol underlying the Marshall shade and deployed by the first crew.

stage. The Shuttle would no longer pursue goals fixed beyond the application of daily or humanitarian needs, but place Earth-based problems firmly in the lap of the space program and in so doing embrace the full gamut of scientific capability, rather than pick and choose the more exciting research areas for performance elsewhere in space. Skylab clearly showed it was time to put away thoughts of an Earth-Moon transportation system and use national economic and human resources on more direct problems; only when the skeleton of capability had been merged with the construction of applications would it be appropriate to turn again to distant goals and far-flung trips of exploration.

NASA Administrator Jim Fletcher summarized feelings: 'In a very real sense, Skylab can be considered a turning point for it possessed many qualities and ingredients that will characterize missions of the future. It has moved the space program from the spectacular into a new phase. It has contributed to an orderly transition from the Apollo era to the Space Shuttle . . . of the 1980s. We have demonstrated that man can perform valuable services in Earth orbit as observer, scientist, engineer and repairman.

'Skylab has given us a wealth of new information about the dynamic processes of the Sun and how they affect us on Earth. We will hear much more about what has been found. . . . We will be living with Skylab achievements for a long time.'

When the last crew had vacated the orbiting cluster, engineers on the ground commanded the vents open to evacuate the interior of oxygen and nitrogen gases. First it was vented down to about 52 mmHg and left to leak from that level; it would take about eight weeks for the interior to become a vacuum. Then the cluster was moved to a gravity-gradient attitude where the station pointed at the Earth with the Multiple Docking Adapter uppermost. The TACS was turned off, the CMG's were shut down and electrical loads were removed from all systems.

During the afternoon of Saturday, 9 February, little more than a day after Carr, Gibson and Pogue returned, the telemetry transmitters were commanded off, effectively killing the station. It had been no mean feat accumulating with a single vehicle more time for men in space than had been accomplished by all previous NASA missions combined: 171 days of habitation, extending manned flight durations from the three weeks of Soyuz 11 to the three months of Skylab 4. One of the last hardware instructions involved a message to the Kennedy Space Center, authorizing Debus' men to roll back the Saturn IB assigned a stand-by role in case of rescue. Only one more conventional manned space flight remained in NASA's calendar: the Apollo-Soyuz Test Project, ASTP, then scheduled for a mid-1975 rendezvous involving American and Russian hardware. Beyond that, only the Shuttle promised a return to manned operations in the wake of redundant expendable launch vehicles of the Saturn class. As Skylab came to an end, plans laid for early Shuttle flights anticipated orbital operations by spring, 1979, a date extended from the earlier 1978 (and original 1977) target by persistent budget cuts. Before that, in 1977, an orbiter vehicle would be released from the upper fuselage of a Boeing 747 to simulate the latter portions of atmospheric descent. There would, consequently, be a planned gap of nearly four years between ASTP and the first Shuttle mission, a longer hiatus than at any time since manned flight began.

On 1 April, 1974, Dr. Debus presided at a formal ground breaking ceremony when construction of a special runway for the Shuttle began at the Kennedy Space Center, north west of the Vehicle Assembly Building. Five months later, Kurt Debus announced his imminent retirement from NASA. Having spent more than twenty years actively engaged in the development of launch techniques at the Florida site, one of the last big names from the old days at Peenemunde was leaving the space agency for whom he had directed the launch of the first American satellite. Before the end of the year he would be replaced by Lee R. Scherer, Director of NASA's Flight Research Center at Edwards, California since 1971, and former manager of NASA's Lunar Orbiter program of five unmanned Moon surveying robots flown in 1966 and 1967.

It was a year of change all round. In December 1973, Bob Gilruth resigned his post as Director of Key Personnel in Washington and the former Manned Spacecraft Center head left NASA. Fifteen years earlier he had been chosen to manage America's first manned space program, Mercury. On 23 April, 1974, Dr. Gilruth was elected to the National Academy of Sciences.

Across the years, change had affected the area from which America sent its most ambitious space missions aloft. Where once the Kennedy Space Center employed more than 26,000 people, it now found work for only 10,000 and prospects were not good that levels would rise much beyond that for with the Shuttle would come rationalization, the specter of unemployment forged by technical innovation, fewer technicians needed in the control centers, fewer new buildings from which to prepare the hardware, and less need for manpower services. Where only five years before the Cape handled a budget of $500 million it now commanded a lowly $219 million, and that figure excluded inflation.

But this was only a sample of conditions reflected throughout the space industry. In real terms, NASA's overall budget was by now down to around one-third the sum it had been a decade before and there were signs of a mini recession in the economy to dissuade proponents from too optimistic a view. NASA fought long and hard to get the Shuttle approved, then only getting permission from Congress because of the professed savings from a reusable transporter, and convinced the Nixon administration of the need to support continued investment in manned flight.

But having seen Apollo blossom from Moon exploration to Earth orbiting Skylab operations, Associate Administrator for Manned Space Flight, Dale D. Myers, tendered his resignation from NASA to return to the company he served before moving to Washington. On 15 March, 1974, Myers left the space agency to become President of Rockwell International's North American Aircraft Operations Group, vacating a post he inherited from George Mueller in early 1970. For a while, Bill Schneider, late of the Skylab program, stood in as Acting Associate Administrator, until, in May 1974, Dr. Fletcher announced the appointment of John F. Yardley to head manned flight development.

As former project engineer for Mercury between 1958 and 1960, launch operations manager for Mercury and Gemini from 1960 to 1964, Gemini technical director from 1964 to 1967, and roving general manager for Skylab since 1967, Yardley was one of the most qualified men in the business. It was his job to oversee the ASTP and Shuttle projects while attempting to secure work on future programs. The next few years were to be some of the more difficult times in NASA history, for denied the general support from Congress it enjoyed in the 1960s, the agency would fight for its very existence amid flagging budgets and decay of political will.

The President approved ASTP only because it supported his foreign policy initiatives toward the Soviet Union. Long time opponent of soft line negotiations, Richard Nixon had just completed one of the most effective and genuinely progressive moves on the international scene. Known for his 'hawkish' attitude about communism and subversive politics in Third World states, he effectively negotiated a cease-fire in Vietnam in the months between the last manned flight to the Moon and the start of Skylab operations.

Less than a year before, in early 1972, Nixon opened Western politics to debate and involvement with China only months before drawing the Soviet Union into an arms agreement aimed at limiting the expansion of strategic weapons. It was within the frame of the SALT-1 environment that Nixon approved the joint docking flight with a Russian Soyuz spacecraft, seeing in this unification of political interests a bringing together of former opponents before a global audience; nothing could more effectively seal world approval, in the eyes of Richard Nixon.

But the man had a certain naivety about clandestine operations more effectively conducted by his immediate predecessors and when the bastion of Presidential protection, J. Edgar Hoover, died in spring 1972, it left the Republican White House totally exposed to self-righteous investigators more concerned with promoting Democratic virtues than national stability; devoid of its dictatorial leader, the FBI was no longer employed for discretionary concealment. Exposed to the full fury of indignant newsmen, Richard Nixon was in the position of carrying the brunt of accusations aimed more at established practice than specific events.

Nevertheless, one particular operation involving lesser powers at the White House flared in the full force of public outrage and the Watergate became more a situation than a building. Wire-taps, bugging, secretly taped conversations, burglary, had all been introduced to the Washington scene by two very efficient Democratic Presidents before Nixon was elected to office. But Nixon happened to be in that place when the bubble burst and his alone was the task of bearing full responsibility for devices and techniques used more effectively, and on a wider scale, than he had seen fit to adopt.

It was as though suspicion erupted suddenly, to wash away in one move corruption nobody wanted to see before but which had been there in reality for nearly fifteen years. In August 1974, Richard Nixon resigned office to be replaced by his own nominee, Gerald Ford. In no other single term of office had a President inspired so large a shift in foreign relations, ending hostilities in South-East Asia, widening the scope of the Western World to embrace China, and effecting the first positive steps toward a tangible detente with Soviet Russia. As for the space program, Nixon's removal had double consequences.

Gone were the days of action and decision, vacated for a period of passive entrenchment; and gone too were hopes of halting a decline in the NASA budget. President Nixon was unable, or unwilling, to honor a promise made in 1972 when Jim Fletcher secured an agreement that NASA's annual budget would not fall below a 'constant level' achieved in fiscal year 1973. Nixon's administration presided over the 1974 and 1975 budgets without keeping to that promise, allowing space agency funding to continue its downward spiral. But without leadership, the years of decay that lay ahead removed all chance of reversing that trend until a change of President would inspire new policy plans.

Yet the lessons from Skylab were not to be completely lost amid concerted efforts to get the Shuttle transportation system into space. In early 1973, the European Space Research Organization (ESRO) concluded an agreement with NASA whereby a consortium of nations would develop and build a so-called Sortie Module, or pressurized laboratory, designed to be carried into orbit by the reusable space-plane. Later called Spacelab, the laboratory would become one of two major projects controlled by the European Space Agency when it was formed from ESRO and the defunct European Launcher Development Organization (ELDO) in 1975.

Spacelab was all that remained of plans for a big 12-man space station leading to large space bases populated by scientists, technologists and engineers. But first, there was one last flight utilizing left-over Apollo hardware: the Apollo Soyuz Test Project.

When Apollo 17 came back from the Moon in December 1972, it was as fitting a time as any to perform a minor adjustment in the management of manned flight programs. Apart from the management shuffles which were expected to follow the completion of Apollo's prime phase, Headquarters restructured lines of responsibility, placing Chester Lee as ASTP program director and Robert O. Aller as his deputy. At Houston, Glynn Lunney replaced Owen Morris as Apollo spacecraft program manager, Arnold D. Aldrich was to be his deputy, with Ken Kleinknecht as director of flight operations; Pete Frank would play the role of lead flight director.

During 1973, while Skylab moved in and out of drama, ASTP kept hundreds of engineers and technicians blissfully ignorant of anything but their own concentrated responsibilities; there was much to do if US and Soviet hardware was to be ready by mid-1975. But now the political commitment had been made, and tentative agreement on the mission profile worked out with the Russians; it was time to dig deep in search of technical solutions to a myriad pitfalls. First, because Soyuz was incapable of reaching altitudes in excess of 225 km the radio coverage possible in such a low path would be limited to an average 17% of each orbit. Something had to be done to improve that because dual operations called for increased communication needs, better television coverage, and adequate data handling rates for the scientific instruments Apollo would carry.

A large communication satellite called ATS-F (ATS-6 after launch in January 1974) would look down on the planet from geosynchronous orbit more than 40,000 km from Earth. Capable of seeing almost one complete hemisphere, use of ATS-6 by Apollo antennae for relaying communication to Houston would increase average cover to 55% of each orbit. The satellite would be stationed over the Pacific.

But the Russians were proving reluctant partners, or so it seemed when schedules and deadlines required delivery of important documentation. And then there was the problem that people with whom NASA officials met to discuss technical or program details were apparently unable to make decisions on the spot, preferring to return for consultations with some higher authority before providing an answer or a comment. Professor Bushuyev, head of the Soviet ASTP team, was noticeably rankled by this need, visibly concerned that things were going slower than he hoped. And then the very content of documentation was seen to have a style and a presentation almost political in content.

At one point NASA refused to accept vague assurances that the catastrophic accident of Soyuz 11 would not result in sudden depressurization during the joint flight, pressing the Soviets to come up with a full technical discussion as to the cause of the accident and corrective procedures worked out since 1971. The space agency wanted to see for itself how safe were procedures developed in the interim. Professor Bushuyev was concerned at this, however, and only succumbed to requests pressed at the highest level. From the information gained, NASA was able to construct a precise sequence of events associated with the Soyuz 11 accident.

But the Russians were apparently unable to keep up with the pace of preparations, falling behind schedule on several occasions. George Low requested from his Soviet counterparts a special mid-term management review to tidy up this and other problems that threatened the program. Glynn Lunney was equally concerned, telling the Russians 'we have experienced a delay in exchange of material of up to nine months . . . even though we have signed minutes committing ourselves to specific dates for these exchanges.' But when George Low was about to depart for Moscow in October to meet with his counterpart, Keldysh, Chester Lee telephoned the NASA Deputy Administrator with news that the Russian scientist was ill and could not attend the mid-term review; Boris N. Petrov had been nominated in his place. Keldysh had received surgery at the hands of Houston heart specialist Michael DeBakey earlier that year but the Russian scientist returned to work before fully recovered and was soon in need of complete rest once more.

By 14 October, George Low and Arnold Frutkin were in Moscow, with Lunney advising them of work performed so far. At the meeting that ensued, the Americans learned the real cause of Soyuz 11's tragic loss. During separation of the orbital module from the descent module prior to re-entry, pyrotechnic charges designed to sever the two structures fired at the same time instead of sequentially as designed. The sudden shock released a seal inside a pressure equalization valve, evacuating the cabin and killing the three men inside. The fact that the Soviets were now prepared to answer technical questions about the accident reflected success achieved with modifications and changes to the basic Soyuz vehicle.

Unmanned test flights of the modified hardware took place during the periods 26 June to 2 July, 1972, and June 15 to 17, 1973, when Cosmos 496 and 573, respectively, were checked out in orbit, each of which weighed 6½ tonnes and was almost identical to the next generation manned Soyuz vehicles scheduled for flight. And then, on 27 September, 1973, just two weeks before the Moscow meeting, cosmonauts Lazarev and Makarov, wearing space suits for the first time on multi-manned Soyuz flight, took off for a two day mission aboard Soyuz 12. It was Russia's first manned mission for nearly 29 months. From this point on, Soyuz flights would be limited to two-man missions, enabling the crew to wear pressure suits in case a similar accident happened again. The Americans were assured that Soviet confidence in their 'new' Soyuz would be heightened by several more manned flights in 1974.

Back in the United States from their fruitful meeting with the Russians, at which agreement was reached to conduct future exchanges on a more timely basis, George Low reported to Congress on preparations for a CSM-only flight should the Russians back out at the last minute or lose all their assigned hardware in pad disasters – the specter of Russian failure that accompanied early appraisal of Soviet activity was still present in the minds of politicians. Chairman Olin Teague of the House Committee on Science and Astronautics was told the astronauts would conduct joint experiments with their Soviet counterparts but that Apollo would also carry several unique experiments to be performed after the joint docking exercise.

By this time the Soviets had named the men who would fly the Russian spacecraft, timing the announcement for the Paris Air Show held in June 1973. Alexei Leonov and Valery Kubasov would fly the prime mission, with Filipchenko and Rukavishnikov selected to fly the second set of hardware if needed. Back-up crew members were Dzhanibekov, Andreyev, Romanenko and Ivanchenko. Leonov had performed the world's first space walk in March 1965, Kubasov had been flight engineer on Soyuz 6, Filipchenko had flown on Soyuz 7, Rukavishnikov was on Soyuz 10; the remainder had yet to make their first flight.

The first real public display of US-Soviet cooperation came when a redundant Apollo was wheeled in to a specially constructed dome to mate with a Soyuz mock-up brought in by the Russians. It was a unique sight, stole the entire show, and was viewed by nearly one half-million people. Stafford, Brand and Slayton had been named to fly the Apollo spacecraft earlier in the year and from that point on Russian became the language to learn at Houston's Johnson Space Center. Not before late 1973, however, would back-up members Bean, Evans and Lousma be free to fully integrate their own tasks with preparations for the ASTP mission.

From spring 1974, the astronauts spent 15 hours a week learning Russian. It was the one part of flight operations where failure could result from an inability to think in the spoken language. Many activities aboard a spacecraft had to be conducted as naturally as walking, sleeping, or eating. If there was doubt about the precise definition of a certain word, or the implication of a statement, catastrophy could result. To accomplish total integration, four language teachers worked with the astronauts, in the classroom, in the gymnasium, and at technical sessions, familiarizing them with the need to assume the role of citizens in whose language they spoke.

Model of the Soviet space station Salyut at the Gagarin Cosmonauts' Training Center.

As astronauts exchanged simulator seats with cosmonauts, and vice versa, the working knowledge of respective systems and spacecraft evolved into a common base from which opposite teams could benefit. But events were not without humor. Equipped with a case of fireworks, Ron Evans entertained Russian and American space men to a traditional Fourth of July celebration, letting off crackers and rockets from their hotel at Star City. When approached by a contingent of policemen, the astronauts engaged in mock combat by firing rockets from a mineral water bottle. When told that it was in commemoration of a much earlier revolution, the policemen went away, bemused but sympathetic.

While flight preparations continued, operations with the new Soviet hardware accelerated. Following the shakedown flight of September, 1973, Soyuz 13 was sent into orbit during the afternoon of 18 December, 1973. Unlike Soyuz 12, it carried solar cell arrays for electrical production; equipped in the configuration of a Salyut space station supply vehicle, its predecessor operated on batteries and was limited to two day operations. Prevented from moving ahead with the Salyut program, Russian engineers were only now recovering from problems with both the Soyuz spacecraft and the space station design.

Salyut 1 had been launched in April, 1971 and served as the target for Soyuz 10, which lasted two days, and the ill fated Soyuz 11, extending Soviet manned flight duration to nearly 24 days. Salyut 2, believed by some to have been a military station equipped with instruments and detectors for observation, reconnaissance, and technology research, was launched on 3 April, 1973. Early reports indicated something drastically wrong had affected Salyut 2 and when no Soyuz was reported to have been launched to man the station, speculation gained momentum. Some said they heard reports that the final stage of the launch vehicle had blown up and that many fragments had been tracked by radar in the same orbital path.

Whatever the cause, Salyut 2 did experience a catastrophic failure eleven days after launch when it was reported to have been disabled, the solar cell arrays to have been wrenched off, and several important antennae and other appendages to have been destroyed. Tumbling uncontrollably in space, the station decayed the following month and came down to Earth near Australia. More than two weeks before that happened, however, a space vehicle officially classed by the Soviets as Cosmos 557, was launched into a Salyut-type orbit. Ascending on 11 May, 1973, it transmitted signals characteristic of the 'civilian' Salyut 1 configuration rather than the 'military' Salyut 2, but it too was unattended and within two weeks had decayed back down through the atmosphere.

In concerted efforts to get under way with a generation of new space station hardware, politically important now Skylab was proving such a success and less than two years to go before the prestigious joint docking flight, Cosmos 613 was sent into orbit on 30 November, 1973, as an unmanned shakedown flight for long duration missions to come. Engineers needed to know how well a spacecraft would survive repeated day-night cycles in orbit around the Earth and sent Cosmos 613, a fully equipped Soyuz, into space for 60 days. At the end of a quiescent period completely powered down, as though it were attached to a space station, controllers activated the vehicle and performed a scheduled retro-fire, modular separation and re-entry, bringing it back to Earth as the last Skylab visit was drawing to a close.

So it was that speculation, rumor, and interest reached new heights when Soyuz 13 ascended on 18 December, 1973, three weeks after the launch of the Cosmos 613 test vehicle. But there was no space station in orbit with which it could dock: Salyut 1 had long since plunged back through the atmosphere; Salyut 2 was down; Cosmos 557 had similarly burned up. From the outset, Soyuz 13 was an independent mission equipped with several important scientific instruments for astrophysical, medical, and Earth resources work. Klimuk and Lebedev were the crew and they remained in space for eight days before returning to a point 200 km southwest of Karaganda in Kazakhstan.

Orion 2, an improved version of Orion 1 carried aboard Salyut 1, was different in that the telescope was secured to the Soyuz exterior rather than being contained within the pressurized interior. Designed by Professor Grigor Gurzadyan of Armenia, it was built to achieve high pointing accuracy through the operation of 13 electric motors. After positioning by hand, the

Cosmonauts Rukavishnikov (foreground) and Filipchenko in the Soyuz → simulator.

Soyuz 15 crewmembers Commander G. V. Sarafanov (right) and Flight Engineer L. S. Dyomin in their spacecraft simulator. Note the switch activation tool, the status annunciators and the extensive interior padding.

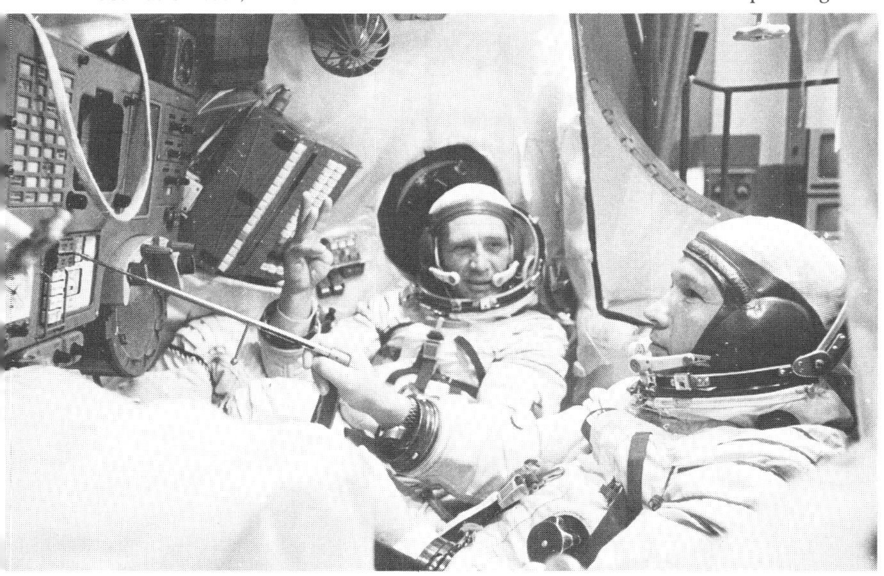

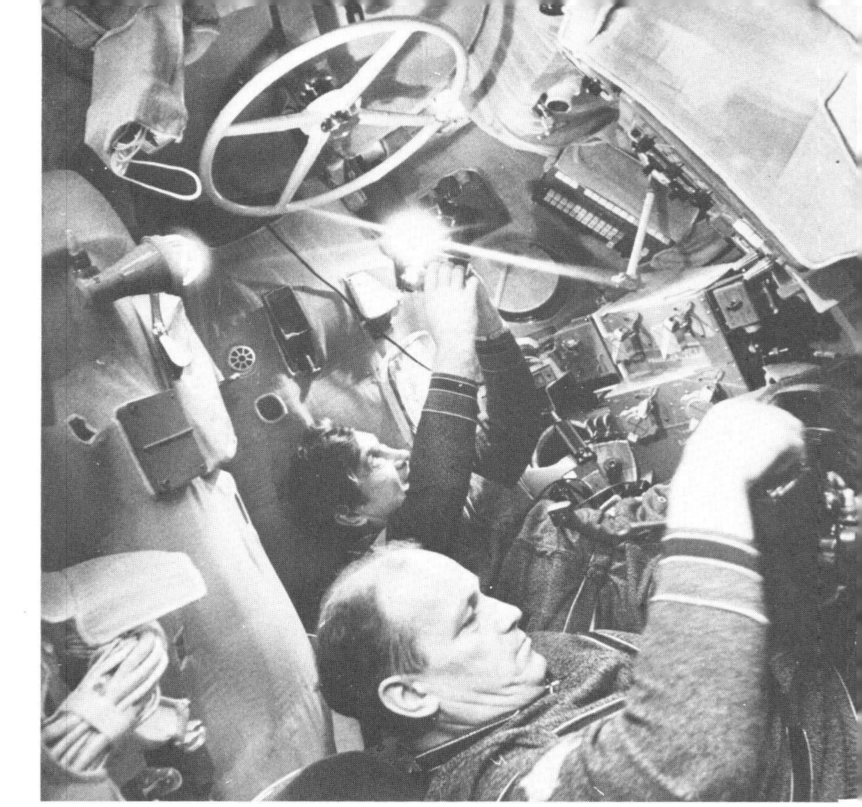

telescope was capable of automatically refining its alignment to within 5 arc-seconds. Called Oasis 2, the biological regeneration device demonstrated the principle of accumulating protein by exchanging biological products. A multispectral Earth resources camera was also used, incorporating nine lens within a triple-film holder. As part of an increasingly cooperative relationship with NASA, George Low provided special film for the Orion 2 experiment, with which the crew obtained more than 10,000 spectrograms of 3,000 stars.

By 1974, Soviet manned space operations were back to the level achieved in 1971. On 25 June, 1974, Salyut 3 ascended to a 219 × 270 km orbit inclined 51.6° to the equator. Skylab had been completed several months before and where NASA left off, the Russians were soon to begin. But operations with the Salyut 3 soon led to speculation that it belonged to the 'military' category like its immediate numerical predecessor. Major changes were made to the exterior appendages, most notably in the form of three solar panels versus the four used with Salyut 1. Salyut 3 carried two wing-like panels and a third of the same size fixed vertically on top of the station. Also, the solar arrays were capable of rotating 180°, obviating the need to move the entire station for the cells to face the Sun. This represented a great improvement on Skylab where the complete assembly had to be moved in attitude.

The interior too was modified, arranged with specific work compartments, and colors were selected for their psychological affect. A tape recorder was provided for private use, as was a chess set and library. A shower was installed in Salyut 3 and although the interior was considerably smaller than Skylab, many of the provisions designed into its US predecessor were duplicated in the Russian station.

The first manned occupation of Salyut 3 began with the launch of Popovich and Artyukhin aboard Soyuz 14 on 3 July. Their spacecraft was of the ferry type, that is it contained equipment canisters in the forward orbital module and used batteries for electrical power rather than solar cells carried on independent flights. Getting into orbit 3,500 km from the 18.5 tonne Salyut 3, Popovich and Artyukhin monitored four orbital corrections, taking over manually about 100 meters away. In the early morning hours of 5 July, the two cosmonauts opened Salyut's hatch and moved inside.

Compared with Skylab, Salyut was quite small, having less than one-third the internal volume, but it was the beginning of a productive use of space stations on a continuous basis. Denied the global communications network characteristic of NASA projects, Russian controllers communicated with their cosmonauts via several tracking ships deployed around the world. With emphasis on ground decisions it was necessary for the Russians to be within contact at frequent intervals. Experiments carried out by Popovich and Artyukhin included a Polinom-2M device designed to study blood circulation, a 'microbiological cultivator', photographic equipment for Earth views, new navigation devices, and an RSS-2 spectrograph for studying molecular components of the upper atmosphere.

The flight lasted less than 16 days, the cosmonauts returning within 2 km of their reported aim point. During the mission, US astronauts in Russia for discussions about the ASTP flight sent their congratulations to the cosmonauts. On 26 August, five weeks after the return of Soyuz 14, Sarafanov and Demin were launched aboard Soyuz 15. It failed to dock with Salyut 3, because, according to Shatalov, it carried an automatic rendezvous and docking system being tested for use on later unmanned tanker vehicles; when a minor technical glitch prevented an automated link-up, the crew were unable to complete the maneuver. The limited life of the battery powered Salyut necessitated its return to Earth, which it did at night on 28 August in a raging blizzard. Special efforts were made to reach the men and they were retrieved from their capsule within minutes of touchdown.

The apparent failure of Soyuz 15 led opponents of ASTP to voice their suspicion about Soviet capabilities. Concern that American astronauts would be exposed to the limited safety levels of Soviet hardware inspired Senator William Proxmire, a veritable thorn in many a technological bush, to question openly the wisdom of ASTP. This led to specialists in the technical press speculating on the performance of Russian spacecraft but ignorance about the precise mechanisms inspired incorrect assumptions. NASA, concerned about the adverse comment, and recognizing that there were indeed several loose ends concerning Russian control mechanisms, pressed for answers which they received to everyone's satisfaction.

Glynn Lunney was told by the Russians that they planned to launch a full dress rehearsal of their side of ASTP before the end of 1974. Preferring not to know anything about it when asked if they would track the spacecraft on details given five days before launch, NASA was unwilling to maintain the secrecy Russia demanded for this information. Nevertheless, it was important for US tracking stations to participate in the test so the Americans told the Russians they would begin tracking only when Moscow announced it publicly. To have deliberately kept quiet about the plan would have contravened a very important element in the NASA charter: total openness and public discussion of every activity conducted by the agency.

The test flight began on 2 December, 1974, with the launch of cosmonauts Filipchenko and Rukavishnikov, the prime back-up crew for ASTP, in Soyuz 16, identical to the spacecraft then being groomed for the joint docking mission. In a flight lasting nearly six days, every major segment of the ASTP mission was successfully simulated, with a special docking device enabling the crew to rehearse the actual link-up. Soyuz 16 was initially placed in an orbit different to that which it would need to rehearse the docking mission, several maneuvers proving that the Soviet spacecraft could accommodate modest errors in the launch vehicle.

For their part, US tracking stations at Bermuda, Tananarive, Antigua, Turk Island, Cape Canaveral, Canton Island, Hawaii, Kwajalein, and Ascension Island satisfactorily simulated American ground operations although Houston did not participate in the rehearsal. Soviet applause for the accomplishments of their two cosmonauts underscored the importance they placed on this test, its satisfactory completion eliminating the final hurdles.

For the previous 3½ months, Salyut 3 had been conducting its own, albeit unmanned, program of scientific (and other) work. On 23 September, just one month after the truncated flight of Soyuz 15, ground controllers commanded separation and ejection of a special recovery pod which duly de-orbited and returned its contents to Earth. Salyut 3 carried a long focal length camera and with a special orientation device was made to point continually at the Earth below, this being possible due to the rotational ability of the electricity-producing solar cell arrays. Moreover, the four cosmonauts associated with its operation were all military men rather than the more normal division between service and civilian personnel.

Nevertheless, by the end of the year, Salyut 3 had performed beyond expectation, exceeding by a handsome margin the initial three-month period for which it was designed. On 24 January, 1975, Salyut 3 was commanded to fire its propulsion system and destructively re-enter Earth's atmosphere over a safe region of the Pacific. By this time, however, its replacement was already in space. Salyut 4, a considerable improvement on its predecessor, ascended on 26 December, 1974, to a 350 km orbit by way of a lower intermediate path before maneuvering to higher altitude. Apart from several new aids provided for the crew in light of experience with earlier stations, Salyut 4 carried an infrared telescope/spectrometer with a 30 cm diameter mirror designed to observe the Earth's atmosphere; it would also be used for the Moon and galactic sources.

Special Emissiya equipment was installed for measuring lines of atomic oxygen in the atmosphere, and biological experiments in an Oasis device involved plant growth and fruit flies. Special exercise equipment was provided, including a treadmill, a bicyle ergometer and a Polinom device like that carried by Salyut 2. The first crew to occupy the station was launched on 10 January, 1975. Cosmonauts Gubarev and Grechko remained in space for 29½ days before returning to Earth having broken the previous Russian manned flight record set up by the ill-fated Soyuz 11 in 1971.

By now all participants in the ASTP mission were in the final stages of preparing hardware and procedures and it was not a good time for either country to suffer an unexpected failure in an ostensibly routine operation. But that was exactly what happened on 5 April, little more than three months before ASTP. In an event known thereafter as 'the 5 April anomaly,' cosmonauts Lazarev and Makarov were launched on what many considered to be an important long duration mission after docking with Salyut. However, minutes off the launch pad the failure of the booster rocket's two stages to separate caused Soyuz to go into an immediate abort, returning the crew on what became the only published suborbital flight of Soviet spacemen; denied the speed needed to achieve orbit, Soyuz separated from the launcher and arched down to the ground in a premature re-entry which put the cosmonauts 1,600 km from the launch site. They landed only 320 km from the Chinese border and were recovered in good health after the unexpected abort.

It was bound to call forth criticism. And it did. Proxmire blasted forth once more – but kept his big guns for the penultimate month before the flight – and got a document from the Central Intelligence Agency quoting an official as having opined the view that the Russians were incapable of monitoring two flights at the same time. That last comment alluded to the flight of Soyuz 18 launched 24 May with Klimuk and Sevastyanov on board.

In a statement issued soon after, Soviet controllers said the spacecraft would probably remain in space until after the planned 15 July launch of the ASTP spacecraft and that it would return to Earth after that mission, implying the simultaneous operation of two spacecraft in two separate missions. The Soviets were keen to show their American counterparts that far from stretching their capabilities, the Russian element of ASTP was, after all, no more to them than an interesting side show and that the main thrust of their manned program would continue apace. As for the failure of the booster rocket on 5 April, that was easy to explain said the Russians.

The rocket used for that particular Soyuz launch was one of a batch now considered obsolete and since the two launch vehicles being prepared for the joint docking mission were of a modified production batch, the same thing could not possibly happen again. But even if it did, they said, it should not concern the Americans since the US was not involved in such an early part of the flight. And it did not. Everybody accepted Bushuyev's explanation. But it had been a gruelling experience for the crew. Travelling at a speed of 19,800 km/hr, the assembly aborted 180 km above Earth and in suddenly decelerating seconds after separation, the Soyuz experienced 14 g on the way down.

Before the successful launch of Soyuz 18, final management reviews were held in Russia. Again, Keldysh was unfit for the meetings, his place taken by Vladimir Kotelnikov who took over his position due to ailing health. On 18 May, less than two months before the launch, George Low, John Yardley, Arnold Frutkin, Glynn Lunney and Walter Kapryan, visited the Soviet launch complex at Baykonur. Professor Bushuyev was their host as the team drove in a van usually reserved for transferring cosmonauts to the launch pad, passing sheds and buildings before moving to the assigned pad for ASTP.

They stopped for a few moments at the monument set up to commemorate the place from where Sputnik 1 ascended more than 16½ years before and from where Yuri Gagarin was launched 3½ years later. It was a historic visit, a rare moment when chief contestants in the game of international space exploration met where it all began. For without Sputnik and Gagarin there would have been no Apollo and no ASTP just weeks hence.

Back in Moscow, George Low and Vladimir Kotelnikov chaired the Flight Readiness Review beginning 22 May. Five working groups had been set up to complete preparations throughout the run up to flight operations. Now, after nearly three years of effort, group heads were in a position to report the state of readiness: V. A. Timchenko on flight operations and mission planning; V. P. Legostayev on guidance control and docking; Bob White on the docking hardware; Boris Niktin on communications and tracking equipment; and Walt Guy on life support functions and crew transfers. It was a vital get-together – the last major meeting of minds prior to lift-off. Nothing must be overlooked.

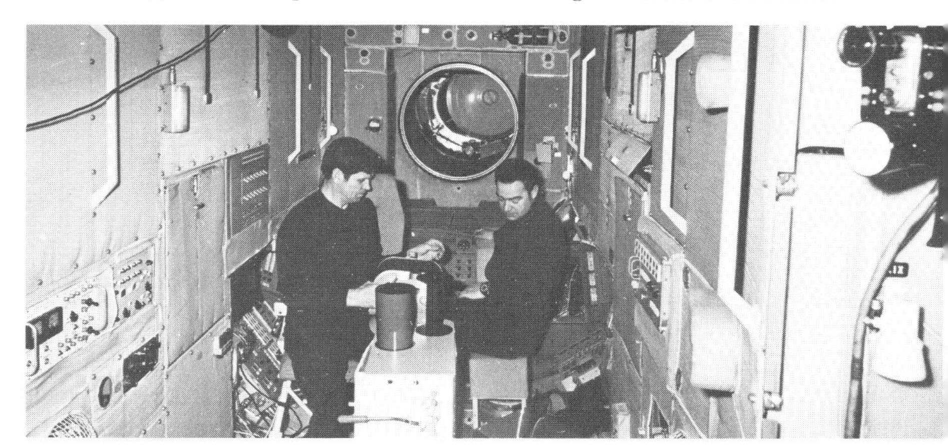

Inside the Salyut 4 mock-up, cosmonauts Gubarev and Grechko reveal the comparatively small size of this production-line facility, smaller than Skylab but the first real step toward a permanent manned station in space.

← Moscow's flight control center during the mission of Soyuz 16, strongly influenced by the design of NASA's Mission Control Center, specially built to serve as both control room and prestiguous show-piece for open coverage the Soviets knew would accrue from the cooperative ASTP mission.

CHAPTER TWENTY-EIGHT

End of the Beginning

The mission plan for ASTP had NASA count down a Saturn IB/Apollo while the Soviets launched their Soyuz spacecraft carrying Leonov and Kubasov. Lift-off time was set for 8:20 am, Cape time, on 15 July, 1975, followed by the Apollo launch at 3:50 pm. Following a series of rendezvous maneuvers similar to those carried out by Skylab astronauts, Stafford, Brand and Slayton in Apollo were to pull alongside Soyuz two days after launch and dock with the Russian spacecraft soon after noon, 17 July. During the next two days, crew transfers would take place via the Docking Module on the front of Apollo with one or two astronauts visiting Soyuz and one cosmonaut visiting Apollo at a time; one astronaut and one cosmonaut would be in Apollo and Soyuz respectively at all times. About 43 hr 48 min later, on 19 July, Apollo would pull free of Soyuz, move away for several combined scientific experiments, and then re-dock, finally separating at about noon on that day.

Less than two days later Soyuz would return to Earth leaving the Command and Service Modules to remain in space for a further five days, conducting experiments before jettisoning the Docking Module and returning to a Pacific splashdown, 24 July. For Apollo, the mission would have lasted 9 days 1½ hours, bringing to an end the last flight with a spacecraft of this type. Compared to other, more ambitious projects, the hardware was simple and rugged, forgiving in the event of unexpected events, and well tested in many simulated environments.

The Russians built a new Soyuz from the ground up for the mission that would unite Soviet and American cosmonauts in orbit for what time would reveal as a unique moment in history. But Apollo hardware was brought out of storage for modifications necessary to equip the American segment. The launcher, Saturn IB AS–210, had been built by early 1967 and placed in storage but the stages were brought out and re-examined for stress corrosion cracks of the type found on the last Skylab launcher, AS–208.

By early 1974 both first and second stages were at the Kennedy Space Center. On 19 February, 1975, inspection of the eight first stage fins revealed a hairline crack in two fin fittings and by the end of the month the other six fins were found to have similar decay. All eight were subsequently replaced and a memo from Glynn Lunney to Professor Busheyev informed the Russians of this action. Spacecraft 111, assigned the role for ASTP, had been completed in March 1967. More than two years later work began to modify the H-series Moon vehicle into an Earth-orbit rendezvous ferry. It was at the Cape by September, 1974.

The changes carried out to CSM-111, and the back-up, CSM-119, were almost identical to modifications for Skylab vehicles. Basically configured with enhanced features brought in as a result of the Apollo 13 accident, CSM-111 had just two SPS propellant tanks, supplementary RCS energy in a Propellant Storage Module comprising four fuel and five oxidizer tanks, and a single SPS helium pressurization tank in the Service Module's center section. But unlike Skylab Apollo vehicles, CSM-111 retained all three fuel cell units fitted to Moon vehicles and the large steerable dish antenna for S-band communication. Like Skylab, ASTP would be an Earth orbit project but, unlike the space station program, would require communication through the large ATS-6 satellite in geosynchronous orbit, hence the antenna.

CSM-111 would carry only 1,198 kg of SPS propellant, a quantity less than that provided for the Service module RCS maneuvering thrusters. Despite much additional equipment necessary to support the unique mission needs of ASTP, and communication systems essential for links through ATS-6, CSM-111 was the lightest manned Apollo to fly. With propellants it weighed only 12.7 tonnes. But Apollo was, of course, not the only payload to be lifted by the Saturn IB and the CSM was necessarily light to reserve a margin below the launch vehicle's 16 tonne lifting capacity so that the Docking Module could be carried beneath the CSM.

The Docking Module, or DM, would provide an airlock facility between Apollo and Soyuz, with docking equipment at each end. Because Apollo was designed to support an atmosphere of pure oxygen at one-third (258 mmHg) sea-level pressure the DM was needed to allow crew transfers to the Soyuz spacecraft which would normally contain a sea-level atmosphere of oxygen-nitrogen; for this mission the Russians agreed to fly with a mixed gas atmosphere at only two-thirds (517 mmHg) pressure. That simplified the transfer. Had a higher pressure differential been present, each crewmember wishing to move through the DM from Soyuz to Apollo would have had to spend considerable time purging nitrogen from his blood.

The module itself comprised a welded, cylindrical, pressure vessel formed from 1.58 cm aluminum, 3.15 meters long and 1.4 meters in diameter. Tapered bulkheads at each end reduced diameter to 80 cm on the Apollo end and 91 cm at the Soyuz end. Both ends had single docking hatches with pressure gauges and an equilization valve. To open or close the circular door a crewmember would rotate six latches simultaneously by moving the handle through 60°. The exterior end faces of the Docking Module supported two different docking units.

On the Apollo end was a conventional drogue to which a probe, pirated from Apollo 14, would be used to draw the CSM on to the Docking Module. Thus fixed rigidly with 12 main docking latches, like similar equipment on Lunar Modules or the front of Skylab's Multiple Docking Adapter, the complete assembly would remain attached throughout the active part of the ASTP mission.

The extreme end of the Docking Module carried a compatible docking unit of the androgynous type evolved through detailed study with the Soviet Union. Both countries agreed the design and fabrication details but each was responsible for respective units. As presented in an earlier chapter, the androgynous concept arose from suggestions that mutual programs conducted by America and Russia could have no better starting point than that based on setting up a space rescue capability. To do that a common docking unit was essential, one which, unlike previous mechanisms, could be active or passive and connect with an identical unit on another spacecraft.

This 'universal' assembly consisted of an extendible guide ring with three petal-like guide plates, three capture latches and

The Soviet docking unit for ASTP inspired by extensive studies into an androgynous concept pioneered by Soviet engineers.

In the clean-room at Rockwell's Downey facility, hardware for the ASTP mission. At right, in foreground, is the US portion of the new docking system, behind it the Docking Module and at left, in background, is the Command Module for the spacecraft the crew would use to reach Soyuz in orbit.

six hydraulic attenuators to absorb the shock of contact. After capture, the guide ring would be retracted to clamp together eight structural latches designed to connect the two vehicles rigidly; the two actions were analogous to 'soft-dock' and 'hard-dock' with Apollo and a Lunar Module. Theoretically, two androgynous docking units of the same type could be installed on two separate spacecraft for satisfactory link-up.

For ASTP, almost identical units were fabricated in respective countries: the American unit, differing only in its cable drive guide rail retraction system versus an electro-mechanical device, was fitted to the Docking Module, while the Russians built their own unit for the front of the Soyuz spacecraft. The Docking Module also supported four spherical tanks holding a total 40.6 kg of gaseous nitrogen and oxygen. On the pad, the DM would contain normal sea-level atmosphere, gradually venting on the way up to one-third sea-level pressure.

Before opening up the tunnel and moving inside, astronauts would evacuate the nitrogen and pump up with oxygen to one-third sea-level pressure. Once inside they could close the hatch and add nitrogen up to two-thirds sea-level pressure compatible with Soyuz on front, thus allowing the forward hatch to be opened and the crew to move inside the Russian spacecraft. A reverse procedure would allow them to get back inside Apollo via the Docking Module. In all, the Docking Module weighed 2.01 tonnes and supported equipment necessary for visual docking procedures. Together, the CSM and Docking Module would be 13.1 meters long and weigh 14.7 tonnes. With Soyuz docked rigidly to the Docking Module, the complete assembly would present a length of about 20 meters and weigh approximately 21 tonnes.

Two Docking Modules were built by Rockwell International's Space Division: DM-1 used for thermal vacuum tests and refurbished to flight status as DM-1A, and DM-2 built as the prime hardware for flight. Structural fabrication of DM-2 began in April, 1973 and by August, 1974, the unit was ready for delivery. It was shipped to the Cape two months later. If needed as back-up hardware, Apollo spacecraft 119 would be installed, with DM-1A, on top of AS-209, the contingency Saturn IB taken over from the Skylab program in February, 1974. This hardware would only be used if the prime vehicle suffered an abort during the ascent, or if the mission was terminated in its early stages.

Supporting communications and data relay from the NASA side were stations at thirteen STDN sites, one ship and three ARIA aircraft. The Soviet network comprised seven ground and two ship stations. Use of ATS-6 would be afforded between approximately 55°W and 125°E, across the Greenwich meridian.

At the Flight Readiness Review, Low and Bushuyev heard from the five working groups established to monitor and implement the various segments of the ASTP objective. Mission planners reported on the maze of alternate flight profiles should something go wrong, on the myriad network of launch windows, and on the abort and contingency plans covering many possible failures. But even if everything went well, the integration of two flight control teams responsible for two spacecraft forming a single assembly in orbit was no mean task. Russian control functions would originate from a new center at Kalinin outside Moscow. Joint flight directors in the two countries would head their respective teams and communicate via seven voice links, two teletype channels, two TV channels, and two air/ground translation channels.

Tests conducted with the flight-rated hardware were satisfying. The Russians were concerned about damage that could occur when selected RCS thrusters impinged upon the Soyuz solar panels so NASA agreed to limit the use of suspect thrusters. Another sensitive area concerned use of the forward firing thrusters when Apollo was close to Soyuz and here too the Americans agreed not to use those thrusters when the two vehicles were close together. That issue had been particularly troublesome when the Russians visited Houston but with agreement reached that the engines would be inhibited within two seconds of docking the Soviets were satisfied.

The Flight Readiness Review was informed that both spacecraft could operate safely in a docked configuration. As for the docking units, they too had come through with flying colors; Vladimir Syromyatnikov's design had been tested repeatedly in compatibility checks in Houston and in Moscow, proving that the two assemblies would satisfactorily link up in space. As for the electrical communications and tracking networks, the two sides agreed that everything here was ready also and that the electromagnetic environments of the two spacecraft were compatible. In January and February the Russians had brought their equipment over to Florida for tests inside Apollo and the Docking Module and in early May the Americans hiked their communications equipment to Moscow for similar tests.

As for environmental compatibility, both sides were concerned about their own particular worries: the Soviets about pressure integrity (remembering the sudden de-pressurization of Soyuz 11), and the Americans about flammability (mindful of the sudden death of the Apollo 1 crew in 1967). Because of the Russian demand for very careful monitoring of minor changes in atmospheric pressure, gauges significantly more sensitive than any previously used in the NASA program were manufactured for the Docking Module. Similarly, the Americans insisted on the Russians fire-proofing everything inside Soyuz. The fact that they had not done this before reflected how much safer it was to fly in a mixed oxygen-nitrogen atmosphere at sea-level pressure and not on any lack of concern they were suspected of exhibiting.

NASA suggested the Russians use material from the United States which had evolved through deep concern for the possibility of a fire on board Apollo in space. The Russians refused and went home to develop their own materials. When it emerged, under the name Lola, it was seen in tests to be superior to the American fabric! The Russians were delighted with this and promptly set about ransacking the interior of their spacecraft, replacing everything with the prestigious Lola.

When Low completed the Flight Readiness Review with his Soviet counterparts he returned with his men to the United States to present a summary to President Ford and the project's most ardent critic, Senator William Proxmire. The latter could do serious damage if unrestrained, for the Russians would not take kindly to, indeed they could not understand, dissident viewpoints publicly expressed from high office. Ford was enthusiastic about the coming flight, but Low had his reservations about the lack of clear control where, in the event of a major malfunction, there was no one man in space or on the ground who had recognized authority over everyone else. It was his only reservation, and one that would not be called to issue.

As for the Flight Readiness Review, the Russians were largely silent, having quietly made sure before that *both* parties were in fact as ready as they claimed. By this time Salyut 4 was inhabited by the cosmonauts from Soyuz 18 and Proxmire was concerned that the Russians might not be in a technical position to control two flights adequately. But with the two flights assigned totally separate control centers, NASA was not at all worried. Many officials did find it rather tiresome that the Soviets wanted to upstage the American side by having their cosmonauts exchange conversation on a mission where only one NASA spacecraft would be in orbit; the Russians would have a manned space station and two cosmonauts in orbit in addition to the two assigned to ASTP.

Three weeks after the Moscow meeting, an internal NASA review was held at the Cape. Representatives from all the concerned NASA facilities were there for a review of the spacecraft and launch vehicle on pad B at complex 39. It was the last flight of Saturn and Apollo, it was the last use of one-shot hardware in the NASA manned flight program, the last splashdown, and the last use of a team set up more than a decade before. And it was a public match of American and Russian technology. It simply had to work, almost better than it had ever done before.

Management personnel were a little concerned about the weather, but July was never a good month at the Cape, although the KSC staff meteorologist Jesse R. Gulick told the review meeting that there was a 77% probability of a clearance for lift-off. Chester Lee went round the table checking with Center personnel, and then heard affirmations of faith from George Low and John Yardley before ordering full steam ahead for launch one month later.

The two stages of AS-210 had been stacked in the Vehicle Assembly Building on 14 January, 1975, and one month later in a separate activity the Docking Module was placed in the adapter shroud on which the Apollo spacecraft was placed by early March. The stacked assembly was then transferred to the launch vehicle and mated to it on 19 March. On 24 March it was rolled to the pad. On the day before the countdown demonstration tests Stafford, Brand and Slayton began a three week 'Flight Crew Health Stabilization Plan,' limiting their contact with germ-ridden personnel. Early next morning, 25 June, CDDT got under way, followed four days later by a 56 hour link-up with the Moscow control center in joint operations aimed at proving the systems.

On 2 July the crew flew to Florida for the CDDT and then returned to Houston for more simulator runs, extra lessons in Russian, and flight document reviews. Two days before blast off, Stafford, Brand and Slayton returned to the Cape.

Meanwhile, on 11 July, the Russians moved their prime hardware out to the launch pad, trundling the horizontal assembly on rails from where it would be raised to a vertical position enclosed within the four main support arms. Two days later, the day the crew arrived at KSC, the back-up launcher was similarly wheeled out from its preparation shed, to another launch pad; that complex was 20 km to the north. A minimum amount of work was to be conducted at the pad, the Russians having selected a checkout procedure that kept the hardware inside protective buildings as late as possible due to severe weather at the Baykonur space base. Both prime and back-up combinations would be moved simultaneously to the T-5 hr mark, the prime set being fuelled at that point ready for launch.

Once filled with liquid oxygen, the Russian booster would have a 'life' of twenty-four hours before corrosion set in and its replacement by the back-up hardware became mandatory. But the Soviets were taking no chances. Their own prestige lay in the balance because for the first time in the space program, Soviet commentators would broadcast live the events associated with countdown and lift-off. Millions of people around the world were watching their TV screens as Leonov and Kubasov left their quarters and were driven to the pad. It was 12:00 noon, Moscow time; in London it was 10:00 am; at the Cape it was 5:00 am; in Houston it was 4:00 am.

At the base of the pad steelwork, Leonov was seen to stop and address the people gathered around, smiling and waving to watchers around the world. Timing respective launches had been difficult. Russian mission rules stipulated that pre-re-entry activities allowed the crew a view of the Sunlit side of the planet and that there should be one full hour of daylight at the end of the nominal mission. American rules required three hours of daylight following launch, in case of abort, and two hours of Sun at the end of a nominal flight. Apart from the staggered interval between launches, necessary to allow the plane of the Soyuz orbit to 'migrate' west, respective clocks in Russia were seven hours ahead of clocks at the Cape. Despite being launched $7\frac{1}{2}$ hours apart, Soyuz was to be sent up at 3:20 pm local time while Apollo would ascend at 3:50 pm, Cape time. This meant that local Sun angles would be almost the same for respective launches, thereby preserving rules which had similar daylight requirements in case of abort. Nevertheless, because of the lag in lift-off times, Leonov and Kubasov would be in orbit even before Stafford, Brand and Slayton were woken.

Just 45 minutes before the launch of Soyuz, two steel arms alongside the rocket were lowered, freeing it of service platforms used to attend the launcher's needs prior to ignition. With less than one minute to go, electrical and propellant umbilicals fell away. From 20 seconds the countdown was fully automatic, and everyone concerned with the flight in Russia and America held their breath as for the first time a live commentary told the world what was happening. It was a truly unique moment, when for this one flight the activities were completely open and public knowledge – as they occurred. From the control center, a clear voice announced events:

'Telemetry registry equipment has been turned on. Also the control, onboard control systems of registration equipment here. Ignition! The engines are powered up. The launch, the booster is off.'

The booster was indeed off. After what seemed like an interminable length of time, fire and smoke thundering from the base of the gleaming white and silver rocket, the four restraint arms fell back away from the ascending structure. 'The flight is proceeding normally. The program maneuver of the booster rockets has been given. The flight is normal, the engine is operating in a stable manner.' At the Cape it was 8:20 am, a real treat for breakfast TV. Minutes later the Moscow announcer told of success: 'The third stage engine has been switched off and the spacecraft Soyuz has now been inserted into orbit. Orbital flight has been initiated. The antennae of the spacecraft are now open and the solar panels . . . are being extended . . . Control over the

Cosmonaut Yeliseyev selected as the Soviet Flight Director for ASTP, momentarily looks up from his console during a joint training session with NASA's Johnson Space Center.

The first man to step outside a spacecraft in orbit, Alexei Leonov (left) would command the flight conceived to begin a new era of cooperation in manned space operations, with his colleague Valeri Kubasov.

Soyuz spacecraft has now been transferred to the Soviet Mission Control Center.'

Unlike NASA operations, launch controllers maintain authority over the assembly until separation of the spacecraft from the last stage in Earth orbit, at which time the Soviet equivalent of Houston's Mission Control takes over. Soon, US controllers would take over the flight of Apollo spacecraft 111 under the baton of flight director Don Puddy. Joint flight directors Jay Honeycutt and Yuri Denisov were also on hand. Until shortly before the Soyuz launch, Neil Hutchinson monitored Soviet operations.

Dignatories were gathering now, drifting in to the Houston Center for the coming events. George Low was there, as was the JSC Director Chris Kraft. In the viewing room too was Max Faget, one of those who began it all. At the Cape, Director Lee Scherer kept himself informed on the countdown and Chester Lee chatted with Walter Kapryan. There were now two built-in holds for the Apollo countdown, designed to accommodate last minute work and to synchronize lift-off precisely with the actual orbit Soyuz was found to be in. As it turned out, the Soviet spacecraft was in an almost perfect path of 186 × 222 km inclined 51.8° to the equator.

The first hold began at T−3 hr 30 min and lasted 54 min 36 sec; the second would begin at T−4 min and last a nominal 5 min 24 sec. Stafford, Brand and Slayton were woken in their private quarters at the Manned Spacecraft Operations Building at 10:10 am, Cape time, just as Soyuz was moving into its second revolution of the Earth. Less than one hour later they sat down with astronauts Young, Evans and Lousma and training officer Dave Bauer for a traditional breakfast of steak and eggs. In CSM-111 several kilometers away on launch complex 39-B, astronaut Bob Crippen set up switch positions and made the spacecraft ready for habitation.

By now the Soyuz crew had equalized the pressure between the descent and orbital vehicles and opened the hatch through into the latter, removing and drying their space suits. In the Launch Control Center at KSC, astronaut Karol Bobko sat at the communicator's console, ready to welcome the flight trio on board. Shortly after 12:30 pm, the astronauts rode an elevator down to the first floor of the MSOB and waddled across to the pad transfer van. It took about fifteen minutes to reach the pad, the three spacemen ascending the umbilical tower at precisely 1:00 pm. Across the other side of the world, Moscow controllers told Leonov and Kubasov they would be required to ride out a correction maneuver at the end of the fourth orbit to trim the path more closely to the desired value.

At the Kennedy Space Center, Tom Stafford slid his bulky frame on to the center couch and then across to the left. He was followed by Slayton while Brand waited near the elevator, and then he too moved in, aided by Crippen. The first words from the spacecraft came as Tom Stafford first hooked up the communication line prior to Slayton's ingress: 'Looks like it's a good day to fly.' That call went out at precisely 1:02 pm. Twenty-three minutes later the hatch was closed and locked.

Now the test conductor, Skip Chauvin, would monitor the pressure integrity checks as they pumped up the cabin to a 60/40 nitrogen-oxygen atmosphere. Just as that was being completed, Soyuz burned its main engine for 7 seconds and nudged into a 192 × 232 km orbit and then settled down into a slow spin, turning at one revolution every two minutes. It would make one more maneuver before being joined by Apollo two days hence. At 2:15 pm, Cape time, the boost protective cover was fastened over the exposed side of the Command Module and fifteen minutes later the closeout crew evacuated the white room. Inside spacecraft 111, the crew were ahead of their checklist. Everything was going just as it should.

On down to the final seconds the clocks ticked away to the last conventional manned flight from Florida's spacecraft:

'Approaching the 30 second mark in our countdown, water pouring on to the flame deflector, now coming onto the deck of the mobile launcher. Everything proceeding smoothly. We'll get a guidance release at the 17 second mark . . . Engine ready light on. 10, 9, 8, 7, 6, 5, 4, 3, 2, engine start sequence, 1, 0, launch commit. We have a lift-off. All engines building up thrust. Moving out, clear the tower.'

STAFFORD 'Roger, tower clear.'
CAPCOM 'Roger, Tom. You got good thrust in all engines. You're right on the money.'
STAFFORD 'Roger, I got roll program started.'

Soviet Ambassador Anatoly Dobrinin (foreground with binoculars), NASA Administrator James Fletcher (right) and test conductor Richard Thornburg watch the Saturn IB as it thunders away from Launch Complex 39 carrying astronauts Stafford, Slayton and Brand toward Leonov and Kubasov aboard their Soyuz spacecraft.

On through max q, the period of maximum dynamic pressure, the old Saturn IB rattled up into the rarefied layers of the atmosphere, the fan tails of flame from eight healthy engines belling out as pressure on the plume decreased. Around the MOCR at Houston, flight director Pete Frank, on duty now for the ascent, went round the room polling his men for staging while Dick Truly maintained a dialogue with the crew. The g forces were building up now as the weight of AS-210 decreased and the engines thrust Apollo on along the programed rails toward space.

At 2 min 20 sec the first stage shut down with the vehicle 58 km high and 64 km out from the Cape. 'Okay, and the light is out on the IV-B,' called Tom Stafford as the indicator light on his control panel extinguished itself at second stage ignition, telling the Commander that all was well. 'And we've got some beautiful sights,' added Stafford. 'Man, I tell you, this is worth waiting sixteen years for,' quipped Deke Slayton, getting his ride into space at last.

For nearly 7½ minutes the S-IVB stage accelerated them to orbital speed, propelling its payload along a flattened trajectory gaining velocity now rather than height. At shutdown, the vehicle was in a 155 × 173 km orbit, a path considerably below that of the Soyuz spacecraft; Apollo was 1,800 km north-east of Florida, Soyuz was about to pass over Belgrade in Yugoslavia. At about that time, the Russian spacecraft reported it had successfully pumped itself down to two-thirds sea-level pressure, the necessary reduction for later events. Lift-off had occurred milliseconds past 3:50 pm Cape time, but in Houston it was 2:50 pm and forty minutes later Stafford declared his crew ready for the transposition and docking maneuver essential to extracting the Docking Module from its hangar on top of the S-IVB stage.

This was the first Saturn IB flight to carry a mission payload in the adapter that supported Apollo; only the Saturn V had the lifting potential for a Lunar Module at that location. Nevertheless, when it came to pulling free from the four petal-shaped doors that held the CSM and turning around to move back in, Tom Stafford's view was blinded by Sunlight. At the moment of separation the four doors had been ejected, exposing the Docking Module sitting proud of the circular base atop the second stage. But it was difficult to see and Stafford nudged close in to about 10 meters. Then, squinting through the COAS that provided an alignment with the docking target, he gently moved the CSM in to soft-capture with the probe. It snagged first time, a perfect docking that some would say was the best ever achieved by an Apollo vehicle. Seconds later the probe retracted, rigidly clamping the two structures together as all twelve docking latches fired on connect with the lip of the collar.

Later, at 6:31 pm Houston time, the Apollo crew performed their first maneuver by using the big SPS engine to put their orbit into a circular path 172 km above the Earth. About 90 minutes later the inert S-IVB stage, now some distance away from Apollo, was deliberately de-orbited by venting residual propellants through the engine bell, bringing it down over the Pacific.

About two hours after the first maneuver, Apollo fired the SPS engine for a second time to set up an elliptical phasing orbit of 174 × 237 km designed to bring the CSM to Soyuz on the latter's 36th revolution of the Earth. Between events, Vance Brand told the Houston controllers that there was a stowaway on board in the form of a small Florida mosquito. They saw it only once, flying around the interior of the Command Module before it succumbed to the reduced oxygen-rich atmosphere. Then there was a minor irritation as the cabin began to overheat but that was soon displaced by a potentially more serious problem.

One of the last activities prior to beginning a rest period required the crew to remove and stow the probe and drogue situated in the tunnel, access to which they gained after first taking out the forward Command Module hatch. This was necessary before sleep because it was felt advisable to stow a special experiment canister in the tunnel. Designed to study the effects of electrophoresis, the equipment canister would release nitrogen which in the following few hours could disturb the oxygen balance in the spacecraft. But after collapsing the probe's support arms, Brand was unable to release the three small capture latches. To do this he would insert a special tool in a connector plate on the probe and turn it 180°. But the tool would not go in, and inspection by torch showed what seemed to be a small connector still bridging the orifice inside.

It was twenty-two minutes past midnight when Tom Stafford informed Houston and flight director Neil Hutchinson told him to get his men to sleep, leave the problem for the next day, but pump up the cabin to 285 mmHg to off-set the effects of additional nitrogen that might bleed from the electrophoresis canister. There was little danger of the probe being stuck, the controllers were confident they knew how to free it. But Soyuz too had a problem. One of the cameras, a black and white television unit, failed, preventing the possibility of pictures of Apollo when the two spacecraft came close together.

When Stafford, Brand and Slayton began their first rest

period, Apollo was trailing Soyuz by about 5,200 km. Having begun their own sleep session several hours earlier, Leonov and Kubasov were up and active by 2:00 am, Houston time, on Wednesday, 16 July, the crew of Apollo rising shortly after 7:00 am. It would be a comparatively passive day occupied with meticulously removing the probe by disassembling it from one end, performing some scientific experiments, and trimming the orbit for rendezvous on Thursday. The Soyuz crew, meanwhile, would circularize their spacecraft's orbit, hold a conversation with the cosmonauts aboard Salyut 4, and send color TV pictures down to Earth.

The Soyuz maneuver came just as the Apollo crew were beginning their second day, an 18 second engine burn that shoved the spacecraft into a path about 229 km above Earth. By mid-morning Brand had the probe out and at 10:46 am Deke Slayton was ready to check the Docking Module. One hour later Houston received television pictures from inside the module while the astronaut carried out some experiments. He was back in Apollo by 1:00 pm. During the checkout, Slayton held up a card inscribed with a familiar question reminiscent of the Mercury days: 'Are you a turtle?' There were loud chuckles in the MOCR.

Shortly after 3:00 pm, Stafford fired the big SPS engine for the third time. That maneuver put Apollo into a 173 × 231 km orbit, a phasing burn designed to set up the relative positions of the two spacecraft for rendezvous next day, now less than 3,100 km apart. At this time the Soyuz crew began their second 'night's' sleep and the Apollo astronauts bedded down shortly after 8:00 pm.

But dramatic events that sent hearts racing on earlier flights haunted the space lanes around planet Earth, for around 3:00 am in the morning a string of master alarms rang in spacecraft 111 in simulation of a similar fault during Apollo 16 more than three years before. It was only a failed diode in the detection logic of the Coupling Data Unit; a deeply buried gremlin grumbling around in the electronics. The crew decided to remain awake, however, and at 8:14 am, Soyuz too was once again communicating with Soviet tracking stations.

It was Thursday, 17 July. The day three Americans and two Russians would shake hands in space for the first time. The only time for many, many years to come. During the night, the Apollo crew slept in various places, locations that would remain their respective choices for the rest of the mission: Stafford in the Command Module, Brand in the Tunnel, Slayton in the Docking Module. But Apollo and Soyuz were now less than 930 km apart and it was time for the first rendezvous maneuver.

Passing over the South Atlantic, the SPS engine fired spacecraft 111 to a new orbit which reduced apogee (the high point) by about 29 km. This allowed Apollo to increase the catch-up rate by dropping to a lower track around Earth and consequently speeding up. Eight minutes later Soyuz was acquired visually through the Apollo sextant and at 8:05 am Deke Slayton managed to get through on VHF:

SLAYTON 'Hello, Valeriy. How are you? Good day, Valeriy.'
KUBASOV 'How are you? Good day.'
SLAYTON 'Excellent.'
LEONOV 'We are very glad. Good morning.'
SLAYTON 'Alexey, I hear you very well. How do you read me?'
LEONOV 'Very good.'

Shortly before 8:30 am, Soyuz switched on the range radar equipment and the two spacecraft were now a mere 222 km apart. At 8:35 am Stafford again fired the main engine, putting the spacecraft into a 195 × 208 km path; Soyuz was waiting in an almost circular orbit at 229 km. Performed in the Pacific dark, this corrective combination burn set up conditions for a coelliptic sequence maneuver just thirty-seven minutes later. Passing on across the Pacific to the coast of South America, Apollo burned into the almost circular path 19 km below Soyuz it would need to begin the terminal phase.

In the Sun's light now, Apollo was right on course. After completing another full revolution of the planet, the NASA vehicle would rendezvous with Soyuz over the South Atlantic and dock just before passing across Europe and North America. But first, the Russian crew were informed of the successful coelliptic burn before TV viewers got a glimpse of the astronauts passing through the Madrid tracking station. In darkness once more, the Apollo men caught sight of Soyuz's flashing lights at 10:00 am, just seventeen minutes before firing the Terminal Phase Initiation burn. That put the spacecraft on course for rendezvous.

TPI had been fired just as the vehicle moved north of Australia. Across on the other side of the Pacific, due west of Chile, spacecraft 111 braked slowly in a series of short RCS thrusts designed to bring Apollo to station-keep with Soyuz. Performed in daylight, both spacecraft swept within view of the ATS-6 communication satellite at 10:55 am, Houston time. In Apollo, the three astronauts had already sealed themselves in the Command Module with the forward hatch to the Docking Module closed and locked. In Soyuz, mindful of the potential accident that could cause the two vehicles to collide, Leonov and Kubasov closed the hatch to the Orbital Module and donned their suits; Stafford, Brand and Slayton would only don their own suits for jettisoning the Docking Module after the joint activities.

From the Descent Vehicle, the cosmonauts waited, lights on the exterior of their spacecraft flashing welcome signs at the American visitors. Through ATS–6, television pictures from Apollo showed the Russian spacecraft only a few hundred meters away. At Houston, officials and dignatories were crowding the viewing room in Mission Control.

Astronauts Allen, Garriott, McCandless, Musgrave, Scott and Schweickart were there, as was another noted explorer, Captain Jacques Cousteau. Former Apollo director Sam Phillips had arrived, and astronaut John Young brought Jim Fletcher and his wife accompanying Ambassador Dobrynin. Caldwell Johnson looked on, as did former JSC Director Bob Gilruth.

It was a significant day, almost exactly six years after Neil Armstrong and Edwin Aldrin left Michael Collins in Moon orbit for a ride down to the lunar surface. Apollo, an epitome of the once strong space race begun by two Presidents, both now dead, who set mankind afire with the burning ambition of ages past, was about to link up with Russia's space ferry. It was to be a different contest from now, the Russians consolidating their plans for manned Earth-orbiting space stations while NASA took time away from manned operations to build a reusable launch vehicle for the 1980s.

As the two spacecraft came within a few meters of each other that morning hour on the Thursday in July, many former officials cast reflective thoughts upon the past fourteen years. But the message from capcom Dick Truly was loud and clear: 'Apollo, Houston, I've got two messages for you: Moscow is GO for docking, Houston is GO for docking. It's up to you guys. Have fun.' The only activity performed by Soyuz was to gently roll 60° for the proper attitude alignment. Slowly, but with the experience of three rendezvous and two docking flights behind him, Tom Stafford nudged the nose of Apollo in toward the Soyuz docking ring. It was a new type of activity, the long Docking Module giving the CSM an extended nose, but it went as well as possible. At 11:09 am, Houston time, exultant calls went between the two spacecraft.

STAFFORD 'Contact!'
LEONOV 'Capture!'
STAFFORD 'We also have capture . . . we have succeeded. Everything is excellent.'
LEONOV 'Soyuz and Apollo are shaking hands now.'
STAFFORD 'We agree.'
LEONOV 'Well done, Tom, it was a good show. We're looking forward now to shaking hands with you on board Soyuz.'
STAFFORD 'Thank you, Alexey. Thank you very much to you and Valeriy.'

It was a triumphant moment, perhaps the single most ebullient time in the MOCR since Neil Armstrong set foot upon the Moon. Glynn Lunney lit up a cigar and spoke to Bushuyev, both men watching each other on the television displays linking Moscow and Houston. The two spacecraft docked on Apollo's 29th orbit, just 1,160 km east of the coast of Portugal. Seventeen minutes later the crew had the Command Module hatch off and the Docking Module hatch open, reporting the presence of a strange smell that temporarily damped spirits at respective control centers. It was traced to a sample left in the materials processing furnace, an experiment installed in the docking module to continue similar tests conducted aboard Skylab.

After some time spent checking out the module and its systems, following which all seemed to be in order, Stafford joined Slayton inside before the hatch was closed. It was the start of a regularized schedule of public relations activity designed to provide equal opportunity for the sense of accomplishment both sides acquired. But it was pre-planned to the letter, a sequence

Artist's rendition of the docked Apollo-Soyuz configuration.

refined in the closing minutes of contact through ATS-6 more than two hours after docking when Houston advised Apollo on what to do after the initial handshake:

'During the first transfer, just after the acquisition of Soviet TV and just after Tom enters the Orbital Module – Deke will still be in the Docking Module – the Soviet leaders would like to pass a message to the crew . . . Immediately after they have finished their message, bring the camera into the Orbital Module and set it up as planned. (Tom) and (Deke) will take positions around the Orbital Module table and the President of the US would like to relay a message to the Commanders of both vehicles. The sequence will be that all will get into position and allow the Commander of the Soyuz to welcome you to the Orbital Module; the President will then speak. (Deke) is requested to give his headset to the Soyuz Commander so the President may speak to the Soyuz Commander.'

With a 'we think we got that' from the crew, Houston prepared for the coming events confident of efficient stage-management. At 2:10 pm the Soyuz crew opened the hatch on the front of their Orbital Module, the Docking Module having been pumped up to an oxygen-nitrogen atmosphere at two-thirds sea-level pressure. Seven minutes later Stafford pulled open the Docking Module hatch, a forward looking TV fixed at the Apollo end looking through into the Soyuz spacecraft. 'Looks like they got a few snakes in there too,' cried Stafford as he saw umbilicals in the Russian spacecraft floating in that un-Earthly state of suspension characteristic of weightlessness; it reminded him of his own 'snakes' on Apollo 10. 'Come in here,' called Stafford as Leonov and Kubasov hovered around the hatch on Soyuz. 'Glad to see you,' said the Russian Commander. Again, the two spacecraft were over Europe.

It was 2:19 pm. The two men shook hands. A few minutes later the Russian communicator at Kalinin read up to the men in space a message from Leonid Brezhnev which said, in part, that 'In the name of the Soviet people and from myself personally, I am congratulating you with a significant event: the first docking of the Soviet spacecraft, Soyuz 19, and the American spacecraft, Apollo.' The Soviet leader went on to affirm that 'The whole world with great attention and delight is observing your responsible work that you are performing in carrying out the scientific experiments.' And of the future, Brezhnev believed 'It could be said that the Soyuz and the Apollo is a prototype of future orbital space stations.'

Minutes after that the Houston controllers set up a communication link with President Ford in the Oval Office at the White House. Grouped around the small table in Soyuz's Orbital Module, the four men were in range of the fixed TV camera. Slayton gave Leonov his headset so the Russian too could receive communication from the US President: 'Gentlemen, let me call to express my very great admiration for your hard work, your total dedication in preparing for this first joint flight. All of us here in Washington, in the United States, send to you our very warmest congratulations for your successful rendezvous and for your docking and we wish you the very best for a successful completion of the remainder of your mission.'

Leonov responded to words of encouragement for future ventures of this type by asserting that 'Mr. President, I'm sure that our joint flight is the beginning for future considerations in space between our countries.' And then Gerald Ford spoke to Vance Brand in the Command Module and then Valeriy Kubasov in Soyuz before signing off. But the ceremonial event was only just beginning. The four crewmen in Soyuz exchanged gifts with one another and Stafford presented Leonov with five American flags, receiving Soviet flags in return. And then it was time for everyone to sit down around the same table in space and eat a meal. While the conversation lessened, officials and dignatories too went off to dine, for it was several hours past noon.

By early evening the first crew visit was over and at 5:47 pm Stafford and Slayton said their farewells and moved back in to the Docking Module, closing the hatches behind them. In the ensuing pressure reduction back down to Apollo's 258 mmHg, Soviet monitoring equipment detected what appeared to be a pressure leak through one of the Docking Module hatches. That delayed Stafford and Slayton moving back to the Command Module but controllers finally agreed to leave the short tunnel between the Docking Module and Soyuz pressurized for the night. If the module was indeed leaking, it would show up in increased tunnel pressure by the morning.

The flight plan called for the crewmembers to begin a scheduled eight hour rest at 6:20 pm. Because of the pressure integrity check, it was 7:36 when the astronauts last spoke to the ground. During the next few hours Walt Guy and his Soviet counterpart V. K. Novikov, monitored the tunnel pressure and concluded that it only appeared to accept a leak from one side or the other because the atmosphere had heated up causing the oxygen and nitrogen gas to expand giving the erroneous impression that gas was leaking in through a hatch.

After five hours of good sleep, Stafford, Brand and Slayton were awake by 2:00 am, Friday, 18 July. Two and one-half hours later, housekeeping chores completed, Stafford accompanied Vance Brand into the Docking Module for the second transfer. At 5:00 am they moved into Soyuz and less than thirty minutes after that Leonov was taken by Stafford back into the Docking Module. At 6:00 am the two Commanders floated into Apollo. For the next

$4\frac{1}{2}$ hours they would remain in this configuration: Leonov, Stafford and Slayton in the Command Module; Brand and Kubasov in Soyuz. It was another well staged show and the actors did their bit to foster a spirit of detente.

First, TV tours of respective space vehicles for the breakfast shows in America, the lunchtime broadcasts in Europe. Then, Kubasov began a commentary as the television cameras showed a view of the Russian landscape below. 'Our country occupies one-sixth of the Earth's surface. Its population is over 250 million people. It consists of 15 Union Republics . . . At the moment we are flying over the place where Volgograd city is. It was called Stalingrad before. In winter 1942–43, German fascist troops were defeated by the Soviet Army here, 330,000 German soldiers and officers were killed and taken prisoner here.'

And then, across in Apollo, Leonov took up the commentary. 'I am now in the Command Module of the Apollo spacecraft. I am here together with Deke Slayton and Tom Stafford. We are flying over the Soviet Union's territory and we are observing everything which is speeding by below us.' After filming a science demonstration for use later in school classrooms showing the behavior of materials in a weightless state, Vance Brand prepared to return to Apollo taking Kubasov with him for the third transfer. Before he left Soyuz, Kubasov formally presented Brand with a gold medal similar to those given to Stafford and Slayton on the first transfer.

After a short meal Stafford moved in to the Docking Module taking Leonov with him, leaving Slayton in Apollo. When the hatch was opened to Soyuz, the two Commanders floated in to the Russian spacecraft while Kubasov and Brand moved back up into the Docking Module, sealing the hatch behind them. By 12:00 midday, the two men had joined Slayton in the Command Module. For the next $2\frac{1}{2}$ hours Stafford and Leonov would be in Soyuz while Brand, Slayton and Kubasov occupied Apollo. Principle activity concerned an open press conference through Moscow and Houston where newsmen got the chance to offer up questions after preliminary statements by respective Commanders.

Leonov showed pictures he had drawn of the American crew, endorsing his well earned reputation as an artist, and Vance Brand followed with a commentary on the view from the window as the docked spacecraft moved across the United States. From the Soyuz TV camera, viewers witnessed the presentation to Stafford of tree seeds to be grown back on Earth and of the joining of a flight medal, separated in two halves until that moment; out of camera shot, a similar performance was taking place in Apollo at the same time.

At the end, shortly before preparing for the final crew transfer, where all astronauts and cosmonauts would be back in their respective spacecraft, Tom Stafford sent an unscheduled message to the Soviet people, speaking in Russian: 'Dear Soviet television viewers, allow me as the representative of the United States of America to transmit to you best regards from the people of the United States. This is a happy time for the whole crew. We're happy, very happy, to receive – to be together here in the first international flight after two years of joint preparation and training.

'We astronauts, and cosmonauts, not only have worked together, but we've become good friends. I'm sure that our joint work – friendship – will continue, even after this flight. I too am sure, dear television viewers, that this flight will open the way to further cooperation and friendship between our two countries. Let the things that went on yesterday in our flight, and today, be a good thing for both of our peoples. Thank you, and good luck.'

Leonov added in English his own words of hope for the future affirming his belief that 'our joint flight is the beginning of very great cooperation in space.'

A few minutes later, at 2:35 pm, Slayton and Kubasov entered the Docking Module. Shortly after 3:00 pm they opened up the hatch to Soyuz and Kubasov was delivered back to his own spacecraft. At 3:49 pm Stafford shook hands with Leonov before drifting across to the Docking Module. Eleven minutes later the hatches were sealed for the last time. By 4:00 pm Stafford and Slayton had re-joined Brand in Apollo. It was over. Both crews now set to clearing up their respective vehicles and little communication ensued. It had been a long day, in fact the events of the preceding 48 hours had been a strain on the five crew members. Nevertheless, the Apollo crew were ready for their scheduled rest session by the flight plan time of 7:20 pm on the Houston clock.

Less than two hours later an alarm in the Command Module sent them scurrying into action. As planned, the crew had purged the spacecraft with pure oxygen by hooking up a pipe to the waste management screen, venting overboard excess atmosphere and traces of nitrogen accumulated from the repeated transfers. But the astronauts forgot to adjust the enrichment pressure setting which inadvertently increased the Command Module to 285 mmHg. Forgetting too that they had left the waste management vent valve open the men went to sleep with the secondary coolant loop still on, initially switched on to remove excess heat which had built up due to lengthy use of the television camera that day.

This unusual combination of circumstances caused the spacecraft to slowly vent its atmosphere to space and when pressure had fallen to 233 mmHg it set the alarms ringing. It was simply a matter of closing the vent and turning the secondary coolant loop off. But thirty minutes after the first alarm they were rudely woken a second time by a noisy communication circuit. Switched to 121.75 MHz, the antennae were picking up landing instructions from a US airport tower!

But the Russians too had received their share of communication problems. Earlier in the flight a controller at the Kalinin center complained of interference from a station at Halifax, Nova Scotia. On hearing this a Canadian reporter covering the flight from Moscow passed word to the station, which had its technicians work to prevent the problem occurring again. The Soviet controller expressed his appreciation by presenting the reporter with a special souvenir of the flight.

Having begun their rest period earlier than the Apollo crew, Leonov and Kubasov were up and working by 1:30 am; Stafford, Brand and Slayton were talking to Houston by 3:30 am. It was Saturday, 19 July – the day the two spacecraft would separate and go their different ways on independent missions. But first, Apollo pulled free from Soyuz at 7:03 am when Slayton unlatched the docking mechanisms and backed off. But in so doing the two vehicles were setting up an important experiment conducted from the Russian vehicle.

By placing the CSM between Soyuz and the Sun, Apollo served as an occulting disc over the bright solar surface to allow an artificial eclipse as viewed by Leonov and Kubasov. The resulting photographs showed the solar corona, normally washed out by brightness from the Sun's surface.

Following the physical separation of Apollo from Soyuz, the Russian crew extended the docking unit on the front of their spacecraft so that it could simulate the active role when Apollo moved in for a re-dock; the CSM was the active vehicle throughout ASTP because Soyuz had limited command and control capability, propellant reserves and sighting alignment along the docking axis. With the compatible docking equipment on Apollo fully retracted, the opposite situation to that for the first docking almost two days before, Deke Slayton occupied the left couch in spacecraft 111. Slowly he approached Soyuz but had difficulty sighting up on the target attached to the Russian spacecraft's Orbital Module because the whole vehicle seemed to be in a different attitude to that planned. Strong light reflections off the target caused Slayton to use more propellant than expected before he finally succeeded in latching up.

Capture came at 7:33 am and within seven minutes the two vehicles were hard docked once more. For the undocking, flyback, and re-docking, Leonov and Kubasov had donned their suits and fastened themselves in the Descent Module just in case an accident punctured the Orbital Module on front. Now safely linked up again, they moved back up into the more spacious Orbital Module and sent to Earth television pictures of their morning meal. But all too soon it was time for the final undocking and a scientific operation where Apollo would fly loops around Soyuz beaming a shaft of light at reflectors on the Russian spacecraft; analysis of the returned beam would determine quantities of atomic oxygen and atomic nitrogen present at more than 200 km.

The Russian crew fastened themselves down in the Descent Module once more and performed the undocking sequence. It was 10:26 am. Apollo began to pull away, backing off for the looping maneuver and although minor problems arose with the reflector, causing a rapid consultation between Houston and Kalinin switching reflectors to a second set on the back of Soyuz, the operation went very well. It was a complex activity and one that resulted in the CSM consuming more thruster propellant that day than on any previous day of manned flight: 290 kg.

Before the final maneuvers of the sequence Houston asked the Apollo crew to switch off the pack of supplementary propellant tanks in the Service Module because they were now depleted to 14%. From this point on the mission would be conducted using the prime tanks built in to the Service Module as part of its basic design.

The combined science activity had kept the Apollo crew busy throughout. Deke Slayton flew the spacecraft while Tom Stafford operated the equipment doors from his position in the lower equipment bay and Vance Brand punched up computer instructions from the right couch. The final maneuver carried Apollo 1 km above and behind Soyuz and a tweak with the RCS thrusters set the spacecraft up for a gradual separation. With each passing hour the two vehicles would drift a further 5 km apart, thus avoiding a collision. At 3:00 pm the Russian spacecraft was back in a spin-stabilized attitude control. Four hours later they began their sleep period followed, at 9:30 pm, by the three astronauts in Apollo.

Sunday, 20 July, was to be a busy day for both teams. Soyuz would perform final experiments and prepare for re-entry next day; Apollo would conduct materials processing tests in the Docking Module and photographic observations of Earth. The CSM supported 28 separate experiments: 21 provided by the United States, 5 comprising joint US-Soviet tasks, and 2 funded by Germany. There had been insufficient money to develop a fully equipped Scientific Instrument Module of the type provided on the last three Apollo Moon missions but a trio of instruments supporting physical science tasks were installed in the Service Module and provided with their respective doors and apertures. The remaining experiments were placed aboard the Command and Docking Modules.

Sunday brought a new day to the Apollo men shortly before 5:00 am, Houston time. The Russian crew had been awake for several hours. At noon, Houston received television pictures from the orbiting Salyut 4 space station hosting cosmonauts Klimuk and Sevastyanov while they chatted with Leonov and Kubasov aboard Soyuz 19. Later, Stafford, Brand and Slayton performed a special measurement check to relate their respective sizes preflight with the effects of weightlessness on the body. This was to calibrate changes to assist in designing seats for the Shuttle.

Unlike conventional couches, Shuttle seats would require the crew to place their feet on aerodynamic control pedals and extensions in body length brought about by the lack of gravity would severely compromise pilot functions during re-entry. It was 10:40 pm when Stafford, Brand and Slayton completed their goodnight call to Houston but the Soyuz crew several hundred kilometers away had been asleep since about 2:00 pm that afternoon. Leonov and Kubasov were awake shortly before midnight for the re-entry preparations that would precede their return to Earth. And through it all, the Apollo crew would sleep soundly on.

Monday, 21 July. Equipment was transferred from the Orbital to the Descent Module aboard Soyuz and at about 1:00 am the crew donned their space suits. Less than one hour later they had sealed themselves up in the Descent Module. Soyuz was now more than 700 km ahead of and slightly below the passive Apollo. Several hours earlier the ground computed instructions for the on-board equipment had been generated at the Kalinin control center and uplinked to the spacecraft. The crew would ride out one of the most publicized re-entries in Soviet space flight history. Live television coverage was planned for domestic and international networks and millions of Russians stood before receivers indulging in this all too rare treat.

It was precisely thirty-five seconds past 5:10 am in Houston when the retro-rocket began to fire. For 3 min 3 sec it continued to burn, slowing Soyuz by 430 km/hr. Nine minutes after the engine shut down the two modules at either end of the Descent Module were pyrotechnically jettisoned. The spacecraft had already descended from 212 km to 154 km and four minutes after that the heat shield began to feel the atmosphere. Soyuz still had 2,000 km to fly when maximum deceleration loads were felt but within a further six minutes range was down to less than 200 km. Shortly thereafter the parachutes were put out at a height of 7 km and Soyuz began a fifteen minute descent to the ground terminating in an event dubbed 'dustdown' by Houston observers.

It was aptly named for after jettisoning the aft heat shield four minutes after deploying parachutes, the Descent Module ignited retro-rockets at the exposed base of the spacecraft to decelerate and cushion the force of impact with a combined thrust of 5.4 tonnes. Recovery for Russian space vehicles appeared as inglorious as splashdowns for NASA spacecraft. With clouds of dust thrown tens of meters into the air, the Descent Module was dragged on to its side before the parachute could be released. From everywhere vehicles and people converged on the truncated spacecraft while helicopters circled transmitting television pictures for the global audience. Three minutes later Kubasov emerged followed seconds later by Leonov.

The spacecraft had landed less than 10 km from the predicted location to the jubilant cheers of controllers and technicians in Houston and the placid acceptance of engineers in the Kalinin control center. Within thirty minutes flight director Pete Frank presented personnel in the Soviet support room with a plaque making them honorary members of the MOCR. By this time communications had begun with Stafford, Brand and Slayton in Apollo, the first call having gone out fifteen minutes after the Soyuz landed at 5:51 am, Houston time. While the crew ate their first meal of the day, the final report was issued from Moscow Mission Control:

'This is the Soviet Mission Control Center. Moscow time is 15:15. ASTP program has been accomplished . . . The flight control directors, in a TV broadcast, have transmitted best wishes for the successful completion of flight to the American crew . . . This completes the transmission of the technical comments by the Soviet Mission Control Center.'

And once again Soviet manned space flight operations descended into camouflage and secrecy. Shortly before 10:00 am that Monday morning, Apollo fired its RCS thrusters in a trim burn designed to set up the orbit for jettisoning the Docking Module two days later. It would play its part in a test to find out the better way of measuring continental movements on Earth. Much of the day was spent enthusing on the view through the windows between experiments and the continual need to operate switches and knobs for general housekeeping duties or periodic servicing of equipment or systems. After getting to bed shortly before midnight the crew were nearly woken at 12:30 am for some switch changes but the flight controllers avoided that by remote commands to the high-gain antenna.

Tuesday, 22 July, was a comparatively quiet day and while wives and families watched TV replays in Mission Control, Stafford, Brand and Slayton conducted more experiments. It was good to have the spacious availability of the Docking Module on Apollo's nose, providing much more room than the Command Module alone.

There were reflective hours in the final days of that last Apollo mission. Before getting the crew to sleep, flight director Neil Hutchinson had the tune 'The First Time I Ever Saw Your Face,' broadcast to the astronauts as he had on Apollo 17 and all three Skylab visits. But it had a special significance. Similar moods had gripped controllers, engineers, technicians and managers twice before in manned space flight.

Mercury, in 1963, had its reflective aura when Gordo Cooper brought the last can home; Jim Lovell and Buzz Aldrin felt the same about Gemini XII in 1966. Nine years on, it was now time for Apollo to bow out, ending nearly two decades of planning, designing and programming for ballistic flights on one-shot rockets. But there was still a full day's work to perform aboard spacecraft 111 when, at 11:15 pm Houston time, the crew said goodnight to Mission Control.

Wednesday, 23 July. A call went up through the Santiago tracking station at 5:47 am and thirty minutes later television equipment was being checked out for the second air-to-ground press conference of the flight. Conducted live from Building 2 at the Johnson Space Center, correspondents had an opportunity to speak direct to the men in space, receiving enthusiastic predictions about the coming decades of space travel. During the early afternoon the crew donned their space suits and prepared to jettison the Docking Module, evacuating the tunnel between the two first to confirm the integrity of the Command Module. At 2:47 pm, the Docking Module was released and after moving some distance away the SPS was fired in a 1 second burst.

Four hours later another SPS burn, this time for 0.9 seconds, stabilized the separation distance at approximately 300 km. The experiment which followed called for a VHF receiver on Apollo to accept signals from the Docking Module, still powered by batteries, and to measure the so-called Doppler effect where the pitch of frequency transmitted by one vehicle changes with relative motion between that and a second vehicle. Because the two structures would be stable with respect to each other at the start of the test, motion put into either would be attributable to mass anomalies in the Earth below – the famed mascons and minicons observed during Moon orbits.

The last conversation before sleep that night went out shortly after 11:00 pm. In the Mission Operations Control Room, the network officer who handled that segment of Alan Shepard's flight fourteen years earlier was on duty routing communications for Stafford, Brand and Slayton.

Thursday, 24 July. The day of the homecoming. Awake by 7:20 am the crew of spacecraft 111 conducted the last experiments before stowing equipment for the return. It was debilitating, having so little room now the Docking Module had gone and all three found it noticeably cramped. During late morning Mission Control received two dozen long-stemmed red roses and a note from a very familiar admirer:

'This Betsy Ross bi-centennial arrangement of my favorite roses is for my country's flags and the greatest Mission Control Center in the world. To honor our great and outstanding Apollo-Soyuz mission and for all our NASA personnel for their dedication to duty, again proving they are the best in their very specialized profession . . . Never before in the world's history has such an historic handshake taken place . . . With all my fondest love forever. Cindy Diane, Keystone, State of Pennsylvania.'

Many dignatories were on hand for this last American splashdown. Administrator Jim Fletcher was in the viewing room on the second floor of the MOCR that had been used to control the first and third manned Apollo flights and all three Skylab visits. With him were Center Directors Scherer, Lucas, and Kraft from Kennedy, Marshall, and Johnson respectively. Also present, Walt Kapryan from KSC, Glynn Lunney who had just been named to vacate his job as head of the US segment of ASTP for administration of Shuttle payload planning, and Chester Lee, Headquarters Director of ASTP In the MOCR, the colorful figure of Gene Kranz with his red, white and blue vest for the bi-centennial celebrations. Capcom Bob Crippen, the man who would team with veteran John Young to fly the first Shuttle mission, talked with the crew of spacecraft 111.

But it would be a return especially tight on reserves. Excessive use of RCS propellant had left the Service Module thrusters with only two-thirds the expected quantity remaining, enough for a 209 km/hr speed change. The SPS engine, with more than the expected quantity of propellant remaining, would have to slow

Cosmonauts Leonov left) and Kubasov minutes after landing back on earth.

Apollo by 208 km/hr. If that engine failed for some reason, the RCS thrusters would have to use their meager propellant stocks to complete the burn. Nevertheless, the burn went exactly as planned when the SPS fired at 3:38 pm in darkness over the Southern Ocean. The engine continued to fire for 7.8 seconds.

Little more than six minutes later the Service Module was pyrotechnically separated from the Command Module. It would follow the manned segment through the atmosphere, breaking up as it went, to fall as debris to the Pacific Ocean. From four minutes prior to re-entry until the spacecraft emerged from communications blackout there would be no contact. On over the Pacific the Command Module descended to complete a total 3,422 revolutions of the Earth by NASA manned spacecraft. But all was not as well as it should have been.

For the re-entry, Brand was in the left couch, Stafford in the center, Slayton in the right seat. With the spacecraft descending at a speed of more than 500 km/hr, rattling around, bumping about, and with a roaring sound as the Command Module passed through the main phase of deceleration, Brand failed to hear a checklist reminder to arm two functions. The switches, called ELS AUTO and ELS LOGIC, would have blown the apex cover when barostats detected an altitude of 7.3 km to start the parachute deployment sequence, simultaneously inhibiting further thruster firings. The switches were to have been thrown when the spacecraft passed through 9.1 km. But they were not.

On noticing the spacecraft was still falling through the 7.3 km height without the expected apex jettison, Slayton told Brand to pull the two manual override switches. Which he did. Three seconds later the apex cover went spinning away and three seconds after that the drogue mortars fired. But no one switched the thrusters off, as they should have done since the operation was being conducted manually, and the spacecraft, swinging wildly under the drogue parachutes, fired thrusters repeatedly in an effort to damp the oscillations. Seeing this thruster activity, Stafford punched the ELS AUTO switch to cut off the propellant flow but residual fuel and oxidizer downstream of the isolation valves continued to pour through the jets.

Seconds later the spacecraft was flooded with a foul gas, Stafford actually saw it swirling in, as propellants from the engines sucked down the pressure relief valve that had opened to admit fresh air and increase pressure, swept in the Command Module. At that, Stafford manually fired out the main parachute six seconds ahead of the auto timer just in case the men passed out and were unable to arrest the 315 km/hr descent rate. At 4:18 pm Houston time, the Command Module struck the water with all three crewmen coughing and spluttering. Brand was in the worse condition, having breathed in a quantity of the noxious fumes, and Slayton was disabled through extreme nausea.

Suddenly, a second after splashdown, the spacecraft flipped over with the astronauts hanging face downwards from their couch straps. Realizing the danger of remaining in the cabin, Stafford slipped his buckles and fell down past the display console on the lower equipment bay which, in the pitching and tossing motion of the external wave action, assumed the position of a floor. Groping for the emergency oxygen he slapped a mask on Brand who was by now unconscious and confirmed that Slayton was holding his own. Seconds later all three were on pure oxygen and Brand quickly regained consciousness.

Having stabilized their condition, Stafford could leave his crew and turn to righting the spacecraft, which he did before waiting the customary three minutes for the bags to inflate. After the spacecraft got back to Stable I Stafford opened the post-landing vent system and unlatched the side hatch to admit fresh air which came flooding into the interior. But nobody outside spacecraft 111 knew anything about all this activity until medical checks about one hour after recovery. The Command Module was on the water only 7.4 km from the planned spot 500 km west of Honolulu. Weather was good and the sea was comparatively calm with low wave action.

Soon, recovery helicopters had swimmers in the water and the crew were transferred to the recovery ship USS *New Orleans* still inside their Command Module as planned. Winched aboard by crane, the only noticeable effect of their experience was a few sore eyes and painful lungs. And then the crew told physicians the story and celebrations on the carrier were cancelled. It turned out that this was the only Apollo mission where parachutes were deployed manually; had the men passed out before activating the switches, they would have been killed on impact with the water.

The astronauts were kept over in Honolulu for a two-week rest and observation period before flying to Houston. The intensive post-flight inspection of each man revealed a small lesion on Slayton's left lung, but surgery proved it to be non-malignant. The celebrations that followed took Stafford, Brand and Slayton to Russia among a series of visits aimed at presenting the achievements of their mission.

A few months after his flight, Tom Stafford left the astronaut corps to resume active service as an Air Force Officer. Deke Slayton was appointed to manage the Shuttle Approach and Landing Test program at Edwards Air Force Base while retaining his position on the list of active astronauts and then as manager of Orbital Flight Tests with the winged transporter. Vance Brand continued as a career astronaut intent on a Shuttle seat during the 1980s.

Throughout the period filled with the first truly international space mission, Russia had in orbit the two-man crew of Soyuz 18 who had docked to the Salyut 4 space station originally launched in December, 1974 and already visited for 29½ days by the crew of Soyuz 17. Klimuk and Sevastyanov docked with Salyut 4 one day after ascending from the Baykonur cosmodrome on 24 May, 1975. Ninety scientific experiments were aboard the manned vehicle orbiting Earth in a 344 × 356 km path.

Activity centered on developing a competent work schedule and in actively pursuing the many scheduled tasks. It was a step toward really long flights planned for the coming years and as such built confidence in the Soyuz-Salyut configuration. They remained in space for 63 days, returning two days after Stafford brought his men back from the ASTP mission. It was the longest Soviet manned space flight, second only to the concluding 84-day Skylab mission. Where America had dominated the first half of the 1970s by advancing the sophistication of Moon exploration and setting dizzy records for long duration flight aboard the roomy Skylab, Russia was now to take over and demonstrate during the second half of the decade variations and adaptations of basic hardware beyond anything they had achieved so far.

Less than four months after the return of Soyuz 18, an unmanned derivative under the Soyuz 20 nomenclature ascended and docked with Salyut 4. For much of that year the Russians had 'leaked' rumors about development of a special tanker version of the Soyuz spacecraft, giving weight to statements from cosmonaut Shatalov that in the near future Soviet cosmonauts would refuel space stations from unmanned supply ships. It all sounded very reminiscent of confident projections recorded during the early 1960s, and not a few in the West debated the validity of such stories. It was certainly true the Russians had been uncommonly cautious.

In twenty-six manned space flights beginning with the late Yuri Gagarin during 1961, the only creditable success had been in demonstrating the introduction of a production-line space station called Salyut. And that only in the last few years. Diverted perhaps from grander goals by technical limitations and the outstanding luck that accompanied Apollo, Russian manned operations had taken a new turn and in the wake of American disinterest in sustaining operations in the interval between one-shot missions and the reusable Shuttle, would press ahead with a coordinated series of missions quite unlike their patchwork efforts hitherto.

Ostensibly a biological flight carrying numerous living species in a test of new environmental control equipment, Soyuz 20 was made to automatically link up with the Salyut 4 station. As time would tell, but as no one knew for sure then, Salyut 4 was not to receive further visitors and the extended docking period of the unmanned Soyuz 20 kept the vehicle in space for 91 days. It was clearly seen then as a precursor checkout for later ferries designed to transport cosmonauts to Salyut stations for stays in excess of 84 days achieved by the last Skylab mission. But other 'biological' flights were being carried out also under the guise of Cosmos designations to remove their affinity to the manned space station program.

Following the return of Soyuz 20 in February, 1976, speculation abounded on the next steps in Soviet manned activity. As each month went by rumor sprang anew with suggestions about new tanker vehicles. Cosmonaut Nikolayev added to the frustration when he was quoted as saying that 'Both Americans and Russians are anxious to find out how long men can live and work in space. We must know how many months . . . maybe six months . . . maybe a year – maybe even more.' And of the cargo carrying derivatives of the basic Soyuz, Nikolayev said that Soviet cosmonauts would extend their stay aboard later Salyuts by receiving unmanned supply vehicles and that 'The 25th Party Congress ordered our cosmonauts to do those things which can benefit the national economy most.' Up to now, Salyut space stations had one docking port at the front, the rear occupied by the propulsion systems employed for attitude and orbit control. The use of unmanned cargo vehicles implied the addition of a second port on developed Salyut variants.

But not immediately, for Salyut 5 which went into orbit on 22 June, 1976, was similar to its predecessor. Exactly two weeks later Boris Volynov and Vitaly Zhdobov flew to Salyut 5 in the Soyuz 21 spacecraft; by late afternoon the next day they had docked and were getting ready to move inside the station. With them went a mandate for extensive scientific observation of the Earth's surface and atmosphere, medical analysis of the effects of long duration space flight and several technical experiments designed to assist with the development of successors.

Among the equipment was a device called 'sphere' in which the cosmonauts could test factory processes like the materials tests conducted by NASA astronauts. The crew also studied aerosols in the Earth's atmosphere and turned to the 'Crystal' equipment designed to study the growth of semiconductor crystals, possibly for commercial exploitation. Like Skylab astronauts, Volynov and Zhdobov operated a teleprinter, called 'Stroka', but this one was a portable device. In biological tests they observed the behavior of fish fertilized in space, the reaction of seeds and plants, and the antics of guppies cavorting in weightlessness.

Like Skylab crews, the cosmonauts spent considerable time studying the Earth, but with a hand-held spectrograph called RSS-2M. In an increasing concern for the resources of their states, Russia was investing heavily in new land management schemes, hydrological development programs and crop inventory; space stations were an important part of the general overview found advantageous from orbit, important adjuncts to the new economic growth sought by Soviet leaders. The crew were given special training in observing likely resource deposits, both oil and natural gas, and in helping to develop new maps of remote areas. In addition, they supported civil engineering projects like the new Baikal-Amur railway by photographing areas through which the work would pass.

Their 49 day mission came to an end in darkness over Kazakhstan at 9:33 pm, Moscow time on 24 August. Three weeks later, on 15 September, 1976, Soyuz 22 lifted off from the Baykonur launch site with cosmonauts Bykovsky and Aksenov. Equipped with a multi-spectral camera developed in East Germany by Karl Zeiss Jena, the spacecraft was not intended to dock with Salyut 5. It remained in space for eight days and returned on 23 September during a morning hour. Unlike its predecessors, designed to dock with Salyut stations, it carried solar cell panels for independent flight.

By now the future shape of Soviet endeavors was beginning to emerge. Already, within a year of the ASTP mission, Soviet officials hinted at the possible flight of international crews from East European countries and others sympathetic to Soviet aims. Little more than three weeks after the independent flight of Soyuz 22, the Soviets launched Zudov and Rozhdestvensky in Soyuz 23. But a fault in the automatic rendezvous and docking equipment prevented the cosmonauts linking as planned with Salyut 5.

Limited to a duration of two days independent flight, the spacecraft was brought down in the middle of a snow storm at 8:46 pm Moscow time one day after launch. Drifting in the strong, blizzard-like, conditions the capsule was set down in Lake Tengiz 195 km south-west of Tselinograd. With temperatures of −15°c the crew were rescued by boats and helicopters, thoroughly shaken by their experience and ready for warm shelter.

Throughout this time, and up to the beginning of 1977, the Salyut 4 station remained in orbit about the Earth. On 3 February, however, Soviet controllers sent commands for the station to reactivate its propulsion system and de-orbit over the Pacific Ocean after making 12,188 revolutions of the Earth in more than two years since launch. Four days later the object of current attention, Salyut 5, got another try at manned habitation when Gorbatko and Glazkov ascended in Soyuz 24.

It was evening when they lifted off into space, their launch window set like all space station missions by the ground tracks of the target vehicle. After successfully accomplishing the docking that escaped the previous manned attempt, the cosmonauts entered Salyut 5 during the morning hours of 9 February. For the next week they thoroughly checked its systems and replaced elements of the onboard computer before setting up comprehensive medical experiments designed to measure heart response. Materials processing experiments followed, as did observation of the Earth's upper atmosphere through an infrared telescope. Before returning to Earth on 25 February, the cosmonauts operated a new systems device designed to change the interior atmosphere by simultaneously opening a vent valve to expell the oxygen and nitrogen while admitting compressed air from storage tanks.

The visit had lasted 18 days, far less than observers speculated it would, and a day after the cosmonauts returned the station jettisoned a recoverable capsule containing research items, an operation previously conducted from Salyut 3. Activities like this usually preceded commands for the station to de-orbit over an ocean area but for most of 1977 Salyut 5 continued to orbit the Earth until, on 8 August, it was brought back down to be destroyed as it struck the atmosphere. It had completed 6,630 revolutions and had been in space for little more than half the duration of its predecessor.

By now the Russians were established on a road toward permanent manned orbiting stations, a goal similar to that proposed by NASA more than a decade before when it sought to turn Apollo Moon-bound hardware into practical applications for the benefit of people on Earth. The failure to impress press, politicians and public adequately with the ultimate value of such endeavor wrote *finis* to full use of redundant hardware for extended surveys from orbit aboard stations like Skylab and the proposed 12-man facility.

But the Russians operated under no such handicap and replaced a democratic accountability to public inspection with a need only to satisfy Ministers and high officials. It was clearly Soviet policy to exploit space technology for the economic and industrial well-being of the Union of Socialist Republics and consistently part of Communist doctrine to carry the banner for mankind deep into the Universe. Yet as economic pressures were exerted by a flagging prospect for growth in the flow of money, Soviet leaders made it clear that extravagant exploration and the human presence of men on the Moon or the planets would have to wait.

If indeed the Soyuz program had its origins in a set of goals similar to those that brought Apollo from prospect to project, perhaps Soyuz-Salyut was, like the American Moon program, a diversion to more mundane objectives, albeit more successful than NASA proposals if only because of the lack of public accountability. The brief hiatus in manned Salyut operations ended seven weeks after the return to destructive impact of Salyut 5, although it had been seven months since the Soyuz 24 crew returned to Earth.

On 29 September, 1977, Salyut 6 lifted into space for 'the purpose of conducting scientific and technical research.' A broad mandate indeed. So far, individual Salyuts had hosted two separate crews at most supporting combined durations no longer than 90 days. By comparison, Skylab hosted three crews totalling 171 days, the last of which was more than three weeks longer than the record Salyut mission. The Soviet station program was aiming far ahead of Skylab objectives and would reach a peak with Salyut 6, the most successful Russian project to date.

Ten days after the launch of Salyut 6, cosmonauts Kovalenok and Ryumin lifted off in Soyuz 25 from the same launch pad used for Sputnik 1 twenty years before. Two days later, on 11 October, they were back on Earth having completely failed to dock with the new station. It was a flop, a dismal failure for what would emerge as a very advanced derivative of the basic Salyut.

Salyut 6 represented a departure from earlier designs in that it carried docking ports at both forward and rear locations. Military Salyuts 2, 3 and 5 had their Soyuz docking ports at the rear while civilian Salyuts 1 and 4 adopted the publicized configuration of a single docking port at the front. The two ports designed into Salyut 6 enabled two Soyuz type vehicles to link up to the station at the same time. Only by careful redesign were the Soviets able to place docking ports at each end. Modifications previously adopted for the military Salyut were mandatory here, as were several new refinements.

On civilian Salyuts designed to accommodate a single Soyuz at the forward port, propulsion systems at the rear completely covered the area, with two main engines in the center and attitude control thrusters on the rear face. Previously, the Soviets employed two different propulsion systems. The main engines used nitric acid and hydrazine in a turbine-driven design while hydrogen peroxide employed for driving the turbines was also used through the attitude thrusters.

To standardize propellant supplies, however, Salyut 6 adopted a pressure fed system making refuelling in space so much simpler. Both main and attitude control engines would use nitrogen tetroxide as the oxidizer and UDMH (unsymmetrical dimethyl hydrazine) as the fuel. Gaseous nitrogen, stored in a separate tank, would pressurize the six fuel and oxidizer tanks for delivery of the fluids to respective motors. Each propellant tank contained an accordion screen pressed flat against the interior wall when filled with liquid. To expel the propellant from the tank, nitrogen would be applied to the other side of the flexible accordion to squeeze out the fluid.

Other changes to Salyut 6 required docking targets and radar equipment at both ends rather than one, handrails for space walking down the outside, a special shower device on the interior, and a more efficient arrangement of controls. Like earlier Salyuts, it had three pivotable solar cell arrays, two diametrically opposed spanning 33 metres and a third on top of the station giving it the appearance of a four-sail windmill with one sail missing. Cosmonaut Feoktistov, a key figure in the development of Salyut 6, said it 'relieved (the crew) of many navigating chores thanks to the economical system of orientation and steering.' Moreover, 'all of the main units of Salyut 6 are controlled from the same switchboard.'

Following the abortive attempt of Kovalenok and Ryumin to board the station after launch on 9 October, preparations got under way to have the next Soyuz attempt a docking at the rear of Salyut 6. Before dawn, on 10 December, 1977, cosmonauts Romanenko and Grechko reached orbit and little more than a day later their Soyuz 26 spacecraft successfully linked up to the aft docking port. Three hours later they moved inside the space station, one of their prime mandates now to see if the forward port and radar equipment would support an equally successful docking; it had been that port Soyuz 25 had been unable to connect to. Cosmonaut Shatalov was not minimizing schedules when he affirmed that the 'working program this time will be rather crowded.' Although Romanenko was making his first flight, flight engineer Grechko had flown 29½ days in the Soyuz 17/Salyut 4 mission.

Following several days' extensive checkout, Grechko opened the forward hatch and performed the first Soviet space walk for nearly eight years. Moving round to the docking equipment he reported its general condition in a 20 minute EVA employing a new suit, more flexible than earlier designs and fully autonomous; Grechko was restrained by a tether, just in case he drifted away. Grechko had, said Shatalov, provided the information that will 'enable us to carry out the whole program of further work.' The forward docking port was declared serviceable.

Soyuz 26 was in every respect a shakedown flight and that fact emerged early in the new year when in 10 January, 1978, cosmonauts Dzhanibekov and Makarov were launched aboard Soyuz 27. A day later it successfully hooked up to the forward docking port, achieving a 'first' in three areas: first four-man station, first manned supply mission to a habited station, first time two Soyuz had docked to a Salyut at the same time. The complete configuration now weighed 32 tonnes and presented a length of about 36.5 meters.

For five days the four cosmonauts worked together, performed experiments, and transferred equipment to the Soyuz 26 spacecraft at the aft port. Seat liners from the recently launched Soyuz 27 were moved across and on 16 January Dzhanibekov and Makarov moved inside the vehicle that brought their colleagues to Salyut 6 and separated from the space station. Leaving their own spacecraft for Romanenko and Grechko, the two visitors came back to Earth in Soyuz 26 after a flight lasting six days during which time they delivered food and supplies. It was another first: cosmonauts returning home in a different spacecraft to that in which they were launched.

Using an MKF-6M multi-spectral camera developed in East Germany, the Salyut 6 crew began an intensive study of the Earth several days later and prepared for the next phase of their greatly expanded program. Dzhanibekov and Makarov had landed back on Earth during the afternoon of 16 January. Less than four days later Russia launched the much publicized transport ship, Progress 1, which docked to the now vacant aft port on Salyut 6 two days later. Romanenko and Grechko were said to have assisted with preparing Salyut for accepting the Progress supply vehicle but when it finally linked up with the manned station it represented clear proof that Russia was now moving visibly forward with Earth orbit activity.

Although the basic technology involved was little more than that demonstrated by Gemini vehicles more than a decade before, adaptation of the standard Soyuz configuration to this unmanned role displayed an innovative capacity hithereto uncharacteristic of the Soviet space program. Progress weighed 7 tonnes of which 2.3 tonnes represented cargo or additional propellant to be piped aboard Salyut. The docking interface on Salyut comprised new fluid transfer equipment adding to the complexity of the basic docking system. If properly connected, Progress would automatically align with connectors on the space station, simplifying the tanker role to one of throwing switches rather than hooking up flexible pipes.

Externally, Progress looked like a standard ferry-type Soyuz, devoid of solar cell panels usually carried on independent missions. But the bell-shaped Descent Module normally used to house two cosmonauts in a pressurized environment was stripped of its heat shield and all life-support functions to serve as a container for the several propellant tanks fulfilling the tanker role. The Orbital Module on front was also unpressurized and equipped with racks and storage bays for containing the 1.3 tonnes of dry supplies to re-equip Salyut. Small items were to be stowed in larger containers attached to the basic frame by quick-release fasteners. Other equipment was bolted to the racks but those too could be quickly liberated for transfer.

The front face of the Orbital Module carried hydraulic and propellant lines for mating with the fixed receptors on Salyut. Propellant carried for transfer was housed in four spherical tanks in the Descent Module, or tanker section, totalling 1 tonne of fuel and oxidizer. On the outside, Progress adopted the same arrangement of fourteen 10 kg thrusters and eight 1 kg thrusters

Cosmonauts Dzhanibekov and Makarov return to earth in Soyuz 26 after visiting Salyut 6.

for basic attitude control with fully automated systems not dissimilar to those employed for a standard Soyuz. Two television cameras and three lights, all on the exterior, assisted with observation from Salyut during rendezvous and simultaneous transmissions of telemetry from Progress to Salyut and the ground enabled both the control center and the cosmonauts to assist with docking if necessary. Progress 1 carried new environmental systems for Salyut, a new set of orientation equipment and a materials processing furnace for the cosmonauts. The unmanned cargo-tanker was capable of an accumulated eight days independent flight and could remain docked to Salyut for a maximum four weeks.

Progress 1 performed rendezvous maneuvers to effect a link-up with Salyut 6 shortly after noon, Moscow time, on 22 January, 1978. Within hours, Romanenko and Grechko had opened the hatches and were unpacking letters and packages from Russia. Additional supplies in the form of special exercise devices designed to maintain physical fitness were in Progress 1, as were a new load of scientific instruments. Two days after docking the crew checked over plumbing lines that would shortly be used to conduct the first orbital refuelling and the day after that they unpacked food and water.

On 2 February the propellant transfer took place with the aid of a specially developed compressor operating on three phase alternating current via an inverter connected to the Salyut's direct current battery system. The compressor was employed to remove nitrogen gas at 20 atmospheres from the Salyut tanks, lowering it to 3 atmospheres so the 8 atmospheres pressure in the Progress tanks would allow the fluids to flow across from tanker to station. Salyut had three oxidizer and three fuel tanks but only two of each were empty, the third in each system providing a reserve. The two fuel and two oxidizer supply tanks in Progress had Salyut topped up by noon on the 3rd.

Like their Skylab counterparts several years before, Salyut cosmonauts found the need to constantly repair and renew ageing systems and components aboard the station. Several hours each day were spent in housekeeping duties and instrument checks. But all was apparently going very well with the first expanded Salyut activity. On 5 February the crew purged the propellant transfer lines of residual fluids by blowing nitrogen down the pipes. And then, in response to a suggestion from one of the flight directors that remaining propellant in Progress' own tanks should be used to modify the Salyut orbit, controllers used the unmanned cargo-tanker as a space tug, another first, igniting its engine to raise the orbit of the docked combination.

During the early morning hours of 6 February, preparations were made for Progress 1 to separate from the aft docking port and at 5:53 am Moscow time it slipped free. After drifting away a distance of approximately 15 km back-up rendezvous equipment was tested and then, its racks packed with redundant equipment, waste containers and other items discarded by the manned Salyut, Progress 1 fired its main engine to bring it back down to Earth. It was destroyed as planned in the atmosphere on 8 February over the Pacific Ocean.

Three days later Romanenko and Grechko exceeded the 63 day space flight of Soyuz 18 two and one-half years before, establishing a new Soviet record, and continued with routine space station experiments. Using the Splav 01 furnace the cosmonauts conducted several materials processing experiments and at other times operated the BST IM infrared telescope designed to study deep galactic sources of interstellar hydrogen. And then, after an official announcement several weeks before, the first truly international space crew ascended from Baykonur aboard Soyuz 28.

Czech cosmonaut Vladimir Remek was taken to the Salyut 6 space station by pilot cosmonaut Alexei Gubarev. Launched on 2 March, it took the spacecraft little more than a day to find the long duration station. Shortly after docking the two cosmonauts moved in to be greeted by Romanenko and Grechko, about to set up a new world space flight duration record. They passed the 84 day Skylab endurance next morning, 4 March, regaining the world record for single flight duration after a lapse of nearly 5 years. The crew drank a toast in cherry juice.

For the first time, a citizen of a third country was in space but not to merely go along for the ride. His specific concern was to operate the Morava smelting test device. Designed by Czech scientists it was further application of the extensive materials processing work being conducted aboard the space station. But it

The activity aboard Salyut 6 involved regular television transmissions. Here, cosmonauts Ivanchenkov (left) and Kovalenok face the camera, July, 1978.

was also opportunity for Moscow to reach out and embrace the territories it occupied, a display of 'sincerity' toward Eastern Bloc states not lost to politicians ever keen to exploit technology for partisan cause.

The visit of Gubarev and Remek ended eight days after launch when they returned to Earth during the afternoon, Moscow time, on 10 March. Almost immediately, Romanenko and Grechko began to prepare for their own return and one week later, after a 96½ day mission, they safely landed 265 km west of Tselinograd on 16 March. They had operated the Salyut 6 station for 13½ weeks.

One day later, NASA announced the names of four two-man crews for Shuttle flights, the first anticipated within the next twelve months: John Young, Robert Crippen; Joe Engle, Richard Truly; Fred Haise, Jack Lousma; Vance Brand, Gordon Fullerton. Although designed to carry a maximum crew of seven, the first five or six flights would be flown by a two man crew during checkout missions designed to explore every facet of the new vehicle. It was a sadly misplaced confidence; the Shuttle would be delayed a further two years at least because of complex financial, technical and management problems. Meanwhile, the Russians demonstrated a continuing exploitation of the Salyut 6 station.

For nearly three months the station remained unattended while controllers periodically inspected the systems by remote command. On 15 June, 1978, the second phase of Salyut 6 activity got under way with the near midnight launch of Soyuz 29 carrying cosmonauts Kovalenok and Ivanchenkov. About 24 hours later they met and docked with the forward port, moving inside the orbiting laboratory several hours later. For several days they reactivated systems and equipment, checked out the condition of the interior and set themselves up for a long stay.

During the evening of 27 June another Soyuz lifted off from Baykonur carrying cosmonauts Klimuk and Hermaszewski, the latter of Polish origin to represent the second crewmember of non-Soviet birth to fly in a Russian spacecraft. Soyuz 30 docked to the aft port on Salyut 6 a day later and for a week the four men performed experiments and tests on board the station. In the early afternoon of 5 July, Klimuk and Hermaszewski landed back on Earth. Two days later, on 7 July, the second Progress supply vehicle was launched and that arrived at the aft docking port on 9 July. It was a busy station and for the next few days the crew unpacked stores and supplies, refuelling from the Progress 2 propellant tanks ten days later.

On 29 July Kovalenok and Ivanchenkov went outside on a 2 hour space walk, removing and replacing materials exposed to the vacuum of space for tests later on Earth. Three days after that, on 2 August, Progress 2 was detached from Salyut 6's aft port – the only one that had the refuelling lines – and was brought back down to destruction in the atmosphere. Progress 2 fell from space on 4 August, a day later the Salyut/Soyuz 29 trajectory was changed by firing the on-board propulsion system, and on the 5th, Progress 3 was launched from Baykonur. Two days later the cargo vehicle docked at the aft port and deputy flight director Viktor Blagov said that it 'is being used to accumulate considerable stocks of food, water and materials.' The spacecraft also brought up mail from friends and relatives – and a guitar for Ivanchenkov!

Unlike its predecessor, which remained docked for 25 days, Progress 3 was separated from Salyut 6 after only ten days although it did not return through the atmosphere until August 24, seven days after undocking. On 26 August, Soyuz 31 carried cosmonauts Bykovsky and Jahn, the latter an East German, into space where they docked with the aft port at Salyut 6 on the 27th. After a stay aboard the station lasting seven days they returned in the Soyuz 29 spacecraft earlier used to carry Kovalenok and Ivanchenkov into orbit and which had been attached to the forward docking port for the past 78 days.

The plan was to have Kovalenok and Ivanchenkov remain in space for considerably longer than the 96½ days notched up by Romanenko and Grechko but engineers were concerned about the condition of the spacecraft after such an extended period. By exchanging their new spacecraft for the Soyuz 29 vehicle, visitors Bykovsky and Jahn were able to leave a reliable taxi for a trip back to Earth. When the Soyuz 29 spacecraft returned on 3 September it left Soyuz 31 docked to the only port equipped for tanker operations. If Salyut was to be refuelled that aft port had to be cleared.

In an unprecedented maneuver carried out during the morning of 7 September, Kovalenok and Ivanchenkov moved inside Soyuz 31, fastened the hatches, and undocked as if they were returning to Earth. Backing off more than 100 meters, the cosmonauts brought Soyuz to a halt, holding their spacecraft steady in this fixed attitude while controllers on the ground commanded the 19 tonne station to pitch through 180°, effectively turning itself over to point its aft docking port in the opposite direction and in so doing present the forward port to Soyuz 31. Kovalenok and Ivanchenkov then simply moved back up to Salyut and slid on to the docking unit once more.

Limited propellant reserves aboard the Soyuz spacecraft prevented the manned vehicle from flying around the station to change ports and necessitated the station itself doing the maneuvering.

On 20 September, the cosmonauts broke the duration record set up by the previous long-term crew aboard Salyut 6 when they notched up 97 days in space. But still the work went on. By the end of the month the crew had taken more than 18,000 pictures of the Earth and accumulated data from the materials processing and infrared telescope equipment. Salyut tasks were not as diverse as those conducted by Skylab astronauts five years earlier, due in part to the comparatively limited volume available inside the station. Soviet efforts in manned space station activity pointed strongly in the direction of medical research, with a priority on studies concerning the response of the human body to very long periods of sustained weightlessness.

Projections concerning the duration of the Kovalenok-Ivanchenkov mission were confounded when further activity developed early on 4 October. But it was not the return of the cosmonauts, rather the launch of yet another unmanned supply vehicle, Progress 4. Docked to Salyut 6's now vacant aft port on the 6th, Progress 4 was unloaded by the crew over the following days. A new tape recorder had been sent up, music tapes were included, and fur boots for legs and feet depleted in blood, a condition due to weightlessness. After refuelling the Salyut tanks, Progress 4 was detached from the space station on 24 October. Two days later it burned up over the Pacific.

Toward the end of the month the cosmonauts began to pack equipment, film, and personal items aboard the Soyuz 31 spacecraft brought up by their last visitors, in anticipation of a return to Earth. They separated from Salyut 6 on 2 November and came home after a flight lasting 139½ days. After detailed examination and analysis the crew were reported to be in a slightly better conditon than Romanenko and Grechko after their 96½ day mission, directly attributable to better exercise sessions and more conditioning in space for the effects of zero g.

So far Salyut 6 had been manned for two continuous periods totalling 236 days, considerably in excess of the 171 days for the three Skylab visits, and received four teams of visiting cosmonauts and four unmanned Progress supply tankers. For nearly four months the station remained dormant. During the interval preparations were accelerated for minor repairs to be conducted by a third long duration team assigned to occupy the facility. Replacement of components and equipment would be essential for any permanent manned station and even this activity was an important part of learning how to live and work for long periods in space. This had been a similar feature of Skylab, more information being accumulated from the limited repair operations carried out than from a station working perfectly all the time. But prior to leaving in November, Kovalenok and Ivanchenkov performed checks which revealed fluctuations with the propellant pressurization system for Salyut's main engine.

Indications were that fuel may have leaked through a damaged accordion membrane separating liquid from pressurizing gas in one of the six propellant tanks. On the ground, extensive analysis of telemetry from the station and of simulated failure in laboratories opened the possibility of repair using a new technique whereby centrifugal force would come to the aid of engineers. By early 1979, more than fifteen months after the launch of Salyut 6, rumor spread that although the station had fulfilled all the original mission objectives it would be put through a third phase of activity beginning soon.

Vladimir Shatalov spoke openly about the possibility of cosmonauts from Bulgaria, Romania, Cuba and Mongolia participating in two-man visits to Salyut 6, extending the provision for international crews pioneered by Remek from Czechoslovakia, Hermeszewski from Poland and Jahn from East Germany. Operating variously between 300 km and 350 km above Earth, Salyut 6 was in a higher orbit than earlier stations of this type and received periodic boosts back to higher altitude by Progress vehicles before they separated. The facility was obviously intended for lengthy use.

During the afternoon of 25 February, 1979, Vladimir Lyakhov and Valeri Ryumin were launched aboard Soyuz 32. They were to put Salyut in order, conduct the planned repairs, and continue research into long duration flight. Fifteen days later Progress 5 was sent up to rendezvous and dock with Salyut 6 at the aft port, Soyuz 32 having linked with the forward station. On 14 March the Progress ferry latched on to the station and two days later Lyakhov and Ryumin worked with controllers on the ground to rectify the apparent leak across the accordion membrane in one of Salyut's three fuel tanks.

Using the station's attitude control motors to spin the complete three-vehicle assembly slowly around its center of mass, the crew induced a centrifugal force which moved the fuel back through the split membrane into the liquid side of the tank. This was then pumped across into one of the other two fuel tanks, leaving the contaminated residue of mixed nitrogen gas and liquid UDMH to be transferred to an empty tank on Progress 5 by the compressor that would normally reduce back-pressure from the refuelling system. After that a propellant vent valve was opened and the damaged tank purged of trace contaminants in a repeated exercise where nitrogen was forced into the tank and through the vent to space, carrying remaining droplets with it.

Every day for a week the crew flushed the tank with nitrogen before closing the vent on 23 March. Four days later the tank, full of nitrogen gas, was isolated from the system. But with two fuel and three oxidizer tanks remaining the station had ample propellant in reserve for maneuvers or attitude control. The temporary application of centrifugal force to perform the initial separation of fluid and gas was typical of the new found tenacity that gripped mission operations personnel. Where once the Soviet engineers would have cautiously avoided using a suspect system, preferring to replace the complete assembly, innovation replaced a seemingly timid approach instilling confidence that seemed in short supply during the first decade of Soviet manned space activity.

Before separating from Salyut, Progress 5 boosted the station to higher altitude by twice using its own propulsion system thereby conserving Salyut propellant. The unmanned supply tanker undocked on 3 April and was returned to destruction in the atmosphere, clearing the aft port for the visit of Salyut's fourth international crew: Nikolai Rukavishnikov and Bulgarian rookie Georgiy Ivanov. But it was not to be.

Launched on 10 April in bad weather and strong winds, Soyuz 33 began the by now familiar rendezvous routine but problems with the primary propulsion engine during the final stages forced the crew to switch to the back-up system and bring the spacecraft down to Earth after only two days, landing in darkness 320 km south-east of Dzhezkazgan. During re-entry Rukavishnikov and Ivanov were subjected to forces of about 9 g versus the normal maximum 4 g on a Soyuz descent, performing only the second ballistic re-entry, the first being the '5 April anomaly' of 1975.

Flight controllers were deeply concerned at the apparent failure to the primary Soyuz propulsion system. Burning nitric acid and hydrazine delivered by turbines driven with hydrogen peroxide, the primary engine facing aft at the center of the Equipment Module delivered a thrust of 417 kg but the back-up engine, feeding 411 kg thrust through two exhaust nozzles, obtained its propellant from the same four tanks carried for the primary system. Rukavishnikov and Ivanov were to have visited Lyakhov and Ryumin aboard Salyut 6 for about one week, leaving their fresh Soyuz vehicle to return in the Soyuz 32 spacecraft docked at the forward port.

On 13 May, about one month after the Soyuz 33 failure, Progress 6 was launched to dock with the aft Salyut port two days later. More propellant was brought up and additional supplies for the orbiting cosmonauts, air to replenish the existing supply, food, new materials for the Splav 01 furnace and sundry items including plant seeds and a fresh stock of light bulbs. During the re-fuelling operation ground controllers monitored and activated propellant transfer equipment before Progress' engine was used to raise the Salyut orbit back up to a height of about 350 km. Meanwhile, on the ground, a parallel effort was under way to bring the next two-man visit to the Salyut crew.

Soyuz 34 was scheduled to fly a Hungarian cosmonaut under the command of a Russian pilot in further conduct of the international flavor added to Soviet space operations. By the end of May, however, a revision to the existing plan brought about by extensive analysis of the propulsion failure on Soyuz 33 grounded further manned visits until modifications to the spacecraft's engine had been proved by unmanned missions.

With Lyakhov and Ryumin already in space, however, a replacement Soyuz was essential. Nobody had kept this type of spacecraft in orbit for the duration these two cosmonauts were scheduled to remain aboard Salyut so it was necessary to replace their ageing Soyuz 32. It had already been exposed to the vacuum of space for three months, longer than any other Soyuz. With Progress 6 still docked to the aft port, the unmanned Soyuz 34 equipped with the modified propulsion unit, lifted from its Baykonur launch pad on the evening of 6 June.

During the morning of 8 June, packed with used equipment and items discarded by the Salyut crew, Progress 6 was undocked to expose the aft docking port. Later that day Soyuz 34 completed its automated rendezvous and linked up to the recently vacated docking unit. But that still left Soyuz 32 on the front port, the spacecraft that would have returned to Earth with the Soyuz 34 crew had that vehicle been manned.

On 13 June, less than a week after Soyuz 34 arrived and Progress 6 was discarded, ground controllers undocked Soyuz 32 and after four revolutions of the Earth commanded it to return, which it did and intact as planned for analysis and examination. Having been in space for 108 days it would provide evidence as to how well the vehicle stood up to sustained exposure. One day later, on 14 June, Lyakhov and Ryumin got into Soyuz 34, backed away from Salyut while the big station turned itself round as it had once before, and then moved in to redock at the exposed forward port.

In what was rapidly emerging as the most active period ever seen in Soviet operations Progress 7 was launched on 28 June, just two weeks after the Soyuz 34 turnaround maneuver, and docked to the aft port two days later. On board were 1.2 tonnes of cargo and 500 kg of propellant. Also carried was a folded radio telescope to be deployed later in the mission, another interesting 'first' for Soviet cosmonauts. Lyakhov and Ryumin had been in space for more than four months and in that time they had been prevented the visitors received by earlier long duration flights; not only because of unmanned tests considered essential to qualify modifications to the propulsion system but also because engineers were completing a new development of the basic Soyuz.

Meanwhile, having unpacked stores and experiments brought up by Progress 7 the previous month, Lyakhov and Ryumin spent 18 July deploying through the aft docking port an extendible 10 meter diameter reflector dish. During early morning the Progress supply tanker was undocked and the port used to push out from the interior the KRT-10 radio telescope reflector. The Russians claimed this was the first deployable structure erected in space during a manned mission but Skylab could arguably lay claim to that 'first' when the parasol shade was put out in May 1973. It was, however, the first erectable radio telescope used in orbit.

For several days the cosmonauts had collected equipment packed in sections within Progress 7 to assemble the parabolic antenna. During the deployment phase, where the device was first pushed out through the aft port, its ribs flush with the central pole, Progress 7 remained close behind Salyut to televise the event and provide ground controllers with a picture of the entire operation. When fully extended through the port, the cosmonauts activated the deployment mechanism from inside, causing the ribs to extend and the reflector to unfurl. It was very large, equal in diameter to the span of the two horizontal solar cell arrays attached to each side of the station.

Two days after putting out the radio telescope reflector, the crew learned from Mission Control that the ground had commanded Progress 7 to re-enter the atmosphere, its work done. On through to the sixth month in space sailed Lyakhov and Ryumin, every day working with the KRT-10, sometimes in conjunction with Soviet ground-based observatories, sometimes alone. They scanned the Milky Way, observed the Sun, recorded data from pulsars and other sources of radio emission. After three weeks the cosmonauts tried to jettison the parabolic antenna, activating from the interior on 9 August a mechanism designed to release the main supports. But for some reason the system failed to work when the mesh of the antenna fouled rods and ribs, entangling the structure in response to unexpected vibrations.

On the ground cosmonauts and engineers worked to perfect procedures whereby Lyakhov and Ryumin would go out on a space walk and use tools already on board to unjam the large reflector and set it free, exposing again the aft docking port which unless vacated would prevent further refuelling options with Progress supply tankers. Using a water tank similar to that at NASA's Marshall Space Flight Center, cosmonauts had the plans and procedures perfected by the middle of August and in a space walk reminiscent of Skylab repair duties, the two spacemen went through the forward airlock on 15 August. Delayed temporarily by a stuck hatch, the men had to wait while the assembly moved through the night side of the Earth; mission rules inhibit Soviet EVA work during night-time passes of the planet.

Attached by a 20 meter line to a point near cosmonaut Lyakhov at the hatch, Ryumin made his way back to the rear of the station. Using pliers from the tool kit, he snipped free mesh entangled round the supports and pushed the freed antenna out into space. It is interesting to note that provision for radio communication during the space walk allowed the cosmonauts to talk to the ground but not to receive word from their Mission Control until they returned to the interior of the space station. However, the $1\frac{1}{2}$ hour space walk had been an outstanding success, proving again how satisfactorily procedures now evolved on Soviet manned missions were compared with the rigidly inflexible nature of flights carried out during the first decade after Gagarin.

During the space walk the cosmonauts retrieved samples left outside by earlier crews, bringing back a cassette of different materials put out prior to launch in September 1977. They also put out other cassettes and samples to be retrieved by later visitors, if such missions were approved.

There was increased interest in the condition of Lyakhov and Ryumin when they got ready to return to Earth aboard the Soyuz 34 spacecraft brought up unmanned for their use at the end of the flight. The men had been in space 175 days when they finally landed back on Earth during the afternoon of 19 August, 1979. They had manned the Salyut space station for almost six months, alone and with several unexpected tasks to perform. This record was more than twice the duration of the longest Skylab flight and brought the total Salyut 6 occupation period to 410 days in three missions spread across nearly two years.

In that time the program generated 18 launches including the station itself, two aborted Soyuz flights, seven manned Soyuz flights, one unmanned Soyuz mission, and seven Progress supply tankers. In addition, fourteen cosmonauts rode Soyuz and visited Salyut for various periods from six days to six months. Two cosmonauts were launched with the intention of docking but never did get to visit Salyut 6, bringing to 16 the number of prime crew assignments associated with the outstandingly successful missions sequence. It far outclassed anything the Russians had done before and demonstrated important new techniques but, above all, a worthy confidence the controllers and managers now appeared to possess.

Unlike earlier programs, Soviet manned space flight had evolved to the level where failures during long duration flights forced controllers to work around the anomaly or for ever fall back from expanded goals. Conservative to a point of criticism, earlier attitudes would never have brought the achievements recorded by Salyut 6. It was a landmark, a true turning point, in operations philosophy. Perhaps it was increased awareness of the responsibility American astronauts were given that persuaded them to bolder steps. They certainly saw this, and expressed trepidation at the thought of cosmonauts given such authority over their spacecraft when, in preparing for the Apollo Soyuz Test Project, Soviet space officials toured US space centers.

Whatever the root cause, developments in Soviet manned flight were given new impetus by the two years of Salyut 6. Before the last Progress tanker departed Salyut 6 while Lyakhov and Ryumin were still on board, the station was pushed to a record altitude for Salyuts. On 3 and 4 July maneuvers were completed that placed the docked assembly in a 399 × 411 km path and when the crew returned to Earth six weeks later that orbit had decayed to 384 × 409 km, preserving options for later re-manning. But would Salyut 6 receive additional visitors? A lot depended on how Lyakhov and Ryumin were seen to have responded to six months in weightlessness.

On their immediate return the cosmonauts remained aboard the Soyuz 34 spacecraft until lifted bodily to recliners designed to prevent their hearts being stressed by having to pump blood against the sudden force of gravity. Not for two days were they allowed to spend lengthy periods in an upright posture. But the report from physicians who remained with them for the following few weeks was one of confidence for flights of even longer duration.

On the first Salyut 6 mission, Romanenko and Grechko were reluctant to perform all the tedious exercises programmed by physicians. The second flight lasting 139½ days gave Kovalenok and Ivanchenkov ample opportunity to adhere to the schedule for physical conditioning. Lyakhov and Ryumin were even more aware of the need for exercise and worked surprisingly well to combat the debilitating affects of weightlessness.

Following several months of inactivity, attention once again focused on Salyut 6 and Russia's manned program plans when, on 16 December, 1979, a new derivative called Soyuz T was launched, albeit unmanned, from Baykonur. Placed initially into a 201 × 232 km orbit, it trailed Salyut 6 by a considerably greater arc than earlier manned Soyuz or unmanned Progress vehicles. Over the next two days, Soyuz T performed several maneuvers unique to this mission until, on 19 December, it automatically docked to the forward port on Salyut 6 to accomplish the eighteenth link-up to this facility.

Soyuz T represented a radical departure from the conventional equipment carried by earlier vehicles of this type. Whereas ferry flights to Salyut used a variant devoid of solar cell arrays and dependent on limited battery power, Soyuz T got back the panels that would allow it to remain in space far longer in the event it failed to dock the first time round. Also, more maneuvering capability was built in to Soyuz T, enabling it to select from several options the trajectory and precise maneuvers it wanted to make in pulling alongside Salyut. Soyuz had been strictly limited in the past by a propulsion system and a control and command capability dating back to the early 1960s; the T derivative had a completely new propulsion system adopting the same nitrogen tetroxide-UDMH propellants first introduced with Salyut 6.

All previous Soyuz models used nitric acid and hydrazine in turbine engines employing pumps run on hydrogen peroxide, also used in the thrusters for attitude control. Now, with the new nitrogen pressure feed system, Soyuz T was compatible with Salyut 6 and, presumably, its successors, improving the opportunity for topping up one vehicle from another in an operation known as propellant crossfeed. Also, pioneering later manned operations, the first Soyuz T employed digital computing equipment for providing the cosmonauts with information on a screen in the pressurized Orbital Module.

Color transmissions from the onboard TV camera were also made possible by improvements to the telemetry and downlink communications equipment. Micro-electronics and other advanced systems aids were now incorporated, bringing Soyuz T to a new level of sophistication that appeared, in principle, to give cosmonauts more control and monitoring capability than they had enjoyed hitherto.

Soyuz T has been described by the Russians as a new manned transportation vehicle, perhaps with a capacity to return to Earth with film, experiments, and equipment from Salyut space stations. Existing Soyuz vehicles have been limited in the weights they can carry back down at the end of a mission and Progress supply tankers, while having ample room for containing much equipment, lack the heat shields necessary for survival through Earth's atmosphere. This one-way, one-shot, use of Progress supply ships is an expensive adjunct to Salyut operations and while details remain uncertain it is possible Soyuz T provides at least a nominal return capability above that exhibited by conventional Soyuz spacecraft.

But more than that, for Soyuz T proved capable shortly after docking with Salyut 6 of boosting the station to higher orbit; hitherto prohibited to all but Progress tankers by limited propellant capacity in the basic Soyuz, that operation is now considerably eased.

After remaining linked to the unmanned Salyut 6 for more than three months, Soyuz T-1 was commanded to undock and separate on 24 March, 1980, finally returning to Earth two days later. One day after that, on 27 March, Progress 8 carried fresh supplies to the big space station. It was the beginning of a fourth long duration flight attempt. With the supply ship still attached to Salyut, Russia's fortieth manned space flight finally got off the ground on 9 April when cosmonauts Popov and Ryumin ascended on the thrust of Soyuz 35's launch vehicle. Valentin Lebedev trained for this flight but was replaced by Ryumin when an injured knee struck him from the flight list.

Little more than a day into their historic, record-breaking, flight, Popov and Ryumin moved across to Salyut 6, having docked at an elapsed time of 26 hours. Their first tasks were to carry out a host of minor repair duties and to complete the transfer of equipment from Progress 8. That unmanned supply ship, locked to the station's aft port, was separated on 26 April and returned to a fiery end in Earth's atmosphere. One day later, Progress 9 was launched with still more supplies and equipment, reaching Salyut 6 and docking to the aft end on 29 April.

Less than a month later it was discarded, days before Soviet cosmonaut Kubasov ascended in Soyuz 36 with Bulgarian cosmonaut Bertalan Farkas on board, late on 26 May. Twenty-five hours later they arrived at Salyut 6, docked to the aft port and joined Popov and Ryumin inside. It was the fourth successful space flight manned by an international crew.

Kubasov and Farkas crawled into the Soyuz 35 spacecraft, docked to Salyut's forward port, to return to Earth during the evening of 3 June, having spent more than six days with the long duration crew. By leaving their own spacecraft for Popov and Ryumin, and returning in the Soyuz that for 55 days had been exposed to the vacuum of space, Kubasov and Farkas left the long duration crew a fresh vehicle in which to return home.

Next day, 4 June, Popov and Ryumin got inside Soyuz 36, backed away from Salyut 6, and held a stable attitude as the massive station slowly cartwheeled around in front of them, presenting in their direction the forward port to which Soyuz was docked, vacating the aft port for later supply flights. A day later, on 5 June, the first manned Soyuz T lifted away from the Tyuratam launch pad with cosmonauts Malyshev and Aksenov on board.

It was an important step in qualifying the re-designed ferry ship for more efficient operations, a major test prior to replacing existing Soyuz vehicles on routine manned missions. As such, Soyuz T-2 was more a qualification and procedures verification flight, than a Salyut 6-dedicated support mission. It lasted just four days, returning to Earth with its two-man crew on 9 June.

During pre-retrofire activity the Orbital Module was separated first, a new procedure for Soyuz, and following re-entry transparent covers placed over the windows, smeared with charred material from the ablated heat shield, were jettisoned to provide for the first time on a Soyuz flight a clear view for the crew after the main heat phase during descent.

It had been an incredible sequence of manned and unmanned missions. In a period of just nine weeks, five flights, three of them manned, had been flown in support of the third Salyut 6 long duration mission. It was time now for Popov and Ryumin to settle down for a long period of research and technical experimentation. By the time Ryumin landed he would have spent only eight of the previous twenty months on Earth. Before that, however, there was much to do.

As the weeks rolled by, Popov and Ryumin set about materials research tasks aimed at showing the feasibility of simple manufacturing techniques in the weightlessness of space, activity similar in principle to that conducted aboard the US Skylab space station. But this fourth long duration mission on Salyut 6 planned to increase significantly the number of sample plates on which materials condensed to a vapour by electron beams were allowed to form for return to earth. Semi-conductor materials were melted down and several unique processes attempted with modest success in the unusual environment.

But throughout the long period of habitation, Salyut itself was constantly in need of attention, minor repairs becoming as integrated an activity with the scientific tasks as had been the same type of operation on Skylab. The attitude control system needed time and attention and a considerable amount of propellant was used up in that process. But this was not as critical a consumable as it had been on the US station, for there was always the progress supply ships to top up the tanks. On 29 June, Progress 10 was launched for an automatic docking to Salyut 6's aft port two days later; it was separated and brought back to destructive re-entry on 19 July.

Exactly four days later, in a rebuff to Romania, Soyuz 37 carried cosmonauts Gorbatko and Tuan into space. Lt. Col. Pham Tuan was from North Vietnam and his assignment, taking precedence over a Warsaw Pact country, represented the product of an overture to Third World communist states. Shortly before midnight the day after launch, Soyuz 37 docked to Salyut 6's aft port, and the two visitors joined Popov and Ryumin inside. Six days later the visitors crawled across to Soyuz 36, leaving their own spacecraft for Popov and Ryumin, and returned to earth.

Throughout August, the cosmonauts continued studies of the earth and performed materials processing and biophysical tests. Frequent use of Progress ferry ships, however, had biased the on-board stores in favour of dry cargo housed in containers and lockers. To conserve propellant, and avoid the necessity for a Progress mission carrying only liquids, the crew put Salyut 6/ Soyuz 37 in a gravity-gradient mode, similar to that tested by Gemini pilots in 1966. Studies of the land and sea below continued and tests were made of the station's back-up propulsion motors, the first time they had been fired in two years. For more than six weeks, Popov and Ryumin circled the globe alone.

On 18 September, shortly after 10:00 pm Moscow Time, Soyuz 38 was launched carrying Soviet cosmonaut Romanenko and Cuban cosmonaut Arnaldo Tamayo Mendez. Little more than 25 hours later they were safely locked to Salyut 6's aft port; Soyuz 37 had been moved to the forward port since the last manned visit. When Romanenko and Mendez returned to earth they used the same spacecraft that carried them to Salyut 6. If Popov and Ryumin were to use Soyuz 37, they would have to return by the end of the third week in October, or contravene a mission rule prohibiting the manned use of a standard Soyuz left in space longer than 90 days.

On 28 September, one day prior to establishing the start of Salyut 6's fourth year in orbit, Progress 11 ascended from its launch pad, reaching the station during the evening hours, Moscow Time, of the 30th. It was time to stock up again and to get ready for return to earth.

On 8 October, the cosmonauts fired the station's main propulsion system to raise and adjust the orbit for manned re-entry via Soyuz 37. Three days later, shortly after noon, Ryumin and Popov were back on the ground, having set up a new manned flight endurance record of 185 days – minus 168 minutes! Cosmonaut Ryumin had beaten his own record breaking flight of 1979 by ten days, accumulating three days short of one full year in space on three separate flights which in itself represented a unique achievement. Both men were in good health, weighing slightly more than when they were launched six months earlier.

In a review of Salyut 6 activity, the mission director said Ryumin and Popov spent about one-quarter of their work time repairing the station, performed more than 50 separate repair tasks, photographed 100 million square kilometers of the earth, shot 3,500 pictures of the ground with the MKF-6 multispectral camera, made a further 1,000 images with the KT-140 camera, and conducted astrophysical studies lasting more than 100 hours.

When Salyut 6 was launched in September, 1977, the Soviet endurance record stood at 62 days, set by Klimuk and Sevastyanov aboard Salyut 4 two years earlier (exceeded as a world record only by the 84-day Skylab flight which ended February, 1974). After three years of sustained operation, involving 28 separate launches and 22 cosmonauts from seven countries, Salyut 6 hosted a total of 12 separate manned missions, successfully received eleven Progress ferry vehicles, failed to receive two aborted flights, and carried for more than three months an unmanned precursor of Russia's latest spaceship, Soyuz T.

Throughout it's more than 1,100 days of continuous operation, Salyut 6 supported four long duration flights, each setting new world endurance records (96, 139, 175 and 185 days, respectively) and eight short duration manned visits. It was, in truth, a program without parallel and one which would point the way to larger and more sophisticated space stations of the future. The epic adventure of Salyut 6 had carried Soviet space time to a staggering 45,969 man-hours, more than twice that accumulated by the US manned flight program.

But it was not the end of Salyut 6, for, while Popov and Ryumin rested in Kislovodsk, three cosmonauts were launched aboard Soyuz T-3 on November 27, 1980. After several days carrying out extensive repairs to Salyut 6, cosmonauts Leonid Kizim, Oleg Makarov and Gennady Strekalov returned to Earth December 10. Only the day before, Progress 11 had been separated from the station's aft port, leaving Salyut 6 stocked, repaired and ready for a further series of manned visits.

As early as 1974 there were strong indications that Russia intended to remain in space for a long time to come, expanding and broadening the base it had so energetically worked to set up for more than a decade. The slow start to major activity (not, it should be said, paralleled in the unmanned space programs where the Russians annually launch three times the number of satellites put up by the West) belies the conviction held by many in Soviet positions of authority that Russia's activity in space is essential to its survival and growth as a major world power. Where once the Russians spoke of manned flights to large space stations and bases on the Moon, they now discuss problems of interplanetary flight and manned expeditions to Mars.

But all that is very much in the future and in the near term, in goals set now on results from Salyut 6, Soviet scientists and engineers are determined to get the best results possible from existing hardware. Flight director Alexei Yeliseyev believes that 'In a sense, we are studying the margin between difficult and impossible' objectives discussed speculatively until the last few years. In combining goals originally set by American proponents of the Air Force Manned Orbiting Laboratory and the NASA Apollo Applications Program, Salyut designers have brought together into a common structure elements of military reconnaissance and civilian research tasks. Two Salyuts were fulfilling a MOL-type role while others operated with a Skylab-style mandate.

A clear indication of emerging Soviet hardware stems from the protracted and cautious effort applied to new designs or variants of existing equipment. For two and one-half years before the launch of Soyuz T in December 1979, vehicles of this type under the Cosmos blanket were tested in Earth orbit on between three and five checkout flights. Similarly, clear evidence exists that the Russians are intent on reducing the cost of their ambitious operations by providing cosmonauts with a reusable transport vehicle like NASA's Shuttle. But unlike its American counterpart, the Soviet Shuttle is believed to be considerably smaller and to be launched on top of a conventional rocket, returning to a controlled re-entry and possibly having wheels for an aircraft style landing on a runway.

Several tests have already been conducted in support of this project and it is believed that the re-entry vehicle is about 10 meters in length with a wing span of 7 meters, or about one-third to one-quarter the size of the NASA Shuttle. As presently conceived, the Raketoplan would carry a crew of four and provide limited storage space for about 500–1,000 kg of cargo. With three propulsion units at the rear and propellant tanks in the sides, Raketoplan is thought to have a delta shaped wing designed to generate lift during flight back through the atmosphere. Unlike the American vehicle of this type, Raketoplan will not be used to replace conventional rockets and its very limited payload capability would not permit it to put out satellites or other space probes, tasks for which the NASA vehicle is specifically designed.

Depending on the precise configuration, Raketoplan may weigh between 7 and 15 tonnes when launched by its conventional booster, the complete assembly standing up to 65 meters tall on the launch pad. Models and full size replicas of Raketoplan have been seen during drop-tests from a Tupolev Tu-95 Bear, an aircraft frequently used in aeronautical research as a carrier-plane for new designs. Soviet interest in reusable manned space vehicles capable of being flown down through the atmosphere to a conventional landing began at the turn of the decade and by 1975 there was abundant evidence that work had progressed well toward definition of the type of vehicle expected to emerge operationally by 1985.

Beyond that, Soviet design teams plan to introduce a reusable rocket stage, thereby achieving maximum economy from expanded use of existing hardware, employed to launch Raketoplan, or its successor, into space. But that may not come before 1990 although it is clearly a distinct goal of Soviet aerospace engineers. Meanwhile, there is much to be accomplished with existing hardware embracing both Salyut 6 and successors that may involve multiple Salyut dockings to expand the size and capability of orbital stations.

The Russians have consistently spoken of the virtues in docking together in orbit several modules of like design proven as independent units. Salyut has been well proven already and refinements in supply ships like Soyuz T point to developments of this kind in the mid-1980s. Several such stations docked together would provide enormous opportunity for research, reconnaissance and the use of manned orbital facilities as jumping off points to the Moon and the planets. In the immediate future, additional crews of mixed nationalities may visit long duration missions utilizing Salyut stations.

One month before Popov and Rumin completed their 185-day flight in October, 1980, French cosmonaut candidates Patrick Baudry and Jean-Loup Chretien began intensive training at the Soviet Star City preparation facility for a mission one of them will fly in 1982 to a Salyut space station. By 1984, cosmonauts from Hungary, Mongolia and Romania will have been taken by Soviet commanders to such stations.

But the concerned interest of Soviet physicians in studying the effects of long duration flight can only have application to planetary missions. A journey to Mars and back would take at least 1½ years with little time for exploration. It is possible that man can survive weightlessness for this time? Evidence from Skylab medical experiments suggests that it may not be.

Careful study of the cardiovascular system shows that the heart and the venous system can be conditioned to accept the unusual environment and that with special exercise routines an astronaut or cosmonaut will not suffer unduly from flights of this duration – provided a carefully balanced exercise program is maintained. And that is vital. As for the vestibular system, initial nausea, vomiting, disorientation, vertigo, and other unpleasant symptoms of an unbalanced middle-ear can be averted. The body quickly adapts.

As for changes to the blood chemistry, red cells are significantly depleted, falling by up to one-fifth in quantity, which appears related to the loss of fluid and a complete halt to red cell production. This may be damaging in the long term, but flights so far have failed to predict a possible deadline. As for muscle atrophy, that can be reduced by exercise of the type used on the last Skylab mission.

But the real problem concerns the loss of bone mineral, suspected of being up to 25% on the two-week flight of Gemini VII in 1965 but recognized later to be a false reading due to errors in measurement methods. Nevertheless, bone mineral is lost at a steady rate and consistently without reduction in the depletion rate over the duration of space flights performed so far. Without calcium, bones become brittle and begin to affect adjacent parts of the anatomy. Bone density was seen to reduce by 8% on the three-month Skylab flight.

Unless some way can be found to halt this decay 'astronauts on long duration flights will need to be protected from demineralization,' according to Consultant Clinical Physiologist Michael W. Whittle of the Oxford Orthopaedic Engineering Centre, on loan to NASA for the Skylab program. 'The skeleton appears to be the only body system in which the effects of weightlessness are neither self-limiting nor preventable.' Quite how immunity to the effects of zero-g can be achieved is difficult to see, unless by stressing the skeleton on a treadmill type exerciser over long periods.

It appears likely that if trends seen after three and six month flights continue for one year man will be unable to spend more than approximately twelve to eighteen months in space without artificial gravity induced by spinning the space station or reconciliation to a destiny beyond the Earth. For beyond eighteen months, the human skeleton may become so brittle that a human being would succumb to a 1-g gravity field and suffer irreparable damage to bones and tissue. Further experiment by Soviet stations may probe that boundary in the years ahead but it would be a crippling irony indeed if the price for journeys to the planets was to be forever condemned to roam the space lanes of the solar system, unable to withstand the crushing force of Earth gravity to which the species has adapted over millions of years.

But it is a biophysical problem, and not perhaps one for the engineers, unless artificial gravity is found essential by spinning stations for centrifugal force. It is an issue likely to dominate Soviet research on long flights around Earth. Not for at least another decade will the United States possess the tools to support such marathon stays in space.

Immediately after completion of the Apollo Soyuz Test Project in June 1975, NASA put aside any plans it retained for further use of Apollo hardware. The big rockets – Saturn V and Saturn IB – would be kept in storage for a few years, just in case they were needed for some unforeseen application, and the spacecraft left over would be held available for possible future use. But it was apparent during 1975 that the Shuttle was the next major commitment for the space-faring aspirants of America and that to allocate funds for other manned ventures between ASTP and the advent of the Shuttle would deny the latter facilities, manpower and finance necessary to get it operational by the end of the decade.

It turned out that this was an optimistic assumption, for the technology involved was necessarily more advanced than some thought, or said, it would be and the space agency was forced to delay the first flight beyond original schedules. However, by dismantling Apollo hardware and failing to invest in any other endeavor capable of long duration manned flight the United States has little opportunity for sustaining the kind of programs begun in the 1960s.

The Shuttle is at most a short-duration transport system, albeit with a capacity to lift 30 tonne loads into a low orbit, and cannot remain in space longer than seven days without additional supplies. A stretched version of the Shuttle may be capable of remaining in orbit for periods of one month but significant advances on existing Soviet systems now being developed call for major new initiatives that seem far off.

To establish a vehicle for truly long duration flight, NASA would have to develop a modular space station for which there seems little prospect before the 1990s; expected to require a development period of at least five years, space money is already committed up to the mid-1980s. But in any event, wide support for such an undertaking would be a prerequisite for Congressional approval and there seems little stimulus for such ostensibly esoteric programs in an age of dwindling economies and depleted energy reserves.

Considerable advantage would befall the nation sufficiently far-sighted to invest in manned Earth orbit stations capable of significantly affecting the way society evolves on the surface of the planet, but few political leaders have the training or the experience to allow them to see and understand the consequences of their apparent disinterest. So the prospects for long duration missions in the future are firmly placed with the ambitions and aspirations of the Soviet Union.

As for manned exploration of the Moon or Mars – the only worlds likely to accommodate human visitors for the next century – here too the prospects are probably loaded in favor of a Russian initiative. If the Soviet Union believed it could achieve political power, prestige or influence by such a venture it would begin such a program at once. As it is, the philosophical and ideological shape of Russian communism converges increasingly upon the view that states within the Soviet envelope have champion rights to exploration and settlement not only on this planet but others too.

The scientific fraternity in Russia is as unwilling to stand apart from suspect politics as were the German technicians and engineers that sought a road to the stars within the rocket research of Hitler's Germany. Political advantage and scientific concurrence combined with motivating forces in Soviet decision-making will inevitably lead to planetary colonization on a grand scale. Perhaps not in the next century but certainly in the one after that. Meanwhile, motivated by profit and loss margins, the United States may well miss out and perchance fall once more into that trap where crash programs are the only solution to diminishing influence.

The single most important factor in the entire history of manned space flight in the two decades since 1960 is that the political machine in America lacks vision and commitment, providing too democratically flexible a response to changing public moods and idiosyncracies. The United States is the most effectively equipped machine for accomplishing massive technical and industrial programs of the type envisioned for the future. But the will and the driving ambition to achieve such things has evaporated under a deluge of commercial interests and self-satisfaction with a seemingly bountiful planet; others with less benevolent philosophies seek to use the new found technology for aims not wholly committed to progress and freedom.

While it was possible once to foresee planetary expeditions and the beginning of a colony on Mars, events of the day threaten to stifle such grand undertakings. For the next several decades the planet will experience a fundamental change in the way man lives and works. In the change from a society dependent on fossil fuels to one more solidly based on solar or other sources of energy there will be little money or manpower for missions to other worlds. But there is perhaps a growing commitment by other countries to participate in ventures they once felt excluded from.

Europe is actively preparing to send its own astronauts into space aboard the NASA Shuttle, probably making the first flight as early as 1983. Having built Spacelab under an agreement signed with NASA in 1973, this manned laboratory designed to remain fixed in the Shuttle's cargo bay will provide ample opportunity for research into materials processing, Earth resource analysis, physics and astronomy. From Spacelab could come the next generation of manned facilities designed to remain in space when the Shuttle returns.

For the moment, the initial activities in manned space flight have been brought to completion by events presaging the new era of reusable Shuttle 'craft. Following several years of valued service to science, the five ALSEP arrays of instruments left by Apollos 12, 14, 15, 16 and 17 at selected sites on the lunar surface were shut down. At the end of September, 1977, engineers switched the equipment off because the money and resources necessary to maintain the flow of data was needed elsewhere; it was a significant footnote to diminishing interest in Earth's nearest neighbor. In all, the five stations had received more than 153,000 commands, operated several years beyond their 12-month design lifetimes, and recorded 12,000 Moonquakes and meteorite impacts. But it was a case of diminishing returns and the comparatively quiescent Moon was giving up decreasing amounts of information so continuous contact with the five ALSEPs was no longer justified.

And then, on 11 July, 1979, another event cancelled hopes of returning to inspect the Skylab space station with a Shuttle vehicle in the 1980s. During the morning hours on that day, Skylab re-entered the atmosphere and broke up over the Indian Ocean, sending debris across the south-western tip of Australia. Its demise had been precipitated by unusually high activity on the Sun which, in turn, increased the density of Earth's outer atmosphere. That had the effect of slowing Skylab down and bringing it back into the denser regions far sooner than anybody predicted. It was, perhaps, a fitting tribute to the end of the reusable age in manned flight. Not for many years would NASA resume space station activity.

The impact of expanded Salyut operations on space policy in the United States is believed by NASA Administrator Robert Frosch to pose a challenge to America and that 'I think people in general would consider it a serious event.' In that way NASA may ultimately secure the funds it now lacks to build a capability similar to the permanently manned space stations sought by Russia. But what of cooperation in the wake of ASTP?

The reason that prospects for continued use of techniques and docking hardware developed for the joint mission in 1975 are lower than they have probably ever been are routed in NASA history and foreign policies of the two nations. For almost two years before the mission was flown, NASA tried to get the Russians talking about following projects after ASTP. They declined to discuss this until the joint mission was over so when Bushuyev and his team visited Houston in November, 1975, everybody wanted to begin serious discussion of a possible Shuttle-Salyut docking early in the 1980s.

America was unwilling to commit Apollo left-overs to further manned missions and expected the Shuttle to be fully operational at the start of the new decade. Bushuyev was reluctant to open negotiations immediately and preferred instead to wait for further deliberations in Moscow. Some people in the United States openly contested the need for such flights, saying that America had already exposed the nerve ends of its technology to Russian inspection during ASTP and that it should not do so again.

In fact, there is little evidence to believe that. The Russians were never given the loan of equipment that would significantly advance their own state of the art and manufacturing techniques are so different that Soviet progress, technological or otherwise, went largely uninfluenced by the exercise. Both sides eagerly sought an inside view of the other's space program, and both learned more than they could have acquired without ASTP. If the Russians gained anything it was seeing for themselves the sincerity of an approach they previously believed to be headstrong and bold.

They learned first hand how engineers, technicians and astronauts when pressed hard to be responsible decision-makers can force the momentum of a major industrial endeavor beyond the pace of its own inertia. For that was really the success of NASA's manned program. Nobody was work-force alone and no one solely management; both did a united job for a common goal outside the capacity of each individual. It was that feeling, of accomplishing more than the sum total of individuals, that bred success.

So when ASTP ended there were plenty of reasons to believe both countries would wish to pursue further negotiations. But then, within a few months of the flight, opposition to Soviet foreign policy and Russia's support for Cuban mercenaries in Africa turned the tide of public opinion away from the Kremlin. Moreover, with Nixon removed from the White House and a caretaker President installed in his place, criticism increased over the deal worked by Henry Kissinger before the SALT I agreement signed in May 1972.

Between 1974 and 1976 it became increasingly apparent that Soviet deployment of new generations of strategic military missiles had been kept back until Washington ratified the treaty. Once secured, open and ambiguous language within the treaty's text was exploited by the Soviets to their own advantage. This created a feeling of betrayal and talks between the two space communities entered a state of limbo. Many parallel agreements swept along with ASTP had been signed, including the flight of US experiments on Soviet biological satelites, using, incidentally, Vostok type vehicles equipped to perform unmanned scientific research. Yet it was not for these reasons alone that the possibility of flying a successor to ASTP evaporated. For NASA itself was adjusting to new roles and commitments.

On 5 June, 1976, George Low ended a long and distinguished career with NASA when he retired to become president of the Rensselaer Polytechnic Institute. Long term advocate of US-Soviet space cooperation, Low was replaced as Deputy Administrator, a post he held for 6½ years, by Dr. Alan M. Lovelace from 2 July, 1976. Only a year before, the space agency restructured its internal profile to reflect more ably the new emphasis on Earth applications rather than lunar and planetary exploration, by abolishing the Office of Manned Space Flight. Set up during the November, 1961, reorganization made necessary by John F. Kennedy's Moon commitment, John Yardley would now hold reign over the Office of Space Flight.

The term 'manned' space flight had become redundant in the wake of transforming such activities into launch systems for unmanned satellites and space probes. All launch vehicles, the manned Shuttle and unmanned Scout, Delta, Atlas, and Titan rockets, would come under Yardley's aegis. He was, in essence, to develop an Earth to orbit transportation system for carrying unmanned satellites, laboratories fixed to the Shuttle cargo bay, and a few planetary robots every few years. It was a very different program from that he worked toward in Mercury, Gemini and Apollo, but it was the only one Congress was willing to pay for.

Major internal shuffles between 1975 and 1977 kept the agency honing its infrastructure for Shuttle operations to begin in the 1980s. Nevertheless, throughout this period, negotiations did move, albeit ponderously, toward a general direction of combined Shuttle-Salyut docking flights. Recently appointed Deputy NASA Administrator, then serving as Acting Administrator, Alan Lovelace and the new head of Russia's Academy of Sciences Anatoly Aleksandrov agreed in May 1977, to work toward this objective. By the end of that year preliminary plans emerged for a Shuttle to carry scientific equipment to a Salyut station 400 km above Earth and to dock for combined operations. The Shuttle would return to Earth within seven days.

A prime advantage in bringing these two together lay in efficient use of Salyut longevity and the heavy payload capacity of Shuttle. Talks assumed the possibility of one day bringing large modules to Salyut equipped for unique scientific tasks carried out by an international crew. However, the change in NASA administration that came with the arrival of a new team at the White House reflected the lack of in-built commitment to international cooperation. Pursued by Hugh Dryden, then Tom Paine and George Low, the incentive to create a joint US-Soviet mission had been a product of people rather than doctrine or national objective.

At the end of 1976, Gerald Ford was firmly beaten in his bid to retain the Presidency by an inexperienced contender fresh from the regional politics of governorship. Jimmy Carter became the thirty-ninth President of the United States. With not only a change in administration but also in party, the responsibility for shaping American domestic and foreign policy produced a democratic stance in conflict with Soviet treatment of dissidents and foreign nationals. There developed during 1977 and 1978 a discontent with issues concerning human rights, agreements signed at Helsinki concerning the basic freedoms of people everywhere under different political systems, and between Carter and the Soviet Praesidium over the strategic intent of Soviet foreign policy.

NASA Administrator Dr. Robert Frosch (left) seen with Johnson Space Center Director Christopher Kraft. Frosch resigned from NASA in January, 1981, verbally criticizing the deflation of NASA's manned flight program; Kraft had been with manned flight activity from the beginning and carried the concept through to reusability and the NASA Space Shuttle, the beginning of a new era where applications took precedence over exploration.

Through it all, negotiators worked long and hard at a SALT-II agreement developed between the super-powers to shape the contour of future arms expansion; Soviet insistence on a hard core of defensive material prevented arms *limitation* from being anything other than a name in the SALT acronym. So once again detente evolved between Russia and America when President Carter learned the ways of international politics and the doubtful advantage of taking unilateral action on arms embargoes, human rights issues, and trade levies.

By 1978 the White House was prepared to accept the forward momentum of discussions about a joint Shuttle-Salyut flight, conducted by NASA and with the Soviet Academy of Sciences. But NASA Administrator Jim Fletcher had resigned 1 May, 1977, and less than two months later Dr. Robert A. Frosch was named by the White House to head the space agency. Frosch was less concerned with international projects than he was with getting the Shuttle operational and smoothing major funding problems even then emerging with the reusable space 'plane. He affirmed that any cooperative flight of the type performed in ASTP 'would have to compete in future budgets with other possibilities for doing science or applications.'

Like Kennedy, Johnson and Nixon before him, President Carter decided to hold a major review of America's space posture and to prepare a report outlining his administration's attitude toward future objectives and the space policy he wished to pursue in the years ahead. For much of 1977 Carter was concerned with other issues, only getting round to a space policy statement more than a year after gaining office. In 1978 Carter made it quite clear that his administration was not in favor of an Apollo-type goal for the future and that any major development would await initial operations with the Shuttle which might, he admitted, lead to new initiatives some time during the coming decade.

Gerald Ford and Jim Fletcher, the President and the Administrator each content with respective caretaker roles, had let the chances slip; Congress was too embroiled with domestic issues, energy crises, and increasing concern over America's weaker military position to bother now about the development of new roles in space.

Frosch allowed talks to proceed on a low level about possible Shuttle-Salyut missions but with chief proponents gone there was little heart for such endeavors. By this time also it was apparent that the Shuttle would not make its initial flight into space during March 1979, as planned. Several delays, not unexpected with such a major technological effort, were forced upon the program as much by inadequate funding during the Fletcher years as any inherent mismanagement during the latter period. What this did was to place in limbo any discussions about dates of possible Shuttle-Salyut options until a major shift in Soviet foreign policy ended all hope of an immediate agreement to go ahead.

Four days before the end of 1979, Soviet infantry divisions poured across the border with Afghanistan bringing to an end a communist regime critical in part of Russian policy. Installed as a puppet government by masters from the Kremlin, the new leaders reflected a not unexpectedly pro-Soviet stance. The response by President Carter was echoed by a majority of free countries around the globe and verbal condemnation led to a halt on all US-Soviet cooperation in science and technology.

Also, increasing concern in Congress about the diminished ability of the United States to develop selective military projects in the framework of SALT II brought a suspicion that Russia was seeking to use negotiations for its own selfish ends. Combined, these changing moods and opinions, backed up by flagrant aggression in the Middle East, inhibited further discussion on a cooperative manned space mission. It may not be possible to resurrect such a flight for at least another decade.

Robert Frosch left NASA on 20 January, 1981, the only Administrator since Glennan – NASA's first boss – not to see a US manned space flight within his tenure of office; in the wake of a chilled dialogue between the super-powers, East-West cooperation was moribund.

And so we come to the end of a period in the history of man when he arose from Earthly bondage and touched the surface of his companion Moon, when for an instant in time it seemed he was indeed writing new chapters in human destiny across the cold expanse of space. As it is, the events of a moment were not to be the precedent for a golden age of exploration remote from continental Earth. And neither should we expect such endeavors again, for many decades may pass before the circumstances and the situations that moved President Kennedy inspire another world leader with similar ambition.

Like all great explorations and discoveries before, the journey to the Moon was neither planned as a liberating crusade for mankind nor as a great new stepping place to the future. To those who call for an end to competition and a halt to causes it is worth a reminder that none of this would have come to pass were it not for the bold pride of a New World people. Without the surge of Apollo, manned flight could well have died in 1960 for it is not clear that Russia would have continued alone along the unchallenged road of grander and more rewarding space missions.

As it is, the legacy of Apollo rides on Shuttle wings designed to make space travel more commonplace than ever before. In 1978, thirty-five new astronauts reported to the Johnson Space Center for training to fly this vehicle in the decade ahead and a further 19 were selected in 1980. But the Shuttle is another story. With the end of the 1970s an age passed into history that will be seen in perspective only when the accomplishments of yesterday have liberated new possibilities for the future.

Gone now are many of the visionary minds that nursed a pioneer group to dizzy heights in space and on the Moon. Shortly after ASTP, Wernher von Braun learned he had cancer and on the last day of 1976 retired from Fairchild. Six months later, on 16 June, 1977, shortly after receiving the US National Medal of Science, Wernher von Braun died. With him went the dreams and ideals of a nascent group from Peenemunde, Germany, whose visions always rested among the stars and planets. Von Braun believed sincerely in the crusading nature of man in space, and who can say that one day his dream will not be fulfilled. In truth it can be said that we went to the Moon not because of our technology but because of our imagination.

Man will probably not descend to the red dust of Mars, or the surface of any other planet, in the lifetime of anyone on Earth today. But there will come a time, and an age dawn, when the emerging purpose of man will be fulfilled in the timeless void. For with him will go the cultures, the beliefs, the sentiments and the creations of ageless history past bringing lessons for tomorrow. No better prelude to that journey can be had than words expressed by Ray Bradbury:

> 'You Jonahs travelling in the belly of a new-made whale,
> You swimmers in the far off sea of Space,
> Blaspheme not against yourself or the frightening
> twins of yourself you find amongst the stars.
> But ask to understand the miracle which is
> Space, Time and Life in the high attics and lost birthing places
> of eternity.
>
> 'Woe to you if you do not find all life most holy
> And coming to lay yourself down cannot say,
> "Oh, Father God, you waken me, I waken thee.
> Immortal We then walk upon the waters
> of Deep Space in the new morn which
> Names itself Forever."'

Project Data

The following tables are compiled to provide the reader with important information in assessing the magnitude, importance and accomplishments of US and Soviet manned space flight programs. It is not intended as an encyclopedic reference and should be read in conjunction with relevant pages of text.

TABLE 1: MERCURY BASIC DATA

The following data refer to a typical spacecraft prepared for a three-orbit mission as flown by MA-6 and MA-7. Data for MA-8 and MA-9 are included, where different, in brackets. Spacecraft weights are typical but specify spacecraft 13 flown by John Glenn.

Physical data:
- Length 7.91 meters on launch pad
  - 3.34 meters without escape rocket
  - 2.92 meters without retro-package
- Diameter 1.89 meters bottom
  - 81.3 cm top
- Weight 1.935 kg launch
  - 1.355 kg orbit
  - 1.131 kg splashdown
  - 1.099 kg recovery

Environmental Control System (ECS):
- Atmosphere 100% oxygen at nominal 258 mmHg
- Storage 2 bottles totalling 3.6 kg (MA-9: 3 bottles totalling 5.4 kg)
- Cooling fluid 17.7 kg water (MA-9: 21.8 kg)
- Drinking water 2.5 kg (MA-9: 4.5 kg)
- Atmosphere cleanser 2 kg lithium hydroxide (MA-8, MA-9: 2.4 kg)
- Odor absorber 0.45 kg activated charcoal (MA-9: 0.091 kg)

Electrical systems
- Batteries 3 × 3.000 watt-hr silver zinc on main 24 volt d-c buses
  - 2 × 1.500 watt-hr silver zinc on 24, 18, 12, 8 & 6 volt d-c standby buses (MA-9: 2 × 3.000 watt-hr batteries)
  - 1 × 1,500 watt-hr silver zinc on 24, 18, 8 & 6 volt d-c isolated buses

Attitude control thrusters

	System A (kg thrust)	System B (kg thrust)	No. of thrusters
Pitch	10.9	1.8 – 10.9	6
Roll	2.7	0.4 – 2.7	6
Yaw	10.9	1.8 – 10.9	6

Propellant: Automotive (ASCS/FBW) 15.8 kg
Manual (MP/RSCS) 11.3 kg
Total: 27.1 kg (MA-9: 33.9 kg) hydrogen peroxide

Attitude control modes

	Attitude thruster system	Electrical power
ASCS	A	d-c & a-c
FBW	A	d-c
MP	B	none
RSCS	B	d-c & a-c

Solid rocket motors

	Number of motors	Nominal thrust (kg)	Burn time (sec)
Escape	1	23.587	1
Tower jettison	1	363	1.5
Posigrade	3	181	1
Retrograde	3	454	10

Parachutes
- drogue 1 × 1.8 metre conical ribbon
- main 1 × 19.2 metre ringsail + identical reserve

ASCS Attitude Stabilization & Control System.
FBW Fly-By-Wire
MP Manual Proportional
RSCS Rate Stabilization & Control System.
d-c direct current
a-c alternating current

Note: RSCS was not installed before MR-4 or for MA-9. Readers should refer to the text for detailed changes.

TABLE 2: MERCURY FLIGHT DATA (unmanned mission)

Mission	Occupant	Date	Spacecraft	Launcher	O.S.D.	Duration	Apogee/ Perigee (km)	Range (km)	max. speed (km/hr)	Max. g	Test objective	Results
LJ-1	—	8/21/59	BP	—	7/59	0:00:20	0.64	0.8	—	—	Max q abort to test launch escape system & recovery equipment	F
Big Joe	—	9/9/59	BP	10-D	8/59	0:13:00	153	2.407	23.909	12	Ballistic test of structures, heat protection & recovery systems	S/F
LJ-6	—	10/4/59	BP	—	—	0:05:10	59.5	127	4.948	5.9	Launch vehicle qualification and capsule aerodynamic evaluation	P
LJ-1A	—	11/4/59	BP	—	—	0:08:11	14.5	17.7	3.254	16.9	Repeat of LJ-1	P
LJ-2	RM (Sam)	12/4/59	BP	—	9/59	0:11:06	85.3	312	7.187	14.8	High altitude abort with primate	S
LJ-1B	RM (Miss Sam)	1/21/60	BP	—	11/59	0:08:35	14.5	19.3	3.254	4.5	Repeat of LJ-1A	S
Beach Abort	—	5/9/60	1	—	—	0:01:16	0.8	1.6	1.571	—	Pad abort to qualify spacecraft structure	S
MA-1	—	7/29/60	4	50-D	11/59	0:03:18	13	9.6	2.737	—	Spacecraft/launch vehicle compatibility during abort	F
LJ-5	—	11/8/60	3	—	12/59	0:02:22	16.2	22.5	2.873	6	Max q abort to qualify launch escape system	F
MR-1	—	11/21/60	2	MR-1	10/59	0:00:02	—	—	—	—	Suborbital qualification of spacecraft and launch vehicle	F
MR-1A	—	12/19/60	2A	MR-3	—	0:15:45	210	378	7.900	12.4	Repeat of MR-1	S
MR-2	C (Ham)	1/31/61	5	MR-2	12/59	0:16:39	253	679	9.426	14.7	Suborbital qualification of abort with primate	S/P
MA-2	—	2/21/61	6	67-D	1/60	0:17:56	183	2.304	21.286	15.9	Repeat of MA-1	S
LJ-5A	—	3/18/61	14	—	—	0:23:48	12.4	29	2.869	8	Repeat of LJ-5	P
MR-BD	—	3/24/61	BP	MR-5	—	0:08:23	183	494	8.244	11	Suborbital test of Redstone modifications	S
MA-3	SM	4/25/61	8	100-D	2/60	0:07:19	7.2	—	1.894	11	Single orbit test of launch vehicle, spacecraft & tracking network	F
LJ-5B	—	4/28/61	14A	—	—	0:05:25	4.5	14.5	2.864	10	Repeat of LJ-5A	S/P
MA-4	SM	9/13/61	8A	88-D	3/60	1:49:20	228.7 × 159.2	41.917	28.205	7.7	Repeat of MA-3 and environmental control system check	P
MS-1	—	11/1/61	—	—	—	0:00:43	—	—	—	—	To launch test satellite for tracking checks	F
MA-5	C (Eros)	11/29/61	9	93-D	3/60	3:20:59	237.2 × 160.1	81.900	28.211	7.7	Three orbit test of life support functions	P/S

LJ: Little Joe
MA: Mercury Atlas
BD: Booster Development

MS: Mercury Scout
RM: Rhesus Monkey
C: Chimpanzee
SM: Simulated Man

BP: Boilerplate
O.S.D: Original Schedule Date (in Jan '59)
F: Failure
S: Success
P: Partial success

Note: Results column refers to total system or launch vehicle/spacecraft
Max q refers to maximum dynamic pressure
Duration is in hr:min:sec

TABLE 3: MERCURY FLIGHT DATA (manned missions)

Mission	Occupant	Age	E.T.	Back-up	Date	Launch time	Spacecraft	Name	S/c weight (kg)	Launcher	O.S.D.	Type	Duration	Orbits	Apogee/ Perigee (km)	Range (km)	Max speed (km/hr)	Max g	Recovery area	Recovery ship
MR-3	Shepard	37	25	Glenn	5/5/61	09:34	7	Freedom 7	1.290	MR-7	1/60	B	0:15:28	–	187.5	487.6	8.262	11	Atlantic	Lake Champlain
MR-4	Grissom	35	27	Glenn	7/21/61	07:20	11	Liberty Bell 7	1.286	MR-8	2/60	B	0:15:37	–	190.4	486	8.317	11.1	Atlantic	Randolph
MA-6	Glenn	30	34	Carpenter	2/20/62	09:47:39	13	Friendship 7	1.355	109-D	4/60	O	4:55:23	3	261 × 161	121.790	28.233	7.7	Atlantic	Noa
MA-7	Carpenter	37	37	Schirra	5/24/62	07:45:16	18	Aurora 7	1.349	107-D	5/60	O	4:56:05	3	268.4 × 160.8	122.340	28.242	7.8	Atlantic	Intrepid
MA-8	Schirra	39	42	Cooper	10/3/62	07:15:11	16	Sigma 7	1.374	113-D	6/60	O	9:13:11	6	282.9 × 161	231.712	28.256	28.1	Pacific	Kearsarge
MA-9	Cooper	36	49	Shepard	5/15/63	08:04:13	20	Faith 7	1.376	130-D	8/60	O	34:19:49	22.5	267 × 161.4	878.946	28.238	7.6	Pacific	Kearsarge

MR: Mercury Redstone
MA: Mercury Atlas
E.T.: Experience Time (since selection as an astronaut in months elapsed duration)

O.S.D. Original Schedule Date (in Jan '59)
B: Ballistic
O: Orbital

Note: Duration is in hr:min:sec
Range for orbital flights equals distance travelled
Age in years at time of flight
Launch time is in local Cape time

TABLE 4: GEMINI BASIC DATA

Physical data:	*Adapter*	*Re-entry Module*
Length	2.286 meters	3.657 meters
Diameter, minimum	2.286 meters	98.2 cm
Diameter, maximum	3.048 meters	2.286 meters
Total length	5.736 meters	

Weight 3.400 kg orbit
2.200 kg splashdown

Environmental control
Atmosphere 100% oxygen at nominal 258 mmHg
Temperature 24°C (49°C during re-entry)
Suit circuit 100% oxygen at 195–217 mmHg
Storage 1 Adapter sphere (47.17 kg) prime containing supercritical oxygen
2 Re-entry Module bottles secondary containing gaseous oxygen
Drinking water 19.1 kg capacity tank in Adapter
Nominal 6.6 kg capacity in Re-entry Module
Fuel cell product water Nominal 19.1 kg capacity tank (also depleted volume in drinking tank)
Atmosphere cleaner lithium hydroxide
Odor absorber Charcoal

Electrical systems
Main 2 fuel cells or combinations of 400 ampere-hr silver zinc batteries
Secondary 4 × 45 ampere-hr silver zinc batteries during re-entry
3 × 15 ampere-hr silver zinc squib batteries for pyrotechnics
Fuel cells (each):
Weight 30.9 kg
Production 1 kW d-c at 23.3–26.5 volts
Size Cylinder 61 cm by 30.5 cm diameter
No. of stacks 3
No. of cells per stack 32

Reactants: Hydrogen and oxygen stored as liquids in separate tanks
Quantities (nominal): 10.5 kg hydrogen at 14.8 kg/cm² pressure (minimum)
82 kg oxygen at 56.2 kg/cm² pressure (minimum)

Attitude control thrusters
OAMS 8 × 11.3 kg thrust engines attached to Adapter
RCS 16 × 11.3 kg thrust engines on Re-entry Module for re-entry only

Translation control thrusters
OAMS 6 × 45.4 kg thrust engines on Adapter
2× 38.6 kg thrust engines on Adapter

Liquid propellant
OAMS quantity Nominal 440 kg N_2O_4/MMH in 3 oxidizer/3 fuel tanks
Pressurant Helium stored at 197 kg/cm² in 2 tanks

Retrorockets
4 × 32.3 cm diameter solid propellant rockets fired sequentially for 5.5 sec each at a nominal thrust of 1.130 kg

Parachutes
Drogue 1 × 3.27 meter conical ribbon
Pilot 1 × 5.58 meter ringsail
Main 1 × 25.6 meter ringsail

d-c direct current
OAMS Orbit Attitude Maneuvering System
RCS Re-entry Control System
N_2O_4 Nitrogen tetroxide
MMH mono-methyl hydrojine

Note Figures refer to typical short-duration spacecraft except reactant quantities denote maximum tanked supplies.
Fuel cell oxygen is obtained from tank employed for environmental control.

TABLE 5: GEMINI FLIGHT DATA (unmanned missions)

Mission	*Occupant*	*Date*	*Launch time*	*Spacecraft*	*S/c weight (kg)*	*Launcher*	*Apogee/ Perigee (km)*	*Splashdown*	*Range*	*Recovery ship*	*Miss distance*	*Recovery time (e.s.t.)*	*Objective*	*Result*
1	—	4/8/64	11:00:01.69	1	3.187	GLV-1	320.3 × 160.3	—	—	—	—	—	Spacecraft/launcher qualification	Success
2	Simulated man	1/19/65	09:03:59.861	2	3.122	GLV-2	171.1	18 min 16 sec GET	3.422	Lake Champlain	62.9 km	10:52	Suborbital systems test	Success

GET: Ground Elapsed Time
GLV: Gemini Launch Vehicle
e.s.t.: Eastern Standard Time
Note: Weights and orbital parameters are as at orbit insertion

TABLE 6: GEMINI FLIGHT DATA (manned missions)

Mission	*Flight crew*	*Ages (yrs)*	*E.T. (mths)*	*Back-up crew*	*Launch date*	*Launch time*	*Spacecraft*	*S/c weight (kg)*	*Launcher*	*Revolution*	*Apogee/ Perigee (km)*	*Highest Apogee (km)*	*Lowest Perigee (km)*
3	Grissom/Young	38/34	71/30	Schirra/Stafford	3/23/65	09:24:00.064	3	3.225	GLV-3	3	224.1 × 161.1	224	158.5
IV	McDivitt/White	35/34	33/33	Borman/Lovell	6/3/65	10:15:59.562	4	3.574	GLV-4	62	281.9 × 162.2	296.1	159.4
V	Cooper/Conrad	38/35	76/35	Armstrong/See	8/21/65	08:59:59.518	5	3.605	GLV-5	120	349.8 × 161.9	349.8	161.8
VI-A	Schirra/Stafford	42/35	80/39	Grissom/Young	12/15/65	08:37:26.471	6	3.546	GLV-6	16	259.3 × 160.9	311.3	160.9
VII	Borman/Lovell	37/37	39/39	White/Collins	12/4/65	14:30:03.702	7	3.663	GLV-7	206	328 × 161.5	327.9	161.4
VIII	Armstrong/Scott	35/33	42/29	Conrad/Gordon	3/16/66	11:41:02.389	8	3.788	GLV-8	7	271.7 × 159.8	298.7	159.8
IX-A	Stafford/Cernan	35/32	45/32	Lovell/Aldrin	6/3/66	08:39:33.335	9	3.668	GLV-9	45	266.7 × 158.7	311.5	158.7
X	Young/Collins	35/35	46/33	Bean/Williams	7/18/66	17:20:26.648	10	3.763	GLV-10	43	268.7 × 159.8	753.3	159.8
XI	Conrad/Gordon	36/36	48/35	Armstrong/Anders	9/12/66	09:42:26.546	11	3.798	GLV-11	44	278.9 × 160.4	1.368.9	160.3
XII	Lovell/Aldrin	38/36	50/37	Cooper/Cernan	11/11/66	15:46:33.419	12	3.763	GLV-12	59	270.6 × 160.7	301.3	160.7

RENDEZVOUS & DOCKING OPERATIONS

Mission	*Target*	*Launcher*	*Launch date*	*Launch time*
VI-A	s/c 7	GLV-7	12/4/65	14:30:03.702
VIII	GATV-5003	TLV-5302	3/16/66	10:00:03.127
IX-A	ATDA	TLV-5304	6/1/66	10:00:02.363
X	GATV-5005	TLV-5305	7/18/66	15:39:46.131
	GATV-5003	TLV-5302	3/16/66	10:00:03.127
XI	GATV-5006	TLV-5306	9/12/66	8:05:01.725
XII	GATV-5001	TLV-5307	11/11/66	14:07:58.688

RENDEZVOUS & DOCKING OPERATIONS

Mission	*Rendezvous revolution*	*Type rendezvous*	*Rendezvous GET*	*Docking GET*	*Docked duration*
3					
IV					
V					
VI-A	4	Co-elliptical	5:56	—	(5:17:29)
VII					
VIII	4	Co-elliptical	5:55	6:33:22	0:41:50
IX-A	3	Co-elliptical	4:15	—	(0:46:00)
	4	Equi-period	6:36	—	(0:39:00)
	12–15	From above	21:42	—	(1:17:00)
X	4	Co-elliptical	5:21	5:52:37	38:47:00
	29	Re-rendezvous	48:00	—	(~3:00:00)
XI	1	First apogee	1:25	1:34:16	48:20:44
XII	3	Co-elliptical	3:46	4:13:53	43:09:24

EXTRA-VEHICULAR ACTIVITY

Mission	*Type*	*Duration*	*Start GET*	*Finish GET*	*Total EVA time*	*Life support*	*Umbilical length*	*Maneuvering unit*	*Participant*
IV	U	0:36	4:18	4:54	0:36	VCM	7.62	HHMU	White
IX-A	U	2:07	49:23	51:30	2:07	ELSS-AMU	7.62	AMU	Cernan
X								—	Collins
	SU	0:49	23:24	24:13		U	0.3	HHMU	Collins
	U	0:39	48:41	49:20		ELSS	15.24	—	—
	EJ	0:01	50:33	50:34	1:29	U	—	HHMU	Gordon
XI	U	0:33	24:02	24:35		ELSS	9.14	—	—
	EJ	0:02	25:37	25:39		U	—	—	Gordon
	SU	2:08	46:07	48:15	2:43	U	0.3	—	Aldrin
XII	SU	2:29	19:29	21:58		U	0.3	HHMU	Aldrin
	U	2:06	42:48	44:54		ELSS	7.62	—	Aldrin
	SU	0:55	66:06	67:01	5:30	U	0.3		

Mission	*Retrofire GET*	*Flight duration*	*Recovery area*	*Recovery ship*	*Miss distance (km)*	*Time from splashdown to: crew recovery*	*s/c recovery*
3	4:33:23	4:52:31	Atlantic 4-1	Intrepid	111.1	71 min	166 min
IV	97:40:01	97:56:12	Atlantic 63-1	Wasp	81.4	57 min	136 min
V	190:27:43	190:55:14	Atlantic 121-1	Lake Champlain	170.3	91 min	235 min
VI-A	25:15:58	25:51:54	Atlantic 17-1	Wasp	12.9	77 min	77 min
VII	329:58:04	330:35:01	Atlantic 207-1	Wasp	11.8	32 min	63 min
VIII	10:04:47	10:41:26	Pacific 7-3	Mason	—	186 min	195 min
IX-A	71:46:44	72:20:50	Atlantic 46-1	Wasp	0.704	53 min	53 min
X	70:10:24	70:46:39	Atlantic 44-1	Guadalcanal	6.2	27 min	54 min
XI	70:41:36	71:17:08	Atlantic 45-1	Guam	4.9	23 min	58 min
XII	93:59:58	94:34:31	Atlantic 60-1A	Wasp	4.8	28 min	67 min

E.T. Experience Time (since selection as an astronaut in months elapsed duration)
GLV: Gemini Launch Vehicle
GATV: Gemini-Agena Target Vehicle
TLV: Target-launch Vehicle
GET: Ground Elapsed Time

EVA: Extra-Vehicular Activity
U: Umbilical
SU: Stand-up
EJ: Equipment Jettison

ELSS: Extravehicular Life Support System
VCM: Ventilation Control Module
AMU: Astronaut Maneuvering Unit
HHMU: Hard Held Maneuvering Unit

Note Rendezvous Revolution column refers to revolution of the chase vehicle
Docked Duration column with brackets indicates duration of station-keeping where no docking occurred
Mission numbers changed from Arabic to Roman numerals from Gemini IV

TABLE 7: APOLLO BASIC DATA

Physical data

	Service Module	Command Module
Length	4.52 meters	3.22 meters
(with SPS nozzle attached)	7.49 meters	
Diameter	3.91 meters	3.91 meters
Loaded weight (typical)	24 tonnes	5.6 tonnes
Splashdown weight (typical)		5.3 tonnes
Total in-flight length	10.1 meters	
Total loaded weight (typical):	29.6 tonnes	
LES length	10.06 meters	
LES weight	4 tonnes	
SLA length	8.53 meters	
SLA diameter top/bottom	3.91/6.6 meters	
Total SLA-CSM-LES length	25 meters	
SLA weight	1.8 tonnes	
Total weight atop launcher plus Lunar Module	52 tonnes	
Total CSM-LM orbital weight (typical)	46.2 tonnes	

Environmental control
Atmosphere 100% oxygen at nominal 258 mmHg
Temperature 21.1°C – 23.8°C
Suit circuit 100% oxygen at 190 mmHg
Storage 2 Service Module tanks containing total 290 kg of liquid oxygen of which 100 kg is for environmental control the balance to fuel cells
1 Command Module surge tank containing 3.7 kg gaseous oxygen for re-entry
Drinking water 15 liters capacity in Command Module from fuel cells
Waste water 26.5 liters capacity in Command Module from suit heat exchanger
Atmosphere cleanser Lithium hydroxide
Odor absorber Activated charcoal

Electrical Systems
Main 3 fuel cells
Secondary 3 × 40 ampere-hr silver zinc batteries during descent
2 × 0.75 ampere-hr silver zinc batteries for pyrotechnics
Fuel cells (each):
Weight 111 kg
Production 1.4 kW d-c at 25-30 volts
Size Cylinder 1.1 meters by 55.9 cm diameter
No. of stacks 1
No of cells 31
Reactants: 2 × 12.7 kg capacity hydrogen tanks at 15.8 kg/cm² (minimum)
2 × 145 kg capacity oxygen tanks at 60.8 kg/cm² (minimum)

Attitude control thrusters
SM reaction control system (RCS) 16 × 45.5 kg thrust engines on 4 SM quads
CM reaction control system (RCS) 12 × 42.2 kg thrust engines around CM

Translation control
SMRCS fired in groups of 2 or 4

Service Propulsion System
1 × 9.3 thrust liquid propellant motor

Liquid propellant
SMRCS quantity Nominal 617 kg N_2O_4/MMH in 8 oxidizer/8 fuel tanks
Pressurant 2.4 kg helium stored at 292 kg/cm² in 4 tanks
SPS quantity 18.4 tonnes N_2O_4/UDMH in 4 tanks
Pressurant 2 helium tanks pressurized to 253 kg/cm²
CMRCS Nominal 122 kg N_2O_4/MMH in 2 oxidizer/2 fuel tanks
Pressurant 0.5 kg helium stored at 292 kg/cm² in 2 tanks

LES solid rocket motors

	Length	Diameter	Thrust	Burn time	Weight
Escape	4.71 meters	66 cm	66.7 tonnes	3 sec	2.132 kg
Pitch control	55.9 cm	22.3 cm	1.1 tonnes	0.5 sec	22.7 kg
Tower jettison	1.41 meters	65.4 cm	14.3 tonnes	1 sec	–

Parachutes
Drogue 2 × 5 meter conical ribbon
Pilot 3 × 2.2 meter ringshot
Main 3 × 25.45 meter ringsail

SPS: Service Propulsion System
LES: Launch Escape System
SLA: Spacecraft-LM Adapter
LM: Lunar Module
CSM: Command & Service Modules
CM: Command Module
SM: Service Module
RCS: Reaction Control System

Note: Data refer to standard Apollo CSM. Apollo 14 carried a third oxygen tank while Apollos 15-17 carried 3 oxygen and 3 hydrogen tanks. The three Skylab CSM carried only 2 fuel cells each while Skylab and ASTP CSM carried only 2 SPS propellant tanks. SPS propellants tank quantities varied from flight to flight.

Physical data:	Descent stage	Ascent stage
Height	3.2 meters	3.76 meters
Diameter	4.2 meters	4.2 meters
Dry weight	1.860 kg (2.760)	2.040 kg (2.130)
Loaded weight	10.000 kg (11.610)	4.670 kg (4.760)
Launch weight	14.670 kg (16.370)	
Total diameter (diagonally across legs)	9.4 meters	

Environmental Control
Atmosphere 100% oxygen at nominal 250 mmHg
Temperature 23.9°C
Suit circuit 100% oxygen at 190 mmHg
Storage 1 (2) Descent stage tank(s) containing 21.7 kg (43.5 kg) gaseous oxygen fitted to Quad 3 (3 & 4) at 192 kg/cm² (189 kg/cm²)
2 Ascent stage tanks each containing 1.1 kg gaseous oxygen at 60 kg/cm²
No. of cabin repressurization 4

Water 1 (2) × 151 kg capacity Descent stage tank(s) fitted to Quad 2 (2 & 4) filled to 75% nitrogen pressurized
2 × 19.27 kg capacity Ascent Stage tanks
Atmosphere cleanser Lithium hydroxide
Odor absorber Activated charcoal
Coolant 11.3 kg of 35% ethylene glycol/65% water solution

Electrical Systems
Main 4 (5) × 400 ampere-hr (415 ampere-hr) silver zinc batteries in forward Descent stage box structure (rear Descent stage box structure)
Secondary 2 × 296 ampere-hr silver zinc batteries in Ascent stage
Power 26-32 volt d-c buses

Attitude control thrusters
Ascent Stage Reaction Control System (RCS) 16 × 45.4 kg thrust engines on 4 quads

Translation control
RCS fired in groups of 2 or 4

Main Propulsion
1 × throttleable 476 kg to 4.477 (580 kg to 4.490 kg) thrust liquid propellant motor: Descent Propulsion System (DPS)
1 × fixed 1.588 kg thrust liquid propellant motor: Ascent Propulsion System (APS)

Liquid propellant
RCS quantity Nominal 287 kg (286 kg) N_2O_4/Aerozene 50 in 2 oxidizer/2 fuel tanks
Pressurant 0.93 kg helium stored at 214 kg/cm² in 2 tanks
DPS quantity Nominal 8.187 kg (8.838 kg) N_2O_4/Aerozene 50 in 2 oxidizer/2 fuel tanks
Pressurant 1 × supercritical helium tank containing 22 kg (23.2 kg) at 109.3 kg/cm² and +240°C
APS quantity Nominal 2.353 kg (2.371 kg) N_2O_4/Aerozene 50 in 1 oxidizer/1 fuel tank
Pressurant 2 × 2.9 kg helium tanks pressurized to 214 kg/cm²

Note: Figures in brackets refer to modified LM for Apollos 15-17 (J-series mission)

TABLE 9: APOLLO FLIGHT DATA (unmanned missions)

Mission	Designation	Date	Time	Launch site	Complex	Spacecraft	Launch vehicle	Launch number	Mission type	Apogee/ Perigee (km)	Range (km)	Recovery Area	Recovery Ship	Test objectives	Result
SA-1	—	10/27/60	13:00:06	ETR	34	—	Saturn I	SA-1	B	137	333	—	—	Live first stage; dummy upper stages. First Saturn	S
SA-2	—	4/25/62	09:00:34	ETR	34	—	Saturn I	SA-2	B	105	80	—	—	Project High Water I	S
SA-3	—	11/16/62	12:45:02	ETR	34	—	Saturn I	SA-3	B	167	211	—	—	Project High Water II	S
SA-4	—	3/28/63	15:11:55	ETR	34	—	Saturn I	SA-4	B	130	352	—	—	Intentional engine shutdown	S
PA-1	—	11/7/63	09:00:01	WSMR	—	BP-6	—	—	B	1.6	1.38	—	—	LES test in simulated pad abort	S
SA-5	—	1/29/64	11:25:01	ETR	37B	—	Saturn I	SA-5	O	262 × 785	—	—	—	Orbital flight test of S-IV second stage	S
Abort test	A-001	5/13/64	05:59:59	WSMR	—	BP-12	Little Joe II	—	B	4.7	3.53	—	—	Verify dynamic shape of LES	S
SA-6	A-101	5/28/64	13:07:00	ETR	37B	BP-13	Saturn I	SA-6	O	182 × 227	—	—	—	Orbital flight test of Saturn/Apollo configuration	S
SA-7	A-102	9/18/64	11:22:43	ETR	37B	BP-15	Saturn I	SA-7	O	185 × 225	—	—	—	Final qualification flight for Saturn I launcher	S
Abort test	A-002	12/8/64	08:00:00	WSMR	—	BP-23	Little Joe II	—	B	4.68	2.3	—	—	Max-q LES test close to guidance limits	S
SA-9	A-103	2/16/65	09:37:03	ETR	37B	BP-16	Saturn I	SA-9	O	496 × 744	—	—	—	BP-16 contained Pegasus meteoroid satellite	S
Abort test	A-003	5/19/65	06:01:04	WSMR	—	BP-22	Little Joe II	—	B	5.94	5.48	—	—	High altitude abort test; launcher broke up	F
SA-8	A-104	5/25/65	15:35:01	ETR	37B	BP-26	Saturn I	SA-8	O	506 × 745	—	—	—	BP-26 contained Pegasus meteoroid satellite	S
PA-2	—	6/29/65	06:00:01	WSMR	—	BP-23A	—	—	B	1.58	2.32	—	—	LES test in simulated pad abort	S
SA-10	A-105	7/30/65	09:00:00	ETR	37B	BP-9	Saturn I	SA-10	O	528 × 531	—	—	—	BP-9 contained Pegasus meteoroid satellite	S
Abort test	A-004	1/20/66	08:17:01	WSMR	—	CSM-002	Little Joe II	—	B	22.6	34.63	—	—	High altitude abort test	S
AS-201	—	2/26/66	11:12:01	ETR	34	CSM-009	Saturn IB	AS-201	B	488	8.472	Atlantic	Boxer	Launch vehicle/spacecraft compatibility	S
AS-203	—	7/5/66	10:53:17	ETR	37B	—	Saturn IB	AS-203	O	185 × 189	—	—	—	Orbital test of S-IVB second stage	S
AS-202	—	8/25/66	13:55:32	ETR	34	CSM-011	Saturn IB	AS-202	B	1.143	28.645	Pacific	Hornet	Suborbital heat shield qualification	S
AS-501	Apollo 4	11/9/67	07:00:01	ETR	39A	CSM-017 LTA-10R	Saturn V	AS-501	O	183 × 187	—	Pacific	Bennington	High apogee test of launch vehicle/spacecraft	S
AS-204	Apollo 5	1/22/68	17:48:08	ETR	37B	LM-1	Saturn IB	AS-204	O	163 × 222	—	—	—	Orbital test of first production Lunar Module	S
AS-502	Apollo 6	4/4/68	07:00:01	ETR	39A	CM-020/ SM-014 LTA-2R	Saturn V	AS-501	O	178 × 367	—	Pacific	Okinawa	High oxygen test of launch vehicle/spacecraft	P

SA: Saturn Apollo
PA: Pad Abort
AS: Apollo Saturn
ETR: Eastern Test Range
WSMR: White Sands Missile Range
BP: Boilerplate
CSM: Command & Service Modules
LTA: Lunar Test Article
LM: Lunar Module
B: Ballistic
O: Orbital
S: Successful
F: Failure
LES: Launch Escape System

Note: Launch times are local times
Apogee/perigee column for ballistic flights indicates maximum altitude
Project High Water released water into upper atmosphere
Max q indicates maximum dynamic pressure

TABLE 10: APOLLO FLIGHT DATA (manned missions)

	Flight Crew					Back-up Crew								
Mission	CDR	CMP	LMP	Ages (yrs)	E.T. (mths)	CDR	CMP	LMP	Launch date	Launch time	CSM	CSM weight (kg)	LM	LM weight (kg)	Docked weight (kg)
Apollo 7	Schirra	Eisele	Cunningham	45/38/36	114/60/60	Stafford	Young	Cernan	10/11/68	11:02:45	101	14.692	—	—	14.692
Apollo 8	Borman	Lovell	Anders	40/40/35	75/75/62	Armstrong	Aldrin	Haise	12/21/68	07:51:00	103	28.897	—	—	28.897
Apollo 9	McDivitt	Scott	Schweickart	39/36/33	78/65/65	Conrad	Gordon	Bean	3/3/69	11:00:00	104	22.028	3	14.525	36.553
Apollo 10	Stafford	Young	Cerman	38/38/35	80/80/67	Cooper	Eisele	Mitchell	5/18/69	12:49:00	106	28.870	4	13.993	42.863
Apollo 11	Armstrong	Collins	Aldrin	38/38/39	82/69/69	Lovell	Anders	Haise	7/16/69	09:32:00	107	28.800	5	15.002	43.862
Apollo 12	Conrad	Gordon	Bean	39/40/37	86/73/73	Scott	Worden	Irwin	11/14/69	11:22:00	108	28.790	6	15.116	43.906
Apollo 13	Lovell	Swigert	Haise	42/38/36	91/48/48	Young	—	Duke	4/11/70	14:13:00	109	28.790	7	15.185	43.975
Apollo 14	Shepard	Roosa	Mitchell	47/37/40	141/57/57	Cernan	Evans	Engle	1/31/71	16:03:00	110	29.230	8	15.277	44.507
Apollo 15	Scott	Worden	Irwin	39/38/41	93/63/63	Gordon	Brand	Schmitt	7/26/71	09:34:00	112	30.343	10	16.434	46.777
Apollo 16	Young	Mattingley	Duke	41/36/36	115/72/72	Haise	Roosa	Mitchell	4/16/72	12:54:00	113	30.358	11	16.429	46.787
Apollo 17	Cernan	Evans	Schmitt	38/39/37	110/80/90	Young	Roosa	Duke	12/7/72	00:33:00	114	30.342	12	16.454	46.796

										Time from splashdown to	
Mission	Launcher	Launch Complex	Lift off weight (kg)	Earth orbit weight (kg)	Initial lunar orbit weight (kg)	Mission type	Revolutions	Flight duration	Recovery area	Recovery ship	Crew recovery	S/c recovery
Apollo 7	AS-205	34	556.904	30.695	—	C	163	260:09:03	Atlantic	Essex	56 min	111 min
Apollo 8	AS-503	39A	2.782.328	128.004	28.498	C′	10	147:00:42	Pacific	Yorktown	88 min	148 min
Apollo 9	AS-504	39A	2.901.704	131.530	—	D	151	241:00:54	Atlantic	Guadalcanal	44 min	132 min
Apollo 10	AS-505	39B	2.908.597	133.788	31.492	F	31	192:03:23	Pacific	Princeton	39 min	96 min
Apollo 11	AS-506	39A	2.902.280	135.104	32.676	G	30	195:18:35	Pacific	Hornet	63 min	187 min
Apollo 12	AS-507	39A	2.941.496	136.105	32.755	G	45	244:36:25	Pacific	Hornet	60 min	111 min
Apollo 13	AS-508	39A	2.912.683	134.476	—	H	—	142:54:41	Pacific	Iwo Jima	45 min	88 min
Apollo 14	AS-509	39A	2.912.335	137.271	32.524	H	34	216:01:57	Pacific	New Orleans	48 min	110 min
Apollo 15	AS-510	39A	2.906.559	140.312	33.803	J	74	295:11:53	Pacific	Okinawa	40 min	94 min
Apollo 16	AS-511	39A	2.921.005	140.042	35.197	J	64	265:51:05	Pacific	Ticonderoga	37 min	99 min
Apollo 17	AS-512	39A	2.923.387	141.138	34.718	J	75	301:51:59	Pacific	Ticonderoga	52 min	93 min

Note E.T. (mths): Time in months between selection as an astronaut and flight date, but note that astronaut may have performed a previous mission.
Revolutions include Earth and Moon missions; Apollo 7 and 9 refer to Earth revolutions.

TABLE 11: APOLLO 7 FLIGHT DATA

Event	Time (GET) hr:min:sec	Burn time (sec)	Resultant orbit apogee/perigee (km)	Period (min)
Lift-off	00:00:00.4			
S-IB inboard engine cutoff	00:02:20.7	143		
S-IB outboard engine cutoff	00:02:24.3	147		
S-IB/S-IVB separation	00:02:25.6			
S-IVB ignition	00:02:27.0			
Escape tower jettison	00:02:46.5			
S-IVB engine cutoff	00:10:16.8	469.8		
Orbital insertion	00:10:26.8		284.6 × 228.3	89.7
Spacecraft/S-IVB separation	02:55:02		309.3 × 232.0	89.99
First phasing mnvr start	03:20:09.9	16.8	305.9 × 231.1	89.95
Second phasing mnvr start	15:52:00.9	17.6	305.0 × 223.7	89.86
Corrective Combination mnvr (SPS)	26:24:55.7	10.0	359.5 × 227.8	90.57
Coelliptic mnvr (SPS)	28:00:56.5	7.8	284.5 × 210.9	89.52
Terminal Phase Imitation	29:16:33		285.4 × 225.2	89.68
Terminal Phase Finalization	29:55:43		298.2 × 226.1	89.82
Separation mnvr start	30:20:00.00	5.4	298.2 × 226.3	89.82
Third SPS test	75:48:00.3	9.0	295.8 × 165.7	89.17
Fourth SPS test	120:43:00.5	0.4	290.2 × 165.0	89.11
Fifth SPS test	165:00:00.5	67.1	452.2 × 165.0	90.77
Sixth SPS test	210:08:00.5	0.5	434.5 × 163.7	90.58
Seventh SPS test	239:06:12.0	7.7	425.6 × 163.9	90.48
Eighth SPS test (de-orbit)	259:39:16.3	11.9	Entry	
Command Module separation	259:43:33.8			
Entry interface	259:53:27			
Enter blackout	259:54:58			
Exit blackout	259:59:46			
Drogue deployment	260:03:23			
Mains deployment	260:04:13			
Splashdown	260:09:03			

S-IB: Saturn IB first stage
SIVB: Saturn IB second stage
mnvr: maneuver
SPS: Service Propulsion System

TABLE 12: APOLLO 9 FLIGHT DATA

Event	Time (GET) hr:min:sec	Burn time (sec)	△V (meters/sec)	Resultant orbit (apogee/perigee, km)
Lift off	00:00:00.7			
CECO	00:02:14.3	141		
OECO	00:02:42.8	169		
S-IC/S-II separation	00:02:43.4			
S-II ignition	00:02:44.2			
S-II skirt separation	00:03:13.5			
LES jettison	00:03:19.0			
S-II cutoff	00:08:56.2			
S-II/S-IVB separation	00:08:57.2			
S-IVB ignition	00:08:57.3			
S-IVB cutoff	00:11:04.7	127.4		
Orbital insertion	00:11:14.7			191.3 × 189.5
Spacecraft/S-IVB separation	02:45:00			
CSM/LM docking	03:02:08			
Spacecraft/S-IVB separation	04:18:00			
Second S-IVB ignition	04:45:47.3	70.4		3,087.0 × 207
First SPS test	05:59:00	5.1	10.4	234.1 × 200.7
Third S-IVB ignition	06:07:19	242.4		Escape
Second SPS test	22:12:03	110.0	259.2	351.5 × 199.5
Third SPS test	25:17:38	281.6	782.6	503.4 × 202.6
Fourth SPS test	28:24:40	28.2	914.5	502.8 × 202.4
First DPS test	49:41:33	369.7	530.1	499.3 × 202.2
Fifth SPS test	54:26:11	43.3	175.6	239.3 × 229.3
Schweickart EVA start (56 min duration)	72:53:00			
Scott EVA start (56 min duration)	72:53:00			
CSM/LM undocking	92:39:30			
CSM separation mnvr ("mini-football")	93:02:53	10.9	1.5	235.6 × 225.6
LM Phasing mnvr (DPS)	93:47:34	18.6	27.6	253.5 × 207
LM Insertion mnvr (DPS)	95:39:07	22.2	13.1	257.2 × 248.2
LM Concentric Sequence Initiation mnvr	96:16:04	30.3	12.2	255.2 × 208.9
LM Constant Delta Height mnvr (APS)	96:58:14	2.9	12.6	215.6 × 207.2
LM Terminal Phase Initiation mnvr	97:57:59	34.7	6.8	232.8 × 208.5
LM Terminal Phase Finalization mnvr	N.A.	N.A.	N.A.	234.5 × 225.4
CSM/LM docking	98:59			
Post jettison CSM separation mnvr	101:32:44	7.2	0.9	235.7 × 224.8
LM APS burn to depletion	101:53:20	350.0	1.643.2	6.939.4 × 230.6
Sixth SPS test	123:25:06	1.29	11.5	222.6 × 195.2
Seventh SPS test	169:38:59	25	199.6	463.4 × 181.1
Eighth SPS test (deorbit)	240:31:14	11.6	99.1	442.2 × −7.8
Splashdown	241:00:54			

CECO: Centre Engine Cutoff (of S-IC Saturn V first stage)
OECO: Outer Engine Cutoff (of S-IC Saturn V first stage)
S-II: Saturn V second stage
LES: Launch Escape System
S-IVB: Saturn V third stage
CSM/LM: Command & Service Modules/Lunar Module
SPS: Science Propulsion System
EVA: Extra Vehicular Activity
DPS: Descent Propulsion System
APS: Ascent Propulsion System
LM: Lunar Module
CSM: Command & Service Modules
N.A.: Not Available

Note △V is the change of speed expressed in meters/sec resulting from the engine burn (to find km/hr multiply by 3.6)

TABLE 13: APOLLO 13 FLIGHT DATA

Event	Time (GET) hr:min:sec	Burn time (sec)	△V (meters/sec)	Propulsion
Lift off	00:00:00.6			
S-IC CECO	00:02:15.2	142		
S-IC OECO	00:02:43.6	171	2.370	S-IC
S-IC/S-II Separation	00:02:44.3			
S-II Ignition	00:02:46.0			
S-II Skirt Separation	00:03:14.3			
Launch Escape System jettison	00:03:20.0			
S-II CECO	00:05:30.6	164.6		
S-II OECO	00:09:52.6	426.6	4.116	S-II
S-II/S-IVB Separation	00:09:53.5			
S-IVB Ignition	00:09:56.9			
S-IVB Cutoff	00:12:29.8	152.9	902	S-IVB
Orbital Insertion	00:12:39.8			
Second S-IVB Ignition	02:35:46.4			
Second S-IVB Cutoff	02:41:37.1	350.7	(3.181)*	S-IVB
Trans-Lunar-Injection	02:41:47.1			
CSM/S-IVB Separation	03:06:39			
CSM-LM/S-IVB Separation	04:01:03			
S-IVB evasive maneuver	04:18:01			
S-IVB lunar impact maneuver	05:59:59			
Mid-Course-Correction No. 2	30:40:50	3.37	7.0	SPS (CSM)
Oxygen tank anomaly	55:54:53			
Free-return burn	61:29:43	30.4	11.5	DPS (LM)
Enter lunar occultation	77:08:35			
Exit lunar occultation	77:33:10			
S-IVB lunar impact	77:56:40			
Pericynthinon + 2-hr maneuver	79:27:39	263.4	262.3	DPS (LM)
Mid-Course Correction No. 5	105:18:32	15.4	2.4	DPS (LM)
Mid-Course Correction No. 7	137:39:49	22.4	0.9	RCS (LM)
Service Module jettison	138:02:06			
Lunar Module jettison	141:30:02			
Entry Interface	142:40:47			
Splashdown	142:54:41			

S-IC: Saturn first stage
S-II: Saturn second stage
S-IVB: Saturn third stage
CECO: Centre Engine Cutoff
OECO: Outer Engine Cutoff
CSM: Command & Service Module
LM: Lunar Module
SPS: Service Propulsion System
DPS: Descent Propulsion System
*: approximate value
RCS: Reaction Control System

TABLE 14: APOLLO LUNAR SURFACE DATA

	Apollo Flight 11	12	14	15	16	17
Stay time (hr:min)	21:36	31:31	33:30	66:54	71:14	74:59
Surface excursions	1	2	2	3	3	3
Duration of excursions (hr:min)	2:40	7:45	9:17	19:08*	20:15	22:04
Experimental weight (kg)	102	166	209	550	563	514
Experiment package	EASEP	ALSEP	ALSEP	ALSEP	ALSEP	ALSEP
Samples returned (kg)	21	34	43	77	94	110
Traverse distance (km)	0.25	2.0	3.3	27.9	27.0	35.0
Traverse mode	W	W	W	LRV	LRV	LRV
In-flight EVA duration	—	—	—	00:38	01:13	01:06
Total EVA time (hr:min)	2:40	7:45	9:17	19:46	21:28	23:10

*: Includes 33 min during a Stand-up EVA (SEVA) on the lunar surface
EVA: Extra Vehicular Activity

ABLE 15: APOLLO LUNAR MISSION DATA

	APOLLO 8				APOLLO 10				APOLLO 11				APOLLO 12			
ent	Start time hr:min:sec	Burn time sec	△V m/sec	Orbit km	Start time hr:min:sec	Burn time sec	△V m/sec	Orbit km	Start time hr:min:sec	Burn time sec	△V m/sec	Orbit km	Start time hr:min:sec	Burn time sec	△V m/sec	Orbit km
C CECO	00:02:05.9	132			00:02:15.2	142			00:02:15.2	142			00:02:15.2	142		
C OECO	00:02:33.8	160	2.304		00:02:41.6	168	2.330		00:02:41.6	168	2.343		00:02:41.7	168	2.352	
I Ignition	00:02:35.2				00:02:43.1				00:02:43.0				00:02:44.2			
I CECO	00:08:44.0	368.8			00:07:40.6	297			00:07:40.6	297.6			00:07:40.7	296.5		
I OECO			4.108		00:09:12.6	389	4.168		00:09:08.2	385	4.182		00:09:12.3	388	4.216	
VB Ignition	00:08:45.0				00:09:13.6				00:09:09.2				00:09:15.6			
VB Cutoff	00:11:25.0	160	980		00:11:43.8	150	879		00:11:39.3	150	858		00:11:33.9	138	815	
bit Insertion	00:11:35.0			191.3 × 181.5	00:11:53.8			190.0 × 184.5	00:11:49.3			192.0 × 190.6	00:11:43.9			189.8 × 185.0
cond S-IVB Ignition	02:50:37.1				02:33:25.1				02:44:16.2				02:47:22.6			
cond S-IVB Cutoff	02:55:55.5	318.4	3.040		02:39:10.0	344.9	3.181		02:50:03.5	347.3	3.182		02:52:03.8	341.2	3.205	
ans-Lunar-Injection	02:56:05.5				02:39:19.9				02:50:13.5				02:53:14			
M/S-IVB Separation	03:20:59.3				03:03				03:17				03:18:04.9			
M/LM Separation	—				03:56				04:17:13				04:13:00.9			
S evasive mnvr	(03:40:01 04:45:01)[a]				04:39:09	2.5	5.7		04:40:01	3.4	6.0		N.P.			
C-1	10:59:59.5	2.4	4.6		—				—				—			
C-2	—				26:32:56.1	6.7	14.9		26:44:58	2.9	6.4		30:52:44.4	9.1	18.8	
uigravisphere	55:38				61:50:50				61:39:55				68:30:22			
C-4	60:59:55	12.8			—				—				—			
nar Orbit Insertion-1	69:12:27.3	246.9	894	312.1 × 110.9	75:55:53	356	908.7	315.6 × 110.4	75:49:49.6	362.1	889.2	312.6 × 113.5	83:25:23.4	352.3	881	312.6 × 115.9
VB fly-by					78:54			(3.245)[b]	78:50:34			(4.334)[b]	85:48			(5.724)[b]
VB impact	—				—				—				—			
nar Orbit Insertion-2	73:35:07	9	41	112.4 × 110.6	80:25:07	13.9	42.2	109.1 × 113.9	80:11:36	17	48.4	121.7 × 99.6	87:48:48.1	16.9	50.3	122.4 × 100.6
ndocking	—				98:22:00				100:12:00				107:54:02.3			
M separation mnvr	—				98:47:16	10.4	0.98	114.6 × 107.4	100:39:50	8.2	0.79	118.0 × 103.3	108:24:36.8	14.4	0.7	117.6 × 104.3
scent Orbit Insertion	—				99:46:00.9	27.4	21.7	113.3 × 15.5	101:36:14.1	29.8	23.3	105.9 × 15.7	109:23:39.9	22.0	22.1	112.2 × 15.0
wered Descent Initiation	—				(100:58:25.2)[c]	40.1	53.8	351.9 × 22.0	102:33:04.4				110:20:38.1			
uchdown	—								102:45:39.9	752	2.065		110:32:36.2	717	2.017	
nar Orbit Plane Change-1	—								—				119:47:13.2	18.2	106.6	115.7 × 106.7
t-off	—				(102:45:00)[d]			352.1 × 21.8	124:22:00.8				142:03:48			
bit insertion	—				102:55:01.4	15.5	67.4	83.9 × 20.7	124:29:20.7	439.9	1.850	83.7 × 16.7	142:10:59.9	423.2	1.846	85.7 × 16.3
ncentric Sequence Initiate	—				103:45:54.6	27.3	13.8	87.4 × 77.4	125:19:34.7	47	15.7	90.0 × 83.9	143:01:51	41.1	13.7	94.4 × 76.9
nstant Delta Height	—				104:43:52.0	3.7	0.94	86.7 × 78.0	126:17:49.6	18.1	6.1	87.0 × 75.7	144:00:02.6	13.0	4.2	82.2 × 74.8
rminal Phase Initiation	—				105:22:55	28.8	7.3	107.4 × 86.7	127:03:30.8	22.8	7.7	113.3 × 81.3	144:36:26	26	8.8	111.5 × 81.1
rminal Phase Finalization	—				106:15				127:39:34.2	28.4	9.6	115.2 × 104.8	145:19:29.3	38	12.2	115.4 × 108.0
M/Ascent stage docking	—				106:22				128:03				145:36:20.2			
cent stage jettison	—				108:43:30				130:10				147:59:32			
M separation mnvr	—				108:43:30	6.9	0.64	117.0 × 101.9	130:30:00	7.1	0.7	115.9 × 101.3	148:04:31	5.4	0.3	114.8 × 106.5
cent Stage de-orbit	—				(108:51:01)[e]	212.9	1.170	110.0 × —	—				149:28:14.8	82.1	59.8	106.9 × −122.8
nar Orbit Plane Change-2	—				—				—				159:04:45.5	19.2	116.4	119.8 × 105.2
ans-Earth Injection	89:19:16.6	203.7	1.066.8		137:36:28	164	1.105		135:23:42	150	999.4		172:27:16.8	130.3	927.2	
C-5	103:59:54	14			—				150:29:54.5	10.8	1.4		188:27:15.8	4.4	0.6	
C-7	—				188:49:56.8	6.54	0.48		—				241:21:59.7	5.7	0.7	
M separation	146:28:48								194:49:19				244:07:20			
try Interface	146:46:12.8				191:48:54				195:03:06				244:22:19			
lashdown	147:00:42				192:03:23				195:18:35				244:36:25			

	APOLLO 14				APOLLO 15				APOLLO 16				APOLLO 17			
ent	Start time hr:min:sec	Burn time sec	△V m/sec	Orbit km	Start time hr :min:sec	Burn time sec	△V m/sec	Orbit km	Start time hr:min:sec	Burn time sec	△V m/sec	Orbit km	Start time hr:min:sec	Burn time sec	△V m/sec	Orbit km
C CECO	00:02:14.7	141			00:02:15.5	142			00:02:17.2	144			00:02:18.8	145		
C OECO	00:02:43.5	170	2.306		00:02:39	166	2.372		00:02:41.2	168	2.369		00:02:40.6	167	2.365	
I Ignition	00:02:44.9				00:02:41.8				00:02:43.6				00:02:43.0			
I CECO	00:07:42.5	297.6			00:07:59	317.2			00:07:41.2	297.6			00:07:40.6	297.6		
I OECO	00:09:18.5	393.6	4.253		00:09:08.5	386.7	4.197		00:09:19.0	395.4	4.199		00:09:19.0	396	4.175	
VB Ignition	00:09:19.6				00:09:09.6				00:09:20.1				00:09:21.1			
VB Cutoff	00:11:39.5	140	823		00:11:34.3	134.7	817		00:11:45.8	145.7	832		00:11:42.2	141.1	860	
bit Insertion	00:11:49.3			188.9 × 183.2	00:11:44.3			171.3 × 169.5	00:11:55.6			175.9 × 166.7	00:11:52.2			170 × 168
cond S-IVB Ignition	02:28:31.4				02:50:02.6				02:33:37				03:12:37			
cond S-IVB Cutoff	02:34:22.4	351	3.154		02:55:53.3	350.7	3.174		02:39:18.9	341.9	3.167		03:18:28	351.0	3.163	
ans-Lunar-Injection	02:34:32.4				02:56:03.3				02:39:28.9				03:18:38			
M/S-IVB Separation	03:02:30				03:22:24				03:04:59				03:42:29			
M/LM Separation	05:47:25				04:18:00				03:59:15				04:45:00			
S evasive mnvr	N.P.				N.P.				N.P.				N.P.			
C-1	—				—				—				—			
C-2	30:36:07	10.1	21.7		28:40:30	0.7	1.6		30:39:01	1.8	3.8		35:30:00	1.7	3.2	
uigravisphere	66:09:01				63:55:20				59:19:45				70:37:45			
C-4	76:58:11	0.6	1.1		73:31:14	0.9	1.6		—				—			
nar Orbit Insertion-1	81:56:40	372.2	921	313.0 × 108.2	78:31:45.9	400.7	20.8	315.0 × 107.2	74:28:28	374.9	854	315.4 × 107.6	86:14:23	393.2	911	314.8 × 97.4
VB fly-by	—				—				—				—			
VB impact	82:37:52				79:24:42				75:08:00				89:39:40			
nar Orbit Insertion-2	86:10:52	20.7	62.7	108.9 × 17.8	(95:56:42)[l]	24.5	65.2	108.3 × 17.0	78:33:45	24.4	63.8	108.3 × 20.2	90:31:37	22.3	60	109.3 × 26.9
docking	103:48:00				(100:39:30)[h]	21.2	0.94	110.9 × 17.8	(96:13:31)[h]	6.8	0.3	109.6 × 19.3	(107:47:56)[j]	3.4	0.3	113.9 × 21.3
M separation mnvr	103:48:00	2.7	0.2	111.5 × 14.4	100:39:30	7.2	0.3	112.8 × 16.8	(102:30:00)[i]	6.8	0.3	110.6 × 20.7	(109:17:29)[f]	3.8	21.5	129.6 × 100.0
scent Orbit Insertion	(105:11:45)[f]	3.8	23.5	118.3 × 103.7	(101:38:58)[f]	3.6	126.5	119.8 × 98.2	(103:21:43)[f]	4.7	24.9	125.9 × 98.3	109:22:42	21.5	2.3	110.4 × 11.5
wered Descent Initiation	108:02:26				104:30:09				104:17:25				110:09:53			
uchdown	108:13:59	693	2.023		104:42:29	740	2.040		104:29:35	731	2.043		110:21:57	725	2.041	
nar Orbit Plane Change-1	117:29:32	18.4	112.9	115.0 × 106.7	165:11:32	18.1	100.8	119.5 × 99.0	169:05:52	7.1	37.8	119.6 × 101.9	179:53:54[k]	20.1	111.5	116.3 × 115.7
t-off	141:45:39				171:37:22				175:31:48				185:21:37			
bit insertion	141:48:51	432.0	1.849	96.5 × 17.0	171:44:38	436.7	1.847	78.7 × 16.7	175:38:55.7	427.7	1.845	74.4 × 14.6	185:28:58	441	1.852	89.8 × 16.9
ncentric Sequence Initiate	(141:56:48)[g]		0.3		—				(175:42:18)[g]		0.3		(185:32:12)[g]	10	3.0	89.8 × 17.4
nstant Delta Height	—				—				—				—			
rminal Phase Initiation	142:30:51				172:29:39	2.5	22.2	119.3 × 71.3	176:26:05	2.5	23.8	118.9 × 74.3	186:15:58	3.2	16.4	119.8 × 89.8
rminal Phase Finalization									177:08:42				187:06			
M/Ascent stage docking	143:33:00				173:35:47				177:41:18				187:37:15			
cent Stage jettison	145:45:00				179:30:14				195:00:12			125.6 × 99.6	191:18:31			117.4 × 114.3
M separation mnvr	145:50:00	7.0	0.4	115.4 × 105.2	179:50:00	12.6	0.6	122.6 × 97.2	195:03:13	6.8	0.6	122.8 × 97.4	191:23:31	12	0.6	118.3 × 113.3
cent Stage de-orbit	147:14:16	75.4	56.7	105.2 × −115.2	181:04:19	86.5	61		N.P.				192:58:14	116	87.2	
nar Orbit Plane Change-2	—				221:20:47	3.3	20.2	140.7 × 100.6	N.P.				N.P.			
ans-Earth Injection	148:36:01	148.1	1.055		223:48:45	141.2	929		200:21:33	162.3	1.027		234:02:09	143.7	928.5	
C-5	165:34:56	2.2	0.15		—				214:35:03	8.0	1.0		—			
C-7	—				291:56:48	24.2	1.7		262:37:21	3.2	0.4		298:38:01	9	0.6	
M separation	215:32:47				294:44:00				265:22:33				301:23:49			
try Interface	215:47:44				294:58:54				265:37:31				301:38:38			
lashdown	216:01:57				295:11:53				265:51:05				301:51:59			

C: Saturn first stage
I: Saturn second stage
V: Saturn third stage
CO: Centre Engine Cutoff
CO: Outer Engine Cutoff
M: Command & Service Modules
S: Service Propulsion System
C: Mid Course Correction
': Delta velocity (change in speed resulting from the maneuver)
'wo small maneuvers performed by the Reactors Control System engines
Closest approach to the Moon
hasing maneuver unique to Apollo 10 (no landing attempt)
eparation of Ascent and Descent stages in flight (unique to Apollo 10)

[e]: Ascent stage firing to eject spacecraft from lunar to solar orbit (unique to Apollo 10)
[f]: CSM circularization maneuver (Descent Orbit Insertion combined with Lunar Orbit Insertion–2)
[g]: Small maneuver to adjust the orbit following ascent
[h]: Combined undocking and separation burn
[i]: Second separation burn
[j]: Combined undocking and separation burn
[k]: CSM trim burn was performed at 178:54:05 (31.3 sec; △V 2.8 m/sec; orbit 124.6 × 115.7 km)
[l]: CSM trim burn
N.P.: Not Performed

Note This table provides data for all flights to lunar orbit but precludes the Apollo 13 mission which was circumlunar for which information is provided elsewhere.
Four trans-lunar and three trans-Earth MCC burn opportunities were provided; MCC-3 and -6 were never used hence they are not shown.

TABLE 16: SKYLAB HARDWARE

Physical Data (in orbit)
Length 36 meters
Weight 90.6 tonnes
Work space 347 cubic meters

Orbital Work Shop (OWS):
Length 14.7 meters
Diameter 6.58 meters
Habitable volume 292 cubic meters
Weight 35.4 tonnes
Living quarters area 35.3 square meters
(Wardroom 9.3
Waste Management Compartment 2.8
Sleep Compartment 6.5
Experiment area 16.7)
Design temperature 15.6–32.2°C
Design pressure 258 mmHg

Airlock Module (AM):
Length 5.3 meters
Diameter 1.67 meters/3.2 meters
Habitable volume 17.3 cubic meters
Weight 22.2 tonnes

Multiple Docking Adapter (MDA):
Length 5.2 meters
Diameter 3.2 meters
Habitable volume 32 cubic meters
Weight 6.3 tonnes

Apollo Telescope Mount (ATM):
Height 4.4 meters
Width 3.4 meters (of structure only)
(31 meters with solar panels deployed)
Weight 11.1 tonnes

Payload Shroud:
Length 16.8 meters
Diameter 6.5 meters
Weight 11.8 tonnes

TABLE 17: SKYLAB MISSIONS DATA

Event	*SL-1*	*SL-2*	*SL-3*	*SL-4*
Prime crew	—	Conrad/Kerwin/Weitz	Bean/Garriott/Lousma	Carr/Gibson/Pogue
Back-up crew	—	Schweickart/Musgrave/McCandless	Brand/Lind/Lenoir	Brand/Lind/Lenoir
Launch date	14 May, 1973	25 May, 1973	28 July, 1973	16 November, 1973
Launch time	13:30	09:00	07:11	09:01
Splashdown date	—	22 June, 1973	25 September, 1973	8 February, 1974
Splashdown time	—	09:49	18:19	11:17
Mission duration (hr:min:sec)	—	672:49:49 (28 days)	1.427:09:04 (59 days)	2.017:15:31 (84 days)
Distance travelled (manned)		18.5 million km	39.3 million km	55.5 million km
Revolutions (manned)		404	858	1.214
EVA 1		3 hr 30 min 7 June	6 hr 29 min 6 Aug	6 hr 33 min 22 Nov
2		1 hr 44 min 19 June	4 hr 30 min 24 Aug	7 hr 1 min 25 Dec
3		—	2 hr 45 min 22 Sept	3 hr 28 min 29 Dec
4		—	—	5 hr 19 min 3 Feb
EVA totals:		5 hr 41 min[1]	13 hr 44 min	22 hr 21 min
EVA crewmembers 1		Conrad/Kerwin	Garriott/Lousma	Gibson/Pogue
2		Conrad/Weitz	Garriott/Lousma	Carr/Pogue
3		—	Bean/Garriot	Carr/Gibson
4		—	—	Carr/Gibson
Docked duration		654 hr 48 min 7 sec	1.416 hr 9 min 42 sec	2.004 hr 32 min 12 sec

Consumable utilization:
Water (at launch/end) 2.722/776 kg
Oxygen " " 2.767/1.254 kg
Nitrogen " " 699/275 kg
TACS " " 36.287/5.664 kg-sec

Notes: [1] Includes 37 min stand-up EVA
[2] All time record

TABLE 18: SKYLAB EXPERIMENT DATA

Experiment category	*SL-2 (hrs/%)**	*SL-3 (hrs/%)**	*SL-4 (hrs/%)**
Solar astronomy	117.2/29.9	305.1/28.2	519/33.2
Earth observations	71.4/18.2	223.5/20.6	274.5/17.6
Student	3.7/0.9	10.8/1.0	14.8/0.9
Astrophysics	36.6/9.4	103.8/9.6	133.8/8.5
Man/systems	12.1/3.1	117.4/10.8	83/5.3
Material science	5.9/1.5	8.4/0.8	15.4/1.0
Life science	143.3/37	312.5/29	366.7/23.5
Kohoutek			156/10.0
	392.2	1081.2	1.563.2

* Indicates percentage of total experiment time devoted to respective categories.

SUMMARY

	SL-2	*SL-3*	*SL-4*
Solar observations:			
Frames returned	28.739	24.942	73.366
Earth observations:			
Frames returned	9.846	16.800	19.400
Magnetic tape	13.716 meters	28.529 meters	30.480 meters

TABLE 19: ASTP HARDWARE

Apollo CSM
(See Apollo Basic Data table)

Docking Module (DM):

Length 3.15 meters
Diameter 1.4 meters
Weight 2.012 kg

Environmental control:
2 spherical tanks = 18.9 kg N_2 stored at 63.28 kg/cm^2
2 spherical tanks = 21.7 kg O_2 stored at 63.28 kg/cm^2
(All 4 tanks 64.31 cm in diameter)

Atmosphere:
258 mm Hg (90% O_2) to 517 mmHg (30% O_2/70%N_2) in orbit
760 mm Hg (21% O_2/79% N_2) on launch pad

Soyuz (see Soyuz Basic Data table)

Docked configurations
CSM/DM length 13.1 meters
weight 14.7 tonnes
CSM/DM/Soyuz length 20 meters
weight 21 tonnes

TABLE 20: ASTP MISSION DATA

Event	*Apollo*	*Soyuz*
Prime crew	Stafford/Brand/Slayton	Leonov/Kubasov
Back-up crew	Bean/Evans/Lousma	Filipchenko/Rukavishnikov
Launch date	15 July, 1975	15 July, 1975
Launch time	19:50:00 GMT	12:20:10 GMT
Initial orbit (km)	154.7 × 173.3	186.3 × 221.9
Soyuz first mnvr		5:19 SGET
Apollo/S-IVB separation	8:44 SGET	
S-IVB evasive mnvr	10:04 SGET	
Apollo circularization	11:11 SGET	
NC-1 phasing burn	13:08:29 SGET	
Circularization burn		24:23:35 SGET
Phasing burn	31:58:00 SGET	
NC-2 phasing burn	48:31:00 SGET	
NCC burn	49:15:05 SGET	
NSR burn	49:52:00 SGET	
TPI	50:57 SGET	
Soft dock	51:49:12 SGET	
Hard dock	51:52:30 SGET	
Stafford/Leonov handshake	54:59:25 SGET	
Undocking	95:43:12 SGET	
Re-docking	96:14 SGET	
2nd undocking	99:07 SGET	
Separation burn	102:27 SGET	
De-orbit burn start		141:50:35 SGET
Landing		142:30:54 SGET
Trim burn	146:37 SGET	
Docking Module jettison	199:25 SGET	
DM-1 mnvr	200:00 SGET	
DM-2 mnvr	204:11:42 SGET	
De-orbit burn	224:37:47 SGET	
Splashdown	224:58:24 SGET	
Flight duration	217:28:24	142:30:54

GMT Greenwich Mean Time

SGET Soyuz Ground Elapsed Time (commencing 12:20:00 GMT. July 15. 1975)

DM Docking Module

TABLE 21: ASTP MANEUVER SUMMARY

Event	*SGET*	*Burn time*	*Velocity charge (m/sec)*	*Orbit (km)*
Soyuz first mnvr	5:19	7	2.8	231.7 × 192.4
Apollo circ[1]	11:11	0.8	5.5	172.2
Apollo NC-1	13:08:29	3.2	20.5	237.4 × 173.9
Soyuz circ[1]	24:23:35	18.5	11.6	229.6
Apollo phasing	31:58	2.4	2.7	230.9 × 172.8
Apollo NC-2	48:31	1.1	7.4	202 × 170.7
Apollo NCC	49:15:05	1.5	10.6	208.2 × 194.6
Apollo NSR	49:52	1.0	6.8	210.6 × 210
Apollo TPI	50:57	0.9	6.7	229.5 × 210
Apollo trim	146:37	37	2.2	223.9 × 216.3
DM-1 mnvr	200:00	1.0	9.5	232.4 × 219.3
DM-2 mnvr	204:11:42	0.9	8.1	223.4 × 211.3

SGET Soyuz Ground Elapsed Time
DM Docking Module

TABLE 22: ASTP CREW EXCHANGE TIMES

Stafford in Soyuz	7 hr 10 min
Slayton in Soyuz	1hr 35 min
Brand in Soyuz	6 hr 30 min
Leonov in Apollo	5 hr 43 min
Kubasov in Apollo	4 hr 57 min

TABLE 23: NASA MANNED FLIGHT PROGRAM COSTS

Program	*$ millions*
Mercury	392
Gemini	1.281
Apollo	25.380
Skylab	2.460
ASTP	218

Note: The above figures reflect total accrued costs and are certified by Frederick L. Dunlap, Director, Budget Operations Division, Office of Associate Administrator/Comptroller, NASA Headquarters, to be the actual program costs following audits and completion of relevant missions, even where the figures are at variance with previously published figures in NASA books and publications.
These costs reflect sums spent on manned operations in addition to the several unmanned test and qualification flights essential to hardware development.

ABLE 24: ASTRONAUT SELECTION *(On active flight list, as at June 1981.)

ame	Born	Selected	Mercury Prime	Mercury Back-up	Gemini Prime	Gemini Back-up	Apollo Prime	Apollo Back-up	Apollo Support Crew	Skylab Prime	Skylab Back-up	Skylab Support Crew	ASTP Prime	ASTP Back-up	ASTP Support Crew
BRAHAMSON, James A. Maj. Gen USAF	19 May, 1933	MOL Group 3													
DAMS, Michael James Maj. USAF	5 May, 1930	MOL Group 1													
LDRIN, Edwin E., Jr. Col. USAF	20 Jan., 1930	NASA Group 3			GT-12	GT-9	11	8							
LLEN, Joseph P. Ph.D	27 June, 1937	NASA Group 6							15						
NDERS, William A.	17 Oct., 1933	NASA Group 3				GT-11	8	11							
RMSTRONG, Neil A.	5 Aug., 1930	USAF Group1/NASA Group 2			GT-8	GT-5/11	11	8							
TWELL, Alfred L. Capt. USAF	1929?	USAF Group 3													
ASSETT, Charles A. Maj. USAF	30 Dec., 1931	USAF Group 3/NASA Group 3													
EAN, Alan L. Capt. USN	15 Mar., 1932	NASA Group 3				GT-10	12	9		SL-3				1	
ENEFIELD, Tommie D.	1929?	USAF Group 3													
OBKO, Karol Joseph	23 Dec., 1937	MOL Group 2/NASA Group 7													1
OCK, Jr. Charles C.	1925?	USAF Group 2													
ORMAN, Frank Col. USAF	14 Mar., 1928	NASA Group 2			GT-7	GT-4	8								
RAND, Vance D.	9 May, 1931	NASA Group 5						15	8/13		SL-3/4		1		
ULL, John S. Lt. Cdr. USN	25 Oct., 1934	NASA Group 5													
ARPENTER, M. Scott Cdr. USN	1 May, 1925	NASA Group 1	MA-7	MA-6											
ARR, Gerald P. Col. USMC	22 Aug., 1933	NASA Group 5							8/12	SL-4					
ERNAN, Eugene A. Cpt. USN	14 Mar., 1934	NASA Group 3			GT-9	GT-12	10/17	7/14							
HAFFEE, Roger B. Lt. Cdr. USN	15 Feb., 1935	NASA Group 3					1								
HAPMAN, Philip K. Sc.D	5 Mar., 1935	NASA Group 6							14/16						
OLLINS, Michael	31 Oct., 1930	USAF Group 3/NASA Group 3			GT-10	GT-7	11								
ONRAD, Charles Jr. Cpt. USN	2 June, 1930	NASA Group 2			GT-5/11	GT-8	12	9		SL-2					
OOPER, L. Gordon Col. USAF	6 Mar., 1927	NASA Group 1	MA-9	MA-8	GT-5	GT-12		10							
REWS, Jr. Albert H.	23 Mar., 1929	USAF Group 2/X-20 Group 1/ MOL Group 1													
RIPPEN, Robert L. Cdr. USN	11 Sept., 1937	MOL Group 2/NASA Group 7										SL-2/3/4			1
UNNINGHAM, Walter	16 Mar., 1932	NASA Group 3					7	1							
UKE, Charles M. Jr. Col. USAF	30 Oct., 1935	NASA Group 5					16	13/17	10						
ISELE, DONN F. Col. USAF	23 June, 1930	NASA Group 3					7	1/10							
NGLAND, Anthony W. Ph.D	15 May, 1942	NASA Group 6							16						
NGLE, Joe H. Col. USAF	26 Aug., 1932	USAF Group 3/NASA Group 5						14	10						
VANS, Ronald E. Cpt. USN	10 Nov., 1933	NASA Group 5					17	14	1/7/11					1	
INLEY, John Lawrence Cpt. USN	22 Dec., 1935	MOL Group 1													
REEMAN, Theodore C. Cpt. USAF	18 Feb., 1930	NASA Group 3													
ULLERTON, Charles G. Lt. Col. USAF	11 Oct., 1936	MOL Group 2/NASA Group 7							14/17						
ARLAND, Neil R. Cpt. USAF	1928?	USAF Group 3													
ARRIOTT, Owen K. Ph.D	22 Nov., 1930	NASA Group 4								SL-3					
IBSON, Edward G. Ph.D	18 Nov., 1936	NASA Group 4							12	SL-4					
IVENS, Jr. Edward G.	5 Jan., 1930	USAF Group 3/NASA Group 5													
LENN, John H., Jr. Col. USMC	18 July, 1931	NASA Group 1	MA-6	MR-3/4											
ORDON, Henry C. Col. USAF	23 Dec., 1925	USAF Group 1/X-20 Group 1						9							
ORDON, Richard F. Jr. Cpt. USN	5 Oct., 1929	NASA Group 3			GT-11	GT-8	12	15							
RAVELINE, Duane E. M.D.	2 Mar., 1931	NASA Group 4													
RISSOM, Virgil I. Lt. Col. USAF	3 Apr., 1926	NASA Group 1	MR-4		GT-3	GT-6	1								
AISE, Fred W. Jr.	14 Nov., 1933	NASA Group 5					13	8/11/16							
ARTSFIELD, Henry W. Jr., Col. USAF	21 Nov., 1933	MOL Group 2/NASA Group 7							16			SL-2/3/4			
ENIZE, Karl G. Ph.D	17 Oct., 1926	NASA Group 6							15						
ERRES, Robert T. Brig. Gen. USAF	1 Dec., 1932	MOL Group 3													
OLMQUEST, Donald L. M.D., Ph.D	7 Apr., 1939	NASA Group 6													
OOVER, Lloyd N. Maj. USAF	1926?	USAF Group 2													
RWIN, James B. Col. USAF	17 Mar., 1930	NASA Group 5					15	12	10						
ERWIN, Joseph P. Cpt USN, M.D	19 Feb., 1932	NASA Group 4								SL-2					
NIGHT, William J. Col. USAF	18 Nov., 1929	USAF Group 1/X-20 Group 1													
NOLLE, Byron F. Maj. USAF	1925?	USAF Group 2													
AWRENCE, Robert H. Jr., Maj. USAF	2 Oct., 1935	MOL Group 3													
AWYER, Richard E. Col. USAF	8 Nov., 1932	MOL Group 1													
ENOIR, William B. Ph.D	14 Mar., 1939	NASA Group 6									SL-3/4				
IND, Don L. Ph.D	18 May, 1930	NASA Group 5									SL-3/4				
LEWELLYN, John A. Ph.D	22 Apr., 1933	NASA Group 6													
OUSMA, Jack R. Lt. Col. USMC	29 Feb., 1936	NASA Group 5							9/13	SL-3				1	
OVELL, James A. Jr., Cpt. USN	25 Mar., 1928	NASA Group 2			GT-7/12	GT-4/9	8/13	11							
ACLEAY, Lachlan, Col. USAF	13 June, 1931	MOL Group 1													
ATTINGLEY, Thomas K. II Cdr. USN	17 Mar., 1936	NASA Group 5					16		8/11/12						
cCANDLESS, Bruce II Cdr. USN	8 June, 1937	NASA Group 5							14		SL-2				
cDIVITT, James A. Brig. Gen. USAF	10 June, 1929	NASA Group 2			GT-4		9								
cINTOSH, Robert H. Cpt. USAF	1926?	USAF Group 2													
ICHEL, F. Curtiss Ph.D	5 June, 1934	NASA Group 4													
ITCHELL, Edgar D. Cpt. USN	17 Sept., 1930	NASA Group 5					14	10/16	9						
USGRAVE, F. Story M.D., Ph.D	19 Aug., 1935	NASA Group 6									SL-2				
EUBECK, Francis G. Col. USAF	11 Apr., 1932	USAF Group 3/MOL Group 1													
'LEARY, Brian T. Ph.D	27 Jan., 1940	NASA Group 6													
VERMYER, Robert F. Lt. Col. USMC	14 July, 1936	MOL Group 2/NASA Group 7							17						1
ARKER, Robert A. Ph.D	14 Dec., 1936	NASA Group 6							15/17						
ETERSON, Donald H. Col. USAF	22 Oct., 1933	MOL Group 3/NASA Group 7							16						
OGUE, William R. Col. USAF	23 Jan., 1930	NASA Group 5							1/7/11/13/14	SL-4					
OGERS, Russell L. Maj. USAF	12 Apr., 1926	USAF Group 1/X-20 Group 1													
OMAN, James A. Cpt. USAF	1927?	USAF Group 3													
OOSA, Stuart A. Col. USAF	16 Aug., 1933	USAF Group 5					14	16/17							
CHIRRA, Walter M. Jr., Cpt USN	12 Mar., 1923	NASA Group 1	MA-8	MA-7	GT-6	GT-3	7	1							
CHMITT, Harrison H. Ph.D	3 July, 1935	NASA Group 4					17	15							
CHWEICKART, Russell L.	25 Oct., 1935	NASA Group 3					9				SL-2				
COTT, David R. Col. USAF	6 June, 1932	NASA Group 3			GT-8		9/15	12							
EE, Elliott J.	23 July, 1927	NASA Group 2				GT-5									
HEPARD, Alan B. Jr. Rear Adm. USN	18 Nov., 1923	NASA Group 1	MR-3	MA-9			14								
LAYTON, Donald K.	1 Mar., 1924	NASA Group 1											1		
MITH, Robert W.	1929?	USAF Group 2													
ORLIE, Donald M.	?	USAF Group 2													
TAFFORD, Thomas P. Lt. Gen. USAF	17 Sept., 1930	NASA Group 2			GT-6/9	GT-3	10	7					1		
WIGERT, John L. Jr.	30 Aug., 1931	NASA Group 5					13	13	7						
AYLOR, James M. Lt. Col. USAF	27 Nov., 1930	MOL Group 1													
HOMPSON, Milton O.	4 May, 1926	USAF Group 1/X-20 Group 1													
HORNTON, William E. M.D.	14 Apr., 1929	NASA Group 6										SL-2/3/4			
RULY, Richard H. Cdr. USN	12 Nov., 1937	MOL Group 1/NASA Group 7										SL-2/3/4			1
WINTING, William T. Cpt. USAF	1929?	USAF Group 2													
HALT, Alfred H. Cpt. USAF	1931?	USAF Group 3													
VEITZ, Paul J. Cpt. USN	25 July, 1932	NASA Group 5							12	SL-2					
VHITE, Edward H. II Lt. Col. USAF	14 Nov., 1930	NASA Group 2			GT-4	GT-7	1								
VILLIAMS, Clifton C. Jr. Maj. USMC	26 Sept., 1932	NASA Group 3				GT-10									
VOOD, James W. Col. USAF	9 Aug., 1924	USAF Group 1/X-20 Group 1													
VORDEN, Alfred M. Col. USAF	7 Feb., 1932	NASA Group 5					15	12	9						
OUNG, John W. Cpt. USN	24 Sept., 1930	NASA Group 2			GT-3/10	GT-6	10/16	7/13/17							

TABLE 25: ASTRONAUT BATCH SELECTION DATES

NASA	Group 1	9 April 1959
USAF	Group 1	15 Mar 1962
USAF	Group 2	20 April 1962
NASA	Group 2	17 Sept 1962
X-20	Group 1	20 Sept 1962
USAF	Group 3	22 Oct 1962
NASA	Group 3	18 Oct 1963
NASA	Group 4	27 June 1965
MOL	Group 1	12 Nov 1965
NASA	Group 5	4 April 1966
MOL	Group 2	17 June 1966
MOL	Group 3	30 June 1967
NASA	Group 6	4 Aug 1967
NASA	Group 7	14 Aug 1969

ASTRONAUTS DECEASED

Adams	15 Nov 1967	X-15 crash
Bassett	28 Feb 1966	air crash
Chaffee	27 Jan 1967	pad fire
Freeman	31 Oct 1964	air crash
Givens	6 June 1967	car crash
Grissom	27 Jan 1967	pad fire
Lawrence	8 Dec 1967	air crash
Rogers	13 Sept 1967	jet accident
See	28 Feb 1966	air crash
Taylor	Sept 1970	air crash
White	27 Jan 1967	pad fire
Williams	15 Oct 1967	air crash

TABLE 27: MANNED FLIGHT LOG

	Mission	Launch date	Duration (days)	Country of origin	Remarks
1	Vostok 1	12 April, 1961	0.075	USSR-1	First manned space flight
2	MR-3	5 May, 1961	0.01	USA-1	First US flight (suborbital)
3	MR-4	21 July, 1961	0.01	USA-2	Suborbital, capsule sank
4	Vostok 2	6 Aug., 1961	1.05	USSR-2	Exceeded 1 day in space
5	MA-6	20 Feb., 1962	0.2	USA-3	First US manned orbital flight
6	MA-7	24 May, 1962	0.2	USA-4	Landed off target
7	Vostok 3	11 Aug., 1962	3.9	USSR-3	Dual flight with Vostok 4
8	Vostok 4	12 Aug., 1962	2.9	USSR-4	Passed with 6.5 km of Vostok 3
9	MA-8	3 Oct., 1962	0.38	USA-5	Employed modified Mercury
10	MA-9	15 May, 1963	1.4	USA-6	First US flight to exceed 1 day
11	Vostok 5	14 June, 1963	4.96	USSR-5	Dual flight with Vostok 6
12	Vostok 6	16 June, 1963	2.9	USSR-6	Carried first woman into space
13	Voskhod 1	12 Oct., 1964	1.07	USSR-7	Carried first three-man crew
14	Voskhod 2	18 Mar., 1965	1.04	USSR-8	Supported first space walk
15	GT-3	23 Mar., 1965	0.2	USA-7	First manned orbit changes
16	GT-4	3 June, 1965	4.08	USA-8	First US space walk
17	GT-5	21 Aug., 1965	7.96	USA-9	First use of electrical fuel cells
18	GT-6A	15 Dec., 1965	1.08	USA-10	Rendezvous with Gemini 7
19	GT-7	4 Dec., 1965	13.8	USA-11	Target for Gemini 6A rendezvous
20	GT-8	16 Mar., 1966	0.4	USA-12	First docking; flight aborted
21	GT-9A	3 June, 1966	3.01	USA-13	Three rendezvous; 2 hr spacewalk
22	GT-10	8 July, 1966	2.95	USA-14	Used propulsion unit on target
23	GT-11	12 Sept., 1966	2.97	USA-15	Tethered exercise with target
24	GT-12	11 Nov., 1966	3.94	USA-16	Successfully resolved EVA problems
25	Soyuz 1	23 April, 1967	1.07	USSR-9	First man to be killed in space
26	Apollo 7	11 Oct., 1968	10.8	USA-17	First US three-man flight
27	Soyuz 3	26 Oct., 1968	3.95	USSR-10	Maneuvered near Soyuz 2
28	Apollo 8	21 Dec., 1968	6.12	USA-18	First flight to Moon orbit
29	Soyuz 4	14 Jan., 1969	2.97	USSR-11	First docking of 2 manned ships
30	Soyuz 5	15 Jan., 1969	3.04	USSR-12	First crew transfer between ships
31	Apollo 9	3 Mar., 1969	10.04	USA-19	First manned test of Moon lander
32	Apollo 10	18 May, 1969	8.0	USA-20	Moon landing rehearsal in lunar orbit
33	Apollo 11	16 July, 1969	8.13	USA-21	First Moon landing (July 20)
34	Soyuz 6	11 Oct., 1969	4.95	USSR-13	Triple flight with Soyuz 7 & 8
35	Soyuz 7	12 Oct., 1969	4.95	USSR-14	Rendezvous target for Soyuz 6 & 8
36	Soyuz 8	13 Oct., 1969	4.95	USSR-15	Rendezvous with Soyuz 7
37	Apollo 12	14 Nov., 1969	10.19	USA-22	Moon landing-2. Deployed instruments
38	Apollo 13	11 April, 1970	5.95	USA-23	Deep-space abort en-route to Moon
39	Soyuz 9	1 June, 1970	17.7	USSR-16	Long duration medical flight
40	Apollo 14	31 Jan., 1971	9.0	USA-24	Moon landing-3. Used hand cart
41	Soyuz 10	22 April, 1971	1.99	USSR-17	Docked with Salyut 1. No transfer
42	Soyuz 11	6 June, 1971	23.7	USSR-18	Occupied Salyut 1. Killed in space
43	Apollo 15	26 July, 1971	12.3	USA-25	Moon landing-4. First use of LRV
44	Apollo 16	16 April, 1972	11.08	USA-26	Moon landing-5. First in lunar highlands
45	Apollo 17	7 Dec., 1972	12.6	USA-27	Moon landing-6. Last Moon mission
46	Skylab 2	25 May, 1973	28.03	USA-28	Occupied Skylab. Repaired workshop
47	Skylab 3	28 July, 1973	59.5	USA-29	Occupied Skylab. Deployed shade
48	Soyuz 12	27 Sept., 1973	1.97	USSR-19	First flight of modified (2 man) Soyuz
49	Skylab 4	16 Nov., 1973	84.05	USA-30	Occupied Skylab. Observed comet
50	Soyuz 13	18 Dec., 1973	7.87	USSR-20	Scientific experiments on lone flight
51	Soyuz 14	3 July, 1974	15.7	USSR-21	Occupied Salyut 3
52	Soyuz 15	26 Aug., 1974	2.0	USSR-22	Failed to dock with Salyut 3
53	Soyuz 16	2 Dec., 1974	5.9	USSR-23	ASTP test precursor
54	Soyuz 17	10 Jan., 1975	29.5	USSR-24	Occupied Salyut 4
55	Soyuz 18A	5 April, 1975	0.01	USSR-25	Abort during ascent
56	Soyuz 18	24 May, 1975	62.97	USSR-26	Occupied Salyut 4
57	Soyuz 19	15 July, 1975	5.94	USSR-27	Target for ASTP Apollo
58	Apollo ASTP	15 July, 1975	9.06	USA-31	Docked with Soyuz 19. Crew transfers
59	Soyuz 21	6 July, 1976	49.1	USSR-28	Occupied Salyut 5
60	Soyuz 22	15 Sept., 1976	7.92	USSR-29	Solo flight
61	Soyuz 23	14 Oct., 1976	2.0	USSR-30	Failed to dock with Salyut 5
62	Soyuz 24	7 Feb., 1977	17.07	USSR-31	Occupied Salyut 5
63	Soyuz 25	9 Oct., 1977	2.0	USSR-32	Failed to dock with Salyut 6
64	Soyuz 26	10 Dec., 1977	96.42*	USSR-33	Occupied Salyut 6
65	Soyuz 27	10 Jan., 1978	6.0*	USSR-34	Occupied Salyut 6
66	Soyuz 28	2 Mar., 1978	7.9	USSR-35	Occupied Salyut 6; Czech co-pilot
67	Soyuz 29	15 June, 1978	139.5*	USSR-36	Occupied Salyut 6
68	Soyuz 30	27 June, 1978	7.97	USSR-37	Occupied Salyut 6; Polish co-pilot
69	Soyuz 31	26 Aug., 1978	67.84*	USSR-38	Occupied Salyut 6; East German co-pilot
70	Soyuz 32	25 Feb., 1979	175.02	USSR-39	Occupied Salyut 6
71	Soyuz 33	10 April, 1979	1.96	USSR-40	Failed to dock with Salyut 6
72	Soyuz 35	9 April, 1980	184.9*	USSR-41	Occupied Salyut 6
73	Soyuz 36	26 May, 1980	7.87**	USSR-42	Occupied Salyut 6; Bulgarian co-pilot
74	Soyuz T-2	5 June, 1980	3.9	USSR-43	Occupied Salyut 6
75	Soyuz 37	23 July, 1980	7.86*	USSR-44	Occupied Salyut 6; N. Vietnamese co-pilot
76	Soyuz 38	18 Sept., 1980	7.88	USSR-45	Occupied Salyut 6; Cuban co-pilot
77	Soyuz T-3	27 Nov., 1980	12.8	USSR-46	Occupied Salyut 6; three man crew
78	Soyuz T-4	12 Mar., 1981	74.8	USSR-47	Occupied Salyut 6; two man crew
79	Soyuz 39	22 Mar., 1981	7.88	USSR-48	Occupied Salyut 6; Mongolian co-pilot

Note: Durations indicate space flight time for crew launched in designated vehicle but Soyuz 27 crew returned in Soyuz 36. Soyuz 31 crew returned in Soyuz 29. Soyuz 32 crew returned in Soyuz 34 (launched unmanned on 6 June, 1979). Soyuz 35 crew returned in Soyuz 37. Soyuz 36 crew returned in Soyuz 35, and Soyuz 37 crew returned in Soyuz 36 spacecraft. Soyuz 33 carried a Bulgarian co-pilot although the mission was a failure. It is useful to note that three-quarters of the 31 US manned flights occurred in the first ten years wheras two thirds

TABLE 26: COMPARISON OF SOVIET SPACECRAFT

VOSTOK	VOSKHOD
Re-entry capsule	
Diameter: 2.3 meters	2.3 meters
Weight: 2.4 tonnes	3 tonnes
Crew: 1	2 or 3
Atmosphere: N_2/O_2 sea-level	N_2/O_2 sea level
Instrument section	
Diameter: 2.42 meters	2.42 meters
Height: 2.2 meters	2.2 meters
Weight: 2.3 tonnes	2.3 tonnes
Complete Assembly	
Diameter (max): 2.42	2.42 meters
Height: 4.3 meters	5 meters
Weight: 4.7 tonnes	5.3 tonnes
Manned Flights: 6	2

SOYUZ orbital module
Diameter: 2.2 meters
Length: 2.65 meters
Weight: 1.2 tonnes

Crew module
Diameter: 2.2 meters
Length: 2 meters
Weight: 2.8 tonnes
Crew: 3/2/3[1]
Atmosphere: N_2/O_2 sea-level
Habitable volume: 10 m³ (including orbital module)

Equipment module
Diameter: 2.2 meters (aft skirt 2.7 meters)
Length: 2.3 meters
Weight: 2.6 tonnes
Solar panel span: 8.37 meters

Combined assembly
Length: 7.1 meters
Weight: 6.6–6.9 tonnes
Propulsion: 2 × 400 kg thrust maneuvering motors
Altitude control: 8 × 1 kg & 4 × 10 kg thrust motors
Manned flights (to end 1980): 38

Notes: 1 Soyuz 1 – 11 capable of 3-man crew; remainder carry 2 cosmonauts but T model from June 1980 could carry 3 and would be alternative to the 2-man model. Also, some Soyuz models employed batteries rather than solar cells for on-board electrical production. Above data is typical but several variations were introduced.

SALYUT Length: 21 meters
Diameter (max): 4.15 meters
Weight: 18.9 tonnes
Atmosphere: N_2/O_2 sea level
Electrical production: 2 or 3 solar cell arrays
Docked Soyuz/Salyut length: 27.2 meters
weight: 25.5 tonnes

TABLE 28: ASTRONAUT CANDIDATES

(As of June, 1981)

Although this book is not concerned with the development or preparation for flight of the NASA Shuttle, the following table lists Pilot and Mission Specialist categories of astronaut recruited as NASA Group 8 and NASA Group 9 in support of Shuttle operations anticipated for the 1980s. Groups 8 and 9 were required to report for duty at the NASA Johnson Space Center, Houston, Texas, on 1 July, 1978, and 7 July, 1980, respectively. Pilot astronauts will be responsible for flying the Shuttle orbiter; Mission Specialists will be responsible for conducting on-orbit operations but not for controlling the orbiter during ascent or descent. Career astronauts from NASA Groups 1 to 7 will continue to serve as experienced pilots from which future crews will be chosen.

Name	Birth	Category	Group
BLAHA, John E. Lt. Col. USAF	26 Aug., 1942	P	9
BOLDEN, Charles F. Maj. USMC	19 Aug., 1946	P	9
BRANDENSTEIN, Daniel C. Lt. Cdr. USN	17 Jan., 1943	P	8
BRIDGES, Roy D. Jr., Lt. Col. USAF	19 July, 1943	P	9
COATS, Michael L. Lt. Cdr. USN	16 Jan., 1946	P	8
COVEY, Richard O. Maj. USAF	1 Aug., 1946	P	8
CREIGHTON, John O. Lt. Cdr. USN	28 April, 1943	P	8
GARDNER, Guy S. Maj. USAF	6 Jan., 1948	P	9
GIBSON, Robert L. Lt. USN	30 Oct., 1946	P	8
GRABE, Ronald J. Maj. USAF	13 June, 1945	P	9
GREGORY, Frederick D. Maj. USAF	7 Jan., 1941	P	8
GRIGGS, Stanley D.	7 Sept., 1939	P	8
HAUCK, Frederick H. Cdr. USN	11 April, 1941	P	8
McBRIDE, Jon A. Lt. Cdr. USN	14 Aug., 1943	P	8
NAGEL, Steven R. Cpt. USAF	27 Oct., 1946	P	8
O'CONNOR, Bryan D. Maj. USMC	6 Sept., 1946	P	9
RICHARDS, Richard N. Lt. Cdr. USN	24 Aug., 1946	P	9
SCOBEE, Francis R. Maj. USAF	19 May, 1939	P	8
SHAW, Brewster H. Jr., Cpt. USAF	16 May, 1945	P	8
SHRIVER, Loren J. Cpt. USAF	23 Sept., 1944	P	8
SMITH, Michael J. Lt. Cdr. USN	30 April, 1945	P	9
WALKER, David M. Lt. Cdr. USN	20 May, 1945	P	8
WILLIAMS, Donald E. Lt. Cdr. USN	13 Feb., 1942	P	8
BAGIAN, James P. M.D.	22 Feb., 1952	MS	9
BLUFORD, Guion S. Jr., Maj. USAF Ph.D	22 Nov., 1942	MS	8
BUCHLI, James F. Cpt. USMC	20 June, 1945	MS	8
CHANG, Franklin R. Ph.D	5 April, 1950	MS	9
CLEAVE, Mary L. Ph.D	5 Feb., 1947	MS	9
DUNBAR, Bonnie J.	3 March, 1949	MS	9
FABIAN, John M. Maj. USAF Ph.D	28 Jan., 1939	MS	8
FISHER, Anna L. M.D.	24 Aug., 1949	MS	8
FISHER, William F. M.D.	1 April, 1946	MS	9
GARDNER, Dale A. Lt. USN	8 Nov., 1948	MS	8
HART, Terry J.	27 Oct., 1946	MS	8
HAWLEY, Steven A. Ph.D	12 Dec., 1951	MS	8
HILMERS, David C. Cpt. USMC	28 Jan., 1950	MS	9
HOFFMAN, Jeffrey A. Ph.D	2 Nov., 1944	MS	8
LEESTMA, David C. Lt. Cdr. USN	6 May, 1949	MS	9
LOUNGE, John M.	28 June, 1946	MS	9
LUCID, Shannon W. Ph.D	14 Jan., 1943	MS	8
McNAIR, Ronald E. Ph.D	21 Oct., 1950	MS	8
MULLANE, Richard M. Cpt. USAF	10 Sept., 1945	MS	8
NELSON, George D. Ph.D	13 July, 1950	MS	8
ONIZUKA, Ellison S. Cpt. USAF	24 June, 1946	MS	8
RESNIK, Judith A. Ph.D	5 April, 1949	MS	8
RIDE, Sally K.	26 May, 1951	MS	8
ROSS, Jerry L. Cpt. USAF	20 Jan., 1948	MS	9
SEDDON, Margaret R. M.D.	8 Nov., 1947	MS	8
SPRING, Sherwood C. Major US Army	3 Sept., 1944	MS	9
SPRINGER, Robert C. Maj. USMC	21 May, 1942	MS	9
STEWART, Robert L. Maj. US Army	13 Aug., 1942	MS	8
SULLIVAN, Kathryn D.	3 Oct., 1951	MS	8
[illegible]	[illegible]	MS	8

TABLE 29: COSMONAUT ASSIGNMENTS

Name	Vostok	Voskhod	Soyuz
AKSENOV, Vladimir V.			22;T-2
ARTYUKHIN, Yuriy P. Lt. Col. AF			14
BELYAEV, Pavel I. Col. NAF		2	
BEREGOVOY, Georgiy T. Maj. Gen. AF			3
BYKOVSKIY, Valeriy F. Col. AF	5		22;31
DEMIN, Col. AF, Lev S.			15
DOBROVOLSKIY, Georgiy T. Lt. Col. AF			11
DZANIBEKOV, Vladimir A. Maj. AF			27;39
FARKAS, Bertalan Lt. Col. AF			36
FEOKTISTOV, Konstantin P.		1	
FILIPCHENKO, Anatoliy V. Col. AF			7
GAGARIN, Yuriy A. Col. AF	1		
GLAZKOV, Yuri Lt. Col. AF			24
GORBATKO, Viktor V. Col. AF			7;24;37
GRECHKO, Georgiy M.			17;26
GUBAREV, Aleksey A. Lt. Col. AF			17;28
GURRAGCHA, Jugderdemidiyn			39
HERMASZEWSKI, Miroslaw Lt. Col. AF			30
IVANOV, Georgi Lt. Col. AF			33
IVANCHENKOV, Aleksander S.			29
JAHN, Sigmund Col. AF			31
KHRUNOV, Yergeniy V. Col. AF			5
KIZIM, Leonid			T-3
KLIMUK, Petr I. Lt. Col. AF			13;18;30
KOMAROV, Vladimir M.		1	1
KOVALENOK, Vladimir Lt. Col. AF			25;29;T-4
KUBASOV, Valeriy N.			6;19;36
LAZAREV, Vasiliy G. Lt. Col. AF			12;18A
LEBEDEV, Valentin V.			13
LEONOV, Aleksey A		1	19
LYAKHOV, Vladimir Lt. Col. AF			32
MAKAROV, Oleg G.			12;18A;27;T-3
MALYSHEV, Yuri			T-2
MENDEZ, Tamayo			38
NIKOLAYEV, Andrijan G. Maj. Gen. AF	3		9
PATSAYEV, Viktor I.			11
POPOV, Leonid			35
POPOVICH, Pavel R.	4		14
REMEK, Vladimir Cpt. AF			28
ROMANENKO, Yuriy V. Maj. AF			19??;26;38
ROZHDESTVENSKY, Valery Lt. Col. AF			23
RUKAVISHNIKOV, Nikolay N.			10;16;33
RYUMIN, Valeriy			25;32;35
SARAFANOV, Gennodiy V. Lt. Col. AF			15
SAVINYKH, Victor			T-4
SEVASTYANOV, Vitaliy I.			9.18
SHATALOV, Vladimir A. Lt. Gen. AF			4;8;10
SHONIN, Georgiy S. Col. AF			6
STREKALOV, Genrodiy			T-3
TERESHKOVA, Valentina V. Col. AF	6		
TITOV, Gherman S. Col. AF	2		
TUAN, Pham Lt. Col. AF			37
VOLKOV, Vladislav N.			7;11
VOLYNOV, Boris V. Col. AF			5;21
YEGOROV, Boris B. Col. AF		1	
YELISEYEV, Aleksey S.			4;5;8;10
ZHOLOBOV, Vitali M. Lt. Col. AF			21
ZUDOV, Vyacheslav Lt. Col. AF			23

Andreyer is recorded as a back-up (BU) because this is the only category he has filled to date (June 1981). Fifty-seven cosmonauts had flown in space up to the end of 1980. Thirty-seven cosmonaut orders are presented because several places are shared where spacecraft carried more than one pilot making his first flight: for instance, Volynov, Khrunov and Yeliseyev shared position as 11th cosmonaut in space.

COSMONAUTS DECEASED

BELYAYEV	10 Jan 1970 Surgical complications
DOBROVOLSKIY	20 June 1071 Died in space
GAGARIN	27 March 1968 Air crash
KOMAROV	24 April 1967 Died in space flight

Index

Entries in italics refer to films, books, reports or articles